3. Early Jurassic, 195 Ma

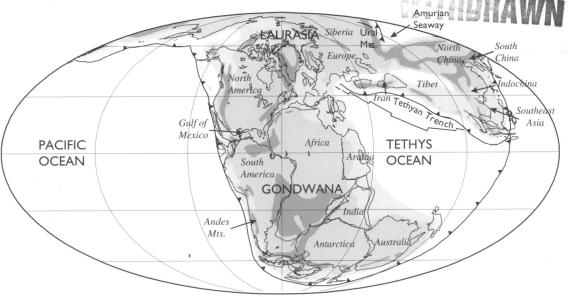

4. Late Permian, 258 Ma

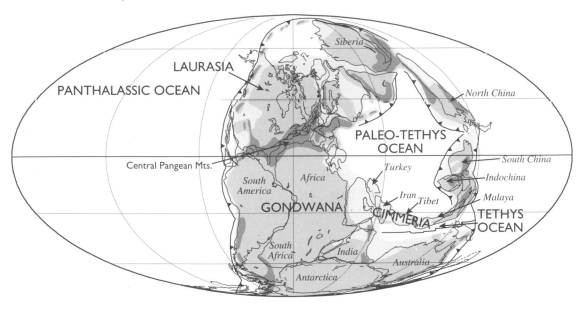

Paleogeographic maps by C. R. Scotese, PALEOMAP Project, University of Texas at Arlington (www.scotese.com)

The Oxford Companion to
The Earth

Editors
Paul L. Hancock[†] and Brian J. Skinner

Associate editor
David L. Dineley

Subject editors
Alastair G. Dawson
K. Vala Ragnarsdottir
Iain S. Stewart

OXFORD
UNIVERSITY PRESS

OXFORD

UNIVERSITY PRESS

Great Clarendon Street, Oxford OX2 6DP

Oxford University Press is a department of the University of Oxford.
It furthers the University's objective of excellence in research, scholarship,
and education by publishing worldwide in

Oxford New York

Athens Auckland Bangkok Bogotá Buenos Aires Calcutta
Cape Town Chennai Dar es Salaam Delhi Florence Hong Kong Istanbul
Karachi Kuala Lumpur Madrid Melbourne Mexico City Mumbai
Nairobi Paris São Paulo Singapore Taipei Tokyo Toronto Warsaw

with associated companies in Berlin Ibadan

Oxford is a registered trade mark of Oxford University Press
in the UK and in certain other countries

Published in the United States
by Oxford University Press Inc., New York

© Oxford University Press, 2000

The moral rights of the authors have been asserted

Database right Oxford University Press (maker)

First published 2000

British Library Cataloguing in Publication Data

Data available

Library of Congress Cataloguing in Publication Data
ISBN 0-19-854039-6

10 9 8 7 6 5 4 3 2 1

Typeset in Minion
by EXPO Holdings, Malaysia
Printed in Great Britain
on acid-free paper by
Biddles Ltd, Guildford & Kings's Lynn

Preface

A true companion should be a person with whom or an object with which one feels comfortable, and to whom, or to which, one can turn for advice and counsel. The *Oxford Companion to the Earth* is designed to be just that: a source of concise, readable, friendly, and stimulating accounts of the many phenomena, processes, and natural materials that make the Earth such a dynamic and fascinating planet. Like other Oxford companions, it is aimed at a wide readership. It is our hope that it will appeal to everyone, but especially to those with a concern for the environment; to those who simply wish to know more about planet Earth; and to students in the Earth sciences. We also hope that the *Oxford Companion to the Earth* will appeal to professional scientists seeking an accessible digest of topics outside their own areas of specialization; to those working in fields concerned with issues of resources and the environment, such as town planners, civil engineers, administrators, and politicians; and to all those, whether in schools or higher education, who are studying or teaching the Earth sciences and related subjects, such as geography and environmental science.

As the foregoing paragraph indicates, the breadth of the *Companion* is wide, and deliberately so; the Oxford University Press decided at the earliest stages of planning that the *Companion* should be concerned with the Earth sciences in their widest sense. We have thus set out to provide information on the atmosphere and the oceans as well as the solid Earth and, indeed, the entire planetary environment. And because we can no longer consider the Earth in isolation, information about other bodies in the Solar System also finds a place here.

In choosing the subjects to be treated, we have laid emphasis on areas that are of particular interest to mankind: geology, including its various applied aspects; solid-Earth geophysics, especially as expressed in surface phenomena; exploration geophysics (vital in the search for oil and minerals); geochemistry (like geophysics, fundamental to the study of the earth); geodesy (of increasing importance in mapping the Earth's surface); palaeontology and palaeobiology; soil science; glaciology; physical oceanography; climatology, palaeoclimatology, and meterology; environmental and resource developments; and the history of the Earth sciences.

The *Companion* contains more than 900 entries, many of which are accompanied by suggestions for further reading. There are numerous cross-references so that the reader can follow up topics of interest. In addition, thematic lists of topics are provided as an aid for readers who wish to look up related entries in order to obtain a general view of a particular subject area. A detailed index is also included so that information contained within the individual entries can readily be located.

The concept of an *Oxford Companion to the Earth* originated with Bruce Wilcock, formerly editor for the Earth sciences in Oxford, and it fell to him to solve the unfortunate and unforeseen problems that plagued the production of the volume. Much of the credit for bringing the project to a successful culmination belongs to Bruce—he has done an extraordinary job.

The initial shaping of the *Companion* was in the hands of Professor Paul Hancock of the University of Bristol. As Editor-in-Chief, Paul Hancock drew up the general outline for the book, and with Bruce Wilcock oversaw much of its early development. The scope of the *Companion* was too wide for any one editor to cover adequately, and Professor Hancock therefore chose four subject editors, who were given responsibility for particular areas: Professor Paul Morgan (Northern Arizona University) for solid-Earth geophysics and applied geophysics; Dr Vala Ragnarsdottir (Bristol University) for geochemistry and environmental and resource development; Dr Iain Stewart (Brunel University) for geomorphology, pedology, and glaciology; Dr Alastair Dawson (Coventry University) for physical oceanography, climatology, paleoclimatology, and meteorology. Paul Hancock himself acted as subject editor for the geological sciences and for the history of geology.

It is not uncommon for a large reference work to have a prolonged, even painful, gestation. The *Oxford Companion to the Earth* has been no exception in this regard. An early setback occurred when Paul Morgan had to withdraw because of pressure of other work. Before standing down, he had, however, given valuable advice and had recruited a number of contributors. His original list of headwords for geophysics has been essentially retained in the final text. On his departure, Paul Hancock took over responsibility for geophysics, adding appreciably to his own editorial burden. A drastic setback was, alas, to follow. Towards the end of 1997 Paul Hancock suffered a marked deterioration in health. He fought on and continued to work on the *Companion*, but by the summer of 1998 it was apparent that he could not continue as Editor-in-Chief. The OUP accordingly appointed him as Consultant Editor, and in this capacity he continued to take a keen interest in the progress of this volume until his untimely death in December

1998. The *Companion to the Earth* is a tribute to the judgement and dedication of Paul Hancock; his name rightly takes first place on the title page. When he had to stand aside, I was asked to take over as Editor-in-Chief and supervise the completion of this project. At the same time, David Dineley, Professor Emeritus at Bristol University, who had already contributed many entries for the *Companion*, was appointed as Associate Editor and was asked to take over editorial responsibility for the completion of geological sciences, geophysics, and the history of geology. David gave invaluable help in many other ways, particularly to Bruce Wilcock during the difficult change-over period; without his efforts, the transfer of editorial work could not have taken place so smoothly.

In referring at length to editorial problems I should not wish it to be thought that the vital contribution made by the contributors was in any way undervalued: that would be far from the truth. This is very much a multi-author book: more than 200 scientists from all over the world have written entries to make the *Companion* a truly international contribution to the wider understanding of the Earth sciences. Our debt to them is great. We are especially conscious of the fact that writing entries for reference books is always a diversion from research activities and that articles of this kind do not count toward the annual quota of publications currently demanded of most academic scientists.

I hope that the *Oxford Companion to the Earth* will prove to be a companion in the truest sense and that it will provide many hours of pleasant and informative browsing as well as being a valuable source of reference.

BRIAN J. SKINNER
New Haven, Connecticut
March 2000

Note to the reader

A *Companion* is a special type of reference book, and specific entries will not be found here for all the terms that would be expected to appear as headwords in an Earth sciences dictionary or an encyclopedia. (There are, on the other hand, numerous entries on topics that would not usually be found in a dictionary, such as *art and the Earth sciences, geoscience and the media,* and *wine and geology,* to quote only three.)

Two aids to navigation are provided to enable the reader to extract the maximum return from the *Companion* with the minimum of trouble. First, there is a comprehensive index at the back of the book which should be consulted if the topic sought does not appear as a headword or a cross-reference in the alphabetical entries. Secondly, those who wish to explore a particular subject area can turn to the thematic lists at the end of the volume. These lists bring together the headwords for entries in specific areas, such as climate and climate change, geological time, or oceanography.

More general information is provided in the appendixes. The geological timescales in Appendix 1, together with the palaeogeographical maps that appear as endpapers, provide the larger framework of time and space into which specific topics can be fitted. Salient facts about the Earth and the Solar System are assembled in Appendix 2. The periodic table of the elements is given in Appendix 3 for ready reference.

Appendix 4 is intended to demystify for the non-scientist the scientific units and notation that are in general use in the Earth sciences, together with the appropriate abbreviations. Tables for converting metric units to Imperial units are also included.

Every effort has been made to contact the original copyright holder for illustrations reproduced in this companion. Oxford University Press will be happy to rectify any omissions in subsequent printings.

Contents

Paul Lewis Hancock (1937–1998)

When Paul Hancock undertook to be Editor-in-Chief of this *Companion* he soon persuaded a group of subject editors and many contributors to join his team. Paul was known by reputation to a wide range of fellow-scientists who had no doubt about the worthiness of the project or of Paul's ability to bring it to a successful conclusion. Sadly, it had in fact to be completed without Paul's guidance, but those friends who have seen it through trust that it bears all the characteristics of a *Companion* and with the imprint of Paul's enthusiasm for the Earth Sciences in the widest sense.

He will have been known to most as a structural geologist, a specialist in neotectonics and one with interests in the field of classical archaeology. To this, add a reputation as a field geologist; it was in the field that he was happiest in his research and in teaching. Pre-eminently a scrupulous observer of the small-scale or mesoscopic structures, he was able to construct the large-scale picture of history. His off-the-cuff analyses and syntheses, given with humour and an enviable command of English, delighted his field colleagues and students alike. Field studies that began upon graduation took him to Wales and southern England, but his interest in brittle fracture in strata soon led him abroad and to larger-scale features. Thus he visited Lebanon and developed a lasting interest in the geology of the Arabian shield; and before long he was also at work in the Pyrenees. The geological history of the Mediterranean region began to exercise him, and he took a special interest in its plate-tectonic development from the old Tethys Ocean. Later he was to find that in this connection Greece and, especially, Turkey offered spectacular neotectonic features and their archaeological sites gave opportunity to both his scientific and historical abilities.

All this experience he put to good service in his lectures and tutorials and in his supervision over the years at the University of Bristol of a succession of British, Arabian, and Turkish graduate students. Equally important was Paul's record in publishing, editing, and contributing to books and journals and his part in the work of several scientific groups and committees. Although Paul would not have taken kindly to being called a 'good committee man', with the rather wordy implications of that phrase, he was a valued member of such bodies, and he was elected to chair several of the most prestigious. Just as the range of his knowledge and experience, the clarity of his argument, and his tact with people fitted him for those tasks, so it did for taking on the job of Editor-in-Chief of this *Companion*. He embarked upon it with gusto, but when he knew he could not complete the task he was anxious to arrange for colleagues to take over. They now hope that he would be pleased with the result and that this is no mere dictionary or encyclopedia, but is a truly good *Companion*.

D. L. D.

Contributors

Inge Aarseth Department of Geology, University of Bergen, Norway

Athol D. Abrahams Department of Geography, State University of New York at Buffalo, USA

C Agee NASA, Houston, Texas, USA

Geoffrey C. Allen Interface Analysis Centre, University of Bristol, UK

John Allen Postgraduate Research Institute for Sedimentology, Reading, UK

Robert J. Allison Department of Geography, University of Durham, UK

Andrew C. Aplin Fossil Fuels and Environmental Geochemistry, University of Newcastle, UK

Graham M. Appleby NERC Space Geodesy Facility, Monks Wood, Huntingdon, UK

Stefán Arnórsson Science Institute, University of Iceland, Reykjavik, Iceland

J. Bahr Department of Geology and Geophysics, University of Wisconsin, Madison, USA

D. K. Bailey Batheaston, UK

Elizabeth H. Bailey Environmental Science, University of Nottingham, UK

J. H. Baker *Formerly:* PRC Environmental Management, Chicago, USA

Victor R. Baker Department of Hydrology and Water Resources, University of Arizona, USA

Keith E. Barber Department of Geography, University of Southampton, UK

Ron D. Barker School of Earth Sciences, University of Birmingham, UK

J. W. Barnes *Formerly:* Earth Sciences, University of Wales, Swansea, UK

Rodey Batiza Department of Geology and Geophysics, University of Hawaii, USA

Douglas I. Benn School of Geography and Geosciences, University of St Andrews, UK

Matthew R. Bennett School of Earth and Environmental Sciences, University of Greenwich, UK

M. J. Benton Department of Earth Sciences, University of Bristol, UK

Richard E. Bevins Department of Geology, National Museums and Galleries of Wales, Cardiff, UK

Paul Bishop Department of Geography and Topographic Science, University of Glasgow, UK

Arthur L. Bloom Department of Geological Sciences, Cornell University, USA

J. D. Blundy Department of Earth Sciences, University of Bristol, UK

John Boardman Environmental Change Unit, University of Oxford, UK

Michael J. Bovis Department of Geography, University of British Columbia, Canada

R. Bradshaw Department of Earth Sciences, University of Bristol, UK

Mark A. Brandon Department of Earth Sciences, The Open University, Milton Keynes, UK

Susan Brantley Geosciences, The Pennsylvania State University, USA

Robin Brett US Geological Survey, Reston, Virginia, USA

C. Bristow Camborne School of Mines, University of Exeter, UK

John P. Brodholt Department of Geological Sciences, University College London, UK

A. Brookes Environmental Rivers Authority, Reading, UK

A. G. Brown Department of Geography, University of Exeter, UK

Judith M. Bunbury Department of Earth Sciences, University of Cambridge, UK

C. Buriks *Formerly:* PRC Environmental Management, Chicago, USA

C. R. Burn Department of Geography and Environmental Studies, Carleton University, Ottawa, Canada

Dave Burnett BMS, Oxford Brookes University, Oxford, UK

E. Burton Department of Geology, Northern Illinois University, DeKalb, USA

D. R. Butler Formerly: Department of Geography, University of North Carolina, Chapel Hill, USA

Ian A. Campbell Department of Earth and Atmospheric Sciences, University of Alberta, Edmonton, Canada

Allan Chapman Faculty of Modern History, University of Oxford, UK

Mark R. Chapman School of Environmental Sciences, University of East Anglia, Norwich, UK

Michèle L. Clarke Centre for Environmental Management, School of Geography, University of Nottingham, UK

Peter Clarke Department of Geomatics, University of Newcastle, UK

Kent C. Condie Department of Earth and Environmental Science, New Mexico Institute of Mining and Technology, Socorro, USA

John E. Costa US Geological Survey, Portland, Oregon, USA

A. Courtice *Formerly:* Department of Geography, University of Sheffield, UK

T. Cousens Department of Civil Engineering, University of Leeds, UK

Andrew B. Cundy School of Ocean and Earth Sciences, Southampton Oceanography Centre, Southampton, UK

John Dalton Department of Earth Sciences, University of Bristol, UK

Alastair G. Dawson Centre for Quaternary Science, Coventry University, UK

Michael Day Department of Geography, University of Wisconsin-Milwaukee, USA

Dirk H. de Boer Department of Geography, University of Saskatchewan, Saskatoon, Canada

Edward Derbyshire Department of Geography, University of London, UK

D. L. Dineley Department of Earth Sciences, University of Bristol, UK

Ronald I. Dorn Arizona State University, Tempe, USA

Ian Douglas Department of Geography, University of Manchester, UK

Julian A. Dowdeswell Bristol Glaciology Centre, School of Geographical Sciences, University of Bristol, UK

Peter Doyle School of Earth and Environmental Sciences, University of Greenwich, UK

Inge Aarseth Department of Geology, University of Bergen, Norway

Frances Drake School of Geography, University of Leeds, UK

D. Drew Department of Geography, Trinity College, Dublin, Eire

S. A. Drury Department of Earth Sciences, The Open University, Milton Keynes, UK

Ernest M. Duebendorfer Department of Geology, Northern Arizona University, Flagstaff, USA

Charles N. Duncan Department of Meteorology, University of Edinburgh, UK

David L. Dunkerley School of Geography and Environmental Science, Monash University, Australia

William M. Dunne Department of Geological Sciences, University of Tennessee, Knoxville, USA

Barbara L. Dutrow Department of Geology and Geophysics, Louisiana State University, Baton Rouge, USA

Geoffrey Eglinton Department of Earth Sciences, University of Bristol, UK

David K. Elliott Department of Geology, Northern Arizona University, Flagstaff, USA

Graham Evans School of Ocean and Earth Sciences, Southampton Oceanography Centre, Southampton, UK

Paul Farrimond Fossil Fuels and Environmental Geochemistry, University of Newcastle, UK

Bruce Fegley Department of Earth and Planetary Sciences, Washington University, St Louis, Missouri, USA

Callum R. Firth Department of Geography and Earth Sciences, Brunel University, UK

D. C. Ford Department of Geography, McMaster University, Hamilton, Ontario, Canada

J. J. Fornós Departament de Ciències de la Terra, Universitat de les Illes Balears, Spain

Ian D. L . Foster School of Natural and Environmental Sciences, Coventry University, UK

Robert O. Fournier US Geological Survey, Portola Valley, California, USA

C. Mary R. Fowler Department of Geology, Royal Holloway, University of London, UK

Martin Frank Institut für Isotopengeologie und Mineralische Rohstoffe, ETH, Zürich, Switzerland

Hugh M. French Departments of Geography and Earth Sciences, University of Ottawa, Canada

John Wm Geissman Department of Earth and Planetary Sciences, University of New Mexico, Alberquerque, USA

Christopher R. German Southampton Oceanography Centre, Southampton, UK

John Gerrard Department of Geography and Environmental Sciences, University of Birmingham, UK

Reto Gieré Department of Earth and Atmospheric Sciences, Purdue University, West Lafayette, USA

David Gillieson School of Tropical Environment Studies and Geography, James Cook University, Cairns, Australia

Sigurdur Reynir Gislason Science Institute, University of Iceland, Reykjavik, Iceland

Andrew S. Goudie School of Geography, University of Oxford, UK

Kenneth J. Gregory Department of Geography, University of Southampton, UK

E. Gruntfest Department of Geography, University of Colorado, Boulder, USA

David Gubbins School of Earth Sciences, University of Leeds, UK

B. A. Haggart School of Earth and Environmental Sciences, University of Greenwich, UK

Martin J. Haigh School of Social Sciences and Law, Oxford Brookes University, Oxford, UK

Ian R. Hall Department of Earth Sciences, University of Cambridge, UK

Anthony Hallam School of Earth Sciences, University of Birmingham, UK

Paul L. Hancock (deceased) Formerly: Department of Earth Sciences, University of Bristol, UK

J. D. Hansom Department of Geography and Topographic Science, University of Glasgow, UK

R. Harré Iffley, Oxford, UK

R. Giles Harrison Department of Meteorology, University of Reading, UK

Stephan Harrison Centre for Quaternary Science, Coventry University, UK

Jane K. Hart Department of Geography, University of Southampton, UK

Malcolm B. Hart Department of Geological Sciences, University of Plymouth, UK

G. B. Haxel US Geological Survey, Flagstaff, Arizona, USA

G Helffrich Department of Earth Sciences, University of Bristol, UK

Denis L. Henshaw Department of Physics, University of Bristol, UK

Stephen Hesselbo Department of Earth Sciences, University of Oxford, UK

L. J. Hickey Peabody Museum of Natural History, Yale University, USA

M. Higgins Formerly: Allott and Lomax Consulting Engineers, Sale, UK

Trevor B. Hoey Department of Geography and Topographic Science, University of Glasgow, UK

J. M. Hooke Department of Geography, University of Portsmouth, UK

Roger LeB. Hooke Deer Isle, Maine, USA

R. J. Huggett School of Geography, University of Manchester, UK

Robert P. Ilchik Geophysical Laboratory, Carnegie Institute of Washington, USA

Allan N. Insole Lower Cheltenham Place, Bristol, UK

Bjorn Jamtveit Department of Geology, University of Oslo, Norway

J. A. A. Jones Institute of Geography and Earth Sciences, University of Wales, Aberystwyth, UK

D. M. Jones Fossil Fuels and Environmental Geochemistry, University of Newcastle, UK

R. L. Jones Centre for Quaternary Science, Coventry University, UK

Tim P. Jones School of Biosciences, University of Wales, Cardiff, UK

P. Kearey Department of Earth Sciences, University of Bristol, UK

E. A. Keller University of California, Santa Barbara, USA

G. R. Keller Department of Geological Sciences, University of Texas at El Paso, USA

Lorcan Kennan Department of Earth Sciences, University of Oxford, UK

Barbara A. Kennedy St Hugh's College, University of Oxford, UK

Christopher J. Keylock Department of Geography, University of Cambridge, UK

J. C. King British Antarctic Survey, Cambridge, UK

Stephen King Formerly: Department of Oceanography, University of Southampton, UK

M. J. Kirkby School of Geography, University of Leeds, UK

Simon J. Knell Department of Museum Studies, University of Leicester, UK

John Knill Newbury, Berkshire, UK

S. C. Kohn Department of Earth Sciences, University of Bristol, UK

M. Krabbendam Australian Crustal Research Centre, Monash University, Australia

Simon Lamb Department of Earth Sciences, University of Oxford, UK

Nicholas Lancaster Desert Research Institute, Reno, USA

L. J. Lane USDA-ARS, Tucson, USA

Steve Larter Fossil Fuels and Environmental Geochemistry, University of Newcastle, UK

A. Lerman Department of Geological Sciences, Northwestern University, Evanston, USA

A. J. Lewis r3 Environmental Technology Ltd, Hicks Lane, Girton, Cambridge, UK

T. Linsey School of Geography, University of Kingston, UK

Richard J. Lisle Department of Earth Sciences, University of Wales, Cardiff, UK

P. Lonsdale Scripps Institution of Oceanography, University of California, San Diego, USA

A. Loy Paltec Ltd, Sheffield, UK

Norman Lynagh Norman Lynagh Weather Consultancy, Chalfont St Giles, UK

G. J. H. McCall Cirencester, Gloucestershire, UK

Danny McCarroll Department of Geography, University of Wales, Swansea, UK

Brian J. McConnell Hydrock Consultants Ltd, Bristol, UK

Christopher McDonald Golder Associates (UK) Ltd, Nottingham, UK

N. MacLeod Palaeontology Department, Natural History Museum, London, UK

John McManus School of Geography and Geosciences, University of St Andrews, UK

Conall Mac Niocaill Department of Earth Sciences, University of Oxford, UK

Judith Maizels Grant Institute of Geology, University of Edinburgh, UK

N. Mann Northwest Leicestershire District Council, Coalville, UK

Larry Mayer Department of Geology, Miami University, Oxford, Ohio, USA

Julian Mayes Environment, Resources and Geographical Studies, School of Life Sciences, University of Surrey Roehampton, UK

Maxwell A. Meju Department of Geology, University of Leicester, UK

A. Mellor Division of Geography and Environmental Management, University of Northumbria, UK

A. Michard Formerly: Laboratoire de Géochimie Isotopique, Centre de Recherches Petrographiques et Geochemiques, CNRS, France

T. Mighall School of Natural and Environmental Sciences, Coventry University, UK

John Milsom Department of Geological Sciences, University of College London, UK

M. Mishra Formerly: PRC Environmental Management, Chicago, USA

N. C. Mitchell Department of Earth Sciences, University of Oxford, UK

David R. Montgomery Department of Geological Sciences, University of Washington, Seattle, USA

Paul Montgomery Department of Geology, The University of Kansas, USA

G. Morteani Lehrstuhl für Angewandte Mineralogie und Geochimie, Technische Universität München, Germany

Bruce W. Mountain Institute of Geological and Nuclear Sciences, Wairakei Research Centre, New Zealand

Duncan Murchison Fossil Fuels and Environmental Geochemistry, University of Newcastle, UK

W. Murphy Department of Geology, University of Portsmouth, UK

John W. Murray School of Ocean and Earth Science, Southampton Oceanography Centre, Southampton, UK

Ted Nield The Geological Society of London, London, UK

E. G. Nisbet Department of Geology, Royal Holloway, University of London, UK

Patrick D. Nunn Department of Geography, The University of the South Pacific, Suva, Fiji

Éric H. Oelkers Laboratoire de Géochimie, Université Paul Sabatier, Toulouse, France

George F. Oertel Department of Ocean, Earth and Atmospheric Sciences, Old Dominion University, Norfolk, Virginia, USA

C. D. Ollier Geology Department, Australian National University, Canberra, Australia

Hamish J. Orr-Ewing Hydrock Consultants Ltd, Bristol, UK

Yoko Ota Yokohama National University, Yokohama, Japan

Lewis A. Owen Department of Earth Sciences, University of California, Riverside, USA

Colin F. Pain CRC LEME, Australian Geological Survey Organisation, Canberra, Australia

M. R. Palmer T. H. Huxley School, Imperial College, London, UK

Jeffrey Park Department of Geology and Geophysics, Yale University, USA

R. G. Park Formerly: School of Earth Sciences and Geography, Keele University, UK

R. John Parkes Department of Geology, University of Bristol, UK

Anthony J. Parsons Department of Geography, University of Leicester, UK

Julian A. Pearce Department of Earth Sciences, University of Wales, Cardiff, UK

R. B. Pearce School of Ocean and Earth Science, Southampton Oceanography Centre, Southampton, UK

Allen Perry Department of Geography, University of Wales, Swansea

C. Pescatore Formerly: OECD Nuclear Energy Agency, Issy Les Moulineaux, France

J. D. Phillips Formerly: Department of Geography and Planning, East Carolina University, Greenville, USA

Nicholas Pinter Department of Geography, Southern Illinois University, Carbondale, USA

G. Postma Department of Geology, Utrecht University, The Netherlands

Monica T. Price Mineral Collections, Oxford University Museum of Natural History, Oxford, UK

Neville J. Price Stamford, Lincolnshire, UK

K. Pye Department of Geology, Royal Holloway, University of London, UK

Dave Quirk Burlington Resources (Irish Sea) Ltd, London, UK

K. Vala Ragnarsdottir Department of Earth Sciences, University of Bristol, UK

V. Rajaram Formerly: PRC Environmental Management, Chicago, USA

Simon R. Randall Ingenica, London, UK

Dhananjay Ravat Department of Geology, Southern Illinois University, Carbondale, USA

Harold G. Reading Department of Earth Sciences, University of Oxford, UK

Stephen H. Richardson Department of Geological Sciences, University of Cape Town, South Africa

J. B. Ritter Department of Geology, Wittenberg University, USA

M. S. Roberts Natural Environment Research Council, Swindon, UK

Neil Roberts Quaternary Environments Research Group, Department of Geographical Sciences, University of Plymouth, UK

E. P. F. Rose Department of Geology, Royal Holloway, University of London, UK

Andrew J. Ross Department of Palaeontology, The Natural History Museum, London, UK

David A. Rothery Department of Earth Sciences, The Open University, Milton Keynes, UK

R. G. Rothwell Challenger Division for Seaßoor Processes, Southampton Oceanography Centre, Southampton, UK

Andrew J. Russell School of Earth Sciences and Geography, Keele University, UK

Paul D. Ryan Department of Geology, National University of Ireland, Galway, Eire

R. W. Sanderson New Malden, Surrey, UK

M. Sato US Geological Survey, Reston, Virginia, USA

Jill S. Schneiderman Department of Geology and Geography, Vassar College, Poughkeepsie, New York, USA

J. C. Schumacher Department of Earth Sciences, University of Bristol, UK

Andrew C. Scott Department of Geology, Royal Holloway, University of London, UK

Mike Searle Department of Earth Sciences, University of Oxford, UK

Roger Searle Department of Geological Sciences, University of Durham, UK

M. J. Selby Department of Earth Sciences, University of Waikato, New Zealand

Richard A. Shakesby Department of Geography, University of Wales, Swansea, UK

D. Sharrock Department of Earth Sciences, University of Oxford, UK

Douglas S. Sherman Department of Geography, University of Southern California, Los Angeles, USA

E. Shock Department of Earth and Planetary Sciences, Washington University, St Louis, Missouri, USA

John F. Shroder Jr Department of Geography and Geology, University of Nebraska at Omaha, USA

H. Catherine W. Skinner Department of Geology and Geophysics, Yale University, USA

Brian J. Skinner Department of Geology and Geophysics, Yale University, USA

Bernard J. Smith School of Geography, Queen's University, Belfast, UK

David E. Smith School of Natural and Environmental Sciences, Coventry University, UK

Michael R. Smith The Energy Network, Chalfont St Giles, UK

Frank D. Stacey CSIRO Exploration and Mining, Pinjarra Hills, Queensland, Australia

R. F. Stallard US Geological Survey, Boulder, Colorado, USA

Iain S. Stewart Department of Geography and Earth Sciences, Brunel University, Middlesex, UK

Stephen Stokes School of Geography, University of Oxford, UK

John Stone Department of Geography and Earth Sciences, Brunel University, UK

Graham Sumner Department of Geography, University of Wales, Lampeter, UK

Diana S. Sutherland Geology Department, University of Leicester, UK

Y. Tardy Institut National Polytechnique de Toulouse, France

K. O. Thomsen Formerly: PRC Environmental Management, Chicago, USA

M. J. Tooley School of Geography, Kingston University, UK

A. S. Trenhaile Earth Sciences, University of Windsor, Canada

Gregory E. Tucker School of Geography, University of Oxford, UK

Jonathan P. Turner School of Earth Sciences, University of Birmingham, UK

Simon D. Turner Department of Geography and Earth Sciences, Brunel University, UK

C. R. Twidale Department of Geology and Geophysics, University of Adelaide, Australia

H. A. Viles Department of Geography, University of Oxford, UK

P. Vincent Department of Geography, University of Lancaster, UK

C. Vita-Finzi Department of Geological Sciences, University College London, UK

John Wainwright Department of Geography, King's College London, UK

Tony Waltham Civil Engineering Department, Nottingham Trent University, UK

Andrew Warren Department of Geography, University College London, UK

R. Washington School of Geography, University of Oxford, UK

Neil A. Wells Geology Department, Kent State University, USA

J. West Department of Earth Sciences, University of Leeds, UK

W. Brian Whalley School of Geography, Queen's University, Belfast, UK

Alfred Whittaker British Geological Survey, Nottingham, UK

Giles S. F. Wiggs Sheffield Centre for International Drylands Research, Department of Geography, University of Sheffield, UK

P. B. Wignall Department of Earth Sciences, University of Leeds, UK

Bruce Wilcock Headington Quarry, Oxford, UK

Ben J. Williamson Department of Earth Sciences, University of Bristol, UK

Ian Wilson English China Clay International Ltd, St Austell, UK

James H. Wittke Department of Geology, Northern Arizona University, Flagstaff, USA

Ellen E. Wohl Department of Earth Resources, Colorado State University, USA

Ivan G. Wong Seismic Hazards Group, URS Greiner Woodward Clyde, Oakland, USA

Alan B. Woodland Mineralogisches Institut, Universität Heidelberg, Germany

Bruce W. D. Yardley School of Earth Sciences, University of Leeds, UK

Robert R. Young Department of Earth and Atmospheric Sciences, University of Alberta, Edmonton, Canada

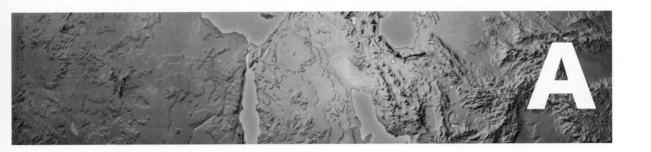

acid rain Acid rain became one of the most emotive environmental issues of the 1970s and 1980s, yet the presence of high concentrations of acids in the urban rainfall of Manchester, in comparison with surrounding rural areas, was identified in a systematic way over a hundred years ago by Angus Smith, an industrial chemist who was employed as the first Inspector of Factories. As a contemporary environmental issue, the acid rain problem came to the forefront of scientific and public awareness in Europe after the first United Nations Environment Conference in Stockholm in 1972 when 'evidence' for a number of detrimental environmental impacts of acid rain was presented. The impacts identified were a rapid increase in the acidity of European rainfall, a parallel increase in the acidity of Swedish rivers and lakes, a decline in fish populations in these rivers and lakes, and a decline in forest growth. In the decades since this conference, the scientific community has endeavoured to evaluate in a systematic way whether there is a sound body of evidence to substantiate these claims. Providing substantive evidence for the effects listed above has proved to be a challenging task for environmental scientists, not least because of the complexity in linking essentially gaseous sources to the environmental effects of acid rain. In consequence, an understanding of the acid rain problem needs to focus upon some of the fundamental principles and concepts involved, of which the first is 'What do we mean by the term "acid rain"?'

Natural rainfall contains many impurities, including dissolved gases, dusts, and salts which it picks up during its passage through the atmosphere to the ground. These constituents can come from a variety of natural sources, which include volcanic and biologically produced gases, sea spray and wind-blown dusts. Impurities in rainfall also come from a variety of polluting sources, such as fossil-fuel combustion and motor vehicles; these impurities include both gaseous and particulate emissions. In order to understand the effect that both natural and polluting substances have on the acidity of rainfall it is essential to start with an examination of the way in which acidity is measured.

Acidity is measured on what is called the pH scale. The scale was developed by chemists at the beginning of the twentieth century to provide a simple way of expressing the concentration of free hydrogen ions (H^+) in a solution. (Free hydrogen

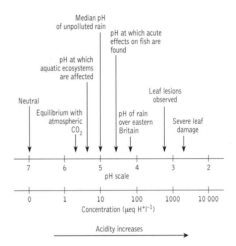

Fig. 1. The relationship between H^+ ion concentration and pH with some observed environmental impacts. (Based on Howells (1995), Fig. 1.2.)

is the hydrogen in a solution which is not part of the water molecule, H_2O.) The pH scale is potentially confusing to the non-chemist, since it is negative and is based upon logarithmic units. This means that the concentration of H^+ increases as the pH value falls. Starting at a neutral pH of 7, where there are no free hydrogen ions, each single unit decrease in pH means that there is ten times more hydrogen in the solution. Between pH 6 and pH 2, therefore, there is a 10 000-fold increase in concentration. The relationship between hydrogen ion concentration and pH is shown in Fig. 1.

One of the most common natural atmospheric gases is carbon dioxide, CO_2, which dissolves in rainwater to form carbonic acid, giving it a pH of around 5.6. The presence of other natural acids can reduce the pH of unpolluted rain to around 5. Dust in rain can change the pH by making it either more or less acidic according to the chemical properties of the dust. While natural rainfall is acidic, the presence of pollutant gases, particularly those containing oxides of nitrogen, sulphur, and chlorine, serves to increase the acidity. In the

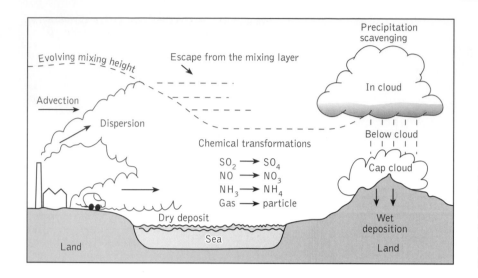

Fig. 2. The origin, dispersal, oxidation, and deposition of acid-forming gases in the environment. (Based on Fowler (1992) In *Air pollution transport, deposition and exposure to ecosystems*, pp. 31–51, (ed. J. R. Barker and D. T. Tingey), Van Nostrand Reinhold, New York.)

industrial regions of North America and Europe the presence of these dissolved gases generally increases the acidity by a factor of 10 (i.e. it produces rainfall with a pH of around 4).

A second fundamental issue in understanding the acid rain problem is to identify the types of pollutant gaseous emissions which produce acid rain and to identify how, and at what rate, acid rain is subsequently produced. As indicated above, the principal pollutant emissions leading to the formation of acid rain are oxides of nitrogen and sulphur. Sulphur is present in fossil fuels and is liberated by their combustion, especially in power generating stations and by the smelting of ores. Nitrogen oxides are generated during the combustion process in the internal combustion engine because of the presence of nitrogen in the air. The degree to which this process occurs depends on the temperature of the reaction. Some nitrogen gases, especially ammonia, are alkaline when dissolved in rain. As ammonia is oxidized, however, it will eventually contribute to acidity. Some 80 per cent of ammonia in the atmosphere is estimated to have come from livestock wastes. Chlorine is a third acid-forming gas, of which 75 per cent in Europe is estimated to come from the burning of fossil fuels. The rate at which sulphur, nitrogen, and chlorine are oxidized in the atmosphere to produce acid rain depends in part on the presence of volatile organic carbons that are also liberated by industrial processes. Furthermore, industry produces considerable quantities of carbon dioxide which influence the amount of carbonic acid present in rainfall. Emissions from point sources such as chimney stacks can be dispersed downwind for hundreds or even thousands of kilometres and are usually retained within an atmospheric layer less than a kilometre thick. The rate at which acids are produced from gases through the oxidation process is fairly slow, and it has been estimated that the conversion rate is only between 1 and 3 per cent per hour. These

gases and oxides are also diluted by as much as 10 000 times in the atmosphere as they disperse.

Environmental scientists have identified a number of ways by which the oxides produced by atmospheric conversion can reach the ground surface without necessarily involving rainfall. The oxides can exist in dusts which reach vegetation surfaces or the ground surface as dry particulate fallout, or as dissolved constituents which may reach the ground surface as wet fallout, which includes not only rain, but snow, mist, and low cloud. In the case of low cloud, acid droplets are deposited on vegetated surfaces by a process called 'occult deposition'. A generalized scheme showing the production, dispersal, chemical transformation, and deposition of acids is shown in Fig. 2.

It has been estimated that the emission of sulphur through the combustion of fossil fuels increased from 1850, when emissions were around 0.5 Mtonne (0.5×10^6 metric tonnes), to 1965, when emissions reached a peak of some 3.5 Mtonne. After 1965, global emissions declined sharply as a result of changing practices in industry and the use of cleaner fuels such as gas and nuclear power. While sulphur emissions have declined, nitrogen emissions have increased, particularly as a result of the increase in road traffic. Although there have been substantial reductions in sulphur emissions, there is little evidence to suggest that this is having a dramatic impact upon the acidity of rainfall.

In trying to link the acidity of rainfall to the acidity of soils and water it is essential to understand the role that soils play in releasing chemicals into solution as a result of natural weathering. Soils are made up of two major components: organic matter, which results from the decomposition of vegetation, and the physically and chemically altered parent material. In soils with a high content of organic matter, and in areas of high rainfall, water draining through the soil becomes more acidic with the release of organic acids. Some parent materials, especially those containing carbonate rocks, can

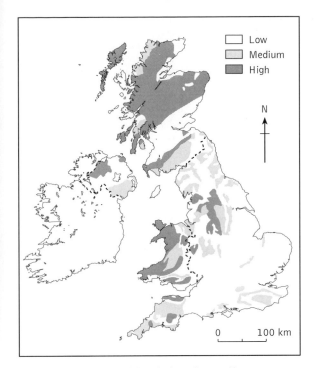

Fig. 3. Areas of the UK which have high, medium, and low susceptibility to the input of acid rain (Based on Foster (1991), Fig. 1.2.)

buffer this acidity and raise the pH of drainage waters. The pathways that water takes through the soil to the rivers are also important because they will determine whether or not the drainage water comes into contact with that part of the soil which can buffer the acidity. The sensitivity of different areas of the UK to acid inputs is highly variable. In regions with high rainfall, acid soils, and hard crystalline rocks, there is a potentially high susceptibility to the input of acid rain with little opportunity to buffer the input (Fig. 3).

The presence of hydrogen ions in rainfall, soil solutions, rivers, and lakes has a secondary impact on the environment through the release of potentially toxic metals, particularly aluminium, which is present in trace amounts in almost all soils. Aluminium is most toxic in the pH range 5–6. It has a number of known effects, such as impairing respiration in fish and possibly reducing the growth rate of phytoplankton communities in fresh water.

Vegetation is important for two reasons. First, it produces leaf litter, which decomposes in the soil to produce organic matter which in turn liberates organic acids. Some tree species especially conifers, are known to produce more organic acids than others. Scientific evidence obtained from studies of lake sediments in North Wales has, however, indicated that acidification began well before the planting of upland forests. It seems, nevertheless, that the rate of acidification increased

after forest plantation. Vegetation also plays an important role in trapping dusts and acid aerosols by occult deposition. The combination of these two processes, in addition to the release of organic acids from leaf surfaces, also serves to increase the acidity of water dripping through the forest canopy (as throughfall: see *hydrological cycle*). Some studies have shown that the pH of throughfall can be at least one unit lower than the acidity of rain as it reaches the canopy surface, that is, the concentration of hydrogen ions is ten times greater. The effect of acidity upon forest growth rates is unproved. Forest decline is not a specific disease with a single cause; many factors, such as acidity, climate change, and the preferential removal of essential elements like magnesium from soils, may all play a part in the decline in forest growth.

The body of scientific evidence accumulated since the 1972 United Nations Stockholm conference has improved our understanding of the acid rain problem. There are, however, many unresolved issues and only recently have scientists begun to turn their attention to the next major phase in acid rain research, which is to find out whether or not the observed chemical and biological changes to the environment are reversible and sustainable. IAN D. L. FOSTER

Further reading
Battarbee, R. W. (1988) *Lake acidification in the United Kingdom.* HMSO, London.

Foster, I. D. L. (1991) *Environmental pollution.* Oxford University Press.

Howells, G. (1995) *Acid rain and acid waters* (2nd edn). Ellis Horwood, London.

Regens, J. L. and R. W. Rycroft (1988) *The acid rain controversy.* University of Pittsburgh Press, Pittsburgh.

Steinberg, C. E. W. and R. F. Wright (eds) (1994) *Acidification of freshwater ecosystems; implications for the future.* John Wiley and Sons, Chichester.

The Quality of Urban Air Group (1993) *Urban air quality in the United Kingdom.* Department of the Environment, Bradford.

active fault An active fault is one that has recently moved in an earthquake. Despite this simple definition, active faults can be hard to identify, because reliable instrumental teleseismic data extend back only a few decades. Records of historical seismicity are far from complete and are often too imprecise as to the location and magnitude of events to allow past earthquakes to be associated with individual faults. Previous movements on an individual fault can sometimes be dated using archaeological evidence or by carbon-dating organic-rich sediments in stream channels offset by the fault (*palaeoseismology*). Even when a history of fault movements exists, it may not be a reliable predictor of future earthquakes on that fault, because earthquake recurrence intervals are not necessarily regular. Whether a given fault ruptures or not depends on the local state of stress in the Earth's crust. When the stress exceeds some threshold, rupture will occur which will relax the stress on the fault, and will increase the stress in some nearby regions and decrease it in others. This will bring neighbouring faults closer to or further from the failure threshold and will thus alter their recurrence intervals.

Faults that have remained active for some time are commonly revealed by the topography but this topographic expression can be very subtle, particularly for low-angle thrust faults, and there may be no fault scarp as such. Recently established faults or older faults that do not reach the surface (blind faults) may not have any topographic expression, especially if erosion rates are high compared with rates of fault growth. As faults grow by repeated earthquakes increasing the fault offset, their effects on the topography will increase with time, as will the lengths of the faults. Around each end of the fault lies the *process zone* into which the fault is propagating as it grows. At first, the deformation in this zone is pervasive, but with repeated earthquakes it will become more localized on to the fault plane and deformation away from the fault will be reduced. Eventually, movement on the fault may cease and it will become inactive. Other faults more optimally aligned to the regional stress field will instead accommodate the deformation.

Longer faults are generally composed of several *fault segments*, each about 10–30 km in length. Segments are often arranged *en echelon*, that is, each one is slightly offset from the line of its neighbours. Not all the segments of a fault will necessarily fail during a single event. Crustal earthquakes smaller than about magnitude 6.5 usually correspond to a single segment; larger crustal events tend to involve several segments and therefore have a more complex pattern of seismic wave radiation. The segmentation of an active fault and its relationship with neighbouring faults has a strong influence on the drainage pattern around the fault. Major river channels are normally diverted round the ends of the dominant, most active faults in the region or between their segments, but stream channels may cross less active faults and be offset by them.

PETER CLARKE

adsorption

adsorption Dissolved constituents in groundwater are attracted to the surfaces of minerals lining the pores through which the water flows. This process is generally referred to as *adsorption*. The dissolved constituents can be of natural origin, e.g. derived from minerals through dissolution by the coexisting water. They can also be of anthropogenic origin, introduced into the environment by a variety of industrial processes.

Dissolved constituents are generally present as charged ions. They have a positive charge (such as the metal cation Pb^{2+}) or a negative charge (e.g. arsenate, $AsO_4{}^{2-}$). Similarly, the mineral surfaces have an electric charge because the metal ions (Me) forming the structure of the minerals are not fully coordinated with oxygen at the surface of the mineral (>Me-O⁻). The surface is then made neutral by a H⁺ ion from the water (>Me–OH). This mineral surface charge is further affected by the acidity (pH) of the groundwater. In general, minerals have a positive charge in water that has high acidity (a high concentration of H⁺ or low pH) and a negative charge in waters that have a low acidity (a low concentration of H⁺ or high pH). The mineral surface generally has a zero charge in neutral waters (intermediate pH). Taking the surface of the mineral feldspar as an example, we can represent its surface as $>MeOH_2{}^+$ at low pH and as >Me-O⁻ at high pH (> denotes the mineral surface). It follows that positively charged metal cations (e.g. Pb^{2+}) adsorb to feldspar at high pH (>Me-OPb⁺) and that negatively charged ions (anions), such as arsenate, adsorb to feldspars at low pH (>Me-OH₂AsO₄⁻). Understanding mineral surfaces and water pH is thus of prime importance for predicting how polluted water will evolve. Adsorbed ions form either strong direct bonds with the surface (inner sphere complex, e.g. Pb^{2+}) or weak electrostatic bonds (e.g. Na⁺). Weaker electrostatic bonding is a result of Na⁺ being firmly coordinated with four water molecules ($Na(H_2O)_4{}^+$) which act as a 'shield' to prevent direct bonding.

K. VALA RAGNARSDOTTIR

aeolian processes

aeolian processes The wind moves large quantities of sediment around the globe. It delivers to the oceans about one-tenth the amount of sediment that is taken by rivers: the North Atlantic in the Trade Wind belt receives $100–400 \times 10^{12}$ grams per year (g yr⁻¹) of dust. In the Sahara, the wind is the dominant erosive agent, removing far more to the surrounding seas than the rivers Niger and Nile in combination, and these are the only significant river outlets. In southern Israel, people can wipe up as much as 0.25 kg m⁻² yr⁻¹ of dust from their balconies, and as much as 8.3 g m⁻² after a single storm. In south-eastern Mongolia there are over 300 dusty occasions in the year, compared with only about 27 in the Negev. In terms of the amount of sediment moved across a standard width, say a kilometre, the wind can move sediment (in sand dunes) at far greater rates than any other process.

Grains are moved when the wind 'shears' the Earth's surface. One process is 'lift' by the Bernoulli effect: because the velocity increases rapidly away from the surface, pressure is lower on the top than the bottom of a particle, helping it to rise. Acceleration of the wind over the protrusion further accentuates lift, but even in total, lift is important only very close to the bed and raises grains only slightly. Shear also causes 'drag', which tends to roll or slide the grain, but even with lift, drag is seldom sufficient to cause much movement. Only when there is also 'bombardment' of the surface by grains already entrained does movement begin in earnest. At the start of movement there is no bombardment, but movement rapidly accelerates as it comes into play.

Not all winds raise sand or dust. Above about 100 micrometres (μm) grain diameter, higher velocities are needed to move larger and larger grains, but below that critical size smaller particles (mostly clays) require greater velocities to mobilize smaller and smaller grains. The reason for this break in behaviour is cohesion, which is more important between fine than coarse grains. In the field, other factors, such as grain sorting, roughness, shape, soil moisture content, algae, and bacteria, have considerable effects on entrainment.

Once in movement, grains travel in four ways. In increasing order of velocity, and affecting increasingly finer sediments, these are *creep, reptation, saltation,* and *suspension.* 'Creep', or surface movement at rates of about 0.005 m s^{-1}, is actually a number of processes: rolling under bombardment; rolling into craters created by bombardment; differential loss or accumulation of different size fractions; preferential movement of coarse particles to the surface when a mixed-size sand is shaken by bombardment; movement of descending grains after they hit the surface and tunnel along just beneath it; consolidation by bombardment; and elevation of grains to positions where they are vulnerable to dislodgement. Creep generally occurs in coarse sands, but can occur even in small pebbles where winds are very strong.

The next two processes occur mostly in sands. *Reptation* is the splashing of grains dislodged by bombardment. Most sand grains in motion at any one time are reptating. *Saltation* is a process in which particles of sediment travel within a few centimetres of the surface, except where they are projected off the lee of a dune, or in high turbulence (see *saltation*). A severe sand storm in the San Joaquin Valley of California, for example, imbedded particles of 23 mm diameter into a wooden telegraph pole at 4.9 m above ground.

Grains in suspension follow the turbulent motion of the air itself. Suspension mostly affects dust, although when winds are particularly turbulent, as in the lee of coastal dunes, sands may also go briefly into suspension (see *dust*).

Wind erosion

Without grains, the wind can accomplish very little, but armed with grains it can be a very effective agent of erosion. Wind erosion is generally ineffective on vegetated or wet surfaces, although between sparse bushes there can be intense localized erosion, and on quickly drying surfaces, such as beaches, the wind can liberate large quantities of sand. These cases aside, wind erosion is most effective in deserts, or on cleared agricultural land.

In deserts (and in some dry periglacial environments), the wind is most effective on sandy river outwash and on loose lake sediments. Here it leaves characteristic residual landforms or 'yardangs', which range in size from small lumps, a few metres across, to the massive 'mega-yardangs' on the Tibesti region in the Sahara, which are ridges many kilometres long and spaced up to 2 km apart. 'Pans', or shallow hollows, seasonally filled with water, range from small indentations to large lakes a few kilometres across, and are formed by wind erosion in conjunction with salt weathering and shoreline erosion. They are characteristic of parts of eastern and western Australia and of South Africa, where they may reach densities of 100 per 100 km^2.

Wind erosion is a much more serious issue on agricultural fields, even in wet parts of the world, such as the English midlands. In the USA it is estimated that each 2 cm of soil eroded reduces crop yield by 6 per cent, and this kind of loss is not unusual. The most notorious occurrence of wind erosion was the 'Dust Bowl' of some parts of the dry Mid-West of the United States in the 1930s. There were, however, even more severe events in the Soviet steppes in the 1950s and 1960s, and even in 1973–4, 50 000 acres of the ploughed land on the Great Plains suffered losses of between 40 and 380 t ha^{-1} yr^{-1}. Various techniques have been introduced to alleviate the problem: windbreaks and strip ploughing across the path of the prevailing wind are the most widely used, but the pressure to produce has meant that some losses are almost unavoidable.

Dunes

Dunes are collections of loose sand built piecemeal by the wind. They can be anything from a few metres across and a few centimetres high to 2 km across and 400 m high. They occur in arid deserts devoid of vegetation and on coasts, where they grow in the presence of vegetation. The processes by which a wind, blowing freely over a desert plain, forms dunes are not entirely clear, but are thought to involve interactions between the plain and the flow of sand in which regular turbulent patterns are set up. Dunes display a number of shapes. The most common is a simple accumulation of sand round a low bush, known as a 'nabkha'. Most larger free dunes can be classified into three types (although there are many intergrades and exceptions): 'transverse' ridges at right angles to the prevailing wind (including 'barchans', some in near-perfect crescents); linear (thought by most geomorphologists to have been formed by two or more common winds in an annual regime blowing from directions at acute angles); and 'network' or 'star' (formed in wind regimes where winds blow from a number of directions throughout the year). Desert dunes are generally accumulated in 'sand seas' and dune fields; the largest is the Rub' al Khali (the 'Empty Quarter') in Saudi Arabia, which forms part of an area of 770 000 km^2 of continuous dunes (bigger than Texas); there are about 50 comparable, if somewhat smaller, sand seas and many thousands of smaller dune fields. Coastal dunes generally form ridges just behind the beach, where plants trap the sand. These ridges may then be eroded by the wind into elliptical hollows ('blowouts') or larger, elongated hollows with accompanying ridges ('parabolic' dunes).

Aeolian processes in the past

The Earth is at present in a calm period; parts of the Pleistocene, particularly the Last Glacial Maximum, were very much windier, and the most spectacular aeolian features are inherited from that time. Examples include the largest desert dunes, mega-yardangs (see above); loess deposits (some 400 m thick) which cover about 10 per cent of the terrestrial Earth, as in large parts of Europe, central Asia and China, and North America, which now form some of the world's most productive soils; and the sand dunes, now fixed by grass and trees, that cover parts of the Sahel and the Kalahari in Africa, a large part of Hungary and central Poland, and about one-third of the state of Nebraska in the USA, to name only a few places. The high winds of the Pleistocene were also the main contributor to the huge thicknesses of dust on the ocean floors, and to dust incorporated into the polar ice caps, where the varving they produced gives some of the best evidence for the climatic fluctuations of the Pleistocene and Holocene.

ANDREW WARREN

Further reading

Livingstone, I. and Warren, A. (1996) *Aeolian geomorphology: an introduction.* Addison-Wesley Longman, Harlow.

Cooke, R. U., Warren, A., and Goudie, A. S. (1992) *Desert geomorphology.* UCL Press, London.

aeromagnetic surveying Aeromagnetic surveying is one of the most cost-effective geophysical methods available for rapidly understanding the complex geology of continental regions. Both fixed-wing aircraft and helicopters are used in this type of surveying. The magnetic effects of aircraft components are minimized either by carefully balancing their effect at the location of the sensors or by towing the sensors in non-magnetic compartments called 'birds' (the corresponding tows in marine surveys are called 'fish'). General aspects of magnetic surveying are discussed in the entry on *geomagnetic measurement: techniques and surveys.*

The technique evolved during the Second World War when the US Navy developed a magnetometer to detect submarines. Over the years, massive data sets have been compiled covering large areas of the continents (for example, North America, Europe, Australia, and large parts of Asia, Africa, and South

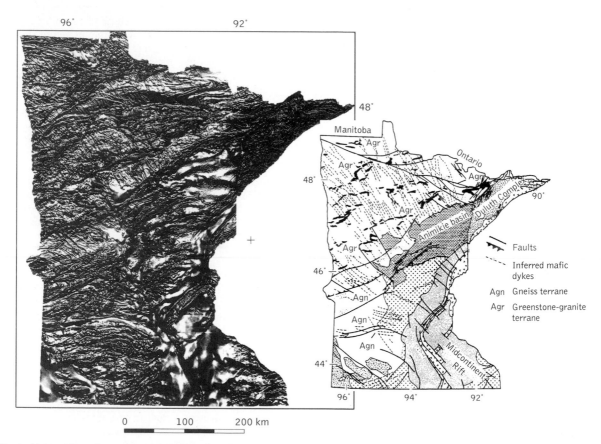

Fig. 1. (a) Aeromagnetic map of the state of Minnesota, USA, acquired with a flight-line spacing of 400 m and a flight elevation of 150 m. (b) Simplified geological map of the Precambrian rocks of Minnesota. The extent of different geological provinces and structural discontinuities (e.g. north-west trending dykes) are readily discernible from the differences in textural pattern of the aeromagnetic data. (Courtesy of V. Chandler and the Minnesota Geological Survey.)

6

America). With the relative ease of positioning afforded by the advent of global positioning systems, aeromagnetic surveying of the oceanic regions is simplified.

For the continents, there is an increasing trend toward maximizing the information content by surveying with small flight-line spacing at low altitude. Such high-resolution aeromagnetic surveys are able to map not only the boundaries of geological units and provinces (Fig. 1), but also aid earthquake research by mapping even small changes in magnetism along near-surface fault zones and igneous dykes. Survey characteristics similar to, or better than, the specifications in Fig. 1 are used in many countries to map the anomalies of interest for the assessment of natural resources, the evaluation of natural hazards, and environmental management.

D. RAVAT

aerosols and climate Aerosols are solid and liquid particles suspended in atmosphere. Their radii range in size from 3 microns (μm) to particles the size of cloud droplets 10 000 times larger. Most aerosols which are able to nucleate cloud droplets are smaller than 0.1 μm and are called *Aitken nuclei*. Larger nuclei, with radii of 0.1–1.0 μm, are 10 000 times less numerous than Aitken nuclei, but constitute nearly half the mass of aerosols.

Aerosols are produced through natural and anthropogenic combustion, decay of organic matter, and volcanic eruptions. Substantial portions arise from the conversion of sulphur dioxide (SO_2) gas to sulphate aerosols, of which more than half is anthropogenic. In continental air, concentrations are approximate 10^6 per litre, with a strong gradient of concentration from a maximum near the surface. Spatially, concentrations are highest over industrial areas (eastern North America, central Europe, and eastern Asia) or where biomass is burned (South America, Africa, and Asia). Water-soluble inorganic species include sulphate, nitrate, and ammonium. The most-studied aerosol is SO_2, which has been characterized by increased emissions since the mid-1950s, following the building of tall stacks which released aerosols above the stable boundary layer. This led to longer residence times and hence higher concentrations. There is also a climatologically controlled seasonal cycle, with global mean aerosol loading reaching a maximum in February and a minimum in October.

Aerosols influence climate to cool the Earth in two ways. Over the oceans, where naturally occurring cloud condensation nuclei (CCN) are less abundant, aerosols act as an important source of CCNs, increasing droplet longevity but also the development of smaller but brighter clouds and hence increased albedo and thus an increase in the amount of solar radiation reflected to space. This role is thought to be critical in the case of marine stratus clouds which cover 25 per cent of the globe; a relative change of CCN of 30 per cent in marine stratus cloud yields a global change in heat balance of 1 watt per square metre (W m^{-2}). This is known as the indirect influence on the radiation budget and remains to be quantified, since present values are merely based on assumed values of anthropogenic CCN enhancement; there are few data on actual CCN populations and poor understanding of what controls production of new CCNs. It is known that the indirect forcing is not simply linearly dependent on the source strengths of aerosol precursor gases such as SO_2, ammonia (NH_3), and nitrogen oxides (NO_x). The way forward lies in describing the evolution of aerosol size distribution in space and time.

The direct effect of aerosols relates to the interaction between the aerosol and radiation, which includes scattering and absorption of solar radiation and absorption of infrared. Aerosols at the surface have a modest effect on long-wave emission through increased absorption of this energy, but may have a substantial effect on the receipt of solar absorption when they overlie a dark surface in a cloud-free area (reflection of as much as 4 W m^{-2}, averaged to 1 W m^{-2} globally). Light-absorbing carbon, for example, has two effects, heating the Earth and heating the atmosphere, the latter changing the vertical temperature distribution and hence stability and convection. Even if the influence on the heat budget is small, this vertical rearrangement might have separate physical consequences. *Clear sky forcing* is the direct backscatter of solar radiation in clear skies. This depends on aerosol mass scatter efficiency and the lifetime of the compound. Overall, aerosols lower the planetary albedo above a high-albedo desert or snow surface, but they increase it over an ocean surface.

The study of aerosol–climate interaction has recently generated much interest since it has been shown through general circulation model (GCM) experiments that a substantial portion of the anthropogenic carbon dioxide-induced (CO_2) warming is hidden by the sulphates. Research may be conveniently divided into theoretical studies (largely GCM experiments) and empirical/historical studies of major volcanic eruptions and their attendant influence on global temperatures. The latter studies are rendered difficult for several reasons, amongst them: the problem of showing a significant relationship with the high natural variability in temperature records; the difficulty of estimating the amount of aerosols emitted in both contemporary and historical eruptions; inaccuracies in assessing the global transport of aerosols; and the difficulty of converting the mass of aerosols to a radiative forcing in W m^{-2}. Nevertheless, reduced volcanic activity after 1914 may have contributed in part to the early twentieth-century warming. New interest has been aroused by the eruption of El Chichon (March 1982) and Mount Pinatubo (June 1991), the latter leading to a global negative forcing of 4 W m^{-2}.

GCMs have offered an alternative tool in studying the effects of aerosols, since these numerical techniques, unlike the real atmosphere, allow one variable to be changed at a time. GCM experiments forced by changing CO_2 alone suggest that the increase in greenhouse gases should have produced a larger warming (0.6–1.3 °C over the last century) than has been observed. The Hadley Centre coupled GCM of the UK Meteorological Office gives a warming of 0.5 °C, with both CO_2 and aerosol forcing, which is close to the observed values. If aerosol forcing is excluded, the increase in modelled temperature is too large. The 1991 eruption of Mount Pinatubo provided a unique opportunity to study the effects of aerosols. Modelling studies undertaken soon after the event predicted a global cooling similar to that which has subsequently been observed, a result which improves the credibility of climate models.

Overall, however, aerosols still pose the largest single uncertainty in calculating the net forcing due to anthropogenic changes in the chemical composition of the atmosphere. Their influence may be vastly different from that of other pollutants, such as CO_2 and ozone, since their residence times are very short, some close to 2 weeks. The aerosol cooling effect could be reduced with a ban on burning of fossil fuels.

R. WASHINGTON

Further reading

Meteorological Office (1991) *Meteorological glossary*. HMSO, London.

Agassiz, Louis (1807–73)

Agassiz, Louis (1807–73) Louis Agassiz (christened Jean Louis Rodolphe), who was to become the foremost of the early proponents of the idea of a recent ice age, was born into the family of a Swiss Calvinist minister. He attended courses at several universities, graduating as a doctor of philosophy at Erlangen and as a doctor of medicine at Munich.

His early scientific leanings were towards zoology and palaeontology and led him to friendship with the famous French comparative anatomist the Baron Georges Cuvier in Paris. Proving a highly intelligent, hardworking, and productive scholar, he was in 1832 appointed Professor of Natural History at the University of Neuchatel. For some time Agassiz had studied both fossil and extant fishes. When only 22 years old he completed a pioneer study of fishes from South America. It was followed by a decade or more of documenting fossil fishes that culminated in his *Recherches sur les Poissons Fossiles* (1833–44). This was not only a description of remains but also a vivid account of the fishes when alive. It was widely acclaimed and it profoundly boosted the study of ancient life.

Agassiz then took up a new interest: the superficial deposits and landscape features of Switzerland and Germany that were attracting attention as being possibly related to a previous much wider extent of the alpine glaciers. This activity culminated in 1840 in his *Études sur les glaciers*, in which he was able to show that Switzerland had recently been covered by a vast ice cap, and from which meltwaters carried far and wide great spreads of sand, gravel, and huge erratic boulders. The thesis brought its author to the immediate notice of European and American geologists.

Two years after a study visit to the USA in 1846, Agassiz accepted a professorship of zoology at Harvard University. Now began an extraordinarily productive period during which he published numerous zoological works, collected specimens in Brazil and California, and set about establishing at Harvard a comprehensive zoological museum. During a period of about 25 years there he had a unique reputation as an inspired teacher. He was very much what today is known as a field man, emphasizing out-of-doors experience and study.

Agassiz was beyond question one of the most able, wise, and well-informed biologists of his day. Although he was a contemporary of Charles Darwin, he seems to have been little influenced by Darwinism. Indeed, he misunderstood parts of Darwin's work on evolution. He nevertheless made a superb contribution to our understanding of biology and the Agassiz Museum of Comparative Zoology is a worthy memorial to his outstanding abilities.

D. L. DINELEY

age and early evolution of the Earth and Solar System

age and early evolution of the Earth and Solar System One of the questions that geologists are most frequently asked is 'How do you *know* the age of the rocks?' The early geologists were aware of the fact that there was a regular order of superposition in sedimentary strata (for example, Nicholas Steno (1638–87), and to William Smith (1769–1839) is attributed the recognition that one can use the fossil assemblages contained in the rocks to identify individual formations, because fossil assemblages are different in rocks of different ages and do not repeat themselves. The early geologists, however, had no idea of the magnitude of the time-spans represented by rock sequences. In the nineteenth century, the definitions of the eras and systems of the geological column were based on fossils and sequences (see *stratigraphy*). Such definitions leant heavily on the work of Sedgwick (1785–1873) and Murchison (1792–1873), but there was still no absolute timescale.

The earliest attempt to determine the age of the Earth was by Bishop Ussher, who in 1654, on the basis of the Scriptures and reportedly the misidentification of a crinoid fossil as an ear of corn, dated the creation at 26 October 4004 BC at the 'sensible hour of 9 a.m.' Buffon in 1749 estimated that at least 75 000 years were required to produce the known fossil-bearing strata. In the 1860s Lord Kelvin used the cooling rate of the Earth to arrive at an estimate of 98 million years, with lower and upper limits of 20 million and 400 million years. John Joly, in 1899, used the saltiness of the sea to arrive at an age of 80 to 90 million years for the Earth. Although we now know that the methods used were flawed, Kelvin's estimate was hailed as 'established by the laws of physics'. Nevertheless, T. C. Chamberlin in the United States argued that it was incorrect and so the laws of physics must be wrong.

The discovery of radioactivity in 1896 supplied a means of calibrating Earth history, and in 1907 Boltwood showed that a figure of at least 400 million years (and possibly 2000 million years) was more correct. Various new methods of radiometric dating have since been developed, based on the decay of elements (for example U–Pb, K–Ar, Rb–Sr, Rh–Os, Sm–Nd, radiocarbon, fission tracks). The accepted age of the Earth rapidly increased to 3.5 billion years and we now accept 4.5 billion years as the likely age of the Earth. Radiometric dating requires no more than the application of a simple equation, but the rock must be impermeable, and have lost or gained neither daughter nor parent isotopes. Nowadays radiometric dating can be effected with great precision using sophisticated instruments, such as the ion probe. As a result, progressive metamorphic changes (caused by heating and pressure), diagenetic changes (caused by burial), and stages in igneous differentiation can be dated in rocks. The entire geological column can in fact be calibrated (Fig. 1).

The development of the Solar System

It is almost universally accepted that the Sun, the planets, their satellites, and the asteroids grew from a cloud of gas and dust, contracting under its own gravity. The cloud originally had some degree of rotation, so, as the centre contracted, the conservation of angular momentum forced the rest of the cloud into a flattened disc, the Solar Nebula, rotating in the same plane as the Sun. This process probably occupied about 10^4 years. Nucleosynthesis had already started by the time this stage was completed and the Sun shone brightly; observations of very young stars support such an evolutionary scenario. Material was then lost to interstellar space and condensation also occurred, producing the solid bodies of the solar system. The density of the nebula decreased because of these processes; heat radiated away more easily and cooling occurred. The planets condensed and aggregated from the nebular material.

The solar abundances of elements (measured spectrometrically) and the abundances in primitive meteorites (measured analytically) enable us to estimate the composition of the original nebula. Most stony meteorites are composed of rounded microscopic lithic particles (*chondrules*), made up of common minerals such as olivine and pyroxene, and these are widely held to be condensation globules, although only in primitive meteorites such as carbonaceous chondrites have they not suffered secondary modification. The chondritic meteorites can be radiometrically dated like igneous rocks, and they are believed to have been formed over a period of 100 000 years or so by condensation 4.5 billion years ago. This is taken as the age of Solar System (and also the age of the Earth).

The favoured scenario for the Solar System requires that bodies further from the Sun will contain progressively more volatile material, and this appears to be the case; there is evidence of water-ice in the outer planets and their satellites. There are two models for the accretion of the planets: slow/heterogeneous and rapid/homogeneous. The answer probably lies between these extremes.

Astronomical evidence of the inclination of orbits favours condensation for the satellites of planets, not capture; but Triton, Jupiter's largest satellite, has a retrograde orbit and might have been captured from somewhere in the outer Solar System; and there may be other captured satellites.

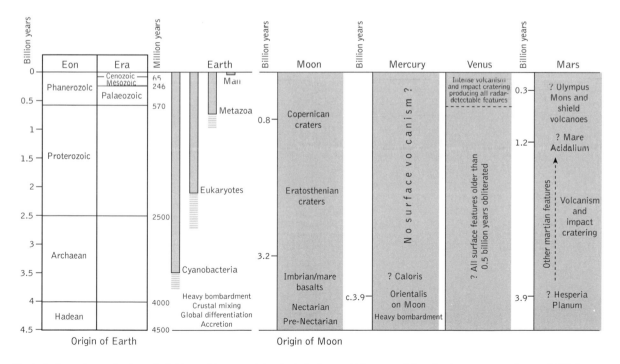

Fig. 1. Radiometric ages for the main divisions of the terrestrial geological column; widely accepted dates for the appearance of terrestrial life forms; and absolute and speculative dates for the Moon, Mercury, Venus, and Mars.

Age calibration for the Moon, Mercury, Venus, and Mars

We have good calibration based on radiometric dating for the lunar surface rocks; and, surprisingly, radiometric dating puts the main cratering events as between 3.92 and 3.17 billion years, and the oldest Pre-Nectarian rocks are dated at 4.17–4.54 billion years. After 3.2 billion years, by which time most of the lunar surface had been formed, the calibration is weakly based. The cratering on the Moon is widely held to be due to impacts.

For Mars, a stratigraphy has been based empirically on crater counts. The oldest rocks of Hesperia Planum are believed to be 3.9 billion years old, various ages suggested down to Mare Acidalium at 1.2 billion years, and an age as young as 300 million years has been suggested for Olympus Mons and the other shield volcanoes. There has, however, been no radiometric dating except on meteorites supposed to have come from Mars and these 'best ages' derived from crater counts may be widely out. There may be a link between the suggested age of Olympus Mons and the shield volcanoes and the age determined radiometrically for the Shergottite meteorites, believed to come from Mars, but determinations on such meteorites provide little in the way of support, or otherwise, for the suggested ages derived from crater counts.

For Venus, again, an empirical stratigraphy has been proposed which is based on the superposition of radar-detected features, but a major surprise is that the global distribution of craters considered to be due to impact cannot be distinguished from a completely random spatial distribution, and the global crater retention age may be no more than 300–550 million years. It is therefore believed that no geological units from the first 80–90 per cent of the history of this planet, heavily influenced by volcanic activity, remain at the surface.

For Mercury, prolifically cratered, it is likely that the cratering history and age of surfaces are not greatly different form those of the Moon, but the suggestion that the post-Orientale (Moon) and post-Caloris (Mercury) surfaces are of the same age (3.8 billion years) is highly speculative in the absence of any radiometric rock dating on Mercury.

For the outer planets and their satellites, all that can be said is that for some of their satellites superposition sequences have been derived.

The Solar System is only a minute fraction of the Universe. The size and age of the expanding Universe is calculated by astronomers on the basis of winking stars called *cepheids*, the nearest of which is 1000–2000 light years away. Recent estimates suggest a figure of 11–12 billion years. This, however, takes one back only to the 'Big Bang'. There is no evidence of what, if anything, preceded the Big Bang.

The early history of the Earth

We can now consider the early history of the Earth after accretion and condensation. There are four strands to this: the early history of the rocks, the origin of the oceans, the origin of that atmosphere, and the origins of life. The Earth evolved into a planet with many of its present properties in a very short time,

geologically speaking: about 500 million years. The oldest known mineral grains in rocks are zircons (for example those 4.10–4.27 billion years old from Mt. Narryer, Western Australia) and the oldest known whole rocks exposed on the Earth's surface—familiar gneisses—are the Acasta gneiss in the Slave Province, Canada, which are about 4.0 billion years old.

There are many models for the 'Hadean' period between 4.5 and 4.0 billion years, of which we have no significant record, but it is reasonable to invoke an intense early bombardment, as for the Moon, and the evolution of a zoned body by gravitational separation—solid inner and liquid outer core, dense and very hot; a molten mantle derived from chondritic silicate material; and a primitive, highly mobile surface layer, perhaps a magma ocean, with small sialic proto-continents segregating as a result of vigorous mantle convection (there are, however, other models for the separation of the sialic crust, such as the one advanced by Grieve, who invokes subsidence of large impact basins and partial melting of basaltic volcanic rocks). Whatever model one espouses for accretion and the early separation of the Earth shells, the early segregation of a proto-water ocean and atmosphere is consistent with the 10–20 per cent water and volatile content of carbonaceous chondrites, the favoured chondritic parent material, but the first atmosphere was reducing or only very weakly oxygenic. The first sialic proto-continents were certainly in place by 4 billion years ago, but continents probably only made up 5–10 per cent of the Earth's surface up to 3.2 billion years ago, when larger areas were cratonized (for example, Kaapvaal, South Africa; Pilbara, Western Australia). The Archaean, up to 2.5 billion years ago, appears to have been characterized by numerous smaller and faster convection cells in the mantle: the larger cells of present-day plate tectonics probably appeared only after that date. The oldest known sediments recorded are metamorphosed sandy sediments (now metaquartzites), iron formations, and clayey sediments (now paragneisses); evaporites (analogous to present-day salt lake and sabkha deposits) 3.484 billion years old are recorded at North Pole in the Pilbara, Western Australia. Banded iron formations are a feature of early sedimentary sequences, the oxygen probably being derived from photosynthetic organisms and being taken up in the oceans at the time when a reducing atmosphere prevailed.

How life originated we do not know, but primitive unicellular cyanobacteria (prokaryotes) are first seen as fossils in rocks 3.5 billion years old (Onverwacht, South Africa; Warrawoona, Western Australia). They had a long innings, for eukaryotic phytoplankton appeared only about 2 billion years ago when the atmosphere became fully oxygenic, and the Metazoa, multicellular animals, appeared about 600 million years ago in the last 100 million years of the Precambrian. The Precambrian has poor fossil preservation, and these established age limits of fossil preservation do not preclude a somewhat earlier existence of these three life forms; possible metazoan fossils are recorded from rocks 1.3 billion years old.

G. J. H. MCCALL

Further reading

Carr, M. H., Saunders, R. S., Strom, R. G., and Wilhelms, D. E. (1984) *The geology of the terrestrial planets*. NASA Scientific and Technical Information Branch, Washington, DC.

Eriksson, K. A. (1995) Crustal growth, surface processes, and atmospheric evolution of the early Earth. In Coward, M. P. and Ries, A. C (eds) *Early Precambrian processes*, pp. 11–25. Geological Society Special Publication No. 95.

McCall, G. J. H. (1996) The early history of the Earth. *Geoscientist*, **6** (1), 10–14.

Rothery, D. A. (1992) *Satellites of the outer planets*. Clarendon Press, Oxford.

Schopf, J. W. (ed.) (1992) *Major events in the history of life*. Jones and Bartlett, Boston.

Van Amdel, T. H. (1994) *New views on an old planet*. Cambridge University Press.

aggregates Aggregates are particles of rock that are used in bound or unbound form as construction material. They consist of particulate rocks that vary in size from sand to pebbles and cobbles. There are two major types of aggregate: natural and processed. *Natural aggregates* are mostly derived from unconsolidated glacial, river, and marine sand and gravel deposits, which are primarily of Quaternary age. They are quarried from surface and subsurface deposits, dug from river and beach deposits, and dredged from offshore sediments. Natural aggregates need minimal processing, such as separation into appropriate grades, to produce the end-use material. *Processed aggregates* are derived from stone, which is quarried or mined and is then crushed to provide the grade required. Crushed aggregates are generally coarse, consisting of rock particles with a diameter of more than 5 mm. They are typically produced from limestone, crystalline rocks (including granite and dolerite), and sandstones. By-products such as blast-furnace slag, broken rubble, and pulverized fly ash from power stations are also used as aggregates. Aggregates are of high economic importance, representing the greatest volume of any geological material extracted in most countries. They are used in the construction industry, either alone or as additives to cement to form concrete.

K. VALA RAGNARSDOTTIR

Further reading

Bennett, M. R. and Doyle, P. (1997) *Environmental geology: geology and the human environment*. John Wiley and Sons, Chichester.

Agricola, Georgius (1494–1555)
Georgius Agricola, the 'father of mineralogy' was born Georg Bauer in Glaucau, Germany but, being educated in the classics and classical languages before studying medicine, used a Latin name in all his writings. After studying at Leipzig and in Italy he returned to Joachimstal in Bohemia as City Physician in that important mining community. There he took a keen interest in minerals, mining, and the miners. In 1534 he moved to another mining town, Chemnitz, where he held various official and civic posts. His books are the first to be devoted to the mining and smelting industries, although he also wrote on many scientific topics.

His first book was a brief introduction to mineralogy and mining. Later the mineralogical part was much expanded in his *De natura fossilium*. Various kinds of rocks, minerals, and fossils were described; volcanic heat was ascribed to the subterranean combustion of coal. Further geological ideas were expressed in *De ortu et causis subterraneorum*, which dealt with such matters as erosion by frost and running water and an attempted classification of ore deposits. Finally, in *De re metallica* he summarized the state of the art of mineral extraction, smelting, and use of metals. It became his best-known work and was enlivened by many excellent woodcut illustrations. This, like his other books, was written in Latin, many of the technical terms being invented by himself. It was to influence both the discussion of geology and mining matters in Europe for the next two centuries.

D. L. DINELEY

air motion, vertical *see* VERTICAL AIR MOTION

air pressure
The atmosphere is made up of a mechanical mixture of gases, which all have mass. Thus a column of air has a weight. If this weight is measured over a unit surface area, it is called *pressure* and may be measured by the force that the column exerts on a column of mercury trapped inside a calibrated tube, otherwise known as a barometer. The unit for measuring air pressure in atmospheric science is typically a millibar (mb) but increasingly the metric equivalent, the hectopascal, hPa, is used (1 mb = 1 hPa). Measurements of air pressure are useful since they relate to the behaviour of the weather. High surface pressure refers to an anomalously large amount of mass above a point with surface values in the range of 1020 to 1040 hPa. Low pressure refers to an anomalously low mass of air above a point, and is characterized by values in the range 970–1010 hPa. Areas of high and low pressure in the mid-latitudes tend to be cellular and a few hundred kilometres in size. High-pressure cells are characterized by air diverging or moving away from the centre of pressure. To compensate, air sinks in the core of the cell, heating adiabatically as it descends. High-pressure areas are thus generally warm and cloud-free. In contrast, areas of low pressure tend to drive airflow towards their centres. This convergence of air leads to ascent, adiabatic cooling, and often rainfall.

Since air pressure refers to the mass of the column of air above a point, it follows that pressure decreases with altitude to zero at the top of the atmosphere. Most weather features occur below the 300 hPa level, which in the mid-latitudes is roughly 8000 m above sea level (corresponding to the highest mountains on Earth). Since air is compressible, the rate of decrease of pressure with altitude is non-linear.

Pressure is inversely related to temperature. Regions which undergo intense solar heating generally experience low pressures (for example, thermal low pressures over Spain and North Africa in July), whereas regions of intense cooling experience high-pressure systems (for example the Siberian High in January). These thermal effects are generally confined to the lowest few kilometres of the atmosphere. A useful meteorological quantity is therefore the thickness of atmosphere between two standard pressure levels (for instance 1000 hPa and 500 hPa), since this reveals the three-dimensional structure of the atmosphere. The thickness is usually expressed in metres, where relatively small values indicate cold air and larger values warmer air.

In countries characterized by marked relief, such as the escarpment of South Africa, a map of surface pressure would simply provide a map of altitude rather than an indication of weather features, since pressure decreases sharply with altitude. In these circumstances it is desirable to plot the altitude (in a unit approximating to metres) at which a certain pressure (such as 850 hPa) occurs. Maps of the topography of the pressure surface are analogous to maps of pressure, where low values indicate low pressure.   R. WASHINGTON

Further reading

Meteorological Office (1991) *Meteorological glossary*. HMSO, London.

Airy, Sir George Biddell (1801–92)
Sir George Airy, who was Astronomer Royal for 46 years, is best known among Earth scientists for his contributions to the theory of isostasy (or 'crustal balance' as it was then termed) and for the experiments he carried out to measure the density of the Earth.

Airy began his career as a mathematician, and he was elected to the Lucasian Chair of Mathematics at Cambridge in 1826. He held that post for only two years before being elected Plumian Professor of Astronomy and Director of the Cambridge Observatory in 1828. His textbook on physical astronomy had been published in 1826. Airy was appointed Astronomer Royal in 1835, and displayed great energy in that role. Among other things he re-equipped the Greenwich observatory with new instruments (which he designed himself), he created a magnetic and meteorological department, and he made important contributions to celestial mechanics. In 1894 he supervised a series of pendulum experiments with the object of measuring the increase in gravity with depth below the surface of the Earth. He was a prolific author, who published works on gravitation and on the downward extension of mountain roots, on the figure of the Earth, and on tides and waves as well as on astronomical subjects. Airy was characterized by thoroughness in his investigations and by his efficiency and strict discipline as an administrator.   D. L. DINELEY

alluvial channels Alluvial channels may be defined as channels formed in sediment transported by flowing water. Many channels are actually formed in a combination of bedrock and alluvium: bedrock outcrops may alternate with thick alluvial fills in a downstream direction; or bedrock may be present beneath a relatively thin alluvial veneer, controlling valley morphology or such channel characteristics as the amplitude of meanders. A segment of an alluvial channel may thus be more specifically defined as a channel with the majority of its length formed in alluvium, and beneath which the depth of alluvium exceeds the depth capable of being mobilized during fairly frequent floods.

Alluvial channels are extremely diverse. The grain size of the alluvium, which may range from clay and silt to boulders greater than 1 m in diameter, interacts with sediment supply, valley gradient, and flow regime to shape the morphology of the alluvial channel. Several classification schemes have been developed for alluvial channel morphology. The classification may focus on channel planform, on bedforms within the channel, on flow characteristics, on sediment load, or on vertical or lateral channel stability, among other characteristics. Recognition of consistent channel patterns facilitates the prediction of river behaviour, as well as river management.

At the smallest spatial and temporal scales, alluvial channels are shaped by the movement of water and sediment within the channel. The geometry and surface roughness of the channel boundaries control the velocity distribution across and along the channel. Channels in which the grain size of the alluvial boundary material is less than about a quarter of the flow depth tend to have predictable velocity gradients, with the highest velocities at the top of the water column and in the centre of the channel. The vertical velocity profile for such channels approximates to a logarithmic curve. Irregularities in the channel planform, such as meandering, may set up strong cross-channel currents; coarser bed alluvium may create low-velocity zones near the channel bed and a steep vertical velocity gradient more like an S-shaped curve than a logarithmic curve.

Turbulence associated with irregular cross-channel or vertical velocity distributions erodes and deposits alluvium from the channel boundaries in a self-enhancing feedback effect that produces bedforms. Bedforms are regular, repetitive variations in the channel bed that, by altering boundary roughness, regulate the expenditure of flow energy. Bedforms common along alluvial channels include step–pool sequences, pool–riffle sequences, the succession of riffles–dunes–plane bed–antidunes, and pebble clusters.

The alluvium that composes bedforms comes both from hill slopes within the drainage basin, and from the channel bed and banks. Bed and bank sediments are frequently eroded, transported, and deposited by the flow, creating a constant exchange of sediment at any one channel cross-section. Alluvial channels tend to scour while the water level is rising during a flood as sediment is entrained from the channel bed by the rapidly increasing flow, and to fill during the falling limb as sediment is again deposited. Equations have been developed to predict accurately the entrainment and transport of sand-sized alluvium as a function of velocity, but such prediction becomes less accurate as grain size increases. For an uneven bed composed of gravel-to boulder-sized particles, the velocity may fluctuate dramatically over small spatial and temporal scales, and the movement of individual particles becomes random. Selective entrainment sometimes occurs; particles that protrude further above the bed and into the flow then move first. At other times, particles of different sizes are mobilized almost synchronously.

The roughness of a channel boundary may also be controlled by biotic factors. Large woody debris (more than 2 m long and 10 cm in diameter) may form individual roughness elements, or may collect in regularly spaced steps that create plunging flow. Beavers create broad step–pool sequences by building dams that temporarily pond the flow. Besides increasing channel-boundary roughness, beaver dams trap sediment and slow the downstream transmission of flood waves. The roots of vegetation growing along the channel banks can substantially increase bank resistance, as well as decreasing the velocity of overbank flows and trapping sediment in transport.

Hydraulic geometry was used by Leopold and Maddock in 1953 to explain systematically the role of the magnitude and temporal distribution of discharge as primary controls on channel geometry. Hydraulic geometry describes the relationships between the dependent variables of channel width, depth, mean flow velocity, slope, and resistance, and the independent controlling variable of discharge. Hydraulic geometry can be used to describe cross-sectional or downstream trends in channel morphology. As a general rule, mean velocity and width: depth ratio both increase downstream along alluvial channels as discharge increases. These trends may be complicated in arid zones or karst terrains, where discharge may decrease downstream as a result of infiltration into the channel bed.

Downstream trends of alluvial channel geometry have also been explained in terms of energy expenditure. In 1960 Leopold and Wolman noted that abrupt discontinuities in the rate of energy expenditure along a channel are less compatible with conditions of balance between discharge and channel geometry than is a more or less continuous or uniform rate of energy loss. Subsequent work indicated that adjustments in the dependent variables in response to changes in discharge tend to be as conservative as possible, and that channel geometry is shaped to minimize total energy expenditure. As a result of this work, alluvial channels are generally considered to be malleable by at least the largest flows, so that channel geometry is shaped by flowing waters to represent the most uniform and efficient expenditure of energy under given conditions of water and sediment discharge.

Hydraulic geometry implies the ability for self-regulation within channels, since a change in discharge will cause a corresponding change in the dependent variables. Hydraulic geometry also implies that channels reach an equilibrium state, in which channel geometry is in balance with the prevailing discharge. In practice, however, different components of a channel may experience different lag times in responding to a change in discharge. In general, bedforms and the channel width: depth ratio are most responsive to change; channel gradient, however, has a longer lag or response time.

It may also be difficult to determine to what level of discharge channel geometry is responding. In 1960, Wolman and Miller proposed that the dominant discharge responsible for transporting the majority of sediment along most alluvial rivers was the 'bankfull discharge' that fills a river channel up to the top of its banks without spilling over on to the flood plain; this typically recurs at least once every 5 years. Subsequent research has demonstrated that as hydrological variability or channel-boundary resistance, or both, increase, the less frequent floods may dominate channel morphology.

Uncertainties about the magnitude and frequency of discharge that dominate channel morphology may also complicate palaeohydrological inferences of former flow regimes, for the geometry of relict or abandoned alluvial channels is used to infer past discharge regimes.

Stanley Schumm defined 'river metamorphosis' as the complete alteration of alluvial channel form as a result of changes in hydroclimatic regime. Metamorphosis may occur over a period of centuries or of decades. In historical times, channel changes have occurred along many of the world's alluvial channels as a result of human interference with flow regimes.

Alluvial channels can also alter their morphology dramatically in the absence of a pronounced change in external controls. A channel reaching an internal threshold may incise independently of changes in discharge or sediment supply. After studying ephemeral alluvial channels in the western United States, Schumm and Parker described a complex response whereby the channels began to incise when gradual aggradation caused the alluvial valley floor to exceed a threshold slope. As the channel incised headward, the increase in sediment supply caused aggradation and braiding in the downstream channel reaches. As incision ceased in the upper reaches, the decrease in sediment supply to the lower reaches triggered a new episode of channel incision. A channel may proceed through two or three such cycles of incision and aggradation before a new balance is established between channel morphology and control variables.   ELLEN E. WOHL

Further reading

Leopold, L. B. and Wolman, M. G. (1957) River channel patterns: braided, meandering, and straight. *US Geological Survey Professional Paper* 282–B, pp. 39–73.

Schumm, S. A. (1985) Patterns of alluvial rivers. *Annual Review of Earth and Planetary Sciences*, **13**, 5–27.

Schumm, S. A. and Parker, R. S. (1973) Implications of complex response of drainage systems for Quaternary alluvial stratigraphy. *Nature*, **243**, 99–100.

Wolman, M. G. and Miller, J. P. (1960) Magnitude and frequency of forces in geomorphic processes. *Journal of Geology*, **68**, 54–74.

alluvial fans and alluvial plains

Alluvial fans and related phenomena are depositional landforms which form a continuum. Alluvial fans are cone-shaped bodies (with areas up to hundreds of square kilometres) of sediment laid down by water-flows and mud-flows where streams emerge from mountains. They radiate from the point of emergence—the *fan apex*—and have a conical shape with concave-up radial and convex-up lateral profiles. They are crossed by radiating channels, which become shallower down-fan to a point—the *intersection point*—where the channel floor and fan surface meet. They may merge to form complex sloping plains—*bajadas*. They are particularly well developed in, but not restricted to, areas of active uplift (e.g. the Basin and Range Province of the USA and the foot of the rising Himalayas and other mountain ranges). Although common in arid or semi-arid areas they are found in all climatic belts. In many areas, fans are relict or palimpsest features produced earlier under different climatic conditions and are today merely being modified by present processes. They form wedges (up to hundreds of metres thick) of sediments with a wide variety of grain-sizes, usually including considerable gravel, deposited by: stream-flows: confined to channels; sheet flows: covering the whole fan surface; and mudflows: acting mainly in the fan head. Changing stream patterns move the locus of deposition and parts of the fan become inactive and coated with vegetation or calcretes, gypcretes, and silicates. Larger-scale climatic changes or tectonic changes in the hinterland produce relative changes in the main agents of deposition and entrenchment of the upper fan (the fan head)—although some geologists claim that this is a natural consequence of fan growth—and in many cases movement of the locus of deposition away from the mountain front. A valuable record of such changes is preserved in the stratigraphy of the fan. 'Alluvial cone' is a term usually applied to fans with slopes greater than 20°. In many cases they contain more mudflow deposits than most fans.

Alluvial plains are relatively flat fluvial-generated surfaces which may be hundreds of kilometres wide and thousands of kilometres long, dipping gently downstream and bordered by bluffs. They may merge downstream to form broad coastal plains with deltas where they reach the sea.

The channels can meander across the whole flood plain or be confined for considerable periods to a narrow belt to produce alluvial ridges. Bedload is confined to the fluvial channels, where it is fashioned into scroll ridges by lateral accretion on point bars at inner bends. Channel abandonment or avulsion, whether for hydrological or tectonic reasons, leads to the construction of new ridges. Sediment escaping from the channel during overbank flooding builds levees bordering the channel, and sheets of sand spread from the channels as crevasse splays. The flat plains bordering the channels—flood-plains—receive mainly suspended sediment during overbank flooding, and aggrade to form accumulations of fine material. Areas remote from the channel in Arctic, humid-temperate, or tropical regions are dominated by organic sedimentation in backswamps to produce peat, or in arid areas, evaporites. Abandonment of a former course through avulsion and meander-loop cut-off produces many lakes.

Some alluvial plains consist of a complex of channels separated by large islands. These anastomosing rivers maintain their courses for long periods, either for tectonic reasons or because of downstream damming by landslides or alluvial cones. The result is vertical aggradation *in situ* with little evidence of lateral movement.

In areas with episodic run-off, the whole plain may become, at periods of high discharge, part of the channel, with the deposition of a complex of shallow gravel and sand channels and bars. Sediment may be stabilized for periods of years or hundreds of years before being again removed. These plains contain less fine-grained sediment than those of meandering and anastomosing systems.

G. EVANS

Further reading

Leopold, L. B., Wolman, M. G., and Miller, J. P. (1964) *Fluvial processes in geomorphology*. W. H. Freeman, San Francisco.

alluvial plains

see ALLUVIAL FANS and ALLUVIAL PLAINS

alpine orogeny

The Alps are an arcuate mountain chain 100–200 km wide extending 1100 km from Nice to Vienna in southern Europe. Peaks over 4 km make them the highest of the circum-Mediterranean mountains. They are part of a chain extending east to the Himalayas formed by Mesozoic–Cenozoic closure of the Tethys Ocean. The Alps, as the world's most-studied mountain range, have contributed fundamentally to our understanding of the orogenic process. The concept of fold nappes was first introduced in 1841 by an Alpine geologist, Escher van der Linth.

The Western and Central Alps (Fig. 1a) have a northern 'external' zone, of low metamorphic grade, containing the Jura, the Molasse Basin, and the Helvetic Nappes. The 'internal' zone, of higher metamorphic grade, comprises the Pennic Nappes. The Southern Alps make up the southern external zone. The foothills of the Jura and the Sub-Alpine Chain comprise folded Mesozoic and early Tertiary strata of the European platform. These are covered by sediments of a mid-Tertiary foredeep, the Molasse Basin, whose southern margin is overridden by the main alpine thrusts in the Pre-Alps. The Helvetic zone, or High Calcareous Alps, contain Jurassic and Cretaceous carbonates, Eocene flysch, and several Variscan basement massifs. Cover has become detached from basement and forms huge north-verging fold nappes (Fig. 1b). Klippe of the highest Helvetic nappes occur in the

Pre-Alps. The Penninic nappes developed in Jurassic and Cretaceous deep-water sediments and ophiolitic rocks are cored by remobilized basement. These are also north vergent except in the southern 'Root Zone', where structures become vertical and late orogenic granitic plutons are emplaced (Fig. 1b). This corresponds to the zone of greatest crustal thickening, where the Moho occurs at 60 km. The Briançonnais, a basement 'high' with only thin Mesozoic cover, occurs in the north and west of this zone. Thick Triassic carbonates and Variscan basement lie in the south-verging nappes of the Southern Alps. A Mesozoic foreland basin forms the Po Plain.

The Eastern Alps differ in structure (Fig. 1a, b). A narrow

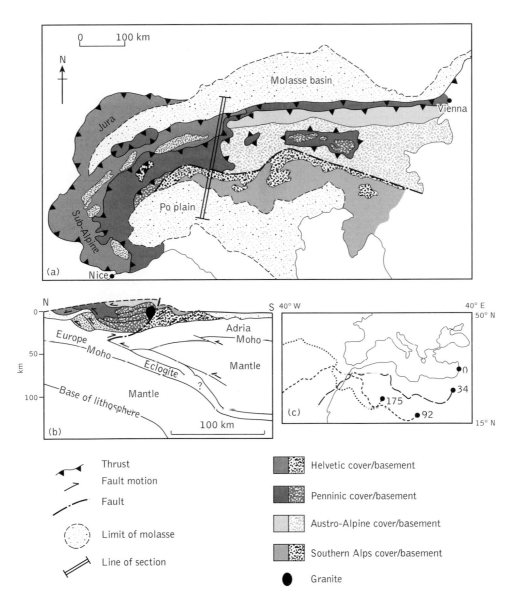

Fig. 1 (a) An outline geological map of the Alps. (b) A speculative cross-section of the Alps. European crust is labelled 'Europe' and Adriatic crust 'Adria'. (c) Maps using present geographic coordinates to show the relative positions of northern Africa with respect to Europe over time. The filled circle (o) corresponds to a point now near Tripoli and the numbers indicate the time (Ma) for each position.

northern external zone of Cretaceous Flysch Nappes (Pennninic?) is structurally overlain by the internal Austro–Alpine Nappes with Triassic carbonates and a Variscan basement. These north-vergent nappes were emplaced over the Penninic Nappes, which are exposed in tectonic windows. They are separated from south-verging nappes of the Southern Alps by a steep zone with late granitoids. The Austro–Alpine and Southern Alps originated on the African margin.

Plate configurations during the Mesozoic–Cenozoic evolution of the Alps can be reconstructed using Atlantic fracture zones and magnetic anomalies (Fig. 1c). The opening of the Central Atlantic Ocean in the Jurassic (175 Ma ago) caused Africa to move eastward with respect to Europe, opening basins on the European margin. The carbonates of the Helvetic Nappes were deposited between structural highs, which later formed basement massifs. The rocks of the Penninic Nappes were deposited in deeper basins, including the Valais Basin separated by the Briançonnais high from the Piemont ocean. Africa moved north-east with respect to Europe in late Cretaceous times (90 Ma), inverting these basins, producing high-pressure metamorphism in subduction zones, depositing early flysch, and causing the collision of the Adriatic promontory of Africa with Europe. The lack of subduction-related igneous complexes indicates that the Piemont Ocean was narrow. Further northward motion of Africa with associated anticlockwise rotation of the Adriatic promontory caused final closure of these basins, deposition of flysch, dextral transcurrent faulting, and formation of a thick north-verging nappe stack by the Eocene (40 Ma) which overrode and flexed the European crust to produce the northern Molasse Basin. The European margin was overthrust by Africa in Oligocene times (30 Ma), producing the steep 'root' zone, south-verging thrusts of the Southern Alps, and a foreland basin in the Po Plain. Late tectonic granites were emplaced in the internal 'root zone' and uplift began at a rate of 2 mm a year, perhaps associated with a loss of a thickened mantle root. The deformation front migrated northwards in the Miocene (20–10 Ma), thrusting the Helvetic Nappes over the southern margin of the Molasse Basin. Rates of uplift reduced to 0.5 mm a year and extensional faults developed in the internal zone as the Africa–Europe convergence rate declined. The Jura was folded in the late Miocene–Pliocene (5 Ma). Reconstruction of the various nappe stacks has lead to estimates of shortening that exceed 200 km across the Alps. The greater degree of shortening in the cover over that in the basement massifs suggests that some continental basement may have been subducted (Fig. 1b).   PAUL D. RYAN

Further reading

Coward, M. P., Detrich, D., and Park, R. G. (1989) *Alpine tectonics*. Geological Society Special Publication, No. 45.

Pfiffner, A. (1992) The Alpine orogeny. In Blundell, D., Freeman, R., and Mueller, S. (eds) *The European Geotraverse*. Cambridge University Press.

alpine pioneers of geology

The western (or Swiss) Alps of Europe hold a special place in the history of geology as well as containing the continent's highest mountains. Their geology consists of gigantic thrust sheets, recumbent overfolds, and slices of basement, deformed and displaced during the collision of the African and European continental plates.

The late eighteenth century saw a large number of Swiss and other students intent upon unravelling the alpine structures, but the dominant figure was Horace-Bénédict de Saussure. His discoveries included much of the Mesozoic and Cenozoic stratigraphy. Then came Arnold Escher of Zürich and Bernhard Studer from Berne who distinguished cover from basement rocks, but who disagreed about the relationships between them. Escher maintained that there were intrusive contacts, while Studer thought that basement and cover had been deformed together. Studer won the day, but Escher gets the credit for recognizing that great overthrusts are present, and that the rocks above the overthrusts have been translated great distances. In the Jura mountains where no basement is seen Jules Thurmann was the first to recognize the box-shaped folds that separate the Mesozoic–Cenozoic rocks from the more-or-less undeformed underlying basement.

By the turn of the century the need was for a structural synthesis to explain the complex distribution of 'normal' and 'inverted' masses of strata, and of basement and post-orogenic formations. The leader of the new generation of geologists was Albert Heim (1849–1937), a Züricher student of Escher who spent his long life investigating the dynamics of mountain-building and wrote a monumental account of the subject, as well as one on glaciers. He was also concerned with the preparation of the geological map of Switzerland. Early on he had studied the structure of the Glarus Alps, but his interpretation was overtaken by that of Marcel Bertrand, who in 1884 used the same data and initiated the nappe (thrust sheet) theory in the Alps. The Pre-Alps, lying to the north of the great Helvetic Alps, were then seen by H. Schardt (1893) as outliers of a pile of nappes that had originated far to the south. It called for earth movements on a previously unheard-of scale. (At about this time great overthrusts were being postulated to account for the deformation observed in the North-west Highlands of Scotland.) The whole style of alpine folding was hotly debated for many years, with the nappe idea gaining support from Maurice Lugeon, then from Edouard Suess, and in 1903 from Heim himself.

Over the next twenty years or so Lugeon and Heim worked on the structure of the Helvetic nappes, while Rudolph Staub unravelled the complexities of the Graubunden. During this time, too, Emile Argand was at work on the Valais Alps and bringing order to bear upon the synthesis of all the various parts of the alpine structures and upon the different styles of deformation within them. The result of all these investigations was the emergence of the Nappe theory, which Heim described in detail in his *Geologie der Schweiz*, and which has been the basis for subsequent refinements, even into the present plate-tectonic age.   D. L. DINELEY

Further reading

Green, M. T. (1982) *Geology in the nineteenth century*. Cornell University Press, Ithaca and London.

alteration of rocks *see* CHEMICAL ALTERATION OF ROCKS

amateurs in geology

For many people the word 'geologist' is apt to conjure up images of whiskery Victorian fossil collectors clad in heavy tweeds (Fig. 1). Some of today's professionals working in the Earth sciences indeed consider that the subject's links with the amateur tradition have not enhanced its status in the eyes of other scientists, such as physicists. Amateurs have nevertheless made important contributions to the geological sciences, and continue to do so even in an era in which highly expensive equipment is regarded as essential for serious research.

Geology can in fact be pursued by amateurs in many different ways and with little investment in scientific equipment. It has been said many times that the best laboratory is the field, where the prime requirement is the observer's sharp eye. As a field science geology attracts many followers, and its attractions can be extended into the laboratory/workshed where specimens may be prepared and studied. Specimens become of greater value as more is known about their provenance. Many amateurs acquire skills as lapidaries, cutting, polishing and investigating specimens of all kinds, as photographers of landscape and geological phenomena, and even as builders of simple seismographic stations.

In a sense, the birth and early progress of geology was due to the enthusiasm and energies of amateurs two hundred years ago. Those who might have been called professionals were the academics who had been appointed to posts in natural philosophy or other disciplines. (The first chair in geology in England was the Woodwardian Chair in the Univeristy of Cambridge, founded in 1728.)

Geology in Britain was particularly fortunate in the number and calibre of the amateurs in the late eighteenth and early nineteenth centuries. James Hutton, the 'Founder of Modern Geology', was a Scottish landowner. Geology was a suitable pastime for country gentlemen, many of whom became inveterate collectors of minerals and fossils. Roderick Murchison gave up his fox-hunting life for that of the field geologist, to the great benefit of historical geology. Another amateur of that time, T. H. de la Beche, was appointed by the government to apply 'geological colouring' to Ordnance Survey sheets and in due course became a leading professional as Director of the Geological Survey of Great Britain. Palaeontology, too, benefited; for example, Dr Gideon Mantell (b. 1790), a Sussex country doctor, made some of the earliest discoveries of dinosaur remains.

The Geological Society of London was founded in 1807 by thirteen amateurs; kindred bodies already existed, the British Mineralogical Society for example. The Geologists' Association, founded in 1858, for professional and amateur alike, has flourished ever since and has regional groups meeting in various parts of Britain. Its activities include field excursions throughout Britain and abroad, and the local recording and sampling of temporary exposures has frequently been carried out by such groups. Recent, too, are geological trails and sites where exposures are made accessible, conserved, and labelled. A code of conduct for field parties and collectors is adopted so that neither landowners nor public are put to inconvenience or risk. Environmental concern is strong.

The trade in geological materials has led to unscrupulous collecting, but for the most part, amateurs resist that kind of activity, and good relations generally exist between them and professional geologists. Journals, newsletters, and other channels advertise geological items for exchange or sale, and fairs are held for the exchange or trade of specimens.

D. L. DINELEY

Amateurs in geology
(Courtesy of *Punch*)

Fig. 1. 'Yes, certainly. Follow the Lower Greensand until you strike the Gault Clay, then turn left along the outcrop of calcareous grit and you'll see Little Gadhurst right there on the chalk escarpment.' The traditional view of the amateur geologist—since dispelled by a new generation of amateurs. (By courtesy of *Punch*.)

amber

Amber is a light, organic substance that is generally yellow or orange in colour and may be transparent or cloudy. It is the fossilized resin from trees that lived millions of years ago Trees produce resin as protection against disease and insect attack. Some trees produce resin in large quantities,

Fig. 1. Piece of Baltic amber containing a cricket, tips of the body and wings of a damselfly, and several flies. Length 73 mm (Phil Crabb (NHM).)

which seeps out of cracks in the bark and flows down the trunk. Resin is very sticky, and insects, other small animals and plant remains are readily trapped (Fig. 1). The resin hardens and becomes incorporated into sediments, where it polymerizes to form amber; any inclusions are usually very well preserved.

Amber is composed of cyclic hydrocarbons known as terpenes; it has a hardness of 2–3 on the Mohs' scale, a specific gravity of 1.04–1.10, and a melting point of 200–380 °C. Amber is generally attractive to look at, warm to the touch, light, and easy to carve. These properties have made it desirable for jewellery since the Stone Age.

Amber is difficult to date and this is mainly done by studying the associated fossils in the sediments from which the amber comes. This, however, indicates only a minimum age because there is no way of knowing how long the amber took to get from the trees to where it was deposited. The associated fossils may give an incorrect age if the amber has been reworked. The oldest amber has been recorded from the Upper Carboniferous. The oldest amber that contains insects comes from the Lower Cretaceous.

Amber occurs in many parts of the world. The largest deposits are from around the Baltic Sea and in the Dominican Republic, which are generally considered to be Upper Eocene–Lower Oligocene and Lower–Middle Miocene in age respectively. Other significant deposits occur in Canada, the USA, Mexico, France, Spain, Germany, Sicily, Romania, Lebanon, Russia, China, Burma, Japan, and Borneo.

The insects and other inclusions that are trapped in amber provide evidence of past biodiversity, ecology, and biogeography. Although most species in amber are extinct, it is often possible to work out how they lived by comparison with their living relatives. There is a bias in what is preserved: many larger animals are strong enough to pull themselves out of sticky resin, whereas small and weak insects are more prone to be entrapped. Furthermore, insects that live on the resin-

producing trees are more prone to entrapment than those that live in habitats away from the trees. Amber is crucial for our knowledge of small extinct insects that would not otherwise be fossilized. Of particular interest are specimens in which two different animals are caught interacting, such as parasites still attached to their host.

In recent years scientists have analysed air bubbles in amber to try to obtain evidence of past atmospheres. However, amber oxidizes, and oxygen in the bubbles would then react with the amber, which would affect the composition of the bubbles. Scientists have also attempted to extract DNA from the insects in amber. Some have claimed success, but independent attempts to replicate the results have so far been unsuccessful, which has led to suggestions that the recovered DNA may be recent contamination rather than ancient materials.

ANDREW J. ROSS

ammonites *see* CEPHALOPODS

amphibians Amphibians are tetrapods that still retain a connection to the water because of their need to lay their eggs in moist conditions. Although modern amphibians, frogs and toads, salamanders, and caecilians (legless amphibians) are familiar organisms today, their Palaeozoic ancestors looked quite different and there are difficulties in establishing relationships between the two groups. The earliest amphibians are known from East Greenland and from Russia. They are exemplified by *Ichthyostega* from the Late Devonian of East Greenland, an animal about 0.5–1 m long that may have breathed air and moved about on land, but retained a fish-like tail and the sensory canal system and gills of its osteolepiform (bony fish) ancestors. These features indicate that it spent most of its time in the water. Sharp conical teeth in the jaws indicate that this animal was a predator, while a number of skeletal adaptations attest to its ability to support its body on land. These include enlarged ribs to support the viscera, and the replacement of the notochord by vertebrae that consisted of a neural arch and a ventral half-ring connected by bony nodules.

A major division of the amphibians occurred in the Early Carboniferous with the development of two groups, the labyrinthodonts and the lepospondyls (Fig. 1). Labyrinthodonts are named for the complex folding of their tooth enamel. Most of them were stocky, short-legged animals up to 2 m in length, with large skulls and jaws lined with sharp conical teeth. They had notches in the back of the skull to accommodate the tympanic membrane (ear-drum). This was joined to the inner ear by a small bone called the stapes, which had been a jaw support (the hyomandibular) in the osteolepiforms. The large body size and short legs suggest that these animals lived mostly in the water, where they adopted a crocodilian life-style. Typical labyrinthodonts of this type are called temnospondyls. This group also includes some forms that developed aquatic characteristics, such as laterally flattened tails and reduced limbs, which indicate a return to a fully aquatic way of life. Temnospondyls were important car-

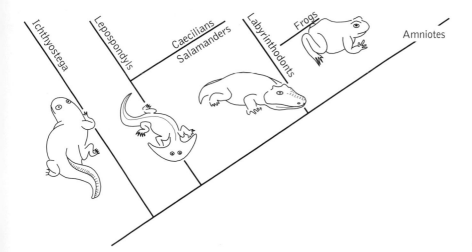

Fig. 1. Cladogram of relationships
for the Amphibia.

nivores during the Carboniferous and Permian and persisted into the Mesozoic. The last examples are known from the Cretaceous of Australia, where they were contemporaries of the dinosaurs and the mammals. The other group of labyrinthodonts was the anthracosaurs, terrestrial amphibians that gave rise to the reptiles in the Early Carboniferous. In anthracosaurs such as *Seymouria* from the Permian of Texas, the skull and dentition are still essentially amphibian while the post-cranial skeleton shows a large number of features characteristic of early reptiles. Whether animals such as this were really amphibians or reptiles ultimately depends on whether they laid amniote (hard-shelled) eggs or not; and the answer to this question is not yet known.

The second major group of Palaeozoic amphibians, the lepospondyls, were mostly rather small animals less than 0.3 m long, in which the teeth lacked the labyrinthodont structure and the skull lacked a notch for the tympanic membrane. They are named for the spool-shaped centrum of bony disc surrounding the notochord that contrasts with the labyrinthodont vertebrae formed of several pieces. This group developed elongated bodies and reduced limbs as an adaptation to a completely aquatic existence. One group, the diplocaulids, developed skulls that were greatly extended laterally, giving them a boomerang shape. Hydrodynamic studies have shown that this shape provided lift and allowed the animal to rise rapidly from its resting-place on the bottom to attack fish swimming above it.

The modern amphibians are separated by a large gap in time and morphology from the major Palaeozoic groups. The earliest frogs and salamanders are found in the Jurassic and look very much like their modern descendents. A possible ancestral frog from the Triassic of Madagascar, *Triadobatrachus*, shows specializations in the ear region similar to those found in some temnospondyl labyrinthodonts, which suggests that they are likely ancestors. On the other hand, it has been argued that salamanders and caecilians were derived from the

microsaurs, a lepospondyl group. If this is correct, then modern amphibians are polyphyletic (derived from more than one ancestor), despite the fact that there appear to be a number of characters linking them, including the development of respiration through the skin, cylindrical vertebral centra, and the presence of pedicellate teeth in which a zone of weakness between base and crown allows them to break easily.

DAVID K. ELLIOTT

Further reading

Carroll, R. L. (1988) *Vertebrate paleontology and evolution.* W. H. Freeman, New York.

Colbert, E. H. and Morales, M. (1991) *Evolution of the vertebrates.* Wiley-Liss, Inc., New York.

Andes of South America The Andes form a sinuous mountain chain, 100–700 km wide and extending along the entire western margin of South America for over 5000 kilometres from 10° N in Venezuela to 55° S in southernmost Chile and Argentina. They pass northwards into the subduction systems of Central America and the Caribbean, and southwards into those of the Scotia Arc and the Antarctic Peninsula. They are conventionally divided into the northern (10° N–10° S), central (10° S–35° S), and southern (35° S–55° S) Andes, separated by marked swings in trend: the Patagonian orocline at *c.* 55° S; the Bolivian orocline at *c.* 18° S, and the Peru–Ecuador orocline at *c.* 5° S. The Andes can be divided into a number of physiographic provinces which parallel the topographic grain, with generally distinct geological histories: the Western Cordillera (Cordillera Occidental or Principal Cordillera) forms a high spine (often rising to over 5 km) of mountains (usually volcanoes) on the western margin; the Eastern Cordillera (Cordillera Oriental) is a rugged region on the eastern margin, usually rising to between 3 and 4.5 km. In the Central Andes, the Altiplano or

Puna is a high plateau or region of more subdued relief between the Eastern and Western Cordilleras; and the Subandean zone or Precordillera, which is the site of active thrusting today, lies in the foothills on the eastern margin. Satellite ranges occur as far as 500 km east of the main Andean range only in central Peru (Shira, Contaya, and Moa uplifts) and in northern Argentina (Sierras Pampeanas). To the east of the Andes, the core of South America, through which most of the major Andean rivers (Orinoco, Amazon, Putumayo) drain, is generally only a few hundred metres above sea level. Mt. Aconcagua (6959 m) at c. 33° S is the highest point in the Andes, but the mountain range as a whole is widest and highest between 12° S and 27° S, forming a mountainous zone nearly 700 km wide, with the Altiplano–Puna plateau on average c. 4000 m high. Overall, the average elevation and width of the Andes decreases more or less symmetrically north and south from central Bolivia.

Plate-tectonic setting

The Andes are a wide zone of convergent continental deformation between the South American plate and oceanic lithosphere of the Nazca and Pacific plates (Fig. 1). Shallow (0–70 km) and deep-focus (70–650 km) earthquakes occur along the entire Andean margin. The shallow earthquakes are predominantly the result of thrust faulting, concentrated offshore in the trench and on the eastern margin of the Andes (Subandean zone or Precordillera). The deep-focus earthquakes define a slab-like zone (Benioff zone) inclined at 30–50° beneath the Andes, reaching depths greater than 600 km in the Central Andes. The Benioff zone lies in oceanic lithosphere that is being subducted at the oceanic trench. This trench runs the length of the Andes, about 75 km offshore and up to 7 km deep off the coast of northern Chile. The relative plate convergence, averaged over the past 3 million years, is about 80 mm per year in a ENE direction. However, at 47° S, near the Taitao peninsula in southern Chile, the active Chile Ridge spreading centre intersects the trench. South of this, the Pacific plate is being subducted beneath the Andes at about 20 mm per year in a more easterly direction, but the southern termination is a sinistral transform. The northern termination of the Andes is dominated by E–W-trending dextral strike-slip motion between the Caribbean Plate and northern Colombia and Venezuela.

Plate convergence is accommodated both by slip at the plate interface in the subduction zone and by shortening in the continental lithosphere of the South American plate, resulting in crustal thickening and mountain building. It has been estimated that during the Cenozoic less than 20 per cent of the total plate convergence between the Nazca and South American plates was absorbed by continental shortening, giving rise to a maximum of about 300–350 km of tectonic shortening in the Central Andes. This has resulted in a crustal thickness ranging from c. 34 km east of the Andes to up to 80 km beneath the Altiplano–Puna. The thickness of the lithosphere also varies markedly, and seems to be relatively thin beneath the Central Andes. The present large-scale topography is almost entirely the result of deformation during the past 50 Ma, in the Cenozoic, and substantial increases in height have occurred in the past 10 Ma.

There are marked variations in the style and amount of deformation along the length of the Andes. Features of the oceanic Nazca plate may have played a role in this, such as a change in age at the trench, between 6° N and 47° S, from Recent (in the north and south) to Palaeocene (at the latitudes of the Central Andes) with a number of prominent ridges that intersect the trench (Nazca Ridge at c. 16° S, Juan Fernandez Ridge at c. 33° S). Two 'flat-slab' regions have been recognized: beneath central Peru (3° S–16° S) and beneath Chile and Argentina (27° S–33° S). In these regions the subducted lithosphere is nearly horizontal in a wide zone before plunging steeply into the Earth's mantle further east. The flat-slab zones also coincide with the presence of outlying ranges far to the east of the Andes.

A volcanic arc running along the western margin of the Andes follows the 90–150-km depth contour of the Benioff zone. However, the 'flat-slab' regions are associated with a gap in the active volcanic arc, although there is evidence for Miocene and older volcanic activity. The gaps in active arc volcanism are often used to divide the arc into northern (NVZ), central (CVZ), and southern (SVZ) volcanic zones. These are the type location for the characteristic intermediate-composition andesitic volcanism, although mafic-rhyolitic volcanism is also common. Widespread ignimbrites, deposited during explosive and ash-rich volcanic eruptions, are an important feature of the CVZ.

Geological evolution

The core of the South American continent is underlain by Proterozoic and older basement, which is the result of episodes of welding of continental blocks during the Pan-African and Brazilian orogenic cycles (700 ± 100 Ma), now separated into two main shield areas, referred to as the Guyanan and Brazilian Shields. The western margin of South America has been the site of continental accretion, crustal growth, and both compressional and extensional deformation throughout the Phanerozoic, situated during the Palaeozoic and early Mesozoic at the edge of the giant landmass of Gondwana. In the Cambro-Ordovician this region seems to have evolved from a zone of continental collision to a passive continental margin associated with the collision and subsequent rifting of the Laurentian (North American) continental mass. It later evolved as a complex subduction margin with associated island-arc systems, rifted continental slivers, and back-arc basins. Thick sequences (up to 10 km) of shelf and continental slope deposits were deposited in the Central Andes from the Cambrian to Devonian. Palaeozoic subduction and accretion also resulted in the amalgamation of various terranes, associated with regional compressive events. Notable terranes include Arequipa terrane in coastal southern Peru and northern Chile; Chilena, Precordillera, and Patagonia terranes in Chile and Argentina. However, since the early Triassic (c. 250 Ma), when the supercontinent of Pangea

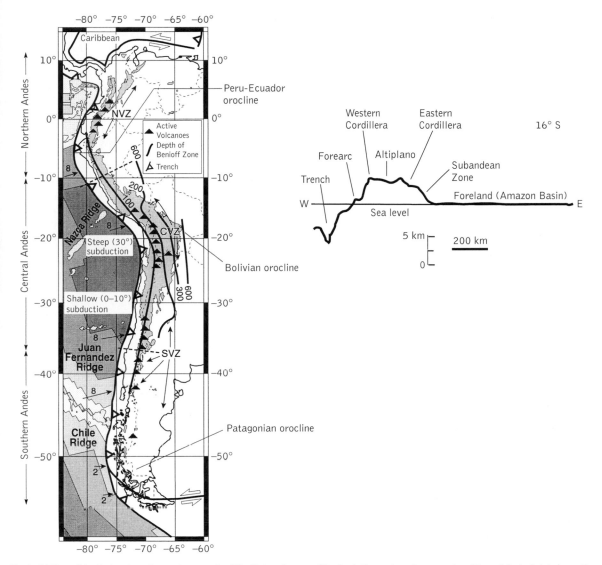

Fig. 1. (a) Map of the Andes along the western margin of South America, showing the principal features. Oceanic lithosphere is being subducted beneath South America at the trench; arrows show direction (with amount in cm per year) of active relative plate convergence. The oceanic plate is shaded according to age, from oldest at the trench (Palaeocene–Eocene) between 10° S and 36° S, to Oligocene, Miocene, Pliocene, and Quaternary at the oceanic ridges. Depth contours (in km) of the Benioff zone show the geometry of the subducted plate beneath South America, with two 'flat-slab' regions in Peru and Chile/Argentina where there is no arc volcanism. The volcanic arc can be divided into northern (NVZ), central (CVZ), and southern (SVZ) volcanic zones. The shaded region on land corresponds to the high parts of the Andes which lie more than 2 km above sea level. (b) Topographic section through the Central Andes at 16° S, showing the main physiographical provinces.

was fully assembled, the central and southern Andes appear to have formed a classic continental-type subduction margin, with eastward subduction of oceanic lithosphere (Nazca and previous oceanic plates), and no further terrane accretion. However, the northern Andes, in Venezuela, Columbia, and Ecuador, have had a more complex history, partly influenced by Caribbean tectonics and relative motion of North and South America. Here, a series of allocthonous terranes (island-arc systems) were accreted in the latest Jurassic to early Cretaceous (Ecuador) and early Tertiary (Columbia).

21

In the Jurassic and Cretaceous there was widespread marine and lacustrine sedimentation along much of the length of the Andes, associated with rifting in forearc and behind arc basins such as the West Peruvian trough at 5°–14° S; the Aimara basin at 16° S–28° S; Neuquen basin near 38° S; and the Megallanes basin south of 47° S, covering much of southern Patagonia. Early Cretaceous back-arc spreading in southern Patagonia was also sufficient to create new ocean floor, which was subsequently obducted in the Middle Cretaceous, forming the Rocas Verdes Ophiolite. Jurassic to Cretaceous magmatic activity, along the entire length of the Andes, was associated with the emplacement of huge granite batholiths which crop out today in the coastal regions of Peru and Chile.

The relative plate motions along the Andean plate boundary are well documented only since the latest Cretaceous (68 Ma), by means of ocean-floor magnetic anomalies. The rate of convergence increases markedly at about 50 Ma, marking the start of the Cenozoic phase of compressional deformation in the Central Andes. The rate of relative plate convergence subsequently decreased in the late Eocene (c. 35 Ma), before increasing markedly again in the latest Oligocene (c. 27 Ma). The two phases of increased relative plate motion correlate well with major phases of compressive deformation in the Andes, sometimes referred to as Incaic (early Cenozoic) and Quechua (Middle–Late Cenozoic). The South American lithosphere, east of the region of mountain-building, has flexed downwards in response to the deformation, resulting in a series of Cenozoic 'foreland' basins which contain a thick sequences, usually less than 5 km thick, of mainly fluvial sediments.

SIMON LAMB and LORCAN KENNAN

Anning, Mary (1799–1847)

In the annals of palaeontology Mary Anning has a secure place for her discovery of the Jurassic marine reptile *Ichthyosaurus*. She was the daughter of a cabinet-maker who sold 'natural curiosities' in the Dorset seaside town of Lyme Regis, and it was from her father that she learned how to collect fossils from the Jurassic rocks exposed in the local cliffs. After his death in 1810 she set out to collect fossils herself, and soon was rewarded with a spectacular success. In 1811 she saw bones projecting from the cliff-face and enterprisingly hired some men to dig out the block in which they were embedded. The skeleton that emerged, about 17 feet (5 m) long, was a sensational discovery. After much debate, it was identified as *Ichthyosaurus platyodon*, a reptile hitherto unknown in Britain. Mary Anning subsequently found specimens of other important vertebrates: *Plesiosaurus* and a pterosaur (*Dimorphodon*) and fish. Her discoveries also included *cephalopods* with their ink-sacs and other soft parts preserved. These finds attracted many distinguished visitors to Lyme Regis, and her death was regretted locally as 'in a pecuniary sense a great loss to the place'.

Mary Anning is commemorated by a stained glass window in Lyme Regis parish church.

D. L. DINELEY

Antarctic climate

The climate of Antarctica is determined by a number of factors, including the continent's geographical location, its topography, and its interaction with the high-latitude circulation of the Southern Hemisphere atmosphere and ocean. In common with the Arctic regions, the surface of the Antarctic continent receives less energy in solar radiation than it loses by infrared cooling. This is because the solar elevation remains low all year round and, for regions south of the Antarctic Circle, the sun does not rise above the horizon for part of the winter. Furthermore, the permanent snow and ice that cover over 97 per cent of the continent and the sea ice that covers up to 20 million square kilometres of the surrounding ocean in late winter both have a high albedo, reflecting up to 85 per cent of the incident solar radiation. The net cooling of the polar regions has to be balanced by the atmospheric transport of heat from lower latitudes. In the Arctic, a significant fraction of this transport is accomplished by the stationary planetary waves that are forced by the major land masses of the Northern Hemisphere. In the Southern Hemisphere, the planetary waves are much weaker and most of the heat transported to Antarctica is carried by the mean north–south circulation of the atmosphere and by transient weather systems.

The East Antarctic ice sheet is a high plateau that reaches a maximum elevation of over 4000 m (Fig. 1). Over the interior of the plateau, the surface slope is generally of the order of 1 in 500 or less, rising to 1 in 100 or greater around the coastal fringes. However, even the small interior slope has a profound effect on the continent's climate as a result of the persistent surface cooling, which generates a nearly permanent surface temperature inversion. The cold, dense air adjacent to the surface accelerates down the surface slope and is turned to the

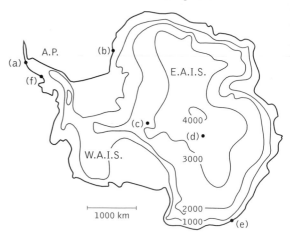

Fig. 1. Map of Antarctica, showing surface elevation contours at 1000 m intervals. A. P., Antarctic Peninsula; W.A.I.S., West Antarctic Ice Sheet; E.A.I.S., East Antarctic Ice Sheet. Key to locations mentioned in the text: (a) Faraday, (b) Halley, (c) South Pole, (d) Vostok, (e) Cape Denison, (f) Wordie Ice Shelf.

left by the action of the Coriolis force. Eventually, the down-slope acceleration is balanced by frictional and Coriolis forces and the resultant flow is known as a *katabatic wind*. Such winds are a characteristic feature of the Antarctic interior. Over the gentle slopes of the plateau, katabatic wind speeds are rarely greater than 5 metres per second (m s⁻¹) and the wind is directed some 30–40° to the left of the local fall line. Because these winds are topographically driven, they exhibit great directional constancy and often appear to be effectively decoupled from disturbances in the free atmosphere. Over the steeper coastal slopes, the katabatic winds accelerate and may converge into coastal valleys, giving rise to extreme wind speeds. For example, at Cape Denison, on the Adélie Land coast, Mawson's 1912–13 expedition recorded a world record annual mean wind speed of 19.4 m s⁻¹ (approx. 43.4 mph) and experienced gale-force winds on all but one of 203 consecutive winter days. Such extremes occur only in a few regions where the local topography favours convergence of the katabatic flow; around much of the Antarctic coast annual mean wind speeds are typically between 5 and 10 m s⁻¹.

The outflow of cold air associated with the katabatic winds is restricted to the lowest few hundred metres of the atmosphere and forms one half of a thermally direct circulation that transports heat to Antarctica to balance the surface radiative cooling. Above the cold outflow, there is a compensating inflow of warmer air that gradually subsides into the katabatic layers. Winds in this upper layer have a weak westerly component in contrast to the easterly component of the near-surface winds. The katabatic winds generated over the Antarctic ice sheets thus exert considerable control over the regional atmospheric circulation, and it is now believed that this influence extends well beyond the Antarctic continent itself.

The annual cycle of air temperature at Antarctic interior stations is markedly different from that at coastal stations (Fig. 2). At the coast, summer mean temperatures are around freezing and the coldest month is usually July or August. On the high plateau, temperatures are much lower as a result of both the high surface elevation and decreased solar heating at these latitudes. A record low temperature of –89.5 °C has been recorded at Vostok station. The most remarkable feature of the temperature records from these interior stations is the lack of any well-defined temperature minimum during the winter months. It is believed that this coreless winter results from the character of the annual cycles of both the surface radiation balance and the atmospheric transport of heat from lower latitudes.

Precipitation over Antarctica falls mostly as snow, although rain can occur during the summer in the Antarctic Peninsula and surrounding islands. The highest precipitation occurs around the coasts of the continent, where weather systems moving in from lower latitudes impinge on the steep coastal slopes, giving annual precipitation totals of 300–400 mm (water equivalent) per year around the East Antarctic coast. Few weather systems penetrate inland on to the high plateau. The air here is very cold and dry, with correspondingly low

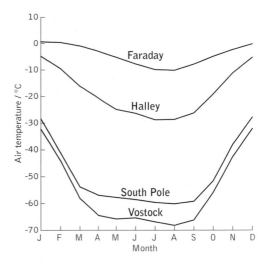

Fig. 2. The annual cycle of air temperature at four Antarctic stations. See Fig. 1 for locations.

precipitation. Much of the area of the East Antarctic plateau is classed as a desert, with an annual precipitation of less than 50 mm (water equivalent). A large proportion of the precipitation over the plateau occurs as a near-continuous fall of ice crystals from a clear sky, a phenomenon known as diamond dust. Although precipitation rates over the interior of Antarctica are very small, the low temperatures mean that melting and evaporation are negligible and the low precipitation is able to maintain the continental ice sheets, which are over 3 km thick in many places.

The weather systems responsible for transporting heat and moisture towards Antarctica have their origins in the midlatitudes of the Southern Hemisphere or in the circumpolar trough of low pressure. This climatological feature marks a storm track that extends around Antarctica at a mean latitude of about 66° S. The latitude of the trough undergoes a semiannual excursion. It is closest to Antarctica in March and October, and the surface pressure in the trough is also lowest during these months. A semi-annual oscillation in surface pressure, reflecting the movement of the circumpolar trough, can be seen in surface pressure records from all high-latitude Southern Hemisphere stations. To the north of the trough, the pressure gradient drives the strong westerly winds experienced over the Southern Ocean; surface winds to the south of the trough are easterly and surface pressure increases towards the continent, indicating the presence of a permanent low-level anticyclone over Antarctica.

In 1995 there were about thirty permanent stations making meteorological and climatological observations in Antarctica, most of them situated on the coast of East Antarctica and in the Antarctic Peninsula. Few of these stations have records extending back beyond the International Geophysical Year of

1957, and little direct information is therefore available on the variability of the Antarctic climate on decadal or longer timescales. Some information can be gleaned from the records kept by the expeditions that visited the continent before permanent bases were established while analysis of Antarctic ice cores provides 'proxy' records of temperature and precipitation extending back over many thousands of years. The longest instrumental records from continental Antarctica come from the Antarctic Peninsula, where there have been permanent stations since 1945. Temperature records from this region show a very high degree of interannual variability superimposed on a warming trend of about 0.6 °C per decade—much larger than trends measured elsewhere in Antarctica or in the middle latitudes of the Southern Hemisphere. This recent warming trend has been confirmed by observations of significant deglaciation in parts of the Antarctic Peninsula, including a reduction of the area of the Wordie Ice Shelf from 2000 km² in 1966 to 700 km² in 1989. However, the region affected by the warming trend appears to be limited, and at present there is little evidence to connect the trend with hemispheric or global warming. Records from East Antarctica are shorter but the climate of this region appears to be much less variable than that of the Antarctic Peninsula, and no significant temperature trends are seen. The high variability of the Antarctic Peninsula climate appears to result from complex interactions between atmosphere, ocean, and sea ice in this region.

The response of the Antarctic climate to global warming caused by an enhanced greenhouse effect is of great interest because increased melting of the ice sheets could contribute to global sea-level rise. Experiments with coupled ocean–atmosphere general circulation models suggest that, while the Arctic regions might be experiencing a warming that is significantly greater than the global average, any warming in the Antarctic will be relatively modest, at least in the short term. This is because the wind-driven overturning circulation of the Southern Ocean, known as the Deacon Cell, is able to sequester large amounts of heat from the atmosphere to the deep ocean, and thus balance any atmospheric warming tendency. Paradoxically, in a slightly warmer climate the Antarctic ice sheets would at first expand, because the warmer atmosphere could hold more moisture, and snowfall over the continent would increase. However, if global warming continued, most of the sea ice around Antarctica would eventually disappear and the ice shelves around the coast would start to melt as warmer ocean waters flowed on to the continental shelf. Once this stage was reached, the West Antarctic Ice sheet could decay quite rapidly, contributing about a metre to global sea-level rise over 500 years.

J. C. KING

Further reading

King, J. C. and Turner, J. (1997) *Antarctic meteorology and climatology.* Cambridge University Press.

King, J. C. (1991) Global warming and Antarctica. *Weather,* **46**, 115–20.

Antarctic ice cores

Three Antarctic ice cores extend back to the last glacial maximum at about 18 000 BP (Fig. 1a). The Byrd Station core was drilled in western Antarctica in 1968 to bedrock at 2163 m depth, and 10 years later the Dome C core was sampled in eastern Antarctica to 905 m depth. The most extensive record, however, comes from Vostok in the central part of eastern Antarctica at an altitude of 3490 m above sea level, where the present mean annual temperature is –55.5 °C.

The first series of drillings (1 Γ and 2 Γ) at the Vostok site began in 1970 and a reached a maximum depth of 950 m in 1974. Low rates of ice accumulation and thick ice make it possible to reach back to the penultimate interglacial and even beyond, in ice that is relatively undisturbed by flow conditions. A third deep core (3 Γ) reached 2083 m in 1982 with a later extension to 2202 m in 1986. Core 4 Γ was started in 1985 and was stopped in 1989 at a depth of 2546 m, where the ice is older than 200 000 BP. A new core, 5 Γ, is currently being sampled.

Obtaining an accurate timescale for the Antarctic ice cores is difficult. Low accumulation rates at Dome C and Vostok mean that the annual isotopic signal, which shows differences in the $^{18}O/^{16}O$ ratio between summer and winter, is not properly formed, so it is not possible to count back using these as annual markers. Two main ways are used to date the cores: comparison of the oxygen isotope profiles from ice with those from deep-sea cores, and using ice-flow models. Correlation between the Dome C and Vostok records was aided because peaks of the isotope ^{10}Be (beryllium) were found in the Vostok core at depths of about 925 and 600 m, corresponding to times of about 60 000 BP and 35 000 BP. The younger peak

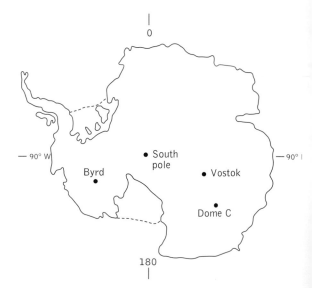

Fig. 1. Map showing location of major deep ice cores in Antarctica. (From Bradley, R. S. (1985) *Quaternary Paleoclimatology.* Allen and Unwin, London. Fig. 5.2.)

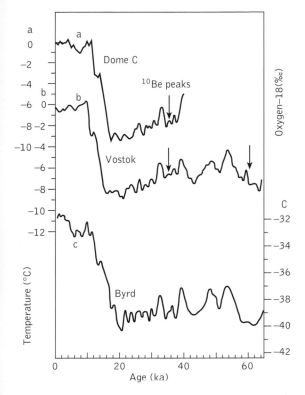

Fig. 2. Comparison of Antarctic ice-core records from Dome C, Vostok, and Byrd over the last 65 000 years. Solid arrows indicate observed [10]Be peaks. (From Jouzel, J. *et al.* (1993) A comparison of deep Antarctic ice cores and their implications for climate between 65 000 and 15 000 years ago. *Quaternary Research*, **31**, 135–50. Fig. 3.)

50 000 and 35 000 BP with temperatures 3–4 °C warmer than the last glacial maximum.

Spectral analysis of the Vostok core clearly shows peaks at 100 000 years, 40 000 years, 23 000, and 19 000 years, which provide support for the theory that Milankovich cycles are the main forcing mechanism of Quaternary climatic change over glacial to interglacial cycles.

The correspondence with the independently dated marine [18]O record is excellent down to 110 000 BP, but before this the records depart slightly and there is a mismatch in age between the two data sets regarding the start and duration of the last interglacial.

Other important studies have been completed at Vostok. Figure 3b shows generalized profiles for temperature, aluminium, sodium, and acidity from the 3 Γ core. The aluminium profile is an indicator of continental input and shows increases during the cold periods. Dust particles also exhibit this property and the pattern can be explained by factors affecting the continental source areas of these impurities, such as increased wind speeds, decreased vegetational cover, increase in arid areas and exposure of continental shelves due to sea-level lowering. In addition, ice accumulation at Vostok tends to decrease during glacial periods by a factor of two, thus concentrating the impurities in any one sample. The sodium profile reflects a source from marine sea salt. Despite the growth of sea ice during glacials, the marine sea salt record appears to increase during glacials, probably reflecting an increase in wind speeds and perhaps a change in the direction of atmospheric circulation. The acidity profile reflects the ejection of aerosols, primarily sulphur, into the stratosphere by explosive volcanism. The mismatch between this profile and the temperature record suggests that there is no long-term correlation between volcanic activity and climate.

Measurements of the main greenhouse gases taken from bubbles of air trapped within the ice are also available from Vostok. Figure 3c shows the variations in CO_2 (carbon dioxide) and CH_4 (methane) over the past 160 000 years, based on measurements taken from the 2083 m deep 3 Γ core. The greenhouse gases can be seen to correlate extremely well with the temperature reconstruction over the whole time period: in fact about 78 per cent of the variation in temperature can be explained by changes in these two greenhouse gases. In more detail, the profiles also suggest that at the maximum of the last glaciation CO_2 levels were 25–30 per cent below Holocene values and CH_4 levels were between 40 and 50 per cent depleted. At the maximum of the last interglacial, both CO_2 and CH_4 also show maxima, but thereafter CH_4 levels seem to decline quite rapidly while CO_2 levels maintain a high level. During the last deglaciation methane concentrations also seem to undergo rapid changes which are not marked by the CO_2 trace.

The levels of CO_2 and CH_4 are primarily controlled by biological production and the rate of abstraction by the various storage reservoirs. The obvious correlation confirms the link between biosphere and climate and implies complex changes

appears to correlate with a similar peak in the Dome C core at 820 m depth.

Figure 2 shows a comparison of the temperature differences from the present day Dome C, Vostok, and Byrd over the last 65 000 years. The cores show remarkable agreement, suggesting that one core may be sufficient to represent climatic changes at the continental scale.

Figure 3a shows the smoothed Vostok temperature profile from core 3 Γ and spans the last 160 000 years. It shows clearly the last glacial cycle, lasting about 100 000 years with an difference of about 6 °C between full glacial and interglacial conditions. Also of note is the fact that the last interglacial, equivalent to Marine Isotope (MI) Stage 5e at about 125 000 BP appears to have been significantly warmer than the Holocene maximum. There are also two peaks of temperature at 100 000 and 80 000 BP which seem to correspond with MI Stages 5c and 5a separated by two troughs. During the last glacial period there seem to have been three temperature minima separated by two double-peaked interstadials at about

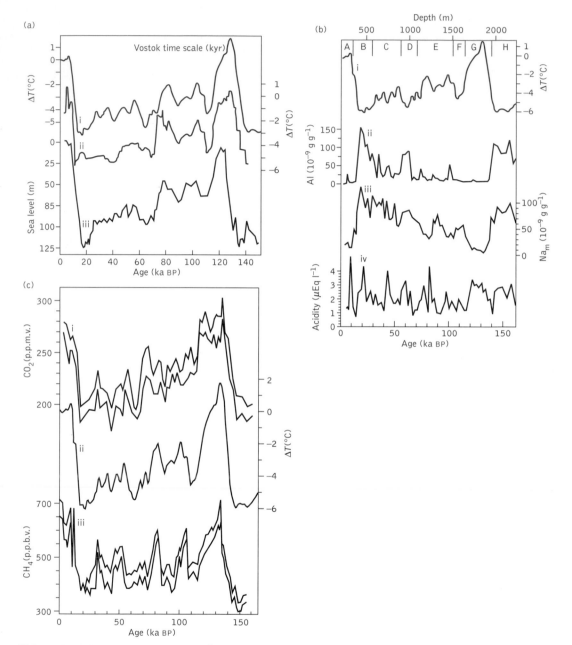

Fig. 3. (a) Time series of: the Vostok temperature record (i); sea surface temperature at the Indian Ocean site MD 84–551 (ii); and the $\delta^{18}O$ SPECMAP record shown as a sea-level curve (iii). (From Lorius, C., Jouzel, J., and Raynaud, D. (1993) Glacials–interglacials in Vostok: climate and greenhouse gases. *Global and Planetary Change* **7**, 131–43, Fig. 3.) (b) Time series of: the Vostok temperature record (i), the aluminium content (ii), the sodium content (iii), and acidity (iv). (From Lorius, C., Jouzel, J., and Raynaud, D. (1992) The ice core record: past archive of the climate and signpost to the future. *Philosophical Transactions of the Royal Society of London*, **B338**, 227–34, Fig. 2.) (c) Time series of: the Vostok carbon dioxide (CO_2) record (i), the temperature record (ii), and the methane (CH_4) record (iii). The temperature is shown as a smoothed record whereas the CO_2 and CH_4 records are shown as error bands. p.p.b.v., parts per billion by volume; p. p. m. v., parts per million by volume. (From Lorius, C., Jouzel, J., and Raynaud, D. (1992) The ice core record: past archive of the climate and signpost to the future. *Philosophical Transactions of the Royal Society of London*, **B338**, 227–34, Fig. 4.)

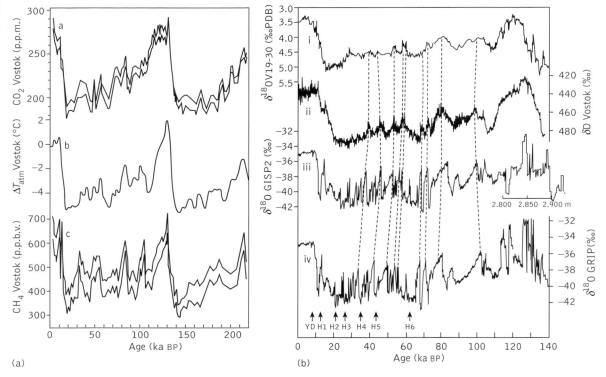

(a)

(b)

Fig. 4. (a) Carbon dioxide (CO_2), methane (CH_4), and temperature using the Vostok core 4 Γ with a timescale back to 22 000 BP. (Source: Jouzel, J. *et al.* (1993) Extending the Vostok ice-core record of palaeoclimate to the penultimate glacial period. *Nature*, **364**, 407–12, Fig. 5.) (b) Time series of: the benthonic $\delta^{18}O$ record from deep-sea core V19–30 (i), the Vostok deuterium profile (ii), the GISP2 $\delta^{18}O$ record (iii), and the GRIP $\delta^{18}O$ record (iv). Heinrich events are shown at the bottom of the plot. (From Jouzel, J. (1994) Ice cores north and south. *Nature*, **372**, 612–13, Fig. 1.)

between climate and sources and sinks of the greenhouse gases. The discovery of this strong correlation has now led to the idea that CO_2 and CH_4 changes have played a part in amplifying the relatively weak Milankovich forcing processes. Because the Milankovich orbital variations have their greatest effect on the northern hemisphere continents, the greenhouse gas amplification may also provide the link between northern and southern hemisphere climatic change. Statistical analysis has shown that about half the temperature change present in the Vostok record can be attributed to amplification by greenhouse gases.

Analysis of the 2546 m long 4 Γ core at Vostok has now been completed, extending the record back through the penultimate glacial period (140 000–200 000 BP) and back into the preceding interglacial. The correlation between the greenhouse gases and temperature (Fig. 4a) is still present but, notably, CO_2 is in phase with temperature at the start of the penultimate glaciation, between 220 000 and 190 000 BP, unlike the period 100 000–70 000 BP. A notable feature of the extended CH_4 results is the absence of oscillations during the

penultimate glacial period. The recalibrated age scale also shows that the duration of the last interglacial at Vostok is about twice as long and that the maximum temperatures were attained some 5000–6000 years before the corresponding features in the deep-sea record. There is debate as to why this should be, involving complicated arguments about validity of timescales. It is possible though that the differences are real and temperatures rose at Vostok some 5000 years before the northern hemisphere ice sheets began to melt.

Comparisons are now being made between Vostok and the detailed GRIP and GISP2 cores from the summit of the Greenland ice sheet. Figure 4b shows correlations between the deep-sea record, Vostok, GISP2 and GRIP over the past 140 000 years. Back to 100 000 years the correlation between GRIP, GISP2, and Vostok is very good. There are differences in detail, however, which may be due to the different response of Antarctica to rapid climatic changes in the North Atlantic. Between 20 000 and 105 000 BP the GRIP and GISP2 cores record 22 warm interstadial events reflecting extremely rapid climatic change in the North Atlantic. At Vostok, however, only nine interstadials are recorded in the same time period. It seems that interstadials occur at Vostok whenever those in Greenland last over 2000 years.

Before 100 000 BP, however, the records do not match well. The GRIP and GISP2 cores show evidence for rapid climatic fluctuations during the last interglacial while the Vostok does not. The 4 Γ core from Vostok at 2546 m is still far from bedrock (which is estimated to be at about 3700 m deep and over 500 000 years old at the base), and so one explanation of the discrepancy could be that ice-flow-induced disturbances have affected the bottom layers of GISP2. In the GRIP core, however, the later part of the last interglacial does not appears to be stratigraphically disturbed. The debate over rapid last interglacial climatic fluctuations continues. If confirmed, these fluctuations pose the interesting question of why our present interglacial should have been so climatically stable until now.

B. A. HAGGART

Further reading

Bradley, R. S. (1985) *Quaternary palaeoclimatology*. Allen and Unwin, London.

Jouzel, J. (1994) Ice cores north and south. *Nature*, **372**, 612–13.

anthropogeomorphology Anthropogeomorphology is the study of the human role in creating landforms and modifying the operation of geomorphological processes such as weathering, erosion, and deposition. The range of the human impact on both forms and process is considerable, and there are very few spheres of human activity which do not, even indirectly, create landforms. It is useful, however, to recognize that some features are produced by direct anthropogenic processes. They tend to be relatively obvious in form and are frequently created deliberately and knowingly. They include landforms produced by construction (e.g. spoil tips, embankments), landforms produced by excavation (e.g. open-cast mines etc.), landforms produced by hydrological interference (e.g. canals) and landforms produced by farming (e.g. terraces). Table 1 lists some of the major anthropogeomorphic processes.

Landforms produced by indirect anthropogenic processes are often less easy to recognize, not least because they tend to involve, not the operation of a new process or processes, but the acceleration of natural processes. They are the result of environmental changes brought about inadvertently by human technology. None the less, it is probably this indirect and inadvertent modification of process and form which is the most crucial aspect of anthropogeomorphology. By removing natural vegetation cover—through the agency of cutting, burning and grazing—humans have accelerated erosion and sedimentation. Sometimes the results will be obvious; for example when major gully systems rapidly develop. Other results may have less immediate effect on landforms but are, nevertheless, of great importance. By other indirect means humans can create subsidence features, trigger mass movements such as landslides, and even influence the occurrence of phenomena such as earthquakes.

Table 1 Major anthropogeomorphic processes

Direct anthropogenic processes
Constructional
    tipping: loose, compacted, molten, graded: moulded,
ploughed,
      terraced

Excavational
    digging, cutting, mining, blasting of cohesive or non-cohesive
      materials
    cratered tramped, churned

Hydrological interference
    flooding, damming, canal construction, dredging, channel
      modification,
    draining,coastal protection

Indirect anthropogenic processes
Acceleration of erosion and sedimentation
    agricultural activity and clearance of vegetation,
      engineering, especially road construction and urbanization
    incidental modifications of hydrological regime

Subsidence: collapse, settling
    mining hydraulic (e.g.groundwater pumping)
    thermokarst (melting of permafrost)

Slope failure: landslide, flow, accelerated creep
    loading
    undercutting
    shaking
    lubrication

Earthquake generation
    loading (reservoirs)
    lubrication (fault plane)

Finally there are situations where, through a lack of understanding of the operation of processes and the links between various processes and phenomena, humans may deliberately and directly alter landforms and processes and thereby set in train a series of events that were not anticipated or desired. There are, for example, many records of attempts to reduce coast erosion by important and expensive engineering solutions, which, far from solving erosion problems, only exacerbated them.

The possibility that the build-up of greenhouse gases (e.g. carbon dioxide, CO_2) in the atmosphere might cause global warming in coming decades has many implications for anthropogeomorphology (Table 2). Increased temperatures will have a direct impact on some landforms, but will also have an indirect effect because of associated changes in precipitation regimes, rates of evapotranspiration, and the distribution and form of vegetation assemblages. Increased sea surface temperatures may change the spread, frequency, and intensity of hurricanes—highly important geomorphological agents. Warmer temperatures will cause sea ice to melt and may lead to the retreat of alpine glaciers and the melting of permafrost (permanently frozen subsoil). The forms of vegetation will change and show latitudinal migration which will also influence the operation of geomorphological

Table 2 Some geomorphological consequences of global warming

HYDROLOGICAL
Increased evapotranspiration loss
Increased percentage of precipitation as rainfall at expense of
winter snowfall
Increased precipitation as snowfall in very high latitudes
Increased risk of cyclones (greater spread, frequency, and intensity)
Changes in state of peat bogs and wetlands
Less vegetational use of water because of increased CO_2 effect on
stomatal closure

VEGETATIONAL CONTROLS
Major changes in latitudinal extent of biomes
Reduction in boreal forest, increase in grassland, etc.
Major changes in altitudinal distribution of vegetation types
(c. 500 m for 3 °C)
Growth enhancement by CO_2 fertilization

CRYOSPHERIC
Permafrost decay, thermokarst, increased thickness of active layer,
instability of slopes, river banks, and shorelines
Glacier melting
Sea ice melting

COASTAL
Inundation of low-lying areas (including wetlands, deltas, reefs)
Accelerated coast recession
Changes in rate of reef growth
Spread of mangrove swamp

AEOLIAN
Increased dust storm activity and dune movement in areas of
moisture deficit

SOIL EROSION
Changes in response to changes in land use, fires, natural
vegetation cover, rainfall erosivity, etc.
Changes resulting from soil erodibility modification (e.g. sodium
and organic contents)

processes. Changes in temperature, precipitation quantities, and the timing and form of precipitation (e.g. whether it falls as rain or snow) will have a whole suite of important hydrological consequences. Some parts of the world might become moister (e.g. high latitudes and some parts of the tropics) while other parts (e.g. the High Plains of the USA) might become drier. The latter would suffer from declines in river flow, reactivation of sand dunes, and increasing frequency of dust storms.

However, among the most important potential future anthropogeomorphological changes are those associated with sea-level change caused by the melting of land ice. Low-lying areas (e.g. salt marshes, mangrove swamps, deltas, coral atolls) would tend to be particularly susceptible, but more generally rising sea levels could promote beach erosion.

ANDREW S. GOUDIE

Further reading

Goudie, A. S. (1999) *The human impact on the natural environment* (5th edn). Blackwell, Oxford.

anticyclone As the name suggests, an anticyclone is the opposite of a cyclone. Whereas cyclones are represented by lower than normal atmospheric pressure at the Earth's surface, anticyclones have higher than normal pressure. Where cyclones are dominated by air moving upward between low-level regions of convergence and upper level regions of divergence, anticyclones are exactly the opposite, with downward motion between upper convergence and lower divergence. Where cyclones have positive vorticity, anticyclones have negative vorticity; that is, they rotate in a clockwise sense in the northern hemisphere, and anti clockwise in the southern hemisphere.

Permanent examples of anticyclones exist in the subtropics, where a belt of anticyclones girdles the world at latitudes between about 20 and 40 degrees. In satellite pictures (Fig. 1) the visible manifestation of an anticyclone is a lack of cloud. Most anticyclones are very large systems, thousands of kilometres across, and deep enough to reach the tropopause. Winds in anticyclones are light; air near the surface therefore moves little and has sufficient time to adopt uniform characteristics. As a result, anticyclones are ideal source regions for air masses.

As a result of the deep downward motion in anticyclones, the air through most of the troposphere becomes much warmer and drier than usual. If temperature is measured at

Fig. 1. An image taken from *Meteosat*, the European geostationary weather satellite, showing anticyclonic cloud-free skies over much of Europe.

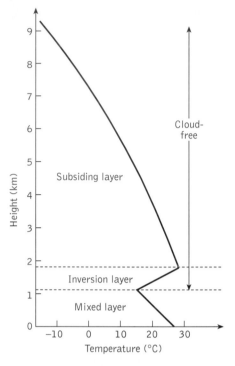

Fig. 2. The vertical variation of temperature through an anticyclone.

to produce convection. The inversion layer exists as the boundary between the downward-moving air in the anticyclone and the turbulent boundary layer.

Between the inversion layer (at about 1 km altitude) and the tropopause (at 10 to 15 km) clouds do not usually develop. However, within the boundary layer shallow clouds can form, particularly if the surface is moist. Shallow layers of cloud often form overnight in anticyclonic regions, but usually evaporate in the morning when heat from the Sun warms the boundary layer. Anticyclones are usually associated with warm, sunny weather and no precipitation, although clear skies during the hours of darkness can lead to very cold nights. Unfortunately, pollution released into the boundary layer is not easily dissipated and cannot be dispersed either above the inversion layer or (because winds are usually light) over large geographical areas. These are the circumstances under which dangerous smog levels are encountered in large cities such as Los Angeles and which in 1953 caused the deaths of thousands in a few days in London.

Anticyclones in middle latitudes are commonly of two types. The first is the ridge of high pressure which exists between depressions moving from west to east steered by the upper-level jet stream. These anticyclones give temporary respite from the wet and windy conditions associated with the

different altitudes, the typical profile through an anticyclone can be separated into three regimes (Fig. 2). The uppermost region, the subsiding region, is dominated by the downward motion; this region is warm and dry and the temperature decreases with height. Below this there is a shallow layer, the inversion layer, in which the downward motion ceases. At the bottom of this layer lies air which has not been moved from upper levels, whereas the air at the top has undergone considerable subsidence. Subsiding air is warmed as it is compressed by ever-increasing pressure; the further it subsides the more it is warmed. The air at the top of the inversion layer has subsided from high altitudes and therefore is warmer by several degrees than the air at the bottom. Although the inversion layer is usually shallow (rarely more than a few hundred metres), it has enormous significance because it is extremely stable. Under average conditions the temperature decreases at a rate of about 6.5 °C per kilometre in the troposphere. In the inversion layer, temperature increases with height, often at a rate of more than 10 °C per km. The great stability of this layer means that it acts as an impenetrable lid on the bottom layer, the mixed boundary layer. This boundary layer is dominated by turbulence caused by friction at the Earth's surface. The mixing in the boundary layer is always present. The mixed layer has a depth of up to about a kilometre, although it is shallower at night when there is no heating of the surface

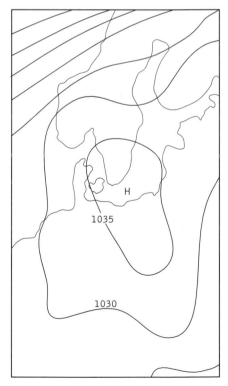

Fig. 3. A typical anticyclone on a weather map. H, high pressure. Figures shows pressures in millibars.

depressions. They are transient features which pass in a day or two. The other type is the 'blocking' anticyclone or 'block'. This is a large quasi-stationary system, similar in structure to a subtropical anticyclone but occurring much further poleward. When these blocks occur they persist for days or even weeks, and because they are located in the latitudes where depressions usually travel from west to east they block the passage of the depressions and force them to divert either north or south of the block (Fig. 3). These blocking anticyclones can provide long periods of stable weather which changes little from day to day, making weather forecasting relatively easy. It is, however, particularly difficult to forecast the onset or the end of a block.

Anticyclonic weather is usually benign. However, the very rapid changes in temperature and humidity in the inversion layer affect the transmission of microwaves used in television and telephone communications. The paths of the transmitted beams are bent to produce 'anomalous propagation', which causes the signals to travel much further than is normal, producing weak or disturbed television signals.

CHARLES N. DUNCAN

Further reading

Ahrens, C. D. (1994) *Meteorology today.* West Publishing Co., St. Paul, Minnesota.

Palmen, E. and Newton, C. W. (1969) *Atmospheric circulation systems.* Academic Press, London.

applied geomorphology Geomorphology has traditionally focused on the study of landforms and on the processes involved in their formation. Applied geomorphology is the practical application of this study to a range of environmental issues, both in terms of current problems and of future prediction. Applied geomorphology provides a strategic tool for informed decision-making in public policy development and in environmental resource management. Key areas of application include specific environmental settings, such as the coastal zone or dryland environments; the impacts of land use and management practice on Earth surface processes; and areas susceptible to natural hazards. Examples of these areas of applied geomorphology are outlined below.

Over 60 per cent of the world's population live in the coastal zone in environments ranging from coral atolls, reclaimed or natural wetlands, dune-backed beaches, and barrier islands to cliff tops. Settlements under threat from coastal erosion and flooding from storm events, sea surges, and rising sea level lobby for protective engineering measures to prevent loss of property, livelihood, and life. Geomorphology has several applications in settings of this type. An understanding of coastal landforms and the processes acting upon them can be used to map areas at risk from cliff failure, beach erosion, and flooding. This approach is of interest to potential developers and the insurance industry and is an important tool in environmental impact assessment (EIA). An understanding of the geomorphology of the coastal zone

can also be used to predict the effects of modifying the coastal system. The installation of groynes, breakwaters, or protective sea walls has knock-on effects on the natural circulation of water and sediment in the near-shore environment. Artificially stabilizing cliffs to prevent erosion may seem the obvious solution for cliff-top dwellers, but a geomorphological evaluation might predict that this approach could starve beaches of the sediment provided by natural cliff fall, with a consequent impact on longshore drift of sediment, and would relocate the focus of erosion further along the coast. The nature of the problem may thus change from cliff failure at one site to beach erosion and subsequent flooding at another. An understanding of the nature and complexity of coastal dynamics is thus an essential component of a coastal-zone management strategy and is important in predicting the future effects on coastal landforms of a rise in sea level.

River-management strategies for flood alleviation have often adopted engineering solutions concentrated in particular river reaches, which are usually in areas of urban development. Reach-specific intervention measures include lining the natural channel with concrete to prevent erosion and bank instability, channel straightening to force flood water to flow rapidly through particular reaches, and flow-control structures such as sluice gates and reservoirs to control water level. These artificial measures are not always successful in preventing flooding and erosion within the river catchment, and natural sections further downstream may be overwhelmed by the river at peak flood. The engineered reaches of rivers often become a sterile landscape because fast-flowing water in a concrete-lined channel, with minimal variation in water depth and channel cross-section, provides a poor habitat for wetland flora and fauna. Geomorphology has been applied to 'river restoration' to recreate an integrated river management strategy within artificially created river systems, maximizing biodiversity while controlling river-flow conditions. Applied geomorphology uses a holistic approach to river response at a catchment-wide scale; the basis here is an understanding of the relationships between river form and process, sediment transport, and the important role of river-bank (riparian) vegetation.

Certain landscapes have specific properties that impinge on our use and development of the environment. In cold environments, the presence of ground ice leads to problems in construction, communication, and housing. In permafrost zones, the ground is permanently frozen except for the upper layers of the soil, which thaw in the summer. The upper soil, known as the active layer, is subject to repetitive cycles of freezing and thawing, making it geomorphologically active. The ground within the active layer will suffer heaving and deformation, disrupting communications and making road construction impracticable. Applied geomorphology can be used in mapping the active layer and ground ice in areas with differing rocks and sediments. This information is then used to evaluate the problems that are likely to affect these areas. Ground heaving depends on the depth of the active layer and the type of sediment present; fine-grained silts present more of a problem than gravels. Additional problems in permafrost areas, as, for example, in some regions of Alaska, occur where structures have suffered dramatic subsidence as a result of heating in buildings. Without appropriate insulation, heat radiates downwards from the building into the ground, thaws the underlying ice, and increases the depth of the active layer, thus effectively changing the structure of the soil. Applied geomorphology is consequently essential in land-use planning and site evaluation, in order to recognize such potential problems as land subsidence, slope instability, invasion of wind-blown sand, and impacts on natural drainage systems.

Land used for agricultural production may suffer from degradation and desertification as a result of soil erosion, landsliding, and over-extraction of water for irrigation. Much agricultural practice focuses on maximizing yield and profit, often using techniques that can be detrimental to the environment, both in the short and the long term. Applied geomorphology uses an understanding of the relationships between surface conditions, climate, vegetation, and soil erosion to advise farmers and politicians on how to improve land management for sustainable use of land and water resources.

Natural hazards such as volcanic eruptions, earthquakes, and mudflows present a significant risk to the population of the surrounding area. Geomorphological mapping can be used to assess the present condition of the landscape and provide a hazard map. The expression of a disaster may result in one settlement having significantly different risk assessment. For example, a volcanic eruption may pose a threat from volcanic ash and lava flows, pyroclastic flows, and bombardment from superheated volcanic bombs or associated hazards such as mudflows, depending on topography, soil cover, type of eruption, and predominant wind direction. This application of geomorphological analysis is of significant interest to the emergency services and the insurance industry.

Applied geomorphology can be used in modelling change to landforms and surface processes. This can include change from human impact on the environment to future prediction of climate change, from short-term El Niño and tropical storm events to longer-term change resulting from greenhouse warming and rising sea levels. In this way, applied geomorphology has a key role in managing the environment to minimize potential degradation of land, water, and natural resources.

MICHÈLE L. CLARKE

Further reading

Cooke, R. U. and Doornkamp, J. C. (1990) *Geomorphology in environmental management* (2nd edn). Oxford University Press.

aquifer 'Aquifer' is the term applied to a geological formation or group of formations with sufficient permeability and water-saturated porosity to transmit and store significant quantities of subsurface water under normal hydraulic gradi-

ents. In this context, 'significant' is usually related to potential water yield from wells or flow to springs. Aquifers can be considered as reservoirs for groundwater resources and thus are also defined, somewhat imprecisely, on the basis of their economic potential.

Geological materials that are generally considered to be good aquifers include unconsolidated sands and gravels, sedimentary rocks such as sandstones or carbonates with abundant primary pores, and sedimentary rocks containing significant secondary porosity resulting from fractures or dissolution, or both. Certain igneous rocks, such as basalt containing abundant cooling fractures or interconnected vesicles, can also be productive aquifers. Geological materials with lower porosity and permeability, such as fractured granitic rock, would not be classified as aquifers in regions where they are overlain by coarse sediments or sedimentary rocks. Such materials of lower permeability could, however, be considered as aquifers in regions where the only other formations were less-fractured crystalline rocks.

Confining units and confined aquifers

Aquifers are distinguished from adjacent formations with lower permeability or storage potential, which are called *confining units*. Aquifers containing the top of the saturated zone, or water-table, are called water-table or unconfined aquifers. Aquifers that are overlain by confining units are known as confined aquifers. The water level in a well drilled into a confined aquifer will rise above the top of the aquifer, in some instances even rising above the land surface so that the well flows freely without pumping. The imaginary surface defined by the stable, or static, water levels measured in wells completed in a confined aquifer is called the *potentiometric surface*. Directions and rates of groundwater flow in a confined aquifer are determined by the slope of the potentiometric surface. Early drilling through a confining unit in the Artois region of France produced flowing wells from the underlying confined aquifer. These flowing wells were subsequently called *artesian* from *Artesium*, the Latin name for the Artois region. Although 'artesian' is still commonly used to refer to flowing wells, according to modern technical usage the terms 'artesian' and 'confined' are synonymous; confined aquifers are therefore equivalent to artesian aquifers.

Geological materials that commonly form confining units include fine-grained sediments such as silts and clays, sedimentary rocks of low permeability, such as shales or evaporites, and relatively unfractured igneous or metamorphic rocks. Confining units are further subdivided by some workers into aquicludes and aquitards. In contrast to an aquifer, an *aquiclude* is not capable of transmitting significant quantities of water. *Aquitards* retard, but do not prevent, flow of groundwater to or from adjacent aquifers. Both aquicludes and aquitards may contain significant amounts of water in storage. During the exploitation of groundwater in basins containing thick sequences of alternating aquifers and confining units, much of the water removed from storage may ultimately come from draining of the confining units. Removal of water from compressible materials such as silt and clay can lead to compaction of the confining unit and subsidence of the land surface. Confining units such as evaporite deposits that neither transmit nor store significant quantities of water are sometimes called *aquifuges*.

Perched aquifers

In cases where large variations in permeability and pore size exist in the sediments or rocks of the unsaturated zone, infiltrating water can sometimes accumulate within and above low-permeability layers such as clays, forming a zone of water-saturated pores above the regional water-table. The upper boundary of this saturated zone is called a *perched water-table* and the saturated zone itself is called a perched aquifer. Perched aquifers may be of limited or large areal extent and they may be stable features or only transient phenomena that develop during periods of intense recharge. Unlike confined or unconfined aquifers, perchd aquifers are rarely reliable and economic reservoirs for water supply. They can, however, be significant to problems of groundwater contamination. Contaminants that infiltrate to a perched aquifer can spread laterally before resuming downward transport to the regional water-table, resulting in a larger area of initial water-table contamination.

Hydrostratigraphic units and regional aquifer systems

Geological formations are frequently divided and combined by stratigraphers into lithostratigraphic or biostratigraphic units on the basis of similarities either in lithological properties or fossil assemblages. The major geological units in a region can also be divided and combined on the basis of similarities in hydrological properties. The resulting units, which can range from low-permeability confining units to highly productive aquifers, are known as hydrostratigraphic units. Important hydrostratigraphic units, some of which extend over very large regions, are commonly given names similar to those assigned to lithostratigraphic units. Examples include the Floridan, a carbonate aquifer in Florida; the Ogalalla aquifer, consisting of sands and gravel that extend from South Dakota to Texas in the United States; and the fissured, highly permeable Chalk aquifer of the United Kingdom.

Patterns of groundwater flow are controlled, in part, by the spatial arrangement of hydrostratigraphic units that make up the regional aquifer system. Early mathematical analyses of regional flow fields were based on assumptions that the aquifer system was essentially homogenous. Later work, facilitated by the advent of computer modelling techniques, demonstrated striking effects on groundwater flow fields caused by contrasts in permeability associated with regional hydrostratigraphy (Fig. 1). A major component of most modern groundwater studies is the identification and characterization of the major hydrostratigraphic units.   J. BAHR

aragonite *see* CALCITE, ARAGONITE, DOLOMITE

Archaean eon

The second of the four eons of Earth history, the Archaean (from the Greek, *beginning*) is the oldest for which a substantial record exists in the rocks of the continental crust. A commonly accepted criterion for establishing the identity of this eon is the presence of life on Earth, although evidence of life may not be obvious from many of the rocks themselves. The record has been much deformed, reconstituted, and obliterated during the subsequent Proterozoic and Phanerozoic eons. Isotopic studies indicate that these two eons amount at least some 2.5 billion years (Ga). The oldest terrestrial materials currently known are detrital zircon crystals from the Jack Hills and Mount Narryer in Western Australia at 4.1 to 4.3 Ga. The Acasta gneiss from the Slave Province of the Canadian Shield is the oldest dated rock at about 4 Ga, and the Isua belt in western Greenland is the oldest large-scale rock succession, estimated at about 3.8 Ga. At the other end of the Archaean, the passage into the Proterozoic is thought of as about 2.5 Ga. By this time as much as 45 per cent of the total history of the Earth had taken place, and the beginning of the Phanerozoic eon was still about 1.9 Ga in the future.

During the Archaean the surface of the Earth underwent much change, together with changes to the atmosphere and hydrosphere. Underlying these changes was the evolution of a cooling pattern in the convecting mantle and the overlying crust by which a form of plate-tectonic activity was developed. The 'rock cycle' operated, and it is believed that the beginning of the eon saw the vigorous and unprecedented production of granitic continental crust. The bulk of this type of crust had been formed by the end of the eon. Environments comparable to those of today can be discerned from parts of the rock record. Nevertheless, evidence for the Archaean world remains very fragmentary, despite recent great improvements in analytical techniques and isotopic dating.

Outcrops of Archaean crust are present on all continents and include the nuclei of several continental shields or cratons (Fig. 1). Southern Africa, Brazil, India, Australia, Antarctica, Russia, China, and North America reveal major areas of Archaean rocks. Early classic studies were also carried out in Finland, Scotland, and Greenland. Many regions bear diverse economic deposits: gold and other metals, asbestos, and radioactive ores. The main Archaean rock units and events of North America, Greenland, and Scotland are set out in Fig. 2.

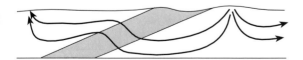

Fig. 1. Groundwater flow paths in a system with complex hydrostratigraphy. The shaded region is a zone of low permeability.

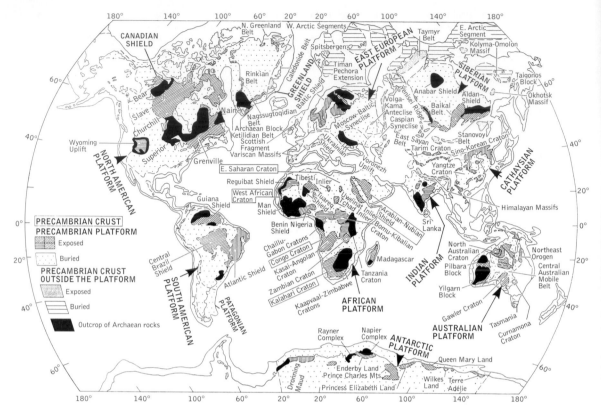

Fig. 1. World map of major outcrops of Archaean rocks (solid shading). Proterozoic rock outcrops of various structural provinces are shown with different shading. The concealed extent of the Precambrian basement of the continents is shown dotted. (After Goodwin (1991).)

Two types of terrain are especially characteristic of Archaean geology: greenstone–granite terrains and granulite–gneiss terrains. They appear to result from orogenic activity, but are not, strictly speaking, linear, and are not comparable to the orogenic belts of later times. The greenstone belts are thick accumulations, sharply defined, mostly composed of ultrabasic lavas and overlying sediments that have undergone several phases of strong deformation. The characteristic metamorphic rock produced is, nevertheless, only of the relatively low-grade greenschist facies. Most typical of the volcanicity is the presence of ultrabasic (ultramafic) lavas and the highly magnesian basic lavas known as komatiites. The latter are regarded as products of partial melting of the underlying mantle at temperatures well above 1700 °C. The sedimentary rocks associated with the greenstones include cherts, banded ironstones, and turbidites, and the whole assemblage has in many places been intruded by later granites. Several greenstone–granite terrains have been ascribed to subduction at continental margins, their evolution occupying several hundred million years. Other volcanic centres were associated with rift basins where tholeiitic basalts were erupted over great areas.

The granulite–gneiss belts contrast with the greenstone belts in their very high grades of metamorphism. They are also characteristically of sedimentary origin, with quartzites, banded ironstone formations, and (dolomitic) limestones. There are enormous intrusions into many of these terrains; layered anorthosites, dyke swarms of basalts, and complexes of tonalite or diorite intrusions also occur. The metamorphism that affected these belts universally accompanied earth movements that warped these crustal accumulations into broad basins and domes and reached very high temperatures. It seems to have attained a peak at about 2.8 Ga: not very long, relatively speaking, before the Archaean may be said to have ended.

The earliest undeniable forms of life are present in rocks of this general age—as in the Onverwacht and Fig-Tree Groups of South Africa, where filamentous bacterium-like microorganisms occur. Algal mats and stromatolites are known from many horizons within the Archaean, commonly well preserved in carbonate strata in Canada, South Africa, and Australia. Many stromatolite beds seem to occur on the margins of greenstone belts, where some may have been asso-

North American Platform less Greenland Shield

Eon	Era	Ga	Orogeny/episode	Salient units and events	Nutak Segment (Nain Province)
		2.5		Huronian supergroup (2.4–2.1 Ga) Proterozoic Cratonization of Archaean provinces. Major crust-forming events (c. 2.6 Ga)	Major deformation, metamorphism, and plutonism
ARCHAEAN	LATE	2.6 2.7 2.9	Kenoran (Algoman, Fiordian)	Granitoid plutonism (2.76–2.65 Ga) Greenstones: Slave Province (2.68–2.65 Ga) Superior Province { (mainly 2.76–2.70 Ga; also 2.85–2.80 Ga; and 3.0–2.9 Ga) Churchill Province: Kaminak (2.7 Ga)	
	MIDDLE	3.0 3.1 3.4	Wanipigowan (Laurentian, Hopedalian)	Prince Albert (2.9 Ga) Granulite-amphibolite metamorphism, tonalitic gneiss, granite (3.1 Ga) Slave Province basement gneiss (3.15 Ga) Beartooth supracrustals (Wyoming) (3.3 Ga)	Granulite metamorphism. Major reactivation of gneiss (3.1 Ga) Anorthosite–gabbro complexes Upernavik supracrustals Saglek dolerite dykes
	EARLY	3.5 3.9	Uivakian (Mortonian)	Morton (Michigan) gneiss (tonalitic) (3.4 + Ga) Pre-Morton supracrustals (Minnesota)	Major crust-forming event Uivak gneiss: deformation, metamorphism (3.5 Ga) Pre-Uivak supracrustals
Pre-Archaean (Hadean)		4.0			

Greenland Shield plus Scottish Shield Fragment

Eon	Era	Ga	Orogeny/episode	Salient units and events	
		2.5		Proterozoic	2.5
ARCHAEAN	LATE	2.6 2.7 2.9	Qorqut	Final cratonization of Archaean block Qorqut granite (2.6–2.5 Ga) Amphibolite-facies metamorphism Late granites	Badcallian granulite-facies metamorphism (2.7 Ga) Lewisian Complex (2.9 Ga), including Scourian metasediments
	MIDDLE	3.0 3.1 3.4	Nuk	Granulites-facies metamorphism (2.8 Ga) Major deformation (nappes) Nuk gneiss and major deformation (3.1–2.9 Ga) Anorthosite (-gabbro) layered complexes	3.04
	EARLY	3.5 3.9	Amitsoq	Malene supracrustals Ameralik dolerite dykes Major crust-forming event Amitsoq gneiss (3.82–3.7 Ga) Isua–Ilasia supracrustals	3.5 3.82
Pre-Archaean (Hadean)		4.0			

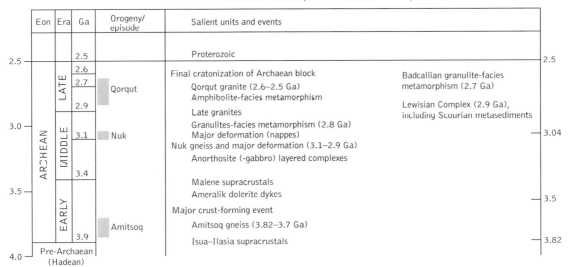

Fig. 2. Archaean units and events in the development of the North American and Greenland platforms and the Scottish Shield fragment. (After Goodwin (1991).)

ciated with hot springs. It is interesting that the Late Archaean stromatolites exhibit a wider variety of dome and pillar-like shapes than do the earlier forms. Graphite deposits also suggest the presence of possible organic material. Geochemical evidence from levels in which stromatolites and other mesosopic evidence of life exists suggests the presence of photosynthetic cyanobacteria and denitrifying bacteria. All in all, the geological evidence for life in the Archaean indicates that single-celled prokaryotes (cells lacking a nucleus) and possibly eukaryotes (cells with a nucleus) were present. It is possible that some archaeobacteria existed close to hydrothermal springs.

Fig. 3. Archaean gneisses (Lewisian Gneiss Formation) near Stoer, north-western Scotland, in the foreground, were highly metamorphosed and peneplaned prior to the deposition of the thick Proterozoic Torridon Sandstone which now remains as a series of relict hills on the Archaean rocks. The time gap between the erosion of the gneisses and the laying down of the Torridon sediments probably amounted to several hundred million years. (Photo British Geological Survey IPR/3-106. © NERC. All rights reserved.)

The Archaean world thus emerges as one with well-defined oceans and continents and a well-developed erosional cycle on the continents. A wide variety of sedimentary environments existed on the continents and the adjacent shallow marine shelves. It is possible that on at least one occasion the continents were clustered into a 'supercontinent'. Vast volcanic eruptions were frequent. Heat loss was rapid but the growth in crustal thickness meant that the rate of plate-tectonic activity would not have been greatly different from today. Volcanic activity on vast scales provided the bulk of the atmosphere and hydrosphere by degassing and the carbon dioxide-rich atmosphere was gradually replaced by an oxygen-rich atmosphere once photosynthesis had provided more than enough oxygen to satisfy the needs of the ironstone formations. A greenhouse effect within the early secondary atmosphere was thereby slowly diminished, and the surface temperature of the planet was not greatly different from that of today. Finally, the global distribution of late Archaean orogenic episodes suggests that it was a time of renewed and heightened vigour, accelerating environmental change prior to the quieter times of the Proterozoic.   D. L. DINELEY

Further reading

Coward, M. P. and Ries, A. C. (eds) (1995) *Early Precambrian processes.* Geological Society Special Publication No. 45.

Goodwin, A. M. (1991) *Precambrian geology: the dynamic evolution of the continental crust.* Academic Press, London.

Nisbet, E.G. (1987) *The young Earth. An introduction to Archaean geology.* Allen and Unwin, Boston.

archaeological geology The study of archaeological sites for data relevant to geological investigations has been much pursued and enhanced in recent decades. They have ranged from ancient hearths in Australia to architectural ruins in the classical sites of the eastern Mediterranean. Clues to the occurrence and magnitude of earthquakes, the nature of local volcanic eruptions, the rise and fall of sea or land levels, floods, and other disruptive events have been gathered and correlated to human history and activities. For example, in the Roman city of Hierapolis in Western Turkey displaced walls and irrigation gutters reflect the city's location astride a newly developing fault zone where shocks occurred many times. At Phalasarna in western Crete a harbour dating from classical times is now several metres above sea level. Its uplift probably came about from movement on an offshore fault, possibly associated with the tectonic plate boundary along which Africa is slowly sliding beneath Eurasia. There many instances like this where the rate of such local or regional neotectonic processes may now be judged.

Clues have been sought also in order to test the validity of legends or folklore. The disappearance of the legendary Atlantis has been linked both to sudden fault-initiated collapse of land areas into the sea and to volcanic explosions. Elsewhere geological processes or events, some involving disaster for human communities, have been revealed from archaeological evidence. One such discovery was of charred skeletons near Pompei which had been smothered by a *nuée ardente* discharge from Vesuvius in AD 79. The instantaneous preservation of so much of that Roman city presented both archaeological and geological problems, but these have been at least partly resolved by experience of *nuées ardentes* eruptions at Mt. St Helen's, and further archaeological work to confirm the enormously high temperatures experienced at Pompei in the AD 79 disaster.

Data of service to archaeological geology may be revealed as a consequence of direct (surface or submarine) observations or may be found as a result of excavations and the employment of geological investigation techniques for primarily archaeological ends. Thus geoarchaeology and archaeological geology can be mutually informative.   D. L. DINELEY

Further reading
McGuire, W. J., Griffiths, D. R., Hancock, P. L., and Stewart, I. S. (eds) (2000) *The archaeology of geological catastrophes.* Geological Society Special Publication No. **171**.

arid-zone hydrology The term 'arid' is used in many ways. In each usage it carries the connotation of words such as *dry*, *desert*, *barren*, and *empty*. Of these, only the first, dry, is generally true of arid zones. Arid areas receive too little rainfall to support dryland agricultural or domestic livestock grazing. In contrast, in semi-arid areas adequate moisture is usually available at some time during the year to produce forage for livestock, and there are some years when dryland crop production may be successful. However, both climates are characterized by extreme variability, with commonly occurring droughts and infrequent periods of flooding.

A first-order classification of arid zones can be made using long-term mean annual precipitation as the criterion. More than a third of the world's land surface is either arid, generally receiving less than 250 mm of annual precipitation, or semi-arid with between 250 mm and 500 mm of annual precipitation. More precise definitions of arid and semi-arid areas are based on climatology and are given in climatic classifications based on precipitation, temperature, and their seasonal distributions. In the remainder of this entry, we use 'arid zones' to mean those areas of the world, excluding the polar deserts, which are arid and semi-arid by the first-order classification given above.

The global distribution and climatic settings of these arid zones are discussed elsewhere (see *deserts* and *drylands*) but, broadly, they are found along two wide belts at approximately 30° latitude north and south of the Equator (Fig. 1). On a more localized scale, however, a combination of terrain and prevailing wind direction can cause rain-shadow effects, resulting in arid zones downwind of major mountain features.

Arid-zone hydrology, literally the study of water in these areas, is conveniently split into surface-water hydrology, dealing with the atmosphere, the land surface, and their inter-

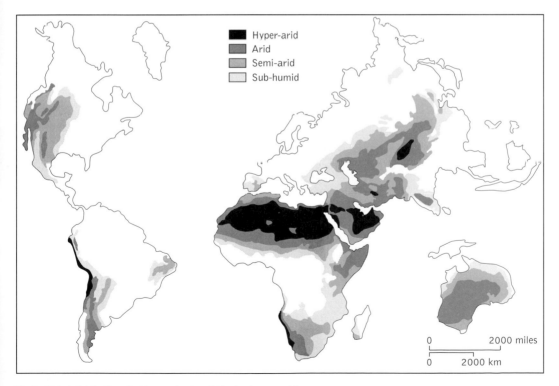

Fig. 1. Global distribution of arid zones (scale valid for land areas only).

actions, and groundwater hydrology, dealing with waters below the surface of the earth (see *subsurface flow and erosion*). Surface water and groundwater are linked in nature, since infiltrated water seeping through the soil can recharge groundwater and groundwater can return to the surface as springs, seeps, and streamflow.

A convenient way to visualize arid-zone hydrology is through the concept of an annual water balance on watersheds. The term 'watershed' (or catchment) means an area above a specified point on a stream channel enclosed by a perimeter of higher ground; the watershed perimeter defines an area where surface run-off will move into the stream or tributaries above the specified point. Precipitation that falls on the land surface can take several paths on its way through the hydrological cycle back to the streams, lakes, rivers, oceans, and atmosphere. An annual water balance is a way of accounting for all the precipitation that falls on a watershed in a year.

If the rate of rainfall is high enough and lasts long enough, surface run-off can occur, resulting in overland flow, streamflow, and, in the extreme, flooding. Rainfall that stays on the surface of rocks, soils, vegetation, litter, etc. can be evaporated directly back into the atmosphere. Rainfall that infiltrates beneath the surface of the soil recharges soil moisture and can then move deeper and recharge groundwater, or can move back into the atmosphere by two routes. Soil moisture can be evaporated directly into the atmosphere. It can also enter plant roots, and through biological processes (transpiration) be released back into the atmosphere. Movement of rainfall back into the atmosphere by these two processes is termed evapotranspiration and simply means the combined processes of evaporation and transpiration. On an annual basis, one accounts for all the precipitation falling on a watershed by stating that the annual change in soil moisture storage in the watershed is equal to the precipitation minus the surface run-off, the evapotranspiration, and the recharge to groundwater. Precipitation falling as snow complicates the water balance as defined in this way, but the principles are the same.

Arid-zone hydrology is thus defined as the study and quantification of components of the water balance. Timescales can be longer (from decades to centuries) or shorter (from seconds to seasons) than a year but the basic principles remain the same. Important features of arid-zone hydrology that differ from hydrology in more humid zones include the key fact that, on an annual timescale, the maximum possible, or potential evapotranspiration, far exceeds the mean annual precipitation, resulting in the dryness associated with arid zones. This dryness, in turn, results in ephemeral stream channels and rivers that are dry for most of the time. Locally, these dry streams are called by a variety of names, for example, wadis (in the Middle East), washes, and arroyos (North America) (see *arroyos*).

Because of the dryness, one might consider hydrology to be less important in arid zones than in more humid areas.

Nothing could be further from the truth. Because water is so scarce, its occurrence and abundance is of primary concern in arid zones: it is here that the links between water and life are most evident.

L. J. LANE

Further reading

Cook, R. U., Warren, A., and Goudie, A. S. (1993) *Desert geomorphology*. UCL Press, London.

Graf, W. L. (1988) *Fluvial processes in dryland rivers*. Springer-Verlag, New York.

Renard, K. G., Lane, L. J., Simanton, J. R., *et al.* (1993) Agricultural impacts in an arid environment: Walnut Gulch studies. *Hydrological Sciences and Technology*, **9** (1–4), 145–90.

Trewartha, G. T. and Horn, L. H. (1980) *An introduction to climate* (5th edn). McGraw-Hill, New York.

arroyos Infrequent, sometimes of high intensity, rainstorms in arid and semi-arid areas can generate large volumes of run-off from the poorly vegetated land surfaces that characterize such regions. This rapid run-off can be highly erosive, detaching and removing soil and rock particles. Where the run-off becomes concentrated, especially in the valleys, deep channel cutting may occur in the unconsolidated valley sediments. These steep-sided drainage courses, which may only occasionally carry streamflow, are known by a variety of names, depending on the region: terms such as 'donga' (southern Africa) and 'wadi' (Arabia and north Africa) are examples. In the south-western USA such channels are sometimes called arroyos—a word derived from the Spanish '*arrugia*' relating to a mine shaft or cutting. Very often, arroyos are simply referred to as gullies.

During the late nineteenth and early twentieth centuries, many valleys in the American west began to be entrenched by arroyos. The introduction of large cattle herds and subsequent overgrazing and depletion of the vegetation cover, causing rapid run-off, was blamed. However, theories of arroyo formation invoking climatic changes have also been proposed. It is known that arroyos existed before human occupancy so that it appears likely that arroyos can occur as a natural part of drainage system development.

IAN A. CAMPBELL

art and the Earth sciences The links between art and the Earth sciences—geology in particular—have not been widely recognized, yet both areas have impinged on each other in a variety of ways. Four main interactions are considered here: Earth sciences phenomena as a source of artistic inspiration; geological illustrations as art; the use of geological materials in art; and the use of geology in the investigation of art objects.

The Earth sciences as a source of artistic inspiration

Many artists have been influenced by phenomena relating to the solid Earth, the atmosphere, and the oceans. Their subjects have been highly varied, ranging from volcanoes, mountain ranges, storms, and other dramatic features to the

Fig. 1. Li K'e-jan: Gorge with coastal village. Ink and paint on paper, 1964.

Fig. 2. Katsushika Hokusai: *Fuji in Clear Weather*. One of the 'thirty-six views of Fuji', 1834–5.

subtleties of atmospheric effects. China and Japan have a long history of landscape painting, dating in China from the ninth century. The Chinese word for landscape is, significantly, *shanshui*, literally a 'mountain-water' picture; the karst scenery of southern China has been a favourite subject for Chinese water-colourists over the centuries (Fig. 1). In Western Europe, landscape painting emerged much later, in the Renaissance, and it was not until the rise of the Romantic movement that a taste for nature in the wild became general. As the western tradition diffused to America and other parts of the world, vast tracts of novel and as yet unspoiled landscape were opened up for a new generation of artists.

Volcanoes have been depicted by many artists, some of whom have approached their subjects geologically as well as pictorially. One such was the German artist Jakob Philip Hackert (1737–1807), who had a close association with the volcanologist Sir William Hamilton; his painting of an eruption of Vesuvius in 1774 is a study of a lava eruption on the flanks of the volcano. At the same period, Joseph Wright of Derby (1734–97), another artist with a keen interest in natural phenomena (although working in a different style from Hackert), painted Vesuvius in eruption while he was in Italy from 1773 to 1775. Mt. Fuji in Japan was the subject of 36 paintings by the celebrated artist Katsushika Hokusai (1760–1849) (Fig. 2). The New Zealand artist Charles Blomfield (1848–1929) similarly became fascinated by the volcanic area of Rotomahana. His atmospheric paintings,

made in the late nineteenth century, of the pink and white terraces formed by hot-spring activity are not only important in their own right but are also a unique record of a landscape that was to be destroyed by an eruption only a few years later.

Mountains have equally—and predictably—been popular subjects for artists. An early example from China is *Summer Mountains* by Ch'ü Ting (active 1023–56). The earliest European paintings in which mountains are primary features (as distinct from merely part of the background) date from the sixteenth century. Albrecht Altdorfer (*c.* 1480–1538) is generally regarded as the first artist to produce pure landscape paintings without figures, although Albrecht Dürer (1471–1528) had earlier made watercolour sketches of Alpine scenery. Ice is a conspicuous feature of much mountain scenery. Paintings and drawings from earlier times have been used to chart the growth and retreat of glaciers (for example, the Grindelwald glaciers in Switzerland) in the past.

In the great flowering of landscape painting that took place in the seventeenth and eighteenth centuries, painters such as Philip James de Loutherbourg (1740–1812) and Caspar David Friedrich (1774–1840) made contributions of geological interest. In America, Thomas Cole (1801–48), who was born in England, became an influential figure as the leader of the Hudson River School, setting out to portray 'the wild and great features of nature: mountainous forests that know not man'. His pupil Frederic Edwin Church (1826–1900) followed him in striving to depict the grandeur of the American landscape. These American artists enjoyed great popularity, and some were influential. Thomas Moran was a member of a government expedition to the Yellowstone Valley in the early 1870s, and the paintings he made in the course of the expedition helped to persuade Congress to make the area a national park—the first in the United States.

In Europe, the Impressionists, with their emphasis on painting *en plein air*, brought a new approach to landscape. In painting the rocks of Belle Isle (1886) Claude Monet (1840–1926) was more concerned with the effects of light than with geological structure. By contrast, the series of paintings that Paul Cézanne (1839–1906) made of the Montagne Sainte-Victoire were as much concerned with structure as with the effects of light. Cézanne's subjects also included the stone quarries in the area. He was not, of course, the first to recognize applied geology as a subject for art. Canaletto had, for example, painted a stonemason's yard in the eighteenth century.

Although some of the leading artists of the twentieth century turned away from landscape, geological subjects were not neglected by others. Two examples may be quoted. Before moving on to abstraction, Piet Mondrian (1872–1944) painted the sand dunes of the Dutch coast. Paul Klee (1879–1940) offered an elemental vision in his *Mountain in winter*. Klee's mountains, pyramidal in form, might suggest to some the tetrahedral silicate structures of rock-forming minerals, but this connection may not have been in the mind of the artist.

The sheer scale of some natural features such as the Grand Canyon presents an artistic challenge that continues to evoke fresh responses. David Hockney (best known as a figure painter) attracted great interest in 1999 with a series of six massive paintings of the Grand Canyon. Large-scale paintings of geological features are not a modern development: James Ward's *Gordale Scar*, with an area of some 14 square metres, dates from 1811–25. A limestone escarpment is here portrayed on a scale more generally associated with historical subjects. On an even larger scale, environmental art modifies the landscape itself. The Bulgarian artist Christo (b. 1935) has, for example, 'packaged' a section of the Australian coast with the object of revealing its basic forms by concealing the details. In another large-scale project, the American artist James Turrell has set out to modify an extinct volcano, the Roden Crater, by excavating chambers and a tunnel to provide a visual experience of varying spatial relationships, the effects of light, and the perception of the sky.

Geological endeavours can also be captured by artists. The work of Thomas Moran has been mentioned above. Another American artist, Peter Hurd (1904–84) travelled with an exploration crew of geologists from Standard Oil through Utah and Wyoming in the 1940s, recording what he saw on canvas. A collection of his paintings has been exhibited at the Philbrook Museum of Art in Tulsa, Oklahoma.

It is not only painters who have been inspired by geological subjects. Photography is also an accepted form of fine art, and geological subjects have been popular with many photographers. In America, for example, Carleton Watkins (1829–1916) photographed mountains, waterfalls, and mining activities. (Many of these photographs were stereoscopic views.) The work of a later generation of photographers, including Ansel Adams, Michael Freeman, and the mountaineer–photographer Chris Bonington, has become well known. Photographs of geological subjects have been used to bring conservation issues to wider notice, as, for example, by the Malvern International Task Force.

Geological photography has not been confined to terrestrial subjects. *Images of the Moon*, published in 1874 by James Nasmyth and James Carpenter, provides an illustration of the problem of representing relief on a two-dimensional surface. Radar images of Venus from the 1991 *Magellan* mission were described in an article in *Nature* by Martin Kemp as 'masterpieces by largely anonymous masters of extraterrestrial landscape painting' that enabled the general public to become 'electronic proxy armchair tourists and aesthetic voyeurs in extraordinary voyages'.

Since the introduction of postage stamps in the mid-nineteenth century stamp designers have on many occasions used geological scenes or objects (Fig. 3). Some of these illustrations may be considered as works of art in their own right. The subjects range from volcanoes to minerals and dinosaurs and other fossils.

The sea and the sky have long been of intense interest to artists. Leonardo da Vinci (1452–1519) produced a series of

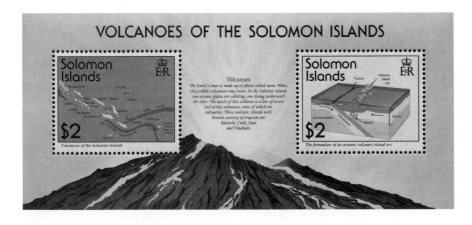

Fig. 3. Postage stamp (minature sheet) showing volcanoes of the Solomon Islands. This is the first stamp showing a plate-tectonic model. (Designed by N. Shewring and A. C. Scott.)

Deluge drawings in which the movement of water is dramatically portrayed. These drawings have been compared to Chinese and Japanese art in their dynamism and convolution. Storms and tempests have continued to be favoured subjects for art: Van Ruisdael (1628–82), Francesco Guardi (1712–93), J. M. W. Turner (1775–1851), Hokusai (1760–1849), and Hiroshige (1797–1888) provide notable examples. The elements are represented in gentler mood in many paintings of all periods, but John Constable (1776–1837) may perhaps be singled out for his cloud studies, in which the sky is the sole subject.

Geological illustrations as art

Geological field drawings were first extensively used by members of the Accademia dei Lincei in Italy in the early seventeenth century. These drawings, collected by Cassiano dal Pozzo for his Paper Museum, which are now in the Royal Collection at Windsor, are important, not only for the history of science, but also as art objects (Fig. 4). Many early geological treatises have been broken up in order to remove the prints, which have then been sold individually. Not only geological drawings and paintings have been treated in this way, but also geological maps.

There has been an increasing interest in geological reconstructions, whether of individual animals or plants or of past time. The fascination with old reconstructions of the past is widespread (a topic that has been discussed by Martin

Fig. 4. Early seventeenth-century drawing of Pliocene fossil wood from the Acquasparta region of Italy drawn as part of a study conducted by Prince Federico Cesi and the Accademia dei Lincei. (Royal Collection © 2000 Her Majesty Queen Elizabeth II.)

Rudwick). These images often gain a wider circulation and can shape public perception of worlds of the past. This can have both positive and negative effects. For the scientist, a reconstruction is at best a working hypothesis, which in some instances may be based upon scant data. This implies that mistakes may be made that will later need to be rectified. The transient nature of reconstructions is not often recognized by those 'not in the know'. This is well illustrated by *Brachiosaurus*, which was shown in the often reproduced illustration by Burian as living in lakes with its head just above the water. Although this interpretation has long since been shown to be incorrect, the illustration continues to be reproduced, so evocative is the original painting. Today a new generation of artists, such as Mark Hallett and John Sibbick, are shaping the public consciousness and their paintings are widely reproduced, even on postage stamps. The visual effect of these reconstructions is illustrated by films such as *Jurassic Park* (in which most of the dinosaurs were in fact of Cretaceous age). More scientific, and no less spectacular, was the 1999 BBC television series *Walking with dinosaurs*, which used animation of a high order to achieve a degree of realism hardly achieved before. Although scientists were closely involved, a large element of conjecture was inevitable. The value of the series in bringing palaeontology to life for a general audience was nevertheless unquestionable. The first episode was watched by an estimated 18.9 million people. Plans for a sequel were being made while showings of the first series were still in progress.

The use of geological materials in art

Geological materials, especially minerals and metals, have been used in art since Prehistoric times. Gold and silver have been widely used in many cultures in the production of art objects. Minerals and gemstones have similarly been used to adorn objects and people in many cultures for many thousands of years. The search for these materials has often gone hand in hand with geological exploration, and the trade in some of them has been of major importance. Amber, for example, is a prized material widely used in jewellery. Amber, a polymerized resin from certain trees, commonly encloses animals and plants. Much of our knowledge of several insect groups comes entirely from amber inclusions, which might not have come to light if it were not for the artistic use of the material. Some pigments, used by the Egyptians and other cultures, were derived from crushed minerals: blue (ultramarine) from lapis lazuli is a well-known example. Fossils are also featured in art as in the use of petrified wood for ornaments or the more general inspiration provided by fossils in painting and sculpture.

The more practical use of stone in architecture is discussed in a separate entry (*see* building stones).

A wide variety of rocks have been used for sculpture. White marble has been classically used but other more unusual materials, such as bitumen have been used (Fig. 5). The physical and chemical properties of geological materials, such as clays, also influence their use. The production of fine

Fig. 5. Drum-shaped bowl with a wild goat protome made from bitumen mastic. Early second millennium BC. Susa, Iraq. (Musée du Louvre, Paris.)

porcelain might be considered as an interaction of geology and art. The use of marbles, agates, and other rocks to produce mosaic images that might be confused with paintings is exemplified by the technique of *pietre dure* (Fig. 6). A factory set up in the Uffizi in 1588 produced table-tops and panels until the nineteenth century.

The application of geology in the investigation of art

Geology has much to offer art historians in the investigation of a wide variety of artistic objects.

Some clays used in pottery have a distinctive chemical signal. Chemical analysis of a pot may thus reveal where it was made. Techniques such as inductively coupled plasma (ICP) have been widely used in this regard. Other techniques, such as atomic absorption (AA), are widely used in the study of metal objects. Geology can similarly help in establishing the provenance of minerals and gemstones. Gemstones of particular regions may, for example, differ in their trace-element composition.

Geology can also help to answer questions about the origin of sculptures. Artefacts such as bitumen carvings from the Middle East have been shown by J. Connan of Elf Petroleum and O. Deschesne of the Louvre Museum to have a distinctive geographical distribution. Organic geochemical analysis, including gas chromatography–mass spectrometry (GC-MS), Rock-Eval, and isotope analysis, has made it possible to identify the sources of the bitumen as well as the relevant trade routes. Geo-detective work of this kind has also been applied to objects such as flint tools. Knowledge of the geology of a region can help archaeologists to identify sources of material and postulate trade between cultures.   ANDREW C. SCOTT

Further reading

Autissier, J. M. (1987) *The world of minerals through postage stamps.* Atelier Saint-Amand-Montrond, France.

Bazarin, K. (1981) *Landscape painting.* Octopus Books, London.

Connan, J. and Deschesne, O. (1996) *Le bitumen à suse.* Collection du Musée du Louvre. Musée du Louvre, Paris.

Grimaldi, D. (1996) *Amber: windows to the past.* American Museum of Natural History and Harry N. Abrams Inc., New York.

Kemp, M. (1998) Kemp's conclusions. [Summarizing *Nature*'s 'Art and science' series.] *Nature*, **39**, 875.

Krafft, M. (1991) *Volcanoes: fire from the Earth.* New Horizons, Thames and Hudson, London.

Malvern International Task Force for Earth Heritage Conservation (1995) *Earth Heritage Conservation.* Joint Nature Conservation Committee, Peterborough.

del Riccio, A. (1996) *Istoria della pietre.* Umberto Allemandi & C. Italy.

Rudwick, M. J. S. (1976) The emergence of a visual language for geological science. *History of Science*, **14**, 149–95.

Fig. 6. Florentine Pietre-Dure Workshop. Tabletop, 1616: polychrome marble and *pietre dure*. (Prado Museum, Madrid.)

Rudwick, M. J. S. (1992) *Scenes from deep time: early pictorial representations of the prehistoric world.* Chicago University Press.

Scott, A. C. (1994–5) Volcanoes on stamps. *Stamp Magazine,* **60** (11), 60–7, **60** (12), 66–9, **61** (1), 67–9.

Tait, H. (1976) *Jewellery through 7000 years.* British Museum, London.

artesian well In 1126, a well was drilled in the province of Artois in north-west France, which produced free-flowing water at the land surface. Flowing wells of this type were also drilled elsewhere in Europe in the seventeenth century and came to be known as *artesian* wells: wells of Artois. Advances in drilling technology in the nineteenth century led to extensive development of flowing artesian wells in many parts of the world. Wells near Paris produced warm water in fountains standing many metres above the land surface. Artesian wells supplied London with water from the Chalk. Water at high pressures from flowing wells in the Dakota aquifer of the United States was used to turn waterwheels and provide power for flour mills. A large region with flowing wells in Australia became known as the Great Artesian Basin.

In modern usage an artesian well is any well in which the water level rises above the top of the confined aquifer to which the well is open. In both flowing and non-flowing artesian wells, the water is under higher pressure than would be created by a column of water extending from the depth of the open well to the top of the aquifer. This excess pressure can be explained by the fact that the recharge area, where water enters the aquifer system, is at a higher elevation than the confined aquifer at the well. In the recharge area the hydraulic head, a measure of the mechanical energy of the groundwater, is determined by the potential energy associated with the elevation of the water-table. As water flows downwards into and within a confined aquifer, friction causes some of the potential energy to be converted into heat. The frictional losses are, however, smaller than the potential energy loss associated with loss of elevation. Much of the potential energy is therefore converted into elastic energy, which is associated with an increase in pressure.

Flowing wells are often assumed to be tapping aquifers that are isolated from shallower groundwater by a confining unit of low permeability. This is not necessarily the case. Flowing wells can occur where there is no confining unit between the open portion of the well and the land surface. All that is required is that the hydraulic head at the depth tapped by the well must be higher than the land surface. This is generally the case in groundwater discharge areas located in topographic 'lows'. Groundwater flow in these areas is upwards, which requires that the hydraulic head must increase with depth since, according to Darcy's law, groundwater flows from a higher to a lower head.

Artesian wells, and confined aquifers in general, were once thought to be virtually inexhaustible sources of water. Continued extraction of water from an artesian well can, however, lead to a decline in pressure, causing a once-flowing well to require pumping. In extreme cases, the water levels can eventually drop below the top of the confining unit. At this point the confined aquifer becomes a water-table aquifer. The extent to which the water level declines depends on the balance between recharge to the aquifer system and the combined discharge through wells and in natural discharge areas.

J. BAHR

arthropods The arthropods are a very large phylum of organisms that are characterized by the presence of a segmented body with a hard exoskeleton and many jointed legs (the name 'Arthropoda' means 'jointed legs'). More species of arthropods have been described than all other kinds of animals put together: almost three-quarters of the million or so known animal species are arthropods. The major groups of modern arthropods are the crustaceans (lobsters, crabs, barnacles, and ostracods) the centipedes, the millipedes, the arachnids (spiders and scorpions), and the insects, which are by far the largest class of all, totalling almost 500 000 species. Important fossil groups are the trilobites, a large group of Palaeozoic marine arthropods of uncertain affinity, and the eurypterids, Palaeozoic marine forms related to the arachnids. It has been suggested that the arthropod body plan might be polyphyletic (i.e. have developed several times), and four phyla are thus sometimes designated: Trilobita, Crustacea, Chelicerata (eurypterids and arachnids), and Uniramia (insects, millipedes, and centipedes).

Because of the adaptability of their body plan, the arthropods have been important and successful organisms since their first appearance in the Early Cambrian. Although they originated in the ocean they have since moved into fresh water, and they were also the first animals to move on to the land and into the air. Despite this, the only groups that are used extensively in geology are the trilobites, which are important biostratigraphic indicators in the Palaeozoic, and the ostracods, which are used throughout the Phanerozoic.

DAVID K. ELLIOTT

asbestos Many aspects of our physical environment pose risks to human health. Contrary to popular belief, asbestos is not, in general, one of them. Despite this fact, various parties continue to make money by exploiting public fears about this material. Furthermore, billions of dollars of state and federal funding continue to be spent to remove asbestos-containing materials from buildings, perhaps unnecessarily.

'Asbestos' is a commercial, rather than mineralogical, term that encompasses six silicate minerals: chrysotile, amosite, fibrous anthophyllite, fibrous tremolite, fibrous actinolite, and crocidolite. All six minerals are referred to as 'asbestos' because they have 'asbestiform habit'; that is, they form bundles of minute fibres. These fibres resist heat and are quite flexible, yet they are chemically and mechanically durable. Asbestos minerals are therefore employed as insulation and to make materials fire-retardant. Commonly, ceiling and floor tiles, pipe insulation, vehicle brake linings, cement, and mortar contain asbestos. The US government characterizes all six forms of asbestos as hazardous and requires its removal from public buildings.

Ninety-five per cent of all asbestos used commercially is chrysotile. This type of asbestos differs fundamentally from the other five kinds. Amosite, fibrous anthophyllite, fibrous tremolite, fibrous actinolite, and crocidolite are amphiboles, double-chain silicates, which observed microscopically look like sharp needles. Most experts agree that crocidolite, which comprises less than 5 per cent of asbestos used in industry, is the only type of asbestos which causes cancer. In particular, it is thought to cause mesothelioma, a cancer of the outer lining of the lung or the abdomen.

When materials that contain any type of asbestos are disturbed or damaged, the fibres can separate and become airborne. If people inhale the fibres, they can cause significant health problems. Asbestosis, first found in naval shipyard workers, is a lung disease caused when asbestos fibres become trapped in lung tissue. Acid produced by a body to destroy the fibres does little damage to the resistant asbestos fibres but scars the lungs, sometimes so severely that they cannot function. However, it seems that only high concentrations of fibres (more than 20 fibres per cubic centimetre of air) inhaled for many years cause this disease. In addition, only the presence of crocidolite among the fibres will cause lung cancer to develop.

This raises the question, should we remove from public buildings asbestos that is in good condition, isolated from the air, and therefore not inhalable?

Individuals who are exposed regularly to high concentrations of asbestos, such as mine, factory, and construction workers, must certainly be protected against the potential health hazards of their work. However, since chrysotile does not cause cancer, and since undisturbed asbestos in good condition cannot be inhaled, the benefits of having all asbestos removed from public buildings are dubious. A number of studies even show that ingestion of asbestos particles that settle in water or on food is not dangerous and that skin contact with asbestos poses no threat to health.

It seems that the health risks associated with minimal exposure to asbestos are tiny. The calculated risk for cancer resulting from smoking is 1 in 5, as compared to that for asbestos-related lung disease, which is 1 in 100 000. The most sensible course for dealing with asbestos in public spaces will include evaluation of the condition and type of asbestos present.

JILL S. SCHNEIDERMAN

aseismic submarine ridges and oceanic plateaux

Aseismic submarine ridges are very large linear features that rise up to 4 km above the surrounding ocean floor and reach 400 km in width and 5000 km in length. The term 'aseismic' has been given to them because they are virtually free of earthquake activity, showing that they are not located near to plate boundaries. Oceanic plateaux are essentially similar features except for their shape. They reach several hundred square kilometres in area and are elevated at least 1000 m above the surrounding sea floor. The two cover, respectively, 25 per cent and 10 per cent of the ocean floor and are therefore extremely significant features of world geomorphology. Frequently the term 'rise' (e.g. Hess Rise) is used for identical features, many 'rises' being, in fact plateaux or ridges. The term rise is better confined to the continental rises of continental margins.

Given the very wide distribution of aseismic submarine highs, it is only to be expected that their origins are quite varied. Some, such as the Rockall Plateau of the North Atlantic, the Agulhas Plateau in the South Atlantic, and the Mascarene Plateau in the western Indian Ocean are isolated fragments of continental crust separated during rifting and sea-floor spreading. They are essentially *microcontinents*.

Of far greater importance, however, are those underlain, not by continental crust but by oceanic volcanic rock. These include the Hawaiian–Emperor chain of the North-west Pacific, the Ninety East Ridge of the Indian Ocean and the Rio Grande 'Rise' and Walvis Ridge on either side of the South Atlantic. In the Hawaiian–Emperor Chain there is a clear progression from high, large islands with active volcanoes at the Hawaiian end to sunken, extinct islands at the northern end. Some earlier explanations proposed that they represented a long deep fracture in the crust along which magma could rise to form volcanic islands, the crack opening progressively southwards. An alternative, now commonly accepted, explanation, associated with the names of J. T. Wilson and W. J. Morgan, was that this lateral migration was the result of movement of the Pacific Plate over a 'fixed' hot-spot.

The twinned Rio Grande 'Rise' and Walvis Ridge are related to the early phases of opening of the South Atlantic; both are composed of oceanic volcanic rocks 100 to 80 Ma old. They developed at a ridge crest in Cretaceous times, their northern and southern margins being formed by major, primary fracture zones. Since that time these two ridges have had a significant influence on the development of the South Atlantic topography and sedimentation, though their history is complicated by the movement of a hotspot, now located beneath Tristan da Cunha.

Inevitably, because aseismic submarine ridges and oceanic plateaux are such voluminous portions of the ocean floor creating a highly irregular morphology, they have an extreme effect on the pattern and nature of subduction and are a major cause of the complexities at consuming plate boundaries (see *island arcs*). This is particularly well seen at the present day where the Hawaiian–Emperor Chain intersects the Kuril–Aleutian island arc.

HAROLD G. READING

assimilation

Assimilation is the process of physical or chemical incorporation of foreign material by a magmatic body that results in a chemical change in the composition of the original liquid. The typical image that this process invokes is that of a magma chamber with bits and pieces of the enclosing country rock or wall rock floating in the magma. If these pieces of wall rock, like ice in a beverage, can melt or, like mud in water, are fine and so widely distributed that they are inseparable, then the resulting liquid will be compositionally different from the original one.

Xenoliths, which are pieces of wall rock, or *xenocrysts*, which are single mineral grains that originate from xenoliths, can commonly be recognized at outcrops of magmatic rocks. The presence of xenoliths or xenocrysts is generally used as evidence that assimilation could have occurred in the magmatic body. However, the existence of foreign material alone is only circumstantial evidence of assimilation, since their mere presence does not prove conclusively that any of the included material has in fact contaminated the primary magma. Resorption textures, which are features that are suggestive of dissolution of xenoliths or xenocrysts, are more persuasive textural evidence for assimilation. A convincing example of a resorption texture would be minerals such as quartz or alkali feldspar with lobate or embayed grain boundaries in basalt. These minerals could not have crystallized at high temperatures in a magma of basaltic composition, which indicates that they represent foreign material, and their lack of crystal faces or angular broken surfaces strongly suggests resorption or dissolution of the minerals.

In the first half of the twentieth century, the cause of variations in igneous rock compositions was an important topic of debate, and the process of assimilation was one of the explanations that enjoyed many advocates. At that time it was thought that superheated magmas could assimilate large amounts of the enclosing country rock as they moved through the crust. The assimilation of carbonate-rich rocks was proposed as a mechanism of desilicification that could explain the origin of silica-deficient alkaline rock suites, and assimilation of quartz-rich pelitic rocks has been invoked to explain the compositional variation between alkaline and sub-alkaline basalt. The main lines of evidence supporting the modification of the major-element compositions through assimilation were field occurrences of xenoliths with compositions that were deemed appropriate to produce the observed range of magma compositions. As the century progressed, knowledge of the heat contents, latent heats, and melting temperatures of rocks and minerals improved. As a result, it became clear that, even under the most favourable conditions, the volume of assimilated country rock relative to the volume of the host magma was far too small to account for the magnitudes of chemical variations in major elements that are observed. The process of assimilation cannot therefore account for large major-element chemical variations, although chemical modification by remelting may locally alter magma compositions at the borders of large magma bodies.

Despite its limited role in modifying major-element compositions, assimilation can significantly affect the con-

centrations of radiogenic isotopes and trace elements in magmas. Among common igneous and sedimentary rock types, the concentrations of these elements can vary by several orders of magnitude. As a consequence, the assimilation of only small amounts of material that is highly enriched in a particular trace element can profoundly alter the composition of a magma that is highly depleted in the same element. Other phenomena related to assimilation are contamination, magma mixing, and the AFC process (simultaneous *a*ssimilation and *f*ractional *c*rystallization).
J. C. SCHUMACHER

asteroids and comets

Asteroids

The asteroids are irregularly shaped rocky or iron-rich bodies, most (but not all) of whose orbits lie between those of Mars and Jupiter. The largest is about 900 km across, and they range downwards in size with apparently no lower limit (Fig. 1). About 5000 are known that exceed a kilometre or so in size. Most are too small to have undergone differentiation, but the few largest ones may have experienced partial melting to the extent of having surface flows of basalt lava very early in the history of the Solar System. The supposed heat source was either the decay of short-lived radioactive isotopes, such as aluminium-26, or electromagnetic induction.

Once thought to be the remnants of a planet that broke apart, the asteroids are now regarded as *planetesimals* that were never able to come together to form a planet because the gravitational disturbance caused by Jupiter made mutual collisions so energetic that fragmentation was just likely as accretion.

A few asteroids have orbits that cross that of the Earth, and collisions with such Earth-crossing asteroids are responsible for many of the impact craters found on the Earth.

20 km

Fig. 1. The asteroid Ida and its tiny satellite asteroid Dactyl imaged by the space probe *Galileo* in 1993 while on its way to Jupiter.

Meteorites

Meteorites are chunks of material that are collected after they fall to Earth. On entry into the Earth's atmosphere, typically at speeds of tens of kilometres per second, the surface of a meteorite is heated to incandescence by friction. Freshly fallen meteorites are thus often recognizable by the black, glassy surface that results, but some incoming bodies break into several fragments before reaching the ground. Only objects more than a few metres across hit the ground fast enough to produce an impact crater. Towards the other end of the scale, most *meteors* are sand-sized objects that are completely vaporized by the heat of their passage into the atmosphere. Their demise is marked by a visible 'shooting star'.

Almost all meteorites are collisional fragments of asteroids, and can be dated radiometrically to about 4.6 billion years (4.6×10^9 years, 4.6 Ga), which represents the birth of the Solar System. Most are stony, and of these the majority consist of *chondrules*, which are globules of silicate minerals, embedded in a finer-grained matrix. These meteorites are therefore referred to as *chondrites*. The abundances of the non-gaseous elements in some chondrites match the elemental abundances in the Sun, and these objects are reckoned to be among the most primitive (least altered) material in the Solar System. Because of their relatively high carbon contents they are known as *carbonaceous chondrites*. Many chondrites contain in their matrices tiny grains whose isotopic composition attests to their pre-solar origin; for example, silicon carbide and diamond believed to have been carried from the atmosphere of giant stars by stellar winds. Other *stony meteorites* are basaltic, having clearly crystallized from a melt, and these provide evidence in favour of early melting events among the larger asteroids. *Iron meteorites* are distinctive in appearance. They consist of alloys of nickel and iron, and are taken to represent samples of the core of a large asteroid that was broken apart by a collision. *Stony-iron meteorites* contain mixtures of the two components, and are interpreted as having originated near the core–mantle boundary of their parent bodies.

A few meteorites have been identified whose ages, textures, and isotopic compositions match those of rocks collected from the Moon. They are evidently fragments of ejecta thrown out by crater-forming events on the Moon at sufficient speed to escape the lunar gravity. Even rarer meteorite types are believed to have come from Mars in the same way. These *martian meteorites* are fine- or coarse-grained basaltic rocks, showing variable amounts of alteration caused by the passage of aqueous fluids. The crystallization ages of most of these meteorites fall in the range 0.2–1.0 Ga, but the oldest sample (ALH84001), on which most of the arguments for life on Mars are based, is much older, having been dated at about 4.3 Ga.

Comets

Comets represent fragments of primitive material from the outer Solar System in the same way that asteroids represent fragments of material from the inner Solar System. Because they were formed further from the Sun, comets are mostly ices of various sorts, with small proportions of rocky and sooty dust. Most are only a few kilometres or tens or kilometres across.

The majority of comets spend most of their time in the Oort cloud, which is a spherical shell some thousand billion kilometres from the Sun (far beyond Neptune's orbit at 4.5×10^9 km). When a comet passes through the inner Solar System, heat from the Sun vaporizes some of the ice, liberating gas and dust to form a tail, shining by reflected sunlight, that may be hundreds of millions of kilometre in length. The tail is pushed outwards from the Sun by the solar wind and radiation pressure, and so the popular conception that a comet's tail streams away behind it is wrong.

The orbit of a comet that passes close to Jupiter or another of the giant planets may become shortened into an ellipse with a period of a few years or hundreds of years. The comet then becomes a periodic comet, such as Halley's comet (Fig. 2), whose successive close passes near to the Sun rapidly deplete its supply of volatiles so that its fate is to become an inert, dusty lump. This being so, the distinction between comets and asteroids is not so clear-cut as it seems at first sight. Many of the small asteroid-like objects orbiting beyond Jupiter may well be most appropriately thought of as dead comets, as may be the small irregularly shaped satellites of the outer planets.

1 million km

Fig. 2. Halley's comet, seen during its 1986 passage through the inner Solar System. Only the escaping gas and dust is visible; at 16 km across, the solid nucleus is far too small to be seen at this scale.

Collisions by comets are responsible for producing the impact craters that are visible on the surfaces of the solid bodies in the Solar System that were not formed by asteroid impacts. Comets may hit giant planets too, as was the case when comet Shoemaker–Levy 9, previously torn into fragments by Jupiter's tidal forces, struck the planet in 1994.

Kuiper-belt objects

Many icy bodies other than Pluto and Charon are now known to circle the Sun beyond the orbit of Neptune in what is known as the Kuiper belt. The largest are only a few hundred kilometres across, and most are much smaller; they are therefore exceptionally difficult to study from Earth.

DAVID A. ROTHERY

Further reading

Beatty, J. K., Peterson, C. C., and Chaikin, A. (eds) (1999) *The new Solar System* (4th edn) Sky Publishing Corporation and Cambridge University Press, Cambridge.

Rothery, D. A. (2000) *Teach Yourself Planets*. Hodder and Stoughton, London.

asthenosphere The athenosphere (derived from the Greek for 'weak sphere') is the relatively weak, ductile layer in the upper mantle immediately underlying the lithosphere. Although solid at normal strain rates, like the rest of the non-lithospheric mantle it can deform slowly by solid-state creep. It has an effective viscosity of between about 10^{19} and 10^{21} Pa s (Pascal seconds), compared with an average of about 10^{21} for the mantle as a whole and 10^{-3} Pa s for water. The lowest viscosities tend to occur under volcanically active areas, and the highest under cratonic areas (which are stable regions within continents).

The top of the asthenosphere is a gradational boundary with the overlying lithosphere, and its depth varies according to the age and temperature of the lithosphere (see *lithosphere*).

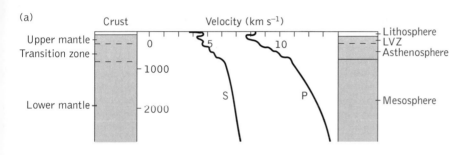

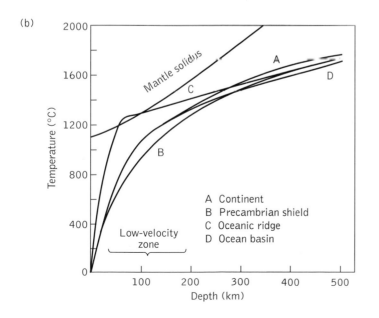

Fig. 1. (a) The P- and S-wave low-velocity zone (LVZ), which provides the primary evidence for a worldwide asthenosphere. (From Kearey and Vine (1996), Fig. 2.16.) (b) Close approach of the geotherm to the mantle solidus at the depth of the low-velocity zone. (After Kearey and Vine (1996), Fig. 2.36.)

The base of the asthenosphere is also gradational, and is not well defined. Some consider that the base occurs around 300–400 km, whereas others consider it to extend to the base of the upper mantle at 670 km.

Evidence for the local existence of an asthenosphere is derived from long-term, regional changes in elevation, such as the postglacial uplift of Fennoscandia. There, historical uplift of about 10 mm per year is occurring as a result of isostatic rebound of the lithosphere following the removal (by melting) of the load represented by the glacial ice cap. When the ice cap first extended over Fennoscandia, its weight depressed the lithosphere, and the underlying asthenosphere gradually flowed away to the sides. As the load is released, the elastically flexed lithosphere 'rebounds' to its pre-deformation level, and the asthenosphere flows back into its former position. The rate of rebound is limited by the viscosity of the athenosphere, and modelling such movements provides estimates of the viscosity.

Although such local studies are important, it is seismology that has demonstrated the global presence of the asthenosphere. This is inferred partly from the almost ubiquitous 'low-velocity zone' for P- and especially S-waves that occurs between about 100 and 300 km below the Earth's surface, and partly from studies of the attenuation of seismic energy. The low-velocity zone (Fig. 1a) occurs almost everywhere, except perhaps under cratonic areas. It consists of a 0.5–1 km s^{-1} reduction in seismic velocity between about 100 and 300 km depth. This velocity reduction is best explained by the close approach of the mantle at this depth to its melting temperature (Fig. 1b), although the actual degree of melt in the asthenosphere is probably very low except under volcanically active areas such as mid-ocean ridge axes, hot spots, and the mantle above subducting slabs. Seismic energy is strongly attenuated in the asthenosphere, in contrast to the lithosphere and the rest of the mantle, further attesting to the near-melting temperature there.

The presence of a worldwide asthenosphere is an important part of the concept of plate tectonics (see *plate tectonics*), allowing the detailed movements of plates to be decoupled from the convection in the underlying mantle. This allows plates to move around relatively freely, interacting only at their edges. This is in strong contrast to pre-plate-tectonic ideas on continental drift, the acceptance of which was hampered by the apparent difficulty of moving continents through a rigid upper mantle.   ROGER SEARLE

Further reading

Kearey, P. and Vine, F. J. (1996) *Global tectonics*. Blackwell Scienctific Publications, Oxford.

astroblemes *see* METEORITE AND COMET IMPACTS ON EARTH

Atlantis The legend of the lost continent or island of Atlantis occurs in the mythologies of many parts of Europe. It is found in the writings of Plato, the Greek philosopher of the fourth century BC. Plato gave a rather full account if its geography and inhabitants and its ultimate fate of being engulfed by the sea. Atlantis, he wrote, had existed beyond, that is, to the west of, the Pillars of Hercules (the Strait of Gibraltar) and was the home of an Aryan people who invaded the Mediterranean, subduing all except the Athenians (Plato was an Athenian). The foundering of Atlantis beneath the ocean was sudden, and little or no trace was said to remain. Celtic folklore regarded Atlantis as a pleasant, romantic place, and in medieval times it was spoken of as possibly the site of the Garden of Eden. Cartographers of the fourteenth and fifteenth centuries and Renaissance writers depicted it as a large island, perhaps part of America, or Scandinavia, or the Canary Islands. All record its end in sudden catastrophe.

Plato was only one of many Mediterranean writers in classical times to dwell upon the civilization that had flourished in Atlantis and also upon the heritage it had bequeathed to the peoples of the eastern Mediterranean in earlier days. There would seem to be some historical basis for this widespread and consistent story of the sudden marine destruction of a previously existing land and a civilized nation with it.

The discovery early in the twentieth century of the Minoan civilization at Knossos in Crete revealed the sudden demise of an early Greek settlement of a highly civilized people. It proved to be a Bronze Age culture, its economy well developed and prosperous but with no defensive fortifications to protect it. The cause of its extinction seems not to have been by conquest from outside. Large buildings had collapsed, and an earthquake was thought to be a possible cause. Once destroyed, the site was not resettled.

About 110 km north of Crete are the closely set islands of Thera and Santorini. They are the remains of an immense volcano, one typical of the kind of volcanicity seen in virtually all the Mediterranean volcanoes, highly explosive. Gigantic explosions occur during what is termed a Plinian eruption, when a blast of upwards-directed gas and dust rises to a great height and then spreads out in a turbulent mass. Torrential rainstorms usually accompany this kind of eruption, causing mud-flows and floods. The dust may settle over very wide areas, hot and toxic, smothering life beneath it. In 1967 excavations on Thera revealed settlement of Late Minoan type, Bronze Age in date and with cultural connections to Crete. The buildings had suffered severe earthquake damage. The ruins lie beneath a thick layer of volcanic ash. The earliest settlement began about 3000–2500 BC. The destruction took place around 1500 BC; 1628 BC is a date favoured by some modern authors; a controversial claim according to others.

Studies on nearby Santorini revealed that a gigantic explosion occurred about 20 000 years ago. The build-up of a new volcanic cone followed, with small Plinian eruptions, and was complete with lush forest and human settlement by about 3000 BC. Later came earthquakes and explosions which blew away most of the island and led to the collapse of a huge central caldera. Gigantic tidal waves, or *tsunamis*, spread out and may have reached shorelines as far as 100–200 km away, wiping out fleets and coastal settlements. The ash cloud probably covered 200 000 square kilometres, darkening the sky and depositing a thick volcanic dust. The date of these events is put at around 1550–1450 BC.

The destruction of Minoan settlements in Thera–Santorini (and even Crete according to some experts) thus occurred as a result of the destruction of the volcanic centre by explosion, associated earthquakes, and their tsunamis. Small eruptions have occurred since and earthquakes in the region are relatively frequent, but nothing comparable has recurred since that devastation. The effects diminished with distance from the centre, but they would have been imprinted upon the folk memories of Greeks, Egyptians, and others locally. It is unlikely that Thera was literally the Atlantis of Plato. He was using folk memory as a basis for his concept of the loss beneath the waves of a civilization not unlike that of the Greeks themselves.   D. L. DINELEY

Further reading

Forsythe, P. F. (1980) *Atlantis, The making of myth.* McGill-Queen's University Press, Montreal; Croom Helm, London.

Luce, J. V. (1970) *The end of Atlantis.* Paladin, London.

atmospheric convergence and divergence *Atmospheric convergence* refers to accumulation of an air mass at a point. This process usually occurs at the surface in the centre of well-developed low-pressure systems, and, in the upper air, along the western limb of Rossby waves. Mass convergence is difficult to measure at the cloud scale, based on updraughts, but values are locally 50 times that of synoptic-scale convergence. Convergence is measured in s^{-1}, with values of 10^{-5} s^{-1} typical for a mid-latitude cyclone. This is achieved through a horizontal gradient of wind speed of roughly 1 m s^{-1} per 100 km, a rate of change that is difficult to measure given the observation network and instrument accuracy. If values of 10^{-5} s^{-1} were maintained throughout the lowest half of the troposphere, an uplift near 5 m s^{-1} would result.

Convergence also results from frictional effects at the surface. If a geostrophic wind, with pressure gradient force balanced by the Coriolis force, were to encounter a rough surface, the wind speed would decrease. Since Coriolis force is proportional to wind speed, this deflecting force would diminish, while the pressure gradient would remain unaltered. The balance of forces is lost and the flow crosses the *isobars* (lines of equal air pressure) towards the low pressure. This frictional inflow leads to an additional component of convergence.

Atmospheric divergence typically occurs where airflow (mass) is moving away from the centre of a pressure system. The air is thus being spread out, stretched, and expanded. This stretching may, however, assume many forms, each at a

characteristic scale. It may result from cross-isobar (*ageostrophic*) flow under frictional effects, from the divergence of *streamlines* (lines of instantaneous air motion), from deceleration of flow, or from the interference of barriers such as mountains.

Divergence typically occurs on a large scale at the surface of large, high-pressure systems such as the Azores High. The spreading out of surface air is compensated by descending air in the core of the pressure system, since outflow must be balanced by vertical motion, a requirement of mass conservation. Subsidence, with typical values of a few centimetres a second, leads to adiabatic warming and cloudless conditions. This link between horizontal divergence and vertical motion is a major cause of weather, atmospheric stability playing a secondary role. Divergence may result from friction with the Earth's surface. Offshore winds, for example, experience a reduction in friction, resulting in acceleration which leads to localized divergence. Divergence may also occur in the upper air as a result of acceleration in a Rossby wave. This promotes uplift and may initiate a surface cyclone. Generally, divergence is well marked either at the surface or upper atmosphere (near 10 km altitude). At a height of roughly 5 km (500 hPa), values of divergence reach a minimum. This is known as the *level of non-divergence.*   R. WASHINGTON

Further reading

Meteorological Office (1991) *Meteorological glossary.* HMSO, London.

atmospheric electricity The atmospheric electrical system is sustained by naturally occurring processes which exchange and release electric charge. Atmospheric electricity is not a recent or unique phenomenon on Earth: fossilized lightning strikes (fulgurites) have been preserved from more than a hundred million years ago, and observations of lightning have been made on other planets. Large-scale natural consequences of lightning include forest fires (which contribute to global carbon dioxide) and the fixation (i.e. chemical reaction) of nitrogen in the air.

Global system

The global atmospheric electrical system consists of a few regions of intense electrical activity and a large area where only small currents flow. Thus at any one time the greater part of the Earth's surface area is only slightly affected by electrical processes. An electric field can be measured at the Earth's surface which is due ultimately to a difference in voltage between the upper atmosphere (at about 80 km) and the surface. Since electric fields dissipate in partial electrical conductors (such as air), the constant presence of this field must be due to continuous replenishment by an atmospheric electrification process. This process is charge separation during more-or-less continuous equatorial thunderstorms.

Figure 1 shows a schematic description of the global electrical processes. Charge separation occurs between the Earth's

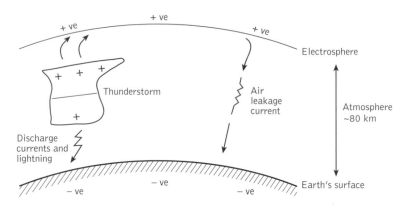

Fig. 1. Schematic diagram of the atmospheric electrical circuit. Global thunderstorms separate charge, which is carried to the Earth's surface and to the upper atmosphere. This establishes the atmospheric electric field, as a result of which a small leakage current flows vertically through the atmosphere.

surface and the upper atmosphere by microscale cloud processes in thunderstorms. Negative charge is generally carried to the surface by leakage currents and lightning strikes beneath clouds. Positive charge carried to the upper atmosphere then spreads over the whole globe. Charge in the upper atmosphere also leaks slowly through the air to the Earth's surface, causing the upper regions of the atmosphere to discharge: the electrical resistance of the entire atmospheric air is effectively a resistor of about 230 ohms. In the absence of any atmospheric charge separation (i.e. if all global thunderstorms were to be switched off), the electric field would decay to significantly in under an hour.

Fair weather conditions

Charges are released and exchanged in air even in the absence of local thunderstorm activity. There are several sources of this free charge, or ionization. Air molecules such as nitrogen, oxygen, and water vapour are split by collisions with high-energy radioactive particles, producing charged fragments or molecular small ions. Small ions may be positive or negatively charged. Another source of ionization (particularly at high altitudes) is cosmic rays (high-energy nuclei). Near the surface, however, the radioactivity released by natural rocks is the dominant source of ionization, and in the boundary layer (the region within about a kilometre of the Earth's surface) the gas radon also causes ionization.

Air ions are highly electrically mobile, and they are therefore greatly influenced by an electric field. Atmospheric air consequently has a slight electrical conductivity, although for everyday purposes it is usually regarded as an insulator. Low conductivity usually indicates a small current flow, but when the total conductivity is calculated over the whole Earth's surface the atmospheric conductivity is quite large, and an appreciable current of about 2000 amperes flows.

Under fair weather conditions (the absence of local thunderstorms), the electric field measured near the surface is generally about 120 volts/metre, that is, the difference in voltage between the Earth's surface and a fixed point one metre above the surface is about 120 volts. At first sight it seems remark-

able that most human existence occurs with such a potential between head and toe; however, only a tiny current flows because of the low conductivity of the air. Measurement of the potential difference is indeed complicated by the smallness of the current: a standard voltmeter is not able to measure the atmospheric field. Electrostatic methods are normally used, such as the field mill, which measures induced electric charge, or an ultra-high-impedance electrometer voltmeter.

The electrical conductivity of air depends on small ion concentration, which is in turn determined by the balance between ion production from radioactivity and ion removal by atmospheric particles such as aerosols (solid particles with diameters between 0.0001 mm and 0.01 mm), and raindrops. Large particle concentrations substantially reduce ion concentrations, and air conductivity is therefore highly affected by particulate pollution. Removal of ions by aerosol particles and raindrops leads to these large particles acquiring the original ions' electric charges. The large particles do not continue charging indefinitely: a raindrop will split apart if the electrical forces exceed the surface tension holding it together, at a critical level of charging known as the Rayleigh limit. Solid particles may emit charges when they become highly charged.

One curious aspect of fair weather atmospheric electricity near to the surface is the electrode effect. This occurs because of negative charge on the Earth's surface, which consequently repels negative ions. In still weather a layer of positive ions may therefore form close to the surface.

Thunderstorms

Thunderstorms offer substantial visible evidence of atmospheric electrical processes. A thundercloud (or cumulonimbus), with a distinctive upper anvil shape, results from air which is moist and unstable rising by convection. This large, deep cloud will eventually lead to lightning, hail, thunder, and heavy rain. Such clouds can exist individually, or as a group of active cells in various levels of development. The three stages of a single thundercloud are a brief cumulus stage (a strong updraft), followed by a longer mature stage (strong updrafts and downdrafts) and a final dissipating stage (weakening downdrafts) of comparable duration with the second stage. A thundercloud has a typical lifetime of about an hour. A full understanding of the electrical processes in a thundercloud has not yet been reached. To be credible, a theory of cloud electrification must explain how an electric field sufficient to produce lightning can be created within tens of minutes. It must also explain observations of an upper region of positive charge in the cloud, a lower region of negative charge, and a small amount of positive charge at the cloud base (see Fig. 1).

Many mechanisms for cloud electrification have been proposed. One major hypothesis depends on transport of external ions leading to regions of different charge in the cloud. The other major hypothesis invokes particle charging interactions of either an inductive or microphysical nature.

The inductive mechanism depends on cloud particles becoming polarized by existing electric fields. If a falling soft-hail (graupel) pellet were to acquire a top negative charge and a bottom positive charge by polarization, collisions beneath the pellet with rising water droplets would carry positive charge aloft on the droplets; negative charge remaining on the hail would be carried downwards. Although this mechanism can account for rapid and substantial cloud electrification, observations have found charges on some hail pellets rather greater than would be expected from polarization alone. These observations have lead to microphysical investigations of the electrical properties of ice. Laboratory studies have found that the sign of the charge exchanged between colliding ice crystals and soft hail pellets depends on temperature. If a riming hail pellet is made to collide with ice crystals, the hail pellet will acquire a positive charge if the temperature is below about −18 °C, and a negative charge if the system is at a higher temperature. The precise reversal temperature varies with many factors, such as the speed at which the pellet falls and the liquid water content. However, because the temperature variations within a thundercloud certainly include the range of laboratory reversal temperatures, this phenomenon does offer an explanation of cloud electrification.

In the colder, upper part of a cloud, (−18 °C typically occurs in about the middle of a cloud, at about 7 km above the surface), falling hail pellets acquire a negative charge, and rising ice crystals are carried upwards, to produce a positive region in the anvil, as observed. The falling hail generates the lower negative region of charge. Additional ice–hail interactions in the lower part of the cloud (which is above the critical reversal temperature) lead to the small positively charged region at the base of the cloud, above the freezing level. Calculations have shown that this process is able to produce electrification at rates comparable with those observed, although some observations remain unexplained.

Thunder and lightning

When sufficient charge has been separated by a storm, the intense local electric fields will eventually cause a lightning discharge to occur. Lightning is a transient electrical discharge several kilometres in length, which may occur within clouds (intracloud), between clouds (intercloud), or between the cloud and the ground. Some discharges can also occur between a cloud and surrounding air, and some can occur *above* a cloud. The most frequent lightning discharge is intracloud. The most dramatic is cloud-to-ground, often seen as forked lightning, which accounts for about 20 per cent of discharges and typically transfers tens of coulombs of negative charge from the cloud.

Forked lightning occurs rapidly, and the structure of an entire path from cloud to ground cannot be resolved by the naked eye. However photographic techniques have been used to investigate the separate stages. A lightning discharge begins with a weak initial discharge or stepped leader, which pursues a tortuous path to the surface, followed by a luminous return stroke.

The stepped leader proceeds in steps of typically 50 m at about $100\,000$ m s^{-1}, with a channel diameter of approximately 5 m. When the leader is close to the surface, streamers move upwards to the leader from several points. If the upward-moving charge reaches the leader, then there is a low-resistance path from the surface to the cloud, and a vigorous return stroke occurs. In a return stroke, currents of tens of thousands of amperes will flow within tens of microseconds, falling to hundreds of amperes sustained for several milliseconds. The lightning discharge may end at this point, or, if there is additional charge in the cloud, an additional dart leader can generate further return strokes using the same ionized channel. There are typically three or four strokes per flash. Lightning discharges also generate electromagnetic energy heard on radio receivers as spherics (crackles with very low frequencies), which may be used to locate distant thunderstorms.

The energy in the return stroke channel gives it a temperature and pressure higher than the surrounding air, so that the channel expands supersonically, generating a cylindrical shock wave which is heard as thunder. Thunder is heard later than its initiating lightning is seen, because the speed of sound in air is slow, compared with the speed of light. It is likely that the base of the channel is the strongest sound source, and is the probable origin of the initial loud bang heard by an observer close to a cloud; at greater distances an observer is more likely to hear a low rumble caused by refraction of the sound. A clap of thunder usually lasts between about one-tenth of a second and two seconds.
R. GILES HARRISON

Further reading

Chalmers, J. A. (1967) *Atmospheric electricity* (2nd edn). Pergamon Press, Oxford.

Saunders, C. P. R. (1988) Thunderstorm electrification. *Weather*, **43**, (9), 318–24.

Williams, E. R. (1988) The electrification of thunderstorms. *Scientific American*, November 1988, 48–65.

atmospheric high pressure Areas of high pressure with central pressure in the range of 1020–45 mb (millibars), occasionally reaching 1060 mb, are known as *anticyclones* or *highs*. The isobars are more or less concentric and are widely spaced around the centre of the high and thus, in contrast to *depressions*, winds are usually light and sometimes rather variable. In general, however, winds normally blow clockwise and outwards in the northern hemisphere (anticlockwise in the southern hemisphere) with a tendency for wind speeds to be greater towards the periphery of the system. High-pressure cells are usually larger (up to 4000 km across), slower moving and more persistent than depressions. Though weather conditions are usually quiet, dry and settled variations can occur over short distances which make forecasting particularly difficult. An example of an area of high pressure can be seen in Fig. 1.

A primary characteristic is the widespread descent or subsidence of air within the *troposphere*, which results in the air being warmed by *adiabatic* compression. This subsidence not only warms the air but produces very low relative humidities; the subsided air is thus frequently cloud-free. However, the descending air, because of convection and turbulent currents, rarely reaches ground level, leaving a cooler moist layer some 500 to 1500 m deep immediately above the surface. The boundary between this cooler air and the subsided air aloft is characterized by a temperature *inversion* (usually referred to as a *subsidence inversion*), which has a profound influence upon whether conditions by limiting the upward movement of convection currents originating at the ground surface.

Types of anticyclonic high-pressure systems

Anticyclones may be categorized as either 'warm' or 'cold' anticyclones. Warm highs result from convergence in the upper troposphere and subsidence beneath, producing relatively warm air throughout the troposphere above the subsidence inversion. The zone of high-pressure cells found in the subtropics provides examples of this type of anticyclone, the Azores high is the main centre in the North Atlantic. The high-level convergence responsible for the formation of these cells takes place beneath the *subtropical jet stream*, which is on the poleward margin of the tropical *Hadley cell*. The resulting highs are slow-moving, produce long spells of fine weather, and are responsible for some of the world's major deserts such as the Sahara and the Kalahari. Warm highs also form in temperate latitudes and are a result of convergence just ahead of a ridge in the *upper westerlies*. They are seen on weather maps either as extensions of the subtropical highs and linked to them by a strong ridge of high pressure, or as persistent

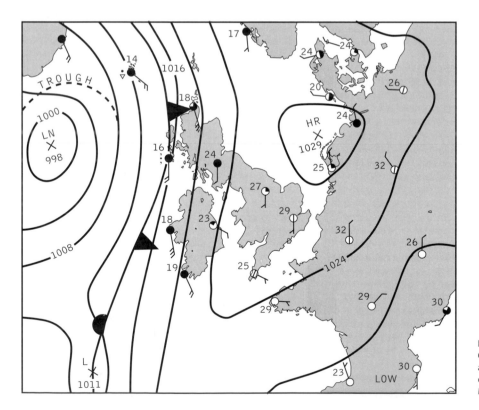

Fig. 1. Synoptic chart for 1200 GMT on 1 August 1990 showing an anticyclone. (Reproduced from data provided by the Meteorological Office.)

'blocks'. Blocking anticyclones block or disrupt the more normal westerly flow and prevent depressions from following their usual routes. When a blocking high develops over north-west Europe, depressions are deflected further north, and sometimes also to the south, and the Azores high is often absent. These developments often lead to anomalous weather conditions in the European and North Atlantic sector. Blocking highs tend to be particularly persistent over Scandinavia, the North Atlantic between 10° and 20° W, Alaska, and the northwest Pacific in the northern hemisphere, and in the Australian and New Zealand sector in the southern hemisphere.

Cold anticyclones are shallow thermal features, often no more than 3000 m deep, which form over cold surfaces as a result of convergence aloft caused by the contraction of the cold low-level air. They may be found over Antarctica and the Arctic Ocean at any time during the year, and in addition form over northern Eurasia, Greenland, and North America during the winter. A classic example is the persistent winter Siberian anticyclone which dominates the weather of this area and sometimes has a central pressure of 1060 mb. Cold anticyclones or ridges of high pressure also migrate from these source regions in the polar air that is found behind depressions bringing short spells of cool but fine weather before the arrival of the next depression; they often approach Britain from the west or north-west.

Weather

Anticyclones are commonly associated with clear skies and light winds. In late autumn and winter these conditions, together with the long nights of this time of year, lead to a rapid cooling of the lowest few hundred metres of the atmosphere owing to long-wave radiation into space. These are favourable conditions for the formation of frost and fog and, because of the weak sun, the fog may persist in places well into or throughout the following day, leading to a spell of raw grey conditions. The subsidence inversion prevents the upward dispersion of pollutants and consequently there is a decline in air quality. If these conditions last for several days, unpleasant smogs may develop, leading to loss of life, especially among the elderly and those who suffer from respiratory problems. Fog is notoriously patchy, and while some areas, especially low-lying regions, may not see the sun for several days, other areas may have unbroken sunshine, resulting in higher diurnal temperature ranges. If the surface layer of air beneath the subsidence inversion is moist then 'anticyclonic gloom' is often a widespread feature. It is characterized by a sky covered by a layer of stratus or stratocumulus which tends to persist because of the weak atmospheric circulation. When this cloud is present, night-time radiational cooling is much reduced and widespread fog is consequently not usually a problem.

In summer, the weather associated with anticyclones is usually fine and cloud-free. During the afternoon small 'fair-weather' cumulus clouds may develop beneath the inversion over the land because of the strong insolation, which causes temperatures to rise rapidly. At night, temperatures fall quickly under the clear skies so that towards dawn short-lived mist or fog patches may form. In coastal areas, especially those adjacent to the chilly North Sea in spring and early summer, sea fog may form where the air has passed over cool waters and is cooled to its dew point. In Fig. 1 an anticyclone situated in the southern North Sea allowed continental air to affect much of the British Isles although a cold front remained close to north-western areas for much of the day. The prolonged sunshine, exceeding 13 hours in many areas away from the cloudy and sometimes wet north-west of Scotland, encouraged temperatures to rise rapidly with maxima in excess of 30 °C in many inland areas of England. However, some anticyclones are cloudy and daytime temperatures do not usually rise much above average for the time of year. The cloud is sometimes the result of a 'decaying' front moving into the circulation of the high; on other occasions it is the result of moist tropical maritime air having been cooled from its passage north over the North Atlantic and reaching the British Isles from the west or north-west.

Persistent blocking anticyclones can have a profound influence upon the weather, producing anomalous months or seasons. The character of the weather depends upon both the precise location of the high and the season. If a blocking high forms over Scandinavia in winter, this area will experience clear, very cold conditions; areas to the south of the centre will be fed very cold polar continental air, which may reach the British Isles. The North Atlantic depressions are steered to the north and to the south of this high and may influence Britain, especially the south, bringing heavy precipitation which in the continental air is likely to be in the form of snow. A situation such as this occurred in early February 1991 when daytime temperatures remained well below freezing for several days in southern Britain with widespread snowfall. The exceptionally cold winter of 1962–3 was largely the result of blocking highs being located either over Scandinavia or between Norway and Iceland, producing very cold easterly winds, severe frosts and significant snowfalls.

Dry spells over Britain (such as the very pronounced drought of 1975–6 and the series of drought episodes during the period 1988–92) are usually the result of persistent blocking anticyclones close to the British Isles. Hot summers are often the result of slow-moving anticyclones situated either over Scandinavia or to the east or south-east of Britain, with very warm and dry continental air being advected across the country. *Cool* but relatively dry summers are caused by high pressure persistently reforming to the west of Ireland.

A general lack of blocking action close to Britain produces cool, moist summers and mild winters as the country is subjected to a changeable zonal westerly airflow bringing a succession of frontal depressions across the country.

JOHN STONE

Further reading

Barry, R. G. (1998) *Atmosphere, weather and climate* (7th edn). Routledge, London.

Young, M. V. (1994) Back to basics: depressions and anticyclones. *Weather*, September 1994, 306–11, and November 1994, 362–70.

atmospheric temperature In the atmosphere, temperature is one of the most sensitive indicators of dynamical and physical processes. It is affected by interactions between the air and the land or ocean, by the radiation received from the Sun and emitted by the atmosphere and the Earth's surface, by chemical interactions (particularly in the upper atmosphere), by changes in state of water from gas to liquid to ice and back again, and by upward and downward motion.

A knowledge of the current temperature in all parts of the atmosphere is crucial to weather forecasting. A massive international effort yields detailed temperature (and wind and pressure) observations every few hours. Because many factors influence the temperature at any given place great care must be taken in making these observations to ensure that they are truly representative of the atmosphere at that location.

Physical meaning of temperature

Temperature is one of the fundamental concepts of physics, like mass, length, and time. Our human senses perceive temperature only when heat is being transferred to or from our bodies; for example, when we stand in front of a roaring fire or we hold an ice cube in our hand. Heat flows from objects which have a higher temperature to objects with a lower temperature; it does not flow in the opposite direction. This is a statement of the second law of thermodynamics.

Heat is a form of energy. All substances, whether solids, gases, or liquids, are composed of molecules moving more or less randomly at different speeds. If we took two solid objects, say the top of a stove and the bottom of a cooking pot, one being much hotter than the other, then the molecules in the hotter object would be moving much faster. When the objects are brought into contact, the faster-moving molecules collide with the slower-moving molecules and transfer some of their energy. The slower-moving molecules in the colder object increase their energy and their temperature. Temperature is a way of measuring the average energy of molecules in a substance. If we pour a glass of hot water and a glass of cold water (each containing the same amount of water) into a jug, then the internal energy of the molecules in the mixture will be the average of the internal energies of the initial two glasses of water—and the temperature of the mixture will be the average of the temperatures of the two initial glasses.

In these examples all the heat transferred to an object results in an increase in the temperature. When this happens the heat is described as 'specific heat'. Another form of heat, 'latent heat', does not immediately result in a temperature change. This applies when materials change from one state to another: for example, from a gas to a liquid (condensation) or from a solid to a liquid (melting). When it rains and a road surface is wet, the temperature of the water on the road and the temperature of the air will be approximately the same. If

heat is supplied, for example by sunshine, then some of the energy supplied to the water will be used to break the molecular bonds that hold liquids together and will enable the molecules to move freely as they do in a gas; the liquid water then evaporates to become water vapour. Because the energy is not used to increase the motion of the molecules, the temperature is not affected during this process: the water changes from a liquid to a gas without changing temperature. The heat used in this process is latent heat. It is latent because the heat is potentially available and will be released when the water vapour condenses at some time in the future to form a liquid. The water which evaporated at the road surface may be carried about in the atmosphere until eventually it becomes part of a cloud and condenses into water drops. When this condensation occurs, the heat originally used to break the molecular bonds is released and warms the air and the water droplets, thus raising their temperature. Latent heat is a very important factor in the atmosphere, for it enables heat received at one location to be released at another location. We can consider the large-scale movement of water vapour carried by winds in the atmosphere to be a large-scale movement of latent heat.

Temperature scales

In common with the other fundamental physical properties, temperature is measured in arbitrary units. To devise a temperature scale, scientists in the eighteenth century chose two situations which were easily reproducible as critical points on which a scale is based. They then divided the interval between these points into intervals called degrees. The two scales still in common use are those named after Fahrenheit and Celsius. Fahrenheit used the boiling point of water as his upper reference point, and the lowest temperature achievable by mixing water, ice, and salt as the lowest. He defined the lower point as 0 degrees and the upper point as 212 degrees (following a temperature scale earlier proposed by Newton). On this scale the freezing point of water lies at 32 degrees. The Celsius scale, for many years called the centigrade scale, uses the freezing point of water as the lower reference point (0 degrees) and the boiling point of water as the upper point (100 degrees). Several other scales were suggested in the eighteenth century but only these two have stood the test of time.

Because temperature is a measure of the motion of molecules there is a theoretical absolute zero temperature at which all molecular motion would cease. A temperature scale can accordingly be devised which has absolute zero as its lower reference point. This scale, known as the Kelvin scale, has the same intervals as the Celsius scale. Zero on the Kelvin scale is −273.15 Celsius. The Kelvin scale is used widely for scientific purposes since it has the advantage that there are no negative values.

Measuring temperature

Thermometers are the most commonly used instrument for measuring temperature, but other instruments are also in regular use for making remote measurements in the atmosphere. Several varieties of thermometer are used in meteo-

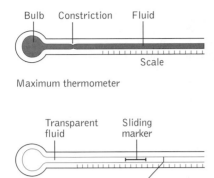

Fig. 1. Maximum and minimum thermometers.

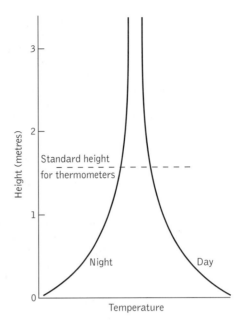

Fig. 2. Temperature varies rapidly near the surface. A standard observing height of 1.5 metres makes it easier to compare measurements made at different locations.

rology. The maximum thermometer is similar to a clinical thermometer used by medical staff. It has a small constriction near the reservoir (Fig. 1), which prevents the fluid from returning once it has expanded; it therefore records the highest temperature reached until it is reset. To record minimum temperatures, some thermometers have a small metallic marker inside the fluid. When the temperature drops, this marker is dragged down the tube by the meniscus of the fluid (Fig. 1). When the temperature rises, the fluid moves along the tube but leaves the marker to indicate the lowest temperature reached.

Glass thermometers are suitable for measuring temperature at the Earth's surface but would be impractical at higher levels. Between the surface and about 30 km altitude the temperature is measured twice daily at about 300 observing stations by using radiosondes. These are instrument packages lifted by balloons filled with hydrogen or helium. The packages have to be very light and the measuring device has to produce an electrical signal which can be transmitted by radio. Most temperature-measuring devices on radiosondes are electrical resistance thermometers. These rely on the fact that the electrical resistance of several materials, for example platinum and certain ceramics, varies with temperature.

Temperature measurements are also made from satellites. Although pictures of weather systems are the most obvious satellite products, many of these satellites also carry instruments designed to measure temperature at various altitudes through the atmosphere at all points along the satellite orbit. This is a very effective way of obtaining observations around the world on a regular basis. The observations made from space are not as accurate or as detailed as those made by radiosondes, but their much broader range makes them invaluable, particularly over the oceans, which are inadequately observed by the radiosonde network. Satellite measurements of temperature are made using a radiometer. Unlike other temperature-measuring devices, the radiometer makes its measurements remotely. It measures the radiation

emitted by molecules in the atmosphere. From the intensity of the radiation it is possible to deduce the temperature of the molecules that were the source of the radiation.

It is vitally important that temperature measurements are made in a way that minimizes errors so that observations made over a tropical ocean or over a polar ice cap are each truly representative of the air at those places. There are many possible sources of error. Factors which can give an erroneously high temperature include sunlight falling directly on the thermometer; observing too close to the ground on a sunny day; insufficient ventilation around the thermometer on a sunny day. Errors leading to a reduction in temperature include exposing thermometers to the sky on a clear night; observing too close to the ground on a clear night; allowing rain to fall on the thermometers. Most of these errors are eliminated by ensuring that thermometers are placed in a well-ventilated box, which is painted white to ensure that it absorbs as little sunlight as possible and is at a height such that the thermometers themselves are exactly 1.5 metres above the ground (Fig. 2). The ground surface should also be grass; the nature of the surface affects the rate at which heat from the Sun is absorbed, and the temperature of the air just above the surface therefore depends on the nature of the surface.

Factors influencing temperature

The temperature at any particular place is influenced by a number of factors: latitude; season; altitude; proximity to a major ocean; time of day; wind direction; present weather conditions. The last three of these control variations in temperature over short periods, of hours to days; the others are important for periods of weeks to months or longer.

On the shorter timescales the diurnal cycle has a major effect. During the day solar radiation is absorbed at the ground and heats the air above it by convection and conduction. Although solar radiation is at its maximum about the middle of the day the ground reacts slowly to the heating and the maximum air temperature lags a few hours behind the maximum radiation. The Earth's surface, in common with all bodies, radiates heat. During the day the incoming heat from the Sun exceeds the radiated heat, except in polar regions, but at night there is a net loss of heat and the air near the surface cools. The cooling process proceeds until the Sun again produces a net input of heat; the minimum in temperature is consequently around dawn. The heat received from the Sun and the heat lost by radiation are, of course, both reduced in cloudy conditions. Local weather systems which determine the direction of the wind influence the day-to-day temperature changes. Air may be directed from warmer regions, from polar regions, or from oceans or continental land masses, each giving different local characteristics.

Seasonal variations exist because the axis about which the Earth spins is tilted in relation to the plane of its orbit around the Sun. This means that during part of the year the southern hemisphere receives more sunlight and six months later the northern hemisphere receives more (Fig. 3). The maximum temperatures in the summer hemisphere are found in a belt between the equator and about 30 degrees, but higher latitudes are also warmer than in the winter hemisphere. The winter hemisphere has no sunlight falling on the polar region and the warm tropical regions are cooler than their summer counterpart.

Since the sunlight falling on tropical regions exceeds that falling on polar regions, there is a variation of temperature with latitude. On a journey in January from the north of Norway (70° N) to southern Spain (40° N), we would experience temperatures from about −12 °C to about 10 °C on average. In North America, the equivalent journey would be from Baffin Island in Canada to New York. The temperatures experienced on this journey would be from about −34 °C to −12 °C. This difference illustrates the effect of the circulation of the Atlantic Ocean. Warm currents are found on the eastern side of the great oceans, and as a result equivalent latitudes in western Europe are considerably warmer than their equivalent on the eastern seaboard of North America. If we travelled away from the coast into the continental land mass, while remaining at the same latitude, we would find that the temperature in January would drop the further we were from the ocean. This is because the thermal capacity of the oceans is very much greater than that of the land. The same amount of heating or cooling will cause a much smaller difference in the ocean temperature than in the land temperature. In summer, the result of this difference is that the continental land masses are heated much more rapidly than the oceans; the coastal regions therefore experience lower temperatures than their inland neighbours.

Altitude is the remaining major factor influencing temperature. Because temperature decreases with height at an average rate of about 6.5 °C for every kilometre it might be expected that, all else being equal, two towns at altitudes one kilometre apart would have average temperatures 6.5 °C apart. This is not the case. Rising up through the atmosphere leaving the land behind is different from rising up a hillside but remaining close to the land. The hillside absorbs solar radiation and is thus warmer than the free atmosphere at the same altitude. The amount of warming depends on the orientation of the hillside. If it is south-facing, then a considerable amount of sunshine may be absorbed, while north-facing slopes can be extremely cold.   CHARLES N. DUNCAN

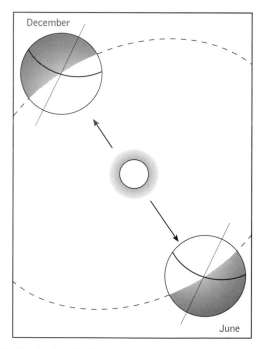

Fig. 3. The tilt of the Earth's axis causes the northern hemisphere to receive more sunshine in June and the southern hemisphere to receive more in December.

Further reading

Wilheit, T. T. (1993) Atmospheric remote sensing by microwave radiometry. *Science*, **262** (5134), 773–4.

Ahrens, C. D. (1994) *Meteorology today*. West Publishing Co., St Paul, Minnesota.

McIlveen, J. F. R. (1986) *Basic Meteorology*. Van Nostrand Reinhold, New York.

avalanches Avalanches may be defined quite simply as the rapid downslope movement of earth, ice, rock, or snow. This broad definition incorporates a variety of phenomena that are typically classified on the basis of the constituent material. However, while debris, ice, and rock avalanches are all important mass-movement processes, attention is focused here upon snow avalanches, which are popularly considered to be synonymous with the term 'avalanche' and are of much greater significance as a natural hazard than other types of avalanche. Examples of major avalanche disasters in the twentieth century include the accident in 1910 at Rogers Pass, British Columbia, Canada, where 62 railway workers were killed while clearing the track of previous avalanche debris; the winter of 1950–1 in Switzerland, which left 279 people dead and 285 injured; and the 1995 avalanche in Flateyri, Iceland, which resulted in 20 fatalities.

Snow avalanches are commonly classified on the basis of release type and free-water content. Loose snow avalanches are released from a localized area or point and in general involve snow only at or near the surface. Small loose avalanches are known as *sluffs*, and in general loose avalanches are smaller than slab avalanches. The latter occur when an approximately rectangular block of snow with a depth of up to the full depth of the snowpack is released and moves downslope. Dry avalanches contain little or no free water at the time of fracture, although it is not unusual to detect some moisture in the snow immediately after deposition. This results from melting induced by frictional heat, which is generated between the avalanche and the sliding surface and by collisions between moving grains.

At the other end of the moisture-content scale, slush avalanches are the flow of a partially or totally saturated snow cover. Wet snow avalanches represent intermediate moisture contents. Wet and slush events are more common towards the end of the winter when temperatures begin to rise. Slush avalanches are particularly common at high latitudes (for example, in Norway), where the sudden return of direct radiation from the Sun promotes intense melting. Wet and slush avalanches commonly incorporate debris from the ground underlying the snowpack; in consequence, they produce dirty deposits that are of interest to geomorphologists studying rates of mountain erosion. It is, however, the dry avalanche, and in particular the dry slab avalanche, that is of primary concern to anyone evaluating the importance of avalanches as a natural hazard.

Dry slab avalanches tend to be released on slopes with angles between 25° and 55°. The majority of slab avalanches have a fracture depth of less than 1 m, although major avalanches will have depths much greater than this: for example, the Flateyri avalanche mentioned above had a maximum fracture depth of 3.7 m. Dry slabs are released when the snowpack fractures, that is, when a catastrophic failure occurs. The state of failure is reached when the downslope component of the weight of the snowpack above the eventual failure plane is approximately equivalent to the shear strength of this failure plane. Thus, the first fracture occurs in a plane approximately parallel to the snow surface. It then propagates both across-slope and upslope. A tensile fracture develops from the bed towards the surface and then spreads laterally, forming the crown of the avalanche, a face perpendicular to the failure plane that is usually clearly observable after the event. Very shortly after the formation of the crown, the flanks and the lower fracture (the *stauchwall*) are created. The slab now begins to move rapidly downslope, usually obliterating the stauchwall in the process.

The most important trigger for natural slab avalanches is the addition of new snow, either by direct precipitation or by wind-driven drifting. Increase in weight due to rainfall is an important cause of wet slab formation. The impact of falling snow from avalanches released above or from cornice collapse is another important natural trigger. Artificial triggers include release by explosives as part of the active control of avalanche-prone slopes and skier loading.

The avalanche failure plane is commonly much weaker than the layers of snow above and below. These weak layers can be remarkably persistent and widespread. A weak layer that formed in mid-November 1996 in British Columbia, Canada was responsible for avalanching in the Rocky, Columbia, and Coast Mountain ranges, including one avalanche 6 months after the layer formed. Weak layers can be formed by a variety of processes. Most weak layers form at or near the surface and are buried by subsequent snowfall. Crusts are an important type of weak layer that can form when water on the surface of the snow (originating from rainfall or melting) freezes. Fresh snow bonds poorly to these layers, promoting instability. If the temperature gradient in the snowpack locally exceeds 10 °C m^{-1} it is possible for snow crystals to change into more angular forms called *faceted crystals*. These crystals bond poorly to one another and also form weak layers. Faceted crystals may also be associated with crusts. An important type of faceted crystal is known as *depth hoar*. These large, cup-shaped crystals can result in deep-seated instability in the snowpack and hence large avalanches.

Various defence strategies are commonly employed to protect people from avalanches. Artificial release can prevent large avalanches from taking place. Supporting structures in the starting zone of the avalanche path help to reduce the chance of release. Fences placed above the starting zone can prevent snow from drifting into leeward basins, thus reducing drift loading. Structures are also used in the runout zone of the avalanche path, where the avalanche begins to decelerate. Deflectors steer the snow away from regions of concern, while dams act as a barrier to the flow. Mounds of earth may also be used to slow down the avalanche and reduce its runout distance. The prediction of avalanches and of their runout distances are important areas of current research aimed at improving the effectiveness of avalanche-protection strategies.

CHRISTOPHER J. KEYLOCK

Further reading

McClung, D. M. and Schaerer, P. A. (1993) *The avalanche handbook*. The Mountaineers, Seattle.

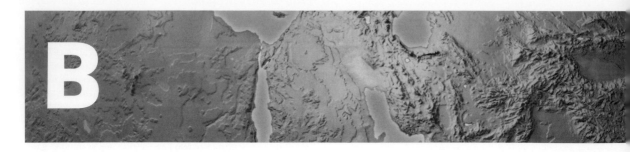

B

bacteria Bacteria are distinguished from all other life forms by their prokaryotic (literally, 'before the nucleus') cell structure. These microscopic organisms, typically 1–5 micrometres (μm) long, are distinguished by the absence of sub-cellular organelles, such as a nucleus, mitochondria, and chloroplasts. These organelles occur in the eukaryotic cells of higher organisms. Bacteria are ancient organisms, being the first to evolve some 3.8 billion years (Ga) ago and they were the sole type of life for 70 per cent of the history of life on Earth. They are not, however, primitive. They are well adapted to their large range of habitats; they are the most numerous organisms on Earth, and their distribution defines the limits of the biosphere.

Although the variety of different bacterial cell types is limited (there are, for example, rods, cocci, spirilla, and filaments), this belies their vast metabolic diversity and their ability to grow under a remarkably wide range of conditions (for example, at –5 to 113 °C, at pH values ranging from 0 to 11, in near-vacuum or at pressures 1000 times greater than atmospheric, and in distilled water or in saturated salt solution).

Different types of bacteria obtain energy in different ways.

(1) Some bacteria obtain energy from photosynthesis, by processes similar to those used by green plants. Blue-green algae, which are actually bacteria, are thought to have been the first to have evolved oxygen-forming photosynthesis; their activity was responsible for the formation of our oxygenated atmosphere and hence the evolution of complex metazoans (plants, animals), which require oxygen.

(2) Some photosynthetic bacteria, however, neither produce nor require oxygen; these *anoxygenic photoautotrophs* represent a more primitive form of photosynthesis. They use reduced compounds such as hydrogen sulphide and ferrous (Fe^{2+}) iron, oxidizing them respectively to sulphur and ferric iron. The anaerobic formation of ferric (Fe^{3+}) iron may have been important for the early formation of geological banded iron deposits.

(3) Like animals, some bacteria can conduct aerobic respiration. These types are termed *aerobic heterotrophs.*

(4) Unlike animals, however, some *heterotrophic bacteria* are not restricted to using oxygen for respiration and for many of these anaerobic bacteria oxygen is poisonous. Other bacteria can use instead the oxygen in other compounds such as nitrates (NO_3^-), sulphate (SO_4^{2-}), carbon dioxide (CO_2), and even from metal oxides (for example, iron and manganese).

(5) Other heterotrophic bacteria do not require any respiratory compound, but instead gain energy from splitting organic compounds into a reduced and an oxidized product, a processes called *fermentation*. Fermentation products are of considerable commercial importance; for example, citric acid is used widely in foods and beverages for flavouring, and itaconic acid is used in the production of acrylic resins.

Heterotrophic bacteria (types (3)–(5) above) are the detrital specialists. Working together with other micro-organisms they degrade the organic compounds from dead plants and animals extremely efficiently and in this way return nutrients that are essential for further photosynthesis. They similarly drive the biogeochemical cycles of the major elements carbon, sulphur, and nitrogen. These microbial processes are optimized in sewage treatment plants, which enable humans to live at high population densities without contaminating each other and their local environment.

(6) Some bacteria specialize in obtaining energy from inorganic rather than organic sources. This unique metabolism includes the oxidation of reduced metals and minerals, generally by using oxygen directly. Relatively little energy is obtained from these reactions, and these bacteria therefore have to process a large amount of material. During mining, minerals are exposed to oxygen. Bacterial oxidation can then be a major problem, especially in abandoned mines. In these circumstances, high concentrations of metals can be produced, together with inorganic acids (such as sulphuric acid), because bacteria oxidize sulphide minerals such as pyrite (FeS_2) to sulphate and ferrous iron. The acid waters are referred to as *acid mine drainage.* Groundwater and local streams can be made very acidic (pH less than 2), which can kill most wildlife. These conditions are, however, optimal for the bacterial 'miners'. Microbial mining reactions can, on the other hand, be turned to commercial advantage to extract metals from low-grade ores. Reduced metals and sulphides are also a major source of energy for bacterial communities at hydrothermal vents at ocean ridges. Reduced hydrothermal fluids are geothermal products, and these bacteria and the animal communities that feed on them are unique ecosystems.

(7) An even stranger inorganic metabolism, inorganic fermentation, is conducted by some bacteria.

In the environment, bacteria tend to work as an interacting team. Although large bacterial populations are commonly present (about 2000 million bacteria per cubic centimetre in soil, for example), a much smaller number may be active. Small environmental changes can produce conditions that are more suitable to a portion of the non-active bacterial population. There can consequently be rapid changes in the bacterial community to maintain efficient processing of energy sources, and hence stable biogeochemical cycles. In addition, bacterial growth rates can be very rapid (the fastest are about one cell division every 20 minutes), providing further opportunity for bacterial populations to adapt to changing conditions. Associated with these high growth rates are high mutation rates, and hence the possibility for genetic modification and adaptation. Consideration of the extensive metabolic activity of bacteria and their capacity to adapt and evolve resulted in the concept of 'microbial infallibility'. This concept implied that bacteria would adapt to degrade any chemicals that were artificially introduced into the environment, such as pesticides and herbicides (xenobiotics, strange to life), and hence few precautions would be required in their use. This proved to be a great oversimplification, because although bacteria can degrade many xenobiotics, some are directly toxic to bacteria and some produce toxic breakdown products; others provide insufficient energy to support degradation.

Bacteria have adapted effectively to inhabit other organisms, both external and internal. They are commonly present in large numbers in the digestive systems of animals, particularly herbivores, where they assist in the breakdown of decay-resistant compounds such as cellulose from plant material. In ruminants, such as cows and sheep, this relationship has developed to such an extent that a separate stomach (the rumen) has evolved to provide space in which large microbial populations can develop and break down the cellulose in grass. The animal does not have the enzymes to break down cellulose, but survives by absorbing the microbial cellulose degradation products and digesting the bacterial cells that are produced. The importance of microbes to cows is demonstrated by the size of its rumen, which is between 100 and 150 litres. A similar symbiosis occurs in the roots of plants, such as peas and beans, where bacteria develop in small nodules. These bacteria are fed by the plant, and in return they supply the plant with ammonia, an essential nutrient. The bacteria obtain this ammonia from nitrogen in air by *nitrogen fixation*, in a process unique to bacteria. Bacterial interactions with higher organisms are not, however, always benevolent, for some bacteria are major pathogens that cause a variety of diseases, some of which can be fatal. Fortunately, micro-organisms, including bacteria, have provided a source of antibotics with which to combat these diseases. Many bacterial pathogens have, however, developed resistance to antibiotics, and some bacterial diseases are now increasing

in their prevalence. For example, tuberculosis, caused by *Mycobacterium tuberculosis*, kills three million people every year.

Molecular genetic analysis has demonstrated two distinct types of prokaryotes: Bacteria (formerly Eubacteria) and Archaea (formerly Archaebacteria). Both of these represent the highest order of life, Domains. All eukaryotes, including plants and animals, exist in a single Domain, Eukarya. Two out of the three Domains of life thus contain exclusively prokaryotic organisms, and this underlines their great diversity. Interestingly, the organelles of eukaryotic cell, mitochondria and chloroplasts, belong to the Domain Bacteria, not Eukarya. This demonstrates that these organelles were originally free-living bacteria that have evolved a stable endosymbiotic relationship with eukaryotic cells. Bacteria are thus an integral component of all of us.   R. JOHN PARKES

Further reading
Madigan, M. T., Martinko, J. M., and Parker, J. (1997) *Brock biology of microorganisms* (8th edn) Prentice Hall, London.

bacterial isotopic fractionation
The number of protons in the atoms of a given element is fixed, but the number of neutrons may vary, giving rise to isotopes of an element which have different atomic weights but the same chemical properties. Some isotopes are unstable and undergo radioactive decay while others are stable. Carbon exists primarily in nature as ^{12}C (6 protons and neutrons), but small amounts of the heavier stable isotope ^{13}C (6 protons and 7 neutrons) and the radioactive isotope ^{14}C (6 protons and 8 neutrons) also occur. Similarly, sulphur exists mainly in the form of ^{32}S, although some occurs as the heavier stable isotope ^{34}S.

Stable isotopic abundances are expressed as parts per thousand (‰) relative to a standard reference material, using the delta notation:

$$\delta X = [R_{sample} \div R_{standard}) - 1] \times 10^3,$$

where X = ^{13}C, ^{34}S, or other elements and $R = ^{13}C/^{12}C$, $^{34}S/^{32}S$, or the appropriate isotopic ratios for other elements.

A sample that is depleted in the isotope of greater mass, relative to the standard, is termed 'light' and will have a negative δ value. By contrast, a sample enriched in the heavier isotope is termed 'heavy' and will have a positive δ value.

Most biological reactions involving carbon, sulphur, and other elements discriminate against heavy isotopes, which means that biological products are 'light'. This provides a clear biological fingerprint in the stable isotopic composition of a compound. For example, in green plant photosynthesis, which incorporates carbon dioxide for the formation of new organic compounds and cells, $^{12}CO_2$ proceeds faster through the enzymatic reactions than $^{13}CO_2$. The carbon in new plant material is thus isotopically 'light', while the remaining CO_2 becomes 'heavy'. This difference makes it possible for organisms to build up carbon compounds from carbon dioxide. The difference thus also provides a means of detecting the presence of living organisms in the geological record. Isotopically 'light' organic carbon occurs in rocks as ancient as 3.5 billion years old, which indicates that life must have evolved earlier than this in order to have produced enough 'light' organic matter to be detectable. Similarly, the 'light' carbon isotopic value of petroleum demonstrates that it was formed from previously living organic material that was modified by burial and heating.

The global distribution of sulphur isotopes also makes it possible to discriminate biological processes from geochemical processes. For example, sulphides can either be of bacterial origin, predominantly from dissimilatory anaerobic sulphate reduction, or of geological origin (for example, from igneous rocks or hydrothermal ore deposits). However, as bacterial sulphate reduction discriminates against the heavy ^{34}S isotope, sulphides from this source will be 'light'. Sulphides of geological origin should show minimum discrimination against the heavy isotope, and hence their δ values should be close to zero. The 'light' sulphur isotope values of sulphides in marine sediments confirm their bacterial origin, while the near-zero δ values of sulphides in igneous rocks confirm their non-biological origin. Most useable sulphur ore deposits are in fact of bacterial rather than volcanic origin.

The isotopic composition of samples from extraterrestrial sources can also help in the search for life on other planets. For example, the isotopic values of sulphides in rocks from the Moon indicate a non-biological origin.   R. JOHN PARKES

Further reading

Madigan, M. T., Martinko, J. M., and Parker, J. (1997) *Brock biology of microorganisms* (8th edn). Prentice Hall, London.

badlands
The word 'badlands' originated in North America, where extensive areas of badlands occur in the north-western Great Plains, such as in Badlands National Monument, South Dakota, and Dinosaur Provincial Park, Alberta. It seems likely that early French explorers in this region translated Plains Indian descriptions of these areas into the phrase '*les mauvaises terres á traverser*' meaning 'land bad (hard) to cross'. In fact, badlands occur in a wide range of environments, and on various materials, from marine silts in valleys of the Canadian Arctic to mine-spoil heaps in New Guinea. Badlands can, therefore, form as purely natural landscapes or as human-created features resulting from unwise agricultural or industrial practices.

Badlands are usually associated with arid or semi-arid regions where occasional intense rainstorms cause highly erosive rainsplash and run-off on exposed, weak rock surfaces. Erosion rates in excess of 20 mm a year have been recorded in some badlands. Relatively impermeable sedimentary rocks, such as mudstones and clay-cemented sandstones, are particularly susceptible to such rapid erosion and

most badlands develop on such materials. Generally, the infertile nature of the rock and the rapid erosion inhibit the establishment of a vegetation cover and the landscape becomes a barren, often spectacular, and seemingly impenetrable maze of deep winding gullies, steep sharp-crested slopes, and an assortment of weirdly shaped hoodoos, rills, and tunnels. The steep slopes are often scarred by a network of closely spaced rills, which form as a result of rainstorm run-off. Rill incision deepens downslope as rills amalgamate into larger channels to form gullies. Wind may assist the processes of badland formation by removing fine grains and causing minor sandblasting effects.

While surface erosion caused by water can produce dramatic landscapes in many badlands, a considerable amount of subsurface erosion in the form of tunnels can also occur. Sedimentary rocks such as mudstones often contain a variety of clay minerals, including swelling clays such as bentonite. The swell–shrink effects of these clays, resulting from alternate wetting and drying, creates deep desiccation cracks which help to channel surface flow underground (see *subsurface flow and erosion*). Extensive subsurface drainage networks may develop, producing an effect that resembles, in some respects, the caves and shafts that form by chemical solution in limestones. In badlands, however, solution processes are minor and tunnel systems are almost entirely erosional in origin as the rock particles are washed away. The tunnels become enlarged and their sides and roofs collapse to form gullies which, in turn, direct flow into newly forming tunnels in a complex interlinked surface and subsurface drainage system.   IAN A. CAMPBELL

Bagnold, R. A. (1896–1990)
Brigadier Ralph Alger Bagnold, born near Devonport, Devon, on 3 April 1896, was a professional soldier who became widely known for his contributions to understanding the physics of sediment transport by wind and water. After serving in Flanders during the First World War, he took an engineering degree at Cambridge University before rejoining the army and being posted to Egypt, where he developed an interest in sand dunes and desert exploration. After retiring from the army in 1935 he undertook laboratory experiments at Imperial College to aid interpretation of his field observations. The results of this work were published in a series of pioneering papers which had a major impact. Bagnold's early work was initially published in 1941 as *The physics of blown sand and desert dunes*; it remains a classic. Recalled to the army in 1939, Bagnold returned to Egypt, where he developed the Long Range Desert Group, a specialist unit which used his knowledge of the desert to harass the enemy with unexpected attacks. He retired again in 1944, and after a short spell as Director of Research for Shell Oil, he returned to his studies of sand transport, extending his interest in wind transport to include sediment movement by running water and waves.

The hallmark of Bagnold's work was a skilful combination of theoretical analysis, laboratory experimentation, and careful field observation. During the later 1940s and 1950s he acted as a consultant in windblown sand control in the Middle East, and, in 1958, he became a consultant to the United States Government. The importance of Bagnold's work is demonstrated by numerous awards and prizes. He remained scientifically active until shortly before his death in May 1990 at the age of 94.   K. PYE

banded iron formations Banded iron formations, or BIFs for short, are an unique type of sedimentary rock in the Earth's geological record, which are as important economically as they are scientifically. They comprise the majority of the world's iron reserves, occurring in vast deposits on every continent, with bulk iron contents locally exceeding 50 per cent by weight. For example, the banded iron formations of the Lake Superior region of North America have been the source of most of the iron ore produced in the United States over the past 120 years and this readily available supply contributed significantly to the rapid industrialization of North America that occurred during this period. The iron formations of north-western Australia and the Transvaal in South Africa occur in layers hundreds of metres thick and are exposed over thousands of square kilometres, providing an important source of income for these countries, as well as an assured supply of iron ore for the foreseeable future.

The oldest known rocks on Earth, from the Isua area of West Greenland, are sedimentary deposits that include a banded iron formation and are about 3.8 Ga (billion years) old. In fact, the vast majority of banded iron formations are very old, having formed prior to 1.8 Ga before the present day. As no modern equivalent is known to exist, the process by which banded iron formations form has been debated by geologists since the late 1800s. However, it is not only their mechanism of formation that makes banded iron formations scientifically intriguing. These very old rocks, with their distinct chemical composition, provide important clues about the early development of the Earth's atmosphere and oceans. As such, banded iron formations record important information that contributes towards unravelling the puzzle of when and how early life developed on Earth. It has even been proposed that biological activity was directly responsible for the formation of banded iron formations.

A banded iron formation is defined as an anomalously iron-rich, chemically precipitated sediment which is usually thinly bedded or laminated and often contains layers of interbedded chert, or microcrystalline silica. These deposits are virtually free of detrital, or terrigenous, sediment. The world's occurrences of iron formations can be subdivided into two groups, based upon their geometric form and their association with other types of sediments: the Algoma-type are lenticular-shaped deposits that are intimately associated with volcanic rocks; and Superior-type iron formations, which are the most common in the rock record, have a large aerial extent on the scale of up to tens of thousands of square kilometres and are associated with sequences of marine sediments such as carbonates, black shales, and quartzites. Banded iron formations can also be characterized by their mineralogy, which reflects their bulk chemistry and the environment in which they were deposited. There are four different mineralogical types:

(1) oxide iron formation, containing haematite, magnetite, and chert;
(2) carbonate iron formation containing siderite, ferrodolomite, and calcite;
(3) silicate iron formation, containing the hydrous iron-silicates greenalite, minnesotaite, stilpnomelane, chlorite, and amphibole; and
(4) sulphide iron formation, containing pyrite.

Over time, the banded iron formations have experienced variable degrees of metamorphism, which has modified the original mineralogy to differing degrees. This makes it difficult to determine unequivocally the exact mineralogy of the original chemical sediment that was deposited.

The chemical composition of banded iron formations is unique and provides useful constraints on the prevailing conditions during their deposition. In addition to iron, they are always rich in silica, indicating that the waters from which the iron formations precipitated were saturated in silica as well as iron. Iron formations have extremely low alumina and titanium contents; elements which are generally associated with extensive land erosion. This characteristic, along with the lack of interlayered detrital sediment, suggests that the banded iron formations were deposited well away from any source of land-derived material, such as a river delta. Banded iron formations themselves also contain very little organic carbon, arguing against a direct link between biological activity and the precipitation of the iron and silica. Their remarkably low manganese and trace metal contents stand in contrast to the manganese-rich nodules currently forming on the deep sea floor. On the other hand, the trace element and rare earth element contents have a 'signature' compatible with formation from waters that were mixtures of normal sea water and hydrothermal fluids, like those forming the ore-rich 'black smokers' located at the mid-ocean ridges. These latter fluids are rich in dissolved metals and other elements through the alteration of the ocean floor by hot percolating fluids.

A further feature of banded iron formations is the chemical nature of iron itself, which can occur in different forms, according to the oxidizing nature of the environment. Under oxidizing conditions, ferric iron is favoured and the most stable iron oxide is haematite (Fe_2O_3). Under more reduced conditions, some or all of the iron can be present as ferrous iron. For example, magnetite (Fe_3O_4) contains a mixture of both ferrous and ferric iron, and many silicate and carbonate minerals contain predominantly ferrous iron (Fe^{2+}). It is important to note that while ferrous iron is relatively soluble in water, ferric iron (Fe^{3+}) has an extremely low solubility; hence the very low iron contents of today's oxidized oceans. In most banded iron formations, the ratio of ferric to ferrous iron is low, reflecting the association of magnetite and ferrous iron-bearing silicate and carbonate minerals along with haematite. As a result, banded iron formations are not as oxidized as one might initially think, simply based upon the occurrence of haematite. Notable exceptions to this are several relatively younger banded iron formations. such as the Rapitan banded iron formation of north-western Canada, which are highly oxidized and have a simple mineralogy of haematite and quartz.

Banded iron formations are so named in part because of their distinctive banded or layered structure, which occurs at various scales from microscopic to macroscopic. It is notable that this layering can have extreme lateral continuity. For example, in parts of the Hamersley iron formation in north-western Australia, individual millimetre-thick chert laminations have been traced for over 300 km, and sequences of layers can be correlated over an area in excess of 50 000 km^2. Such fine-scale layering and extensive continuity point to a quiescent environment of deposition for the iron formation. There are other occurrences, however, where layering is locally disrupted and sedimentary structures like ripple marks, scours, and channels are recognizable, together with oolitic and peloidal textures. These features are indicative of deposition in a high-energy environment where currents vigorously rework the sea-bed sediment.

The distribution of banded iron formations in the geological record is limited to an early period in the Earth's history. Radiometric dating reveals that banded iron formations were primarily deposited during the Archaean (>2.5 Ga old) through the Early Proterozoic (between 2.5 and 1.6 Ga) eras, their greatest development occurring between 2.6 and 1.8 Ga ago. After about 1.8 Ga ago, there was essentially no deposition of banded iron formations, except for a slight resurgence of deposition that occurred between 800 and 600 Ma ago. These younger deposits, including the Rapitan iron formation in north-western Canada, have a distinctly different character in comparison with the older banded iron formations, suggesting that they formed under different environmental conditions. Since 600 Ma ago, no true banded iron formations have been deposited.

Origins and formations of BIFS

Hypotheses to explain the origins of banded iron formations have been many and varied; some early geologists even proposed that they had crystallized from a magma. Although their sedimentary origin through precipitation has long been agreed, the mechanism of iron deposition and the type of environment in which they formed are still the subject of debate. Any hypothesis must adequately account for a number of features, namely: the layering, which exhibits continuity over large distances; the variation of mineralogical type; and the limitation of their formation prior to 1.8 Ga ago. In addition,

there must be a plausible source and mechanism for the transport and precipitation of large quantities of iron and silica.

It is generally agreed that the iron and silica must have been in solution in order to account for the widespread deposition of banded iron formations. This implies that the iron was in the ferrous state, and it then follows that the amount of free oxygen was much lower than it is in today's oceans and atmosphere, if it was present at all. There are two conceivable sources for the iron and silica. One possibility is that they were the product of weathering of rocks exposed on land. A lack of atmospheric oxygen would facilitate the transport of iron to the oceans by rivers. Another likely source is from volcanic or hydrothermal activity. Indeed, the Algoma-type banded iron formations, which are most common in the Archaean, have volcanic rocks associated with them. The trace-element contents of banded iron formations also point to a hydrothermal source. However, although hydrothermal activity appears to have supplied much of the iron (and silica), other geochemical considerations suggest that river run-off also made a contribution.

The mechanism of precipitation of the iron has exercised the imagination of many geologists. There is general agreement that oxidation and changes in acidity (pH) are necessary to cause iron precipitation and to form the different mineralogical types of iron formations. It is also considered that precipitation of silica may have been more or less continuous, since without the presence of micro-organisms that take silica out of the water to make shells and skeletons the ocean would have been always close to silica saturation. One proposal calls upon evaporation to cause iron precipitation. Such a model entails deposition in a basin with restricted circulation or even a setting of playa-lake type in which the lake periodically dries up. In this case, the microscopic laminations and macroscopic layering would be related to daily and seasonal fluctuations in temperature. A difficulty with this model is that there is often no geological evidence for such a restricted basin, and rocks associated with banded iron formation tend to show marine affinities rather than the characteristics of lake deposits. A second model envisages a direct role of biological activity in the precipitation of iron. Here, primitive algae provide the necessary oxygen (through photosynthesis) to oxidize the iron locally. Support for this hypothesis comes from the presence of iron in biologically produced organic pigments and the observation of iron oxides precipitating directly on some modern bacteria. The layering in the iron formations would then be due to daily and seasonal variations in photosynthetic production of oxygen. Major drawbacks to a direct link between biological activity and the precipitation of iron come from the absence of fossils and the general lack of organic carbon in the iron formations themselves, particularly those of the oxide type. A further implication of both the models so far described is that iron formations were deposited in quite shallow water, where evaporation was effective and where photosynthesis could occur. However, carbonate sediments associated with banded iron formations contain fossils that are indicative of shallow-water conditions and biological activity; and these sediments are remarkably poor in iron. This suggests that although the amount of atmospheric oxygen may have been minimal, the surface waters were oxidizing enough to preclude the presence of significant amounts of dissolved iron.

Another hypothesis is based upon the mixing of reduced, iron-rich waters with oxidized water to cause precipitation of iron. This involves the notion that the early oceans were stratified into two layers by contrasts in density and chemical composition, effectively isolating them from each other. A relatively thin upper layer, containing some oxygen resulting from photosynthetic activity near the surface, would overlie relatively reduced waters that had some input from hydrothermal sources on the ocean floor. Iron precipitation would occur in regions where upwelling currents brought the deep, iron-rich waters up on to the continental shelves to mix with the relatively oxidizing surface waters. Upwelling currents occur on a regional scale in today's oceans, as, for example, along the western coast of South America. Such currents could be more effective during periods of rising sea level. The precipitated iron-bearing minerals would settle to the ocean bottom and accumulate in more or less cyclic layers related to daily, monthly, or seasonal variations in the strength of the upwelling currents, and possibly in the supply of oxygen from photosynthetic activity. Although this mixing hypothesis requires a biological input of oxygen, it is indirect since iron precipitation generally occurs away from the location of biological activity.

The different mineralogical types of banded iron formation formed in response to differing degrees of acidity and oxidation and the supply of organic material, reflecting distinct environments of deposition. Sulphide-type iron formations, which are predominantly black shales or cherts, have relatively high organic carbon contents, indicating that they were deposited close to the site of biological activity and, therefore, in shallow protected basins where upwelling water periodically penetrated. Carbonate-type iron formations have a lower content of organic carbon and contain sedimentary structures that indicate deposition in shallow water, but further from the site of biological activity. Oxide-type iron formations, with their generally undisturbed fine-scale layering and low organic carbon contents, indicate a quiet, deep-sea environment. Some oxide-type BIFs can, however, possess sedimentary structures indicative of a high-energy environment with current action, implying formation in shallow waters. It is likely that the oxide-type iron formations can be deposited in a variety of water depths and that the transition between the oxide and carbonate types may be controlled by chemical factors, namely acidity and the supply of organic material.

The formation of the younger banded iron formations that were deposited between 800 and 600 Ma ago requires a different explanation, since they have a different mineralogy and there is ample geological evidence for an oxidizing atmosphere at this time. The intimate association of these deposits with glacial sediments suggests a cause and effect. It

has been proposed that extensive ice sheets covered much of the Earth's surface during this period in its history, essentially isolating the oceans from the atmosphere. Other evidence in the geological record also supports the notion of extensive glaciation. This could have led to stagnant reducing conditions and the progressive build-up of dissolved ferrous iron in the oceans. Subsequent melting of the ice sheets would have restored water circulation patterns and caused oxidation and precipitation of the iron from solution, producing haematite-rich deposits.

Evidence gleaned from the study of banded iron formations has made important contributions towards understanding the early evolution of the Earth. The earliest occurrence of a banded iron formation, deposited 3.8 Ga ago, indicates that oceans had already developed by this time. It seems that the early oceans were stratified, with a thin oxygenated surface layer overlying reduced, deep-ocean water. The amount of oxygen in the ocean surface layer was probably very low relative to that in today's oceans, but it indicates the establishment of oxygen-producing micro-organisms as early as 3.8 Ga ago. Some oxygen could also have been present in the atmosphere, but the amounts must have been very low and possibly transient. It is likely that locking some of the biologically produced oxygen into the precipitating iron-bearing minerals in banded iron formations by oxidation helped to limit the oxygen content of the early atmosphere. Conversely, the banded iron formations owe their existence to photosynthetic micro-organisms that supplied the oxygen necessary for the oxidation of the dissolved ferrous iron in the oceans. By comparison with current conditions, there must have been a point in time in the past when the atmosphere and oceans changed from being dominantly reducing to being dominantly oxidizing. This would represent the time when the rate of supply of biologically produced oxygen overwhelmed the rate at which the oxygen could be consumed through various oxidation reactions. The rather abrupt end to the formation of banded iron formations about 1.8 Ga ago could coincide with this transition. In fact, there is other corroborating evidence in the geological record that points to 1.8 Ga ago as being an important time in the evolution of the Earth's hydrosphere and atmosphere. Except for an anomalous period between 800 and 600 million years ago, the oceans have since remained oxygenated and, as a result, contain only small amounts of dissolved iron.

ALAN B. WOODLAND

Barrell, Joseph (1869–1919)

Barrell, Joseph (1869–1919) As one who was interested in the broad philosophical aspects of geology as well as its field discipline and in mining engineering, Joseph Barrell exercised a wide influence in geological thought and teaching in America during the early decades of the twentieth century. He was trained at Lehigh and Yale universities and in 1903 he returned to a professorship at Yale.

Barrell's approach to geology was essentially analytical and quantitative and this led him to consider the rates of geological processes, both internal and external. He was interested in isostasy and the strength of the Earth's crust and their effects upon sedimentation and the geological record. To Barrell we owe the concept of the depositional *base level*. It had become apparent from the use of radiometric dating that the rates of sedimentation for ancient strata were much slower than previously believed, yet many of these beds showed signs of rapid accumulation. It could be shown that these successions were heavily invested with small breaks in sedimentation—diastems. These were brought about by fluctuations in base level, and the fluctuations could be influenced by many factors, acting intermittently or cyclically, although there are exceptions. Barrell emphasized the discontinuous nature of deposition and the control of deposition by subsidence of the depositional basin.

Despite his short life, Barrell's graduate teaching had a profound effect upon the progress of academic geology in America.   D. L. DINELEY

basalt 'Basalt' is a term widely used and abused. It has been employed both as a specific rock name, designating a narrow range of compositions (46–50 per cent silicon dioxide, (SiO_2)), and as a general term for almost any dark, fine-grained igneous rock, more particularly for those with the minerals plagioclase feldspar and pyroxene. Not all rocks of basaltic composition are, however, dark and fine-grained, nor are all dark, fine-grained, igneous rocks of basaltic composition. While the term 'basalt' remains a useful field name, it should be borne in mind that whole books have been written on the finer details of the sub-classification of rocks of basaltic composition. In practice many basalts also contain easily visible crystals (phenocrysts) of feldspar, pyroxene, olivine, or amphibole.

Basalts are the most common type of rock on Earth; they are also abundant on the moon and on other planets in the solar system. Almost the entire ocean floor is made of basalt, which was, and still is, erupted at mid-ocean ridges. These mid-ocean ridge basalts (MORB for short), in common with some other types of basalt, are erupted as a result of the divergence of tectonic plates. As two plates pull apart, the Earth's mantle rises to fill the gap, is decompressed, and melts to form magma. As the melt reaches the surface of the Earth it cools to form basalt.

Most MORBs are of what is known as tholeiitic composition; i.e. they have free silica and contain calcium-poor pyroxene. The basalts of oceanic islands (OIBs) are also tholeiitic, but vary widely in composition. Of special interest are the komatiites, which are all more than 2.5 billion years old. They are rich in magnesium and contain large olivine crystals. Komatiites were evidently erupted at higher temperatures than other volcanic rocks, and it is likely that they were generated by plumes in the mantle.

Extensive flows of basalt are present on the Moon, where they form the dark areas known as the maria. These lunar basalts are in general similar to those found on Earth, although they are much older (all the available samples have of ages of more than 3000 million years), but their chemical composition is more variable and they contain some minerals not known on Earth. A notable feature of the lunar basalts is the complete absence of water.   JUDITH M. BUNBURY

base level Base level refers to the elevation below which downcutting by a stream is not possible. A local base level may be created by a resistant material through which the stream cannot incise as rapidly as it is able to upstream and downstream. Local base levels may be formed by resistant bedrock, accumulations of large woody debris, beaver ponds, or such artificial construction as dams, grade-control structures, or weirs. The ultimate base level for any stream, however, is the elevation of the reservoir (ocean, lake, playa) or larger channel into which that stream flows.

In 1948, J. H. Mackin described an equilibrium condition for streams along which controlling variables and base level remained constant long enough for the stream to develop a longitudinal profile that is concave upward, with the gradient decreasing in the downstream direction. In such a graded stream, slope is adjusted to provide, with available discharge and prevailing channel characteristics, the velocity necessary to transport the sediment load supplied from the drainage basin. If base level changes, the stream will adjust, whether by incising (for falling base level), by aggrading (for rising base level) so as to maintain a graded profile, or by altering its channel pattern, width, or roughness.   ELLEN E. WOHL

Further reading

Mackin, J. H. (1948) Concept of the graded river. *Geological Society of America Bulletin*, **59**, 463–512.

Schumm, S. A. (1993) River response to baselevel change: implications for sequence stratigraphy. *Journal of Geology*, **101**, 279–94.

basins, drainage *see* DRAINAGE BASINS

basins, ocean *see* OCEAN BASINS

basins, sedimentary *see* SEDIMENTARY BASINS

bauxites Bauxites are aluminous ores containing more than about 50 per cent of Al_2O_3, formed under humid tropical climates by the weathering of aluminium-rich minerals. The parent materials can be igneous or metamorphic rocks (all basic or moderately acid) as well as sedimentary rocks: shales marls, limestones, or sandstones (rich in clay minerals but relatively poor in quartz). Bauxites are lateritic materials either remaining *in situ* (lateritic bauxites) or located in karstic areas within which limestones have been dissolved in ground and surface waters (karstic bauxites). They are chiefly composed of aluminium (gibbsite, $Al_2O_3 \cdot 3H_2O$; boehmite or diaspore, $AlOOH$), but may contain significant amounts of iron (goethite, FeOOH or haematite, Fe_2O_3) and small amounts of silica (kaolinite, $Al_2O_3 \cdot 2SiO_2 \cdot 2H_2O$ or quartz, SiO_2) and sometimes titanium (anatase, TiO_2). Three major classes of bauxite are distinguished. (1) Orthobauxites are gibbsitic, of massive structure, red in colour, relatively rich in iron, and overlain by an iron-rich hardcap; (2) metabauxites are orthobauxites subsequently transformed under arid conditions; they are boehmitic, white in colour, poor in iron, exhibit a pisolitic structure, and are underlain by kaolinitic nodular ferricrete; (3) cryptobauxites are orthobauxites that evolved under very humid conditions and are covered by a kaolinitic soft horizon of a biological origin. Bauxites are good continental indicators of palaeoclimatic changes.

Y. TARDY

Beaches Beaches consist of accumulations of unconsolidated sand or shingle deposits, or both, along the shoreline. The form of a beach is largely controlled by wave energy and particle size, and its shape can change rapidly in response to short-term variations in wave climate. The form of the beach is thus able to maintain a dynamic equilibrium with the prevailing environmental conditions.

Although many people consider the beach to extend from low water of spring tides to the limit of wave action, this is in error since this definition does not contain the sediments in the offshore zone, which are still moved by wave processes. Perhaps a better definition is provided by Komar, who describes the beach (littoral zone) as the zone in which coastal sediments are affected by wave processes. The zone thus extends from water depths of 10–20 m below low-tide level to the landward limit of storm waves.

In general, two major beach profiles tend to be recognized: steep/storm profiles and shallow/swell profiles. These are, however, really end-members within a continuum. The gradients associated with these profiles also vary between sand and shingle beaches, the latter being generally steeper. The profile usually consists of a series of ridges and troughs that occur both above and below the water level (Fig. 1). These features tend to be mobile, moving offshore and onshore. Steep profiles usually consist of a marked landward ridge, usually

Fig. 2. A bay-headed beach: Luce Bay, Scotland. Note the steep berm developed along the shoreline, which is fronted by a low-gradient sandy beach that is covered at high tide.

referred to as the *berm*, which forms the landward limit of wave swash. In contrast, shallow beaches rarely have a berm developed; rather, a bar forms just below the low-tide level. This bar is referred to as the *long-shore bar*. The shape of the profile is controlled by two factors: waves and sediment type. In terms of waves, spilling breakers tend to be associated with shallow gradients and the landward movement of material, whereas plunging/surging breakers are commonly associated with steeper gradients and the seaward movement of material. In terms of sediment type, the coarser the material, the steeper the gradient. This relationship has often been linked to the associated percolation rate. With shingle, the water tends to sink into the beach and hence there is little backwash. The material is consequently not combed down and the gradient is steep. In contrast, on sandy beaches percolation is limited. Backwash is thus more pronounced and the process maintains a lower gradient. In many instances a combination of sediments is present. Hence along the shoreline there is a well-developed, steep shingle ridge, whilst seaward of this is a low-gradient sandy beach (Fig. 2). In such circumstances the day-to-day evolution of the profile may vary with tidal movements. At low tide, the low sandy beach promotes spilling breakers and thus landward movement of beach material. In contrast, at high tide, the steep shingle beach produces plunging breakers against the berm causing the seaward movement of material. Clearly, contrasting processes will also occur during storms, when the characteristics of the waves change.

Viewed from the air, most beaches are curved in outline and many of them contain a series of regularly spaced secondary curved features. In general, the smaller the feature the shorter its residence time within the environment. For example, cusps (small crescentic forms on the upper beach) may last only for a matter of hours, whereas accumulations developed between headlands may remain for several hundred years.

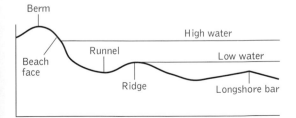

Fig. 1. A beach cross-profile showing the various geomorphological features.

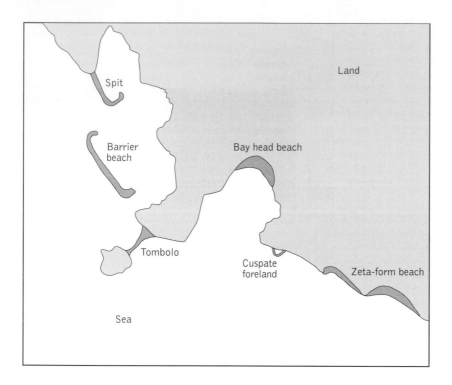

Fig. 3. Schematic diagram of a variety of beach landforms.

A wide variety of beach features have been described. More detail is provided in the books by Pethick and Carter. The various forms include:

Bay-head/pocket beach: an accumulation of sand and/or shingle located along the shoreline between two headlands (Figs. 2 and 3).

Zeta-form or fish-hook beach: the width of the beach increases down-drift to terminate against a headland. The feature is then usually repeated downdrift of this headland (Fig. 3).

Spits: beach, which is attached to the coastline at only one point (Fig. 3; *see spits*).

Cuspate foreland, where the beach leaves the coastline and runs seaward at an angle before returning to meet the coast once more (Fig. 3); e.g. Dungerness, Sussex.

Tombolo: beach, which joins an island to the mainland (Fig. 3).

Barrier beach: a sand or shingle accumulation completely detached from the main shoreline (Fig. 3); e.g. The Bar, Nairn, Scotland.

The form of the beach is largely dependent upon the balance between wave energy and availability of sediment. In general, the waves will transport material along the shore until an equilibrium form is developed (sediment arriving in the area is balanced by material being eroded). Clearly, however, the nature of the feature varies on a day-to-day basis with changing wave energy. CALLUM R. FIRTH

Further reading

Carter, R. W. G. (1988) *Coastal environments: an introduction to the physical, ecological and cultural systems of coastlines.* Academic Press, London.

Komar, P. D. (1976) *Beach processes and sedimentation* Prentice-Hall, Englewood Cliffs, New Jersey.

Pethick, J. (1984) *An introduction to coastal geomorphology.* Edward Arnold, London.

Viles, H. and Spencer, T. (1995) *Coastal problems: geomorphology, ecology and society at the coast.* Edward Arnold, London.

beachrock Beachrock is a lithified sediment which forms as a result of cementation, usually by calcium carbonate, in the beach zone of mainly tropical and subtropical coasts. Similar cementation also occurs within some marine intertidal flat sediments and storm ridges which lie above the normal high water limit. Examples of lake-shore, estuarine, and lagoon beachrock are also known. In the majority of beachrocks the detrital sediment grains remain in contact with one another; that is, the sediment maintains a clast-supported fabric, while between 10 and 100 per cent of the intergranular porosity is filled by precipitated calcium carbonate.

The size distribution of the detrital sediments varies widely, ranging from relatively clean, well-sorted sands to poorly sorted conglomerates and muddy sands. Volcaniclastic grains are common on volcanic islands (for example the Canary Islands); quartz, feldspars, and lithic grains dominate on many continental mainland shores (such as Egypt, Israel, and Crete); while carbonate grains, including ooids, coral fragments, shell debris and algae, are predominant on beaches with relatively low rates of clastic sediment supply (for example the Bahamas and the Florida Keys). Aragonite ($CaCO_3$) and magnesium calcite (containing 10–14 mol per cent $MgCO_3$) are the two most important cementing minerals, particularly in tropical and subtropical beachrocks. Occasionally low-magnesium calcite, dolomite, siderite, and other minerals also occur as major components of the cement. These minerals tend be more common in lacustrine, estuarine, and lagoonal settings.

Most marine beachrocks form below the surface of the beach, in the zone affected by fluctuating groundwater. As long ago as 1851, J. D. Dana was one of the first scientists to suggest that tropical marine beachrock forms as a result of evaporation of sea water at low tide, a view endorsed by many subsequent authors. An alternative mechanism involves lithification by magnesium calcite, aragonite, or dolomite brought about by the subsurface mixing of sea water with fresh (or brackish) waters draining from the land. A few beachrocks, composed mainly of low-magnesium calcite, have also apparently formed entirely as a result of evaporation of land-derived fresh waters rich in calcium carbonate.

Although beachrock formation can be an entirely inorganic process, microbial activity and degradation of organic matter, including upward movement of microbial degradation products (such as carbon dioxide and methane) from greater depths, often play a part. Analysis of the stable carbon and oxygen isotope composition of the cements, together with the nature of trace and rare-earth elements, is often useful in identifying the relative importance of different carbonate sources.

A wide variety of cement textures has been described from beachrocks. Textures developed in the relatively free-draining vadose zone above the lowest level of the beach groundwater table are often dominated by *meniscus* cements, formed by precipitation during evaporation of water films held at points of grain contact, and *pendant* cements, formed by evaporation of moisture held by gravity on the underside of grains. Cements formed by precipitation from supersaturated fluids below the lowest level of the beach groundwater table often form *rim* cements which encircle the grains, or *blocky* cements which completely fill the pore spaces. In practice, many beachrocks show more than one type of cement fabric, because of spatial and temporal fluctuations in groundwater levels. Partial or complete replacement of detrital grains, composed of both carbonate and siliciclastic material, is common.

The cementation of beachrock can be very rapid by geological standards. Beachrocks on Pacific atolls which contain military artefacts have reportedly formed within less than seven years. Reports have also indicated that on some Indian Ocean islands it is possible to make an annual harvest of beachrock for use as a building stone. In other instances,

however, beachrocks have been shown by uranium-series and radiocarbon dating to be several thousands to hundreds of thousands of years old. Fossil beachrock has been used widely as a palaeo-sea-level indicator, and lateral variations in the elevation of beach rocks can provide useful information about recent tectonic deformation associated with earthquakes and volcanic activity. Accurate radiometric dating of beachrocks can be difficult, however, owing to diagenetic changes which involve exchange of radioelements between the cements and ambient fluids, possibly leading to under or overestimation of the true ages.

Beachrock is often exposed at the surface on shores where the overlying sediment has been eroded, or where the beachrock has been exposed subaerially by a fall in relative sea level. In such instances the upper surface of the beachrock may exhibit a variety of karstic weathering features, including miniature dolines (sink-holes) and pinnacles. Exposure to rainwater, sea spray, and endolithic organisms may also bring about changes in the rock texture and mineralogical composition.

K. PYE

Further reading

Scoffin, T. P. and Stoddart, D. R. (1983) Beachrock and intertidal cements. In Goudie, A. S. and Pye, K., (eds) *Chemical sediments and geomorphology*, pp. 401–25. Academic Press, London.

bearing capacity

Bearing capacity is a general term used in the design of foundations for engineered structures to define the ability of a rock or soil immediately below the foundation to bear the stress that will be placed upon it when the structure is complete. Bearing capacity is defined in more specific terms: *ultimate bearing capacity* (q_f) is the limit of stress which can be imposed upon a soil or rock without exceeding the strength of the material and thus causing it to fail and, consequently, to move and thus cause the foundation and the structure based on that foundation to fail. The value of q_f for soil or rock increases with depth because of the constraint produced by the weight of the increased thickness of soil or rock material adjacent to it.

Foundation engineers take account of the probability that the ultimate bearing capacity of the material might not be known accurately enough because of deficiencies in sampling or strength testing and because the nature and properties of geological materials can change over small lateral and vertical distances. Foundation engineers also take account of the consequences of inadequate design, which could lead to catastrophic failure of the structure. To allow for such possible inaccuracies and their consequences, a factor of safety (F) is employed. Its value is usually between 1.1 (for a road foundation, for example) and 3 (for a dam). The ultimate bearing capacity q_f is then divided by the chosen factor of safety to give the *safe bearing capacity* (q_s). Thus, $q_s = q_f/F$. The safe bearing capacity q_s is then used in the design calculations for the foundation. Using the safe bearing capacity in the calculations should ensure that the strength of the geological material will not be exceeded and it will therefore not fail, but because of the imposed stress it will deform. The effect of that deformation is that the structure will settle. The amount and rate of settlement has then to be considered to ensure that the structure can accommodate the settlement without serious damage. If there is a possibility that the deformation of the underlying geological material that is to bear the weight of the structure will be too great under pressure equal to q_s, then the deformation must be reduced by changing the size or shape of the foundation so that the stress is reduced sufficiently to keep the deformation and settlement within acceptable limits. The bearing pressure that takes into account of the demands of the built structure as well as the strength of the geological material under the foundation is termed the *allowable bearing pressure* (q_a), because that is the limit of stress (pressure) that is allowable if deformation, and consequently settlement, are to be kept within the limits that the structure can accommodate. Thus q_a will be equal to or less than q_s.

J. WEST

beginnings of geological thought

Many ancient peoples had models of the Earth and its place in the Universe. To the ancient Babylonians the Earth was a hollow mountain supported and surrounded by the ocean. The realm of the dead lay beneath the ground: the Sun, Moon and stars all moved across a solid canopy above. To the ancient Egyptians a quartet of deities was responsible for the universe: Keb was a reclining god, the earth, covered with vegetation; the skies were a gracefully bent goddess supported by the god of the air; and the sun god sailed across the heavens into the night and death. A Hindu version was that the Earth was held up by four elephants standing on a giant tortoise, which in turn stood upon a cobra. These concepts were religious rather than scientific, but did nevertheless attempt to account for the Earth, Sun, Moon, stars, earthquakes, and seasons. They were discarded only as information improved and the need for different models increased or because of changes of religion or culture.

It is difficult to understand how many of these models originated, but myth and folk memory clearly played a part. It is possible that logical thought operated at the level of human experience of the immediate environment but ceased when the magnitude of the unknown, as distinct from the known universe, and the passage of time was contemplated. Here the supernatural was invoked, as it still is by much of mankind.

The Greeks and later the Arabs were led to group and classify objects and phenomena, and as the conquest of materials developed they enlarged the categories and transformations that operated in their universe. Aristotle (384–322 BC) was impressed with the transformations between different states of matter and between different substances. Each of his four elements, fire, earth, air, and water, possessed two of four basic properties: hot, cold, wet, and dry. Cold and wet in combination gave earth, and so on. Other early writers developed alternative versions. They were concerned in the debate about the nature of matter rather than with forces in nature. Empedocles, however, saw the need for forces in nature that brought about material changes; he argued for two, love to bring about unifications, and hate which brought about disunity. Centuries later, the alchemists held to the doctrine of the two contraries, in which opposing qualities exemplified by particular substances were said to come into contact deep within the Earth to produce different metals. Throughout these centuries the knowledge of how and where different substances such as the metals and precious stones occur had grown and the properties (real and imaginary) of such materials were recorded at some length.

The sixteenth century in Europe saw the flourishing of a new method of handling the mass of information by them available to philosophers, alchemists, and artisans alike. Francis Bacon (1561–1626) Lord Chancellor of England, philosopher, and man of letters, is regarded as the father of the scientific method of enquiry. He was not a practical scientist but saw how knowledge could be codified and tested. His influence upon those following him was great. He divided physics from metaphysics; natural philosophy became the search for the expression in nature of different forms (objects). It took some time for this to have effect upon geological debate. Meanwhile continental scholars were busy collecting and classifying all manner of geological and biological objects. The great change to come was that the ancient values and magical properties of objects were discarded. The mineralogists in particular were beginning to treat Earth materials as scientific objects, no longer imbued with supernatural or biological qualities. Fossils were to present problems for rather longer.

In the early seventeenth century the Italian Ulyssis Aldrovandi used the terms *geologia* and *fossillibus*; and the word *geologia* first appears in English in a work on minerals by Lovell (1661). It was not until nearly 150 years later that the word *fossil* was used in the modern sense.

D. L. DINELEY

Further reading

Adams, F. D. (1938) *The birth and development of the geological sciences.* Dover Publications, New York.

Crombie, A. C. (1994) *Styles of scientific thinking in the European tradition* (3 vols). Duckworth, London.

Lindberg, D. C. (1992) *The beginnings of Western science: the European tradition in philosophical, religious, and institutional context, 600 BC to AD 1450.* University of Chicago Press.

Beloussov, V. V. (1907–90)

Vladimir Vladimirovich E. Beloussov was a senior Earth scientist in the Soviet Union and a protagonist of alternatives to the theory of ocean-floor spreading and continental drift/plate tectonics during the mid-twentieth century when debate was so vigorous. He was an Academician and member of the Institute of Physics of the

Earth in Moscow. He was also a Foreign Member of the Geological Society of London whom he addressed in 1966 on 'Modern concepts of the structure and development of the Earth's crust and the upper mantle of continents' and, later, on the work of Soviet expeditions to East Africa. Beloussov remained opposed to the theory of ocean-floor spreading and offered detailed alternatives. His view of the nature and structure of the crust and mantle was complex, and he maintained that processes were widespread and active by which hot basic and ultrabasic magmas transform continental crust to oceanic crust by destroying and dissolving the former. These processes, he held, were most active at the edges of continents. As late as 1971 he was offering possible relationships between magmatisim and tectogenesis (major tectonic events) that differed very much from the views commonly held in the West.

Beloussov refused to believe in the regular magnetic striping of the ocean floor about the oceanic ridges. He thought of it as an expression of a layering of basalts and as occurring only as 'rather irregular scattered patches'. Vertical movements he favoured; great horizontal movements of entire continents he would not accept. Beloussov nevertheless made significant contributions to Earth physics, and was highly regarded in the Soviet Union and elsewhere. He led three Soviet expeditions to the Rift Valley area of East Africa in the 1960s and concentrated on studies of deep continental structure and the mantle. He was President of the Upper Mantle Commission of the International Union of Geodesy and Geophysics in 1960, and at the Union's Thirteenth General Assembly in Finland he called for a new international project that could follow the then recently concluded International Geophysical Year. The title of this project was to be 'The upper mantle and its influence on development of the Earth's crust'. It was a non-starter.

During his career Beloussov was appointed to many prestigious positions in Soviet academic circles, and was a member of several international scientific bodies. Although his views have not found subsequent favour, he possessed great intellectual powers, originality, and drive. His influence upon the development of the Earth sciences in the Soviet Union after the Second World War was very important.

D. L. DINELEY

Benioff H. (1899–1968)

A *Benioff zone* is an active seismic zone that dips below island-arc systems to depths of 200 k m and more. It is named after Hugo Benioff, who was for many years Professor of Geophysics at the California Institute of Technology, Pasadena. In 1946 he described the zone on the basis of the location and depth of earthquake foci in a belt about 50 km thick which dips under the arcs at about 33° from the vertical towards the continents to a depth of about 300 km, whereupon it descends at a steeper dip of about 60° to a depth of 700 km.

Hugo Benioff was born in California in 1899 and spent most of his life there, graduating from the California Institute of Technology and subsequently becoming a faculty member.

He was employed at the Mount Wilson Observatory from 1917 to 1921 and was associated with the Carnegie Institute in Washington from 1923 onwards, by which time he was committed to the physics of the Earth. Seismology was the field in which he made his most significant contributions.

His research into locating the foci of earthquakes was well known when it was interrupted by the outbreak of the Second World War. He was taken on as a research engineer by the Submarine Signal Co. and was called upon to join various government wartime research bodies. In particular his seismological skills were in demand. In July 1945, Benioff, using data from about ten seismic stations in the south-western USA, determined the time and place of a strange seismic event in the New Mexico desert; it was the world's first, and highly secret, atomic explosion. Seismic detection of such explosions became a high priority in American defence research, and for some years Benioff was a member of the Air Force Office of Scientific Research, and later a consultant to the National Science Foundation. He was also Chairman of a consultative board of the California Department of Water Resources. He was universally acknowledged as the master craftsman of seismometer design and was, in that seismically troubled State, well aware of the importance of his science to the public safety and good.

Benioff retired in 1964, honoured by academic bodies in America and elsewhere, and died in California in 1968.

D. L. DINELEY

Benioff zone *see* SUBDUCTION ZONES

Berzelius, Jöns Jacob (1779–1848)

The Swede J. J. Berzelius, 'the greatest chemist of his day', made important contributions to the geological sciences and especially to mineralogy. He first studied medicine and was elected to a chair at the Carolinska Medico-Chirurgical Institute in Stockholm in 1807. From there he moved in 1811 to take on the management of the Academy of Agriculture, where he became interested in mineralogy. As a chemist, his work on electrochemical theory and the nature of the atom was fundamental, and he introduced the use of symbols for the elements. In a publication of 1818 he gave analyses of up to 2000 simple and compound substances; he discovered the elements selenium, cerium, and thorium; he was the first to isolate niobium and silicon; and he determined the atomic weight of at least 12 elements.

In 1811, when he became interested in mineralogy, the classification of minerals was essentially on the basis of their physical properties, although as early as 1742 Wallerius had suggested that ultimately classification must be based on chemistry. Berzelius saw mineralogy very much as a branch of chemistry; he recognized silicates as a distinct group of minerals and (*c.* 1815) he produced a chemical classification which, with modifications, is still in use today. With a pupil, Mitscherlich, he introduced the concepts of polymorphism and isomorphism.

His geological activities involved travel in Scandinavia and central Europe. At Karlsbad he debated the origin of volcanoes with Goethe and, we are told, showed Goethe that his ideas on the subject were wrong. The raised beaches of Scandinavia he thought to be related to the fall of sea level as the Earth contracted on cooling.   D. L. DINELEY

BIFs *see* BANDED IRON FORMATIONS

biodegradation Biodegradation is the decomposition of organic compounds mediated by micro-organisms. This involves the oxidation of organic matter under both aerobic and anaerobic conditions by a variety of different micro-organisms. Bacteria and fungi are largely responsible for the breakdown of organic matter on Earth. They thus play a fundamental role in driving the biogeochemical cycles of elements at the Earth's surface. The degradation of natural organic compounds is very efficient; little survives to be buried in soils and sediments. In the absence of oxygen, biodegradation can be more restricted than in its presence, leading to greater preservation.

Biodegradation is harnessed in sewage treatment and in landfills to treat domestic and industrial waste. It can also be used to produce fuels, such as methane, from cheap or surplus agricultural crops. Conversely, the biodegradation of food and related products has to be prevented by killing micro-organisms (for example, by means of high temperatures during canning, or by irradiation), or slowing down their activity (for example, by freezing, refrigeration, drying, or preservatives). On a larger scale, biodegradation has to be prevented in the oil industry (in reservoirs, storage tanks, and pipes) and other industries, since bacterial activity can degrade the product and corrode and block pipes with biofilms. In these situations large quantities of biocides are used for bacterial control.

Micro-organisms can also degrade a range of toxic xenobiotic compounds (synthetic chemicals, which do not occur naturally) such as herbicides and pesticides. Some xenobiotics are, however, resistant to biodegradation and thus can accumulate to toxic levels in the environment. Micro-organisms are not invincible, despite their diverse metabolism and ability for metabolic evolution.

Related to biodegradation is biodeterioration, in which microbes do not significantly degrade a compound or matrix, but cause more subtle changes which results in its devaluation (e.g. odours in food and drinks, staining of surfaces, corrosion of metals, minerals, and concrete).

R. JOHN PARKES

Further reading

Atlas, R. M. and Bartha, R. (1998) *Microbial ecology: fundamentals and applications* (4th edn). Benjamin/Cummings, Menlo Park, California.

biogenic weathering Over geological time-spans (more than 10 000 years) silicate weathering reactions have controlled the movement of carbon dioxide between the atmosphere and the oceans, and during the past 400 million years the animals and plants of a region (the biota) have played an important role in the process. The weathering of calcium and magnesium silicates is the primary 'sink' for atmospheric carbon dioxide (CO_2) over geological time. James Lovelock considers that amplification of weathering caused by the appearance of plants and animals in the Precambrian almost certainly caused a decrease in atmospheric CO_2 levels. Although the cause and effect linkages are fairly straightforward, the precise magnitude of the shift from abiotic to organically mediated weathering is uncertain. It has been argued that the effect of biotic enhancement is to increase weathering by at least one order and perhaps more than three orders of magnitude. This uncertainty about biotic enhancement is a critical problem in models of early climate, since large-scale shifts in atmospheric CO_2 levels would probably, because of the greenhouse effect, result in temperature excursions. A minor biotic enhancement of weathering would imply that global temperature changed relatively little with the colonization of land by animals and plants. A high biotic weathering enhancement would, on the other hand, point to a substantial lowering of temperature. For example, if biotic weathering were ten (or a hundred) times faster than abiotic weathering, then the Earth would be 15 °C (or 30 °C) warmer than at present. The order-of-magnitude uncertainty in the biotic effect thus makes it difficult to model global habitability over geological time. It is thus important to determine the order of magnitude of the acceleration effect on biogenic weathering by soil microorganisms, lichens and vascular plants. This is, however, not a straightforward question to answer: the factors that control the weathering of the Earth's crust are complex, often coupled, and as a result are understood at the field scale only in a semiquantitative sense. Soils rich in organic matter commonly have high CO_2 pressures and contain abundant organic acids, and are commonly warmer than soils that are not; soils exposed to heavy rainfall commonly have high organic activity. All these conditions would result in accelerated weathering. Weathering rates increase with temperature, ambient moisture, and organic activity, but the derived dependencies are somewhat approximate because no two weathering basins are the same in a mineralogical, hydrological, or biological sense. Measurements of abiotic versus biogenic weathering suggest that abiotic weathering is proportional to rainfall and that biogenic weathering is far more sensitive to rainfall. Abiotic weathering is therefore limited by the availability of weathering solutions, and the weathering is further enhanced in the presence of the biota because of the effects of three possible factors: the ability of organic acids to form complexes; the extended retention of moisture in the pore walls because of the presence of a microflora; and the effect of the biota in increasing the surface area that is being weathered.   K. VALA RAGNARSDOTTIR

biogeochemistry Biogeochemistry refers to the coupling of biological, chemical, and geological processes that together drive the major chemical cycles on the Earth's surface. Life plays a major role in this cycling, and there are few chemical reactions on the surface of the Earth that are not affected. The chief biological catalysts of these cycles tend to be micro-organisms, and in particular, bacteria. Micro-organisms do not have the same visual impact as large animals and plants, but the power of these small microscopic organisms is in their great metabolic diversity. For 70 per cent of the history of life on Earth (before 3.8 billion years (Ga) ago) bacteria were the only type of organism present and it is their activity that has shaped the surface chemistry of our planet, including the atmosphere and thus the climate. Indeed, their development of oxygen-producing photosynthesis both dramatically changed the atmosphere and made possible the subsequent development of metazoans. Today the atmosphere still provides a clear signature of active biogeochemical cycles in the form of a mixture of oxidized gases (e.g. oxygen, O_2) and reduced gases (e.g. methane, CH_4) that is far from chemical equilibrium. This chemical disequilibrium is produced by life and would provide clear evidence for life and active biogeochemical cycles on other planets.

Biological activity accelerates the speed of natural chemical reactions, such as rock weathering. For example, the weathering of pyrite is increased up to 1 million times. Biological activity also makes possible a range of unique reactions that could not otherwise take place. The importance of oxygen from photosynthesis has already been mentioned, but oxygen is only a waste product. The production of new organic compounds from atmospheric carbon dioxide (CO_2), and thus the formation of new cells and an energy source, are the crucial products for photosynthetic organisms. This photosynthetic 'fixation' of CO_2 traps energy from sunlight and is considered to be the base of all ecosystems on Earth, since herbivores, and indirectly carnivores, rely on it for their energy supply. However, after the death of these organisms their carbon and nutrients must be recycled or photosynthetic production would eventually cease because of lack of nutrients. The decomposition of organic matter is therefore just as important as its production through photosynthesis in maintaining life on Earth. These two processes drive the biological carbon cycle and the cycles of associated elements (for example, nitrogen, sulphur, and phosphorus—key elements in biomolecules). In addition, if carbon dioxide were not returned to the atmosphere through decomposition, the greenhouse effect of carbon dioxide would decrease and the planet would start to freeze (in about 4–10 years).

Other reactions unique to life, except at extremely high temperatures and pressures, include nitrogen fixation, nitrification and denitrification, sulphate and sulphur reduction, and methane formation. These processes have a direct effect on element cycles and can have major global impacts. For example, most commercial sulphur deposits are of biological origin, and the majority of atmospheric methane (which is 60 times more powerful as a greenhouse gas than CO_2) is of bacterial, not volcanic, origin.

A small amount of biologically produced organic matter leaks from the biosphere and becomes buried and preserved in the geosphere. Over geological time, this accumulates in vast quantities, making the geosphere the largest store for most elements on Earth. For example, vast amounts of organic matter are stored in sedimentary rocks, some of it in the form of fossil fuels. Naturally, this material would eventually be returned to the surface by the geological processes of mountain building and volcanic eruptions, so completing the element cycle. These processes, however, are very slow, taking approximately 300 million years. In contrast, the biological cycle is extremely fast (taking years to tens of years), and it is the interplay between the slowly turned over but vast geological reservoirs and the rapidly turned over but small biological reservoirs that controls the global biogeochemical cycles. There is evidence that biological interaction with the geological reservoirs has kept the climate stable for life despite changes in heat output, volcanic and tectonic activity, and solar radiation during the development of the Earth.   R. JOHN PARKES

Further reading

Schlesinger, W. H. (1997) *Biogeochemistry: an analysis of global change.* Academic Press, London.

biogeomorphology 'Biogeomorphology' is a term coined in the 1970s to describe the interrelations between *geomorphology* (that is, landforms and the processes that shape them) and plants, animals, and micro-organisms. Organisms have an important effect on many geomorphological processes; they can, in some circumstances, directly produce landforms, and in turn their *ecology*, distribution, and growth patterns are affected by geomorphological processes and landform assemblages. Biogeomorphology encompasses all such interactions and provides a useful link between ecologists and geomorphologists. The specific study of the role of animals in landscape modification has become termed 'zoogeomorphology'.

The term 'biogeomorphology' appears to have developed from the work of James Knox, an American fluvial geomorphologist, who wrote about the 'biogeomorphic response' of vegetation and sediment yield to changes in climate and human impacts in a paper on valley alluviation in Wisconsin he published in 1972. Other papers dating from this time also referred to various links between sediment systems and vegetation in a range of different environments, such as coastal barrier islands, using terms such as 'biogeomorphological'. In the late 1980s and early 1990s several edited volumes appeared with the word 'biogeomorphology' somewhere in their titles, or with the concept of biogeomorphology at their core. These books, some of them derived from conferences, were largely made up of contributions from Anglo-American and Japanese geomorphologists and, to a lesser degree, ecologists.

Biogeomorphological interests are, however, much older and can be traced back to the work of many nineteenth-century natural historians and scientists, such as Charles Lyell. In his *Principles of geology*, published in 1835, Lyell talked about the many roles of plants in geomorphological processes, or as he put it, 'the superficial modifications caused directly by the agency of organic beings'. There are many wonderful nineteenth-century descriptions of esoteric biogeomorphological processes, such as the production of depressions by wallowing African game animals and the role of spiders in boring into rock surfaces. For much of the twentieth-century, geomorphologists, especially those working in the United States and Britain, ignored or played down the role of plants, animals, and micro-organisms in shaping the land surface. Large-scale landform development studies, which focused on the evolution of landscapes over millennia, seemed to have no place for the consideration of biological influences. Biogeomorphology was reduced in many minds to descriptions of curious and exceptional features. It was, perhaps, a combination of new technological developments, an increased interest by geomorphologists in ideas coming out of biology rather than physics, and a move towards looking at shorter-term, smaller-scale issues in geomorphology, which encouraged the redevelopment of biogeomorphological interests from the 1970s onwards. In Europe, links between ecology and geomorphology have always been much stronger, reflecting the very different intellectual traditions there.

Organic influences on geomorphological processes range from the spectacular but spatially limited influences of large mammals on erosion (such as grizzly bears in Canada) to the more subtle, but almost ubiquitous role of rock-surface micro-organisms in weathering and crust formation. Many such processes have been studied by geomorphologists, leading to a fragmentary but steadily growing body of knowledge on how fast such processes operate and how important they are in relation to other geomorphological processes. For example, many studies have been made on the role lichens play in rock weathering in arctic–alpine, desert, Mediterranean, and temperate-zone environments, and estimates have been made of the rate of lichen weathering activity.

Plants, animals, and micro-organisms can also be the major controlling factor in the creation of some entire landforms. Thus, mima mounds, an often debated feature of prairie grasslands, are sometimes created by gopher burrowing; and many organisms dwelling on rocky coasts create their own micro-landforms in the form of burrows in which they then live. Beavers create often spectacular landforms in the form of beaver dams, and also have a huge impact on the entire fluvial geomorphology of some areas through the creation of such dams. Tufa, which is a friable deposit of calcium carbonate, accumulates in some streams and log dams commonly provide the chief nuclei for such deposition, which is also often largely mediated by micro-organisms such as cyanobacteria.

The influences discussed above are certainly not just a one-way affair, however. Geomorphology exerts an equally important influence on the development of biological communities, with, for example, landslides producing new surfaces for plant colonization and complex topography producing subtle spatial variations in soil and vegetation communities. In many cases, clear two-way interactions can be seen, and it remains a challenge for biogeomorphologists to understand these and the influences upon them. For example, sediment movements and vegetation growth on hill slopes are intimately related, as has been shown by the work of the British geomorphologist John Thornes and his co-workers in Mediterranean scrub environments.

The study of biogeomorphology is not a separate sphere of environmental science, but very much an integral and integrating part of both geomorphology and ecology. To date there have many field-based empirical studies of biogeomorphological interactions, but little in the way of sophisticated modelling. The major exception to this statement has been on studies of hill-slopes with, for example, the prolific modelling work of Mike Kirkby from the University of Leeds, England, who has developed a whole suite of models relating plant growth, sediment transport, and hill-slope hydrology (see *landscape modelling*). Biogeomorphological work is now being applied to environmental problems and issues. For example, the response of different coastal areas to future sea-level rise will depend very much on their biogeomorphological characteristics, and especially the interactions between vegetation and sediment dynamics. Biogeomorphology is certainly a fertile area for future research.   H. A. VILES

Further reading

Butler, D. (1995) *Zoogeomorphology: animals as geomorphic agents*. Cambridge University Press.

Viles, H. (ed.) (1988) *Biogeomorphology*. Basil Blackwell, Oxford.

biomarkers (chemical fossils)

Chemical fossils, biological markers, or biomarkers, are the wide variety of individual organic carbon compounds (biomolecules) derived from living organisms and found in trace quantities in fossils and sediments. The most commonly used term is *biomarker*. Complex mixtures of biomarker hydrocarbons are major constituents of petroleums; for example, phytane, a branched-chain, saturated hydrocarbon with 20 carbon atoms in the molecule, derived ultimately from the chlorophyll of marine algae, can reach 1–2 per cent of an oil.

Biomolecules made by living organisms are small, as in cholesterol, medium-sized like hormones, or very large as in structural proteins and DNA. The vast assemblages of different molecules making up a living organism can give a form of molecular signature of that life for the fossil record, provided that some molecules survive decay and diagenesis, either intact or so little modified that we can recognize them. We now know that some biomolecules and their breakdown products persist at levels of parts per million or parts per thousand in fossils and sediments for thousands, sometimes millions, of years. However, most of the carbonaceous organic matter in the Earth's crust is insoluble amorphous debris, which is very difficult to analyse. The initial record in recent sediments is very close to that of the biochemistry of contemporary living organisms with small amounts of fairly well preserved DNA, protein, and carbohydrates still detectable. Most of this record is erased within a few tens of years, principally by microbial action. The process, however, is selective, patchy, and incomplete. The end result is that relatively easily biodegraded compounds and biopolymers survive in small amounts in aquatic sediments that have not been subjected to deep burial and geothermal heating. Biomarkers are extractable from sediments and fossils by organic solvents. Those that have been most studied are of the lipid (fat) type, although certain pigments (e.g. metal porphyrins) are also referred to as biomarkers. Biologically characteristic distributions of carbon compounds are seen in extracts of sediments from the present back to Archean times, nearly 4 billion years ago. The contemporary biota uses enzyme systems inherited more or less unchanged from the Archean. Biomolecules can provide clues to their origin and subsequent history throughout the time they persist, whether in their original or in altered form. The biolipid record goes back into the Archean, even to the earliest microscopic and isotopic indications of life at 3 to 4 billion years ago.

Information of this kind has opened up new avenues of research in archaeology and the Earth sciences, for example in studies of past climatic change. Molecular information proves of enormous value in correlating petroleum bodies and their source rocks, the assessment of thermal history, and in stratigraphical, production, and refining considerations.

GEOFFREY EGLINTON

Further reading

Logan, G. A., Collins, M. J., and Eglinton, G. (1991) In Allison, P. A. and Briggs, D. E. G. (eds) *Taphonomy: releasing data locked in the fossil record.* Plenum Press, New York.

biosphere The biosphere is the part of our planet that is inhabited by life. It consists of three zones: the *lithosphere*, comprising the Earth's crust, together with accumulated soils and sediments; the *hydrosphere*, that part of the lithosphere covered in water, including groundwater; and the *atmosphere*, the gaseous envelope surrounding the Earth. Life on Earth is extremely ancient, having evolved before 3.8 billion years (Ga) ago. It has played, and still plays, a major role in the development of our planet and its climate.

The greatest biomass, and largest variety, of living organisms inhabit the low-altitude surface lithosphere and upper hydrosphere. Most organisms rely on photosynthesis, and thus on solar energy, for their existence, either directly (plants and some bacteria) or indirectly (animals and most micro-organisms). Life continues to greater depths and higher altitudes away from this surface layer. However, the further away from this layer the more reduced the variety, and density, of

organisms becomes—as, for example, when ascending mountains. Numerically, bacteria are the dominant form of life (total $4–6 \times 10^{30}$ cells); they also represent a major component of the total biomass on Earth. This is because they can grow under a wide range of conditions (for example, at −5 to 113 °C, at pH 0 to 11, in near-vacuum or at pressures 1000 times greater than atmospheric, and in distilled water or in saturated salt solution). They thus greatly extend the biosphere.

In the upper atmosphere life becomes very restricted; some, indeed, question whether this is a true habitat. Nevertheless, some viable bacterial cells and bacterial and fungal spores are present. Growth and multiplication of micro-organisms here is limited by lack of nutrients and water; and solar radiation can be lethal.

The hydrosphere contains living organisms throughout, although the greatest abundance of organisms is in the near-surface photic zone, where light penetrates. Below this, life is adapted to live in dark, high-pressure conditions and depends primarily on dead organisms from above for food. At the sea floor, which may be as much as 11 km below sea level, life is sparse and is dominated by invertebrates that feed on falling particles and on the sediments, which contain large bacterial populations. The upper sediment layers can contain 10 000 million (10^{10}) bacterial per millilitre. Below the depth of oxygen penetration life is restricted to anaerobic bacteria. Although they decrease in numbers from the sediment surface, bacteria have still been detected at 850 m below the sea floor (the deepest sample analysed) and even in the rocks beneath.

Where new crust is being formed at ocean ridges, unique ecosystems exist as oases of life at hydrothermal vents. The expulsion of hot magma results in deep circulation of sea water, which reacts with the hot crust and is chemically altered before being expelled as 'black smokers', at temperatures up to 350 °C, or as 'white smokers' at lower temperatures. The fluids are rich in hydrogen sulphide, methane, and dissolved reduced metals; for example, Fe^{2+} in the presence of oxygen in sea water provides a chemical energy source for high-temperature bacteria. This energy is used to produce new organic compounds form carbon dioxide, and thus enables bacteria to grow and divide in the absence of light. Some bacteria do not even need oxygen and thus are considered to be completely independent of photosynthesis, using geothermal rather than solar energy. The mass of bacteria around vents can be so great that they provide food, locally, for vast numbers of clams, shrimps, and tube worms.

In the lithosphere, soils and sediments contain mainly invertebrates and micro-organisms. But bacteria also live deep, to at least 3.5 km depth, in the cracks of subsurface rocks, in aquifers, in salt and mineral mines, and even in oil reservoirs. Temperature alone should not be a limiting factor until between 5 and 10 km depth. Some of these bacterial populations obtain their energy from subsurface geochemical or geothermal processes and thus, like hydrothermal bacteria, may be independent of the surface biosphere and the sunlight that drives it.   R. JOHN PARKES

Further reading

Postgate, J. R. (1994) *The outer reaches of life*. Cambridge University Press.

biostratigraphy Biostratigraphy is that aspect of stratigraphy which is concerned with the sequential subdivision of rock strata on the basis of their fossil content. Biostratigraphy had its origins in the late eighteenth century, when William Smith, a civil engineer, deduced that the same strata invari-

ably occurred in consistent sequences and included unique assemblages of fossils. Its establishment as a discrete field of study is considered to have followed from the mid-nineteenth century work of Albert Oppel on Jurassic strata. The term 'biostratigraphy' was used first by Dollo in 1910. Most of its current concepts had been formulated by the start of the twentieth century and have subsequently become more subtle.

By the 1950s confusion had arisen over the use of geological terms referring to strata and to time. As a result, an international commission chaired by Hollis Hedberg was established to formulate principles and to agree practices for stratigraphy. The work of this commission resulted in the publication of the *International Stratigraphic Guide* in 1976.

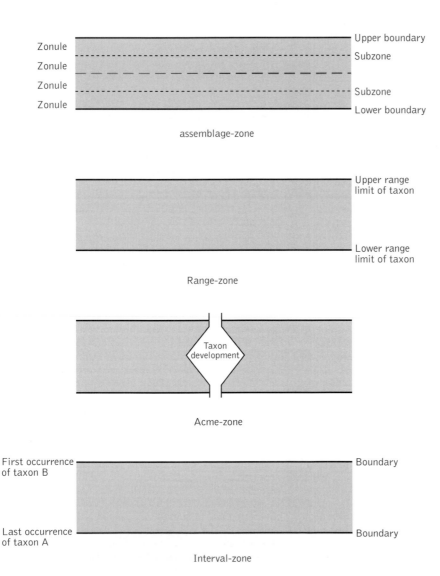

Fig. 1. Types of biozone. (a) assemblage-zone, (b) range zone, (c) acme-zone, (d) interval zone. (After H. D. Hedberg (ed.) (1976).)

This states that biostratigraphical classification aims to organize rock strata systematically into named units based upon the content and distribution of fossils. All rocks possess lithostratigraphical and chronostratigraphical features. However, as considerable tracts of the Earth's rocks are devoid of fossils, they lack biostratigraphical character. Furthermore, biostratigraphical units are different from other stratigraphical units because they are based on varied, separated remains scattered in uneven quantities within rocks.

The fundamental biostratigraphical unit is a *biozone*—the general term for any kind of unit used in calibration and correlation. Fossils that are diagnostic of biozones do not occur everywhere that relevant rocks are found. The areas of distribution of fossils vary; some are limited, others extensive. Widely distributed, abundant, and easily recognizable remains of former living organisms that occur over a narrow span of geological time are called *index fossils*. These fossils are normally of species, the narrowest major classificatory grouping of organisms, and basic to biostratigraphical investigations. However, other members of the taxonomic hierarchy (genera and families in particular) are also used for biostratigraphical purposes. A taxon (plural *taxa*) refers to any biological category, whatever its classificatory rank, and is a widely used biostratigraphical term. The lateral limits of a biozone are determined by the distribution of the taxa by which it is defined; its vertical limits by the persistence of these taxa over time. Surfaces of biostratigraphical change or distinctive character (zone boundaries) are called *biohorizons*. Horizons (biohorizons) are increasingly recognized in modern stratigraphical practice. If a number of biozones possess common features they may be grouped into *superzones*. Biozones may be divided into *subzones* to demonstrate finer detail. Division of subzones into *zonules* is the ultimate expression of such patterns.

Four main formal types of biozone can be designated (Fig. 1). An *assemblage-zone* is characterized by certain association of fossil taxa. Its biohorizons delimit this characteristic assemblage. Boundaries can be defined either by eye, or by computerized numerical methods (such as cluster analysis, ordination, and principal components analysis) in order to avoid human bias. A constituent taxon may be distributed beyond the boundaries of an assemblage-zone. The written definition of a zone is by the scientific (Latin) names of two or more of the principal taxa of which it is composed; for example, *Betula–Pinus–Corylus* pollen assemblage-zone. A *stratotype* (i.e. reference stratal section) should also be designated in order that the zone can be identified at other sites. Assemblage-zones are of most use for local correlation, and their constituent taxa can provide important information on past environmental conditions. A *range-zone* is a body of strata that represents the horizontal and vertical range of a specified taxon. An *acme-zone* is characterized by the maximum development, abundance or frequency of occurrence of a taxon but not its entire range. An *interval-zone* occurs between two clearly defined biostratigraphical horizons. It may include

nothing diagnostic biostratigraphically, and is normally named after a characteristic of one of its boundaries, such as the appearance or disappearance of a taxon.

A biozone should also be formally described when it has been defined. Important in this description are its name and type locality (stratotype), its kind and rank, diagnostic fossils, associated lithological and chronostratigraphical units, boundaries, thickness and lateral extent, age and correlation, and facies (individual characteristics) and climatic significance.

Informal biostratigraphical units may also be defined; for example, the 'reindeer stratum'. Also informal is the use of letters or numbers, or both, for biozones in place of formal names. Unlike names, letters and numbers express position in a sequence. They also facilitate communication between geologists, some of whom may be unfamiliar with Latin nomenclature.

R. L. JONES

Further reading

Blatt, H. Berry, W. B. N., and Brande, S. (1991) *Principles of stratigraphic analysis*. Blackwell Scientific Publications, Boston, Mass.

Hancock, J. M. (1977) The historic development of concepts of biostratigraphic correlation. In Kauffman, E. G. and Hazel, J. E. (eds) *Concepts and methods of biostratigraphy*, pp. 3–22. Dowden, Hutchinson and Ross, Stroudsburg.

Hedberg, H. D. (ed.) (1976) *International stratigraphic guide: a guide to stratigraphic classification, terminology and procedure*. John Wiley and Sons, New York.

birds Birds, together with bats and pterosaurs, form the three groups of terrestrial vertebrates that developed flight independently. Each of them met the structural requirements for flight in a slightly different way. In the birds the development of feathers for use as a flight membrane is the major unique derived character (autapomorphy).

The most famous bird, and probably the most famous fossil, is *Archaeopteryx*. This was first discovered in the Late Jurassic Solnhofen Limestone of Bavaria in 1860, and since then a total of only six specimens have been found. *Archaeopteryx* is generally recognized as being the earliest fossil bird and, as such, it provides a considerable amount of information towards an understanding of the origin of birds and the development of flight. Birds have a number of specializations of the skeleton that are related to flight, particularly enlargement of the sternum (breastbone) for the attachment of flight muscles and the fusion of the clavicles (collar bones) to form a V-shaped furcula (wishbone). *Archaeopteryx* possesses a furcula, but its sternum is not enlarged, which suggests a limited flight ability. In birds the fingers have become fused and act only as supports for the wing feathers, whereas in *Archaeopteryx* the fingers are still distinct and bear functional claws. All birds, both fossil and modern, have reduced the tail to a mass of fused vertebrae termed the pygostyle. *Archaeopteryx*, however, retained a long tail. It also retained socketed teeth and a pelvis that is structured like that of carnivorous dinosaurs. In fact *Archaeopteryx* has so many characters in common with carnivorous dinosaurs that without its flight feathers (preserved as impressions in the fine-grained Solnhofen Limestone) it would be identified as one. The relationship between birds and dinosaurs was identified more than a century ago by T. H. Huxley, and this view has been strongly substantiated in recent years by cladistic analysis. Such analyses have shown that some features considered to be typically avian, such as the furcula, first appeared in carnivorous dinosaurs. Feathers, although a major character of birds, probably developed initially from epidermal scales as insulation and were only later adapted for flight, a view substantiated by recent Chinese discoveries of dinosaurs with feathers.

There are two main views about the way in which flight developed in birds. The arboreal theory is the traditionally favoured view and holds that birds started by gliding from trees and then progressed to flapping flight. A more recent cursorial theory suggests that flight developed in small terrestrial animals that ran and jumped into the air, gaining flight ability gradually as their wings developed. Although it is certainly easier to gain speed by dropping from a tree than running along the ground, the cursorial theory does fit better with the dinosaurian ancestry of birds. More evidence is needed to provide a resolution of these conflicting theories.

Little is known of Cretaceous birds, although Early Cretaceous birds discovered in Spain, South America, and China and called enantiornithines show a full complement of avian features such as a pygostyle and enlarged sternum. Late Cretaceous birds are represented by open-water forms such as *Hesperornis*, which was loon-like, and the tern-like *Ichthyornis*. None of these are members of the modern bird orders, most of which appear in the Eocene or later and led to the numerous and highly adapted organisms that we know today.

DAVID K. ELLIOTT

bivalves The phylum Mollusca is extremely diverse and contains a number of classes that at first sight can appear to be so different as to be unrelated. One of these is the bivalves (sometimes called pelecypods or lamellibranchs), an important group of bivalved molluscs familiar to all from the numerous shells that litter beaches. Bivalves are aquatic suspension-feeders, inhabiting a variety of infaunal and epifaunal habitats (i.e. living in and on the bottom) and are particularly characterized by their ability to burrow, some of them even into rock and wood. About 50 000 species have been described, of which about 10 000 are modern. It is their adaptability that has made them so successful today, although their geological history goes back into the Early Cambrian.

All bivalves have two shells or valves (Fig. 1a) that are usually virtual mirror images of each other about the commissure or edge along which the valves are in contact. The valves are connected to each other by a ligament which forms a hinge and is instrumental in opening the valves. Inside the hinge area the valves feature a set of teeth and sockets, the dentition, which is extremely variable in morphology and ensures a tight fit when the valves close. In addition the valves may show large internal scars from the attachments of the adductor muscles that operate to close the valves, and a linear scar parallel to the margin of the valves termed the pallial line. This is the line along which the muscles of the inner muscular part of the mantle are attached, the mantle being a sheet of tissue that forms the inner lining of the shell and secretes it. The mantle encloses the mantle cavity, and within this lie the gills, elongated ciliated feeding structures termed the palps, and the stomach. In burrowing forms a muscular foot projects outside the valves, and a siphon formed from the mantle projects to the surface (Fig. 1b). The presence of the siphon can be recognized by a deflection in the pallial line termed the pallial sinus.

Bivalve classification has always been a difficult problem because shell shape is often closely related to life habits, and important taxonomic characters are often present in the soft parts which do not get preserved. Virtually every organ system and hard-part structure has been used as the basis for grouping, and for many years a classification based on the gills was used. This has now been superseded by a multi-character

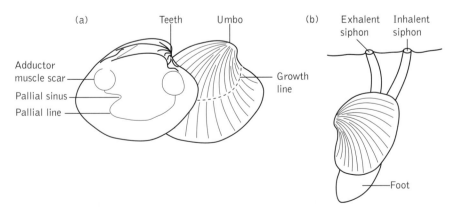

(a) Teeth Umbo — Adductor muscle scar — Pallial sinus — Pallial line — Growth line

(b) Exhalent siphon Inhalent siphon — Foot

Fig. 1. (a) Hard-part morphology of a typical bivalve; (b) burrowing bivalve in life position.

classification in which hard-part structures, particularly the dentition, are important.

The shape and general morphology of bivalve shells directly reflects their mode of life; hence our understanding of the ways in which modern bivalves live enables us to make inferences about the way in which fossil forms lived. Many bivalves are shallow or deep burrowers, using the muscular foot to pull themselves through the sediment while rocking motions of the valves also help the digging process. Deep burrowers generally have elongate streamlined shells that enable them to move rapidly, while shallow burrowers tend not to have elongate shells. These infaunal bivalves live within the sediment for protection and remain connected to the surface by their siphons. Attached epifaunal bivalves are also common; they may be attached by a mass of threads, the byssus, as in the mussel, or be cemented down as in oysters. An extinct group of Mesozoic bivalves, the rudistids, became closely adapted to a cemented mode of life: one valve became conical while the other formed a lid. In some areas these animals were numerous enough to form reefs. In this way they parallel some Late Palaeozoic brachiopods and, remotely, even the corals. Swimming is a way of life adopted by some modern bivalves, such as the scallops, which are free-lying on the sea floor. Rapid opening and closing of the valves by the large adductor muscle allows the animal to eject water forcefully from the mantle cavity and can move it erratically just above the sea floor. This is an exhausting activity for the animal and it is used only to escape predators such as starfish. Some bivalves are adapted for life in hard substrates, boring into rock and wood by rocking and scraping actions of the valves and, in some instances, by using corrosive secretions.

Bivalves first occur in the Early Cambrian of Australia, North America, Denmark, and Siberia. These are extremely small (1 mm) and show similarities to an extinct molluscan class, the rostroconchs, which appear to have been ancestral to them. The rostroconchs were bivalve-like but possessed no hinge; the bivalves seem to have developed from them by reducing mineral deposition and increasing the deposition of ligamentous tissue in the hinge area. During the Ordovician the major groups of bivalves appeared and burrowing and bysally attached forms developed during a rapid burst of adaptive radiation. Although bivalves were less important than brachiopods (which are also bivalved benthonic suspension-feeders) during the Palaeozoic they were able to displace them in the early Mesozoic after the Permo–Triassic extinction event. This was mainly due to their ability to exploit infaunal habitats as a result of their development of a muscular foot and siphons formed by fusion of the mantle edges. At this time they appear to have first extensively colonized intertidal habitats—environments that they inhabit to the present day.

The bivalves have a long time-range and their biostratigraphic utility is limited. They are, however, used locally in Pennsylvanian (Late Carboniferous) coal measures, and in the Upper Cretaceous.   DAVID K. ELLIOTT

Further reading

Boardman, R. S., Cheetham, A. H., and Rowell A. J. (eds) (1987) *Fossil invertebrates*. Blackwell Scientific Publications, Oxford.

Morton, J. E. (1967) *Molluscs*. Hutchinson, London.

black shales Although many sedimentary rocks are coloured various shades of grey, truly black examples are rare: coal and black shales are the only two common types. Both these rocks owe their colour to a high concentration of organic matter; a content that makes them the most economically important of all rock types as they are the source of all the world's fossil fuels. The organic matter of black shales tends to comprise a high proportion of complex organic molecules derived from plankton which, when buried deeply enough (typically to depths of 3–4 km), break down to form oil; black shales are therefore often petroleum source rocks.

As well as their economic importance, black shales are also celebrated for their exquisite fossil content. Black shales tend to accumulate slowly in deep, poorly oxygenated, stagnant seas and lakes. These conditions are unfavourable for bottom-living creatures but they have little affect on the animals swimming around in the upper water column. On death, such animals (e.g. ammonites, fish, whales, and marine reptiles) sink to the sea floor where they are buried and fossilized in beautiful detail: fish scales and bones are commonly preserved in place, undisturbed by scavengers, and even skin impressions are not uncommon.   P. B. WIGNALL

black smokers and white smokers At the mid-oceanic ridges (MOR) sea water is heated and circulated at depth, forming localized convective systems similar to geothermal systems on land (see *geothermal energy*). The bedrocks are basaltic and are very similar in composition to sea-water-based geothermal systems in south-west Iceland. The main differences are that the MOR systems are located at depths of about 3000 m and are hence under a hydrostatic pressure of about 300 bar, whereas geothermal systems in Iceland are at 1 atm (or about 1 bar pressure) at the Earth's surface. This difference causes MOR geothermal systems to be rich in hydrogen sulphide gas (H_2S); the pH of the circulating water is thus low (about 3).

The first hydrothermal vents to be discovered on mid-ocean ridges were found near the Galopagos Islands in the Pacific in 1979. It was not until 1985 that hydrothermal vents were found on the Mid-Atlantic Ridge. The vents were first discovered with the submarine *Alvin*, owned by the US government. As seen from *Alvin*, the vents were of two different types, one black, the other white. The vents have since been referred to as 'black smokers' and 'white smokers' (Fig. 1).

Black smokers have fluid temperatures of more than 360 °C, whereas white smoker fluids are generally lower in temperature (260–300 °C). Once the vent water enters the sea, the hot fluid mixes with sea water and the dissolved constituents rise because of the buoyancy resulting from the higher temperature. These anomalies are generally referred to

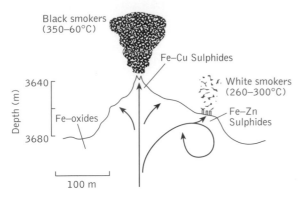

Fig. 1. Schematic cross-section of the active TAG mound on the Mid-Atlantic Ridge. Arrows indicate subsurface fluid flows. (After Humphris, S. H. *et al.* (1995) *Nature*, **377**, 713–16.)

as *plumes*. They can nowadays easily be traced by temperature and chemical anomalies. The latter consist of trace concentrations of metals such as manganese (Mn) or a rare isotope of helium (^{3}He). The plumes are typically about 40 km wide and 1 km high (Fig. 2). Tens of such systems have now been discovered around the globe.

Active hydrothermal vents typically have diameters of 50–200 m, and some are over 20 m high. These structures are generally referred to as *mounds*. In general, mounds have both black and white smokers. The chemistry of the sea water has been altered by reaction with basalts at depth. As mentioned above, the pH of the water is low (3–4) and it also has elevated concentrations of hydrogen sulphide gas (H_2S) as well as the metals manganese (Mn), iron (Fe), copper (Cu), zinc (Zn), and to a lesser extent cobalt (Co), lead (Pb), and cadmium (Cd).

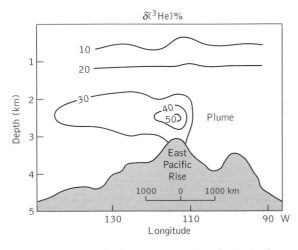

Fig. 2. Distribution of δ^3He in sea water at 15° S on the East Pacific Rise in the Pacific Ocean. (After Lupton, J. E. and Craig, H. (1981) *Science*, 214, 13–18.)

The so-called TAG hydrothermal mound on the Mid-Atlantic Ridge (26° 08′ N) has been studied by researchers from the Woods Hole Oceanographic Institute in the USA. The field was drilled in order to decipher the structure of mounds. The TAG mound is typical of mounds found on the MOR in having both black and white smokers. The top of the mound is composed of massive pyrite breccia. At depth, first anhydrite and then amorphous silica are also found. At still greater depth, the basalt wall-rock is rich in silica; the basalts then become rich in the alteration mineral chlorite. The mound itself is a 20-m cone with several black smoker chimney clusters (Fig. 3). The minerals that form these chimneys are primarily pyrite (FeS_2), with some chalcopyrite ($CuFeS_2$) and anhydrite ($CaSO_4$). The fluids have a relatively low concentration of zinc (45 μmol l^{-1}). White smokers are located at the edge of the mound. The minerals forming the vents are primarily pyrite and sphalerite (ZnS). The fluids have elevated concentrations of zinc (300–400 μmol l^{-1}) in comparison with those of the black smokers. The fluid is thought to evolve, in that primary black smoker fluid flows through the mound, precipitating metal sulphides (e.g. FeS_2). This reaction causes the pH to decrease and dissolve the sphalerite. The white smokers thus represent secondary fluids that are rich in zinc. This process can be represented by the following chemical reactions:

$$Fe^{2+} + 2HS^- = FeS_2 + H^+$$
$$ZnS + H^+ = Zn^{2+} + HS^-.$$

Anhydrite ($CaSO_4$) is precipitated when the hot vent fluid (at a temperature of more than 150 °C) mixes with sulphate-rich sea water. This is because anhydrite is unusual in having retrograde solubility (its solubility decreases as the temperature rises).

In the hot sea-water plume above the vents metals are found in the form of dissolved constituents and small particles. Manganese remains in the plume for up to two weeks (at concentrations up to 14 nmol l^{-1}), whereas other metals such as iron, cobalt, copper, and zinc are precipitated from the plume at plume height (Fig. 4). 70 per cent of the metals are precipitated as sulphides and the remainder as oxides (e.g. haematite; Fe_2O_3) and oxyhydroxides (e.g. goethite; FeOOH) and hydroxides (limonite; $Fe(OH)_3$).

The TAG mound contains 1–2 weight per cent (wt per cent) of copper. The black smoker deposits contain 700 parts per million (ppm) of zinc and 250 ppb (parts per billion) of gold, whereas the white smoker deposits contain 1–4 wt per cent of zinc and 3 ppm of gold. The TAG mound has been compared to ancient massive sulphide deposits such as the deposits in Cyprus, Oman, and Newfoundland. The total amount of copper in the TAG mound is about 3.9 × 10^6 tonnes: more or less identical to that of the Cyprus deposit (3 × 10^6 t). The major difference between the Cyprus and the TAG metal deposits is that no anhydrite is found in Cyprus whereas the TAG mound has about 10^5 m^3 of that mineral. This is due to the retrograde solubility of the anhydrite, which has been dissolved out by cold waters that have percolated

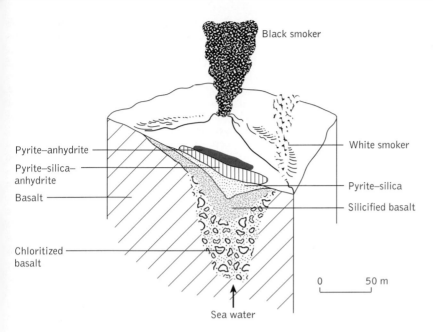

Fig. 3. Block diagram of the TAG mound on the Mid-Atlantic Ridge. (After Humphris, S. E. *et al.* (1995) *Nature*, **377**, 713–16.)

through the Cyprus formation over a long period of time. This has caused the Cyprus mound, to collapse and form a breccia. The only other difference is that the TAG mound contains amorphous silica (SiO_2), whereas the Cyprus mound has jasper—a microcrystalline form of silica, SiO_2. Jasper is a typical ageing product of amorphous silica.

The fallout from plumes forms sediments that are very rich in iron hydroxides. They are found up to tens of kilometres away from the vents (Fig. 4). The vent fluid chemistry affects the global elemental cycle. The plumes are also hosts to lush populations of gigantic clams and worms. (see *biosphere*)

A. MICHARD and K. VALA RAGNARSDOTTIR

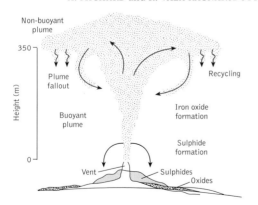

Fig. 4. Chemical and physical dynamics of the TAG buoyant plume. (After Rudnicki, M. D. (1995) In *Hydrothermal vents and processes.* Geological Society Special Publication No. 87.

Blackett, P. M. S. (Baron Blackett of Chelsea) (1897–1974)

Patrick Maynard Stuart Blackett was born in London. He trained for and entered the Royal Navy for the duration of the First World War. He then entered the University of Cambridge to read general studies before taking up physics. Blackett was to remain at Cambridge until 1933, working on nuclear physics in the Cavendish Laboratory and winning high esteem for his work. In 1933 he moved to Birkbeck College in London, where he began work in earnest on cosmic rays. Four years later he transferred to Manchester University's Chair of Physics, where he followed W. L. Bragg. Blackett transformed the Department of Physics into a modern institution concentrating on cosmic ray research. Here he also developed his interests in astronomy generally and in astrophysics. It was while considering the behaviour of the magnetic fields of the Sun and the Earth that be became interested in geomagnetism and palaeomagnetism. The need for an improved magnetometer prompted Blackett to design and produce one, and the need for a non-magnetic laboratory building for geomagnetic observations led him to build one at Jodrell Bank, where the university had its astronomical observatory.

At the approach of war and during the war years Blackett was greatly involved with government committees concerned with defence research and he was a member of the Committee on Atomic Energy.

In 1953 he returned to London, to the Chair of Physics at Imperial College, where he continued his work on palaeomagnetism, influencing a number of younger workers who were to apply this research to problems of palaeogeography and continental drift. The significance of the results that were then materializing soon became known, and 'polar wandering' became a new topic of relevance to the general debate on continental drift. Blackett was instrumental in organizing the Royal Society's Discussion Meeting on continental drift in 1964, and he was very influential in changing some of the antagonistic attitudes towards the idea. This was one of the last meetings at which geophysicists voiced objections to drift. Blackett was pressing in his call for further geophysical research.

In addition to these endeavours, Blackett served in several capacities for university and government bodies, as adviser or committee member, and he was always interested in the academic and scientific development of the developing nations, especially India. He was elected President of the Royal Society in 1965, and was created a Life Peer in 1967; he died in 1974. The great debt the Earth science community owes to Blackett acrues not only for his pioneer work and influence in palaeomagnetism, but for his persuasive powers in bringing researchers in physics and the Earth sciences together. His influence was most important in the great debates about sea-floor spreading and continental drift.   D. L. DINELEY

blockfields and blockstreams Blockfields (other names are given below) are large, sheet-like expanses of weathered blocks (usually greater than 0.25 m in size) with low surface gradients (less than 10°) covering bedrock. They are generally found in arctic or alpine regions. No source of the rock debris (such as a cliff) is seen, although the slope may rise to a ridge crest. The blocks themselves are usually angular and are often thought to be the result of 'mechanical' weathering processes. Weathering may make the blocks less angular with time. The surface appearance of open-work blocks belies the internal composition, which generally includes fine material down to silt size. On steeper slopes, fines may be removed by running water and it has been suggested that the steeper the slope, the larger the surface blocks and the greater the sizes of the fines.

The terminology is somewhat complex, being used for a variety of features: alternatives to 'blockfields' are *felsenmeer*, *blockmeer*, and *stone fields*. A useful distinction is between *autochthonous blockfields*, where the material is derived *in situ*, and *allochthonous forms*, where the material has been derived from elsewhere (e. g. till). Nomenclature is made still more intricate when the elongate forms, generally running down from mountain ridge crests, are considered. Such features are at a higher angle (perhaps up to 35°) than the fields and are usually called *blockstreams* (*blockstrome*). Note, however, that some blockstream features have been called blockfields and that some of the forms called felsenmeer are, in fact, the elongate 'stream' forms. The 'stone runs' of the Falkland Islands are blockstreams. Some forms have even been called 'rock glaciers' but a visual distinction can usually be made between

blockstreams and rock glaciers. The Russian distinction between *kurums* (blockstreams) and rock glaciers (*khemenigletscheri*) is not always appreciated. Blockstreams, like rock glaciers, commonly show indistinct flow features and surface clasts with a preferred orientation, but rock glaciers are topographically more distinctive than blockstreams. Movement of blockfield and, especially, blockstream material is generally thought of as being related to soil creep rather than the ice creep found in rock glaciers.

Blockfields are generally considered to be mountain plateau or ridge-related features, but lowland forms have been described from Scandinavia. The forms found on high mountain ridges (blockslopes) are generally at steeper angles than blockfields *sensu stricto*. They may be related genetically to the blockstreams found in present-day temperate latitudes such as the Appalachian mountains.

A distinction has been made between angular blocks produced in the Pleistocene and rounded blocks produced by chemical weathering in pre-Pleistocene times. Three basic models of formation have been proposed:

(1) formation of blockfields in a periglacial environment;
(2) chemical weathering and the formation of blockfields by corestones of residual (i. e. pre-Pleistocene) material; and
(3) a combination of older (Tertiary) deep weathering and frost action during Quaternary glaciations.

The mechanical weathering apparent on blockfields is usually considered to be the only mechanism of formation, indicating a cold-climate origin. The blockfields of the Appalachian mountains of the eastern USA south of the glacial border are considered to be 'palaeoperiglacial'. Work in northern Norway indicates a Tertiary age for blockfields which are now being revealed as plateau ice masses retreat. The clay minerals and copious fines reported suggest that blockfields were produced by chemical weathering in a Mediterranean-type climate. The virtually flat, autochthonous blockfield has prevented washing away of the fines. Not all blockfields may be so old; several authors provide convincing evidence for a Pleistocene origin. Although blockfields may appear to be relatively insignificant features, their importance is considerable when linked to the formation and occurrence of other features such as tors and the possible protection given by ice where blockfields have been preserved under glaciers.   W. BRIAN WHALLEY

Further reading

Rea, B. R., Whalley, W. B., Rainey, M. M., and Gordon, J. E. (1996) Blockfields, old or new? Evidence and implications from some plateaus in northern Norway. *Geomorphology*, **15**, 109–21.

White, S. E. (1976) Rock glaciers and blockfields, review and new data. *Quaternary Research*, **6**, 77–97.

body waves *see* SEISMIC BODY WAVES

bogs *see* PEATLANDS AND BOGS

Bowen, Norman Levi (1887–1956) One of the foremost investigators of igneous rocks in the twentieth century, N. L. Bowen was born in Ontario, Canada. His career as the great pioneer of experimental petrology was at the Geophysical Laboratory of the Carnegie Institution in Washington, DC.

Bowen studied for a short while with the Norwegian petrologist J. H. L. Vogt and at the Geophysical Laboratory under another Canadian, A. L. Daly. His classical studies of rock-forming minerals began with the investigation of the relationships of the various plagioclase feldspars and then those of the iron–magnesium silicates (mafic minerals) found in igneous rocks. He soon discovered the so-called continuous and discontinuous reaction series to which all these minerals belong. He was able to demonstrate how, from basaltic magma, a succession of rocks of progressively more siliceous composition could evolve, terminating in granite. This has become one of the fundamental concepts of modern geology. Bowen was also a principal investigator of the role of water in magma, and of reactions between different rocks at high temperatures and pressures.

Bowen's great work *The evolution of the igneous rocks* (1928) has been one of the cornerstones of modern petrology. He went on to investigate many silicate mineral systems and the role of volatiles, especially water, in the formation of different crystalline rocks. Bowen took part in the great debates of the mid-twentieth century on the origin of granite and the problems of metamorphism. Much of is research was of necessity in the laboratory, but he never lost sight of the importance of field work. His studies of the physico-chemical bases of geological processes have inspired many geologists worldwide.

D. L. DINELEY

brachiopods The brachiopods are a phylum of benthonic marine organisms in which the soft parts are enclosed in a pair of hinged calcareous valves. Although similar to bivalves in this and in their filter-feeding habit, they are in fact not closely related to them. In brachiopods the two valves are different in size but symmetrical about a median plane, whereas in bivalves they are equal in size but inequilateral. Brachiopods are very abundant in the fossil record, particularly in Palaeozoic shallow-water marine deposits, and range from the Cambrian to the Present. They are not, however, as abundant today as they were in the past, and of the 3000 described genera only 100 are modern. Their organization reflects their fixed mode of life; they have no locomotory devices and no highly developed sense-organs. Their only protection from predators or other dangers, apart from being inconspicuous, is to close their valves. Like most other sessile or attached organisms they feed by filtering nutritive particles from sea water. Most fossil brachiopods appear to have been limited to the same way of life and their basic anatomical and physiological organization has remained unchanged throughout their geological history. Within these limits, however, the brachiopods display a remarkable diversity of form and habit.

Morphology

The brachiopod shell is usually anchored to the sea floor by a muscular stalk, the pedicle, which projects through the pedicle opening in the posterior end of one valve (Fig. 1). This pedicle valve is ventral in relation to the organs of the animal and is thus sometimes called the ventral valve. The other dorsal valve is frequently termed the brachial valve because of the presence of the brachidium, a calcareous internal structure that supports the lophophore or food-gathering apparatus. The two valves fit closely along the commissure when closed, giving protection to the tissues of the organism. The hinge mechanism at the posterior end of the valves consists of interlocking teeth and sockets. Internal muscles control valve movement; adductor muscles connect the valves and operate to close them; didductor muscles which operate around the hinge, can open them. Hinges may be strophic, in which case the hinge axis is coincident with the edge of the valve, or non-strophic in which case the edges of the shell are curved. Between the hinge and the initial growing area of the valve, the beak, lies an area termed the interarea. This is the site of the pedicle opening in the pedicle valve; in some brachiopods this opening may be represented

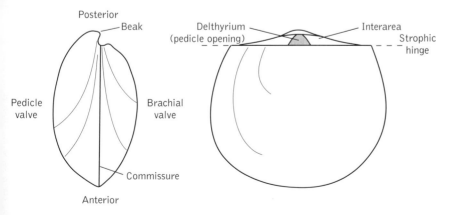

Fig. 1. Hard-part morphology of a typical brachiopod.

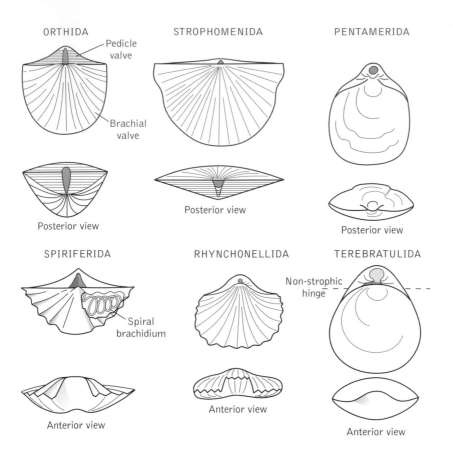

ORTHIDA
Pedicle valve
Brachial valve
Posterior view

STROPHOMENIDA
Posterior view

PENTAMERIDA
Posterior view

SPIRIFERIDA
Spiral brachidium
Anterior view

RHYNCHONELLIDA
Anterior view

TEREBRATULIDA
Non-strophic hinge
Anterior view

Fig. 2. Typical members of the main orders of brachiopods, illustrating the general character of the shells. In each case the anterior or posterior and brachial views are shown.

by a broad triangular notch termed the delthyrium, through which the pedicle projects.

Taxonomy

Brachiopods have long been divided into two main classes: the Inarticulata, in which there are no teeth and sockets and the pedicle projects between the valves, and the Articulata, which possess a functional hinge and a pedicle opening. More recently, however, the inarticulates have been further subdivided and a new class, the Lingulata, has been recognized. These are brachiopods in which the valves are chitinophosphatic rather than being calcareous, as in all other brachiopods, probably because of their early adaptation to coastal environments with extreme salinity changes that most marine organisms cannot tolerate. They include the well-known genus *Lingula*, which has remained essentially unchanged since the Cambrian. Lingulates are the only known burrowing brachiopods. They use the two valves in rotary and scissoring motions to dig down anteriorly and then follow a U-shaped trail, coming to the surface in feeding position. Although they were important initially, lingulates and inarticulates form only 5 per

cent of the known brachiopod genera, and it is the articulates that are most important and abundant. They are divided into six orders based on features of the gross morphology (Fig. 2). The Orthida were the earliest; they are biconvex forms with non-strophic hinges, large interareas, and delthyrium. These were followed by the Strophomenida, with convex pedicle and concave brachial valves, and strophic hinge-lines. The Pentamerida were biconvex with incurved beaks and were characterized by an internal muscle platform, the spondylium. The Rhynchonellida are small non-strophic forms, typically with strongly ribbed valves. The Spiriferida also had strongly ribbed valves but were strophic and possessed spiral brachidia. Finally, the Terebratulida are biconvex and non-strophic. The differences in morphology are in all cases related closely to the mode of life of the organisms.

Evolutionary history

The earliest brachiopods appeared in the Early Cambrian and include abundant lingulates and the first articulates, the orthids. This fauna became more abundant and diverse through the Cambrian and pentamerids are first seen in

Upper Cambrian rocks. All the classes became more diverse in the Ordovician, although the articulates were greatly at an advantage and assumed dominance from the inarticulates and lingulates at this time. The inarticulates and lingulates declined from this point and by the Devonian were reduced to a few stable stocks, which have survived almost unchanged to the present day. The articulates achieved their most significant evolutionary radiation during the Ordovician, adding the strophomenids, rhynchonellids, and spiriferids. The earlier orders reduced somewhat during the Silurian and, towards the end of the period, the last new order, the terebratulids, was added. The Devonian was chiefly a period of expansion, but major extinctions of invertebrate faunas at the end of the Devonian took their toll and the pentamerids were lost. The later Palaeozoic was a time of expansion, particularly for the strophomenids, since they were able to colonize quasi-infaunal habitats. However, the severe extinctions of the Late Permian affected the brachiopods as they did all other invertebrate groups, and by the end of the Jurassic the only articulate orders left were the terebratulids and the rhynchonellids, both of which are still extant. The decline of brachiopods in the Mesozoic is probably related to competition with bivalves. Once the strophomenids had become extinct the bivalves were able to colonize habitats within the sediment (infaunal), which were thenceforth denied to brachiopods. These habitats provided the bivalves with protection from predators, particularly starfish, while the brachiopods remained vulnerable.

Ecology

Brachiopods are normally part of the sessile epifauna and as such they show many and varied adaptations to different types of benthonic existence. Many were attached through life by the pedicle, either directly to hard substrates or into soft sediments by the division of the end of the pedicle into numerous threads. It has been demonstrated that some brachiopods with long thin pedicles would act like kites, 'flying' in water currents. This would result in a region of reversed flow round the anterior edge that would provide them with feeding currents. Some forms attached themselves directly to the substrate by the pedicle valve. In some species this was a juvenile development; once the brachiopod became large enough it would break off and become free-lying. Attached forms are prominent members of reef associations in the Permian of Texas. Here conical forms known as richthofeniids attached themselves to the hard substrate by long tubular spines, forming an interlacing network that contributed to the reef fabric. They may have generated currents by flapping the small brachial valve. Many brachiopods had a closed pedicle opening and must have been free-lying on the sea floor. This resulted in adaptations such as dish-like valves to spread weight on unstable substrates and radiating spines that supported the animal. In the spiriferids the interareas were developed as broad stable platforms, and in some species the shell was thickened in the basal area to provide additional stability.

Burrowing forms are rare; only the lingulids exploited this way of life; many of the strophomenids, however, particularly the productids, lived a partially buried or quasi-infaunal existence, held in place by strong spines. Short spines were present on the brachial valves to hold sediment and prevent it being winnowed away; the animals were thus effectively hidden from predators.

Communities and distribution

Brachiopods are found in recurring associations commonly composed of several species. These appear to be characteristic of particular environments and are presumably related to physical parameters such as temperature, water depth, salinity, and substrate type. Work on the Ordovician of Wales in 1983 by Martin Lockley of the University of Colorado has demonstrated the presence of eight palaeocommunities that persisted until the Silurian. He found that coarse sediments were inhabited by coarse-ribbed orthids, while the fine sediments supported an assemblage of inarticulates and small strophomenids. A similar picture is seen in the Silurian, with five communities inhabiting the same area and forming concentric belts parallel to what was then the shoreline. This seems to indicate a relationship to water depth, but factors that vary with depth, such as temperature, pressure, turbulence, or food supply, may be more important.

On a global basis the distribution of brachiopods is best known from the Palaeozoic, where they are most abundant, and is similar to the distributions of trilobites and graptolites. In the Ordovician an early fauna of lingulates and inarticulates occupied the polar regions while a separate fauna of orthids inhabited the more equatorial areas. The closer approach of the continental areas during movements of the lithospheric plates resulted in a lessening of provincialism during the Ordovician and this was further reduced by the expansion of polar ice-caps at the end of the Ordovician, which severely affected shallow-water marine faunas. Silurian faunas were generally more cosmopolitan, but by the Devonian continental movements had once more resulted in the development of five provinces.

Brachiopods are generally too long-ranging in the Mesozoic to be of more than local stratigraphic use. They are, however, useful in the Palaeozoic, despite their facies dependence and provinciality. They have proved to be useful, together with trilobites, for zoning the shallow-water shelly sequences of the Ordovician and Silurian, where they provide, in some instances, a greater degree of accuracy than can be provided by the standard graptolite zones. DAVID K. ELLIOTT

Further reading

Clarkson, E. N. K. (1993). *Invertebrate palaeontology and evolution.* Chapman and Hall, London.

Richardson, J. R. (1986). Brachiopods. *Scientific American,* September 1986, pp. 100–6.

Rudwick, M. J. S. (1970). *Living and fossil brachiopods.* Hutchinson, London.

braided rivers Braided rivers have a distinctive divided channel form. The simplest braided pattern is where a single channel divides into two channels round a bar, and rejoins at the downstream end of the bar. Large braided rivers can have 20 or more channels at any one location. Braided rivers are mainly found in glaciated mountainous regions where channel slopes are steep. As a consequence, braided rivers are usually gravel-bedded, although braiding can develop in sand- or silt-bed streams, particularly where the sediment load is high (the Brahmaputra, for example).

Braided rivers were traditionally regarded as a distinct type of river channel pattern, but more recent work has recognized that there are many rivers with patterns that are intermediate between braided and meandering. These *wandering* rivers are part of a continuum of pattern types from straight, through meandering and wandering, to braided. The bars and individual braid channels are modified frequently by river flow, changing in both size and position. Where the channels flow round stable, usually vegetated, islands, the individual channels are more stable through time and the pattern is referred to as *anastomosing*.

Because of the settings in which it is usually found, braiding has been explained with reference to steep slopes, highly variable water discharge, a predominance of bedload, and high rates of sediment supply to the river. It is now accepted that these variables are related to each other, and that the requirements for braiding are high stream energy and sediment input from hill slopes, tributary streams, or glaciers. Work on the gravel-bed rivers of New Zealand has suggested that the pattern of a particular river depends on the ratio between the erodibility of the river banks and the rate of supply of sediment from upstream. Braiding occurs when the banks are easily eroded and wide shallow channels thus form. Meandering is favoured when the banks are more resistant to erosion, producing channels which are deep and relatively narrow. Wandering channels form under intermediate conditions.

The presence of mid-channel bars round which channels diverge creates sedimentological characteristics and hydraulic conditions which are peculiar to braided rivers. Of particular importance are channel confluences, where scour holes develop. These are important for river morphology because they control the downstream transfer of bed sediment. They are important locations for the deposition of placer deposits. Braided river bars have low elevation and usually consist of coarse sediment at their upstream end which becomes finer downstream. The downstream tail of bars are important locations for silt- and clay-sized sediment deposition.

Because of their association with high rates of sediment supply, braided rivers are commonly aggradational, and many sandstones and conglomerates were intially deposited in braided river environments. Mineral and oil reserves from braided river deposits are frequently exploited, and this has stimulated interest in present-day braided river processes. However, the unstable nature of these rivers creates particular difficulties where their management is required as, for example, in Iceland and New Zealand where large braided rivers present a serious flood hazard and threaten structures.

TREVOR B. HOEY

Further reading

Best, J. L. and Bristow, C. S. (eds) (1993) *Braided rivers*. Geological Society Special Publication No. 75.

brines Brines are highly saline aqueous fluids, even more saline than sea water, with concentrations of total dissolved solids often in the order of hundreds of thousands of parts per million (milligrams per kilogram). Their ionic strength (I) is greater than 1.0 moles per kilogram.

Brines are formed by a number of processes. Perhaps the most famous brines in the world are those found in the Dead Sea in Israel, an inland lake which is so salty that bathers float in its waters. This is an example of a brine formed by evaporation when water molecules escape from the lake to the atmosphere and the solutes are concentrated in the residual solution. Another example is Mono Lake in California, which receives rainwater and snow melt from the surrounding mountains but has no outlet. Both of these lakes are important tourist attractions and are examples of geological phenomena that are also important economic resources. When brines such as these become supersaturated with their solutes, mineral salts will precipitate out of solution as the solubility product of each salt is passed. The least soluble minerals such as calcium carbonate (calcite, $CaCO_3$) precipitate first, usually followed by calcium sulphate (gypsum, $CaSO_4 \cdot 2H_2O$), sodium chloride (halite, common rock salt, NaCl), potassium chloride (sylvite, KCl), and magnesium salts. Thick sequences (thousands of metres) of evaporates were deposited over Europe during the Permian and Triassic as a result of mineral salt precipitation from huge land-locked seas, forming valuable economic resources which are still mined in north-west Europe today. Re-dissolution of such deposits will result in brine production.

Connate or formation waters are formed from fluids trapped in spaces between particles in sediments. During diagenesis and burial some fluids are squeezed out and will escape, but some will remain trapped in isolated pockets and form oilfield brines. These will have remained in contact with the rock at elevated temperatures and pressures for millions of years and will be highly saline as a result of extensive interaction between water and rock. Pore fluids play an important role in metamorphism and in the formation of economically valuable ore deposits. The presence of fluids in the sedimentary pile can reduce the effective pressure across grain boundaries and lead to hydraulic fracturing, providing a pathway for fluid flow and a void for mineral deposition. Magmatic brines are produced during the final stages of the crystallization of a magma (e.g. granite) as fluids and other non-compatible components are excluded from the crystallizing magmatic melt. These brines will interact with the country rock at the intrusion-country-rock interface to form skarns, or vein deposits if they are able to penetrate into the country rock.

Hot saline aqueous fluids are a necessary component in the formation of hydrothermal ore deposits. Metals such as lead, zinc, and copper are much more readily transported in saline fluids and in the presence of ligands such as chloride (Cl^-) and bisulphide (HS^-) with which they form complexes. For example divalent copper (Cu^{2+}) forms the complexes $CuCl^+$ and $CuCl_2^0$; this dramatically increases the concentration of copper in solution and hence the possibility of copper transport. Complex stability is sensitive to fluid temperature, pH, salinity, redox conditions, and pressure in the system. When one or more of these parameters changes to lower the solubility of the mineral phase with respect of the fluid phase, mineral deposits may be formed.

Very small inclusions of these fluids can be trapped in a crystal during its growth. These fluid inclusions can be analysed to indicate the temperature at which the mineral formed as well as estimates of its salinity.   A. Y. LEWIS

Bryozoans The bryozoans, sometimes also called the Ectoprocta, are a phylum of colonial, mostly marine, organisms. The individuals or zooids are physically connected and form a calcareous skeleton, the zooarium, that is commonly similar in appearance to the structures formed by other colonial organisms such as corals and hydrozoans. The individual zooids are microscopic but the colonies may be up to 1 m across. The individuals are organized as functional and morphological units capable of feeding and carrying out other functions such as reproduction and digestion. The feeding zooids use retractile tentacles, called the lophophore, to filter feed and have a U-shaped gut for digestion. The Bryozoa are often grouped with the phylum Brachiopoda in the 'Superphylum' Lophophorata because of the presence of a lophophore in both.

Bryozoans are common today in shallow-water marine environments and were common in the past; at least 3500 living and 15 000 fossil species are known. The fossil record of the phylum extends over 500 million years, from the Ordovician to the present day, and during most of that time they were widely distributed and abundant. As bryozoan genera tend to be long-lived and facies-controlled they have little stratigraphic utility. They have, however, been important sediment binders in reefs through geological time and, as the morphology of a colony is related to current strength, they are useful in palaeoenvironmental reconstructions.

DAVID K. ELLIOTT

Buckland, William (1784–1856) Fossils fascinated William Buckland, even as a child in Dorset, England. At Corpus Christi College, Oxford, in 1809 he was both ordained and elected Fellow. He was then made Reader in Mineralogy and elected to the Geological Society in 1813. Buckland's interests in fossils led him to study both vertebrates, especially mammals, and invertebrates, particularly molluscs. The vertebrates were an interest he had in common with the anatomist Baron Cuvier whom he visited in Paris several times.

In his study of Quaternary cave faunas he regarded a knowledge of the living representatives of the fossils as of great importance. His researches included the origins of the caves and the various deposits they contained, and the nature of flood and river deposits. As a churchman he contributed to the debate about the Noachian flood, without identifying it specifically in the superficial deposits he examined. He did, however, hold that there had been great inundations in prehistoric times which had spread the 'diluvium' far and wide. He was a plutonist and something of a catastrophist, since he believed that sudden great events had recently occurred to remove the fauna seen in the deposits we now know as of Pleistocene age. He was not opposed to Louis Agassiz's theory of an ice age and the previous great spread of glaciers. In fact, he described glacial deposits and glacially smoothed rock surfaces to the Geological Society when there was still much doubt about the former extent of highland ice.

Buckland's contribution to geology was not so much in the discoveries he made as in his attempt to redefine nature and in his methods of geological explanation. He was instrumental in establishing the Museum of Practical Geology in London. In 1845 Buckland was appointed Dean of Westminster, and thereafter was no longer able to devote time to geology.   D. L. DINELEY

Further reading

Rupke, N. A. (1983) *The great chain of history. William Buckland and the English School of Geology (1814–1845)*. Clarendon Press, Oxford.

building stones It is probably true to say that all the rocks that are accessible to humans have been used for constructional purposes at one time or another. In general, the stones used for building are those that have sufficient strength and rigidity to support free-standing structures, and are amenable to the technical and organizational resources of the culture wishing to use them. The innumerable varieties of rock are not uniformly distributed over the surface of the Earth, and any one culture will consequently have only a limited range with which to work. Stone, at least in its natural state, is a bulky product of low value, which until modern times was rarely transported great distances. This put tight constraints on the types of building that could be erected, and led to culturally distinctive forms of architecture, ranging from the comfortable oolitic limestone villages of the English Cotswolds to the sandstone and granite temples of the Nile Valley.

Building in stone has an extremely long history. It developed to satisfy several requirements: better protection from the weather and from enemies; longevity of the structure; and prestige for the builders. Building is essentially a craft industry, and knowledge of the reaction of stone to the environment has developed by experience over the generations. Rocks are relatively strong when placed under compression, as in traditional block-on-block construction, and even material as weak as sun-dried mud is capable of supporting multi-storey buildings, such as are common in Arabia and North Africa. When subjected to tensional (stretching) forces, however, all rocks are weak. Because of this fact the distance that can be bridged by a single slab of stone used, for example, as a door lintel, is rarely more than three or four metres, and flat ceilings of stone are almost unheard of. (This limitation is apparent in the temples of ancient Egypt.) The arch was developed to overcome this problem, which it does by smoothly diverting vertical stresses around the open space. In short, the traditional building stands by harnessing gravitational forces to balance the structure.

Of the 2000 or so mineral species currently known to science, only about a dozen are important as structural components of building stones. Quantitatively the most important are quartz, feldspars, micas, the clay minerals, pyroxenes, amphiboles, and the carbonate minerals (calcite and dolomite). Of lesser general importance are two sulphate minerals, gypsum and anhydrite, and iron hydroxides ('limonite' and haematite). Individual rock types are usually composed of no more than four essential minerals, and it is the relative proportions and interrelations of these, together with a degree of post-formational alteration that gives the variety of appearance and physical characteristics in the building stones that are available.

Rocks, although durable, are not indestructible. They have each formed under restricted environmental conditions and when subjected to a different set of conditions may undergo mineralogical alteration. Stone in buildings has been removed from a relatively stable environment within the Earth's crust and has been exposed to the atmosphere, to the effects of changes in temperature, to the weather in general, and to man-made pollutants. The result is that the stone suffers from the adverse effects of weathering or decay. This process can be rapid, as measured on a human timescale.

Classification of building stones

The stone trade has its own nomenclature, which in several respects is at variance with geological usage. For commercial purposes, building stones are usually divided into three lithological groups, which are scientifically ambiguous, although based on geological classification.

(1) Igneous and metamorphic rocks, also referred to as 'granites' or 'crystalline' rocks.
(2) Sedimentary rocks: limestones, sandstones, and 'slates'.
(3) 'Marbles': rocks of any type, regardless of origin, that are used for decorative rather than for structural purposes.

It should be appreciated that rocks show continuous ranges of compositions and structures. The classification of a building stone within the three groups listed above depends much on the use to which it is to be put.

Igneous and metamorphic rocks. Rocks that result from the crystallization of molten magma within the Earth's crust, or from surface extrusions of lava, are called 'igneous', those produced by molecular reorganization in the solid state are 'metamorphic'. Both types are products of crystallization in the mass.

In general, igneous and metamorphic rocks are composed of silicate minerals, among which quartz, feldspars, and ferromagnesian minerals, such as the micas and pyroxenes, predominate. The individual crystal grains that comprise these rocks have grown by molecular accretion, and the resultant interlocking structure is commonly extremely strong when the crystals are randomly orientated. When crystallization has occurred within an environment of uneven pressure, elongate or flat crystals tend to form, with parallel orientations that result in planes of weakness within the rock. Schists and slates are formed by pressure-induced metamorphism, generally from muddy sediments in which recrystallization of the mud has resulted in a fabric of parallel aligned flakes of clay or mica. Such weakening characteristics can be exploited in the production of thin sheets for roofing (slates) or paving.

Sedimentary rocks. Sedimentary rocks are formed by the deposition of detritus and solutes derived from the chemical and physical breakdown, as a result of weathering, of pre-existing rocks of all kinds. All sedimentary rocks have a layered structure, or bedding, produced by the successive settlement of material. The beds represent the sea floor or other surface upon which successive layers are deposited. Many sedimentary rocks also display joints, at right-angles to the bedding. The presence of closely spaced joints can greatly reduce the value of a stone for building purposes. Both bedding planes and joints are planes of weakness exploited by quarriers, and control the maximum block size that can be obtained from a stratum. They may be very thin (laminations) or several metres in thickness. Mechanical deposition, by water or wind produces clastic rocks composed of broken fragments, generally of silicate minerals and small pieces of rock: sandstones and conglomerates. The physicochemical and biochemical concentration of dissolved salts results in limestones and alabaster.

Sedimentary rocks can be thought of as being composed of three parts: (1) the fabric of relatively coarse sand grains, granules, and shell fragments; (2) an original fine-grained component or matrix of clay minerals; and (3) a mineral cement introduced after deposition. It is the relative proportions and interactions of these three components which determine the usefulness of a rock. The mineral cement, which controls the strength and weather resistance of the stone, is especially important.

Marbles. In scientific parlance, marbles are limestones that have suffered metamorphic recrystallization, but common usage applies the term to any rock that may be used for decorative rather than for structural purposes. An unmetamorphosed limestone with an attractive appearance may thus be called a 'marble' by the stonemason.

A great variety of rocks have been used for building. In the past the choice was usually, as indicated above, confined to what was readily available in the locality. Limestone was used for many of the buildings of classical Greece. Tufa was among the materials used in ancient Rome. Andesite was used for Borobudur, in central Java, the world's largest Buddhist monument, built in 778–856. The Maya pyramids of Yukatan were built of limestone. The Jaisalmer Citadel in north–west India, dating from the sixteenth century, is built of sandstone. In the absence of other sources of building stone, glacial erratics have been extensively used in Finland and northern Poland.

The structural and decorative uses of building stones are usually distinct. The stones used for decoration are usually more expensive than structural stones, and in many instances are applied as thin slabs. The Taj Mahal in India, the Cathedral of San Marco in Venice, and (a modern illustration) Canary Wharf in the London Docklands are among many examples of decorative stonework that could be cited. Sedimentary, igneous, and metamorphic rocks have all been used for decorative purposes. (One igneous rock, a syenite from the Oslo region of Norway, was used so extensively in London that it was nicknamed 'public-housite' by metropolitan geologists.)

Durability of building stones

Rocks are natural materials, and they can vary significantly within relatively short distances. A illustration of this fact was the choice in 1839 of a Permian dolomite for the construction of the Houses of Parliament in London. In the event it became necessary to obtain most of the stone from another quarry, where it was extracted unselectively from a thickness of more than 30 metres. The result was a disaster, and in 1927 it was decided to restore the building using a sandstone at a cost of more than a million pounds—a huge sum in those days.

The observation of some very simple rules can help to increase significantly the effective life of building stone. For example, sedimentary rocks will resist erosion much better if they are put in place with their bedding planes horizontal—that is, in their natural position.

One of the unforeseen effects of the Industrial Revolution has been the need to use complicated measurements of physical characteristics in engineering calculations. This fact, combined with the desire of architects to build gravity-defying structures, has made it increasingly necessary to understand the materials used in construction. Modern buildings are commonly built of cast reinforced concrete, and the exterior is clad with thin slabs of stone, or some other material, to give protection from the weather. Instead of being supported on four or five sides by adjacent blocks, as in traditional buildings, the cladding slabs are held by a small number of brackets which grip into slots cut into their thickness. Among the industrial nations, such organizations as the British Standards Institute and the American Society for Testing Materials have sought to develop test regimes of general application. For building stones the results have been generally disappointing, and no unified testing system has yet proved entirely satisfactory. At best, it might be possible to relate a combination of factors such as porosity and saturation coefficient to the known past behaviour of a limited range of building stones, but the results should not be extrapolated to other stones without careful consideration.

The results of failure are dramatically illustrated by a high-rise building in Chicago, where the stone used for cladding reacted badly to exposure to the elements and fell off in large quantities: a serious hazard to life as well as a commercial loss of considerable magnitude.

Problems of supply

Suppliers, especially those in industrialized countries, are finding it difficult to meet the demand for stone for building and other purposes as existing quarries become worked out. Possible sites are few, and there is—understandably—strong resistance to the opening of new quarries. Large 'super-quarries' in remote locations that have suitable anchorage for large ships have been opened up in such places as the west of Scotland, Newfoundland, and the Yucutan peninsula of Mexico. Other large quarries have also been developed at inland sites, even in the United Kingdom.

R. W. SANDERSON

Further reading

Burton, M. (ed.) (1999) *Designing with stone*. Ealing Publications Ltd, Maidenhead.

Clifton-Taylor, A. (1987) *The pattern of English building* (4th edn). Faber and Faber, London.

Hund, R. (ed.) (1990) *Dimension stones of the world*. Marble Institute of America, Farmington, Michigan.

Smith, M. R. (ed.) (1999) *Stone: building stone, rock fill and armourstone in construction*. Geological Society Engineering Geology Special Publication No. 16.

Bullard, Sir Edward (Crisp) (1907–80)

Edward Bullard (1907–80) began his career as a physicist, but in 1931 joined the then new Department of Geodesy and Geophysics at Cambridge. He was attracted to some of the major aspects of the physics of the Earth and spent two years conducting gravity surveys across the Rift Valleys of East Africa. He found marked gravity deficiencies in the vicinity of the rifts and proposed new models to account for them, but the origins were still questioned. He was also active at sea, where prior to the outbreak of war in 1939 he made seismic measurement from trawlers. This work in fact initiated the geophysical study of the ocean floor, later to be so important an area of research in the Earth sciences.

After war service with the Navy, Bullard returned to his research on marine geophysics, seismic measurements, and studies on gravity. He also carried out research on the Earth's magnetic field, and at the same time as Walter Elsasser in the USA, he put forward the idea of a self-sustaining dynamo as its source.

The early 1950s were occupied with study of heat flow through the floor of the North Atlantic, during which he successfully designed new equipment for measuring this and for retrieving samples from the floor for conductivity studies. By the end of that decade he was deeply interested in the role of heat flow in continental drift and in the nature of the deep ocean floor where heat anomalies were apparent. This was later to be justified by the recognition of the mid-oceanic ridges and submarine volcanicity. His influence upon the Cambridge school of geophysics and, later, the development of the idea of ocean floor spreading was profound. He was knighted in 1953.   D. L. DINELEY

Burgess Shale fauna The Burgess Shale is a lens-shaped mass of muddy sediments in the basinal shales of the Middle Cambrian Stephen Formation near Field, British Columbia. It has become exceptionally well known because of the extensive soft-bodied fossil fauna that it contains, making it an important example of a conservation Lagerstätten or accumulation of unusually well-preserved fossils. The importance of such accumulations cannot be denied: recent work on the history of lineages has shown that 20 per cent of major groups are known exclusively from their presence in the three great Palaeozoic Lagerstätten: the Burgess Shale, the Devonian Hunsrückschiefer, and the Carboniferous Mazon Creek fauna.

The Burgess Shale fauna was discovered in 1909 by Charles Walcott, then head of the U.S. Geological Survey and Secretary of the Smithsonian Institution, while mapping in the Canadian Rockies. He opened up a small quarry and collected for the next eight years, eventually amassing a collection of 80 000 specimens at the Smithsonian Institution. Administrative duties prevented him from publishing thorough descriptions of the fauna, and that work was taken up in the 1960s by Harry Whittington of Cambridge University and his students Simon Conway Morris, also of Cambridge, and Derek Briggs of Bristol University. Through their work a clear picture is now developing of the range of organisms present and the way in which the accumulation was formed. The Burgess Shale was

deposited in relatively deep water seaward of an enormous algal reef (Fig. 1a). The reef had a vertical face hundreds of metres high and the organisms lived in and on the muds that accumulated at its base. Patches of mud slumped downslope periodically, carrying with them the carcasses of dead organisms together with live inhabitants of the sediment, and all these remains were deposited together when the mud settled out. The animals appear to have been carried into an anaerobic (low in oxygen) environment that inhibited decay and were preserved as flattened films in which the organic material has been replaced by calcium aluminosilicates. In order to study the compressed specimens, layers are carefully removed by hand using needles, and reconstructions are then built up. As the fossils consist of black films on black shale, photographs are taken with ultraviolet light, or with the specimens under water or ethyl alcohol to exploit the reflectivity of the fossils.

The fauna itself includes over 120 species, representing major groups such as arthropods, molluscs, brachiopods, cnidarians, polychaetes (bristle worms), priapulid worms, echinoderms, chordates, and many forms that cannot be attributed to known phyla. The species are not equally abundant: some are represented by thousands of individuals, others by only a few. The largest proportion (37 per cent of the organisms) are arthropods, which are extremely abundant, both in numbers and diversity. The trilobites include specimens of *Olenoides* in which the appendages are preserved, a rare circumstance providing valuable information on these structures. Although the non-trilobite arthropods were originally classed together, subsequent work has shown that a great diversity of groups is present, including early representatives of crustaceans, phyllocarids, merostomes, and other forms with puzzling combinations of characters that make their evolutionary relationships difficult to understand. The polychaetes are represented by five genera and are, therefore, only minor constituents of the fauna; the priapulids, however, appear to have been important infaunal carnivores. One significant organism is *Pikaia*, which appears to show the notochord and chevron-shaped muscle blocks that characterize the phylum Chordata and is thus the first-known representative of the phylum that includes the vertebrates. The most interesting parts of the fauna are those organisms that do not fit into known phyla. Of these, *Wiwaxia* was a hemispherical animal that seems to show affinity with molluscs, although it is covered by most unmollusc-like scales and spines. *Opabinia* is superficially arthropod-like with a long segmented body; each segment, however, had a flexible lateral lobe rather than jointed legs and the head bore five eyes on short stalks, and a long flexible process armed with a terminal claw. The aptly named *Hallucigenia* was originally reconstructed as an animal that walked on seven pairs of spines, but in later reconstructions it was turned upside down and reinterpreted as an onychophoran, a group similar to both annelid worms and arthropods. The largest animal in the fauna, *Anomalocaris* (Fig. 1b), is also of unknown affinity. The large segmented appendages, mouth with cutting teeth, and

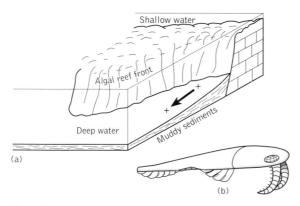

Fig. 1. (a) Reconstruction of the environment in which the Burgess Shale fauna lived and was preserved. Large arrow shows the direction of transport of organisms from the environment in which they lived to that in which they were preserved. (b) Reconstruction of the large predator *Anomalocaris*.

segmented lobed body show that it was clearly adapted for life as a predator on active benthonic invertebrates such as trilobites.

The significance of the Burgess Shale fauna lies in its ability to show us the diversity of organisms during the 'Cambrian explosion', the initial period of development of invertebrate phyla during which many new body plans were developed. Evolution since then has progressed by the refinement of the body plans that survived. In addition, an analysis of the feeding habits of the organisms within this fauna show that the fundamental trophic structure of marine organisms had already been established at this time. The study of this fauna has also led palaeontologists into a new dimension of evolutionary thinking. When Walcott initially described the organisms, he 'shoehorned' them into established categories, but later work by Whittington and his colleagues showed how much greater a diversity of body plans was present, and this observation enabled them to document the explosion of diversity that occurred at the beginning of the Phanerozoic.

DAVID K. ELLIOTT

Further reading

Gould, S. J. (1989) *Wonderful life.* W. W. Norton and Company, New York.

Conway Morris, S. (1998) *The crucible of creation.* Oxford University Press.

Whittington, H. B. (1985) *The Burgess Shale.* Yale University Press, New Haven.

calcite, aragonite, dolomite Calcite, aragonite, and dolomite are the most important and widespread of the carbonate minerals, of which at least 60 are known. The essential unit of carbonate composition is the CO_3^{2-} anion linked by a variety of cations, especially calcium and magnesium.

Calcite and aragonite are polymorphs of calcium carbonate ($CaCO_3$); they have different structures and properties and are stable under different conditions. They make up, about 50 per cent each, the skeletal material of many invertebrate groups, but with time the aragonite changes into calcite. Both can be precipitated from water, either through the agency of organisms such as algae or by inorganic processes. The resulting limestones are ultimately composed mainly of calcite and are important sedimentary rocks. Calcite is a common cement of clastic sedimentary rocks and occurs widely in mineral veins.

Both calcite and aragonite can form under metamorphic conditions. Pure calcite limestones are stable to high temperatures and pressures—the mineral simply recrystallizing, with a change in texture. Aragonite is an important constituent of so-called 'blue schists'.

Dolomite is a double carbonate of calcium and magnesium: $CaMg(CO_3)_2$; there is some solid solution with calcite. Extensive bodies are believed to have formed on the sea floor by the reaction of calcite or aragonite with magnesium in the sea water; more irregular and restricted occurrences developed from magnesium-rich groundwaters reacting with already-formed limestone bodies.

Calcite is perhaps best known for its property of extreme double refraction, for its use as a polarizer in older petrological microscopes, and for the large number of crystal habits it assumes; 224 have been noted in Scotland alone.

R. BRADSHAW

calcrete see PALAEOSOLS, DURICRUST, CALCRETE, SILCRETE, GYPCRETE

Caledonides The Caledonides are a linear orogen continuously exposed over some 6600 km from the Southern Appalachians to Svalbard in the Arctic circle. This orogen was formed by the closure of two oceans during the Silurian period: Iapetus, which separated Laurentia from Baltica–Avalonia; and the Tournquist sea, which separated Baltica from Avalonia (Fig. 1). It is a classic 'Wilson-cycle' orogen; the North Atlantic ocean reopened in early Tertiary times along the line of Iapetus. Associated rifting and magmatic underplating produced the present topography, which exceeds 2 km in east Greenland and Scandinavia.

The Caledonides can be divided into five segments, four developed on continental margins and one within the Iapetus Ocean (Fig. 1). The Laurentian margin has Neo-Proterozoic rift basins overlain by Cambro-Ordovician carbonate shelf sequences. Mid-Ordovician foreland-directed thrust sheets, not recorded in Greenland, produced during dextral convergence, contain ophiolitic or island-arc lithologies. Silurian folding and thrusting, associated with sinistral strike-slip, produced foreland basins in the western Appalachians. The Baltic margin has Neo-Proterozoic rift basins and Cambro–Ordovician shelf sequences with evidence of early Ordovician arc collision and continental subduction during the Silurian to produce a classic thin-skinned nappe pile and foreland basins in north-west Europe. The Avalonian margin and the German–Polish Caledonides have a basement recording Neo-Proterozoic volcanism and orogeny with Cambrian-to-Silurian shelf sequences and local evidence of Ordovician arc accretion. The shelf sequences of Laurentian, Baltica, and Avalonia all contain markedly different Ordovician fossil assemblages and were probably separated by large oceans. The Mauritanides of north-western Africa formed on the Gondwanan margin and are correlated with rocks in Nova Scotia. They show geological similarities with the Avalonian margin but were deformed during the Devonian and Carboniferous, as were the Southern Appalachians of the Laurentian margin. Finally, a Central Belt comprises Ordovician and Silurian igneous and sedimentary rocks characteristic of island-arc systems similar to those in the western Pacific today. Subduction-related calc-alkali volcanism ceased by late-Ordovician times along the Avalonian margin, but continued into the Silurian on the Laurentian margin.

Deep seismic profiles show marked changes in structure along the orogen. In the Southern Appalachians thin-skinned nappes emplaced during the Carboniferous Alleghanian orogeny override the Laurentian foreland by some 240 km (Fig. 1). In the Northern Appalachians, the Irish and the

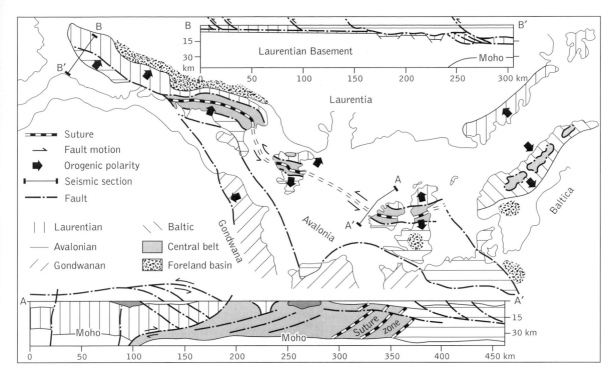

Fig. 1. The Caledonian orogen: palaeogeographical map and seismic sections.

British Caledonides a narrower thin-skinned thrust belt gives way to a central thick-skinned fold belt in which a zone of north-dipping crustal reflectivity marks the Silurian suture zone that separates Ordovician faunal provinces (Fig. 1). The Greenland–Scandinavian Caledonides represent a paired thrust belt with subduction of the Baltic margin beneath Laurentia and thrusting of the Central Belt rocks on to Baltica. The Polish Caledonides show little evidence for substantial crustal thickening.

The Caledonides are generally of low metamorphic grade, higher-grade rocks being restricted to lower levels in the nappe piles. Strike swings are attributed to ancient continental promontories rather than indenter tectonics. The orogeny did not, as in the Himalayas or European Variscides, involve substantial deformation of the continental cratons. The Caledonides are marked by the emplacement of extensive late-orogenic granites, whose ascent was usually controlled by major strike-slip faults.

The tectonic evolution of the Caledonides began with the break-up (750–600 Ma) of a late Precambrian supercontinent, Rodinia. Laurentian and Baltica drifted north away from Gondwanan land masses, opening a wide Iapetus ocean. Palaeomagnetic evidence suggests that Baltica rotated through 180° during the Cambrian to early Ordovician and interacted with newly established subduction systems in Iapetus. Oceanic arcs collided during the Ordovician (470 Ma) with the margins of Laurentian and Avalonia, which was probably still attached to Gondwana. Subduction polarity reversal followed these collisions and produced active continental margins on the south of Laurentia and the north of Gondwana. Avalonia rifted from Gondwana at this time and move rapidly northwards, overriding the Iapetus mid-oceanic ridge by the late Ordovician (455 Ma) and ending subduction-related magmatism along this margin. Subduction beneath Laurentia destroyed the remaining oceanic crust, causing collisions with Avalonia and Baltica during the Silurian (430–400 Ma). The Tournquist sea also closed at this time, perhaps by subduction beneath eastern Avalonia. Gondwana collided with the southern Avalonian margin and the Southern Appalachians during the Carboniferous (300 Ma) as a southern Rheic ocean closed during the Variscan orogeny.

PAUL D. RYAN

Further reading

Gayer, R. A. (ed.) (1985) *The tectonic evolution of the Caledonide–Appalachian Orogen.* Vieweg und Sohn, Wiesbaden.

Gee, D. G. and Sturt, B. A. (eds) (1985) *The Caledonide Orogen: Scandinavia and related areas.* John Wiley and Sons, New York.

Williams, H. (ed.) (1995) *Geology of the Appalachian–Caledonian Orogen in Canada and Greenland.* Geological Survey of Canada, Geology of Canada No. 6.

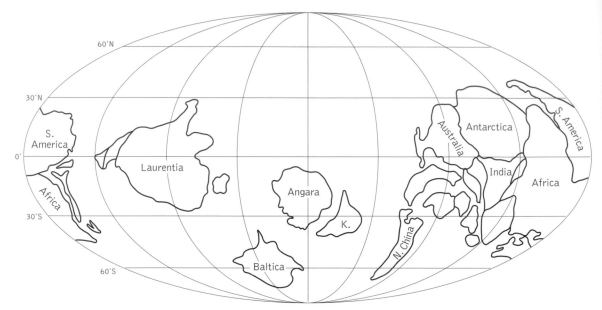

Fig. 1. The world at the end of Early Cambrian time. All the continental blocks are to some extent rotated from their present orientations. South America lies in south polar latitudes. North America (Laurentia) is rotated so that Alaska is its northernmost point. Southern European basement units are grouped with the Gondwana fragments to the south-east. The Kazakhstan craton has yet to take shape. (After Scotese.)

Cambrian The Cambrian System is named after Wales (Cambria) and, being the oldest of the three Early (Lower) Palaeozoic systems, constitutes the first of the Phanerozoic systems of rocks. Much of the fascination that this system holds for geologists is on account of its fossils, which represent the first great successes of invertebrate animals. It was formed over a time-span from about 450 Ma to around 495 Ma, and during this time the major patterns of invertebrate diversity were established.

The first mention of the name Cambrian was in 1835 by Adam Sedgwick and R. I. Murchison, both of whom were pioneer explorers of the older rocks of Wales and the Welsh Borderlands. Sedgwick, who was the first professor of geology at the University of Cambridge, took particular interest in recording the rock succession between Anglesey and the Berwyn Hills of North Wales, and in similar formations in the north of England. He was able to discover the general sequence of these sparsely fossiliferous and rather deformed rocks and referred to them as 'Cambrian'. The base of the system was never clearly defined, but it was generally taken to be where the first abundant shelly fossils occur. As for the top, it too was a matter of some disagreement. The ultimately bitter dispute that arose a few years later between Sedgwick and Murchison as to the boundary between the Sedgwick's Cambrian and Murchison's lowest Silurian was finally resolved by Charles Lapworth in 1879 with the creation of the Ordovician system. Lapworth offered a compromise, but it was a wise and practical one.

Many suggestions have been made about the best level for the base of the system. The inception of the Phanerozoic Eon is seen as marked by a revolution in the organic world with the first great radiation of new marine invertebrate groups to be 'written' into the geological record. Many of these creatures possessed little or no preservable hard tissues, and fossils are thus rare. It has always been difficult to designate the fossils that can widely be found and identified, as the earliest that are essentially and distinctively Cambrian. The base of the system has consequently been hard to define. Since Sedgwick's day and the number of known early Cambrian and late Precambrian fossils has greatly increased, but this has not made the problem much easier to solve. The matter has been studied by working group of the International Commission on Stratigraphy, and a level at or near the base of the Tommotian Stage with its many shelly microfossils is favoured. A stratotype section and point in eastern Newfoundland appear to satisfy the requirements. Other important comparative sections are in China and Siberia. How readily the geological community accepts and uses this choice remains to be seen during the next decade or so.

The Cambrian System is divided into three series. The lower is subdivided into four stages, the middle into two, and the upper into three. As originally defined, these were basically lithostratigraphical divisions, but palaeontological—or rather biostratigraphical—criteria now prevail. Cambrian biostratigraphy has from the outset been largely based upon the distribution of the trilobites (a group of extinct marine arthropods). These fossils, present above the basal Tommotian

Stage, have proved to be useful in inter-regional, and even in some intercontinental, correlations. Trilobite faunas, nevertheless, exhibit some provincialism and cannot be used for global correlation. Other marine groups, such as sponges, archaeocyathids, echinoderms, brachiopods, and some primitive molluscs are known from parts of the Cambrian, while problematic fossils of unknown affinities or uncertain taxonomic status are locally common, especially as microfossils.

An exceptional, though not unique, view of Middle Cambrian marine life is afforded by the Burgess Shale fauna of British Columbia. There, about a hundred genera of mainly soft-bodied creatures have been exceptionally and exquisitely preserved. They indicate a more taxonomically diverse and ecologically versatile assemblage of animals than is apparent elsewhere. The fauna also includes some taxa surviving from the Precambrian Ediacarian stage. The Burgess Shale fossils, and others from China and Greenland, record the major adaptive radiations that were in progress at the time—the 'Cambrian explosion' of the invertebrates as it is often called. There is much debate about the extent to which the Cambrian biota spread throughout the breadth and the depth of the oceans, but there seems to be little doubt that the continental seas were effectively exploited, at least within the tropical and temperate latitudes.

There is no convincing direct evidence of life in Cambrian terrestrial or freshwater environments, but primitive forms such as algae and bacteria were undoubtedly present in soils and fresh waters. They were pioneering the spread of life into the continental parts of the world, and in the following Ordovician period this spread was to gather pace.

Cambrian palaeogeography was characterized by considerable continental movement. Much of the Proterozoic supercontinental mass persisted in the lower latitudes, but the continental blocks of Baltica, Laurentia, Siberia, Kazakhstania, and China were separated from it and from each other by narrow oceans, spread out for the most part within the tropics (Fig. 1). As time progressed the Gondwana supercontinent rotated clockwise and swung into a south polar location.

The effects of this, together with those of climate and depositional environment, were to produce several regionally distinct faunas throughout the period. The Lower Cambrian archaeocyathids, for example, seem to be confined to Siberia and the nearby margin of Gondwana at first, but later spread to include Laurentia and Eastern Gondwana. Trilobite provinciality was manifest throughout the entire period. To what extent it was controlled by climate and geography is debated, but it must have been considerable.

Cambrian sedimentation was profoundly influenced by rising sea-level. Lower Cambrian successions almost everywhere are dominantly clastic and transgressive. Commonly they rest unconformably upon Precambrian rocks and the derivation of new sediment from a thick regolith (superficial unconsolidated rock) gave rise to extensive spreads of quartzites across the shallowly flooded floors of the continental margins. Carbonate precipitation (in the form of limestones and dolomites) had been facilitated in the Precambrian by the relatively higher propor-

tion of atmospheric carbon dioxide. Then, during the Precambrian–Cambrian transition, world atmospheric conditions went from 'ice house' to 'greenhouse'. Such changes were to take place several times later in earth history. Carbonate-secreting organisms spread rapidly across the cratonic margins and other basins, giving rise to a new major phase of carbonate deposition. Plate-tectonic activity with much oceanic volcanism was more conspicuous during Cambrian times, so that calcium carbonate precipitation was again favoured, especially in northern and central Laurentia and Siberia. Evaporites are found associated with the shallow-marine areas of Siberia and northwestern Gondwana of early Cambrian time.

The active continental margins of Laurentia, western Baltica, and parts of eastern and western Gondwana were the sites of deposition of thick turbidite sequences and, later, volcanic rocks. Mountain-building activity within the Cambrian period seems to have been limited, but a major event in Scotland and mild tectonism in Iberia are known to have occurred. The long time during which the continental interiors had been subaerially weathered and lowered by erosion was succeeded by one of progressive flooding and coverage by shallow seas. The shallow-marine realm for the expansion of life's activities was continually expanding, and the opportunity it afforded for new arrivals was not neglected. By the end of the period the tectonic unrest had died out and a quiet passage into the Ordovician ensued.

Late Cambrian time also witnessed a lowering of sea level, and the idea has been put forward that cool, oxygen-poor waters from the deep levels of the oceans spread repeatedly up on to the continental shelves. These chills brought about a series of closely spaced marine extinctions, and the deposition of anoxic black shales. Trilobites, conodonts, and brachiopods suffered most during these crises; the archaeocyathids had suddenly vanished at the end of the Early Cambrian epoch, together with the olenellid group of trilobites.

Cambrian magnetostratigraphy appears to be remarkably simple overall. It begins with a span of mainly reverse polarization to 530 Ma, followed by a normal episode (to about 526 Ma) and then mainly reverse polarization to the end of the period. If geomagnetic activity is the result of changes within the core and mantle, its Cambrian low level seems to accord with the low level of plate-tectonic events; but this is only speculation at the moment.   D. L. DINELEY

Further reading

Brasier, M. D. (1992) Background to the Cambrian explosion. *Journal of the Geological Society of London*, **149**, 585–7. [This is the first of a thematic set of papers on the Precambrian–Cambrian boundary in this part of the *Journal*.]

Cocks, L. R. M. (1978) Cambrian. In McKerrow, W. S. (ed.) *The ecology of fossils*. Duckworth, London.

Holland, C. (ed.) (1974) *Cambrian of the world*. Wiley Interscience, New York.

Whittington, H. B. (1985) *The Burgess Shale*. Yale University Press, New Haven.

Canada, geological exploration *see* GEOLOGICAL EXPLORATION OF THE NORTH AMERICAN WEST

carbon cycles Carbon is one of the fundamental building blocks of the Earth. Most life forms on Earth consist of organic carbon, while inorganic carbon may dominate the visible physical environment. What makes this element particularly interesting is its involvement in a continuous and complex exchange between the atmosphere, ocean, and land as a result of energy flux originating from the Sun. The increase in atmospheric carbon dioxide (CO_2) by 26 per cent since the Industrial Revolution, largely attributed to the burning of fossil fuels, and the links that this increase may have with climate change, have led to an emphasis on understanding the carbon cycle. To quantify this cycle requires an understanding of the functioning of the major components of the planet.

The processes controlling the flux of carbon are, by themselves, fairly simple. Three types may be identified: (1) the major energy-transforming reactions of life—assimilation and dissimilation in photosynthesis and respiration—which cycle some 10^{11} metric tons of carbon each year; (2) simple physical exchange of carbon dioxide; and (3) dissolution and precipitation (deposition) of carbonate compounds, resulting in the formation of sedimentary rocks such as limestone and dolomite. These are essentially balanced processes, although precipitation was dominant during the past. In aquatic systems these processes are two orders of magnitude slower than assimilation and dissimilation. These processes occur within and between three major carbon pools: the ocean, the atmosphere, and the terrestrial system. Figure 1 shows the sizes and fluxes of carbon between the major carbon reservoirs on timescales of up to 10^5 years. Each is now discussed in turn.

The main forms of carbon in the atmosphere are carbon dioxide, methane, and carbon monoxide of which carbon dioxide and methane have long residence times. Climate and the carbon cycle have undergone huge, seemingly contemporaneous changes (Fig. 2). One of the remaining questions in climate research is which system, carbon cycling or climate, has forced the change. Over the past 200 years humans appear to have perturbed the carbon cycle. Since carbon dioxide serves as a greenhouse gas, trapping long-wave radiation emitted by the Earth in the atmosphere, it is possible that temperature increases in the atmosphere over the past decades have occurred because of this increase in atmospheric carbon.

The ocean contains over 60 times more carbon than the atmosphere, making it the largest pool of the planet's mobile carbon. Within this pool, dissolved inorganic carbon is the most common form. Smaller pools, but with higher turnover,

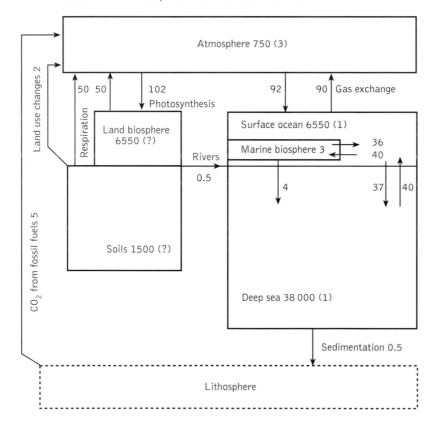

Fig. 1. Carbon pools in gigatonnes of carbon (GtC) and fluxes in GtC yr^{-1}.

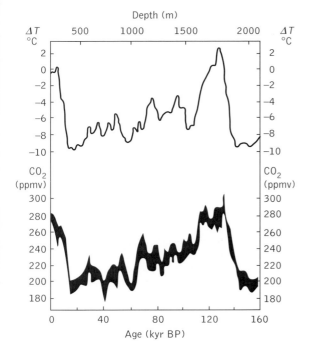

Fig. 2. Variation of temperature and carbon dioxide (CO_2) over the past 160 000 years.

occur in the marine biosphere as dissolved organic carbon. Primary production in the surface layer is some 30–40 per cent of terrestrial vegetation. Processes operative in this large pool may be considered as two pumps: the solubility pump, which transfers CO_2 to the deep ocean, and the biological pump.

The ocean may be divided into three layers: the surface layer (75 m), which is well mixed, overlies the thermocline (to 1 km), which is a stagnant zone stabilized by increasing density and decreasing temperature, the bottom layer is the deep ocean. Cold, saline surface water sinks, spreading out at deeper levels. Such descending waters, most common in polar regions, transport dissolved surface CO_2 to the deep ocean, where it remains trapped for hundreds or thousands of years. The efficiency of this solubility pump depends on CO_2 being dissolved in surface waters; this in turn, depends on the difference in the pressure of CO_2 in the sea water and air. More gas may be dissolved in colder water. Carbon dioxide thus leaves the ocean in the tropics and enters in the polar regions, such as the North Atlantic and Antarctic ocean. Elsewhere, such transport occurs through convergence within the subtropical gyres and vertical diffusion between the thermocline and the subtropical waters. This process, however, is very slow; it takes hundreds or thousands of years for surface waters to penetrate below the mixed layer.

The biological pump occurs through photosynthesis, although most of the atmospheric carbon fixed in this way is respired in the surface ocean within days to months. An estimated 10 per cent precipitates out to be oxidized in the deep ocean. This carbon remains at depth for centuries.

Nitrogen and phosphorous act to limit primary production. Upwelling in the ocean brings nutrient-rich water to the surface, a process which promotes phytoplankton growth. A definite seasonal cycle is attached to this occurrence, the stabilizaton of the water column due to surface warming and reduced turbulence in spring, together with increased sunlight, causes an explosive growth of phytoplankton populations. Wind-driven upwelling regimes, such as that off the west coast of southern Africa, display an interesting ocean–atmosphere coupling which may be changed if the climate is perturbed as a result of anthropogenic changes in the carbon cycle.

Most of the Earth's carbon resides in sedimentary rocks, such as limestone, but since this is largely a captive source, we shall concentrate on three sub-pools: living biomass, litter, and soil carbon. A distinction between herbaceous and woody plants, owing to differences in turnover rates, is often made within the living biomass component. Boreal, temperate, and tropical forests are regarded as the most important carbon reservoirs in this pool. Recycling occurs through photosynthesis, respiration, and decomposition.

Living vegetation contains roughly the same amount of carbon as is stored in the atmosphere, while dead biomass on land consists of twice as much carbon as the atmosphere. The carbon absorbed by photosynthesis of land plants amounts annually to 100 gigatonnes of carbon (GtC) (gross primary productivity). About half this amount is returned to the atmosphere by the autotrophic respiration of plants. The remainder, termed 'net primary production', amounting to 60 GtC, is transformed into organic carbon and built into plant tissue. Some of the dead plant carbon is transformed to soil organic carbon, which is oxidized very slowly.

One of the most important current research questions centres on plant response to changes in atmospheric carbon, and is part of the so-called indirect perturbation problem. The *fertilization effect*, where plants grow faster under enhanced atmospheric CO_2, provided they are not limited by nutrients or water, is thought to be the dominant result. Under water stress and enhanced CO_2, stomata remain open for shorter periods, leading to reduced water loss owing to evapotranspiration. It is not clear whether increased CO_2 in the atmosphere results in increased terrestrial carbon storage or simply a faster turnover rate. Indeed, some studies suggest that plants acclimatize to the increased CO_2. The terrestrial biosphere forms an important component of flux under climate change. Increased aridity, for example, may lead to forest dieback, increased incidence of fire, and a change in the total carbon stored on the landmass of the planet.

Models of the global carbon cycle have been developed to help quantify the carbon cycle and to assess its sensitivity to change. In most models, pools of carbon are represented as boxes (as in Fig. 1) and the fluxes between them are represented as simple first-order flux dynamics. Tracers with dynamics similar to carbon have been used to calibrate the carbon exchanges. The most important tracers are the stable carbon isotope ^{13}C, and the radioactive isotope ^{14}C. Changes in the ratio of $^{14}C/^{13}C$ before and after nuclear bomb tests provide one means of tracing the dynamics of the cycle. Tree rings indicate past changes in this ratio, a decrease in ratio up to 1950 resulting from the flux of carbon from fossil fuels (radiocarbon-free CO_2) into the atmosphere.

The calibrated models range from box diffusion models developed in the 1970s to high-resolution numerical models based on the general circulation models of the atmosphere and ocean, which treat the planet as a three-dimensional structure. Flux rates depend on local properties such as ecosystem structure, soil composition, land use, agricultural practices, or, in the ocean, on the rate of upwelling or deep-water production.

Perhaps the most intriguing element of the carbon cycle concerns the 'missing sink', whereby 1.5 GtC per year during the decade of the 1980s is unaccounted for in the total budget. This missing sink, which is thought to be a northern hemisphere effect, is probably related to fertilization. Other effects such as eutrophication, may be important. Part of this problem is the question of how much of the flux of CO_2 into the atmosphere has remained in the atmosphere? The observational record from Mauna Lao since the International Geophysical Year in 1957/8 shows an increase, but what is lost in this record is the increased flux sequestered by the ocean and the terrestrial biospheric carbon pools. Changes in atmospheric carbon with time depend on the amount released from fossil fuel burning, the amount in or out of the ocean, and the net biomass term due to land use changes, biomass burning, etc. Changes in atmospheric CO_2 over time are known. Fossil fuel burning can be estimated by energy consumption figures, but terrestrial and ocean fluxes are more difficult to estimate. Four approaches exist to estimate their value: (1) carbon modelling; (2) direct measurement of ocean–atmosphere fluxes and extrapolation; (3) direct measurement of surface– atmosphere fluxes extrapolation; and (4) tracer studies. The most accurate method is thought to be modelling, with the ocean being the better quantified component. R. WASHINGTON

Further reading

Schlesinger, W. H. (1991) *Biogeochemistry: an analysis of global change*. Academic Press, London.

carbonate platforms, reefs, and carbonate mounds
Carbonate platforms, reefs, and mounds form in relatively shallow-water areas of the sea where there is extensive deposition of carbonate undiluted by siliciclastic (sandy and muddy) sediments. There may be various reasons for the absence of siliciclastic sediments: the area of deposition may be remote from possible sources of such sediments; the source areas may be of low relief; or intervening deep water may isolate the source area from the area of deposition.

Some carbonate platforms have persisted over long periods of geological time and have provided sites for the accumulation of thick and extensive sequences of limestones.

Today such platforms are relatively narrow (e.g. the Florida, Belize, and Yucatán shelves), but in the past, during periods of high eustatic (worldwide sea level, huge areas of the continents were covered by shallow epeiric shelf seas. The slowly subsiding passive continental margins were ideal sites for the accumulation of thick carbonate sequences. Also, fragmentation of continental crust, resulting from tectonic movements (e.g. in Triassic–Jurassic times in the Mediterranean region) produced isolated fault-bounded blocks remote from sources of siliciclastic detritus. Other small platforms developed on volcanic seamounts.

Changes in sea-level, caused either by eustatic or tectonic changes, have resulted in the development of long-persisting platforms showing complicated cyclical changes. These were at times exposed, with the development of palaeosols, the introduction of siliciclastic detritus, or the development of shallow-water evaporites. They have been explained by repeated normal sedimentary progradation on a subsiding platform, eustatic changes, tectonic changes, or combinations of all three. However, the widespread extent and correlatable pattern of such cycles suggests that long-term climatic controls or tectonic process related to large-scale plate tectonic movements are more likely causes.

Carbonate platforms sometimes aggrade on slowly sinking continental crust and extend seawards by shedding carbonate sediment into flanking deep-water areas. In contrast, platforms sometimes sink so that shallow-water carbonate sedimentation

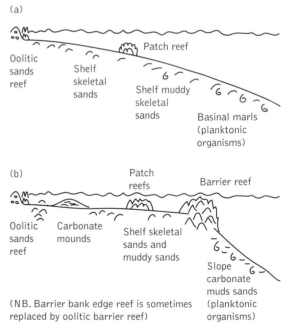

(NB. Barrier bank edge reef is sometimes replaced by oolitic barrier reef)

Fig. 1. Carbonate platforms. Cross-sections of a typical ramp (a) and rimmed shelf (b).

ceases and becomes cloaked with carbonate sediments composed solely of planktonic or nektonic remains.

Carbonate platforms have been broadly classified into *carbonate ramps* and *rimmed shelves*. A carbonate ramp (Fig. 1a) is a gently sloping surface extending seawards from the coastline towards deeper water. There is no marked break in slope between shallow and deep water (e.g. the Southern Arabian Gulf). Carbonate sands, often oolitic, accumulate along the inner part of the ramp adjacent to the continent, and may be accompanied by fringing reefs. Other isolated reefs develop over the shallower part of the ramp on tectonic highs. Shallow-water skeletal sands and muddy sands cover the ramp and gradually pass seawards into finer carbonate mud, commonly containing an increasing proportion of siliciclastic detritus to form marl.

Rimmed shelves are carbonate platforms with a relatively sharp break at their transition into deep water (Fig. 1b). They are subject to high-energy wave and current activity along their margins. Oolitic sand accumulation forms barrier islands along the rim (as in the Bahamas). Elsewhere, coral and other reef-building organisms have built extensive reefs along the rim (the Great Barrier Reef is an example) capped with small islands, composed of reef debris. The shelves may contain isolated reefs and are covered with carbonate sands, muddy sands, and carbonate muds.

Reefs are biologically constructed bodies which form positive upstanding features on the contemporary sea floor. They may form linear features along the margins of rimmed shelves or may occur as small features on the platforms of rimmed shelves and ramps, as fringing reefs along mainland shelves of carbonate platforms around volcanic seamounts, or sometimes as isolated features in deep water. They form massive features in the geological record which contrast markedly with adjacent bedded sediments.

Reefs consist of a framework of carbonate-secreting organisms. Today corals are dominant. In the past, stromatolites, calcareous algae, calcareous sponges (stromatoporoids), and rudist bivalves have been important reef builders. The framework is inhabited by a wide variety of organisms (bivalves and echinoderms for example), which attack the framework and but also add to the skeletal accumulation. The skeletal debris produced by physical and biological destruction (fills in the skeletal framework, and the whole mass becomes encrusted by various skeletal organisms; it is also cemented to form a rigid wave-resistant structure. Waves and currents attack the edges of the reef and sediment is piled up on the reef crest to form small islands as well as being deposited either bankwards or seawards into deeper water.

Modern reefs are composed of what are known as hermatypic corals, which are restricted to waters with temperatures of 10–36 °C and salinities of 22–40 ‰ (per mil). They contain symbiotic micro-organisms—*zooxanthellae*—which can survive only in the photic zone (the zone that is reached by sunlight); reefs thus extend only to depths of 80–100 m. The diversity of the corals and their development are restricted by availability of light, and hence are affected by the presence of suspended sediment. Whether similar restrictions affected reef-building organisms in the past is unknown because of the lack of modern counterparts.

Related to reefs are so-called carbonate mounds. These are positive features on the contemporary sea floor produced by biological processes. They form massive structures that contrast with adjacent bedded deposits when found in the geological record. Unlike reefs, they are not rigid structures but have formed by the trapping and binding action of algae, grasses, and other organisms. In some ancient examples the exact nature of the binding organism is problematic. They mounds are normally formed on carbonate platforms away from their exposed edges, but they occasionally form in deep water.

The abundance of reefs and mounds and their fauna have varied throughout geological history. Stromatolitic mounds are known in the late Archaean and became widespread in the Proterozoic when they formed barriers and patch reefs; some possibly contained calcareous algae. Throughout the Early Palaeozoic they contained increasing amounts of skeletal organisms such as sponges, algae, and primitive corals. Palaeozoic reefs were best developed in Ordovician–Devonian times, when they were formed by stromatoporoids (calcareous sponges). Mounds with abundant bryozoa, calcareous algae, and sponges were characteristics of the Carboniferous–Middle Triassic.

Reefs like those of modern seas developed in the Late Triassic. They were dominated by scleractinian corals, calcareous algae, and calcareous sponges. The reefs of the Late Jurassic were composed of sponges and algae. During the Cretaceous, rudist bivalves became important mound-formers, and in the Late Cretaceous rudist coral reefs developed. Since the beginning of the Cenozoic, reefs have become dominated by scleractinian corals and calcareous algae.   G. EVANS

Further reading

Tucker, M. E. and Wright, P. V. (1990) *Carbonate sedimentology*. Blackwell Scientific Publications, Oxford.

carbonatites Volcanism is dominated by the eruption of silicate melts, as might be expected from the abundance of silicon in the Earth's mantle and crust (which contain about 66 per cent of silicon by weight). Carbon dioxide (CO_2) is one of the important gases emitted by volcanoes, but the amount of CO_2 is tiny compared with the melts and ashes, and molten carbonate might not therefore be expected to occur. Carbonate eruptions are, however, the only important exception to silicate magmatism. This unexpected composition, and the nature of the activity, although minute in amount, afford special insights into Earth processes.

Igneous rocks containing over 50 per cent of carbonate are defined as carbonatites. More than 350 carbonatite intrusions are known at present; most of them consist of calcium carbonate. Less common are magnesio-carbonatites (with calcium and magnesium) and ferro-carbonatites (with calcium, magnesium, and iron). All three types may occur in the same complexes, most of which are sub-volcanic intrusions. Surface eruptions of ashes and lavas (rare) are also preserved; the list of discoveries is growing rapidly with new research interest. In northern Tanzania, a volcano built largely of alkali silicates is currently erupting sodium, calcium, and potassium carbonates as lava and ash.

Many carbonatites form parts of alkali-rich silicate magmatic complexes. The first discoveries of carbonatites were in such complexes, and this led to a fierce controversy in the early half of the twentieth century as to whether or not carbonatites were merely sedimentary limestones that had been mobilized by the heat from silicate magmas moving through the Earth's crust. It is now clear from their chemistry, however, that the ultimate source of carbonatites must lie below the crust, in the Earth's mantle. Most of the known carbonatite bodies are small, with areas in the region of 1 km^2; none has yet been recorded that is larger than 20 km^2. In age they range from mid-Precambrian (2000 Ma) to the present, but most are younger than 150 million years (150 Ma). Nearly all are located in the continental plates, very commonly in the rift zones of the stable interiors; about half the recorded occurrences are in Africa.

An outstanding feature of carbonatites is that they contain relatively large amounts of elements that are otherwise rare in the outer layers of the Earth. Elements such as phosphorus, normally found in trace amounts, reach percentage levels. This exotic aspect of carbonatite chemistry (as well as a similar global distribution) is shared by kimberlites, and both are widely regarded as small-volume partial melt extracts from the mantle—a process in which uncommon elements would be concentrated. Kimberlites also contain carbonate, and with increasing amounts of carbonates grade into carbonatite. A feature of carbonatite intrusions is that the surrounding rocks show intense chemical alteration, marked by new sodium and potassium minerals replacing the primary ones. This distinctive alkali metasomatism is known as *fenitization*. (It was first described from the intrusion at Fen, southern Norway.)

A further aspect is gradually emerging as more data reveal patterns in the ages of carbonatite eruption. Not only is the activity repeated in the same areas, over periods extending back into the Precambrian, but repeated episodes take place at the same time at widely separated locations. These episodes, occurring within plates, coincide with plate-collision events elsewhere and with other major igneous activity worldwide. To explain carbonatite activity may thus call for radical re-thinking about mantle processes and geodynamics.

D. K. BAILEY

Carboniferous
The Carboniferous system of rocks takes pride of place as the first stratigraphical system to be established formally. Early in the nineteenth century, long before W. D. Conybeare and W. Phillips coined the term 'Carboniferous' (1822), the Coal Measures, the Millstone Grit, the Mountain Limestone and the Old Red Sandstone had been recognized and shown as important formations on the several pioneer geological maps of Britain. The word *Carboniferous* seemed appropriate for rocks that occurred commonly together and several of which contained bands of coal (Latin *carbo*, coal). Later, the Old Red Sandstone was dropped from the quartet, but the others were perceived as related by their fossil content. Today the Carboniferous period is held to have begun around 355 Ma and to have lasted for 165 million years. It was a time of great advances in the history of life and one in which geographical change was immense. Perhaps we should not be surprised that so much changed: 165 million years is no small stretch of time, even of geological time.

Interest in this system of rocks has always been intense on account of the great reserves of hydrocarbons it contains. These fossil fuels lay at the core of the exploitation of natural resources that arose with the industrial revolution. Coal was a prodigious source of energy and materials for the production of iron and steel and for much of the nineteenth-century chemical industry. In the second half of the twentieth century, oil and gas from Carboniferous sources have largely replaced coal. Fossil fuels are formed, impounded, and located very largely on the basis of regional stratigraphy and where this is extensive, as in north-western Europe and eastern North American, great natural wealth results. It is no wonder, then, that the Carboniferous system has been the subject of a great deal of attention as well as academic interest.

The Carboniferous is unusual in that it is today divided into two Subsystems (Upper and Lower) and five series. A previous division of the system into a lower part, the Dinantian, and an upper, the Silesian, was based upon the succession in western Europe. These two divisions have not been retained in the global stratigraphic column, where the Upper and Lower Subsystems are not exactly their equivalents. Alternatives, though not exact equivalents, to the upper three series exist in the former USSR. Twenty-five stages have been named in western Europe; rather more have been erected in what is now Russia. Regional differences are also embodied in the American divisions of the system. Despite the general international usage of the European subdivisions, the American are almost exclusively used in North American literature.

Mississippian was the term first applied in 1870 by Alexander Winchell (1824–1891) to the limestone lying beneath the coal-bearing strata in the Mississippi valley. In 1891 Henry Shaler Williams (1847–1918) established the correlation of these strata with the Lower Carboniferous of Europe. For the overlying coal-bearing series he proposed the name *Pennsylvanian*, after the State in which they had become best known. Because there is a widespread unconformity between these two formations, they were for some time given the status of geological systems. It was not until 1953 that the United States Geological Survey adopted these terms, and it has since been shown that the Pennsylvanian boundary is younger than the Lower–Upper Carboniferous boundary in Europe. The International Commission on Stratigraphy accepts that the Carboniferous System is inconveniently large and proposes dividing it into two subsystems. The exact stratigraphical level

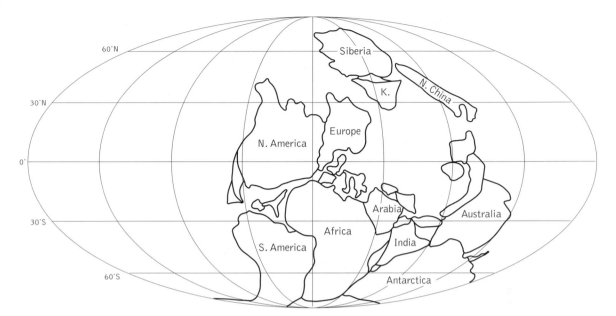

Fig. 1. The world in Early Late (Namurian) Carboniferous time, about 330 Ma ago. The continents are gathered into a large southern mass, Gondwanaland, consisting of South America, Africa, Arabia, India, Australia, and South-East Asia with Antarctica. To the north lies the North America–Europe block, Laurussia; Siberia, Kazakhstania (K), and North China lie to the north-east. Gondwanaland is by far the biggest land mass, stretching from the equator to the South Pole, and in the southern latitudes it experiences repeated glaciation.

at which a boundary is to be drawn between the two is not yet decided, but it might be close to the traditional level of the Mississippian–Pennsylvanian break.

Correlation within the Carboniferous has been developed to a relatively high level of precision on the basis of biozones using conodonts, brachiopods, corals, foraminifera, cephalopods, and brachiopods in the marine facies and macroplants and palynomorphs (spores and pollen) and non-marine bivalves in the continental facies. There are many horizons that can be dated on the basis of isotopic analysis. The base of the system has now been defined in the pelagic (open water) marine facies at the excellent section at La Serre in southern France. The base of the bed in which the conodont *Siphonodella sulcata* first appears (bed 89) and which is within the lineage from *Siphonodella praesulcata* to *Siphonodella sulcata* is the point at which the boundary is drawn. The pelagic facies is of world-wide distribution, and auxiliary stratotypes have been selected at Nanbiancun in South China and Hasselbachtal in Germany. This level is as near to the base of the classic cephalopod zone of *Gattendorfia* as can be found.

Maps of the Carboniferous world show us the gradual drawing together of three great continental masses and associated smaller islands, continuing the process which had got under way during the preceding Devonian period (Fig. 1). *Laurussia* is the name given to the continental grouping of North America, Greenland, and southern and western Europe. Associated with these ancient stable cratons were new

sedimentary basins at their margins (and also adjacent to the old Caldonian orogenic belt). Across the middle of Laurussia ran the Equator, while its northern margin lay just beyond 40° N. Not far off to the north-east lay the craton of Siberia, known as Angaraland, almost half as big as Laurussia. To the south of these continents was Gondwanaland, an agglomeration of today's southern continents and India. It stretched from close to Central America to its north-eastern Australian extremity in the southern tropics and into the south polar latitudes. Plate tectonic movements now combined to suture Laurussia and western Gondwanaland together. Orogenic activity was concentrated along the southern margin of Laurussia where the small land masses were driving against the continental foundation and deforming the sedimentary basins as they progressed. Seas closed, uplands emerged, volcanic activity took place, and the entire continental margin grew by the addition of these new terranes.

On the western margin of Laurussia and the Andean sector of Gondwanaland, similar crustal activity was adding new strips of upland to the continent. Elsewhere around Gondwanaland comparable accretionary processes were enlarging the continental margins. Granitic batholiths were slowly emplaced in many of these areas. By the end of the Carboniferous period the effect of all this plate-tectonic activity was twofold: to draw all the continental fragments into the supercontinent Pangaea, and to increase the volume of the ocean basins.

Almost everywhere the upper part of the Carboniferous

system shows cyclic sedimentation, and finally a pronounced regression of the sea accompanied by the spread of continental sedimentation. In a great swathe of country from the western USA to eastern Canada, the British Isles, and across Europe into the Donetz basin there were repeated sudden episodes of rising sea level followed by longer phases of deltaic and continental deposition. The culmination of each cycle was the establishment of a tropical forest, on the floor of which accumulated the plant debris that was to become almost a quarter of the world's total coal resources. In the southern parts of the USA, coal-bearing cycles of sedimentation gave rise to widespread but geographically variable units, known as *cyclothems*.

Long-continued, and sometimes fierce, argument has concerned the underlying cause of this repetitive process. Repeated transgression by the sea was followed by sand and clay build-up and the establishment of a widespread and dense vegetation cover. Such vigorous forest growth required a hot, humid climate. This seems to have been experienced throughout much of the tropics and to the north in North China, Kazkhstania, and Siberia (parts of Angaraland). Thus climate, local tectonics, and the global rise and fall of sea level have all been considered as prime causes of cyclothems. All three must have had an influence, but the view today is that sea-level changes in response to continental ice growth and decay in southern Gondwanaland was the prime mover.

Carboniferous plants in Europe show little trace of strong seasonal growth. This implies that they grew under constantly warm rather than seasonal conditions. In Angaraland the plants do have well-developed seasonal rings, showing that the winter months were cold. Coals occurring in Gondwanaland also show the effect of a seasonal climate, cool rather than warm. Elsewhere in the tropics, both north and south of the equator, evaporites were being precipitated in shallow arms and coastal reaches of the sea. Carboniferous evaporites occur today in the western USA, the Arctic, adjacent to the Ural mountains, and in central Asia, as well as in parts of Brazil and North Africa. Here, aridity was the norm, and the landscapes around the salt basins were probably clothed only in very sparse vegetation, if any.

The ultimate in cooler global conditions was attained in the far south. Glaciation is recorded by widespread boulder clays or tillites in Gondwanaland south of latitude 60°, which persisted from the Early Carboniferous into the Early Permian. The ice sheets grew to enormous sizes in South America, South Africa, India, and Antarctica, in places almost reaching the middle latitudes. They had a history of repeated growth and decline, and their effect upon sea level was to keep it constantly moving up or down as the ice volume changed. From time to time, and as the continents moved, the centre of ice accumulation shifted, but the maximum spread of the ice seems to have been late in the Carboniferous.

Life in the seas and oceans was prolific in Carboniferous time, especially in the shallow waters of the tropics. Bottom-dwelling faunas were dominated by brachiopods, corals, and echinoderms. The productid brachiopods included some of the largest

forms ever to exist, while many other kinds persisting from the Devonian were also very successful. The crinoids were locally sufficiently abundant to produce limestone deposits extending over thousands of square kilometres. Rugose and tabulate corals gave rise to reef structures in many regions, usually with the help of stromatoporoids and calcareous algae, which were also reef-builders in their own right. Many reefs were of great size and exercised some control over local sedimentation. Bryozoa, bivalves, and gastropods, together with a small number of trilobites, added to benthonic populations. The fusulinids were among the foremost of the foraminifera and were sufficiently abundant in late Carboniferous time to produce extensive sheets of foraminiferal limestone in North America. Among the plankton and nekton, the cephalopods were conspicuous predators. The goniatite ammonoids evolved rapidly and are of great biostratigraphic importance; the nautiloids, however, were in decline. Fish included many elasmobranchs as well as osteichthyes such as palaeoniscoids, dipnoi, selachians, and crossopterygians, together with the ancient acanthodians. Many of these groups produced animals of a relatively large size, but the giant placoderm fish that had been present in late Devonian seas were now extinct.

On land the vascular plants produced diverse and luxuriant floras with lycophytes, sphenophytes, ferns, seed ferns *Cordaites*, and, eventually, the first of the conifers. They were present virtually throughout the full range of climatic zones. Three botanical provinces have been recognized: in Laurussia, China, and Gondwanaland. The tropical Laurussian coastal coal forests were dominated by the lepidodendrales and the sphenophytes. Elsewhere a more xerophytic vegetation was present across the land. The forests afforded habitats for many aquatic invertebrates, fish, and amphibia. Terrestrial invertebrates were mostly arthropods, but land and fresh-water gastropods also existed. The insects rapidly evolved many very large flying species; many of them were clearly carnivores, but their mode of life remains uncertain. The base of the ecological pyramid, however, was occupied by the multifarious plants, and it was highly productive.

The vertebrates, too, were commonly very large, adapting to new habitats by the evolution of new forms. Early Carboniferous tetrapods are known only from the coastal land of Laurussia from Iowa to Germany. Most are amphibians, but the earliest reptiles are recorded in Visean rocks. Late Carboniferous tetrapods have been discovered over a wider area, and a great increase in reptile diversity occurs. Not being tied by the necessity of a watery habitat for reproduction, the reptiles were able to enter the drier environments and to become much more active and mobile. The terrestrial habitats were undergoing wide geographical expansion at this time in both tropical and temperate latitudes.

Towards the end of the Carboniferous period the growing proximity of the continents to one another brought about marked changes in climate and crustal deformation of the now impacting continents. The coal forest regimes came to an end with the uplift of the land surface and with growing aridity. As the early students of the Palaeozoic rocks in Britain had perceived, the Carboniferous system is sandwiched between two systems with very prominent and extensive continental facies, the Devonian Old Red Sandstone and the Permian New Red Sandstone, and it constitutes in time a cycle of geological events and processes.   D. L. DINELEY

Further reading

Dias, C. M., Granados, L. F., Wager, R. F., and Winkler-Prins, C. F. (eds) (1983, 1985) *The Carboniferous of the world*, Vols I and II. IUGS Publications 16 and 20. Instituto Geológico y Minero de España and Empresa Nacional Adaro de Investigaciones Mineras, S. A., Madrid.

Ramsbottom, W. H. C. (1978) Carboniferous. In McKerrow, W. S. (ed.) *The ecology of fossils*, pp. 146–83. Duckworth, London.

catastrophic geomorphology

Processes that fashion the landscape act continually and gradually over thousands of years. On rare occasions, they act briefly with great force and suddenness. These rare events occur in small, medium, and large landscapes. Landslides are sudden and relatively rare events in the development of a single hillside, but they are common events in long-term regional landscape evolution. Catastrophic geomorphology is chiefly about catastrophic events that affect regional (medium and large) landscapes.

Most regional geomorphological catastrophes involve the sudden release of large volumes of water. Releasing mechanisms can be terrestrial and extraterrestrial in origin. Massive undersea landslides and the breakage of ice or sediment dams impounding large lakes are terrestrial processes that lead to release of water. Bombardment by asteroids and comets is an extraterrestrial process that may release huge quantities of water.

Undersea landslides of Lanai, Hawaii, may have caused a series of enormous waves to have swept over the Hawaiian Islands and along the Australian coast some 100 000 years ago. Evidence for the passage of this wave sequence comes from gravel deposits, probably produced by wave action, that blanket coastal slopes on the island of Lanai (*see tsunamis*). Other evidence comes from the eastern seaboard of Australia, where coastal sand barriers in southern New South Wales were almost utterly destroyed and coastal abrasion ramps were eroded by wave action to at least 15 metres above present sea level. Earthquakes along the outer coast of Washington State appear to have created tsunamis that, at least six times in the past 7000 years, buried well-vegetated lowlands under sheets of sand.

Several cases of catastrophic flooding after dam breakage are known. Examples from North America include the Spokane Flood, which occurred between 13 000 and 18 000 years ago; the Lake Bonneville Flood, which occurred about 15 000 years ago; and the catastrophic drainage of glacial Lake Agassiz through a north-western outlet after the breaching of a drainage divide about 9900 years ago.

'Superwaves' created by asteroids or comets crashing into the ocean may have caused 'superfloods' in lowland landscapes around continental margins. Good sedimentary evidence now exists for the production of superwaves by such impact events. Some 65 million years ago, an object with a diameter of about 10 km appears to have struck the Yucatán Peninsula, Mexico. The impact seems to have triggered massive earthquakes and the collapse of soft sediments down nearby continental slopes. In turn, the earthquakes and submarine landslides appear to have generated massive tsunamis that scoured sediments from the sea floor and rushed over surrounding lowlands, including the Gulf of Mexico, depositing a jumble of fine and coarse sediments.   R. J. HUGGETT

Further reading

Atwater, B. F. (1987) Evidence for great Holocene earthquakes along the outer coast of Washington State. *Science*, **236**, 942–36.

Huggett, R. J. (1989) *Cataclysms and Earth history: the development of diluvialism*. Clarendon Press, Oxford.

Huggett, R. J. (1997) *Environmental change: the evolving ecosphere*. Routledge, London.

Swinburne, N. (1993) It came from outer space. *New Scientist*, **137** (1861): 28–32.

catastrophism and uniformitarianism

The terms *catastrophism* and *uniformitarianism* refer to concepts and ensuing arguments that preoccupied Earth scientists about two hundred years ago, and which to some extent recur from time to time as our perception of the nature of the geological record changes. Until the early nineteenth century the origin of the crystalline rocks that we know as volcanic or igneous was regarded by some European scholars as being from precipitation in a primeval ocean. The concept has been called 'neptunism', but as new knowledge about volcanic activity and the immense heat within the Earth became generally available the Neptunists had to retreat before the criticisms of the Plutonists with their new (modern) views on the origin of igneous materials. This may have cleared away some of the obstacles to understanding the nature of the crust of the Earth and the processes that have produced it, but geological argument remained fierce and widespread. The new dramatic aspect of Earth history inferred from igneous and structural features, and especially from unconformities, required radically novel explanation. It was soon forthcoming, but new arguments were to break out, as described below.

Catastrophism was the early nineteenth-century line of reasoning that ascribed major changes in the physical environment to sudden, violent, and short-lived episodes or *events*. It was advocated most notably by the French baron Georges Cuvier (1769–1832), the foremost anatomist of the day, who sought to explain the unconformities separating rocks containing very different kinds of fossils. He had, in fact, little field experience outside the Paris basin, but worked with such geologists as A. Brongniart and J. B. Elie de Beaumont who had demonstrated the marine–freshwater cycles in the Tertiary (Cenozoic) rocks there. Cuvier also believed that the entire history of the Earth could be encompassed in

75 000 years, and he rejected his contemporary James Hutton's view that gradual and constant action by geomorphic processes was sufficient to account for great environmental changes. Cuvier's fellow-countryman the Chevalier de Lamarck (1744–1829), a fellow-student of the living world, thought that gradual changes in the physical environment were the cause of changes in animals and plants, with, in time, a progressive 'improvement' of living things. That idea, too, was firmly rejected by Cuvier. Elie de Beaumont went on to develop the idea that catastrophic, sudden, but infrequent upheavals of mountain ranges were responsible for the environmental changes and the destruction of much of the contemporary biota. This catastrophism was, on the evidence available, an attractive idea, but there was no discussion of the source of the energy to bring these upheavals about, nor any explanation of why and where they occurred as they did.

Here we turn to Hutton's great contribution to geology. Uniformitarianism has a place in every textbook of Earth science, for it has been regarded as the closest thing to a fundamental 'law' that geologists have within their discipline. It has also been referred to as the principal of *uniformity*, and as perhaps one of geology's major contributions to science. Students of Earth science in Hutton's day saw the need to explain even the simplest geological phenomena in the light of what they knew of contemporary natural processes, and supernatural or divine causes were discounted. This was eventually developed by the Scot (and Plutonist) James Hutton as a critical tenet in his *Theory of the Earth*, published in 1795. From then on it has become part of the lore of geology, for rather than being adopted as a law, it has been interpreted in so many different ways as to be in disrepute. It has been expressed variously in four common, but not necessarily equivalent, forms:

(1) As another famous Scot, Sir Archibald Geikie, put it, the present is the key to the past.
(2) Former changes of the Earth's surface may be explained by reference to processes now in operation.
(3) Earth history may be deciphered in terms of present observations on the assumption that the laws of physics do not change with time.
(4) Further to (3) above, events and processes in the geological past have proceeded at the same rate and in the same manner as they do today.

Since Hutton's day, and even more since the publication of Charles Lyell's *Principles of geology, being an attempt to explain the former changes of the Earth's surface by reference to causes now in operation* (1830–3), these postulations have been debated and denied. Hutton himself denied that any sort of supernatural agency has been involved in Earth history or in the functioning of the Earth as a machine or system. He seems, however, to have espoused the view that we should seek to explain the past in terms of causes and processes now existing or known today. It seems an acceptable and simple way of going about Earth history, and Lyell, one of the most influential geologists of all time, tended to agree. The principle has been termed 'actualism'.

Geological changes brought about by the continuing activity of observable processes require very long periods of time. Hutton's recognition of geological time as being orders of magnitude longer than had previously been considered was one of his most perceptive achievements. Nevertheless, the principle of uniformitarianism that he propounded has now been rejected on a number of grounds. Two considerations lead most obviously to this rejection. The first is that many factors have been changing gradually with time; as examples, heat flow from the Earth has been diminishing since the planet first formed; the composition of the atmosphere has changed several times; and in pre-Silurian times there was virtually no terrestrial life. We know that both atmosphere and terrestrial life are important in bringing about environmental or geomorphic changes. The second objection stresses that the present provides few or only vague indications of the infrequent, but sudden, major changes and events that must have taken place in the past. More than one geologist has emphasized that big geological events are rare, small ones are common. It has even been suggested that we need the concept of 'neocatastrophism' to cover the recognition of rare violent events (such as giant meteor impacts) taking place in addition to the continuing activity of 'normal processes'. Derek Ager likened the incidence of geological catastrophes to the life of a soldier: long periods of boredom and short periods of terror. The American biologist S. Jay Gould (1977) preferred to use the term 'punctuationalism' for long periods of tranquility interrupted by brief moments of profound change.

Table 1. Major extinction events

Extinction episode	Major animal groups affected	Percentage of extinct families
Late Cretaceous	Ammonites, belemnites, rudist bivalves, corals, echinoids, bryozoans, sponges, planktonic foraminifera, dinosaurs, marine reptiles, pterosaurs	26
Late Triassic	Ammonites, brachiopods, conodonts, reptiles, fish	35
Late Permian	Ammonites, rugose corals, trilobites, blastoids, inadunate, flexibiliate and camerate crinoids, productid brachiopods, fusulinid foraminifera, bryozoans, reptiles	50
Late Devonian	Corals, stromatoporodis, trilobites, ammonoids, bryozoans, brachiopods, fish	30
Late Ordovician	Trilobites, brachiopods, crinoids, echinoids	
Late Cambrian	Trilobites, sponges, gastropods	52

(After N. D. Newell)

A summary review of the catastrophes or crises that have taken place during the history of life on Earth reveals that no single repetitive process can be shown to be solely responsible for sudden drastic reduction of animal and plant numbers and diversity. In total, these events number about eighteen, but rather less severe crises are much more numerous—and are that much more difficult to detect in the geological record. It was the Late Permian and late Cretaceous crises that first impressed the early geologists, but those listed in Table 1 have attracted the most attention so far. Smaller events, when smaller particular groups were wiped out, appear now to have been more numerous, but the causes of these extinctions remain as uncertain as those of the bigger mass mortalities.

Geologists have been able to witness and record very few catastrophes on anything more than the local scale: the scenarios put forward to account for sudden great changes in the past are based upon little or no experience of the processes involved. So to explain these events it is necessary to examine the entire geological record. And there is another possible dimension to consider: from a careful study of the timing of major extinctions and others, less obvious, some palaeo-biologists have claimed that it is not random but shows a periodicity of about 30 million years (Ma).

The more widely discussed and more plausible causes of these catastrophes include the following:

extraterrestrial: bolide impact, cosmic ray flare; volcanic activity; geomagnetic reversal; changes of sea level, salinity changes, anoxic events in the oceans; climatic changes: glacial episodes, increasing aridity; disturbance of the ecological balance; and disease or geochemical poisoning.

Each of these has, in theory, been shown to be a possible agent for disaster, but to what extent do they lend themselves, singly or in combination, to a periodicity scenario? What phenomena that might be the ultimate cause(s) are both periodic and global in their influence?

A sharp fall in sea-level seems to be the only likely cause to occur at or shortly before most mass extinctions, and like many of the other possibilities, it may be linked to sudden violent or widespread volcanic activity, changes in the volume of the ocean basins, or to continental collisions and resulting Earth movements (orogeny). In the eyes of many writers, all these stem from the behaviour of the mantle deep below the crust, for it is there that plate motions and volcanic activity originate. Moreover, it has been argued that there is a periodicity of about 30 million years in the frequency of reversals of the Earth's magnetic field, which itself is related to dynamics within the core and mantle. To the mantle, too, may be traced the causes of exceptional phases of volcanicity, and it seems likely that this holds for the entire Phanerozoic Eon or longer.

Early Earth history differs significantly from that of more geologically recent times; some aspects of, say, the Archaean cannot be explained entirely by the processes known from the Phanerozoic record. The changing nature of the interior of the planet during the 4500 or so million years of its existence remains obscure, but it was presumably unaffected by events elsewhere in the Solar System.

Nevertheless, we are becoming more aware of the influence of cyclic changes in the behaviour of the Earth in its motion around the sun (Milankovich cycles), and the idea that other astronomical cycles may have influenced the history of life on Earth has been widely discussed. Some palaeobiologists have claimed to see a regularity rather than a randomness in the timing of extinctions precisely on account of such influences, and even on account of bodies beyond the Solar System. They invoke phantom stars, as yet unobserved planets, and passage of our Solar System through the spiral arms of the Galaxy. Much is hypothesis rather than fact and some of these propositions seem to verge upon the fanciful. Most palaeobiologists and geologists today, taking the cautious approach, favour combinations of volcanic activity, climatic change, and lowering of sea level as providing sufficient environmental stress to bring about crisis in the biosphere. Meteor or comet impact while stress was mounting would perhaps give the final push to an impending mass extinction.

Meteor impacts are not in fact very rare events, geologically speaking, and their influence in early Earth history was probably important. It seems, however, wise to look carefully at terrestrial processes as prime causes of catastrophes, even on the grandest scale. Uniformitarianism in its original sense is no longer tenable; and we acknowledge terrestrial agencies beyond present experience that offer better models of Earth history and greater insight into particular events. This *neocatastrophism* may serve in the explanation of mass extinctions, but where does uniformitarianism stand today in providing us with the key to the past—to the evolution of the Earth during all those long spells of time between the 'catastrophes'?

The advent of the paradigm of plate tectonics opened up great attractions as an explanation of how the Earth itself works, that is, how the geological cycle has operated throughout a great part of geological time. Research suggests that plate tectonics has been in operation for the past 4000 million years. In spite of the fact that heat production was so much greater in early Precambrian time than it is now, the tectonic and geochemical processes that produced the early Archaean rocks were not fundamentally different from those we see today. Brian Windley, a British authority on Precambrian geology, reviewed the evidence and concluded that, allowing for greater heat production in early Precambrian times, the basic processes responsible for the geology of continents and ocean basins have not changed significantly throughout the ensuing time. They could also have brought about the conditions of stress which gave rise to mass extinctions. With this concept, he suggests, those great protagonists of uniformitarianism, Hutton and Lyell, would have agreed.

D. L. DINELEY

Further reading

Adams, F. D. (1938) *The birth and development of the geological sciences.* Dover Publications, New York.

Albritton, C. C. (1989) *Catastrophic episodes in Earth history*. Chapman and Hall, London.

Berggren, W. A. and Van Couvering, J. A. (eds) (1984) *Catastrophes and Earth history*. Princeton University Press.

Gould, S. J. (1990) *Time's arrow, time's cycle: myth and metaphor in the discovery of geological time*. Penguin Books, Harmondsworth.

Stanley, S. M. (1987) *Extinction*. Scientific American Library, 20. Scientific American Books, New York.

Windley, B. F. (1993) Uniformitarianism today: plate tectonics is the key to the past. *Journal of the Geological Society of London*, 150, 7–19.

caves A cave is a natural cavity in bedrock which acts as a conduit for water flow between input points, such as sinking streams or soil percolation water, and output points, such as springs or seepages. This three-dimensional complex of surface closed depressions, subterranean conduits, caves, and springs is known as karst terrain. Once the developing cave conduit has a diameter larger than 5–15 mm, the basic form and hydraulics do not change much, although the diameter can be as much as 30 m in the case of caves in New Guinea or Sarawak. This minimum diameter threshold allows turbulent flow, optimizing both the solution of rock and the effective transport of sediment through the conduit. These conduits are principally formed by the dissolution of the rock in acidified water, most commonly circulating in limestone and other carbonate rocks, such as dolomite, but dissolution also occurs in evaporites such as gypsum and halite, and in silicates such as sandstone and quartzite. Caves can also develop by other processes: by weathering, the hydraulic action of waves, tectonic movements, and glacial melt water, and the evacuation of molten rock in lava flows.

The longest caves known are found in flat-lying limestones or gypsum and have been the scene of systematic exploration and mapping over many decades. Most of the world's known deep caves are in Europe, especially in the younger mountain ranges of western and central Europe. But cave exploration in the tropics is starting to yield very significant caves, especially in China, Mexico, New Guinea, and Sarawak. The five longest and deepest caves are listed in Table 1.

All the caves listed in Table 1 are active, that is they are still enlarging as a result of dissolution by acidified water. In limestone caves, the main source of this acidity is carbon dioxide dissolved to form weak carbonic acid in meteoric waters, although sulphuric acid may be important in thermal waters. The rate of dissolution depends on limestone purity, water velocity, the concentration of dissolved gas, and whether the chemical system is open (a water–air interface is present in the conduit) or closed (no water–air interface is present in a flooded conduit). The karst drainage system may be divided into a number of zones, each of which has distinctive hydraulic, chemical, and hydrological characteristics (Fig. 1). These zones are not mutually exclusive, and under flood conditions one zone may temporarily gain the properties of the one below it in

Table 1. Longest and deepest known caves

Cave	Country	Length/depth (m)
Longest caves		
1. Mammoth, Cave System	USA	563 270
2. Optimisticheskaya	Ukraine	208 000
3. Jewel Cave	USA	193 200
4. Holloch	Switzerland	175 150
5. Lechuguilla Cave	USA	161 900
Deepest caves		
1. Lamprechtsofen–Vogelschacht	Austria	1632
2. Gouffre Mirolda/Lucien Bouclier	France	1610
3. Réseau Jean Bernard	France	1602
4. Torca del Cerro	Spain	1589
5. Pantyukhinskaya	Georgia	1508

the sequence. The uppermost zone, or *epikarst*, contains the surface and soil as well as the subcutaneous zone of weathered rock and enlarged fissures. In this zone percolating water picks up carbon dioxide from root respiration and soil bacteria and becomes acidified. Below this is the *endokarst*, the zone in which most active conduits are found. This is usually divided into the *vadose* or unsaturated zone (free air surfaces) and the *phreatic* or saturated zone (water-filled with no free air surfaces). The upper portion of the vadose zone, where discrete threads of water in the subcutaneous zone join to form percolation streams, is often termed the *percolation zone*. In recognition of the transient boundaries between some zones, the lowest portion of the vadose zone within the range of fluctuations in flood-water level is the *epiphreatic* zone (or temporary phreatic). Below the vadose zone is the permanently water-filled *phreatic* zone. Here tubular conduits are still actively forming and may rise or fall tens of metres along the line of least resistance—the *hydraulic gradient*—from the input points or stream sinks to the outputs at the springs. There are three types of water flow in karst drainage systems: conduit, fissure, and diffuse. The water flowing in a cave stream is typical of very rapid *conduit flow*. Water dripping from straws and stalactites in the roof is obeying less rapid *fissure flow* carried by the fractures, including joints, in the rock mass. Water seeping through the intergranular pores of the rock and emerging as wet patches on the roof is governed by slow *diffuse flow*. The magnitude and relative importance of each type in a karst hydrological system will depend on the porosity and fissure density in the host rock.

Many caves have been abandoned by the waters that formed them and are now inactive, except for the slow processes of ceiling and wall collapse and infilling by cave deposits (known as *speleothems*), forming features such as stalactites and stalagmites. Many of these inactive caves are now perched hundreds of metres above valley floors, owing to incision (cutting down by streams). These drained caves offer Earth scientists a library of environmental history which parallels and complements the information to be gained from ice and deep-sea cores and lake deposits. In particular, we can examine the relationships

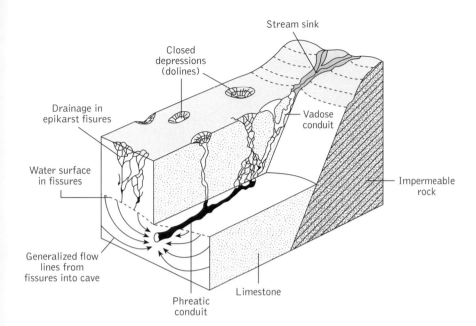

Stream sink

Closed
depressions
(dolines)

Drainage in
epikarst fisures

Vadose
conduit

Water surface
in fissures

Impermeable
rock

Generalized flow
lines from
fissures into cave

Limestone

Phreatic
conduit

Fig. 1. Diagrammatic
representation of a simple karst
drainage system.

between geology, landscape denudation, and cave development, and can date cave levels using radiometric techniques such as uranium series disequilibrium.

Mammoth Cave in Kentucky, USA, is currently the longest cave in the world, and displays the role of geology and landscape denudation in determining the shape and orientation of cave passages. Its development and hydrology have been studied by Arthur Palmer and Jim Quinlan over many years. The cave system is formed in limestones of Mississippian (Early Carboniferous) age which dip gently to the north-west. The limestone forms a broad plain of low relief, the Pennyroyal Plateau, punctuated by closed depressions, or *dolines*, and dissected by several sinking streams draining to the Green River. The border of the limestone plain is an upland composed of limestone ridges capped by sandstones and other insoluble rocks. The Mammoth Cave system lies under these ridges, occupying a sequence of limestones 110 m thick. The Ste Geneviève Limestone in the middle of the sequence contains most of the cave passages.

Most canyons and tubular conduits in Mammoth Cave are highly concordant with the bedding of the limestones. Thus the main entrance passage follows the same beds of the Ste Geneviève Limestone for several kilometres. Passages tend to run parallel to the strata because there are very few joints and they rarely intersect more than one bed. In the south-eastern area of Mammoth Cave abrupt changes in passage level are associated with faults and major joints. The three most common passage types in Mammoth Cave are canyons, low-gradient tubular passages, and vertical shafts (Fig. 2). The tubular passages are generally elliptical, reflecting the control

of bedding. These are of phreatic origin, trending along the strike of the beds, and their sinuosity is controlled by local variations in the orientation of the controlling bedding plane. Former siphons in the passages are often visible. The vadose canyons are much modified by collapse but can be seen to drain down-dip. The numerous vertical shafts in the cave are controlled by major joint intersections, with inflowing water perched on bedding planes or relatively resistant beds.

A phreatic tube named 'Cleaveland Avenue' in Mammoth Cave illustrates geological control very well. It is about 1500 m long with almost no slope (4 m km[-1]), and the limestone beds exposed in its walls show little variation along its length. Broad bends in the passage are caused by two gentle folds that push the tube to the west on the nose of an anticline and to the east into the trough of a syncline. The concentration of large passages at certain levels is a combination of stratigraphic control and the progressive downcutting of the Green River. Thus the passages cluster at three distinct elevations: 180, 168, and 152 m above sea level. These passages were backflooded by the river at the time when their levels were accordant, leaving the underground equivalents of flood plains and terraces. These deposits, and the stalagmites that cap them, have have been dated by a combination of palaeomagnetism and uranium series dating methods, and indicate that the cave has formed over the whole of the Quaternary period.

If the dissolution process has operated efficiently through time, then extensive caves will be found in a limestone massif. If that process has been severely interrupted by glaciation, aridity, or sea-level change, then these effects will be reflected

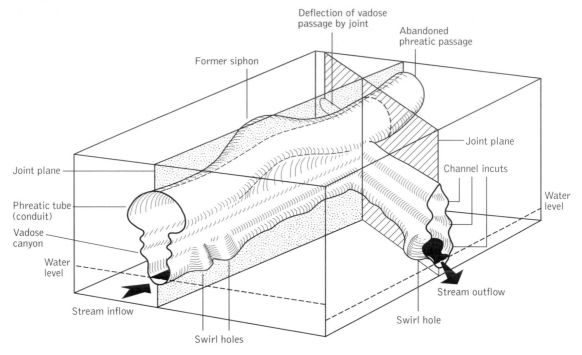

Former siphon

Deflection of vadose
passage by joint

Abandoned
phreatic passage

Joint plane

Joint plane

Channel incuts

Phreatic tube
(conduit)

Water
level

Vadose
canyon

Water
level

Stream outflow

Stream inflow

Swirl hole

Swirl holes

Fig. 2. Typical passage forms in Mammoth Cave, Kentucky (from Gillieson 1996, Fig. 3.23).

in cave morphology. Thus the extensive caves of the arid Nullarbor Plain, Australia, probably formed as large phreatic tubes under conditions of greater rainfall in the Tertiary. Deepening of the caves has been enhanced by lowering of sea level in the Pleistocene, and today there are extensive flooded tunnels, explored by divers for just over 6000 m. In regions where climatic change has been minimal, such as the humid tropics, variation in cave development may reflect regional uplift patterns. In the karst towers of Guilin, China there are caves at various levels—right up to the summits—which have been abandoned by their streams as the valleys have incised into rapidly rising plateaux. Finally, the fluctuations in sea level during the Quaternary have also produced cave development below the present mean sea level, such as the Blue Holes of the Bahamas and the Great Barrier Reef.

Once the products of surface and underground processes enter the cave system, they are likely to be preserved with minimal alteration for tens of millennia, perhaps even millions of years. In the near-constant temperature and humidity of the cave, weathering processes are reduced in intensity compared with the surface environment. Caves can thus be regarded as natural museums in which evidence of past climate, geomorphic processes, vegetation, animals, and people can be found.

Caves overrun by glaciers can accumulate valuable records of the ice ages. Paul Williams of the University of Auckland,

New Zealand has studied the record of glacial advances preserved within Aurora Cave, Fiordland. The overflow from the remote Lake Orbell sinks into this steep active cave system to emerge at a spring on the shore of Lake Te Anau, lying in a deep glacial trough. The cave began to form at least 230 000 years ago, and in it sequences of glaciofluvial gravels are interbedded with dateable flowstones and stalagmites. In the past 230 000 years seven glacial advances have filled the valleys with hundreds of metres of ice and brought gravels into the cave; the last glaciation is dated at about 19 000 years ago. This provides direct evidence lacking from surface deposits, which have been eroded by successive glaciations. Older deposits near by hold the promise of extending the glacial record of New Zealand well back into the Middle or Early Pleistocene.   DAVID GILLIESON

Further reading

Ford, D. C. and Williams, P. W. (1989) *Karst geomorphology and hydrology.* Unwin Hyman, London.

Gillieson, D. (1996) *Caves: processes, development and management.* Blackwell Scientific Publications, Oxford.

Jennings, J. N. (1985) *Karst geomorphology.* Blackwell Scientific Publications, Oxford.

White, W. B. (1988) *Geomorphology and hydrology of karst terrains.* Oxford University Press, New York.

Cenozoic The Cenozoic, Cainozoic, or Kainozoic (Greek: recent life) Era covers approximately the last 65 million years of Earth history, and includes those periods which have been known as the Tertiary and the Quaternary. The word *Kainozoic* was introduced by John Phillips in 1840 and as *Cainozoic* the following year. The spelling most commonly used today is that given in the heading. The term was used to signify the time equivalent to the Tertiary strata of the Paris Basin as described by Alexandre Brongniart some years previously. The conspicuous mollusc and vertebrate faunas of those rocks clearly set them apart from the Chalk and other older formations below, and the sudden change to forms of life very much more akin to those of today impressed the early geologist.

During Cenozoic times most continents have continued to move northwards, dispersing from their positions at the end of the Mesozoic era. The closure of the Tethyan Ocean was completed with the collision of Africa, Arabia and India against the southern margin of Eurasia. The Mediterranean and other adjacent seas are the last remnants of the former Tethys ocean; the mountains there preserve many of the rocks that were on the floors of the surrounding seas. The convergence that caused the uplift of the mountain ranges continues to the present day. These topographic changes have in turn been responsible for climatic changes across the face of Asia from Turkey to China, cutting off wind-borne moisture and giving rise to vast deserts.

The North and South Atlantic oceans have continued to open and this has been influential in the general lowering of temperatures throughout the era. During Pleistocene times this resulted in a glacial climate and widespread continental glaciations in high latitudes.

Throughout the Cenozoic the mammals have adapted to almost every environment to become the most successful and conspicuous members of the tetrapods. The birds similarly have achieved enormous success while the teleost fishes have dominated the seas and fresh waters.   D. L. DINELEY

Further reading

Hailwood, E. A. and Kidd, R. B. (eds) (1993) High resolution stratigraphy. Geological Society Special Publication No. 70.

Pomerol, C. (1982) The Cenozoic Era: Tertiary and Quaternary. (Trans. D and E. E. Humphries.) Ellis Horwood, Chichester.

Savage, R. J. G. and Long, M. R. (1986) Mammal evolution, an illustrated guide. British Museum (Natural History), London.

Stanley, S. M. (1993) Exploring Earth and life through time. W. H. Freeman, Oxford.

cephalopods Cephalopods are the most highly evolved members of the phylum Mollusca. They are marine animals that are mostly fast-moving predators and they are named for the close union of the head and foot (Greek *kephale*, head and Latin *pes*, foot), the latter organ having been elaborated into a series of tentacles that surround the mouth. Modern members of the class include the squids, octopus, and *Nautilus*, while belemnoids and ammonites are important fossil groups. Modern cephalopods include within their numbers the largest, fastest, and most intelligent marine invertebrates; the giant squid *Architeuthis princeps*, for instance, reaches lengths of 22 m; they are, however, relatively insignificant when numbers of species are considered. Only 730 species exist in our oceans today compared to the approximately 10 500 extinct species that are known. Cephalopods have a long geological history, ranging from the Cambrian to the present day. They were particularly abundant during the Mesozoic, and the ammonites of that period are particularly valuable as zonal indicators.

The molluscan body is characteristically divided into four components; the head, foot, visceral mass, and mantle. As in the other classes of molluscs the cephalopods have a mantle cavity that lies behind the head. In this class the mantle cavity has, however, been adapted as a propulsive organ. Water enters around the edge of the mantle but is ejected only through a flexible tube, the hyponome, producing a strong propulsive jet. This system is particularly well developed in the squids, which with their long streamlined bodies are the only invertebrates capable of competing with fish as fast efficient predators in the oceans. All cephalopods have tentacles surrounding the head. In *Nautilus*, which is a primitive form, they are small and numerous; higher cephalopods have eight to ten muscular arms with suckers. Squids have eight short arms and two long ones, whereas octopus have eight arms of equal length; both use the arms for catching active prey such as fish and crustaceans. Modern cephalopods have very well-developed brains and sense organs, particularly the eyes (which show a similar design to the vertebrate eye). The main feature that enabled cephalopods to become capable of sustained rapid swimming was the development of a buoyant shell. This is an important feature of the ammonites and *Nautilus*, where it is a large coiled and chambered structure; it has, however, been reduced and adapted in the modern squids and is lost entirely in the octopus. One feature of major taxonomic importance in the shell is the suture, which is the line along which the walls between the chambers meet the main shell wall.

Taxonomy

The numerous shells of fossil cephalopods present in rocks as far back as the Late Cambrian attest to the importance of this group in the past. Using this abundant material as a basis, it has been possible to erect a classification that recognized a division into three broad groups: the Nautiloidea, extending back to the Late Cambrian and characterized by an external shell that may be straight, curved, or coiled and with a simple suture; the Ammonoidea, also with an external shell but with a complex suture, and found from the Early Devonian to the Late Cretaceous; and the Coleoidea, which have an internal shell and are known from the Early Carboniferous.

Nautiloidea

The modern cephalopod *Nautilus* (Fig. 1a) appears to be very similar in design to the shelled cephalopods of the past and

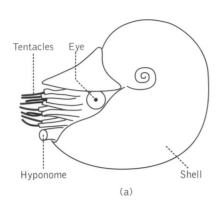

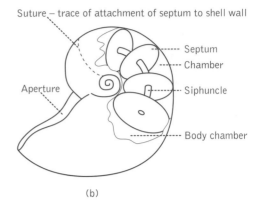

Fig. 1. *Nautilus*, showing (a) the living animal, and (b) the morphology of the shell.

has been studied extensively for this reason. It consists of two parts, the body and the shell that encloses it. The soft body is surrounded by the mantle and can be completely accommodated within the final chamber of the shell. Surrounding the mouth are about ninety tentacles. These do not have suckers but are very adhesive and are used to catch and hold prey and draw it towards the mouth, where a beak, like that of a parrot, is used to cut it up. On the lower part of the animal is the hyponome, a funnel through which it can eject water from its mantle cavity to provide a propulsive force. Sideways movements of this funnel can be used to change direction. The shell is a smooth, thin, and light planar-spiral. It is held above the animal and opens to the front. Internally the shell is divided into about thirty chambers, increasing in size towards the most recent, and separated by thin walls or septa, which meet the main shell walls along a gently curving line termed the suture. Most recent work on *Nautilus* has been carried out by Peter Ward of the University of Washington, who has shown that the rate of chamber formation diminishes with age. Early chambers are formed at two-week intervals, but in the adult the formation of a new chamber can take up to three months. The chambers are filled with gas, which makes the shell buoyant, and are connected with each other and the body of the animal by a fleshy tube, the siphuncle. Ward has shown that *Nautilus* strives to maintain steady non-fluctuating neutral buoyancy during its life by secreting liquid into the chambers or withdrawing it, using the siphuncle, to compensate for changes in body weight. These processes are too slow to power vertical movements, which the animal accomplishes by active swimming.

Nautilus is restricted to the Indo-Pacific area and is a predator that is active at night. During the day the animals rest on the bottom at depths up to 500 m. As the light starts to fail they move up to the reef, a journey that may take several hours. Predators such as fish that might otherwise be a threat are asleep at this time and the *Nautilus* are able to feed through the night on crustaceans and their moulted exoskeletons, apparently a favourite food. As morning approaches they move back over the reef edge and descend to the bottom, where they will spend the day. They are slow swimmers and appear to find their prey by touch; their sight is poor.

Nautiloids are the first known cephalopods, represented by small curving shells in the Late Cambrian of China. An explosive radiation within the group during the Early Ordovician produced the many straight and curved shell forms that were important into the Silurian. Many of these were of large size, some of the straight forms reaching lengths of 9–10 m. These endoceratoids gave rise to several groups, most notably in the Late Silurian or Early Devonian to the Nautilida, which includes the modern *Nautilus*. Most of the Palaeozoic forms became extinct by the end of the Permian, only the Nautilida surviving until the present day, although their relatives the ammonoids were extremely successful during the Mesozoic.

Ammonoidea

Ammonoids, and particularly the Mesozoic forms known as ammonites, are abundant and well-known fossils frequently collected because of the beauty of their shells. They were once known as 'Ammon's Horns' from a fancied resemblance of the coiled shells to the coiled horns of the Egyptian god Ammon, who was represented by a ram's head. The common name 'ammonite' is derived from this. Ammonoids evolved from straight-shelled forms during the Palaeozoic and developed the typical planar-spiral coiled form during the Early Devonian.

Ammonoid shells differ from those of other cephalopods in having a complex suture (Fig. 2a). This is due to the fact that the septa became increasingly frilled towards the point of attachment with the shell. Early ammonoids or goniatites may have simple zig-zag sutures, but they become increasingly complex from the Triassic to the Cretaceous. These sutures are often represented graphically, one side being drawn starting from the venter, or outer edge of the shell, and proceeding round to the dorsum or inner edge. The direction of the shell

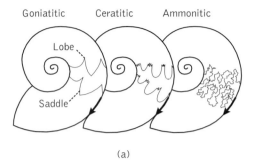

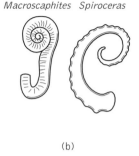

Fig. 2. (a) Three ammonoids showing goniatitic, ceratitic, and ammonitic suture patterns; (b) Cretaceous heteromorph ammonites.

opening or aperture is indicated in the diagram by an arrow, and inflections in the suture that point in that direction are termed saddles; those that point in the opposite direction are termed lobes. As the sutures are important in classifying ammonoids, these diagrams are useful tools in developing an understanding of relationships. It is generally considered that ammonoids achieved buoyancy in much the same way as the modern *Nautilus*, and that the increased complexity of the septa might have developed to improve the strength of the shell so that it would resist implosion at depth. Calculations of shell strength suggest that many ammonoids had shells that were similar in strength to that of the modern *Nautilus*, although the shell material was thinner, while others had weaker shells and must have been restricted to shallow waters.

Many modern cephalopods show size differences between the sexes or sexual dimorphism, and this has been demonstrated in ammonites also. It was noted by the Polish palaeontologist H. Makowski in 1963 that Jurassic ammonites of the genus *Quenstedtoceras* from one locality consistently showed two adult forms, one of which was larger and had more whorls or turns to the shell. The earlier whorls of both forms were exactly the same but the later whorls of the larger form showed a different ornament. This same relationship has since been shown to occur in many ammonites, and it is thought that, by analogy with modern forms, the microconchs, or smaller shells, were the males while the macroconchs, or larger shells, were the females.

Little is known about predation on modern *Nautilus* beyond the fact that turtles and sea perch will feed on them. It is clear that the shell provides little defence from the attacks of powerful vertebrate predators. Evidence of predation on ammonoids comes from shell damage or the presence of shell material in the stomach contents of other organisms. Healed damage to shells of Early Jurassic ammonites has been attributed to attacks by fish; shells of various ammonites have been found in pellets derived from plesiosaurs, which suggests that they were preyed on by these marine reptiles. Erle Kauffmann of the University of Colorado published in 1960 a study of a large ammonite of the genus *Placenticeras* that had been bitten a number of times by a mosasaur, a large marine lizard. The mosasaur had clearly moved the ammonite around in its mouth several times before devouring it, apparently indicat-

ing that the mosasaurs hunted ammonites methodically and knew exactly how to handle them.

The earliest ammonoids appeared during the Early Devonian and were derived from a group of straight-shelled nautiloids. The sutures rapidly became more complex leading to the 'goniatitic' condition, which characterized a large group of ammonoids that were important during the Carboniferous and the Permian. This group almost became extinct at the end of the Devonian, only one genus surviving to give rise to the Carboniferous radiation. Almost all the goniatites became extinct at the Permo–Triassic boundary. The few survivors again gave rise to an explosive radiation during the Triassic, in this case of a group known as the ceratites in which the basic goniatitic suture had increased in complexity. From the ceratites the basic ammonite stock arose during the Triassic; these ammonites, of the Order Phylloceratida, gave rise to the very diverse ammonites that occur throughout the Jurassic and Cretaceous. This pattern of short-lived groups diversifying rapidly and then being replaced by offshoots from an ancestral stock is termed iterative evolution.

Ammonites became extinct at the end of the Cretaceous after a slow decline that started in the early Late Cretaceous and was probably due to adverse environmental conditions related to a series of marine regressions. During the Late Cretaceous a group of ammonites developed with shells that deviated from the normal planar-spiral pattern. These are termed heteromorphs (Fig. 2b). At one time their sometimes bizarre shapes were thought to indicate racial senility of the lineage. More recent study has shown that this view is incorrect and that they were in fact highly specialized forms adapted to a variety of environments. Their asymmetric shapes would have made locomotion by jet propulsion impossible and they would have led a predominantly benthonic life as scavengers, or even filter-feeders.

Coleoidea

The coleoids include modern squids, cuttlefish, octopuses, and the argonauts or paper-nautilus, together with extinct groups such as the belemnites. In these animals the body structure is broadly comparable to that seen in *Nautilus*; the shell, however, is internal or, in the case of the octopus, has been lost entirely. In the cuttlefish *Sepia* the internal shell is

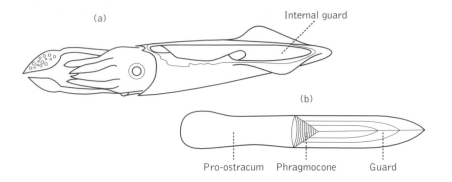

Fig. 3. Belemnites, showing (a) a reconstruction of the living animal with the internal guard, and (b) the morphology of the hard parts.

oval and consists of a series of closely spaced chalky partitions (the cuttlefish-bone). The animal is able to control its buoyancy by pumping liquid in and out of the chambers between the partitions, a system very similar to that found in *Nautilus*. In addition many squids use a system in which ammonium chloride is stored in the tissues to provide buoyancy. This system freed them from the depth limitations imposed by buoyant shells, and to this day the deeper parts of the ocean are extensively colonized by squid.

The earliest coleoids are known from the Mississippian (Early Carboniferous) of North America and appear to have been derived from straight-shelled nautiloids. By the Triassic the well-known belemnites had appeared (Fig. 3a). These are characterized by a solid calcite rod or guard, which has developed over the initial part of the chambered shell or phragmocone, which has itself been reduced to a small conical structure (Fig. 3b). A flat expanded tongue extended forward (the pro-ostracum) and presumably protected the anterior part of the body. Like the ammonites, the belemnites expanded and diversified through the Jurassic and Cretaceous and then dwindled, although the belemnites continued into the early Tertiary before becoming extinct.   DAVID K. ELLIOTT

Further reading

Lehmann, U. (1981) *The ammonites: their life and their world.* Cambridge University Press.

Morton, J. E. (1967) *Molluscs.* Hutchinson, London.

Ward, P. D. (1988) *In search of the Nautilus.* Simon and Schuster, New York.

CFBs *see* CONTINENTAL FLOOD BASALTS

chalk In geology the word 'chalk' is used in two senses: for a type of limestone; and (with a capital C) for a stratigraphical formation of Upper Cretaceous age that is notably exposed in the white cliffs of south-east England. (Blackboard 'chalk' is nowadays not chalk at all: it is in fact gypsum, calcium sulphate.) Chalk in the first of these senses is a very pure, white, fine-grained silty carbonate mudstone. It is a rock-type that is particularly well developed in the upper Cretaceous of north-western Europe and the western USA, but is also found elsewhere.

Maurice Black of Cambridge showed that chalk consists mainly of two sizes of particles, 0.5–4 micrometres (μm), and 10–100 μm. The finer size consists mainly of tablet-shaped plates of calcite, which are the remains of marine algae. In life, the alga consisted of a sphere (coccosphere) of overlapping rings (coccoliths) or rings with spines (rhabdoliths) of small plates. Although complete coccoliths are sometimes preserved, they have usually disintegrated into their constituent plates. The coarser fraction of chalk consists mainly of the skeletal debris of foraminifera, calcispheres, bivalve fragments, echinoid plates, and bryozoan, ostracod, and coral debris. Chalk contains a benthonic macrofauna of sponges, brachiopods, molluscs, and echinoids, as well as necktonic ammonites and belemnites.

Chalk is composed dominantly of calcite and was deposited originally as this mineral. It thus underwent much less change during diagenesis than limestones originally composed of aragonite and high-magnesium calcite. Chalk has subsequently retained much of its original porosity and typically has porosities of 35 to 47 per cent; even at depths of 1500–2000 m porosities are still 2–25 per cent. Chalk has a characteristically low permeability (6–12 millidarcies) owing to the very small intraparticle pores. It sometimes contains fine-grained clays which are important in the Lower Chalk, where marls or alternations of marl and chalk are common.

Chalk is well stratified, although its homogeneity makes this less obvious than in other rocks. It is more visible when there are beds rich in clays, shell layers, or flints nodules. In some instances chalk has accumulated in biohermal mounds (metres high and kilometres long) and shows evidence of contemporary slumping and deposition by dense gravity flows. It is commonly intensely bioturbated (disturbed by organisms) and has complex burrow patterns mainly infilled by precipitated silica (flint). Prominent well-cemented layers, termed *hardgrounds*, occur in the otherwise weakly cemented rock; these are named Chalk Rock, Top Rock, etc. They are the result of early cementation on the sea floor and are analogous to cemented crusts on the floor of modern carbonate seas. They mark diastems or pauses in sedimentation when the sea-floor sediment was cemented, bored, encrusted with epifauna, and commonly coated with phosphate and glauconite. Courses of flint nodules, parallel to the bedding, developed around original concentrations of siliceous remains or organic matter or were formed by precipitation in burrows.

The Chalk was deposited in a marine environment at water depths between 100 and 600 m, according to the evidence of the faunas and their state of preservation. Modern analogues are rare, although coccolith muds are known to be forming on the shelves of Yucatán. Sea-level was high in the late Cretaceous, and the relatively small land areas supplied little siliciclastic detritus. Because of the scarcity of marginal deposits there is little definite proof of the actual climate during deposition.

In spite of its low permeability chalk is a good aquifer, for water moves through an intense network of joints; and it sometimes acts as a petroleum reservoir. The Chalk is, in its lower part, extensively quarried for cement.   G. EVANS

Further reading

Hancock, J. M. (1975) The petrology of the Chalk. *Proceedings of the Geologists' Association.* **86**, (4), 499–536.

Chamberlin, T. C. (1843–1928)

The American educator and geologist Thomas Crowder Chamberlin was born in Illinois, graduated from Beloit College in Wisconsin in 1866, and returned to the college as professor of geology in 1873. In 1876 he became chief geologist in the state geological survey, but moved on again to be chief of the glacial geology division of the US Geological Survey. He had been able to study the glaciers of Switzerland in 1878. In 1894 he went to Greenland as geologist to the Peary relief expedition. At that time he was head of the department of geology at the University of Chicago, where he remained until he retired in 1919.

His early works concerned the geology of Wisconsin State and glacial geology, and he continued writing throughout his active life upon several important, even fundamental, themes in geology. His greatest contribution, however, was his planetesimal hypothesis of the origin of the planets. This suggested that a passing star drew the surface of the Sun into eruptions and raised great tidal bulges upon it. There followed the drawing out of two great curved arms of gaseous matter, like the solar prominences that are known today, but on a far greater scale. Most of the mass of these arms fell back on to the sun, but an appreciable quantity remained in orbit, gradually to condense from the gaseous state into small solid lumps, planetesimals. These in turn joined together by collisions, with the smaller gravitating to the larger, eventually to produce the planets. In time the space within the planetary orbits was swept clean. The idea of planetary accretion from cold matter was subsequently to be developed by several other geologists and cosmologists.

Later, many objections were raised to this hypothesis, but the principle of accretion has remained attractive. In the 1940s and 1950s several new hypotheses invoked condensation from a nebula, cloud, or disc of cold matter. All owed something to Chamberlin's original idea.   D. L. DINELEY

channelization *see* RIVER CHANNELIZATION

chemical alteration of rocks Alteration is the result of chemical reactions that transform the composition of a rock into a more stable configuration. These reactions can occur (1) in the solid state (2) by a coupling of mineral dissolution and precipitation reactions, or (3) through the partial melting of a rock. Such mineralogical changes are a fundamental process in geology and are the basis of a wide variety of geological phenomena including chemical weathering, diagenesis, metamorphism, metasomatism, and hydrothermal alteration, the last sometimes accompanied by ore deposition. The scale of alteration can vary in extent from less than a centimetre to many tens of kilometres. In the broadest sense, alteration is a process by which the mineralogical composition of a rock is changed by physical or chemical means, or both.

The most widespread physical alteration of a rock is brought about by a significant change in temperature, pressure, or differential stress. The deep burial of a lithological unit is associated with increases in temperature and pressure that can lead to alteration on a regional scale as part of metamorphism. For example, as a sedimentary rock, originally formed at temperatures no greater than l00 °C, is buried it experiences significant increases in temperature and pressure. The original mineral assemblage of the rock becomes unstable as a result of these changes in temperature and pressure. This instability triggers a sequence of chemical reactions to form a mineral assemblage that is more stable at the higher temperature and pressure conditions to which the rock is now subjected. In such cases, although the mineral composition has been changed, the overall chemical composition of the rock may not be altered; in this situation the alteration is termed *isochemical*. Similarly, the local emplacement of a high temperature intrusion can locally alter the mineralogical composition of the surrounding country rock, forming a contact metamorphic aureole.

The chemical alteration of a rock is commonly caused by its interaction with a moving, chemically incompatible fluid that is present at contacts between minerals. The resulting alteration is related to the origin of the reactive fluid. Common types of chemical alteration include: (1) reaction among rocks and water percolating down from the surface leads to *supergene alteration*; (2) reactions between rocks and ascending fluids yield *hypogene alteration*; (3) reactions between rocks and rainwater at the Earth's surface cause *chemical weathering*; and (4) interaction of country rock with hot fluids originating from cooling intrusions results in *hydrothermal alteration*. In each instance the altered mineral assemblage forms as the product of reactions among one or more of the minerals in the rock and the coexisting fluid, leading to changes in the bulk composition of the altered rock. The rate and extent of this alteration depends on the temperature, quantity, composition, and velocity of the fluids present within the rocks. The fluid serves both as a reactant in the alteration process and as a transporter of chemical reactants and products as dissolved species. Rapidly moving fluids can readily move large quantities of dissolved chemical species, yet stagnant fluids can move significant quantities of dissolved chemical species by diffusion.   ERIC H. OELKERS

chemical fossils *see* BIOMARKERS (CHEMICAL FOSSILS)

chemical properties of soils *see* SOILS

chemistry of geothermal waters and hydro-thermal alteration
Rainwater and snow always contain small amounts of dissolved solids, which are largely derived from sea-water spray and aerosols. Aerosols are small suspended particles in the air that are washed down with the precipitation. Sea-water spray forms when the wind rips small droplets off wave crests. When transported into the air the droplets may evaporate, leaving the dissolved solids floating in the air to form aerosols.

Rainwater and melted snow are undersaturated with all common rock-forming minerals. Thus, upon contact with the soil and bedrock this water starts to dissolve the minerals, and in this way its content of dissolved solids increases. Additionally, carbon dioxide and organic acids may be added to the water from the soil, where they form by decay of organic matter. Dissolution of the rock and soil by rainwater seeping into the ground is generally sufficient to saturate it with some minerals, that will consequently tend to be deposited from solution.

The process of rock dissolution and secondary mineral deposition that occurs in the near-surface environment is collectively termed *chemical weathering*. When these chemical processes occur deep in the ground at elevated temperatures and pressures, they are termed *hydrothermal alteration*. Weathering and hydrothermal alteration invariably lead to changes in the mineralogy of the soil and rock, and sometimes also to changes in their chemical composition.

The process of weathering and hydrothermal alteration is often termed 'hydrogen-ion metasomatism' because it may be viewed as a chemical reaction between acids and bases. The aqueous solution acts as an acid and the rock as a base. Many silicates, which are outstandingly important rock-forming minerals, behave as quite strong bases when dissolving in water. Their dissolution thus entails consumption of hydrogen ions. Various cations are simultaneously released into solution. The reactivity of natural water, that is, its tendency to dissolve the rock minerals, depends on the supply of acids to this water. In most natural waters the most important acid is carbonic acid (i.e. dissolved carbon dioxide). In geothermal waters, hydrogen sulphide, silicic acid, and boric acid may also be important. Waters containing acids in rather high concentrations are relatively reactive and have a relatively high hydrogen ion content (low pH). By contrast, waters with a low content of dissolved acids, and in contact with very reactive rocks, attain low hydrogen ion concentrations (high pH) through the dissolution of rocks.

The chemical composition of geothermal waters is extremely variable. The content of dissolved solids ranges from a few hundreds of milligrams per litre to as much as 30 per cent. The main factors affecting the composition are temperature and the chemical composition of the rocks through which the water flows. However, infiltration of sea-water into the bedrock and degassing of a magma heat source to many geothermal systems may also contribute to the chemical composition of geothermal waters. Although all common rocks are mostly composed of various oxide and silicate minerals, they contain a small but variable content of soluble salts. The amount of these soluble salts in the rock has a profound effect on the concentration of dissolved solids in the water. Certain rocks, such as evaporite sediments, are solely composed of soluble salts. Water associated with such rocks is hypersaline and referred to as brine.

Studies of the chemical composition of the water in many drilled geothermal fields around the world, together with the alteration mineralogy, indicate that chemical equilibrium is closely approached between these minerals and the aqueous solution, at least if temperatures exceed 50–100 °C.

Hydrothermal alteration has been studied extensively in active geothermal systems, which have been drilled for the purpose of exploiting geothermal resources, and also in fossil systems of this kind that have been exposed by erosion. Temperatures increase with depth in the drilled areas and may reach as much as 350 °C; even more in a few instances. The hydrothermal minerals in active geothermal systems typically display a depth-zonal distribution, suggesting that each of these minerals, or an assemblage of minerals, is stable over a certain range of temperature. Other minerals, such as quartz, may be stable over almost the whole temperature range. In fossil geothermal systems the zonal distribution of the hydrothermal minerals defines an aureole around intrusive formations that are considered to represent the heat source for the fossil geothermal system. In some geothermal systems total reconstitution of the primary rock minerals by hydrothermal minerals may have occurred; alteration in others is limited. The age of the system, the contact area between water and rock, and the degree of instability of the primary rock minerals determine the extent of the alteration.

The solubility of many of hydrothermal minerals, such as quartz, is temperature-dependent (Fig. 1). Geothermal waters reside for a considerable time, probably years, at depth in geothermal reservoirs, or long enough to come close to chemical equilibrium with the hydrothermal minerals. The water may cool extensively in upflow zones between hot springs and the reservoir, either through conductive heat loss to the wall rock or by boiling. The residence time in the upflow is relatively short and is often not sufficient to allow the water to re-equilibrate much as it cools. In other words, the chemical composition of the water emerging in hot springs is very much the same as that in the reservoir. This has been made use of for geothermal exploration. By collecting samples of hot-spring waters and analysing them, (e.g. for silica) one can predict

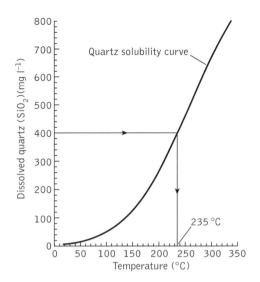

Fig. 1. Temperature dependence of quartz solubility as determined by Fournier and Potter (1982, *Geochemica Cosmochimica Acta*, vol. 46).

reservoir temperature (Fig. 1). For example, if the analysis of a water sample yields a silica concentration of 400 mg per litre (mg l^{-1}), the reservoir temperature can be predicted to be 235 °C; a significant conclusion considering that one of the most important exploitation characteristics of a geothermal reservoir is its temperature.

The quartz geothermometer was the first to be developed for geothermal exploration. Many other geothermometers have since been developed, particularly those that are based on the aqueous concentration ratios of various cations, such sodium/potassium, sodium/lithium, and magnesium/potassium.

In many geothermal fields, particularly, those located on high ground, the groundwater table is so low that no hot-water springs exist; only fumaroles (steam vents) are present. The steam contains various gases, including carbon dioxide, hydrogen sulphide (easily recognized by its notorious smell), hydrogen, and nitrogen. Gas geothermometers have been developed for geothermal exploration that relate the gas content of steam in fumaroles to temperature in the underlying geothermal reservoir where the steam originated.

STEFÁN ARNÓRSSON

Further reading

Ellis, A. J. and Mahon, W. A. J. (1977) *Chemistry and geothermal systems*. Academic Press, New York.

Nicholson, K. (1993) *Geothermal fluids*. Springer-Verlag, Berlin.

chemostratigraphy Chemostratigraphy uses inorganic geochemical data to characterize and correlate strata. It is a technique that has made great progress in recent years and it is applied mostly to sequences of siliciclastic sedimentary rocks (rocks such as sandstones which are made up of silicate

fragments) that are biostratigraphically barren. The composition of these sequences is controlled largely by the nature of their source regions, and especially by the composition of the rocks denuded there. The strata in these sequences may extend, and be identified by their geochemical signatures, over wide areas. Their geochemical signatures may involve major element concentrations and ratios; trace-element concentrations may also serve to identify specific horizons and 'events', as in event stratigraphy.

The introduction of rapid and cost-effective analytical methods (e.g. IC (isotope-concentration) mass spectrometry) in the 1980s greatly boosted the potential of chemostratigraphy for more routine use in correlating biostratigraphically barren sections for the petroleum industry and for academic research. Even where a chemostratigraphy has been identified over distances of as much as 500 km it can be used to identify sediment source areas and pathways. In thick sequences of Devonian black shales in several parts of the world individual widespread geochemically identifiable bands a few centimetres thick have proved valuable in regional correlation, and the famous iridium-rich horizon at the top of the Cretaceous seems to be global in extent.

D. L. DINELEY

china clay China clay, also known as kaolin, is a white commercial clay consisting predominantly of the mineral kaolinite, a hydrated aluminosilicate. The name 'kaolin' is derived from the village of Gaoling in Jiangxi province, China, where the white clay was mined. The nearby Jingdezhen potteries used the kaolin to create their fine white porcelain. The composition of Chinese procelain was identified by Europeans in the eighteenth century and deposits of kaolin were sought in Europe. In 1746 William Cookworthy, a Plymouth chemist, recognized the occurrence of kaolin in Cornwall, in southwest England, and subsequently manufactured porcelain from Cornish china clay. The deposits of china clay in Cornwall and Devon rapidly attracted the attention of the Staffordshire potters, who until the early nineteenth century used Cornish and Devonian kaolin almost exclusively in the production of fine ceramics.

China clay deposits have now been found throughout the world and new deposits are still being discovered. Kaolin deposits may be classified as primary or sedimentary. Primary, or residual, deposits were formed by the alteration *in situ* of the parent rock, which may have been igneous, metamorphic, or sedimentary, by volcanic, hydrothermal, and weathering processes. Sedimentary, or secondary, kaolins are derived from the erosion of pre-existing deposits and the subsequent transport and deposition of the clay. In Cornwall and Devon the kaolinite is derived from the late-stage magmatic or hydrothermal decomposition of feldspars within granite. It is separated from the host granite by washing it out with high-pressure water hoses, a process known as monitoring. The kaolin content rarely exceeds 20 per cent of the altered granite, but the depth of kaolinization extends in many places down to 300 metres. The best-known sedimentary kaolin deposits are from Georgia, in the USA. These were formed from the erosion of deeply weathered crystalline rocks in the Piedmont Plateau and deposited along Georgia's Fall line during the Cretaceous and Tertiary periods. Here the kaolin is found in lenses, often up to 20 metres thick and with a high percentage of kaolinite, around 80–90 per cent.

The world market for high-quality kaolin is about 25 million tonnes per annum. The USA is the largest producer with 9 million tonnes, and south-west England the second with 3 million tonnes. Other important producers are Germany, France, Ukraine, China, Czech Republic, New Zealand, Brazil, Spain, Indonesia, and Australia.

Kaolin today is used in making paper, plastics, rubber, paints, fibreglass, ceramics, some foods, sunscreen lotion, and many other products. Almost 80 per cent of the kaolin produced in Georgia, Devon, and Cornwall is used in filling and coating paper. Filler clays are so called because they fill the gaps between wood fibres in the papermaking process. The addition of these fillers also improves the strength, smoothness, brightness, and opacity of the paper. Coating clays may be applied to this base paper to impart a glossy surface suitable for high-quality printing of illustrations. Additional uses of kaolin continue to be developed; they include a wide range of new applications in the paper industry, ranging from low-cost pulp extenders to high-opacity fillers and high-gloss and high-brightness coatings.

IAN WILSON

chordates Chordates are interesting to us because they include our own taxonomic group, the vertebrates. The phylum as a whole is characterized by the presence of a longitudinal cartilaginous stiffening rod (the notochord), a single tubular dorsal nerve cord, and perforations in the pharynx comparable to gill slits. On the basis of these and additional characters we divide the phylum into three subphyla. The urochordates or tunicates are sessile (attached to the bottom) marine filter-feeders, and rather sponge-like as adults. The larvae, however, are tadpole-shaped and mobile and, unlike the adults, show the full complement of chordate characters. The cephalochordates include only about twenty species of two genera of organisms commonly called amphioxus. They are small fish-like organisms (about 7 cm long) in which the chordate characters are retained in the adult. They live in sandy coastal sediments and filter the interstitial waters. The vertebrates are characterized by the presence of a bony skeleton and a brain. Although they retain the chordate characters as adults gill slits are present only in the embryonic stages of land vertebrates.

Study of types of cell division shows that chordates are most closely related to echinoderms, and it is currently accepted that echinoderms and lower chordates diverged from vertebrate ancestors no later than the Early Cambrian. Although fossil chordates (other than vertebrates) are rare, a fossil cephalochordate *Pikaia* is known from the Middle Cambrian Burgess Shale, and vertebrates have now been reported from the Middle Cambrian of Chengjiang, southern China.

DAVID K. ELLIOTT

chromites Chromite ($FeCr_2O_4$) is the only ore mineral of chromium (Cr) and is found, usually in small amounts, in mafic and ultramafic rocks (igneous rocks containing a large proportion of olivine and pyroxene). Chromite deposits are of two main types:

(1) 'Magmatic sediment layers' in layered basic intrusions where the chromite ores appear to have segregated during crystallization of the magma body, e.g. Bushveld complex, South Africa. These layers vary from a few centimetres to several metres in thickness and often extend laterally for considerable distances. Many show features analogous to sedimentary rocks. Occurrences are isolated, but are of economic significance; approximately 97 per cent of the world's chromite reserves occur in such layered deposits in South Africa and Zimbabwe.

(2) 'Podiform chromites' are found in peridotite or serpentinized masses associated with mountain belts (sometimes termed Alpine-type), e.g. the Perm mining district in the Ural Mountains and deposits in Turkey and Albania. Such deposits are found in highly unstable tectonic environments. The deposits are believed to have formed initially within the oceanic lithosphere and to have been incorporated subsequently into the continental crust during mountain building.

ELIZABETH H. BAILEY

chronostratigraphy *see* STRATIGRAPHY

Clarke, Frank Wigglesworth (1847–1931) An

American chemist with an intense interest in mineralology, F. W. Clarke is best known for his compendium *Data of geochemistry*, which first appeared in 1908 as *US Geological Survey Bulletin* 330. Subsequent editions appeared in 1911, 1916, 1920, and 1924 as *Bulletins* 491, 616, 695, and 770 respectively.

After professional appointments in chemistry at Howard University and the University of Cincinnati, Clarke joined the US Geological Survey in 1883. During his academic years Clarke's research was directed to the precise determinations of atomic weights and of other fundamental physical and chemical constants. When he joined the US Geological Survey as Chief Chemist, he began to produce a steady stream of papers on the compositions and origins of minerals, rocks, waters, and other natural materials. From this body of analytical data he prepared *Data of geochemistry*, the first such comprehensive compilation; it became one of the foundations on which the modern field of geochemistry has been built.

By naming one of its two medals for Clarke (the other is named for V. M. Goldschmidt), the Geochemical Society recognized the seminal role of Clarke's studies in the rise of the study of the chemistry of the Earth. BRIAN J. SKINNER

classical times and the Earth sciences Ideas about

the nature of the Earth, if not the Universe, which we might call scientific, since they were founded upon direct observations, have reached us only sparingly from the classical world. The Greeks from Aristotle onwards have left little; the Romans even less. This is despite the voluminous nature of their literature. The list of Greeks writing on aspects of geology begins with Thales of Miletus (c. 636–546 BC), who recognized that alluvial deposits at the mouths of rivers were water-transported and that the action of water was important in bringing about environmental changes. Those who followed may have subscribed to several themes that appear commonly in ancient Greek work, namely: a central fire existed within the Earth; the relative positions of land and sea have in places changed with time; fossil remains occurring in strata may show where sea previously existed, but may be due to a special force acting upon the rock. Most of these Greeks lived in the fourth century BC. Aristotle himself (384–322 BC) wrote upon a wide range of topics in describing the physical world, but no treatise of his on rocks, fossils, or minerals survives. One of his pupils, Theophrastus, compiled a work *Concerning stones* which survived and shows that there was a great deal of mineralogical knowledge among the miners and quarrymen of the day. This book served as a guide for almost 2000 years.

It seems that several of the earlier philosophers had concluded that the Earth is a globe. In and about 250 BC, Eratosthenes, a Greek in Alexandria, determined the Earth's dimensions by geometry. He arrived at a circumference of about 40 000 km, which is not far from the true value. Other, later Greek writers recorded a great deal of information about the location and nature of ores and other interesting or useful mineral and rock-types within the Mediterranean–Middle East region. The presence of tin in Britain was known to them. Descriptions of mountains, rivers, and various landforms were discussed. This work was continued by a number of Roman writers.

Of the Romans, the earliest and most sapient was Lucretius (99–55 BC), who was certainly one of the greatest writers of Roman times. He was concerned with the origins and behaviour of subterranean water, caverns, springs and rivers, volcanic eruptions, and earthquakes. He was interested in the occurrence and distribution of metallic ores and other minerals, and took time to discuss the 'exhalations' that are present in mines to the detriment of the miners' health. In typical Roman fashion, he was concerned with how materials could be used, and he devoted some attention to mining and smelting practices. The role of climate and drainage in modifying landscape was also of interest to him. Lucretius believed that the existence of subterranean winds gave rise to volcanism and earthquakes. His knowledge of geological phenomena seems to have been enormous for his time, and his lines of reasoning are similarly impressive.

A fellow roman, Vitruvius (a military engineer at the time of Julius Caesar and Augustus), was also a writer of distinction, but was interested principally in the kinds and distribution of natural resources that were available to the Roman world. He noted many local building materials and the means of utilizing them. The account of the famous mortar and concrete that the Romans used under water came from his writings. His observations on volcanic materials and products were extensive, accurate, and perceptive. Vitruvius was fascinated by all the indications of great temperatures beneath the land around Vesuvius, and he recorded the presence of asphalt in the lands east of the Roman Empire.

Completing the Roman succession of writers on Earth science topics were the two Plinys, uncle and nephew. Pliny the Elder (AD 23–79) was born to a wealthy family and spent his life largely in imperial service, which gave him opportunity to visit many countries around the Mediterranean. His *Natural history* appeared about AD 77. It is a large work, and includes not only much simple factual data, but also a lot of myth, hearsay, and total misinformation. The range of geological subjects is wide: minerals, rocks, precious stones, and the means of working them. Much of the description is very detailed. The whole survived to form part of the foundation of knowledge from which the Renaissance grew. Pliny the Elder lost his life while visiting Vesuvius during an eruption. His nephew, Pliny the Younger, recorded the event. His interest in the natural world has been overshadowed by his extensive writings and correspondence on state, social, and commercial matters.

We have little or no knowledge that any of the classical writers, save Vitruvius or Pliny the Elder, was intent on examining geological phenomena at first hand. The extent to which they possessed maps and mine plans is doubtful. None of them appears to have set out to identify the geological components or structure of a region, or to propose a theory of the Earth in the modern sense. The Greeks were more intent upon framing untestable hypotheses; the Romans were interested in Earth science for what they could get out of it in a purely pragmatic sense.

D. L. DINELEY

Further reading

Adams, F. D. (1938). *The birth and development of the geological sciences.* Dover Publications, New York.

Crombie, A. C. (1994) *Styles of scientific thinking in the European tradition* (3 vols). Duckworth, London.

Lindberg, D. C. (1992) *The beginnings of Western science: the European scientific tradition in philosophical, religious, and institutional context, 600* BC *to* AD *1450.* University of Chicago Press.

clastic dykes

Clastic dykes are intrusive bodies of sediment that cut the normal sedimentary layering. They originate in two very different ways. Some come from below; these are termed *injection clastic dykes*. Others have been filled from above; these are *neptunean dykes*. Associated with injection dykes are sills (tabular bodies parallel with the bedding) that at outcrop frequently appear very similar to beds deposited at the surface by normal sedimentary processes.

Dykes may be composed of almost any material. The most common is sand, but dykes of mud, silt, asphalt, and bituminous sediment occur. On the shores of Lake Huron, conglomeratic quartzite dykes with granite pebbles 15 cm in diameter are known from the Precambrian.

Clastic dykes are produced by the injection of sand from an underlying source during a short-lived post-depositional event when both the buried source and the host layer were in a weakened condition. The event may be a seismic shock, the sudden emplacement of a slumped mass flow from above, or the impact of storm waves.

Neptunean dykes are a characteristic feature of limestones, because such rocks are easily hardened at the surface to become fractured and dissolved so that very large solution cavities can form. These are not only filled by overlying sediment, but may be inhabited by special organisms that colonize these submarine fissures: for example, dwarf ammonites in the Jurassic reefs of Sicily.

HAROLD G. READING

clays are one of mankind's oldest raw materials, used for making pottery since Neolithic times. The plasticity of clays enables them to be shaped; firing then converts them into rigid, impermeable articles which can be used in cooking and other purposes. Modern society uses a wide range of clays, ranging from common clays used for brick manufacture to speciality clays used to make high-grade paper and plastics. Clays are also important components of most natural soils.

In engineering or soil science a clay particle is usually defined as bring less than 2 or 4 microns (0.002 or 0.004 mm), although many geological formations described as clays contain a substantial proportion of particles larger than 4 microns.

Sedimentary clays are common and are usually laid down in relatively still water: below the range of wave action or in lakes, for example. When buried by later sediments, clays can undergo changes in their mineralogy, a process known as diagenesis. Further increases in temperature and pressure will convert a clay into a shale and then a slate. Because of this, clay becomes progressively less common in older geological periods and is almost never found in Precambrian formations.

Clays are also be formed by hydrous fluids or gases altering harder rocks such as igneous rocks or slates. The altering fluids can be meteoric water derived from the surface (weathering), or hot water or gas associated with magmatic activity (hydrothermal alteration). Both processes usually involve the removal of silica and alkalis, and the addition of water.

The dominant minerals in a clay are usually layer silicates and fine silica (quartz), together with smaller amounts of iron sulphides and oxides, titanium minerals, various carbonates, and organic matter. The layer silicates (phyllosilicates) can be classified into various groups, according to their chemical composition and the layer structure of their crystal lattice. The layers are composed of various combinations of tetrahedral silica sheets and octahedral hydrated aluminium oxide sheets, frequently with appreciable amounts of potassium, calcium, magnesium, sodium, and iron. Some phyllosilicate particles are composite alternations of different clay minerals. When mica occurs in fine-grained form in a clay it is usually

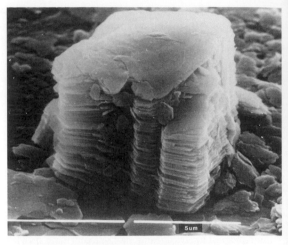

Fig. 1. Scanning electron microscope image of a kaolin crystal, showing the layered structure. The crystal is about 10 microns (0.01 mm) in width. (Illustration by courtesy of IMERYS.)

referred to as *illite* or *clay mica* and is similar in composition to muscovite, but there is usually some substitution of hydroxyl ions into the lattice. There are many other clay minerals, including chlorite, vermiculite, and various chain silicates such as attapulgite and pyrophyllite, some of which are valuable industrial materials.

Techniques commonly used to study clay minerals include X-ray diffraction for the crystal lattice structure, X-ray fluorescence spectrometry for chemical composition, and scanning electron microscopy for the morphology of clay particles.

The *kaolin* group of clay minerals includes kaolinite and halloysite, both of which are hydrated aluminium silicates. Kaolinite has a simple layer structure (see Fig. 1); a kaolinite layer is composed of one tetrahedral sheet combined with one octahedral sheet. Clays predominantly composed of kaolinite are commercially known as 'kaolin'. The china clays of southwest England are 'primary' kaolins formed *in situ* from granites by a combination of hydrothermal and weathering action. Elsewhere in the world important deposits of sedimentary kaolin of late Cretaceous–early Tertiary age are found in Georgia, USA; others, of Neogene age, are in the Amazon Basin. The principal use for these kaolins is in the manufacture of paper, both as a filler and as a coating material to provide a receptive surface for printing ink. Other uses include ceramics, and as a filler in paint, rubber, and plastics.

Ball clays are sedimentary rocks, usually laid down in fresh water, which are composed of a special type of kaolinite known as *b-axis disordered*, together with clay mica (illite), which gives the clay good plasticity and strength, making it particularly suitable for ceramics. Fireclays have a similar mineral composition to ball clays, but are generally less pure and lack plasticity and strength. Large quantities of fireclay were formerly used in the iron and steel industry, but nowadays their main use is for making bricks and sanitary ware. Halloysite has the same basic layer configuration as kaolinite, but the layers are rolled up into scrolls. Deposits of halloysite are valued for use in high-quality ceramics, such as porcelain.

Another important group of layer silicates are the *smectites*, many of which have the property of absorbing water by expanding their layer structure. Their chemical composition is variable, owing to much substitution in the crystal lattice and to the presence of variable amounts of cations such as calcium and sodium between the layers. The purer deposits of smectite are known as bentonites and are valued for use in various applications such as drilling mud, iron ore pelletizing and foundry use, and in civil engineering, as well as for clarifying liquids used in the food and drink industry. Less pure deposits of smectite are known as fuller's earths and were originally used for removing oil and grease from wool and cloth. Nowadays their main application is as an absorbent (cat litter). Traditional sources of high-quality bentonite include Wyoming, USA and Mediterranean islands such as Milos. Fuller's earth deposits are widespread, but the fuller's earth near Bath, which gave rise to the term in the late eighteenth century, is no longer exploited.

Common clays are usually mixtures of clay minerals such as illite, smectite, and kaolinite, together with fine silica and other minor constituents. Interlayered clay minerals (see above) also frequently occur. Formations such as the Oxford Clay or the London Clay typify this type of clay. Although widespread, these clay formations are important economically, for they provide the basic material for brickmaking and for heavy clay products such as sewer pipes and clay floor tiles.

COLIN BRISTOW

cleavage and other tectonic foliations in rocks

Although tectonic foliations are best displayed in thin sections, which enable the viewer to see through rock specimens, and in outcrops at the Earth's surface, these small-scale structures result primarily from tectonic deformation on a much larger scale. These structures generally occur as products of mountain-building.

A foliation is a pervasive set of parallel surfaces or zones in a rock. This broad definition encompasses features such as sedimentary bedding, igneous cumulate layering, and tectonic foliations (Fig. 1). Narrowing our focus, tectonic foliations develop from differential stress during deformation with metamorphism. The original grain arrangement or rock fabric changes by developing four fabric elements (Fig. 1): compositional banding such as alternating mafic and felsic layers in a gneiss, parallel discontinuities such as pressure-solution surfaces in a cleaved rock, elongate grains such as deformed quartz grains in a schist, or inequant grains such as aligned mica grains in a schist. The parallelism of these elements constitutes the foliation.

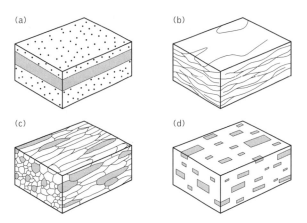

Fig. 1. Fabric elements that define foliations. In each case the foliation is parallel to the top of the block. (a) compositional banding; (b) parallel discontinuities; (c) elongate mineral grains; (d) inequant mineral grains.

A cleaved rock has fabric domains of two types: cleavage zones, where the fabric elements and metamorphic minerals form the foliation; and *microlithons*, which mostly or entirely preserve the original mineralogy and fabric. Cleavage is either continuous (Fig. 2) or spaced. Continuously cleaved rocks have cleavage zones so close together that their spacing is not detectable, or all fabric elements are parallel to the foliation. A common rock with this type of foliation is slate, which splits along an almost innumerable number of parallel planes. Essentially, continuously cleaved rocks lack microlithons, whereas a rock with spaced cleavage has discernible microlithons.

Many workers view schistosity as a coarse continuous cleavage with visible grain size and all fabric elements parallel to the foliation. This description does not do justice to the appearance of these beautiful rocks. They have parallel bright shiny mica

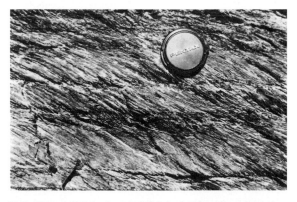

Fig. 2. Fine continuous cleavage deforming bedding. The cleavage is at an angle of about 30° to the bedding, which is here horizontal.

wrapping round large grains of pink garnet, brown staurolite, or blue kyanite. In contrast to cleaved rocks or schists, gneisses contain differentiated layers. The primary fabric element is compositional banding rather than parallel discontinuities or long axes of grains. The alternating bands of mafic and felsic minerals are a striking and distinctive feature of all gneisses.

A spectrum of foliations from cleavage to differential layering exists because the structures develop under different metamorphic conditions. The transition from cleavage to gneiss represents a progression in metamorphic intensity during deformation. If an unmetamorphosed mudstone is progressively metamorphosed and deformed, the grain size and mineral segregation in the evolving foliated rock increase until deformation ceases. Increases in conditions such as temperature, pressure, and fluid content drive this change.

Foliated rocks are common folded, and the folds and foliations will have simple geometric relationships if they were formed at the same time. The foliation is generally parallel or subparallel to the axial plane of the host fold, but it can also converge through the fold core. These differences in geometry indicate different sequences of deformation. For example, divergent cleavage geometry can indicate that shear planes parallel to the layers in the rock were rotated by the formation of cleavage during folding. Parallel cleavage geometry can indicate a fold that was modified by the later formation of foliation. Not all foliations and folds are, however, of the same age. Rocks that have been regionally metamorphosed can exhibit several phases of folds and foliations that were produced by successive phases of deformation during mountain-building. In such cases, younger folds will deflect an older foliation, or a younger cleavage will transect older folds without there being a consistent geometrical relationship between them.   WILLIAM M. DUNNE

Further reading

Twiss, R. J. and Moores, E. M. (1992) *Structural geology.* W. H. Freeman, New York.

cleavage in minerals Cleavage is the property of a mineral to split along smooth, well-defined planes that are planes with a high density of atoms and are parallel to possible faces in the crystal. Cleavage arises because the bonds between the atoms are less strong across the planes. Graphite, for example, has a sheet structure with its carbon atoms strongly bonded within the planes but with only feeble residual forces between the planes; as a result, it has a perfect cleavage.

Cleavage, defined as imperfect, distinct, perfect, or eminent, according to its quality, usually gives rise to smooth, glistening faces (in contrast to the external faces of crystals, which may be dull, etched, or uneven). The number of cleavage planes in a mineral depends on the symmetry of the structure. The cleavage may consist of either one form or set of planes (as, for example, in fluorite (4 planes, octahedral), galena (3, cubic), andalusite (2, prismatic), gypsum (1, pinacoidal), or there may be several sets in one mineral (as in barite with

one set of two planes (prisms) and two sets of one each (pinacoids), i.e. four cleavages in total. Some minerals (e.g. quartz) have no cleavage but a fracture which is irregular. The orientation of cleavage planes is defined by using Miller indices.

Cleavage may give useful clues to identification, as in the case of the pyroxenes and amphiboles. These rock-forming minerals are similar chemically and in colour and have two cleavages, which in the pyroxenes intersect at an angle of 87° but in the amphiboles at 124°.   R. BRADSHAW

cliffs and rock slopes Many of the most spectacular landforms on Earth are associated with high mountain chains, towering cliffs, and extensive, steep rock outcrops. Although the slopes that develop in lithified, jointed rock masses vary in form, they include a number of regularly occurring components (see *hill slopes*). At the crest of the slope a free face is often present, where rock *in situ* forms a steep, near-vertical break between ground at the top and base of the cliff. The free face is usually skirted by debris or *talus*, consisting of blocks of rock that have become detached from above. Talus will vary in size and shape, according to the properties of the rock and the joint patterns. As a rule, loose material transported from the upper to lower parts of the talus slope decreases in size. Talus can take the form of a sheet of debris where rockfall activity has been more or less uniform along a cliff. Talus cones can develop if falling debris is funnelled or local activity rates increase above mean conditions. At the toe of the slope a rock *pediment* can be found. Pediments are low-gradient ground surfaces, seldom of more than 2° to 3°, which extend from a cliff front away into the surrounding landscape, in some instances over many kilometres (see *pediments*). The presence of each of the three rock-slope components depends on the environmental setting. At the coast, where the processes of basal erosion are active, rock debris is likely to be removed by the sea. In these situations the free face can extend to sea level and talus slopes will be absent. Along scarp fronts in tectonically stable, arid environments, all three components are usually found along a slope profile.

The rate at which rock slopes change is varied because of the many controls on development that operate at different rates. The most important natural controls on cliffs and rock slopes can be considered in three groups: climate, weathering, and geology. Climate influences the failure mechanism of a jointed rock mass, rates of rock disintegration, changes in form, and rates of slope development. In wet environments, high-intensity rainstorms may cause a sudden increase in water moving through discontinuities in a rock mass, destabilizing the slope and generating failure. In cold, polar regions, frost and ice may widen joints and promote disintegration of blocks of rock. The weight of snow and ice at the top of a slope may add a load to the rock mass which creates a change in stress and leads to failure. Weathering promotes the breakdown of rock, either as the detachment of material from outcrops *in situ* or through the disintegration of blocks previously separated from a rock outcrop. Weathering may be physical, chemical, or biological, depending on the environment, climate, and rock properties. Physical breakdown of rocks results in progressively smaller fragments without change to the mineral matrix. Chemical disintegration alters the original mineral composition. Biological change includes both physical and chemical alteration and is usually confined to the uppermost few metres of rock. In tropical areas, chemical weathering will be significant because of the high temperature and humidity, particularly in rock materials that have a susceptible mineral matrix. In the arid zone, physical weathering is usually the main agent of rock disintegration. Thermal expansion and contraction of rock occurs between day and night time as temperatures fluctuate, generating sheeting of the outer layers of blocks and the sudden splitting of boulders.

Geological factors have great control on the form and development of rock slopes. At the largest scale, regional tectonics and structures, such as fault systems, will influence cliff patterns. Earthquakes and relative movements across plate margins have probably triggered more large rockfalls and slope changes than any other single cause. A good comparison can be seen in the juxtaposed countries of Australia and New Zealand. Australia has an ancient landscape, far removed from plate margins and with little tectonic activity. Combined with an arid climate and slow weathering rates, the slopes of areas such as the Kimberley Plateau in Western Australia are smooth, curvaceous, and reflect development over tens of thousands of years. In contrast, the South Island of New Zealand sits astride a collision zone between the Pacific and Indian–Australian plates. The product is the Southern Alps. Uplift rates exceed 10 mm a year, precipitation is intense, weathering rapid, and denudation rates high. The topography is rugged and complex, prone to rockfalls and avalanches. The slopes and cliffs reflect a highly unstable and dynamic landscape. At a local level, particularly at individual sites, the main geological controls are the properties of intact rocks and structural discontinuities. The stronger the rock, the greater its competence and the steeper the slopes that are likely to develop. The significance of inherent rock strength is, however, commonly eclipsed by the properties of eclipsed by joints and bedding planes. The orientation, width, spacing, and continuity of joints play an important role in slope form and evolution. Widely spaced joints, which dip out of a slope and have a clay fill, promote the movement of one block over another. The consequence will be an increased likelihood of failure, a reduced slope angle, more talus, and enhanced slope retreat rates. The dip (inclination) of the bedding planes frequently controls rock-slope form and stability. If the bedding is horizontal, slope failure will be rare, the rock slope will be almost vertical, and cliffs may develop. Inclined bedding will reduce the angle of hillslope stability and slope gradient. The likelihood of failure will also increase; bedding that dips out of the slope will lead to the least favourable situation. The interaction of intact rock properties and discontinuity characteristics in controlling the form and development of slopes is

often examined using rock-mass strength classification techniques. Properties of site rock materials are synthesized in a semi-quantitative format to explain differences in slopes and the evolution of features such as glaciated valleys, bornhardts (large inselbergs) and scarp fronts.

It is the failure of rock that generates the most rapid change in slope profiles and determines the relative dominance of the main rock-slope components of free-face, talus slope, and pediment. The type of failure is greatly influenced by the intensity and inclination of discontinuities. In the most simple case, failure entails detachment of individual blocks of rock. The precise failure mechanism relates to the orientation of joint sets, the dimensions of the block that becomes detached, and the shape of the resulting space or rock scar (Fig. 1). For slides and wedge failures, orientation of joints is the determining factor. The failure of a number of blocks is usually termed a *fall*, with debris accumulating as a talus slope below the zone of rock detachment. More complex movements develop where failure generates a number of displaced blocks. Topples are a good example: a mode of failure in which blocks overturn and a column of rocks can shift. The precise nature of the displacement depends on local conditions, including the size and shape of individual boulders.

Large movements involve masses of material and the controls on failure are somewhat different (see *landslides*). Not only are the properties of individual joints or blocks significant, but factors such as the location and dynamics of the water-table and relative relief become important. The proportions of solid rock material, water, and air will vary, affecting the velocity of movement and destructive force of a rockfall. Rock debris becomes fluidized by water and air which becomes trapped in the moving mass. If a cushion of air forms at the base of the moving material, velocities can be great. An extreme form of movement is a rock avalanche, where large volumes of debris move considerable distances, over short time periods. Rock avalanches are associated with steep slopes, well-jointed rock, a weathering environment which reduces material strength, and travel distances which allow moving rock to descend at high velocities. The result is rapid change in slope form, the displacement of many thousands of tonnes of material, and a major natural hazard. Examples include the rock slide on the flanks of Mount Cook, New Zealand, in 1991, and the failure on Mount Huascaran, Peru, in 1970. The Mount Cook event reduced the 3764 m peak of New Zealand's highest summit by 10 m. An estimated 2.5 million m^3 of rock travelled 699 m horizontally and involved a vertical descent of 2700 m. During the Huascaran event, a mass of rock and ice moved at velocities exceeding 97 m s^{-1}, destroying the town of Yungay and killing at least 17 000 people. As the two examples confirm, the largest failures and most rapid rock-slope development rates are associated with steep, mountain environments and with active tectonics. Rock avalanches with particularly large volumes have been called *sturszstroms* and are major processes of erosion and rock-slope development in high mountains. In

119

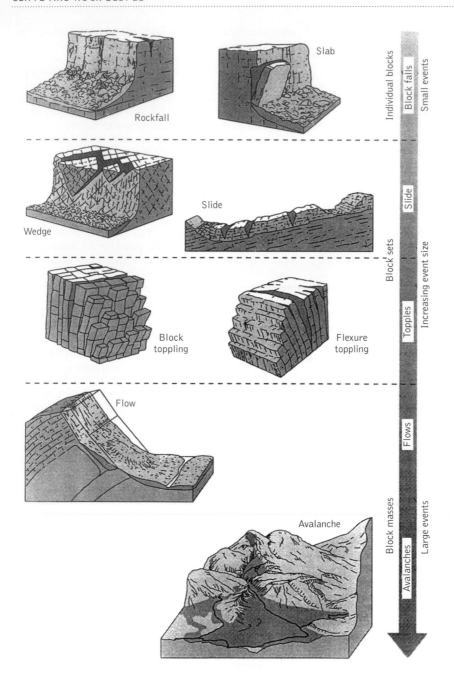

Fig. 1 Types of failure mechanism found in rock slopes and cliffs.

contrast, individual block detachment can occur from any rock slope and may reflect no more than the passage of time and the natural course of events, with weathering gradually widening a joint and a block suddenly falling away from the cliff as a consequence.

As rock-slope failures occur and form changes, there may be a gradual process of slope retreat. Two alternative suggestions have been proposed for rock-slope development. The first, parallel retreat, is a mechanism by which overall slope shape is maintained but the location of the cliff front recedes through time. Cliffs forming the margins of rifts are thought to have developed in this way, with retreat occurring since the Late Cretaceous in Precambrian to Palaeozoic rocks. An example is Fish River Canyon, Namibia, where a complex sequence of rock slopes and cliffs has retreated up to 10 km from the line of the river. The alternative model involves down-wearing of the landscape, with a gradual decline in mean slope angle and the relaxation or lowering of the terrain. Profiles eventually reach equilibrium with a thin talus cover, termed *Richter slope*. Thereafter the slope weathers uniformly, maintaining its angle but reducing in size. Good examples are found in cold environments such as the Transantarctic Mountains and the Koettlitz Valley, Antarctica. Recent developments in computing and the application of advanced mathematical modelling now permit the use of rock-mass property quantitative data to examine rock-slope evolution and link controls on development, mechanisms of failure, and changes in cliff form. ROBERT J. ALLISON

Further reading

Duff, D. (ed.) (1994) *Holmes' principles of physical geology* (4th edn). Chapman and Hall, London.

Selby, M. J. (1993) *Hillslope materials and processes* (2nd edn). Oxford University Press.

Summerfield, M. A. (1991) *Global geomorphology*. Longman, Harlow.

CLIMAP Project The CLIMAP (Climate: Long Range Mapping and Prediction) project was established in the 1970s to investigate global climatic change over the past million years. The principal objectives of this programme were the identification and examination of the key factors responsible for past climatic change and the exploration of the mechanisms that control the complex interactions between ocean–ice–atmosphere processes. One of the strategic goals of the CLIMAP project was to reconstruct the surface conditions of the Earth at the time of the last glacial maximum (LGM) around 18 000 years ago.

The main body of CLIMAP research concentrated on deciphering the records of past climatic history preserved within deep-sea sediments. This study of deep-sea material was advantageous for a number of reasons: (i) the global collection of deep-sea cores and the near-continuous palaeoclimatic records which they provide enable comparisons to be made at regional and global scales for a selected time interval; (ii) quantitative proxy measurements of past oceanographic conditions, in particular sea surface temperature, can be calculated from population counts of various planktonic microfossil groups using transfer functions determined through multivariate statistical techniques; and (iii) oxygen isotope data, which record global ice volume, and other sedimentological properties from deep-sea cores, together with radiocarbon dating, provide the detailed and global-scale stratigraphic constraints required for accurate time-slice reconstruction.

In order to achieve a comprehensive global reconstruction, it was necessary to assimilate additional geological datasets relating to other important aspects of the climate system. The massive build-up of the global ice sheets during the LGM dramatically altered the geography and nature of the land masses and considerably lowered glacial sea level. Indicators of former sea levels, vegetation patterns, lake levels, and glaciers record the impact of these changes on glacial terrestrial environments and yield additional information about the climate of the ice-age world.

CLIMAP concluded that the surface ocean temperature of the ice-age world was, on average, cooler by 2.3 °C. However, this average value concealed complex and striking regional patterns. For instance, thermal changes of around 10 °C accompanied the southward displacement of polar and subpolar water masses in the mid- to high-latitude North Atlantic, whereas the position and temperature of the tropical and subtropical Atlantic water masses remained relatively stable. This equatorward movement of the polar frontal systems at the LGM resulted in more marked latitudinal thermal gradients and a general intensification of oceanic circulation.

CLIMAP's results provide another important scientific contribution as the geological boundary conditions identified in the LGM reconstruction (the extent and elevation of permanent ice, global sea surface temperatures, continental geography, albedo patterns) have been incorporated into general circulation models which simulate the dynamic processes that control the climate of the Earth in both its ice-free and glaciated state.

A similar investigation was carried out for the last interglacial, some 122 000 years BP, when the global ice volume was comparable to present values. This CLIMAP reconstruction indicated that temperatures during the last interglacial were not significantly different from those of today.

MARK R. CHAPMAN

Further reading

CLIMAP Project Members (1976) The surface of the ice-age Earth. *Science*, **191**, 1131–7.

climate change and deep water formation

During the past million years the Earth's climate has varied on two broad timescales. There have been regular interglacial–glacial cycles over thousands of years and irregular, decadal changes within these cycles. Astronomically driven changes in the intensity of the seasons might periodically trigger climate change. It is, however, increasingly clear that other factors result in the rapid changes recorded in ice-core records, most probably variations in deep water formation.

Modern oceanic deep water masses

A deep water mass is a body of water, in contact with the sea floor, which is identifiable by its distinct combination of physical and chemical characteristics. The properties most used in identifying deep water masses are potential temperature and salinity, since away from the sea surface these can be changed only by mixing with other water masses. At present, deep water is formed in the North Atlantic and to a much smaller degree in the Weddell continental shelf, in the Atlantic sector of Antarctica. These regions produce two very distinct water masses, the southward-flowing, 'warm' (2 °C), saline (>34.9‰, i.e. more than 34.9 parts per thousand) North Atlantic Deep Water (NADW), and the northward-flowing, colder (<0 °C), less saline (>34.7‰) Antarctic Bottom Water (AABW).

Each winter in the Norwegian–Greenland sea northward-flowing water at approximately 800 m depth rises to the surface because of wind stress. This water is chilled from 10 °C to 2 °C, which, coupled with its already high salinity, results in a water mass dense enough to sink to the ocean floor, producing NADW. The winter injection of saline water from the Arctic Ocean increases the density of the NADW. NADW ponds up north of the ridge between Greenland and Scotland (which forms a major barrier to deep water flow below 400 m), until it intermittently overflows, cascading into the North Atlantic. About 20 per cent of NADW is formed in the Labrador Sea and mixes with the NADW travelling west-ward from the Norwegian Sea. The formation of NADW gives off heat equal to 30 per cent of the yearly direct input of solar energy to the surface of the North Atlantic (5×10^{21} cal); this heat is the reason for western Europe's mild winters.

NADW flows into the Pacific, via the Indian Ocean and the Antarctic Circumpolar Current. It slowly upwells, becoming shallower, reaching the surface in the North Pacific and returning to the North Atlantic as part of the upper warm water circulation. This is termed the *thermohaline circulation*. The global thermohaline circulation appears to be self-sustaining. In the North Atlantic, NADW outflow is balanced by warm surface inflow. Owing to the low salinity of sea surface waters in the North Pacific no deep water is formed at present in this region. The low salinity is due to the upwelling of cold deep water with low evaporation rates and the input of fresh water evaporated from the North Atlantic.

AABW forms by (1) the seasonal removal of water to marine ice sheets, which increases sea-water salinity, and (2) the surface cooling of sea water in pockets of ocean enclosed by sea ice termed *polynyas*. The largest polynya recorded is 1000 km by 350 km. In the Weddell Sea cold-core eddies up to 28 km across form cold-water chimneys siphoning water down into the AABW. AABW occurs in all the Atlantic, Indian, and Pacific ocean basins. It is the densest water mass found in the oceans, and in the Atlantic it travels northwards below the NADW at a depth greater than 2.5 km. Variations in NADW and AABW properties are largely a reflection of mixing between the two water masses.

Proxies of deep water masses

As deep waters travel from their regions of formation they become enriched in total dissolved organic matter, because of the continual 'rain' of detritus from the photic zone. This is termed the *ageing effect*. This enrichment results in a water mass that (1) is more acidic and therefore more corrosive to

$CaCO_3$; (2) has lower $\delta^{13}C$ values, because $\delta^{12}C$ is preferentially fixed in organic matter ($\delta^{13}C$ and $\delta^{12}C$, the delta values for the isotopes carbon-13 and carbon-12 respectively, are measures of the relative differences between the values for a sample and a standard); and (3) has an increased phosphorus content; the Atlantic deep water contains on average about half as much phosphate and nitrate as deep water in the Pacific Ocean, while the Circumpolar Deep Water (CPDW) in the southern ocean reflects a mixture of the two.

Changes in deep water distribution over time are reflected in the sedimentary record. The $CaCO_3$ content and degree of dissolution of calcite fossils varies, reflecting the corrosiveness of the benthic water mass. In the Atlantic glacial stages, calcite dissolution often increased, suggesting the presence of older deep water. Epifaunal benthic foraminifera have been used to trace past water masses by using the levels of carbon stable isotopes and cadmium incorporated in their shells. Certain species are known to incorporate $\delta^{13}C$ and cadmium into their shells in equilibrium with the surrounding sea water. Other species form shells at a constant disequilibrium to the surrounding sea water. These can also be analysed and a correction factor can be applied to obtain a value for the original sea-water composition. Cadmium is important, for in the modern ocean it shows a strong correlation with phosphorus and nitrate in sea water. Both proxies indicate, in general, an increase in the ageing effect in the NADW during glacial periods. The $^{14}C/^{12}C$ ratio within calcite shells can be used as a proxy for the age of the surrounding water when the shell calcified. Variations between $^{14}C/^{12}C$ ratios for coexisting planktonic and benthic foraminifera are used to date the benthic water mass. In the tropical Pacific the age difference today is 1600 years, whereas in the tropical Atlantic it is 350 years. It appears that during the glacial period the deep water mass in the Atlantic was at least twice this age. Sediment grain-size variations, the distribution of volcanic ash fragments, and deep-sea channels have also been used to track changes in deep water distribution and directions of bottom-water flow.

Past changes in deep water formation

Using a variety of geochemical proxies it is clear that the thermohaline circulation has more than one stable mode. There are two timescales of variability in NADW production: variations on a glacial-to-interglacial scale, and short rapid changes within glacial-interglacial periods (for example the Younger Dryas). NADW production may be a stable system for millennia and then undergo a rapid transition to a new state on the time scale of a decade. It appears that for at least the past 1.5 Ma glacial stages have had significantly less NADW production compared to interglacial periods. During the last glacial maximum the production of NADW was greatly reduced. It occurred in a region south of Iceland, and the water mass formed was convected to a shallower depth, not reaching the sea floor.

The most likely cause of the larger time-scale variability in NADW production are changes in the North America ice sheet, possibly driven by astronomical processes. These variation are on a 23 000-year cycle and therefore appear to be linked to the Earth's precessional cycle. Atmospheric general circulation models show that the North America ice sheet directly controls the sea-surface temperature of the North Atlantic. Strong cold winds, generated on the ice sheet's northern flanks, blow out across the ocean, cooling the surface waters. Lower sea-surface temperatures result in less evaporation and therefore less saline, less dense surface waters. Modelling has shown that a slight decrease in surface-water salinity could markedly decrease deep-water formation in the Norwegian Sea within a few decades. There is a strong correlation between sea-surface temperature (based largely on planktonic foraminiferal assemblages) and rates of NADW production on a glacial–interglacial time scale. A growth in ice sheets, causing a southerly shift in the polar front, might also inhibit the inflow of saline surface water from the South Atlantic which is required for NADW production.

Modelling has indicated that changes in wind direction along the Marginal Ice Zone (MIZ), may also be important in controlling the rate of NADW production. At present the dominant wind direction during winter along the MIZ is north-east, resulting in northward Ekman transport of water and ice. As ice drag on the underlying water is greater than that due to surface water drag, water is being removed to the north faster than it is being replaced from the south along the MIZ. Water therefore upwells along the MIZ to remedy the imbalance. This water is substantially cooled and sinks to form NADW. During glacial periods, because of the large North America ice sheet, the wind direction is westerly. This results in a southward Ekman transport, a build-up of water along the MIZ, and downwelling of water. This water is not, however, dense enough to sink to the sea floor.

Short timescale variations

The Younger Dyras represents a period of almost complete return to glacial conditions at 11 000–10 000 radiocarbon years BP in the North Atlantic during the last deglaciation. This has been related to injections of melt water from the North American ice sheets. During the Younger Dryas it appears that meltwaters ponded up on the southern edge of the ice sheets forming Lake Agassiz. This appears to have subsequently overflowed into the North Atlantic in a series of complex catastrophic events. This would result in a less dense 'freshwater cap' in the North Atlantic, which would favour stable stratification of the ocean. Modern observations have shown that a decrease in salinity of more than 1‰ in the surface waters is sufficient to suppress NADW formation.

There have been at least three other periods of rapid cooling of the North Atlantic climate since 14 500 BP. These are all believed to have been caused by injections of melt water, reducing the rate of NADW production. The point of entry of the melt water appears to have varied between the Gulf of Mexico, via the Mississippi, and the Arctic Ocean via the Siberian river system.

The termination of the Younger Dryas was extremely rapid, possibly taking less than a decade. Greenland ice cores indicate a rapid increase in snow accumulation at the termination. Modelling suggests that a switching on of NADW production will result in increased advection poleward of warm surface waters. This is a suggested cause for the end of the Younger Dryas and the associated increased input of atmospheric moisture over the North Atlantic.

Late Quaternary sediments in the North Atlantic contain clear horizons of ice-rafted debris. These are termed 'Heinrich events'; they represent decadal-length events of atmospheric and sea-surface cooling, reduction in the flux of planktonic foraminifera, lower sea-surface salinities, and brief, but exceptionally large, discharges of icebergs. These phenomena are the result of the slow thickening of the North America ice sheet on frozen sediment (Binge phase), giving way to rapid ice-stream surges as the water-saturated sediment defrosts (Purge phase). The injection of large amounts of ice and melt water again forms a 'freshwater cap', reducing the rate of NADW production and resulting in colder north Atlantic temperatures. The effect of Heinrich events is widespread and is seen in widely separated places: in pollen records from Florida and ice-core records from the Andes. Heinrich events and their effect on NADW production may well be a major cause of rapid climate changes within interglacial periods.

Global effects

Prolonged droughts in the Sahel and tropical Mexico during the past 14 000 years coincided with major injections of fresh water into the North Atlantic. A reduction in the production of NADW would slow the thermohaline circulation, and less water would be drawn into the North Atlantic, resulting in colder surface waters and a decrease in the rate of evaporation. This would in turn cause reduced precipitation on surrounding land masses. Between 1968 and 1982 colder, fresher waters in the Norwegian and Labrador seas, indicative of decreased NADW production, have been correlated with decreased precipitation over West Africa as the tropical rain belt shifted southward. A more sluggish thermohaline circulation would also reduce the strength of upwelling in the equatorial oceans. This would affect the rate of moisture input into the world's great tropical convective systems, resulting in more arid conditions at high latitudes.

Near-contemporaneous climate variations recorded in the Greenland and Antarctic ice cores are believed to have been driven by variations in NADW input into the Southern Ocean. Reduced NADW input would lower Antarctic surface temperatures because of decreased heat release from ocean to air, amplified by increased albedo due to ice growth. In addition, a northward shift of the Antarctic climate would reduce the poleward heat transport through the atmosphere. Modelling has, however, suggested that increased input of NADW could cool the southern hemisphere, as a larger volume of warm surface/intermediate water must be removed to replace that incorporated in NADW. The input of NADW could also displace warm intermediate waters northward in the Southern Ocean.

Deep water and CO_2 variations

Carbon dioxide (CO_2) is a major greenhouse gas, and variations in the levels of atmospheric CO_2 have a profound effect on the Earth's climate. Substantial changes in atmospheric CO_2 between glacial–interglacial periods are recorded in ice cores. These changes are at least partly driven by variations in the rate of NADW production. NADW is known from present-day measurements to have a very low alkalinity: approximately 2.3 10^{-3} eq/kg, as compared with 2.4–2.45 10^{-3} eq/kg for all other deep-water masses. Alkalinity in the oceans may be defined as the combined concentration of bicarbonate and carbonate ions. In modern oceans the CO_2 partial pressure, $p(CO_2)$, is controlled by the alkalinity; higher alkalinity relates to increased $p(CO_2)$. In the Atlantic, changes in the deep water $p(CO_2)$ are largely a result of mixing of low-alkalinity NADW and high-alkalinity AABW. The NADW flows into the CPDW, part of which subsequently upwells in the Antarctic, forming the southern low-latitude surface waters. There is little production of $CaCO_3$ by organisms within these surface waters, which could have regulated the alkalinity of the sea water. Changes in CPDW alkalinity are therefore almost directly controlled by the input of NADW.

CO_2 is transported from warm surface waters to cold surface waters. An increase in the potential of the cold high-latitude surface waters as a sink for CO_2 would therefore have an influence far beyond that indicated from its area. Ocean box models have shown that the polar oceans control the $p(CO_2)$ in warm surface waters and therefore also atmospheric CO_2 levels. Changes in the alkalinity and therefore $p(CO_2)$ of the southern low-latitude surface waters may have been involved in the decrease in CO_2 and cooling observed during glacial stages. During glacial stages NADW still reached the Antarctic, but at much reduced levels. Decreased NADW input into the CPDW resulted in surface waters with an increased $p(CO_2)$. It is important to note that changes in the input of NADW to the Southern Ocean can explain less than half the variation in atmospheric CO_2. The remainder is most likely driven by increases in high-latitude surface water productivity and variations in total ocean nutrient distribution. The use of boron isotopes ($^{11}B/^{10}B$) and barium concentrations in the calcite of foraminifera as a proxy of sea-water pH many well increase our knowledge of this topic. At present the Ba/Ca ratio of NADW is 1.8, as compared with the value of 4.5 for the more acidic Pacific Deep Water.

During the first pulse of deglaciation at 15 000 years BP it appears that there was a major flushing of the North Atlantic deep-sea basins with oxygen-rich water as NADW production was reinvigorated. This oxidized previously deposited organic matter, resulting in the generation of CO_2 and dissolution of carbonate *in situ*. The effect of this on NADW, CPDW, southern low-latitude surface water $p(CO_2)$, and atmospheric CO_2 is still uncertain.

Neogene glacial development

The development of the southern ice sheet and the resulting steepening in the global thermal gradient is related to deep water formation. NADW was not initiated until the late Miocene. Deep water during the early Miocene was formed in the low latitudes of the Tethys Sea through to the Indian Ocean and appears to have been warmer and more saline than NADW. Warmer deep water upwelled at high latitudes, resulting in warmer surface waters in the Southern Ocean. There was consequently a large amount of evaporation next to the cold land masses. More water in the atmosphere resulted in greater snowfall and is believed to be a cause of the development of the East Antarctic Ice Sheet (EAIS) during the middle Miocene. As the atmospheric temperature gradient increased, boundaries between climate zones strengthened, increasing the aridification of mid-latitude continental regions (Australia, Africa, and the Americas), and grasslands developed, providing an environment for the evolution of grazing mammals.

Another feature of an increased temperature gradient is stronger oceanic circulation. An increase in upwelling strength during the middle Miocene resulted in greater sea-surface productivity, which is recorded in extensive deposits of organic-rich sediments in the Pacific; for example the diatom-rich Monterey formation in California. This resulted in the removal of CO_2 from the atmosphere into sediments. A positive feedback loop was set up with a lowering of temperatures, since the removal of this major greenhouse gas was initially driven by cooling due to EAIS growth. The loop appears to have been broken when the nutrients available in the oceans were used up. This appears to have been a crucial step, cooling the Earth sufficiently for late Neogene glaciation. In general the Neogene unipolar ice sheet system was considerably less sensitive to variations in deep water temperatures than the late Quaternary bipolar ice sheet system.

Palaeocene benthic extinctions of deep-water fauna might be related to variations in deep-water masses. A switch from high-latitude cold, relatively fresh, deep waters to low-latitude warmer, saline deep water would result in a deep-sea oxygen deficiency, causing the remarkably rapid extinctions recorded in benthic foraminiferal assemblages, which took place in less than 3000 years. Even older deep-sea organic-rich sediments may reflect changes in the preservation potential of carbon caused by changes in deep water circulation. These sediments, acting as a sink for CO_2, may have had a major influence on the levels of atmospheric CO_2 and subsequent global climate change.

STEPHEN KING

Further reading

Broecker, W. S. and Denton, G. H. (1990) What drives glacial cycles. *Scientific American* **262** (1), 42–50.

Duplessy, J. C., Shackleton, N. J., Fairbanks, R. G., Labeyrie, L., Oppo, D., and Kallel, N. (1988) Deep water source variations during the last climate cycle and their impact on the global deep water circulation. *Paleoceanography* **3**, 343–60.

climate change and lake levels

Lakes occupy a mere 1 per cent of the Earth's land area, but they are of great importance in reconstructing its past environments. Just as mountains are areas of net erosion and sediment loss, so lakes are subject to net accumulation and sediment gain. The resulting lacustrine sediments preserve a record of past ecological conditions and geomorphic processes, which can be studied by techniques such as pollen analysis and magnetic measurements. Indeed, the build-up of sediment means that most lakes are relatively ephemeral features, as they eventually fill up and become terra firma.

As well as acting as a natural 'archive' for the surrounding landscape, lakes can also provide a valuable record of regional climate change via study of fluctuations in their water level and salinity. While all lakes respond to changes in water balance, the most responsive and easily quantified are those which are hydrologically closed. These 'non-outlet' lakes adjust their surface area and water depth dynamically in response to changing inputs in the form of rainfall and river flow, and to changing losses in the form of evaporation. Among the most important non-outlet lake basins are the Caspian Sea, Lake Eyre in Australia, and the Great Salt Lake in North America. Climatically induced fluctuations in these and many other smaller lakes have been documented in recent decades through historical records and remotely sensed imagery. Lake Chad, for example, has shrunk dramatically in area during the late twentieth century. This shrinkage has been associated with the Sahel droughts of the 1970s and 1980s (Fig. 1a). Similarly (although for different reasons), the surface area of the Aral Sea has shrunk by 40 per cent since 1960, with disastrous ecological and human consequences.

Lake hydrology

The water balance of any lake is determined by the inputs, namely precipitation onto the lake surface, inflowing streams, surface run-off and subsurface inflow, e.g. springs; and the outputs, namely evaporation from the lake surface, outflowing streams or rivers, and subsurface outflow, e.g. sink holes. If a lake is in hydrological equilibrium, inputs and outputs will balance. In a hypothetical case where groundwater exchanges are negligible, then the variable which adjusts to maintain equilibrium is surface outflow. As a lake water surplus increases, so the outflow discharge becomes greater to compensate for it. It is possible, however, that there will be no water surplus from the lake, in which case there will be no surface discharge and the lake will be hydrologically closed.

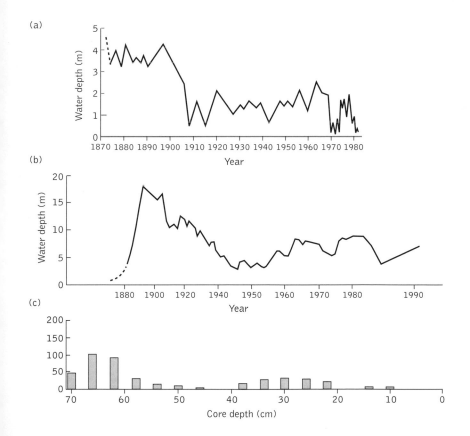

Fig. 1. Historical lake-level curves for (a) Lake Chad (Sahel) and (b) Lake Naivasha (Kenya); (c) abundance of *Microchironomus deribae* in a short core from Oloidien Bay, Naivasha. (Data from D. Verschuren (1994) *Journal of Paleolimnology* **10**, 253–63.)

In a lake without an outlet, hydrological equilibrium is maintained, not by changes in outflow discharge, but through adjustment of the area of the lake and hence by net evaporation loss. Although the surface area is the important hydrological parameter for a lake, it will co-vary with lake water level. The former can be calculated from the latter by means of the area–depth hypsometric curve determined from the individual morphometry of each lake. It is this fundamental hydrological relationship which links climate to fluctuations in lake level, and enables the past water balance and climate to be reconstructed from palaeolimnological data (i.e. data from ancient lakes).

In reality, few lakes fulfil the assumptions set out above in acting like giant 'rain gauges', particularly because of groundwater flows. A lake can be closed in terms of surface hydrology, but may not be isolated from subsurface inflows and outflows. Indeed some lakes, such as those in the Corangamite region of western Victoria in Australia, are largely fed by subsurface inflows and act essentially as groundwater 'windows'.

Lake levels and salinity

When a lake moves from being open to closed, it usually also changes from being freshwater to saline. Instead of being washed away down the outflow stream, solutes (salts) are retained within the lake water and are progressively concentrated by evaporation. As lake levels fall, the remaining water initially becomes brackish, later fully saline, and eventually hyper-saline like the Dead Sea. For this reason most arid-zone lakes are salt lakes. If, on the other hand, the level of a lake rises to the point of overflowing, salts are then flushed out of the system and the lake water becomes fresh. Past salinities are recorded in the chemical and isotopic composition of lake sediments. In a lake with an outflow, sediments such as freshwater diatomite or *gyttja* (organic mud) will be deposited, but if a lake moves from positive to negative water balance and the water becomes chemically concentrated, carbonates of various types will be laid down. The subsequent chemical evolution normally follows one of two main pathways, depending on the initial composition of the water. One pathway leads towards carbonate being the dominant anion and the eventual precipitation of salts such as trona (a form of hydrous sodium carbonate); the other leads towards chloride or sulphate dominance, or both, and the formation of compounds such as gypsum. The presence of these kinds of evaporite in a sedimentary sequence would be indicative of a hyper-saline, 'playa'-type lake environment. Evaporative concentration also influences the stable isotope ratio of the sediments, and, in particular, the oxygen-isotope ratio has proved valuable in palaeolimnological reconstruction.

This alternation between fresh and saline conditions not only influences the character of the sediments deposited in the lake, but also largely controls the organisms living within it. When preserved after death in the lake-bottom sediments, their remains can be used to reconstruct past salinities and water depths, and hence also climate. Among the most useful of these indicator organisms are diatoms, ostracods, and molluscs. Figures 1(b) and 1(c) show a historically reconstructed water-level curve from Lake Naivasha in Kenya, together with the abundance of head capsules of a salt-tolerant chironomid (midge larva) from a short sediment core taken in Oloidien Bay. This arm of the main lake is isolated and saline during periods of low water level, as occurred before 1887 and again in the middle decades of the twentieth century, and at these times this chironomid species was relatively abundant. Much recent research on biological indicators has gone on to use statistical methods to reconstruct past lake salinity, especially from diatoms.

Where the regional bedrock is highly permeable, salts may not be retained in the lake, and in this case water-level fluctuations will not be accompanied by significant changes in lake salinity. This is true, for example, of the groundwater-fed lakes of the Parkers Prairie Sand Plain in Minnesota and those of the limestone massifs in Morocco's Middle Atlas mountains. At these sites, fluctuations in water depth rather than in salinity provide the main record of past changes in water balance and climate. Former high lake levels are also indicated by geomorphological evidence, notably as shoreline terraces left above the present-day water surface.

Lake-based climate reconstruction

Various methods have been used to calculate past precipitation levels for individual lake basins. Two of the principal methods are the simple water balance and combined water- and energy-balance approaches. In the first of these, the former lake area is used to calculate total evaporation losses from the water surface, which can then be balanced against the input of precipitation plus runoff from the catchment. The chief assumption that this approach makes is that past temperatures—the primary control over evaporation rates—are known. In the second approach, the evaporation term is eliminated between the water- and energy-balance equations, but it is necessary to estimate certain surface properties such as albedo and the Bowen ratio (a function of vapour pressure and temperature, which is used in the assessment of evaporation). These properties are related to the former vegetation. Both these and area-based approaches have been used to provide estimates of past precipitation in regions such as inter-tropical Africa. They indicate an average annual increase in rainfall of at least 250 mm for the Sahara, East Africa, and South Asia about 9000 (radiocarbon) years ago.

It was originally believed that the pluvial (wet) conditions responsible for climatic changes in lower latitudes matched up with the glacial (or cold) periods of high latitudes—and in a few subtropical areas, notably the American south-west, this appears to have been the case (see *pluvial lakes*). However, for many other regions, particularly those within the tropics, radiocarbon dating has now shown that the last wet period occurred during the Late Pleistocene and the first half of the Holocene (between 12 500 and 5000 years ago). In areas such as the Sahel, lakes which today are saline or dry contained fresh water during the early Holocene, and were able to support rich and varied aquatic faunas. The increased wetness and higher lake levels of the early Holocene appear to have been brought about by an intensification and northward displacement of the Indian Ocean monsoonal circulation, itself related to orbital forcing associated with Milankovich cycles.

Because of their typically rapid response time, lake levels also provide important palaeoclimatic data on shorter-term 'Sub-Milankovich' timescales. Many of the more detailed lake-level records show significant fluctuations lasting between 100 and 1000 years. These mark abrupt climatic events produced by phases of major explosive volcanism, reorganization of oceanic circulation, or similar factors. At least one of these events, the Younger Dryas, appears to have been global in nature, and is marked in the tropics by a sharp fall in lake levels, indicating a period of climatic aridity. Lakes—or at least their remains in the form of lacustrine deposits—therefore provide valuable evidence for the changing climate and hydrology of the world's dry lands over a hierarchy of timescales during the late Quaternary.   NEIL ROBERTS

Further reading

Almendinger, J. E. (1993) A groundwater model to explain past lake levels at Parkers Prairie, Minnesota, USA. *The Holocene*, 3, 105–15.

Street-Perrott, F. A. and Harrison, S. P. (1985) Lake levels and climate reconstruction. In Hecht, A. D. (ed.) *Paleoclimate analysis and modeling*, pp. 291–340. John Wiley and Sons, New York.

Street-Perrott, F. A. and Roberts, N. (1983) Fluctuations in closed-basin lakes as an indicator of past atmospheric circulation patterns. In Street-Perrott, A., Beran, M., and Ratcliffe, R. A. S. (eds) *Conference on variations in the global water budget*, pp. 331–45. Reidel, Dordrecht.

climate change and laminated sediments

Laminated sediments can often be used to infer short-term (e.g. seasonal to annual) climatic or oceanographic fluctuations, since each lamina is comprised of sediment deposited under unique conditions. For example, annual winter rains lead to the deposition of silty laminae in the Santa Barbara Basin, off the coast of California. A first step in deciphering past climate records from laminated sediments is thus to identify the sedimentary components of successive laminae. This can be done *in situ* by using data from sediment traps. Sediment traps continuously collect descending sediment (e.g. biogenic or terrigenous material), collection periods sometimes being for a year or more. Over such periods of time, any seasonal changes in sedimentation patterns will be observed, such as increasing silt deposition due to increased rainfall or glacial discharge, or abundant diatoms due to algal blooms. Such observations can be compared with the laminated sedimentary record observed directly from the sediments. Advances in scanning electron microscope (SEM) back-scattered electron imaging techniques make it possible to examine micron-scale lamina variations in rock sections; these can provide detailed records of sub-annual deposition. Data from both sediment-trap studies and SEM examination of core intervals can be compared with his-torical climate records. The interpretations provided by such studies greatly aid both the study of laminated sediments that predate historical records and the understanding of longer-term variations in palaeoclimate and palaeoceanography.

Many studies of hand or microscopic specimens analyse specific features of laminated sedimentary sections; for example, colour or grey-scale changes. The occurrence through time of any variations can be documented, and their statistical significance investigated. The periodicity of any significant frequencies is in some instances related to climatic or oceanographic phenomena; for example, periodicities between 2 and 7 years correspond to the periodicity of the El Niño–Southern Oscillation (ENSO).

Laminated sediments occur in various depositional environments, but primarily where water is poorly oxygenated at the sea or lake bed, that is, where water is either dysaerobic (oxygen-poor) or anoxic (no oxygen present at all). Under these conditions, the activity of benthonic organisms is severely or entirely curtailed, and individual laminae are preserved. Typical locations where this occurs are: (1) basins with limited deep water replacement; for example, Californian Borderland Basins; (2) areas of intense upwelling, which experience high organic-carbon sedimentation rates owing to increased primary productivity, for example, the Gulf of California, and off the coast of Peru; and (3) lakes, where stratification of the water column often prevents oxygenation of the deeper waters; for example, Arctic lakes and high alpine lakes. More recently, deep-sea laminated diatom oozes have been recovered from the Pacific and Atlantic Oceans, making it possible to investigate short time-scale oceanographical processes such as Miocene–Pliocene El Niño-like events in the eastern equatorial Pacific. In the ancient sedimentary record, laminated sediments are preserved in marine sediments from periods of ocean anoxia, and similarly from freshwater and saline lakes, providing insights into ancient climate and ocean change.   R. B. PEARCE

Further reading

Kemp, A. E. S. (ed.) (1996) *Palaeoclimatology and palaeoceanography from laminated sediments*. Geological Society Special Publication No. 116.

climate change and palaeovolcanism

Scientists have long known that major explosive volcanic eruptions, which produce large quantities of ash, often result in cooling of global climate. Such eruptions are often associated with the injection of large quantities of volcanic ash into the stratosphere, where it can persist for several years. Dust injected into the troposphere, by contrast, is soon washed out by rain. The ash that reaches the stratosphere, however, absorbs incoming solar radiation and is heated.

It is generally believed that large low-latitude eruptions are likely to cause climate cooling on a global scale, whereas those that take place at higher latitudes tend to cause cooling within one particular hemisphere. Professor Hubert Lamb attempted to compare the possible climatic effects of the veils of ash caused by different historical eruptions. His 'dust veil index' (DVI) has a reference value of 1000, based on the violent eruption in 1883 of the island of Krakatoa in Indonesia. Table 1 lists the major volcanic eruptions that occurred between 1680 and 1970; the DVI value attributed by Lamb to each eruption is also given. A par-

ticularly well-known and catastrophic event took place during 1815 when the eruption of Mount Tambora in Indonesia resulted in the establishment of a dust veil over much of Europe and led the poet Byron to write 'Time of Darkness'.

A question of great scientific interest has long been whether or not extremely large volcanic eruptions in the past have been capable of inducing global climate cooling that would be sufficient to cause the growth of glaciers and ice sheets. Several clues are found in the geological record, where there is clear evidence of several exceptionally explosive volcanic

Table 1 Major volcanic eruptions during the past 300 years
The associated Dust Veil Index value according to Lamb is shown for each eruption. (After Lamb 1982.)

Year	Volcano	DVI
1680	Krakatoa, Indonesia	400
1680	Tongkoko, Celebes	1000
1693	Hekla, Iceland	100
1693	Serua, Molucca Is.	500
1694	Amboina, Molucca Is.	≥250
1694	'Celebes'	≥400
1694	Gunung Api, Molucca Is.	400

Note: If the low temperatures prevailing in England, as well as Iceland and a wide surrounding region, over the years 1694–8 were representative of a world-wide anomaly of about the same amount, and provided their departure from the temperatures prevailing in the immediately preceding and following years were entirely due to volcanic dust, the total DVI for 1694–8 should be 3000–3500.

Year	Volcano	DVI
1707	Vesuvius, Italy	150
1707	Santorini	250
1707	Fujiyama, Japan	350
1712	Miyakeyama, Japan	200
1717	Vesuvius, Italy	100
1717	Kirishima Yama, Japan	200
1721	Katla, Iceland	250
1730	Roung, Java	300
1744	Cotopaxi, Ecuador	300
1752	Little Sunda Is., possibly Tambora	1000
1754	Taal, Luzon, Philippines	300
1755	Katla, Iceland	400
1759	Jorullo, Mexico	300
1760	Makjan, Moluccas	250
1763	Molucca Is.'	600(?)
1766	Hekla, Iceland	200
1766	Mayon Luzon, Philippines	2300(?)
1768	Cotopaxi, Ecuador	900
1772	Gunung Papandayan, Java	250
1775	Pacaya, Guatemala	1000(?)
1779	Sakurashima, Japan	450
1783	Eldeyjar, off Iceland	
	Laki and Skaptar Jö kull, Iceland	700
1783	Asama, Japan	300
	Total veil 1783:	1000
1786	Pavlov, Alaska	150
1795	Pogrumnoy, Umanak Is., Aleutians	300
1796	Bogoslov, Aleutians	100

Year	Volcano	DVI
1799	Fuego, Guatemala	600
1803	Cotopaxi, Ecuador	1100(?)
1807	Various, including	
−10	Gunung Merapi, Java	(?)
	and Säo Jorge, Azores	(?)
	Total veil. 1907–10:	1500(?)
1811	Sabrina, Azores	200
1812	Soufriére, St Vincent	300
1812	Awu, Great Sangihe, Celebes	300
1813	Vesuvius, Italy	100
1814	Mayon, Luzon, Philippines	300
1815	Tambora, Sumbawa, Indonesia	3000
1821	Eyjafjallajökull, Iceland	100
1822	Galunggung, Java	500
1826	Kelud, Java	300
1831	Giulia or Graham's Island	200
1831	Pichincha, Ecuador	(?)
1831	Babuyan, Philippines	300
1831	Barbados	(?)
1835	Coseguina, Nicaragua	4000
1845	Hekla, Iceland	250
1846	Armagora, South Pacific	1000
1852	Gunung Api, Banda, Moluccas	200
1856	Cotopaxi, Ecuador	700
1861	Makjan, Moluccas	800
1875	Askja, Iceland	300
1878	Ghaie, New Ireland, Bismarck	
	Archipelago	possibly 1250
1883	Krakatoa, Indonesia	1000
1888	Bandai San, Japan	250
1888	Ritter Is., Bismarck	
	Archipelago	250
1902	Mont Pelée, Martinique	100
1902	Soufriére, St Vincent	300
1902	Santa Maria, Guatemala	600
	Total veil 1902:	about 1000
1907	Shtyubelya Sopka Ksudatch,	
	Kamchatka	150
1912	Katmal, Alaska	150
1963	Mt Agung (Gunung Agung), Bali	800
1966	Awu, Great Sangihe, Celebes	150–200
1968	Fernandina, Galapagos	50–100
1970	Deception Is.	(200)

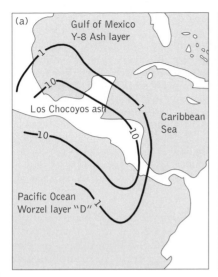

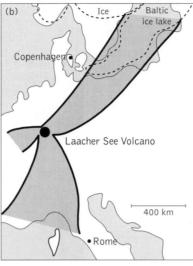

Fig. 1. (a) Contour map showing distribution and thickness (in cm) of ash layer associated with Atitlan eruption; (b) distribution of the Laacher See Tephra. (From Dawson 1992.)

eruptions that have taken place in the last 100 000 years. Each eruption had a magnitude much greater than those associated with volcanic eruptions during the Holocene (i.e. the past 10 000 years).

Evidence for past volcanic eruptions is not only present on the land surface, where volcanic ash layers and basalt deposits testify to the occurrence of former volcanic activity. Numerous layers of volcanic ash are also present within sediments on the floors of the world's oceans and some of these bear witness to the former occurrence of exceptionally large eruptions on land. Evidence of former volcanic activity is also present within ice cores taken from the Greenland and Antarctic ice sheets, within which layers of volcanic ash are present. However, the ice cores also contain an 'invisible' record of past volcanic activity because changes in the acidity of the ice indicate the occurrence of former volcanic eruptions. This is a consequence of explosive volcanic eruptions being commonly associated with the production of dilute hydrochloric and sulphuric acids, which result in acid precipitation. It has long been known, for example, that change in the acidity composition of ice in the Greenland ice cores provides a surrogate record of past volcanic activity in neighbouring Iceland.

One of the most explosive volcanic eruptions that has taken place during the last 100 000 years is recorded in layers of volcanic ash that are present in ocean sediments in the Gulf of Mexico and the eastern equatorial Pacific Ocean. The most widespread ash horizon in this area occurs across a large area of the eastern Pacific, and the same ash layer has been observed in sediments on the floor of the North Atlantic Ocean, in the Gulf of Mexico, and in the Caribbean Sea (Fig. 1a). This ash layer, known as the Los Chocoyos Ash, occurs across an ocean area of approximately six million square kilometres; the source of the eruption appears to have been the Lake Atitlan area of the Guatemalan Highlands. Scientific analysis of marine microfossils in some of the sediment cores in which this ash layer was present appears to indicate the occurrence of a period of climate cooling at approximately 84 000 years before the present (BP). It has also been noted from oxygen isotope studies of Greenland ice cores that there is a pronounced 'spike' in the ice record that appears to indicate a major period of global ice accumulation at this time. At present, research is under way to define more precisely the relative timing of this period of ice accumulation on the continents and the date of the volcanic eruption.

The explosive event of greatest magnitude that took place during the past million years was the eruption of Toba, northern Sumatra (Fig. 2). This eruption dwarfs, both in scale and magnitude, all other volcanic eruptions during the Quaternary and may have profoundly influenced global climate. Initial accounts of the eruption noted that the area of the volcanic caldera at Toba, now occupied by a lake, had an area of approximately 3000 km². The Toba ash layer is present throughout much of the Indian Ocean and is also present on the Indian subcontinent. Detailed dating of the ash layer indicates that this eruption took place at approximately 75 000 years BP. Calculations of the rate of discharge of volcanic magma from the volcano appear to indicate that the column height of the eruption was in the region of 50 to 80 km. If this is true, the Toba eruption would have been associated with the injection of immense volumes of volcanic ash into the stratosphere, and possibly also into the overlying mesosphere.

The great significance of the Toba eruption is that the event seems to have taken place at about the same time as a major episode of rapid and widespread ice-sheet growth throughout the northern hemisphere. It is tempting to speculate that rapid northern hemisphere glaciation at 75 000 years ago was

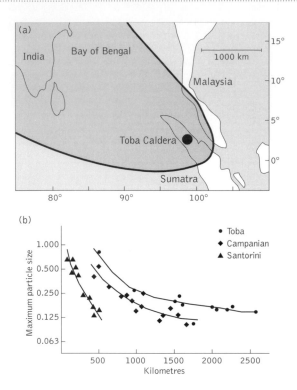

Fig. 2. (a) Distribution of the tephra plume associated with the Toba eruption derived from deep-sea core evidence; (b) plot of maximum particle size with distance as related to three volcanic eruptions of high magnitude. (From Dawson 1992.)

place in the Laacher See area of northern Germany at approximately 11 000 years BP (Fig. 1b). In these eruptions volcanic ash was deposited over at least 170 000 km^2 of the northern European lowlands and also across southern Scandinavia. In Iceland, an extremely explosive volcanic eruption appears to have taken place at approximately 10 600 years BP. This eruption was associated with the deposition of volcanic ash across the North Atlantic Ocean, and also across Northern Europe: the ash is widely recognized in the ocean sediments in the North Atlantic. The source of the eruption is believed to have been Mount Katla in southern Iceland, where there was an explosive eruption from underneath a pre-existing ice cap. The climatic significance of these various extremely large volcanic eruptions in the Northern Hemisphere is, at present, unknown.

There is a school of thought that maintains that certain environmental changes during the last Ice Age may have influenced the timing of individual volcanic eruptions. For example it has been argued that the redistribution of water associated with the growth and decay of ice sheets during this period may have given rise to both hydroisostatic and glacioisostatic processes that triggered volcanism. Crustal adjustments would have been most active along plate margins and the intersections of major lineaments (i.e. faults). In this way the reactivation of faults might have triggered subcritical magma bodies through the mechanism of magma mixing. It has been argued on various grounds that the rapid melting of ice sheets might frequently have led to both faulting and volcanic activity. It has been proposed that the very rapid ice thinning during the melting of the last ice sheets could have led to considerable changes in the stress field in the bedrocks and that the fault zones activated by these processes might later have acted as conduits for ascending magma. It has in addition been argued that ice-sheet melting might have been associated with an increased thermal gradient in the Earth's crust as well as causing increased pressures within individual volcanic magma chambers that led, on occasions, to explosive eruptions. It is certainly true that several very large eruptions took place at the time of the melting of the last ice sheets. For example, the eruption at Mount Katla (see above) in Iceland took place from beneath a rapidly thinning ice mass. Similarly the very explosive eruptions in the Glacier Peak range were associated with rapid ice sheet thinning. Other authors have shown that very large eruptions took place on Iceland during the last interglacial period and that this was in some way related to melting of the Icelandic ice sheet. Certain authors have described a scenario in which the build-up of individual ice sheets results in the downwarping of the upper part of the crust, which subsequently causes a reduction in the amount of volcanic activity. It has in addition been argued that rapid changes in ocean water pressure can also influence the amount of volcanic activity that takes place. Thus, a fall in relative sea level in a particular area might be accompanied by an increase in volcanism, whereas a rise in sea level would have the opposite effect.

triggered by this eruption and that the eruption provided a critical threshold for the initiation of widespread glaciation at a time when, according to Milankovich reconstructions, there was also a marked decline in summer solar radiation.

Some scientists have argued that whereas a single highly explosive volcanic eruption in the past may have been responsible for the initiation of a period of global climate cooling, it is also possible that similar periods of climate cooling may have been caused by a series of volcanic eruptions that took place at various locations throughout the world during a relatively short time interval (10 to 1000 years). For example, between 13 000 and 11 000 (radiocarbon) years BP there appears to have been a series of major eruptions in the north-western USA, Germany, and Iceland. In the Cascade range of the north-western USA, two major volcanic eruptions took place in the Glacier Peak region between approximately 12 750 and 11 250 years BP. These eruptions were associated with the transport of ash for a distance of at least 1000 km east of the volcanoes and the deposition of ash to a thickness of 2 to 3 m as far as 30 km downwind. In Germany a period of explosive volcanism took

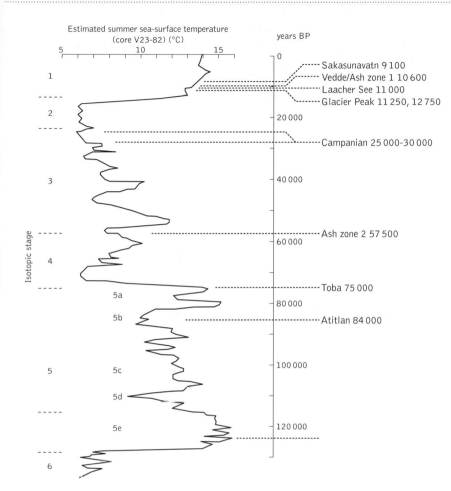

Estimated summer sea-surface temperature
(core V23-82) (°C)

years BP

Isotopic stage

------- Sakasunavatn 9100
------- Vedde/Ash zone 1 10600
------- Laacher See 11000
------- Glacier Peak 11250, 12750

------- Campanian 25000–30000

------- Ash zone 2 57500

------- Toba 75000

------- Atitlan 84000

Fig. 3. Timing of major volcanic eruptions in relation to a reconstruction of Late Quaternary summer sea-surface temperature changes for the North Atlantic. The oxygen isotope stages are also shown. No correlation between eruptions and sea-surface temperatures is implied. (From Dawson 1992.)

The interrelationships between explosive volcanic activity and climate are a priority area in the study and understanding of climate change. Most computer models that attempt to predict future patterns in climate change do so on the assumption that volcanic activity has a negligible impact on climate. Yet the examples of Toba and Atitlan demonstrate that there can indeed be a relationship between extremely large explosive eruptions and climate (Fig. 3). Scientists who investigate past, present, and future climate change ignore volcanoes at their peril.   ALASTAIR G. DAWSON

Further reading

Dawson, A. G. (1992) *Ice Age Earth: Late Quaternary geology and climate*. Routledge, London.

Kemp, D. D. (1990) *Global environmental issues: a climatological approach*. Routledge, London.

Lamb, H. H. (1982) *Climate history and the modern world*. Methuen, London.

climate models Climate models are an attempt to simulate and predict changes to the climate system. The climate system consists of five components, each of which is a system in itself. The components are the atmosphere, the hydrosphere (all liquid water), the cryosphere (all frozen water), the biosphere (living things) and the lithosphere. In essence the climate is driven by the balance between the incoming solar (short-wave) radiation and the outgoing terrestrial (thermal and infrared) radiation. The amount of solar radiation received at the tropics is far more than at the poles. This results in a temperature imbalance. Atmospheric and oceanic circulations consequently work to redistribute heat from the Equator to the poles. These circulations interact with the other components of the climate system through processes which transfer energy, mass, and momentum and this then determines the Earth's climate.

The climate is in dynamic equilibrium, that is, over a sufficient period of time the incoming and outgoing radiation are in global balance. Should there be changes (forcings) in either the amount of incoming or outgoing radiation, then the climate system will respond to that change by moving to a new equilibrium position. The climate will not move instantaneously to the new equilibrium climate; there will be a transition period during which there is a transient climate. The forcings which cause changes can be external to the system, such as changes in the amount of solar radiation reaching the Earth due to changes in the orbital configuration, or they can be internal. These internal forcings result from feedback within the climate system. Climate models show that increased carbon dioxide from human activity will result in a temperature increase, often referred to as global warming. This increase in temperature will lead to more water vapour in the atmosphere, and, as water vapour is also a greenhouse gas, there will be a further increase in temperature. As this reinforces the original change it is called a positive feedback. There are many examples of this type of feedback within the climate system. Negative feedback, which will counteract the original change, is less common.

Each component of the climate system takes a different amount of time to respond to a change in the radiation regime. The atmosphere takes only a few days to respond, but the surface layers of the ocean take weeks to months, and the deep circulation takes centuries. Temporal scales are as important to climate modelling as spatial scales, for they will determine which components and which processes need to be considered in a climate model. An example of this point is the lithosphere, which through processes such as mountain building can alter the climate. The processes occur, however, on such long timescales that, except for studies concerned with geological climate change, the lithosphere can be regarded as constant.

Analogue models

The term 'climate model' is most often used to describe computer-based models of the climate. However, past climates can be used as analogue models of future climates. These have been used to predict the consequence of global warming. There is strong evidence that during several periods in the geological past carbon dioxide levels in the atmosphere were far greater than they are now. By using data derived from geological sources it is possible to reconstruct temperature and precipitation at those times and hence build up a picture of the climate. These models can then be used as analogues for a climate perturbed by anthropogenically increased greenhouse gases. The mid-Holocene optimum (5–6 thousand years BP), The Last Interglacial (125–30 thousand years BP), and the Pliocene (3–4 million years BP), have been used as analogues for 2000, 2025, and 2050 respectively. The chief criticism of analogue models is that the forcing mechanisms in the past may not mimic those of the future. Other criticisms are that relatively local geological data are being used to represent a large area, and that the proxy data are imprecisely dated; there can also be disagreement over the reconstructed climate.

Computer-based models

Computer-based climate models attempt to predict the climate from basic physical principles. These models can be used to examine present, past, and future climates. The difference between the present-day climate simulated by the model and the climate produced by some change in the forcing is the simulated climate change. The simplest of computer-based climate models are energy balance models. These models divide the Earth into latitude zones. As latitude is the only dimension used, these are one-dimensional models. For each latitude zone the model computes the incoming and outgoing energy. The equations that govern the model are then all written in terms of the variable to be predicted, usually surface air temperature. Energy balance models of this type came to prominence in the late 1960s, when they were used to assess the sensitivity of the climate to a change in the amount of incoming solar radiation. The models predicted that a relatively small decrease in the amount of radiation reaching the Earth could lead to total glaciation. At the time when these results were published, temperatures in the northern hemisphere were on a downward trend. This led to speculation that the Earth could be heading for a new 'ice age'. The extreme sensitivity turned out, however, to be due to the simple way in which the model represented the climate and did not present a true picture of the climate response. Energy-balance models have since increased in sophistication and are still in use today, mostly to study climates in the geological past.

The first type of model to be used, the three-dimensional atmospheric general circulation model (AGCM), is the most complex. Models of this type emerged from numerical weather-prediction models which were developed in the years after the Second World War. In the 1960s, oceanic general circulation models were developed. Since then many other components of the climate system, such as sea ice and the biosphere, have been modelled. All these models have to be coupled together in order to simulate the climate. Until the mid 1990s most large-scale climate change experiments were restricted to AGCMs, and the following discussion reflects that fact.

As the name suggests, AGCMs attempt to predict the general circulation of the atmosphere from first principles. The basic laws of physics that govern the motion of the atmosphere are embodied in what are called the fundamental equations. They are the conservation of momentum (Newton's second law of motion), the conservation of mass (the continuity equation), the conservation of energy (the first law of thermodynamics), the equation of state (the ideal gas law); there is also a fundamental equation for moisture. This group of equations relates the basic variables (pressure, wind velocity, temperature, and humidity) in both space and time. Given a set of values for the variables at some starting time (the initial conditions), it should be possible to solve the equations for some future time and obtain new values for the variables. The equations are, however, extremely complex, and in order to solve them approximations have to be derived. This can be achieved in two ways, leading to two types of GCM. One is called a Cartesian grid GCM; the other a spectral GCM. The former works on a grid akin to the latitude–longitude grid which appears on maps of the Earth.

The latter is much more difficult to visualize, for it represents the fundamental equations as wave forms. Both types of models typically have a horizontal grid spacing of between 250 to 800 km in both latitude and longitude. The third dimension, the vertical, usually has 10 to 20 levels to describe it. It might be thought that a sensible choice for the vertical coordinate would be pressure. This, however, leads to some interesting problems; grid points could change height and could even appear inside a mountain. To avoid these problems a method was devised called the sigma coordinate system. In this system the actual pressure at a point is divided by the surface pressure at the point vertically below it. The lowest sigma level follows the contours of the Earth's surface exactly.

Although the fundamental equations can predict the dynamics of the atmosphere, there are other physical processes that also have to be modelled, such as radiation, clouds, and surface exchanges. These processes occur on spatial scales far smaller than the grid spacing of the climate model and they are not predictable from first principles. Instead they are parameterized; that is, the equations to predict these processes are simplified and formulated as functions of the variables from the fundamental equations. There are good reasons for parameterization. First, it would be impossible computationally to predict everything from first principle. Secondly, we often do not know in enough detail the underlying principles of the processes, and so the equations have to be simplified in some way. Most parameterizations are based on sound empirical evidence or are approximations of more complex functions. An example of the latter is the parameterization of the incoming and outgoing radiation streams. For both Cartesian grid models and spectral models the calculation of these parameterizations is completed on a Cartesian grid. Spectral models therefore have to transform from one grid type to the other.

To a large extent oceanic general circulation models (OGCMs) are based on the same fundamental equations as for the atmosphere, as both are fluids. There are, however, some important differences. The atmosphere responds much more quickly to changes in the energy balance than the ocean; OGCMs therefore need to consider far longer time-scales than AGCMs. In contrast, the spatial scales of OGCMs need to be much smaller than AGCMs, for important motion occurs at lower spatial scales. The radiation treatment is less complex than for the atmosphere. The ocean is affected far more by the ocean bottom than the atmosphere is by the Earth's surface. Sea ice, by its very nature, is very different from the atmosphere and the ocean, and consequently so are the models. To a large extent sea-ice models concentrate on predicting whether sea ice exists or not at a particular location and time, rather than determining its properties as in the atmospheric models.

Once a climate model has been constructed, it needs to be validated. This is achieved by comparing the simulated present-day climate with observed data. Validation studies have revealed that general circulation models can predict on a global scale the features of the present-day climate. It should, however, be borne in mind that in the earlier studies sea-surface temperatures and sea ice were often constrained to have realistic values only. Predictions of regional-scale features are widely recognized as being poor. These large-scale climate models have been extensively used to investigate global warming. The studies suggest a 2–5 °C temperature rise in the global average. Such equilibrium studies are rather false, as the climate will reach equilibrium only long after the amount of carbon dioxide has reached some constant level. Initially most of the studies were equilibrium studies based on an instantaneous doubling of carbon dioxide. They are, however, computationally efficient, for they do not require the deep circulation of the ocean to be considered. Furthermore, the results from different climate models can easily be compared. Nevertheless atmospheric and oceanic general circulation models, as well as sea-ice models, are increasingly being coupled to predict transient climate changes. These studies allow for gradually increasing levels of carbon dioxide as well as deep ocean circulation. It should be mentioned that GCMs are not without their critics, who argue that the sensitivity of the climate to changes in carbon dioxide is overestimated.

In an effort to understand more about the regional changes that might occur with global warming, the results from large-scale climate model studies, such as those discussed above, are being used as initial conditions for regional-scale model studies. Also, as the need for policy decisions related to global warming increases, models are being formulated which include a climate model together with various socio-economic models in an effort to model future emissions of greenhouse gases and the consequent climate change.

FRANCES DRAKE

Further reading

Houghton, J. T. (1994) *Global warming: the complete briefing.* Lion Publishing.

climatic trends, recent *see* RECENT CLIMATE CHANGES

Cloos, Hans (1885–1951)

Hans Cloos is best remembered for his work on structural geology, and in particular the tectonics of granites and other plutonic rocks, but he also made contributions to geomorphology. He was born in Magdeburg in Germany, and began a training in architecture before he turned to geology. On leaving the University of Freiburg, he went to South-West Africa, where he mapped plutonic rocks and began the research that was to be a significant part of his life's work. Apart from a spell working in Indonesia for an oil company and service with the German army in the First World War, the rest of his career was in German universities, first at Marburg, then at Breslau, and finally at the Geologisches-Palaeontologisches Institut of Bonn University, where he was professor from 1925 until 1941. While at Bonn he was the editor of the journal *Geologisches Rundschau*.

In the late 1920s Cloos began a series of experiments in which he used sand and clay to model the deformation of rocks, reproducing folds, faults, and fracture systems. Perhaps the best known of these experiments was the simulation of the Rhine graben.

As well as a textbook, *Einführung in die Geologie*, Cloos published an autobiography, *Gespräch mit der Erde*, which appeared in an English translation as *Conversation with the Earth* in 1954.   D. L. DINELEY

clouds In 1803 Luke Howard developed a classification system for naming different cloud types. Many of the names that he gave to clouds are still in use by surface observers today. There are ten distinguishable types, called genera, which have a Latin name and a recognized abbreviation. The identification of a cloud as of a particular type is based on a subtle blend of its appearance and altitude. The basic cloud forms are cumulus, which are heaped clouds; stratus, which are layer clouds; and cirrus, which are wispy. If the cloud is raining then the term 'nimbus' is used, as in cumulonimbus and nimbostratus. Further extensions to the genera names, called species, provide a fuller description. Apart from some notable exceptions in the stratosphere, clouds form in the lowest layer of the atmosphere, that is, in the troposphere. Although clouds can form at any altitude in the troposphere, they are usually assigned to one of three levels, low, middle, or high, which further aids classification. The exact height of these groups varies with latitude and season. Some clouds, such as cumulonimbus, develop vertically beyond one level. This often leads to a fourth group being defined: those clouds with vertical extent. Table 1 lists the ten genera together with their abbreviation and height of the cloud bases above sea level.

The appearance that clouds take is determined by the mechanism which forms them. Clouds form when water vapour, which is always present in the atmosphere, condenses out. Condensation occurs when the air is cooled to such a point that the air is saturated with respect to water vapour. Further cooling results in liquid water or ice. Condensation in clean air is very difficult to achieve, and air can easily become supersaturated under such conditions. Most condensation takes place on aerosols which are hygroscopic (water-absorbing). Naturally occurring aerosols are abundant in the Earth's atmosphere, for example in the form of dust and salt from sea spray. Pollution from human activity also contributes suitable aerosols for cloud formation. Cooling of the air to produce clouds can occur in a variety of ways. The most common is when air rises; the pressure decreases, and so the air expands, doing work and expending energy and therefore cooling. Small parcels of air rising because the atmosphere is unstable result in cumulus clouds. Clouds can result when air is forced to rise over mountains. Large-scale uplift, as seen at a front (the intersection of two air masses with different characteristics) creates predominantly stratus clouds. Other cooling mechanisms, such as warm moist air being cooled when it comes into contact with a cold surface, can result in fog. Lower clouds are predominantly composed of liquid water droplets, but high clouds such as cirrus are composed entirely of ice crystals.

Clouds, being white, are highly reflective and they reflect the incoming short-wave radiation from the Sun. They thus have a cooling effect on the Earth, the cloud albedo effect. However, because clouds are composed of water, which is highly efficient at absorbing long-wave radiation, clouds also have a greenhouse effect; that is, they cause the Earth's surface to become warmer by stopping outgoing long-wave radiation from escaping into space. Clouds therefore have a strong control on the radiation budget of the Earth; and as it is this budget which ultimately determines the Earth's climate, the radiative effect of clouds is of immense importance to climate studies. Low- and middle-level clouds are thought to cool the Earth's surface—that is, their albedo effect dominates—but for high cloud it is believed that the greenhouse effect dominates. The effect of clouds on the radiation budget is determined not only by their gross properties, such as their amount and height, but also by their microphysical properties, such as the size of the droplets in the clouds and their water content.

Surface observers throughout the world record the type, height, and amount of cloud covering the sky. The amount of sky covered is typically measured in oktas, eighths of sky cover, a non-linear scale. Although height can be measured by instrumental methods, it is often an estimate. All measures of cloud parameters observed at the surface are subjective and subject to human error. Moreover, human surface observers are unable to provide complete Earth coverage. Satellite observations of clouds, which began in the 1960s, potentially offer an objective measure of cloud parameters over the entire globe. Although trained meteorologists can, most successfully, interpret satellite cloud images, the volume of satellite data and the need for objectivity require computer analysis. The interpretation of digital satellite data through computer algorithms is, however, complex and not without its own problems. Total cloud amount, as a percentage of ground cover, is relatively easy to assess, but estimation of the cloud type and

---

Table 1. Cloud genera

Low clouds

Height of cloud bases: all regions below 2 km

Cloud types: stratus (St), stratocumulus (Sc)

Cloud types with vertical extent; all have bases in the low-level group:

nimbostratus (Ns), cumulus (Cu), cumulonimbus (Cb)

Middle clouds

Height of cloud bases: tropical region, 2–8 km; temperate region, 2–7 km; polar region, 2–4 km

Cloud types: altostratus (As), altocumulus (Ac)

High clouds

Height of cloud bases: tropical region, 6–18 km; temperate region, 5–13 km; polar region, 3–8 km

Cloud types: cirrus (Ci), cirrostratus (Cs), cirrocumulus (Cc)

the amount of each cloud type has proved more difficult. The whole idea of cloud types as discussed earlier is in fact rather misleading when referring to satellite data: satellite-retrieved cloud parameters are essentially a measure of the cloud effects on the radiation reaching the satellite sensor. Satellite-retrieved cloud parameters do not therefore readily equate to cloud parameters observed at the surface. We thus have two incompatible sets of data. Satellite data nevertheless hold the promise of being able to provide far more information than merely the gross physical properties of clouds.

FRANCES DRAKE

Further reading

Bohren, C. F. (1987) *Clouds in a glass of beer: simple experiments in atmospheric physics.* John Wiley and Sons, New York.

Scorer, R. (1986) *Cloud investigation by satellite.* Ellis Horwood, Chichester.

coal Coal is an important fossil fuel. As with 'oil', 'coal' is an umbrella term encompassing a wide range of fuels varying considerably in their physical, chemical, and technological properties. Although solid, and thus less convenient to handle and less versatile in usage than oil, coal continues to be a diverse energy source. Although the use of coal for gasification has diminished in many economies, if not disappeared completely, coal remains a major resource to meet growing energy demands, particularly in emerging economies.

Once a land flora was extensively established some 400 million years (Ma) ago, the potential for the formation of substantial coal deposits existed. Coals have formed at all latitudes, but the greatest volume has accumulated at middle latitudes. Major coal-forming events in the Earth's history are episodic, but only one such event has widely affected both hemispheres simultaneously. This took place during the Carboniferous and Permian periods, reaching its acme c. 300 Ma ago. In the Mesozoic era (249 to 65 Ma ago), there were several less extensive periods of coal formation. Throughout the Cenozoic (from 65 Ma ago to the present time), further coal-forming episodes have occurred and still continue, particularly in the Far East.

Older coals are usually more valuable as energy sources than younger coals because of their lengthy maturation in the Earth's crust. Their longer history has the disadvantage that the extensive depositional areas of which they were once part have been disrupted and fragmented by structural movements and, when exposed at surface, they have been weathered and eroded, leaving isolated coalfields.

Formation of coal-forming peats

Most economic humic coal deposits are *autochthonous*; that is, the peats from which the coals formed largely accumulated *in situ* from the fallen debris of trees and plants with relatively little movement of the dead vegetation. The *allochthonous* coals are a smaller group, forming chiefly from finely degraded vegetation, which was transported into lakes and ponds, principally by water, but also by wind. These organic muds form sapropels (cannels, algal cannels, and boghead coals), which are usually thin, of limited extent, and may occur within or on top of normal humic coals. Parts of some humic coal seams may be allochthonous.

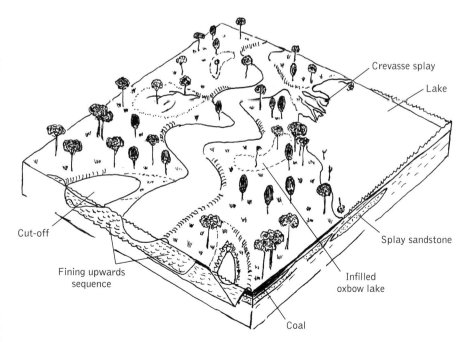

Fig. 1 Coal peats accumulating in low-lying swamps beside a meandering river (from McCabe, P. J. (1984) *International Association of Sedimentologists*, **7**, 13–42).

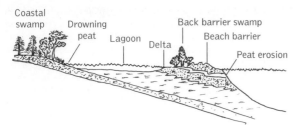

Fig. 2 Section across a coastline undergoing a marine transgression (from McCabe, P. J. (1984) *International Association of Sedimentologists*, **7**, 13–42).

An essential requirement for economic coals is that at no time in their history should they be heavily contaminated by mineral matter. A long-established view is that coal-forming peats accumulated in swamps on delta surfaces or in coastal swamps (Figs. 1, 2). Some coals may have formed in such environments, but they would not be 'clean' coals. Delta surfaces and coasts receive an influx of clastic material at times of flooding and during storms. The mixing of this mineral matter with the organic debris would be more likely to give rise either to high-ash coals or to carbonacous shales or mudstones, but certainly not to coals low in mineral matter.

'Clean' autochthonous peat swamps in sedimentary basins are of three types: 'floating', which form only thin peats, 'low-lying', and 'raised' (Fig. 3); those of the second and third types form thick high-quality peats. The peats may pass into one another, according to the depositional conditions. Active clastic deposition can occur where the swamps develop, but all the peats are low in mineral matter. In the floating swamp, the top of the peat mat is above water level; where the swamp is low-lying, covering pre-existing topography, clastic material and the peat are deposited at different times; and in raised swamps the peats lie above flood levels and are not contaminated by water-borne minerals; they also possess high acidity which may leach out any minerals present.

For peat accumulation to proceed satisfactorily, organic production and decay must roughly balance. Both are controlled by climate, particularly by temperature and humidity. Decay can be halted or retarded only if the vegetation accumulates under oxygen-free, stagnant conditions below the water table, which must be at the surface of the sediment. If the water table

rises at an appropriate rate, peat growth will keep pace with this rise. The water table may fluctuate; the peat will then either decay because of exposure to the air or it may temporarily be drowned. Fluctuations of this kind occur throughout peat accumulation, but when subsidence becomes rapid and continues, the drowned peat will then be covered by clastic sediments. Most major coalfields have formed by drowning of the peats after each has had a period of successful growth.

Composition of coals

In the field and under the microscope most coals display heterogeneity governed by their initial compositions and the biochemical histories of the peats from which they formed. While sapropelic coals are massive and fine-grained, humic coals are banded, the bands being composed of differing proportions and sizes of constituents that reflect the changing character and depositional conditions within the peat. Plant components range from the massive parts of trees, shrubs, and plants to small fragments of tissues, seed and spore coats, cuticles, and sometimes algae and highly resistant plant impregnations. After deposition these components offer differing degrees of resistance to influences such as oxidation and microbiological activity, which continue to operate (the latter especially), until the peat is buried so deeply that they are no longer effective. Thus, not only do the initial compositions of the peats vary, but the peats are subsequently modified during what is termed the 'biochemical stage of coal formation'.

The plant components in peat are transformed into 'macerals' (broadly analogous to the minerals of inorganic rocks), which fall into three groups.

(1) Vitrinite, the dominant constituent of humic coals and formed from the lignin–cellulose complexes of trunks, branches, roots, and other plant organs along with degraded plant and humic-peat material, some of which forms gels;
(2) Liptinite (exinite) formed of the waxy resistant parts of plants, spore and seed coats, resins, waxes, tannins, algae, and degradation tissues;
(3) Inertinite, much of which is material oxidized either before or after incorporation into the peat, with a similar origin to those components forming the vitrinite group, and also other material of fungal origin, particles, and granules arising from redeposition of inertinite and degradation of liptinites.

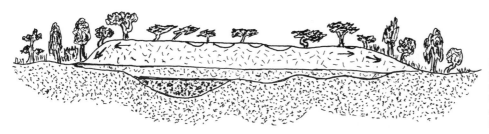

Fig. 3 Peat formation taking place in a raised mire (swamp) (from McCabe, P. J. (1984) *International Association of Sedimentologists*, **7**, 13–42).

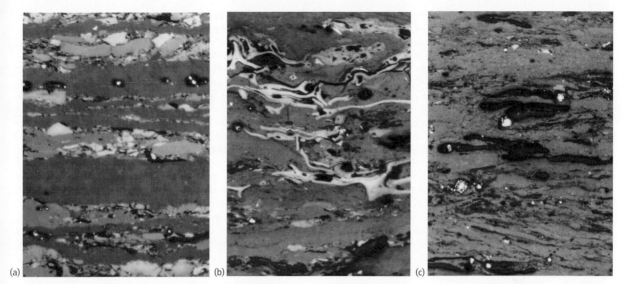

(a) (b) (c)

Fig. 4 Photomicrographs of humic coal in reflected light using oil-immersion objectives: (a) Principally vitrinite (medium grey bands and inertinite (light grey to white); (b) Fusinite (white) and vitrinite (medium grey), both derived from similar plant tissues, but each with a different biochemical history, and a few spores (dark grey); (c) Spores and other liptinitic material (dark grey) in a vitrinitic gel (medium grey) with some high-reflecting pyrite globules. All photomicrographs: 30 mm represents *c.* 0.1 mm

Minerals form a fourth group of constituents, of which mineral sulphides and carbonates are the principal components. The differing proportions of these constituent groups govern the properties and technological behaviour of any coal. Figure 4(a)–(c) illustrates typical appearances of a humic coal under the microscope. The highly reflecting material is inertinite, with some mineral matter (pyrite in Fig. 4a); the medium-reflecting constituent is vitrinite, and the dark grey constituents are liptinite, mainly spores.

The petrographic compositions of coals and their technological potential are best assessed by light microscopy. The macerals present, and their proportions, are determined together with that fraction of contaminating mineral matter which is not too finely divided to be detected by light microscopy. From an economic viewpoint, coals are categorized chemically most effectively by proximate and ultimate analyses. Proximate analysis defines the moisture content of a coal, its ash content, and its yield of volatile matter (which is directly related to the gas

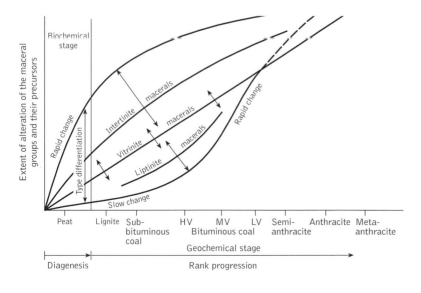

Fig. 5 Variation in the rates of coalification of the different maceral groups related to different coal rank stages. HV, high volatile; MV, medium volatile; LV, low volatile (from Smith G. C. and Cook A. C. (1980) *Fuel*, **59**, 641–6).

137

potential of the coal), and fixed carbon. Ultimate analysis yields the elementary composition of a coal: its carbon, hydrogen, oxygen, nitrogen, and sulphur contents and other properties. The calorific value and the coking potential of the coal are decided by other tests.

Figure 5 summarizes qualitatively what happens to an accumulating peat and then later when the peat is transformed into coal. The properties of any coal are directly related to its original composition and to its biochemical and geochemical histories. Biochemical activity occurs throughout the peat stage when coal types are differentiated, including a range of humic-coal types and the sapropelic coals: the cannels, algal cannels, and boghead coals. Once all biochemical activity has ceased, the 'geochemical stage of coal formation' begins. The coal is subjected only to physical influences: rise of temperature, length of exposure to increased temperature, raised overburden pressure, and sometimes stress. With increasing length of geological time, as the organic mass is exposed to raised temperatures, coals pass through a series of coalification (rank) stages: lignite, sub-bituminous coal up to meta-anthracite. Table 1 illustrates how some important properties

of coals vary with rising rank. The coal depicted in Fig. 4 belongs to the high-volatile bituminous group.

Reserves and consumption of coal

Proven reserves of coal are usually taken as those quantities that geological knowledge and current engineering technology indicate can be recovered from known deposits. Table 2 shows the world distribution of major reserves in 1997, dominated by the Asia–Pacific region. Only in this region do the reserves of anthracite and bituminous coal exceed the reserves of sub-bituminous coal and lignite.

All regions show interest in their reserves, but the greatest concentration is focused on the length of time the reserves will last. This is estimated by dividing the known reserves remaining at the end of a year by the production in that year. The calculation gives the 'Reserves/Production (R/P) ratio', representing the number of years the reserves will last if production continues at that level. Figure 6 shows R/P ratios for oil, natural gas, and coal for three major regions. The dominance of coal reserves over those of oil and natural gas at current production rates is striking.

Table 1 Variation of some important coal properties with rank.

Rank (USA)	Bed moisture (%)	Volatile matter d.a.f. (%)	Carbon d.a.f. (%)	Hydrogen d.a.f. (%)	Reflectance (%) Vitrinite	Calorific value (kcal/kg)
Peat		70			0.2	
	ca.75	65	ca.60			
Lignite		60			0.3	
	ca.35	55				4000
Sub-bituminous C / B / A	ca.25	50	ca.71		0.4	5000
	ca.8–10	45	ca.77		0.5 / 0.6	7000
High-volatile bituminous C / B / A		40				
		35			1.0	
Medium-volatile bituminous		30	ca.87	5.5	1.2	8650
		25			1.4	
Low-volatile bituminous		20			1.6	
		15			1.8 / 2.0	
Semi-anthracite		10				
			ca.91	4.0		8650
Anthracite		5		2.5	3.0 / 4.0	
Meta-anthracite				1.5		

d.a.f.-dry ash free

Table 2 Distribution of major coal reserves in 1997.

Region	A/B	SB/L	Total	R/P
	thousand million tonnes			years
North America	111.9	138.5	250.4	233
Europe	59.1	97.6	156.7	191
Former Soviet Union	104.0	137.0	241.0	>500
Asia/Pacific	178.2	133.3	311.5	146
Total World	519.4	512.3	1031.6	219

A/B, Anthracite/Bituminous coal
SB/L, Sub-bituminous coal/Lignite
R/P, Reserves/Production ratio

Coal production is closely related to consumption: most coal is consumed in the region in which it is produced. The balance between production and consumption is roughly equal in most regions, but North America is a net coal exporter while Europe is a net importer.

The future for coal

Table 3 illustrates the pattern of world energy consumption over the period 1972–97, when consumption increased by over 60 per cent. The largest increase (*c.* 90 per cent) was in natural gas. Consumption of oil rose by only some 36 per cent, but in total substantially exceeded consumption of any other individual fuel. Usage of coal increased by approximately 44 per cent, and was the second largest fuel consumption over the period.

Table 3 Variation in world consumption of primary energy (million tonnes of oil equivalent) over 25 years.

Fuel	1972	1977	1982	1987	1992	1997
Oil	2591	3000	2773	2955	3182	3409
Natural gas	1045	1182	1318	1573	1818	2000
Nuclear energy	24	136	245	455	545	636
Hydroelectric	91	118	136	182	182	218
Coal	1545	1727	1864	2182	2145	2309

Consumption of energy varies from region to region. In the Asia/Pacific region the largest usage in 1997 was of coal; in the Former Soviet Union it was of natural gas; but elsewhere oil was dominantly consumed. Table 3 and Fig. 6 illustrate the implications of these patterns of usage upon the world reserves' position. Even allowing for substantial future discoveries of oil and natural gas, but discounting a dramatic development of new technologies, which would diminish or replace the use of oil, or a reversal of attitudes to large-scale expansion in nuclear-energy programmes, use of coal as an energy source must increase apace, particularly in developing countries with few other fossil-fuel reserves. The greatest demand for coal will be for electricity generation, but coal will continue to feature in conversion processes such as gasification, liquefaction, and pyrolysis, and in the production of metallurgical coke.   DUNCAN MURCHISON

coastal dunes Coastal dunes are deposits of wind-blown sediment that occur next to large bodies of water. With few exceptions, these dunes are formed when sand is blown landward from a sandy beach and is accumulated where there is some degree of protection from waves and currents. As such, coastal dunes may form in marine, lacustrine, or fluvial environments. Most of the dune forms that develop in coastal systems are similar to those found in arid environments (see *desert dunes*): for example, barchanoid, coppice, longitudinal, star, and transverse dunes. It has been argued that the foredune is the only distinctive coastal dune form, because its geometry is a fundamental response to interactions between nearshore processes, wind, sediments, and vegetation.

Foredunes occur at locations far enough from the water for wind-blown sediments to be able to accumulate at a rate that exceeds, at least temporarily, their rate of removal by waves and currents. Foredune formation is usually initiated when vegetation or debris slows or stops landward-blowing sand. Incipient dunes, usually coppice dunes, may form within hours as an accumulation grows. These dunes afford a favorable habitat for many salt-tolerant plant species, such as *Ammophila* or *Spinifex*. As the density of the plant community increases, the ability to trap sand also increases, and the rate of dune growth is increased. If the nascent dunes are not destroyed by nearshore erosion, they will continue to grow until they merge into a foredune proper. In many environ-

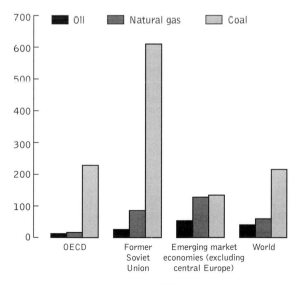

Fig. 6 Reserves to production ratios (*R/P*) for oil, natural gas, and coal for different regions at the end of 1997 (from the British Petroleum Review of World Energy (1998).)

ments, especially in the mid-latitudes, vigorous plant colonization is juxtaposed with erosion by nearshore processes. This results in a near-linear foredune, usually aligned transverse to the prevailing winds, with an erosion-steepened stoss slope and a densely vegetated crest and lee slope. A scarped windward slope is the aspect most characteristic of coastal dunes.

In prograding coastal systems, individual foredune ridges may be stranded as new ridges form in front of them. This may result in the formation of dune-ridge plains, which are common along many sediment-rich coastlines. In eroding or transgressing systems, destabilization of the foredune system facilitates movement of sand inland. This is often manifested in the formation of blowouts through the foredune and the subsequent development of inland-migrating parabolic dunes. Winds are accelerated through blowouts, increasing sand transport rates, causing additional scour in the gap, and creating a depositional lobe downwind. The edges of the blowout remain anchored by vegetation, the back wall of the lobe migrates downwind, and lengthy sidewalls maintain the integrity of the dune. Sand transport landward of the foredunes effectively removes sediment from the active coast and provides material for the formation of any of the other typical dune forms.

Coastal dunes are a resource that requires careful management. In many environments, these dunes represent a critical habitat. They may serve as important sources of fresh water. Along many coasts, the sediments stored in dunes represent a significant fraction of the material in the littoral system. During periods of beach erosion, the dunes act as a sediment reservoir to feed the nearshore system. During periods of elevated water levels, linear foredune systems act as barriers to flooding. For these reasons, coastal dunes are often socially valuable landforms.   DOUGLAS J. SHERMAN

coastal management Identification of the best way towards sustainable use of the coastal zone is an area of common concern and there is little doubt that coastal zone management (CZM) is now a major issue worldwide. However, not only is there a spectrum of perceptions attached to what is meant by the term coastal management but there is also a wide range of policies adopted by various governments aimed at achieving it. For example, from a coastal engineering perspective, coastal management is frequently taken to mean shoreline management and thus principally deals with the ways in which the effects of coastal erosion may be mitigated by the use of engineered solutions. At the other extreme, the OECD has defined integrated coastal zone management (ICZM) as the management of the coastal zone as a whole, with respect to local, regional and national goals. This is a much wider perspective and implies a focus on the interactions between the wide range of activities and resource demands that occur within the coastal zone and those that occur outside at a variety of scales. Operationally this means that ICZM must seek to integrate the various goals of envi-

ronmental protection, economic development, pollution control, tourism development, defence requirements, shipping and port management, coastal erosion, and flooding, amongst others, within different parts of a particular coastal zone. Simultaneously, ICZM should also ensure the consonance of such local goals with wider regional and national goals that might be involved, for example, in the strategic siting of nuclear facilities and defence establishments, or of industrial developments judged to be in the national interest. The policies adopted by governments range from no formal provision at all, through simple extensions of the inland countryside planning provisions to encompass the coast, to full-blown coastal zone management programmes where national legislation places local implementation into context.

Definitions

One of the problems in formulating a CZM policy lies in the definition of the coastal zone itself. The term 'coastal' relates to the land–sea interface along two axes. The shore-parallel axis is non-controversial from a scientific viewpoint since it is generally continuous, although where it crosses national or even regional boundaries there arise political and administrative problems associated with co-ordination of policies. More debate exists regarding coastal limits along the axis perpendicular to the coast. For example, it could be argued that since the sediments and water reaching the coast come from inland upland areas, then the inland coastal limit is the main mountain divide! Similarly, the seaward limit could be argued to be the maximum reach of national jurisdiction (e.g. the 200 mile limit). This point has been demonstrated in Chesapeake Bay in the United States, where the watershed includes portions of six states and so extends well beyond any conventional definition of the coastal zone. Soil erosion and upstream pollution along the upland rivers have a major influence on the water quality of Chesapeake Bay. However, most definitions rely in some measure on the extent of tidal influence and include the land–air–sea interface around the continents and islands, and extending from the inland limit of tidal or windblown sea or sand influence to the outer edge of the continental shelf. Figure 1 shows a representation of the coastal management boundaries and plethora of government acts relating to the coastal zone of New Zealand in 1991. The difficulty of definition lies in the fact that what constitutes the coastal zone depends on the purpose in hand and so the limits of the zone will vary depending on the nature of the problem. The inland coastal zone limit has been set by some to include the associated aquatic ecosystems and tributaries that drain into an estuary up to the height of migration of spawning fish or the height of tidal influence, whichever is the higher.

Although there are strong reasons for basing the definition of the coastal boundary on scientific criteria, such as the inputs and outputs of a coastal sediment cell or the limits of an ecosystem, there are many operational difficulties that first need to be resolved, including a marked lack of coincidence between the physical ecosystem boundaries and the political

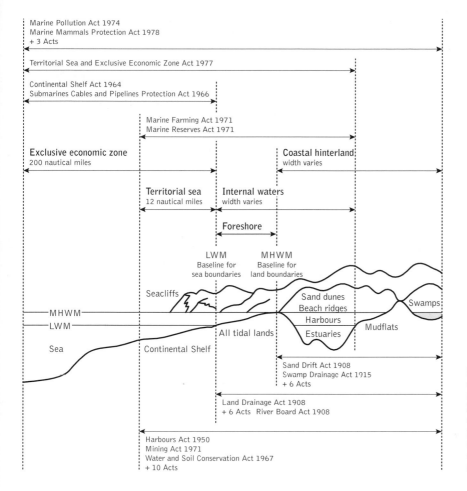

Marine Pollution Act 1974
Marine Mammals Protection Act 1978
+ 3 Acts

Territorial Sea and Exclusive Economic Zone Act 1977

Continental Shelf Act 1964
Submarines Cables and Pipelines Protection Act 1966

Marine Farming Act 1971
Marine Reserves Act 1971

Exclusive economic zone
200 nautical miles

Coastal hinterland
width varies

Territorial sea
12 nautical miles

Internal waters
width varies

Foreshore

LWM
Baseline for
sea boundaries

MHWM
Baseline for
land boundaries

Seacliffs

Sand dunes
Beach ridges

Swamps

MHWM

Harbours

LWM

Mudflats

Sea

All tidal lands

Estuaries

Continental Shelf

Sand Drift Act 1908
Swamp Drainage Act 1915
+ 6 Acts

Land Drainage Act 1908
+ 6 Acts   River Board Act 1908

Harbours Act 1950
Mining Act 1971
Water and Soil Conservation Act 1967
+ 10 Acts

Fig. 1. Boundaries of the coastal zone and the range of government acts affecting the coastline of New Zealand in 1991. Although now superseded in New Zealand, the plethora of boundaries shown is a common feature of the CZM of other countries. (From Hansom and Kirk, 1991.)

and social boundaries of administrative authorities; a lack of integration between the administrative units responsible for the management; and the fact that the area defined may be much too large and diverse to be treated as a coastal zone management unit, thus requiring any integration to be focused on the extent to which activities impact on coastal waters. Since the legal powers of local authorities often extend only to either mean low or high water mark, in order to operate outside this range the local authority must be integrated with the relevant national authorities. In 1993, the OECD highlighted two overriding requirements for effective coastal management: the extent of the coastal zone needs to be broadly based enough to include most of the relationships between economic and natural systems; and a high level of integration is needed between the various local, regional, and national administrative authorities involved in management.

Resources

At the core of ICZM lies an acknowledgement that coastal resources possess several characteristics important to their sustainable management. Sustainable management must take into account the fact that coastal resources, such as wetlands, sand dunes, beaches, wading birds, fin fish, and shellfish, to name but a few, behave as systems that interact with each other. These resources have also been identified as having multiple uses with a complex matrix of complimentaries and conflicts that need careful resolution and compromise to achieve sustainable development. However, coastal resources also deliver private goods via the market in addition to the delivery of public goods from which nobody can be excluded, the overuse of which poses dangers to sustainability. The OECD noted that the delivery of these resources by private and public providers requires co-ordination.

CZM programmes

CZM can be approached from a negative perspective in that coastal development is controlled through legislation that limits options, together with disincentives such as the withholding of planning permission for developments that are deemed to be inappropriate. Positive programmes allow coastal development within carefully constructed strategies that impose certain levels of environmental control and management. It has been suggested that most CZM in developed countries is negative, whereas positive programmes can be found in developing countries that are keen to maximize the foreign exchange potential of coastal tourist developments. Some programmes adopt positive elements embedded within a negative framework by setting aside zones for development. This analysis forms the basis for the two main approaches to CZM: the UK approach (used widely throughout the world) which is negative and reactive largely in the wake of changes and development applications, and the US approach which has positive elements embedded within negative and thus involves a proactive strategy.

The US model has adopted a central stance on CZM policy, where national objectives are enshrined in legislation aimed at guiding individual CZM plans along a uniform set of principles that ensure integration of a wide range of coastal uses. The US CZM Act 1972 (later amended several times) was the first CZM plan aimed at fully integrated management of the coast. Coastal states need federal approval of their programmes in order to attract substantial funding for development and administration of CZM plans. Thus, although the aim was to achieve national policy objectives for coastal management through a voluntary partnership between federal and state government, few coastal states could afford not to become involved. Central to this model is an acceptance of the coastal zone as an entity, the need to define the spatial limits within which permissible land and water uses are possible, and the need for legislation to control these uses (Fig. 2). The CZM Act has been generally regarded as a success, since it requires state coasts to be managed according to a set of national rules which provide direction towards sustainable coastal development but which also allow a fair degree of flexibility in how this might be achieved. Though not without its problems, some regard the result to have been a 'quantum leap' over the 1972 institutional capacity for coastal planning and management in the USA.

The approach to CZM adopted in the UK has evolved within the framework of existing administration and planning legislation. Several government departments have coastal responsibility and interests but none of these has traditionally led the process and the result has been a vacuum of policy that has been filled in the past by voluntary organizations such as the National Trust, who have managed their coastal properties along nationally agreed guidelines. The lack of any national policy on matters of coastal management has resulted in the emergence of an *ad hoc* system of coastal decision making, in which little or no regard has been taken of the negative impacts of developments on other parts of the coastal system. Since the late 1990s, however, the Ministry of Agriculture, Fisheries, and Food (MAFF) has taken a more prominent role in the development of coastal planning, and shoreline management plans (SMPs) are being developed to

1 Boundaries of the coastal zone and critical areas
2 Intergovernmental co-ordination
3 Areas of special concern for economic, social, historical and environmental reasons
4 Public participation
5 Public beach access
6 Beach erosion control
7 Management of major land and water uses including those of regional and national interest
8 Monitoring nearshore waters for pollution
9 Environmental impact statement co-ordination and funding

Fig. 2. Major components of a successful CZM programme (from Hansom (1998), by permission of Cambridge University Press.)

deal with coastal defences and erosion, and estuary management plans (EMPs) to deal with rivers and their exits. SMPs are constituted to reflect the natural sediment cells around the coast, and any coastal defence schemes must be consistent with a strategic plan in order to be eligible for grant aid. Thus there is now a clear financial imperative for coastal authorities to produce SMPs. The situation surrounding CZM is less clear and the present situation is that the UK government does not expect all stretches of the coast to be subject to this approach. This represents a restatement of existing policies with no national coastal objectives that would provide the policy overview required to ensure consonance within and between individual coastal management plans. The emerging model of CZM in England and Wales revolves around local authorities and agencies working together on a voluntary basis to produce CMPs that seek to better integrate the use of the coastal zone. However, the plans are non-mandatory and may be paralleled or overlapped by the proposed SMPs and EMPs.

It is clear that most of the research and experience so far gained in ICZM favours an holistic approach that clearly specifies the limits of the coastal zone and employs government legislation to guide coastal use towards sustainability, since without a strategic overview provided by top-down government legislation, the aims of coherent coastal management are devalued. This may inevitably sacrifice local issues to satisfy national aims, but regional policies are much more restricted in their ability to identify or even see the national issues and so may conflict with national sustainability.

J. D. HANSOM

Further reading

Beatley, T., Brower, D. J., and Schwab, A. K. (1994) *An introduction to coastal zone management.* Island Press, Washington.

Brahtz, J. F. P. (1972) *Coastal zone management: multiple use with conservation.* John Wiley and Sons, New York.

Carter, R. W. G. (1988) *Coastal environments.* Academic Press, London.

Godschalk, D. R. and Cousins, K. (1985) Coastal management: planning on the edge. *Journal of the American Planning Association*, **51**, 263–5.

Hansom, J. D. and Kirk, R. L. (1991) Change on the coast: the need for management. In Johnston, T. R. R. and Flenley, J. R. L. (eds) *Aspects of environmental change*, pp. 49–62. Massey University Press, Palmerston, New Zealand.

Hansom, J. D. (1995) Managing the Scottish coast. *Scottish Geographical Magazine*, **111**, (3), 190–2.

OECD (1993) *Coastal zone management, integrated policies.* Organization for Economic Co-operation and Development, Paris.

coasts and coastal processes
In his book on coastal environments, published in 1988, R. W. G. Carter defined the coastal zone as 'that space in which terrestrial environments influence marine (or lacustrine) environments and vice versa'.

The coastal zone is of variable width and may also change in time. Delimitation of zonal boundaries is not normally possible, more often such limits are marked by an environmental gradient or transition. At any one locality, the coastal zone may be characterized according to physical, biological or cultural criteria. These need not, and in fact rarely do, coincide. In more simple terms, the coastal zone is often taken to be that area where land, water, and air meet or the zone in which wave processes have an active affect upon the Earth's crust. It is thus important to realize that the coastal zone includes areas both above and below sea level.

The coastal zone has always been important to society, for it provides food and was once a focus for transport and industrial development. In recent years, the emphasis has shifted towards leisure. The importance of this zone is highlighted by the fact that the United Nations estimate that some 66 per cent of the world's population live within a few kilometres of the coast. The coastal zone is thus under pressure, and a good understanding of the processes associated with it is essential to society.

Classification
Various attempts have been made to classify coastal landscapes, but it has been recognized that none of them is ideal. On the one hand there are the genetic classifications based on the origin of the coastal landforms (e.g. shorelines of emergence or submergence); on the other there are descriptive classifications based on the form of the landscape (e.g. fjord, delta, cliffed, coral, barrier island, mangrove). A widely used classification was developed in the 1950s by Valentin, who made a fundamental distinction between advancing and retreating coasts (Fig. 1). His classification recognized that

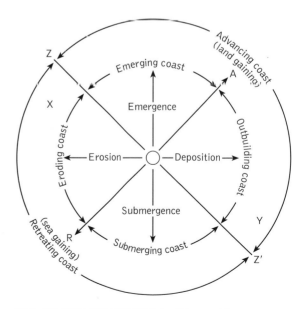

Fig. 1. Valentin's scheme (1952) for classifying coastal areas.

Table 1. Classification of world coastlines (after Valentin (1952))

(A) Coasts that have advanced

    (1) as a result of emergence

        (a) emerged sea-floor coasts (e.g. Hudson Bay)

    (2) as a result of organic deposition

        (b) phytogenic (formed by vegetation): mangrove coasts (e.g. north coast of Western Australia)

        (c) zoogenic (formed by fauna): coral coasts (e.g. Tahiti)

    (3) as a result of morganic deposition

        (d) marine deposition where tides are weak: lagoon-barrier and dune-ridge coasts (e.g. Gulf coast of USA)

        (e) marine deposition where tides are strong: tide-flat and barrier-island coasts (e.g. Holland)

        (f) fluvial deposition: delta coasts (e.g. Rhône, Niger, Mississippi)

(B) Coasts that have retreated

    (1) as a result of submergence of glaciated landforms

        (g) confined glacial erosion: fiord-skerry coasts (e.g. Norway)

        (h) unconfined glacial erosion fiard-skerry coasts (e.g. Baltic Findland)

        (i) glacial deposition: morainic coasts (e.g. Baltic Denmark)

    (2) as a result of submergence of fluvially eroded landforms

        (j) on young fold structures: embayed upland coasts (e.g. Greece)

        (k) on old fold structures: ria coasts (e.g. south-west England)

        (l) on horizontal structures: embayed plateau coasts (e.g. Red Sea)

    (3) as a result of marine erosion

        (m) cliffed coasts (e.g. the English Channel)

advance may be due to coastal emergence and/or progradation by deposition, whereas retreat is due to coastal submergence and/or retrogradation by erosion (Table 1).

A more recent classification based on a tectonic framework was produced by Inman and Nordstrom in 1971. This scheme (Fig. 2) recognizes broad distinctions which are influenced by tectonics (mountain coast, narrow shelf, broad shelf) and morphologic form (headlands and bays, coastal plain, deltas, reefs, glaciated regions) and it accepts that more one than factor may influence a region.

Coastal processes

The geomorphic form and evolution of coasts is largely controlled by six major factors: waves, tides, offshore topography, bedrock geology, sediment supply, and sea-level changes. However, in the recent past society has been having an increasing impact on the system.

Waves

Waves are undulations of the water surface produced by winds blowing over it. The turbulent flow of air across the water surface sets up pressure variations which initiate the waves. Once in place the waves help to disturb the air flow, and as a consequence continue to grow while the air flow continues. The final size of the wave depends upon the wind speed, how long the wind continued to flow, and the distance over which the wind blew (fetch). In enclosed basins (Irish Sea, North Sea) fetch is usually of vital importance, since it will determine the magnitude of the wave and hence the energy it can expend in sediment movement and erosion.

Waves provide a means of transmitting energy through water with relatively small displacement of the water particles in the direction of energy flow. The water particles move in an orbital fashion, and the magnitude of movement decreases with depth. In deep water the waves move unimpeded through the liquid, but as they move into shallow coastal waters (where the water depth is less than half the wavelength) they begin to interact with the sea floor. In general, water movement at depth becomes retarded and as a result the wave starts to slow down. As the wave slows down there is a decrease in wavelength (the distance from one wave crest to the next) and an increase in wave height (the distance between the wave crest and the trough). During this process waves approaching at an angle to the shoreline tend to become realigned (refracted) so that their angle to the shoreline is reduced. The orbital movements within the wave become more elliptical and the shoreward movement of the water increases until it is greater than the wave velocity. At this point, the orbital motion can no longer be completed and so the wave front collapses (breaks) sending a rush of water (swash) on to the shore. The breakers are said to

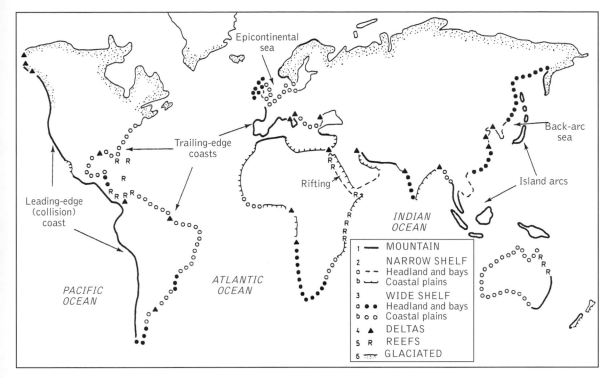

Fig. 2. The tectonic classification of the world's coastlines. (After Inman and Nordstrom (1971).)

plunge (Fig. 3) when the wave crest curves over and collapses with a crash, or to *spill* (Fig. 4) when the crest of the wave flows more gently down the wave front to produce a rush of water on to the shoreline. The water then withdraws (the backwash) either as undertow (sheetflow near the sea bed) or in localized currents known as rip currents. In simple terms, spilling (constructive) breakers tend to be associated with the transport of sediments onshore, whereas plunging (destruc-

tive/ storm) breakers are associated with the movement of material offshore. Along the open coast waves are generally the principal source of energy and so they do most of the work. The waves arriving at the coast will thus determine its form and how it evolves. What must be remembered is that the waves are controlled by wind direction and air speed and, as a result, they can vary from day to day. The direction of wave approach will also vary and, consequently, sediment move-

Fig. 3. Plunging waves breaking as they near the shore, Hawaii.

Fig. 4. Spilling waves breaking as they near the shore, Aberystwyth, Wales.

145

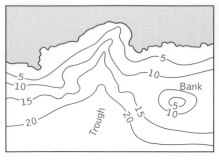 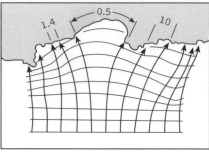

Fig. 5. Wave refraction patterns associated with an embayed coastline. Depths are shown in metres. The lines drawn at right angles to the wave crest are known as orthoganals. The dispersion and concentration of the orthoganals indicate the variations in wave energy along the shoreline. Note that within the bay the wave energy is dispersed (0.5); along the headland it becomes enhanced (1.4)

ment in the coastal zone can be complex, moving onshore, offshore, and alongshore. Often most of the work is achieved during storm (high-energy) events in which more sediment is transported in a few hours than during the preceding months of low-energy conditions.

Tides

Tides are movements of water bodies (oceans or seas) set up by the gravitational effects of the Sun and Moon in relation to the Earth. They are important in the coastal zone since they lead to regular changes in water level along the coast. In general, as the Earth rotates the gravitational attraction of the moon results in a highstand of water on the Earth's crust directly beneath the moon and on the opposite side of the earth. When the Sun and Moon are in alignment, the gravitational forces are greater and a higher (spring) tide is produced, but when they are aligned at right angles to each other the gravitational forces are reduced and a low tide (neap) results. The form of the tidal wave is also dependant upon the size and shape of the ocean basin. As a consequence, some water bodies (for example, Mediterranean Sea, the Black Sea; many Atlantic and Pacific shorelines) experience only very small tidal variations (less than 2 m). In other areas, in particular estuaries and embayments, the tidal variations become accentuated, resulting in spring tidal ranges in excess of 4 m (for example, the Bristol Channel (UK), 12.3 m; the Bay of Fundy, Canada, 15.2 m).

Contrasts in tidal range have implications in coastal geomorphology. A large tidal range produces a broad intertidal zone, and waves breaking against the high-tide line thus have much reduced energy, having crossed the shallow shore zone. In such circumstances, the waves will break along the high-tide line for only a short period of time. Wave action then has a limited opportunity to modify the shoreline. Such areas are often characterized by extensive salt marshes and mud flats. In contrast, where the tidal range is limited, waves constantly break along the same section of the shoreline and thus the potential to develop coastal wetlands is limited.

The movement of water associated with tides can also result in the formation of tidal currents. Along the open shoreline the currents associated with the rising (flow) tide and the falling (ebb) tide often counteract each other. In enclosed settings, such as estuaries, the tidal movements often

set up currents which can transport sediment. For example, in some estuaries the ebb tide is often strongest within the central channel, whereas the flood tide is stronger along the shoreline. In this way, currents are set up which tend to move sediment up the estuary close to the shore and towards the open sea near the centre of the embayment.

Offshore topography

As a wave nears the coast in starts to interact with the sea floor. The topography of the offshore zone can thus influence how the wave energy is dissipated along the shoreline. When a wave moves into an embayment, the wave at the centre is in deeper water than those at the edge. As a consequence, the wave in the centre moves more rapidly and the wave front becomes refracted, tending to become aligned parallel to the shoreline (Fig. 5). The result is that wave energy tends to become dispersed in bays and concentrated on headlands. The offshore topography will thus determine the distribution of wave energy along the coast for a given direction of wave arrival. If the offshore topography is shallow, there is greater opportunity for the waves to interact with the sea bed and hence to become fully refracted. If the offshore zone is instead steep and narrow, the waves have little time to interact with the sea bed and thus refraction is minimal.

Bedrock geology and sediment supply

The evolution of the coastal zone is clearly controlled by the bedrock geology, both in terms of rock strength and structure. In general terms, 'hard-rock' coasts (granite, quartzite) evolve more slowly than 'soft-rock' coastlines (clays, tills). It is noteworthy, however, that processes of coastal erosion are also controlled by the climatic setting of the region (*see* rock platforms). Within erosional coastal settings wave processes tend to exploit zones of weakness, and the structural form of the bedrock geology can thus influence the form of the shoreline. Lines of weakness (e.g. faults) tend to undergo accelerated erosion, and as a consequence bays and headlands develop.

The nature of coastal sediments and their rate of supply will also influence coastal evolution. On open coasts where wave processes dominate, clastic sediments (sands to cobbles) will characterize the coastal zone. It is only in sheltered locations (embayments, estuaries) that energy levels are low enough for fine sediments (silts and clays) to be deposited. Where the rate of sediment delivery is high, the coastline will tend to prograde (build seaward). Where sediment supply is limited in relation to available wave energy, erosion will dominate.

Relative sea-level changes

The level of the water surface varies over time as a result of changes in ocean volume (*see* Quaternary sea-level changes) or movements of the Earth's crust. Such changes clearly affect the zone over which coastal processes operate, and thus influence the long-term evolution of the coastal zone. The impact that changes in sea level have upon a coastline are complex and are dependent upon other factors, such as sediment supply, coastal topography, and wave climate. Changes in sea level can certainly result in changes in the tidal range since they will alter the shape of an ocean basin. A detailed account of the impact that sea-level changes can have upon coastal systems is provided by R. W. G. Carter and C. D. Woodroffe in their book on coastal evolution.

Impact of society

In the recent past, society has increasingly had an impact upon coastal processes. Many of the structures (jetties, harbours, sea walls, breakwaters) built by society disrupt the natural coastal processes and consequently result in erosion and deposition. A classic example is provided by the breakwater at Santa Barbara, California, built in 1930 to protect the harbour. This structure produced a barrier to sediment transport which resulted in the accumulation of material updrift of the barrier, within 7 years a spit began to build across the harbour mouth. In contrast, downdrift of the structure the lack of sediment being moved along the coast resulted in erosion of the shoreline over a distance of some 40 km. Additional examples of the impact that society can have upon the shoreline are given in Carter's book on coastal environments.

CALLUM R. FIRTH

Further reading

Carter, R. W. G. (1988) *Coastal environments: an introduction to the physical, ecological and cultural systems of coastlines.* Academic Press, London.

Carter, R. W. G. and Woodroffe C. D. (1994) *Coastal evolution: Late Quaternary shoreline morphodynamics.* Cambridge University Press.

Inman, D. and Nordstrom, C. (1971) On the tectonic and morphological classification of coasts. *Journal of Geology*, 79, 1–21.

COHMAP Project
The Cooperative Holocene Mapping Project (COHMAP) is an interdisciplinary research project with a general aim of increasing our understanding of the relationship between climate and environmental changes during the last 18 000 years. A description of the aims of the project, the research design, and some of the results were published in *Science* in 1988.

A unique dual approach was adopted by COHMAP by combining knowledge of past climate and environmental changes, as derived from the interpretation of field and analytical data, with the results of an atmospheric general circulation model (GCM) used to simulate and model past climate. Palaeoenvironmental information, such as stratigraphic pollen data, evidence for past lake levels, and marine microfossils, was collected from several hundred sites distributed throughout the world. This information was then used to reconstruct a spatial and temporal sequence of palaeoclimatic changes on a global scale. With the development of high-powered computers that can handle large volumes of data, it is now possible to construct numerical models of the atmospheric general circulations (GCMs). Most GCMs are used to predict future changes in climate as the atmospheric concentration of the greenhouse gases changes. The COHMAP GCM was designed to simulate climatic changes in the past, and certain boundary conditions were therefore built into the model. These include values of insolation according to Milankovich cycles, sea-surface temperature estimates based on foraminiferal assemblages, areas and heights of ice sheets, and estimates of atmospheric carbon dioxide and aerosols. The project was thus designed specifically to test two independent data sets, the palaeoenvironmental data set and the climate model, in the endeavour to reconstruct and explain the interaction between past changes in climate and the environment.

Initially, the project concentrated on reconstructing the climatic and environmental conditions at seven specific periods of time: 18, 15, 12, 9, 6, and 3 Ka (thousand years ago) and the present day. Results from the model have enabled climatologists and Quaternary scientists to begin to confirm links between climatic and environmental changes. For example, two COHMAP members, Kutzbach and Otto-Bliesner, showed that variations in the Earth's orbital elements increased the intensity of mid-latitude summer insolation 9000 years ago, which resulted in enhanced monsoonal rainfall and the expansion of northern African lakes. A comparison between climatic conditions as simulated by the COHMAP GCM with the distribution of the spruce tree, as indicated by pollen records, shows a fair to good agreement, indicating that temperature and precipitation have influenced the geographical distribution of spruce during the past 18 000 years. These examples illustrate how the COHMAP project has simulated the patterns of the movements of weather systems, the expansion and contraction of ice sheets, and temperature and precipitation levels and then compared the results with changes in the distribution of vegetation or lake levels.

The COHMAP project has been particularly useful because it illustrates that a combination of field and analytical data combined with model simulations can be a powerful tool for revealing features that appear to have determined the nature of past climatic conditions or environmental systems, or both. For example, Wright and Bartlein have found that the presence of a large ice sheet altered the pattern of atmospheric circulation in

the northern hemisphere, and these features seem to be able to explain palaeoclimatic patterns inferred from fossil data. The results of the project are important, not only because they confirm the broad agreement between the model and the palaeoenvironmental evidence, but also because they show up the discrepancies. This is extremely useful because it identifies gaps in our understanding and areas for future research. As the accuracy and quality of both the model simulations of climate and the field and analytical data improve, advances in our understanding of the relationships between past climates and the environment should follow.
                                                    T. MIGHALL

Further reading

COHMAP members (1988) Climatic changes of the last 18 000 years: observations and model simulations. *Science*, 241, 1043–52.

collisional orogeny Orogeny forms mountains by thickening the continental crust. Collisional orogenies are caused by the convergence of two plates of continental lithosphere when the intervening oceanic lithosphere is destroyed by subduction. Continental lithosphere cannot be subducted wholesale, for it is buoyant with respect to the underlying asthenosphere; further convergence is therefore accommodated by lithospheric thickening, which produces a mountain belt and converts the lower continental crust to denser metamorphic phases that allows partial subduction of continental lithosphere. The formation of a mountain belt is the more important process here. The subducting slab is attached to one plate and descends below the other, giving an inherent asymmetry to collisional orogens (Fig. 1). Collision is driven by the consequent slab pull or by ridge push from oceans elsewhere on either plate. Crustal thickening and orogeny above a

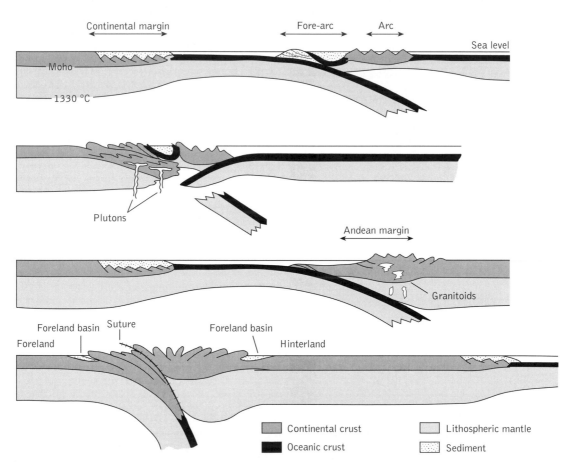

Fig. 1. Evolution of an arc–continent collisional orogeny: (a) the fore-arc, comprising an accretionary prism, an ophiolite, and a fore-arc basin, and the arc override the passive continental margin; (b) collision is followed by 'subduction flip'. (c) Evolution of a continent–continent collisional orogeny in which an Andean margin overrides a passive continental margin; (d) subduction ceases when all lithosphere within the ocean has been destroyed. The base of the lithosphere is shown at 1330 °C.

long-lived subduction zone, as in the Andes, are also caused by the addition of andesitic or granitic material, or both (Fig. 1c).

Two fundamental types of continental lithosphere are involved in collisional orogeny in an island arc bordered on both sides by oceanic lithosphere and normal continental lithosphere (Fig. 1a). There are therefore three fundamental types of collisional orogeny: arc–arc, as in the Moluccan Sea; arc–continent, as in Taiwan; and continent–continent, as in the Alps or the Himalayas. This might give the impression that collisional orogenies have a simple structure. However,

they show great variability and complexity owing to the nature of the continental margins, the width of the ocean destroyed, the number and polarity of subduction systems, and the presence of microcontinents and oceanic plateaus.

Lithospheric thickening associated with collision is caused by foreland-verging thrusts in the upper crust, by ductile deformation and folding in the lower crust, and by flow beneath the Moho that develops a mantle root (Fig. 2c). Continental lithosphere is about 2.5 per cent less dense than the asthenosphere. An increase of 40 km in its thickness will therefore produce an uplift of 1 km. Structural and geophysical data

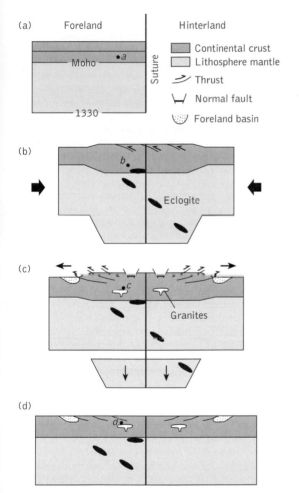

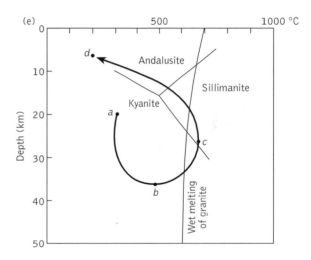

Fig. 2. Highly schematic diagram showing the tectonic and thermal evolution of a continent–continent collisional orogen. The *polarity* of the orogen is the same as in Fig. 1, and strike-slip motion parallel to the orogen is not shown. (a) The lithosphere of the foreland plate before collision; a reference point, *a*, at 20 km depth is shown. (b) The lithosphere of both plates after collision. Eclogites may form on the subducting plate or at the base of the thickened crust. Point *a* has now

moved to *b*. (c) The start of the collapse phase, in which the heavy lithospheric root is being removed mechanically. It may also be removed thermally. Point *a* has now moved to *c*. (d) The end of the collapse phase; point *a* now lies at *d*. (e) Depth–temperature diagram showing the thermal evolution of the orogen. Points *a*, *b*, *c*, and *d* correspond to those shown in (a)–(d) above.

indicate that in many mountain belts the lithosphere has thickened by a factor of 2, giving mountains 3 km high for lithosphere initially 120 km thick. This factor was as high as 3 in the Siluro–Devonian Caledonides of western Norway and as low as 1.4 in the Caledonides of the British Isles. Thickening stops when there is a change in the configuration of the plates driving the collision or when the potential energy of the mountains is sufficient to stop further convergence. Metamorphic transformation of lower crustal rocks to denser garnet–granulite or eclogite–facies assemblages reduces the buoyancy of the continental lithosphere. In some orogenies this has allowed thickening by a factor of more than 2 without producing excessively high mountains. Lithospheric thickening generally takes a few tens of millions of years in continent–continent collision zones. In the Alps, convergence began in the mid-Cretaceous and continues today; the slow northward motion of Africa has taken 90 Ma to thicken the rifted Mesozoic southern European margin. Convergence was assisted by subduction of eclogitized European lower crust. Pressure increases with thickening in continent–continent collisions (Fig. 2e). The increased thickness of rocks relatively rich in radiogenic heat–producing elements (potassium, uranium, and thorium) causes a slow rise in temperaturethat peaks after some 20–30 Ma to produce moderate-temperature, moderate-pressure Barrovian metamorphic assemblages (Fig. 2e).

In arc collision events, the oceanic lithosphere attached to the oceanward side of the arc is compressed and thickened by the collision. It is denser than the asthenosphere and it will therefore subduct. This produces a change in subduction polarity, allowing the back-arc oceanic lithosphere to be destroyed by subduction beneath the continent. The subducting slab attached to the continental margin will detach, and shortening will then most probably cease (Fig. 1b). These 'subduction flip' events are, therefore, rapid, taking less than 10 Ma, as in the Bismark Ranges of Papua New Guinea or the Grampian orogeny of the Caledonides. Thickening and uplift are due to the overthrusting of the arc and fore-arc upon the passive continental margin. The rapid nature of this event means that there is not enough time for radiogenic heating of the orogenic pile. However, the detachment of the slab of oceanic lithosphere during subduction flip is accompanied by voluminous plutonism that advects heat into the orogen causing Barrovian metamorphism (Fig. 1b).

The thick, dense mantle root beneath a continent–continent collision zone is neither thermally nor mechanically stable (Fig. 2b, c). Its removal greatly increases the buoyancy of the orogen, leading to a rapid uplift, or 'morphogenic' phase (Fig. 2c). This can produce mountains up to 5 km high which spread laterally as in the 700-km-wide Tibetan Plateau. In an arc-collision event, buoyancy is gained by the detachment of the heavy subducting slab from the continental margin causing rapid uplift (Fig. 1b). The increased gravitational potential of the mountains will cause large fold nappes to be thrust over the foreland of the orogen and produce foreland basins filled with molasse. Such nappes are thrust both towards the foreland and the hinterland (Fig. 1d, 2c), widening the collision zone, as in the Alps. and causing rapid uplift of metamorphic rocks (Fig. 2c, d, e). These rocks do not have time to cool, but pressure is dropping as they are uplifted, and the earlier Barrovian metamorphism is followed by a later moderate-temperature, low-pressure, Buchan event (point c in Fig. 2e). The thinning of the orogen reduces buoyancy and leads to subsidence or 'collapse' of the mountain belt (Fig. 2d), sometimes below sea level, as in the Alboran Sea in the core of the Betic–Atlas orogen. This event is also associated with production of late orogenic granite magma (Fig. 2e, point c). Orogenic collapse is relatively rapid; for example it took some 20–30 Ma in the Variscan orogeny of Europe, reducing the height of mountains much faster than erosion.

Many mountain belts record several collisional events. The Himalayas resulted from the closure of the 5000-km-wide Tethyan Ocean during the late Mesozoic–early Tertiary. The Caledonides have a record of subduction throughout Ordovician and Silurian times, suggesting that a wide ocean was destroyed. These events inevitably involved the production and collision of island arcs before final closure of the ocean. Volcaniclastic flysch sediments are deposited during these events. In long orogens these events will be diachronous; for example, in Taiwan, the Luzon Arc is colliding obliquely with Asia. They may also be missing along strike; for example, the Ordovician arc-collision event recorded in the British Isles is not present in the Caledonides of Greenland. The Alps were formed by convergence in extended continental crust and closure of ocean basins so small that they could not produce significant island arcs; there was therefore no early arc-collision event there.

The lines marking the closure of oceans are called *sutures* (Fig. 1d) and are usually marked by ophiolitic remnants (e.g. the Indus–Tsangpo suture of the Himalayas), by subduction-related metamorphism (e.g. the Variscan suture), by a marked change in fossil fauna and flora, and by a fundamental change in the geophysical properties of the crust (e.g. the Iapetus suture in Britain). Recognition of these sutures is critical to the understanding of orogens that are older than the modern ocean floor and cannot be reconstructed using the principles of plate tectonics; compare, for example, the models presented for the Variscan and Alpine orogenies.

It is extremely unlikely that convergence will be orthogonal to the strike of the orogen. Thickening is therefore likely to be associated with strike-slip displacements and the orogeny will undergo transpression. Different plates colliding with one margin at different times can produce changes in shear sense. For example, Ordovician arc collision on the Laurentian margin of the Caledonides was associated with dextral motion, but later Silurian continent–continent collision was associated with sinistral motion. During the collapse phase this motion became transtensional and opened strike-slip basins within the orogen or allowed emplacement of large granite batholiths such as the Donegal batholith of Ireland.

Many collisional orogenies follow the Wilson Cycle, in which oceans open along the lines of earlier mountain belts whose lithosphere is weaker than that of the foreland. The time interval between closure and re-opening can be up to 400 Ma. The relative weakness of the orogen cannot therefore be attributed to thermal effects of collision, for the lithosphere would have cooled after such a long interval. It may be due to the presence of partially subducted continental eclogites within the mantle (Fig. 2d), which not only produce radiogenic heat but are weaker than the mantle they displace.

Collisional orogens form at continental margins and are therefore long and commonly linear. A linear form could reflect the original geometry of the margin, or be superimposed by shear, parallel to the orogen during collision. Arcuate sectors in orogen can occur because one plate, usually a microcontinent, acted as an indentor, as in the Iberian Variscides, or because of original embayments in the continental margin, as in the Appalachians. The role of indentors in the Himalayan orogeny is controversial and two models are proposed. One model suggests that India indented Asia extruding material eastwards along strike-slip faults. The other argues that Asian lithosphere behaved as a viscous slab, allowing the Indian collision to produce a wide zone of deformation; the strike-slip faults are related to dextral transpression along the eastern margin. Orogens with complex geometries can form around remnant oceans, such as those marginal to the Mediterranean, as a result of the interference of several plate processes such as microcontinent or arc collision, subduction roll-back, and back-arc spreading. Final closure may remove much of this complexity, producing a deceptively simple linear orogen.   PAUL D. RYAN

Further reading

Dewey, J. F. and Bird, J. M. (1970) Mountain belts and the new global tectonics. *Journal of Geophysical Research*, 75, 2625–47.

Windley, B. F. (1995) *The evolving continents.* John Wiley and Sons, Chichester.

Burg, J.-P. and Ford, M. (1997) *Orogeny through time.* Geological Society Special Publication No. 121.

comets *see* ASTEROIDS AND COMETS

compaction and consolidation of soil
Ideally, soil can be considered as a two- or three-phase system consisting of soil particles and water or air, or both. Dry soils and fully saturated soils have two phases and partially saturated soils have all three phases. When a soil is compressed it becomes more dense by rearranging its particles in a tighter packing arrangement, but in order to do this, air or water must be expelled. *Compaction* is the densification of soil by the expulsion of air, with negligible change in the water content of the sample; *consolidation* is densification by the expulsion of water.

Compaction reduces the permeability of the soil and increases its shear strength and bearing capacity. The degree of compaction is measured in terms of the dry density of the soil, i.e. the mass of solids per unit volume of soil, and for a given compactive effort there is a maximum dry density that can be achieved. When material is to be used to form an embankment or dam or to raise ground levels it is important to attain a dense condition. In order to accomplish this, loose soil material is generally placed in layers of between 75 and 450 mm thickness. Each layer is then compacted with a specified compactive effort using specialized heavy machinery such as rollers and rammers, to expel as much air as possible.

In such operations, it is necessary to determine the suitability of available material. If it is either too dry or too wet it may be unsuitable and the maximum dry density will not be achieved. The amount of water contained is critical to ensure effective compaction of a soil for a given compactive effort.

A dry soil will not compact well and reach low dry densities because it is without the lubricating action of any water. Higher dry densities are obtained in the presence of some water; the soil particles can then move past each other and air is easily expelled. If, however, the water content is too high, the water starts to occupy space that would otherwise be occupied by soil particles, resulting in a lower dry density. Laboratory tests can determine the maximum dry density and the optimum moisture content for a given compactive effort. Greater compactive efforts will give higher maximum dry densities at lower optimum moisture contents. In practice, not all the air can be expelled from a soil, which may typically contain 5–10 per cent of air voids.

Consolidation is the process of applying a load to a saturated soil in order to squeeze out some of the pore water. In soils of high permeability this is immediate; in soils of low permeability, however, drainage, and therefore consolidation, is slow.

A common analogy of consolidation is that of a frictionless piston pushing against a spring in a cylinder full of water (Fig. 1). If a load is applied to the piston, the spring does not compress since all the load is supported by increased water pressure. If a valve is opened so that the water can escape slowly (to represent the low permeability of the soil), then the load is gradually transferred to the spring and it compresses until it finally supports the entire load and water pressure dissipates. A stage during this process can be represented by an equation fundamental to soil mechanics:

$$\sigma = \sigma' + u,$$

where σ is the total stress (i.e. the applied load), σ' is the effective stress (i.e. that supported by the soil skeleton), and u is the pore water pressure.

Such conditions may apply, for instance, when a building is constructed on a saturated soil of low permeability. If it is constructed quickly, the load will be supported initially by an increase in the pore-water pressure. With time, the building

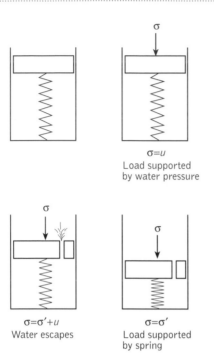

Fig. 1. Consolidation analogy showing a cylinder in which a load (σ) is gradually transferred from the water pressure (u) to a spring representing soil skeleton (σ').

$\sigma = u$
Load supported
by water pressure

$\sigma = \sigma' + u$
Water escapes

$\sigma = \sigma'$
Load supported
by spring

may settle as some of the pore water is squeezed out and the ground beneath consolidates, gradually transferring the load to the soil particles. The settlement resulting from dissipation of pore-water pressure occurs at a rate that depends on the permeability of the soil and is termed *primary consolidation*. Further settlement or secondary consolidation may then occur in the long term as the soil particles adjust and rearrange themselves to the applied load.

In such instances, before the construction of the building the consolidation characteristics of the underlying soil would need to be ascertained from laboratory tests. These tests would consolidate a soil sample under given loads by measured amounts to obtain two important soil parameters: the coefficient of volume compressibility (m_v), which is a measure of the amount by which a soil will compress when loaded and allowed to consolidate; and the coefficient of consolidation (c_v), which indicates the rate of compression and the time period over which settlement will take place. The permeability of the sample can be shown to be proportional to the product of m_v and c_v.

CHRISTOPHER McDONALD

conglomerate and gravel *see* GRAVEL AND CONGLO-MERATE

conodonts Conodonts are an extinct group of marine organisms whose preserved skeletal parts consist of tiny phosphatic tooth-like elements. These microfossils are found worldwide and have been recovered from sediments ranging in age from the Late Proterozoic to the Upper Triassic. They are constructed of laminations of the mineral apatite that accreted around a growth centre; in the earliest forms the laminations were laid down internally, but in the Cambrian the growth surfaces became external, indicating that the elements were embedded in secretory tissue. Although it is now recognized that the elements formed an apparatus composed of a series of bilaterally symmetrical elements of different shape, originally each different element was given a separate name (a form-generic approach). Elements are categorized morphologically into one of three major groups: coniform, simple and conical; ramiform, in which a main cusp is flanked by lateral processes bearing denticles; or pectiniform, platforms with lateral denticle-bearing processes.

Only more recently have generally accepted conodont animals been recognized in the fossil record, despite the long history of study of the conodont elements. Although previous examples of soft parts related to conodont elements were known, the extent to which they represented conodont animals or conodont predators had been debated. In 1982, however, convincing conodont body structures were reported from the Mississippian (Early Carboniferous) of Scotland by Derek Briggs of the University of Bristol and his colleagues. These are elongate, worm-like animals, about 4 cm in length with impressions of a terminal tail fin and midline, and an almost complete assemblage of conodont elements in the head area. Although it has generally been accepted that the

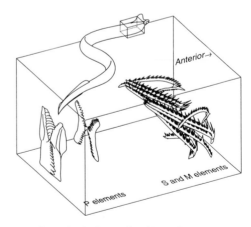

Fig. 1. Conodont animal soft parts found in Scotland, N. America and S. Africa now indicate that the microfossil conodonts were assembled as a complex feeding apparatus in the 'head' region of the animal. These extinct animals may have been related to modern agnatha and cephalochordates. (Reprinted from *Tree*, vol. 11, Aldridge, R. J. and Purnell, M. A., The conodont controversies, p. 466, 1996, with permission from Elsevier Science).

unique structure of conodont elements indicates that they belonged to a separate phylum, it has recently been suggested by Moya Smith of Guy's Hospital (London) that morphological details of the Scottish conodont animal, together with microscopic details of the hard tissue, indicate a close affinity with vertebrates.

The association of conodonts with pelagic (free-swimming) organisms such as fish and cephalopods suggests that they are broadly adapted to life within benthonic (bottom-dwelling) and pelagic environments. They are rare in shallow-water environments and absent from deep basinal environments, but are abundant in sediments containing organisms adapted to normal marine conditions in which nutrient levels were high. They also seem to have been temperature-controlled, since they occur in abundance in zones parallel to the Equator but are absent from cold-water environments. Although little is known about their method of feeding, it is conjectured that the apparatus may have been used in food-gathering, either to strain microplankton from the water or as an aid in grasping prey.

The principal geological application of conodonts is in biostratigraphy. Because of the morphological changes that occur through time an extensive range of morphotypes is available, which has made it possible to develop a sequence of 140 zones through the Palaeozoic and Triassic. This has world-wide utility and is one of the most widely applicable biostratigraphic zonations in invertebrate palaeontology. A second important application is the use of a colour alteration index (CAI) that recognizes the degree of heating at depth in the Earth's crust by means of the colour change shown by a conodont element. As petroleum does not occur in areas of intense geothermal activity, a map of CAI contours can help to delineate areas of low or high petroleum potential.

DAVID K. ELLIOTT

conservation, geological *see* EARTH-HERITAGE CONSERVATION

consolidation of soil *see* COMPACTION AND CONSOLIDATION OF SOIL

contact metamorphism (thermal metamorphism)
Probably the earliest understanding of the causes of metamorphism came from the study of *contact-metamorphic aureoles*, that is, metamorphic rocks that occur around an igneous intrusion, and have been produced by the transformation of pre-existing sedimentary or metamorphic rocks which still persist at a distance from the igneous contact. Most of the classic metamorphic aureoles studied in the nineteenth century are associated with granitic plutons emplaced late in the history of an orogenic belt; hence the metamorphic assemblages produced by magmatic heating are superimposed on earlier, regional metamorphic assemblages. One of the distinctive features of most contact-metamorphic rocks is that mineral growth is not accompanied by deformation. As a result, contact metamorphic minerals grow in random inter-

locking patterns giving rise to *hornfels*, a tough rock with no direction along which it will split preferentially. This texture is often found in rocks that appear to retain a pre-existing schistosity, because this can remain picked out as a fine compositional segregation, irrespective of subsequent recrystallization. One of the best-known rock types of contact metamorphism is *spotted slate*, a rock found in the outer parts of an aureole hosted by low-grade slaty regionally metamorphosed rocks. It has conspicuous spots, up to a few millimetres in diameter, which overprint the slaty fabric and are, or were, *poikiloblasts* (large inclusion-riddled crystals) of cordierite. In many instances the cordierite proves under the microscope to have been extensively altered to a fine-grained hydrous products known as *pinite*.

Most typical contact-metamorphic aureoles formed in the upper half of the crust, at pressures less than about 4 kbar. The effects of magma emplacement at greater depth are likely to be associated with more widespread heating of a regional character. For example, many granulite-facies terrains reflect additional magmatic heat contributions to areas of regional metamorphism. Because aureoles form at shallow depths, they are characterized by minerals stable at low pressures, typical of the hornfels facies (see *metamorphism, metamorphic facies, and metamorphic rocks*). Most typical are the assemblages of pelitic rocks in which cordierite is widespread but garnet generally rare, while andalusite normally occurs in place of kyanite. So strong is this association that when kyanite was first reported from the aureole of the Main Donegal Granite by W. S. Pitcher, an eminent reviewer condescended to declare that the mineral identification must have been incorrect, since it was well known that kyanite could not form during contact metamorphism.

Contact-metamorphic aureoles show marked zoning in their assemblages, and the rocks nearest the contact may have experienced sufficiently high temperatures for the onset of partial melting. This is seen in one of the best-studied modern examples, the Ballachulish aureole in Scotland. The strong variation in metamorphic grade across the zones in an aureole is indicative of conductive heating, and a number of workers have used simple models of conductive heating to estimate the duration of contact metamorphism and the pattern of heating. Simple calculations suggest that the maximum temperature in the aureole should not exceed the midpoint between the temperature of the magma and the initial temperature of the country rocks. Higher temperatures are, however, readily achieved if the pluton represents a conduit through which magma passed to an overlying volcano.

The simplicity of the classic contact-metamorphic aureole is somewhat misleading; there are a number of important aspects of contact metamorphism that remain obscure or at least controversial. One of the most elementary difficulties, alluded to above, is where to draw the distinction between regional and contact metamorphism. In areas of voluminous magmatism, low-pressure, high-temperature metamorphism commonly occurs beyond the immediate proximity of the igneous bodies, has zones that are not concentric about the pluton margins, and is typically accompanied by deformation. (north-eastern Maine is an excellent example.) This is much more a problem of classification that of understanding. It is clear that where magmatism results in extensive volcanism, as well as forming plutonic rocks, the distribution of metamorphism will reflect the total magma flux rather than the volume of pluton remaining. Furthermore, prograde metamorphism can result in weakening and superplastic behaviour of the rocks affected, as fluid pressure rises, minerals are weakened, and the grain-boundary structure is transformed by breakdown and growth. Modern work on stresses in the crust has demonstrated that large areas of apparently stable continent are in fact stressed close to the point of failure, and so deformation can be triggered as readily by weakening of the rocks as by increased stress. It is not therefore surprising that, if a sufficiently large volume of rock is weakened by thermal metamorphism, it will begin to deform in response to the ambient stress regime.

One of the most curious features of contact metamorphism is the places where it does not occur. Despite the fact that basic magmas are emplaced at higher temperatures than granitic ones, basaltic intrusions emplaced into sedimentary rocks at shallow levels commonly lack contact aureoles, or display extreme melting effects only close to the contact. It appears that at shallow levels the properties of water and the nature of the host rock to the intrusion play an important role in determining whether contact metamorphism, in the normally accepted sense, will take place. It is much easier to understand this by studying what is happening today around volcanoes where contact metamorphism might reasonably be expected to be occurring at depth. Many volcanoes have areas of geothermal activity on their flanks, and exploitation for geothermal energy means that their thermal structure has often been investigated by drilling. Frequently, temperature increases with depth in a way that is close to the boiling point–depth curve of water. A steep increase in temperature over a distance of a few hundred metres below the water-table is followed by rather uniform temperatures over a long vertical distance, typical of fluid convection. R. O. Fournier has pointed out in the context of the Yellowstone geothermal area in the USA that, when the temperature in the geothermal field is taken into account, there must be a very steep increase in temperature beneath it towards the magma body that remains at depth. These two thermal regimes correspond to an upper zone of relatively cool permeable rocks (at up to 400 °C), with extensive circulation of hot water, and a deeper zone in which the steep temperature gradient indicates conductive heat loss and is typical of a thermal aureole. This is illustrated in Fig. 1; indeed direct evidence for such a structure has been reported from deep drilling at the Larderello geothermal field in Italy.

BRUCE W. D. YARDLEY

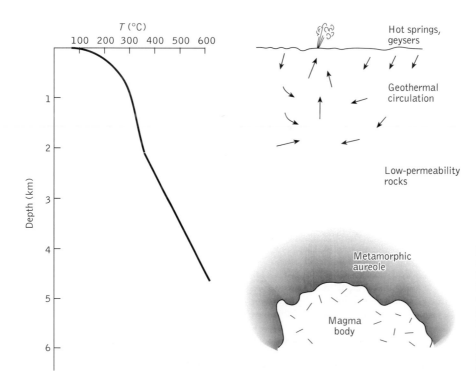

Fig. 1. Schematic representation of the relationship between a magma body emplaced at depth, the metamorphic aureole that develops around it, and geothermal circulation in volcanic rocks near the surface. The temperature profile is keyed to the geological cartoon and shows the temperature variation with depth above the pluton.

Further reading

Lux, D. R., De Yoreo, J. J., Guidotti, C. V., and Decker, E. R. (1986) Role of plutonism in low-pressure metamorphic belt formation. *Nature*, **323**, 794–7.

Voll, G., Töpel, J., Pattison, D. R. M., and Seifert, F. (eds) (1991) *Equilibrium and kinetics in contact metamorphism*. Springer-Verlag, Berlin.

containment walls in environmental management

Vertical containment walls have been used for several decades to provide low-permeability barriers to depths of over 30 m below ground surface. Typical applications of these walls include containing sanitary and hazardous waste landfill contaminants, sealing dams and dykes, enclosing oil and chemical tank farms to prevent releases, dewatering structural excavations, and hydraulically isolating lagoons.

Containment walls, also known as 'cutoff walls' or 'slurry walls', are constructed either of soil–bentonite or of cement–bentonite slurry. For most applications, the soil–bentonite slurry provides the most economical and impermeable wall. Containment walls can be constructed using several methods according to depth and permeability requirements. Slurry walls are built by excavating a narrow trench 0.5 to 1 m wide while pumping in the slurry and maintaining its level at or near the top of the trench during excavation. The excavated soil is mixed with additional clay (the quantity is determined from tests) and bentonite slurry to a consistency similar to that of wet concrete. The mixture is used to fill the trench and sets to form a low-permeability containment wall. The bottom of the trench is set 30 cm to 1 m into an underlying low-permeability formation. The wall, in conjunction with the low-permeability formation, provides containment of the material within the wall.

Containment walls can be constructed *in situ* by using other techniques such as deep soil mixing (DSM). The DSM system uses a set of leads supported by a crane that guides a series of mixing shafts each 1 m in diameter. Three to four shafts are advanced vertically into the soil while bentonite slurry is injected through their hollow stems. A combination of auger flights and mixing blades along the shafts lift the *in situ* soil and mix it with the slurry. Continuity of the wall is ensured by using multiple sets of shafts overlapping the previously drilled columns. Containment walls typically reuse most of the excavated material, and the quantity of bentonite used varies from 5 to 10 per cent. Cement–bentonite slurry walls are used where self-hardening, stability-enhancing walls are required.

Permeability is the critical design parameter for containment walls; it varies from 10^{-8} m s^{-1} to 10^{-11} m s^{-1}. The permeability is reduced as the natural clay content or bentonite content increases. The percentage of coarse-grained materials in the soil affects the strength and compressibility of the wall. The strength increases as the percentage of coarse-grained material increases. As the water content in the soil–bentonite

slurry reaches equilibrium with that of the surrounding soil, the strength of the wall is approximately equal to the strength of the surrounding soil.

The development of high-density polyethylene (HDPE) products led to the installation of vertical HDPE liners underground without excavation. Since 1991, some manufacturers in the USA have installed HDPE containment walls to depths of about 10 m below ground without excavation. These walls consist of a plastic barrier of HDPE liner panels locked together with a watertight joint.   V. RAJARAM

contaminant transport

Contaminants are dissolved constituents in surface waters (rivers, lakes, and seas) and groundwaters that originate from direct or indirect anthropogenic sources. The contaminants can be individual metal ions (e.g. lead, Pb^{2+}) or complex organic molecules (e.g. the pesticide DDT). The amount of the contaminant in water is governed by the solubility of a metal with respect to a mineral phase or the ability of an organic molecule to dissolve in water. For example, water, a dipolar fluid, can dissolve slightly polar organic molecules (e.g. benzene) more easily than non-polar molecules (e.g. oil). Once the contaminants have entered a watercourse, they are transported at the same rate as the water flow unless retardation occurs because of transport, chemical, or biological processes (see *pollution control*). A contaminant that is transported with the water without any chemical or biological control (e.g. chloride (Cl^-) and the solvent trichloroethylene) is affected only by dilution processes such as diffusion and advection.

Dilution processes

Two processes are the main cause of diluting dissolved constituents in groundwater: *diffusion* and *hydraulic advection/dispersion*. Diffusion occurs in the absence of any bulk hydraulic movement caused by the kinetic energy of solutes. Hydraulic dispersion (mechanical dispersion) occurs by mechanical mixing during fluid flow caused by motion of the fluid. In groundwater studies, the term 'hydrodynamic dispersion' covers diffusion and hydraulic dispersion.

Diffusion

Diffusion (self-diffusion, molecular diffusion, ionic diffusion) occurs when a solute moves under the influence of thermal–kinetic energy in the direction of its concentration gradient. Diffusion continues until the concentration gradient is zero. In technical terms, the mass flux F (in kg m^{-2} s^{-1}) of a diffusion substance passing through a given cross-section per unit time is equal to the concentration gradient dC/dL, where C (in kg m^{-3}) represents the solution concentration and L the distance (in metres) multiplied by D (in m^2 s^{-1}), the diffusion coefficient. This relationship is known as *Fick's first law*.

The diffusion coefficient, D, for major ions (Na^+, K^+, Mg^{2+}, Ca^{2+}, Cl^-, HCO_3^-, and SO_4^{2-}) in groundwater (at 25 °C) varies from 1×10^{-9} to 2×10^{-9} m^2 s^{-1}. D is temperature-dependent, and is lower by about 50 per cent at 5 °C (the temperature of most groundwaters) than at 25 °C. In porous media the *apparent diffusion coefficient (D*)* is smaller than D in water because of the presence of particles and adsorption. D^* is thus generally lower than D by a factor of 0.5–0.01.

Using Fick's first law and the equation of continuity, we can relate the concentration of a diffusion substance to space and time. The change in concentration with respect to time (dC/dt) is equal to D^* multiplied by the second derivative of the change in concentration with respect to distance, L (d^2C/d^2L), the second derivative of the concentration gradient. This is *Fick's second law*. A calculation using Fick's second law gives the change in concentration with respect to time and distance and is often plotted as the dimensionless ratio C_i/C_o where C_i represents the concentration at time t and distance x and C_o is the initial concentration. Diffusion is a relatively slow process. The diffusion time for a relative concentration, C_i/C_o, of 0.01 is, for example, 500 years if we assume that the apparent diffusion coefficient, D^*, is 15×10^{-10} [m^2 s^{-1}].

Diffusion will cause a solute to spread away (be transported) from the place where it is introduced into a porous medium, even in the absence of groundwater flow. Figure 1 shows the distribution of a solute introduced at concentration C_o, at time t_o, over an interval $(x - a)$ to $(x + a)$. At succeeding times, t_1 and t_2, the solute has spread out, resulting in a lower concentration over the space interval $(x - a)$ to $(x + a)$ but increasing concentrations outside this interval. The solute concentration follows a normal or Gaussian distribution.

In zones of active groundwater flow the effects of diffusion are usually masked by the effects of the bulk water movement. In low-permeability deposits such as clay or shale, in which the groundwater velocities are small, diffusion over periods of geological time can, however, have a strong influence on the spatial distribution of dissolved constituents.

Transport by advection

Dissolved solids are carried along with the flowing groundwater. This process is called *advective transport*, or *convection*. The amount of solute being transported is a function of its

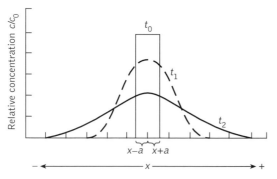

Fig. 1 The spreading of a single injection of solute as a result of diffusion. At time t_o the initial concentration was C_o.

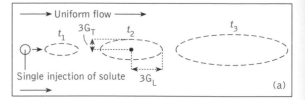

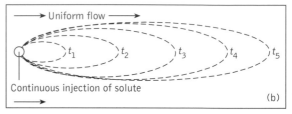

Fig. 2 The spreading of a solute in a uniform flow field in an isotropic formulation: (a) single injection of solute; (b) continuous infection of solute. (After Freeze, R. A. and Cherry, J. A. (1979).)

concentration in the groundwater and the quantity of groundwater that is flowing.

Mathematically, the one-dimensional mass flux, (F_L) due to advection is equal to the average linear velocity V_L (from Darcy's law: see *groundwater*) multiplied by the product of, the effective porosity n_e, and the concentration C of the dissolved solids: $V_L n_e C$.

The one-dimensional advective transport or the change in concentration with respect to time (dC/dt) is equal to the average linear velocity times the change in concentration with respect to distance (dC/dL). A solution of the advective transport equation yields a sharp concentration front. On the advancing side of the front, the concentration is equal to that of the invading groundwater, whereas on the other side of the front it is unchanged from the background value. This is known as *plug flow* (Fig. 2a). All the pore fluid is replaced by the invading solute front.

Mechanical dispersion

Groundwater moves at rates that are both greater and less than the average linear velocity. This is caused by three effects (Fig. 3): (1) differences in pore size; (2) tortuosity, e.g. the branching and interfingering of pore channels; and differences in the velocity of the water across the pores because of drag caused by roughness of the pore surfaces.

Because all the water flowing in a porous medium is not travelling at the same velocity, mixing occurs along the flow path. This mixing is called *mechanical dispersion*, and it results in dilution of the solute at the advancing edge of the flow. The mixing that occurs in the direction of the flow path is called *longitutional dispersion* (D_L).

An advancing solute front will also tend to spread in directions that are normal (perpendicular) to the direction of flow, because at the pore scale the flow paths can diverge, as shown

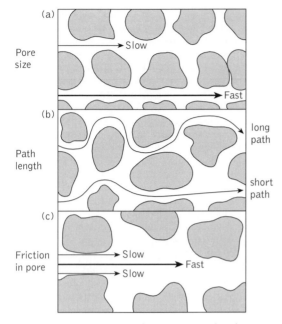

Fig. 3 Factors causing dispersion. (After Fetter, C. W. (1988) Applied hydrogeology. Macmillan, New York.)

in Fig. 3. The result of this mixing in directions normal to the flow path is *transverse dispersion* (D_T).

Hydrodynamic dispersion

If we assume that mechanical dispersion can be described by Fick's law for diffusion and that the amount of mechanical dispersion is a function of the average linear velocity, we can then define the coefficient of mechanical dispersion to be the dynamic dispersivity (α) multiplied by the average linear velocity (V_L). Since the process of molecular diffusion cannot be separated from mechanical dispersion in flowing groundwater, the two are combined to define a parameter called the *hydrodynamic dispersion coefficient, D*. D is thus equal to the coefficient of mechanical dispersion plus the apparent diffusion coefficient. The hydrodynamic diffusion coefficient can be defined for longitudinal flow (L, principal direction) and transverse flow (T, perpendicular to L). In simple terms,

the transverse dispersion is close to being one-tenth of the longitudinal dispersion.

Figure 4 shows the effect of diffusion and mechanical dispersion on the relative concentration (C/C_o) of a solute acting as a tracer that has been injected into a porous medium under one-dimensional flow conditions. A mass of solute is instantaneously introduced into the aquifer at time t_o over the interval $x = k + a$. The resulting initial concentration is C_o. The advecting groundwater carries the mass of solute with it. In the process the solute slug spreads out, so that the maximum concentration decreases with time, as shown for times t_1 and t_2. The diffusional model of hydrodynamic dispersion pre-dicts that the concentration curves will have a Gaussian distribution.

Figure 2 shows this dispersion in two dimensions (one-dimensional flow (horizontal), two-dimensional dispersion, L and T). This is based on an experiment in which a non-reactive tracer was injected into isotropic sand. The tracer is transported along the flow path, spreading in both the L and T directions. The total mass of the tracer does not change, but with time the mass occupies an increasing volume of the porous medium. Fig. 2a shows the spread of the tracer for a point source (e.g. accidental spillage); Fig. 2b depicts the spreading of the tracer for a continuous source (e.g. a landfill, a leaking underground tank).

The advection–dispersion equation

By applying the law of conservation of mass we can write an overall relation that takes account of fluid flow and mass flux:

$$\begin{bmatrix} \text{Net change} \\ \text{of mass of} \\ \text{solute in} \\ \text{volume} \end{bmatrix} = \begin{bmatrix} \text{Flux of} \\ \text{solute out} \\ \text{of the} \\ \text{volume} \end{bmatrix} - \begin{bmatrix} \text{Flux of} \\ \text{solute} \\ \text{into the} \\ \text{volume} \end{bmatrix} \pm \begin{bmatrix} \text{Loss or gain} \\ \text{of solute} \\ \text{mass due to} \\ \text{reactions or} \\ \text{adsorption} \end{bmatrix}$$

The physical processes that control the flux into and out of the elemental volume are advection and hydrodynamic dispersion. The loss or gain of solute mass in the elemental volume can occur as a result of chemical or biochemical reactions or radioactive decay.

The relation above can be put into mathematical form for solute transport in saturated porous media by assuming the law of conservation of mass and that the porous media is homogeneous, isotropic, saturated, a steady-state flow, and that Darcy's law applies. Using these assumptions, the flow is

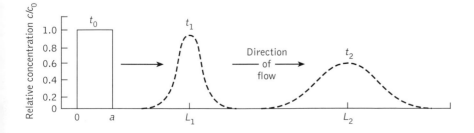

Fig. 4 Transport and spreading of a single injection of solute by advection and dispersion. (After Fetter, C. W. (1993).)

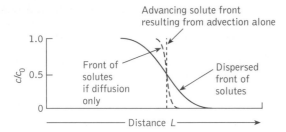

Fig. 5 Diagram showing the contributions made by molecular diffusion and mechanical dispersion to the spread of a solute concentration front. (After Fetter, C. W. (1988) Applied hydrology. Macmillan, New York.)

described by the average linear velocity which carries the dissolved substances by advection, and dissolved constituents are nonreactive. A solution to the advection–dispersion equation (in three dimensions) can be obtained by assuming that the contamination is caused by an instantaneous slug at a point source. The maximum concentration distribution of the contaminant mass $(M = C_o, V_o)$ at time $t(C(x,y,z,t))$ is then as shown in Fig. 2a. Fig. 5 shows the tracer front if diffusion only takes place, the dispersed tracer front, and a vertical dashed line which represents an advancing solute front resulting from advection alone.

These equations can describe the flow of contaminants in an L–T plane for uniform flow for an instantaneous point source. A continuous tracer feed takes the configuration shown in Fig. 2b. For most practical purposes models for one-dimensional flow (longitudinal) and two-dimensional dispersion (D_L and D_T) can adequately describe the transport of contaminants in groundwater systems. The advection–dispersion equation for continuous tracer feed (e.g. landfill) under such conditions can involve *reactive constituents*. Results from using this solution are shown in Fig. 2b. The figures shows a 'contamination plume' in $L–T$ space (L is the longitudinal direction of flow and T the transverse direction perpendicular to L) as a function of time.   K. VALA RAGNARSDOTTIR

Further reading

Fetter, C. W. (1993) *Contaminant hydrogeology*. Macmillan, New York.

Freeze, R. A. and Cherry, J. A. (1979) *Groundwater*. Prentice-Hall, Englewood Cliffs, N.Y.

Continental drift

Continental drift describes the process whereby continents move laterally over the surface of Earth in the course of geological time. Early versions of the theory, current during the first half of the twentieth century, generally viewed the continents as somehow moving through a sea of mantle. This concept never found widespread acceptance, and in particular was opposed by geophysicists who considered the mantle, on seismic evidence, to be too strong for such motion to occur. This problem was resolved with the discov-

ery of thermally activated solid-state creep and the development of plate tectonics in the 1960s. In consequence, the phenomenon of continental movement is now widely accepted. However, because of its association with a discredited mechanism, the term 'continental drift' is now eschewed by some in favour of 'plate tectonics'. It is nevertheless still useful as a term for the phenomenon of continental movement as opposed to the theory of its mechanism.

Although a number of early scholars had noted the matching shapes of the coastline flanking the Atlantic Ocean and had pondered on the origins of their congruence, it appears to have been the American glaciologist Frank Taylor who first suggested, in 1910, that the continents move about on the surface of the Earth. In 1912 the German meteorologist Alfred Wegener proposed his hypothesis of continental drift, and produced a variety of evidence in its favour. Wegener's theory received its greatest support in the southern hemisphere, where some of the consequences of continental drift after the break-up of Gondwanaland are best displayed. The South African geologist Alexander du Toit was one of its leading proponents.

There is a wide variety of geological evidence for continental drift. Perhaps the most direct is evidence of climatic variation with time, especially where it can be shown that it was not part of a global change but that different patterns of variation occurred in different continents.

A number of lines of evidence suggest that continents that are now widely separated were once joined. These include the matching of rock types, climatic regions, ancient mountain belts, and fossils across continental margins that were once contiguous. An important example is a Permo-Carboniferous glaciation which occurred approximately 300 million years (Ma) ago and affected parts of all the southern continents, leaving characteristic deposits such as tillites. When these continents are reconstructed in their pre-drift configuration, not only do the boundaries of the glaciation match across continental boundaries, but also the total extent of the glaciation forms a roughly circular area, approximately centred on the ancient pole position.

Further important evidence for continental drift comes from palaeontology. For example, some species in now separated land masses can be seen to have followed identical evolutionary paths until the time of continental break-up, and have subsequently diverged. In many instances species, particularly terrestrial ones, inhabit a restricted region bounded by climatic or geographic barriers such as seas or mountains, and one can track the break-up or collision of continents by the fossil evidence of species either becoming isolated by the creation of new barriers or freed to expand by the removal of barriers. Many good examples of the latter effect are provided by the migration of species between Asia and Australia as these two continents gradually collided.

Despite the now compelling nature of much of the geological evidence for continental drift, the most unequivocal proof of the movement of continents now surely comes from

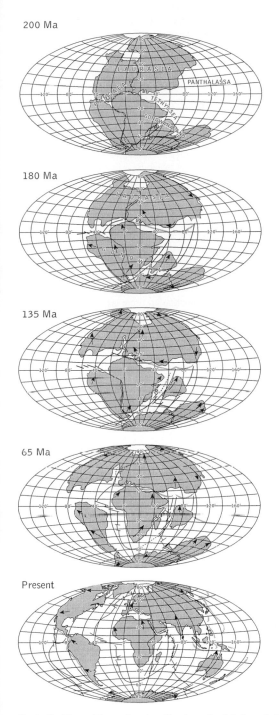

200 Ma

180 Ma

135 Ma

65 Ma

Present

Fig. 1. Positions of the continents, with schematic plate boundaries, over the past 200 million years (Ma). (After F. Press and R. Siever (1986) *Earth*, Fig, 20–17. W. H. Freeman, New York.)

palaeomagnetism. This branch of geophysics studies the magnitude and direction of the ancient magnetic field. Several natural mechanisms can permanently magnetize rocks with a magnetic field parallel to the Earth's field at the time. The direction of this natural remanent magnetism can be determined from oriented samples, and the position of the ancient magnetic pole can be inferred.

When these ancient pole positions are plotted, it is found that those for different continents are often widely separated. The only way of explaining such results is by assuming that the continents have moved in relation to each other and in relation to the Earth's magnetic field. Moreover, if pole positions from samples in the same continent but for different epochs are plotted, it is found that they do not coincide but seem to move in what is known as an apparent polar wander (APW) curve. APW curves from different continents are generally different, although some may share matching sections (indicating that the continents concerned were joined and moved together for that period) separated by non-matching ones (indicating periods of drift).

Finally, for about the past 200 Ma there is widespread evidence of plate movements preserved in the sea-floor. The linear magnetic anomalies produced by sea-floor spreading are especially important. These mark the ancient positions of plate boundaries, and from them detailed reconstructions of plate (and hence continental) motions can be inferred.

A combination of palaeomagnetic and geological evidence has made it possible to trace the history of continental drift back to the Precambrian, although the degree of detail diminishes for earlier times (Fig. 1). Wegener himself suggested that in Permo-Carboniferous time all the continents were joined in one supercontinent called Pangaea, although du Toit recognized a northern supercontinent of Laurasia (North America, Greenland, Europe, and Asia) and a southern supercontinent of Gondwanaland (South America, Africa, India, Australia, and Antarctica). Pangaea began to break apart during the Mesozoic, and this break-up continues today. Pangaea itself was, however, formed from the earlier convergence of Laurasia and Gondwanaland in the Silurian, and Laurasia itself was formed from a number of earlier continental blocks. The recurring episodes of continental break-up, drift, and collision have been called the Wilson cycle after J. Tuzo Wilson, one of the pioneers of plate tectonics. It is likely that there have been a number of such cycles, each lasting a few hundred million years, in the history of the Earth.   ROGER SEARLE

Further reading

Kearey, P. and Vine, F. J. (1996) *Global tectonics*. Blackwell Science, Oxford.

continental drift and palaeomagnetism *see* PALAEOMAGNETISM AND CONTINENTAL DRIFT

continental flood basalts Continental flood basalts (CFBs) are the largest volumes of basalts that are found on the continents, having volumes of thousands of cubic kilometres; some flows, such as those of the Deccan in India and the Columbia River in the north-western United States, extend for hundreds of kilometres. The Snake River Plain in the north-western USA is the only area in which flood basalts are currently erupting, but it is thought that volcanic emissions from larger earlier eruptions might have been sufficient to change global climate, perhaps even causing a 'nuclear winter'.

CFBs are generally of tholeiitic composition, but show signs of contamination by crustal material.

Modern theories ascribe the formation of CFBs to melting of the Earth's mantle as a direct consequence of a continent being pulled apart over a plume of hot material rising from deep in the mantle. This hypothesis is supported by the existence of CFB provinces in Brazil and South Africa, which can be closed matched. The eruptions that produced these basalts took place before the opening of the South Atlantic in Cretaceous times.   JUDITH M. BUNBURY

continental margins Continental margins mark the transition between thin, dense, oceanic crust and the thicker, lighter, chemically different continental or transitional crust. If it were not for our landlubber's bias in terminology, they could equally well be termed ocean margins. However, this simple definition obscures the fact that there are different types of continental margin, a point made by E. Suess as early as 1885 when he differentiated between Atlantic margins as steadily sinking regions that accumulate thick sequences of sediments, and Pacific margins that are, on the whole, rising, and are associated with volcanism, folding, faulting, and other mountain-building processes. Thus for many years continental margins were classified into two types: Atlantic and Pacific. The Atlantic, or passive, type has a stable continental block on the landward side that has been little deformed since Palaeozoic times. It lacks earthquakes and widespread volcanism. The Pacific, or active, type is associated with a trench, volcanism, active mountain-building, and earthquakes.

As our understanding of global tectonics developed during the 1960s, it became apparent that the theory of plate tectonics offered at least a partial explanation as to why these two types of margin existed: passive margins bordered diverging oceans where new oceanic crust was being created, while active margins lay above or close to subduction zones where oceanic crust was being lost. True though this simple division is, it is only part of the story and obscures major differences within each type. In particular, there is a third type, the transcurrent margin, previously referred to as active, which has elements of both types: although it is seismically very active, it lacks significant volcanicity.

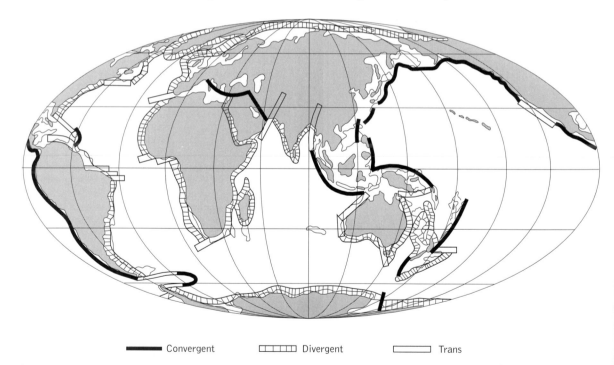

Fig. 1 Present-day distribution of convergent (= Pacific), divergent (= passive = Atlantic), and transcurrent (= translational = transform) types of continental margin. (After Emery, K. O. (1980) *Bulletin of the American Association of Petroleum Geologists*, **64**, 297–315; modified by Kennett, J. P. *Marine geology*. Prentice-Hall, Englewood Cliffs, N.J.)

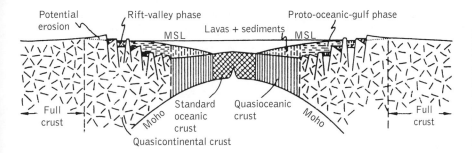

Fig. 2 Post-rift, early drift phase in the development of an immature Atlantic-type, passive continental margin, as exemplified by the present Red Sea. (From Ingersoll, R. V. and Busby, C. J. (eds) (1995) *Tectonics of sedimentary basins*. Blackwell Science, Cambridge, Massachusetts.)

Continental margins are now divided into three types:

(1) A passive, aseismic Atlantic type that sits upon a plate above the transition between oceanic and continental crust. It is on a divergent plate but does not mark a plate boundary; this lies at the adjacent mid-ocean ridge (see *mid-ocean ridges*). However, within this broad definition there are many different types, reflecting primarily the age of the margin (immature or mature) and the type and amount of sediment supplied from the neighbouring continent.

(2) An active, seismic Pacific type that marks a convergent plate boundary. However, not all convergent boundaries are continental margins. At intra-oceanic island arcs, oceanic crust is subducted under oceanic crust (see *island arcs*) and there is a spectrum from that type of oceanic arc, through those where oceanic crust is subducted beneath crust transitional between continental and oceanic and those subducted under isolated continental islands (e. g. Japan), to those that genuinely mark the margin of continents (e.g. Central and South America).

(3) Transcurrent (translational, transform) margins, where movement between plates is lateral, seismicity is high, and volcanic activity limited (e.g. the coast of California).

Passive continental margins almost entirely encircle the Atlantic and Indian Oceans; even the Red Sea has passive margins. Exceptions are the Antillean and Scotia arcs in the Atlantic and the Burma–Andaman–Indonesian island arc along the north-eastern side of the Indian Ocean and some transcurrent margins indicated by major fracture zones (Fig. 1).

The present-day Red Sea exemplifies the early stages of continental break-up (Fig. 2). Situated at the northern end of the East African Rift system (see *extensional tectonics on the continents*), it forms one arm of a triple junction, the third arm being the Gulf of Aden. In the Red Sea a sequence of first arching, then compound, segmented rifting and finally separation of continental plates has occurred. The early stages of arching and rifting were accompanied by basaltic igneous activity. Drainage was away from the rift because of the elevations of its margins, and sedimentation within it has been limited to fault-bounded troughs where very complex pat-

terns of sedimentation occur because of an alternate pattern of asymmetrical extensional faults linked by transfer zones. In this arid climate evaporites form; alluvial fans lie at the base of footwall scarps and carbonate reefs grow.

The significance of the present-day Red Sea and Gulf of Aden is that they show the kind of structural and sedimentary patterns that underlie more mature passive margins, such as much of the central Atlantic, and those that have aborted to be preserved today under younger basinal sediments, as have the hydrocarbon-bearing Mesozoic sediments of the North Sea.

If the early drifting continues and the continents become separated by a widening ocean, mature passive continental margins develop. These can be divided into a shelf, a slope, and a rise, the shelf and slope being collectively known as the continental terrace (Fig. 3).

The shelf lies between the coast that forms the seaward extension of the adjacent continent and the shelf break or shelf edge, along which there is a marked increase of gradient from about 1 : 1000 to more than 1 : 40. The depth of the shelf break varies, but averages 130 m.

The continental slope is relatively narrow (less than 200 km), extending from depths of 100–200 m down to 1500–3500 m. Its important feature is its steepness. Its profile may show anything from a rather gentle concave upward curve to a series of complex scarps and sediment-filled basins. Although sedimentary draping of the slope may take place, sediments are rather unstable and large masses of sediment may intermittently slump downslope to give a confused pattern of rotational slump faults.

The continental rise has a gentle gradient (1 : 100 to 1 : 700) that diminishes oceanward. Its width may vary from 100 to 1000 km as it passes into the more-or-less flat abyssal plain (see *ocean basins*). It is primarily a depositional feature and is today the receptacle for most of the sediments coming off the American, African, and European continents, as well as from those rivers of Asia that flow into the Indian Ocean. The bulk of the sediment is transported from shallower water by sporadic turbidity currents that construct large sandy deep-sea fans that may extend out onto the abyssal plains.

In the Atlantic Ocean another process of sedimentation is of equal importance in the construction of the continental

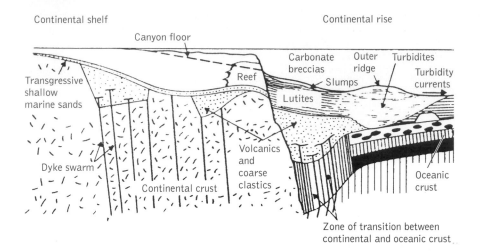

Fig. 3 Late-drift, mature Atlantic-type passive continental margin. (After Dewey, J. F. and Bird, J. M. (1970) *Journal of Geophysical Research*, **75**, 2625–47.)

slope and rise. This is due to the very complex pattern of semi-permanent ocean bottom currents, which are the deep-water expression of oceanic thermohaline currents. Deep-ocean bottom water is formed by cooling and sinking of surface water at high latitudes. In the North Atlantic, for example, these cold, dense bottom waters flow south from the Arctic, banked up against the western margin by the deflecting effect of the Coriolis force, which is always to the right in the northern hemisphere. Thus an undercurrent, called the Western Boundary Undercurrent, travels along the continental slope and rise. Since it parallels the contours, this type of current has been known as a 'contour current' and the deposits that form from such currents are called contourites, as opposed to turbidites for the deposits of turbidity currents. Contourites are made up of fine sand, silt, and mud. This contrasts with the very broad range of grain sizes that can make up a turbidite, the composition of which depends primarily on the nature of its source. Though they are semipermanent, contour currents fluctuate considerably, but in the western North Atlantic they may attain velocities of 10–20 cm s^{-1} and even exceed 100 cm s^{-1}.

The nature of passive continental margins varies not only with age but also because of a number of other factors: seg-mentation, sediment supply, climate, and sea-level changes.

At first sight continental margins appear to run for thousands of kilometres along the side of the ocean. However, they are in fact broken into segments by transverse fractures and faults, just as the underlying rifts and mid-ocean ridges are (see *mid-ocean ridges*). Some of these fractures are fundamental faults that penetrate the continent; others originated as small transform faults at the mid-ocean ridges or as transfer faults in the original continental rift system. Even the western margin of the North Atlantic, which at the surface appears to be a relatively continuous feature, has been shown, by seismic profiles and drilling, to be composed of a number of separate sedimentary basins filled by 8–18 km of sediment, separated by fracture zones, some of which penetrate the continent.

Where sediment supply is enormous, as off the Mississippi, the Niger, the Indus, and the Ganges–Brahmaputra rivers, the margin is dominated by great piles of sediment deposited in deltaic–submarine fan complexes. Thicknesses in the Gulf of Mexico margin reach 16 km, over 7 km having been deposited locally during the Pleistocene alone. The high rate of sedimentation leads to sediment instability and the development of a whole range of synsedimentary deformational processes, including growth faults and shale diapirs (Fig. 4).

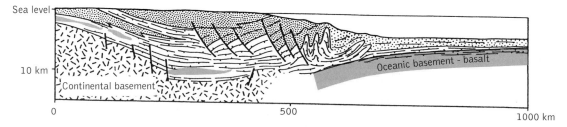

Fig. 4 Atlantic-type passive continental margin forming in front of a major delta, displaying synsedimentary tectonic deformation. (After Beck, R. H. and Lehner, P. (1974) *Bulletin of the American Association of Petroleum Geologists*, **58**, 376–95.)

In contrast to the Gulf of Mexico, the eastern margin of the North American continent is relatively starved of continental sediment, but, as has been shown above, sedimentation is dominated by contour currents flowing southwards. These contribute substantially to the thicknesses of sediment along the margin. Because these thermohaline contour currents are dependent on cold polar waters descending in the Arctic, there have been considerable fluctuations in their effectiveness in the past. During interglacial periods they were much less important and sedimentation at the continental margin would have been much reduced.

Some continental margins are more less completely starved of terrigenous sediment, whether supplied from the neighbouring continent or from along its length. This is exemplified today by the Blake Plateau, a continental margin 850 m deep and 300 km wide that has been kept clear of sediment since the early Tertiary when the Gulf Stream first swept northwards from the Straits of Florida. Virtually no sediments have accumulated since the Cretaceous; the Mesozoic was a time when huge carbonate build-ups were formed. Such sedimentation can be seen today in the Bahamas, which have provided us with the classical area for the study of modern carbonate platforms.

The Bahamas developed when the shelf was cut off from the mainland by the deep Florida Straits. They built up by carbonate growth as a series of isolated shallow-water platforms surrounded by deep water. As with all isolated reefs, the facies distribution is primarily dependent on the prevailing wind and storm directions.

There are, however, other types of carbonate continental margins. Some, such as the so-called rimmed shelf, have a pronounced break of slope between the shallow-water platform and deep water. Present-day examples are the Great Barrier Reef off the eastern coast of Australia, the South Florida shelf, and the Belize shelf on the Caribbean coast of Central America. The rim is formed by a nearly continuous line of barrier reefs or skeletal-oolite sand shoals or both. Behind these rims lie extensive lagoons. Carbonate ramps, such as the Yucatan coast of Mexico in the Gulf of Mexico, have a gently sloping surface (less than 1°) along which there is a gradual passage from nearshore carbonate sands to deeper-water muds. There is no offshore barrier or lagoon.

Another effect of oceanic currents on the build-up of continental margins is the upwelling of nutrient-rich ocean waters. The upwelling of phosphate-rich waters associated with the nutrient-rich waters off south-west Africa and off the convergent margin of Peru is of great economic importance. It provides a large and very significant proportion of sediment along such margins where terrigenous material is lacking because of the very arid nature of the adjacent continent, owing to its position on a west-facing coast in the high-pressure desert belt.

Eustatic and relative sea-level changes are the main factors in controlling the detailed pattern of sedimentation, though not the broad outlines. The effects of sea-level changes are extremely complex (see *sequence stratigraphy*). Although there are many exceptions, resulting from local circumstances, in general a relative fall in sea level on a clastic continental margin induces an increase in sedimentation, whereas a relative rise in sea level reduces the sediment supply to the ocean as the coastline is moved landward. This contrasts with carbonate systems, where the carbonate sediment is produced within the photic zone at or just below sea level within the general area of sediment accumulation. The optimum time for sediment to be manufactured and supplied is thus during a rise in sea level.

An important feature of modern passive continental margins that should never be ignored is the presence of underlying salts and the diapiric structures that they produce (Fig. 5). Because salt basins abounded during the early stages of rifting, the continental margins of West Africa, especially off Gabon, Angola, and Morocco as far north as off southern Spain, off South America, and in the Gulf of Mexico show domination of the structure and overlying basin patterns by the upward and lateral movement of salt domes, although shale diapirs may locally be of equal importance.

Convergent continental margins are best seen along the eastern side of the Pacific, at the Andean margin (Fig. 6) off Central America, and off the North American states of Oregon and Washington. Many of the processes involved are discussed under *island arcs*. Here, however, no backarc basins developed, and the volcanic arc stayed well within old continental crust. Compressional forces are dominant. As with all such belts, the Andes themselves are segmented into five zones, only three of which have active volcanoes. The divisions into volcanic and non-volcanic zones are related to dif-

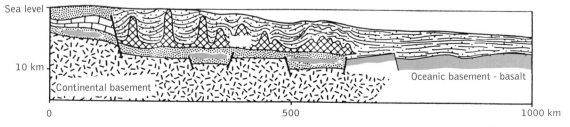

Fig. 5 Atlantic-type passive continental margin, showing deformation caused by diapiric movement of an underlying salt layer. (After Beck, R. H. and Lehner, P. (1974) *Bulletin of the American Association of Petroleum Geologists*, **58**, 376–95.)

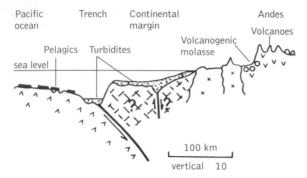

Fig. 6 Andean-type convergent continental margin. Note that continental basement may lie immediately above the subduction zone.

ferences in the nature of the subducting Pacific plate, in particular the dip of the subduction zone. Where the dip is about 30° there are active volcanoes, but where the dip is much lower (5–10°) volcanic rocks are absent. A feature that is almost entirely absent from the Andean margin (except at the very southern end of Chile) is a forearc basin and trench slope break. Trenches are relatively narrow and small, and are close to the continental mainland. In part this is because sediment from the Andes and the interior of the continent is transported eastwards towards the Atlantic Ocean, but principally it is because the climate is dry and therefore little sediment is transported into the Pacific Ocean. In contrast, off Central America and off Oregon in the north-western USA, accretionary prisms are very well developed. This is a consequence of the heavy rainfalls in these two regions. Thus the nature of the margin, and whether an accretionary prism develops or not, may depend on the rainfall.

Transcurrent continental margins are extraordinarily complex and it is almost impossible to make general statements about them. The best-studied example today is the west coast of North America, from California into northern Mexico (Fig. 7). Here a belt thousands of kilometres long and up to 500 km wide, of which the San Andreas fault system is only a minor part, is being subjected to dextral (right-handed) translation as the Pacific plate moves northwards relative to the North American continent.

This broad belt includes, in its northern part, compressional and thrust-faulted mountains; basins, both lacustrine

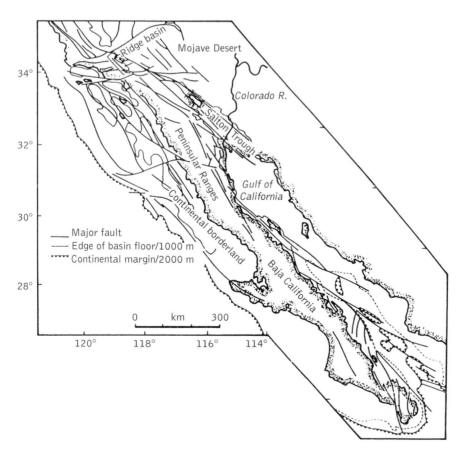

Fig. 7 Transcurrent, translational continental margin between the Pacific and North American plates. (Based on several sources.)

and marine; and a wide area of continental crust between the true shelf, which here is very narrow, and the continental margin, which is taken at the 2000 m isobath, known as the continental borderland (Fig. 7). Within this area are some 20 strike-slip basins and intervening highs or islands. The basins are commonly 1000 m, even 2000 m, deep, and within them are innumerable deep-sea fans, which have been studied extensively as models of deep-sea sedimentation.

In the southern part of this transcurrent margin the degree of regional extension increases so that oceanic crust is created as the Gulf of California opens. At its northern end is the Salton Trough, which lies 110 m below sea level. It has considerable volcanic and intrusive magmatic activity, due probably to an incipient spreading centre beneath. These features are, in fact, seen in the Gulf of California itself, where new oceanic crust is forming at mid-ocean ridges. Thus the gulf represents a piece of ocean separated from the main Pacific by the equally large slice of continental crust of Baja California.

Other transcurrent margins occur along the north coast of the bulge of South America and its counterpart on the west coast of Africa, where plate motion in this central Atlantic area was primarily along fracture zones, rather than by spreading centres. The southern margin of South Africa is another transcurrent margin. In this case sedimentation is dominated by an oceanic current, the south-westward flowing Agulhas Current that impinges on the shelf. It was from this margin that the Falklands plateau became detached and, during translation, was rotated 180°, thus demonstrating the complex tectonic histories of such margins.

HAROLD G. READING

Further reading

Burk, C. A. and Drake, C.L. (eds) (1974) *The geology of continental margins.* Springer-Verlag, New York.

Edwards, J. D. and Santogrossi, P. A. (eds) (1989) Divergent/passive margin basins. *Memoir of the American Association of Petroleum Geologists* No. 48.

Watkins, J. S. and Drake, C. L. (eds) (1982) Studies in continental margin geology. *Memoir of the American Association of Petroleum Geologists* No. 34.

continental rifts
A rift is essentially an elongate down-faulted block or graben, usually associated with a pronounced linear depression. However, major continental rifts comprise complex systems of faults in zones extending for many hundreds or even thousands of kilometres. Rifts typically also exhibit characteristic seismic and volcanic features, and contain thick sedimentary deposits. The primary structures of rift zones are steep faults of extensional origin, and the process of horizontal extension is fundamental to the origin of all rifts, although the amount of extension may be quite small: about 10 per cent in many instances.

All active rifts exhibit anomalous crustal and upper-mantle profiles, which are interpreted as the result of lithosphere thinning, and share many of the characteristics of ocean

ridges. For example, both are underlain by a region of low-density mantle (termed the rift 'pillow') beneath the central part of the structure, correlating with a region of higher than normal heat flow.

Associated volcanicity is highly variable, both in type and quantity. Some rifts exhibit no volcanicity, whereas others are highly active and are surrounded by vast lava fields.

Some rifts form part of the network of currently active plate boundaries; others appear to link with that network but project far into a plate interior; yet other examples are situated entirely within a plate, far from any plate margin. Some past rifts have evolved into oceans (for example, the precursor rifts of the present Atlantic Ocean), whereas others have existed over long periods of geological time without opening to form oceans (e.g. the East African rift system). Rifts that join a successfully opened ocean are termed 'failed arms'. These usually join the successful branch of the system at an

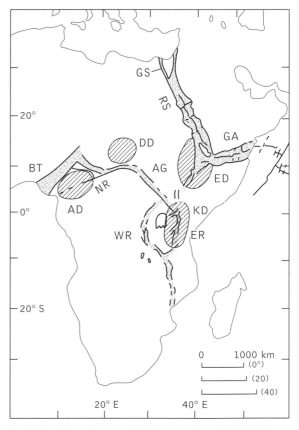

Fig. 1. The African rift system. AD, Adamaoua dome; AG, Abu Gabra rift; BT, Benue trough; DD, Darfur dome; ED, Ethiopian dome; ER, East African rift, eastern branch; GA, Gulf of Aden; GS, Gulf of Suez rift; KD, Kenya dome; NR, Ngaoundere rift; RS, Red Sea; WR, East African rift, western branch. (After Girdler, R. W. and Darracott, B. W. (1972) Comments on the Earth sciences. *Geophysics,* **2** (5), 7–15.)

angle to form the third arm of a 'triple junction'. The Benue Trough of West Africa, at the western end of the African rift system, is a good example of a failed-arm rift (Fig. 1).

Three examples illustrate the diversity of natural continental rift systems: the African and Rhine rifts and the Basin and Range Province of the western USA.

The African rift system

The well-known rifts of East Africa are part of a much larger, continent-wide system that extends across Central Africa to link up with the Atlantic Ocean to the west on one side, and embraces the Red Sea and Gulf of Aden rifts to the east on the other side (Fig. 1). To the south, the two main branches of the East African system (the Western and Eastern rifts) join and continue southwards to meet the Indian Ocean at Beira in Mozambique. This tripartite rift system thus divides Africa into three segments. The Red Sea and Gulf of Aden are both parts of the active plate-boundary network, forming the eastern margin of the African plate. They link up to the north with the plate boundary that runs through the Mediterranean and to the south with the Indian Ocean ridge system. The other two branches connect with the now inactive continental margins of the Atlantic and Indian oceans respectively.

Associated with the rift system are four major dome-shaped uplifts that have been centres of volcanicity for the past 10 to 20 million years (Ma), since the Miocene period, in Ethiopia, Kenya, Darfur, in the western Sudan, and Adamaoua in Cameroon. These domal uplifts appear to be the focal points for the rift volcanism, and it is in these areas that the lithosphere has undergone the greatest thinning.

The Eastern, or Kenya, rift crosses the Kenya dome, which is elliptical in plan, and about 1000 km in extent along its longer axis, parallel to the rift axis. The dome rose to a mean height of 1400 m during its maximum development in the early Pleistocene period, about 1.5 Ma ago.

The major faults of the Kenya rift zone (Fig. 2) define a complex branching graben system with an overall N–S trend, although individual faults and graben segments generally strike NNW–SSE or NNE–SSW. The conspicuous central graben (the 'Gregory Rift') traverses the elliptical uplift, and at its northern and southern ends is replaced by less well-defined broad depressions. The main graben is 60–70 km wide and 750 km long, and is bounded by steep normal faults arranged *en echelon*. Between the ends of adjacent *en echelon* faults, sloping ramps descend from the marginal plateaux to the rift floor.

The major fault escarpments bounding the rift are up to 2000 m in height, and the floor of the rift valley is filled by a layer of sediment that is up to 2 km in thickness, implying a total displacement on these major faults of up to 4 km. The faults, which dip steeply towards the graben floor, are extensional in origin, and at least 5 km of crustal extension is required to explain the displacements on the visible faults. The rift floor is, however, cut by abundant minor faults parallel to the graben walls, and the amount of extension should perhaps be doubled to 10 km to take account of these concealed faults.

Activity on the Kenya rift appears to have commenced during the Miocene period, 10–20 Ma ago, with flexuring of the western rift flank accompanied by extensive volcanicity. The first major faulting developed in the early Pliocene, about 5 Ma ago. The magmas produced were predominantly of alkaline type, rich in potassium, characteristic of rift volcanicity in general. Further north, in Ethiopia, the rift system widens into a broad fault zone (Fig. 1) that is extending at a rate of about 3–5 mm per year and is associated with much more voluminous volcanicity.

The East African rift system is linked across Central Africa to two other domal uplifts, Darfur and Adamaoua, via the NW–SE Abu Gabra rift in the western Sudan, and the NE–SW Ngaoundere rift in Cameroon (Fig. 1). These two rifts are sediment-filled troughs, now largely inactive, which saw their maximum development in the Cretaceous period, more than 100 Ma ago. The Ngaoundere rift is linked in turn with the Benue trough in Nigeria, which is a failed arm connecting to the mid-Atlantic rift in Cretaceous times, prior to the opening of the South Atlantic Ocean.

On the other side of Africa is the Gulf of Aden, which is floored by oceanic crust formed during the past 10 Ma and connects with the Indian Ocean ridge system (Fig. 1). The Gulf of Aden continues northwards into the Red Sea rift, also floored by oceanic crust, which has opened at an average rate of about 2 cm per year over the past 3–4 m.y. However, prior to their oceanic opening stage, these two structures were sites of active subsidence and volcanism from the Cretaceous period onward. At its northern end, the Red Sea rift branches into two further rifts: the Gulf of Suez and Dead Sea rifts, which connect in turn with the convergent plate boundary through the Mediterranean. The Gulf of Suez rift is non-volcanic and consists of a central trough flanked by fault blocks tilted at angles of between 5° and 35° away from the rift axis. There is no evidence of any previous doming, and the fault geometry indicates that the rift formed as a result of horizontal extension perpendicular to the rift axis. Estimates of the amount of extension vary from 20 to 50 per cent over the past 25 to 30 Ma.

The African rift system as a whole thus consists of a network of rather disparate structures, originating at different times, parts of which are no longer active. Some sectors of the rift system are intimately associated with domal uplifts and have been or are now the focus of extensive volcanicity; others are volcanically inactive and exhibit only extensional faulting. There appear to have been three main phases of movement on the network. The initial phase seems to have been in the early Cretaceous period, about 130 Ma ago, and was probably connected with the opening of the South Atlantic Ocean via the Benue trough. At that time, a continuous series of sedimentary basins crossed Africa, linking the Atlantic and Indian Oceans. The second phase of activity commenced about 35–40 Ma ago with the initiation of the Red Sea rift. The third phase commenced in the late Miocene period, between 5 and 10 Ma ago, with the development of the East African rifts on

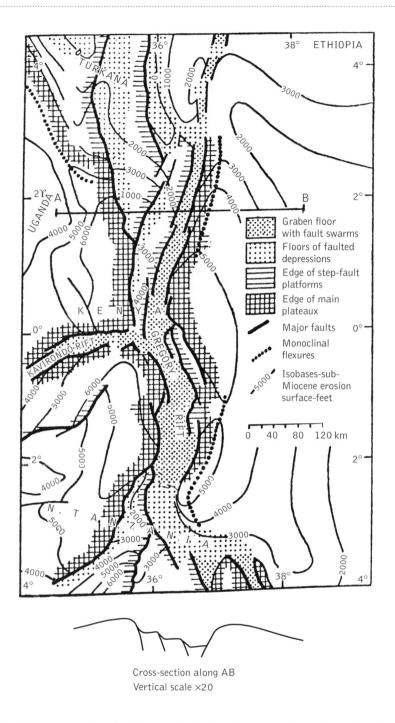

Fig. 2. The Kenya sector of the East African rift system, showing fault pattern and elevated rift shoulders. Contours indicate amount of crustal uplift in feet. (After Baker, B. H. and Wohlenberg, J. (1971) *Nature*, **229**, 538–42.)

Cross-section along AB
Vertical scale ×20

the site of the volcanically active Kenya and Ethiopian domes. At the same time, the Red Sea and Gulf of Aden rifts commenced their opening stages and extended northwards to the Gulf of Suez rift and southwards into the east African system to form a continuous active rift dividing Africa from Arabia and isolating Somalia from the rest of Africa (Fig. 1).

167

The Rhine rift

Because of its situation at the heart of Western Europe, the Rhine rift is one of the best known and most extensively studied examples of currently active rift systems. The rift extends from the Alpine front at Basle for a distance of 440 km, in a NNE–SSW direction, as far as Kassel, but the main branch turns north-westwards at Mainz, following the Rhine river for 300 km to Arnhem, where it disappears under recent deposits (Fig. 3). It is thought to connect northwards with the graben system underlying the North Sea basin.

The fault pattern in the southern sector of the rift consists of a branching system of faults that curve from NE–SW outside the rift into a N–S orientation within the rift—a pattern indicative of a sinistral (left-lateral) shear along this portion of the rift (Fig. 3). At many localities, however, normal faults with an originally extensional geometry are overprinted by horizontal striations, indicating two successive phases of movement with different directions. The northern

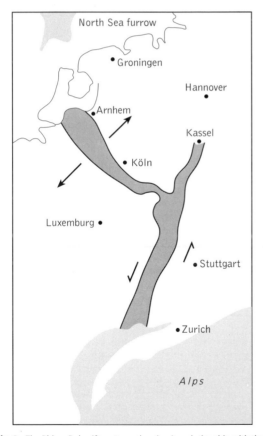

Fig. 3. The Rhine–Ruhr rift system, showing its relationship with the Alpine front and the North Sea Furrow. Note NE–SW extension in the northern branch and left-lateral shear on the southern. (After Illies, J. H. and Greiner, G. (1978) *Bulletin of the Geological Society of America*, **89**, 770–82.)

sector of the rift system is characterized by abundant steep faults, parallel to the NNW–SSE trend of the rift, formed under ENE–WSW extension (Fig. 3). There is no continuous graben feature in this part of the system.

Like the African examples, the Rhine rift is associated with volcanism and with a regional upwarp that forms raised borders to the rift. Examination of pebbles from conglomerate deposits within the rift fill has shown that uplift of the rift borders evolved simultaneously with the main stages of graben subsidence between about 100 and 35 Ma ago. Geophysical evidence indicates that the rift is underlain by a thin crust that correlates with a zone of higher than normal heat flow.

When the rift system was initiated, between about 100 and about 80 Ma ago during the late Cretaceous period, the movement pattern must have involved extension in an approximately E–W direction (to explain extension in both rift branches). This activity may be related to a general extension that characterized northern Europe before the formation of the North Atlantic Ocean between 90 and 55 Ma ago, but could also be related more directly to the initiation of compressive deformation in the Alps to the south which took place about 80 Ma ago.

Rifting and subsidence continued through the Eocene and Oligocene periods, reaching a climax about 35 Ma ago, but activity decreased during the Miocene period as Alpine convergence showed down and was replaced by uplift. About 3 Ma ago, during the mid-Pliocene period, the rift was reactivated but with a different movement direction, that is, left-lateral shear along the southern rift and ENE–WSW extension along the northern (Fig. 3). The present stress field has been thoroughly investigated; the pattern indicates a general NW–SE compression both in the Alps and across Western Europe generally, which is compatible with the current movement pattern.

The Basin and Range province

The Basin and Range province is located in the western part of North America, extending from the Gulf of California in northern Mexico through the western USA and dying out in southern British Columbia—a distance of over 3000 km (Fig. 4). In Nevada and adjoining parts of Utah, California, and Oregon, the province is about 1000 km wide, but narrows to the north and south. Unlike the previous two examples, the Basin and Range province is not a well-defined rift, but a broad extensional zone containing many individual rifts (for example, the Rio Grande rift) and basins.

The province is situated within a broad elevated plateau between 1 and 2 km in height, and includes a large region of internal drainage: the Great Basin. The southern end of this extensional province links through the Gulf of California with the East Pacific ocean ridge; the northern end dies out east of the volcanically active Cascades range. This belt of volcanoes is situated about 300 km east of the active subduction zone along the west coast of Oregon and Washington.

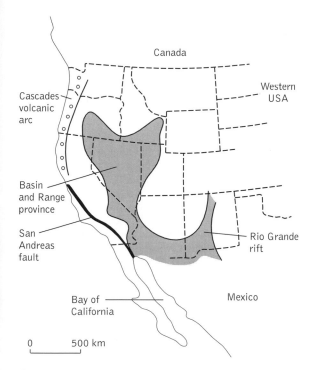

Fig. 4. The Basin and Range province and the Rio Grande rift, in relation to the Cascades volcanic arc and the San Andreas fault. Black arrows represent the current WNW–ESE extension direction; white arrows show the older extension direction. (After Zoback, M. L., Anderson, R. E., and Thompson, G. A. (1981) *Philosophical Transactions of the Royal Society*, **A300**, 407–34.)

The main part of the province, in Nevada and the adjoining states, is seismically active and, although largely non-volcanic, exhibits a high heat flow (around three times the normal level). It consists of a linear topography of elongate fault-bounded ranges separated by basins filled with Cenozoic to recent sediments. The blocks are spaced 25–35 km apart and the intervening basins are 10–20 km wide. The block structure is controlled by a system of steep faults in response to a WNW–ESE extensional stress field. The amount of extension has been estimated at about 20 per cent during the past 10–13 Ma, but much larger extensions are thought to have taken place during an earlier phase of extension more than 20 Ma ago.

The earlier phase of extension was accompanied by extensive volcanicity, and the faults of that period show evidence of much larger movements; in many instances they have been rotated to low angles by crustal stretching. These low-angle faults form the roofs of large areas of crystalline basement, known as core complexes, that have been brought to the surface as a result of the stretching and thinning of the crust. Some of these structures have been imaged in seismic

reflection profiles and appear to descend to depths of 15–20 km. Estimates of the amount of extensional strain vary from 50 per cent to as much as 300 per cent—very much higher than anything seen in the African or Rhine systems.

The origin of the extensional province is controversial. It has been ascribed by some to 'back-arc' extension on the upper plate of the subduction zone that bordered the coast of North America throughout much of the Mesozoic era. Reorganization of the plate structure in the Pacific Ocean 40–50 Ma ago resulted in the progressive cessation of subduction-related volcanicity along the southern sector of the western American continental margin, to be replaced by the San Andreas fault. This resulted in a diachronous 'switching off' of the subduction-related magmatism commencing about 20 Ma ago in the south and gradually moving northwards to the present end of the Cascades volcanic arc in Oregon. It is thus more difficult to link the present extensional field in the southern and central Basin and Range directly to subduction. Others have ascribed the origin of the province to the effects of a deep mantle plume, originally situated beneath the Pacific Ocean, gradually overridden by North America as the latter moved westwards, and now situated beneath the region of extensive volcanicity at the northern end of the province.

Genesis of continental rifts

There have traditionally been two schools of thought on the genesis of continental rifts: one ascribing the initiation of rifts to the effect of mantle plumes, the other to crustal extension. In the plume-driven model, a series of volcanic domes is formed, associated with thinning of the lithosphere, resulting in crustal extension and eventual collapse of the central parts of the domes. The rift segments thus formed then connect up to form a continuous rift. This model was proposed by James Burke and John Dewey in 1973 to explain the formation of oceans. The second model ascribes the formation of rifts primarily to crustal extension arising from major plate-boundary forces. In this model the doming and volcanicity are secondary effects of the lithospheric thinning produced by the extension.

From the examples described above, it is clear that two types of active rift exist: one characterized by doming and volcanicity, the other solely by extensional faulting. Both types are found in the same system (compare the East African with the Gulf of Suez rifts, for example.) The volcanic examples are, however, often preceded by extensional basins, as in the Rhine rift, and it is hard to escape the conclusion that both mechanisms play a part in the origin of rift zones. It appears to be necessary for regional (plate-scale) extension to exist before rifts can 'take off' to become oceans, as in the case of the Gulf of Aden and the Atlantic. It seems unlikely that those rifts, such as the East African, that are clearly plume-related but lie within a compressional stress field, will evolve into major extensional features or oceans. Regional extension would seem to be a necessary, but perhaps not a sufficient, condition for major rift systems to develop. When the African rift system was first formed, probably during the Cretaceous period, the whole African plate may have lain in an extensional stress field, but the process of rift formation was probably aided, certainly along the course of the future South Atlantic Ocean, by strategically placed mantle plumes.

R. G. PARK

Further reading

Park, R. G. (1988) *Geological structures and moving plates*, Chapter 4. Blackie, Glasgow.

continental shelf geology The continental shelf includes the gently sloping area from the shoreline to the shelf break where the inclination of sea floor increases with passage to the continental slope. The continental shelves represent about 8 per cent of the area of the oceans and have an area of 29×10^6 km^2. The majority are shallower than 130 m but some reach depths of 550 m. Continental shelves are generally narrow along convergent plate margins (as around the Pacific Ocean), where they border deep trenches, and wide along intraplate passive margins (as around much of the Atlantic Ocean) where they border a wide continental slope/rise.

Shelves are the most accessible marine areas and they provide rich economic resources such as oil and gas, minerals (aggregate, sand, heavy minerals rich in rare elements such as titanium and chromium, diamonds) while the contemporary sea floor provides the habitat for fish and shellfish which are important food resources for mankind. Their accessibility has led to abuse by man. Marshes and shallow areas are cordoned off to 'reclaim' land, which in reality means destruction of marine habitat (almost universally but on a major scale in The Netherlands). Soft sediment is hydraulically removed by civil engineers to build up the adjacent land and leads to the formation of marine deserts in the source areas (for instance along parts of the nearshore of the United Arab Emirates). Continental shelves are also used as a dumping ground for wastes of various kinds (sewage sludge, munitions), and the overlying water is often polluted with industrial effluents and untreated sewage. All these activities are detrimental to the environment. and the marine biota, especially in coastal areas. They have serious consequences for the sustainability of food resources derived from shallow seas.

Continental shelves are the more ephemeral parts of the marine realm because of oscillations of climate and sea level. During the Last Glacial Maximum (LGM) 18 000 years ago, large areas of modern continental shelves were exposed subaerially. Those at high latitudes were ice-covered, and in areas bordered by mountains they are commonly cut by deep ice-gouged valleys that now form fiords, for example, along the margins of the Norwegian–Greenland Sea. There are also large quantities of morainal deposits (as on the shelf off Newfoundland). In middle to low latitudes, the shelves were crossed by rivers that disgorged their sediment load on to what is now the outer shelf and the adjacent continental slope.

Sea level has risen by about 130 m since the LGM. As a consequence, the lower parts of river valleys have become flooded. River-borne sediment has then invariably been deposited close to the point where the river enters the sea. Where the rate of input is low, the flooded valley forms a *ria* (as along the west coast of Spain). Where the rate of input is moderate, the result is an estuary largely infilled with sediment and bordered by intertidal flats and marshes (as with most rivers entering the North Sea). Where the rate of input is high, a delta progrades across the shelf (as in the Mississippi in the Gulf of Mexico). Since the sediment is then ponded close to the shore, the shelves are largely starved of sediment. The deposits are thus out of equilibrium with present sea level and are partially or entirely relict, relating to deposition in the early stages of the marine transgression. Such deposits cover some 70 per cent of modern shelves. In some instances, former alluvial deposits associated with a previous lower sea level remain exposed on the sea floor. Deposits of this kind off Africa yield alluvial diamonds. These sediments are sucked from the sea floor in order to extract the diamonds. Elsewhere, sand and gravel are extracted in the same way for use in the building industry (as off southern England). Sands containing heavy minerals have been extracted from beaches (as at Redondo Beach, California). The process of extraction has serious consequences for marine life, not only on the sea floor, but also through putting sediment into suspension in the water column, which affects the primary productivity of the plankton.

In some instances, the absence of detrital input has led to the accumulation of carbonate deposits formed from the shells of marine organisms. In middle to high latitudes such deposits lack aragonite mud; they are formed of bioclasts of molluscs, barnacles, bryozoans, and foraminifera and are termed *heterozoan*. The rates of accumulation are a few centimetres per thousand years. In low-latitude arid climates, such sediments commonly contain aragonite mud; the deposits are formed of bioclasts of molluscs, hermatypic (reef-forming) corals, and calcareous algae and are termed *chlorozoan* because the hermatypic corals and calcareous algae are light-dependent. The rates of formation here are considerably greater than those in middle to high latitudes. In some instances, large protected shelf lagoons have developed where reef growth on the shelf has formed a barrier that isolates the inner shelf from severe wave attack (for example, the lagoon behind the Great Barrier Reef). Fine-grained sediments accumulate in such settings.

The shelves along passive margins were initiated by rifting associated with the divergence of two lithospheric plates. After the rift phase, during which half-graben basins formed (and were infilled with syn-rift, mainly clastic, sediments), such areas underwent subsidence caused partly through thermal cooling of the lithosphere and partly through loading caused by the deposited sediment. In areas well nourished by clastic sediment (the east coast of the USA), the sediments reach a thickness in excess of 10 km. In areas starved of clastic sediment, subsidence is much less and, although the syn-rift sediments may be clastic, the post-rift sediments are at least partly calcareous (as in the Western Approaches to the English Channel and the Celtic Sea). In the North Atlantic, the initiation of rifting took place during the Triassic. Wide epicontinental shelves may also be underlain by basins that were initiated by rifting (such as that of the North Sea). In this case there are thick pre-Permian shallow marine and continental deposits which formed prior to rifting. Permo-Triassic rifting led to the initiation of the North Sea basin. Deep-water deposits rich in organic matter formed in the central graben, and these provided a source for oil which has accumulated in Jurassic–Palaeogene reservoirs now exploited in off-shore oilfields. In the southern North Sea, natural gas derived from Carboniferous sources has accumulated in Permian sandstones.

Pacific shelves are situated in areas of active tectonism resulting from the convergence of two lithospheric plates. The oceanic plate plunges beneath the continental plate along the line of the trench, which marks the zone of subduction. The shelf may suffer uplift during tectonism and, in very shallow areas, this leads in some instances to emergence as land (as off Alaska). There may also be mass failure of the slope and part of the shelf as a result of oversteepening through compressional tectonics and of build-up sediment. Shelf deposits are then transported either as a fluidized flow or *en masse* into deeper water. The sediments are partly clastic and partly volcaniclastic because volcanism is active along such margins.

On the eastern sides of oceans where the coast lies to the left of the coast-parallel prevailing wind, Ekman transport of the surface waters is away from the coast (as off Peru). This results in upwelling: the ascent of nutrient-rich water from the thermocline or deeper. These regions of high productivity promote the development of plankton, which feeds planktivorous fish such as anchovies. The latter produce faecal pellets rich in phosphate, which in turn may lead to the accumulation of phosphate-rich sediments. If the supply of organic material to the sea floor is very high, there may be temporary or permanent anoxia of the bottom waters. Under such conditions biogenic carbonate may be replaced by phosphorite (as off the coast of South Africa beneath the Benguela Current).

There is popular concern that global warming may cause sea level to rise above its present level. Geologists are aware that sea level has changed constantly though time so, regardless of whether or not global warming is taking place, there is no reason to suppose that the level of the sea will stay at its present position. What are the consequences of a rise? Clearly the effects will be felt mainly in very shallow areas, but the consequences will not uniform because different ecosystems and organisms respond to change in different ways. Deltas would be more affected by an increase in sea level than an increase in temperature, whereas for corals the opposite is true (increased temperature leads to a higher incidence of disease). Sea level has, however, risen about 130 m in the past 18 000 years and the faunas and environments have managed to adjust to this rapid change. There is therefore no cause to be pessimistic about future changes to the natural environment, even though there may be severe consequences for those countries lying close to present sea level.

JOHN W. MURRAY

Further reading

Kennett, J. (1982) *Marine geology*. Prentice Hall, New Jersey.

Seibold, E. and Berger, W. H. (1996) *The sea floor: an introduction to marine geology* (3rd edn). Springer-Verlag, Berlin.

controlled-source electromagnetic mapping

Electromagnetic mapping is a geophysical technique for mapping subsurface structure and composition by way of variations in the electrical conductivity of rocks. It relies on the principles of electromagnetic induction, whereby (1) a changing magnetic field will induce a potential difference across a conductor, which can drive a current, and (2) an electrical current produces a magnetic field. Because it works through induced fields, there is no need for direct coupling to the ground, as is needed for electrical resistivity surveying. The method is therefore generally faster and the equipment is more portable; it is even suitable for airborne applications. Moreover, controlled-source electromagnetic mapping gives the user control over the frequency of the applied signal and over the geometry of the transmitting and receiving antennas, both of which can be chosen to enhance the method's response to particular types and depths of target.

There are two main methods: the *frequency-domain* and *time-domain* methods. The principle of the frequency-domain method is illustrated in Fig. 1. An oscillating current, usually with a frequency of a few thousand hertz (Hz: cycles per second), is generated in a transmitting coil. This current sets up a magnetic field (called the primary field) which oscillates in phase with the transmitter current. The primary field spreads out in space and can penetrate the ground, although the depth of penetration is limited, being less for higher signal frequencies and for higher ground conductivities; it is therefore important to choose an appropriate frequency for the survey. The primary field will induce a varying voltage in any electrical conductor it encounters, such as a conductive orebody. This induced voltage drives a secondary oscillating current in the conductive body, also at the same frequency as the primary, but with a phase difference that depends on the electrical properties of the conductor. The secondary current generates a secondary oscillating magnetic field, which can be detected at the surface by a receiving antenna. In general, the combined effect of the primary and secondary fields at the receiver is detected, but some instruments are designed so that the effect of the primary can be removed. In any case, as the instrument is moved over the survey area, a varying signal will indicate the presence of variations in ground conductivity.

In the time-domain method, a constant primary current is applied, and this is suddenly switched off. The primary field produced then decays with time, and this time variation induces a secondary field, the effect of which is again detected at the surface.

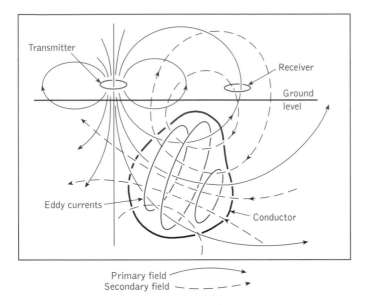

Fig. 1. Principle of controlled-source electromagnetic mapping. A transmitting coil carries an oscillating electric current. This produces a primary magnetic field which penetrates the ground. This field will induce secondary currents in any underground conducting body, which will generate a secondary magnetic field. A receiving antenna at the surface detects the combined effect of the primary and secondary magnetic fields. Variations in this combined field indicate changes in the subsurface conductivity. (Modified after J. M. Reynolds (1997) *An introduction to applied and environmental geophysics*. John Wiley and Sons, Chichester.)

Controlled-source electromagnetic instruments typically have ground penetrations ranging up to a few tens of metres. Greater penetration requires the use of very low frequencies, which are difficult to produce in portable transmitters, although some applications using large, semi-portable antennas hundreds of metres across are known. The more usual instrument consists of two coils (transmitter and receiver), either mounted on a common platform or else deployed separately. Smaller ones can easily be carried by one person; those with a separate transmitter and receiver may require two operators.

Details of both the amplitude and phase of the recorded signal can give information on the position, shape, size, and electrical properties of the conducting body, although as with most geophysical methods there is some inherent ambiguity; it can, for example, be difficult to isolate the effects of increasing conductivity and increasing body thickness. However, one of the advantages of the method is that it has quite good horizontal resolution, and is capable of pinpointing isolated conducting bodies which are hard to resolve using direct current (DC) resistivity.

The method can be applied to any survey where variations in electrical conductivity are expected. A common application is in the search for metallic ores, many of which have high electrical conductivity compared with most common rock types. In environmental work, the method is useful for detecting buried non-ferrous metals, which are non-magnetic. Another important field of applications is the investigation of groundwater. It can be used for detecting depth to the water table (though DC resistivity does this at least as well), detecting water-filled fissures and aquifers, and mapping variations in groundwater conductivity caused by varying levels of dis-

solved contaminants. This last areas is becoming increasingly important in investigating environmental contamination.

As well as mapping particular targets such as orebodies and contaminant plumes, controlled-source electromagnetic methods can be used to map geological structures. The method is particularly useful for mapping variations in thickness of layers of constant conductivity; for example, of a surficial conducting clay layer, or of relatively conductive soil over permafrost.

A relatively recent development in controlled-source electromagnetic mapping is *ground-penetrating radar* (GPR). Suitable short-pulsed radar signals, usually at frequencies of around 100 MHz, are generated in a portable transmitter, which is usually mounted on a towed sledge and directed vertically down. These pulses can penetrate depths of up to a few tens of metres in normal soil or rock. The method was originally developed for sounding through ice, where penetrations of hundreds or thousands of metres are possible.

Radar pulses in the ground are reflected at interfaces where the electrical properties (the conductivity and refractive index or electrical permittivity) change, and these reflected signals are detected at the same antenna that is used for transmission. When the instrument is moved along a traverse, many closely spaced traces can be stacked to reveal a map of subsurface structure, in almost exactly the same way as is done in the seismic reflection method, in which formation boundaries and possibly even individual beds can be seen. Indeed, many of the processing and display techniques used in seismic reflection can also be applied to GPR data. GPR is finding important applications in, for example, engineering-site surveys and environmental investigations.   ROGER SEARLE

Further reading

Milsom, J. (1996) *Field geophysics* (2nd edn). John Wiley and Sons, Chichester.

controlled-source seismology Seismological methods for determining Earth structure are often classified as being either active or passive in nature. Passive methods involve waiting for an earthquake to occur and provide the seismic source for recording. In contrast, controlled-source seismology refers to active methods where the experimenter provides the source by an explosion or a mechanical device such as a hammer, a weight that is dropped, or a vibrator. Active methods can, in turn, be divided into two basic classifications. The first is seismic reflection profiling, which is a clearly defined approach in which the goal is to produce an image of the subsurface in which structures can be seen directly, in the way that an X-ray image reveals features inside an object. This technique is discussed in more detail elsewhere (see *deep seismic reflection profiling* and *seismic exploration methods*). Other active seismic methods infer seismic velocities and the presence of discontinuities in velocity and structure (such as faults) using a variety of approaches that analyse the arrival times and sometimes the shape of seismic waves travelling along different paths through the Earth. As a consequence of advances in seismic instrumentation and national programmes increasing the number of instruments available, there have been many recent developments in these techniques that do not directly image the Earth. A convenient aspect of the theoretical basis for virtually all active-source techniques is that they are largely independent of scale. Thus, detailed studies to address environmental problems and regional studies to determine deep Earth structure employ the same basic techniques of analysis.

Controlled-source seismology is not a precisely defined term and, as discussed above, could be considered to encompass all active-source techniques. The common practice is, however, to associate this term with large-scale investigations to determine deep Earth structure and large offsets (of at least tens of kilometres) between the sources and receivers. I exclude tomographic approaches which may have these characteristics but which tend to form a distinct line of investigation on their own (see *seismic tomography*).

Most introductory texts use the term 'refraction techniques' to describe what is often thought of as controlled-source seismology. If one assumes an Earth composed of constant-velocity layers separated by planar interfaces, simple equations can be derived that can be solved for the velocities of each layer and the geometry of the boundaries between each layer (see *seismic waves, principles*). This approach is used in many small-scale applications, such as the search for the top of the water table. As practised in large-scale studies, the refraction technique classically focused on determining the thickness of the crust and velocity variations in the crust and upper mantle. Early refraction studies employed a few dozen seismographs so that spacing between recordings was large and resolution was low. However, in modern studies, the distinction between refraction and reflection techniques is gradually disappearing because it is becoming possible to record the entire wave field generated by a seismic source.

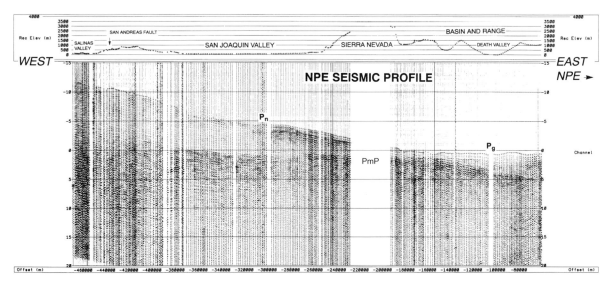

Fig. 1. Example of a long-range seismic profile recorded across the western USA. The profile is 450 km long. P_n is the phase travelling in the mantle as a refraction. P_g travels in the upper crust as a refraction and the apparent continuation of P_g to the west is P_mP, reflections from the Moho.

Modern studies of lithospheric structure usually employ hundreds of portable seismographs of portable seismographs, tens of seismic sources, intervals between recording stations of less than 1 km, digital data processing, and sophisticated computer modelling schemes capable of handling very complex Earth structure. Formal inversion techniques and synthetic seismogram modelling (see *synthetic seismograms*) to match observed waveforms are also employed regularly. Although a modern experiment may be referred to as a refraction study, the actual seismic waves recorded and interpreted are often mostly wide-angle reflections. An example of data and an initial interpretation of Earth structure from a recent seismic experiment are shown in Fig. 1. The non-proliferation experiment (NPE) was a large chemical explosion which was detonated in Nevada in 1993. Its purpose was to provide data for efforts to discriminate between nuclear blasts and mine blasts. About 600 seismographs were deployed eastward from this explosion along a line which crossed Death Valley, the Sierra Nevada range, and California's Great Valley. The data show strong P_n arrivals from the upper mantle and strong reflections from the Moho (Fig. 2).

Controlled-source seismology has made a major contribution to our understanding of the Earth's crust and upper mantle. This approach generally produces our best constraints on the thickness of the crust (depth to the Moho) and the velocity of the upper mantle (P_n velocity). These two values have become measurable quantities, understood and used by a large segment of the geoscience community because of the insight they provide in terms of plate tectonic processes. Areas of extension (see *grabens and rift valleys*) are associated with crustal thinning, and the amount of thinning reflects the amount of extension. On the other hand, large compressional mountain belts, such as the Alps and Himalayas, have thick crust as a result of crustal-scale thrust faulting. Upper mantle velocities are an indication of the tectonic and heat flow regime of an area, and a linear relationship between heat flow and P_n velocity has been shown to exist. Tectonically active areas tend to have slow P_n velocities and high heat flow. In addition, global and regional compilations of controlled-source seismology results are showing indications of crustal structure variations that reveal differences that might even correlate with crustal age.

The lower crust has been the focus of much interest in recent studies. The lower crust has often been found to be highly reflective. This has been particularly true in extended areas and may indicate a fabric imparted by ductile flow. In some areas, the lower crust has been found to have unusually high velocities, suggesting that it may have been modified by magmatism.

Still other long-range results are discovering interfaces in the upper mantle which may be related to flow, as suggested by the presence of anisotropy (see *seismic anistropy*). A major development in controlled-source seismology has been the release of ultra-long seismic profiles from the former Soviet Union which employed peaceful nuclear explosions to probe several hundreds of kilometres into the mantle. These results will probably never be duplicated. The deeper we look into the Earth, the more heterogeneity we see. We can thus look forward to many discoveries from controlled-source seismology.   G. R. KELLER

Further reading

Blundell, D., Freeman, R., and Mueller, St. (1992) *A continent revealed, the European geotraverse.* Cambridge University Press. (With maps and database on CD-ROM.)

Meissner, R. (1986) *The continental crust, a geophysical approach.* Academic Press, San Diego.

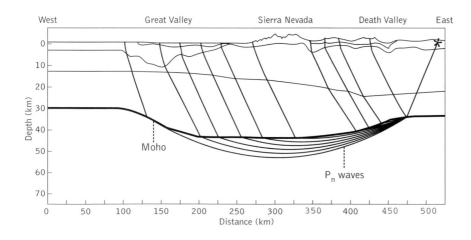

Fig. 2. Example of ray tracing for the P_n phase of the NPE explosion.

controversies, geological *see* GEOLOGICAL CONTRO-
VERSIES

convergent plate margins Convergent margins are one
of the three types of plate boundary envisaged in plate tectonics, and are arguably the most complex of the three. They
occur where two lithospheric plates are colliding, and their
detailed structure depends on the natures of the two plates
involved.

There are three possibilities: ocean–ocean, ocean–
continent, and continent–continent. The differences between
them arise as a result of the differences in strength and
density of oceanic and continental lithosphere. Oceanic
lithosphere is relatively strong and has almost the same
density as the underlying asthenosphere. At a convergent
boundary it therefore tends to bend rather than break, and is
relatively easily subducted. Continental lithosphere, on the
other hand, is lighter and therefore difficult to subduct. It is
also relatively weak compared with oceanic lithosphere, and
its yield strength is less than the force needed to subduct it.
Continental lithosphere will therefore tend to break up rather
than be subducted. (A consequence of this is that continental
rocks, once created, generally remain at the surface of the
Earth; so the area of continental rocks tends to increase with
time.)

The simplest of the three margins is the ocean–ocean
margin (Fig. 1a). Both plates are strong and can bend elastic-

ally with only minor fracturing. Either is relatively easy to
subduct, and a subduction zone is readily formed. Which of
the two plates actually becomes the subducting one may
depend on detailed differences in density (the older plate
being the denser) and perhaps also on details of the local
stress regime and mantle convection pattern. Once subduction begins, a classic ocean–ocean subduction zone is
established, exemplified by those of the western Pacific such as
the Mariana Trench (between the Pacific and Philippine
plates) or the Tonga–Kermadec Trench (between the Pacific
and Indo-Australian plates). The plate boundary is a narrow
zone lying in a deep-sea trench, and can often be localized to a
single, active décollement fault. In plan view, earthquake epicentres associated with the subduction zone may occupy a
band several hundred kilometres wide, but in three dimensions the earthquake foci are seen to be mostly confined to a
narrow, dipping Benioff zone that lies along and within the
subducting plate.

The next simplest convergent boundary is ocean–
continent. Here, too, the oceanic plate is readily subducted,
and the continental one simply overrides it (Fig. 1b). Such
convergent margins are exemplified by those of the eastern
Pacific between the South American and Nazca plates. The
subducted side is similar to that in an ocean–ocean margin;
again there is a narrow plate boundary zone and a narrow
dipping zone of earthquakes. However, because of the weakness of continental lithosphere, the continental plate
may undergo various forms of deformation, including both
extensional and compressional faulting and folding.

Continent–continent convergent margins are the most
complex. They are exemplified by the great fold-mountain
belts and associated regions, such as the Himalayan mountains and Tibetan plateau of the India–Asia continental convergence. Neither plate is readily subducted, and initially it is
probable that both are strongly deformed. At least one plate
becomes shortened and thickened through thrust faulting and
folding, and the other is underthrust at a shallow angle,
though without being subducted (Fig. 1c). As convergence
continues, the thickening lithosphere rises; large, regional
blocks may be extruded sideways between the converging
plates, and eventually the margin may become so thick and
high that the weak continental crust begins to collapse under
its own weight. Although the boundary between the two
plates may be localized along a single thrust front, active
deformation goes on over a broad zone many hundreds or
even thousands of kilometres wide, including important
zones of strike-slip and tensional deformation, as well as
thrust faulting and folding (Fig. 2).   ROGER SEARLE

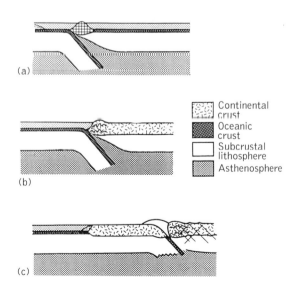

Fig. 1. Simplified cross-sections through (a) ocean–ocean,
(b) ocean–continent, and (c) continent–continent convergent
boundaries. (Modified from Kearey and Vine (1996), Fig. 9.1.)

Continental
crust
Oceanic
crust
Subcrustal
lithosphere
Asthenosphere

Further reading

Kearey, P. and Vine F. J. (1996) *Global tectonics*. Blackwell Science,
Oxford.

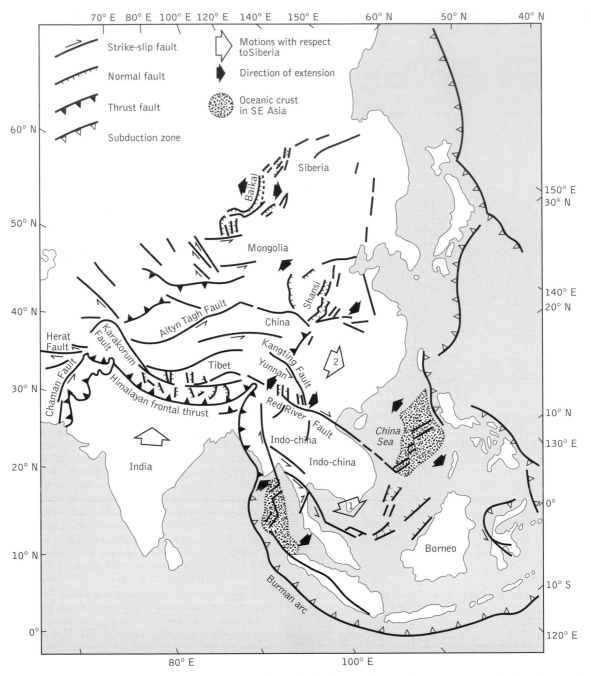

Fig. 2. Simplified map of the complex deformation zone associated with the Himalayan continent–continent convergent margin. (From Kearey and Vine (1996), Fig. 9.14.)

Cope, Edward Drinker (1840–97) One of the last great zoologist–explorers of America, E. D. Cope was also one of the most prolific of its scientific authors and editors. Of Quaker origins, he developed a keen interest in fossils and natural history in his early youth, graduated from college, and spent the Civil War years studying in Europe. There he visited all the major natural history collections that he could, supported by family wealth. His return at the age of 24 was to a college professorship in zoology in Philadelphia. Cope gave that up to go fossil collecting in the western territories.

In Kansas he made spectacular discoveries of Cretaceous marine reptiles, toothed birds, and fish and of early Cenozoic bizarre giant mammals. Cope became a leading authority on the Cenozoic strata and vertebrates throughout the world. His private collections of vertebrate material became enormous.

Cope was continually publishing descriptions and discussions of his (and others') discoveries and was editor of the *American Naturalist*. Some of his more than 1200 publications were large lavishly illustrated monographs, and his reputation as a comparative anatomist was international. His views on Lamarck's evolutionary ideas—'Neolamarckism'—did not, however, find favour. Cope's capacity for controversy was great and was at its apogee in his bitter rivalry with O. C. Marsh in collecting dinosaurs from the West. This so consumed his better judgement that it became a national scandal. Despite this, he was one of the most profound and prolific of America's palaeontologists in the nineteenth century.   D. L. DINELEY

coral reefs Coral reefs are masses of calcareous rock deposited by living organisms (not all of them corals), representatives of which commonly inhabit reef tops a little below the low-tide level of the ocean. At present, the principal hermatypic (or reef-building) organisms are the calcifying rhodophytes (red algae), molluscs, sponges, polychaetes, and cnidarians. The latter include the corals; the most important contributors to the growth of coral reefs are scleractinian corals, together with a few octocorallians and hydrocorallians (Table 1). The Scleractinia are an order of corals, mostly colonial, which have a calcareous external skeleton with radial partitions or septa. The Octocorallia form fan-shaped colonies with interconnecting branches. The Hydrocorallina also have calcareous skeletons.

Reef-building corals are animals containing soft parts called polyps that contain symbiotic algae. One of the major constituents of a polyp is calcium bicarbonate, which is broken down by the coral into calcium carbonate and carbonic acid. The latter is then broken down into carbon dioxide and water. The carbon dioxide is taken by the algae which, through photosynthesis, produce metabolites that feed the polyp. The process of extracting carbonic acid from calcium bicarbonate entails the precipitation of calcium carbonate, a process that is responsible for the construction of coral reefs:

$$Ca(HCO_3)_2 \rightleftharpoons CaCO_3 + \qquad H_2CO_3 \rightleftharpoons CO_2 + H_2O.$$

| Calcium bicarbonate | calcium Carbonate (precipitated) | carbonic acid | carbon water dioxide (removed by algae) |

The whole process of coral-reef build-up is therefore driven by the demand for carbon dioxide of the algae which live in symbiosis with corals. The calcium carbonate produced in the reaction above is precipitated around the polyp, where it forms a variety of calcareous structures that become incorporated into the reef upon which younger corals will ultimately grow.

Table 1. Biological classification of corals.
The principal reef-building corals are the Madreporaria (scleractinian corals), together with a few hydrocorallians and octocorallians.

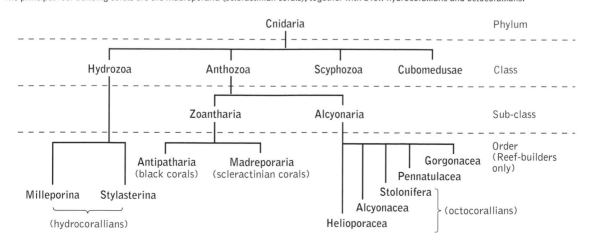

Most corals live only within the photic zone, that is, the upper 20 to 30 metres of the ocean into which enough light can penetrate for their symbiotic algae to photosynthesize. Reef-building or hermatypic corals live only in tropical seas, where temperature, salinity, and lack of turbid water are conducive to their existence.

So much emphasis has been placed on large, readily visible corals in reef-building that their role may have been overestimated. The role of calcifying rhodophytes (the crustose coralline algae which build impressive algal ridges and trottoirs (narrow features) along many tropical reef edges today) has been increasingly upgraded in importance during the past thirty years. Unlike corals, such algae are not confined to tropical waters but are involved in organic reef construction elsewhere. The paucity of corals in many emerged Quaternary reefs supports the view that the importance of corals as reef-builders may have been overestimated; the 1945 account by Harry Ladd and J. E. Hoffmeister of the limestones of the high Lau islands in the South Pacific is a good example. These geologists demonstrated that most (of the few) corals in Lau limestones were not in growth positions and had not therefore contributed significantly to building up the reefs.

Whatever the relative contributions of various groups of organisms to reef construction—a contribution which is unlikely to be globally uniform—reefs remain among the largest organic structures on Earth: the Great Barrier Reef of Australia can, for example, be seen from the Moon.

All reef builders require an initial surface upon which to grow, together with favourable oceanographic conditions. At the end of the Last Glacial (Würm), sea level rose from about 120 m below its present position. This postglacial rise in sealevel provided ideal situations for upward reef growth (or re-growth). Warming ocean waters increased nutrients from enhanced organic activity and allowed reefs to become established on the flanks of larger edifices (see *oceanic islands*). Subsequent upward growth was associated with a rise in sea level: the reef top needed to remain within the photic zone for the reef to survive. The resulting reef types have been characterized as 'keep-up', 'catch-up', and 'give-up' reefs. In the Pacific, for example, most reefs were probably catch-up reefs but a few were keep-up reefs and can be used for the accurate calibration of postglacial changes in sea level. To judge from the many submarine banks, some of which are known to be submerged reefs, in the Indian Ocean and around the Bahamas in the Caribbean, many reefs were also unable to grow upwards at the same rate as sea level rose, and have thus been submerged.

The absence of late Holocene emerged reefs in many places has given rise to the belief that contemporary sea level in these areas never exceeded its present level—as it is known to have done elsewhere. For the western Pacific, this view is largely incorrect, since the reefs in the areas where no emerged Holocene reef is found today are either occupied by catch-up reefs or by keep-up reefs which have been planed down to sea level since their emergence.

Since sea level stabilized around the middle to late Holocene, most coral reefs have extended laterally rather than vertically. This has meant a change in the dominant coral genera on many reefs, which in turn has led to a facies change in reef material. Studies of reef facies have made it possible to distinguish fossil reefs that grew vertically from those that grew horizontally. In consequence, a much clearer relationship between reef growth and sea-level change has been deduced. This understanding is important when the ability of many reefs to respond to future sea-level rise by growing upwards is assessed; many reefs will require a major species change before they can grow upwards because the branching corals that create a reef framework are generally not as abundant as they were during the period of postglacial rise in sea level.

Older fossil reefs, emerged and submerged, have been successfully used to detect Quaternary environmental changes. Methods have included oxygen-isotope analyses, which enable ocean palaeotemperatures to be known, and various techniques for determining past ocean productivity levels, which can be linked to palaeotemperature, ocean sediment levels, and other oceanographic variables. Drilling and dating of fossil reefs have allowed chronologies of reef build-up, tectonic and eustatic (sea-level) change to become precisely known. Reconstruction of Cenozoic sea-level changes has been made possible by drilling of Midway and other atolls in the North Pacific.

A classification of reefs, based on their geographical relationship to the land masses from which they have grown, is still appropriate. *Fringing reefs* are those that fringe the coast of a landmass. They are often ephemeral, young, and may be highly localized in occurrence along a particular coast. They are usually characterized by an outer reef edge capped by an algal ridge, a broad reef flat, and a sand-floored 'boat channel' close to the shore. Within late Holocene times, most fringing reefs have been growing seawards. Many fringing reefs grow along shores which are protected by barrier reefs and are thus characterized by organisms that are best adapted to low wave-energy conditions. *Barrier reefs* occur at greater distances from the shore than fringing reefs and are commonly separated from it by a wide deep lagoon. Barrier reefs tend to be broader, older, and more continuous than fringing reefs; the Beqa barrier reef of Fiji stretches unbroken for more than 37 km; that off Mayotte in the Indian Ocean for around 18 km. The largest barrier-reef system in the world is the Great Barrier Reef, which extends 2300 km along the east Australian coast, usually tens of kilometres offshore. *Atoll reefs* (or ring reefs) rise from submerged volcanic foundations and often support small islands (*motu*) of wave-borne detritus, sometimes armoured with beachrock or containing conglomerate platforms (*pakakota*) or phosphate rock which cause them to resist wave erosion. Atoll reefs are essentially indistinguishable in form and species composition from barrier reefs except that they are confined to the flanks of submerged oceanic islands, whereas barrier reefs may also flank continents.

Fig. 1. Miocene reef limestone at Qalimare, Sigatoka valley, VitiLevu, Fiji, unconformably overlying basement rocks.

Many ancient (fossil) reefs show similar facies patterns to modern reefs. Some of the most closely-studied exhumed reefs are the Permian reefs of west Texas, the Devonian reefs of western Canada, Europe, and Australia, and the Triassic reefs of the European alpine province. Facies variations are also the cause of significant variations in the depth and thickness of freshwater lenses in high limestone islands like Niue. The Miocene reef limestone in Fiji illustrated in Fig. 1 shows evidence of a contemporary hiatus accompanied by *subaerial diagenesis*.

In his 1842 classic book *Structure and distribution of coral reefs*, Charles Darwin outlined the way in which coral reefs could grow upwards from submerging foundations. From this, it became clear that fringing reefs might be succeeded by barrier reefs and thence by atoll reefs (Fig. 2). Later writers, particularly W. M. Davis, took this to mean that the three types were stages in an evolutionary succession and could thus be used to infer the stage of development that a particular reef had reached. Although partly a response to theories put forward which had questioned at the beginning of the twentieth century Darwin's ideas, Davis's views proved too inflexible and have since been challenged, principally by those concerned to incorporate the effects of sea-level changes in any explanatory framework of reef types.

Reefs in many parts of the world are currently under severe stress for a number of reasons. Direct human impacts include physical damage, pollution, and sedimentation. The rise of ocean-water temperature in many places has combined with

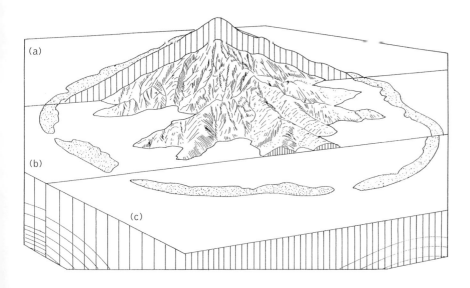

Fig. 2. Darwin's theory of atoll formation through submergence, a volcanic island with a fringing reef (a) will become converted to an island with an embayed coast and a barrier reef (b), and finally an atoll reef (c). (From Nunn (1994), after Davis (1913).)

other sources of stress to produce coral bleaching, a phenomenon in which polyps eject their symbiotic algae, resulting in coral death. The effects of predatory organisms, particularly *Acanthaster*, can be devastating in the short term, although perhaps important in long-term regeneration. Reef damage resulting from storm surges, particularly those associated with tropical cyclones (hurricanes) and tsunamis, and from earthquakes can be catastrophic.

Most authorities are agreed that, should sea level rise in the future, reefs will be able to make optimal responses in many places only if stress levels are reduced. There is consequently an urgent pragmatic need for reef conservation apart from a desire to protect this unique ecosystem. Should sea level rise and reefs be unable to respond to exert the same degree of protection along landward coasts that they do now, many of these coasts will be subject to greatly increased wave attack and erosion, particularly during storms.   PATRICK D. NUNN

Further reading

Darwin, C. R. (1842) *Structure and distribution of coral reefs.* Smith, Elder, London.

Guilcher, A. (1988) *Coral reef geomorphology.* John Wiley and Sons, New York.

Jones, O. A. and Endean, R. (eds) (1973) *Biology and geology of coral reefs*, Vol. 1. Academic Press, New York.

Nunn, P. D. (1994) *Oceanic islands*. Blackwell, Oxford.

Nunn, P. D. (1999) *Environment change in the Pacific Basin*. John Wiley and Sons, Chichester.

Wiens, H. J. (1962) *Atoll environment and ecology*. Yale University Press, New Haven.

coral terraces

coral terraces A coral terrace is a special type of marine terrace, comprising a staircase-like set of subhorizontal terrace surfaces, each separated by a former sea cliff. The base of each sea cliff, defined by the inner landward edge of the terrace surface, is a good marker of past sea level, to the extent that the location and elevation of these former shorelines record sea-level and tectonic fluctuations during Quaternary times. The age of emergence of a coral terrace can be determined by uranium-series dating of uncrystallized coral samples and other dating methods. By using these data, former sea-level elevations and the magnitudes of past changes in sea level can be quantified.

The formation of coral terraces is interpreted as the product of approximately uniform long-term uplift superimposed on eustatic changes in sea level. Coral terraces are thus well developed on tectonically active coastlines close to convergent plate boundaries, such as at the Huon Peninsula of Papua New Guinea, Barbados Island in the West Indies, and Ryukyu Islands of south-western Japan. Past sea-level elevations deduced from coral terraces at Huon Peninsula have been combined with oxygen isotope data to establish global sea-level histories that are widely used for Quaternary studies.

Furthermore, rates of uplift estimated from coral-terrace data may reveal significant regional and local differences in tectonic behaviour. For example, the coral terrace corresponding to the last interglacial maximum of about 125 ka ago (oxygen isotope stage 5e) occurs at an elevation of 400 m at Huon Peninsula and at 220 m at Kikai Island near Ryukyu trench. This difference indicates greater tectonic uplift in the Huon Peninsula area, although both areas are characterized by relatively rapid uplift. At both Huon Peninsula and Kikai island, successive former sea levels younger than isotope stage 5e (stages 5c, 5a, and 3) are recorded as a series of lower terraces. This enables rates of uplift in both areas to be tracked through time.

Holocene coral terraces fringing the older terraces along rapidly uplifting coasts are usually composite features, each comprising a series of small terraces. At Huon Peninsula, for example, the Holocene coral terrace reaches up to 25 m in elevation but it is subdivided into a maximum of seven discrete steps, indicating rapid but intermittent uplift over a short time-span. Such small terraces can also be found in late Pleistocene terrace profiles at Huon Peninsula, indicating that a similar tendency operated in earlier times. These multiple small terraces are interpreted as evidence for repeated major jerks of the coast during earthquakes, and can thus be used for reconstructions of palaeoseismicity.   YOKO OTA

Further reading

Chappell, J. (1974) Geology of coral terraces at Huon Peninsula, Papua New Guinea; a study of Quaternary tectonic movements and sea level changes. *Geological Society of America Bulletin*, **85**, 553–70.

Chappell, J. and Shackleton, N. J. (1986) Oxygen isotopes and sea level. *Nature*, **324**, 137–40.

Chappell, J., Omura, A., Esat, T., McCulloch, M., Pandolfi, J., Ota, Y., and Pillaus, B. (1996) Reconciliation of late Quaternary sea levels derived from coral terraces at Huon Peninsula with deep sea oxygen isotope records. *Earth and Planetary Science Letters*, **141**, 227–36.

Ota, Y. and Omura, A. (1992) Contrasting styles and rates of tectonic uplift of coral reef terraces at Ryukyu and Daito Islands, southwestern Japan. *Quaternary International*, **15/16**, 17–29.

Ota, Y. and Chappell, J. (1996) Late Quaternary coseismic uplift events at Huon Peninsula, Papua New Guinea, deduced from coral terrace data. *Journal of Geophysical Research*, **101B**, 6071–82.

corals and related fossils

corals and related fossils The corals, jellyfish, sea anemones, and hydroids are all members of the phylum Cnidaria, a group of fairly simple aquatic organisms of both solitary and colonial habit. They differ from higher invertebrate groups in lacking cells organized into organs, and in their radial symmetry and the possession of stinging cells or cnidoblasts, which give the phylum its name. The basic structure of the animal is a simple sac in which a central mouth

(a)

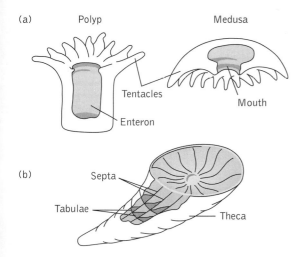

Fig. 1. Cnidarian morphology: (a) polyp and medusoid stages showing major cnidarian features; (b) coral skeleton showing major structural features.

surrounded by tentacles leads to a closed stomach or enteron (Fig. 1a); sea anemones exhibit this form. They are passive predators, catching in their tentacles small organisms that blunder into them. Their fossil record is uneven, because many groups do not have a skeleton; corals, however, secrete a massive calcareous skeleton and because of this are the most important fossil group.

Cnidarians show an alternation of generations in which two body forms, the polyp and the medusa alternate (Fig. 1a). The medusae reproduce sexually, giving rise to the polyps, which reproduce asexually by budding off medusae, and so on. All cnidarian groups are modified versions of one or other of these, and three major classes are recognized. The Hydrozoa are marine and freshwater cnidaria in which the polyp stage dominates; however, they only rarely produce a massive calcareous skeleton and are therefore of only marginal importance in the fossil record. The Scyphozoa, or jellyfish, are an exclusively marine class in which the medusoid stage dominates. Although they do not secrete a skeleton they do occur in the fossil record and are known from as far back as the Proterozoic, where they occur in the Ediacara Fauna. Included within this class by some authorities are the extinct conularids whose conical, four-sided, chitinous skeletons occur in marine rocks from the Cambrian to Triassic. The major class is the Anthozoa. This class includes the corals and encompasses a group in which there is no trace of the medusoid form. This is the largest group today, with about 6000 living species, and is also the largest fossil group with about 2300 genera known from the Proterozoic Ediacara Fauna to the Present.

The main anthozoan subclass, which includes the corals, is the Zoantharia. These organisms live in most marine environ-

ments, from tide pools to depths of 6000m or more. They are characterized by the presence of a calcareous skeleton that consists of a cylinder or cone, secreted by the polyp, which inhabits the upper part (Fig. 1b). The base of the polyp forms radial folds that secrete ridges, the septa, that are extended upwards as the polyp grows and are joined at the outer ends by the theca or skeletal wall. Horizontal floors, or tabulae, may be laid across the theca periodically as the animal grows.

The earliest corals are the Tabulata, colonial forms with conspicuous tabulae but poorly developed septa, which appear in the Early Ordovician. By the Middle Ordovician they are common and diverse elements of the marine fauna and are joined by the Rugosa, a group both solitary and colonial in which the septa are grouped in four quadrants. The Tabulata maintained numerical superiority until the Devonian, when a burst of activity brought the Rugosa level. The Late Devonian saw a wholesale extinction of coral faunas from which the Rugosa recovered with a further proliferation of families in the Mississippian (Early Carboniferous). Both became extinct by the Early Triassic, at which time they were replaced by the modern corals, the Scleractinia, which appear to have risen independently from soft-bodied ancestors. During the early Mesozoic the Scleractinia steadily increased in numbers, diversity, and distribution culminating in extensive reef development in the Late Jurassic. Additional periods of widespread reef development occurred in the Cretaceous and the Late Tertiary.

The development of reefs is caused in part by the colonial form of many corals. This is a successful strategy allowing rapid growth without the dangers of larval and immature stages; and with the ability to grow laterally allowing them to take advantage of whatever space is available in areas of high competition. Modern reef corals also take advantage of a symbiotic relationship with zooxanthellae, algae that live within their tissues. These hermatypic corals are restricted to shallow tropical waters, where they exist to 90 m depth in waters at 25–29 °C. This beneficial arrangement allows them to build massive fringing reefs, barrier reefs, and atolls, representing successive stages in the growth of coral around a sinking volcanic island, coral growth keeping pace with subsidence. The ahermatypic forms are less restricted environmentally and are found down to 6000 m and in temperatures at low as 1 °C.

Corals are generally too long-ranged to be useful zonally, although they have been used in Europe in the Early Carboniferous, where A. Vaughan established a zonal scheme in 1905 based on corals and brachiopods. They have, however, been useful in geochronometry. Many rugose and scleractinian corals show fine daily growth increments, which are often grouped in monthly and yearly annulations by growth constrictions. Study of Palaeozoic corals by Colin Scrutton of the University of Durham (England) has indicated that there were an average of 400 days in the Devonian year, a figure that corresponds well with astronomical estimates of the slowing of the Earth's rotation due to tidal friction. DAVID K. ELLIOTT

Further reading

Clarkson, E. N. K. (1993). *Invertebrate palaeontology and evolution.* Chapman and Hall, London.

Muscatine, L. and Lenhoff, H. M. (1974). *Coelenterate biology.* Academic Press, New York.

Coriolis effect or Coriolis force

The Coriolis effect, often referred to as the Coriolis acceleration or force, produces an apparent deflection of airflow relative to the Earth's surface. For horizontal airflow this deflection is to the right in the northern hemisphere and to the left in the southern hemisphere. The degree of deflection is zero at the geographical Equator, and at a maximum at either pole. The Coriolis acceleration (or force) is an important determinant in the geostrophic wind.

Understanding the Coriolis effect centres on an appreciation of the consequence of the Earth's rotation for air motion within the atmosphere: atmospheric air motion must be considered in relation to the Earth's surface. The Earth rotates on its axis (which passes through the North and South Poles) from west to east approximately once every 24 hours. Viewed from a point over the North Pole, this rotation is anticlockwise. Viewed from a point over the South Pole, the rotation is clockwise. However, points on the Earth's surface near the Equator have a speed of movement relative to space in excess of 1600 kilometres per hour, whereas, points at either pole simply rotate once per day about the axis.

These facts have two important consequences. First air moving across the Earth's surface will carry with it an 'inherited' velocity which can be greater or less than that of the Earth's surface at its new location. In either hemisphere air moving from lower to higher latitudes will be moving over areas where the surface speed is decreased. In the anticlockwise-rotating northern hemisphere, and viewed from the surface, there will thus be an apparent deflection to the right, because the air sample will now be moving faster relative to the surface beneath. Air moving in the reverse direction will be moving over areas where the surface speed is increased, so that, again, there will be an apparent deflection to the right. For the clockwise-rotating southern hemisphere the effect is mirrored, so that deflection is to the left.

Second, the Earth is nearly spherical, so that the orientation of its axis of rotation relative to air motion must be also considered. Exactly on the Equator horizontal airflow (i.e. that tangential to the Earth's surface) is normal (perpendicular) to the 'sense' of the Earth's spin, so that the effect of the rotation on airflow is zero. At the poles horizontal airflow is in exactly the same 'sense' as the Earth's spin, so that the effect of rotation is at a maximum. A similar, but contrasting, argument may be applied to air moving vertically relative to the Earth's surface, so that the Coriolis effect applies to all air motion in the atmosphere, and may be resolved into its vertical and horizontal components, more commonly the latter.

In the general case for horizontal motion, the Coriolis acceleration, a, is proportional to the sine of the latitude, Φ, and the velocity of airflow, V: $a = 2\Omega V \sin \Phi$, where Ω is the angular velocity of the Earth's rotation.   GRAHAM SUMNER

cosmochemistry

Cosmochemistry is concerned with experimental, observational, and theoretical studies of the chemical, isotopic, and mineralogical composition of extraterrestrial materials such as cometary dust particles (e.g., the *Giotto* mission to comet P/Halley), interplanetary dust particles (known as IDPs) collected using high-flying aircraft), lunar samples, and meteorites. The seminal work by Harold Urey, Hans Suess, and Harrison Brown on the chemical processes involved in the origin and evolution of the Solar System and the abundances of the elements led to the emergence of cosmochemistry as a separate sub-discipline in the late 1940s.

Historically, two major areas in cosmochemistry research have been: (1) the determination of the solar-system abundances of the elements, and (2) the chemical behaviour of the elements in a solar composition (i.e. hydrogen-rich) environment. These two topics are interwoven because the observed elemental abundances in primitive meteorites (chondrites) are generally correlated with the volatility of the elements, or their compounds, in material of solar composition.

Before proceeding some commonly used terms need to be defined. Chondrites are stony meteorites that contain small, melted beads known as chondrules and finer-grained material known as matrix. Some chondrites also contain calcium, aluminium-rich inclusions (CAIs), which are dominated by calcium, aluminium, and titanium-bearing refractory oxides and silicates. Observational studies indicate that the chondrules, mineral fragments, inclusions, and matrix in chondrites formed in the solar nebula and have been little altered since that time by planetary processes (e.g., aqueous alteration, igneous differentiation). The chondrites are divided into three groups; carbonaceous, ordinary, and enstatite chondrites, on the basis of their major element composition and mineralogy. These three major classes are further subdivided into different petrographic types. The most primitive chondrites, in the sense of most closely reproducing the elemental abundances in the photosphere of the Sun, are the CI (or CI) carbonaceous chondrites.

Solar abundances of the elements

Figure 1 shows the relative abundances of the 20 most abundant elements in the Sun and solar nebula. These abundances are primarily derived from chemical analyses of chondrites and astronomical observations of the elemental abundances in the Sun. To a very good approximation, the abundances of most elements in the CI carbonaceous chondrites and in the Sun are identical. The few exceptions to this generalization are: light elements such as lithium, which are destroyed by thermonuclear reactions in the Sun; atmophile elements such

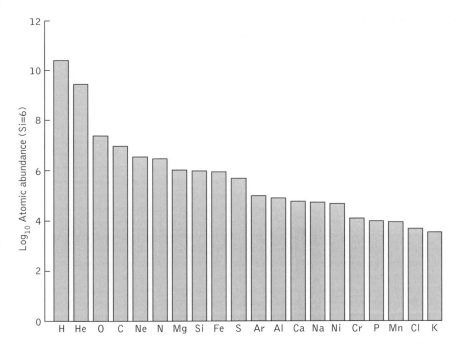

Fig. 1. The relative abundances of the 20 most abundant elements in solar composition material according to Anders and Grevesse (1989), *Geochinica et Cosmochinica Acta*, **53**, 197–214. The abundances are normalized to Si = 10^6 atoms.

as hydrogen, oxygen, carbon, nitrogen, and the noble gases, which are incompletely condensed in meteorites; and some rare elements such as mercury, germanium, lead, and tungsten, which are difficult to analyse in the Sun or in meteorites. To a lesser extent there is also a good correspondence between elemental abundances in the Sun and in all chondrites. This close relationship has led cosmochemists to believe that chondrites are relatively unaltered samples of material that condensed in the solar nebula. Over 30 years of meteorite studies and chemical equilibrium modelling of nebular chemistry demonstrate that evidence of nebular chemical reactions is preserved, with varying degrees of alteration by subsequent processes such as thermal metamorphism and aqueous alteration, in chondritic meteorites.

Cosmochemical behaviour of the elements

Chemical analyses of chondrites and interplanetary dust particles, and thermochemical equilibrium calculations of chemistry in hydrogen-rich solar material, show that the elemental fractionations in chondrites and IDPs are primarily the result of volatility controlled processes such as gas → solid condensation and solid → gas evaporation in the solar nebula. The occurrence of some characteristic mineral morphologies, such as enstatite ($MgSiO_3$) whiskers with screw dislocations, in some IDPs shows that gas → solid condensation was directly responsible for growing some minerals.

Cosmochemists therefore classify the elements according to their chemical behaviour in hydrogen-rich solar material. Refractory elements are the first elements to condense from solar composition gas. Both lithophiles (preferentially found in oxides or silicates, or both) and siderophiles (preferentially found in metals) with low vapour pressures, or that form compounds with low vapour pressures, fall into this category. The condensation of iron metal alloy and magnesian silicates ($MgSiO_3$, enstatite, and Mg_2SiO_4, forsterite) divides the refractory elements from the moderately volatile elements. In turn, troilite (FeS) condensation divides the moderately and highly volatile elements. Finally, water-ice condensation separates the highly volatile elements (e.g., lead, indium, bismuth, and thallium) from the atmophile elements (hydrogen, carbon, nitrogen, and the noble gases). Table 1 summarizes the major condensation reactions in solar composition material.

Refractory lithophiles include the alkaline earths (e.g. calcium and magnesium) the lanthanides (rare earth elements, or REE), the actinides, aluminium, and elements in groups 3b (scandium, yttrium), 4b (titanium, zirconium, and hafnium), and 5b (vanadium, niobium, and tantalum) of the periodic table. The refractory siderophiles are the platinum-group metals (except palladium), molybdenum, tungsten, and rhenium. As shown in Table 1, the refractory lithophiles and siderophiles constitute about 1 per cent by mass of the total condensable material (rock + ices) in the solar nebula. Exten-

Table 1. Condensation sequence for the 20 most abundant elements in the solar nebula (10^{-4} bar total pressure)

Reaction	Temperature (K)	Fraction condensed (mass %)
Ca, Al oxides, and silicates condense	1670–1530	1.1
Fe alloy condenses	1337	9.1
Forsterite (Mg_2SiO_4) condenses	1340	20.4
Schreibersite (Fe_3P) forms	1151	20.5
Alkali feldspar (Na, K)$AlSi_3O_8$ forms	1000–970	20.9
Sodalite [$Na_4(AlSiO_4)_3Cl$] forms	863	20.9
Troilite (FeS) forms	719	23.4
Magnetite (Fe_3O_4) forms	400	24.0
Hydrated silicates (talc, serpentine) form (?)	~280	—
H_2O ice condenses	180	50.9
Ar, Kr, and Xe clathrate hydrates condense	74–50	51.5
CO and N_2 ices condense	~20	100.0

sive studies of the chemical composition of stony meteorites show that these refractory elements behave as a group in most meteorites: that is, their abundances in different types of meteorites are either enriched or depleted by about the same factor. Large enrichments, which are 20 times solar elemental abundances on average, of the refractory lithophile and siderophile elements are found in CAIs in the Allende meteorite and other carbonaceous chondrites. The CAIs have a mineralogy dominated by minerals rich in calcium, aluminium, and titanium such as hibonite, $CaAl_{12}O_{19}$, melilite, a solid-solution of gehlenite, $Ca_2Al_2SiO_7$, and åkermanite, $Ca_2MgSi_2O_7$, spinel, $MgAl_2O_4$, and perovskite, $CaTiO_3$.

Table 1 gives the 50 per cent condensation temperatures for metallic iron alloy and forsterite (Mg_2SiO_4). Metallic iron and magnesian silicates account for most of the rocky material in solar composition matter. The large excess of molecular hydrogen (H_2) in solar gas leads to extremely low oxygen fugacities and the FeO content of the magnesian silicates is insignificant until low temperatures of about 400–600 K. At these temperatures olivine and pyroxene solid solutions containing ~20 mole per cent of fayalite (Fe_2SiO_4) and ferrosilite ($FeSiO_3$) are predicted to form and any remaining iron metal is predicted to form magnetite at a pressure-independent temperature of about 400 K. However, slow solid-state diffusion at 400–600 K may inhibit solid–solid reactions such as the formation of FeO-rich silicates over the estimated 10^5–10^7 year lifetime of the solar nebula.

The moderately volatile elements have condensation temperatures intermediate between those of the major elements iron, magnesium, and silicon and troilite, FeS. The elements in this group are geochemically diverse and include sodium,

potassium, rubidium, chromium, manganese, copper, silver, gold, zinc, boron, gallium, phosphorus, arsenic, antimony, sulphur, selenium, tellurium, fluorine, and chlorine. The highly volatile elements condense at temperatures below 719 K, where troilite forms. These elements include mercury, bromine, cadmium, indium, thallium, lead, and bismuth. The condensation chemistry of many of the moderately and highly volatile elements is not well known because of uncertainties in the relevant thermodynamic data.

As Fig. 1 shows, hydrogen is the most abundant element and H_2 is therefore the most abundant gas in material of solar composition. At sufficiently high temperatures, dissociation to atomic H occurs. However, the phase boundary where abundances of H_2 and H are equal is at lower pressures and higher temperatures than those expected in the solar nebula. H_2 remains in the gas until temperatures of about 5 K, where it will condense out as solid hydrogen. It is unlikely that temperatures as low as this were ever reached in the solar nebula.

About 0.1 per cent of all hydrogen condenses out as water-ice at temperatures of 150–250 K, depending on the total pressure. Hydrated silicates such as serpentine and talc are also predicted to form by reactions between anhydrous silicate grains and water vapour in the nebular gas at temperatures below 300 K at 10^{-4} bar. However, although they are thermodynamically favourable, these reactions probably did not occur in the solar nebula because the vapour phase hydration of rock in a near-vacuum is a very slow process. Theoretical studies of hydration kinetics in the solar nebula and petrographic studies of water-bearing chondrites both suggest that the production of hydrated minerals occurred on the meteorite parent bodies. It is thus very likely that water-ice is the first H-bearing condensate to form.

Carbon chemistry is significantly more complex. To a good first approximation, carbon monoxide (CO) is the dominant carbon gas at high temperatures and low pressures and methane (CH_4) is the dominant carbon gas at low temperatures and high pressures in solar composition material. The two gases are converted by the net thermochemical reaction: $CO(g) + 3H_2(g) = CH_4(g) + H_2O(g)$. (The symbol g denotes the gaseous state). Increasing the H_2 pressure (essentially the total pressure in solar material) or decreasing the temperature drives this reaction to the right and yields more CH_4. The CO-CH_4 boundary is in the region of 600 K at 10^{-4} bar total pressure. CO is more abundant at higher temperatures, and CH_4 is more abundant at lower temperatures.

As first noted by Urey and later quantified by Lewis and Prinn, the kinetics of the CO→CH_4 conversion may be so slow under the pressure and temperate conditions expected in the solar nebula that CO cannot be converted to CH_4 within the lifetime of the nebula. An exception to this occurs in the giant protoplanetary subnebulae, which are higher-density environments that are predicted to exist around Jupiter and the other gas giant planets during their formation. The CO→CH_4 conversion is predicted to take place in these environments.

At low temperatures in the outer solar nebula and the giant protoplanetary subnebulae, CO and CH_4 may react with water-ice to form the clathrate hydrates $CO \cdot 6H_2O(s)$ and $CH_4 \cdot 6H_2O(s)$ (The symbol s denotes the solid state.) (Clathrates are solids in which one chemical component is enclosed in the structure of another, as if in a cage.) The formation of these clathrate hydrates requires sufficiently rapid diffusion of CO or CH_4 through the water-ice crystal lattice. Theoretical models, which use experimentally determined activation energies for clathrate formation, predict that CH_4 clathrate hydrate can form in the giant protoplanetary subnebulae but that CO clathrate hydrate cannot form in the lower-density environment of the outer solar nebula.

The most important features of nitrogen chemistry are that N_2 is the major nitrogen gas at high temperatures and low pressures while NH_3 is the major nitrogen gas at low temperatures and high pressures. The two species are converted by the reaction $N_2 + 3H_2 = 2NH_3$, which is analogous to the reaction which converts CO and CH_4. Reduction of N_2 to NH_3 is also predicted to be kinetically inhibited in the solar nebula and to be both thermodynamically favoured and kinetically facile in the giant protoplanetary subnebulae. This is true even when the possible catalytic effects of grains of metallic iron are taken into account. Thus, N_2 is predicted to be the dominant nitrogen gas throughout the solar nebula, and NH_3 is predicted to be the dominant nitrogen gas throughout the giant protoplanetary subnebulae.

At low temperatures in the outer solar nebula, $N_2 \cdot 6H_2O(s)$ becomes thermodynamically stable, but its formation is probably inhibited by two factors. One is the limited availability of water-ice, which may already be totally consumed by reactions to form other hydrates and clathrates. The other is the expected kinetic inhibition of clathrate hydrate formation in the outer solar nebula. In this case, N_2, like CO, will not condense until temperatures of about 20 K (at 10^{-4} bar pressure) are reached, where the solid ices form. On the other hand, $NH_3 \cdot H_2O$ formation is predicted in the giant protoplanetary subnebulae, because it is both thermodynamically favoured and kinetically facile.

The noble gases helium, neon, argon, krypton, and xenon display fairly simple chemistry in material of solar composition. All are present in the gas as the monatomic elements and argon, krypton, and xenon undergo condensation to either ices or clathrate hydrates at sufficiently low temperatures. Condensation of the pure ices will occur at slightly lower temperatures than condensation of the clathrate hydrates. The formation of these species, like the clathrates of CO and N_2, may, however, be kinetically inhibited. Temperatures of about 20 K (at 10^{-4} bar pressure) are required for the quantitative condensation of argon, krypton, and xenon as pure ices. Neither helium nor neon will condense out of the gas because temperatures of 5 k or below are required for this to happen.

BRUCE FEGLEY

Further reading

Kerridge, J. F. and Matthews, M. S. (eds) (1988) *Meteorites and the early Solar System.* University of Arizona Press, Tucson.

Lewis, J. S. and Prinn, R. G. (1984) *Planets and their atmospheres: origin and evolution.* Academic Press, New York.

Weaver, H. A. and Danly, L. (eds) (1989) *The formation and evolution of planetary systems.* Cambridge University Press.

cratons The term *craton* has been used for many years for the broad central parts of continents that are affected only by *epeirogenic* movements. These areas thus contrast with mobile eugeosynclinal, *orogenic* regions. Cratons are essentially rather stable, continental, shield areas that have a basement of Precambrian rocks. They make up the cores of present-day continents in both North and South America, Australia, Russia, Fennoscandia, and Africa, where there are several cratons separated by more mobile belts. Within them *intracratonic basins* may develop that are characterized by very slow subsidence.

The nature of sedimentation is particularly important, both on the cratons and within their basins. The absence of faulting and of any rapid vertical uplift leads to a shortage of sediment. At the same time the very slow subsidence of the basins means that very little accommodation space is generated in which the sediments can collect. Instead of being swallowed up, as in a typical rift valley, the sediments are spread widely over the basin or anywhere that is slightly lower than the uplifted area.

Many of these basins develop as lakes, known as 'sag' lakes to contrast them with lakes that form in rift valleys. Lake Chad in North Africa is a typical example. It is very shallow, just a few metres deep, yet, over very short periods of time it can extend and contract its margin by hundreds of kilometres as rainfall fluctuates. Lake Eyre in Australia and the Great Lakes of North America are other examples. All these lakes are characterized by a lack of sediment from the surrounding rivers.

A major feature of ancient cratons is the presence of extensive 'sheet' sandstones of great textural and compositional maturity, some of which yield almost pure orthoquartzites. These orthoquartzites cover very large areas of cratons: the Nubian Sandstones of North Africa, the Roraima Sandstone of Brazil and Venezuela, the sandstones of North Norway around the Baltic Shield, and the extensive late Precambrian and Cambro-Ordovician Sandstones of north-central USA.

One feature of such sandstones is that their depositional environments are not easy to establish. Even the most experienced sedimentologists have difficulty in deciding whether they have been formed in braided rivers, as shoreface-beaches, or as shelf sand bars, especially when fossils are absent or sparse, as in the Precambrian.

A second, but equally important, feature of cratons, is the existence of unconformities that can also be traced over vast distances. In North America these 'interregional' unconformities have long been used for correlation purposes. It is significant that these ideas were developed by L. L. Sloss and his colleague W. C. Krumbein, living and working on the North American craton at Northwestern University in Illinois. They used the word 'sequence' for the major unconformity-bounded masses of strata.

Since there were essentially no tectonic movements and only a very limited sediment supply, discussion as to the causes of the extensive unconformities and sedimentary packets revolved around whether episodic sedimentation was caused by minor epeirogenic movement or by sea-level changes, a controversy that still continues. It was also no coincidence that when the Exxon Production Research Group developed their ideas on sequence stratigraphy the driving force should be Peter Vail, a student of Krumbein and Sloss.

HAROLD G. READING

Creationism Creationism, or what since the 1960s has tended to call itself 'Creation Science', is historically a product of American fundamentalist religion. Though it rejects, or is guarded about, many of the chronologies of modern science, such as those used in cosmology, the particular thrust of its disagreement with the rest of contemporary science lies in its wholesale repudiation of Darwinian biology and the theory of evolution. Creationism considers the Creation narrative presented in the first books of *Genesis* in the Holy Bible to be a complete and adequate account of the origins of the natural world. And in particular, Creationists find the evolution of man from lower animal forms to be especially repugnant. To most Creationists, all life, and especially mankind, was created in the Garden of Eden in a single Divine act extending over six days, traditionally dated to 4004 BC: a date derived from the genealogies of descent from Adam and Eve found in various books of the Bible. Creationists generally argue that geological fossils are not of great antiquity, but are the preserved remnants of creatures that failed to get into Noah's Ark as described in *Genesis*, chapter 7.

Yet orthodox science, even when practised by ordained scientists, had come to question the literalist calculation of the 4004 BC date long before Charles Darwin published his *Origin of Species* in 1859. William Buckland in Oxford and Adam Sedgwick in Cambridge—who were professors of geology in their respective universities, and canons of English cathedrals—had come to accept by 1820 that, while Adam and Eve might have been created in 4004 BC, the universe, the globe, and the geological strata, rich as they were in extinct fossil forms, were immeasurably older. The Victorian geologists argued that the Bible never mentioned extinct forms because ichthyosauri and similar creatures did not possess immortal souls, and were not therefore of interest to the writer of *Genesis*, which is pre-eminently concerned with the spiritual

history of mankind. Some twentieth-century Creationists, such as John William Dawson and George Frederick Wright, have however been willing to countenance variations on these ideas, in which an ancient and possibly cataclysmic Earth history pre-dated Adam and Eve.

Although the descent of the human race from lower forms, which Darwin had implied in his *Origin* and discussed explicitly in his *Descent of Man* (1871), undoubtedly challenged the historical and spiritual status of Adam and Eve, many devout scientists, such as Asa Grey of Harvard from the 1860s, were able to develop a reconciliation between the new sciences of geology and orthodox Christian theology. Evolution, after all, indicated an active God who had formed a universe of vastly greater wonder than that resulting from a single fiat of creation.

It must further be remembered that Creationists refused to recognize not only much of late nineteenth-century science, but also the 'higher criticism' of Biblical texts developed by contemporary philologists and textual scholars, mainly in Germany. This new scholarship, while in no way denying the divine truth and inspiration behind the Bible, none the less acknowledged that the book itself was to some extent a human literary composition, containing textual contradictions along with mythological and allegorical components.

It was *The Fundamentals*, initiated by the preacher A. C. Dixon, a series of twelve booklets published in America between 1905 and 1915, which fired the first popular salvos against both higher criticism, Darwinism, and a non-literalist understanding of *Genesis*. The heartland of Creationist influence for most of the twentieth century was, and has remained, the staunchly Protestant American South, Mid-West, and West. Creationism is, moreover, very much of a product of a movement within Protestant Christianity which draws its spiritual authority not from apostolic or sacramental traditions within the historical church, but solely from the Bible as expounded by a preaching ministry. Consequently, anything which challenges the literal authority of Scripture strikes to the very heart of the faith. Roman Catholics and Orthodox, Anglican, and other Christian denominations which are not based wholly on a literal understanding of the Scriptures are therefore relatively untouched by Creationism.

Fundamentalism and Creationism also grew out of an American radical political tradition which aimed to change society by legislation: abolition of slavery, prohibition of alcohol, and, after 1919, against evolution. This tradition was exemplified in the notorious Scopes trial at Dayton, Tennessee, USA, in 1926, in which the radical fundamentalist politician and lawyer William Jennings Bryan used the force of law to attempt to prevent the teaching of evolution in state schools. Indeed, much of the Creationist controversy in America since 1919 has focused on attempts to control what was taught in public schools and colleges.

In spite of its strict Biblical base, Creationism has none the less never lacked diversity of opinion within its own ranks. In 1954, for instance, Bernard Ramm's *The Christian View of Science and Scripture* advocated a less rigid understanding of the Genesis creation narrative in which God had developed and perfected the pre-human globe over millions of years, only to provoke John C. Whitcombe's and Henry M. Morris's *The Genesis Flood* (1961), which reasserted the young Earth and the Biblical deluge as a primary geological agent.

In the early 1960s the Creation Research Society came into being, and figures like Whitcombe, Morris, Walter E. Lammerts, and Duane Gish attempted to develop a 'creation science', much of which was, and still is, aimed at discrediting evolution and demonstrating the historical reality of Noah's Flood.

Although now possessing considerable media resources, and always trying to assert its scientific credentials, Creationism remains culturally linked to American fundamentalism. Its theology is still overwhelmingly concerned with preserving the literal authority of *Genesis*, while its geology and biology cannot run where the evidence leads, but must always be capable of a precise reconciliation with the Bible. And although Creationism has tried to put down roots in Canada, Australasia, Great Britain, and other Protestant countries, it is rejected not only by modern scientists, but also by the generality of present-day Christians, to whom the literal accuracy of the ancient *Genesis* narrative is not necessarily an article of faith.

ALLAN CHAPMAN

Further reading

Rupke, N. A. (1983) *The great chain of history: William Buckland and the English school of geology.* Clarendon Press, Oxford.

Larson, E. J. (1989) *Trial and error: the American controversy over evolution and creation.* Oxford University Press, New York.

Numbers, R. L. (1993) *The Creationists. The evolution of scientific creationism.* University of California Press, Berkeley.

creep *see* HILL-SLOPE CREEP

Cretaceous The third and youngest system of Mesozoic rocks takes its name from the Latin *creta*, meaning chalk. The word *Cretaceous* first appeared in an early book on the geology of England by W. D. Conybeare and William Phillips in 1822. It was based on the term *Terrain Cretacé* which had recently been used by J. J. d'Omalius d'Halloy in France. Lasting for about 80 million years, the Cretaceous period terminated around 65 Ma ago with the ending of the Mesozoic Era and the occurrence of a major extinction event. Thus it is one of the longest of the Phanerozoic periods. It is customarily divided into two rather than three epochs, the formations being known as Upper and Lower Cretaceous. There are 12 stages, and 13 biozones based on ammonoid cephalopods are recognized. Biozones are also established in the Chalk (Upper Cretaceous) on the basis of foraminifera and coccolith assemblages. The Cretaceous system is very widely distributed about the world; great areas of the continental cratons are covered by the chalky limestones that mark the extent of the late Cretaceous very high sea level. In North America the *Zuni*

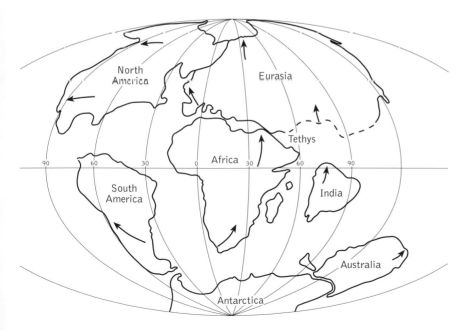

Fig. 1. The world in mid-Cretaceous time. The Atlantic Ocean has yet to open northwards from Spain, and the South Atlantic is confined to south of the Brazilian–West African shield. India is about to move northwards, while Africa begins to rotate anticlockwise. These movements will result in the disappearance of the Tethyan Ocean. Australia remains in contact with the pole-centred Antarctic continent.

synthem (sequence) represents a major transgression from the margins of the craton towards the interior, and equivalent transgressions are found on other continents. It lasted from Mid-Jurassic time until the Paleocene.

Cretaceous palaeogeography was primarily influenced by the continuing dispersal of the continental fragments of Pangaea. Indeed, it seems that during this period the rate of ocean floor spreading increased. Africa became detached from South America and a seaway between the North and South Atlantic oceans opened up in the later part of the Early Cretaceous. The North Atlantic also was opening relatively fast (Fig. 1). The Bay of Biscay came into existence late in the period, as perhaps did the Rockall Trough. Tethys now ceased to exist as an open ocean with the movement of Africa and Arabia northward against the southern margin of Eurasia. The Alpine–Himalayan orogenic episode had begun. Madagascar became well separated from Africa and, probably before the end of the Early Cretaceous, India was detached from Antarctica and headed northward to make contact with the Lhasa Block by the middle of Late Cretaceous time. Orogeny on this margin was beginning in Kashmir and northern Afghanistan as the Tethyan Ocean closed. Accretion to the western margin of North America was operating as the Kula plate 'docked' against the mountains there, and then slid northwards through some 20 degrees of latitude.

The vigorous ocean-floor spreading of Cretaceous time is one aspect of the igneous activity that also affected the continents. There were also very large and complex intrusions emplaced in the Cordillera of South America, western USA, and Canada; the famous Boulder Batholith of Colorado is one of these. Associated with these intrusions were outpourings of andesitic and more acidic lavas and fine ashes. Some of the ashes are important as bentonite ash bands, which are of enormous geographic extent and yield good isotope late Cretaceous dates. High in the Canadian arctic and in Greenland tiny kimberlite pipes and small dykes were intruded as Greenland rifted away from the Canadian Arctic Islands. In India, the famous Cretaceous Deccan Traps are tholeiitic basalts, over 1200 m thick and attaining some 5000 00 km² in area. In South America the Parana basalts cover an area of 12 000 000 km² and seem to be associated with fissures near to the continental margin, which was under tensional stress as the south Atlantic spreading developed. On the opposing stable continent, Africa, kimberlite pipes were being punched through the crust in Angola.

As we noted above, global sea level rose throughout the period, beginning at a level rather lower than today's average and rising to a point in the Late Cretaceous where only 18 per cent of the Earth's surface remained as land, compared with 28 per cent today. A sharp fall began shortly before the end of the Cretaceous (Fig. 2).

Cretaceous climates seem to have been more varied and seasonal than those earlier in the Mesozoic. Clear evidence of a more conspicuous temperature gradient between high and low latitudes is provided by fossil plants. A cold climate flora and ice-rafted pebbles occur in Alaska and arctic Canada.

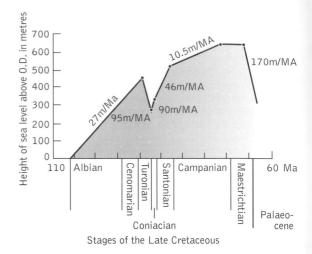

Fig. 2. Late Cretaceous times saw world sea level rise very high, until near the end of the period it suddenly began the downward shift that continued into the Cenozoic era. The graph is based upon one by Hancock and Kauffman (1977). This remarkable trend throughout most of the Late Cretaceous is thought to have been due to the great growth of the mid-oceanic basalt ridges and perhaps the development of 'hot-spot' bulges in the ocean floor.

Nevertheless there are indications that these were the products of temporary spells of colder climate. In detail, it seems that the cold snaps were more pronounced in the Early Cretaceous and the notably warmer phases were about equally distributed between the Early and Late Cretaceous. The Milankovitch climatic cycles found so clearly preserved in Quaternary deposits are also present in Cretaceous sediments, and presumably indicate similar short-term fluctuations of climate.

One of the fascinations of the Cretaceous system is the great variety of sediments that were deposited during that period. We are able to interpret the majority of them in the light of our knowledge of modern sedimentary environments. For many years the origin of the chalky limestones were, however, a puzzle, solved only since the invention of the electron microscope. They are the products of coccolith accumulation. Coccoliths are the extremely small calcareous parts of microscopic marine algae. In the Late Cretaceous they were extremely abundant in the warm clear seas that flooded the continents. As much as 90 per cent of the rock is commonly made of coccolithic debris. Other remarkable and typically Cretaceous limestones are the massive shelly limestones formed of the heavy rudist bivalves. The limestones are predominantly white, but are at several levels interrupted by think black shale bands. These are carbon-rich deposits, formed during brief phases of oxygen depletion in the bottom waters of the oceans. They are most common in the lower part of the Upper Cretaceous. Dark sediments are also characteristic of the main basin of the Early Cretaceous Atlantic. Most of the Cretaceous seas were organically highly productive: it is estimated that more than half of the world's oil and gas comes from reservoirs in Cretaceous rocks.

Cretaceous palaeontology is very rich, with well-preserved plants, invertebrates, and vertebrates from a wide range of ancient habitats. The marine coccolith-bearing algae have been mentioned; on land a new floral feature was the spectacular rise of the angiosperms, so that by mid-Late Cretaceous time they were dominant in many of the world's floras. They seem to have differentiated into three main floral provinces.

Marine communities also seem to have established themselves within geographical provinces. The Boreal province persisted from the Jurassic but was less marked, the Tethyan province was well populated with characteristic benthos. Molluscan faunas are rich, but with declining populations of cephalopods. Ultimately the ammonites, belemnites, inoceramid bivalves, and rudists became extinct. Among the microfaunal elements, the planktonic foraminifera suffered almost complete annihilation, and catastrophic extinction of almost all coccolithic algae also occurred at the end of the period.

The vertebrates of the Cretaceous made many evolutionary advances, but towards the end of the period the dinosaurs, pterosaurs, and giant marine reptiles died out. Dinosaur populations in a wide range of habitats were large and included both large and small herbivors and carnivores. They appear to have been present on most continents—even at high latitudes—during this period. The pterosaurs produced the largest-ever flying animals, while the birds achieved success in the air and as swimmers. Mammals remained small in size but increased in variety over their Jurassic ancestors.

The progressive disappearance of so many forms of life towards the end of the Cretaceous period has been the subject of intense research and debate. Several arguments have been put forward to account for the demise of creatures both on land and in the sea. They include arguments about late Cretaceous climatic change, the impact of some extraterrestrial object, cosmic radiation, and prodigious volcanic activity. The general trend of opinion is towards a combination of several of these factors operating over a short but decisive interval of time to accomplish a colossal extinction event.

All but about a dozen genera of dinosaurs were extinct before the end of the period, and their final days were seemingly under climatic stress. It had been getting progressively colder and drier during the last 10 Ma of Cretaceous time as sea-level fell. This change in climate must have affected the vegetation, including food for the dinosaurs. Added to this there is the possibility that the herbivorous mammals, being smaller but more prolific breeders and competing for food, were able to survive on seeds, nuts, and the new vegetation that was insufficient for the larger reptiles.

The impact scenario is based upon the occurrence of considerable widespread geochemical and mineralogical evidence of a large bolide striking the Earth. An event of this kind would create a huge crater, send out shock waves, and create tremendous atmospheric disturbance. Traces of a possible end-Cretaceous impact crater perhaps 300 km in diameter have been found in the Yucatán area of Mexico. Critically, and apart from local blast effects, immense volumes of dust would have been carried high, far, and wide, screening out the sun's heat and creating a cold snap lasting several years. Added to this would have been chemical consequences, the generation of poisonous hydrogen cyanide and nitrous oxide in the atmosphere. Late Cretacous volcanism is known to have been on an unusually large scale, again with the injection of huge quantities of dust into the atmosphere and acidification of the air by aerosols. Almost certainly this would have attacked the ozone layer to admit larger doses of harmful ultraviolet radiation.

Other cosmic agencies have also been called in as possible sources of additional doses of radiation. Whatever the actual cause, any combination of these factors would seriously have affected the global environment for some time. The end of the period was marked by a major crisis in the history of life, the like of which had not been seen since the end of the Permian period, 185 Ma earlier.

That the end of the Cretaceous period marked a turning-point in the history of life was first recognized by the early geologists and palaeontologists in Europe, who noted that the widespread and conspicuous Chalk is overlain unconformably in Cenozoic sands and clays. The unconformity itself represents a period of uplift and non-deposition, short but immensely significant. Only in certain areas, such as parts of Denmark and Italy, is the sequence continuous; and it was in the Italian succession that a geochemical anomaly—a relatively high proportion of the element iridium in a clay band at the top of the Cretaceous—was discovered in the 1970s. The discovery led to the hypothesis of a cosmic impact and to the successful search for this or a similar anomaly elsewhere in the world. The discovery of the Yucatán crater seems to provide the site from which the Cretaceous catastrophe originated. The original Catastrophists would be gratified at this development.

D. L. DINELEY

Further reading

Kennedy, W. J. (1978) Cretaceous. In McKerrow, W. S. (ed.) *The ecology of fossils*, pp. 280–322. Duckworth, London.

Moullade, J. and Nairn, A. (1978) *The Phanerozoic geology of the world. II. The Mesozoic A.* Elsevier Scientific Publications, Amsterdam.

Moullade, J. and Nairn, A. (1983) *The Phanerozoic geology of the world. II. The Mesozoic B.* Elsevier Scientific Publications, Amsterdam.

Stanley, S. M. (1987) *Extinction*. Scientific American Books, New York.

Croll, James (1821–90)

James Croll was born in Scotland. After his early years spent as a millwright, carpenter, shopkeeper, and insurance salesman, Croll, aged 36, moved to

Glasgow, where he wrote his manuscript 'The philosophy of theism'. Shortly afterwards, he was employed at the Andersonian College and Museum in Glasgow, where he developed an interest in physics. In 1864 he started to become extremely interested in geology—in particular with the controversies surrounding the Glacial Epoch—and began to study the influence of the Earth's orbit on climate change. He formed the view that changes in the eccentricity of the Earth's orbit may have played a large part in determining the timing of ice ages. His first paper on this subject appeared in the *Philosophical Magazine* of August 1864. Croll attempted to calculate changes in the orbital eccentricity of the Earth for the last 3 million years and concluded that there was almost certainly a correlation between the timing of ice ages and times in the history of the Earth when its orbit was markedly elongate. He subsequently attempted to estimate variations in the intensity of radiation reaching the Earth's surface during each season. Furthermore, he was the first scientist to recognize the importance of the albedo of snow and ice surfaces in promoting glaciation.

Croll presented carefully reasoned arguments that explained the occurrence of ice ages as a result of astronomical changes in the nature of the Earth's orbit. He maintained that, as a result of changes in orbital eccentricity, the occurrence of periods of glaciation would have alternated between the northern and southern hemispheres. Croll also demonstrated an awareness of the importance of oceanic circulation in initiating glaciation. He pointed out that the growth of ice sheets in the northern hemisphere would always be associated with the displacement of ocean currents towards lower latitudes. In his later years, he was employed by the Geological Survey of Scotland, and in 1875 he published the influential textbook *Climate and time*. He was subsequently of elected a Fellow of the Royal Society.   ALASTAIR G. DAWSON

cross-sections Maps and cross-sections are a most fundamental source of geological data, and a large part of an Earth scientist's training entails their preparation and interpretation. Maps and cross-sections are both two-dimensional representations of the spatial distribution of one, or a combination of several, particular data sets, such as rock type and structure. Maps are drawn parallel to the Earth's surface (more properly that of the geoid), whereas cross-sections are generally normal to the surface.

Largely because of the emphasis that oil and gas exploration places on understanding subsurface geological structure, most recent work on cross-sections has concentrated on improving *structural cross-sections*. They display geological structure within a frame whose horizontal and vertical scales measure distance. Many structural cross-sections, especially those showing the deep structure of a region, will not have equal horizontal and vertical scales. For example, equal vertical and horizontal scales on a 1 : 1 00 000 scale cross-section through the crustal thickness of Western Europe, from north-

ern Norway to southern Italy, would make the section 380 cm long, but only 35 mm deep. Consequently, vertical exaggeration is often employed to make a cross-section understandable. Remember, however, that vertical exaggeration will artificially increase the inclination of surfaces in the cross-section compared with how they would appear on a true-scale cross-section.

A second type of cross-section in widespread use is the *chronostratigraphic cross-section*, or *Wheeler diagram*, which plots distance horizontally against geological time vertically. Thus, instead of showing the geological structure along the chosen line of section, chronostratigraphic sections simply display the presence or absence of stratigraphic units. They are particularly effective where extensive, but subtle, unconformities cause widespread hiatuses, or breaks in the rock record. In the central North Sea, the economic viability of fairly small oilfields depends on chiefly on the geologists' ability to determine accurately from sections the potential volume of the porous reservoir units. Chronostratigraphic sections have been used to correlate palaeontologically dated sandstones, on occasion demonstrating that sandstones thought previously to be isolated are in fact connected together.

Where possible, the line of a structural cross-section should be chosen so that it is roughly at right-angles to the strike of geological structures, such as folds and faults. This has several advantages. It will mean that the cross-section depicts the real, rather than the apparent, inclination of non-horizontal surfaces; it will also mean that plane strain can be assumed. Plane strain expresses the idea that there has been no movement of rock into, or out of, the plane of the cross-section during deformation, and hence, that volume has been preserved.

Drawing a cross-section always entails a certain amount of geological interpretation. The credibility of this interpretation will depend on the degree to which it is independently constrained by additional geological criteria. The cross-section should be based on a high-quality map of the surface geology, preferably supplemented by a series of maps produced at different levels within the sub-surface. Ideally, boreholes drilled along the line of cross-section provide essential information on the depth of key marker horizons whose structure the cross-section depicts.

In the late 1960s, Canadian geologists exploring for oil and gas in the foothills of the Rocky Mountains began testing the geological credibility of their structurally complex cross-sections by unravelling them, or restoring them. They developed the notion of the *balanced cross-section* in which the credibility of a structural interpretation is improved significantly if the section can be sequentially restored to its undeformed state (Fig. 1). Balanced cross-sections work best in thrust belts of mountain ranges, where the rocks are commonly well exposed, and therefore accessible for detailed field mapping. Further, the *layer-cake stratigraphy* that characterizes sedimentary successions in many thrust belts, in which beds exhibit consistent thicknesses over large areas, means

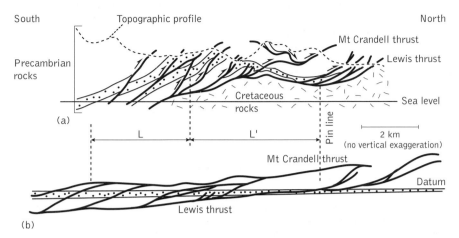

Fig. 1. (a) Balanced cross-section through a duplex structure in the Lewis Thrust Sheet, near the Canada–USA border. The section depicts complex thrust geometry within a Precambrian succession thrust over Cretaceous rocks (stippled). The dotted ornament highlights the principal marker horizon used during the restoration of this section. (b) Restored version of the balanced cross-section in (a). In order to represent the geometry of the full structure before it experienced recent erosion, the structure has been interpreted and restored above the level of the present topographic profile. *L* and *L′* denote, respectively, the restored (*L* + *L′*) and deformed (*L′*) length of a part of the cross-section. Thus, by comparing the balanced (deformed) and restored (unravelled) cross-sections, we can measure directly the amount of crustal shortening accommodated by the thrusting (3.5 km, or 44 per cent shortening of the original undeformed stratigraphy between *L* and *L′*).

that the geometry of the units that make up the undeformed *template* is considerably simpler to deform and restore.

At its simplest, restoration requires only that the beds being restored can be traced to one end of the cross-section, the *pin line*, where they lie in their original vertical and horizontal position relative to one another. Subsequently, the deformed cross section is unravelled in the reverse order to that in which the structures formed, starting at the pin line. A restorable, or balanced, cross-section is one in which restoration can be accomplished without needing to introduce, or explain away, deficit and excess areas. In theory, comparison of the length of the restored and deformed cross-sections through a thrust belt enables a direct assessment of the amount of shortening, expressed either as a percentage of the length of the initial restored section, or as the difference in kilometres between the length of the restored and deformed sections. Typical shortening values are 150 km in the Pyrenees, 400 km in the European Alps, and more than 500 km in the Himalayas.

Work in the Appalachians has demonstrated the degree to which the implicit assumption, that shortening is accomplished by displacement on mappable fault planes, can lead to severe underestimates of the amount of shortening. This work records small-scale structures such as cleavage and stylolites, accommodating as much as 25 per cent of the total shortening along a line of cross-section. On the tens to hundreds of kilometres scale of many conventional balanced cross-sections, such structures are usually more or less completely neglected.

The increasing use of computers and digital data storage throughout the geosciences has speeded up the process of producing maps and cross-sections, and has improved considerably the choice of presentation formats. Most exciting, however, is the way in which computers are being used to build three-dimensionally balanced cross-sections, thus obviating the requirement of conventional cross-section construction to demonstrate plane strain. Using algorithms borrowed from fluid dynamics, the latest restoration procedures literally 'flow' detached cover successions over three-dimensional, ramp-flat fault systems. These produce balanced cross-sections comprising ramp anticlines and rollover anticlines whose geometry is determined by the curvature of the ramp, the depth to the *décollement*, and the deformational characteristics of the hanging wall succession undergoing deformation. *Isostatic balancing* of cross-sections is a further development that has proved particularly effective in producing reliably constrained cross-sections across extensional basins, such as the North Sea. Here, the deformed section is not only in geometrical equilibrium with its restored equivalent, but the restoration procedure also maintains isostatic equilibrium by taking into account the isostatic consequences of thickening or thinning the lithosphere during deformation.　　　　JONATHAN P. TURNER

crustaceans The crustaceans are a class of the phylum Arthropoda, a large and important phylum that includes animals with a segmented body, hard exoskeleton, and many jointed legs. The name 'Crustacea' was originally used to designate animals with a hard but flexible 'crust', but as this applies to most arthropods we now distinguish crustaceans as arthropods that breathe by means of gills and have two pairs of antennae (crabs, lobsters, barnacles, and ostracods). The first crustaceans are known from the Early Cambrian and were forms in which a bivalved carapace almost covered the body. These phyllocarids might have been ancestral to the more advanced shrimp- or lobster-like forms which appeared in the Late Devonian, and have become very important since then, particularly in the oceans.

Most crustaceans have little geological value despite their abundance; however, one group that does is the Ostracoda, small laterally compressed crustaceans enclosed within a protective shell which is formed of two chitinous or calcareous valves that are hinged above the dorsal region of the body. These animals are usually about a millimetre in length and are filter-feeders, consuming micro-organisms stirred up by their appendages. They mostly live in aquatic environments and are benthonic (bottom-dwelling) or planktonic (floating) organisms. The abundance of their shells in marine sediments makes them particularly useful for biozonation and also as indicators of palaeosalinities (ancient salinities). They have a long and well-documented fossil record from the Early Cambrian to the Present.　　　DAVID K. ELLIOTT

crustal composition and recycling
The *crust* is a rigid outer layer of the Earth above a discontinuity in seismic wave velocities known as the *Mohorovičić discontinuity* or simply the *Moho*, which separates the crust from the underlying mantle (Fig. 1). The crust is really the top compositional layer of the lithospheric plates that move about on the Earth's surface. There are two major crustal divisions: oceanic and continental. Oceanic crust ranges from 5 to 15 km thick and comprises 59 per cent of the total crust by area; continental crust, which ranges from 30 to 80 km, comprises 79 per cent of the total crust by volume. Islands, island-arcs, and continental margins are examples of transitional crust with thicknesses of 15–30 km. Although both oceanic and continental crust are recycled into the mantle in subduction zones, oceanic crust is recycled relatively quickly, such that the oldest oceanic crust is only about 200 million years old. On the other hand, at least a few minor remnants of continental crust date to about 4 billion years in age, and the average age of the continents is about 2 billion years.

Geologically, the Earth's crust is very diverse and can be divided into 11 crustal types, each with similar characteristics. *Shields* are deeply eroded segments of the continental crust that have very little relief and have remained tectonically stable for long periods, mainly since Precambrian time. If shields are covered with several kilometres of mainly flat-lying sedimentary rocks they are known as *platforms*. *Orogens* are long, curved belts of deformed rocks that have been elevated into mountain ranges when continents collided with each other. *Continental rifts* are fault-bounded valleys, such as the Rio Grande rift in New Mexico, and the Rhine rift in Europe, and they are formed by tensional forces in the crust. *Volcanic islands* are produced by active volcanoes on the sea floor, and *volcanic arcs* are volcanic chains formed above subduction zones, where oceanic plates return to the mantle (Fig. 1). *Trenches* are the deepest parts of the oceans and mark the beginning of subduction zones. *Ocean basins* comprise most of the Earth's surface and are composed of a thin sediment cover over baratic lavas. *Ocean ridges* are linear rift systems in oceanic crust where new crust is constantly being produced by magmas derived from hot, upwelling mantle. Lastly, *marginal-sea basins* are small ocean basins between island-arcs, and *inland-sea basins* are basins, such as the Caspian Sea and the Gulf of Mexico, that are partially or completely surrounded by continents.

Crustal structure is determined from the velocities of seismic waves that pass through the crust or are reflected by bodies of rock within it. Using evidence from seismic wave velocities, the oceanic crust can be divided into two or three layers, in order of increasing depth: the sediment layer (0–1 km thick), the basement layer (mainly basalt lavas)

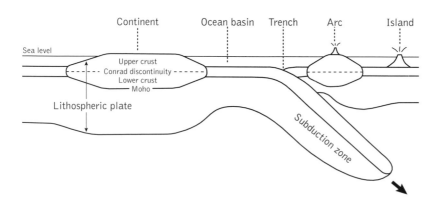

Fig. 1. Generalized cross-section of the lithosphere, showing continental and oceanic crust phenomena.

(0.7–2.0 km thick), and the oceanic layer (sheeted basalt dykes uncertain by gabbons) (3–7 km thick). Some parts of the continental crust can be described in terms of an upper layer (10–20 km thick) and a lower layer (15–25 km thick), and may or may not have a layer of sedimentary rocks on top of the upper layer. The boundary between the upper and lower layers is known as the *Conrad discontinuity* (Fig. 1).

How do we estimate the composition of the crust?

Because most the Earth's crust is buried and not available for sampling, geologists have had to come up with indirect methods of estimating its composition. One of the earliest methods was based on chemical analysis of glacial clays or other fine-grained sediments, which were assumed to be representative of large portions of the continents. Some scientists have used the compositions and abundances of rocks in exposed crust to estimate the composition of its hidden parts. Also helpful are seismic wave velocities measured in rocks in the laboratory compared to velocity distributions observed in the crust. Perhaps the most definitive evidence comes from uplifted and exposed segments of both oceanic and continental crust, where geologists in the field can look at the lower crust. Also useful are fragments of the lower continental crust brought to the surface during volcanic eruptions.

Seismic wave velocities

Seismic wave velocities measured in the laboratory at pressures corresponding to depths of 1–15 km in the continents are consistent with the presence of large amounts of granitic rock at these depths. At depths in the continents of greater than 15 km, measured seismic wave velocities suggest that *granulites*, which are metamorphic rocks formed at high pressures, are most widespread. However, the seismic wave velocities are not sensitive enough to tell us very much about the chemical composition of these granulites.

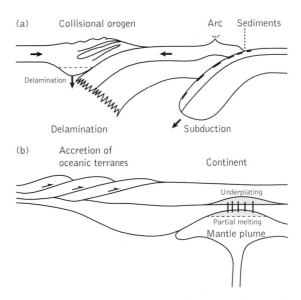

Fig. 2. Cartoons showing (a) recycling and (b) growth mechanisms of continents.

Shields

Widespread sampling and chemical analysis of rocks exposed in shields have provided an extensive database for estimation of both the mineralogical and chemical composition of the upper part of the continental crust. Why are rocks formed at depth now exposed in shields? As rocks are buried, primarily during continental collisions (Fig. 2), their mineral assemblages change during a process known as *metamorphism.* Later, these rocks return to the surface as the crust is uplifted and material is removed from the top by erosion (i. e. *exhumation*). Studies of metamorphic rocks exposed in

Table 1. Average chemical composition of continental and oceanic crust

| | Continental crust | | | | | Oceanic |
	Granodiorite	Upper crust	Lower crust	Total crust	Andesite	crust
	1	2	3	4	5	6
SiO$_2$	66.8	65.5	49.2	57.4	58.8	49.6
Al$_2$O$_3$	16.0	15.0	15.0	15.0	17.2	16.8
TiO$_2$	0.5	0.5	1.5	1.0	0.9	1.5
FeO	4.3	4.3	13.0	8.7	7.3	8.8
MgO	1.8	2.2	7.8	5.0	3.5	7.2
CaO	3.9	4.2	10.4	7.3	6.9	11.8
Na$_2$O	3.8	3.6	2.2	2.9	3.5	2.7
K$_2$O	2.8	3.3	0.5	1.9	1.7	0.2

Chemical analyses of rocks are expressed as weight per cent of the oxides. Note that both the continental and oceanic crust are composed principally of SiO$_2$ and Al$_2$O$_3$ and that they differ chiefly in their contents of CaO and K$_2$O.

shields suggest that the upper layer of the continental crust is composed of about 80 per cent granitic rocks and 20 per cent volcanic and sedimentary rocks, and that collectively this upper crust has a chemical composition similar to a common granitic rock known as granodiorite (Table 1).

Fine-grained sediments

As continents are uplifted, they are weathered and eroded, the denuded material being carried to the oceans as suspended matter in rivers and the air. This suspended matter is deposited in the oceans forming *sediments*. During transport, the weathered materials are well mixed, especially the fine-grained particles, and thus fine-grained sediments may give us some sort of weighted average for the composition of the upper continental crust. This is especially true for elements like the rare earths and thorium, which are relatively insoluble in natural waters. The remarkable uniformity of rare-earth element distributions in fine-grained sediments compared to their great variability in crustal rocks attests to the efficiency of mixing during erosion and deposition. Estimates of upper continental crustal composition based on the chemical composition of fine-grained sediments agree well with the results from shields, again suggesting a composition similar to that of granodiorite.

Crustal sections

During continental collision (Fig. 2), portions of the continental crust are deeply buried in orogens and, later, elevated again as the crust seeks a new level of equilibrium. During this process, continuous sections of the deep continental crust may be brought to the surface, generally along thrust faults. From these sections we see that the lower crust is composed chiefly of granulites of basaltic composition and the upper crust is a mixture of granitic, volcanic, and sedimentary rocks, with granitic rocks dominating. More than anything else, the crustal sections indicate considerable variations in rock types and chemical composition, both laterally and vertically in the continental crust.

Crustal xenoliths

Crustal *xenoliths* are fragments of the deep crust that are brought to the Earth's surface during volcanic eruptions. They generally range from less than 1 cm to about 25 cm in size. From studying the chemical composition of minerals in xenoliths, and from laboratory experimental data on the temperatures and pressures at which these minerals form, we can estimate the depth in the crust from which xenoliths have come. In some volcanic fields, such as the Navajo field in the south-western United States, xenoliths come from many different crustal depths, and are very helpful in reconstructing a cross-section of the crust. Lower crustal xenoliths from volcanic rocks erupted on the continents are quite diverse and indicate that the deep crust is heterogeneous. Granitic rocks and granulites dominate in deep crustal xenoliths. Xenoliths from depths greater than 20 km are chiefly basaltic granulites, which suggests that the lower part of the continental crust is basaltic in composition.

Ophiolites

Fragments of oceanic crust may be thrust on to continents, either during subduction or when arcs or continents collide. These fragments, known as *ophiolites*, appear to have formed at ocean ridges mainly in marginal-sea basins, and they may later be uplifted to the surface where they can be studied and sampled. By chemically analysing samples from ophiolites, together with volcanic samples dredged from the ocean floor, we can estimate the composition of the oceanic crust.

Composition of the crust

The average chemical composition of the upper continental crust is well known from widespread sampling of shields, from chemical studies of fine-grained sediments, and from exposed crustal sections. As shown in Table 1, the upper crust is enriched in silica (SiO_2) and potash (K_2O); these oxides are contained chiefly in the minerals quartz and feldspar, which are abundant in the upper crust. Although the upper continental crust is very heterogeneous, its average composition is similar to that of granodiorite (compare columns 1 and 2, Table 1). The composition of the lower continental crust is less well constrained. Crustal sections and deep crustal xenoliths suggest that a large part of the lower crust is basaltic in overall composition (Table 1, column 3). An estimate of total continent composition based on a one-to-one mixture of upper and lower continental crust is given Table 1, column 4. This bulk composition is similar to that of the volcanic rock andesite (compare columns 4 and 5).

An estimate of the major element composition of oceanic crust is given in Table 1, column 6. It is based on chemical analyses of ophiolites and data from dredge-and-core samples that have penetrated the basement (basaltic) layer of oceanic crust. Deep-sea sediments, which contribute less than 5 per cent to the oceanic crustal composition, are ignored in the estimate. The average composition of the oceanic crust, in striking contrast to continental crust, is that of a basalt that is greatly depleted in potash (note: only 0.2 per cent K_2O).

Chemical elements can be grouped according to their preference for liquid or solid phases. Those elements that are strongly partitioned into a liquid phase are known as *incompatible elements* and include such elements as potassium, thorium, and barium. Those elements that prefer to stay in solids are referred to as *compatible elements*, and include such elements as iron and nickel. In Fig. 3, the distributions of many elements in the crust according to their compatibility are shown. All element concentrations are normalized to the primitive composition of the Earth's mantle so differences are easily seen on the graph. When partial melting occurs in the mantle, the most incompatible elements, such as rubidium and uranium, are strongly concentrated in the melt, which because it is less dense than the surrounding mantle, rises towards the Earth's surface bringing incompatible elements from the mantle to the crust. Moderately incompatible elements, such as zirconium and titanium, and compatible elements remain chiefly in the mantle during partial melting.

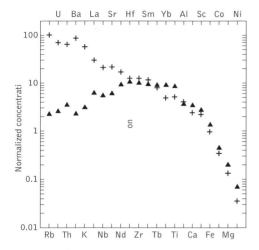

Fig. 3. Distribution of elements in continental (+) and oceanic (▲) crust. Element concentrations are divided by the concentrations in the primitive Earth's mantle and the corresponding normalized concentrations are plotted on the vertical axis. Elements are arranged on the horizontal axis from most incompatible at the left to most compatible at the right.

This mantle, now impoverished in incompatible elements, is often referred to as *depleted mantle*. Because incompatible elements are greatly reduced in abundance in depleted mantle, the next time the depleted mantle is partially melted the magmas inherit this depletion in incompatible elements; for example, the oceanic crust, which is derived from depleted mantle (Fig. 3). The complementary incompatible element distributions in the continental and oceanic crust shown in Fig. 3 suggest a two-stage model for extraction of the Earth's crust from the mantle. During the first stage, continental crust is extracted and, during the second, oceanic crust is extracted from the depleted mantle left behind from the first stage. However, we know that both continental and oceanic crust are continually generated in the mantle, so in reality a simple two-stage model will not work. Both types of crust are continually being extracted from and returned to the mantle, but in such a way that their strikingly different incompatible element distributions are maintained. This means that continental crust cannot be recycled into the depleted mantle from which oceanic crust is derived, and hence that the Earth's mantle is composed of two or more 'geochemical reservoirs' that do not appreciably mix with each other through time.

Growth of continents

The distributions of isotopic ages of igneous rocks in the continental crust shows that rocks of certain ages (for instances 2.7 and 2.0 billion years old) are more widespread than other ages, and that crust older than 3 billion years is of very limited extent. Although the peaks in crustal age have been interpreted by some to reflect episodic continental growth, this is only so if continental rocks are not recycled into the mantle. It is more

accurate to interpret these ages in terms of the preservation rate of continental crust, which is equal to the growth rate (i. e. the rate at which new crust is extracted from the mantle) minus its recycling rate back into the mantle. Today we recognize three major mechanisms by which continents grow: terrane collision, magma underplating, and subduction-related magmatism. Oceanic terranes, such as island-arcs, submarine plateaus, and islands, frequently collide with continents (Fig. 2). During such collisions, oceanic terranes are thrust on to and become part of the continental crust. As an example, about 25 per cent of the growth of North America occurred in the past 100 million years by terrane collisions in western North America, and almost all of Alaska comprises such terranes. Continents can also grow by underplating of basaltic magma that comes from mantle plumes. *Mantle plumes* are buoyant masses of mantle that have risen from great depths in the Earth, probably from just above the core boundary. As they reach the base of the lithospheric plates they begin to melt, producing basaltic magmas, which can rise and underplate continents or erupt at the surface producing flood basalts like those of the Deccan Plateau in India. If they come up beneath oceans, they may erupt and form submarine plateaus, like the vast Ontong–Java plateau in the south-western Pacific. Continents can also grow above subduction zones that have developed along continental margins, such as in the Andes in South America. Here, magmas derived from the mantle are added to the margins of continent, both by intrusion of granites at depth and by eruption of volcanics at the surface.

Continental materials are recycled into the mantle by two major mechanisms (Fig. 2). As continents are uplifted, they are eroded and carried to the oceans where they are deposited as sediments, and some of these sediments may be subducted back into the mantle. The second recycling process, known as *delamination*, is poorly understood. When continents collide, the crust greatly thickens in response to the large compressive forces and, as the lower crust bends downward, changes in mineralogy occur in response to elevated temperature and pressure. The newly formed minerals include garnet, which has a density greater than the crust, and hence the root zones of the collisional orogens can become gravitationally unstable and sink into the mantle by delamination (Fig. 2).

Thus, the amount of continent formed during any particular period of time in the geological past is determined by a delicate balance between the rate at which new crust is extracted from the mantle and the rate at which old crust is returned to the mantle. Because the recycling rate has probably decreased with time, paralleling the cooling of the Earth, the average preservation rate of continental crust has increased, and hence there is more young crust that has survived to the present than there is old crust.

KENT C. CONDIE

Further reading

Condie, K. C. (1997) *Plate tectonics and crustal evolution* (4th edn). Butterworth–Heinemann, Oxford.

Taylor, S. R. and McLennan, S. M. (1985) *The continental crust: its composition and evolution*. Blackwell Science, London.

crustal movements and deformation styles

Most crustal movements and accompanying deformation owe their origin to plate motions, which in turn are driven by thermal and gravitational processes within the Earth. The mechanisms of deformation in rocks are a complex function of forces applied to a rock volume and the ambient physical and chemical conditions. The most important of these properties are mineralogy, temperature, pressure, fluids, and strain. Brittle deformation is favoured at shallow crustal levels at low temperatures (<300 °C) and pressures. It involves loss of cohesion of rock at the grain scale. Ductile deformation is favoured at higher temperatures and pressures, and comprises several grain-scale mechanisms including pressure solution, grain boundary creep and sliding, and dislocation creep. Both brittle and ductile processes can occur simultaneously in a rock; however, one mechanism commonly dominates.

In the absence of anomalous heat sources and at a typical geothermal gradient (20–30 °C km⁻¹), the transition between brittle and ductile deformation occurs at depths between 10 and 15 km. This depth corresponds to the base of the seismogenic zone for large intracontinental faults. Recognition of deformational features formed at middle to lower crustal levels is important because mountain belts commonly expose rocks which, before uplift and denudation, were deformed at depths of 10–25 km. Cores of mountain belts therefore provide a 'window' into the physical and chemical conditions that prevail at middle and lower crustal levels during deformation. In summary, the type of deformation observed in rocks presently exposed at the Earth's surface is a function of many factors, not the least of which is the level of exposure.

The following discussion of crustal movements and rock deformation is placed in the context of plate tectonic settings and processes. Particular attention is paid to styles of deformation, both as a function of position within tectonic setting and the crustal level of deformation.

Convergent plate boundaries

Convergence of tectonic plates has produced some of the most spectacular topographic and geological features present on Earth, including volcanic chains and large mountain belts. The deformational features that develop at convergent plate boundaries depend on whether a dense oceanic plate sinks beneath a buoyant continental plate (non-collisional convergence) or whether two continental masses collide (collisional convergence).

Ocean–continent (non-collisional) convergence

When oceanic and continental plates converge, the denser oceanic plate sinks beneath the continental plate in a process called *subduction*. This process has profound effects on the overriding continental plate. The most visible product of subduction is the development of a magmatic arc. The magmatic arc separates an oceanward, or fore-arc area, from a continentward, or back-arc area. Below, the deformational styles that occur in different parts of a non-collisional plate boundary are discussed (Fig. 1). Not all components are equally developed in these zones.

The most conspicuous outboard structure is the accretionary wedge that lies just continentward of the ocean trench, the bathymetric manifestation of the subduction zone. The accretionary wedge is an accumulation of material derived from the descending slab. Slivers of oceanic crust and sediment deposited on the slab may be too buoyant to subduct. These materials can be scraped off the slab and accreted to the leading edge of the overriding continental plate. Accretionary wedges parallel the trench, may be emergent or below sea level, and may be in excess of 100 km wide and 15 km thick. Accretionary wedges are characterized by

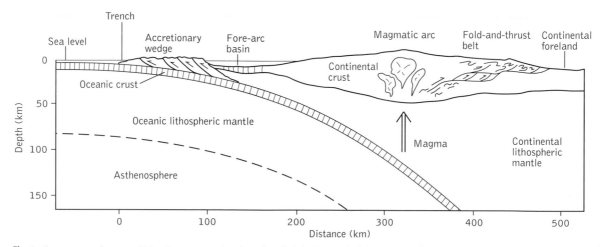

Fig. 1. Components of a non-collisional convergent plate boundary (subduction zone). (After Hamilton (1988) In Ernst W. G. (ed.) *Metamorphism and crustal evolution of the western United States*, Vol. 7, pp. 1–40. Prentice-Hall, Englewood Cliffs, New Jersey.)

intense shortening, and the structural geometry and kinematics of accretionary wedges are characteristic of fold-and-thrust belts. Seismic reflection profiles show that the internal structure of most wedges consists of 20–30° continentward-dipping, imbricate thrust sheets that are thrust toward the trench. Modern accretionary wedges include western Taiwan, the eastern Aleutian Islands, and Barbados.

A fore-arc basin may intervene between the magmatic arc and the accretionary wedge. This basin collects sediment from both the wedge and the arc. Deformation is absent or minor within the fore-arc basin during subduction but these sediments may be deformed during subsequent arc–continent or continent–continent collision.

The back-arc region may be under compression, under tension, or in a neutral state of stress. The origin of tensional forces in the back-arc regions of some convergent zones is poorly understood and controversial. It is clear that many back-arc regions are marked by extensional tectonism and volcanism in both oceanic and continental settings. Examples of the latter include the Basin and Range province of western North America (however, here at least some extension initiated within a broad zone of arc magmatism) and the Cretaceous Rocas Verdes complex of southern Chile. Studies of earthquake data in the south-western Pacific show that strike-slip movement may be significant in back-arc regions. Sedimentary and volcanic rocks, and associated oceanic crust, may be deformed by contractional deformation during subsequent collisional orogenesis.

Perhaps the most conspicuous and spectacular deformation in convergent zones occurs where the back-arc region is under regional compression or basal shear traction from low-angle subduction. In this case, a zone of intense contractional deformation, known as a fold-and-thrust belt may develop. Material is moved from the magmatic arc region toward the continent. The sense of tectonic transport of upper crustal rocks relative to the undeformed craton is opposite to that of the overriding plate with respect to the oceanic plate.

Fold and thrust belts may be shortened by 50 per cent; that is, a region formerly 200 km wide is 100 km wide after deformation. Towards the continent, in the foreland, shortening is accommodated at shallow crustal levels by folding and thrust faulting of sedimentary strata that overlie uninvolved crystalline basement. Thrust faults generally follow interfaces between sedimentary strata for great distances (flats) and locally cut stratigraphically upsection (ramps) in the direction of tectonic transport. Large-scale folds form over ramps and at the ends of 'blind' thrusts that do not reach the surface. Upper-crustal stratal shortening in the foreland may be balanced by ductile shortening of the deeper crust toward the arc in the hinterland. Hinterland regions of many thrust belts record true crustal shortening that is manifested by major, thrust-sense, ductile shear zones that cut pervasively folded rocks.

Continent–continent (collisional) convergence

Subducting oceanic plates can contain seamounts, island arcs, or continents that approach the subduction zone as convergence between plates continues. These features are thicker and more buoyant than oceanic crust, and therefore cannot be completely subducted. The result is a collision between arc systems, an arc and a continent, or two continents. This results in the formation of a suture zone between the colliding bodies. Suture zones are characterized by folds and thrust faults at shallow crustal levels and penetratively deformed crustal blocks bounded by reverse-sense ductile shear zones at deeper levels. Deformation may extend for hundreds or even thousands of miles away from the suture zone and may be markedly heterogeneous.

Perhaps the most spectacular example of continental collision is the Early Tertiary to present-day collision of the Indian Plate with the Eurasian Plate to produce the Himalayas and the Tibetan Plateau. Collision was preceded by subduction of Indian oceanic lithosphere beneath Asia, producing a magmatic arc on the leading edge of the Asian Plate. As a consequence, the Asian Plate was warmer and more buoyant than the Indian Plate. This circumstance resulted in incipient subduction of Indian continental crust when it encountered the trench.

Folding and thrust faulting is localized along a band, 200–300 km wide, north of the main frontal thrust of the Himalayas. However, deformation associated with the collision is manifested far inboard (up to 2500 km) of the suture, leading some researchers to suggest that up to 2000 km of Indian continental crust has been subducted beneath Asia. Major strike-slip faults in China and Mongolia appear to accommodate north–south shortening associated with the collision and to facilitate eastward and south-eastward movement or 'escape' of wedge-shaped blocks of the Eurasian continent. In China and Siberia, east–west extension is manifested by the formation of extensive graben systems such as the Baikal rift.

Underthrusting of India beneath Asia doubled the thickness of the continental crust beneath the Himalayas and Tibetan Plateau. Several independent workers have recently documented the presence of large-scale extensional structures that accommodate north–south extension in the Tibetan Plateau. The driving force for extension in the Tibetan Plateau is attributed to gravitational instability induced by overthickened crust. These extensional structures are coeval with contractional structures, but they occur within different parts of the orogen and at different crustal levels. Thus, a complex array of structures, including thrust, normal, and strike-slip faults can be produced during collisional orogenesis, depending upon crustal level and location relative to the suture zone.

Divergent plate boundaries

Tensional forces within the Earth's lithosphere can produce continental rifts—elongate fault-bounded depressions which may ultimately result in the formation of new oceanic crust between diverging continental plates. Almost all regions undergoing extension are also sites of magmatic activity; however, the relationship between magmatism and extension remains controversial. If the extensional process is arrested before development of oceanic crust, a continental rift is produced. If rifting results in formation of a new ocean basin, passive plate margins will form on the rift-facing sides of the diverging plates.

Crustal extension at shallow crustal levels is accommodated by normal faults that range in dip from less than 30° to more than 70°. Among the best-studied regions that offer insight into the rift-to-drift process is the Red Sea–Gulf of Aden–East African rift system. The East African rift is in the incipient stage of rift development, whereas the Gulf of Aden is the locus of active continental separation. Initially, limited crustal extension is accommodated by moderately to steeply dipping normal faults, oriented approximately, but not necessarily exactly, parallel to the rift axis. These structures may be discontinuous along the strike and relay displacement between *en échelon* segments. Typically, a master set of normal faults will form the boundary of one side of the rift, creating a half-graben. Along strike, this set is replaced by oppositely dipping normal faults that form the boundary of the other side of the rift. The intervening region is one of intermeshed normal faults, called an accommodation zone. With increasing extension and differential lateral movement of crustal blocks, strike-slip transfer faults form to link areas undergoing coeval extension.

A different style of rifting is manifested in the Basin and Range province of western North America, and the process may also be occurring today in the Tibetan Plateau. In these areas, crustal extension is distributed over a region up to 800 km wide. Extension is accommodated by both high- and low-angle normal faults. The latter appear to be instrumental in localizing areas of extreme upper-crustal extension which is heterogeneous in space, time, and magnitude. Although the

reason for the two different styles of extensional regimes is not known, it has been suggested that broad regional extension (Basin and Range style) is favoured where body forces dominate (e.g. overthickened crust). More localized rifts form due to traction forces at the base of the lithosphere that may be induced by asthenospheric flow.

The manner in which crustal extension is accommodated in the crust and mantle is controversial (Fig. 2). One model invokes deformation by *pure shear strain*, in which extension is accomplished by uniform ductile stretching and consequent thinning of the middle and lower crust. The upper brittle crust fails under tension, creating a series of fault blocks. In this model, there is little or no lateral translation between the hanging-wall and the footwall. Any given vertical crustal column is simply thinned by a combination of brittle and ductile extension at different crustal levels. The other model involves *simple shear*, in which extension is accommodated by lateral translation of hanging-wall blocks relative to the footwall along an extensional detachment fault. Detachment faults may root into a zone of mid-crustal fluid flow. As much as 60 km of translation has been suggested for some detachment faults that appear to have been active at angles to the surface of less than 20°. Detachment faults can also initiate at high angles and rotate to lower angles by footwall rebound mechanisms.

Perhaps the most spectacular manifestations of extensional process are metamorphic core complexes. Metamorphic core complexes can occur on the scale of a mountain range and have been documented in the Basin and Range province, the Tibetan Plateau, the eastern Mediterranean Ocean, and the Alps. Metamorphic core complexes consist of a central 'core' of metamorphic and plutonic rocks. These rocks may have resided at middle crustal levels just before extension. The

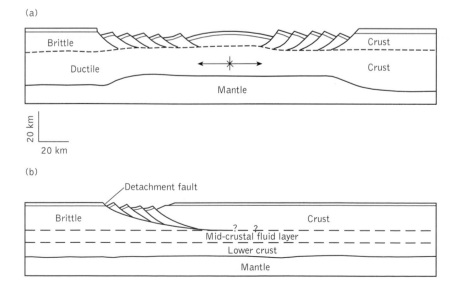

(a)

(b)

Fig. 2. Two models for continental extension: (a) pure shear model; (b) simple shear model. Modified after Lister *et al.* (1986, *Geology*) and Wernicke (1992, *The geology of North America*).

crystalline rocks are separated by a detachment fault from an upper plate that has been highly extended along brittle normal faults, most of which dip in the same direction. The upper plate may contain rocks of any age but always contains sedimentary and/or volcanic rocks that were deposited at the surface prior to, or during the early stages of, extension. The detachment fault therefore juxtaposes surficial rocks against middle crustal rocks, implying an omission of up to 15 km or more of crust. The crystalline rocks in the lower plate of the detachment typically contain mylonite zones that may be up to several kilometres thick. Sense of shear within the mylonite zones is consistent within any given core complex and is kinematically compatible with the movement sense of most of the brittle faults in the hanging-wall.

Transform plate boundaries

Transform plate boundaries are strike-slip faults that link other plate boundaries such as spreading centres, subduction zones, or other transforms. Because most transform boundaries occur in oceanic crust, much of the deformation produced by these structures is obscured. The effects of purely horizontal motion in oceanic crust may be relatively minor. However, in continental settings, transform boundaries may be complex systems of strike-slip faults tens or hundreds of kilometres wide that produce profound deformational effects.

Typically, lateral motion on an individual strike-slip fault will die out along strike and be transferred to an *en échelon* segment. Stepping (sometimes called jogging) of faults in this manner may produce local zones of extension or shortening in the region of overlap. The result is that many major strike-slip boundaries are characterized by metre- to kilometre-scale 'pull-apart' basins or uplifts ('push-up' structures). These effects may be accentuated where the relative plate motion vector is not exactly parallel to the transform boundary zone.

Where there is a component of convergence between plates along a boundary that is mainly a transform one, shortening structures develop in addition to structures normally associated with strike-slip deformation. Slightly oblique convergence between plates is not accommodated by oblique motion along faults, but rather, deformation is partitioned into a shortening component and a horizontal slip component. Shortening is manifested by the development of folds and thrust faults; the horizontal slip is accommodated by strike-slip faults. A modern example of such a transpressional plate margin is the San Andreas fault system in California. Plate motion reorganizations about 4 million years ago resulted in slight convergence between the Pacific and North American plates. As a consequence, a fold-and-thrust belt developed within the Coast Ranges and these structures appear to be functioning completely independently of the San Andreas fault.

Where there is a component of divergence between plates along a transform boundary, extensional structures will develop in a zone of transtension. Normal and strike-slip faults will operate in a kinematically co-ordinated fashion to produce extensional basins, as in the Gulf of California region.

ERNEST M. DUEBENDORFER

Further reading

Moores, E. M. and Twiss R. J. (1995) *Tectonics*. W. H. Freeman, New York.

crystallographic systematics

Symmetry

Virtually all minerals are crystalline with a regular atomic structure, but in many instances with a poor external form. Under favourable conditions of growth, however, the internal regularity may result in well-formed crystals in which a regularity of external form reflects the internal regularity. These crystals are rarely perfect; some faces may be distorted or even totally suppressed, and some crystals may be single-ended (uniterminal), one end being keyed-in to the side of a cavity in which they grew. Doubly terminated crystals may show a certain symmetry in the distribution of their faces. Crystal symmetry is related to four types of symmetry operation (Fig. 1):

(1) Reflection across a mirror plane which divides the crystal into two halves that are mirror-images of one another.
(2) Rotation about an axis through the centre which brings the crystal into an identical position 2, 3, 4, or 6 times after a complete rotation of 360°. (A 5-fold axis is not possible, nor is one higher than 6-fold.) These axes are, respectively, diads (2), triads (3), tetrads (4), and hexads (6).
(3) 'Roto-inversion' about a point. This entails rotation about an axis followed by inversion across the centre.
(4) Inversion across the centre, so that for every face there is a parallel face diametrically opposed to it.

Table 1 shows that some crystals have a high degree of symmetry while others have none. On the basis of symmetry crystals can be assigned to one of seven *systems*, each with a certain minimum symmetry. Each system can be further subdivided into *classes*, of which there are 32, each with its own characteristic symmetry. Many crystal classes are represented by a unique geometrical body.

Steno (1669) found that the external angles between the corresponding faces of quartz crystals were constant. This relationship was refined by Romé de Lisle (1779) and defined as the 'law of the constancy of interfacial angles', which states that in all crystals of the same substance the angles between corresponding faces have a constant value. The law applies irrespective of distortions in the crystal. Nowadays the angle is taken as that between the normals (perpendiculars) to the faces and is measured roughly by a contact goniometer or more accurately by an optical goniometer. In minerals which have similar compositions and structures the value of the interfacial angle may be used to identify specific minerals.

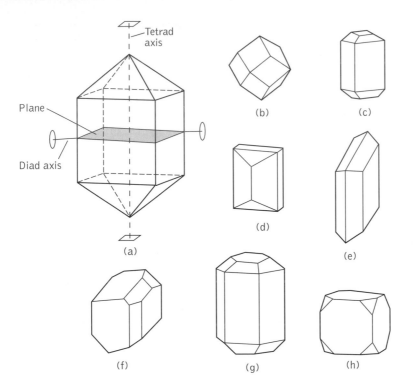

Fig. 1. Symmetry: (a) elements of symmetry; (b–g) crystals from the cubic, tetragonal, orthorhombic, monoclinic, triclinic, and hexagonal systems; (h) combination of cube and octahedron.

Forms

Well-formed crystals may have several faces that are similar in shape and orientation with respect to the crystallographic axes. Such a group of equivalent faces or assemblage of faces constitutes a *form* which can be generated by the symmetry elements if one face is given. A *general form* results when the given face bears no special relationship to the symmetry elements and can be acted upon by all of them. General forms have a wide variety of shapes and number of faces; in the most symmetrical of the classes in the cubic system, for example, the resultant form has 48 faces (hexoctahedron), whereas in the least symmetrical class of the triclinic system the form is a single-faced one (pedion). There is only one general form in each class. A *special form* is developed if the given face bears a special relationship to the symmetry elements; if, for example, it is parallel to or perpendicular to an

Table 1 The crystal systems and their symmetry

System	Minimum axial symmetry	Maximum symmetry			classes
mum		Number of			
		Axes	Planes	Centre	
Cubic	4 triads	3–4, 4–3, 6–2	9	C	5
Tetragonal	1 tetrad	1–4, 4–2	5	C	7
Orthorhombic	3 diads	3–2	3	C	3
Monoclinic	1 diad	1–2	3	C	3
Triclinic	None	None	None	C	2
Hexagonal	1 hexad	1–6, 6–2	7	C	7
Trigonal	1 triad	1–3, 3–2	3	C	5

Note. In column 3 (Maximum symmetry, axes), the pairs of figures denote numbers and types of axes of symmetry: thus, for example, the tetragonal system has one fourfold symmetry axis and four twofold axes.

Table 2 Parametral intercepts and interfacial angles

System	Parametral intercepts	Interfacial angles
Cubic	$a = b = c$	$\alpha = \beta = \gamma = 90°$
Tetragonal	$a = b \neq c$	$\alpha = \beta = \gamma = 90°$
Orthorhombic	$a \neq b \neq c$	$\alpha = \beta = \gamma = 90°$
Monoclinic	$a \neq b \neq c$	$\alpha = \gamma = 90°, \beta > 90°$
Triclinic	$a \neq b \neq c$	$\alpha \neq \beta \neq \gamma \neq 90°$
Hexagonal		$x \wedge y = y \wedge x = u \wedge x = 120°$
Trigonal	$a^1 = a^2 = a^3 \neq c$	$x, y, u, z = 90°$

These are *open forms*; examples are the pinacoid with two parallel faces and the orthorhombic prism with four faces.

Crystallographic axes

In order to specify the orientation of planes (faces) in space it is necessary to relate them to a set of non-co-planar axes, here called the *crystallographic axes*, which are chosen according to a set of rules. If possible they should be parallel to symmetry axes, at right angles to each other, and parallel to actual or possible edges in the crystal. They cannot be chosen unambiguously from the external form. The three axes intersect in a point, have positive and negative ends and are lettered x, y, z (sometimes u in addition). The angles between the axes are: $y \wedge z = \alpha$, $x \wedge z = \beta$, $x \wedge y = \gamma$ (see Table 2 and Fig. 2a).

Miller indices

The *law of rational indices* enunciated by Haüy in 1801 states that the intercepts made by any face in a crystal are simple rational numbers by reference to the intercepts made by the *parametral plane*. In the early days of crystallography the parametral plane was chosen arbitrarily, but because of the development of X-ray crystallography it can now be chosen by reference to the edges of the unit cell. Using the *xyz* axial notation the intercepts made by the parametral on these axes are lettered *abc*. The intercepts made by any other face can now be related to these unit intercepts.

axis or perpendicular to a plane. In the cubic system the cube, with only six faces, is special to a tetrad axis and to four planes of symmetry.

The growth of some crystals is such that only one form is developed: the cube and hexoctahedron in the cubic system, the bipyramid in the orthorhombic and hexagonal systems, and the trigonal trapezohedron in quartz are examples, but many crystals develop combinations of forms so that many faces result. Some forms, e.g. the rhombic dodecahedron, totally enclose space and are *closed forms*; others do not enclose space and therefore cannot exist alone in a crystal.

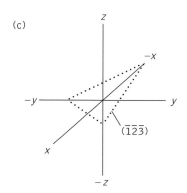

(a)

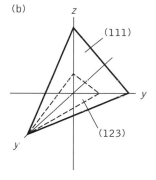

(b)

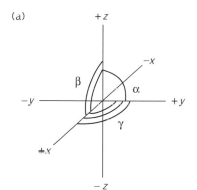

(c)

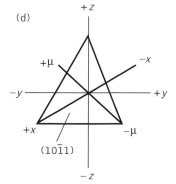

(d)

Fig. 2. (a) Crystallographic axes and angles: (b) parametral plane (111) (thick lines) making intercepts *a*, *b*, *c* on axes; face (123) (dashed) making intercepts 1*a*, ½*b*, ⅓*c*; (c) face ($\bar{1}\bar{2}\bar{3}$) cutting negative axes at 1*a*, ½*b*, ⅓*c*; (d) hexagonal axes and parametral plane (10$\bar{1}$1).

In Fig. 2b the parametral plane (shown in bold lines) makes intercepts a, b, c on the crystallographic axes. A second face (dashed lines) cuts the axes at $1a$, $\frac{1}{2}b$, $\frac{1}{3}c$; these multipliers, 1, $\frac{1}{2}$, $\frac{1}{3}$, of the unit intercepts are the *Weiss symbols*, but it is more convenient to use their reciprocals to give the *Miller indices*; the index of the new face is thus 123, written (123). A third face (dotted) makes intercepts of $1a$, $\frac{1}{2}b$, $\frac{1}{3}c$ on the negative ends of the axes; its Miller index is thus ($\bar{1}23$).

A face parallel to an axis cuts it at infinity; so the Weiss symbol is ∞ and the Miller index is 0. A face with an index of (100) thus cuts the positive end of the x axis and is parallel to the other two. The index of the parametral is clearly (111). It is important to note that it is not the magnitude of the intercept that is critical but its relation to the unit intercept. It will be apparent that the larger the index the smaller the intercept and the larger the intercept the smaller the index.

The structure of the mineral determines that in the cubic system the parametral (111) cuts off equal intercepts on the three axes and is thus an equilateral triangle, but in the tetragonal system (111) is isosceles, and in the orthorhombic systems ($11\bar{1}$) is scalene.

In the hexagonal and trigonal systems, in order to fit with the internal structure and hence the external morphology the z axis is chosen parallel to vertical edges and there are now three horizontal axes, x, y, u, which are at 120° to one another and at 90° to z. The parametral cannot cut off equal intercepts on all three axis: its index may be ($10\bar{1}1$) with the first three digits adding up to zero.

The index of a form is written {100} or {321}, indicating all the faces that can be generated from the face (100) or (321) by the symmetry elements of the particular class. In the holosymmetric class of the cubic system, {100} gives six faces (the cube); in the tetragonal system four faces (prism); in the monoclinic system two faces (pinacoid), and in the lowest symmetry of the triclinic system only one face (pedion). In the holosymmetric class of the cubic system {321} gives 48 faces (hexoctahedron), but in the triclinic system only one face (pedion).   R. BRADSHAW

Further reading

Bishop, A. C. (1967) *An outline of crystal morphology*. Hutchinson, London.

crystals and crystal growth
Crystals are solid bodies bounded by naturally formed plane faces with a regularity of external form which reflects the regular arrangement of the constituent atoms.

Externally, crystals are rarely perfectly developed; some faces are usually distorted or even suppressed, or have striations or etch pits. Internally, what appears to be a perfect crystal may have millions of defects in a specimen a few millimetres across. This is perhaps not surprising when a garnet 10 mm in diameter consists of at least 10^{20} atoms. What is believed to be the biggest crystal ever discovered, a microcline 50 m $\times$ 36 m $\times$ 14 m calculated to weigh 15 900 tonnes, might

have 10^{35} atoms. Under tightly controlled conditions of growth in the laboratory crystals of much greater perfection are grown.

The growth of a particular crystal depends on the chemistry of the local environment, the presence of fluids, the degree of supersaturation, and the temperature and pressure conditions. Well-formed crystals can grow from a melt (e.g. orthoclase feldspar from a granite magma; quartz from aqueous solution in, say, a hydrothermal vein; garnet in a solid rock during metamorphism; gypsum in a clay deposit). In all cases small groups of atoms are forming and dispersing but as the physical conditions change or the chemical concentration increases these groups will grow larger and reach a critical size, at which stage a new mineral can nucleate and the atoms assume regular positions. In ionic crystals every cation is completely surrounded by anions, and vice versa. The numbers are determined by the ionic charges and the relative size of the ions; the total crystal will be electrically neutral. Slow growth with few nuclei will lead to large crystals; rapid growth with a high level of supersaturation will produce many small crystals.

The bonds of the atoms on the margins of the crystal will be unsatisfied and will latch on to appropriate atoms in the vicinity. Once a crystal face begins to grow, it is comparatively easy for a layer of atoms to be completed, but it is much more difficult to initiate a new layer. However, spiral growth means that there is no new layer to initiate and the face continues to grow spirally. These spirals on the faces of crystals, particularly those growing from aqueous solution, are readily visible under a microscope.

Even though the atomic structure remains the same, the presence of impurities in the growing medium and variations in the physical conditions affect the growth of particular crystal faces, resulting in a variety of external forms. Thus, for one structure many different habits may result—at least 200 in the case of calcite.

Already-formed, small 'foreign' minerals and globules of fluid may adhere to the growing face and then become trapped by the further addition of atoms. These inclusions provide valuable information about the environment in which the mineral grew.

Crystals may grow as composite solids in two or more parts, the individual parts being related in a quite specific geometrical way according to the nature of the lattice. These twins can be of contact or interpenetrating type and may be simple or repeated. A plagioclase feldspar crystal may have 200 or more individuals, all closely related.

The large microcline crystal has already been mentioned. It may not in fact be a single crystal. Other large specimens include garnet, beryl, and phlogopite crystals, weighing 38, 370, and 334 tonnes, respectively.   R. BRADSHAW

Curie temperature (Curie isotherm)
The forces that cause magnetic ordering in ferromagnetic materials can be overcome by raising the temperature of such materials; the increased temperature changes the intermolecular spacing

and thereby weakens the forces holding the molecules together. The temperature at which the magnetic ordering in ferromagnetic substances vanishes is known as the *Néel temperature*. In ferrimagnetic substances (a subclass of ferromagnetic materials), this same temperature is referred to as the *Curie temperature* (T_C).

The *Curie isotherm* is a temperature surface in the Earth's subsurface connecting depths at which Curie temperatures are reached. Curie temperatures for natural magnetic minerals range from below 0 °C to over 1000 °C. The most important magnetic minerals encountered in the Earth's lithosphere have lower Curie temperatures. These magnetic minerals belong to two solid–solution series: the titanomagnetites, for which T_c ranges from about 200 °C to 580 °C; and the titanohaematites, with values of T_C up to 680 °C. Since these Curie temperatures fall in the range of temperatures generally inferred for the deeper parts of Earth's crust and the uppermost mantle, magnetic anomaly data are capable of yielding useful information about the depth of the Curie isotherm and, in turn, the dominant magnetic mineralogy and the lithologies at that depth—provided that the magnetic survey has been carried out with appropriate survey parameters (for example, 10 km observation elevation and spacing, and hundreds of kilometres of continuous flight-lines).

DHANANJAY RAVAT

Cuvier, Georges (Baron) (1769–1832)

One of the great names in natural history, Georges Léopold Chrétien Frédéric Dagobert Cuvier, was a Frenchman of prodigious attainments and international importance. Of poor family but exceptional intellectual ability, he managed to attend university at the age of 15, and to be appointed assistant at the Museum of Natural History in Paris in 1795.

His life was dominated by the care and study of his zoological collections at the museum, and the creation and enhancement of French teaching institutions. He carried on an enormous international scientific correspondence and travelled widely.

Napoleon recognized his abilities, promoting him to positions where he could act for the benefit of French science and education. At that time excavations near Paris and elsewhere in Europe were yielding the remains of numerous fossil mammals. Many of these were sent directly to Cuvier or were described to him by correspondents. The results of his palaeontological studies were published in two great works, *Recherches sur les ossements fossiles de quadrupèdes* (Paris, 1815) and the *Discours sur les révolutions de la surface du globe*.

Cuvier held that the structure of an animal is, of necessity, in harmony with its mode of life. He was a believer in divine providence and also an out-and-out catastrophist. This, and his classification of the animal kingdom, was propounded in his famous four-volume work, *Le règne animal distribué d'après son organisation* (1817).

Raised to the French peerage in 1831, he died after a brief illness the following year.

D. L. DINELEY

cycles, geochemical *see* GEOCHEMICAL CYCLES

cyclicity The idea that there are cycles of events in Earth history is an old one. Even when attempting to reconcile geological observation with biblical teaching, scholars saw cycles in the process of creation. With the recognition of 'deep' geological time and with the advent of the uniformitarian approach to earth history the smooth flow of geological history seemed to have proceeded from one state or condition to another several times over. Such cycles were terminated by 'catastrophes' and were followed by a well-defined series of evolutionary happenings. In another sense it was recognized that geological processes of construction and destruction are continually at work in what is known as the *geological cycle*. This concept is discussed in a separate entry.

Another view of the geological cycle is that embodied in the movement of matter through the crust from an igneous origin, via weathering and erosion, to deposition as a sediment followed by burial, metamorphism, and remobilization as a magma, whereupon it is transferred once more to the outer part of the crust. Different chemical elements have different pathways through the crust and through environments at the Earth's surface which may be referred to as *geochemical cycles*.

One particular cog within the geological cycle may be designated the *cycle of erosion* (or *geographical cycle*), a concept which originated with the American geomorphologist W. M. Davis in 1899. Here the sequence of processes and landforms which exists between the tectonic uplift of an area and its reduction to a peneplain or erosion surface close to base level is believed to have occurred many times in certain regions. Each cycle has stages of youth, maturity, and old age recognizable from their characteristic landforms.

Stratigraphy reveals that throughout the Phanerozoic there has been an unceasing movement of the strandline of the sea: transgression has followed regression without cease. These oscillations can have cumulative effects over tens of millions of years, gradually pushing major spreads of sediment from the continental margin towards the cratonic heartland. They have been called 'Sloss sequences' after the geologist who first documented them in the USA; they are also known as *synthems* (Fig. 1). These synthems of sedimentation are revealed in North and South America, the Russian Platform, and Siberia. There is a range of small-, medium-, and large-scale unconformities preserved in each continent, and some of them can be matched from one land mass to another. The Carboniferous and Permian systems in particular show a closely similar pattern of transgression–regression cycles in most continents.

An American, Peter Vail, and his colleagues have identified several orders of these cycles in stratigraphy. The first-order cycles (*supercycles*) span hundreds of millions of years; second-order cycles (synthems) last tens of millions of years; third-order cycles span a few million years; fourth-order cycles (*mesothems*) range from 600 000 to 3 million years; the fifth-order cycles (the classic cyclothems of the Late Palaeo-

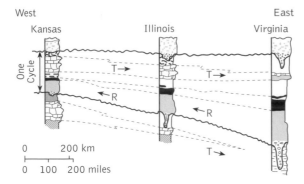

West   East

Kansas   Illinois   Virginia

Fig. 1. A Pennsylvanian (Upper Carboniferous) coal-bearing cycle in the eastern USA, showing characteristic lithologies and regional variations. In the east (Virginia) it is about 100 m thick and accumulated in the rapidly subsiding Appalachian basin; to the west it was spread across the relatively stable interior (cratonic interior). The transgressing sea and seashore moved progressively northwards and eastwards. During regressive intervals they withdrew in the opposite directions. T, Transgression; R, regression.

zoic) run for 5000 to 300 000 years; and the six-order cycles range from 50 000 to 130 000 years (Fig. 1). Smaller still are the *Milankovich cycles* of around 28 000 years.

First-order cycles have been postulated in the USA by A. G. Fischer and may be related to major plate movements. It has been suggested that the Earth has since late Proterozoic times gone through two supercycles 300–7 million years long, moving between 'ice house' and 'greenhouse' states.

Second-order cycles may be driven by cycles of fast and slow mantle convection which regulate sea-floor spreading on a scale of 30 to 60 million years. Periods of Phanerozoic fast spreading are correlated with those of high magnetic field activity, which is also related to the vigour of mantle convection. When mantle activity is reduced, spreading is slower and there is a drop in sea level; magnetic polarity can be frequently reversed. There is much argument about the reality of these cycles, seemingly because they reveal an overall eustatic nature but are subject to a great deal of local tectonic 'noise'.

Interpreting the past 200 million years of Earth history in terms of such cycles, Fischer and a colleague have suggested that cyclic changes occurred between two distinct conditions or episodes. In their simplest form, these are as follows. *Polytaxic episodes* are seen as characterized by rising sea level, higher and more uniform oceanic temperatures, high organic diversity, and continuous open-sea deposition; sea-floor oxygen values were low and heavier carbon isotope values prevailed in marine organic matter. *Oligotaxic episodes* have lower marine temperatures with steeper latitudinal and vertical temperature gradients. The climate is more equable. Sea level falls and there is a lowering of diversity in the pelagic communities. It is suggested that eight cyclic alternations, each lasting about 32 million years, can be discerned in the record as far back as the Triassic period. Today we are within an oligotaxic phase.

This model has aroused both interest and detailed criticism, and it remains to be seen how well it can be applied to pre-Mesozoic stratigraphy. There are very many variables to be considered, and more data are needed to complete the assessment.

Eustatic sea-level changes of less than a few million years duration are usually held to be the consequences of changes in global ice volume. Warmer and cooler periods of climate control the growth and melting of the ice and hence sea-level rise and fall. If the present ice caps were to melt the oceans would rise by some 40 to 50 m: enough to bring about changes comparable to those of the late Palaeozoic, though not of the Jurassic and Cretaceous, when no ice caps existed.

The glacial-control idea has been advanced to explain the origin of the numerous cycles of sedimentation that are present in the Late Palaeozoic of the Appalachian and Mid-West areas of the USA. These cyclothems begin in the late Mississippian (Early Carboniferous) and persist up to the Early Permian. In all, there are more than fifty cycles. The lowest few consist of simple triplets of sandstone, shale, and limestone, a few metres thick, but remarkably uniform over hundreds of square kilometres. The clastic material was brought into the region by numerous delta streams flowing from south-eastern Canada; the carbonate was deposited in an open clear sea with an abundance of living organisms. In the Pennsylvanian (Late Carboniferous) times that followed, the cyclothems became thicker and more variable; they also contain persistent seams of coal and black shale. Each of these cyclothems rests upon an erosion surface and begins with coarse conglomeratic and cross-bedded sandstone upon which rests fossil soil or underclay, and then a seam of coal. There are many local variations, and filled channels that cut through these strata indicate local erosional events. The full extent of the Pennsylvanian cyclothems seems to have reached well into the area of the present Appalachian Mountains and west beyond the present Mississippi River. These cyclothems are known in great detail from the evidence gathered from mines and boreholes. They are seen to be dominated by sandy strata in the east and by carbonates and fine clastic sediments to the west. Their formation appears to be controlled by two influences: rapid transgressions and regressions superimposed upon a more intermediate-term warping of the craton (stable region). The debate about the cause of the short-term cycles continues, but the glacial sea-level control hypothesis is generally accepted as the basic mode. This control, emanating from the Gondwanaland glaciations, also affected Carboniferous sedimentation in Eurasia. Within the paralic (coastal swamp) coal measures there are cycles similar to those in North America, with coals extending over thousands of square kilometres and rare marine limestone and shale bands similarly widespread. Here the Earth movements of the developing Variscan orogeny imposed variations on the rates and areas of sedimentation.

Almost the smallest and briefest of cycles might also be climatically controlled by means of the Milankovich cycle

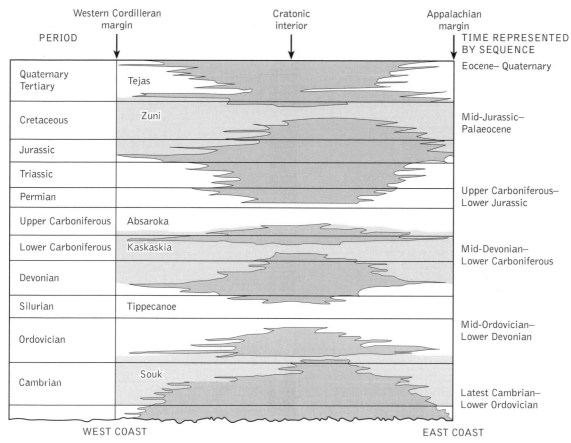

PERIOD — Western Cordilleran margin — Cratonic interior — Appalachian margin — TIME REPRESENTED BY SEQUENCE

PERIOD		
Quaternary Tertiary	Tejas	Eocene– Quaternary
Cretaceous	Zuni	Mid-Jurassic– Palaeocene
Jurassic		
Triassic		
Permian		Upper Carboniferous– Lower Jurassic
Upper Carboniferous	Absaroka	
Lower Carboniferous	Kaskaskia	Mid-Devonian– Lower Carboniferous
Devonian		
Silurian	Tippecanoe	
Ordovician		Mid-Ordovician– Lower Devonian
Cambrian	Souk	Latest Cambrian– Lower Ordovician

WEST COAST EAST COAST

Fig. 2. Major unconformity-bounded sequences (synthems) have been distinguished in North America as reflections of global cycles of sea-level rise and fall. They were first named, after native American tribes, by L. L. Sloss. Each cycle rests upon an unconformity which becomes progressively younger towards the continental interior (left in figure). The closely stippled area represents significant gaps in the record and it is apparent that the cycles are not of the same duration, though the rates of movement of sea-level change do not differ greatly with time. Regressional movement is thought to be generally somewhat faster than transgressional.

mechanism. This relates to the Earth's orbital geometry and its control over the amount of the Sun's energy that reaches the Earth's surface. Sedimentary cycles from many parts of the stratigraphic column as far back as the Devonian have been attributed to the Milankovich mechanism. Detailed examination of the Chalk in western Europe, and of its equivalent formations in the USA, reveals great numbers of thin beds of alternating higher and lower carbonate content. They are thought to have been deposited under the climatic influence of Milankovich cycles. Jurassic rocks also show an abundance of similar small cyclic units in Europe.

Even finer cyclic or rhythmic alternations of sediment types may be found in great numbers locally. Laminae about or less than a millimetre thick are the norm. Where deposition was in periglacial lakes the laminae (*varves*) are paired, finer and coarser, and represent the annual deposition of the winter and summer seasons. Comparable laminations of carbonate sediment are known in the Jurassic, especially in south-eastern France and in the Atlantic, where deposition was far from land and under the open ocean. The laminae occur in 'bundles' distinguished by their sizes and carbonate content, and are thought to include both annual and cyclic climatic units, perhaps influenced by solar (sun-spot) cycles.

The recognition of cyclic sedimentary sections is now commonplace and, increasingly, these are being subjected to spectral analysis. The purpose of this is to establish the regularity of the cycles and any patterns that they might show. A possible connection with climatic control may be indicated, and some workers have endeavoured to use the technique to reveal the actual time taken for the section to form. This becomes high-resolution stratigraphy at its most sophisticated.

D. L. DINELEY

Further reading

Butcher, S. S., Orians, G. H., *et al.* (1992) *Global biogeochemical cycles*. Academic Press, London.

Einsele, G., Ricken, W., and Seilacher, A. (eds) (1991) *Cycles and events in stratigraphy*. Springer-Verlag, Berlin.

Fischer, A. G. (1984) Two Phanerozoic supercycles. In Berggren W. A. and Van Couvering, J. A. (eds) *Catastrophes in Earth history*, pp. 97–104. National Academic Press, Washington, D.C.

Fischer, A. G. (1986) Climatic rhythms recorded in strata. *Annual Revues of Earth and Planetary Sciences*, **14**, 351–76.

Gould, S. J. (1987) *Time's arrow, time's cycle. Myth and metaphor in the discovery of geological time*. Harvard University Press, Boston, Mass.

Vail, P. R., Mitchum, R. M., Jr., and Thompson, S. III. (1977) Global cycles of relative changes of sea level. *American Association of Petroleum Geologists Memoir* **26**, 83–98.

cyclone The term 'cyclone' refers to large-scale rotating weather systems which rotate with positive *vorticity*. Two major types of cyclones are tropical cyclones and extra-tropical cyclones. The latter are also known as depressions. The mechanisms of formation and the structures of these two forms are so different that they should be considered separately. It is unfortunate that they share the same name.

Tropical cyclones

Tropical cyclones are also known as hurricanes and typhoons. They develop over tropical oceans and can produce extremely heavy rainfall and devastating winds with sustained wind speeds sometimes in excess of 100 metres per second (m s⁻¹). Satellite pictures (Fig. 1) reveal a striking circular symmetry in tropical cyclones with a small (50 km diameter) cloud-free 'eye'. The preferred regions for hurricane development are oceans where the winds are light, the humidity is high, and the surface water temperature is high (usually over 26 °C) over an extensive area. Since these conditions exist in some places for only part of the year, there is a tropical cyclone 'season', from June to November, in the tropical north Atlantic and Pacific.

The conditions required for tropical cyclones are also suitable conditions for thunderstorms, deep convective clouds with strong updraughts. Thunderstorms can be organized into a tropical cyclone in the presence of low-level convergence. The winds converging in one particular region increase the large-scale rotation in an anti clockwise sense in the northern hemisphere and in a clockwise sense in the southern hemisphere. Because this rotation is an important factor in the development of tropical cyclones, and because no rotation is imparted at the Equator, no development of tropical cyclones is found within about 5 degrees of the Equator.

As the air converges, the thunderstorms become more organized and closer together. Huge amounts of water evaporating from the warm ocean surface are carried aloft in the bands of thunderstorms. As the air rises it cools, and the water condenses releasing latent heat. The latent heat released

greatly enhances the buoyancy of the air, producing even stronger updraughts, which in turn draw in more converging air at the base. There is positive feedback as more warm moist air being drawn into the base of the clouds produces even stronger updraughts. At the tropopause the air spreads out in bands moving away from the centre of the cyclone. When this divergence of air at high levels exceeds the convergence of air at low levels, the surface pressure drops, forming a low-pressure centre around which the air circulates and converges, bringing in even more warm moist air to feed the cyclone.

The clear 'eye' at the centre of the cyclone is formed by air subsiding as it emerges from the intense eye-wall thunderstorms. This small region of subsiding air warms as it descends through the atmosphere, inhibiting the growth of clouds and producing a clear eye. The fiercest storms are those in the eye-wall cloud with the highest rainfall rate, up to 250 mm per hour, and the strongest winds, up to 100 ms⁻¹.

Tropical cyclones move somewhat erratically, making it difficult to predict their exact course. As they move to areas where the sea is not so warm, their energy source is removed and they dissipate.

Extra-tropical cyclones

Extra-tropical cyclones are the middle latitude tropospheric circulation systems also known as depressions. The life cycle of a depression is often described by the polar front approach, in which the depression is seen as a disturbance which grows and modifies the front as it develops. Another approach is used here to illustrate the three-dimensional structure and the

Fig. 1. Satellite image of a vigorous hurricane with a cloud-free 'eye' at the centre.

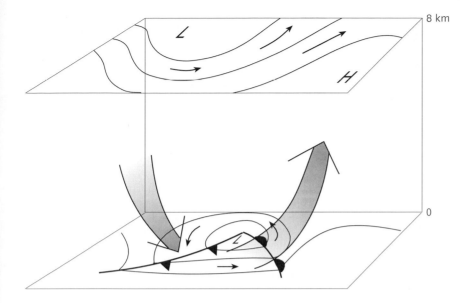

8 km

0

Fig. 2. The structure of a developing depression at the surface and in the upper troposphere. L, low pressure, H, high pressure.

development mechanism by which an extra-tropical cyclone develops. Fig. 2 shows a wave depression at its most vigorous stage of development. The surface weather map shows an open warm sector with a deepening low-pressure centre. The low-pressure centre is situated below a region in the upper troposphere where a wave or trough of low pressure is lying slightly behind the surface low. This is exactly the configuration required to enable a low-pressure centre to deepen.

At the upper level there is a region of divergence. The air moving around the trough moves more slowly at the curved section and faster as the flow becomes straighter. The result of this is that some air is removed from the column of air above the low-pressure centre. However, near the surface the flow is not exactly around the centre of low pressure but slightly towards the centre. This is a region of convergence where air is added to the column. At this stage in the development of the depression the divergence at the upper level is greater than the low-level convergence, and so the net effect is to remove air from the air column and thus reduce the surface pressure. At an earlier stage the development was initiated when the upper level trough, which is moving faster than the low-level system, approached the region of the polar front and the initial divergence produced a low-pressure centre.

The circulation of the wave depression is in an anticlockwise sense. (In the southern hemisphere the circulation would be clockwise, but the cold air would be to the south and so the mechanism would be the same.) The warm air at low levels in the warm sector is lifted above the warm front. This applies not only to air at the surface, but to all the air through a substantial depth. At the same time the cold air moving southwards behind the low-pressure centre is losing height. The net effect of warm air rising and cold air sinking is to decrease the 'centre of gravity' of the system because the cold air is more dense than the warm air. By lowering the centre of gravity, some potential energy is removed from the system and converted to kinetic energy—the energy of motion. (When we release an object from a height and let it fall, we are converting potential energy into kinetic energy). In the wave depression the kinetic energy is manifested by the strength of the winds in the circulating system. Because it is these winds that are moving the warm air upwards and the cold air downwards, the process accelerates and feeds on itself. This unstable situation, known as baroclinic instability, continues until the warm air is lifted from the surface into the upper troposphere.

By moving warm, less dense, air into the column ahead of the upper trough the pressure ahead of the upper trough is reduced. Conversely, the cold air introduced behind the trough increases the pressure: so the trough minimum is moved from east to west. This means that the region of divergence is now no longer above the centre of low pressure at the surface. In fact, there is neither divergence nor convergence at upper levels, but convergence remains at the surface. The effect is thus to increase the surface pressure and complete the last phase in the life of the depression.   CHARLES N. DUNCAN

Further reading

Ahrens, C. D. (1994) *Meteorology today.* West Publishing Co., St. Paul, Minnesota.

Palmen, E. and Newton C. W. (1969) *Atmospheric circulation systems.* Academic Press, London.

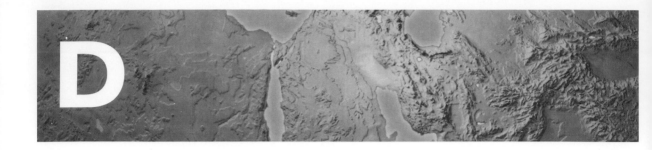

D

Daly, Reginald Aldworth (1871–1957)

Originally trained as a mathematician, R. A. Daly has been recognized as one of the first to see the need for geology to be put on a more quantitative and experimental basis. Born in Canada, Daly followed a career centred on Harvard University, where he was a professor in the Department of Geology from 1912 until 1942. The range of his research was wide. His early work was in petrology, where he studied a variety of topics relating to the composition and origin of igneous rocks. In 1910 he put forward his glacial control hypothesis to account for the origin of coral reefs and atolls. (This hypothesis is not today regarded as generally valid.) In the 1920s he was among the few scientists in the USA who supported the idea of continental drift. The mechanism that he envisaged was one in which continents resting on a molten mantle slid downwards under the influence of gravity. In 1932, together with physicists and astronomers at Harvard, he instituted a research programme on problems concerned with the interior of the Earth.

Daly was among the first to realize that measuring the response of the Earth's crust to loading by continental ice sheets could be used to obtain information about the crust and the underlying mantle. He was also the first to suggest (in 1936) that the submarine canyons that form remarkable features of the continental shelves could be have been eroded by turbidity currents of the kind seen in lakes.   D. L. DINELEY

dams

A dam is an engineering structure built to block a valley or other low-lying area for the purpose of storing water. The water may be required for a variety of purposes: to regulate the flow of a river, for irrigation, or for the generation of hydroelectric power.

Dams in narrow valleys are usually made of concrete; those in wide valleys are usually built with rock or soil obtained close to the construction site; these are called *earth dams*. *Concrete dams* are constructed specifically to suit the geological conditions at the site. They must resist the forces imposed by the stored water, which tend to push the dam downstream. This tendency is enhanced by upward forces on the structure resulting from the difference in water levels between the upstream (heel) side of the dam and the downstream (toe) side. These forces can be resisted by making a concrete dam very heavy (a *mass concrete* or *gravity dam*), by supporting the downstream face of the dam with buttresses, or by curving the dam in the horizontal plane (an *arch dam*). The effect of the curvature is to direct some of the force laterally, where it can be resisted by the rocks of the valley sides. The most elaborate concrete dams are curved in both horizontal and vertical planes (*cupola dams*). An *earth dam* is essentially an impermeable membrane (made of anything from clay to steel plate) held in place and protected from erosion by shoulders of rock or soil, or both, obtained locally.   J. WEST

Dana, James Dwight (1813–95)

For several generations of geologists the name of J. D. Dana was associated with the concept of the geosyncline, an elongated downwarp in the Earth's crust in which a great volume of sediments accumulated. (In his 1873 paper, Dana in fact used the term 'geosynclinal' rather than 'geosyncline'.) He regarded the accumulation of the sediments as a consequence rather than as a cause of the downwarp, which he ascribed to contraction of the Earth's crust as a result of cooling. Geosynclinal interpretations prevailed until the 1960s, but have since been replaced by the concepts of plate tectonics.

More than half of Dana's published work is on geological subjects, but he was a mineralogist and a zoologist as well as a geologist. His first published work was on Vesuvius, and he later studied the volcanism of the Hawaiian islands. From 1838 until 1842 he was a member of a U.S. Government expedition that circumnavigated the globe. After that he was at Yale University. In 1846 he maintained that the continents and oceans had never changed places and that the Earth's general framework was essentially stable. In the same year he proposed that all the Moon's craters had been formed by internal activity rather than by impact. Neither view has stood the test of time—but that is in the nature of science. Dana also published work on coral reefs and coral islands, the Taconic orogeny, and metamorphism.   D. L. DINELEY

Darwin, Charles (1809–82)

Although he is known worldwide as a biologist, Charles Robert Darwin had a keen interest in geology throughout his life. His contributions to the science were not inconsiderable and his work on evolution has had a profound influence, not only on palaeontology but also on much of Western thought. He went up to read medicine at the University of Edinburgh in 1825. It was not a success, and 3 years later he entered Christ's College, Cambridge. He was, however, not an outstanding academic success there either, and on modern criteria would have been unlikely to have obtained a research grant. Among his friends at Cambridge was Adam Sedgwick, the Woodwardian Professor of Geology.

Darwin was attracted to join the voyage of HMS *Beagle* as scientist; its voyage from 1831 to 1836 took him to many countries of the South Atlantic and the Pacific. His geological observations there were astute and detailed in a number of different fields, even including an accurate analysis of axial-plane cleavage; he collected a wealth of geological specimens. Much of this was figured in his publications on return to Britain. In *The structure and distribution of coral reefs* he advanced a theory of reef formation that has remained largely intact ever since.

Darwin's time in southern South America allowed him to collect fossil mammals and molluscs which demonstrated the identity of the zoological province there throughout Cenozoic time. He also made volcanological observations in the Andes, and he noted seismic events and structural features in the terrains of the Pacific coast. There is evidence in his unpublished writings that he was intrigued by the distribution of continents and of the orogenic features exhibited by many of their margins. Thoughts of some form of continental drift or mobility may have crossed his mind.

On returning to Britain he seems to have restricted his palaeontological writing to monographic work on cirripedes (barnacles). He was in almost continual poor health, and when his writings were the source of controversy he refrained from taking part in the arguments. The *Origin of species*, begun in 1856, was half completed when Darwin received a manuscript from Alfred Russel Wallace which he found to duplicate his own views and findings. His geologist friend Charles Lyell was instrumental in arranging a joint presentation by Darwin and Wallace to the Linnean Society to the satisfaction of all parties. The book (1250 copies) sold out on the day of publication.

The ensuing controversy about the origin of species and, further, the origins of mankind, was long lasting and intense, but Darwin's work was championed by, among others, Thomas Henry Huxley ('Darwin's bulldog'). It provided an immense stimulus to western scientific thought. Darwin's reputation as a biologist greatly overshadows his renown as a geologist, but recent research shows that he had ideas that seem to foreshadow modern theories of tectonics and orogeny. He died in April 1882, and was buried in Westminster Abbey.   D. L. DINELEY

Darwinism *see* EVOLUTION (DARWINISM AND NEO-DARWINISM)

dating rocks absolutely The term 'absolute age' has been used to mean a geological age (either of a mineral, a rock or an event) measured in years, as distinct from an age that is merely relative, that is, expressed in relation to other formations or events. In practice, 'absolute age' has generally been a synonym for 'radiometric age', although it includes other methods of dating rocks (varve analysis and tree-ring dating or dendrochronology).

The application of stratigraphical principles in the early nineteenth century made it possible to establish rock successions and to correlate beds in different places by means of the fossils contained in them. In this way a stratigraphical column was gradually built up, but the relationships thus established were only relative: it was not possible to determine the age of any particular formation in absolute terms—that is, in numbers of years.

By the mid-nineteenth century most geologists accepted that the total time represented by the rock succession was immense ('deep time'), but it was still not possible to assign an absolute to any event in the Earth's history.

Attempts were made in the second half of the nineteenth century to obtain absolute ages for at least part of the geological column. Various methods were used, including rates of sedimentation or denudation, the salinity of the oceans, and glacial theory. The results were inconclusive; too many assumptions had to be made and there were too many uncertainties in the calculations.

In 1863 Lord Kelvin, an eminent physicist of the time, calculated the age of the Earth by assuming that it was originally in a molten state and had since cooled gradually. (A calculation of this kind had in fact been attempted in the eighteenth century by the Comte de Buffon.) Kelvin's figures were in the view of many geologists inadequate, and a long debate followed. Kelvin was eventually shown to have been wrong because his assumptions were incorrect.

Radiometric dating

In the early years of the twentieth century, radiometric age determinations (strictly speaking, isotopic age determinations) pioneered by B. B. Boltwood in the USA and Arthur Holmes in Britain, finally gave useful ages for rocks and minerals. It was not, however, until the 1950s, when better analytical techniques using mass spectrometers were available, that radiometric dating came into general use.

We now know that the Earth is about 4600 million years (Ma) old. The oldest known rocks are dated at about 3800 Ma; zircons in ancient sediments in Western Australia have yielded an average age of 4200 Ma, which must represent the age of a portion of continental crust that was destroyed long ago. No less importantly, we now have an abundance of radiometric dates for rocks of all ages.

The term 'absolute age' is open to question. Arthur Holmes was indeed one of those who objected to it as long ago as the 1960s, describing it as a 'meaningless term'. In certain circumstances radiometric ages can be regarded as absolute, but, apart from the uncertainties that are inherent in any laboratory method, the ages obtained can be ambiguous if there is the possibility that thermal events may have reset the 'radiometric clock'. 'Radiometric age' is accordingly considered by many to be a more suitable term.

Radiometric age determinations can be applied to a wide range of rocks over time-spans ranging from thousands of years to the oldest rocks in the Solar System. Other methods of arriving at absolute ages are of much more limited extent and are applicable only to the youngest rocks. Analysis of varves (banded sediments in lakes that show annual variation) can be used to date Pleistocene sediments. Tree-ring dating (dendrochronology) is restricted to ages up to about 8000 years.

D. L. DINELEY

Further reading

Hawkesworth, C. J. (1992) Geological time. In Brown, G., Hawkesworth, C., and Wilson C. (eds), *Understanding the Earth* (2nd edn), pp. 132–44. Cambridge University Press.

Lamb, S. and Sington, D. (1998) *Earth story: the shaping of our world*, pp. 11–31. BBC Books, London.

See also: *age and early history of the Earth and Solar System; geological time; Holmes, A.; isotopic dating; stratigraphy*

Davis, W. M. (1850–1934) and landscape evolution

The influence of William Morris Davis on geomorphology in the first half of the twentieth century was greater than that of any other geomorphologist; by the end of the second half, Davisian geomorphology was regarded with contempt by many. Seldom has a scientist risen so high in his lifetime to have fallen so low soon thereafter.

Davis was born in 1850 and began his academic career as an instructor at Harvard University in 1878. In 1882, with the threatened termination of that position, he began to publish with vigour. By 1881 he had published three papers. By the end of 1883 he had another twenty to his credit. This dramatic increase is an early example of 'publish or perish'. He was promoted to Assistant Professor of Physical Geography in 1885. In 1886 he published 27 more papers. His famous 'geographic cycle', first presented at the meetings of the American Association for the Advancement of Science in 1884, was published in full in 1899.

Davis's 'geographic cycle of erosion' was the first big modern idea on the formation of landscapes. He recognized that landscapes are the result of structure, process, and stage, but he emphasized only stage. He proposed that landscapes evolve through a series of developmental stages as rivers erode into rapidly uplifted terrain. He envisaged three main stages of sequential denudation, which he named 'youth', 'maturity', and 'old age'. Davis thought that an initial planar surface was uplifted rapidly so that erosion had little time to act upon it until uplift was complete. He supposed that in youth the rivers eroded rapidly down to base level and then in maturity eroded laterally. Finally, in old age, as the hillslope angles

declined, the surface would be eroded down to a rolling peneplain ('almost a plain'), upon which resistant rocks could form an erosional residual that he termed a 'monadnock', after the mountain of that name in New Hampshire. Renewed rapid uplift would start downward erosion again, in the process of rejuvenation, to complete the cycle. This highly theoretical scheme of landscape development seemed so logical, and so self-evident when expressed in his clear and straightforward fashion that it became immensely popular in the first half of the twentieth century.

Davis, we now know, relied upon a number of basic but flawed assumptions in constructing his erosion cycle. For example, certain landforms were associated with his normal cycle, rather than the climate change that in fact produced them. Little attention was paid to the details of the underlying rocks. He associated renewed downcutting with changes in base level caused only by uplift, rather than with a multiplicity of other causes such as changes in climate, in load, in sea-level, and gulley-head cutting. Furthermore, we now know that uplift occurs with wide temporal and spatial patterns. No peneplains are known anywhere in the world. Widespread surfaces of erosion, or unconformities, are common in the stratigraphical record and indicate that landscapes must be eroded down in time, but perhaps not in the way that Davis thought. A major problem with the Davis legacy was that denudation chronologies constructed using his method had little relation to the real geological history of a region. Many people even failed to differentiate between erosional and depositional landforms.

In response to criticisms, Davis later reformulated his ideas into humid, arid, glacial, and marine cycles of erosion. Other scientists established cycles of karst formation, periglacial activity, and slope failures. His presentations using block diagrams and drawings were profound influences that are with us still. Today, however, few serious geomorphologists use the theoretical ideas of Davis to explain the evolution of landforms. An unfortunate result has been the neglected study of the general long-term evolution of landscapes, a field of research that is only now starting to re-emerge (see **landscape evolution**).

Davis was promoted to full professor at Harvard in 1889, and in 1890 he was appointed to the Chair of Physical Geography. Throughout the 1890s, he constantly advanced the cause of geography. He defined physical geography as the physical, rational, and explanatory study of the present-day features of the Earth that enter into relation between all life, particularly humanity. Physical geology, on the other hand, he saw as concerned with the processes of the past. He defined geology as composed of past geographies, as well as a study of the past in the light of the present, whereas geography was a study of the present in the light of the past. He championed the development of human geography, and maintained that it would benefit from the same explanatory treatment that he advocated for landforms. In 1904, he founded the Association of American Geographers, which is now one of the chief organizations of geographers in the world. Ironically, human geography did indeed rise in importance as a result of his efforts, ultimately eclipsing his own brand of man–land relationships and the science of physiography that he advocated but hardly ever practised. That subject is in fact no longer even taught, and has been largely replaced by process geomorphology.

Davis tirelessly promoted his erosion cycle and conducted a sustained campaign to advertise his theories. He travelled widely throughout Asia, Africa, and Europe, and lectured at major universities in Europe and America. He was promoted to the Sturgis-Hooper Chair of Geology at Harvard University in 1899, a post that he held until his resignation in 1912. After his retirement he lectured widely at universities, predominantly in the western United States, and continued his research and writing. He then began to modify his views, to do more fieldwork, and to become more inductive in his approaches. He never changed his attitude to his cycle of erosion, but he did say that it was only a mental aid or just one model. He developed and refined his simple cycle until it became highly complicated, and more elastic. He became far less dogmatic in his later years, and even finally accepted the competing theory of parallel retreat of slopes. His most positive influence on students was arguably during his 70s and 80s. As an octogenarian, he still managed to write 33 papers on such diverse topics as the origin of caves, desert landforms, marine terraces, and the classification of lakes. His total lifetime production of 615 papers was impressive in its breadth and scope. He died in 1934, only a few days before his 84th birthday.

W. M. Davis so dominated his field that progress in geomorphology was retarded by his efforts. While he was alive, and for several decades thereafter, one either taught Davisian models of landform evolution, or one left the field; no alternatives were available. The Davisian legacy is with us still; its greater points have largely been subsumed by younger, better, and more explanatory models, but its influence is still present in the literature, if greatly modified.   JOHN F. SHRODER, JR.

debris flows
A debris flow is a gravity-induced, rapidly moving body of sediment particles, water, and/or air. Such flows are an intermediate stage between landsliding and water flooding, but debris flows originate when poorly sorted rock and soil debris are mobilized from hillslopes and channels by the addition of moisture. These conditions exist in a variety of settings, including mountainous areas in arid and semi-arid, arctic, and humid regions. The exact mechanism by which landslides change into debris flows is uncertain, but the transformation from a solid, rigid soil mass to a viscous fluid may occur as a landslide changes from a close-packed and dense soil structure to an open-packed structure, accompanied by an increase in pore volume. Incorporation of moisture then transforms the sliding mass into a flowing, viscous fluid.

Debris flow resemble wet concrete, and when it flows usually follows pre-existing drainage ways, but debris flows can also move down hillslopes and across unobstructed alluvial fans in almost any direction. Flows may appear as a series of waves or surges with periods ranging from a few seconds to several hours. Flow velocities can be very fast, from 1 to 20 m per second, and are controlled by the characteristics of the sediment in the flow (size, shape, sorting) and by the topography (channel slope, width, shape, and sinuosity).

Physical properties of debris flows vary widely. The density of clear water is 1.0 g ml^{-1}, but during floods, when large amounts of sediment are being moved, the density of stream flow is typically 1.01 to 1.3 g ml^{-1}. Measured densities of debris flows range from about 1.40 to 2.53 g cm^3, and of common rocks from about 2.7 to 3.0 gm/cm^3. Large and small pieces of rock can thus nearly float in debris flows because they are so similar in density. The high density of debris flows imparts an internal strength to the material that must be overcome before the material will begin to flow. This is very different from sediment-free water, which has no internal strength. Debris flows have been known to transport boulders weighing 30–40 tonnes for tens of kilometres, and during the passage of flows, witnesses have reported ground shaking and loud roaring and rumbling noises. Debris flows are capable of exerting enormous impact forces on objects in their path. Buildings have been destroyed, and large trees snapped off.

Debris flows can flow for many kilometres beyond their source areas, and tend to stop upon reaching areas with relatively low gradients or areas of decreased confinement, such as alluvial fans at the mouths of small watersheds or canyons. At Mount St. Helens in 1980, when glacial ice within a huge debris avalanche that accompanied the major eruption melted it mobilized over 150 tonnes of sediment into an extraordinary debris flow that moved over 60 kilometres from the volcano. This flow deposited so much sediment in the Columbia River that shipping lanes were closed until dredging eventually cleared the channels. Debris-flow deposits from side tributaries are the origin of most of the large rapids in deep canyons such as the Grand Canyon in the United States. The exact mechanism by which debris flows stop flowing is uncertain. Lateral spreading might result in the thickness or depth of a flow to decrease below the minimum required for flow movement to continue; the escape of pore fluids such as water, clay, and fine silt might result in an increase in internal friction. Because debris flows have a finite strength, their deposits have unique characteristics. At the distal and marginal edges of flows, lobes and levées with steep fronts and surface concentrations of large boulders commonly occur. These landforms are characteristic of debris flows and can be preserved for many years. In cross-section, the deposits consist of unsorted pebbles, cobbles, and boulders in a matrix of fine-grained debris. Bedding, characteristic of river-laid sediments, is absent in debris-flow deposits.

Damage and loss of life from debris flows can be mitigated by four general kinds of remedial measures: (1) identification and avoidance; (2) control of grading, clearing, and drainage; (3) protective structures; and (4) warnings and evacuations.

Because of their elevation above floodplains, alluvial fans have long been favoured sites for development. Unfortunately, mitigating procedures and identification of risk areas for debris flows are poorly developed in comparison with those for water floods. Dangerous areas cannot be identified systematically and consistently, nor can reliable data be obtained on frequency of inundation. As a general rule, the bottoms and mouths of small, steep ravines that originate in steep, hilly or mountainous terrain (especially volcanic areas), or in areas of historic and prehistoric debris flows, should be considered as potential debris flow areas and avoided.

It is generally believed that erosion by debris flows can be reduced by strict controls of land use, grading, and drainage. On artificial slopes, this could include limiting the height of slopes, properly compacting fills, and ensuring that drainage is channeled away from potential source areas. Devegetation by wildfires or overgrazing in source areas generally increases the likelihood of debris flows.

The construction of protective barriers to stop, slow, or divert debris flows may be necessary if avoidance of hazardous areas is not possible. Channelling of debris flows is usually ineffective because channels can quickly become choked with sediment, allowing subsequent surges to overflow the channel and flow in different directions. Closely spaced trees can be quite effective in stopping boulders and other large debris. Structural fences of steel and reinforced concrete, steel cable nets, debris fences, and sediment barriers can be effective in stopping or separating large boulders from debris flows. Large reservoirs to trap and store debris upstream of developments have been successful in many locations. Because debris flows frequently originate from sudden landslides in remote locations and travel at high speeds, it is difficult to provide direct warnings. Ground shaking and loud noises may provide a short warning, and sensors and tripwires installed in upstream locations can detect the passage of debris flows and enable alerts to be issued. Longer-term warnings may consist of identification of minimum precipitation thresholds for slope failures in debris-flow prone areas. Despite the expenditure of large sums of money on protective and warning devices, debris flows will probably continue to reap a large toll in property and lives throughout the world.

JOHN E. COSTA

Further reading

Johnson, A. M. with contributions by J. R. Rodine (1984) Debris flow. In Brunsden, D. and Prior, D. B. (eds) *Slope instability* pp. 257–361. John Wiley and Sons, Chichester.

Pierson, T. C. and Costa, J. E. (1987) A rheologic classification of subaerial sediment–water flows. In Costa, J. E. and Wieczorek, G. F. (eds) *Debris flows/avalanches: process, recognition, and mitigation*, pp. 1–12. Geological Society of America, Reviews in Engineering Geology, vol. VII.

deep-sea geomorphology
Deep-sea geomorphology entails studying the shape of the deep-sea floor: measuring it, mapping it, and explaining it. To many research workers, the

term 'deep-sea' means anything deeper than the continental shelves, the fringes of the continents that were exposed as dry land by the 120-m drop in sea level that took place during Pleistocene glacial periods. Here, however, we are concerned principally with the geomorphology of oceanic crust, the material that accretes at oceanic spreading centres and is modified by tectonic and volcanic processes, and by sedimentation, on the floor of the ocean basins. Our definition all but ignores, therefore, the submerged sides of the continents, that is, the continental slopes, even though they commonly plunge to depths of several kilometres, and have features such as submarine canyons that have been targets of much geomorphological research.

Oceanic crust has a shorter, simpler history than continental crust, and at the largest scale, which is concerned with regional differences in elevation, its geomorphology is better understood. The oceanic spreading centres at which the crust originates have relatively uniform depths (mostly at 2500–3500 m below sea level), and cooling of the lithosphere as the crust spreads away causes it to subside at a predictable rate, which is proportional to the square root of its age. This tendency towards steadily increasing depth with increasing age and increasing distance from the spreading centre, the phenomenon that creates mid-ocean ridges, is counteracted by the disposition of sediments, causing shoaling of the sea floor. The rate at which sediment builds up is much more variable than the rate of thermal subsidence, but because the continents are the principal sources of sediment, the ocean floor commonly shoals towards the continental margins, as the upper surface of a thickening sediment wedge. Striking exceptions to this pattern occur where active subduction zones drag oceanic crust down into marginal trenches that may be twice as deep as the adjacent basin. Even in some of these situations, accumulation of land-derived sediment may be fast enough to overcome the rapid tectonic sinking of the crust, so that an uphill slope towards the continent is maintained (as at the Cascadia margin off the north-western United States).

The gross patterns of regional depth variations and the presence of mid-ocean ridges and marginal trenches were established during the nineteenth century by widely spaced wire-log soundings. They could not, however, be explained until the development in the 1960s of the theory of plate tectonics and techniques for measuring sediment thickness. Pursuing deep-sea geomorphology at the same spatial scales as those of most subaerial studies, and with the same goals of defining and explaining the relief-forming processes, also had to wait for technological developments of the mid-twentieth century.

Measuring the relief of the ocean floor, expressed as the varying depth below sea level, is the chore known as bathymetry. For most of the continental surface, government agencies have published highly accurate topographical maps that can serve as the foundation, or even the sole data source, for morphological studies. When governments have conducted detailed bathymetric surveys of the ocean floor, however, the products have often been classified as military secrets. Fortunately, in recent years civilian scientists have acquired new tools that produce bathymetric maps comparable in accuracy and resolution to the best 1:100 000 topographic sheets available to continental geologists, at survey rates of 5000 to 10 000 km^2/day. These new tools are multibeam sonars. They are mounted in the hulls of survey ships, and take echosoundings every few seconds to measure the depth of the seafloor, not only directly beneath the ship, but at many (45–75) points on either side of the ship's track. Shipboard computers log the location and depth of every point. These digital data are immediately machine contoured to yield a bathymetric swath 10–25 km wide along the ship's track. The data are also amenable to statistical manipulation, providing morphometric indices such as roughness and mean slope angle. Sonar surveying is the counterpart of photogrammetric mapping of the land surface. In the deep sea, because of the poor transmission of light in sea water, photography (or visual observation from submersibles) is effective only for close-up examination of tiny areas.

The first step in converting bathymetry into geomorphology is establishing whether the sea-floor relief is caused by roughness of the igneous crust or by a variable thickness of overlying sediment. This step also helps to determine whether interpretation of the relief requires the techniques of structural geomorphology (which is concerned with how tectonic deformation and volcanism create and modify sea-floor relief) or of sedimentary geomorphology (which is concerned with the ways in which different types of ocean currents have shaped the sea floor). Bathymetric sonars which record the strengths of sea-floor echoes as well as their ranges can identify suspected areas of bare-rock 'outcrop'. Many of these, on closer examination, prove to be largely smothered with the rockfall talus that is readily produced by failure of the weak, porous lavas at the top of the igneous crust. Where the igneous crust is buried by sediments and sedimentary rock, its relief can be imaged by acoustic or seismic profiling, using low-frequency sound that partially penetrates the sediment and is reflected from the igneous basement, as well as from the sea floor. The relief of the igneous crust, whether exposed or buried, is created almost exclusively by tectonic and volcanic processes; erosion of basement rocks, which rapidly modifies and degrades subaerial tectonic and volcanic landforms, is of minor significance in the deep sea, where the principal modifying agent is deposition of sediment. In some instances, very even sediment accumulation causes sea-floor relief to mimic that of the underlying basement, so that the morphology is inherited from pre-existing (and buried) tectonic landforms. At the other extreme, deposition of sediment from deep-sea currents can be highly localized, producing a new relief unrelated to that of the underlying basement rocks. Huge areas, the so-called 'abyssal hill terrains' that cover more than 50 per cent of the ocean floor, lie between these two extremes: the tectonic basement relief still influences the shape of the sea floor, but has been modified; it has commonly, for example, been gradually smoothed by preferential deposition in structural depressions.

Studies of the structural processes responsible for sea floor topography have been concentrated at plate boundary zones, the principal sites of intense and frequent tectonic and volcanic activity. The crests of mid-ocean ridges, where structures are created that may have persistent effects on sea-floor relief for many millions of years, are especially important. Here, along axes of sea-floor spreading, volcanic intrusion and extrusion associated with extensional faulting and tectonic tilting produce volcanic rift zones. Where a high rate of intrusion maintains a permanent lens of magma in the crust, the rift zones occupy the crest of axial ridges 2 to 10 km wide; elsewhere (generally at slower spreading axes) the rift zones occupy the floors of axial rift valleys. At the margins of the zones of crustal accretion a diminished rate of tectonic activity is dominated by continued extensional faulting in response to crustal stretching, producing closely spaced fault scarps parallel to the spreading axis. This fault-block terrain, which subsequently becomes blanketed with sediment, is the foundation of the corrugated 'abyssal hill' topography of much of the ocean floor. The other important elements of what has been called the 'plate fabric' are cross-cutting structural lineations created at lateral offsets of spreading axes. The larger examples of these offsets have transform faults parallel to the relative motion of the separating oceanic plates. Crust that spreads past them acquires lineated relief that is orthogonal to the abyssal hills and forms so-called 'fracture zones' on the flanks of mid-ocean ridges. The azimuths of abyssal hills and fracture zones preserve a record of changes in the direction of relative plate motion.

Where oceanic crust approaches a subduction plate boundary it is also subject to intense relief-creating tectonic activity. Stretching of the upper surface of the lithosphere as it bends down into the subduction zone causes renewed extensional faulting of the igneous crust. This in turn either rejuvenates the abyssal-hill faults inherited from the margins of the crustal accretion zone, or, if those faults are unsuitably oriented, creates a new set of fault scarps parallel to the trench. Large tectonic landforms can form rapidly in these settings; for example, some rift valleys that originated less than 10 000 years ago in western Pacific trenches are now more than 1000 m deep. In many trenches the upper crustal layers are unconsolidated sediments that respond to stretching by monoclinal folding rather than by faulting. More intense folding of these layers by the compressive stresses of plate convergence may occur in the trench axis, producing a characteristic morphology of arcuate fold ridges on top of a prism of sediment scraped off the subducting oceanic lithosphere.

Studies of the geomorphological effects of the sedimentary processes of deposition and erosion have also concentrated on the most active sites, those shaped by fast bottom currents. There have been few investigations of such morphologically significant but exceedingly slow processes as the downslope creep that contributes to the smoothing of abyssal hills. Direct study of turbidity currents, one of the most energetic and morphologically significant current types, has also proved difficult. Turbidity currents (downslope flows driven by the density difference between turbid and clear sea water) are generally episodic with long recurrence intervals, and can be catastrophic. Most of the information on their speeds, and on the shear stresses they exert on their beds, comes from modelling studies rather than from measurement *in situ*. Where turbidity currents are confined in downslope channels, their erosive effects, which maintain the channels, may be the most impressive. Some of the channels, with meandering and braided patterns resembling those of large alluvial rivers, extend for thousands of kilometres. Deposition from turbidity currents that spill out of their channels has built deep-sea fans ranging in area from a few square kilometres to the Bengal Fan's several million square kilometres. The distal parts of deep-sea fans merge into extraordinarily flat and even more extensive abyssal plains (with gradients of less than 1 in 1000). Obliteration of basement relief, replacing it with a much smoother surface (except at the steep, sometimes undercut, walls of channels) is a characteristic effect of turbidity-current activity.

Thermohaline bottom currents, which are part of the general circulation of the oceans, also entrain large amounts of sediment and shape the sea floor by maintaining fields of bedforms and depositing huge sediment drifts. They are amenable to direct measurement, and recording their velocities and near-bed patterns is an important step in process-oriented geomorphological studies. A problem, however, in examining the relationship between an observed landform and a measured current is that bottom currents are known to fluctuate over seasonal and much longer timescales (for example, many of them have been more intense during recent glacial periods). Thus, the present relief may not be in equilibrium with the present current regime. This is especially true of the largest depositional landforms, huge drifts of current-sculpted sediment like the Blake Outer Ridge of the western North Atlantic. This ridge, which is more than 2 km high and 500 km long, is the product of several million years of changing interactions between the deep Gulf Stream and southward-flowing currents.

Sometimes a change in these interactions and in the amount of sediment supplied to the bottom water causes the effect of bottom currents to change from depositional to erosional. Much of the surface of the Blake Outer Ridge, and of many other large sediment drifts, has large fields of erosional furrows, troughs a few metres deep and several hundred metres apart that are cut by filaments of faster flow within the bottom boundary layer of steady bottom currents. The characteristic depositional landform of sediment drifts is the mud-wave, forming fields of very linear corrugations 50–200 m high and 5–10 km apart. On sandy sea floors, sand waves and ripples very similar to those of shallow-water settings are common.

If even the smaller-scale deep-sea bedforms are not in equilibrium with present currents, as seems to be commonly the case, then measuring flows around them may reveal more about how the landforms affect the currents, rather than how the currents create sea-floor relief. However, flow measurements also provide opportunities of using the sea-floor patterns to infer different current regimes that may have prevailed in the past. In this way, the sedimentary geomorphology of the deep sea can contribute to studies of global oceanographic change.   P. LONSDALE

Further reading

Heezen, B. C. and Hollister, C. D. (1971) *The face of the deep.* Oxford University Press, New York.

deep-sea trenches Deep-sea trenches are the surface expressions of subduction zones, and mark the sites of convergent boundaries, where an oceanic plate is overriden either by another oceanic plate or by a continental plate. Most trenches occur around the rim of the Pacific Oceans, although they also occur in the north-east Indian Ocean (Java Trench), the Atlantic between the Caribbean and American plates (Puerto Rico Trench and Antilles subduction zone) and between the Scotia and Antarctic plates (South Sandwich Trench). The deepest point on the surface of the Earth occurs in Challenger Deep (11 034 m) at the bottom of the Mariana Trench in the western Pacific.

As seen on a map, trenches are narrow, elongate, and usually arcuate. They may be thousands of kilometres long with widths of about 100 km. In cross-section they present an asymmetric V-shaped profile, the subducted plate dipping gently down towards the axis of the trench, while the side of the overriding plate is generally steeper (Fig. 1). Typical slopes are in the range 5–50°, but steeper slopes can occur locally. The floor of the trench is usually flat, having been covered by sediments. The relative size of the sedimented floor will depend on the balance between sediment supply from the surrounding area and removal by subduction or scraping off on to the overriding plate. Maximum depths (both absolute and relative to the surrounding ocean floor) vary considerably, although most are 2–4 km deeper than their surroundings.

Because oceanic lithosphere has a very similar density to that of the underlying asthenosphere, an oceanic plate converging with another plate is easily subducted; that is, it is deflected downwards into the mantle. The deepest point of the trench generally occurs at the boundary between the two plates. The vertical cross-sections of trenches are well described by the deformation of a thin elastic–plastic shell overlying a viscous substratum (Fig. 1). Such flexural calculations also predict the observed 'outer rise', a low rise a few hundred metres high lying some 120–150 km seaward of the trench axis on the subducting plate.

One of the most noticeable characteristics of most trenches is their arcuate plan, convex toward the subducting plate. An

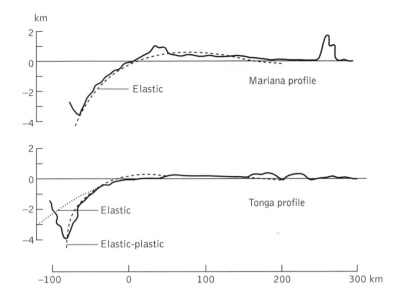

Fig. 1. Cross-sectional topographic profiles through the Mariana and Tonga deep-sea trenches (solid lines) compared with theoretical profiles calculated on the basis of the deflection of thin elastic or elasto-plastic plates overlying a viscous asthenosphere (broken lines). (After Kearey and Vine, Fig. 8. 11.)

excellent example is the Aleutian Trench in the north Pacific. It is explicable in the way a spherical cap of lithosphere must deform when it is bent at a subduction zone, changing from convex upwards on the sea floor to concave upwards in the subducting slab. It is exactly comparable to the way the surface of a ping-pong or tennis ball can be indented. The trench effectively follows the boundary between the convex and concave parts, which delineates a circular arc on the surface.   ROGER SEARLE

Further reading

Kearey, P. and Vine, F. J. (1996) *Global tectonics*. Blackwell Science, Oxford.

deep seismic reflection profiling For over 50 years, images of the subsurface produced by recording seismic energy reflected from velocity discontinuities in the Earth's upper crust have been the primary tool in the exploration work of petroleum companies (see *seismic exploration methods*). The expenditure (about a billion pounds a year) on acquiring seismic reflection data has fuelled technological advances that have made this technique an important driving force in the development of computer hardware and software. In employing this technique, the goal is to produce an image that is so clear that its geological interpretation will be readily evident. There are also subtleties in the data, such as amplitude variations with distance from the source, which can reveal details about the physical properties of a region of the image. Unlike other active source methods, the reflection technique is configured to record seismic waves (usually P waves; see *seismic body waves*) at receivers whose distances from the source are small in relation to the reflecting target.

Many repetitive sources are used and many hundreds of groups of receivers are deployed. This approach results in a great deal of multiplicity, which is used to suppress noise and enhance the usually weak reflected signals and to infer variations in velocity. A knowledge of velocities is needed to form the image, and velocity variations can help to identify the types of rock present and their physical properties, such as their porosity. An individual seismogram on a typical seismic reflection record section (Fig. 1) is in fact the sum of many thousands of individual seismic pulses that have travelled through the Earth.

The technology involved in the reflection technique comes at a high price. Each kilometre of data costs thousands of pounds for data recorded on land. A marine vessel equipped for conducting reflection surveys is a technological wonder capable of determining its position with great precision as well as recording huge volumes of data. Once deployed, such a vessel tows the seismic sources and a streamer containing the receivers, and it is thus very efficient. The unit cost of acquiring data is consequently reduced.

Once the data have been recorded, the task of organizing and manipulating the data to produce an image is still considerable. The data-processing and analysis entailed are greater than for other seismic methods. The computations required are, in fact, some of the most intensive in any scientific endeavour. Once a satisfactory image has been obtained, the task of modelling and interpreting the data is also very computer-intensive.

Geophysicists interested in deep Earth structure and processes recognized the value of seismic reflection data early in the evolution of the method, but academic researchers were frustrated for many years because they could not publish proprietary data from the petroleum industry and could not

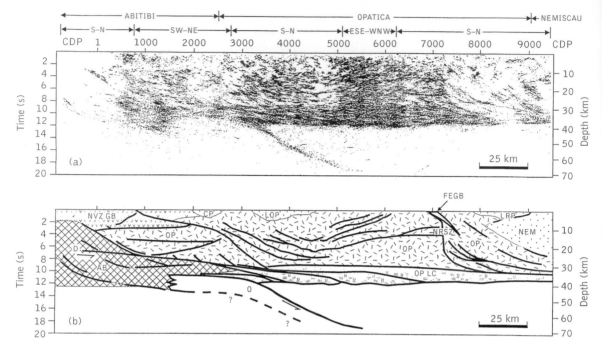

Fig. 1. Migrated seismic image of ancient crust in the Superior Province, Canada, showing the collision zone between the Abitibi subprovince on the south (left) and the Opatica belt on the north (right). The seismic profile was recorded in an approximately north–south direction; the exact orientation of each section is shown at the top of the figure. The common depth points (CDP) are shown for location purposes. AB, Abitibi crust; CP, Canet pluton; D, fault dividing the Abitibi crust; GB, greenstone belts forming the upper Abitibi crust; LOP, Lac Quescapis pluton; LRP, Lac Rodayer pluton; NEM, Nemiscau crust; NRSZ, Nottaway River shear zone; NVZ GB, greenstone belts forming the upper Abitibi crust; O, relic of oceanic slab; OP, Opatica crust; OPLC, Opatica lower crust. For interpretation, see text. (From Calvert, A. J., Sawyer, E. W., Davis, W. J., and Ludden, J. N. (1995) *Nature* **375**, 670–4. Image provided by the LITHOPROBE Secretariat.)

afford to acquire their own data. From the 1970s , however, deep seismic reflection data began to be gathered by academic groups on a regular basis. This advance was led by the Consortium for Deep Continental Reflection Profiling (COCORP), which was funded by the National Science Foundation in the USA. By the 1980s, many other countries were funding the collection of deep seismic reflection data for academic research.

The images produced by these efforts have provided many insights into the structure and evolution of the Earth's crust. On occasion, the uppermost mantle has also been imaged. These contributions are too numerous to discuss in detail. However, one example is that the effects of compression in the lithosphere have been shown to take many forms. In the southern Appalachian Mountains of the eastern USA, reflection data indicate that crystalline basement rocks have been transported hundreds of kilometers inward from the edge of the continent as result of the collision of Africa and North America about 300 million years ago. In the Alps, the convergence of Africa and Europe caused virtually the entire crusts of the two plates to be stacked on top of each other. Elsewhere, relatively simple crustal-scale faults have accom-

modated compression by uplifting large blocks of the crystalline basement. An example of an excellent image is shown in Fig. 1, which illustrates a section of the ancient crust of the Canadian Shield near Hudson Bay that was imaged as part of a LITHOPROBE study. LITHOPROBE is a highly successful Canadian effort to understand the lithosphere and is based on a series of deep seismic reflection profiles. Figure 1 shows the oldest subduction zone yet recognized (about 2.7 billion years old). The interpretation of this image is that the Abitibi crust on the south subducted beneath the Opatica crust to the north. The crust–mantle boundary (Moho) occurs at a depth of about 40 km, where the reflections stop. The dipping reflections extending into the mantle to a depth of almost 70 km are interpreted as revealing a relic of an oceanic slab.

Another example of the use of deep seismic reflection data is in rift zones where extension has greatly modified the crust. Reflection profiles have shown that most rift basins are asymmetrical, indicating that one bounding fault is dominant. They have also shown that the modification of the crust by magmatism is extremely variable, ranging from nil to the replacement of virtually the entire pre-existing crust.

G. R. KELLER

Further reading

Blundell, D., Freeman, R., and Mueller, St. (1992) *A continent revealed: the European Geotraverse.* Cambridge University Press. [With maps and database on CD-Rom.]

Meissner, R. (1986) *The continental crust: a geophysical approach.* Academic Press, San Diego.

Meissner, R., Brown, L., Durbaum, H.-J., Franke, W., Fuchs, K., and Seifert, F. (1991) *Continental lithosphere: deep seismic reflections.* American Geophysical Union, Geodynamics Series, Vol. 22.

deep-water sediments

Vast quantities of siliciclastic sediments (muds, sands, etc.), and lesser amounts of carbonate sediments, escape seawards across the continental shelves into deep water to interfinger with the pelagic sediments of the open sea. Siliciclastic sediments form a fringe around the continents and infill small marginal basins.

Fine-grained mud is spread seawards by oceanic and wind–driven currents. Mud stirred up by storms on the shelves forms dilute clouds—nepheloid clouds, which flow down the slope, where the mud is spread along the continental margins by geostrophic currents flowing parallel to the slope.

Sand and coarser sediment also escapes seaward. However, in contrast to mud, it mainly moves seaward along submarine canyons which cut the shelf. Many of these canyons are supplied directly from rivers and their associated deltas, but elsewhere, where the shelf is narrow, canyons are filled with sand moving alongshore under the action of waves in the nearshore zone. During times of low eustatic sea-level large quantities of sediment have escaped in this way; by contrast, in times of higher sea-level, such as today, many rivers do not supply sediment directly to the canyons. The sediment is then largely retained on the shelves, except where they are very narrow.

The sediment that escapes down the canyons is re-sedimented. It is deposited in the canyons, and then when it is disturbed moves downslope as sediment gravity-flows. These are water–sediment mixtures that move under the action of gravity. They are of various types, which form a continuum. Turbidity currents are high-density flows in which the sediment is supported by the upward component of fluid turbulence. Because of their relatively high density, turbidity currents can travel great distances. Less important types of flow are grain flows, in which the sediment is supported by interparticle collision, and fluidized flows, in which the grains are supported by the upward flow of escaping water. Large debris flows also occur when the sediment is supported in a muddy matrix as it moves downslope.

Turbidity currents and related processes spread coarser sediment seawards to produce large submarine fans (Fig. 1), which have many features in common with the alluvial fans of continents. Some are huge features that stretch over thousands of square kilometres, such as the fan of the Ganges–Brahmaputra delta. Elsewhere, as in the basins off California,

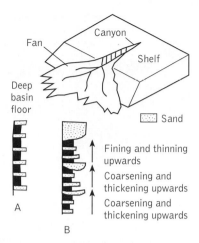

Fig. 1. Typical sequence of sediments of (a) basin plain (b) deep-sea submarine-fan. (Sketched from Walker and Mutti (1972), American Petroleum Geology and Geoscience Canada.)

they are smaller but dominate basinal sedimentation. The sediments deposited by individual turbidity flows show upward-fining graded bedding, with a typical sequence of sedimentary structures known as a Bouma sequence. They contain displaced remains of shallow-water organisms, which contrast with the interlayered fine-grained mud which normally has deeper–water pelagic or nektonic faunas.

The sediments of the fans becomes finer to seaward. When they build out (prograde) into deeper water they produce upward-coarsening sequences with subsidiary upward-fining sequences in which individual channels have become filled (Fig. 1a). Some individual fans have more complex stratigraphics owing to the movement or abandonment of channels. On the adjacent deep-water areas fine-grained siliciclastic sediment (muds) accumulates, either from dilute turbidity currents originating from the canyons, or from suspended sediment in the water column above. The fine-grained sediment accumulating on the continental slope between the canyons is sometimes mobilized to form large slumps or debris flows, particularly on active margins.

Thick sequences of interbedded sandstones and mudstones/shales which contain sands with graded bedding and displaced faunas, with planktonic and nektonic faunas in the interbedded mudstones, are common features in the geological column. In addition, they show well-marked bottom-structures, which have not been described from modern deposits, such as groove and flute casts on their undersides. These structures have been produced by the cut and fill of the erosional scours in the underlying muds. These are useful to geologists in elucidating the directions of palaeoslope and source areas. They show vertical sequences of upward-coarsening and sometimes upward-fining sequences with infilled upward–fining channels. Particularly good examples

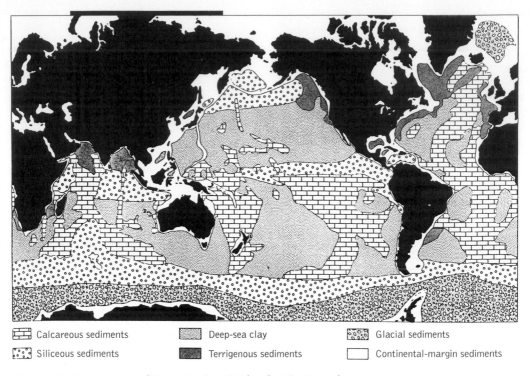

Calcareous sediments   Deep-sea clay   Glacial sediments

Siliceous sediments   Terrigenous sediments   Continental-margin sediments

Fig. 2. Distribution of sediments in oceans. (After Davis and Gorsline (1976) Academic Press.)

are known from the Mesozoic and Cenozoic rocks of Spain, the Cenozoic rocks of California, the Cretaceous and Cenozoic rocks of the Alps (Flysch), and the Palaeozoic rocks of most continents.

Large parts of the deeper-water areas of many basins and the ocean are beyond the reach of the siliciclastic sediment derived from the continent by turbidity currents, or they are protected from it by deep-sea trenches and basins which trap the sediment close to the continent. The only land–derived sediment that can reach these areas is transported by ocean currents, wind, or icebergs which calve-off from the ice sheets and then melt. Ice-transported sediment is common in high latitudes (Fig. 2).

The sediment that reaches the sea floor—pelagic sediment—is dominantly supplied by skeletal remains of planktonic and nektonic organisms living in the surface waters. Because of the great depth, benthonic organisms are unimportant contributors to this sediment although there are few places where their effects, such as bioturbation, are completely absent. The term *hemipelagic sediment* is used when there is still appreciable terrestrial mud, but pelagic components dominate.

Vast quantities of calcium carbonate are extracted from sea water by phytoplankton such as coccolithophores (calcite tests), zooplankton such as foraminifers (calcite tests), and

planktonic molluscs such as pteropods (aragonite tests), as well as other less important organisms. Foraminifera are the most important, and vast areas of the ocean floors are covered with a mantle of fine-grained foraminiferal ooze (Fig. 2). Whereas surface waters are saturated with calcium carbonate, the deep cold oceanic waters are undersaturated; most of the remains of organisms are thus dissolved before they reach the sea floor. The level at which this solution increases and becomes noticeable is the so-called 'lysocline'. At greater depths the rate of supply is exceeded by the rate of solution. The limit is known as the carbonate compensation depth (CCD): sediments below this depth are devoid of calcium carbonate. The calcite tests of foraminifera and coccolithophores survive to greater depth than the aragonite tests of pteropods. The remains of pteropods are consequently found only on the upper parts of seamounts or the shallower parts of the sea floor. The depth of the CCD varies across oceans and has also varied during geological time owing to varying oceanographic conditions. (At present it is approximately 3500–4500 m in the Pacific). Silica is also extracted from sea water by phytoplankton such as diatoms and silicoflagellates, by zooplankton such as radiolarians, and by siliceous sponges.

Although diatoms are universal in fresh and sea water, they achieve greater importance in high-latitude oceanic areas, as well as in some continental margin basins where they form diatomaceous ooze (Fig. 2), composed mainly of opal. Radiolarians form significant deposits only in low latitudes, where the ocean floor is covered with radiolarian ooze (Fig. 2).

The deepest parts of the ocean are remote from sources of continental detritus except wind-borne dust and, in high latitudes, detritus from icebergs. In these areas, which are too deep for the accumulation of organic oozes, the ocean floor is covered with a chocolate-brown to red clay termed *red clay* (Fig. 2). This is very fine -grained clay composed of clay minerals, quartz silt, some volcanic minerals, partly of windblown origin, and other minerals produced by submarine weathering of volcanic rocks. It also includes some minerals (e.g. zeolites) which have grown authigenically (i.e. *in situ*) in the sediments. It contains large amounts of iron and manganese hydroxides. Sedimentation is very slow and the deposits contain unusual components, such as earbones of whales, shark's teeth, and cosmic spherules from extra-terrestrial sources.

Crusts of iron-manganese oxides and manganese nodules occur, particularly in red clays. They originate in part from precipitation from sea water, solution and redeposition within the sediment, and from hydrothermal and magmatic waters. They are usually rich in a variety of metals such as zinc, cobalt, and nickel, particularly when formed on spreading ridges.

Sediments equivalent to modern-day pelagic sediments are found in the geological record, where they occur as fine-grained foraminiferal limestones, radiolarian and diatomaceous-cherts, and red-mudstones with manganese nodules. They are commonly associated with pillow-lavas and other igneous rocks typical of the oceanic crust. The term *Steinmann Trinity* was introduced by Bailey and MaCallien in 1960 to describe the association of radiolarite, serpentinite (altered ultrabasic rock), and greenstones (slightly metamorphosed basalts and gabbros) as a tribute to Steinmann, who first suggested their oceanic origin. Particularly good examples of such sequences, which would now be called ophiolites, are known from the mountains of Cyprus, the Cretaceous of Timor, the Middle Eocene of Barbados, and also in the rocks of the Alps and other mountain chains.

Organic matter is present only in small amounts in most deep-water sediments, for it is recycled in the water column before it reaches the bottom. However, in some marginal basins with restricted circulation (e.g. the Santa Barbara Basin of California and the Cariaco Trench), the bottom waters have become anoxic (oxygen-free) and organic matter descending from the surface water is consequently preserved, commonly as annual laminations of planktonic origin alternating with normal siliciclastic clay layers. Elsewhere (e.g. in the Black Sea) density stratification has prevented oxygenation and allowed preservation of organic matter. The development of basins with restricted circulation appears to have extended over vast areas at particular intervals. Widespread organic-rich black shales are found in the Cretaceous (Late Barremian–Albian, Cenomanian–Turonian boundary, Coniacean–Santonian) and the Miocene, and are also known in Palaeozoic rocks. Considerable debate has arisen over their origin, and many explanations have been given. However, increased upwelling and hence increased productivity due to the more vigorous circulation coupled with the presence of isolated basins which restricted access to oxygen-rich waters appears to be the likely cause.

Much controversy exists over the depth of deposition of evaporites. Although there are no places where extensive deep-water evaporites are forming today, the character of many ancient deposits, including those in the Devonian of Canada and the Zechstein of Europe, indicates that this was not always true. These deposits are often beautifully laminated sediments in which individual laminae can be traced over wide areas; and they are associated with slumped beds and turbidites. The rate of subsidence required to accommodate the great thickness of evaporite in many basins would have been too great to be feasible, according to present day ideas about basin subsidence, if they had been deposited in shallow water basins.

G. EVANS

Further reading

Heezen, B. C. and Hollister, C. D. (1971) *The face of the deep*. Oxford University Press, New York.

Walker, R. G. (1992) Turbidites and submarine fans. In Walker, R. G., and James, N. P. (eds), *Facies models*, pp. 239–64. Geological Association of Canada.

deforestation and landscape

Deforestation is associated with a spectrum of Earth surface changes, including accelerated soil erosion, increased landslide activity, sediment pollution, changes in fluvial geomorphology, and changes in the hydrological, biogeochemical, and climatic regime. Deforestation is counted among the most important environmental crises facing our planet, not least because of its role in reducing biodiversity, increasing global warming, and expanding deserts.

Deforestation is also a hallmark of the human impact on the environment. Most humans live in degraded forest landscapes like those of western Europe, China, and India. These societies would not wish to change their situation. Their landscapes demonstrate that when the trauma of forest conversion is past, most former forest-lands may be managed sustainably and productively by agricultural husbandry.

Current estimates of deforestation rates are alarming. The FAO Forest Resource Assessment suggests that the world's forest covers 3454 million hectares, a little more than a half of which lies in the developing nations. In the 5 years to 1995, the world forest gained 8.8 million ha and lost 65.1 million ha. The area of plantations in the developing world has doubled since 1980 and now reaches 80 million ha. In 1997 the FAO suggested that 15 million ha of tropical forest are lost annually. About 50 per cent of the surviving tropical rainforests grow in the Amazon Basin.

Deforestation is a pejorative term that has been applied to a wide range of forest conversion activities. A few of these are permanent but many are temporary. They range from cyclical forest harvesting and long-rotation forest–fallow agriculture through forest thinning and canopy degradation. The popular concept of deforestation has been created by images of forest burning and clear felling for the creation of arable or grazing land. Landscapes at the front line of forest conversion undergo very dramatic environmental changes. Forest harvesting, official and unofficial, can create major problems. Forest roads are notorious sources of accelerated run-off, sediments, and landslides. Timber-extraction techniques, especially those that employ skid runs or slide logs off site, may accelerate erosion and run-off. Research on steep slopes in the Tatra Mountains found that for each cubic metre of timber harvested, an equivalent volume of forest soil was lost. Sensitive forest management can eliminate such problems.

Forests are resilient systems. Given time, they will regenerate and reclaim their land, even if they are not actively replanted. In many cases, deforested land has to be kept deforested by the regular suppression of regeneration, often through grazing and burning. An outer zone of fire-resistant species characterizes most forest islands in Africa's savannah regions. The last remnant of prairie in Illinois is preserved by regular burning. The uplands of Wales and other European landscapes are kept deforested by overgrazing sheep.

Most deforestation is not achieved by sudden or dramatic clearances, but by slow, progressive forest degradation. Mature forests are replaced by immature forests through accelerated recycling in a stressed long-rotation forest–fallow agricultural system (shifting cultivation, *jhum*, *taungya*, etc.). Forest edges are nibbled away by timber harvesting and agricultural extension, and forests are broken into patches by road construction. Air photographs from both the Himalaya and Amazon show how forest clearance spreads to either side of new roadways. However, even in Amazonia, clearance is not a one-way process; for every 3 ha cleared, perhaps one is reclaimed by forest regeneration.

Landslides

Deforestation has its greatest geomorphological impact in mountainous regions. In steeplands, many rock and debris slopes have evolved to a condition called *self-organized critically*, where they are vulnerable to very small disturbances. Tectonic uplift and river downcutting have produced hill slopes that are close to the margins of their stability. In geotechnical terms, these slopes have a '*factor of safety*' close to unity. This means that the forces that preserve the slope and resist failure, such as the strength of the rock or soil, are in balance with those that encourage its failure, such as gravity. Trees are employed increasingly by bioengineers in the stabilization of steep slopes, because their roots have a tensile strength that can be equivalent to, or greater than, steel. In forest environments, tree roots provide a significant component in the capacity of a hillside to resist failure. Research in Alaska's declining yellow cedar forests suggests that tree roots

can be as important a control on landsliding as pore-water pressure on shallow soils. However, the bulk of the landslide activity that follows deforestation entails the generation of shallow slumps of the deep soils and debris covers that forests preserve on even quite steep slopes. The process is affected by slope angle. Studies at Taranaki, New Zealand, find that 28° is the threshold for post-deforestation landslides and that most occur on slopes steeper than 32°. Average surface lowering was 0.2 m in 10 years.

Soil loss and sediment yield

Deforestation is associated with dramatic increases in soil loss and sediment yield. A review of about 80 studies of surface erosion in natural forest and tree-crop systems finds median soil loss in natural forests to run at about 0.3 t ha^{-1} yr^{-1} (range 0.03–6.2 t ha^{-1} yr^{-1}). Plantations and tree crops with ground cover/mulch suffer losses of about 0.6–0.8 t ha^{-1} yr^{-1} (range 0.02–6.2 t ha^{-1} yr^{-1}). Studies of the cropping phase under long-rotation forest–fallow suggest soil losses of around 2.8 t ha^{-1} yr^{-1} (range 0.4–70.0 t ha^{-1} yr^{-1}). Conversion to arable cropping can lead to much higher rates. The degree depends upon the slope angle, the land husbandry, and the erodibility of ancillary features such as tracks and ditches. Well-managed agricultural lands may have soil losses comparable with those under natural forest. After deforestation, soil losses rise to measurable levels. In Brazil's Rio Grande de Sul, conversion of 90 per cent of the forest allowed soil losses to climb to a regional level of 7 t ha^{-1} yr^{-1} and was followed by the emergence of surface flow pathways. On steep slopes, the effect may be more dramatic: cleared sites in the western Ghats, India, may shed 120 t ha^{-1} in a single season. However, erosion rates subsequently decline rapidly. In the Ghats, erosion from sites cultivated to pepper was less than 3.5 t ha^{-1} yr^{-1}, and in Rio Grande de Sul, erosion became too small to record after conversion to no-till agriculture. In the western Himalaya, comparison of bedload sediments trapped from parallel streams draining steep, 1 km^2 microcatchments found that the sediment loads from undisturbed forest were 5–7 times smaller than those from deforested areas covered by grass and scrub. The depths of soils on the deforested areas were significantly smaller and there were larger patches of exposed bedrock. On poorly managed agricultural steeplands and sites recently cleared of ground cover, huge soil losses can be reported; for example, from tens to hundreds of tonnes per hectare per year, including a maximum of around 500 t ha^{-1} yr^{-1} for fields planted to onions, tilled up-and-down slope in Java. Professional soil conservationists consider recently cultivated, up-and-down slope, bare fallow to be the most erosive condition for an arable field. The losses from the same land fallowed under forest may be anything up to three orders of magnitude smaller, depending on the slope angle, soil type, climatic regime, and scale of the study. It is axiomatic that the amount of soil loss recorded depends less on how much is mobilized than where the loss is recorded. The larger the size of the catchment, the greater the chance that mobilized sediment will come to rest before it is recorded. Sediment delivery ratios, which link mobilization to export in basins of different sizes, range from 0.6 for fields to less than 0.006 in basins larger than 1000 km².

Hydrological impact

Hydrological impacts of deforestation are diverse and have so far had no measurable effect on the record from the world's largest forest basin, the Amazon. However, human activity, including deforestation, is thought to have an important impact on sediment yields from the major rivers of South and East Asia.

Where soils are deep, through-flow is the dominant supplier of storm flow. It is unusual for rainfall to exceed the infiltration capacity of forest soils, except where there has been compaction, but saturated overflow is not unusual. Annual evapotranspiration in tropical, moist, lowland forests ranges up to 1500 mm yr⁻¹ with transpiration accounting for a maximum of 1045 mm yr⁻¹. In most environments, these figures are very much lower. However, when forests are removed, a great deal more water remains in the environment and the total discharge of affected streams tends to increase

(Fig. 1). In the tropics, increases of water yield equivalent to 110–825 mm yr⁻¹ are reported in the year following clearance. What happens next depends on local conditions and the degree to which evaporation supplants evapotranspiration. If more rainwater is converted into near-surface run-off, if soil storage is lost through erosion, and if the groundwater receives less water through infiltration, then run-off increases and so does the height of the mean annual flood. Sometimes, the duration of flow is also reduced. However, where the forest is replaced by vegetation, which is equally effective in soil protection or in permitting infiltration of rainwater, then volume and duration of river run-off may increase. Indeed, local deforestation is employed by some nations, such as Israel, as part of a programme of water harvesting. The role of deforestation in causing the very largest floods is controversial because the capacity of forest geosystems to restrict run-off is both variable and limited.

Channel response

Increased soil losses may have an important impact on stream channels. In the Himalaya and its Siwalik Fringe, deforesta-

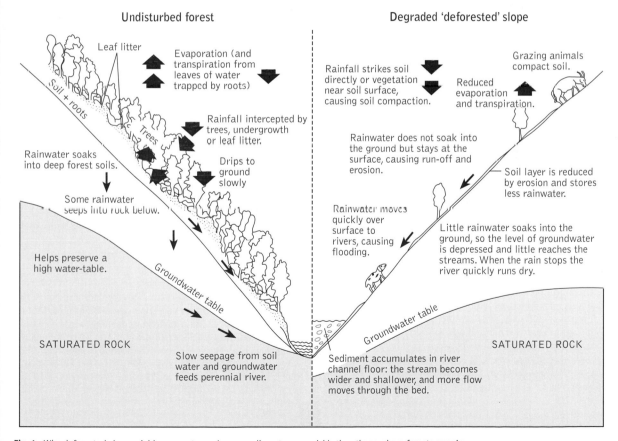

Fig. 1. Why deforested slopes yield more water and more sediment more quickly than those where forests remain.

tion and development are associated with changing the dominant processes in stream channels from incision to aggradation. Affected streams may be converted from surface channels to bedload channels. They tend to become wider, sometimes braided, shallower and less sinuous. Surface flows frequently decrease and become restricted to flood conditions. There have been dramatic increases in stream width and it is likely, but not yet proved, that this has led to further undercutting and destabilization of hill slopes. Elsewhere, in situations where the sediment supply to the rivers is only slightly higher, the increased run-off, especially the enhanced severity of the mean annual flood, that accompanies deforestation, together with reduced vegetative protection and greater compaction of the soil surface, may lead to exactly the reverse effect. In semi-arid environments, deforestation often leads to stream-channel trenching, decreases in the width/depth ratio, and increased channel sinuosity. The impact of the affected channels depends upon the balance and the character of the local change in the supply of water and sediment.

Other factors

Deforestation affects many aspects of the environment that are beyond the concern of this brief account. Forests, especially tropical forests, are major, and largely unexplored, reserves of biodiversity. Deforestation effects a transformation of the ecology of affected lands and rivers. It has major impacts on biogeochemical processes. Soil-formation processes are entirely transformed. Deforestation has major impacts on climate at various scales, from micro to macro. Large forests, especially the Amazon rain forest, are thought to play a key role in the geophysiological regulation of the Earth's atmosphere and climate. Computer simulations have indicated that the complete transformation of the Amazon rainforest to pasture would increase regional temperature by 2.5 °C with consequent impacts on evapotranspiration. Forests are major sinks of CO_2 and forestation plays an important role in the remediation of global warming.

A cautionary note

Forests play many roles and some are part of very complex environmental systems. It is dangerous to generalize about the interactions between forests and the environment. This does not mean that deforestation is not associated with accelerated soil loss, landsliding, river-channel aggradation, with desertification, or with increased severity of floods. Very commonly, it is. Equally, it is not necessarily. The reality is sometimes very different. For example, the presence of trees does not automatically mean reduced erosion. Forest research has found positive correlations between tree height and rainfall erodibility. It takes a fall of 4 m for a raindrop to achieve its maximum velocity. Many forest canopies exceed this height. Many leaves have drip tips that create raindrops of large size that are consequently more erosive. Research in the Carpathian Mountains has found a positive correlation between tree density and erosion. The reason is that sheep cluster for shelter beneath trees, degrading the ground vegeta-

tion cover, compacting the soil, and so encouraging run-off and erosion. In the Himalaya, there is a positive correlation between forests and landslides because development has cleared the more stable slopes and forests persist only on the steepest and least stable slopes. In the same area, there is a positive association between forests and the perennial flow of springs, which persists despite the fact that trees return a large amount of rainwater to the atmosphere as evapotranspiration. The reason is that the trees preserve deep, moist soils that encourage infiltration to groundwater and thence to supply springs.

Reforestation

The environmental impacts of reforestation are not the reverse of those due to deforestation. Many of the impacts of deforestation occur because of the loss of the forest soil and litter layers. New forests mine the environment for nutrients. They accelerate weathering and fracture rocks for anchorage. There are relatively few studies of the long-term environmental impact of current reforestation programmes. The new plantings may not, however, have the same geomorphological impact as the forests that are removed. In Nepal, new forests tend to be planted on relatively gentle slopes, whereas deforested areas are more often steeplands. Recent findings from reforestation research, however, demonstrate that, against expectation, trees (in this case cultivated on steep banks in the Balkans) reduce erosion by up to four times, even in the absence of soil and litter layers. In Japan, reforestation has resulted in increased stream flow, despite the increased losses to evapotranspiration, by encouraging deep seepage.

MARTIN J. HAIGH

Further reading

Bruijnzeel, L. A. (1990) *Hydrology of moist tropical forests and effects of conversion: A state of knowledge review*. UNESCO/IHP, Paris.

FAO (1997) *Sustainable development dimensions: Environment policy planning and management*, Special Agenda 21 Progress Report, chs 10–15, FAO, Geneva.

Haigh, M. J., Krecek, J., Rajwar, G. S., and Kilmartin, M. P. (eds)(1998) *Headwaters: water resources and soil conservation*. Balkema, Rotterdam.

Hewlett, J. (1982) *Principles of forest hydrology*. University of Georgia, Athens, Georgia.

deformation

The deformation of geomaterials, that is, rocks and soils, is of considerable importance, both in understanding geological processes and for the construction industry. Before discussing deformation we first need to consider the fundamental concepts of stress and strain.

Stress and strain

Stresses are forces acting on materials that tend to change the dimensions of those materials. When a material is distorted by stresses it is said to be strained. A stress is measured in terms of force per unit area; a strain is the ratio of an elongation or a deflection to an original dimension (see *stress analysis*).

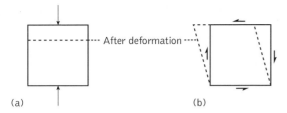

Fig. 1. Deformation under normal and shear stresses. (a) A normal stress applied to the face of a cube of material results in a decrease in its height; (b) a shear stress applied to the faces of a cube results in a shear strain.

The stress state in a geomaterial can be altered in various ways: for example, by changes in external loading caused by the construction of buildings or by the removal of surface layers during erosion. When this happens, the material will respond by straining; that is, its shape and size will change to reach a new equilibrium under the changed stress state—unless the change is great enough to cause the material to fail altogether. If a state of failure is reached, the resulting strains are usually very large and are difficult to predict.

In general, a material is acted on by normal (compressive or tensional) stresses and shear stresses. A normal stress applied to the upper face of a cube of material will produce a linear strain and will result in a decrease in the height of the cube (Fig. 1a). A shear stress applied to one face of the cube will produce a shear strain and will result in distortion of the type illustrated in Fig. 1b. Engineers also refer to volumetric strain, which is the ratio of the change in volume to the original volume.

The deformation of the mass can be determined by finding the strains, multiplying the strains by the appropriate lengths, and summing the effects.

Materials, including soils and rocks, can deform in various ways. The two broad modes of behaviour are *elastic* and *plastic*.

Elastic deformation

In elastic behaviour deformations are recoverable: when the load is removed, the material will return to its original shape and size. If a material is isotropic and perfectly elastic, its behaviour is linear, that is, strain is proportional to stress. Such a material can be characterized by simple moduli (ratios of stress to strain): Young's modulus, E, and Poisson's ratio, v. (Isotropic materials have the same properties in all directions.) Young's modulus and Poisson's ratio for intact samples can be determined from compression tests; a cylindrical sample of the rock or soil, which may be subjected to an all-round confining stress, is loaded axially, and the vertical and horizontal changes in dimensions are measured (Fig. 2). If the material does not change in volume on loading, then v is equal to 0.5; if the material increases in volume on shearing (dilation), v is greater than 0.5.

As well as Young's modulus and Poisson's ratio, the *bulk modulus* (K) and the *shear modulus* (G) are often used to characterize geomaterials.

For a perfectly elastic material the cumulative effects of changes in loading can be found by simply summing the effects of each change. A complex change in conditions can be analysed as a series of simpler changes. This is known as *superposition*.

Elastic materials may be non-linear. The stiffness of the material will then vary according to the conditions. This is illustrated in Fig. 2. In these cases it is more difficult to characterize the material by simple moduli, although for many problems the behaviour of the geomaterial can be approximately represented by a linear response.

Plastic deformation

If when a material is loaded and then unloaded it retains a permanent deformation, it is said to have undergone plastic deformation. Three simplified stress–strain curves are shown in Fig. 3. Materials generally show elastic behaviour, shown here as perfectly elastic, up to a *yield point*, at which point both elastic and plastic strains occur. If the material is perfectly plastic (Fig. 3a), then large, indeterminate strains occur when it yields. If a material shows strain-hardening plastic behaviour, as it strains after yielding it becomes stiffer (Fig. 3b). When the load is removed, the elastic strains are recovered and permanent plastic strains remain. As the material strain-hardens, the yield stress increases. If it is deformed enough, the material will fail, that is, suffer large, sudden deformations. A good

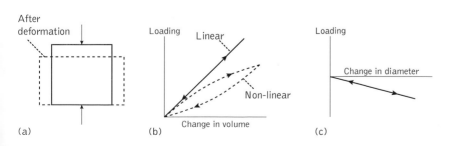

Fig. 2. Unixial compression test. (a) An axial load is applied to a cylindrical sample of the material under test and the change in height and the change in radius are measured. (b) Plot of change in volume against loading for linear and non-linear responses; arrows indicate direction of loading. (c) Plot of change in length against change in diameter; arrows indicate direction of loading.

223

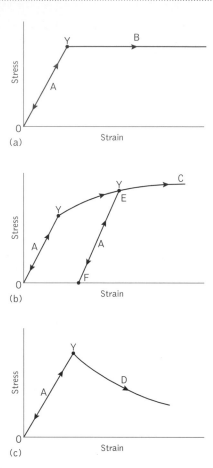

Fig. 3. Typical elastic–plastic stress–strain curves (stress against strain): (a) elastic—perfectly plastic behaviour; (b) elastic—strain-hardening plastic behaviour; (c) Elastic—strain-softening plastic behaviour; the plastic strain increment and the elastic strain increment, are shown for a sample loaded from O to E. A, elastic behaviour; B, perfectly plastic behaviour; C, strain-hardening plastic behaviour; D, strain-hardening plastic behaviour; Y, yield point.

example of a strain-hardening material is copper. Initially it is quite easy to bend a copper bar, but this becomes more difficult as it is bent to and fro and the metal strain-hardens. Eventually the bar will fracture and break, that is, fail. Figure 3c shows a strain-softening material, which after yielding becomes weaker. Such materials are very difficult to model.

To model plastic materials it is necessary to define *yield criteria*, that is, those combinations of stress at which plastic deformations begin, a *strain-hardening* relationship for suitable materials, which defines the change of the yield criteria as the material strains, and a '*flow*' relationship which links plastic strain increments with stresses. This can be complex

for soils and rocks, although quite simple models exist for some metals.

If a material is plastic, then the effect of a change in loading cannot be found by summing the effects of individual load components, as can be done for elastic materials; the stress history of a plastic material is important because it affects its current yield stresses and other properties.

Creep and other processes

In addition to elastic and plastic deformations, which occur relatively quickly in response to changes in loading, geomaterials can deform as a result of other processes. One of these is creep, where continual movement occurs over an extended time period under a constant load. This can be thought of as a viscous flow. A common example is the slow flow of bitumen down a sloping surface.

Deformation of fine sedimentary deposits (such as clays) can be caused by a change in loading, usually vertical, which results in a change in the pressure in the water contained in pores in the soil. As a result water will tend to flow out of, or into, the soil, which will then decrease or increase in volume with time. This process of *consolidation* or *swelling* may be very slow. It is nevertheless significant on geological timescales and can affect the construction of buildings on such materials.

The response of soils and rocks is not constant but will depend on conditions; for example, the general stress level on the material and whether it is constrained. A granite buried deep in the Earth's crust may deform plastically and flow into tunnels cut into the rock, whereas on the surface the same material may be very stiff and deform elastically until it fails in a brittle fashion. In a similar fashion, a soft clay will generally behave plastically, whereas an over-consolidated clay (for example, London clay) will be stiff until it reaches its yield point, when it will break up into blocks.

The deformation of large masses of rock and soil is heavily influenced by the structure of the mass. Rock masses commonly contain joint systems which cut through the mass, separating it into blocks. The blocks may be large or small, and the joints may be smooth and planar or rough and complex. The joints can be extremely thin or wide and filled with altered rock material. Therefore, the response of the mass to a change in loading, for example to a cutting being excavated or a tunnel being driven, may well be determined by the way in which blocks of intact rock move and slide over each other. In attempting to predict, or understand, the deformation of a soil or rock mass the structure of the mass must therefore be understood and appropriate field and laboratory tests undertaken to determine its properties.

Values of Young's modulus for rock masses and typical construction materials vary widely. Typical values are 30 000–

40 000 MN m^{-2} (meganewtons per square metre) for granite, 35 000–60 000 MN m^{-2} for slate, 7000–20 000 MN m^{-2} for sandstone, 5000–17 000 MN m^{-2} for clay, 200 000 MN m^{-2} for steel, and 25 000 MN m^{-2} for concrete.

Additional complications arise from the fact that many geomaterials are anisotropic and non-homogeneous; that is, their properties vary with direction at any point and also from point to point. At any particular site there are likely to be various layers of soil and rock, and these layers will generally be of variable thickness and geometry, according to their geological history. Sufficient field and laboratory testing is therefore required if a reasonably accurate model of the problem area is to be developed and the characteristics of the constituent materials determined. Techniques for investigation in the field include geophysical methods (such as seismic, resistivity, and magnetic methods), drilling boreholes to identify strata and obtain samples, and tests made *in situ* (with pressure meters, for example).   T. COUSENS

deformation bands *see* SHEAR ZONES, DEFORMATION BANDS, AND KINK BANDS

deformation, crustal *see* CRUSTAL MOVEMENTS AND DEFORMATION STYLES

deformation of rocks
Buried flat-lying sediments beneath a river plain can be elevated into a mountain belt by the processes of folding and faulting. Deep within the mountain belt, a metamorphic rock becomes a locus for movement, flowing so intensely from shear that the rock is changed. This change extends down to the scale of the crystal lattices of the minerals that constitute the rock. Later as the mountain belt collapses perhaps under its own weight, normal faults grow, fracturing the rocks and juxtaposing folded sediment against sheared metamorphic rock. These events are all examples of deformation: the process by which rocks move and alter in response to tectonic stress. Types of movement or *kinematic behaviour* include distortion, which changes the shape of rocks; rotation, which changes the orientation of rocks; and translation, which changes the position of rocks. A commonly used measure of distortion is strain, which is sometimes incorrectly assumed to represent the entire deformation of a rock.

Deformation leaves its imprint on geological structures at all scales. For example, large structures include the trace of the San Andreas fault through the countryside of California, the scarp faces of normal faults framing the rift valleys of East Africa, and individual folds which define bulbous linear mountains in the Zagros Mountains. Yet, despite producing structures that range from mountains to defects in mineral lattices, almost all deformation processes or mechanisms operate at the scale of mineral grains and their atomic lattices. The larger-scale structures are simply the cumulative result of these fine-scale processes operating over a large volume of rock for extended periods of time.

Deformation mechanisms involve either fracture or flow. Differentiating between these two types of processes can be difficult. A single structure may display both fracture and flow effects as either a function of position or time. For example, a large vertical fault contains fracture-related structures in the upper crust, but flow-related structures where it is expressed by a shear zone in the lower crust. Alternatively, rocks initially shortened by cleavage formation during flow may then shorten by fracture-related thrusting as deformation progresses. Another problem is that some processes are viewed as having elements of both. For example, *cataclasis* is a fracture-related process that changes a fault from a simple pair of moving surfaces to a zone containing deformed, poorly sorted, fractured, fine-grained material by converting country rock to fault rock. The conversion is by grain-size reduction, grain-boundary sliding, and microfracturing driven by the work done to overcome the frictional resistance to fault movement. These processes involve fracturing, but some research workers wish to view the product at regional rather than grain scale, and interpret cataclasis as a flow mechanism. Such disagreements will no doubt continue for a long time among structural geologists.

Fracture produces clean breaks or discontinuities where a rock loses cohesion. The common motions are rotation and translation across the discontinuities. The most common structures are joints, where the rock dilates across the fracture. Joints can provide pathways for fluids such as oil or groundwater. The most important structures formed by fracturing are faults, where the rock slips parallel to the walls of the new fracture. Large faults form mountain belts, create basins during crustal extension, form plate boundaries, and are sites of almost all major shallow earthquakes.

Although flow, like fracturing, may produce discontinuities, it preserves rock cohesion and deformation is continuous. Flow causes rotation and translation, but it can also be marked by tremendous distortions. Common structures that are the result of flow include folds, foliations, lineations, and shear zones. Large folds may form individual mountains such as in the Zagros, Appalachian, or Rocky mountain belts; or they can decorate the entire sides of mountains, as in the Helvetic nappes of the Swiss Alps. Folds can also create sites for hydrocarbon accumulation.

All flow mechanisms entail a change in some combination of grain size, mineralogy, and rock chemistry. The type of flow is a function of such parameters as temperature, pressure, strain rate, and fluid content. For example, pressure solution, where grains dissolve at grain contacts in response to tectonic stress, operates at low temperatures (commonly less than 250 °C). Fluids are prevalent in rocks, and the small-scale products of pressure-solution deformation include pitted grains and sutured solution surfaces. An important result of this process is rock cleavage and the formation of stylolites.

At intermediate temperatures (commonly 250–500 °C), fluids are much less abundant, and stress drives intragranular mechanisms that operate within mineral lattices. Three mechanisms are common under these conditions. Diffusion is the process in which atoms leave their lattice sites and migrate to other positions within the same lattice or other lattices. Twinning of crystals occurs where the lattice kinks and collapses like an accordion. Dislocation glide happens when lattice bonds break and heal during slip on crystallographic slip surfaces in the grains. Although slip occurs on a surface during glide, it is not a fracture process, because healing of broken bonds after translation prevents the formation of discontinuities and preserves lattice cohesion. These processes produce strained grains with folded or twisted atomic lattices, or even grains that are compartmentalized into regions (i.e. subgrains) of common lattice orientation. They also increase the number of defects or imperfections in the lattice as an artefact of the deformation.

At elevated temperatures (commonly more than 500 °C), intragranular processes such as grain-boundary formation and migration become active. Some grains become bigger, while others disappear, as a function of lattice orientations, stress orientations, accumulated defects, and other conditions. These processes produce recrystallization, replacing the original texture and even the composition of the rock during the formation of a schist, gneiss, or mylonite. All flow mechanisms may produce a shape fabric in which the mineral grains of the rock have parallel long axes. The intragranular processes may also produce a fabric in the rock where all the same minerals have commonly oriented mineral lattices, producing a crystallographic-preferred orientation. The production of these fabrics is a result of flow mechanisms operating pervasively through the rock, unlike fracture mechanisms, where deformation concentrates at discontinuities.

Deformation mechanisms compete. Changing conditions such as temperature, pressure, fluid pressure, or grain size determine which mechanism is most efficient, and hence, dominant during deformation. The dominant mechanism or mechanisms will dictate which structures develop in the rock.

WILLIAM M. DUNNE

Further reading

Hancock, P. L. (ed.) (1994) *Continental deformation*. Pergamon Press, Oxford.

Twiss, R. J. and Moores, E. M. (1992) *Structural geology*. W. H. Freeman, New York.

De la Beche, H. T. (1796–1855)

De la Beche was the originator and first director of the Geological Survey of Great Britain, which was officially recognized in 1835. A Londoner, and educated for the army, he returned to London at the end of the Napoleonic wars in 1815. There his private income enabled him to take up the new science of geology. He became a Fellow of the Geological Society of London, and 2 years later was elected to the Royal Society, aged 23. De la Beche was an enthusiast for field geology and travelled widely in Europe.

He was something of a catastrophist rather than uniformitarianist, although he accepted granite as igneous, and he grasped the significance of fossils in stratigraphy. In 1831 his *Manual of geology* was an immediate success in Britain and abroad, and he was by then well known in the geological community. He became Secretary and then a vice-president of the Geological Society. In 1832 he was given a commission to colour geologically the one inch to a mile Ordnance Survey maps of south-west England.

In this task he made rapid progress; the maps were clear, excellently coloured and covered an area of about 965 square miles, very many of which De la Beche had himself examined. An explanatory memoir appeared in 1839. Officers of the Geological Society meanwhile lobbied for the work to be extended and for De la Beche to continue in his appointment. This was agreed in 1835 and the *Report on the geology of Cornwall, Devon and west Somerset* was the first memoir of the Geological Survey. Part of the new appointment was to set up a Museum of Practical Geology, the Royal School of Mines, and the Mining Record Office. De la Beche set to work to good effect but in the 1850s his health failed and he died in 1855. All his enterprises expanded and prospered.

D. L. DINELEY

deltaic sedimentary deposits

Deltas are important sites of deposition in modern seas, and ancient deltaic sequences are common in the geological column. Fluvio-marine deltas develop where large drainage systems terminate in marine basins and smaller fluvio-lacustrine deltas build into lakes. All major deltas develop in tectonic depressions, and large modern deltas show evidence of long-continued subsidence (e.g. those of the Mississippi and Niger).

The subaerial delta-top or delta plain often covers thousands of square kilometres; the subaqueous prodelta tens to thousands of square kilometres on the adjacent shelf; and associated submarine fans hundreds of thousands to millions of square kilometres on the adjacent ocean floor. The morphological development of a delta and the resulting importance of the various sedimentary environments of deposition are controlled by the relative importance of fluvial, wave, and river processes.

Fluvial-dominated deltas have elongate bodies of channel sands enclosed in inter-channel muds and peats with their crevasse splays of sand produced when the river escapes from its channel in times of flood. *Wave-dominated deltas* have wide beach-dune barriers which build seawards to produce sheet-sands as well as having channel sands. *Tidally dominated deltas* have complex channel-sand bodies with intervening intertidal deposits and peats. In all cases these pass seawards into more uniform subaqueous muds.

The seaward growth of a delta—progradation—produces a vertical sequence of deposits (Fig. 1) that comprises tens to hundreds of metres of upward coarsening sediments, which also show an upward increase of complexity of sedimentary structures and facies, and an upward decrease in abundance and diversity of marine fauna; the latter reflect increasing

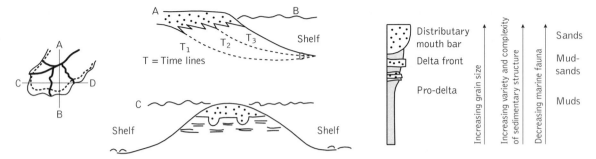

Fig. 1. Cross-sections of a typical prograding delta and resulting sedimentary succession. The schematic map of the delta shows the section lines A–B and C–D.

fluvial influence (Fig. 1). However, deltas rarely build continually seawards; constructional phases alternate with destructional phases at any one location. Channels are constantly abandoned and sites of deposition become erosional. The coastline retreats and the deposits of the constructional phases are trimmed and become overlain by a thin layer of marine deposits. Repeated channel changes produce a series of overlapping lenses 10 to 100 m thick and many tens of kilometres across which form a complex stratigraphy of repeated truncated upward coarsening cycles.

Deltaic accumulations are extremely unstable. The rapid deposition of fine-grained sediment on the pro-delta slope produces mudflows and slumps. Synsedimentary faults, i.e. growth faults, with associated roll-over anticlines and horst and graben structures, develop in the advancing deltaic wedge. Muds flow upwards into areas of lower stress to produce mud-diapirs; and gas generated from sedimentary organic matter escapes upwards and disrupts the sediments.

The relationship between the various deltaic facies depends upon the relative importance of the rates of fluvial sediment supply, sea-level changes, and tectonic subsidence. Slowly rising, stable, or falling relative sea level allows the delta to build upward-coarsening sequences during progradation. Even though vast quantities of sediment reach the coast, if the relative rise of sea level is relatively fast (whether because of eustatic sea-level rise or tectonic subsidence), the coastline will either remain stable and the delta will build up (i.e. aggrade) or it will retreat, and the delta will retrograde. In the former case, sediments that would be in a prograding delta normally lie above one another, build up by its side; in the latter case, the normal sequence will be reversed, and deeper-water sediments come to overlie shallow-water sediments.

Ancient deltaic sediments are of considerable economic importance: peat accumulations on the delta plain produce coal and gas on burial, and pro-delta muds are a source of oil and gas. Porous sands and complex synsedimentary structures provide reservoir rocks and traps, respectively, for hydrocarbons, as in the Jurassic Brent Group of the North Sea.

G. EVANS

Further reading

Elliott, T. (1986) Deltas. In Reading, H. G. (ed.) *Sedimentary environments and facies* (2nd edn), pp. 113–54. Blackwell Scientific Publications, Oxford.

Whateley, M. K. G. and Pickering, K. T. (1989) *Deltas: sites and traps for fossil fuels.* Geological Society of London Special Publication No. 41.

deltas A delta constitutes both the alluvial tract of land at or near the river mouth (*delta plain*), the adjacent coastal zone (*delta front*) that is reworked and shaped by waves, and the offshore area below storm-wave base (*prodelta*) that is still influenced by the deposition of most fine-grained, river-derived material. Its morphology is the result of the river interacting with the receiving basin. Figure 1 shows how delta morphology is defined by fluvial processes (fluvial regime) on the one hand, and basinal processes (basinal regime) on the other. The fluvial regime is controlled by hinterland characteristics, such as

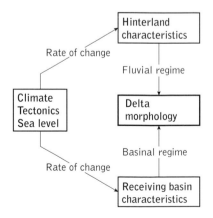

Fig. 1. Conceptual framework for comparative studies of delta morphology.

227

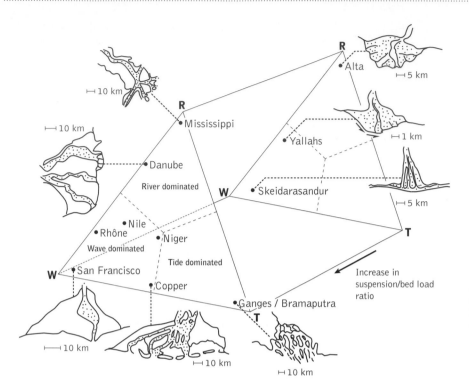

Fig. 2. Classification of modern deltas based on the dominant process of sediment dispersal and on the prevailing grain size (synthesized from classifications put forward by W. E. Galloway in 1975 and by G. J. Orton and H. G. Reading in 1993).

drainage area, geology (e.g. lithological variation), and topographical relief of the catchment, which are all subject to continuous change. The rates at which these changes take place depend on the rate of variation in climate, tectonic, and sea-level change. For instance, periods with increased run-off at times of climate deterioration, or periods of tectonic uplift of the hinterland, will lead to an increase in sediment supply to the delta due to increased erosion rates in the hinterland. Other factors like vegetation and human modification also modify the input from the fluvial regime. The basinal regime is controlled by the geographical setting, shape, depth, and size of the receiving basin. These characteristics define its dynamics in terms of available tide and wave energy, but other variables, like the density of the basin water relative to the river water, also play a role in shaping the morphology of the delta.

Fluvial regime

The control of the fluvial regime on delta architecture is best shown by deltas prograding in low-energy basins, i.e. in basins where wave and tidal influence can be neglected. High gradient (>0.5°), closely spaced, and highly mobile distributary channels in the delta plain feed the delta front essentially as a line source (a more or less uniform supply of sediment along the delta front), which results in a typical, fan-shape delta-plain morphology (Alta delta, Fig. 2). Deltas built by these high-gradient streams are generally small in size and coarse grained,

consisting predominantly of sand and gravel. A special example is the *fan delta*, a delta that is fed by a sandy to gravelly alluvial fan. In contrast, low-gradient (>0.5°), widely spaced, and relatively stable distributary channels in the delta plain feed the delta front essentially as a point source, which results in a typical lobate-shaped delta-plain morphology (*birdfoot delta*). The latter delta type is generally fine grained and develops at the coastal margins of extensive lowland areas. A good example is the Mississippi delta. The regular splitting and shifting of distributaries in the delta plain is caused by bars that are formed at the river mouth. There, the velocity of the river is checked by the standing water of the basin, causing deceleration of the river current, thereby reducing its transport capacity and promoting sedimentation. The deposited sediment causes an obstruction in the river mouth in the form of a bar which is coarse upstream and gradually fining downstream. As these *mouthbars* grow in size, they will deflect and split the flow, causing the river to bifurcate. This process will eventually lead to the characteristic lobate pattern of a fine-grained, *fluvially dominated delta*. Bar characteristics, such as width/length ratio, depend strongly on river-mouth processes, which are controlled by density contrasts between river and basin water, by the channel- to basin-depth ratio at the river mouth, and by the bed load to total load ratio. In the case of a line source, bars also develop but will coalesce to create a uniform delta front. The steepness of the delta front in low-energy basins depends mostly on the initial basin relief (e.g. absence or presence of a coastal shelf) at the river outlet, and the fluvial regime. Low-gradient delta fronts are often associated with rivers with a high suspension load to total load ratio. Examples include the many large, fine-grained deltas. Deltas with steep-gradient delta fronts (so-called *Gilbert-type deltas* named after G. K. Gilbert who, in 1885, described these deltas from Lake Bonneville) are often sandy to gravelly and associated with rivers with a low suspension load to total load ratio.

Basinal regime

The role of the basinal regime in creating delta morphology can be pictured as the relative contribution of wave and tide processes to river processes. Often a triangular diagram is used to distinguish between fluvial-, wave- and tide-dominated delta systems (front triangle of Fig. 2). *Wave-dominated deltas* tend to have straight delta fronts, because the wave energy that causes the sediment redistribution along the delta front is evenly distributed along the entire coast line. Examples include the Nile and the Rhône deltas which both have delta plains with a shape similar to the Greek capital letter D, with the apex of the delta pointing upstream. The characteristic 'delta' shape of wave-dominated deltas was first recognized by Herodotes from the Nile delta plain in the fifth century BC, and the term *delta* has since been used to denote all types of deltas. *Tide-dominated deltas* are characterized by tidal sand ridges that are oriented at a high angle to the coastline. Examples include the fine sandy Ganges/Brahmaputra delta and the Klang delta. The ridges develop best if the ratio of tidal to fluvial discharge is high. The importance of the type of distributary system is visualized in Fig. 2 by giving the ternary diagram a third dimension with the prevailing grain size.

G. POSTMA

dendroclimatology
Dendroclimatology is the reconstruction of aspects of past climate based on the analysis of tree-ring widths. The technique was pioneered in the early twentieth century by the American astronomer A. E. Douglass. His studies have been expanded considerably in the 1980s and 1990s by workers at the Laboratory of Tree Ring Research in the University of Arizona at Tuscon.

The technique is based on the recognition and counting of annual growth layers in trees. Such tree rings are typically most clearly pronounced in species from temperate forests and are usually poorly defined or absent in tropical species. The growth increments in temperate varieties consist of relatively large and widely spaced thin-walled cells (early wood), followed by more densely packed, thick-walled cells (late wood). The average width of an annual tree ring is controlled both by intrinsic factors (tree species and age) and extrinsic factors (including soil nutrient and soil moisture status, sunshine, precipitation, temperature, wind speed, humidity, seasonality). Additionally, the occurrence of forest fires and of periods of extensive frost or drought conditions results in specific stress to tree health, which is frequently recorded in the annual rings.

The longest chronologies established extend back for periods in excess of 9000 years. Such long chronologies have been based on a variety of tree species, including *Quercus* (Oak) and *Picea* (Spruce) in Europe, *Pinus longaeva* (Bristle Cone Pine, formerly *Pinus aristata*) in the western United States, and *Agathis Australis* (Kauri) in New Zealand.

Tree-ring width is commonly utilized by observations of interannual variability or variations in wood density. Year-to-year variability is most pronounced where climate limits growth (termed sensitive environments). Dendroclimatic reconstructions require the precise age assessment of each ring. This is achieved by cross-matching (or cross-dating) many sets of individual rings for the same species, collected in close geographical proximity, and preferably extending forward to a living tree. Once cross-dated, a sequence must be standardized in order to correct for the effects of variations in ring width associated with increasing tree age.

Climatic reconstruction is achieved by a variety of statistical methods which regress ring-width variations against one or more climatic parameters (e.g., temperature, precipitation, sunshine hours) measured for the same area over the span of instrumental records. Demonstration of significant correlations between ring width and climate then makes it possible to extend the relationships beyond the time range of the instrumental records. It is common to generate the tree-ring–climate associations over only a portion of the period of instrumental records, and then to use the remaining period as a test set to demonstrate the degree of agreement between actual climate and that inferred from ring-width variations.

More recent studies have investigated the use of ratios of stable isotopes of oxygen and hydrogen as a means of directly estimating past temperatures. An additional, highly significant aspect of tree rings has been the calibration of the radiocarbon timescale, which involves correction for variations in the radiocarbon production rate in the atmosphere.

STEPHEN STOKES

Further reading
Lowe, J. J. and Walker, M. J. C. (1984) *Reconstructing Quaternary environments*. Longman, Harlow.

density distribution within the Earth
Our estimates of the total mass of the Earth, M, and therefore its mean density, rely on the measurement of gravity. Outside the Earth, at a distance r from the centre, this is, by Newton's law,

$$g = GM/r^2 \qquad (1)$$

with small corrections for ellipticity and, if measurements are made on the surface itself, also for rotation. G is known as the gravitational constant. Observations of the orbital accelerations of satellites have yielded values of the product GM with 9-figure accuracy. G is, however, by far the least well known of the fundamental constants of nature, and this limits our knowledge of M to about 1 part in 8000. Using the present best estimate of G, obtained at the National Bureau of

Standards in the USA, the mean density of the Earth is 5515 kg m^{-3} with an uncertainty of one in the last figure.

The mass of the Earth, obtained in this way, constrains the estimates of density at all levels. If we imagine the value of G to be revised downwards by 1 per cent (a hypothesis that seemed just possible in the 1980s), then M and the densities at all levels would be revised upwards by the same factor. A similar upward revision of the elasticities of the deep Earth materials would leave gravity at all depths, the observed seismic velocities, and the frequencies of the Earth's free oscillation all unaffected. The mean density is thus a scaling factor for density throughout the Earth.

Another whole-Earth property that any model of the Earth's density structure must match is the moment of inertia (see *moment of inertia and precession*). It was first used in Earth model studies in the 1930s by K. E. Bullen, whose models became the reference standards for several decades. The axial moment of inertia, C, is related to the mass, M, and the equatorial radius, a, by a numerical coefficient which is known much more precisely than the mass or mean density:

$$C = 0.330695 \, Ma^2. \qquad (2)$$

If a model of the Earth's density profile were to be rescaled to match a new value of M (perhaps arising from a new and better measurement of G) in Eqn (2), then the numerical coefficient would be unaffected. It is therefore a scaling factor with a role that is rather different from that of M. The mass controls the absolute densities in a model; C/Ma^2 gives a control on relative densities at different levels.

Seismological observations

Although they are not entirely distinct, there are three kinds of information derived from seismology that are used to obtain earth models: body waves, surface waves, and free oscillations give complementary data. The earliest Earth model studies used the travel times of body waves through the interior to obtain the variations in velocity of compressional (P) and shear (S) waves. Major boundaries show up as discontinuities or as abnormally rapid increases in velocity with depth (Fig. 1). These boundaries separate the regions within which the velocity increases can be explained as due to increasing pressure on homogeneous material. Body-wave speeds give the ratios of elastic moduli to density. Density is not obtained independently but must be deduced from addi-

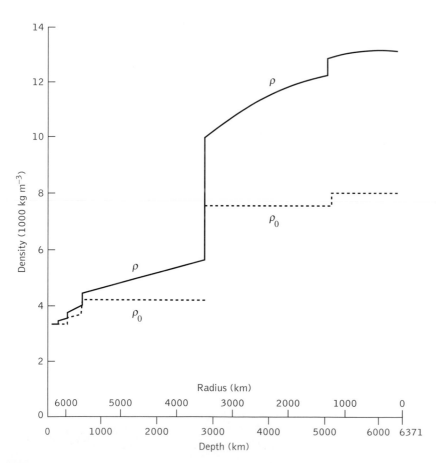

Fig. 1. Density profile of the PREM Earth model (solid line) with theoretical extrapolation to zero pressure and low temperature (broken line). (From Stacey (1992).)

tional data. Before free oscillations were observed this meant total mass, moment of inertia, and high-pressure (shock wave) observations on properties of likely minerals.

Surface waves are guided by the layered structure near the surface of the Earth. The amplitude of wave motion is virtually restricted to a depth of a half or a third of a wavelength and falls off exponentially at greater depths. The longer wavelengths (lower frequencies) thus sample greater depth ranges. Since the (body wave) speeds generally increase with depth, the surface waves of longer wavelength sample more of the high-velocity material and so travel faster. They show strong dispersion (velocity variation with wavelength or frequency), unlike the body waves which are almost non-dispersive. But like the body waves, surface waves give ratios of elastic moduli to density. They give information about near-surface layering and are important in indicating lateral variations in structure, but give no independent density information.

Free oscillations are literally the ringing of the Earth like a large bell after it has been struck (by an earthquake). There are two basic types of free oscillation: toroidal, or torsional, and spheroidal. The simplest torsional oscillation is a twisting motion between two hemispheres. No radial motion or change in shape occurs and no independent information about density is conveyed. The spheroidal oscillations entail radial motion, causing small changes in the Earth's shape, and so gravity contributes to the restoring forces and hence the mode frequencies. This means that density enters the equations of motion in two ways, both as a ratio with elasticity (as in the body and surface waves) and coupled with gravity. The spheroidal mode frequencies thus give an independent measure of the density structure, but one that is still subject to the total mass of the Earth as a scaling factor.

Variation of density with depth

The most widely used Earth model is known by its acronym, PREM (Preliminary Reference Earth Model). It was published in 1981 by A. M. Dziewonski of Harvard and D. L. Anderson of the California Institute of Technology at the request of an international committee that was set up in response to a perceived need for a generally accepted reference model. A very large number of data points were used in preparing the model, the free oscillation frequencies being particularly important. The word 'preliminary' in the name implies that replacement by improved models was expected, but this is certainly not happening quickly, essentially because the changes would be small and, for most purposes, would not justify the effort of producing a new model.

PREM specifies density and the P- and S-wave velocities as polynomial functions of radius in each of a number of radius ranges. It is a spherically averaged model, except in the outermost part, where different continental and oceanic structures are recognized. The density profile is plotted in Fig. 1. Gravity and pressure, which are uniquely determined by the density profile, are plotted in Fig. 2.

Calculation of zero pressure densities requires a finite strain theory, so called because conventional elasticity theory applies only to infinitesimal strains. For pressure ranges that are not very small compared with elastic moduli, the moduli cannot be treated as constant but increase with pressure. In fact they increase much more strongly than does density, which is why the seismic velocities increase with pressure. The value for the bulk modulus (usually called *incompressibility* in geophysics) in the inner core is about eight times the zero pressure value for iron. Compression becomes progressively more difficult with increasing pressure. Numerous empirical formulae have been fitted to Earth model tabulations to account for this. Although there is no really satisfactory theory of finite strain, the differences between alternative calculations are slight and the extrapolated zero pressure densities are well constrained. These are shown as the broken lines in Fig. 1.

Temperature as well as pressure increases with depth in the Earth. The deep interior minerals are therefore thermally expanded and the temperature profile of the Earth acts to oppose the increase in pressure with depth. For much of the Earth a temperature gradient of about 7 °C per kilometre would suffice to cancel the effect of pressure on density. This gradient is exceeded by a factor of three or so in the upper crust, but in the rest of the Earth, except for a thin layer at the core–mantle boundary, the gradient is smaller by a factor of about 20 and the density variation by self compression is completely dominant. Nevertheless, when we interpret zero pressure densities the temperature effect is important. The plot in Fig. 1 gives the low-pressure densities of cooled material, with allowance for thermal expansion.

One of the targets of fundamental Earth science is to determine the deep interior composition. The zero pressure densities give a basis for initial guesses. In the shallowest layer of the mantle, where minerals have familiar low-pressure forms, the matching of possible compositions to the observed density is straightforward. At greater depths high-pressure phases of closer atomic packing appear, and these may be observable in laboratory measurements only in very high-pressure apparatus. High-pressure experiments on likely deep-Earth compositions are therefore vital to the interpretation of the density data.

The core

The abundance of iron in meteorites and in the solar atmosphere made it the obvious candidate material for the Earth's dense core, as soon as the core was identified by seismologists. For this reason the high-pressure properties of iron have received considerable attention. F. Birch, who pioneered finite strain theory, first pointed out that the core density was less than that of pure iron at core pressures. How much less depends on our still insecure knowledge of the high-pressure phase diagram of iron and on temperature, but it may be as much as 10 per cent. There is little doubt that the core composition is dominated by iron, but there are divergent views on the light ingredients that reduce the density to the observed value. Oxygen, sulphur, carbon, silicon, and hydrogen all have their advocates. It is probable that these elements and others are all present to some extent. There is almost

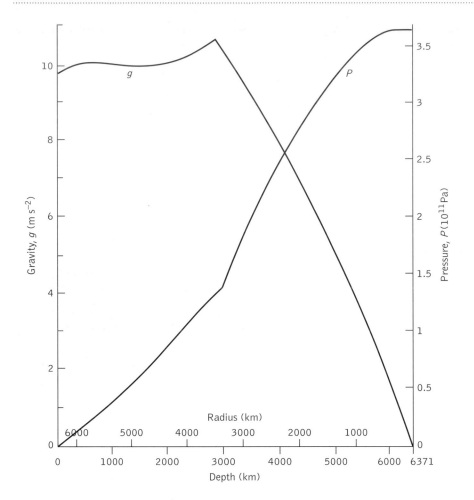

Fig. 2. Variations in gravity (g) and pressure (ρ) with radius for the density profile in Fig. 1.

certainly some nickel and other transition metals, but they do not reduce the density.

The boundary between the liquid outer core and the solid inner core is marked by a density jump of about 650 kg m^{-3}. It is not very precisely observed, but certainly exceeds the density increment due to solidification of a homogeneous alloy at core pressure, which is estimated to be about 200 kg m^{-3}. The difference must be accounted for by a difference in composition. The inner core contains less of the light ingredients (but they are not entirely absent). Since the Earth is cooling slowly, the inner core must be growing by progressive solidification, and there is therefore a continuing process of rejection of the light ingredients by the solid. This leaves an excess of the light ingredients at the bottom of the outer core. The process of upward mixing into the whole outer core releases gravitational energy, which is believed to be the dominant (or sole) source of power for the geomagnetic dynamo.

The outer core is liquid and so lacks a defined crystal structure, but the inner core is solid with a crystal structure that is still in doubt. Under ordinary laboratory conditions iron has a body-centred cubic structure (the alpha phase), but this is stable only at low pressures and is not relevant to the core. Even at zero pressure a face-centred cubic structure (the gamma phase) appears at high temperatures, and this is more strongly favoured at higher pressures. However, at low temperatures increasing pressure causes alpha iron to transform to the denser hexagonal close-packed (epsilon) form. Until recently it was believed that this replaced also the higher-temperature gamma form when core pressures were approached. There is now evidence of further phase transitions, so that the solid phase of iron at core pressures is unclear. In any case the lighter core ingredients probably modify the phase diagram. However, the observation that the inner core is seismically anisotropic, with P-waves travelling faster from pole to pole than across the Equator, is easier to explain if the inner core is composed of a phase with an anisotropic crystal structure, such as epsilon iron. It is on the basis of a comparison of the core with epsilon-iron that the density decrement of the core relative to pure iron is believed to be close to 10 per cent.

The lower mantle

By convention the lower mantle extends from a depth of about 670 km to the outer boundary of the core at 2890 km depth. In Fig. 1 it is seen as a region in which density progressively increases with depth but without discontinuities and over most of the depth range appears to be homogeneous in composition and phase. There are heterogeneities in a layer just above the core, known in the jargon of geophysics as D″ (dee-double-primed), and probably also in the 100 km or so just below 670 km, but these do not show in the figure. The significance of the selection of 670 km as the upper boundary of the lower mantle is that this marks the deepest of the series of phase transitions by which upper mantle minerals are converted by high pressure to the close-packed crystal structures characteristic of deep mantle mineralogy.

In 1976 an experiment at the Australian National University by L. G. Liu, using the then new technique of compressing very small specimens between diamond anvils, showed that when important upper mantle minerals are heated at lower mantle pressures the predominant product is a ferro-magnesian silicate, $(Mg, Fe)SiO_3$, with a perovskite structure. This is a crystal structure first found for the relatively rare mineral perovskite, $CaTiO_3$. A second mineral, $(Mg, Fe)O$, also appeared. This is usually referred to as magnesio-wustite, wustite being FeO, but ferro-periclase might have been a better name as the magnesium is dominant. In a subsequent experiment in the same laboratory, S. E. Kesson and J. D. Fitzgerald examined very small inclusions in diamonds that were believed to have originated in the lower mantle. They inferred that the Fe/Mg ratios in lower mantle minerals are about 1:20 in the silicate perovskite and about 1:7 in the magnesio-wustite; that is, iron favours the magnesio-wustite.

With these compositions of silicate perovskite and magnesio-wustite we know the zero pressure densities and elasticities of what are believed to be the two dominant minerals in the lower mantle (Table 1).

Table 1 Zero pressure densities and elasticities for lower-mantle minerals

	Silicate perovskite $(Mg_{0.95}Fe_{0.05}) SiO_3$	Magnesio-wustite $(Mg_{0.88}Fe_{0.12}) O$
Density ($P = 0, T = 0$)	4163 kg m⁻³	3880 kg m⁻³
Incompressibility ($P = 0, T = 0$)	263 GPa	162 GPa

Data from L.G. Lin, S. E. Kesson and J. D. Fitzgerald

These values can be compared with the extrapolated zero pressure properties of the lower mantle, but remembering that the lower mantle values still refer to a high temperature, although at zero pressure.

Lower mantle ($P = 0$)   Density   3977 kg m⁻³

                          Incompressibility   205 GPa

Knowing how density and incompressibility vary with temperature, we can find a mix of the two minerals and a temperature that match the extrapolated properties of the lower mantle. We find about 75 per cent perovskite, 25 per cent magnesio-wustite, and a temperature of about 1800 °C. This temperature estimate allows us to calculate the range of actual lower mantle temperatures, 2300 to 2700 °C, that would be obtained by compressing the mixture to lower mantle densities starting from 1800 °C. These estimates must of course be treated with caution: the mineralogy of the lower mantle is probably not quite as simple as is assumed here.

Densities of the crust and upper mantle

The upper part of the crust has been sampled by numerous drill cores and is found to be very variable in composition and density. Sedimentary rocks are generally much less dense than the deeper igneous material, but they represent only a thin veneer on the Earth and are not very significant for the total density structure of the planet. The two representative types of igneous rock in the crust are granite (with a density of about 2700 kg m⁻³) and basalt (density about 2900 kg m⁻³). These are both much less dense than the uppermost mantle (3370 kg m⁻³), in which the lighter continental rocks are effectively floating, as revealed by gravity observations. Between the uppermost mantle and the 670-km boundary the mantle minerals undergo a succession of phase transitions to progressively denser structures, as shown in Fig. 1. High-pressure experiments have now identified these intermediate phases as well as the lower mantle minerals.

Lateral density variations

Lateral heterogeneity of the crust is very obvious, especially in the division of the Earth's surface into continents and oceans. Differences in structure between continental and oceanic areas are seen to persist to depths of at least 100 to 200 km, but the variations diminish with depth. In the deep mantle the lateral variations appear to be very small, but they exist at all depths down to the core. At the core–mantle boundary itself there is a layer of quite strong heterogeneity. Seismologists are studying the lateral structure of the mantle by techniques known collectively as tomography, and are obtaining a three-dimensional picture of the mantle. Tomography indicates that there are variations in the velocities of seismic waves within the mantle. The interpretation of these variations in terms of densities is still a matter of contention, but a particularly interesting aspect of this work is that lateral density variations mean buoyancy variations; they must therefore be related to the convective pattern that causes the tectonic processes that make the surface of the Earth such an interesting place.   FRANK D. STACEY

Further reading

Bullen, K. E. (1975) *The Earth's density*. Chapman and Hall, London.

Stacey, F. D. (1992) *Physics of the Earth* (3rd edn). Brookfield Press, Brisbane.

denudation *see* EROSION

desert dunes
Deserts are often imagined to be blanketed in sand dunes, but although almost half of the Australian deserts are sand covered, globally only about 20 per cent of drylands actually reflect the classical image of blowing sand sheets and dunes. Indeed, only 1 per cent of desert areas in the Americas are composed of sand deposits. Despite this, where sand deposits do occur they commonly exist as extensive sand seas (or *ergs*) and so form an impressive large-scale landscape. The nature of these sand seas is highly variable. Many (about 60 per cent) are dune-covered, the type and form of dunes being controlled by variations in sand supply and wind direction. Others may be dune-free and consist of low-profile sand sheets often with some vegetative cover. The chief areas of mobile and stabilized dunes are shown in Fig. 1.

Dune types
Dune sand is generally derived from sources where it has been concentrated by marine or fluvial action (e.g. coastal beaches and dry river valleys) and then transported by desert winds to regions of aeolian accumulation. These winds are not necessarily stronger than winds in humid regions but are more effective in transporting sand because of the lack of surface vegetation in drylands. Sand-sized material (0.04–1.0 mm) also has a low shear velocity threshold and is preferentially transported when compared to larger or smaller grain sizes. Unvegetated (or free) dunes can be classified in relation to their morphology and formative wind characteristics.

Barchan dunes are characterized by their crescentic plan form, with the dune moving in the directions of the horns, as shown in Fig. 2. They are found in regions of limited sand supply and unidirectional winds, resulting in relatively small and highly mobile stores of sand migrating in a downwind direction. Sizes are highly variable, ranging from 1 m in height and 5 m in width to over 30 m high and 300 m wide. Their speed of movement is related to their size; smaller dunes can move at rates of over 40 m a year. Barchan dunes are found within aeolian sand transport pathways and are common in dryland coastal regions such as the Gulf States,

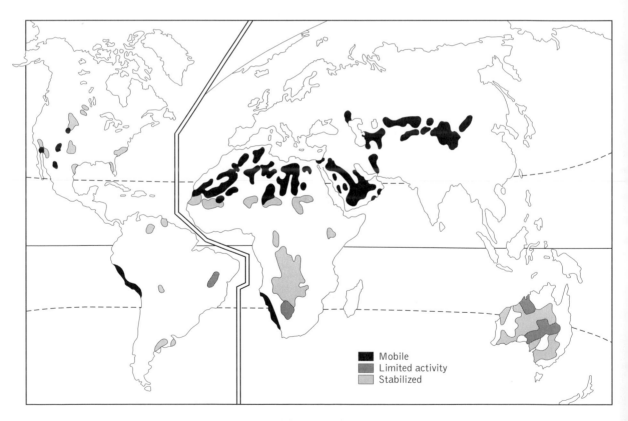

Fig. 1. The global distribution of mobile and stabilized sand dunes and sand sheets. (After Thomas (1997).)

Fig. 2. A 10-m-high barchan sand dune in eastern Oman. Dune movement is from right to left.

North Africa, Namibia, Peru, and California. Transverse (or barchanoid) ridges have a higher sand supply than barchans but also exist in unidirectional winds. In morphology they resemble a series of barchan dunes in which the horns have coalesced into a ridge. Transverse ridges are a very common dune form and can be found throughout the North African, Middle Eastern, and Chinese sand seas.

Linear dunes form in regions with bidirectional wind regimes and relatively limited sand supply. They take the form of linear ridges, between 20 m and 200 m high and over 1 km apart, with steep flanks and a sharp crest. Dunes such as these tend to be accumulating forms, trapping windblown sand from two wind directions and slowly accumulating and extending linearly in the resultant wind direction. They are common in the Namib sand sea. Vegetated varieties occur in Australia and the Kalahari. With complex wind regimes and high sand supply, star dunes may form, with a number of 'arms' of sand radiating from a central peak. Each 'arm' corresponds to a different wind direction. These are accumulating forms with no lateral movement. Star dunes can reach heights of 300–400 m. They are common in the central Saharan sand seas and also in Namibia, China, and Iran.

The classification given above can be complicated by such factors as vegetation and topography. Examples are provided by parabolic dunes and echo dunes. Parabolic dunes are similar in plan-form to barchans but have lower relief, with the horns fixed in position by vegetation. The central part of the dune moves freely downwind and the horns face upwind (in contrast to barchans). Echo dunes accumulate in the lee of topographic barriers where the lee-side airflow may oscillate above the surface, providing regions of low wind velocity where sand masses may accumulate. Owing to their semistabilized nature, parabolic, echo, and vegetated linear dunes may have been fixed in position for thousands of years, providing a useful sedimentary sequence for environmental reconstruction.

Dune initiation and dynamics

The processes by which sand dunes are formed are still little understood. It is clear that a reduction in the shear velocity (u^*) of the wind is required so that grains fall out of the transporting wind and are deposited at the surface. Obstacles ranging from large hills to small shrubs may initiate deposition in their lee, and small hollows in the surface may encourage flow expansion and subsequent deposition. Reduction in shear velocity may also occur because of subtle reductions in the aerodynamic roughness of the desert surface caused by changes in surface grain size. Once a patch of sand has been formed it may grow into a dune by trapping more sand and developing a slip face. As the wind progresses up the windward slope of the dune form, streamlines are compressed by the dune body, causing an acceleration of flow towards the crest, as shown in Fig. 3. This accelerated wind erodes sand from the windward slope and deposits it beyond the dune brink. The lee slope is characterized by low wind velocities in a reverse-flow vortex where sand is deposited on the slip face, which lies at an angle of 32–4°, the natural angle of repose of sand. With erosion on the windward slope and deposition in the lee, the dune body moves in a downwind direction. As the dune develops, an equilibrium is achieved between the angle of the windward slope, the height of the dune, the degree of airflow acceleration, and hence the rates of erosion and deposition on the windward and lee slopes.

Desert dune management and control

Active dunes may pose a hazard to human activity where the blowing sand comes into contact with communication lines, buildings installations, or oases. In semi-arid areas, dunes which are stabilized by vegetation (with more than 10–15 per cent vegetation cover) may also cause problems at a local scale where they become re-activated because of the removal of vegetation owing to farming pressures, drought, or fire. In

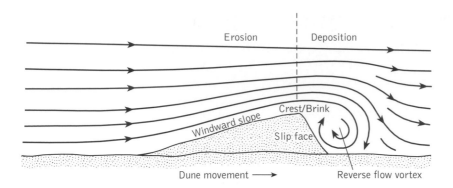

Fig. 3. The compression of streamlines on the windward slope of a barchan dune leading to windflow acceleration and erosion. Deposition occurs on the slip face downwind of the dune brink.

these cases careful management and control of dunes is required. In the semi-arid context, dunes can revegetate quite naturally and quickly once the initial disturbance has been removed. Grazing management in these instances is quite simple, although the effects of a longer-term change in climate may be more serious. Actively mobile dunes in arid areas can be more problematic, especially in the Middle East where infrastructure for the oil industry is commonly placed within or across actively moving dunefields. In these situations the dunes may be controlled with the use of fences or they may be removed with a long-term and work-intensive programme. The simplest and most effective solution, however, is to use careful planning and avoid hazardous areas completely.   GILES F. S. WIGGS

Further reading

Cooke, R. U., Warren, A., and Goudie, A. S. (1993) *Desert geomorphology*. UCL Press, London

Lancaster, N. (1995) *Geomorphology of desert dunes*. Routledge, New York.

Thomas. D. S. G. (ed.) (1997) *Arid zone geomorphology: process, form and change in drylands* (2nd edn). John Wiley and Sons, Chichester.

desert pavement Unusual features abound in dry landscapes. One is the desert pavement. This is a veneer of stones lying at the ground surface, in some instances so tightly packed together that there is almost no soil visible. Indeed, the surface takes on the appearance of being finely paved or cobbled with stone (Fig. 1).

The stones in a desert pavement are usually about halfway embedded in the soil surface, with only their upper surfaces exposed, and are fairly tightly held in place. More rarely, the stones may be loose. Below the stones, there is finer material that may be silty, dry, and powdery, or of heavier clay texture. This material itself contains few, if any, stones. Vascular plants are generally absent in areas of pavement, but the sheltered environment at the interface between the stones and the finer material beneath may be extensively colonized by hypolithic algae, especially beneath translucent stones.

Desert pavement is significant in landscape behaviour because it acts as an efficient seal, preventing ready entry of rainwater into the material below. It also provides a relatively smooth surface across which the surplus water flows, and which facilitates wind transportation of materials. The limited entry of water between the stones entraps air beneath, and the resulting positive pressure deforms the moist soil to form a distinctive vesicular zone containing abundant spherical pores up to 1 mm or so in diameter.

The stones in a pavement are usually coated with a desert varnish of materials deposited from water (see *rock coatings*). The older the pavement, the more developed and thicker is the coating. This can be used as a means of determining the relative ages of pavements within an area.

Desert pavement can arise in several ways. Wind and water erosion may in places gradually sift out and carry away finer materials from an earlier mixed deposit, leaving more and more of the immovable stones lying at the surface. Eventually, so little fine material is exposed that winnowing ceases, and a stable pavement, formed as a lag deposit, results.

Dry landscapes are periodically subject to blowing dust, which can settle across favourable sites as a growing surface mantle. Dust is washed from the upper surfaces of stones to accumulation sites between them (and beneath any overhanging edges). Upon drying out, layers of these fine materials crack, and newly settled dusts both fall and are washed into these openings, which may be too narrow to receive stones. The wind-blown materials thus work their way under the stones. Desert pavements may develop in this way during multiple episodes of deposition and in climatic conditions different from those of today. The pavement stones are not a lag deposit left by erosion, but rather are from the original landscape below the dust, and have been kept at the surface by the downward infiltration of finer grains as they accumulate. This mechanism for pavement development affords an explanation of the rarity of stones in the finer sub-pavement material.

The development of a nearly complete desert pavement may effectively terminate the developmental process in either of the mechanisms just outlined. Further loss of fine material is prevented by a dense pavement, while dust accretion is also

Fig. 1. Embedded desert pavement of varnished siliceous stones in the Sturt Stony Desert near Innamincka, northern South Australia.

restricted by rapid and nearly complete run-off of rain. Thus, well-developed desert pavement may be an indicator that the surface on which it lies has been stabilized. Landscape activity may then switch to other areas, perhaps to the stream channels that must carry away the water shed from the pavement surfaces. DAVID L. DUNKERLEY

desert sedimentary deposits Today's deserts occur mainly either side of the Equator, between latitudes 10° and 30° in belts of high aridity associated with area of persistent high pressure. They are in addition found in the arid interiors of some continents, on the leeward sides of mountain ranges in rain shadows, and along coastal areas where the presence of cold oceanic water prevents transference of moisture to a neighbouring land-mass.

Most desert surfaces are erosional: they are formed of bare rock (*hamada*) or deflated stony plain (*serirs* or *reg*). However, large volumes of sediments accumulate in alluvial fans, aeolian sand seas (*ergs*), and in playa lakes or sabkhas.

Long rainless periods accompanied by diurnal temperature changes of large magnitude, together with crystallization of salts drawn to the surface by capillary action, combine to cover barren mountainous areas with a cloak of broken rock and mineral debris. Periods of violent rainfall, although rare, produce short intervals of intense run-off. Sediment-laden streams carve deep, steep-sided valleys which, where they debouch into adjacent lowlands, produce alluvial fans. These have the shape of a segment of a cone, with a profile that is convex upwards as seen across the fan and concave upwards along the fan profile. Surface slopes vary from 5° to 25°, but

are usually less than 10°, and everywhere decrease away from mountain fronts. Adjacent fans may merge to form a broad sloping surface, known as a *bajada*, at the foot of a mountain range. They are crossed by a diverging pattern of shallow channels; streams flowing through the channels deposit cross-bedded imbricate gravels, sandy gravels, and sands which are lenticular, and commonly form down-fan longitudinal bodies. During extreme floods streams escape from the channels and deposit sheet-flow deposits, which consist of low-amplitude bars of clast-supported gravels and planar and cross-stratified sands in shallow channels over the whole fan surface. Where the source produces little fine material, open-work gravels—sieve deposits—are formed, which are often infilled by a later infiltration of finer sediment. The streams are sometimes so overloaded with sediment that they degenerate into viscous sediment gravity flows—mudflows—which may be confined to channels but, particularly in the upper fan spread over the fan surface. They produce matrix-supported conglomerates (paraconglomerates) that commonly show inversed grading with large boulders at the top of the deposit.

During intervening dry periods, winds scour the fan surfaces and sandblast exposed gravel clasts to produce smooth faceted *ventifacts*. Water drawn to the surface by capillary action precipitates iron and manganese crusts ('desert varnish') on the surface of the pebbles. Capillary movement of waters leads to precipitation of crusts of calcium carbonate (calcrete) or gypsum, (gypcrete). Such accumulations which develop in surface sediments mark hiatuses of deposition.

Apart from scattered plant roots, skeletal remains of terrestrial organisms, and some burrows, fossil remains are rare in

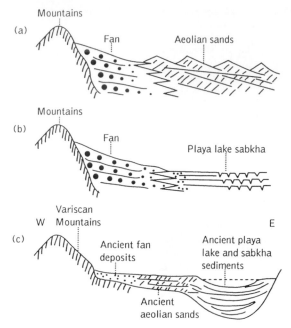

Fig. 1. Various types of association of desert sediments (a) Fan-aeolian sands; (b) Fan-playa; (c) section of Rotliegende sediments across southern North Sea. (Sketched from Glennie (1972), *Bulletin of the American Association of Petroleum Geologists*, **56**, 1048–71.)

desert deposits. Repeated flooding produces a thick wedge of coarse-grained deposits termed fanglomerates, which extend from the mountains into the adjacent lowlands. The sediments become finer and show increased roundness from the top to the bottom of the fan. Tectonic changes in the source area produce cyclic sequences (10–100 m thick) that coarsen upwards or sequences that become finer upwards where faulting has caused retreat of the mountain margin.

Although they cover only approximately 20 per cent of the world's deserts and are rare in some, aeolian sands are perhaps regarded as the most typical desert sediments. The sand seas or ergs cover thousands of square kilometres, and in places the sand cover may be hundreds of metres thick.

Wind winnows fine sediment from rocky, stony areas and the surfaces of alluvial fans to leave a gravel lag. The material of silt-sand size is transported in suspension to settle in playa lakes or sabkhas, or form loessic deposits on desert fringes.

Sand is transported by rolling, creep, saltation, and in suspension to form fine to medium, well-sorted well-rounded sands fashioned into a variety of bedforms. These grade from ripples (with heights ranging from centimetres to metres and wavelengths up to metres), dunes (with heights up to metres, wavelengths of hundreds of metres) and huge bedforms: draas (with heights up to hundreds of metres and wavelengths measured in kilometres) with superimposed dunes. Various

dune forms develop, according to the wind regime and availability of sand.

Barchan, barchanoid dunes, and transverse dunes develop under winds with a low directional variability and show considerable lateral movement. Linear or seif dunes occur in areas of more variable winds and show less lateral movement; star-shaped dunes develop in areas of very variable winds and show little lateral movement and considerable vertical accretion.

Migration of dunes and deposition on their lee flanks produces cross-stratified sands up to 10 m or more in scale with moderately high angles of dip, and some superimposed ripple cross-stratification. Cross-stratification in barchan, barchanoid, and transverse dunes has a variable down-wind unimodal pattern. In linear dunes laminae dip away from the main axes to give a bimodal pattern, and in star dunes there is a poly-modal pattern. The various patterns of cross-stratification have been used to differentiate various forms of dunes in ancient deposits.

As various bedforms migrate over one another, they produce a variety of erosional surfaces between sets or groups of cross strata; these are termed 'bounding surfaces'. First-order bounding surfaces are formed where draas migrate over older aeolian sands; second-order bounding surfaces form where dunes move over the flanks of draas or one another; and third-order bounding surface are surfaces that cut across sets of cross-strata and are formed by erosion or changes in local wind direction.

In some deserts, salty flats (sabkhas) or ephemeral lakes or playas with marginal sabkhas occur. Wind-blown dust settles in these areas. They are sometimes fed directly by surface flow from adjacent fans or mainly by surface flow.

Evaporitic minerals (carbonates, gypsum, halite, borates, etc.) are precipitated from evaporating waters. The sediments show polygonal desiccation cracks and tepee structures—polygonal patterns of ridges formed by the expansive force of crystallization of the various salts.

As Fig. 1 shows, the general relationships between various facies—alluvial fan, aeolian sands and playa deposits—depend on the tectonic setting. In many fault-bounded desert basins (e.g. the western USA), fans pass directly into playa lake sediments with little dune sand. In other deserts (e.g. Arabia) alluvial fans are bounded by wide dune fields with few playas.

Desert sediments are well known in the geological column. Although they are commonly red or buff in colour with low contents of organic matter, their coloration is mainly of diagenetic origin and is not exclusive to these deposits. It cannot therefore alone be used as an environmental indicator of arid conditions. The Permian deposits of the North Sea show the development of alluvial fans, aeolian sands, and playa lakes in a most striking way (Fig. 1c). G. EVANS

Further reading

Glennie, K. W. (1970) *Desert sedimentary environments*. Elsevier, Amsterdam.

desertification Desertification is the name given to a complex of interrelated human and natural processes that lead to environmental degradation in drylands. It was first used in 1949 by the French scientist Aubréville to explain environmental degradation in sub-humid tropical Africa, caused mainly by human impacts. Since that time, numerous individuals and organizations have produced different definitions of the phenomenon. As defined by the United Nations Environment Program in 1992, desertification is 'land degradation in arid, semi-arid, and dry sub-humid areas caused by adverse human impact'. Land degradation is the reduction or loss of potential productivity and biological potential by adverse changes in soil characteristics and/or soil loss by wind and/or water erosion. These changes are such that natural recovery is not possible or takes so long that in, human terms, it represents a permanent change.

Approximately 40 per cent of the Earth's land surface, encompassing a population of around 850 million people, is susceptible to desertification. Seasonal rainfall regimes with a pronounced dry season, an annual excess of evapotranspiration over precipitation, and fragile ecosystems create an environment that can be degraded by population pressure, inappropriate technology, and poor land-use practices. Drylands are regions that often experience drought (periods when rainfall is 80 per cent or less of normal) that last for years or decades (e.g. the Sahel region of Africa in the late 1960s and 1970s). Droughts reduce biomass and increase the potential for wind and water erosion of cultivated lands. At the same time, human pressure on the land through overgrazing, overcultivation, and gathering of fuelwood may reduce the ability of the natural system to withstand drought or recover from it.

Estimates of the rates of desertification have varied considerably. In the 1980s, the United Nations estimated that 20–27 million hectares were being lost to productive use each year. More recent studies have suggested that the much of the evidence for progressive and permanent degradation (the so-called 'advancing desert' phenomenon) is incorrect. Simplistic analyses of changes in vegetation cover probably confused natural temporal and spatial variability with long-term degradation caused by human activities. Many dryland ecosystems are very resilient, and vegetation cover recovers rapidly in periods of good rainfall. For example, the Sahara desert expands and contracts as rainfall varies from year to year. Soils, however, can degrade progressively and permanently. Using this approach, the United Nations Environment Program estimates that 1036 million hectares or 20 per cent of susceptible drylands have experienced degradation (Fig. 1). Of this, only 4 per cent is strongly or extremely degraded. Water erosion is the main process, affecting 48 per cent of the area, with wind erosion affecting 39 per cent. Physical and chemical changes (e.g. salinization, waterlogging) affect a further 14 per cent of susceptible areas.

NICHOLAS LANCASTER

Further reading

Thomas, D. S. (1993) Sandstorm in a teacup? Understanding desertification. *Geographical Journal*, **159**, (3) 318–31.

UNEP (1992) *World atlas of desertification.* Edward Arnold, London.

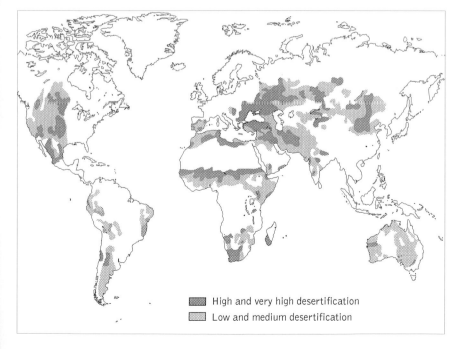

High and very high desertification
Low and medium desertification

Fig. 1. Map of the extent of desertification in the world's drylands (after UNEP 1992).

deserts Deserts are regions characterized by very low annual rainfall (usually less than 300 mm), sparse vegetation, extensive areas of bare, rocky mountains and plateau, and alluvial plains. Sand dunes cover less than a third of desert regions. Many deserts are regions of high temperature (the hot or tropical deserts), but some polar regions (including most of the Antarctic continent) are also classified as deserts.

A generally accepted definition of a desert is a region in which mean annual potential evapotranspiration (Etp) exceeds mean annual precipitation (P) by a factor of two or more. A world map of desert regions (Fig. 1) identifies three main classes of aridity: (1) hyperarid ($P/Etp < 0.03$); (2) arid ($0.03 < P/Etp > 0.20$); and (3) semi-arid ($0.20 < P/Etp > 0.50$). Defined in this way, deserts cover approximately one-third of the Earth's land surface.

Causes of deserts

Desert climates are characterized by low humidity (except in cool, foggy coastal deserts like the Namib), a high daily range of temperatures, and precipitation that is highly variable in time and space. The most extensive deserts lie astride the tropics. Descending, dry stable air masses in the subtropical anticyclonic belts maintain arid conditions throughout the year. The effects of stable air masses are reinforced by large land masses. Long distances to continental interiors restrict the influence of moist oceanic air masses in summer, as in the central Asian and African deserts. In winter, large continental areas develop strong high-pressure cells, reducing the influence of frontal systems. Mountain barriers block rain-bearing winds and create rain-shadow areas in their lee,

especially in the Great Basin Desert of North America and in central Asia, where the Himalaya prevent penetration of the south-west monsoon to the Gobi and Takla Makan deserts. Deserts located on the west coast of South America and southern Africa (Atacama, Namib) owe their hyperarid climates to the influence of cold oceanic currents offshore. These reinforce the subsidence-induced stability of the atmosphere by cooling surface air masses and creating a strong temperature inversion.

Desert landforms

Although no landforms or geomorphological processes are unique to deserts, certain characteristics of desert environments have a significant effect on the operation of the major processes of weathering, erosion, transport, and deposition. A sparse vegetation cover with a high percentage of bare ground results in rapid run-off of water when intense rainfall does occur, and enhances the ability of the wind to erode and transport sand and dust (silt- and clay-sized sediment). Sand accumulates in areas of lower wind velocity and transport capacity to form dunefields and sand seas comprised primarily of crescentic, linear, and star dunes, with sand sheets in marginal areas. Dune form is governed by the availability of sand and the variation in wind direction from season to season. Dust storms transport fine-grained material away from desert regions to be deposited in ocean sediments and desert margin soils. Some may reach the polar ice caps.

The excess of evaporation over precipitation gives rise to physical or mechanical, rather than chemical, weathering of rocks, and to upward movement of soil moisture and near-

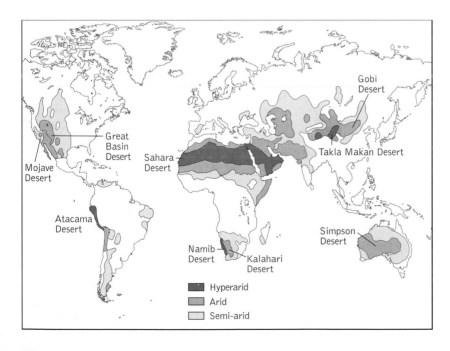

Fig. 1. Map of the global extent of hyperarid, arid, and semi-arid regions.

(a)

(b)

Fig. 2. (a) Landscape of a representative shield desert: the Namib. Note the abundance of low-relief rocky surfaces and sand dunes in the foreground. (b) Landscape of a high-relief, tectonically active desert region: Death Valley; California. Note the steep mountain slopes, alluvial fan at the mountain front, and salt-encrusted playa in the centre of the valley.

Table 1. Proportions of landform types in deserts (percentage of area covered) (from Cooke *et al.* 1993)

Landform type	South-western USA	Sahara
Desert mountains	38.1	43.0
Playas	1.1	1.0
Desert flats	20.5	10.0
Bedrock fields	0.7	10.0
Regions bordering through flowing rivers	1.2	1.0
Ephemeral streams	3.6	1.0
Alluvial fans	31.4	1.0
Sand dunes	0.6	28.0
Badlands	2.6	2.0
Volcanic fields	0.2	3.0

surface groundwater. As a result, water-soluble salts (principally sodium chloride, calcium carbonate, and calcium sulphate) accumulate in desert soils, forming calcic and gypsic horizons in the subsoil. Insolation weathering and salt weathering dominate processes of rock breakdown. On a regional scale, lack of water gives rise to internal drainage and thus to playas and salt lakes.

The character of desert landforms is also affected strongly by the regional geological and tectonic environment. Two end-member models can be recognized (Table 1). The tectonically stable Old World shield deserts, such as those in the Arabian Peninsula, Australia, and southern Africa (Fig. 2a), are characterized by low relief and extensive rocky plains and isolated hills (*inselbergs*), with up to 30 per cents of the land surface occupied by areas of sand dunes or sand seas. Variants of this landscape depend on whether the bedrock is sedimentary, as in the northern and eastern Sahara, or crystalline, as in much of Australia and Namibia. At the other end of the spectrum are the high-relief deserts of the tectonically active areas of the Atacama, the Basin and Range Province of western North America, and parts of central Asia (Fig. 2b). These deserts have only rare and small sand-dune areas, but extensive areas of desert mountains, alluvial fans, and internal drainage.

Deserts in the past

The distribution of desert climates has changed significantly over geological time. Many of the modern areas of aridity originated during late Tertiary times (especially the Mid- to Late Miocene) as the 'modern' climate and geography of subtropical regions developed. Aridity in north Africa and Asia intensified with the uplift of the Himalaya and the formation of the Tibetan Plateau, blocking the penetration of the monsoon to central Asia and reinforcing the tropical easterly jet stream that brings dry, stable air masses to the Sahara. Narrowing of the Tethys seaway increased the con-

tinentality of northern Africa, culminating in the isolation of the Mediterranean in the Miocene and the first signs of true aridity in the Sahara. The southern African deserts owe their origins to the development of the Antarctic ice sheet and cooling of the Southern Ocean, which led in turn to the formation of the Benguela Current and its upwelling system. Australian deserts developed as the continent 'drifted' northward to reach its present latitude in the Miocene. In North America, uplift of the Sierra Nevada and the Transverse Ranges of southern California created a barrier to penetration of moist air masses from the Pacific, giving rise to the Great Basin and Mojave deserts.

During the Quaternary period, climatic changes associated with glacial–interglacial cycles at high latitudes resulted in changes in the extent of deserts and the intensity of aridity. The core hyperarid areas, such as the central Sahara and Namib, were relatively unaffected; instead, desert margin areas (the Sahel and Kalahari) were affect most. For example, there is no evidence to suggest that the Namib has experienced any climate wetter than semi-arid at any time during the Quaternary. By contrast, degraded dune systems, now covered by savannah vegetation, occur in sub-humid areas adjacent to the southern Sahara, the Kalahari, and in Australia, and indicate a expansion of arid conditions during glacial periods. In North America, dunes in the semi-arid High Plains were also mobile as little as 1000 years ago in periods of extended drought. Conversely, lake and spring deposits, as well as archaeological and faunal evidence, show that many parts of the Sahara experienced humid conditions in the period 9000–6000 years ago.

In the more distant past, the rock record of ancient aeolian (i.e. wind-blown) sandstone and evaporite deposits is evidence for extensive desert conditions during the Permian (the Rotliegendes Sandstone of the North Sea Basin) and during the late Late Carboniferous, Permian, and Jurassic (e.g. the Weber, Navajo, and Entrada sandstones of western North America). The Old Red Sandstone (Devonian) of Britain is also thought to have been deposited under somewhat arid conditions.   NICHOLAS LANCASTER

Further reading

Cooke, R. U., Goudie, A. S., and Warren, A. (1993) *Desert geomorphology.* UCL Press, London.

Thomas, D. S. G. (ed.) (1977) *Arid zone geomorphology.* John Wiley and Sons, Chichester.

Devonian The Devonian System was first recognized as a major Palaeozoic stratigraphical unit in its own right by Adam Sedgwick and R. I. Murchison in 1839. Their researchers in south-west England had convinced them that it was younger than the Silurian System of Wales and the Welsh Borderland and at the same time older than the Carboniferous Limestone. Its correlation with the Old Red Sandstone of England and Wales was also recognized. The system is conventionally divided into three series and seven stages on the basis of its marine faunas. Devonian rocks occur on all the continents, cropping out over some 77 500 00 square kilometres. The period lasted from around 410 Ma to 355 Ma.

The biostratigraphy of Devonian marine facies is based upon conodonts or, in certain facies, upon ammonoids. The continental non-marine facies is divisible on the basis of its vertebrate faunas and macroplant or spore assemblages. The base of the system has by international agreement on the recommendation of the International Commission on Stratigraphy (ICS) been defined at the base of the Lochkovian Stage, which is also the base of the graptolite zone of *Monograptus uniformis uniformis* and of the conodont *Icriodus woschmidti*. The global stratotype section and point is at Klonk in the Czech Republic, and in 1970 was the first such point to be designated formally by the ICS.

Other regional stages based on marine biostratigraphy have been used in the USA, the former USSR, and China, but the chronostratigraphy advocated by the ICS is now increasingly used.

The Devonian continents were gathered into groupings inherited from the early Palaeozoic, with a major Gondwana land-mass in the southern hemisphere and four separate continents in the equatorial and north temperate latitudes (Fig. 1). Laurentia (N. America) and Baltica (Europe) became sutured together in late Silurian time to give the single conti-nental block, Laurussia. This was separated from Angaraland (Siberia) to the north by a narrow ocean, and from Kazakhstania by a somewhat wider one to the east. China remained isolated at the eastern end of this group of continental blocks. Orogenic and mountainous chains existed along the Caledonide–Appalachian belt from northern Greenland to the south-eastern USA, and in the Ural region. Minor uplifts occurred in western North America. There were pronounced upland belts in Kazakhstania, Siberia, and China, where volcanic activity also persisted. Orogeny continued spasmodically in many of these regions throughout the period. In Gondwanaland the western seaboard of South America from Colombia to southern Argentina was the setting for much volcanic and tectonic activity. At the other end of the supercontinent, eastern Australia experienced similar events. Shallow epeiric seas flooded great areas of the continents, especially in mid- and later Devonian times. In the equatorial regions this allowed the spread of carbonate sedimentation and, locally, the precipitation of thick evaporites. Reef growth took place in many areas of western Canada, eastern central USA, western and central Europe, and in Russia, China, and parts of Australia.

Clastic sedimentation around the continental margins was generally at a gentle pace, but with notable exceptions in the vicinity of orogenic uplifts. Here, thick clastic wedges soon built up, giving rise to the continental facies, the Old Red Sandstone, as in the North Atlantic area, Arctic Canada, and

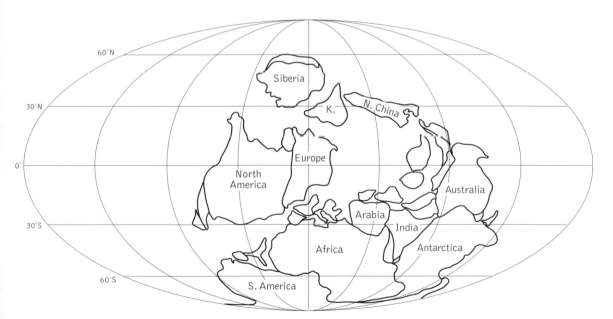

Fig. 1. The world during the Givetian epoch, that is, during the later part of Middle Devonian time, about 380 Ma. At this juncture the main land-masses were gathered into a major group constituting the southern supercontinent of Gondwanaland. To the north of this the continents of Laurussia (North America and northwestern Europe), Siberia (Angaraland), and Kazakhstania (K) and North China lay in a great arc. The central ocean between all these periodically flooded the edges of the continents, eventually producing one of the major marine transgressions of Phanerozoic time. In the south polar area glaciations occurred over parts of South America. (After Scotese and McKerrow.)

parts of China and eastern Australia. During late Devonian time black shale events occurred, induced by phases of anoxia in the shallow epeiric waters. Eustatic changes of sea level have left widespread effects, and several cycles of transgressive–regressive movement of the strandline are well documented in North America, Europe, and Asia.

There is abundant evidence of Devonian continental glaciation in southern South America, but the extent, volume, and age of the ice mass is uncertain. While palaeontological and sedimentological evidence points to tropical, even monsoonal, climates in the equatorial regions, there are also indications of cool- or cold-water animal communities in South America. To what extent the glacial phase or phases contributed to eustatic changes is conjectural, but it was probably only on a minor scale. Ocean-ridge volcanism and tectonics may have been more significant.

Life in the equatorial shallow waters was prolific to the extent of providing organic matter for the accumulation of large volumes of oil and gas. Bottom-dwelling communities were widespread and diverse. Brachiopods reached their heyday during this period and there were also many few families of both the rugose and the tabulate corals. Corals and stromatoporoids were, with the algae, builders of chains of reefs in parts of North America, Europe, North Africa, many parts of Asia, and Australia. Reefs and the waters about them offered habitats to many new bivalves and other molluscs, a somewhat declining number of trilobite families, and new groups of crinoids. Some calcareous foraminifera now began to make their appearance in vast numbers. Among the larger pelagic animals none were more successful than the ammonoids. Beginning with small uncoiled forms in the early Devonian, their rapid radiation give rise to widespread quickly evolving forms. The tiny conical shells known as dacryoconarids were also abundant and evolving rapidly at this time, but only a few dendroid graptoloids remained from the Silurian faunas. The ostracodes included both marine and freshwater forms, and the scorpion-like eurypterids were now formidable members of the freshwater communities.

The conodont vertebrate animals reached the zenith of their existence during the Devonian, with species living in several different marine environments. Their pelagic existence and rapid evolution make them ideal for biostratigraphy and global correlation. During this period the agnathan and gnathostome fishes, too, gave rise to many distinctive groups: the armoured ostracoderm and placoderms, the bony fishes, and elasmobranchs to populate marine and freshwater environments. Towards the end of the period the first tetrapod amphibia had evolved from advanced bony fish. Their terrestrial mode of life was possible in the great swamp forests of middle and later Devonian times. At the start of the period the only vascular (land) plants were primitive psilophytes, but in Middle Devonian time these disappeared and were replaced by the 'Archaeopteris flora' with fern-like leaves and spore-bearing organs. The later forests also included horsetail ferns, seed ferns, and lycopods. Within these forests there were also

numbers of insects, including some with the power of flight.

Towards the end of the period a crisis occurred that affected many marine animal communities. On the other hand, the forests and the freshwater and terrestrial communities do not seem to have been affected. Among the marine groups to suffer heavily were certain brachiopods, trilobites, conodonts, and corals. There seems to be evidence that a sudden catastrophic event, such as a meteorite impact, caused immediate significant changes, but elsewhere the extinction seems to have taken some time. The scenario proposed is like that offered for the Cretacous extinction, but is less convincing. An alternative and tentative explanation of this culling of the marine biota is that glaciation in the south polar area caused sufficient cooling of the surface ocean waters to bring about the collapse of the reef communities. Both processes may have taken place.

D. L. DINELEY

Further reading

Dineley, D. L. (1984) *Aspects of a stratigraphic system: the Devonian.* Macmillan, London.

McMillan, N. J., Embry, A. F., and Glass, D. J. (eds) (1988) *Devonian of the world. Proceedings of the Second International Symposium on the Devonian System, Calgary, Canada.* Canadian Society of Petroleum Geologists.

diachronism The concept of diachronism is easy to grasp but is commonly difficult to demonstrate. *Lithostratigraphic* (i.e. rock) *units* may not correspond to *chronostratigraphic* (i.e. time–rock) *units*, but can show a progressive change of age from one place to another. They may thus be said to be time-transgressive. Fossils normally provide the critical data for determining the age of the rock units; they also give information about the facies of the rock body in question. A typical example of diachronism would be a sequence of beds deposited in a shallow sea that gradually encroaches (transgresses) upon the land. As the sea advances, so do the belts of sedimentary (facies) deposition on the sea floor, each (facies) more distant from the shore advancing over its inshore neighbour as time goes on. For example, the various Lower Palaeozoic and Devonian formations on the stable interior platform of North America have presented many problems of correlation largely because the different facies are diachronous. Similar lithologies in this area contain very different fossils, the inheritance of a constantly migrating set of facies belts developed over a long period of geological time. In the Triassic rocks of South Wales and the west of England, a local formation known as the Dolomitic Conglomerate passes laterally into rocks of both the Sherwood Sandstone Group and the overlying Mercia Mudstone Group. It is regarded as a diachronous facies of rocks marginal to highlands that existed at that time (and in part still do exist). Likewise, in southern England some sands of Early to Middle Jurassic age are markedly diachronous over a distance of about 160 km from north to south (Fig. 1).

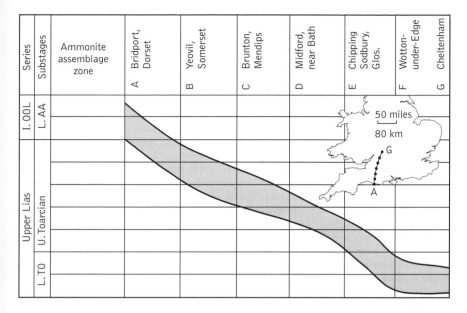

Fig. 1. A diachronous facies in the Lower to Middle Jurassic rocks of southern England. Here the Bridport, Yeovil, Midford, and Cotswold Sands units are parts of a facies that becomes progressively younger southwards. It reflects the continuing southward shift of a depositional environment during a time-span of perhaps 10 Ma or more. I. OOL, Inferior oolite; L. AA, Lower Aalenian; L. TO, Lower Toarcian.

This progressive migration of sedimentary facies is common and led the German geologist Johannes Walther to note in 1894 that the lateral succession of facies is frequently found to be repeated vertically. In other words, Walther's 'law' suggests that vertical stratigraphy may represent lateral transgression; the rock units and facies are diachronous. When used with caution, this is a useful tool that enables stratigraphers to limit possible interpretations of lithological successions. The need for good fossil data for correlation remains critical. D. L. DINELEY

diagenesis The term 'diagenesis' is attributed to von Gumbel who coined it at the end of the nineteenth century. It includes all processes which affect sediments after their deposition and during burial, excluding the effects of tectonism and metamorphism. Diagenesis usually, but not always, results in the production of an indurated or lithified (i.e. hard) rock. Diagenesis grades imperceptibly into metamorphism: the division between them is taken at the point where the first metamorphic index mineral appears.

Diagenesis is controlled by several factors: the composition and grain size of the original sediment; the rate of deposition and environment of deposition; the rate of burial and physical and chemical changes during burial. It is also influenced by the character of adjacent sediments. Diagenetic changes are both physical and chemical. Water is expelled as the component grains are packed together by overburden pressure on burial. Expelled water escapes into more porous adjacent sediments where it may cause solution or precipitate mineral cements according to the chemical conditions.

Compaction is the main physical process. Organic sediments such as peats, which originally contain more water than

sediment, are reduced to a fraction of their original thickness. Fine-grained siliciclastic muds which may contain up to 80 per cent water when originally deposited, retain only about 10 per cent of it after burial. During compaction of siliciclastic muds the flaky clay particles are realigned to be parallel to each other and impart a fissility (a direction of easy splitting) to the rock, which is commonly called shale. Sands and coarser sediments, which can contain up to 50 per cent water on deposition, undergo minor compaction. Individual grains are rotated slightly and may slip and even sometimes fracture as they are forced into a tighter packing. However, this produces only minor changes in porosity. Miocene sands in the Gulf Coast of the USA retain porosities of around 30 per cent at depths of 3226 m. Where sands contain interstial mud or grains of mud or fine-grained rocks they may undergo greater changes of porosity. The softer grains are smeared between more resistant grains to produce a matrix and hence reduce the pore space.

The character of the chemical changes undergone by sediments during diagenesis is controlled by the activity of ions in the pore-water solution between the grains, its Eh (redox potential) and pH as well as the temperature and pressure. These changes can take place in several different environments. For example, they may occur in the continental realm above the water table—the vadose zone; or in the permanently shallow saturated zone—the phreatic zone, in either freshwater or marine conditions. They may also occur in the zone of deep burial.

Although the climate of the environment of deposition controls the pathways of early chemical diagenesis in terrestrial and coastal situations, the effect of surface water lessens during burial. The pore waters become more reactive as the solubility of various mineral phases increases with increasing temperature and pressure.

Organic matter in sediments is very reactive to diagenetic change. The fairly large polymers of primary organic matter are degraded by bacteria during burial and converted to monomers to form the complex organic compound kerogen as well as releasing biogenic gas (methane and carbon dioxide). As temperature increases during burial, kerogen breaks down to yield oil and wet gas—the process known as *catagenesis*—and ultimately it passes into the zone of *metagenesis*, where only dry gas is released.

The humic organic matter of peat is also altered on burial, water and volatiles being lost during the process of *coalification*. The carbon content of the organic matter increases, that is, it is said to increase in rank. The parent material, peat, is converted to soft brown coal and then to hard brown coal and ultimately via bituminous coal to anthracite under the influence of increased pressure and temperature, methane gas being produced during the process.

Evaporites undergo changes on burial owing to reactions between interstitial brines and previously deposited salts to produce new suites of minerals. Fine-grained siliceous deposits originally composed dominantly of opal lose water and recrystallize as chalcedony, and then eventually quartz. Fine-grained carbonates are replaced so that they consist entirely of calcite and can suffer further diagenesis and may be replaced by dolomite.

Fine-grained siliclastic deposits are also prone to diagenetic change. The clay mineral kaolinite can develop early in acidic conditions but on burial changes to other polymorphs which are, however, destroyed at higher temperatures or can be converted to other clay minerals as the water chemistry changes. Smectites and vermiculites develop in alkaline waters during early diagenesis, as well as minerals such as attapulgite (authigenic minerals). They may be converted to kaolinite if they are brought in contact with acidic pore waters. However, usually during burial they become dehydrated and transformed to mixed-layer clays, and then to illite or chlorite according to the pore-water chemistry. Illite and chlorites form more stable polymorphs during diagenesis in the presence of alkaline pore waters, and the amount of the two minerals increases with time. It is thus difficult to ascertain the nature of the original or early diagenetic clay in Palaeozoic rocks, but this becomes increasingly easier in Mesozoic and Cenozoic rocks.

The coarser grains of siliclastic sediments usually survive chemical diagenetic changes with only minor modification, although some minerals such as feldspars are sometimes broken down to clay and carbonates. In carbonate deposits individual grains are either recrystallized or replaced by calcite when they were originally composed of aragonite or high-magnesium calcite. Grains become coated with a precipitated mineral cement consisting mainly of quartz, carbonates, iron oxides, and clay. Some of this cement is of local origin but most of it originates from water expelled from adjacent mud-stones. In some instances complex sequences of cements are present which were deposited at various times during diagenesis and have different origins. Studies using staining techniques, cathode-luminescence, ion-probes, and isotopic analyses have helped greatly in unravelling the details of cementation histories.

In carbonates (mainly limestones), cementation may occur very early while the sediment is exposed on the sea floor, but continues during burial, often with several episodes of cementation. Carbonates are very easily altered. Both coarser and fine-grained carbonates may be replaced by silica (chert), phosphate and dolomite at various times during their diagenetic history.

Deep burial sometimes leads to pressure solution at the points of contact of grains in siliciclastic and carbonate sediments. This produces sutured interpenetrative contacts between grains, and these may form thin extensive sheets of concentrations of insoluble material called stylolite seams. Whereas most diagenetic changes produce a denser indurated deposit, sometimes, solutions driven out of adjacent sediments can lead to solution of earlier cements and the production of high-porosity zones in the subsurface.

The cumulative affect of diagenesis is to modify the porosity and permeability of sediments and thus their economic potential.

G. EVANS

Further reading

Burley, S. D., Kantarowicz, S. D., and Waugh, B. (1985) Clastic diagenesis. In Brenchley, P. J. and Williams, B. P. J. (eds) *Sedimentology: recent developments and applied aspects*, pp. 189–228. Blackwell Scientific Publications, Oxford.

Dickson, J. A. D. (1985) Diagenesis of shallow-marine carbonates. In Brenchley, P. J. and Williams, B. P. J. (eds) *Sedimentology: recent developments and applied aspects*, pp. 173–188. Blackwell Scientific Publications, Oxford.

Leeder, M. R. (1982) *Sedimentology, process and product*. George Allen and Unwin, London.

diagenesis in deep-sea sediments The organic material that rains down on to the sea floor is buried and is oxidized (or broken down) by bacteria present in the sediment. The most efficient way for the bacteria to break down organic material is to use oxygen as an oxidizing agent (aerobic respiration). In truly pelagic deep-sea sediments where the supply of organic material and the rate of burial are low, diagenesis does not generally continue beyond this stage. Where the rate of organic carbon supply or burial is higher, the oxygen concentration in the sediment interstitial water can fall to such a low level that diagenesis must proceed using secondary oxidants (anaerobic respiration). Oxidants are consumed (or reduced) in the order nitrate ~ manganese oxides > iron oxides > sulphate. This produces a vertical sequence of 'diagenetic zones' in the sediment. In hemipelagic deep-sea sediments where the organic carbon content and sedimentation rate are higher, diagenesis can proceed to the iron oxide reduction stage. In deep trenches, where overlying waters might be oxygen free, reduction of sulphate and in some cases methane production occurs.

During early diagenesis, elements are mobilized into solution and so are able to migrate through the interstitial waters. These may be reprecipitated and incorporated into authigenic materials, such as glauconite, various carbonates, and iron and manganese oxides. Alternatively, the element can escape from the sediment entirely and be released into the overlying water column.

As the sediment is buried more deeply the sedimentary material begins to dewater under the effects of compaction, and primary cementation of sediment can then begin.

Deep burial

During deep burial, dewatering and compaction continue, and pore waters are modified further by reactions with clay minerals, dissolution of unstable material, and precipitation of authigenic minerals. The sediment is cemented by material derived from pore waters and grain dissolution. Deep burial processes operate over tens of millions of years and affect sediments to depths of around 10 km. At greater depths reactions occur at elevated temperature and pressure and are termed metamorphic.   ANDREW B. CUNDY

Further reading

Chester R. (1990) *Marine geochemistry*, pp. 468–528. Chapman and Hall, London.

Tucker M. E. (1991) *Sedimentary petrology. An introduction to the origin of sedimentary rocks*. Blackwell Scientific Publications, Oxford.

diamonds Diamonds are for ever, or almost so. Most diamonds crystallize as the stable form of pure carbon in the deep subcontinental mantle, and are subsequently carried to the surface by rare volcanic eruptions of rocks known as kimberlites and lamproites. Once at the surface, diamond should change to graphite, but fortunately the conversion rate is negligible.

Diamonds sometimes contain minute inclusions of the minerals garnet, olivine, and pyroxene, which indicate formation in two major mantle rock-types, peridotite and eclogite. The inclusions crystallize at the same time as the host diamond and incorporate trace elements such as samarium and neodymium, which may be used for radiometric dating. Peridotitic diamonds are older (up to 3200 Ma) and eclogitic diamonds younger (e.g. 1000–1600 Ma). Both types may be stored in old continental mantle (at depths of 150 to 200 km) for long periods of geological time before being transported to the surface by kimberlites or lamproites of various ages (e.g. 20–1200 Ma).

Diamonds are mined from lamproite and kimberlite pipes, as well as from secondary 'placer' concentrations in river and marine gravels. While diamond is the hardest substance known and its status as a gemstone is legendary, it is not indestructible. Flawed diamonds are readily pulverized for use as abrasives.   STEPHEN H. RICHARDSON

diapirs and diapirism Intrusive bodies of buoyantly upwelling rock, most commonly halite (rock salt), are a common feature of many sedimentary basins. These are *diapirs*, and the mechanism by which they move laterally and vertically in the subsurface is termed *diapirism*. Diapirs may also be composed of magma, and they share many of the same structural features and mechanisms of formation as the salt diapirs dealt with here.

One of the most impressive manifestations of salt diapirism is where diapirs emerge at the present surface as a glacier of pure salt, moving at rates of between ten and one hundred metres per thousand years. Perhaps the world's most spectacular modern salt glaciers are those of the remote Great Kavir, in the Zagros Mountains of central Iran. In the westernmost Pyrenees of northern Spain, a row of Tertiary diapirs, comprising salt of Triassic age, lines up along a major fault zone. They contain kilometre-scale blocks of igneous basement rocks, plucked by the rising diapir from the footwall of the fault. Both the Iranian and Spanish examples occur in dry climates where the low rainfall means that, in spite of its solubility, much of the salt is preserved.

A universal feature of salt diapirs is the way in which they move and deform internally by ductile, as opposed to brittle, processes. Essentially, this means that over geological time-scales, diapirs flow through the sub-surface. Thus, their internal structure is dominated by highly plastic folds; faults and fractures are virtually absent. Furthermore, experimentation has shown that salt requires only one part of water in ten thousand for its material properties to change so that it softens to behave as a highly viscous fluid. This finding was backed up by a surprising observation from one of the Iranian glaciers which flowed only following rainfall, but at a rate 100 000 times faster than was predicted by experiments on dry salt.

The detailed study of salt diapirs has come about largely because of their importance for the search of oil and gas in the sedimentary successions in which they occur. Salt diapirs usually comprise an impermeable barrier to the migration of fluids, thereby forming a highly effective seal, against which oil and gas may become entrapped. Additionally, their upwelling can cause potentially hydrocarbon-trapping faults and folds to develop in the overlying strata. However, permeability barriers may also inhibit the migration of oil and gas from the source rock to the hydrocarbon reservoir. Also, halite has a very high sonic velocity (speed of sound waves in the rock) of about 4500 metres per second (m s^{-1}), as compared with average clastic sedimentary rocks at between 2500 and 3500 m s^{-1}. This difference leads to strong refraction of seismic rays at the contact of a salt diapir. In consequence, even simply shaped diapirs can produce complex seismic ray paths, resulting in seismic profiles that are difficult to interpret.

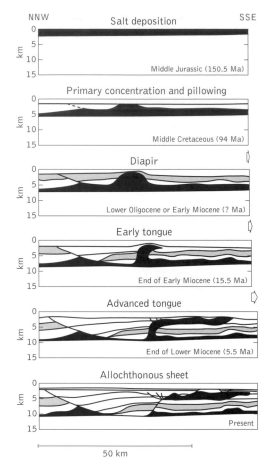

NNW   Salt deposition   SSE

Middle Jurassic (150.5 Ma)

Primary concentration and pillowing

Middle Cretaceous (94 Ma)

Diapir

Lower Oligocene or Early Miocene (? Ma)

Early tongue

End of Early Miocene (15.5 Ma)

Advanced tongue

End of Lower Miocene (5.5 Ma)

Allochthonous sheet

Present

50 km

Fig. 1. Reconstruction of the evolutionary stages of development, between the Middle Jurassic and the Recent, of an allochthonous (i. e. far-travelled) salt sheet from the Gulf of Mexico. Note that the area of the salt (shown in black) cannot be balanced within the plane of the cross-section. The left-hand end of the section is fixed, most of the extension being accommodated by the SSE-dipping normal growth fault. (Modified from Wu, Bally, and Cramez (1990).)

Despite the technical difficulties that salt poses to those interpreting seismic profiles, it was the wealth of new seismic data, particularly from the oil-rich passive margin basins where sedimentary salt is common, that led to the first appreciation of the huge size and abundance of far-travelled (allochthonous) salt sheets. In the Gulf of Mexico basin, off the Mississippi delta, Tertiary salt tectonics controls structure development, facies distribution, and hydrocarbon accumulation. Middle Jurassic salt has pierced through a sedimentary cover more than 5 km thick (Fig. 1) at rates varying from 10 to 2000 m per million years. An extensive geometric range of salt diapirs has been distinguished, the main types being the mushroom-like stocks, which can become completely

detached from their parent salt bed (Fig. 1), linear walls, and lobe-like tongues. Individual salt tongues are as long as 80 km and up to 7 km thick. Similarly, a large variety of structures in the sedimentary cover are also recognized. Rim synclines develop in response to the combination of withdrawal of salt from areas adjacent to a diapir, and dragging of the limb of the syncline against the rising diapir.

The density of halite, the most abundant constituent of naturally occurring salt deposits, is 2.2 (g cm^{-2}). Salt density does not change with burial depth because, unlike clastic sediment, rock salt is more or less uncompactable. This means that, taking freshly deposited sediment, and progressively burying and compacting it, rock salt will become less dense than the surrounding sediments at a depth of about 450–900 m. A buried bed of salt in the sub-surface will therefore become increasingly unstable below depths of a kilometre owing to the buoyancy of the salt relative to its encasing sediment. This buoyancy provides the principal driving force behind diapirism; but for diapirism to be initiated, an external trigger is required. For example, a normal fault zone, possibly with listric faults detaching on the unstable salt bed, will thin the overburden, thereby locally reducing the mass of the overlying rock column resisting the buoyant force of the salt. Thus, not only do diapirs induce faulting in the cover because of the ballooning effect of the growing diapir, but faulting in the cover may likewise induce diapirism.

JONATHAN P. TURNER

diastrophism *see* EARTH MOVEMENTS (DIASTROPHISM)

diatoms *see* FORAMINIFERA AND OTHER UNICELLULAR MICROFOSSILS

Dietz, Robert Sinclair (1914–95)

In the course of a long and very productive scientific career Robert Dietz made many contributions, but two were of especial importance and ensure that his name will long be remembered. Born and educated in the United States, Dietz was guided by a childhood interest in mineralogy to study geology at the University of Illinois. There he came under the influence of Professor Francis P. Shepard (1897–1985), who had recently started his epochal studies of the submarine canyons off the coast of California. Dietz made marine geology the main focus of his life's work, and from his studies came one of the key forward steps in the development of plate tectonics. In June 1961 Dietz published a paper in *Nature* entitled 'Continent and ocean basin evolution by spreading of the seafloor'. The paper proposed that new sea floor is created along mid-ocean ridges and then moves laterally away from the ridge. In the paper Dietz introduced the now familiar term 'sea-floor spreading'. The paper attracted an enormous amount of attention. Unknown to Dietz, Harry H. Hess (1906–69) of Princeton University had, in 1960, written and circulated—but not yet published—a preprint suggesting the same idea. The concept of sea-floor spreading, as propounded by Hess and Dietz, provided a reasonable mechanism that explained some of the dynamic difficulties with the idea of continental drift. Sea-floor spreading quickly became a key step in the development of the plate-tectonic paradigm later in the 1960s.

A boyhood interest in astronomy led Dietz to ponder the origin of the surface features of the Moon: he even proposed the topic for his Ph.D thesis at the University of Illinois, but the proposal was rejected. Nevertheless, he wrote a paper on the meteoritic origin of the Moon's surface features which was published in the *Journal of Geology* in 1946. He also entertained the idea that the Kentland structure in Indiana, long considered a crypto-volcanic feature, might be the eroded remnant of a meteoric impact site. When he visited Kentland, he saw and measured the orientation of shatter cones and found that the cones indicated a fracturing force from above, not from below as would be the case for a volcanic effect. Subsequent work at the Flynn Creek and Wells Creek Basin structures in Tennessee proved that shatter cones from an impact were present there too. Over the years Dietz examined and demonstrated an impact origin for many features in various parts of the world.

Among the many structures that Dietz was the first to recognize as having a probable impact origin is the Sudbury igneous complex in Ontario, Canada. Host to one of the great nickel resources of the world, the origin of the deformed ring-structure of the Sudbury Complex had long presented a puzzle to geologists. Dietz's discovery of shatter cones and other features suggestive of an ancient impact origin was not immediately accepted by all, but with time and close study of the evidence, almost all geologists now agree with his conclusions.

BRIAN J. SKINNER

diffusion and dispersion in groundwater flow

Dissolved constituents, or solutes, that are introduced to groundwater in a small volume of an aquifer are usually observed to spread throughout large volumes of the aquifer as they migrate in the direction of groundwater flow. The general process of spreading is known as dispersion. While dispersion reduces contaminant concentrations in the aquifer, a potentially positive effect, it also results in larger total volumes of contaminated water to remove and treat when an aquifer is cleaned up. Molecular diffusion, driven by concentration gradients, is one contributor to the observed dispersion. Variations in the microscopic velocity within a given pore, and between pores of different sizes, lead to further 'mechanical' dispersion, a process that is generally more significant than diffusion except where groundwater velocities are very small. Observed dispersion at the scale of laboratory columns can often be explained by expected rates of molecular diffusion and mechanical dispersion. In aquifers, however, dispersion tends to occur at much higher rates than would be predicted from laboratory experiments. Enhanced dispersion at the field scale is primarily the result of larger-scale variations in groundwater velocity resulting from variations in permeability in the aquifer. Development of improved models to predict field-scale dispersion is a continuing area of research.

J. BAHR

dinosaur hunters In the early nineteenth century William Buckland of Oxford coined the name *Megalosaurus* for a large Jurassic fossil carnivore. Soon afterwards a Sussex family doctor, Gideon Mantell, discovered and named *Iguanodon*. Like Buckland, he thought he had a giant lizard to deal with. By 1841 enough Mesozoic bones and teeth had been collected for Richard Owen, a young anatomist, to use the term 'Dinosauria' for these ancient reptiles.

By the middle of the nineteenth century Mesozoic bones and teeth were known from many parts of Europe and from a few localities in eastern North America. Then from the American west came reports of huge bones that drew the attention of two very energetic zoologists in the east, Othniel Charles Marsh (1831–99) and Edward Drinker Cope (1840–97). In competition, they set out to collect for their institutions as many of the new spectacular fossils as they could. Their teams shipped back east many hundreds of tons of material excavated in Colorado, Montana, Wyoming, and Utah. Their rivalry reached positively dangerous levels, but the fossils were becoming famous. Marsh, a professor at Yale with a private fortune behind him, collected for the Peabody Museum at Yale. Marsh was eventually appointed palaeontologist to the new US Geological Survey in 1881. His rival Cope, from Philadelphia, was an adventurous biologist intent upon the natural history of the west. He was also editor of the *American Naturalist* and himself published over 1200 books and papers.

Marsh and Cope were followed by others who had been their students and assistants and who were to make additional discoveries. Henry Fairfield Osborne (b. 1857) and William Berryman Scott were disciples of Cope. Others were W. D. Matthew and Barnum Brown, Earl Douglass and Charles W. Gilmour, all of whom continued excavating in the west and adding new giants to the collections.

The first Canadian dinosaurs were found in Saskatchewan by G. M. Dawson of the Geological Survey of Canada, and in 1883 his assistant J. B. Tyrrell found bones in the Red Deer River valley in Alberta. Another member of the Canadian Survey, Lawrence M. Lamb, reached Alberta in 1897. His discoveries initiated the Red Deer River dinosaur rush that lasted until 1917. Here the Sternbergs, father and sons, and Barnum Brown collected great numbers of the latest Cretacous dinosaurs.

Meanwhile in Europe, the German Friedrich von Huene (b. 1875), a renowned traveller and scholar-collector of dinosaurs at the University of Tübingen, made discoveries in many parts of Europe, the Americas, Africa, and Asia. He was still active in his nineties. A rival of von Huene was the Baron Nopsca von Felsö-Szilvás, a Hungarian eccentric genius who committed suicide in 1933. Prior to the First World War, Werner Janensch of the Berlin Museum used large numbers of natives to collect huge dinosaurs in Tanganyika, East Africa.

Important discoveries were made in the Gobi desert of Mongolia by American expeditions in 1922, 1923, and 1925, led by Roy Chapman Andrews. They included eggs and baby dinosaurs. Parties from Soviet Russia, led by I. Efremov, visited Mongolia in 1946–49. They collected giant carnivore and gigantic duck-billed dinosaur remains. Communist China has been active in dinosaur research, and numbers of Middle Jurassic giants have been found in Chengde and other areas.   D. L. DINELEY

Further reading
Colbert, E. H. (1968) *Men and dinosaurs, the search in field and laboratory.* Dutton, New York.

dinosaurs The name 'dinosaur' ('terrible reptile') was first coined by Sir Richard Owen, a British anatomist, in 1841. He used the term to denote the large reptiles *Megalosaurus* and *Iguanodon* that had been collected and described in the 1820s. A considerable amount of work has been done on dinosaurs since then and we now recognize them as a monophyletic (ancestor and all descendants) group of archosaurian reptiles that are closely related to the flying reptiles, the pterosaurs (Fig. 1). Dinosaurs are characterized by having erect limbs and a pelvis that incorporates at least five vertebrae, characters (using that word in its biological sense) that are related to their active terrestrial lifestyle. The earliest dinosaurs are known from the late Middle Triassic of South America and have been studied and described by Paul Sereno from the University of Chicago. Forms such as *Herrerasaurus* and *Stau-*

rikosaurus were small bipedal (moving on two legs) carnivores showing a variety of advanced characters that quickly allowed them to gain ascendancy over the then dominant terrestrial vertebrates, the mammal-like reptiles. In these animals the hind leg was elongated and the foot was functionally three-toed and digitigrade, that is, the animal stood on its toes. The acetabulum (hip-socket) was partly open and the upper rim was buttressed to help support the body on the vertical legs. In addition, the two outer fingers were reduced and the thumb had an offset to bring it into opposition with the other fingers, thus providing a grasping hand. During the Late Triassic, the two major groups of dinosaurs developed. They are separated by characteristics of the structure of the pelvis (Fig. 2). In the Saurischia (lizard-hipped) the pubis points forwards, while in the Ornithischia (bird-hipped) the pubis has rotated backwards to lie parallel to the ischium.

Ornithischia
The ornithischian dinosaurs were an extremely diverse group of terrestrial herbivores. Apart from the structure of the pelvis they characteristically had an additional bone at the front of the upper jaw, the predentary, which bore no teeth but acted like a beak for cropping vegetation. Their teeth were rather leaf-like in shape, similar to those of the modern *Iguana*, and they had ossified tendons running along the back and tail. This group includes such familiar forms as the stegosaurs

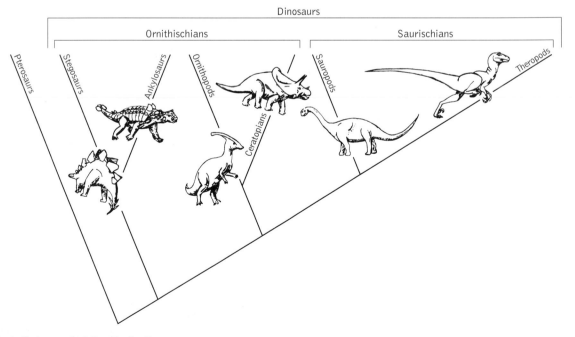

Fig. 1. Cladogram of relationships for dinosaurs.

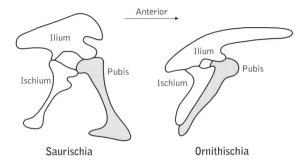

Fig. 2. Pelvic structure of the two main dinosaur groups.

(plated dinosaurs), ankylosaurs (armoured dinosaurs), ornithopods (bird-footed dinosaurs), and ceratopids (horned dinosaurs) (Fig. 1). They were all large animals, between 3 and 9 m long. Whereas the ornithopods were bipedal the other groups were secondarily quadrupedal.

A number of interesting structural adaptations were developed within the ornithischians. The ornithopods included the hadrosaurs or duck-billed dinosaurs, in which the front of the mouth was enlarged to form a broad flat beak and a variety of bony crests were developed on top of the head. The crests were formed by enlargement of the nasal bones so that the nasal passages extended through them. Although at one time it was thought that the crests might have been an adaptation for storing air while swimming and diving, it is now thought that they provided a resonating chamber for the production of loud calls. Stegosaurs bore a double row of diamond-shaped plates along their back and a pair of spikes at the tip of the tail. Although the spikes were clearly defensive, the plates were probably for thermoregulation, enabling the animal to control its internal temperature. Ankylosaurs were broad, short-legged dinosaurs, covered by a mosaic of defensive bony plates while the tail bore a large bony club. Because of this defensive armour they have been called 'dinosaurian tanks'. In the ceratopians, posterior skull bones extended back to form a large neck, while most also had horns on the nose and above the eyes. *Triceratops* is a typical example of these forms. The extension of the skull bones provided additional area for the attachment of jaw muscles, and together with the extensive battery of teeth made these animals extremely efficient at shearing and grinding tough vegetation.

Saurischia

The saurischian dinosaurs include both carnivores, the theropods, and herbivores, the sauropods. The theropods were all bipedal and ranged in length from less than 1 m up to 12 m. The smaller carnivores, such as *Coelophysis*, had small heads, long necks, long arms with grasping hands, and long hind limbs. They were lightly built and were probably opportunistic feeders, eating insects as well as small vertebrates. At the other end of the scale, forms such as *Tyrannosaurus* were the largest land predators that have ever lived. They possessed

very large heads with serrated, blade-like teeth; the hind limbs were long and powerful, but the forelimbs were very much reduced and in some forms the hand had only two fingers. They were probably adapted as ambush predators of large herbivores and were capable of moving very quickly for short distances. A third group of theropods includes *Deinonychus*, a medium-sized predator in which one hind claw was greatly enlarged for use as an offensive weapon. It is from this group that the birds are thought to have developed.

The sauropods include the largest land animals that have ever existed; they may have reached 24 m in length. They had small heads with weak teeth at the front of the mouth, long necks and tails, and short barrel-shaped bodies supported on massive vertical limbs. At one time they were considered to have been swamp-dwellers, too massive to have been able to move on land, but further analysis has shown that they were terrestrial animals adapted to browse in high trees. As there were no crushing teeth in the mouth, vegetation must have been swallowed and then crushed in a gizzard similar to that found in many birds. The presence of gastroliths (gizzard stones) in the rib cages of some specimens shows that this view is correct.

Dinosaur metabolism

It was long considered that dinosaurs had an essentially reptilian metabolism; that is, they controlled their internal body temperature by activities such as moving into or out of the sun. This is termed ectothermy and differs from the mammalian system in which control is internal or endothermic. In accordance with this view, dinosaurs were often reconstructed as large sprawling lizards with activity levels similar to those of modern reptiles. In 1970 John Ostrom of Yale and Robert Bakker of Harvard University suggested that dinosaur metabolisms may have been more mammalian, thus initiating an argument that still continues. Bakker pointed out that dinosaurs' skeletons show that they had upright postures and gaits similar to those of modern endotherms. In particular, the combination of long hind limbs and flexible limb joints, together with long rigid tails for balance suggests that bipedal dinosaurs were fast and agile. The bone structure of dinosaurs is also similar to that seen in modern endotherms and unlike that of ectotherms. More recent evidence includes the fact that dinosaurs are found in localities much further north or south than ectotherms inhabit now, again suggesting that they were endothermic. Bakker also suggested that fossil predator–prey ratios supported endothermy in dinosaurs. These ratios are based on the fact that endothermic predators consume more than ectothermic predators in order to support their higher metabolic rates, and hence there are fewer of them in proportion to prey organisms. Although some analyses suggest that dinosaurs showed an endothermic ratio, these have been challenged as unreliable because of problems such as preservational biases. In fact, the case for endothermy in all dinosaurs is still not proved, although it is generally accepted for small carnivorous dinosaurs. Other dinosaurs, particularly large herbivores, would have been more efficient as ectotherms because the temperatures of their large bodies would not have fluctuated easily, and they would have needed less food to support an ectothermic metabolism.

Dinosaur behaviour

Information about the behaviour of dinosaurs has been gleaned from trace fossils, mainly footprints, analysis of various skeletal structures, and the presence of eggs and nests. These suggest that many dinosaurs exhibited group behaviour like that seen in modern large mammals rather than the essentially solitary behaviour observed in modern reptiles. Trackways of dinosaurs commonly show multiple tracks moving in the same direction, indicating herding behaviour. Many finds of dinosaur eggs and nests, particularly those made in Montana by Jack Horner of the Museum of the Rockies in Bozeman, Montana, show that the young must have been cared for by the adults for an extended period, for they hatched out at a stage at which they were not self-sufficient. This indicates co-operative group behaviour of the adults at group nesting grounds on a level similar to that seen in birds, and far beyond that seen in any modern reptiles. Finally, the presence of possible display structures, particularly crests on hadrosaur skulls, suggests sociality among groups of dinosaurs, for such structures would have aided the recognition of potential mates or opponents within a social group.

Dinosaur extinction

Despite their success during the Mesozoic, the dinosaurs became extinct at the end of the Cretaceous. How this happened has become the focus of a highly charged scientific debate. The dinosaurs were only one of the groups that became extinct at the end of the Cretaceous; also hard-hit were the unicellular Foraminifera, the ammonites, and marine reptiles such as the mosasaurs and plesiosaurs. On land the pterosaurs and some mammalian groups became extinct, but many terrestrial organisms were unaffected. This rather selective extinction has been explained by changing climatic conditions, particularly a gradual cooling and drying of the climate during the Late Cretaceous. William Clemens of the University of California at Berkeley has pointed out that in the rare cases where there is a good record of dinosaurs up to the Cretaceous–Tertiary boundary it shows a gradual reduction in diversity that would tend to support the climatic model. However, more recent work by Louis and Walter Alvarez, also from Berkeley, has demonstrated the presence of a clay layer at the boundary in many localities. This contains high levels of elements that are usually rare at the Earth's surface, particularly iridium, together with droplets of shocked quartz which indicate a high-speed impact. They have interpreted this as evidence the Earth was hit by a large extraterrestrial body or bolide, and that the impact generated large amounts of dust, causing global darkness and ultimately extinction of many groups of organisms. Although this appears to be a simple and elegant solution to the problem, it has been criticized for a lack of connection between such a catastrophic event and the selective pattern of extinctions that

is seen. No generally accepted scenario has yet been proposed and the debate will certainly continue for some time— although it has been pointed out that as the dinosaurs survived as birds perhaps they did not really become extinct.

DAVID K. ELLIOTT

Further reading

Lucas, S. G. (2000) *Dinosaurs: the textbook.* McGraw-Hill, Boston.

Norman, D. (1985) *The illustrated encyclopedia of dinosaurs.* Crescent Books Ltd, New York.

dispersion *see* DIFFUSION AND DISPERSION IN GROUND-WATER FLOW

disposal methods *see* WASTE DISPOSAL METHODS

divergent plate margins Divergent plate margins occur where two tectonic plates are separating. Most current instances are along the mid-ocean ridges, where new oceanic plate material is being accreted through sea-floor spreading, although in a few instances such 'oceanic' spreading centres occur on land, such as in Iceland and the Afar Depression of Ethiopia.

Some divergent boundaries also occur through active continental rifting, although at the present epoch these are rare. The best example of a boundary of this type is the East African Rift Valley, where it is just possible to demonstrate statistically significant differences in plate motion between the Nubian (western Africa) and Somalian (eastern Africa) plates. There are also areas, most notably the Woodlark Basin of the Solomon Sea in the south-west Pacific, where an oceanic spreading centre is actively propagating into the continental crust of Papua New Guinea.

The geology of mid-ocean ridge spreading centres and continental rifts is described in separate entries. Here we focus on the geometry and kinematics of divergent margins at mid-ocean ridges.

Oceanic spreading centres

Plate-tectonic theory does not require divergent boundaries to be orthogonal (at right angles) to the direction of plate separation, although in most instances oceanic spreading centres are orthogonal at the local level. Nevertheless, regionally they are often far from orthogonal. Divergent plate margins inherit at least their initial shapes from those acquired during the initial break-up of a plate. In the case of continental break-up in particular, the initial shape may be controlled strongly by the accidents of the pre-existing continental structure and geology, so that large sections of the plate boundary are highly oblique to the new spreading direction. An obvious example is the almost semicircular boundary between North America and West Africa along the Mid-Atlantic Ridge between the Azores and the Equator.

Ridge–ridge transform faults

Locally, near-orthogonal plate boundaries can be reconciled with such regionally oblique ones by the existence of numerous offsets in the ridge. In classical plate-tectonic theory such offsets are accomplished by transform faults, and an oblique divergent margin would typically have a staircase pattern of orthogonal spreading centres separated and offset by ridge–ridge transforms.

Detailed, high-resolution surveying in the 1970s and 1980s demonstrated that true transform faults containing through-going strike-slip faults are quite rare. They occur only when the ridge–ridge offset is longer than about 30 km, and tend to be fairly widely spaced along a ridge: typically perhaps 1000 km apart. For example, there may only be one true transform fault, at Charlie-Gibbs Fracture Zone, 53° N on the Mid-Atlantic Ridge between Iceland and the Azores; in the 2500 km length of ridge south-west of the Azores there are four transforms: Hayes, Oceanographer, Atlantis, and Kane. Similar frequencies occur on other mid-ocean ridges.

These major transform faults tend to persist throughout the lives of ocean basins, and are probably inherited from the original continental break-up. In the equatorial Atlantic, for example, the large east–west offset of the Mid-Atlantic Ridge between the North and South Atlantic is accompanied by some of the longest and closest-spaced transforms in the world, and their fossil fracture-zone trails can be traced back to the continental margins, where they coincide with significant structural weaknesses and offsets on land.

Non-transform offsets

Between the major transform faults are sections of spreading centres many hundreds of kilometres long but still typically trending obliquely to the spreading direction. These ridge sections are offset at shorter intervals, but by non-transform offsets (sometimes called non-transform discontinuities) rather than by transform faults. The ridge segments between non-transform offsets tend to be almost orthogonal to the spreading direction.

Non-transform offsets are typically oblique offsets of spreading centres by distances of 25 km or less. Most such offsets, except the very smallest, are associated with along-axis deeps, and leave off-axis traces that are either trails of aligned, but isolated, depressions or continuous valleys, similar to

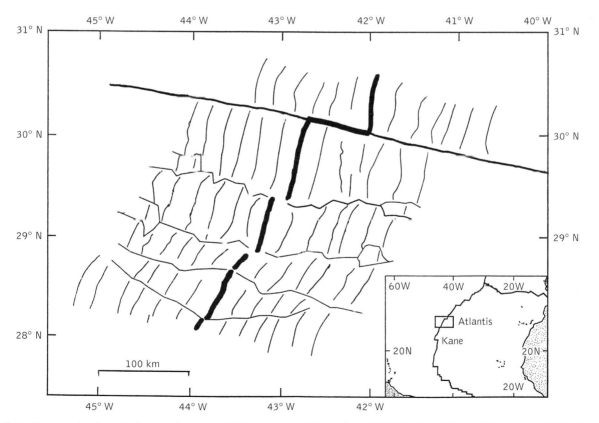

Fig. 1. Plate boundary (heavy line), magnetic isochrons (light, subvertical lines) and traces of a transform and several non-transform offsets on part of the slow-spreading Mid-Atlantic Ridge. (Modified after R. C. Searle *et al.* (1998) *Earth and Planetary Science Letters*, **154** (1–4), 167–83.)

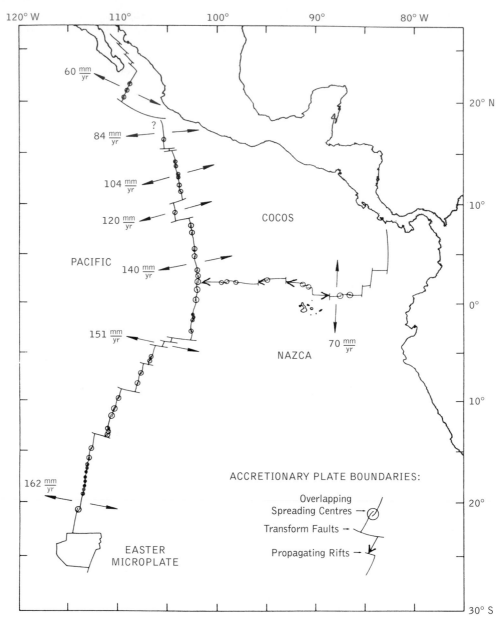

Fig. 2. Plate boundaries of the fast-spreading Pacific–Cocos and Pacific–Nazca and the medium-spreading Cocos–Nazca plate boundaries in the Eastern Pacific. Lines cross-cutting the general trend of the plate boundary at right angles are transform faults; arrow-heads mark those marking the off-axis traces of true transform faults. Non-transform offsets also tend to be associated with relatively high gravity anomalies, indicating that they are associated with thin crust. These non-transform offsets are spaced out along the ridge axis with separations of around 50 km (Fig. 1). propagating rifts, and other discontinuities in the boundaries are non-transform offsets, mostly overlapping spreading centres. (Modified from Macdonald K. C. *et al.* (1986) *Journal of Geophysical Research*, **91**, 10501–10.)

The actual shape of the offset in map view may either be a short section of oblique plate boundary, an offset *en echelon*, or a slight overlapping of the adjacent spreading centres. Normal faults, which elsewhere tend to run normal to the spreading direction, curve obliquely into non-transform offsets, particularly in the cold, strong axial lithosphere of slow-spreading ridges, such as the Mid-Atlantic Ridge, where they may reach angles of up to 45° to the spreading direction. At fast-spreading ridges, such as the East Pacific Rise, overlapping spreading centres are well developed, typically overlapping by about three times the length of the offset (Fig. 2). Although most overlapping spreading centres are less than 25 km across, some have grown to several hundred kilometres in diameter to become oceanic microplates.

Both transform faults and non-transform offsets can change their length of offset by asymmetric accretion. Although sea-floor spreading is usually approximately symmetrical if averaged over periods of ten of millions of years, on a timescale of a few million years it can be locally asymmetrical by up to about 30 per cent. If a non-transform offset grows to exceed the critical length of about 30 km, a permanent through-going strike-slip fault may develop and will become a transform fault. Similarly, if a transform fault shrinks below the critical length, it may turn into a non-transform offset; if it continues to shrink, the offset may be eliminated altogether or its sense may even be reversed.

Propagating rifts

Just as the lengths of non-transform offsets may change with time, so may their positions migrate along the ridge axis (although the same is not true of transform faults, which are quite stable against along-axis migration). It is quite common for most non-transform offsets to migrate to some extent, and as they do so their off-axis trails make V-shaped wakes (Figs 1 and 2).

Along-axis migration of ridge offsets is most spectacular at medium-spreading ridges, such as the Galapagos Spreading Centre (between the Cocos and Nazca plates), the Juan de Fuca Ridge (between the Pacific and Juan de Fuca plates in the north-east Pacific), and the Australian–Antarctica Rise (south of Australia). Here, large (approximately 25 km) non-transform offsets can migrate along the ridge axis with a constant direction and speed for millions of years.

Such features are called *propagating rifts*. They leave spectacular V-shaped wakes called 'pseudofaults', and in their passage they transfer narrow strips of lithosphere from one plate to another. They appear to be driven by pressure differences derived from regional along-axis topographic gradients (often associated with hot spots lying above mantle plumes), and may be an important mechanism in realigning divergent plate margins.

Ridge segmentation

The discovery of non-transform offsets was critical in understanding the structure and accretionary processes of mid-ocean ridges. Ridges had previously been largely thought of as two-dimensional structures with little variation along their length between widely spaced transform faults. Indeed, one model of mid-ocean ridge magma-chambers was called the 'infinite onion', referring to its supposed cross-sectional shape and infinite along-strike extent. Now, however, it can be seen that there is significant structure and variation in the along-axis direction as well. It has become common to refer to the segmentation of ridges, where individual segments are the lengths of spreading centre (about 50 km long) between non-transform or transform offsets. These segments are now believed to be the individual building blocks of the oceanic lithosphere.

Gravity and seismic refraction studies have shown that segments tend to have the thickest crust (about 8 km at slow-spreading ridges) at their centres and thin crust (about 3 km) at their ends. They thus produce ribbons of alternately thick and thin crust aligned along the spreading direction. This pattern indicates that more melt is delivered to the centres of segments than to their ends, although the precise controls on this 'melt focusing' are still not completely understood. Melt focusing has an important effect on the architecture of the oceanic lithosphere. The larger melt input at the centres of segments gives them a thicker crust and greater heat input than at their ends. Both effects tend to produce a weaker lithosphere, and the effects of this can be seen in variations of faulting, and possibly volcanic activity, along segments.

ROGER SEARLE

Further reading

Macdonald, K. C., Scheirer, D. S., and Carbotte, S. M. (1991) Mid-ocean ridges: discontinuities, segments and giant cracks. *Science*, 253, 986–94.

Sempéré, J.-C., Purdy, G. M., and Schouten, H. (1990) Segmentation of the Mid-Atlantic Ridge between 24° N and 30° 40′ N. *Nature*, 344, 427–31.

DNA (the ultimate biomarker)
The most remarkable of biomolecules is the blueprint of life, DNA, which in principle, has the potential as a biomarker to answer questions about the evolution of species. Also remarkable is the fact that recognizable portions of this apparently delicate type of molecule have been detected in archeological finds and certain fossils. However, early claims for DNA in Jurassic dinosaur bones are now discounted.

The DNA molecule is comprised of specific sequences of thousands to millions of nucleotide bases strung together like beads on a chain. The highly sensitive and precise amplification of target nucleotide sequences by the polymerase chain reaction (PCR) is the uniquely important search capability made possible by this structure. However, finding enough intact pieces of DNA of the sequence being sought is a major challenge in a fossil, since contaminants usually far outweigh the minuscule amounts of ancient DNA. Nevertheless, it is apparent that fragments of a few hundred base pairs are regularly retrievable by using the PCR method of amplification on buried plants, bones, and other remains up to a few tens of thousands of years old, but only where preservation of organic matter has been good, owing to exceptional burial conditions, such as low temperatures, extreme dessication, and enclosure in resins, crystals, or anoxic sediments. Replication has not been achieved for any of the specimens that are millions of years old, not even the DNA from insects trapped in amber, which some researchers had expected to be an optimal medium for preservation.

DNA may be the most remarkable molecular survivor, but it is by no means the most durable. Structural proteins usually last longer and components of woody tissue, plant and insect coats, fats, oils, and waxes may persist as useful biomarkers into geological time. Sediments on lake bottoms and sea beds are far richer in biomarker molecules than we might suspect. Some bacteria slowly consume these molecules in the most extraordinary circumstances, a kilometre deep in the ocean floor muds, and possibly even within rock salt crystals. If so, these crystals may harbour some of the most ancient life on the planet, an intriguing possibility which is under active study.

GEOFFREY EGLINTON

Further reading

Herrmann, B. and Hummel, S. (eds) (1994) *Ancient DNA*. Springer-Verlag, Berlin.

dolines *see* SINKHOLES

dolomite

The rock dolomite, or *dolostone* as it is sometimes called, is closely related to limestone, but unlike limestone it is composed of more than 50 per cent of the rhombohedral carbonate mineral dolomite $CaMg (CO_3)_2$, rather than calcite or aragonite, which are calcium carbonate, $CaCO_3$.

The origin of dolomite has intrigued geologists since its discovery in 1791 by Deodat de Dolomieu in the dolomites of Italy, both the mineral and the mountains being named after the man. The understanding of dolomite was hampered because modern deposits of dolomite were unknown until the 1950s and were not extensively studied until the 1960s, when it was found forming in intertidal and supratidal deposits of the Arabian Gulf and in the Bahamas and Florida.

Primary dolomite is sometimes precipitated directly from lake or shallow lagoon marine water or from interstitial water in deep marine sediments; in the geological record it is sometimes found as a primary precipitate associated with marine sediments and as cement. Most dolomite appears, however, to have originated by reaction of earlier-formed calcium carbonate with magnesium in solution: either in brines with an Mg/Ca ratio of 10 or more, or in brackish waters with a Ca/Mg ratio of about 5. In modern environments this is achieved in brines by the removal of calcium by organisms and the precipitation of aragonite and gypsum during evaporation. In brackish waters, the mixing of marine and fresh waters produces the same result.

Dolomitization (a diagenetic process) can take place at various times and in various settings during the history of a carbonate sediment. In an arid environment, flooding of supratidal flats introduces marine waters that are subsequently concentrated by evaporation to form brines. After the precipitation of gypsum, these brines have an Mg/Ca ratio greater than 10, and dolomitize aragonitic muds: this process is *evaporative dolomitization*. These dolomites are usually associated with stromatolitic bedding, gypsum or anhydrite, mudcracks, and cavities produced in drying cyanobacterial mats: 'bird's-eye' or 'fenestral' structure.

Brines concentrated in coastal lagoons sink and move seawards in the subsurface to react with the carbonate sediments of barriers and nearshore areas: this is *reflux dolomitization*. Elsewhere, mixing of fresh and marine ground waters produces a brackish water with an Mg/Ca ratio of about 5, which on contact with carbonate sediments can cause dolomitization: this is *mixing dolomitization* or so-called *Dorag dolomitization* after the geologist who first suggested this mechanism. After burial, carbonate sediments can be dolomitized by Mg-rich waters, expelled by compaction from adjacent siliciclastic deposits: this is *burial dolomitization*. Even after lithification and deformation, limestones sometimes experience dolomitization. Fluids exploit unconformities, faults, joints, and other permeable pathways to produce dolomites which cut across the original stratification of the rock: *tectonic dolomitization*.

Dolomites are more common in Precambrian than in Phanerozoic rocks. This has been attributed to the higher partial pressure of CO_2 in the Precambrian atmosphere and the higher Mg/Ca ratio and lower SO_4 content of Precambrian sea water, as compared with the values of today.

Mole-by-mole replacement of calcium carbonate by dolomite produces a decrease in volume and hence an increase in porosity (approximately 13%). When this replacement occurs in soft sediment much of the increased porosity is lost by subsequent compaction, but when indurated sediment is replaced, the increased porosity will be preserved. Dolomite is important as a raw material for the manufacture of refractories. Also because of their increased porosity, dolomites often act as hosts for hydrocarbons and for lead–zinc and other ores.

G. EVANS

Further reading

Purser, B., Tucker, M. E., and Zenger, D. (1994) *Dolomites*. International Association of Sedimentologists, Special Publication No. 21. Blackwell Scientific Publications, Oxford.

Tucker, M. E. and Wright, V. P. (1990) *Carbonate sedimentology*. Blackwell Scientific Publications, Oxford.

drainage basins

The drainage basin is a unit that integrates the atmosphere, the geosphere, and the hydrosphere. The basin is that area technically delimited on the land (the surface of the geosphere) by the watershed, the imaginary line passing through the highest points that separate land draining to one river from land draining to adjacent rivers. The drainage basin thus delimited is the area that receives, collects, and concentrates water falling from the atmosphere, together with groundwater emerging from springs in the geosphere, to give river discharge (Fig. 1). The drainage basin is therefore a unit or system that is central to the hydrosphere.

The idea that the drainage basin is the unit of the Earth's surface that concentrates run-off to produce river flow is derived from the work of Perrault in the seventeenth century on the basin of the Seine in France. Perrault calculated that approximately one-sixth of the water or precipitation falling over the river basin of the Seine actually flowed down the river. It had previously been thought that rivers emerged on the land surface fed from underground sources; the relationship between the supply of flow from groundwater and rainfall over the basin was not appreciated. It was not until 1945 that the drainage basin was acknowledged to be the fundamental unit in hydrology.

The study of water movements in the hydrosphere is fundamental to hydrology. The water balance equation relates precipitation (P) to discharge or run-off (Q), evapotranspiration (ET), and changes in storage (S). The equation is given as

$$P = Q + ET \pm S.$$

The equation needs to be applied to a standard unit of land area on the surface of the geosphere, and the drainage basin provides that standard unit. In the study of geomorphology, the drainage basin is also acknowledged to be a fundamental unit, central to the way in which the land surface changes in those areas of the world influenced by running water.

In studying a drainage basin unit it is necessary to relate input from precipitation to output in the form of river discharge. The input of rainfall from a particular rainstorm is described by a 'hyetograph', which is the graph relating rainfall amount to time. The output of water as river discharge is described by a hydrograph. A hydrograph is simply a graph of the variation in the discharge (volume of water per unit time) over time. For a specific drainage basin it is therefore possible to demonstrate how a particular rain storm is translated into a particular hydrograph of river discharge: the drainage basin is the transfer function responsible for the conversion of rainfall into a particular type and amount of river discharge. The drainage basin also acts rather like a conveyor belt for material in solution (solutes) and for sediment, which is why it is a fundamental geomorphological unit. Water moving through the drainage basin can convey material in solution (as solutes), or mechanical particles in suspension (suspended sediment) or by being rolled or jumped (saltated) along the

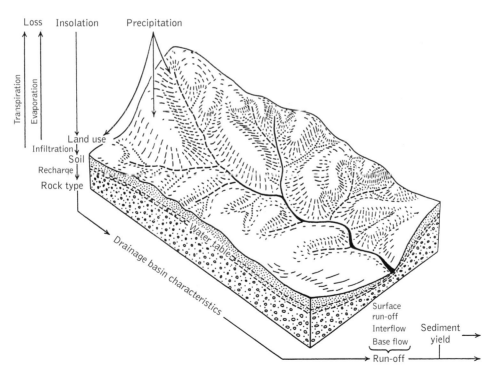

Fig. 1. A geomorphological view of a drainage basin. It depicts the way in which drainage-basin characteristics (form) influence the transformation of input (precipitation-losses) into output of run-off and sediment yield. The dynamic nature of the drainage network is incorporated by representing perennial (solid), intermittent (dashed), and ephemeral (dotted) streams.

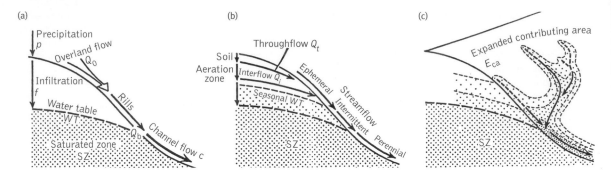

Fig. 2. Various views of drainage-basin dynamics explained in the text: (a) the Horton overland flow model; (b) the throughflow model; (c) some characteristics of the contributing area. (Based on Gregory and Walling (1973).)

bed of a river as bedload. The volume of material moved as solutes, suspended sediment, or bedload can be totalled for each year and dividing this total volume by the area of the drainage basin gives an indication of the rate at which erosion is taking place, represented in terms of the depth of erosion from the drainage basin as a whole. In fact, as explained below, the sediment is derived from particular parts of the drainage basin rather than from the entire surface.

Input–output studies simply view the drainage basin as a uniform unit and relate the input from precipitation to the output as river discharge. It is necessary, however, to know how the processes within the drainage basin work and how precipitation is turned into river flow (Fig. 2). Originally it was thought that there were two major types of water flow throughout the drainage basin. First, surface run-off or overland flow is water that flows over the slopes of the drainage basin and is then concentrated into river channels. Secondly, water that infiltrates into the soil, then through the rock and down to the water table leads to discharge of groundwater from springs and seepages producing base flow. In 1945 the view of R. E. Horton, then a leading hydrologist in the United States, was thus that rivers are sustained by two types of river flow: 'overland flow' and 'base flow'. In the 1960s it was realized that there are other types of flow that occur between that flowing over the surface and that derived from below the water table. These include 'throughflow', which is the lateral flow of water through the soil, and 'interflow' which is the lateral flow of water through the unsaturated zone above the water table. In addition there is 'pipeflow', the flow of water through small pipes in the soil, pipes which can vary from several centimetres to several metres in diameter. It is now appreciated that in any one drainage basin there is a range of types of routes that water can follow from the moment that it hits the surface of the drainage basin to the time when it reaches the river channel. It can move over the land surface as overland flow or it can infiltrate down to the water table to take decades or even centuries before it emerges as water from

springs. In addition there are a number of other routes through the soil as matrix flow, throughflow, interflow, and pipeflow, as mentioned above. Many of these flow types can either be saturated, when the water table rises locally to higher levels, or unsaturated, when lateral flow is determined by the existence of relatively impermeable layers in the soil or in the unsaturated zone of the weathered bedrock.

In the same way that different types of water flow have been recognized, attention has been given to the sources of sediment and solutes (Fig. 3). When precipitation hits the surface of the drainage basin it already contains solutes derived from the atmosphere and from vegetation; for example, water flowing over leaves, branches, or trunks of trees collects solutes which are dissolved in the rainwater by the time it reaches the surface of the drainage basin. As water passes over the surface or along the various routes mentioned above, then more solutes are obtained. It is the water flowing over the land surface, through pipes or in open channels, that obtains sediment derived either from the channel banks or from the river bed, and then transports it as suspended sediment. Bedload is the coarser material which is picked up only when the discharges are high and which can be rolled or bounced along the bed of the channel. The relevant significance of bedload, suspended load, and solute load varies significantly from one drainage basin to another according to the characteristics of the drainage basin, including its size.

It is now appreciated that drainage basins are dynamic in that the network of rivers and streams in the basin expands and contracts. This network is analogous to the circulation of blood in the human body, because the network of streams and rivers is equally vital to the dynamics of the basin. When the network of stream and river channels in the drainage basin is viewed in plan form it is possible to characterize the channels into one of three types (Fig. 1). First, there are those channels that carry water at all times throughout the year; these are described as *perennial* streams. Secondly, there are those channels that have water flowing in them seasonally (for example,

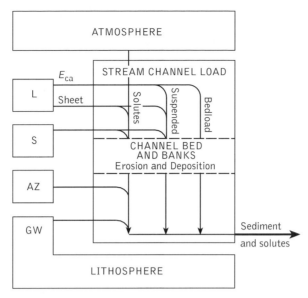

Fig. 3. The major components of sediment and solute movement in the drainage basin are here shown conceptually. Sediments and solutes are visualized as coming from four main sources: the land surface (L), the soil (S), the underlying aeration zone (AZ), and groundwater (GW); they can be derived from the land surface by expansion of the channel network (E_{ca}) or by sheet flow. The land surface and the channel bed and banks can supply all three components (bedload, suspended load, and solutes) to the stream channel load, whereas the other sources provide two or one components. The stream channel load can be supplemented by solutes from the atmosphere, and additional changes occur during erosion and deposition along the channel bed and banks. (From Gregory and Walling (1973).)

during the wet season of a seasonal climate); these are described as *intermittent* channels. At the extremes of the drainage network of river channels there is a third type, which is described as *ephemeral* and consists of channels that have flowing water in them only during or immediately after rainstorms. The functions of river channels will also vary as the saturation level rises, so that along some valley floors saturation overland flow may occur when the local water table rises to the surface. It is therefore possible to envisage any drainage basin as possessing a network of channels that expands during and immediately after rainstorms and then contracts rather more slowly after a storm event has ended. There are, of course, considerable variations in the character and density of drainage networks from one area to another; the highest densities of channel that have ever been recorded, expressed as drainage densities in kilometres per square kilometre, have values of over one thousand on gullied landfill sites. Drainage densities in temperate landscapes tend to be between one and two kilometres per square kilometre (km km^{-2}), whereas in tropical areas densities can rise to as much as 10 or 20 km km^{-2}. It has been found that the density of perennial streams relates directly to the available moisture, which is defined as precipitation minus evapotranspiration. However, the density of the total network, including the perennial, intermittent, and ephemeral channels, reflects the relationship between the intensity of precipitation and the degree of resis-

tance which that precipitation meets when it reaches the surface. The highest densities of ephemeral channels can therefore occur in desert or semi-desert areas where there are very occasional but intense rainstorms combined with a very small amount of vegetation cover to resist the formation of channels.

The relationship between the density of the drainage network, the precipitation input, and the vegetation cover is one example of the way in which the characteristics of a drainage basin control the translation of precipitation input to the output of water and sediment. A drainage basin has four major groups of characteristics: characteristics of rock type, of soil type, of vegetation and land use, and of topography. Each of these has an effect on the way in which the drainage basin acts as a transfer function; for example rocks that are permeable tend to have basins with higher proportions of groundwater flow than of surface run-off or quick flow. Similarly, soils that are permeable tend to facilitate subsurface flow including interflow and throughflow, rather than surface run-off. Vegetation and land use can also encourage infiltration where vegetation cover is dense and continuous. Topographic characteristics include four major categories: the size of the basin, the drainage network (classically expressed as drainage density mentioned above), the relief aspects of the basin, and the shape of the basin. Relief is significant because the greater the relief the higher the slope and the faster the movement of water through the basin. The shape of a drainage basin is significant because the more pear-shaped a basin is, the more efficient it becomes in concentrating precipitation into run-off. Various quantitative indices have, therefore, been invented to express the separate characteristics of the drainage basin to make it possible to construct simple equations relating the precipitation input, the drainage basin characteristics, and the output of discharge. In particular, if data can be obtained from at least ten or twenty different basins it is possible to produce equations giving values of different types of run-off, as the dependent variable, in terms of drainage basin characteristics, precipitation, and other climatic characteristics as independent variables. There are various ways in which discharge can be expressed for such statistical (often multiple-regression) models. It is therefore possible for discharge (as described in the form of a hydrograph) from a single basin to be expressed in terms of instantaneous values of discharge, average values for, say, a period of a year, or to be expressed by an index value that reflects either high or low values of discharge with a particular frequency. A particularly important index in relation to the management of drainage basins is the use of the flood-frequency analysis, which establishes the flood value with a particular recurrence interval for a specified basin.

The development of measures of drainage-basin characteristics for a particular basin might imply that characteristics within the basin are uniform, which is not the case. Along the course of a river channel there are considerable variations in the character of the channel, and a whole range of types of river channel can be established according to the clarity of definition of the channel. A way of characterizing the clarity of channel definition is by its roughness. Various equations developed by hydraulic engineers link the velocity of water flow to the characteristics of the river channel cross-section, usually in terms of the hydraulic radius, R, the cross-sectional area divided by the perimeter, the slope of the channel, s, and the estimated roughness, n. One such equation, proposed by the engineer Robert Manning, has the form:

$$V = \frac{R^{2/3} s^{1/2}}{n}$$

Such flow equations enable estimates of velocity, and therefore of discharge, to be made (by multiplying velocity by cross-sectional area) for locations along rivers for which there are no continuous discharge records. A particular way in which river channels vary in both morphology and process is in the river-channel pattern, which is the pattern of the river channel as seen from the air. Two major types of river-channel pattern have been distinguished. 'Single-thread' channels include all unitary river channels that are straight or meandering; 'multi-thread' channels are composed of several channels and include braided channels and anastomosing channels.

Using the knowledge of the mechanics of the drainage basin it has been possible to build quantitative models of drainage-basin processes for particular planning purposes

such as flood prediction. Models were originally thought of as 'back-box' models, whereby links were established between input and output without adequate knowledge of what actually occurred within the drainage basin. It is now more usual to resort to what are described as 'grey-box' models, which to some extent reflect a knowledge and understanding of the processes operating in the drainage basin that are responsible for the conversion of precipitation into river flow. It will be some time before we can move to a 'white-box' model that is based on a full and complete understanding of all the processes operating within the basin. Although the processes are known and understood in terms of the way in which they operate in different parts of the basin, it is very difficult to link them all together in a quantitative model.

The development of an effective quantitative computer model of the operation of a particular basin is obviously a major way in which management of drainage basins can be informed. The management of drainage basins can focus upon the incidence of flooding, the provision of water supply, the use of the basin for effluent disposal and industrial purposes, the use of the basin for recreational and leisure purposes, and the control of erosion in relation to land-use management and river-channel management. In the case of erosion control, developments occurred in a number of basins in the United States in response to erosion problems: the work of the Tennessee Valley Authority in the 1930s was very prominent in this regard. A number of slogans were developed, including 'Erosion begins at the top of the hill' and 'Stop the little raindrops where they fall', to press home the message that erosion control in part of the basin was necessary to prevent large-scale erosion taking place elsewhere, perhaps with severe consequences such as the removal of topsoil and extensive gullying, which could not then be reversed. Other ways in which basin management has been applied relate to flood prediction and flood control. Good models of the ways in which floods are generated in a basin make it possible for advance warning to be given of the likelihood of floods occurring and to predict how floods may be generated as flow passes downstream. Flood models and flood-warning systems are in operation in major river basins such as the Mississippi in the United States and the Murrumbidgee in Australia. Four basic strategies are available for flood-prevention measures: doing nothing; providing relief measures, such as tax and financial benefits; introducing land-use zoning (in which the most vulnerable and expensive land uses are kept furthest away from the river); and undertaking structural solutions, which entail either channelization of the river or the institution of small or large dams for flood control. Water supply in drainage basins is provided either by direct abstraction from rivers or by impoundment, which requires the construction of reservoirs. After water has been used for domestic and industrial purposes, it is inevitably returned to the river. To avoid pollution, there must therefore be controls on the condition of the water that is returned. Pollution was certainly a feature of major European rivers

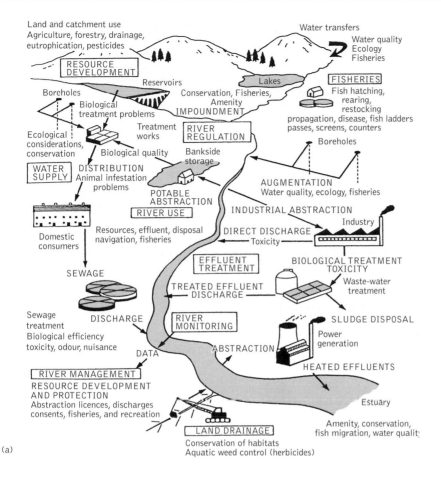

(a)

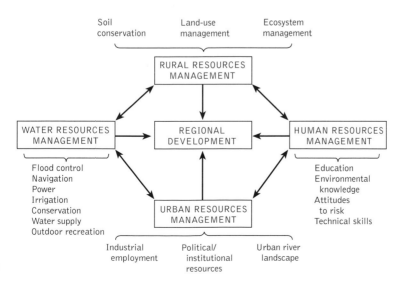

(b)

Fig. 4. Different perspectives of the management of water within the drainage basin: (a) a water manager's view of the river basin; (b) the complex of projects involved in river-basin development. (From Newson (1992).)

before legislation was introduced in the 1960s to regulate the quality of the water returned to them.

It is clear that there are many potential uses for the water in a drainage-basin system (Fig. 4). Leisure use of waterways and river systems, for example, includes water sports, bathing, and access to rivers and streams; various uses related to agriculture and irrigation require specialized types of water supply. All these uses can affect the hydrological cycle and the way in which the drainage basin operates, particularly the amount of, and the rate at which, water and sediment are conveyed through the drainage basin. It has therefore been necessary to devise ways in which the management of the drainage basin reflects the fact that all these components are interrelated: integrated basin management was developed as a way of signifying the need to manage the basin with an awareness of the impact of the range of activities operating within a specific basin. As a number of approaches to integrated basin management have been devised, they have increasingly tended to reflect the fact that the activities that are dominant and produce problems in one basin are not those that are dominant in another, and also the fact that particular disciplines tend to be associated with their own particular emphasis in drainage-basin management. Thus, an engineering approach may tend to focus on a structural solution to river problems, and an ecological approach may emphasize the impact upon the river ecology, whereas a geomorphological approach may reflect an awareness of the impact upon the flood plains and on the drainage system as a whole. In the light of the differences from one basin to another and from one disciplineto another, there have now been many calls for a holistic approach that seeks to acknowledge all the ways in which the drainage basin and its processes are affected by human activity and proposing a method of management that is most similar to natural processes and involves the least amount of disturbance. Whereas initial approaches to a particular problem within the drainage basin were perhaps driven by the idea that 'technology can fix it', it has subsequently transpired that such technological fixes could have significant implications for other parts of the drainage basin; for example, immediately downstream. It is therefore being realized that it is better to work with the river rather then against it and wherever possible to imitate nature rather than to superimpose a system unlike that which occurs naturally in the drainage basin. It is now generally accepted that a knowledge of the drainage basin and how it works is important, not only to understand this aspect of the operation of nature, but also to underpin ways in which management of drainage basins is constructed.

KENNETH J. GREGORY

Further reading

Downs, P. W., Gregory, K. J., and Brookes, A. (1991) How integrated is river basin management? *Environment Management*, 15, 299–309.

Gregory, K. J. and Walling, D. E. (1973) *Drainage basin form and process*. Edward Arnold, London.

Newson, M. (1992) *Land, water and development*. Routledge, London.

Walling, D. E. (1987) Hydrological processes. In Clark M. J., Gregory K. J., and Gurnell A. M. (eds) *Horizons in physical geography*. Macmillan, London.

droughts During the 25-year period between 1968 and 1992 there were 446 droughts worldwide, resulting in 1.8 million deaths; a further 1474 million persons were affected or made homeless by drought and famine (global data from the *World Disasters Report* prepared by the International Federation of the Red Cross and Red Crescent Societies). With the exception of civil strife, which accounted for around 2.5 million deaths, 90 000 injuries, and 139.7 million affected and homeless people, droughts caused more social disruption over this 25-year period than any other natural or non-natural disaster. The geographical distribution of droughts is not uniform: 62.8 per cent of droughts over the same 25-year period occurred in Africa, 18.6 per cent in Asia, 11.8 per cent in America, and 3.4 per cent in each of Europe and Oceania.

Numerous attempts have been made to define droughts on the basis of a range of physical and socio-economic criteria. For example, *permanent droughts* are characteristic of the arid regions of the world, which make up about a third of the world's land surface (Fig. 1). In regions subject to permanent drought, society has adjusted to a lack of rainfall and utilizes a range of alternative strategies in order to provide a reliable water supply. These include the exploitation of deep groundwater (as in Australia, Libya, and Tunisia), the use of desalination (Kuwait, Saudi Arabia, and Iran), abstraction from rivers fed by lakes, rainfall, and snowmelt in areas with more reliable precipitation (the Nile, Tigris, Euphrates, Murrumbidgee), and the development of large-scale water storage reservoirs (Ataturk Dam, Turkey; Aswan High Dam, Egypt; Bhakra Dam, North India). It is interesting to note that the completion of the Ataturk Dam in southern Turkey in 1993 has substantially reduced the quantity of water reaching Iraq through the Tigris and Euphrates river systems; this fact well illustrates the international political dimension to drought amelioration.

Seasonal droughts occur in many areas on the margins of arid regions where there is a strong seasonality in the precipitation regime and in equatorial regions dominated by monsoon climates. Ironically, many of the monsoon regions that experience seasonal drought also experience seasonal flooding.

Contingent drought occurs when lower than average rainfall conditions persist for months or even years in environments where the expectation of rainfall is high and in which an alternative water supply infrastructure has not been developed. Contrary to popular belief, contingent droughts are not limited in extent to the developing world. Examples include the dustbowl years of the 1930s in the USA, the European drought of 1976, the Sahelian droughts of North Africa in the mid and late 1980s, and the south-east Australian drought of the 1990s.

An *agricultural drought*, sometimes referred to as an *invisible drought*, is one in which rainfall appears to be adequate to

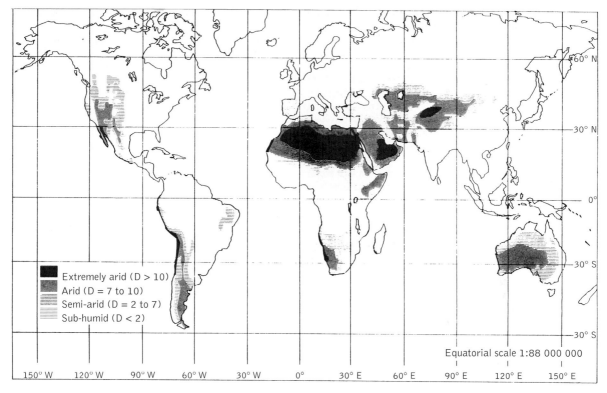

Fig. 1 The world distribution of drylands based upon the dryness index of Budyko (1977). The dryness ratio (D) is calculated from the number of times the mean net radiation at the Earth's surface (R) in a year can evaporate the mean precipitation (P) and the amount of heat required to evaporate a unit volume of water (latent heat of vaporization, L). The dryness ratio is derived from $D = R/L \times P$. (From Middleton (1991), Fig. 1.2.)

sustain agricultural production but evaporation rates are higher than normal, causing plant stress and a reduction in crop yield. It is rarely detected as a result of an examination of rainfall statistics alone, because account needs to be taken of the *water balance*, which determines the relationship between rainfall, run-off, and evaporation (see *hydrological cycle*).

Physiological drought is a form of drought in which plants suffer from an excess concentration of salt in the soil. It is common in irrigated dry-land areas. Although the soil contains sufficient quantities of water, the high salinity prevents the plants from taking up enough water for proper growth of the crop.

Different criteria are applicable to different parts of the world. This is partly because the expectation of rainfall and the natural variability in rainfall vary in different geographical regions. The spatial pattern in rainfall is mirrored by statistical variability in river runoff, which increases significantly between latitudes of 25° and 45° north and south of the Equator (Fig. 2). In general terms, the greater the rainfall that is expected in a region, the shorter the time interval without rain which is used to define a drought. In the UK, for example, an absolute drought is a period

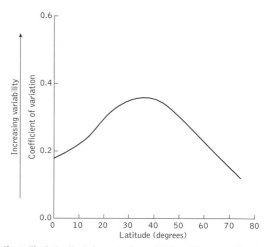

Fig. 2 The latitudinal change in the variability of river run-off at the surface of the Earth. (Based on an original diagram in McMahon T. A. (1982) *Hydrological characteristics of selected rivers of the world*. UNESCO Technical Documents in Hydrology. UNESCO, Paris.)

of at least 15 consecutive days during which no more than 0.2 mm of rainfall is recorded in any one 24-hour period. In arid and semi-arid environments, by contrast, a period of several months may elapse before a drought is officially declared.

Droughts are not abnormal phenomena; they are a normal component of contemporary climates in many regions of the world. They often cause severe disruption to agriculture and to human society generally, but they are a perfectly normal aspect of many climatic regions, especially in dry-lands where rainfall is unreliable and sporadic. In all the definitions given above, drought has the same effects: plants die, vegetation is reduced, and soil is either blown away or washed away by subsequent high-intensity rainfall (Fig. 3). In most dry areas, natural vegetation is well adapted to drought conditions, and seeds may lay dormant for several years until enough rain falls for germination.

Numerous attempts have been made to explain why droughts occur. In part, a meteorological explanation is appropriate. In middle latitudes, belts of west-travelling cyclones or depressions bring rain to areas of hundreds of square kilometres. In arid areas, the rainfall is caused by convectional instability through heating of the ground surface. Here, areas as small as a few square kilometres may receive rainfall while the surrounding areas remain dry. Since the worlds dry-lands are found in regions where air masses are stable, few rain-bearing systems develop. Research has shown that there is a relationship between rainfall in the Sahel region of Africa and the surface temperature of the tropical parts of the Atlantic Ocean. When surface temperatures are low, droughts are more frequent—perhaps because less evaporation occurs from a cooler body of water. The reason for a cooler sea surface has not, however, been established. The link between climate and the oceans has been shown to be important in other areas. For example, in the Eastern Pacific Ocean a large area of equatorial water is occasionally much warmer than usual in December. This so-called El Niño event appears to be associated with droughts as far apart as Australia and the Sahel.

It is also important that droughts are seen in the context of climate change. They occur repeatedly in historical rainfall records spanning at least a hundred years. Some research workers have been tempted to argue that the persistent droughts in the Sahel in the 1970s and 1980s are indicative of longer-term shifts in the world's climate (Fig. 4). Others have argued that these periods of droughts are not abnormal.

In comparison with other types of natural hazard, such as floods, hurricanes, earthquakes, and tsunami waves, droughts are often characterized by their widespread geographical impact, their longevity, their lack of physical impact other than on vegetation, and their dramatic impact on human population displacement. Droughts are often, and incorrectly, directly associated with the problem of desertification. Desertification is the reduction or destruction of the biological potential of the

Fig. 3 Evidence of desertification in New South Wales, Australia, December 1994. (Author's photograph.)

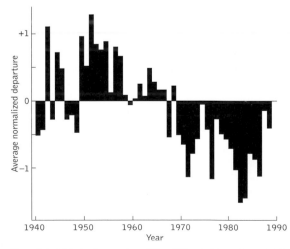

Fig. 4 Rainfall index for a region of West Africa, 1940–90. (From Smith (1992).)

land that can ultimately lead to desert-like conditions. While climate change and drought may exacerbate the problem, many research workers believe that desertification is a human-induced problem caused by overgrazing, the cutting of vegetation for fuel, the burning of animal manure, and the increased pressure of rising human populations on smaller areas of land leading to non-sustainable development.

The most usual long-term defence against drought has been the construction of dams and reservoirs for the artificial storage and transfers of water supplies. This reliance on high-cost technological solutions is demonstrated by the remarkable increase in river regulation, particularly between 1945 and 1970. By the late 1970s, over 40 per cent of the stable river flow in Europe, North America, and Africa was regulated by reservoirs. Although reservoirs often provide a reliable source of water under extreme drought conditions, a number of problems have arisen. First, many of the world's major rivers cross international boundaries. Disputes have consequently arisen over headwater impoundment and abstraction which have been resolved only by international treaty. Secondly, the life expectancies of many reservoirs in the dry-land areas of the world are proving shorter than was originally predicted. For example, the Mohammed V reservoir of north-east Morocco has been silting up at rates approaching 1 m per year since its construction in 1984. A survey of the 21 reservoirs currently operating in Morocco has shown that all have now lost over 20 per cent of their original storage capacity, and some have lost over 70 per cent. Thirdly, in many rural areas, such as the Sahel, which have been suffering severe droughts over the past twenty years, large-scale technological solutions are not available, and 95 per cent of agriculture is rain-fed and not irrigated. This leaves fewer options for mitigating drought than in irrigated areas, where it is possible to ration water and set priorities on its use. Some scientists have argued that one of the greatest social effects of droughts has been caused by the change from traditional to cash-crop agriculture in the Sahel, which has been prompted by a need to earn foreign exchange. This has seen an increase of the area of land under continual cropping, which in many instances has led to declining fertility and the displacement of traditional pastoralists. The pastoralists had developed diverse stocking management practices in keeping camels, cattle, sheep, and goats, all of which have different grazing habits and water requirements. This allowed a range of responses to drought conditions, including free movement to new pastures or the option of selling animals at market.

The greatest single disaster caused by drought is undoubtedly famine. Although severe droughts occur in many developed countries, far fewer deaths are recorded than in less developed countries. Famine appears to be almost endemic in some of the poorer countries of sub-Saharan Africa. Such disasters are multi-causal and often include an element of civil war and unrest. The international response is usually one of aid, the distribution of which is often made difficult by poor communications and civil unrest. Since sub-Saharan Africa has 29 of the 36 World's poorest countries, social scientists

have argued that the people of the region are locked into an economic system that obliges them to produce food they do not consume and to purchase goods they do not produce. Some have argued that colonialism and the dominance of 'Western' trading systems have reduced the ability of sub-Saharan Africa to cope with variations in their physical and social environments. The worst-hit groups are undoubtedly the poorest members of society—the landless and jobless, the women and children, who cannot ensure the security of their own food supplies.

IAN D. L. FOSTER

Further reading

Chen, M. A. (1991) *Coping with seasonality and drought.* Sage Publications, New Delhi.

Dawson, A. G. (1991) *Global climate change.* Oxford University Press.

Middleton, N. (1991) *Desertification.* Oxford University Press.

Smith, K. (1992) *Environmental hazards: assessing risk and reducing disaster.* Routledge, London.

Somerville, C. M. (1986) *Drought and aid in the Sahel.* Westview, Boulder.

World Disasters Report (1994) International Federation of Red Cross and Red Crescent Societies, Geneva, Switzerland.

drylands The term 'drylands' is easy to understand, but rather more difficult to define. Intuitively, it implies a shortage of water, but the amount of rainfall is a poor measure of water shortage. In consequence, there have been many attempts to produce definitions of this term, or the related term 'aridity'. One that has gained wide acceptance is that produced on behalf of UNESCO in 1977 (Fig. 1), which is based upon Thornthwaite's index of moisture availability (effectively the ratio of precipitation to potential evapotranspiration), together with information on soils, biology, and vegetation of drylands. Useful as this is, it begs the fundamental question, 'How far does this climatic definition of drylands describe a distinctive landscape?'. The short answer is that in part it does, and in part it does not.

It is a characteristic of drylands that they experience great climatic variability at all scales. At the daily scale, fluctuations of temperature are most significant and are typically very similar to the annual range of mean daily temperature. This daily temperature fluctuation is particularly important for weathering processes, both physical and chemical.

On the longer timescale, the variability in precipitation is most marked. The significance of this variability is to enhance the landscape-forming importance of rare extreme events, which probably have a greater role in drylands than in any other environment.

Precipitation not only is temporally variable; it is characteristically spatially variable also, and relatively small areas are likely to be affected by high-magnitude precipitation events. Erosion and mobilization of sediment are, therefore, characteristically both episodic and patchy, in contrast to more humid environments where these processes are more integrated.

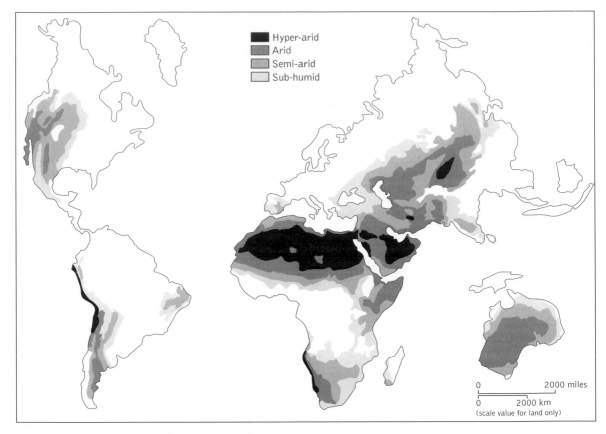

Fig. 1. Global distribution of drylands. (After UNESCO 1977.)

The present climate of drylands is, however, only part of the story. Many writers have commented on the persistence of landscape features in drylands. Williams described them in 1994 as 'excellent geological, geomorphic and archaeological museums'. The landscape of each dryland area reflects its own climatic history as much as it does its present climate. Despite the similar climatic setting of many of today's drylands, their climatic history is very diverse.

This far-from-simple influence of climate on the geomorphology of drylands is further complicated by the diversity of their geological inheritance. Mabbutt, in his book *Desert landforms*, grouped the world's drylands into two broad types on the basis of major geological structure and gross relief, namely those of the shields and platforms and those of the tectonically more active zones (mountain-and-basin deserts). Within these gross morphostructural classes, Mabbutt distinguished different physiographical settings, most notably the uplands and the piedmonts, and emphasized that different geological settings and consequent major relief produce different physiographical responses to aridity. A consequence is that different drylands contain different suites of landforms. For example, whereas alluvial fans and bajadas occupy more than 30 per cent of the American south-west but sand dunes occupy less than 1 per cent, in the Sahara dunes cover almost 30 per cent of the area and fans and bajadas about 1 per cent.

The drylands of the world are far from looking all alike. Their differences arise from their different climatic and geological histories; their similarities arise from their common susceptibility to the same landscape-forming processes and events.

ANTHONY J. PARSONS

Further reading

Abrahams, A. D. and Parsons, A. J. (eds) (1994) *Geomorphology of desert environments.* Chapman and Hall, London.

Cooke, R. U., Warren, A., and Goudie, A. S. (1993) *Desert geomorphology.* UCL Press, London.

Mabbutt, J. A. (1976) *Desert landforms.* ANU Press, Canberra.

duricrusts *see* PALAEOSOLS, DURICRUSTS, CALCRETE, SILCRETE, GYPCRETE

du Toit, Alex (1878–1948) Alex du Toit was a South African with a wide and detailed knowledge of his own country and of other parts of the southern hemisphere, and who was in consequence a disciple of Alfred Wegener, the originator of the theory of continental drift. In particular, du Toit was struck by the remarkable similarity of the geology of southern Africa and the eastern parts of South America. When Wegener published his book *The origins of continents and oceans* in 1915 it had met with a great deal of hostile criticism, especially from geophysicists who denied that such movement was possible. Some geologists, too, found it unacceptable. To du Toit the concept was attractive, and in 1927 he published a lengthy article of the support that the structural, stratigraphical, and radio-isotope data from the South Atlantic countries gave to the Wegener thesis.

Wegener read du Toit's paper, and in the fourth edition of his book he referred to the evidence it offered in support of the drift idea. Then in 1937 du Toit published his own book *Our wandering continents*, which contained small improvements and modifications of the original scenario. Despite the continuing hostility of many geologists, this contribution to the debate aroused much interest. Du Toit differed from Wegener in that he proposed two supercontinents: a northern or equatorial Laurasia and a southern, polar Gondwanaland. Du Toit's book was the last major contribution to support Wegener's thesis before the outbreak of the Second World War. When peace returned, interest was rekindled. A symposium on the subject was held in New York in 1949, but by then du Toit had died in South Africa.

D. L. DINELEY

dust *Dust* can be loosely defined as a suspension of solid particles in air, or a deposit of such particles which have been transported and deposited by the wind. Dust particles which are transported in suspension in the Earth's atmosphere are mostly smaller than 100 micrometres (μm) and a high proportion is less than 20 μm. This is because grains larger than 20 μm have relatively high settling velocities and tend to fall back quickly to the ground surface except under conditions of very strong turbulent winds.

Atmospheric dust originates from several different sources, including volcanic eruptions, industrial emissions, and outer space (cosmic dust), but wind deflation of particles from surface sediments and soils is quantitatively most important. At the present day the most significant sources of dust are arid and semi-arid regions, particularly those affected by periodic droughts and human activities such as cultivation, overgrazing, and construction. The Sahara desert and the deserts of northern China provide the two most important individual source regions for dusts. They produce dust plumes which extend thousands of kilometres over the equatorial Atlantic and North Pacific, respectively. During cold stages of the Quaternary, very large amounts of dust were also generated by wind erosion of extensive fluvioglacial outwash deposits, notably in North America, Europe, and Siberia.

Many different types of wind system are involved in dust transport, including small-scale dust devils (whirlwinds), mountain and valley winds, trade winds, thunderstorm downdrafts, monsoonal winds, and winds associated with mid-latitude depressions. Localized dust blowing may be caused by any set of conditions that produces steep thermal or pressure gradients. However, long-range transport of dust usually requires the dust to be lifted to relatively high levels in the troposphere, where it becomes incorporated in fast-moving upper-level wind systems.

Dust transport is of considerable geological, geochemical, and biological importance. As much as two-thirds of the sea-floor sediments in parts of the North Pacific consist of deposited dust derived from the Asian deserts, while in north-central China terrestrial accumulations of dust, known as loess, exceed 300m in thickness. These deposits form some of the most fertile soils in the world and support a population of more than 600 million people. The iron present in windblown dust provides an important control on the levels of biological activity in remote oceanic areas, and as such might exert a significant indirect influence on the global carbon budget, concentrations of atmospheric carbon dioxide (CO_2) and climate. Deflated soil dust and volcanic dust can also influence climate directly through their effect on the proportion of light reflected by the atmosphere (its albedo) and the Earth's radiation budget, particularly if the dust reaches the stratosphere, where residence times can be of the order of several years.

In areas of frequent dust-storm activity, blowing dust poses a significant hazard to human health and results in significant economic costs. High atmospheric dust loadings are associated with high incidences of respiratory diseases and a range of bacterial and viral infections. Reduced visibility caused by blowing dust is a significant cause of vehicle and aviation accidents, while the list of adverse economic effects includes contamination of drinking water supplies; damage to crops, engines, and electronic equipment; and interference to radio transmissions and telecommunications. In the more developed parts of the world where dust storms are a hazard, such as the south-western United States and south-eastern Australia, dust-storm warning systems have been set up in an attempt to minimize their negative impacts. Dust transport and deposition may, however, also have positive effects, including the neutralization of acid rain, the addition of mineral nutrients to soils, and the formation of stabilizing surface crusts in areas of active sand dunes.

K. PYE

Further reading

Pye, K. (1987) *Aeolian dust and dust deposits.* Academic Press, London.

Dutton, Clarence Edward (1841–1912)

One of the first geophysicists in the United States, C. E. Dutton was educated at Yale University and served in Union Army during the American Civil War. In 1875, he began a ten-year study of the great plateau region of the southern-western United States, first as a member of the United States Geological and Geographical Survey of the Territories, under the directorship of J. W. Powell, then, after 1879, as a member of the United States Geological Survey.

His studies of the high plateau of Utah and the geological history of the Grand Canyon district led Dutton to ponder the causes of elevations and subsidences of restricted areas of the Earth's crust. He reasoned that unloading of the land by erosion and loading of the sea floor by deposition of sediment must result in a force that tends to push the loaded sea bottom inward on the unloaded land. For this counterbalance between depression and elevation, Dutton invented the term *isostasy*.

Dutton subsequently studied volcanoes in Hawaii, Oregon, and California. From its similarity to the calderas of Hawaiian volcanoes, he recognized the caldera origin of Crater Lake in Oregon. After his studies of the earthquake of 31 August 1886 that devastated Charleston, South Carolina, Dutton devised new methods of calculating the depth of earthquake foci and for ascertaining the velocities of propagation of earthquake waves.   BRIAN J. SKINNER

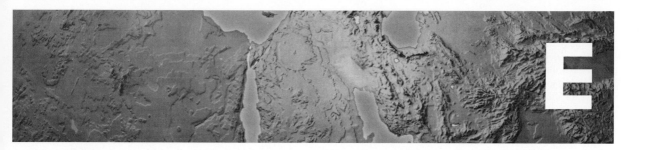

early vertebrates and fish

early vertebrates and fish Vertebrates are distinguished from the other members of the phylum Chordata by the presence of a brain, specialized paired sensory organs (hearing, sight, smell), and an internal skeleton. The earliest vertebrate fossils have recently been reported from the Middle Cambrian of Chengjiang, southern China; however, they are not well known until the Middle Ordovician. These organisms had a cartilaginous internal skeleton and a bony exoskeleton. Bone may have developed initially as a protective armour, perhaps as a defence against invertebrate predators such as eurypterids, or it might have developed as a storage site for calcium and phosphate, both of which are necessary for muscle activity and metabolic processes.

Agnathans

The earliest fossil vertebrates belong to a group called the Agnatha or jawless vertebrates. They are represented now only by the lampreys, eel-like forms that are parasites on fish, and the hagfish, also eel-like but feeding on dead or dying animals. Ordovician vertebrates occur in the Harding Sandstone of North America and were first reported by Charles Walcott of the US Geological Survey in 1892. More recently they have been discovered in rocks of similar age in Australia and Bolivia. Work on these new forms by Pierre-Yves Gagnier of the Natural History Museum in Paris, and reassessment of the North American material by David Elliott of Northern Arizona University, has shown that these animals were about

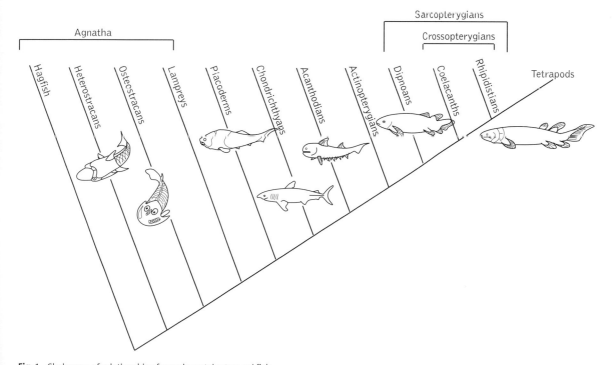

Fig. 1. Cladogram of relationships for early vertebrates and fish.

15 cm long with numerous separate gill openings and no paired or dorsal fins. All of them were found in shallow-marine rocks, and it is generally accepted that the vertebrates originated in marine conditions, a view supported by the recent Chinese discoveries.

Related agnathans with external armour occur widely from the Late Silurian to the Devonian, but they all became extinct before the start of the Carboniferous. There are two major groups, one of these, the osteostracans ('bone-shelled') had a broad triangular bony shield covering the front of the body (Fig. 1). The eyes were close-set on the top of the shield and between them was a nasal opening and an opening for the pineal organ, a light-sensitive structure. Also in the mid-line was an elongated area covered by small polygonal plates, which together with two similar areas, one on each side of the head, constituted the main vibration receptors for the animal. The mouth and the gill opening were below the head-shield, and this together with the shape of the body indicates that these animals were bottom-dwellers. Paired pectoral fins and a dorsal fin were present, and the caudal fin was asymmetrical with a large upper lobe. These fish were neither fast nor agile swimmers, but probably wriggled along the bottom sucking up organic-rich material.

The second major group of agnathans, the heterostracans ('different shelled'), were animals in which the head and front of the body were covered by bony plates. There were no paired or dorsal fins, although some forms developed bony plates extending out from the body to act as control surfaces (Fig. 1). They had no median nostril. A pineal eye was present, although not usually open to the surface. The water from the gills exited through a single branchial opening on each side of the body. A related group, the thelodonts, were covered by small bony elements. Although they are rarely well preserved as complete animals, their bony elements are sometimes very abundant in marine sediments and they have been used extensively in the biostratigraphic classification of the Late Silurian and Devonian.

Early jawed fishes

The development of jaws was a major step forward in the evolution of vertebrates, opening up many new possibilities in feeding and defence. Evidence from embryology and comparative anatomy shows that jaws developed from an anterior pair of gill arches (bony supports within the gills) that moved forward and became supports for the edges of the mouth. At the same time the upper part of the next gill arch (the hyoid arch) also became enlarged and formed a brace (the hyomandibular) between the back of the jaws and the brain-case. There are, however, differences between the gill arches of modern jawless and jawed fishes, indicating that neither group could have been developed from the other and that they must have had a common ancestor far back in vertebrate history.

The first jawed fish in the fossil record are the acanthodians ('spiny ones'), which first occur in the Late Silurian. These generally small fish were laterally compressed and were characterized by the presence of long spines in front of the two dorsal fins and the pectoral fins, and in a double row along the underside of the body. Some had small sharp teeth and fed on small invertebrates and fish; some of the later forms had no teeth and filtered suspended particles. The acanthodians became extinct in the Permian and are thought to be related to the modern bony fish.

The placoderms ('plate-skinned') are the second group of early jawed vertebrates and have a purely Devonian distribution. These fish were characterized by bony armour over the skull and front of the body, with a joint between the two areas. They developed a great diversity of body types, particularly within the major group, the arthrodires, some of which were predators up to 6 m long. Another placoderm group, the antiarchs, had a long box-like armour and a pair of arm-like pectoral appendages that may have helped them to move across the bottom. Although in the past they have been thought to show a relationship to sharks or to bony fish, they are now considered to be an entirely separate group.

Chondrichthyans

The advanced jawed fish the chondrichthyans (Fig. 1) first appeared in the Late Silurian. Their internal skeleton is composed of cartilage rather than bone and does not preserve well, so their fossil record consists mostly of teeth, fin spines, and scale-like dermal denticles, which are ossified. The teeth appear to have been formed from dermal denticles that migrated to the edge of the jaw and developed there. In the sharks, a method of tooth replacement was developed in which teeth migrated from the inner to the outer edge of the jaw, becoming functional and then being lost. Because of this replacement system, one shark may produce 20 000 teeth during its lifetime. Well-preserved specimens of the shark *Cladoselache* from the Late Devonian Cleveland Shale of Ohio show that this animal was an active predator with a streamlined body form. The pectoral fins were broad-based where they were attached to the body, which suggests less flexibility than in modern forms; the mouth and eyes were also further forward than is the case in modern sharks. Sharks of this type persisted until the Triassic, when they were replaced by the hybodonts. The hybodonts were dominant in the oceans until the Early Cretaceous, when the ancestors of the modern sharks appeared. The main advances seen are in the jaws, which became shorter and more curved, allowing for a more powerful bite. In addition, the upper jaws lost their contact with the brain-case, allowing the jaws to be protruded during biting. The snout was also developed in many sharks to accommodate the very sensitive electroreceptor system. Sharks include the largest modern fish predator, the Great White Shark, as well as the largest fish, the Whale Shark, which may reach 18 m in length. A separate development of animals adapted to life on the bottom produced the skates and rays, in which the pectoral fins enlarged to take over the main propulsive function. In addition, the chimaerids or ratfish represent a separate radiation of deep-water fish in which large crushing toothplates are fused to the brain-case.

Osteichthyans

The osteichthyans or bony fish also appeared in the Late Silurian, together with the chondrichthyans and the early jawed vertebrates. They were streamlined, laterally compressed fish with a bony skeleton and thick bony scales. The gill openings were covered by a large plate, the operculum, and the head was covered by interlocking bony plates. Paired pectoral and pelvic fins and a large dorsal fin provided stability and control, while a heterocercal tail (in which the vertebral column projects into the upper lobe) provided the motive power. At the time of their first appearance the osteichthyans had already separated into two groups, the actinopterygians ('ray-finned') which includes most modern fish, and the sarcopterygians ('lobe-finned') which include the ancestors of the tetrapods, or land vertebrates.

The actinopterygians have only one dorsal fin and all the fins are supported by numerous fin rays controlled by muscles inside the body wall. Although they were initially a rather conservative group of visually oriented, active predators, they began in the Early Mesozoic a series of modifications that resulted in the enormous range of forms that we see today. Thinning of the scales, strengthening of the vertebral column, the forward movement of the pelvic fins, and the development of a homocercal, or symmetrical, caudal fin resulted in faster and more manoeuvrable animals. At the same time, shortening of the jaw and uncoupling of the upper jaw from the skull made possible a new feeding mechanism in which rapid opening of the mouth caused suction, pulling food into the mouth. These advances culminated in the modern bony fish, or teleosts, which show a tremendous diversity of mouth parts and feeding specializations. Although about 23 000 species of teleosts are alive today, there are also some examples of the earlier actinopterygians still living. The sturgeon is an example of the initial radiation of Palaeozoic chondrosteans, while the gar pikes are representatives of the initial radiation of neopterygians in the Mesozoic.

The sarcopterygians include the lungfish, together with the crossopterygians, which are composed of coelacanths and the extinct rhipidistians. These forms had two dorsal fins rather than one, and all the fins were supported by an internal skeleton and musculature. The stout fins may have been useful in propelling them across the bottom. The lungfish use lungs as accessory breathing organs, and during droughts modern forms can survive for several years in burrows. Massive crushing teeth on the upper and lower jaws are used to feed on shelled molluscs. The lungfish were extremely conservative and show little change through their evolutionary history. The crossopterygians, however, developed a diverse group of medium-sized freshwater predators. They were thought to have become extinct in the Cretaceous until, in 1948, a coelacanth was caught off the coast of East Africa. Rhipidistians did become extinct in the Early Permian, but at some time in the Devonian the osteolepiform rhipidistians had given rise to the first land animals, the amphibians.   DAVID K. ELLIOTT

Further reading

Carroll, R. L. (1988) *Vertebrate paleontology and evolution.* W. H. Freeman, New York.

Colbert, E. H. and Morales, M. (1991) *Evolution of the vertebrates.* Wiley-Liss, Inc., New York.

Earth, age of *see* AGE AND EARLY EVOLUTION OF THE EARTH AND SOLAR SYSTEM

Earth movements (diastrophism)

'Earth movements' is a term used in the older literature to describe larger-scale crustal deformation, that is, deformation other than the movements that produce individual faults or folds. Thus, Arthur Holmes, in his celebrated textbook *Physical geology*, first published in 1944, described under the heading of 'Earth movements' the effects of vertical movements involved in uplift or depression of the Earth's surface to create plateaux, mountain ranges, or basins. He also included the buckling of strata into folds produced by horizontal compressive forces, but did not speculate on their origin. At that time it was generally thought that crustal deformation was primarily due to vertical movements: continental drift was then a controversial topic.

The term 'diastrophism' also belongs to the older literature. It was first used by J. Milne in 1881, and was held by G. K. Gilbert in 1890 to include both orogenic (mountain-building) and epeirogenic (vertical) movements, but it was subsequently applied (as, for example, by J. W. Powell in 1895) more particularly to epeirogenic movements. Orogenic or mountain-building movements are today ascribed to relative plate motion, and the terms 'Earth movements' and 'diastrophism' are generally absent from the current literature.   R. G. PARK

Further reading

Duff, D. (ed.) (1993) *Holmes' Principles of physical geology* (4th edn). Chapman and Hall, London.

Earth orientation

Earth orientation is defined as the instantaneous angular relationship between an Earth-fixed reference frame and an inertial, external reference frame. Intimately linked with this concept is that of Earth rotation or rate of change of orientation over timescales of from thousands of years to a few hours. To first approximation, the Earth in its annual orbit about the Sun is inclined such that the angle, the obliquity, between the Earth's equator and the orbital plane, the ecliptic, is approximately 23.4 degrees, and the Earth's rotational axis is fixed in inertial space. The Earth spins about this axis in a period of 24 hours, or 86 400 seconds. The study and observation of Earth orientation and the development of a theoretical understanding of the processes affecting Earth rotation seek to improve the parameters of this model so that orientation at any instant may be determined at the centimetre level of accuracy.

The Earth is an oblate spheroid, with polar diameter some 45 km less than the equatorial diameter. The gravitational attraction of the Sun and Moon on the resultant equatorial bulge causes the rotational axis to precess smoothly about the pole of the ecliptic with a period of some 25 600 years. Superimposed on this smooth precession is a somewhat irregular small-amplitude motion (nutation) at periods of from a few days to 18.6 years, due to the gravitational attractions both of the Sun and of the Moon during its complex orbit about the centre of gravity of the Earth–Moon system. The ecliptic itself moves slowly in inertial space, owing to the gravitational attraction of the planets on the Earth, such that the obliquity changes very slowly with time.

In addition to the precession and nutation of the Earth's rotational axis, the Earth itself undergoes a wobble such that the crust wanders about the rotational axis with a combination of two periods and amplitudes. One motion, called the *Chandler wobble*, occurs at the natural resonant frequency of the Earth and its oceans, having a period of about 435 days and amplitude on the surface of the Earth of about 6 m. The second, annual, wobble is understood to be atmospherically excited and has amplitude about 3 m. Together these wobbles are referred to as *polar motion*.

The rotational period of the Earth (or length-of-day, LOD) is not constant, varying from the standard 86 400 seconds by several milliseconds over a range of periods, as outlined here.

The tidal interaction between the Moon and the Earth is such that the Moon in its orbit exerts a retarding torque on the spinning Earth and slows its rotation thus causing a linear increase in LOD, of about 2 milliseconds per century. To conserve angular momentum within the Earth–Moon system, the Moon responds by increasing its orbital distance from the Earth at a rate of some 3 cm each year. In addition, the tides raised both in the oceans and in the solid Earth by the Moon and Sun cause changes in angular momentum of the Earth, with resultant short-periodic changes in LOD of up to one millisecond. Interactions between the Earth's core and solid mantle are believed to be responsible for changes in LOD of several milliseconds, over periods measured in decades. Exchanges of angular momentum between the atmosphere and the rugged surface of the Earth on annual timescales can cause periodic changes in LOD of up to one or two milliseconds in a matter of a few weeks.

As a result of these mechanisms, on average over recent years the LOD has been about 2 milliseconds longer than the standard day as kept by uniform atomic clocks, the basis for the worldwide timescale, UTC. This excess accumulates day by day, so that after about 18 months the timescale as determined by monitoring the rotation of the Earth, which is referred to as UT1, loses about one second as compared with UTC. By international agreement, one-second leap seconds are periodically introduced into UTC, in this case to delay its

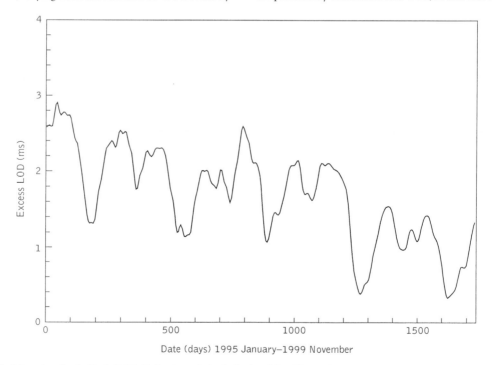

Fig. 1. Length of day values for the Earth (1995–9) showing variation in day length in milliseconds.

progress and keep civil timekeeping in step with the rotation of the Earth.

Earth orientation is monitored on a routine basis by world-wide networks of observational facilities and analysis centres. There are two classes of technique, astronomical and space geodetic. They contribute complementary and necessary duplicate information leading to very accurate, rapid determination of all the components of Earth orientation for a variety of operational and research activities.

A network of radio telescopes makes interferometric observations of compact extra-galactic radio sources whose positions define an inertial system with respect to which precession, nutation, polar motion and the difference between atomic time and UT1 are determined.

A network of optical telescopes makes laser range measurements with centimetric accuracy to Earth-orbiting satellites and in a few cases to the Moon. The observations are analyzed to determine both the orbital motions of the satellites and the Earth orientation parameters polar motion and LOD. Similar analyses of the navigational signals from the satellites of the US Global Positioning System are also used routinely to determine components of Earth orientation.

Figure 1 shows a series of LOD values for five years, determined from satellite laser range measurements and expressed as excess LOD in milliseconds with respect to the standard day of 86 400 seconds. For clarity in the presentation, the predictable, short-period variations in LOD caused by the tidal interaction of the Moon have been removed from the data. Clearly seen are the annual, atmospherically driven variations in LOD, of amplitude between one and two milliseconds. Also apparent is the trend towards smaller values of excess LOD. If this trend continues, the introduction of leap seconds into UTC will become less frequent.　　GRAHAM M. APPLEBY

Further reading

McCarthy, D. (ed.) (1996) *International Earth Rotation Service Technical Note* 21: *IERS Conventions*. Observatoire de Paris, 61 Avenue de l'Observatoire, Paris.

Munk, W. H. and Macdonald, G. J. F. (1960) *The rotation of the Earth*. Cambridge University Press, New York.

Earth resources and management

The Earth is a repository of materials which have diverse application for human life. The Earth's geological resources include water, building materials (for houses), metals (for cars), oil (for petrol, plastics, and fabrics), coal (for energy and heating), and gas (for cooking and heating). A *geological resource* is defined as a naturally occurring solid, liquid, or gas that is known to exist or thought to exist in or on the Earth's crust in concentrations that make extraction economically feasible, either at present or at some time in the future. (Note that this definition includes deposits that are thought to exist.) A *reserve*, on the other hand, is the portion of an identified resource that can be extracted economically using current technology. In the short term, resource management focuses on reserves. Environmental considerations are of increasing importance and have the potential to reduce available reserves by altering the economics of exploitation. An important consideration is thus to determine whether the deposit is economic by taking into account the price in relation to the production cost, as well as supply and demand.

Economic resources include geological commodities (economic goods) such as gemstones, fossil fuels, water, and aggregates. The cost of production for an economic resource includes capital costs and operational costs. Capital costs include the cost of exploration to identify reserves, the cost of mining rights, planning and political costs, the cost of the mining and processing infrastructure, investment in the transport infrastructure, and the cost of quarrying or mine aftercare (restoration). Operational costs include the production of unit volume of a marketable product, which is influenced by the quality and extent of the reserve, the cost of removing the commodity from the ground, the cost of extracting it from the rock, and the size of the operation. In general, the bigger the operation the lower the cost per unit of output.

Supply and demand define the price obtained for the mined commodity. *Supply* refers to the quantity of a commodity that suppliers are prepared to sell at a given price. *Demand*, on the other hand, is the quantity of a commodity a consumer is prepared to buy at a given price. In a modern free-market economy the supply, demand, and price of a product are thus interwoven. Some commodities are described as being *elastic*; others as *inelastic*. Inelastic commodities include water, which is essential for life: a price increase will hardly affect demand. For elastic commodities, on the other hand, such as gemstones, small price increases result in a large reduction in demand. An equilibrium price is reached where the supply and demand curves intersect. Changes in demand can result from increasing consumer spending power, changes in consumer priorities, and the availability of cheaper substitutes. For example, the price of copper has in the past been affected by an increase in demand which increased the commodity price. This increased incentive for suppliers to produce copper led to exploration and the introduction of new supplies. For their part, consumers looked for alternatives (for example, aluminium). This in turn led to oversupply of copper and the price fell. The chief influences on the price of copper in the twentieth century were the two world wars, the Wall Street crash in 1930, and the oil crisis in the 1970s. For a commodity such as copper it is also important to consider the grade (concentration) of the ore. It is not economic to mine the ore if the grade is low, and that in turn causes the reserves to increase.

In the past some reserves have had their prices fixed by large organizations. These prices were then not subject to a free-market economy. Such price fixes were applied by OPEC (the Organization of Petroleum Exporting Countries), which was formed in the 1960s by ten Middle Eastern countries that had at the time 60 per cent of oil production and 90 per cent of world exports. OPEC chose an artificially high oil price, which increased the value of world reserves, including those in the North Sea. This caused the 1991 OPEC world exports to drop to 38 per cent.

Effective resource management requires consideration of the sustainability of the resource and the environmental impact associated with its exploitation. Resources are either sustainable (renewable through natural processes) or non-sustainable (finite). An example of a sustainable resource is water. Non-sustainable resources include the majority of geological resources, such as fossil fuels and ore deposits. There are two perspectives on the future availability of mineral resources. One is the so-called *Ricardian paradigm*, which predicts that when reserves are consumed growth in demand and technological evolution will allow the exploitation of lower-grade ore. Examples commonly given are limestone, aggregates, and iron ore. On the other hand, the *Malthusian paradigm* proclaims that economically viable resources are finite and are consumed at exponential rates and that there exists a technological barrier to exploitation of low-grade ores.

Technological alternatives have a very important impact on commodity prices. For example, the use of optical fibre instead of copper wire for information distribution in

Germany has saved 200 000 tonnes of copper which is approximately 18 per cent of German production. Recycling also reduces the demand on primary reserves.

Environmental impact also needs to be minimized and managed. Working towards this goal is probably the most important contribution that an Earth scientist of the twenty-first century can make to the sustainable management of the environment as a whole. Some examples of the world's reserves are given in Table 1. It is notable that several commonly used metals are predicated to last only into the first half of the present century.

Water is the Earth's most important resource. Without it, there would be no life on the planet. There are 1410×10^6 km^3 of water on Earth. However, over 97 per cent of this water is saline and in the oceans, and 2 per cent is locked in ice and glaciers. Only 0.3 per cent of the Earth's water is shallow groundwater, 0.01 per cent is lake water, and 0.0001 per cent river water. Less than 1 per cent of the water on Earth is therefore available for drinking. Unfortunately, supplies of fresh water are unevenly distributed across the globe. For example, Iceland has about 558 times more water per head of population than Belgium, and 20 000 times more water per head than Egypt. It is therefore not surprising that the provision of fresh drinking water is a major preoccupation of many third-

Table 1. Earth resources: world reserves (1990 figures)

Commercial energy Commodity	Resource	Reserve
Coal		
bituminous	1 212 852	521 413 million tonnes
lignite	743 193	517 770 million tonnes
Crude oil		134 792 million tonnes
Natural gas		128 852 billion cubic metres
Uranium		(<$80/kg) 1 410 040 tonnes
		(<$130/kg) 673 670 tonnes
Hydroelectric power (installed)		624 044 MW

Ores and metals Commodity	Reserve (million tonnes)	Life index* (years)
Bauxite	23 000	22
Cadmium	0.54	
Copper	310	33
Lead	63	18
Mercury	0.13	43
Nickel	47	51
Tin	8	45
Zinc	140	20
Iron ore	150 000	161

*Resource-based life index is generally 100 per cent higher.

world and developing countries and that water supply has become a major factor in territorial disputes. It is estimated that at least 1.2 billion people on Earth lack a satisfactory or safe water supply.   K. VALA RAGNARSDOTTIR

Further reading

Bennett, M. R. and Doyle, P. (1997). *Environmental geology. Geology and the human environment*. John Wiley and Sons, New York.

World Resources Institute (1994) *World resources 1994–1995. A guide to the global environment*. Oxford University Press.

Earth sciences The Earth sciences are those scientific disciplines which help in understanding how the planet works and how it has reached its present state. They may be thought of as primarily those that reveal the complex systems which regulate the Earth's behaviour and history (Fig. 1). They progress today largely by research by teams of scientists from different basic disciplines, working on major topics or problems together, though the role of the lone scientist is not diminished. Advanced technological methods are nowadays supplemented by those of mathematics, statistics, and information handling.

During the nineteenth century the Earth sciences greatly stimulated discussion about man's place in nature. The uniformitarian view that 'the present is the key to the past' was widely adopted. There is a strong historical component to the Earth sciences: one of the major contributions to knowledge is the recognition of the magnitude of geological or 'deep' time. Another concerns the major role of life in the evolution of Earth's crust and atmosphere throughout the last 3.5 billion years. Urgent and ongoing themes relate to the conservation of natural resources and to the rapid natural changes that influence the future of mankind. The Earth sciences now also have a bearing on the new fields of planetology and space exploration.   D. L. DINELEY

Earth sciences in Classical times *see* CLASSICAL TIMES AND THE EARTH SCIENCES

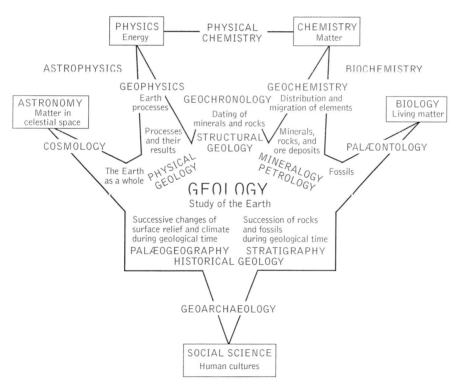

Fig. 1. In his classic *Principles of physical geology* in 1944 Arthur Holmes depicted the subdivisions of geology and their relationships to the other sciences. These subdivisions are today often referred to as the solid Earth sciences; their sister disciplines, meteorology and oceanography are also Earth sciences. The use of geological techniques and reasoning in archaeology has made great progress since Holmes drew his original figure. Other recent special developments such as palynology and sedimentology can be included under existing headings (palaeontology, petrology, etc.).

Earth sciences and information technology *see* INFORMATION TECHNOLOGY AND THE EARTH SCIENCES

Earth sciences and myths *see* MYTHS AND THE EARTH SCIENCES

Earth structure
The interior of the Earth is inaccessible: all that we know about it has had to be deduced from our knowledge of the rocks accessible to us and from the shape and physical properties of the Earth as a whole, and from the results of geophysical and geochemical experiment.

Our desire to understand the Earth goes far back into history, to the Hebrews, the Ancient Greeks, and the Chinese, but our understanding of the internal structure of the Earth really starts with the Victorian physicists (e.g. Rayleigh and Rutherford) and then Sir Harold Jeffreys with his classic book *The Earth*, first published in 1924, who laid the foundations for modern geoscience. With the very rapid advances in equipment, and particularly in computer technology, our knowledge of the details of the fine structure and workings of the Earth's interior has improved greatly since the 1980s.

The Earth is, in the broadest sense, a series of concentric spherical shells, each shell having distinct physical or chemical properties (Fig. 1, Table 1). The outermost, and thinnest, shell is the *crust*. Then, descending into the interior, the next shell is the *mantle*, which extends to a depth of 2891 km. This is subdivided into two: the upper mantle and the lower mantle. Finally at the centre of the Earth is the *core*. This also is subdivided into two: the outer core and, the innermost sphere, the inner core.

The crust

The rocks exposed at the surface of the Earth are part of the crust. The crust is a thin layer of silica-rich rocks which have been derived from the underlying mantle by melting and subsequent metamorphic or erosional processes,—or, in places, both. Crustal rocks can thus be broadly classified as igneous, metamorphic, or sedimentary, according to their individual histories. The crust is comprised of rocks of all ages from the oldest Hadean (more than 4000 Ma old) to the youngest modern lavas.

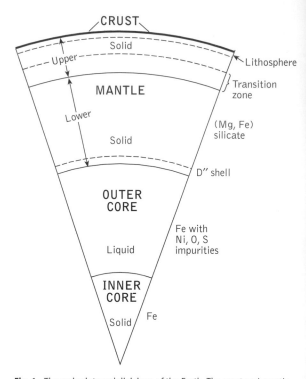

Fig. 1. The major internal divisions of the Earth. The crust and mantle are silicates and the core is predominantly iron. The crust and mantle are solid, the outer core is liquid and the inner core solid.

The crust that makes up the continents differs in origin, structure, and composition from the crust beneath the oceans. On average the continental crust is of granitoid composition, whereas the oceanic crust is basaltic. Beneath the continents the crust has an average thickness of about 38 km. However, in areas of continental extension, such as the Rhine Graben in Germany and the East African Rift, the crust is locally thinner by 10–15 km than the crust on either side of the rift zone. Beneath features such as the Andes, the Alps, the Himalayas, or Tibet, the continental crust is much thicker

Table 1 Volume, mass, and density of the Earth

	Depth (km)	Volume ($10^{18}m^3$)	Volume (% of total)	Mass ($10^{21}kg$)	Mass (% of total)	Density ($10^3 kg\ m^{-3}$)
Crust	0–Moho	10	0.9	28	0.5	2.60–2.90
Upper mantle	Moho–670	297	27.4	1064	17.8	3.35–3.99
Lower mantle	670–2891	600	55.4	2940	49.2	4.38–5.56
Outer core	2891–5150	169	15.6	1841	30.8	9.90–12.16
Inner core	5150–6371	8	0.7	102	1.7	12.76–13.08
Whole Earth	—	1083	100	5975	100	

than normal, often far exceeding 50 km. In some respects the continental crust can be regarded as the light 'scum' that floats on the Earth's denser mantle, in the same way that icebergs or logs float on water. The total volume of continental crust has increased through Earth history. The regions termed 'cratons' which now form the centres of several of the continents (e.g. North America) have crust dating from the Archaean (pre-2500 Ma). Younger rocks then surround and overlie these ancient cratons as new material has been accreted to the continent over time. Models of 'crustal growth rate' (which are based on isotopic ratios) indicate that the continental crust formed gradually through much of the Archaean with an increased growth rate in the late Archaean, since when there has been a gradual increase. Over 70 per cent of the present surficial area of the continents was formed more than 450 Ma ago. The continual processes of erosion and deposition of sediments mean that a good deal of 'recycling' takes place in crustal rocks. There is a continuing loss of some sediment into the mantle at subduction zones, although the 'conveyor-belt' system of plate tectonics ensures that most sediment is added to the accretionary wedge of the overriding plate. New crust in the form of volcanic rocks derived directly from partial melting of the mantle is continually added to the continents.

In contrast to the continental crust, the oceanic crust is young, thin, and chemically magnesium-rich. All the oceanic crust has been formed since the Jurassic, and only fragments of mid-Jurassic crust remain. The average thickness of the oceanic crust is 7 km. Oceanic crust is formed as a result of decompression melting in the mantle at shallow depths beneath the mid-ocean ridges. As a result, the oceanic crust is basaltic and is uniform in composition. Oceanic basalts are generally termed MORB (Mid-Ocean Ridge Basalts). Some of the rising magma that forms the oceanic crust is erupted at the sea bed, but much more solidifies without erupting, to form the characteristic stratified layering of the oceanic crust. The broad lithological layering is inferred from seismic P-wave velocities. Layers 2 and 3 are commonly subdivided in terms of the details of lithology and physical properties (Table 2).

The uppermost parts of the crust have been sampled directly by drilling. The international Ocean Drilling Program (ODP), a major co-operative programme of drilling in oceanic regions, has provided detailed information on the fine structure of the oceanic crust and has answered many questions about the details of the formation of oceanic regions. On the continents there are just two deep boreholes that penetrate to mid-crustal levels: one in Germany (KTB) and the other on the Kola peninsula in Russia. Despite the scarcity of drill information, a variety of geophysical techniques are used to great advantage to determine the gross overall structure of the continental crust in different tectonic regions, as well as some of its fine structure. Gravity surveys enable models of possible underlying density structures to be established; and as the density of rocks is broadly dependent

Table 2 Gross layering of the oceanic crust

	Lithology	Average thickness (km)	Seismic P-wave velocity (km s^{-1})
Layer 1	sediments (variable)	about 0.5 (depends upon age)	about 2
Layer 2	basalts fractured lavas	2.1 ± 0.6	2.5–6.6
Layer 3	dykes, gabbro, cumulates	5.0 ±0.8	6.6–7.6

upon their composition, gravity measurements can be used to infer lithology. Electrical and magnetic surveys enable models of the electrical and magnetic properties of the crust and uppermost mantle to be determined. Mineral composition, porosity, and permeability are additional factors controlling the electrical conductivity and magnetic susceptibility of rocks. However, seismic methods provide the most detailed and unique images of the structure of the crust (both continental and oceanic).

Seismic reflection profiling can yield images of the fine structure over small areas. It is the mainstay of the oil industry's search for hydrocarbons in sedimentary rocks. The method is used to image the structure of the crystalline crust and on occasion even the uppermost mantle. National programmes, such as COCORP in the USA, Lithoprobe in Canada, and BIRPS in the UK, have been very successful. Seismic reflection profiling provides images of any structures or features that send back a reflection. Reflections arise only when there is a change in seismic velocity or density. Reflections should thus be recorded from a sediment–igneous contact or from the base of a pluton, but totally homogenous material, such as a salt diapir or the interior of a granite pluton, will be reflection-free. Wide-angle seismic reflection profiling and seismic refraction are the other controlled-source seismic methods that are used to image the crust and uppermost mantle. While these methods do not generally provide such fine detail as reflection profiling, they have the significant advantage that they yield the seismic velocity structure. The velocity–depth structural models can then be interpreted in terms of lithology. The continental crust is not stratified like the oceanic crust and so does not have a characteristic seismic velocity structure. Nevertheless, the uppermost 10 km or so of the crystalline crust beneath the sedimentary cover generally has a P-wave velocity of 6.0–6.3 km s^{-1}, and beneath that the velocity is normally in excess of 6.5 km s^{-1}. According to the tectonic history of an area and its complexity, there may be low-velocity zones in the crust, or lower crustal material with a velocity in excess of 7 km s^{-1}.

The boundary between the crust and the underlying mantle is called the Mohorovičić discontinuity (abbreviated to Moho). It is named after Andrya Mohorovičić , who first

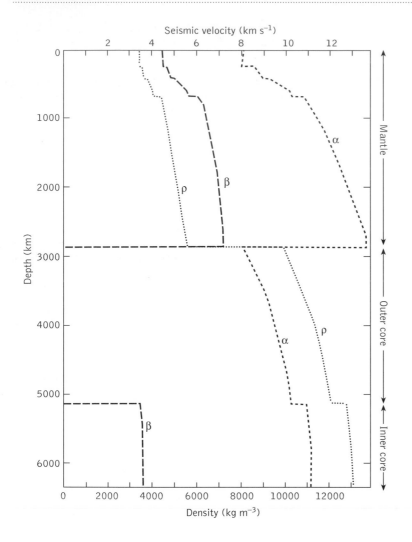

Fig. 2. The PREM (Preliminary Reference Earth Model) seismic velocity and density model for the Earth. This model was determined from a joint inversion of the travel–time and distance data for earthquakes as well as the free oscillation periods of the Earth, its mass and moment of inertia. α, P-wave velocity; β, S-wave velocity; ρ, density.

delineated the boundary in 1909. The normal seismic P-wave velocity of the uppermost mantle is 8.1 km s^{-1}, but considerable variation is observed locally. In regions where the mantle is hotter, such as along the axis of the mid-ocean ridge system, the seismic velocity is reduced. In contrast, cold dense regions can have elevated seismic velocities.

The mantle and deep interior

The controlled-source seismic methods used to determine crustal and shallow mantle structures are not suitable for determining seismic velocities deep within the Earth. Instead, methods utilizing earthquakes as the energy source and networks of seismic recording stations are used to calculate the travel times of seismic waves. These travel times are then used to calculate the variation of seismic velocities with depth in the Earth. Seismic velocities in the mantle are also determined

by using the dispersion of surface waves (i.e. the change in the character of the wave with increasing distance, which is due to the fact that waves of different frequency travel with different velocities). Figure 2 shows the seismic velocity structure for the whole Earth. This was determined using travel-time data, as well as the periods of the Earth's free oscillations and its mass and moment of inertia.

The Earth can also be classified by the way in which heat is transferred through it. The lithospheric plates are the Earth's outermost near-rigid, cool 'skin'. These are the plates that move about on the Earth's surface and along whose edges much of the seismicity and volcanism occur. Conduction is the main mechanism of heat transfer through the lithosphere. As the plates are about 100 km thick, the lithosphere comprises both the crust and the uppermost part of the mantle. The mantle beneath the lithosphere is hotter and, although behaving as a solid on a short timescale, is able to flow on a geological timescale. This means that convection is the mechanism for heat transfer through the sub-lithospheric mantle. There has been considerable scientific debates as to whether or not the upper and lower mantle convect as two separate systems. That they may be separate is suggested by four lines of argument: (a) the increase in seismic velocities and density at 670 km; (b) the maximum depth of earthquakes at subduction zones is 670 km; (c) some subducting plates break off at 670 km; and (d) some geochemical models suggest that the upper mantle is depleted (e.g. by partial melting and extraction of crust) and has been separate from the lower mantle for much of the Earth's history. Detailed three-dimensional seismic images of the velocity structures in the mantle indicate, however, that some plates are subducted into the lower mantle, thus indicating that the mantle may not be completely stratified.

Both the seismic P-wave and S-wave velocities increase with depth through the mantle. The P-wave velocity increases from a value of 8.1 km s^{-1} at the top of the mantle to 13.7 km s^{-1} at the core–mantle boundary (CMB). There are, however, several major irregularities superimposed on the steady increase of both seismic velocities and density with depth. There is a low-velocity zone for S-waves in the upper mantle which extends to depths of approximately 220 km. This low-velocity zone, which has been well defined by surface wave dispersion data, is generally known as the asthenosphere (from *asthenia*, Greek 'weak' or 'sick'). Beneath the asthenosphere, the seismic velocities and density increase steadily down to 400 km depth. There, and again at 670 km depth, all increase sharply. The velocities then increase steadily though the lower mantle to the basal 200 km of the mantle, where the rate of increase is much reduced. This is also a region that produces increased scatter in the amplitude and travel times of seismic waves, indicating that it is a zone of considerable inhomogeneity.

Experimental work on olivine, (Mg, Fe)$_2$SiO$_4$, a major constituent mineral of the mantle, has shown that its atoms undergo a phase change at pressures equivalent to depths of about 400 and 670 km. Phase changes like these do not involve any change in mineral composition; rather the atoms are reorganized into more closely packed crystalline lattice structures. Between about 390 and 450 km, the olivine is changed into a spinel structure, which results in a 10 per cent density increase. This change from olivine to spinel is exothermic, which means that the change involves a release of energy in the form of heat. The other major mantle minerals, the pyroxenes, also undergo a phase change at these depths, to garnet. The change that occurs at about 670 km is from spinel to post-spinel structures, perovskite and magnesium oxide. This change also results in a density increase of 10 per cent, but it is endothermic, which means that heat is required for the change to take place. It is thought that these phase changes may control the cessation of seismicity at 670 km depth as well as maintaining the pattern of convection in the mantle as two separate systems: upper mantle and lower mantle. The

convection regime may undergo periodic flushing events when large volumes of accumulated upper mantle material descend from the base of the upper mantle into the lower mantle. The base of the lower mantle may in effect be a graveyard for subducted plates.

Some old terminology remains in describing the mantle. The whole region between 400 and 670 km depth is called the mantle transition zone. Beneath this, from 670 to 2700 km depth, the mantle is sometimes referred to as the D′ shell. The basal 200 km of the mantle (from 2700 to 2900 km depth) is called the D″ shell. The reduction of the seismic velocity gradient in the D″ shell, and its heterogeneity, may be due to the shell acting as a thermal boundary layer through which heat is conducted rather than convected as it is through the rest of the mantle. It may also be in part caused by vigorous chemical interaction between the silicate mantle and the iron core.

The core

The core was discovered by R. D. Oldham in 1906 and was accurately delineated as being at 2900 km depth by Beno Gutenberg in 1912. The core–mantle boundary, or CMB, is also known as the Gutenberg discontinuity. The core is physically and chemically distinct from the mantle. In composition it is predominantly iron with small amounts of other elements. Work on tides enabled Sir Harold Jeffreys to establish in 1926 that the outer core must be fluid. A decade later, in 1936, Inge Lehmann (1888–1993) was able to show that there was a solid inner core at the very centre of the Earth. She did this by using seismic energy from an earthquake in New Zealand that was recorded in Europe after having passed through the centre of the Earth. The outer core–inner core boundary is called the Lehmann discontinuity in her honour.

At the CMB the P-wave velocity drops from 13.71 km s^{-1} to 8.06 km s^{-1} and the S-wave velocity drops from 7.26 km s^{-1} to zero, while the density increases from 5567 to 9903 kg m^{-3}. This is consistent with the outer core being liquid (S-waves cannot be transmitted through a liquid). Within the outer core the P-wave velocity increases steadily, reaching 10.36 km s^{-1} at the outer core–inner core boundary. At that boundary the P-wave velocity increases from 10.36 to 11.03 km s^{-1}, the S-wave velocity from zero to 3.50 km s^{-1}, and the density from 12 166 to 12 764 kg m^{-3}. The P-wave velocity, S-wave velocity, and density then increase only slightly through the inner core, reaching values of 11.26 km s^{-1}, 3.66 km s^{-1}, and 13 088 kg m^{-3} respectively at the centre of the Earth.

The composition of the core is hard to verify: there are no core samples to be studied. Instead we have to rely on ingenuity and analogue. The relative abundances of elements in the Sun and in meteorites indicate that the core should be predominantly iron, and in bulk could approximate to Fe$_2$O with a small proportion of nickel. The seismic velocity and density structure, together with experiments conducted in the laboratory at high pressure and temperature, imply that the inner core may be almost pure iron. The outer core is an iron alloy with about 10 per cent of lighter elements, the most likely candidates being oxygen, sulphur, nickel, and silicon. Experiments have shown that liquid iron and iron alloys react strongly with solid iron and magnesium silicates. The seismic complexity of the CMB can thus be explained in terms of the chemical reactions taking place there. This may be the most chemically active part of the planet.

That the outer core is liquid and the inner core solid is consistent with all seismological observations as well as studies on tides and the Earth's rotation, which both require a liquid core. The liquid outer core is the source of the Earth's magnetic field. It acts as a giant spherical dynamo, in which less dense rising convection currents of liquid iron also carry electric currents. The interaction of these electric currents with the Earth's magnetic field then results in an enhancement of that magnetic field. This is called a self-exciting dynamo, and can occur in the Earth only because the outer core is liquid, convecting, and, being iron-rich, conducts electricity.

C. MARY R. FOWLER

Further reading

Bolt, B. A. (1991) *Inside the Earth: evidence from earthquakes.* Freeman, San Francisco.

Brown, G. C. and Mussett, A. E. (1993) *The inaccessible Earth* (2nd edn). Chapman and Hall, London.

Claerbout, J. F. (1985) *Imaging the Earth's interior.* Blackwell Scientific Publications, Oxford.

Fowler, C. M. R. (1990) *The solid Earth: an introduction to global geophysics.* Cambridge University Press. (2nd edn, 2000.)

Grand, S. P., van der Hilst, R. D., and Widiyantoro, S. (1997) Global seismic tomography: a snapshot of convection in the Earth. *GSA Today,* Geological Society of America, **7**, No. 4, 1–7.

Dziewonski, A. M. and Anderson, D. L. (1981) Preliminary Reference Earth Model. *Phys. Earth Planet. Inter,* **25**, 297–356.

Jeanloz, R. (1983) The Earth's core. *Scientific American,* **249**, 56–65.

McKenzie, D. P. (1983). The Earth's mantle. *Scientific American,* **249**, 66–113.

Earth tides

The gravitational attraction between the Earth and Moon keeps them in orbit about their common centre of mass, which is a point within the Earth 4670 km from the centre. Both the Earth and the Moon are nearly spherical, which means that the total attractive force between them is nearly the same as if they were point masses at their centres. Thus the centripetal force on each is just that required to keep point masses at their centres in orbit about one another. The side of the Earth nearer to the Moon experiences a stronger gravitational pull than is needed to keep it in orbit with the rest of the Earth. It is therefore pulled towards the Moon. Conversely, on the far side of the Earth the gravitational force of the Moon is weaker than the average. The result is a pattern of tidal force pulling the Earth into a prolate ellipsoidal form aligned with the Earth–Moon axis, as shown in Fig. 1.

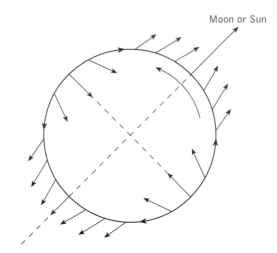

Moon or Sun

Fig. 1. The pattern of the tidal forces at the surface of the Earth resulting from the gravity of the Moon or the Sun.

The tide due to the Sun is smaller than the lunar tide, by a factor of 0.46: the effect of the Sun's much greater mass is more than compensated by its distance. The total tide is therefore a superposition of the lunar and solar tidal bulges, with a relative alignment that changes progressively through the lunar month. At new and full moon, when the Sun and Moon are in the same direction or opposite one another, the two tidal components are added and maximum (spring) tides are observed. Weaker (neap) tides occur at half-moon, when the sun and Moon are 90° apart.

The observed tide is a result of the rotation of the Earth within the envelope of the two deformations, which remain fixed in orientation relative to the Moon and Sun. In most places the lunar semidiurnal tide is dominant, which means a tidal period of 12.42 hours, with an amplitude that oscillates semi-monthly as a beat with the 12-hour solar tide. Misalignment of the Equator with the orbits gives a prominent diurnal tide at some latitudes, and there are other periods in tidal records arising from the orbital ellipticities and precession of the lunar orbit.

The *equilibrium tide* is the sum of the two prolate ellipsoidal elongations in the directions of the Moon and Sun. The solid Earth is deformed elastically in this manner, but the motion, about 0.4 m, can be observed only with sensitive instruments. The marine tide is more obvious, but is much more complicated. Even in the deep oceans the natural speed of a tidal wave around the Earth is much less than the speed of the Earth's rotation, and the ocean consequently responds to tidal forces with a phase lag of nearly 90°. This means that high tides appear where lows would be expected for an equilibrium tide. The tidal bulge observed by analysis of satellite orbits is the sum of the solid Earth and marine tides, and so is smaller than would be observed with the solid Earth alone.

The term *Earth tide* sometimes refers to this total tide, but historically it has meant the solid Earth tide.

The marine tide is locally very variable. Large tidal lags and resonant amplifications occur in shallow areas, particularly in bays and estuaries. Tidal movements in shallow water are believed to account for most of the dissipation of tidal energy, which globally amounts to 3×10^{12} watts. A very small fraction of this is due to imperfect elasticity of the solid Earth.

As seen by satellites, the energy dissipation appears as a 2.9° lag in the orientation of the globally averaged lunar tidal bulge relative to the Earth–Moon axis. Since the bulge is caused by the Moon itself, the misalignment produces a torque in the direction that would pull the bulge back into line. This torque acts as a brake on the Earth's rotation, causing the length of the day to increase gradually. The combined effect of the lunar and solar tidal torques is to increase the length of the day by 24 microseconds each year. Remembering that the Earth is about 4.5×10^9 years old, it is evident that over geological time tidal friction has had a major effect: 650 million years ago there were 400 days per year, and when the Earth was very young the day was probably only 6 to 8 hours long.

The lunar torque on the Earth is balanced by a torque on the Moon's orbit, causing the radius of the orbit to increase by 3.7 cm per year. The angular momentum lost by the Earth's axial rotation appears in the lunar orbit, the total angular momentum being conserved. The Moon was much closer early in the life of the Earth. It may also have been rotating, but now presents a constant face to the Earth because the friction of the large tide raised in the Moon by the Earth has completely stopped its relative rotation.   FRANK D. STACEY

Further reading

Darwin, G. H. (1898) *The tides and kindred phenomena in the solar system.* (Reprinted 1962). W. H. Freeman, San Francisco.

Stacey, F. D. (1992) *Physics of the Earth* (3rd edn), pp. 115–33. Brookfield Press, Brisbane.

earthfills

'Earthfill' is the general term used to describe the end result where soil or rock is moved and placed, generally in a controlled way, to form an engineered structure, such as an earthfill dam, a highway embankment, a spoil tip, or a platform on which a building is to be constructed. The performance of the placed materials, for example their strength, compressibility, and permeability, depends on the placed density, which in turn depends on the compactive effort and type of compaction. For soil, it depends also on the water content.

Compaction results in a reduction in the volume of air in the soil or rock and an increase in the dry density (the dry weight of soil divided by the volume of soil) with an associated decrease in the voids ratio (the volume of voids divided by the volume of solids). For a given compactive effort a typical response is for the dry density of the soil to increase to a maximum value and then fall as the moisture content increases. A simple explanation for this is that at low moisture contents the soil is stiff and has low 'workability', that is, it is difficult to mould or deform. Increasing the amount of water 'lubricates' the soil, making it easy for the soil particles to move in relation to each other. At high water contents the water effectively fills the voids within the soil and is not forced out during compaction, which is a rapid loading technique, and so the dry density decrease. Tests, such as the Proctor compaction test, can be carried out in the laboratory to assess the suitability of a soil for compaction, but field trials are generally needed to select the best compaction method and machinery. While it is sometimes possible to adjust the field moisture content of the soil for optimum compaction of fine-grained soils, such as clays, this can be very difficult.

Soils are generally placed in 'lifts' of a certain height depending on the soil type, each layer being compacted before the next layer is placed. The process can be controlled and monitored either by checking the 'as-placed' water content and density or by checking performance criteria, e.g. strength using a penetrometer. Compaction is carried out using various types of machinery according to the soil type and the volume to be compacted. For very large volumes purpose-built machinery may be purchased.

For many earthfill structures the cost of transporting the fill, given its unit weight, is a major consideration and it is normal to minimize the haul distance (the distance from the point of excavation to the point of placing) by using the nearest practicable soil, provided that an acceptable performance can be obtained. The local availability of materials can have a significant effect on earthfill structures such as dams; where supplies of certain soil types are limited zoned construction may be adopted to make the best use of specific soil types, such as low-permeability clays for the core of the dam.   T. COUSENS

Earth-heritage conservation

Earth-heritage conservation encompasses the protection of the physical resources of the planet from adverse development, misuse, or destruction. The Earth's physical resources are directly accessible only at its surface, and for our present purposes can be considered in terms of two basic components: landforms, comprising natural physical features of the landscape, which are primarily the subject of geomorphology, and rock outcrops, representing rock units that are present, but not necessarily exposed, at the surface, and are primarily the subject matter of geology. Rocks may be exposed both by naturally occurring and by artificial means. The built environment is also fundamentally an expression of the geological and geomorphological resource. All these features are under threat from development and mismanagement, and their protection comes under the general heading of Earth-heritage conservation.

The basis for Earth-heritage conservation

In common with other aspects of conservation, the basis for Earth-heritage conservation relies upon four basic principles:

(1) that the resource should be conserved for its own sake; (2) that the resource provides a basis for economic exploitation; (3) that the resource provides a basis for research, training, and education; and (4) that the resource has an aesthetic or cultural value (Fig. 1).

The first of these principles is most commonly understood with reference to wildlife conservation, but unlike wildlife, Earth-heritage features usually convey an appearance and impression of solidity and permanence that is at odds with their general vulnerability. It is for this reason that Earth-heritage conservation is often viewed as a poor relation of wildlife conservation, which is concerned with seemingly fragile resources. The second and third principles are interlinked and are associated with exploitation, both through economic exploitation, and with use in research and training, which are ultimately important for human progress. Finally, the Earth's physical resources provide inspirational and aesthetic assets, which materially contribute to the spiritual and emotional well-being of people.

The basis for the protection of the Earth-heritage resource is effective management, which is the key to protecting vulnerable sites and features and enhancing their value to the user. Management seeks to achieve a balance between creating a heritage resource and depleting that resource through development (Fig. 1). Effective management is primarily a function of four components: assessment, awareness, protection, and enhancement. These components are discussed below.

Assessment and awareness

Assessment of the value of Earth-heritage features is primarily subjective. It rests on the assumption that some areas of the Earth's surface are more important than others on aesthetic, cultural, or scientific grounds, and leads to the protection of areas of various sizes. Whereas no one would doubt the importance of Antarctica as 'the last great wilderness' and as a continent of great scenic and scientific importance, the scientific reasons for protecting more typical sites each a few square metres in area, such as a single rock exposure, or an

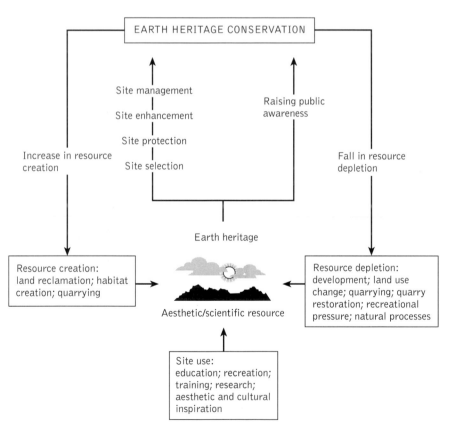

Fig. 1. Earth-heritage conservation: a balance between creation of a heritage resource through effective management, and the depletion of that resource through development and adverse land use. The basis for protection is largely exploitative for research, for education, or simply for recreational use. (From Bennett, M. R. and Doyle, P. (1996) *Environmental geology.* John Wiley and Sons, Chichester.)

important landform, may be a source of puzzlement to the general public. It is for this reason that promoting awareness of the value and fragility of Earth-heritage features is a most important aspect of successful conservation (Fig. 1). Unfortunately, there is a lack of public awareness of the ultimate fragility of geological features, and this has hampered their effective conservation. For example, in Great Britain, where 80 per cent of the population live in urban areas, awareness of the decline in the quality of the Earth-heritage resource in country areas is low, and there is consequently no national mood in favour of effective conservation.

Protection and enhancement

The protection of the Earth-heritage resource is dependent on the definition of an appropriate conservation strategy. In some cases the choice is akin to the difference between preservation, or 'moth-balling' in an unchanged state, and conservation, or protection with a recognition that the resource is bound by nature to change with time. Preservation is most commonly associated with the protection of historic monuments in a state close to suspended animation, while conservation is most commonly practised in ecological and other wildlife sites where some active management is necessary to prevent encroachment by unwanted flora or fauna.

Landscape areas selected on the basis of their scenic beauty and cultural associations are mostly managed using both preservation and conservation principles. In essence, active conservation management is necessary to maintain the character of the environment and to prevent unwanted intrusions, while specific components within that landscape will be subject to stringent preservation orders.

Geological or geomorphological features selected on the basis of their scientific importance are also managed with reference to preservation and conservation principles, but the application of these principles is governed by the nature of the scientific interest. Here, the distinction is between those sites or features that are representative of a finite resource—*integrity sites*—and those that are representative of a much greater resource that is otherwise obscured elsewhere. These are referred to as *exposure sites*, since they generally include outcrops in mountain sides, coastal cliffs, and quarries that display rocks that are not otherwise commonly exposed at the surface.

Integrity sites include those with rare fossil or mineral assemblages and those that cannot be reproduced and are unique. These include most landform sites. Exposure sites include most geological sites that have been selected for conservation because they represent the typical expression of a particular geological unit. The conservation principle applied at integrity sites is to maintain the integrity of the site. This is clearly akin to the concept of preservation. At exposure sites, effective management is needed to keep the exposure open, but if the site were threatened it could in theory be duplicated by selecting an alternative, or by creating a new exposure of the same rock unit, which is otherwise plentiful underground.

Effective management and conservation also call for the development of enhancement techniques. These can vary from the promotion of greater awareness, through better access, to physical actions such as, for example, the clearance of unwanted vegetation or dumped rubbish. Site enhancement is often necessary at the small scale, particularly at exposure sites, where the overriding conservation principle is to maintain the exposure. Here, particularly where the material is unconsolidated and prone to movement, periodic cleaning and clearance of the most important faces is continuing need.

Threats to Earth heritage

Threats to Earth-heritage sites may be classified according to their potential impact in impairing the quality of the specific areas of the physical landscape. Their mitigation depends on the scale of the problem, the area to be conserved, and the rationale behind its conservation: whether aesthetic, cultural, or scientific.

Aesthetic and cultural sites are commonly associated with large-scale land-use areas such as national parks. They can be threatened by large-scale industrial or housing developments; by road or similar modes of transport; by large-scale construction works; by large-scale mineral extraction, particularly from newly opened quarries; by waste management schemes that significantly alter the landscape; by recreational pressure from visitors in large numbers and by their activities, such as walking, mountain biking, and using four-wheel-drive vehicles; all of which cause erosion; and by incremental erosion resulting from small-scale operations, such as new building works, changes in agricultural practices, and so on.

The management of most sites selected on scientific criteria depends on whether they are deemed to be integrity or exposure sites. The main threats to both types of site are associated with large-scale development, and depend on the setting of the site; whether, for example, it is situated on a coastal cliff or within a disused quarry. The main threats are from industrial, road, and other major developments, which obscure exposure and destroy integrity through foundation and other construction works; from coastal defence schemes, which obscure exposure and damage the integrity of coastal sites; from waste disposal, since many exposure sites are situated in former quarries which are commonly targeted as landfill voids, and waste disposal is a damaging activity that ultimately leads to loss of exposure; and from quarrying, which although beneficial in creating new exposures, is particularly damaging to integrity sites such as cave systems, static landforms, and exceptionally rare mineral and fossil sites. Fossil and mineral collecting, whether for scientific, commercial, or recreational reasons, is rarely a major threat to all but the most sensitive sites, although many countries forbid collecting in sensitive areas under the banner of the preservation of national heritage.

Countering threats

Countering threats to sites depends largely upon awareness and effective legislation. An enhanced awareness of the value of the Earth's geological and geomorphological heritage provides the best possible counter to most of the threats listed above. This is demonstrated most effectively in its counterpart, wildlife conservation, which can muster a powerful 'green' lobby in defence of threatened biological sites. The 'green' lobby is also efficient in the protection of large-scale landscape areas, as in the fight for the protection of Antarctica and other major wilderness areas, as well as in the protection of the National Parks of countries in many parts of the world. It is currently less effective in protecting rock outcrops or landforms where the rationale for their conservation is purely scientific. Here the need for effective legislation is strong. Methods of control range from heavy fines for site damage to consultation over operations that are considered to be potentially damaging to a site.   PETER DOYLE

Further reading

O'Halloran, D., Green, C., Harley, M., Stanley, M., and Knill, J. (eds) (1994) *Geological and landscape conservation*. Geological Society, London.

Wilson, R. C. L. (ed.), Doyle, P., Easterbrook, G., Reid, E., and Skipsey, E. (1994) *Earth heritage conservation*. Geological Society, London.

earthquake hazards and prediction Many of the world's greatest natural disasters have been due to earthquakes (Table 1). In part, this is because earthquakes, like hurricanes and storms, can generate severe effects over large areas. The hazards from earthquakes can be classified into two

Table 1 Partial list of world's most significant earthquakes[1]

Date	Location	Magnitude[2]	Fatalities
1290 September 27	Chihli, China	Not known	100 000
1531 January 26	Lisbon, Portugal	6.9	30 000
1556 January 23	Shensi, China	Not known	830 000
1737 October 11	Calcutta, India	Not known	300 000
1755 November 1	Lisbon, Portugal	8.5	70 000
1811 December 16	New Madrid, Missouri, USA	8.2	Uncertain but probably
1812 January 23	New Madrid, Missouri, USA	8.1	less than 100 in all three
1812 February 7	New Madrid, Missouri, USA	8.3	earthquakes
1819 June 16	Kutch, India	7.8	1500
1857 January 9	Fort Tejon, California, USA	8	2
1897 June 12	Assam, India	8.7	1500
1899 September 10	Yakutat Bay, Alaska, USA	8.0	None known
1906 April 18	San Francisco, California, USA	7.9	2500
1908 December 28	Messina, Italy	7.5	120 000
1920 December 16	Kansu, China	8.5	180 000
1923 September 1	Tokyo, Japan	8.2	143 000
1939 December 27	Erzincan, Turkey	8.0	23 000
1960 May 22	Chile	9.5	5700
1964 March 27	Prince William Sound, Alaska, USA	9.2	131
1970 May 31	Offshore, Northern Peru	7.9	66 000
1976 February 4	Guatemala City, Guatemala	7.5	23 000
1976 July 27	Tangshan, China	7.7	650 000 (?)
1985 September 19	Michoacan, Mexico	8.1	10 000
1988 December 17	Spitak, Armenia	6.9	25 000
1990 June 10	Iran	7.7	40 000
1999 August 17	Izmit, Turkey	7.4	>17 000
1999 September 21	Chi Chi, Taiwan	7.6	2400

[1] Magnitudes cited are in several different scales and there may be a range of values for each earthquake (see *Earthquake seismology* for discussion).

[2] Earthquake data compiled and modified from Bolt (1993).

Fig. 1. Surface rupture along the Lost River fault after the 1983 Borah Peak earthquake, magnitude (M) 6.8, in the western United States. Borah Peak is seen in the background. The maximum vertical displacement along this normal fault in Idaho was about 2.5 m. Note the displaced concrete conduit, showing that a small component of horizontal movement also occurred during this earthquake.

categories, based on whether they are primary or secondary effects. Primary hazards include ground shaking, surface fault rupture, and uplift or subsidence. Liquefaction, landslides, and water waves such as tsunamis and seiches are secondary effects because they are either caused by strong ground shaking or, in the case of tsunamis, *coseismic* (i.e. at the same time as the earthquake) elevation changes (uplift or subsidence).

Ground shaking is the result of seismic waves (see *earthquake mechanisms and plate tectonics*) reaching the Earth's surface. This is the most damaging of all earthquake hazards because of its far-reaching effects. Ground shaking or ground motions are a function of earthquake size and mechanism, distance from the causative fault, the attenuating properties of the Earth along the path of travel of the seismic waves, and the near-surface geological conditions beneath the location of the observer (see *strong-motion seismology*).

Surface fault rupture or surface faulting occurs when an earthquake is generated along a fault that is expressed at the Earth's surface (Fig. 1). As a result, movement generated at depth along the fault is propagated upward, resulting in displacement of the ground. For example, in the 1992, magnitude (M) 7.3, Landers, California earthquake, the maximum displacement along the fault observed at the ground surface was about 6 m horizontally (see *earthquake seismology* for discussion of magnitude). Any structure situated along a fault which has a surface expression is subject to damage in a future earthquake. In the United States, some state laws require that certain facilities should not be built along active fault zones. Not all faults reach the Earth's surface, including the greatest faults—the megathrusts located in subduction zones (see *earthquake mechanisms and plate tectonics*).

Significant coseismic elevation changes can also occur in large earthquakes; these are generally associated with large subduction zone megathrust earthquakes. For example, in the great 1964 Alaskan earthquake, M 9.2, portions of Montague Island were uplifted 11 m on the upthrown side of the megathrust fault and about 2 m on Prince William Island which is located on the downthrown side. Because of the coincidence of extensive coastlines along the Pacific Rim and the subduction zones within the 'Ring of Fire' (see *earthquake mechanisms and plate tectonics*), this hazard can be significant over a large part of the Earth. Also, because these elevation changes are generally long term in nature relative to sea level, they can lead to permanent loss of function of coastal facilities, such as harbours.

A possible secondary and dramatic effect of coseismic elevation changes are tsunamis. These giant sea-waves are generated by sudden displacement of the sea floor or by submarine landslides. In the open ocean, tsunamis have small wave heights but they can travel up to speeds of 500–800 km s^{-1}. As the waves approach the coast, the water depth decreases, resulting in increases in wave heights up to 20 m. In the 1964 Alaskan earthquake, the resulting tsunami caused extensive damage along the Pacific coast of Canada and western USA and Hawaii, and killed more than 100 people. Another type of water wave, a *seiche*, is generated in an enclosed body of water such as a lake. Earthquakes can induce seiches, such as was the case in the 1959, M 7.3, Hebgen Lake, Montana earthquake in the western United States.

When water-saturated sandy and/or silty soil is subjected to strong earthquake ground shaking, a phenomenon called liquefaction can occur. Shaking realigns the soil particles, resulting in an increase in pore pressure and a decrease in strength. Soils in such a state acquire a degree of mobility sufficient to permit deformation. In extreme cases, the soil particles become suspended in groundwater and the deposit reacts as a fluid giving rise to sand or mud volcanoes. There have been spectacular examples of damage due to liquefaction as it often occurs in large earthquakes.

Earthquake-induced landslides are like other types of slope failure, except they are triggered by strong ground shaking. One of the most catastrophic landslides to ever occur took place in the 1970 Peru earthquake (Table 1). The landslide was actually a large debris avalanche which originated from a peak of the Nevado Huascaran volcano, travelled at a speed of about 320 km s^{-1}, and destroyed two villages. The death toll was more than 18 000 lives.

One of the most important applications of the Earth sciences today is the evaluation of earthquake hazards. Such evaluations generally comprise two major steps: the identification and characterization of the sources of the earthquake and the assessment of the associated hazards. The former requires a multidisciplinary approach, utilizing not only seismology but also other fields of geophysics and, most importantly, geology and geomorphology. Possibly the most important breakthrough in our understanding of earthquake processes has been due to the relatively new field of palaeoseismology. Palaeoseismic studies consist of geological investigations of prehistoric earthquakes through evaluation of their impacts on the environment, such as surface faulting, tsunami deposits, liquefaction features, and buried marshes.

The assessment of specific earthquake hazards has involved seismologists, geologists, and earthquake engineers. Strong earthquake ground shaking has been the subject of greatest attention and studies (see *strong-motion seismology*).

Earthquake prediction

Motivated by the desire to mitigate the often disastrous effects and hazards of earthquakes, the early prediction of such events has long been attempted. However, modern attempts at earthquake prediction were not initiated until about the 1960s, and such efforts were generally concentrated in Japan, China, the former USSR, and, to a lesser extent, the United States.

Earthquake prediction has revolved around the search for earthquake precursors, those physical indicators that might signal the future occurrence of a seismic event. Precursors might include unusual changes in:

(1) seismicity in the region of an impending event;
(2) crustal deformation of the ground, including horizontal and vertical land changes, water-well fluctuations, and changes in the Earth's seismic velocities;
(3) changes in the Earth's gravity, magnetic, or electrical fields;
(4) emission of gases such as radon;
(5) unusual animal behaviour.

All or most of these effects might be related to strain and stress changes in the Earth before earthquake.

Earthquake prediction involves answering the three basic questions (see *earthquake seismology*) of where, when, and how big. Thus, earthquake precursors can be classified on the timescale at which they may be useful. The classifications include: (1) long-term or up to a few hundred years; (2) intermediate-term or up to several years; and (3) short-term or on the order of days to months. A successful example of a long-term precursor is based on the concept of a 'seismic gap'. First applied to the world's subduction zones, a *seismic gap* is a region where large earthquakes have occurred in the past but none in past decades or longer and where any seismicity appears to be nearly absent. Eventually these gaps are filled by large earthquakes and their aftershocks. The concept has been applied successfully in the circum-Pacific belt of subduction zones.

Although of considerable value, the weakness in long-term precursors is that the uncertainties in predicting the time of a future earthquake can be large, on the order of years or decades. Herein lies the basic problem in present earthquake prediction efforts. Great strides have been made in predicting the place and the size of future earthquakes. The problem of predicting the specific time of an event has, for the most part, remained totally unsolved. No reliable intermediate or short-term precursors have been identified to date. This is not to say there have not been successes. Probably the most spectacular example was the short-term prediction of the 1975, M 7.3, Haicheng earthquake by the Chinese, based on earthquake foreshocks, unusual animal behaviour, and a few other precursors. By evacuating Haicheng, probably ten of thousands of lives were saved. Unfortunately, this event was followed a year later by the catastrophic 1976, M 7.7, Tangshan, China earthquake which took up to 650 000 lives with no apparent warning and no prediction.

In the United States, efforts in earthquake prediction accelerated in the 1970s as a result of reported successes by Soviet seismologists who had observed changes in travel times and velocities of seismic waves as they passed through the source region of an impending earthquake. In general, however, US scientists failed to observe similar anomalies. Recently, the only significant US effort has been focused on an experiment along a section of the San Andreas fault near the town of Parkfield, California. At this location, an M 6 earthquake had been predicted to occur at some time in the time frame 1988 ± 5 years, based on an apparent regularity of previous earthquakes. Although this event has not yet occurred, efforts are continuing to monitor the proposed epicentral area with a wide variety of instrumentation, including seismometers, water-well gauges, tiltmeters, levelling surveys, triangulation and trilateration arrays, radon gas monitors, gravimeters, and magnetometers. Currently, efforts in the United States are focused on long-term forecasting, specifying the location and

size of future earthquakes and their probability of occurrence within a time period of several decades. IVAN G. WONG

Further reading

Bolt, B. A. (1993) *Earthquakes.* W. H. Freeman, New York.

Brumbaugh, D. S. (1999) *Earthquakes, Science and Society.* Prentice Hall, New Jersey.

Reiter, L. (1990) *Earthquakes hazard analysis.* Columbia University Press, New York.

earthquake mechanisms and plate tectonics

Since the dawn of man and his first experiences with earthquakes, the question of their origin has been asked. However, the responses were by and large mystical until about the past millennium. The modern concept of earthquakes, their causes and mechanisms, had its beginning in the 1880s. G. K. Gilbert, an American geologist, was one of the first to suggest that earthquakes were the result of displacement along geological faults. In his observations of the Wasatch Mountains in north-central Utah in the western United States, Gilbert suggested that the mountains were the result of upwards incremental displacements (earthquakes) along a range-bounding fault that moved due to the constant accumulation of strain. In the century prior to this, it was believed that earthquakes were the result of magmatic processes or some form of

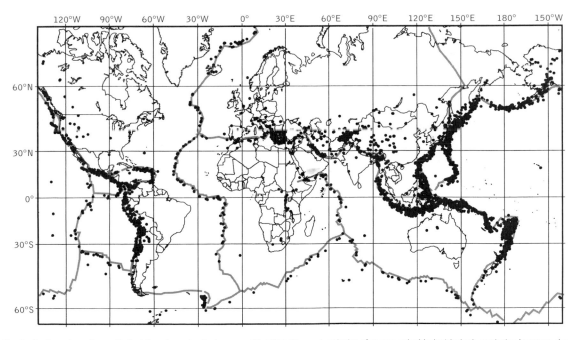

Fig. 1. Earthquakes of magnitude 4.0 and greater that occurred in 1994. The vast majority of events coincided with the boundaries between the major tectonic plates. (Seismicity data courtesy of Stuart Koyanagi and Waverly Person, US National Earthquake Information Center.)

geologically generated explosions, and that faulting simply accompanied earthquakes rather than being their cause.

In 1910, H. F. Reid suggested that earthquakes were the result of a phenomenon called 'elastic rebound', based on his observations of the great 1906, magnitude (M) 8, San Francisco (California) earthquake. This theory states that an earthquake is generated by rupture or sudden displacement along a fault when it has been strained beyond its elastic strength. In the process of strain accumulation, the opposing sides of the fault are stressed until failure occurs, that is, sudden displacement takes place, and then they rebound back to an unstrained position. The result of each cycle of strain accumulation along a fault generates an earthquake. The elastic rebound theory has become the accepted model for the generation of most, but not all, earthquakes. Some types of volcanic earthquakes and deep earthquakes may have different mechanisms. Also, displacement along faults does not necessarily always result in earthquakes. Slow displacement, or creep, is an aseismic (i.e. non-seismic) process that has been observed along faults worldwide.

Most earthquakes are essentially the product of tectonic stresses which are generated at the boundaries of the Earth's tectonic plates. Such tectonic earthquakes, as we shall refer to them, can range in size from magnitudes less than zero, resulting from fault slippage of a few centimetres, to the largest events, M greater than 9, where fault displacements are on the order of many metres. The size of an earthquake is not only a function of the amount of displacement but also the area of the fault plane that ruptures (see *earthquake seismology*). Hence the larger the rupture area, the larger the earthquake. An M 7 earthquake ruptures a fault area of about 1000 km^2 or about 50 km long and 20 km wide.

The majority of the seismic energy released in the world is from earthquakes occurring along the plate boundaries, particularly around the Pacific Rim or the so-called 'Ring of Fire' (Fig. 1). Specifically, the Earth's greatest earthquakes are the result of incremental plate movement within subduction zones. The largest known earthquake was the 1960, M 9.5, earthquake that occurred along the South American subduction zone off the coast of Chile (see *earthquake hazards and prediction* and *earthquake seismology*). Earthquakes that occur along or in the vicinity of the plate boundaries are called interplate earthquakes, in contrast to intraplate earthquakes that occur in the interiors of the tectonic plates (see *intraplate seismicity*). Intraplate earthquakes seldom exceed M 8 in size.

Earthquakes can result from very rapid displacement on all types of faults. Faults can be classified into three general types based on their sense of displacement: normal, reverse or thrust, and strike-slip. The great subduction zone earthquakes are generated along the great thrust faults that constitute the boundaries between downgoing and overriding tectonic plates. Major strike-slip faults, such as the San Andreas fault, 1300 km long, which separates the Pacific and North American plates in California, and the North Anatolian fault

in Turkey, 1100 km long, have also ruptured repeatedly in earthquakes as large as M 8. In extensional tectonic regimes where the Earth's crust is being extended or stretched, normal faulting has generated large earthquakes but not quite as large as those along subduction zones. In the interior of the western United States, normal-faulting earthquakes have created the horst and graben landscape of the Basin and Range province and the Rio Grande rift.

Most of the world's earthquakes occurring outside the subduction zones have their origins in the Earth's upper crust. At these relatively shallow depths, of generally less than 15–20 km, the temperatures are usually low enough (less than about 350 ° ± 100 °C) for sudden displacement to occur along faults cutting brittle rocks. This form of displacement is often called 'stick-slip'; that is, strain builds up in the rocks next to the fault but there is no slip for some time because the rocks on either side of the fault are stuck together by features such as asperities. Eventually the resistance to sliding is overcome by shear stresses acting along the fault—giving a very rapid increment of slip. At greater temperatures (except for uppermost mantle rock, where the limiting temperature is thought to be about 650 ± 100 °C), faults will behave plastically and deform by aseismic creep instead of brittle failure. The deepest earthquakes in the world occur at depths of 600–700 km within subducting tectonic plates. Because the temperatures at these depths are extremely high, the mechanism by which these earthquakes are generated is not well understood.

Earthquakes often occur in sequences. The principal and largest event in a sequence is called the main shock. Foreshocks precede the main shock, and aftershocks can occur for tens of years after the main shock. Many tens of thousands of aftershocks occurred for several years after the 1964, M 9.2, Prince William Sound, Alaska earthquake, with several events as large as M 6. Aftershocks are generally the result of the release of residual stress in areas not ruptured in the main shock. Aftershocks need not, however, occur on the same fault as the main shock. For example, the 1992, M 7.3, Landers, California earthquake was followed 3 hours later by the M 6.2, Bear Lake earthquake, the source of which was an unnamed fault located 30–40 km to the west of the Landers zone.

During the process of fault slip and, hence, earthquake generation, tectonic strain energy is expended by the crushing of rock within the fault zone, production of heat, and a release of a small percentage of energy as seismic waves. The point on the fault plane where the earthquake rupture is initiated is called the focus or hypocentre. The point on the Earth's surface directly above the hypocentre is the epicentre. The focal depth of an earthquake is the depth of the hypocentre below the Earth's surface.

Seismic waves can be classified into three basic types: compressional or primary (P) waves, shear or secondary (S) waves, both which are body waves, and surface waves. P- and S- waves are called body waves because they can travel through the interior of a body such as the Earth. The P-wave, the fastest wave (i.e. it has the highest velocity) results in particles in the medium in which it is travelling, moving in a longitudinal direction (the same direction as the movement of the wave). S-waves generate transverse particle motion and they generally have velocities that are about a half to two-thirds that of the P-wave velocity.

In contrast to body waves, surface waves are confined to the outer layers of the Earth. There are two types of surface waves: Love and Rayleigh waves. The Love waves has a particle motion, which, like the shear wave, is transverse to the direction of propagation but with no vertical motion. Rayleigh waves have an elliptical and retrograde particle motion confined to the vertical plane in the direction of propagation. Deep earthquakes generally do not generate surface waves.

At the frequencies of most engineered structures, most of the seismic energy radiated as seismic waves is contained in the S-wave. It is for this reason, and also the fact that S-waves

approaching the Earth's surface result in horizontal ground movement, that most of the ground shaking damage from earthquakes is due to S-waves (see *strong-motion seismology*).

IVAN G. WONG

Further reading

Bolt, B. A. (1993) *Earthquakes*. W. H. Freeman, New York.

Brumbaugh, D. S. (1999) *Earthquakes, science and society*. Prentice Hall, New Jersey.

Fowler, C. M. R. (1990) *The solid Earth*. Cambridge University Press.

Yeats, R., Sieh, K., and Allen, C. (1997) *The geology of earthquakes*. Oxford University Press, New York.

earthquake seismology The first recorded accounts of earthquakes date back to 2000 BC. However, the era of earthquake seismology as a modern science began in the late 1800s with the installation of sensitive instruments called seismo-

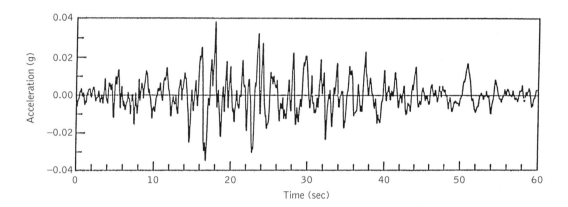

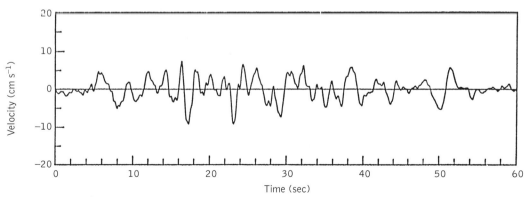

Fig. 1. Seismograms of the 1985, M_w 8, Michoacan, Mexico earthquake as recorded at a site in Mexico City. A seismogram is a record of the ground movement, usually in terms of velocity as a function of time, but it can also be in units of acceleration or displacement. Although this earthquake occurred more than 300 km away, along the Pacific coast of Mexico, it killed more than 10 000 people in Mexico City, principally because of the collapse of high-rise buildings.

graphs. The key questions of scientific enquiry (what, where, how, and how big?) have been recurring themes in earthquake seismology probably ever since humans first experienced this natural phenomenon and suffered its consequences.

In the attempt to answer these basic questions, three principal tools have been developed and used by seismologists. All three techniques require a seismogram or recording of the seismic waves generated by an earthquake (Fig. 1). These seismic waves are recorded as ground motions by seismographs at the Earth's surface.

When an earthquake is recorded by many seismographs arranged in a spatial network, its origin time, location, focal depth, and sometimes focal mechanism can be calculated. The intent of the first seismographic networks was indeed to determine the locations of, and to eventually evaluate the sources of, earthquakes. Such locations consist of a position defined in some horizontal coordinate system (e.g. latitude and longitude) and depth with respect to a datum, usually the Earth's surface. Earthquake origin times or the time of the rupture initiation of the causative fault can also be computed. The Worldwide Network of Standard Seismograph Stations (WWNSS) in the early 1960s provided the first opportunity to locate and quantify the size of earthquakes in a relatively uniform manner on a global scale (see *Worldwide Standardized Seismographic Network*).

Locating an earthquake is actually a sophisticated form of triangulation. Present-day location schemes commonly use an iterative least-squares approach (e. g. Geiger's method). Based on the arrival times of seismic waves (usually the initial compressional (P) wave and sometimes later phases such as the shear (S) wave; Fig. 1), the observed travel times to a trial location are compared against theoretical travel times based on a velocity model of the Earth. The differences between these times to all stations are then minimized by converging upon an event location where the sum of the square of the travel time residuals is minimized. The use of computers has made such iterative techniques more efficient.

Prior to the development of the idea of earthquake magnitude, the size of an earthquake could only be measured crudely by the intensity of its effects. The first intensity scale was developed by M. S. de Rossi and F. A. Forel in the 1880s. The most widely used scale today is a modified version of one originally developed by G. Mercalli. This 12-value scale is shown in Table 1. By taking an intensity scale as a basis, maps showing the distribution of reported intensities for a particular earthquake can be developed, and contours encompassing areas of similar intensity can be drawn. These isoseismal lines clearly indicate that generally the larger the earthquake, the larger the felt area and that intensities decrease away from the epicentre, that is, the point on the Earth's surface above the focus of the earthquake. Robert Mallet's detailed study of the 1857 Naples earthquake resulted in the first known isoseismal map. Figure 2 shows an isoseismal map for the destructive 1994 Northridge, California earthquake, magnitude 6.7.

Table 1. Abridged modified Mercalli intensity scale

I	Not felt except by a few under favourable circumstances (RF I)[1]	VIII	Damage slight in specially designed structures; considerable in ordinary substantial buildings with partial collapse; great in poorly built structures. Panel walls thrown out of frame structures. Fall of chimneys, factory stacks, columns, monuments, walls. Heavy furniture overturned. Sand and mud ejected in small amounts. Changes in well-water levels. Persons driving cars disturbed (RF VIII + to IX)
II	Felt only by a few persons at rest, especially on upper floors of buildings. Delicately suspended objects may swing (RF I–II)		
III	Felt quite noticeably indoors, especially on upper floor of buildings, but many people do not recognize it as an earthquake. Standing motor cars may rock slightly. Vibration like the passing of a truck. Duration estimated (RF III)	IX	Damage considerable in specially designed structures; well-designed frame structures thrown out of plumb; great in substantial buildings; with partial collapse. Buildings shifted off foundations. Ground cracked conspicuously. Underground pipes broken (RF IX +)
IV	Felt indoors by many, outdoors by few during the day. Some awakened at night. Dishes, windows, doors disturbed; walls make creaking sound. Sensation like heavy truck striking the building. Standing motor cars rocked noticeably (RF IV–V)	X	Some well-built structures destroyed; most masonry and frame structures destroyed with foundations; ground badly cracked. Rails bent. Landslides considerable from river banks and steep slopes. Shifted sand and mud. Water splashed, slopped over banks (RF X)
V	Felt by nearly everyone, many awakened. Some dishes, windows, and other fragile objects broken; cracked plaster in a few places; unstable objects overturned. Disturbances of trees, poles, and other tall objects sometimes noticed. Pendulum clocks may stop (RF V–VI)	XI	Few, if any, [masonry] structures remain standing. Bridges destroyed. Broad fissures in ground. Underground pipelines completely out of service. Earth slumps and land slips in soft ground. Rails bent greatly
VI	Felt by all, many frightened and run outdoors. Some heavy furniture moved; a few instances of fallen plaster and damaged chimneys. Damage slight (RF VI–VII)	XII	Damage total. Waves seen on ground surface. Lines of sight and level distorted. Objects thrown into the air
VII	Everybody runs outdoors. Damage negligible in buildings of good design and construction; slight to moderate in well-built ordinary structures; considerable in poorly built or badly designed structures; some chimneys broken. Noticed by persons driving cars (RF VIII)		

[1] Equivalent Rossi–Forel (RF) intensities.

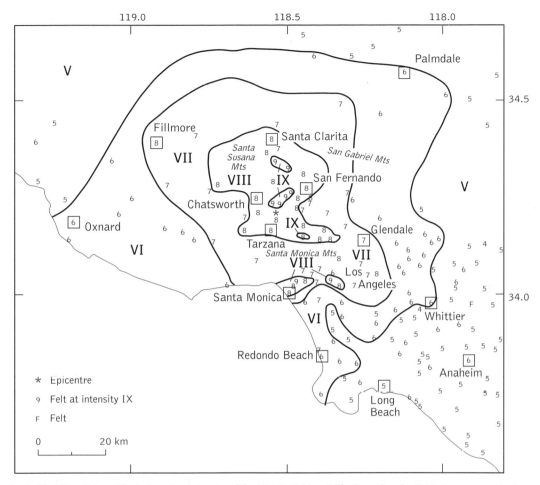

Fig. 2. Modified Mercalli intensities in the epicentral region of the 1994 Northridge, California earthquake, M 6.7. Roman numerals give average intensities within isoseismal contours. (Illustration courtesy of James Dewey, US Geological Survey.)

In the early 1930s, Charles Richter, using a particular type of instrument called the Wood–Anderson seismograph, developed the local magnitude (M_L) scale for southern Californian earthquakes. This was a monumental step in earthquake seismology because it allowed for the first time a precise quantification of the size of a seismic event based on instrumental recordings.

Because M_L values were based on the amplitude of the largest wave recorded on a seismogram and were thus simple to calculate, the scale rapidly became a worldwide standard. Since then, several other magnitude scales, such as surface wave (M_s) and body-wave magnitudes (m_b), have come into use. Up to now, however, the M_L scale has been the most commonly used magnitude measure. The moment magnitude scale (M_w) has increasingly become the scale of choice among seismologists because it is based on seismic moment and is

the best measure of earthquake size. The seismic moment of an earthquake is a function of the area of the fault that ruptures, the average displacement on the fault, and the shear modulus, a parameter that is related to the rigidity of the rocks in the fault zone. The units of seismic moment are dyne.cm (g.cm² s⁻²). In Table 1 of *Earthquake hazards and prediction*, some of the world's largest earthquakes ever recorded, their locations, magnitudes, and some of their reported effects are listed.

One of the most significant advances in earthquake seismology has been the development of the technique of focal mechanisms (also referred to as fault-plane solutions) by the Japanese seismologists T. Shida and H. Nakano, and Perry Byerly of the University of California at Berkeley in the early 1900s. Focal mechanism studies provided the first insights into the source processes of seismic events.

Fig. 3. Principal types of earthquake focal mechanisms. The dark quadrants represent compressional first motions (ground up). The remaining quadrants are for dilatational first motions (ground down). P (pressure) and T (tension) axes are also shown.

The point source representation of an earthquake (a shear failure in rock; see *earthquake mechanisms and plate tectonics*) is that of a double-force couple consisting of two opposing force couples having no net force or torque. The P-wave radiation of such a double-couple results in a pattern of alternating quadrants of compressional and dilatational first motions; S-wave polarizations also show a quadrantal distribution for this type of source. The two orthogonal nodal planes separating these quadrants represent the fault plane and an imaginary plane called the auxiliary plane. When the first motion pattern over the focal sphere is plotted on a stereographic or equal-area projection, as is done for focal mechanisms, the orientations of the nodal planes and the type of faulting can be determined, as can two important axes, the pressure (P) and tension (T) axes of the causative stress field. These two axes approximate the directions of the maximum and minimum principal stresses, respectively. Figure 3 illustrates focal mechanisms for the three principal types of faulting: normal, reverse or thrust, and strike-slip (see **earthquake mechanism and plate tectonics**). Combinations of these types, such as oblique faulting, are depicted by variations of these principal types of focal mechanisms.

More sophisticated techniques to analyse the source and rupture process of an earthquake were developed in the 1990s. Parameters such as seismic moment, stress drop (the difference between the initial and final stress acting on the fault resulting in an earthquake), average displacement across the fault, and rupture dimensions can now be calculated, based on an assumed model of the source such as the circular crack model developed by James Brune in 1970. More recently, inversion techniques utilizing seismograms have provided an even more detailed view of the rupture process by deducing the varied distribution of slip on the fault plane.

Using these basic tools, earthquake seismology has evolved into several fields of study focused on various aspects of earthquakes and their effects, including for example: their source processes (see *earthquake mechanisms and plate tectonics*); seismic wave propagation and Earth structure; seismicity and seismotectonics, the geographic distribution of earthquakes and their relationship to geological structures; and strong motion and seismic hazards (see *strong-motion seismology* and *earthquake hazards and prediction*).

Much of what has been learned about the internal structure and composition of the Earth has come from the analysis of seismograms. As seismic waves propagate through the Earth, their velocities vary, they attenuate at different rates as a function of the material they are travelling through, and they are reflected and refracted off compositional and structural boundaries. These effects can be deciphered from seismograms of recorded earthquakes (or explosions) to develop models of the Earth's velocity structure and hence its geological structure. For example, the analysis of a specific reflected phase recorded on seismograms led to the discovery of the Earth's outer core at a depth of 2900 km by R. D. Oldham and Beno Gutenberg at the beginning of the twentieth century.

Not only can the gross structure of the Earth be imaged, but also its finer structure, particularly within the crust. Using mathematical techniques, the travel times of earthquakes are inverted to develop an image of the continental crust. In volcanic areas, inversion has been successful in characterizing underlying magma chambers. Other approaches using artificial sources of seismic waves, such as explosions, and analysis of their reflected and refracted behaviour have been used extensively by the petroleum industry in its search for oil.

Seismicity studies attempt to characterize the spatial distribution of earthquakes, their temporal behaviour, and their sources, particularly as they pertain to geological structures such as faults. Evaluating the seismicity of a region is not only important in terms of understanding what role earthquakes might have in the tectonic deformation of a region and development of geological structures, but is vitally important in terms of addressing the hazard from identified seismic sources such as faults.

IVAN G. WONG

Further reading

Bolt, B. A. (1993) *Earthquakes*. W. H. Freeman, New York.

Brumbaugh, D. S. (1999) *Earthquakes, science and society*. Prentice Hall, New Jersey.

earthquakes, volcanogenic *see* VOLCANOGENIC EARTHQUAKES

ecclesiastical geology The study of the geology of church masonry is an application of what might be called mural geology, which can be useful to architectural historians and archaeologists, and interesting to the geologist. Recent years have seen collaboration between the disciplines in research on standing buildings as well as in archaeological excavations. For example, historic buildings have often undergone successive phases of modification and repair, and a geological study can help to identify different batches of masonry. Moreover, sources of building stone are seldom recorded, but they are important to those concerned with restoration, and to the historian.

A comprehensive study requires an accurate survey of the masonry, producing stone-by-stone drawings of each elevation. Above 2 or 3 metres, the team will need scaffolding. For some buildings photogrammetry may provide useful drawings, but individual stones are not always clear; in any case, the geologist usually needs to get up there. Identification is done by means of a hand-lens, occasionally with the aid of a small tool for scraping, but obviously not a hammer! The drawings are then annotated to indicate recognizable varieties of the rock-types present, and other materials such as brick. It is also useful to note whether a stone has been reddened by burning. Identification of the exterior fabric is generally aided by the weathering of the stone. (Lichens can sometimes actually be useful, different species indicating limestone, ironstone, or the presence of dolomite). Interior masonry is more

difficult, with smooth cut surfaces, poor lighting, and often the use, at some time, of limewash.

A building that has yielded many of its secrets by this kind of investigation is the ancient church of All Saints', Brixworth, in Northamptonshire. The survey was directed by David Parsons for the Brixworth Archaeological Research Committee. Although founded as a monastery *c.* AD 680, none of the present fabric pre-dates the eighth century. The geologist W. J. Arkell once described it as a 'museum of rock-types'. Among the 30 different types of stone recognized are varieties of diorite, granite, bedded tuff, sandstones, limestones, and tufa. When plotted in the form of a mural geological map, the various assemblages reveal phases in the development of the

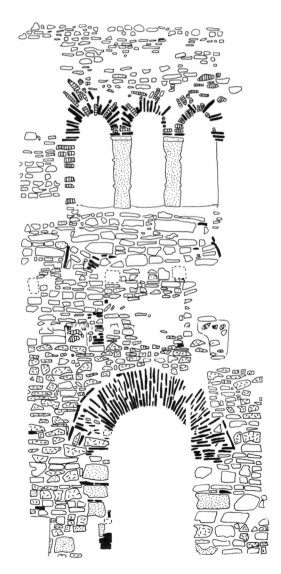

Brick

Tufa

Jurassic limestone

Varieties of Northampton Sand

Non local sandstone, Triassic & Carboniferous

Igneous & associated Leicestershire rocks

1 m

Fig. 1. Part of a 'geological map' of the west nave wall of All Saints' Church, Brixworth, Northamptonshire. This is the west side of the wall, as seen inside the tower. Note the stratification of the masonry (eighth–ninth century); the trace of the gallery doorway above the main west door, blocked with limestone when the triple arch was inserted (*c.* eleventh century); the introduction of tufa at this stage (also used to fill former beam holes). (Reproduced with permission from Sutherland and Parsons (1984), *Journal of the British Archaeological Association*.)

293

church. Around the building, except in the stair turret, a motley exotic collection of igneous rocks, Triassic sandstones, Jurassic limestone, and brick occurs in the masonry up to about 4 m. Above this, except in the tower and stair turret, the fabric appears more uniform in local Northampton Sand sandstone and ironstone, although the 10 varieties recognized enable a more detailed analysis, and features such as putlog holes come to light. (These mark the position of scaffold poles used in construction.)

The petrological survey has demonstrated graphically the modifications introduced with the addition of the stair turret (c. eleventh century). The lowest, mixed masonry of the tower was not built up, like the nave, in Northampton Sand, but in the same limestone now brought in for the stair turret (was the tower perhaps only a single storey porch before this, or had it, maybe, a wooden superstructure?) Higher up, tower and turret are similarly stratified, with incoming Northampton Sand and now the conspicuous use of tufa, not previously used in the church. Tufa was employed particularly for the spiral staircase in the turret, but also for quoins and window casing, interspersed with Northampton Sand, in the upper tower; similar deployment of tufa is seen at the east end, in the Saxon remains of the polygonal apse. The delightful triple arch inserted in the west nave wall cuts an early arched doorway above the lower stage of the tower. The door was blocked in limestone (as used in the turret and this stage of the tower) and part of its brick heading was perhaps re-used, with tufa, for the new triple arch head (Fig. 1).

The task of locating possible sources of building stone requires close examination of the petrology, and comparison with known rock types, bearing in mind the material available locally. Many early churches incorporated re-used Roman material, assumed to have been nearby as, for instance, in York. For Brixworth, surprisingly, some was apparently brought 35 km, overland; the igneous varieties match those of Roman ruins (still extant) in Leicester, which could also have provided the large quantities of brick. The random distribution of burnt stone, too, throughout the fabric at Brixworth

shows that many later consignments of stone were also reclaimed, rather than newly quarried material.

Most rural parish churches, however, are built of rubblestone from the nearest geological source, and indeed can often provide us with information on the local geology; towers (when added later, as were spires) may be of ashlar using different stone, as are the carved windows and doorways. The dressed stone also tends to come from the nearest suitable source, though it was sometimes transported several kilometres, and certain sources became well known. 'Barnack Rag', a distinctive spar-cemented shelly oolite from the Lincolnshire Limestone at Barnack near Stamford was obtainable in very large blocks and can be recognized, for example, with typical ribbed cross-bedding, in the quoins and decorative stonework of the Anglo-Saxon tower at Earls Barton 50 km to the south, as well as at Barnack itself and many medieval churches, including Peterborough Cathedral. The quarries were owned by the abbey at Peterborough, which controlled the distribution of the stone by waterways to the abbeys of the Fens. Though the quarries remain as grassy 'hills and holes', no section can be seen today. The Jurassic outcrop across England has provided many varieties of stone—some, including the various Bathonian oolites from around Bath, used extensively, others, like the Northampton Sand sandstone, with more local distribution. In Yorkshire, geologist John Senior has traced a limestone used in medieval churches around Malton to ancient quarries in the Oxfordian Hildenley Limestone in the Howardian Hills; it is a very fine-grained white limestone containing the microfossil *Rhaxella perforata*. North and west of the Jurassic outcrop in Britain, ecclesiastical buildings reflect the whole spectrum of older geological sources, and to the east the Cretaceous has furnished local sandstones and flint.

In areas lacking suitable stone is was sometimes imported by sea. Portland Limestone was loaded into boats on the Dorset coast for Wren's churches in London. Long before this, Quarr stone from the Oligocene Bembridge Limestone was brought from the Isle of Wight for use in several of the churches of Kent. More detective work, by Bernard Worssam

Fig. 2. St Martin's Church, Canterbury: south wall of the (?) seventh-century nave. Calcaire Grossier (Eocene) and Marquise Oolite (Jurassic) were imported from France; there is much re-used Roman brick. (From Worssam and Tatton-Brown, in Parsons 1990, illustration 20.)

and Tim Tatton-Brown, has identified this hard, shelly gastropod limestone in the eleventh-century church at Brook, near Ashford, and in small amounts in Canterbury Cathedral. For the cathedral, great quantities of Caen stone were brought from western Normandy, and also another Middle Jurassic limestone from near Boulogne, the *Oolithe de Marquise*, which includes fine-grained, yellowish oolite with a 'millet-seed' structure and a coarser variety. A further French limestone, the *Calcaire Grossier* from the Eocene north or north-east of Paris is found in some Kentish churches: a whitish-grey, finely granular limestone, not oolitic, but containing sections of serpulid worm tubes (*Ditrupa*) and foraminifera. Some of these stone types are, however, also known to have been imported to Kent by the Romans, and at St Martin's in Canterbury, for example (Fig. 2), the presence of Roman brick indicates again that building materials could be recycled.

DIANA S. SUTHERLAND

Further reading

Clifton-Taylor, A. (1972) *The pattern of English building*. Faber and Faber, London.

Parsons, D. (ed.) (1990) *Stone quarrying and building in England AD 43–1545*. Phillimore/Royal Archaeological Institute, Chichester.

Parsons, D. (1996) Brixworth, Northamptonshire. In Turner, J. (ed.) *Dictionary of art*, Vol. 4, pp. 829–30. Macmillan, London.

Rodwell, W. (1989) *Book of church archaeology* (2nd edn). Batsford/English Heritage, London.

Sutherland, D. S. and Parsons, D. (1984) The petrological contribution to the survey of All Saints' Church, Brixworth, Northamptonshire: an interim study. *Journal of the British Archaeological Association*, CXXXVII, 45–64.

Wenham, L. P., Hall, R. A., Briden, C. M., and Stocker, D. A. (1987) St Mary Bishophill Junior and St Mary Castlegate. *The Archaeology of York*, Vol. 8, Fasc. 2. Council for British Archaeology. (Petrological analysis by P. C. Buckland.)

echinoderms The phylum Echinodermata is moderately successful at present with five classes and about 6000 living species. They are complex invertebrates that are common in a

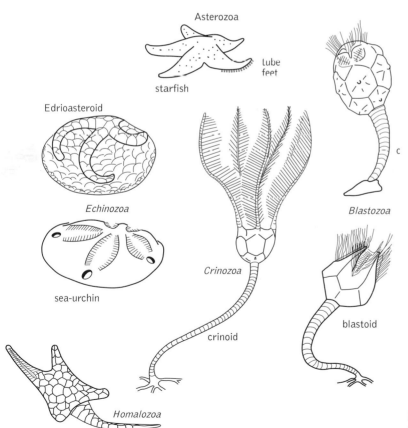

Fig. 1. Typical members of the main subphyla showing the major features of echinoderms.

variety of marine habitats, and include forms such as the starfish, brittle-stars, sea urchins, sea cucumbers, and crinoids. Because most echinoderms have a calcite skeleton composed of distinctively shaped plates and spines they are potentially preservable, and in fact they have a long fossil record extending back at least into the Cambrian. Up to 20 classes are represented in this rich record and as many as 12 or 13 may have occurred together. Echinoderms of all kinds are characterized by the presence of a five-rayed or pentameral symmetry, which is most obvious in the starfish and brittle-stars but is present in a sometimes modified form in the other classes. All echinoderms also possess a skeleton composed of calcite plates which are normally spiny and give the phylum its name (Greek *echinos*, spiny, and *derma*, skin). In addition they posses an internal water-vascular system consisting of tubes and bladders, which hydraulically controls the movement of the tube feet, blind-ended tubular projections that are involved in locomotion, respiration, and feeding.

Taxonomy

Echinoderms have traditionally been divided into five subphyla of fossil and modern forms, based on major features of the skeleton (Fig. 1). However, it has been suggested recently by Andrew Smith of the Natural History Museum in London that there is a natural division between attached (pelmatozoan) and free-living (eleutherozoan) echinoderms and that this should be recognized in the classification. Of the five subphyla of the traditional scheme, the first is the Echinozoa, which are usually globose or discoidal in form. These include three classes: the sea urchins and the sea cucumbers, both of which have a record going back to the Ordovician, and an extinct group, the edrioasteroids, which have a Cambrian to Mississippian (Early Carboniferous) record. The first two classes are mobile, travelling by means of their tube-feet; the edrioasteroids were sessile (attached) benthonic (bottom-dwelling) forms. The second subphylum, the Asterozoa, includes the starfish and brittle-stars. These are mobile predators that have a fossil record reaching back to the Ordovician. The Crinozoa or sea lilies are an important group of attached filter-feeders that have been part of the marine benthos since the Middle Cambrian. The Blastozoa includes several groups of stalked echinoderms, all of which are now extinct. The blastoids (Silurian and Permian) and cystoids (Cambrian to Devonian) were both groups in which complex respiratory systems were developed. Finally, the Homalozoa were a group of peculiar echinoderms that were calcite-plated but had no regular symmetry. They have been the subject of some controversy; some workers led by Richard Jefferies of the Natural History Museum consider them to have been closer to the chordates than the echinoderms and refer to them as calcichordates. They occur from the Cambrian to the Silurian.

Ecology

Fossil and living echinoderms have occupied a number of the different ecological niches available to benthonic marine invertebrates. During the Palaeozoic the majority of echinoderms were suspension-feeders, trapping small organisms that were settling out of the water column or being moved along the bottom by currents. Most of the echinoderms involved in this way of life were attached to the bottom by a long holdfast or stem that enabled them to lift their feeding appendages well above the sea floor. Crinoids, blastoids and cystoids used this way of life and are termed high-level epifaunal suspension-feeders. In contrast, low-level epifaunal suspension feeders, such as the edrioasteroids, had feeding appendages raised only a few centimetres above the sea floor and fed on bottom currents. This preponderance of suspension-feeders was probably related to the high productivity of micro-organisms in the Palaeozoic oceans, the lack of predators, and the virtual absence of infaunal organisms that might disturb the substrate on which the echinoderms lived.

Infaunal-detritus or deposit-feeders became more important in the Mesozoic. These animals browsed through the organic debris on the ocean floor, ingesting large amounts of sediment to extract nutritious particles. Many sea cucumbers occupied this niche, and other echinozoans such as the heart-shaped or irregular sea urchins became moderate to deep-burrowing forms. In addition, many echinoderms developed as mobile herbivores or predators during the Mesozoic. This lifestyle is found particularly in sea urchins, which developed strong mouthparts that enabled them to graze on algae or prey on other benthonic organisms. Starfish developed the tactic of pulling bivalve shells open and inserting their digestive stomach, and have continued as important predators on molluscs to the present day. These changes in the Mesozoic appear to have been related to a decrease in primary productivity in the oceans and the rise of efficient predators such as the fish.

Only a few other ways of life are found in the echinoderms. Some sea cucumbers are able to float or swim, and a few stalkless crinoids are also capable of swimming for short periods. Echinoderms are not, however, adapted for fast movement and would be easy prey, and so this way of life is a minor one.

Evolutionary history

The history of echinoderms is characterized by three main developmental stages, two of which are separated by a major extinction event. The first of these was the initial radiation that took place in the Early Cambrian to Middle Ordovician. This was followed by a period of stability dominated by stemmed forms and lasting until the Permo–Triassic extinction event. The final stage is the re-expansion that took place in the Mesozoic and saw the rise of mobile and burrowing forms.

The initial radiation of echinoderms may have started in the latest Precambrian, where the puzzling fossil *Tribrachidium* occurs. This shows some similarity to the later edrioasteroids, but the lack of any preserved skeleton makes its relationship to echinoderms somewhat problematic. By the Middle Ordovician this radiation had produced all 20 echinoderm classes known in the fossil record, although by that point five had already become extinct. One feature that is seen in this development is that new classes tend to develop suddenly in the record without obvious intermediate forms, suggesting either that they were initially soft-bodied and appeared in the fossil record only when the skeleton developed, or that evolution of new classes was indeed very rapid. Members of the Early Cambrian classes tend to have rather simple morphology with short stems, poorly developed pentameral symmetry, and numerous plates; in addition, diversity was low with only one or two genera occurring together. In contrast, by the Middle Ordovician advanced features, such as longer and more elaborate feeding structures and better-organized pentameral symmetry, had developed, and the blastozoans had developed complex respiratory structures. Diversity at this time was much greater with five to twenty echinoderm genera occurring together.

The period of stability that occurred in the Late Palaeozoic is dominated by crinoids and is characterized by the reduction in numbers of classes and the increase in diversity of those that remained. The number of classes dropped from 17 in the Middle Ordovician (the largest number to occur together) to 11 in the Middle Devonian and 6 in the Permian. This 'weeding-out' process was probably driven by the rise of efficient predators and a drop in primary productivity in the oceans; however, the echinoderms were also severely affected by the major extinctions of marine invertebrates that took place at the end of the Permian. Crinoids almost became extinct at this point and the blastoid record does in fact end here. Sea cucumbers, sea urchins, starfish, and brittle-stars all survived the extinctions, although none of them showed very great diversity during the Permian and Triassic.

The five classes slowly recovered during the early Mesozoic, but the crinoids never regained their dominant position, and the sea urchins became the major group by virtue of their movement into a burrowing life-habit. Although starfish, brittle-stars, sea cucumbers, and crinoids have apparently only a moderate amount of fossil diversity in the Mesozoic and Cenozoic, this may be due in part to the fact that their skeletons are composed of loosely sutured plates or ossicles that tend to disarticulate when the animal dies. The preservation of identifiable animals as fossils is thus rare.

Biostratigraphy

Echinoderms are restricted in their use as zonal indicators by the fact that so many of them were sessile, benthonic forms dependent on larvae for their distribution. Although evidence exists for transatlantic dispersal of larvae in some modern genera, there is little evidence for this in the Palaeozoic, despite the then close continental juxtaposition of North America and Africa and western Europe. Only a small number of widespread species of echinoderms are known, in particular a number of stemless and presumably pelagic crinoids that developed worldwide distributions in the Silurian and the Late Creta-

ceous. Isolated plates of echinoderms can be extremely abundant in marine limestones, forming up to 90 per cent of the rock in some instances. Some work has been done on crinoid and sea cucumber fragments to determine their biostratigraphic utility; those of sea cucumbers in particular appear to have considerable potential. More work needs to be done, however, before these fossils can be widely used.

DAVID K. ELLIOTT

Further reading

Boardman, R. S., Cheetham, A. H., and Rowell, A. J. (eds) (1987) *Fossil invertebrates*. Blackwell Scientific Publications, Oxford.

Broadhead, T. W. and Waters, J. A. (eds) (1980) Echinoderms, notes for a short course. *University of Tennessee Department of Geologic Sciences, Studies in Geology* **3**.

Nichols, D. (1962) *Echinoderms*. Hutchinson, London.

eclogite Eclogite is a distinctive, dense, green rock, composed principally of pyroxene and garnet and with no plagioclase. It has the bulk chemical composition of basalt or gabbro, and in some instances textural evidence shows it to have formed from progressive metamorphism of such rocks. Eclogite can also be produced by the primary crystallization of basaltic magmas at upper mantle pressures, but it appears that most eclogites are a product of metamorphism at similarly high pressures. Under these conditions, plagioclase is unstable: the notional albite component is present in eclogite as jadeite in solid solution in the pyroxene (which is omphacite), and as minor quartz, while the notional anorthite component occurs primarily as grossular in garnet, but sometimes also as zoisite. Other common minor constituents of eclogite include kyanite, orthopyroxene, rutile, amphibole, pyrite, and white mica.

There are three main settings in which eclogites are found: as xenoliths in kimberlite or basalt (as at Oahu crater, Hawaii), as bands or lenses in high-grade gneiss terranes (as in west Norway or the Dabie Mountains of central China), and as bands or isolated blocks associated with blueschists (for example in the Cyclades). Mineral chemistry indicates significant temperature differences between these modes of occurrence, with xenoliths representing the highest temperatures and bodies in blueschist terranes the lowest.

The origin of eclogites, especially those from gneiss terranes, has been controversial for many years. While some geologists have favoured a primary origin by crystallization of basalt magmas in the upper mantle, others have argued that these eclogites originate at sub-solidus temperatures by metamorphism of basalt or gabbro. Although eclogites in west Norway and other gneiss terranes were seen for many years as tectonically emplaced slices of mantle, modern work has largely moved away from this view. Some eclogite bodies retain in part the textures of precursors formed at lower pressure; for example, eclogite assemblages may develop only locally in shear zones, because it appears that deformation or ingress of water, or both, is needed to catalyse the reactions that convert dry granulite or gabbro to eclogite. Another argument for tectonic emplacement from the mantle was that the host gneisses themselves recorded no comparable history of high-pressure metamorphism. However, in 1982 C. A. Heinrich showed that, whereas retrogression of eclogite requires influx of water to proceed, high-pressure gneiss assemblages undergo spontaneous and pervasive dehydration as they are uplifted, and are thus much less likely to retain any high-pressure relicts by the time they reach the surface. Nevertheless, some evidence of high-pressure metamorphism has been reported from plagioclase-bearing gneisses associated with eclogists. For a long time, battle was joined over the issue of the pressure (and hence depth) required for eclogites to form, with the 'crustal eclogite lobby' consistently claiming lower formation pressures than the 'mantle school'. This dispute was finally resolved by the discovery in 1984 by D. C. Smith of tiny inclusions of coesite in pyroxene from a Norwegian eclogite. Coesite has subsequently been reported from a number of eclogite terranes worldwide, although it is more normally hosted by garnet. As a result, a distinct group of ultra-high-pressure eclogites is now recognized, formed at pressures of around 30 kbar or higher, and corresponding to depths of burial of at least 100 km. Interestingly, many of these eclogites appear to have originated in the crust, rather than in the mantle.

Some occurrences of eclogite in blueschist terranes are closely associated with blueschists of broadly similar metabasite compositions, and several suggestions have been made as to the origin of the association. In the Cyclades, eclogite appears to correspond to relatively massive gabbro bodies, while metavolcanic rocks have blueschist assemblages. Restricted access of water to the gabbros may thus have favoured production of the more anhydrous eclogite assemblage, but there are chemical differences in Mg/Fe and silica content that may also control the assemblages.

BRUCE W. D. YARDLEY

Further reading

Carswell, D. A. (ed.) (1990) *Eclogite facies rocks.* Blackie, Glasgow.

economic deposits, history of the search for *see*
HISTORY OF THE SEARCH FOR ECONOMIC DEPOSITS

economic geology
Rocks and unconsolidated surface deposits that contain some valuable commodity in a high enough concentration to be mined for a profit are known as mineral deposits. This applies whether or not they have, or have not, been mined. Such deposits are rare. The study of mineral deposits in order to elucidate how they were formed is recognized as a separate discipline in geological sciences that is known as *economic geology*. The chemical and physical processes that created these unusual deposits do not differ from those responsible for the formation of other economically uninteresting rocks. However, economic geology endeavours to discover what special combination of these processes

occurred to facilitate the genesis of mineral deposits. In the past, the discipline had made great progress based on studies of the structural geology, mineralogy, and global distribution of particular types of deposit. More recently, advances in the geological sciences have provided further valuable tools in the study of mineral deposits. For example, the theory of plate tectonics has enabled economic geologists to explain and predict the distribution of particular types of deposit in terms of processes occuring at colliding and spreading plates. In geochemistry, developments such as the ability to measure the isotopic composition of certain elements, the understanding of how minerals react with high-temperature fluids and with each other, and the accurate analysis of minute concentrations of important trace elements, have resulted in great progress in the study of mineral deposits. The ultimate goal of economic geology is to provide geologists with explanations of how mineral deposits are formed in order to provide a prospecting tool for the location of resources as yet undiscovered.

BRUCE W. MOUNTAIN

El Niño
During the nineteenth century Peruvian fishermen were aware of a warm water current that flowed southwards along the coast. Because it usually started after Christmas, *Los Dias del Niño*, the days of the (Christ) Child, they termed it *Corriente del Niño* or the El Niño current. The tropical marine conditions that El Niño brings are in great contrast to the habitual coolness caused by upwelling of cold water from Antarctica along the western South American coast. The annual El Niño current heralds summer rains over the coastal desert and Andean Mountains, bringing much-needed run-off for the irrigation of crops. The fishing industry also has to adapt from pursuing schooling fishes of the cool, dense, nutrient-rich upwelling water to foraging predator species of the less productive and less saline warmer waters.

At irregular intervals the El Niño current is markedly stronger than normal and carries the warmer, less saline water much further south. Scientists now use the term El Niño to refer to these irregular, stronger events, whose effects can last for more than a year and have important repercussions for weather and climate, and for social and economic well-being, throughout the Pacific basin and even further afield. The first written evidence for an El Niño event (though not named as such) is found in the campaign diaries of Pizarro in 1525–6. El Niño events seem to have occurred at intervals of three to four years, with the strongest type occurring at intervals of more than 20 years. During the twentieth century there were nine strong and 16 moderate El Niño events.

The atmospheric equivalent of El Niño is termed the Southern Oscillation (SO). The Southern Oscillation is a large-scale atmospheric pressure change over the southeastern Pacific and Indian Ocean. It was first documented in 1932 by Sir Gilbert Walker, who as Director-General of Observatories in India was studying the causes of variation in the Indian Monsoon. During the 'high phase' of the Southern Oscillation, strong tropical convection causes air to ascend

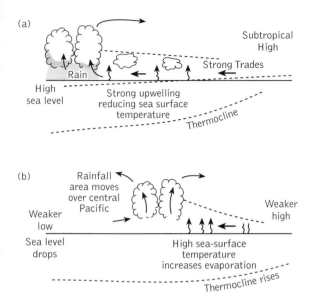

(a)

Subtropical
High

Strong Trades

Rain

High
sea level

Strong upwelling
reducing sea surface
temperature

Thermocline

(b)

Rainfall
area moves
over central
Pacific

Weaker
high

Weaker
low

Sea level
drops

High sea-surface
temperature
increases evaporation

Thermocline rises

Fig. 1. (a) Generalized cross-section of atmospheric and oceanic features along the equator during the normal phase of the Southern Oscillation. (b) Generalized cross-section of atmospheric and oceanic features along the equator during the El Niño phase of the Southern Oscillation. (From Bigg (1990), Figs 1 and 2.)

over the western Pacific and Indonesia, producing low pressure which draws in strong trade winds over the ocean surface from the south-east. The air rises and returns at higher altitudes before descending and causing high pressure in the south-eastern Pacific (Fig. 1a). During a 'low phase' the pressure pattern is reversed as the zone of low pressure associated with tropical convection moves to the middle and eastern Pacific, resulting in weaker than normal trade winds (Fig. 1b).

The Southern Oscillation Index (SOI) is a measure of the pressure difference between Darwin in Australia and Tahiti (Fig. 2). Troughs in the index correspond to El Niño events when pressure is low in Tahiti and high in the western Pacific, whereas peaks in the Southern Oscillation Index correspond to an accentuated 'normal' phase (now termed La Niña) characterized by strong trade winds. Scientists were unaware of the link between the Southern Oscillation and El Niño until 1957, when a strong correlation was noted between the SOI and sea surface temperatures off Peru (Fig. 2).

The Norwegian scientist Jacob Bjerknes was the first to suggest a mechanism for linking these atmospheric and oceanic phenomena. A simplified structure of the tropical ocean can be thought of as comprising a layer of less dense, fresher, warmer water overlying a layer of denser, colder, more saline water, the two separated by a zone of rapidly changing density and temperature known as the thermocline. During 'normal' conditions, the west-blowing trade winds move the upper warmer layer towards the western Pacific, where it increases in thickness such that the thermocline can be at depths of 150 to 200 m, leaving only a shallow upper layer and

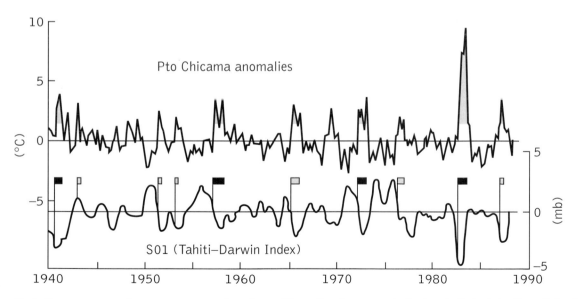

Fig. 2. Upper curve, sea-surface temperature anomalies at Puerto Chicama, Peru; lower curve, the Southern Oscillation Index. The staff of each flag marks the start of an El Niño event, the length of the flag shows its duration. Solid flags, severe or very severe El Niño; tinted flags, weak or medium El Niño events. (From Enfield (1989), Fig. 3.)

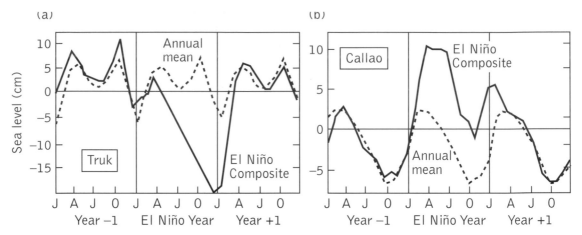

Fig. 3. (a) Changes in sea level during El Niño and non-El Niño years at Truk Island in the central Pacific. Note the single large fall in sea level as the western Pacific is 'drained' during an El Niño event. (b) The same variables plotted for Callao, Peru. (From Enfield (1989), Fig. 8.)

thermocline (30–50 m) in the east. At the same time the west-blowing trade winds and the Earth's rotation combine to produce a surface movement of water away from the Equator, both to north and south. This in turn induces upwelling of cold water from below, and an elongated tongue of cool surface water is produced. The cool tongue does not extend to the western Pacific, because here the trade winds are weaker and the thermocline is deeper. When the trade winds relax at the start of an El Niño event, the upwelling ceases and the upper warm water moves eastwards, bringing with it the zone of low pressure and strong atmospheric convection.

The relaxation of the trade winds initiates a large-scale oceanic wave termed a Kelvin wave, which travels from west to east in the equatorial region at maximum speeds of 2.8 m s⁻¹, crossing the basin in one to two months, lowering the thermocline and raising sea level as it passes. Once a Kelvin wave reaches the eastern side of the Pacific ocean, some of the wave energy may also be translated into coastal Kelvin waves which travel north and south parallel to the coast of the Americas, taking up to three to four months to reach high latitudes. In regions away from the Equator the principal large-scale wave form is a westward-travelling Rossby wave, which can be produced by reflection of Kelvin or coastal Kelvin waves. The speed of a Rossby wave decreases with distance from the Equator, and so El Niño effects that are carried into middle and high latitudes may persist for months and even years. At the latitude of Japan, for instance, a Rossby wave reflected westwards across the Pacific would take several years to cross the basin. Research has indicated that after particularly strong El Niño events, such as in 1982–3, the effects on sea surface temperature and, by implication, weather conditions have been found at higher latitudes up to 11 years later.

A typical El Niño event seems to be presaged by a period of stronger than normal trade winds. In the onset phase, which usually begins in the early part of the year, there is a westerly (east-blowing) component to the wind in the western and central equatorial Pacific. In April, peak sea-surface temperatures are noted off Peru and Ecuador corresponding to the arrival of the oceanic Kelvin wave. Sea level also rises because of the lower atmospheric pressure and the arrival of warmer surface water from the west. Figure 3 shows the change in the sea-level signature during an El Niño event at two locations: Truk Island in the Western Pacific, and Callao, Peru. High sea-surface temperatures can last until the following boreal spring of year 2, perhaps with another small peak during the winter before collapsing at the end of the sequence.

The effects of an El Niño event can be global. The eastward movement of the low-pressure convection area results in drought conditions in Australia, Indonesia, Africa, and India. The reverse occurs in the central Pacific, where there is intense rainfall and hurricane initiation. Western South America experiences heavy rainfall, resulting in flooding and erosion, and the effects can be felt in North America, where the change induced in the pressure patterns results in frontal systems bringing high rainfall further south to the western and southern United States (Fig. 4).

The 1982–3 El Niño was exceptional in its severity and was undoubtedly the strongest of the twentieth century; it was also one of the best recorded so far. There were record-breaking amounts of rainfall in Peru and Ecuador in early 1983, resulting in record amounts of run-off, flooding, avalanches, and erosion. Translation of the effects took place to higher latitudes with concomitant flooding and costal erosion. Further afield there was a record drought in Australia. New types of disturbance were noted to the marine biota on a global scale. For instance, the deepening of the thermocline and the associated fall in nutrients within the surface waters of the eastern Pacific resulted in the devastation

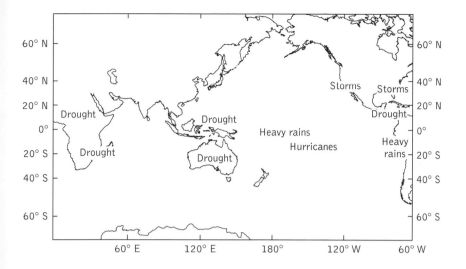

Fig. 4. The global meteorological impacts of El Niño. (From Bigg (1990), Fig. 4.)

of many coral reefs which had experienced almost uninterrupted growth for several centuries. The loss of the nutrients led to a decrease in the zooplankton, which resulted in fish and squid mortality, reproductive failure in marine birds, and food shortages for penguins, cormorants, seals, and sea lions. In addition, rough seas and high sea levels made feeding very difficult for near-shore feeders such as marine iguanas, and kelp colonies were devastated.

Historical studies over longer timescales using a wide range of approaches suggests that El Niño events can be detected using such proxy evidence as changes in tree ring widths in the south-west of the United States and Mexico; Nile River flood data; tropical and subtropical ice cores; coral growth records; fishery catch records; and from a wide range of evidence derived from marine and lake sediments.

It is now evident that El Niño teleconnections (atmospheric and oceanic responses away from the equatorial Pacific) indicate that the El Niño phenomenon is the largest source of interannual climatic variability on the global scale. Attempts at predicting the onset and severity of El Niño events have not so far been satisfactory. Much more work will have to be done by ocean modellers and palaeoclimatologists before the factors behind the variability are understood.

B. A. HAGGART

Further reading

Bigg, G. R. (1990) El Niño and the Southern Oscillation. *Weather* 45, 2–8.

Enfield, D. B. (1989) El Niño, past and present. *Reviews of Geophysics* 27, 159–87.

elastic wave propagation Observations of the waves that are propagated from a seismic source such as an earthquake or an explosion led early workers to recognize that these waves represent elastic behaviour on the part of the Earth. Hooke's law is based on the observation that, for many materials, there is a linear relationship between stress (force per unit area) and strain (deformation). Materials that obey Hooke's law when stressed are defined as being elastic (see *seismic waves: principles*). Most of the materials that make up the solid Earth are elastic, at least when the level of stress is low. A perfectly elastic body would immediately return to its original shape after an applied stress was removed, and no energy would have been lost through breakage of the material or conversion to heat. The oscillations of a spring after a force stretching it is released provide an analogy for the elastic waves that are propagated through the Earth when a seismic source is triggered. These waves travel from the source like the waves generated when a stone is dropped into a quiet pond, but in three dimensions. One of these waves is simply a sound wave. The amplitude of the waves diminishes with distance, since a given packet of energy is spread over a larger and larger area as the wave spreads out. In addition, discontinuities in seismic velocity scatter some of the energy, and some energy is lost by conversion to heat. The patterns and amplitudes of arrivals observed by seismographs far from the source are used to infer the elastic properties and structure of the Earth as well as the nature of the seismic source itself. G. R. KELLER

electrical techniques in geophysics The electrical techniques used in geophysics include some of the most important tools employed in the geophysical exploration of the shallow subsurface. The normal depth range that is investigated is from a few centimetres to a few hundred metres. These techniques are thus particularly applicable in archaeological, engineering, groundwater, and mineral investigations.

The simplest technique is the self-potential (SP), which entails the measurement on the ground surface of voltages related to currents that flow naturally below the surface. These currents are largely produced by electrochemical reactions and fluid flow in the subsurface, and the measurable voltages can be up to several hundred millivolts. Natural voltages are measured simply by using a sensitive voltmeter connected to two non-polarizing electrodes pushed into the soil. Mapping the voltage pattern on the surface of the Earth can be used to trace leachate plumes from landfills, leaks from dams, and the extent of mineral deposits.

By far the most important of the electrical techniques is the investigation of the electrical resistivity of the Earth by measuring the voltages that arise when an electrical current is applied to the ground. In all electrical surveys the object is to determine the rock resistivity, which is one of its basic physical properties. In this context the resistivity of a rock means its resistance to the flow of an electrical current. Although most common rock-forming minerals, such as quartz (in sandstone) and calcite (in limestone), are insulators, most rocks are porous. Electrical charge is then carried by ions in solution in the porewater. Clay is also able to conduct electricity well. As a result, the principal factors that govern the resistivity of rocks are the quantity of groundwater that is present, the salinity of the groundwater, and the amount of clay present.

The resistivities of common rocks vary widely; typical values (in Ohm-metres) are: clay, 5–50 Ωm; sand, 50–300 Ωm; gravel, 200–1000 Ωm; sandstone, 100–400 Ωm; limestone, 500–5000 Ωm; igneous and metamorphic rocks, more than 1000 Ωm.

The resistance of a volume of geomaterial is measured using a four-electrode system in which a low-frequency current, I, is injected into the ground through a metal, steel, or copper rod pushed into the soil, and the current is returned through a second electrode placed some distance away. The voltage, V, is measured across an additional pair of electrodes. The resistance is the ratio of the voltage to the current. When this ratio is multiplied by a factor that takes into account the

relative spacing between the electrodes, the product is the *apparent resistivity*.

In order to unravel a geological structure, a series of measurements must be made using electrodes at different spacings and positions.

In *horizontal traversing*, the spacing between the four electrodes is fixed; the whole array is moved along a line, and measurements are repeated at intervals. This is a simple technique for locating areas of particular interest. For example, traversing techniques might be employed to locate fracture zones (low resistivity) in a limestone (high resistivity), where there is soil or alluvium concealing the solid geology (Fig. 1). Interpretation is generally qualitative, the results being tested by drilling.

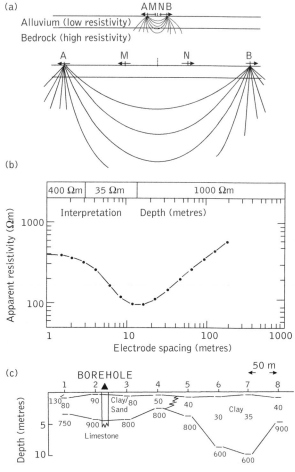

Fig. 2. Vertical electrical sounding. (a) Current electrodes (*A*, *B*) and voltage electrodes (*M*, *N*) are expanded about a centre point so that current penetrates progressively deeper; (b) sounding curve and interpretation; (c) geoelectrical section with formation resistivities in Ωm.

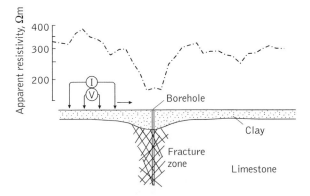

Fig. 1. Principle of horizontal traversing.

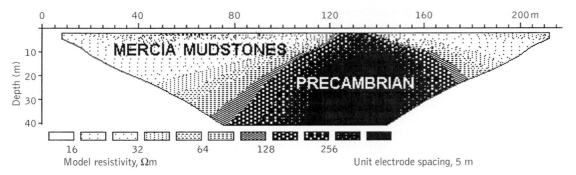

Fig. 3. Electrical image across Precambrian pinnacle, Leicestershire, England.

Vertical electrical sounding (VES, electrical drilling) is used to investigate the vertical variation of resistivity and might, for example, be used to determine the depth to bedrock or thickness of a gravel resource. The centre point of the four electrodes is fixed and the electrode spacing is increased between measurements (Fig. 2a). In this way current is forced to flow to increasingly greater depth with the result that the resistivity measurement also responds to changes with depth.

A sounding curve, produced by plotting resistivity against electrode spacing (Fig. 2b), is interpreted quantitatively to derive thicknesses and resistivities of subsurface layers. The results from a series of soundings are used to produce a geoelectrical section (Fig. 2c). The technique has important applications in groundwater, extractive mineral, and engineering investigations to determine geological structures that are essentially horizontal.

Electrical imaging (tomography) is a later development and incorporates a combination of both traversing and sounding. As the process calls for a large number of measurements, the survey is normally carried out under computer control and the collection of data is automatic. The data are subjected to computer inversion procedures to produce an image of the variation of rock resistivity along the line of the survey (Fig. 3). Electrical imaging is particularly useful in areas of complex geological structure where conventional techniques are unsuitable. By using a knowledge of the resistivities of the common rocks, the images can be interpreted in geological terms.

Techniques are now being developed that utilize the small variations in resistivity that occur when the frequency of the applied current is varied over the range 0.001–1000 Hz. These *spectral-induced polarization* or *complex resistivity* techniques respond to the presence of disseminated sulphide mineralization and are sensitive to the electrochemical effects caused by the differences in the chemistry of minerals and fluids. These techniques hold the promise of being able to differentiate low levels of impurities in investigations of contaminated land.   RON D. BARKER

electromagnetic mapping, controlled-source

see CONTROLLED-SOURCE ELECTROMAGNETIC MAPPING

electromagnetic methods in applied geophysics

Varying currents are associated with varying magnetic fields, and varying magnetic fields induce currents in conductive materials. These currents in turn produce *secondary* alternating magnetic fields which can be detected at the ground surface. These facts form the basis of electromagnetic methods in applied geophysics. Broadly, the methods fall into two main categories. In *continuous wave (CW)* methods, the inducing current has a sinusoidal form and a fixed frequency, which is generally in the range from a few hundred to a few thousand Herz (Hz). Some use is also made of communications transmissions (particularly military 'VLF' transmissions in the 15–25 kHz range) and of natural fields (magnetotellurics), which cover a very wide range of frequencies. In *transient e.m. (TEM)*, current is induced by the collapse in magnetic field which occurs when current flow in a circuit is terminated abruptly. The TEM method is fundamentally multi-frequency and provides information which can be obtained from CW systems only by operating at a number of discrete frequencies. Another advantage of TEM is that measurements are made at times when no current flows in the source circuits, whereas CW measurements of secondary magnetic fields are made against primary-field backgrounds which are generally much stronger.

Various configurations of local electromagnetic source have been used, including very long grounded wires, large rectangular loops, (sometimes enclosing the area being surveyed and sometimes offset to one side of it), and small portable loops with diameters of about 10 cm. Small loops, which produce radiated fields approximating to those produced by alternating magnetic dipoles, are usually ferrite-cored to enhance the magnetic effects. Receiver coils are almost always small loops. Measurements may be made, very crudely, by measuring the dips of the resultants of the primary and secondary magnetic fields or in more sophisticated ways by measuring amplitudes and phases of various field components. Measurements of field directions and phase shifts make it possible to distinguish primary from secondary fields, and to estimate conductivity parameters for subsurface conductors.

One great advantage of electromagnetic methods is that they can be used from aircraft, making it possible to cover large areas rapidly. Most airborne CW systems now use multiple receiving coils with different orientations, and some use multiple transmitting coils. Coils may be mounted in the nose and at the tail of the survey aircraft, or at the wingtips, or in tubular fibreglass 'birds' towed beneath helicopters. Whatever the arrangement, corrections must be made for changes in separation and, more importantly, relative orientation of the receiver and transmitter coils caused by vibration and flexure. The importance of these corrections can be appreciated from the fact that, whereas a secondary field measured on the ground may exceed 10 per cent of the primary field, significant airborne secondary fields may amount to no more than a few parts per million.

Although electromagnetic methods were primarily developed for mineral search, relying on the very high conductivities of massive sulphide deposits, they are increasingly being applied to regional conductivity mapping for hydrogeological and environmental purposes. The use of repeat airborne surveys to monitor salinization of irrigated agricultural lands has been pioneered in Australia using both CW and TEM methods.　　　JOHN MILSOM

elemental associations and ore minerals and allied deposits
Each chemical element or group of elements is characterized by distinct properties which result from element-specific electronic and nuclear configurations. The existence of differences in chemical, physical, and thermodynamic properties among different elements plays a key role during many geological and biological processes and leads to significant fractionation or separation of elements. These processes, in turn, are a function of the temperature, pressure, and chemical composition of the local environment. As a result of fractionation the chemical elements and their isotopes are not homogeneously distributed in the Earth. Instead, elements with similar properties tend to form associations, and are usually separated from elements that exhibit contrasting properties. Under certain circumstances geochemical fractionation leads to economic enrichment of metallic ores and industrial minerals. Recognition of these elemental associations and their environments of formation is the basis for geochemical exploration.

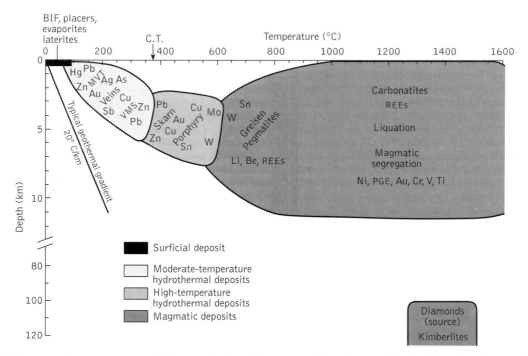

Fig. 1. Pressure and temperature ranges of various ore-forming environments, and the major commodities found in each environment. Note that in comparison to the typical geothermal gradient, ore-forming environments represent thermal, as well as elemental, anomalies. BIF, banded iron formation; C.T., critical temperature of water; MVT, sediment-hosted base metal deposits; REEs, rare-earth elements; PGE, platinum group elements; VMS, volcanogenic massive sulphide deposits.)

The formation of unusual concentrations of a group of elements generally involves a mobile phase, that is, a magma or a fluid; this is because in the crystalline state the elemental composition of any portion of the crust is essentially fixed. Changes in bulk composition require a phase that is capable of dissolving and transporting material. Magmas and fluids become enriched in certain elements and depleted in others because solubility is a function of the chemical characteristics of the individual elements and the fluid or magma. Mineral deposits form where new minerals are precipitated from magmas or fluids in response to changes in pressure, temperature, and chemical environment.

Mineral deposits occur in a variety of environments in the Earth's crust (Fig. 1). In the high-temperature environments of the lower crust and upper mantle, basic and ultrabasic magmatism and release of volatiles dominate the processes of ore formation. In the cooler middle and upper crust, felsic magmatism and aqueous metasomatism are most important. At the low temperatures of the Earth's surface and hydrosphere, biological activity and hydrological processes give rise to additional types of concentrations. These various processes lead to the formation of elemental associations that are characteristic of the various geological environments. Thus, for convenience, economically valuable deposits of ores and industrial minerals are classified according to the processes and temperature of their formation: (1) magmatic; (2) high-temperature hydrothermal; (3) moderate-temperature hydrothermal; and (4) low-temperature surficial.

Mineral deposits of an individual element need not be restricted to a single geological setting. Elemental associations may, however, be indicative of a specific setting because the geochemical processes leading to enrichment are different for different environments. Gold, for example, is found in association with platinum group elements (PGE) in the high temperatures of the magmatic environment. In the hydrothermal environment, gold occurs with copper in high-temperature porphyry deposits, and with arsenic, antimony, and mercury in the moderate-temperature vein environment. And finally, in the low-temperature surficial environment of placer deposits, gold can be associated with uranium.

Magmatic environment

The formation of several types of deposits can be ascribed to processes that are mainly magmatic, that is, to processes related to the cooling and solidification of magmas. During the crystallization of minerals from a cooling melt, some elements exhibit a preference for the solid phase (compatible elements), whereas others tend to remain in the liquid phase (incompatible elements). As crystallization proceeds, the incompatible elements will become concentrated in the melt. They will eventually be removed either by crystallization of late-stage minerals that can accommodate them in their structure (as in pegmatite deposits), or by fractionation into a fluid phase that either accumulates in the upper parts of the magmatic system (greisen deposits) or escapes into the surrounding country rocks (hydrothermal deposits). This magmatic process, termed fractional crystallization, also promotes the accumulation of compatible elements in the early-formed minerals (e.g., chromite, olivine) which settle at the bottom of the magma chamber because of their relatively high density. Economic deposits can also form from liquation, the process in which two immiscible liquids separate from a single magma. This separation creates significant fractionation during the redistribution of the chemical elements between the two liquids. The nature of a magmatic ore deposit is largely determined by the source and the chemical composition of the magma from which it formed.

Deposits related to mantle sources. Some ore deposits occur in alkalic ultrabasic rocks derived from the Earth's mantle. Because of their deep-seated origin, these rocks contain typical mantle minerals or elemental associations that are unusual for crustal rocks.

Kimberlites and other alkalic ultrabasic rocks are the primary source for diamonds of both gem and industrial quality. Extremely high pressures and temperatures are required to stabilize diamond rather than graphite, and all natural diamonds found at the Earth's surface must therefore be ultimately derived from the upper mantle. Kimberlites originate at depths of approximately 100–300 km below the surface. They ascend through the lower crust and intrude the upper crust, where they deposit the diamonds in volcanic pipes. Examples of such deposits occur at Kimberley (South Africa) and in various kimberlite pipes in Northern Lesotho and Eastern Siberia (Russia).

Carbonatites are magmatic carbonate rocks that are usually associated with alkaline igneous complexes. Originally derived from mantle rocks through partial melting, alkaline magmas can generate a volatile-rich carbonate melt through exsolution, wall-rock reactions, or fractional crystallization during their ascent. Carbonate melts are generally enriched in incompatible elements which can be concentrated into economic deposits. Carbonatite deposits are exploited primarily for lead, zinc, niobium, rare earth elements (REE), uranium, thorium, barium, strontium, and fluorine. Examples of these resources occur at Phalaborwa (South Africa), Khibina and Kovdor (Russia), and Mountain Pass (California, USA).

Magmatic segregation deposits (orthomagmatic). These deposits are formed from basic or ultrabasic magmas by fractional crystallization or liquation. Fractional crystallization is usually important in plutonic rocks, whereas liquation can be significant in both plutonic and volcanic environments.

Cumulate rocks produced by fractional crystallization are characterized by alternating layers rich in olivine, pyroxene, and plagioclase. Crystallization of oxides can generate chromium-, iron-, vanadium-, or titanium-rich layers of economic importance. Ore layers, a few millimetres to more than a metre in thickness, are commonly found in the lower parts of large igneous complexes, and can be traced laterally for several kilometres. These deposits also produce large quantities of PGE and some gold. The Bushveld (South

Africa), the Great Dyke (Zimbabwe), and Tellnes (Norway) are places where such deposits are exploited.

Liquation deposits, an important source of nickel, copper, iron, and PGE, are formed when a sulphide liquid separates from a mixed sulphide–silicate magma. Sulphide droplets coalesce, sink, and accumulate at the base of the intrusion or lava flow. These sulphide liquids have a strong chemical affinity for iron, nickel, copper, and PGE, and concentrate these metals in relation to the silicate magma. The resulting massive orebody is usually overlain by a zone with subordinate silicate minerals enclosed in a network of sulphides (disseminated ore). This zone, in turn, is overlain by sulphide-poor rocks which grade into unmineralized ultrabasic or basic rocks towards the upper part of the intrusion or lava flow. At Sudbury (Ontario, Canada), Eastern Goldfields (Western Australia), and Noril'sk (Siberia) there are examples of these deposits.

Pegmatite deposits. Pegmatites are late-stage, very coarse-grained igneous rocks, usually of granitic composition. Several types of pegmatites have been recognized, but only those associated with felsic intrusions at higher crustal levels are economically important. Many of these pegmatites are characterized by high concentrations of sodium and potassium, of volatiles such as boron, fluorine, and water, and of incompatible elements (lithium, beryllium, carbon, phosphorus, rubidium, caesium, tin, molybdenum, tungsten, niobium, tantalum, zirconium, REE, thorium, and uranium. This elemental association represents the final stage of fractional crystallization of granitic magmas. Most of these elements were present only in trace concentrations in the original magmas, but became enriched in the residual, water-rich melt from which pegmatites crystallized. The inner portions of pegmatites contain large crystals, which can be up to several metres long. Pegmatites are mined for gemstones (beryl, topaz, and tourmaline), industrial minerals (feldspar, muscovite, lepidolite, and spodumene), or metallic constituents. Examples of pegmatite deposits occur at Bancroft (Ontario, Canada) and Bikita (Zimbabwe).

Greisen deposits. A greisen is a quartz–mica aggregate containing significant amounts of cassiterite, wolframite, fluorite, topaz, rutile, and tourmaline. This rock type develops at the upper contacts of granite intrusions, where late-stage hydrothermal solutions rich in fluorine have intensely altered the granites. Fluorine plays an important role in the mobilization, transport, and deposition of high-field-strength elements such as tin, tungsten, and titanium. Greisen deposits are mined particularly for the production of tin and tungsten. These deposits represent a transition from the magmatic to the high-temperature hydrothermal environment. Deposits of this type are found in Cornwall (England) and the Erzgebirge (Germany and the Czech Republic).

High-temperature hydrothermal environment

High-temperature hydrothermal processes are related to the emplacement and cooling of magmas, and are of primary importance in the distribution and concentration of elements in the upper levels of the Earth's crust. These processes are responsible for the genesis of several types of ore deposit of enormous economic value.

Porphyry deposits. These deposits are intimately associated with the porphyric parts of intermediate to acid plutons in orogenic belts. They are principally mined for copper, molybdenum, and tin, but contain subordinate amounts of lead, zinc, silver, and gold. These metals are constituents of ore minerals that occur with quartz in a complex network of veinlets (stockworks) or are scattered throughout the host-rocks (disseminated mineralization). Both styles of mineralization are commonly found in the same deposit. Porphyry copper and porphyry molybdenum deposits are usually extremely large (up to 10^9 tonnes), whereas porphyry tin deposits (mainly stockworks) are smaller.

Porphyry deposits form during the cooling of magmas when the crystallization of anhydrous minerals leads to an increase in the volatile content of a melt and a concomitant increase in vapour pressure. This results in boiling and separation of a hydrothermal fluid from the magma, and often causes brecciation of the surrounding, solidified rock. The fluid is enriched in chlorine, which enables it to leach metals from the igneous rock and transport them outwards through the new fractures. The fluid is also enriched in the sulphur required for precipitation of the metals as sulphides at the lower temperatures found at the site of the future orebody. Chuquicamata (Chile) and Bingham (Utah, USA) are important porphyry deposits.

Skarn deposits. Skarns are important sources of tungsten, copper, iron, molybdenum, zinc, tin, lead, and graphite. They result from the high-temperature replacement of carbonate rocks adjacent to porphyry deposits. The fluids and ore-forming elements are derived mainly from the intrusion. The orebodies are commonly irregular in shape and may terminate abruptly at structural discontinuities. In skarns, unlike most other hydrothermal deposits, the mineral assemblage depends on the composition of the invaded country rocks, and not simply on temperature, pressure, and fluid composition. Skarns are generally zoned, with anhydrous minerals (typically garnet, pyroxene, and wollastonite) closest to the intrusion, followed outwards by hydrous minerals (e.g. amphibole and chlorite). Ore minerals include magnetite, scheelite, wolframite, cassiterite, and graphite, or base-metal sulphides, which usually occur in the outer parts of a skarn deposit. This zonal sequence reflects differences in temperature and in the properties of the metals in solution. The presence of chlorine-rich minerals in many deposits and the composition of fluid inclusions trapped within skarn minerals document the importance of metal–chloride complexes in ore-forming fluids, particularly at higher temperatures. Among examples of such deposits are those at Sarbai (Kazakhstan), Mary Kathleen (Queensland, Australia), and King Island (Bass Strait, Australia).

Moderate-temperature hydrothermal environment

Slightly further from intrusions and in some metamorphic environments, several types of ore deposits are formed by the circulation of moderate-temperature hydrothermal fluids. Fluid circulation is driven by local thermal anomalies, such as that produced by emplacement of a nearby intrusion. These deposits generally form below the critical temperature of water (374 °C), and aqueous complexes with anions other than Cl$^-$ can be important for metal mobility.

Volcanogenic massive sulphide deposits. These deposits are composed predominantly of iron sulphides (pyrite, pyrrhotite), and lesser amounts of copper, zinc, and lead sulphides. Economically important contents of gold and silver are present in some orebodies. These sulphide ores can grade laterally into massive oxide ores with increasing magnetite or hematite content. Massive sulphide deposits occur as groups of lenticular or sheet-like orebodies at the interface between volcanic and sedimentary rocks or the sea floor. The formation of such deposits has been observed at submarine hydrothermal vents ('black smokers') on the present-day ocean floor adjacent to many mid-ocean spreading ridges: base metal sulphides precipitate when hot (~350 °C), generally reducing, and acid fluids rich in hydrogen sulphide, are expelled from 'chimneys' into the cold sea water. The hydrothermal fluids appear to be sea water which has been heated and enriched in metals by reaction with hot rocks below the sea floor. Beneath the orebodies there is usually a stockwork which represents a system of feeder channels along faults and cracks through which the ore-forming fluids ascended to the main deposit. At Troodos (Cyprus) and Kuroko (Japan) there are examples of these deposits.

Vein deposits. Vein deposits of precious and base metals form in the upper few kilometres of the Earth's crust, where brittle faults and fractures can remain open for long periods of time and provide channel-ways for hydrothermal fluids migrating upwards from hotter regions. Occasionally these fluids reach the surface and form hot springs (as, for example at Yellowstone, USA). Vein deposits are often found in volcanic rocks overlying large plutons, but they are also present in sedimentary and metamorphic rocks. Cooling magmas or metamorphism provide the heat that drives fluid circulation, but stable isotope measurements indicate that the water in these hydrothermal fluids is derived from the surface (i.e., it is meteoric water).

Vein deposits represent some of the highest concentrations of precious metals which result in very rich 'bonanza' ores. In addition to the more valuable silver and gold, common constituents of these ores are copper, zinc, lead, arsenic, antimony, bismuth, mercury, and tin. Vein deposits are smaller in size than porphyry deposits because the ore is restricted to narrow zones along faults. Because they form near the surface, these deposits are often eroded before they can be mined. Where gold was present in large amounts, it can be redeposited in placers. The Comstock Lode (USA) and deposits at Yellowknife (Canada) and Potosi (Bolivia) are examples of such occurrences.

Sediment-hosted base-metal deposits. Some important sources of lead and zinc form through the expulsion of large volumes of water from the pore space of sediments as pressure and temperature increase during burial and compaction in large, subsiding basins. Overlying impermeable rocks prevent this water from moving upward, and force it to move laterally, sometimes for hundreds of kilometres. These fluids become very saline, and dissolve lead and zinc which are present in trace concentrations in the strata. The metals precipitate where these fluids encounter organic-rich sedimentary rocks due to reduction of the metals and associated oxidation of the organics. Other elements commonly associated with these deposits are fluorine and barium. Some deposits of uranium form in a similar manner. Examples of these deposits occur at Viburnum Trend (Missouri, USA), Upper Silesia (Poland), and Grants (New Mexico, USA). They are generally referred to as rollfront deposits.

Low-temperature surficial environment

Ore deposits can also result from low-temperature geochemical and sedimentary processes at, or near, the Earth's surface. In addition to the geochemical processes discussed above, biological activity can play an important role during ore formation. Moreover, deposits related to clastic sedimentation (placer deposits) demonstrate that typical elemental associations can also be generated by mechanical processes.

Deposits related to chemical sedimentation. Valuable deposits of many different elements can be produced by chemical precipitation from oceans, lakes, and pore fluids. The precipitation of metalliferous sediments depends on a variety of factors, including pH, Eh, temperature, availability of metals, and biological activity (mainly bacteria and algae). Metals are derived from terrestrial rocks and from submarine volcanic sources. Several different elemental associations are found in such sedimentary deposits, each being indicative of a specific geological environment.

(1) Banded iron formations (BIF) are iron-rich sediments consisting of laminated silica (chert) layers alternating with layers of iron minerals (haematite, magnetite, or pyrite). These important deposits occur as stratigraphical units several hundreds of metres thick that are hundreds of kilometres in lateral extent. They were formed mainly in the relatively shallow waters of continental shelves by chemical or biochemical precipitation. The source of these huge amounts of iron has not yet been established with certainty. Most of the world's BIF deposits were deposited during the late Archaean and early Proterozoic (2600–1800 Ma), but precipitation of BIF is not restricted to this period. Major BIF deposits occur in the Lake Superior region (North America) and at Hamersley (Western Australia).

(2) Manganese is mined principally from shallow marine or estuarine sedimentary environments, where it is derived from terrestrial source rocks. Significant, but unmined, enrichments occur also in deep-sea environments where

manganese (present as nodules) appears to be derived from submarine volcanic activity; these manganese nodules contain significant amounts of copper, nickel, and cobalt. Important manganese deposits occur at Nikopol (Ukraine) and Chiatura (Georgia).

(3) Stratiform base-metal sulphide deposits can form in shallow, reducing, and euxinic environments (lagoons or stagnant basins similar to the present Black Sea) or in sabkhas (tidal marshes in arid regions). In such deposits early diagenetic sulphides of copper, lead, and zinc are disseminated throughout a matrix of alternating layers of carbonate, clay and organic matter. The European Kupferschiefer and the Zambian Copperbelt are good examples of such deposits.

(4) Phosphate-rich sedimentary rocks (phosphorites), exploited mainly for fertilizers, develop on continental shelves by the chemical precipitation of apatite. Precipitation occurs when cold, upwelling and landward-directed sea water becomes supersaturated upon mixing with warmer shelf waters. Phosphorites also contain large amounts of biogenic phosphate resulting from the thriving biological activity that develops as a result of upwelling, nutrient-rich waters. Morocco, Florida (USA), and Peru contain major phosphate deposits.

(5) Evaporites form by desiccation of ocean or lake water in arid climates, and commonly occur in rift zones and depressions. Marine evaporites represent the solutes of sea water, and have a rather uniform chemical and mineralogical composition; they constitute the major source of salt (NaCl, KCl), gypsum, anhydrite, sulphur, magnesium, bromine, iodine, and strontium. Lacustrine evaporites have more diverse compositions which are dependant on the local surroundings; individual deposits provide various commodities including nitrates (Chile), boron compounds (California, Turkey), lithium (Utah), and sodium carbonate (East African Rift).

Placer deposits. Placer deposits represent important accumulations of gold, diamonds, and other heavy minerals produced by mechanical sedimentary processes. Heavy minerals have a high density which makes it possible for gravity separation and concentration to take place by the action of irregularly flowing water. In addition, placer minerals must be chemically resistant and mechanically durable to survive exposure, erosion, dissolution, and abrasion during transport from their source to the site of deposition. Minerals found in placers include gold, ilmenite, rutile, cassiterite, zircon, monazite, xenotime, diamond, ruby, and sapphire, and contribute significantly to the world's needs in gemstones, industrial minerals, and metals (gold, titanium, tin, zirconium, thorium, uranium, REE, PGE). Placer deposits are found in meandering streams (for example, the cassiterite placers of Malaysia and Brazil; and the diamond gravels of South Africa), and along the shorelines of major oceans (beach placers). The shoreline deposits occur mainly within the trade wind zones; here strong waves and longshore currents lead to their development, particularly within the tidal zone, as in the beach sands of Australia and India and the gem gravels of Namibia. Most placer deposits are of Cenozoic age, but older fossil placers are also of great importance (e.g. the Proterozoic gold–uraninite deposits of the Witwatersrand Goldfields, South Africa).

Residual deposits. These deposits are formed in regions with intense chemical weathering typical of tropical areas. In these climates, characterized by high rainfall and high temperatures, most rock types decompose rapidly, yielding soils from which all readily soluble components have been removed. These soils, called *laterites*, are composed mainly of iron and aluminium, but they sometimes contain other valuable metals such as nickel, cobalt, and gold. The chemical composition of a laterite, and therefore its economic importance, depends on several factors, including the original rock composition and the pH of the infiltrating water, which is strongly influenced by vegetation. *Bauxites* (laterites consisting of almost pure aluminium hydroxides) develop on rocks with a low initial iron content or from which iron has been removed, and represent the primary source for aluminium. Residual nickel deposits are commonly found in association with intensely weathered peridotites which had high initial nickel contents. Residual deposits occur at Sangaredi (Guinea); in Central Jamaica; at Boddington (Western Australia); and in New Caledonia.

Deposits formed by supergene enrichment. These deposits represent enrichments of ore minerals caused by surface waters that percolate downwards through an existing sulphide-rich orebody. These waters oxidize primary ore minerals (e.g., pyrite and chalcopyrite), yielding insoluble iron hydroxides and solvents that dissolve other minerals. During this process valuable elements such as copper, zinc, or gold are leached from the upper part of an orebody and carried downwards until the solution reaches more reducing conditions (e.g. the groundwater table). Here, the metals are reprecipitated as new, secondary ore minerals which replace primary ones. The upper, leached part of the orebody forms a barren residuum of iron hydroxide, commonly known as a gossan or 'iron hat', which overlies a zone of supergene enrichment that has significantly higher copper, zinc, or silver contents than the underlying primary orebody. Examples of such deposits occur at Butte (Montana, USA) and Escondida (Chile).

RETO GIERÉ AND ROBERT P. ILCHIK

Further reading

Evans, A. M. (1993) *Ore geology and industrial minerals: an introduction.* Blackwell Scientific Publications, Oxford.

Edwards, R. and Atkinson, K. (1986) *Ore deposit geology.* Chapman and Hall, London.

energy budget of the earth To understand the thermal and tectonic history of the Earth it is necessary to recognize the enormous release of gravitational energy that took place in the process of its formation. This accretion energy dwarfs all other energy sources in the entire life of the Earth, including radiogenic heat, to which the high temperature of the interior is sometimes erroneously attributed. A comparison of the important contributions to global energy is given in Table 1, in which the first entry is the total gravitational energy released by the accretion of the Earth and segregation to its present density structure. The following four items are components of this total accretion energy. The reason for identifying them separately is that they are derived from processes that were not simultaneous. They therefore have a bearing on our understanding of the evolution of the Earth.

Of the total accretion energy, 90 per cent would be accounted for by the accumulation of originally widely dispersed material into a homogeneous sphere with the mass and radius of the Earth. The remaining 10 per cent is due to the concentration of mass towards the centre and, of this, 80 per cent was due to the separation of the dense core. The progressive growth of the solid inner core by freezing the material from the liquid outer core and the release of gravitational energy by separation of the light crust at the surface are geophysically important phenomena, but appear as minor items in Table 1.

The estimate of the total radiogenic heat given in Table 1 is based on the abundances of the four thermally important isotopes of the elements uranium, thorium, and potassium. The present rate of release of this heat is about 30×10^{12} Watts (W). At the time of formation of the Earth, 4.6×10^9 years ago, these elements would have been producing four times as much heat, and the average over the subsequent life of the Earth is twice the present value. It is possible that the early Earth incorporated also some short-lived isotopes that contributed to its heating. They cannot, however, affect the conclusion that radioactivity provides only a topping-up of the Earth's internal heat. The fact that the Earth is hot inside is due to the early gravitational energy release. This is emphasized by a comparison of the total radiogenic heat produced in the Earth during its lifetime with present stored heat, which is twice as great.

Rotational energy is dissipated by the tides. Angular momentum is conserved by expansions of the Earth–Moon and Earth–Sun orbits, but rather little of the energy lost by the Earth's slowing rotation goes into the orbits. Most is dissipated

in the Earth. The sea is, however, the principal sink of tidal energy and only about 5 per cent of it heats the deep interior.

Much of the gravitational energy would have been radiated away during the accretion process itself. The fraction retained as heat would have depended somewhat on the speed of this process. Core separation would, however, have required the Earth to be hot and substantially complete. It has even been suggested that core formation was delayed by up to 100 million years. Almost all the gravitational energy of core formation would thus have appeared as heat, and it would have sufficed to raise the temperature of an already hot Earth by a further 3000 °C. This is the source of the residual heat that the Earth is still losing at a rate of about 11×10^{12} W.

Cooling of the Earth

Our current best estimate of the total heat lost through the Earth's surface is 44×10^{12} W. With an estimated 29×10^{12} W of radiogenic heat with 2.5×10^{12} W of gravitational energy due to thermal contraction and 1.5×10^{12} W of gravitational separation, the deficiency is 11×10^{12} W. This represents a net loss by cooling of the mantle at about 70 °C per billion years, with slower cooling of the core. The inevitability of this continuing net loss of heat can be seen by considering the mechanism by which the heat is transported to the surface from the interior.

The mantle is undergoing thermal convection. Hot material rises, being replaced by cooler, sinking material, resulting in a net upward transfer of heat. The rate of convection is controlled by the viscosity or stiffness of the mantle material, and this depends very strongly on temperature. The temperature-dependence of mantle viscosity has a stabilizing effect on the convection. If convection proceeds too fast, then the mantle cools too fast, becoming stiffer and so slowing the convection, allowing the heat sources to catch up again. Conversely, if the convection were to stop or even slow down, the mantle would heat up and soften, accelerating the convection. But the principal continuing source of heat, being due to radioactive decay, is decreasing with time. The self-stabilizing effect of the temperature-dependent viscosity therefore requires that the convected heat flux must also be slowing down; and since this is controlled by temperature, the temperature must be falling.

Mechanical energy and earthquakes

The mantle of the Earth behaves as a heat engine. The process of transferring heat from the hot interior to the cool surface generates mechanical power with an efficiency that is readily calculated from the thermal properties of the mantle materials. It is about 15 per cent, which means that the mechanical power generated is this fraction of the convected heat flux. With a mantle heat flux of 32×10^{12} W, the total mechanical power driving global tectonics is about 4.8×10^{12} W. This power is used in deforming the solid mantle material; that is, the convection itself consumes the power that it generates. In doing so it heats up the deformed material, but this is not an independent heat source; it is part of the mantle heat that is converted to mechanical energy and back again.

Table 1. Contributions to global energy

Gravitational energy of accretion	
Present density structure	2.49×10^{32} J
Uniform Earth	2.33×10^{32} J
Core separation	1.61×10^{31} J
Inner core formation	8.3×10^{28} J
Separation of crust	7.6×10^{28} J
Radiogenic heat	8.0×10^{30} J
Residual (stored) heat	1.8×10^{31} J
Tidal dissipation	2 to 3×10^{30} J
Present rotational energy	2.1×10^{29} J

The speed of convection and the mantle stresses involved, including the stresses apparent in earthquakes, are directly related to this mechanical power. If we consider a block of deformable material and apply stresses to its faces, the power that is applied to it is the product of stress, strain rate, and volume. This principle is a variant of the familiar equation 'work = force × distance'. It can be applied directly to the convective deformation of the Earth's mantle. Observations of the motions of the surface plates indicate the rate, and we know the volume involved. With the thermodynamic value of total power, we can then estimate the average magnitude of tectonic stress. The answer is about 5 MPa (50 bar), with an uncertainty factor of 2 because of the variability of flow rates in the mantle. This value falls neatly into the range of stresses deduced from the elastic waves radiated by earthquakes, which are usually in the range 1 to 10 MPa. Since the stress release in earthquakes is the best indicator that we have of the general magnitude of tectonic stress, we have a satisfying coincidence of numbers to demonstrate that tectonics, including earthquakes, are driven by thermal convection.

The annual average energy release by earthquakes is about 4×10^{17} joules (J), corresponding to a mean power of 1.3×10^{10} W. This is about 0.25 per cent of the total tectonic power. Earthquakes are localized irregularities in the tectonic motion of the mantle and crust, and through most of the mantle the motion is aseismic. Occasional very large earthquakes release more energy than the annual average. The largest well-recorded shock occurred in Chile in May 1960, with an estimated energy of 1.6×10^{19} J.

Core energy and the geomagnetic field

The magnetic field of the Earth is produced by electric currents in the fluid, metallic outer core, driven by turbulent convection motion. The energy source for this motion must be identified with the slow cooling of the Earth. Radiogenic heat in the core is discounted because none of the thermally important elements, uranium, thorium, or potassium, separates into the iron in melting experiments on mixtures of iron and mantle-type silicates. These elements are completely absent from iron meteorites.

Thermal convective cooling of the core is possible in principle but appears not to be very effective, primarily because the high thermal conductivity of the core causes a conductive loss of heat, amounting to about 3.7×10^{12} W; only the core heat flux in excess of this would cause convection and that only with an efficiency of 12 per cent. It is not in any case, likely that the core heat flux is greater than this. S.I. Braginsky, then in Moscow, first pointed out that cooling of the core would cause the solid inner core to grow by the solidification on to it of outer core material but that the solid does not have precisely the same composition as the liquid. Light solutes in the outer core (which may be some mixture of oxygen, silicon, sulphur, carbon, and even hydrogen) are rejected by the solid and so remain in the fluid at the boundary as an excess of the light components. Convective mixing of this excess into the whole outer core releases the gravitational energy of inner-core formation listed in Table 1. The present rate is estimated to be 3×10^{11} W, at least half of which, 1.5×10^{11} W, is available to drive the geomagnetic dynamo.

The kinetic energy of convective motion is very small, even in the fluid core where the motion is a million times faster than in the mantle. The convective energy in the core is converted directly to magnetic energy by the dynamo action. The total magnetic energy is about 10^{22} J. This energy is continuously lost by the resistive heating of the core by the electric currents. We can thus identify the power input, 1.5×10^{11} W, with the rate of loss and so calculate the time that it would take an unmaintained field to disappear,

$$10^{22} \text{ J}/1.5 \times 10^{11} \text{ W} = 6.7 \times 10^{11} \text{ seconds} = 2000 \text{ years}.$$

This is a rough but reasonable estimate of the characteristic time-scale for changes in the field.

The surface energy balance

The average flux of heat through the Earth's surface from the interior is 0.086 W m^{-2}. The corresponding temperature gradient in the crust, about 25 °C/km, is very noticeable in deep mines and boreholes, but at the surface itself this heat is completely insignificant. The power in the solar radiation is 1367 W m^{-2}, or 342 W m^{-2} when averaged over the whole surface of the Earth, including the dark side. The internal heat has no effect on climate, except indirectly when it causes volcanic eruptions.

A look into the future

To the extent that radioactivity maintains the internal heat that drives the tectonic engine, it has been likened to a spring-driven clock that is winding down. But the analogy is very imperfect. The energy of a spring has a definite end-point, but radioactive decay does not. Moreover, as the figures in Table 1 show, the Earth's stored heat is twice as great as the total radiogenic heat release since its formation; it is six times as great as the radiogenic heat release from now until the end of time. Radioactivity slows the cooling of the Earth, but the stored energy source will suffice to maintain convection and all the tectonic processes for at least another 10^{10} years; and in less time than that the Earth will be engulfed by a profligate expansion of the dying Sun.   FRANK D. STACEY

Further reading

Stacey, F. D. (1992) *Physics of the Earth* (3rd edn). Brookfield Press, Brisbane.

engineering geology Engineering geology is the application of principles and methods of geology to the purposes of civil engineering projects. Engineering geologists (i.e. those with a first degree in geology) and geotechnical engineers (i.e. those with a first degree in civil engineering) work closely together to form a comprehensive ground engineering team for the investigation, design, and construction of major infrastructure projects. Engineering geologists are involved in the study of raw materials, assessing the stability of natural and man-made slopes, foundations for buildings, excavations for cuttings, the construction of dams and embankments, the stability and construction of mines and tunnels, the containment and disposal of waste materials, and the location of aggregates and building stone.

In the twentieth century, engineering geology has become increasingly scientific as the use of natural soil and rock as construction materials has been understood and the impact of ground conditions on modern projects realized. Ground conditions are extremely variable but may nearly always be accommodated by engineering design, provided that they are correctly assessed and understood. Engineering geology is now firmly established within the curriculum of geology, applied geology, and civil engineering degree courses.

Engineering geologists have usually been trained in general geology to first degree level before a more specialized second degree. They are depended upon by other members of the industry to apply their geological skills to detailing aspects of the feasibility, design, and construction of projects. Typically, their work may involve a variety of activities:

A 'desk study' that describes the search through records and other literature relevant to the geology of an area. This may entail the study of topographical maps, aerial photographs, local historical archives, geological maps, scientific literature, and information from local authorities and geological societies.

Field reconnaissance to assess the geomorphology and geology of the site. Such a preliminary assessment may outline areas of landslipping, springs, or fault lines and may highlight areas for specific investigation.

A field investigation to observe, describe, and identify rocks and soils exposed in trial pits and trenches or in boreholes. Tests may be undertaken *in situ*, such as water injection tests to assess permeability or bearing capacity tests for foundations. Subsurface geology may also be investigated using geophysical methods.

Undertake a programme of laboratory testing to determine parameters such as moisture content, soil or rock classification, strength, consolidation and compaction behaviour, permeability, soil and rock chemistry.

A continual assessment of the geology exposed during construction, e.g. as a tunnel boring progresses or foundations are dug, or as part of a quality-control operation to ensure that rocks and soils being used as construction materials are suitable.

The measurement or prediction of the movement or dispersion of contamination within the ground where these are controlled by the geology or hydrogeology.

The assessment, during and after construction, of any instrumentation that has been installed, or the monitoring of the groundwater regime or ground movements.

CHRISTOPHER McDONALD

enhanced greenhouse effect The term 'enhanced greenhouse effect' has been used to describe anthropogenic impacts on the Earth's atmosphere which influence the global heat budget. It is generally suggested that these impacts will result in an increase in global temperatures and, as a result, the term 'global warming' has also been used to describe the phenomenon.

The Earth's atmosphere is naturally warmed as a result of its interaction with incoming and outgoing radiation. The Earth receives energy from the Sun, which mainly emits it at the short-wave (ultraviolet) end of the spectrum. This energy is readily transmitted through the atmosphere unaltered to heat the Earth's surface during the day. The Earth's surface, heated by the sun's energy, subsequently re-radiates this heat back into space as long-wave (infrared) energy. Certain components of the atmosphere (water vapour, carbon dioxide, methane, ozone) absorb some of this energy, and, as a result, the temperature of the atmosphere is raised. The energy is subsequently radiated from the atmosphere to the Earth's surface or out into space. After a short while the atmosphere achieves a balance, in which it receives and loses the same amount of heat, the trapped heat fluctuating only in response to variations in the levels of carbon dioxide, water vapour, ozone, and methane. As a result of this process it is estimated that the atmosphere's average temperature is raised from −17 °C to +15 °C.

This natural process of atmospheric warming (and hence the term 'greenhouse effect') has been likened to the way in which glass in greenhouses allows short-wave radiation into the structure but stops long-wave radiation from being transmitted out. The temperature in the greenhouse thus rises. It must, however, be remembered that the glass also prevents the air in a greenhouse from mixing with the colder air outside. Such a barrier is not present in the natural system. Many scientists would thus argue that the greenhouse is an inappropriate model and the term 'greenhouse effect' is consequently a rather poor one. It nevertheless continues to be widely used, particularly by the media and the general public.

Study of air samples trapped in polar ice has indicated that levels of carbon dioxide in the atmosphere have varied in the past. During Quaternary glaciations carbon dioxide levels ranged between 180 and 200 parts per million by volume (ppmv); during the warm interglacials they rose to 275 ppmv. The latter figure is taken as being a good estimate of carbon dioxide concentrations prior to the Industrial Revolution. Since the level of carbon dioxide in the atmosphere will control the amount of heat trapped, it is concluded that atmospheric temperatures will have fluctuated in response to these variations. It is now believed that human activities may be increasing the concentrations of this and other 'greenhouse' gases, thus enhancing the effect.

Most attention has been paid to the human impact on carbon dioxide concentrations, and the possible consequences of such changes on atmospheric temperature. Detailed measurements of carbon dioxide concentrations have been made only since 1957, but levels appear to have risen from around 290 ppmv in the 1880s to 353 ppmv in the 1990s. Much of this rise has been attributed to human activity. For example, the burning of fossil fuels (coal, oil), which are largely carbon-based, results in the generation of carbon dioxide, which is released into the atmosphere. Similarly, the clearing of vegetation usually ends with the material being burnt, which directly increases the amount of carbon dioxide in the atmosphere. Such clearing also reduces the rate of photosynthesis, which extracts carbon from the atmosphere, and it increases the rate of oxidation of carbon from the newly exposed soil. At present, clearing processes are believed to generate between 5 and 20 per cent of the anthropogenic generated carbon dioxide, but in the recent past (agricultural settlement of North America) they may have contributed 50 per cent of anthropogenically produced carbon dioxide. At present it is estimated that human activities put some six billion tonnes of carbon dioxide into the atmosphere each year. Much of this is taken up by natural sinks (oceans, terrestrial ecosystems). The detailed processes involved that operates in such sinks, and the full range of natural sinks are, however, rather poorly understood and it is recognized that less carbon dioxide appears to have remained within the atmosphere than current understanding of the systems would suggest.

Measurements indicate that concentrations of carbon dioxide in the atmosphere have increased rapidly in recent years (by 2 to 4 ppmv per year). If the present rate of increase continues, the concentrations will be double those of pre-industrial levels by the year 2075. It must, however, be recognized that such predictions are uncertain, since they depend on the rate at which fossil fuels and natural ecosystems are exploited, and on the measures taken to reduce such inputs.

Since carbon dioxide is an agent known to promote atmospheric warming, it has been concluded that increased concentrations should result in increased atmospheric temperatures. The extent of potential warming has usually been determined by using computerized General Circulation Models. The carbon dioxide concentration is raised within the model, which is then allowed to run until an equilibrium is reached. Models utilized by the International Panel on Climate Change (IPCC) indicate that a doubling of carbon dioxide concentrations would raise global temperatures by between 1.5 and 4.5 °C, with a best estimate 2.5 °C. These estimates mask wide-scale variations at a regional level, since in high latitudes temperatures are predicted to rise by as much as 8–12 °C in winter but by such less during the summer. Similarly, low-latitude areas are predicted to show only minor increases in temperature (1 °C). The models also make predictions relating to changes in precipitation. For example, much of Europe and the USA will become drier, while Canada becomes wetter in the winter and drier in the summer. Confidence in these predictions is, however, limited.

The predictions from most of the General Circulation Models are based solely upon changes in the carbon dioxide content of the atmosphere, and they thus fail to recognize that the concentrations of other greenhouse gases are also being altered. As a result the accuracy of the models is questionable and it is possible that the greenhouse effect may be enhanced to a greater extent.

Methane is principally produced through the anaerobic decay of organic matter and its concentration within the atmosphere is relatively low at 1.72 ppmv. Methane is, however, a far more effective greenhouse gas, being 21 more times more effective than carbon dioxide. It has been estimated that levels of methane are increasing at about 0.8–1.0 % per year, primarily as a result of agricultural practices. By far the largest source of anthropogenically produced methane are the world's rice paddies, which provide a suitable anaerobic environment when flooded. Other major sources include domestic animals, which produce considerable amounts of methane in their digestive systems, and biomass burning, which releases methane as the land is cleared for cultivation. Methane is also produced by the burning of coal and from the utilization of natural gas, as well as from the decay of organic wastes.

Another natural greenhouse gas is nitrous oxide, which is produced by the denitrification of soils. It has been suggested that levels have been increasing because of increased use of nitrogen fertilizers and from the burning of fossil fuels. The IPCC assessment has, however, indicated that the sources and sinks of this gas are at present very difficult to estimate, and, as a result, its possible contribution to the enhanced greenhouse effect can not be determined accurately.

In recent years, society has introduced new greenhouse gases into the atmosphere, namely CFCs (chlorofluorocarbons) and other halocarbons. These gases are released from aerosols, refrigeration units, insulating foams, and industrial plants. They are now recognized as one of the principal causes for the destruction of the ozone layer. They are, however, also very effective greenhouse gases. For example, when compared to carbon dioxide, the anthropogenically produced gas CFC-11 is some 120 00 times more effective in trapping heat in the atmosphere. The atmospheric concentration of these gases is very low, ranging between 2 and 484 parts per trillion by volume (pptv) but until recently they have have been increasing at a very rapid rate (4–15 per cent per annum). Although society has started to reduce rates of emission, in response to the ozone hole problems, such gases are estimated to have a long residence times in the atmosphere, in some instances up to 400 years. As a result they will continue to contribute to the enhanced greenhouse effect.

It is now generally accepted that concentrations of greenhouse gases have increased in the recent past, mainly as the result of human activities, and they are likely to increase in the future. The impact of such changes on the global heat budget remains uncertain. It is recognized that the atmospheric system is characterized by a wide range of feedback effects which could both enhance and reduce the greenhouse effect. At the same time it is recognized that many of the processes involved in the removal of greenhouse gases from the atmosphere are poorly understood. The General Circulation Models that have been used to estimate the possible impacts of an enhanced greenhouse effect also have their limitations. In the main they are based solely on changes in carbon dioxide and thus ignore the contribution of other gases. At the same time, the models have not effectively incorporated the ocean system, although this is essential if a full understanding of the global heat budget is to be achieved. At present there is no consensus about the possible impact of increased concentrations in greenhouse gases. Most scientists consider that warming will take places, but the likely rates and extent of warming, particularly at a regional level, remain poorly defined.

Some scientists have suggested that global warming has already begun (0.5 °C in the past hundred years). Most, however, have concluded that any changes that have taken place lie within the bounds of natural climate fluctuations, and thus the enhanced greenhouse effect remains unproved. Society has so far done little to reduce the possible effects, and it seems likely that while the levels of uncertainty surrounding this issue remain high, few changes will occur. If the models are correct, the enhanced greenhouse effect should be noticeable within the next twenty years.   CALLUM R. FIRTH

Further reading

Kemp, D. D. (1995) *Global environmental issues: a climatological approach* (2nd edn). Routledge, London.

environmental geology

Environmental geology is the application of geological techniques to analyse and monitor environmental impacts to the Earth. The term was first applied in the 1970s to described the activities of geologists engaged in exploiting commercial reserves of fuel and minerals and in minimizing volcanic, earthquake, and other cataclysmic disturbances to the urban and industrial infrastructure. One current activity of environmental geologists is the study and correction of pollution and waste-disposal problems.

The exploitation of fuels and minerals has always been a major activity of the environmental geologist. Since the energy crisis, a greater emphasis has also been placed on the discovery of new fossil fuel sources by economic geologists. Traditional materials such as precious metals and stones, industrial metals such as iron and copper, and building materials such as stone, sand, and gravel are still in demand. A bigger challenge today is to determine the environmental effects of these activities and to reclaim disturbed areas.

Volcanic and earthquake activity are responsible for disturbing some urban and industrial areas. Where activity of this kind occurs, geologists monitor it and conduct research to improve predictive methods. The geologist's expertise in these areas can be used by the planner, engineer, and architect to construct buildings and communication networks that are resistant to geological hazards.

The study of geology is interdisciplinary. Mathematics, physics, chemistry, and biology have always been keys to the study of geology and have gained even more importance in the application of geology to environmental problems, particularly in the study of the geological aspects of pollution and waste disposal. Geologists contribute to the formulation of environmental policy and the development of environmental regulations to protect geological media and groundwater resources.

Selecting sites for the location of waste-disposal facilities, designing waste-disposal facilities such as landfills and incinerators, and characterizing waste sites are other aspects of environmental geology. For example, to select a site for a waste-disposal facility, the environmental geologist defines geological stability and then identifies the geological characteristics required by the design of the disposal facility. Environmental geologists also characterize hazardous waste sites by determining the extent of contamination in the soils, groundwater, and other environmental media. This information is necessary for the engineer to design the remediation system.   K. O. THOMSEN

environmental toxicology

Environmental toxicology is the qualitative and quantitative study of the adverse or toxic effects of contaminants and other anthropogenic materials on living organisms. Toxic effects may be either lethal, causing death, or sublethal. Sublethal effects include changes in the reproduction, development, growth, pathology, physiology, and behaviour of organisms. The toxicity of a contaminant depends on the physical and chemical properties of the contaminant, the physical, chemical, and biological properties of the ecosystem studied, and the sources and rate of input of the contaminant into the ecosystem.

Environmental toxicologists study the potential of a contaminant to have harmful effects on living organisms. For any particular contaminant, this potential depends on its concentration and the duration of exposure. Toxicity tests are used to evaluate the adverse effects of contaminants on living organisms using standardized, reproducible conditions that enable comparison to be made with other contaminants that have been tested. After any chemical spill, bioassays are performed using the chemical in questions and various species of animals to assess the effects of the spilled material on the organisms. Toxicity depends on both abiotic factors such as pH, temperature, and organic matter, and biotic factors such as the species, life stage, size, and health of the organism. Chemical mixtures may demonstrate a toxicity greater than expected (synergistic) or less than expected (antagonistic) according to the effects of exposure on each individual contaminant in the mixture.

Many toxicological studies were performed after the *Exxon Valdez* oil spill in 1989 to evaluate the effects of crude oil on the organisms living in Prince William Sound (Alaska). As in other environments, the interaction of chemical, physical, and biological components had to be considered there. Polluted environments may be aquatic (freshwater, estuarine, or marine) or terrestrial, and each environment has its own mechanisms for responding to pollution.

Toxicology also entails the study of the distribution, transformation, transport, and fate of contaminants in plants and animals. Studies conducted after spills determine the horizontal and vertical extent of the contaminant and whether the original material has been transformed into other chemicals (metabolites) that might be more lethal. Environmental toxicologists also study the transport of contaminants through water currents, air currents, movement through porous sediments into groundwater, and volatilization. Toxicologists determine the fate of these chemicals, which can include ingestion by animals and eventual movement up the food chain, chemical uptake through the roots of plants that are then eaten by various animals and man, or absorption or adsorption by sediments ingested by bottom-feeding animals.

J. H. BAKER

Eocene Originally coined by Charles Lyell (1833), the term 'Eocene' was at first used for the oldest of the Cenozoic epochs. The subsequent discovery of older Cenozoic rocks (the 'Paléocene' of Wilhelm Schimper, 1874) relegates it to the second place with a time-span of 15 million years. It is suceeded by the Oligocene at 33.7 Ma. It was originally distinguished by Lyell on the basis of its molluscan fauna in the Paris basin, with 3.3 per cent of living species present. Four stages are present, the last being represented by two different facies developments, the Priabonian and Latdorfian.

Plate-tectonic activity saw a northward motion of the continents and the closure of the Tethyan region with the main Alpine orogenic event towards the end of the epoch. Sea-floor spreading also brought about the split between Greenland and Scandinavia. Australia and Antarctica remained joined at the Tasman Rise but were moving northwards. In the seas, the calcareous nannofossils were still in evidence but in shallow waters the benthonic nummulite and other larger foraminifera became very abundant. Mammals became the dominant tetrapods, replacing those of the Paleocene and Mesozoic. By 40 Ma there were advanced whales in the seas. In the lower latitudes, broadleaved forests were widespread; conifers dominated at higher latitudes and the grasses continued to diversify and reach new territory.

Palaeomagnetic activity was considerable, with reversals of polarity for about 50 per cent of the time. The imprint of this behaviour of the geomagnetic field on igneous rocks in particular has been useful in the correlation of ocean floor basalts in many parts of the world.

Eocene formations are important source or reservoir rocks for petroleum or gas in several parts of the world. The tropical seas of the time were highly productive in hydrocarbon materials, and the stratigraphy and structures of areas of continental shelf around Europe and North America have been responsible for many oilfields. The Near and Middle East have also seen the benefit of Eocene formations yielding abundant hydrocarbons.

D. L. DINELEY

Further reading

Pomerol, C. (1982) *The Cenozoic Era: Tertiary and Quaternary.* (Trans. D. and E. E. Humphies.) Ellis Horwood, Chichester.

Pomerol, C. and Premoli-Silva, I. (eds) (1986) *Terminal Eocene events.* Elsevier Science Publishers, Amsterdam.

Schaal, S. and Ziegler, W. (eds) (1992) *Messel: an insight into the history and life of the earth.* Oxford University Press.

epeirogeny Some very large areas of the continental crust, the so-called cratons, are remarkably stable, undergoing relatively slow or only slight tectonic movements without significant folding or fracture. These gentle movements have, in the past, been called *epeirogenic*, in contrast to those *orogenic* areas where more intense tectonic movements take place to produce faults and folds and a generally mountainous topography.

The term was introduced by G. K. Gilbert in the late nineteenth century.

HAROLD G. READING

erosion Erosion is the process that moves material resulting from the breakdown, or weathering, of bedrock. Weathering itself may be distinguished into two broad categories of process: physical and chemical (see *weathering*). Physical weathering results from the mechanical action of environmental agents on bedrock, including wind, moving water, freezing and thawing of water, crystallization and dissolution of salts, expansion of roots, digging by animals, and fire. Glacial erosion is perhaps the most powerful manifestation of the mechanical breakdown of bedrock. Sometimes, physical erosion acts alone, for example during the erosion of barely consolidated bedrock. Chemical weathering entails the chemical breakdown of bedrock minerals by water; it can also operate alone, as when carbonate or evaporite deposits dissolve. For much of the Earth's surface, however, physical and chemical weathering act in concert. When liquid water interacts with siliceous bedrock, chemical weathering breaks down some of the silicate minerals in the bedrock, converting these to solutes plus clays and aluminium and iron sesquioxides.

Measuring rates

Most weathered materials are removed and transported by water and wind. The discharge of material transported by rivers is normally measured at fixed gauging sites. The material discharge divided by the area of the gauged catchment gives the yield. The yield does not equate directly to erosion or weathering rates because of mass storage during transport downslope and through river channels. Instead, yields are said to equate to rates of denudation, a measure of the removal of material, thus generally leading to a reduction of elevation and relief. Possible sites of storage include (1) colluvium or sediment deposited at the base of hillslopes before even enter-

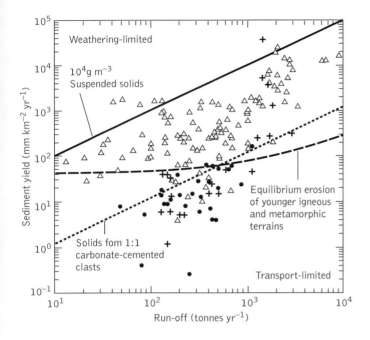

Fig. 1. Sediment yields of rivers that drain to the ocean from a global data set prepared by Milliman and Syvitski (1992). Their classification has been simplified into three classes: rivers with mountainous headwaters (open triangles), rivers with uplands less than 1000 m (plus signs), and rivers draining coastal plains and piedmont (solid circles). Two model trends cross the middle of the graph, roughly separating a region labelled 'weathering-limited' from a region labelled 'transport-limited'. The curved trend represents a model of equilibrium erosion of younger igneous and metamorphic terrains (Stallard 1995b). Presumably, a model developed for erosion of older igneous and metamorphic terrains would fall significantly below this trend (Stallard 1995b). The straight trend is for the weathering of 1:1 carbonate-cemented rocks (Stallard 1995b). This trend would shift up or down depending on the ratio of clasts to carbonate cement. Sediment yields that plot well above these trends may represent either human-induced erosion of weakly consolidated sediment, loose volcanic material, or tectonized bedrock. Low yields reflect non-erosion and soil development, sediment trapping in flat landscapes, or interception of sediment by dams.

ing a channel, (2) alluvium or sediment deposited by fluvial activity within a channel or on adjacent floodplains, and (3) lacustrine sediments deposited in lakes. Various techniques are used to measure local rates of physical erosion, such as erosion pins, erosional troughs, and isotopic techniques. Particles moved by wind do not conveniently pass by a single gauging point, and intermediate storage in dunes and sediment sheets is almost always important. Thus aeolian denudation rates must be synthesized from many local measurements.

In rivers, the concentrations of individual solutes normally decrease with increasing discharge, whereas suspended sediment concentrations and bedload movement normally increase. Consequently, solute transport is dominated by typical flow conditions, and solid transport by discharge events. Good estimates of sediment yield from a watershed, or catchment area, thus require years of measurement. Many rigorous studies have been undertaken to determine sediment yields from large watersheds, including those on sedimentary substrates. The mass of sediment in motion is large, $30–100 \times 10^{15}$ g year^{-1}, with approximately 20×10^{15} g year^{-1} being transferred to the ocean (Fig. 1). Still larger quantities of sediment are probably deposited on land. Storage of solutes on land appears minimal, and about 9×10^{15} g year^{-1} are transferred to the ocean.

Natural erosion in tropical to temperate settings

Under natural conditions, the supply of terrigenous sediments from a watershed is largely controlled by its topographic relief, its bedrock geology, and its Pleistocene climatic history.

The fundamental division is between steepland and flatland terrains. Runoff follows in importance, whereas temperature appears less significant. The erosion of Pleistocene glacial and periglacial deposits continues to affect the solute and solid loads of many rivers. Finally, given the same general rock type, topography, runoff, and temperature, rates of chemical weathering of younger igneous rocks, such as those in island arcs, are roughly twice rates in old cratonic settings.

Feedback among the chemical and mechanical processes that erode developing soils strongly influences both the rate of erosion and the composition of solid and dissolved erosion products. This interaction can be examined in a simple framework, the division of erosional regimes into two types: weathering-limited and transport-limited. Typically, soils develop from the weathered and partially weathered mineral grains, while solutes are transported away by surface and ground waters. Loose soil material can then be eroded by physical processes. For weathering-limited erosion, the supply of loose weathering products is controlled by weathering rate. Soils tend to be thin. Most rock types are more vulnerable than young crystalline igneous bedrock to chemical and physical erosion, and weakly cemented sediments, such as some shales, and unconsolidated glacial deposits, especially loess, contribute exceptional quantities of sediments to rivers. For transport-limited erosion, weathering supplies loose solid materials at rates that exceed the capacity of transport processes. As soil thickens, it develops an internal structure which in turn reduces the rate of interaction between water and fresh minerals, thereby reducing weathering rates. Rates may drop exponentially with soil thickness. Weathering rates can be low in the absence of soil, because water fails to make prolonged contact with fresh minerals. Millions of years may be required to reach equilibrium.

Vegetation has two primary roles in weathering and erosion. First, it reduces the power of physical erosion processes. Plants protect and anchor loose material with their roots, while litter and low ground cover protect soil from raindrop impact and surface wash. This allows thicker soils to develop and stabilizes vulnerable bedrock. Physical erosion usually accelerates after loss of vegetation by fire, cutting, cropping, or grazing. Second, plants produce chemicals that promote the breakdown of bedrock and soil minerals. Some are released through roots, perhaps with the help of symbiotic

root fungi (*mycorrhyzae*). Chemicals include an array of complexing agents, organic acids, and carbonic acid. Other chemicals, such as carbonic and nitric acids, are generated during the decay of organic matter. Experiments indicate substantially greater rates of chemical weathering under well-developed plant cover. Soils with plant cover and active soil organisms appear to have greater infiltration rates of water, perhaps allowing more interaction with bedrock minerals. On slopes where soils tend to be thin owing to weathering-limited erosion, plants probably greatly increase erosion rates because of their chemical effects. In transport-limited settings, with thick soils, the influence of plants may be less; carbon dioxide, however, is mobile throughout the soil profile.

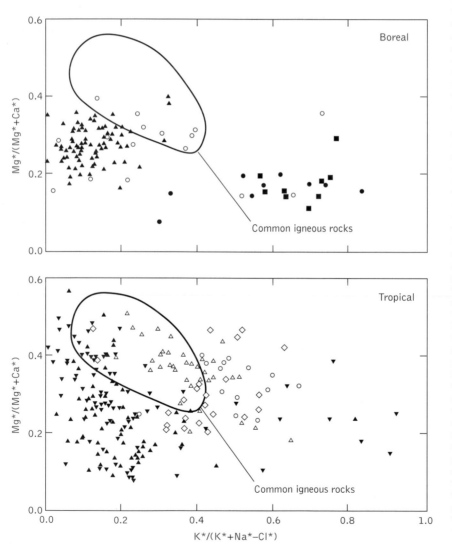

Fig. 2. Representation of major cation fractionation during weathering of different types of landscapes. Asterisks are used to show that sea salt contributions to the cation concentration have been removed; the remainder presumably represents bedrock-derived material. Solid triangles represent mountainous region; upward-pointing triangles represent continental settings, and downward-pointing triangles represent island-arc settings. Solid squares represent a recently glaciated mountainous landscape; solid circles represent subglacial waters. Open triangles represent hilly to mountainous shield terrains. In the upper diagram, open circles represent flat shield and sedimentary landscapes. In the lower diagram, open circles represent flat shields, and open diamonds represent flat sedimentary landscapes. The oval field represents complete leaching of common igneous rocks. In mountainous terrains, rivers appear to have an excess of Ca* over Mg* and of Na* over K*. Hilly to mountainous shield terrains plot in the oval, presumably representing complete leaching and equilibrium erosion of these settings. Some scatter to larger K*/(K* + Na* − Cl*) ratios in the tropical graph reflects the difficulty of making cyclic-salt corrections in rivers near the ocean. In addition, disequilibrium weathering of potassium-bearing minerals in sediments contributes. The scattering to larger K*/(K* + Na* − Cl*) ratios in the boreal plot is characteristic of glaciated and recently glaciated terrains. (Adapted from Stallard (1995a).)

The interaction of vegetation with erosional processes promotes styles of erosion where mass movement is episodic. Two types of cycles seem to be especially important. One is a *landslide (soil–avalanche) cycle* that may be a dominant form of mass wasting in many humid regions. The other is a *fire–flood erosional cycle*. In the landslide cycle, soil is anchored by vegetation. The soil profile eventually thickens to the point where heavy rainfall or an earthquake destabilizes the slope and causes soil avalanches. This normally involves the entire profile above a hard *saprolite* (partially weathered rock). Vegetation re-establishes, and the soil profile begins to thicken, starting the cycle once again. In eastern Puerto Rico, the recurrence interval is roughly 10 000 years. In a fire–flood sequence, vegetation anchors loose soil. After sufficient burnable material accumulates in and upon the soil, a hot fire can occur. Such burns commonly transform soils chemically into a hydrophobic form. Instead of infiltrating, heavy rains run off, thereby eroding the soil and scouring channels. Debris torrents and landslides develop. Vegetation re-establishes and the cycle starts once again. The fire–flood sequence requires a threshold accumulation of soil. Thus, the recurrence intervals of the fire–flood sequence are much shorter, probably by several hundred years. Because these processes involve loose material and act minimally on hard saprolite, they cause a characteristic partitioning of major cations in stream waters. Sodium and calcium are preferentially leached from the hard saprolite, and the sodium-to-potassium ratio in water (after correcting for sea salt) exceeds the bedrock ratio, as does the calcium to magnesium ratio.

For landscapes in which chemical weathering represents the rate-limiting step for physical erosion, the rates of physical erosion (solid yields) can be predicted from chemical erosion (Fig. 2). A hypothetical starting point for analysing physical erosion in a watershed is to assume that the watershed is in geomorphic equilibrium. This is a state in which the statistical measures of a landscape (hypsography, river-channel density, soil types and thicknesses, vegetation types, disturbance types and frequencies) do not change through time. Most importantly, equilibrium requires that there is no net thickening of soils or accumulation of sediments within the landscape. In this context, we can define trends for equilibrium erosion for geologically young, crystalline, igneous bedrock and carbonate-cemented bedrock. When sediment yields for watersheds are compared with run-off, the predicted equilibrium–erosion trends separate high-sediment-yield rivers that drain steeplands from low-sediment-yield rivers that drain flatlands. Thus, these trends approximate to the transition between weathering-limited erosion in steeplands and transport-limited erosion in flatlands.

Sediment yields of flatland watersheds do not seem to depend on watershed area, whereas yields in mountainous watersheds generally decrease with increasing watershed area. The yields of the world's largest rivers are only a few-fold greater than those of flatland watersheds. Several factors are involved in the apparent decrease in yields with increasing

catchment area. In uplands, sediments accumulate as colluvium at the base of hillslopes or in other areas of reduced steepness, or they deposit as alluvium in lower-order streams, without having ever entered a large river channel. Alluvial accumulation is especially significant in flatter regions. Large watersheds tend to have extensive flatland areas. These settings are transport-limited and either fail to contribute sediment through physical erosion or act as depositional areas, where sediments accumulate as alluvium in river channels and adjacent floodways, and as a mix of clastic and organic deposits in lakes and wetlands.

Glacial and cold climate weathering and erosion

Glacial erosion is the most powerful form of physical erosion. Perhaps hundreds of metres of rock were removed during the Pleistocene by continental ice sheets from the Laurentide region and smaller ice sheets in Asia and Europe. Glaciers have also deeply sculpted mountainous areas worldwide. Ice caps still cover much of Antarctica and Greenland. The intensity of glacial erosion is much greater when liquid water is present at the ice base, a condition referred to as a *warm-based glacier*. Glaciers lacking liquid water near the base, *cold-based glaciers*, sometimes erode rather little. For some regions with extensive Pleistocene glacial sediments, sediment yields of watersheds increase with increasing area, an indication that Pleistocene sediments have been eroded out of the smaller basins but are still moving downstream in larger watersheds.

Chemical weathering in cold, wet environments contrasts in notable ways from that of warmer settings. Glacierization and freeze-thaw processes fracture bedrock, exposing fresh mineral surfaces. In effect, the bedrock-to-water ratio in these settings is much greater than for warmer settings. Because of this, the weathering of less abundant but easily weathered minerals in siliceous bedrock, especially carbonates and sulphides, contributes disproportionally to solutes in surface waters. Subglacial waters are especially rich in potassium relative to sodium, perhaps because the physical breakdown of micas releases large quantities of potassium.

Human effects

Much of the Earth's land surface is now occupied by people and has been cleared for timber, agriculture, and urbanization. Although land clearing increases rates of physical erosion, the resulting changes in rates of physical erosion have, in general, not been well documented. Agriculture, civil engineering, and mining mobilize vast quantities of soils, unconsolidated sediment, and bedrock. Engineering activities, such as road building and building construction, may, in concert, move more solids over the Earth's surface than all ordinary natural processes combined. The hydraulic architecture of the land has been transformed to serve agriculture, water supplies, land reclamation, navigation, and power generation. Few rivers lack dams and reservoirs; large rivers are confined to their channels by levees; normal floodplains are converted to paddylands and aquiculture; water is mined;

irrigation raises soil moisture and creates artificial distributary systems; and so forth. These changes confound the quantification of sediment dispersal from the uplands to the ocean.

Human activities have accelerated physical erosion over equilibrium levels. In flat coastal plains and piedmonts, yields appear to increase five- to ten-fold with agriculture, whereas in loess terrains 100-fold increases are typical. The greatest yields are in the smallest watersheds. In tiny watersheds in the upper Mississippi River valley, yields approaching $10\ 000\ \mathrm{g\ m^{-2}\ year^{-1}}$ have been reported.

Research directions

Much of the process-level research relating biogeochemical processes to erosion is routinely undertaken in small research watersheds. Most of these are on igneous and metamorphic rock to minimize groundwater losses. In contrast, most of the world's chemical and physical erosion is from sedimentary rocks, and these are mostly shales. Moreover, sampling for most small watersheds has been from periodic (interval-based) sampling, thereby ignoring a major characteristic of such watersheds, that sediment moves during discharge events. Finally, most research watersheds are not large enough to examine the effects of sediment storage during transport. Thus, many opportunities remain for well-designed process-level studies of weathering and erosion.

Perhaps one of the most fascinating research directions in the study of local erosion is the analysis of cosmogenic radionuclides produced *in situ*. These are created when cosmic-ray neutrons and muons strike atoms within mineral grains in a soil profile. Some of the important isotopes are ^{10}Be, ^{26}Al, and ^{14}C. The concentration of these isotopes depends on erosion rate and irradiation intensity. The latter depends on altitude and latitude and depth below the ground surface. Two situations are easily modelled. One is the erosion event, where it is assumed that a surface was formed at some time in the past, and that there was no subsequent erosion. Isotope concentrations are used to date this event. Settings locally dominated by transport-limited erosion would be suited to this approach. The second situation is to assume that erosion is steady, as might be expected in an erosion-limited setting. The ground-surface isotope concentration equates, for a broad range of erosion rates, to the erosion rate. These methods average erosion over thousands of years.

R. F. STALLARD

Further reading

Milliman, J. D. and Syvitski, J. P. M. (1992) Geomorphic tectonic control of sediment discharge to the ocean—the importance of small mountainous rivers. *Journal of Geology*, **100**, 525–44.

Stallard, R. F. (1995a) Relating chemical and physical erosion. *Reviews in Mineralogy*, **31**, 543–64.

Stallard, R. F. (1995b) Tectonic, environmental, and human aspects of weathering and erosion—a global review using a steady-state perspective. *Annual Review of Earth and Planetary Sciences*, **23**, 11–39.

erosion surfaces Erosion (planation) surfaces are cut by erosion across varied geological structures and rock types, and have the general form of plains. W. M. Davis proposed that areas erode to an erosion surface, or *peneplain*, formed by slope decline. Penck and King proposed that slopes evolve by parallel retreat to form pediments that coalesce to make *pediplains*. These workers thought that landscapes consist of a series of peneplains or pediplains. Indeed, until the 1950s geomorphology was largely concerned with erosion surfaces. Geomorphologists explained landscapes in terms of cycles of erosion and successions of erosion surfaces or partial erosion surfaces. Other terms are *etchplains* and *panplains*. A non-genetic word is *palaeoplain*.

In the 1960s, work on erosion surfaces was strongly attacked in favour of process studies and systems theory. Some workers even doubted that erosion surfaces existed. Nevertheless, planation surfaces are real. Their existence is evident in many geological sections where the amplitude of topography is trivial compared with the size of folds, faults, and intrusions in the bedrock. They are also clear on the ground, and on continents like Australia and Africa they dominate the landscape.

It is difficult to explain how plains really formed. Moreover, there are many parts of the Earth's surface where only one palaeoplain can definitely be recognized. As C. D. Ollier has pointed out, palaeoplain development takes place on the same timescales as plate tectonics and biological evolution; so in most parts of the Earth there has probably not been enough time for more than one or two erosion surfaces to develop. In south-eastern Australia, for example, the palaeoplain first described by E. S. Hills is preserved along much of the Great Divide, and is probably Mesozoic in age. In other places, for example western South America, there are a number of erosion surfaces, probably because of the greater rates of uplift there.

Erosion surfaces are sometimes dissected into remnants. Observation of these remnants, and early enthusiasm for erosion surfaces, probably caused many phantom surfaces to be reported, although bevelled cuestas are real. At its extreme, accordant summits may be taken as evidence of a former erosion surface (*Gipfelflur*), but how far such interpretations can be carried is problematic.

Many unconformities in the geological record separate older, folded rocks from younger, nearly horizontal rocks. These unconformities are buried surfaces, some formed by terrestrial, others by marine, planation. Where the younger rocks have been eroded away, the exposed unconformity becomes an exhumed erosion surface. In the Kimberley region of Western Australia the removal of a thin cover sequence resulted in the modern land surface corresponding very closely to the Precambrian erosion surface.

Major rivers developed on erosion surfaces show remarkable persistence in the landscape. In Uganda and eastern Australia some rivers on the palaeoplain pre-date continental rifting. In Australia this is true even of rivers east of the Great

Divide that were reversed and now flow into the Tasman Sea instead of to the north-west. COLIN F. PAIN

Further reading
Ollier, C. D. (1991) *Ancient landforms*. Belhaven Press, Cambridge, Mass.

erosion, wind *see* WIND EROSION

escape tectonics
The term 'escape tectonics' refers to the lateral extrusion of fault-bounded geological blocks as a result of compression. This extrusion can occur on all scales, ranging from small-scale faults, with only centimetres or metres of displacement, to large-scale crustal faults, with hundreds of kilometres of displacement, such as the Red River Fault in Asia (Fig. 1). The classic example of escape tectonics

has long been held to be the eastward extrusion of parts of south-east Asia as a response to the collision and northward motion of India into Asia. Initial collision of India with Asia occurred about 50 million years ago; geophysical evidence, derived from the pattern of magnetic stripes in the Indian Ocean, indicates that the northward motion of India, with respect to Asia, slowed from about 100 mm a year to about 50 mm a year at this time. This lower rate continues to the present day. As a result, almost 2500 km of convergence between the two has occurred since collision, shortening and thickening the crust to produce the Himalayan mountain chain and the Tibetan Plateau.

Estimates of the amount of crustal mass beneath the Tibetan Plateau are, however, about 30 per cent too little to account for all the shortening that must have taken place. In addition, many of the large earthquakes that affect the region

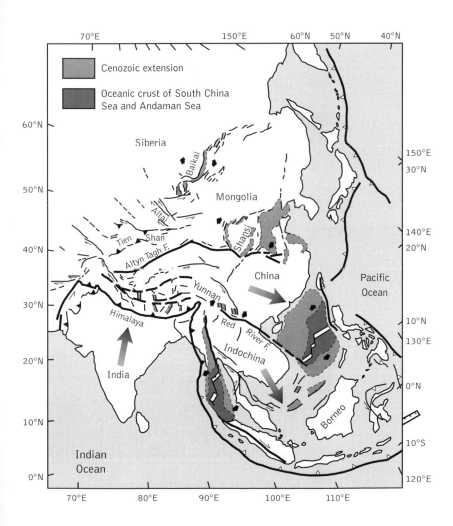

Fig. 1. Large-scale escape tectonics in south-east Asia. The northward motion and collision of India with Asia is proposed to have driven the lateral extrusion of parts of Indochina and China. The lateral escape of this large crustal block may also be responsible for the opening of the South China and Andaman seas. (Modified from Tapponier *et al.* (1982).)

319

occur, not at the boundary between the Indian and Asian plates, but within the Asian plate. In the 1970s a number of workers drew on these lines of evidence, and on the recognition of large-scale strike-slip faults (hundreds of kilometres in length) in south-east Asia from then newly available satellite imagery. As a result, they proposed that the balance of the missing material from the India–Asia collision must have been extruded or 'escaped' eastwards (Fig. 1). The escape of this material is thought to have been accommodated by movements on the large strike-slip faults that cut across south-east Asia, and is thought to be responsible for much of the Cenozoic deformation of the region. Supporting evidence for this extrusion model was obtained from a series of laboratory experiments using a rigid indentor (representing India) and plasticine (representing Asia) which produced a series of 'slip lines' of material motion, generated by the indentor penetrating into the Asian plate, in a pattern strikingly similar to the pattern of faults observed in south-east Asia.

More recent work, however, has raised some concerns about the validity of applying the concept of 'escape tectonics' to the large-scale Cenozoic evolution of south-east Asia. The extrusion of south-east Asia requires that the continental fragments behave and deform as rigid (or semi-rigid) plates. While such an assumption may be valid over short timescales, over the longer timescale of the India–Asia collision and the Cenozoic deformation of south-east Asia continents are thought to behave in a more viscous manner and 'flow' rather than act as rigid blocks. Computer models of India–Asia collision, which treat Asia as being viscous rather than rigid, indicate that there is very little effect on the tectonics of the region beyond the length of the collision zone. Studies of slip rates on active faults indicate that continental blocks within Tibet are indeed moving eastward at rates of some 10–30 mm per year, as predicted by the extrusion model. These measurements, however, provide only a snapshot of the rates at a particular time, and few of the measured faults have been continuously active since the time of collision. The cumulative motions measured on the larger faults in the area, from the offsets of key well-dated rock units, while large, fall short of the total predicted by a model of 'pure' escape tectonics.

A definitive test of such large-scale models of escape tectonics is not currently possible, given that much more information needs to be gathered on the motion histories of the large crustal-scale faults. It can be stated, however, that some extrusion does indeed appear to occur. It may well be that escape tectonics on a short timescale may be part of a longer-term and larger-scale process of viscous flow of continental material.

CONALL MAC NIOCAILL

Further reading

Royden, L. H. and Burchfiel, B. C. (1997) The Tibetan Plateau and surrounding regions. In van der Pluijm, B. A. and Marshak, S. (eds) *Earth structure: an introduction to structural geology and tectonics*, pp. 416–23. WCB/McGraw-Hill, New York.

Tapponier, P., Peltzer, G., Le Bain, A. Y., Armijo, R., and Cobbold, P. (1982) Propagating extrusion tectonics in Asia. New insight from simple experiments with plasticine. *Geology*, 10, 611–16.

escarpment An escarpment is a long, relatively continuous line of cliffs or steep slopes that divides the landscape into two level or gently sloping surfaces. Of French origin, the term 'escarpment' is often used synonymously with 'scarp', although the latter is more often used to refer to smaller-scale features and/or to features of tectonic origin.

Escarpments can originate through faulting, differential erosion, or by a combination of the two. A 'fault scarp' is formed by vertical motion along a fault, where one block of land is displaced upward relative to an adjacent block (see *fault scarp*). Fault scarps are common along the edges of fault-block mountains, such as those in the western United States and the eastern African rift valleys, where valley floors are sometimes downfaulted by as much 2.5 km. A 'fault-line scarp' is an escarpment formed where ancient faulting has juxtaposed rocks of contrasting resistance to erosion.

Escarpments can also form by differential erosion, where erosion or faulting has exposed an erodible rock layer lying beneath a more resistant one. The resistant rock layer (typically sandstone, limestone, or basalt) will often act as a 'caprock', supporting the top of a cliff face. An escarpment thus formed can wear back over time, in a process known as 'escarpment retreat', as the less-resistant materials (typically shale or clay) are weathered and eroded from beneath the caprock, causing its collapse. Such caprock-supported escarpments are common in arid regions containing layered sedimentary rocks. In the Colorado Plateau region of the western United States, for example, escarpment retreat is responsible for forming the dramatic 'butte-and-mesa' topography.

Fig. 1. Part of the Tuwayq escarpment in central Arabia. Jurassic limestones form the cap to this escarpment and overlie shales. (Photograph by P. L. Hancock.)

Long, continuous escarpments formed by undercutting of a resistant limestone layer can be found in the deserts of Arabia and the Gulf of Suez region in Egypt (Fig. 1). Low rates of weathering in these arid settings promote escarpment retreat by enhancing the contrast between different rock types. However, caprock-supported escarpments can also form in humid regions; the Niagara escarpment, over which the eponymous Falls spill, is one example. In some cases, escarpments occur where there is no obvious lithologic or tectonic origin. One such example is the Blue Ridge escarpment of western Virginia and North Carolina in the eastern United States, which is sculpted from apparently homogeneous metamorphic rocks.

Escarpment retreat has long been an important problem in geomorphology. The geologist L. C. King believed that escarpment retreat was an ubiquitous erosional process on Earth. As an alternative to William Morris Davis's 'geographical cycle', King proposed a model of landscape evolution based on the concept of cyclic episodes of escarpment backwearing. King's 'pediplanation' model retained Davis's view that landscape evolution is characterized by brief periods of rapid tectonic uplift separated by long intervals of erosion. But where Davis considered landscape erosion to occur by a process of more or less uniform relief reduction, leading ultimately to a nearly level surface (or 'peneplain'), King instead proposed that erosion occurs primarily through escarpment retreat, which would leave behind 'pediplains' formed at the foot of a retreating escarpment. Each episode of uplift would lead to the creation of new escarpments along the shoreline or fault lines, which would then wear back to form a new set of pediplains sloping gently toward the coast. Multiple episodes of uplift and erosion over millions of years would create a staircase-like series of escarpments and pediplains, each of a different age.

King's theory was certainly influenced by the landscape of his homeland, South Africa. One of the largest escarpments on Earth is the Great Escarpment of southern Africa, which rings the sub-continent 100 or more kilometres from the coastline, separating the interior Kalahari Plateau from the coastal plains. The highest elevations in southern Africa are found along or near the rim of the Great Escarpment, which rises to over 3 km in elevation in the Natal region. Escarpments similar to the Great Escarpment occur along the edges of many continents, including eastern Brazil, western India (the Western Ghats), and eastern Australia. Many geomorphologists now believe that these 'great escarpments' originated along newly formed continental margins during the early break-up of the supercontinent Pangaea, and have since been eroded back to their present inland positions (sometimes hundreds of kilometres from the original rift zone near the coast). It has been speculated that the long-term survival of these 'great escarpments' may be due in part to coastal isostatic rebound as material is eroded from the escarpment face.

Even more impressive escarpments can be found on the planet Mars, where erosional processes have presumably been much less vigorous than those on Earth. The Valles Marineris, a vast, steep-walled, equatorial canyon that dwarfs Earth's Grand Canyon, is bounded by escarpments up to 6 km high.

GREGORY E. TUCKER

Eskola, Pentti Eelis (1883–1964)

It is to Pentti Eskola that we owe the concept of metamorphic facies. He was born in south-west Finland, the second child of a farming family, and his scientific career was spent almost entirely within the University of Helsinki. After graduating in chemistry there he went on to a doctorate at Freiburg. His early research was on reactions in the solid state, and he moved to the Department of Geology at Helsinki University in order to apply chemical methods to the study of minerals and rocks. He was Professor of Geology and Mineralogy at Helsinki from 1924 until he retired in 1953.

Eskola's geological research spanned a range of topics in petrology and mineralogy, including eclogites, the synthesis of minerals at high pressures, silicate equilibria, the origin of granites, mantled gneiss domes, metamorphic differentiation, and the Precambrian rocks of Finland. His classic work on metamorphic facies, which was based on a comparison of mineral assemblages in Norway and Finland, appeared in 1920 after he had spent a year in Oslo with the Norwegian geochemist W. M. Goldschmidt. In 1921 he also worked in the Geophysical Laboratory in Washington, and in 1922 he did fieldwork with the Geological Survey of Canada. He received many honours and awards during his lifetime, and on his death was given a state funeral.

D. L. DINELEY

estuaries

An estuary is a transitional zone between a river and the sea as represented by a shallow marine shelf or platform. Dominated by tidal currents and by interactions between sediment-laden salt and river waters, it is a generally narrowing, elongated inlet which reaches across a coastal alluvial plain or inward along a river valley as far as the upper limit of tidal rise. Estuaries experience strong spatio-temporal gradients in many important environmental factors, and consequently are affected by a rich variety of processes while displaying an exceptional range of morphological, sedimentological, and biological responses. Although among the most productive of biological systems, estuaries are for these same reasons stressful environments for organisms. Under appropriate conditions—a rising sea level seems particularly favourable— estuarine deposits have become preserved in the rock record, where they may now retain fossil fuels. Over the last few centuries in particular, estuaries have provided the people of many of the world's cities and towns with their nearest experience of a natural habitat. Yet the quality and functioning of estuaries can be irreparably damaged by poorly judged land-claim, ill-regulated industrial development, and inadequate waste management. Contaminants and pollutants become concentrated and trapped in estuaries, because of the natural processes operating there. Estuarine salt marshes form part of the natural system of flood defence, but require careful management if they are to remain effective.

Setting

The character of an estuary as a coastal landscape feature largely depends on the more recent geological history of its surroundings. Today's estuaries occupy landscapes influenced by major oscillations of global sea level, as well as by slow vertical crustal movements of tectonic origin. At higher latitudes, glacio-hydro-isostatic adjustments and direct glacial action also played their parts. All estuaries as we see them today represent a lengthy integration of the actions of numerous processes.

As global sea level most recently moved upward, many estuaries occupy drowned river valleys, some narrow and bold but others hardly noticeable. Good examples of boldly framed estuaries are provided by the Columbia River, on the tectonically active North American West Coast, and by the Severn in south-west Britain and the Gironde in France. By contrast, estuaries on the North American East Coast fill many-branched drowned valleys of subdued relief. Here, modern sedimentation is adding to a thick, offlapping sequence of later Mesozoic and Cenozoic coastal-plain and shallow-marine deposits. Estuaries in drowned valleys are to varying degrees constrained or 'rock-bound'. Because of resistant hills and valley sides, their shape in plan departs to some degree from the funnel-like form otherwise developed, and there is little room for the normal marginal environments.

Estuaries of alluvial type, unconstrained by bedrock features, arise where sediment is sufficiently abundant for wide valley-flats and coastal plains to have formed. Here, the inner estuary is a sequence of tidal meanders, cuspate giving way seaward to sinuous, which lead to an outer part with the shape of an approximately exponentially expanding funnel, as in the Alligator Rivers of the Northern Territory, Australia, and the Ord River estuary, Western Australia. Each of these presents the full set of estuarine environments permitted by the prevailing tidal and climatic regimes.

Hydraulics

That an estuary is a place where river and sea water mix is the main determinant of its morphological, sedimentary, and biological features. The mixing is neither uniform nor steady. It is non-uniform because the ratio of salt to fresh water increases as the estuary grows in cross-section seaward. The mixing is unsteady, because of variations in the height and strength of the tide, especially on semidiurnal, diurnal, and spring–neap scales, and because of seasonal and storm-related changes of river discharge. Thus different mixing states can exist in an estuary at any one moment over space and at different times at the same place. Averaged tidally, however, estuaries are of three mixing types (Fig. 1): salt-wedged, partially mixed, and well-mixed.

Salt-wedge estuaries typically are microtidal, the characteristic tidal range being below 2 m. In a salt-wedge estuary, the river water, because of its lower density, rides over a wedge of salt water which moves short distances up and down the estuary in response to the weak flood and ebb of the tide. The stream of river water erodes the sharp top of the salt wedge or *halocline*, and there is a consequent upward mixing into it. Replacement sea water enters the estuary as a weak landward current. Normally, relatively fresh water spreads far and wide over the sea from the mouth of a salt-wedge estuary, with off-shore waves accomplishing the final mixing.

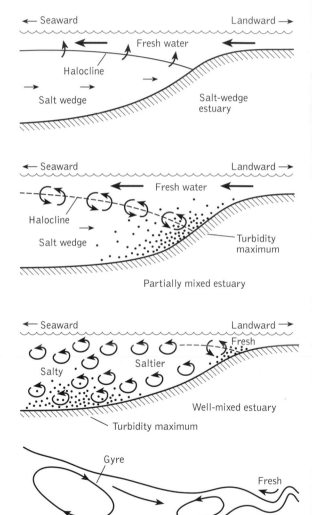

Fig. 1. Schematic tidally averaged water circulations in salt-wedge, partially mixed, and well-mixed estuaries. The well-mixed estuary is shown in plan as well as longitudinal profile.

A partially mixed estuary normally has a mesotidal regime, typified by a range of 2–4 m. The tidal currents are now quite strong and the whole water mass in the estuary moves first landward and then seaward with a tidal periodicity. Because of strong shear the currents are very turbulent. The halocline is now a broad vertical zone through which salt water mixes upward as fresh water mixes downward. The tidally averaged inflow of salt water to replace what is mixed out of the bottom layer is relatively vigorous and there is a correspondingly much enhanced discharge of brackish water through the surface layer. A seaward increase in the salinity of both the surface and bottom layers is assured by this mechanism of mixing.

A tidal range in excess of 4 m gives an estuary a macrotidal or hypertidal regime and invariably makes it well-mixed. The vigorous tidal currents are sufficient to preclude any vertical salinity stratification, creating an effectively vertically homogeneous water mass. Only a gradual longitudinal salinity gradient can be sustained. Some lateral variations in salinity and water velocity arise, however, and horizontal circulations tend to develop on a tidal scale. There can be a landward drift of water along one bank but a seaward drift along the other; an axial inward drift balanced by an outward drift along each bank; or in the largest estuaries a complex set of large horizontal gyres.

Sediment supply and dispersal

Estuaries receive sediment from many sources. Some bed-material is supplied by rivers, but tracer studies show that a part, and in some cases nearly all, of the sand originates offshore. Although the river catchments are the primary providers of most suspended fine sediment, which becomes flocculated in the saline environment, in many estuaries there is a vigorous recycling of mud because of alternating erosion and deposition at the bed and banks. Well-mixed estuaries are particularly turbid. The suspended mud in the hypertidal Severn Estuary, for example, in mass equals many times the annual fluvial supply. Some mud reaches estuaries from seaward, having been scoured from neighbouring coastal cliffs or the sea bed. Cliffs within rock-bound estuaries can be local sources of coarse debris, which can form beaches.

A complex interaction between the tidal flow and sediment erosion–deposition creates in partially and well-mixed estuaries a *turbidity maximum* (Fig. 1). Here the greatest concentration of suspended mud can be one to two orders of magnitude more than in most of the estuary. Because the turbidity maximum is a dynamic phenomenon, it varies in strength and position on tidal periodicities and also with river discharge, as research in the Seine Estuary, France, has demonstrated.

Sedimentary processes, environments, and facies

Sediment erosion and deposition roughly balance on a tidal scale in estuaries, implying that the tidal streams have much the same speed everywhere. This ensures that estuaries, which are blind-ended, must progressively widen and deepen seaward in order for the tidal flows to meet the requirements of hydraulic continuity. Because the 'tidal wave' is reflected in an estuary, on account of the blind end, high and low tide are times of slack or sluggish water, whereas the tidal streams peak in speed at about mid-tide. The vertical distribution of sedimentary environments and facies in an estuary is controlled by this pattern of currents.

Suspended mud settles out from the low-velocity waters associated with high tide, creating tidal marshes and mudflats traversed by dendritic to net-like creek-systems in the upper part of the intertidal zone on the margins of estuaries. Tidal marshes trap silt brought in by the tides high enough to drown them, but also obtain sediment, which may be preserved as peat, from the salt-tolerant plants (halophytes) growing there. The salt marshes of high and middle latitudes, populated by herbaceous and some shrubby plants, are replaced in the tropics by *mangals*, where shrubs and long-lived trees, particularly species of mangrove, predominate. Tidal marshes support dense population of invertebrates. The lower limit of halophyte growth, about the level of neap high tides, is controlled by predation on seeds and seedlings as well as by the salt tolerance of adult plants. Mudflats lie lower in the upper intertidal zone than marshes and are smooth, almost level surfaces across which tidal creeks meander. Lacking macroscopic plants, but with many algae, they support huge, low-diversity populations of filter- and surface-feeding invertebrates, preyed on particularly by birds.

Sandy or gravelly beaches replace tidal marshes and mudflats in the upper intertidal zone in the outer parts of large estuaries and at the foot of bedrock cliffs, wherever the fetch at high tide and exposure can allow powerful wind-waves. On exposed coasts, the wind can blow sand inland from the beaches to form belts of supratidal aeolian dunes, so long as a sufficiently protective vegetation cover is unable to form. Together with salt marshes, these dunes also contribute to the natural flood defences of estuaries, protecting low-lying alluvial areas to landward. Strong mid-tide flows create unstable sand shoals, sand flats, and channels in the subtidal and low intertidal zones of estuaries, for example, the Severn Estuary and Bay of Fundy. Tidal ripples and dunes decorate these shoals, which internally are mainly cross-stratified, with some parallel laminations formed by peak currents. Few organisms can tolerate these shifting substrates, but transported shells and waterlogged wood and plant debris are common.

Slack water occurs again around low tide. Not only do gravels arise subtidally, but mud can accumulate permanently at these depths, especially where the turbidity is high. Associated bodies of *flud mud*—partly settled but not truly deposited—feature in the subtidal zones of many partially and well-mixed estuaries.

JOHN ALLEN

Further reading

Adam, P. (1990) *Saltmarsh ecology*. Cambridge University Press.

Allen, J. R. L. (1993) An introduction to estuarine lithosomes and their controls. *Sedimentology Review*, **1**, 123–8.

Dalrymple, R. W., Zaitlin, B. A., and Boyd, D. (1992) Estuarine facies models: conceptual basis and stratigraphic implications. *Journal of Sedimentary Petrology*, **62**, 1030–46.

Dyer, K. R. (1994) Estuarine sediment transport and deposition. In (Pye, K. (ed.) *Sediment transport and depositional processes*, pp. 193–218. Blackwell Scientific Publications, Oxford.

McLusky, D. S. (1981) *The estuarine ecosystem*. Blackie, Glasgow.

evaporation Evaporation is the process by which a liquid is transformed into a gas. The most common example in the natural world is the evaporation of water from the oceans, lakes, and rivers into the atmosphere. The main difference between a liquid and a gas is that in a gas the molecules are free to move anywhere. Each molecule moves until it collides with another molecule, after which it moves off in another direction. In a liquid the forces of attraction between the molecules bind them together. The binding is not as strong as in a solid, such as ice, but it is strong enough to keep almost all the molecules together as a liquid. The molecules do not all have the same energy; some move faster than others.

A few of the molecules in a liquid have sufficient energy to escape from the binding forces that keep them joined to the other liquid molecules. When these molecules leave the liquid surface they have entered the gaseous state, and are then free to move anywhere. On the other hand, some molecules in the gas, while moving about randomly, will collide with a liquid surface. When this happens, they will become bound to the liquid by the attractive forces, and these molecules will have changed from the gaseous to the liquid state. The net effect of these two processes is evaporation if more molecules are leaving the liquid state to become gas than the other way round. If more molecules are leaving the gas and becoming liquid then the process is *condensation*.

A state of *saturation* is reached when the molecules are leaving the liquid state at the same rate as molecules are joining the liquid from the gas. Once the gas has become saturated, the only way to increase the number of molecules leaving the liquid is to supply more energy to increase the energy of the molecules. This can be done by heating the liquid, producing thermal energy. (Heat energy which is used to provide energy for evaporation rather than to increase the temperature is called the *latent heat* of evaporation.) After heating, a new equilibrium level can be reached in which the gas is saturated; that is, the rates at which molecules leave and rejoin the liquid are the same. However, the number of molecules in the gaseous state is now greater than before heating because they have more energy and more molecules have been able to escape the attractive bonds of the liquid. This shows that the state of saturation is temperature-dependent. At higher temperatures gases are capable of supporting more of the molecules from the liquid. This explains why condensation occurs, for example, on windows at night, as the temperature drops. Air, which initially may not have been saturated with water, cools and eventually reaches a temperature at which it is saturated with water. If the air cools below this point it is supersaturated with water, and condensation occurs to restore the balance.

Evaporation and condensation are major factors in the hydrological cycle. Although only about 0.001 per cent of all the water on Earth is in the atmosphere, there is a never-ending cycle of evaporation from the surface, condensation into cloud droplets, and eventually the fall of raindrops and snowflakes returning the water to the surface. Averaged over the entire globe, the annual amount of precipitation would be about a metre deep. By comparison, the total depth of all the atmospheric water vapour at any one time is equivalent to a liquid depth of about 25 mm. To produce the required amount of rain in one year, water must be recycled round the hydrological cycle approximately every nine days.

CHARLES N. DUNCAN

Further reading
Ahrens, C. D. (1994) *Meteorology today*. West Publishing Co., New York.

McIlveen, J. F. R. (1986) *Basic meteorology*. Van Nostrand Reinhold, New York.

evaporites Evaporite deposits are sediments of economic importance that are formed by chemical precipitation from saturated marine and continental waters. The main minerals of evaporites are halite (sodium chloride, $NaCl$), anhydrite (calcium sulphate ($CaSO_4$), and gypsum (hydrated calcium sulphate, $CaSO_4$. $2H_2O$). Today, evaporites mainly form in mid-latitudes, but they also develop in other arid areas. Evaporites are thus important indicators of palaeolatitude. Because of their solubility, they are uncommon at the Earth's surface, but they underlie large areas of the continents and marginal marine basins.

Complete evaporation of a depth of 1000 m of sea water would produce about 15 m of evaporites. The great thicknesses of evaporites found in many ancient basins could thus have been formed only by repeated incursions of sea water into basins. Figure 1 shows one model for these conditions based on present-day observations of halite forming in the Gulf of Kara Bogaz in the Caspian Sea. The cyclical nature of evaporite sequences supports this conclusion. Theoretically, evaporation of sea water should produce a regular sequence: calcium and magnesium carbonates (*calcite*, $CaCO_3$ and *dolomite*, $CaMg (CO_3)_2$; calcium sulphate (as *gypsum* and *anhydrite*); sodium chloride (*halite*); and potassium and magnesium salts (*kainite*, $KMgSO_4$ Cl. $3H_2O$; *carnallite*, $KMgCl_3$. $6H_2O$; *sylvite*, KCl; and *kieserite*, $MgSO_4$. H_2O) respectively. Although some basins contain complete evaporite sequences, many ancient sequences display anomalous amounts of calcite and dolomite with gypsum and anhydrite. This suggests that the evaporating brine underwent only partial desiccation, and after precipitating the more insoluble salts sank to the basin floor and returned seaward.

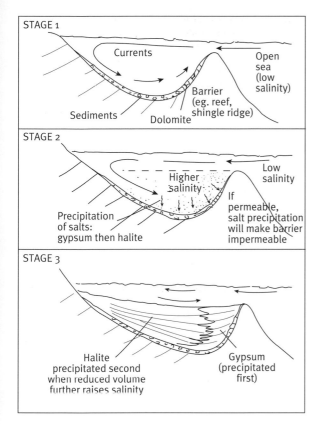

Fig. 1. Model for evaporite formation in a barred basin. This model is based on observations in the Gulf of Kara Bogaz in the Caspian Sea, where halite is forming at the present day.

There has been much controversy about the depth of water in which evaporite sequences in the stratigraphical record were formed. Because many of them are finely laminated, it was earlier assumed that they had formed in deep water. However, most modern evaporites form in shallow-water marine lagoons or on wide coastal salt flats (*sabkhas*), and this prompted some geologists to argue that all evaporites were of shallow-water origin. Detailed studies of ancient evaporites reveal that they show the same range of sedimentary structures and environments as more normal marine sediments: subaerially exposed supratidal and intertidal flats; current- and wave-dominated shallow marine areas; and deep marine areas with laminated sediments and turbidites. Precise comparison with modern basins is difficult because the density of the brines in evaporitic basins would modify the effects of currents and waves, and the high salinity would prevent colonization by bottom fauna. Hence, laminated sediments may have been preserved in shallower water than is normal today.

Continental evaporites are less common than marine evaporites. They form today in arid, land-locked interior basins by evaporation of water flowing from surrounding rocks, and in some instances from hydrothermal springs. The waters are more variable in composition than marine waters and precipitate a greater variety of minerals.

Evaporites are readily dissolved and may therefore migrate and recrystallize in other beds. The cavities left behind may collapse to form breccias or may be replaced by other minerals (usually calcite). This process can be recognized by the shape of the cavity or the growth structures of the replacement mineral or minerals.

Gypsum changes to anhydrite ($CaSO_4$), according to salinity and temperature. The calcium sulphate may thus be in either form, depending on the geological conditions and history. As more water is evaporated in an arid sabkha, the large gypsum crystals are replaced by finer anhydrite within the original shape of the crystal. If this continually occurs, large nodules of anhydrite develop and the sediment between them is squeezed into thin stringers, giving a 'chicken-wire' appearance that is commonly seen in evaporite rock sequences. In sabkhas gypsum surrounded by sediment grains forms 'desert roses'. A variety of large twinned crystal forms of gypsum grow on the floors of evaporating lagoons and lakes including fans and needles. In contrast, beds of very fine gypsum and anhydrite, usually laminated, form in deeper waters and may be subject to reworking, as happens to other sediments.

Halite also exhibits a variety of forms, depending on the environment. If it is formed in evaporated lakes, it can show desiccation structures such as cracks, polygons, and tepees (fold-like structures resembling tepee tents). Trace levels of bromine substituted for the chlorine in halite can indicate the source and environmental conditions during formation. For example, non-marine halite is low in bromine. Marine halite with increasing bromine reflects a salinity increase and indicates little influx of sea water and no loss of brine, whereas a decrease in bromine content indicates that there has been a significant influx of sea water or that the magnesium salts (with high levels of bromine) have been removed. Because of their buoyancy, halite beds may deform or rise and drag surrounding rocks, producing salt domes (diapirs) or act as fill or lubrication in geological faults.

Economic importance

Evaporites are important raw materials. Gypsum is used primarily in the plaster-making industry. The soft variety, alabaster, is used for sculpting. Halite is used in the chemical industries as a source of sodium and chlorine. Sylvite is particularly valuable as a source of potassium for agricultural fertilizers.

There are many other minerals, such as potash (sylvite, potassium chloride), that can form under certain evaporating conditions, especially in inland salt lakes. They are generally rare, but can be of economic importance.

Evaporites also form source rocks for hydrocarbons and act as impermeable seals—cap rocks—over permeable reservoir rocks. They flow under differential loading and when subjected to such loading or tectonic stress can produce diapiric structures, many of which have associated traps for hydrocarbons. They commonly form horizons (*décollement surfaces*) along which the overlying sediments detach themselves from the underlying strata and slide during thrusting and folding. The ease with which evaporites flow and dissolve in fresh water poses problems for civil engineers.

G. EVANS and N. MANN

Further reading

Borchert, H. and Muir, R. O. (1964) *Salt deposits — the origin, metamorphism and deformation of evaporites*. Van Nostrand Reinhold, London.

Schreiber, B. C. (1988) *Evaporites and hydrocarbons*. Columbia University Press, New York.

Selley, R. C. (1988) *Applied sedimentology*. Academic Press, London.

event stratigraphy Event stratigraphy is a comparative latecomer to the Earth sciences, but is proving to be a useful and interesting concept. Extraordinary or singular widespread geological events of very short duration may leave an impression in the stratigraphical record. Fixing the order and dates of these happenings in the stratigraphical column (chronostratigraphy) gives us event stratigraphy. Sudden extinctions or appearances of new species of fossils also may constitute 'events', and event stratigraphy is important in our understanding of the history of life. The catastrophes postulated by geologists early in the last century are happenings of this kind, as are the meteor impacts now regarded as possibly responsible for certain mass extinctions. In dating these events, detailed correlation and detailed (high-resolution) biostratigraphy is essential. With increasingly detailed knowledge of the stratigraphical column, many more events are now proposed than ever the Catastrophists had in mind. Many global events recognized today were, however, on a much less spectacular scale than those pinpointed so long ago.

What may be described as *event deposits* include flash-flood conglomerates and the current-induced sediments: *inundites*, *tempestites*, and *turbidites*, the products of mass flow. Then there are the *seismites*, the products *in situ* of earthquake shock deformation. Volcanic action may produce widely distributed ashfalls and other distinctive deposits that are the results of sudden, shortlived, or singular, commonly explosive events.

Geochemical events are recognized where there are sudden, brief anomalies in the chemical composition of a succession of sediments. For the most part, they can be detected only when detailed sampling and analysis of the rocks have been made. They may involve trace elements, or *isotope perturbations* of carbon, oxygen, strontium, and sulphur. Under some conditions amino acids are now revealed as possible geochemical event markers. The anomalies usually are revealed as sharp 'spikes' on the spectrographs produced from the analyses. Graphs representing values of these substances drawn for stratigraphically sampled sections can provide regional—if not global—correlation. Sharp anomalies may also be widespread in distribution and constant in stratigraphical position. They constitute event markers that can coincide with other evidence to indicate a break in the pre-existing regime. The black shale bands within the upper Cretaceous seem to denote sudden brief episodes of ocean anoxia, and are correspondingly widespread. Other evidence of singular events may include bands of microtectites, the microscopic glass spheres that are generated at the point of a meteoritic impact and scattered far afield at a single instant.

Biostratigraphical events are detected where there is a sudden hiatus in the progress of a lineage through the strata. This is possible only where there is an abundance of fossils to give a clear lineage, preserved in an unbroken section of sediment, preferably where no facies change occurs. We know many such events in the ammonoid-bearing rocks of the European Mesozoic, even where no lithological break has been suspected.

There has been a tendency to regard many of the major extinction events as the result of meteorite impact, but the matter remains one of debate. The most intensely studied event in this connection is that at the Cretaceous–Tertiary boundary. At that horizon geochemical anomalies were noted in the late 1970s in sections in Italy, North America, and subsequently many other parts of the world. The element iridium occurs in abnormal quantities throughout a 5-cm layer. The proportion is consistent with that found in meteoritic material but not otherwise common. Walter Alvarez and his colleagues in the USA proposed in 1980 that the sudden rise in the amount of iridium present results from the outfall from a giant meteorite impact. This hypothesis is supported by the subsequent discovery of what may have been a 300-km-wide crater at Chicxulub in the Yucatán Peninsula of Mexico and megabreccias in Haiti that appear to be of the right age and size.

Within many basins of deposition there are large numbers of various kinds of event markers. They are put to use in what is known as *high-resolution event stratigraphy*. From this technique it may be possible to calculate the rate and manner in which the basin was filled.

D. L. DINELEY

Further reading

Einsele, G., Ricken, W., and Seilacher, A. (eds) (1991) *Cycles and events in stratigraphy*. Springer-Verlag, Berlin.

Geldsetzer, H. H. J. and Nowlan, G. S. (eds) (1993) Event markers in Earth history. *Palaeogeography, Palaeoclimatology, Palaeoecology*, **104**.

Sharpton, V. L. and Ward, P. D. (eds) (1990) Global catastrophes in Earth history; an interdisciplinary conference on impacts, volcanism, and mass mortality. *Geological Society of America Special Paper* No. 247.

evolution (Darwinism and neo-Darwinism)

Evolution can be defined as a sustained change in the gene frequencies of populations over the generations, producing new species. Although it remains a theory, evolution links together observations made by scientists in many different fields: geologists, biologists, palaeontologists, and geneticists. Evolutionary theory is generally linked with the name of Charles Darwin and hence is often termed 'Darwinism' (or 'neo-Darwinism' with the addition of modern genetic information). There were, however, many pre-Darwinian scientists who recognized that organisms had changed over time and that new species had developed from pre-existing ones. One of the best known of these was the French naturalist Jean Baptiste Lamarck, who in 1809 put forward the theory of inheritance of acquired traits as a mechanism to explain evolutionary change. Although he recognized that animals became better adapted to their environment through time, his concept of the retention of characteristics acquired during the life of the organism is no longer supported. In his classic example, giraffes acquired long necks by stretching to reach leaves higher on the tree; the length gained during the parents' lifetime was then passed on to the offspring. It can very easily be shown, however, that changes such as increase in muscle size or loss of limbs are not inherited. Darwin put forward a testable theory for evolutionary change that has stood the test of time and has provided a basis for further development.

Darwin proposed his main ideas on biological evolution in 1859 in his book *On the origin of species by means of natural selection, or the preservation of favoured races in the struggle for life.* These ideas were based to a great extent on his observations while on the voyage of H. M. S. Beagle (1831–6), and were particularly influenced by his visit to the Galápagos Islands in the Pacific Ocean. He subsequently added much corroborative evidence from detailed observations of artificial breeding as well as of fossil and living species, but seems to have been galvanized into publication only when another biologist, A. R. Wallace, formulated the same idea independently. Darwin's discovery is commonly called the 'theory of natural selection'; its main points are as follows. First, animal species reproduce more rapidly than is necessary to maintain their numbers, but despite this their numbers tend to remain stable. Secondly, there is a 'struggle for existence' within and between species in which individuals are competing for resources such as food and living space. Thirdly, all animals vary within a species and this variation can be inherited. Finally, in the struggle for existence those organisms best fitted to the environment will survive and their characteristics will be passed on to future generations, while those that are not well suited will be weeded out. By these processes favourable characteristics are accumulated and lead to the separation of species. Although this process seems to be logical, Darwin was ignorant of the mechanisms by which characters are passed from generation to generation and of the ways in which new characters appear.

Heredity and variation

Darwin believed that all the parts of an organism secrete minute granules that circulate throughout the body and finally accumulate in the reproductive organs ready to be passed on. In 1866 Gregor Mendel published the results of his experimental work on breeding and developed a theory of heredity. Mendel had experimented with varieties of garden peas and had tracked the appearance and disappearance of characters such as seed and flower colour during cross-fertilization experiments involving thousands of plants. In a typical experiment he cross-fertilized plants grown from green seeds with plants grown from yellow seeds and found that the resulting hybrid seeds were all yellow. He then planted those seeds, allowed the resulting plants to self-fertilize, and found that the seeds produced from those plants were both yellow and green. Of 8023 seeds in the second generation, 6022 were yellow and 2001 were green, giving a ratio of 3:1. Mendel was able to demonstrate from experiments such as this that inheritance is particulate and not blending, and also that there is a clear difference between the appearance of an organism (its phenotype) and its genetic constitution (its genotype).

Mendel's work lay unnoticed until the early 1900s, when other scientists independently rediscovered the gene as the basic unit of heredity. Genes are molecular blueprints that contain instructions on how to replicate the cell and how it should function. The genes occur on strands called chromosomes which occur in pairs within the nucleus of the cell. The number of pairs is constant within a species. Humans have 23 pairs of chromosomes containing a total of about 100 000 genes. In sexually reproducing organisms several variants of a particular character (e.g. eye colour) can coexist within a single species. These homologous genes are termed 'alleles'. As each species normally possesses a large number of such alleles, the number of potential combinations of different characteristics is very great. In the production of gametes or sex cells a process called meiosis results in the splitting of each pair of chromosomes so that only one of each pair is present in each gamete, which is termed 'haploid'. In addition, recombination or shuffling of the genes during meiosis results in a different subset of genes being passed on each time. This in turn results in a greater amount of variation on which evolution can operate. The haploid gametes will combine ultimately to form diploid cells (with the full complement of chromosomes). This explains Mendel's theory, which was proposed before anything was known about chromosomes. He guessed that each individual has a double dose of factors, that they segregate independently and at random in the gametes, and that they recombine in the next generation.

The resolution of the problem of how information was held in the chromosomes and passed on was provided in 1953 by Francis Crick and James Watson, then of Cambridge University. They proposed a model for the structure of DNA (deoxyribonucleic acid) molecules, which are the main component of chromosomes, that showed how they could have a

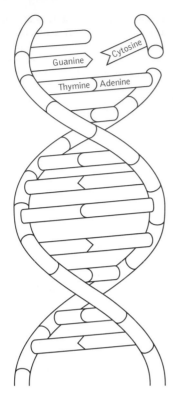

Fig. 1. The structure of the DNA molecule. This shows the spiral side pieces and the interlocking cross pieces formed of the bases guanine, cytosine, thymine, and adenine.

genetic function. This model (Fig. 1) shows the molecule to be a twisted spiral ladder, or double helix, in which the uprights are long chains of alternate sugars and phosphates while the rungs are formed of two bases each. The base pairs are of two kinds only, adenine bonding with thymine, and cytosine with guanine, but they can be oriented either way round and can occur in any sequence. This provides a means of coding information by the sequence of bases along the side chains. It also provides a means of replication of the molecule in which the bonds between the bases separate and the double helix 'unzips' allowing a new chain to be synthesized next to each of the old chains, which act as templates. Ribonucleic acid (RNA) reads the message on the DNA (transcription) and then acts to synthesize the relevant proteins (translation).

Domesticated animals and plants exhibit a tremendous range of variability, which is also potentially available in wild species. This variation is acted on by natural selection and much of it comes from the new combination of genes produced during meiosis. There must, however, be a source of new variations if radically new features are to develop. This lies in mutations, which are changes that occur in the genotype. These may be small-scale changes (point mutations) in

which base pairs are lost or gained (deletion or insertion) or replaced by a different one (substitutions) during transcription. Large-scale or chromosome mutations involve changes in large segments of DNA, segments of chromosomes or even whole chromosomes, and tend to have large-scale effects on the phenotype. Although these mutations have been attributed to damage to or mistranscription of DNA, it has now been shown that 'jumping genes' or mobile genetic elements can also be involved. Change that results in an advantageous character will spread through the population only if it is recurrent, and thus constantly strengthened, and if the population is small enough for it to be spread fairly quickly.

Natural selection

Natural selection operates to eliminate unsuccessful forms and allow those that function most efficiently to survive, and it acts upon the phenotype. The success of individuals depends on their reaction to their environment; their genetic make-up is hence vitally important. The carriers of advantageous alleles will gradually outnumber other individuals in the population because they are better fitted to their environment. If, however, the environment should change, selection will operate to bring forward a different variant. A recent example of this can be seen in colour changes in the British peppered moth (*Biston betularia*), which is nocturnal, resting during the day on surfaces such as tree trunks. The typical form is pale with dark speckling, which makes it almost invisible on lichen-covered tree trunks, while the occasional dark forms in the population stand out and are rapidly predated by birds. In the 1700s and early 1800s the populations were all light coloured; dark forms first appeared in the mid-1800s, and by 1900 98 per cent of moths collected near Manchester were black. This time period correlates with the most rapid increase in the human population in that area, and in the quantity of coal burned, which resulted in pollution that darkened the tree trunks. Birds acted as the selective agent; they removed the dark variants when the trees were lichen-covered, but removed the light variants when the tree trunks were blackened. This shows both the importance of environmental change in selection and also the extreme rapidity with which mutations can spread through a population. However, although selection is seen as a primary agent in evolution it does not act to produce ideal genotypes but only those that are best fitted at the time, given the numerous constraints that are constantly in operation.

Microevolution and macroevolution

The two main modes of evolution are termed microevolution and macroevolution. Macroevolution concerns evolution above the species level and the origin of major groups; microevolution deals with evolution at the species level, and particularly with the development of new species. Darwin had viewed the development of new species as occurring slowly, by a gradual shift of characters within populations, so that a

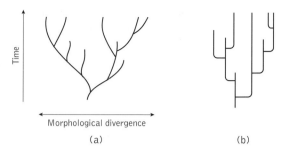

Fig. 2. Branching evolution in (a) gradual and (b) punctuated patterns.

transition from one species to another took place (Fig. 2). A number of examples from the fossil record were advanced to support this view, particularly that of the horse (in which changes to the feet and the jaws and teeth seem to have progressed in one direction through much of the Cenozoic). More recently, Peter Sheldon of the Open University has shown gradual change in eight lineages of trilobites over a period of three million years in the Ordovician of Wales. This process of phyletic gradualism is, however, rarely seen in the fossil record, where the common picture is of stable populations separated by abrupt morphological breaks (Fig. 2). Although this may partly be due to the incompleteness of the fossil record, it appears to be a real phenomenon and in 1972 was named 'punctuated equilibrium' by Stephen Gould of Harvard and Niles Eldredge of the American Museum of Natural History. One explanation of how this takes place, termed 'allopatric speciation', suggests that reproductive isolating mechanisms create a barrier to gene flow. Allopatric or geographical isolation could result when the normal range of a population of organisms is reduced or fragmented. Small parts of the population can then become separated in peripheral isolates, and if the population is small it can rapidly become modified, particularly if it is adapting to a new environment. After a few generations the peripheral population may have changed enough to become reproductively isolated from the parent population. In the fossil record this will be seen as a period of stasis in which the parent population occurs, followed by a rapid morphological change as the small population is first isolated from it and then competitively displaces it. Examples of this are rare in the fossil record, but Peter Williamson of Bristol and Harvard universities published in 1981 an account of punctuated equilibrium in molluscs in the Late Tertiary sediments of Lake Turkana, East Africa. It is also considered that such punctuated changes might have taken place as a result of heterochrony, i.e. a change in the timing of appearance or rate of development of ancestral characters. As an example, the rate of development can be affected so that the descendant will continue to grow, producing a giant form, or cessation of growth can occur earlier than normal, producing a dwarf. This has the advantage of being essentially instantaneous and thus producing the means of rapid change.

Macroevolution concerns evolution at levels above the species, and it is in this area that palaeontology may be able to make the greatest contribution. As the resolution of the fossil record is generally quite coarse, several thousand generations having lived and died during the period in which a few millimetres of sediment have been deposited, palaeontological studies are best at resolving major evolutionary events. In macroevolution the unit of selection is the species, and net evolutionary changes are seen within a plexus of related species rather than within a species, as in microevolution. Processes that operate at this level include allometry, which concerns change in relative growth of different parts of the body, resulting in different proportions. Abrupt and major changes such as the sudden development of new taxa at the base of the Cambrian are difficult to understand. They were explained in the 1940s by the concept of 'hopeful monsters': organisms that exhibited major mutations and could be the progenitors of entirely new taxa. Such views are no longer supported. Heterochrony might, however, provide an answer to the question of how major changes occur. A change in timing of developmental events could produce a descendant that was quite different from its parent, but was a viable organism. A means of creating a new higher-level taxon could thus be developed. Euan Clarkson of Edinburgh University has shown that the development of trilobite eyes could have occurred by just such a mechanism (see *trilobites*). Heterochrony has also been considered as being responsible for the development of the fish-like ancestral vertebrates from primitive chordate ancestors, and for the transition of vertebrates from aquatic to terrestrial environments.   DAVID K. ELLIOTT

Further reading

Mckinney, M. L. (1993) *Evolution of life: processes, patterns and prospects*. Prentice Hall, New York.

Patterson, C. (1978) *Evolution*. British Museum (Natural History), London.

Ewing, W. M. (1906–74)   William Maurice Ewing was a Texan who took to the sea and over a period of forty years or more made marine geology and geophysics of prime importance in the Earth sciences. A physicist by training, he was the first to carry out reflection and refraction seismic traverses across the continental shelf (1935). He had attended a symposium on the applications of geology to the ocean basins and margins and was invited to direct the geophysical investigations aboard a US Coastguard Service ship off the coast of Virginia. His design of equipment and survey methods in deep waters was masterly, and he showed that 3000 m or more of sediment had accumulated at the continental edge.

Ewing was to develop gravity and magnetic survey techniques for use in the oceans and was engaged in this work up until the Second World War, when he was to work on antisubmarine research for the US National Defense Research Board. In 1949 he became head of the marine geology and geophysics unit at the new Lamont Geological Observatory

(Columbia University, N. Y.—later the Lamont–Doherty Geological Observatory). Seismic work on land was also initiated, but it was at sea that he was continually at work with his team. With Bruce Heezen, for example, he confirmed Daly's ideas about turbidity-current erosion of submarine canyons on the Grand Banks. His survey work in the Atlantic revealed the great spread of sediments over the ocean plains, and he initiated coring and sampling programmes.

A topic that Ewing was anxious to initiate in the mid-1950s was deep drilling in the ocean to penetrate the basalt ocean floor and reach the Mohorovicić discontinuity. The project never materialized, but in 1968, with the marine drilling ship *Glomar Challenger* as his base at sea, Ewing began the first of many investigations of the Joint Oceanographic Institutions Deep Earth Sampling (JOIDES) project. This drilled at 31 sites in the Atlantic and at 53 sites in the Pacific to obtain cores of basalt to test the ocean-spreading hypothesis. The

results were spectacular and confirmatory. JOIDES and its successor organization's cruises and programmes have continued to this day, providing data of the widest significance to the Earth sciences.

Ewing died in 1974, a man who had by his enormous energy and intellect established marine geology and geophysics as a vital part of the Earth sciences. He often worked an 18-hour day at sea, took part in the operations no matter what the weather, and took no vacation, it is said, for 30 years.

D. L. DINELEY

exotic terrane An exotic terrane is a terrane that has been transported some considerable distance from its original location prior to its amalgamation to a continental margin, or another terrane. (The spelling 'terrane' is used to avoid confusion with 'terrain', which is a geographical term). Terranes are defined as being three-dimensional geological blocks that

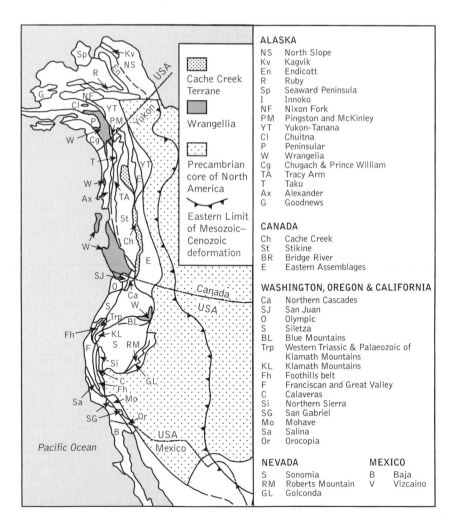

ALASKA

NS North Slope
Kv Kagvik
En Endicott
R   Ruby
Sp Seaward Peninsula
I   Innoko
NF Nixon Fork
PM Pingston and McKinley
YT Yukon-Tanana
Cl Chuitna
P   Peninsular
W   Wrangelia
Cg Chugach & Prince William
TA Tracy Arm
T   Taku
Ax Alexander
G   Goodnews

CANADA

Ch Cache Creek
St Stikine
BR Bridge River
E   Eastern Assemblages

WASHINGTON, OREGON & CALIFORNIA

Ca   Northern Cascades
SJ   San Juan
O   Olympic
S   Siletza
BL   Blue Mountains
Trp Western Triassic & Palaeozoic of Klamath Mountains
KL   Klamath Mountains
Fh   Foothills belt
F   Franciscan and Great Valley
C   Calaveras
Si   Northern Sierra
SG   San Gabriel
Mo   Mohave
Sa   Salina
Or   Orocopia

NEVADA

S   Sonomia
RM   Roberts Mountain
GL   Golconda

MEXICO

B   Baja
V   Vizcaino

Fig. 1. Simplified tectonic map of the Cordillera of western North America, illustrating the various terranes that have accreted to the North American margin during the Mesozoic. Not all of these may have been transported very far along the margin from their original locations. The current locations of the far-travelled terranes referred to in the text, the Wrangellia and Cache Creek Terranes, are marked by grey shading and a coarse stipple respectively. (Redrawn from Coney *et al.* (1980).)

preserve a coherent internal stratigraphy, be it sedimentological, metamorphic, or igneous, marking a geological history that differs from that of its surrounding terranes or continental margin. It could, therefore, be said that all terranes are 'exotic' to each other. The term is, however, usually restricted to those terranes that can be shown to have undergone large lateral displacements with respect to the surrounding rocks. By definition all terranes are fault-bounded. In practice, however, the margins of a terrane can often only be identified by discordances in stratigraphy or structural fabric, or both, and the terrane-bounding fault may only be inferred. As a result it is possible to misidentify as a terrane a geological block bordered by rapid lateral facies changes rather than by a bounding fault or faults.

The recognition of exotic terranes in the geological record arose primarily from attempts to unravel the complex sequence of Mesozoic and Cenozoic deformation events in the Cordillera of western North America (Fig. 1). The Cordillera comprise a vast mosaic of more than 50 terranes, each with a geological history different from that of its neighbours. These are referred to as 'suspect terranes' because their palaeogeographic setting with respect to North America is unknown for much of the Phanerozoic. Detailed investigations of the fossils contained in the rock sequences in a number of these terranes, in combination with palaeolatitudes derived from palaeomagnetic analyses, have identified several terranes that have travelled large distances with respect to the North American continent and, thus, constitute 'exotic terranes'.

Two particular examples from the Cordillera of North America serve to illustrate the lateral mobility of such terranes: the Wrangellia and Cache Creek terranes (Fig. 1). Disaggregated fragments of the Wrangellia terrane, preserving a coherent internal geological history, are now found over a length of some 2000 km in the Cordillera, stretching from Alaska to Oregon. The terrane seems to have originated as an oceanic plateau or seamount out in the Pacific ocean which subsequently collided with the North American margin as a result of subduction of an oceanic plate underneath North America in Jurassic and Cretaceous times. Palaeomagnetic data from identical Upper Triassic basalt units along the exposed length of the terrane indicate that the rocks were formed at a latitude of some 11°. If they had formed in their present position with respect to the North American continent, and allowing for the plate motion of North America since the Triassic, they would have been extruded at a latitude of 34° N. The inescapable conclusion of these analyses is that the Wrangellia terrane must have moved northwards by some 23° of latitude (approximately 2500 km) with respect to North America since the Upper Triassic. Astonishing as this conclusion may seem, it is equally possible that if they formed at 23° S (palaeomagnetic data cannot distinguish between the northern and southern hemispheres without a continuous sequence of palaeolatitudes going back from the present to the time period in question) the terrane has been moved by upwards of 4500 km with respect to North America. This

motion must have occurred between the Upper Triassic and Late Cretaceous, for the terrane is known to have collided with the North American margin at this time. The Cache Creek terrane, on the other hand, also contains a sequence of Palaeozoic to Upper Triassic rocks, with evidence of subduction-zone deformation, containing large limestone blocks that contain equatorial fossils. This terrane is now smeared out along some 1000 km of the Cordillera from the Yukon to Washington State. The conclusion again is that the Cache Creek Terrane must have travelled large distances with respect to North America prior to its accretion and dismemberment along the margin.

The large lateral motion of these exotic terranes along the western North American margin and their subsequent accretion is linked to the oblique convergence of the North American Plate with the various Pacific plates, including some that have long since been completely subducted. The subduction of any oceanic plate beneath a continent is likely to carry edifices of the seafloor, such as seamounts, oceanic islands, and oceanic plateaux, into the subduction zone, where they will collide with the continental margin. It is thus almost certain that all mountain belts contain exotic terranes.

CONALL MAC NIOCAILL

Further reading

Coney, P. J., Jones, D. L., and Monger, J. W. H. (1980) Cordilleran suspect terranes. *Nature*, **288**, 329–33.

Howell, D. G. (1989) *Tectonics of suspect terranes: mountain building and continental growth*. Chapman and Hall, London.

experimental petrology Experimental petrology is a branch of Earth science that attempts to reproduce, in the laboratory, the conditions under which igneous and metamorphic rocks are formed at great depth within the Earth where temperatures and pressures are extremely high. For example, diamonds form at depths of about 150 km and temperatures over 1300 °C. Experimental petrologists grow samples of minerals, observe the melting behaviour of different compositions of rock, examine the movement of trace elements between minerals, and, in short, investigate many processes of geological interest.

Experimentalists use very small rock samples, often no more than a millimetre across. A tiny sample is used so that it will equilibrate quickly at the temperature and pressure of interest, and can quickly be cooled to 'freeze in' the experimental results. If the samples were large, long periods of time would be required for the sample to reach equilibrium, i.e., the point at which, if the samples remained indefinitely at the same temperature and pressure indefinitely no further change would take place. Processes that take thousands or even millions of years to occur in the Earth can be simulated, by using small enough samples, in only days or weeks in the laboratory.

Most of the work done before the 1950s was at atmospheric pressure. Notable contributions to the study of silicate systems were made by N. L. Bowen and his colleagues at the Geophysical Laboratory at Washington, D. C. Experiments at high pressure and temperature, corresponding to conditions at depth in the crust, became possible with the introduction in the late 1940s of the 'Tuttle bomb': essentially a thick metal test-tube closed by a screw cap. The 'bomb' was heated in a furnace and a fluid pumped in to raise the pressure. Higher pressures were achieved in the 1960s with the piston-cylinder apparatus, heated internally and using one or more pistons to compress the sample. The diamond-anvil apparatus carried the range of pressures still higher, with the advantage that the specimen could be observed during the experiment through the diamond anvil. (With the classical methods it was possible to examine the experimental charge only after it had been chilled and prepared for observation under a petrological microscope.) A laser beam is used to heat the sample. For the highest pressures, brute force is applied in the form of the shock-wave apparatus, in which a projectile is fired at the sample.

The equipment now available makes it possible to simulate conditions deep in the mantle. Experimental petrology is adding much to our understanding of the interior of the Earth and the way in which it melts to produce the material from which igneous rocks are derived.

JUDITH M. BUNBURY

exploration geophysics Exploration geophysics is the application of geophysical techniques to mineral exploration. A range of methods is available, usually grouped according to the physical property to which they respond (Table 1). The choice of the appropriate method for each application is therefore important. In some instances, geophysical methods cannot be used to sense a deposit directly, but are used indirectly to detect characteristic features that may be associated with it, or just to determine the geological structure of the prospect. It is therefore common for a mixture of methods to be applied. Cost is important, and so cheaper methods tend to be used for preliminary surveys and more expensive ones kept for later refinement.

Many of the methods have been adapted for airborne, marine, and borehole surveys. Many are finding increasing application in environmental surveys. ROGER SEARLE

exposure *see* OUTCROP, EXPOSURE

extensional tectonics on the continents Continental extension is manifested in three main ways: by long, narrow rift valleys, such as the East African rifts; by very broad areas of extension (highly extended terranes), such as the Basin and Range province of south-western USA; and by

Table 1. Methods of exploration geophysics

Method	Sensed property	Relative cost	Typical targets
Gravity	Density, mass	Moderate	Regional and local structure
			Ore bodies with large density contrasts (e.g. chromite, salt)
			Estimating total reserves
			Detecting cavities
Magnetic	Magnetic susceptibility; natural remanent magnetization	Cheap	Regional and local structure
			Magnetic ore bodies (e.g. magnetite)
			Ferrous metal deposits
Electromagnetic	Electrical conductivity	Cheap to moderate	Regional and local structure
			Conducting ore bodies
			Groundwater exploration
			Contaminated groundwater
Electrical	Electrical conductivity; electrical potential	Moderate	Local structure
			Conducting (or resistive) and electrically charged ores;
			Groundwater and water table
			Contaminated groundwater
Radioactive	Natural or induced radioactivity	Moderate	Radioactive minerals
			Clay and other lithologies
Seismic refraction and reflection	Seismic velocity; acoustic impedance	High	Regional and local structure
			Hydrocarbons

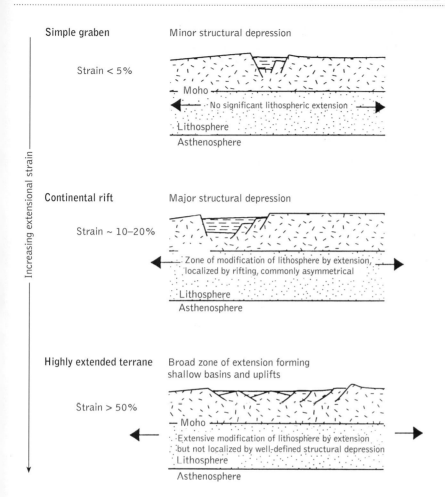

Increasing extensional strain

Simple graben

Strain < 5%

Minor structural depression

— Moho
No significant lithospheric extension
Lithosphere
Asthenosphere

Continental rift

Strain ~ 10–20%

Major structural depression

Zone of modification of lithosphere by extension, localized by rifting, commonly asymmetrical
Lithosphere
Asthenosphere

Highly extended terrane

Strain > 50%

Broad zone of extension forming shallow basins and uplifts

— Moho
Extensive modification of lithosphere by extension but not localized by well-defined structural depression
Lithosphere
Asthenosphere

Fig. 1. Major styles of continental extension, from a simple graben where there is no lithospheric extension and strain is less than 5 per cent through a continental rift with stress of 10–20 per cent to a highly extended terrane such as the Basin and Range province. Notice that rifts are asymmetrical in cross-section. (From Olsen, K. H. and Morgan, P. (1995) In Olsen, K. H. (ed.) *Continental rifts: evolution,structure and tectonics*, Elsevier, Amsterdam.)

extensive plateau basalts (see *continental flood basalts*). Yet recognition that large parts of the continents are undergoing extension rather than compression took a long time to establish and had to await geophysical, especially gravimetric, studies in the middle decades of the twentieth century to confirm the relationship between rift valleys and extension. The main reason why most structural geologists in the nineteenth century and the first half of the twentieth century were reluctant to believe this relationship was that they had been conditioned to take a cooling, shrinking Earth for granted and therefore saw no problem with interpreting both folded mountain belts and rifts as being zones of compressional shortening.

Studies of continental rifts began with the Rhine graben, *Graben* being a German term for ditch or trench, on the doorstep of Vienna University where the great Alpine geologist, E. Suess, taught (see *Suess, Edouard*). In his treatises on world geology and on the geology of rift valleys (1891) he drew attention to a worldwide system of rifts, including the

African rift system that runs from the Zambesi via the Afar and Red Sea into Palestine. Inspired by these Austrian global syntheses, a Scotsman, J. W. Gregory, undertook thorough detailed fieldwork of the East African rift, publishing his results in two large tomes in 1896 and 1921.

Throughout this time, and well into the 1930s, even the most distinguished geologists believed in a compressional origin. However, gradually during the 1950s compressional mechanisms gave way to extensional ones as the causes of rifts, especially as data accumulated to show that oceanic mid-ocean ridges and their central rifts were the result of ocean-floor spreading (see *mid-ocean ridges*). In particular, the link between the continental rifts of East Africa and the initial stages of ocean-floor spreading as seen in the Red Sea and Gulf of Aden seemed to fit into one consecutive sequence that continued into the creation of new oceans. The only question that seemed to arise was that the East African rift appeared to be quite an old structure and that, while it clearly was a zone of extension, significant lateral movement did not appear to have taken place. It is apparent now that, although some continental rifts do become sites of future oceans, others do not and either abort, to become the so-called *aulacogens* (aborted rifts) such as the Benue trough in the bight of west Africa or the North Sea, or persist for at least tens of millions of years as the major rift systems of today.

Today everyone agrees that the rift systems of the Rhine, East Africa, Rio Grande, and Baikal are zones of extension (Fig. 1), yet it is also clear that the amount of horizontal extension they have undergone is relatively small, being less than 10 km for the Baikal rift, 4.5–5 km for the Upper Rhine graben, and 8–10 km for the Kenya rift. They are also commonly associated with large domal uplifts and have an underlying lithosphere that, like mid-ocean ridges, is abnormally thin, and invaded by low-velocity, high-temperature material. Magmatic activity is strongly alkaline.

Initially, ideas and models of rifts envisaged a simple symmetrical basin deepening towards the centre, but over the past 20 years, from detailed mapping studies and geophysical profiles in the Gulf of Suez, the East African rift, the Rhine graben, and Baikal rift, it has become apparent that this is not only wrong but very misleading if one is to understand properly the patterns of sedimentation, location of volcanoes, and structural style. Rift valleys are, in fact, segmented into a series of asymmetrical basins or half grabens that alternatively face opposite ways (Fig. 2). Thus at any one point in a rift basin, one side is a major fault scarp; the opposite margin is a more gently faulted monocline. Between each oppositely dipping half graben is a complex zone of oblique strike-slip faulting, termed an accommodation zone. In the non-volcanic Lake Tanganyika these divide the lake into a number of sub-basins.

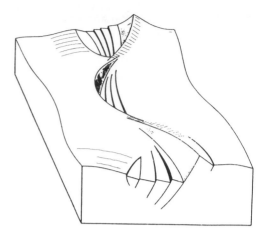

Fig. 2. Sketch showing alternation of half graben, facing opposite ways, linked by accommodation zones.

In the Gregory rift of western Kenya the major volcanoes (Elgon, Kenya, and Kilimanjaro), situated 100–150 km to the east and west of the rift axis, alternate either side of the rift, each being located close to the abrupt edge of the escarpment.

Once it was established that rifts are zones of both extension and high heat flow, the debate turned to whether the initial cause of rifting was an underlying heat source, such as a convection plume in the mantle or a hot-spot, the rifting then being a *consequence* of expansion, uplift, and doming brought

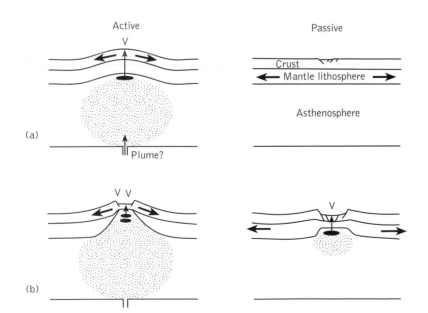

Fig. 3. Idealized diagrams comparing the active and passive hypothesis for initiation of continental rifting, showing (a) the initial stage and (b) subsequent stage of development. (From Bott, M. H. P. (1995) In Olsen, K. H. (ed.) *Continental rifts: evolution, structure and tectonics*, Elsevier, Amsterdam.)

about by the thermal activity (an *active* system), or whether the magmatic activity was merely the result of the regional or local extensional stresses causing failure of the continental lithosphere and consequent development of an anomalous underlying mantle, allowing magma to come to the surface (a *passive* system) (Fig. 3). Although it is clear that all rift systems require both an element of extension and an underlying heat source, some can be differentiated into those that are more active (the northern East African rift) and those that are more passive (Rhine graben and southern East African rift). The Baikal rift has features of both types. One also has to remember that, as with all global tectonic features, particularly on continents, the actual location and the orientation of such features depends largely on the position of older tectonic lines such as lineaments and sutures.

Active rifts are characterized by more voluminous volcanic activity, higher rates of crustal extension, mildly alkaline basalts, and a bimodal distribution of basic and acid magma types. Active rifting requires the presence of an upwelling convective plume at the base of the lithosphere prior to crustal extension. This process may result in convective thinning from below, either thermally (due to increased heat input displacing the solidus upward) or mechanically (due to removal of material from the base of the lithosphere). Lithospheric thinning should cause uplift of several kilometres but little extension of near-surface rocks. Active rifts should be floored by erosional unconformities, and initial centrifugal drainage patterns should result in clastic sediment starvation within the rift but abundant volcanigenic and volcaniclastic sediment instead.

Since passive rifts have only localized centres of alkaline volcanism they are characterized by small volumes of eroded products, low rates of crustal extension, discontinuous volcanic activity, and a wide spectrum of basaltic and more differentiated magma compositions. Unconformities do not form and early sagging diverts drainage into centripetal or axial patterns. Basaltic volcanism should occur late, following crustal extension.

One possible hypothesis for the formation of some passive rifts, such as the Rhine graben and the Baikal rift, is that they are the consequence of nearby, contemporaneous collisional tectonics that has imposed a large-scale compressional regime in the vicinity. This is the idea of indentation tectonics, where continental-collision-related rifts are thought to be produced by secondary tension set up at high angles to a zone of compression. Such rifts are also known as impactogens.

Earlier models of rifting, such as the well-known McKenzie model, invoked symmetrical, pure-shear extension to explain crustal thickness, subsidence histories, and gravity profiles of extended rifts and basins. However, as it became evident that rifts were asymmetrical and new geophysical data became available, models incorporated low-angle detachment faults that descended even into the asthenosphere (Fig. 4). In the Wernicke model of crustal extensional faulting, large, very low-angle normal asymmetrical faults in the upper crust pass

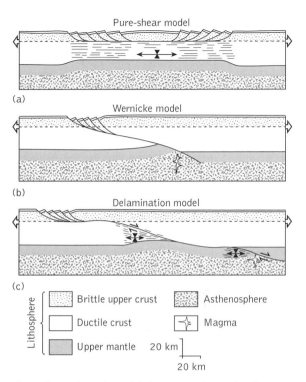

Fig. 4. Three end-member models for continental extension. (From Lister, G.S., *et al.* (1986) *Geology*, **14**, 246–50.)

down into a detachment surface which allows extension of the ductile part of the lithosphere to be offset from the surface structure. An alternative model includes delamination of the lithosphere with crustal detachment at the brittle–ductile transition at the Moho.

At the surface, rifts are indicated by wide valleys, 40–100 km across, filled by young alluvial and lacustrine sediments and bordered by steep escarpments, rising in some cases a few hundred metres, in other cases 3000–4000 m above the valley floor. In the East African rift system the degree of volcanism within the rift varies substantially. In rifts such as the Gregory rift, north of 1° N, volcanoes are so abundant that the rift is filled by volcanoes and volcaniclastic sediments. Other parts of the rift, such as the Western rift, have volcanoes at sporadic intervals along the rift, so that the volcanoes separate a series of lakes. In the southern part of the rift, where volcanism is essentially absent, very large lakes such as Lake Malawi and the 1400 m deep Lake Tanganyika form.

East African lakes traverse the whole range of climates from the arid desert belt in the north to the wet tropics in the south. In the north, where volcanic activity dominates the rift and neighbouring areas, we have alkaline, soda-rich lakes, especially Lakes Magadi and Natron, which have given their names to unique forms of sodium silicates (magadiite) and an authigenic zeolite (natrolite), as well as a whole range of mineral assemblages that have yielded models for saline, alkaline, ephemeral closed-basin lakes and for the formation of sodium carbonates and so-called Magadi-type cherts in ancient rocks.

The East African rift system also contributed to knowledge because of the abundance of ash falls. These not only serve as the most reliable and short-lived time markers but preserve fossils, including footprints and soft bodies, in places where fossils do not normally form, especially in the surrounding savannahs. Of particular importance is the chemistry of the ash, which in Eastern Africa is low in silica and high in lime. East African volcanoes are famous for their carbonatites, and most of the eruptive magmas were rich in alkalis and natro-carbonatite. When falling into water this ash instantly cements as sodium carbonate (trona) and, in this alkaline environment, bone apatite and calcium carbonate fossils are relatively insoluble. In addition, the fine-grained matrix preserves footprints and moulds of soft bodies, including the muscles of birds and even the tongue of a lizard. The calcium carbonate will also calcify and coat or fill the bones. It is no coincidence that East Africa is the domain of so many beautifully preserved continental animals (particularly primates and other mammals, and reptiles), especially soft-bodies ones.

Although not as long as the East African rift system (about 2000 km), the Baikal rift system is just as spectacular in other ways. It is associated with an approximately 1500 km long domal uplift that today is only part of a much larger uplifted region, the Mongolian plateau. It contains some 15 individual topographic depressions bounded by fault scarps related to half graben. In places, some 8000 m of sediment accumulated, 3000 m having been penetrated by drilling. It also has the deepest (1700 m) and largest lake in the world, containing 20 per cent of the world's fresh water. Since it is also probably the oldest lake and has been cut off for millions of years from other freshwater lakes, a unique fauna has evolved. Volcanic rocks, ranging in composition from tholeiite to alkaline basalts occur, the tholeiites being older; but the volume of volcanic rocks (5000–6000 km³) is small compared with other rift systems, nearly two orders of magnitude less than in the East African rift system. It had two stages of evolution: an earlier Oligocene, 30–35 Ma, or perhaps even earlier, and a later stage from the Pliocene, 4 Ma (4×10^6 years), to the present.

Whether the Baikal rift has an active or passive origin has been much debated. An active origin is suggested because volcanism and uplift appear to have preceded rifting, an active plume passing over a large area. On the other hand, the very small volume of volcanic rock and its location between two distinctive tectonic blocks are arguments for a more passive origin and an interpretation as an impactogen.

The very broad areas of continental extension known as 'highly extended terranes' may be less discussed than rifts, but, in terms of the amount of extension involved, they are significantly more important. One problem is that most such terranes are found today below sea level or under thick piles of basinal sedimentary rocks. The one area where such a terrane is not only known today, but has been studied extensively in excellent exposures, is the Basin and Range

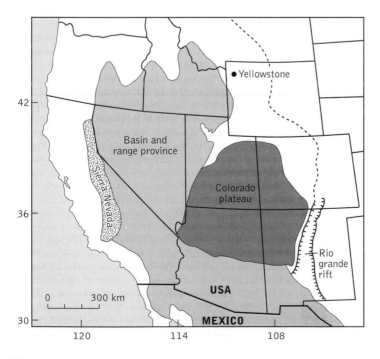

Fig. 5. Map showing the extent of the Basin and Range province, which extends some 1000 km further south into northern Mexico, the Colorado plateau, and Rio Grande rift. (After Parsons, T. (1995) In Olsen, K. H. (ed.) *Continental rifts: evolution, structure and tectonics*, Elsevier, Amsterdam.)

province that extends across much of the western USA and into Mexico, and is made up of hundreds of basins of varying depth, age, and orientation (Fig. 5). The name derives from the characteristic pattern of alternating basins and ranges, resulting from a particular type of extensional block. At its widest point it is 900 km across. Estimates of the amount of crustal extension range between 50 and 100 per cent, with local extremes of perhaps 300 per cent or 10 per cent. Thus extension is perhaps an order of magnitude greater than in rifts and spread over a much wider area. Volcanism is abundant and diverse in this extensional terrain. Basaltic and acidic rocks predominate, but intermediate types also occur. Some magmatic activity is concentrated along basement lineaments and other volcanism appears to be related to rifting in a similar way to that of major rift valleys. The basement of the province is continental but thinner than normal, less than 30 km thick, and heat flow is high. The problem, therefore, of its origin is why should the province be so high, and why should it be so wide? Extension should lead to isostatic subsidence. There are no clear answers to these questions, but it is evident that the entire western third of North America is elevated above sea level, regardless of whether it was extended or not. This is generally thought to be the result of long-term subduction of the Pacific Plate under the North American Plate, leading to backarc extensional stresses in the Basin and Range province.

Almost all the rifts and highly extended terranes discussed so far have some degree of magmatic activity associated with them. However, the quantity of extrusive rocks pales into insignificance when compared to the vast outpourings of basaltic lavas associated with continental flood basalt provinces. They compare closely with oceanic flood basalts that make up many oceanic ridges, plateaux, and sea mounts. Continental flood basalt provinces are generally vast laterally extensive volcanic plateaux which are commonly associated with regional extensional stress fields, as evidenced by dyke and vent alignments (Fig. 6). They differ from rifts in that they lack any tectonic depressions. However, they may be closely associated with major rifts and highly extended terranes. For examples, the Ethiopian Flood Basalt province embraces the Afar–Yemen triple-rift junction, where the East African rift joins the Red Sea and Gulf of Aden sea-floor spreading centres. The Jurassic–Cretaceous basalts of South American, southern Africa, and Antarctica were associated with the break-up of the great Gondwana continent. One of the most remarkable features of these continental flood basalts provinces is the scale and rate of lava expulsion. The Parana basalts of Brazil and the Deccan Traps of India covered more than 1 million km². The Karoo field of southern Africa was twice as large and reached 9 km in thickness. The youngest example is the Columbia River plateau of north-western USA that lies just to the north of the Basin and Range province. It was formed by a plume or mantle hot-spot that is now located under the currently active Yellowstone Park region.

HAROLD G. READING

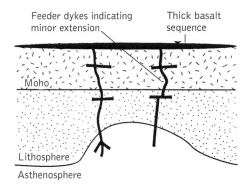

Fig. 6. Model for continental flood basalts. (From Olsen, K. H. and Morgan, P. (1995) In Olsen, K. H. (ed.) *Continental rifts: evolution, structure and tectonics*, Elsevier, Amsterdam.)

Further reading

Busby, C. J. and Ingersoll, C. V. (eds) (1995) *Tectonics of sedimentary basins*. Blackwell Science, Cambridge, Mass.

Coward, M. P., Dewey, J. F., and Hancock, P. L. (eds) (1987) *Continental extensional tectonics*. Geological Society Special Publication No. 28.

Frostick, L. E., Renaut, R. W., Reid, I., and Tiercelin, J.-J. (eds) (1986) *Sedimentation in the African Rifts*. Geological Society Special Publication No. 25.

Olsen, K. H. (ed.) (1995) *Continental rifts: evolution, structure, tectonics*. Developments in Geotectonics 25. Elsevier, Amsterdam.

extinctions and mass extinctions

Extinction of species is a continuous process, and evidence of its occurrence abounds in the fossil record. It has been estimated that marine species persist for about four million years, which translates into an overall loss of about two or three species each year. This is considered to be the background extinction rate and it is balanced by speciation events that result in the development of new species. Mass extinctions are events during which the rate of extinction rises dramatically above this background rate. A number of these have occurred during Phanerozoic times, i.e. since the end of the Precambrian, about 600 million years ago. Five major events have for some time been recognized: the end-Ordovician; late Devonian (Frasnian–-Fammenian); end-Permian; end-Triassic; and end-Cretaceous events. These are best seen in the record of shallow-water marine organisms; in each event at least 40 per cent of the genera were eliminated. By using statistical methods it has been possible to estimate that at least 65 per cent of species became extinct at each of these events, 77 per cent being eliminated at the end-Cretaceous event and 95 per cent at the end-Permian event. Research by David Raup and Jack Sepkoski at the University of Chicago has demonstrated the presence of a

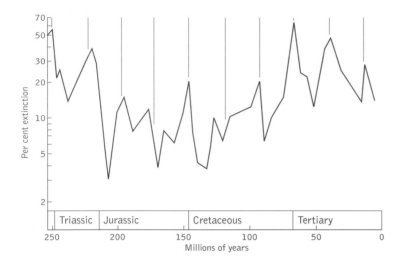

Fig. 1. Extinction rates for marine animal families, showing 12 peaks of extinction since the Late Permian. The periodicity (roughly 26 million years) is indicated by the vertical lines.

number of other such mass extinctions and has shown that they occur with a regular periodicity of about 26 million years (Fig. 1). This periodicity has become a controversial topic because of its association with ideas about possible extraterrestrial causes of such events.

Causes of mass extinction

Attempts to explain the causes of mass extinctions have traditionally centered on terrestrial phenomena such as sea-level changes, climatic changes, or volcanism. Sea level has shown regular fluctuation on a global level during the Phanerozoic (Fig. 2). The term eustatic was coined for this by the Austrian geologist Edward Suess at the turn of the century. These eustatic changes appear to be related to the melting or formation of polar ice caps or to major tectonic events, such as continental splitting or collision and the uplift and subsidence of ocean ridges. Extinction events appear mostly to be correlated with periods of marine regression. (i.e. retreat of the sea from the land when the sea level was low). The reason for this appears to be that a withdrawal of the ocean leaves a much smaller habitat area for shallow-water marine organisms. This leads to increased crowding and competition, and ultimately to an increased extinction rate. Reduction of large terrestrial vertebrates during these regressions, as happened during the end-Permian, Triassic, and Cretaceous events, might be related to increased annual seasonality caused by the loss of the ameliorating influence of the shallow epicontinental seas. It has also been shown that some extinctions are related to transgressive events (rising sea level), possibly resulting from the spread of anoxic waters across epicontinental areas, and represented by extensive deposits of black shales. Climatic changes seem to be generally correlated with eustatic events, and the evidence implicating temperatures as the main cause of extinctions seems to be weak. For example, the most important extinction event, the end-Permian, occurred at a

time of temperature amelioration marked by the disappearance of the Gondwana ice sheet.

Volcanism has been presented as a possible cause for the Cretaceous–Tertiary boundary extinctions. The lavas forming the present Deccan flood basalts of northern India were erupting at that time and would have produced large quantities of volatile emissions that could have resulted in global cooling, depletion of the ozone layer and changes in ocean chemistry. However, no evidence exists as yet for the involvement of volcanicity in other extinction events.

Although various extraterrestrial causes for mass extinction events have been suggested in the past, these ideas have gained greater credence since the publication in the early 1980s of work by Louis and Walter Alvarez of the University of California at Berkeley, who ascribed the Cretaceous–Tertiary boundary extinction event to the effects of the impact of a large bolide (extraterrestrial object) in the Caribbean region. The impact of such a large object, estimated to have been 10 km in diameter, is estimated to have resulted in some months of darkness caused by the global dust cloud that was generated. This would have halted photosynthesis and would thus have resulted in the collapse of both terrestrial and marine food chains. Although cold would initially have accompanied the darkness, greenhouse effects and global warming would follow as atmospheric gases and water vapour trapped infrared energy radiating from the Earth. Physical evidence for an impact rests on the presence in boundary layers of high concentrations of iridium and other elements that are generally rare at the Earth's surface but abundant in asteroids. In addition these layers often contain shocked quartz grains which are otherwise known only from impact craters and nuclear test sites, and microtektites, which are glassy droplets formed by bolide impacts. Soot particles are also present in some localities, which suggests that extensive wildfires may have raged across the continents. Although the evidence for

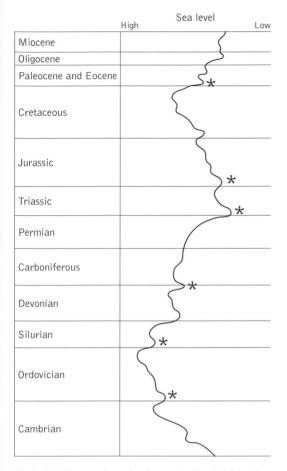

Sea level

High   Low

| Miocene |
| Oligocene |
| Paleocene and Eocene |
| Cretaceous |
| Jurassic |
| Triassic |
| Permian |
| Carboniferous |
| Devonian |
| Silurian |
| Ordovician |
| Cambrian |

Fig. 2. The Phanerozoic sea-level curve, showing the six marine mass extinctions (indicated by asterisks) that appear to be associated with changes in sea levels.

extraterrestrial impacts at the other major extinction events is slight, this causal factor has been linked with the regular periodicity of extinctions demonstrated by Raup and Sepkoski. It has consequently been suggested that perturbation of the Oort cloud of comets by the regular passage of a large planetary body as yet unidentified (Planet X, or Nemesis the death star) would result in increased asteroid impacts and extinction events.

Mass extinction events

The first mass extinction event that can be recognized in the fossil record occurred in the middle Vendian about 650 million years ago. Although its study is hampered by the paucity of macrofossils, it is clear that micro-organisms such as acritarchs (resting stages of planktonic, eukaryotic marine algae) underwent a severe decline. This extinction event has been linked to climatic cooling related to the Varangian glaciation, which occurred during the Lower Vendian.

The extinction in the Late Ordovician was a major event in which 22 per cent of marine families became extinct; graptolites and corals were particularly hard hit. As there were two main pulses of extinction and no iridium anomaly is known, an extraterrestrial cause seems unlikely. Changes in sea level and temperature have been cited as likely causal factors. In addition oceanic overturn might have brought biologically toxic bottom waters to the surface during periods of climatic change.

The end-Devonian (Frasnian–Fammenian) event had a catastrophic effect on brachiopods, which lost about 86 per cent of genera, and on reef-building organisms such as corals and stromatoporoids. Shallow-water faunas were most severely effected; only 4 per cent of shallow-water species survived, but 40 per cent of deeper-water species survived; and cool-water faunas also survived better. This has been linked with a significant drop in global temperatures, of unknown cause, during this time period.

The end-Permian event was the most severe of Phanerozoic time: it resulted in the extinction of up to 95 per cent of all marine invertebrate species. Taxa that became extinct include rugose and tabulate corals, trilobites, goniatites, and many groups of crinoids, bryozoans, brachiopods and foraminifera. On land amphibians and therapsids (mammal-like reptiles) were both badly affected, while vascular plant diversity dropped by 50 per cent. Although extraterrestrial causes have been suggested, no iridium anomaly is present. The most likely explanation is climatic instability caused by continental amalgamation and the simultaneous occurrence of marine regressions, which would have resulted in trophic (nutritional) disruptions on a major scale.

The end-Triassic event was much less severe but still resulted in major reductions in ammonoids, brachiopods, and marine reptiles in the oceans. On land there was a major faunal turnover in which labyrinthodont amphibians, early reptile groups, and mammal-like reptiles died out and were replaced by archosaurs, lepidosaurs, and mammals. Again, no evidence of an impact event is present and the extinctions are generally correlated with widespread marine regressions.

The Cretaceous–Tertiary boundary (K–T) mass extinction has been hotly debated, largely because of the bolide impact hypothesis of Louis and Walter Alvarez. Although the broad pattern of extinctions is known for marine organisms, the detailed picture is known only for planktonic foraminifera and calcareous nannoplankton. Study of the ranges of these micro-organisms shows that the extinctions of foraminifera occur over an extended period of time, starting well before and finishing well after the boundary. Cretaceous species of calcareous nannoplankton likewise survived across the boundary and became extinct some tens of thousands of years later. The macrofaunal record as yet shows insufficient resolution, although it is evident that brachiopods clearly suffered badly across the boundary and were replaced almost entirely by new species in the earliest Tertiary. Although much has been made of the extinction of ammonites at the end of the Cretaceous, there are too few ammonite-bearing sections to show if this was gradual or abrupt. On land, the evidence for a dramatic increase in fern species just above the boundary suggests the presence of wildfires, for ferns are usually the first plants to recolonize an area devastated in this fashion. However, in many sections a return to the Cretaceous vegetation is seen above the fern 'spike', indicating little extinction. Among the vertebrates a picture of gradual change is seen for mammals, with drastic reductions occurring only in the marsupials. The boundary also does not seem to have been a barrier for turtles, crocodiles, lizards, and snakes, all of which came through virtually unscathed. The dinosaurs did become extinct, and much argument has centred on whether or not their extinction was abrupt or occurred after a slow decline. In this context it must be noted that there is only one area where a dinosaur-bearing sedimentary transition across the Cretaceous–Tertiary boundary can examined; this is in Alberta and in the north-western USA. Records of dinosaurs in this area during the later part of the Cretaceous show a gradual decline in diversity with a drop from thirty to seven genera over the last eight million years of Cretaceous time. Although explanations of the extinction of dinosaurs have ranged from mammals eating their eggs, through terminal allergies caused by the rise of flowering plants, to the current ideas about bolide impacts, the answer is probably related to climate. A major regression of the oceans occurred at this point, resulting in a drop in mean annual temperatures and an increase in seasonality. The bolide impact would have dealt the *coup de grâce* to taxa that were already declining.

The major extinction event that occurred at the end of the Pleistocene primarily affected large terrestrial mammals. In North America, 33 genera of large mammals were lost; 46 were lost in South America, and 13 in Europe. In North America these included mammoths and mastodon, giant ground sloths, and glyptodonts. These extinctions coincided with a shift from a more equable to a more seasonal climate as a consequence of the end of the last glacial stage, thus suggesting a climatic cause. However, they also coincided with the arrival of humans in North America, which suggests that overpredation by these new and skilful hunters might have been the cause.

DAVID K. ELLIOTT

Further reading

Elliott, D. K. (ed.) (1986) *Dynamics of extinction.* John Wiley and Sons, New York.

Nitecki, M. H. (ed.) (1984) *Extinctions.* University of Chicago Press.

Raup, D. M. (1991) *Extinction: bad genes or bad luck?* W. W. Norton, New York.

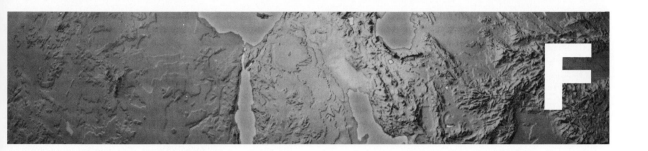

facies The term 'facies' was first used in 1838 by the Swiss geologist Amanz Gressly (1814–65) in reference to Jurassic rocks in the Jura Mountains. He found that the individual strata there change their characteristics laterally: they are not uniform throughout their extent and undergo facies change. Emile Haug (1907) gave the definition 'a facies is the sum of lithologic and palaeontologic characteristics of a deposit in a given place'. This useful concept immediately caught on and was used in a great many different ways. Today the word is usually qualified by a descriptive term (for example, 'sandy facies', 'graptolitic facies'), or by a genetic term (for example, 'deltaic facies', 'pelagic facies', etc.). The usage is also extended to cover associations and packets of rock units rather than single strata.

Several meanings have been attached to the word 'facies'. It may refer to the petrology of a sediment: *petrofacies* e.g. quartz sandstone petrofacies; or it may refer to the lithology of the deposit: *lithofacies*, e.g. cross-bedded sandstone facies; to the inferred mode of deposition: e.g. *aeolian facies*; to the included fauna: *biofacies* e.g. algal-limestone facies; or to a deposit formed with a particular tectonic background: *tectofacies* e.g. stable-shelf sandstone. Alternatively, it can be used to describe the inferred environment if deposition: e.g. *estuarine facies*. Particular facies may together form a characteristic *facies association* or may succeed one another to form a characteristic *facies sequence*.

Facies analysis is an important tool in the reconstruction of ancient environments of deposition and hence of palaeogeography. It has also proved invaluable in the prediction and exploitation of economic deposits.

The boundaries between facies, either in the vertical or the horizontal direction, may be sharp or indistinct, depending to some extent upon how the facies are defined. Many writers have used definitions so vague as to bring the concept into disrepute; others have used them as synonymous with lithological units (rock bodies).

Sedimentary facies commonly occur in associations typical of specific depositional environments, and the environments occur next to one another in sequence. This led Johannes Walther in 1894 to declare that facies that occur in conformable vertical successions also occur in laterally adjacent environments. This has been referred to as Walther's 'law of correlation of facies'.

Since the studies by the Finnish geologist P. Eskola (1915), the term *facies* has also been used to denote metamorphic rocks containing similar assemblages of minerals (for example, 'greenschist facies', 'amphibolite facies') that are interpreted as having formed under comparable conditions of temperature and pressure at depth in the Earth's crust.

D. L. DINELEY and G. EVANS

Further reading
Reading, H. G. (ed.) (1996) *Sedimentary environments: processes, facies, and stratigraphy* (3rd edn). Blackwell Scientific Publications, Oxford.
Schoch, R. M. (1989) *Stratigraphy: principles and methods*. Van Nostrand Reinhold, New York.

fault rocks Fault rocks are a texturally distinct category of rocks that occur smeared along fault planes. They form in response to friction-generated heating and localized crushing during faulting. Their importance comes from the fact that they provide unique insights into the temperature–pressure conditions during faulting, and the age of faulting. The dating of hydrothermal muscovites within fault rocks associated with crustal-scale normal faults in the Nepal Himalayas has made it possible to relate the timing of uplift during orogenesis to the subsequent collapse of the elevated topography. Fault rocks can also be of considerable economic significance. In deltaic settings, as in Nigeria and Sarawak on the island of Borneo, many oilfields are characterized by large anticlines bounded by faults that appear to juxtapose permeable rock types one each side of the fault. However, impermeable muds smeared along these fault planes mean that, in spite of the absence of an obvious oil trap, the muds can seal significant accumulations of oil.

A study of thrust faults in the Outer Hebrides of Scotland led Richard Sibson to classify fault rocks according to their grain size, fabric, and the extent to which they underwent recrystallization during their formation. In particular, he recognized two principal categories of fault rocks—cataclasites and mylonites—which he suggested are respectively diagnostic of faulting above and below the brittle–ductile transition. Cataclasites exhibit fragmented, brecciated textures, whereas mylonites are recrystallized, extremely hard metamorphic rocks with penetrative strain fabrics.

JONATHAN P. TURNER

fault scarp A fault scarp is a type of escarpment that arises from repeated earthquake displacement on a fault in the Earth's crust. Strictly speaking, the term 'fault scarp' refers to the scarp formed along the intersection of a fault plane with the Earth's surface and should only be used in the case of a fault that reaches ground level. However, the term is also used loosely of an escarpment that follows the approximate surface line corresponding to a fault that is active at depth. The latter should properly be called a *fault-line scarp*.

For a true fault scarp to form, displacement at the hypo-centre of the earthquake (the focus of the earthquake, typically 10–20 km beneath the surface) must be sufficiently large to propagate to the surface. Because displacement across the fault plane decreases away from the focus, this condition is generally met only for faults associated with earthquakes of magnitude 6 or greater, in extensional or strike-slip settings. Each fault movement in an earthquake results in a sudden increase in the size of the scarp, by an amount ranging from a few centimetres for a magnitude 5.5–6 event to several metres for a magnitude 7.5–8 earthquake. The scarp may also grow more gradually as creep occurs on the fault plane in the hours and days following the earthquake. The freshly exposed portion of the scarp is formed of shattered fragments (fault breccia) from the rocks on either side of the fault and is usually highly polished. This breccia is easily eroded, and the polished surface usually persists for only a few metres above the base of the scarp.

Thrust faulting occurs on shallow dipping faults and does not commonly produce a true fault scarp, because the tip of the upper, overthrust block tends to collapse, leaving a sinuous scarp that do not coincide with the fault plane. Strike-slip faulting may result in a scarp if there is a significant extensional or thrust component of motion, or if rock-types with different weathering characteristics are juxtaposed by faulting. Failing this, both active and inactive faults are often marked by a *fault-line valley*, a line of preferential erosion along the fault.

Fault scarps are generally visible only on faults that have been active within the past few tens to hundreds of thousand years. The competing effects of fault motion, erosion, and sediment deposition all affect the form of the scarp and the latter two will eventually eliminate it after movement on the fault ceases. If the rates of erosion of the rocks either side of the fault are known or can be assumed to be negligible, and the sediments deposited contain material that can be dated, then the magnitude of the scarp can be used to determine the age at which the scarp began to form and the long-term rate of movement on the fault. Additionally it may be possible to calculate the displacement in individual earthquakes and the recurrence interval of such events. More information about the relative significance and rates of movement of neighbouring faults can be gleaned by observing their effects on the drainage pattern in the area. Both major and minor faults influence the directions of river channels, and even minor faults can have *wind gaps*, low points on the scarp where a stream once crossed the fault but has now been diverted because of the changed topography.   PETER CLARKE

faults and faulting A fault is a fracture, or a zone of several fractures, across which movement has taken place, as shown by the fact that a reference marker of some kind has been offset. As such, faults differ from joints, which exhibit no measurable displacement. In the upper crust, above a depth of about 15 km, fault displacement is the principal means by which energy generated at plate margins is dissipated. Understanding the geometrical expression of faults, and the mechanisms by which they develop, is crucial to a number of key needs of modern society, such as the exploitation of fault-related energy and mineral reserves, and the prediction of earthquake hazards.

Classification of faults

Two main categories of fault geometry are recognized: *dip-slip* and *strike-slip*. Dip-slip faults are those faults on which the slip vector is parallel to the dip of the fault, whereas strike-slip faults have a near-horizontal slip vector. Dip-slip faults fall into two further subdivisions (Fig. 1). In normal faults the rock overlying a dip-slip fault plane, the *hanging-wall*, moves down relative to the underlying *footwall*. Conversely, *reverse faults* push hanging-wall rocks up, relative to the footwall.

(a)

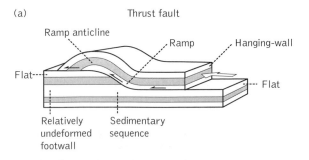

(b)

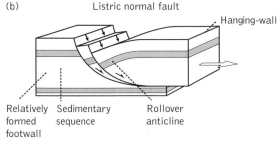

Fig. 1. Typical features of linked, thin-skin thrust and reverse faults. (Modified from Powell (1992).)

Normal faults are the principal structure by which the upper crust undergoes an overall lengthening, or extension. By placing relatively older, deeper rocks on younger footwall rocks, however, reverse faults cause shortening. A vertical profile, such as a borehole, through a reverse fault will therefore encounter the same vertical succession of rocks twice, in both the hanging-wall and the footwall, whereas a borehole through a normal fault will record apparently thinned, or even 'missing', rock units across the fault plane (Fig. 1).

Strike-slip faults are also subdivided according to the relative displacement between each side of the fault. If we imagine ourselves standing on one side of a strike-slip fault that cuts a linear marker, the fault is termed *dextral* or *sinistral*, according to whether the marker is offset to our right or left, respectively.

Linked fault systems

In practice, individual exposures of faults seldom display perfect dip-slip or strike-slip geometry. Slip-parallel striae, termed *slickenside lineations*, usually indicate oblique slip, in which displacement is sub-vertical or sub-horizontal. Furthermore, dip-slip and strike-slip faults commonly occur in the same setting, forming part of a *linked fault system*. In cross-section, these consist of a staircase-like profile of sub-horizontal *flats*, and steeper *ramps*, which cut through the footwall, detaching a relatively thin-skinned slice of hanging-wall rocks. The idea that disparate faults with different senses of slip often belong to a single, linked fault system spawned a major new research theme in the 1970s and 1980s. Such studies explored the range of patterns of linked fault systems, and the way in which a hanging-wall succession deforms internally in response to its movement over variously oriented ramps and along flats in the underlying fault system. These studies are based on data obtained from fieldwork, especially from mapping reverse fault systems in the Canadian Rockies, the Appalachians, and the European Alps, and from seismic reflection profiles imaging normal fault systems in the sub-surface.

A more-or-less universal observation from linked fault systems is that the constituent faults have a *listric*, or curviplanar cross-sectional profile (Fig. 1). Listric fault systems are particularly widespread at fairly shallow depths within sedimentary basins. They form most commonly in well-layered rock sequences, comprising a variety of strong and weak horizons, when gravity is important as a driving force. At the surface, the ramps of listric dip-slip faults will commonly attain angles of between 30° and 60° to the horizontal, but their dip will become gentler with depth, before they become detached along a flat.

Flats commonly consist of a particularly weak rock-type, such as halite (rock salt) or a mud containing trapped, high-pressure water. The combination of the relative weakness of these low-strength, *décollement* (detachment) horizons and the entrapment of high-pressure fluid makes them an ideal décollement zone because they minimize the frictional resistance to movement of the hanging wall over the footwall. Off the Caribbean island of Barbados, where a major reverse fault

system is actively deforming the sea floor, mud volcanoes record the advance of a flat up to 10 km in front of the ridge marking the line of most intensive hanging-wall deformation. Provided the frictional resistance to movement along a flat is minimal, the hanging-wall will slide passively over the footwall, and little or no internal deformation will occur.

Faulting and folding

Two main fold structures are associated with deformation above ramps, *rollover anticlines* and *ramp anticlines* (Fig. 1). Rollover anticlines form where a relatively steeply inclined ramp in a listric normal fault system flattens with increasing depth to connect with a sub-horizontal flat. Gradual displacement along the fault system will, theoretically, open an expanding, unfilled gap between the footwall and the hanging-wall. In practice, however, such gaps are never observed as the hanging wall responds by collapsing into the gap. The result is a monoclinal rollover anticline, its tightness increasing progressively with continued fault displacement. Excellent examples of rollover anticlines are associated with listric fault systems affecting oil- and gas-bearing sandstones in Tertiary sedimentary sequences in the Gulf of Mexico and the Niger delta.

Ramp anticlines record a comparable process to the formation of rollovers, but here applied to listric reverse, or *thrust*, faults. In the case of thrusts, an anticline will always form above an underlying ramp: the empirical rule states that the angle between a ramp and the geological marker horizons it cuts must be approximately maintained. Thus, if thrust ramps attain angles of around 30° to the horizontal (and hence 30° to a marker, such as sedimentary bedding), then subsequent movement of the hangingwall up the ramp and on to a flat must be accompanied by folding as the bedding flexes to maintain its original angle with the flat. Continued displacement along the thrust will result in the increasing separation of the ramp anticline from the ramp above which it originally formed. Alternatively, a new fault may develop which cuts into the former footwall, thereby generating a closely spaced, or *imbricate*, array of faults. According to whether or not they connect at both the foot and the top of the ramp, such structures are termed, respectively, *duplex* or *imbricate fans* (Fig. 2). Studies of the detailed history of internal deformation of hanging-wall rocks that have experienced movement over multiple ramps reveals complex sequences of repeated bending and unbending. For this reason, ramp anticlines are commonly referred to as *fault-bend folds*. A drive across any of the world's currently or recently active thrust belts—the foothills of the Himalaya, or the Jura Mountains of the European Alps—demonstrates spectacularly the role of fault-bend folding in generating their topography. These regions are characterized by linear ranges of hills and valleys, the hills marking the ramp anticlines, the valleys and plateaux expressing the flats between thrust ramps. Ramp anticlines in thrust belts in Colombia, Papua New Guinea, and particularly the Zagros Range in Iran, have yielded some of the largest single oil accumulations in the world.

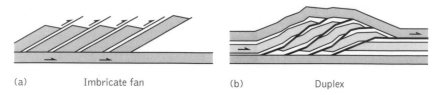

Fig. 2. Schematic cross-section showing the typical geometries of an imbricate fan and a duplex. The sections emphasize the way that duplexes are bounded by a fault, both above and below, whereas an imbricate fan is unconstrained by an overlying fault. Note that these structures are not confined to thrust belts, but can also occur in extensional settings.

Strike-slip faulting

So far, discussion of linked fault systems has concentrated exclusively on dip-slip settings. However, many of the characteristic features of linked fault systems—their staircase profile, comprising variously orientated ramps and flats, the interconnection between disparate fault types, and the intimate relation between faulting and folding—are seen also in areas undergoing strike-slip faulting—with the exception, that is, that the ramps and flats are turned on end so that the staircase profile is seen, not in cross-section, but in plan view (Fig. 3). Strike-slip fault systems are therefore dominated by long straight fault segments along which the adjacent sides of the fault move passively past one another. Deformation is largely confined to the offsets of these straight segments, at *bends* along the fault (Fig. 3). Two types of bend are recognized: *transtensional bends* (Fig. 3), where the stresses are locally tensional, and compressional *transpressional bends*. Seismic reflection profiles through transtensile and transpressive bends reveal dendritic arrangements of normal and reverse dip-slip faults respectively, termed *flower structures*. Rocks adjacent to a major strand of an active strike-slip fault commonly display evidence of repeated episodes of reverse faulting, and normal faulting, as they pass through multiple zones of transpression and transtension.

Strike-slip faults are the most long-lived of all fault systems, sometimes remaining active for 100 Ma, over which time they can accumulate as much as 5000 km displacement. The only structures capable of absorbing such enormous displacements are the plate margins. Among the best-known examples of these trans-lithospheric, or *transform*, faults are the dextral San Andreas and Alpine faults in California and New Zealand respectively.

Growth faulting

One of the most exciting developments in the study of linked fault systems has been the realization that faulting and sedimentation are commonly synchronous processes. In the case of displacement along a listric normal fault, the development of a rollover anticline will cause a topographic depression, itself a focus of further sediment accumulation. Consequently, sediment will accumulate preferentially in the hanging-wall of an active normal fault, thereby generating a thicker sedimentary succession. This is *growth faulting* and it can lead to dramatic changes in sedimentary thicknesses on each side of a fault, the thicker sedimentary succession sitting always on the downthrown side of a growth faults. Furthermore, growth faulting will lead to variable inclination of growth strata; the older, stratigraphically deeper beds will dip more steeply into a listric normal growth fault because they have experienced a longer period of fault displacement and bed rotation.

Growth faulting is also an important process in active thrust belts, in which thicker successions are predicted to accumulate in the footwalls, where extra space is continuously being created by hanging-wall uplift. By the same token,

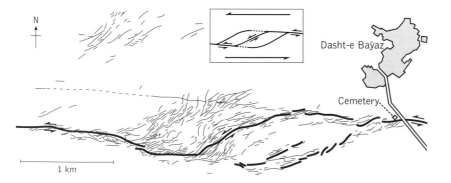

Fig. 3. Map of mainly normal fault segments in a small zone of intensive deformation associated with a transtensional bend that broke during a destructive earthquake in 1968 on the sinistral Dasht-e-Bayaz strike-slip fault, Iran. Major fault strands are shown in bold, subsidiary faults and fractures are lighter. Note the comparison between the plan view strike-slip geometry depicted here, and the dip-slip duplex illustrated in Fig. 2. (Modified from Tchalenko and Ambraseys (1970).)

actively amplifying ramp anticlines experience syntectonic erosion of their crests, thereby providing a ready source of sediment. Using seismic reflection profiles, analysis of growth strata associated with modern thrust faults in the Philippines, Venezuela, California, and Oklahoma provides a promising means of assessing earthquake hazard. Here, dating of the growth strata yields a direct estimate of rates of fault-bend folding during the past 8 Ma. or so, the most rapid rates corresponding to the more earthquake-prone fault segments.

Crustal-scale faulting and earthquakes

Despite the undoubted importance of listric faults, deep geophysical surveys have demonstrated that the largest-scale faults which control the overall shape, position, and long-term development of sedimentary basins are, in fact, more commonly planar. Some planar faults have been shown to extend as deeply as 30–40 km, to the base of the crust, sometimes actually offsetting the Moho itself. In the northern North Sea, planar normal faults define many of the largest oilfields, such as the Brent field. Field-based studies of planar normal fault systems liken them to the behaviour of books on a bookshelf. With increasing extension, or shear, vertically stacked books will progressively rotate, attaining ever-decreasing angles to the horizontal. Examples mapped in the highly extended Basin and Range province of Nevada show early formed, highly rotated planar normal faults now orientated at near-horizontal attitudes.

Earthquakes occur during episodes of sudden fault movement and they express the catastrophic release of energy which builds up within a volume of rock immediately adjacent to a geologically stressed fault plane. The existence of earthquakes is perhaps the most graphic evidence we have for the ability of the upper crust to store energy elastically. Analysis of global earthquake data shows that the overwhelming majority of earthquakes of small to medium magnitude have their epicentres within the top 10–15 km of the crust. Below 15 km, where temperatures begin to exceed 250–300 °C, the brittle upper crust changes its physical properties and begins to deform in a more ductile fashion. This is the *brittle–ductile transition*, and it explains why few earthquakes are recorded from deeper than this seemingly arbitrary depth: beneath the brittle–ductile transition, only small quanta of elastic energy can accumulate along an active fault segment before it deforms plastically, and harmlessly. Studies of magnificently exposed planar normal fault systems in the Aegean region of Greece reveal a remarkable consistency in the dimensions of continental fault systems. Active faults are inclined at angles of between 40°–60° to the horizontal, and they are spaced 20–100 km apart, with individual faults attaining lengths of between 15–20 km. Such relative consistency has led researchers to suggest that it is the fairly uniform deformational characteristics of the brittle upper crust that exert the chief control over the dimensions of continental fault systems.

Why in general do large-magnitude earthquakes not occur above the brittle–ductile transition? This is due to the way in which the upper crust is criss-crossed by billions of pre-existing faults and fractures. From the scale of millimetres to hundreds of kilometres, all these fractures represent planes of weakness, each with the potential to be reactivated time and time again. In the upper crust, therefore, elastic energy will not usually build up for long before the magnitude of stress is sufficient to overcome the frictional resistance to reactivation of a pre-existing fracture, and only a small earthquake will result.

Slipping and creeping

Cumulative displacement on large fault systems can be of the order of thousands of kilometres in the case of strike-slip fault systems and hundreds of kilometres for dip-slip fault systems. However, measurement of *fault scarps*, the surface expression of recently ruptured fault planes, indicates that the amount of slip that accompanies even a large-magnitude earthquake is usually less than a metre. This introduces the notion of *co-seismic slip* and *aseismic creep* along faults, defined as fault displacement occurring respectively during and after an earthquake. Sporadic episodes of co-seismic slip take place at rates of between 100 mm and a metre per second, whereas more-or-less continuous aseismic creep, as the name suggests, occurs as slowly as 1 to 100 mm a year. Despite the considerably more dramatic and hazardous manifestation of faulting during co-seismic slip, it is actually responsible for generating as little as a millionth of the total length of a fault. Aseismic creep also has the effect of smoothing the spatial variation in displacement over the area of a fault plane. Most faults exhibit a smooth decrease in cumulative displacement, from a maximum at the centre of the fault to zero at the fault tip.

Faults and plate tectonics

Is there a systematic distribution of the different fault types according to plate-tectonic setting? Using earthquake data to map the global distribution of faulting, it has long been appreciated that most faulting is restricted to linear zones of intense activity, with few faults in between. These are, respectively, the plate margins and intra-plate regions. The dynamics of plate movement exerts the principal control over the global distribution of horizontal stress. Three main stress domains can be discriminated, each with characteristic types of faulting. Normal faults occur where the least principal, or tensile, stress is horizontal, and the greatest principal, or compressive, stress is vertical. Reverse and thrust faults are found in regions where the least (but not tensile) stress is vertical and compressive stress is horizontal. Lastly, strike-slip faults occur where both the compressive and tensile stresses are horizontal, and the intermediate stress direction, which is compressive is vertical. It is easy to see, therefore, why the main plate tectonic settings of reverse faults are the world's major mountain belts, where strong horizontal compression is generated in response to the collision of adjacent plates. Conversely, normal faults are commonest in proto-oceanic environments of continental splitting, where plate divergence generates strong horizontal tension. Strike-slip faults, however, characterize those plate margins where adjacent plates are moving past one another, neither colliding nor splitting apart. JONATHAN P. TURNER

345

Further reading

Bolt, B. A. (1993) *Earthquakes*. W. H. Freeman, New York.

Hancock, P. L. (ed.) (1994) *Continental deformation*. Pergamon Press, Oxford.

Twiss, R. J. and Moores, E. M. (1992) *Structural geology*. W. H. Freeman, New York.

faunal provinces Faunal provinces are large geographical areas within which the organisms, even perhaps up to the family level, have a distinct identity. Provinces may be separated by sharp or gradational boundries, and in some cases the fauna may be endemic, that is, not found outside a particular province. As marine invertebrates are the commonest fossils, most interest in past faunal provinces concentrates on the marine realm and particularly on the continental shelves. The main modern zoogeographical regions are controlled particularly by temperature and, therefore, by latitude so that we can delineate tropical shelf, warm temperate, cold temperate, and polar regions. Within these areas there are further subdivisions termed realms whose boundaries are controlled by the presence of warm or cold currents or the position of land masses. Thus the Isthmus of Panama separates two different tropical shelf faunas, so that of the 517 species on the Caribbean side and the 805 species on the Pacific side only 24 are held in common. The cold Humboldt current that sweeps up the western side of South America likewise restricts the tropical fauna to a few degrees south of the equator.

Similar faunal provinces can be detected in the fossil record and can help to corroborate ideas about continental movements through time, as well as providing information about palaeotemperatures and even the pattern of ancient oceanic current systems. The distribution of rugose corals in the Devonian of North America has been described in terms of faunal realms. During the Early Devonian two realms were separated by a ridge extending through the centre of the continent. Through the Middle and into the Late Devonian, however, a steady mixing of the faunas showed that the influence of the barrier was waning until it became inundated towards the end of the Devonian. Separation and amalgamation of faunal realms has been instrumental in tracing the movements of Gondwana (southern hemisphere) continents during the Late Palaeozoic and early Mesozoic. The mammal-like reptile *Lystrosaurus* is found in Antarctica, India, and South Africa but clearly could not have travelled across oceans to establish such a range. In the nineteenth century hypothetical land bridges were postulated to explain distributions such as this that we now know were caused by plate motions. The opening and closing of ocean basins can be timed by the increasing similarity or dissimilarity of faunas on the opposing coastlines. In the early Palaeozoic, North America and Europe were separated by the Iapetus ocean. During the Ordovician, only planktonic animals such as graptolites were able to establish themselves on both sides; during the Silurian benthonic organisms with swimming larvae such as brachiopods were able to bridge the gap; and by the Devonian the northern part of the ocean had closed and even freshwater fish occurred in both faunas. Accreted terranes, that is, portions of present continents that have come from other former continents or oceans, can also be recognized because they contain fossils from different faunal provinces. For instance, the Avalon terrane, which occurs on the east coast of North America, contains fossils otherwise known only from the Palaeozoic of western Europe. The explanation for their distribution is that after the Iapetus ocean closed in the late Palaeozoic it opened again a little further to the east, leaving part of the western shore of Europe attached to North America.

DAVID K. ELLIOTT

figure of the Earth The 'figure of the Earth' means the 'shape' of its gravitational field. Mean sea level closely follows this shape in oceanic areas, and we can extend the concept to continents by imagining narrow canals connected with the oceans. The water level would then follow a surface called the *geoid*. This is the surface of constant gravitational potential that most closely matches sea level. The figure of the Earth is the form of this geoidal surface.

The geoid is much closer to a smooth ellipsoid than is the Earth itself. The presence of continents and ocean basins does not show on a contour map of the geoid, even after the ellipticity, or equatorial bulge, is subtracted to make the smaller wrinkles more obvious. The continents are isostatically balanced: that is, they appear to be floating on the denser mantle. They are not related in any direct or simple way to the geoid undulations, which have deeper origins.

The dominant departure of the Earth's figure from a sphere is the ellipticity or polar flattening. This is the result of a balance between gravity and the centrifugal effect of rotation. The equatorial radius is about 21 km greater than the polar radius, about one part in 300, and all other geoid undulations are many hundred times smaller. The greatest departure of the geoid from ellipsoidal form is a 105-m low south of India. This is very small compared with undulations of the land surfaces or the 5-km mean difference in level between continents and ocean floors.

The combination of ellipticity and rotation gives a variation of gravity with latitude. Gravity at the poles is 0.53 per cent greater than its equatorial value. Most of this difference (0.35 per cent) is due to the centrifugal force itself, but the balance (0.18 per cent) is due to the elliptical form of the Earth's mass.

If the Earth were uniform in density then the equatorial flattening would be one part in 230. The fact that the observed flattening is much less is due to the concentration of mass towards the centre, where the centrifugal effect is relatively less important. Moreover, the denser a planet is the more effective is gravity in pulling it towards a spherical shape against the centrifugal distortion. This means that the inner layers of the Earth are less elliptical than is the surface and, as a consequence, their contribution to the ellipticity of the geoid is reduced.

The observed flattening of the Earth (1/298.26) is slightly greater than the equilibrium (hydrostatic) value that would be observed if the Earth were fluid (1/299.63). There are several reasons for the difference, which amounts to a 100 m excess equatorial bulge. The two most important reasons are : (1) the Earth tends to adjust its orientation to favour equatorial locations for denser material (geoid highs). This gives the lowest rotational energy for fixed angular momentum; (2) the polar regions were depressed by ice loads during the most recent series of ice ages and are still rebounding towards isostatic balance.   FRANK D. STACEY

figured stones *see* MEDIEVAL MINERALOGY AND FIGURED STONES

fiords (fjords) Fiords are relatively long and fairly narrow sea inlets, straight or curved in plan, and with steep sides that descend to considerable depths. Major fiords, such as the Sognefjord, Norway, can be up to 200 km long, 5 km wide, and 1300 m deep, although depths may decrease to around 200 m at shallow 'thresholds' within a fiord and at its entrance. Fiords are found in many glaciated or formerly glaciated coastlines across both hemispheres; around the shores of the Antarctic peninsula and Greenland, and along the Atlantic and Pacific seaboards of Canada, Alaska, Iceland, Norway, Sweden, Scotland, southern Chile, Tasmania, and the South Island of New Zealand. They are young (immature) features, eroded by glaciers during the multiple glaciations of the past 2 million years (the Quaternary). The glaciers developed in mountainous areas at high latitudes and flowed down to sea level through fluvial valleys. These preglacial valleys were eroded along zones of weakness in the bedrock, and thus both the main fiords and their tributaries mirror the structural bedrock geology by following the trend of bedrock lineaments.

Glaciers themselves erode in three ways. (1) Water at the base of the glacier (at the 'pressure melting point') is forced into joints and, as pressure decreases, it refreezes, causing fractions of the rocks to be wedged loose; this process is called 'plucking'. (2) Rock fragments frozen to the base of the ice may move along with the ice, grinding into the underlying bedrock in a process called 'scouring'. (3) Melt water under pressure may find its way to the base of the glacier, where it picks up loose material that is used in further grinding of the bedrock, often making potholes and canyons. This process is called 'glacio-fluvial erosion'.

Glacial erosion may change V-shaped fluvial valleys into U-shaped glacial valleys. The glaciers move in the direction of the surface slope, so that at the glacier base both ice and water can move uphill in relation to the basement. Thick glaciers can even erode below sea level. With their erosion, U-shaped glacial valleys become progressively deeper, and direct more

and more ice into 'ice streams'. At the confluence of valley glaciers, glacier velocity, and thus also their erosional power, is increased. During interglacial periods the steep, unstable U-shaped valley sides are subject to mass movements such as rock falls and large rock avalanches, often exploiting structural weaknesses in the bedrock. The result is the gradual widening of the valleys, which in turn makes room for larger glaciers in the following ice age.

During deglaciation, glaciers 'calve' into the sea and, in valleys where glaciers had earlier eroded below sea level, fiords develop as rising seas flood the inlets, forcing calving glaciers to retreat up-valley. Along tributary valleys, 'hanging fiords' form, being shallower than the main fiords because of the reduced erosional power of the smaller glaciers that occupied them. Shallower zones or 'thresholds' within the inner portions of the main fiord usually result from thick submarine moraines, deposited during short-lived episodes of climatic deterioration, but the outermost threshold that limits its entrance is generally a bedrock sill. These sills mark areas at the coast where low relief makes it possible for a glacier to spread out and thus lose its erosional power. Despite their impressive depths, major fiords such as the Trondheimsfjord, Norway, may contain sediments up to 600 m thick, which are mainly the depositional record of deglaciation.

INGE AARSETH

fire as a geomorphological agent In many environments, particularly drought-prone ones, the frequency of fires is high (once in 1–100 years). Fire can cause rock weathering by exfoliation and even cracking across boulders through the differential expansion of surface and subsurface, coarse-grained rocks being most susceptible. The amount of rock affected by exfoliation can be substantial. For instance, fire in the Blue Mountains, Australia caused the loss of 2–6 kg m^{-2}, involving half the original sandstone surface.

Fire is also important geomorphologically through its alteration of the vegetation and litter cover, soil properties, and hydrology. Loss of the thermal insulation of the vegetative cover is particularly important in permafrost terrain. Deep permafrost thawing may result in irregular surface subsidence and thermokarst pits and mounds.

Fire leaves the soil more susceptible to a range of erosion processes. Tree and shrub mortality causes reduced rooting strength. Evapotranspiration is reduced, causing increased surface moisture availability. Snow hydrology and soil properties (including reduced aggregate stability and enhanced soil hydrophobicity (water repellency)) can be altered, with important erosional consequences. Hydrophobicity has been recorded in dry, unburned soils but may be intensified by fire. Burning causes organic matter to be volatilized, some of which descends steep temperature gradients in the upper few centimetres of soil and condenses, forming a hydrophobic layer. The degree of repellency depends on temperature-gradient steepness, soil moisture, and physical properties. The result is reduced infiltration and increased overland flow, aided by less litter, raindrop compaction, and pores blocked by fine soil particles (on a wettable soil surface), and sometimes soil fusing.

These fire-induced changes often cause soil erosion peaks several orders of magnitude above rates typical of unburned sites. Soil losses decline sharply within the first 1–2 years, but may take many years to recover completely. In some areas, more frequent, shallow, rapid mass-movement processes have been attributed to fire-induced losses of rooting strength and higher groundwater levels. Forest fire can increase snow avalanche activity on steep slopes through the reduced anchoring effect of the vegetation. On slopes affected by permafrost, melting of the ice matrix may liquefy sediments and promote mudflows and slumping. Other fire-enhanced erosion processes include gravity flow of dry soil particles, surface creep, wind erosion, needle-ice formation, and rill, gully, and sheet-wash action. Where the soil surface is hydrophobic, rainsplash detachment of particles can be enhanced through dry, repellent soil being easily detached even from beneath a water film.

Increased overland flow and sediment losses on hillslopes after fires can cause higher stream discharge and sediment yields. Sediment lags and storages are often altered, with the response of sediment transport varying spatially and temporally. Organic stream load may also be affected.

Overall, the effectiveness of fire in promoting accelerated erosion varies with the intensity and frequency of fire and landscape erosion potential, with frequent, intense fires on steep slopes causing most erosion. On such landscapes, fire disturbance may account for much of the long-term sediment yield.

RICHARD A. SHAKESBY

fish *see* EARLY VERTEBRATES AND FISH

flash floods
Flash floods are a significant threat to lives and property in many parts of the world. Unlike 'slow-rise' floods, they result from too much water in too little time. The amount of lead time varies from place to place. In some localities 6 hours' warning can be given, but in arid, mountainous, or canyon regions only half an hour's notice is possible. Loss of life generally presents a larger problem than property losses. In 1996, for instance, more than 800 people were killed in flash flooding in South Africa, Morocco, and Afghanistan.

Flash floods occur in both rural and urban areas. Steep topography, sparse vegetation, and infrequent but intense thunderstorms typify many flash-flood hazard areas. Flash floods on alluvial fans are an increased threat as the population living in hazardous areas continues to rise. Urban environments, where vegetation has been removed, where bridges and culverts constrict flow, and where building and paving have added greatly to impermeable surfaces, also present an increasingly serious flash-flood problem. Urban drainage problems compound the effects of severe thunderstorms with catastrophic results. Many of these storms, if they occurred in uninhabited areas, would pass without any notice or impacts. Levee breaks, dam breaks, and ice jams can also result in serious flash floods.

Two flash-flood case studies
Rapid City, lying at the foot of the Black Hills in South Dakota (USA), has developed along Rapid Creek. On 9 June 1972, heavy rain began to fall just after 6 p.m. Up to 15 inches (38 cm) of rain fell in less than 6 hours. The spillway at Pactola Dam, upstream, plugged up with cars and house debris, causing the water level to rise 3.6 m (12 feet). A dam then failed at about the same time as the natural flood crest occurred, unleashing a torrent of water on the city. The death toll was 238 people.

On 31 July 1976 the Big Thompson Canyon, near Estes Park, Colorado (USA), was filled with residents and visitors. That Saturday night a flash flood ravaged the canyon, and 140 people lost their lives. Heavy rain fell over an area of 70 square miles (180 km^2) in the central portion of the Big Thompson watershed between 6.30 and 11.00 p.m. The most intense rainfall, between 12 and 14 inches (30 and 36 cm), fell on slopes in the western end of the canyon. (The flood washed out all stream and rain gauges, and accurate measurements were thus not possible.) If people had climbed to higher ground, rather than driving their cars or staying in their homes, fewer lives would have been lost.

Reducing the threat: warning systems and advances in forecasting
Structural control measures are not effective because of the local nature of flash floods and the short lead times. Because of the threat to humans, most efforts at mitigation focus instead on the development of detection and response warning systems. The burgeoning number of ALERT (automated local emergency response in real time) systems has been the most dramatic type of hazard reduction effort since 1980. Successful warning systems must include detection and response elements, but often only detection is provided. Detection alone does not save lives unless mechanisms are in place to warn the populations at risk.

ALERT systems are also being used extensively for water-quality monitoring, water-supply decision making, fine-weather forecasting, and air-quality monitoring. ALERT use for reservoir management and watering of urban golf courses saves cities millions of dollars a year and pays for the cost of the system. Internet capability enables officials and flood-plain residents to monitor gauge conditions 24 hours a day in some places, such as Maricopa County, Arizona. Knowing the conditions in real time is a step towards a warning but it is not an effective mechanism for ensuring that people are notified and take appropriate action in a timely fashion.

Flash-flood forecasting is becoming more reliable. Doppler radar and geographical information systems, often combined with ALERT data, help to pinpoint the severe storms. However, because of the impacts of fires, local land-use patterns, and the limitations of the technology, some of the worst flash floods are not recognized until after the damage has occurred.

In the USA and in Europe most flash-flood deaths occur in cars. People underestimate the power of water. Flowing water less than a metre deep can carry away a car. People often feel a false sense of security in their cars. For instance, in 1996 in Pennsylvania, police closed flooded roads. Then they stood in the water and gave 50 tickets to motorists who tried to drive through the barriers on the closed roads. If flash-flood deaths are to be reduced, motorists in particular must be educated to heed warnings and interpret environmental cues (the length of a storm, the soil moisture, and the sound of the river), which often indicate when to climb to high ground. Nevertheless, some flash-flood victims die downstream of the rain, in 'dry' areas.

Technological advancements in forecasting make it more likely that major flash floods will be predicted in advance, but recent research on the seasonality of flash floods has made prediction more complicated, and forecasting rare events, especially when they take place outside 'rainy' seasons, is extremely difficult. In the 1990s, for example, summer flash floods took their toll on tourists: 79 people were swept to their deaths by a raging river in Spain in 1996, and 11 tourists died in a narrow canyon in Arizona in 1997. In this respect, hydrological science and meteorology are still far from being able to predict, before the rains come, which thunderstorms will turn out to be the worst killers.

E. GRUNTFEST

Further reading

Gruntfest, E. and Huber, C. J. (1991) Toward a comprehensive national assessment of flash flooding in the United States. *Episodes*, **14**(1), 26–35.

floods Physically, a flood is defined as a high level of water which overtops either a natural or an artificial river bank, or rises above normal levels or sea defences in coastal environments. Unless such an event poses a threat to human life or property, however, it cannot properly be considered as an environmental hazard.

The physical causes of floods are highly variable. Figure 1 distinguishes between coastal and river floods. River floods can be caused by a number of atmospheric, seismic, or technological hazards and some scientists have argued that deforestation, particularly in large areas of the Himalayas, has increased the likelihood of flooding in recent years. The flood hazard may be exacerbated by various human activities, particularly urbanization (see *hydrological cycle*) and by the construction of flood defences. Urban areas, for example, increase flooding for four major reasons: (1) the creation of impermeable surfaces, which cause more surface runoff; (2) the presence of smooth surfaces, which increase flow velocities; (3) the constriction of natural channels by bridges and other structures, which can increase the depth of floods; and (4) increasing urbanization leading to the construction of urban storm drains, many of which cannot now cope with the increasing peak flows. In the UK alone, over 3000 sewer failures are reported annually. Many of these lead to the surging of water through the nearest point of escape in the sewer system.

Many scientists have argued that the construction of embankments and levees as flood defences provides a false sense of security for those people occupying surrounding low-lying land. In its natural state, the river floodplain provides an opportunity for the dissipation of energy and the attenuation of flood peaks. Construction of levees and embankments prevents the floodplain from performing this function and transfers the problem further downstream to areas which were not subject to flooding. Once started, there seems little option other than to continue constructing flood defences and levees. A dramatic example of this problem is to be found in the lower reaches of the Yellow River in China, near the city of Jinan. Levee building, once started, has continued over the past fifty years because of rapid sedimentation on the bed of the river has been too rapid for dredging to keep pace with it. The river bed currently stands between 6 and 10 m above the level of the surrounding floodplain. Once levees are breached, the severity of flooding and the longevity of the flood event is dramatically increased because water cannot return easily to

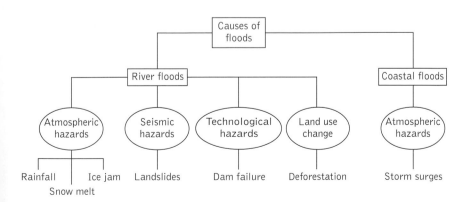

Fig. 1. Physical causes of floods in relation to other environmental hazards. (Atmospheric hazards resulting from high rainfall are the most important contribution to flood hazard.) (Modified from Smith (1992) Figure 11.3.)

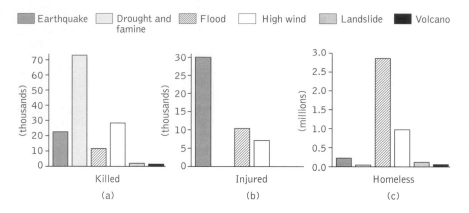

Earthquake ▪ Drought and famine □ Flood ▨ High wind □ Landslide ▩ Volcano ■

Fig. 2. The average number of people killed (a), injured (b), and made homeless (c) resulting from a range of natural hazards over the 25-year period 1968–92. (Data from *World Disasters Report* (1994), Table 3.)

the channel. This problem was graphically illustrated during the 1993 Mississippi floods.

Flooding is the most widespread of all geographical hazards. There were significantly more reports (1302) of flooding worldwide over the 25-year period 1968–92 than any other natural hazard with the exception of high winds (1494), yet in general floods cause less loss of life than earthquakes, drought and famine, and high winds (Fig. 2a). Although there are fewer mortalities, floods are responsible for a high incidence of injuries (Fig. 2b) and have a significant impact upon homelessness in comparison with other natural hazards (Fig. 2c). In most cases, floods are of limited geographical extent. In England and Wales and in Australia, for example, less than 2 per cent of the population live in flood-prone areas. In the United States, over 10 per cent of the population is at risk from flooding, particularly along the main floodplain of the Mississippi river. In contrast to the relatively small land areas subject to flooding in most countries of the world, the 1988 floods in Bangladesh inundated almost 50 per cent of the country to a depth of one metre. Nearly 1500 people died as a result of this flood alone.

While it is possible to generalize about the social impact and consequences of flooding by means of the figures given above, flooding is a complex subject because of the various causes of the flood hazard and human responses to it. Some particular areas are at greater risk of flooding than others. For example, populations occupying low-lying parts of active floodplains and estuaries, small basins subject to flash floods, areas below unsafe and inadequate dam structures, low-lying inland shorelines, and alluvial fan environments are all at high risk. In general, the nature of the flood hazard is different from other natural hazards (see *droughts*). With limited exceptions, floods tend to affect relatively small geographical areas and are short-lived, usually lasting only days to weeks. Population displacement is usually localized but structural damage to housing, communications, and power supplies is often severe: the estimated damage caused by flooding worldwide between 1988 and 1992 was in excess of US$8.5 billion. Damage is a function not only of the depth of water, particularly in river floods, but also of the velocity of the river

and the amount of silt it carries: more severe damage is caused by high-velocity silt-laden rivers.

Despite increased global investment in flood control, flood losses continue to increase. This is true even when allowance is made for inflation. There would appear to be only two plausible explanations for this trend in recent years. First, that there has been an increase in the magnitude of floods caused by changes in climate. There is evidence to suggest that damaging floods in Eastern Australia, for example, have increased since 1945 in comparison with the previous hundred years or so. The second explanation, and one that by consensus seems more likely, is that the increased risk has been driven by greater floodplain occupancy putting more people at risk to the flood hazard.

The responses to flooding have evolved in what hydrologists have recently referred to as *the structural era* (1930s–60s), *the unified floodplain management era* (1960s–80s), and *the post-flood mitigation era* (1980s–). The structural era is a period in which reliance was almost totally placed in engineering solutions through the construction of reservoirs and levees and through channel improvements (usually widening and straightening). The unified floodplain management era is characterized by the use of a combination of mitigation measures which in part utilized structural approaches but is dominated by the use of non-structural approaches, which included: (1) improved flood warning, through the use of weather radar, rainfall and flood level telemetry, and computerized flood forecasting models; (2) land use planning, where high-value developments in flood-prone areas are controlled and regulated through the planning system; and (3) increased use of insurance to offset the costs of clean-up to national and local government. These measures have often proved to be less successful than anticipated; many of them rely on government-run rather than private schemes to be effective.

The post-flood mitigation era is one which is still emerging, but it includes measures taken in the post-flood period to effect better control or minimize risk in the future. A good example of such an approach is provided by an economic study of Soldiers Grove in the 1980s. Soldiers Grove is a small settlement located on the Kickapoo river in south-west Wisconsin, USA, which was severely damaged by a series of

Dynamic pressures

Unsafe conditions
produced by

Floods:
hazard types

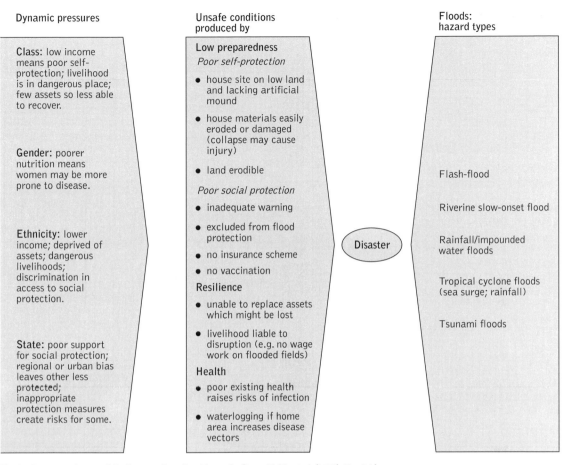

Class: low income
means poor self-
protection; livelihood
is in dangerous place;
few assets so less able
to recover.

Gender: poorer
nutrition means
women may be more
prone to disease.

Ethnicity: lower
income; deprived of
assets; dangerous
livelihoods;
discrimination in
access to social
protection.

State: poor support
for social protection;
regional or urban bias
leaves other less
protected;
inappropriate
protection measures
create risks for some.

Low preparedness
Poor self-protection

- house site on low land
  and lacking artificial
  mound

- house materials easily
  eroded or damaged
  (collapse may cause
  injury)

- land erodible

Poor social protection

- inadequate warning

- excluded from flood
  protection

- no insurance scheme

- no vaccination

Resilience

- unable to replace assets
  which might be lost

- livelihood liable to
  disruption (e.g. no wage
  work on flooded fields)

Health

- poor existing health
  raises risks of infection

- waterlogging if home
  area increases disease
  vectors

Disaster

Flash-flood

Riverine slow-onset flood

Rainfall/impounded
water floods

Tropical cyclone floods
(sea surge; rainfall)

Tsunami floods

Fig. 3. Pressures that result in disasters from flood hazards. (From Blakie *et al.* (1994), Fig. 6.1.)

floods in the 1970s. Here, two post-flood solutions were considered; the construction of a flood storage reservoir with associated river channel levees; and the relocation of the urban population away from the flood-prone area. The cost of the two schemes was roughly equal and other benefits accrued as a result of relocation of the urban population.

While many of the schemes identified above may be applicable to small urban populations, it is not surprising that many of them are not appropriate in countries where flooding can cover a high percentage of the land area. In less developed countries, disaster aid, which includes technical assistance as well as disaster relief, is an essential factor in flood mitigation. After the 1988 floods in Bangladesh, the UN Development Programme commissioned a range of studies by overseas experts, who recommended an increased reliance on the construction of embankments along the major rivers. The scheme had an estimated cost of US $6 billion with an annual maintenance cost of US $600 million. On grounds of cost alone this scheme is unlikely to be implemented. Even then, the

social costs have not been taken into account, for both fish farmers and jute farmers benefit from the spread of the monsoon floods in normal seasons. The proposed scheme would also be unable to cope with the most extreme floods and might therefore increase the hazard during these events.

Social scientists have argued that flood hazards impart a variable impact on people according to vulnerability patterns which are generated by the socio-economic system in which they live. This is seen as a function of class relations and the structures of domination within the society, which determine levels of ownership, the means of production, and the ability of individuals to respond to the flood hazard. Figure 3 attempts to summarize the main factors which could turn flood hazards into potential disasters. In common with droughts, vulnerability to the flood hazard in less developed countries is essentially a function of strengths and weaknesses in the society. The homeless, women, and children appear to be those groups in society exposed to the greatest risk.

IAN D. L. FOSTER

Further reading

Blakie, P., Cannon, T., Davis, I., and Wisner, B. (1994) *At risk: natural hazards, people's vulnerability, and disasters*. Routledge, London.

Burton, I., Kates, R. W., and White, F. G. (1993) *The environment as hazard*. Guildford Press, New York.

Penning-Rowsell, E. C., Parker, D. J., and Harding, D. M. (1986) *Floods and drainage. British Policies for hazard reduction, agricultural improvement and wetland conservation*. Allen and Unwin, London.

Smith, K (1992) *Environmental Hazards*. Routledge, London.

World Disasters Report (1994) International Federation of Red Cross and Red Crescent Societies, Geneva, Switzerland.

fluid inclusions

Fluid inclusions are bubbles of liquid trapped inside crystals, and were among the first new features to be discovered with the introduction of microscopic studies of thin sections of rocks. A seminal paper by H. C. Sorby, published in the *Quarterly Journal of the Geological Society* in 1858, recognized many of the major inclusion types and showed how useful information could be obtained from them. Strictly speaking liquid, vapour, or supercritical fluid (e.g. methane) may be present in fluid inclusions. Most workers would include melt inclusions in igneous minerals in the same general category, on the grounds that they were liquid when trapped in the host crystal.

Fluid inclusions are particularly common in vein minerals; myriad fine inclusions give rise to the milky appearance of much vein quartz. However, fluid inclusions occur in a wide range of rock-types, including diagenetic overgrowths in sediments, veins in active geothermal systems, hydrothermal ore deposits, metamorphic rocks and veins, granites, pegmatites, and some other types of igneous rock. Melt inclusions are particularly common in volcanic phenocrysts, especially in andesites. In addition to quartz, fluid inclusions can occur in carbonates, gypsum, halite, fluorite, feldspar, garnet, pyroxene, amphibole, tourmaline, olivine, and sphalerite; indeed in most minerals except phyllosilicates. Fluid inclusions have also been studied in opaque ore minerals, using infra red microscopy.

Most fluid inclusions are small, only a few microns to tens of microns in diameter, but large inclusions visible to the naked eye are known. Fluid inclusions can be trapped in crystals as they grow (*primary inclusions*), in which case they often define zones reflecting the morphology of the growing crystal, or can form during the resealing of cracks in a pre-existing crystal (*secondary inclusions*). Water is the most common fluid present, but carbon dioxide, methane, heavier hydrocarbons, and nitrogen can also occur. If the fluid carried a large dissolved load, material may precipitate as *daughter crystals* during cooling to surface conditions. Halite is the most common daughter mineral, but many other salts and some silicate minerals also occur in the same way.

The value of fluid inclusions is twofold: they preserve samples of the fluids present during mineral-forming processes and, in favourable circumstances, they also preserve

the density of these fluids, from which the pressure or temperature of mineral growth can be estimated. Rocks formed in the deep crust normally contain multiple generations of fluid inclusions, formed at different times and under different conditions. This makes it difficult to relate specific inclusions to particular events in the history of the rock.

The approximate composition of fluid inclusion contents is normally determined by microthermometry, that is, by studying their freezing behaviour (in thin, doubly polished slices) using a microscope with a heating–freezing stage. Carbon dioxide occurs in liquid form in fluid inclusions, reflecting a significant internal pressure, and its purity can be estimated from the melting point. Likewise, the salinity of aqueous inclusions is determined from the temperature at which ice melts. More precise chemical information is obtained by laser Raman spectroscopy, for measurement of gas compositions of individual inclusions, or by crushing and leaching of bulk samples for cation and anion analysis. Elemental analysis of individual inclusions is now being developed using laser and ion beam techniques.

Fluid inclusion densities are also measured by microthermometry. The hole in the host crystal that the fluid occupies has an almost constant volume, and so the density of the filling remains unchanged during uplift and cooling, unless the inclusion leaks. If an inclusion is originally trapped at the pressure–temperature conditions of point A shown in Fig. 1, then on cooling, the internal pressure of the fluid contents may vary along a line of constant density, or *isochore*, irrespective of the external pressure acting on the rock. At

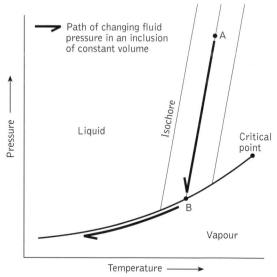

Fig. 1. Pressure–temperature diagram to show the changes in fluid pressure within an inclusion of constant volume, formed at the conditions of point *A*, as it cools to surface temperature.

point B, the isochore intersects the liquid–vapour curve of the fluid. With further cooling, a vapour bubble develops in the inclusion, and the internal pressure lies along the liquid–vapour curve. As the liquid portion becomes denser, the proportion of vapour to liquid increases. In microthermometric studies, this history is reversed under the microscope, and the temperature of homogenization, that is, where the vapour bubble vanishes, is measured. At this temperature, the density of liquid coexisting with vapour (which is well known for a range of fluids) corresponds to the density of the inclusion filling as a whole, and hence an isochore can be constructed which passes through the P–T conditions at which the inclusion formed, or last leaked. Synthetic fluid inclusions, made under known condition of composition, pressure, and temperature, have been used to obtain primary data on the volume of fluids over a wide range of physical conditions.

BRUCE W. D. YARDLEY

Further reading

Roedder, E. (1984) Fluid inclusions. *Reviews in Mineralogy*, 12, Mineralogical Society of America, Washington D. C.

fluids in the Earth One dictionary definition of a fluid is 'a substance that flows freely'. This definition is, however, more general than is commonly used in the Earth sciences, for it would include silicate melts, or even the solid mantle, since it too flows, albeit extremely slowly. Some authors use the term 'fluids' for common gases and liquids (such as water, water vapour, carbon dioxide, etc.); others restrict the noun 'fluid' to volatile substances at temperatures above their critical temperatures. Regardless of the terminology, fluids are understood to be distinct from the denser silicate melts (and of course the solid silicates) and generally to have compositions in common with substances that are gases or liquids at the Earth's surface. Obvious examples are water, methane, carbon dioxide, carbon monoxide, and sulphur dioxide.

Although there are geological systems in which free fluid phases exist or have existed in the past, there are also many processes and systems in which fluids are important but do not form their own separate phase. Rather they are incorporated into minerals and melts, either making distinct minerals or present as defects. While it is then not strictly correct to describe them as fluids (since they are not capable of flowing freely), this is commonly done. References are often made to the water content of a magma, even though there is no water in the sense of a free water phase, nor may there be any H_2O molecules in the magma. This common usage comes about because when heated to above saturation, oxygen and hydrogen in the melt or mineral which exist as OH groups can dehydrate to form bubbles of water. It is more strictly correct to refer to these components as volatiles rather than fluids, but the terms are commonly used synonymously.

The chemical composition of fluids in the Earth

The predominant chemical components of most fluids in the Earth are carbon, hydrogen, oxygen, and sulphur. According to the conditions (temperature, pressure, and oxygen partial pressure) and composition, these elements will form molecular species such as CO, CO_2, CH_4, H_2, H_2S, SO_2, etc. In addition they are able to dissolve large concentrations of many elements, particularly at higher temperatures. At the low temperatures and pressures of the Earth's surface, water will readily dissolve large amount of salts, such as sodium chloride ($NaCl$), but the solubility of metals is extremely low. At high temperatures (500 °C or more), however, metal concentrations in water–$NaCl$ fluids can reach 1 per cent by weight (1 wt%) or more. Clearly such fluids have been important in ore formation.

How much fluid is there, and where is it?

Taking the broader definition of a fluid as being the volatile component of the systems rather than a free fluid phase, estimates for the fluid content of the whole Earth are very variable and are susceptible to large uncertainties. An analysis of meteorites suggests that the water content of the whole Earth could be as high as 2 or 3 per cent by weight and the carbon dioxide (CO_2) content even higher. These amounts are far greater than estimates for the current volatile content of the Earth. The reason for the discrepancy is twofold. First, if, as is suggested, the Earth was hit by a Mars-sized object early in its history, then a large proportion of the volatile fluids would have been lost to space during this collision. Secondly, the Earth has been continually degassing since its formation, although much of this happened early and the rate is now far slower. This means that we cannot rely on meteorites; instead we have to calculate the total Earth fluid budget by summing up the different reservoirs: the hydrosphere, crust, mantle, and core. It should be recognized that estimates made this way can differ significantly, primarily because of uncertainties in the estimates for the mantle and, in the case of carbon and sulphur, the core.

Estimates for the hydrogen content of the whole Earth are around 30 ppm by weight, which amounts to an equivalent of about 300 ppm (parts per million) water. This is much more water than is in the oceans alone, with much of it contained in the Earth's crust and mantle, and perhaps also the core. In environments very near the surface, water concentrations in aquifers can reach 50 to 60 per cent by volume, but this rapidly falls off with depth since the pressure forces the fluid out of the pores. At greater depths and higher temperatures, the water reacts with the host rock to create hydrous phases such as clays, serpentines, amphiboles, and micas. In the Earth's mantle the water content is thought to be in the region of 100 ppm. This is estimated from the water content of mid-ocean-ridge basalts (MORBs), which typically contain about 3000 ppm water, and assuming that MORBs come from about 10 per cent melting of the upper mantle. Some rare MORBs, however, contain significantly more water. The water in the mantle may be contained either in high pressure hydrous phases, or alternatively as small hydrous defects in normally anhydrous minerals. Olivine, the most abundant mineral in the upper mantle, can contain 1000 ppm water, and perhaps significantly more. Wadsleyite, a dense magnesium silicate thought to exist in the Earth's transition zone (400 to 550 km depth), is capable of containing up to 3 wt% water. This has led to suggestions that this part of the Earth's interior might be a large reservoir for water.

Estimating the amount of carbon in the whole Earth is also problematic, and estimates range from 30 to 450 ppm by weight. When converted to CO_2 this is about 100 to 1500 ppm. One of the problems in estimating the carbon content of the Earth is that there may be considerable amounts in the core. If we consider mole per cent instead of weight per cent, the concentrations of carbon and hydrogen in the Earth are similar. The bulk of CO_2 in the near surface is contained in carbonate rocks, primarily limestones. Estimates for mantle CO_2 abundances are more difficult to obtain than for water, since MORBs almost always show evidence of CO_2 loss. Nevertheless, estimates of mantle CO_2 are about 300 ppm. Mantle CO_2 may be primarily contained in the high-pressure magnesium carbonate phase, magnesite ($MgCO_3$), although it is very rare to find carbonates in mantle xenoliths. Another likely possibility is that the majority of the mantle carbon is in the form of diamond (C).

As with carbon, sulphur is also an element that could be in the core. Estimates for the amount in the core range from 0 to 9 wt% and have a strong influence on estimates for whole Earth concentrations. In the silicate part of the Earth (the crust and mantle), sulphur concentrations are about 250 ppm, with similar amounts in the crust and mantle. This is about 1000 ppm (sulphur dioxide (SO_2)). Sulphur in the mantle is assumed to be in the form of iron–nickel sulphides. These are fairly widespread in mantle xenoliths.

The effect on the physical properties of minerals and melts

The presence of small amounts of fluid species such as H_2O and, to a lesser extent, CO_2 can have significant effects on many physical properties of minerals and melts. Concentrations of water as low as 100 ppm can increase creep rates in minerals and melt by more than two orders of magnitude. Hydrous phases can be greater than 10 per cent more compressible than their anhydrous equivalents. Seismic velocities are lowered and attenuation is also increased by the presence of water. Similarly, electrical conductivity can be strongly increased or decreased by small amounts of water. Perhaps one of the most important effects is that melting temperatures of rocks and minerals are strongly affected by water, with liquidus temperatures being lowered by many hundreds of degrees. This is an extremely important effect and has implications in many different environments.

Fluids in metamorphism

The majority of metamorphic reactions typically involve fluids in some way or another. This can be either by adding fluids, as in hydrothermal alteration, or by releasing fluids during decarbonation or dehydration reactions. Apart from changing the equilibrium phases in the rock, fluids can also change the chemical composition by adding and taking away chemical components.

Fluids in the lower crust

Magnetotellurie studies of the electrical conductivity of the lower crust show that it is anomalously conducting. This has led to the suggestion that the lower crust is saturated with a film of brine. There is, however, an alternative view which argues that the lower crust cannot be wet and that we must look for another mechanism for the high conductivity. This alternative view argues that at the temperatures and pressures of the lower crust, any free fluid would react with the rock to form hydrous mineral phases.

Fluids in volcanism

Fluids dissolved in magmas are the main cause of explosive volcanism. If the fluid reaches saturation in the magma, it then begins to exsolve. If this happens slowly, the gas bubbles will slowly migrate through the magma and degas. If, on the other hand, the gas exsolves quickly, the rapid and very large increase in the volume of the fluid as it enters the gas phase can generate tremendous mechanical energy and cause the eruption to be explosive. This can come about if the pressure in the magma chamber is suddenly released or if the magma has a high viscosity that inhibits migration of the bubbles. Water release from volcanic eruptions through geologic time has produced the Earth's oceans and atmosphere.

Fluids and earthquakes

For an earthquake to occur, the stress has to exceed the strength of the rock. One of the ways in which this can be achieved is if the fault is lubricated by a fluid. This has the effect of decreasing the friction. Dehydration reactions release water and increase the total volume of the rock, thereby pushing apart the grains and removing some of the strong frictional forces.

Dehydration reactions may also be necessary for deep subduction-zone earthquakes to occur. Many earthquakes in subduction zones occur at depths at which the pressures are so large that the friction between two sides of a fault will stop the fault from sliding; the stress should then be released by creep rather than by faulting. It has been suggested that release of water during the dehydration of serpentine can effectively lubricate a fault.

Water in subduction zones

It is generally accepted that water plays a crucial role in the generation of melting and volcanism beneath island arcs, where cold oceanic lithosphere is subducted into the mantle. The altered basaltic layer in the ocean crust typically contains about 3 per cent H_2O; as the lithosphere subducts and heats up, the hydrous phases in the slab dehydrate and eventually release their fluids. These fluids migrate into the overlaying dry mantle, which causes the liquidus temperature of these rocks to be lowered. At some point some of the mantle melts and migrates to the surface, where it forms island-arc volcanoes. There is some evidence that the amount of melting is correlated with the amount of water coming from the slab. Without the water, the cold subducting slab would simply cool the surrounding mantle and there would be no arc volcanism.

Recycling of fluids into the mantle

Mid-ocean ridge basalts show clear evidence for small amounts of fluids being degassed from the mantle. Although subduction zones return fluids to the mantle, much of that fluid then returns to the surface in island-arc volcanism. The question whether any fluid continues past the melting zone into the deeper mantle remains contentious. It has been suggested that in subduction zones where old, cold lithosphere is being subducted, the mantle near the slab would be too cold to melt and water might either react with it to form hydrous phases or be incorporated as defects in mantle olivines or pyroxenes. It might then be carried down to the transition zone, where it could readily be incorporated into wadsleyite, and then deeper, into the lower mantle.   JOHN P. BRODHOLT

fluvial sediments Rivers are the avenues through which most of the detritus produced by the denudation of the continents reaches the sea. However, large volumes of sediment are deposited in the fluvial environment and become preserved as part of the geological record. These deposits are sometimes very thick and cover vast areas.

Generally, sediments become finer and coarser grains become more rounded away from their sources. Less resistant rocks and minerals are broken down both by chemical and by physical processes during their transport to the sea, so that the deposits become increasingly composed of components which are stable under surface conditions—an effect that is more pronounced in tropical regions than in temperate regimes.

The river channels that transport sediment vary in character because of differences in slope, discharge (amount and frequency), and type (amount and size) of the load. Channel morphology is the most important factor in determining the geometry of the resulting fluvial sediments. Straight channels are relatively rare in nature and, when found, occur in areas of low slopes where there is an abundance of fine-grained, suspended sediment. Meandering channels (Fig. 1a) develop with low slopes, steady low discharge, and relatively fine-grained loads, although they can be found in rivers with coarser mixed loads (both bedloads and suspended loads). Anastomosing channels (Fig. 1b) consist of channels of varying sinuosity, which are separated by islands as a consequence of commonly intersecting each other. They develop with variable discharge, loads, and slopes, but when base level rises as a result of various causes. Braided channels (Fig. 1c) develop in areas of steep slopes, high and spasmodic discharge, and relatively coarse loads (dominance of bedload). The last-mentioned are also favoured by easily erodible banks and hence also by a lack of vegetation. Braided channels were more important in earlier periods of the Earth's history (pre-Devonian), before the continents were clothed in vegetation.

Meandering rivers are floored with gravelly sands and sands which are usually fashioned into large-scale three-dimensional ripples (dunes) which extend up the flanks of the inner bends of the channel (*point bars*) to produce trough cross-stratified sands or gravelly sands. Interbedded with these forms are plane-bedded sands produced at high flow stages, and tabular cross-stratified sands or gravelly sands formed by the movement of large-scale two-dimensional ripples across the point bar. In the upper part of the point bar, where the sediments are finer, small-scale ripple cross-stratification, often covered with mud drapes, and flaser bedding are common.

Lateral migration of the channels leads to growth of point bars to produce an upward fining sequence of sediments with a decreasing scale of sedimentary structure (Fig. 1a). Variations on this model in the upstream and mid-bend portions of point bars result in sequences of sand and gravelly sediments which may show no marked vertical sequence in grain size, or even a coarsening-upwards sequence. In rivers with more episodic flow, where the flood

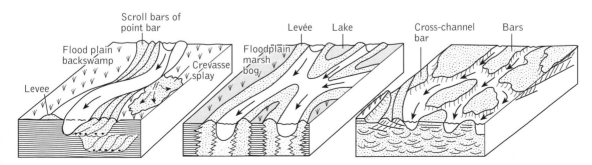

Fig. 1. Sediments of various types of rivers: (a) meandering; (b) anastomosing; (c) braided. (Sketched from Allen (1964) *J. Sed. Pet.*; Smith and Smith (1980), *J. Sed. Pet.*; Smith and Cant (1982) American Association of Petroleum Geologists.)

waters sometimes cut across the bar to develop chute cut-offs, other variations occur in the vertical sequence, and large-scale cross-stratification may develop on the upper parts of the point bar.

Overbank flooding, when bankfull discharge has been exceeded, leads to the deposition of sands and muds as raised banks or levées and the deposition of finer sediment over the entire flood plain. Floods sometimes cut across l!evees to produce crevasses, and pour sandy lobes of sediment on to the flood plain—*crevasse splays*. Cut-offs across meander bends produce isolated channel shapes—ox-bow lakes—which are ultimately filled by overbank flooding. In humid areas, flood plains may contain lakes and usually have a dense cover of vegetation, which develops into peats and ultimately into coals. In arid areas, where there is an excess of evaporation over precipitation, soluble salts are concentrated in the upper layers of flood-plain sediments to produce calcareous concentrations and stringers (calcrete or caliche, gypcrete, silcrete, and concentrations of other salts). Whereas the flood-plain deposits tend to be buff or brown in arid or semi-arid environments, they are grey in humid areas, because of the presence of organic matter.

Anastomosing rivers are those with relatively fixed channels. Within the channels flat-bedded gravels and trough and planar cross-stratified sands accumulate. The channels enclose islands on which fine-grained levée and flood-plain sediments are deposited. In humid areas thick vegetation gives rise to peat deposits (ultimately coals). Owing to rising base levels because of downstream obstructions, or possibly tectonic subsidence, the channels and the enclosed islands accrete vertically (*aggrade*) with only small amounts of lateral movement.

Braided or low-sinuosity rivers contain less suspended load sediments than meandering streams and are mainly composed of bedload material. They show less well-developed cyclicity than the deposits of meandering rivers and are composed of units that lie in shallow-shaped scours (Fig. 1c). In the upper reaches of such rivers (i.e. proximal positions) and on alluvial fans, subject to stream flows, are low-relief bars which consist mainly of horizontally bedded, clast-supported, imbricate gravels. Occasional trough and tabular cross-stratification is produced by migration of large-scale ripples (dunes and bars). Cross-stratified gravel and sand wedges are deposited on the sides of the bars, around cores of flat-bedded gravel. Flow velocities are generally very similar throughout the channel, and little segregation of grain sizes occurs. These bars generally show poor cyclicity.

Rivers with mixed loads develop bars with more relief than in the previous type. The deposits consist of sequences crudely fining upwards, beginning with horizontally bedded or trough cross-stratified gravels of the channels where they are covered with large-scale rippled sands which pass upwards into cross-stratified gravelly sand, and into tabular cross-stratified sands or parallel-laminated sands formed on the bars. These coarser deposits are capped by fine-grained rippled sands and muds formed on the bar-crests. Movement

of a channel may leave a complete vertical sequence in the geological record.

Sand-dominated rivers containing some gravel show sequences fining upwards. Channels are commonly floored with gravel, with large-scale ripples producing trough and tabular cross-stratified sands. Bars migrate downstream, producing tabular cross-stratified sands, and these merge to give sand flats covered by rippled and planar-bedded sands developed during flood stages. Fining-upward cycles develop with a great divergence of the direction of the cross-stratification as bars move across the main channel.

Sand-rich, exceptionally broad, shallow rivers with steadier discharge than other types have sinuous, linguoid, or straight-crested bars, and flood plains on to which coarse-grained sediment is often driven. The channel floors are covered with sinuous dunes, which form trough cross-stratified sands and gravels. Constantly moving bars deposit sets of tabular cross-stratified sands at their leading edges (1–3 m). These bars may stack up to produce extensive sand flats. Where channels become filled, or on the tops of bars, smaller-scale cross-stratification develops as a result of the migration of large-scale two-dimensional ripples; and, in places, horizontally bedded sands and muds accumulate by vertical accretion on islands and the flood plain.

Rivers that are subject to flash floods and are rich in sand with little or no gravel show little relief between bars and channels. The river sweeps over the whole valley, resulting in little vertical segregation of sediment of different grain size. Deposits consist mainly of plane-bedded sands with only minor amounts of cross-stratified and climbing-ripple cross-stratified sands.

Fluvial sediments lack a marine fauna. Faunal remains are generally sparse, but fossil plant fragments are found in channel sediments. Remains of freshwater fossils, many of them shelly, are preserved in fine-grained flood-plain deposits. Thick accumulations of plant debris are commonly associated with these deposits.

Cross-stratification shows an orientation down the palaeo-slope with a variable spread of current vectors and modes depending upon channel type. There may be occasional reversals of palaeoflow direction resulting from local conditions, but these are rare.

Channel deposits form sand bodies of variable geometry. Shoestring-type sands develop in meandering and anastomosing river channel systems enclosed by fine-grained over-bank flood-plain sediments. Sheet sands stacked in channel-like scours develop in low-sinuosity and braided river channel systems. Migration on avulsion of channels will lead to the production of cyclical systems.

Fluviatile sediments form sheets or wedges of sediment and elongate, ribbon-like infilling troughs in the underlying rocks. They lie unconformably or disconformably on the underlying older rocks and thin out towards the source, or pass into alluvial-fan deposits. They pass basinwards into deltaic and nearshore marine or lacustrine sediments. Thick sequences of cyclical fluviatile sediments have been preserved during many period of the Earth's history, and fluviatile sediments form a substantial part of the preserved rock records of some systems; for example, the Old Red Sandstone fluvial facies of the Devonian, and the Palaeozoic and Mesozoic Nubian sands of North Africa.

Fluvial-channel sands are potentially good reservoir rocks. Channel sands embedded in their own alluvium or cut into older rocks can be important traps for oil and gas if the sediments are covered with a good seal, such as clay. In coalfields, channel sandstones can be a great nuisance and hazard as they can contain explosive gas and water which causes flooding.

Some fluvial-channel deposits contain important *placer deposits*, including cassiterite, gold, or diamonds, that is, concentrations of unusually dense minerals that are deposited at the same time as coarse grains or pebbles. Ores of uranium and other elements can also be precipitated from chemical solutions moving through porous, permeable channel sands.

G. EVANS

Further reading

Collinson, J. D. (1986) Alluvial sediments. In Reading, H. G. (ed.) *Sedimentary environments and facies* (2nd edn), pp. 20–62. Blackwell Scientific Publications, Oxford.

Collinson, J. D. and Lewis, J. (1983) *Modern and ancient fluvial systems*. International Association of Sedimentologists, Special Publication No. 6.

flysch *Flysch* is a dialect word used in the Alpine Oberland to describe shaly rocks with intercalated sandstones. It probably derives from the tendency of the rocks to slip down hillsides. It was introduced into the geological literature by Studer in 1827 and was later used by Boussac in 1909 as a stratigraphical term for a sequence of Late Cretaceous–Tertiary rocks deposited in the sea during the growth and deformation of the Alpine mountain chains, but before the main climax of the earth movements.

The term is now widely used outside Switzerland for rocks of a recurrent facies consisting of sandstones or calcarenitic limestones (clastic limestones) of uniform thickness over large areas interbedded with shales deposited in a tectonic setting similar to that of the Alpine flysch. The Chinese geologist Kenneth Hsu, working in Zürich, recommended in 1970 that when the term is used in this way it should be written with a small *f*—*flysch*—to distinguish it from the Swiss term *Flysch*, which has stratigraphical overtones.

The sandstones or calcarenitic limestones have graded bedding and bottom structures, and generally contain shallow-water fossils that have been transported a long way from where the animals lived. These characteristics, together with the associated trace fossils and nektonic–planktonic fossils of the intercalated shales, indicate deposition by turbidity currents into a deep-water environment. Although the term is sometimes used without reference to tectonic setting, some have argued that it should then be replaced by 'flyschlike' of 'flyschoid'.

G. EVANS

folds and folding

An interest in geology may be sparked by a first contact with the products of geological processes, and the resulting sense of aesthetic pleasure may lead to a lifetime's study of the Earth. A well-exposed fold can trigger this process. Something about the appearance of a fold appeals to those who have a taste for geometry, lines, or symmetry (Figs 1 and 2). Besides, how can something as 'hard' and 'rigid' as rock possibly fold?

For a rock to be folded, it must have a set of parallel surfaces and preferably layering. The deflection of this layering defines the folds (Figs 1, 2, and 3). Because most folds look different from one other, a few basic geometric terms are necessary to distinguish fold appearances (Fig. 3). Determining fold appearance in outcrops serves two essential roles. First, the geometry and style of small folds can reflect these aspects of larger associated folds of the same age. These larger folds are usually poorly exposed, and so their appearance, size, and location are therefore inferred from their smaller or *parasitic* brethren, which commonly are better exposed. A knowledge of larger folds is important because by virtue of their size they contribute more to rock deformation and the geometry of rock layers in the Earth.

Secondly, appearance can indicate the origin of a fold, although a particular fold morphology can be the product of more than one possible folding history. The preservation of geological features as end products that lack clear evidence for their formative history poses a constant problem in geology. As with fold geometry and origin, the final appearance of a geological feature offers geoscientists many opportunities to leap to conclusions about its formation. Many times, they subsequently discover, often with the 'aid' of their colleagues, that the rocks lack sufficient evidence to prove

Fig. 1. Folded schists, Swiss Alps. (Dark circular object is a camera lens cap.)

Fig. 2. Refolded schists, Swiss Alps. (Dark circular object is a lens cap.)

their interpretation conclusively. This problem is analogous to attempting to deduce how a book was written, while having only the final printed copy before you as a source of data.

Let us establish the vocabulary of a few basic geometric terms (Fig. 3). The zone of maximum curvature in the fold is the *hinge zone* or fold nose. The portions of layer that connect the hinge zones of adjacent folds are *fold limbs*. The inner arc of a fold is its *core*. The two types of fold are *antiforms* that fold up and over (*A*, Fig. 3), and *synforms* that fold down and under (*S*, Fig. 3). The *axial surface* is an imaginary surface that connects the hinge zones of the various folded layers and commonly bisects the angle between adjacent fold limbs. The shape of a fold is a function of the thickness of a layer with respect to a referent; for example a thickness measured at right angles to the surface of a layer (Fig. 3) or the thickness of the axial surface. If the thickness of a particular layer is constant normal (perpendicular) to the layer, the shape of the fold is *parallel*. If the thickness is constant parallel to the axial surface, the shape is *similar*.

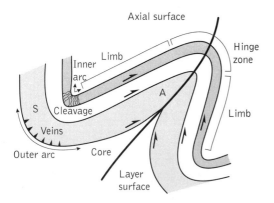

Fig. 3. Fold appearance and kinematic behaviour.

We can now apply this vocabulary to Fig. 1 to see what can be gained. The metamorphic rocks from the Swiss Alps illustrated contain a series of asymmetric folds with limbs of differing length and axial surfaces tilted to the right. The asymmetry is an important interpretative tool. For example, these folds can be parasitic to larger unexposed folds. Because the antiforms close to the right and the axial surfaces tilt to the right, the asymmetry indicates that the host antiform hinge zone is to the right of the photograph and the host synform hinge zone is to the left. This information provides a key for a field geologist who is constructing a geological map. The map shows the surface distribution or outcrop of the rocks, and large map-scale folds strongly control outcrop patterns. The importance to non-geologists of determining rock outcrop in an area includes: locating weak rock types that favour mass wasting; finding sources of resistant rocks for road metal or building; locating the surface outcrops of aquifers and aquicludes that control the distribution of groundwater; and finding natural resources such as coal horizons or subsurface traps of hydrocarbons.

Another topic on which the fold asymmetry of Fig. 1 can shed some light is the regional deformation pattern. The asymmetry can indicate that during Alpine mountain building, rocks moved from left to right, as seen in the photograph. Small-scale data of this kind gathered from a large number of outcrops across the Alps, provide a basis for interpreting the overall horizontal and vertical displacements of the rocks that formed the mountains. In general, the basic pattern of rock motion in this region is one of shortening by rocks being displaced horizontally in one dominant direction. In the Alps, data from widely distributed outcrops of asymmetrical folds indicate that the mountains formed by Italy (part of the African plate) being driven north into Europe with concurrent vertical uplift. Another feature of Fig. 1 is the fold shapes of the layers. Light-coloured layers are mostly parallel in shape, whereas medium to dark grey layers are more similar in shape. This difference can indicate that the darker layers sustained more flow during fold formation.

A long narrow fold with parallel or isoclinal limbs closes to the right in Fig. 2. The fold is neither an antiform nor a synform; because it is lying on its side, it is called *recumbent*. Such folds are common in high-grade metamorphic rocks in the cores of mountain belts. They result from rocks travelling large horizontal distances by flow, including folding. The left-hand side of Fig. 2 illustrates another common feature in such settings, where two younger folds *refold* or *overprint* an older fold. They deform both the fold and its axial surface, producing a zigzag interference pattern. The key observation from this pattern is that the rocks were deformed more than once. The rule that younger structures deform older structures is the essence of interpreting the relative ages of multiple deformations. This is shown in Fig. 2, where younger folds deform older folds. The angular relationships between axial surfaces and hinge zones of different fold phases are used to classify the types of refolding and to provide insight into changes of regional shortening direction during mountain-building.

The various types of information listed above also have a role in determining how a fold developed. When one looks at rocks on the ground surface, it is very difficult to believe that anything so rigid could fold. However, almost all folds in rocks form in the subsurface region at depths in the crust of about 1 to 45 km. Buried rocks in the crust are subjected to elevated temperatures, pressures, and fluid pressures that trigger deformation mechanisms, allowing tectonic forces to drive folding processes.

Three processes produce folds through processes called buckling, bending, and passive folding (Fig. 4). During buckling, compression is applied parallel to the layers, which are of differing strengths (Fig. 4b). The compression causes shear stresses at a buckling instability, deflecting layers and hence initiating a fold. Most of the small and medium-sized folds seen in exposures of sedimentary and low-grade metamorphic rocks are buckle folds.

In contrast, bending folds develop when compression is normal to layering (Fig. 4b). The most common types of bending folds form above salt domes and as fault-bend folds. Salt rises into domes because it is much less dense and much weaker than most rocks. When a more dense rock buries salt, a gravitational instability exists because less dense materials are more stable when they are above more dense substances. So, the salt tends to escape vertically by doming and bending the layers above it as a tight, almost isoclinal fold of large proportions. Fault-bend folds occur when rocks bend during translation through a change in fault dip.

The third process, passive folding, unlike the first two, produces folds as an artefact because layering has no mechanical significance (Fig. 4c). The rock layers are not mechanically active; they merely record passively the deformation by changing shape. This type of process is particularly common in thick sequences of fine-grained sedimentary rocks and high-grade metamorphic rocks. These sequences lack the strength and density contrasts which aid buckling and bending.

The three folding processes can operate simultaneously in adjacent rock bodies, or sequentially in the same rock sequence if deformation conditions change. The processes do not always yield folds with distinctly different appearances. As a result we cannot make a simple correlation between appearance and process. The focus here will be on aspects of the development of buckle folds.

Two common kinematic behaviours, which are combinations of distortion, rotation, and translation, occur during buckling. One is *flexural slip* or *flow*, in which layers in the fold limbs move parallel to layering and each other. With flexural slip, this movement is restricted to slip on layer surfaces. An analogous behaviour is to flex (fold) a pack of cards. The cards arch by sliding over one another. If you have formed an antiform, the cards on top slide toward the hinge zone relative to those underneath. The same is true in a natural fold where layers slip from synforms to antiforms. Flexural flow distorts layers by layer-parallel shear, rather than by slip at layer boundaries as with flexural slip.

This ability to slip or flow parallel to the layering greatly weakens the rock, and hence makes folding much easier to initiate. Again, an analogy shows this result. Take two telephone books, cut off their bindings, remove their hard covers,

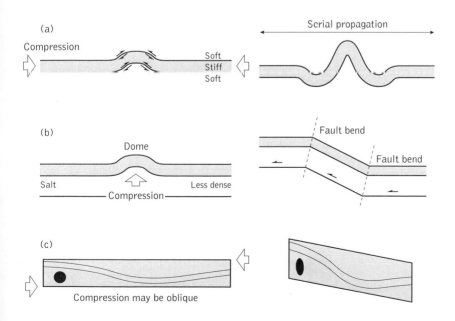

Fig. 4. Three fold-forming processes: (a) buckling; (b) bending; (c) passive folding.

and glue all the pages of one book together. If you flex (fold) these two 'books', the one with unglued pages will fold easily because the pages will slide on each other. The glued 'book', however, requires much more force because it is a single mechanical unit with no ability to slip.

The glued telephone 'book' would most probably fold by the process of *tangential longitudinal strain*. The behaviour is exactly the opposite of flexural slip because deformation is concentrated in the hinge zone as distortion and is not distributed through the fold limbs as slip. It is common in thick, strong rock layers or layers that are unable to slide. The hinge collapses by extending the outer arc of the layer and contracting the inner arc (synform *S* in Fig. 3). Veins, intragranular processes such as twinning, dislocation glide, and grain-boundary migration, or normal faults achieve the outer-arc extension. Inner-arc contraction can be by cleavage development, parasitic folding, reverse faulting, or intragranular processes. An interesting aspect of this behaviour is that the distortion pattern changes from the inner arc to the outer arc: the strain progresses from decreasing contraction to no deformation at a neutral point to increasing extension. The strain is thus positionally dependent within the hinge zone.

When examining an exposure of buckle folds, it is tempting to think that they form simultaneously. Yet, when rock layers compress parallel to layering, folding becomes a race between instabilities and later, fold amplification. Any rock sequence contains many imperfections that act as buckling instabilities, such as nonplanar layer surfaces, differentially cemented layer surfaces, or heterogeneously distributed minerals such as mica or clay minerals. Each instability generates a fold of a particular wavelength, which is a measure of fold width much like the wavelength of ocean waves. A layer has a dominant wavelength during buckling that grows more efficiently as a function of layer thickness and strength. Those instabilities that yield wavelengths closest to the dominant one grow much faster and becomes the visible folds. This selection process is enhanced by folds growing even faster when they amplify to a critical size, as would be the case for dominant folds.

Another feature of this growth process is that an amplifying fold will deflect adjacent layering and initiate two more folds (Fig. 3). Thus, folds grow serially along a layer from the initial buckling instability as they are also amplifying. Such a growth sequence is also possible during bending folds when one salt dome triggers an adjacent one, but it is absent during passive folding.

WILLIAM M. DUNNE

Further reading

Price, N. J. and Cosgrove, J. W. (1990) *Analysis of geological structures.* Cambridge University Press.

Ramsay, J. G. and Huber, M. I. (1987) *The techniques of modern structural geology*, Vol. 2: *Folds and fractures*. Academic Press, London.

foliation (tectonic) *see* CLEAVAGE AND OTHER TECTONIC FOLIATIONS IN ROCKS

foraminifera and other unicellular microfossils

The biological kingdom Protista contains all those relatively simple organisms which consist of a single cell containing a nucleus and other internal structures. Their dimensions range from a micron to several centimetres. Most protists occur as single cells but some occur as loose aggregates of several cells. Many forms are capable of photosynthesis and hence resemble plants in the way that they obtain energy. Others ingest other living organisms from their surroundings and thus are animal-like. A third group is capable of obtaining energy by both methods. This diverse group of organisms is subdivided into a number of groups but, since most lack any form of skeleton, only a few have a fossil record and, of these, only the coccolithophorids, dinoflagellates, diatoms, foraminifera, and radiolarians are important.

Coccolithophorids (Coccolithophorida) are an important group of planktonic marine protists which are capable of photosynthesis. They are spherical or oval and generally less than 20 micrometres (μm) in diameter. The living cell possesses a gelatinous sheath in which are embedded calcareous plates termed coccoliths. As an individual cell grows older, the gelatinous sheath becomes calcified and the plates become rigidly fused. The older coccoliths are gradually shed from the sheath and sink to the ocean floor. Hence, fossil coccolithophorids are generally seen as isolated plates, although entire skeletons are sometimes found. The plates are minute (2 to 30 μm) and variously shaped (Fig. 1). The coccolithophorids range in age from Triassic to Recent, and form a major constituent of Mesozoic and Tertiary chalks. Their extremely small size, great abundance in many deep-water deposits, and widespread geographic distribution make them valuable in biostratigraphic correlation. They can also be used to infer oceanic palaeotemperatures and salinities.

Diatoms (Bacillariophyceae) are a group of photosynthesizing protists characterized by a siliceous cell wall consisting of two elements (valves) with one valve overlapping the other like the lid on a box (Fig. 1). The cell size varies from 5 to 2000 μm. The members of one group of mainly marine diatoms are circular with a radial symmetry (centric diatoms). Those of a second group, mostly confined to freshwater environments, are elliptical with bilateral symmetry (pennate diatoms). Cells usually occur singly but some forms are colonial or filamentous. The cell wall is usually delicately ornamented and perforated by minute holes. The oldest known diatoms are found in the Jurassic. From the Cretaceous onwards, diatom ooze containing more than 30 per cent diatom skeletons has been deposited in high-latitude areas of the deep ocean. Diatoms are valuable in biostratigraphic correlation and, because many fossil diatoms have living relatives, they are also valuable in palaeoenvironmental work.

Dinoflagellates are globular protists with two flagella (thread-like structures) of unequal length. Their size range is from about 5 to 2000 μm. Dinoflagellates exist in two forms. In the motile stage, the organism propels itself through the water by active beating of the flagella. In this form, the organ-

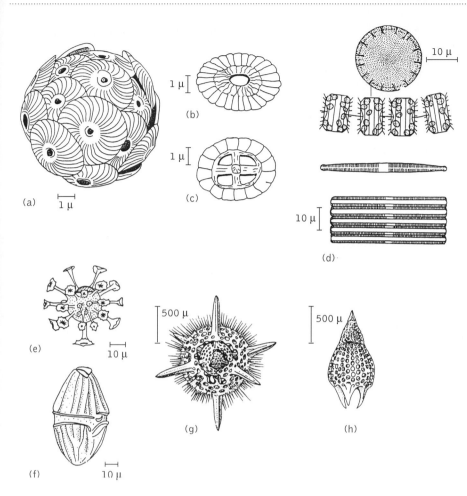

Fig. 1. Coccoliths (a) Recent coccolithophorids: *Cyclococcolithina*; (b) *Pseudoemiliania*; (c) *Prediscosphaera*. Diatoms: (d) centric diatom *Thalassiosira*, valve view and lateral view of colony; (e) pennate diatom *Fragilaria*, valve view and lateral view of colony. Dinoflagellates: (f) *Hystrichosphaeridium*; (g) *Dinogymnium*. Radiolaria: (h) *Actinomma*; (i) *Podocyrtis*. (Adapted from Brasier (1980).)

ism may either have a flexible cell wall or it may have a rigid armour of cellulose plates. Some forms bear simple processes. Whenever conditions are unsuitable, the organism becomes dormant and enters the cyst stage. Only the cysts are preserved in the fossil record. They comprise a cellulose skeleton with a variety of spines or processes (Fig. 1). Dinoflagellates first appear in the Cambrian. Modern forms occur not only in marine environments but also in quasi-marine situations, such as estuaries, and in some large lakes. The dinoflagellates are useful in both biostratigraphy and palaeoenvironmental studies.

Radiolarians (radiolaria) are a group of marine protists characterized by a perforated central chitinous or membranous capsule which contains the bulk of the protoplasm and the nucleus. They range in size from 100 to 2000 μm. The central capsule is rarely found fossilized but they possess a delicate skeleton of opaline silica or strontium sulphate which may be preserved. This skeleton is a lattice of variable morphology made up of needles (spicules), bars, and spines (Fig. 1).

Radiolaria are relatively uncommon as fossils, but occur sporadically from the middle Cambrian onwards. All modern forms are marine, most being free-floating forms occurring either at the surface or close to the sea bed. Radiolarians are used in biostratigraphic correlation of oceanic sediments and are particularly useful in cases where calcareous microfossils have been subject to dissolution. Fossil Radiolaria occur in a wide range of sediments, but of particular interest is radiolarian ooze and its presumed fossil equivalent, radiolarian chert. Radiolarian ooze is a relatively rare deep ocean sediment which consists of more than 30 per cent radiolarian skeletons. It is formed in equatorial deep-sea areas where the water is rich in dissolved carbon dioxide and is readily able to dissolve any calcium carbonate present. This occurs at water depths of about 4500 m (the carbonate compensation depth) and more. A few fossil examples of deposits of this type, termed radiolarian cherts, are known.

Foraminifera (technically Foraminiferida, but variously termed foraminiferids, foraminiferans, or forams) are the most important protists from a palaeontological point of view. They are amoeboid organisms whose cell is protected by a test (or protective shell) comprising one to many chambers. The test may have one or more large openings or apertures. The protoplasm is extruded as pseudopods via the aperture and through any perforations that may be present in the test. The pseudopods form a complex network which is used to capture food (bacteria, other protists). On the basis of test structure and composition, five subdivisions of the group can be recognized. The most primitive forms, the Allogromiina, secrete a thin non-rigid test composed of a chitin-like, organic material termed tectin. Similar material occurs as a thin lining within the chambers of most other foraminifera. Some Allogromiina tests may be covered in loosely attached grains of sediment. The second foraminiferan group, the Textulariina, includes all those forms whose test is composed of grains of organic and mineral matter (e.g. sand grains, diatoms, other foraminifera, sponge spicules) bound together by an organic, calcareous, or ferric oxide cement. These are called agglutinated tests. While some species are very selective in terms of the size, texture, or composition of the grains they use, others are fairly catholic.

Foraminifera with calcareous tests are by far the most abundant forms and fall into three suborders, each possessing a different wall structure. Fusulinina have tests which consist of minute crystals of calcite. These crystals are all approximately the same size and are closely packed together. In thin sections under the microscope it can be seen that the calcite crystals are arranged either randomly, producing a granular appearance, or aligned at right angles to the surface of the test, giving the wall a fibrous appearance. In most fusulinine foraminifera, granular and pseudo-fibrous layers of micro-crystalline calcite are combined to produce a multi-layered test wall.

The Miliolina have tests consisting of tiny needles of high-magnesium calcite. The calcite crystals are randomly oriented except adjacent to the outer and inner surface of the test, where a layer constructed of horizontally or vertically arranged needles occurs. This arrangement gives milioline tests a porcellaneous appearance. They are distinctively milky white when viewed in reflected light and amber in transmitted light.

The final group, the Rotaliina, possess tests which appear hyaline (glassy) when viewed in reflected light and grey to clear in transmitted light. However, the clarity is not always apparent, being obscured by a thick test wall, by a large number of fine perforations, or by the surface of the test being ornamented by spines. In thin section, the calcite crystals in rotaliine walls have a variety of arrangements including monocrystalline, granular, and radial crystalline.

The walls of many foraminifera are perforated by narrow straight or branched pores. In life, these pores allow the transfer of food particles from the external to the internal cytoplasm. Radial pores are characteristic of the Rotaliina but they also occur in some of the more complex Textulariina and Fusulinina. Where they are present, these pores give the test wall a pseudo-radial or pseudo-fibrous appearance in thin section.

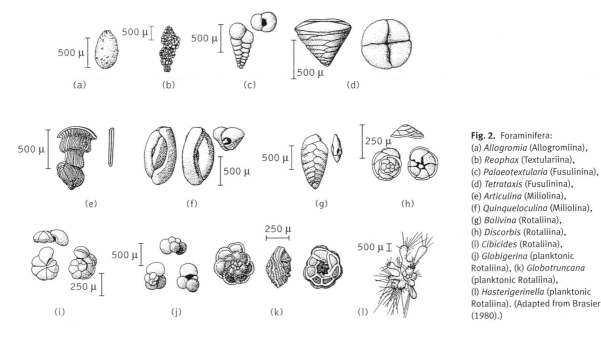

Fig. 2. Foraminifera:
(a) *Allogromia* (Allogromiina),
(b) *Reophax* (Textulariina),
(c) *Palaeotextularia* (Fusulinina),
(d) *Tetrataxis* (Fusulinina),
(e) *Articulina* (Miliolina),
(f) *Quinqueloculina* (Miliolina),
(g) *Bolivina* (Rotaliina),
(h) *Discorbis* (Rotaliina),
(i) *Cibicides* (Rotaliina),
(j) *Globigerina* (planktonic Rotaliina), (k) *Globotruncana* (planktonic Rotaliina),
(l) *Hasterigerinella* (planktonic Rotaliina). (Adapted from Brasier (1980).)

Forms range from a single, non-chambered flask- or sac-shape to very complex, chambered discs (Fig. 2). In multi-chambered forms, the chambers may be arranged in a great variety of modes including linear, planispiral, and trocho-spiral patterns. The chambers are connected together via pores or foramina in the dividing walls. It is from this characteristic that the group name is derived. Most foraminifera are small, generally less than 1 mm across. However, some forms are considerably larger; a few measure as much as 100 mm in diameter.

The earliest foraminifera occur in Cambrian rocks, where a few simple tubular membranous forms are recorded. Agglutinated tests first occur in Ordovician rocks, and textulariines dominate foraminiferal assemblages in the early Palaeozoic. Coiled tests appeared in the Silurian but chambering did not arise until the Devonian. Although forms with calcareous tests first evolved in the Ordovician, they remained relatively uncommon until the Devonian. There followed a general trend in many lineages towards increasing complexity both of the wall structure and chamber arrangements. The first planktonic species evolved in the Jurassic, and all belong to the Rotaliina.

Foraminifera occupy a wide range of marine habitats, from estuaries and tidal flats to the deep ocean. The great majority are benthic, some being free-living while others live attached to objects. Characteristic suites of species occur in each environment, although an individual species may be found in more than one habitat. Foraminifera are therefore extensively used as environmental indicators, providing information on substrates, palaeosalinities, water depths, palaeotemperatures, etc. Their small size and relative abundance makes them particularly useful in this respect. A few forms (about fifty modern species) have adapted to a planktonic existence by increasing the horizontal dimensions of the test, reducing the thickness of the test, increasing pore size, developing supplementary apertures, or by enlargement of the aperture. Although there are only a few planktonic species, they are extremely abundant in modern oceans, and much of the deep ocean floor is covered with an ooze largely composed of their tests. Planktonic foraminifera can be used to obtain data relating to surface water temperatures and oceanic circulation patterns.

The small size and abundance of foraminifera in many marine sediments make them useful in biostratigraphy. Planktonic forms have been intensively studied and some species can be used for stratigraphical correlation on a near worldwide scale. Benthic species are less useful, although they can be used for local correlation.　　　　ALLAN N. INSOLE

Further reading

Bignot, G. (1985) *Elements of micropalaeontology*. Graham and Trotman, London.

Brasier, M. D. (1980) *Microfossils*. Allen and Unwin, London.

Jenkins, G. and Murray, J. W. (1981) *Stratigraphical atlas of fossil Foraminifera*. Ellis Horwood Ltd, Chichester.

Lord, A. R. (1982) *A stratigraphic index of calcareous nannofossils*. Ellis Horwood, Chichester.

foreland, foreland basin, and hinterland In its strictest sense, the term 'foreland' is assigned to the down-going or subducted plate of a collisional mountain belt (orogen). It differs from *hinterland*, which describes the over-riding plate of an orogen. For example, the Pyrenean mountain belt formed in response to the subduction of the Spanish plate beneath France. In relation to the Pyrenees, therefore, the Spanish plate is referred to as the foreland plate, France being part of the hinterland plate.

At the Earth's surface, a universal response to orogenic thickening is for the mountain belt to rise isostatically. Isostatic uplift at a geologically gradual rate creates much of the topography that we commonly associate with mountain belts, whose most active periods have often long since ceased. An integral part of mountain-belt anatomy, however, is seen also in the flat-lying rocks of regions adjacent to all the world's major collisional mountain belts. Lithosphere adjacent to the mountain belt is capable of supporting the extra mass of the thickened, and therefore heavier, mountain belt by down-bending, or flexure. These areas of flexing lithosphere are *foreland basins* or foredeeps. Good examples are the Ganges Plains of northern India, adjacent to the Himalaya, the Cumberland and Allegheny Plateaux of the Appalachians, and the Aquitaine and Ebro valleys to the north and south of the Pyrenees, respectively.

Another common setting of foreland basins is the regions hindward of oceanic island arcs, which are the surface expressions of linear zones of compression where oceanic crust is being subducted beneath continental crust. These are *retroarc* foreland basins, also subsiding in response to thrust faulting and lithospheric thickening related to subduction of oceanic plates. Good examples of retroarc foreland basins occur around the western margin of the shrinking Pacific Ocean, where the Pacific plate is being subducted beneath Asia.

Numerical models show how the stiffness, or elastic thickness, of the lithosphere exerts the chief control over the width and depth of foreland basins: the greater the elastic thickness, and therefore stiffness, of the lithosphere, the wider but shallower will be the resultant foreland basin. The curvature of foreland basins in plan view provides a simple method of assessing elastic thickness, which, in turn, appears to be determined by the length of time since the lithosphere was last thinned and heated as part of a passive margin. Thus, the markedly curved west Alpine foreland basin comprises relatively unstiff lithosphere with a small elastic thickness of 25 km, owing to the fact that it has been relatively recently heated, some 250 Ma ago. The much less curved Himalayan foreland basin, however, comprises ancient Indian Shield lithosphere last heated some 1000 Ma ago, thereby leading to an elastic thickness of 90 km.

Perhaps more than in any other geological setting, the coupling of foreland basin development with the evolution of thrust belts has promoted an understanding of relationships between plate collision, uplift, basin subsidence, and sedimentation. An increasingly well-documented process which emphasizes these interrelations is the incorporation of early foreland basin deposits in subsequent deformation, as the adjacent thrust belt advances into its own foreland basin, in a process known as *cannibalization*.

JONATHAN P. TURNER

fossil plants

The study of fossil plants, palaeobotany, it not only of interest in itself, but can be applied to solving a wide range of biological and geological problems. Palaeobotany is concerned with the study of macroscopic plants, but can also include palynology—the study of spores, pollen, and other microscopic plants remains.

The nature of fossil plants

Fossilization processes: taphonomy

Plants differ from animals in a number of important respects so far as fossilization is concerned: they tend to live in erosional rather than depositional sites; they are entirely organic (except for some microscopic algae such as diatoms and coccoliths); they can show an alternation of generations, one being more easily fossilized than the other; and they readily fall apart. The size of the plant may also play an important role in controlling the nature of fossilization: plants range in height from less than 1 cm to over 100 m.

Many plants have a number of distinct organs: roots, stems or trunks, leaves, fertile parts. On death these organs separate and will usually be found as individual fossils. In life also, plants may shed leaves, seeds, and other organs. In consequence, only isolated organs are commonly encountered as fossils. Palaeobotanists are thus rarely, if ever, faced with a whole plant. Even today, relatively few plants have been reconstructed for which all the organs are known.

Because the fossil record is fragmentary, different parts of the same plant are given different names. For example, the large tree-like Carboniferous plant known as *Lepidodendron* has a rhizome underground called *Stigmaria*, leaves known as *Cyperites*, cones known as *Lepidostrobus*, and so on.

To complicate matters even further, the chemistry of the various plant organs also controls their preservation. Wood, for example, is composed of cellulose and lignin, each of which decays at a different rate. Leaves similarly have cellulosic tissues that readily decay but have an outer cuticular layer that may be more resistant.

Plant fossils accumulate in a wide variety of environments and can be found in a wide range of rock-types. They may be buried in place by inundation of sediment or even by lavas or pyroclastic flows. Fossil forests are commonly found entombed by volcanic rocks (e.g. the early Carboniferous forests on Arran, at Weeklaw, Scotland; the Jurassic forest in Patagonia, Argentina; and the Tertiary forest of Yellowstone National Park, USA). In other instances they have simply been entombed by clastic sediment from river flooding (e.g. Joggins, Nova Scotia) (Fig. 1a). Wetland settings, such as bogs and swamps, provide areas where plant material can accumu-

late under reduced oxygen conditions so that the decay of the organic material is slowed. Production of plant material then exceeds decay, leading to an organic accumulation in the form of peat.

Waters charged with minerals may infiltrate buried plants and calcite, silica, or pyrite may be precipitated in the empty cells spaces, 'entombing' the plants. 'Permineralization' is the term used where the original organic cell walls of the plant remain (Fig. 1b). The organic cell walls may subsequently be replaced by minerals (Fig. 1g). A good example of this type of petrification is the Triassic fossil forest of Arizona.

Plants can be transported and buried in sites far from where they lived, to become fossils. A fossil leaf may retain some of its organic component in the form of a coaly layer. An impression of a leaf can also be preserved in the rock; it is possible for all the organic material to be removed at any stage after burial, leaving a simple impression (Fig. 1h). This form of preservation may look spectacular but it is seldom particularly informative to the palaeobotanist.

Some plants may have been subjected to wildfire, and incomplete combustion produces charcoal which is relatively inert and has an enhanced preservation potential (Fig. 1d). Fossil charcoal (fusain) characteristically shows excellent preservation of three dimensional anatomy (Fig. 1e, f).

Palaeobiology of plants

The reconstruction of plant fossils is important for the interpretation of the biology of the plants. Investigation of the morphology and anatomy of fertile structures is needed to interpret the reproductive biology of the plants, which may be important for the understanding of their ecology. For example, vegetative structures may yield data on the life habit of a plant: whether it was free-standing, a scrambling plant, or a climber; investigations on seeds may yield data on their mode of dispersal.

Palaeoecology and reconstruction of ecosystems

An important aspect of palaeobotanical research is to unravel the palaeoecology of plant fossils to use the results in reconstructing ecosystems. Palaeoecological data are gathered in the field, where the occurrence of plants, *in situ* and drifted, can be documented, together with aspects of their completeness, preservation, and abundance. Interpretation of the biology of the plants, their transport and burial history, together with data on their preservation, can all help in palaeoecological reconstructions. The ecology of plants in wetland mires (peat-forming systems) has been widely studied using both microfossils and mesofossils.

Plant biology can also give clues to ecology. Plants with numerous small seeds are usually colonizers, whereas those with larger seeds usually live in more mature communities. The evolution of trees in the late Devonian opened up many new ecological niches. Likewise the evolution of the seed enable plants to thrive in drier environments. It is not until the late Carboniferous that we have evidence of upland

Fig. 1. Fossil specimens illustrating various parts of plants and modes of preservation:

(a) Upright sandstone-filled lycophyte trunk *in situ*; Upper Carboniferous, Joggins, Nova Scotia, Canada.

(b) Transmitted light micrograph of peel of coal ball (plants preserved as permineralizations by calcium carbonate) showing transverse sections of the seed-fern (pteridosperm) *Lyginopteris*; Upper Carboniferous, Lancashire, England (×2).

(c) Lycophyte megaspore, *Lagenicula*; Lower Carboniferous, Foulden, Scotland (×50).

(d) Fusain (fossil charcoal); Middle Jurassic, Scalby Formation, Long Nab, Yorkshire, England (×1/2).

(e) Scanning electron micrograph of gymnosperm wood charcoal (fusain); Lower Cretaceous, Nova Scotia, Canada (×50).

(f) Scanning electron micrograph, *Scandianthus*, preserved as charcoal (fusain); Mid-Cretaceous, Scania, Sweden (×30).

(g) Cut and polished section through conifer cone of *Araucaria mirabilis* preserved as a silica petrification; Late Jurassic, Cerro Cuadrado petrified forest, Patagonia, Argentina (×2).

(h) Angiosperm leaf (leguminosaceous leaflet) showing marginal insect damage; Eocene, Tennessee, USA (×2).

(i) Sandstone cast of pteridosperm seed (*Trigonocarpus*) showing position of arthropod boring; Upper Carboniferous, England (×1).

vegetation. Palaeoecological data on plants, together with evidence of plant–animal interactions, may be integrated with data from animals and sedimentary environments to reconstruct an ecosystem.

Plant evolution

Palaeobotany is also concerned with the evolution of plants. Most studies now consider relationships between plants, and the use of cladistic analysis has become widespread. Traditionally palaeobotanists had sought ancestor–descendant relationships, for example, the evolution of the conifers from the cordaites (an extinct group found in rocks of Carboniferous and Permian age), but more recently cladistic analysis has shown that these two groups are sister groups and that the conifers may be a polyphyletic group.

The evolution of plant strategies or organs has been the main focus of some research. Other research has been concerned with the evolution of a plant family or genus: for example, the evolution of the flowering plant family the leguminoseae or the genus *Ginkgo*.

Plant–animal interactions

Plants form the base of most food chains. Plants may be eaten dead, as part of the decaying litter, as by many arthropods, or alive, as by leaf-feeding insects (especially the caterpillars of butterflies and moths). It is not only invertebrates that rely on a regular supply of plant food but also vertebrates such as grazing and browsing mammals.

Such plant–animal interactions may be of three types: feeding, dispersal, and shelter. These interactions can be recognized from the study of both plant and animal fossils. Interactions can be revealed by damage to plants (Fig. 1h, i); for example, chewed leaves, bored wood, or the occurrence of specially evolved features such as nectaries in flowers, glandular hairs, and thorns on stems and leaves. Plant–animal interactions have, however, rarely been studied using fossil material. Damage to plants by feeding arthropods is only rarely recorded, because many collectors throw away 'damaged' specimens. In particular, leaves showing evidence of feeding (chewed leaves, mines and galls) become increasingly common from the Carboniferous to the Recent; Tertiary floras show a marked increase in numbers and diversity of specimens. A number of evolved features in flowers from the Cretaceous onwards have been linked to insect pollination.

The origin of life

Studies on the Precambrian fossil record have focused attention of the origin and early evolution of life. Simple, single-celled micro-organisms including bacteria, cyanobacteria, and green algae have a considerable Precambrian fossil record. Studies of the diverse fossil biota from the 1800 Ma Gunflint Chert in Canada alerted researchers to the possibility of Precambrian fossils. Rocks more than 3000 Ma old in Australia have since yielded fossils, pushing further back the origin of life. Early fossils are all simple-celled prokaryotes such as bacteria and cyanobacteria. The evolution of eukaryotic algae with a discrete nucleus made possible the evolution of sex and more rapid diversification; it also led to the evolution of multicellular organisms.

The uses of fossil plants

Fossil fuels: their origin, use, and exploitation

Plant fossils provide the basis for most fossil fuels (coal, oil, and gas), and palaeobotany helps in all aspects of our understanding of these fuels. In terrestrial wetlands plants may accumulate in waterlogged conditions and the anaerobic environment may prevent decay and promote the formation of peat. In aquatic settings, organic matter transported into the environment or living in the environment (e.g. algae, including planktonic algae in the sea) can accumulate in anaerobic settings. On burial and with increasing temperature, the organic materials alters and produces oil and gas. Peats will *in situ* change successively to lignite, to bituminous coal, and eventually to anthracite. Gas, and in some instances oil, will be generated from such deposits. In organic-rich sediments oil and gas are generated from hydrogen-rich organic particles (e.g. algae, spores or pollen, cuticles, etc).

Because much of the plant material remains in coals, it is possible to isolate chemically identifiable plant parts such as spores and cuticles. Palaeobotanical data can be used to interpret the ecology and environment of the original peat deposit and to determine sequential changes that have taken place during the development of the peat. These data may help in dating coal seams and in their correlation, and can even be used to identify coals sold on the international market.

Plant fossils can be extracted from oil and gas source rocks, again helping not only in their dating but also providing palaeoenvironmental data and information on the petroleum potential. The chemical or physical alteration of the organic matter may can help with maturity assessment (i.e., whether the oil or gas 'window' has been reached).

Pollen, spores and marine phytoplankton are widely used to date sediments. When oil migrates from its source to a trap it may during that journey pick up pollen and spores from the rocks through which is passes. A study of this material can aid studies of fluid migration.

Past atmospheres

The study of fossil plants has contributed to our understanding of the evolution of past atmosphere, and in particular of the changes in oxygen and carbon dioxide (CO_2). Oxygen and carbon dioxide levels are controlled by the activity of plants. The evolution of plants that photosynthesize caused major changes in the biosphere–atmosphere cycle. Green plants utilize sunlight to extract carbon from the atmosphere and (through photosynthesis) to incorporate the carbon into their organic 'skeletons'. Oxygen is released as a by-product of photosynthesis. Oxygen concentrations in the atmosphere began to rise significantly in Precambrian times. Shortly after the spread of plants on to the land during the Devonian period, oxygen concentrations reached their present levels. We know that oxygen levels must have been between 15 and 35 per cent through the period from the Devonian to the Recent because of evidence from the charcoal record of wildfire from the latest Devonian onwards.

Carbon dioxide is also controlled by the global carbon cycle. When they decay, plants release carbon dioxide (CO_2) into the atmosphere. However, if sufficient carbon is held back in the lithosphere, in sediments or peats, then the overall CO_2 levels will decline. This can become self-perpetuating, because low CO_2 levels lead to global cooling. (CO_2 is an important greenhouse gas.) This in turn can lead to the formation of ice caps, a lowering of sea level, and a spread of land vegetation, increasing carbon draw-down. There have been many fluctuations in CO_2 levels during the Earth's history. The stomata (pores) on plant leaves vary with CO_2 concentrations in the atmosphere, and changes in stomatal density can act as a proxy in interpreting ancient CO_2 levels in the atmosphere.

Past climates

Fossil plants have been widely used to interpret past climates. Five main approaches have been used.

Nearest living relative (NLR) In Tertiary plant assemblages the climatic ranges of the nearest living relative (either at generic or family level) can be used to make broad interpretations of past climates. The method has the disadvantage that it assumes that the present climate range of a particular taxon is the same as it was in the past.

Leaf physiognomy Several aspects of the shape and size of angiosperm leaves are related to climate. For example, leaves in the tropics are generally over 10 cm long, are tough and evergreen, and have entire margins and drip-tips. In contrast, temperate leaves are often less than 10 cm long, are deciduous, and have interrupted margins and no drip-tips. Calculations can be made using data from leaf assemblages to extract a climate record.

Growth rings In the tropics plants and trees grow continuously, their wood consequently shows no interruption of growth. Wildfires can cause fire scars, and in tropical fire-prone environments these may be regular. In temperate latitudes, because of the variation of light throughout the year, there are interruptions of growth which produce distinctive growth rings. These appear yearly and hence provide the basis for dendrochronology. The size of any growth ring and its regularity depend not only on latitude and hence the length of the growing season, but also on temperature; the largest rings form in warmer climates. Indices such as mean sensitivity are used to categorize growth rings in fossil plants for palaeoclimate analysis. Growth rings are known in fossil woods from the Devonian onwards, and trees from polar areas have been shown to have had strong growth rings in the Mesozoic.

Fossil charcoal Charcoal (as fusain) occurs widely in post-Devonian sediments and coals and forms as the result of wildfire (Fig. 1d). Wildfires can occur in many climatic zones, but they are most prevalent where there is a build-up of fuel and where there are periods of dryness and frequent lightning strikes.

Isotopes In fixing their carbon from carbon dioxide in the atmosphere, plants may use a variety of processes. Each of these incorporates different amounts of the stable carbon isotopes ^{12}C and ^{13}C. The most common photosynthetic cycle, known as C3, is common to most plants. Some plants that are found in drier environments and in highly stressed environments such as salt marshes use the C4 metabolic pathway. C4 plants characteristically produce carbon with a lighter $\delta^{13}C$ isotopic signature. Studies of on the isotopic composition of soil carbon have been used to interpret climatic changes by means from C3 to C4 plants in the profile.

Biogeography and plate movements

Plant distribution is broadly controlled by climate. Climate is in turn controlled not only by latitude but also by height above sea level. Mountain ranges and oceans both provide barriers for the migration of plants. The movement of continental plates and the changing distribution of oceans and mountains, together with changing climates, have all contributed to the formation of distinct biogeographical regions on Earth today. The occurrence of distinctive palaeobiogeographical regions is recognized as a feature since the early evolution of life on land.

The study of such provinciality may provide useful data on the positions of former continents. In early Permian times the gymnosperm *Glossopteris* lived in the southern hemisphere temperate zone. Fossils of *Glossopteris* are now found widely in South America, South Africa, Antarctica, Australia, and India. This reinforces the idea that these continents were once joined together into a large southern continental landmass called Gondwana.

The development of plant palaeobiogeography through the late Palaeozoic has been widely studied, and increasing regionalization is seen as continents move and split up. Equally, when previously separate continental blocks with different floras (e.g. the North and South China blocks) collide, their floras intermix.

Dating

Plant fossils have been widely used for dating rocks. Macrofossil plants have traditionally been used for dating non-marine clastic sequences. Fossil plants zonations have been erected for the Upper Carboniferous and have been used to correlate coal measures sequences across Euramerica.

Microfossil plants have been applied to biostratigraphical problems since the 1970s. Spores and pollen occur widely and abundantly in a variety of sediments and environments. Several types of zonation scheme have been used, but the concurrent-range biozone defined by the simultaneous presence of two or more species) is the most common. Of particular use is the fact the spores and pollen can be obtained from a variety of facies (including clastic rocks and even volcanic ashes). Spores and pollen also have the advantage of being transported into marine environments. In some cases instances (e.g. in the late Devonian and Carboniferous), megaspores have been used successfully to date and correlate sequences. In marine sediments marine phytoplankton (floating plants) are widely used in biostratigraphy. Palaeozoic sequences yield abundant and diverse acritarchs (organic-welled microplankton of unknown affinity). Mesozoic and Tertiary sequences yield diverse dinoflagellates (a class of uni-cellular algae), which have been particularly useful in cor-relating the marine Jurassic rocks of the North Sea. Other plant microfossils, including diatoms and coccoliths (remains of unicellular algae) have been found useful for dating in Cretaceous and younger sediments.

Palaeosols

Fossil soils are known as *palaeosols*. Plant roots leave distinctive traces in sediments: the activity of plants rooting on a rock sub-strate has the effect of altering the rock physically and chemi-cally. Organic decay releases acids that chemically attack rock and produce a soil. Study of the texture and zonation of palaeo-sols can yield data on the temperature and humidity of the environment.

Provenance studies

In some instances smaller plant fragments and organs may be reworked and redeposited into younger rocks. This commonly happens with palynomorphs (organic microfossils), including megaspores, with coal particles and with fossil charcoal (fusain). The erosion and transport history of the early Ter-tiary Thanet beds of southern England were deduced from the presence of megaspores of various ages, including some from the Carboniferous and Mesozoic.

Techniques in the study of fossil plants

Collection

Fossil plants can be found in a wide variety of terrestrial and marine rocks, not only sedimentary but also igneous. Methods of collection vary according to the nature of their preservation and whether qualitative or quantitative data are required. Most geologists are familiar with the occurrence of plant compression fossils in bedded sedimentary rocks. Both the part and the counterpart of the fossil need to be collected, since they will yield complementary data. Quantitative collection may require the counting of specimens per bed or the excavation of a uniform bedding area. Palynological samples may be just rock samples in which no macroscopic plants are visible.

Fossil wood is generally easy to find and collect. Per-mineralized plants (other than large pieces of wood) are often difficult to recognize in the field. Stems, leaves and, fructifications may not look well preserved in the field, and as they occur in cemented rocks that do not break along bedding planes they can easily be missed. Even volcanic ashes and some lavas may contain such material. For example, the Lower Car-boniferous green volcanic ashes and agglomerates from Oxroad Bay in East Lothian, Scotland contain abundant anatomically preserved plants that were overlooked by those studying the

geology. With such material it is necessary to collect large blocks of material for subsequent treatment in the laboratory.

Coals can be sampled in a variety of ways, either by taking a single-channel sample of the coal or by taking more detailed samples. The choice will depend on the thickness of the coal and the nature of the investigation. Some studies require continu-ous sampling of every lithology; others require regular spaced samples.

Isolated plant compressions are also commonly found, as are permineralized plants, in marine sediments, even in some instances associated with ammonoids. These are often over-looked.

Fusain (fossil charcoal) is rarely collected, for it is often mistakenly believed to be poorly preserved. Under the scan-ning electron microscope blocks containing charcoal com-monly show exquisite preservation (Fig. 1e).

Extraction

Serious study of plant fossils may require extensive preparation. Any parts of the specimen buried in the rock are carefully dug away with a fine mounted needle under a microscope using a technique known as *dégaugement*. Organic material can be removed and treated chemically for microscopic study. Per-mineralized fragments can be embedded in resin and sectioned or peeled. Bulk maceration (physical and chemical treatment) can be used to break down the rock matrix and release plant tissues. This is widely used for the study of plant mesofossils and microfossils. Plant mesofossils (in the size range 180 μm to 5 mm) include small seeds, megaspores (Fig. 1c), and plant cuticle. Such material has been referred to as 'fossil tea leaves'.

Permineralizations and petrifications

Petrifications, where no organic material remains, are usually prepared as thin sections or polished and studied under reflected light. For permineralized plants, the peel technique is useful; this provides a section of the original plant material embedded in a cellulose acetate sheet (Fig. 1b).

Microscopic studies

Fossil plants are studied using a wide variety of microscopic techniques. For many initial investigations, a low-power binocular microscope may be all that is required. For more detailed investigations of macrofossil, mesofossil, or micro-fossil plants there are many techniques that can be employed, ranging from traditional high-power transmitted microscopy, using normal or UV light, to more specialist microscopes such as laser scanning microscopy, scanning acoustic microscopy, infrared microscopy, and cathodoluminescence microscopy. Some relatively simple techniques, such as the use of polar-ized light to study macroscopic and compression fossils or the immersion in liquids, are also useful.

More powerful electron microscope techniques are widely used. These include scanning electron microscopy (which is now possible for large specimens by using an environmental chamber) and transmission electron microscopy, in which details of the plant ultrastructure can be observed.

Geochemical studies

Plant fossils have been subjected to a number of chemical studies, on the organic material itself, on the permineralizing cements, or on the enclosing sediments.

Organic geochemical studies Organic compounds extracted from fossil plants can be studied using such techniques as gas chromotography–mass spectrometry. The compounds identified include biomarkers: compounds that can be related to identifiable biological chemical 'species'. This technique is particularly useful for interpreting the source of oil. More resistant plant tissues containing macromolecules can be studied using a range of techniques, including spectroscopic methods such as solid-state ^{13}C nuclear magnetic resonance and fourier transform micro-infrared spectroscopy and by destructive techniques such as flash pyrolysis–gas chromotography–mass spectrometry. Information has been obtained on the chemistry of sporopollenin, the walls of spores and pollen, on wood chemistry, on fossil cuticles, and on algal cell walls. Studies on both modern and fossil plants are helping to interpret changes that occur in the organic structure of the plant tissues during burial diagenesis. As indicated above, isotopic studies are also undertaken. Claims have been made for the occurrence of DNA in several fossil plant specimens. The records are sparse and none has yet been repeated by independent laboratories. There is thus still doubt as to whether such proteins can survive more than a few thousand years.

Inorganic geochemical studies Enclosing sediment and permineralizing cements can be subjected to geochemical analysis to help understand their origin. Destructive geochemical methods are commonly combined with petrographic methods (thin sections, stained sections, cathodoluminescence) to interpret the history of rock cementation. Chemical analysis may include the use of energy-dispersive X-ray analysis (EDAX) linked to a scanning electron microscope or such techniques as X-ray spectroscopy or inductively coupled plasma mass-spectroscopy (ICP), which yields data on element abundances. More recently, stable isotopes of oxygen and carbon have been used to interpret the origin and history of cementation of calcareous and siliceous plant permineralizations.

Geochemical analysis of the enclosing sediment has been used to 'fingerprint' plant horizons with the aim of identifying the origin of loose specimens, whether from the field or in old museum collections.

Experiments

Useful data have been obtained from taphonomic experiments, which include studying the transport and burial of plants in modern environmental settings and in laboratory experiments using flume and settling tanks. The natural landscape has also been used for fire experiments. Charring experiments have been undertaken under controlled laboratory conditions. Plant decay has been subjected to controlled experiments both in the laboratory and under 'natural' conditions.

Diagenetic processes have also been studied experimentally from compression experiments using artificial and natural materials; in addition there have been attempts to permineralize and petrify plants.

The occurrence of fossil plants

Facies distribution of fossil plants

Fossil plants occur in a wide variety of terrestrial clastic rocks. They are also equally common in some limestones, volcanic rocks, and several marine facies. It is not only the spectacular plants that are worth collecting; less 'promising' fragments can yield abundant data of high quality.

In general, rocks that are grey, black, or green yield organic material, whereas brown and red rocks do not, the plant material being oxidized. Even in red rocks some plant impressions may be found. Also as a rule coarser clastic rocks contain fewer fossil plants then finer clastic rocks, but large logs and even small charcoal fragments can be found in such rocks. Likewise, while most lavas and ashes do not contain plant material, others do and they may be spectacularly preserved. Marine rocks commonly contain plant fossils but they are outnumbered by more common and spectacular shelly invertebrate fossils. Some rocks are made up entirely of plant fossils: for example, coals, diatomites, and even chalk, made up of the coccoliths, algae that secrete calcium carbonate.

Stratigraphical distribution of plants and plant groups

Microscopic algae are known from some of the oldest rocks on Earth. Indeed, many hundreds of Precambrian localities are now known to yield plant fossils. Macroscopic terrestrial plants are not known until the late Silurian, and it was in the early Devonian that vascular plants diversified. Early land plants were small, some only a few centimetres high, and did not grow more than a metre tall. All early plants reproduced by spores. These early plants belong to a group known as Psilophyta, which is probably a group of simple plants, some of which, but not all, were descendants of a common ancestor (paraphyletic). The Lycophyta (clubmosses such as the living *Lycopodium* and *Selaginella*) also evolved in the early Devonian (Fig. 2a). The Mid–Late Devonian saw the evolution of the tree habit, with the evolution of secondary growth.

The seed habit evolved among plants in the late Devonian in the gymnosperms. Seed-ferns or pteridosperms evolved rapidly in the Carboniferous; the first conifers are known from the Upper Carboniferous (Fig. 2b). The Carboniferous saw the spread and diversification of three groups of spore-bearing plants: the lycophytes, the sphenophytes (horsetails), and the ferns. The Lycophytes became particularly important in tropical peat-forming areas and were particularly well represented by the tree-like forms such as *Lepidodendron*. The sphenophytes were also represented by small herbaceous forms as well by as arborescent taxa such as *Calamites*. Ferns diversified in the Carboniferous, but much of the fern-like foliage found in the clastic rocks associated in the coals belongs to pteridosperms or seed-ferns such as *Neuropteris* and *Alethopteris*.

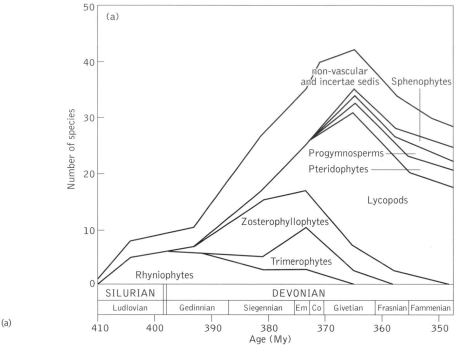

(a)

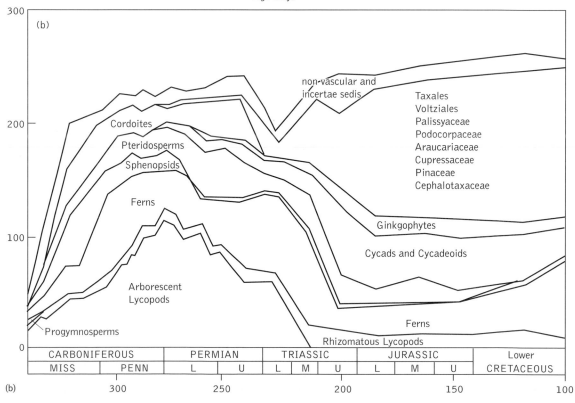

(b)

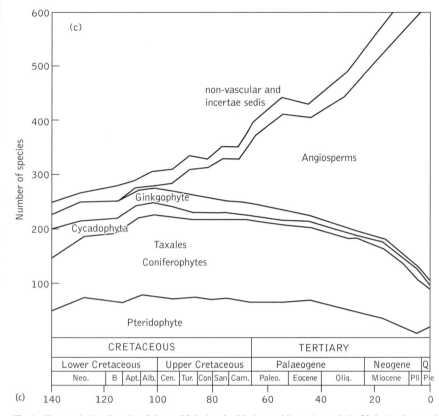

Fig. 2. Changes in the diversity of plants: (a) during the Silurian and Devonian periods; (b) during the Late Palaeozoic and Mesozoic eras; (c) during the Mesozoic and Cenozoic eras. (From Niklas, K. J., Tiffany, B. H., and Knoll, A. H. In Valentine, J. W. (ed.) *Phanerozoic diversity patterns: profiles in macroevolution.* Princeton University Press, New Jersey.)

Major climate changes at the end of the Carboniferous caused the extinction of many of the spore-bearing groups (Fig. 2b). Seed plants evolved rapidly during the Permian and late Mesozoic (Fig. 2c). Conifers continued to expand both geographically and ecologically, and they also diversified. Cycads and the extinct bennettites (cycadeoids) became important, and in some areas dominant, elements of the vegetation. Seed-bearing glossopterids characterize the Permian of the southern continents. The Triassic saw the continued evolution of seed-plants with the first occurrence of *Ginkgo* (a living genus) and also the Peltasperms.

A major change in terrestrial vegetation took place in the Cretaceous with the evolution of the angiosperms, or flowering plants (Fig. 2c). Some of these plants are wind-pollinated; others are insect-pollinated, and the close interaction between the angiosperms and insects has often been cited for the success of both groups.

By the end of the Cretaceous, several groups of plants, such as the bennettites, peltasperms, and pteridosperms, had become extinct. The Tertiary saw the rapid diversification of the angiosperms (Fig. 2c). Grasses did not evolve until the mid-Tertiary, and grasslands were consequently not abundant until Mid–Late Tertiary times.   ANDREW C. SCOTT

Further reading

Cuneo, N. R. and Taylor, T. N. (eds) (1996) Palaebiology of fossil plants: new insights and perspectives. *Review of Palaeobotany and Palynology*, **90**, 155–373.

Jones, T. P. and Rowe, N. P. (eds) (1999) *Fossil plants and spores: modern techniques.* Geological Society, London.

Scott, A. C. and Calder, J. (1994) Carboniferous fossil forests. *Geology Today*, **9**, 230–4.

Stewart, W. N. and Rothwell, G. W. (1993) *Palaeobotany and the evolution of plants* (2nd edn). Cambridge University Press.

Taylor, T. N. and Taylor, E. L. (1993) *The biology and evolution of fossil plants.* Prentice Hall, New Jersey.

fossils and folklore The commonly symmetrical shapes and similarities of fossils to parts of living organisms have caught the eye of men throughout time and in most parts of the world. Many resemble other organic forms or structures. In some instances the fossils are associated with particular kinds of rock or with specific localities. Fossils have been found associated with the remains of early humans, some being present in burial furnishings or as adornments. Cultures throughout the world have ascribed special properties to particular types of fossil, and these objects are well known in European folklore that goes back at least to the Middle Ages.

In folklore many such fossils have been identified as supernatural or exotic objects and bear fanciful names. Some may be accorded talisman status, promoting good luck or boosting the possessor's virtues, etc. In Britain, some of the earliest systematic collecting of data on the topic is found in the work of Dr Robert Plot, the first keeper of the Ashmolean Museum in the University of Oxford.

Coiled shells, commonly of ammonoid cephalopods, nautiloids, or gastropods, have been given colloquial names such as rams' horns, snakestones, serpentstones, and conger eels. The bullet-shaped solid guards of belemnites have been referred to as thunderbolts, devil's fingers, or even St Peter's fingers. In Yorkshire they are known as scaur pencils, while in Scandinavia they are gnome's lights or candles. Screwstones are the casts of the interiors of spirally coiled gastropods from the Jurassic of southern England.

Very common in the Lower Lias (Jurassic) of Britain, the bivalve mollusc *Gryphaea*, with its characteristic layered curved shape, is known as the devil's toenail.

The columnal plates of crinoids, common in many Palaeozoic limestones, are locally weathered out of the matrix singly or in groups, and may be threaded together through the central hole. They have been called fairy money, St Cuthbert's beads, and, in Germany, St Boniface's pennies.

Columnals of the Mesozoic crinoid *Pentacrinites* have a star-like shape and are known as star-stones. Fossil echinoderms have been called shepherd's crowns, fairy loaves, and even thunderbolts. Others have been dubbed chalk eggs.

Many of the objects mentioned above have been credited with beneficial or medicinal properties, and prescribed in one form or another as curatives or aphrodisiacs. In China fossil bones (dragon bones) have been used for centuries, usually ground to powder, in various prescriptions and for similar purposes.

D. L. DINELEY

Further reading

Bassett, M. G. (1982) '*Formed stones', folklore and fossils*. Geological Series No. 1, National Museum of Wales, Cardiff.

Edwards, W. N. (1967) *The early history of palaeontology*. British Museum (Natural History), London.

fossils and fossilization
Fossils (Latin *fodere*, to dig up) are the remains of once-living organisms preserved in rocks or sediments. The term originally referred to anything dug out of the ground, whether animal or mineral and, therefore, included gems and minerals, archaeological remains, and even strange-shaped nodules; however, the measuring of the word has gradually become more restricted. We now generally divide fossils into two main types: body fossils, which are the remains of hard parts or more rarely, soft parts of organisms, and trace fossils, which are the traces of biological activity, such as burrows and tracks and trails. The record of fossils goes back 3400 million years to preserved single-celled organisms (bacteria and blue-green algae). Complex organisms or metazoans do not appear in the fossil record until about 700 million years ago when the Ediacara fauna has an almost worldwide distribution. Hard parts were not commonly present in organisms until about 544 million years ago at the beginning of the Cambrian Period, when there was a tremendous development of invertebrate organisms often called 'the Cambrian Explosion'. Since then a diverse assemblage of invertebrates and vertebrates is available throughout Phanerozoic time for study by palaeontologists.

Fossilization is a rare occurrence, and it has been estimated that of the more than one million living species only 10 per cent are likely to be preserved as fossils. It must be appreciated, therefore, that the fossil record is exceedingly incomplete and biased. The natural preservation of an organism, when it does occur, is dependent on a number of factors, of which the main ones are: the composition and structure of the organism; its abundance; the sedimentary environment in which it lives; and what post-depositional changes take place.

All organisms are composed of delicate tissues known as soft parts, but many also have more resistant tissues referred to as hard parts. The hard parts may be mineralized, as in the shells of bivalves, or composed of organic material, such as the chitin that makes up the exoskeleton of arthropods. Although in general the possession of mineralized hard parts is a prerequisite for preservation, soft-part preservation can occur and some entire fossil groups, such as the graptolites, have only organic hard parts. Another important feature is the density of the hard parts, particularly from the point of view of the ratio of organic to inorganic material; in general a low proportion of organic material favours fossilization.

The numerical abundance of organisms is generally considered to be important, since, all other things being equal, abundant organisms would seem to be more likely to be preserved than rare ones. It has, however, been shown in a number of instances that the relative abundance of an organism in the fossil record does not accurately reflect its abundance in the living population. This differential preservation is frequently related to the preservability of the hard tissues.

The most common sedimentary rocks are those deposited in shallow-marine environments, and because of this the majority of fossils are of organisms that lived in those environments. Deep-water sediments are less commonly exposed, and in consequence fossils of organisms that lived in those environments are rarely found. Terrestrial organisms generally lived in areas of non-sedimentation and thus stood little chance of preservation. In addition, the type of sediment will have an effect on preservation: coarse sediments, indicating high-energy conditions, are unlikely to contain fossils, whereas fine-grained sediments are more conducive to preservation. Sediments, while directly preserving organisms by enclosing them, can also produce indirect fossils by creating impressions of surfaces that are termed moulds. These may be impressions of the outer surface, or of the inner in which case they are often called steinkerns (Fig. 1).

Post-depositional changes are important because both biochemical and chemical processes act upon the remains of the organism to preserve or destroy it. Groundwater percolating through sediments can dissolve enclosed hard parts, removing all trace of the incipient fossil, though if an external mould is left this might be filled later by other minerals to form a cast (Fig. 1). Minerals can also be deposited in the pore spaces of hard parts such as bones in a process called permineralization, thus strengthening them and improving the chances of preservation. A mineral may also be completely replaced by another mineral; for example, calcium carbonate shells can be replaced by silica or pyrite (Fig. 1). Another process that destroys details in a fossil, although it will improve the preservation potential, is recrystallization. In this process less stable minerals forming hard parts are transformed through time to more stable forms, as when gastropod shells formed of aragonite are transformed into calcite.

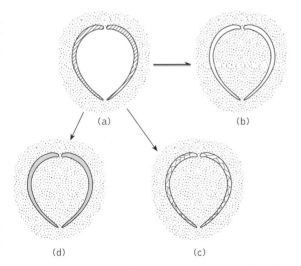

Fig. 1. Possible processes of fossilization of a bivalve shell: (a) original shell; (b) shell dissolved and moulds and steinkern developed; (c) recrystallization; (d) replacement.

Fossils are the main phenomena studied by palaeontologists to gain a better understanding of evolution, extinctions, palaeoecology, and many other topics. Preservation is of vital importance to the acquisition of this information and it is important that it should be understood.   DAVID K. ELLIOTT

Further reading

Briggs, D. E. G. and Crowther, P. R. (eds) (1990) *Palaeobiology: a synthesis*. Blackwell Scientific Publications, Oxford.

Clarkson, E. N. K. (1993) *Invertebrate palaeontology and evolution*. Chapman and Hall, London.

fractals in Earth science

The term 'fractal' is informally applied to phenomena or objects which manifest some type of structure or organization which is related to itself across scale, or which exhibits scale invariance. In the late 1960s B. Mandelbrot, of the IBM Thomas J. Watson Research Center, established the importance of 'fractional dimension' in the context of measuring the length of the British coastline, and coined the term 'fractal'. The length of the coastline, and of many other objects, is dependent upon the scale at which one measures. A simple way to visualize this fact is to imagine making the measurements with a ruler. If you use a very long ruler, several kilometres long, the measured length appears to be smaller than if measured with a smaller ruler, one say, several metres long. The apparent change in length of the coastline with the scale of measurement is an important characteristic of fractals. For special objects, sometimes called 'monster curves' by mathematicians, the apparent length approaches infinity as the ruler gets infinitely short.

Earth scientists commonly use an object, such as a hammer or camera lens, as a scale in their photographs. One reason for this practice is that many geological phenomena, such as landforms, require an object of known size in order to define the scale; without the hammer, it would be hard to determine their size simply from their shape. In a prolific stream of ideas, Mandelbrot suggested that fractals are a significant and useful alternative description of the geometry of many natural phenomena. In general a 'fractal is a shape made of parts similar to the whole in some way', an attribute called self-similarity. This shape can refer, for example, to an outline of a physical object, a theoretical geometric object, or even patterns produced by the functioning of some process. The fractal represents an important concept for the Earth sciences because it shows how phenomena can have systematic relations to scale. The application of fractal concepts is here outlined briefly with reference to geomorphological ideas.

In geomorphology, fractals occur in many contexts: to describe or quantify the shapes of landforms, patterns of stream networks, or the rhythm of geomorphic processes and their effects. Landscape roughness provides an example of how fractals may be applied to geomorphology. In the context of fractals we are interested to know at what scale a landscape exhibits roughness of a certain magnitude. A landscape may appear rough at one scale and smooth at another, and these scale relations could be the result of processes operating on the landscape but at different scales. Because fractals provide tools that quantify the shape of landforms, they have been used in morphometric analysis. The scale at which particular signals appear in a landscape, tectonic or climatic for example, can be studied using the method of fractal science. Fractals are also used to analyse temporal patterns in the records of surficial processes, including flood records, earthquakes and tectonic uplift, precipitation, and sedimentation.

LARRY MAYER

Further reading

Jürgens, H., Peitgen, H-O., Saupe, D., and Zahlten, C. (1990) *Fractals, an animated discussion: A film and video.* W. H. Freeman, New York.

Mandelbrot B. (1967) How long is the coast of Britain? Statistical self-similarity and fractional dimension. *Science,* **156,** 636–8.

Snow, R. S. and Mayer, L. (eds) (1992) Fractals in geomorphology. Special Issue. *Geomorphology,* **5.**

fraud in geology The Earth sciences seem, on the whole, to have suffered relatively little from the antics of fraudsters. In early days, students, quarrymen, and other pranksters sought to confound the professors or pundits for fun rather than for financial gain. Fossils, in particular, have been prone to imitation or 'improvement', In 1726 Professor Johann Beringer at Würzburg was deceived by his students, who knew of his interest in fossils from the Muschelkalk limestones. They prepared a number of 'fossils' by moulding various forms of living or imaginary creatures in clay. Baked to resemble stone, these objects were scattered at the fossiliferous localities where Beringer would find them. Many of them were extremely fanciful, even with Hebrew characters included. Beringer was duly deceived to the extent of publishing a treatise with illustrations, and even refuting the possibility that these could be fraudulent objects. One day he found a fragment with his own name upon it, and the hoax dawned upon him. In great chagrin he spent his savings trying to buy up the whole edition. It is said that it shortened his days.

It is not unknown for quarrymen and others to join together the different parts of fossils to manufacture a 'perfect' or 'complete' specimen; the instance of the tail of one species being carefully joined to the head of another has occurred many times. Instances of carved fossils with soft and imaginary parts delicately delineated are numerous, ammonoids being particularly favoured with the addition of snake-like heads.

In early Victorian time a notorious Dr Albert Koch exhibited in several European cities a giant skeleton of ancient whale bones as a sea-serpent (as he had on previously occasions, using other fossil material). He was denounced wherever he set up his show by the indignant but highly competent Dr Gideon Mantell. Koch had to move on.

The Piltdown forgery, which was exposed in 1953, was more serious. Fragments of a purported human skull had been collected from river gravels at Piltdown in Sussex (England),

and in 1912 Arthur Smith Woodward of the British Museum (Natural History) and Charles Dawson announced the spectacular discovery of a remote human ancestral form (*Eoanthropus dawsoni*). There were also artefacts and animal bones to accompany the humanoid. At various times different authorities expressed doubts about the compatibility of the several skull fragments and teeth and about the provenance of the stone tools and animal bones. Nevertheless, it was accepted for many years that the Piltdown finds did indicate an early tool-using hominid in Britain where none was ever indicated previously. Refined analytical techniques eventually established that the skull and teeth were a mixture of recent human and ape fragments, abraded and stained to appear ancient. A bone tool was found to have been carved by a sharp metallic edge. Thus there never was a Piltdown hominid. It is thought that the Piltdown hoax may have begun as a joke, but it got out of hand. The leading figures in the debate, including anatomists such as G. Elliot Smith and Arthur Keith, were all deceased before the forgery was detected.

Fraud on a large scale has been detected more recently in the publications of an Indian professor, V. J. Gupta. He described fossils said to have been discovered by himself at various localities in the Himalayas. Many of the fossils were of considerable biostratigraphical significance, and a spurious stratigraphical record with numerous international correlations was set up on the basis of the identifications. Illustrations of the fossils were found to resemble those used by other authors elsewhere; the localities proved fictitious. Several papers were published by Gupta and one or other of several different, mainly Western, authors, who in some instances were unaware that their names had been used. The principal sleuth in the affair was an Australian, Professor John Talent. Despite strenuous denial of fraud, the published refutations of Gupta's work have come from responsible sources and the field evidence that contradicts his misleading ideas is convincing.

Fraudulent presentation of gemstones, ores and other samples in order to attract investment in mining ventures has undoubtedly occurred many times. Petroleum deposits have also been 'proved' on the basis of fraudulent records and samples. Very large sums of money have been involved on occasion, but today evidence of commercially important discoveries has to be corroborated by data from different sources. The possibilities for fraud are therefore fewer.

D. L. DINELEY

Further reading

Weiner, J. S. (1955) *The Piltdown forgery.* Oxford University Press, London.

free oscillations After a large earthquake the Earth rings like a bell. The earthquake sets up standing waves that can persist for weeks, each with a characteristic oscillation period measured in minutes. These are called *free oscillations* because they continue without any forcing, unlike tides for example, which are continuously driven by the Moon's gravitational

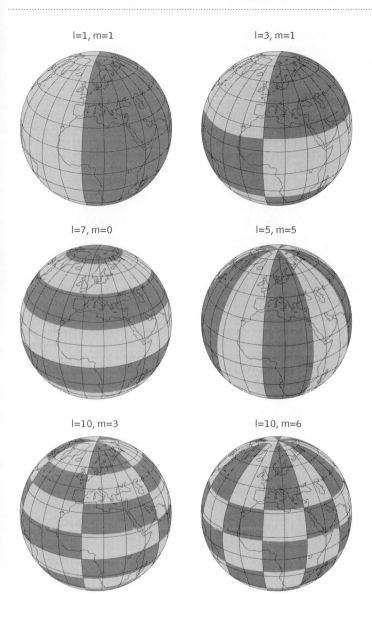

l=1, m=1　　　l=3, m=1

l=7, m=0　　　l=5, m=5

l=10, m=3　　　l=10, m=6

Fig. 1. Free oscillations with various angular and azimuthal order numbers *l* and *m*. Nodal lines divide the sphere into light and dark regions. For spheroidal modes, the dark regions move outwards while the lighter regions move inwards.

attraction. Since the discovery of free oscillations in 1960, the periods of many hundreds have been measured precisely. They depend on details of the Earth's internal structure, and consequently have told us much about the Earth's interior.

Plucking a guitar string sets up standing waves on the string with an integral number of wavelengths between the fixed ends. The standing wave with the longest wavelength is called the *fundamental*; the *overtone number* keeps count of the number of half-wavelengths. Points where the string remains stationary are called *nodes*. The frequency (pitch) is determined by the wavelength, which is altered by changing the length of the string using the frets. Free oscillations of a sphere are three-dimensional standing waves and require three overtone numbers. The nodes are surfaces lying on spheres (constant radius), cones (constant latitude), and meridians (constant longitude): see Fig. 1. The *radial overtone number (n)* counts the number of spherical nodal surfaces and the *angular order number (l)* counts the rest. The *azimuthal order number (m)* counts those on meridians, leaving *l–m* on cones of constant latitude. The 'wavelength', and therefore the frequency, depends on *n* and *l*; *m* merely keeps track of the distribution of nodal surfaces with respect to geographic north. Many free oscillations therefore have the same frequency and are said to be *degenerate*. The theory for a uniform sphere was worked out around the end of the nineteenth century. In 1882 H. Lamb showed there were two types of free oscillation, called *spheroidal* and *torsional*, the latter having purely horizontal motion with no compression. In 1911 A. E. E. Love determined a period of 60 minutes for the Earth's slowest oscillation.

Observation of the waves was held up by the technical difficulty of recording long periods: rapid vibrations are much easier to detect (they make cups rattle on shelves, for example), but periods of minutes to an hour require a very stable instrument. By 1952 H. Benioff was detecting long-period oscillations with his strainmeter, but positive identification had to await the very large Chilean earthquake of May 1960. Several research groups announced measurements at the 1960 meeting of the International Association of Seismology and Physics of the Earth's Interior (IASPEI) in Helsinki, heralding a new era of long-period seismology. Some of the frequencies observed with strainmeters were missing from gravimeter records: these were the torsional oscillations, which do not register on a gravimeter because their motion is purely horizontal.

Over a thousand modes, with periods ranging from 56 minutes down to less than 40 seconds, have been identified using the present network of long-period seismometers and gravimeters. Any record of a moderate-sized earthquake may be transformed from time to frequency to reveal a comb of peaks at each free oscillation frequency (Fig. 2). The frequencies are in close agreement with theoretical predictions calculated using detailed models of the Earth's internal structure.

In 1975, F. Gilbert and A. M. Dziewonski used frequencies of 1044 free oscillations to refine our understanding of the internal structure still further, including a demonstration of the solidity of the Earth's inner core. The Earth is not quite a perfect sphere, and departures from spherical symmetry split the degeneracy: each single frequency peak becomes a 'fine structure' of several closely spaced peaks. The fine structure can be explored using high-resolution seismograms. Splitting studies have revealed a large-scale departure from spherical symmetry in the transition zone of the mantle.

The peaks in Fig. 2 are fundamentals: they have no nodes in radius. Fundamentals are set off by shallow earthquakes; radial overtones require a deep earthquake source and are relatively rare. On 9 June 1994 the largest deep earthquake in recorded history occurred beneath Bolivia. Although it caused little damage, it was felt as far away as Canada. It registered on the entire global network of modern instruments, including many temporary field stations, and has provided data for the identification of new free oscillations and their fine structure.

DAVID GUBBINS

Further reading

Dahlen, F. A. and Tromp, J. (1998) *Free oscillations of the Earth.* Princeton University Press.

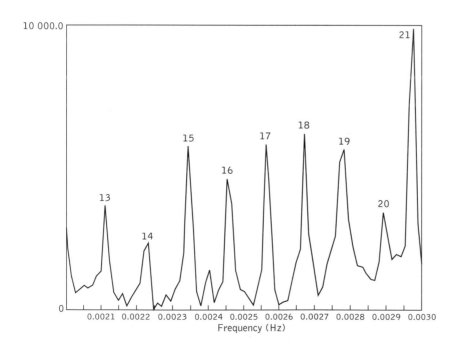

Fig. 2. Shallow earthquake in Mexico recorded on a gravimeter at station CMO in College, Alaska. Time has been converted to frequency to show the fundamental modes. Angular order number is shown.

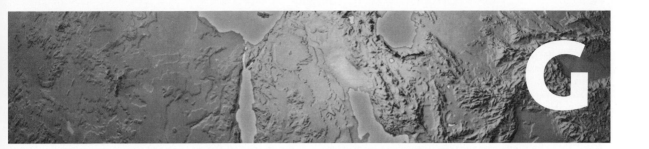

Gaia Interactions between the geosphere, biosphere, and atmosphere have been studied for at least the past 40 years, but, until the late 1970s, conventional wisdom has been dominated by the view that life exists only because material conditions on Earth happen to be just right. The idea that these interactions can be self-regulating is new. *Gaia*, the Greek goddess, was the name suggested to James Lovelock by the novelist William Golding to express the idea that life defines the material conditions needed for its survival, and that it makes sure they stay there. They hypothesis first grew from attempts to explain why conditions on Earth differed so markedly from those on its 'dead' neighbours Venus and Mars. In collaboration with the microbiologists Lynn Margulis, Lovelock became intrigued by several observations that appeared to suggest that conditions ideal for supporting life had been maintained on Earth against odds as unlikely as 'surviving unscathed a drive blindfold through rush-hour traffic':

1. For at least the past 2 billion years, the Earth's atmosphere has been maintained in a state of profound chemical disequilibrium, in which incompatible gases such as oxygen and methane coexist. Indeed, it was while working at NASA on the *Viking* mission to Mars that Lovelock, in collaboration with the philosopher Dian Hitchcock, first hypothesized the unlikeliness of there being life on Mars because of the chemical equilibrium of its atmosphere. By their reckoning, the gases of Mars and Venus are like the exhaust gases from an internal combustion engine in which all the useful energy is spent.

2. The Earth appears to have maintained a surface temperature of between 10 °C and 30 °C, ideal for the sustenance of life, throughout the past 3.5 billion years. This is in spite of the fact that Earth now receives between 1.4 and 3.3 times more energy than it did at the time of its formation.

3. The apparent indispensability of certain gases that are peculiar to Earth led protagonists of the Gaia hypothesis to ask: 'What purpose does constituent X serve in the atmosphere?' Ammonia, for example, is present in trace quantities, yet it is considered to be essential in maintaining soils at a pH of around eight, that is, optimal for sustaining life. Compared to the atmospheres of Mars and Venus, in which carbon dioxide is concentrated at levels of more than 95 per cent, carbon dioxide makes up only 0.03 per cent of the Earth's atmosphere, where it is essential in driving photosynthesis. However, the atmosphere is particularly susceptible to increases in carbon dioxide—too much triggers a devastating 'greenhouse effect' in which planetary temperature would soon rise above 30 °C and all life would die. Methane is more or less absent from the Martian and Venusian atmospheres, but its presence in minute but ubiquitous concentrations of 1.7 parts per million on Earth is crucial to the support of life by maintaining atmospheric oxygen levels.

4. Considering the current input of salt to the sea from the land, it would take only 80 million years for the oceans to reach their present level of salinity. In fact, the salinity of the world's oceans has been maintained at a more or less constant 3.4 per cent. The importance of 'managing' salinity is illustrated by the fact that new organisms are capable of surviving in salinities greater than 6 per cent.

In 1981, W. Ford Doolittle published an important critique of Gaia theory in which he wondered 'how does Gaia know if she is too cold or too hot, and how does she instruct the biosphere to behave accordingly? From his perspective, Doolittle was uncomfortable with the idea that Gaia seemed to require a teleological capacity for foresight and planning in the biota. Strenuous efforts were now required to understand mechanisms by which the planet might self-regulate. The result was Daisyworld, a numerical simulation of an ecosystem comprising, at first, just two species of daisy—black and white.

Daisyworld is a bleak and cloudless planet, with a constant and low concentration of greenhouse gases, that orbits a star not unlike our Sun in which solar luminosity is increasing. The mean surface temperature of Daisyworld is therefore determined by the balance between radiant energy received from the star and energy radiated back to space from the planet, according to the reflectivity of its surface (albedo). Both species of daisy survive at temperatures of between 5 °C and 40 °C, their optimal temperature being 22.5 °C. The model examines changes in the relative abundance of the black and white daisies in response only to changing temperature, itself controlled largely by whether the high-albedo white daisies, or the low-albedo black daisies, dominate.

During the first 'growing season', as temperature rises above 5 °C, the white daisies are disadvantaged because, by reflecting sunlight, they make the planet too cool. As Daisyworld's surface temperature creeps above 22.5 °C, the black daisies are increasingly disadvantaged because they absorb too much energy and, consequently, the planet overheats. Between them, the daisies act as a negative feedback mechanism stabilizing their environment. It is only because of the inexorable rise in solar luminosity that the daisies are eventually no longer able to regulate the temperature, and they all die. Increasingly sophisticated simulations of Daisyworlds, some with many more species of daisies, some incorporating 'rabbits' that feed on the daisies, themselves predated by 'foxes', all demonstrate a powerful tendency for the planet to regulate temperature at a level most conducive to the survival of its biota.

Self-regulation, or homeostasis, has thus become the essential defining property of the modern Gaia theory. Restated, it now proposes that living organisms and their material environment are tightly coupled. The coupled system is a superorganism, and as it evolves there emerges the ability to regulate climate and chemistry. By analogy to the impact that physiology had in bringing together microbiology and biochemistry with medicine, geophysiology has become the new trans-disciplinary environment in which planetary-scale feedback mechanisms are investigated. Geophysiology, an approach to Earth science first advocated by the pioneering geologist James Hutton more than 200 years ago, imposes no teleological demands on the biota. Homeostasis arises as a natural consequence of biota–environment interactions.

Geophysiologists are only just beginning to glimpse mechanisms by which homeostasis might proceed. Physiological analogues abound, such as the homeostatic regulation of blood glucose level by the hormones glucagon, which stimulates glucose production, and insulin, which stimulates glucose utilization. Faced with acutely high blood glucose levels, the body will generate large quantities of insulin that eventually decline to normal levels as the glucose anomaly is dissipated. On a global scale, comparable 'acute' crises in the Earth system are exemplified by the relatively sudden decline in atmospheric carbon dioxide levels that followed the formation of the Himalayan mountains. By drawing down and fixing atmospheric carbon dioxide as calcium carbonate, or limestone, the weathering of calcium silicate rocks acts as a huge negative greenhouse effect. The Himalaya are therefore implicated fundamentally in global cooling in the Tertiary period. Furthermore, the uplift of the Tibetan plateau, the largest elevated area of the continents, and the subsequent initiation of the monsoon, has had a profound effect on global climate circulation. The manifold responses of the Earth system to Himalayan mountain-building are yet to be understood fully.

Most geochemists agree on the important role of land vegetation in promoting chemical weathering, thereby leading to the drawing down of atmospheric carbon dioxide and a reduction of the greenhouse effect. However, perhaps the most convincing proof of the control exerted by biota over climate arose from the observation, first made from satellite data, of a possible connection between cloud cover over the oceans and lush, oceanic algal blooms. Nearly all species of oceanic algae produce dimethyl sulphide (DMS) as a by-product of a reaction by which they protect themselves from the saltiness of the sea. Some of the DMS is released into the air where it is oxidized to form microscopic particles of methane sulphonate. These particles constitute the principal cloud condensation nuclei: without them, clouds cannot form. DMS production therefore appears to control cloud cover, and hence the Earth's albedo, in a comparable way to the daisies on Daisyworld.

The importance of DMS and atmospheric carbon dioxide concentration in controlling global climate is supported by analyses of ice cores from Antarctica. Perhaps not surprisingly, they reveal that recent glacial periods coincide with unusually high abundances of methane sulphonate and low carbon dioxide, suggesting that the low glacial temperatures are promoted by high percentage cloud cover (high albedo) and low greenhouse effect. More ominous, however, is the recent analysis of the effects of temperature change on the feedbacks induced by changes in the surface distribution of marine algae and land plants. During the rising temperatures of the interglacials, such as we are now experiencing, the negative feedback mechanisms of both marine algae and land plants are increasingly disabled. As global mean temperature rises above 20 °C, the marine and terrestrial ecosystems are in positive feedback, thereby amplifying further increase in temperature. No one is claiming that the Daisyworld simulations come close to representing the true complexity of the Earth system. Nevertheless, Lovelock and his co-worker, Lee Kump, emphasize that these models do serve to warn of the dangers of the anthropogenic addition of greenhouse gases to the atmosphere, and the destruction of natural ecosystems, at a time when the geophysiological system may be at its least effective, and when the consequences of these actions may be amplified by positive feedback.

JONATHAN P. TURNER

gas hydrates A hydrate, or clathrate, is a solid, ice-like substance composed of rigid cages of water molecules enclosing gas molecules. Low temperatures (4–6 °C), high pressures (above 50 atmospheres), and gas concentrations exceeding solubility are necessary for their formation. The two components are not chemically bonded, but are linked by van der Waals forces.

Naturally occurring gas hydrates are present in polar continental regions and continental margin marine sediments. In marine sediments hydrate layers usually occur below the sea floor, where their presence can be detected by seismic reflections parallel to the topography of the sea bed. The gas component may be ethane, propane, butane, carbon dioxide, or hydrogen sulphide, but is most commonly methane. One volume of methane hydrate may contain 164 volumes of methane gas. Hydrate deposits generally contain biogenic methane, generated from bacterial anaerobic degradation of organic matter, rather than thermogenic methane derived from the breakdown of organic matter at high temperatures.

Gas hydrates have global significance as a potential energy resource. Most natural gas on Earth is in the form of hydrates, and estimates indicate that methane hydrates contain twice the carbon contained in all known fossil-fuel deposits. Hydrate breakdown contributes to global warming, since methane is a 'greenhouse gas'. Geological evidence indicates that rapid falls in sea level and warming of the deep ocean have occurred in the past, destabilizing the gas hydrate and releasing methane, which can cause global warming. Gas hydrate dissociation is also linked to submarine landslides, which can generate tsunamis.   R. JOHN PARKES

gastropods Gastropods are one class of the phylum Mollusca, a large phylum that contains such diverse organisms as the bivalves, octopus, and ammonites. The gastropods are the largest class of molluscs with about 1650 genera, and are by far the most varied, occurring in marine, freshwater, and terrestrial environments and occupying a variety of habitats. They normally inhabit a coiled shell and move by means of a muscular foot: however, in the slugs and marine nudibranchs the shell has been lost. Although gastropods are at their most successful today, they have a long geological history stretching back into the Cambrian.

Snails are perhaps the most characteristic gastropods. A helically coiled shell contains most of the organs, while the head and foot project outside (Fig. 1). The head bears the main sense organs, the eyes and tentacles, and also the mouth, which contains the radula, a rasping structure consisting of a tongue-like band with many small teeth. In some forms the radula is used to scrape algae from rocks, but in predatory gastropods it is used to bore holes into the shells of other molluscs. The foot is a muscular structure used to creep over hard substrates, propulsion being provided by waves of muscular contraction that pass along it. Both the head and foot can be pulled inside the shell for protection. The shell is secreted by the mantle, which lines it and also encloses the mantle cavity within which are the gills, the osphradium (a chemo-sensory organ), and openings of the digestive glands. The mantle cavity opens at the shell aperture, and may be partly drawn out to form a siphon whose function is to draw water into the gill and osphradium. In terrestrial gastropods the gill is lost and the mantle cavity has evolved into a vascularized lung.

Gastropod classification is largely based on the anatomy of living forms, fossils being placed by comparison of the morphology of their shells. A major characteristic of gastropods is the torsion of their bodies: the posterior part of the mantle cavity is rotated laterally and anteriorly so that it comes to lie above the head. This torsion is evident in the phylogeny of gastropods as they develop from bilaterally symmetrical ancestors with simple conical shells. Detorsion also took place later in forms such as slugs, which have lost their shells and reverted to a bilateral symmetry. The classification of gastropods is based chiefly on this feature. There are three subclasses: the Prosobranchia, which show full torsion, are marine, freshwater, and terrestrial gastropods that are generally herbivorous and include about 50 per cent of the members of the class; the Opisthobranchia are detorted gastropods in which the shell is absent or concealed in the mantle (e.g. nudibranchs); and the Pulmonata, which are also detorted but retain a spiral shell, have a mantle cavity converted to a vascularized lung, since they are terrestrial forms.

Unequivocal gastropods first occur in the Early Cambrian where Bellerophontacean gastropods with shells coiled in one plane are found. Problematic simple conical shells are known from the Early Cambrian; there is little agreement about their relationships, although it is probable that the earliest forms

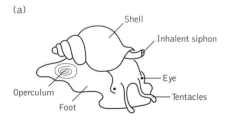

(a)

Shell

Inhalent siphon

Eye

Tentacles

Operculum

Foot

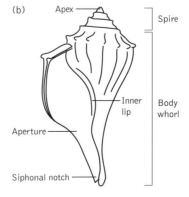

(b)

Apex

Spire

Inner lip

Body whorl

Aperture

Siphonal notch

Fig. 1. (a) Major structures of a typical gastropod. (b) Gastropod shell morphology.

developed from a monoplacophoran ancestor that initiated torsion by its ability to withdraw deep into a narrow shell. This ability may have developed independently in several lineages; if so, the gastropods could be polyphyletic (developed from several different ancestors). The first helically coiled shells appeared by the end of the Cambrian and development continued until the Early Carboniferous (Mississippian), by which time the gastropod faunas were very diverse and included the first non-marine forms. The gastropods were affected by the Permian–Triassic extinctions, as were all marine invertebrates, but they continued to evolve through the Mesozoic, the pulmonates making a successful transition to land at this time.

Gastropods have been little used in biostratigraphy because of the relatively restricted range of their shell morphology; they have, however, proved useful in evolutionary studies. In the late 1970s work by Peter Williamson of the University of Bristol on the molluscan faunas of the Tertiary sediments of Lake Turkana in Kenya provided fine-scaled evolutionary data that supported the punctuated equilibrium (evolution in steps) model of speciation. Periodic regressions of the lake left peripheral isolates (isolated groups of organisms) in which major changes took place under the influence of environmental stress, contrasting with the slow rate of change in the main lake. Short bursts of rapid speciation thus alternated with long periods of stasis in the fossil record.   DAVID K. ELLIOTT

Further reading

Boardman, R. S., Cheetham, A. H., and Rowell A. J. (eds) (1987) *Fossil invertebrates*. Blackwell Scientific Publications, Oxford.

Morton, J. E. (1967) *Molluscs*. Hutchinson, London.

Geikie, Sir Archibald (1835–1924)

Although his education was in the classics and literature, Archibald Geikie exhibited a keen enthusiasm for geology in his youth, and his early geological observations in Scotland were noticed by the then Director-General of the Geological Survey of Great Britain, Sir Roderick Murchison. As a result, Geikie joined the Survey in 1855 at the age of 20. He became Director for Scotland in 1867 and was Director-General from 1882 until he retired in 1901. During his time at the Survey he established a programme of mapping and research priorities that has remained largely unchanged (though greatly enlarged) until today. From 1871 to 1881 he was also Professor of Geology and Mineralogy at the University of Edinburgh.

Geikie made notable contributions to geology in several fields: the study of glacial deposits, erosion by rivers, igneous petrology, and stratigraphy. The work he did on volcanic rocks is generally regarded as his most important contribution; his 1888 memoir on the history of volcanic action during the Tertiary era in Britain and his two-volume book *Ancient volcanoes of Great Britain* are classics. His works on the glacial drift of Scotland and on the Old Red Sandstone of western Europe was also important. A celebrated conflict took place in the 1860s between Geikie and other geologists on the one hand and Lord Kelvin (William Thomson) on the other over estimates for the age of the Earth. (Kelvin, from physical arguments, set a maximum age of for the Earth that was less than a quarter of what was envisaged by the geologists. It was subsequently shown that Kelvin had not taken account of the radioactive heat generated within the Earth.)

Geikie was one of the first people to write on the history of geology. Among other books, he also wrote a volume on the scenery of Scotland (1865), which he illustrated himself, a textbook of geology (1882), and an autobiography (1924). His many administrative activites included extended service to learned societies and to education. He was knighted in 1891 and received the Order of Merit in 1913.   D. L. DINELEY

gemstones Gemstones are minerals esteemed for their qualities of beauty, durability, and rarity. Beauty is generally perceived as a good clear colour and transparency, as shown by fine emeralds or rubies. Beauty may also result from a play of colours or an attractive sheen, as in diamond and precious opal. The dispersion or 'fire' of a colourless diamond is due to the splitting of white light into its spectral colours as it is reflected inside the stone (Fig. 1). In precious opal, the minute spheres of silica which make up its internal structure act as a diffraction grating, splitting light into its spectral colours. 'Star stones', which include some sapphires and rubies, show asterism, a star-shaped reflection of light from orientated fine needle-shaped inclusions in the stone. Reflection of light from parallel inclusions or a fibrous structure, known as chatoyancy, is shown best by the cat's-eye variety of chrysoberyl.

Although soft or fragile minerals can make exquisite cut stones for the collector, the ability to withstand chemical attack and the abrasion of everyday use is essential for gem-

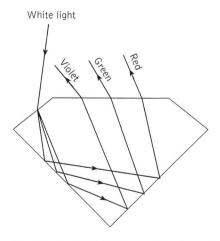

Fig. 1. The dispersion of white light into its spectral colours when it is reflected inside a brilliant-cut diamond.

stones used in jewellery. When measured on Mohs' scale of hardness (a scale of ten points from the softest, talc (1), to the hardest, diamond (10)), most gemstones measure at least 7 and so are at least as hard as quartz. Even hard gemstones may be vulnerable to abrasion. The brittleness of zircon is often revealed by chipping on its facet edges, while topaz will readily split along its perfect basal cleavage.

The most highly prized gems are those rare in nature; the larger and more flawless the stone, the greater its rarity and hence the higher the value attributed to it. The quality of diamond in particular and of gemstones in general is assessed according to the 'four C's': colour, clarity, cut, and carat weight (5 carats = 1 gram). The world's largest faceted diamond is the yellow 'Golden Jubilee', unveiled in 1995 and weighing 545.67 carats.

Gems and gem deposits

Gem minerals are formed in a very wide range of geological environments. Igneous hosts include volcanic pipes and basalt flows, but particularly important are pegmatites. Formed from slow-cooling volatile-rich magma, these coarse-grained rocks are the source of beryl, corundum, tourmaline, topaz, and many other precious stones. Garnet, emerald, jade, and lapis lazuli are among the gemstones created by metamorphic processes. All the more durable gems may be found concentrated in alluvial and beach deposits following erosion of their host rocks. Discovery of these gem gravels has often preceded the search for the source rock and its subsequent exploitation.

The following accounts of the main precious and semi-precious stones give only a few of their most important sources; Robert Webster expands on this subject in his book (revised by Read). New gem deposits continue to be discovered, particularly in the countries of the former USSR, China, and remoter areas of Asia and Africa.

Diamond is composed of pure carbon and is the hardest material known to man. Familiar in its colourless form, it also occurs in a wide range of strong colours including blue, green, pink, and yellow. These rare coloured stones are highly treasured; their coloration comes from traces of elements such as boron and nitrogen or from structural flaws in the crystal lattice. Diamond is obtained from volcanic pipes composed of kimberlite or lamproite, rocks that are found only in cratons, very old stable areas of the Earth's crust. The deposits of South Africa are well known, but most diamond today comes from Australia, Botswana, Zaire, and Russia.

Volcanic rocks in parts of Hungary (now Slovakia) were the main source of precious opal (hydrated silicon dioxide) from Roman times until the discovery of the Australian opal fields in the nineteenth century. Australian opal is formed by the circulation of low-temperature silica-rich water in sedimentary rocks. It occurs as nodules and thin bands and sometimes replaces organic material in marine fossils. Australia yields the bulk of the world's supply of opal, the remainder coming mainly from Mexico and Brazil.

Gem varieties of beryl (beryllium aluminium silicate) include emerald (bright green), aquamarine (blue-green), heliodor (yellow), goshenite (colourless), and morganite (pink). Beryl is derived mainly from granite pegmatites and alluvial deposits, although gem deposits of emerald are also found in schists, gneisses, and hydrothermal veins. By far the finest emeralds come from Colombia.

Ruby and sapphire are gem varieties of corundum (aluminium oxide). The name 'sapphire' is given to all colours excluding red and includes green, yellow, and colourless stones as well as the more familiar blue. The finest rubies are of metamorphic origin, occurring in marbles and gem gravels in the Mogok area of Burma. Sapphires are also found in Burma, where they occur in syenites and pegmatites. Sapphire and ruby are obtained from basaltic rocks, pegmatites, and limestones in several countries of Asia, including Thailand, Cambodia, Vietnam, India, Sri Lanka, and, more recently, China. Australia and some countries in Africa and America also produce corundum of gem quality.

Cat's-eye (cymophane) and alexandrite are rare and much sought after varieties of chrysoberyl (beryllium aluminium oxide), a mineral often found in the same gem deposits as ruby and sapphire, together with spinel (magnesium aluminium oxide) and zircon (zirconium silicate).

The finest cat's-eyes are honey-coloured and come from the gem gravels of Sri Lanka. Alexandrites from Sri Lanka, Brazil, Zimbabwe, Madagascar, and other modern sources rarely show the extraordinary colour change of this mineral so well as the old Russian gems. It appears red under tungsten lighting and green by daylight because it contains trace amounts of chromium, the same element that colours rubies red and emeralds green.

Spinel may be blue, purple, pink, or red, while the colour range of zircon includes yellow, green, red, orange, and brown. The high refractive index of zircon gives it a particularly bright lustre.

Gem-quality olivine (magnesium iron silicate), known as peridot, is an oily olive-green colour. Although olivine is an important constituent of many mafic igneous rocks, large crystals are rare. St John's Island (Zebirget) in the Red Sea yielded the peridot brought to Europe by crusaders in the Middle Ages. Today, some of the finest crystals come from Pakistan and Brazil.

Brazil is the main source of gem topaz (an aluminium silicate fluoride in which hydroxyl ions replace some of the fluorine ions). Topaz can be colourless, blue, and rarely pink as well as the more familiar yellow. The colour range of tourmaline (a group of complex borosilicates) is exceptionally wide, and reflects its considerable variation in composition. Single crystals of two or more different colours are not uncommon, making unusual multicoloured faceted stones.

Amethyst, rock crystal, citrine, rose quartz, and smoky quartz are all varieties of quartz (silicon dioxide). Most gem quartz comes from pegmatite deposits worldwide, but especially from Brazil, Madagascar, and Namibia. Large amounts of amethyst are also obtained from geodes in volcanic lavas in Brazil and Uruguay. Crocidolite (blue asbestos) replaced by quartz, known as tiger's-eye, comes mainly from South Africa and Australia.

Garnets are a structurally related group of silicate minerals of various compositions. They include the more familiar almandine (deep pink) and pyrope (red) as well as grossular (pink, green, colourless, and the orange variety known as hessonite), spessartine (orange-red), and the demantoid variety of andradite (bright green). Most garnets form as a result of metamorphic processes.

Metamorphic environments also yield some of the finest gemstones for carving: jade and lapis lazuli. The term 'jade' is applied to two different materials, jadeite and nephrite. Jadeite (sodium aluminium silicate) is the rarer, harder, and more highly prized form. Its colour range is diverse and includes the prized emerald-green imperial jade. Most jadeite has come from Burma and Guatemala. Nephrite jade is the compact variety of the closely related minerals tremolite and actinolite (calcium magnesium iron aluminium silicate). It ranges in colour from mid- and pale green to white and it is obtained from Burma, China, and elsewhere. Nephrite jade from New Zealand has long been used by the Maori people to make tools and weapons.

Lapis lazuli is an ornamental rock formed by the regional metamorphism of limestone; it has blue lazurite (sodium calcium aluminosilicate sulphate), white calcite, and pyrite as its main constituents. Most lapis lazuli today comes from Afghanistan, a source since antiquity, although it is also obtained from the Chile, Russia, and elsewhere.

The term 'gemstone' also encompasses some materials of organic origin. These include jet (fossil wood), amber (fossil tree resin), coral, pearl, and ivory.

The cutting of gemstones

Translucent and opaque stones, and those showing asterism or chatoyancy, are usually cut in the cabochon style, with a polished domed surface. Other gems are faceted with flat faces, a process that requires great skill and precision to maximize both the beauty and the size of the finished stone.

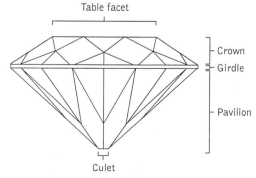

Fig. 2. The parts of a gemstone.

The process takes place in three stages. First the crystal is sawn using a blade impregnated with diamond powder to remove flawed sections and form the initial shape. The faces are then ground on a lap, a fast-revolving horizontal metal disc charged with a mixture of abrasive and oil or water. The stone is mounted on a dop-stick and held against the lap at precise angles to grind each face in turn, first the table facet, then the crown facets, and lastly the pavilion facets (Fig. 2). Finally each face is polished on the lap using various polishing agents.

Because of their extreme hardness and high value, diamonds are prepared for faceting by slightly different methods. Where possible, flawed sections are removed and larger crystals cut into smaller pieces with minimal wastage by splitting the crystal along natural cleavage planes. First, a notch or kerf is cut using a laser or another diamond. A metal blade is set in the kerf and this is tapped to split the stone. Any cutting needed in other directions is carried out using either a very thin diamond-charged saw blade or a laser, the latter being very much faster. The stone is next mounted on a lathe and, using a second diamond, the points of the crystal are ground away to form a round girdle, a process known as bruting. The individual faces are then ground and polished on a lap using diamond powder as an abrasive. In the past, each stage of diamond cutting was carried out by hand, but the process has become increasingly automated in recent years, benefiting from computer technologies.

The style of cut of any gemstone will depend on the shape and size of the crystal, the intensity of colour, and on other optical properties of the mineral. Styles range from variants of the traditional rose, brilliant, and step cuts more commonly used in jewellery (Fig. 3) to a multitude of modern designs, some of striking appearance.

The brilliant cut was reputedly introduced by Vicenzio Peruzzi in the seventeenth century; it was refined by Marcel Tolkowsky, who published his design in 1914. It has 58 facets, the proportions and angles chosen to optimize the fire and brilliance of a diamond or any stone of high optical dispersion. The zircon cut has an extra set of facets to reduce the loss of light through the pavilion. The step (or trap) cut is particularly well suited to coloured stones, the depth of the gem regulating the intensity of the stone's colour. The truncated corners of the emerald cut give the stone an octagonal outline and help to reduce risk of abrasion. The scissor or cross cut is a form of step cut with triangular faces. Small stones step cut in an elongate rectangular shape are known as baguettes. The benefits of two cuts are obtained for stones of high refractive index and good colour with a mixed cut of brilliant crown and step cut pavilion.

Enhanced, simulant, and synthetic gemstones

The enhancement of natural gemstones to give a better colour or clarity has a long history. In his *Natural History*, Pliny the Elder (AD 23–79) describe the oiling of 'smaragdi' (emeralds and other green stones) to disguise flaws and enhance their

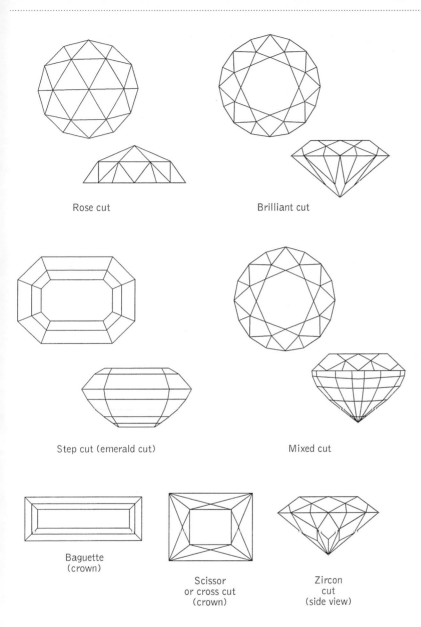

Rose cut

Brilliant cut

Step cut (emerald cut)

Mixed cut

Baguette
(crown)

Scissor
or cross cut
(crown)

Zircon
cut
(side view)

Fig. 3. Some traditional styles for faceted gemstones.

colour; today oils, waxes, plastics, and resins are widely used. Flaws in diamonds may be burnt out with lasers, while natural fractures in diamonds and other gems can be filled with a glassy material. As the materials applied are less durable than the mineral itself and may degrade in an unsightly way, such treatments do not enhance the value or desirability of the stone, and their detection is important.

Traditional methods of colour enhancement include painting or applying foil to the backs of stones, and the dyeing or bleaching of porous gems such as agate. Heating under con-

trolled conditions or irradiating a mineral using X-rays, neutrons, gamma rays, or other energy sources will effect colour changes in many gems. Sometimes these processes simulate the effects of natural heating and radiation within the Earth, but in others they produce colours unknown in nature. The blue zircon seen in jewellery stores is obtained only by heating brown crystals. Similarly, citrine, though uncommon in nature, is widely available in the form of heat-treated amethyst. The colour of sapphires and other gems may be deepened by controlled heating. Much smoky quartz is produced by irradiating rock crystal, and irradiation generates the intense blue colour of some topaz gems. Even diamonds may be irradiated to alter or enhance their colour. The regulations relating to disclosure of gemstone enhancements vary from country to country.

The imitation of gemstones by less costly materials also has a long history. Coloured glasses, known as pastes, have been widely used as crude imitations of many gemstones, and synthetic spinel is manufactured in a variety of colours for the same purpose. The quest for a good simulant of diamond has led to the production of a number of materials matching ever more closely its optical and physical properties. These include colourless zircon, strontium titanate, yttrium aluminium garnet (YAG), colourless sapphire, cubic zirconia, and, most recently, synthetic moissanite (colourless silicon carbide).

Doublets and triplets are composite stones consisting of two or three layers of natural or manufactured gem materials cemented together. Some composites such as precious opal protected between layers of quartz or glass are sold as such. Others, typically a coloured glass base and natural gem top, are intended to deceive, and may be difficult to detect when set in jewellery.

The best simulants of natural gemstones are synthetic materials manufactured under laboratory conditions, identical in both chemical composition and crystal structure to their more costly natural counterparts. They can be distinguished by small variations in their physical properties. Synthetic ruby, sapphire, spinel, emerald, opal, and turquoise are commonly encountered, but synthesis of diamond for gem cutting has so far been very limited.

The mechanisms for enhancement of gemstones are explained in the book by Nassau, and accounts of many enhanced, simulant, and synthetic stones are given in O'Donoghue's book.

Identifying gemstones

Only non-destructive testing can be used to identify a gemstone and detect simulant and synthetic stones. Methods include visual examination, study of the absorption spectrum, and measurement of refractive indices, specific gravity, and thermal conductivities. Read's *Gemmology* (1999) explains these and other tests in detail.

Close visual examination of a stone using a hand lens or microscope will reveal any microscopic inclusions of crystals, liquids, and gases, the shapes and characters of which may not only assist with identification of the stone, but may also show whether it is natural or synthetic, and in some cases may indicate its geographical origins. Zoning or patchiness of colour distribution may also provide useful clues. Examination may also reveal a double image of the pavilion facets seen though the table facet, an indication that the stone is not diamond, garnet, spinel, or any other mineral belonging to the cubic system of crystal symmetry.

Minerals of the cubic system are isotropic, have just one refractive index, and do not polarize light passing through them. Those of the other six crystal systems are doubly refrac-

tive. They polarize light passing through them, the polarized rays having different refractive indices. For most gemstones the refractive index can be measured with a gemmologist's refractometer, but the refractive indices of diamond, some zircons, and certain other gem materials are too high to be measured with most of these instruments.

The emerging polarized rays from a doubly refracting stone may be of different colours, a property known as pleochroism. This can be observed with a dichroscope, which splits the emergent rays and shows them side by side. Not all doubly refracting stones show pleochroism but for those that do, the effect can be striking. Cut stones of iolite, the blue gem variety of cordierite (magnesium iron aluminium silicate), show a marked pleochroism of yellow, light blue, and dark violet-blue, distinguishing them from sapphire, which shows a dark blue–pale greenish-blue pleochroism.

The absorption spectrum of gemstones in the visible light region of the electromagnetic spectrum is observed through a spectroscope, and appears as fine dark bands. The positions of these bands provide information about the chemical composition of the stone and may also indicate whether the stone is synthetic or artificially modified. Any fluorescence under ultraviolet light may also provide clues to the identity and status of a gemstone.

The specific gravity (the ratio of its weight to the weight of an equal volume of pure water at atmospheric pressure and a temperature of 4°C) of an unmounted gemstone can be gauged using an accurate balance and applying Archimedes' principle. It may also be estimated by immersing the stone in a series of liquids of known specific gravities, to see whether it sinks, remains suspended, or floats.

None of the tests mentioned above can identify a diamond in jewellery with absolute certainty. One of the most effective tests is measurement of thermal conductivity using a small electronic probe. Conductivity is particularly high for diamond but low for nearly all its simulants.

MONICA T. PRICE

Further reading

Harlow, G. E. (ed.) (1998) *The nature of diamonds*. Cambridge University Press.

Nassau, K. (1994) *Gemstone enhancement* (2nd edn). Butterworth-Heinemann, London.

O'Donoghue, M. (1997) *Synthetic, imitation and treated gemstones*. Butterworth-Heinemann, London.

Read, P. G. (1999) *Gemmology* (2nd edn). Butterworth-Heinemann, London.

Webster, R. (1994) *Gems: their sources, descriptions and identification* (5th edn revised by P. G. Read). Butterworth-Heinemann, London.

general circulation of the atmosphere 'The

general circulation' is the term given to the way in which the atmosphere circulates air that carries with it moisture and heat. Day-to-day changes are not significant in the general circulation, which takes account of the long-term average motion of air. However, because the circulation is usually different in different seasons it is also possible to define different average circulations for each season.

The most important question to try to answer is, 'Why does the air in the atmosphere move at all?' Is the motion random or is there a reason for it? If there were no atmosphere, as for example on Mercury, then the parts of the Earth that are most exposed to the Sun would be hottest and those that get least sunlight would be coldest. Because the Earth is almost spherical and, over the course of a year, the average position of the Sun is over the Equator, we would expect equatorial regions to receive most sunlight and be the warmest. Near the poles the Sun is never overhead, and here the Sun's rays are spread over a wider area and the heating effect is less. This is what we observe. However, if the atmosphere, and also the oceans, were not circulating, equatorial regions would be far too hot to support life as we know it, and polar regions would be much colder than they are now. The atmosphere takes the enormous input of heat around the tropics and tends to redistribute it to the cooler parts of the planet. In the process the cooler air is brought to the tropics to be heated. The oceans also play a major role in this circulation. The effect of the general circulation is to remove large temperature differences.

The largest differences in temperature are between the equatorial regions and the poles. These differences between different latitudes are much larger than differences between longitudes. It is conventional, therefore, to describe the general circulation in terms of what happens at different latitudes and to ignore, to some extent, the variations which occur at any particular latitude around the world.

Some of the earliest attempts to describe the general circulation of the atmosphere were undertaken in the days of sailing ships when it was vitally important to know the direction and strength of the prevailing wind in different regions of the globe. One of the simplest models of the circulation, although it is incorrect, highlights several important points. If air is heated in the tropics it will become more buoyant and will rise. Cooling air in polar regions will sink as it becomes more dense. A circulation could be established if the cool air from the poles were to flow equatorwards near the Earth's surface while the warm air from the tropics flowed polewards at high levels (Fig. 1a). This circulation would satisfy the requirement of transferring heat from the equator to other parts of the planet. If such a circulation existed what would be the wind pattern observed at the Earth's surface? If we ignore the fact that the Earth is rotating, the flow at the Earth's surface would be from the poles to the Equator. However, the Earth is rotating and the effect of that rotation is that for observers on the Earth there appears to be a component of the wind at right angles to the direction of motion. In the northern hemisphere this means that a wind from the north pole blowing southward would appear to be deflected towards the west. The prevailing wind would be a north-easterly. (Wind directions are referred to by the direction from which the wind comes.) In the southern hemisphere the prevailing wind would be a south-easterly. Such a uniform distribution of winds is not observed. However, the wind pattern postulated above is close to what is observed in the trade wind regions between about 30° north and south of the equator. This part of the circulation is known as the Hadley cell, after George Hadley, an English meteorologist who first proposed it in 1735.

The theoretical circulation described above, with a single cell in each hemisphere, was later replaced by a three-cell explanation which included a Hadley cell in the tropics with a similar cell in polar regions and a cell circulating in the opposite direction in mid-latitudes (30° to 65°) (Fig. 1b). This had the advantage that it explained the prevailing wind pattern at the surface quite well. However, it became clear later, when observations of the upper atmosphere were carried out, that winds at upper levels were not correctly described in the mid-latitude region. A full explanation of the way in which heat is transferred in mid-latitudes was not produced until the twentieth century. It was then realized that the depressions and anticyclones that predominate in this region can be described as almost horizontal waves. These waves transport warm air towards the poles on the eastern side of depressions and transport cold air towards the Equator on the western side of depressions. In fact, the waves are not exactly horizontal, for the poleward-moving warm air is also rising slowly while the equatorward-moving cold air is sinking (Fig. 1c).

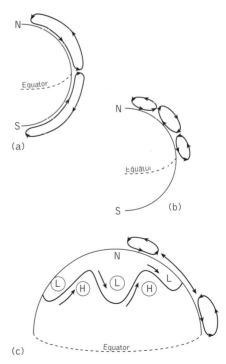

Fig. 1. Three models of the general circulation: (a) a simple single cell in each hemisphere; (b) a more realistic three-cell circulation; (c) an accurate model showing the role of waves in middle latitudes. N, North Pole; S, South Pole; H, high pressure; L, low pressure.

We can now consider in detail what happens in the two main parts of the circulation. These are the Hadley cells, which cover about half the Earth's surface, and the mid-latitude waves, which cover more than 40 per cent of the surface. The Hadley cells have rising air near the Equator and sinking air near 30° north and south. There is one cell in the northern hemisphere and another in the southern hemisphere. In the spring and autumn each of these cells is approximately symmetrical. During the summer in one hemisphere, and in the winter in the other, the Sun is not directly above the Equator and heating is greater in the tropics in the summer hemisphere. The boundary between the two Hadley cells then moves into the summer hemisphere and the Hadley cell in that hemisphere is stronger. This boundary is characterized by rising air which is converging near the surface from a north-easterly direction to the north and a south-easterly direction to the south. As the air rises at this boundary, it cools and clouds form as water condenses. The broken band of clouds that is seen on satellite images in this area is called the ITCZ (Inter-Tropical Convergence Zone) (Fig. 2). Very heavy rainfall is produced at the ITCZ because water vapour has been transported into this region from the tropical belts to the north and south. The relatively dry air at upper levels in the Hadley cell moves away from the ITCZ, and also eastward owing to the Earth's rotation. The air subsides in the region about 20 to 30 degrees north and south of the ITCZ. Because subsiding air is becoming warmer and more stable, clouds do not form in this region, which is sometimes called the sub-tropical anticyclone belt. The subsidence of the air in this subtropical region is very slow in comparison with the rapidly rising air in the ITCZ. Clear skies in this region mean that more sunshine can reach the surface than over the cloudy ITCZ, and the highest temperatures at the surface are usually observed in this area. These high temperatures also enhance the evaporation of water from the ocean surface, and this moisture-laden air is carried towards the ITCZ in the low-level flow of the Hadley circulation.

The mid-latitude circulation takes warm air from low levels in the subtropics and transports it almost horizontally poleward. The warm air is carried poleward between depressions and anticyclones on the eastern flank of the depressions. As it moves it rises slowly at a rate of about a kilometre upwards for every 200 or 300 km horizontally. This slow rising motion cools the air, and condensation occurs to produce clouds and precipitation over large areas. On the other flank of the depressions cold air is moving from high levels at high latitudes and slowly subsiding as it is carried towards the subtropics.

The general circulation succeeds in transporting heat from the warmest parts of the Earth to the cooler parts. In so doing it also transports moisture from places where evaporation is dominant to places where precipitation is greatest. There are also significant components to the general circulation, such as monsoons, that occur on smaller scales than the phenomena described here.

CHARLES N. DUNCAN

Further reading

Hunten, D. M. (1993) Atmospheric evolution of the terrestrial planet. *Science*, **259** (5097), 915–20.

Kasting, J. F. (1993) Earth's early atmosphere. *Science*, **259** (5097), 920–6.

Palmen, E. and Newton C. W. (1969) *Atmospheric circulation systems*. Academic Press, London.

Fig. 2. An image from the European weather satellite, *Meteosat*, showing the deep clouds at the ITCZ, the cloud-free subtropical anticyclonic belts, and the waves of cloud in mid-latitudes.

geoarchaeology In its baldest sense, geoarchaeology is the application of geology to archaeological interpretation. Any implication that archaeologists would otherwise be innocent of geological procedure is of course unwarranted: many of their techniques were forged in the nineteenth century by geologists. Notable among these techniques are stratigraphy, with its emphasis on superposition of strata, and palaeontological dating, whereby the age of a horizon can be estimated by reference to its included fauna. Moreover, several pioneering prehistorians have been geologists by training. But the evergrowing complexity of archaeology sometimes subordinates Earth science (together with botany and other disciplines) to such services as the dating of selected strata, whereas geoarchaeology strives to fuse the many ways of evaluating the human record.

The issues to be resolved range from the grander puzzles of human evolution and speciation to parochial matters of subsistence and trade. The tasks of the geoarchaeologist amount at one extreme to the assessment of accessibility and at the other to fact-finding expressed in millimetres of rainfall or perhaps the quality and availability of minerals and grazing land. The language, once brashly deterministic, is nowadays permissive: fertility and remoteness are human constructs, and nature may circumscribe but rarely prescribes.

The techniques on which geoarchaeology draws likewise go beyond conventional geology into the physical and biological sciences. The dating of settlements and soils nowadays may use radiocarbon and other isotopes, pollen, and palaeo-magnetism; their interpretation may refer to ecology, genetics, and seismology; their fate may invoke climatic change and thus astronomical as well as atmospheric variables. But at the core of geoarchaeology is the analysis of sediments, minerals, and topography. Again, although the evidence may come primarily from a restricted excavation (as when a change in the local topography is under review) or from far afield (for example if shifts in sea level or volcanic eruptions are under suspicion), geoarchaeology is primarily concerned with specific sites and their changing settings.

The success of geoarchaeology thus hinges on well-defined, primarily archaeological questions and an eclectic array of answers from the geosciences. Classic examples include the role of the Bering Straits in the peopling of America, the extent to which Greece and Rome enjoyed a more productive landscape than the present, and the significance of silting by the Indus in the decline of Mohenjodaro.

It follows that geoarchaeology is not parasitic on Earth science but rather its partner: geomorphology, seismology, and neotectonics are among the fields that draw on the human record for dates and insights. The study of the Earth's changing magnetic field gained new depth from analysis of the magnetism of ancient pottery kilns; the short instrumental earthquake record has been greatly extended by the evidence gleaned from ruins and inscriptions; crustal upheavals are being pieced together from submerged or disturbed human settlements and structures; and the extinction of mammals and flightless birds can be followed from midden to midden across archipelagos and continents.

C. VITA-FINZI

Further reading

Rapp, G., Jr and Hill, C. L. (1998) *Geoarchaeology*. Yale University Press, New Haven.

Herz, N. and Garrison, E. G. (1998) *Geological methods for archaeogeology*. Oxford University Press, New York.

geochemical analysis Geochemistry is the study of the composition of geological materials and the behaviour of individual elements during geological processes. Geochemical analysis is now a vital tool in most fields of geological and environmental research. It is used, for example, in studies of water, soil, and air quality, of formation of rocks and minerals, of fossilization mechanisms, of metal accumulation in

organisms from contaminated water and soil, and in determining the suitability of materials for civil engineering purposes. Some of the most commonly available geochemical techniques are described below.

Electron probe microanalysis (EPMA)

EPMA is a non-destructive technique for the analysis of polished and carbon-coated specimens of minerals, glasses, and synthetic materials. It can also be used to give an indication of the elements present in organic or unpolished or uncoated materials. Analyses can be made at individual points (usually 1 to 40 μm (micrometres) in diameter) or over small areas (usually less than 1 mm^2) of the sample surface to produce geochemical maps. Almost every element of the periodic table can be determined, with the exception of those with low atomic numbers, mainly hydrogen, lithium, and beryllium, and for many instruments boron, carbon, nitrogen, and oxygen. Specimens are introduced into the sample vacuum chamber and viewed under high magnification to select areas of interest. A beam of electrons is fired at the surface of the sample, producing X-rays with energies and wavelengths specific to the elements present. The instrument is calibrated by analysing standard samples with known compositions. Among the most useful applications of EPMA is in the study of subtle chemical zoning in minerals by element mapping. Zoning of this kind, which is often invisible using light microscopy, can yield information on crystallization conditions and changes in the chemistry of a magma or hydrothermal fluid.

X-ray fluorescence spectrometry (XRF)

XRF is a routine technique for the determination of major elements and many trace elements in rocks and minerals, at concentrations from 1 or 2 ppm (parts per million) to 100 per cent. Samples are usually prepared as glass discs for major element analyses, by fusing the sample powder with a known proportion of a commercially available flux, or as pressed powder pellets for trace-element analyses, made by mixing the sample powder with a binding agent, then pressing the mixture into a compact disc with a smooth upper surface. The sample surface is irradiated with primary X-rays, producing secondary X-rays with energies and wavelengths characteristic of the elements present. The concentration of the elements is determined by comparing the intensity of the various energy or wavelength peaks with those produced by standard samples of known composition. One of the most common uses of XRF is in the geochemical analysis of suites of igneous, metamorphic, and sedimentary rocks in studies of crustal and mantle evolution.

Inductively coupled plasma–atomic emission spectrometry (ICP–AES)

ICP–AES is used for the rapid measurement of major elements and a wide variety of trace elements in waters and sample solutions. Solid materials are powdered and homogenized, and a known amount is dissolved, using a combination of high-purity acids, and diluted to produce a sample solution. This is nebulized (sprayed) into a high-temperature argon plasma, where it is vaporized and ionized to yield radiation characteristic of the elements contained in the sample and with intensities proportional to their concentration. The intensity of the signal is converted into element concentration by comparison with a calibration curve. Detection limits are often less than 0.005 wt % (weight per cent) in the sample. Almost all elements of the periodic table can be measured, although for some elements, or those at low concentrations, other techniques may be more appropriate. Some of the most useful applications of ICP–AES are in the analysis of natural waters for transition metals or other potentially contaminating elements, not ideally determined by ICP–MS (see below).

Inductively coupled plasma–mass spectrometry (ICP–MS)
ICP–MS is used for the rapid, quantitative determination of a variety of elements at low concentrations. As for ICP-AES, sample solutions are introduced into an argon plasma, although ions are measured rather than photons. For the analysis of small volumes of solid materials, it is also possible to use a high-intensity laser beam to ablate holes 10 to 100 μm diameter (by variable depth) in the surface of a polished sample. For the analysis of solutions, detection limits are very low (below ppb (parts per billion) in solution) but are higher for laser ablation techniques (between 10 ppb and 1 ppm in the solid). Most elements in the periodic table can be determined although the best results are obtained for elements with an atomic mass greater than 80, and the method is not generally appropriate for major elements. Examples of the use of ICP–MS are the determination of rare-earth elements in magmatic rocks for studies of their origin and evolution and the analysis of road dusts, vegetation, waters, soils, and animal tissues for potentially harmful elements such as lead.

The interpretation of geochemical data
Data for major elements are generally reported in the form of oxides as weight per cent (e.g. silica, SiO_2 wt. per cent) and trace elements in parts per million (ppm) or micrograms per gram (μg g^{-1}). An important consideration in the interpretation of data is analytical uncertainty, particularly when evaluating the existence or significance of small variations in composition. Some of the most common sources of uncertainty are sample contamination, incomplete sample digestion (some minerals are extremely resistant to acids), interferences between elements (for example in EPMA, where X-ray energies for two elements may overlap), and poor instrument calibration. The extent of these problems is best limited at the time of analysis by taking necessary precautions and through discussion with experienced laboratory staff. They are, however, impossible to eliminate completely and an indication of analytical uncertainty should therefore be reported with the data, ideally as 'error bars' on geochemical diagrams. Uncertainties can be estimated from the regular analysis of reference materials of known composition and

from the analysis of two or more samples. To help in interpretation, data are often imported into graphing and statistical computer programmes. A wide variety of diagrams can be produced to compare the data with those from previous studies and to demonstrate trends or new interpretations.

BEN J. WILLIAMSON and TIM P. JONES

Further reading
Gill, R. (1997) *Modern analytical geochemistry: an introduction to quantitative chemical analysis techniques for Earth, environmental and materials scientists.* Longman, Harlow.

Rollinson, H. (1993) *Using geochemical data: evaluation, presentation, interpretation.* Longman, Harlow.

geochemical anomalies At many locations within and on the Earth concentrations of some element or set of elements are found which are much higher than the background concentrations normally encountered in the host material. The host can be any naturally occurring substance such as rock, soil, a stream or lake sediment, glacial debris, vegetation, or water, and the abnormal concentration is referred to as a *geochemical anomaly*. Geochemical anomalies are the result of the concentration of an element by some mechanism into a smaller volume of material when compared to the source of the element. Such types of anomaly are best represented by ore deposits. The processes leading to their formation collectively cause what is known as *primary dispersion*. Alternatively, an anomaly can be the result of a mechanism or mechanisms which remobilize a pre-existing primary concentration, creating a halo of lower but elevated concentration over a wider area and volume of material. This process is known as *secondary dispersion*. These two types of geochemical anomaly are the focus of geochemical mineral exploration because they provide much larger targets when geologists are searching for undiscovered deposits of valuable elements.

The Earth is a constantly changing system, and chemical elements within and on its surface are always on the move. At great depths, where temperatures and pressures are very high, rocks melt. In the process, some elements are concentrated in the magma while others are left behind. As these magmas ascend, the decrease in pressure and temperature causes them to crystallize. The minerals that first appear may tend to settle from the magma and form highly anomalous concentrations of some elements. An example of this is the formation of chromium deposits from ultramafic magmas composed almost entirely of ferromagnesian minerals. Crystallizing magmas may also release aqueous fluids as they solidify, and in the process some elements will tend to concentrate in the fluid. These fluids can then ascend through networks of fractures in the overlying rock. As they ascend and cool, the dissolved elements will be precipitated in the fractures and in the surrounding rocks. Cold fluids may also descend into the Earth, become heated, and at the same time react with rocks extracting certain elements. If these fluids then ascend and cool, they may also generate anomalous concentrations. These mechanisms are the source of most of the world's deposits of copper, lead, zinc, silver, gold, and many other metals. The processes of melting, crystallization, dissolution, and precipitation are the chief mechanisms for the formation of geochemical anomalies within the Earth and on its surface, commonly known as *hypogene* or *primary mineralization*

On the human timescale, rocks seem to be permanent, unchanging features. However, over geological timescales, they are constantly being uplifted and eroded. As this occurs, primary anomalous concentrations of elements are exposed to surficial physical and chemical processes. These result in secondary dispersion, creating geochemical anomalies in surficial materials. These anomalies are of great interest to exploration geochemists and environmental scientists.

Dispersion by physical processes is driven by gravity, whether due to simple downward movement, transport by water, or transport by glacial ice. On sloping ground, soil can slip downwards at an imperceptibly slow rate by a process known as *lateral creep*. If the soil is composed of weathered rock material containing an anomalous concentration of some element, there is a tendency for the primary anomaly to become distorted and spread out in the direction of movement. Rapid downward movements of loose rock or soil, such as occur on talus slopes or in landslides, will also tend to displace or completely detach the anomaly away from the source.

The most common process that causes secondary dispersion by physical means is transport by flowing water. The material transported can range from large boulders to fine clays, the amount and size being dependent on the speed and turbulence of the water. During flooding caused by heavy rainfall and spring runoff, streams can transport large fragments of rock from their banks. These are progressively broken down into smaller fragments and are sorted as the water slows until only fine clay material is transported in suspension in slow-moving streams and rivers. Seasonal variation in water level and velocity can result in further reworking of earlier deposits. It is evident that transport and sorting by moving water of a relatively restricted primary geochemical anomaly, can lead to a complex and much larger secondary anomaly.

Secondary dispersion by physical means can also be caused by the effects of continental and mountain glaciation. In the northern regions of North America, Europe, and Asia, the land has been periodically covered by thick sheets of glacial ice. As the ice moves it can erode or excavate large quantities of the rock and unconsolidated debris on which it rests. This material is carried along with the ice and is abraded as the glacier advances. Along the margins of glaciers, large quantities of sediments are transported and reworked by melting ice. Glacial action can thus serve to generate large secondary geochemical anomalies from mineralized bedrocks. Exploration geochemists need to take great care when attempting to interpret geochemical data from areas that have undergone extensive glaciation.

Secondary dispersion by chemical means can also serve to generate geochemical anomalies. Many ore minerals, and in particular sulphide minerals, are not stable in surficial environments. If they are exposed to surface waters they tend to break down, or oxidize, releasing their contained metals into the drainage system. Materials transported by physical means also undergo chemical reactions, causing them to disintegrate or be transformed to new minerals. During these processes, elements can be dissolved into the water and carried further downstream, where they can become concentrated in the sediments or retained in the water. All elements are soluble to some degree in surficial waters. The amount transported is dependent on the chemical mobility of the element in question and the composition of the water in which it is dissolved. Relatively immobile elements will be precipitated close to the source and can lead to relatively restricted stream sediment anomalies; other elements may be carried farther from the source and form widespread geochemical anomalies that are difficult to detect.

The important feature of geochemical anomalies formed by secondary dispersion, whether by physical or chemical processes, is that they are usually much larger than the source from which they were generated. In consequence, they are used extensively in the field of mineral exploration in the search for undiscovered mineral deposits as well as in environmental geochemistry to detect the sources of contaminants in surface waters.

BRUCE W. MOUNTAIN

geochemical cycles A great variety of processes on Earth are cyclical in nature: in them, materials are transformed from their original state into other forms and eventually return to their original state. A fundamental picture of a global cycle of Earth materials is the so-called rock cycle that represents a pathway from the molten magma in the interior crystallizing into rocks that make the oceanic and continental crust; the breaking up or erosion of the crustal surface by the physical and chemical agents of the atmosphere and running water; transport of the eroded and dissolved materials to the ocean where they are deposited on the oceanic floor; and subsequent return of oceanic sediments by crustal subduction into the deeper regions of the Earth where they can melt, completing the cycle to the original magmatic state.

Any cycle has a characteristic time that it takes to complete, and the very large cycle of materials travelling from the interior of the Earth to its surface and back into the interior takes a geologically long time, measurable in hundreds of millions to more than a billion years. This is a long time even on the scale of the Earth's age, which is about 4.5 billion years, or of the age of the oldest sediments deposited in water, about 3.8 billion years.

Although geochemical cycles, as the name implies, encompass the processes regulating the chemical composition and material balances of the solid, liquid, and gaseous parts of the Earth, they are also closely tied to many other physical, geological, or biological processes. The geochemical cycles are controlled to a varying extent by such major factors as the configuration and elevation of the continents, volcanic emanations, spreading of the ocean floor, flow of water on the Earth's surface and in the subsurface, vegetation cover of the land, biological production on land and in waters, and climate.

The distinction between the cycling processes at, or near, the Earth's surface and those in the Earth's deeper interior found its way into the scientific terminology, which distinguishes between exogenic and endogenic cycles. The historically older concept of epicycles (small cycles on a bigger cycle) has not found use in the geological literature. Early discussions of the cyclical nature of processes on the continents and in the oceans go back to James Hutton's description in 1785 of sediment erosion on land, transport to the ocean, accumulation on the ocean floor, hardening, and subsequent rising to be eroded again. Perhaps it was not a small feat of geological thinking in the late eighteenth century to deduce a picture of the global cycle of sedimentation that included processes not immediately observable on the surface.

The water cycle

The global water cycle (Fig. 1) is one of the major drivers of the geochemical cycles on the Earth's surface, in which water is both a means of transport and a substance that reacts chemically with minerals and gases in the Earth's crust and the atmosphere. Professor C. Bryan Gregor, of Dayton University, wrote in 1988 in a prologue to the volume

Chemical cycles in the evolution of the Earth that one of the best of the cyclical processes that are not immediately observable as cycles is the global cycle of water: if the rivers flowed only one way, as an observer might simplistically conclude, without the water coming back somehow, then the global sea level must be rising continuously. Such a rise, at an extreme, would have been about a metre per decade, a change that could not have remained unnoticed in historical times. Since the last glaciation maximum about 18 000 years ago, melting glaciers have added water to the oceans, raising the sea level a total of some 120 metres. The rate of sea-level rise was not uniform during this period, but it translates into an average of only several centimetres per decade.

The oceans are by far the greatest reservoir of liquid water on the Earth's surface and they are the main source of water vapour in the atmosphere. Water transport from the ocean surface to the atmosphere is driven primarily by heat, and water vapour is a major greenhouse gas in the atmosphere that makes the Earth's climate much warmer than might have been expected from only the solar radiation received by the Earth. Condensation of water vapour in the atmosphere over the oceans returns it directly to its source, and its condensation over land accounts for surface flow, recharge of underground water aquifers, and plant and animal life. Glaciers and ice sheets grew in the past because the climatic conditions made the atmospheric

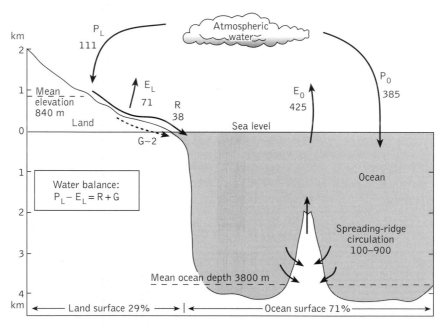

Fig. 1. Global water cycle. Vertical scale shows the mean elevation of land (840 m), depth of the oceanic ridges (approx. 2000 m), and mean depth of the ocean (3800 m). Horizontally not to scale. Arrows show the main water flows, in thousands of cubic metres per year (10^3 m^3 yr^{-1}). In the ocean: evaporation (E_O), precipitation (P_O), and circulation in the spreading zones. On land: evaporation and transpiration by plants (E_L), precipitation (P_L), river and surface runoff to the ocean (R), and groundwater flow (G).

precipitation accumulate as ice in the colder regions. Today, about 2 per cent of the volume of water in the oceans is locked in the Arctic and Antarctic ice. Ocean water circulates through the spreading zones of the oceanic lithosphere, where new material is added to the ocean floor and where water reacts with rocks and melt at high temperatures. This is an important mechanism of chemical transport between the lithosphere and the ocean.

The sodium cycle

Sodium is one of the major constituents of crustal minerals, sediments, and ocean waters. The geochemical cycle of sodium is also an example of how our thinking about the cyclical nature of geological processes has evolved since the earlier part of the nineteenth century. Sodium and chlorine are the two most abundant elements in solution in ocean water, and salt is the main mineral (halite, of composition NaCl) that precipitates by evaporation of sea water. However, areas of strong evaporation and salt precipitation near sea shores are scattered around the world, and we cannot easily observe mineral-growth reactions on the ocean floor where sodium is removed from solution into clay minerals. The earlier thinking, based on the processes and pathways more easily observable at the time, was that sodium accumulated continuously in the ocean, where it was brought by river flow from land; the age of the ocean could therefore, it was thought, be determined from the mass of sodium in the ocean and its annual import by rivers. Then, as the reasoning went, the ratio of the mass of sodium in the ocean to the mass brought by rivers every year would give the length of time needed to build up the amount in the ocean or, in other words, the age of the ocean. This figure, estimated by various authors at various times, is about 65–110 million years: much too short for the age of the oceans on Earth, where marine life began not less than 3.8 billion years ago and highly developed organisms lived in the oceans this side of 600 million years ago. This estimate, however, is not the age of the ocean, as it was once thought, but it is the time needed to replenish all the ocean-water sodium by river flow from the continents. In general, when the mass of an element in some reservoir (such as the mass of sodium in solution in ocean water) does not change with time or, allowing for inflows and outflows, it is in a steady state, then the amount brought in over some period of time must be balanced by the same amount removed from that reservoir. Therefore dividing the mass in the reservoir by the rate of input gives the length of time needed to renew the reservoir mass; this time is also called the *residence time*.

In reality, the picture of the geochemical cycle of sodium is considerably more complex. Starting with the continents, sodium is a constituent of crustal rocks, such as granite, where it occurs in a common rock-forming mineral, albite, composed of sodium, aluminium, silicon, and oxygen ($NaAlSi_3O_8$); it is subsequently leached into waters when crustal rocks become exposed to, and react with, surface and ground waters; it is transported in solution by rivers to the ocean. Rivers, however, carry sodium in solution from three sources. Some of it comes from the spray from ocean surface carrying sea salts on to the land, where they are ultimately washed back into the ocean; some of it comes from dissolution of old salt deposits that formed from ocean water in the course of the geological history of the Earth; and some of it comes from the dissolution of aluminosilicate minerals, such as albite, in the continental crust. Thus a good part of the sodium carried by rivers had already been in the ocean and it is being recycled on the Earth's surface between the ocean and the land.

Sodium is removed from ocean water by chemical reactions into sedimentary minerals on the ocean floor, into the basalt in the spreading zones, and by precipitation of mineral halite in the regions of strong evaporation; ultimately, sodium returns from the ocean and oceanic sediments into the deeper crust of the Earth. The longest paths in this process are the travel times from the sediment into the Earth's interior by subduction of the ocean floor and the time during which sodium resides in crustal rocks before it becomes exposed to chemical reactions with waters near the Earth's surface.

The carbon cycle and its couplings

The geochemical cycle of carbon (Fig. 2) is a fascinating story of the unique conditions on our planet during most of its history. Carbon dioxide, water, and other volatile substances are believed to have been degassed from the Earth in its early history of formation and cooling. Subsequently, two main processes were responsible for removal of carbon from the atmosphere of the early Earth: one was the deposition of limestones, made of calcium carbonate minerals, from ocean water by combinations of inorganic precipitation and extraction by organisms secreting calcium carbonate in their tissues or skeletons; the other was the process of photosynthesis that converts carbon dioxide (with nitrogen, phosphorus, and sulphur) to organic matter, releasing free oxygen in the process. Since the 1970s, James Lovelock has written extensively in the context of his 'Gaia' model of the Earth about the modification of the chemical composition of the Earth's atmosphere by living organisms and, in particular, their role in the removal of carbon dioxide to its present levels. Professor Robert A. Berner, of Yale University, has calculated the levels of carbon dioxide in the Earth's atmosphere, in periods of the past 500 million years, to have been up to 18 times higher than in modern times. Over the long term, carbon dioxide returns to the atmosphere through decomposition of limestones subducted from the ocean floor into the hotter interior, and through volcanic emissions in the oceans and on the continents. Behind this process is a chemical reaction between calcium carbonate and silicon dioxide minerals, occurring at high temperatures, which produces carbon dioxide and a calcium silicate mineral phase, known as the Urey reaction and named after Harold C. Urey, of the University of California, who proposed it.

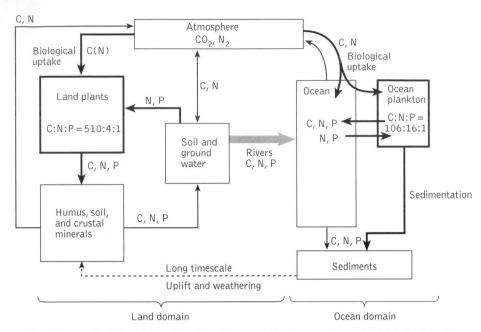

Fig. 2. Diagram of the main reservoirs of the carbon cycle, and the nitrogen and phosphorus cycles coupled to it in the atmosphere, on land, and in the ocean. Heavy lines show the biological reservoirs, their C : N : P atomic ratios, and fluxes of carbon, nitrogen, and phosphorus associated with them.

The mass of carbon stored in limestones and in organic matter in sediments is very great: about 2000 times greater than all the carbon in ocean water and atmosphere combined. Today's land plants contain approximately the same amount of carbon as the atmosphere (within 20 per cent) on which they depend for carbon dioxide. This balance indicates that changes in the global distribution of vegetation on land, possibly caused by natural factors or human activities, may strongly affect the mass of carbon dioxide in the atmosphere. The mass of carbon in photosynthetic plants living in the surface layer of the ocean is only a small fraction of its mass in land plants, but because of a much faster growth of oceanic plants the flows of carbon dioxide from the atmosphere to the plant reservoirs on land and in the surface ocean waters are not too different.

Every molecule of carbon dioxide converted to organic matter by photosynthesis releases one molecule of oxygen. In fact, the occurrence of oxygen gas in the atmosphere is a direct outcome of the activity of living organisms in photosynthesizing their organic matter from carbon dioxide and water. When plants die, their carbon is ultimately oxidized back to carbon dioxide. This process, however, is not 100 per cent efficient, and a small fraction of organic matter escapes oxidation by storage in sediments. All the organic carbon preserved in sediments represents an amount of oxygen produced that is about 30 times greater than the present-day oxygen in the atmosphere. The efficient recycling of the atmospheric carbon and oxygen on the geological timescale prevented a complete drain of the atmospheric carbon dioxide and accumulation of oxygen to very high, environmentally improbable concentrations.

The connections between geochemical inorganic and biological processes in the global cycles were well recognized by the early 1920s, when Alfred J. Lotka, then at the Johns Hopkins University, wrote a book, *Elements of mathematical biology*, in which some of the chapters were devoted to the carbon dioxide, nitrogen, and phosphorus cycles. The conditions of life on Earth are embedded in the geochemical cycles of these chemical elements, which are among the main building blocks of organic matter. (Other main constituents of organic matter are hydrogen and oxygen in water, and sulphur.) The mechanism that essentially connects the geochemical and living worlds is that of chemical reduction and oxidation reactions: carbon, nitrogen, and sulphur are chemically reduced in the process of photosynthesis and formation of living organic matter; they are oxidized when organic matter respires or decomposes, returning them to the environment.

The biological reservoirs of plants on land and in the ocean maintain close coupling between the geochemical cycles of the four life-essential elements: carbon, nitrogen, phosphorus, and sulphur. In the process of photosynthesis, land and aquatic plants take these elements in certain nearly fixed proportions (known after their discoverer as the Redfield ratios) from different environments. For example, carbon dioxide is taken from the atmosphere, and phosphorus comes from dissolution of minerals in crustal rocks and its release into waters from dead organic matter in soils and sediments. For every atom of phosphorus available, land plants take up 500-800 atoms of carbon; in water, 106 atoms of carbon go into the formation of organic matter with every atom of phosphorus. The low abundance of phosphorus in the form of the phosphate-ion in natural waters makes its availability critically important to land and aquatic plants. Decomposition of organic matter supplies most of the phosphorus needed on a short timescale, but the slower process of leaching from crustal and soil minerals is the ultimate source of new supply of phosphorus to the biosphere.

For the past two to three centuries, the geochemical cycles of the life-essential elements have been perturbed by industrial and agricultural activities on a global scale. Observations of the effects of these perturbations are generally focused on the rising concentrations of greenhouse gases such as carbon dioxide, methane, and nitrous oxide in the atmosphere, and their future effects on the global climate. However, the coupling between the geochemical cycles ensures that global changes are unlikely to be confined only to the atmosphere, but will affect the processes and flows of materials interacting within the Earth's surface system and its major reservoirs of the atmosphere, land, biota, and waters.
                                                                A. LERMAN

geochemical differentiation Chemical differentiation occurs whenever coexisting phases with different chemical compositions become physically segregated from each other. Examples of chemical differentiation include fractional crystallization in magma chambers, formation of slags in blast furnaces, and partial melting of crust or mantle rocks. Chemical differentiation is the principal process of chemical diversification within the Earth, and has led to its chemically layered structure.

Differentiation is a consequence of the chemical and physical properties of coexisting phases, such as crystals, liquids, and gases, in multicomponent systems. The partitioning of chemical components between these phases depends on the pressure, temperature, and bulk composition of the system. For thermodynamic reasons elements are rarely partitioned equally between phases. This leads to the coexistence of phases with quite different chemical, and hence physical, properties. Thus, a basaltic magma at high temperature may comprise crystals of olivine (rich in magnesium and nickel), spinel (rich in chromium), augite (rich in calcium and magnesium), and plagioclase (rich in calcium and aluminium) in equilibrium with a silicate liquid or melt, the composition of which is distinct from that of the crystals. If the crystals become physically separated from the melt, they will take with them those components in which they are rich. Conversely, elements excluded from the crystals (for example, silicon, potassium, uranium, and thorium) will become concentrated in the melt. Thus, with increasing crystallization the melt becomes progressively enriched in silica (SiO_2) and depleted in magnesium oxide (MgO). This process, which is widespread in magmatic systems, is called *fractional crystallization*. An analogous process occurs during the melting of rocks. For example, beneath mid-ocean ridges the magnesium-rich peridotite mantle melts by approximately 10 per cent to produce a basaltic liquid enriched in silicon, aluminium, and alkalis. Segregation and ascent of this material gives rise to the oceanic crust, which is chemically distinct from its parent peridotite. This process is called *fractional melting*. Further differentiation of basalt, either by fractional crystallization or by solidification and subsequent melting, can generate a diversity of rocks, richer in silica, aluminium, and alkalis, of the type that characterize the continental crust. The Earth's core was formed early in the history of the planet by an analogous differentiation process in which a dense iron-rich metal phase (solid or liquid) was segregated downwards from a residual silicate mantle. Degassing of volatiles from this early Earth during cooling played a key role in generating the atmosphere and hydrosphere.

Physical segregation is an essential feature of chemical differentiation. In many instances segregation is driven by gravity acting on phases of different density. Thus, in magma chambers dense ferromagnesian minerals may sink to form cumulate rocks at the base of the chamber, whereas in partial melting a lower-density melt may percolate upwards. The rate of phase separation governs the extent to which phases can chemically re-equilibrate with each other, and hence the efficiency of differentiation. For example, rapid extraction of mantle melts along fractures produces more efficient chemical differentiation than slow porous flow. Phase segregation is in reality a complex fluid-mechanical process that is affected by viscosity, density, chemical diffusivity, and physical form. Other examples of physical processes that involve liquids and that may give rise to chemical differentiation include flow segregation, for example in dykes and sills; expulsion of evolved liquids by filter pressing; separation of immiscible liquids, particularly in carbonatite or ferrobasaltic systems; and thermally driven Soret diffusion (in which a concentration gradient is produced by a temperature gradient). Chemical differentiation may also occur in the solid state, for instance in slowly cooled rocks and minerals, where exsolution of one crystal phase from another can occur, or in the gaseous state, for example as magma degasses beneath a volcano. In a natural system, several differentiation processes may operate in tandem.
                                                                J. D. BLUNDY

geochemical distribution The chemical elements and their isotopes are not homogeneously distributed in the Earth (or in the other planets). As a result of their characteristic properties, the geochemical behaviour of each element or group of elements is distinct, leading to significant fractionations between fluids, melts, and solids (minerals or rocks).

The following parameters play a major role in the fractionation, and thus the distribution, of elements: (1) physical conditions during geochemical processes (e.g. temperature, pressure, presence or absence of volatiles); (2) properties of solids, melts, fluids, and their constituents (chemical, physical, thermodynamic, and structural properties); and (3) kinetic factors (e.g. nucleation and growth rate of crystals, reaction rate). To some extent biological activity may also exert controls on the distribution of elements and isotopes (e.g. production of biogenic sediments, reduction of sulphate).

Because of geochemical fractionation, elements with similar electronic configurations, and thus properties (e.g. rare earth elements), tend to be associated with one another, but are usually separated from elements that exhibit different properties (e.g. copper and zinc). Many elemental associations are characteristic of certain geological environments (e.g. lithium, beryllium, boron, fluorine, and alkali elements in pegmatites; silver in massive sulphide deposits). Recognition of typical elemental associations is therefore crucial in geochemical exploration.   RETO GIERÉ

geochemical indicator species

The search for undiscovered economic deposits of essential commodities has led to the development of numerous methods of geochemical exploration that are based on the recognition of abnormal metal concentrations in surface materials. Other unusual characteristics of these materials, due to the presence of subsurface mineral deposits, are also important guides to hidden mineralization. One type of naturally occurring surface material that has become useful in geochemical exploration is vegetation. Plants can be used as guides to mineralization in two ways. First, plant material can become enriched in metals taken up from soils that are enriched in these metals. This leads to a biogeochemical anomaly, which can be detected by the collection and analysis of plant material. Second, the effects of elevated metal concentrations may lead to visible differences in plant characteristics or distribution. An example is the common occurrence of dwarfism and sparse vegetation on soils developed from ultramafic (silica-poor) rocks. This may be due to the relatively high concentrations of nickel, copper, and chromium in these soils. If elevated concentrations of a particular metal lead to an unusual characteristic in a plant species, this species is a potential indicator of elevated concentrations in the soil or underlying bedrock. The indicator species can either be observed directly on the ground by exploration geochemists or detected indirectly in aerial photographs or satellite images.

All plants require water and carbon dioxide, as well as many common elements, such as potassium, nitrogen, phosphorus, and magnesium, in order to generate the organic compounds necessary for their growth. Most plants also require small amounts of other elements, such as copper, zinc, iron, and less common metals to generate specialized organic molecules. Some species even require high soil concentrations of particular elements. For example, the calamine violet grows only where the soil is highly enriched in zinc, and the copper flower grows in areas of copper mineralization. Metals are present in the soil as dissolved ions in soil water, as component parts of soil minerals or organic compounds, or as adsorbed ions on the surface of soil minerals or organic compounds. Plant roots penetrate into the soil or fractures in the underlying bedrock, sometimes to depths of up to 50 metres. Local acidic conditions immediately near the plant roots result in the release of metals from the soil. These metals may be concentrated in the root structure or they may migrate to the upper parts of the plant, such as leaves and seeds. If the soil or bedrock is enriched in a particular metal or group of metals, these may become enriched in some part of the plant. Plants growing over potentially mineralized rocks thus provide a valuable tool for geochemical exploration.

Although most plants are able to absorb selectively the particular compounds they require, a plant may not be able to cope if an excessive amount of an element is present in the soil. Many metals, for example, have adverse effects on plants if they are concentrated to high levels; the result is plant deformities, dwarfism, or death of the plant. Although other environmental factors, such as variable drainage or local variations in the amount of sunlight, lead to normal plant distributions, excess metal concentrations in soils may create some specialized occurrences. In this way, the presence of particular indicator species or of groups of indicator species provides a clue to the location of possible hidden mineralization.

BRUCE W. MOUNTAIN

geochemical mineral exploration

Geochemical exploration, also known as geochemical prospecting and exploration geochemistry, is the search for economic deposits of minerals or petroleum by detection of abnormal concentrations of chemical elements or hydrocarbons in surficial materials such as soils, waters, and plants. For convenience, geochemical exploration is usually divided into two areas of specialization: (1) the search for metallic minerals deposits, and (2) the search for accumulations of crude oil and natural gas. The object of the search is the same in each case—the discovery of some dispersion of chemical elements or hydrocarbon compounds at levels sufficiently above normal to be called a geochemical anomaly. It is hoped that the anomaly might indicate the presence of mineralization or hydrocarbon accumulations at depth.

Prospecting for metallic ores

Geochemical prospecting for buried ore deposits is an ancient technique. For thousands of years prospectors have sought iron and copper stains on rocks as possible indicators of mineralization below; the stains are geochemical anomalies that arise as a result of the interaction of the atmosphere and rainwater with a sulphide mineral deposit, and the dispersion of the oxidized products so formed. Detectable amounts of gold in stream sediments are geochemical anomalies. Panning for gold in stream sediments is one of the most ancient and successful methods of prospecting.

Modern methods of geochemical exploration came into practice in the 1930s in Russia and in the 1940s in North America. What brought about the change from traditional methods such as the panning of stream sediments was the development of rapid and accurate means of making chemical analyses in the parts per million range so that anomalies below the visibly detectable could be located.

Kinds of geochemical surveys

Surveys may be either reconnaissance or detailed. The object of a reconnaissance survey is the evaluation of an area of hundreds or even thousands of square kilometres. Only small sample densities are feasible with large areas; a typical example would be one sample per square kilometre. Reconnaissance surveys are usually carried out by sampling stream sediments and their purpose is not so much to locate a specific mineral deposit as to assess the likelihood that mineralization might be present in the region sampled and that further, more detailed exploration might be warranted.

Detailed surveys utilize closely spaced samples over an area of a few square kilometres, and individual samples may be as close as two or three metres apart. The object of a detailed survey is the outlining of a specific anomaly and thereby the location of a specific deposit, or the possible extension of a known deposit.

Soil surveys are one of the most widely used methods of geochemical exploration for the detailed, local assessment of an anomaly is a soil survey. The method works because weathering and leaching of buried deposits can release anomalous concentrations of heavy metals to the soil and ground water. The released heavy metals spread outwards and create a dispersion halo in the soil which is much larger than the deposit itself.

In general, it is the A and B soil horizons (especially the B) that are most effective for soil surveys.

Rock surveys (also known as *lithogeochemical* or *bedrock surveys*) entail the sampling of unweathered bedrock, usually by drilling. The object is to outline host rocks that are favourable for mineralization. Rock surveys are most effective for reconnaissance surveys.

Stream-sediment surveys are used most commonly for reconnaissance surveys. Stream sediment surveys are conducted by sampling the sediments of drainage basins because, if sampled correctly, a stream sediment is an ideal composite of the materials in the drainage basin lying upstream from the sample site. Panning for gold is an example of a stream sediment survey.

Water surveys involve the sampling of surface or groundwater. Higher concentrations of chemical elements tend to be found in groundwater and for this reason groundwater surveys are preferred in most cases to surface-water tests. Groundwater surveys are often carried out in conjunction with soil surveys and are most effective for detailed surveys. Surface waters tend to be very dilute because they lose much of their dissolved load by adsorption on fine-grained clays in stream sediments.

Biogeochemical surveys are used in cases where vegetation can be used as a test medium. Geobotanical studies have been used since ancient times as a prospecting tool. It has long been known that specific plants or plant communities are indicative of a high concentration of a given chemical element in the soil. It is also true that high concentrations of certain trace elements can cause the malformation of leaves and add colourings in plants. Such clues have long been used as possible indications of buried mineralization. Modern biogeochemical surveys entail the sampling and chemical analysis of plants for unusual concentrations of elements. The system works because the roots of some trees can reach as deep as 50 metres. A large volume of soil can thus be sampled by analyzing appropriate parts of a tree, such as twigs, needles, leaves, or bark. Biogeochemical surveys are less certain than soil or groundwater samples because factors such as the age of a plant, the local climate, and the part of the plant sampled can affect the readings obtained.

Gas surveys are useful to aid in the location of buried ore deposits. These can be found through the detection of gases such as sulphur dioxide, hydrogen sulphide, and vapours of mercury, iodine, and radon. Some gases must be measured in the soil but others, such as mercury vapour, can be detected up to 100 metres above the ground surface and can therefore be measured using airborne techniques.

Effectiveness of geochemical exploration

Geochemical survey techniques are generally carried out in combination with other methods of exploration, such as geophysical techniques. It is often difficult, therefore, to say what fraction of a successful discovery is due to geochemistry. Despite the caveats, modern geochemical exploration has been used extensively for over half a century, and the discoveries of a number of deposits in Canada, Australia, Mexico, and Africa can be attributed in part to geochemistry.

Geochemical prospecting for petroleum

Two different techniques have been successfully used in geochemical prospecting for oil and gas. The first, which has been widely used and is very informative, in the detection and analysis of organic matter encountered during subsurface drilling. Subsurface geochemistry is only an indirect geochemical method of prospecting for petroleum in that it provides a regional indication of the petroleum potential of a sedimentary basin. Organic matter analysis is analogous to rock geochemistry in the search for metallic ores.

The second technique is the detection of surface geochemical anomalies. Surface geochemistry works because the overlying rock that seals a petroleum accumulation is never completely impermeable, and the upward migration of small amounts of petroleum liquids and gases inevitably occur. A surface geochemical anomaly is the end result of petroleum leakage. Detection of the anomaly, either on land or beneath the sea, generally involves the detection and measurements of hydrocarbon gases in the soil or sea water. An anomaly may also be detected through non-hydrocarbon gases, generated as a result of soil–petroleum reactions, or through the release of radon, helium, or even halogen gases.

Many of the techniques used in geochemical prospecting are also used in monitoring and assessing the effects of environmental pollution, such as slow leaks of petroleum, and the accidental dispersal of heavy metals into streams as a result of the accidental spillage of mining waste.   BRIAN J. SKINNER

Further reading

Kovalensky, A. L. (1987) *Biogeochemical exploration for mineral deposits.* VNU Science Press, Utrecht, Netherlands.

Levinson, A. A. (1974) *Introduction to exploration geochemistry* (2nd edn). Applied Publishing, Wilmette, Illinois.

Tedesco, A. A. (1995) *Surface geochemistry in petroleum exploration.* Chapman and Hall, New York.

geochemical mobility The geochemical mobility of elements is primarily important because it controls their transport, and thus their availability to take part in geochemical reactions. Such reactions control the Earth's mineralogy, which is affected by plate tectonics, earthquakes, volcanism, and the development of metal ore bodies, and the chemistry of surface waters, groundwaters, and the atmosphere, which has effects on regulation of climate and the distribution of life on Earth, and the availability of the nutrients on which all life depends.

The majority of chemical transport in geological processes occurs in the liquid phase, although volatile components can be transported in the gas phase. Gaseous material (e.g. from volcanic degassing) is far more mobile than material in the liquid phase because gases can migrate rapidly. The term 'sorption' applies to the processes of precipitation, adsorption, and complexation (formation of a complex ion) which can semi-permanently remove substances from solution, hence reducing their mobility. The mobility of many inorganic elements, especially the 'major' elements such as calcium, sodium, magnesium, and potassium, is controlled by equilibria with solid phases in which they are found. Such elements are removed from solution by precipitation when a certain concentration in solution is exceeded, but may be returned to solution by dissolution. Trace inorganic elements, such as heavy metal pollutants and many organic chemicals, can be removed from solution by adsorption to the surface of solid phases such as minerals and organic matter. This involves an electrical interaction with the surface, and can lead to direct bonding with it. Formation of complexes with oppositely charged entities in solution can reduce the mobility of a metal in solution if the resulting complex is more strongly adsorbed than the metal alone. The geochemical mobility of a substance can be increased if the resulting complex is less strongly adsorbed than the uncomplexed constituents. Adsorption on to very fine amorphous mineral phases held in suspension (known as *colloids*) can also increase the mobility of many substances. Once adsorbed on colloidal material, a substance is no longer affected by processes that could otherwise reduce its mobility. It may then be transported at the same rate as the water in which it is suspended.

Precipitation, adsorption, and the formation of complex ions are all strongly influenced by the Eh of a system (how oxidizing it is), and its pH (its acidity). These factors influence the speciation (chemical form) of substances in solution, and the speciation and charge of solid surfaces. An element may be highly soluble under acidic, reducing conditions but very insoluble under alkaline, oxidizing conditions, because of changes in speciation. The solubility of solid phases varies as a function of pH and Eh, and solubility commonly rises with temperature. Positively and negatively charged species are most strongly adsorbed under alkaline and acidic conditions respectively, because surfaces are generally negatively charged under alkaline conditions but positively charged under acidic conditions. Microbial activity exerts a strong control on the geochemical mobility of many substances. Microbes exist in a wide range of environments, and they can control the rate and extent of many reactions by mediating the processes described above.

The reactions that affect geochemical mobility can be described in terms of chemical thermodynamics. Enough equilibrium and kinetic data are now available to allow limited computer modelling of geological systems. For instance, we can predict the migration of contaminants in aquifers and the formation of metal orebodies.

SIMON R. RANDALL

geochemical residence times Residence times are a representation of the average time that a substance (e.g. a chemical element or sedimentary particle) remains in a reservoir before it is removed into another reservoir or is transformed into another species. Residence times are applied to many areas of the Earth sciences; for example, the residence time of atmospheric pollutants in the troposphere, dissolved elements in the oceans, or sediments in a basin. The measurement of residence times requires a knowledge of the flux of a species into the reservoir, the amount of the species within the reservoir, and the flux of the species out of the reservoir. The residence time is then defined as:

$$\frac{\text{Amount of species in reservoir}}{\text{Flux of species into reservoir}} =$$
$$= \frac{\text{Amount of species in reservoir}}{\text{Flux of species out of reservoir}}$$

If the two estimates of the residence time are not equal, then the system is not in a steady state and the amount of the species in the reservoir must be changing. Frequently, however, a precise measurement of all these parameters is not possible so that residence times are often estimated on the basis that the system is in a steady state.   M. R. PALMER

geochemistry, history of *see* HISTORY OF GEOCHEMISTRY

geochemistry of lakes

The sources of dissolved solids in lake water are in many respects those of dissolved solids in rivers. The chemistry of the two parts of a freshwater lake, the *epilimnion* and the *hypolimnion* (see *lakes*), is affected by biological processes and by circulation.

The process of *photosynthesis* and *respiration* in lakes can be represented by the following reaction:

$$106\ CO_2 + 16\ NO_3^- + HPO_4^{2-} + 122\ H_2O + 18\ H^+ (+ \text{trace elements and energy}) \rightarrow C_{106}\ H_{263}\ O_{110}\ N_{16}\ P_1 + 138\ O_2.$$

Photosynthesis is the combination of carbon dioxide and nutrients (nitrogen, N, and phosphorus, P), aided by trace elements and solar energy to produce organic matter and oxygen (the reaction being driven to the right). The process of *aerobic respiration* involves the breakdown of organic matter and the consumption of oxygen but also the release of nutrients, carbon dioxide, protons, trace elements, and energy (the reaction being driven to the left). There is more photosynthesis than respiration in the epilimnion. The excess photosynthesis in surface waters manifests itself in the precipitation of dead organic matter to the bottom of the lake. However, in the hypolimnion (deep water) there is a net respiration, which causes the oxygen concentration to decline in the water but the nitrogen, phosphorus, proton, and carbon dioxide concentrations to increase according to the reaction for photosynthesis. Prolonged isolation of deep waters from the atmosphere (with no overturn) results in dramatic changes in water quality. The deep water becomes anoxic, and bacterially mediated chemical reactions, such as the reduction of nitrate, sulphate, iron, and manganese, and the formation of methane can take place.

Lakes are classified as *oligotrophic* or *eutrophic* according to whether their concentration of plant nutrients or their productivity of organic matter. *Trophic* means concerned with nutrition, and oligotrophic lakes are poorly fed; that is, they have a low concentration of nutrient elements such as nitrogen and phosphorus. On a geological timescale, lakes are short-lived. Streams bring sediment in to the lake and the accumulation of organic deposits may cause shallow lakes to change to bogs or swamps. Eventually they may become dry land. Relatively long-lived lakes are those located in deep basins or in arid regions where the drainage will integrate only very slowly. *Eutrophication* is the process in which a lake gradually fills in with organic-rich sediments, eventually becoming a swamp and then disappearing. Humans have greatly accelerated the process by artificially enriching lakes with too many nutrients or with an excess of organic matter, which can result in oxygen-depleted bottom waters. This process has been called *cultural eutrophication*.

Of the nutrient elements needed for photosynthesis (shown in the equation above), hydrogen and oxygen are readily available and carbon is generally supplied from the atmosphere, but nitrogen and phosphorus are not always available. They are the *limiting nutrients*. Relatively small amounts of nitrogen and phosphorus can produce relatively large amounts of organic matter. According to the equation for photosynthesis, only 7 g of nitrogen and 1 g of phosphorus are required to synthesize 100 g (dry weight) of algae. If the mass ratio of nitrogen to phosphorus is greater than 7 in a particular water, phosphorus is the limiting nutrient; conversely, if the ratio is smaller than 7, nitrogen is the limiting nutrient. However, nitrogen deficiencies (N/P ratios smaller than 7) can be made up by the development of blue-green algae, which are capable of fixing nitrogen from the atmosphere. Phosphorous is thus usually the limiting nutrient.

The primary sources of phosphorus and nitrogen in lakes are direct rainfall and snowfall on the lake itself and runoff from the surrounding drainage area. In oligotrophic lakes most of the phosphorus in the runoff comes from rock weathering and soil transport. However, in areas influenced by humans there are additional sources of phosphorus, including agricultural runoff containing phosphorus from fertilizer and animal waste, and sewage containing phosphorus from human waste, detergents, and industrial waste. The pollution of lakes has been shown to be proportional to population density per lake volume and energy consumption per capita.

In past decades dilute freshwater lakes and streams in southern Scandinavia, south-eastern Canada, and the north-eastern United States have been acidified by acid precipitation. These lakes and their drainage areas are underlain by weathering-resistant igneous and metamorphic rocks or non-calcareous sandstones and have thin, patchy, acid soils. In contrast, other areas that are also receiving acid precipitation, but have limestone and calcareous sandstone, contain lakes whose pH values are essentially unaffected by acid precipitation. The rapid dissolution of calcium carbonate in the rocks is able to neutralize the acid precipitation. In addition to permanently acidified lakes, there are also episodic decreases in pH in dilute, but usually non-acid lakes, caused by runoff of acid snow melt in the spring. Acid pollution accumulates in the snow in the winter and is preferentially released by the first snow melt in the spring.

Saline and *alkaline lakes* are common in arid to semi-arid climates. Arid conditions, however, do not always produce saline lakes. The necessary conditions for saline lake formation and persistence are: (1) the outflow of water must be restricted, as it is in hydrologically closed basin; (2) evaporation must exceed inflow during the initial stages; and (3) for persistence, the inflow must be sufficient to sustain a standing body of water. Lakes are saline and alkaline because salts derived from the surrounding rocks are brought into the lake in solution and evaporation causes the proportion of dissolved salts to increase steadily. Eventually, with continued evaporation, saturation is reached with respect to soluble minerals, and they are precipitated, resulting in *chemical fractionation* of the remaining water. Among the first minerals to be precipitated are calcium and magnesium carbonates. The pH of a saline lake can rise to values greater than 10 if the concentration of bicarbonate (HCO_3^-) in inflow waters exceeds that necessary to precipitate all the Ca and Mg ions.

Bicarbonate can build up with continued evaporation, causing the reaction

$$H^+ + HCO_3^- \rightarrow H_2O + CO_2$$

to be driven to the right, resulting in the loss of hydrogen ions and a loss of CO_2 to the atmosphere. In other words, the pH of the remaining water increases. An example of a saline but non-alkaline lake is the Great Salt Lake in the United States; in contrast, Lake Magadi in Kenya is saline and alkaline.   SIGURDUR REYNIR GISLASON

Further reading

Lerman, A., Gat, J., and Imboden, D. (eds) (1995) *Physics and chemistry of lakes* (2nd ed). Springer-Verlag, New York.

geodesy and geodetic measurements Geodesy,

the science of precise measurement of the Earth's shape and gravitational field, has at its root many of the methods employed by the local surveyor or cartographer, but is distinguished from these by the scale of interest: in general geodesy refers to the analysis of observations made over distances at which the curvature of the Earth becomes significant. Modern geodesy utilizes surface-to-surface and ground-based astronomical observations, and increasingly also data from signals transmitted and received by artificial Earth-orbiting satellites, to quantify the form of our planet.

Geodesy and gravity

Physical geodesy is that branch of the subject concerned with the measurement of the Earth's gravity field and the *geoid*, the surface of equal gravitational potential that in the oceans is close to mean sea level. On the planetary scale the geoid is approximated by an ellipsoid of revolution about its minor axis with a degree of flattening of around 1 part in 300 caused primarily by the Earth's rotation. Permanent deviations of the geoid from an ellipsoid occur as a result of non-uniform mass distribution within the Earth, and also temporarily over the oceans by up to 1 m in response to winds, tides, and ocean currents. Both types of perturbation may be of interest to Earth scientists, but to geodesists the term 'geoid' refers to the stable long-term features of the equipotential surface. Geoids are expressed in terms of the *geoid–ellipsoid separation* or *geoid height*, the perpendicular distance to the geoid from a given *reference ellipsoid*. Historically, different countries tended to use their own reference ellipsoids with differing origins and major and minor axes, chosen so as to minimize the geoid height in the region of interest, but the increasing use of space geodetic data makes it more appropriate nowadays to use a common global reference ellipsoid such as WGS-84 or GRS-80 with its origin at the centre of mass of the Earth. In this case, geoid heights may reach several tens of metres.

The geoid can be derived from gravity measurements at known positions over land or sea, or more directly by using precise radar altimetry readings taken from low-orbiting satellites over the oceans. The latter method has proved particularly efficient in determining the short-wavelength components of the geoid. A satellite orbiting at an altitude of 100 km and emitting microwave radar pulses with a beam width of 1° will illuminate patches of the sea surface with radii of about 3–5 km. The radar altimeter measurement is affected by the roughness of the ocean surface and by atmospheric refraction, and so the accuracy of the measurements is limited to approximately 0.2 m, comparable with the precision to which the satellite's position can be tracked. In contrast, obtaining a *gravimetric geoid* from terrestrial gravity measurements is much more complicated and error-prone. The spatial resolution of gravimetric geoids is limited to length-scales smaller than the area over which gravity observations are given, and the observations themselves must be subjected to elaborate corrections to remove the effects of matter situated above the points at which gravity was measured.

Terrestrial gravity measurements can be made in an absolute sense by measuring the time for a pellet to fall a fixed distance in an evacuated chamber, or in a relative sense by using mechanical force-balance instruments. Relative gravimeters can be of the static variety, in which the change in gravity from the nominal value displaces a mass on a spring by an amount proportional to the change, or the more accurate astatic variety in which the displacement for a given gravity perturbation is much larger. The mean value of the acceleration due to gravity at the Earth's surface is approximately 9.8 m s^{-2}. Gravity anomalies are commonly expressed in units of gal (named after Galileo; 1 gal = 0.01 m s^{-2}) or gravity units (1 gu = 10^{-6} m s^{-2} = 0.1 mgal). The precisions of modern astatic relative gravimeters are of the order of 0.1–0.01 mgal but are subject to long-term drift in excess of 0.05 mgal/month, whereas absolute gravimeters can reach accuracies of 0.001 mgal and do not suffer from temporal drift. Seaborne gravity observations are less accurate than those on land, first because of the pitch and roll of the ship, which cause unwanted accelerations of the gravimeter (although gyroscopic stabilizing platforms can limit these), and secondly because of the *Eötvös effect*. This effect is caused by the east–west component of the ship's velocity, which modifies the centrifugal acceleration due to the normal rotation of the Earth about its axis. An uncertainty in ship speed of 1 km hr^{-1} results in an error in the Eötvös correction of up to 4 mgal at the Equator and less at higher latitudes. With modern GPS navigation systems (see below) providing ship velocity, accuracies one order of magnitude better than this can easily be achieved. Airborne gravity measurements also require an Eötvös correction and are significantly degraded by vibration and acceleration of the aircraft. Another problem with airborne measurements is that the aircraft's altitude must be known to within 1 m to attain gravimetric precision of the order of 1 mgal.

Positions and reference frames

Fundamental to positional geodesy is the reference frame in which the resulting coordinates are expressed. Early marine navigators could compute their astronomic latitude relatively easily, by observing the highest and lowest angles above the horizon attained by the Sun and other stars. Astronomic longitude is a slightly trickier problem because an accurate determination relies on the ability to ascertain the local time with respect to the time at the origin of longitude (by modern convention the Greenwich meridian). Until accurate chronometers were developed in the eighteenth century, this could be achieved only by means of complex calculations based on the eclipses of stars and planetary bodies. Although astronomical measurements can give independent positions at any point on the Earth, their disadvantage is that the observations are taken with respect to the local vertical (the direction of the gravitational force, perpendicular to the geoid). Because the geoid is not a perfect ellipsoid and undulates gently, the directions of the vertical in two places can in fact be parallel, resulting in identical astronomical coordinates for the two locations. Correction must be made for this deflection of the vertical to give geodetic coordinates (latitude, longitude, and height above the ellipsoid); these coordinates, being relative to a given ellipsoid, are unique. The parameters of the ellipsoid used must then be specified together with the coordinates.

Because, as described above, many countries used their own ellipsoidal parameters (major and minor axis radii and location of the centre) to achieve as close as possible a match between the geoid and ellipsoid in their region, ellipsoidal coordinates are not universal. An alternative way of describing the position of a point is by using Cartesian geocentric coordinates with respect to three mutually perpendicular axes with their origin at the centre of mass of the Earth. The Z-axis is chosen to coincide with the Earth's rotation axis, the X-axis intersects the Equator at the Greenwich meridian, and the Y-axis is at 90° E. The rotation axis changes continuously in a way that is not entirely predictable in the short term as a result of day-to-day and seasonal redistribution of mass within the atmosphere and oceans; it also changes more predictably in the longer term as a result of the *Chandler wobble* with a period of 14 months and as a result of *nutation* of the Earth (small oscillations of its axis) with a period of 18.6 years. In practice, therefore, the Z-axis is defined by the mean rotation axis over a fixed interval. The X- and Y-axes are no longer defined directly by observations at Greenwich, but in an implied sense through the choice of coordinates for several sites around the globe.

When dealing with astronomical and satellite measurements it is necessary to consider not just the Earth-fixed reference frame referred to above but also a suitable *celestial reference frame* in which the distant stars are (approximately) fixed and the equations of motion of satellites can be written without including any external rotational terms (an *inertial frame*). In practice such a frame is not perfectly realizable, because even the most distant stars exhibit small *proper motions* with respect to each other, and because it is not yet possible to model accurately all the resistive forces acting on near-Earth satellites. According to the problem in question, it may be more appropriate to consider an Earth-centred or Sun-centred (*barycentric*, fixed to the centre of mass of the Solar System) reference frame. In either case, the system is defined with respect to the ecliptic, the plane of the Earth's orbit around the Sun, and the Earth's Equator. The line of intersection of these two planes marks the equinoxes, the points at which the Sun apparently crosses the Equator as it passes from one hemisphere to the other. The vernal equinox (also known as the ascending node), where the Sun passes from the southern to the northern hemisphere, is chosen as the fundamental axis of the reference frame. Like the Earth's rotation axis, the actual Equator and ecliptic vary with time, and so the celestial reference frame is defined using mean values of these entities, observed over a fixed period.

Terrestrial geodetic measurements

Geodetic positioning by surface-to-surface observations has in some respects been superseded by space geodesy (in particular GPS, see below), but for small- and medium-scale positioning it often remains the best method. Also, many geophysical studies, in particular those of Earth rotation and lithospheric deformation, benefit from the longer history of terrestrial geodetic measurements. The first true geodetic measurements, dating back to the time of ancient Greece, were astronomical in nature. Eratosthenes (a contemporary of Archimedes) derived a remarkably accurate value for the Earth's circumference by comparing the zenith angle of the noonday sun at Alexandria with that at Aswan. Advances in timekeeping, telescopes, star catalogues, and the modelling of atmospheric refraction of light mean that astronomical positions this century are accurate to within 0.3″ of latitude or longitude, an absolute positional accuracy of 10 m, which is insufficient for survey purposes. The difference in position between two nearby points can, however, be measured more precisely: the azimuth from one to the other can be measured to within 0.5″ of arc, corresponding to 25 mm over a 10-km baseline. Many modern survey networks have such azimuth observations to provide orientation control.

Triangulation, first proposed by the sixteenth-century Dutch geodesist Gemma Frisius, was until the development of satellite geodesy the main method by which regional geodetic networks were measured. In its simplest form, triangulation relies on the fact that if the angles at the vertices of a set of abutting triangles are known, and also the length and orientation of one side of a triangle, then the positions of all the other vertices can be calculated using simple trigonometric relations. This method obviously requires that the vertices are intervisible, which is why triangulation pillars are usually sited on the tops of mountains or on tall buildings. The angle measurements are made with a theodolite, which is essentially a telescope free to rotate in the horizontal and vertical planes

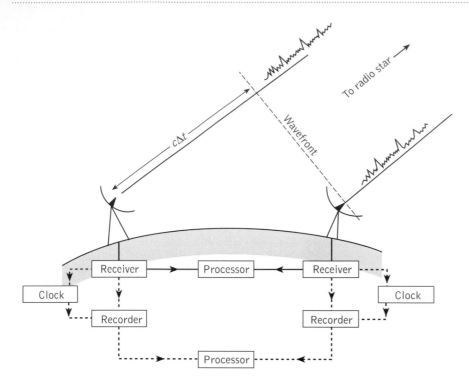

Fig. 1. Schematic illustration of radio interferometry. The component of the baseline between the radio telescopes in the direction towards the star is proportional to the time difference Δt between the arrival of a wavefront at the two telescopes, multiplied by the speed of light c. In older, conventional, short-baseline systems, the signals are transmitted immediately by cables (solid arrows) to the processing unit. In VLBI, the signals are time-tagged and recorded, then later transferred to the processing unit (dashed arrows). (From Lambeck (1988).)

and with graduated circles so that the angle can be read off to within up to 0.1″. Atmospheric refraction of light is a significant source of error, and to counter this several sets of measurements are taken, often at night when the temperature structure of the atmosphere is more stable.

Historically, the main advantage of triangulation over traverse surveys (in which successive bearings and distances are used to proceed from the known point to each unknown point in turn) was that there is a need for only one distance measurement in the entire network, although many networks had more than one to counteract errors. Distance measurements were previously made with Invar tape, designed to have a low coefficient of thermal expansion, but this method entailed laborious corrections for gravitational sag of the tape and the underlying topography between the two ends of the baseline. Since the 1950s *electronic distance measurement* (EDM) systems, using microwave radar or visible light, have made it relatively easy to measure the distance between intervisible points. Atmospheric refraction errors are mitigated by using two frequencies of light, because at optical frequencies the atmosphere is dispersive (i.e. it delays different signals by an amount depending on their wavelength, which can later be compensated). In this way the distance-dependent errors of EDM can be limited to about 1 ppm. Using EDM, *trilateration*, which is similar to triangulation but uses distance measurements instead of angles, becomes feasible.

The vertical component of position can be measured either by *vertical triangulation (trigonometric levelling)*, which is severely limited by atmospheric refraction errors because the density of the atmosphere changes rapidly upwards, or more accurately by spirit levelling. The difference in height between successive points a few tens of metres apart is measured using a telescopic level and graduated rod and the process is repeated along a traverse or loop. Accuracies of up to 0.1 mm per kilometre of traverse can be achieved.

At the length-scales involved in geodesy, ordinary plane trigonometry cannot be used and the computations must be performed for a curved reference surface. This is a major reason why an appropriate ellipsoid must be chosen to approximate the geoid in the region. All these terrestrial methods are affected by the deflection of the vertical explained above, and this must be estimated either independently or, as part of the complete geodetic problem, together with the positions of the points. Because trigonometric and spirit levelling give the *orthometric height* (relative to the geoid), not the ellipsoidal height, the difference between the two (i.e. the geoid height or *geoid–ellipsoid separation*) must also be computed.

Space geodesy and satellite positioning

Soon after the discovery of extraterrestrial radio sources in the 1930s, it was realized that these signals could be used to deter-

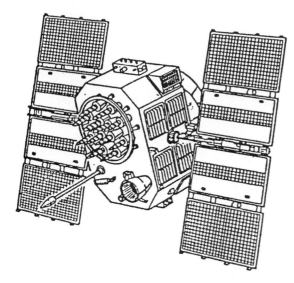

Fig. 2. A Navstar Block II GPS satellite. The large panels to either side generate electricity from solar power, and rotate so as always to face the Sun. The various antennae on the satellite transmit the GPS signal, and also other military communications. (From Fliegel *et al.* (1992), *Journal of Geophysical Research*, **97** (B1), 554–68.)

mine the distances between radio telescopes. Use of this technique of *very-long-baseline interferometry* (VLBI) for geodesy did not come to full fruition until the late 1960s, when precise atomic clocks made it possible to record and time-tag independently the signals received at widely separated points (Fig. 1). The time difference between the arrivals of a wavefront at each telescope is proportional to the component in the direction towards the radio star of the baseline joing the two telescope. Measurements are made in the S- and X-bands (2–8 GHz). At these frequencies, signal delay caused by charged particles in the ionosphere is a problem which can be overcome by measuring at more than one frequency because the effect is dispersive. More significant is the delay caused by the troposphere, in particular tropospheric water vapour, which is non-dispersive and must be estimated using models of pressure, temperature, and humidity variation with altitude. With these corrections, modern VLBI position measurements are accurate to within a few centimetres. Because the signals observed come from distant stars, the system is particularly sensitive to small variations in the Earth's rotation rate and axis, but it has also been used to measure plate-tectonic and solid-Earth tidal movements.

Space geodesy has now expanded to include the use of radio and visible-light signals from various artificial satellites since they were first placed in orbit in the late 1950s. Conceptually, the simplest method is that of *satellite laser ranging* (SLR), in which the two-way travel time is measured of a laser pulse transmitted from a ground station and reflected from a satellite back to the observatory. Although the travel time can

be ascertained to within 0.1 ns, the accuracy of the system is limited by atmospheric refraction errors, which can reach several nanoseconds, and also by the need for accurate modelling of the low-Earth satellite orbits (about 1000 km altitude), which are affected by atmospheric drag and short-wavelength variations of the Earth's gravity field. Fortunately, at visible light frequencies (unlike microwave frequencies), the signal-delaying effect of tropospheric water vapour is minimal and the 'dry' component can readily be modelled. SLR systems yield positional accuracies at least as good as those obtained with VLBI, but can more easily be transported, so SLR has been used to measure global and regional tectonic motions.

Early microwave radio satellite systems involved the tracking of the Doppler shift of signals received from low-orbit satellites. Corrections similar to those from VLBI for tropospheric and ionospheric signal delay must be used, although the frequencies are lower (150 and 400 MHz for early Doppler systems). Because of the low altitude of the orbits (each satellite was visible for only a short time during an orbit), many tens of orbits, over a period of several days, must be tracked before an observer position accurate to 1–2 m can be determined.

The various Doppler satellite have now been superseded by the *Global Positioning System* (GPS) operated by the US Department of Defense. Unlike previous systems, GPS is designed to provide both long-term geodetic-quality coordinates and also instantaneous positions accurate enough for navigation. This is possible because the system consist of 24 satellites in high orbits (20 000 km altitude), such that at least four satellites are visible to a ground-based observer at any time. Each satellite (Fig. 2) contains a precise atomic clock and transmits two L-band microwave carrier signals (on 1.2 and 1.6 GHz), both of which are modulated with codes defining the time at which the signal left the satellite. Observation of four satellites allows the receiver to compute the arrival time of the signal as well as the three components of position, thus removing the need for an accurate clock in the receiver. There are two codes: the C/A (Coarse/Acquisition) code, on the lower frequency only, and the shorter-wavelength P (Precise) code on both frequencies. Each satellite also transmits information about its current orbit. The C/A code enables the range to each satellite to be measured with a precision of 10 m, but errors are deliberately introduced into the satellite clock and orbit information so as to reduce civilian coordinate accuracy to 100–150 m. The P-code is intended for military use and is further encrypted, but some GPS receivers can bypass this encryption and give instantaneous positional accuracy of 10–15 m using the P-code. An enhanced method, *differential GPS*, uses corrections to the satellite ranges which are calculated at a nearby receiver of known position and transmitted by radio to the roving receiver. Precisions of better than 5 m are possible for rovers that are close enough to the base station for the tropospheric, ionospheric, orbital, and deliberate errors to be common to the two receivers. This method is increasingly used for vehicle navigation and surveying. For accurate positioning, the phase of the GPS carrier wave recorded at two or more receivers can be combined in an interferometric manner (*relative GPS*) that is similar to the procedure used for VLBI observations. Using several hours of measurements, accuracies of 5 mm can be achieved, and precisions of a few centimetres can be realized in seconds even for moving receivers (*kinematic GPS*). As with VLBI, ionospheric and tropospheric errors must be counteracted using dual-frequency observations and theoretical models of the atmosphere. In addition, the satellite orbits must be re-computed after the event using data from a global tracking network, because the orbit information broadcast in real time is based on an imprecise extrapolation of previous orbits. Because GPS receivers are readily portable and give precise results, relative GPS has become an important tool in the study of plate-tectonic deformation and earthquakes as well as for precise navigation and land surveying.

PETER CLARKE

Further reading

Bomford, G. (1980) *Geodesy* (4th edn). Clarendon Press, Oxford.

Lambeck, K. (1988) *Geophysical geodesy*. Clarendon Press, Oxford.

Leick, A. (1990) *GPS satellite surveying*. Wiley Interscience, New York.

Smith, J. R. (1988) *Basic geodesy: an introduction to the history and concepts of modern geodesy without mathematics*. Landmark Enterprises, Rancho Cordova, California.

geodesy, planetary *see* PLANETARY GEODESY

geodynamics Geodynamics deals with geological change and its processes, particularly in the area of structural geology, although it may also encompass sedimentation, erosion, volcanism, fluid flow, heat transfer, etc. Topics range in scale from global (e.g. mantle convection, plate tectonics) through regional (mountain-building, basin formation, lithospheric flexure) to local (fault movements, folding, magma intrusion and extrusion). The field is, however, restricted to macroscopic structures, and so would not normally cover microscopic structure or individual rock grains. Nevertheless, microscopic and even atomic-scale processes may be directly relevant to geodynamic behaviour; for example, the effect of dislocation creep in determining the macroscopic rheology of rocks.

Geodynamics applies physical theory to dynamic geological phenomena, and is usually characterized by mathematical or numerical modelling and quantitative prediction. At the scale of local deformation, it may be contrasted with rock mechanics, which is mainly concerned with static structural phenomena or the outcomes of dynamic processes.   ROGER SEARLE

geoecology Geoecology (or landscape ecology) combines geography (the landscape) with ecology (the interaction of life and its environment). Geoecologists (or landscape ecolo-gists) see the landscape as a non-uniform and non-random (heterogeneous) mosaic. The landscape mosaic consists of three basic elements: patches, corridors, and background matrices. These landscape elements are made of individual plants (trees, shrubs, herbs) and small buildings. Patches are fairly uniform (homogeneous) areas that differ from their surroundings. Woods, fields, ponds, rock outcrops, and houses are all patches. Corridors are strips of land that differ from the land to either side. They may interconnect to form networks. Roads, hedgerows, and rivers are corridors. Background matrices are the background ecosystems or land-use types in which patches and corridors are set. Examples are deciduous forest and areas of arable cultivation.

Geoecology seeks to understand how landscape mosaics are created and how they change. Three mechanisms are primarily responsible for creating landscape mosaics: substrate heterogeneity, natural disturbance, and human disturbance. Substrate heterogeneity includes the distribution of hills and valleys, of wet spots and dry spots, of different rock types, and of different soil types. Natural disturbance is caused by physical agents (wind, running water, fire) and by biological agents (burrowing animals, grazing animals, pests, and pathogens). Human disturbance arises through such activities as ploughing, forest clearance, mining, quarrying, road building, and city building. Biological and ecological processes usually modify the landscape mosaics created by these mechanisms.

Change in landscape mosaics results from interactions between landscape elements. The interactions occur as natural materials (water, gases, and solids), human-made materials (such as fertilizers and cars), and organisms (live animals and plants, seeds, spores, and so on) move from one patch, corridor, or background matrix to another. Such movements cause the landscape mosaic to change. In turn, change in the landscape mosaic influences the movements of materials and organisms. Change in the landscape mosaic is thus self-perpetuating.

Traditionally, landscape ecologists have confined their interest to landscape elements no smaller than about 10 m^2 and whole landscapes no bigger than about 10 000 km^2. The trend is now to study a much wider range of landscape units. A hierarchy of landscape units is recognized running from small (microscale), through medium (mesoscale) and large (macroscale), to very large (megascale). Small-scale landscapes are a few square metres in extent. Medium-scale landscapes have areas up to about 10 000 km^2. Each main type is influenced by the same macroclimate, by similar geomorphology and soils, and by a similar set of disturbance regimes. Large-scale landscapes have areas up to about 1 000 000 km^2, which is about the size of Ireland. Very-large-scale landscapes are regions more than 1 000 000 km^2 in extent. They include continents and the entire land surface of the Earth.   R. J. HUGGETT

Further reading

Forman, R. T. T. (1995) *Land mosaics: the ecology of landscapes and regions*. Cambridge University Press.

Huggett, R. J. (1995) *Geoecology: an evolutionary approach.* Routledge, London.

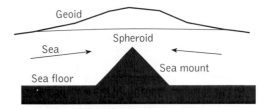

Fig. 1. The relationship between the best-fit or reference spheroid and the actual sea-level surface, or geoid. The dense sea mount has attracted the surrounding sea water gravitationally, causing it to pile up and thus raise the geoid above the sea mount.

geoid The geoid is defined as a gravitational equipotential surface that coincides with mean sea level. Any massive body in or around the Earth has a gravitational potential (a measure of the work that has to be done to move it against the gravitational force) that is due to the Earth's gravitational field. This potential varies in inverse proportion to the distance from the centre of mass of the Earth. We know that the potential energy of a body is greater on top of a hill than at its foot (that is why bodies fall under gravity); so gravitational potential energy increases away from the Earth.

It follows that if the Earth were spherically symmetrical and were not spinning, any given equipotential surface would be a sphere, and one could imagine successive equipotential surfaces of steadily increasing potential energy extending away from the Earth like the skins of an onion. Since sea water is free to flow in the gravitational field, it will flow (all else being equal) until every point in its surface is at the same minimum potential energy. This is the mean sea-level surface, and is also clearly a gravitational equipotential. The geoid is simply this one among the infinite number of 'skins' that coincides with mean sea level. Although the geoid is defined in terms of sea level, since it is a continuous surface one can determine mathematically its height over land, and can think of this as the level that water would reach in a network of canals connected directly to the sea.

In fact the Earth is a slightly flattened sphere (technically, an oblate spheroid) with its equatorial radius some 21 km longer than the polar one. This flattening, combined with the spin of the Earth, modifies the gravitational potential somewhat, so that the geoid is also slightly flattened. To a good approximation (about 1 part in 60 000), the geoid is an oblate spheroid whose major axis is about 0.3 per cent longer than the minor one. Cartographers, surveyors, and geodecists normally use a perfect oblate spheroid, which is a best fit to the actual geoid, as the datum for height measurements on the Earth. Nevertheless, there are important departures of the actual geoid from this simple shape.

Geoid anomalies

These departures, or geoid 'anomalies', are the result of the non-uniform distribution of mass within the Earth. A simple way to picture this is to imagine a uniformly deep ocean in which there is an isolated underwater mountain (a sea mount; Fig. 1). In the absence of the sea mount, the sea surface would have a uniform height everywhere. However, the sea mount, being denser than water, exerts a gravitational attraction on the surrounding water, which therefore tends to pile up above the sea mount, raising the geoid at that point. Another way of looking at this is to realize that the Earth's gravitational field is slightly enhanced over the sea mount because of its greater density, so the same gravitational attraction is found at a slightly greater distance from the Earth's centre.

The increase in geoid height over a large sea mount can be several tens of centimetres, but large, deep-seated mass anomalies in the Earth can distort it by much more. Figure 2 shows a map of geoid anomalies (measured relative to the reference spheroid), which range up to about 100 m (just south of India). Such large anomalies must reflect planet-scale variations in mass distribution, perhaps arising from thermally induced density variations due to mantle convection, or from substantial topography on the core–mantle boundary.

Determining the geoid

The foregoing description shows that there is an intimate relationship between the height of the geoid, the Earth's gravitational attraction, and the distribution of mass within the Earth. There is, in fact, a unique mathematical relationship between the geoid and the gravity field, although the relationship between either and the mass distribution is not unique (because a variety of mass distributions can give rise to a given gravitational field). Nevertheless, the shape of the geoid does place important constraints on the mass distribution. Moreover, since the local vertical (as shown by a plumb-line or spirit-level) is defined as the direction of the gravitational attraction, which is always perpendicular to the geoid, geoid anomalies cause 'deflections of the vertical' (relative to an astronomical framework), which are of vital practical importance for surveyors. Determining the geoid is therefore an important objective in geodesy and geophysics.

In the pre-satellite era the geoid was determined from the gravity field, which entailed measuring gravitational attraction or deflections of the vertical (or both) at a large number of locations around the Earth. Accuracy and resolution were limited by the number and distribution of observations that were feasible. Since the advent of artificial satellites, geoid determinations have become easier, more accurate, and of higher resolution. There are basically two methods. In the first (and earliest), satellites are tracked precisely from ground stations, and the geoid inferred from small fluctuations in their measured orbits.

More recently, sea level can be measured directly by a radar-altimeter mounted on a satellite. It is important to remove the effects of waves, storms, ocean currents, and other similar effects, but ultimately this method gives a precision of a few

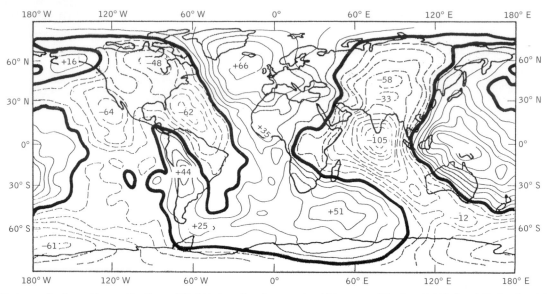

Fig. 2. A map of major geoid anomalies (the geoid height above the reference spheroid) in metres. (After Bott, M .H. P. (1982) *The interior of the Earth* (2nd edn), Fig. 1.2. Edward Arnold, London.)

centimetres in geoid height. Although they cannot be used over land, such measurements have yielded unprecedented high-resolution coverage of the geoid over the sea. Along-track resolution can be as good as a few kilometres; cross-track resolution is limited by track spacing, which can be of the order of 10 km in the best cases. This method has been used in recent years to provide very high-resolution maps of the gravity field over the oceans. Since the main source of short-wavelength gravity anomalies there is the varying topography of the sea bed, the method has revealed this topography in unprecedented detail.   ROGER SEARLE

Further reading

Heiskanen, W. and Moritz, H. (1967) *Physical geodesy.* W. H. Freeman, New York.

Sandwell, D. T. and Smith, W. H. F. (1997) Marine gravity anomaly from Geosat and ERS 1 satellite altimetry. *Journal of Geophysical Research*, **102**, (B5), 10 039–54.

geological controversies
Geology emerged as a distinctive science only about two centuries ago. From its earliest days the subject has been associated with controversy. Focusing on controversy can be illuminating because issues tend to become dramatized and the underlying assumptions and attitudes of the protagonists are often brought out into the open. Furthermore, attention is concentrated on matters most critical to growth and development of a given subject. Attention is here confined to four major controversies, which are considered briefly in historical sequence.

Nature versus neptunists and plutonists

In interpreting his rock sequence from 'primitive' igneous and metamorphic rocks to unconsolidated sediments, the pioneer German geologist Abraham Gottlob Werner adopted in the late eighteenth century the widely accepted view that initially the Earth had been enveloped by a primeval ocean covering even the highest mountains. All the older rocks we see today, even igneous ones such as granite, he perceived as being chemical precipitates from this ocean. The emphasis upon the role of water in Werner's theory led to its being called *neptunist*. A major conflict was, however, to arise about the origin of basalt. Werner could not deny, of course, the existence of volcanoes, but he restricted their activity to very recent times, and related them to local melting by the combustion of underlying seams of coal. This interpretation was progressively undermined by detailed researches in the Auvergne of France and elsewhere. More fundamental opposition to the neptunist scheme was provided by the recognition that granite was not necessarily primordial. For example, James Hutton (the original plutonist) recognized intrusive granite veins in Scotland, thereby conclusively establishing that some granite, at least, was younger than the enveloping country rock. The first two decades of the nineteenth century witnessed progressive defections from the neptunist camp, and plutonism became firmly established.

Catastrophists versus uniformitarians

The terms 'catastrophism' and 'uniformitarianism' were coined in 1832 by William Whewell, but 'catastrophism' fails to do justice to the cluster of beliefs that characterized the opposition to Charles Lyell's doctrine, because what can be called 'directionalism' was also an important component of their system. 'Uniformitarianism', as outlined by Lyell, is both a system and a method. The term has frequently been used as an exact equivalent of the continental term 'actualism', and refers to the study of present-day processes as a means of interpreting past events. It is, however, perfectly possible to use actualistic methods and come to 'catastrophist' conclusions.

The leading 'catastrophists' of the early nineteenth century were the Frenchmen Georges Cuvier and Leonce Elie de Beaumont. Basing his ideas primarily on his stratigraphic work in the Tertiary of the Paris Basin, Cuvier invoked a succession of catastrophes that not only disrupted strata and caused dramatic changes in relative sea level but resulted in mass extinctions of fauna. Elie de Beaumont followed Cuvier in arguing that folded and tilted strata implied sudden disturbance, and that one was not entitled to extrapolate to such 'catastrophic' phenomena from the manifestly slow and gradual 'causes now in operation'.

A lively debate was engaged in England in 1820s and 1830s. Leading catastrophists such as William Buckland and Adam Sedgwick promoted the so-called diluvial theory, which accounted for many geological phenomena by the action of the biblical flood. Although the diluvial theory was quickly abandoned, a stubborn belief persisted that Lyell had overstated his case, and that there was indeed a *direction* to Earth history rather than a steady-state condition of the sort favoured by Lyell. This was established most clearly from the fossil record, which indicated a kind of organic progression to more complex forms culminating in Man, a fact that was later explained by Darwin's theory of evolution by natural selection.

The age of the Earth

By the middle of the nineteenth century the belief that the Earth was a mere 6000 years old was adhered to only by biblical fundamentalists. While the consensus of geologists, following Lyell, was that the Earth was immensely older than that, there was a general reluctance to undertake even approximate estimates of age. The first serious attempt was made in 1860 by John Phillips, who adopted the cumulative thickness of strata as the best available measure of geological time, using the best current estimates of sedimentation rates. He arrived at an age estimate of nearly 96 million years for the formation of the Earth's crust, which strongly challenged the Lyellian notion of virtually unlimited time.

Only a few years later the Scottish physicist William Thomson, later elevated to the peerage as Lord Kelvin, attempted an estimate on a completely different basis. This was made on the widely held assumption that the Earth was originally a hot, molten sphere that had cooled gradually, with the heat, derived ultimately from gravitational energy, being transmitted solely by conduction. His best estimate of the age was 98 million years, a figure remarkably close to that of

Phillips. Kelvin's work was initially well received by geologists. Darwin became very concerned, however, about the limited time allowed for evolution, but 'Darwin's bulldog' Thomas Henry Huxley questioned the validity of Kelvin's underlying assumptions.

A growing opposition to Kelvin emerged progressively among geologists, which became especially sharp when he became increasingly dogmatic, reducing his estimate by 1897 to a mere 24 million years. The American geologist Thomas Chamberlin speculated that there might be sources of energy locked up in atoms, of which nineteenth-century physicists were completely unaware. The discovery of radioactivity in 1896 confirmed this prescient thought, and within a few years Kelvin's assumptions had been completely undermined. With radiometric dating well established, it became generally agreed early this century that the Earth must be several thousand million years old.

Continental drift

By the end of nineteenth century a consensus had emerged among geologists that the Earth had both cooled and contracted in volume through time, and many attempted to explain orogenic belts as the consequence of this contraction. The Europeans tended to believe that extensive sectors of ocean were underlain by subsided continent. Isostatic theory had more influence among the Americans, who were impressed by the fact that gravity surveys appeared to indicate that the oceans were underlain by denser crust than the continents, implying that they were permanent features whose underlying crust was not interchangeable with continents. Both groups denied the possibility of any significant lateral movement of continental masses through the oceans, a phenomenon that would have been completely inconsistent with the stabilist Earth model.

Although he was not the first person to propose such lateral movement, the notion of continental drift is irrevocably associated with the name of the German meteorologist and geophysicist Alfred Wegener, because he was the first to put forward substantial evidence for a coherent and logically argued hypothesis that took account of a wide variety of natural phenomena. He challenged the cooling, contracting-Earth model on a number of grounds. It was not clear from the contraction hypothesis why the shrinkage 'wrinkles' represented by fold mountains were not distributed uniformly rather than being confined to narrow zones. Further, some basic assumptions about the Earth's supposed cooling had been undermined by the discovery of widespread radioactivity in rocks, leading to the production of considerable amounts of heat acting in opposition to thermal loss into space by radiation.

Wegener postulated that, commencing in the Mesozoic and continuing up to the present, a huge supercontinent, Pangaea, had rifted and the fragmented components had moved apart, creating the Atlantic and Indian Oceans. During the westward drift of the Americas, the western Cordilleran ranges had been produced by compression, as had the Alps and Himalayas as

Africa and India had converged with Eurasia. He deployed a variety of arguments in support of his hypothesis, citing a number of notable geological facts, such as matching orogenic belts on the two sides of the Atlantic, and close resemblances among fossils of the southern continents implying former land connections. There was in addition convincing evidence for Late Palaeozoic ice sheets in South America, southern Africa, Australia, and especially India that was not consistent with the present dispersed continental configuration. Wegener also put forward a number of geophysical arguments but was unable to propose a plausible mechanism for drift.

The initial reaction to Wegener's hypothesis was not uniformly hostile, but it became increasingly so in the years between the two world wars. The most formidable opposition came from certain geophysicists, who insisted that the Earth possessed too great a strength for continents to migrate across its surface. These scientists ridiculed Wegener's proposed mechanism for drift. Wegener nevertheless had some distinguished supporters, most notably Emile Argand, Alexander du Toit, and Arthur Holmes, the last of whom provided a plausible mechanism for drift involving convection in the mantle. Despite their efforts, supporters of continental drift were generally dismissed as cranks, and by the middle of the twentieth century the hypothesis had been almost totally rejected by Earth scientists. The situation was radically transformed by research developments after the Second World War, most notably in the study of rock magnetism, and a vastly increased knowledge of what underlies the oceans. These developments led to a posthumous vindication of Wegener. The theory of plate tectonics, an outcome of continental drift, was put forward in the late 1960s and generally accepted by the Earth sciences community within a few years.

ANTHONY HALLAM

Further reading

Hallam, A. (1973) *A revolution in the Earth sciences: from continental drift to plate tectonics.* Oxford University Press.

Hallam, A. (1989) *Great geological controversies* (2nd edn). Oxford University Press.

geological exploration of the North American West

The USA

From the earliest times of European colonization the West fascinated those who had settled or who had been born along the Atlantic seaboard. The scientific exploration of the North American continent was begun by the Lewis and Clark expedition (1804–6). Sent by President Jefferson to find a north-west passage across the continent by travelling via the great Missouri and Columbia rivers, they were instructed to report on the 'soil and face of the country', and this they very effectively did. They were well equipped and well suited for the task and reached the Pacific ocean before returning to St Louis and Washington. They established the topographic outlines of the continental interior and showed that such exploration

could be very successful. Several other government-sponsored expeditions soon followed. Striking west from the Missouri up the Platte River, Z. M. Pike in 1806–7 crossed the Great Plains to the foothills of the Rocky Mountains, and in 1819–20 S. H. Long followed the Arkansas River west to the 'Great American Desert'. By this time there were many individual explorers and trappers, the 'mountain men', penetrating beyond the central plains into the Rockies, the Great Basin, and the western ranges to the Pacific. Jedediah Smith, Joseph T. Walker, and Captain Benjamin de Bonneville, to name only the most famous, were widely travelled and observant, covering ground from the present Canadian border south to the Mojave Desert and the Colorado River. Although not professional geologists, these remarkable men produced valuable accounts of the terrain, its wildlife, and natives.

Growing population and the quest for land had prompted westward development to the edge of the great plains; now many private parties set out for the west coast, which was known to be habitable. In 1841 there began a series of expeditions by the US Corps of Topographical Engineers to survey the intervening lands. They were led west from the Rockies by Lieutenant John C. Fremont and from the Pacific inland by Captain Charles Wilkes. They successfully managed their expeditions as scientific teams and professional surveyors. The extent, variety, and mineral wealth of the mountainous lands west of the Rockies was becoming apparent.

With the expansion westwards, the 1850s saw railway route-finding surveys striking out from Minnesota, Iowa, and Arkansas. These afforded tremendous opportunities for the gathering of geological information during the course of engineering work, and the mid-century gold rushes to California focused public attention to the extent that several states west of the Mississippi River set up their own geological surveys. National surveys had begun as early as 1860 and now took in enormous swathes of country. The names of the leaders, Hayden, King, Powell, and Wheeler, are famous in the geology of the west. Professional geologists were among them, and Clarence King became the first Director of the US Geological Survey in 1879. Mining operations followed on prospecting by increasing numbers of geologically trained personel. By 1870 the unique volcanic nature of the ground around the Yellowstone River was recognized and the movement to establish the first US National Park began.

Canada

Canada is the world's second largest country, with an area of 9 971 500 km², most of it being north of the 49th parallel of latitude and extending to the high Arctic. It is vast, cold, and empty, and about half of it lies on the Canadian (Precambrian) Shield. The Prairies and the Rockies are continuations of the American plains and Cordilleras. Early explorers travelled mostly by canoe. The open water season was limited to less than six months. Minerals were generally less of an attraction than furs to early explorers, and settlements in Hudson's Bay were the starting points for many explorations. Samuel Hearne's journey to the northern ocean (1769–72) from

Hudson's Bay crossed the Coppermine country. John Richardson in 1821 travelled from the north coast southwards noting the rocks and minerals *en route*. Others looked for a land-based north-west passage or for the lost Franklin expedition in the 1850s, but the geology of these parts was scarcely touched upon. J. B. Tyrrell (1858–1957) was a widely travelled geologist who saw much of the Precambrian terrain between the mountains and Hudson's Bay in the nineteenth century.

During Tyrell's time the Geological Survey of Canada began systematic examination of the country west of the Great Lakes and Hudson's Bay. Robert Bell's Geological Survey of Canada party mapped a total of 2735 km in track surveys north and north-west of Lake Winnipeg in 1879. Between the two world wars aircraft were first employed to supply field parties, and in 1955 helicopters were much used in the first comprehensive mapping project in the Arctic Islands.   D. L. DINELEY

Further reading

Goetzmann, W. H. (1966) *Exploration and empire: the explorer and scientist in the winning of the American West.* Knopf, New York

Zaslow, M. (1975) *Reading the rocks—the story of the Geological Survey of Canada 1842–1972.* Macmillan Company of Canada and Canada Department of Energy, Mines and Resources, Ottawa.

geological exploration of the polar regions

The two polar regions together occupy about one-seventh of the land area of the globe, but are very different from one another. The Arctic is an ocean basin with peripheral continents, whereas the Antarctic is continental. The north polar domain has been explored and developed by Canada and the USA, by the British and Scandinavians, and by the Russians. The search for fish, whale oil, and furs and a trade route to China has been succeeded by exploration for mineral resources and more recently by strategic military occupation. Antarctica has been visited only by expeditions with primarily scientific or other non commercial objectives. By international agreement, mining and other resource-based activities are not carried out there. Unique geological, glaciological, and meteorological studies continue there from purpose-built bases around the edge of Antarctica or at the South Pole.

After Martin Frobisher collected iron ore (believing it to contain precious metals) from Baffin Island, Canada, in the 1570s, little sustained attention was paid to geology there until the Alaskan–Yukon gold rush (1896–9). The Geological Survey of Canada had been established in 1842 and the US Geological Survey in 1879, but their concerns were primarily with more southerly terrain. In the twentieth century geological surveys over land and sea areas by national institutions and international consortia of oil companies have covered most of the western half of the Arctic basin. The work of the Geological Survey of Greenland (founded 1876) has followed on that of private expedition begun in 1913. Both national geological surveys in North America have been supportive of private expeditions, as have been the Arctic Institute of North America and the Canadian Polar Continental Shelf Project. For Spitsbergen Norsk Polarinstitutet has been the agency through which much geological research (British, Scandinavian, Polish, and Russian) has been assisted or carried out.

The Soviet Union set up its Arctic Institute and, later, its Institute of the Geology of the Arctic in Leningrad (St Petersburg, 1947) to supervise work throughout its national territories.

Antarctic geology has been pursued by many national expedition during the twentieth century; Argentina, Chile, Australia, and New Zealand have taken part in continuing programmes of research. France, the Scandinavian countries, the USSR, and Japan have also contributed, while Australia, the USA, and Britain have a continuing commitment to geological work from permanent research stations. The British Falkland Islands Dependencies Survey has been superseded by the British Antarctic Survey with its bases on the Antarctic Peninsula. All these ventures have benefited from the advance of late-twentieth century technologies of cold-climate travel and living. Centres such as the Scott Polar Research Institute in Cambridge (UK), as well as those mentioned above, maintain geological libraries and data banks. A boost was given in 1957–8 to research in Antarctica when, under the auspices of the International Geophysical Year, 12 countries set up bases there for programmes of geomagnetic, geophysical, and other research. Many of these bases have been kept in operation since the official end of the IGY. An international Scientific Committee for Antarctic Research (SCAR) has been effective since then in promoting cooperation on scientific topics and arranging conferences and publications.   D. L. DINELEY

Further reading

Raasch, G. O. (ed.) (1961) *Geology of the Arctic.* (2 vols) University of Toronto Press.

Adie, J. R. (ed.) (1972) *Antarctic geology and geophysics.* (2 vols) Universitetsforlaget, Oslo.

geological humour

The Earth sciences have lent themselves to many different forms of humour since the early nineteenth century, and even the 'lying stones of Dr Beringer' (manufactured fossils, 1726) may be regarded as a joke on the part of his students—although it was not so by Beringer himself. This was a case of humour of malicious kind being combined with all-too-obvious fraud. It has not been common. For the most part the subjects of geological jokes have been ideas or facts rather than persons, although geological personalities have often been caricatured. One of the earliest was James Hutton, so depicted by John Kay in Edinburgh in 1838; another was William Buckland exaggeratedly fitted out for a field study of glaciation and with two hand specimens, one 'scratched by a glacier thirty-three thousand three hundred and thirty three years before the creation, a second scratched by a cart wheel on Waterloo Bridge the day before yesterday' (T. Sopwith). These are British examples, but others could be quoted from other countries, especially the USA and France.

407

Humour in Earth sciences poetry and prose has taken many guises. There have been essays and other sundry prose forms in which geology or geologists, or both, have been lampooned by authors from Mark Twain onwards. Spoof reports on fictitious geological matters and even planetary geology have appeared from time to time. Some have revealed their intent to amuse rather than to confuse the reader only at the last moment. A long-continuing series of commentaries on the geological community and its activities, with many humorous asides, published over many years was 'The geologic column' by R. L. Bates in *Geotimes*, the monthly magazine of the American Geological Institute.

Limericks and other rhyming forms of humorous verse are innumerable, and not a few are not only very funny but also scurrilous or simply unprintable. Longer contributions, both serious and ribald, have appeared in several anthologies of geological verse. Several famous Earth scientists have been the authors of some of these.

Geological howlers, culled from university examination scripts, have often been quoted and were even compiled in a small book by the Geological Society of Glasgow (1980). Short exhortations have in recent years appeared as printed bumper stickers or window stickers on cars. 'Stop continental drift' is a typical example. Hats and T-shirts follow suit; 'Reunite Gondwanaland' and similar mottoes have proved popular with students.

D. L. DINELEY

Further reading

Craig, G. Y. and Jones, E. J. (eds) (1982) *A geological miscellany*. Orbital Press, Oxford.

Rhodes, F. H. T. and Stone, R. O. (eds) (1981) *Language of the Earth*. Pergamon Press, New York and Oxford.

geological ideas and the Renaissance in Europe

see RENAISSANCE IN EUROPE AND GEOLOGICAL IDEAS

geological maps and map-making

A geological map shows the rocks that would be seen at the Earth's surface if all the soils were removed. The various rocks are distinguished either by colours or by black and white patterns (Fig. 1a). The reliability of such maps depends on the extent to which the rocks crop out, on the ability of geologists to interpret what happens beneath the soil between one rock outcrop and another, and on the geologists' knowledge of the rocks they are dealing with; different types of rock were formed differently and occur in different ways.

The attitudes and structures of rocks are indicated on a map by symbols. Using the information provided by the map, cross-sections can then be drawn to show how the rocks continue in depth (Fig. 1b). Cross-sections, are however, merely projections, and their reliability depends on the reliability of the surface mapping. In some instances, subsurface information from drill holes, mine workings, and geophysical surveys provide supporting factual information. There is, however, a

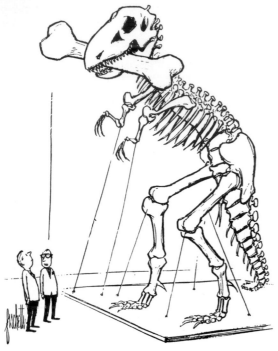

'It seemed logical. We had one bone left over.'
(Courtesy of *Punch*)

Fig. 1. An example of geological humour.

Cartoons depicting cave-men with dinosaurs, mammoths, or other extinct beasts in an enormous variety of humorous situations have been very popular and feature in almost every anthology. They have also been made the subject of highly successful film cartoons, such as *The Flintstones*. Dinosaurs have figured very frequently in the wake of renewed scientific and public interest in them. Their huge size, the reputed ferocity of some, and their bizarre skeletons lend them readily to humorous situations, whether it be in museum exhibit or alive in some Mesozoic scenario. Lesser creatures such as fish and amphibians about to give rise to higher tetrapods have been subject also to much mirth. At the other end of the vertebrate scale, hominoids have similarly provided the butt for many a joke, while Darwin himself has not escaped completely the humorists' pen or pencil. Perhaps most ventures of this kind are never published; a remarkable Victorian example is *Geology familiarly illustrated*. This is a single 12 ft long, hand-coloured pull-out sheet depicting a landscape with cartoon figures, illustrating terms such as puddingstone, Old Red (Sandstone) and intrusion, by C. M. Webster, produced in London about 1840. Today its humour seems quaint, schoolboyish, a far cry from modern jokes, but such is the changeable nature of humour, and in 1840 geology was a popular new science.

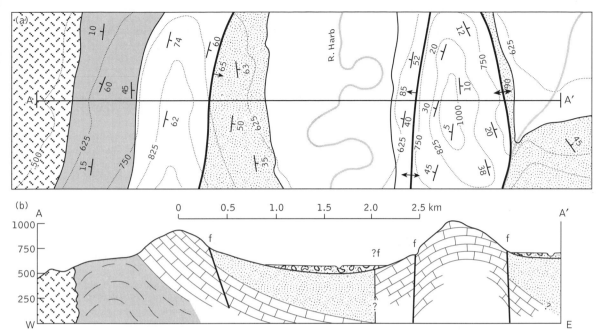

Fig. 1. (a) Portion of a geological map showing two limestone ridges (no patterns) with associated shales (dotted patterns) and a granite intrusion in the west (dashed pattern). Thick alluvium covers the valley floors. The rocks are faulted, and the dip (inclination) of the strata is shown in degrees. (b) A cross-section drawn on the line A–A′ showing an interpretation of the structure in depth. In the section the limestones are depicted by a brick-like ornament.

limit to how far geology can be extended downwards with any certainty: at best, perhaps no more than 4 or 5 km, usually much less. Below that is conjecture, but at least it is educated conjecture, based on a knowledge of how rocks behave.

There are many reasons for making geological maps. The most obvious is to aid the search for water, oil, and minerals, for they are all associated with specific types of rock and rock structures. Planning and engineering projects also require the information provided by geological maps: the construction of dams, tunnels, and high-rise buildings are just a few examples. Geological maps are in fact essential to economic development and also for the protection of the environment, such as in limiting the spread of industrial pollutants. It is to be regretted that this is not always appreciated by the powers that be. There are academic reasons too—we would like to know how our planet was formed, its history, and, possibly, even its future. We might even eventually learn to predict earthquakes and volcanic eruptions in time to warn the people who may be affected.

Because there are so many uses, there are many different types of geological map, produced at many different scales. Reconnaissance maps of geologically poorly known regions—and they still exist—may first be made at 1 : 250 000 or at even smaller scales. Maps for more general use are more likely to be surveyed at 1 : 50 000. In well-developed areas, geological surveying may be 1 : 10 000 or even larger scales, whereas maps for mining and engineering purposes are commonly made at 1 : 500, or larger still. Of course, a map made at one scale in the field may be reproduced at a smaller scale for publication: thus the 1 : 50 000 geological map series of Great Britain was complied from field information gathered at 1 : 10 000.

Geological map-making

During the past 50 years or so, geological map-making has changed radically. Before then, geologists took topographic maps into the field and covered the ground on foot. In some underdeveloped regions they even had to make their own topographic maps, often by plane-tabling, because of the total lack of any reliable base maps. In the field they systematically identified and recorded the rocks they encountered, measured their attitudes, traced the contacts between different rock types, and took rock samples for laboratory examination. If the rocks being investigated were fossiliferous, they would identify the fossils they found and take specimens for further examination and comparison with fossils from elsewhere; rocks can be dated by fossil evidence and correlated with rocks in other regions, often very far afield.

In the later 1940s aerial photographs began to become more available. By the early 1950s excellent topographic maps were being produced from these photographs, thus solving some of the geologists' problems. In addition, methods of photogeological interpretation that had been developed by oil companies were now being adopted more generally. These were methods by which 'stereo' pairs of aerial photographs were viewed through a stereoscope to give a three-dimensional image of the ground below, which was then systematically examined. Many geological features—often invisible on the ground—could be seen, even under soil cover. Changes in tones and textures indicated different rock types, fine erosional patterns distinguished softer rocks from harder, and changes in vegetation showed that plant life has preferences for soils covering one type of rock rather than another. Contacts between rock groups, fault lines, and major joints in rocks could also be traced on photographs, often far more easily than on the ground. Geologists still had to work on the ground to confirm what they saw on the photographs, and of course they could not measure the dip of strata, identify a rock properly, or collect fossils, except in the field; but photogeology proved to be of enormous benefit in geological interpretation. In addition, aerial photographs made locating oneself on a map a great deal easier.

Other aids to geological mapping were also developing at much the same time. Hand-held magnetometers could distinguish magnetite-bearing rocks from those that were not magnetic. Thus the Karroo dolerites, important aquifers in southern Africa, could be easily traced despite soil cover. Subsurface mapping was aided by geophysical methods of many types. The systematic measurement of gravity could outline the shape of large buried masses of less dense material, such as salt domes, while seismic surveys could penetrate to considerable depths by recording the reflection and refraction from deep geological boundaries of waves produced by small surface explosions.

Airborne geophysical methods were also developed. Recording magnetometers flown over large areas could distinguish rapidly between magnetic and non-magnetic rock groups. A scintillometer could make a radiometric survey at the same time which would distinguish potassic from sodic granites. Such methods were particularly useful in covering large areas quickly, especially in mineral surveys. The term 'remote sensing' was coined for such methods.

Since the 1970s satellites have also played their part in geological map-making. Satellite images (resembling colour photographs, some taken in 'false colour' extending beyond the visible spectrum) can often show major structures in astonishing detail. Satellites can also help a field geologist who has a hand-held GPS (Global Positioning System) instrument. GPS is in theory capable of giving a position to within 10 m or better, but the Pentagon restricts civilians to a 'coarse acquisition', accurate only to between 50 and 100 m. This can still be useful in the field.   J. W. BARNES

Further reading

Barnes, J. W. (1995) *Basic geological mapping* (3rd edn). John Wiley and Sons, Chichester.

Maltman, A. (1990) *Geological maps: an introduction.* Open University Press, Milton Keynes.

geological museums *see* MUSEUMS AND GEOLOGY

geological societies The variety of geological societies in various parts of the world is wide indeed, and the following notes provide no more than a few selected snapshots.

Societies concerned specifically with geology did not exist until early in the nineteenth century. The Geological Society of London, established in 1807—in the middle of the 'Heroic Age of Geology'—was the first, and in many respects typical, providing a forum for scientific discussion, publishing journals, and maintaining a library. Its founder members included prominent members of the short-lived British Mineralogical Society (founded in 1799), who decided to transfer their activities to the more comprehensive science of geology. Before this, the emerging geological sciences in western Europe were under the umbrella of the general scientific societies and national academies, such as the Royal Society of London and the Académie des Sciences in Paris. In England, for example, the Royal Society had included geological topics in its meetings and publications (a paper on 'a peculiar lead-ore of Germany' appeared in the first issue of its *Philosophical Transactions* in the 1660s, and important papers on such subjects as volcanism and vertebrate palaeontology were to follow). The Society had also assembled a collection of geological specimens. The Society of Arts, founded in 1754, had promoted mineralogical maps and had supported William Smith financially. At a later stage, the Royal Institution (founded in 1791) also played a significant role. Outside London, geology featured in the activities of the Lunar Society of Birmingham, which flourished in the 1770s. Further north, the Literary and Philosophical Society of Newcastle upon Tyne, founded in 1793, had concerned itself with geology, and especially with mining; and many geological papers appeared in the *Transactions of the Royal Society of Edinburgh*.

The Geological Society of London was granted a Royal Charter in 1825 (but apparently never sought royal patronage). Its founding members were a group of 'gentleman amateurs': the professionalization of science was, of course, a later development. Like other scientific societies of the time, its meetings were social as well as scientific occasions. Members would dine first and then, fortified with bumpers of claret, would move on to the reading and discussion of papers. (In later years the sequence was reversed. Whether the quality of the discussion was thereby enhanced or diminished remains a matter for speculation.) Until it was reconstructed in the 1970s the Society's meeting room was notable for the 'parliamentary' arrangement of its seating, with two sets of benches facing each other: an appropriate setting, perhaps, for the great debates that took place there in the nineteenth century. (Some of the Fellows of that time were in fact also Members of Parliament.)

The relatively small number of active members at that early period have been described as a family or an élite. If a family, it was one without women. Lady guests were first admitted to the Society's meetings in 1860, but this arrangement lasted only until 1862. The first paper by a female author (Maria Graham) was published in 1824, but it was not until 1907 that women were admitted as Associates, and they had to wait until 1919 before they could become Fellows of the Society in their own right.

Publication of papers read before the Society began in 1811 with the first volume of the *Transactions of the Geological Society*, which appeared in quarto. The *Proceedings* followed in 1827 and the *Quarterly Journal* in 1845. Delays in publication and disappointing sales led to the abandonment of the *Transactions*, of which the last part appeared in 1856.

Like the Royal Society before it, the Geological Society built up a collection of mineral specimens and fossils. The expenses of curation, the need for space for the library, and the realization that it was no longer necessary for the Society to maintain a collection of its own, resulted in the disposal of the specimens to the Museum of Practical Geology in 1911.

The Geological Society of London did not remain alone on the stage for long. In the British Isles, the Royal Geological Society of Cornwall was founded in 1814, and its *Transactions* first appeared in 1818. That Society, which has concerned itself very much with mining and mineralogy, is still in existence and has retained its collections and museum. A singular, but ephemeral, development was the formation of the Wernerian Natural History Society of Edinburgh, which broke away from the Royal Society of Edinburgh in 1808 but atrophied in the 1830s as the theories of A. G. Werner became discredited. (Most of the initial members of the Geological Society of London were in fact Wernerians, but the London Society did not have the fatal disadvantages of having nailed its colours to any particular doctrinal mast.)

Outside Britain, geological societies soon sprang up. Russia was quick off the mark. The foundations had been laid in the eighteenth century by Catherine the Great, who had invited many foreign scientists, including geologists, to visit Russia. The Mineralogical Society that was set up in St Petersburg in 1817 covered geology in the wider sense as well as mineralogy, and played an important part in the development of the geological sciences in Russia. In western Europe, the Société Géologique de France was founded in 1830 and the Deutsche Geologische Gesellschaft in 1848. Across the Atlantic, the Geological Society of America was founded in 1888. The pattern of the modern scientific society with its own publications and its own premises (in many instances housing a library as well as offices), with a salaried staff working under the direction of honorary officers and committees, was quickly established. Today there are geological societies in practically every developed country, and in most of the developing countries.

Developments in Russia and China were, not surprisingly, rather different. In 1918 Lenin took over the Academy of Sciences that had been set up by Peter the Great in 1725 and reorganized it in sections, of which a Geological–Geographical Section was one. Under Lenin's reorganization the Academy combined the functions of a scientific society with the direction of research, the emphasis being heavily on the latter activity. Centralization is also evident in China, where the learned scientific societies belong to the All-China Federation of Scientific Societies. The Geological Society of China, based on the Library of Geology in Beijing, was founded in 1922. The Chinese societies for palaeontology, oceanography and limnology, and geophysics are later creations, dating from the late 1940s.

Once what may be termed the general geological societies were established it was not long before more specialized ones appeared. An early arrival was the Palaeontographical Society, set up in 1847 to publish plates of fossils from British formations. The Palaeontological Association, with wider aims, did not emerge until over a century later, in 1957. Mineralogy provided a focus for numerous societies, some of which combined mineralogy with related subjects such as petrology. The Mineralogical Society was set up in London in 1876; the Société Minéralogique de France in 1878.

The applied aspects of the Earth sciences were another fruitful field, which flourished as the various areas of the subject became increasingly professional: the Royal Meteorological Society (1850) and the Institution of Mining and Metallurgy (1892) can be instanced here. The main development was in the twentieth century, when such areas as exploration geophysics and petroleum geology became of great economic importance. The American Association of Petroleum Geologists, founded in 1917, has in fact become the world's largest geological association.

The American Geophysical Union, founded in 1919, looked after the 'pure' aspects of geophysics and had established itself as the leading society in that field. The Society of Exploration Geophysicists, also based in the USA, was formed in 1930 and became numerically a very large society, even by American standards. In the UK, both the Geological Society and the Royal Astronomical Society (founded in 1820) had legitimate concerns with geophysics. The unseemly prospect of having two learned societies competing for the same territory was avoided by the formation in the late 1970s of the Joint Association for Geophysics, in which both societies participate.

In some countries geological societies (*sensu lato*) have taken on such functions as the validation of qualifications and the establishment of codes of conduct for their members. In the UK, for example, the Geological Society has since 1990, when it merged with the Institute of Geologists, functioned as a professional body as well as a learned society and has awarded the titled Chartered Geologist to Fellows with suitable qualifications and experience. Elsewhere, as for instance in France, specific organizations, such as the Union Française des Géologues and the corresponding French regional organizations, all formed in the 1960s, have been created for these purposes. Similar functions are performed in the USA by the American Institute of Professional Geologists (the AIPG).

A further twentieth-century development has been the formation of supranational bodies such as the European Union of Geosciences. Its biennial meetings in Strasbourg are major events in the geological calendar and the programmes include papers on most aspects of the Earth sciences, including palaeontology. On the professional side, there is the European Federation of Geologists, which was formed in 1980 with the object of facilitating free movement of professional geologists within the European Community. The Federation represents more than 70 000 geologists in sixteen countries and awards the title European Geologist (EurGeol) to members of national associations who have reached a high level of training and experience. It maintains working groups on education and training, environmental engineering, mineral resources, and land-use planning.

The final word on all these activities will perhaps one day be pronounced by the History of Earth Sciences Society, founded in 1981.   B. WILCOCK

geological surveys A knowledge of the areal distribution of rocks and minerals occurring within a region has long been recognized as having economic significance. Several ancient civilizations recorded carefully the whereabouts of deposits of ore minerals, useful or decorative stones, and the like. During the Renaissance, maps with geological data were produced in Italy and central Europe. Relatively modern maps, firmly linked to topography, date from the seventeenth century, and within the hundred years following their introduction maps specifically showing outcrop geology were produced in increasing numbers and complexity in Europe. With the Industrial Revolution came a greatly increased demand for raw materials, especially coal and iron ore. Geological prospecting for entirely commercial ends was accompanied by great intellectual progress. Access to geological information was seen to be necessary for the good of all. For some years after 1794, the Board of Agriculture in Britain had published a series of county maps of soils and exposed rocks. With the new discipline of stratigraphy to hand, geological maps of both practical and academic value could be, and were, produced by surveyors and scholars towards the end of the eighteenth century.

The British Geological Survey came into being in 1835 as the Geological Ordnance Survey, and is generally regarded as the oldest national geological survey in the world. Its establishment came after T. H. De la Beche (1796–1855), Secretary of the Geological Society of London, had been appointed to add colouring to Ordnance Survey maps of south-west England, in order to portray the geology. His work was greatly admired, and in 1835 he received a commission to continue, with the aid of two geologically minded members of the Ord-

nance Survey, to carry out a geological survey of Cornwall. From then on the work continued in earnest: the first memoir was published in 1839 and the surveying extended into South Wales. A prime objective was to produce a map on the scale of one inch to the mile for the entire British Isles. Mining, quarrying, and agriculture were also dealt with in some detail.

Hereafter the importance of the British Geological Survey was not in question and it has evolved into a modern survey and research institution upon which many other countries have modelled their own national geological surveys. France instituted its own geological survey in the 1830s (Service de la Carte géologique de la France) and in 1842 William Edmond Logan, who had been of much assistance to De la Beche, left for Canada to found the Geological Survey of Canada. By the beginning of the twentieth century the Geological Survey of Canada was sending ships north into Hudson Bay and the Canadian Arctic Islands. Also in North America, several of the United States set up their own state surveys of geology and natural resources, and in 1889 the US Geological Survey (USGS) was established. Its budget for its first year was US$100 000. Similar government bodies were created in various countries of Europe and in other British possessions overseas during the nineteenth century. The Geological Survey of India, for example, was set up in 1851 and achieved much within its first few decades. British colonial possessions in various parts of the world were surveyed by the Colonial Geological Survey, the services being administered later by the Overseas Geological Survey (OGS). In 1966 the OGS was combined with the Geological Survey of Great Britain as the Institute of Geological Sciences. In 1984 the title was once again changed, this time to the British Geological Survey.

The enormous territories of Imperial Russia, the later the Soviet Union, received the attention of various official government survey organizations. The Ministry of Mining was at first responsible for official mapping. The Ministry was a direct descendant of the 'Stone Department' set up in 1584, which under Peter the Great became the Department of Mining in 1700 and was renamed in 1729 as the Ministry of Mining. It was not until 1833 that a survey organization as such, the 'Geological Committee', was set up. Under the auspices of the Committee large areas of the country were mapped. A reorganization in 1929 resulted in the breaking-up of the Geological Committee and the formation of an Institute for Geological Maps. This arrangement was found to be inefficient, and a new organization, the Central Scientific Research Geological Prospecting Institute (ZNIGRI) was created. The ZNIGRI was in turn reorganized in 1938 as the All-Union Scientific Research Geological Institute (ZNIGRI-VSEGEI).

In China, systematic geological mapping is the responsibility of the Institute of Geology, one of several institutes under the control of the Ministry of Geology. The Institute, which was created in 1954, has its headquarters near Peking. The Chinese Academy of Sciences, which is concerned with basic research, has also undertaken regional surveys.

Even the distant continent of Antarctica has received great attention from the national governments that administer territories there. Although international agreement prohibits industrial development in Antarctica, a natural interest in discovering what resources exist is keen. The British Antarctic Survey, for example, maintains a vigorous geological survey programme and the USGS is similarly active.

New techniques have increasingly been adopted for improving and extending geological surveying everywhere. In particular, geophysical exploration methods and aerial photography (later to become remote sensing) have been utilized. The eventual use of satellites, and a number of remote sensing techniques deployed from them, has enabled more widespread and even global data to be gathered. Even the business of drawing maps has been vastly improved and speeded up by the use of computers and geographical information systems.

These information systems and advancing high-speed digital data transmission are already revolutionizing the operations of geological surveys; the exploration of ever-deeper parts of the crust is a good example. New mineral deposits are an obvious target, but the need for safe storage of hazardous wastes becomes increasingly urgent. A Deep Geology Unit was established in the Institute of Geological Sciences in 1977, and continues its exploration of British territory. Earthquakes, which are persisting hazards in many densely populated parts of the world, originate at depth within the crust. Some might be predicted from seismic observatories, but improving knowledge of the state of local or regional crustal stress increases our ability to foretell these shocks, as does surveying visible expressions of earthquakes, such as fault scarps. Most geological survey organizations include units concerned with this aspect of monitoring.

In addition to producing maps of both outcrop and sub-surface geology, geophysical characteristics, and mineralogical and geochemical maps, geological surveying has, since the 1950s, been extended to the continental shelves adjacent to land masses. This activity has been driven by the discovery of petroleum and gas in these marine regions. Today there is close cooperation between petroleum companies and national geological surveys.

Immediate and future activities for many geological surveys are concerned with environmental planning, linked closely with civic authorities and industry. Disposal of radioactive waste poses geological problems. Other natural hazard predictions and assessments appear to be increasingly necessary in heavily populated and urbanized areas. In 1993 the USGS had a budget of US$35 000 000 for this field of activity, but this has since been significantly decreased, an unwise development according to many commentators.

Geological survey organizations throughout the world account for a large portion of published Earth science data, ranging from geochemical maps to mineral statistics and to micropalaeontological taxonomy. Most surveys have close links with national industries and also provide data and advice to the public on request, as did the first geological survey in 1835. The costs of such services have risen greatly, especially with the use of new technologies, but they remain a very small part of gross national expenditures. Nevertheless, in recent years, governments have directed their surveys to seek private contract work with industry or for other government departments or organizations. This imposes restrictions upon the ability to pursue research of a purely scientific nature, and even to continue with a basic mapping programme. Undertaking work for private concerns is far from the original purpose of a geological survey such as that of Britain; in so doing, its ability to provide a public service must to some degree be impaired. Yet that original aim must be preserved in a modern technological society. No national geological survey foresees the end of its mission, and its demise would be a matter for regret.   D. L. DINELEY

Further reading

Bailey, Sir E. B. (1952) *Geological Survey of Great Britain.* George Allen and Unwin, London.

Wilson, H. E. (1985) *Down to Earth. One hundred and fifty years of the British Geological Survey.* Scottish Academic Press, Edinburgh.

Winch, K. L. (ed.) (1976) *International maps and atlases in print* (2nd edn). Bowker, London.

Wood, D. N., Hardy, J. E., and Harvey, A. P. (eds) (1989) *Information sources in the Earth sciences* (2nd edn). Bowker-Saur, London.

geological time Time, to a geologist, is a familiar dimension. Often, it is easier to assess the time history of a rock sequence than to travel horizontally, to identify environments and their facies that coexisted at a single time. While to a physicist the fourth dimension is something that depends on theory, to a geologist time is intuitive and is obvious from the rocks.

The scale of geological time is now known to be immense: over 4500 million years (4.5 Ga) of Earth history to be comprehended. There is danger here: geologists dealing with the earlier part of that history can fall into the trap of regarding 10 or even 100 million years (Ma) as a short time, in which little of importance has happened, forgetting that many important events in geology are very sudden: a flood, an earthquake, a volcanic eruption, or a meteorite impact; and it is the effects of sudden large-scale events that are commonly preserved. But there is advantage too in the vast scale: long-flow processes such as the formation and destruction of ocean basins, or the secular cooling of the Earth, can be investigated.

The concept of geological time

That the Earth is old is not a new idea. In the ancient world, both Egyptian and Sanskrit writers held the view that the world itself is ancient, and the Roman author Lucretius also discussed the long past and distant future. The Christian

413

Main divisions of geological time

Eon	Era	Period	Epoch	Age (Ma)
Phanerozoic	Cenozoic	Quaternary	Holocene	
				0.01
			Pleistocene[1]	
				1.8
		Neogene[2]	Pliocene	
				5.3
			Miocene	
				23.8
		Palaeogene[2]	Oligocene	
				33.7
			Eocene	
				54.8
			Palaeocene	
				65.0
	Mesozoic	Cretaceous	L / E	
				142
		Jurassic	L / M / E	
				205.7
		Trias	L / M / E	
				248.2
	Palaeozoic	Permian	L / E	
				290
		Carboniferous — Pennsylvanian[3]	L	
				323
		Carboniferous — Mississippian[3]	E	
				354
		Devonian	L / M / E	
				417
		Silurian	L / E	
				443
		Ordovician	L / M / E	
				495
		Cambrian[4]	L / M / E	
				545
Precambrian[7]	Proterozoic[5]			
				2500
	Archaean			
				4000
	Hadean[6]			
				4560

(Tertiary spans Neogene[2] and Palaeogene[2])

This table, which shows only the main divisions of geological time as generally agreed*, is not to scale. (For example, Precambrian time (the Hadean, Archaean, and Proterozoic eons) represents about seven-eighths of geological time from the formation of the Earth to the present.) More detailed schemes, showing the names for the major Phanerozoic divisions and isotopic ages for the entire timescale, including the Precambrian, will be found in Appendix 1 (pp. 1119–20).

The letters E, M, and L denote 'Early', 'Middle', and 'Late' respectively.

The dates shown here for the Phanerozoic eon are those given by F. Gradstein and J. Ogg (1996) A Phanerozoic time scale *Episodes*, **19**, (1–2), (3–5).

[1] Opinion on the duration of the Pleistocene is divided between advocates of a 'shorter' chronology in which the Pleistocene lasted for about 1.6 to 1.8 Ma, and a 'longer' timescale of *c.* 2.5 to 2.6 Ma.

[2] The Palaeogene and the Neogene periods were together originally categorized as a period or sub-era, the *Tertiary*, and this term is still in general use.

[3] In America the Carboniferous period is divided into two sub-periods, the Pennsylvanian and the Mississippian, as shown.

[4] The base of the Cambrian period is important in that it is also the base of the Palaeozoic era, and the time value of this has been contested in recent years. The range of 540–60 Ma seems to be in accord with most of the evidence.

[5] The Proterozoic eon is commonly divided into Neoproterozoic, Mesoproterozoic, and Palaeoproterozoic, each with a duration of several hundred million years.

[6] The Hadean eon is not recognized in all schemes. The boundary between the Archaean and the Hadean is placed by some at the origin of life on Earth and by others at the oldest coherent rock (as opposed to individual minerals). The former criterion is unlikely to be detected in any rock; the latter may be found as exploration continues and analytical techniques improve.

[7] The term 'Precambrian' is used informally for the timespan represented by the Hadean, Archaean, and Proterozoic eons.

* A new edition of the International Stratigraphic Chart compiled by Jürgen Remane, A. Fuare-Muret, and G. S. Odin was published by the Division of Earth Sciences, UNESCO, in 2000 for the International Commission on Stratigraphy with the collaboration of all subdivisions of the ICS. The age values given in this chart differ by a few million years from those of Gradstein and Ogg (1996), but these small refinements do not significantly change the magnitudes given here.

western world has been influenced by the Bible and varying interpretations of its constituent books. Augustine, arguably the greatest of the early Church fathers, for his writings profoundly influenced later Christian thought, clearly pointed out in his commentary on Genesis that God's 'vision is outside time': the evening and morning of the six days of creation described in Genesis were not in his view to be regarded as solar 24-hour days, for the Sun itself was not created until the fourth day. Augustine held that our present historical existence comprises the seventh day, a view consistent with Hebrews 4, 7–9. The modern 'creationist' opinion that the world was created in six 24-hour days is thus not supported by traditional theology or necessarily by biblical texts. Indeed, scholars generally affirm that it is in conflict with them. Archbishop Ussher's celebrated calculation in the seventeenth century that the world began in 4004 BC is, if it has any relevance, only a minimum estimate: the Bible leaves open the question of the age of the Earth and, as Augustine clearly points out, time itself is a created dimension. The geologist, whose concepts can range freely through time, has in this power a slight hint of eternity, in which all dimensions exist.

The study of geological time in natural science, as opposed to metaphysics, began early. In the Bible, Job discusses erosion, and Psalm 104 (which is closely similar to a monotheistic hymn ascribed to the Egyptian Pharoah Akhenaton) has a clear vision of geological processes. Modern geology, with its emphasis on uniformity, reflects both the Hebrew and Roman (Lucretian) heritage of ideas: James Hutton and Charles Lyell, who pioneered the subject, saw no vestige of a beginning and no prospect of an end to Earth history, but more recently geologists have come to discover clear vestiges of that beginning, and can predict certain prospects of the distant end.

Since the eighteenth-century Age of Enlightenment, stratigraphical methodology has been applied. In 1759 the Italian Giovanni Arduino distinguished three successive categories for rocks, and consequently for geological time: Primary, Secondary, and Tertiary. Later, geological *systems* were defined for rocks, with corresponding *periods* for geological time, initially with the notion that they were divided by massive catastrophies. The systems and periods described individually in this book which form the basis of modern Phanerozoic timescales (see Fig. 1, Appendix 1) were all distinguished by the end of the nineteenth century: the Cambrian by Adam Sedgwick in 1835 on the basis of rocks in Wales; the Ordovician by Charles Lapworth in 1879, also in Wales; the Silurian by R. I. Murchison in 1835 in Wales and the Welsh borderland; the Devonian by Murchison and Sedgwick jointly in 1839 in Devon; the Carboniferous by W. D. Conybeare and William Phillips in 1822 in England and Wales; the Permian by Murchison in 1841 in Imperial Russia; the Trias by F. von Alberti in 1834 in Germany; the Jurassic by A. von Humboldt in 1795 in the Jura mountains of France and Switzerland; the Cretaceous by J. J. d'Omalius d'Halloy in 1822 on the basis of the European Chalk. Arduino's terms 'Primary' and 'Secondary'

became obsolete, but 'Tertiary' was maintained, and in 1833 Charles Lyell initiated its subdivision (into Eocene, Miocene, and Pliocene) on the basis of increasing percentages of Recent species in fossil mollusc faunas. Beyrich distinguished the Oligocene in 1854, and Schimper the Palaeocene in 1874. The Quaternary was named by Morlot in 1854. Thus, by the turn of the century the Phanerozoic was fairly well chronicled. Study of older rocks had hardly begun.

The timescale

The geological timescale is now divided into a hierarchy of units. The largest divisions of geological time are the *eons* (or aeons). No consistent terminology as yet exists for the first 4 Ga of Earth history, but the names of four eons are in current common usage:

(1) *The Hadean* (named from Hades, the underworld of Greek mythology) starts at the time of accretion of the planet, about 4.56 Ga ago, and includes the probable great impact of a Mars-sized object 4.45 Ga ago that may have led to the ejection of fragments which coalesced to form the Moon. The Hadean–Archaean boundary is as yet ill-defined. We here take this boundary as the time of the last common ancestor to life, perhaps about 4 Ga ago (but not better dated than in the interval 4.4–3.7 Ga). At present, the oldest terrestrial materials known are zircon crystals from the Jack Hills and from Mount Narryer, Western Australia, some nearly 4.3 Ga old. The Acasta gneiss, from the Slave Province, Northwest Territories, Canada, is the oldest known rock, about 4 Ga old. The oldest large-scale rock succession is the Isua belt, in western Greenland, about 3.8 Ga old, widely accepted as of Archaean age.

(2) *The Archaean* (from Greek: 'beginning') (about 4.0 Ga to 2.5 Ga) is the time of the beginning of the history of life. During this eon primitive microbes (bacteria and archaea) colonized the planet and changed the composition of the atmosphere and the redox (oxidation–reduction) state of the Earth's surface. The Archaean–Proterozoic is another disputed boundary. Currently it is placed arbitrarily at 2.5 Ga, but this usage breaks international stratigraphical practice, that definition should be 'in the rock'. It might be defined at a location in the Hamersley succession in Western Australia, or by the cooling age of the Great Dyke in Zimbabwe, both of which are dated at about 2.5 Ga.

(3) *The Proterozoic* (from Greek: 'first life') marks the dominance in the geological record of eukarya, organisms with more complex cell structure than the primitive bacteria and archaea. Oxygen-rich sedimentary facies became widespread after 2.2 Ga, and possible metazoa after about 1 Ga. The Proterozoic–Phanerozoic boundary is not disputed in location (Newfoundland), but its age (arguably 545–570 Ma) is not well known.

(4) *The Phanerozoic* (from Greek: 'apparent life') is characterized by a fossil record in which some organisms possess a mineralized skeleton, and so have relatively high preservation potential.

The Phanerozoic is divided into three *eras*: Palaeozoic, Mesozoic, and Cenozoic—names given by John Phillips in 1840–1
on the basis of their differing fossil biotas (Greek 'ancient,
middle, and recent life' respectively). Seven eras are now
commonly used to subdivide the earlier eons.

The late Proterozoic and the Phanerozoic have been divided
into *periods*. Those of the Phanerozoic are named above; their
sequence is remembered by some students through the
mnemonic: 'Camels Ordinarily Sit Down Carefully, Perhaps
Their Joints Creak Tremendously Quietly' (although the
Tertiary and Quaternary are more correctly regarded as sub-
eras of the Cenozoic, as shown in Fig. 1).

Periods are in turn subdivided into *epochs*, and epochs into
stages or more correctly for time units, *ages*.

Relative time

Until the early twentieth century geologists had little idea of
the absolute ages of rocks, but—at least in the Phanerozoic—
had an excellent understanding of the relative ages of rock
units. Chronostratigraphical units (sometimes known as
time-stratigraphical units to distinguish them from rock-
stratigraphical units, whose definition is based on rock type
alone) were distinguished and arranged in sequence with
increasing precision. The chronostratigraphical scale now
provides a global standard by which geologists can correlate
local rock sequences. Major boundaries are being increasingly
defined by markers in stratotypes: rock sequences agreed by
the International Commission on Stratigraphy to serve
as the primary reference points for regional or global correlation.

Correlation was and still is mainly by the use of fossils
(biostratigraphy). William Smith and others at the turn of the
eighteenth and nineteenth centuries demonstrated the validity
of the method, which is based on changes in the fossil record
through geological time, and it has been increasingly refined
subsequently. Macrofossils (especially trilobites, graptolites,
and ammonoids) eventually provided high-resolution biozonation and correlation for appropriate parts of the Phanerozoic, supplemented more recently by the use of microfossils
(notably foraminifera, nannoplankton, radiolaria, dinoflagellates, and conodonts).

Correlation may now also be effected by a wide variety of
other means, such as event stratigraphy (based on sudden
major events, e.g. volcanic eruptions, which can define time
planes); magnetostratigraphy (based on changes in the Earth's
magnetic field through time); chemostratigraphy (based on
changes in the global chemical environment, especially of the
oceans); and changes in global climate.

'Absolute' time

In the nineteenth century, attempts were made to date the age
of the Earth and its rocks using a variety of techniques, for
example, by calculating the length of time necessary for sediments to accumulate, or for the oceans to reach their present
salinity. Although lacking in accuracy, these techniques did

indicate that the Earth was of great age—arguably hundreds
or thousands of million of years. In contrast, the physicist
William Thomson (Lord Kelvin) calculated a relatively young
age (20–40 Ma) for the Earth, assuming that the planet was a
cooling conducting body (but not modelling an Earth that
convects). In the early twentieth century Lord Rutherford
suggested that naturally occurring radioactive elements might
be used to date rocks, and that radioactivity contributes to the
Earth's heat, providing the basis to confirm that our planet is
indeed very old. Arthur Holmes and others subsequently constructed timescales based on isotopic dating, and these have
been developed into the high-resolution scales available today.

Isotopic dating is based on the observation that isotopes of
certain elements are unstable: a 'parent' isotope decays to
produce a 'daughter' isotope at a constant rate (often expressed
as a 'half-life'—the period of time required to reduce the
parent by one-half). Rocks and minerals incorporate such isotopes at the time they form. By measuring the proportion of
parent and daughter isotopes present now, and making
certain assumptions about the amount of the daughter
isotope initially present, from a known rate of decay it is
sometimes possible to calculate the time when the rock or
mineral was formed in absolute terms, provided that the
system is 'closed' (i.e. that no parent or daughter isotopes have
been added or removed), or allowance is made for any such
anomalies.

Isotopes commonly used for dating younger rocks are those
produced continuously, notably carbon-14 (which decays to
nitrogen-14, with a half-life of 5730 years), uranium-235
(which decays to lead-207 with a half-life of 704 Ma),
uranium-238 (which decays to lead-206 with a half-life of
4469 Ma), and thorium-232 (which decays to lead-208 with a
half-life of 14 010 Ma). Isotopes used for dating older rocks
are not produced continuously; they include potassium-40
(which decays to argon-40 with a half-life of 11 930 Ma),
rubidium-87 (which decays to strontium-87 with a half-life of
48 800 Ma), and samarium-147 (which decays to neodymium-
143 with a half-life of 106 000 Ma). By the mid-twentieth
century isotopic data had provided dates, in million of years,
for the main divisions of the Phanerozoic succession, for
which only relative ages had previously been known. The
resulting timescale is being continually refined. Appendix 1b
shows changes in widely accepted interpretations between
1989 and 1996.

In rocks older than the Cambrian (eight-ninths of the
world's history), isotopic dating is becoming increasingly
significant. Dating based on zircon crystals is estimated
now to provide ages accurate to within a million years, making
correlation nearly as precise as that permitted by fossils in
Phanerozoic sequences. The various isotopic techniques all
have specific uses and problems, but nevertheless now serve to
elucidate not only ages of igneous events but also ages of uplift
and cooling in igneous and metamorphic terrains.

In the Quaternary (arguably the last 1.6 or 2.6 Ma of Earth
history), rocks are too young to apply some isotopic dating

techniques used for the Precambrian. A wide range of other methods is used. However, some numerical methods (such as carbon-14, argon-40/argon-39, and uranium-series isotope analyses) can be used to produce quantitative estimates of age (and uncertainty) on a ratio or absolute scale. Several other methods measure systematic changes in a parameter (such as lichen growth) with time, but the rate of change must be calibrated by use of another technique. In addition, relative methods (such as amino acid racemization) provide ages in order of rank, but since long-term rates of change in this category are difficult to calculate they require independent chronological controls to convert them to calibrated ages. There are also correlated age methods, but these only provide evidence of age equivalence, as in oxygen-16/18 ratios in fossil foraminifera from deep sea cores, and so must be used in conjunction with other corroborative evidence to relate them to the timescale. Additional techniques that are being increasingly applied to provide numerical or calibrated ages in the Quaternary are represented by fission-track, luminescence, and electron-spin resonance dating.

The term 'absolute' has commonly been used for ages derived from numerical (usually isotopic) methods, but it should be avoided, since there are inherent uncertainties in all such methods that may invalidate the derived ages.

The record

The geological record, though varied and extensive, is, even at its best, a poor sample of geological time. Most of the record is of major but short-lived events, such as volcanic eruptions or turbidity currents. Even in the best parts of the record, for example in slowly deposited sediments, the time captured in the rock is only a small sample. In the history of tectonic or metamorphic events, only peak events are usually well recorded, and the intervening periods, when significant but lesser events may have been frequent, may not be recorded at all.

The record is thus a partial and biased sample of geological time; worse, we do not know what we miss. From this is reconstructed the history of our planet.

E. G. NISBET and E. P. F. ROSE

Further reading

Dalrymple, G. Brent (1991) *The age of the Earth*. Stanford University Press.

Harland, W. B., Armstrong, R. L., Cox, A. V., Graig, L. E., Smith, A. G., and Smith, D. G. (1990) *A geologic time scale 1989*. Cambridge University Press.

Rogers, J. J. W. (1993) *A history of the Earth*. Cambridge University Press.

geology and art *see* ART AND THE EARTH SCIENCES

geology and other Earth sciences

The word 'geology' is derived from two Greek words (*geō, logiā*) and translates literally as the study or science of the Earth. The word was coined in medieval Latin in the eighth century by the Northumbrian monastic scholar and historian Bede ('The Venerable Bede') to mean the study of things earthly (as opposed to the study of things godly (i. e. theology). In the late eighteenth century it acquired something like its modern meaning of scientific observation of the Earth, its structure, composition, and history. Subsequently, various related disciplines emerged, such as palaeontology, the study of fossils; mineralogy, the study of minerals; geomorphology, the study of landscapes; geochemistry, the study of the behaviour of chemical elements within the Earth and the associated analytical techniques; and geophysics, the investigation of the Earth's magnetic, electrical, and gravitational fields, and their use, along with the propagation of seismic waves, to probe the planet's interior. This left the meaning of 'geology' impoverished in scope, with a geologist being defined as someone who investigated the Earth primarily by seeking to interpret rocks and the settings in which they formed.

The term 'Earth sciences' came into vogue during the 1960s and 1970s as an inclusive expression that was intended to embrace geology and all its related disciplines. Geologists, geophysicists, and geochemists often restrict the meaning of 'Earth sciences' pretty much to just those three disciplines, but in other contexts (as in this *Companion*) the term is much wider in scope, embracing even such fields as geodesy (the science of measuring and mapping the surface of the Earth), oceanography (the study of the oceans), meteorology (the study of the atmosphere), and sometimes even anything that could be described as geography.

The term 'planetary sciences' is used to describe the application of the techniques of 'Earth sciences' (in either the broader or the narrower sense) for the study of planets other than the Earth, or of planets in general. It is ironic that the more modern invention 'Earth sciences' should have proved inadequate to the task of describing the extension of the science beyond our home planet, and that people are beginning to use 'geology' to cover any or all related disciplines employed on any Earth-like body. For example 'lunar geology' and 'martian geology' have now ousted their more abstruse equivalents 'selenology' and 'areology'. This is perfectly justifiable, because in the original Greek 'geo' can carry the connotation of any Earth-like body rather than the Earth alone.

DAVID A. ROTHERY

Further reading

Rothery, D. A. (1997) *Teach yourself geology*. Hodder and Stoughton, London.

geomagnetic measurement: techniques and surveys

Geomagnetic measurements owe their beginning to an uncommon rock: lodestone. In the first century AD, Chinese people knew that when a spoon cut from lodestone is spun, it comes to rest every time in nearly the same direction. The compass needle and a knowledge of *magnetic declination*, the angle between the magnetic north and geographic north, were developments from this knowledge; they were described by Shen Kua in 1088. Figure 1 shows the local magnetic coordinate system. In 1581, Robert Norman, a London instrument maker, invented the dip circle by suspending a compass needle on a horizontal pivot. It thus became possible to record *magnetic inclination*, the angle in a vertical plane between the Earth's surface and the direction of the Earth's magnetic field. However, the mysteries of the Earth's magnetism began to unfold only when William Gilbert took the novel approach of systematically experimenting with a terella, a model Earth made of lodestone. His enquiries into the properties of lodestone, perhaps spurred by speculations about its possible medicinal use, resulted in the first modern scientific book, *De Magnete* (1600). Later, Gilbert became physician to Queen Elizabeth I and King James I. Although the concept of the strength of Earth's magnetic field was known qualitatively to Gilbert, it was Baron von Humboldt who discovered (*c.* 1798) that the field strength at a point was inversely proportional to the square of the period of oscillation of a dip circle compass needle.

Perhaps the most important contributions to geomagnetism were made by Carl Friedrich Gauss (1777–1855). Among his many scientific contributions to the field, he standardized the local magnetic coordinate system to perform his measurements (Fig. 1); developed experiments in collaboration with W. Weber to measure magnetic moment and horizontal field intensity using two magnets in certain configurations (named after him as Gauss-A and Gauss-B positions) and the method of least squares that he had devel-

oped himself; and represented the intensity and magnetic moment in mass, length, and time units. He also developed the spherical harmonic analysis method so that local magnetic field parameters distributed over the globe could be expressed collectively by a mathematical equation. Later developments in the field of geomagnetic measurements resulted from the relations between electricity and magnetism established by Michael Faraday (1791–1867) and James Clerk Maxwell (1831–79). These relations, in turn, led to the sophisticated instrumentation, surveying techniques, and technologies of the twentieth century (for example, ground-, air-, ship-, and satellite-borne magnetometers).

Today, the objectives of a survey govern the use of specific magnetic instruments, surveying parameters (for example, observation elevation and data spacing), and the coordinate system used for the analysis of the data. One can measure the relative variation in the intensity of the magnetic field to a very high precision in any given direction using vector instruments (e.g. a fluxgate magnetometer); one can also measure the total intensity of the field using scalar instruments (traditionally, the proton precession or, commonly, the alkali vapour magnetometer). For geological interpretation, data from three orthogonally directed vector instruments are equivalent to obtaining the scalar data, and vice versa. A configuration of vector instruments, however, is advantageous when studying the magnetic effects of dynamic current sources such as those found in the Earth's outer core and ionosphere and magnetosphere. The sensitivity of these instruments usually ranges between 0.1 and 1 nT (nanotesla, also known as gamma) and is adequate for most investigations. (For comparison, at the Earth's surface, the magnetic effects of geological sources range from a few nanoteslas to thousands of nanoteslas in aeromagnetic surveys. The Earth's main field ranges between ~24 000 nT and ~66 000 nT, from the magnetic equator to the magnetic poles respectively.) Sampling rates and portability dictate the choice of an instrument in a specific application. Instruments with low sampling rates are not adequate for fast-moving platforms.

In general, magnetic observations made anywhere in the vicinity of the Earth contain information from the Earth's internally generated field (the main field), the external fields from the ionosphere and the magnetosphere (see *geomagnetism, external fields*), the fields from geological sources, and sometimes from cultural (artificial) sources such as ferromagnetic objects and power lines. According to the application, the magnetic field due to the source of interest is retained for further analysis; the fields from other sources are first subtracted, if possible, by modelling those phenomena and by signal-processing techniques. The magnetic field retained for geological analysis is generally referred to as the *magnetic anomaly*: it is a deviation of observations (corrected for external fields) from the 'normal' or the background intensity of the local field. Magnetic anomalies are caused by relative variations in the quantity and type of magnetic minerals in geological sources and formations (primarily the

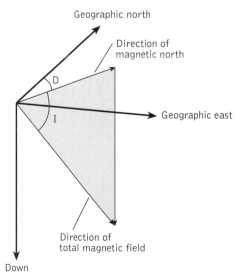

Fig. 1. The local geomagnetic coordinate system: D, magnetic declination; I, magnetic inclination. Several different coordinate systems are used in the analyses of geomagnetic data.

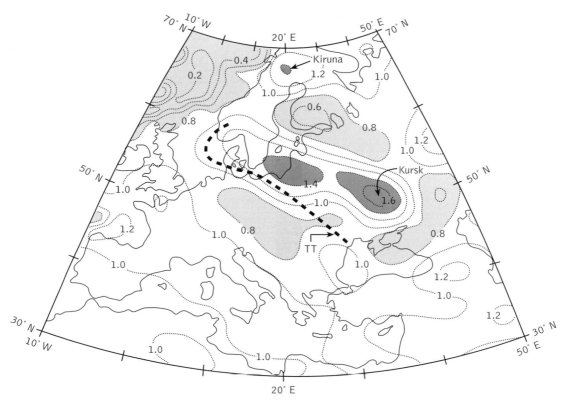

Fig. 2. Magnetic susceptibility map of Europe derived from Magsat data (by courtesy of M. Purucker and R. A. Langel). Many of the susceptibility differences in this map reflect large-scale variations in the magnetic properties of rocks formed during the distinctly different tectonic events that built the European continent. TT, the Tornquist–Teisseyre zone which separates the younger, thinner, and hotter crust of Palaeozoic western Europe from the older, thicker and cooler Precambrian crust to its north-east; Kursk and Kiruna are two of the prominent satellite altitude magnetic anomalies of Europe.

mineral magnetite (Fe₃O₄), but also haematite (Fe₂O₃) and pyrrhotite (FeS)) and are thus extremely helpful in providing geological information about the subsurface. Since rocks are magnetized by the ambient magnetic field, the anomalies also record the reversals of the Earth's magnetic field. The anomalies are therefore quite useful in deciphering the nature and evolution of geological provinces. Indeed, it was the high–low–high–low pattern of magnetic anomalies observed on the ocean floor that simultaneously led L. W. Morley and F. J. Vine and D. H. Matthews in the early 1960s to unambiguously confirm the hypotheses of sea-floor spreading and geomagnetic reversals.

The surveying parameters (such as data spacing and elevation) also depend on the objectives of the survey. For geological surveys, the desired spacing between measurements (data spacing) depends on what is necessary to characterize completely the anomaly of interest. (This sounds like a circular argument, but to design an optimal survey one must estimate the shortest spatial dimension of the anomaly of interest.) A magnetic anomaly due to any source decreases in strength with increasing distance from the source, on the other hand, with increasing distance (for example, a higher elevation), the anomaly broadens. These relationships result in a useful 'rule of thumb' that the line spacing should be roughly equal to the vertical distance between the sources of interest and the sensor. Ground surveys over small areas and satellite magnetic surveys can achieve this line spacing without much difficulty; indeed, there is an increasing trend in aeromagnetic surveying to achieve such spacing (see *aeromagnetic surveying* for a further discussion and an aeromagnetic map of Minnesota, USA).

The gradual temporal changes of the Earth's main field do not pose difficulties in isolating magnetic anomalies due to geological sources as long as the main field is known at the time of the surveys. (The International Geomagnetic Reference Field (IGRF) is updated every 5 years.) However, external magnetic fields change over the time-span of the surveys and the magnetic variations due to these fields must be removed from the data by tracking the temporal magnetic variations at fixed locations using continually recording base-station magnetometers. This procedure is adequate for ground, ship-borne, and aeromagnetic surveys because temporal variations due to external fields have similar manifestations in both base-station and moving magnetometer sensors (because of their vicinity). In mapping the magnetic anomalies from space, however, empirical models of the magnetic fields due to ionospheric and magnetospheric effects must be derived simultaneously from the satellite data themselves. To analyse ionospheric and magnetospheric effects, it is necessary to use coordinate systems that organize the current systems optimally in time or space, or both. Some examples are: local time, magnetic (dip) latitude, geomagnetic (dipole) latitude, magnetic local time, and even the geomagnetic field lines. Such processing efforts have led to determinations of satellite magnetic anomalies of the Earth. Once the anomalies are compiled, they can be transformed into regional magnetization or magnetic susceptibility variations which aid in the understanding of the evolution of the Earth's lithosphere (Fig. 2). With lower-altitude, longer-duration satellite magnetic missions in the future, a significant increase in our knowledge of the geological processes of the upper litho-

sphere, the outer core, and the ionospheric current systems will be achieved during the twenty-first century.

DHANANJAY RAVAT

Further reading

Merrill, R. T. and McElhinny, M. W. (1983) *The Earth's magnetic field: its history, origin and planetary perspective.* Academic Press, London.

geomagnetism: external fields

Not only does the Earth have its self-generated internal magnetic field (see *magnetic field, origin of the internal field*); it is also surrounded by magnetic fields formed outside the Earth. These fields are called external fields. External fields originate as a result of complex conditions created in Earth's vicinity by the interactions of solar wind, solar radiation, the Earth's internal field, and the Earth's atmosphere. Extremely high temperatures prevail in the Sun's *corona*, a region immediately above the surface of the Sun. In the corona, protons and electrons of solar hydrogen can exist separately as charged particles in an ionized state known as *plasma*. Because of their very high speeds, these charged particles defy the Sun's gravitational field and continuously break away from the corona in a radially

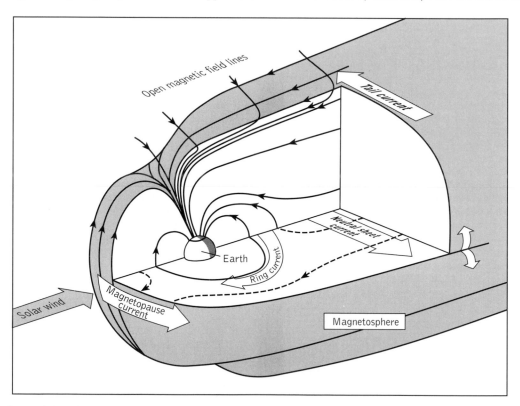

Fig. 1. The morphology of the Earth's magnetosphere. Arrows show the direction of various currents. (Based on L. J. Lanzerotti, R. A. Langel, and A. D. Chave (1993) Geomagnetism. In *Encyclopedia of applied physics*, Vol. 7, pp. 109–23. VCH Publishers, Inc.)

outward manner. The constant bombardment of the charged particles into space, with velocities of some 400 km/s is called *solar wind*. The solar wind is deflected around obstacles in its path—obstacles such as planets, comets, and other objects in the solar system. (The first clues to the existence of the solar wind were in fact provided by the tails of comets, which always point away from the Sun.) But because the solar wind is made up of charged particles, the form of the internally generated magnetic field of a planet is significantly distorted by the solar wind. The interaction between the two encloses and compresses the planetary magnetic field on the side of the Sun (in the planet–Sun axis) and drags it further out on the side away from the Sun. This vast region in which the geomagnetic field is enclosed is called a *magnetosphere* (Fig. 1). Each planet that has its own internally generated magnetic field has its own magnetosphere. Within the magnetosphere, trapped charged particles (in a region known as *Van Allen belts*) result in one of the most prominent magnetospheric currents, the ring current. Its magnetic effect in the near-Earth environment can be large, measured in hundreds of nanoteslas (nT) (see *geomagnetic measurement: techniques and surveys*).

Planets with atmospheres have regions surrounding them called *ionospheres*. The presence of the Earth's ionosphere makes worldwide short-wave communications possible because the radio waves are bounced off this region. The *ionosphere* is produced by bombardment of solar ultraviolet photons on the atmosphere. The photons can break apart, or ionize, molecules and atoms of the atmosphere into protons and electrons, producing plasma. The ionized particles of the plasma are significantly affected by the Earth's magnetic field and the atmospheric circulations caused by solar heat and the Earth's rotation, which cause them to flow in various forms of circulation patterns (in the bulk sense). In 1882, Stewart suggested that the flow of these charged particles creates electrical currents and, consequently, the magnetic fields in the ionosphere (see *magnetic field (origin of the Earth's internal field)*). During normal solar activity, two magnetic effects from currents in the ionosphere are prominent in the equatorial and mid-latitude regions. One is called *Sq* (solar quiet) and the other is *equatorial electrojet* (Fig. 2); both are strongest during a local noontime. The Sq current has its centres near 30° magnetic (dip) latitude (see *geomagnetic measurement: techniques and surveys*) in both hemisphere, flowing counterclockwise in the northern and clockwise in the southern hemisphere. Its magnetic effects range up to about 40 nT near the Earth's surface. The equatorial electroject, on the other hand, flows eastward during the daytime in a narrow swath of latitudes along the magnetic (dip) equator; it is also restricted in altitude, close to about 105 km above the Earth. Its magnetic effects at the Earth's surface range up to about 100 nT.

Clinging tightly to the superhighway of magnetic field lines, the charged particles in the solar wind are accelerated by the Earth's magnetic field, attaining speeds of over 60 000 km/s during solar storms. Their descent into the Earth's polar ionosphere is marked by colourful shimmering curtains of

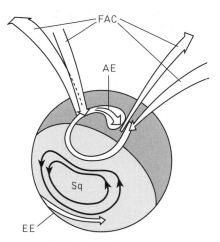

Fig. 2. Main features of the Earth's ionosphere. FAC, field-aligned currents; AE, auroral electrojet; EE, equatorial electrojet; Sq, solar quiet time variations. (Based on L. J. Lanzerotti, R. A. Langel, and A. D. Chave (1993) Geomagnetism. In *Encyclopedia of applied physics*, Vol. 7, pp. 109–23. VCH Publishers, Inc.)

light called the *aurora*. Aurorae are by far the most spectacular of geomagnetic phenomena. When the charged particles of the solar wind hit the atmosphere, various gases present in the atmosphere heat up, split apart molecules, and ionize atoms. Eventually, the atoms cool down, deionize (i.e. gain some electrons back), and in the process emit photons (in a process almost opposite to the one that causes ionization in the equatorial region). According to their energy, the charged particles penetrate to different levels in the atmosphere, ionize different gases, and hence produce the different colours of the aurorae. In addition to aurorae, complex interactions between the magnetosphere and the ionosphere take place over the polar regions. Streams of charged particles produce *field-aligned* (or *Birkeland*) *currents*. Coupling between the ionosphere and the atmospheric circulation around the poles also produces an intense current called the *auroral electrojet*, with magnetic effects at the Earth's surface as high as 1000 nT.   D. RAVAT

Further reading

Consolmagno, G. J. and Schaefer, M. W. (1994) *Worlds apart: a textbook in planetary sciences*. Prentice Hall, New Jersey.

geomagnetism: main field, secular variation, and westward drift
The magnetic field that originates in the Earth's outer core (see *magnetic field (origin of the Earth's internal field)*) is known as the *main field*. Roughly 90 per cent of the main field at the Earth's surface can be approximated as a field produced by a magnetic dipole; hence that part the main field is known as the *dipole field*. The part that cannot be expressed in this way is known as the *non-*

dipole field. At the Earth's surface, the intensity of the main field varies from roughly 24 000 to 66 000 nanoteslas (nT), from the magnetic equator to near the geomagnetic poles, respectively. (The term *geomagnetic* is generally reserved for a mathematical dipole model of the main field, whereas the term *magnetic* usually represents local field conditions and includes the non-dipole field and the effect of geological sources.) Observations of changing magnetic declination made over a period of 50 years in London convinced H. Gellibrand (*c.* 1634) that the main field changes not only in space, but also over time. These temporal changes in the characteristics of the main field (for example, intensity, inclination, and declination) are known as the *secular variation* (or time variation) of the main field. The Earth's rotation plays a significant role in maintaining the orientation of the geomagnetic dipole close to the axis of rotation (see *magnetic field (origin of the Earth's internal field)*). In the past 400 years, based on historical magnetic charts, the dipole axis has moved westward at a rate of about 0.80° per year. However, the most remarkable westward drift is displayed by some features of the non-dipole field. Historical magnetic charts and archaeomagnetic data (especially the archaeomagnetic inclination data that systematically deviate from that of the dipole field) going back about 2000 years show that some features of the non-dipole part of the field move westward at rates of between 0.3° and 0.6° per year. The remaining features of the non-dipole field are more stationary. However, both parts of the non-dipole field change in amplitude over time. The dipole field also varies in strength with time. The strength of the ancient field is inferred from the use of various intensity measurement techniques on archaeomagnetic samples (such as pottery, bricks, and kilns) and sequentially formed rock units (for example, lake sediments). A large number of such studies show that the strength of the Earth's field for the past 10 000 years has varied from about 0.8 to 1.5 times the strength of the present geomagnetic field. (The dipole moment is presently about 8×10^{22} A m^2.) However, measurements going back further to roughly 50 000 years exhibit a much smaller value for the field strength (about half the present value), excluding perhaps a few episodes when the magnetic field was extremely strong (reaching about 5–6 times its present strength). In general, all these different types of secular variations, their rates and their peculiarities, constrain models of fluid motions in the outer core and the generation of the Earth's magnetic field.

DHANANJAY RAVAT

Further reading

Merrill, R. T. and McElhinny, M. W. (1983) *The Earth's magnetic field: its history, origin and planetary perspective.* Academic Press, London.

geomagnetism: polarity reversals
In contrast to the periodicity displayed by the Sun's magnetic field in its reversals (which correspond with a sunspot cycle of approximately 11 years), the Earth's magnetic field appears to reverse its polarity chaotically. The two states of the Earth's magnetic field are termed 'normal' (corresponding to the present direction of the field) and 'reverse' (the direction opposite to the present field) and their alignment close to the Earth's axis of rotation testifies to the influence that rotation has on generation of the field (see *magnetic field (origin of the Earth's internal field)*). The interval between reversals can be as short as 100 000 years (the average is about 500 000 years), but there have been long periods of tens of millions of years in both polarity states without any reversals (for example, Cretaceous normal and Permo-Carboniferous reverse superchrons, i.e. major time intervals). Records of the direction of the past magnetic field preserved in oceanic sediments indicate that a reversal of polarity takes roughly 1000–10 000 years. However, Coe and his co-workers have found evidence from cooling lavas that at least some reversals might have occurred within a matter of weeks. In addition to reversals, some lake sediments and lavas record a significant change with time in the orientation of the Earth's dipolar magnetic field. These large changes in the orientation of the Earth's field are called *geomagnetic excursions*; it is possible that these could be aborted reversals.

Past behaviour of the Earth's magnetic field can only be traced through the ancient characteristics of the field preserved in rocks. It has been known since the middle of nineteenth century, through the work of Melloni, that volcanic rocks can acquire permanent magnetization in the direction of the Earth's ambient magnetic field. But the possibility that the Earth's magnetic field may have reversed its direction at various times in the geological past was tested by David and by Brunhes (by studies of clays baked by lavas) and by Mercanton (by a global sampling test on the reversed directions) only at the beginning of the twentieth century. In the 1920s, Matuyama calculated the locations of geomagnetic poles from careful magnetic measurements made on a number of dated volcanic basalt samples, and demonstrated that the field may have reversed as recently as the Pleistocene. As palaeomagnetic evidence for the reversals and their timing was slowly building in the 1950s through the work of Runcorn, Irving, Rutten, and Cox, it had to survive the blow of the Nagata–Akimoto–Uyeda discovery of mineralogical self-reversal, displayed by the Mt. Haruna dacite, and Néel's theoretical explanations substantiating self-reversals. (Self-reversal does in fact occur in rocks but is a rare phenomenon.) At about this time, theoreticians working on models of the origin of the Earth's magnetic field were also able to simulate reversals with dynamo models (see *magnetic field.*) Also, in the context of oceanic research, R. S. Dietz and H. Hess were synthesizing ideas about the evolution of ocean basins through the process of sea-floor spreading (see *sea-floor spreading*). Thus, the time was ripe for a far-reaching idea that revolutionized the Earth sciences.

As shipborne and airborne magnetic anomaly surveys were being compiled during the 1950s and 1960s by the Scripps Institute of Oceanography in the eastern Pacific and the Geological Survey of Canada in the northern Atlantic, respect-

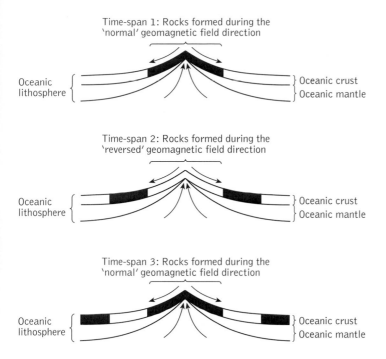

Time-span 1: Rocks formed during the 'normal' geomagnetic field direction

Oceanic lithosphere { Oceanic crust / Oceanic mantle

Time-span 2: Rocks formed during the 'reversed' geomagnetic field direction

Oceanic lithosphere { Oceanic crust / Oceanic mantle

Time-span 3: Rocks formed during the 'normal' geomagnetic field direction

Oceanic lithosphere { Oceanic crust / Oceanic mantle

Fig. 1. Schematic diagram showing how magnetic stripes of the mid-oceanic ridges can be explained by the process of sea-floor spreading and reversals of the Earth's magnetic field. See text for details.

ively, it became obvious that the magnetic make-up of ocean floors was different from that of the continents: the oceans had alternating tracts of magnetic highs and lows paralleling their ridge axes. F. Vine and D. H. Matthews of Cambridge University in England were at the same time interpreting similar magnetic anomalies in the Indian Ocean. They, and L. Morley (a Canadian geomagnetist working independently) synthesized, through the process of sea-floor spreading, the evidence related to reversals of the Earth's magnetic field and the stripes of magnetic highs and lows on the ocean floor. Their explanation was remarkably simple: they proposed that the magnetic pattern of the ocean floor was derived from processes occurring at the mid-ocean ridges. The new volcanic rocks at the ridges acquire a permanent magnetization in the ambient direction of the Earth's magnetic field as they cool through their Curie temperature (see *Curie temperature (Curie isotherm)*), and the process of sea-floor spreading splits apart these rocks at the ridges and moves them in opposite directions away from the ridges (Fig. 1). Thus, the mid-ocean ridges are like magnetic heads of a tape recorder, recording the ambient Earth's magnetic field and its reversals in the new oceanic crust and spooling it out toward the older parts of the ocean. Not only was this synthesis more plausible than the other alternatives that could explain the magnetic stripes on the ocean floor, but it could also predict the kind of new information one might look for to prove the model by drilling into areas of adjacent magnetic stripes—namely, the rock

types, their geochemical ages, their magnetization directions, and, of course, the data had to maintain symmetry across ridges. These ideas proved remarkably contagious, bringing about the plate tectonics revolution.   DHANANJAY RAVAT

Further reading

Cox, A. (1973) *Plate tectonics and geomagnetic reversals.* W. H. Freeman, San Francisco.

Merrill, R. T. and McElhinny, M. W. (1983) *The Earth's magnetic field; its history, origin and planetary perspective.* Academic Press, London.

geomatics *Geomatics* is a term that encompasses the various scientific disciplines relating to the collection, analysis, management, and presentation of spatial data from the Earth. These data may be related to environmental, physical, geographical, social, or engineering phenomena; one key aspect of geomatics as opposed to the former term, surveying, is that geomatics spans many subject areas. Alternatively, the positional information may be viewed as an end in itself, as in the case of vehicle navigation. The means of data collection include space geodesy, terrestrial land surveying, mining and engineering surveying, geophysical surveying, hydrography, remote sensing, and photogrammetry. Data management involves not only the techniques necessary to process and interpret these spatial datasets individually, but also the construction of computer databases to assimilate several coordinated datasets (Geographical Information Systems or GIS), and the development of algorithms to extract useful information from the GIS. The data can be presented in the form of traditional cartography or as a statistical presentation, but are nowadays commonly interactive and computer-based. In this case, the relationships between differing spatial datasets can be explored to the full.   PETER CLARKE

geomicrobiology
Geomicrobiology is the study of micro-organisms as agents of geochemical change. Geochemical transformations control the chemical composition of the Earth, its minerals, rocks, and atmosphere, and hence its climate. Near the Earth's surface micro-organisms drive most geochemical transformations, and they thus have a fundamental effect on the surface chemistry and geology of our planet. These biogeochemical processes drive the cycles of elements, in which the near-surface biological cycles rapidly turn over small reservoirs and interact with the much larger reservoirs of the geosphere, which turnover very slowly (about every 300 million years). This interaction has lead to the concept of two separate spheres, the *biosphere* and the *geosphere*. In reality, there is great overlap between these systems and their processes. As this overlap is largely concerned with micro-organisms, this has lead to the development of geomicrobiology. For example, micro-organisms, and particularly bacteria, can exist at considerable depths (kilometres) in the geosphere; geothermal activity can provide energy for microbial populations that are independent of photosynthesis and hence of the surface biosphere; microbes can both produce and consume minerals; subsurface bacterial activity can be enhanced in petroleum reservoirs and ore deposits; and micro-organisms have been responsible for major changes in near-surface chemistry during the development of the Earth.

Evolution of bacteria: stromatolites, photosynthesis

Bacteria were the first organisms to evolve and were the sole inhabitants of out planet for 70 per cent of the history of life on Earth. They evolved some 3.8 million years (Ga) ago, and there is clear fossil evidence for diverse cellular forms by 3.5 Ga. At the same time, significant amounts of isotopically 'light', sedimentary organic carbon were formed and existed at concentrations similar to those that occur today. The biomass of these primitive bacterial communities was thus comparable to that of modern-day ecosystems. Complex microbial mat communities soon evolved which interacted with minerals by trapping sediment within layers of bacterial filaments and precipitating calcium carbonate. This activity resulted in the formation of characteristic laminated mound-shaped structures: *stromatolites*, which are now well preserved as rocks. These structures grew and coalesced into large reefs in shallow coastal lagoons and intertidal zones. By 2.5 Ga stromatolites were everywhere: in Australia, North America, Spitsbergen, and Africa. They are now rare, but can still be found in the warm, shallow waters of the Gulf of California and Western Australia and in hot springs. Here, a combination of warmth and high salinity restricts burrowing and grazing by metazoans. The activity of these organisms, which are widespread in other sedimentary environments, disturbs the bacterial mat community, preventing the formation of stromatolites.

Although ancient stromatolites probably contained photosynthetic bacteria, these would have been very different from those in modern stromatolites. The early Earth was then still anoxic (contained no oxygen), and hence the photosynthetic bacteria would have to have been anoxygenic (neither requiring nor producing oxygen). Anoxygenic photosynthetic bacteria still exist today. They use compounds such as hydrogen sulphide or elements such as sulphur and ferrous iron (Fe II), which would have been common on the early Earth, instead of water (H_2O) for photosynthesis. They also readily form microbial mats with other bacteria, producing an interacting community (Fig. 1).

In surface sediments anoxygenic photosynthetic bacteria live on top of a layer of anaerobic bacteria (sulphate-reducing bacteria) which bring about the decay of organic matter, producing the hydrogen sulphide the anoxygenic photosynthetic bacteria require. In turn the anoxygenic photosynthetic bacteria provide organic matter, such as dead cells, and sulphate for the anaerobic bacteria. As a result, both carbon and sulphur are cycled between these two types of bacteria (Fig. 1). This demonstrates the ability of coupled bacterial processes to drive chemical cycles and provides a possible model for an ancient biogeochemical sulphur and carbon cycle.

About 2.7 to 2.2 Ga ago, some anoxygenic photosynthetic bacteria, probably a cyanobacteria (also called blue-green algae, but they are bacteria), developed the ability to use water (H_2O) for its reducing power in photosynthesis. This is a more complicated form of photosynthesis that involves another photosystem, but it enabled these bacteria to gain access to an almost limitless supply of electrons from water,

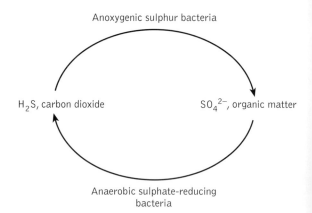

Anoxygenic sulphur bacteria

H_2S, carbon dioxide SO_4^{2-}, organic matter

Anaerobic sulphate-reducing bacteria

Fig. 1. Simple sulphur and carbon cycle between two anaerobic microbial mat bacteria.

and hence these organisms proliferated. However, their waste product, oxygen, was extremely reactive in the anoxic and reducing early Earth and was rapidly removed in the oxidation of compounds such as hydrogen sulphide, ferrous iron (Fe II), and other reduced metals. Huge amounts of reduced compounds were oxidized by oxygen before oxygen concentrations began to increase in the atmosphere around 2 Ga ago. This delay in the oxygenation of the atmosphere was very fortunate for the evolution of higher forms of life, since it provided time for anaerobic bacteria to evolve biochemical mechanisms to cope with and then exploit toxic oxygen. This evolutionary development resulted in the first aerobic organisms.

Many bacteria were not able to adapt to oxygen and are still killed even by traces of it. These *obligate anaerobes* inhabit the wide range of anoxic environments which still exist: subsurface soils and sediments, waterlogged soils, some lakes, fjords, and so on. In addition, many aerobic bacteria are so efficient that they rapidly remove oxygen, creating anaerobic micro-habitats just a few millimetres away from a fully oxygenated environment (for example, within soil crumbs, biofilms, and bacterial colonies). Increased atmospheric oxygen also made possible the formation of an ozone shield that filtered out harmful ultraviolet (UV) radiation, allowing the further spread of oxygenic photosynthetic bacteria. It has been suggested that the consequent rapid increases in oxygen triggered the Cambrian 'explosion' in metazoans (540–510 Ga). This event also provided new habitats for bacteria, both on these organisms and inside them; for anaerobes, these new habitats included the digestive systems of metazoans.

Bacteria and ore deposits: 'black smokers'

Bacteria have thus had a profound effect on the development of the surface chemistry of the Earth, and they continue to do so by driving the biogeochemical cycles of elements. In addition, bacteria also have a profound effect on the formation of minerals and rocks. For example, the oxidation of ferrous iron (Fe II) by bacterially produced oxygen, over a period of a billion years, resulted in extensive banded iron formations (BIFs), which account for 90 per cent of the world's iron reserves. The majority of commercial sulphur deposits are similarly the result of bacterial sulphate reduction (Fig. 1).

Other extensive ore deposits are formed by hydrothermal activity (for example, copper, zinc, iron, and even gold) at mid-ocean ridges where new ocean crust is formed. The cooling crust cracks and draws down sea water, which reacts with hot basaltic rocks, extracting metals into hydrothermal fluids. These reduced hydrothermal fluids, at 270–380 °C, discharge into oxic (oxygen-containing) sea water at 4 °C and precipitate metals, resulting in black plumes called 'smokers'. In the presence of oxygen the reduced metals and sulphides provide a unique energy source for bacteria. Dense bacterial populations (10^8–10^9 cells per millilitre) exist close to the vents and even in the mineral chimneys. It is clearly to their advantage to be as close to the smoker as possible, and *hyperthermophiles* (bacteria that grow at temperatures above 80 °C), exist to exploit this unique energy source. *Pyrolobus fumarii* is the current record holder, with maximum and minimum temperatures for growth of 113 °C and 90 °C respectively. Its energy source is hydrogen from the hydrothermal fluid.

The high bacterial populations around the vents form the base of a unique ecosystem of suspension-feeding benthic organisms. These represent biological oases in the deep sea, where large organisms are normally scarce. This unique ecosystem obtains its energy from a geothermal source rather than from sunlight, and bacteria are here the primary producers of organic matter. This food web is taken to the extreme in the *Pogonophora* tube worms that are found at some vents. These organisms have no mouth, gut, or anus, but have a large, spongy tissue, the trophosome, which represents half the worm. The trophosome is filled with bacteria (*chemolithoautotrophic bacteria*) that use hydrogen sulphide from the vent and oxygen from sea water, both supplied by the worm, for energy. In this symbiotic relationship the worm obtains its energy from the organic matter produced by the bacteria, including dead bacterial cells. This is a very short and surprising food chain:

$$H_2S + CO_2 \rightarrow \text{Bacterial carbon} \rightarrow \text{Worm}$$

There are nevertheless similarities here to animals with which we are more familiar: ruminants, such as cows and sheep, which do not eat grass directly but bacterial products. In these animals large microbial populations occur in a special digestive organ, the rumen, and the microbes break down the cellulose in grass. The ruminant absorbs the cellulose breakdown products for energy and digests bacterial cells for essential proteins and vitamins.

Bacteria in the subsurface region

Hyperthermophiles (both Bacteria and Archaea) are the most ancient group of bacteria on Earth. Molecular phylogenetic analysis demonstrates that they cluster at the base of the universal tree of life. This suggests that the first bacteria were hyperthermophiles and that life may have evolved in hydrothermal systems, including hot springs. Bacteria, as well as occurring in hydrothermal vent fluids and chimneys, are present in the surrounding rocks and sediments, where they may be stimulated by subsurface lateral flow of hydrothermal fluids and sea water. Hence, there may be even larger bacterial populations beneath vent fields, possibly living at even higher temperatures. Research has certainly demonstrated the presence of a deep bacterial biosphere in other environments, such as aquifers, Cretaceous shales, terrestrial basalts and granites, marine sediments and the rocks beneath them, gas hydrate deposits and oil reservoirs, salt and mineral mines, and glaciers. Bacteria have been found to about 3 km depth, and will probably be found even deeper, and as old as Permian in age (250–90 Ma) in salt mines. The size of their populations could be considerable, and there is clear evidence that they are active, and not just surviving, in these deep formations. For example, in a North Sea oil reservoir 3–16 kg of hyperthermophile cells are produced each day with the production fluid. It has been estimated that over 90 per cent of all bacteria on Earth live in the subsurface region. This has led to speculation that bacteria may originally have evolved in the subsurface. In this region they would have been safe from late meteoritic and cometary impacts after the accretion of the Earth. (These impacts could have heated the surface to sterilizing temperatures.) These subsurface bacterial populations would subsequently have inoculated the surface, just as hyperthermophiles from oil reservoirs do today.

Where, however, do these subsurface bacteria obtain energy when they are so far from the surface and from the photosynthetic organisms that have been thought to be the base for all life on Earth? Hydrothermal and geothermal activity is one source already discussed above. In addition, some bacteria can probably survive on buried organic matter from the surface for long periods of time, by growing only very slowly (for example, one division every 1000 to 2000 years); active bacterial populations are known to occur in Cretaceous shales laid down approximately 90 Ma ago. As organic matter is buried even deeper it will be heated by the thermal gradient of the Earth (c. 30 °C per km), and under some circumstances the organic matter will be made so reactive that it breaks down to methane gas and other hydrocarbons forming fossil fuels. There is some evidence that this may also result in the stimulation of deep bacteria. So far as temperature is concerned, there is no reason why bacteria should not be active even into oil 'window' (100–150 °C, the temperature at which petroleum is formed). The involvement of bacteria in deep oil formation would certainly explain the occurrence of bacteria in some oil reservoirs. Bacteria can also degrade components of crude oil, even anaerobically, and under certain conditions they produce methane (CH_4), a valuable natural gas, in a processes analogous to the physical 'cracking' of hydrocarbons.

A deep bacterial energy source that is completely independent of photosynthesis, since neither organic matter or oxygen is required, occurs at a depth of 1.5 km in the Columbia River basalts, which are igneous rocks and contain little organic matter. Below the surface, conditions are anaerobic, but active microbial populations are present which produce methane in such quantities that natural gas was commercially exploited early in the twentieth century. The energy for this methane formation comes from rock-weathering reactions that form hydrogen (H_2), which is then used by autotrophic bacteria to produce new cells and also, as a waste product, methane gas. An interacting bacterial community is present called 'SLIME' (Subsurface Chemo Lithoautotrophic Microbial Ecosystem) that might be a model for a deep biosphere on other planets, without the necessity for surface life. For example, basalt, liquid water, and bicarbonate are all believed to be present in the subsurface on Mars.

Basalt weathering:
$$FeO + H_2O \rightarrow H_2 + FeO_{3/2}$$
SLIME Community:

Methanogenesis	$4H_2 + CO_2 \rightarrow CH_4 + 2H_2O$
Acetogenesis	$4H_2 + 2HCO_3 \rightarrow CH_3COOH + 4H_2O$
Sulphate reduction	$4H_2 + SO_4^{2-} + H^- \rightarrow HS^- + 4H_2O$

Similar weathering reactions might also be involved in providing energy for bacteria in oceanic basalts, which occur beneath sedimentary layers. With an estimated volume of 10^{18} m^3, oceanic basalts represent the largest habitat on Earth. Bacteria have been found at a depth of 237 m in the basaltic rock layer, where they are associated with weathering channels and pits.

Bacteria and minerals

Minerals are of great importance to micro-organisms. They are (a) substrates to which micro-organisms can attach themselves; (b) they are the origin of many dissolved constituents essential for the metabolism of micro-organisms (NH_4^+, K^+, Mg^{2+}, Co^{2+}, Cu^{2+}, Fe^{3+}, Mn^{2+}, Ni^{2+}, Zn^{2+}, and other trace cations, the oxyanions PO_4^{3-}, SO_4^{2-}, and less abundant anions); and (c) for lithoautotrophs (rock-eaters), they are the source of their energy. This, combined with the need to cope with a range of other metals that can be toxic (such as Ag^+, Cd^{2+}, Hg^{2+}, Pb^{2+}), demonstrates the extensive interaction that microbes can have with cations and anions and the minerals from which they are derived. As well as colonizing and dissolving minerals, micro-organisms can also produce minerals such as carbonates and dolomites, gypsum, metal sulphides, uraninite, apatite, siderite, iron oxide, magnetite, manganese oxides (birnessite, vernadite, buserite), silica, and mixed iron–aluminium silicates.

The formation of these minerals by micro-organisms may be direct or microbially controlled, such as the production of magnetite during the respiratory reduction of ferric iron (III) by sulphate-reducing bacteria; or indirect and microbially induced, such as the formation of pyrite as a consequence of bacterial sulphate reduction producing hydrogen sulphide (H_2S) in the presence of iron. A characteristic of bacterially produced minerals is that they are extremely fine-grained. For the palaeontologist, this is particularly valuable in providing accurate replication during fossilization. The activity of anaerobic bacteria is particularly associated with mineral formation. Their activity has been associated with the rapid replacement of soft tissue by minerals (within weeks) in the formation of exceptionally preserved fossils. Hence, paradoxically, the formation of exceptional fossils requires elevated and not restricted bacterial activity.

A particularly impressive bacterially controlled mineral is produced by *magnetotactic* bacteria, which can detect magnetic fields. These motile bacteria are ubiquitous in aquatic habitats. Their highest concentrations are at the sediment–water interface, where the transition between oxic and anoxic conditions occurs. They produce either magnetite (Fe_3O_4) or greigite

(Fe_3S_4) in a magnetosome (a magnetic organ), which enables them to align themselves with the geomagnetic field so that they can swim towards their sediment habitat. When magneto-tactic bacteria die, their magnetosomes can be preserved in the sediment and provide an excellent paleomagnetic record.

In contrast, microbial mineral dissolution drives rock weathering. Microbial weathering rates are particularly elevated when the mineral is being used as an energy source, because relatively little energy is usually obtained from the reaction and large amounts of the mineral thus have to be oxidized. An important example of this type of weathering is the oxidation of metal sulphides (for example, pyrite (iron), chalcopyrite (iron and copper), chalcocite (copper), arsenopyrite (arsenic and iron), galena (lead), and sphalerite (zinc)). The reaction is normally an aerobic one (Me represents a metal):

$$MeS_2 + 15/4O_2 + 7/2H_2O \rightarrow Me(OH)_3 + 2H_2SO_4$$

This is one of the most acid-producing biologically mediated processes known. It can result in the local formation of very low pH values (around pH 2). Bacteria like *Thiobacillus ferro-oxidans* can increase the rate of this reaction by up to a million times. They grow best in these acid conditions and are thus well adapted to their exotic lifestyle. These 'microbial miners' are used commercially to extract metals from low-grade ores of metals such as copper, uranium, and gold, for which normal processing would be uneconomic. However, these acid producers can also cause considerable damage to underground concrete pipes and other structures. In addition, sulphate-reducing bacteria of the SLIME community (see above) that use hydrogen cause corrosion of metals.

It is clear that bacterial processes extend into the geosphere to a much greater extent than had previously been considered possible. Research in this area has uncovered unique deep biospheres which are fuelled, not by sunlight, but by geosphere processes. Similarly, several subsurface processes previously thought to be abiological have been shown to involve bacteria. These results have profound implications for our understanding of life on Earth, and possibly on other planets.

R. JOHN PARKES

Further reading

Banfield, J. F. and Nealson, K. H. (1997) Geomicrobiology: interactions between microbes and minerals. *Reviews in Mineralogy*, 35.

geomorphological equilibrium
Geomorphological equilibrium has been defined in a variety of often contradictory ways, but they generally boil down to one or both of two key concepts: invariance and adjustment.

Invariant equilibria occur where landforms or Earth-surface processes either do not change (static or stationary states), or where a generally constant form is maintained even though the geomorphological system may change (steady states). Examples might include a stream which has cut down to its base level, so that no further entrenchment will occur. A steady-state example would be a concavo-convex slope profile with a concave lower portion and a convex upper slope typical of humid climates. While erosion, deposition, and mass movement continue to operate, the basic slope form is maintained.

Adjustment forms of geomorphological equilibrium exist where landforms or landscapes have had sufficient time to adjust to environmental constraints (for instance, climate, tectonic movements, sea-level change) so that the landforms and surface processes are characteristic of, or at least reflect, those constraints. Thus a river which has been cutting down during a low stand of sea level would be considered to be in equilibrium with a subsequent rise in sea level if cutting down ceased and aggradation commenced. Or, an equilibrium soil landscape in a desert would include pedogeomorphic features characteristic of arid environments (such as desert pavements or accumulations of salts and carbonates) rather than features reflecting previous, wetter climates.

In geomorphology there is no general agreement, and considerable controversy, over the precise definition of geomorphological equilibrium, the extent to which landforms and surface processes tend toward equilibria, and the role of equilibrium states in landscape evolution. Equilibrium concepts first came into wide use in geomorphology as an alternative to the cyclic theories of landscape evolution of William Morris Davis. Although an antithesis between 'historical' and 'equilibrium' approaches to landform study may have been more apparent than real, such was the general perception by Earth scientists through much of the late twentieth century. J. T. Hack of the United States Geological Survey promulgated the concept of dynamic equilibrium as a philosophical basis for landscape analysis. The landscape and the processes moulding it are part of an open system in a steady state where every slope and every form is adjusted to every other. Thus, in Hack's conceptual model, landscapes tended toward an equilibrium state in which change was continual, but slope forms and relief were maintained over time. Because of disturbance, individual landforms and landscapes might not be in equilibrium but they could be expected to tend toward an equilibrium state.

Hack and others promoting dynamic, process-based approaches to geomorphology traced the origin of the dynamic equilibrium concept to G. K. Gilbert, a highly respected geologist and physical geographer who worked in the American West for the United States Geological Survey and whose best-known works were published in the late 1800s. Gilbert described equilibrium as a balance of forces by which (for example) less resistant rocks are worn away more rapidly than harder rocks, which are left more prominent. The differentiation continues until an equilibrium is reached through the 'law of declivities' (slope gradients). When the ratio of erosive action as dependent on slope gradients becomes equal to the ratio of resistances as dependent on rock characteristics, there is 'equality of action'.

Gilbert's equilibrium concept is distinctly geomorphological in origin. It describes internal processes within a system, is scale-dependent and is focused on mass (rather than energy) flux, and embodies form. Unfortunately, later writings by others mixed this geomorphological concept with broadly analogous, but sometimes inappropriate, concepts of equilibrium derived from physics, mathematics, and general systems theory. The confusion was exacerbated by the fact that Hack's influential 1960 paper used 'dynamic equilibrium' to describe a steady-state equilibrium.

Some geomorphologists have more recently advocated abandoning the equilibrium concept, or at least its terminology. Others have argued for more precise and restricted definitions. Reconsideration of the equilibrium concept and its definition in geomorphology reveals four emerging trends:

(1) Input–output equilibria. A geomorphological relationship is in equilibrium if there is a constant relationship between mass and energy inputs and outputs or form, across the entire range of the inputs, within some consensual degree of variability. Equilibrium would thus be defined quantitatively, and could be a property only of specific mass or energy fluxes, not of whole geomorphological systems.

(2) Mass–flux equilibria. Equilibrium exists if the amount of material removed from a unit of the land surface per unit time is equal to the amount supplied during the same period. *Dynamic equilibrium* applies only to the relationship between process rates. *Steady-state equilibrium* exists when the relationship is time-independent.

(3) *Ad hoc* definitions. Some geomorphologists argue that a single, generally agreed definition is unlikely, but that equilibrium concepts are too useful and ingrained to be discarded. Therefore, the term must be defined or described at each usage. For example, a geomorphologist referring to an equilibrium regolith thickness would take care to define it as the thickness which would exist when the rate of production of weathered debris is approximately equal to the rate of removal by erosion and mass wasting.

(4) Complex equilibria. The idea of a geomorphological system proceeding inexorably (though perhaps interrupted by perturbations) toward a single stable end state is too narrow to describe the rich variety of trends in landscape evolution. In particular, some landforms and landscapes are non-equilibrium and do not tend toward any particular stable end-state. This is distinguished from disequilibrium systems that are tending toward, but have not yet reached, a steady state. Other geomorphological systems may have multiple equilibria, rather than a single stable end-state, and those may be simple (e.g. a steady-state slope form) or more complex (e.g. a mosaic of erosional, depositional, and transportational zones). Finally, some geomorphological systems may be stable (equilibria being restored or maintained by, for example, negative feedback after perturbation) or unstable (disturbances tending to persist or grow, resulting in disequilibrium or non-equilibrium).

J. D. PHILLIPS

Further reading

Ahnert, F. (1994) Equilibrium, scale, and inheritance in geomorphology. *Geomorphology*, **11**, 125–40.

Gilbert, G. K. (1877) *Report on the geology of the Henry Mountains*. US Geographical and Geological Survey of the Rocky Mountain Region. US Government Printing Office, Washington, DC.

Howard, A. D. (1988) Equilibrium models in geomorphology. In Anderson, M. G. (ed.) *Modelling geomorphological systems*, pp. 49–72. John Wiley and Sons, Chichester.

Phillips, J. D. (1992) The end of equilibrium? *Geomorphology*, **5**, 195–201.

Thorn, C. E. and Welford, M. R. (1994) The equilibrium concept in geomorphology. *Annals of the Association of American Geographers*, **84**, 666–96.

Thornes, J. B. (1983) Evolutionary geomorphology. *Geography*, **68**, 225–35.

geomorphological systems The Earth's landscapes are typically characterized by the presence of many components with a dense network of linkages, such that each component affects, and is affected by, a number of the others. Some (reductionist) approaches to geomorphology cope with this complexity by disaggregating the web of interactions to isolate specific 'process–response' relationships. Other (holistic) approaches strive to deal directly with the web of interactions as a whole; hence the study of geomorphological systems.

Systems are sets of interconnected parts which function together as complex wholes. In *geomorphological (geomorphic) systems* the parts are landforms, Earth-surface processes, and environmental factors which influence or control processes and forms. The interconnections between the parts involve transformations, storage, flows, and cycles of mass and energy.

We may distinguish between systems-oriented concepts of, or approaches to, geomorphology, and the study of geomorphological systems themselves. The former is a perspective which recognizes the value of considering multiple components and interrelationships in landscapes to explain geomorphological phenomena, and is quite common. The latter is focused on a holistic understanding of system-level behaviour, and is less common.

A systems-oriented perspective on the study of Earth-surface processes and landforms is characterized by:

(1) concern with interactions among landscape elements in addition to the behaviour of individual elements;

(2) consideration of the mutual adjustments of landforms, surface processes, and environmental controls; and

(3) an emphasis on dynamic change rather than description, classification, or dating.

These characteristics are an implicit part of much of the dynamic, process-based geomorphology of recent decades.

The study of geomorphological systems subsumes the three elements of the systems perspective described above. It differs from the implicit, nearly pervasive systems perspective of mainstream process geomorphology in two major respects. First, understanding dynamic behaviours, interactions among landscape elements, and mutual adjustments is the primary goal of the geomorphologist, rather than aspects which are of interest only to the extent that they are helpful in understanding a particular feature or process. The concern with dynamics, interactions, and mutual adjustments is thus explicit rather than implicit. Secondly, the study of geomorphological systems is concerned with the evolution of whole systems rather than one or more system components. The approach is thus holistic, and disaggregation is seen as a tool for understanding system-level behaviour. This contrasts with reductionist process geomorphology, where geomorphological systems provide contexts or boundary conditions for the details of interest.

Explicit applications of systems theory also characterize the study of geomorphological systems. These fall into two general classes. Statistical–mechanical forms of systems theory, such as general systems theory, rely on an essentially probabilistic approach and assume that the number of elements is large and the network of relationships too complex to follow individually in detail. Non-linear dynamical systems theory assumes that the geomorphologist has at least a general idea of the functional relationships between system elements, typically expressed in the form of equations or interaction matrices.

A systems-oriented perspective in geomorphology can be traced at back at least to G. K. Gilbert's work with the United States Geological Survey in the 1870s. The study of geomorphological systems, including explicit applications of statistical–mechanical forms of systems theory, can be traced to Arthur Strahler of Columbia University, who in turn influenced Richard Chorley of Cambridge University. Strahler and Chorley, along with the Austrian Adrian Scheidegger, Luna B. Leopold, and Walter Langbein of the United States Geological Survey, argued for and demonstrated the applications of systems theory to geomorphological problems. While they had limited success in fostering formal applications of systems theory, this and related work between the early 1950s and early 1970s was instrumental in engendering the pervasive systems-oriented perspective in process geomorphology.

Geomorphological systems are open and dissipative. The former means that they exchange matter and energy with the environment; the latter that they are governed by irreversible processes which result in the dissipation of energy (generally in form of friction or turbulence). Thus a geomorphological system dissipates energy from external sources (such as sunlight, tectonic uplift, and precipitation) to maintain itself. This is sometimes called a far-from-equilibrium system, but in this case the reference is to thermodynamic equilibrium, which is quite different from geomorphological concepts of equilibrium (see *geomorphological equilibrium*). The conceptual model of open, dissipative geomorphic systems accords well with intuitive notions of landscape evolution. Geomorphological processes are indeed generally irreversible: weathered rocks do not reconstitute; landslides do not move back uphill; glacially carved mountains are not re-formed; water and sediment only rarely and locally move back upstream. Likewise, the vast majority of landscapes are far from the thermodynamic equilibrium of a featureless plain. Energy moving through geomorphological systems is dissipated to such an extent that energy fluxes are notoriously difficult to measure; mass fluxes are usually measured instead.

Many studies of geomorphological systems have focused on the nature of their equilibrium states (see *geomorphological equilibrium*) and the processes and evolutionary pathways by which landscapes and geomorphological systems tend towards equilibrium. Until relatively recently, the (often tacit) assumption was that there was a single steady state, adjusted to climate, tectonic activity, and other controls, towards which landscapes and landforms tended. One example is the concave-upward longitudinal profile of stream channels; deviations from this 'equilibrium' norm are ascribed to disturbances such as tectonic uplift, or to insufficient time for the stream to achieve equilibrium. More recently, however, studies of geomorphological systems have recognized that a monotonic trend towards a single stable equilibrium, or a situation where a system either is or is not in equilibrium, is one of only many possible modes of landscape evolution. Geomorphological systems may be *non-equilibrium* (they do not tend toward any particular stable steady/state) as well as *disequilibrium* (tending toward a stable end state, but not yet there) or equilibrium systems. Topographic evolution is sometimes observed to result in increasingly irregular and, complex terrain; this is one manifestation of a non-equilibrium geomorphological system. Further, there may be multiple equilibria for any given set of environmental controls; equilibrium states may be stable or unstable; and there may be many ways for a geomorphological system to achieve a given equilibrium. In a stream channel, for example, there are many different combinations of adjustments of width, depth, shape, roughness, and slope which can accommodate a given change in imposed flow, and numerous mechanisms by which any given combination can be achieved.

The study of thresholds and non-linearity has also been critical to the study of geomorphological systems. A *threshold* is the point at which a fundamental change in system response occurs: for example, at a critical threshold of shear stress resulting from the weight of ice accumulation, a glacier begins moving by plastic deformation. Considerable insight into system behaviour can be gained by identifying important thresholds. Major advances in the understanding of erosion, sediment transport, and deposition by water, air, and ice, and of slope stability and mass movements have come as a result of identifying critical thresholds of force or power versus resistance. Thresholds have also been examined at broader spatial and temporal scales. These are often in the form of rate comparisons, such as comparing the rate of sea-level rise with the rate of sediment infilling of an estuary, or the rate of uplift with the rate of denudation of a mountain range.

Thresholds are ubiquitous in geomorphology; their presence indicates that geomorphic systems are non-linear: the outputs of matter and/or energy are not proportional to the inputs across the entire range of inputs. This has led to increasing applications of *non-linear dynamical systems* (NDS) theory to geomorphology. NDS theory in geomorphology treats geomorphic systems as complex, non-linear, dynamic systems which may be far from thermodynamic equilibrium and which may exhibit self-organization (for example, the formation of evenly spaced, nearly uniform ripples on sand dunes or stream beds; or of patterned ground in periglacial landscapes). NDS theory also more easily accommodates non-equilibrium geomorphological systems. Further, some geomorphological systems are characterized by disproportionate responses, whereby small perturbations or minor differences in initial conditions lead to disproportionately larger responses or changes as the system evolves. Topographic irregularities in hill slopes may grow larger, for example, rather than being smoothed out; or minuscule variations in rock or sediment may give rise to an increasingly variable soil cover (in spatial terms) over time. NDS approaches are well equipped to explain phenomena of this kind.

Another common theme in geomorphological systems is the response to disturbance, particularly the role of mutual adjustments and feedbacks. *Mutual adjustments* are particularly important in relationships between form and process, as in the interactions between waves, which shape beaches, and the beaches themselves, which help determine the characteristics of the incoming waves. *Feedbacks* are the mechanisms, or links between system components, by which a geomorphic system responds to changes or disturbances. For example, uplift increases elevation differentials, which in turn results in steeper slopes. Erosion rates increase in response to the steeper slopes, with a tendency toward redistributing mass from higher to lower points, reducing elevation differences. Feedbacks may be negative, as in the previous example, and work to reduce or mitigate changes in the system. They may also be positive, and reinforce change. Enlargement of a nivation hollow (a depression created by processes around snow patches) allows the hollow to capture more snow the following winter, and thus further increase the size of the hollow size, to give one example. Most geomorphological systems exhibit numerous feedback links between system components, both positive and negative.

Few geomorphologists would argue in favour of the uniqueness of specific landforms and landscapes, or the place- and time-dependence of many geomorphic phenomena. However, the study of geomorphological systems also involves the search for universality. This is based on the belief that there are widely applicable laws governing the responses and evolution of broad classes of Earth-surface systems. An example is the use of maximum and minimum values of energy to explain geomorphic evolution, such as the argument that stream channels are adjusted so as to minimize energy dissipation rates of flowing water. An important inspiration for this work is the observation of similar forms in a variety of environments. The alternating pattern of 'noses' and 'hollows' along the sides of valleys; crescentic-shaped sand dunes; and the veins-in-a-leaf pattern of stream channel networks are examples. All occur in a variety of climatic and geological settings, and over a range of spatial scales, suggesting common governing principles which are not place- or time-dependent.

Traditionally, most studies of geomorphological systems have focused on landforms and surface processes, and have treated other environmental factors as part of the environment, context, or source of inputs to the geomorphological system. More recently, there has been increasing interest in broader-scale studies, explicitly linking landforms and surface processes with hydrological, climatological, and biological phenomena. A traditional study of soil erosion in an overgrazed grassland would have treated the erosion as the response of a geomorphic system to a perturbation, and would have focused on the interactions and mutual adjustments between run-off, soil properties, topography, and sediment fluxes. Today, such a study would be more likely to include the geomorphic variables together with ecological and climatological variables in a single model. The study of geomorphological systems is thus likely to become increasingly intertwined with, and indistinct from, the study of hydrological, pedological, and climate systems and with ecosystems. Accordingly, more general terms such as Earth-surface systems are increasingly coming into use.   J. D. PHILLIPS

Further reading

Chorley, R. J. and Kennedy, B. A. (1971) *Physical geography: a systems approach.* Prentice Hall, London.

Coates, D. and Vitek, J. (eds) (1980) *Thresholds in geomorphology.* Allen and Unwin, London.

Huggett, R. J. (1991) *Climate, Earth processes, and Earth history.* Springer, Berlin.

Phillips, J. D. and Renwick, W. H. (1992) *Geomorphic systems.* Elsevier, Amsterdam.

geomorphology People had speculated on the forces and mechanisms that had created the natural landscape around them long before the term 'geomorphology' was introduced in the 1880s to describe this preoccupation. Essentially, geomorphology can be considered as the science concerned with the form of the land surface and the processes that create it. Although the term is derived from the Greek terms *geo*, meaning 'Earth', *morph*, meaning 'form', and *logos*, meaning 'discourse', recent geomorphological research has extended the scope of the discipline beyond its traditional confines, taking in the study of submarine features (see *deep-sea geomorphology*), and, more controversially, of planetary landscapes (see *planetary geomorphology*). Such widening scope reflects the recognition that landscapes of any type provide testing grounds for making inferences and observations about fundamental natural processes. In this way, geomorphology has the potential, as yet only partially realized, of making significant contributions to other areas of knowledge. The location of the land surface at the interface of the Earth's lithosphere, atmosphere, hydrosphere, and biosphere means that geomorphology is closely related to a range of other disciplines. It depends on the application of basic principles of physics, chemistry, biology, and mathematics to natural systems, and overlaps with allied research fields such as geology, hydrology, glaciology, engineering, statistics, climatology, soil science, computer modelling, aerodynamics, and even space science.

An important current area of focus for geomorphological research is the relationship between landforms and the processes currently acting on them. But many landforms cannot be fully explained by the nature and intensity of present-day geomorphic processes, and so an important aspect of geomorphology is to consider how past events may have helped shape the landscape. To paraphase Stanley Schumm, one of the twentieth century's most influential geomorphologists, the landscape is a physical system with a history. To appreciate it, geomorphologists must develop both general physical theories of landscape development as well as more specific theories concerning the historical evolution of particular landscapes. In this last respect, then, geomorphology is also very much a historical science, and in the early years of geomorphological research, during the late nineteenth and early twentieth century, it was this aspect that dominated the discipline, led by W. M. Davis's long-term and largely deductive models of landform development. During this period, much of geomorphologists' efforts were directed at classifying the surface of the Earth into various descriptive units. Interest in these largely descriptive accounts of regional landscapes waned mid-way through the twentieth century, and, by the 1950s and 1960s geomorphologists began instead to view landscapes as complex, dynamic systems. Ushering in the modern era of 'process geomorphology', geomorphologists sought to collect empirical evidence as to the nature and rate of landscape change through time, generally based on field measurement of geomorphic processes. So what of the future directions in geomorphology?

Currently, geomorphology spans a greater diversity of themes than at any time in its history, driven by technological advances that provide an ever-increasing array of research techniques and by emerging interdisciplinary collaborations that offer new and exciting challenges. While current major research trends will undoubtedly continue, advanced by improved empirical datasets and expanded theories, it is likely that future geomorphological studies will increasingly address important societal issues. One of the main challenges is to assess the impact of future global environmental change on landforms and on the landscape. Global warming, as a consequence of the increasing concentration of certain gases

(carbon dioxide, methane, CFCs, etc.) in the atmosphere, is likely to have direct and indirect impact on landforms and on the nature and intensity of processes involved in their formation. As the British geomorphologist Andrew Goudie points out, to deal effectively with global changes like greenhouse warming requires more research into rates and mechanisms of likely geomorphological response; the frequency of tropical cyclones, the speed and degree of permafrost degradation, the response of glaciers and ice caps, the extent of sea-level rise, the reaction of beaches to rising sea levels, and the state of wetlands, deltas, and coral reefs. Such geomorphological phenomena are dynamic and highly variable physical systems in which changes are continually occurring, and long-term studies are therefore needed to characterize their inherent variability and assess their sensitivity to change. With humans becoming one of the Earth's major modifiers, human-altered landscape systems are likely to be a prime focus of future research, tackling environmental issues such as deforestation and pollution. In turn, land-surface processes pose a suite of threats to human resources, through the extreme action of soil erosion, slope instability, and river and coastal flooding, and with humans increasingly impinging on the natural environment, hazards research will be of critical importance in most fields of geomorphology.

At the same time that predictive models of short-term landform change are increasingly being used to assess the likely landscape response of human activities, there are signs that the early interests of geomorphology are being rekindled. Thus, satellite imagery is allowing regional landscapes to be viewed at manageable scales, and new techniques, such as the application of cosmogenic isotopes, are opening up the possibility of quantifying rates of landscape processes on timescales of up to 100 million years or more. It is now recognized that significant components of the Earth's land surface are of considerable antiquity and that their survival is contrary to the expectation for conventional models of landscape development, necessitating a fresh look at explanations of landscape longevity. There are now geophysical models to underpin studies of long-term landscape evolution, and there is growing interest among the geophysics community in the potential of geomorphological studies to provide insights into tectonic mechanisms: hence the burgeoning research field of tectonic geomorphology (see *tectonic geomorphology*). The continual tectonic and geomorphological modification of the Earth's surface means that the occasional action of catastrophic processes, such as impact cratering, leaves little imprint. The more stable planetary landscapes of the Moon, Mars, and other bodies in the solar system offer an opportunity to examine the effect and incidence of these rare events in landform genesis. In short, the technological and methodological advances of the past few decades, many of them supported by the increasing application of computer modelling, mean that geomorphology at last has the promise of bridging the gap between surface processes and long-term landscape development.

IAIN S. STEWART

geomorphology, applied *see* APPLIED GEOMORPHOLOGY

geomorphology, fluvial *see* DEEP-SEA GEOMORPHOLOGY

geomorphology, global *see* GLOBAL GEOMORPHOLOGY

geomorphology, planetary *see* PLANETARY GEO-MORPHOLOGY

geomorphology, soil *see* SOIL GEOMORPHOLOGY

geophysics, electrical methods *see* ELECTRICAL TECHNIQUES IN GEOPHYSICS

geophysics, electromagnetic methods *see* ELECTROMAGNETIC METHODS IN APPLIED GEOPHYSICS

geophysics, exploration *see* EXPLORATION GEOPHYSICS

geophysics, history of *see* HISTORY OF GEOPHYSICS

geoscience in the media How much geoscience is in the news? Information on the level of coverage gained by different disciplines is hard to find, and much existing research is out of date. Calls from UK journalists received at the Media Resource Service run by the Novartis Foundation in London between 1985 (when the MRS was launched) and December 1998, broke down as follows (in percentages): health and medicine; 39.2; psychology and social sciences; 13.6; life sciences: 10.6; environment: 4.2; physical sciences: 5.6; industry: 3.1; space and military: 2.3; Earth sciences: 1.8; energy: 1.1; other: 18.1. (Source: Media Research Service)

The Novartis Foundation's MRS responds to unprompted inquiries from journalists. Experience of operating a similar system for putting journalists in touch with UK university experts has shown that the motivation for such enquiries comes from items that are in the news in any case, on which the journalist wants expert comment. This gives enquiries a distinct news-desk bias, greatly favouring direct human-interest subjects such as medicine, biology, and the social sciences.

Patterns of coverage have undoubtedly changed since 1989, when Anders Hansen (University of Leicester) analysed the UK national press and found that Earth science accounted for only 0.8 per cent of total coverage over its two-month sample. Interest in environmental change, which has brought increased interest in natural hazards of all kinds, including those within the realm of classical geology, has shifted the emphasis of science coverage. Also, the limits of Earth science have broadened out as classical geology (the science of things you can hit with a hammer) has become part of what is now seen as 'Earth system science'.

The general impression is that, taking in dinosaurs, earthquakes, volcanoes, and other natural hazards like tsunamis whose cause is geological, Earth science often commands much more than 0.8 per cent or 1.8 per cent of science coverage. For example, during the 1998 week-long meeting of the British Association Meeting (the UK equivalent of the Ameri-

can Association for the Advancement of Science), geological stories commanded about 30 per cent of all the science news generated (counting column centimetres). This would be a considerable proportion for any discipline, but it is vast when one realizes that Earth science made up only about 5 to 8 per cent of the total programme.

Science journalists are aware of the huge public interest in aspects of geoscience, and are eager to write about it *when it is available*. Outside the major *sciencefests* like the BAAS and AAAS, however, levels of coverage slide back because there are fewer people pushing the stories. Earth science is mostly small-scale and dispersed, whereas big science and medicine have bigger, more cash-rich installations with more money to devote to telling the world how worthwhile they are. As high consumers of public money, it is very much in their interests to do this.

Fortunately not all sciences are equal in press terms. Volcanoes and dinosaurs are inherently more accessible than subatomic particles. This is why geoscience punches above weight once a suitable story comes to be promoted. Experience at the Geological Society of London in placing geological stories has been a happy 100 per cent record of success (at the time of writing!). In two years, not one story that the Society has attempted to promote from its publications and conferences has failed to gain coverage. The constraint on the process is the availability of suitable material.

Geoscientists, like most scientists and many other committed professionals, think that the media give them a raw deal. Those with experience of the press often find it unnerving because careful scientific caveats are left out, and the story may develop a headline of which they disapprove. Scientists who have worked with press officers and been successful are often discouraged by the time and effort required.

Many scientists of all disciplines want to be in the media spotlight, but want to be there only on their own terms. Their failure to get in on those terms generates much futile chagrin. Scientists believe that the media are in the education business, but the truth is that they are part of the entertainment industry. This misapprehension is especially strong in the UK, where the spirit of Lord Reith, the redoubtable first Director-General of the British Broadcasting Corporation, still walks abroad (at least among highbrow consumers).

For these reasons much misguided effort is expended by scientific groups in asking how those in the media might change their ways to suit science better and give it a better press. This is not likely to succeed. The true road to greater coverage lies through providing the media with what they want. Improving the quality of coverage (which most scientists think necessary) will come about only through the diligent cultivation of professional press relations, and through more scientists learning to be journalists.

Documentary TV programmes

Like those working in other media, TV producers have generally looked to the more sinister aspects of the Earth's behaviour to provide material for documentary programmes. Volcanic eruptions, earthquakes, floods, tsunamis, and tornadoes are undeniably photogenic and awe-inspiring; climatic change and other threats to the environment, if less immediately dramatic, also pose obvious threats to life and property and can thus gain the viewers' attention.

A wider interest in the Earth sciences has more recently been evident in British television and radio programmes. In 1998, for example, BBC TV put out two geological series. The flagship was *Earth Story*, broadcast on Sunday evenings in eight parts. This was a lavish, presenter-led documentary of the old school, reminiscent of those 'voyages of personal discovery' exemplified by Kenneth Clarke's *Civilization* and Jacob Bronowski's *The Ascent of Man*. Aubrey Manning, a retired professor of natural history from Edinburgh, was an inspired choice to front the series. Not being a geologist allowed him to play the role of explorer, newly returned from an expedition to the land of the geologists, eager to introduce the world to the wonders he had found there. David Sington was the producer for *Earth Story*; Simon Lamb of the University of Oxford was the series consultant.

Series consultants are the relatively unsung heroes of TV documentaries. Anna Grayson fulfilled this role, and that of presenter, in the BBC's other 1998 TV geology series, *The Essential Guide to Rocks*. This was an attempt to present Earth science to young people, and unlike *Earth Story* it was aimed at those with short attention spans. Each 30-minute programme was in magazine format with four slots, each a self-contained story. There was no attempt here to create a structured course, although this was done in the book accompanying the series. The presenters for *The Essential Guide to Rocks* were Ray Mears (a survival expert, chosen, it seems, for high recognition with the target audience), Duncan Kopp (a geologist and also series researcher), and Kate Humble, who, like Professor Manning, took the role of the perpetually amazed *ingénue*: 'Being down in a deep mine in Wales and finding real gold made me feel like I was an excited ten-year-old again', she said.

In 1999 the BBC World Service broadcast a series entitled *Earthworks* for its massive 50 million audience. This series of 12- to 15-minute programmes traced the entire history of the Earth, from Big Bang to eventual supernova destruction. The producer was Merilyn Harris and the presenter (again) Anna Grayson. Many geologists were called upon for help in producing the series. Anna Grayson commented, 'None of this could have been achieved without all the scientists who have been so helpful, polite, and have given up their time to communicate to wider public.' An encouraging postscript to this story is that both Simon Lamb (*Earth Story*) and Anna Grayson (*Essential Guide*) subsequently won media prizes for their work.

Geoscientists in films and television

Most people make their minds up about others very quickly. Stereotyping is easy and comfortable. It confirms prejudices and saves time. It is also, as Stephen Jay Gould reminds us, 'untrue, but culturally powerful'. Scientist characters in Hollywood films demonstrate, however, that fundamental changes are taking place in those stereotypes, with geoscientists leading the way. These changes are significant because they mark a changing attitude to science among the public.

In drama, time is of the essence. Characters must be established quickly with the audience. Quick, easy images are the stuff of TV news, advertising, and public relations. Stories must be tellable in quick, simple images. In film portrayals of scientists, we see what screenwriters believe will be recognizable by a mass audience. To those interested in the image of scientists, progress would consist of a change in this stereotype. This is what is happening.

The sexy, dashing whip-wielding archaeologist Indiana Jones, played by Harrison Ford, is based on a real-life character: the palaeontologist Roy Chapman Andrews, discoverer of the dinosaur *Oviraptor*. Indiana Jones is unusual in portraying the academic as action-man. Hitherto, we were much more familiar with the egghead—scientist or not—as an arrogant, unworldly, megalomaniac obsessive (with a beautiful daughter to provide romantic interest for the hero action-man). This ancient dramatic tableau can be traced back to *The Tempest* and probably beyond. But with Indiana Jones we saw the beginning of a reaction. Increasing audience sophistication is part of the reason.

The screenwriter's challenge is to take recognizable character traits that might be a cliché elsewhere, and transplant them on to figures in the plot who would not normally be expected to display them.

Spielberg's *Jurassic Park* presents three principal scientists: Jeff Goldblum's theoretical mathematician, Sam Neil's vertebrate palaeontologist, and Laura Derne's palaeobotanist. All the main scientist characters in this picture come over well. Derne is strong-willed, independent, feminist and sexy. She is everything, in fact, because she also wants marriage and children. Neil is dedicated—perhaps a bit too educated—but is also intuitive, a superb communicator, and, above all, knowledgeable about dinosaurs, Goldblum is weird, roguish, and cool.

These characters have failings, but these are not numerous and do not include the old stereotypes of arrogance, selfishness, greed, or malevolence. Those traits are reserved for Attenborough, the aggrieved computer nerd on his staff, and the hateful accountant.

Geoscientists-as-saviours appeared in the 1997 volcano movie, *Dante's Peak* (Universal Pictures). In classic *Jaws* format, the film opens with a cosy community about to get the shock of its life from the Force of Nature on its doorstep: in this case, a large volcano in the Cascade Mountains of the western USA. The vested business interests and the local council, with the exception of feisty single mum Mayoress (Linda Hamilton), conspire to prevent the necessary action being taken until it is too late.

To the rescue comes the US Geological Survey in the form of a geologist, Harry Dalton (Pierce Brosnan). At this point we understand why the mayoral character has to be a single female. Intuition, with which Dalton is richly endowed, quickly tells him that the time for careful surveys is past. He is about to arrange an evacuation when his boss arrives. He, Paul Dreyfus (Charles Hallahan), is thoughtful and cautious.

Dreyfus is distressed that his impetuous subordinate is about to panic the natives on little more evidence than his senses and a few asphyxiated squirrels. A bad experience, years ago when he too was young and impetuous (but not, we suspect, as handsome as Mr Brosnan), urges him to caution. He is right, of course; but Brosnan is righter.

The drama unfolds as one would expect. Brosnan is vindicated. The citizens are evacuated just in time. Paul Dreyfus, despite being man enough to admit his error, pays the ultimate penalty, as is dramatically right. Brosnan and Mayoress survive by hiding in a mine, and all ends as it should.

This is a very encouraging film for geoscientists. The young, pretty, student-like whizz-kids who work for crusty, overweight Paul are bright and attractive. Paul is hamstrung by protocol and procedures. He isn't In Touch With The Earth; but he is neither evil nor incompetent, and admits his mistake manfully. Conversely, Harvey Dalton uses his senses. He tastes things. He feels for the flora and fauna. These empathic traits are the anti-cliché devices that the scriptwriters have attached to him. He is also extremely handsome. Gorgeous, athletic scientists are undoubtedly a novelty.

The public has become better informed about science. One of the results is an increasing awareness of the difference between science and technology. From Mary Shelley on, tampering with Nature has always been a major theme in science fiction. As we move into the next century, however, the public is shifting its view. Scientists can be cool and brilliant, in touch with their feelings and brim-full of 'emotional intelligence'. Villains are now much more likely to be the money- or power-mad, or those who thoughtlessly *apply* science. The heat seems to have come off those who are merely curious about Nature's workings.

This explains why geoscientists are in the vanguard. Geoscientists can more easily be cast as sensitive seers than physicists, chemists, or geneticists. Moreover, by being associated with the open air and fieldwork, they can take on some of the clichéd but healthy characteristics usually associated on film with oilmen and lumberjacks.

The battle for the rehabilitation of science in the public mind is not over. But these new fictional representations of science and its practitioners show us that it has moved into a distinctly new phase.

The people in the media

Many first-rate science journalists are not scientifically trained, but few would disagree that first-hand knowledge of science's methods and culture is a great help. Competition in journalism is intense, and so the ability to specialize provides the aspiring reporter with a distinct advantage. A degree in geoscience provides an ideal background for a science jour-

nalist: the study of the Earth brings all the sciences together, giving the graduate an unusually broad education.

The traditional route into newspaper journalism has been to start as a junior on local papers taking professional qualifications along the way, like an apprentice. Most science journalists tend, however, to be recruited from specialist journals.

Many science journalists work independently. They either go freelance after a period as 'staffers', or begin writing as a sideline to their main occupation, gradually allowing their journalistic work to grow until it can provide them with a living. With recent developments in IT and communications, working from home is increasingly attractive.   TED NIELD

Further reading

Pollock, J. and Steven, D. (1997) *Now for the science bit— concentrate! Communicating science.* River Path Associates, 5 Old Road, Wimborne, Dorset BH21 1EJ.

Scanlon, E., Whitelegg, E., and Yates, S. (eds)(1999) *Communicating science: Reader 2: contexts and channels.* Routledge, London.

Anon. (1997) *So you want to be a science writer?* (2nd edn). Association of British Science Writers, 23 Savile Row, London W1X 2NB.

geosphere The term 'geosphere' refers to that part of the solid Earth on which humans live and depend upon for most of their food, fuels, and mineral resources. Thus the geosphere encompasses the upper part of the continental crust of the Earth (made primarily of granite), basalts and other products of igneous activity; sediments deposited during the Earth's long history; and the uppermost weathered layer of the sediments and crustal rocks, called the regolith. The weathered minerals of the regolith, together with an admixture of organic matter and water, make up the soil.

The totality of living organisms is the *biosphere*, although this term is also used to denote the environment inhabited by living organisms. Water, in its three forms, liquid, vapour, and ice, makes up the hydrosphere. The geosphere, atmosphere, hydrosphere, and biosphere are clearly not only in physical contact one with the other, but they interact among themselves, and changes in any one of them affect the others to varying degrees. In the system formed by these four major spheres, any two contiguous spheres exchange materials. There are six pairs of two-way exchanges (Fig. 1). For example, the geosphere provides physical habitat and nutrient substances to the biosphere and receives back some of the nutrient elements in organic matter; it exchanges gases with the atmosphere, reacts with them chemically, and receives atmospheric precipitation; the geosphere's interaction with the ocean is heavily weighted by transport from land, but it has two-way interactions with continental and atmospheric waters. The nature and magnitude of the exchanges between

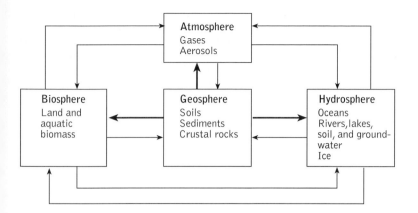

Fig. 1. Schematic representation of the geosphere and its interaction with other spheres of the Earth's surface. Thick arrows indicate direct impacts of the geosphere and human perturbations on the contiguous spheres, as explained in more detail in the text. Transport of materials from the geosphere affects the content of the other spheres and two-way flows between them.

the geosphere and other Earth's surficial spheres have changed considerably in historical times, as the world's human population has grown and its technology has developed. These have become important geological factors and will remain so for the foreseeable future.

There are two main reasons for the geosphere's importance in the global environment:

(1) The mass of the geosphere and the diversity of its chemical and mineral composition are greater than those of the other spheres, making the geosphere the ultimate source of many materials. The elevation of the land surface determines the direction of primary transport from the land to the ocean.

(2) Human industrial activities occur on land, deriving most of the needed fuels and raw materials from the Earth's sediments and crust, and their agricultural activities affect primarily the land vegetation and soils. These activities, originating in the geosphere, perturb, through their products, the hydrosphere, biosphere, and atmosphere with which the geosphere exchanges materials.

The first point, concerning the mass of the geosphere, can be illustrated by the following orders of magnitude. The mass of all the sediments has been estimated as 3×10^{21} kg; with a layer of the crystalline continental and oceanic crust down to 2 km depth, the total mass is about 6×10^{21} kg. The hydrosphere holds 1.4×10^{21} kg of water (97.5 per cent of it in the oceans), but only about 5×10^{19} kg of dissolved solids. The mass of the atmosphere is 5×10^{18} kg. The mass of the biosphere is contained mostly in land plants, with the animals accounting for about 1% of the total and marine life for an even smaller fraction. The biosphere consists of six main elements: carbon, nitrogen, phosphorus, sulphur, oxygen, and hydrogen. The mass of dry organic matter in the biosphere is about 1.5×10^{15} kg. In other words, the hydrosphere is about one-fifth to one-quarter of the geosphere mass, as defined above, and its mass of dissolved solids is about one-hundredth of the mass of the geosphere. The atmosphere is one-thousandth and the biosphere about one millionth of the of the geosphere mass.

The geosphere of the past was far from immutable, and changes in such features as elevation of the continents, the mass of ice, the land surface area, and its vegetation cover were, over periods of geological time, much greater than the changes attributable to human technological activities.

Plant life gains physical support in the weathered surface of land, where crevasses, cracks, or joints between rock particles provide growth space for simple plants or for roots of vascular plants. Organic acids generated by plants and bacteria, or formed in the process of decomposition of dead organic matter, dissolve soil minerals and leach their chemical constituents into water. Decomposition of dead organic matter in soils releases such nutrient elements as nitrogen and phosphorus, which are essential to plant growth. The leached components of crustal rocks and sediments, and products of decomposition of organic matter, are carried into groundwaters and by rivers to the oceans. Some of the products of decomposition of organic matter return to the atmosphere as carbon dioxide, and oxides of nitrogen and sulphur. In the opposite direction, molecular nitrogen of the atmosphere is biologically fixed on land; some of the nitrogen and sulphur oxides are deposited as particles and in rain on land; and carbonate dioxide was, until the start of the human perturbation, one of the main gases that gave rainwater its acidity. Acid waters reacting with soil and rock-forming minerals are chemically neutralized to a greater or lesser extent, according to the minerals with which they react: limestones, reacting faster with percolating water, neutralize the acids and dissolve, contributing calcium and bicarbonate ions to solution; igneous rocks, in general, react more slowly and their ability to buffer or neutralize acid waters is not as great over comparable periods of time.

The present industrial age is characterized by rates of global population growth that are significantly higher than in the historical past; by intensive extraction of organic and inorganic materials from the geosphere; and by land-use activities. The beginning of human perturbation of the geosphere is somewhat loosely placed at about 300 years ago, near the year 1700, and it is clearly associated with the 'industrial age'. Human agricultural activity, the cutting of forests, and burning of wood are undoubtedly much older, but their scale has much increased with the growing population, its agricultural practices, and with industrial development. Agricultural activity converts natural land to farmland (cropland), and it is part of a broader class of land-use activities, including such processes as deforestation, replanting, soil erosion, and, in general, changes in the residence time of carbon on land. At present, the cultivated area of the world is about 13 per cent of the continental surface, and 64 per cent of land is under natural vegetation cover; the remaining 23 per cent comprises ice, deserts, and non-vegetated land. Cultivation of land, deforestation, construction, and associated land-use activities increase the rate of land erosion by water and wind. Soil erosion in the United States has been estimated to average 1.1 kg of soil per square metre per year ($1.1 \text{ kg m}^{-2} \text{ year}^{-1}$), affecting a total agricultural area of about 18 per cent of the country or close to 1.7 million square kilometres. This rate of erosion is equivalent to the loss of a layer of soil about 4 cm thick in 100 years, and it is much higher than the longer-term rates of land denudation estimated from the sediments transported by rivers on individual continents and for the world as a whole; the world mean denudation rate is about $0.2 \text{ kg m}^{-2} \text{ year}^{-1}$, equivalent to the removal of about 0.7 cm of continental surface in 100 years. It is likely that soil erosion rates of agricultural and cultivated land are higher because much of the eroded material is not immediately carried to the oceans but is stored elsewhere on land and in river valleys.

Deforestation of the Earth's surface, particularly of the tropical forests, is (after the burning of fossil fuels) the second most important source of carbon dioxide emissions from land to the atmosphere. Agricultural and other land-use activities are, however, also responsible for a greater transport of inorganic and organic materials from land to the oceanic coastal zone, and they have also led to a negative feedback—a faster recycling of organic matter in soils and greater release of the nutrient elements nitrogen and phosphorus that promote fertilization and plant growth. Land-use activities are also the likely causes of stronger emissions to the atmosphere of such gases as nitrous oxide and methane, the latter associated with bacterial activity in rice fields and the digestive systems of cattle.

The historical path to the useable energy sources did not lead to the biggest natural source, solar radiation, but to fossil organic materials occurring in the geosphere as coal, petroleum, and natural gas. Combustion of fossil fuels in the past three centuries has been the main cause of the increase, by nearly 30 per cent, in the carbon dioxide content of the Earth's atmosphere, from 275 to 350 p.p.m.v. (parts per million by volume). Unrelated to fossil-fuel burning and land-use activities in an earlier period, since the last glaciation maximum about 18 000 years ago to the beginning of the industrial age, atmospheric carbon dioxide rose by a greater amount, from 180 to 275 p.p.m.v.

The burning of fossil fuels releases to the atmosphere not only climate-affecting, greenhouse gases, but also gases that make acid rain. Coal and petroleum contain nitrogen and sulphur, and their oxides formed in combustion together with carbon dioxide are emitted to the atmosphere. Further photochemical oxidation of nitrogen and sulphur oxides converts them into compounds that make nitric and sulphuric acid when they dissolve in water. Dissolution in rain droplets becomes acid rain, and deposition in aerosol particles on land makes acidic waters on the surface. As nitric and sulphuric acids are much stronger than a solution containing dissolved carbon dioxide at its present concentrations in the atmosphere, waters in certain locations have become progressively more acidic during the past several decades. As in pre-industrial times, acid waters are not always, and not everywhere, neutralized in the geosphere, and they may remain acidic for long periods of time. Acid precipitation has been linked to forest damage in industrialized countries, and acidification of lakes has been said to reduce the diversity of ecosystems because of the poor acid tolerance of species of fish and other aquatic organisms. An additional adverse effect of acid waters on the biosphere may be related to the stronger leaching of such metals as aluminium from crustal rocks and their retention in acidic waters at concentrations toxic to some organisms.

Our technological culture is essentially an iron culture: global production of metallic iron from iron-ore minerals exceeds by far the production of other metals. However, the greatest mass of natural materials used in the technological society are crystalline rocks and sediments, used in construction under such names as stone, gravel, and sand. The annual consumption rates per person, in the USA in 1989, were reported as 7700 kg of stone, sand, and gravel, and about 750 kg of such mineral materials as iron, phosphate, gypsum, salt, and other metals. Large as these numbers seem when viewed per person, the total per unit of the country's land area is $0.23 \text{ kg m}^{-2} \text{ year}^{-1}$, representing an average extracted mass of crustal materials that is considerably smaller than the mass of soil removed by erosion, cited above as $1.1 \text{ kg m}^{-2} \text{ year}^{-1}$.

Materials extracted from the geosphere are converted into forms used for different purposes and subsequently disposed as wastes. This is very much a one-way process, and recycling of Earth's materials from industrial wastes is limited on a global scale.

There is a strong coupling between present-day perturbations of the geosphere and the density of human populations. In industrialized countries, for which statistical data

are available, such processes as energy production from fossil fuels, application of chemical fertilizers to agricultural land, and generation of industrial, agricultural, and municipal wastes (measured as mass per unit of land area per year) correlate with the population density: countries with higher population density produce, in general, more energy, use more fertilizers, and generate more wastes per unit of their land area. This is a further indication of the role of humankind as an increasingly important geological agent acting in the geosphere. As the geosphere does not absorb or store all the products of human perturbations, these are passed on to the oceanic and land waters, the biosphere, and the atmosphere in progressively greater amounts and with uncertain consequences for the future direction of change in the global environment.
A. LERMAN

geothermal energy Hot springs and volcanic eruptions indicated to ancient man that the interior of the Earth must be hot. In most parts of the world deep drilling has revealed that temperatures increase with depth by 20–100 °C for every kilometre, sometimes more. The highest thermal gradients are observed in areas of active volcanism. Hot springs and fumaroles are most abundant in these areas. They represent discharge from underground bodies of hot water, and sometimes steam, that are termed geothermal systems or geothermal reservoirs. In most instances the source of supply of the heat is cooling molten rock in a magma chamber. The rocks in the uppermost 2–4 km of the Earth's crust are generally porous and the pores are filled with water (groundwater). When a magma is intruded into the crust it heats up its surroundings, particularly those above its roof, including the groundwater. When heated, the groundwater expands and, as a results, it tends to rise. The rising hot groundwater is replaced by cooler groundwater from the sides, which in turn gains heat over the magma heat source. In this way a circulation (convection) of groundwater is established. It is this circulation that brings about cooling of the magma by extracting heat from it. Thus, a part of the Earth's interior heat is brought to the surface, first by rising magma and, secondly by geothermal water and steam.

A volcano, particularly one that develops a caldera by collapse of its summit, typically forms above a shallow magma chamber. The magma chamber is fed by a larger magma reservoir at the base of the crust or in the mantle. Thus, geothermal reservoirs are usually, but not at all always, found in the vicinity of active volcanoes. The ages of individual volcanoes and their associated magma chambers are usually in the range of 0.1 to 1 million years, and so are the ages of individual geothermal systems.

Not all geothermal systems have a magmatic heat source. The temperature of groundwater always increases with depth in harmony with the increased temperature in the crust. Hotter groundwater underlying colder and, therefore, denser groundwater represents an unstable condition. If the permeability of the rock is sufficiently high, and the density difference of the hot and cold groundwater is large enough, the groundwater body will start convecting in much the same way as happens above magmatic intrusions. In some areas of rapid sedimentation, as, for example, north of the Gulf of Mexico, the porosity of the sediments and the temperatures at deep levels may be high enough to make hot-water geothermal reservoirs exploitable through deep drilling although no convective groundwater systems exist.

Surface geothermal manifestations include hot steaming ground, fumaroles, mud pools, and warm to boiling hot springs. The ground surrounding some hot springs may be intensely altered, forming colourful deposits, such as native sulphur (yellow) and haematite (red). Periodic explosive boiling in the feeder pipes of hot springs leads to geyser action, in which a jet of water and steam can be thrown as much as 50–60 m into the air. The most famous geyser in the world today is Old Faithful in the Yellowstone National Park in Wyoming, USA.

The primary requisites for the formation of geothermal systems, i.e. a heat source (either magma or hot rock) and sufficiently permeable rocks, are mostly found in areas of active volcanism. Such areas generally coincide with areas of active Earth movements, as witnessed by earthquakes. Hot springs and geothermal energy sources can also occur in association with earthquake faults, even in areas free of active volcanoes (e.g. Dixie Valley, Nevada, USA and the Menderes graben, western Turkey). The earth movements create and maintain rock permeability by fracturing of the rock. Open fractures do not extend to great depths; they may be not much more than about 4 km deep in most places. Under the load of the overburden, particularly when temperatures are high, the rock yields, behaving more as a plastic material than a brittle one, with the result that pores and fractures cannot remain open. Thus in any specific area the base of a geothermal system coincides with the lowest level of permeable rocks.

Geothermal waters generally contain more dissolved gases and solids than ordinary groundwaters. The source of supply for these dissolved substances is mostly the rock in contact with the water. Some substances, particularly those forming volatile compounds, at least at high temperatures (such as carbon dioxide, boron, and hydrochloric acid), may be partly derived from the magmatic heat source through its degassing. Although 99 per cent of most common rocks consists of relatively few elements (ten to be exact), these rocks host a large number of trace elements, some of which form soluble salts, such as chlorine and boron. Many trace elements once transferred to the water, remain in solution and, as a result, become concentrated in the water. Most major elements, on the other hand, are removed from solution through deposition of hydrothermal minerals. The solubility of many minerals in water increases with temperature. Chemical reaction rates are also enhanced by rising temperature. This, together with the long residence time (tens to thousands of years) of geothermal waters underground, explains their higher content of dissolved substances in comparison with cold groundwaters, particularly of those elements forming soluble salts. The content of soluble salts varies enormously between rock types. It is highest in evaporative and marine sedimentary rocks but lowest in basalts. Geothermal waters associated with the former types of rocks, such as the Salton Sea field in southern California, are very salty (30 per cent dissolved solids, or almost ten times the salt content of sea water), whereas those in basaltic terrain, such as the Krafla geothermal system in Iceland, are very dilute (0.1 per cent dissolved solids). Hot brines, such as those in Salton Sea, are relatively rich in many metals, including lead, zinc, manganese, copper, and silver. These metals are relatively mobile in hot, saline waters because they complex with chloride, the main soluble salt-forming anion. The Salton Sea brine is considered to represent a modern ore-forming fluid. Studies of this brine, and other similar ones, are therefore not only of interest to those engaged in developing geothermal resources but also to those involved in mining of metalliferous deposits.

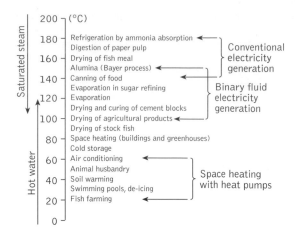

Fig. 1. Modified Lindal diagram: temperature requirements of geothermal fluids for various uses. (From Líndal (1973), *Geothermal Energy*, **12**, 135.)

(a)

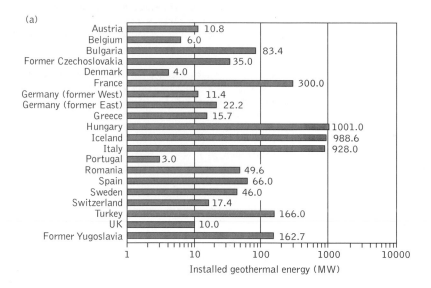

(b)

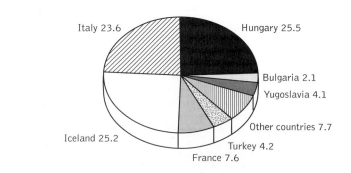

Fig. 2. Installed geothermal energy in European countries: (a) in megawatts (MW) and (b) as a percentage of 3926.8 MW. (From Hurtig (1992) *Proceedings of the International Conference on Industrial Uses of Geothermal Energy*, Reykjavik.)

Modern use of geothermal energy is based on drilling for hot water and steam. The deepest geothermal drillholes approach 4 km, but a common depth is in the region of 2 km. The highest temperatures measured in wells are close to 400 °C, on the northern periphery of the Geysers field in California and at Nesjavellir in Iceland. Temperatures commonly approach or exceed 300 °C. The fluid in some drilled geothermal reservoirs is largely steam. They are then said to be vapour dominated. In others the fluid is mostly water. Such reservoirs are said to be liquid dominated. Vapour-dominated reservoirs are much less common than the liquid-dominated ones, the ratio probably being 1 to 20. Examples of the former include the Geysers field in California, Larderello in Italy, and Kamojang on the island of Java in Indonesia. Examples of liquid-dominated reservoirs include Wairakei in New Zealand, Tongonan in the Philippines, and Cerro Prieto in Mexico.

The technical possibilities of exploiting geothermal fluids for various uses are largely dictated by the temperature of the reservoir (Fig. 1). The boiling point of water is pressure-dependent. It is 100 °C at atmospheric pressure but it increases with pressure, being, for example, 300 °C at 85 bars (equivalent to about 1000 m depth). When water above 100 °C ascends a well it will begin to boil when the pressure created by the load of the overlying fluid equals the boiling point pressure. Because of this, steam can be produced from liquid water geothermal reservoirs. However, reservoir temperatures well in excess of 200 °C are desirable if steam at 140–190 °C is to be produced for electrical power generation (Fig. 1).

Drilling for geothermal water and steam employs mostly the same technology as drilling for oil. For conventional turbines, about 2 kilograms per second (kg s^{-1}) of steam are required to generate 1 megawatt (MW) of electricity. Many geothermal wells yield about 10 kg s^{-1} of steam and occasionally as much as 40 kg s^{-1}. The cost of drilling a typical geothermal well (2000 m deep with a diameter of 50 cm at the top and 20 cm at the bottom) is in the region of US$1.5 million.

Geothermal energy is extensively used in many countries, including China, Hungary, Iceland, Italy, Japan, Mexico, New Zealand, the Philippines, and the USA (California). Figure 2 depicts usage in Europe by country. Table 1 provides information on the worldwide use of geothermal energy for power generation and direct use. The most important category is the generation of electric power from geothermal steam. The steam is used in much the same way as steam produced from water by burning coal or oil. Its pressure is used to rotate a turbine connected to a generator that, in turn, produces electricity by converting the mechanical work carried out by the flowing steam into electrical energy. On a worldwide scale the electrical energy produced by geothermal power is very small. Thus, for example, the integrated production capacity of the world's geothermal power plants in 1997 (8240 MW) was only about 7 per cent of annual electric energy demand in Great Britain. In 1970, power generation by geothermal steam was only 675 MW, showing that development of this natural resource has been growing rapidly. Other uses of

Table 1. Use of geothermal energy in the world by country in 1997

Country	Electric production		Direct use	
	Power MW[1]	Annual use GWh[2]	Power MW[1]	Annual use GWh[2]
Algeria			1	5
Argentina	0.7	3.5		
Australia	0.4	0.8		
Austria			21.1	84
Belgium			3.9	19
Bosnia-Herzegov			33	230
Bulgaria			95	346
Canada			3	13
China	32	175	1914	4717
Costa Rica	120	447		
Croatia			11	50
Czech Republic			2	15.4
Denmark			3.2	15
El Salvador	105	486		
France	4	24	309	1359
Georgia			245	2145
Germany			307	806
Greece			22.6	37.3
Guatemala	5			
Hungary			750	3286
Iceland	140	375	1443	5878
Indonesia	590	4385		
Ireland			0.7	1
Israel			42	332
Italy	768	3762	314	1026
Japan	530	3530	1159	7500
Kenya	45	390		
Macedonia			75	151
Mexico	743	5682	28	74
New Zealand	345	2900	264	1837
Nicaragua	70	250		
Philippines	1848	8000		
Poland			44	144
Portugal	11	52	0.8	6.5
Romania	2	?	137	528
Russia	11	25	210	673
Serbia			86	670
Slovakia			75	375
Slovenia			34	217
Sweden			47	351
Switzerland			190	420
Thailand	0.3	2	2	8
Tunisia			70	350
Turkey	20	71	160	1232
Ukraine			12	92
United States	2850	14600	1905	3971
Europe	936	4309	4368	20505
America	3883	21529	1908	3984
Asia	3031	16092	3075	12225
Oceania	345	2901	264	1837
Africa	45	390	71	355
Total	8240	45220	9686	38906

[1] Megawatts. The number are valid for 1998 and are based on information from the International Geothermal Association (http://www.demon.co.uk/geosci/igahome.html). [2] Gigawatthours. (From Fridleifsson and Stefánsson (1998).)

geothermal energy, which totalled 9686 MW thermal in 1997, are for greenhouse farming, space heating, balneology, and various industrial uses. Balneological use is the oldest. For example the Romans used hot spring waters extensively for bathing in many circum-Mediterranean countries, including even Britain. Industrial uses include paper production in New Zealand and drying of diatomaceous sediments in Iceland. Geothermal water and steam have also been used as sources of various chemicals. The classic example includes production of boron from geothermal steam in Larderello, near Pisa in Italy, although this production ceased some years ago. Common salt and carbon dioxide gas are today produced from brine and steam at the Reykjanes geothermal field in Iceland.

Only a very minor part of the Earth's heat is conveyed to the surface by rising magma and convecting groundwater. Most of it is transferred conductively. By far the larger part of the Earth's heat is therefore to be found in hot rock at depth and not in shallow crustal magma chambers or geothermal systems. This fact has stimulated projects aimed at developing techniques to extract and exploit economically the heat stored in rocks at shallow depths in the crust although the rock may be quite impermeable. The technology so far developed entails drilling a pair of appropriately spaced wells in an area of high thermal gradients. In order to create permeability the rock at depth is hydrofractured by pumping water (commonly containing sand to keep cracks open) under high pressure into one of the wells until the rock ruptures. Water is subsequently injected and it flows through the rock to the other well. Hot water is thus produced as the injected water gains heat from the hot rock through which it flows. Projects of this kind have been termed hot dry rock projects. An example is furnished by a drilling into one of the Cornish granites in south-west England. Exploitation of thermal energy from hot dry rock does not yet seem to be economically attractive, although it is technological achievable. Future projects may drill into fractured, and therefore permeable, rocks, although they do not host a convecting geothermal fluid.

Geothermal energy is preferred to fossil fuel (coal and oil) from an environmental point of view: much less carbon dioxide and sulphur are discharged into the atmosphere for every generated unit of electricity when using geothermal steam as compared to fossil fuel. The advantage of geothermal energy is even more pronounced when it is being used directly, that is, as a source of heat, as is the case with space heating and greenhouse farming. However, geothermal water and steam commonly contain appreciable amounts of some substances that are harmful to the environment, such as sulphur, arsenic, boron, and mercury. Exploitation of geothermal resources may thus create local pollution problems. In order to reduce this pollution as much as possible, a technique has been developed during the past 20 years by which the used geothermal water is reinjected into the ground through wells drilled specifically for this purpose. In the Paris basin, for example, hot water pumped from deep wells is being exploited for house heating. The water in the wells is saline and thus chemically undesirable for surface disposal. After passing through heat exchangers to extract the heat it is therefore injected back into the ground through injection wells.

STEFÁN ARNÓRSSON

Further reading

Armstead, H. C. H. (1978) *Geothermal energy*. John Wiley and Sons, New York.

Rinehart, J. S. (1980) *Geysers and geothermal energy*. Springer-Verlag, New York.

geothermal waters, chemistry *see* CHEMISTRY OF GEOTHERMAL WATERS AND HYDROTHERMAL ALTERATION

giant planets Our Solar System contains four giant planets which are so massive (Table 1) that their solid interiors lie deep within enormous liquid and gaseous envelopes that deny them Earth-like characteristics. Their atmospheres are composed mostly of hydrogen with some helium, but the remaining fraction of a percentage that forms the cloud layers prevents us from seeing deeper. The topmost cloud layer on Jupiter and Saturn is of ammonia, whereas the clouds are formed mostly of methane on Uranus and Neptune.

In Jupiter and Saturn gaseous matter persists to depths of tens of thousands of kilometres, until pressure is sufficient to compress hydrogen into a metallic state. Below this there is believed to be a water-rich layer (possibly a high-pressure form of ice) surrounding a central rocky core of several Earth-masses. Pressure within Uranus and Neptune is insufficient for metallic hydrogen to exist; instead their gaseous shells probably directly overlie the water–ice interiors, and their rocky centres are somewhat smaller.

We can study only the topmost part of the atmosphere of these planets directly. Atmospheric circulation is controlled more strongly by heat escaping from their interiors and by the rapid planetary spin than on the Earth, where the main driving force is solar heating. Major storm systems have been observed on all four giant planets, but only on Jupiter do they persist for more than about a year. The Great Red Spot

Table 1. The giant planets.

Name	Distance from Sun (millions of kilometres)	Diameter (km)	Mass relative to Earth*	Density (tonnes m^{-3})
Jupiter	778.3	142.8	317.8	1.33
Saturn	1427	120.0	569	0.69
Uranus	2870	51.2	14.5	1.29
Neptune	4497	48.6	102	1.64

*The mass of the Earth is 5.98×10^{24} kg.

10 000 km

Fig. 1. Jupiter's Great Red Spot seen in two images recorded 9 hours apart. The spot is an atmospheric vortex rotating anticlockwise. Changes are most noticeable near the outer edge of the system, where wind speeds are up to 150 m s⁻¹.

(Fig. 1) is the most famous of these, and has persisted since the seventeenth century. All four giant planets have ring systems, made of dust and boulder sized icy debris, but only Saturn's ring is wide and spectacular. Most ring-forming material probably originates from collisions of cometary material with satellites of these planets.

Extrasolar planets have been detected around several nearby stars; those found so far they have all been giant planets in the Jupiter class. This is not surprising, because it is only the really large planets that we can expect to detect with current technology. It seems likely that smaller Earth-like planetary bodies must exist in abundance around other stars.

DAVID A. ROTHERY

Further reading

Rothery, D. A. (2000) *Teach Yourself Planets*. Hodder and Stoughton, London.

Beatty, J. K., Peterson, C. C., and Chaikin, A. (1999) *The new Solar System*, (4th edn). Sky Publishing Corporation and Cambridge University Press, Cambridge.

Gilbert, G. K. (1843–1918) Grove Karl Gilbert is widely regarded as the most accomplished American geologist. After early experience with the Ohio, Wheeler, and Powell Surveys, he was one of the six original geologists hired in 1879 to the US Geological Survey. By 1888, he had risen to Chief Geologist, and his professional career was fully linked to the Survey. Gilbert received many honours in his own time, including the distinction of being the only person ever to serve twice as president of the Geological Society of America.

Gilbert is best known for the exemplary quality of his scientific monographs. As a member of the Powell Survey, he produced the 150-page *Report on the geology of the Henry Mountains*. In his structural analysis of the Henry Mountain laccoliths, Gilbert used quantitative mechanical procedures that anticipated the geodynamical modelling of today. Similarly, his analysis of erosional processes relied upon principles of energy and equilibrium in the operation of streams. These themes were continued in later works. His 1890 *Lake Bonneville* (US Geological Survey Monograph No. 1) was one of the consummate works of nineteenth-century geology. Not only did he describe the dynamics of lake shoreline formation; he showed how warping of the potential surface evidenced in the shorelines could be interpreted in terms of Earth's crustal dynamics. Later works, *The transportation of debris by running water* (1914) and *hydraulic-mining debris of the Sierra Nevada* (1917), are classics of fluvial geomorphology.

Gilbert believed that 'the inculcation of scientific method' was best achieved by example. In an 1885 Presidential Address to the American Society of Naturalists in Boston, he emphasized the role of hypothesis generation in geology. He proposed that scientific investigators, whom he contrasted with 'theorists', generated hypotheses via a creative process of reasoning via analogy. In a 1896 paper entitled 'The origin of hypotheses, illustrated by a discussion of a topographic problem', Gilbert expanded upon these views, equating the method of hypotheses to the method of science. In their quest to determine the antecedent causes of phenomena in evidence via their consequences, geologists must first make a kind of guess, to frame a hypothesis, that must then be evaluated for comparison of its deduced consequences against newly discovered facts. Although this pragmatic view of scientific method was shared by contemporaries such as T. C. Chamberlain, it was ignored by twentieth-century philosophers of science, who dismissed the 'logic of discovery' as unworthy of serious logical inquiry. Nevertheless, the respect accorded to Gilbert's methodology by practising Earth scientists continues to stand in sharp contrast to their general apathy in regard to the ideas of modern philosophers.

VICTOR R. BAKER

glacial landforms When studying ancient landscapes we are often left with an array of landforms and sediments having few or no known modern analogues. Such is the case with landscapes left from Pleistocene glaciation. For many of the landforms and sediments found in glaciated regions there are no equivalents currently being formed, even in regions

Fig. 1. Woodworth Glacier and forefront, Alaska (Bradforth Washburn). A former ice position is illustrated by the moraine and transition from esker to fan deposition. The flutings are located behind boulders and are composed of till.

submerged under deep, relatively fast-moving ice. Around many modern glaciers, the close proximity of some landforms and sediments to their parental ice bodies enables us to be confident in making inferences about their construction (Fig. 1). However, these landforms represent only a limited range of those formed during the Pleistocene.

In attempting to explain older landscapes, we must use our knowledge of processes to make inferences about landform development, because nobody has ever directly observed the actual processes. With this knowledge we can build hypotheses that can be tested by predicting the expected results of different processes. If a prediction does not account for the observations, then the hypothesis fails and must either be abandoned or modified.

In general, geomorphological or landscape features can occur as the result of three processes: (1) erosion of pre-existing rock or sediment; (2) deformation of a previously existing landform; or (3) deposition of a new landform. In the case of depositional landforms, the composing sediment should reflect the processes of accretion as beds are added to form the final product. Where the landform is thought to be the result of deformation of a pre-existing landform, the internal sediment or rock should be deformed. Examples are seen where deformed rocks occur in fold-belt mountains. Finally, if the landform was eroded from previously existing material, the older sediment or rock within it does not reflect the method of formation.

Glacial landforms

The origins of many glacial landforms near modern glaciers can be confidently reconstructed (Fig. 1). The long sand and gravel bodies called eskers were formed in subglacial cave-like tunnels where sand and gravel were deposited by flowing water. Moraines were deposited in front of, beside, and below the glacier during 'stillstand' phases when the melt rates were roughly equal to the flow from the glacier source. The ice acted as a conveyer belt, delivering unsorted, mixed glacial debris, or 'till', to its margins. As melt exceeded supply, the glacier snout retreated, leaving a relatively flat-lying till plain in its former region of occupation. Small-scale flutings were formed as glacial debris was squeezed into long linear grooves in the ice as it passed over an obstacle. Here, both hard obstacles and lee-side till fills should be found. These observations give coherent results that record a range of processes and resultant landform/sediment assemblages.

However, a number of glacial landforms, such as drumlins, large-scale flutings, rogens, and hummocky topography, have no modern analogues. The forms differ from those found close to modern glaciers. Two hypotheses for the formation of these non-analogous landforms dominate current thought. One invokes glacial deformation of saturated sediment as the major formative process. This explanation implies that the saturated substrate was deformed into different landforms as ice flowed over it. This theory infers that sediment contained within a given landform will reflect its accretion or moulding as a slurry was moulded into shape. There are, however, many examples of glacial landforms that do not exhibit the landform/sediment packages predicted by the subglacial deformation theory. Also, in areas of ice flowing over deforming sediment, as in parts of Antarctica, landforms similar to those formed during the Pleistocene are not currently being developed.

The other hypothesis postulates that turbulent melt-water sheetfloods beneath the Pleistocene ice sheets were the major agent of change. The resultant landform/sediment assemblages should, therefore, reflect erosion or deposition, or both, by turbulent flow. Again, these processes are not occurring at this scale in the modern world, but we can use a dual approach to test the theory. On one hand, our knowledge of fluid flow shows that vortices of extremely high velocity can be set up in sheet flows. The vortices interact differently, according to their angle relative to the surface they encounter, whether it be near-parallel to vertical (Fig. 2). The agent must have been fluid in order for these processes to act, because the viscosity of ice is too great to allow turbulent flow. Examples

of erosion in turbulent fluids can be found in the Channeled Scablands of Washington State, USA, and in many desert regions (Fig. 3). The Channeled Scabland contains a range of features similar to those formed in glaciated regions, but the Scabland examples are thought to be the result of short-lived, turbulent, outburst sheet flows across parts of the state. Linear erosional features (called *yardangs*) found in deserts are the result of long-term erosion of wind blowing in a consistent direction. As the wind blows, horseshoe vortices form, and, over long periods of time, these produce numerous lineations at different scales.

The existence of Pleistocene subglacial water flows is inferred through a range of features. Scour marks from vortices of varying sizes have been observed in granite rocks in glaciated areas of southern Canada (Fig. 4). They have also been observed in granite on the continental shelf of Antarctica in a region that is now submerged, but was glaciated during the Pleistocene. These features could not have been formed by ice deformation of a substrate, because granite cannot be deformed by ice.

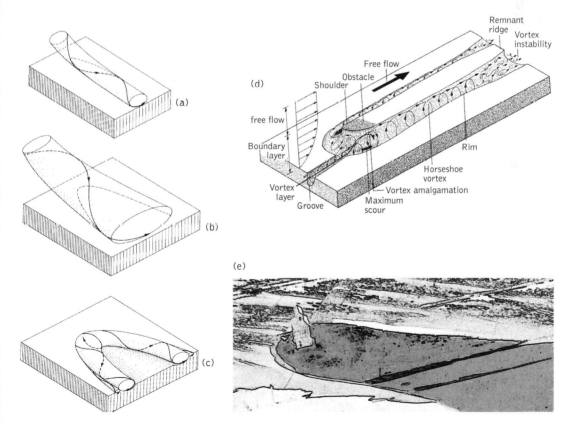

Fig. 2. (a–c) Turbulence within sheet flows can erode bedforms that vary according to the different possible shapes and/or angles of contact to the ground; (d) when sheet flows encounter an obstacle or uneven bed, horseshoe vortices are often set up, forming elongate streamlined erosional remnants; (e) both air and water can form turbulent vortices, but not ice.

Fig. 3. (a) Erosional remnants in the Channeled Scabland of Washington State, USA, that were formed by erosion of water sheetfloods. (b) Yardangs, Cero Yesera, Peru, with smaller intervening flutes formed by wind.

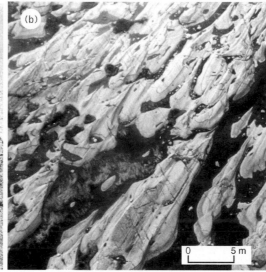

Fig. 4. Turbulent vortices are effective erosional agents that erode their forms into granite (a) and consolidated sedimentary rock (b) at greatly differing scales.

Vortices may change in orientation from linear to transverse, so that the same process results in different landforms. Under ice, vortices can affect both the ground and the base of the ice. Varying flow conditions can cause vortices to erode at higher velocities, or to deposit material in cavities as flow wanes. Deposition into eroded cavities has also been noted to create landforms called 'cavity-fill drumlins'.

Drumlins and giant flutings–remnant ridges have been topics of debate for decades. They are streamlined hills and troughs, and are commonly used to infer past flow directions of ice (Fig. 5). The 'classic' drumlins contain many types of material, including undeformed glacial sediment and bedrock. Such drumlins could not have been formed by

deformation, or all the sediment within them would have been deformed. The common element, exterior form, indicates that erosion by turbulent vortices was responsible, much as in the formation of streamlined hills in the Channeled Scabland and for aeolian yardangs. Giant flutings–remnant ridges generally mirror the drumlins' formation, except for the small-scale type found near modern glaciers (Fig. 1). Cavity-fill drumlins contain bedded sediment that indicates rapid deposition into subglacial cavities during the waning stages of a catastrophic subglacial sheetflow. Continued reduction in flow may result in channelized flows that erode parts of some drumlins. Later, eskers are considered to form in smaller tunnel-channels.

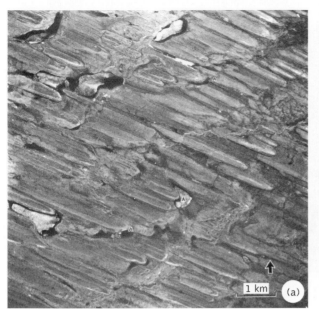

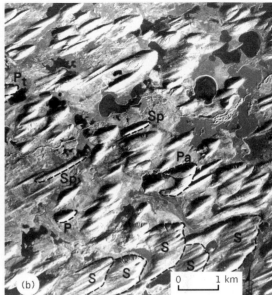

Fig. 5. Drumlins: (a) of the conventional erosional type; (b) of cavity-fill type.

Rogens, commonly called rogen moraine, resemble the transverse erosion marks that are occasionally found eroded upwards into river ice (Fig. 6). They have long been regarded as the product of smearing of underlying, pre-existing sediment. However, they frequently contain undeformed sand and gravel units, indicating that the smearing explanation alone does not work. A more recent view is that both erosion and deposition by water gave them their final form. In this alternative view, subglacial cavities are carved out by transverse 'rolling' styles of vortices that enlarge the cavities and consequently cause a local decrease in flow velocity, thereby permitting sand and gravel to be deposited. Thus, like cavity-fill drumlins, rogens can be viewed as preferred sites of deposition during the waning stages of flow.

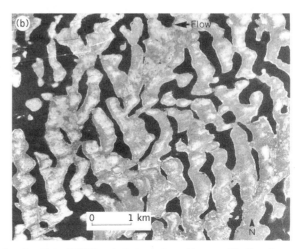

Fig. 6. (a) Transverse erosional marks cut upwards into river ice; (b) rogens from Boyd Lake, NWT, Canada that were formed by a combination of erosion and cavity filling.

Fig. 7. Hummocks cut into undeformed glaciolacustrine sediment by turbulence within subglacial sheet floods, near Maple Creek, Saskatchewan, Canada.

Hummocks in Pleistocene landscapes have usually been interpreted as moraines—comparable to those of many modern glacial systems. Recent studies indicate, however, that a range of material exists within them, including undeformed glacial and nonglacial sediment (Fig. 7). The external form, therefore, is often unrelated to the sediment contained within hummocks, indicating that the hummocks are largely erosional features. Again, in such cases, the best explanation for the mechanism of formation is by turbulent vortices in water.

Varieties of subglacial megaflood landscape features are arranged in a predictable topographic continuum (Fig. 8),

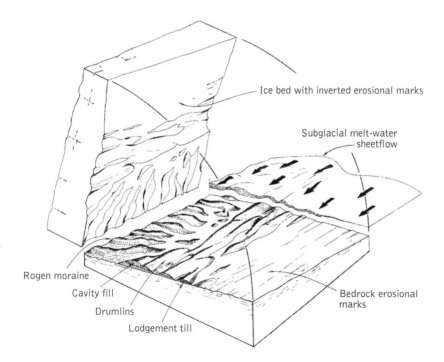

Ice bed with inverted erosional marks

Subglacial melt-water sheetflow

Rogen moraine

Cavity fill

Drumlins

Lodgement till

Bedrock erosional marks

Fig. 8. A model of subglacial landforms produced by subglacial sheet floods. The various landforms are either eroded into previously existing sediment, or are formed by ice-bed cavity erosion and filling.

suggesting that relative topographic position probably exerted some control as pressure of overlying ice and fluid sheet flow volumes varied. In higher areas, faster fluid flows occurred as smaller gaps between the ice and bed led to elongate vortices, whereas in lower areas less energetic vortices prevailed. While these ideas remain controversial, the discovery of large subglacial reservoirs under the Antarctic ice sheet, and increasing evidence of equivalent systems under the Laurentide ice sheet, are stimulating re-examinations of some long-held paradigms in geomorphology. ROBERT R. YOUNG

Further reading

Dowdeswell, J. A. and Siegert, M. J. (1999) The dimensions and topographic setting of Antarctic subglacial lakes and implications for large-scale water storage beneath continental ice sheets. *Geological Society of America Bulletin*, **111**, 254–63.

Munro, M. and Shaw, J. (1997) Erosional origin of hummocky terrain in south-central Alberta, Canada. *Geology*, **25**, 1027–30.

Shaw, J. (1996) A meltwater model for Laurentide subglacial landscapes. In McCann, B. and Ford, D. C. (eds) *Geomorphology sans frontières*. John Wiley and Sons, Chichester.

Shipp, S., Anderson, J., and Domack, E. (1999) Late Pleistocene–Holocene retreat of the West Antarctic Ice-Sheet system in the Ross Sea: Part I—Geophysical results. *Geological Society of America Bulletin*, **111**, 1486–516.

glacial rivers

Glacial rivers are distinguished by four main characteristics:

(1) their regular seasonal and diurnal cyclical fluctuations;
(2) their liability to numerous irregular and catastrophic flood events associated with sudden drainage of ice-dammed lakes;
(3) their high sediment loads, especially of suspended sediment; and
(4) their tendency to form extensive and unstable braided channel systems.

Cyclical fluctuations of melt-water run-off

Most glacial rivers are characterized by strong seasonal variations in melt-water discharge (except in equatorial glacierized basins) where discharge fluctuations closely follow the annual temperature curve. Thus, during winter there is little melt-water flow, for the catchment remains frozen and the only source of melting is from subglacial geothermal heat flux. Spring melt-water flows begin with the break-up of river ice followed in early midsummer by a rapid rise in snow-melt run-off. Further seasonal warming leads to melting of glacier ice, which generates the maximum melt-season run-off into glacial rivers because flows are now composed both of melt-water derived from ablation (melting) and rainfall inputs, and from stored water, all of which run off rapidly through fully expanded englacial routeways. By late summer, supplies of stored water are exhausted, and, as temperatures begin to decrease, discharge declines in the form of a recession curve. Finally, as autumn comes to an end, the winter freeze-back begins and flows approach their seasonal minimum.

Most glaciers also exhibit a strong diurnal cycle in run-off, reflecting diurnal temperature fluctuations, and thereby generating the equivalent of a complete flood cycle every 24 hours. The timing of the peak daily flow largely depends on the transit time for water to flow through the glacier system from different parts of the ablation zone. Hence, peak flows occur earlier in the day towards the end of the melt-season, when the tunnel systems are most fully developed, and in rivers draining small glaciers.

Irregular and catastrophic run-off events

Superimposed on the regular seasonal and diurnal cycles of run-off are two other populations of flood events. First, floods generated by summer or autumn rainstorm events which, when combined with relatively warm conditions, may form the largest floods of the melt season. Secondly, floods generated by sudden failure of an ice dam which is ponding up melt waters in an ice-marginal lake. These floods are known as 'glacier burst' or 'jökulhlaup' floods in which discharges commonly exceed 'normal' ablation-related flows by up to several orders of magnitude (see *jökulhlaups*). For example, Michael Church, working in Baffin Island, found that jökulhlaup flows from sudden lake drainage reach *c.* 200 m^3 s^{-1}, compared with more normal diurnal peak flows of *c.* 20 m^3 s^{-1}, and accounted for over 10 per cent of total melt-season run-off. By contrast, Helgi Björnsson, studying jökulhlaups from the Grimsvötn area of the Vatnajökull ice cap in southern Iceland, which result from subglacial geothermal activity, found that they can reach magnitudes of 40 000 m^3 s^{-1} (e.g. in 1922), compared with more normal run-off flows of *c.* 400 m^3 s^{-1}. Some of the largest floods on Earth have been jökulhlaups. For example, during deglaciation of the Pleistocene ice sheets in North America, 'Glacial Lake Missoula', which impounded up to 2500 km^3 of water at the edge of the Cordilleran ice sheet in Washington State, drained catastrophically through failure of an ice dam to generate peak river flows of 21 × 10^6 m^3 s^{-1}. Similarly cataclysmic floods occurred in the Altay Mountains in Siberia, where Late Pleistocene glacial lakes drained catastrophically to generate peak flows of 18 million m^3 s^{-1}.

High sediment loads

Glacial rivers are characterized by extremely high sediment loads. Suspended sediment loads generally exceed 1000 p.p.m. (parts per million) and 400 metric tons km^{-2} of catchment area per annum. However, the actual amounts also depend on the resistance of the bedrock to glacial and glacio-fluvial erosion. In ancient crystalline shield areas, for example, suspended loads of glacial rivers are significantly lower than those of young, mountainous volcanic regions. For example,

Church's measurements of suspended sediment concentrations of melt waters in Baffin Island indicated peak amounts of 1060 p.p.m., compared with 39 000 p.p.m. recorded by Kazimier Klimek from melt waters on Skeidararsandur in southern Iceland. Such high volumes of sediment are derived both from (a) a variety of sediment transport zones within the glacier and (b) non-glacial sources, such as mountain slopes, fans, avalanches, and rain- and snow-fed tributary streams. One study from the Hilda Glacier in Alberta, Canadian Rockies, for example, estimated that 6 per cent of suspended loads were derived from supraglacial and englacial debris, 47 per cent from subglacial sources, and a further 47 per cent from materials lining the banks of the proglacial rivers. Sediment sources are thus highly variable over space and time, the former reflecting the geographical distribution of sediment pathways to the proglacial environment, and the latter reflecting seasonal, diurnal, and episodic variations in run-off and availability of sediments. For example, jökulhlaup flows have been found to account for between 25 and 75 per cent of the total annual sediment transport from glacial catchments. Nummedal and his colleagues have estimated that a 2-week jökulhlaup on Skeidarasandur could move as much sand as would otherwise be transported in 70–80 years of normal melt-water flows.

Braided channel patterns

Proglacial river systems are widely characterized by the development of braided channel patterns. These result from the high degree of channel instability associated with numerous and repeated flood cycles that act to weaken the non-cohesive channel bank deposits. These are made highly erodible by the absence of binding vegetation and the paucity of silts and clays, which are washed away by daily high flows. Channel widening through repeated bank collapse leads to shallowing of the flow and erosion of meander bends, so that many glacial rivers characteristically form shallow channels of low sinuosity. High volumes of sediment laden melt waters flowing through shallow channels can lead to large-scale aggradation. In mountainous catchments aggradation is confined between valley walls to form a 'valley train' or 'valley sandur', while river systems that are unconfined can form vast 'outwash' plains or sandurs. Modern sandur plains are located typically in piedmont zones, especially those which extend seawards from young glaciated mountain ranges on to coastal plains, such as those of southern Iceland, south-western Alaska, and western Greenland. Pleistocene (ice-age) sandurs were more extensive including, for example, the Canterbury Plains of South Island, New Zealand; the Patagonian plains of southern Argentina; and those bounding much of the North American and Eurasian ice sheets. Huge glacial rivers crossed these plains, forming the ancestors of such major drainage routeways as the Mississippi and the St Lawrence, the Rhine and the Danube, the Ob and the Yenisei, many subject to catastrophic jökulhlaup flooding and drainage diversions during each episode of glacial fluctuation and ice-sheet decay.

JUDITH MAIZELS

Further reading

Baker, V. R. (1973) Paleohydrology and sedimentology of Lake Missoula flooding in Eastern Washington. *Geological Society of America Special Paper*, **144**.

Church, M. and Gilbert, R. (1975) Proglacial fluvial and lacustrine environments. In Jopling, A. V. and McDonald, B. C. (eds) *Glaciofluvial and glaciolacustrine sedimentation*, pp. 22–100. Society of Economic Paleontologists and Mineralogists Special Publication No. 23, Tulsa, Oklahoma.

Maizels, J. K. (1995) Sediments and landforms of modern proglacial terrestrial environments. In Menzies, J. (ed.) *Modern glacial environments. processes, dynamics and sediments*, pp. 365–416. Butterworth-Heinemann, Oxford.

glacial sediments Ice covers only a relatively small part (10 per cent) of the Earth's surface today. However, in the past it has extended over vast areas. In the last few million years the Quaternary icesheets covered 32 per cent of the surface; and during earlier geological history many continents that are now remote from polar regions were in the reach of ice sheets.

Ice moving from polar regions or high ground becomes armed with a load of frost-shattered rock and sediment of an enormous range of grain-sizes. As it spreads over the landscape it removes soil, weak superficial deposits, and sometimes large slabs of bedrock. It modifies the landscape: valleys are deepened and straightened. Hardrock surfaces are moulded into characteristic forms such as *roches moutonées*, which are left striated and grooved. Both these phenomena make it possible to the determine the direction and sense of movement of former ice. Subsequent melting leaves a sheet of *till* (*boulder clay* or *diamictite*) with boulders (erratics) derived from the areas crossed by the ice. Beyond the limits of the ice advance vast areas are cloaked with water-laid sand and gravel and wind-transported sediment.

Ice moving actively over its bedrock, even when retreating, deposits a layer of dense ill-sorted till with a great range of grain-size, and composed of broken and crushed rock debris (*lodgement till*). Individual clasts show a marked elongation in the direction of flow. The deposits are often lenticular with many erosional breaks due to variations in local ice flow. They are cut by channelized bodies of sand and gravel deposited in courses cut by fast-flowing high-pressure subglacial streams into the underlying bedrock. The infill of other channels entirely within the ice is left as upstanding elongate ridges (*eskers*) of cross-bedded gravels and sands with collapse structures produced when the confining ice walls melt. Elsewhere, small mounds (*kames*) are formed of cross-stratified sand deposited by streams from the ice.

When ice which is inactive and frozen to the bedrock melts, it leaves an irregular hummocky mass of less densely packed till, sometimes moulded into elongate mounds (*drumlins*) without orientated clasts (*ablation till*). When this becomes exposed, as the ice melts, it flows into the depressions to form complexly folded *flow-till*. Ridges of till (moraine) mark the position of former edges of ice sheets. These have sometimes been highly deformed to produce alpine-type folds and thrusts (*push moraines*) by ice that readvanced.

Huge volumes of sediment-laden water are released by glaciers during seasonal melting. Fast-flowing streams issue from the glacier to produce complex braided fluvial plains (*sandurs*) (Fig. 1a). These are composed of horizontally bedded gravels and plane and cross-bedded sands in a complex of channels and bars which generally fine away from the ice edge. Near the contact with the ice they are usually deposited on unmelted ice, and when this melts become deposited as highly contorted masses of sand and gravel. Where outwash sands and gravels reach the coast, sandy deltas form and the sediment is reworked and fashioned into beach-barrier complexes by waves as in ice-free areas. They enclose swamps and lagoons in which peat develops and may produce cold-climate coals if preserved. The coastal sediments show interesting features produced by freeze-thaw and distortion produced by pack-ice impinging on the shoreline that are not found in other climatic regimes.

Winds blowing over the unvegetated surface of the outwash plains winnow out the finer sediment and leave gravel lags on which the gravel clasts are faceted by sand-blasting (*venti-*

facts). Extensive sheets of cross-stratified aeolian sands form (Fig. 1a) with disturbances and slumps resulting from freeze-thaw of the interstitial water (e.g. the *cover sands* deposited in Holland during the last Quaternary glaciation).

Finer sediment is transported great distances from the edges of the former ice-sheets to blanket the surrounding areas in fine sandy-silt and sand (*loess*) (Fig. 1a). This covers hundreds of thousands of square kilometres ($800\,000$ km^2 in the USA) and may be hundreds of metres thick (300 m in China).

Till, cover sands, and all other sediments deposited by ice, as well as those of the bordering regions (*periglacial regions*) are commonly disturbed by freeze-thaw processes: ice wedges fracture the surface; melting of groundwater produces fluidization and expansion during freezing, which produces intense deformation (*cryoturbation*); mobilization and flow down slopes forms ill-sorted gravity-flow sediments; and repeated freeze and thaw sorts the various grain sizes to form stone *polygons* and *stripes*.

Lakes are very abundant in glacial terranes. Some of them occupy valleys overdeepened by ice erosion; others valleys blocked by ice or till; some occur as depressions on the uneven till surface left by retreating ice. Where they are in contact with the ice (*proglacial lakes*) meltwater builds rapidly deposited deltas of coarsely bedded gravel and sand with many liquefaction structures (Fig. 1b). Sediment is transported as density currents by the cold dense heavily laden waters and deposited as graded sands (turbidites). These are interbedded with fine-grained muds which settle during more tranquil periods. The coarser sediment grades into well-laminated deposits in deeper water. Similar laminated deposits are formed in more distal *periglacial lakes*. The deposits consist of couplets of silt and clay layers with sharp boundaries (the couplet being called a *varve*). The silty layers are in some instances graded or multiple owing to deposition by either single or repeated events. The darker clay layers may also show grading. The couplets are produced by the introduction and deposition of silt during summer melting followed by slow deposition of fine suspended clay during winter, often under an ice cover.

The continental shelves surrounding glaciers are cloaked with till deposited by ice rafting from the adjacent land. They are commonly cut by erosional surfaces with gravel concentrates produced by storm-generated currents. Muds with isolated clasts (*dropstones*) are interbedded with these deposits together with carbonate skeletal debris of benthonic organisms formed during periods of limited glacial supply. On the outer shelves the sediments may consist mainly of tests of planktonic foraminifera and diatoms.

The shelf sediments pass across the shelf edge into a complex of submarine fans, which apart from the greater abundance of coarse detritus of striated or faceted clasts resemble similar deposits in other climatic zones (Fig. 1c). These deposits grade seawards into normal pelagic sediments with turbidities, occasional lenses of sand or gravel, or isolated dropstones released from melting icebergs (Fig. 1d).

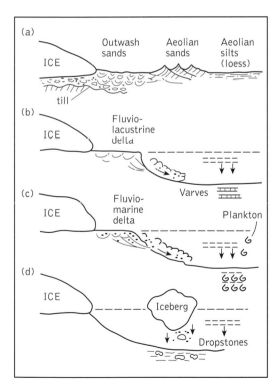

Fig. 1. Glacial sedimentary environments: (a) terrestrial sedimentary pattern; (b) lacustrine sedimentary pattern; (c) marine sedimentary pattern; (d) marine sedimentary pattern with icebergs. (Based on Edwards (1978) Blackwell Scientific Publications, Oxford.)

As well as being abundant over continental areas covered by Quaternary glaciations, sediments and features of glacial origin are found elsewhere within the Phanerozoic sedimentary record. Glacial deposits are particularly well known in the Pre-cambrian, Ordovician, and Permo-Carboniferous. They include striated and moulded rock surfaces with scattered boulders, thin sheets of ill-sorted indurated till (*tillite*), and associated fluvial and eolian sandstones, and fine-grained laminated sandstones-mudstones with dropstones.

G. EVANS

Further reading

Edwards, M. (1996) Glacial environments. In Reading, H. G. (ed.) *Sedimentary environments* (3rd edn), pp. 445–70. Blackwell Scientific Publications, Oxford.

Eyles, N. and Miall, A. D. (1984) Glacial facies. In Walker, R. G. (ed.) *Facies models*, pp. 15–38. Geoscience Canada Reprint Series 1. Association of Canada, Toronto.

glacier mass balance and climate The relationship between glaciers and climate has been studied for many years. Its basis lies in the observed correlation of glacier advances with climatic deterioration and glacier retreat with climatic amelioration. However, this relationship is not as straightforward as it at first seems.

It is useful to think of glaciers as systems, with zones of inputs and outputs of mass and energy and zones of storage (Fig. 1). Ice is obviously the most significant component of glacier mass. Mass is added to a glacier through a number of processes. The most important of these is through the accumulation of snowfall in the upper reaches of the glacier (called the accumulation zone). Over time and with the increased pressure as further snowfalls are added, this snow turns into ice and is incorporated into the glacier. Hence the glacier gains mass. Other accumulation processes include avalanches, the freezing of basal melt water and the addition of rime. Towards the snout of the glacier, mass is lost from the system mainly through melting but also by other processes such as calving (the loss of large ice masses from the glacier into bodies of water) deflation (removal of ice by wind), and sublimation (the phase change of ice directly to water vapour in cold, dry environments). These processes are collectively termed ablation. The difference between accumulation and ablation for a glacier over one year is termed the net mass balance. The mass balance thus describes the 'health' of a glacier.

When a glacier displays positive mass balance, accumulation of mass is greater than ablation. This can be achieved by an increase in snowfall, for instance, or by a decrease in ablation. The glacier will respond to such changes in many ways but will tend to thicken and advance its snout. Conversely, under conditions of negative mass balance (when ablation is greater than accumulation) the glacier will lose mass by retreat of its snout or by lowering and thinning of its surface. Negative mass balance may also have a climatic cause, such as an increase in temperature (which increases the melting component of ablation) or a decrease in total precipitation (which reduces accumulation).

Zones of accumulation and ablation can therefore be seen to have a spatial dimension, accumulation processes being most effective at higher altitudes and ablation becoming predominant near the snout of the glacier. In addition, these processes will also vary temporally; during the winter the glacier may accumulate mass over its whole surface, whereas during the summer ablation processes may take over. Ablation is usually greatest at the snout and falls to zero at higher altitudes. Accumulation is greatest at higher altitudes and falls to zero at lower altitudes. The zones of net accumulation and net ablation are thus separated by a point where accumulation and ablation are both zero; this is called the equilibrium line.

It is clear then that glaciers will respond to climatic variations by changes in input and output. The speed with which a glacier will respond to a climatic stimulus is called the response or relaxation time and is defined as the time taken for the glacier to reach equilibrium after a change in input. The response time tends to be longer for large glaciers than for small ones owing to the larger amount of ice flux through the system that is required to initiate changes in the terminus.

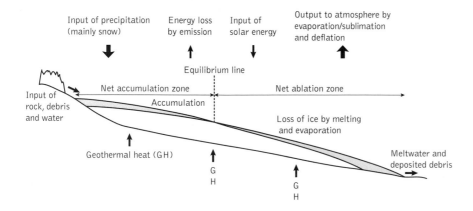

Fig. 1. Simplified diagram showing the main inputs and outputs of mass and energy of a non-calving glacier. (Adapted from Mannion, A. M., Mitchel, C. W., Parry, M., and Townshend, J. R. G. (1986) *Processes in physical geography*. Longman, Harlow.)

As a result, if changes in inputs occur often, some large glaciers may never achieve equilibrium. The shape of a glacier may also affect the response time through its influence on the distribution the area of a glacier with altitude.

The position of a glacier snout is generally more responsive to climatic changes than the glacier surface; it may fluctuate significantly on a yearly timescale while the surface height remains relatively constant. In addition, the positions of glacier snouts respond more sensitively to periods of negative mass balance (by showing retreat) than periods of positive mass balance (when the snout may be stationary but the thickness of the glacier increases). This is because increased ice flux from the accumulation zone may take a number of years to affect the position of the snout (yet will lead to a thickening of the glacier), while increased ablation will make its effects felt first at the terminus of the glacier.

Many morphological variables must be analysed to explain the behaviour of glaciers; these include aspect and relief. Climatic controls include temperature, precipitation, and continentality (Fig. 2).

Temperature. Although glaciers receive heat from a number of sources, including geothermal heat and frictional heat at the glacier bed, solar radiation has the biggest impact on ablation. Melting of the ice takes place in two ways. Short-wave radiation from the Sun and long-wave radiation from the valley sides, from water vapour, and from material lying on the glacier is affected by the reflectivity of the glacier surface (albedo). This is greatest when the glacier is white (usually in early summer before the winter snow cover has melted) and least in late summer when ablation is therefore most significant. Secondly, conduction of heat to the glacier surface from the air and condensation of water vapour are important variables causing ablation and are magnified by high winds.

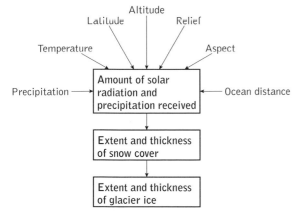

Fig. 2. Geographical variables controlling the extent and thickness of glacier ice. These may operate and change over different timescales. (Adapted from Sugden and John (1976).)

As a result, there seems to be a close relationship between glacier ablation and mean summer temperature.

Precipitation. Knowledge of total yearly precipitation for an area in which glaciers are present does not necessarily give an accurate estimate of glacier cover or behaviour. On low-latitude glaciers, for instance, much, or all, of the precipitation may fall as rain during the summer months and run off the glacier and out of the system as melt water. This precipitation is glaciologically ineffective, contributing little to the mass balance of the glacier. As a result, glaciers in areas of very high precipitation (e.g. the west coast of Canada, the South Island of New Zealand, and the Patagonian icefields in South America) may have very high accumulation rates, but these are offset by equally high ablation. In consequence, these glaciers fluctuate wildly in response to changes in these variables.

On the other hand, in extremely arid and cold environments (northern Greenland and central Antarctica, for instance) glacier ice is very stable and long-lived, because most (if not all) of the precipitation falls as snow and, by contributing directly to glacier mass, is glaciologically effective.

Continentality. Distance to the nearest ocean (or continentality) has been shown to be a significant variable accounting for the amount of precipitation received by glaciers. Continentality also affects the temperature regime experienced by the glacier. Glaciers in coastal locations tend to have higher accumulation and ablation rates (if they are in mass balance) than glaciers in continental interiors. This is because air from the oceans is moist, having picked up water vapour through evaporation, resulting in high levels of snowfall over the glaciers. In extreme continental areas precipitation may be very low but, owing to the low temperatures, may be preserved as snow.

Although the glacier–climate relationship has been demonstrated from many mountainous regions of the world, there are some glaciers which behave asynchronously to climatic inputs. One of the most important factors which disrupt this glacier–climate signal is when calving makes up a significant component of the ablation process.

When ablation is dominated by glacier melting, retreat or advance of the snout can be seen to be related to climatic factors. Calving, on the other hand, involves the large-scale and often catastrophic release of ice blocks from the glacier snout into water. This process is very rarely climate-driven; it is affected more by factors such as water depth at the calving front and topographic factors such as valley width. In addition, it is now known that the calving speeds of glaciers vary, according to whether the glacier ends in fresh water or tide water; freshwater glaciers are more stable and calve more slowly. The reasons for this behaviour are, as yet, unknown.

Precipitation seems to be the forcing mechanisms which accounts for the climatic component in calving glacier dynamics. This is because ablation is dominated by calving and not by melting; hence this suppresses the importance of temperature. In areas where glacier snouts are both land-based and end in water they may display asynchronous responses to climatic change, with land-based glaciers responding to variations in temperature and calving glacier fluctuations being controlled by precipitation and topographic factors.

Although the links between climate and glacier mass balance are evident, the study of climate change through knowledge of glacier oscillations is fraught with difficulties. For instance, there are only a few glaciers for which we have detailed mass balance records for periods of more than a few years. In addition, in most cases the relative contributions to the mass balance equation of non-climatic processes such as avalanches and calving are not known. As a consequence, it is often extremely difficult to extract the climatic signal from the fluctuation behaviour of glaciers.

However, in the absence of long-term climatic records in many remote parts of the world, glacier fluctuations provide us with the only record of climate change. This 'proxy record' allows us to reconstruct past climates and to monitor present-day climatic fluctuations. Correct interpretation of the glacio-climatic record depends on a proper understanding of the response of glaciers to variations in climatic inputs.

STEPHAN HARRISON

Further reading

Sugden, D. E. and John, B. S. (1976) *Glaciers and landscape.* Edward Arnold, London.

glaciers and glaciology
A glacier will form whenever a body of snow accumulates, compacts, and turns to ice. If large enough, this body of ice will flow, thus forming a glacier. Glaciology is the study of glaciers. Glaciers range in size from small valley glaciers, such as those in the European Alps, to large ice sheets, such as that in Antarctica which today contains 91 per cent of all the glacier ice on Earth. In the past, during the Quaternary Ice Age, large ice sheets covered much of northern Europe and North America and the landscape of these regions is dominated by glacial landforms and glacial sediments.

Glaciers can be classified in terms of a combination of topography and size. There are two broad categories: those glaciers that are constrained by topography, and those that are not. Ice sheets and ice caps are unconstrained by topography and are sufficiently large to drown much, if not all, of the landscape over which they flow. An ice sheet is simply a large ice cap (>50 000 km²). Ice sheets are composed of several smaller elements. In the centre there may be one or more domes forming the central topography of the ice sheet. Ice flows from these domes (ice divides) towards the ice-sheet margin, either as a broad sheet or in well-defined outlet glaciers or *ice streams*. Where ice sheets terminate in water they may form *ice shelves*, which are floating slabs of glacier ice. In Antarctica 7 per cent of the ice volume is due to ice shelves, the largest of which is the Ross Ice Shelf over which polar explorers such as Shackleton, Scott, and Amundsen travelled to the South Pole.

Three main types of glacier are constrained by topography: the ice fields, valley glaciers, and cirque glaciers. An *ice field* is an interconnected network of valley glaciers and snow fields, while a *cirque glacier* is a small, discrete body of ice within an armchair-shaped depression.

Glacier mass balance and flow

Glacier formation and size is controlled by the balance between the accumulation of snow or ice and its loss through ablation. Ablation includes losses due to evaporation (sublimation), melting, and the breaking off (calving) of icebergs where glaciers terminate in water. A glacier forms whenever the accumulation of snow/ice exceeds ablation over a sustained period of time. Once a glacier is established, this balance will also determine glacier size and is known as a glacier's *mass balance*. If glacier mass balance is positive, that is, if there is more accumulation than ablation, then the glacier will grow in size to cover a greater area. If, however, the mass balance is negative (that is, if there is more ablation than accumulation), then the glacier will shrink and ultimately disappear. Changes in mass balance need to be sustained over a number of years to have an impact on the size of a glacier.

Accumulation and ablation do not occur uniformly over the surface of the glacier, and it is this that causes glaciers to flow. Snowfall, and therefore accumulation, increases with altitude and is greatest high up on the glacier. In contrast, ablation decreases with altitude, as it gets colder and less snow/ice melts. A glacier can be divided, therefore, into an upper accumulation zone, where accumulation exceeds ablation, and a lower ablation zone, where ablation is greater than accumulation (Fig. 1a). Where these two zones meet there is a balance between ablation and accumulation, a point on the glacier surface known as the *equilibrium* or *firn line*. This imbalance in accumulation and ablation over the surface of a glacier drives glacier flow. The addition of snow and ice in the accumulation zone and its loss in the ablation zone cause the glacier's downslope or long profile to steepen, increasing the stress on the ice crystals within it (Fig. 1b). Above a critical level of stress, ice crystals will deform or creep past one another. In this way, ice is transferred from the accumulation zone to the ablation zone to counteract the increasing glacier slope and associated build-up of shear stress.

The degree of imbalance between accumulation and ablation across the glacier determines the velocity of ice flow. The greater the difference between accumulation and ablation across the glacier, the greater the amount of ice that must be transferred to maintain an equilibrium glacier slope, and consequently the faster the glacier will flow. The difference in accumulation and ablation across a glacier is known as the *mass balance gradient*. A glacier with a high mass balance gradient (that is, a large amount of both accumulation and ablation) will flow faster than a glacier with a low mass balance gradient. Warm, wet, maritime climates give high mass balance gradients and have fast-flowing glaciers. The

(a)

Ice sheet

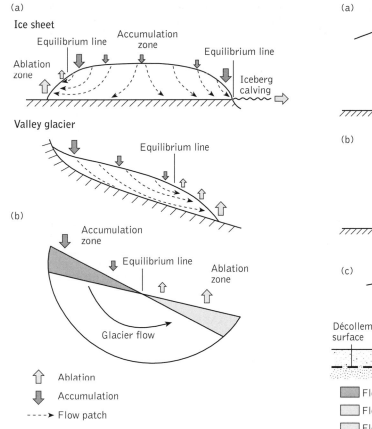

(b)

⇧ Ablation

⬇ Accumulation

----➤ Flow patch

Fig. 1. (a) Schematic diagrams of an ice sheet and a valley glacier, showing the location of the equilibrium line, the accumulation zone, and the ablation zone. Principal flow paths are also shown. (b) Idealized valley glacier, showing glacier flow in response to the uneven distribution of accumulation and ablation on a glacier surface. Accumulation at the top of the glacier and ablation at the bottom causes a steepening of the glacier profile, resulting in ice deformation and mass transfer from the accumulation to the ablation zone.

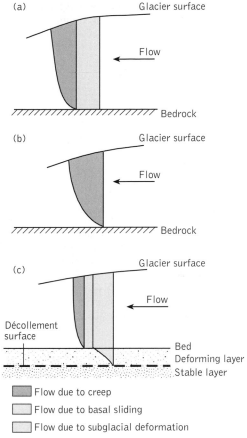

Flow due to creep

Flow due to basal sliding

Flow due to subglacial deformation

Fig. 2. (a) Mechanisim of glacier flow for a warm-based glacier resting on bedrock. The solid vertical line on the right-hand side is deformed by glacier flow. (b) Mechanism of glacier flow for a cold-based glacier resting on bedrock. (c) Mechanism of glacier flow for a warm-based glacier resting on deformable sediment.

Franz Josef glacier in New Zealand is a good example, with a velocity of 300 m year^{-1}. Dry, continental climates, in contrast, are associated with slower-moving glaciers; for example the Meserve Glacier in Antarctica flows at only 3 m year^{-1}.

Glacier flow is not restricted to ice deformation or creep, but may also occur through internal deformation, basal sliding, and subglacial deformation (Fig. 2). Internal deformation is the folding, faulting, and thrusting of ice which causes displacement and flow. Basal sliding may occur by one of two processes: enhanced basal creep and regelation slip. At a glacier bed the pressure of ice pushing against basal obstacles cause the ice to deform or creep at a faster rate (enhanced basal creep). Similarly, this pressure against obstacles may cause ice close to its melting point to melt. The resulting melt water

flows round the obstacle to freeze again as regelation ice downstream of the obstacle where the pressure is lower (regelation slip). As a result of both these processes, ice may slide or move rapidly over its bed. Glaciers can also flow by subglacial deformation. When glaciers flow over unfrozen sediment they may cause this sediment to deform beneath the weight of the ice. The deformation occurs when the water pressure in the pores or spaces between the sediment grains increases sufficiently to reduce the internal friction of the sediment. This allows the grains to move or flow relative to one another as a slurry-like mass. In response to the shearing force imposed by the overriding glacier, this slurry forms a continuously deforming layer on which the glacier moves. This process may be dramatic. In the case of the Icelandic glacier Breiðhamerkurjökull, 90 per cent of the glacier flow is due to subglacial deformation. The principal factor controlling which combination of flow processes—creep, basal sliding, or subglacial deformation—operates beneath a given glacier is the temperature of the basal ice, its basal thermal regime.

Basal thermal regime

The *basal thermal regime* is the temperature of the ice at the glacier bed. Some glaciers are frozen to their beds. No melt water is present at the ice–bed interface and such processes as basal sliding and subglacial deformation cannot therefore operate. Such glaciers are known as cold glaciers. In contrast, other glaciers are composed of warm ice, where basal ice is constantly melting and the ice–bed interface is therefore lubricated with melt water. Basal sliding and subglacial deformation may therefore both operate. A glacier composed of warm ice has a much greater potential for fast flow and therefore greater potential to modify its bed by erosion, producing glacial landforms, than one which is frozen to it. Basal thermal regime is therefore one of the most important controls on the geomorphological impact of a glacier, since it controls the processes of erosion, transport, and deposition.

Not only does basal thermal regime vary between glaciers, but it may also vary within a particular ice body. In some cases the glacier margin may be cold and frozen to the bed while the middle of the glacier is warm and melting, such glaciers are referred to as *polythermal.*

The temperature at the base of a glacier is determined by the balance between (1) the heat generated at the base of the glacier by friction and from geothermal sources and (2) the temperature gradient within the overlying the ice. The temperature gradient within the overlying ice is critical; if it is positive, heat is conducted away from the glacier bed and it may be frozen. If the gradient is negative, heat is not conducted away and may build up to cause basal melting. This temperature gradient is determined by ice thickness, climate, mass balance, and the flux of cold and warm ice through the glacier.

In warm temperature regions most glaciers are warm-based, while in cold, continental regions, such as Antarctica, they may be completely cold-based. In practice, the pattern of basal thermal regime within a glacier is complex and varies through time with changes in glacier geometry. The pattern of basal thermal regime will control the patterns of erosion and deposition within a glacier and therefore the regional character of the glacial landscape produced.

MATTHEW R. BENNETT

Further reading

Bennett, M. R. and Glasser, N. F. (1996). *Glacial geology: ice sheets and landforms.* John Wiley and Sons, Chichester.

glaciotectonics Glaciotectonic deformation may be defined as 'structural deformation as a direct result of glacier movement or loading' (INQUA Work Group on Glacier Tectonics

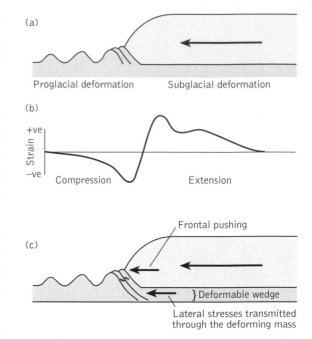

Fig. 1. (a) A schematic diagram showing the position of subglacial and proglacial glaciotectonic deformation. (b) The theoretical pattern of longitudinal strain. (c) Schematic diagram of the forces producing proglacial deformation, frontal pushing and compressive stresses transmitted through a subglacial deforming layer.

1988). This topic has received a great deal of study in recent years because glaciotectonic processes have been related to glacial dynamics (Fig. 1). There are two types of glaciotectonic deformation formed by the action of a moving glacier:

1. *Subglacial deformation*, which takes place beneath the glacier and is characterized by simple shear and extensional tectonics, i.e. attenuated folds, boudins, and augens, and results in the formation of deforming bed till and/or flutes and drumlins;
2. *Proglacial deformation*, which takes place at the glacier margin and is characterized by pure shear and compressional tectonics, i.e. open folds, thrusts, and nappes. This results in the formation of push moraines.

Additionally, deformation also occurs within the glacial environment as a result of gravitation instabilities associated with stagnant ice and is known as *dead-ice tectonics*. Within the progracial area, features that are associated with melting ice form, such as ice-collapse structures in the outwash plain and debris-flow mobilizations of till. In the subglacial environment, 'crevasse infill' structures are formed when saturated tills are extruded up through crevasses on to the glacier surface.

Subglacial glaciotectonic deformation

Many research workers have shown that when a glacier moves over an unconsolidated bed there is a coupling between the glacier and the underlying bed. This leads to an increase in the velocity or a decrease in the slope angle of the glacier, or both, and the deformation of the sediments below. This deformation occurs because the water produced at the base of a temperature glacier cannot drain away well enough to prevent the build-up of pore-water pressures and the associated reduction in the strength of the subglacial sediment.

Subglacial deformation has been recorded beneath a number of modern glaciers, in particular Breiðhamerkurjökull, in Iceland, and Ice Stream B, in western Antarctica. It has also been inferred from sedimentological studies of the deposits from several neoglacial surging glaciers and Pleistocene glaciers (where the ice moved over unconsolidated rocks associated with the European (and British) and Laurentide (North American) ice sheets).

Subglacial deformation occurs within a subglacial shear zone. This is a thin layer undergoing high ductile shear in which rocks are broken and ground up by mechanical processes (cataclastic mylonization). The resultant till (a deforming bed or deformation till), is a primary till formed from a combination of both deformation and deposition.

At relatively low shear strains, deformation is apparent from the slight deformation of strain markers, such as the overturning of ice-wedge casts. At higher shear strains, criteria for the recognition of subglacial deformation include folds, tectonic laminations, boudins, augen-like features, and rotated clasts and boulder pavements. At very high shear strain (or sites with homogeneous bedrock), the deformation processes result in the homogenization of the till, and so macroscale structures may not be visible. Instead, criteria such as a specific till fabric (low strength associated with a thick deforming layer, high strength associated with a constrained deforming layer), specific micromorphology structures (e.g. rotation of clasts, shears), and a high fractal dimension can be used.

It has also been argued that subglacial streamline bedforms (flutes and drumlins) are a product of subglacial deformation. These features form because of the presence of more competent masses (or cores) within the deforming layer, which act as obstacles to flow.

Proglacial glaciotectonic deformation

Proglacial glaciotectonic deformation is generally characterized by large-scale compressional folds and thrusts. These are very common and are associated both with modern and with Pleistocene glaciers.

Proglacial deformation spans a continuum of compressive deformation styles, from folding (ductile deformation) to stacking (brittle deformation), both of local and of far-travelled sediment or rock masses, and the final result depends upon the rheology and competence of the bed and the behaviour of the ice sheet.

The usual result of the proglacial deformation processes is to produce a topographic ridge transverse to the ice margin, called a push moraine. There is often a basin up-glacier from which the material of the ridge has been removed. These features do not, however, always have a topographic expression because many proglacial structures have been subsequently overridden by ice and so have become incorporated into drift deposits.

Thrust blocks (within push moraines) can be modelled rather like nappes in hard-rock terrains; they move along incompetent rock units or planes of weakness resulting from high pore-water pressures, and are formed by lateral push, gravity gliding, or gravity spreading. Some push moraines form simply by lateral push, and need not necessarily be associated with the deforming bed. In contrast, where a deformable bed is present, the push moraine forms by a combination of frontal pushing and stresses transmitted through a subglacial deformable wedge (the latter being a direct result of the gravity spreading of the ice sheet). Small-scale versions of this phenomenon are common in modern glacial environments associated with deforming beds. Here the push moraine is composed of subglacial till that has been moved out into the foreland while the slower-moving sediment remains behind in the form of flutes.

Conclusions

There have been many studies of glaciotectonics in recent years because it has been shown that glaciotectonic deformation is a fundamental process within the glacial environment. It is not 'special' or unusual, but occurs in association with glacier movement, deposition, and deformation. Very few modern-day glaciers do not have push moraines currently being formed at their margins. Evidence for deformation (both on a macro- and microscale) within tills is similarly very common, and is now being reported regularly. Future research in glaciotectonics is, however, needed in the field of thermal properties, and the geotechnical behaviour of tills and their relationship to ice dynamics.   J. K. HART

Further reading

Hart, J. K. and Martinez, K. (1997) *Glacial analysis*. CD-ROM, Routledge, London.

Wateren, van der F. M. (1995) Processes of glaciotectonism. In Menzies, J. (1995) *Modern glacial environments: processes, dynamics and sediments*, pp. 309–33. Butterworth-Heinemann, Oxford.

glass (natural and industrial)
Glass is an amorphous material which lacks the structure that characterizes crystals; glass is formed when molten materials are cooled too rapidly for crystallization to occur. Natural silica-rich liquids readily form glasses, and acid volcanic activity frequently generates the glassy rock known as obsidian, which has a distinctive black appearance and conchoidal fracture.

Basic magmas must be cooled much more rapidly if glasses are to be formed. In consequence, the only basaltic glasses known have been quenched by contact with water (e.g. rims on pillow lavas) or air (spatter). Glasses formed by meteorite impact are also known. Because of their tendency to crystallize (devitrify), most natural terrestrial glasses are geologically young. The presence of glass 4 billion years old on the Moon indicates that water is crucial to the rate of devitrification.

Any liquid can be quenched to form a glass, and technologically important glasses are very varied in composition. The most familiar type of glass used in windows and bottles is made from sand, sodium carbonate (obtained from salt), and limestone, but specialist glasses can contain components as diverse as boron, lead, rare-earth elements, and fluorine.

The colours in industrial glass are usually due to small additions of impurities. These impurities can dissolve either as ions (e.g. Co^{2+} (blue), Cr^{3+} (green) and Ni^{2+} (yellow or purple)), single atoms, or tiny particles (e.g. gold or copper, which give a ruby colour).   S. C. KOHN

global geomorphology Global geomorphology refers to the study of regional topography and processes that shape regional topography over geological time. Geomorphology at these scales reflects the sum total of local processes acting upon the landscape. During the past century of research, geomorphologists have altered their approach, from emphasizing time-dependent evolution of regional landforms, based on W. M. Davis's *Geographical Cycle*, to focusing on local time-independent process–response studies. Davis's model, with its elegant progression of landscapes from 'youth' to 'maturity' to 'old age', is often blamed for leading generations of geomorphologists on fruitless searches for low-relief peneplains. In retrospect, the greater disservice of the Geographical Cycle may have been that the backlash against it caused additional generations of researchers to eschew regional synthesis and turn blind eyes to the role of time in geomorphological systems. More recently, however, researchers have begun to utilize new techniques to re-examine regional topography and long-term landscape evolution (see *landscape evolution*).

Examining landscape at a global scale (Fig. 1), we see that the Earth's topography reflects the operation and interaction of three major subsystems: tectonics, surface processes, and climate. The Earth's area–altitude distribution is distinctly bimodal, with one mode near sea level and the second at depths of roughly 4–6 km. This bimodal distribution reflects the tectonic differentiation of oceanic and continental crust. Tectonics is responsible for the most dramatic contrasts in the long-wave length topography visible at this scale, including the deepest spots on the surface—the oceanic trenches—and the highest peaks. The most dramatic manifestations of tectonic activity, at least to our human, land-dwelling eyes, are the Earth's mountain ranges. Most mountain ranges are intimately associated with the boundaries of the mobile lithospheric plates. Mountain building can be viewed as the tectonic addition of mass to the crust, through either compression or magmatism.

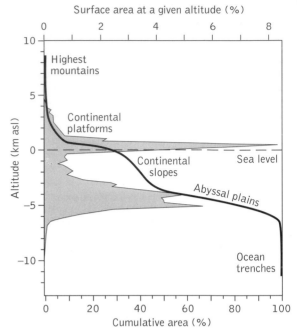

Fig. 1. Area–altitude distribution of the surface of the Earth. For a given altitude, both the percentage area (shaded plot) and the cumulative area higher than that altitude (bold line) are shown. The most striking feature of this distribution is that the Earth's topography is bimodal, representing the differences between continental and oceanic crust. (Modified from E. A. Keller and N. Pinter (1996) *Active tectonics*, Fig. 9.2, p. 292, John Wiley and Sons, New York; and P. J. Wylie (1976) *The way the earth works*. Fig. 3–11, p. 38, John Wiley and Sons, New York.)

At one time or another during the eons of geological time, tectonic activity has acted upon all of the Earth's continental crust (to some, this action defines continental crust and explains its origin). If mountain building has been so pervasive in the past, why then do the high-altitude areas in Fig. 1 represent such a minuscule portion of the continents? The answer, of course, is that erosion acts over time to remove mass, transport it, and eventually deposit that material close to sea level. The low- to mid-altitude areas of the Earth generally represent regions where geomorphological processes have predominated for long periods of time; hundreds of millions of years or longer in some areas. In addition, geomorphological processes are responsible for generating relief. Whereas tectonic activity may supply the thickened crust and the high surface altitudes, most of the Earth's dramatic scenery, from sea cliffs to continental escarpments to deep mountain valleys, is cut by erosional processes.

Climate is also central to the creation and evolution of regional-scale topography. Solar insolation and atmospheric and oceanic circulation determine the patterns of temperature and precipitation across the Earth's surface, which in turn determine the distribution and rates of virtually all geomorphological processes. Under otherwise equal conditions, temperature will determine whether glacial or fluvial processes will predominate in a region, and precipitation rates will determine whether a limestone outcrop is the most resistant unit in a landscape or the least. Climatic conditions also control the rates at which most geomorphological processes operate. For example, precipitation rates tend to increase and average temperatures decrease with increasing altitude. Measurements suggest that, as a result, rates of erosion tend to be greater in high-altitude drainage basins. This effect is also partially the result of the fact that high-altitude, mountainous regions have increased topographic relief, and taller and steeper slopes result in faster rates of gravity-driven processes. Integrated over time, the increased rates of erosion at high altitudes mean that growing mountain ranges tend to be self-destructive—the higher they grow, the more potent becomes the force of climate-driven erosion in wearing those mountains away.

The 'cybernetic' model of regional landscape

The new model of regional topography and long-term landscape evolution has been called 'cybernetic', and not because the model implies that electronic or mechanical forces are at work. The term comes from the Greek '*kybernetes*', which refers to piloting or steering, as in a ship. 'Cybernetic' refers to the command-and-control elements in an integrated and self-regulating system. The model of mountain building is called 'cybernetic' because it emphasizes the strong interactions and feedback between tectonics, surface processes, and climate in the geomorphic system. Changes in any one of these three sub-systems seem to be capable of causing major changes elsewhere in the system.

One of the most significant feedback mechanisms at work is isostasy. This refers to the processes and mechanisms by which topography at the surface is supported by the buoyancy of the crust beneath. As a result of isostasy, surface processes appear to be capable of guiding or even accelerating tectonic processes deep within the crust. The removal of mass at the surface by regional erosion causes an upward force of isostatic uplift as the lithosphere acts to re-establish buoyant equilibrium. The rate of isostatic uplift is always less than the rate of denudation so that, in the absence of any additional tectonic uplift, the average altitude of topography can only decline in response to the combined effects of erosion and isostasy.

The significance of isostatic uplift becomes most apparent when one considers those few regions where it is essentially absent; for example, the Tibetan Plateau. Continental collision of India with southern Asia has uplifted the plateau to a mean altitude above 5000 m, and some researchers have suggested that this is as high as the Tibetan Plateau can go. With little precipitation and no large-scale redistribution of mass by erosion—because of internal drainage—the mass of the plateau may act as a 'gravitational cork', blocking any additional tectonic uplift. But how can 5000 m be a gravitational limit when the Himalayas just to the south have peaks of nearly 9000 m? The average altitude throughout the Himalayas is about equal to that of the Tibetan Plateau, with the high peaks matched by an equal area of deeply incised gorges. The difference between the Tibetan Plateau and the Himalayas, it is suggested, is that the latter is characterized by through-going river systems that cut the deep valleys and promulgate erosion through the region. Isostatic compensation to erosional mass removal apparently unplugs the gravitation cork and allows tectonic uplift to continue.

Evolution of regional topography over time

Recent recognition of the significance and impacts of the linkages between tectonics, surface processes, and climate has led to greatly improved qualitative and quantitative models of long-term landscape evolution. W. M. Davis's Geographical Cycle was based on the flawed assumption that tectonic construction occurs in brief, powerful spasms and is followed by the much slower effects of erosion. Geologists have since learned that tectonic activity may persist for long periods of geological time, that erosional denudation may be just as rapid as tectonic uplift, and that the two sets of processes almost always act simultaneously.

Japanese researchers were among the first to formulate a comprehensive model of landscape evolution rich in feedback (Fig. 2). This model bears a superficial resemblance to the Geographical Cycle in that it, too, suggests a three-stage progression. During the first, or 'developing stage', uplift rates exceed denudation rates so that average altitudes increase over time. In Japan, low-relief surfaces of Tertiary age diminish in extent as altitudes increase, relief increases, and erosion rates rise. During the second, or 'culminating' stage, uplift rates and denudation rates are regionally balanced. If tectonic activity is sufficiently long-lived, and if climatic conditions allow erosion rates that are high enough, then the topography may enter a period of dynamic equilibrium with no change in regional average altitude. The third, or 'declining' stage could commence with either a reduction in tectonic uplift or a sustained increase in denudation. A landscape in decline would be characterized by declines in average altitude and lithospheric thickness—rapid at first but gradually decelerating—as both the high-standing topography and the associated isostatic root were gradually consumed by erosion over time.

Numerical models of regional topography based on a strongly coupled tectonic–geomorphological–climatic system have yielded strikingly realistic simulations of landscape evolution under various conditions. Perhaps for the first time, researchers are bridging the gap between process-based empiricism and regional synthesis. Future research is likely to add additional degrees of detail to these models, incorporating such complexities as lithological heterogeneity and climatic variability over time.

NICHOLAS PINTER

457

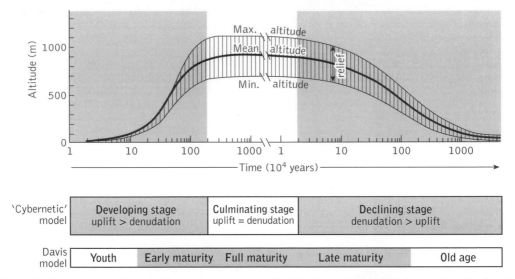

Fig. 2. Model of landscape evolution due to simultaneous tectonic construction and erosional destruction. According to this model, the most important factor controlling regional average altitude change is the balance between tectonic uplift and climate-driven denudation.

(Modified from H. Ohmori (1985) A comparison of the Davisian scheme and landform development by concurrent tectonics and denudation. *Bulletin of the Department of Geography, University of Tokyo*, **17**, 19–28, Fig. 4.)

Further reading

Keller, E. A. and Pinter, N. (1996) *Active tectonics: earthquakes and landscape.* Prentice Hall, Upper Saddle River, NJ.

Ohmori, H. (1985) A comparison of the Davisian scheme and landform development by concurrent tectonics and denudation. *Bulletin of the Department of Geography, University of Tokyo*, **17**, 19–28.

Pinter, N. and Brandon, M. T. (1997) How erosion builds mountains. *Scientific American*, **276**, (4), 74–9.

Global Positioning System (GPS)
The Global Positioning System (GPS) is operated by the US Department of Defense and can be used worldwide for both navigation and accurate positioning. The system consists of 24 satellites in high (20 000 km altitude) orbits, positioned so that at least four satellites are always visible to an observer at any point on the Earth with a good view of the sky (Fig. 1). Each satellite contains a precise atomic clock and transmits two L-band microwave radio signals (on 1.2 and 1.6 GHz), both of which are modulated with codes defining the time at which the signal left the satellite. Observations from four satellites allow the receiver to compute the precise arrival time of the signal as well as the three components of position, thus removing the need for an atomic clock in the receiver. If the receiver altitude is known (e.g. on sea craft), only three satellites are required.

There are two codes: the C/A (Coarse/Acquisition) code on the higher frequency only, and the shorter-wavelength P

(Precise) code on both frequencies. Each satellite also transmits its current orbital parameters and clock error. The C/A code enables the range to each satellite to be measured with a precision of 10 m. However, errors (known as 'Selective Availability' or SA) are deliberately introduced into the satellite clock and orbit information so as to reduce civilian positional accuracy to 100–150 m. The P-code is intended for military use and is further encrypted (by 'Anti-Spoofing' or AS), but

Fig. 1. GPS satellites orbit the Earth at 20 000 km altitude in six orbital planes, with four satellites in each plane, so that a receiver on the ground can obtain signals from at least four satellites at all times.

some GPS receivers can bypass this encryption and give instantaneous positional accuracy of 10–15 m using the P-code. An enhanced C/A code method, *differential GPS* or *DGPS*, uses corrections to the satellite ranges that are calculated using a nearby receiver at a precisely known location and transmitted by radio to roving receivers. Precisions of 1–5 m are possible for rovers that are close enough to the base station for the atmospheric, orbital, and deliberate (SA) errors to be common to the two receivers. This method is often used for vehicle navigation and lower-precision surveys. A further development, *wide-area dGPS*, involves corrections interpolated between an extensive network of base stations, so that the region of useful accuracy is enlarged.

For more precise positioning, observations of the GPS carrier wave phase recorded at two or more receivers can be combined in an interferometric manner (*relative GPS*) analogous to VLBI radio astronomy. Using several hours of measurements, accuracies of 5 mm can be achieved, and precisions of a few centimetres can be realized in seconds even for moving receivers (*kinematic GPS*) close to a fixed station. Ionospheric and tropospheric errors must be counteracted using dual-frequency observations and theoretical models of the atmosphere. For the highest precision, the orbit information must be later improved using data from a global network of receivers because the parameters broadcast in real time, based on an extrapolation of previous orbits, are too inaccurate. Because GPS receivers are readily portable and give such precise results, relative GPS has become an important tool for Earth scientists studying plate-tectonic deformation, post-glacial rebound, and earthquakes.

PETER CLARKE

Further reading

Hofmann-Wellenhof, B., Lichtenegger, H., and Collins, J. (1997) *Global Positioning System theory and practice* (4th edn). Springer-Verlag, Vienna.

Leick, A. (1990) *GPS satellite surveying.* Wiley Interscience, New York.

Smith, J. R. (1988) *Basic geodesy: an introduction to the history and concepts of modern geodesy without mathematics.* Landmark Enterprises, Rancho Cordova, California.

global warming

'Global warming' is the term applied to increasing average global temperature, popularly associated with the enhanced greenhouse effect. Since the mid-1980s, the prospect of warming has become reality. Global temperature series, combining observations from land and sea surface, have shown that temperatures have risen by between 0.5 and 0.8 °C over the past 100 years, the warmest year being 1998, followed by 1997, 1995, and 1999. This warming has coincided with advances in computer modelling of the enhanced greenhouse effect. This is the process by which long-wave (outgoing) radiation emitted from the surface of the Earth is absorbed by greenhouse gases, primarily carbon dioxide, methane, nitrous oxide, various chlorofluorocarbons, tropo-

spheric ozone, and water vapour. The atmospheric life of greenhouse gases means that emissions in the late-twentieth century will still be making a contribution to the greenhouse effect at the end of the twenty-first century.

Climate modelling is an attempt to simulate the effects of changing atmospheric processes, including changes in the radiation balance caused by greenhouse gases. The equilibrium (eventual) response to a doubling of the pre-Industrial Revolution greenhouse effect is expected to result in a warming of 2–4 °C by the late twenty-first century.

Since the late 1980s, research has been focused through the Intergovernmental Panel on Climate Change (IPCC). The rate at which greenhouse gases concentrations will increase and the sensitivity of the climate system to such change are key uncertainties, together with various forms of positive and negative feedback (e.g. those associated with changes in cloud cover). In the early 1990s, transient (time-dependent) models started to incorporate more fully the thermal inertia of oceans, indicating possible rates of warming. These indicate an average warming of 1 to 1.4 °C by about 2040. This is likely to be exceeded in the interiors of the larger continents (especially in winter at high latitudes), whilst some sea surfaces and adjacent coastal regions may have little or no warming.

The uncertain outcome and consequences of climate change in the twenty-first century accounts for the urgency of securing international policy responses. The Climate Convention signed after the 1992 Earth Summit require signatories to reduce greenhouse gas emissions to 1990 levels by 2000, with further reductions being sought by 1997. We are, however, still likely to have committed the Earth to a warming over the next 50 years two to three times that of the last 150 years. Additional uncertainty arises from the response to warming of other components of the climate system, notably any changes of the atmospheric circulation on the distribution of rainfall (*see recent climate changes*). Global warming can thus be seen as one of the most far-reaching environmental problems of our time.

JULIAN MAYES

Further reading

Ferguson, H. L. and Jager, J. (eds) (1991) *Climate change: science, impacts and policy.* Proceedings of the 2nd World Climate Conference. Cambridge University Press.

Houghton, J. T., Jenkins, G. T., and Ephraums, J. J. (eds) (1990) *Climate change: the IPPC scientific assessment.* Cambridge University Press.

Houghton, J. T., Callander, B. A., and Varney, S. K. (eds) (1992) *Climate change 1992: the supplementary report to the IPPC scientific assessment.* Cambridge University Press.

Watson, R. T., Zinyowera, M. C., and Moss, R. H. (eds) (1996) *Climate change 1995. Impacts, adaptations and mitigation of climate change: scientific–technical anlayses. Contribution of Working Group II to the (2nd) Assessment Report of the Intergovernment Panel on Climate Change.* Cambridge University Press.

Parry, M. L. (ed.) (1991) *The potential effects of climate change in the United Kingdom*. UK Climate Change Impacts Review Group, 1st Report. HMSO, Norwich.

gneiss *see* HIGH-GRADE GNEISS TERRAIN

gold, silver, and copper deposits

Gold, silver, and copper were the first metals used by humans. Under suitable geological conditions each of these metals can be found in the native, or elemental state, and thus can be gathered for use without having to resort to smelting. Our ancestors had learned to work these metals into useful and ornamental objects by at least seven to eight thousand years ago.

Gold and silver are present in the Earth's crust at extremely low levels of concentration. The average geochemical abundance of silver is approximately 80 parts per billion (ppb), and gold is even lower at about 2.5 ppb. Various geological processes can lead to the localized concentrations of gold and silver that are called ore deposits. Gold must reach a concentration of about 1 part per million (ppm) for mining to be considered, and silver must reach a level of 50 to 60 ppm to be profitably recovered. The several circumstances under which such concentrations can occur in nature determine the various kinds of gold and silver deposits.

Gold deposits

Gold occurs primarily as the native metal, but it almost always has some silver admixed by atomic substitution. If the silver content is 20 per cent or more by weight, the name *electrum* is used. Gold also forms telluride minerals, such as calaverite ($AuTe_2$), but tellurides are rare. Although much of the gold that is mined is recovered from mines worked specifically for gold, it is also present in small amounts in many kinds of deposits, and as a result a great deal of gold is also recovered as a by-product from the mining of copper, lead, and zinc.

Gold deposits can be divided into *lode* and *placer* deposits. Lode deposits are *primary*: that is they are emplaced in the hard rocks of the Earth's crust. Placer deposits are *secondary*: that is, they are derived from lode deposits and are found in the zone of weathering and sediment accumulation at the Earth's surface.

The formation of lode deposits requires the collection and transport of gold in hot, aqueous solutions moving through the crust. Although gold is a noble metal (that is, it is chemically unreactive), it can form soluble halide complexes, thiosulphate complexes, and other chemical complexes; it is in these complexed forms that it is transported in solution during ore-forming processes.

Ore-forming solutions can either be released by crystallizing magmas, or they can evolve from deeply circulating meteoric waters. As the solutions move through fractures in the crust, the gold carried in solution can be precipitated as a result of cooling, boiling, or chemical reactions with enclosing rocks.

Many of the most ancient gold mines, and some of the more famous ones from modern times, were simple gold-bearing quartz–pyrite veins in faults and other fractures in rocks. Examples of gold–quartz–pyrite vein deposits are the Mother Lode in California, the Hollinger deposit in Ontario, Canada, and the Bendigo and Ballarat deposits of Victoria, Australia.

In the classification of mineral deposits proposed by Waldemar Lindgren (1860–1939), deposits that are formed under conditions of low to intermediate temperatures (150–350 °C) and low-pressure, near-surface conditions are called *epithermal deposits*. They are of two general kinds. First, and most common, are *epithermal deposits* formed as a result of subaerial volcanism of the kind associated with convergent margins of tectonic plates. Such deposits tend not to extend to great depths but to be exceptionally rich and deserving of the term *bonanza*. Examples are Cripple Creek, Colorado; Goldfield, Nevada; and El Oro, Mexico.

A second kind of epithermal deposit has been recognized only in relatively recent times because the gold is present in tiny grains. The deposits are accordingly referred to as 'invisible gold deposits', or 'no-see-em' deposits. Examples of such deposits are Carlin, Cortez, Jerritt Canyon, and Getchell, all in Nevada. The deposits, which are of relatively low grade and are mined on a very large scale, were formed from relatively low-temperature solutions (250 °C or less) that circulated through large volumes of fractured rock and deposited the tiny grains of very finely dispersed gold.

Because gold is inert to chemical action by rainwater and air, or nearly so, grains of gold persist and concentrate in the soil as other minerals are weathered away. When the products of weathering are removed by flowing water and redeposited as sediments, gold can become concentrated in placer deposits. Gold is malleable and does not fracture as it is tumbled about in running water. With a density of 15 to 19 g cm^{-3}, depending on the silver content, gold becomes concentrated by flowing water as less dense grains of sand (density 2.6–3.0 g cm^{-3}) are winnowed out. Tens of thousands of placer gold deposits have been discovered in various parts of the world. Indeed, placer gold is so widespread that it has been won on every continent and in almost every country.

It was surely placer gold that was first exploited by our ancestors, and an important amount of gold is still recovered today from placers in Alaska, the Yukon in Canada, Russia, and elsewhere. More important than young placers, however, are the vast palaeoplacers, estimated to be about 2.7 billion years old, of the Witwatersrand in South Africa. The source of the gold in the Witwatersrand placers is a geological puzzle, but the size and importance of the deposits are not in question. It is estimated that about 40 per cent of all the gold ever mined has come from the Witwatersrand mines, and even though production is now declining, South Africa is still the world's leading gold producer.

Silver deposits

Unlike gold, silver forms many minerals by combining with sulphur arsenic, antimony, and other chemical elements.

Native silver is a relative rarity and most of the world's silver is recovered from minerals such as argentite (Ag₂S) and proustite (Ag₃AgS₃), or from tetrahedrite (Cu₁₂Sb₄S₁₃), in which silver replaces some of the copper in the atomic structure.

Like gold, a great deal of silver is recovered as a by-product of bulk mining of copper, lead, zinc, and gold—indeed far more silver is produced as a by-product than is produced from what are primarily silver mines. Deposits that are mined largely or entirely for their silver content are mostly epithermal veins. The bonanza silver deposits of Potosi, Bolivia; Guanajuato, Mexico; Creede, Colorado; and Tonopah, Nevada, are classic bonanza-type epithermal deposits. The Andean volcanic province, together with the Mexican and western North American volcanic provinces, are host to the world's largest and richest epithermal silver deposits, and more than half the silver ever mined has come from the region. Indeed, silver was essentially a rare metal in most parts of the world until the Spanish conquistadors invaded South and Central America and silver started to flow from the vast silver stocks of the Incas and Aztecs.   BRIAN J. SKINNER

Goldschmidt, Victor Moritz (1888–1947)

V. M. Goldschmidt, a pioneer geochemist, was born in Switzerland but at the age of thirteen moved to Oslo and later took Norwegian citizenship. After studying chemistry, mineralogy, and geology and completing his doctorate in Oslo he was appointed to the staff in 1911 and three years later was made a full professor and director of the Mineralogical Institute. From 1929 until 1935 he was professor and head of the Mineralogical Institute at Göttingen but returned to Oslo in 1935. He was imprisoned by the Nazis but escaped to England and returned to Norway in 1945. Goldschmidt's early work was on the metamorphic rocks of south Norway. He showed that in contact metamorphism of a variety of sedimentary rocks the mineralogy of the ten hornfelses produced was controlled by composition and temperatures up to at least 450 °C but increasing with depth. Resulting from this was his concept of the Mineralogical Phase Rule, which states that the number of mineral phases in stable equilibrium in a rock is equal to, or sometimes less than, the number of chemical components.

Later in the Stavanger area Goldschmidt showed that metamorphism of argillaceous rocks by granitic and similar intrusions, together with the addition of materials from the intrusions, resulted in the formation of gneisses—a process of metasomatism, in which he recognized four principal types.

As a result of geochemical and crystal structural work on some 200 compounds of 75 elements he was able to elucidate laws of geochemical distribution and to produce the first tables of atomic and ionic radii, ionic charges, and interatomic distances. He was also able to relate the hardness of crystals to their structure. For these nine volumes of the *Geochemische Verteilungsgesetze der Elemente* (the last one published after his return to Oslo in 1935) he was editor and principal contributor. Arising from this work he proposed a

geochemical classification of elements according to their affinity with iron, sulphur, or silica. *Geochemistry*, edited after Goldschmidt's death by A. Muir, was published in 1954.

R. BRADSHAW

Gondwanaland The Austrian geologist Eduard Suess gave the name 'Gondwanaland' to a former continent embracing the Late Palaeozoic Gondwana fauna common to its components. The term Gondwanaland has in its general usage been subsequently extended to embrace Australia, South America, and Antarctica as well. 'Gondwana' (originally the name of a region in eastern India) is actually the more correct term, since it means 'land'. Gondwanaland is the southern component of Wegener's supercontinent Pangaea.

ANTHONY HALLAM

GPS *see* GLOBAL POSITIONING SYSTEM

Grabau, Amadeus William (1870–1946)

A. W. Grabau was one of the most influential geologists of the first half of the twentieth century. Born in Wisconsin to parents of German heritage, he expanded a childhood interest in natural history through training in palaeontology from the Massachusetts Institute of Technology (BS, 1896), and Harvard University (Ph.D., 1900). From that training he structured two careers, one in the United States, the other in China.

Grabau commenced teaching at Rensselaer Polytechnic Institute in Troy, New York, then in 1903 moved to Columbia University, where he taught both palaeontology and stratigraphy. He was impressed by the work of the German geologist Johannes Walter (1860–1937), and dedicated his book, *Principles of stratigraphy*, published in 1913, to Walter. Ideas and concepts in the book met with vigorous opposition, but today the book is recognized as a classic that was far ahead of its time. Grabau focused his book on sedimentation by using present-day depositional environments as guides to the interpretation of past sedimentary environments and to their fossil assemblages. In so doing, he laid the groundwork for the field now called palaeoecology. Where Grabau's work raised the strongest opposition was in his expansion of Walter's idea of the sedimentary facies concept—that lateral changes occur in both the lithological and biological record within a sedimentary formation. The principal opponent to the facies concept was E. O. Ulrich (1857–1944), a geologist with the US Geological Survey, who argued that local hiatuses and unconformities separate strata of constant lithology. Nearly 25 years were to pass before the geological profession recognized fallacies of interpretation in the field evidence on which Ulrich based his ideas, and then embrace the sedimentary facies concept as articulated by Grabau. In the meantime, a dramatic upheaval occurred in Grabau's life, and this upheaval led to his second career.

In 1919, after the end of the First World War, Grabau abruptly left Columbia. The circumstances of his departure have never been made fully public, but they are rumoured to have involved charges that he was propounding pro-German ideas. He accepted an invitation to serve as Professor of Palaeontology at the National University in Peking (today's Beijing), and Chief Palaeontologist of the Geological Survey of China. Grabau spent the rest of his working life in China, and his impact on Chinese geology was massive. He taught and supervised the work of an entire generation of Chinese stratigraphers and palaeontologists. At the same time he continued his forward thinking on sedimentary processes and stratigraphy, developing ideas that presage concepts embodied in today's sequence stratigraphy.

Grabau was interned by the Japanese during the Sino-Japanese war and he died a few months after the liberation of China. His grave is on the campus of Peking University, marked by a white marble tombstone, engraved in Chinese and English.　　　　　　　　　　　　BRIAN J. SKINNER

grabens and rift valleys A *graben* (German, a ditch) is an elongate block of crustal rocks that has been thrown down by normal faults dipping towards it on one or both sides (Fig. 1). A graben of regional extent is usually called a *rift*, whereas the term *rift valley* is generally used to describe a fault-controlled topographic trench irrespective of the character of the faults. Perhaps the most famous rift valley that is

not a graben is the Dead Sea rift valley, which coincides with the outcrop of an intracontinental transform fault on which there is horizontal (i.e. strike-slip) motion. Many other famous rift valleys, such as those of East Africa (the cradle of mankind), the Rhine, and Lake Baikal (the site of the largest volume of fresh water on the Earth), are, however, rifts in the strict sense. The uplifted block between two grabens is called a *horst* (German, an eyrie). In this account only grabens and rift valleys within the continental crust of the Earth are discussed. *Oceanic rifts* are treated in the entry on *mid-ocean ridges*.

Grabens in cross-section

The traditional view of a graben is that it is a structure framed by two inwards-dipping normal faults on which displacements are roughly equal (Fig. 1a). The tops of the plateaux on either side of such *whole-grabens* slope away from the rift valley. Although some small grabens are whole-grabens, most grabens are *half-grabens* bounded by a major normal fault on only one side (Fig. 1b). The subsidiary fault on the opposite side of a half-graben is generally a product of local strains imposed when the hangingwall block (i.e. the block above the fault) slides down the main fault (Fig. 1c).

There has been much discussion by geologists and geophysicists about the shapes in profile of graben-bounding faults. Seismological data analysed by James Jackson and

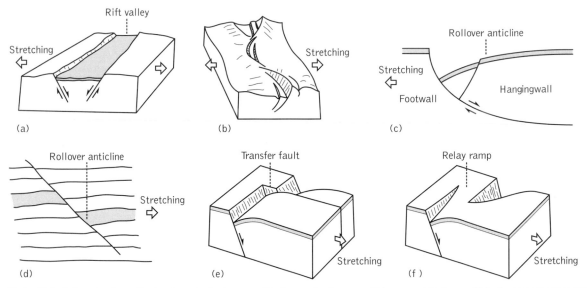

Fig. 1. (a) A whole-graben framed by two inwards-dipping normal faults of roughly equal status along its entire length. Note that the plateaux on either side of the rift slope away from it. (b) A series of half-grabens framed by a single major normal fault that switches sides as the graben is traced along its length. (c) A half-graben bounded by a listric normal fault which has given rise to a rollover anticline and a subsidiary fault in its hangingwall. (d) A rollover anticline resulting from variations in the amount of displacement along a plane normal fault. (e) A transfer fault at a step-over within a normal fault zone. (f) A relay ramp at a step-over within a normal fault zone (b) modified from L. E. Frostick and I. Reid (1987) *Geology Today*, **3** (4) 122–6; (c) modified from A. D. Gibbs (1983) *Journal of Structural Geology*, **5** (2) 153–60; (d) modified from A. Roberts and G. Yielding (1992) in *New concepts in continental deformation*, pp. 223–50. Pergamon Press, Oxford.

others after large earthquakes on major normal faults suggest that they are plane surfaces maintaining a roughly constant dip of about 45° down to the base of the brittle crust, at 10–20 km beneath the Earth's surface. By contrast, other geophysical and geological evidence, much favoured by Alan Gibbs, suggests that some major normal faults are concave up, that is, they are *listric* in shape. Another piece of evidence which is often used to argue that some normal faults are listric is the presence of a *rollover anticline* in the hangingwall of the fault (Fig. 1c). Such rollovers form as a result of the back-rotation of the hangingwall block during slip down the curved fault surface. The presence of a rollover, however, is not by itself diagnostic of a normal fault being listric; displacement along a plane fault can also give rise to one (Fig. 1d).

Grabens in plan

In the same way that our ideas about the profile shapes of grabens have changed, so have our notions about their shapes in plan. For example, rather than thinking that both boundary faults are of roughly equal status along the length of a graben (Fig. 1a), it is now known, largely as a result of work in the East African rift system, that the dominant fault commonly switches from one side of a graben to the other (Fig. 1b). Most graben-boundary faults are segmented, each segment being fairly short and offset from its neighbour across a step-over. Linkage at a step-over is via either a *transfer fault* (Fig. 1e) or a *relay ramp* (Fig. 1f). Transfer faults share some characteristics

in common with the strike-slip faults that link segments of oceanic spreading ridges. During an earthquake on a major normal fault it is common for rupture to occur on only a few adjacent segments; the other segments may fail in earthquakes tens or hundreds of years later.

Isolated grabens and graben fields

Some grabens, such as the Rhine graben (Fig. 2a), are isolated structures but many belong to graben fields containing numerous nearly parallel structures. Two of the world's best known graben fields are the Aegean extensional province in the eastern Mediterranean region (Fig. 2b) and the Basin and Range province in the western USA (Fig. 2c). Death Valley is the most famous half-graben in the Basin and Range province. In western Turkey, the Menderes valley, which follows a large half-graben in the Aegean province, has, since Greek and Roman times, been an important route between eastern and western civilizations. Incidentally, the Menderes River, the ancient River Meander, has given the world its name to describe any markedly sinuous river or stream. Displacements on faults in both the Aegean region and the Basin and Range province have caused many damaging earthquakes both in the past (e.g. Sparta, 464 BC) and in modern times (Dixie Valley, 1954). Other important graben fields are hidden from view beneath the seas surrounding the continents. Although we cannot see them in the field, such hidden grabens are of great importance, as is explained in the next section.

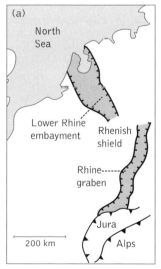
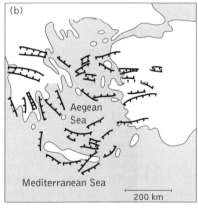
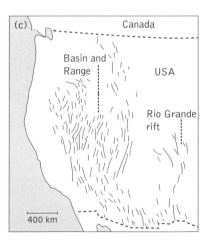

Fig. 2. (a) The Rhine graben system: an isolated graben cutting 'stable' Europe north of the Alps. (b) Normal faults in the graben field of the Aegean extensional province. (c) Normal faults in the graben fields of the Basin and Range province and the Rio Grande rift, western USA. (a) After J. H. Illies (1981) *Tectonophysics*, **73**, 249–66; (b) after A. A. Barka and P. L. Hancock (1984) In Dixon, J. E. and Robertson, A. H. F. (eds) *The geological evolution of the eastern Mediterranean*, Special Publication of the Geological Society No. 17, pp. 763–74; (c) after M. L. Zoback *et al.* (1981) *Philosophical Transactions of the Royal Society of London*, **3**, 407–34.)

Grabens as sites of sedimentary basins

Rift valleys expressing whole-grabens above sea level are relatively starved of sediment because the catchment areas of any streams draining into them are mainly restricted to their flanking escarpments (see *fault scarp*). Because there is commonly a long slope down to a half-graben, they are generally sites of sedimentary basins, the greatest thickness of sediment accumulating close to the boundary fault. Many rifts beneath the continental shelves are outstandingly important sedimentary basins; some, such as the North Sea, are now major oil and gas fields. At the same time as many rifts were being filled with sediments there were displacements on the rift-bounding faults (see *growth fault*).

Origins of grabens

Grabens are a reflection of the horizontal stretching of the brittle upper part of the Earth's crust as a result of plate motions or underlying mantle processes. Grabens within continents occur where: (1) there was rifting before the actual or failed opening of an ocean, (2) a convection-current cell formed in the mantle above a subducting slab, (3) collision between two continental plates split the undeformed rocks ahead of the collision zone, (4) a high mountainous region spread out laterally under its own weight, or (5) the foreland to a mountain belt subsided and was stretched as a consequence of the load imposed by the mountains.

P. L. HANCOCK

Further reading

Coward, M. P., Dewey, J. F., and Hancock, P. L. (eds) (1987) *Continental extensional tectonics*. Geological Society of London Special Publication No. 28.

Frostick, L. and Reid, I. (1987) A new look at rifts. *Geology Today*, 3, 122–6.

Park, R. G. (1988) *Geological structures and moving plates*. Blackie, Glasgow.

gradient wind One of the simplest concepts of airflow is the geostrophic wind, where the *pressure gradient force* (the driving force arising from the difference in pressure between a high- and a low-pressure area) is balanced by the *Coriolis force* (the deflection arising from the Earth's rotation). Friction is ignored and the lines of equal pressure, or *isobars*, are assumed to be straight. The existence of high- and low-pressure cells in the atmosphere means that isobars are generally curved rather than straight. Wind direction remains parallel to the isobars above the frictional layer of a few hundred metres; a centrifugal force must therefore be present to maintain the balanced flow. The centrifugal force is related to the square of the velocity of the flow and inversely proportional to the radius of curvature. The three-way balance between centrifugal force, Coriolos force, and the pressure-gradient force is called the *gradient flow*. In a high-pressure cell, the pressure-gradient force and the centrifugal force act in the same direction, which is radially outward. In a low-pressure cell, centrifugal force acts radially outward but the pressure gradient force radially inward.

This difference between low and high pressures is of significance in determining the wind speed in pressure systems with a radius smaller than 500 km. In low-pressure systems the speed of the gradient wind is slower at a point than that of the geostrophic wind. In other words, the curved isobars introduced an additional force which acts to slow down the flow compared with the speed that would be realized if the isobars were parallel, but with equal distances between them (equal pressure gradient force). In the case of a high-pressure cell, the speed of the gradient wind is greater than that of the geostrophic wind. In practice two factors diminish this effect: (1) the pressure gradient in a low-pressure system is much larger than that in a high-pressure system; and (2) the centrifugal acceleration is typically fairly small. The gradient wind is important in fast, curved flow (hence high centrifugal acceleration) found in upper-air westerly waves with short wavelengths. Typically, the flow alternates between cyclonic and anticyclonic curvature. Thus the wind speed does not remain constant through the wave, but increases through the anticyclonic curvature (supergeostrophic flow) by as much as 20 per cent and slows down where the flow is cyclonically curved (subgeostrophic flow). Areas where the wind increases in speed cause upper-air divergence, rising air, and disturbed weather. Areas where the flow slows down lead to upper-air convergence, sinking air and clearer, more stable weather.

Similarly, if the wind speed in a hurricane is calculated on the basis of the pressure gradient, values of 500 m s^{-1} result. If a gradient wind is assumed to exist, speeds nearer 75 m s^{-1} can be calculated. These are closer to the typical observed values.

R. WASHINGTON

Further reading

Meteorological Office (1991) *Meteorological glossary*. HMSO Publications, London.

granite Granites are intrusive igneous rocks that contain large amounts of silica (SiO_2). They contain minerals such as feldspar, which is white or pink, quartz, which is generally colourless, and either biotite, a flaky black mineral, amphibole, a black mineral, or both.

Granitic magma is thought to be generated in two ways: either by the melting of an existing rock or by the differentiation of a basaltic magma by fractional crystallization. In the process of fractional crystallization, denser magnesium-rich and silica-poor minerals crystallize and are separated from a melt, leaving it richer in silica. If a large volume of crystals is lost, the composition of the melt will be sufficiently modified to form a granite.

Perhaps the most spectacular granite bodies are those of the cordilleras of North and South America: a series of batholiths extending from the Coast Range in Canada to Chile. Some of these plutons are of vast extent; the coastal batholith of Peru, for example, is more than 2400 km long and up to 65 km wide.

Granites are not only strong and durable but also decorative; they are thus popular as facing stone for buildings, or as restaurant counters. They are particularly popular for the decoration of banks that wish to suggest financial durability. One of the most popular British granites for such purposes is the Shap granite, quarried on the Shap Fells in Cumbria, which is easily recognized by the large (2–3 cm) pink feldspar crystals in it.   JUDITH M. BUNBURY

granite–greenstone terrains

Much of the Earth's very old crust regions (of Archaean age), such as the Canadian Shield, large areas of Western Australia, and the Kaapvaal and Zimbabwe cratons in southern Africa, consists of granite–greenstone terrain. Such terrains have two main components: a number of batholiths of granite or granite-gneiss, each several tens of kilometres across, with relatively narrow greenstone belts wrapped around and between them. The greenstone belts consist largely of metamorphosed basalts (sometimes including recognizable pillow lavas) and some metamorphosed sediments such as shales and greywackes.

A common interpretation is that the granites represent small elements of the earliest continental crust and greenstones are the compressed remains of narrow ocean-like basins that formerly separated them. In some instances, however, the granites can be seen to intrude the greenstones. Whatever their origins, granite–greenstone terrains appear to have become stable cratons by Proterozoic times, and have remained so until the present day. Many are unconformably overlain by sedimentary or volcanic cover of Proterozoic age.

DAVID A. ROTHERY

granite landforms

Granite is an 'acid' (silica-rich) plutonic rock, consisting of minerals that react with moisture to a greater or lesser degree. Contact with water is thus a critical factor in the weathering and erosion of the rock, and hence in the development of granite landforms. Granite consists of interlocking crystals, and is of low porosity and permeability. But it is characteristically well-jointed, with orthogonal systems and sheeting fractures well and widely developed. Fracture patterns are critical, because the interaction of granite with meteoric and shallow groundwaters, initially circulating along fracture planes, determines morphology. Alteration of the bedrock produces a regolith (soil), the base of which is known as the weathering front. Many granite forms, including representatives of such major features as plains and residual massifs (inselbergs, Fig. 1) and hills, as well as minor forms such as boulders, basins, and gutters, are initiated at the weathering front, or, in the case of corestone boulders, within the weathered mantle, as a result of the exploitation of weaknesses in the country rock by groundwater. As the regolith is stripped, irregularities developed at the weathering front are exposed as etch forms. Other minor weathering features, however, are wholly surficial in origin, and a few, both major and minor, reflect the activity of tectonic forces.

Areally, plains are the most characteristic granite landform. Long-continued weathering has produced weathering fronts of remarkable smoothness and etch plains of essentially no relief, as, for example, in the Bushmanland Surface of Namaqualand and Namibia, and the Meekatharra plain, in the interior of Western Australia. Elsewhere, weathered granite has been eroded to form plains of rolling relief, as in the northern Transvaal, the southern Yilgarn of Western Australia, and Eyre Peninsula in southern South Australia. Bordering uplands, the granite has been eroded to form remarkably smooth, gently inclined surfaces carrying a mantle of weathered detritus and known as mantled pediments; these, however, with the stripping of the cover material, are converted to rock pediments or platforms.

Boulders are perhaps the most common positive relief form associated with granitic outcrops. Corestones are formed beneath the surface, within the regolith. Regoliths with corestones attain considerable thicknesses, showing that in some areas weathering has outpaced erosion, but elsewhere erosion has proceeded more rapidly than the alteration of the rock, and the matrix has been removed, leaving the corestones at the surface as boulders.

Massive compartments are delineated by fractures of the regional orthogonal system, though rhomboidal patterns are also common. Such massive compartments offer few avenues for water penetration but, like corestones, they evidently become rounded, either as a result of weathering being guided by upward-curving sheeting fractures, or as a result of the preferential weathering of the edges and corners of the resistant masses to form domical hills or bornhardts. The many geometrical variations of bornhardts (e.g. 'whalebacks', 'elephant rocks', 'sugarloafs') reflect fracture spacing and the distribution and intensity of weathering. Block- and boulder-strewn nubbins (or knolls) and castle koppies (and the tors of south-western England) apparently evolve through the further weathering, in the subsurface, of incipient bornhardts. Where vertical fractures are prominent, needle-like forms and towers are developed, as in the Organ Mountains of southern New Mexico, and the eastern Sierra Nevada of California, around Mt. Whitney.

Some workers maintain that inselberg forms are the last remnants of denudation of the surrounding area by long-continued scarp retreat, but there is much contrary evidence. It is clear that some bornhardts are exposed intrusive stocks, and that others are upfaulted blocks. In addition, the presence of domical forms, or nascent bornhardts, in artificial excavations shows beyond doubt that some bornhardts, at any rate, are initiated by differential weathering at the base of the regolith. Stepped inselbergs and the manifest subsurface origin of many of the minor forms developed on bornhardts, the occurrence of bornhardts in valley floors and valley-side slopes as well as on high plains, and the fact that they are not confined to any one climatic zone all sustain the suggestion

465

Fig. 1. (a) Inselberg landscape in granite, in monsoonal western Pilbara, north of Western Australia. (b) Inselberg massif in granite gneiss, in arid central Namibia.

that most bornhardts are two-stage forms. They are comparable to boulders, except that whereas the latter are commonly detached, bornhardts retain a physical connection with the solid bedrock below. Pillars a few metres high but clearly still in physical continuity with the bedrock provide miniature examples of this model.

In many shield areas inselbergs stand in isolation above the level of extensive, almost featureless, plains. Some, like the Groot Spitzkoppe of central Namibia, are huge, but many are minute in comparison with the extent of the surrounding plains. These small inselbergs reflect long-continued subsurface weathering beneath plains of low relief and the reduction of even the most resistant compartments to small dimensions. Massifs reflect their large initial size, low fracture density, and reinforcement effects, or a combination of these factors.

Although some minor features, such as rock basins, gutters, and grooves (or *Rillen*), undoubtedly evolve on exposed surfaces, other examples of these forms, as well as flared slopes and scarp-foot depressions are, like the residual hills, boulders, or platforms on which they are developed, initiated at the weathering front. Some become diversified after exposure, so that basins, for example, give rise to pits, pans, armchairs, and cylinders according to local slope and structure. Others, such as rock doughnuts, rock levees, and fonts, evolve as a result of partial exposure of the bedrock. Blocks and boulders protect the rock beneath, which results in their standing on low tables or plinths. A few minor granite landforms are of tectonic origin, for crustal stress has given rise to an assemblage of minor neotectonic features, including A-tents or pop-ups, vertical and horizontal wedges, and displaced slabs and blocks. In addition, because crystals in strain are more susceptible to water attack than those in equilibrium, linear zones of strain have been preferentially weathered and eroded to produce clefts or slots.

Thus, in considerable measure, granite landforms reflect the susceptibility to attack by moisture of the minerals that constitute granite, and the access of water to the rock mass. Fractures are significant in this respect, but time, the duration of attack, is also critical, and rock stresses, remnant and contemporary, are increasingly recognized as finding expression in the landforms of granitic terrains.   C. R. TWIDALE

Further reading

Godard, A. (1979) *Pays et paysages du granite*. Presses Universitaires de France, Paris.

Twidale, C. R. (1982) *Granite landforms*. Elsevier, Amsterdam.

Vidal Romani, J. R. and Twidale, C. R. (1998) *Formas y Paisajes Graníticos*. University of Coruña Press, Coruña.

Wilhelmy, H. (1958) *Klimamorphologie der Massengesteine*. Westermanns, Brunswick.

granulite and the granulite facies

The term 'granulite' was formerly used to refer to a wide range of siliceous metamorphic rocks with a medium- to coarse-grained granoblastic texture (one with equidimensional grains) and lacking a marked schistosity. Since the 1970s, however, it has come to be used exclusively for silicate rocks that have experienced granulite-facies metamorphism. Thus the former use of the term 'granulite' for lower-grade psammitic rocks is now obsolete. True granulites are of very varied composition, and their lack of a prominent schistosity arises because micas are scarce or absent in the granulite facies, reflecting the intense dehydration that occurs in this, the metamorphic facies formed at the highest temperatures (see *metamorphism, metamorphic facies, and metamorphic rocks*).

There has been much controversy over the origin of granulites, and in particular the cause of their intense dehydration. One, probably always predominant, view has been that granulites are residual rocks that have experienced at least incipient partial melting. Not only has any pore water originally present dissolved in a melt phase, but the temperatures of the granulite facies are high enough for melting to proceed

directly by breakdown of hydrous minerals such as biotite, providing a water component for the melt. Granulites can display reaction textures compatible with such a process and appear migmatitic on a larger scale. On the other hand, not all granulites are obviously migmatitic, and some do not show the depletion in incompatible elements that the melting model would indicate. The recognition by Jaques Touret in the 1970s that fluid inclusions in granulite are commonly dominated by carbon dioxide led to an alternative model for granulite formation. This model invoked streaming of carbon dioxide from the mantle through the lower crust, driving dehydration through lowering the partial pressure of water by dilution.

In the course of the ensuing debate, many points of importance have emerged. One of the arguments in favour of the carbon dioxide streaming model was that the geothermometer temperatures for granulite often overlap with those of amphibolite-facies rocks. However, it is now recognized that this is largely the result of cation exchange by diffusion, which means that simple cation-exchange geothermometers often preserve a blocking temperature, rather than the true temperature of formation, of rocks of very high grade. Indeed, further work has identified a number of distinctive assemblages that require exceptionally high temperatures to be stabilized (e.g. sapphirine with quartz) and appear to typify a distinct type of ultra-high-temperature granulite. On the other hand, another compelling argument for the carbon dioxide model is that some granulites (generally charnockites, i.e. rocks with orthopyroxene and potassium-feldspar) can be seen to have been produced very locally from apparently lower-grade wall rocks in certain localities, forming zones of replacement that may be just a few centimetres in width, usually flanking veins. Such charnockites formed *in situ* may contain much more carbon dioxide in fluid inclusions that the adjacent parent rocks, and this is a rather effective rebuttal to the claim that the carbon dioxide fluids of many granulites arise because water in an original mixed H_2O–CO_2 fluid is selectively dissolved in melt.

At the time of writing, there is increasing consensus about granulites, although many points remain to be resolved. The distinction between granulites produced during slow cooling of deeply emplaced plutons and those formed by progressive metamorphism and dehydration of lower-grade rocks is far from straightforward, and the former mechanism accounts for the apparently anomalous lack of melting features of certain bodies. Likewise, release of carbon dioxide from crystallizing magmas is a much more potent means of introducing it more or less pervasively into lower crustal rocks than long-range migration as a separate fluid, because of its poor wetting properties. This process, and the presence of marbles in some CO_2-rich granulite terranes, probably accounts for the well-documented examples of carbon dioxide involvement in granulites. Most examples of extensive prograde granulite-facies metamorphism appear to be clearly related to melting, and efficient removal of melt accounts for the observation that migmatite features are usually sparse in comparison with those of amphibolite-facies migmatites, from which the melt was unable to migrate without freezing. In some instances, there is clear evidence for the accompanying emplacement of mantle-derived mafic magmas, providing a source for the anomalous heating.   BRUCE W. D. YARDLEY

Further reading

Harley, S. L. (1998) On the occurrence and characterization of ultrahigh-temperature crustal metamorphism. In Treloar, P. J. and O'Brien, P. J. (eds) *What drives metamorphism and metamorphic reactions?*, pp. 81–107. Geological Society, Special Publication 138.

Touret, J. L. (ed.)(1985) *The deep Proterozoic crust*. Reidel, Dordrecht.

graptolites The graptolites were a group of colonial marine organisms that lived from the Cambrian to the Pennsylvanian (Late Carboniferous). The colonies ranged in size from 5 mm to 1 m in length and were linear or complexly branched. They are generally preserved as carbon films, stick-like and minutely serrated along the edges like tiny saw-blades. This gives them a resemblance to writing on rock surfaces, and gave rise to their name (*graptos*, Greek 'written'). Although for many years their origins were somewhat enigmatic they are now considered to be part of the phylum Hemichordata and allied therefore to the pterobranchs, a modern group of similar appearance. Their abundance in Palaeozoic black shales, together with their rapid and distinct shape changes through time, makes them very useful in Palaeozoic biostratigraphy.

Although graptolites are usually preserved as carbon films that provide little morphological detail, they are sometimes found uncrushed and unaltered, particularly in limestones, from which they can be freed by acid etching. From these specimens a considerable amount can be deduced about the growth and structure of graptolites. The members of the colony secreted a hard skeleton or rhabdosome which originated as a conical sicula housing the original zooid produced by sexual reproduction (Fig. 1a). The colony appears to have developed by the asexual budding off of new individuals, each of which formed a theca, a tubular structure forming a linear overlapping sequence. Each sequence of thecae forms a stipe, or branch, and colonies varied in the numbers of stipes that they possessed. Although no preserved soft parts are known it is assumed that the zooids could extend themselves from the thecae and that they possessed food-gathering apparatus of some kind (Fig. 1b). The material forming the rhabdosome occurs in two layers; the inner fusellar tissue consists of interlocking half-rings, while the outer cortical tissue is composed of a series of bandage-like overlapping sheets. This laminated structure must have given the rhabdosome a great deal of strength while maintaining flexibility.

The shape of the thecae can vary considerably, particularly around the aperture or main opening where ornate projec-

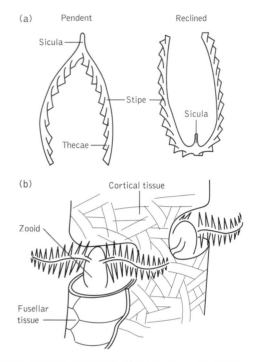

Fig. 1. Graptolite morphology. (a) Two planktonic graptoloid rhabdosomes showing major morphological features. (b) Detail of part of the rhabdosome showing zooids and cortical and fusellar tissue.

tions may develop. The overall shape of the colony also shows great diversity and, as an easily identifiable feature, is very useful in biostratigraphy. Colonies in which the stipes and the thecal apertures are directed down from the sicula are termed pendent, while those in which the stipes are at right angles to the sicula are termed horizontal. Reclined rhabdosomes are ones in which the stipes are inclined upwards, and those in which the stipes have joined to form a linear structure are termed scandent.

Graptolites are separated into orders that correspond to modes of life. The Dendroidea were bush-like forms that were probably attached to the sea bed like modern seaweed; they were important Cambrian forms but then dwindled to their extinction in the Late Carboniferous (Pennsylvanian). The Graptoloidea were planktonic and extremely numerous from the Ordovician to the Early Devonian; these are the forms generally thought of as 'graptolites'. In addition there are four orders of encrusting forms that are of minor importance. Only the planktonic graptoloids are sufficiently numerous and widespread to have biostratigraphic value; major shape changes of the rhabdosome have been used to develop precise correlation schemes. In the Early Ordovician, many-branched planktonic forms developed from the attached dendroids.

These early forms branched dichotomously and produced a diverse group of eight-, four-, or two-branched forms that were generally pendent. From these developed the two-stiped forms ('tuning-fork' graptolites) that dominated the Late Ordovician. These were pendent to reclined, and even scandent. At the base of the Silurian this fauna was replaced by the monograptids, scandent forms in which the thecae were arranged along one side of the stipe only. These were generally straight but also included forms that were curved or spiral or had lateral arms.

Graptolites are distributed in patterns similar to those seen for modern planktonic organisms. They are richest in rocks formed on the outer parts of continental shelves, where upwelling of nutrients along shelf edges would have provided abundant food resources. They also show vertical distribution; those that are most widespread were surface planktonic forms, while those that occur only in shelf-margin and shelf-slope deposits probably lived at some depth within the water column. A number of the planktonic forms display features related to floating. These include long spines projecting from the rhabdosome or loosely woven mesh-like structures surrounding it. Vane-like or bubble-like structures present particularly on scandent forms may have aided flotation if they were gas-filled. Other graptoloids with helical coiling may have been able to spin if the zooids could have generated enough activity with their tentacles, and it is possible that they might have migrated diurnally within the water column. During some parts of their faunal history the graptolites occupy faunal provinces, separate areas characterized by particular graptolite taxa. These appear to have been temperature-controlled, with graptolites preferentially inhabiting warm-water areas.

DAVID K. ELLIOTT

Further reading

Clarkson, E. N. K. (1993) *Invertebrate palaeontology and evolution.* Chapman and Hall, London.

gravel and conglomerate
Gravel and its indurated (i.e. hardened) equivalent conglomerate is a sediment (or sedimentary rock) composed of more than 50 per cent of rounded fragments (clasts) with diameters greater than 2.0 mm. The name is, however, commonly applied to deposits with as little as 10 per cent clasts.

When deposited by currents and waves the clasts form a supportive framework enclosing voids (open-work gravel) which when infilled with fine sand and then cemented is termed *orthoconglomerate*. The clasts can be of one lithological type (*oligomict conglomerate*) or of several types (*petromict conglomerate*). When the clasts have been derived from immediately underlying rocks in the basin of conglomerate deposition the term *intra-formational conglomerate* is used; when the clasts have come from distant sources the deposit is called an *extraformational conglomerate*. *Paraconglomerate* is a term used for deposits which are not true conglomerates but contain gravel clasts enclosed in a matrix of sand and mud and were not deposited by traction currents but by sedimentary gravity flows and other processes. The terms *diamictite* or *pebbly-mudstone* have also been used for this type of deposit. Glacial tills (boulder clays) and their ancient equivalents tillites are of this type. *Tilloid* is a term used for till-like deposits of doubtful but probably non-glacial origin. Although widely used in provenance studies, these deposits are quickly modified by transport: clasts are rounded and the softer rock fragments are lost after several kilometres of transport. They thus contain relatively larger proportions of hard components, such as vein quartz and quartzite, than the original source area. Conglomerates and gravels are generally only crudely stratified and commonly occur as lenses or as channel-fills in finer deposits. The individual clasts are commonly imbricated (i.e. they lie parallel to each other) and they dip upstream. This relationship is used in palaeocurrent studies for determining the directions of former currents. Conglomerates undergo relatively minor diagenetic changes, such as recrystallization of any clay matrix and precipitation of mineral cement. Pitting at the points where clasts are in contact can be observed in some conglomerates when high pressures have caused solution to occur at their contacts.

Petromict orthoconglomerates and subsidiary paraconglomerates are the principal components of huge alluvial fans along the feet of mountain chains. Thick wedges of such ancient fan deposits mark the locations of modern or old mountain belts. They usually wedge out laterally over short distances into fluvial, lacustrine, or marine deposits. The repeated occurrence of conglomerate wedges is often taken as evidence of tectonic movements in the source area.

Gravels are common in rivers, they cover old river terraces, and they infill buried channels beneath many river valleys in ancient fluvial sedimentary sequences. Conglomerates commonly occur as thin basal beds within cyclic sedimentary sequences, within which they represent the lateral migration of channels. Gravel beaches are common features of many coastlines, and thin layers of gravel commonly cover raised beaches and other surfaces of marine planation. They form relict deposits on continental shelves and sometimes thin contemporary deposits—*gravel lags* where currents are sufficiently strong. Conglomerates form thin layers above unconformities, where they represent the beginning of a new cycle of sedimentation; that is, they are basal conglomerate. Matrix-supported gravels form slumped masses along modern continental margins and occur interbedded with ancient slope and basin sediments. Matrix-supported gravels also form extensive sheets of till which blanket vast areas in former Quaternary glaciated regions. Similar deposits that accumulated during earlier glaciations occur at various levels in the geological column.

Both gravels and conglomerates are valuable raw materials for the construction industry. They are also important aquifers and are hosts to some mineral deposits, especially gold and tin.

G. EVANS

Further reading

Pettijohn, E. J. (1975) *Sedimentary rocks*. Harper and Row, New York.

gravel resources *see* SAND AND GRAVEL RESOURCES

gravitational gliding

Gravitational gliding (often simply called *sliding*) refers to the overall downslope movement of a coherent rock mass which maintains its coherency during gliding. It contrasts, therefore, with *gravitational spreading*, which describes the gravity-driven thinning and radial collapse of a relatively incoherent rock mass. As a mass wasting process (commonly called landsliding) on steep hillsides, where gliding usually occurs above a *listric*, rotational detachment, gravitational gliding can have catastrophic economic and social consequences.

Gravitational gliding, whether on the scale of a landslide or on the scale of tens of kilometres, often generates superficial deformation in which compressional, reverse faults at the downslope end of a structure are balanced by normal faults at the extensional, upslope end. In the past, gravitational gliding was considered by many to be the chief force driving thrust sheets in mountain belts. However, the absence of upslope, extensional structures, and the coupling of most thrust sheets with deeper, older fault systems, indicates that gravity is much less important than plate-generated compressional stress in driving thrust sheets.

One of the principal settings in which gravitational gliding occurs is at passive margins, where sediment accumulates at the shelf edge before sliding down to the basin floor. For example, off the south-east coast of Africa, a 20 000 km³ volume of coherent strata, 750 km long and 106 km wide, moved *en masse* approximately 12 km down the continental slope little more than 1 Ma ago. This movement was triggered by small earthquakes.

JONATHAN P. TURNER

gravity measurement and interpretation of anomalies

Measurements of gravity field are used to determine the distribution of rock units in the subsurface but, because gravity is the force that drives the Earth's tectonic engine, they also provide information about some of the most fundamental of tectonic mechanisms. Measurements are made at all scales, from detailed gravity meter surveys using station intervals of as little as 5 m, to observations of perturbations of satellite orbits which define gravity variations with wavelengths of thousands of kilometres.

The main problems in the use of gravity methods arise from the very small sizes of the effects being measured. The Earth's field averages about 9.8 m s^{-2}, but the convenient unit for Earth observations is only one ten-millionth of this, the μm s^{-2} or gravity unit (g.u.). An older unit, the milligal (mGal), equal to 10 g.u., is also still widely used. From the poles to the equator, the gravity field at sea level has a range of about 50 000 g.u. (from 9.78 m s^{-2} to 9.83 m s^{-2}), so that a rather large person might weigh a pound less at the equator than at the poles (and even less on top of a suitably high mountain). This change arises partly from the increase in the Earth's radius towards the equator and partly from the accompanying increase in the effect of the Earth's rotation. Geologically significant changes can amount to no more than a few tenths of a gravity unit, and instruments for gravity surveys must therefore operate to precisions of the order of 10 parts per billion of the total field.

Gravity was first measured with any degree of accuracy using pendulums, and these continued to be used for special purposes until well into the second half of the twentieth century. For survey use, pendulums were supplanted early in the twentieth century by torsion balances, which measured gradients rather than field magnitudes, but the major advance came in the 1930s with the introduction of gravity meters based on spring-balance principles. These accurate and portable instruments, which enabled readings to be obtained in less than a minute (a day was typical for a torsion-balance measurement), led to a rapid increase in the numbers of gravity surveys, and in the areas covered. As with many other geophysical advances, the impetus came from the oil industry, since torsion balances had proved the ability of gravity measurements, under favourable circumstances, to locate salt domes.

Spring-balance gravity meters continue to be the mainstay of gravity surveys but suffer from all the disadvantages associated with mechanical systems. They are temperature-sensitive, because the elastic properties of springs are temperature-dependent, and rather delicate. Sensitivity is enhanced by use of multiple-spring astatic systems, which are more vulnerable than single springs to breakage or simple tangling. Gravity meters also require calibration, and even then measure only differences in gravity field between two points rather than absolute field values. Surveys must therefore be based on networks of readings, ultimately linking back to one or more points of known (or arbitrarily assigned) absolute gravity. In calculating field differences, allowances have to be made not only for instrument drift but also for background variations (known as *tidal effects*) caused by the changing positions of the Sun and Moon. Fortunately these are predictable and are also small (generally less than 1 g.u.).

Although originally developed for land surveys, spring-balance meters have been adapted for use on ships and, more recently, in aircraft. Since gravity fields are accelerations, vehicle movements are important sources of 'noise', but their effect can be minimized by heavy damping, coupled with actual measurement of transient accelerations using short-period accelerometers. A more difficult problem is presented by the change produced by the movement of the vehicle in the effective rate of rotation of the gravity meter about the Earth's spin axis. The very precise navigation required to eliminate this *Eötvös effect* in airborne work became available only with the advent of global positioning satellites.

To achieve absolute measurements of gravity field which are accurate to 0.1 g.u. or better, observations are now made of corner cubes in free fall in evacuated 'co-falling' carriages. Corrections have to be made for such things as the frictional and buoyancy effects of the residual gases in the chamber, and for microseismic activity, which accelerates the entire instrument. Measurements with a repeatability of better than 0.1 g.u. can now be made routinely in a few hours with instruments that are reasonably portable, but it is usual for each determination to span a period of at least one full daily tidal cycle. The cost of an absolute instrument, at about US$250 000, is some ten times the cost of a spring-balance gravity meter. The current IGSN71 system of worldwide base stations is therefore based on only a few high-precision absolute determinations controlling a network in which the vast majority of values have been determined by gravity meter links.

Artificial satellites have brought new ways of measuring gravity fields. Almost as soon as the first satellite launches had been achieved, changes in gravity field with wavelengths of the order of thousands of kilometres were being defined by observations of very small orbital irregularities (perturbations). More recently, radar satellites have measured sea-surface elevations averaged over 'footprints' of a few square kilometres with precisions of the order of a few centimetres, and so have defined the *geoid*, the surface of constant gravitational potential energy which coincides with mean sea level. Because the differences between this surface and the Earth ellipsoid (the best second-order approximation to the shape of the Earth) are caused by local gravity variations, altimeter readings can be inverted to obtain the gravity field. This technique has now mapped marine gravity anomalies with wavelengths down to about 10 km in all the world's oceans except the areas within 10° of latitude of the North and South poles, which are not covered by the present generation of radar satellites.

Gravity reductions

Before geological use can be made of gravity measurements, the numerous and in some cases very large effects of masses which are not of direct geological interest must be eliminated. The most important of these is the gross background effect of the Earth, for which allowance is made by subtracting the normal sea-level gravity from the absolute gravity value. This leaves a remainder which is strongly dependent on elevation. Two opposing effects are involved, these being the decrease in gravity due to the increased distance of the observation point from the Earth's centre of mass, and the increase due to the gravitational pull of the rock material between this point and sea level. The 'free-air' correction for the first of these effects depends only on the radius and mass of the Earth, and its application leads to a quantity known as the *free-air gravity* or *free-air anomaly*. The second correction depends on both the shape of the rock mass and its density, and is usually approached in two stages. As a first step, the gravity station is

assumed to be located on the upper surface of an infinite horizontal plane, and the topography is approximated by a slab of material (the *Bouguer slab*) of uniform density and equal in thickness to the height of the gravity station. The field due to this slab is directly proportional to density and thickness. The free-air effect is approximately equal to 3 g.u. m^{-1} and the *Bouguer effect*, for normal rock densities, is about 1 g.u. m^{-1}. Subtraction of the Bouguer effect from the free-air gravity produces a quantity known as the simple *Bouguer gravity* or simple *Bouguer anomaly*.

The infinite slab is obviously a very poor approximation to the real topography, and although its gravity effect is usually a reasonable approximation to the gravity effect of this topography, further corrections are needed in many instances. Regardless of whether the deviation from the Bouguer slab take the form of masses above the elevation of the station or valleys below, the corrections must be added to the simple Bouguer gravity, producing a quantity known as the *extended Bouguer gravity*.

Isostasy

A final correction that is sometimes applied to gravity data before interpretation allows for the inability of Earth materials to support large loads for long periods without deforming. Mountain masses must therefore be supported, as icebergs are supported, by mass deficiencies at depth. Conversely, near-surface mass deficiencies, such as the water layers in seas and lakes, are balanced by deeper mass excesses. The deep compensation, whether positive or negative, produces gravity fields that oppose the gravity fields from the near-surface sources. Thus, free-air gravity values across the deep Tertiary sedimentary basin in the northern North Sea are generally close to zero, since the low gravity due to the low-density sediments is almost exactly cancelled by the high gravity due to the deep compensation.

Where study of compensation effects is not the primary objective of a gravity survey, it may be convenient to attempt to remove them before interpretation and calculate '*isostatic gravity*' (isostatic anomalies). This is usually necessary only where, as in the vicinity of mountain chains and deep marine basins, the effects are likely to be very large. Even in such areas, corrected maps can be misleading since the gravity field at the surface depends on the way in which compensation is implemented, and this is not usually known. The problem was recognized even in the nineteenth century, when the astronomer Airy and Archdeacon Pratt independently suggested different compensation mechanisms. According to Airy, surface masses are supported by the deepening of an interface at which there is an abrupt increase in density (Fig. 1). This interface is now usually identified with the seismological Moho (Mohorovičić discontinuity), and Airy isostasy (or the Airy–Heiskänen modification which allows the crust to have a finite rigidity) is now thought to apply in most continental areas. Pratt's scheme, first suggested to explain observations in the Himalayas, supposes surface

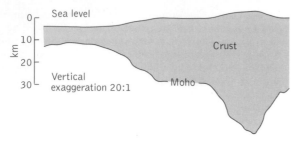

Fig. 1. Airy isostasy. Mass balance is maintained because density increases across the Moho, which rises under the ocean basins and is depressed beneath mountain ranges. The density contrast is somewhere in the vicinity of 0.4 Mg m⁻³, implying relief on the Moho about four times as large as the relief of the sea floor and six and a half times the relief of the subaerial topography.

masses to be supported by density reductions throughout columns of material extending from the surface down to some fixed depth within the Earth (Fig. 2). This form of isostasy seems to hold beneath mid-ocean ridges, which are high because they are hot and therefore of low density, and also, in part, beneath continental extensional regions such as the Ethiopian highlands.

Gravity interpretation

Gravity (and magnetic) fields can be described mathematically by a potential energy function and are therefore sometimes referred to as potential fields. It is a truism of geophysical interpretation that any pattern of variation in a potential field can be produced by an infinite number of possible distributions of source material. This rather depressing fact might seem to make the gravity method useless as a

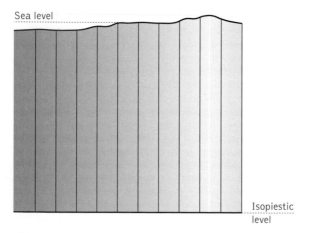

Fig. 2. Pratt isostasy. The densities of entire columns of rock above the isopiestic level vary with surface elevation. In the example shown, which is based on the same topographic profile as Fig. 1, dark-shaded columns are denser than light-shaded columns.

geological tool, but it is also the case that many of the theoretically possible source distributions will, in practice, be impossible (e.g. will require densities corresponding to no known geological material, or structures that contravene surface geological information) and many others will differ only trivially from each other. Although there is always a residual ambiguity, the gravity method can, in many cases, be used to decide between two quite different geological hypotheses. Thus, in using gravity to look for caves in limestone, the two competing hypotheses for any given area are that it is, or is not, underlain by a concealed void. Gravity can tell which of these hypotheses is correct, even though ambiguities will remain as to the exact size, shape, and depth of any cavity present.

In theory, there is one item of information that can be deduced unambiguously from the results of a gravity survey, this being the total mass (or mass deficit) which produces a given anomaly. According to the Gauss relationship, the integral over any closed surface of the scaler product of gravity field and surface area (the *gravitational flux*) is proportional to the total mass enclosed by the surface. It is not, of course, normally possible to carry out a gravity survey over a closed surface, but if a survey is extended over a sufficiently wide area, half the total flux produced by a mass anomaly will be measured and half will be directed downwards and be unobservable. It is thus only necessary to sum the measured flux to calculate the amount by which the mass in a high- (or low-) density body exceeds (or falls short of) the mass of an equivalent volume of country rock. Although there are obviously practical difficulties in deciding the background fields against which gravity anomalies should be estimated, the method provides a valuable technique for quantifying such things as the sizes of cavities and the tonnages of massive sulphide orebodies.

Where the shapes and depths of anomaly sources are important, gravity data are usually interpreted by the method of forward modelling, in which the gravity field of a subsurface model prepared using all available geological information is compared with the field actually observed. The model is then modified, within the limits set by the geological constraints, until a satisfactory level of agreement is reached between calculation and observation. The final model is not, of course, necessarily correct, since other mass distributions will exist that satisfy all the conditions, but it is at least a possible solution. In most instances the modifications to the model are performed interactively on a computer screen which shows both the model and the gravity data. Because the human brain is rather poor at visualizing mass distributions in three dimensions (and also because three-dimensional mass distributions may take an inordinate time to input into a computer), 'two-dimensional' (2-D) approximations are used in which the geology is assumed not to vary at right angles to the line of profile and section. The 2-D approximation is generally adequate provided that the strike length of the anomaly is at least three times as great as the cross-strike width.

An approach to three-dimensional modelling which preserves most of the simplicity of the 2-D approach is provided by $2^1/_2$-D models, which extend unchanged for some defined distance from the line of section and then terminate abruptly. Bodies which are completely offset from the line of profile can be included by inserting negative strike lengths in the equations. Where even this approach is unsatisfactory, the full complexity of three-dimensional modelling must be faced.

Computer programs are also available for gravity field inversion, implying that solutions can be calculated directly from the gravity measurements. Strictly, potential field ambiguity makes this impossible and the programs necessarily rely on initial models which are then modified automatically. Different initial models produce different final solutions. Two inversion methods are in general use. In *non-linear optimization*, the computer keeps track of the changes made in the model and consistently moves towards minimizing the differences between observed and calculated fields, whereas in *Monte Carlo methods* the model is changed at random and all models which provide a pre-specified level of agreement are recorded. Optimization methods produce only a single answer, which may correspond to some local rather than absolute minimum in the difference function, whereas a single 'run' with a Monte Carlo program produces a set of solutions. The latter can provide a graphic picture of the range of solutions possible within the constraints provided jointly by the gravity field and the available geological information.

Computer modelling has now generally replaced interpretation based on the anomalies due to very simple bodies, but the anomaly produced by a point mass or uniform sphere is still occasionally used. The *half-width* (half the width at half the maximum amplitude) of the anomaly due to such a source is equal to three-quarters of the depth to the centre of mass. However, bodies which are laterally more extensive can produce very similar anomalies and, using this rule, will be interpreted as lying deeper than is actually the case. For extended bodies, it can be useful to remember that a 1 km Bouguer slab which is 0.1 Mg m^{-3} (tonne m^{-3}) denser than its surroundings produces a gravity field of approximately 40 g.u. Fields due to slabs with other densities and thicknesses can be obtained as simple ratios.

Gravity and geology

The magnitudes of anomalies produced by real geological bodies vary widely, although within limits set by the operations of isostasy. Beyond a certain point, the Earth itself will deform to eliminate, or at least reduce, abnormal gravity fields. Isostasy is not generally a factor in interpreting the negative anomalies, which may amount to only one gravity unit, produced by cavities, nor the positive anomalies, of up to perhaps ten g.u., which may be associated with massive sulphide orebodies. Salt domes, which owe their existence to gravitational forces, may cause negative anomalies of similar magnitude. Positive anomalies of several thousand gravity units have been observed over some of the world's major ophiolite belts (former oceanic crust). The highest known free-air gravity at sea level (+4000 g.u.) occurs over the Bonin forearc in the western Pacific, which can be regarded as an ophiolite waiting to be emplaced. Strong negative free-air gravity values are observed over oceanic trenches, while Bouguer gravity values are strongly negative in major mountain ranges (to below –6000 g.u. in the Himalaya region).

Gravity anomalies associated with sedimentary basins provide guides not merely to the distribution of sediments but also to the genesis of basins. Isostatic equilibrium is maintained at all times in classical models of extensional basin development and, because the compensating mass lies immediately beneath the main sediment depocentre, the net

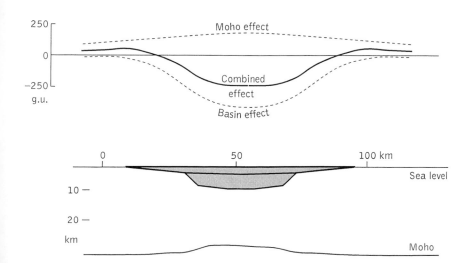

Fig. 3. Gravity field over an extensional basin with full Airy isostatic compensation and consisting of a central rift overlain by a later and broader thermal sag basin. The broader the basin, the more exact will be the cancellation of the low field due to the sediments by the high field due to the elevated Moho.

473

gravitational effect is generally small, although usually still negative (Fig. 3). Deltaic wedges, on the other hand, are commonly associated with gravity highs because, although the sediments making up the wedge have comparatively low densities, they are on average at least twice as dense as the water they displace (Fig. 4). Unless compensation occurs point-for-point at shallow depths and virtually instantaneously, significant positive effects will persist.

Basins in the forelands just beyond mountain belts are characterized by large negative anomalies because they are produced by load masses which are offset laterally from the depocentres. Because the support of the offset loads is only partially local, isostatically corrected gravity is generally positive in the load region although Bouguer gravity is usually negative (Fig. 5). Over subsea fold-and-thrust belts associated with offshore foreland basins, free-air gravity is also usually strongly positive, and effect accentuated in some cases by the presence of a dense subducted slab beneath the belt.

The very-long-wavelength gravity anomalies that perturb satellite orbits are generally thought to originate from sources

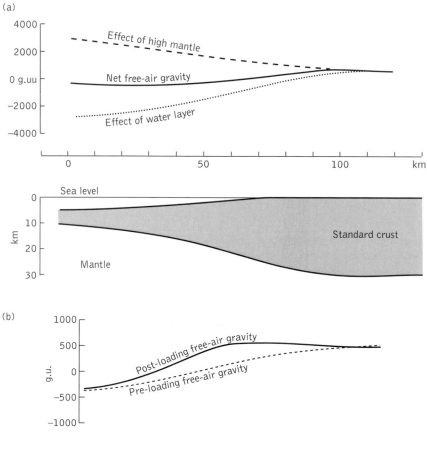

Fig. 4. Gravity effect of a delta. Deposition of deltaic sediments on the continental margin shown in (a) constitutes a net transfer of mass into the delta region, despite the depression of both the underlying basement and the Moho. As a result, the presence of these low-density sediments (b) leads to an increase in gravity field.

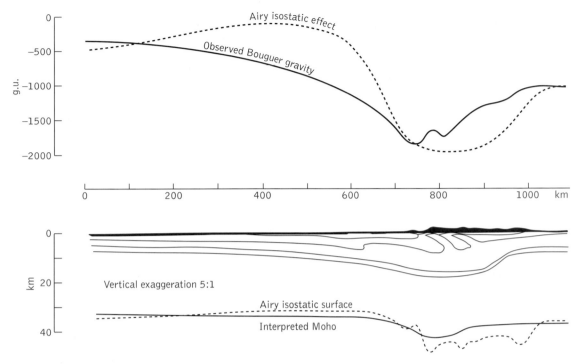

Fig. 5. Foreland basin. The fold and thrust belt beneath the mountains is indicated by a few large, conventionalized thrusts (oblique lines show these imaginary faults). The mass of the mountains is supported not only by the buoyancy of the thick crust beneath them but also by lithosphere rigidity which bends down the crust ahead of the thrust belt.

As a result, there is a considerable difference between the gravity effect that would be produced by simple Airy isostatic compensation and the Bouguer gravity actually observed. An isostatic correction based on the Airy curve would therefore over-correct in the mountains and under-correct in the foreland.

below the base of the lithosphere, both because the wavelengths involved are at least ten times greater than lithosphere thicknesses and because there is no consistent correlation with any one type of surface environment. There are, however, striking correlations with regions of high seismic velocity deep within the asthenosphere, and it is becoming widely accepted that high regional gravity fields mark the locations of unassimilated subducted lithosphere at depths of 600 km or more.   JOHN MILSOM

Further reading

Fowler, C. R. M. (1990) *The solid Earth*, pp. 160–90. Cambridge University Press.

greenhouse effect *see* ENHANCED GREENHOUSE EFFECT

Greenland ice cores At 12.30 on 12 July 1992 the forty or so field scientists working on the Greenland Ice Core Project (GRIP) funded by the European Science Foundation, celebrated as the drill bit penetrated ice over 200 000 years old and finally reached bedrock at a depth of 3028.8 m. The GRIP core was located at Summit, some 3208 m above sea level situ-

ated on the ice divide at the highest part of the ice sheet in central Greenland (Fig. 1). The project entailed a great deal of logistical and technical expertise. The base camp had been constructed in 1989; drilling started the following year and took three field seasons to complete. Over 1300 individual core segments 2.3 m long and 100 mm in diameter were recovered by an electromechanical drill 11.5 m long. (Less than 1 m of core was lost in total during the drilling process.) The cores were partly analysed in an underground laboratory 50 m long and were kept on site before being transferred to a purpose-built storage facility in Copenhagen, where they will be available for further study.

Almost exactly a year later, on July 1993, the Greenland Ice Sheet Project 2 (GISP2), funded by the United States, completed a five-year drilling programme at a site only 28 km to the west of the GRIP core (Fig. 1). The GISP2 core consisted of segments up to 6 m in length and 13.2 cm in diameter. At 3053.44 m (not including the 1.5 m drilled into bedrock), it was the deepest ice core ever recovered.

The completion of the deep GRIP and GISP2 cores heralded a new era in palaeoenvironmental research. They provided the most detailed, high-resolution data on climatic

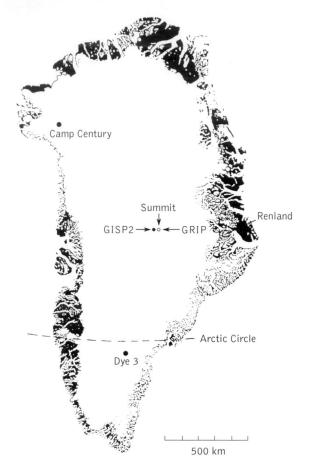

Fig. 1. Map of Greenland showing location of deep ice cores. (From Peel, D. A. (1994) The Greenland ice-core project (GRIP): reducing uncertainties in climate change? *NERC News* **29**, 26–9. Fig. 1.)

analysis (particularly deuterium and ^{18}O), analysis of physical features (such as melt horizons and particulate matter), and chemical and electrical variations. However, the Camp Century, Renland, and Dye 3 cores had important limitations. The Camp Century core was located at an altitude of 1885 m above sea level (some 1345 m below GRIP) at a considerable distance from the ice divide. Because of ice flow (Fig. 2), the deeper the ice, the further it will have travelled and the greater the potential for a disturbed stratigraphy. Renland also had this problem, lying east of the ice divide. The Dye 3 core was located about 100 km from the ice divide south of the Arctic Circle in a zone of summer melting. Melting introduces an additional problem because the meltwater absorbs soluble gases from the atmosphere before refreezing, and this alters the chemical composition of the ice. Allowances have therefore to be made for these factors in interpreting the data from these cores.

At only one point on the Greenland ice sheet, at Summit, is there potential for recovering a complete undisturbed record with no horizontal movement or melting problems. Even here, however, there is the possibility that at lower levels the great pressure of ice might have squeezed out part of the record. It is also possible that even the location of the ice divide itself may have changed through time. This is why it is so important to have two high-resolution deep ice cores, GRIP and GISP2, so near to one another. By comparing the two cores it is possible to assess the amount of disruption to each record caused by ice flow.

Much work still remains to be completed on GRIP and GISP2, but some remarkable results have already been published. Perhaps the most amazing is shown by the oxygen isotope record (Fig. 2), which gives a reliable estimate of the prevailing air temperature over the ice sheet when it was deposited. These data, together with the chemical and electrical signatures, show that during the past 10 000 years (i.e. during the Holocene) the climate has been relatively stable. Most striking, however, is the evidence of rapid climatic change; for instance the temperature rise of 7 °C that marked the most recent climatic shift at the end of the Younger Dryas some 11 700 years ago (Fig. 2) seems to have taken only 50 years.

There is also remarkable evidence for twenty or so large-scale, rapid, and short-lived cyclical fluctuations from warm interstadial to cold glacial conditions within what is generally termed the last glacial period, between 20 000 and 100 000 years ago. Eleven of these interstadials are shown in Fig. 2.

These warm–cold oscillations, termed 'Dansgaard–Oeschger' events (after two prominent ice-core scientists), had previously been observed near to bedrock in the Camp Century, Renland, and Dye 3 cores but it was considered a possibility that these changes might have been caused by disturbed stratigraphy. The almost exact correlation of the [18]O profiles between the Summit, Dye 3, Camp Century, and Renland ice cores shows, however, that these inferred rapid changes in temperature were real (Fig. 2).

change yet produced, and have yielded information that is invaluable for unravelling past climate history, for testing models and theories of climatic change, and for predicting the future rate and direction of climate change. But GRIP and GISP2 are not the first deep cores available from Greenland. Why are they considered to be so important?

In 1966 the USA Cold Regions Research Engineering Laboratory completed the first deep ice core at Camp Century in north-west Greenland, where the ice was 1390 m thick. Further deep cores were subsequently completed at Renland (1981) in eastern Greenland and at the Dye 3 radar station (1987) in southern Greenland (Fig. 1).

Ice cores potentially contain a wealth of information on atmospheric composition, circulation, and pollution, volcanic activity, climatic change, and changes in solar output. The main techniques used in studying them include stable isotope

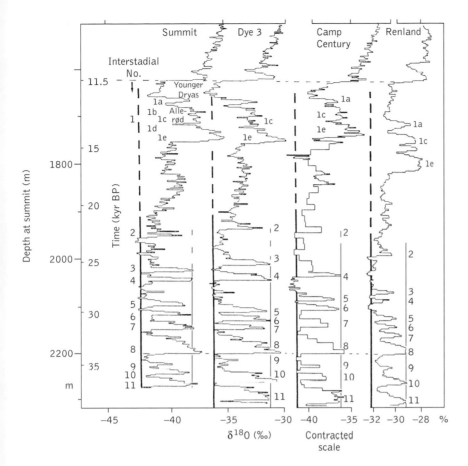

Fig. 2. ^{18}O profiles from four Greenland ice cores over the last 40 000 years plotted on a linear depth scale. Significant mid- and late-glacial warm interstadials are numbered from 1 to 11. (From Johnsen, S. J. and Clausen, H. B., et al. (1992) Irregular glacial interstadials recorded in a new Greenland ice core. *Nature* **359**, 311–13, Fig. 2.)

These events seem to have occurred at irregular intervals. Each comprised a rapid warming phase and a longer, step-like cooling phase. Indeed, the abrupt warming to interstadial conditions probably reflects temperature rises of 5–8 degrees C within a few decades—which is more than half the total glacial-to-interglacial temperature difference.

It is believed that the driving force behind these cyclical events is probably connected with the changing strength of and/or direction of the North Atlantic warm surface ocean current associated with changing amounts of sea ice and deep water formation. The Greenland climate is greatly influenced by the amount of heat released by the sea. The direction and amount of this heat release is closely linked to the global oceanic thermohaline circulation, which acts rather like a conveyor belt. In the North Atlantic the thermohaline circulation is driven by the formation of cold, dense, saline water which sinks and travels southwards, inducing a return flow of surface warm water. When the 'conveyor' is working, warm surface water is introduced into the North Atlantic. It seems

likely that the rapid changes seen in the Greenland ice cores relate to the switching on and off of this conveyor.

The mechanism that switched the conveyor off may in part be due to the release of huge 'armadas' of icebergs from the huge Laurentide ice sheet which spread across the northern Atlantic ocean. The result would be a rapid cooling of the surface water and the creation of a freshwater 'lid'. This would have the effect of reducing surface salinity and hence density, and would result in the thermohaline circulation being switched off. Evidence for these 'armadas', known as Heinrich events, comes from the analysis of deep ocean cores in the North Atlantic (Fig. 3). Layers in these cores have been found to contain ice-rafted debris from Canada and low numbers of foraminifera, which reflect low oceanic productivity. The timing of Heinrich events H1 to H6 is shown in Fig. 4.

The Heinrich events seem to be spaced at intervals of about 10 000 years and occur during the coldest of a series of Dansgaard–Oeschger phases. The apparent regularity of these cycles may suggest a link with the Earth's precessional cycle.

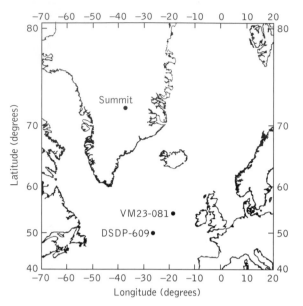

Fig. 3. Map showing locations of the Summit ice core and the deep-sea cores used to construct Fig. 4. (From Bond *et al.* (1993), Fig. 1.)

However, others argue it might be due to the internal dynamics of the Laurentide ice sheet, alternately building up and discharging large amounts of ice. The fact that similar and synchronous discharge events have now been recorded from the Scandinavian ice sheet make it more likely that a factor external to the ice sheets is responsible, since it would be unlikely that the internal dynamics of two dissimilar ice sheets would be in harmony.

The GRIP and GISP2 cores also penetrated through the previous interglacial (Eemian) and revealed further evidence for climatic instability. In the GRIP core the Eemian period comprises about 80 m thickness of ice some 163 to 238 m above bedrock. Only in the first part of the interglacial, equivalent to Marine Isotope Stage 5e, do temperature appear to reach interglacial warmth. Even within this stage there seem to be many short-lived returns to interstadial conditions (5e2, 5e4; Fig. 5) interspersed with warmer episodes (5e1, 5e3, 5e5; Fig. 5). And even within these warmer episodes there are very short-lived changes; for instance, within warm period 5e1 there is a cooling event (event 1) which lasted about 70 years, during which temperatures dropped from interglacial to mid-glacial and back again (Fig. 5).

The Eemian at its warmest was on average 2 degrees C warmer than the peak of our present interglacial and has been

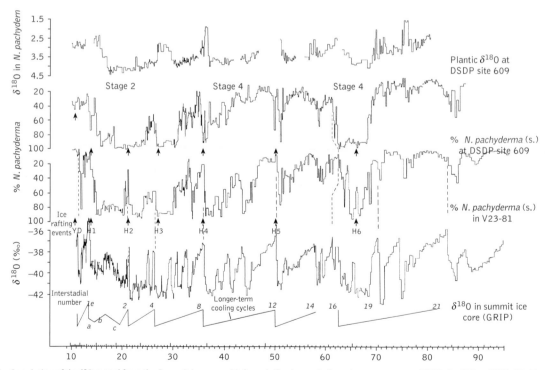

Fig. 4. Correlation of the ¹⁸O record from the Summit ice core with foraminiferal records from deep ocean cores DSDP site 609 and V23–81. Also shown are interstadials in the ice-core record numbered 1 to 21, longer-term cycles (termed Dansgaard–Oeschger cycles), and Heinrich events H1 to H6. (From Bond *et al.* (1993) Fig. 3.)

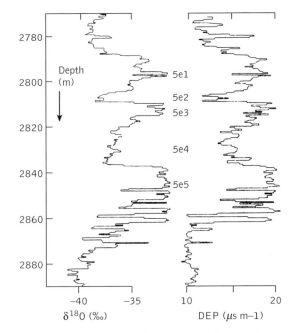

5e1
5e2
5e3
5e4
5e5

Depth (m)

2780
2800
2820
2840
2860
2880

−40   −35
$\delta^{18}O$ (‰)

10   20
DEP ($\mu s\ m{-}1$)

Fig. 5. Detailed profiles of $\delta^{18}O$, electrical conductivity and Ca^{2+} through the Eemian section of the GRIP core (Marine Isotope Stage 5e) showing rapid fluctuations in climate. (From GRIP members (1993) Climate instability during the last interglacial period recorded in the GRIP ice core. *Nature* **364**, 203–7, Fig. 2.)

used as a model for possible climates in a future warmer world. These abrupt shifts within the Eemian have not been found in other cores. If valid, they may indicate the possibility of future rapid climate changes on a scale not seen before during the present interglacial.

There is some debate over these Eemian fluctuations because they occur lower in the core and may be the result of disturbed stratigraphy. Indeed comparisons of the GRIP and GISP2 cores using ^{18}O stratigraphy and electrical conductivity confirms excellent agreement apart from the bottom 10 per cent of the cores. Below about 2700 m, from interstadial 22 and through the Eemian, there is a breakdown in the correlation between the two cores. This could mean that either or both the GRIP and GISP2 cores have disturbed stratigraphy.

There are two ways in which the stratigraphy might be disturbed: through large-scale overturn folds and through small-scale squeezing out of some layers (a process comparable to boudinage in rocks). One method of detecting disturbance is to look at the dip of cloudy bands within the ice. Cloudy bands consist of microbubbles that form round microparticles of dust, particularly in the colder ice, once the core has been retrieved from great depths and the pressure is relaxed. The microparticles tend to be deposited in early summer and therefore show the position of former ice sur-

faces. While there is some evidence for disturbance in the GRIP core between 2900 and 2954 m, this is below the Eemian layers.

On the smaller scale, boudinage might result in the squeezing out of some layers and thickening of others, but this would only stretch some parts of the timescale and contract others; the broad outline of the timescale would still be valid and it could not change an originally stable ^{18}O record into an unstable one.

The evidence for rapid Eemian temperature changes is not found elsewhere. There could be three possible reasons for this: (1) disturbed stratigraphy (thought unlikely by GRIP scientists); (2) the evidence observed may represent changes of limited geographical extent that are consequently not found in other areas; and (3) the changes might have occurred outside the North Atlantic region but the effects were damped because of low sensitivity or time resolution of the techniques applied in other areas. The search is now on for comparable abrupt events in the oceanic and continental records, including North Atlantic ocean cores and north-west European lake and peat deposits, and a more detailed reanalysis of the Vostok ice core. Plans are also being made to drill new ice cores in Antarctica. It is hoped that the European Project for Ice Coring in Antarctica (EPICA) programme will focus on the periods of rapid climatic change through several glacial cycles.

B. A. HAGGART

Further reading

Bond, G., Broecker, W., Johnsen, S., McManus, J., Labeyrie, L., Jouzel, J., and Bonami, G. (1993) Correlations between climate records from North Atlantic sediments and Greenland ice. *Nature*, 365, 143–7.

greenstone terrains *see* GRANITE–GREENSTONE TERRAINS

ground subsidence Any downward movement of the ground surface may be referred to as subsidence, and it can occur on a variety of scales. Regions hundreds of kilometres across may subside very slowly through tectonic movements, while a single building may be destroyed by a localized collapse over a natural ground cavity.

Large-scale crustal subsidence is essential to the continued activity of sedimentary basins and the creation of new generations of sedimentary rocks. Thick sequences of shallow water sediments, including the thousands of metres of coal measures, can only have accumulated in subsiding basins. This style of tectonic subsidence can occur either in a tension zone where the Earth's crust is 'necking' and thinning, or in the compressive subduction zone of a convergent plate boundary. One of the reasons why London needs its Thames flood barrier is its annual 2 mm of subsidence resulting from the extension and crustal sag of the North Sea basin.

Within sedimentary basins, ground subsidence also occurs through compaction of clay-rich sediments as they are lithified to form denser sedimentary rocks. The combined surface effect of this compaction and the crustal sag can be significant. The Po delta in Italy has a thick sediment sequence which is both sagging and compacting; the combined process is a minor but unstoppable component of the long-term subsidence of Venice.

Among the common rock materials, clay is unique in that it is compressible; it can be compacted because plastic deformation of its mineral structure allows its high initial porosity to be greatly reduced at modest pressures. Ground subsidence can be caused by clay compaction under the weight of a single building; the Leaning Tower of Pisa is the most conspicuous example. The lean to the south has been caused by differential subsidence as the clay beneath has compacted where its porosity has been reduced by pressure from the tower. It has been stabilized by drilling small-diameter holes beneath the north side of the tower. These artificially increase the clay porosity, and their slow collapse causes compaction and ground subsidence, partially correcting the inclination of the tower.

Clay also compacts when its water content is reduced, because the pore-water pressure is a significant force in keeping the clay grains apart and supporting a superimposed load. Extensive damage to buildings was caused by localized ground subsidence when the near-surface layers of the London Clay were desiccated in the series of dry summers that started in 1976. Subsidence also occurs, on a much larger scale, where groundwater abstraction from wells causes loss of pore-water pressure in clays within buried sediment sequences. Water is pumped only from sand aquifers, but the loss of water pressure affects both the sands and any inter-bedded clays, and the clays then compact. This is the main cause for increased subsidence and flooding of Venice in the twentieth century and also for the 9 m of ground subsidence in parts of Mexico City. At both sites the subsidence has now almost stopped because groundwater pumping has been greatly reduced by strict legislation.

Peat is the one natural material which compacts more than clay when its water content is reduced. The fenlands of eastern England were originally marshland, but have been turned into rich farmland by efficient drainage. The lowered water tables have, however, caused many metres of ground subsidence on the peat, and large areas now lie below sea level; all land drainage is now pumped up into river channels, which are maintained between high levees.

Limestone regions with karstic landscapes have a reputation for ground subsidence and collapse, related to their underground rivers and caves. Some rocky and steep-sided dolines and potholes have been enlarged by progressive collapse of the limestone and retreat of their exposed rock walls. Cavern collapse is rare and only a few limestone gorges have been formed in this way.

Subsidence sinkholes are diagnostic landforms of karst, which form in unconsolidated soils or drift deposits overlying cavernous limestone. These are created by ground subsidence, where surface soils or sediments are washed down into fissures and caves in the limestone, but they do not involve any failure of the normally strong limestone. The down-washing is normally by infiltrating rainwater, or leakage from a stream, lake, or drainage channel, and this process of sediment movement may be referred to as *suffosion*. The thousands of 'shakeholes' in the karst of the Yorkshire dales are subsidence sinkholes, formed where the glacial till has been washed into limestone fissures. These are youthful landforms because the till is only about 12 000 years old. Rapid subsidence of new sink-holes can be a hazard in populated karst regions, especially where excessive pumping causes a water table decline and therefore increases suffosion. Florida usually loses two or three houses a year where subsidence sink-holes develop in the thick soils overlying its buried and over-abstracted limestone aquifer; most of these losses occur after rainstorms or where drains have broken.

Ground subsidence may occur on any other rock types in which cavities can form; these include chalk, gypsum, and basalt, but the subsidences are generally very localized. Rock salt is so highly soluble in groundwater that it can cause rapid ground subsidence. The 'meres' of the Cheshire plain are all post-glacial lakes in subsidence hollows formed by dissolution of the salt which is at rock head beneath the glacial and glacio-fluvial drift. Modern dissolution is most active where 'brine streams' of groundwater flow along the salt rock-head. These create linear subsidences, which generally extend for a few kilometres and can subside at rates of many metres per century; lakes suddenly formed in them are known locally as 'flashes'.

Mining subsidence can cause serious damage to structures by its local ground movements, but is only rarely on a scale significant in terms of landforms. The inevitable consequence of modern longwall mining is a shallow bowl of ground subsidence over the panel where all the coal has been removed. Many lakes have been formed in the English coalfields where valley floors have been given a locally reversed gradient out of a subsidence bowl. Block caving is another modern method of total extraction mining, and this causes inevitable surface collapse on a grand scale. The 'Big Holes' left on the South African diamond pipes are hundreds of metres deep, and their floors continue to subside as the rock is dug away from below. Older methods of mining were mainly by partial extraction, leaving rock pillars for roof support; these decay over time, and frequently cause destructive ground subsidences many years after the mines have closed.

Mining is only the most obvious of the ground subsidence processes induced by man. Peat and clay subsidences are greatest where groundwater is pumped artificially, most new collapses of subsidence sinkholes are caused by drain breaks, and salt subsidences are greatly accelerated by brine pumping. Tectonic subsidence and some of the movements on clay are wholly natural, but society causes most of its own problems in ground subsidence.
TONY WALTHAM

Further reading

Waltham, A. C. (1989) *Ground subsidence*. Blackie, Glasgow.

groundwater Groundwater comprises water that exists beneath the land surface, held within openings, or pores, of soils and geological formations. According to most classical definitions, this term refers exclusively to water occurring at or beneath a surface, known as the water-table (see *water-table*). Below the water-table, in a region known as the saturated zone, pores are completely filled with water (Fig. 1). Other definitions of groundwater include water held in the unsaturated zone, also known as the soil moisture or vadose zone, which is located above the water-table. Within the unsaturated zone, pores may contain both water and air. This broader definition makes sense when one recognizes that fluids above and below the water-table are intimately connected, being part of a continuous flow field that moves water through the subsurface portion of the hydrological cycle. Although the following discussion will focus primarily on water below the water-table, consideration of water in the unsaturated zone, as well as water in other components of the hydrological cycle, is essential to understanding this important resource.

Groundwater as a resource

Quantitative evaluation of groundwater resources requires information on volumes of water stored in the subsurface, on rates at which groundwater can flow to wells or other extraction points, and on rates at which the resource can be replenished by transfer from other components of the hydrological cycle. The potential importance of groundwater as a resource is indicated by estimates of the world water balance. According to most estimates, saline oceans and seas account for 94 per cent to more than 97 per cent of the world's water, while another 2 per cent is held in glaciers and polar ice caps. Although groundwater constitutes only a few per cent by volume of the total water at or near the Earth's surface, estimated volumes of groundwater are much greater than the combined volumes in lakes and reservoirs, stream channels,

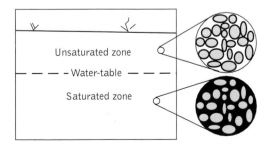

Fig. 1. Pores between grains in saturated zone are filled with water (black), while in the unsaturated zone they contain both water and air.

and freshwater wetlands. Thus, it becomes apparent that groundwater makes up the majority of the world's utilizable freshwater resources, 95 per cent or more by volume.

Storage and flow properties of geological materials

The amount of water stored in the saturated zone depends on the volume of pore space. This volume is quantified as the *porosity*, defined as the volume of pores divided by the total volume of the material (see *permeability and porosity*). The porosity of typical geological materials can vary from less than 0.01 in relatively unfractured igneous or metamorphic rocks to more than 0.5 in silt or clay soils and sediments. High porosities are also found in some limestones and other carbonate rocks in which pores created by depositional processes or fracturing have been enlarged by dissolution of carbonate minerals. The wide range of porosity in geological materials translates into a wide range of water storage potential in the subsurface. Because porosity is determined to a large extent by the type of material, geological maps provide a useful starting point for many evaluations of groundwater resources.

The rate at which water can move through the pore space of geological materials is obviously related to porosity, since the greater the volume of pores, the greater the open area through which water can flow. However, the ease with which water can move through a porous material, quantified by the property known as permeability, depends not only on the porosity, but also on the size of pores (see *permeability and porosity*). Frictional resistance between water and solids where they are in contact at pore walls converts the mechanical energy of flowing water to heat, thereby limiting the rate of flow. For a given porosity, there is more contact area, and hence greater frictional resistance, for materials with smaller pores. In soils and sediments, the sizes of pores vary with sediment size and sorting. The effect of pore size on permeability dominates that of porosity. For example, a clay soil (particle size /< 0.04 mm) with a porosity of 0.5 typically has a permeability many orders of magnitude smaller than a sand (particle size 0.06–2 mm) with a porosity of only 0.3. Permeability also depends on the interconnectedness of the pore network. Relatively isolated pores, such as those created by gas bubbles in cooling lava, can make up a significant portion of the total porosity but will contribute little to flow. In groundwater resource evaluation it is often useful to distinguish between total porosity and the effective porosity, which includes only the well-connected pore space.

Geological formations with sufficient porosity and permeability to supply water to wells are classified as aquifers (see *aquifer*). Formations through which very little flow can occur under normal conditions have been called *aquicludes*, a term suggesting the exclusion of water. Because geological materials are rarely (if ever) completely impermeable, many workers prefer to refer to low-permeability geological materials as *aquitards*, a term that implies retardation of flow rather than exclusion of water. In practice, the distinction between aquifers and aquitards is somewhat subjective, depending on the rates of flow required to meet water supply needs and on the range of permeability and porosities of the local formations.

Rates of replenishment and use

Rates of long-term groundwater use that can be sustained in a given region depend not only on the presence and extent of aquifers, but also on the rates of replenishment, or recharge, through infiltration of rainwater, snow-melt, or other surface water (see *recharge and the hydrological cycle*). In most lakes and streams, water has a residence time of the order of weeks to tens of years. Groundwater, in contrast, may remain in aquifers for hundreds to thousands of years before discharging into the ocean, lakes, streams, or wetlands, or returning to the atmosphere through evaporation or uptake and transpiration by plants. In many arid parts of the world, existing aquifers contain water that infiltrated during periods of wetter climate thousands of years ago. Current rates of recharge in these areas may be very limited. Extraction of groundwater in arid regions is often referred to as mining because such use is similar to exploitation of a non-renewable mineral resource.

Even in moderately humid areas, where current recharge rates are not negligible, rates of groundwater abstraction for irrigation or other intensive uses can exceed natural rates of recharge. One consequence of excessive rates of groundwater use is a significant decline in the water-table. This occurs as water drains from pores, causing portions of a saturated aquifer to be converted into part of the unsaturated zone. As the water-table drops, wells must be deepened. Pumping costs increase because of the increased energy required to bring water to the surface from greater depths. In coastal areas, a decline in the water-table can also induce flow of saline groundwater from the formation beneath the ocean or sea toward wells on land. Salt-water intrusion into freshwater aquifers has long been recognized as a potentially serious source of groundwater contamination. This problem was first addressed by theoretical analyses in the late 1800s and continues to be a topic of research and practical interest.

Groundwater hydraulics

Quantitative estimation of the rates of recharge to and discharge from aquifers requires an understanding of the physics of groundwater flow (see *groundwater flow rate*). The empirical relationship governing rates of groundwater flow, and more generally the flow of fluids through any porous material, was first demonstrated experimentally by Henry Darcy, an engineer working in Dijon, France in the 1850s. This relationship, which has come to be known as Darcy's law, states that groundwater flow rates are proportional to permeability and to the change along the flow path of a property called hydraulic head. Hydraulic head is a measure of the mechanical energy of the groundwater and can be evaluated by measuring the level to which groundwater rises in an open well. Groundwater flows from areas of high head to areas of

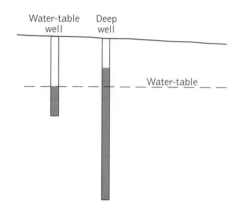

Fig. 2. Water (tint) rises exactly to the water-table in a shallow well, while in a deeper well the water level may be above (or below) the water-table.

lower head, in other words from areas of higher to lower energy.

Measurements in a number of wells are required to map the distribution of hydraulic head within an aquifer. Water in wells open at the water-table will rise to the position of the water-table, while deeper wells may have water levels that are lower or higher than the water-table at the same location (Fig. 2). A contour map of the water-table shows the distribution of shallow hydraulic head and, therefore, provides an indication of the horizontal direction of shallow groundwater flow. Water-table elevations measured relative to a reference elevation, such as sea level, are highest in regional areas of groundwater recharge and are lowest in areas where groundwater discharges to a surface water body or is removed from the subsurface by evapotranspiration.

Darcy's law also provides the basis for predicting the response of water levels in an aquifer to pumping, a problem that has been studied extensively in the field of well hydraulics.

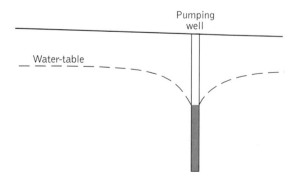

Fig. 3. The water-table surrounding a pumping well exhibits a cone-shaped depression.

Removal of water from a well generates a decline of water levels in a pattern that resembles a downward-pointing cone centred on the well. This water-level pattern is called a cone of depression (Fig. 3). The rate at which the cone of depression spreads away from a well when pumping begins depends on the storage properties and permeability of the aquifer. Pumping tests, which entail monitoring water levels in a pumped well and in nearby observation wells, are commonly used, not only to evaluate the water-supply potential of a particular well, but also to quantify aquifer properties so that groundwater flow rates under other conditions can be estimated.

Groundwater contamination

In recent years, the emphasis of groundwater studies in many parts of the world has shifted from questions of quantity to problems of quality. In addition to contamination associated with natural sources of salinity, such as salt-water intrusion in coastal areas, threats to the quality of groundwater as a source of drinking water result from a variety of chemicals that have been introduced to the subsurface, intentionally or unintentionally, through human activities. Common sources of groundwater contamination include applications of agricultural fertilizers and pesticides, land disposal of municipal solid and sewage wastes, and spills or leaks of fuel and industrial solvents during transport or storage. Because of the long residence times typical of most bodies of groundwater, contaminated aquifers are not readily restored by natural processes.

Attempts to restore water quality of contamination aquifers by pumping are often met with limited success, partly because of the large volumes of water that must be pumped to remove groundwater contaminated even by relatively small, localized sources. Diffusion of dissolved chemicals can spread contaminants originating from a restricted source area, such as a spill, through large volumes of groundwater. Flow of contaminated groundwater through complex pore networks and, on a larger scale, through geological materials of varying porosity and permeability, causes even greater mixing of contaminated and uncontaminated groundwater. In aquifers, flow-induced mixing usually dominates diffusion in the observed spreading, or dispersion, of contaminated groundwater (see *diffusion and dispersion in groundwater flow*). As a result of dispersion and of the low solubility in water of many contaminants, relatively small sources can generate very long 'plumes' of contaminated groundwater. For example, less than 100 litres of organic solvent at a site near Denver, USA generated a plume containing over 4 billion litres of contaminated groundwater that extended over a distance of more than 5 km.

Transport of groundwater contaminants is also affected by the rate and extent of chemical reactions between the dissolved contaminants and the aquifer solids or absorption to the solid surfaces. One possible consequence of chemical reactions is a decreased velocity, or retardation, of contaminants relative to the velocity of the groundwater (see *retardation in groundwater*). Retardation of contaminants necessitates removing even greater volumes of water in order to clean up an aquifer.

Because of the difficulty of completely removing contaminated groundwater from the subsurface region, limiting the continued migration of contaminants may be a more realistic objective at many sites. Containment techniques applicable to shallow contaminant sources include a variety of passive barriers such as surface or subsurface drains and slurry walls, vertical trenches that have been filled with a low-permeability mixture of clay and other sediment. Containment of deeper plumes generally requires the installation of wells that can be pumped to alter the natural ground-water flow field. The design of these systems is based on the principles of well hydraulics and an understanding of the physical and chemical processes that affect the transport of contaminants. Effective containment also requires adequate characterization of the nature and subsurface distribution of contaminants, which can be highly dependent on the subsurface distribution of porosity and permeability.

While much current research is devoted to the development of more efficient methods for removing and containing contaminated groundwater, there is increasing emphasis on identifying microbial and geochemical mechanisms that can break down or permanently immobilize contaminants within aquifers. Techniques that enhance natural degradation or attenuation of contaminants or provide new mechanism for subsurface treatment might, in the long run, provide the best chance of remedying contaminated aquifers.

Groundwater as a geological agent

The flow of groundwater through geological materials also has important effects on a variety of geological processes that shape the land surface and control the distribution of mineral and energy resources. In the absence of water, the entire stress resulting from the weight of overlying material or from tectonic forces must be borne by the solid phases of sediments or rocks. When groundwater fills the pores, however, stress in any direction is reduced to an effective stress equal to the total stress minus the fluid pressure. Changes in effective stress can cause changes in the shear strength of materials, the property that controls the resistance to fracture and deformation. Triggering of landslides and earthquakes has been attributed to changes in fluid pressure of groundwater.

Changes in fluid pressure can be induced by natural events such as heavy rains, or by human activities such as subsurface injection of liquid wastes. Fracturing of rocks and movement along faults can change the permeability of geological formation, generating potential feedbacks between groundwater flow, fluid pressure, and deformation. These feedbacks can become even more complex in *geothermal systems*, where groundwater circulation is driven in part by a heat source at depth. Hot groundwater and steam generated from groundwater in geothermal areas can provide energy for heating and generation of electricity.

Dissolution and precipitation of minerals and cement by flowing groundwater are important processes leading to low-temperature, or diagenetic, alteration of sediments and sedimentary rocks. This alteration can cause significant changes in porosity and permeability. An extreme case is the dissolution of carbonate rocks to produce caves and the distinctive irregular landscape of sinkholes and steep hills known as karst. Less extreme cases of groundwater-related porosity changes can alter the properties of water supply aquifers and create reservoirs or traps for petroleum resources.

Groundwater flow on a regional scale can transport dissolved metals for long distances and is thought to be responsible for the creation of a wide variety of mineral resources, such as stratabound lead–zinc deposits. Migration of petroleum, either dissolved in groundwater or as a separate immiscible fluid, can also be strongly affected by the groundwater flow field and may accompany the transport of dissolved metals.

Groundwater and engineering problems

The presence of groundwater at shallow depths can create problems for a variety of human activities. Obvious examples are the excavation of building foundation and tunnels that extend below the water-table. The prediction of water infiltration rates and construction of suitable drainage systems or low-permeability barriers requires an understanding of groundwater hydraulics and of the porosity and permeability of the local geological formations. Seepage of water through foundation rocks beneath a dam or through the dam itself can limit reservoir storage and, in some instances, can undermine the stability of the dam. The 1976 failure of the Teton Dam in Idaho, USA, which killed 14 people, resulted from reservoir water that infiltrated through fractures in the dam foundation and then reached the erodible silt core of the earth-fill structure. Even more destructive was the 1963 landslide induced by high fluid pressures in rocks surrounding the reservoir of the Vaiont Dam in Italy. Water was displaced and overtopped the dam as debris from the landslide filled a large area of the reservoir. Some 3000 people died in the resulting flood. Consideration of the effects of reservoir filling on pore pressures and groundwater flow might have helped to prevent both these disasters.

As noted above, excessive abstraction of groundwater can limit the quantity and quality of the water resource. In some geological settings, excessive pumping of groundwater can also create problems for engineering structures by inducing subsidence, or lowering, of the land surface. Compaction of sediments occurs naturally over geological time in areas of active deposition. This compaction can be accelerated when fluid pressures are lowered, increasing the effective stress on the sediments. These effects are particularly pronounced in sedimentary basins containing large thicknesses of compressible clays. Pumping from permeable sands and gravels in the same basins causes a delayed drainage of the clays, which then compact. The resulting subsidence is mostly irreversible. Large areas of Mexico City have subsided up to 7 m over the past century. Subsidence can cause building foundations to crack and buried utility lines to rupture. Venice has subsided by only about 15 cm, but this is sufficient to have greatly increased the frequency of flooding in the city. As the connection between subsidence and groundwater withdrawals became recognized, some areas, such as the Santa Clara Valley in California, began to develop alternative water sources and systems of artificial recharge in attempts to limit the potential for future subsidence.   J. BAHR

Further reading

Domenico, P. A. and Schwartz, F. W. (1990) *Physical and chemical hydrogeology.* John Wiley and Sons, New York.

Freeze, R. A. and Cherry, J. A. (1979) *Groundwater.* Prentice Hall, Engelwood Cliffs, NJ.

groundwater chemistry The geochemistry of groundwaters is governed by reactions between water and the minerals present in the bedrock. The minerals that comprise the crust of the Earth are primarily composed of silicon, aluminium, and oxygen, with varying amount of akalis (e.g. potassium, K; sodium, Na) and alkaline earth elements (e.g. calcium, Ca; magnesium, Mg). These minerals are collectively referred to as aluminosilicates. The most common aluminosilicate minerals found in the Earth's crust are the feldspars ($KAlSi_3O_8$, $NaAlSi_3O_8$, and $CaAl_2Si_2O_8$) common in granites. The only other mineral types commonly found in groundwater reservoirs are carbonates: calcite ($CaCO_3$) and dolomite ($CaMg(CO_3)_2$). Once rainwater percolates into the Earth's crust it reacts with the minerals forming the groundwater reservoirs. The composition of the water thus reflects the composition of the bedrock. There are therefore two main types: granite and carbonate groundwaters. The major cations in each groundwater type are shown in Table 1. The less common basaltic waters are shown for comparison. Rivers reflect the composition of the groundwater of the area they drain (see *rivers*).

Table 1. Analysis of groundwaters (p.p.m.)

Bedrock/ Element	Carbonate (USA)	Granite (Cornwall)	Basalt (Iceland)
SiO_2(aq.)	8.9	5.4	14.6
Ca^{2+}	48.0	8.3	2.5
Mg^{2+}	5.8	2.0	0.5
Na^+	4.0	15.4	7.5
K^+	0.7	2.3	0.4
$Al(OH)_4^-$	–	0.3	–
HCO_3^-	168.0	8.8	17.1
SO_4^{2-}	6.4	11.5	3.7
Cl^-	4.8	24	4.0
Total	246.6	78.0	50.3
pH	7.5	6.0	8.8

Granite groundwaters are dominated by positively charged (cations) Na^+ and Ca^{2+}. In calcium carbonate bedrocks the major cations is Ca^{2+}, whereas Na^+ dominates in basaltic bedrocks. In all waters silicon (Si) is present as an uncharged ion ($SiO_2(aq.)$). The dominant anions (negatively charged ions) are HCO_3^- in carbonate water, with varying amounts of the anions SO_4^{2-} and Cl^- in all three water types. The pH of granitic waters is generally low (pH 5–6). Carbonate waters have an intermediate pH value of 7.5. In contrast, waters in basaltic reservoirs have a pH of 8–9.

Carbonate-rich waters are generally referred to as being 'hard'. This is due to the fact when they are heated scale is precipitated which has the composition of calcite, $CaCO_3$. This occurs because carbonates have the unusual property of having a solubility which decreases with increasing temperature (retrograde). On the other hand, silicates, which in this respect resemble common salt and sugar, have a higher solubility as temperature rises (prograde). Dissolved constituents of waters from granite and basaltic bedrocks therefore increase with temperature.

The concentrations of inorganic constituents in aqueous solutions are thus controlled by chemical equilibria which can be described in terms of the *solubilities* of given minerals. The reactions between minerals and water involve the transfer of protons (H^+)). They are referred to as *hydrolysis reactions*.

Interaction between minerals and water

The majority of reactions between minerals and water are hydrolysis reactions. Two common hydrolysis reactions include the breakdown of calcite and feldspar. When writing these reactions we have to remember that water (H_2O) is in equilibrium with the ions H^+ and OH^- ($H_2O = H^+ + OH^-$). Therefore H_2O is often not shown in the equations for geochemical reactions.

The hydrolysis of calcite can be represented by the equation

$$CaCO_3 + H^+ - Ca^{2+} + HCO_3^-.$$

The dissolution of calcite thus entails the exchange of H^+ from water for Ca^{2+} from the calcite. Similarly, the hydrolysis of feldspar (unstable at the Earth's surface) can be represented by the formation of the clay mineral kaolinite ($Al_2Si_2O_5(OH)_4$, which is stable at the Earth's surface):

$$2NaAlSi_3O_8(s) + 2H^+ + H_2O = Al_2Si_2O_5(OH)_4(s) + 2Na^+ + 4SiO_2 \text{ (quartz)}.$$

Thus, when feldspar breaks down, kaolinite and quartz are formed. The reaction entails the exchange of H^+ for Na^+. Note that aluminium (Al) is preserved within the kaolinite. This is reflected in water compositions where the amount of dissolved aluminium (as $Al(OH)_4^-$) is in general very low or negligible (Table 1). The feldspar–kaolinite transformation can be observed in weathered granitic plutons such as those in Cornwall, which are mined for their rich kaolinite component.

We can sum up by saying that the composition and pH of drinking water are governed by the composition of the bedrock of the groundwater reservoir.

Waters that are rich in dissolved constituents are often referred to as 'mineral waters'. When they contain calcium carbonate gas (CO_2) they are described as carbonated. The gaseous waters come from pressurized carbonate bedrocks. At high pressure, the CO_2 is dissolved in the water. When the pressure is released, the carbonate gas has a lower solubility in the water; it can then exsolve from the water and form visible bubbles. This process can be readily observed when a bottle of carbonated water is opened. The bottle was filled under a pressure of more than one atmosphere. While the bottle remains closed, only a few bubbles can be seen. When the bottle is opened and the pressure released, many bubbles of CO_2 become visible.   K. VALA RAGNARSDOTTIR

groundwater flow rate In hydrology, the volume of water flowing through a unit area of aquifer per unit of time is the groundwater flow rate. A scientific understanding of groundwater flow, and more generally of fluid flow in porous media, was first established in the 1800s through sand-column experiments performed by the French engineer Henry Darcy. From his results, Darcy deduced an empirical relationship between flow rates and the physical parameters of his experimental system. This relationship has come to be known as Darcy's law and can be summarized as follows: the rate of flow through a unit area of porous material is proportional to the permeability and to the hydraulic gradient. The hydraulic gradient is defined as the head change, a measure of mechanical energy of the water, divided by the distance over which the head change occurs. The permeability is a property of the porous material. The flow rate per unit area is known by a variety of names, including specific discharge and Darcy flux. The specific discharge has units of length divided by time, and for this reason it is sometimes called the Darcy velocity or seepage velocity. This terminology is misleading, however, because the microscopic velocity within a pore is much larger in magnitude than the flow rate through an area of porous material containing both pores and solids.

Specific discharge is the flow rate of interest when attempting to quantify potential production rates of wells or regional fluxes for water supply purposes. When evaluating rates of contaminant migration in groundwater, the specific discharge may be of secondary interest compared to the rates at which contaminants move from a source area to some critical location such as a well. In such cases, a more useful measure is the advective, or average linear, velocity. The advective velocity is determined by dividing the macroscopically observed travel distance of a tracer, measured along a straight line rather than along the true tortuous path through the pores, by the travel time. The advective velocity can also be estimated by dividing the specific discharge by the 'effective' porosity, which excludes dead-end or disconnected pores that do not contribute significantly to groundwater flow.

Movement of a contaminant or other dissolved substance by groundwater flow is the process known as advection. Local variations in velocity lead to spreading, or dispersion, of contaminants around the average position that would be predicted using the advective velocity. Not all dissolved substances, or solutes, migrate at an average rate equal to the advective velocity. Large solutes may be excluded from small pores. Because the true velocity in the smallest pores can be much lower than the velocity in the larger pores, these excluded solutes have macroscopic velocities that are greater than the average linear velocity of water. Other solutes interact with aquifer solids through physicochemical processes known as sorption. *Sorption* can reduce the effective macroscopic velocity of reactive solutes, a phenomenon known as retardation.

J. BAHR

growth fault

A growth fault is a type of fault on which there were displacements at the same time as the sediments on either side of the fault were accumulating. Most growth faults are normal faults because such faults cause the basins in which sediments are deposited to subside. A growth fault is characterized by preserving greater vertical thicknesses of sedimentary horizons on the side of the fault that has been thrown down. Where a rollover anticline (see *grabens and rift valleys*) has formed in the hangingwall of a normal fault, strata commonly show a noticeable thickening as they are followed towards the fault on its downthrown side. Where a rollover is absent, sedimentary formations on both sides of the fault are of tabular shape in profile but of contrasting thickness.

The idea that deposition and deformation, including faulting, can occur at the same time is not entirely new; neither is the notion that deformation and denudation (i.e. weathering plus erosion) can also be synchronous processes. However, the realization that deposition, deformation, and denudation often act on a relatively small area at the same time is more recent. An older view, before the 1960s, was that sedimentary rocks were initially deposited and then deformed, perhaps to form a mountain range, before finally being denuded to a peneplain. This view, that there was an orderly sequence of events, almost became an axiom; one that was frequently reflected in geology courses taught at universities and colleges.

In certain areas it is possible to show that some sedimentary formations were deposited before normal faulting, some during faulting, and others after faulting. The thicknesses of pre-faulting formations will be the same on either side of a fault, but they will be displaced vertically, relative to each other. Those of the same age as the movements on a fault will again be displaced but will be of different thickness on either side of the fault. Formations that were deposited after faulting will not be cut by the fault.

Determining whether there was growth faulting during the deposition of a formation not only adds to our understanding of the geological history of an area but can also be of practical value. Suppose, for example, that a formation cut by a growth fault contains valuable resources, such as oil or gas. The explo-ration geologist or geophysicist searching for these resources would be able to estimate that greater reserves would be preserved on the downthrow side of the fault, in contrast to its upthrow side. Such apparently subtle variations in formation thickness may make all the difference to whether a deposit is, or is not, of commercial value. The attention that was given to the detailed geometry of growth faults in the Gulf of Mexico, the North Sea, and the Niger Delta helped the exploitation of oil and gas fields in these areas.

PAUL L. HANCOCK

Gulf Stream

The Gulf Stream is a western boundary current and is part of the North Atlantic Gyre (Fig. 1). The name 'Gulf Stream' is commonly applied to the entire northern branch of the gyre, but the current is in fact made up of at least three individual parts: the Florida Current, which flows from the Caribbean to approximately 37° N, the Gulf Stream proper between 37° N to approximately 40° W, and then the extension of the current towards Europe, which is the North Atlantic Drift. The velocity of the Gulf Stream is strongest just south of Cape Hatteras and can reach 2.5 m s^{-1}. At this point the current is constrained to the continental shelf and limited to approximately 800 m in depth. Meanders in the current are common but only of approximately 50 km in amplitude. Just north of Cape Hatteras the current leaves the shelf and moves over the deeper water of the North Atlantic, slowing to velocities that are typically 0.7 m s^{-1}. The northern boundary of the current usually corresponds to a sharp gradient in temperature with cold water on the northern side. The southern boundary is typically more diffuse.

The slowing down of the current as it moves away from Cape Hatteras and the lack of topographic steering allows meanders of up to 350 km in amplitude to develop. These large meanders can break from the main current to form eddies called Gulf Stream rings, which can typically be 200 km in diameter. Because of the temperature gradient on either side of the current, the meanders can form rings with either cold or warm water in their centres ('cold-core' or 'warm-core') and can extend to full ocean depth. The eddies can be re-absorbed by the current or can exist independently for up to a year, and at any one time there may be up to ten rings in existence. Similar rings have been observed forming in other western boundary currents such as the Agulhas and the Kuroshio (see *ocean currents*).

The reason for the existence of the Gulf Stream and the North Atlantic Gyre is principally Ekman drift (see *ocean currents*) coupled with the latitudinal variation of the Coriolis force. The prevailing wind field across the North Atlantic is zonal (see *Walker circulation*), being from the west above approximately 30° N and to the east south of this towards the Equator. These prevailing winds generate currents which are deflected to the right by the rotation of the Earth and result in a general inflow of water into the subtropic convergence (that is the centre of the gyre). The pool of water which develops in the centre causes a decrease in vorticity, and because of the

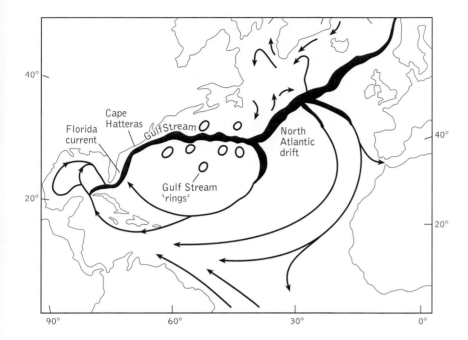

Fig. 1. The approximate position of the Gulf Stream, showing Gulf Stream rings.

conservation of angular momentum an anticyclonic gyre develops across the basin. The intensification of the western boundary of the gyre (i.e. the formation of the Gulf Stream) results from the fact that the Coriolis force changes with latitude, coupled with the conservation of vorticity across the whole gyre.   MARK A. BRANDON

Further reading

Apel, J. R. (1987) *Principles of ocean physics.* International Geophysics Series, Vol. 38. Academic Press, London.

gullies and gullying

gullies and gullying Although 'gully' can refer to clefts in sea cliffs and natural trenches eroded in deep-sea fans, it principally connotes recently eroded channels in soil or sediment, too deep to be crossed by wheeled vehicles or eliminated by ploughing. In this sense, gullies are intermediate between rills (tiny entrenched ephemeral rivulets) and arroyos or gulches (much larger incised stream beds). Gullies tend to be deep and long relative to their width, although not as long as valleys and incised rivers, and they are typically steep-sided with a steep headscarp or waterfall at the upper end. Most are dug in loose earth by ephemeral run-off from storms or melt water, although perennial flow, erosion by debris flows, and incision into coarse gravel or bedrock are possible.

Beyond this, local combinations of various processes of initiation and growth, together with the diverse constraints of substrate, vegetative cover, and climate, mean that gullies are quite variable and may have distinct regional differences. Thus the literature contains many redefinitions and also many local names for unusual gullies, such as *donga, vocaroca, ramp,* and *lavaka.* Among the variations, gullies may be continuous or discontinuous, depending on whether or not the gully has become connected at grade to the main drainage system, and they can grow downward from mid-slope positions, or from the hill-toe up. Valley-side gullies represent a new or future extension of a drainage network into hillsides, whereas valley-floor gullies represent an incision by the drainage on the valley floor.

Gullies are considered to be diagnostic of recently disrupted or non-equilibrium landscapes. Probably the most dramatic and common causes of gullies arise from human misuse of the land, especially overgrazing, too-frequent fires, deforestation, ploughing downslope rather than across it, failure to terrace fields, and poor management of run-off from paths, roads, and construction sites. However, the ultimate causes of gullies also include:

(1) relative lowering of local or regional base level, including tectonic uplift of the headwater area, subsidence of the lowland region, a fall in sea level, drying or drainage of a lake, or erosive breaching of a sill or a natural dam;

(2) increased run-off, which might result from increased rainfall, increased seasonality or storminess, a particularly severe storm, soil compaction or plugging, or decreased vegetation (which can be caused by aridification and natural fires, as well as the human and animal activities mentioned earlier);

Fig. 1. Gullies in central Madagascar. (a) An elongate and dendritic gully system formed predominantly by incision of a drainage network. (b) The lobate mid-slope gully is not linked to valley-floor drainage but has expanded upslope as a result of sapping of the headscarp once its base intersected groundwater seeping downhill. The middle concavity is an old gully now stabilized, overgrown, and smoothed over, whereas the rightmost gully has started to fill in. (c) Severe gullying, probably resulting from climate change exacerbated by fires and overgrazing. A change from infiltration and chemical weathering to surficial run-off has caused extensive destruction of the previous landscape of smooth and very deeply weathered lateritic hills.

(3) slope collapse, including: (a) mass movements ranging from deep slumps to soil slips (possibly triggered by earthquakes or undercuts by springs, rivers, and artificial excavations); and (b) collapse of pipes, which are natural subsurface passageways through soils, enlarged by rapid drainage along burrows, root voids, interconnected soil pores, and the like.

Thus, the causes of gullying range from entirely natural to totally cultural.

Growth processes include crumbling, spalling (surface exfoliation) and collapse of walls, direct attack by rain drops, slope wash, incision of the floor and undercutting, and collapse of walls by channelled run-off or mass flows, collapse of feeder pipes, and sapping due to seepage of groundwater out of the bases of the walls and headscarps.

The shape and position of a gully usually reflect its dominant growth process or processes, and in some cases its immediate cause (Fig. 1). For example, collapse of pipes, downcutting along stream beds, and headward retreat by springs or waterfalls create very long gullies; whereas growth by seepage and sapping can lead to a lobate shape with a broad and arcuate headscarp. Incision by slope wash and early channels (i.e. exaggerated rilling) results in gullies that lie in the floors of hillside hollows, re-entrants, and valleys, below well-defined areas of collection of drainage; whereas seepage and sapping need not reflect surface topography at all, and can even grow backward through a ridge crest, if subsurface plumbing supplies groundwater from the far hillside. Gullies that have grown along tracks and paths may extend perfectly along the crests of hill spurs, if these provide the easiest routes up the hill.

Because many gullies represent readjustment of a landscape to a new equilibrium after some sort of threshold of resistance to erosion has finally been exceeded, they may grow astonishingly quickly once they have been initiated. For the same reason, they are not easily stopped until they have extended to the upslope limit of concentrated or rapid flow, or have regraded to the new base level, or both. As a result, gullies are easier to prevent than to cure, at least until the impetus for growth has been exhausted, at which time the gully will begin to stabilize and to heal itself. (Very long gullies may be healing at the downstream end while the head end is still expanding.) Thus, preventing or curing a gully requires understanding of its stage of growth and its specific proximate causes and growth processes. For example, if the gully is growing by headward retreat of a waterfall, then upslope management of run-off is required, such as diversion ditches and regrading the channel floor with loose broken stone ('rip-rap'). If unchannelled slope wash is the problem, then the best solutions would be to promote infiltration relative to run-off by contour-ploughing of the watershed, diverting flow into blind trenches, and planting trees and other vegetation. In contrast, to cure a gully that has grown largely by seepage and sapping, diversion ditches are irrelevant and increased infiltration is undesirable; one should instead stabilize the gully floor and rebury the zone of seepage. In lateritic terrain, where the soil surface hardens in the sun but the underlying saprolite becomes increasingly moist and soft downward (at least until the transition into fresh bedrock), ploughing or otherwise breaking the surface and destroying vegetative cover may actually make a bad situation much worse.   NEIL A. WELLS

Further reading

Higgins, C. G. (1990) Gully development. In Higgins, C. G. and Coates, D. R. (eds) *Groundwater geomorphology*. Geological Society of America Special Paper 252, pp. 18–58.

Harvey, M. D., Watson, C. C., and Schumm, S. A. (1985) *Gully erosion*. US Department of the Interior, Bureau of Land Management, Technical Note 366.

NWSCA (National Water and Soil Conservation Authority) (1985) Soil erosion in New Zealand. *Soil Water*, **21**, (4), supplement.

Wells, N. A. and Andriamihaja, B. (1993) The initiation and growth of gullies in Madagascar: are humans to blame? *Geomorphology*, **8**, 1–46.

Gutenberg, Beno (1889–1960)

Important contributions to our understanding of the interior of the Earth were made by Beno Gutenberg, who was born at Darmstadt, Germany, and became Professor of Geophysics at the University of Frankfurt-am-Main in 1926. In that year he published the seismic evidence of the low-velocity layer in what is now known as the upper part of the mantle, having been exercised since his early work at Strasburg by the need to explain the apparent passage of earthquake tremors within the Earth. The following year (1927), intrigued by Alfred Wegener's idea of continental drift, he proposed that the Moon had been torn from the Pacific and that what had been a single continent was now spread as several fragments under the influence of the Earth's rotation.

By then Gutenberg had to give time to the (Jewish) family business as well as to his research, and Germany was about to succumb to Nazism. In 1930 he was attracted to a professorship at the California Institute of Technology. He was also appointed Director of the Seismological Laboratory at the Carnegie Institution.

By studying the passage of earthquake waves through the interior of the Earth, Gutenberg was able to calculate the diameter of the core. The seismic boundary between the mantle and the core is now know as the Gutenberg discontinuity. In California he worked with C. F. Richter. They drew up proposals for making seismological observations more accurate by taking into account the ellipsoid shape of the Earth. Their book *The seismicity of the Earth* became a standard work. After the Second World War Gutenberg served on government bodies concerned with the use of seismology to determine the sites of atomic explosions.   D. L. DINELEY

guyots During the Second World War, H. H. Hess, who later, together with R. S. Dietz instigated the theory of sea-floor spreading, served with the US Navy in the Pacific. He took advantage of the opportunity to record echo-sounding profiles across parts of the Pacific and discovered several flat-topped submarine mountains or seamounts. These tablemounts he called *guyots* after Arnold Guyot, a Swiss geographer of the eighteenth century. Hess hypothesized, correctly as it turned out, that these tablemounts were ancient islands that had become submerged to depths of 1 to 2 km.

This discovery led to a number of expeditions in the Pacific in the decade after 1945 that established by dredging and drilling that guyots were truly submerged volcanic islands. The causes proposed for their origin were, however, numerous, ranging from the subsidence of individual volcanoes to subsidence of the whole sea floor and including a major sea-level rise. In truth, their relatively small size (4–10 km across) and the discovery of evidence for shallow water in the way of subaerial erosion, truncated surfaces, wave and tidal action, corals, and sediments (e.g. coals) above a volcanic base shows that most are submarine volcanoes that have had their tops truncated and have then been submerged.   HAROLD G. READING

gypcrete *see* PALAEOSOLS, DURICRUSTS, CALCRETE, SILCRETE, GYPCRETE

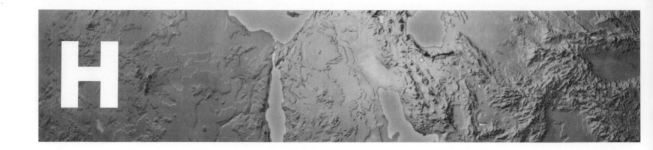

Hadean eon

Hadean eon The term 'Hadean' (after the Greek mythical underworld, *Hades*) was coined by the American geologist Preston Cloud for the earliest eon of geological time from the formation of the planet to the inception of the Archaean eon at about 4 Ga (4000 million years ago). This earliest eon has also been called the Priscoan. There are no known rocks from this time persisting at the surface of the Earth, nor is there any tangible material evidence by which to define the Hadean–Archaean boundary. It has been suggested that, to be consistent with the criteria used for the definition of the other eons of Earth history, reference should be made to the history of life on the planet. One way of meeting this criterion would be to designate it as the time preceding the appearance of life on Earth at around 4.0 Ga (between 4.4 and 3.8 Ga).

Fierce meteoritic bombardment may have continued throughout this eon, adding both solid rock and water to the accreted material of the early Earth. During this time the Earth began to differentiate into the concentric shells that have influenced its behaviour ever since. A solid crust was formed, and in due course degassing and heat loss reached the point when oceans began to accumulate. This process may have been interrupted by the impact of a giant meteorite, which ejected molten chunks of the planet to form the Moon. Such degassed atmosphere (and hydrosphere) as there was would have been blown away by the solar wind and the intense meteoritic onslaught. Eventually the bombardment declined, although volcanic activity continued at a very high level, adding water, carbon dioxide, and other gases to a new atmosphere. Much of this would have been from volcanoes on the forerunners of mid-oceanic ridges, which at this time may have risen above sea level. A continuing gradual dehydration of the Earth's mantle may by then have begun to drown the ridges and to flood the surface of the planet. Heat loss from the Earth now took place more through the oceans into the atmosphere, rather than directly into the atmosphere itself.

Temperatures in the seas sank to a level at which the persistence of life in its earliest form was possible. Debate about the processes and probable stages in development of self-replicating carbon-based structures covers many scenarios, including the injection of such molecules from outer space. The advent of life set in train new processes, both chemical and physical, that were to result in progressively greater changes to the surface of the Earth with each succeeding eon. Once these biological processes had begun, it may be said, the Archaean eon had started. D. L. DINELEY

Further reading

Coward, M. P. and Ries, A. C. (eds) (1995) *Early Precambrian processes.* Geological Society Special Publication No. 95.

Hall, James, Jr (1811–98) James Hall, Jr became recognized as America's foremost palaeontologist, whose life was devoted to collecting and describing fossils and to promoting state institutions concerned with the Earth sciences. He spent most of his life in New York State, joining the New York Geological Survey in 1836 as a curator of invertebrate fossils collected by Ebenezer Emmons and others. This resulted in the beautifully illustrated 13 volumes of *The Palaeontology of New York State*. In 1843 Hall was appointed State Palaeontologist.

So energetic was Hall in building up the collections that by 1857 he was in need of, and built, a new laboratory in Albany, the state capital. He then made a compelling plea for a state museum, with the result that he was appointed as its first curator in 1865, and in 1871 became the first director of what was soon to be the New York State Museum. Hall's experience and expertise in establishing this kind of institution led to six other states seeking his help in setting up their own equivalent bodies. They became productive, and thriving models for new Earth science centres in other states. Hall's achievements as scientist and administrator won him many honours.

Hall was the author of over 300 scientific works, some of them very large. In 1845 he described the Mesozoic fossils collected by the (J. C.) Frémont expedition to the south-western territories. Nothing like them was known in the east, and they aroused much scientific interest. Hall was interested in the environments in which his Palaeozoic fossils had existed when alive, and he was aware of the great thicknesses of Palaeozoic rocks in the Appalachian area as compared with equivalent strata in the mid-west. His concept (1857) of a great downfold or depression filling with sediment at the edge of a continent and then isostatically uplifted to produce the Appalachian fold ranges, was the first suggestion of an Appalachian geosyncline. Over 60 years passed before Stille's geosynclinal model was proposed. D. L. DINELEY

Hall, Sir James (1761–1832)

A geologist and chemist, James Hall has been recognized as the first experimental petrologist. After leaving Cambridge (without a degree) he espoused the Uniformitarian views of James Hutton and set out to put to the test the objections made by the Werner and his fellow-Neptunists. He began by examining rocks in Scotland and on the European mainland, where he combined his geological work with architectural studies. On returning to Scotland in 1785 he began his experiments on the melting and crystallization of rocks. He was able to demonstrate that if basalt was fused and then cooled slowly it became crystalline and did not form a glassy material. Powdered calcite (calcium carbonate), when heated under pressure, yielded a crystalline marble, not 'quicklime' (calcium oxide). Experiments on melting a mixture of crushed feldspar and quartz led to the suggestion that slow cooling resulted in a product very similar to the material that was originally melted: slow crystallization explains the texture of granites—a novel concept in 1790. These results, which Hall published in a series of memoirs (1790–1805), provided critical evidence in favour of the Vulcanist (Plutonist) interpretation.

Hall also studied boulder deposits and striated rock surfaces and concluded—erroneously—that they were formed by floods of sea water sweeping across country.

Subsequent work showed these phenomena to be of glacial origin. For five years (1807–12) he was a Member of Parliament for a constituency in Cornwall. D. L. DINELEY

Harker, Alfred (1859–1939)

When Alfred Harker entered St John's College, Cambridge as an undergraduate in 1878 he could not have foreseen that he would remain there for 61 years. During that interval he became one of the outstanding figures of British petrology. Harker began life at Cambridge reading physics, but soon transferred his affections to geology and was appointed a demonstrator at the Sedgwick Museum in 1884. He worked primarily on the volcanic and hypabyssal igneous rocks of the older Palaeozoic rocks in North Wales and the English Lake District. From 1895 to 1905 he worked with the Geological Survey in Scotland and commenced investigations of the Tertiary igneous province in the Hebrides.

Harker's writings were many and ranged over such subjects as diverse as slaty cleavage, metamorphic rocks, plutons, and Pleistocene glaciation in the Isle of Skye. He worked out in detail the extrusive–hypabyssal–plutonic sequences in the Tertiary volcanic centres of the Inner Hebrides, matching field relationships with petrographic characters and producing lucid, beautifully illustrated accounts.

He published three remarkably successful textbooks. *Petrology for students* first appeared in 1895 and further editions were still forthcoming 60 years later. His *Natural history of the igneous rocks* came out in 1909 and was not to be superseded until Bowen's work on the reaction series of rock-forming minerals was published, some 20 years later. His book *Metamorphism*, published in 1932, similarly was acclaimed as the most authoritative work on the subject up to that time. It, too, was to stand at the peak of the subject for many years.

These superb texts were in keeping with Harker's standards of teaching and research, which were widely recognized. He was honoured accordingly by the Geological Society and was elected a Fellow of the Royal Society. He died in 1939. D. L. DINELEY

Haüy, René-Just (1743–1822)

Haüy dominated mineralogy and crystallography for over 25 years from 1800 onwards and has been styled the 'father of crystallography'. He studied physics and mineralogy, and in 1802 was appointed Professor of Mineralogy in the Museum of Natural History and later in the University of Paris; he was an Honorary Canon of Notre-Dame.

In 1784 he published his *Essai d'une théorie sur la structure des cristaux*, in which he postulated that crystals are made up of small fundamental units which he called *molécules intégrantes*. He arrived at this from a study of calcite crystals of various habits and noted that the cleavage rhombohedra were always the same. In elegant drawings he showed that all the forms of calcite could be generated by suitable stacking of the units. The problem is that not all crystals, quartz for example, cleave as perfectly as calcite but have irregular fractures.

Haüy regularized the use of crystallographic axes and formulated the law of rational indices, which states that the intercepts made by faces on the axes are small rational multiples of those made by a parametral face.

In his *Traité de Minéralogie* (1801) he classified minerals by using Linnaean principles, emphasizing physical properties and the *molécules intégrantes*. Many minerals were described and figured with reference to their chemistry. Some chemical analyses of a fair accuracy were included. Haüy named and described crystal forms, appreciated the nature of twinned crystals, and discussed the pyroelectricity of tourmaline. A second edition with more data but few other changes appeared in 1822. R. BRADSHAW

heat flow in the Earth

From its molten outer core to its temperate surface, the temperature of the Earth varies by several thousand degrees. The heat of the Earth's interior accumulated by means of a poorly understood combination of mechanisms: (1) potential energy acquired as the planet accreted from infalling meteorites and asteroidal fragments, (2) the conversion of gravitational to thermal energy as metallic iron segregated from silicate rock and sank to form the planetary core, and (3) the energy released by the continuing decay of radioactive elements within rocks and the core.

Internal heat has been gradually escaping to outer space during the Earth's history, impeded by the thermal insulation of the planet's outer layers of silicate rock. This thermal energy is transported primarily by two mechanisms: conduction and advection.

In conductive heat flow, thermal energy flows from warmer to cooler materials. The energy flux is proportional to the temperature difference $\Delta T = T_1 - T_2$ over a distance interval ΔZ, where T_1 and T_2 are the temperatures of the warmer and cooler materials respectively. For any particular material, there is a proportionality constant κ that scales $(\Delta T/\Delta Z)$. This constant, κ, is known as the *thermal conductivity*. The thermal conductivity of silicate rock is very low, typically 10 times smaller than metal sulphide minerals like pyrite (FeS_2) and 100 times smaller than pure metals like copper (Cu). A steady conductive heat flow near the Earth's surface is maintained by a gradual increase of temperature with depth, roughly one degree Kelvin for every 50 metres near the outer surface. The conductive vertical heat flow $Q = \kappa(\Delta T/\Delta Z)$ can be estimated by measuring temperatures at two depths Z_1 and Z_2 within a borehole, and estimating the thermal conductivity κ from the rock-types within the interval $\Delta Z = Z_1 - Z_2$.

Because conduction in silicate rocks is inefficient, thermal energy in the Earth is commonly advected, that is, transported by the motion of material. Heat can be advected by water that percolates along fissures or through porous rock. Rainwater in volcanic areas can descend as far as several kilometres into the crust, later rising to the surface in hot springs or geysers. A spectacular form of heat advection occurs when molten rock, or magma, erupts from a volcano. Although the silicate rock of the Earth's mantle and crust is predominantly solid, on geological timescales it can flow in response to buoyancy forces. Positive and negative variations in temperature cause rocks to expand and contract. Just like a hot-air balloon, but much more slowly, heated rock expands, experiences a proportional decrease in its density, and tends to rise within the planetary interior. Similarly, cooled rock contracts, experiences an increase in density, and tends to sink. The Earth's mantle advects heat upward by a continuous exchange of rising warmer rock for sinking cooler rock, in the process called *convection*.

How do geophysicists determine whether the mantle convects heat from the molten-iron core to the surface crust, rather than conducting heat through motionless rock? Consider a sizable volume of heated rock. The tendency of heated rock to become buoyant is scaled by its coefficient of thermal expansion α. Typical values of α for silicate rocks imply expansion by two parts in 1000 for every 100 °K increase. Buoyancy is also favoured if the temperature difference ΔT between the top and bottom of the mantle is large. Average surface temperatures are close to 290 °K. The temperature at the base of the mantle must be higher than the melting point of iron in the underlying core. This melting point is imperfectly known, but is likely to exceed 3500 °K. The buoyancy forces must overcome the resistance of the rock to flow,

as determined by its viscosity η. Buoyancy forces must also overcome the tendency of heat to escape by conduction from the buoyant rock, as determined by its thermal conductivity κ. The total size of the (potentially) convecting region, measured by a length scale, is also important, because opposing flows of warmer and cooler rock are difficult to maintain if the system is of limited extent. The balance of parameters can be combined into a single dimensionless ratio, called the Rayleigh number Ra, that governs whether overall heat flow is convective or conductive. Above a critical value of Ra, which can be estimated using model experiments in the laboratory, a system will experience a convective 'instability' and be subject to advective heat transport. By the mid-twentieth century, geophysicists had determined that the Earth's mantle was likely to advect heat from the core, because its Rayleigh number far exceeds the convective threshold. In fact, convective instabilities are probable on regional scales within the mantle, not just for the system as a whole.

Conduction still plays a large role in the Earth's heat flow, mainly in the top and bottom 100–200 km of the mantle, roughly 10 per cent of its 2900 km thickness. The conductive zones form boundary layers where heat is absorbed from the underlying core and, through the crust, released to the overlying atmosphere. The upper boundary layer is known as the lithosphere. The lower boundary layer has been named D″ by seismologists. It is characterized by unusual wave propagation behaviour (scattering, for example) that suggests a structural complexity comparable to that of the lithosphere. Substantial increases in temperature occur with depth in the boundary layers, enhancing the heat flow through them. Between the boundary layers, however, the near-absence of conductive heat transport is reflected in a weaker 'adiabatic' temperature profile, along which the increase in temperature with depth is associated with compression of the rock by increasing overburden pressure, not an increase in its intrinsic heat content.

The famous 1944 textbook *Principles of physical geology* by Arthur Holmes, written prior to the advent of plate tectonic theory, illustrated the advection of mantle heat as a steady migration of silicate rock within a convection cell, in which regions of upwelling warmer material were matched by symmetrical regions of downwelling cooler material. The real mantle is more complex. Its high Rayleigh number encourages a sporadic progression of upwellings and downwellings in both space and time. If Earth history could be accelerated to fit within a typical television advertisement, the planet's underlying mantle flow would look quite turbulent. The viscosity of silicate rock depends strongly on temperature; cooler regions therefore tend to be stiffer, and warmer regions more prone to flow. As a result, the surface of the Earth has, as it cooled, formed stiff tectonic plates (the lithosphere) that are subject to brittle fracture (which is expressed as earthquakes) along their mutual boundaries. When these cooling plates founder and sink, they descend as sheets, or 'slabs', through the upper mantle at ocean trenches, forming earthquake-prone Benioff–Wadati zones. Upwelling mantle rock is less viscous, typically rising as localized 'plumes' or 'hot spots'. The collision of a plume with the lithosphere gives rise to volcanic islands in the world's oceans, as well as progressions of volcanoes on the continents. Although plumes probably represent most of the advective heat that is transported from the core–mantle boundary to the surface boundary layer, most surface volcanism occurs elsewhere, along the weak boundaries of the thick surface plates. One of the most prominent hot spots has formed the Hawaiian Island chain, which can be traced across the north-west Pacific Ocean as a crooked line of extinct underwater volcanoes. The motion of the hot spot is only apparent. The tectonic plate that underlies most of the Pacific Ocean has instead been sliding across the mantle, at a speeds of 5 to 10 cm per year, for more than 100 million years. By contrast, the mid-Atlantic Ridge has hovered above the hot spot that formed Iceland for a comparable length of time without significant drift.

The lower part of the mantle is estimated to be 10–30 times more viscous than its upper layers, presumably because of mineral phase changes from low- to high-pressure crystal structures at a depth of about 670 km. The increase in viscosity with depth probably forms a barrier to downward convective flow, causing some slabs to stagnate at a depth of 600–750 km. The high-viscosity lower mantle probably discourages the lateral drift of upwelling plumes from the core–mantle boundary. The hot spots therefore form a fixed reference frame on which the motions of the surface plates can be tracked. Computer models of convection that incorporate large Rayleigh numbers and temperature-dependent viscosity can reproduce the gross features of observed mantle convection, such as stiff surface boundary layers, downgoing slabs, and rising plumes. Computer simulations also indicate that cool slabs that stagnate at 600–700 km depth are occasionally flushed catastrophically into the deeper mantle. Such overturns of mantle rock may help to explain episodes of widespread intense volcanism in Earth history.

The pattern of heat flow at the Earth's surface is governed partly by mantle convection beneath the lithosphere, partly by the location of plate boundaries, and partly by the concentration of radioactive elements in the crust. Geologists have identified several dozen active hot spots, each a centre of present-day volcanism. The heat flow at plate boundaries depends on their character. At transcurrent boundaries where plates slide past each other, such as the San Andreas Fault in California, surface volcanism is absent or minimal. Where plates spread apart, mantle rock at an estimated 1550 °K rises to fill the gap and undergoes pressure-release melting. The magma rises by porous flow to form new oceanic crust along linear ridges on the sea floor, as well as linear volcanic zones on land like the East African Rift. Tectonic plates develop at mid-ocean ridges, because the viscosity of mantle rock increases substantially once temperatures fall below 1350 °K. New lithosphere thickens and stiffens as it cools. Much of the near-surface cooling is accomplished by sea water, which circulates within the oceanic crust. Theoretical calculations

predict, and measurements confirm, that the heat flow of a new tectonic plate decreases inversely with the square root of its age. The thermal buoyancy of the cooling lithosphere is reflected in the elevated sea-floor topography of mid-ocean ridges, whose crests are about 2 km above the abyssal sea floor. The near-constant depth of the abyssal sea floor indicates that the lithosphere thickens to roughly 100 km in 70 million years, but then ceases to grow. Small-scale convective instabilities cause any additional thickness of cooled rock to peel off and sink into the mantle.

At boundaries where plates converge, one plate typically underthrusts the other and sinks into the mantle, forming a deep oceanic trench in its wake. As it descends, volatile compounds, such as water (H_2O), carbon dioxide (CO_2), and sulphur dioxide (SO_2) are released from the descending slab and catalyse melting in the surrounding mantle. Rising magma forms lines of volcanoes in the overriding plate at a predictable distance from the trench. If the overriding plate is oceanic, an arc of islands is formed; for example, the Tonga–Fiji island arc north of New Zealand. If the overriding plate is continental, the volcanic arc expresses itself as an extended mountain range such as the Andes of South America.

The total conductive heat flux at the Earth's surface is estimated to be 32×10^{12} W (32 billion billion watts). Heat flow from circulating water is estimated to add 10×10^{12} W to the conductive heat flux, giving a total of 42×10^{12} W. This massive outpouring of thermal energy can be put in perspective by noting that the average conductive heat flux is only 0.06 W m^{-2}. The conductive heat flux from an area the size of a football field would, according to this average, power three 100-W light bulbs.

On the continents, heat flow tends to be greater where the surface is underlain by younger rocks. In contrast to the oceans, the main factor in this trend is the age-dependent concentration of radioactive elements in the granitic rocks of the continental crust. Oceanic crust is thinner than continental crust and basaltic in composition, with lower concentrations of radioactive elements. In continental crustal rocks formed over a billion years ago, much of the uranium (U), thorium (Th), and the radioactive isotope ^{40}K of potassium has decayed, either to lead (Pb) or, in the case of ^{40}K, argon (Ar). Radioactive decay in old rocks therefore contributes less thermal energy to the crust. The large radiogenic heat production of continental crustal rocks compensates for a low conductive heat flow from the underlying continental mantle, which is estimated to be 0.02–0.025 W m^{-2} on average. The greater thickness and age of continental lithosphere (much of it more than 200 km thick and older than a billion years) as compared with oceanic lithosphere (which is no thicker than 100 km and no older than 200 million years) may be the root cause of this heat-flow deficit. Continental lithosphere may simply have cooled for a longer time. However, why small-scale convective instabilities should not cause the base of the continental lithosphere to founder and sink, thereby limiting its growth, is thus far not understood well.

JEFFREY PARK

Further reading

Fowler, C. M. R. (1990) *The solid Earth: an introduction to global geophysics.* Cambridge University Press.

Sigurdsson, H. (1999) *Melting the Earth: the history of ideas on volcanic eruptions.* Oxford University Press.

Heezen, Bruce (1923–77)

Bruce Heezen was born in Iowa, the son of a farmer. He later studied at the University of Iowa, where he developed a particular interest in palaeontology. Shortly after the end of the Second World War, he was introduced to Maurice Ewing and soon after he joined an exploration of the submerged continental margin of the east coast of the United States. He later spent several months on a marine exploration of the Mid-Atlantic Ridge. He was the first scientist to draw attention to the fact that the Mid-Atlantic Ridge coincided with a zone of earthquake activity. He suggested that the crust along the axis of this ridge is also stretched at right-angles to the axis and he proposed that these displacements were accommodated by the creation of crust at ridge axes, and that the history of continental displacements was recorded in the sea-floor sediments and rocks.

Heezen was the first scientist to produce (in 1956) a physiographic diagram of the North Atlantic, and in the subsequent two decades he was responsible for the production of a series of physiographic maps of all of the world's oceans. This collection of maps is probably the most widely distributed set of sea-floor maps available at the present time. Heezen, one of the foremost scientists who have studied the science of the sea floor, published more than 300 academic papers and two books. He received awards from many scientific societies for his fundamental contributions to marine geology and the understanding of the Earth. He died, aged 54, in June 1977.

ALASTAIR G. DAWSON

Hercynian orogeny *see* VARISCAN OROGENY

Hess, Harry Hammond (1906–69)

The American geologist H. H. Hess is perhaps best known for his work on the geology and geophysics of the ocean floors. As a young man he accompanied Vening Meinesz on a submarine cruise to make gravity measurements in the Caribbean, and later did much of his research in that region. By developing echo-sounding submarine survey techniques during and after the Second World War he made significant contributions to the bathymeytry of the Pacific Ocean and coined the word 'guyot' for the submerged flat-topped sea mounts there. He also investigated the gravity anomalies beneath the submarine trenches of the Pacific.

At Princeton University he followed his classic work on pyroxenes and ultrabasic rocks, and their role in the development of the crust, by developing a model in which ocean-floor basalts were generated along rifts in the mid-ocean ridges. There upwelling of the converting mantle led to extrusion and lateral spreading of the ocean floor. In this work he

was joined by the Canadian J. Tuzo Wilson, and it was to be tested and supported by many other investigators on both sides of the Atlantic in the late 1960s and 1970s. The 'plate tectonics' concept took shape from these beginnings.

Hess was an able administrator as well as a brilliant scientist and played a leading role in the ambitious Mohole Project, initiated in 1957 and abandoned in 1966. This had the aim of drilling through the ocean floor to reach the Earth's mantle. It was superseded by the Deep Sea Drilling Project. Hess was eventually the Chairman of the Space Science Board of the National Academy of Sciences and was for a time the principal non-government adviser on the scientific objectives of planetary exploration.

D. L. DINELEY

heterochrony in palaeontology

Heterochrony ('different time') refers to a change in the timing or rate of development of characters relative to those same events in the ancestors. All evolution involves change, but heterochrony emphasizes change in patterns of individual development rather than in adult stages only. The recognition that the development of an individual (ontogeny) is related to its evolutionary history (phylogeny) was formalized in the nineteenth century by Ernst Haeckel, who stated that 'ontogeny recapitulates phylogeny'. This means that the juvenile forms of an organism repeat the evolutionary stages of their ancestors, as for example the fish-like stages in the embryos of mammals. This situation in which the descendent passes morphologically beyond its ancestor is now termed 'peramorphosis'. Many examples of exceptions to this rule were, however, found; and the biologist Walter Garstang noted in the 1920s that the evolution of many groups of organisms must have involved the retention of juvenile characters into the adult stage, or paedomorphosis. Stephen J. Gould of Harvard University has shown the development of paedomorphosis in Mickey Mouse as drawn by cartoonists over a 50-year period, with larger head, larger eyes, and shorter arms (all juvenile characters) becoming more apparent in later versions.

Three main kinds of change can be seen in the characteristics of an organism: change in rate of growth, beginning, or end of growth. As each change can be either an increase or decrease, a total of six kinds of heterochrony are possible: neoteny (slower rate), acceleration (faster rate), predisplacement (earlier start of growth), postdisplacement (later start), progenesis (earlier end of growth), and hypermorphosis (later end). Thus, paedomorphosis could be caused by neoteny, or by post-displacement or progenesis.

Examples of heterochrony are known from the fossil record both at and above the species level. At the species level, for instance, studies of Cenozoic brachiopods from Australia and New Zealand show a paedomorphocline in which adults in the Pliocene show features found in juveniles in the Palaeocene. In this case the change appears to be related to the adaptations from deep to shallow water, particularly in the size of the pedicle foramen. This was large in the deep-water juveniles, reflecting the large pedicle needed to confer stability, and was also large in the shallow-water adults living in turbulent conditions. At a higher level, it is considered that vertebrates might have arisen from the pelagic larva of a tunicate-like organism by paedomorphosis, causing the retention of the larval characters into the adult phase. It has also been suggested that birds may be paedomorphic theropods (carnivorous dinosaurs). Birds have a number of characters that could be seen as juvenile in relation to theropods, including large orbits and brain-cases and retarded dental development. It has also been suggested that feathers were present on juvenile theropods. In this case the processes would probably have been neoteny and post-displacement. Although these examples are ones in which the processes appear to be operating in one direction, it is more common for mosaic heterochrony to occur, in which opposite heterochronic processes produce some characters that are paedomorphic but others that are peramorphic. DAVID K. ELLIOTT

high-grade gneiss terrain

'High-grade gneiss terrain' is the name traditionally given to one of the two main types of Archaean crust, the other being the 'granite–greenstone terrain'. High-grade gneiss terrains are characterized by metamorphic rocks of amphibolite to granulite facies and dominated by gneissose igneous material with a wide range of composition, but containing a large proportion of granodiorites and tonalites. Gneisses of sedimentary and volcanic origin have also been recognized in these terrains but these generally form only a minor component. These rocks are typically highly deformed, which accounts for their gneissose appearance, and ductile shear zones ranging from centimetres to many tens of kilometres across are characteristic.

Although it was once thought that high-grade gneiss terrains represented a fundamentally different kind of Archaean crust from the granite–greenstone type, it is now realized that the two kinds of terrain represent only different levels of exposure. In the Superior province of Canada, for example, which contains perhaps the best-known examples, it has been shown that high-grade gneiss-type crust underlies the granite–greenstone terrain at depths of around 10 km, and is in several places upthrust to the surface along what are interpreted as major collisional suture zones. These high-grade rocks are indistinguishable from the high-grade gneiss terrains exposed, for example, in the nearby Godthaab craton of south Greenland, long considered to be a type example of the high-grade gneiss terrain. R. G. PARK

high-latitude ocean currents

The ocean circulation in high-latitude waters has been determined only relatively recently by a somewhat different method from that employed for the more temperate ocean currents. The principal difficulty in studying the polar currents is a consequence of the fact that for some or all of the year, large portions of the oceans are ice-covered and there is still a very small database. Whereas the positions of the Gulf Stream and the Agulhas (see *ocean currents*) have historically been determined by trading ships on passage, the general circulation of the Weddell Sea and Arctic Ocean were initially determined using more dramatic means. More so than other oceans, the general circulation of polar waters has been determined by temperature and salinity analysis (see *ocean water*) of relatively few measurements.

Northern high-latitude ocean currents

The Arctic Ocean is a deep basin almost completely surrounded by continental land masses with just one deep entry and exit called Fram Strait. The Arctic Ocean is permanently covered in ice up to 5 m thick, and it was this ice cover which made the circulation difficult to determine. Towards the end of the nineteenth century several renowned scientists began to believe that there might be a current from the Bering Strait across the north pole to Fram Strait because of the discovery of certain items of ship wreckage and Siberian fir trees on the coast of Greenland. This hypothesis was not tested until Fridtjof Nansen led the carefully designed *Fram* (Norwegian for 'forward') to the Arctic Ocean, where she drifted across the basin from 1893 until 1896 in what we now know as the Transpolar Drift (Fig. 1).

Exploration of the Arctic Basin really began in earnest in the 1930s with the drift of the *Sedov* and the start of the Russian North Pole annual drifting ice stations, which began in 1937 and continued until 1990. At any one time there were up to three 'North Pole' camps on the ice, each studying various physical and biological processes. Position data from these drifting ice camps and the addition of modern satellite-tracked drifting buoys deployed on the surface of the ice have revealed that the speed of the Transpolar Drift can vary between 10 to 40 mm s^{-1}. The other main feature of the Arctic Ocean circulation is a large slow anticyclonic circulation over the Canadian Basin called the Beaufort Gyre. At cold temperatures, salinity has a strong effect on controlling the density of sea water. Coupled with the Arctic halocline this means that most of the transport in the Beaufort Gyre (approximately 80 per cent) is in the upper 300 m of the gyre. The two ocean currents have a significant effect on the ice conditions within the Arctic. The Transpolar Drift moves a large volume of ice from the Siberian Coast towards Fram Strait. The open water caused by the removal of the ice allows significant ice growth and salt rejection and may contribute to the maintenance of the halocline. The Beaufort Gyre is responsible for the piling up ('ridging') of large volumes of ice on the north Greenland and Canadian coasts, making the mean ice thickness in these places up to as much as 8 m. There is still much we do not know about the circulation within the Arctic Ocean; satellite-tracked drifting buoys have revealed that the Beaufort gyre has been known to slow and stop for periods of several months. Many large research programmes to investigate these features are planned.

As the transpolar drift exits through Fram Strait it is responsible for transporting large quantities of sea ice into

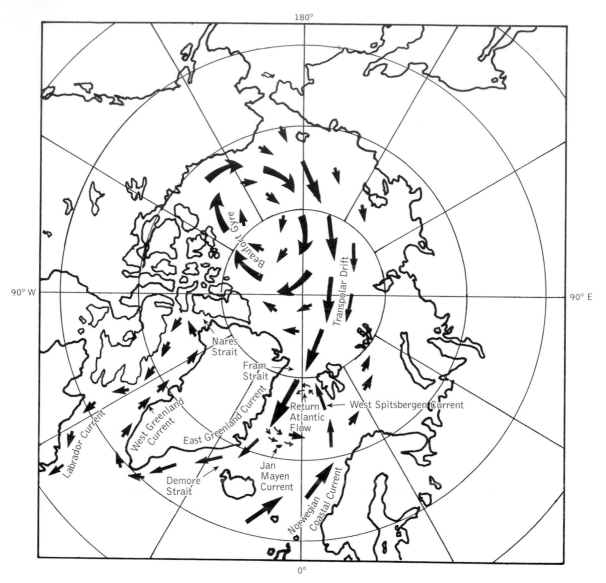

Fig. 1. Polar projection of the Arctic region showing the major ocean currents.

what is called the Nordic Basin. This ice travels down the east coast of Greenland within the important East Greenland Current, which is one of the major sources of water entering the Nordic Basin. Again it was Nansen who first correctly suggested, and then proved with data from the *Fram*, that the origin of the flow was due to the rotation of the Earth and the operation of buoyancy forces along the coast.

The other major current of the Nordic Basin is the remnant of the North Atlantic drift, which becomes the Norwegian Current along the coast of Norway, and then the West Spits-

bergen Current at roughly 78° N. The flow of the warm, salty Atlantic-derived waters is complex at Fram Strait. A certain fraction enters the Arctic Ocean. Another fraction of this water turns westwards and then southwards at approximately 79 °N to join the flow of the East Greenland Current in what is called the Return Atlantic Flow.

The two source waters of the cold fresh Arctic water and the warm salty Atlantic water form a cyclonic gyre which is closed in its southern section at approximately 72 °N by the eastward-flowing Jan Mayen Current. This current is cold and

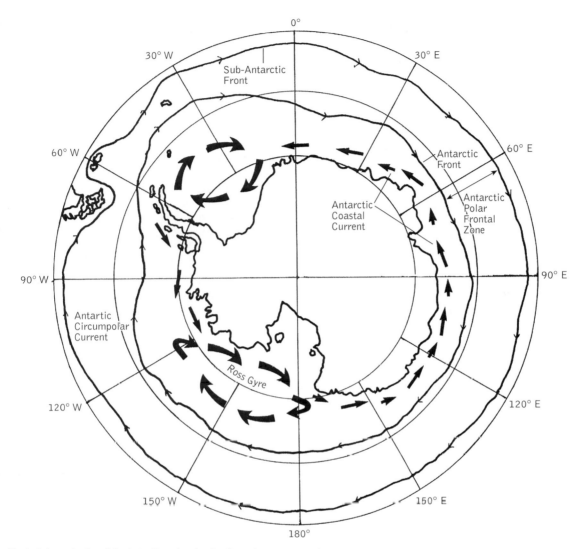

Fig. 2. Polar projection of the Antarctic region showing the major ocean currents.

relatively fresh. In winter it is covered by a famous ice feature called the 'Odden', which may be important in the driving of the ocean conveyor belt (see *sea ice and climate*).

The cold East Greenland Current leaves the Nordic Basin through the Denmark Strait and turns north at the southern tip of Greenland to become the West Greenland Current. As this current heads north, it collects icebergs from some of the most active glaciers in the world. At Nares Strait the current then turns south again to become the Labrador Current. Past Newfoundland, the current, now iceberg-laden, can be a major hazard to shipping.

Southern high-latitude ocean currents

The current structure in the southern high latitudes is less complex than in the northern high latitudes because of the absence of land. Antarctica is bounded on all sides by deep water and the most significant current in the high latitudes is the Antarctic Circumpolar Current (ACC) or West Wind Drift, which lies between 40° and 60 °S (Fig. 2). The current flows around the Antarctic continent in a clockwise direction uninterrupted by land with surface speeds that can range between 0.5 and 1.5 m s⁻¹, and can be considered the ocean analogue of the atmospheric jet stream. The passage of the

ACC around Antarctica is heavily determined by topography, but this itself is not well defined in the high latitudes because of the sea ice cover. The ACC is especially strong below the Agulhas current and in the Drake Passage, where recent measurements have determined the transport as 130×10^6 m³s⁻¹ with a uncertainty of roughly 10 per cent; the instantaneous flow may, however, differ by as much as 20 per cent from this figure. The ACC derives from the Ekman transport, which is to the left of the prevailing wind (the 'Roaring Forties') and raises the sea surface to the north. The surface slope generates a current which is stable and balanced by the Coriolis force.

The ACC delineates two major frontal regions: at the northern boundary the Sub-Antarctic Front, and at the southern boundary the Antarctic Front. The fronts are highly variable and variations in position of 100 km in 10 days have been observed. Larger meanders in the fronts can form eddies similar to those formed in the Gulf Stream. The greater of the two fronts is the Antarctic Front (southern), which separates the warmer surface waters of northern origin from the cold waters of southern origin. This front is historically identified as the Antarctic Convergence, as water is converging on this region, but recent measurements have shown that the ACC between the two fronts is zonal and complex in structure throughout its north–south extent and there is a series of convergences. Consequently the region is now termed the Antarctic Polar Frontal Zone. It is important for biological processes: deep water rich in nutrients is brought to the surface, resulting in an area of high localized primary production. The primary production is then grazed by zooplankton, and in particular by krill, which is a staple food for many species of birds, seals, fish, squid, and whales.

Close to the Antarctic continent a narrow and westward-flowing current at approximately 65° S is called the Antarctic Coastal Current or occasionally the East Wind Drift. This current has not been well studied but it has a speed of typically 0.1 m s⁻¹. The current is not, however, continuous around Antarctica and it is absorbed in the two large gyre systems of the Weddell Sea and the Ross Sea.

The Weddell Sea is one of the few areas of the Antarctic to retain a permanent ice cover. A gyre was first suspected from the drift of the *Deutschland*, which was trapped in the southeast Weddell Sea in 1911 and drifted for nine months before escaping from the pack ice close to the Antarctic Peninsula. A worse fate struck Shackleton's 'Imperial Trans-Antarctic Expedition' in the Weddell Sea when the *Endurance* was beset just off the coast of the Filchner ice shelf in January 1915 and sank in November of the same year. Shackleton and his men drifted within the northward portion of the Weddell gyre until April 1916, when they landed on Elephant Island at the edge of the peninsula and went on to complete one of the most famous self rescues in polar history. The existence and continuity of both the Weddell Sea and Ross Sea gyres have been subsequently confirmed from hydrographic measurements and satellite-tracked drifting buoys. The region at the north edge

of the Antarctic Peninsula marks the Weddell– Scotia Confluence, a region which has been identified as being biologically extremely important.   MARK A. BRANDON

Further reading

Smith, W. O. (ed.) (1990) *Polar oceanography. Part A, Physical science.* Academic Press, San Diego.

Dunbar, M. J. (ed.) (1977) *Polar oceans.* Arctic Institute of North America, Calgary.

high-latitude tropospheric circulation

The high latitudes of the northern and the southern hemispheres are very different geographically. In the northern hemisphere the polar region is an ocean basin which is almost completely enclosed by surrounding major land masses. Much of the ocean basin is covered by floating sea ice. In the southern hemisphere the polar region is occupied by the Antarctic continent, much of which is at an altitude greater than 3000 m. This land mass is completely surrounded by the Southern Ocean and in middle latitudes of the southern hemisphere there is very little land. As a consequence of these very different geographical configurations, the tropospheric circulations at high latitudes in the two hemispheres are also very different.

Southern hemisphere

The continental landmass of Antarctica is almost entirely ice-covered and the ice surface is mostly at a high altitude. Air temperatures at the surface are very low throughout the year, but particularly so in the dark winter months. From the edge of the continent, northwards across the Southern Ocean to mid-latitudes, there is a very marked temperature gradient. This temperature gradient provides an environment which forces the continual development of mobile low-pressure systems in a band circling the continent. These generally move in a south-easterly direction, and each one typically has a life span of several days. Monthly mean surface pressure charts show a band of low pressure encircling Antarctica between 60° S and 70° S. Figure 1 is an example, showing the average mean sea-level pressure for July.

In the low-pressure belt circling Antarctica the weather is constantly very changeable and is often stormy. In many ways the weather patterns in this belt are similar to those in the northern fringes of the temperate latitudes in the North Atlantic and North Pacific Oceans. The main differences is that the blocking anticyclones which are common in the northern hemisphere are much less common in the southern hemisphere. When they do form they are normally short-lived. The reason for this is that the southern hemisphere has no substantial land masses in middle latitudes and the weather patterns are totally dominated by the mobile maritime low-pressure systems.

Polarwards of the belt of low pressure the average mean sea-level pressure is somewhat higher. The active low-pressure systems surrounding the continent do not normally extend

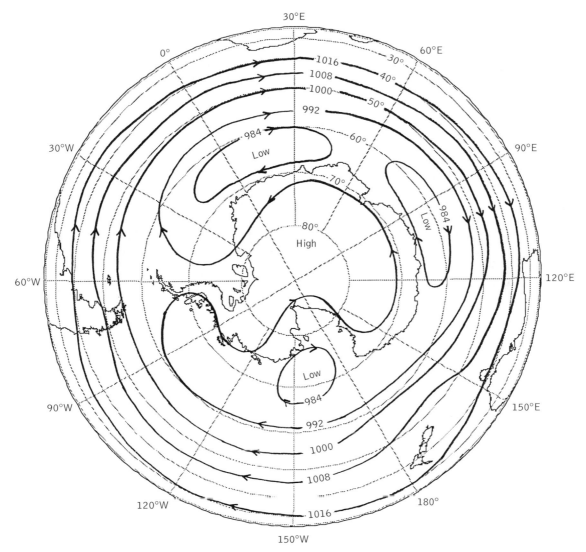

Fig. 1 Average mean sea-level atmospheric pressure (millibars) in the southern hemisphere in July (mid-winter).

their influence much into the continent itself. Over the Antarctic Plateau the tropospheric air is generally subsiding, and there is thus little cloud or precipitation. Indeed, the very small amounts of precipitation categorize Antarctica as a desert. Precipitation does occur around the coasts on the fringes of the low-pressure systems but inland, conventional precipitation is rare. There is occasional precipitation inland in the form of ice crystals. This occurs when relatively moist maritime air is carried inland at levels above the surface. This air cools significantly to become supersaturated with respect to ice, and some of the moisture precipitates out in the form of ice crystals. The great blizzards, for which Antarctica is notorious, are caused by blowing snow rather than by falling snow.

One very notable feature of the lower troposphere over the Antarctic Plateau is the very marked surface temperature inversion that exists throughout the year, although it is most pronounced in winter. This temperature inversion means that at heights of a few hundred metres above the surface the temperature is normally significantly higher than at the surface. In extreme conditions in winter the temperature difference has been known to exceed 30 °C. One effect of this marked temperature inversion is that winds at the lowest levels, below the inversion, are normally completely decoupled from the

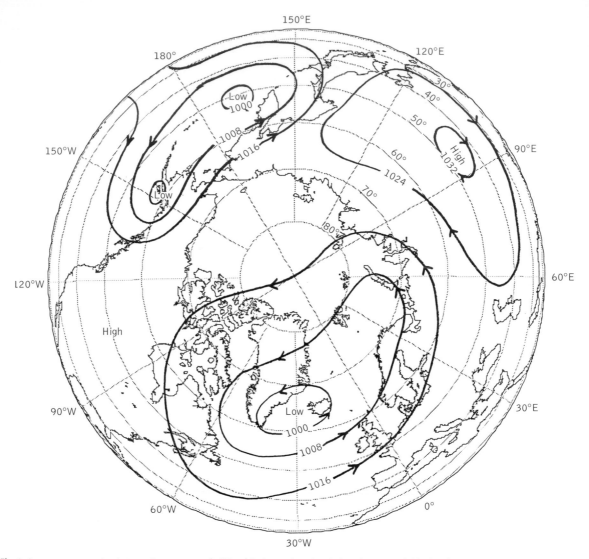

Fig. 2 Average mean sea-level atmosphere pressure (millibars) in the northern hemisphere in January (mid-winter).

winds above the inversion. Above the inversion the winds follow the normal relationship between pressure gradient and wind. Below the inversion the winds are largely driven by gravity, with the cold air draining down the slopes. These winds are known as *katabatic winds*. In some localities the shape of the land can cause winds of gale force or stronger to blow for days or even weeks at a time. At the top of the inversion there can be very marked changes in wind direction and wind speed over only a very small vertical distance. This is a very important factor for aircraft pilots to take into consideration during take-off and landing or when engaged in low-level

flying. A sudden decrease in head wind or increase in tail wind can cause an aircraft to stall.

Over the Antarctic Plateau there is normally very low humidity throughout the troposphere. This results in largely clear skies allowing maximum radiation cooling in the winter months. Winters are very long with constantly very low temperatures. The lowest temperature every recorded (–89 °C) on the surface of the Earth was at the Russian base, Vostok, which is at an altitude of 3400 m. Summers are short and temperatures are slow to rise to their midsummer maximum and quick to fall again subsequently.

Northern hemisphere

Although the northern hemisphere polar region is an ocean basin mostly surrounded by land masses, much of the sea surface is ice-covered for most of the year. The main exceptions to this are the Norwegian Sea and the Barents Sea, which are ice-free in summer as far north as 80° N. Even in winter the ice edge is often north of 70° N, especially in the Norwegian Sea. The coastal fringes all around the Arctic Ocean have ice-free stretches in summer but freeze over in winter. Because of the extensive ice cover in winter the region has more of a land environment in that season than a marine environment. The great difference compared to the polar regions of the southern hemisphere is that the surface is mostly at or close to sea level. The only high elevations are found on the peripheral land masses, notably Greenland.

In the middle and upper levels of the troposphere the circulation is dominated by a cold-cored polar vortex throughout the year. This is at its most pronounced during the winter months. Strong westerly winds circle the globe in middle latitudes around this vortex. This basic westerly flow is modified by the land masses, notably over North America, where the Rocky Mountains cause a semi-permanent upper atmosphere ridge resulting in a corresponding upper atmosphere trough downstream over eastern North America.

Minor short-wave troughs moving in the westerly upper air flow provide the dynamics conducive to the development of surface low-pressure systems. Conditions are particularly suited to such cyclogenesis in winter off the east coasts of Asia and North America where horizontal temperature gradients are greatest. These lows typically track north-eastwards, reaching maximum intensity in the vicinity of the Aleutian Islands in the Pacific Ocean and in the vicinity of Iceland in the Atlantic Ocean. Figure 2 shows the average mean sea-level pressure for January. This shows clearly that the main low-pressure areas of middle latitudes in winter are to be found over the seas, while the land masses are dominated by areas of high pressure. This juxtaposition of high-pressure and low-pressure systems in similar latitudes causes much more meridional airflow than is normally experienced in similar latitudes in the southern hemisphere. As a result of this, in the northern hemisphere, polar air in the lower troposphere is frequently advected into middle latitudes and, occasionally, even as far as subtropical latitudes. These outbreaks of polar air can occur at any longitude but are particularly common and intense along the eastern fringes of Asia and North America. The outbreaks of polar air to middle latitudes are normally balanced by intrusions of mild maritime air from middle latitudes into the Arctic and polar regions. The preferred regions for this to happen are in the Bering Sea area and, particularly, through the gap between Greenland and Scandinavia.

It is not uncommon for the Atlantic and Pacific low-pressure systems to continue their north-easterly track during the decaying phase of their life cycle and move completely into the polar regions. By this stage they are normally rela-tively weak systems, having moved away from the regions of marked horizontal temperature gradient that caused their formation. The circulation is then entirely of cold polar air, and there is little precipitation associated with such lows when they reach the polar areas. The very low annual precipitation totals categorize the northern polar regions as a desert area: the very cold air cannot hold enough water vapour to permit substantial precipitation.

In contrast to situations dominated by low pressure, the northern polar regions also experience spells of very strong high-pressure domination in winter, often linked to major middle-latitude high-pressure systems over Asia or North America. In extreme cases the surface atmospheric pressure in the polar regions can reach or exceed 1050 millibars.

In the summer months the upper-air polar vortex is much weaker and the persistent cold high-pressure systems are absent from the middle-latitude land masses. As a result, there is much less interaction between the lower troposphere air masses of the polar regions and middle latitudes. During this season the polar vortex of the middle and upper troposphere is also reflected in the lower troposphere, with relatively low surface pressures being typical of the polar regions.

NORMAN LYNAGH

Further reading

King, J. C. and Turner, J. (1997) *Antarctic meteorology and climatology*. Cambridge University Press.

Sater, J. E., Ronhovde, A. G., and Van Allen, L. C. (1971) *Arctic environment and resources*. The Arctic Institute of North America, Washington, DC.

hill-slope creep In landscape studies, the term 'creep' covers all slow downslope movements of unconsolidated material or of soft rock that is easily deformed under pressure. The rate of downslope movement through creep is typically very small, rarely more than 1–2 cm per year, but it is the ubiquity of this tendency, affecting almost any hill slope with an appreciable gradient, that makes it a significant geomorphological process. In fact, the general process of creep can arise from several independent mechanisms. The first, continuous creep, occurs by the slow plastic flow of clay-rich material, typically unconsolidated soils, shales, and clays. The key prerequisite for continuous creep is an easily deformable substrate, an attribute that depends on clay content, moisture content, the thickness and bulk density of the regolith or rock layer, and the angle of slope. Where all these are moderate to high, overburden pressure from overlying strata (caprock) or constructions (walls, power poles, buildings, etc.) can readily induce creep. This in turn may lead to the slumping or cambering of the slope, the bulging of valley sides, and the downslope curvature of surficial strata.

Creep is often referred to as 'soil creep', but strictly this refers to two distinct mechanisms that primarily operate within a regolith environment, although they may also affect loose rock material (talus creep). The mechanisms are creep by expansion and contraction and creep by swelling and shrinking, but they are often referred to together as 'heave'. Expansion and contraction generally occur through the action of freeze–thaw, where the expansion of water, present as soil moisture, upon freezing causes a consequent increase in soil volume and, in turn, in the bulging of the soil surface perpendicularly outwards from the slope. On thawing, the regolith material sinks vertically downwards under the direction of gravity. This results in a net downslope movement of individual soil particles, although cohesion between the particles usually retards a pure vertical return drop. Soil particles at the surface are affected most by this tendency; the propensity for heave and its efficacy decrease with increasing depth, reaching zero at the lower boundary of the soil layer that has been frozen. Over a period of time, because repeated freezing and thawing affect surface layers more than deeper portions of the regolith, creep rates are disproportionately greater in the near-surface zone. Expansion and contraction creep can also occur in soils containing clay minerals that are capable of swelling, such as montmorillonite. Alternating wetting and drying out of clay minerals causes a smaller net downslope movement that results from freeze–thaw because shrinking may draw a clay particle back upslope before it is moved downslope during drying. Both mechanisms are extremely widespread on the Earth's surface. Their action is commonly considered to be manifest as flights of narrow steps, called 'terracettes', on steep, grass-covered slopes (Fig. 1).

Because expansion and contraction depend on the direction of gravity and not its strength, creep processes are also likely to affect other planetary landscapes. Thus, on the Moon and Mars large temperature fluctuations at the surface (+130 °C to –150 °C on the Moon; 15 °C to –85 °C on Mars) give rise to alternating expansion and contraction of the surface material.   IAIN S. STEWART

Fig. 1 Terracettes affecting a chalkland hillslope in Wiltshire, southern England. Although these flights of narrow steps have been attributed to preferential trampling by animals, they are widely viewed as the expression of discontinuous downslope creep of the soil.

hill slopes Hill slopes occupy most of the land surface with the exception of terraces and plains formed by river deposits. Even extensive surfaces such as the plateaus of Africa and Great Plains of North America are largely comprised of hill slopes of low angle between barely definable interfluves (i.e. drainage divides) and valley floors. The most spectacular hill slopes are the sides of deep gorges and the great rock cliffs of high mountain chains.

Hill-slope forms in landscapes with soil cover

Form is the shape of the ground surface that can be identified by consideration of its profile-form with respect to a vertical plane (cross-sectional view) and its plan-form (map view) normal to the profile and therefore in a horizontal plane parallel to the contours. Profile- and plan-form together provide a three-dimensional image and record of hill-slope forms which are made up of a combination of planar, concave, and convex segments (Fig. 1). Shapes of hill-slopes may be identified from very detailed maps with closely spaced contours but, in most parts of the world, such maps are not available and detailed measurements must be made in the field with the aid of theodolites and survey levels. Measurements of hill-slope form are usually undertaken to identify the association of form with the processes of erosion and deposition acting on each slope unit, with the soil and vegetation characteristics, and with actual or planned use of land.

Convex slope segments commonly occur on the upper parts of slopes, near the drainage divide, as a result of soil creep and rainsplash erosion. Rainsplash is most effective in sparsely vegetated areas and soil creep in well-vegetated areas. The principles applying to convex slopes were first stated by G. K. Gilbert, who showed that soil eroded from the upper part of a slope has to pass each point below it and consequently that the volume of soil being moved increases with distance from the divide. If the transport rate for soil creep and rainsplash is proportional to the slope angle, then the slope angle must also increase with distance from the slope crest.

Concave slope segments usually form towards the bases of slopes where wash processes (see *overland flow*) transport and deposit sediment, or rolling rock fragments accumulate as a talus deposit below a cliff; with wash the discharge increases downslope, and on talus slopes boulders with the greatest kinetic energy roll beyond the depositional zones of those with less energy.

As discharges increase downslope, because of the greater collection area for wash, water velocity can be maintained on lower slope angles. Particle sizes of sediment being transported tend to decrease downslope as a result of weathering and abrasion, and any given flow can carry a larger load of fine than coarse material. There is thus an excess transporting capacity in the flow and the slope angle can reduce while the power to transport is maintained. Such slopes may be called transportational slopes, developed as the transporting capacity of the wash attains an equilibrium with the load of sedi-

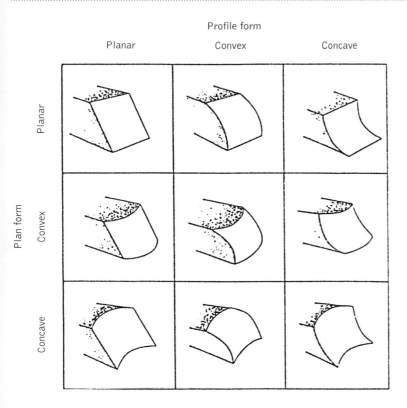

Profile form

Planar Convex Concave

Plan form

Planar

Convex

Concave

Fig. 1. The nine possible shapes for hill-slope units. (Based on Ruhe (1975), as classified by Parsons 1988.)

ment carried by the water. Depositional slopes are concave because large sediment particles are dropped by the water flow before finer particles. Concave wash slopes are commonly developed in semi-arid areas where surface wash is the dominant process. Concave segments of footslopes can also be created by deposition of soil and other debris below a slide plane of a landslide.

Straight slope segments develop where the dominant erosional processes are shallow landsliding with slide planes parallel to the ground surface, and where talus slopes are formed by rock debris from rolling particles with a limited size range.

Compound slope profiles are very common. On any slope with a soil cover, several processes may operate and there may be zones of erosion, of transport of sediment, and of deposition of the transported material. A hypothetical nine-unit model of slope segments is shown in Fig. 2. This model does not imply that all nine units will occur in every landscape, nor that the units will be arranged in the order shown. A slope with several outcrops of horizontal hard rock units, for example, may be undercut at the base so that, starting at the crest, it may have units, 1, 2, 3, 4, 5, 4, 5, 8, 9. A slope without rock units may consist of units 1, 2, 3, 5, 6, 8, 9.

The use of the nine-unit model assists in the recognition of the relationship between hill-slope form and the dominant process. However, caution is required: soil creep may not be the only cause of upper slope convexity, especially if the upper slope units are of fractured rock overlying a mud–rock unit which is deforming, with opening of joints, so that rock blocks are slowly migrating downslope—a process called *cambering*. Furthermore, the current landscape may contain slope units which are relict from former environmental conditions, such as glacial, periglacial, or subhumid climates in an area that is now under temperate forest.

Hill slopes formed on bare rock

These normally have forms controlled by the resistance of the rock, not by the erosional processes acting on it (see *cliffs*). This is because slope processes have limited erosional power compared with the resistance of the rock. The important exceptions are those slopes that are undercut by rivers or waves and those that have forms inherited from tectonic processes, for example fault-line scraps, or by structural controls, such as granite domes.

Hill slopes that are controlled by rock resistance are called *strength equilibrium slopes*. They owe their form to the effects of the following eight features: strength of the intact rock (free of fractures); state of weathering of the rock; the spacing of joints and fissures; the dip of the fissures with respect to the cliff face; the aperture, or width, of the fissures; the continuity of the fissures; the amount of infill of the fissures with weak

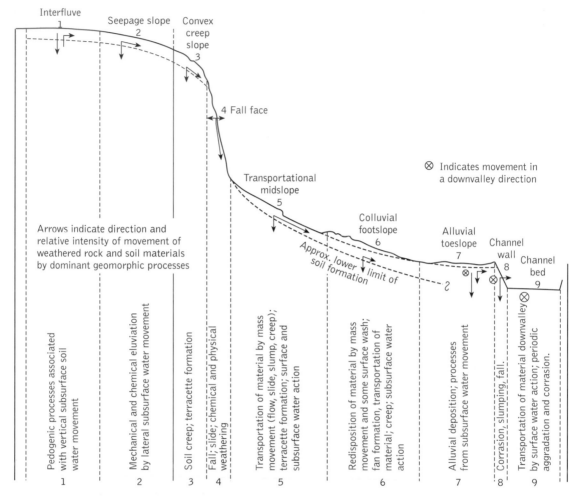

Fig. 2. The hypothetical nine-unit land-surface model. (After Dalrymple *et al.* (1968) A hypothetical nine-unit landsurface model. *Zeitschrift für Geomorphologie* **12**, 60–76, with permission from Gebrüder Borntraeger.)

soil; and the amount of water within the fissures. Approximately 70 per cent of the total strength of the rock mass is due to the nature of the joints and fissures, and only 30 per cent to the strength of intact rock and its state of weathering.

Very steep cliffs are usually formed on rocks with widely spaced joints which are tightly closed and which either dip back into the slope or are approximately horizontal. More gently inclined slopes usually occur on rocks with very closely spaced open joints, or very weak rock which weathers rapidly, releasing particles that can be removed by wash or shallow landslides. Many rock slopes have complex forms with varied segments of different inclination. This variety of inclination usually reflects variations in rock-mass strength properties.

Hill-slope deposits

Hill-slope deposits and their layering (stratigraphy) are the best indicators of past history and development of the slopes. Deposits formed on hill slopes are called *colluvium* where they are fine-grained, and *taluvium* if they are of coarse-grained rock debris. Exposures through such deposits are commonly the most reliable indicators of past processes. Deposition at the base of a slope is an indication of erosion on upper slopes, and changes in erosional intensity may be indicated by the presence of distinctive layers in colluvium and by buried soils. The lowermost deposit in an accumulation is the indicator of the oldest event, or period, in the depositional record.

Many processes of erosion and deposition occur too frequently for soil profiles to develop distinct horizons in colluvium; rather, there is a continuing process of gradual deposition with weakly developed bedding and horizon formation. Rare but catastrophic processes, such as landslides, may produce thicker deposits of unsorted debris, possibly containing such datable organic material as logs, and in the interval before the next major depositional event the colluvium may weather and develop a vegetation cover and a soil profile. The colluvium is, consequently, an indicator of the magnitude and frequency of the major land-forming processes.

In addition to magnitude and frequency, colluvium may preserve indicators of past environments which are fundamentally different from those operating in the modern environment. In many parts of Europe, colluvium has preserved a record of processes of the last glacial period and the warming at the beginning of the Holocene. Where this record is present, the basal deposit in the colluvial footslope is a rock rubble of periglacial origin, with weak soil development; above that may be a unit with finer-grained sediment formed in a warm, moist climate with landsliding and soil creep on upper slopes, with upper colluvial deposits of thick soils formed under a forest cover. Human settlement from the Bronze Age onwards then became the dominant influence on upper slopes and hence on the nature of the colluvium of the footslope.

Hill-slope evolution

In the long term, hill-slope evolution has rarely left evidence that can be used to test the various hypotheses that have been put forward, but there are some conditions under which an initial slope form of known age may be reasonably assumed and subsequent forms dated in a sequence: fault-scarps; riverside bluffs; bluffs of old lake-shorelines; spoil heaps of coal mines; and marine terraces. These features present opportunities for studies of the development of hill slopes in sediments in weak rock. Cliffs in resistant rock have been studied where infill at the base of the cliff has progressively protected cliff units from coastal erosion, and high walls of open-pit mines of various ages since they were abandoned have provided information on rock-slope evolution.

Soil-covered slopes show a decrease in slope angle, over time, of the mid-slope segments, with an increase in the extent of the convex crest and in the concave segment at the slope base. Rock slopes progressively develop forms that tend increasingly towards equilibrium with the strength of the rock mass of each rock unit of the exposure, but there may also be other dominant factors in rock-slope development, such as structural controls, units of weak rock dominated by erosional processes, or deposition of talus covers. Hypotheses of long-term development of hill slopes are not directly testable.

M. J. SELBY

Further reading

Parsons, A. J. (1988) *Hillslope form.* Routledge, London.

Ruhe, R. V. (1975) *Geomorphology.* Houghton Mifflin, Boston.

Selby, M. J. (1993) *Hillslope materials and processes* (2nd edn). Oxford University Press.

Himalayan tectonics

The Alpine–Himalayan mountain belt formed mainly during the Cenozoic era as a result of the collision between the northern Laurasian landmass and Gondwana continental fragments to the south. The ocean that separated these two continents during the Mesozoic, called Tethys, was born during the Permian period some 250 million years ago after the breakup of the supercontinent Pangea. Several smaller continental fragments within the Tethyan ocean are now represented by the Cimmerian terranes stretching from central Turkey eastwards to China. The main suture zone dividing the two continents is represented by a line of remnant ophiolites (ancient oceanic crust and mantle), deep-sea sediments, and mélanges, occasionally with high-pressure blueschists, stretching along the Bitlis and Zagros mountains of eastern Turkey–Iran to Oman and round the northern margin of the Indian plate.

The Himalaya is the world's highest mountain range. It has been forming since the closure of Tethys some 50 million years ago and the subsequent indentation of India northwards into central Asia. The actual zone of collision is the Indus suture zone, which runs east–west for about 2500 km north of the Himalaya across northern Pakistan, the Ladakh region of India, and south Tibet. After the collision of India with Asia, crustal shortening and thickening began, and thrust-related shearing and folding affected the entire northern continental margin of the Indian plate. Spectacularly folded sedimentary rocks now form the mountains of the Zanskar–Spiti range immediately north of the High Himalaya. Remnant ophiolite complexes emplaced on the northern continental margin of India are still preserved at Spontang in Ladakh and in southern Tibet.

About 25 million years ago thrusting propagated southwards to the High Himalaya, which suffered extreme amounts of crustal shortening, thickening, and regional metamorphism. These middle and lower crustal rocks are bounded by a crustal-scale thrust fault along the base of the slab (Main Central Thrust) and a large-scale north-dipping normal fault zone (North Himalayan fault or South Tibetan Detachment) along the top of the slab (Fig. 1). Both these structures were active simultaneously about 20–18 million years ago. The thickening of these Indian plate rocks of the High Himalaya resulted in an increase of temperature and pressure, causing new metamorphic mineral assemblages to form. The resulting schists and gneisses contain garnet, micas, staurolite, kyanite, and sillimanite. At the highest temperatures, about 750 °C, the gneisses began to melt, forming migmatites (rocks consisting of two components—a granitic melt and a remnant gneiss), and finally distinctive leucogranites. These leucogranites contain the minerals tourmaline, muscovite, and garnet as well as quartz and feldspars. They form many of the highest peaks of the Himalaya today, including Kangchenjunga,

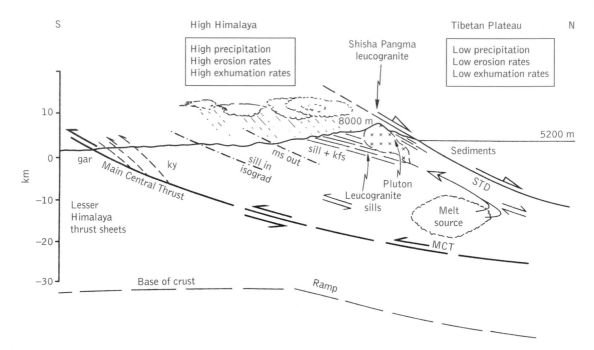

S   High Himalaya   Tibetan Plateau   N

Shisha Pangma leucogranite

High precipitation
High erosion rates
High exhumation rates

Low precipitation
Low erosion rates
Low exhumation rates

8000 m

5200 m

sill + kfs

gar   ky   ms out   sill in isograd   Sediments

Main Central Thrust   STD

Lesser Himalaya thrust sheets

Pluton Leucogranite sills

Melt source

MCT

Base of crust   Ramp

km   10   0   −10   −20   −30

Fig. 1. Schematic scaled cross-section of the Langtang-Shisha Pangma High Himalaya, showing the distribution of the garnet (gar), kyanite (ky), and sillimanite (sill) metamorphic zones, isograds, and relationships between the Main Central Thrust (MCT) and South Tibetan Detachment (STD). kfs, potassium feldspar; ms, muscovite. The topographic rise of the High Himalaya was directly related to motion along these two major structures, resulting in a southward-thrusting and extruding orogenic wedge which was being actively uplifted by underthrusting. The rise of the Himalaya provided a topographical barrier to the northward-directed monsoon winds that are the cause of the high rainfall along the southern slopes of the Himalaya. High precipitation along the Himalaya resulted in increased mechanical and chemical weathering and increased rates of erosion. (From Searle, M. P. (1997), *Journal of Geology*, **105**, 295–317.)

Jannu, Makalu, Nuptse, Shisha Pangma, and Manaslu in Nepal and Changabang, Shivling, and the Bhagirathi peaks in India. These granites lie at the top of the High Himalayan slab immediately beneath the South Tibetan Detachment normal fault. On Mount Everest, for example, the base of the mountain consists of gneisses and leucogranite sheets and dykes, whereas the upper part, including the famous 'Yellow Band', consists of unmetamorphosed Palaeozoic sediments that were deposited under the ocean. The normal fault that separates these rocks cuts across the south face of Everest and dips at low angles to the north along the Rongbuk glacier under the Tibetan Plateau.

During the past 15 to 20 million years, southward-directed thrusting spread to the Lesser Himalaya, and active thrusts now bound the southern margin of the Himalaya along the Salt Ranges in Pakistan and the Siwalik hills in India and Nepal. Two molasse basins accumulating erosional debris shed off the rising mountains were formed, the Siwalik basin along the southern margin of the Himalaya and the Indus basin, along the trace of the Indus suture zone, north of the Himalaya. Some late-state reactivation of earlier structures has resulted in north-directed thrusting along the Indus

suture zone. The highest erosion and exhumation (rock uplift) rates today occur at the two syntaxes (bends in the orogenic belt) at the far western end of the Himalaya (Nanga Parbat in Pakistan) and the far eastern end (Namche Barwa in south-east Tibet).

Since the collision of India into Asia about 50 million years ago India has continued its northward indentation into Asia. Some 1000 km of crustal shortening may have occurred in the Indian plate rocks across the whole Himalaya, while the Tibetan plateau has also been shortening by about the same amount, resulting in a doubling of crustal thickness to around 70 km. Erosion rates in Tibet are, however, very low; most of the plateau surface consists of sedimentary rocks, which have never been deeply buried, or volcanic rocks. The Tibetan plateau is the largest uplifted part of the Earth today, occupying an area of some two million square kilometres at an average elevation of just over 5000 metres, stretching from the northern margin of the High Himalaya north to the Kun Lun–Altyn Tagh mountains. Models proposed to accommodate the convergence between India and Asia, north of the Himalaya include homogeneous north–south shortening and thickening of the plateau and continental escape or extrusion of Tibet eastwards away form the indenting Indian plate. The present thick crust of the Tibetan plateau is bounded by large-scale strike-slip faults, notably the dextral Karakoram fault and the sinistral Altyn Tagh fault.

Fault-plane solutions of earthquakes in Asia show that the present-day kinematics are dominated by thrust faulting along the Himalaya, but Tibet shows a combination of east–west extension on normal faults and stike-slip faulting. Whereas the Himalaya are still actively rising as India continues its northward penetration, the Tibetan plateau is now in a state of collapse and shows crustal extension in a series of graben structures aligned north–south. It has been proposed that the uplift of a feature as large and high as the Tibetan plateau caused perturbations in the global climate, particularly in initiation of the south-west monsoon which draws warm, moist air in from the Indian ocean and sweeps northwards across the Indian subcontinent to stop abruptly at the Himalayan crest.

The Alpine–Himalayan belt continues eastward along the Indo–Burman ranges into the complex geology of south-east Asia and Indonesia. This belt shows some of the most spectacular evidence for plate tectonics, and for processes of continental deformation at plate boundaries and within the plates, on this planet.   MIKE SEARLE

Further reading

Windley, B. F. (1995) *The evolving continents* (3rd edn). John Wiley and Sons, Chichester.

Himalayan–Tibetan uplift and global climate change

During the late 1980s and early 1990s much attention was focused on the potential role of late Cenozoic uplift of the Himalayas and Tibetan plateau in driving global cooling and the growth of large continental ice sheets during Late Tertiary and Quaternary times. The formation of the Himalayas and the Tibetan plateau is thought to have altered global atmospheric circulations and to have changed the concentrations of gases in the atmosphere as a result of changes in the biogeochemical cycles associated with the increased weathering of newly exposed rock surfaces, which in turn led to global cooling.

There is unequivocal evidence to show that the Indian and Asian continental plates collided between 54 and 49 Ma (million years) ago. Since then, India has continued to move northwards into Asia at a rate of between 40 and 50 mm a year^{-1}. This has resulted in folding and thrusting of the Indian crust in the Himalayas and southern Tibet. The homogeneous deformation of Tibet entailed shortening and doubling the thickness of the crust, elevating the Himalayas and Tibetan plateau to an average height of between 5000 to 5500 m above sea level. The tectonic forces resulting from the northward movement of India were not, however, in the opinion of some geologists great enough to support the thickened crust, and the plateau began to spread under its own weight. Currently, the plateau is spreading, extending east–west on normal faults, by about 1 per cent of its extent every million years.

This rise in elevation may have altered climate in a number of ways. First, the rising plateau would have deflected and blocked regional air systems, causing jet streams to meander, and facilitating the southward migration of cold polar air into key locations in the northern hemisphere, particularly North America and north-west Europe, the sites of ice sheet development. This in turn would have affected atmospheric circulations on a global scale. Secondly, the elevated region would have enhanced temperature-driven atmospheric flows as higher and lower atmospheric pressure systems developed over the plateau during the winter and summer, respectively. This would have intensified the Indian monsoon, leading to heavier rainfall along the frontal ranges of the Himalayas, while the aridity across the Tibet plateau and central Asia would have increased as northward-moving monsoon winds lost their moisture as they were forced over the Himalayas. In Central Asia, the increased aridity would have led to higher temperatures during the summers and extremely low temperatures during the winters. In addition, the plateau would have cooled as it rose higher into the atmosphere, and glaciers would have been able to develop. These glaciers would have provided positive feedback, reflecting incoming solar radiation and leading to further cooling. Thirdly, the greater availability of new exposed rock surfaces and detritus produced by denudation processes would have allowed more chemical weathering to take place. During chemical weathering atmospheric carbon dioxide (CO_2) reacts with rock-forming silicate minerals to produce bicarbonates, which are washed into the oceans where they eventually form carbonate rocks. Because CO_2 is an important greenhouse gas, helping to warm the atmosphere, a decrease in the amount of atmosphere CO_2 would lead to global cooling. It has been argued that this process resulted in decreased CO_2 over the past 40 Ma leading to an icehouse effect and the onset of the Quaternary ice age.

Throughout the Cenozoic, other major tectonic events also took place, which could have forced climate change. Continental displacements led to changes in the configurations of the oceans, and seaways opened and closed. These would have given rise to major changes in global and regional oceanic circulations, altering the exchange of heat throughout the globe. The rate of sea-floor spreading also decreased during early Tertiary times, reducing the supply of CO_2 from volcanic eruptions to the atmosphere. As sea-floor spreading was reduced, the oceanic spreading ridges cooled and subsided, lowering sea levels. This produced more land available to be weathered and allowed more plants to grow, storing CO_2 as organic carbon, further reducing the amounts of atmospheric CO_2. The importance of each of these processes, and of mountain uplift, in causing global cooling throughout the Tertiary, and ultimately the onset of the Quaternary ice age, is difficult to prove and assess.

A variety of proxy data have been presented to demonstrate the connection between Tibetan uplift and climate change on regional and global scales. The ratio of oxygen isotopes

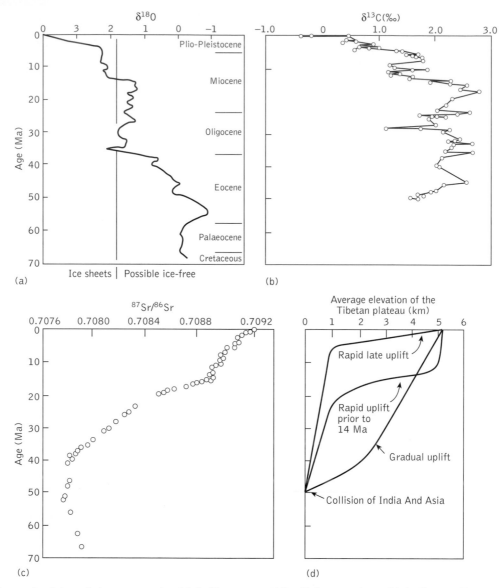

Fig. 1. A comparison between isotope curves and models for Tibetan uplift (a) Simplified compilation of $\delta^{18}O$ measurements from deep-sea cores in the Atlantic Ocean; (b) $\delta^{13}C$ curve for marine carbonates over the past 70 Ma; (c) $^{87}Sr/^{86}Sr$ curve for marine carbonates for the past 70 Ma; (d) the contrasting models for Tibetan uplift. Note the broad correlation between the isotope curves (see text for discussion) and between the model for rapid Tibetan uplift before 14 Ma ago. (a, b, and c adapted from Raymo and Ruddiman (1992).)

($^{18}O/^{16}O$) in the calcareous shells of foraminifera (unicellular marine organisms) preserved in marine sediments essentially records past ice-sheet volumes and is a proxy for ocean temperatures. The oxygen isotope record for the past 55 Ma shows a long-term increase in $\delta^{18}O$ values ($\delta^{18}O$ expresses the changes in the $^{18}O/^{16}O$ ratio), indicating progressive cooling,

punctuated by times of rapid cooling (Fig. 1a). The first major increase in $\delta^{18}O$ values (indicating cooling) occurred about 35 Ma ago and probably represents the initiation of the growth of the Antarctic ice sheet. This was followed by 20 Ma of cooling. Further increases in $\delta^{18}O$ values occurred during mid-Miocene times (approximately 15–13 Ma ago) and late

Pliocene times (approximately 4–2 Ma ago); these increases probably represent further growth of Antarctic ice and the build-up of the northern hemisphere ice sheets, respectively. Tills, moraines, ice-rafted sediments and palaeontological evidence provide additional support for this cooling trend and the onset of glaciation in the Antarctic at about 35 Ma ago and in the Northern Hemisphere at approximately 4.5 and 2.5 Ma ago.

$^{87}Sr/^{86}Sr$ isotope ratios recorded in marine carbonates provide an important proxy for estimating the rates of chemical weathering. A marked increase in $^{87}Sr/^{86}Sr$ ratios after Late Eocene times (approximately 38 Ma ago: Fig. 1c) may be attributed to increased weathering or a change in the type of rock being weathered (erosion of rocks supplies radiogenic strontium from the land) or a decrease in sea-floor hydrothermic supply. The latter is controlled by changes in the rates of sea-floor spreading. Because sea-floor spreading rates changed little during the past 30–40 Ma, the changes in the $^{87}Sr/^{76}Sr$ ratios may be attributed to increased weathering. Today, the Ganges–Bramaputra river systems have unusually high $^{87}Sr/^{86}Sr$ isotope ratios, and it is therefore argued that the late Cenozoic $^{87}Sr/^{86}Sr$ ratios may be attributed to enhanced weathering in the Himalayas. Others argue, however, that the increased ratios may reflect enhanced supply by glacial erosion associated with the growth of the Antarctic ice sheet.

Such an increase in weathering would deplete the atmosphere of all its CO_2 within a few million years. Negative feedback must therefore operate to return CO_2 to the atmosphere. One possible mechanism is related to imbalances in the organic carbon subcycle. $^{13}C/^{12}C$ ratios in marine carbonates reflect changes in the size of the organic carbon reservoir. If, for example, the burial of organic matter decreases, then $\delta^{13}C$ would decrease (a decrease in the $^{13}C/^{12}C$ ratio). $\delta^{13}C$ values in marine carbonates showed a marked decrease during the Cenozoic (Fig. 1c), indicating that carbon burial also decreased. This in turn could have caused additional CO_2 to be released into the oceans and atmosphere. Because oxygen is necessary to oxidize organic matter and release CO_2, the decrease in buried organic carbon may reflect an increase in dissolved oxygen concentrations in the oceans. Oxygen solubility increases with decreasing temperature. The dissolution of oxygen into the oceans would therefore have increased as the oceans began to cool. This would lead to the oxidation of organic matter and a greater release of CO_2 into the atmosphere. The residual times of CO_2 (less than 1 Ma) and O_2 (c. 10 Ma) in the atmosphere would not lead to a perfect balance of these processes. Rather, the balance of the organic subcycle would become less effective with time. This may have led to more intense forcing of cooling in late Cenozoic times as CO_2 became progressively depleted. Other possible negative feedback mechanisms that could release CO_2 into the atmosphere include the precipitation of silicate minerals in the deep ocean, and sea-floor weathering of basalt, but these processes are poorly understood.

It is difficult to assess the relative importance of all these complex processes in driving climate change. Attempts have been made, however, to model the evolution of Cenozoic climate using computer models of the Earth's atmosphere (General Circulation Models: GCM). Using GCMs, American Earth scientists have investigated the atmospheric changes that would be associated with the uplift of the Himalayas and Tibetan plateau. Their initial results have shown that many of the global changes in precipitation and temperature that were calculated by the models are consistent with those observed in the geological record over the past 40 Ma. Additional experiments have extended the GCMs to examine the sensitivity of the Indian monsoon. The results showed that strong monsoons, similar to those of today, could have been induced by strong solar forcing when the elevation was only half its present height. The question then arises as to when was Tibet high enough to initiate the monsoons, and was this coincident with the proxy evidence for global cooling and increased monsoon activity at around 7 to 8 Ma ago?

The timing of uplift is one of the major uncertainties in attributing uplift to climate change. With regard to the timing of uplift, there are three contrasting schools of thought. Primarily on the basis of mammalian fauna, palaeokarst, and geomorphology, the first school advocates very rapid late Pliocene–Pleistocene uplift. These changes in fauna, palaeokarst, and geomorphology could equally, however, be attributed to climate change. The second school suggests that uplift has occurred rather gradually since Early Eocene times, with substantial elevations being reached by late Eocene times. This is based on lithological variations in lacustrine rocks from eastern China, but there is a large degree of uncertainty with this hypothesis.

The third school, and the one whose view is most generally accepted, considers that rapid uplift occurred after about 25 Ma ago, and that the plateau attained its present elevation by about 14 to 15 Ma ago. This is based on the assumption that the present elevation and extensional deformation of the plateau probably resulted from uplift owing to convective thinning of the underlying lithospheric mantle. Post mid-Miocene lavas from the northern part of the Tibetan plateau, derived from lithospheric mantle, owe their origins to the thinning of the lithospheric mantle. These lavas have been dated at about 13 Ma. Furthermore, extensional faulting in the Thakkola graben in Nepal, which is the result of the extensional collapse of the plateau, has been dated at 14 Ma. These examples support the view that maximum elevation must have been achieved prior to 14 Ma. A reassessment of the sediments deposited on the Indo-Gangetic Plain and the Bengal fan has shown that there was a reduction in sediment supply from the Himalayas 8 Ma ago. This could be attributed to reduced tectonic activity after 8 Ma, further supporting the view that plateau uplift occurred more than about 8 Ma ago. Alternatively, reduced denudation could be related to major vegetational changes which also occurred at about 8 Ma ago. These uplift dates are essential for modelling and understanding the role of uplift on climate change.

An alternative view on uplift and climate change proposes that the Quaternary glaciations in the Himalayas could have enhanced uplift. This theory suggest that when the Earth's surface is evenly eroded, and is therefore unloaded, a corresponding readjustment will occur raising the crust to approximately 5/6th of its original height. If, however, the crust was not evenly eroded, for example by glaciers which deeply incised into the landscape, but the same amount of material was removed, as was the case with even erosion, then the rebound would help produce higher mountain peaks. This process may have helped produce the high peaks in the Himalayas. While this process may provide an alternative mechanism for uplift, the geological data suggest, however, that much of the Himalayas obtained their height before the onset of major glaciation. Unfortunately, however, there are no reliable dates for the oldest glaciations in the Himalayas and Tibet, and the extent of glaciation is poorly constrained. It is difficult therefore to evaluate the connection between glaciation and uplift.

To conclude, it is not possible to unequivocally attribute late Cenozoic climate change to uplift of the Himalayas and Tibetan plateau. Researchers are only beginning to try to understand the complex processes and feedback mechanisms that drove the Late Cenozoic climate change. More proxy data for climate change in the Himalayas and Tibetan regions, and elsewhere in the world, are needed to substantiate the timing and nature of tectonic and climatic changes.   LEWIS A. OWEN

Further reading

Raymo, M. E. and Ruddiman, W. F. (1992) Tectonic forcing of late Cenozoic climate. *Nature*, **359**, 117–22.

Ruddiman, W. F. and Kutzbach, J. E. (1991) Plateau uplift and climatic change. *Scientific American*, **264**, (3) 42–51.

Ruddiman, W. F. (1997) *Tectonic uplift and climate change.* Plenum Press, New York.

hinterland   *see*   FORELAND, FORELAND BASIN, AND HINTERLAND

historical geology   Geology is a historical science in that it is concerned with the evolution of the planet. Pre-Archaean history (say, before 3500 million years (Ma) ago) is largely speculative; but subsequent events and Earth history are interpreted on the basis of an understanding of rocks, minerals, fossils, and geological processes giving rise to phenomena such as tectonic structures, and the immensity of the time entailed. Three principal ordering activities are involved: correlation, recognition of the sequence of events, and the measurement of time (*geochronology*).

We assume that the laws of physics have remained constant throughout time and that the general motions within the Solar System have not changed greatly with time. It is also assumed that there has been a continuous loss of heat from the Sun and the other bodies within the Solar System since it came into being. Before these fundamentals were grasped, theories of the origin of the Earth and its history were based upon myth and belief. Biblical teaching restricted enquiry and argument, but during the Renaissance a number of scholars began to make observations which did not conform to Scripture. Niels Steensen (1638–86, usually known as Nicolaus Steno), observing stratified fossiliferous rocks in Italy, postulated that the original sediments were deposited on a horizontal base, that they formed once-continuous sheets subsequently eroded and deformed, and that each layer is necessarily younger than the one upon which it rests. This last concept is known as the 'law of superposition'. Steno also postulated that the fossils present are the remains of creatures living when the sediments were accumulating. These were revolutionary ideas and were not at all popular with the Church at the time.

Although the word 'stratigraphy' did not come into use until the nineteenth century, Steno had laid a foundation for this branch of the Earth sciences that fairly soon became widely appreciated. He had made careful observations establishing the order in which strata occurred, and on the basis of this stratigraphy he proposed a sequence of causative processes and events—a small but significant essay in historical geology. His lead was soon followed in Britain and northern and central Europe. Early eighteenth-century geological argument in Europe identified lithological categories of rocks which were said to occur everywhere in regular succession. The formation of these deposits was thought to have been within the time allowed by Scripture, the latest having taken place since Noah's survival of the Deluge. These developments led to the eventual rise of the *Neptunist school*, which advocated precipitation from a primeval ocean to produce the ancient crystalline rocks. This view was soon challenged by the *Plutonists*, who emphasized the importance of terrestrial heat and volcanic activity in the formation of crystalline rocks. Among the latter, the influential James Hutton (1726–97) explained rocks he examined in Scotland by volcanic action and by sedimentary processes that continue today. He advocated the principle of uniformitarianism and, with it, the view that the Earth must be immensely old, showing 'no vestige of a beginning, no prospect of an end'. The 6000 years of biblical times were grossly inadequate for Earth history.

It was not long before the use of fossils to define relative geological age was realized, but this did not solve the problem of measuring geological time. Despite many attempts, the solution was approached only once the principles of radioactivity and isotope analysis had been mastered. These have permitted not only the estimation of the age of different formations and events in Earth history, but also the age of the Earth and the rates at which geological processes proceed.

Today's model of the Earth is a set of interacting systems which have evolved continuously since the planet became a discrete and self-contained unit. It acknowledges our limited understanding of these systems and also the possibility that factors as yet not recognized may on occasion have been important throughout the life of the planet. Secular changes in the Earth's orbit and precession are now recognized as possibly important influences upon world climate throughout much of geological time, and secular changes in solar radiation and sunspot cycles may have been similarly important. Increased understanding of the other terrestrial planets in the Solar System has some bearing upon modern interpretations of early Earth history. Attention is now also focused upon the nature, behaviour, and evolution of the inner shells of the Earth, the mantle and core, since these will have provided many (literally) underlying influences upon events at the surface.

Life itself is seen not so much as a superficial chemical bloom but as an important factor in the geological cycle since early Proterozoic time, around 2000 million or more years ago. Its cumulative effects have made their mark upon the evolving lithosphere to the point where calcium carbonate production, giving rise to limestones, is a major part of the sediment budget. The atmosphere has also been profoundly changed by organic activities throughout geological time, as numerous geochemical studies show us. The evolution of lithosphere, atmosphere, and global geography is, however, no longer thought of as a continuous unchanging and smooth process. Historical geology views this process as one with many sudden unique or recurrent events, some of which may have triggered great geological, geographical, and palaeobiological changes.   D. L. DINELEY

Further reading

Berry, W. B. N. (1987) *Growth of a prehistoric timescale* (2nd edn). Blackwell Scientific Publications, Palo Alto, California.

Gould, S. J. (1987) *Time's arrow, time's cycle*. Harvard University Press, Cambridge, Massachusetts.

history of geochemistry Geochemistry deals with the distribution and migration of the chemical elements within the Earth and, as such, is distinct from mineralogy and petrology, which have a history of going back to classical times. This branch of Earth science has largely developed during the twentieth century, though the word itself was first used by the Swiss chemist, C. F. Schönbein, in 1813. Thus the discovery of the different elements marks episodes in the history of the subject. Perhaps it all began with Lavoisier, who recognized some 31 elements in 1789. Before the end of the eighteenth century another eight were known. The nineteenth century saw the virtual completion of the periodic table of D. I. Mendeleev. Some short-lived radioactive elements were soon to be added. Thus for about a hundred years geochemical data had been largely the by-products of mineralogical investigations of the outermost parts of the crust of the Earth.

In 1884, the US Geological Survey appointed a Chief Chemist, F. W. Clarke, and set up a laboratory to investigate the chemistry of the planet. It was to be concerned with analyses of all kinds of materials sent in by the Survey's field officers, and it amassed a wealth of sample data. With the establishment of the Geophysical Laboratory by the Carnegie Institution in Washington in 1904, the principles of physical chemistry were to be applied to the study of geological processes, especially the evolution of the rock-forming minerals.

A few years later V. M. Goldschmidt at the University of Oslo applied the phase rule to mineralogical changes brought about during contact metamorphism of sedimentary rocks. His subsequent studies in metamorphism all showed that metamorphic changes could be interpreted in terms of the principles of chemical equilibrium. Meanwhile X-ray crystallography had been initiated by M. T. F. von Laue in Germany, although it took some time to make an impact on geochemistry. Spectrographic methods were also developed. Geochemists were interested in the concept of the geochemical cycle and began to speculate on the nature of the matter in the interior of the Earth and on the chemical evolution of the planet. In the USSR a school of geochemistry was developed energetically by V. I. Vernadsky and his successors, such as A. P. Vinogradov. Their work was largely directed towards the search for and exploration of mineral resources.

After the Second World War, there was a surge of interest in radioactivity within the Earth, and new methods of analysis and instrumentation, such as the electron probe, were invented. Two of the most significant and rapidly expanding fields of research today are organic geochemistry and biogeochemistry. Organic materials are found in many sedimentary rocks of all ages from the Archaean onwards. Organic geochemical research has been concerned primarily with the origin and evolution of living materials and their fate after burial. Much research has been of interest to, and sponsored by, the petroleum industry, and many facets of the research are as significant to stratigraphy and basin analysis as to palaeobiology   D. L. DINELEY

history of geophysics Geophysics employs the principles and methods of physics in the study of the character and history of the Earth. Its modern offshoot, *geophysical prospecting* or *exploration geophysics*, applies these principles and methods to the search for natural resources to obtain information about subsurface structure for engineering purposes. Archaeological information, especially the location of buried remains, is also obtainable by these means.

As early as AD 132 Chang Heng set up a seismic instrument in China to indicate not only when an earthquake occurred but also the direction of the first shock waves. During the Renaissance, Leonardo da Vinci (1452–1519) acquired a considerable knowledge of geology, the Earth's gravity field, and wave propagation and reflection. It was William Gilbert (1540–1603) who founded the sciences of magnetism and electricity and noted that the magnet dipped at an angle dependent on latitude. Slightly later, in Italy, Galileo Galilei (1564–1642) developed the correct formula for the motion of a pendulum, so important in the study of gravity. A Dutchman, Christian Huygens (1629–95), was able to explain the behaviour of waves impinging upon an interface, and also diffraction. Sir Isaac Newton (1642–1727) produced the concept of calculus, the basis laws of motion, and explained the true cause of ocean tides.

Experiments and instruments were developed in the eighteenth century and enabled Pierre Bouguer (1698–1758) to compare the regional mass of the Earth with the local mass of mountains by pendulum experiments. The Bouguer correction is used today for gravity measurements at an elevation above a datum. In 1798 an early Woodwardian Professor of Geology at Cambridge, John Michell, described a method for measuring a gravity field by a 'torsion balance', and in the same year Charles Coulomb built a similar device to study magnetic and electrical attraction. During the nineteenth century much progress resulted from the invention of the Eötvös torsion balance and the setting up of global observational networks. Magnetic stations were established in many places throughout the world and gravity was measured in a wide variety of topographical settings.

Seismic observations on a network basis had to await the invention of accurate seismometers. The first such instruments were made in 1880 and seismographs were installed at the University of California in 1887. A co-ordinated network of 27 seismic stations was soon operating on a global basis and in 1903 the International Seismological Association was formed. Then began a period of intense seismological study during which discontinuities deep within the Earth were discovered by A. Mohorovicić (the Moho discontinuity) and Beno Gutenberg (the Gutenberg discontinuity).

During the two world wars location by acoustic and seismic means achieved some success on land and at sea. Between the wars the search for natural resources became intense, using gravity, seismic, electric, and magnetic methods. The petroleum industry in particular invested heavily in these advances, being largely concerned with seismic surveying. After the Second World War progress continued with the aid of microelectronics and computers. Airborne and marine patrols during the 'cold war' period were equipped with advanced means of locating nuclear explosions, and remote sensing was developed. The rise of plate tectonic concepts after the Second World War was largely a consequence of the exploration of the deep oceans (again, defence needs were paramount), using increasingly sophisticated instruments.　　D. L. DINELEY

Further reading

Bates, C. C., Gaskell, T. F., and Rice, R. B. (1982) *Geophysics in the affairs of man*. Pergamon Press, Oxford.

history of plate tectonics *see* PLATE TECTONICS: THE HISTORY OF A PARADIGM

history of the search for economic deposits

Economic deposits, as considered here, include the metallic ores and other industrial minerals, building materials, and fossil fuels. Man's need for an increasing range of them has risen dramatically during the twentieth century, but his earliest needs were possibly for rock types that gave sharp cutting tools. Natural glasses, flint, and other siliceous rocks were used throughout the stone ages, and prehistoric workshops and mining sites are known in many parts of the world. Native copper was probably the first metal to be fashioned into tools and weapons, as in the Middle East, the Americas, and Asia during the Bronze Age. Copper was in use in Egypt around 7000 BC, but gold and possibly native silver were collected even earlier.

By the time of the Roman Empire, iron and other metals were in common use; Roman mining methods were locally very efficient. In the western part of the Empire the Romans discovered all the principal mining regions, and no significant major discovery has since been made there for almost two thousand years.

Three main methods of exploration for metallic (ore) mineral deposits have been used over the years: geological, geophysical, and, more recently, geochemical. They can also be applied to the search for stratiform (layered) and superficial deposits. In addition, there has always been the use of the divining rod. Divining existed in ancient times; Agricola (1494–1555) referred to its use in mining and in the search for water. More reliable methods, developed in recent decades, include remote sensing and aerial surveys.

Geological prospecting has been the means of discovering resources of many kinds, though it has been carried out on a scientific basis only for the past two hundred years. It brings together information to build a geological map; that is, a picture or model of the resource sought, from outcrop data and frequently nowadays also from borehole data and cored samples. Geological sampling has been extended to the sea floor over the continental shelves during the nineteenth century, and to the oceanic basins since the Second World War. The discovery of manganese and other nodules at depth has prompted proposals for oceanic dredging of metal-bearing materials.

From the beginning, the British Geological Survey has been required to record data of use to, and provided by, mining undertakings. Most other surveys have the same commitment. Mining geologists, petroleum geologists, and hydrogeologists are highly qualified specialists; degrees in mining geology have been given for about 150 years at various universities. Before this, mining academies, as in central Europe, provided a training largely on the engineering side.

Geophysical prospecting has become increasingly important over the past hundred years as better instruments have been developed. The oldest techniques are those of magnetic and gravity prospecting, in which the operator looks for regional anomalies within the global magnetic or gravitational field. In the eighteenth century the simple magnetic compass was used to detect buried iron-ore bodies in Europe, and shortly afterwards in North America too. Several improvements were made in the late nineteenth century and again during the twentieth century, enabling the surveyors to operate with greater speed and to make magnetic surveys from the air.

Gravity survey, using pendulums and the torsion-balance type of gravity meter are used for detecting certain metallic bodies, but are most frequently employed in prospecting for potential oil-bearing structures at depth. Seismic surveying methods, again primarily for oil or gas-bearing structures, have been developed during the twentieth century. A seismic survey was used in Texas to locate a salt dome in 1919. By 1926 seismic survey methods had discovered an oilfield for the first time in the southern USA. Today, wide areas of country and sea floor are surveyed by both reflection and refraction methods, and can yield seismic 'profiles' that reach depths of several thousand metres. Radio, radar, and sonic methods of survey are also now employed to detect discontinuities or surfaces at depth.

Electrical survey methods were developed early in the twentieth century to locate orebodies, especially *metallic sulphides*, at depth. Between the two world wars large areas of Canada and Australia were surveyed by private and governmental organizations. Airborne surveys were possible and techniques for borehole surveys became very sophisticated with the use of microprocessor technology.

In the mid-1930s radioactivity surveys became possible with new instruments, and since the Second World War they have been important in the search for fossil fuels, evaporites, and uranium deposits.

Remote sensing, using a variety of spectra, has achieved a major role in broad aerial surveys since the mid-1960s. It has proved the location and identification of many sedimentary deposits, especially in desert areas.

Geochemical prospecting, developed largely since the 1950s, involves taking samples of soils and other superficial deposits, (and in some cases vegetation) for rapid microanalysis. Maps built up to show the distribution of selected elements may offer clues to the whereabouts of subsurface concentrations of economic value. Computer techniques for handling data now enable these surveys to proceed at high speed.   D. L. DINELEY

Further reading

Reedman, J. H. (1979) *Techniques in mineral exploration*. Applied Science Publishers, London.

Robinson, E. S. and Cahit, C. (1988) *Basic exploration geophysics*. John Wiley and Sons, New York.

Townsend, J. R. G. (1981) *Terrain analysis and remote sensing*. George Allen and Unwin, London.

Watson, J. (1983) *Geology and Man: an introduction to applied Earth science*. George Allen and Unwin, London.

Holmes, Arthur (1890–1965)

Arthur Holmes was one of the most distinguished British geologists of the twentieth century. He carried out some outstanding research in geochemistry, made an early but highly significant contribution to geochronometry and the determination of the age of the Earth, proposed a mechanism to drive continental drift, and wrote a best-selling and innovative textbook, which ran to four editions.

In 1907 Holmes entered Imperial College to read physics, and then stayed on for a further degree in geology. Field experience in Mozambique followed and gave him an early taste for ancient rocks, the genesis of igneous materials, and physical geology. He served as geologist to an oil company in Burma before returning to the University of Durham as professor of geology. From 1943 to 1956 he was Regius Professor at Edinburgh.

In 1913 Holmes published a small book, *Age of the Earth*, which discussed the pioneer radiometric analyses that he had worked on. It gave a series of age estimates for points in the stratigraphical column that were to be unchallenged for a long time. Holmes showed that Kelvin's estimate of the age of the Earth was too small and that an age of at least 1600 Ma was needed. His was, in fact, the first modern scale based upon analyses of isotopes of uranium, thorium, and lead. At this time, too, the contribution of radiogenic heat within the Earth was recognized as a most important factor in the planet's history.

He also continued with his work on igneous rocks, in particular, granites, for which he came to favour an origin in 'granitization' (i.e. the transformation *in situ* of country rocks to granite by the introduction and removal of constituents on a large scale). During the years around the Second World War this was a topic of fierce debate, but Holmes's views were never seriously questioned. He also published contributions on the igneous rocks of Uganda and the rift valleys of East Africa.

Wegener's hypothesis of continental drift induced Holmes to propose a revolutionary model in which mantle convection played a vital part. The first edition of his *Principles of physical geology* (1944) showed how well Holmes appreciated that fundamental geophysics was of service to historical geology and to tectonics. In fact, this immediately successful volume has become a British classic.

During his lifetime Arthur Holmes was much honoured by the geological community in Britain and abroad.

D. L. DINELEY

Holocene/Recent

The Holocene is the later of the two epochs of the Quaternary Period, the earlier being the Pleistocene. The Recent Series was proposed for post-glacial deposits by Charles Lyell in 1833, but the definition of the base of the series is vague at best. It was suggested that deposits formed since the arrival of man should be designated as Recent. The term is synonymous with Holocene, which was the term agreed for the post-Pleistocene by the International Geological Congress of 1885, but the means by which the base of this division could be recognized were still far from certain.

After exhaustive discussion relating to glacial and post-glacial successions in many parts of the world, general agreement now favours the selection of criteria that can be recognized without reference to specific local conditions. There are several, but nevertheless common consent is given to the selection of an age in years. This most recent epoch is held to date from 10 000 years ago to the present moment. It is a span of time in which climatic fluctuation has continued and the natural environment has come under increasing pressure from mankind.

The general warming of world climate has led to the diminution of ice caps and glaciers, and as a consequence sea level has been rising. This, the Flandrian transgression, began to take effect about the beginning of the Holocene. During this time, levels might have risen as much as 170 m. The Flandrian transgression is thought to have reached its climax about 6700 years ago with sea level a few metres above the present.

Climatic changes are reflected in the preserved record of plant species present in the deposits, and pollens contribute substantially to this. In Britain, the pollens indicate cooling and increased precipitation after the so-called climatic optimum around 5000 BP. Botany, too, helps to establish a quantitative means of dating Holocene and Recent events; the discipline is known as dendrochronology, or tree-ring analysis. The technique involves the counting and measurement of tree-ring thicknesses in ancient timbers. They reflect the growing (i.e. climatic) conditions prevailing year by year as the trees grew, and they tend to show very similar patterns over spans of time even in different species. European dendrochronology now goes back over 11 000 years before the present, and research in Germany has corroborated tree-ring dates with radiocarbon and stable isotope analyses.

In Scandinavia and North America, sediments from several recent and existing lakes receive an annual layer (of sediment) with the spring thaw. It can be shown that these layers, rhythmites (or varves), have been accumulating for a long time, and provide a kind of calendar dating back throughout the Holocene.

A chronology unique to the Quaternary, and especially important in the Holocene, concerns the succession of stone artefact cultures. Stone tools, commonly of flint, from the Holocene date back to the Mesolithic (Maglemosian–Tardenoisian) in Europe. After the Mesolithic, and beginning about 8000 years ago, came the Neolithic, during which the rise in sea level came to an end. Human cultures now spread from the subtropical regions into the steppes and temperate climes where the grasses could be selected to provide abundant food. Cereal collecting soon gave way to cereal cultivation and the domestication of sheep and cattle. Pastoral nomads evolved their special niches in both the Old World and the New by about 5000 years ago, and the Bronze Age dawned in the Near East. Migration increased in pace and the world human population was in the realm of tens of millions.

The process of change continues and we are at some pains to understand the mechanisms by which it proceeds. The future comfort and prosperity of the human race depends to large measure upon ability to contend with inevitable natural changes—which may be drastic—and also to ensure that human activities do not further detract from the planet's amenities. The Holocene epoch is a short one, so far: the changes it has seen have been minor by geological standards; no catastrophic rare global event has occurred so far. There are many pitfalls in the process of prediction. UNESCO and other agencies are establishing work in natural disaster reduction and in monitoring the effects of global geological and biological change. Understanding the past may be of some assistance in understanding the present. Understanding what may lie ahead is fraught with even greater difficulties.

D. L. DINELEY

Further reading

Calder, N. (1984) *Timescale: an atlas of the fourth dimension.* Chatto and Windus, London.

Taylor, J. (1978) Present day. In McKerrow, W. S. (ed.) *The ecology of fossils.* Duckworth, London.

Hooke, Robert (1635–1703) The experimental philosopher Robert Hooke was active in many areas of science and made a number of significant contributions to the Earth sciences. He was the son of a clergyman, but because of his poor health was not educated for the Church, although he was able to take a degree at Oxford. He was closely associated with the Royal Society, holding for many years the office of Curator of Experiments as well as being a Fellow.

Hooke has been described as hurrying from one enquiry to another with 'brilliant but inconclusive results'. He used a pendulum to measure the force of gravity, and his subsequent analysis of the movements of the Earth and Moon round the Sun provided the basis for the understanding of planetary movements. Among a host of other activities he observed comets, investigated the relation of barometric pressure to changes in the weather, discoursed on the effects of earthquakes, and put forward a correct interpretations of the nature of fossils and the succession of living things on the Earth. His *Discourse on Earthquakes* (1668) emphasized the striking results of earthquakes in modifying the surface of the ground and the Earth on a large scale—an aspect that was to influence Lyell's interest and investigations 180 years later. His experiments on elasticity led to the formulation of what is known today as Hooke's law, which describes the relation between stress and strain for a perfectly elastic material and has applications in structural geology.

D. L. DINELEY

hot springs A spring is called 'hot' or 'thermal' if it has a temperature noticeably above the mean annual air temperature in the locality. In Europe the general practice has been to apply the term 'thermal' only to springs having temperatures greater than 20 °C.

Hot springs are found throughout the world wherever there are permeable geological structures that (1) channel meteoric water (water derived from rain or snow) into the ground where heating occurs, and (2) channel the heated water back out of the ground at a rate that is fast enough to prevent all the thermal energy from being dissipated by conduction into the surrounding rock during upward flow. Most hot springs with temperatures above 60–70 °C occur in volcanic belts, or where the continental crust has been thinned and faulted as a result of geological processes that pull the crust apart, as in the Basin and Range province of the western United States.

The driving force for the discharge of hot springs is the bouyancy that results from the lower density of hot water compared to cold water. The temperature attained by meteoric water during deep convective flow depends on the Earth's thermal gradient where the flow occurs, and the depth of its circulation. Temperature normally increases by about 25° to 30 °C per kilometer of depth, and most hot spring waters attain temperatures less than 100 °C in the course of circulation to depths of 2 or 3 kilometres. Heating to more than 100 °C can result either from deeper circulation along faults in non-volcanic regions, or by relatively shallow circulation that brings water into contact with anomalously hot rock in volcanic belts. Drilling in search of geothermal resources in volcanic regions has encountered groundwaters that have been heated to temperatures as high as 350 °C at depths as shallow as 2 km. This is near the expected upper temperature limit for circulating meteoric waters because faulting generally does not occur in the continental crust where temperatures are greater than 350–370 °C. Repeated faulting appears to be essential for keeping open channels of flow that otherwise would become clogged by minerals that are deposited from circulating hot waters.

Where the circulating waters attain underground temperatures higher than 100 °C and the rate of upward flow is relatively fast, boiling will occur in response to decompression (the decrease in the weight of the overlying column of water as the upward-flowing water rises to shallower depths). A mixture of boiling water and steam will then be discharged at the Earth's surface. In some places this discharge occurs episodically as geyser eruptions. The depth at which the boiling is initiated in the upward flowing water, and the ratio of steam to water in the fluid that is discharged by a boiling spring, increase in proportion to the maximum temperature attained by the deeply circulating water. The maximum temperature exhibited by a hot spring is independent of the water to steam ratio; it is controlled by the boiling point of water at the elevation where the discharge occurs. At sea level pure water boils at 100 °C. There is a decrease in the boiling temperature by 1 °C for each 303 m increase in altitude. On the other hand, hot springs that issue on to the floors of lakes, or at the bottom of the ocean, can have temperatures much greater than 100 °C because the boiling temperature at the point of discharge is determined by the weight of the overlying column of cold lake or ocean water. Vent temperatures

as high as 360 °C have been measured at hot springs deep in the ocean where volcanic activity occurs along plate boundary spreading centres. These springs are called 'black smokers' because of the dark colour imparted to their water by precipitating minerals.

The compositions of hot spring waters are determined by the temperatures and types of rocks encountered during underground flow, and, in volcanic regions, by the degree of mixing with waters and gases given off from crystallizing magmas. The types of minerals deposited by hot springs reflect these conditions. Massive travertine (calcium carbonate) deposits occur where hot spring waters issue from limestone after having attained a temperature no higher than about 100–200 °C. Siliceous sinters are deposited where hot spring waters flow relatively quickly to the surface after having attained a temperature greater than about 180 °C. Acidic springs and 'mud pots' occur in volcanic regions where upward-flowing steam rich in hydrogen sulphide condenses in perched bodies of groundwater at or near the Earth's surface.

ROBERT O. FOURNIER

Further reading

Waring, G. A. (Revised by R. R. Blankenship and R. Bentall) (1965) *Thermal springs of the United States and other countries of the world.* US Geological Society Professional Paper 492.

humour, geological *see* GEOLOGICAL HUMOUR

Hutton, James (1726–97)

The founder of modern geology, James Hutton, came of an Edinburgh family, and attended the universities of Edinburgh, Paris, and Leiden, where he qualified in medicine in 1749 before returning to Edinburgh. He did not practise but took up farming about 40 miles from the city, travelling in Europe from time to time to observe the best farming methods. In 1768 he sold the now prosperous farm and returned to Edinburgh to devote his time to chemistry and geology. He was active in the Edinburgh Philosophical Society (forerunner of the Royal Society of Edinburgh), and he travelled about Britain keenly noting all the geology he could.

By the time he published *Theory of the Earth* in 1795 he had been actively interested in geology for about 30 years. Hutton had in 1783 and 1788 already firmly established the concept of the geological cycle and an immense length of time for its operation. The book set out to discuss the length of time that a habitable world had existed, what changes it had undergone, and what might overtake the present state of affairs. Religion and science he kept quite separate. Hutton discounted the biblical version and divine intervention; he found time endless and he was impressed with the role of heat in Earth history. Hutton's own field observations had made him a convinced plutonist, rejecting the idea that crystalline rocks were precipitated from an ancient ocean. Neither was he a catastrophist, but he sought to account for the constant change proceeding on the surface of the Earth by the processes observable today. He held that igneous rocks originated at great depth and in regions of great pressure.

The basis of Hutton's work was observation, but he had read widely the literature available. Nevertheless, he was strongly independent of mind and, like his friend James Hall, believed that experimental research was a powerful aid to geological problem solving. The two friends published jointly on intrusive granite in the new *Royal Society of Edinburgh Transactions*.

James Hutton died in 1797, leaving one of the most influential books on the Earth sciences, yet it was not an easy book to read. Only when his friend John Playfair produced his *Illustrations of the Huttonian theory of the Earth* in 1802 was its significance widely appreciated.   D. L. DINELEY

hydrological cycle The hydrological cycle describes the various routes by which water moves from the atmosphere and oceans to the land surface and returns to the oceans via surface and subsurface pathways (runoff). It is the central concept in hydrology, and hydrologists strive both to understand and to predict the fate of water on the land at a variety of spatial scales.

At the global scale, 97.2 per cent of water resides in the oceans. Of the remaining 2.8 per cent, which is made up of both fresh and brackish water, 77 per cent is locked up in ice sheets and glaciers, 22 per cent is stored beneath the Earth's surface as groundwater, and the remaining 1 per cent is stored in the atmosphere, rivers, lakes, inland seas, and soil. The hydrological cycle is driven by the evaporation of water from both land and ocean surfaces and the lateral movement of moisture-laden air to the continents. Some 419 cubic kilometres (km^3) of water are evaporated from the oceans every year, and about 69 km^3 of water are evaporated from the continents. Precipitation on the continents amounts to about 106 km^3 and run-off is about 37 km^3. The oceans receive proportionally less rainfall (382 km^3) than they supply to the atmosphere by evaporation, and the continents therefore receive proportionally more than they supply to the atmosphere by evaporation. The inputs to and outputs from the continents are in balance. The following expression shows this balance as a simple equation (quantities are in km^3):

$$\text{Evaporation} + \text{Run-off} = \text{Precipitation}.$$
$$69 \quad + \quad 37 \quad = \quad 106$$

This is referred to by hydrologists as the *water balance*. From this expression we can calculate the run-off ratio (the ratio of run-off to precipitation) which, at the global scale, is 35 per cent. As we shall see later, the water balance varies significantly at the surface of the Earth because of regional differences in precipitation and evaporation.

At the regional scale, there are significant differences in the amount of precipitation received by the continents. This results from variations in the global movement of moisture-laden air masses. Two regions receive more than the average annual global precipitation of about 85–90 cm. These are the low-latitude equatorial regions (e.g. Amazonia and Indonesia) and the west coasts of middle-latitude continents (e.g. Western Europe, Alaska, and Western Canada). There are two regions which receive less than the annual average global precipitation. These include areas where semi-permanent high-pressure systems develop (e.g. the arid and semi-arid regions of the south-west United States and the Saharan region of North Africa) and the continental interiors of high-latitude regions (e.g. central Siberia, central Canada); the low-temperature air masses in these regions have a very low moisture content.

The global inequalities in the distribution of precipitation and in evaporation have a profound effect on the water balance at the regional scale. Table 1 gives annual precipitation, evaporation, and run-off data for three geographical regions of Eastern Europe. These data show that there is an unequal distribution of precipitation at the regional scale but, more significantly, also show us that the run-off ratio varies with latitude. In general, high run-off ratios (over 50 per cent) are experienced in cool high-latitude regions whereas low run-off ratios are found in the relatively warmer low-latitude regions.

At the scale of the catchment or drainage-basin, the hydrological cycle comprises one major input, six stores, seven transfer processes, and three outputs (Fig. 1).

Catchment inputs; precipitation. Precipitation occurs in a variety of forms, including fog, drizzle, rain, sleet, hail, and snow. The most important input in high latitudes and at high altitudes is snow. Snow will often accumulate and can eventually compact to form glaciers and ice caps. In this case, the input is released slowly into the hydrological cycle and the output via river run-off is controlled by temperature. The melting of ice and snow reach a seasonal maximum in the summer months, and river discharge is high at this time of the year. At the daily timescale, the highest run-off is recorded in the late afternoon following a rise in temperature. The most common form of precipitation outside high latitudes and high altitudes is rain. Hydrologists are concerned with the amount of rain received by a drainage basin (usually expressed as rainfall depth in millimetres or centimetres) and the intensity at which it falls, measured in millimetres per hour (mm hr^{-1}). Rainfall volume and intensity are important

Table 1. Annual water balance of three geographical regions of Eastern Europe

	Precipitation (cm)	Evaporation (cm)	Run-off (cm)	Run-off ratio (%)
Tundra	45	11	34	76
Mixed forest	58	37	21	36
Semi-desert	20	19	1	5

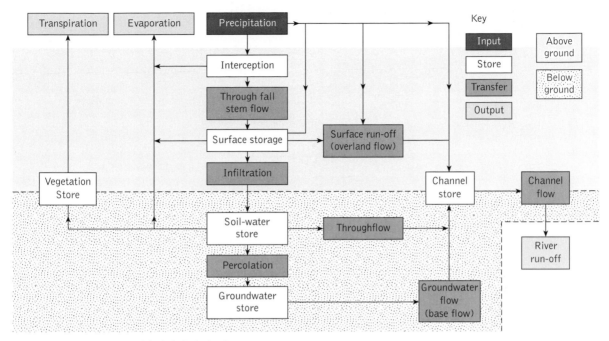

Fig. 1. The major components of the hydrological cycle.

in controlling the output from the catchment. High-intensity storms are usually associated with convectional instability, which causes thunderstorms. These are often local in nature but can lead to excessive amounts of precipitation in local areas and can cause severe flooding. Many of these storms occur in hot arid and tropical environments. Since the soils of arid environments are often thin and compacted, flooding is exacerbated by the inability of the catchment soils to absorb the high input. In tropical areas, high-intensity rainfalls are commonly associated with landslides, mudflows, and other mass-movement hazards.

Catchment stores. Water moving through the hydrological cycle can be stored in one or six locations (Fig. 1). Above the ground surface are the interception, vegetation, surface, and channel stores. Below the ground surface are the soil and groundwater stores. Vegetation acts both passively and actively in storing water. Water in the interception store is held on the canopy surface, which includes the leaves, stems, and branches of growing vegetation. In the vegetation is the water held at any one time *within* the vegetation. Vegetation is usually replenished directly from the soil-water store. The quantitative importance of the interception and vegetation stores depends on the type of vegetation. In general, forests have a higher storage capacity than crops and grasslands, although there are significant differences in the volume stored, depending, for example, on the age and species of the trees. The species is particularly important in relation to the seasonal variation in storage volume, since conifers retain a leaf cover

throughout the year, whereas deciduous forests lose their leaves in the winter. Recent studies in Europe have shown that oak forests can store as much as 20 mm per month whereas Douglas Fir forests can store more than 40 mm per month. Seasonality in vegetation and interception storage is also found in cropped agricultural systems. The importance of canopy interception increases in areas where rainfall occurs as light frequent showers. In mixed deciduous forests, 60–70 per cent of incoming precipitation can be stored in the canopy for storms of 1–2 hours duration. This decreases to less than 20 per cent in storms of over 12 hours duration. Snow can also be stored in substantial quantities in the canopy, and its transfer to the ground is largely controlled by wind and exposure.

The surface store accounts for the volume of water which can be stored in micro-scale hollows in the ground surface, which, although they are most apparent in cultivated landscapes, are also known to occur naturally. Any regular or irregular depression in the ground surface can prevent water flowing downslope towards the river channel. These depressions include plough furrows running at right angles to the dominant slope direction or irregularities left after harrowing. The presence of shallow depressions in the ground surface allows time for water to percolate into the soil and reduces the volume and speed of flow across the slope. The importance of small depressions in allowing the soil-water store to be recharged has long been appreciated in arid and semi-arid environments, where contour ploughing and bunding (creation of embankments) are commonly used as a means of sustaining rain-fed agriculture by inducing soil-water recharge. Although we usually assume that surface storage is a small-scale phenomenon, it may also be of significance at the larger scale, where water may enter the soil and groundwater stores from deflation hollows (depressions left from the activity of wind in arid environments) and kettle holes (depressions left by the melting of stagnant ice during deglaciation).

The channel store measures the amount of water held within the river channel at any particular instant. Although recognized by hydrologists as one type of store, it is a relatively minor component in the hydrological cycle.

Beneath the ground surface are the two major storage components of soil and groundwater. Precipitation enters the ground surface and moves downwards through the soil towards the water table, which marks the upper limit of the zone of saturation (the groundwater zone). Water is held against the influence of gravity in the soil zone because of the attraction of water molecules for each other and for fine soil particles. There are three types of water held within a soil. In dry soils, a thin film of *hygroscopic water* is retained around individual soil particles. It is effectively unavailable to plants, which cannot overcome the surface tensions involved in order to release the water from the soil. Other water molecules link up with the hygroscopic water and are held less strongly in the soil because the water–water bond is less strong than the soil–water bond. This is known as *capillary water*. The third type of water is not held by the soil but moves slowly through it under the influence of gravity. This is *gravitational water*, which drains slowly downwards towards the groundwater zone. A soil that has its full quota of capillary water is said to be at field capacity (i.e. it contains the maximum amount of water that a soil can hold against the influence of gravity). The amount of water that can be stored in the subsurface zone depends upon the size and arrangement of particles. A soil composed predominantly of clay and fine silt particles, for example, will have a higher storage capacity (porosity) than a sandy soil. In the soil zone the actual amount of water stored will depend on the porosity of the soil and the amount lost by evaporation, drainage, and plant uptake. Soils therefore usually hold less than their maximum potential. In contrast, groundwater is permanently saturated and the porosity is controlled by the void spaces in the rock. These voids include the intergranular spaces in the intact rock as well as the fractures and joints that develop as a result of earth movement. Groundwater can occur at great depths in the subsurface zone and in near-surface horizons where impermeable strata prevent the downward movement of water. In this latter case, a perched water table will develop.

Catchment transfers. Water transfer through the hydrological cycle is controlled through seven mechanisms. Three of these, throughfall and stemflow, surface run-off, and channel flow occur above the ground surface; a fourth, infiltration, controls the transfer of water from the ground surface to the subsurface zone; and the remaining three, percolation,

throughflow, and groundwater flow, occur beneath the ground surface (Fig. 1).

Throughfall and stemflow are terms that describe the movement of water between the interception store and the ground surface. Throughfall refers to the free fall of water to the ground after it has collected on vegetation surfaces. This water commonly has a much larger drop size than rain and is therefore more destructive when it hits the ground. Stem flow describes the flow of water along branches and down the trunks and stems of growing vegetation to reach the ground surface at the base of the plant. Stemflow and throughfall transfers therefore play an important role in modifying the spatial distribution and intensity characteristics of rainfall reaching the ground surface.

Surface run-off, sometimes called overland flow, is an important transfer mechanism in the hydrological cycle. As more water is transferred to river channels by this process, so the magnitude of a flood from the same volume of rainfall increases. This is partly because the velocity of water across the ground surface is much higher than through the soil. Surface run-off velocities of over 0.14 metres per second $(m\ s^{-1})$ have been observed in comparison with lateral water velocities through the soil of only 1 to 80 m $s^{-1} \times 10^{-6}$. The amount of rainfall that becomes surface run-off is controlled by one of two mechanisms. Where the intensity of rainfall is higher than the rate at which water can enter the ground (the infiltration rate), surface run-off will occur. This is a common mechanism by which surface run-off is generated in arid and semi-arid areas, where rainfall intensities are often high and infiltration rates are low. In humid environments, the soil may become saturated and rainfall cannot enter the ground surface. The British geomorphologist M. J. Kirby has used the analogy of a bottle to distinguish between these processes. We either have a bottle with a very narrow neck that fills slowly or a bottle that is already full. In either case, the net result is the generation of surface run-off.

Infiltration refers to the rate at which water enters into the soil profile. The rate of infiltration is clearly of significance to the generation of surface run-off as described above, and also to the recharge of the soil and groundwater stores. Of greatest significance in controlling infiltration are the nature and properties of the soil and the presence or absence of a vegetation cover. Soils that have recently been ploughed or have a high-density vegetation cover generally have a high infiltration rate. Grassland that has been compacted by the trampling of grazing animals or bare soils that have been beaten by high-intensity rainfall generally have low infiltration rates. The size of soil particles is also important. Soils with even small quantities of clay can form impermeable crusts owing to high-intensity rainfall. In arid environments, the presence of biofilms (thin algal crusts) may reduce infiltration substantially.

Percolation refers to the vertical movement of water through the soil profile, whereas throughflow (or interflow) refers to the downslope transfer of water through the soil towards river channels. Percolation occurs through the soil matrix (the main mass of the soil) and through small cracks and spaces that are created by burrowing animals, tree roots, or the presence of certain clay minerals that expand and contract upon wetting and drying. If the upper soil profile contains gravitational water with the soil at or near saturation, water will move downwards under the influence of gravity at a rate that is controlled by the characteristics of the soil. Where capillary water is held, movement will usually be from a zone of high moisture content to a zone of low moisture content. Because this movement is controlled by the capillary tension forces that hold water in the soil, it is more precise to say that water moves from a zone of low capillary tension to a zone of high capillary tension. This movement can occur both upwards and downwards within the soil profile.

Throughflow occurs when there are significant changes in the density of different layers within the soil horizon. High-density impermeable layers are commonly found within soil horizons that allow the build-up of a saturated wedge of water above the impermeable layer. On a slope, water will move laterally under the influence of gravity. The rate of water movement is again controlled by the number and size of the void spaces within the soil. Although the term 'throughflow' is generally used to describe the lateral movement of water through the soil matrix, in some environments small pipes and tunnels develop at the junction between a permeable and impermeable layer in which water can move more rapidly than through the soil matrix. Pipes and tunnels, in which water velocities approach that of overland flow, are common in areas of peat and also in areas with fine loess (wind-blown) silts, where tunnels of up to 4 m in diameter may develop.

Groundwater flow, like throughflow, is the slow downslope movement of saturated water under the influence of gravity within the saturated zone. Rates of movement are variable but are usually very slow. Sandstones typically have water velocities of between 0.28×10^{-6} and 0.28×10^{-2} m s^{-1}, whereas in closely jointed limestones velocities may be as high as 0.14 m s^{-1}. In areas where the water table is confined between impermeable strata and the groundwater recharge area is at a higher elevation some distance away, water may be forced to the ground surface under pressure. This is artesian water, which is often a commercially important source for water supply.

Channel flow is the most efficient means of removing water from a catchment. At high water discharge, velocities can exceed of 3 m s^{-1}. The velocity of flow is controlled by the gradient of the channel, the shape of the channel, and the roughness caused by the presence of stones and vegetation. In both urban and rural areas, artificial drainage frequently increases the length of open channels and pipes. While these serve an important function in improving soil drainage for cultivation and in preventing flooding in urban areas, they are also perceived as posing an increased risk of flooding in downstream areas.

Catchment outputs. There are three catchment outputs, evaporation, transpiration, and run-off (Fig. 1). Evaporation is an output from the interception, soil surface, and soil water stores. Transpiration is an output from the vegetation store, usually via the stomata on the leaves of living vegetation. The vegetation stores can tap water directly from the soil water store. To an extent, transpiration rates are regulated by the biosphere. Evaporation and transpiration rates are primarily controlled by a number of other factors, not least of which is the availability of a moisture supply. Supply limitation is of little importance in controlling evaporation from open water bodies, but is of great significance in relation to evaporation from bare soils, from the interception store, and from the vegetation store. The most significant factors controlling evaporation and transpiration losses are the amount of incident solar radiation (the energy which drives evaporation and transpiration) and the vapour pressure of the air relative to saturation. This controls the capacity of the air to absorb more moisture from the surrounding environment. In arid and semi-arid environments, evaporation dominates the output from the hydrological cycle. Run-off describes the loss of water from the catchment. It is measured by hydrologists at a wide range of timescales from hours to years. Annual run-off is controlled by the balance between the input of precipitation and the combined outputs of evaporation and transpiration giving rise to differences in the run-off ratio described above. At shorter timescales, run-off is controlled by differences in the magnitude of the storage components within the hydrological cycle and by the transfer processes between the catchment and river channel.

Of major significance in hydrological research is the direct and indirect impact of human activity on the various storages and transfer processes illustrated in Fig. 1. The process of urbanization, for example, reduces the capacity of the interception and vegetation stores, increases the efficiency of surface run-off and channel flow transfers, and consequently increases run-off at the hourly and annual timescales. Forest clearance and cultivation have similar effects on the natural hydrological cycle, but pose additional threats to the environment by making large quantities of sediment available for erosion.

IAN D. L. FOSTER

Further reading

Newson, M. (1992) *Land, water and development.* Routledge, London.

Petts, G. E. and Foster, I. D. L. (1990) *Rivers and landscape.* Arnold, London.

Shaw, E. M. (1994) *Hydrology in practice* (3rd edn). Chapman and Hall, London.

Ward, R. C. and Robinson, M. (1990) *Principles of hydrology* (3rd edn). McGraw-Hill, Maidenhead.

hydrology A synthesis of many disciplines, hydrology is the science of water, its properties, its circulation, and its distribution in Earth and atmospheric systems. In the Earth sciences, hydrology is studied at various scales, ranging from hill slopes ($10^0–10^3$ m^2) to drainage basins ($10^0–10^6$ km^2). The hydrology

(a)

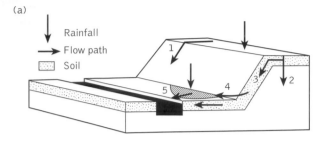

Rainfall

Flow path

Soil

(b)

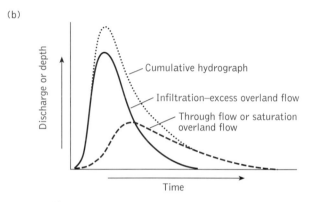

Discharge or depth

— Cumulative hydrograph

— Infiltration–excess overland flow

— Through flow or saturation
overland flow

Time

Fig. 1. (a) Possible flow paths of rainfall on hill slopes, including (1) infiltration–excess overland flow; (2) groundwater recharge; (3) through flow; (4) return flow; and (5) saturation overland flow. (b) Cumulative flow hydrograph and its relation to contributions from multiple flow processes.

of hill slopes is a combination of surface flow processes, which include infiltration–excess overland flow and saturation overland flow (see *overland flow*) and shallow subsurface flow processes (see *subsurface flow and erosion*) (Fig. la). Soil infiltration capacity, the rate at which soils absorb rainfall, determines whether water flows on the surface or through the subsurface. Infiltration–excess overland flow occurs when intensity of rainfall exceeds the capacity of the soil to absorb water. If rainfall intensity does not exceed infiltration capacity, then through flow or saturation overland flow occurs (Fig. la). Saturation overland flow is the combination of return flow (subsurface flow that returns to the surface downslope because of saturation) and direct rainfall on to areas of subsurface saturation. Throughflow travels entirely through the subsurface, typically along the interface of a less permeable soil horizon. For a given hill slope, the dominant run-off process is therefore regulated by infiltration capacity, which distributes rainfall between surface and subsurface hydrological systems. Infiltration capacity and the flow process that dominates on a slope are a function of several variables, including climate and vegetation, rainfall characteristics, soil thickness, slope morphology, and human disturbance.

The hydrology of drainage basins is a combination of hill-slope flow processes, groundwater flow processes, and channelized flow processes in surface streams (see *drainage basins*). Hydrographs, plots of either flow discharge or depth versus time, are the cumulative record of discharge by the various flow processes on the hill slopes that comprise a drainage basin (Fig. 1b). The discharge characteristics of a drainage basin, such as the peak and total discharge and the time-to-peak discharge, depend on the flow process as well as on the characteristics of the drainage basin such as the shape of the basin and the drainage density. The flow process is critical to the shape of the hydrograph because of differences in flow velocity between surface flow paths, including channelized and non-channelized flow, and subsurface flow paths. Hydrographs dominated by infiltration–excess overland flow tend to have higher peak discharges and faster times-to-peak discharge than hydrographs dominated by saturation overland flow. In addition, the recession limbs of saturation overland flow hydrographs are generally less steep and of longer duration because of the lag in response of the return flow or throughflow contribution to discharge.

Hydrologists' understanding of the conditions under which the different flow processes dominate and their characteristic hydrographs, has wide-ranging applications in the Earth sciences, from flood forecasting and routing to the evolution of drainage basins. Accurate flood forecasts require an understanding of the response of dominant flow processes to rainfall events of varying intensities and durations, superimposed on seasonal differences in soil moisture and vegetation, among other variables. If we consider the hydrograph characteristics and their relation to flow process, as described above, determinations of peak and total discharge and time-to-peak discharge from storm characteristics and basin parameters are especially critical for evaluating potential human impacts as well as the integrity of flood-control structures within the impacted drainage basin. Hydrological routing is used to predict downstream changes in the hydrograph.

The role of hydrology is being incorporated increasingly in models of landscape evolution. The frequency, magnitude, and spatial extent of surface and subsurface flow processes also affect the evolution of drainage basins. Drainage networks develop in response to surface flow processes and are therefore limited in extent to the portion of the basin that experiences overland flow. The role that flow process plays in drainage-basin evolution is best expressed by considering the hydrological response of a hill slope and drainage basin to climatic change. Change from a semi-arid to an arid climate is associated with increased dominance of desert scrub communities over grassland communities. In the scrub community, rainfall is preferentially infiltrated in the vicinity of scrub vegetation, while interscrub areas are more susceptible to soil erosion by infiltration–excess overland flow. Spatially heterogeneous soil moisture and soil degradation, associated with the semi-arid to arid climate change, further stress remnant grassland vegetation and decrease soil infiltration capacity over time. Consequently, the dominant flow process would shift from throughflow or saturation overland flow to infiltration–excess overland flow. Aggradation in alluvial and fluvial systems has been temporally linked to reductions in effective precipitation during the late Quaternary and explained as the geomorphological response to newly imposed run–off infiltration balances. The increased dominance of infiltration–excess overland flow results in increased peak run-off in the basin hill-slope and channel systems, exceeding stream power thresholds for sediment entrainment, and would result in increased sediment supply and delivery to downstream depositional sites.

J. B. RITTER

Further reading

Dunne, T. (1978) Field studies of hillslope flow processes. In Kirkby M. J. (ed.) *Hillslope hydrology*, pp. 227–93. John Wiley and Sons, New York.

Gregory, K. J. and Walling, D. E. (1973) *Drainage basin form and process*. Halsted Press, New York.

hydrothermal solutions *Hydrothermal* means, simply, hot water, but in geological parlance the term is restricted to hot subsurface waters. If hot subsurface water contains dissolved materials—as most heated subsurface waters do—it is said to be a *hydrothermal solution*. Research has shown that hydrothermal solutions have the capacity to dissolve, transport, and redistribute minerals within the Earth's crust, and, under suitable circumstances, to scavenge and concentrate valuable metals, such as gold, silver, copper zinc, and tin into rich mineral deposits. Indeed, most mineral deposits mined for their metallic contents were formed through the agency of hydrothermal solutions and are referred to as hydrothermal deposits.

Origins of hydrothermal solutions

The water in a hydrothermal solution can come from any of four principal sources, and is commonly a mix from more than one source. The four kinds of water are groundwater, formation water, metamorphic water, and magmatic water.

(1) Groundwaters are meteoric in origin and can be either fresh or saline (as in sea water). When groundwater comes in contact with hot igneous rock, or penetrates deeply enough to be heated by the geothermal gradient, a hydrothermal solution is the result.

(2) Formation waters are waters trapped between particles of buried sediment, within the strata of a sedimentary basin. Compaction slowly expels formation water, which then rises and mixes with groundwater.

(3) Metamorphic waters are formed by the dehydration of hydrous minerals such as clays, micas, and amphiboles as a result of high temperature and high pressure deep within the crust.

(4) Magmatic waters are formed from the vapors released by a magma that is cooling and crystallizing to an igneous rock.

There are two sources for the metals and other constituents dissolved in hydrothermal solutions: the trace amounts of minerals present in all common rocks, and magmas.

When a magma crystallizes, most of the chemical elements present are incorporated into minerals such as feldspar, mica, amphibole, and pyroxene. Some trace elements do not easily fit into the atomic structures of the crystallizing minerals and become concentrated in the magmatic vapour that is released by the magma. As such vapours cool, hydrothermal solutions of magmatic origin are formed; they are commonly enriched in copper, lead, zinc, gold, silver, tin, tungsten, and molybdenum (Table 1).

Table 1 Compositions of some hydrothermal solutions

Element	Amounts (parts per million)		
	1	2	3
Cl	155 000	157 000	158 200
Na	50 400	76 140	59 500
Ca	28 000	19 708	36 400
K	17 500	409	538
Sr	400	636	1 110
Ba	235	Not det.	61
Li	215	8	–
Rb	135	1.0	–
Cs	14	–	–
Mg	54	3 080	1 730
B	390	–	–
Br	120	526	870
I	18	32	–
F	15	–	–
NH_4^+	409	–	39
HCO_3^-	>150	32	–
H_2S	16	0	–
SO_4^{2-}	5	309	310
Fe	2 290	14	298
Mn	1 400	46.5	–
Zn	540	3.0	300
Pb	102	9.2	80
Cu	8	1.4	–
Temperature	320 °C	>80 °C	

1. Salton Sea Geothermal field, California. After Muffler, L. J. P. and White, D. E. (1969) *Bulletin of the Geological Society of America*, **80**, 157–82.

2. Well in aquifer No. 9, Cheleken Pennisula, USSR, now Kazakhstan. After Lebedev, L. M. and Nikitina, I. B. (1968) *Doklady of the Academy of Sciences of the USSR, Earth sciences sections*. [English translation] American Geological Institute, Washington. **183**, 180–2.

3. Oilfield brine, Gaddis Farms D–1 well, lower Rodessa reservoir, central Mississippi, 11 000 ft. After Carpenter, A. B., Trout, M. L., and Pickett, E. E. (1974) *Economic Geology*, **69**, 1191–206.

Any heated water, regardless of origin, will react chemically with the rocks with which it is in contact. If a heated water is even slightly saline, it is an effective solvent for any metals present in trace amounts in all sediments and rocks. By reaction, a hot, saline solution of groundwater, formation, or metamorphic origin will slowly evolve into a hydrothermal solution that contains metals of the same kinds as those carried by solutions of magmatic origin.

Compositions of hydrothermal solutions

Many hydrothermal solutions have been encountered by drillers seeking oil or other commodities; three examples are shown in Table 1. The solutions encountered are all chloride-rich, and this explains how significant amounts of heavy metals can be carried in solution. Minerals such as galena (PbS), sphalerite (ZnS), and metallic gold (Au) have exceedingly low solubilities in water. However, in saline solutions, chloride complexes such as $(AuCl_4)^{2-}$, and $(PbCl_4)^{2-}$ can form and thereby raise the solubility even though sulphur ions (S^{2-}) might also be present in solution.

Deposition from hydrothermal solutions

The origins of hydrothermal solutions, the sources of the dissolved metallic constituents, and the means of solution and transport of heavy metals in solution, despite the very low solubility of metallic sulphur minerals, have become clear as a result of research in the second half of the twentieth century. Less clear, however, are the mechanisms whereby solutions deposit their dissolved loads and form mineral deposits.

Hydrothermal solutions range in temperature from about 120 °C to as high as 600 °C; and the higher the temperature, the more material the solution can dissolve. When a hydrothermal solution cools, therefore, minerals will be precipitated from the solution. However, the chemical nature of a solution—the acidity and its capacity to serve as an oxidizing or reducing agent—is also affected by temperature. Because the stabilities of chloride complexes are dependent on acidity and the reduction—oxidation capacity of a solution, the chemical effects of cooling on the solubility of a hydrothermal solution may be more pronounced than simply the temperature effect of cooling.

Other factors beside temperature change can influence the chemical nature of a solution, and thereby cause precipitation. For example, an acid solution that reacts with a limestone will become less acid; a metal-rich solution may mix with another water mass, such as a brine containing dissolved hydrogen sulfide (H_2S); or a solution might be reduced by organic matter in a sedimentary rock. All such reactions, of which the three cited are selected examples among a large number of possible reactions, can cause precipitation, and the formation of a mineral-rich deposit. It is difficult in most cases to determine exactly which reactions were most important in the formation of a specific hydrothermal deposit.   BRIAN J. SKINNER

hypabyssal igneous rocks Hypabyssal igneous rocks are defined as those rocks that lie between the two categories plutonic and volcanic. Plutonic rocks form in large igneous intrusions, deep in the Earth's crust, and are consequently coarse-grained (lentil-sized grains and larger). Volcanic rocks, by contrast, are those that were erupted on or near the surface of the Earth and are fine-grained (grains not visible to the naked eye), having cooled quickly. There is a continuous range of sizes and depths of intrusions between plutonic and volcanic and it is here that the category 'hypabyssal' lies. The boundaries of the group of hypabyssal rocks are thus blurred and the term has been dropped by some geologists.

In practice, the term 'hypabyssal' is used to encompass dykes and sills (Fig. 1), which are parallel-sided, sheet-like igneous intrusions that form as a result of dilation, that is, a rock mass splits to allow magma to force its way through the country rock (i.e. the rock through which the magma is flowing) without otherwise deforming it. Dykes and sills propagate as the 'hydrostatic' pressure of magma flowing through them splits the country rock in the direction of least resistance i.e. at right angles to the least principal stress. During the initial intrusion of the dyke, magma chills on to the sides of the conduit to form a chill margin, but if there is prolonged transport of magma the chill margin and even part of the country rock may be melted. Where oceanic crust is forming at the mid-ocean ridges, the new crust consists entirely of dykes and each successive intrusion is inserted through the centre of its predecessor. This type of intrusive body is called a sheeted dyke complex and is characterized by multiple half-dykes, each of which has a single chill margin.

The word 'dyke' ('dike' in American usage) is often used to describe a sheet intrusion that cuts across strata (sedimentary beds), in contrast to a sill which was intruded parallel to the strata. Most dykes were intruded vertically, whereas sills were intruded horizontally. Dykes range in thickness from a matter of centimetres to over 100 m but are generally about a metre

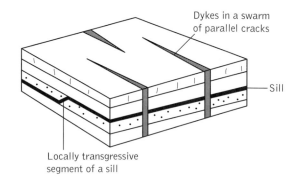

Fig. 1. Block diagram showing vertical dykes cutting horizontal sedimentary strata and a sill that was intruded into the sediments after they were deposited.

thick. Smaller dyke-like intrusions (less than 10 cm thick) are known as igneous veins.

Confusion between the terms 'hypabyssal' and 'plutonic' arises when a large igneous body is formed by the intrusion of a dyke or a sill. For example, the Palisades sill of New Jersey in the USA is over 300 m thick and is unusually large. It shows many features such as chilled margins, cumulates, and pegmatites that, had its lower contact been obscured, would lead to its having been classified as plutonic. Sills can also be of large areal extent. Those of the Karroo, in South Africa, extend over an area of more than 500 000 km².

Dykes, belonging to dyke swarms comprising many parallel and closely spaced dykes commonly propagate along planes of weakness that are fractures formed at right angles to the least pressure and parallel to the regional direction of the greatest horizontal pressure. The orientation of such a regional stress field can be perturbed by the parent igneous intrusion itself and, in extreme cases, radial and concentric dyke swarms can be produced. Dykes may travel large distances under the ground before they break the surface as a fissure eruption. The Cleveland dyke, part of the Skye dyke swarm, extended from the Isle of Skye on the west coast to Scotland, 400 km across mainland Britain to Yorkshire. The surface of the planet Venus also shows linear features that are interpreted as dykes and might express the pattern of stresses on the planet. Sills are less common than dykes, because for magma to occupy a horizontal sheet all the overlying strata have to be lifted up; in contrast, the crust has merely to extend horizontally for a dyke to form. In the north of England, the Whin (hard) sill underlies an area of at least 6000 square kilometres.

JUDITH M. BUNBURY

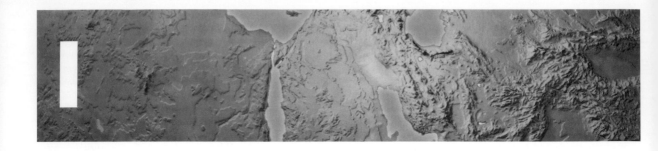

Iapetus Iapetus is the name given to a Pacific-style ocean that existed from the Late Precambrian to the mid-Silurian period (around 600–420 million years ago). It separated several continental masses by a distance of up to 5000 km. To the north was the continent of Laurentia (present-day North America, Greenland, and northern Britain), while to the south lay Gondwanaland (now South America, Africa, most of Asia, Australia, Antarctica and most of Europe) and to the east was Baltica (Scandinavia). The configuration of these former continents is shown in Fig. 1.

Earth scientists are now confident of the existence of Iapetus, although the exact timing of its stages of development is more controversial. Several lines of evidence are used to infer the presence and position of the ocean, such as data from fossils, geophysical studies, and field geology.

Fossil evidence is provided by a phenomenon known as *provinciality*. Large oceans can act as barriers to the inter-

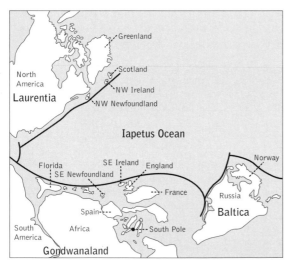

Fig. 1. Idealized plate reconstruction of the Iapetus Ocean and surrounding continents during the Early Ordovician. (After Prigmore, J. K., Butler, A. J., and Woodcock, N. H. (1997) Rifting during separation of Eastern Avalonia from Gondwana: evidence from subsidence analysis. *Geology* **25**, 203–6.)

action of different species of animals living on the continental shelf. For instance, it is difficult for a shallow marine species on one side of an ocean to mix with any similar organisms on the other side, and hence they evolve separately. Thus, fossils found in Cambrian and Ordovician rocks on the northern side of the Iapetus Ocean, as in north-west Newfoundland, are different from those found on the southern side, such as in south-east Newfoundland. These fossils include graptolites, trilobites, and various types of microfossils visible only under the microscope.

Other evidence for Iapetus comes from rocks of similar age that occur in close association with each other but have very different physical and magnetic characters, indicating that they were not originally formed in the same place. For example, in Ballantrae in Scotland there are unusual rocks interpreted as representing oceanic crust in an area where there is no longer an ocean.

Present estimates place the formation of the Iapetus Ocean in the Late Precambrian (between 600 and 550 million years ago). During this time plate-tectonic processes caused the rifting of an old supercontinent, the pieces of which (Laurentia, Gondwanaland, and Baltica) drifted apart as oceanic crust was created at mid-oceanic spreading centres. The ocean continued to expand until mid-Ordovician times.

No ocean lasts for ever, and the continuous processes of plate tectonics that originally formed Iapetus were also responsible for its eventual destruction. During the Ordovician, oceanic crust at the edges of Iapetus began to slide, or subduct, under Laurentia, Gondwanaland, and Baltica, causing the ocean to shrink slowly and the continents to drift back together (Fig. 2). One of the initial consequences of the shrinking of Iapetus was a gradual decline in the provinciality of fossils. Organisms that earlier had been separated by thousands of kilometres of water were able to cross the ever-decreasing distance between the margins of the ocean. A gradual mixing of fossil species is thus seen in progressively younger rocks.

The climactic result of the closure of the Iapetus was the eventual collision of the opposing continents about 420 million years ago in the events known as the Caledonian and Appalachian orogenies. As Laurentia, Baltica, and the northern part of Gondwanaland approached each other, marine sediments and volcanic islands at the margins of these continents

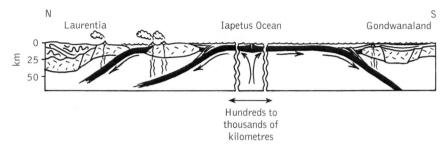

Fig. 2. Idealized cross-section through the Iapetus Ocean and its continental margins during the Early Ordovician. Black shading indicates oceanic crust, stippling indicates crystalline continental crust. (After Watson, J. and Dunning, F. W. (1979) Basement-cover relations in the British Caledonides. In Harris, A. L., Holland, C. H., and Leake, B. E. (eds) *The Caledonides of the British Isles; reviewed.* Geological Society Special Publication No. 8, pp. 67–91.

began to be pushed together, causing them to be deformed by folding and faulting and to become altered (metamorphosed) by the effects of pressure and heat. Thus, the continents slowly ploughed into each other to form mountain belts of deformed, metamorphosed Lower Palaeozoic rocks, within which is a record of the Iapetus Ocean. These mountain belts are known as the Caledonides and despite their great age, are still expressed today as, for example, the Appalachians in North America and the Highlands of Scotland. The mountain belt was later broken up by the rifting events that led to the formation of the North Sea and the North Atlantic ocean.

The join, or suture, between the old continents can today be located with the aid of geophysical techniques. Seismic reflection profiles, which provide an image of the structure of the crust by deep echo soundings, indicate that the suture line between Laurentia and the northern part of Gondwanaland stretches from the border between England and Scotland, through the centre of Ireland, across Newfoundland and down the east coast of the USA. If today you travel from the south-eastern to the north-western coast of Ireland, it is worth remembering that to have made this journey 500 million years ago would have required a 5000 km sea crossing across the Iapetus Ocean.   DAVE BURNETT and DAVE QUIRK

ice-age aridity Observation of the current and past extent of aeolian sand dunes (see Fig. 1) indicates that during the past they have expanded considerably. The current distribution of active sand seas is for the most part restricted to geographical areas close to 30° north and south of the Equator (in the vicinity of the descending limbs of Hadley circulation cells), in the inland areas of continents, and in regions influenced by cold ocean currents. These areas are characterized by precipitation levels which do not exceed 150 mm per annum. In the past they were larger, occupying over 50 per cent of the continental land area between 30° N and 30° S. The timing of desert expansion and its relation to global climate cycles has been a subject of great controversy.

Early theories generated late in the nineteenth century by the American geomorphologist G. K. Gilbert correlated cold (glacial) periods with times of pluvial (wet) conditions in arid regions. He studied former lake levels and their relationship to glacial moraines in the Great Salt Lake and Wasatch Mountains area of the south-western United States and concluded that high lake levels correlated with maximum advances of mountain glaciers.

In the mid-1970s this theory was brought into question by three separate lines of evidence. First, radiocarbon dating of some desert and desert-associated sediments suggested that during glacial times the sand dunes had been active. The southern extent of the Sahara, for example, was displaced southwards by some 500 km. Secondly, evidence from lakes in tropical and North Africa indicated that they attained their lowest levels during the Last Glacial Maximum, about 18 000 years ago. Thirdly, dust and aeolian sand input into ocean cores off the coasts adjacent to large deserts were at a maximum during glacial times. Maxima of dust deposition during glacial periods have also subsequently been recognized in polar ice cores. A global correlation between high-latitude glacial expansion and low-mid-latitude aridity was thus established.

It is now realized that cold dry conditions experienced in low-latitude desert regions during glacial times are mainly related to a weakening of the African and Asian monsoonal rainfall regime. The monsoon rains are driven primarily by a strong summer temperature gradient between the warm continental areas and the colder adjacent deserts. Weakening of the land–sea summer temperature differential reduces the intensity of monsoon rains and thus leads to the expansion of deserts. This trend may be further enhanced if the wind-driven upwelling of relatively cold waters is reduced in adjacent oceans. The main influence on summer temperatures is the solar insolation input, which is controlled over glacial–interglacial timescales by the Milankovich or orbital cycles.

Additional evidence of ice-age aridity has been obtained from the analysis of extensive loess sequences in Europe and

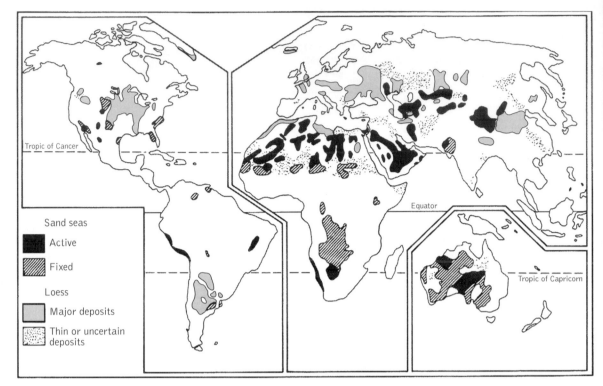

Fig. 1. Present and former extent of major desert areas. (After Thomas, D. S. G. (ed.)(1989) *Arid zone geomorphology: process, form and change in drylands.* John Wiley and Sons, Chichester.)

Asia (Fig. 1). Loess is a fine-grained aeolian sediment, typically liberated by the grinding action of glaciers, which may be transported great distances and indicates both windiness and aridity. Periods of loess deposition are found to correlate with glacial periods when loess profiles are compared to palaeo-climatic evidence from deep-sea cores.

An additional factor which works to enhance ice-age aridity relates to the lowering of sea levels associated with glacial periods. Lower sea levels expose in many coastal areas large expanses of continental shelf (in places over 200 km wide), which result in considerably larger glacial continental areas. The effect of this continentality is that any moisture-laiden air masses travelling inland are likely to have lost much of their moisture by the time they have reached inland continental areas. Pole-to-Equator (meridional) temperature gradients are also at a maximum during ice ages, and this results in the intensitification of trade-wind circulation. Intensified trade winds encourage evaporation.

Over shorter durations, changes in ocean thermohaline circulation and sea-surface temperatures may also have a significant influence, as has been most clearly indicated by a correlation between North Atlantic and Pacific sea-surface temperature anomalies and drought conditions in the Sahel.

Anomalies in the general patterns outlined above do occur. An especially significant example of this is the lake basins examined by G. K. Gilbert in the Great Salt Lakes region. Lakes Bonneville and Lahontan were in this area, reaching a maximum level at about the time of the last glacial maximum. The regional evidence for pluvial conditions at that time is explained by the occurrence of increased rains associated with a south displaced limb of the westerly polar jet, and subsequent increases in the occurrence of winter cyclonic storm activity. The jet was displaced for a period of time when the Laurentide ice sheet reached its maximum late Pleistocene extent. Retreat of the ice sheet during late glacial times has resulted in the return and reconnection of the jet stream limbs in more northern latitudes.

The most recent period of Late Quaternary maximum wetness in arid and semi-arid regions is broadly coincident with the maximum of Northern Hemisphere isolation which occurred approximately 9000 years ago. This was a period of intensive monsoonal rainfall.   STEPHEN STOKES

Further reading

Dawson, A. G. (1992) *Ice Age Earth: Late Quaternary geology and climate.* Routledge, London.

ice-age theories The form of recent climatic change is becoming increasingly clear from a wide range of geological evidence. Three times in the last 800 million years the Earth has been subject to glaciation; during the Late Precambrian, the Permian and Carboniferous, and, most recently, during the Quaternary. Our understanding of the Quaternary is the most detailed because the evidence is more abundant, the range of geological techniques available is larger, and the potential for resolution of events in time and space is greater. Analysis of deep-sea cores has shown that within the last 2 million years or so of the Quaternary ice age there were over 60 isotopic stages reflecting major changes in global climate. Any attempt at explaining ice ages must therefore address two problems: first, why do ice ages start and, secondly, why were there such major alterations of climate during the Quaternary?

Taking the second question first, it seems to be accepted that Milankovich cycles are the main forcing mechanism behind the alterations of climate during the Quaternary on timescales of tens of thousands of years. In addition, recent work on ice cores from Greenland has suggested that other mechanisms, such as the turning on and off of the oceanic thermohaline circulation, may also have a profound effect on climatic change, especially on short timescales. But Milankovich cycles have operated throughout Earth history, much of which has been ice-free, so they cannot be invoked as a mechanism to initiate ice ages.

In order to answer the question 'How do ice ages start?' it is first necessary to review the history of the transition from an ice-free to an ice-bound world. There is evidence from oxygen isotopes at Deep Sea Drilling Program sites that ice began to build up in Antarctica about 36 million years (Ma) ago during the Eocene, following some 20 Ma of cooling (Fig. 1a). Further marked decreases in temperature during the Middle Miocene at 15 Ma probably reflect increased ice growth in Antarctica, while the decrease in the late Pliocene is probably due to ice build-up on the continents of the northern hemisphere.

Dealing with the Late Pliocene cooling in more detail, recent oxygen isotope analysis of deep-sea cores suggests that there was a period of enhanced ice volume between 4.6 and 4.3 Ma, a long-term warming trend from 4.1 to 3.7 Ma, followed by a distinct cooling trend that was initiated at 3.5 Ma and which progressed through to the initiation of the large-scale northern hemisphere glaciation, known as the Quaternary ice age, after 2.85 Ma.

Many theories have been put forward to explain the initiation of the Quaternary ice ages. These include: plate motions (continental drift), uplift of the Earth's crust, changes in atmospheric circulation, variations in the gaseous composition of the atmosphere, volcanic activity, and changes in oceanic circulation.

The theory that displacements of continents over the poles might explain the initiation of glaciation is an accepted one, especially for glaciation during the Permian and Carboniferous. However, over the past 100 Ma there has been negligible continental displacement towards the poles, and this cannot explain the observed cooling since the Early Eocene, some 55 Ma ago, which eventually led to the Quaternary ice ages.

A second theory suggests that land uplift, notably of the Tibetan plateau, may have been sufficient to initiate glaciation. The Tibetan plateau is one of the most imposing large-scale topographic features on Earth. It has an average elevation of over 5 km and an area half the size of the United States. It was formed by the collision of the Indo-Australian plate with the Asian plate which began during the Middle Eocene (52–44 Ma ago) and which continues today.

The Tibetan uplift has had two main effects on atmospheric circulation. First, the modern Tibetan plateau is so high and broad that it affects atmospheric circulation on a global scale. High mountain ranges, especially those with a large north–south frontage, act as barriers to east–west circulation and force the atmospheric circulation into poleward (meridional) diversions. The large ranges of the Himalayas and also the Rockies of North America have the effect of diverting upper atmospheric circulation in a meridional sense and are probably the cause of large-scale standing Rossby waves in the upper atmosphere. Rossby waves are responsible, amongst other things, for bringing moist air from the warm south-west Atlantic towards the colder continents adjacent to the north-east Atlantic, an important source of moisture for the build-up of ice sheets.

Secondly, the Tibetan plateau is a major factor driving the Indian monsoon circulation. Because of the height of the plateau, the Sun's rays have less atmosphere to pass through. In summer the plateau heats up quickly, air rises and causes intense low pressure at the surface which draws in the monsoon winds from the south and east.

However, work by climatic modellers suggests that the development of a Tibetan plateau does not have any appreciable effect on high-latitude summer temperatures, which exert an effective control on ice sheets by melting the previous winter's snow. This suggests, therefore, that the creation of high mountain ranges in Tibet and North America is not enough by itself to initiate the growth of large terrestrial ice sheets. In 1899, T. C. Chamberlin put forward a theory linking land uplift to changes in the gaseous composition of the atmosphere that would be sufficient to initiate global cooling.

As we know today to our cost, carbon dioxide (CO_2) levels in the atmosphere influence global temperature. On short timescales the main sources of CO_2 are from the burning of organic matter, from volcanic eruptions, and from changes in the rate at which CO_2 is absorbed and stored in the oceans as carbonate rock. On timescales longer than a million years, however, the level of CO_2 in the atmosphere is controlled by the balance between the rate of input of CO_2 (which largely comes from the Earth's interior and is released through volcanic activity) and the rate of loss of CO_2 (which is largely achieved by the chemical weathering of silicate rocks which absorb CO_2 from the atmosphere as part of the chemical reaction).

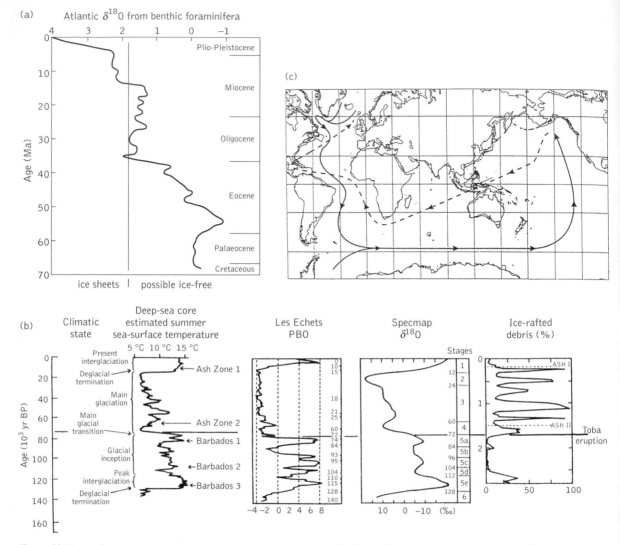

Fig. 1 (a) Climatic change since the Eocene. Oxygen isotope measurements for the Atlantic ocean during the past 70 million years. The data are compiled from ten Deep Sea Drilling Program sites. (From Raymo, M. E. and Ruddiman, W. F. (1992) Tectonic forcing of late Cenozoic climate. *Nature*, **359**, 117–22, Fig. 1.) (b) Climatic indicators for the last 140 000 years. Horizontal solid bold line shows age of Toba eruption. PBO, Palaeobioclimate Operator, a measure of the 'best possible' climate profile. Higher PBO values indicate a warmer climate (From Rampino, M. R. and Self, S. (1993) Climate-volcanism feedback and the Toba eruption of 74 000 years ago. *Quaternary Research* **40**, 269–80) (c) The global thermohaline 'conveyor belt'. Deep water is shown by solid lines, very general surface transport by dashed line. (From Hay, W. W. (1993) The role of polar deep water formation in global climate change. *Annual Review of Earth and Planetary Sciences*, **21**, 227–54, Fig. 7.)

Maureen Raymo of the Massachusetts Institute of Technology and Bill Ruddiman of the University of Virginia suggest that Tibetan uplift both raised and exposed silicate rocks to chemical weathering, thereby abstracting CO_2 from the atmosphere. At the same time the enhanced monsoon climates produced by the Tibetan plateau would flush away the weathering products down the steep slopes into the rivers, and continually expose a fresh supply of silicate minerals for renewed chemical weathering.

One requirement of this theory, however, is for there to be a mechanism for adding to the levels of CO_2—for if the process were to run unchecked, the atmosphere would lose all its CO_2 within a few million years. Researchers suggest that this can be achieved by, first, a decrease in the amount of carbon being buried, resulting in more CO_2 being available, and secondly, by a decrease in the rate of carbonate precipitation on the ocean floor (which would abstract CO_2) relative to the rate of silicate deposition (which does not abstract CO_2). This hypothesis has become increasingly popular in recent years and is seen by many as providing a convincing mechanism for inducing the downward climatic spiral which began in the Miocene, continued through to ice build-up on Antarctica, and ended with the Quaternary ice age.

Volcanic activity has also been promoted as a possible cause of glaciation. The geological record suggests that Quaternary explosive volcanism was about four times higher on average than that during the Miocene and Pliocene. Explosive volcanism can eject large amounts of dust and debris into the stratosphere which can have a cooling effect. However, the main climatic effects are probably caused by sulphur, ejected into the stratosphere, which rapidly converts to sulphuric acid, an aerosol which cools the troposphere by back-scattering incoming solar radiation.

Recent studies suggest that dust does not remain in the atmosphere for long after a volcanic event and that volcanic aerosols do not have a long-lasting effect on climate. It is thought, for instance, that the 1815 eruption of Tambora led to an average fall of 0.7 °C in the northern hemisphere the following year but that temperature soon recovered. Similar short-term effects have been noted from the 1980 eruption of Mount St Helens. However, the events noted above are comparatively small scale. By contrast, the eruption of Toba in Sumatra some 73 500 years ago was the largest known explosive event in the Quaternary. The eruption seems to have coincided with a period of rapid global ice accumulation at the start of the last major cold period (Fig. 1b). It is suggested that the eruption could have led to a cooling of about 12 °C in average summer temperatures for 2–3 years in the areas of growth of the Laurentide ice sheet which built up over northern North America. It is not impossible, therefore, that a very large volcanic eruption in conjunction with other climatic factors might be capable of initiating an ice age.

Changes in ocean circulation have also been put forward as a possible mechanism for initiating an ice age. The closure of the gap between the Atlantic and Pacific oceans in the Central American region which took place gradually between 13 and 1.8 Ma is seen as a significant event, mostly for its effects on oceanic circulation in the North Atlantic.

One major determinant of climatic change during the Quaternary both on long and short timescales has been the strength of the thermohaline circulation (Fig. 1c). In the North Atlantic cold, dense saline water is formed which flows southwards at depth, eventually reaching all the world's oceans. A return flow of warmer water is induced which greatly influences temperatures in the North Atlantic and adjoining continents. One major requirement of the thermohaline circulation is the formation of dense saline water. When the Panama gap was open and the Pacific and Atlantic were joined, it is suggested that more water flowed from the Pacific to the Atlantic than vice versa, limiting the build-up of salinity in the North Atlantic. Once the gap closed, the Gulf Stream flow increased, producing a region of high evaporation off eastern North America and an increase of salinity in the North Atlantic.

There is also evidence that during the Miocene a sill surrounding the Arctic Ocean, known as the Greenland–Scotland Ridge, subsided, allowing more cold polar water to escape into the North Atlantic. As the salinity of the North Atlantic grew and as outflow of cold polar water increased, so the thermohaline circulation increased in vigour, providing the mild winter temperatures and large amounts of moisture to the North Atlantic, which are prerequisites to the build-up of the large continental ice caps on the adjacent cold continents.

It therefore seems likely that the cause of the Quaternary ice age is a very complex question to answer. The climatic cooling induced by Tibetan uplift and consequent decrease in atmospheric CO_2 may have started the climatic downturn in the Miocene, but the closing of the Central American isthmus seems to have been the trigger for the build-up of ice in the northern hemisphere. This then provides the opportunity for Milankovich cycles to become the pacemaker of the ice ages, causing alternating glacials and interglacials.   B. A. HAGGART

Further reading

Andersen, B. G. and Borns, H. W. (1994) *The ice age world.* Scandinavian University Press.

Imbrie, J. and Imbrie, K. P. (1979) *Ice ages: solving the mystery.* Erslow Publishers, Short Hills, New Jersey.

ice ages Repeated fluctuations between warm (interglacial) and cold (ice age or glacial) climates have characterized the palaeoclimate of the past 2.5 million years (the Quaternary). Other geological periods when ice ages occurred include the Permo-Carboniferous (250 million years ago), the Cambrian (650 million years ago), and two occasions during the Proterozoic (900 and 1250 million years ago). The Quaternary glaciations were initially thought to have occurred only four times, interrupted by widespread interglacial conditions. However, detailed chemical and other evidence from deep-sea cores now indicates that more than twenty such climatic shifts have occurred during the past 2.5 million years.

Evidence of former extensive glaciations is recorded in high-latitude ice cores and in ocean sediments by evidence for dramatic cooling and increases in the input of continental dust. At mid-latitudes a variety of sedimentary, ecological, and morphological evidence indicates the occurrence of ice in terrain that is currently non-glaciated. At low latitudes evidence for ice-age environments is scarce, but is generally suggestive of extensive desiccation and desert expansion, caused by a combination of reduced evaporation from cooler ocean surfaces and stronger trade winds (see *ice-age aridity*).

Ice sheets appear to grow and decay in response to broad-scale changes in solar insolation at high northern latitudes. A reduction in solar heating in these latitudes can result in year-round snow cover which would reinforce cooling by reducing the surface albedo; in doing so it would establish a positive (cooling) feedback. The retreat of ice masses appears more complex and may be caused by insolation-related, positive (warming) feedbacks, including moisture starvation, sheet instability induced by a rise in sea level, and changes in the abundance of greenhouse gases. Expansion and retreat of ice sheets appears to have been globally synchronous, despite the non-synchronous distribution of insolation. This is thought to be related to the significant role of intermediate and deep ocean circulation in redistributing heat. Abrupt shifts in climate during glacial–interglacial transitions have been documented. They seem to occur more rapidly than would be expected from insolation-driven forcing and they are assumed to be related to changing ocean–atmosphere and ocean–land interactions.

The last glacial period culminated approximately 18 000 years ago. In addition to dramatically changing a vast range of continental environments, it caused the sea level to be lowered by approximately 120 m. During this time it is estimated that the average annual temperature was 5–10 degrees C cooler and that ice covered nearly one-third of the land area. Large portions of Western Europe and North America were covered by ice masses several kilometres in thickness. These ice masses, termed the Fennoscandian (Europe) and Laurentide and Cordilleran (North America) ice sheets, caused dramatic re-organizations in mid-latitude Rossby Wave configuration.   STEPHEN STOKES

Further reading

Dawson, A. G. (1992) *Ice age Earth: Late Quaternary geology and climate*. Routledge, London.

Lowe, J. J. and Walker, M. J. C. (1984) *Reconstructing Quaternary environments*. Longman, Harlow.

ice sheets and climate Ice sheets and ice caps are dome-shaped ice masses overlying relatively flat, but not necessarily low-lying, areas of land. The dome is typically convex in shape and ice flows radially from the centre outwards. The difference between ice sheets and ice caps is one of scale; the latter being less than 50 000 km^2 in area. As a result, although a number of ice caps are recognized (e.g. in Patagonia, Iceland, and the Canadian Arctic), only two ice sheets exist at the moment. These are the Greenland ice sheet (1 726 400 km^2 in area) and the Antarctic ice sheet (12 535 000 km^2), both of which are at high latitudes (between 60 and 82° N and 65 and 90° S respectively).

Ice sheets display a number of climatic characteristics. First, they are characterized by extreme cold. For example, the mean annual temperature in Antarctica varies from about −10 °C in coastal locations to almost −60 °C on the summit of the ice

sheet at 4200 m altitude (see Fig. 1). Winter temperatures are lower; down to −70 °C or below in the interior. This makes Antarctica the coldest place on Earth with a recorded minimum of −88.3 °C at Vostok on 24 August 1960. Greenland is less cold, but mean annual temperatures approaching −20 °C are not uncommon and mean winter temperatures on the north coast fall to −45 °C.

Secondly, precipitation levels are generally low. The centre of the East Antarctic ice sheet receives less than 50 mm per year, as do ice-free enclaves such as Victoria Land. Precipitation increases towards the coast, where totals of 200–600 mm per year are usual. Most precipitation falls during the winter months. The Greenland ice sheet shows similar precipitation characteristics: the centre and north are driest (around 140 mm per year) and the south coast receives the highest totals of about 1000 mm, although levels peak during the summer.

Thirdly, winds are an important feature of ice-sheet climates. Extreme cold caused by radiation cooling causes the development of a pronounced temperature inversion layer some 100–200 m thick over the ice. This, in conjunction with the convex profile of the surface, forces air to flow downhill towards the coasts to form katabatic winds, which may be diverted slightly by Coriolis effects. These winds may be almost constant in their strength and direction. Parts of East Antarctica are probably the windiest places on Earth as a result of these katabatic winds, with recorded mean annual wind speeds of 19.4 metres per second. Wind speeds are further enhanced by the lack of friction over the ice surface.

There are a number of reasons for these distinct climatic characteristics of ice sheets. The intense cold can be partly explained by the low level of net solar radiation received at high latitudes owing to the low angle of the Sun (although in midsummer the poles receive more solar radiation than anywhere else on Earth). In addition, the albedo (reflectivity) of the surface is very high and up to 90 per cent of the solar radiation may be re-radiated and unable to heat the surface. This process is helped by the lack of natural and anthropogenic pollutants in the atmosphere which would otherwise trap outgoing long-wave (heat) radiation. The high altitudes of the ice sheets (more than half of Antarctica is above 2000 m) are also an important factor: the air is cooled by adiabatic processes at the rate of about 1 degree C per 100 m altitude.

Blocking, deep high-pressure air masses (anticyclones) are a characteristic feature of the ice sheets, especially in winter, and the clear skies that accompany them explain to a large extent the low levels of precipitation. On the fringes of the ice sheets (for example in the Sub-Antarctic islands and on South Georgia) low-pressure weather systems (cyclones) travelling along the boundary of the atmospheric polar front bring wet, cloudy, and stormy conditions. Maximum precipitation values in Antarctica are found a short way inland at an altitude of about 1600 m, where orographic precipitation is greatest. Inland, precipitation is very much lower.

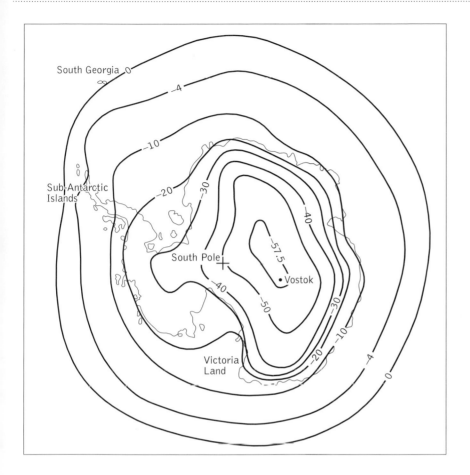

Fig. 1. Mean annual surface temperatures in Antarctica and other locations mentioned in the text. (Modified from John and Sugden (1975) Coastal geomorphology of high latitudes. *Progress in Geography*, **7**, 53–132.)

It can be seen therefore that a number of factors combine to perpetuate cold conditions over the ice sheets. The annual deficit of radiant energy has to be balanced by additions of heat from lower latitudes; otherwise the ice sheets would become progressively colder. Air masses from lower, warmer, latitudes transport sensible heat to the ice sheets, while latent heat is released by the conversion of water vapour to precipitation. Warm ocean currents, especially the North Atlantic Drift from the Gulf of Mexico, also bring heat to the seas surrounding the Greenland ice sheet but are of less importance in warming the Antarctic because currents circumnavigate the continent.

STEPHAN HARRISON

ice sheets and landscape evolution The surface of the Earth has been profoundly modified by the action of ice sheets. On several occasions during the Quaternary period of the past 2 million years, ice sheets and glaciers covered approximately one-third of the Earth's surface, and parts of every continent bear the imprint of glaciation. Ice sheets modify the landscape in two principal ways. First, they act as powerful agents of erosion, scouring out and removing large volumes of rock and sediment; and second, this eroded material is laid down by glacial deposition, either beneath the ice sheet, along the ice edge, or beyond the ice margins. Ice sheets therefore erode new landscapes, or bury old land surfaces beneath new glacial landforms. In some circumstances, ice sheets can also protect the underlying land surface from erosion or deposition, preserving ancient landscapes beneath their frozen covering.

The most important control on erosion and deposition beneath an ice sheet is the temperature at the base, in particular whether the ice is at or below the local melting point (see *glaciers and glaciology*). If the ice is below the melting point (which may be below 0 °C if the ice is under high pressure), it will be frozen to the underlying land surface and the base of the ice sheet will be immobile. (The high adhesive strength of cold ice can easily be appreciated while clearing a

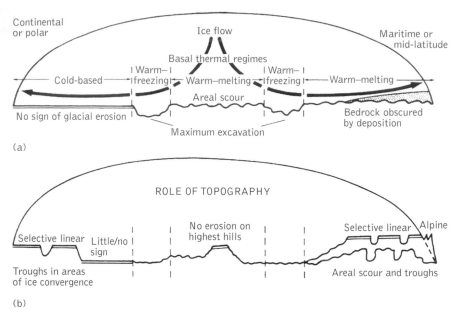

(a)

(b)

Fig. 1. A model of erosion by ice sheets. The left-hand side of the diagram represents polar or continental conditions; the right-hand side represents mid-latitude or maritime conditions. (a) Thermal regimes and idealized erosional effects; (b) schematic effect of topography, showing how uplands experience minimal erosion or selective linear erosion. Massifs protruding above the surface of the ice sheet near the margin are sculpted into alpine landscapes. In cold-based zones, troughs will form in areas of flow convergence. (Redrawn from Chorley, R. J., Schumm, S. A., and Sugden, D. E. (1984) *Geomorphology*. Methuen, London.)

car windscreen on a frosty morning.) In contrast, if the basal ice is at the melting point the ice sheet can slide over a thin film of meltwater at the bed, and significant erosion and deposition can occur as rock fragments are dragged along beneath the ice. Warm-based conditions tend to occur below thick, rapidly moving ice, or where the surface climate is relatively mild, whereas cold-based conditions are encouraged

beneath thin, sluggish ice in extremely cold environments. Warm- and cold-based conditions can occur simultaneously below different parts of the same ice sheet, depending on variations in ice thickness, flow rates, and climate. Glaciated landscapes can thus provide important insights into patterns of freezing and melting conditions at the base of former ice sheets.

Fig. 2. Glacial trough, Gjende, central Norway. During several stages of the Quaternary a major stream occupied this trough. Ice flow was towards the viewer.

Landscapes of glacial erosion

Five basic landscapes of glacial erosion were identified by Professor David Sugden of the University of Edinburgh. Models of erosion by ice sheets are illustrated in figure 1. First, *landscapes of areal scouring* are extensive tracts of subglacially eroded bedrock, consisting of hundreds of overdeepened rock basins and intervening rock knobs. In Scotland, such terrain is known as *knock and lochain* topography, from the Gaelic words *cnoc*, meaning 'knoll' and *lochain*, meaning 'small lake'. The widespread evidence for abrasion and plucking in such landscapes indicates that they develop below extensive areas of sliding, warm-based ice. Spectacular examples of areally scoured landscapes occur in north-west Scotland, especially in Sutherland and the Isle of Lewis.

Second, in *landscapes of selective linear erosion*, glacial erosion is confined to deep, linear troughs incised into unmodified plateau surfaces. Such landscapes develop where warm-based ice streams occur between areas of slowly moving or cold-based ice. This situation is common in parts of Greenland and Antarctica today, where ice streams drain the greater part of the ice that accumulates on the great inland snowfields (Fig 2). Third, *landscapes of little or no glacial erosion* consist of essentially unmodified preglacial landscapes that have survived one or more periods of glacial occupancy. Sometimes the only indication that such landscapes have been glaciated at all is the presence of rare erratic boulders, or evidence for glaciation in surrounding regions. Landscapes of little or no glacial erosion persist beneath cold-based ice where sliding is non-existent or insignificant. Examples include Buchan in north-east Scotland and parts of the Canadian Arctic islands, where Tertiary landforms and sediments have been preserved despite repeated occupancy by glacier ice during the Quaternary.

The fourth and fifth landscape types are *alpine landscapes*, consisting of branching networks of troughs separated by ridges; and *cirque landscapes*, in which separate cirques (or corries) are set in an upland massif. These landscapes develop in mountainous terrain, where ice cover is insufficient to bury summits and ridges. Mature alpine landscapes exhibit many of the 'classic' features of glaciation, including troughs, hanging valleys, truncated spurs, and narrow arêtes rising to narrow rock peaks. These are most spectacularly developed in high mountain environments, such as the Himalayas, Rockies, Andes, and the Alps of New Zealand and Europe (Fig. 3). Cirque landscapes reflect more restricted glacial conditions than alpine landscapes.

Landscapes of glacial deposition

Landscapes of glacial deposition tend to develop in the outer parts of ice sheets, where melting ice releases its sediment load or deforming sediment accumulates at the base of the ice. Depositional landscapes also predominate in areas underlain by weak sedimentary rocks, which are easily reworked by glacial action. Examples include the spectacular drumlin fields of lowland Britain, and the ice-pushed hills of northern Europe and the northern Prairies of North America.

Fig. 3. Glaciated alpine landscape, central Andes, Chile, showing characteristic features.

Continent-wide patterns of glacial landscapes reflect long-term influences of pre-existing topography and geology, glacier and ice-sheet morphology, basal temperatures, and ice-flow rates, and systematic variations in these factors can explain many major landscape patterns, particularly in the mid-latitudes. A glance at the map of Canada, for example, reveals a broad arc of large lakes, from the Great Bear Lake to Lake Ontario, which is thought to reflect the distribution of former warm-based zones below the Laurentide Ice Sheet. In the west, the fiordlands and alpine terrain of British Columbia attest to vigorous glaciation of high-relief mountains in a snowy, maritime climate. In the north-east, the plateaux and troughs of Labrador and Baffin Island record the selective action of ice streams in a colder, more arid environment. In contrast, depositional landscapes predominate in southern Alberta and Saskatchewan, where soft sedimentary rocks crop out near the southern limit of the ice sheet.

Glacial landscapes develop cumulatively over multiple glacial cycles, and are therefore more likely to relate to 'average' glacial coverage during the Quaternary than to 'instantaneous' ice configurations at glacier maxima. This average ice cover, which lies about halfway between present-day and ice-age maximum conditions, represents the principal control on much Quaternary landscape evolution, not only beneath the ice sheets themselves, but also the development of erosional shore platforms, tropical atolls (because of the correlation between ice cover and global sea level), and long-term sediment yields to the ocean basins.

DOUGLAS I. BENN

Further reading

Benn, D. I. and Evans, D. J. A. (1998) *Glaciers and glaciation*, Chapter 9. Arnold, London.

Bennett, M. R. and Glasser, N. F. (1966) *Glacial geology: ice sheets and landforms.* John Wiley and Sons, Chichester.

ichnology *see* TRACE FOSSILS AND ICHNOLOGY

ichthyosaurs
Ichthyosaurs (fish-reptiles) were among the most specialized of the marine reptiles and show very well the changes that had to be made in animals that became secondarily aquatic. The reptile lungs were retained, requiring the animals to breathe at the surface, as whales do now; the legs and feet were transformed into paddles; and tail fins similar to those of fish were developed to propel the animals through the water. As the reproductive strategy of laying eggs in water had already been abandoned by reptiles, those aquatic reptiles that could not come out on land to lay eggs developed live birth.

Ichthyosaurs first appeared in the Early Triassic as fully fledged aquatic animals with no obvious ancestral forms. It is, however, clear that their ancestry must lie among the diapsids, the major group of reptiles that also includes the crocodiles, dinosaurs, lizards, and snakes. These most specialized of aquatic reptiles were essentially like large fish in their way of life and show similar specializations for fast swimming. The typical genus *Ichthyosaurus* was a fish-like animal up to 3 m in length with a large elongated head and enormous eyes. The jaws were long and narrow and lined with small sharp teeth, and the nostrils were positioned on top of the head above the eyes. The head merged with the body to produce a tapering streamlined shape, indicating that this animal was a powerful swimmer. Well-preserved fossils from marine black shales sometimes shows the body outline of these animals, and from this material it has been possible to reconstruct the large dorsal fin and the tail, which was a large lunate structure like that of a tuna. The four limbs had been altered to paddles by a shortening of the limb bones and a great increase in the number of finger and toe bones, which formed flattened hexagonal elements pressed closely to each other. It is clear that the animal moved by lateral undulations of the body, as a fish does, and used its paddles to steer with. The ichthyosaurs were active hunters, tracking their prey by sight, and stomach contents show that they fed on squids and probably also on marine molluscs such as the ammonites. Some specimens from black shales in Germany have unborn embryos within the body cavities of adults, showing that these animals bore living young.

The largest ichthyosaurs occur in the Triassic; *Shonisaurus* from Nevada reaches a length of almost 15 m. Later ichthyosaurs were smaller but were very widespread, occurring in all the Mesozoic oceans. Two main groups of ichthyosaurs can be distinguished; the latipinnate (wide paddle) and longipinnate (long paddle) forms. The latipinnate forms were abundant during the Jurassic but became extinct before the end of that period; longipinnate forms lasted into the Late Cretaceous. Despite their specializations for a marine existence, this entire group became extinct at the end of the Creataceous, as did the other main group of marine reptiles, the plesiosaurs, and the dinosaurs, as well as many invertebrate groups.

DAVID K. ELLIOTT

igneous melt compositions (silicate and carbonate)
The term 'melt' is widely used within Earth sciences to specify a liquid produced by a melting process within the Earth and is in many respects equivalent to the term 'magma' which may consolidate to form an igneous rock. Magma erupted at the Earth's surface is known as lava. The study of magmas and melts, their origin, composition, and physical properties, lies at the heart of igneous petrology and petrogenesis, although the term 'melt' is also used to describe liquids produced by melting processes that occur as a result of extreme metamorphism.

Except for rare cases where quenched melts are preserved as glasses in mantle xenoliths, the composition of melts produced within the Earth cannot be directly ascertained and must be inferred by several means. First, we can study the consolidated magmas and lavas exposed at the Earth's surface. These are overwhelmingly silicate-based, with silica (SiO_2) contents ranging from 35 to 75 per cent by weight, alumina (Al_2O_3) from 12 to 18 per cent, Na_2O from 2.5 to 8.0 per cent, K_2O from 0.5 to 10 per cent, iron oxides (FeO and Fe_2O_3) in the range 2 to 14 per cent, CaO from 0.4 to 13 per cent and MgO from 0.1 to 15 per cent in the vast majority of silicate igneous rocks. Next in abundance, although rare in comparison to silicate rocks, are carbonate-based magmas and lavas known as carbonatites, which have little (less than 15 per cent by weight) or no silica content. There is only one presently active carbonate volcano on Earth, Oldoinyo Lengai in Tanzania, East Africa. This one occurrence can be contrasted with the 500–600 active silicate lava volcanoes around the world. The composition of the lava erupting from Oldoinyo Lengai differs considerably from geologically ancient carbonatites because it has, approximately, an Na_2O content of 30 per cent; CaO, 12 per cent; K_2O, 10 per cent; with 27 per cent CO_2 and significant amounts (greater than 4 per cent) of chlorine, fluorine, and sulphur. This composition contrasts sharply with ancient carbonatites (about 500 in the world), which have CaO contents up to 55 per cent, MgO up to 20 per cent, and K_2O and Na_2O contents less than 1 per cent. It is interesting to note that it is since the 1960s that carbonatites found at the Earth's surface were accepted as being magmatic. The observation of carbonate lava erupting from Oldoinyo Lengai in Tanzania together with the experimental demonstration that a melt rich in calcium carbonate ($CaCO_3$) could exist at the surface, silenced critics of the magmatic origin who had proposed that carbonatites were products of hydrothermal replacement at the surface. Other types of magma, such as sulphide and iron oxide liquids, are extremely rare indeed and are not considered here. It should, however, be pointed out that early in the Earth's history a melt composed mainly of iron with some nickel separated from the bulk silicate Earth, giving rise to the core. Although both silicate and carbonate igneous rocks found at the surface give us an indication of the range of melt compositions that may be produced within the Earth, many of these rocks have undergone significant modification through processes of igneous *differentiation* and contamination, and thus are not direct samples of melt produced within the Earth's crust and mantle which are termed primary melts.

A second way by which we can constrain melt compositions is through knowledge of the chemical composition and mineralogy of the Earth's upper mantle, because this is thought to be the source region for the majority of magmas. We know from studies of xenoliths (fragments of the upper mantle brought to the surface in kimberlites and alkali basalts), and also from constraints imposed by seismology and chemical data from extraterrestrial bodies (such as meteorites) that the Earth's upper mantle is mainly composed of FeO, MgO, and SiO_2 (approximately 90 per cent by weight) with some 5 to 10 per cent of CaO, Na_2O, and Al_2O_3, the remainder being minor and trace elements plus volatiles. In mantle xenoliths the two most common minerals are olivine (a Mg–Fe silicate) and orthopyroxene (a Mg–Fe silicate), followed by clinopyroxene (a Ca–Mg silicate) and an aluminous-bearing phase (spinel or garnet). Thus, in considering silicate melt compositions in the upper mantle it would seem advisable to begin by investigating the melting products of an assemblage consisting of the main mantle minerals: olivine, orthopyroxene, clinopyroxene, and spinel or garnet. This approach entails subjecting the assemblage to pressures and temperatures equivalent to those within the upper mantle. Such laboratory experiments have been carried out for over 30 years and have revealed that a broad spectrum of silicate melts can be generated by partial melting of normal mantle materials. For example in the simple mantle system CaO–MgO– Al_2O_3–SiO_2 (CMAS) at 1 atmosphere (Earth's surface), partial melting of this simplified mantle produces a basaltic melt in equilibrium with olivine, orthopyroxene, clinopyroxene, and plagioclase feldspar (Ca–Na–Al silicate). At 15 kbar (equivalent to about 45 km depth) the melt has much less silica and is in equilibrium with olivine, orthopyroxene, clinopyroxene, and spinel (Mg–Al oxide). At even higher pressures of 30 kbar (equivalent to about 90 km depth) the melt, although still basaltic in composition, is even poorer in silica and is in equilibrium with olivine, orthopyroxene, clinopyroxene, and garnet (Mg–Al silicate). These experiments demonstrate quite clearly that with increasing pressure the silica content of a basaltic-type melt decreases, but

that the MgO content of the melt increases. Such experiments on simple analogous mantle systems are extremely important for establishing primary melting relationships among the main mantle minerals, and lead us to ask vital questions concerning melt generation within the upper mantle. For example, in the natural mantle system can the range in basaltic magma compositions seen at the Earth's surface be generated by the partial melting of an mantle assemblage at a variety of pressures in a similar manner to that described above for the CMAS system?

One must, however, be cautious in directly applying experimental results on simple systems to the variety of igneous rocks observed at the Earth's surface because the upper mantle is chemically much more complex than CMAS. Of course we can, and do, conduct experiments on assemblages chemically akin to those found in nature. Alternatively, we can take samples of igneous rocks and subject them to the pressures and temperatures of the upper mantle so that they melt, and we can then determine the composition of this melt and the minerals that coexist with it. Experiments on natural systems are difficult to undertake and sometimes difficult to interpret. However, these experiments can be very useful for distinguishing those igneous rocks found at the surface that have been modified by igneous differentiation (secondary magmas), from those that have not (primary magmas). For example, a basic prerequisite for a primary magma produced by partial melting of the upper mantle is that it is in equilibrium with the mantle assemblage at pressures and temperatures which correspond to the depth of segregation of the melt from the upper mantle.

Together with the experiments on simple systems, these more complex experiments reveal that the range in melt compositions that can be generated in the upper mantle is dependent on the bulk chemical composition and mineralogy of the source, pressure, temperature, and the degree of partial melting. For example, melt compositions ranging from alkali basalt to tholeiitic basalt can be produced by partial melting of the upper mantle, while more silica-rich magmas of granitic, rhyolitic, compositions may be produced by partial melting of the continental crust. The effects of depth and degree of melting are well illustrated by terrestrial basaltic magma types, which can be divided into three broad groups: basalts, komatiites, and picrites. From experimental studies it has been shown that tholeiitic basalts can be generated by 20–30 per cent partial melting of either depleted or enriched lherzolite (olivine + clinopyroxene + orthopyroxene) at 15–20 kbar (45–60 km depth). By 'depleted' is meant lower amounts of aluminium, calcium, titanium, sodium, and potassium, and an increased Mg/(Mg + Fe) ratio relative to average mantle composition. Picritic melts can be generated by 20–30 per cent partial melting at higher pressures, and alkali basalt by less than 20 per cent partial melting of enriched sources at pressures greater than 10 kbar (30 km depth). Komatiites can be generated either by 30–40 per cent partial melting of a depleted lherzolite or by 40–50 per cent partial melting of an enriched lherzolite. Although partial melting of lherzolitic mantle can produce a range of basaltic magmas, it is also the case that basalts may be the end products of fractional crystallisation of primary picritic magmas. For example, a comparison of experimental data with tholeiitic basalts from Kilauea and Mauna Loa volcanoes in Hawaii, shows that olivine fractionation from a primary picritic magma at pressure can produce part of the observed trend in composition. Both these tholeiites and basalts erupted at mid-ocean ridges show evidence for extensive fractionation at shallow depths.

It was mentioned above that small amounts of volatiles are present in the mantle. We know this from: (1) volatile emissions that commonly accompany volcanic eruptions; (2) the ubiquitous presence of CO_2-rich fluid inclusions in mineral in mantle xenoliths and (3) the occurrence of volatile-bearing minerals (e.g. carbonates (CO_2) and amphiboles (H_2O)) in xenoliths of mantle origin. From such lines of evidence it is known that the most abundant volatile in the upper mantle is water, followed by carbon dioxide, with minor amounts of chlorine, fluorine, and sulphur. Although water and carbon dioxide occur only in minor amounts in the upper mantle (100–400 parts per million and 50–300 parts per million, respectively), they have significant effects on the melting relationships of mantle assemblages. The presence of either volatile will lower the melting temperature of the assemblage and the melting products of a carbonated and/or hydrous mantle will be different from volatile-free mantle. The presence of volatiles in the melt will also effect the physical properties (e.g. viscosity) of that melt.

In this respect it is important to conduct experiments on mantle assemblages in the presence of carbon dioxide and water. For example, experiments in both simple (e.g. $CaO-MgO-SiO_2-CO_2$) and natural mantle systems have shown that carbonate melts can be produced by small amounts of partial melting of carbonated upper mantle at depths greater than approximately 75 km. These melts have low silica contents (less than 10 per cent) and are dominated by calcium and magnesium carbonate. As for silicate melt compositions, source composition, pressure, temperature, and extent of melting will have a strong influence on the melt composition. From experimental studies of the melting products of carbonated and hydrous upper mantle, it is believed that water plays a vital role in the petrogenesis of subduction-zone magmas, while carbon dioxide is influential in the petrogenesis of kimberlites, carbonatites, alkali-rich silica-poor rocks such as nephelinites and melilitites, all of which are characteristic of continental intraplate magmatism.

Igneous rocks like carbonatites illustrate one of the major problems facing the igneous geologist in that many carbonatites have had their compositions modified by secondary processes (e.g. fractional crystallisation, contamination) and there are few unequivocal examples of primary carbonatite magmas. The same can be said for silicate magmas with primary mantle magmas such as picrites being comparatively rare at the Earth's surface. As for so many cases in the Earth sciences, the igneous geologist has to work back in time using all available field, geochemical, and experimental data to construct a viable petrogenetic history for any particularly igneous rock.   JOHN DALTON

Further reading

Wilson, M. (1989) *Igneous petrogenesis: a global tectonic approach.* Chapman and Hall, London.

igneous rock classification

Igneous rocks are those that form from molten rock (magma), itself a product of the melting of parts of the Earth's interior. Igneous rocks display a variety of compositions, colours, grain sizes, and other attributes. Much scientific effort has focused on how igneous rocks have formed but, because it is difficult to observe and measure them at the relevant temperature and pressures, the search for the explanation of igneous rock formation has centred upon the rocks themselves. The study of the way in which rocks are formed is known as *petrogenesis*. Research on igneous rocks has included the identification of the minerals and the determination of the concentrations of the chemical elements in them. Each type of study has produced its own system of nomenclature and, hence, classification. These classifications place the rocks into pigeonholes which, although useful, tend to conceal the fact that there is a continuum of rock compositions.

The proliferation of igneous rock names that were based upon the names of the localities at which type specimens of the rocks were found introduced unnecessary confusion. The Subcommission on the Systematics of Igneous Rocks of the International Union of Geological Sciences (IUGS) has a glossary of more than 1500 rock terms, from 'A-type granite' and 'abessedite' to 'zobtenite' and 'zutterite' (an erroneous spelling of 'rutterite'). The subcommision has made a strong recommendation that the naming of rocks should be simplified and standardized in accordance with the guidelines laid out in their book (see Further reading).

Acid–basic

One simple classification is based upon the silica (SiO_2) content of the rock. Rocks with less than 45 per cent by weight SiO_2 are termed 'ultrabasic'; 45–52 per cent SiO_2, 'basic'; 52–66 per cent 'intermediate'; and more than 66 per cent 'acidic'. The terms 'acidic' and 'basic' are not intended to express the idea that the rocks are acidic or alkaline in the customary sense; they are employed merely to describe the silica content of the rock. Rocks that are rich in silica ('acid' or silicic) are commonly (with the notable exception of obsidian, a black volcanic glass) *leucocratic* (pale in colour), while 'basic' rocks are darker, that is *melanocratic*. The terms *leucocratic* and *melanocratic* may be replaced by *felsic* and *mafic*, respectively. The simple categories 'ultrabasic', 'basic', 'intermediate', and 'acidic' can be subdivided in a number of ways by consideration of grain size, concentrations of other elements, or the mineral content of the rock.

Increasing silica (SiO_2) content ⟶

	Ultrabasic	Basic	Intermediate		Acid
			plagioclase > K-spar	K-spar > plagioclase	
Fine-grained or glassy (grains invisible to naked eye)	(Rare)	Basalt	Andesite	Trachyte	Rhyolite Obsidian
Medium- to fine-grained (grains visible to naked eye; ⩽ 1 mm in diameter)	(Rare)	Dolerite	Microdiorite	Microsyenite	Microgranite
Coarse- to medium-grained (⩾ 1 mm in diameter)	Peridotite	Gabbro	Diorite	Syenite	Granite

(left axis: Increasing grain size)

Fig. 1 . A traditional classification of igneous rocks based on silica content ('ultrabasic' to 'acid') and grain size.

Grain-size classification

Rocks of identical composition can be distinguished from each other on the basis of their grain size. Figure 1 shows how rocks in each compositional group can be additionally subdivided in this way. Rocks of smaller grain size are those that have cooled quickly; they tend to be associated with volcanoes and near-surface intrusions, coarser-grained rocks are most likely to be plutonic (formed within the crust) in origin. This method of classification by composition and grain size can be enhanced by the addition of qualifiers; for example, a gabbro in which olivine is prominent is called an olivine gabbro, or a rhyolite that is strongly banded may be called a banded rhyolite.

The methods of classification detailed above are extremely useful for the naming of rocks in the field. Although a field geologist cannot necessarily identify all the minerals in the rock, and almost certainly cannot calculate the silica content, the clues yielded by colour, grain size, and readily identifiable minerals allow a field name based upon the system to be assigned. Laboratory analysis of the concentrations of numerous elements in igneous rocks makes it possible to define smaller and more specific categories that can, for example, hinge upon the relative concentrations of potassium and sodium. Some other methods of rock analysis and classification are summarized below.

Although the terms 'acidic', 'intermediate', 'basic', and 'ultrabasic' are potentially misleading from the viewpoint of the chemical composition of a rock, their use persists because they are well-established words and because all geologists understand their true meaning.

CIPW-norm

An igneous rock of given chemical composition (of elements) usually contains crystals of the same minerals (crystalline molecules) as other rocks of that composition. Figure 2 shows the minerals that are associated with the various silica contents of igneous rocks. This led to the use of the CIPW-norm in the classification of rocks. The CIPW-norm was developed in 1902 by four eminent petrologists who also left the initial letters of their surnames (Cross, Iddings, Pirrson, and Washington) to posterity. Using their knowledge of coarse grained rocks and the results of chemical analysis, they were able to reconstitute their samples theoretically as if they were coarse-grained. This operation allowed rocks of different grain size to be compared in terms of their bulk chemistry with recourse to only a few simple rules and some elementary, but tedious, arithmetic. Comparison of the normative composition of basalt, which is very fine grained, with gabbro, which is coarse grained, shows that the two are identical in CIPW-norm terms. This implies that, apart from their cooling history, they have a similar petrogenesis. Modem techniques for the determination of bulk chemistry have made analyses widely available for comparison with one another, and the CIPW-norm as an intermediary has become largely redundant. However, as many rock analyses in the literature have been classified in terms of the CIPW-norm, it is still a standard method of classification.

Rock analysis

In the past, the concentrations of major elements (those that make up more than 1 per cent by weight) of the rock were determined by wet chemistry, a long and tedious process which had to be performed by specialists. Modern analytical tools include X-ray fluorescence, a technique by which the quantities of major and trace elements (those that make up less than 1 per cent by weight) of a rock can be determined. Analysis for major and trace elements is thus now relatively

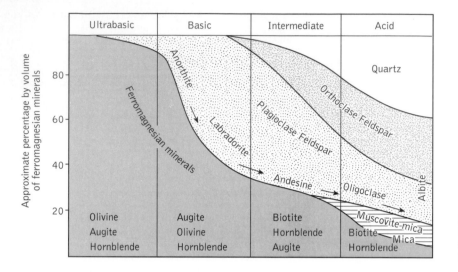

Fig. 2 . Proportions of minerals in igneous rocks classified according to where they are located on the 'ultrabasic' to

Harker plots

fast and can be performed by less expert chemists, provided that the equipment is available. This change in analytical procedure has led to a change of emphasis in rock nomenclature, now that there are enough analyses to consider the mean and variance of the rock compositions. A sample may include rocks that fall into a number of different named categories, even though analyses cluster as though they were the result of the same geological process or set of processes.

As a means of addressing these new data sets, the use of element–element plots (Harker diagrams), a method first used in 1909 by Alfred Harker, has become widespread (Fig. 3). After plotting, the rocks can be divided and assigned names according to where they plot on the graph. An enormous variety of different indices has also been used to split groups of rocks more finely and thus assign names to the groups. The user should ascertain, before applying such an

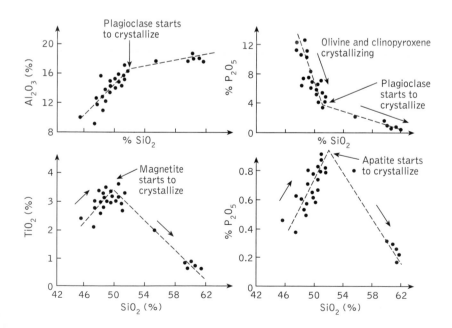

Fig.

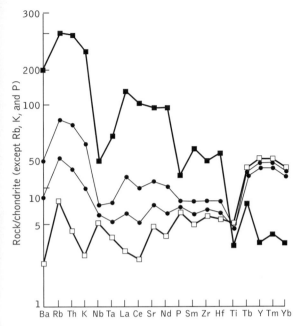

Fig. 4 . Spider diagram illustrating normalized element compositions in a range of igneous rocks.

index plucked from the literature, what the index represents and what he or she expects from it.

Spider diagrams

The spider diagram (Fig. 4) has become a popular tool in the analysis of trace-element compositions. Spider diagrams overcome the difficulty of plotting elements whose concentrations vary over several orders of magnitude by normalizing the elemental concentrations to a carefully chosed standard known as a norm. Common normalization standards include the concentrations of trace elements in chondrites (meteorites that are thought to have the same composition as the Earth) and MORB, the mean composition of normal (as opposed to enriched) mid-ocean ridge basalts. The compositions relative to the norm are plotted against the element names and a comparison of rocks or groups of rocks is made. It should be borne in mind that normalized values express the relative concentrations of elements to the norm and that the actual values of the elemental concentrations have been obscured.

Rare-earth elements (the elements lanthanum (La) to lutetium (Lu) are also normalized and plotted in this way (Fig. 5). Rare-earth elements (REE) are particularly useful in petrogenetic studies because they are chemically similar to one another and they are *incompatible*; when a small amount of melting occurs they move from the solid phase into the melt. Incompatible elements are hence very mobile and can move with extremely small melt fractions. This fact accounts for the extremely variable concentrations of these elements in rocks and the plethora of rock names that they have spawned. Incompatible elements are much invoked in the study of melting, particularly where small degrees of melting have occurred.

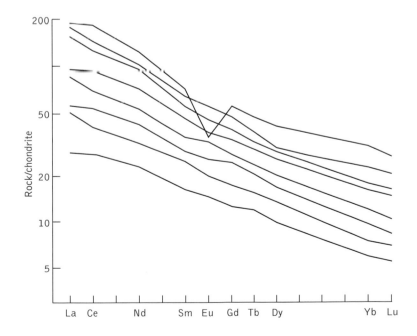

Fig.

The naming of rocks began with field names that were assigned so that a rock, or group of rocks, could be labelled easily in the field. As methods of analysis proliferated, and hence the criteria that could be used to discriminate between rocks became more numerous, so the range of names has snowballed. Now that modern analytical techniques can determine the concentrations of almost every element in the periodic table, there is a movement away from intensive naming of rocks towards a statistical treatment of multi-element data. Erstwhile names such as kulaite (a particular type of volcanic rock from the village of Kula in western Turkey) have been subsumed by much broader categories like 'alkali-basalt'. In time, more and more use will be made of statistical analysis of data to describe rocks that fall into a few named categories. Indeed, it is essential to replace a proliferation of rock names with a sound knowledge of the petrogenetic origin of the rocks but, at the same time, to avoid generating, in their place, too many complicated 'black-box' diagrams of one index against another. Modern petrogenetic studies are deeply involved in the study of the different environments in which igneous rocks occur, and in modelling the processes that lead to the range of rock types and compositions observed.　　　　JUDITH M. BUNBURY

Further reading

le Maitre, R. W. (ed.) (1989) *A classification of igneous rocks and glossary of terms, recommendations of the International Union of Geological Sciences Subcommission on the Systematics of Igneous Rocks.* Blackwell Scientific Publications, Oxford.

Kennet, P. and Ross, C. A. (1983) *Igneous rocks.* Longman Group Resources Unit, York.

Wilson, M. (1989) *Igneous petrogenesis: a global tectonic approach.* Unwin Hyman, London.

industrial minerals
Industrial minerals are defined as non-metallic, non-fuel minerals which can be exploited commercially. Minerals have been used as a natural resource by humans ever since flint axes were first worked in the East African rift valley over two million years ago. Since then, we have used minerals to ever-increasing degrees in order to develop our complex present-day society.

Ironically, the first natural resource to be exploited, the mineral quartz, forms the basis of the industry that has dramatically transformed society in modern times, namely the communications industry. Manufactured quartz crystals are used as oscillators in all types of radios, telephones, and satellites, and the advent of the personal computer was dependent on development of the microchip, etched on a silicon wafer grown from artificial silicon, itself generated from melted-down naturally occurring quartz of high purity.

From high-tech to low-tech applications, minerals form the basis for our survival. Our homes are built from many different industrial minerals, including not only natural stone but also clays for bricks and tiles, slates for roofs, limestone for

cement, silica sands for glass, and gypsum for plasterboard for walling.

A wide variety of industrial minerals are used as *aggregates* in the construction industry. These range from *limestone* and *dolomite*, which are used for roadbases and sub-bases, but principally for concrete manufacture, and harder igneous rocks for road construction and ballast.

Various salts, typically of evaporitic origin, are extracted worldwide, such as rock salt (sodium chloride), in the Cheshire Basin, potassium salts in North Yorkshire, and boron salts in the USA, Turkey, and South America. Salt is used by the chemical industry in the manufacture of a wide range of products, from pesticides to soaps and detergents. More visibly it is spread on roads to minimize the formation of ice. Potassium salts, such as sylvite (potassium chloride) are used to produce fertilizers for the agricultural industry, while boron salts, especially borax, are a basic resource for the glass industry.

Zeolites are a group of minerals with hollow crystal lattices that make them useful as industrial sieves. Solutions passed through a filter bed of zeolite minerals undergo ion exchange as particular ions or molecules become trapped within the crystal lattice. Zeolites, both natural and synthetic, therefore have wide industrial applications. In the petroleum industry they are used to 'crack' oil; in the nuclear industry ion exchange makes it possible to remove soluble radioactive elements from low-level radioactive waste solutions; and in the detergent industry zeolites are added to soap powder to remove calcium ions from 'hard' water, a process that leads to better lathering qualities.

Clay minerals find a wide range of applications. *Kaolinite*, extracted extensively in the south-west of England, is used to coat paper, as a filler in plastics and paint, and also in the ceramic industry; ball clay is also used in the ceramic industry; and bentonite is used in the formulation of oilfield drilling fluids, in the production of cat litter, and in the generation of hydraulic barriers, particularly for sites of landfill waste disposal.　　　　RICHARD E. BEVINS

information technology and the Earth sciences
As a result of a strong applied tradition the Earth sciences have had a long history of interdisciplinarity and the assimilation of new technologies. The use of sonar after the Second World War to map the sea floor and the use of stable isotopes to study ancient climates are but two examples of technology transfer into the field of Earth science that precipitated conceptual, as well as practical, revolutions in understanding the history of our planet. On the cusp of the twenty-first century Earth science finds itself, once again, at the forefront of another interdisciplinary technological revolution; the electronic communications or information technology (IT) revolution.

Apart from the (now) commonly recognized advantages of electronic mail to speed communication between research collaborators and between students and teachers, the IT revolution will fundamentally change the way in which all science is conducted. It will do so, not by providing tools that enable scientists to study the history of our planet in new ways *per se*, but by ensuring that information about such tools is distributed to the widest possible audience, and (even more importantly) by making it possible to gain access to large amounts of highly specified information from remote locations. The IT revolution will also allow the results of research to be more effectively communicated to the scientific community and to the general culture quickly, easily, and inexpensively; and because of these factors it will force the professional Earth science community to reorganize its activities in the creation, management, and interpretation of data. While this IT revolution will be experienced by most scientific disciplines, it will have a disproportionate effect on the Earth sciences because of the popular nature of their subject-matter and their place at the heart of many scientific controversies.

Databases

Like all the sciences, Earth science is becoming increasingly specialized. However, cutting against this trend is the need for specialists and students to place their observations within a more general context and to make efficient and correct use of data generated by other groups (both within the Earth science community and elsewhere). Moreover, as the total number of scientists increases, the rate of data acquisition and publication has increased to the point where it is impossible for individuals to keep track of all significant developments in any but the most narrow research programmes. Coping with this flood of information will be one of the first great challenges for the Earth sciences IT revolution.

At present the most popular way of resolving this problem lies in the creation of searchable databases and metadatabases, usually linked to the World-Wide Web (WWW), that can direct individuals to the specific information they need, maintain the quality and integrity of that information, and present it in a form (e.g., as graphs, tables, and figures) that can be understood and used by clients. To take a simple example, the German GEOMAR Research Centre for Marine Geosciences has made its Ocean Drilling Stratigraphic Network (ODSN) database available in the form of a WWW-accessible software interface that enables scientists quickly to construct accurate base maps of tectonic plate configurations for a wide variety of times and regions.

This trend towards using IT to make the results of important research available to the scientific and lay communities is being encouraged by both government and private research foundations as the normal outcome of Earth science research projects. The first generation of Earth science IT research products has been favourably received by researchers and educators alike. As the number and quality of these products improve, and as the Earth science community educates itself to take advantage of them, access to the Internet and WWW will become as indispensable as access to a library for research and education.

Electronic publishing

The reason why Earth scientists are finding it more and more difficult to keep up with the data produced by their increasingly specialized colleagues is that the number of relevant publications is exploding as specialization proceeds. This is compounded by the interdisciplinary nature of much Earth science and the fact that scientific publication in general is undergoing a phase of increased diversification. Add to this the (now) constant constraint of diminishing library acquisition budgets and rising journal subscription prices (especially for institutions), and the result is decreased access to relevant information from traditional print sources. The need to maintain access to scientific publications will be addressed by another aspect of the IT revolution, electronic publishing.

In recent years the WWW has demonstrated its ability to be used as an efficient publishing medium. Virtually all major western business and cultural institutions boast a 'web page' that growing numbers of people regard as their first port of call for both general and specific information. From here it is but a small step to the creation of formal scientific journals that have no print-based equivalents. While few fully electronic Earth science journals exist at the time of writing (*Palaeontologia Electronica* is one example), their advantages are widely appreciated within the Earth science community. In addition, documents published electronically can form the basis for on-line discussions among specialists, the text of which can be included in the body of the document. This will, in effect, turn previously static scientific documents into dynamic, living documents whose value can grow with time.

Several difficulties remain before electronic publications becomes commonplace in the Earth sciences. These include concerns about permanence; the lack of library facilities, personnel, and resources to manage electronic documents; and both professional and nomenclatural formalisms that currently exclude electronic publications from being considered valid for certain types of publishing purposes. These issues are not unique to the Earth sciences and are progressively succumbing to the demand for increased numbers of electronic publications from scientists, professional societies, librarians, funding institutions, and the general public.

Organizational impacts

Over and above the changes it is causing in the creation, management, and dissemination of data and interpretations, the IT revolution is also forcing virtually all institutions involved in the creation or use of Earth science data to re-examine their position with the organizational structure of the field. The days when even large institutions devoted to the study of the Earth could provide support for all their own scientific activities are largely over. In the petroleum industry, for instance, many research and technical services that were performed 'in house' in the 1970s, have now been outsourced. The IT revolution will reinforce this trend, since it abolishes many of the physical constraints that were formerly imposed. Moreover the infrastructure and personnel requirements that underpin the creation of Earth science IT products and databases will force the community to locate and support specialized activities in certain types of institutions (for example, databases in museums or commercial providers, publications in electronic libraries) to which all have access instead of continuing to duplicate these resources themselves. If managed correctly, this re-organization will provide an opportunity for the development of further expertise and increased interaction between the various sectors of the Earth science community (academic, museum, commercial, government). It will also require some adjustment of roles in each sector.

The perceived need, tools, incentive, and resources for a major IT-driven revolution in the Earth sciences are thus all in place Even just a few years ago such a step could not have been taken because of the limitations imposed by desktop computer capabilities and by costs. Thanks to the continuing development of computer technology, these limitations have been overcome, and the transfer to large-scale, high-capacity networks will continue to eliminate the remaining technological barriers. More Earth scientists (and students) already have easier access to computers, the Internet, and the WWW than they do to high-quality Earth science libraries. This revolution will dramatically improve the utility of data, the skills of scientists, and the relevance of Earth science to the larger issues of society and culture.   N. MACLEOD

inlier, outlier The terms 'inlier' and 'outlier' each describe a situation where a outcrop of one type of rock appears at the surface completely surrounded by another. This is usually a result of erosion having removed part of an overlying rock unit from an underlying unit. An inlier is then an isolated region of underlying rock exposed by erosion of the overlying units, whereas an outlier is an isolated remnant of overlying rock that has survived the erosion that has stripped it away everywhere else to reveal the underlying rocks. In regions of flat-lying or gently dipping strata, inliers tend to occur in valley bottoms whereas the hilltops may be occupied by outliers. Where the strata have been strongly folded prior to erosion these simple topographic relationships do not necessarily apply, and it is possible to have an inlier preserved in the core of an anticline occupying high ground or an outlier preserved in the core of a syncline occupying low ground. Inliers and outliers can also be generated by fault movements followed by erosion. For example, if an older unit is thrust over a younger unit erosion can isolate part of the thrust sheet, leaving an outlier of older rock surrounded by younger rock (a reversal of the usual age relationship). Similarly erosion can expose a window through the thrust sheet to reveal an inlier of younger rock surrounded by older rock. Fault movements by which blocks of terrain move up or down relative to each other, as when horsts or grabens are formed, can also produce inliers and outliers.   DAVID A. ROTHERY

insects Insects are invertebrates belonging to the phylum Arthropoda, which is composed of animals with a segmented body, many jointed appendages, and a chitinous exoskeleton (an external skeleton made of chitin, a hard brown substance). They occupy first place among animals in terms of biological success: almost 75 per cent of all known animals are arthropods (750 000 species), and of these almost 500 000 are insects. Insects form an easily distinguishable natural group of organisms in which the body is divided into a head with jointed appendages adapted as mouthparts, a thorax with legs and two pairs of wings, and an abdomen. They are separated into major groups based on gross morphology; Palaeoptera have wings that cannot be folded when the animal is at rest (e.g. dragonflies), while Neoptera have wings that can be folded. In addition, the palaeopterans have immature forms that are voracious aquatic predators, quite unlike the adults, while neuropterans have immature forms that are either smaller versions of the adults or are initially worm-like and then pass through a pupa stage in which they are transformed into adults.

The fossil record of land arthropods is scanty, like that of all land animals, and hence does not necessarily give a true picture of numbers of species or time of earliest appearance. Although almost 500 000 species of insect have been described, only a few thousand fossil insect species are known; most of these are from a few special regions where fossilization took place under unusual conditions. Baltic amber deposits from the Tertiary are well known; deposits of Cretaceous age are also known from Alaska and New Jersey. The Late Carboniferous (Pennsylvanian) Mazon Creek Fauna from Illinois is an important Palaeozoic locality and contains over a hundred insect species, almost half of the animals recorded from there. Insects are also known from trace fossils; brood chambers of mining bees are known from the Eocene of Wyoming and the Upper Cretaceous of Arizona; insect trails have been reported from the Permian of Germany; and insect-cut leaves are known from the Eocene of Kentucky.

Insects probably developed in the Devonian from an unknown arthropod ancestor. The earliest described insects are flightless; they occur in Scottish Devonian rocks, and are similar to modern springtails. The first important evolutionary event was the development of wings, which must have taken place in the Early Carboniferous (Mississippian) although currently no Mississippian insects are known. By the Late Carboniferous (Pennsylvanian) large flying insects had developed, including dragon-flies with wingspans up to 750 mm, the largest insects known. The ability to fold the wings close to the body developed during the Late Palaeozoic, and allowed insects to escape predators by running along the ground and hiding in crevices. The development of a pupal stage probably took place at about the same time; the first examples are found in the Permian. This development avoided competition between larval and adult forms for food sources and also provided for overwintering in the pupal stage. The last major development was the co-evolution of insects and flowering plants, which took place in the Late Mesozoic. Although the record is poor it is clear that insect pollinators such as moths and butterflies, which appear in the Cretaceous, arose in concert with the flowering plants, while the appearance at the same time of fleas marks their development together with their hosts, the mammals.

DAVID K. ELLIOTT

International Union of the Earth Sciences (IUGS)
A comparative latecomer among international scientific unions, the International Union of Geological Sciences (IUGS) is the senior body of geoscientists. Several such unions operate within the general framework of the International Council of Scientific Unions (ICSU), under the auspices of UNESCO in Paris. IUGS was set up following discussions at international geological congresses and at UNESCO in the 1960s and met for the first time in Rome in 1963. It is a voluntary, non-profit-making organization, aiming to facilitate and encourage study of the geological sciences, and to facilitate international co-operation and continuity in geology and related sciences, and to promote the International Geological Congress (IGC). Its council is composed of one delegate per member country, of which there were 112 in 1998; of these, 28 were inactive. Its research activities are embodied in commissions, advisory boards and committees, and 28 other affiliated organizations. The costs of these activities are large and financial difficulties are aggravated by the tardiness of member countries in paying their subscriptions. Day-to-day work is carried out by the office of the Secretary-General, currently at the US Geological Survey, Reston, Virginia, USA. The Union sponsors meetings and publishes several series of reports and papers.

The International Geological Congress meets once every 4 years, in a different location each time, and is the largest such geological gathering, with many symposia, workshops, exhibitions, and excursions as well as a very full programme of lectures in all areas of the geosciences. It publishes abstracts of papers, general congress proceedings, and guides, as well as special volumes of papers. Attendance numbers several thousand scientists.

Among the continuing endeavours of IUGS are the several commissions, such as those on experimental petrology at high pressures and temperatures, on marine geology, on systematics in petrology, and on tectonics. The largest body within IUGS is the International Commission on Stratigraphy (ICS) and its subcommissions, working groups, and committees. It is the only organization concerned solely with stratigraphy on the global scale. Its aims are to clarify and coordinate principles of stratigraphical procedure and to produce a single unified nomenclature for a standard stratigraphical scale. This, in turn, is a means of documenting unambiguously global events with reference to a scale established by biostratigraphy and all other means available for correlation. It

seeks to achieve the precise definition of systems, series, and stages by the use of fossils within the Phanerozoic rocks. To do this it seeks the international recognition of and agreement to the use of Global Stratotype Sections and Points (GSSP), each a unique point in the rock section and marking a defined point in geological time. Each system receives the attention of a subcommission made up of stratigraphers specializing in that system. First to be considered have been the definitions of the base of each system (and hence the top of the underlying one). Not every one has yet been finally agreed but progress is made each year. With the whole world to consider, and the long traditions of some of the developed nations exposed to change, the choice of some boundary stratotypes has been hotly contested. In the case of the Devonian system, which was the first to acquire a GSSP for the base, the series are now also defined, as are most of the stages. The level of the Devonian–Carboniferous boundary presented a number of problems worldwide, and a working group studied sections in several countries before selecting a GSSP in France and auxiliary stratotypes in China and Germany.

There is a long history behind the ICS; most of the early international geological congresses had symposia and meetings to debate and decide stratigraphical procedures and nomenclature. At the nineteenth IGC in Algiers in 1952 the present commission was set up and in 1965 it was recognized by IUGS. Today the commission consists of its officers and the chairmen of the subcommissions; each subcommission may have as many as 50 or more members, titular or corresponding. Field or business meetings, or both, are held annually if possible. The commission has a small fund to assist the business of conducting meetings and publication of results and decisions. Recommendations about boundary levels are submitted from the subcommissions and working groups to the commission for approval. Subsequently they are passed to the council of IUGS for ratification. An official journal of IUGS, *Episodes*, contains news releases, progress, proceedings, and decisions of IUGS.

Another joint venture, the International Geological Correlation Programme (IGCP) was set up in 1972 at the twenty-fourth International Geological Congress in Montreal, Canada, to promote international projects designed by co-operating scientists. International liaison, co-operation and communication have, from the outset, been regarded as important as the scientific objective. Special areas for favour were defined: better methods for finding and assessing resources, a better understanding of the way geological processes and events affect the environment and human activities, standardization of terminology and procedures in research. The programme operates as a collaboration between IUGS and UNESCO, with a scientific board to receive, assess, and respond to proposals and reports from project leaders. The scientific board is made up of active scientists who give their time to the programme. Funds are allocated annually to leaders of recommended projects to provide for communication and travel by project members; funds are not directly applied to actual research. Most projects, selected by peer review, receive funds for 5 years. In 1972 20 projects were funded; in 1994 the number was 54, including 10 accepted that year, from 34 different countries.

The International Geosphere/Biosphere Programme (IGBP) was set up in 1986 in response to the urgently perceived rapid and potentially catastrophic changes in the global life-support system. It seeks to promote observations and communication across all the geological and biological disciplines on a grand scale to advance our understanding of progressive changes in the environment of the human species on Earth—and to make predictions about the long-term future.

The International Lithosphere Program (ILP), organized by the Inter-Union Commission on the Lithosphere, was founded in 1980 by ICSU at the request of the International Union of Geodesy and Geophysics (IUGG) and IUGS and it is in the tradition of the earlier Upper Mantle Project and the Geodynamics Project. It seeks to investigate all aspects of the lithosphere, and its origin and evolution prior to 200 Ma. Special topics within the programme include great earthquakes of the late Holocene, mid-oceanic ridges, reflection survey of the continental crust, and the geodynamics of deep internal processes. Subduction-related magmatism and palaeomaps of continental reconstructions are also targeted. Much of the investigation involves high technology and concomitant high expense, especially where multidisciplinary approaches to a research topic are adopted. ILP seeks to initiate and promote the best use possible of available resources internationally in pursuing such research. Its practical aspects include the prediction and mitigation of natural hazards, especially in developing countries. The programme operates through a number of working groups and co-ordinating committees.

With the closer integration of the affairs of the European Community, a European Union of Geoscientists has been formed, and liaison between national societies has flourished. Similar improvements in international contacts have been achieved on other continents, the American Geophysical Union being a good example.   D. L. DINELEY

Further reading

Fuchs, K. (1990) The International Lithosphere Programme. *Episodes*, **13**, 239–46.

Fyfe, W. S. (1990) The International Geosphere/Biosphere Programme and global change: an anthropocentric or an ecocentric future? A personal view. *Episodes*, **13**, 100–2.

Naldrett, A. J. (1990) International Geological Correlation Programme: an example of collaborative geoscience. *Episodes*, **13**, 22–7.

Skinner, B. J. (1992) Scientific highlights of two decades of international cooperation at the grassroots level. *Episodes*, **15**, 200–3.

intraplate seismicity The Earth's most seismically active regions and the sites of its largest earthquakes coincide with the boundaries between the tectonic plates (see *earthquake mechanisms and plate tectonics*). However, the vast majority of the Earth's surface and continents lie within the interiors of these tectonic plates, and these 'intraplate' regions can also exhibit significant earthquake activity. To a large extent, such intraplate seismicity is in response to tectonic forces as they propagate from the plate boundaries into the interiors of the plates. Examples of tectonically and seismically active intraplate/intracontinental areas include the intermountain region of the western USA (e.g. Nevada and Utah) and much of China.

Stable intraplate regions include the central and eastern portions of the United States and Canada, and Australia. Although these regions are relatively distant from the plate boundaries, large earthquakes can and have occurred. A significant example is the 1811–12 New Madrid, Missouri earthquakes, which occurred near the Mississippi River in the central United States. Over the course of two months, three earthquakes of approximate moment magnitude (M_w) 8 (see *earthquake seismology*) occurred. If repeated today, they would undoubtedly result in the worst natural catastrophe in the history of the United States. Other notable and damaging intraplate earthquakes have been the 1819, M_w 7.8, Kutch, India; 1886, M_w 7.6, Charleston, South Carolina, USA; 1918, M_w 7.4, Nanao, China; the 1933, M_w 7.7, Baffin Bay, Canada; and 1935, M_w 7.1, Libya earthquakes.

A widely accepted model for the occurrence of earthquakes in stable intraplate settings requires the presence of suitably oriented pre-existing zones of weakness in the Earth's crust, such as faults. When such structures are oriented favourably with respect to the tectonic stress field, they are subject to reactivation and hence may generate earthquakes. A favourable orientation occurs when the fault is aligned to within 10–50° of the direction of the maximum principal stress. The pre-existence of such zones of weakness is required, in large part, because of the lower levels of tectonic stress present in intraplate regions. The physical properties that control the strength of the fault zone are also important in determining whether it will be reactivated and capable of producing earthquakes.

A continental rift beneath the New Madrid seismic zone (Fig. 1), which was the source of the 1811–12 New Madrid earthquakes, is an example of a reactivated pre-existing zone of weakness. The rift, about 70 km wide and 200 km long, is an intracontinental graben-like structure of Precambrian age. It may have formed originally along a boundary separating basement rocks of different composition. Bounding the rift, and probably contained within it, are crustal faults, which are the sources of the New Madrid seismicity. Igneous plutons have also intruded through this zone of weakness in the Earth's crust. Although the rift was probably the product of crustal extension, the earthquakes observed today are the result of strike-slip and reverse faulting resulting from compressive tectonic stresses at present affecting the region. The nature of these stresses, which reflect the modern-day stress field, indicates that the rift is currently being reactivated. The rift beneath the New Madrid Zone is buried by several

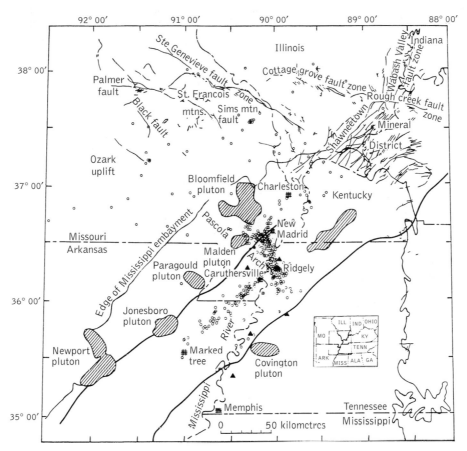

Fig. 1. The New Madrid seismic zone. Crustal faults (fine solid lines), earthquake epicentres (open circles), igneous plutons (shaded areas), and rift boundaries (heavy solid lines) are shown. (From Hamilton, R. M. and Zoback, M. D. (1982) Tectonic features of the New Madrid seismic zone from seismic reflection profiles. In F. A. McKeown and L. C. Pakiser (eds) Investigations of the New Madrid, Missouri, earthquake region, *U.S. Geological Survey Professional Paper* **1236**, pp. 55–82).

kilometres of alluvial sediments, and investigations to characterize its true nature, such as traditional geological approaches, have thus been hindered or prevented. For the majority of significant intraplate earthquakes, the identity and characteristics of their seismic sources remain unknown.

IVAN G. WONG

intraplate tectonics Since the development of the theory of plate tectonics in the 1960s it has been recognized that this theory, although a very good approximation to the behaviour of much of the Earth's lithosphere, cannot be strictly applied to some regions. The boundaries of some plates are not sharp but are spread out over hundreds of kilometres, and the interiors of several plates have significant deformation. The existence of this intraplate deformation is revealed by the global

pattern of recent shallow earthquakes (Fig. 1). Deep earthquakes (below 30 km depth) which are associated with the lower parts of subduction zones are not included in this map. Although the majority of earthquakes occur at the boundaries of plates (in particular the 'ring of fire' surrounding the Pacific Ocean), a significant number of earthquakes, including many damaging ones, occur elsewhere. Other evidence for the existence of intraplate deformation comes from the direct observation of crustal movements by geodetic means.

Continental intraplate tectonics

Many parts of the continental lithosphere do not deform. These *cratons* or *continental shields* are parts of the continental lithosphere, usually well within the interior of the plates, that have been unaffected by plate-boundary or mantle-plume

545

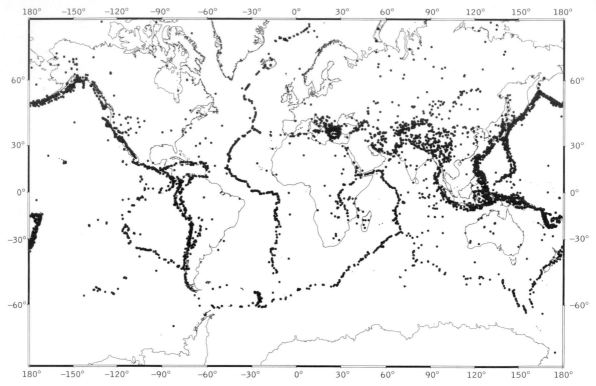

Fig. 1. Crustal earthquakes of magnitude 5 and above that have occurred in the past 30 years indicate both the well-defined plate boundaries, such as the mid-Atlantic ridge running approximately north–south down the centre of the plot, and also zones of widespread intraplate deformation such as the Basin and Range region of the western USA.

activity for tens to hundreds of millions of years. As a result these regions, for example, central Canada and Siberia, have cooled at depth, thickening the lithosphere, and in consequence are relatively strong and resistant to deformation. Surrounding the cratons are areas of younger continental lithosphere, which are warmer and therefore weaker and more easily deformed. The most significant intraplate deformation takes place within the Alpine–Himalayan belt, a band of intense seismic activity running from Albania and Greece through the Caucasus and Iran to Tibet and south-east Asia. The width of the zone varies from a few hundred kilometres in Greece to over 1000 km in Tibet. The Eurasian plate to the north and the African, Arabian and Indian plates to the south are converging at up to 40 mm yr^{-1}, and so the predominant style of deformation is thrusting. Strike-slip and extensional faulting are, however, also observed. With the exception of smaller regions, such as central Turkey and the Tarim basin in Tibet, which appear to be rigid and have sharp plate-like boundaries, the plate motion in the Alpine–Himalayan belt is accommodated pervasively throughout the zone. A contrasting situation exists in the Basin and Range of the Western USA, where extensional intraplate deformation takes place

despite the existence of an abrupt strike-slip plate boundary to the west.

Although the continental part of the African plate is mostly regarded as cratonic, extension is occurring across the East African Rift, which runs south from the Red Sea to roughly the northern border of South Africa. The situation here is that the African craton has been heated from underneath by a plume of hot, upwardly convecting mantle material. This has resulted in uplift of the continental lithosphere, which has been weakened and has eventually collapsed under its own weight and riven apart. In the Red Sea and the Gulf of Aden, rifting has taken place and a new plate boundary has been formed at which oceanic crust is being created, but the southern arm has not rifted completely. This style of deformation, in which continental rifting does not always lead to the formation of a plate boundary but still results in significant deformation, is widespread throughout geological history and is responsible for many of the old sedimentary basins in which much of the Earth's oil and gas reserves are found.

A third style of continental intraplate tectonics is represented by large, isolated earthquakes such as the 1811–12 New Madrid (eastern USA) events of about magnitude 8 and the even larger 1755 Lisbon (Portugal) earthquake. Because they are infrequent, such events are not well understood. Although each one results in a significant relative displacement of parts of the Earth's crust, how such events lead to long-term tectonic motion is unclear. Smaller isolated intraplate earthquakes also exist, but their contribution to tectonic movement is negligible.

Oceanic intraplate tectonics

Because the oceanic lithosphere is made up of olivine-rich material, in contrast to the quartz- and feldspar-rich continental lithosphere, the oceanic plates are (at a given temperature) much stronger and less likely to break than the continents. The outcome of this is that wide zones of deformation within the oceanic lithosphere near to plate boundaries simply do not occur. The main intraplate tectonic activity in the oceans takes the form of *hot-spot volcanism*. At the hot-spot, a rising convective plume of hot mantle material causes melting in the upper asthenosphere. Much of the melt never reaches the surface, but some is erupted to form a submarine volcano which may eventually grow into an island. Because the plates are moving in relation to the hot-spots, over a long period of time a chain of oceanic volcanoes will be formed. Only those close to the plume will be active; older volcanoes further away will erode and subside as the plate cools, forming a chain of seamounts. A change of direction of a seamount chain is the result of a change in the relative movement of the plume and the oceanic plate. A good example of this is the Hawaii–Emperor chain in the Pacific Ocean, which has a pronounced kink corresponding to a change in relative plate velocity 43 Ma ago.

Although intraplate volcanism is a sign of tectonic activity, it does not result in large-scale deformation of the lithosphere. The only current evidence of significant intraplate oceanic deformation comes from the equatorial Indian Ocean within the Indo-Australian plate. Here, the existence of widespread seismicity and folding of sea floor rocks indicates that deformation at a rate less than 1 cm year^{-1}, less than half of that at the slowest-spreading current mid-ocean ridge, has taken place during the past 10–20 Ma. Geodetic measurements demonstrate that this intraplate movement is an ongoing phenomenon.   PETER CLARKE

intrusion types and intrusive igneous rocks Intrusive igneous rocks are formed where molten rock (magma) cools and solidifies without reaching the surface of the Earth to form a volcano. Large intrusive bodies cool slowly and are coarse-grained, whereas small intrusive bodies can cool quickly to form fine-grained igneous rocks. The coarse-grained rocks of large intrusive bodies are known as *plutonic rocks*. The edges of large intrusive bodies that have come into contact with the cooler surrounding country rock are, however, fine-grained. This fine-grained margin is known as a *chilled margin*. Intrusive igneous rocks, by definition, solidify beneath the Earth's surface

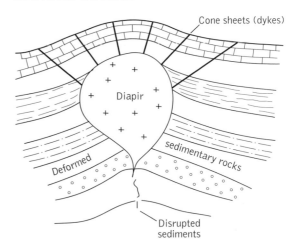

Fig. 1. Cross-section though an imaginary diapir of actively emplaced plutonic igneous rock deforming formerly horizontal sedimentary rocks.

and are seen at the surface only when uplift and erosion have unroofed them during exhumation.

Intrusion types

There are two main methods by which igneous rocks intrude other rocks: active and passive intrusion. Actively intruding rocks deform the rocks around them to create a space for themselves. Figure 1 shows a rising diapir of buoyant molten rock that has forced its way through the surrounding sedimentary rocks, deforming those above into a dome, and rupturing the lower strata. The diagram also shows the way in which this type of deformation produces fractures in the rocks through which the intrusion is passing. Both radial and concentric dyke swarms can be formed. A section through some concentric cone sheets is illustrated in Fig. 1. A variation of diapirism is the formation of a *lopolith*, a tree-like igneous intrusion formed where the magma has forced its way not only upwards but also along planes of weakness, such as bedding planes (Fig. 2). The Duluth gabbro in Minnesota is a classic example.

Passively intruding rocks are accommodated as the country rocks are broken, without being otherwise deformed, to make the necessary space. Figure 3 shows one mode of passive intrusion. Faults, which are usually circular in plan and steeply inclined outwards, allow a cylindrical block of material to separate and subside. Magma rises into the space created above the subsiding block and begins to solidify as a cauldron-shaped intrusion. This type of passive intrusion is called *cauldron subsidence*. It can occur repeatedly so that numerous concentric circular intrusions are formed. Classic examples are found in Yellowstone Park (Mount Holmes) and in Glen Coe in Scotland.

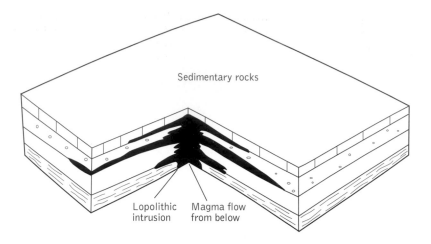

Fig. 2. Structure of a lopolith intruding flat-lying sedimentary rocks.

Not all passively intruded igneous bodies replace a large block of subsided rock; the igneous body may also make space for itself by splitting off small blocks from the top of the magma chamber or by the intrusion of multiple dykes, side by side (Fig. 4). Notable examples of parallel dyke systems are found in the Tertiary intrusions of the west of Scotland, in Colorado (Spanish Peak), and in Montana (the Crazy Mountains). In the former case, liberated blocks sink into, and can be melted by, the magma, which, being more buoyant than the overlying rock, rises to fill the space above. The removal of individual blocks is called *stoping*. The blocks are called *stoped blocks*; pieces of rock that have not fallen away and remain hanging from the roof of the magma chamber are *roof pendants*. When erosion cuts into an igneous body of this type, stoped blocks can in places be seen preserved within the solid magma. Roof pendants may be preserved as islands of country rock in the middle of the roof of an exposed igneous intrusion.

A dyke is a parallel-sided, sheet-like, igneous intrusive body that cuts vertically through the country rock regardless of its type) and is commonly either perpendicular to sedimentary layers (strata) or vertical. If the sheet intrusion is emplaced along the interface between two beds and is horizontal, it is known as a *sill*. Well-known examples of sills are those of the Karroo region of South Africa, which extend over areas of thousands of square kilometres, and the Whin sill in northern England, along which much of Hadrian's Wall was built. The rock of which the Whin sill is formed is more resistant to erosion than the surrounding sedimentary rocks and consequently forms a prominent natural feature. Dykes and sills may range from tens of metres thick down to centimetres thick, but intrusive igneous bodies less than a centimetre or so thick are more often refered to as *igneous veins*.

Dykes and sills make space for themselves by pushing the rocks on either side apart because the magmatic pressure

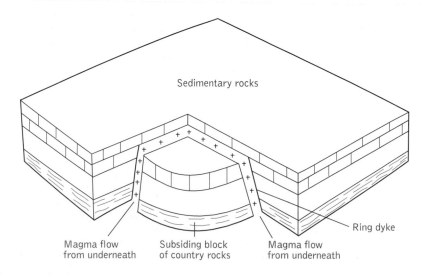

Fig. 3. Passive emplacement of an igneous rock body leading to cauldron subsidence.

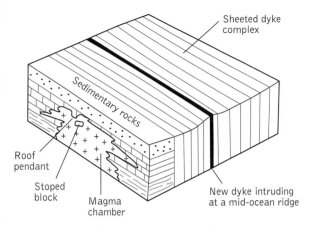

Fig. 4. A sheeted dyke complex at a mid-ocean ridge.

exceeds the inward pressure on the walls of the dyke or sill. This process, which is analogous to hydraulic fracturing, causes the rock mass to dilate. Multiple dykes may cause large, cumulative amounts of dilation, in some instances reaching as much as 100 per cent. This means that no original rock remains and the whole of the section being examined is dyke material. Dilation is one means by which an igneous body can be accommodated. The whole of the ocean floor has been formed by means of dilation. In this case, new dykes are added in the middle of the ocean, meaning that each dyke only has one chill margin, on the outside. Younger dykes are emplaced into the centre of the most recent dyke to be formed in the area, splitting the older dyke into two thinner dykes, each with one chill margin, and themselves solidifying to form a two-sided dyke. This type of intrusion formed of numerous dykes is called a *sheeted dyke complex*.

Passive intrusion requires brittle fracturing of the country rock, in contrast to active intrusion which entails some ductile deformation. Brittle fracturing is said to have occurred when the country rock has been broken, whereas ductile deformation occurs when a rock has behaved as a plastic solid or a viscous liquid. Whether the stresses set up by an intruding magma body result in ductile or in brittle deformation of the country rock depends principally on temperature, the pressure, and the rate at which the stress is applied. The term 'country rock' is used in igneous petrology to describe any rock into which an igneous rock intrudes, irrespective of whether it is brittle or ductile. In general, the upper part of the crust is more susceptible to brittle fracture, while magmas deeper within the crust are more likely to cause ductile deformation of the country rock. In some tectonic environments, the deformational environment set up by the tectonic activity may favour one or other type of behaviour. In County Donegal, Ireland, a number of granites have been intruded at the same crustal level and would be expected to have similar intrusion styles. However, the complicated tectonics of the region at the time of intrusion led to a mixture of active and passive styles of intrusion. Those granites intruded into areas of

tectonic extension (where brittle fracture occurs more easily) had a tendency to be of the passive type, whereas the granites in areas of compressive tectonics generally exhibit active intrusion by ductile deformation. Passive intrusions at high levels within the crust are thought to be fed by lower-level magma chambers that have been actively intruded.

Contact relations

In an area where there have been multiple igneous intrusions their relative ages can be deduced from the contact relations. The fine-grained selvage called a chilled margin, formed in a younger rock in contact with an older rock, has already been mentioned. Other contact relations that are used to infer relative age are dykes and veins that cut across each other or features of an older rock. (An exception to this is back veining, see below.)

Back veining

Back veining is a feature of some igneous contacts that occurs where molten rock of high melting temperature (gabbro, for example) intrudes a rock of lower melting temperature, for example granite. In this situation the gabbro will intrude the granite at about 1300 °C, forming dykes and veins of gabbro which are then chilled. However, granite, which melts at about 800 °C, will still be melted by the heat of the gabbro after the gabbro has been chilled. The molten granite then intrudes the recently formed gabbro. At a normal igneous contact, the relative dates of the rocks can be deduced from their contact relations. The second rock to be intruded has veins into earlier rocks. Where back veining has occurred, each rock looks as though it has intruded the other. It can thus be concluded that the rock of lower melting temperature was the first to be emplaced.

Intrusive igneous rocks

The grain size of an igneous rock depends upon the rate of cooling. Slow cooling gives time for the nucleation and growth of crystals; fast cooling forces the molten rock to freeze into a fine-grained mass or a glass. Small intrusions, such as narrow dykes and sills, are usually fine grained, while larger dykes and other intrusions are characterized by their larger grain size. Crystals form when molten rock begins to cool in a magma chamber. Material intruded as dykes from the magma chamber after crystallization has begun will form *porphyritic dykes* (i.e. dykes with easily visible crystals, some up to a few centimetres across). Many plutonic rocks have names that are distinct from those of their fine-grained compositional equivalents. Gabbros, dolerites, and basalts differ in grain size but are of the same chemical composition.

Most parent magmas, that is magmas whose composition has not been modified since the time of melt generation, are roughly gabbroic in composition. If a gabbroic melt crystallizes completely to a coarse-grained igneous rock, it is called a *gabbro*. Rocks that are both poorer and richer in silica than a gabbro are formed during the process of fractional crystallization, in which silica-poor material is crystallized and drops out of the magma, whose composition becomes more evolved and richer in silica. If the crystals in the magma chamber are denser than the liquid of the melt, they sink; others may remain in suspension, while some plagioclase crystals, which are of low density, are thought to float, forming rafts at the top of the magma chamber. Crystals that fall to the bottom of the magma chamber are usually those that are rich in magnesium and poor in silica, such as olivine and pyroxene. The crystals settle to form a *cumulate*, which is an ultrabasic rock (that is, it contains less than 45 per cent SiO_2 by weight).

Cumulates, being deposited by the liquids of the magma chamber, commonly display sedimentary structures that are the result of fluid dynamic processes acting upon the piles of crystals. Sedimentary features typical of this type of rock are graded bedding, developed when larger, heavier crystals were deposited at the bottom of a layer, and channel deposits, where a current of magma flowed across the surface of the crystal pile, forming deposits like those in a river channel. Ultrabasic cumulates are common in rock masses that appear to represent the remains of magma chambers.

The altered remains of igneous cumulates that form a large proportion of an ophiolite have been interpreted as the result of crystal fractionation during ocean-floor formation. The melt that remains after fractional crystallization is richer in silica than the parent magma. The formation of some very silica-rich granites has been attributed to the process of fractional crystallization.

During crystallization, melt may escape and be erupted at the surface, or new melt may arrive in the magma chamber, mixing with the melt already present and altering its composition back towards the composition of the parent magma. Through the cycle of fractional crystallization and replenishment, layered igneous intrusions may develop, in which each layer represents the products of crystallization of the parent magma, followed by the crystallization of the more evolved magma. The Skaergaard intrusion of east Greenland is a famous example of a layered ultrabasic intrusion showing chilled margins and coarse-grained crystalline layers.

The composition of a magma may also be changed by the addition (assimilation) of new material that has been melted from the sides of the intrusion. Assimilation entails the loss of heat from the magma in order to melt the contaminating rocks. It therefore seems unlikely that large amounts of contamination can occur in this manner without the solidification of the parent magma.

Granites

Granites are pale-coloured, silica-rich, intrusive igneous rocks. Current theory ascribes granitic magmas to two possible sources: some granites are interpreted as the result of partial melting of the crust to form a granitic melt; others are seen as the result of extreme fractional crystallization which removes mafic (dark-coloured, silica-poor) components from the melt. Large quantities of granite are known to be formed in the mountain ranges that result from continental plate collision

where they intrude the cores of the mountains as a granite batholith. Along the subducting margin of the Pacific Ocean are found the batholith belts of western North and South America. The largest of these is the coastal batholith of Peru, which is over 1600 km long and 60 km wide. One such batholith in England is the batholith of Devon, Cornwall, and the Isles of Scilly. The moorlands of the area are each apophyses rising from the top of a large batholith at depth. The roofs and edges of these granites are associated with the mineral and metal wealth of Cornwall, which has drawn traders and miners to the area for over 4000 years. The Phoenicians are known to have traded in the area and the Beaker People, who were associated with mining operations in continental Europe, also found their way to the west of England.

Contact metamorphism

Country rock intruded by magma is subjected to sudden heating, usually at low pressures. The mineralogical and textural alteration in the country rocks resulting from this heating by igneous bodies is called *contact metamorphism*. Contact metamorphism often produces hardening of the country rock, which is then known as a *hornfels*. New mineral growth is particularly obvious in shales that are rich in aluminium. The new minerals that grow in the dark shales include spots of the mineral cordierite ($Al_3(Mg,Fe^{2+})_2 [Si_5AlO_{18}]$), white needles of sillimanite (Al_2SiO_5), and crystals of andalusite (Al_2SiO_5) and, where high temperatures have been sustained, garnet (e.g. almandine ($Fe^{3+}_2Al_2Si_3O_{12}$)) may grow.

JUDITH M. BUNBURY

inverse methods Data analysis, whether in economics or the Earth sciences, entails the use of inverse methods. Inverse methods provide a set of techniques that can be used to reduce various observations or measurements into information about the world around us. Perhaps the simplest application of inverse methods is the fitting of a straight line to a suite of data. At the other extreme, a more sophisticated application is seismic tomography, the mapping of seismic velocity structure in the Earth.

The aim of most data analysis in the Earth sciences is to model an observed physical system in terms of model parameters—measurable characteristics of the system—constrained by experimental data. The problem as outlined incorporates three components: model parameters, observational data, and a relationship describing how the model parameters relate to the observed data. The model parameters completely describe the physical system. In addition we must consider how we represent these parameters; whether, for example, discretely or continuously, exactly or with associated error bounds. The observational data are the raw material from which we aim to extend our understanding of the physical system. The forward model defines the relationship between the observed data and the model parameters. There is an infinite variety of models, each a different representation of the same system. We aim to find the simplest model that explains the largest percentage of the available data. What follows is an introductory discussion on some concepts used in general inverse methods.

Parameterization

Model parameters and data can be defined continuously or discretely. Discrete data might include the mass of the Earth, or the travel time of a P phase, while continuous data might include a velocity gradient across the mantle. Most data are presented in discrete form for computational purposes: computers handle digital (discrete) information rather than analogue (continuous) information.

The physical system can be parameterized in many ways. For example, consider the velocity structure of the Earth. We can represent the velocity structure as horizontal layers of constant velocity. This is the starting point for many seismic refraction experiments used in the search for oil. Alternatively, large regions of the Earth can be parameterized into a number of cells, like a Rubik's cube, in which each cell has a constant velocity. The correct parameterization is implied by the type of observational data collected.

Further optimization is possible and often essential. The forward problem can be a linear or non-linear relationship. Mathematically solving linear problems is more straightforward. Thus we aim to find a linear relationship, or if this is not possible, approximate a non-linear relationship with a simpler linear relationship. For example, the travel time t of a seismic arrival depends inversely on the velocity v in the Earth ($t = x/v$, where x is the distance travelled by the ray). However, by defining the slowness s (equal to the inverse of velocity, $1/v$) we have a linear rather than non-linear forward model ($t = xs$), which can be solved more easily.

There are methods in the inversion tool kit that can be used to solve non-linear inverse problems within a certain framework. For example, perturbation theory enables complex non-linear systems to be solved on the assumption that in responding to small changes the system behaves in a linear fashion. A solution may then be found by making small perturbations to a starting model and minimizing the residuals (the error terms) between the data predicted by the new model and the experimental data. We assume that the starting model is close enough to the real solution for the inversion to converge to the appropriate solution. However, most non-linear problems cannot be treated in this fashion, and ideally we aim to 'linearize' the inverse problem.

The forward problem

We represent the N experimental data as an N-dimensional vector d and the M model parameters describing the physical system as an M-dimensional vector m. The forward model relating d and m usually takes the form of one or more formulae that d and m are expected to satisfy. These equations are often non-linear and difficult to solve, and we usually simplify the problem by linearizing. In this case, we can define an ($N \times M$) matrix G containing the linearized equations relating d and m:

$$d = G\,m$$

This matrix equation represents N algebraic equations. We then seek to solve the resulting set of linear equations, a basic problem in linear algebra.

$$d_1 = G_{11}M_1 + G_{12}M_2 + \ldots + G_{1M}M_M$$

$$d_2 = G_{21}M_1 + G_{22}M_2 + \ldots + G_{2M}M_M$$

$$\vdots$$

$$d_N = G_{N1}M_1 + G_{N2}M_2 + \ldots + G_{NM}M_M$$

Ideally, the inverse problem consists of N independent linear equations. This depends on the number of independent observations and model parameters. Three possibilities exist. If there are fewer observations than model parameters the inversion will in mathematical terms be poorly constrained or under-determined: no solution can be found without additional information. If the number of observations is about the same as the number of model parameters, the problem is said to be well posed: the model is well described by the parameters and there are enough data for a minimum in the space of all available models to be found. Finally, if the number of observations greatly exceeds the number of model parameters, the problem is over-determined.

The effect of noise (unwanted fluctuations accompanying the data) means that we can never exactly replicate our data. In consequence, finding exact solutions of inverse problems is rarely possible. Typically we want numerical values of the model parameters but we are usually forced to compromise between the information we would like to retrieve and what is obtainable from the actual data set. Instead of an exact solution we search for an 'answer', which may take the form of estimates of the model parameters, bounding values, or weighted averages of model parameters.

Most problems fall between the categories described above. These mixed-determined problems are composed of a combination of over-determined and under-determined problems. For example, in seismic travel-time tomography the data are travel times related to the velocity along the ray path. The model parameterization may be in terms of cells in the region of interest. The problem is at once over-determined, where many rays criss-cross in a single cell, and under-determined, where no rays pass through a particular cell. There will therefore be parts of the solution that are well resolved, corresponding to a high density of rays and thus data, and parts that are unresolved, corresponding to 'holes' in the data coverage.

The inverse problem

We want to solve for m given the linear problem posed in the form $d = Gm$ as defined above. Because of experimental errors, observational data will not fit the model exactly, and so we introduce a residual or error term e so that our equation becomes $d = Gm + e$. The best way to obtain a unique solution for the model parameters is to minimize the residuals. Several methods exist for doing this. For example, we can define a norm, a measure of the length of the vector e. These length measures are called L_N-norms where the length is

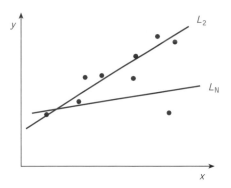

Fig. 1. Deciding on the measure used to define goodness-of-fit depends on the errors in the data and acceptable errors in the model parameters. In the example shown here, L_2-norm is appropriate. The higher-order L_N-norm is too sensitive to the outlier. In some cases, outliers are significant and so must be included in the calculation of the solution.

Fig. 2. A simple representation of a model space, the region containing all possible models that exist. The residual between a model and the experimental data is plotted on the y-axis. In the example shown, there are two local minima and a global minimum. There is no way of differentiating between a local minimum and a global minimum. However, various techniques exist for preferentially finding the global minimum.

defined as e to the power N. The simplest of these is the L_1-norm $(d - Gm)^1$, the linear distance between the predicted and observed data. The L_2-norm $(d - Gm)^2$, is more commonly known as the sum of least squares. The higher the order of the norm used, the greater significance outliers have for the final solution. If we are confident of our data, outliers must contain information about the solution. Using a norm of higher order (Fig. 1) increases the weighting of outliers in the inversion accordingly.

Although inverse methods provide a powerful tool for investigating physical systems, care is needed to ensure that the solutions found are the best possible. There may be several local minima that represent convergent solutions to the inverse problem that are nonetheless distinct from the global minimum we are searching for (Fig. 2). There is no means of telling that the solution found is not a local minimum without finding another smaller minimum. We must therefore also consider how to search the model space sufficiently well to be sure the solution we find is the global minimum. Many techniques exist. For example, we can perform a random search of the model space (the Monte Carlo method). In seismic tomography it is usual to use a starting model that is assumed to be already close to the global minimum, so that if a solution is found it must be the global minimum.

The formulation of an inverse problem is such that many models exist, although the impression is that the solution is a unique well-constrained model. Inversion models should be interpreted within the boundaries imposed by the context of the problem. Inverse methods can then continue to provide invaluable insight into physical systems in the Earth sciences.

D. SHARROCK

inversion tectonics Inversion tectonics describes the process by which formerly extensional, normal faults are reactivated in compression (*positive inversion*), and conversely,

reverse faults are reactivated in extension (*negative inversion*). The effect of positive inversion is to generate uplift, with the result that initially low-lying areas become topographically high. Inversion is a particularly common process at the margins of collisional mountain belts, where passive margins are subsequently incorporated in thrust belts; and along strike-slip fault systems, where rocks can experience repeated episodes of *transpression* and *transtension*.

Inversion can both enhance and reduce the oil and gas prospects for a sedimentary basin. It results in the formation of new structures capable of trapping oil and gas accumulations, many of which are expressed at the present surface as small hills. Surface mapping of these inversion-related anticlines has proved highly successful in finding oil and gas in eastern Sumatra. However, reactivation of faults may breach overlying horizons that seal an oil or gas accumulation, thereby destroying the effectiveness of the trap. Additionally, inversion can mean that reservoir and source rocks have formerly been buried too deeply to have retained significant porosity and organic carbon content. Because it is important to ascertain the maximum depth attained by a succession before its inversion, oil-exploration geologists have evolved specialized techniques to help quantify the magnitude of inversion. Vitrinite reflectance data provide an estimate of maximum burial depth by optical assessment of the thermal alteration of coaly material. Porosity studies compare the actual with the expected exponential decrease in the porosity of sedimentary rocks with depth. Abrupt jumps in this porosity profile correspond to episodes of inversion by an amount proportional to the size of the jump. Fission-track analysis looks at the intensity of spontaneous radiation damage, in the form of fission tracks in common sedimentary minerals such as apatite and zircon. Because the radiation damage occurs at known rates and temperatures, inversion events will be recorded by detectable changes in the rate of formation of fission tracks.

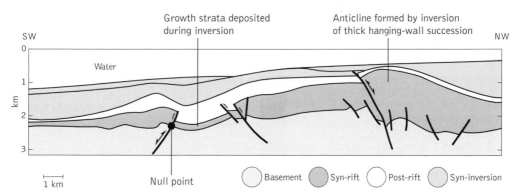

Fig. 1. Cross-section based on a seismic profile through part of the Sunda Arc in Indonesia. Note how the original, pre-inversion downthrow of faults can be recognized by the pronounced thickening of syn-rift strata in their hanging walls. Double arrowheads depict those faults which have experienced both initial normal movement and subsequent inversion. Modified from Letouzey (1990).

The effects of inversion can be subtle and its recognition requires well-constrained structural cross-sections supplemented, ideally, by seismic reflection profiles. Figure 1 illustrates the two principal observations that allow inversion to be diagnosed. First, thicker packages of *syn-rift* strata will occur on the upthrown sides of positively inverted faults. Second, the initially smooth increase in fault displacement, from a fault tip towards its centre, will be disturbed. Inversion is essentially instantaneous, imposing a more-or-less fixed increment of displacement on all parts of the fault plane undergoing inversion. Thus, a positively inverted normal fault will show the greatest reverse displacement where the initial phase of normal displacement was least.

In the Alpine foothills of Provence, Cretaceous syn-rift rocks deposited in the hanging walls of normal growth faults are typically ten times thicker than the equivalent footwall successions. Horizontal compression during the Palaeogene formation of the Alps subsequently inverted the normal faults, thereby placing the thickest units on the upthrown sides of faults whose sense of displacement is manifestly reverse. Seismic reflection profiles from geological settings comparable to the Alps, and analogue models of inversion using accurately scaled sandbox experiments, allow predictions to be made as to the most likely configuration at depth of the Provence examples. Such profiles reveal so-called *null points*. A null point is the point along an inverted fault at which the sense of offset of strata each side of the fault changes from normal to reverse (Fig. 1). The position of the null point along a fault provides an indication of the magnitude of the inversion event. With increasing amounts of positive inversion, the null point will migrate progressively deeper, down the fault plane, until the fault is said to have been totally inverted, so that it exhibits a reverse sense of offset along its full length.

Research has tended to concentrate on positive inversion, largely because of its special economic implications. This does not mean, however, that negative inversion is less common, or indeed less important as a geological process. Recent work in regions of lithospheric thickening has demonstrated how, after the relaxation of horizontal compression, the lithosphere may often be too thick to support itself. In north-west Scotland, for example, the lithosphere thickened during Caledonian orogenesis began to collapse under the influence of gravity from Early Devonian times. This collapse was accommodated, at least partly, by negative inversion of Caledonian thrust faults, such as the crustal-scale Moine thrust.

Clearly, not all faults will be equally susceptible to inversion, which may vary over small areas, often within individual sedimentary basins. The main factors governing susceptibility are the steepness of faults prior to their inversion, the orientation of faults relative to the stress doing the inverting, and the frictional resistance to reactivating the faults, as determined by, for example, fluid pressure along the fault plane. Positive inversion may be likened to squeezing a sponge in which the liberation of pore fluid causes local fluid pressure to build up along fault planes, thereby reducing their frictional resistance to reactivation. Faults formed during the initial phase of pre-inversion faulting are sometimes modified in an effort to lessen their steepness, and thereby minimize the resistance to inversion. In south-west Wales, for example, a slice of Precambrian crystalline rock, originally sited in the footwall of a major normal fault, has been sheared off by a relatively low-angle, 'footwall shortcut fault', such that it is now incorporated in the hanging wall of an inversion-generated thrust fault.

JONATHAN P. TURNER

invertebrates The kingdom Animalia, comprising over a million described species, has traditionally been divided into those animals that have backbones and those that do not; the vertebrates and the invertebrates. This is an artificial distinction and reflects our anthropocentric viewpoint; however, in the past the invertebrates were regarded as a formal taxonomic group of high rank. The artificiality of this arrangement becomes clear once the range of organisms included within the invertebrates is considered. No single character unites the group and the invertebrates range enormously in size, structural diversity, and in adaptations to different modes of existence. Some of the groups have common phylogenetic origins while others are only remotely related. Some are actually more closely related to vertebrates than they are to invertebrates.

Despite the fact that the word 'invertebrates' is currently used mainly as a term of convenience in biology and palaeontology, there are advantages in studying invertebrate organisms. Collectively they comprise much of the fossil record, just as they comprise much of the diversity of life now. The evolution of any one group is tied inextricably to that of the other invertebrates around it and to its changing environment. The evolutionary processes and environments of the past will thus be better understood if the invertebrates are studied as a whole.   DAVID K. ELLIOTT

ionosphere The upper levels of the atmosphere, above about 80 km, act as an absorber for harmful *ultraviolet radiation* and high-energy particles emitted by the Sun. Although some ultraviolet radiation passes through his level to be absorbed by ozone in the stratosphere, the most harmful radiation is absorbed at altitudes above 80 km. As this stream of particles and radiation from the Sun collides with atmospheric molecules, it causes some electrons to be split from their atoms and molecules. The resulting particles are ions. Positive ions are atoms and molecules that have had electrons removed, leaving them with a net positive electrical charge. Negative ions are those that have absorbed the free electrons, producing a net negative charge.

The ionosphere is the layer extending from about 80 km to the outer reaches of the atmosphere that includes charged particles. Since the air at these levels is electrically conductive it reflects some radio waves. Long wavelengths, for example those used in AM radio, are reflected at lower levels than shorter wavelengths, for example, the short-wave transmissions used by 'ham' radio operators. Very short wavelengths, such as those used for television and FM radio, are not reflected at all. The reflection of radio waves means that AM radio reception is possible at great distances from the transmitter.

The Aurora Borealis, or Northern Lights (Aurora Australis in the southern hemisphere) is also a manifestation of the ionosphere. It is produced when atoms and molecules which are excited (that is, put into a state of higher energy) by collision with particles from the Sun decay to their original state by emitting light.   CHARLES N. DUNCAN

Further reading
Slanger, T. G. (1994) Energetic molecular oxygen in the atmosphere. *Science*, 265 (5180), 1817–18.

iron–nickel deposits Nickel and iron are important elements used in a variety of everyday materials. Both are important constituents of stainless steel, where nickel, for example, gives the steel strength and anti-corrosion properties.

Two main types of iron–nickel deposits exist: ores related to laterites and those associated with mafic and ultramafic rock types in a wide range of tectonic settings. Each type of deposit is discussed below.

Lateritic ores
Laterites are formed in subtropical climates where silicate minerals are broken down leaving behind soils composed of insoluble oxides such as aluminium oxide (Al_2O_3) and iron oxy(hydr)oxides ($FeOOH$, Fe_2O_3). Dissolved silica and other alkali metals such as potassium, sodium, and calcium are leached from the rock into the underlying water-table, where they may subsequently react to form secondary minerals.

Where laterites form above ultramafic igneous rocks the nickel and iron contained within the minerals that comprise the rocks may be concentrated to form economic deposits. The climate and topography of the region in which laterite formation occurs will control the grade of ore produced. In tropical climates weathering of the silicate minerals is too rapid for new silicate minerals to be formed at depth, and in temperate climates weathering is too slow to generate an economic deposit. Subtropical regions therefore yield the richest lateritic nickel deposits. In areas where topography is steep, weathered material may be removed before nickel has been released from the underlying formations and had an opportunity to form new nickel-rich minerals. Important lateritic nickel deposits are found in New Caledonia, Indonesia, and Cuba, all of which are associated with island arc volcanism and plate margins. The typical structure of a nickel-rich laterite is shown in Fig. 1.

Nickel sulphide ores
Nickel sulphide ores can be subdivided into two main types: those associated with dunites/peridotites in Archaean Greenstone belts (e.g. Kambalda and Agnew deposits, Australia) and those associated with layered igneous intrusions (e.g. the Bushveld complex, South Africa). The most productive nickel ore deposit in the world is in Sudbury, Ontario (Canada). The geological interpretation of this deposit has been a source of controversy, but it is now generally agreed that it was formed as a result of an asteroid impact that triggered magmatic activity.

Sulphide ores associated with peridotites occur within flows of ultramafic lavas classified as komatiites, which usually contain elevated magnesium concentrations. The nickel sulphide ores are found at the base of the sequence of lava flows. The main ore mineral is pentlandite ($(Fe,Ni)_9S_8$); other common minerals include pyrrhotite ($Fe_{1-x}S$), chalcopyrite ($CuFeS_2$), pyrite (FeS_2), magnetite (Fe_3O_4), and iron-rich chromite ($FeCr_2O_4$). The ores are thought to have formed by

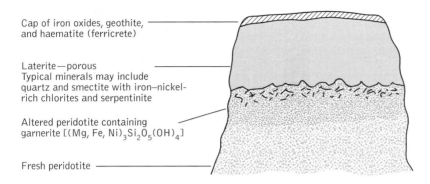

Cap of iron oxides, geothite, and haematite (ferricrete)

Laterite—porous
Typical minerals may include quartz and smectite with iron–nickel-rich chlorites and serpentinite

Altered peridotite containing garnerite [$(Mg, Fe, Ni)_3Si_2O_5(OH)_4$]

Fresh peridotite

Fig. 1. Structure of a nickel-rich laterite.

the segregation of an immiscible sulphide-rich liquid from an ultrabasic magma, but the precise mechanism of the process is not fully understood.

Ores associated with intrusive dunites tend to occur in lens-shaped bodies in which the primary mineral is olivine or its breakdown products. One theory proposes that these deposits are the feeders for nickel sulphide deposits in komatiitic formations. Sulphide mineralization in these deposits tends to be disseminated throughout the body and is dominated by pyrrhotite and pentlandite phases. The origin of the deposits is considered to be similar to that of the volcanic–peridotitic ores, where segregation from an ultrabasic magma appears the most likely scenario. The disseminated nature of the deposits indicates, however, that segregation probably took place at a later stage than for the komatiitic-associated ores.   ELIZABETH H. BAILEY

island arcs Island arcs have long been recognized as tectonically active belts of intense seismic activity containing a chain or arc of active volcanoes. At the turn of the nineteenth century, W. J. Sollas drew attention to the correspondence of the arc-like form of the Aleutians/Alaskan Peninsula, the East Indies (Indonesia), and several mountain chains to a series of great circles, and C. Lapworth discussed the 'Volcanic Girdle of the Pacific' (better known as the Pacific 'Ring of Fire') as a continuous 'septum' separating 'plates' with different histories and thicknesses. The deepest parts of the oceans, the elongate deep-sea trenches, were located on the oceanward side of these arcs. As the nature of the ring of fire was examined, it was realized that a line, called the *andesite line* could be drawn around the Pacific outside which andesites occurred (named after their type area in the Andes) and inside which basalts predominated.

Little could, however, be done to elucidate the origin of island arcs until geophysical data were acquired. The first, in the 1930s, to tell us something about the nature of the internal structure of the oceans was the evidence obtained by the Dutch geophysicist F. A. Vening Meinesz, who showed that an extensive belt of large negative isostatic gravity anomalies ran

parallel to the island arc. A surprising feature of this discovery was that the centre of this belt ran to the island side of the trench axis, not under the trench itself. it is therefore not related to the present topography but to some deeper root; and, since all surface evidence in trenches and island arcs shows that they are under tension, it was considered as proof that there must be some downward stress that was dragging down a mass of material of relatively low density. It gave rise to the concept of the *tectogene*, which involved the vertical down-warping of oceanic crust, either by compressional forces or the drag of converging convection currents. It was not, however, until 1949, when H. Benioff showed that earthquake epicentres became progressively deeper as one went from the ocean side of the trench to the volcanic arc, that the idea of a relatively simple steeply dipping thrust plane extending from near the trench to a depth of as much as 700 km was clearly established.

Thus, by the 1950s substantial geophysical data had been acquired around the Pacific, off Indonesia, and in the Caribbean to suggest that large slabs might be dragged down beneath island arcs along subduction zones, which are commonly known as Benioff zones.

It was nevertheless not until 1968 that the next significant advance was made. The establishment of the hypothesis of ocean-floor spreading in the 1960s had shown that new lithosphere was being continuously created (see *mid-ocean ridges*), and it was realized that unless the Earth was expanding (and at that time some geoscientists seriously argued in favour of this) an equal amount of lithosphere must be lost, and that this must happen at the subduction zones. As the slab of oceanic lithosphere goes down, it melts partially at about 150–200 km depth, giving birth to magmas that rise and are extruded in volcanoes located 150–200 km from the axis of the trench.

A subduction zone is identified by seismic foci, the seismic activity being concentrated on the upper surface of the down-going slab of lithosphere. The seismic activity defines the 'seismic plane' of the subduction zone, which may be up to 20–30 km wide. Subduction zones dip mostly at angles between 30° and 70°, but individual subduction zones dip more steeply with depth. The dip of the slab is related inversely to the velocity of convergence at the trench, and is a function of the time since the initiation of subduction. Because the downgoing slab of lithosphere is heavier than the plastic asthenosphere below, it tends to sink passively; and the older the lithosphere, the steeper the dip.

It is important to remember that the term 'island arc' is commonly used as a synonym for 'volcanic arc', yet the two terms are not quite the same. Volcanic arcs include all volcanically active belts located above a subduction zone, whether they are situated as islands in the middle of oceans or on continents, as along the west coasts of Central and South America. Island arcs include only those separated from the land by a stretch of water.

There is therefore a spectrum of island-arc types. Some are truly intraoceanic, being situated entirely within the oceans (Marianas, New Hebrides, Solomons, and Tonga in the Pacific; the Antilles and Scotia arcs in the Atlantic). Others are separated from major continents by small ocean or marginal basins with a crust that is intermediate between continental and oceanic (Andaman islands, Banda, Japan, Kuril, and Sulawesi). The Aleutian arc passes laterally into a continental 'Cordilleran-type' fold belt on the North American continent. At the extreme end of the spectrum are those arcs built against continental crust, such as the Burmese and Sumatra/Java portions of the Burmese–Andaman–Indonesian arc, and finally the Andean chain, where the volcanic belt is located entirely within the continent and is not therefore an island arc. The age also varies. Some are very young: less than 10 Ma (10×10^6 years). Others are much older, with histories that go back at least to the Tertiary, if not to the Cretaceous.

The exposed magmatic island arc is only one of a number of features or tectonic zones that extend from the trench at the oceanward end to the marginal or back-arc basin on the continental side (Fig. 1).

Trenches are the deepest features of ocean basins, with depths ranging from 7000 to almost 11 000 m, the deepest being the Mariana and Tonga trenches. Most deep-sea trenches in the Pacific are floored by normal basaltic oceanic crust overlain by pelagic sediments and ash. This relatively thin sedimentary layer is easily subducted under the overriding plate. This simple subduction pattern may, however, be disturbed because, in some instances the trench may receive a substantial amount of terrigenous sediment derived from a continent. Although this feature is most important in the fill of deep-sea trenches immediately adjacent to continents (e.g. the Andean-type subduction zones of Central and South America), it may also be important where the trench is fed from one end, as off the Andaman islands. In addition, enormous volumes of sediment accumulate on ocean floor as deep-sea fans. The largest of these, the Bengal Fan, has a volume of 4 million km³, and this is being swept eastwards towards the Burmese–Andaman–Indonesian arc. The other complication is due to the fact that ocean floors are not simple, smooth features of a constant 5000–6000 m depth, but are composed of a variety of topographic features, such as oceanic plateaux, ridges, sea mounts, and guyots (flat-topped submarine mountains) that may rise thousands of metres above the ocean floor and have areas that range from a few square kilometres to thousands of square kilometres. When these irregularities reach the deep-sea trench and come up against the overriding plate they may either be subducted and incorporated into the downgoing slab or scraped off the oceanic plate and incorporated in the overlying wedge.

In this way an accretionary prism or wedge develops that has a ridge at the surface, known as the outer arc ridge or trench-slope break, behind which is the fore-arc basin. This prism comprises a thrust stack of scraped-off sediments from the trench, which forms by tectonic accretion. The sediments can be examined today in the Mentawai islands west of the main Indonesian islands and in Barbados, a low-lying nonvolcanic island east of the Lesser Antillean volcanic chain. An important feature of the accretionary prism is that while each sedimentary succession gets younger upwards, there is a decrease in depositional and metamorphic age as one goes from the earliest-formed wedge on the volcanic arc side towards the toe on the ocean side, as successively younger slices of oceanic sediments and floor are scraped off. It is in fact this successive 'underplating' that progressively raises the older slices and causes the elevation of the fore-arc region. A striking feature of some accretionary prisms, recognized in Barbados for many years, has been the abundance of mud diapirs (see *mud and sand volcanoes*) formed by high-pressure fluids within the unconsolidated hemipelagic muds and biogenically generated methane that are carried down

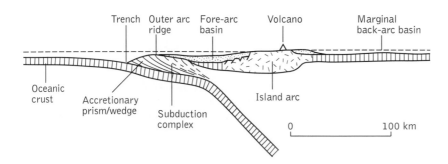

Fig. 1. Cross-section through an intra-oceanic island arc.

beneath the accretionary prism so that they become over-pressured and rise through the prism. The surface of the accretionary prism is not a simple one. Because it was formed as a result of the successive accretion of individual slabs of crust, the surface displays a rugged and irregular sea-floor morphology governed by numerous tectonic ridges that form by folding and dislocation. As each successive thrust fault propagates oceanward through the wedge of uplifted trench sediments, a deformation front will separate the undisturbed trench environment from the disrupted accretionary prism, and this structural boundary migrates seaward with time. Simultaneously, however, older, higher thrusts may be reactivated. Thus this constantly reactivated complex and prism has innumerable small sedimentary basins known as trench-slope basins.

Between the volcanic arc and the accretionary prism is the fore-arc basin, which may be as much as 100 km wide. Typically such basins are filled by immature volcaniclastic sediments derived from the erosion of the volcanic arc. Sedimentation in fore-arc basins can, however, be quite complex and almost any type of sediment can be deposited.

For example, in the very extensive fore-arc basin west of the Burma–Sumatra–Java volcanic arc and east of the Andaman–Nicobar–Mentawai islands outer arc, sediments range from fluvial–deltaic in the north, fed by the Burmese rivers, to deep-water turbidites and shelf sediments further south.

The volcanic suites of island arcs vary according to the thickness and composition of the overriding lithosphere. At true intraoceanic islands, mantle-derived magmas are relatively unimpeded in their ascent, resulting in eruption of very fluid, often aphyric (microcrystalline and without phenocrysts) tholeiitic basalts and basaltic andesites. As the arc develops, the thickening arc massif depresses the oceanic crustal layer. Once the crust has thickened to some 20 km, it may start to act as a density filter, and primary magma may become ponded in a series of interconnected high-level magma chambers. The upward ascent of magma, particularly under the centre of the arc, is a slow and fitful affair, and differentiation processes generate calc-alkaline magmas, which contain relatively large amounts of calcium (CaO) in relation to alkalis (Na_2O and K_2O), and more intermediate magmas such as andesite. In the Japanese arc it appears that erupted

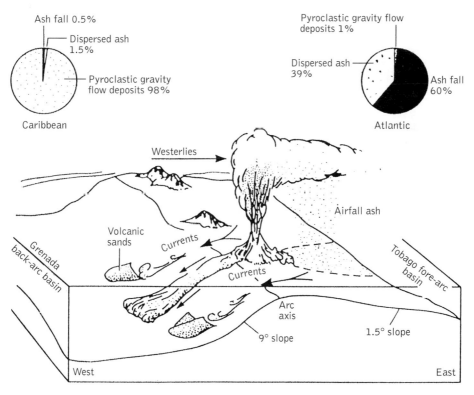

Fig. 2. Schematic illustration of the Lesser Antilles volcanic arc, showing the distribution of volcanigenic deposits and main processes responsible for the asymmetric redistribution of sediment. Along the 400 km length of the Lesser Antilles volcanic arc some 500 km³ of volcanic material has been produced in the past 100 000 years. Only 20 per cent of this material remains in the volcanic arc; 80 per cent has been redistributed. (From Sigurdsson *et al.* (1980) *Journal of Geology.*)

557

magmas become more alkaline away from the trench, apparently reflecting the increasing depth to the source of the magmas along the Benioff zone. Not all arcs follow this simple pattern, however, probably because of tectonic inhomogeneities in the overriding plate, variable depths of partial melting, and the type of lithosphere subducted.

Oceanic volcanic arcs are surrounded by large volcaniclastic aprons, kilometres thick, whose volume may far exceed that of the volcanoes. Although the foundation of the arc may be dominated by lavas, most of the apron consists of pyroclastic fragments generated by explosive volcanic activity in shallow water or on land, or reworked volcaniclastic rocks. As the submarine slopes of arc-related volcanoes are steep, there is great seismic activity and sedimentation is rapid resedimentation by slumping, sliding, and turbidity currents is common. Unfortunately, the sediments immediately adjacent to island arcs are not easily penetrated by deep-sea drilling, and we mostly know about such sediments from their uplifted equivalents on land, either in modern island arcs such as the New Hebrides and Japan or in their ancient equivalents such as the Ordovician of North Wales and in the English Lake District.

Some of the more interesting results that have come from the study of both modern (e.g. the Lesser Antilles, New Hebrides) and ancient examples have shown that:

(1) the distribution of sediments around an arc is usually asymmetric, owing to the prevailing wind patterns, different arc slopes, and ocean currents (as in the Lesser Antilles where westerly winds blow most of the ash into the Atlantic) (Fig. 2);

(2) as the island arc develops, older and more mafic volcaniclastics, including scoria, hyaloclastites (formed when hot lava comes into contact with water or wet sediment), and shards are succeeded by products of more differentiated magmas, including abundant pyroclastic material; dormancy of the island arc can result in blanketing of the volcaniclastic apron with silty epiclastic (redeposited) turbidites;

(3) the position of the island or individual volcanoes can migrate, often oceanward towards the trench or along the arc;

(4) as the volcano grows into shallow water and emerges, eruptions become more explosive, with ash dispersed over greater distances from the volcano;

(5) sedimentary processes continually sort volcaniclastic fragments by grain size and density into a proximal coarse-grained facies of pillow breccias, debris-avalanche, and lahar (mudflow) deposits; a medial debris-flow facies; and a distal facies consisting of thin distal turbidites and fallout ashes; and

(6) the rates of arc volcanism and back-arc spreading can vary with periods of continuous arc volcanism and reduced spreading favouring progradation of the volcaniclastic apron.

As island arcs develop, enlarge, and become more mature, as in Japan and the North Island of New Zealand, terrestrial sediments and plants abound, and lagoons and lakes develop, especially within the caldera of the volcanoes. Uplift may reveal the plutonic core of the arc, to expose calc-alkaline, predominantly dioritic to granodioritic plutons. Because island arcs are zones of overall extension, rift valleys develop, some with a component of strike-slip, as in Japan or the Taupo rift of the North Island of New Zealand.

Although the type of mineralization found within an island arc varies according to its age, the characteristic class of ores are syngenetic massive sulphides, the so-called Kuroko ores named after their type locality in Japan. These contain pyritic zinc–lead–copper as well as silver and gold. They are closely associated with marine pyroclastic rhyolitic domes and calderas and were probably deposited from submarine brines in or on the flanks of local sedimentary basins.

A striking feature of the western Pacific Ocean is the enormous area covered by a large and complex pattern of basins that lie behind the volcanic arcs and are marginal to the continent. These marginal basins have been a source of controversy ever since it was realized that their crusts, while usually having thicknesses close to that of continental crust, have seismic velocities closer to those of oceanic crust. Most marginal basins are now known to be old ocean floor trapped behind an island arc and are recognized not only in the western Pacific but also in the Andaman sea behind the Burmese–Indonesian volcanic arc, and behind the Antillean and Scotia arcs. They range in age from very young back-arc basins that have developed within oceanic crust relatively recently (intraoceanic back-arc basins) to those mature basins adjacent to continents, such as the Japan Sea, which is inactive at present (continental back-arc basins).

We now know that these oceanic back-arc basins are formed by crustal extension producing first rifts and then new ocean crust by sea-floor spreading, similar to that which forms mid-ocean ridges (see *mid-ocean ridges*). The extensional origin of these basins is indicated by the presence of normal faults, high heat flow, and, in some instances, magnetic linear anomalies about a central rift. Back-arc spreading occurs particularly when older, colder lithosphere is being subducted and where the subduction zone dips steeply. Although there is no consensus as to the precise mechanism for the spreading, it is probably a consequence of several mechanisms, such as migration oceanward of the trench and volcanic arc over the subducted plate, or magma intrusion and upwelling due either to thermal upwelling of a magma diapir or to heat generated by a plume or by the subducted slab. HAROLD G. READING

Further reading

Busby, C. J. and Ingersoll, C. V. (1995) *Tectonics of sedimentary basins*. Blackwell Science, Cambridge, Massachusetts.

Fischer, R. V. and Smith, G. A. (eds) (1991) *Sedimentation in volcanic settings*. Special Publication No. 45, Society of Economic Paleontologists and Mineralogists, Tulsa.

Orton, G. J. (1996) Volcanic environments. In Reading, H. G. (ed) *Sedimentary environments: processes, facies and stratigraphy*, pp. 485–567. Blackwell Scientific Publications, Oxford.

islands, oceanic *see* OCEANIC ISLANDS

isostasy
Isostasy (a word derived from the Greek for 'equal standing') is essentially the principle of hydrostatic equilibrium applied to the Earth. In its simplest form, it considers rigid blocks of the Earth (usually taken to be the crust) to be buoyantly supported in an underlying fluid medium (usually taken as the mantle) and free to move vertically. These blocks will then move until their weight is exactly balanced by their buoyancy, at which point they are said to be 'in isostatic equilibrium'.

The concept was originated by French surveyors in the eighteenth century working around the Andes, who noted that the observed gravitational attraction of the mountains was less than that predicted. They inferred the presence of a low-density 'root' that balances the excess weight of the mountain range and also reduces its gravitational attraction. The theory was further developed by the English geodesists Pratt and Airy in the nineteenth century, who have given their names to two forms of the theory.

In both versions, it is assumed that different crustal blocks have different thicknesses. Pratt's hypothesis (Fig. 1a) assumes that the bases of the floating blocks are at a constant depth and that their densities vary inversely with their thickness, so that the total weight of each column, or pressure at its base, is

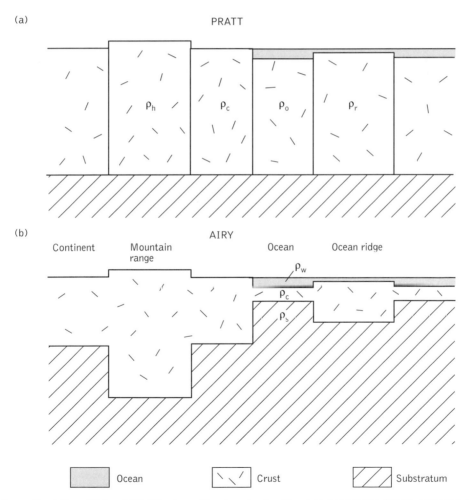

Fig. 1. Varying forms of isostatic equilibrium. (a) Pratt isostasy achieved in variable thickness and density blocks with a constant basal depth; (b) Airy isostasy achieved with blocks of variable thickness and constant density with variable basal depths. (Modified after Bott, M. H. P. (1982) *The Interior of the Earth, its structure, constitution and evolution*, Fig. 2.14. Edward Arnold, London.)

the same everywhere. Nowadays the requirement for constant-depth bases seems arbitrary and contrary to much observation, although some real situations (e.g. the variations in depth, thickness, and density of oceanic lithosphere with age) approximate to the Pratt mechanism.

In Airy's mechanism, the surface layer has a constant density but variable thickness (Fig. 1b). Blocks move so that at a certain constant depth (at or below the deepest base) the weights and pressures are again equalized. The tops of the thicker blocks will then float higher, while their bases sink deeper into the substratum, forming, for example, mountains and their roots. Airy's mechanism is more generally applicable than Pratt's, and is a reasonable model for variations in continental crust. However, it should be realized that both the Airy and Pratt mechanism are particular cases with simplifying assumptions, neither of which is required by the general idea of isostasy. In general, both the thickness and density of blocks vary, and there is usually no reason to assume that their bases lie at a given depth.

The concept of isostasy was originated well before modern ideas of a lithosphere and asthenosphere. When the Mohorovičić discontinuity, separating the light crust from the denser mantle, was discovered by seismology, it was natural for geologists to equate the crust with the solid, floating layer of isostasy and the mantle with the fluid substratum. However, it is now clear that the rheology of the lithosphere is more complex than this. In many applications it is appropriate to consider the lithosphere (crust and uppermost mantle) as the floating layer, while the asthenosphere provides the 'fluid' substratum.

In fact, isostasy is not confined to lithospheric-scale blocks, but will apply at any scale where rigid blocks overlie ductile layers. For example, in many places there is a weak, ductile layer in the mid- to lower crust, and fault blocks in the rigid, upper crust can move in this ductile layer to reach equilibrium. It is also often unrealistic to think of free-moving blocks bounded by simple crustal- or lithospheric-scale faults. A more realistic model contains a strong but elastic lithosphere which can flex under applied loads, so that its equilibrium does not depend simply on local hydrostatic equilibrium; instead, the loads are distributed by the strength of the plate and compensated regionally by its flexure.

In any form of the theory, the depth at which pressures are equalized is called the 'compensation depth'. For 'pointwise' isostatic mechanisms, such as Airy's or Pratt's, if the thickness and densities of the blocks or layers are known, it is a matter of simple arithmetic to calculate the expected topography for bodies in isostatic equilibrium. In fact, if one assumes isostatic compensation, then given any two of surface topography, crustal thickness, or crustal density, the third can be calculated. Alternatively, if all three are known, one can estimate the degree to which the observed topography is in isostatic equilibrium. For regional or flexural isostasy this can still be done, but the calculations are more complex.

Areas that have thicker or lower-density roots than are required for equilibrium are said to be 'overcompensated', and those that are too thin or too dense are 'undercompensated'. In either case, the lack of isostatic equilibrium indicates that the topography must be supported either by the finite strength of the lithosphere or by dynamic forces (as, for example, in a tectonically active area).

Many regional-scale geological processes tend to disturb isostatic equilibrium. For example, erosion makes blocks thinner or reduces their weight, so that a mountain being eroded will tend to rise to maintain equilibrium. In contrast, the deposition of sediments represents an added load, so that sedimentary basins tend to sink. Extension of the crust to form rift valleys thins it, while compression in mountain-building thickens it. Loading and unloading by glaciers also affects isostatic equilibrium. In all these cases, the lithosphere will move in order to maintain or regain isostatic equilibrium, and the ductile asthenosphere will flow in response.

Isostatic forces are thus of major importance in controlling the topography of the Earth's surface. For the geologist reconstructing ancient topography (for example, the throw on a fault or the depth of a sedimentary basin or continental margin), it is vital to correct for varying loads and their isostatic effects.

ROGER SEARLE

Further reading

Kearey, P. and Vine, F. J. (1996) *Global tectonics*. Blackwell Science, Oxford.

isotope geochemistry
Isotope geochemistry has provided the basis for some of the most important advances in the Earth sciences. Studies of the distribution of radiogenic isotopes in rocks and minerals are the only accurate and precise way we have of measuring the passage of time in geological processes. This has implications for fields as diverse as dating the age of the Earth to measuring rates of evolutionary change in planktonic microfossils. Radiogenic and stable isotopes can also be used to trace the sources of elements in processes such as crust–mantle interactions and the factors that control the composition of sea water. The temperature-dependence of stable isotope fractionation can also be used to recover information about low-temperature phenomena, such as climate change between glacial and interglacial intervals as well as processes at higher temperatures, such as the crystallization of granite magmas. In short, there is almost no field of geology that has not been touched by the study of natural isotope variations.

General principles

The nucleus of an atom is made up of neutrons and protons. By definition, all atoms of a particular element have the same number of protons, but most elements have atoms with differing numbers of neutrons. These different types of atom of the same element are termed *isotopes*. For example, carbon has three isotopes of interest to Earth scientists: carbon-12 (^{12}C), carbon-13 (^{13}C), and carbon-14(^{14}C). All these isotopes contain six protons in the nucleus, but each has a different number of neutrons. For example, the nucleus of ^{12}C also contains six neutrons, to give a total mass number of 12. Of these three carbon isotopes, ^{12}C and ^{13}C do not undergo radioactive decay, and are termed *stable isotopes*. In contrast, ^{14}C is unstable and decays to ^{14}N (nitrogen-14). The process by which unstable isotopes spontaneously decay to other isotopes (which may themselves be unstable) is called *radioactivity*. During radioactive decay a parent isotope decays by one of several different mechanisms to form a daughter isotope of a different atom. The isotopes that are formed from radioactive decay are termed *radiogenic isotopes*. The variations measured in the isotopic composition of elements in different reservoirs on the Earth are due to two separate processes. Different amounts of radiogenic isotopes arise from variations in the parent/ daughter ratios of different reservoirs and the passage of time, which leads to the steady addition of radiogenic isotopes to a reservoir. Isotope variations can also arise among stable isotope compositions if one isotope is preferentially incorporated in one of two phases that interact with one another. These two processes form a natural division in isotope geology between radiogenic isotope studies and stable isotope studies.

Radiogenic isotopes

There are many natural radioactive isotopes. Those of special interest to geological studies are $^{40}K/^{40}Ar$ (potassium–argon), $^{87}Rb/^{87}Sr$ (rubidium–strontium), $^{147}Sm/^{143}Nd$ (samarium–neodymium), $^{232}Th/^{208}Pb$ (thorium–lead), $^{235}U/^{207}Pb$, and $^{238}U/^{206}Pb$ (uranium–lead). Details of these systems are given in Table 1. The half-life of a radiogenic isotope is the time it takes for half a given number of atoms of the parent isotope (e.g. ^{40}K) to decay to the daughter isotope (e.g. ^{40}Ar).

Each of the isotope systems has slightly different applications in the Earth sciences. Isotopic dating is among the most important (see *isotopic dating*). The essential provisions for this dating technique to be valid are that the minerals were formed at the same time in equilibrium with one another and that they have not gained or lost any rubidium or strontium since they were formed. Various isotope systems can be used for dating igneous rocks, but differences in their geochemical behaviour mean that they are applied to different dating problems according to how abundant the parent or daughter

Table 1 Natural radioactive isotopes of special geological interest

Isotope system	Half-life ($t_{1/2}$)	Isotope ratio of interest
$^{40}K \rightarrow {}^{40}Ar$	1.40×10^9 years	$^{40}Ar/^{36}Ar$
$^{87}Rb \rightarrow {}^{87}Sr$	4.89×10^{10} years	$^{87}Sr/^{86}Sr$
$^{147}Sm \rightarrow {}^{143}Nd$	1.06×10^{11} years	$^{143}Nd/^{144}Nd$
$^{232}Th \rightarrow {}^{208}Pb$	1.40×10^{10} years	$^{208}Pb/^{204}Pb$
$^{235}U \rightarrow {}^{207}Pb$	7.04×10^8 years	$^{207}Pb/^{204}Pb$
$^{238}U \rightarrow {}^{206}Pb$	4.47×10^9 years	$^{206}Pb/^{204}Pb$

Table 2 Stable isotopes of special geological interest

Isotopes	Isotope ratios	Notation
^1H, ^2H	$^2H/^1H$	δD
^{10}B, ^{11}B	$^{11}B/^{10}B$	$\delta^{11}B$
^{12}C, ^{13}C	$^{13}C/^{12}C$	$\delta^{13}C$
^{16}O, ^{17}O, ^{18}O	$^{18}O/^{16}O$	$\delta^{18}O$
^{32}S, ^{33}S, ^{34}S, ^{36}S	$^{34}S/^{32}S$	$\delta^{34}S$

isotopes are in the rocks of interest and how susceptible the isotope system is to varying degrees of metamorphic alteration.

Different reservoirs in the Earth have evolved different radiogenic isotope compositions over time as a result of varying parent/daughter ratios. For example, rubidium is preferentially partitioned into the crust relative to strontium. The Rb/Sr ratio of continental crust is consequently higher than that of the mantle, and on average the $^{87}Sr/^{86}Sr$ ratios of crustal rocks are higher than those of rocks derived from the mantle.

Stable isotopes

Variations in the stable isotope composition of an element can arise during several different kinds of chemical reactions and physical processes. This is a consequence of the fact that certain thermodynamic properties of molecules are dependent on the masses of their component atoms. In practice, this stable isotope fractionation is significant only in isotopes with masses of less than 40. The major stable isotopes of interest in geological applications are shown in Table 2.

Stable isotope variations are generally referred to in terms of the del-notation, rather than by their isotope ratios. For example, oxygen isotope compositions are given by:

$$\delta^{18}O = \left[\left(\frac{^{18}O/^{16}O_{sample}}{^{18}O/^{16}O_{standard}} \right) - 1 \right] \times 10^3$$

Hence, the more positive the $\delta^{18}O$ value, the heavier the oxygen isotopic composition; that is, the higher the $^{18}O/^{16}O$. The extent of stable isotope fractionation between two different species is generally dependent on temperature in a relatively simple way, the magnitude being greater at lower temperatures. Other general rules of stable isotope fractionation are: the light isotope of an element is preferentially partitioned into reduced species (e.g. the $\delta^{34}S$ of SO_4^{2-} is heavier than that of coexisting H_2S); biota tend to enrich the light isotope of an element (e.g. the $\delta^{13}C$ values of plants are lighter than that of atmospheric CO_2); and the light isotope of an element will favour the heavier cation in two minerals formed in equilibrium (e.g. the $\delta^{34}S$ of galena, PbS, is lighter than that of coexisting sphalerite, ZnS).

Other applications of isotope geochemistry in the Earth sciences

The stable isotope composition of rocks and minerals derived from the mantle (e.g. mid-ocean ridge basalts (MORB) and ocean island basalts (OIB)) do not generally show large variations; this is because the high temperatures in the mantle preclude significant stable isotope fractionation. However, radiogenic isotopes (mostly, strontium (Sr), neodymium (Nb), and lead (Pb)) show considerable variations that indicate that the mantle is not a homogeneous body, but contains reservoirs with different isotope compositions that reflect processes within the mantle and exchange of material between the crust and mantle. Some portions of the mantle have isotope compositions similar to those expected of a homogeneous Earth that has not undergone any differentiation. In contrast, some mantle-derived rocks have isotope compositions that indicate that they have had substantial portions of crust extracted from them. Other portions of the mantle have isotope compositions that appear to reflect recycling of continental material back into the mantle or movement of metasomatic fluids within the mantle, or both.

Destructive plate margins such as island arcs are major sites of interaction between the crust and the mantle. In some instances they have stable isotope compositions (particularly oxygen and boron) and radiogenic isotope composition (mostly strontium and lead) that extend outside the variation shown by MORB and OIB, indicating that they contain an additional source of some elements. The isotope composition of this additional source is similar to that of the altered oceanic crust and sediments that are carried down by the subducting plate beneath the island arc. Hence, isotope studies provide some of the clearest evidence that at least some subducted material is driven off the subducted slab (either by melting or dehydration reactions) and incorporated into the overlying volcanic rocks.

The nature and the growth and evolution of the continental crust is one of the most fundamental aspects of the study of the Earth. There has been some disagreement over whether virtually all the continental crust formed very early in the Early's history and has remained relatively constant in size since that time, or whether it has been growing at a more or less constant rate since the Earth formed. Isotope studies have provided partial answers to this question through the direct dating of portions of the continental crust (primarily using U–Pb dating of zircons) and by determining the average age of sediments weathered from the continents (from the Sm–Nd isotope system). These studies appear to show that there have been periods in Earth history when there has been more rapid continental growth (particularly around 2.7 Ga (2.7×10^9 years)) and that the continental crust has remained relatively constant in size over the past 2 Ga.

Granites comprise a significant portion of the continental crust. Isotopes (and other geochemical tracers) indicate that a high proportion of granites have compositions that are intermediate between two end members: I- and S-types. The S-type end members tend to have isotope compositions (in particular for oxygen and strontium) that are very similar to old continental crust and clastic sediments. It is therefore thought that S-type granites are derived from melting of these materials. In contrast, I-type end members have isotope compositions that are closer to those of mantle-derived rocks. They are accordingly thought to represent melts derived from the upper mantle without significant subsequent interaction with the crust.

Many types of ore deposits are thought to have formed from the circulation of aqueous fluids through heated rocks. During this process chemical and isotopic exchange takes place between the reservoirs and is recorded in the isotope composition of hydrothermal minerals. Isotope studies (particularly of boron, carbon, sulphur, strontium, and lead) yield information about the source of the elements in the mineral deposit (for example, whether an element was derived from sea water or from rocks). Oxygen and hydrogen isotopes are especially useful in determining the origin of the fluids and the hydrology of the ore deposit (for example, whether the fluids were sea water, brines, or meteoric water, and how extensive the reactions were between the fluids and the rock).

When plankton precipitate calcium carbonate shells the oxygen isotope composition of their shells records the temperature and isotopic composition of the sea water in which they grew (which is largely dependent on ice volume). When they die, their shells accumulate in sediments at the sea floor. Analyses of microfossils from shells of different ages can thus be used to reconstruct the climatic history of the oceans. These studies have revealed regular variations in the Earth's climate, and it is apparent that our climate is linked to changes in the Earth's orbit and in the tilt of its axis of rotation (Milankovich cycles).

M. R. PALMER

Further reading

Faure, G. (1986) *Principles of isotope geology* (2nd edn). John Wiley and Sons, New York.

isotopic dating The possibility of determining the ages of rocks and minerals by using radioactive isotopes was recognized early in the twentieth century by scientists such as Arthur Holmes in the UK and B. B. Boltwood in the United States, but the widespread application of these methods was not possible until the 1950s, when precise analytical methods became available. The special merit of radiometric dating is that it provides ages in years (usually in millions of years), as compared with most of the other methods available to the geologist, which yield only relative ages.

Isotopic dating depends on measuring variations in the abundance of naturally occurring isotopes. Each chemical element is characterized by its number of protons, but also has isotopes which differ in the number of neutrons in the atomic nucleus. An isotope can be stable or unstable (radioactive). An unstable isotope (a *parent* or *radioactive isotope*) decays into an isotope of another element (a *daughter* or *radiogenic isotope*), which itself may be stable or unstable. In the process of decay, a radioactive element emits various forms of energy (α, β, or γ radiation or spontaneous fission). A decay series of unstable isotopes ends when a stable isotope is reached: for example, the stable lead isotope ^{206}Pb (lead-206) ends the decay series that starts with ^{238}U (uranium-238). The rates of decay and the half-lives of unstable isotopes vary greatly but they are accurately known. This makes it possible to date rocks and minerals over a wide range of geological ages.

A simple equation describes the exponential decay of a radioactive parent isotope and is the basis for every kind of dating involving radioactive and radiogenic isotopes:

$$N = N_0 e^{-\lambda t},$$

where N is the number of atoms of a radioactive parent isotope present at any time t, N_0 is the number of parent atoms that were present when the number of daughter atoms was 0 ($t = 0$), and λ is the decay constant, which is specific for any particular isotope. Solving this equation for t gives

$$t = (1/\lambda) \ln(N/N_0),$$

or expressing t as a function of the number of atoms of the radiogenic daughter isotope D produced,

$$t = (1/\lambda) \ln((D + N)/N)$$

with $D = N_0 - N$. In dating, concentrations of isotopes and isotopic ratios are measured to determine the time that has elapsed since the rock or mineral was formed. Many different techniques are employed to determine the concentrations of isotopes or their respective ratios. Concentrations of isotopes with short half-lives can be accurately measured by the radiation they emit, even when their concentrations are relatively low. The development of mass spectrometers that can also be coupled to particle accelerators has brought particularly significant gains in precision.

The radiometric age obtained by isotopic dating of a rock or a specific mineral within a rock may date various geological processes, such as the formation of magmatic and volcanic rocks, metamorphic alterations, and sedimentary or diagenetic events. Most ages are *closure ages*, which correspond to the time when diffusion of the radiogenic isotope out of a mineral ceased. This would have occurred at a given 'blocking temperature', when the mineral became closed to exchange with the surrounding material. This point 'sets the clock' for isotope dating and corresponds to the age determined. A basic requirement for successful isotopic dating is that the rock or mineral has been a closed system since its formation. If at some point in the history of a rock or mineral the system had become open again (for example, because of metamorphic processes) the clock would be disturbed or reset. This means that the date obtained always corresponds to the last resetting of the clock. Different minerals within the same rock sample can thus give different ages if, for example, one of the minerals was thermally less stable and became open as a consequence of reheating.

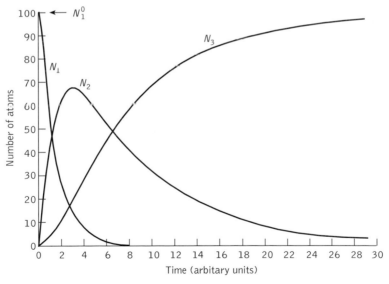

Fig. 1. Radioactive decay and growth of an arbitrary three-isotope system, such as would be observed for uranium-series decay. The half-life of the parent isotope N_1 is five times shorter than that of the daughter isotope N_2 and daughter isotope N_3 is stable. Uranium-series disequilibrium dating methods are based on the measurement of differences between the amounts of parent and daughter isotopes.

The period of time to be measured cannot be too short or too long in relation to the half-lives of the isotopes because the measurement of the concentration of the isotope with the lower concentration (either daughter or parent) limits the precision of the dating. (Adapted from Kaplan (1955) *Nuclear physics*. Addison-Wesley, Reading, Mass.)

563

If a mineral has been a closed system since the point of its formation, the simplest case for dating is a parent–daughter isotope system in which the content of the daughter isotope was zero when the system became closed. In such favourable cases as U/Pb (uranium–lead) dating of zircons, which do not initially contain lead, or ^{87}Rb/^{87}Sr (rubidium–strontium) dating of muscovite micas without initial common strontium, ages may be calculated from the determination of parent and daughter isotope abundances alone. This approach is also valid for fossil Pleistocene corals because the calcium carbonate of the corals contains relatively high concentrations of uranium but virtually no thorium when it is precipitated by corals. The uranium starts to decay and the *radioactive disequilibrium* between the parent ^{238}U (uranium-238) and the daughter ^{230}Th (thorium-230) can be used to derive very precise datings for events that took place during the past 500 000 years. (In radioactive *equilibrium* the amounts of decaying parent and shorter-lived daughter isotopes are the same; see Fig. 1.) Estimates of the extent and timing of past sea-level changes can, for example, be inferred from datings of corals that grew on fossil beach terraces during periods of maximum of sea-level. Speleothems (stalactites, stalagmites, and other chemical deposits in caves and veins calcites from caves can also be dated by this method. To achieve the required precision, the minerals used for dating generally have to be separated and the elements of interest have to be chemically purified before measurement. Ion-probe techniques are

an exception in that they allow direct measurement and dating of single specific mineral grains or even subzones of grains such as zircons or monazites.

In most other instances concentrations of the daughter isotopes are present when the system becomes closed; for example, ^{143}Nd (neodymium-143) initially present in a rock or mineral that is to be dated using the ^{147}Sm/^{143}Nd (samarium–neodymium) scheme. Age determinations from parent–daughter isotope systems therefore have to be corrected for the initial content of the daughter isotope. This is normally done by developing *isochrons*. The ratios of the respective parent and daughter isotopes to a stable isotope of the daughter element are determined in various minerals or in a suite of well-homogenized whole-rock samples and are plotted against each other. If the system has been closed and the initial ratio between daughter isotope and stable isotope was the same for all compounds, the data points form a straight line (the *isochron*); an example is shown in Fig. 2. The intercept of the isochron with the Y-axis gives the initial ratio between daughter and stable isotope in the homogeneous reservoir from which the minerals precipitated, while the slope of the isochron corresponds to the age. Suitable isotope systems are now available for almost every type of rock, ranging from magmatic and metamorphic rocks to sedimentary rocks. Parent–daughter isotope couples which can be applied for dating using the isochron technique include ^{147}Sm/^{143}Nd (samarium-147–neodymium-143), ^{238}U/^{206}Pb

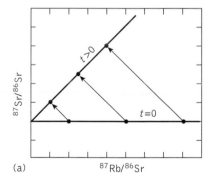

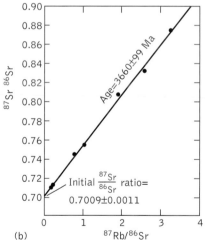

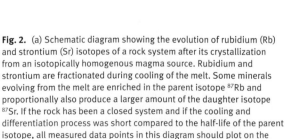

Fig. 2. (a) Schematic diagram showing the evolution of rubidium (Rb) and strontium (Sr) isotopes of a rock system after its crystallization from an isotopically homogenous magma source. Rubidium and strontium are fractionated during cooling of the melt. Some minerals evolving from the melt are enriched in the parent isotope ^{87}Rb and proportionally also produce a larger amount of the daughter isotope ^{87}Sr. If the rock has been a closed system and if the cooling and differentiation process was short compared to the half-life of the parent isotope, all measured data points in this diagram should plot on the

isochron, the slope of which corresponds to the age of the differentiation of the melt. (From Faure 1986). (b) Whole-rock Rb/Sr isochron for the Amîtsoq gneiss (Greenland). In this case the age defined by the isochron represents the metamorphism of originally igneous source rocks of granitic composition. (Data from Moorbath, S. *et al.* (1972) Further rubidium–strontium age determinations on the very early Precambrian rocks of the Goodthaab district, West Greenland. *Nature*, **240**, 78–82.)

(uranium-238–lead-206), $^{87}Rb/^{87}Sr$ (rubidium-87–strontium-87), $^{187}Re/^{187}Os$ (rhenium-187–osmium-187), and $^{176}Lu/^{176}Hf$ (lutetium-176–hafnium-176). The long half-lives (low decay constants) of these isotope systems also make them suitable for dating meteorites, which are thought to represent undifferentiated material of the Solar System, by calculating isochrons of various minerals of the meteorites. Most meteorites are found to be 4.56 Ga (billion years old, and this age has been interpreted as corresponding to the age of the Solar System. The oldest rocks on Earth dated so far are gneisses from Greenland, which yield U/Pb (uranium/lead) ages of up to 3.8 Ga. Applying an ion-probe mass spectrometer, Compston and his colleagues at the Australian National University studied zircon grains in a quartzite from Western Australia which gave even higher U/Pb (uranium–lead) ages, between 4.1 and 4.2 Ga. These point to the existence of chemically evolved rocks of this age.

For the uranium–lead dating method, two radioactive uranium isotopes (^{238}U and ^{235}U) that decay into the stable isotopes ^{206}Pb and ^{207}Pb are available. When the daughter to parent ratios of these isotopes are plotted against each other on a graph, the result is not a straight line but a curve. Their combination offers the possibility of dating single minerals by two different isotope systems of the same elements. These are considered reliable when they both give the same ages (*concordant ages*), but it is also possible to take later losses or gains of uranium or lead into account when the ages are *discordant.*

Although no older rocks are at present known, isotopic dating methods exist that could be used to gain information on the timing of processes such as the formation of the Earth's atmosphere, which took place in the early Solar System prior to 4.2 Ga: that is, within a few tens or hundreds of millions of years after the accretion of the Earth. Daughter isotopes of *extinct isotopes* can be used for this purpose. Examples are the iodine isotope ^{129}I (iodine-129) which had a half-life of 17 Myr (17 million years) and decayed into ^{129}Xe (xenon-129), and the plutonium isotope ^{244}Pu (plutonium-244), which had a half-life of 83 Myr and decayed into ^{136}Xe (xenon-136). The presence of radiogenic ^{129}Xe and ^{136}Xe in the atmosphere today shows that there must have been an atmosphere that was able to retain xenon at a time as early as 50 to 100 Myr after the formation of the Earth.

Another frequently used absolute isotope dating method that is suitable not only for magmatic or metamorphic rocks but also for certain sedimentary rocks is based on the isotope couple $^{40}K/^{40}Ar$ (potassium-40–argon-40), in which the daughter isotope ^{40}Ar is a gas. The argon present in the rock can be extracted and measured by a number of techniques. If the potassium-bearing mineral or rock has been a closed system for the noble gas argon, the amount ^{40}Ar is proportional to the amount of potassium and the time that has elapsed since closure of the system. The total half-life of ^{40}K amounts to 1.25 Ga. This makes possible a wide range of dating applications, from ages of a few million years back to the Precambrian, because potassium is present in most rock-forming minerals,

some of which represent closed systems for argon. Not only basalts, granites, and volcanic tuffs, but also shales and glauconite-bearing sediments can be dated by this method.

Cosmogenic isotopes

Isotopes generated by the interaction of cosmic rays with gases or solid matter on the Earth can also be applied for dating purposes. The most widely used cosmogenic isotope for absolute dating is carbon-14 (^{14}C), which is exchanged with living organisms as $^{14}CO_2$. The clock for dating is set when the organism dies and stops exchanging. The initial $^{14}C/^{12}C$ ratio is well known, and the age of an organism can thus be very precisely determined from its present $^{14}C/^{12}C$ ratio. Carbon-14 is used to date biogenic carbonate in such organisms as foraminifera (zooplankton) or corals, organic material in sediments, or fossil wood back to about 35 000 years BP (before the present). Dating with ^{14}C is not reliable beyond that because after five to six half-lives (in this case one half-life is 5730 years) it cannot be measured accurately. Short-lived isotopes (with half-lives of several years) such as ^{7}Be (beryllium-7) introduced from the atmosphere can be used to date mixing processes in the ocean. A different dating approach using cosmogenic isotopes such as ^{10}Be (beryllium-10), ^{26}Al (aluminium-26), ^{36}Cl (chlorine-36), ^{32}Si (silicon-32), or ^{39}Ar (argon-39) is based on their specific production rates *in situ* in minerals which are exposed to cosmic radiation. The quantity of a cosmogenic isotope present, for example in a quartz sample found on top of a moraine, is ideally proportional to the duration of its exposure to cosmic radiation. The age of the retreat of the glacier that left the moraine can thus be determined, provided that the reaction cross-section of the corresponding mineral (i.e. how much of a cosmogenic isotope is produced per volume of mineral) is accurately known. Such *exposure ages* can also serve to estimate the erosion rates of landscapes or the rates of lateral movement along strike-slip faults by dating river terraces cut by faults and measuring the amount of their displacement. Exposure ages have further applications in cosmochemistry to date the exposure times of lunar rocks or the travel times of meteorites after their expulsion from larger parent bodies.

Relative isotopic dating methods

In addition to the 'absolute' dating methods considered above, there are also various indirect isotope dating methods, which are mainly used to date sedimentary sequences. These methods are based on a reference data set with a well-constrained age model, derived, for example, from absolute dating methods. The sample record to be dated is compared with the reference data set. On Phanerozoic timescales the best-documented indirect isotopic dating method uses variations of the $^{87}Sr/^{86}Sr$ ratio measured in biogenic carbonates, which are supposed to have recorded the strontium isotope ratio of ambient sea water. The strontium isotope record shows variations that have been mainly related to orogenic events and changes in weathering regimes.

On Tertiary and Pleistocene timescales, the $^{18}O/^{16}O$ ratio (both isotopes are stable) in carbonate shells of marine zooplankton is one of the most important relative dating methods for marine sediments. During the Tertiary sub-era, the oxygen isotope composition of deep water underwent drastic and well-dated changes caused by the onset of formation of cold deep-water masses and the build-up of the first ice sheets in Antarctica, which caused a global fractionation of oxygen isotopes. During the Pleistocene the oxygen isotope composition of the deep water changed on glacial–interglacial timescales when, in addition, huge amounts of isotopically light oxygen were stored in the ice sheets during glacial periods and related temperature fluctuations enhanced the difference between the glacial and interglacial oxygen-isotope signature of the ocean water. Profiles of relative oxygen-isotope variations measured on sediment cores with unknown sedimentation rates are compared to the reference data set and can then be translated into absolute ages using a model of well-constrained astronomical forcing of the reference data sets. Other examples of relative isotopic dating are the deuterium (^2H) variations in ice cores and the osmium isotope record of marine sediments. These relative methods can be combined with dating methods based on the decay of radioactive isotopes such as ^{230}Th (thorium-230) or cosmogenic ^{10}Be (beryllium-10) and ^{26}Al (aluminium-26). These isotopes are introduced into the ocean at a nearly constant rate and are particle-reactive, which causes their rates of deposition in marine sediments to be constant to a first approximation. Concentration profiles of ^{230}Th or ^{10}Be in marine detrital or chemical sediments such as ferromanganese crusts show an exponential decrease in downcore profiles as a function of sedimentation rates or growth rates, and can therefore be used for dating.

Indirect isotopic dating techniques

Some dating techniques are not based on the isotopes themselves but on the traces their decay has left; for example, in a mineral. In the fission-track method for uranium the tracks of destruction caused by the spontaneous fission products of uranium in a mineral are counted. Knowing the rate of spontaneous fission, ages may be derived from a number of assumptions. Similar damage is caused by the alpha recoil, which is a trace of the energy released during the alpha decay of uranium and thorium isotopes. MARTIN FRANK

Further reading

Faure, G. (1986) *Principles of isotope geology.* (2nd edn). John Wiley and Sons, New York.

Dickin, A. P. (1995) *Radiogenic isotope geology.* Cambridge University Press.

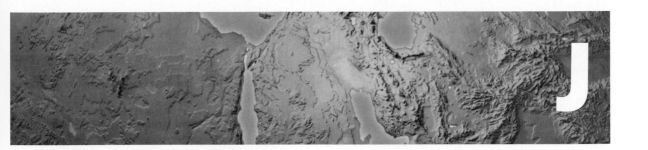

Jamieson, Thomas (1829–1913)

Thomas Francis Jamieson was born in Aberdeen on 1 April 1829, the son of a jeweller. He was educated at Aberdeen Grammar School and Aberdeen University, subsequently becoming factor of the Ellon Castle Estate.

Jamieson developed an interest in geology early in life and corresponded with both Sir Roderick Murchison and Charles Darwin. He was elected a Fellow of the Geological Society of London in 1862. His early papers largely concerned petrology, but he is best known for his contributions to Quaternary science. His discourse on the ice-worn rocks of Scotland (1862) helped to establish the glacial theory in Britain. His interpretation of the Parallel Roads of Glen Roy (1863) as the shorelines of ice-dammed lakes ran contrary to Darwin's explanation of those features as formed during changes in sea level.

Jamieson is, however, remembered mostly for his contributions to sea-level studies. It was he who put forward the view in 1865 that the marine sediments in the Forth Valley, now raised above present sea level, had been deposited while the area lay at a low level, having been depressed beneath the weight of ice and then elevated after the ice had melted. This was effectively the first expression of the concept of glacio-isostasy. In subsequent years he elaborated this view, ultimately conflicting with the orthodox opinion then promulgated by the Geological Survey of Scotland, which was that raised shorelines in that country lay at elevations of 25, 50, or 100 feet (7.6, 15, or 30.5 m). In five papers published between 1882 and 1908, he effectively demonstrated the concept of differential uplift of isostatically raised shorelines, although his views for Scotland were not to be accepted until the 1960s.

Jamieson died in 1913. During his lifetime he made arguably some of the greatest contributions to the development of Quaternary geology.   DAVID E. SMITH

Jeffreys, Sir Harold (1891–1989)

The British geophysicist Harold Jeffreys made important contributions to our knowledge of the Earth's interior. His book *The Earth: its origin, history and physical constitution*, first published in 1924, became a classic text.

Jeffreys spent most of his working life in Cambridge, except for a few years at the Meteorological Office. He was Plumian Professor of Astronomy and Experimental Philosophy from 1946 until 1958.

By studying earthquake waves Jeffreys showed that the Earth's core is liquid, at least in part. He also found a marked difference between what would we today call the upper and lower mantle. With K. E. Bullen he analysed earthquake records to produce curves showing travel times for seismic waves—an essential tool for geophysicists. Using seismic records of an explosion in a chemical factory Jeffreys and Dorothy Winch were able to estimate the velocities of seismic waves in the crust and to show that it must be multi-layered. This was the first experimental use of an explosion in seismology.

Jeffreys did much research on other aspects of the physics of the Earth's interior. His long-held opposition to continental drift was based on physical arguments. He also did important work on the dynamics of the Solar System and the Earth and on the application of fluid dynamics to the atmosphere and the oceans.   D. L. DINELEY

jet streams

Wind speeds normally increase with height above the Earth's surface, often reaching a maximum close to the tropopause. The strength of these upper winds is not uniform over wide areas; in certain regions the wind becomes concentrated into narrow cores of maximum wind speed which are known as jet streams. They are often thousands of kilometres in length, hundreds of kilometres in width, and some 2 to 4 kilometres in depth with the wind speed typically between 40 and 80 metres per second. In some areas air rises into the jet, while in other locations air sinks beneath it.

There are three principal global jet streams in the troposphere. The polar front jet stream normally reaches a maximum beneath the tropopause at a height of about 10 or 11 km between 40 and 60 degrees latitude. It is the core of a broad westerly flow formed by the thermal gradient between the warm tropical and the cold polar regions. The polar front jet stream is closely associated with polar front depressions. The westerly subtropical jet stream is found at a height of about 12 km between 25 and 30 degrees latitude at the poleward limit of the tropical Hadley cell. It is much more persistent than the meandering and sometimes discontinuous polar front jet stream. The easterly equatorial jet stream occurs at a height of about 12 km and is especially well formed over the Indian subcontinent during the summer monsoon.   JOHN STONE

joints and jointing The word 'joint' has been used by quarrymen and miners since the eighteenth century to express the idea that the parts of a rock mass are joined together across fractures. More correctly, a formerly intact rock mass is now split by joints, rather than being joined by them. Although the word 'joint' is used by most geologists, the structure could equally well be called a crack. In contrast to faults, along which rock masses have slid, joints are fractures on which it is not possible to see any evidence for there having been sliding (i.e. shear) along the fracture. A vein is similar to a joint but differs from it in that the crack is filled by minerals deposited from water that formerly flowed through it.

Joints, the most abundant structures caused by Earth movements in sedimentary rocks, are a result of the widespread cracking of rocks in the uppermost 10 km of the Earth's crust, which is brittle. Individual joints are often only a few centimetres apart, and their lengths commonly range from less than a metre to a few tens of metres. Understanding the intricate geometric patterns made by joints and knowing how they were formed is especially worth while because joints commonly influence the underground flow of fluids, such as oil and water, and they can lead to many exposed rock masses being unstable. Joints also profoundly influence the shape of the landscape by controlling, for example, the layout of valleys and ridges and the shapes of rock outcrops. The study of joints allows geologists to determine directions of pressure (stress) that acted on rocks in the past, or are still affecting them at the present day.

A *joint set* is a group of parallel joints, and a *joint system* consists of two or more symmetrically arranged sets (Fig. 1). A *joint spectrum* is a suite of neighbouring joints that are not perfectly parallel but show a continuum of orientations up to a maximum angle of about 45°. It is rare for the members of a natural joint spectrum to intersect at a common node as has happened at the outcrop in northern Spain shown in Fig. 2. A joint that is a plane or nearly plane surface and belongs to a set or spectrum is called a *systematic joint*. Most systematic joints are of tectonic origin, that is, they are related to deep-seated Earth movements. By contrast, *non-systematic joints*, which are mostly irregular surfaces linking systematic joints (Fig. 1), are of superficial origin. They are products of processes such as weathering and erosion which act on a rock mass after it has been exposed at the Earth's surface.

The trace of a joint that is younger than one of its neighbours will abut, or terminate, at it. This allows one to deduce the order of formation of joints exposed in an outcrop. The abutting relationship arises because a younger joint propagating through intact rock is unable to jump over the space

Fig. 1.   The nearly straight traces of three sets of systematic joints exposed on a limestone bedding plane in central Arabia. One set trends parallel to the 25-cm rule, one at right-angles to it, and the third from top left to bottom right of the photograph. The three joint sets define the joint system at this outcrop. Note that some joints intersect in T-shaped junctions where a younger joint abuts an older one. Where joint traces cross each other making + shapes it is not possible to decide which is the older joint. Joints in the two sets at right-angles to each other are extension fractures. The most irregular joint traces are those of non-systematic joints.

between the walls of an older joint (Fig. 1). The buffering of younger joints by older ones means that most older joints are of greater dimensions than younger ones. If joint traces cross each other it is not possible to determine which is the older joint; all that can be said is that the walls of one of them must have been welded together before the other joint propagated across it.

Genetic classification of joints

The majority of joints are *extension fractures* belonging to single sets of cracks that formed perpendicular to a direction of least stress, commonly a tension (Fig. 3). Some stress fields can also give rise to two sets of joints of roughly the same age. Such *conjugate joint sets* (Fig. 3) consist of fractures inclined to the directions of greatest and least stress. Some conjugate joint sets enclose an angle of about 60° with each other and can be interpreted as *shear fractures*, that is, fractures along

Fig. 2. A spectrum of hybrid-shear joints intersecting at a common node in Miocene limestones near Tudela, northern Spain.

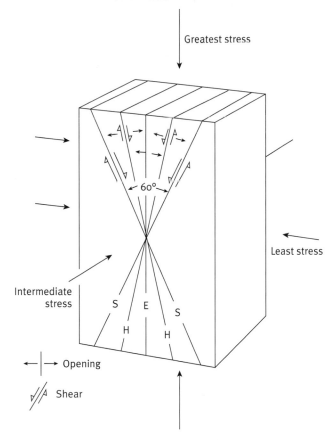

Greatest stress

Least stress

Intermediate stress

60°

S E S

H H

←|→ Opening

//◢ Shear

Fig. 3. Relationships between principal stress directions and classes of fracture. On extension fractures [E] there is opening, on shear fractures [S] there is shear, and on hybrid-shear fractures [H] there is both opening and shear.

which there was slip, but of such small magnitude that it is not detectable by eye in the field. More commonly, conjugate joints enclose an angle of less than 60° about the direction of greatest pressure. Because these joints show some characteristics in common with extension fractures and some characteristics in common with shear fractures they are often called *hybrid-shear fractures*. Most of the joints in the spectrum shown in Fig. 2 are examples of hybrid-shear joints.

Origins of joints

In 1985, Terry Engelder, who had been studying joints in the Palaeozoic rocks of the Appalachian Plateau in the northeastern USA, concluded that there are four main types of joints in the Plateau and other areas of simple structure.

(1) *Tectonic joints* form at depth in the Earth's crust and before rocks are uplifted and denuded. The initiation and propagation of tectonic joints are a response to the combined action of stresses of tectonic origin and of water in the crack that has been pumped up to abnormally high pressures as a result of the sediments having been tectonically compacted during Earth movements. Such high water pressure in a small crack-like flaw in a rock enables it to open up and increase in size, becoming a joint in the process.

(2) *Hydraulic joints* also form at depth, before uplift, and as a consequence of the influence of high water pressures; but, unlike tectonic joints, the cause of the high water pressure is solely related to compaction of sedimentary origin.

(3) *Unloading joints* are formed close to the Earth's surface after the uplift and exhumation of a sedimentary rock sequence. Although unloading joints are formed at shallow depths beneath the Earth's surface, their orientations are controlled by the directions of deep-seated tectonic stresses at the time of their formation. The idea that some joints are caused by unloading after uplift is not a new one: in 1959 Neville Price had introduced the notion to explain the widespread occurrence of joints in otherwise 'undeformed' rocks.

(4) *Release joints* are also formed in near-surface rocks after uplift and exhumation, but their orientations are largely determined by pre-existing structures. For example, some release joints open up along cleavage planes dating from a time when the rocks were being folded at a depth of a few kilometres beneath the Earth's surface.

Exposed *neotectonic joints* are a special category of unloading joints formed in a stress field that has persisted with little or no change of orientation until the present day. Neotectonic joints are useful structures for inferring the directions of present-day stresses where they are not known from geophysical measurements (Fig. 4). Understanding present-day stresses is important because if their orientations and magnitudes are known it is possible to reduce earthquake hazard and predict the underground flow of fluids.

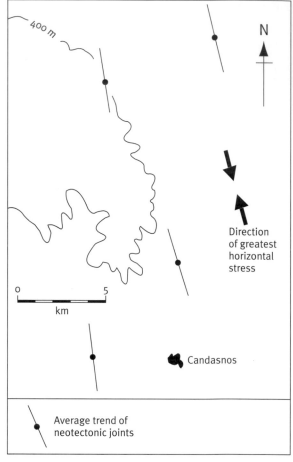

Fig. 4. Strike of neotectonic joints in the Candasnos area of northern Spain, with the direction of the present-day greatest horizontal stress determined from independent geophysical evidence superimposed on the map. Note that the strike of the joints is subparallel to the direction of compression. (Much modified from Hancock (1991), *Philosophical Transactions of the Royal Society of London*).

Many of the mechanisms leading to the formation of regionally extensive joints sets in areas of simple structure, such as the Appalachian Plateau, also give rise to joints in structurally more complex terrains, but some joints in such settings are of more local origin, as is explained below.

Relationships between joints and other structures

Many studies of joints have concentrated on their relationship to other structures. An important question to be answered is: are some types of joints associated with particular structures? The answer to this question is a provisional 'yes'. For example, Jonathan Turner and Paul Hancock found that in the thrusted and folded rocks of the south-western Spanish Pyrenees joints trending at right-angles to thrusts are restricted to the rocks

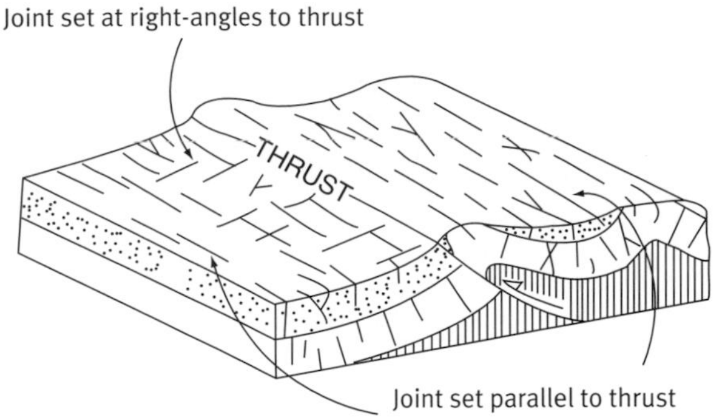

Joint set at right-angles to thrust

THRUST

Joint set parallel to thrust

Fig. 5. Joint sets symmetrically arranged about thrust faults and folds in the south-west Pyrenees. Note that the dip-joints, at right-angles to the thrust, are restricted to the rocks immediately beneath the thrust, whereas the strike-joints, parallel to the thrust, occur both above and below it. (Modified from Turner and Hancock (1990), *Journal of Structural Geology*.)

immediately beneath them, whereas joints trending parallel to thrusts occur both above and below thrusts (Fig. 5). These Pyrenean joints, and many others in deformed sedimentary rocks elsewhere, are also symmetrically arranged about the folds containing them.

In addition to wanting to understand geometrical relationships between joints and other structures, geologists have also attempted to determine the relative timing of jointing with respect to folding. Some workers have claimed that joints are early structures that are formed before beds are folded. A powerful argument in favour of this view is that many joints appear to have been rotated with strata during folding. However, where joints cut cleavage planes that are of the same age as the folds, the joints must be younger than the folds.

Joints in igneous rocks

Because they are products of tectonic stresses, many of the systematic joints cutting igneous rocks are similar to those in sedimentary and metamorphic rocks. In addition, some igneous rocks also contain joints related to their cooling history. The best known and most aesthetically pleasing of these joints are *columnar joints* bounding long polygonal columns. The columns may be straight or curved, and analysis of their shapes enables successive cooling surfaces to be identified. Columnar joints are best developed in sills and dykes, volcanic vents, and former lava lakes. Some joints in major intrusions, especially those near the margins of plutons, are also thought to be related to stresses generated during cooling or the closing stages of the intrusion of the igneous rock. PAUL L. HANCOCK

Further reading

Price, N. J. and Cosgrove, J. (1990) *Analysis of geological structures*. Cambridge University Press.

Suppe, J. (1985) *Principles of structural geology*. Prentice Hall, New York.

jökulhlaup *Jökulhlaup*, the Icelandic term for a glacier-burst, is used worldwide to describe sudden flood-releases of melt water from glaciers and ice sheets. The phenomenon is relatively common in actively glaciated environments, where it can pose a considerable hazard.

Jökulhlaups are generated by a variety of mechanisms, including volcanic activity and the sudden drainage of stored glacial melt water. Volcanic eruptions underneath glaciers cause the rapid melting of glacier ice, releasing large volumes of melt water and erupted material. Volcanic jökulhlaups are best documented in Iceland. The 1918 eruption of the volcano Katla underneath the Myrdalsjökull, Iceland, generated a peak discharge of 1.6×10^6 m^3 s^{-1} within 4–5 hours with depths of 70 m. This jökulhlaup deposited enough sediment to extend the nearby coastline by 4 km.

Storage of melt water either in glacier marginal lakes or as water bodies on, within, or underneath the glacier is relatively common. Glacier margin collapse may also result in the storage and sudden release of melt water. Glacial moraine ridges impounding lakes often fail rapidly, generating jökulhlaups. Many ice-dammed lakes drain and refill regularly with a periodicity ranging from several months to decades. The frequency of jökulhlaups is controlled by the size of the lake basin, the refill rate, and the mechanism of dam failure.

Ice-dammed lake drainage can be triggered by several mechanisms:

(1) flotation of the ice dam once the lake reaches nine-tenths of the ice dam thickness;

(2) over-topping of the glacier dam accompanied by sudden downcutting;

(3) plastic deformation of the ice dam when ice-dammed lake depths exceed 200 m;

(4) pressure variations within the glacier-drainage network; and

(5) catastrophic ice dam failure.

Once drainage is initiated, tunnels conveying lake water through the glacier enlarge exponentially through melting, allowing progressively more water to drain. This sequence produces the distinctively steady exponential rising limb of many jökulhlaup hydrographs.

Jökulhlaups vary in duration from several hours to several weeks or even months. Volcanic jökulhlaups can reach peak flows exceeding 10^6 m^3 s^{-1} within as little as 4 hours. In contrast, a jökulhlaup draining from an ice-dammed lake may take months to reach a similar discharge and would involve a much greater volume of water. It is thought that over 2000 km^3 of water drained from the Late Pleistocene glacial Lake Missoula in less than 4 days, producing flow velocities up to 30 m s^{-1} and a peak discharge of 17×10^6 m^3 s^{-1}. Volcanic jökulhlaups are commonly 'hyperconcentrated' or heavily laden with erupted debris, giving the flow the consistency of wet cement. As jökulhlaups have higher discharges than flows generated by other mechanisms, sedimentary structures are usually well preserved. Large-scale jökulhlaup bars can extend for tens of kilometres, and can have relief amplitudes of up to 100 m; they commonly support large areas of gravel and boulder dunes reaching heights of up to 15 m.

Jökulhlaups play an important role in transporting sediment away from glaciers and ice sheets as well as creating distinctive fluvial landforms and deposits. The impact of jökulhlaups from vast Late Pleistocene ice-dammed lakes in North America and Siberia extended well beyond the immediate glacier margin into marine and climatic systems.

ANDREW J. RUSSELL

Further reading

Maizels, J. K. and Russell, A. J. (1992) Quaternary perspectives on jökulhlaup prediction. In Gray, J. M. (ed.) Applications of Quaternary research, *Quaternary Proceedings*, pp. 133–53.

Jovian planets *see* GIANT PLANETS

Jurassic
The splendidly fossiliferous limestones and shales of the Jura Mountains of Switzerland attracted the attention of the early scientists of Europe long before their true nature was realized. They eventually drew the attention of the traveller and scientist, Alexander von Humboldt (1769–1859) who in 1795 had written about them, using the term *Jura-Kalkstein*. Somewhat later, when Leopold von Buch came to study them in the 1830s, their correlation with similar fossiliferous beds in England and Germany became apparent to him and he used Humboldt's term in establishing a formal system of rocks for all three areas (1839). The name was immediately used by the many amateur and professional geologists who then were beginning to make maps of the fossilerous formations in the various parts of Europe. The ammonite fossils were everywhere an immediate clue to the age and correlation of the rocks of the Jurassic system. They are distinct from those of the marine Triassic rocks beneath and from the ammonites of the Cretaceous system above.

The Jurassic is the second of the three Mesozoic systems, and was formed over a span of about 70 million years prior to the commencement of the Cretaceous period (135 Ma). The system is divided into 11 stages and over 70 ammonite zones in Britain. Virtually all the stages are named after localities in western Europe where the system is extensively developed in neritic (shallow-water) facies. The German palaeontologist Albert Oppel realized in the mid-nineteenth century that ammonites give by far the best means of correlation within the Jurassic. They are widespread and evolved rapidly; only within a few nearshore or brackish-water facies are they rare or absent. Oppel's detailed zonal scheme for the Jurassic has provided the basis for our modern zonation. Moreover, sub-zones can be defined within many of the Jurassic ammonite zones, and even smaller biostratigraphic units (horizons) characterized by the presence of a short-lived species have been identified. There are some 44 of these in the Middle Jurassic Aalenian and Bajocian Stages in western Europe. This is biostratigraphy on a very detailed scale. Where alternative zonal schemes have been erected—using foraminifera or coccolith microfossils, for example—the species used have longer time ranges than the ammonites. None of the Triassic super-families of ammonites lived on into the Jurassic, but three new superfamilies appeared in the Early Jurassic. Significant extinction events of ammonites occurred in the late Early Jurassic, at the Middle–Late Jurassic boundary, and at the end of the period. Such was the palaeogeography of the Jurassic that the biota was distributed across a number of palaeoecological provinces; and this provinciality poses a number of problems. This is apparent in the uppermost stage, known variously as the Tithonian, Portlandian, and Volgian in different parts of the Eurasia. They are not exactly equivalent, and the Volgian has a boreal fauna, different from the *Tethyan* (Alpine–Himalayan) faunas of the other two. In the absence of ammonites, the dinoflagellates may provide marine correlation, while spores and pollens serve best in non-marine facies.

Many stratigraphers have regarded this system in western Europe as reflecting a full cycle of marine deposition, beginning with a marine transgression and ending with a sharp regression. Detailed study, however, has shown that this is an oversimplification: several important transgressions follow on from phases of uplift and erosion, despite the then relative stability of much of the western European continental area.

The palaeogeography of the Jurassic period was one of change, with the break-up and dispersal of the rifted fragments of Pangaea now well under way (Fig. 1). The two principal continental masses, Laurasia and Gondwanaland, became increasingly separated by the equatorial Tethyan ocean. During the latter half of the period, north-western Africa parted from North America to create a narrow ocean, the forerunner of the Atlantic. Somewhat later the separation of South America, Africa, and Antarctica began, while ocean floor spreading in the proto-Pacific Ocean began to move small land masses on to the margins of Asia and the Americas. Plate-tectonic motions thus appear in the Jurassic period to have been more important in bringing about geographical change, with accompanying rise in sea level and climatic amelioration, than they had been in the Triassic period. Volcanic activity took place on an enormous scale in various parts of the world. Dyke-swarms were intruded in many regions of Africa. The lavas of the Karroo in southern Africa covered perhaps $2\,000\,000$ km^2 to a thickness approaching 9 km. Lavas are also present in a zone running across the centre of Antarctica.

One of the most tectonically active regions of the Mesozoic world was the great belt of island arcs, deep troughs, and volcanoes stretching along the western margin of the Americas from Alaska to Panama and southwards to southern Chile. There was almost continuous crustal unrest and volcanism here, and at depth beneath the newly rising mountains great batholiths of granite and granodiorite were intruded. Volcanic rocks of enormous thickness and deep-seated

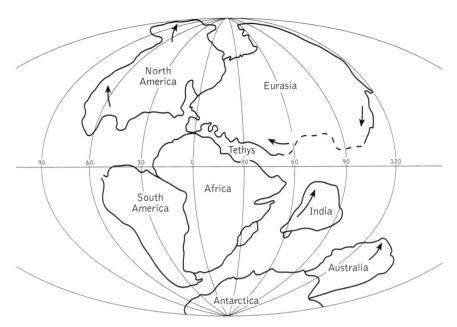

Fig. 1. The Jurassic world saw the continuation of the dispersal of parts of the old Pangaea supercontinent. Laurasia was beginning to split apart with the growth of new ocean floor between North America and south-western Europe. The Gondwanaland blocks were now parting on account of the growth of new ocean floor between the southern tip of South America and eastern Africa–Arabia to the north, and Antarctica, Australia, and India to the south an east. In particular, India embarked upon its spectacular motion northwards towards the southern margin of Laurasia. The Tethyan Ocean was the site of much crustal unrest, with several microcontinents moving between the giant continents to north and south. Most of the continental margins were progressively and repeatedly flooded by eustatic rises in sea level. (Based on an early map by Dietz and Holden.)

igneous intrusions from this period have created much of the geology of the Peruvian Andes. Subduction of the Pacific crustal plates beneath the margins of the North American and South American plates was responsible, and the process has continued ever since.

It was a time of slow, impersistent but inevitable rise in sea level, with up to a quarter of the present continental area flooded in Late Jurassic time. The period began with sea level at a relatively low stand at around 14 per cent of the continental area under water and rather less at the end. Many geologists see this period as exhibiting a great eustatic cycle within which many smaller cycles developed. Only towards the very end of the period did sea level begin to decline rather sharply. This Jurassic inundation was very extensive in Laurasia, but most of North America and Gondwanaland was not greatly affected. Nevertheless, a narrow sea spread down the eastern flank of the present Rockies during the mid-part of the period. Much of eastern Asia persisted as an upland area. Palaeomagnetic polarity was very mixed throughout the Jurassic, but with an interval of relatively steady normal polarity in mid-period.

The climate seems to have been warm and equable; with no polar icecaps and with an equatorial ocean, conditions existed to give tropical climates well up into the present north temperate latitudes. The continental interiors, distant from the sea, were generally low-lying and arid.

Marine life was prolific with a widespread development of both calcareous benthonic organisms and vigorous free-moving forms. Reefs of corals, algae, bryozoa, and sponges are known, while the bivalvia were the almost universally dominant stock of the benthos elswhere. Gastropods, belemnites, echinoids, foraminifera, and ostracods were important in many deposits. A complex picture of developing faunal provinces emerges from studies of the Jurassic shallow-water facies. It appears to depend upon many factors, including climate and water circulation. Many areas of very strong salinity have been demonstrated. Open-water planktonic and nektonic faunas were dominated by the cephalopods. The ammonites were by far the most prolific, with many fast-evolving lineages. They have proved to be most useful in - biostratigraphy because of their robust preservation, rapid evolution, and widespread distribution. This serves to illustrate the remarkable change in cephalopod fortunes following the late Triassic extinction, when only one family survived into the Jurassic. Marine vertebrates included the large dolphin-like ichthyosaurs and the long-necked plesiosaurs, marine crocodiles, turtles, and other reptiles. So high was local organic productivity during Jurassic times that many sedimentary basins became rich in petroleum hydrocarbons.

On land, there was a rich flora in the more rainy regions to provide swamp and forest coverage. Gymnosperms, including conifers, gingkoes, and cycads, were very numerous, as were ferns and horsetails. From the Jurassic forests are derived extensive coal deposits, some of the seams being 5 m or more in thickness. They are largely confined to the eastern parts of Laurasia and Gondwanaland, where a climate for long-continuing forest growth persisted into the Cretaceous. Palaeobotanical studies suggest that the temperature gradient from low to high latitudes was significantly less than today in the northern hemisphere.

Tetrapod populations were overwhelmingly of reptiles. The Triassic thecodonts had given rise to the dinosaurs and pterosaurs in the late Triassic, and these now gave rise to an extraordinary range of adaptations to new habitats. Bipedal and quadrupedal herbivors achieved considerable size, while predatory types also are well represented by active bipedal forms. By late Jurassic time the pigeon-sized *Archaeopteryx* had evolved and true birds were soon to follow. The mammals remained small and relatively subordinate, but the insectivors and omnivors were undoubtedly diverse and numerous. Among the fishes the holosteans, with their thick enamelled scales, were the most prolific. Some were of very large size. The sharks included large species that preyed upon the schools of cephalopods, and shell-feeding forms that snatched large bivalves from the sea floor.

The natural resources of the Jurassic system are very varied and extensive. They include all manner of construction materials, mineral ores, and non-metallics associated with igneous and sedimentary formations, fossil fuels, and other strategic materials.

D. L. DINELEY

Further reading

Arkell, W. J. (1956) *Jurassic geology of the world.* Oliver and Boyd, Edinburgh.

Hallam, A. (1975) *Jurassic environments.* Cambridge University Press.

Hallam, A. (1984) Continental humid and arid zones in the Jurassic and Cretaceous. *Palaegeography, Palaeoclimatology, Palaeoecology,* **46**, 195–223.

Sellwood, B. W. (1978) Jurassic. In McKerrow, W. S. (ed.) *The ecology of fossils.* Duckworth, London.

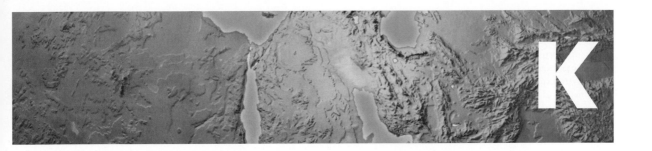

karren 'Karren' is the term used to describe the microsolutional features and sculpturing that form on exposed bedrock surfaces of soluble rocks. Karren is a German term (*lapies* in French, *lapiaz* in Spanish) which has come into general usage in English in past decades and which is derived from the old High German word char = kar.

The small-scale karren features, usually a few millimetres to a few metres in size, are the most widespread karstic landform. The karren shows narrow channels or furrows, generally separated by sharp, knife-like ridges or crests, that are formed on the surface of limestone by the solvent action of rainwater (Fig. 1). Some of them have microscopic solutional and

Fig. 1. (a) General view of karren developed in bedrock limestones at Escorca (Mallorca Island, Spain) (b) Close-up view of karstic solutional runnels at Escorca (Mallorca Island, Spain).

biokarstic features related to highly specific limestone weathering processes. Others remain hidden under the soil or buried by clastic sediments, as happens with the typical subcutaneous karren features.

The occurrence of karren is widespread wherever karstic rocks are exposed, occurring in recently glaciated areas or in tropical and Mediterranean environments. Their universal distribution and morphologies make them very characteristic of karst regions.

Karren features that occur in bare carbonate, gypsum, and salt karst rocks or under soil cover have been formed by the chemical and physical erosive action of water. The solutional mechanisms that produce these features may be due to direct rainfall, sheet wash, channelized flow, or percolating flow. Thus their growth and development remain basically under the control of atmospheric precipitation. Their morphology and genesis illustrate the different controls and variations in the solution processes.

Karren morphologies are especially well developed when pure and homogeneous rocks display a low primary permeability and are affected by widely and well-developed joints. As a result of karren development the fissures (*grikes*) are widened and channel edges are eroded. This promotes catchment reduction and residual rock ridges may be broken up by mechanical weathering. Thus ultimately karren can be considered a self-destroying process.

We can consider two basic types of karren: circular plan forms, developed on horizontal rock surfaces (solutional hollows—*kamenitzas*), and linear forms, developed principally where the initial rock surface is sloping. In the first case, water can accumulate and stand for some time and *kamenitza* development can be accelerated by biologically produced carbon dioxide. Saturated water in the hollows must be removed periodically and replaced by fresh and 'aggressive' water. *Kamenitzas* range from a few centimetres in width and depth to a metre in depth and several metres in width. In the second case, the rainwater will run down the steepest slope. Should some degree of concentration of flow occur, then a channel will gradually be etched into the rock. Water is in contact with the rock for only a brief time and biological carbon dioxide is available to the water only if channelling occurs beneath the soil cover. These linear forms may be fracture controlled (*Splitkarren*—joints or stylolites, *Kluftkarren*—major joint- or fault-guided) or hydrodynamically controlled, giving 'gravitomorphic' solution channels. Among these, *Rillenkarren* and *Rinnenkarren* (solution runnels) are more common. They are razor-shaped, finely chiselled, solution runnels from 1 to 2 cm wide and deep, 20 cm long, up to 50 cm wide and deep and 20 m long or even longer, forming downward-oriented (gravitational) patterns along bare rock slopes (Fig. 1). These features develop preferentially on well-cemented, hard, and compact rocks.

The superposition and complex intermixing of karren morphologies are possible, giving rise to more complex polygenic forms. In addition, in the coastal 'littoral' zone the effect of boring and grazing marine organisms, as well as salt weathering, may produce a special case, called *littoral karren*.

Scientific interest in karren features is based not only on the processes that lead to such morphologies, but also on the fact that they preserve valuable information about the climate of the recent past, since they are the result of the weathering of calcareous and evaporitic rocks by meteoric waters as well as other meteoric and biological phenomena.   J. J. FORNÓS

Further reading

Bögli, A. (1980) *Karst hydrology and physical speleology.* Springer-Verlag, Berlin.

Ford, D. and Williams, P. (1989) *Karst geomorphology and hydrology.* Unwin Hyman, London.

Fornós, J. J. and Ginés A. (eds) (1996) *Karren landforms.* Servei de Publicacions, Universitat de les Illes Balears, Palma de Mallorca.

karst The term 'karst' is used to describe an area of limestone or other highly soluble rock in which the landforms are of dominantly solutional origin and in which the drainage is underground in solutionally enlarged fissures and conduits (caves). The word *karst* is derived from the Kras, a region is Slovenia, some 400 km^2 in extent, which was one of the first areas of well-developed limestone landforms to be investigated by scientists. The term has affinities with the Greek word *chalis*, the Latin *calx*, and the Italian *carso*. *Karst* has now become the generic term for all landscapes dominated by the distinctive suite of surface and subterranean features described above, even though the landscape of the Slovenian type-area is not particularly representative of other karst areas elsewhere.

The degree to which soluble rocks have been karstified varies greatly from place to place, ranging from areas such as parts of southern China or Indonesia, where almost all the landforms are understandable in terms of karstic processes (*holokarst*), to areas of limestone in which other geomorphological processes have been at least as important as dissolution in creating landforms. For example, *fluviokarsts* are the products of both rivers and karstic forces. Figure 1 shows the global distribution of carbonate rocks, most of which exhibit karstification at least to some extent.

Karst areas are relatively uncommon in Africa and Australasia but widespread in Europe (more than 12 per cent of the land surface) and in south-east Asia.

Karst landforms are developed on highly soluble rocks such as halite and gypsum, but their most widespread occurrence is where carbonate rocks, and especially pure limestones (calcium carbonate), crop out at the surface.

The geomorphological evolution of limestone terrains is uncomplicated in comparison with other rock types. A single process, chemical solution of the limestone by acidified rainwater, operates on a rock-type composed overwhelmingly of a single mineral (calcite). Thus virtually all of the rock is dissolved and then transported away as invisible solute load in a

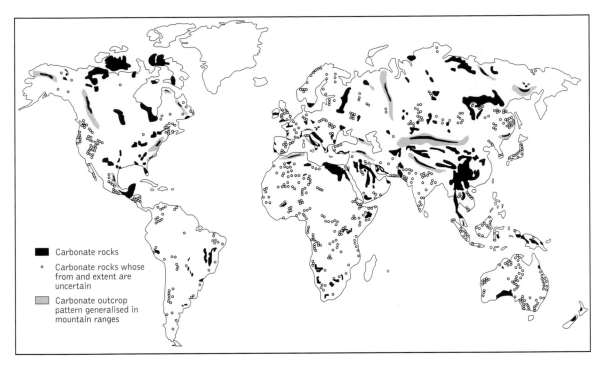

Fig. 1. The global distribution of outcrops of carbonate rocks, most of which exhibit karstification, at least to some extent. (After Ford and Williams (1989).)

Carbonate rocks

Carbonate rocks whose from and extent are uncertain

Carbonate outcrop pattern generalised in mountain ranges

single operation. Little all no insoluble material accumulates to blanket the rock surface and to require mass-movement processes to move it downslope, as is the case with most other rock-types.

Humans have settled in karst regions since prehistoric times, apparently often in preference to other apparently more attractive areas. The barren terrain, patchy soils and vegetation, and the difficulties encountered in obtaining reliable supplies of surface water have not prevented the creation of high civilizations, such as those of meso-America and south China, in karstic terrains. Nowadays some 25 per cent of the world's population obtains its water supply from karst supplies, despite the high risk of contamination to which karst water is liable. Karst regions include some of the world's most spectacular scenic attractions, such as the Stone Forest of China, the travertine terraces of Pammukale in Turkey, and the great caves of Carlsbad and Mammoth in the USA. Increasingly these are becoming major, global tourist attractions and their integrity is accordingly under threat.

Most erosion of limestone is the result of the action of the dissolving action of dilute carbonic acid in rainwater on the limestone. Rainwater is slightly acidified by equilibrating with atmospheric carbon dioxide, but a much more significant source of carbon dioxide, and hence carbonic acid, is from soil-air dissolved by rainwater as it percolates through to the underlying limestone bedrock. Soil-air concentrations of carbon dioxide are commonly several orders of magnitude higher than atmospheric levels, and so therefore are carbonic acid concentrations in soil water.

In turn, the more acidic waters are able to dissolve larger amounts of limestone, and hence the most rapid and best development of karst landscapes is in areas which have abundant rain and thick soils rich in carbon dioxide: the humid tropics, for example. The character of the limestone is also important in the development of karst. Massive, very pure limestones karstify best and an essential prerequisite is that the main body of rock is impermeable. Thus, water can filter underground only through secondary openings in the rock such as joints and bedding planes. Such openings are spaced some distance apart in most limestones and these are the features that may become selectively enlarged into characteristic karstic landforms such as dolines, caves, swallow-holes, and gorges.

Water, ultimately derived from rainfall that is not evaporated (runoff), is essential to the formation of karsts, and almost all the major and minor landforms (Table 1) are readily understandable in terms of erosion or deposition of limestone by running or ponded water.

When an area of limestone is first exposed to external weathering agencies, perhaps after the removal of overlying

Table 1 A classification of karst landforms by size (micro to regional) and by location (surface or subterranean)

Scale of landform	Positive-surface landforms	Negative-surface landforms	Positive underground	Negative underground
Small scale (millimetres to metres)	Spitzkarren (ridges and pinnacles)	Karren pits, runnels and cockles	Calcite deposits (speleothems)	Micro-fissures
Medium scale (metres to hundreds of metres)	Cone and tower karst, tufa dams	Stream sinks enclosed hollows (dolines, sinkholes uvalas)	Calcite deposits (speleothems)	Caves
Large scale (100 m to kilometres)	Karst plateaux, plains, and cliffs	Dry valleys, gorges, poljes	None and cave	Major caves and cave chambers

non-carbonate rocks, the secondary openings in the rock will commonly be hairline fissures, able to engulf only seepages of rainwater. Initially a surface drainage system of streams is therefore likely to develop, with valleys being excavated as on non-limestones rocks, although in limestone regions the valleys are often gorge-like with steep sides. With time, progressively more of the runoff is able to sink underground as the openings are gradually enlarged by solution. Surface rivers will carry smaller and smaller flows and eventually may be active only during especially wet weather. Finally, the valleys become wholly dry and completely lose their original hydrological function. If the stream that formed the valley originated on adjacent impermeable rocks, a perennial stream sink (swallow-hole) may develop at the contact between the two rock-types. The surface stream will continue to deepen its valley upstream of the sink, and in time a reverse gradient or even a cliff will develop immediately down-valley of the point of engulfment forming a *blind valley* (Fig. 2b).

In the floor of the dry valley and at other locations between valleys, points at which rainwater is funnelled into the ground from the immediate surroundings will experience greater solutional erosion and bedrock hollows will develop. Some of these may enlarge greatly by capturing the water feeding adjacent hollows and become fully formed enclosed depressions or basins termed *dolines* or *sinkholes* (USA), ranging in diameter from a few metres to several kilometres and with depths of up to several hundreds of metres. Figure 2 illustrates the decay of a river valley into enclosed depressions.

The extent to which a river (fluvial) system develops varies greatly between karst regions, but ultimately the process of karstification will destroy the surface drainage system and replace it with a landscape of enclosed basins of all sizes which serve to conduit runoff underground as efficiently and expeditiously as possible. These dolines are the equivalent of small valleys in a fluvial system, the waters from them uniting underground into the major stream draining the area. If river systems do survive it is usually because they have cut deep gorges and thus occupy the lowest parts of the karst region. Such rivers may be fed by the water of karst springs. The Causse plateaux of south-central France are an example of wholly karstic uplands separated by rivers such as the Tarn, flowing in deep gorges and draining the upland karsts.

The surface of a mature karst area seems chaotic, with hollows and hillocks of every size, little or no level ground, barren and often steep slopes. The 'sense' of the landscape is not obvious from above because the controls are underground. The rivers that erode and transport away the weathered limestone are in cave systems emerging from the large springs that characterize the margins of karst regions.

Caves are the equivalent of surface streams, but instead of cutting valleys cave streams dissolve an increasingly large tunnel through the rock, the stream eventually occupying only the lowest part of the passage. Cave streams may deposit insoluble sediments brought from the surface, but more typically underground deposits are of calcium carbonate reprecipitated from solution as stalactites, stalagmites, and other *speleothem* forms.

Karst landscapes are developed wherever soluble carbonate rocks outcrop and where surplus rainfall is available to dissolve the limestone. Karst landscapes are comparatively poorly developed in dry and cold areas which lack both water and carbon dioxide in soils. The best development is found in the wet, warm limestones of the tropics. Areas such as Indonesia and adjacent areas of south-east Asia, many of the Caribbean islands, much of central America, and southern China exhibit some of the world's finest karst landscapes, the product of extended karstic erosion under optimal conditions.

The landforms of karst range from tiny solutional etching and pits on exposed rock surfaces (*karren*) to flat-floored enclosed basins several hundreds of square kilometres in area enclosed by steep slopes some hundreds of metres high (*poljes*). However, a predominantly solutional origin for landforms on whatever scale is the essence of karsts.

D. DREW

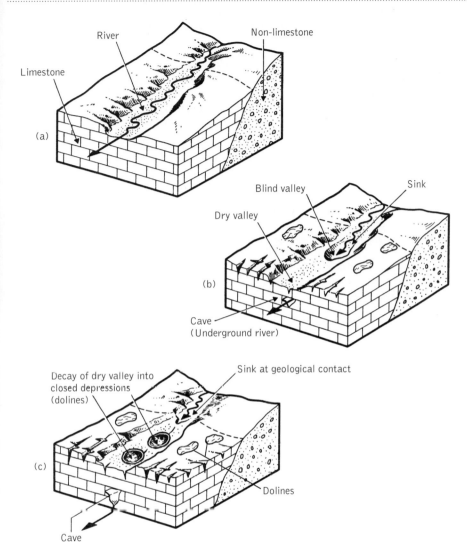

Fig. 2. The progressive decay of a surface drainage system as karstification proceeds, with the establishment of a swallow-hole at the contact between limestone and non-limestone rocks and the replacement of a valley by dolines. (After Drew, D. (1985); original drawing by J. Hanwell.)

Kelvin (Lord), *see* THOMSON, WILLIAM (LORD KELVIN)

kerogen, *see* ORGANIC MATTER (KEROGEN)

kimberlites Diamonds have been prized since antiquity, but until 1870 they had been recovered only as detrital grains from gravels and conglomerates. In that year, an area rich in diamonds was found in the north-eastern Cape Province, South Africa, and subsequent diggings revealed an underlying volcanic breccia containing diamonds. The town that rapidly grew up there was called Kimberley, after the then Colonial Secretary, and the new rock was named 'kimberlite'. For more than a hundred years, until the recognition of diamoniferous lamproite (a related rock) in Western Australia, kimberlite seemed the only primary source of diamonds. Both kimberlite and lamproite form small intrusions, typically vents filled with a breccia of wall-rock fragments in an ultramafic matrix (mica–peridotite in the case of kimberlite). Compared with other volcanic rocks, kimberlite is minute in terms of volume, but on the other hand its scientific value is immense because it has been erupted at high speed from depths in excess of 150 km (a region in which diamond is stable) and has sampled all the wall rocks in its path. In consequence, kimberlite provides the deepest samples of the planet, and is a unique source of information about deep Earth composition and processes. Until recently, diamond extraction methods meant that the rock fragments in the kimberlite were separated and left as dumps of mine waste. These dumps provided a rich scientific resource when their content of deep samples was recognized. Research has increased dramatically over the past 25 years (since companies have allowed wider access to mines and samples) to the extent that it has been found necessary to have special kimberlite conferences for the regular exchange of new knowledge.

Diamondiferous kimberlite is now known on all the continents, where it is confined to the oldest shield segments, the cratonic nucleii. The eruptions range in age from early Precambrian to Recent, with a notable display in Cretaceous times in Africa and South America. Kimberlites without diamonds are found outside the oldest cratons. No kimberlites have been recorded in the ocean basins, from which it may be inferred that thick, cool lithosphere is a prerequisite for the generation of diamondiferous kimberlite. Most kimberlite pipes are irregular in surface form, with areas of a few hectares; the largest, in Tanzania, has an area of 1.4 km², and is also one of the few examples where subaerial volcanic ashes are preserved. (Most kimberlite pipes are relatively shallow subvolcanic vents.) In many instances the vent flares towards the surface, so that its profile can be described as 'carrot-shaped'. This form results from the eruption of a volcanic breccia charged with gas that is expanding very rapidly with falling pressure, especially nearing the Earth's surface. Calculations give an average speed through the lithosphere of about 70 km h⁻¹, with near-surface velocities greater than the speed of sound. Such velocities, with rapid cooling resulting from gas expansion, explain not only the survival of diamond far outside its stability field, but, most importantly, provide rock and mineral samples from the deep mantle that retain the 'frozen' conditions of their formation. From the detailed mineral chemistry it is possible to estimate the mantle pressure and temperature at the time of eruption, and because the samples come from all depths to about 250 km it is possible from a range of specimens to find the Earth's geothermal gradient at that locality when the kimberlite erupted. The geothermal gradients from kimberlites of different ages and from different continents are all similar and are close to the calculated shield geotherm based on surface heat flow. This indicates that regions of this kind are a requirement for kimberlite activity.

Diamonds are constantly yielding fresh information about the Earth's interior, because they contain inclusions of melts and minerals. Some varieties of diamond have been found to contain minerals that may have come from the lower mantle, possibly even from the core–mantle boundary (at 2750 km). The ages of the diamonds are also interesting, because they are commonly much older than the host kimberlite; some, indeed, are as old as the oldest known rocks on Earth, showing that many, if not all, were picked up by the kimberlite during its passage to the surface.

As implied by the name mica-peridotite, kimberlite is olivine-rich, as might be expected of material erupting from the mantle, but it is difficult to specify the composition closely because of the high content of accidental fragments, ranging in size down to finely comminuted debris. Kimberlite may be thought of as composed largely of olivine with variable amounts of other minerals, most notably mica and carbonate, because these exemplify the most striking departure from other peridotites, which are normally deficient in volatile elements. Such elements, together with others having large ionic radii and high field strengths, are excluded by the crystal structure of typical mantle minerals such as olivine. The source of these elements and the process by which they were concentrated are two of the prevailing questions about kimberlite. At the very least, any melt must represent a minute fraction of the mantle from which it formed.   D. K. BAILEY

King, L. C. (1907–89)

Born in London, Lester King was educated in New Zealand. His name is associated with the analysis and explanation of African landscapes, and with the resultant model of landscape evolution involving scarp recession. In southern Africa he discovered landform assemblages that were incomprehensible in terms of the then conventional models of landscape evolution. King noted that scarp form and inclination are similar regardless of the degree of dissection of the landscape. He concluded that valley-side slopes developed a maximum inclination commensurate with stability and were then worn back. As scarps retreated, pediments were formed at their bases by sheet wash, eventually coalescing to form pediplains. King insisted that scarp retreat and pedimentation were active wherever running water was primarily responsible for shaping the surface. He applied his model in various contexts, interpreting inselbergs as the last remnants surviving after scarp retreat, and identifying in Africa three major planation surfaces and associated cycles of scarp retreat and pedimentation, of Mesozoic, Early–Mid Tertiary, and later Cenozoic ages. Many of King's ideas have been modified or disproved, but the scarp retreat model is valid in some contexts, as is his threefold scheme of major planation surfaces. He challenged conventional wisdom, and he thought globally. He was a major figure in twentieth-century geomorphology.   C. R. TWIDALE

kink bands, *see* SHEAR ZONES, DEFORMATION BANDS, AND KINK BANDS

Kuenen, Ph. H. (1902–76)

The Dutch sedimentary petrologist and marine geologist Philip Kuenen is best known for his research on turbidity currents and their deposits. His celebrated paper on turbidity currents as a cause of graded bedding, written jointly with C. I Migliorini, was published in the *Journal of Geology* in 1950. In it they brought together observations made in the field with experimental work in the laboratory to interpret flysch deposits (marine sediments of a special type, typically found in the Alps) in the stratigraphical record as having been deposited by turbidity (or density) currents: masses of water carrying a heavy load of material and travelling down undersea slopes with turbulent motion rather like a mud flow at the surface. Many more papers on the subject were written by Kuenen himself.

Kuenen was born in Dundee (his mother was Scottish, his father Dutch), but left Scotland at the age of five. After a doctorate at the University of Leiden, he joined the Snellius expedition to Indonesia in 1929–30 and his interest was aroused in marine geology. In 1934 he moved to the University of Groningen, becoming Rector in 1960. He became well known among English-speaking geologists for his book *Marine geology*, published in 1950, which became a standard work.

Kuenen's geological interests were wide. He also studied, among other things, the abrasion of sand grains, the origin of beach cusps, the formation of ptygmatic folds, and eustatic changes in sea level.   D. L. DINELEY

Kuiper, G. P. (1905–73)

The important contributions made to planetary astronomy by G. P. Kuiper were also of great significance for the geological sciences. Kuiper was born in the Netherlands and studied astronomy at the University of Leiden. In 1933 he moved to the United States and was subsequently known as Gerard Peter rather than as Gerrit Pieter Kuiper. In America he worked first at the Lick Observatory in California, and subsequently at Harvard, Chicago, and the University of Arizona, where he established the multidisciplinary Lunar and Planetary Laboratory.

It was Kuiper who first discovered (using spectroscopic methods) that Titan, Saturn's largest moon, has an atmosphere. Among Earth scientists he is perhaps best known for his theory of the origin of the Solar System, in which he postulated a 'reservoir' of icy planetismals beyond the orbit of Pluto. His research provided evidence in support of the impact theory of crater formation.

Kuiper studied the surface of the Moon and produced lunar atlases. He was also the editor or co-editor of substantial publications on the atmospheres of the Earth and the planets, the Solar System, and stars and stellar systems. The Kuiper Belt is named after him.   D. L. DINELEY

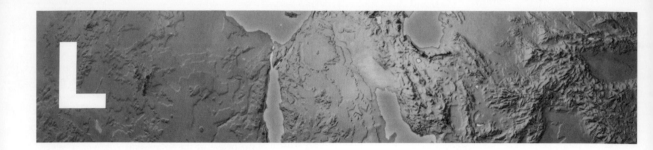

lagoons

lagoons 'Lagoon' is a term given to a shallow water body that is connected with the sea or a large lake through a narrow entrance. At landscape scales lagoons have elongate trends that are parallel to the coastline; lagoons of small drainage basins may, however, be symmetrical or even elongate in the coast-normal direction. By their nature, lagoons are bounded by at least two land masses. The mainland forms a coastline along the landward side of the lagoon. The outer-bounding coastline is generally a narrow strip of land or a chain of barrier islands. Lagoons are similar to estuaries in that they are enclosed water bodies landward of a sea (see *estuaries*). However, coastal lagoons are generally not associated with large river systems, but rather they occur along coastlines between large river systems. As such, the mainland shorelines of coastal lagoons are often influenced by a number of low-order drainage basins. The magnitude of run-off from the mainland margin is generally relatively low compared to that of a major river system, and the salinities of coastal lagoon are close to those found in adjacent nearshore waters. The amount of run-off from the mainland is only one of many factors affecting lagoon salinity. Lagoon size, exchange with the ocean, and rates of evaporation are other important factors causing lagoonal salinities to range from fresh to hypersaline.

Since flow in most coastal lagoons is not river-driven, circulation is primarily influenced by inlet tides and wind stresses on the lagoon surface. Shallow lagoons experience a great deal of sediment resuspension by waves, and wave-borne turbidity tends to cloud the water column. Deep lagoons tend to be less turbid, although tidal channels which have strong currents may produce enriched bands of turbid water.

Tidal inlets are the main conduits of exchange between the ocean and the lagoon, and have the greatest influence on circulation and flushing. The ebb and flood of the lagoonal tidal prism is a principal process affecting the flushing of the lagoonal basin. The total volume (capacity) of the lagoon with respect to the tidal prism therefore plays a major role in the exchange of water with the marine environment (see also *estuaries*).

Lagoon formation occurs when a portion of the mainland coast is separated from the sea by an outer or 'barrier' land mass. Migration of spits and submergence of back barriers are the two principal ways in which lagoons form. The lateral migration of spits from coastal headlands produces a new coastline seaward of the original mainland coastline. This process occurs in wave-dominated regions where sand sources are plentiful and littoral transport is large. The floors of spit-formed lagoons are characteristically smooth because they are inherited from the wave-smoothed sea floor. The floors of spit-formed lagoons tend to slope seaward, the greater depths being along the landward sides of the barriers.

The most common mode of origin of lagoons is by submergence of the land behind beach and dune landscapes. This method of lagoon formation is dependent on a rise in sea level, and requires only that the landwards migration of the mainland–lagoon shoreline is faster than that of the outer barrier. Although local subsidence of land surfaces can produce relative rises in sea level, global fluctuations in sea level have occurred many times during the Earth's history. These fluctuations are responses to changes in water volume stored in the oceans. Glaciation and deglaciation of the continents are important mechanisms for changing the volume of water in ocean basins. Since the last glacial lowstand, about 18 000 years ago, sea level has risen over 100 m, submerging the continental margin in its path. The patterns of local landscape submergence are strongly influenced by the erosional maturity of the land surface. In consequence, the floors of coastal lagoons are commonly very irregular and take on the character of the former land surface (antecedent surface) prior to submergence.

Once formed, the evolution of a coastal lagoon is dependent on the relative migration rates of the inner and outer coastlines and the rate of sediment influx into the lagoonal basin. During a period of rising sea level the size of a lagoonal basin is rarely constant because the inner and outer shorelines migrate landwards at different rates. If both shorelines were to remain stationary, then the sedimentary evolution of a lagoon would depend on the rate of sediment accumulating in the lagoon versus the size (or capacity) of the lagoon. When the inner shoreline migrates landward faster than the outer shoreline, then the basin capacity increases. However, when the outer shore migrates landward faster than the inner shoreline, the size and capacity of the lagoonal basin is diminished. Migration of the *inner* shoreline is primarily a passive process caused by submergence of an antecedent surface. The rate of migration is controlled by landscape slope. Shorelines of gently sloping landscapes migrate more rapidly than shorelines on steep slopes. The landward migration of the *outer* shore requires the movement or rollover of the outer-barrier landmass. The rollover process takes place in response to intermittent overwashes of sand from the beach to the back side of a barrier. Rollover is more prevalent at low-profile barriers than at barriers with multiple sets of dune ridges.

The rate of basin fill by sediment accumulation is controlled by: (1) run-off from the mainland; (2) flooding of sandy sediment through the tidal inlet; (3) overwash of sandy sediment across the barriers; and (4) the settling and accumulation of fine-grained sediment in protected areas of the lagoon. The pattern of fill is strongly influenced by the character of the antecedent surface, bathymetric shielding, and the type of circulation. In general, lagoonal sediments tend to be coarser textured near inlets, along the outer margins of the basin near barriers, and at exposed wave-stressed tidal flats. Fine-grained sediments accumulate in protected bays and fringe marshes which characteristically occur along the boundaries of temperate lagoons.

GEORGE F. OERTEL

Further reading

Ward L. G. and Ashley, G. M. (1989) Physical processes and sedimentology of siliclastic-dominated lagoonal systems. *Marine Geology*, 88 (special issue).

lake (lacustrine) sediments Lakes show a great range in size and depth, from small local ponds to bodies of water covering thousands of square kilometres. Many large lakes originate in tectonic depressions (rift basins) (for example, Lake Tanganyika and Lake Baikal) or major crustal sags (for example, Lake Chad and Lake Eyre), whereas others may form in extinct volcanoes, be produced by lava flows, or are the results of glaciation. Many small lakes are produced by changes in river courses or landslips. Lakes show a bewildering variety of sediments, reflecting local topography, climate, and availability of sediment.

Sediment may be supplied to lakes by rivers, wind, and ice. Wave- and wind-driven currents distribute the incoming sediment load in a similar way to that along marine coastlines. When surface winds drive water to one side of a lake they produce return flows along the bottom of the lake. When the water of lakes is significantly disturbed it oscillates to produce a slopping motion, the phenomenon being known as a *seiche*. Incoming surface water, depending upon its temperature and/or sediment load may spread over the surface, sink and flow at a level of similar density, or sink to the bottom to produce bottom currents (*turbidity currents*) which can deposit graded sands far from the point of entry (Fig. 1a).

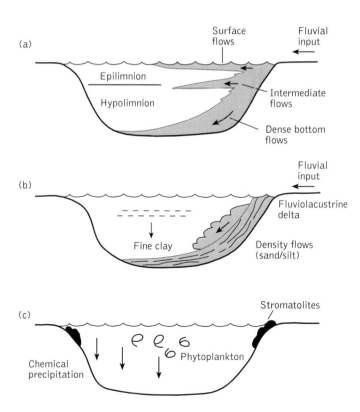

Fig. 1. (a) General zonation and processes in lakes. (b) Processes and sediments in lakes with abundant supply of terrestrial sediment. (c) Processes and sediments in lakes with dominant carbonate sediment and little influx of terrestrial sediment. (Sketched from data in Hutchinson (1957) *A treatise on limnology*, Wiley; Reeves (1968) *Introduction to paleolimnology*, Elsevier; and Matter and Tucker (1978) *Modern and ancient lake sediments*, International Association of Sedimentologists, Special Publication No. 2, Blackwell.)

Lake waters show a well-marked density stratification. The upper, warm waters are usually well mixed (the *epilimnion*) and these overlie denser, colder waters (the *hypolimnion*). The upper water is usually oxygenated, whereas the lower layer becomes anoxic due to the exhaustion of oxygen by the oxidation of organic matter descending from the surface. Whereas in temperate lakes, surface cooling in winter leads to overturn of the waters and oxygenation of the bottom waters (*oligotrophic lakes*); in low-latitude lakes, where there is little variation in seasonal temperatures, the bottom waters remain without any replenishment of oxygen (*eutrophic lakes*). Lakes receive water by precipitation, surface flow, and, particularly in arid areas, by subsurface flow from surrounding rocks. Also, in some areas (rift valleys) they receive hydrothermal magmatic waters from the subsurface. Lake water has a much greater variation in composition than sea water, and hence there is a greater variation in the resulting chemical precipitates.

Lakes experience much more frequent and faster changes in water level than seas. As a consequence, their nearshore and marginal environments suffer exposure and fast transgressions and regressions, and have a complicated stratigraphy when preserved in the geological record.

Fluviolacustrine deltas (Fig. 1b), constructed by incoming rivers are very similar in structure and stratigraphy to fluviomarine deltas. Sands pass lakewards into fine-grained sediments. Turbid, cold waters are often sufficiently dense to plunge beneath the surface waters and flow down the slope to form subaqueous fans with channels and levées as seen in fluviomarine deltas. The resulting sediments are composed of graded and plane-bedded sands, sometimes with cross-stratification in the channels, wavy laminated silts in the levées, and graded silts and laminated muds on the more distal (i.e. distant), deeper, lake floor. Subaqueous slumps occur on deltas (e.g. Lake Brienz, Switzerland) and they may also occur around the margins, to produce gravity-flow deposits. Seasonal floods frequently produce graded sands and silts over lake floors. In glacial terrains coarse-grained, so-called Gilbert-type deltas of gravels and sands are common, and these pass in deeper water into varved sediments.

A great variety of carbonate sediments are present in lakes draining carbonate-rich terrains (Fig. 1c). Laminated, fine-grained carbonate mud and marl, commonly interlaminated with fine-grained organic layers produced by planktonic remains descending from the surface, are common bottom sediments of temperate lakes.

In shallow water in arid areas, oolitic sands are common along the shorelines of some lakes (e.g. Great Salt Lake, Utah). Elsewhere, plants such as charophytes coat their stems with calcium carbonate and produce calcified oogonia which accumulate on the lake floor to produce *chara marls* when mixed with siliciclastic detritus. Various particles become coated with films of cyanobacteria which trap and precipitate calcium carbonate to produce coated grains (*oncoids*) as found in marine lagoons. Also, cyanobacterial mats cloak sediment and rocks to produce *stromatolitic coatings*. Large mounds, often several metres high and tens of square metres in area (*tufa mounds*) are produced by bacteria that trap or precipitate calcium carbonate or gypsum issuing from springs. Dolomite is also precipitated in lake sediments in Lake Baikal, Deep Spring Lake, California; and other lakes in Texas.

Lake deposits are often very rich in organic matter. It may be derived from surrounding vegetation and be composed of the debris of land plants. Also, diatoms and dinoflagellates living in lake waters produce vast quantities of organic matter in large temperate lakes, while green algae produce most of the organic matter in hypersaline lakes. Large amounts of organic matter are found in low-latitude lakes that do not experience a seasonal overturn and remain eutrophic. The organic-rich bottom deposits (*sapropels or gyttja*) are rich in oil-producing organic compounds. In vegetated areas, lakes are often colonized and invaded by marginal plants and become filled with peat.

In arid areas a rich variety of chemical deposits is formed by subsurface water seeping from adjacent rocks into the depression (*playa*). Many of these lakes are ephemeral and dry out in summer to produce bare salt-covered flats or *sabkhas*. In some lakes in western USA, carbonate occurs with gypsum and halite. Sulphate-rich waters in bitter lakes produce *mirabilite* ($Na_2SO_4.10H_2O$) which alters on exposure to *thenardite* (Na_2SO_4) both along the margins of the lake and in its deeper parts. In alkaline lakes, *natron* ($Na_2CO_3.10H_2$)) and *trona* ($NaHCO_3.2H_2O$) are found in the United States, Africa, and South America. In lakes in the East African rift valley, volcanic emanations supply high concentrations of sodium carbonate, and reaction between these products and volcanic detritus produces unusual mineral suites. *Trona* ($Na_2CO_3.HCO_3.2H_2O$) forms layers interbedded with detrital clays and volcanic detritus in which there is formation *in situ* of zeolites. The rare minerals *magadiite* ($NaSi_7O_{13}(OH)_3.3H_2O$) and kenyaite ($NaS_{11}O_{20.5}(OH)_4.3H_2O$) and associated cherts are thought to form by leaching of the latter. In other lakes, borates are deposited as the rare mineral *borax* ($Na_2B_4O_7H_2O$). Searles Lake in California produces 90 per cent of the world's boron from such a deposit. Other more unusual deposits are the diatomaceous-rich sediments found in temperate and high-latitude lakes, and iron and manganese crusts and iron hydroxides and oxides of some high-latitude lakes (e.g. Scandinavia).

Thick deposits of ancient lake sediments are known in many parts of the geological column. The Eocene Green River Formation is one of the most famous, but ancient lake deposits are known in the Devonian and Triassic of Europe and North America. They are extremely valuable economically, both for their rare evaporitic salts and as important sources of hydrocarbons.

G. EVANS

Further reading

Galloway, W. E. and Hobday, D. K. (1983) *Terrigenous clastic deposition systems, applications to petroleum, coal, and uranium exploration*, pp. 185–201. Springer-Verlag, New York.

Matter, A. and Tucker, M. (eds) (1978) *Modern and ancient lake sediments*. International Association of Sedimentologists, Special Publication 2.

Reeves, C. C. (1968) *Introduction to paleolimnology*. Elsevier, Amsterdam.

lake levels and climate change *see* CLIMATE CHANGE AND LAKE LEVELS

lakes A lake is a standing body of fresh or saline water on the land surface, into which and from which rivers may flow. Although lakes originate in many ways, they may be as prolific today as at any time in the geological past. This is attributable to the legacy of the Pleistocene glaciation during which ice scraped across the land surface, accentuating hollows, picking out weak rocks, and depressing great areas of the Earth's crust to entrap vast volumes of water, as in the Great Lakes of North America. However, glacial activity is by no means the only process whereby lakes are formed. Small lakes may form as parts of river channels become separated from the main flow channel during flood. Other ephemeral lakes develop in volcanic craters or collapsed caldera systems. These are destined to be lost during subsequent volcanism. Most lakes will become filled with sediments in due course, but cores show that some of the largest, such as Lakes Baikal and Tanganyika, are at least 20 million years old. Anthropogenic lakes, reservoirs for water storage or for power generation, abound in both the developed and developing worlds, and many of these are filling so rapidly with sediment that their useful life is gauged as a few decades.

Large lakes may develop as rivers become blocked when mountain belts rise across their former routes, as with the Tibetan rivers encountering the rising Himalayas. In major rift valleys lowering of the graben floor creates linear depressions, as in East Africa, but the uplift of the marginal horsts may equally block pre-existing rivers draining across the area to create lakes such as Lake Kyoga. Some such rift valley floors may fall below sea level, as in the Dead Sea, so that their lakes are below sea level. Downfaulting away from rift systems may also create substantial basins isolated from the surrounding areas and within which streams flow to the low points where water accumulates. Such waters carry salts in solution and become increasingly saline as the water evaporates from the lake surface; the result is that salts are precipitated on to the bed of the lake. Short-lived saline lakes or playas are particularly common at the margins of many deserts, where waters from flash floods become ponded up, but evaporate away in time (see *playas*). In other arid areas, where the wind blows particles away from the floors of the hollows, large deflation hollows (e.g. Quattara) may form whose floor level is defined by the presence of the water table below surface. The capillary action of the water serves to retain the sand particles, and water seeps may form. Many oasis lakes are of this origin.

Whatever their mode of origin, lake basins contain bodies of water which are dynamic in that they respond to solar energy input to give warmth to the waters, and to the volumes of water being carried to and from the basin. The influent waters carry not only solute loads, but also transport sediments both in suspension and along the bed of the streams. At the points of entry of streams deltas may build into the water body, and where these extend to the opposite shore they subdivide the lake. Factors such as wind activity, which creates surface waves capable of eroding shorelines, and the generation of 'seiches', which cause the water mass to oscillate within the basin, contribute to the creation of a range of associated, often submerged, geomorphological features.

The solar energy input is related to the position of the lake on the Earth's surface. In tropical areas there is always substantial heat input, but in temperate or sub-arctic zones the lake waters may be very cold or frozen for part of the year when solar energy input is low, increasing to a thermal peak in the summer months. Under such conditions the density of the water mass varies as the water temperature changes and the lake waters become layered. The warmer surface waters are less dense than those at depth, and form a distinct layer which responds readily to wind-induced current generation and wave formation. The lower water mass remains cool and little of the energy input from sun or wind penetrates across the thermocline, a barrier across which water temperatures may fall by as much as 6 °C, marking a horizon at which settling suspended sediment may be held during its descent to the lake bed. Many fish species feed at the thermocline, although its level falls as the summer progresses. With the first frosts of approaching winter, the surface waters are chilled, increasing in density, and they begin to sink into and below the summer bottom layer. The turnover of water leads to drastic changes

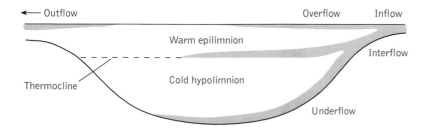

Fig. 1. Motion of influent waters reaching a lake basin during the summer period of thermal stratification. Locations of the overflow, interflow, and underflow are dependent on temperature.

585

in the colour of the water if materials from the lake bed are carried into the upper waters. The position at which influent water penetrates the lake water depends on the relative densities of the water masses. In summer, influent waters are commonly warm and enter the uppermost waters (overflow), but cooler rainwater falling at night may sink to the level of the thermocline before moving out into the basin (interflow) (Fig. 1). Cold rain derived from thunderstorms may hug the bed as it flows along the basin floor (underflow). Although it is possible to calculate residence times for the waters entering and leaving a lake system using influent and effluent figures, the relative importance of the overflows, interflows, and underflows is the defining factor for the geomorphological behaviour of the lake.

The behaviour of chemicals entering the water with the influent rivers or directly from the atmosphere is complex, being dependent on the chemicals concerned, their relative solubilities, and the pH and Eh of the systems concerned. This fascinating topic is discussed in the entry on the geochemistry of lakes.

Casual observers may regard the beds of lakes as being inherently rather dull with little variety of sediment texture throughout. The belief that the floor is normally flat may be derived from memories of dry lake-bed salt flats used for high-speed vehicle testing. Few lakes have this characteristic, and many have substantial deltas ending in steep slopes where streams enter the water body. In others there are submerged, often broad, benches and platforms which French and German workers have shown to be related to a widely recognized period of low water levels at about 4500 years BP. At this stage the vegetation of the shallower parts of the lake floor was replaced by emergent species before a return to wet conditions again raised water levels. Many ancient lake-dwellings in Scotland ('crannogs') and elsewhere have their foundations on this now submerged surface or on rocky outcrops protruding above it. It appears that as lake levels rose slowly the people built upwards, strengthened their access causeways, and continued to occupy the site. Archaeologists have shown that some were occupied for over 3000 years before being abandoned in the fifteenth century.

Steep-sided glacially formed lakes occupying deep U-profile valleys, as in northern Switzerland, show slopes at angles close to those of the angle of stability of the sediments. Silts and muds are known to slide from these slopes into the deeper parts of the basin, sometimes creating small density flows in which the additional sediment load in the sliding waters provides the power enabling the flow to extend for considerable distances from the originating collapse. The atmospherically controlled seiching phenomenon which causes very long-period waves to become established, both at the water surface and also deeper at the thermocline, may exert large enough shear stresses to help in dislodging this material.

In ice-marginal lakes, notably those in Norway, where the active ice alternates between retreat and advance, the lake basin is subject to very rapid sedimentation. Some of the sediments settling to the bed are contributed directly by melt waters, but much is also derived from calving of the glacier. The upward-calving ice from the bed of the glacier is usually heavily charged with morainic matter. The bergs, driven towards the lake outlet by katabatic winds, become stranded as they move into shallower waters. Here melting continues, with accompanying release of sediment to the bed, and a broad shallow shelf is commonly created, with an erratic-strewn beach lining the shore. The finer sediments are frequently moved by the small tsunami-like waves created as the upward calving of the glacier ice continues. During advance phases the deposited morainic and layered bottom sediments become pushed by the base of the ice, and sequences of push-moraines may form beside the glacier margin during retreat dominance.

In the deeper ice-marginal lakes where sedimentary deposits may be preserved, seismic stratigraphic methods can distinguish between morainic material and proglacial deposits, thereby making it possible to reconstruct the glacial history of the lake. At favoured sites (such as Loch Muick in Scotland), the submerged terminal moraines from the last phase of glacial retreat can be recognized. As well as shedding light on past climate changes, lake sediments are sensitive indicators of recent environmental variability. Sediments of the lake floor commonly show alternating coarse and fine layers, which are sometimes attributed to varving from summer–winter cycles of change. It is known that in some lakes several varve successions develop during one year in response to freezing and refreezing in spring.

As most well-established lakes have been in existence for at least several thousand years, the coastal geomorphology of their shorelines has long since reached a mature condition. Pebble beaches may show growth and retreat of longshore ridges and bars in response to wind-induced wave action, but although some will also show development of spits where coastline directions change abruptly, few show more exciting features of coastal erosion. That these existed in the past can be demonstrated from examination of the shores of recent man-made reservoirs, where within 5 years substantial cliff systems form on glacial tills, and within 30 years complex systems of caves, stacks, and geos are seen to have developed on similar shore materials exposed to maximum wave activity.

JOHN McMANUS

Further reading

Gierlowski-Kordesch, E. and Kelts, K. (1994) *Global geological record of lake basins*, Vol. 1. Cambridge University Press.

Lerman, A., Imboden, D., and Gat, J. (eds) (1995) *Physics and chemistry of lakes*. (2nd edn). Springer-Verlag, Berlin.

McManus, J. and Duck, R. W. (eds) (1993) *Geomorphology and sedimentology of lakes and reservoirs*. John Wiley and Sons, Chichester.

Lamarck, Jean-Baptiste Pierre Antoine de Monet (1744–1829)

The French naturalist Lamarck had a significant influence on palaeontology by virtue of his theory of the inheritance of acquired characteristics. An aristocrat by birth, his education was in botany and his first appointment was as 'Botanist to the King'. In 1793 he was appointed to a professorship of zoology at the Musée Nationale d'Histoire Naturelle, based on the Jardin du Roi (which later became the Jardin des Plantes under the Republic). He established several taxonomic classes that still form part of today's scheme of zoological classification, but his teachings were based more on the ideas of Aristotle than those of Linnaeus. He recognized what today is regarded as ecology and pondered on the nature of species and the causes of extinctions.

Lamarck's evolutionary theory was expounded in his *Philosophie Zoologique*, published in 1809. He argued that structural changes in organisms were caused by changes in the environment, which made it necessary for the organisms to adopt new habits. Thus, he believed that the long neck of the giraffe was the result of stretching upwards to compete for the higher foliage. An individual that had thus stretched its neck passed on this characteristic to its progeny; and so on through many generations. Lamarck even included Man in his theory of descent, despite the fact that no human or pre-human fossils were known at that time. Lamarck's theory was eventually supplanted by Darwin's theory of evolution, which did not require acquired characteristics to be inherited.

D. L. DINELEY

Lamb, H. H. (1913–97) The climatologist Hubert Horace
Lamb was born in Bedford, England. After studying geography at the University of Cambridge he joined the Meteorological Office, and in 1939 was undertaking research on upper winds and preparing weather maps for the transatlantic route. He spent most of the Second World War in Ireland, where he was in charge of meteorological training courses and was also responsible for forecasting the weather for transatlantic flights. He returned to the Meteorological Office in 1946 and was soon posted to the Forecasting Research Branch to study medium- and long-range forecasting. He subsequently worked in the Climatology Division, where he made his first investigations of historical climate data. While there, he published several papers describing the relations between variations in sea-surface temperatures and variations in climate.

When his retirement from the Meteorological Office approached in the 1960s, Lamb was offered the task of establishing the Climatic Research Unit at the University of East Anglia. He remained there until his final retirement in 1977. Under his direction the Unit became internationally recognized as one of the foremost centres of the subject in the world. Lamb himself was well known for his many detailed scientific papers on climate research. He was also the author of *Climate: present, past and future* (Volume 1 of which was published by Methuen in 1972 and Volume 2 in 1977) as well as his later book *Historic storms of the North Sea, British Isles and north-west Europe* (Cambridge University Press, 1991).

ALASTAIR G. DAWSON

laminated sediments and climate change *see*
CLIMATE CHANGE AND LAMINATED SEDIMENTS

landfill techniques A landfill is a site where material
('waste') discarded by individuals and organizations is disposed of because there are no technically and economically viable methods for its reuse, recycling, or treatment. Waste in landfills decays in physical, chemical, and biochemical processes that may eventually render it environmentally harmless.

Early landfills and their problems
Early landfills were simply piles of waste disposed of on the ground, generally in topographically depressed areas. Natural depressions, such as valleys, and artificial depressions, such as abandoned quarries and surface mines, were mostly chosen as sites for landfills because of their container-like shape. However, because of their topographically depressed nature, landfills at such sites received surface water runoff from surrounding areas, and groundwater infiltrated into them from their sides and bottom. Infiltration of surface water and groundwater into a landfill results in water passing through waste in the landfill, which frequently picks up contaminants from the waste. Such contaminated water is called 'leachate', and the process that transfers contaminants from waste to water is called 'leaching'. Leachate from a landfill can contaminate various environmental media, such as groundwater, surface water, soil, and air, that come in contact with it. Exposure to leachate may also be hazardous to people, animals, and plants.

In addition to generating leachate, water passing through landfilled waste helps micro-organisms in the landfill biochemically to degrade organic matter in the waste by providing moisture and conveying nutrients to the microorganisms. Biodegradation of organic matter in a landfill ultimately produces a mixture of gases generally referred to as 'landfill gas', or 'LFG'. The LFG is typically 40 to 65 per cent methane by volume, the remainder being carbon dioxide. Traces of hydrogen sulphide and other gases can also be presented, according to the nature of the waste. LFG poses fire and explosion hazards because of its methane content.

The location of early landfills in topographically depressed areas resulted in the generation of large volumes of leachate that often contaminated the environment. In addition, the design, construction, and operation of early landfills allowed both vertical and lateral movement, or 'migration' of LFG. Upward migration of LFG caused fires on landfill surfaces and posed explosion hazards to crews and equipment operating on the landfills, as well as to aircraft flying overhead. Downward migration of LFG threatened the quality of groundwater because of the methane and hydrogen sulphide in the gas. Lateral migration of LFG to nearby farmland damaged crops, and its migration to the basements of buildings posed explosion hazards. In addition, the method of operating early landfills provided shelter and food for disease-carrying creatures, or vectors, such as birds, rats, flies, and mosquitoes.

of denudation of the land surfaces since these dated materials were deposited or formed are lower than had been previously thought, especially in areas of low tectonic activity or those that did not experience Quaternary glaciation.

Powerful computer-based numerical models have been developed that link lithospheric and asthenospheric tectonic processes with surficial processes of weathering and erosion (generally in a plate-tectonic context). These models have provided significant insights into long-term landscape evolution. They have been able to simulate the topography observed along passive continental margins, for example, highlighting the relative importance of tectonic processes and surficial processes (weathering and erosion) in generating topography. They have also shown how denudational and flexural isostatic processes modify topography. These models are being used to elucidate landscape evolution at the temporal and spatial scales considered by Davis, Penck, and King, but we are not yet able to integrate all aspects of landscape evolution into a grand theory. At this stage, we probably know more about rates of landscape evolution and the important roles of passive and active tectonics than about the precise forms through which the landscape passes during its evolution.   PAUL BISHOP

Further reading

Summerfield, M. A. (1991) *Global geomorphology*, Chapter 18. Longman, Harlow.

landscape modelling
Because most landscapes change only very slowly, the erosion of valleys, for example, typically taking periods of 100 000 years or more, it is exceptional to be able to observe landscape evolution directly. Models provide an essential means of extrapolating measurements taken over a few years, or at most a century, to the geological time spans over which landscapes develop. There are difficulties in this extrapolation, because of progressive changes in the form of the land and substantial climate change over these periods, but the principles of landscape modelling are reasonably well established.

Models may be either qualitative or quantitative. W. M. Davis developed the concept of the geographical cycle of erosion in the 1890s, which provided a descriptive model of the sequence by which hillslopes and rivers decline in gradient over time in the absence of significant tectonic uplift, eventually becoming low-relief peneplains, which could later be uplifted to form erosion surfaces. Walter Penck in the 1920s and Lester King in the 1950s proposed that hillslopes normally evolved through parallel retreat of steep slopes, although Penck attributed the retreat primarily to a balance with tectonic activity, whereas King recognized parallel retreat as characteristic of steep semi-arid landscapes. John Hack in 1960 suggested that landscapes were generally in a balance between denudation and either tectonic uplift or fluvial downcutting, so that the landscape was essentially in a state of dynamic equilibrium. Although these models were essentially qualitative, they contain important concepts which are also relevant in the context of quantitative models, particularly the ideas of slope decline, slope retreat, and equilibrium.

One of the earliest quantitative models was set out in the Revd Osmund Fisher's 1866 paper 'On the retreat of a chalk cliff'. G. K. Gilbert, working in the Henry Mountains of Utah in the 1870s, also formulated a set of principles for badland slopes, which are seen to be related to current slope models. In the 1950s, Bakker and de Heux made further progress on modelling the retreat of a cliff face to form a talus slope at its foot. In the 1950s too, Frank Ahnert was among the first to begin development of slope models in a recognized current form. Initially most slope models were for the evolution of a single slope profile and, before the widespread use of computers, many models were based on seeking mathematical solutions rather than using numerical simulation. Ahnert, in the 1970s, was again a leader in developing three-dimensional models of landscape evolution, and today slope profile and three-dimensional models coexist, both relying mainly on computer simulation, although there is also considerable value in scale models of landscapes, especially for fluvial and coastal geomorphology. Slope profile models allow more complex processes to be simulated, while three-dimensional models provide a basis for studying the integration of stream and slope processes to create a complete landscape.

Principles of current simulation models

Most, although not all, current slope models are built from four basic building blocks: (1) a framework consisting of a sediment budget equation; (2) one or more 'process laws'; (3) boundary conditions linking the model to the remainder of the landscape; and (4) an assumed initial form from which the landscape evolves. These create a soluble system, which can be described formally by a set of partial differential equations.

The sediment budget equation essentially takes the form:

$$\text{Input} - \text{Output} = \text{Net increase in storage,}$$

which is familiar in hydrology. For landscape models, the inputs and outputs are sediment and solute transport rates, and the net increase in storage represents an addition of sediment, that is deposition, and a net decrease (negative increase) erosion, or denudation (where relevant it should include any net addition or removal by wind transport). The sediment budget may be applied to many kinds of landscape unit, but is most commonly applied to sections of a slope profile (Fig. 1), although it may also be applied to parts of drainage basins (e.g. hillslopes and valley bottoms) or whole basins. A flow strip, which outlines a slope profile, lies between neighbouring lines of greatest slope (flow lines), and may have a variable width if there is convergence or divergence of the flow lines.

The sediment budget equation is undoubtedly the best established relationship in slope modelling. It not only plays a vital role in maintaining the integrity of models (by ensuring

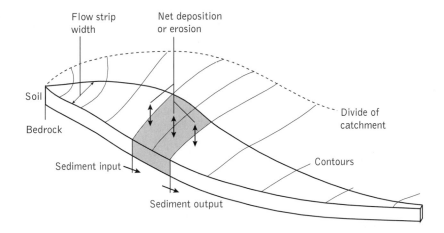

Fig. 1. Sketch of a flow strip with its catchment. It is bounded by lines of steepest descent, which are orthogonal to the contours. Strip width varies with convergence or divergence of flow lines. Terms in the storage equation are shown for the shaded section of the flow strip.

that all sediment is accounted for), but also provides the formal link between spatial rates of change (in sediment transport rates) and temporal rates of change (in elevation), which is at the methodological heart of slope modelling.

Process laws describe the observed empirical dependence of sediment and solute transport capacities (the maximum rate of steady sediment transport) on the landscape topography, usually through slope gradient, s, and distance from the divide, x (which can be generalized to flow strip area per unit strip width). Several process laws have been generalized to the form:

$$\text{Transport capacity} \propto x^m s^n$$

The distance (x) term indirectly represents the effect of the collecting area for water at any point. The gradient (s) term reflects the influence of gravity. For processes such as soil creep, rainsplash, or solifluction, the transport capacity is (approximately) directly proportional to gradient, and independent of flow (and therefore collecting area). The exponents m and n are therefore 0 and 1. For soil erosion (rillwash) processes, acting in rills or small gullies, the effect of overland flow is much more important, and the exponents are $m = 1$–3; $n = 1$–2. For a process such as solution, which depends directly on flow and hardly at all on gravity, $m = 1$; $n = 0$.

Actual sediment transport may not take place at the full sediment transport capacity, although in simple slope model, this is usually assumed to be the case. More generally, the rate of pick-up of sediment may be related to the 'excess capacity' principle. This states that where sediment is carried at less than its full capacity, the rate of pick-up is proportional to the excess capacity, that is the difference between the capacity and the actual sediment transport rate. This difference is particularly important in rivers, where different grain sizes are transported selectively, with different excess capacities for different grain size classes. Coarse material is normally carried at its full (rather low) capacity rate, while fine material may be carried

at far below capacity, limited by the supply of suitable material upstream.

Boundary conditions may take many forms. For a simple slope profile, the upper boundary is most simply represented as a divide between two valleys, at which the sediment transport is zero. The base of a slope profile is usually at a river bank. The simplest way to represent this boundary is to define how its elevation changes over time: it may either be fixed (the simplest case) or gradually changing in response to the behaviour of the river, and/or its interaction with the sediment delivered by the hill slope.

The fourth and last essential component of a slope model is a set of initial conditions. For a slope profile, it is necessary to assume a profile form at the start of the period to be simulated. This may be chosen as an arbitrary form, such as a plateau or a slope of uniform gradient, it may be based on independent evidence of a previous landscape form, or it may use the surveyed form to begin the simulation from the present.

Figure 2 shows a sequence of slope profile forms simulated for a combination of rillwash and rainsplash processes, beginning from a plateau, with a divide at the top (left) and a fixed stream at the base. The compound process law used in this example is a sum of two terms for rainsplash and rillwash respectively.

The simulated slope forms are initially strongly convex, because of the incision of a ravine into the plateau but, over time, the profiles become convexo-concave. As time goes on, the influence of the initial form becomes progressively weaker, and the profile becomes largely characteristic of the processes forming it. This may be regarded as a useful property, in that direct evidence of past forms is generally rather speculative, but it also follows that it may not be possible to reconstruct former landscapes from the present. This property, for distinct initial profiles to evolve convergently, is known as 'equifinality'.

Calibration of the models can be partially achieved by comparing the process laws with short-term field measurements

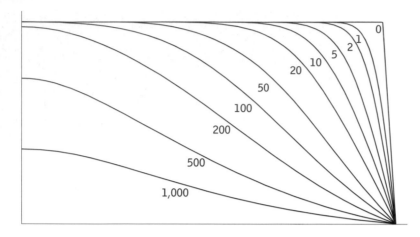

Fig. 2. Modelled evolution of a slope profile under a combination of rainsplash and rillwash. The top of the profile is a symmetrical divide, and the base a river which remains at fixed elevation. Numbers show times in thousands of years.

of process rates, but it must be recognized that climate has changed, and process rates with it, so that extrapolation of present rates further back than the mid-Holocene may be highly speculative. Formal validation of slope models is, therefore, largely impracticable at present, even though the compatibility of known processes and observed forms may be highly suggestive.

Among the problems encountered, there is considerable error in assessing many of the process rate parameters, and many of the processes are inherently non-linear. There is, therefore, the strong expectation that there are some situations of chaotic sensitivity to initial conditions, and an inherent unpredictability in the landscape evolution. Perhaps the best example is the difficulty in predicting exactly where channel heads will form, even though it is possible to predict the overall density of the drainage network.

Special cases

In a number of cases, simplified assumptions within slope models help to give an overview, and link back to the older qualitative models described above. For example it may be assumed that the landscape is uplifted tectonically at a rate exactly equal to the denudation rate. This matches Hack's steady 'dynamic equilibrium' assumptions, and it can be shown that for the simple single process law above the steady-state equation shows convexity for certain conditions and concavity otherwise. For the compound process law of Fig. 2 (for rainsplash plus rillwash), the steady-state profile is convexo-concave, and the convexity has a width corresponding exactly to the area where the first (rainsplash) term is dominant. Similarly the base of the slope is concave in the zone where the second (rillwash) term is dominant.

A second kind of simplifying assumption is to hold the bottom point of the slope profile at a fixed elevation (base level). For many process laws, including the examples given here, the slope profile can be shown to evolve towards a 'characteristic form' in which the denudation at every point is

strictly proportional to its elevation above the base level. This assumption, and its consequences, correspond closely to the Davisian model, with the characteristic form corresponding to 'maturity' of the profiles, and the final form to a 'peneplain'. Although less amenable to a simple analysis than the dynamic equilibrium above, it may be seen from Fig. 2 that, for the assumption of fixed base level, the profile due to rainsplash and rillwash is also convexo-concave, although the width of the convexity is slightly narrower than for dynamic equilibrium.

Developments in slope modelling

Beyond the simple slope models described here, there are many significant developments which are gradually reducing the limitations of the simple forms. The main growth areas are in three-dimensional models; in establishing explicit links between process rates and climate; in the enrichment of models with an understanding of soil and vegetation dynamics; and in attempting to reconcile models at different scales. These areas represent the current research frontiers in landscape modelling.

One important growth area is the extension of models from profiles to three-dimensional landscapes. The main difference in principle is that the distance from the divide (x above) needs to be replaced with drainage area per unit flow strip width, and this changes dynamically as the landscape evolves, and must be re-computed periodically. Although three-dimensional models make greater computing demands, these can now be met, allowing investigation of the dynamic interaction between stream and hillslope processes. This determines not only stream long profiles, which determine the evolution of each slope base, but also the extension or contraction of the entire stream network, through the position of channel heads. Large-area three-dimensional models are also potentially valuable tools for exploring the response of landscapes to regional patterns of tectonic movements, and the isostatic uplift resulting from unloading due to erosion.

Although simple process laws are related to distance from divide (or area) and gradient, these laws are based on an understanding of the physical processes, and implicitly sum rates over the frequency distribution of rainfall and other climatic events. Increased knowledge of each process provides a means of building up the aggregate process laws from physical and chemical understanding of single events. This approach has the potential of making an explicit link from climate to process rates. As knowledge of past climates improves, it may become possible to simulate slope evolution reliably through periods of major past and future climate change.

One element in linking process rates and climate lies in a better understanding of the links with vegetation and soils. Vegetation cover and organic soil are highly effective in reducing erosion; stripping by erosion changes soil properties; where surface wash is important, there is strongly selective transport amongst the grain sizes moving on the surface; an understanding of these processes therefore plays another major part in mediating the impact of climate on process rates.

The last major growth area is in attempts to make workable models at a wide range of spatial scales and timescales. In assessing priorities for mitigating soil degradation, there is a need to understand area distribution at regional to continental scales. There is also a need to link slope evolution and erosion rates to the very broad scales of general circulation models for climate. Similarly, there is a need to understand processes down to the time scale of a single storm, and up to the Pleistocene or the Tertiary. In changing time and/or space scales, it is essential, although inherently difficult, to ensure that similar principles are applied at each scale, that models remain compatible as scales are changed, and that dominant process at each scale are properly represented.   M. J. KIRKBY

landscape sensitivity

Unlike people, all landscapes are not created equal. The fascinating variety of landforms that seduces many geomorphologists to their calling also gives rise to a tremendous range in landscape response to natural and anthropogenic change. Landscape sensitivity may be considered to measure the degree to which a landscape will respond to a unit change in geomorphological forcing. Landscape sensitivity therefore characterizes the intrinsic susceptibility of a landscape to geomorphological change. Such susceptibility is a function of the nature of the change and the nature, distribution, and juxtaposition of spatially variable landscape characteristics, such as soils, topography, and bedrock lithology and structure.

Variability in the intrinsic sensitivity of a landscape to different geomorphological processes, and the importance of being able to relate specific processes to specific places, is readily illustrated through the example of shallow landsliding. Common sense suggests that steep slopes are more prone to landsliding than gentle slopes. Other factors that influence the potential for shallow landsliding are drainage area (which influences how wet a site is), soil strength properties (i.e. cohesion and friction angle), and the effective cohesion provided by the root strength of vegetation. The sensitivity of a landscape to loss of root strength (such as follows a timber harvest) varies greatly according to topographic position and soil properties. Steep, convergent topography with cohesionless soils is the most sensitive location, whereas low-gradient, divergent topography with highly cohesive soils is least sensitive. Hence, the overall sensitivity of a landscape to changes in processes that influence shallow landsliding will depend upon the specific topography and soils present in the landscape.

Similarly, the juxtaposition of elements within a landscape can strongly influence landscape sensitivity to particular processes. Consider further the example of shallow landsliding discussed above. The net downstream impact of a change in land use that accelerates landsliding, and thereby increases the sediment supply in the headwaters of a drainage basin, can depend upon the nature of the channel system between the input and the site of interest. A channel system with abundant sediment-storage elements distributed along its length may rapidly damp out the effects of a pulsed increase in sediment supply. In contrast, a channel system with little buffering capacity for storage of sediment will rapidly deliver increased sediment loads to downstream environments. Landscape sensitivity must be evaluated in relation to particular processes operating in the context of a particular landscape.

Landscape sensitivity also varies with different types of external forcing on landscape processes. The sensitivity of a landscape may be very different for changes in climate, the nature of the ground surface (e.g. changes in impervious area or in soil properties), or in boundary conditions (resulting, for example, from dam construction or the removal of flow obstructions, such as large logs from streams). Again, the sensitivity of the landscape will vary according to the influence of the change on hydrogeomorphological processes that govern the erosion, transport, and deposition of sediment. Landscapes are perhaps most sensitive to changes in those processes that dominate their morphology and dynamics.

A key aspect of landscape sensitivity is resistance to change, or the ability of a landscape to resist or absorb impulses of change. Landscape resilience can differ over different timescales and for systems that encompass different spatial scales. Gradual change and the accumulated effects of individually small changes in landscape processes can eventually cause a large landscape response. Or a landscape could be very sensitive to minor changes, but the response might also be minor. In contrast, catastrophic changes can result in rapid pulses of dramatic change, even in relatively insensitive landscapes. Moreover, a system could be very sensitive to large changes, but also quite resilient and recover quickly from even catastrophic disturbance. Finally, the scale of the system can influence its sensitivity to change. Small systems, for example, may equilibrate rapidly with changes in hydrogeomorphological forcing, and hence be quite sensitive to anthropogenic change. Conversely, large systems may respond slowly and the landscape response may integrate environmental changes over longer time periods.

Although landscape sensitivity is difficult to quantify because of the huge range of possible processes and intrinsic variability in landscape attributes, the concept is important for understanding natural landscape dynamics and for planning and evaluating land use. Humans have long recognized the disruptive effects of natural disturbance (albeit under the guise of natural disasters), but our cultures have not been very good at recognizing the sensitivity of the landscape to our actions. This has had unfortunate consequences, for failure to gauge landscape sensitivity to forest clearing and agricultural practices contributed to the decline of many once-great civilizations.

In the context of watershed management, landscape sensitivity reflects the intrinsic capability and capacity of a landscape to sustain a given land use. Hence, landscape sensitivity is a function of what is done where on a landscape. Although land use has traditionally been determined primarily on the basis of land ownership and short-term economic interests, the recent formalization of methodologies for conducting watershed analysis provides a framework for holistically examining how a landscape functions as a system and thereby how better to match human actions to the ability of a landscape to endure or sustain particular actions. Landscape sensitivity is not something that can be measured precisely and accurately, but the concept is useful in examining the interaction of humans and their environment.

DAVID R. MONTGOMERY

Further reading

Carter, V. G. and Dale, T. (1974) *Topsoil and civilization.* University of Oklahoma Press, Norman.

Goudie, A. (1986) *The human impact on the natural environment.* MIT Press, Cambridge, Mass.

Marsh, G. P. (1864) *Man and nature; or physical geography as modified by human action.* Charles Scribner, New York.

Thomas. D. S. G. and Allison, R. J. (1993) *Landscape sensitivity.* John Wiley and Sons, Chichester.

Turner, B. L., II, Clark, W. C., Kates, R. W., Richards, J. F., Mathews, J. T., and Meyer, W. B. (eds) (1990) *The Earth as transformed by human action.* Cambridge University Press.

landslides
Landslides are phenomena in which soil or rock moves down a slope under the action of gravity. The nature of movement may be falling, flow, sliding, or toppling. Landslides generally have distinct lateral boundaries (Fig. 1), and are therefore readily identifiable as such. They are distinct from soil creep, which, while satisfying the conditions of downslope movement, does not entail motion on a discrete shear surface. All landslides are therefore forms of mass movement, but not all mass movements are landslides.

When the strength of a soil or a rock is insufficient to support the stresses acting on the slope, a landslide will develop. The strength of soils and rocks can be described in terms of

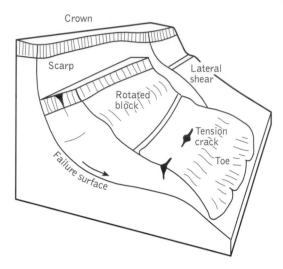

Fig. 1. The general geomorphology of a rotational landslide.

the cement and the bonds between grains (*cohesion*) and the interlocking of the grains through their surface roughness (*friction*). Clays have high cohesion but low frictional strength, whereas sands have low cohesion but higher frictional strengths. Rocks can be high in both characteristics and tend to slide along pre-existing failure planes such as joints. The biggest single influence on the occurrence of landslides is the presence of water. Landslides tend to happen as a result of high moisture contents which result in high water pressures in the soil pores. These high pore-water pressures reduce the contact between grains, thus reducing the frictional strength of the soil.

Falls

The simplest type of landslide is a fall, which can occur either in soils or in rocks. Falls are generally thought of as rock falls. They happen when steep slopes and inherent weaknesses in the rock or soil allow a block of the material to detach and fall to the ground. There may be some degree of bouncing and rolling, but the initial movement is that of a falling object. Very large rock falls can trap enough air to create a cushion, enabling a landslide to travel great distances at high speeds. This type of landslide, known as a *rock avalanche*, has resulted in large numbers of casualties. The 1970 Huascaran rock avalanche in Peru caused the deaths of at least 17 000 people in the towns of Yungay and Ranrahirca. It is believed that this landslide moved at speeds of over 320 km h^{-1} (200 miles per hour). If a mix of soil and rock is involved, the movement is referred to as a *debris fall*. Rock avalanches can be recognized by the long distances they travel, their great lengths compared to their breadths, and the fact that they may move so fast that they can even travel over small hills and up valley sides. Rock falls range in size from single blocks of less than 1 m in diameter to multiple blocks with volumes of many thousands of cubic metres.

Slides

Those landslides that occur through sliding on a discrete surface are known as *rock slides, debris slides*, or *earth slides*. Mudslides, the third of these groups, are a specific form of slide in which a fluidized mass moves on a shear plane. Unlike falls, slides move through shearing on a discrete surface. The movement can be either simple translation, in which the moving mass slides on a planar shear surface, or rotation, in which the mass slides on a semicircular shear plane. Rotational movement results in the sliding block becoming tilted as well as rotated. Slides are probably the most complex and varied group of landslides that can involve both soil or rocks. In soils there will not usually be a pre-existing weakness, but in rocks there normally will be. Such weaknesses can be present as fractures in the mass of the rock, or as weaker beds, such as mudstones.

Flows

In earth, debris, or rock flows there is movement of saturated soil material, often at very rapid rates (see *debris flows*). Earth or debris flows occur where soil material becomes saturated and oozes, sometimes at high speeds, downslope. Rock flows tend to occur through the saturation of weak, often clay-rich, layers in the ground. When these begin to move, they carry the overlying, more rigid, rocks with them. Flow slides of any variety tend to be the result of high fluid pressures. While these fluid pressures are normally water pressures, flow slides can be triggered by high air pressures. During the Kanto Province earthquake in China in 1920 large flow slides occurred in loess, a silty soil deposit. These landslides are believed to have resulted from the collapse of the delicate loess soil structure, which was caused in turn by earthquake shaking. When this occurred the air could not escape from the open pore structure, and the air pressure in the silt reduced the frictional contact. Huge flow slides then took place which resulted in over 20 000 fatalities.

Topples

Topples represent a variety of landslide that occurs when rock or soil material becomes detached from an exposed face. Rather than a simple falling motion, a topple will involve some form of outward rotational motion from the rock or soil mass. Toppling generally occurs when blocks are tall in comparison with their width. Rock topples, for example, are common in rock masses that show columnar jointing. Small toppling failures have occurred from the crags on which Stirling and Edinburgh castles sit, and they posed a hazard to the general public before work was undertaken to reduce the risk.

Recognition of landslides

Landslides can be recognized by their geomorphological features, both from satellite images and aerial photographs as well as from detailed ground investigation work. Active landslides tend to show exposed soil or rock, well-defined boundaries, egress of water from various points of contact, and a readily identifiable source and accumulation zone for the sliding material. In relic landslides, the shape of the source and accumulation zones may be visible, but these will normally be muted by vegetation cover. Relic landslides can, however, be differentiated from the surrounding area because of their shape and their vegetation, which is often found to contrast with that of stable ground. This is due to frequent differences in soil moisture conditions on and off the sliding mass. The geomorphological features associated with landslides in soils are shown in Fig. 1. The sides of landslides will be marked by lateral shear surfaces, while tension cracks and minor scarps may occur near the head of the landslide and along the body of the sliding mass. The soil or rock from the landslide accumulates in what is called the 'toe' of the landslide.

Landslides are three-dimensional features. In addition to the geomorphology that can be used to define a landslide in plan view, landslides can also be clearly defined in cross-section. The basal shear surfaces can be defined through drilling programmes designed to assess the three-dimensional shape of the landslide. An understanding of the geometry of the sliding mass, a knowledge of the strength of the materials, and a good assessment of the groundwater conditions can be used to form an estimate of the hazard posed by a landslide.

Landslides as extreme events

Landslides can be triggered by a variety of events. Coastal landslides, such as Black Ven in Dorset, can be triggered, or speeded up, by changing rates of coastal erosion. Landslides may also be initiated by extreme events. Cyclone Bola resulted in thousands of landslides in New Zealand in 1986, and the Loma Prieta earthquake in California in 1989 caused many large landslides. In large-magnitude earthquakes (magnitude 6.7 or greater) more than half the fatalities are the result of landslides triggered by ground shaking. Volcanic activity can also make slopes unstable. An example was the collapse of the east flank of Mount St Helens in 1980. After a period of volcanic inflation, the slopes became so steep that they were no longer stable. An earthquake triggered the landslide from the side of the mountain at 8.32 a.m. on 18 May. Once this slope had collapsed it allowed the release of the magma from the volcano, and the famous eruption followed.

These large-magnitude events are normally infrequent, but when they do occur they can be very damaging. It is such catastrophic slope failures which are generally thought of when landslides are mentioned, although the smaller, less newsworthy types are much more common.   W. MURPHY

Further reading

Crozier, M. J. (1986) *Landslides: causes, consequences and environment*. Croom Helm, London.

Lapworth, Charles (1842–1920)

Charles Lapworth's most renowned contribution to geology was the founding, in 1879, of the Ordovician system in the stratigraphical column. He was, however, also responsible for improving our understanding of two other important aspects of the geology of Britain, namely the structure and biostratigraphy of the southern uplands of Scotland, and the Caledonian structures of the north-west Highlands. In addition, his studies on graptolites won him wide acclaim, as did his work on the geology of the English Midlands.

Lapworth was born in Faringdon, Berkshire. He took up a post as schoolmaster in Scotland in 1864. During the years there he devoted his spare time to the geology of the deformed but fossiliferous rocks of the southern uplands, and to probing the mysteries of graptolite anatomy and lineages. In the southern uplands he showed that what had been thought of as an enormously thick sequence was in fact a series of folds with repetitions. His field determination of the graptolite zones there has stood the test of time; the International Commission on Stratigraphy returned to his section at Dob's Linn (Dumfries and Galloway) for its Geological Stratotype Section and Point (GSSP) for the base of the Silurian system. In 1881 Lapworth was appointed to the newly created chair of geology at Mason College in Birmingham and there continued his studies of graptolite successions, and of Palaeozoic structures in Scotland. He showed that the great dislocation beneath the Moine Schists was an overthrust, but left open the question of what the original nature of the Moines had been. From Birmingham he also mapped in great detail the complex Harlech Dome in North Wales.

British and North American geologists (but not the Geological Survey of Great Britain) soon adopted his Ordovician system as resolving the Sedgwick–Murchison controversy as to whether these strata belonged to the Cambrian or Silurian systems. Lapworth continued his graptolite work throughout his life, and was a notable teacher and writer. He received the Geological Society's premier award, the Wollaston Medal, in 1899, and also received a Royal Medal of the Royal Society.

D. L. DINELEY

laterites

Laterites are products of tropical weathering, formed *in situ* on old continental shields, regions that are tectonically stable and relatively well preserved from mechanical erosion. Lateritic profiles can be up to several tens of metres thick, and have generally been formed over periods of tens of millions of years. They are widely distributed and include bauxites, ferricretes (concretionary haematitic iron duricursts), latosols *sensu lato* (loose microgranular red soils), and lithomarges (soft kaolinitic saprolites—chemically decomposed rock). Lateritic profiles are mature: except for resistant minerals, such as quartz (SiO_2), which is usually abundant, almost all the primary minerals are completely hydrolysed. Alkaline and alkaline-earth elements are almost entirely leached out. Silica is generally partly, and sometimes totally, removed, but iron, aluminium, and manganese tend to accumulate in the profiles. The secondary minerals that are formed are oxides such as haematite (Fe_2O_3) or pyrolusite (MnO_2); oxyhydroxides such as goethite ($Fe_2O_3.H_2O$), boehmite ($Al_2O_3.H_2O$), or manganite ($Mn_2O_3.H_2O$); hydroxides such as gibbsite ($Al_2O_3.3H_2O$) or ferrihydrite ($Fe_2O_3,3H_2O$); and abundant clay minerals belonging exclusively to the kaolinite group ($Al_2O_3.2SiO_2.2H_2O$). Many lateritic profiles are concretionary; the petrographic features and soil organization are controlled by hydration–dehydration and silication–desilication processes. Most laterites are polygenetic, having been formed and transformed under various types of climatic successions. Laterites thus serve as continental archives and are excellent indicators of change in palaeoclimatic or palaeogeographical conditions.

Y. TARDY

Laurasia

Rather than accepting Wegener's single supercontinent Pangaea, the South African geologist Alexander du Toit preferred to argue for northern and southern supercontinents (Laurasia and Gondwanaland, respectively) which had been more-or-less separated since the Late Palaeozoic by the seaway Tethys, which was not broken up until the Tertiary as Africa and India drove northwards towards Eurasia. Laurasia thus comprises North America and Eurasia.

ANTHONY HALLAM

lava

Molten rock at the surface of the Earth is called lava. Lavas vary in composition: when they are rich in silica (SiO_2) they are highly viscous (sticky); when poor in silica, they have low viscosity. Some lavas contain large amounts of gas; others are gas-free. As lavas erupt the pressure in the magma is reduced and, in the manner of a bottle of fizzy drink (soda) being opened suddenly, the gas bubbles out. Different combinations of viscosity and gasiness lead to a wide variety of behaviours on eruption. Very viscous, gas-rich lavas erupt explosively, as at Mt. St Helen's, because the gas bubbles cannot escape quickly enough and the lava is blown apart. Low-viscosity, gas-poor lavas erupt quietly and flow away in thin sheets. Viscous lavas form thick, slow-moving flows; less viscous lavas form thin faster-moving flows. The surface of a lava flow cools quickly in contact with the air and solidifies rapidly. Continued movement of the lava underneath this surface may cause it to break into blocks, or the surface may become wrinkled and folded to form a ropy texture. Blocky and ropy lavas are often known as *aa-aa* and *pa-hoe-hoe*, onomatopoeic Hawaiian words.

JUDITH M. BUNBURY

lead–zinc deposits

Lead and zinc are commonly found together in many types of ore deposit. They may constitute the metals of primary interest or may be produced as by-products of the mining of other metals, in particular, copper, silver, and gold. The sulphides of lead and zinc (galena and sphalerite) are important ore minerals that are found in volcanogenic massive sulphide deposits, where they occur with those of copper. Galena and sphalerite are also commonly found in skarn (contact-metamorphic) and epithermal (shallow vein) deposits, where they are co-deposited with ore minerals of copper, gold, silver, tungsten, and tin. Although these types of deposit contain high concentrations of the two metals, there are types of deposit that are mined almost exclusively for their lead and zinc contents. Carbonate-hosted lead–zinc deposits occur in ancient reef deposits that have undergone partial dissolution by hydrothermal solutions. The open spaces thus created become filled with high concentrations of lead and zinc sulphides, together with minor amounts of copper and cadmium. The metals are transported by heated aqueous fluids derived from adjacent sedimentary deposits, which migrate into the porous reef horizons. The name 'Mississippi-Valley type' has been applied to these deposits because of important occurrences in that area of the United States. Sedimentary exhalative deposits also host lead and zinc as their primary ore metals. These deposits are comprised of fine layers of galena and sphalerite, together with pyrite, pyrrhotite, barite, and minor chalcopyrite interbedded with deep-water sedimentary rocks. Important examples are found at Mt Isa and McArthur River, Australia, and Sullivan, Canada.

BRUCE W. MOUNTAIN

Leakey, L. S. B. (1903–72)

Louis Seymour Bazett Leakey was born in Kenya and spent his early years largely in company with Kikuyu children. There he acquired his lifelong interest in East African landscape and wildlife. At the age of 16 he entered school in England and, later, St John's College, Cambridge. In 1923 he joined an expedition to Tanganyika and between 1926 and 1935 he led four highly successful expeditions to East Africa, studying ancient lake levels and prehistoric cultures.

In 1937 he returned to Kenya to study the Kikuyu people, but on the outbreak of the Second World War he entered government service. In 1945 Leakey was appointed curator of the Coryndon Museum in Nairobi, making it eventually one of the finest museums on the African continent. During this time he developed with his wife, Mary, new excavation techniques for palaeontological and occupation sites. Leakey, already well versed in human and hominid palaeontology, had established the presence of forms of *Ramapithecus* and of *Dryapithecus* in the Kenyan Miocene. Now he became much sought after to advise on the study of sites elsewhere in Africa. Discoveries with Mary Leakey and others of a series of australopithecine and hominid fossils and artefacts in Olduvai Gorge established the African origin of mankind and the unique importance of the region in palaeoanthropology and human evolution.

Leakey initiated the first Pan-African Congress on Prehistory at Nairobi in 1947, and in 1962 he became the first director of the Centre for Prehistory and Palaeontology there. Its facilities have been used by many expeditions and distinguished specialists, including the Leakeys' son Richard, in continuation of Louis Leakey's research.

D. L. DINELEY

Lehmann, Inge (1888–1993)

Inge Lehmann was a Danish seismologist who spent her professional career maintaining seismic observatories in Denmark and Greenland for the Danish Geodetic Institute. During this time she used seismograms from these and other European observatories to study how earthquake waves interact with the deep structure of the Earth. In 1936 Lehmann published evidence for the existence of an inner core within the liquid iron–nickel outer core. By timing the arrivals of pressure waves from distant earthquakes that penetrated the core, Lehmann determined that the waves appeared to be refracted by a sharp increase in wave speed over 1000 km from the Earth's centre. In the 1940s Francis Birch and Keith Bullen hypothesized that Lehmann's observations indicated the conversion of molten iron to a solid by the extreme pressures of the planetary interior. Evidence that the inner core is truly solid was not definitive until the 1970s and 1980s, when subtle vibrational resonances of the inner core were observed by Adam Dziewonski, Freeman Gilbert, and Guy Masters, and were shown to include shear, as well as pressure, motions.

Inge Lehmann was remarkable for the number of important contributions she made after her formal retirement in 1952. As a visiting scientist in the United States she reaped the data harvest from a new generation of seismic observatories, installed during the Cold War to monitor underground nuclear explosions. From such explosion data, she was one of the first researchers to identify a deficit in seismic velocity under the mantle immediately beneath the western USA. This deficit is now known to be associated with the elevated topography of the region. In 1966, at the age of 78, she published evidence for a subtle discontinuity in wave speed at a depth of 220 km within the rock of the Earth's mantle. Now recognized worldwide as the Lehmann discontinuity, the exact nature of this internal feature is still under debate.

JEFFREY PARK

Lehmann, Johann Gottlob (1719–67)

One of the early pioneers in unravelling the stratigraphical column was Johann Gottlob Lehmann, a teacher of mining and mineralogy in Berlin. He had visited many of the mining regions of central Europe and acquired at first hand a knowledge of not only the minerals and their modes of occurrence but also the different rock formations in which they were present. These he described in several large and influential volumes together with his classifications of the rocks themselves and the terrains to which they give rise (1756).

Lehmann recognized the way in which rocks may be layered and he coined the term 'strata' for them. His classification of 'strata' was threefold. The Primary rocks are crystalline, unfossil-

603

iferous, and of chemical origin precipitated from water at the Creation. The Secondary formations are conspicuously layered, fossiliferous, and deposited from the waters of the Noachian flood. The Alluvial deposits, laid down since then, are loosely consolidated, largely superficial, and locally accompanied by volcanic rocks. Lehmann also distinguished three kinds of terrain (mountains) typified by these formations. The oldest (*Urgebirge*), dating from the time of the Creation, are of massive crystalline rocks, relatively uniform in composition. Surrounding and covering the flanks of these are the Secondary (*Flötzgebirge*) terrains composed of well-stratified sedimentary rocks, remaining from Noah's Flood. Last are the *Aufgeschwemmtgebirge*, lands composed of alluvial beds covering parts of the other two and hence most recent. In 1761 Lehmann moved to St Petersburg at the invitation of Catherine the Great to take up the posts of Professor of Chemistry and Director of the Imperial Museum. There unfortunately he died in an accidental explosion in his laboratory.   D. L. DINELEY

Libby, Willard (1908–80)

The use of radioactive carbon, carbon-14 (^{14}C) for dating carbon-bearing artefacts and fossils fills a vital time-gap in the range of dating tools available for the measurement of geological time. The person who researched and demonstrated the method was Willard Libby; for originating and developing the ^{14}C dating method, he won the Nobel Prize in Chemistry in 1960.

Born and raised in farming communities in Colorado and California, Libby discovered science, and in particular chemistry, at the University of California, Berkeley. When the United States entered the Second World War, Libby, then at Princeton University, was invited to join the Manhattan Project in order to carry out research into the separation of uranium isotopes by gaseous diffusion. The key discoveries leading to a successful separation method were made by Libby and a co-worker. After the war, Libby joined the faculty of the University of Chicago and there began his study of ^{14}C. He demonstrated that ^{14}C in the atmosphere is formed through the action of cosmic rays on nitrogen-14 (^{14}N), that the ^{14}C quickly oxidizes to carbon dioxide and is rapidly and uniformly mixed throughout the atmosphere and, through solution, into sea water. A key step in the research on ^{14}C as a dating tool was the demonstration that through photosynthesis, respiration, and food ingestion, all the carbon in plants and animals is continually exchanged with the carbon in air, food, and water, and therefore that all living matter has the same percentage of ^{14}C as that in the atmosphere and the ocean. When a plant or animal dies, exchange ceases and the amount of ^{14}C decreases steadily through radioactive decay. Libby having successfully studied all aspects of the method, then developed rapid and accurate ways to measure the ^{14}C content of a sample, from which he could calculate the time since the plant or animal died. The method is accurate for samples as old as 50 000 years. The first radioactive carbon dates were published in 1949.   BRIAN J. SKINNER

lichenometry

Lichenometry is a rock-surface dating technique that relies on measurements of the diameter of lichens. The technique was once described as akin to attempting to establish the age of a man by measuring his waistline, but despite such disparaging comments it remains widely used to date, in particular, glacial landforms such as moraines. It is particularly useful in arctic and alpine areas beyond the treeline, where other dating techniques are difficult to apply. The lichens most commonly used belong to the genus *Rhizocarpon*, although a range of others has been used.

The standard approach is to use surfaces of known age to establish a 'lichenometric dating curve' which relates surface age to the size of the single largest or, more commonly, the mean of the five largest lichens. Surfaces whose age is commonly known include glacial moraines dated by using historical evidence such as paintings and photographs, made ground such as mine spoil, or even gravestones. Surfaces of unknown age are then searched and the single or mean largest lichen value is compared with the curve to obtain an age value. The age range of the technique depends on the growth rate of the lichens, but most studies concentrate on the past few hundred years. Lichenometric dating curves are curvilinear, and so the older the age estimate the less accurate it is likely to be.

The implicit reasoning behind the standard approach to lichenometry is that the largest lichens on a surface are likely to have been among the first colonizers and the fastest growers. A more direct approach to lichenometry, which might support this assumption and also make dating possible in areas where there are few surfaces of known age, entails measuring the growth of individual lichens over a number of years and so producing a growth curve. The small amount of work that has been done so far on 'direct lichenometry' suggests that these 'growth curves' are very different from 'lichenometric dating curves' from the same areas. It is, therefore, all the more surprising that the standard approach seems to work so well.

A serious limitation of both the standard and direct approaches to lichenometry is that they produce age estimates without clearly defined confidence limits. Techniques have recently been developed which attempt to solve this problem by constructing dating curves using very large samples of lichens rather that just the few largest individuals. The most common approach is to use the largest individual lichen on each of a large sample of boulders. Confidence limits can then be placed round the dating curve and, more importantly, around the mean lichen diameter obtained from the surface to be dated. Where these new approaches are based on very large samples, they also open the possibility of dating diachronous surfaces, such as avalanche or talus deposits where boulders accumulate over time. There is potential, therefore, for lichenometry to be used to date periods of enhanced rockfall or avalanche activity related to seismicity or climate change. Furthermore, lichenometry also has the potential for dating rock surfaces recently exposed as a result of increments of slip on earthquake faults. Such an application would aid in calculating the recurrence interval of earthquakes.

DANNY McCARROLL

life, origin of *see* ORIGIN OF LIFE: GEOLOGICAL CONSTRAINTS

limestone pavements Pavement-like exposures of hard limestone strata are one of the most recognizable karst landforms. For the past fifty years or so it has been widely accepted that these pavements are simply outcrops of more or less horizontal limestone that have been scoured by erosion (Fig. 1). By far the most important erosional agent is ice, and for this reason most pavements are found within the limits of the last major glaciation affecting much of northern Europe, Canada, and the northern USA (Devensian–Würm–Wisconsin cold stages). In these areas, limestone pavements are thus properly described as being *glacio-karstic* in origin.

With the withdrawal of the ice, the scoured pavement surfaces are thought to have been covered by peat or wind-blown loess, under which a variety of solutional grooves (karren) developed in postglacial times. The solution of vertical joint sets into grikes characteristically fragments the pavement surface into blocks known as clints. Beyond the limit of the last ice cover, the limestone outcrops were not swept clean of their covering of interglacial/periglacial rubble and pavements are absent.

More recently, it has been recognized that the origin of limestone pavements found on outcrops of Lower Carboniferous (Dinantian) outcrops, such as those in northern England and western Ireland, might be a little more complex. We know that these limestones were laid down in a shallow shelf sea that was periodically exposed as sea levels fell worldwide owing to the build-up of ice in a former glaciation. The exposed surfaces were covered by volcanic dust (wayboard clay) and invaded by

Fig. 1. Limestone pavements on the Burren, Co. Clare, Ireland.

land plants. Karst processes must have been very active in the root zone and the exposed limy sediment became hardened into a tough calcrete. The sea then flooded the shelf once more to complete the cycle. About thirty such cycles are known in the Lower Carboniferous sequence, each lasting for about 30 000 years. The calcreted zones in these limestones are relatively poorly jointed, and as the Quaternary glaciers scoured the outcrops, highly jointed strata were removed to expose the poorly jointed, resistant calcrete horizons. In this sense, limestone pavements on Dinantian ourcrops can be said to be glacially exhumed, palaeokarstic features. In some rare instances palaeokarstic pits and volcanic clays have been preserved, supporting this theory.

P. VINCENT

Further reading

Ford, D. and Williams, P. (1989) *Karst geomorphology and geology.* Chapman and Hall, London.

Vincent, P. (1995) Limestone pavements in the British Isles: a review. *Geographical Journal*, **161**, 265–74.

limestones Limestones are sedimentary rocks that contain more than 50 per cent of calcium carbonate, mainly in the form of the mineral calcite, but also as aragonite, which has the same chemical composition but a different crystal structure. After mudstones and sandstones, limestones are the next most abundant type of sedimentary rock. They extend over huge areas of the continents and the continental shelves, and many mountain chains are dominated by them. Limestones are relatively soluble in meteoric water (water derived from the atmosphere) and they give rise to distinctive landscapes, called karst landscapes, as a result of solution at the surface and for hundreds of metres underground.

Modern shallow-water calcium carbonate sediments are initially composed of the orthorhombic mineral aragonite, together with the rhombohedral minerals calcite and high-magnesium calcite. Ancient indurated (hardened) limestones are composed of calcite. Aragonite and calcite have each been the dominant carbonate phase originally deposited at various times throughout geological history, and all limestones do not originate from what was once thought to be a dominantly aragonitic sediment. These temporal changes in mineralogy are considered to be due to changes in the partial pressure of carbon dioxide and the calcium : magnesium ratio of sea water, which in turn are a reflection of deep-seated crustal processes. Limestone and their modern carbonate sediment equivalents exhibit the same variety of grain sizes, textures, and sedimentary structures as siliciclastic deposits and, in addition, others not exhibited by siliciclastic sediments. The old terms used to describe siliciclastics have, in some classifications, been retained for carbonates, with the addition of the prefix 'calci', for example, gravels are calcirudites, sands are calcarenites, and muds are calcilutites.

Carbonate sediments and limestones are composed of grains and crystals produced by biological and chemical extraction of calcium carbonate from solution. In contrast to siliciclastic deposits, they are normally deposited close to their site of formation within the basin of deposition; that is they are autochthonous, and are rarely transported great distances from their site of origin. The various components which combine together to form modern carbonate sediments and limestones are of diverse origins and sizes.

Gravel-grade material is generally composed of whole disarticulated or broken skeletal fragments together with sand-grade material of whole, disarticulated or disaggregated and broken skeletal debris. Such sediments can contain fragments of early cemented limestones of local origin, known as *intraclasts*, or of extra-basinal origin, that is, *extraclasts*. Sand-sized carbonate sediments commonly consist of sub-spherical or spherical grains with cores consisting of carbonate or siliciclastic grains coated with concentric layers of precipitated carbonate. Such spherical grains are known as *ooids*, and the common name for a rock consisting of ooids is an *oolite*. Gravel-grade sediment clasts may have coatings of fine carbonate bound by a cyanobacterial film. The resulting particles are called *oncoids*. Some sand-sized carbonate grains composed of fine-grained carbonate are faecal pellets produced by sediment-ingesting organisms. Grains with a similar structure and form are, however, produced by cyanobacteria and fungi boring into skeletal debris or ooids, destroying the structure, and then filling their pores with fine-grained precipitated carbonate. When the origin of such grains is uncertain, the term *peloid* is used; *pellet* is retained if the origin is certain. Individual grains of all other types are sometimes bound together by sticky coating of cyanobacteria or cement to form composite grains which behave hydrodynamically as a single particle—*grapestones*.

Mud-grade carbonate is composed of whole tests of small organisms, disarticulated, disaggregated skeletal debris, and broken and finely ground skeletal debris, together with precipitated crystals. The exact origin of this material, particularly when it has been recrystallized in ancient limestones, is often very difficult or impossible to ascertain, although precipitated muds usually have a smaller variety of grain sizes than muds produced by other processes.

Carbonate sediments display the same geomorphological depositional landforms (beach-dune barriers, etc.) as siliciclastic deposits. They can also form wave-resistant structures as a result of binding by skeletal organisms (in reefs or accretionary mounds). The term *biostrome* is often used to describe these organically produced units when they are tabular or sheet-like, and the term *bioherm* is used when they form positive elevations on the sea floor.

Unlike siliciclastic deposits, carbonate sediments of different geological age are commonly unlike each other because of the change in flora and fauna over time, some disappearing and others appearing. Furthermore, the mineralogy of individual groups of organisms has varied with time.

Carbonate sediments, and their ancient equivalents, that is limestone, form in a wide variety of sedimentary environments, extending from soils on older rocks, or river floodplains (limestone nodules in such settings being called *calcrete*) to lakes to a variety of coastal and marine environments. However, unlike siliciclastic deposits carbonates are unable to accumulate in very deep water because they are dissolved by carbon dioxide-rich bottom waters at the so-called carbonate compensation depth (CCD), where the rate of solution of calcium carbonate equals the rate of supply from overlying water. Below this depth carbonate sediment cannot accumulate.

Marine carbonates are by far the most abundant, both today and in the geological record. As they are the result of extraction of calcium carbonate from sea water, they are most abundant in the low-latitude saturated and supersaturated waters of the tropics where there is greater biological productivity and diversity, and where evaporation enhances the chemical precipitation of calcium carbonate.

Carbonates are also important contributors to shelf sediments in high-latitude areas and become dominant whenever the supply of siliciclastic sediment from adjacent land is limited. Important accumulations of shelf carbonates are found in high latitudes, as, for example, off Scotland and Ireland, and also off southern Australia; their importance has become increasingly recognized over the past few decades.

In warm subtropical and tropical waters of moderate to high salinity, ooids form in turbulent coastal waters and accumulate in shallow water or form beach-dune barriers. The nearshore shelf is dominated by remains of benthonic skeletal organisms, together with some peloids and pellets; there are also variable qualities of carbonate mud, produced by breakdown of skeletal grains as well as chemical precipitation. The remains of planktonic organisms increase away from the shoreline and become dominant in the deeper waters of adjacent marine basins. Hermatypic (reef-building) corals form fringing reefs along coastlines, patch reefs on the shelves, and sometimes barrier reefs along the shelf edge.

The barriers commonly enclose lagoons where pelletal (and peloidal), skeletal, and grapestone sands and carbonate muds accumulate. This association of carbonate sediments forms the so-called chloralgal association of Lees and Buller. Where lagoons have high salinities, hermatypic corals disappear and green algae are rare or unimportant; this is a chlorozoan association. In temperate latitudes, where hermatypic corals and green algae are absent, the sediments are dominated by the remains of molluscs, foraminifera, and red algae, which can form beach-dune barriers similar to those in tropical areas: the foramol association. Ooids and grapestones are, however, absent, although pellets are present but are less important. Carbonate mud is also less important, forming only by skeletal breakdown and not being chemically precipitated.

Limestones and their present-day equivalents are classified using the nature of the component framework grains and their texture. The classifications introduced by Folk in 1959 and 1962, by Dunham in 1962, and the modifications of Dunham's classification by Embry and Klovan in 1971 have

Fig. 1. Types of limestone sediments: (a) oolitic grainstone; (b) pelletal or peloidal grainstone; (c) skeletal grainstone; (d) skeletal packstone; (e) carbonate wackestone; (f) floatstone; (g) rudstone; (h) carbonate mudstone; (i) bafflestone; (j) boundstone; (k) framestone; and (l) crystalline carbonate. (Drawn from James (1992, Geological Association of Canada); and Dunham (1962, modified by Embrey and Klovan, 1971).)

superseded all process classifications. Dunham's classification, with its later modifications, is more easily applied to both modern and ancient carbonates than is that of Folk. According to Dunham (Fig. 1), carbonates composed of a supporting framework of grains of various types separated by voids which later may have carbonate cement precipitated in them are termed *carbonate grainstones* (Fig. 1a, b, c), or *packstones* if they contain some interstitial mud (Fig. 1d). In contrast, sediments having a carbonate-mud matrix are termed *carbonate wackestones* (Fig. 1e) if they contain more than 10 per cent of individual grains, or *carbonate mudstones* (Fig. 1h) if the grain content is less than 10 percent. In 1971 Embry and Klovan added two further terms: *carbonate floatstones* (Fig. 1f) for sediments containing more than 10 percent of the grains having a grain size greater than 2 mm diameter supported by a mud matrix; and *carbonate rudstones* (Fig. 1g) for those with more than 10 percent of the grains having a grain size greater than 2 mm diameter but which are grain supported. Sediments composed of a crystalline mosaic of doubtful, probably diagenetic (post-depositional), origin are merely termed *crystalline carbonates* (Fig. 1l)

The terms *bafflestone*, *boundstone*, and *framestone* are used where the sediments are bound together by marine grasses, encrusting organisms, or a rigid structure such as a reef, respectively (Fig. 1i, j, k). Grainstones are characteristic of high-energy conditions and mudstones of low-energy, less turbulent conditions, with the intermediate types of sediment forming a continuum between these extreme conditions.

A common facies pattern seen in modern seas and in ancient environments from a shoreline out into a deep marine basin is:

(1) lagoonal pelletal (and peloidal) and skeletal sands and muds (peloidal–skeletal packstones, wackestones, and mudstones);

(2) beach-dune barrier oolitic and/or skeletal sands (oolitic–skeletal grainstones) with or without reef sediments (coral–algal boundstones);

(3) nearshore shelf skeletal sands, muddy sands, and muds (skeletal packstones, wackestones, and mudstones) with patch reefs (coral boundstones) and sometimes barrier reefs; and

(4) deep-water carbonate muds with planktonic faunas (carbonate mudstones), sometimes mixed with increasing amounts of siliciclastic mud (marls).

Calcium carbonate is a relatively soluble compound and undergoes a wide variety of diagenetic changes after deposition. Grains soon become bound together by the precipitation of fringes of cement around them on beaches, to form beachrock and also on the sea floor at almost all depths where sediment deposition is slow. This produces hard crusts which are bored by bacteria and fungi and support an epifauna. Similar phenomena found in the geological record are termed *hardgrounds*.

At the present day, cements formed at an early stage are aragonite and high-magnesium calcite in marine sediments, and normal low-magnesium calcite in freshwater deposits and deep-water sediments. During burial the early cements recrystallize, a process called *neomorphism*, to calcite, and are then overlain by later cements, sometimes produced by solution of the less stable mineral aragonite, by carbonate released by pressure solution at point contacts of adjacent grains, or by solutions moving from adjacent deposits. Fine-grained carbonates recrystallize to form a mosaic of calcite crystals. Aragonitic framework components are usually leached and infilled with a mosaic of calcite. Rarely do aragonitic grains retain their original structure, as they must go through a solution phase before being replaced by calcite.

The diagenetic processes generally produce a rock with less permeability and porosity than the earlier deposit. Occasionally, however, solutions may pass through the sediment and dissolve earlier cements to produce a sediment with high porosity and permeability. Late diagenesis generally greatly reduces both porosity and permeability.

Carbonate sediments are commonly replaced by other minerals; for example, dolomitization can occur at various times in the history of a deposit. Both phosphatized carbonates and silicified carbonates are also common.

Carbonates and limestones are important as raw materials for the building industry (building stone, cement) and for agriculture (lime). Because of their porosity and permeability (which may be primary or may be secondarily produced by waters moving over grains and along fractures) many limestones are important host rocks for water, oil, and gas, for example, in many of the oilfields of the Middle East. They can also act as hosts for mineral deposits, in particular lead and zinc ores.

G. EVANS

Further reading

Bathurst, R. G. (1971) *Carbonate sediments and their diagenesis.* Elsevier Publishing Company, Amsterdam.

Tucker, M. E. and Wright, V. P. (1990) *Carbonate sedimentology.* Blackwell Scientific Publications, Oxford.

Lindgren, Waldemar (1860–1939) Swedish-born,
educated in both his native country and at the Royal Mining Academy of Freiburg, Saxony, Waldemar Lindgren gained fame for his scientific studies in the United States of America. His years in Freiburg were times of intense debate concerning the genesis of mineral deposits. Opinions divided over the origins of the waters that deposited veins of ore minerals; one group favoured a meteoric origin, the other a magmatic origin. Lindgren listened to the debate and sought to resolve it when he came to the United States in 1883.

Lindgren's early years in the United States were in the employ of the US Geological Survey, in which position he had the opportunity to travel widely and examine many of the mineral deposits newly discovered in the American West. He realized from what he saw that the evidence favoured a magmatic involvement in the genesis of most deposits. While not the originator of the theory of magmatic hydrothermal origin for ore deposits, Lindgren became its most thoughtful supporter. He developed a depth–temperature classification for

mineral deposits based on the geothermometry of mineral assemblages and on mineral textures; in a modified form, his classification continues to be used today. Lindgren was also the first to articulate the concept of metallogenic provinces, a hypothesis that has gained much support from the theory of plate tectonics.

In 1912 Lindgren resigned from the US Geological Survey and joined the faculty of the Massachusetts Institute of Technology, where he continued teaching until 1933, at the age of 73. During his teaching years Lindgren wrote a highly influential text on the genesis of mineral deposits. Through four editions, the last in 1933, Lindgren's *Mineral deposits* was the standard reference and teaching text for several generations of economic geologists.

BRIAN J. SKINNER

lineaments Crustal lineaments are major 'fundamental' faults
or fault zones which have had a lasting influence on the geological evolution of the continental lithosphere. Many were initiated in the early Proterozoic, or even in the Archaean, and have since been periodically reactivated; some are still active at the present day. Many extend for a thousand or more kilometres, and although approximately linear, they may have widths of tens or hundreds of kilometres. They may be marked by a single strand of intense deformation or fracturing, or may consist of a complex geometrical array of faults and shear zones. Their presence is expressed in a variety of ways. Some are the loci for present-day intracontinental seismicity; others have associated gravity and magnetic anomalies and gradients, and many have a marked topographic expression and can be recognized on satellite imagery.

Some lineaments not only subdivide the continental crust but may extend into oceanic crust where they form major primary transform faults. Many show a major phase of strike-slip deformation and may form links between widely spaced thrust and extensional zones previously regarded as unrelated structural features.

During reactivation, fault lineaments may episodically be changed from thrust to extensional or strike-slip features. Lineaments are also potential conduits for the ascent of magmas and for the upward and downward migration of water; they consequently exert a major control on the location of ore deposits.

HAROLD G. READING

Further reading

Reading, H. G., Watterson, L. J., and White, S. H. (eds̊) (1986) Major crustal lineaments and their influence on the geological history of the continental lithosphere. *Philosophical Transactions of the Royal Society*, Ser. A., **317**, No. 1539.

lineation 'Lineation' is a term used in structural geology for a
repeated or penetrative linear structure in a rock mass. A lineation may be present on discrete planes or developed throughout the body of the rock; in the latter case the rock is called an L tectonite. Most lineations are associated with the

processes of deformation and metamorphism; they are less common in igneous and sedimentary rocks. Any discussion of lineations requires reference axes or planes, as illustrated in Fig. 1a. The tectonic axis a is the direction of tectonic transport 'up' the cleavage; b is parallel to the fold axis; and c is perpendicular to the ab plane, which in cleaved rocks is the plane of slaty cleavage. Deformation may be multi-phase (that is, it may take place in more than one episode), in which case the successive phases are designated D_1, D_2, etc., yielding folds F_1, F_2, etc., planar structures S_1, S_2, etc., and linear structures L_1, L_2, etc. S_0 is used for the original surface, which is usually a bedding surface. In the investigation of deformed rocks it is important to relate the linear and planar features to the appropriate phases of deformation in order to be able to elucidate the geological history.

Lineations take a variety of forms, as shown in Fig. 1b: the intersection of two planar surfaces, elongation of minerals or mineral aggregates, elongation or rodding of pebbles, deformed ooids and reduction spots, oriented fold axes, mullions and boudins, grooves or mineral growths on fault surfaces, preferred orientation of larger crystals (phenocrysts) in igneous intrusions; variety of sole marks, and primary current lineations in sedimentary rocks.

During deformation at low temperatures rock layers may become flexed and the upper ones may slip over the lower and ride up towards the hinge zone of the flexure in the same way that playing cards will move over each other when a pack of cards is flexed. This bedding-plane slip produces *slickensides*, which may be in the form of grooves or elongation of minerals or mineral aggregates or growth of new minerals such as quartz or calcite. This *elongation lineation* is perpendicular to the fold axis in the direction of slip.

With low to moderate temperatures and increasing deformation, a planar closely spaced cleavage develops parallel to the axial plane (ab) of the fold, although in some instances the cleavage fans and is not strictly parallel to the axial surface. The intersection of the cleavage on the bedding and of the bedding on the cleavage (see Fig. 1) gives an *intersection lineation* which is parallel to the fold axis (b). In slates a *mineral stretching lineation* commonly develops 'up' the cleavage plane and at right-angles to the fold axis; if there are porphyroblasts of, say, pyrite in the slate, then new grains of quartz may grow in their 'shadow' also 'up' the cleavage. In some slates, especially in purple ones, there may be green spots which were originally spherical reduction spots but are now triaxial ellipsoids whose eccentricity is a measure of the strain the rocks have undergone. The maximum elongation which gives a type of lineation is in a direction perpendicular to the fold axis and parallel to (a), the transport direction; the maximum flattening is at right-angles to the cleavage. Ooids in a limestone with a composition similar to the matrix are deformed in a similar manner to the reduction spots and give similar lineations.

Pebbles and even boulders deformed under even higher temperature and pressure, when both fragments and matrix become more ductile, undergo considerable elongation with

the long axes of the pebbles most often parallel to the fold axis, a 'b' lineation. There is some debate about how there can be such a high degree of stretching parallel to the fold axis. In

(a)

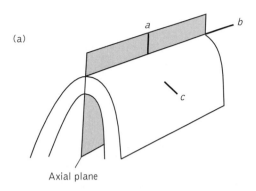

Axial plane

(b)

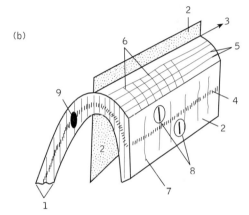

(c)

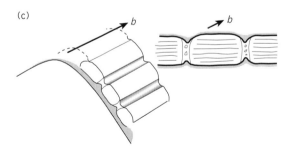

Fig. 1. (a) Tectonic axes a, b, c and axial plane in a simple anticline. (b) Diagram to show a number of lineations in folded and cleaved rocks. 1, bedding S_0; 2, axial-plane cleavage S_1; 3, fold axis b; 4, bedding on cleavage intersection L_1; 5, cleavage on bedding L_1; 6, Slickensides or mineral elongation lineation; 7, 'down-dip' mineral elongation on cleavage plane L; 8, reduction spot on cleavage plane elongate parallel to a; 9, reduction spot flattened in cleavage; (c) Boudins parallel to b with incompetent material squeezing into necks.

these high-temperature conditions new inequant minerals such as hornblende, tourmaline, and sillimanite grow with their long axes along the fold axis; in a later deformation event some of these may be fractured across their length and pulled apart along their length. Further evidence of multiphase folding is given by crenulation or microfolding with hinge zones a centimetre or so apart which crinkle an earlier cleavage or schistosity. These small-scale folds are parallel to a later phase of folding.

On the limbs of folds affecting alternations of competent and incompetent beds where there has been considerable layer-parallel extension, segmentation and extension of the competent beds results, with the incompetent beds squeezing into the spaces between and new minerals growing in the gaps with an elongation parallel to the direction of stretching. The result is a set of long, parallel sausage-shaped bodies called *boudins*, elongated in the direction of the fold axis (Fig. 1c); *mullions* have a similar orientation and shape with deformed bounding surfaces which have not fractured.

Slickensides as defined above but associated with faults are usually not penetrative but are restricted to specific planes. If shearing along thrust planes was intense, penetrative lineations may extend through a considerable thickness of rock. The direction of these lineations indicates the relative movements of the rocks involved.   R. BRADSHAW

Further reading

Hobbs, B. E., Means, W. D., and Williams, P. F. (1976) *An outline of structural geology*. John Wiley and Sons, New York.

liquefaction Liquefaction is a condition in which a soil or sediment behaves as a fluid. This is the result of strong vibration and is normally related to a moderate to large earthquake. The process of liquefaction normally occurs only in sands and silty sands, although recent evidence suggests that gravels may also liquefy. The actual process of liquefaction is complex. Soils are composed of a mineral phase and a fluid phase. During an earthquake the soil mass may begin to compact. When this happens, the water in the pore spaces between grains exerts a higher pressure on the mineral particles. It has also been shown that the passage of seismic waves, which are a transient stress in their own right, also has an effect on these forces, referred to as *pore-water pressures*. As these pore-water pressures increase, the grains are forced apart, there is less frictional contact between sand grains, and the shear strength of the soil decreases. When the soil loses all its shear strength, liquefaction has occurred. This process is therefore important in soils that gain their strength from internal friction (i.e. grain-to-grain contact). Liquefaction does not occur in clay-rich sediments because the strength of clays generally comes from interparticle bonds.

The susceptibility of soils to liquefaction is controlled by a number of factors. These can be divided into two groups: the ground conditions, and the characteristics of the earthquake ground motions. The ground conditions that give rise to a high liquefaction susceptibility are low density, water saturation, and a silty sand particle-size distribution. The distribution of grain sizes is critical, and it is possible to identify a range of particle sizes at which liquefaction is likely to occur. Dense sands, with low water contents, are less likely to liquefy. The geological history of the soil is important, since river-deposited sands are more likely to liquefy than glacial deposits because of their differences in density.

In general, the larger the earthquake, the greater the possibility that liquefaction will occur. The duration of shaking is, however, a critical factor. A long duration of seismic shaking allows for the gradual increase in pore-water pressures as each stress pulse passes through the soil. The greater the stress pulse (measured as a ground acceleration), the greater increase in pore-water pressure. A long duration of shaking with a high ground acceleration is therefore more likely to cause liquefaction than a short duration of shaking with low ground acceleration. This would normally require an earthquake of magnitude greater than 5.

The engineering effects of liquefaction can be profound. The complete loss of bearing strength had serious consequences during the 1964 Niigata earthquake in Japan. The Kwangishicho apartment block in Niigata itself was tilted through an angle of over 40° as a result of liquefaction around the foundations, although the buildings themselves did not suffer structural failure. Buried objects, such as septic tanks, have been found to float to the surface in liquefied sediments. Liquefaction can also result in a lateral spread, or flow slide, which can occur on slopes that are normally too low for landslides to occur. This effect was seen during the 1964 Alaska earthquake in which major landslides occurred at Turnagain Heights and Government Hill owing to liquefaction of underlying sand layers.   W. MURPHY

literature and geology Reference to geology in literature may occur in several ways. It may be by way of setting a scene, a topography, a location; it may be in connection with a resource or other geological materials (gold is a favourite); it may refer to a natural disaster; it may relate to the characters involved.

Many ancient writings are concerned with the nature of the physical world, its place in the cosmos, and the natural hazards it inflicts upon mankind. Most of them reveal little observation and much myth, but they are of no less value because of that. Astronomical observations in some other ancient writings are, on the other hand, detailed and extensive. Reference to ancient Egyptian, Hindu, and other scripts, and the writings of scholars in classical and other periods of history is made elsewhere in this volume. The Bible has many passages which attracted the attention of the early geologists. Most notable is the story of the Deluge. There are many indications of widespread floods in the Quaternary geology of Mesopotamia and in the Black Sea basin which have exercised scholars on the subject of Noah's adventure.

The crossing of the Red Sea by the fleeing Israelites has similarly prompted speculation about tsunamis and sudden violent earth movements.

Other natural disasters have been part of the scenarios of novels such as Bulwer Lytton's *The last days of Pompeii* (1834). There, the sudden eruption of Vesuvius in AD 79 provided a suitable background climax to the misdemeanours of local society.

At a later period, the permeation of geological ideas into literature is exemplified in Thomas Hardy's novel *A Pair of Blue Eyes*, published in 1873. In a dramatic episode, Henry Knight, 'a fair geologist' in Hardy's description, is left clinging to a cliff-face with the sea far below him. He sees 'a creature with eyes', a trilobite, standing out in low relief from the rock, and in his perilous situation reflects on the myriad forms of life that had existed in the millions of years between the present and the time when the fossil was still a living creature.

The American humorist Mark Twain, in his *Life on the Mississippi* (1883), refers to the meanders in the course of that great river and comes to some amusing conclusions about the rates of fluvial processes. He knew some of the jargon and was happy to use it in a carefree way.

Descriptions of spectacular topography as a setting for narrative or history have been provided by many modern writers. The novels of Zane Grey, which were among the first of recount adventures in the American West, contain much accurate topographical detail. In more recent times the tradition has been continued by Louis Lamour. T. E. Lawrence was aware of the importance of arid landscape as a war setting in his *The Seven Pillars of Wisdom*. Among modern novelists, Hammond Innes and Neville Shute have set the action in remote regions described with a keen eye for topographical, and even geological, detail.

Dr Tony Cooper of the British Geological Survey has suggested that the famous episode at the beginning of *Alice in Wonderland* in which Alice falls down a rabbit-hole may have been inspired by a geological event. In 1834 a collapse in gypsum beds at Ripon left a hole more than 18 metres deep near a house occupied by a clergyman with whom Lewis Carroll (Charles Dodgson) was well acquainted. Tony Cooper thinks it probable that Lewis Carroll, who grew up in Ripon, would have heard about this and other similar events in the area. The collapse resulted from the solution of gypsum (calcium sulphate) by underground water.

Few geologists have themselves turned to writing poetry or prose other than doggerel or autobiography. Hugh Miller (1802–56), best known for his book *The Old Red Sandstone*, is a notable exception. This is perhaps surprising: a geologist's life is one of action in the field, which may bring encounters with exotic places and wildlife, danger as well as natural beauty, and possibly hostile natives. In modern times, several North American Earth scientists, F. J. Pettijohn , G. G. Simpson, and J. Tuzo Wilson among them, have produced highly readable accounts of their travels or exploits.

Science fiction would seem to be a suitable genre for the subject, but in fact there are few such novels, apart from some notable early exceptions. Jules Verne's novels *A journey to the centre of the Earth* (1864) and *Twenty-thousand leagues under the sea* (1873), published originally in France, refer knowledgeably to geological phenomena and were highly successful in their day. Conan Doyle's *The Lost World* (1912) described both geologists and dinosaurs in an Amazonian forest setting and was a splendid example of adventure writing of its time. The resurrection of dinosaurs from geologically preserved dinosaur DNA was the subject of Michael Crichton's *Jurassic Park*, which was spectacularly adapted as an epoch-making film in 1992.

Earth science heroes are rare in literature. In John Fowles's novel *The French lieutenant's woman* the hero is a young amateur palaeontologist visiting Lyme Regis in Dorset to collect fossils from the Mesozoic rocks in the cliffs. Is there perhaps an echo here of Hardy's Henry Knight?

Non-technical anthologies of writings on the Earth sciences are rather few—at least in English. The notable exception is the *Language of the Earth*, edited by F. H. T. Rhodes and R. O. Stone (1981). It includes prose items from the time of Goethe to the 1970s; its selection of poetry extends back to the Book of Job, in which Chapter 28 concerns the valuable or precious materials man has mined from the Earth.

The surprising source of an article on 'Palaeontology in literature' by Archie Lamont, a palaeontologist and Scottish Nationalist, was *The Quarry Manager's Journal* for 1947 (vol. 30, January and March). Lamont found more to quote from poetry than from prose and referred to many of the better-known poets and to others less often read. It may be time for an anthology of Earth science poems—or at least verses—to be published. There is much from which to choose: Milton, Shelley, Tennyson, and Wordsworth are among the many poets who have touched upon geological themes. Many of the spectacular advances in the geological realm in recent years have inspired palaeontologists and their kin to compose but not to publish formally some quite respectable lines. The advent to continental drift and plate tectonics as major concepts in modern geological thinking also have prompted several well-known geologists to compose poetry.

The popularization of science and of its history has achieved excellence by an increasing number of authors in many countries in recent decades. In Britain the Earth sciences have been fortunate in attracting such writers as Nigel Calder, Ron Redfern, and John Gribbin, while in the USA Stephen Jay Gould has produced a stream of remarkable essays. Writing for television science programmes was the spur that eventually led to some of these essays or books, and there is no doubt that radio, television, film, and the popular science press today all encourage writers of ability and interest in the Earth sciences. Their influence on literature may be at second hand but is none the less important.   D. L. DINELEY

611

Further reading

Craig, G. Y. and Jones, E. J. (1982) *A geological miscellany*. Orbital Press, Oxford.

Hazen, R. M. (ed.) (1982) *The poetry of geology*. Allen and Unwin, London.

Rhodes, F. H. T. and Stone, R. O. (1981) *Language of the Earth*. Pergamon Press, New York.

lithification *see* DIAGENESIS

lithosphere Plate tectonics relies on the concept of a rigid lithosphere. Although the term 'crust' was used originally in geology to denote such a rigid layer, the modern concept of the lithosphere generally includes the crust *and* the uppermost part of the upper mantle (Fig. 1). Whereas the base of the crust is essentially a compositional boundary between mafic and ultramafic rocks, characterized by a rapid increase in seismic P-wave velocity to more than 8 km s^{-1} (the Mohorovičić discontinuity), the lithosphere is defined rheologically as the strong, elastic layer at the surface of the Earth, overlying the weak, ductile asthenosphere. The base of the lithosphere is marked by the transition from brittle to ductile deformation, and is probably gradational. The lithosphere, as defined in this way, is sometimes referred to as a 'mechanical boundary layer'.

Because the lithosphere is rigid, it cannot convect as the deeper mantle does. Heat therefore passes through it mainly by conduction, which is a less effective mechanism than convection. In consequence, the geothermal gradient in the lithosphere is considerably higher than that in the rest of the mantle, and there is a large temperature difference across it. One can therefore also think of the lithosphere as a thermal boundary layer.

There are several ways of estimating the thickness of the lithosphere. Because it is a rheological boundary, and rheological properties depend, *inter alia*, on strain rate, the thickness also depends on strain rate.

One of the most common ways of estimating lithospheric thickness is to use seismology. Since this depends on the elastic deformation of rocks at frequencies of about 0.01–1 Hz, it is effectively a very high strain-rate method. Seismology shows that the velocity of seismic waves generally increases with depth in the upper mantle. However, in most places there is a small decrease in velocity starting at depths of around 100 km, which is thought to mark the transition from lithosphere to asthenosphere. The thickness of the lithosphere defined on this basis varies between different geological provinces. It is greatest under the cratonic centres of continents, where it may possibly reach around 300 km, and it is least, probably about 4 km, under mid-ocean ridges where new lithosphere is formed.

Lithospheric thickness generally increases with age since its last major tectonic reactivation. The lithosphere–asthenosphere transition should roughly follow the isotherm corresponding to the brittle–ductile transition. For oceanic lithosphere the depths of isotherms are determined by conductive cooling since its creation at a mid-ocean ridge. These depths can be computed and are approximately proportional to the square root of the age, as therefore is the lithospheric thickness. The seismically defined oceanic lithosphere increases in thickness from just a few kilometres at zero age to about 100 km at ages of over 100 Ma (million years). These depths correspond approximately to the 1000 °C isotherm.

Another method of estimating lithospheric thickness is by measuring how much it flexes under applied loads. Geophysicists have shown that the lithosphere can be modelled successfully as a thin elastic plate overlying a viscous substratum. The magnitude and wavelength of the flexure depend on the

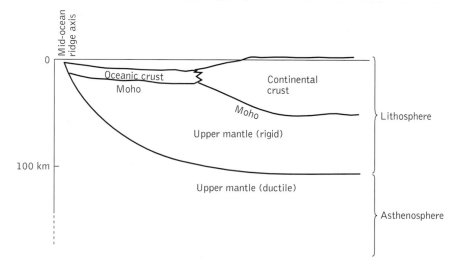

Fig. 1. Cross-section from a mid-ocean ridge axis to a continental interior, showing the relationship between crust, mantle, lithosphere, and asthenosphere. Vertical exaggeration approximately 20 : 1.

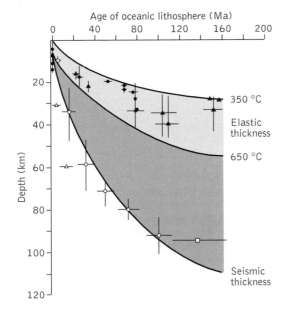

Fig. 2. Estimates of the seismic and flexural elastic thickness of oceanic lithosphere compared with the predicted depths of selected isotherms. (Modified from Kearey and Vine (1996), Fig. 2. 38.)

applied load and on the thickness of the plate. A variety of loads can be modelled, including sea mounts, ocean islands, subducting slabs, and the isostatic forces acting on the dipping boundaries of fault blocks. Such studies also show an increase of lithospheric thickness with age, but the thicknesses inferred are approximately one-third of those determined seismically. This is because flexural adjustments take place over long timescales of around 10^5 years or more, and at these low strain rates the brittle–ductile transition occurs at lower temperatures. In fact, the elastic thickness estimates from flexural studies fall between the 350 °C and 650 ° C isotherms (Fig. 2).

The fact that the lithosphere comprises a significant part of the upper mantle has an important effect on its rheology. Oceanic lithosphere is typically over 90 per cent mantle; its strength, dominated by the mineral olivine, is high, and its density is also relatively high. Continental lithosphere may be only 60 per cent mantle, and with significant quantities of typically crustal minerals such as feldspar and quartz it is much weaker, and relatively light. Thus, if two lithospheric plates are in collision and one is oceanic, the oceanic one can resist major deformation and be relatively easily overridden to form a subduction zone. However, if two continental plates are in collision, both are too light to sink easily, but they are also weak enough to be relatively easily folded or faulted. Thus continental collision zones tend to be wide and complex, containing many individual fault blocks as well as large fold mountains. ROGER SEARLE

Further reading

Kearey, P. and Vine, F. J. (1996) *Global tectonics*. Blackwell Science, Oxford.

Little Ice Age Beginning at some time in the thirteenth and fourteenth centuries and lasting until the end of the nineteenth century was a period of generally colder climate informally termed the Little Ice Age. Its effects can be detected over most parts of the globe and it represents the most recent of a series of colder episodes lasting several hundreds of years which have occurred during the past 11 500 years since the end of the last ice age.

The special interest of the Little Ice Age is that its later parts lie within the range of instrumental measurement and the period as a whole is susceptible to analysis by a wide range of geological, biological, and historical techniques.

Prior to the Little Ice Age conditions were warmer; the northern sea routes had been free of pack ice and Norse settlements were established in western and eastern Greenland in the 980s AD. Throughout Europe there was an extension of agriculture; vineyards were established some 300–500 km further north, cereal crops were grown at higher altitudes and latitudes, and glaciers were in recession.

During the Little Ice Age, however, average temperatures were depressed by 1–2 °C. This fall in temperature affected different areas at different times, but the overall result was increased hardship for human groups, plants, and animals, and an increase in geomorphic and meteorological activity, such as glacier growth, flooding, landsliding, avalanches, and storminess. At present a fall of 1 °C in summer temperature in Iceland results in a 15 per cent reduction in crop yield. A prolonged period of decreased temperatures, perhaps up to double this figure, would have produced severe difficulty for the inhabitants of more marginal areas in less technologically advanced times than those of today.

In Greenland, for instance, conditions became more difficult after AD 1200. Increasing storminess and sea-ice cover made voyages between Norway, Iceland, and Greenland more troublesome. The Western Settlement was abandoned in mysterious circumstances about 1350 and the larger Eastern Settlement died out in the late 1400s. Evidence from the Crête ice core in central Greenland suggests that this abandonment took place when temperatures were depressed by up to 2 °C for long periods compared to the time of the original settlement. Archaeological evidence also shows evidence for climatic deterioration. Clothing from bodies buried before about 1350 at Norse settlements in Greenland was penetrated by tree roots, whereas those bodies buried between 1350 and 1500 were in a remarkably good state of preservation owing to the extension of permanently frozen ground.

The disappearance of the Greenland settlements under conditions of climatic stress is of wider interest to us because it is a rare example of failure in a developed European society. There is no doubt that this failure was due largely to Greenland's climatic marginality, but isolation and lack of adaptability brought about by social and economic factors may have also played a part.

The Crête ice-core record can also be used to reconstruct temperatures for southern Greenland during the Little Ice Age. Most striking is the fact that the temperature record between 1200 and 1850 shows a series of fluctuation superimposed on the generally cooler trend, yet there is some evidence for the maximum cold to have been divided between two periods—just before 1400 and between 1750 and 1900. The later part of the record appears to agree well with temperature reconstructions from Iceland and central England based on temperature measurements and other proxy information (Fig. 1a).

These temperature records are in accord with other geological and historical evidence. In Norway, outlet glaciers from the Jostedalsbreen ice cap, such as Nigardsbreen and Austerdalsbreen, advanced over formerly cultivated farmland and reached their maximum extent between 1740 and 1750. Tax inventories throughout this period show evidence for a

decrease in wealth and also physical damage, both directly by glacier ice and also by flooding, rockfall, and landslides (Fig. 1b). Since 1750 the precision with which the retreating outlet glacier positions can be mapped increases with time as Jostedalsbreen was visited by increasingly numbers of travellers, climbers, and latterly scientists, who recorded their observations (Fig. 2a).

In the Swiss Alps, the Grindelwald glaciers have also been subject to intense analysis. Oil paintings, drawings, prints and photographs, maps and literature, as well as direct measurement since 1880 have been used to chart the changing positions of the front of the Lower Grindelwald glacier between 1590 and 1970 (Fig. 1d). Between 1590 and 1640 the ice front of the Lower Grindelwald seems to have been situated some 600 m forward of a prominent rock band called the Unteren Schopf. This rock band, a prominent marker and easily detected in drawings and sketches, remained covered until about 1870 although it was almost uncovered by a

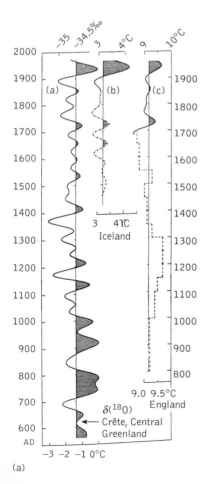

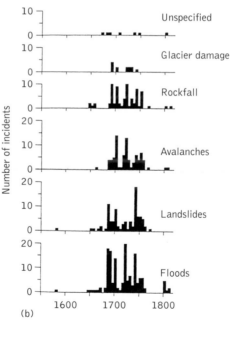

Fig. 1 (a) Reconstructed temperatures for southern Greenland based on the Crête ice core compared to those from Iceland and central England. The Icelandic temperatures are based on measurement back to about 1850 and on sea-ice data prior to this. The central England data set is based on measurement back to 1698 and historical and botanical data prior to this date. (Source: Bradley, R. S. (1985) *Quaternary Paleoclimatology*, Allen and Unwin, London, Fig. 5.24.) (b) The incidence of geomorphic activity such as glacier damage, rockfall, avalanches, landslides, and floods as shown by requests for tax relief in parishes near the Jostedalsbreen ice cap in southern Norway. (Source: Grove, J. (1988) *The Little Ice Age*, Methuen, London, Fig. 12.2.)

(a)

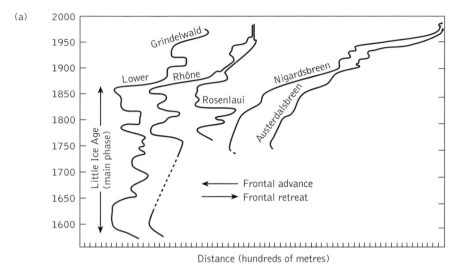

Distance (hundreds of metres)

(b)

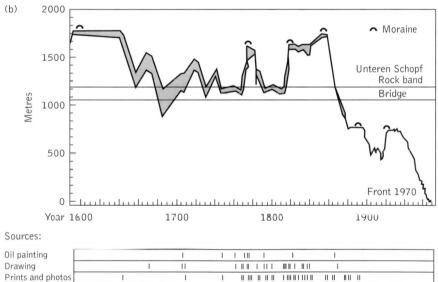

Fig. 2 (a) Time–distance diagram of glacier margin fluctuations in the Swiss Alps and in southern Norway. The termination of the Little Ice Age after 1850–1950 is clearly shown. (Source: Broeker, W. S. and Denton, G. H. (1989) The role of ocean–atmosphere reorganizations in glacial cycles. *Geochimica et Cosmochimica Acta*, **53**, 2465–501.) (b) Frontal positions of the ice margin of the Lower Grindelwald glacier between 1590 and 1970. (Source: Grove, J. (1988) *The Little Ice Age*, Methuen, London, Fig. 6.4.)

retreat phase in the late seventeenth century. Between 1770 and 1780 another advance left a moraine, since eroded, which appears in several oil paintings of the time. There was a further retreat in 1794 followed by a readvance lasting until about 1822. Another advance brought the ice front forward until 1856, after which it has retreated, albeit with fluctuations, to the present. The total amount of retreat between 1856 and 1970 was about 1800 m (Fig. 2b).

Glacier margin fluctuations derived from a wide range of techniques can therefore provide a useful line of evidence for assessing the timing and scale of climatic change during the Little Ice Age, despite the difficulties of assessing lag effects between climatic change and changes in ice front position.

Iceland is a rich source area for historical information regarding Little Ice Age climatic fluctuations, because factors such as civil unrest and war have not complicated interpretations. Grass for haymaking is the main crop and its yield is directly affected by temperature, both in the growing season and during the previous winter, length of snow cover, and duration of frosts. Severe climate restricts yields, causing a reduction in the livestock population as animals either die or are killed for meat. A survey of primary sources has shown that in 24 out of the 36 years between 1730 and 1766 livestock died of cold and hunger.

Fishing records are also a rich historical source. The staple fish and main export of Icelanders since the fourteenth century has been cod. However, cod cannot survive for long in waters colder than 2 °C and are abundant only at higher temperatures up to 13 °C. During cold periods cod has to migrate south. Until the 1890s Icelanders used to fish from open boats. The restricted range of the boats, increased storminess and presence of sea ice, coupled with the southward migration of cod, meant greatly diminished catches during colder periods of the Little Ice Age. The fishing industry failed during most of the seventeenth and eighteenth centuries, this failure being most marked between 1685 and 1704 and 1744–59.

Presence of sea ice around Iceland can also be used as a climatic indicator. In 'normal' conditions the north coast will be ice free, with the ice edge situated 100 km to the north. In severe years the sea ice extends along the north coast, and in extremely severe years it reaches the south coast. By surveying historical records, ice incidence and therefore temperature can be estimated back to the time of first settlement. Figure 3 shows a record extending back to 1600 which is in good agree-

ment with information derived from farming and fishery records.

The cause of the Little Ice Age is subject to much debate. The glacial evidence suggests that it was a global phenomenon, the latest of a series of periods of colder climate lasting several hundreds of years within the past 10 000 years. Within the Little Ice Age itself, there seems to be a consistency in the timing of the main ice advances within Europe, dating from 1600–10, 1690–1700, 1770–80, 1815–25, 1845–55, 1870–80, and 1920–30. The causal mechanism or mechanisms must therefore be capable of producing such a pattern within the postglacial and within the Little Ice Age itself. Prime contenders at present are variations in the deep-water circulation of the oceans, volcanism, and changes in solar output.

Today, colder more saline water forms in the northern North Atlantic, sinks, and travels south eventually to reach all the world's oceans. This in turn brings in a surface return flow of warmer water. Small changes in the rate of this circulation would affect the amount of warmer water reaching the North Atlantic and would produce climatic fluctuations on the scale of the Little Ice Age.

There is also abundant evidence for increased volcanic activity during the Little Ice Age. Volcanic dust and debris ejected into the stratosphere, together with aerosols, are known to have the effect of reducing temperatures at the Earth's surface. The Crête ice core provides a detailed record of volcanism since about AD 533 and shows a lack of major erup-

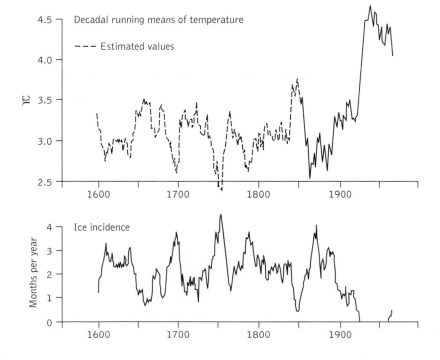

Fig. 3 Sea ice incidence and estimated temperatures for Iceland since AD 1600. (Source: Grove, J. (1988) *The Little Ice Age*, Methuen, London, Fig. 2.4.)

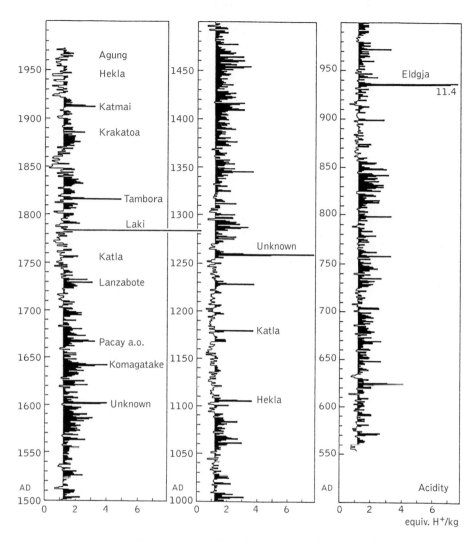

Fig. 4 Volcanically induced acidity layers in the Crête ice core, central Greenland between AD 533 and 1972. (Source: Hammer, C. U., Clausen, H. B., and Dansgaard, W. (1981) Past volcanism and climatic revealed by Greenland ice cores. *Journal of Volcanology and Geothermal Research*, **11**, 3–11.)

tions between AD 1260 and 1600 and an increase during the Little Ice Age (Fig. 4). The main point of debate is whether the volcanic effect could be sustained for long enough to produce the Little Ice Age and the climatic variation observed within it.

Finally, it is now also known that the Sun undergoes periodic changes in output of 1–2 per cent which could easily translate to the drop of 2 °C needed to account for the Little Ice Age. Before the recent instrumental measurement of solar output a good substitute measure was the number of sunspots observed. Numbers of sunspots are reduced during periods of low solar output. Periods of reduced sunspots, known as the Spörer Minimum (1450–1534) and the Maunder Minimum (1645–1715), did occur during the Little Ice Age. However, the correlation is not perfect and reduced sunspot numbers did not cover the whole period.

The cause of the Little Ice Age and its fluctuations may include elements of all three of these mechanisms. A complete answer must await a better understanding of how our climatic system operates.   B. A. HAGGART

Further reading

Grove, J. M. (1988) *The Little Ice Age*. Methuen, London.

littoral sedimentary deposits *see* SHORELINE (LITTORAL) SEDIMENTARY DEPOSITS

loess Loess is a terrestrial sediment formed by the accumulation of wind-blown dust. It is mainly composed of silt-size particles, most being in the size range 0.01–0.06 mm. Loess has been accumulating on the land surface for several million years, but accumulation increased notably at the beginning of the Quaternary period (i. e. since about 2.6 million years ago), the greatest accumulation rates coinciding with the extremely cold and arid conditions during the glaciations. Loess is found in many parts of the globe, but the thickest and most extensive cover occurs in northern China, middle Asia (Tajikistan, Uzbekistan, Kazakhstan), central and western Europe, central North America, the *pampa* and sub-Andean basins of South America, and New Zealand. The thickness of the loess and its average grain size decrease with distance from the source of the dust. Both display a close relationship to prevailing wind directions, reflecting both present and former atmospheric circulation patterns.

The mineralogy of loess is rather simple: quartz, often reaching 60 per cent by weight, is the principal mineral in most deposits, with feldspars and micas as important ancillary minerals. Also commonly present is calcium carbonate, which, because of its solubility, varies considerably with variations in climate; for example, 16–30 per cent by weight in the humid south-eastern parts of the North China loess belt, falling to as little as 7 per cent on the margins of the sand deserts to the west. The mineralogy of loess commonly differs considerably from that of the underlying bedrock, further demonstrating its derivation from a distant source.

As an air-fall deposit, loess is massive in structure, any evidence of bedding, and the inclusion of coarser grains such as pebbles, being indicative of post-depositional reworking. Such reworked loess is referred to as a *loess-like* or *loessic* deposit; for example, loessic alluvium (when reworked by water). An additional characteristic of loess, also arising from its wind-lain origin, is its low bulk density (high porosity). The generally low moisture contents in loess in the continental interiors keep any intergranular water at high tensions. This, together with the cementing action of the scattered calcium carbonate, is a source of some strength and characteristic sub-vertical joint systems so that, despite its low mechanical strength, loess commonly maintains steep, cliff-like slopes. For many centuries, the Chinese have taken advantage of such 'dry-strength' by excavating cave houses that are warm in winter and cool in summer. These same properties make loess a friable and so a readily cultivated soil capable of high crop yields: in fact, loess is a major component of the soils of the world's 'breadbasket' regions. At the same time, the characteristic high porosity and weak cementation render loess subject to instantaneous collapse (hydrocollapsibility) when groundwater conditions approach saturation. Large-scale landslides (specifically flowslides) and mass flowage of loess are frequently triggered by heavy, seasonal rainstorms (including the summer monsoonal rains) in the drylands of Asia and elsewhere. This gives rise to some of the highest known rates of soil erosion (around 20 000 tonnes km^{-2} $year^{-1}$ in parts of the Loess Plateau of north-central China). Essentially dry loess may also suffer collapse as a result of earthquake shock, a common phenomenon in central and eastern Asia. Such failures are a major geological hazard with a costly annual toll in lives, livelihood, and infrastructure, as well as in delivering large pulses of sediments into some of the world's greatest rivers and so directly affecting their flow regimes and flood behaviour.

EDWARD DERBYSHIRE

Further reading

Pye, K. (1987) *Aeolian dust and dust deposits*. Academic Press, London.

loess deposition and palaeoclimate

Since the 1970s, the analysis of oxygen isotope ratios ($^{18}O/^{16}O$) in foraminifera in deep-sea cores, particularly some from the equatorial Pacific, has provided an unprecedented proxy record of climate change throughout Quaternary times. In essence, increased concentrations of ^{18}O in foraminifera (unicellular marine organisms) within the deep-sea sediment reflect colder times of lower sea levels and larger ice sheets. The variations in $^{18}O/^{16}O$ ratios within foraminifera in deep-sea sediments, which have been continuously deposited throughout Quaternary times, show a complex pattern of climate change. This pattern corresponds well to theoretical cycles (Milankovich cycles) of change in the distribution of incoming solar radiation across the Earth's surface. This is the result of variations in the orientation of Earth's axis and its orbit around the Sun. The analysis of oxygen isotopes therefore shows that climate change was much more complex than had previously been thought. This stimulated the need to look for detailed proxy data for climate change on land surfaces. Unfortunately, subaerial erosion has made the continental sedimentary record very incomplete and discontinuous. The most complete terrestrial sedimentary record is provided by thick loess sequences. Much attention has therefore been given to trying to reconstruct palaeoclimates from loess.

Loess is a wind-deposited sediment made up of angular quartz and feldspar grains with diameters of between 20 and 60 microns (silt size). This sediment is essentially unlaminated; structurally it is held together by the point-to-point contacts between silt grains, although clays may form bridges between the silt grains or carbonate cements may provide support. Loess is loose and friable and has a high porosity, commonly more than half its bulk volume. It is capable of supporting substantial structures, but when wet the contacts between grains break and the loess may collapse or flow. In some deposits described as loess, fine sand and clay fractions may be more dominant and careful examination often reveals laminations. A precise definition of loess is therefore difficult.

Two models of evolution exist for the origin of the loessic silt. The first suggests that silt was formed by the grinding of rock beneath ice sheets during glacial times. The silts were washed out of the glaciers and deposited on outwash plains, where they were deflated by wind action and carried away from the margins of the ice and subsequently redeposited. The loess deposits of Europe and North America are explained in this fashion. The second school of thought advocates a desert origin for loess. In this hypothesis, the silts form by aeolian abrasion and attrition of sand grains and by rock-weathering processes. Winds blow the silts from the deserts and deposit them in adjacent regions. The loess of central China is explained in this way. Substantial quantities of silt can also be produced by other processes, such as fluvial abrasion and attrition, rock-fall processes, and freeze–thaw (cryogenic) weathering. In support of the two main mechanisms loess deposits have finer grain size away from their glacial or desert sources, for example, the Chinese loess fines westwards away from its source, the Gobi Desert. Characteristic heavy-mineral assemblages also enable the source of the silt to be identified. There is much debate over the exact mechanism of silt deposition, but in the main it can be regarded as a general fallout process.

The critical point, however, is that the processes of formation, transportation, and deposition of silt are strongly controlled by climatic conditions. During cold glacial times, desert regions become more arid and the production of silt increases and its transportation is enhanced. In glaciated areas glacial grinding increases and wind systems are intensified, producing more silt and transporting the sediment further away. Adjacent to both regions the deposition of loess increases. The grain-size characteristics and mineralogy of the loess should reflect climatic conditions. It is generally considered that the deposition of loess has been an almost continuous process for about 2.5 Ma (million years), but the rate of deposition has been highly variable, reflecting past climate conditions. Subsequent erosion of loess, however, makes a continuous record of deposition rare, and the stratigraphy is often difficult to interpret. In addition, during interglacial times soil-forming processes dominate and silt deposition is greatly reduced. As a result, many loess successions have distinct horizons representing ancient soils (palaeosols). The structure and mineralogy of these provides particularly valuable information about past climatic conditions. Loess and palaeosols have been particularly helpful in reconstructing the fluctuations of the desert margins and the intensity of past monsoons.

The thickest loess occurs in central China on the Loess Plateau, where it reaches a thickness of about 330 m near Lanzhou. Very little reworked and continuous deposition appears to have been the rule on the Loess Plateau. The Chinese loess sequences therefore provide the most complete terrestrial record of climate change for the Quaternary. The loess overlies sedimentary rocks of Late Pliocene age, the 'Red Clay Formation', and palaeomagnetic work has established an age for the basal contact of the loess of 2.48 Ma. It may be no coincidence that the onset of loess deposition is approximately contemporaneous with the start of the Quaternary Ice Age. This may be because the build-up of ice sheets and the intense rapid global climate change may have resulted in loess-forming processes becoming more dominant. It is also argued that the uplift of the Tibetan plateau during late Tertiary may have lead to increased aridity in central Asia and the onset of loess deposition. Most geologists however, believe that the Tibetan plateau attained its height about 14 Ma ago and that the loess cannot be attributed to tectonics (see *Himalayan–Tibetan uplift and global climate change*).

Up to 32 identifiable palaeosols (S1 to S32, S1 being the youngest) alternating with loess units (L1 to L33, L1 being the youngest) have been identified in the Chinese loess (Fig. 1a). These are thought to indicate cold dry periods, with high rates of loess deposition, and warm wet periods, with lower rates of deposition and the formation of soils.

At the time of deposition, magnetic minerals within the loess align themselves parallel to the Earth's magnetic field. The Earth's magnetic field fluctuates and reverses periodically. The nature and timing of these changes are reasonably well known, and the measurement of the orientation of these magnetic minerals within the loess therefore enable to loess to be dated. Such palaeomagnetic dating studies have been undertaken on several key sections (see Fig. 1b). Luminescence dating techniques, which measure the remnant radiation left in mineral grains after they were exposed to light, has provided an additional calibration of the loess successions. Accurate dating is critical to apply multi-proxy measures of climate to the loess–palaeosol successions. Many techniques have been applied, including particle size analysis, magnetic susceptibility, sediment fabric, micromorphology, mineralogy, palaeontology (mainly molluscan fauna, but also vertebrates and flora), carbonate content, organic carbon, and extractable iron.

Magnetic susceptibility (MS) has been used to detect palaeoclimatic variations, to correlate palaeosols and to correlate between the oxygen isotope record. MS is generally higher in palaeosols than in loess. Although the reasons for this variability are not fully understood, it may be attributed to the enrichment of detrital magnetic minerals in soil during interglacials owing to concentration by decalcification and soil compaction processes. It may also be the result of subaerial deposition of ultrafine magnetic minerals from distant sources whose concentrations are diluted during the higher rates of silt deposition associated with cold times. Alternatively, the MS may be the result of the formation *in situ* of magnetic minerals by soil-forming processes. On the basis of the types of magnetic minerals present within the loess, it has been suggested that the formation *in situ* of magnetic minerals by soil-forming processes is the most important control of the MS. The MS may therefore be broadly considered to be a function of palaeoprecipitation. One of the most intensive MS studies on loess was undertaken at three key sections on the Loess Plateau (Xifeng, Luochuan: Fig. 1). The sediments at

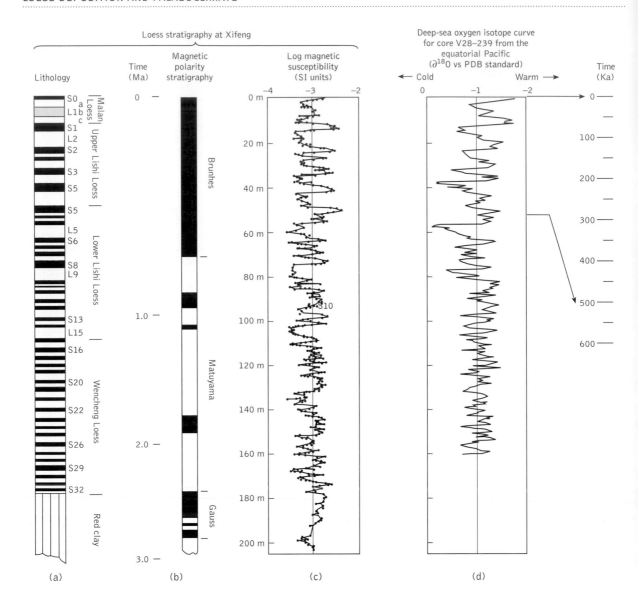

Loess stratigraphy at Xifeng

Deep-sea oxygen isotope curve
for core V28–239 from the
equatorial Pacific
(∂^{18}O vs PDB standard)

(a)　　　　　　(b)　　　　　　(c)　　　　　　(d)

these locations represent a time-span of about 2.5 Ma. The combined magnetic susceptibility results from these sections showed a general agreement with the astronomically tuned oxygen isotope deep-sea chronology in the upper part of the succession, but less agreement prior to 0.5 Ma (Fig. 1). Gross correlations between other sections throughout the Loess Plateau, for example at Lui Jia Po and Baoji near Xian in the humid warm south and at Lanzhou in the semi-arid west. Minor variations are more difficult to correlate, probably because of regional variations in climate and soil-forming processes. Broad correlations have also been made between

the MS results and aeolian sediment present in deep-sea cores from the Pacific Ocean. High MS values correspond well with high concentrations of aeolian sediment in deep-sea cores (Fig. 1e, f, and g). This probably indicates glacial times when stronger westerly winds carried sediment from China into the Pacific Ocean. A better understanding of the controls of MS will help to refine its use as a detailed proxy measure of climate change.

In the Chinese loess, the median grain size is essentially a measure of the vigour of the north-westerly (winter) monsoon. Coarse median values probably represent cold and

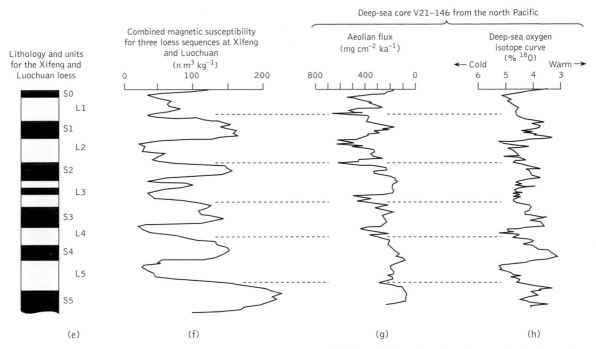

Fig. 1. Summary of the loess–palaeosol stratigraphy and magnetic susceptibility (MS) of selected loess sequences in China in comparison with deep-sea oxygen isotope curves and aeolian flux in the Pacific Ocean. (a), (b), and (c) compare the loess stratigraphy and MS at Xifeng with the deep-sea oxygen isotope curve from the equatorial Pacific Ocean. Note the close similarity between the curves. The age of the loess–palaeosol sequence at Xifeng was obtained by palaeomagnetic dating (b); (e) and (f) show the combined stratigraphy for three sections at Xifeng and Luochuan, and (g) and (h) show the aeolian flux and oxygen isotope curve for a deep-sea core from the north Pacific Ocean. Note the close correlation between loess, cold times, decreased MS values, and increased aeolian flux. (Adapted from Xiuming Liu *et al.* (1992), Kukla *et al.* (1990), and Hovan *et al.* (1989).)

dry glacial times. Some loess horizons are particularly sandy, such as the L9 and L15 sandy loess units (Fig. 1a), and these are thought to represent extensive advances of the desert margins. The median grain sizes match closely with the MS, but recent results show a much more complex pattern than that obtained from the MS analysis. This technique has great potential for detailed interpretation of past climate.

Study of the microscopic structures (the micromorphology), the clay mineralogy, organic carbon, and faunal excrements within palaeosols has helped to determine the nature of soil-forming processes across the plateau and between palaeosols of different ages. The abundant molluscan fauna within the loess is also being used to help provide information on past humidity and temperature. Early results show that the abundance of mollusca closely parallels the MS, further supporting the idea that these were times of thermal and humidity maxima.

Loess successions have also been studied in detail elsewhere in the world, but the record is less complete and more complex than in the thick loess of China. In the 'Palouse' region of Eastern Washington, USA, for example, the loess

sequence reaches 75 m in thickness and contains 19 or more palaeosols. They possibly record 1.5 to 2 Ma of deposition, but the record is discontinuous, with the initiation of cycles of loess deposition controlled by giant glacial floods from the Laurentide (North American) ice sheet during glacial times. The loess sequences in Kashmir and Europe are discontinuous and fragmented, yet quite convincing correlations have been made, mainly on the basis of the MS, with the oxygen isotope deep-sea chronology.

A record of palaeoclimates derived from the interpretation of the loess–palaeosol sequences is becoming greatly improved through higher resolution of the data and a better understanding of the mechanics involved in controlling regional climate change and orbital forcing. Subtle changes in grain size, MS, mineralogy, micromorphology, and other characteristics of loess provide great potential for reconstructing past climate change on the continents.   LEWIS A. OWEN

Further reading

Hovan, S. A., Rea, D. K., Pisias, N. G., and Shackleton, N. J. (1989) A direct link between the China loess and marine records: aeolian flux to the north Pacific. *Nature*, **340**, 296–8.

Kukla, G., Heller, F., Lui, X. M., Xu, T. C., Lui, T. S., and An, Z. S. (1988) Pleistocene climates in China dated by magnetic susceptibility. *Geology*, **16**, 811–14.

Rutter, N. (1992) Presidential address, XIII INQUA Congress 1991: Chinese loess and global change. *Quaternary Science Reviews*, **11**, 275–81.

Xiuming Lui, Shaw, J., Lui, Tungheng, Heller, F., and Baoyin, Yuan (1992) Magnetic mineralogy of Chinese loess and its significance. *Geophysical Journal International*, **108**, 301–8.

low-latitude tropospheric circulation The circulation of air in low latitudes is taken here as referring to circulation in the region of the globe where the Sun is within 10° of zenith at least once a year. Most textbooks lead one to believe that the structure and functioning of the atmosphere in this zone is simple, with little variability in the day-to-day wind flow. Interestingly, there is as yet still no simple theoretical framework similar to that which has been developed for the mid-latitudes (i.e. quasi-geostrophic theory) that can be used to provide an overall understanding of the tropical atmosphere. Instead, largely because of problems of scale, the search for an integrated mathematical framework for the tropics can be regarded as one of great remaining problems of global atmospheric circulation.

One of the simplest criteria for evaluating the large-scale structure of the atmosphere is the distribution of pressure. The surface pressure at 30° N or S is higher than the pressure near the Equator. This drives the surface easterlies (easterly because Coriolis force deflects the flow) or trade winds towards the Equator. Evaporation removes heat and water from the oceans and transports it equatorwards. Consequently, the near-surface air flow is both warm and moist. Convergence and forced ascent near the Equator lead to heavy precipitation at the Intertropical Convergence Zone (ITCZ) or Intertropical Confluence (ITC). The release of latent heat and the convergence of sensible heat flux both help to drive strong upwards motion. Latent and internal energy are here converted into potential energy and exported towards the pole in the upper air. This export exceeds the sum of the flow of latent and internal energy towards the Equator in the lower branch, giving a small net flow of energy towards the pole. As a result of the constraints of the angular momentum balance, the meridional cell is not a particularly efficient means of polewards energy transport.

Ascent in the ITCZ is balanced by descent in the subtropics. This is evidenced by subtropical highs, intense subsidence, clear skies and deserts (e.g. Namib, Karoo, Kalahari, Sahara, Atacama). Diurnal temperatures range by up to 50 °C, because of intense solar radiation by day and loss of terrestrial

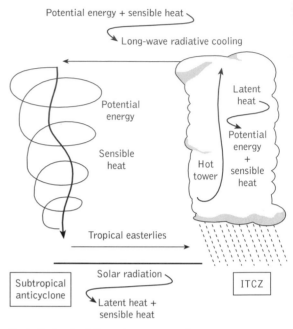

Fig. 1. Hadley circulation and energy conversions.

radiation by night in a dry, cloud-free atmosphere.

Overall, the low latitudes are characterized by a lack of temperature gradients away from land/sea boundaries. This, in turn, means that the atmosphere is driven by energy sources very different from those of the mid-latitudes, whose overlying atmosphere relies on available potential energy derived from the tilting of the atmosphere, itself brought about through temperature contrasts. Instead, the large-scale overturning in the tropical atmosphere is maintained largely by the release of latent heat.

These overturning thermally driven cells in the low latitudes of both hemispheres are known as the *Hadley cells* (Fig. 1), in honour of George Hadley (1685–1768), who assumed that such a circulation extended from the Equator to the poles. The northern cell dominates from November to March, the southern cell from May to September, with rapid transitions between April and October.

Many of the characteristics of the tropical atmosphere come about because of the small value of the Coriolis parameter, or, put differently, the small effect of the Earth's rotation, which allows air to flow more or less directly from a region of high pressure to one of low pressure. Regions of different pressure and temperature are thus cancelled out efficiently. By contrast, in the mid-latitudes the large effect of the Earth's rotation causes much greater deflection of the wind; in consequence, airflow rotates around regions of low pressure or high pressure, essentially maintaining differences in pressure and temperature.

Unfortunately, while this overview of the low-latitude

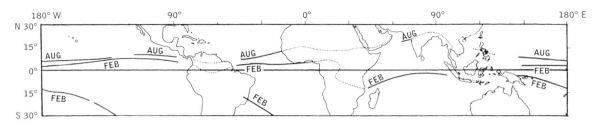

Fig. 2. Position of the Intertropical Convergence Zone (ITCZ) in summer and winter; continuous lines, Intertropical Confluence (ITC) cloud bands; dotted lines, Intertropical Front (ITF) over land. (After Barry and Chorley (1993).)

atmosphere provides the most convenient insight into broad-scale functioning, it also conceals features of the tropical atmosphere which make it both interesting and complicated. For example, the simple model of a large-scale ascent in the ITCZ at the Equator is not consistent with either observation or theory. The large-scale tropical atmosphere is stable above approximately 5 km. Large-scale ascent to the top of the atmosphere would thus have to take place through a stable layer or, in other words, up the energy gradient. This would in effect mean that the upper troposphere would be cooled by the ascent, a situation which would not satisfy the heat imbalance of the tropical zone. Riehl and Malkus showed that ascent needs to occur within clouds, in the so-called hot towers, where the air is saturated. The warmer air could still arrive at the top of the atmosphere by travelling inside the cloud conduits. The ITCZ thus occurs as a collection of cloud clusters rather than as a large ascending branch. At any one time, 1500 to 5000 clouds or hot towers would be necessary to account for the required heat transport of the tropics.

The outcome is that in order to understand and integrate the dynamics of the global tropics, one must really understand and integrate the dynamics at the cloud scale. This introduces one of the most difficult scale problems imaginable in the study of fluids. With this context in mind, we can consider some of the features of the tropical atmosphere, including the ITCZ, easterly waves, tropical cyclones, the monsoons, and the interannual variability of the tropics.

Averaged over time, the ITCZ appears as one of a line of deep convective clouds, for example in the Atlantic and Pacific Oceans between 5 and 10° N. On a day-to-day basis, however, the feature appears as a discrete but transient cloud cluster several hundreds of kilometres in diameter separated by regions of clear skies. Over the continents, the ITCZ clusters are much larger and less zonally constrained, essentially spreading out over the heated land mass. Three equatorial regions of maximum cloudiness are apparent: South America, Africa, and Indonesia. The last-mentioned is an equatorial location, with shallow seas and relatively small land masses, giving rise to the region of greatest rainfall. In fact this region provides a centre for strong thermal forcing of the rest of the tropical atmosphere.

The flat Amazon Basin, oriented towards the moisture-bearing easterlies from the Atlantic, receives more than 2 m of rainfall a year. The moisture is prevented from flowing out over the Pacific by the large barrier of the Andes. Convection leads to release of latent heat and the warming of the lower troposphere, which, in turn, lowers the pressure and increases the influx of surface moisture. Maximum rainfall occurs during the early months of the year and a minimum in August, when precipitation occurs over Central America.

Seasonal shifts in the ITCZ are largest over the land, because of the very small effective heat capacities of land surfaces (Fig. 2). The most dramatic seasonal movement in the mass of convection is centred over Indonesia, with both north–south and east–west components. During the summer of the southern hemisphere, the convection extends southwards over the south Pacific Ocean and is identifiable as far away as 30° S, 140° W. This feature is called the South Pacific Convergence Zone (SPCZ). During the northern summer the SPCZ is retracted back across the dateline, with Indonesian convection extending north-westward into the Bay of Bengal, where it becomes connected with the convection associated with the Asian summer monsoon.

Embedded within the tropical easterly flow are weak equatorial wave disturbances driven by latent heat release, which propagate westward at speeds of 8–10 m s^{-1} along the ITCZ. The cloud bands are separated by 3000–4000 km with characteristic periods of 4–5 days. The largest disturbances and the highest precipitation tend to occur when mid-tropospheric temperatures are warmer than average (although usually by less than 1 °C). These disturbances involve an interaction in which large-scale convergence at low levels moistens and destabilizes the environment so that small-scale thermals can easily reach the level of free convection and produce deep cumulus clouds. The clouds then act as a large-scale heat source that drives the secondary circulation responsible for low-level convergence. This is true for most parts of the tropics, but in North Africa waves feed from an easterly jet which is driven largely by temperature contrast between the Sahara and equatorial regions.

These systems have the role of converting potential (derived from release of latent heat and ascent) to kinetic energy. This is worth emphasizing, since mid-latitude westerly waves retain potential energy with little conversion to kinetic energy. At times, easterly waves develop into disturbances with closed isobars, which may follow a continuum of intensification from

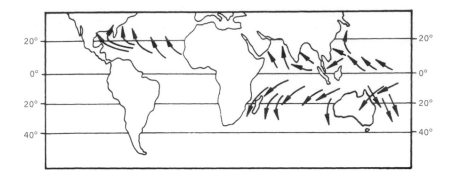

Fig. 3. Principal tropical cyclone tracks. (After Barry and Chorley (1993).)

easterly storms to tropical storms (winds of more than 19 m s^{-1} at 10 m), culminating in tropical cyclones (sea-level pressure of 950 hPa or less, pressure gradients of 1 hPa km^{-1}, winds in excess of 33 m s^{-1} at 10 m level), which are the most powerful weather disturbance on Earth. Roughly 80 occur per year, forming over oceans where surface temperatures exceed 27 °C over an area of 5° of latitude square (Fig. 3).

Mature tropical cyclones consist of a dense white ring of cumulonimbus and cirrus extending a couple of hundred kilometres out from the centre above a ring of updraft surrounding a central eye. In the eye, pressures drop so low that the stratosphere sinks to the surface; the sinking and warming air is therefore clear. The spirals, arms of cloud stretching hundreds of kilometres outward, often equatorwards, from the central structure, comprise long, curving lines of cumulonimbus.

The warm core of the tropical cyclone sets it apart from other tropical disturbances, which are cool-cored (though penetrated by the hot towers discussed earlier). A warming of the core by 6 °C compared with the surrounding air mass, for example, would lead to a drop in surface pressure of the tropical cyclone of about 50 hPa. Positive feedbacks between frictional convergence in the surface layer supply moisture to the base of the storm. Cumulonimbus convection then strengthens the convergence and hence total growth.

The central pressure of the storm usually rises when it reaches land, which cuts off the moisture supply and therefore energy supply in the form of latent heat. Many storms then travel westwards and then polewards. An interesting scientific advance has been the seasonal forecasting of tropical cyclone frequency in the Atlantic on the basis of sea-surface temperatures, the state of the El Niño–Southern Oscillation (ENSO) event, the quasi-biennial oscillation in the stratosphere, and the rainfall in the Sahel.

While tropical cyclones represent one of the smaller, yet most intense types of tropical disturbance, the monsoons, with a near-global influence, are among the largest. This seasonally reversing circulation system is driven by temperature contrast between land and sea. Land has a different thermal response because the heated layer is thin and its heat capacity is small. Solar radiation thus raises the temperature much

more quickly over land. Surface warming leads to cumulus convection and release of latent heat, producing warm temperatures throughout the troposphere. A pressure difference sets in between the land and the sea to drive a direct thermal circulation with low-level convergence and upper-level divergence over the land. The positive correlation between vertical motion and the temperature field points to the role of the monsoons in converting eddy potential energy to eddy kinetic energy, which is frictionally dissipated.

One of the largest meridional seasonal shifts occurs in the Indian–Asian regions and is associated with the monsoon. As a result, India experiences warm, wet summers and cool, dry winters. The monsoon usually becomes established over the Indian subcontinent by late June. Southerly flow into the monsoon is a continuation of the south-easterly trades of the southern hemisphere, except that they are deflected to the right by the Coriolis force in the northern hemisphere to become south-westerlies. These winds evaporate moisture off the Arabian Sea, which reaches temperatures of 29 °C during the boreal summer. Slow-moving monsoon depressions lead to enhanced rainfall. A tropical easterly jet stream is a semi-permanent feature of the region in summer at latitudes of 15° N, developing in response to the heating gradient between the warmer north and the cooler equatorial region (a reversal of the norm). This jet is thus driven by a thermal wind mechanism.

The monsoon begins to withdraw in September and is absent by November. The easterly jet is replaced by a narrow westerly subtropical jet stream on the southern edge of the Himalaya. In winter the circulation reverses to low-level offland flow as the land cools more quickly than the oceans. Cooling is particularly marked over the Himalaya. High pressure and dry, descending air occurs over the land and rain falls over the ocean. North-westerly cold surges of air occur during winter from the vast pool of air north and east of the Himalaya.

The tropical atmosphere is characterized by large interannual variability in pressure and rainfall. This has been shown to be forced to a large degree by variability in sea-surface temperature; the ENSO mechanism is the best-known example. A unique insight into this behaviour has been derived from the Hadley Centre Climate of the Twentieth Century experiments, in which a general circulation model was forced by observed sea-surface temperatures for the period 1900 to 1993. The model was run several times, but each run had different starting conditions. The degree to which each of the runs produced different interannual variability determined the degree to which the atmosphere was controlled solely by varying boundary conditions. Early results showed that the tropics respond directly to sea-surface temperature forcing, whereas the resulting interannual variability in pressure, temperature, or rainfall in the mid-latitudes is extremely sensitive to starting conditions. The mid-latitudes appear therefore to be more chaotic. An important consequence of this tendency of the tropical atmosphere is that seasonal rainfall may be forecast several months ahead. The forecasts are based on the state of the slowly evolving sea-surface temperatures.   R. WASHINGTON

Further reading

Barry, R. G. and Chorley, R. (1993) *Atmosphere, weather and climate*. Routledge, London.

Burroughs, W. J. (1991) *Watching the world's weather*. Cambridge University Press.

Meteorological Office (1991) *Meteorological glossary*. HMSO, London.

Lyell, Charles (1797–1875)

Charles Lyell was born into a well-to-do Scots family in Forfarshire in the year of James Hutton's death, and was to become one of the foremost British geologists of all time. Between 1816 and 1819 he was at Oxford reading law. There he attended lectures in geology by Dr William Buckland, who greatly influenced the rise of geology in England and aroused Lyell's enthusiasm to the point that he practised law for only a couple of years before turning to a life in geology. Being well-heeled, Lyell was able to travel extensively in Europe as well as in Britain. He became a fellow of the Linnean and Geological Societies, corresponded with many of the foremost geologists of the day, and was elected to the Royal Society in 1826.

These were days in which the catastrophists were still ascendant, but by 1825 Lyell was convinced of the validity of a form of uniformitarianism as providing the grand explanation of the visible condition of the crust of the Earth. In 1830 he published the first volume of his great work, *The principles of geology*. Two more volumes were published by 1833. Lyell's visits to France and Italy led him to propose a threefold stratigraphical division of the Tertiary rocks there on the basis of the proportion of recent to extinct species of mollusc present. His terms Eocene, Miocene and Pliocene for these are still in use in modified form.

Lyell's *Elements of geology* (1838) was to become a standard work for stratigraphy for many years. He published a stream of contributions in several journals, and was keen to see and study as much as he could in order to explain geological phenomena. His travels in Europe and his two visits to North America helped him to visualize ancient environments in which geological formations such as coal measures were

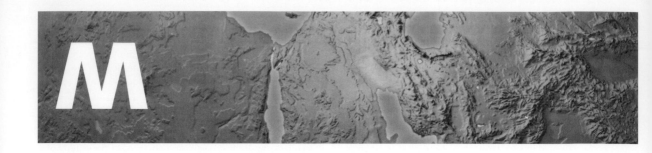

magma and magmatism

magma and magmatism Magma is molten rock that has not yet reached the surface of the Earth. Magmatism is the set of processes that lead to the generation of magmas and their emplacement. Although volcanoes are the surface expression of magmatism, a magma, once it has reached the surface of the Earth, is called a *lava*.

Much progress has been made in understanding the processes that generate magmas, and models have been developed for the composition and volumes of magma whose products are observed on Earth. The development of plate-tectonic theory in the late 1960s led to the recognition that certain types of volcanic activity are associated with specific plate-tectonic environments, and this has in turn led to refined models of melt generation. Melting under these conditions produces a range of parent magmas of different chemical composition. The compositions of volcanic rocks at the surface of the Earth reflect the composition of the parent magma and the modifying effects of the processes occurring between the source region and the point of eruption. The composition of a parent magma can be modified by crystallization or by contamination by the rocks through which it passes on its way to the surface.

The outer part of the solid Earth consists of a thin (30–150 km) solid *lithosphere*, composed of tectonic plates, 'floating' on the *asthenosphere*, which is the upper part of the mantle. Although the mantle is a crystalline solid, over millions of years it convects, carrying heat from the Earth's core to its surface; the lithosphere can be thought of as a lid, insulating the mantle from space. Experiments in the Department of Applied Mathematics and Theoretical Physics at Cambridge University have shown that the convection in the 3000 km thick mantle can be modelled by a 3 cm thick layer of golden syrup, heated from underneath. The syrup layer has the advantage of being thinner, cooler, and less viscous than the mantle and it therefore convects much more quickly. The syrup in these experiments forms polygonal convection cells with hot, rising plumes at the centre and cold, sinking sheets at the edges. Analogous structures are observed in the mantle: the hot, rising plumes are at about 1550 °C and the cold, sinking sheets are identified with tectonic plates sinking into the mantle (i.e. subducting). Heat is continually being lost from the Earth, which is slowly cooling. Heat is continually being supplied to the mantle from two sources: the radio-active decay of elements and the heat of crystallization of the Earth's core.

Figure 1 shows the thermal structure of the lithosphere and mantle. The mantle, because it is convecting, is of roughly constant temperature, while the lithosphere acts as a conducting lid. The temperature profile across the two layers is called the geotherm. Material in the mantle that becomes unusually cold sinks; material that has been heated by the Earth's core rises as a hot plume. If the mantle is dry, it melts at the dry solidus and if volatiles (such as water and carbon dioxide) are present it melts at the wet solidus. The solidus temperature is the temperature at which the first droplets of melt are formed. Figure 1 shows that there are three main ways in which the

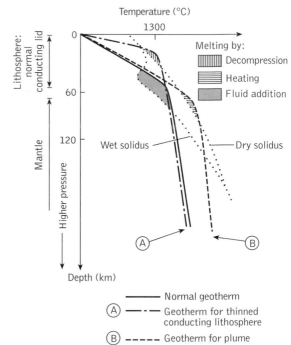

Fig. 1. Thermal structure of the lithosphere and mantle, illustrating the principal causes of melting. D, decompression; F, addition of fluid; H. heating.

mantle can be melted: by a reduction in pressure, by an increase in temperature, and changing from the dry solidus to the wet solidus by adding volatiles to the melt.

Pressure is reduced when the lithosphere is stretched and consequently thinned, leading to a steepening of the geothermal gradient (A in Fig. 1). The geotherm now crosses the dry solidus and melting occurs. The temperature of a mantle plume is around 1550 °C. An increased mantle temperature also causes the geotherm to cross the solidus and generate a melt (B). It can be seen from Fig. 1 that the wet solidus corresponds to lower temperatures than the dry solidus, so that melting may be caused by the addition of volatiles. Melting is also aided by any combination of these three processes: decompression, heating, or the addition of volatiles.

The composition of the melt produced depends on the composition of the material that is melting and the conditions under which it melts. Thus certain compositions of melt are related to specific plate tectonic environments. Figure 2 shows the plate tectonic environments in which magmatism is most likely to occur. Examples of some of these types of magmatism and the processes that generate them are discussed below.

Decompression melting

Decompression melting is melting caused by a reduction in pressure during tectonic extension. As the lithosphere is stretched horizontally it is thinned as the top surface sinks and the bottom surface rises. The mantle flows to fill what would otherwise be a space at the base of the lithosphere, and is decompressed. Where there has been only a little stretching, only a little melting occurs (e.g. in rift valleys). Large amounts of stretching result in the lithosphere becoming so attenuated that it breaks and separates completely. The volcanic rocks that are generated at the breaks form mid-ocean ridges: they solidify to form new crust called oceanic crust. The most common rocks on the Earth are the volcanic rocks formed at

mid-ocean ridges. These rocks are basaltic in composition and are called *mid-ocean ridge basalts* (MORBs). A mid-ocean ridge is a place where two tectonic plates are diverging; as they move apart, decompression melting of the mantle occurs to form basalts that erupt to form new sea floor. Most of this new oceanic crust eventually sinks back into the mantle at a subduction zone and is recycled. The process by which MORB is generated is remarkably constant: it is a melt generated from the mantle by continuous extension of the oceanic crust. The composition and thickness of the oceanic crust is correspondingly constant. Oceanic crust has a thickness of only about 7 km.

Continental rift valleys express much smaller amounts of extension. If there is extreme extension of a rift, it will ultimately become an ocean, as for example, the present-day Red Sea. The volumes of magma generated in rifts are correspondingly smaller than in oceans and their compositions represent much smaller degrees of melting. These smaller volumes of melt are rich in elements such as sodium and potassium (alkali metals), which are the first elements to move into the melt; the resulting rocks are called alkali basalts.

Melting by heating

Melting by heating occurs when a mantle plume rises beneath the lithosphere. The centre of the rising plume is very hot and, where it reaches the top of the mantle, the pressure is low enough for it to melt. The melts thus generated rise through the lithosphere and eventually erupt at the surface. Volcanic activity of this type is often called *hot-spot volcanism*. Where a plume rises under an oceanic plate the volcanic rocks formed from the magma generated form a sea mount (a mountain on the sea floor). An example of a hot-spot sea mount is the island of Hawaii in the Pacific Ocean. There has been so much melting beneath Hawaii that the sea mount has become taller than the ocean is deep and it emerges as an island. The oceanic plate upon which Hawaii is being built is moving slowly over

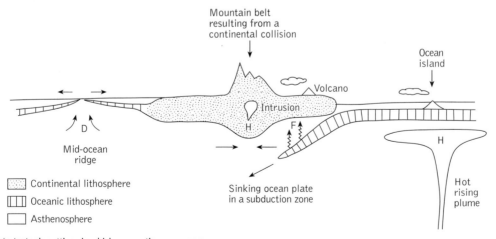

Fig. 2. Plate-tectonic settings in which magmatism can occur.

the plume, producing a trail of islands across the ocean known as the Emperor–Hawaiian chain. The dog-leg in this chain of islands shows that the plate has changed its direction of motion since the hot-spot originated (Fig. 3). Most hot-spot volcanoes of this type erupt *ocean island basalts* (OIB).

Because the continental crust is much thicker than the oceanic crust, plumes cannot rise to areas of low pressure beneath them without colliding with the base of the lithosphere; continental hot-spot activity is thus rare. However, the hot springs and volcanic activity of Yellowstone Park in the western USA, related to the 'Yellowstone plume', are an expression of such activity.

The most common cause of melting by heating is a plume, but during metamorphism (change in the mineral composition of a rock through the application of heat and pressure) part of the rock may be locally melted and escape. This process of partial melting is called *anatexis*; once an anatectic melt has escaped from the parent rock, it behaves like an igneous melt.

Melting by the addition of volatiles

Under normal conditions volatiles (mainly water and carbon dioxide) are present only in small quantities in the mantle

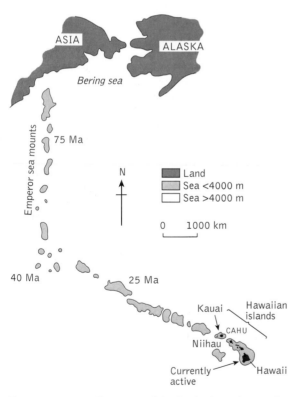

Fig. 3. Emperor–Hawaii sea-mount chain, showing increasing age of sea mounts to the north-west and west as a consequence of a change in motion of the Pacific plate.

(less than 0.5 per cent). However, in a subduction zone, where an oceanic plate sinks beneath another plate, it takes water down with it into the mantle. The water in the descending plate is trapped in hydrous (i.e. water-bearing) minerals at the top of the oceanic crust. As the crust is heated, the minerals break down and water is released, rising into the wedge of mantle above the plume and causing it to melt. This process generates large volumes of melt by wet melting and forms chains of volcanoes whose composition is andesitic. Examples of chains of andesitic volcanoes occur in the Andes, and in parts of the western Cordillera of North America, both areas of volcanism being related to the subducting eastern margin of the Pacific Ocean. Where one oceanic plate is subducted beneath another, the volcanism can form a chain of islands; Japan is an example. This type of magmatism is very complicated and is still not fully understood.

Mixed melting

All the examples given above involve one melting process. There are also examples of tectonic environments where more than one melting process contributes to the formation of a group of volcanic rocks. Figure 4 shows the possible combinations of melt types, and gives examples of places where they occur.

In Iceland the mid-ocean ridge and a plume are acting together to cause very large amounts of melting (heating and decompression). Without the presence of the plume, the crust would be of normal oceanic thickness, that is about 7 km thick; but the additional heat of the plume means that sufficient melting has occurred for an abnormally thick (30 km) oceanic crust to have formed. This young, thick crust is very buoyant and rises above the level of the ocean surface. It is known that the Iceland plume has been coincident with the Mid-Atlantic Ridge since the ocean began to open. The perennial presence of the plume is reflected in a ridge of thickened crust that joins Iceland to the British mainland. Although when a plume rises under an oceanic plate, melting is restricted to the hottest part of the centre of the plume, it is thought that hot material spreads out in a mushroom-shaped head at the base of the lithosphere, thus raising the temperature of the mantle over a much wider area.

Even though the continental crust is much thicker than the oceanic crust, decompression melting in the presence of a hot plume generates the aptly named continental flood basalts (CFB). These are the largest volcanic effusions on the continents. The most recently active continental flood basalt provinces are those in the USA, which include the Columbia River plateau (active from 17 to 6 million years ago) and the Snake River plain (active from 17 million years ago until the present) provinces. The activity of the Deccan Traps in India during the time of transition from the Cretaceous to the Tertiary is considered by some geologists to have been so catastrophic that it has been proposed as a cause of extinction of the dinosaurs.

In southern Africa, melts that are very small in volume have been generated by wet melting at the base of the lithosphere,

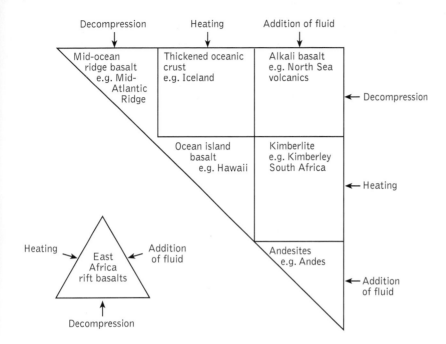

Fig. 4. Magma types related to two or three causes of melting.

enhanced by the presence of a hot plume. These rocks, called kimberlites, came from very deep in the lower lithosphere and were so rich in volatiles that they bubbled up through the crust quickly enough to preserve diamonds picked up from their source region at a depth of about 150 km. Kimberlites are named after one of the most diamond-rich sites in Kimberley, South Africa.

The volcanic rocks of the East African Rift Valley appear to be the product of all three melting processes combined. The African lithosphere is thick (150 km), and there is much debate about the number and location of plumes in the area, but it seems certain that there is at least one plume under the region. The thick African crust displays a low geothermal gradient which inhibits decompression melting. Even though melting would not normally be expected because of the small amounts of extension, the combination of extension, volatiles, and the plume has produced alkali basalts. These basalts, being the result of all three types of melting, have exotic compositions (Fig. 4).

Modification of magma

Melts are less dense than the materials in their source regions; so, as soon as melting occurs, they start to rise. Melts move by percolation and as blobs of melt (*diapirs*). *Percolation* is the movement of small amounts of melt in tiny canals along grain boundaries. Work by Mark Spiegelmann in the United States suggests that diapirs can be interpreted in terms of a special kind of wave called a solitary wave. The waves of this type take the form of increased porosity in a sphere of several kilo-

metres radius. However they move, rising magmas eventually come to a region where they no longer have positive buoyancy and they then cease to ascend. In some instances, stationary magmas solidify *in situ*. As they cool, dense minerals usually crystallize out which sink to the bottom of the magma body. This process is called fractional crystallization. As the density of the remaining melt is lower, it can again rise in the crust. The density of the magma is progressively lowered because it loses magnesium and becomes richer in silica. For example, a gabbroic melt rising through the crust will pond in a magma chamber when it reaches neutral buoyancy. As the magma chamber cools, crystals of olivine form and sink to the bottom of the magma chamber and collect to form a peridotite. The magma, now lighter having lost olivine, continues to rise and can eventually erupt as a basalt enriched in silica.

The composition of magmas is also modified by crustal contamination. Many of the crustal rocks with which a magma comes into contact are silica-rich. Because they have lower melting temperatures than the magma, they are melted on contact with it and can be assimilated, changing its composition.   JUDITH M. BUNBURY

Further reading
Wilson, M. (1989) *Igneous petrogenesis: a global tectonic approach.* Unwin Hyman, London.

magma production *see* PARTIAL MELTING AND MAGMA PRODUCTION

magnetic field (origin of the Earth's internal field)

Although most of us have used a magnetic compass to orient ourselves, we do not often think in terms of the source of the energy that powers the compass needle. This energy is provided by the Earth's magnetic field. The origin of the field is a complex subject which has enthralled a large number of the best scientists for the past four centuries; Albert Einstein even referred to it once as one of the three most important unsolved problems in physics. Certain aspects of how the field is generated are, however, now fairly well understood. Earth's magnetic field originates in the fluid outer core and is generated by electrical currents resulting from convective fluid motions of the highly conducting iron–nickel alloy of which the core is made. Even though the fluid motions in the core are quite complex, the magnetic field at the Earth's surface (roughly 3000 km away from the core) is nearly as simple as the dipolar magnetic field of a simple bar magnet (Fig. 1). Unfortunately, this simplicity inhibits our ability to deduce the precise mechanisms that could generate such fields; neither the exact mechanism that generates the field nor the patterns of the convection cells are known precisely. As a result, scientists who investigate the physical origin of the field have to rely on peculiarities that the field exhibits, as well as its gross features. In general, any reasonable physical model of the field must explain the secular variations of the field as well as its capability to reverse itself on a timescale of thousands of years (see *geomagnetism: main field, secular variation, and westward drift; geomagnetism: polarity reversals*). Under these constraints, the overall mechanism that generates the field is thought to be similar to an ordinary electrical dynamo. A dynamo converts mechanical energy from a moving electrical conductor into electromagnetic energy and thus generates current. But the fluid motions in the highly conducting outer core could not, by themselves, generate the electric currents in the outer core and the consequent magnetic field. An initial (seed) magnetic field is required to start a dynamo. For the Earth, this seed magnetic field may have been supplied by the Sun when the metallic iron–nickel core first formed within the Earth. Since that time the geometry of the convection cells and also the characteristics of the field might have changed as the inner core began solidify and grow in size. When all the core has solidified, the internally generated magnetic field will cease to exist (as on the present-day Moon).

The magnetic field cannot, however, be sustained simply by the presence of highly conducting fluid in the outer core. (Because of the high temperature, this fluid is as mobile as is water on the Earth's surface.) Thermo-gravitational convection acts to sustain the field: if the convection is turned off and the fluid remains static, the field decays, the inferred characteristic decay time being about 10 000 years. The direction of the field comes from the Earth's rotation, which shapes the patterns of the convection. Because of the Earth's rotation, the convecting fluid experiences the Coriolis effect, this causes clockwise circulation in the downwelling limbs of the convecting fluid in the northern hemisphere, and counterclockwise circulation in the southern hemisphere (Fig. 1).

It is clear that the magnetic field has its origin in electrical currents in the outer core and some sort of dynamo-like process resulting from interactions of conductive fluids, gravity, Earth's rotation, temperature differences between the upper and lower boundaries of the outer core, convective fluid motions, etc. Developing physical models of such a dynamo is, however, extremely complex. A mathematical treatment requires simultaneously dealing with about ten formidable equations containing physical parameters whose values are very uncertain. Despite their complexity, dynamo models are attractive because magnetic fields can be modelled which more or less parallel the Earth's axis of rotation, as well as reversals of the field.

There are many dynamo models and it is not possible even to mention all of them here; the following short list is intended to give only an overview. In 1919, J. Larmor suggested a rotating disc dynamo model to explain the origin of the solar and the Earth's magnetic field (Fig. 2a). In a disc dynamo, an initial magnetic field induces in a rotating conductive disc an outwardly directed electric current which can be tapped with an assembly of a brush and a wire. The wire can be wound around the axis of the disc to reinforce the initial field. Reversals in the direction of the magnetic field cannot occur in the single-disc dynamo, but chaotic reversals, such as are observed on the Earth, can occur in a coupled double-disc dynamo (Fig. 2b), as was discussed by Rikitake in the late 1950s. (Because reversals of the Earth's magnetic field were firmly established only in the late 1950s,

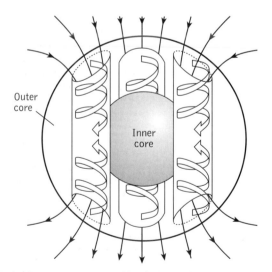

Fig. 1. The Earth's magnetic field has its origin in convection in the outer core of the Earth; the exact configuration is still under investigation. After Bloxham and Gubbins (1989).

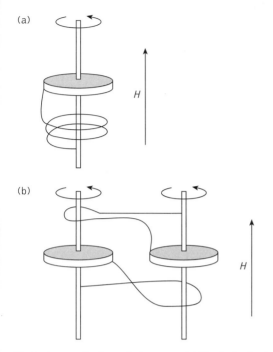

(a)

(b)

H

Fig. 2. An example of an early dynamo model. The single-disc dynamo of Larmor (a) is able to reinforce an initial magnetic field, but a coupled double-disc dynamo (b) is also able to replicate the chaotic reversals observed in nature. H magnetic field.

has also been inferred by J. Bloxham and D. Gubbins (Fig. 1) by mathematically projecting the observed magnetic field at the Earth's surface to the surface of the core–mantle boundary.

Geomagnetists deciphering the mysteries of the origin of the Earth's magnetic field constrain their models with snapshots of the magnetic field obtained from a variety of sources: satellites such as Magsat (*c.* 1980), which measure the near-Earth magnetic field; the POGO series satellites flown in the mid-1960s; surface geomagnetic observatories scattered round the world; and historical navigational charts (going back to the time of Halley); they also use information about the strength and direction of the field preserved in archaeological artefacts such as baked pottery and kilns (over thousands of years) and rocks formed over geological timescales (millions to hundreds of millions of years). As mentioned above, models of the physical processes in the outer core must be able to replicate the nature and form of variations that are observed (for example, periods of frequent reversals, periods of quiescence in reversals, rate of change of dipole moment, rate of secular variation of the dipole field, westward drift of the non-dipole field, etc.). DHANANJAY RAVAT

Further reading

Bloxham, J. and Gubbins, D. (1989) The evolution of the Earth's magnetic field. *Scientific American*, **255**, 68–75.

Merrill, R. T. and McElhinny, M. W. (1983) *The Earth's magnetic field: its history, origin and planetary perspective*. Academic Press, London.

Larmor had no reason to propose a reversing disc dynamo model.) An important transition from the mechanical analogues to the self-sustaining magnetic field involving fluid motions was made in the 1940s by W. M. Elsasser in the United States and E. C. Bullard in England. This led to the first testable numerical computations of the Earth's magnetic field. In 1955, E. N. Parker was able to show heuristically that by assuming a certain velocity field for the fluid motions in the outer core and the frozen-influx concept of H. Alfvén (see *magnetohydrodynamic waves in the Earth*), one can reinforce the original magnetic field (an $\alpha\omega$ dynamo). In a kinematic model such as this, the magnetic field may continue to grow indefinitely. But the magnitude of the field can be maintained, and even reversals can be produced, by calling upon instabilities created by preferentially located cyclonic activity in the outer core (Levy–Parker models). Instabilities can also play an important role in the magnetohydrodynamic models proposed by S. I. Braginsky, where magnetic forces (Lorentz forces) become important in shaping the fluid motions. In another set of models, called hydromagnetic models, a velocity field, is derived by appropriately coupling the dynamo equations. In the hydromagnetic models constructed by F. H. Busse in the 1970s, convection can occur in columns parallel to the rotation axis. A columnar convection pattern

magnetic pole The north and south magnetic poles are two special locations on the Earth's surface near the geographic poles, where a horizontally pivoted compass needle shows a vertical orientation with respect to the Earth's surface (see *geomagnetism: main field, secular variation, and westward drift* for the difference between the terms 'magnetic' and 'geomagnetic'). There are in fact many places on the Earth where a compass needle may show a vertical magnetic inclination because of the effect of strong local geological or cultural sources; thus, to qualify for a magnetic pole, the orientation of the compass needle must gradually become vertical over a very long distance. Magnetic poles have no particular scientific significance. Finding their location has, however, been historically important as a test of human endurance in Arctic and Antarctic climates—perhaps also as symbols of the human spirit of exploration of the unknown. On the other hand, the locations of the geomagnetic poles (the places where the axis of the geocentric magnetic dipole that best fits the main field intersect the Earth's surface) and the geomagnetic latitudes are considerably more important because many ionospheric phenomena in polar regions are related to geomagnetic latitudes and field lines. Magnetic latitudes are, however, scientifically more critical in the equatorial–mid-latitude regions. DHANANJAY RAVAT

magnetohydrodynamic waves in the Earth (MHD waves)

Magnetohydrodynamic waves (*Alfvén waves*) are produced by coupling forces between the magnetic field and highly conductive fluids. To understand magnetohydrodynamic waves, let us first consider purely magnetic waves. If we replace the usual strings of a guitar with strings made up of a permanently magnetized material and pluck one of the strings, we can observe the physical vibration of the string. Because the string is magnetized, the magnetic field surrounding it will also vibrate with the string. A magnetometer placed near the guitar will be able to detect the changes in the intensity of the magnetic field with time that are created by the oscillations of the string. If we use a continuously recording, fluxgate-type magnetometer (see *geomagnetic measurement*), the measurements of the magnetic field intensity will vary systematically in time, with a period corresponding to the mechanical vibration frequency of the string. We can thus show that magnetic waves can be generated by the vibration of a magnetized string.

Let us imagine that we could further replace the strings with the magnetic lines of force (the field lines) themselves, but could retain their elastic properties. Also, to pluck a string, let us squirt jets either of charged particles or of highly conductive fluid across the string. The permanently magnetized string that we first considered vibrates when plucked, and so would the imaginary elastic field line when the fluid was squirted across it. Once again, we could deduce with fluxgate magnetometers that waves were generated within our imaginary string (the field line) and that they travelled most efficiently along the string. These are Alfvén waves.

The outer core of the Earth is not permanently magnetized because it is at a temperature above the *Curie temperature* (the temperature below which a material can become magnetic) of its constituents, a molten alloy of iron and nickel. The electrical conductivity of this liquid is high enough to conduct the electric currents that generate the Earth's magnetic field (see *magnetic field*). Physical motion of the fluids (i.e. hydrodynamics) can 'carry' the magnetic field along. This ability of a highly conducting material to carry the magnetic field along with it is known as the frozen-in flux model, the subject of H. Alfvén's Nobel-prize-winning theorem. The complicated oscillatory motions of the conducting liquid are influenced by the Earth's rotation and the thermo-compositional convection in the outer core (leading to more complex Magnetic–Archimedean–Coriolis (MAC) waves, so named by S.I. Braginsky in the 1960s). In addition to MAC waves, there are also other kinds of MHD waves. In the guitar-string experiment we needed continuously recorded magnetic data to determine the frequency of vibration of the string. Similarly, scientists who study magnetic processes in the core need at least a semi-continuous record of variations in the Earth's magnetic field. Such data come mainly from magnetic observations and from certain palaeomagnetic investigations. Cyclic variations in the observed values can help in evaluating physical models of the fluid motion in the outer core; with

sufficient data, it may perhaps be possible to elucidate the origin of the Earth's main field.

DHANANJAY RAVAT

magnetostratigraphy

Magnetostratigraphy is the application of the chronology of reversals in polarity of the geomagnetic field to the study of the stratigraphy of layered materials (i.e. sedimentary and volcanic rocks). An essential criterion for accurate magnetostratigraphic work is a reliable timescale for the reversal chronology of the Earth's magnetic field: hence the development of the geomagnetic polarity timescale (GPTS) (Fig. 1), in itself a remarkable story. Early work in palaeomagnetism demonstrated that some rocks were magnetized in a direction opposite to that of the present field. For example, a north-seeking compass needle, as we think of it today, would have pointed to the south pole. One explanation required that the geomagnetic field reversed itself at least once, if not several times, in the geologic past. Alternatively, it was argued that at least some rock types are capable of self-reversal, where the magnetization acquired is opposite to the field applied. The possibility that the geomagnetic field itself does reverse has important consequences for processes in the core and lower mantle and, as noted in the 1920s, is testable because, worldwide, rocks of the same age should have the same polarity. Research provided convincing evidence that the geomagnetic field has reversed its polarity several times in the past, and a crude GPTS was born.

The discovery of the relationship between alternating stripes of ocean floor that were normally or reversely magnetized and reversals in the geomagnetic timescale lead to the platetectonics paradigm and provided the opportunity for substantial refinements in the GPTS. Although first-order estimates of age durations of polarity magnetozones could be provided by date on marine magnetic anomalies, the absolute ages of magnetozones are based on independent geochronologic information (e.g. dating of lava flows just younger than and just older than a specific reversal) from the terrestrial record, and are constantly being refined. More recently, the combination of high-precision $^{40}Ar/^{39}Ar$ isotopic age spectra and magnetic polarity data for igneous rocks has led to improved estimates of ages of parts of the polarity timescale, particularly for the Cenozoic. For example, the age of the most recent complete reversal, from the reversed polarity Matuyama magnetozone to the present normal polarity Brunhes magnetozone, has been revised from 730 000 years to about 780 000 years. Further research will significantly improve the chronological framework provided by the geomagnetic polarity timescale and thus estimates of ages of sequences of rocks by magnetostratigraphy and geological processes based on the polarity timescale. In addition, detailed palaeomagnetic records from some sequences of lava flows and recent sediments have revealed the morphology of the transitional part of a reversal. Rather than an actual 'flipping' of the main dipole, the short transitional period (estimated to be about a few thousand years in duration) is characterized by a decay of the main dipole to an almost non-existent state, and domination by smaller, non-dipole components.

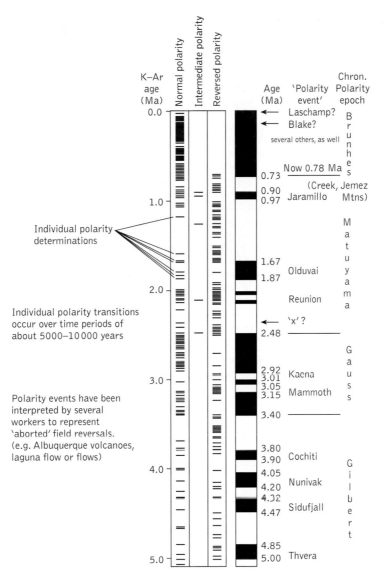

Fig. 1. Part of the geomagnetic polarity timescale (GPTS) for the latest Cenozoic. In the timescale, black refers to a period of time (magnetozone) where the Earth's field is dominantly normal polarity; white to a magnetozone where the field is dominantly of reverse polarity. At the present, that is, in the Brunhes normal polarity magnetozone, for example, there is evidence for at least five different and independent short-lived, unsuccessful field reversals, the most recent of which was about 30 000–40 000 years ago. This geomagnetic polarity timescale was compiled by Ed Mankinen and G. Brent Dalrymple of the US Geological Survey in 1979. The ages of many of the magnetozone boundaries have been refined since this compilation. For example, the age of the Matuyama to Brunhes polarity reversal is now considered to be about 780 000 years.

Magnetostratigraphic studies have several goals, among which are:

(1) defining the polarity history of the Earth, in particular prior to about 160 million years, the age of the oldest oceanic lithosphere yielding interpretable marine magnetic anomaly data;

(2) time correlation of different sections of strata; and

(3) determination of absolute ages of overall sections of strata where independent information (e.g. isotopic age data) is available.

For any magnetostratigraphic study, the approach should be no different from that of conventional palaeomagnetism, with the caveat that the sampling strategy is to obtain a polarity record of a rock section that is as complete as possible. Several factors, including the amount of time missing in the rock sequence, the spacing of individual measurements, and the fidelity of the process by which magnetization is acquired, make magnetostratigraphic studies challenging and complicated. At each stratigraphic level, the quality of the magnetization signal and hence the reliability of the polarity determination must be documented. Furthermore, it must be demonstrated that the rocks possess a primary magnetization. This is relatively straightforward for sequences of lava flows but is more difficult in sedimentary rocks. As an example, a generally high remanence intensity and overall abundance in the geologic record have made haematite-cemented detrital sedimentary rocks ('red beds') a focus of much magnetostratigraphic as well as palaeomagnetic research. Perhaps by default, these rocks have also been the focus of considerable controversy that bears on the interpretation of magnetostratigraphic data. At the centre of this debate is the carrier and age of the magnetization characteristic of the rocks. Rock-magnetic data indicate that the magnetization is typically carried by haematite, but the

haematite could occur as detrital grains, whose magnetic moments aligned themselves with the ambient field during deposition, or as the interstitial cement, which could have been precipitated at any time after deposition of the sediment. If a consensus is emerging, it is that the cement in many red beds is acquired relatively early after sedimentation; many red beds are thus considered to yield magnetostratigraphies of reasonably high fidelity.   JOHN WM GEISSMAN

Further reading

Butler, R. F. (1992) *Paleomagnetism: magnetic domains to geologic terranes*. Blackwell Scientific Publications, London.

Tarling, D. H. (1983) *Palaeomagnetism*. Chapman and Hall, London.

magnetotelluric prospecting Naturally occurring variations in the Earth's magnetic field induce eddy currents in the Earth that are detectable as electric (or *telluric*) field variations on the surface. The magnetotelluric (MT) method is an electromagnetic (EM) technique for determining the resistivity distribution of the subsurface from measurements of natural time-varying magnetic and electric fields at the surface of the

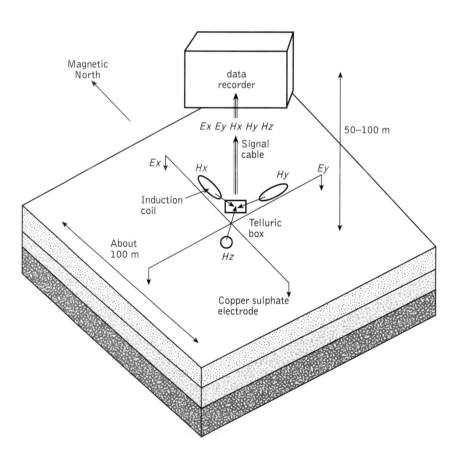

Fig. 1. Typical field set-up for magnetotelluric depth sounding.

Earth. The ratio of the horizontal electric field to the orthogonal horizontal magnetic field (termed the *EM impedance*), measured at a number of frequencies, gives Earth resistivity as a function of frequency or period, resulting in a form of depth sounding. This is a wide-band depth-sounding technique; the frequencies utilized range from about 10^4 to 10^{-6} hertz (Hz). The method is suitable for shallow as well as deep geological investigations. The depth of investigation in MT is a function of subsurface resistivity and frequency (or the inverse, period) of the EM signals. The depth of sounding can be roughly related to frequency by the use of the *skin depth*, which is defined as

$$\delta \approx 0.5 \text{ (resistivity/frequency)}^{0.5} \text{ km,}$$

where resistivity is expressed in ohmmetre and frequency in hertz. MT can probe great depths (up to 600 km) by utilizing sufficiently long periodicities.

Natural time-variations of the Earth's magnetic field of frequencies greater than 1 Hz are of atmospheric origin and are caused by global thunderstorm activities. Some thunderstorm energy is converted into EM fields which are propagated with slight attenuation in the Earth–ionosphere interspace as a guided wave. At a large distance from the source this is a plane wave of variable frequency. The MT signals of frequencies lower than 1 Hz are of magnetospheric origin (that is, they result from the interaction of solar wind and the Earth's magnetic field).

In the *scalar* MT method, the electric field in one horizontal direction and the magnetic field in an orthogonal horizontal direction are measured simultaneously at a surface

location. The *tensor* MT method is more popular. This requires simultaneous observation of two orthogonal horizontal (usually north–south and east–west) components of the magnetic field (Hx, Hy) and the electric field (Ex, Ey) and the vertical magnetic field component (Hz) at a surface location (Fig. 1). Magnetic sensors are squid magnetometers or induction coils consisting of several hundred turns of fine copper wire wound over a ferrite core or mu-metal of high magnetic permeability. Electric sensors are non-polarizing electrodes: commonly copper rods inserted in supersaturated copper sulphate solution or lead rods in supersaturated lead chloride solution in porous pots.

Data processing entails converting the field records into EM impedance (Z) data and determining apparent resistivity and phase information from Z for the various observational frequencies. The EM wave impedance is obtained as $Z = E/H$. This is a complex quantity (it has amplitude and phase). Since two mutually perpendicular horizontal components of E and H are measured in the field, Z can be defined for any given direction in *tensor* MT. For example, the electric field in the north–south direction and the magnetic field in the east–west direction can be used to define the impedance $Zxy = Ex/Hy$, while the electric field in the east–west direction and the orthogonal magnetic field component can be used to define the impedance $Zyx = Ey/Hx$. Zxy and Zyx are equal if the subsurface is homogeneous and isotropic. If the ground is heterogeneous, both impedances differ in magnitude and phase, as in the example presented in Fig. 2. In general, the magnetotelluric impedance is a tensor whose values depend on the direction of measurement. The components of the impedance tensor can be rotated mathematically to attain maximum values in the direction of linear conductive subsurface structures (and minimum values in the perpendicular direction). In MT parlance, the direction in which the impedance tensor elements are maximized during rotation is termed the *geoelectric strike* or *azimuth* of the impedance tensor. It usually coincides with the trend of a major geological feature.

The magnitude of Z is used to define an interpretative quantity called the *apparent resistivity*, ρ_a, which for a given EM wave frequency or period (T) is obtained as, $\rho_a = 0.2 T|Z|^2$.

Apparent resistivity is in ohm.m (symbolically written as Ωm) when the wave period T is in seconds (T is the reciprocal of frequency in hertz), E is in mV/km, and H is in nanotesla or gamma.

The phase of Z (symbolically denoted by θ) is obtained as the ratio of the imaginary to real parts of the complex impedance, i. e. $\theta = imag Z/real Z$ and is in degrees.

The relationship between the vertical magnetic field Hz and the horizontal magnetic components can also be used to estimate the geoelectric strike (and to define directions of current concentrations or induction arrows in MT parlance). The differences in the strike directions determined by impedance tensor rotation and vertical–horizontal magnetic field relationships serve as indicators of the three-dimensionality of subsurface conductivity structures.

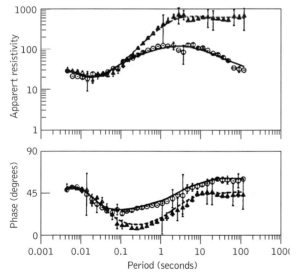

Fig. 2. Example of single-station MT apparent resistivity and phase sounding curves for two orthogonal directions. The data recorded in the east–west and north–south directions are shown by the triangular and round symbols respectively.

Mathematical modelling techniques exist for interpreting MT data (commonly apparent resistivities and phases). When the subsurface consists of horizontally stratified rock sequences, the resistivity varies significantly only with depth and the resistivity structure is said to have a one-dimensional (1-D) physical property distribution. The parameters sought in 1-D interpretation are layer resistivities and thicknesses, or alternatively conductivity–thickness products. When the geological sequences are not horizontally stratified, the resistivity structure is described as being three-dimensional (3-D) and may vary in all directions. There are, however, many situations where tectonic processes have imparted strong structural lineations in subsurface materials (e.g. major faults, angular unconformities, and discordant intrusive bodies) so that there is a well-defined strike. The gross geoelectric structure is then taken to be approximately 2-D (i.e. it changes only across strike and at depth). 1-D, 2-D, and 3-D data interpretations are carried out by interactive forward or automatic inverse numerical modeling on computers.

Geological and environmental applications

Mineral, hydrocarbon, and geothermal resources and contaminant dispersal processes in the subsurface are all genetically related by water and heat. The physical property most affected by changes in subsurface fluid content and temperature is electrical resistivity. Hence, MT can be used in routine exploration for natural resources and in environmental investigations.

The geological basis for employing MT in hydrocarbon exploration is the fact that prospective basins tend to have successions of source or cap rocks (dominantly shaly formations) and reservoir rocks (typically sandy formations and fractured crystalline rocks such as limestones and dolomites), which have contrasting electrical properties: shales and other fine clastic rocks are highly conductive (generally <30 Ωm), whereas sandstones and carbonate rocks are highly resistive (usually >50 Ωm). In a typical hydrocarbon trap, oil and gas are insulators and rest on highly conductive brines; such a trap may sometimes provide good targets for the MT method especially in occurrences at shallow depth. The MT method also has good potential for imaging deep sedimentary basins, since a broad depth range (approximately 10 m to 600 km) can be achieved by selecting appropriate wave frequencies or periodicities. A frequency bandwidth of about 10^{-4} to 10^3 Hz (i.e., 10^{-3}–10^4 seconds) is used in typical basin-evaluation studies. Basin exploration entails the determination of structural and stratigraphical variations down-dip. This can be achieved by performing depth soundings over a network of stations in the prospective region. The interpretative model will define the thickness of the sedimentary cover, the approximate distribution of the resistive and conductive members of the sedimentary sequence, and hence the possible location of reservoir and source or cap rocks.

The MT method also finds application in studies of large-scale geological features such as crust and mantle structure, the structure of mountain belts, and the structure of major rift valleys. It can also be used for imaging ancient and modern volcanoes.   MAXWELL A. MEJU

Further reading

Cagniard, L. (1953) Basic theory of the magnetotelluric method of geophysical prospecting. *Geophysics*, **18**, 605–35.

Vozoff, K. (1972) The magnetotelluric method in the exploration of sedimentary basins. *Geophysics*, **37**, 98–141.

major elements A major element is often defined as an element that occurs in the Earth's crust in an average concentration greater than 1.0 wt %. This definition restricts the number of elements to eight: oxygen, silicon, aluminium, iron, calcium, magnesium, sodium, and potassium in decreasing abundance. Other elements are, however, also fairly abundant; and the definition therefore is commonly extended to include titanium, manganese, hydrogen, and phosphorus.

Major element data are useful in several ways, ranging from simple characterization of rock types to quantitative assessment of geochemical processes. One of the principal purposes for obtaining major element data is the classification of rocks, particularly of igneous rocks. The characterization of volcanic rocks is often based on whole-rock major element data, whereas the nature and abundance of the minerals present in a given rock provide the basis for a widely used nomenclature of plutonic rocks. This difference is due to the fact that most volcanic rocks are microcrystalline or partly glassy, and thus it is more difficult to classify them by their mineralogical composition. In order to characterize volcanic rocks, major element data are used in a variety of classification schemes that use discrimination diagrams based on oxide or cation contents (in weight % or mole % respectively), or on the so-called norm calculations, which yield, in conjunction with several assumptions, a theoretical mineral content derived from the chemical analysis of the rock. Major element data are of only limited importance in the classification of sedimentary rocks (e.g. some clastic sediments), but may help in determining the original nature of a metamorphic rock by comparing its composition to that of present-day igneous or sedimentary rocks. This approach is, however, valid only if the metamorphic rock has not been altered or weathered, and if it can be shown that the metamorphism was isochemical. It may be difficult to demonstrate this, because several major elements (particularly potassium, sodium, and hydrogen) have been shown to be easily removed from, or added to, a rock during metamorphism (e.g. dehydration of sedimentary rocks or metasomatism). In both igneous and metamorphic rocks the major element content plays an essential role in constraining the possible mineral assemblages as well as the mineral compositions at a given pressure and temperature. Major element data are commonly displayed graphically in variation diagrams, which are invaluable in unravelling possible correlations and trends between different elements in a data set. From existing interrelationships it is often possible to infer and model quantitatively geochemical processes such as the mixing of two or more components, fractional crystallization, partial melting, assimilation of country rocks by an ascending melt, or element mobility. The use of variation diagrams is not restricted to major elements. They are particularly powerful tools when both major and trace element data are included.

Oxygen and hydrogen play an important role in modern geochemistry, because their stable isotopes may be fractionated as a consequence of mass differences between the isotopes. Determination of isotope ratios ($^{18}O/^{16}O$, D/H) in minerals and rocks thus provides insight into a variety of processes including reaction mechanisms, diffusion, evaporation, and fluid–rock interaction. In addition, the stable isotopes of oxygen and hydrogen are used as palaeothermometers and as tracers to determine the source of an element. Only one major element, potassium, is used for radiometric dating of rocks (the K–Ar method).

Major element concentrations (traditionally expressed as oxide wt %; O not analysed) are commonly determined by X-ray fluorescence analysis. The determination of the hydrogen (or H_2O) content and the distinction between FeO and Fe_2O_3 require other analytical methods, such as conventional wet-chemical, colorimetric, or spectroscopic techniques.

RETO GIERÉ

major fossil groups: their origins

The major fossil groups are those that have resistant hard parts and are thus preservable. They are not representative of the entire range of organisms that have lived, and many important taxa have left no fossil record. The major groups represented are invertebrates, an informal division of the kingdom Animalia that ranges from some organisms almost as simple as protists to highly complex animals such as insects. The number of invertebrate phyla is probably close to fifty, of which only eight have a significant fossil record. In addition the phylum Chordata (which includes the vertebrates) also has a significant fossil record.

A major problem in palaeontology is the origin of invertebrate phyla from a presumed eukaryotic (unicellular) ancestor at some time in the Precambrian. Evidence for a solution to this problem comes from the fossil record and also from the comparative anatomy and embryology of living representatives. Comparative anatomy shows that the invertebrates can be divided into groups by the way in which the coelomic cavities (cavities between the endoderm and ectoderm) formed. If this scheme is taken as a basis, the most primitive group consists of Protista, Porifera (sponges), and Cnidaria (corals). From these developed three other groups: Chordata and Echinodermata (starfish, etc.); Bryozoa and Brachiopoda; and Arthropoda, Mollusca, and Annelida (worms). The sequence of development of the mouth and anus proceeds in opposite directions in the Chordata–echinoderm group (deuterostomes) and the arthropod–annelid–mollusc group (protostomes), while the bryozoan–brachiopod group (lophophorates) shows affinities with both.

The fossil record of the earliest metazoans consists of trace fossils that can be recognized at the base of the Ediacaran interval about 680 million years ago. These show a change through the interval from simple shallow horizontal burrows

to more complex horizontal and vertical burrows, and finally to arthropod traces. A fauna of complex metazoans found in the Ediacaran fauna consists of soft-bodied organisms of unusual design (the Vendozoa) that may not be directly related to the development of any later taxa. At the base of the Cambrian, a new fauna of animals with hard parts appears. This Tommotian fauna contains sponges, brachiopods, primitive molluscs, the calcareous archaeocyathids, and other organisms of unknown affinity. Shortly after this, the 'Cambrian Explosion' marked the rapid development and diversification of invertebrate taxa.

On the basis of this information it has been proposed that the earliest metazoans were probably wormlike organisms, similar to the larvae of modern cnidarians and developed from flagellate protists. From this form the cnidarians and poriferans developed; and after the development of a coelom gave more rigidity to the body, the wormlike ancestral form was able to live infaunally and leave burrowing traces. A necessary response to the requirements of locomotion is segmentation. Once this had appeared, the protostomes could develop. Deuterostomes and lophophorates did not develop full segmentation, and the lophophorates might have developed directly from filter-feeding worms. Deuterostomes were a later development, for echinoderms are not present until Early Cambrian times, and chordates, which probably developed from a swimming ancestor, are not present until the Middle Cambrian.

DAVID K. ELLIOTT

mammal-like reptiles

The mammal-like reptiles together with the mammals form a monophyletic group (ancestor and all its descendants) termed the synapsids. They are all characterized by having a single skull opening behind the orbit. The earliest synapsids appeared in rocks of Late Carboniferous age at the same time as the first members of the main reptile group, the diapsids. Throughout their history the synapsids were quadrupedal (four-footed) terrestrial organisms that showed steady improvement both in their locomotor ability and the differentiation and specialization of their teeth. These advances helped them to become the dominant terrestrial carnivores and herbivores of Permian and Early Triassic times. They consist of two successive groups, the pelycosaurs that occur from the Late Carboniferous through the Permian, and the therapsids that occur through the Permian and Triassic and into the Jurassic, and which gave rise to the mammals.

The pelycosaurs (Fig. 1) were initially all carnivorous, but they subsequently developed herbivorous forms. From relatively small ancestors, pelycosaurs developed into animals up to 3.5 m in length. Reconstructions of their skeletons show that the massive front part of the body was supported on sprawling front limbs while the hind limbs had a greater range of movement although they were still sprawling. The ankles were poorly defined and could not provide much forward thrust as the feet were extended laterally. The spine was stiffened to allow the movement of the hind limbs to be converted into forward thrust with little lateral body movement. This

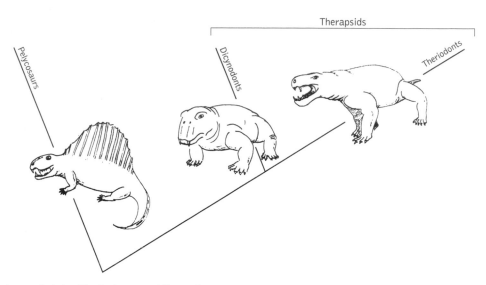

Therapsids

Pelycosaurs

Dicynodonts

Theriodonts

Fig. 1. Cladogram of relationships for the mammal-like reptiles.

method of progression circumvented a problem that constrains the endurance of many quadrupedal reptiles. David Carrier of the University of Michigan has shown that the twisting of the body that occurs as most amphibians and reptiles run compresses each lung alternately and prevents the animal from breathing while running. Lizards, therefore, progress in short rushes with frequent stops to breathe. The stiff spine and sprawling gait of the pelycosaurs, although resulting in a more ponderous progression, was actually an improvement, allowing them to keep running for longer periods.

A characteristic pelycosaur was *Dimetrodon*, a large predator from the Permian of Texas. This animal possessed tall neural spines supporting a web of skin forming a 'sail' along its back. Living reptiles are ectothermic, that is, they need external sources of heat to regulate their body temperature. Because of this their internal temperature can fluctuate with the external temperature. To counteract this, living reptiles bask in the sun and orient their bodies for maximum heat absorption. This behaviour suggests that *Dimetrodon* and related forms with similar 'sails' may have used them as surfaces to pick up heat rapidly when needed, or to radiate it when they became too warm. The pelycosaurs included large herbivores such as *Edaphosaurus* that also possessed large sails supported by neural spines.

Drier conditions in the Late Permian that resulted from the amalgamation of the continental blocks contributed to the extinction of the pelycosaurs. However, a second group of synapsids, the therapsids (Fig. 1), evolved from them in the Late Permian and became important Triassic tetrapods consisting of several carnivorous and herbivorous lineages. One group of therapsids, the dicynodonts, were herbivores with a worldwide distribution. They were particularly abundant in the Late

Permian, where they constitute 90 per cent of the therapsid diversity. They are characterized by a short snout and the loss of almost all their teeth, which were replaced by a turtle-like beak used for cropping vegetation. Ranging from rat- to cow-sized, and with specializations for a variety of modes of herbivory, their success was due, at least in part, to their ability to replace the cutting surface of the jaws continuously, an improvement on the intermittent tooth replacement of most reptiles. Another group of therapsids, the dinocephalians, included both carnivorous and herbivorous forms, some of which grew to very large size and had large flanges and crests projecting from their skulls. Herbert Barghusen from the University of Chicago has suggested that these were used in head-butting behaviour to establish dominance within a group.

The theriodonts include the ancestors of the mammals. They were specialist carnivores, some quite large, in which the teeth were well differentiated and the jaw musculature had been rearranged to provide a more precise occlusion of the teeth. This resulted in a much more efficient biting and chewing mechanism, the presence of which suggests that these animals had a high metabolic rate, although probably not up to mammalian grade; they are sometimes reconstructed with mammalian characteristics such as hair, for which there is no fossil evidence. The major skeletal character that separates them is the presence of a typical reptilian jaw articulation between the quadrate and the articular. This is lost in mammals, in which the joint moves to the dentary and the squamosal.

Although the synapsids were dominant in the Permian and in the Early Triassic, they were replaced during the Triassic by the diapsid archosaurs, particularly the dinosaurs. These were animals that had developed bipedality (moving on their hind limbs), an advance that eliminated the special problems of breathing while running that are inherent in quadrupedal reptiles, and one that led to their dominance through most of the Mesozoic while the mammal-like reptiles dwindled.

D. K. ELLIOTT

Further reading

Carroll, R. L. (1988) *Vertebrate paleontology and evolution.* W. H. Freeman, New York.

Colbert, E. H. and Morales, M. (1991) *Evolution of the vertebrates.* Wiley-Liss, New York.

mammals Mammals are not a diverse group, and yet they have achieved a dominant position in many ecosystems. There are estimated to be 4000 species of mammals alive today, a low number compared to the insects or flowering plants, and indeed only two-thirds of the current diversity of reptiles. The success of the mammals is measured rather in their wide adaptability, and by the sheer abundance of some species, such as humans and rats.

Class Mammalia is divided into three unequal groups, the Subclasses Monotremata, Marsupialia (or Metatheria), and Placentalia (or Eutheria). The monotremes, the duck-billed platypus and the echidnas of Australasia, are characterized by the fact that they lay eggs, the retention of a primitive reptilian character. Marsupials, such as the opossums of the Americas and the kangaroos, koalas, and wombats of Australasia, generally give birth to tiny immature young which then complete their developments in a pouch. The placental mammals, by far the largest group, and consisting of all other mammals from mice to elephants, aardvarks to zebras, and bats to whales, give birth to well-developed young which have been nourished by means of a placenta in the maternal womb.

The success of the mammals: chewing and homeostasis

It is impossible to point to a single feature of mammals that can explain their great success. Mammals are distinguished from their reptilian forebears by a number of anatomical and physiological characters, and some of these appear to have led to the great adaptability of the group.

Perhaps mammals owe their success to their teeth. Unlike reptiles, amphibians, and fishes, which have rather uniform teeth, mammals have differentiated teeth, snipping incisors at the front, pointed piercing canines to the sides, and broad crushing and shearing cheek teeth (premolars and molars) behind (Fig. 1). Tooth differentiation allows mammals to process their food efficiently: less is wasted and it is cut up or ground into fragments before it is swallowed. The jaws and teeth are modified in other ways.

Unlike reptiles, mammals generally have only two sets of teeth during their lives, a milk set and an adult set. Reptiles retain the primitive style in which teeth are inserted continuously as old ones wear out. This might seem to be a better system than ours; indeed, continuous tooth replacement means no visits to the dentist. However, mammals are forced to minimize the amount of tooth replacement because of the precise occlusion (fitting) between the upper and lower cheek teeth. The benefits of close interlocking between the crests (cusps) and troughs of the premolars and molars, the fact that this arrangement allows mammals to chew and comminute their food, thereby aiding digestion, far outweigh the advantages of being able to shed worn-out teeth.

Reptiles cannot chew; their jaws generally merely open and shut in a single orientation, rather like a simple hinge. Mammals have complex rotatory and sliding jaw joints, and the mandibles (lower jaw) may move sideways and backwards and forwards over considerable distances. Parents and nannies used to urge children to chew their food a hundred times before swallowing; this injunction was intended to instil good manners and to prevent indigestion, but chewing means that the digestive processes begin even while the food is in a mammal's mouth. Reptiles and birds gulp their food down and hence lose much nutritive value.

The mobile jaw joint and differentiated teeth of mammals have been cited as a major reason for the success of the group as a whole, and as an explanation for their wide diversification. Mammals have a broad range of diets, including fish, shellfish, plankton, flying insects, ants, worms, snails, other land vertebrates, grass, leaves, roots, wood, fruit, seeds, and nectar. The teeth are highly variable and specifically adapted for the diet: needle-like piercers in insectivores, filtering curtains in plankton-eating whales, and broad ever-growing crushers in grasseaters (Fig. 1). Some anteaters even have no teeth.

The second major adaptive complex of mammals concerns their control of body temperature. Mammals, independently of birds, evolved endothermy, commonly called warm-bloodedness. In other words, mammals produce a great deal of heat within their bodies by physiological means: indeed, it is estimated that fully nine-tenths of what we eat is devoted to temperature control (which is why a crocodile consumes one-tenth of what a human does). The heat production is closely monitored to maintain a precisely constant body temperature. In other words, mammals are endothermic ('internal heat') homeotherms ('constant temperature'). Different species of mammals maintain different temperature set points, but these are generally in the range 35–40 °C, and small deviations may be fatal. The adaptive advantage of such constancy of body temperature is in terms of biochemical functioning; mammalian enzymes have evolved to act at peak efficiency at the normal set-point temperature, whereas reptilian enzymes must be able to cope with wide fluctuations in body temperature, and they can rarely function ideally. The maintenance of constant body temperatures involves a complex array of homeostatic (feedback) mechanisms: when you are too hot, you sweat, you turn red (blood flows in the skin layers), and you seek a cool place; when you are cold, you seek a warm place, you shiver, your peripheral blood vessels close off, and your body generates internal heat as fast as it can.

The advantages of endothermic homeothermy are seen in the wide range of habitats occupied by mammals, wider than for any other group of animals of similar diversity. Mammals

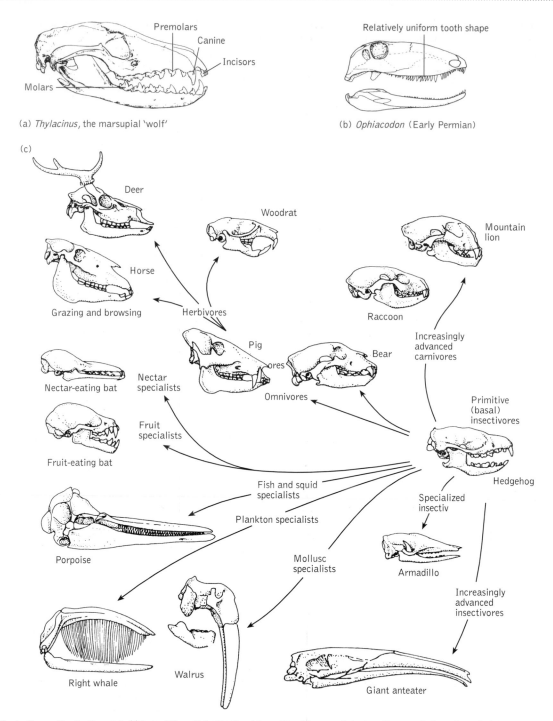

(a) *Thylacinus*, the marsupial 'wolf'

(b) *Ophiacodon* (Early Permian)

Premolars

Canine

Incisors

Molars

Relatively uniform tooth shape

(c)

Deer

Woodrat

Mountain lion

Horse

Grazing and browsing

Herbivores

Raccoon

Increasingly advanced carnivores

Nectar-eating bat

Nectar specialists

Pig

Bear

Fruit specialists

Fruit-eating bat

Omnivores

Primitive (basal) insectivores

Fish and squid specialists

Hedgehog

Plankton specialists

Specialized insectiv

Porpoise

Mollusc specialists

Armadillo

Increasingly advanced insectivores

Right whale

Walrus

Giant anteater

Fig. 1. Mammal teeth. Mammals (a) have differentiated teeth, while reptiles (b) generally have uniform teeth throughout the jaws. Mammalian teeth have diversified into a wide range of forms (c) which relate to diet. (Based on several sources.)

uniquely live from the Equator to the poles, in the depths of the oceans, underground, and in the treetops and skies. Mammals operate at night and during the day.

The third major reason for the success of the mammals is clearly their relatively large brains. In proportion to body size, mammals have brains about ten times the size of reptilian brains. Large parts of the brain are, admittedly, devoted to the homeostatic control of body temperature, but the bulk of the new brain volume is concerned with intelligent functioning. Many mammals live complex lives, for example in trees, and they must learn how to move around and find a diverse array of food. Most mammals are more able to solve problems than reptiles.

The large brain and intelligence of mammals requires a learning period after birth. Reptiles generally abandon their nests after the eggs have been laid, but mammals devote long periods to feeding and caring for their young. Indeed, reptiles typically produce large numbers of eggs, perhaps 20–50, and their evolutionary strategy is to invest in quantity rather than quality (r-selection strategy). Mammals do the opposite, producing few young, typically 1–8, and devoting effort to protecting them while they grow to adulthood (K-selection strategy). The large brain and need to learn require parental care. Parental care includes feeding, and a characteristic feature of mammals is that the mothers feed their young on milk produced by mammary glands (mammae, hence, mammal, and indeed 'mama').

The origin of the mammals

Teeth, warm-bloodness, and brains: which came first? All three adaptive complexes seem to have evolved in parallel in the ancestors of the mammals, but warm-bloodness may have come last. The ancestry of mammals can be traced back to the mid-Carboniferous, when the great vertebrate group, the Amniota arose. Amniotes are the fully land-living vertebrates, including reptiles, birds, and mammals, that broke away from the dependence on water shown by the amphibians. The Amniota divided early on into three lineages, the Anapsida (leading to turtles), the Diapsida (leading to lizards, crocodiles, and birds), and the Synapsida (leading to mammals).

The pre-mammalian synapsids, often called mammal-like reptiles, showed hints of mammalian characters from the start. Tooth differentiation was well advanced by the Late Permian. In addition, brain expansion was also under way. Reptilian skulls are typically in two parts, a large outer portion that supports the jaws and sensory organs, and a small braincase fitted fairly loosely inside at the back. The best analogy is a shoebox with a cigarette packet placed inside at one end. Permian and Triassic mammal-like reptiles showed various stages in a process of brain enlargement. The bones surrounding the brain moved outwards, and fused with the outer skull bones. In the end, in mammals, the brain dominates the back half or more of the skull, and the sense organs are concentrated in front.

Mammals arose from within a major clade of synapsids termed the Cynodontia. Cynodonts included a number of lineages in the Triassic, many of which probably looked very mammal-like (Fig. 2). The teeth were fully differentiated, the braincase bulged at the back of the head, and many of them walked in an upright manner, instead of the primitive sprawling gait of other reptiles. Later cynodonts may have had hair, providing an insulating layer, which is a necessary part of warm-bloodedness. These animals still laid eggs, and indeed all Mesozoic mammals probably did. The oldest mammal is Adelobasileus from the Late Triassic of Texas, and the transition from mammal-like reptile to mammal may seem rather subtle.

The key feature that is traditionally taken to define the Class Mammalia is the full expression of the mammalian jaw joint (Fig. 2). In reptiles, the jaw hinges between the articular bone at the back of the lower jaw, and the quadrate. In mammals, the jaw joint has shifted bodily to the dentary in the lower jaw and the squamosal. This extraordinary shift did not occur in one dramatic step: some later cynodonts had both joints close together and in operation at the same time. The old articular–quadrate joint of the reptiles did not disappear, but it changed function and became part of the chain of three ear ossicles ('little bones') of mammals. Reptiles have a single ear ossicle, the stapes or stirrup bone. Mammals have three, the stapes, the malleus (hammer), and incus (anvil), the last two being the old reptilian jaw joint still conserved deep within our ear. Ears and jaw joints are still closely associated: you can 'hear' yourself chewing.

Mammalian evolution

The oldest mammals (Fig. 2) originated at the same time as the dinosaurs. However, the dinosaurs rose to dominate the Earth for the remainder of the Mesozoic, a span of some 155 million years. During this time, the mammals radiated and evolved, but they could not make the breakthrough to becoming large or to diversifying their modes of like. Most were rat- or mouse-sized, and they filled nocturnal niches, mainly as insectivores. One group, the multituberculates ('many cusps'), were highly successful gnawing herbivores, and they outlasted the Mesozoic, surviving until the Oligocene. The multituberculates had gnawing incisors, a deep cutting premolar, and broad multi-cusped molars.

The monotremes, marsupials, and placentals arose in the Early Cretaceous. These modern groups were relatively rare during the Cretaceous; the early monotremes are known only by isolated specimens from Australia and South America, and the marsupials and placentals from North and South America, with some superbly preserved placentals from Central Asia. Even these mammals of modern aspect, although diversifying to some extent, did not exceed cat size, and they would have been unimpressive when seen beside the larger and more abundant dinosaurs with which they shared their habitats.

As is well known, the dinosaurs died out 65 million years ago, at a time of numerous other extinctions at the Cretaceous–Tertiary (K–T) boundary. Several groups of Mesozoic mammals also disappeared, and the marsupials nearly died out. Little is known of the evolution of monotremes, beyond the fact that

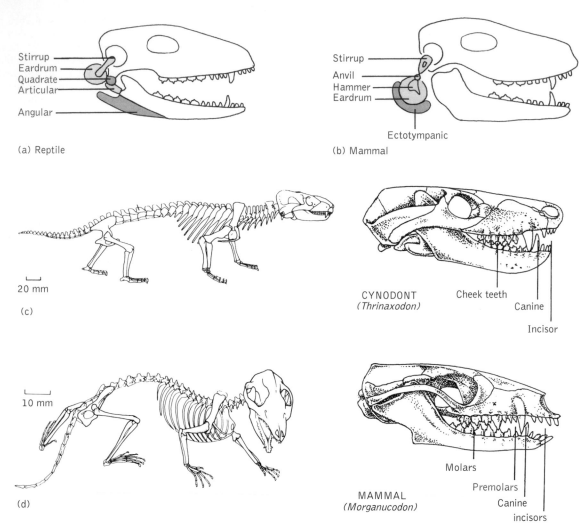

(a) Reptile

(b) Mammal

20 mm

(c)

CYNODONT
(*Thrinaxodon*)

Cheek teeth

Canine

Incisor

10 mm

(d)

MAMMAL
(*Morganucodon*)

Molars

Premolars

Canine

incisors

Fig. 2. The mammal jaw joint. The mammalian jaw joint evolved from the standard reptilian condition (a), where the joint is between the articular and quadrate, through an intermediate condition in many cynodonts, to the mammalian dentary-squamosal joint (b). The early cynodonts, such as *Thrinaxodon* (c), were mammal-like in many respects in comparison to the oldest mammals, such as *Morganucodon* (d). (Based on Kemp 1982.)

they appear to have existed always at low diversity, and are essentially restricted to Australasia.

The marsupials that survived the K–T event radiated in North and South America. They diversified particularly successfully in South America, giving rise in time to a wide diversity of carnivores, from opossums to marsupial sabre-toothed cats and dogs. At some time in the Eocene marsupials traversed Antarctica to reach Australasia. As Gondwana broke up, the link was broken, and the Australasian marsupials evolved in isolation, giving rise to the modern groups of koalas, bandicoots, wombats, and kangaroos. From North America, marsupials migrated to Europe, and from there to Central Asia and Africa. Those radiations were short-lived, and marsupials died out all over the northern hemisphere by Mid-Tertiary times. The South American marsupials radiated back into Central and North America when the Panamanian land bridge was formed 3 million years ago, but many of the more dramatic South American forms eventually died out at the end of the Pleistocene.

The placentals seemingly diversified through the K–T mass extinction, and the modern orders virtually all originated within the first 10 million years of the Tertiary, essentially

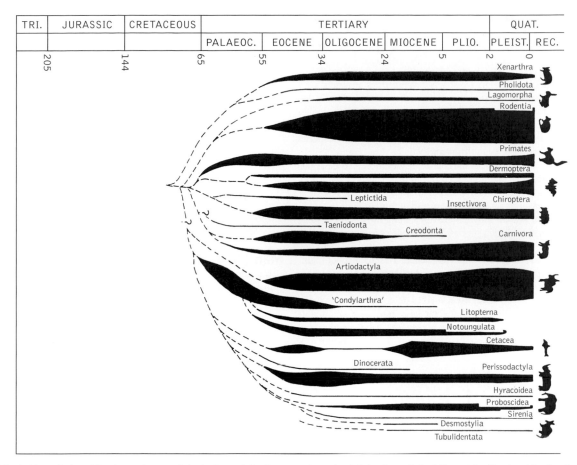

TRI.	JURASSIC	CRETACEOUS	TERTIARY					QUAT.	
			PALAEOC.	EOCENE	OLIGOCENE	MIOCENE	PLIO.	PLEIST.	REC.
205	144	65	55	34	24	5	2	0	

Xenarthra
Pholidota
Lagomorpha
Rodentia
Primates
Dermoptera
Leptictida
Insectivora
Chiroptera
Taeniodonta
Creodonta
Carnivora
Artiodactyla
'Condylarthra'
Litopterna
Notoungulata
Cetacea
Dinocerata
Perissodactyla
Hyracoidea
Proboscidea
Sirenia
Desmostylia
Tubulidentata

Fig. 3. The radiation of the placental mammals in the Cenozoic is often treated as a classic 'adaptive radiation'. Placental radiated rapidly after the extinction of the dinosaurs, and the modern diversity of form was established within the first 10 million years of the Tertiary. (Based on Gingerich 1984.)

during the Palaeocene and Early Eocene. The oldest bats, whales, proboscideans (elephants), perissodactyls (horses, rhinos), artiodactyls (cattle, pigs, camels), primates (monkeys, humans), insectivores (shrews, hedgehogs), rodents (rats, squirrels), and carnivores (cats dogs) are known from this early phase of radiation. In addition, there were numerous other major groups of Palaeocene and Eocene mammals, some plant-eaters, others meat-eaters, that radiated, some successfully, but died out during mid-Tertiary times, as the modern orders became established.

The radiation of the placental mammals (Fig. 3) is often chosen as a classic example of an adaptive radiation, in other words, a rapid diversification of a clade that is apparently driven by, or facilitated by, some adaptation or adaptations. To a large extent, the radiation was opportunistic: the extinction of the dinosaurs vacated large volumes of ecospace on land, and at first it was not clear that the mammals would

prevail. In certain parts of the world, huge hunting birds or fast-running crocodiles dominated the Palaeogene scene. The radiation can be dissected into several phases: (1) the opening of ecospace; (2) the initial diversification when new species seem to have arisen rapidly and with minimal competition from others; (3) a later phase when the ecospace was filling up, and more stable food chains and competitive interactions became established; and (4) a final phase when many groups died out and modern ecosystems became established.

Modern terrestrial mammalian ecosystems perhaps owe their shape to two further major environmental changes that occurred in the Mid- to Late Cenozoic. During the Oligocene and Miocene, climates worldwide became more arid, and areas that had formerly been covered by lush tropical and sub-tropical forests opened up and became grasslands. Great savannahs spread over much of the Americas, Europe, Asia and Africa. Mammals that had formerly lived secretive lives in

643

the forests either died out or adapted. Horses adapted: they had previously been small, terrier-sized animals that fed on leaves, and they evolved into taller, long-legged animals that fed on grass.

The second major event was the Pleistocene ice ages. In the northern hemisphere, ice sheets several times advanced over much of North America, as well as northern Eurasia. Pluvial and arid climatic changes affected the whole world. Land mammals underwent numerous large-scale migrations as climates changed, and some groups died out. Other ice-adapted species evolved, such as mammoths and woolly rhinos, and perhaps Neanderthal humans. At the end of the Pleistocene, some 11 000 years ago, large mammals died out in many parts of the world, perhaps as a result of environmental changes, perhaps partly as a result of the rapid worldwide spread of human hunters.

Current human activities are threatening the futures of some endemic mammalian species, for example certain primates that occur in small populations in Madagascar, the African apes, pandas, elephants, and other specialists. The majority of mammalian species, however, seem to be highly adaptable and rather immune to such dramatic environmental modification.

Mammals can never rival insects or microbes in terms of abundance or diversity, but their large body size and adaptability have made them highly successful over the past 65 million years.   M. J. BENTON

Further reading

Benton, M. J. (1991) *The rise of the mammals.* Apple Press, London: Crescent Books, New York.

Benton, M. J. (1997) *Vertebrate palaeontology* (2nd edn). Blackwell Science, Oxford.

Carroll, R. L. (1987) *Vertebrate palaeontology and evolution.* W. H. Freeman, San Francisco.

Gingerich, P. D. (1984) Primate evolution. In Gingerich, P. D. and Badgley, C. E. (eds) *Mammals. Notes for a short course,* pp. 167–81. University of Tennessee, Knoxville.

Kemp, T. S. (1982) *Mammal-like reptiles and the origin of mammals.* Academic Press, London.

Pough, F. H., Janis, C. M., and Heiser, J. B. (1989) *Vertebrate life* (5th edn). Prentice Hall, New Jersey.

Savage, R. J. G. and Long, M. R. (1986) *Mammal evolution.* British Museum (Natural History), London.

Man (hominids)
Hominidae, the mammalian family that includes modern humans and their immediately extinct ancestors, contains four genera: *Ardipithecus, Australopithecus, Paranthropus,* and *Homo.* The family is distinguished from the African apes (their closest living relatives) by a number of anatomical and behavioural traits. The most important of these are: (1) possession of a relatively large brain in which

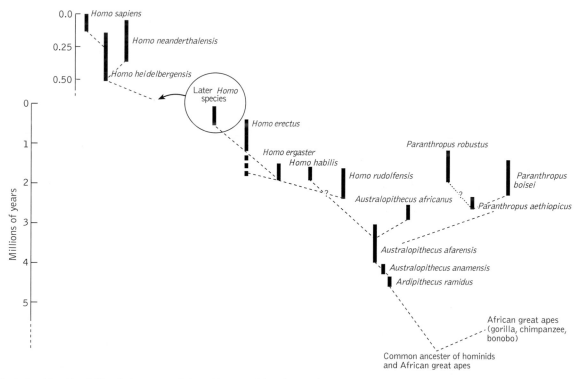

Fig. 1. Provisional hominid evolutionary tree. (Adapted from Wood (1994) *Nature.*)

the frontal and occipital lobes are especially enlarged, allowing more complex behaviour; (2) adoption of a bipedal gait, which involves a distinctive complex of skeletal features; and (3) a slow rate of postnatal growth and development, favouring complex social organization. The adoption of bipedalism as the main mode of locomotion was the fundamental anatomical and behavioural change that led to the divergence of the hominid lineage from that of the African great apes.

In recent decades, the volume of fossil material relating to the evolution of the hominids has greatly increased The modern multidisciplinary approach has begun to produce a clearer picture, not only of the hominids themselves, but also of the environments they inhabited. New and more precise dating techniques are making it possible to analyse how the hominids evolved anatomically and technologically against a background of palaeoenvironmental change. The intensive study of modern apes has established a whole set of links between anatomy and behaviour which can be used to interpret the fossil material. New biomolecular techniques have also added another avenue of research. While a number of areas of uncertainty remain, it is now apparent that the pattern of hominid evolution was more complex than had been thought (Fig. 1).

The most primitive hominid currently known, *Ardipithecus ramidus*, was found in a 4.4 million-year-old (Pliocene) horizon in Ethiopia. The pelvis and leg bones of *A. ramidus* are unknown, but cranial evidence hints that it was to some extent a biped. The key hominid locomotor adaptation, and consequently the divergence from the African ape lineage, therefore took place prior to 4.4 million years ago. This agrees with biomolecular data which supports a Late Miocene date (about 6 million years ago (Ma)) for the last common ancestor of the hominids and African apes. *Ardipithecus ramidus* may have been ancestral to all later hominids. New, but as yet undescribed, material should help clarify both relationships of this form with *Australopithecus* species and its mode of locomotion.

The ancestry of *Ardipithecus* is unknown and, without direct evidence, there has been considerable speculation on how and why the ancestral hominids became bipedal. The most popular hypothesis suggests that the shift in locomotor repertoire was promoted by an environmental change involving the break-up of the equatorial forest into discontinuous stands of trees. This might have forced the essentially arboreal proto-hominids to travel across open ground from one food patch to another. The available evidence indicates that *Ardipithecus ramidus* inhabited woodland environments. However, it is now beginning to appear more likely that bipedal locomotion evolved while the ancestral hominids inhabited wooded areas and not as an adaptation to open environments. Postural bipedalism, similar to that displayed by chimpanzees, could possibly have originated in arboreal or terrestrial feeding situations in wooded environments and was subsequently modified to locomotor bipedality. Whatever the cause, the adoption of bipedalism had profound behavioural implications. Bipedalism pre-adapted

the early hominids to life in more open environments in a number of ways. These included freeing the hands for other activities, increasing the visual horizon for a ground-dwelling animal in flat country, providing an energetically efficient means of terrestrial travel between widely scattered food resources, and minimizing the area of the body exposed to solar radiation when the sun was overhead, thereby reducing the thermal stress experienced in open country.

The genus *Australopithecus* contains three species, all from African Pliocene sites: *A. anamensis*, known only from Kenya in beds between about 3.9 and 4.2 Ma old: *A. afarensis*, found in Ethiopia, Tanzania, and possibly Chad in horizons between 3.9 and 3.0 Ma old; and *A. africanus*, which inhabited South Africa between 3.2 and 2.3 Ma ago. All three forms are characterized by a chimpanzee-like skull with a large face attached to a relatively small cranium (Fig. 2). The cranial capacity (a measurement of the volume of the brain) was very similar to that of a modern chimpanzee. All three species were relatively slender and lightly built animals, standing between 1.0 and 1.5 m high. *Australopithecus* species exhibit a high degree of sexual dimorphism; among several morphological differences, the males are considerably larger than the females. *Australopithecus* postcranial skeletons demonstrate that these forms were upright bipeds. A trackway preserved in volcanic ash 3.2 Ma old at Laetoli in Tanzania confirms that australopithecines were bipedal striding animals with a foot similar to that of modern humans rather than that of the apes. However, they also had relatively long arms, and this, together with some details of the forelimb anatomy, indicates that they may have spent a substantial amount of time climbing in trees. *Australopithecus* species lived in a variety of habitats, ranging from forested areas to relatively open environments.

The three species of *Paranthropus* (*P. aethiopicus, P. robustus,* and *P. boisei*) are considered to represent a specialized group of hominids. *Paranthropus aethiopicus* and *P. boisei* are east African forms, while *P. robustus* occurs in southern Africa. They were small-bodied animals, ranging in height from 1.1 to 1.4 m, but their long bones were more robust than those of *Australopithecus*. They are distinguished by the presence of pronounced facial buttressing and cranial cresting, an extremely robust mandible, and massive cheek teeth (Fig. 2). These characteristics form a unique functional complex which was an adaptation to powerful or repetitious chewing. It is believed that these animals ate a wide range of plant foods, including nuts, seeds, and fibrous material. The postcranial skeleton appears to have been similar to that of *Australopithecus* and they also exhibit the same high degree of sexual dimorphism. *Paranthropus* diverged from the *Australopithecus* lineage prior to 2.6 Ma ago and survived until about 1.0 Ma ago.

Homo, the genus to which modern humans belong, appears in the east African fossil record about 2.4 Ma ago. Its origin was in some way related to a massive cooling of the world's climate which began about 2.6 Ma ago. The current model suggests that, as the east African climate became colder and more arid, the vegetation became sparser and this compelled

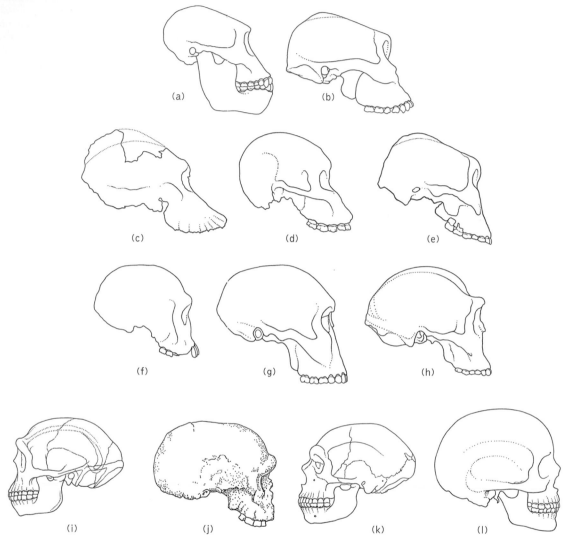

Fig. 2. Lateral views of hominid crania: (a) *Australopithecus afarensis*; (b) *Australopithecus africanus*; (c) *Paranthropus aethiopicus*; (d) *Paranthropus robustus*; (e) *Paranthropus boisei*; (f) *Homo habilis*; (g) *Homo rudolfensis*; (h) *Homo ergaster*; (i) *Homo erectus*; (j) *Homo heidelbergensis*; (k) *Homo neanderthalensis*; (l) *Homo sapiens*.

(Adapted from Aeillo and Dean (1990) *An introduction to human evolutionary anatomy*, Academic Press, except (j), adapted from Stringer and Gamble (1993) *In search of the Neanderthals*, Thames and Hudson.)

the ancestral *Homo* to become more mobile and consume a more varied diet. The defining characteristics of the genus are in a state of flux but it appears to contain relatively large-bodied forms with a large cranial capacity relative to the body size, thick cranial roof bones, and a parabolic dental arcade. The increase in cranial capacity was the second key event in hominid evolution, providing our ancestors with greater behavioural, and technological innovation. Some authorities recognize as many as seven species: *Homo rudolfensis,*

H. habilis, H. ergaster, H. erectus, H. neanderthalensis, and *H. sapiens.*

The oldest species, *Homo rudolfensis*, occurs in eastern and southern Africa in horizons dated at between 2.4 and 1.8 Ma. It exhibits some traits resembling *Australopithecus*, including the skull shape (Fig. 2), but the cranial capacity is relatively large and the brain had a fully *Homo*-like pattern. Like *H. rudolfensis*, *H. habilis* occurs in eastern and southern Africa but at sites dating between 1.9 and 1.5 Ma old. The cranial shape (Fig. 2) in this species resembles that of later *Homo* species but the cranial capacity is smaller than that of *H. rudolfensis*. The postcranial skeleton of both species is poorly known but appears to have been rather *Australopithecus*-like.

The earliest stone implements found so far in eastern Africa are about 2.5 Ma old and it is usually assumed that the appearances of *Homo* and artefacts are linked. It is probable that the appearance of lithic artefacts was preceded by a long period in which plants were used as raw materials for tools, similar to the digging sticks and twig probes employed by chimpanzees today. The earliest tool kit, termed the Oldowan, is characterized by choppers, crude implements fashioned from cobbles by striking off flakes from either side of the stone to create a sharp edge. Analysis of archaeological sites, which are essentially associations of artefacts and mammalian bones, supports the view that, while meat had begun to be used as a regular source of food, the hominids were scavenging carnivore kills.

About 1.9 Ma ago, the third species in the genus, *Homo erectus*, evolved, probably in eastern Africa. This had a long, low-vaulted and thick-walled skull with a massive, straight brow ridge (Fig. 2). The brain was larger and the face less projecting than in the earlier forms. This species was similar in stature to modern humans, and sexual dimorphism was much less than that exhibited by *Australopithecus* and *Paranthropus* species. Some authorities suggest that the earliest material is sufficiently different from later specimens to warrant allocation in a separate species, *H. ergaster*, but this conclusion is not universally accepted. The appearance of *H. erectus* has been linked to tectonic uplift and consequent climatic shifts in eastern Africa, but the evidence is not entirely convincing. Shortly after its appearance, this species migrated out of Africa, probably before 1.8 Ma ago; its remains are found in Israel, Georgia, China, Vietnam, and Java. While fossils of this species do not occur in Europe, archaeological finds suggest that it periodically inhabited the Mediterranean region. Its appearance outside Africa marks the third key event in hominid evolution. Surprisingly, in spite of its wide geographical distribution, *H. erectus* remained little changed for over a million years. Associated with some, but not all, *H. erectus* sites is a new, larger, and more sophisticated tool kit, the Acheulian, the characteristic implement of which is the large hand axe. Archaeological data indicate that this species was a hunter–gatherer. Some late *H. erectus* groups were able to control fire and possibly cooked some of their food.

In the late Early Pleistocene, a form intermediate between *H. erectus* and *H. sapiens*, sometimes called archaic *H. sapiens* or *H. heidelbergensis*, is found in Africa, Asia, and Europe. Regional differences can be detected amongst the available fossil sample and it is probable that a number of separate species may eventually be recognized. The details of the transition from *H. erectus* remain unclear and its date appears to have varied from region to region, ranging from about 800 000 to about 300 000 years ago. All the transitional forms retain the low forehead of *H. erectus*, but the thick brow ridges

curve over the eye sockets, the rear of the skull is more rounded, and the cranial capacity is larger (Fig. 2). Some populations of this group adapted to cool temperate conditions and they were clearly able to cope with and exploit a great diversity of habitats. The archaeological materials found in association with this species also exhibit variation, the tool kit being modified to suit local requirements. The first evidence for the construction of shelters appears at this time.

In Europe and western Asia about 300 000 years ago, *Homo heidelbergensis* evolved into a distinctive form, *H. neanderthalensis*. This species was heavier and more muscular than modern man, but, contrary to popular opinion, it had a completely upright stance and a physiology close to our own. Cranially, its main distinguishing feature was a long skull with a flat top, resembling *H. erectus* (Fig. 2). The cranial capacity was, on average, actually larger than that of modern man. The distinctive anatomy of the Neanderthals appears to be an adaptation to the glacial climate experienced intermittently in its geographic range during the Late Pleistocene. Associated with the Neanderthals is the Mousterian tool kit, comprising smaller and more refined implements than the Acheulian. The Neanderthals buried their dead, sometimes with grave goods, implying that they had some form of ritual or spiritual aspect to their life.

The place, time, and cause of the emergence of *Homo sapiens*, the fourth and final critical step in hominid evolution, remains controversial. On balance the available evidence supports the hypothesis that our species evolved over 100 000 years ago, either in southern or eastern Africa, and subsequently migrated to other parts of the world. It had reached parts of the Middle East by 90 000 years ago, China by 67 000 years ago, and Australia by 40 000 years ago. It did not enter Europe until after 35 000 years ago, after which it either displaced or replaced Neanderthals. Possibly it did not possess the requisite technology to deal with the climatic conditions in this area until this late date. Early *H. sapiens* is associated with a sophisticated stone and bone technology, exhibiting a large number of local variants.

ALLAN N. INSOLE

Further reading

Aeillo, L. and Dean, C. (1990) *An introduction to human evolutionary anatomy*. Academic Press, London.

Gamble, C. (1993) *Timewalkers: the prehistory of global colonization*. Harvard University Press, Cambridge, Mass.

Jones, S., Martin, R., and Pilbeam, D. (1992) *The Cambridge encyclopaedia of human evolution*. Cambridge University Press.

Leakey, R. and Lewin, R. (1992) *Origins reconsidered: in search of what makes us human*. Little, Brown, and Co., Boston, Mass.

Stringer, C. and Gamble, C. (1993) *In search of the Neanderthals*. Thames and Hudson, London.

Tattersall, I. (1995) *The fossil trail: how we know what we think we know about human evolution*. Oxford University Press.

manganese nodules When HMS *Challenger* made its epochal voyage of oceanographic research and discovery under the auspices of the Royal Society and the British Admiralty, scientists aboard made many unusual discoveries. One of the least expected discoveries was the presence of manganese-rich nodules sitting on the sea bed in the deepest parts of the ocean. During the voyage, from December 1872 to May 1876, 133 sites were sampled by dredging the sea floor. Many of the dredge hauls brought up curious black nodules that ranged from pea-sized to grapefruit-sized, but averaged about 3 cm in diameter. The *Challenger* scientists analysed the nodules and found them to be composed of the oxides of manganese and iron, and also to contain small amounts of copper, nickel, cobalt, and other minerals.

Manganese nodules were treated as little more than scientific curiosities until the period of intense oceanographic research that started in the years following the Second World War. Ocean-going geologists confirmed the *Challenger* observations and quickly discovered that vast areas of the sea floor are literally peppered with nodules, some of which have combined contents of copper, cobalt, and nickel of several per cent by weight. It was soon realized that the metals in nodules are a potentially vast mineral resource, and by the mid-1960s commercial interest in the nodules had developed. Several ways of mining the nodules were devised, and ways of processing them in order to recover their metal contents were developed. Commercial attention given to the economic potential of manganese nodules in the 1960s and 1970s raised government concerns in many countries and was a major factor in starting the discussions that led to the 1982 Convention of the Law of the Sea. The Law calls for an equal sharing of all marine mineral resources among both developed and developing nations. Commercial mining and processing of nodules has still not commenced (1999), but many experts anticipate that it will not be far into the twenty-first century before mining does start, most likely in the Pacific Ocean.

Although the nodules are generally called manganese nodules, following the terminology introduced by the *Challenger* scientists, it would be more accurate to describe them as ferromanganese nodules because they contain both manganese and iron compounds. Some of the iron is present as the mineral goethite, but much is amorphous ferric hydroxide. The principal manganese compound is δ-MnO_2, but the manganese minerals todorokite and birnessite are also commonly present, together with amorphous manganese dioxide. Copper, cobalt, and nickel are not present in the nodules as separate minerals. In part they are incorporated in the manganese minerals, and in part they are absorbed on the surface of the amorphous manganese dioxide. Internally, the nodules are layered, like the layers of an onion, and at the centre of each nodule is a small nucleus of foreign matter around which growth occurred. The nuclei are varied and include hardened fragments of deep-sea clay, diatoms, radiolaria, chips of basalt, sharks' teeth, and even whale ear-bones. Nodular growth is thought to be very slow. Radiometric dating of growth bands is not precise but different methods give concordant ages suggesting growth rates of a few millimetres in a million years. The slow growth rate is thought to enhance the absorption of heavy metals like copper and cobalt from sea water.

Closely related in origin to manganese nodules are the ferromanganese incrustations found on exposed rocks of the mid-ocean ridge, seamounts, and other places in the ocean where bare rock is exposed. Up to 20 cm thick, the encrustations are similar in composition to nodules, and like the nodules they sometimes contain potentially valuable concentrations of copper, nickel, and cobalt.

The compositions of both nodules and encrustations vary widely from place to place. The highest concentrations of copper, nickel, and cobalt seem to occur in the regions of the ocean where the rate of sedimentation is least and the nodule growth rate is slowest. These conditions occur on the abyssal plains of the central Pacific Ocean where the underlying sediments are red clays.

The origin of the nodules and the encrustations has given rise to much debate. Although the involvement of microscopic organisms in the precipitations of the manganese and iron-rich layers has sometimes been suggested, no clear proof of biological mediation has yet been demonstrated. Most scientists who have studied nodules now support an inorganic origin for them, with oxidation of iron and manganese in solution as the cause of precipitation.

There are two principal ways by which manganese, iron, and the other metals found in nodules reach the sea: by weathering of terrestrial rocks followed by river transport to the sea, and by deep-sea volcanism, principally along the mid-ocean ridges. The pathways followed by the metals once they enter the sea are less certain. Many nodules contain evidence of dissolution on the under surface where the nodule was in contact with sediment. This indicates that nodules may encounter reducing conditions in the sediments. Under such circumstances, iron and manganese would be reduced and go into solution. From such evidence it is hypothesized that the sediment pile is the main reservoir of manganese and iron and that there is a slow, upward flux, through the sediments, of dissolved manganese and iron. On reaching the sea floor the dissolved species are oxidized and precipitated on growing nodules.

BRIAN J. SKINNER

mangroves 'Mangrove' is the general term given to a variety of halophytic (salt-tolerant) trees and shrubs found in low-energy intertidal areas. While some of these species can tolerate relatively cool temperatures, the major mangrove forests are found in the tropics and subtropics, where the mean winter water temperature is over 20 °C. Like the salt marshes found in higher latitudes, mangroves grow best in low-energy, sheltered environments such as estuaries, river deltas, and less-exposed open coast areas. The extent of an individual mangrove forest is largely dependent on coastal relief: the largest expanses of mangrove forest are found along gently sloping shorelines such as those found on major river deltas. The Sundarbans, for instance, which is the largest continuous area of mangrove forest in the world, has developed on the extensive delta of the Ganges, Brahmaputra, and Meghna rivers. Tidal range is also important, since a high tidal range will flood a greater area of coast and allow a correspondingly bigger (and often more diverse) mangrove forest to develop.

The fine sediment that is characteristic of the low-energy environments in which mangroves grow is frequently oxygen-poor (or even oxygen-free). This is a result of the small grain size of clay and silt particles, which maximises water retention at low tide and so limits the penetration of oxygen into the sediment, and to the high quantities of organic material in the sediment, the breakdown of which depletes the sediment of oxygen. Owing to the oxygen-poor nature of the sediments (and to provide anchorage in the muddy substrate) mangrove trees and shrubs have extensive root systems that may be aerial (in the form of prop roots) or have numerous breathing roots (pneumatophores) which extend up above the mud surface. Once established, mangrove vegetation promotes rapid vertical build-up of sediments by the trapping of suspended sediments by the extensive above-ground root network.

Longer-term mangrove forest development is intimately linked to changes in relative sea-level. Under a stable, or falling, sea-level, mangroves will tend to prograde or expand seawards, as long as the supply of sediment is sufficient to allow the growth of new mangrove areas. Mangroves in deltaic areas, however, are often subject to a rising relative sea level, owing to land subsidence caused by sediment compaction (a process in which delta sediments compact under the weight of overlying sediment). Provided that the sediment supply is sufficient, the mangrove substrate can still build up rapidly, offsetting this rise in sea level. Where sea-level rise is more rapid or sediment supply is lower, mangroves will undergo erosion and landward migration. Over shorter timescales, mangrove development can be dramatically influenced by tropical cyclone activity. Because of their dominantly tropical and subtropical location, mangrove forests are often subject to damage from severe tropical storms. Trees may suffer direct breakage and leaf-loss, and extensive erosion of the muddy mangrove substrate may occur. In areas prone to regular tropical cyclones, the frequency of these storms may limit the extent to which the mangrove ecosystem can develop.

ANDREW B. CUNDY

mantle convection, plumes, viscosity, and dynamics

The theory of plate tectonics drastically changed our understanding of the dynamics of the Earth's interior. Until the theory of plate tectonics was developed, the primary means of transporting heat from the Earth's interior to the surface was thought to be by conduction. Rock is an extremely poor conductor of heat, and the observed heat flow at the Earth's surface was therefore attributed either to conductive heat flow from the near-surface or to radiogenic sources.

Conduction, unlike convection, requires no movement of material, and so the dynamics of the deep interior of the Earth would then have little effect on the observed surface processes and was of only academic interest.

In plate tectonics, large rigid plates float on a fluid plastic mantle. These plates move about the surface of the Earth at speeds of up to 10 cm per year. The motion of the plates has to be driven in some way. Furthermore, if mantle rocks are able to flow, then the transport of heat can be achieved more efficiently by convection. Heat can be transported to the Earth's surface from deep inside the Earth. This leads to the concept of a dynamic mantle composed of 'solid' rock that can behave as a viscous fluid over geological timescales. On a geological timescale, lithospheric plates behave as elastic solids and the mantle as a viscous fluid. If the motion of the plates is controlled by mantle dynamics, the Earth's deep interior holds the key to understanding the evolution of much of the Earth's surface.

The viscosity of the mantle

The mantle is known to flow like a fluid over geological timescales. We quantify the ability of a material to flow under pressure by its viscosity. The viscosity is defined as the ratio of shear stress to strain rate, and is thus a measure of the internal friction of a fluid: the more viscous a fluid, the more resistant it is to flow. Measuring mantle viscosity is not a trivial problem. Direct measurements are not possible, and the viscosity must therefore be inferred from surface processes.

Geologists in Finland and Scandinavia noticed that the sea-level appeared to be falling in Scandinavia: dating terraced beaches found in Scandinavia and Finland showed that during the past 10 000 years the sea level had dropped while the global trend was for a rise in sea level. This was due to the rapid removal of a large mass (the retreating Fennoscandian ice sheet which covered the region during the last Ice Age), resulting in the lightened lithospheric plate rising. This process, known as *isostatic rebound*, provides evidence for a flowing viscous mantle.

The rate of isostatic rebound enables the viscosity of the mantle to be estimated. Relatively small loads such as the Fennoscandian ice sheet yield estimates for the viscosity of the uppermost mantle. The rapid removal of much larger loads is required to estimate the viscosity of the lowermost mantle. Using these observations, the viscosity of the mantle is estimated to be reasonably constant at between 10^{20}–10^{21} Pascal seconds (Pa s) in the upper mantle and 10^{21}–10^{23} Pa s in the lower mantle (compared to the viscosity of water which is 10^{-3} Pa s). Although the viscosity of the mantle is high, it is still low enough for the mantle to flow over geological timescales. Using these viscosities, it is estimated that the mantle can overturn completely by convection once every 100 million years.

Studying mantle dynamics

Seismologists have succeeded in describing the deep structure of the Earth in some detail. We know that the lithospheric plates are on average 20 km thick in oceanic areas and 80 km thick in continental areas. The mantle extends from the base of the lithosphere to the D″ layer (pronounced 'dee double prime'), which acts as a narrow (about 200 km) buffer between the mantle and the core, at a depth of 2700 km. On a global scale the mantle can be thought of as having two distinct regions: the upper mantle contains many scatterers of seismic energy; the lower mantle is relatively homogeneous, with little scattering or anomalous refractors of seismic energy. The upper mantle and the lower mantle are separated by a weak seismic discontinuity (a jump in seismic velocity) at a depth of about 670 km. This weak seismic boundary also marks the lower boundary of observed seismic activity: no earthquakes have been recorded below this depth.

There are several weak seismic discontinuities in the mantle, and the nature of these discontinuities provides useful constraints on its dynamics. Seismic discontinuities may be a result either of compositional changes, where the main mineral constituents differ across the boundary, or of phase changes, where the composition stays the same but as a result of changes in temperature or pressure the crystallographic structure, and hence the physical properties, of the minerals change. The 670-km discontinuity is thought to be due to a phase change in the mainly iron and magnesium silicates that make up the mantle. There is a 'gradient' associated with a phase change that acts as a physical barrier to the flow of material across the boundary, which is similar to the contact between oil and water.

Seismic tomography provides a means of imaging the mantle using differences in seismic velocity or attenuation. These seismic properties can be related to more fundamental material properties such as temperature and density. The images produced by seismic tomography have become increasingly detailed, and it is now believed that descending slabs and even mantle plumes can be imaged with some confidence.

Understanding mantle dynamics requires detailed knowledge about the structure and composition of the Earth's deep interior. Unfortunately, seismology alone is not sufficient to understand mantle dynamics. Seismology tells us only about physical properties such as seismic velocity, attenuation, and, more indirectly, temperature. In order to understand how these properties relate to mantle dynamics we need to know something about the composition of the mantle. Mineral physics and laboratory experiments at high temperatures or pressures provide useful constraints on the probable structure of mantle minerals. The composition of these minerals is known from a handful of outcrops containing inclusions of mantle material. From geochemical studies of the radioactive isotopes that are present in volatiles released from volcanic eruptions the evolutionary history of the mantle can be pieced together.

Mantle convection

There is a strong correlation between the distribution of seismicity and mantle dynamics. Figure 1 shows a map of global seismicity. This pattern of seismicity, which can be interpreted as a map of dynamic surface features, owes a lot to mantle dynamics. Mid-ocean ridges bisect many of the world's oceans; the geography of the plates is intrinsically effected by mantle plumes; plate motions (and hence collisions) are controlled by mantle convection. All these processes depend intrinsically on mantle dynamics.

Oceanic crust is created at mid-ocean ridges: huge volumes of lava are extruded, which subsequently cool and thicken to form the oceanic crust. Remote diving vessels have observed 'black smokers': vents belching out the volatile components of the lava. Geochemical analysis of radioactive isotopes released by these vents and deposited at mid-ocean ridges can reveal the evolutionary history of the magma source. We have a fair idea of the relative abundance of radioactive isotopes such as helium-4 (^{4}He) when the Earth was formed. We also know their half-lives and can thus determine their relative abundances at the present day, assuming that the isotopes have not been depleted in some way.

A magma source will first lose its volatile content by degassing. If the measured ^{3}He/^{4}He ratios are close to predicted present-day relative abundances, this indicates that the magma source has never before been tapped. The ^{3}He/^{4}He ratios measured at mid-ocean ridges are depleted, which is indicative of previous degassing of the magma source. On the other hand, oceanic island basalts deposited by volcanoes associated with mantle hot-spots show different ^{3}He/^{4}He ratios, which are in line with those predicted for primordial concentrations. This is indicative of an unmixed source that has not undergone degassing. The volatiles in the hot-spot source have not 'partitioned out' or been mixed with depleted mantle material.

This evidence suggests that mid-ocean ridge basalts and ocean island basalts are supplied from chemically distinct regions of the mantle. This observation, together with the different observed seismic characteristics of the upper and lower mantle, led many scientists to suggest that mantle convection occurred in two separate layers with little movement of material between the layers. The 670-km discontinuity was thought to provide a barrier to movement across the boundary, effectively splitting the mantle into two separate convection regimes. Other evidence includes laboratory work on convection. In the preferred convection patterns conforming to the boundary conditions present in the mantle the aspect ratio (the ratio of cell depth to cell width) of convection cells is about one. For two-layer convection, the cells would be about 700 km by 700 km in size, which fits neatly with the spatial distribution of surface features, such as mid-ocean ridges and subduction zones that are thought to be associated with downwelling or upwelling mantle material (see Fig. 1). For layered convection the subducting slabs must settle at 670 km, the accepted boundary between two layers of convection cells.

Evidence for whole-mantle convection includes calculations of heat flow. Without complete mixing of the mantle

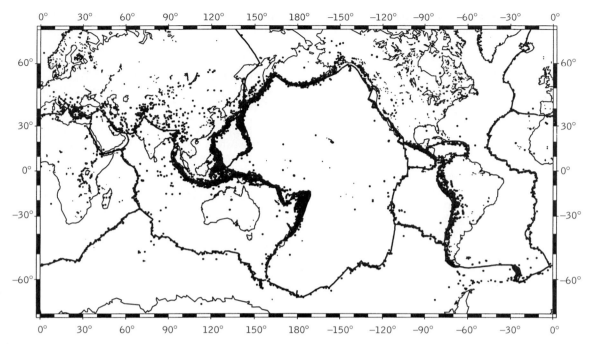

Fig. 1. Map of global seismicity, illustrating the effect of mantle dynamics on the surface processes of the Earth. The map shows large earthquakes reported in the Harvard earthquake catalogue (black dots). No depth dependence is shown; although most earthquakes occur in the lithosphere, earthquakes have been recorded at depths of 670 km that are associated with lithospheric slabs descending into the mantle. Seismicity, or earthquake distribution, maps show dynamic regions near the surface of the Earth, most of which are strongly influenced by mantle dynamics. The plate boundaries (thick black lines) have been superimposed on the map. Much of the observed seismicity is linked to global tectonics controlled by mantle dynamics. Mid ocean ridges, such as the North Atlantic ridge running the length of the Atlantic Ocean, drive the formation of oceanic crust, moving the continents apart at rates of up to 10 cm a year. Lithospheric plates are subducted into the mantle along convergent margins, such as those around the Pacific rim. Hot-spots, such as Hawaii, occur where mantle plumes break through the lithosphere.

the Earth would simply overheat. Also, it is argued that plumes originating deep in the mantle would not be the long-lasting stationary features that we observe if layered convection were the primary method of heat transport through the mantle.

Evidence from seismic tomography experiments shows slabs of subducted lithospheric plates descending through the 670-km discontinuity deep into the lower mantle. In addition, the latest attempts to image plumes show plumes emanating from deep in the lower mantle, possibly from the D″ layer. This suggests that the mantle convects as a whole, although the geochemists now require an explanation for the existence of pockets of unmixed mantle material. Perhaps the key to a more complete description of mantle dynamics lies with understanding the mechanics of mantle plumes.

Mantle plumes

Mantle plumes are perhaps the most interesting features in the mantle. They originate deep in the Earth and yet have a dramatic influence on surface features. Elucidating the origin of plumes and understanding why they are such long-lasting

stationary features in the mantle will hold the key to a more complete understanding of mantle dynamics.

Hot-spots are the result of plumes of molten rock rising through the mantle and impacting on the lithosphere, where in effect they burn a hole through the lithosphere and form volcanic islands. These islands are large topographic features: to rise above sea level in the Pacific Ocean, Hawaii must be at least 4 km high. The plumes show considerable longevity and are thought to be stationary over their life cycles. Evidence for this can be seen in the chains of volcanic islands associated with a single hot-spot. The trail of extinct volcanic islands shows the speed and direction in which the overlying lithospheric plate has travelled. The largest of these island chains is the Emperor chain culminating in the volcanic island cluster of Hawaii. This process is ongoing and today a new volcano can be seen off the main flank of Hawaii. In about 10 000 years' time this will break surface and form a new volcanic island, extending the chain.

Evidence for much larger mantle plumes can be seen in India. The Deccan traps are a series of lava beds hundreds of metres thick that were once thought to have been deposited over many millions of years. However, palaeomagnetic measurements tell a different story. The direction of the geomagnetic field is frozen into the lavas as they cool below the Curie temperature. The Earth's geomagnetic field changes with time and is known to reverse approximately once every million years. When the directions of the palaeomagnetic fields frozen into the lavas of the Deccan traps were measured, instead of a record of many geomagnetic reversals only two different palaeomagnetic directions were observed. Given the frequency of geomagnetic reversals this meant that the Deccan traps were deposited in only about a million years. This is an incredible volume of lava to be deposited in such a short space of geological time; it would be enough to cover the entire United States to a depth of almost 1 km.

The plume located underneath Iceland almost certainly resulted in the break-up of the supercontinent Pangea into what are now Western Europe and North America. The match between the coastlines of the Americas and Europe and Africa is almost perfect, except for the island of Iceland. This is because Iceland, the world's largest volcanic island, is located on top of the super-plume that resulted in the break-up of Pangea. Iceland is a segment of mid-ocean ridge domed up above sea level by the plume.

Large-scale eruptions such as those that resulted in the Deccan traps and Iceland are one end of a spectrum. The other end is represented by small island chains or isolated hot-spots, which indicate continuous sources of magma from deep in the mantle: a pulsating plume, rather than one large plume head. Super-plumes can obviously rapidly alter the Earth's surface. For example, the timing of the Deccan traps is coincident with a major mass extinction: the Cretaceous extinction in which many species disappeared, including the dinosaurs.

Mantle dynamics are certainly more complex than many current models suggest. It is now time to move away from simple stratified or whole-layer convection models to a more dynamic model of convection incorporating both models of convection at different times. We are aware of dramatic events in the Earth's history, and have already correlated some, such as the mass extinction in the Cretaceous and the break-up of Pangea, with the arrival of super-plumes at the surface. Possible scenarios for time-variant convection models include quiescent periods of two-layer convection interspersed with short periods of rapid mixing and whole-mantle convection. For example, subducted slabs may 'pool' at the 670-km discontinuity before eventually breaking through the 670-km barrier and surging into the lower mantle, entraining upper mantle material with them. The role of the D″ layer in mantle dynamics is still unclear. It has been suggested that mantle plumes are formed in the D″ layer and rise to the surface from the base of the lower mantle. On the other hand, some believe that the D″ layer is a 'graveyard' of subducted slabs. It is clear that there are many exciting discoveries to be made before our understanding of mantle dynamics is complete.

D. SHARROCK

Further reading

C. M. R. Fowler (1990) *The solid Earth: an introduction to global geophysics.* Cambridge University Press.

mantle and core composition Together, the mantle and core make up more than 99 per cent of the solid Earth's volume (Fig. 1). The mantle occupies the shell directly below the crust and extends to a depth of approximately 2890 km. It is thought to be composed primarily of silicate minerals, some of which are in high-pressure, close-packed structures. Beneath the mantle is the Earth's core. It consists of two parts, an outer molten shell with radius of 3480 km and a thickness of 2259 km, and a solid inner core with a radius of 1221 km. The core makes up about 32 per cent of the Earth's mass and is probably composed of an iron alloy.

There are still some significant gaps in our knowledge of the chemical composition of the mantle and core. The main reason for this is that we have never collected a sample from either of these shells. Instead, our knowledge of mantle and

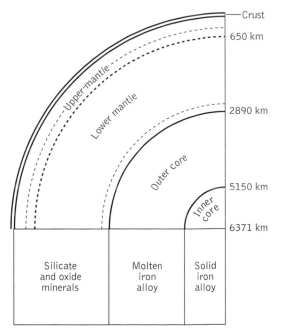

Fig. 1. Cross-sectional slice of the Earth showing the major regions of the mantle and core. The lower, horizontal panel gives the primary compositional constituents of these regions. The only indisputable compositional boundaries in the Earth are between the crust and the upper mantle, and between lower mantle and the outer core. Other compositional boundaries may exist between the outer and inner core at 5150 km depth, and between the upper and lower mantle at 650 km depth. The crust is represented by a narrow band on the outermost portion of the cross-section; its thickness varies from *c.* 5 km beneath the ocean basins to as much as *c.* 70 km beneath continents.

core compositions is derived through several lines of indirect evidence.

Much of what is known of the constitution of the Earth comes from observations of seismic body waves (compressional waves and shear waves) that travel through its interior. The velocities of the waves depend on the physical properties of rocks. By recording large amounts of seismic data, good estimates of the Earth's radial and lateral seismic velocities can be obtained. These can then be compared and matched with high-pressure velocities of minerals and rocks measured in the laboratory. Unfortunately, knowledge of the body wave velocities alone does not offer a unique solution for the rock types or chemical make-up of the mantle and core. Moreover, data on the properties of Earth materials at deep mantle and core pressures are sparse and very difficult to produce. Hence the uncertainty in determining the types of minerals and rocks through which the seismic waves are passing becomes greater with increasing depth.

Chemical equilibrium studies also shed light on the compositions of the mantle and core. High-pressure, high-temperature experiments can be designed to examine the element exchange reactions that occur between different minerals and between minerals and magmas under conditions found in the Earth's interior. These experiments yield estimates of the probable distribution of elements in the mantle and core.

Another source of information on the chemical composition of the Earth's interior is provided by analyses of peridotite xenoliths and oceanic basalts. These rocks are believed to have originated in the upper mantle and have been transported to the Earth's surface in a relatively unaltered condition. Peridotites and basalts yield the most reliable information about the composition of the mantle down to about 300 km, but it is possible that peridotites exist at deeper levels in the mantle; indeed it is often assumed that the whole mantle is of peridotitic composition. However, truly pristine samples that preserve the mineralogy of the high-pressure and high-temperature conditions of the deeper mantle have yet to be discovered. There are two possible explanations for the absence of rocks at the Earth's surface formed at great depths. First, it is imaginable that convective transport in the mantle is slow enough for high-pressure minerals, conveyed from great depths, to convert to stable, low-pressure forms by the time they reach the surface. Secondly, the mantle circulation pattern may be one of layered convection, with the deeper convection cell isolated from the shallow, subcrustal ones. Accordingly, material from the deep cells would rarely, if ever, reach the Earth's surface. For the material that constitutes the core, this is almost certainly the case. The mantle and the core are thought to have entirely separate and distinct convective regimes. Furthermore, at the core–mantle boundary, the core is nearly twice as dense as the overlying mantle. Such a large density contrast would be a major impediment to large 'core xenoliths' being swept up into the mantle by rising convection currents. If some minor amount of the core has ever been entrained in the mantle, it has not reached the Earth's surface in a recognizable form; core xenoliths have never been documented.

If the core has never been sampled and core xenoliths do not exist, how then can we conclude that it is probably composed of an iron alloy? Seismic velocities tell us that the core must be much denser than the mantle, but how do we know it is not made of some other dense materials such as lead, uranium, or gold? The answer comes from a fourth line of evidence on the deep-Earth composition: cosmic abundances of the elements deduced from analyses of chondritic meteorites and the solar photosphere. We know that the core makes up almost a third of the Earth by mass, and iron is the only dense element that is abundant enough in the solar system to supply this large reserve. Similar mass-balance arguments can be made to deduce the budget of major elements, such as silicon, magnesium, iron, aluminium, and calcium, that should reside in the lower mantle.

Mantle composition

One of the major controversies concerning the nature of the mantle is whether it is of a uniform composition (homogeneous) throughout or whether there are large regions of distinctly different composition (heterogeneous). Many Earth scientists subdivide the mantle into two parts: an upper and a lower mantle. The upper mantle extends from the base of the crust to a depth of approximately 650 km. The lower mantle continues from 650 km down to the core–mantle boundary, located at approximately 2890 km depth. Between 400 and 650 km lies the transition zone, a region of the upper mantle where seismic velocities increase relatively rapidly. In particular, at approximately 400 and 650 km abrupt seismic velocity discontinuities or jumps are observed. In a homogeneous mantle composed of peridotite, these seismic discontinuities can be attributed primarily to the transformation of the mineral olivine ($[Mg,Fe]_2SiO_4$) to closer-packed, denser, high-pressure forms. At approximately 400 km, olivine undergoes an isochemical phase transformation to the denser mineral, wadsleyite (also referred to as beta-phase), and at about 520 km depth wadsleyite is converted to the denser, cubic-structured, ringwoodite (or gamma spinel). Near 650 km depth, ringwoodite breaks down into two different crystalline phases, perovskite ($[Mg,Fe]SiO_3$) and magnesiowüstite ($[Mg,Fe]O$). Perovskite and magnesiowüstite are thought to be the major constituents of the lower mantle (Fig. 2).

A. E. Ringwood's 'pyrolite' is perhaps the most widely accepted composition for a homogeneous mantle. Pyrolite is not an actual mantle rock like a peridotite xenolith. Instead, it is a hypothetical mantle composition consisting of a mixture of 83 per cent harzburgite and 18 per cent basalt. The mineralogy of pyrolite is 57 per cent forsteritic olivine, 17 per cent enstatitic pyroxene, 12 per cent diopsidic pyroxene, and 14 per cent pyropic garnet. Pyrolite can be thought of as an analogue for the original or primordial mantle composition that existed after planetary accretion and core formation. It is supposed

Mantle mineralogy

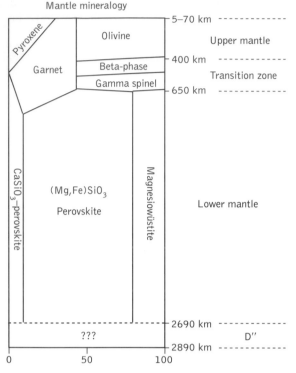

Fig. 2. Volume per cent mineralogy with depth assuming a homogeneous pyrolite composition for the mantle. In this model the shallowest portion of the upper mantle consists of 57 per cent olivine, 29 per cent pyroxene, and 14 per cent garnet. The 400-km seismic discontinuity is attributed to the polymorphic phase transition of olivine to beta-phase (wadsleyite). The stability of garnet is at a maximum in the upper mantle transition zone, while pyroxene stability decreases with depth in the mantle. The upper mantle–lower mantle boundary at 650 km is interpreted as the depth range at which gamma spinel and garnet convert to $(Mg,Fe)SiO_3$-perovskite + magnesiowüstite. In layered mantle models the 650-km discontinuity is also the location of a compositional boundary. A lower mantle with higher SiO_2 content than the upper mantle will have more than 75 per cent $(Mg,Fe)SiO_3$-perovskite. The composition and mineralogy of the D″ region is unknown.

that basaltic magmas were extracted from the primordial pyrolite mantle to form the crust. According to the model, the shallowest part of the mantle has been continually depleted of a basaltic component over geological time, while the remainder of the mantle is undepleted pyrolite.

In contrast to the pyrolite model, D. L. Anderson and others contend that the mantle is compositionally layered. In most layered models the 650 km discontinuity is not only the site of a major mineralogical phase transformation, but it is also a compositional boundary. Geochemical layering also implies layered convection in the mantle, with little or no mass transport across the 650 km discontinuity. These models

usually assume that most of the upper mantle is an olivine-rich peridotite similar in composition to pyrolite, and that the lower mantle is enriched in $(Mg,Fe)SiO_3$-perovskite. This means that the lower mantle has a higher proportion of SiO_2 than does the upper mantle. A compelling reason for favouring an SiO_2-rich lower mantle comes from cosmochemical evidence. A mantle composed solely of pyrolite is deficient in SiO_2 relative to chondritic meteorites. If the missing SiO_2 is actually sequestered in the lower mantle, then the bulk composition of whole mantle can be matched with chondritic elemental abundances. How could the mantle become layered? One possibility is that the mantle became differentiated during the high-temperature, early stages of Earth evolution. Differentiation would have been enhanced if deep-level melting or a planet-wide magma ocean existed. Upon cooling, the mantle could have solidified like a giant layered intrusion.

Lateral heterogeneities in mantle composition might also exist. Evidence for this idea stems from seismic tomography. Tomography creates three-dimensional images of the Earth's interior and can distinguish between regions within which rock properties are seismically 'fast' or 'slow'. Fast and slow velocity regions are both radially and laterally distributed throughout the mantle. One interpretation is that the 'fast' regions are simply colder downwellings and that slow regions are hotter upwellings in a compositionally homogeneous mantle. The alternative possibility is that some regions are relatively fast or slow because of compositional differences. An interesting possibility is that primordial differentiation formed a compositionally layered mantle, and that whole-mantle convection has, over the past four billion years, been working to erase and homogenize the compositional layering.

Related to the question of mantle convection and composition is the nature of oceanic lithosphere (slab) subduction. One possible fate for subducted slabs is that they sink through the entire mantle and ultimately reach the core–mantle boundary. If this occurs, the whole mantle should be affected to some degree by the consumption of oceanic crust. At any given time, there will be a number of slabs transiting at various depths in the mantle. The base of the mantle may be a slab 'graveyard' with a composition that is different from the rest of the mantle. This idea is consistent with the seismically observed D″ (D-double-prime) region that occupies approximately the bottom 200 km of the mantle. However, if slabs do reach the core–mantle boundary, they might also have a short residence time there. Chemical assimilation and convective recycling may incorporate them back into the whole mantle. Hence, a slab graveyard is by no means the only interpretation of D″. Other ideas include: (1) D″ is a chemical reaction zone between the mantle and core; (2) D″ is the remnant of layering formed during primordial differentiation; and (3) the seismic velocity change at D″ is caused by a yet unknown phase transformation. Another idea is that there might exist a large-enough density and/or viscosity contrast at the 650-km discontinuity physically to impede subducted slab penetration. In this view of subduction, only the upper mantle is affected by this process. Here the slab graveyard would be located on top of the 650 km discontinuity and much of the upper mantle transition zone would be compositionally distinct from the regions above and below it.

Isotopic compositions of oceanic basalts suggest that there are at least two distinct chemical reservoirs in the mantle. In general, mid-ocean ridge basalts have high $^{143}Nd/^{144}Nd$, low $^{87}Sr/^{86}Sr$ and $^{206}Pb/^{204}Pb$, indicating that they originate from a part of the mantle that is relatively depleted in incompatible elements. In contrast, many basalts from oceanic islands have low $^{143}Nd/^{144}Nd$, high $^{87}Sr/^{86}Sr$ and $^{206}Pb/^{204}Pb$, as would be expected from a mantle source region enriched in incompatible elements. But there are many exceptions to the depleted and enriched isotopic ratio relationships just described. When the whole spectrum of oceanic basalts is considered, it is not possible to account for all the observed isotopic ratios by simply mixing one enriched and one depleted reservoir. It has therefore been proposed that as many as five different basalt source regions exist in the mantle. Although isotope ratios argue strongly for a heterogeneous mantle, perhaps on several different scales, they are not particularly good in giving precise information about the petrological, spatial, or physical nature of the source region. One major complication in interpreting isotopic signatures of basalts is the role of crustal contamination and metasomatism. For instance, a mantle source region could actually be 'depleted', but if basalts originating there become enriched by incorporation of incompatible elements when they pond in the crust, then the source region would appear to be 'enriched'. Another difficulty is distinguishing between enriched mantle reservoirs that are primordial and those that might have formed by crustal subduction. As is the case with all the methods discussed here, the diagnostic power of isotope geochemistry is most effective when combined with other independent lines of evidence.

Basalts and peridotites, although volumetrically most prevalent, are by no means the only rock types that originate in the mantle. Kimberlites are among the most compositionally and isotopically 'enriched' mantle rocks. Their occurrence is rare, both spatially and temporally. They are found almost exclusively as pipe-like intrusions, or diameters in continental crust. Modern kimberlites may be non-existent; many have radiometric gases that place their formation in the Cretaceous. Kimberlites can be diamond-bearing and originate at depths greater than 150 km. They reach the Earth's surface in violent eruptions, often carrying along fragments of peridotite mantle. Kimberlites have especially high concentrations of potassium, water, and carbon dioxide, attesting to the notion that the mantle can be extremely heterogeneous from place to place.

A small number of kimberlites contain eclogite xenoliths. Eclogites consist primarily of diopsidic pyroxene and pyropic garnet and, in contrast to mantle peridotites, they are not rich in olivine. Deep-source eclogites may have two types of origin. Basalt, when metamorphosed at high pressure, trans-

forms into eclogite. Hence, some eclogite xenoliths may have originated as subducted oceanic crust. Another possibility is that some eclogite is primordial and that it formed during the crystallization of an early magma ocean. One way to think of primordial eclogites is that they represent a 'crustal' component that has been trapped in the mantle. The upper mantle transition zone has been one of the proposed locations for primordial eclogites. Petrological constraints cannot rule out the possibility that eclogites are the mantle source rock for basaltic volcanism. However, the paucity of eclogite xenoliths compared to the widespread occurrence of peridotites makes them a less popular candidate.

The theory of plate tectonics is unequivocal about the need to subduct oceanic lithosphere into the Earth's interior. The fate of the sedimentary veneer that covers the ocean floor is, however, more uncertain. Some models propose that during subduction most or all of the sedimentary layer is scraped off and accreted on to the overriding plate. Another view is that a significant portion of the sediments are subducted. Beryllium-10 has been used as an isotopic tracer to confirm that some of the sedimentary material reaching a trench is indeed transported down in the mantle during subduction. The average composition of the material will vary, and will depend on the local sources of sediment. It could be rich in silica, feldspar, carbonate, and hydrous phases. Subduction of sediments might be an important factor in lowering the temperature at which the mantle melts, thus contributing to arc magmatism. The degree to which the sedimentary component is recycled back up into the crust during subduction melting has yet to be resolved. If the process is not completely efficient, then some sediments may be retained in the mantle for geologically significant lengths of time.

Core composition

The most simplistic view of the Earth's core is that it is composed mostly of metallic iron, the inner core being pure iron in a hexagonal close-packed (hcp) crystalline form, and the outer core being molten iron with approximately 10 per cent by weight of a 'light' element such as oxygen or sulphur. This view is satisfactory from a geophysical perspective in that it accounts for the core density estimated from the mean density and moment of inertia of the Earth, as well as from seismic body wave velocities. If, however, we use the crust and the mantle as a guide, the Earth's core is probably anything but nearly pure or of simple composition. The main problem with determining the actual core composition is caused by its remoteness, the absence of core xenoliths, and the difficulty of replicating the extreme pressures and temperatures of the core in the laboratory.

Most current ideas on the composition of the core have been strongly influenced by the iron–nickel meteorite analogy. In the earlier part of the twentieth century, the idea was developed that meteorites were remnants of a single, disrupted planet. As a result, the meteoritic Earth hypothesis was formulated to give insight on the Earth's internal composition. The hypothesis held that the stony meteorites represented the disrupted planet's mantle, and iron–nickel meteorites the core. It followed then that the Earth's core must also be similar in composition to iron or iron–nickel meteorites. By the 1960s the meteoritic Earth hypothesis had fallen out of favour because it had been discovered that meteorites did not originate from a single parent planet. In particular, oxygen isotope studies showed that many parent bodies must have existed. Moreover, petrological investigations revealed that some meteorites experienced igneous processing while others have remained relatively unaltered since the formation of the solar nebula. Iron meteorites were found to have formed at low pressure, in small, possibly asteroidal-sized bodies. Their genesis and nature, therefore, may be completely unrelated to that of the Earth's core.

A more fruitful approach to deducing the core composition is achieved by referring to the abundances of the elements in the most primitive, unaltered meteorite type: carbonaceous chondrites. Carbonaceous chondrites are the oldest objects dated by radiometric methods and are most likely remnants from the formation of the solar system. In addition, they have concentrations of non-volatile elements that closely match solar abundances. Carbonaceous chondrites are the starting point for a bulk earth composition in numerous geochemical and geophysical models. They are comprised primarily of silicate minerals, a carbon-rich matrix, and approximately 30–40 per cent iron in oxidized, sulphide, and metallic form. The high abundance of iron in carbonaceous chondrites is the most convincing evidence to support the premise that it is the primary dense element in the Earth's core. Additional dense elements alloyed with iron are not required to satisfy geophysical constraints; however, the chondritic abundance of nickel and its tendency to form metallic compounds supports the likelihood that nickel is in the core at concentrations of 5–15 per cent. Geophysical data do indicate that the core is not dense enough to be composed of pure metallic iron or an iron–nickel alloy. A better match for the estimated core density is obtained by diluting the liquid outer core with 8–12 weight per cent of a 'light' element or elements. Hydrogen is the most abundant light element in the Solar System, but because of its high volatility, much of the Earth's original hydrogen budget may have been lost to space under the high-temperature conditions of planet formation, and only small amounts may be left in the core. Oxygen and sulphur are more likely candidates for being the major light core elements. Oxygen is often favoured because of its ubiquity in minerals that make up the crust and mantle. Furthermore, it has been proposed that bonding in iron oxide changes from ionic to metallic at very high pressure, thus enhancing the solubility of oxygen in iron alloys under core conditions. Sulphur is favoured as a major core component because it readily forms stable sulphide compounds with iron, at both low and high pressure. Sulphur, on the other hand, is relatively volatile at low pressure and some of the Earth's original chondritic abundance of sulphur may also have been lost during accre-

tion. At present there are no experimental data on the combined solubility of oxygen and sulphur in molten iron alloy at outer core pressures. Until such data exist, notions on the precise make-up of the outer core remain speculative. Other light elements that are commonly proposed for the outer core are silicon, carbon, helium, and nitrogen. It is probable that every naturally occurring element is present in the outer core at least at trace levels.

The inner core is a relatively small region, making up only 5 per cent of the mass of the whole core, and for this reason it is difficult to estimate its density. Seismic studies have revealed that the inner core is markedly anisotropic with respect to compressional wave velocities. Some geophysical models assume the inner core to be composed of epsilon (ε) iron, a crystalline, hexagonally close-packed compound that is stable at very high pressures and temperatures. Crystallographic alignment of these hexagonal ε-iron crystals has been proposed as a mechanism producing the inner core seismic anisotropy. On the other hand, it is conceivable that the inner core is similar in composition to the outer core, with several per cent nickel and various light elements. At present there are no thermodynamic data on the stability of ε-iron diluted with other chemical species. It therefore is possible that yet-unknown high-pressure crystalline phases are present in the inner core.

There are at least two explanations for the existence of a molten outer core and a solid inner core. First, the primordial core could have been completely molten and homogeneous. If the pressure versus temperature melting curve for the core alloy is steeper than the core geotherm, then the first solid to form upon cooling would be at the deepest level. Hence, as the Earth cooled through time, the inner core grew at the expense of the molten outer core. Given enough time, the core will become completely solid. In this scenario the inner and outer cores are likely to have similar, but not identical chemical compositions. The determining factor will be the partitioning behaviour of elements between liquid and solid iron alloy. Conventional wisdom based on element partitioning at low pressures assumes that the 'light' elements will have an affinity for the molten outer core over the solid inner core. This assumption requires testing by future high-pressure experiments. The second explanation for a solid plus liquid core is that the inner and outer cores were originally formed as distinctly different regions and are still separated by a compositional boundary. Such a configuration could have a nearly pure iron inner core with a high melting temperature and a light-element-diluted outer core with a comparatively low melting temperature. In this model, the primordial inner and outer cores could have originally been out of chemical equilibrium. With time, however, chemical potential would tend to drive the exchange of mass across the inner core–outer core boundary. The evolution towards chemical equilibration could have caused the purer inner core to shrink through time as it was contaminated and consumed by the molten outer core.

Are the mantle and core in chemical equilibrium? This is one of the great unanswered questions about the composition of the Earth's interior. If low-pressure experimental data on silicate–iron alloy phase equilibria are good indicators, then the answer is most likely to be 'no'. The upper mantle does not, at least, bear the chemical signature of a silicate body that is in contact with a large mass of iron alloy. Evidence for this is the relatively high concentration of oxidized siderophile elements, such as ferrous iron (Fe^{2+}), nickel (Ni^{2+}), and cobalt (Co^{2+}), in peridotite xenoliths. The presence of an iron core in chemical equilibrium with the upper mantle would cause a large proportion of these species, through oxidation–reduction reactions, to change to metallic iron, nickel, and cobalt in lower valence states. This does not rule out the possibility that the lower mantle is in chemical equilibrium with the core. It would of course require that the upper mantle and lower mantle were isolated from each other and were compositionally different. We also cannot rule out the possibility that mantle–core equilibration is a very slow process compared to the age of the Earth and at present the chemical reaction is far from complete. The alternative view is that the very high pressures and temperatures present at the core–mantle boundary have a dramatic effect on silicate–iron alloy phase equilibria. In some cases, element partitioning would even be reversed from what is observed near the Earth's surface. If this is true, then conventional wisdom from the low-pressure experimental data would be a poor guide for determining mantle–core chemical equilibrium.

C. AGEE

Further reading

Anderson, D. L. (1989) *Theory of the Earth*. Blackwell Scientific Publications, Oxford.

Jeanloz, R. (1990) Nature of the Earth's core. *Annual Reviews of Earth and Planetary Sciences*, 18, 357–86.

Ringwood, A. E. (1975) *Composition and petrology of the Earth's mantle*. McGraw-Hill, New York.

marble 'Marble' is a general term for any kind of limestone or other carbonate rock that has been metamorphosed. The heat and/or pressure experienced during metamorphism causes the calcium carbonate to recrystallize into a tightly interlocking mass of calcite crystals. The resulting rock is often pure white, and because it takes a polish well marble is a favoured material for ornamental building stone or sculpture. For example, the Taj Mahal, the Parthenon, and Michelangelo's statue of David are all made of marble. However, some stone described by builders or masons as marble is not really marble in the geological sense, but is just some variety of unmetamorphosed limestone that can be cut and polished.

Swirls of colour within some true marble indicate impurities within the original limestone. Marble produced by contact metamorphism of an impure limestone adjacent to an igneous intrusion may contain such minerals as olivine and garnet.

DAVID A. ROTHERY

marine erosion *see* COASTS AND COASTAL PROCESSES

marine geology The oceans cover 71 per cent (approximately 360 million km^2) of the Earth's surface. Marine geology is the study of the nature and history of the earth beneath the sea. The origins of marine geology lie in the development of submarine telegraphy in the latter half of the nineteenth century. For the first time it became imperative to find out the depth and nature of the sea bed, knowledge essential for routing and operating submarine telegraph cables. Early cable laying ships were the first to make routine soundings and recorded the shoal topography of the Mid-Atlantic Ridge between America and Europe. Increasing interest in the oceans led the British Admiralty to sponsor the first large-scale truly oceanographic expedition and during 1872–6 HMS *Challenger* circumnavigated the globe making the first systematic collection of oceanographic and geological data, ultimately visiting all the oceans and travelling over 100 000 km. However, our knowledge of the shape and structure of the sea floor, and the processes that shape it, progressed only modestly from the late nineteenth century until the Second World War. The development of the sediment corer was a major innovation, allowing the recovery of sediment sequences from the sea floor and analysis of the geological history contained within them. Several geophysical techniques for investigating the sea bed and its substructure, such as sonar, gravity, magnetic, and seismic methods, which had largely been experimental in the 1920s and 1930s, were developed during the Second World War and became major geological surveying tools during the subsequent great expansion of oceanographic research, which continues to the present day. In the late 1960s, these tools played a major role in revolutionizing the Earth sciences by providing evidence for plate tectonics. For the first time, the major features of the oceans basins could be explained as the edges of plates, where new crust is created (at mid-ocean ridges) and destroyed (at ocean trenches).

Today, marine geologists use a variety of tools to investigate the ocean floor. These are largely either imaging techniques (geophysical mapping and profiling or observation using cameras) or direct sampling.

The depth of the ocean floor is conventionally measured using echo-sounders, in which depth is determined by measuring the time between the emission of a sound pulse and the detection of the echo from the sea floor. If the speed of sound in the water is known, then depth can be calculated. However, modern survey vessels commonly use swath bathymetry systems that use a fan of sound beams to produce highly accurate contoured strips of sea floor whose width may be up to twice the water depth. Acoustic plan-view images of the sea floor can also be obtained using sidescan sonar, in which an array of transducers is used to illuminate the sea floor with sound on either side of the survey vehicle. The sound is backscattered from the sea bed according to the topography, sediment type, and small-scale

Fig. 1. Sidescan sonar mosaic of the sea floor in the eastern Gulf of Mexico (analogous to an aerial view). The straight lines running obliquely across the picture are ships' track lines and are spaced about 30 km apart. The bright lobate areas in the lower part of the picture are debris flows (mudslide deposits) derived from the eastern flank of the Mississippi fan. They are brighter because they contain coarser-grained sediment and have a rougher surface than the darker flatter muddy sea floor to the north. One debris flow overrides an earlier one (lower right), which itself overrides a bright sinuous deep-sea channel (upper right).

Fig. 2. A corer being recovered on a research ship. Sediment cores are the primary source of raw data for determining the geological history of the oceans and recent climate change.

bottom roughness. The resulting acoustic image provides the equivalent of aerial photographs of the sea floor (Fig. 1). Swath bathymetry (which provides topographic maps) and sidescan sonar (which provides 'aerial' views) can be combined using computational tools such as geographical information systems (GIS) to provide detailed geological information. High-resolution views of the sea floor can be obtained by fixing sidescan sonar apparatus to deep towed instrument platforms that operate within a few hundred metres of the sea floor. Although deep-towed sidescan sonar will have a higher resolution than near-surface towed vehicles, the width of coverage will be narrower and the survey speed much slower. A range of instruments can be attached to deep-towed vehicles for high-resolution work. These can include geochemical 'sniffers' which can locate hydrothermal plumes and hence proximity to hydrothermal vent fields, where metal-rich, often superheated, water exits from the sea bed after having passed through oceanic crust in circulation systems driven by magmatic heat.

The substructure of the sea floor can also be investigated using sound waves, by using a frequency of operation that allows the sound to penetrate the sea floor and to be reflected back from interfaces between different types of sediment or rock. This forms the principle of seismic reflection profiling, in which a picture of the subsurface structure is built up along the track of the ship. Surveys using closely spaced track lines can be used to build up three-dimensional pictures of the geological structure beneath the sea floor.

The sea floor can be directly sampled using dredges, corers, and grabs (Fig. 2). Dredges are metal frames attached to chain-mail type bags, which, when dragged over rocky topography, break off pieces of rock which are collected in the bag. Corers recover continuous sequences of sub-sea floor sediment. There are various types. Most cores taken today are box cores or piston cores. Box corers are lowered on to the sea floor and recover shallow blocks of sea bed, preserving the sediment–water interface. A piston corer consists of a tubular barrel driven into the sediment by a large weighted head. The barrel contains a piston which reduces the internal friction, effectively sucking the sediment into the barrel. Piston corers can take cores up to 50 m in length. Analysis of cores recovered by this method has revolutionized our understanding of recent climate change.

The ability to core long intervals beneath the ocean floor became a reality with the birth of the Deep Sea Drilling Project (DSDP) in 1968. This project was conceived by scientists from American oceanographic research centres, who set up a parent organization, Joint Oceanographic Institutions for Deep Earth Sampling (JOIDES), and aimed to investigate Earth history by drilling the deep ocean floor using the dynamically positioned drill-ship *Glomar Challenger*. On its first cruise cores were recovered from 140 metres below the sea floor in the Gulf of Mexico and contained traces of oil, providing the first direct evidence that oil occurred in ocean depths deeper than the continental shelves. Later the same year, *Glomar Challenger* drilled a transect of holes across the flanks of the Mid-Atlantic Ridge between Dakar, Senegal, and Rio de Janeiro, Brazil. The recovered cores conclusively showed that the age of the Earth's crust directly beneath the overlying sediment pile increases systematically with distance from the mid-ocean ridge crest, validating the theory of plate tectonics and continental drift. On DSDP Leg 13, gypsum, a mineral produced by the evaporation of sea water (an evaporite) was recovered from a drill site in 2000 m water depth in the western Mediterranean, in sediments 5 million years old. Gypsum is typically found in lagoons along arid coasts. The only reasonable explanation for this is that the Gibraltar gateway to the Atlantic became closed and the Mediterranean Sea evaporated to dryness, resulting in a gigantic desert basin over 2 km deep between Europe and North Africa, although there were periodic marine incursions. Some 700 000 years later, the Gibraltar gateway subsided a little and the connection to the Atlantic was fully re-established, restoring marine conditions to the Mediterranean Basin.

Such was the success of DSDP that in 1975 an international phase was initiated. Research organizations in Britain, France, West Germany, the USSR, and Japan participated in the project. The International Phase of Ocean Drilling (IPOD) was superseded by a new international research programme: the Ocean Drilling Program (ODP), which commenced operations at sea in 1985 with a new drill-ship, *JOIDES Resolution*, which had improved capabilities. In 1997, ODP provided striking evidence for what was behind one of the great mysteries of Earth history: the extinction of the dinosaurs and many other organisms 65 million years ago. Geologists on land have recovered continental cores from this time containing iridium (an element abundant in extraterrestrial material), shocked quartz and soot, probably derived from global wildfires. This pointed to global devastation, probably caused by an asteroid, estimated to be 10 km across, crashing into the Earth at a location that has since been identified as the Yucátan Peninsula in Mexico. On ODP Leg 171B, *JOIDES Resolution* recovered a core from this time interval at a drill site 560 km east of Florida and 2000 km from the now-buried Yucátan impact crater. These sediments of 65 million years ago contained a layer 15 cm thick of ejecta, which included glassy spherules condensed from the vapour cloud produced by the impact and dust and ash produced by the resulting fireball. Today, the Ocean Drilling Program continues to be the world's leading international marine geological research program and comprises a partnership of 22 countries.

Marine geology is one discipline of oceanography but it under pins the other disciplines in that the record of the sea floor reflects the chemical and physical evolution of the oceans above. Study of the deep sea floor and the record it preserves is therefore essential to understanding Earth history and the oceans themselves.

R. G. ROTHWELL

Further reading

Open University Course Team (1998) *The ocean basins: their structure and evolution* (2nd edn). Butterworth-Heinemann, Oxford.

Seibold, E. and Berger, W. H. (1996) *The sea floor: an introduction to marine geology* (3rd edn). Springer-Verlag, Berlin.

Summerhayes, C. P. and Thorpe, S. A. (1996) *Oceanography—an illustrated guide*. Manson Publishing, London. (Chapters 8, 9, 10, and 20 in particular.)

Marsh, Othniel Charles (1831–99)

O. C. Marsh was pre-eminent as a collector of vertebrate fossils and one of the founders of American vertebrate palaeontology. After a baccalaureate from Yale College (1860) and study in New Haven and Germany, he was appointed by Yale to the first professorship of palaeontology in the Americas. With a bequest from his uncle, the financier George Peabody, who had founded Yale's Peabody Museum of Natural History in 1866, Marsh established a highly successful programme of expeditions to the Western Territories to collect from the newly discovered Mesozoic and Tertiary formations. From 1882 to 1892 he served as the first vertebrate palaeontologist of the US Geological Survey.

Marsh's systematic collecting in the western United States, which began in 1868, provided direct corroboration of Darwin's theory of evolution. He discovered Cretaceous toothed birds and recognized them as a link to dinosaurs, completed the horse evolution sequence of Thomas Huxley, elucidated the evolutionary sequence of the camel, and described the first known American Mesozoic mammals and archaic groups of ungulate mammals. Marsh emphasized the collection of complete skeletons, which enabled him to publish the first such reconstructions of Mesozoic birds, marine reptiles, and dinosaurs. He and his team developed the use of modern jacketing techniques for collecting their material. His descriptive monographs aroused immense interest and have become classics of the literature.

Marsh was in fierce competition with E. D. Cope in his search for and description of dinosaurs and other vertebrates from the West. Their mutual antagonism eventually reached the proportions of a national scandal, but both were most extraordinarily productive scientists.

L. J. HICKEY

mass extinctions *see* EXTINCTIONS AND MASS EXTINCTIONS

massive sulphide deposits

Massive sulphides are rocks made up primarily of sulphide minerals. There is no accepted definition of the term 'massive sulphide', but it is most commonly used in describing strata-bound (and often stratiform) deposits in sedimentary or volcanic settings formed from the flow of heated fluids (usually sea water) through sediments or igneous rocks, or both, the circulation being driven by associated volcanic activity. Massive sulphides range in age from Archaean to those forming today. Although they form a minor part of the Earth's crust, massive sulphides have received considerable attention throughout history. This interest has arisen mainly because they are a major source of many metals (in particular lead, zinc, copper, and silver) that have been critical in human technological development and are vital to a modern industrial economy. Exploration continues for new ore bodies, and considerable effort has been expended to develop models for the formation of massive sulphides in the hope that these will aid in the search for new reserves. Massive sulphide deposits are also of interest for the record they preserve of the magmatic and hydrothermal processes that control the composition of the Earth's crust, oceans, and atmosphere.

Generalized genetic model

Although various types of massive sulphide deposits have been described, they share some basic similarities in their paragenesis. Cold aqueous fluid (commonly sea water) is drawn down through sediments or igneous rocks and its temperature is raised by an underlying heat source. This heat source is usually a relatively shallow magma chamber or a recent igneous intrusion. As the cold water is heated to become a hydrothermal fluid, any dissolved sulphate that it contains is reduced to sulphide or precipitated as anhydrite (calcium sulphate). The fluid also becomes depleted in magnesium, and this causes a drop in pH. The resultant hot, acidic fluid reacts with the solid phase through which it is flowing. Various elements are leached from the rock and dissolved as complexes (predominantly chloride complexes, with smaller amounts of sulphide complexes). This modified hydrothermal fluid rapidly reaches equilibrium with an assemblage of secondary minerals. The hot hydrothermal fluid becomes less dense and flows upwards. As it nears the sea floor it cools. This causes precipitation of minerals such as pyrite (FeS_2), chalcopyrite ($FeCuS_2$), sphalerite (ZnS), and galena (PbS) that form massive sulphide deposits, together with sulphates (anhydrite ($CaSO_4$) and barite ($BaSO_4$)) as the hydrothermal fluid mixes with sea water entrained into the system. The exact mineralogy depends on the physico-chemical nature of the hydrothermal fluids and the extent of sea-water mixing. If, as usually happens, the rising hydrothermal fluids exit to the sea floor, they may then rise hundreds of metres into the overlying water column. Here they can form precipitates that can be carried many kilometres by bottom currents. Alternatively, if they are sufficiently dense, the exiting hydrothermal fluids may form pools of metal-rich brines similar to those observed in the Red Sea today. The modern oceans are generally well oxygenated, and the particulate plumes are consequently dominated by iron and manganese oxides. In the Archaean oceans the plumes would have consisted largely of sulphide minerals.

Idealized volcanogenic and sediment-hosted massive sulphide deposits are illustrated in Fig. 1. The deposits consists of a zoned body, in which the central core of iron-rich sulphides (pyrite or pyrrhotite, or both and a copper-rich sulphide (chalcopyrite) is overlain by more zinc- and lead-rich sulphides (sphalerite and galena) and is underlain by an alteration zone containing veins ('stringers') of iron and copper sulphides in a matrix of hydrothermal silicate minerals. This zone of alteration becomes progressively less distinct from the surrounding ('country') rocks as one moves laterally away from the deposit. The sulphides may be overlain by, or partially intermingled with, a cap of sulphate minerals. The deposit is usually mantled by iron and manganese oxides ('ochres and umbers') and iron-rich cherts that become thinner away from the deposit; these may extend for several kilometres before grading into the background sediments. In deposits that have been uplifted and exposed on land, the upper sulphides and oxides are invariably oxidized and weathered to a variety of secondary minerals ('gossan') by circulating meteoric water. During this process some metals, such as gold, may become enriched in the weathered zones.

Volcanic-hosted, or volcanogenic, massive sulphide deposits can be defined as deposits formed from circulation of hydrothermal fluids through volcanic rocks without involvement of significant amounts of non-volcanic sediments. Hence, these deposits contain only species leached from igneous rocks (except for a few elements derived from the circulating sea water). Sediment-hosted massive sulphide deposits are also generally associated with igneous activity, which provides the heat source for hydrothermal circulation. They differ, however, from volcanogenic deposits in that the hydrothermal fluids have also flowed through sediments overlying the igneous rocks. Hence, the hydrothermal fluids and resultant massive sulphide deposits may contain species derived from the sediments, as well as from the igneous rocks.

Volcanic-hosted deposits

Massive sulphide deposits are currently forming from hydrothermal systems at a large number of sites on mid-ocean ridges. The largest of these yet discovered is the TAG (Transatlantic Geotraverse) site at 26° N on the Mid-Atlantic Ridge at a depth of 3500 m below sea level. Detailed sampling by submersibles, surface ships, and deep-ocean drilling has provided much information about this site. *Black smokers*, debouching fluids at 350 °C into the overlying sea water sit atop a mound of hydrothermal minerals, including sulphides, sulphates, oxides, and silicates. The mound is 200 m in diameter and 50 m high and contains approximately 5 million tonnes of hydrothermal minerals. Radiometric dating suggests that hydrothermal activity has been intermittent at the site for about 50 000 years. There are also older hydrothermal deposits in the TAG area that are heavily weathered, the massive sulphides being altered to oxides and eroded by bottom-water

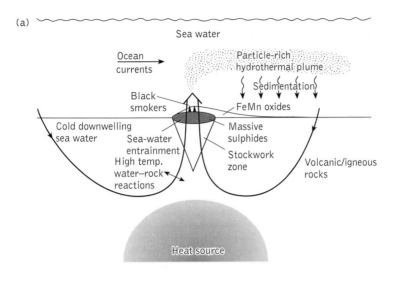

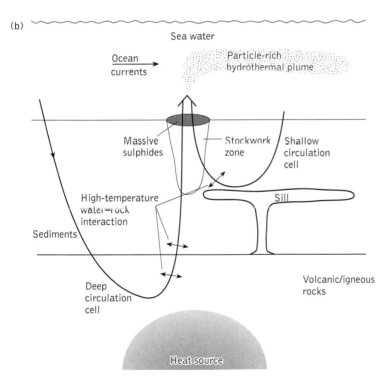

Fig. 1. Idealized forms of (a) a volcanogenic massive sulphide deposit and (b) a sediment-hosted massive sulphide deposit.

currents and tectonic activity. It is likely that most massive sulphides formed on mid-ocean ridges are destroyed within a few million years after hydrothermal activity ceases.

The closest analogues in the geological record to modern hydrothermal systems are the massive sulphides found in ophiolite belts throughout the world (e.g. Cyprus, Oman, the Urals). These deposits are generally small (few are larger than

20 million tonnes and most are smaller than a million tonnes) and are most commonly mined for copper, although some contain significant amounts of gold in gossans. The commonly accepted interpretation of these deposits is that they are located in portions of uplifted oceanic crust that formed in the same manner as modern, mid-ocean ridge hydrothermal systems. Massive sulphide deposits in ophiolite settings are commonly overlain by later flows of basalt pillow lavas that prevent them from being destroyed.

Kuroko-type massive sulphide deposits are also found in environments with modern tectonic equivalents, thought to have formed in back-arc and interplate settings. They differ from ophiolite-hosted deposits in a number of ways. They are associated with felsic volcanism and they were emplaced in relatively shallow seas (shallow enough for steam explosions). They have a higher proportion of galena and sphalerite in the massive sulphides and smaller amounts of chalcopyrite and pyrite than is found in ophiolite settings, and their sulphate minerals include a higher proportion of barite relative to anhydrite. The deposits are generally overlain by mudstones or tuffs, or both. Kuroko deposits tend to be larger than ophiolite-hosted deposits, with orebodies up to 75 million tonnes. Kuroko deposits were first described in Japan (*Kuroko* means 'black ore' in Japanese), but similar types of deposits have been identified in Fiji, Turkey, and the Philippines.

The largest volcanogenic massive sulphide deposits are found in greenstone belts in Archaean cratons, such as those in South Africa and Canada. They are not the products of presently active tectonic processes, but the general form of the deposits is similar to other volcanogenic massive sulphides. The greenstone-hosted deposits formed in shallow water in association with basic and more silicic rocks; they have an underlying hydrothermally altered zone and contain abundant copper, zinc, and lead sulphides. The greenstone-hosted deposits differ in that they lack a cap of sulphate minerals and they are overlain by later volcanic rocks rather than clastic sediments. The major difference is, however, one of scale. For example, the Kidd Creek deposit (in the Atitibi greenstone belt of south-eastern Canada) contains over 250 million tonnes of ore, and there are several other deposits containing in excess of 100 million tonnes of ore. The size of the greenstone-hosted deposits may be related to the greater intensity and duration of igneous activity in the Archaean.

Sediment-hosted deposits

There are several places in the oceans (e.g. Guaymas Basin, Gulf of California) where spreading centres are buried beneath sediments shed from the nearby continents. Hydrothermal circulation in these environments differs from sediment-starved mid-ocean ridge sites. Instead of the fluids flowing directly into sea water from the oceanic crust, they must first pass through an overlying layer of sediments. When the fluids do reach the overlying sea water, their chemistry is substantially altered. They have lower dissolved metal and

sulphide concentrations due to precipitation of massive sulphides in the sediments and the chemical signature of the fluids provides clear evidence of leaching of elements from the sediments by the hydrothermal fluids.

The heat source that powered hydrothermal circulation that formed sediment-hosted massive sulphides is not always exposed, but fluid inclusion and mineralogical studies indicate that these deposits were the products of high-temperature (around 350 °C) fluids. Sediment-hosted massive sulphides have a wider range in mineralogy than volcanic-hosted deposits, reflecting the variation in the composition of sediments (e.g. arenites, carbonates, evaporites, etc.). In terms of their economic geology, sediment-hosted massive sulphide deposits tend to have higher concentrations of lead, zinc, and silver and relatively smaller quantities of copper and iron than volcanic-hosted deposits. Two of the largest sediment-hosted massive sulphide deposits are the Sullivan (Canada) and the Broken Hill (Australia) deposits. The Sullivan orebody comprises 160 million tonnes of massive and bedded sulphides, consisting mainly of sphalerite, galena, and pyrrhotite. The deposit is located in a thick sequence of continentally derived turbidites which were intruded by gabbro sills shortly after the sediments were deposited. These sills may have provided the heat source for hydrothermal circulation. The massive sulphides formed at, or close to, the sediment–sea-water interface, possibly in a brine pool. The deposit is overlain by turbidites that were altered to an albite-rich assemblage in the waning stages of hydrothermal activity. The funnel-shaped alteration zone underlying the sulphides extends to a depth of least 1500 m and is unusual in containing very high concentrations of tourmaline. The Broken Hill district comprises 180 million tonnes of lead- and zinc-rich orebodies and has been a major source of these metals in the world. Unlike Sullivan, it has been extensively metamorphosed, so that many of the original features of the ore bodies are obscured. However, it also appears to be the product of hydrothermal activity within a thick sediment column. Both Sullivan and Broken Hill are thought to have formed in early continental rifting environments.

The factors required to generate a large orebody include: a source of metals, a means of transporting large amounts of metals in solution, and a means of trapping precipitates from these hydrothermal fluids before they are dispersed or destroyed. An intriguing feature of both Broken Hill and Sullivan is that there is evidence for the involvement of non-marine evaporites in their formation, and these evaporites may have been the key to the genesis of these two very large, rich ore bodies. Evaporites contain high concentrations of chloride and other anions that are readily dissolved by circulating hydrothermal fluids and greatly enhance the solubility of metals in solution through formation of dissolved complexes. A modern example is provided by the Salton Sea geothermal field in southern California, where deep hydrothermal fluids interact with non-marine evaporites to produce very high dissolved concentrations of ore-forming metals and actively deposit iron–zinc–lead–copper sulphide minerals. Non-marine evaporites are relatively common in early rifting environments. Their presence may be very important in the formation of giant sediment-hosted massive sulphide deposits.

M. R. PALMER

Further reading

Guilbert, J. M. and Park, C. F. (1986) *The geology of ore deposits.* W. H. Freeman, New York.

Matthews, Drummond Hoyle (1931–97) D. H.

Matthews was one of the British geophysicists who contributed most significantly to the development of plate tectonic theory and later to the study of the deeper parts of the continental crust. After National Service in the Royal Navy and graduating at Cambridge, he joined the Falklands Islands Dependencies Survey (now the British Antarctic Survey). He was engaged in field work in the Antarctic from 1955 to 1957 and while there became interested in the concept of continental drift. At that time the concept was regarded rather as geological heresy in Europe and North America, but the supporting geological evidence for the idea was more in favour in the southern hemisphere. Matthews returned to Cambridge in 1958 to study basalts dredged from the deep sea-floor of the eastern North Atlantic.

This work was extended in a series of marine geophysics research cruises into the North Atlantic and Indian Oceans and the Mediterranean. The Indian Ocean yielded data that led him with F. J. Vine in 1963 to explain the magnetic stripes that cover the ocean floor as injections at the mid-oceanic ridges and by the periodic reversals of the Earth's magnetic field. From this developed the paradigm of the impermanence of the ocean floor and the migration of continents, a concept adroitly and rapidly developed by Matthews and his graduate students.

In the 1970s Matthews and D. J. Blundell initiated the British Institutions Reflection Profiling Syndicate (BIRPS) to study deep crustal continental seismology. This highly successful project carried out seismic profiling on and around the British Isles, reaching depths of over 50 miles, and it revealed much of the deep foundations upon which the structures seen on the geological map are founded.

D. L. DINELEY

meanders and meandering rivers 'Meandering' is

the term used to describe the plan form of a river channel whose course winds smoothly through a series of bends. This type of pattern is usually distinguished from that of straight or braided (i.e. divided) channels. Meandering is a very common feature of natural river channels, but the morphology and stability of meanders varies.

In river meanders, the channel curves alternately right and left about the down-valley direction. Banks are steep on the outside of the bend, where erosion generally takes place, and are gently sloping on the inside where deposition normally occurs. Sediment deposited on the inside of the bend forms a *point bar*. The channel also varies in cross-sectional and longitudinal form through a meander: there is usually a shallowing (riffle) at the entrance or inflexion point of the loop and a deeper part (pool) on the outside of the bend in the apex. The plan form of a meander is frequently characterized as a smooth, symmetrical loop, often as a sine-generated form. Research has, however, shown that asymmetric and compound forms are very common.

The size of meanders, measured by the meander wavelength, is closely related to the amount of flow (discharge) of the river. The shape of meanders can be measured by the ratio of wavelength to meander breadth (the cross-valley dimension of the loop), or the ratio of the size of a circle that would fit the loop (radius of curvature) to the channel width. The degree of meandering for a reach can be described by the sinuosity, which is the ratio of the channel length to the straight line distance between two points on the channel. The shape and sinuosity of meanders is influenced by the erodibility of the materials of the banks and by the slope or energy of the river.

Why rivers meander is still not fully understood. Theories of origin include those associated with the distribution and dissipation of energy, those associated with helical flow, and those associated with bank erosion, sediment load, and deposition, but the consensus is that meandering is associated with inherent flow characteristics and intrinsic instabilities in the turbulent flow of water against a mobile channel boundary.

Early experiments showed that, under constant conditions, meanders tend to develop a stable, equilibrium form. Classically, river meanders have been represented as migrating in the downvalley direction while retaining their equilibrium form. Irregularities of form and occurrence of cutoffs were thought to be caused by variations in the bank material. Some river meanders have very low rates of erosion and are now stable, though they must have been active at some time in the past to have created the meandering pattern. Other meanders exhibit much more active processes, resulting in movement of the channel over time. Some of these meanders migrate in an orderly manner, but others have been found to evolve in form, with a tendency to become more asymmetric and compound in pattern over time. Cutoffs can occur by intersection of these evolving meanders. It is still debated whether such evolution is inherent to meanders or whether it is a response to changes in environmental conditions, such as the amount of flow in the river. Rates of movement and development on some active meandering rivers are such that change can be detected over periods of a century or so, for example by using earlier and later maps and aerial photographs.

The rate of change in a meandering river reflects the relative strengths of force and resistance in the bed and banks. The forces are influenced not only by the amount of water and the gradient of the reach but also by the pattern of currents within the bend. The pattern of currents was originally represented as a spiral or helicoidal flow, but more complex cells of circulation have since been identified which can influence the distribution of erosion and deposition. Bank erosion can take place by a number of mechanisms, but it occurs mainly by undercutting of the bank through direct removal of material by the river, after which blocks of material fall from the bank. Banks composed of silts, sands, and gravels are much more easily eroded than those with a high clay content or composed of bedrock. Deposition takes place in the meander where the energy of flow is insufficient to carry the size or amount of sediment in transport. Coarser material is deposited first. A decrease in sediment size is commonly evident downstream and across the bar. The form and stability of meanders is also influenced by the amount and size of sediment transported into a bend from upstream.

The consequence of meander migration and flooding is to create a zone of alluvial sediments bordering the river. The composition of this floodplain reflects the history of development of the river, its sediment load, and its mobility. Thus, traces of old channels (some of them oxbow lakes) are commonly found in the floodplains of meandering rivers. In some instances, meanders may not be developing entirely in recent alluvium but may be constrained by river terraces or the valley sides. A confined style of meandering then develops. Meanders have also been found to develop in bedrock by vertical incision of a meandering channel, usually over a much longer period of time (incised meanders). Some of these features are inherited from very different environmental conditions in the past.

The mobility of many meandering channels can lead to practical problems, particularly the undermining of structures built adjacent to the channel. The modern trend in channel management is to recognize this natural behaviour of rivers and to accommodate it. Many attempts have been made to model meander movement, both through computer simulations based on mathematical theory and from physical models. These are becoming increasingly sophisticated, and prediction of meander movement may become a reality in the near future.   J. M. HOOKE

Further reading

Schumm, S. A. (1977) *The fluvial system*. John Wiley and Sons, New York.

Leopold, L. B. (1994) *A view of the river*. Harvard University Press, Cambridge, Mass.

medical geology The relationship between the Earth's surface that we humans inhabit and our health is under debate. The fact that a continuum and indelible link exists is not in doubt. We have obtained food, water, and shelter since *Homo* arrived, but in the twentieth century we have learned that disease as well as health may by derived from our environment.

The geochemical distribution and biochemical availability of the elements that are required for human existence are not uniformly distributed over the Earth's surface. For example,

low concentrations of iodine (I) characterize the soils and rocks at high elevations and in limestone terrains. This is a natural global phenomenon. Medical acumen and geostatistical and epidemiological investigations have identified iodine as an essential nutrient. The thick necks that were depicted in ancient Chinese scrolls, and the cretinism found in mountainous regions, are now recognized as symptoms of the endemic disease goitre. Reduction, but unfortunately not eradication, of this preventable malady is now possible through the use of iodine-enriched table salt and oils.

Fluorine (F), another element that is a constituent of some minerals, is now added to drinking water to minimize the development of dental caries, especially in children. Apart from the beneficial effects of maintaining a healthy oral cavity to aid mastication and minimize pain, it is probable that ingestion of fluorine in small amounts (parts per million) over a lifetime will stave off osteoporosis, or at least serve to preserve the mineral materials in the skeleton in old age. It was the recognition of a connection between high natural fluorine concentrations (100 ppm) in the drinking waters of certain localities in Oklahoma and India and over-abundant calcium phosphate mineral deposition in the skeleton that most clearly illustrates the essential and continuing basic interactions between geology, geochemistry, medicine, and biochemistry. The fluorine effect, fully researched, led to applications aimed at reduction, if not prevention, of disease.

However, in the past half-century we have become very aware that the extraction of mineral materials that increase our comfort and standard of living may also be potential health hazards. Most of the health effects to which we have been alerted are maximized in occupational settings where high doses of natural products such as particulates (asbestos, for example) or gases (radon) have been implicated as disease-causing agents. We do not yet fully understand the mechanisms of action of these factors in the human body, but the cooperation of geologists, who investigate the mineralogy and crystal chemistry of such materials, and their distribution in the environment, with medical practitioners and researchers will be essential if we wish to benefit future populations.

Radon is a naturally occurring colourless, odourless gas that is emitted from rocks containing minerals rich in the transuranic elements. The occupational health effects, in particular lung cancer, suffered by some European coal-miners who mine such rocks were ascribed to radiation, but may equally well have been induced by smoking. Granites that underlie portions of the north-east of the United States (New England) are known to contain minerals that emit radon. Recent epidemiological studies that measured environmental exposure (the average was less than 4 picocuries for the region) were not able to demonstrate an association between the incidence of lung cancer and sites where radon concentrations (possible doses) were elevated.

A similar conundrum exists for asbestos. The asbestos miners of 25 to 50 years ago who were exposed to high doses of particulates may now have asbestosis, but not one case of this fibrotic pneumoconiosal disease has been unequivocally attributed to environmental exposure. In environmental situations, whether at home, in commercial buildings, or in schools, the measured number of fibre particles per cubic centimetre of air (about 0.001) is much too low to be significant, even if we assume accumulation over many years. The potential for a variety of diseases, especially lung cancers, in workers receiving high doses is obvious, but it is hard to evaluate the contribution made by personal choices, lifestyle, or habits to susceptibility to harm from any naturally occurring materials.

Occupational and environmental health studies undertaken to establish a connection between hazards and disease and to estimate exposures must rely on data that are difficult to obtain. Cancer is likely to have a long gestation period (lag-times of more than 20 years are usual) and the individuals, both affected and unaffected, who might have had similar exposures cannot be tracked adequately because people are more mobile today. Even though the technology exists to measure minute quantities of potentially hazardous materials, the determinations available are often too few and far between, or not necessarily appropriate to the particular site or situation we wish to evaluate.

As we contemplate an increase in world population and an ageing population, it becomes apparent that evaluating long-term exposure to natural materials in our environment makes cooperation and coordinated study of geology and medicine essential. The intertwining of these areas of knowledge should enable us to continue to improve health and combat disease, and contribute to better living conditions for all people.   H. CATHERINE W. SKINNER

medicval mineralogy and figured stones

It is true to say that during the Middle Ages no progress was made in mineralogy in Europe. What Aristotle (384–322 BC) had written in ancient Greece, classifying the minerals known then, remained unchallenged and unimproved into the nineteenth century. Not until well into the Renaissance were minerals and fossils denied occult, healing, or other supernatural powers.

Several theories were extant during medieval times to account for the varieties of 'stones' and their properties. Many of them refer to the 'life', a kind of inorganic force, that these objects may have had. Other ideas concerned the 'petrific seed', which had the power to condense water into mineral form, or to interact with emanations from deep in the Earth to produce mineral matter. Then came the theory of the 'lapidifying juice', a fluid that was thought to circulate through the crust of the Earth, the sea, and the atmosphere. This fluid might, it was believed, cause the development of mineral matter in the Earth or elsewhere, even in the human body—to produce kidney or other stones. Gems were though to have been produced from a simple crystallization of this fluid. Some writers regarded 'stones' as being either male or female and capable of producing young. Structures and markings, real or perhaps fanciful, were thought to indicate biological affinities and to be indicative of specific connections with living things. A further concept was that certain animals and plants produce gems or 'stones' in certain parts of their bodies, and that these, again, had special properties.

Healing or restorative powers were attributed to many minerals and rocks. Cures for a number of diseases, poisoning, or even mental conditions could, it was said, be made by administering mineral matter in one way or another. Several ancient compendia nevertheless list accurately the physical and chemical characteristics of many common minerals or rock types.

It was not until the sixteenth century that many of these notions were committed to print, but it is clear that many of the writers were then referring to beliefs held for many generations, if not centuries. Only at this time, also, do illustrations of minerals and fossils as we know them appear. The term 'glossopetra' was by then in use for objects of specific mineral character, such as colour, shape and symmetry, density, etc. They were held to be capable of growth while within the ground or while submerged in water, where they would acquire their characteristics. Clearly some of these objects were fossils, and to some were attached histories of their origin; devil's toe-nails, thunderbolts, and other fanciful names were applied to them. Some, such as the 'thunder stones' were, in fact, stone axes or other implements from prehistoric ages, and not uncommon in parts of Europe. The recognition of the true nature of such objects had to wait until the Renaissance, when scholars had access to, or made for themselves, collections of rocks, minerals, and fossils from many sources.   D. L. DINELEY

Further reading

Adams, F. D. (1938) *The birth and development of the geological sciences*. Dover Publications, New York.

Medieval Warm Period

The Medieval Warm Period (MWP) is the name given to the warm period observed in many parts of the world between 900 and 1200 AD approximately. The peak of this period is sometimes referred to as the 'Little Optimum', distinguishing it from the longer 'Postglacial Climatic Optimum' that lasted from the fifth to the second millennium BC.

Climate reconstruction is possible from two main sources: historical archives and dating techniques. Knowledge of the climatic history and its effects upon people owes much to the independent archival research of the British climatologist Hubert Lamb and the French historian Le Roy Ladurie. The most useful sources include agricultural records and accounts—proxy indicators of climate as a potentially destabilizing influence on society.

Archival research in and around Europe

Scrutiny of archive sources by Hubert Lamb in the 1960s revealed how climate affected human prosperity in medieval Europe. The Norse colonization of Greenland and Iceland—itself probably made possible only by the moderation in climate—provided a legacy of records in a climatically sensitive zone. Indeed, high latitudes provide the most emphatic evidence for increasing temperature. This is usually attributable to variations of sea ice; for most of the eleventh and twelfth centuries the waters around Iceland and southern Greenland were largely clear of ice.

The discovery early in the twentieth century of plant roots and Norse burial grounds in subsequently frozen soil suggests that temperatures were 2–4 °C higher than in the twentieth century. A chronology of annual temperature for Iceland, derived from sea-ice records, indicates a peak of warmth around 1100 AD, a finding confirmed by evidence of agricultural prosperity: the largest surviving foundations of farmhouses are of Norse origin.

Central European vineyards have been found at elevations more than 200 m higher than today, indicating that summers were up to 1.5 °C warmer. Vineyards spread northwards as well as upwards; they were quite widely distributed over southern England. This implies a freedom from late spring frosts, and summers that were relatively dry and sunny. This is confirmed by estimates of seasonal temperature and rainfall compiled by Lamb, who estimated the mean temperature for July and August in central England from 1150 to 1300 at 16.3 °C compared with 15.8 °C for 1900–50.

Apart from obvious problems of exaggeration, Lamb had to make adjustment for changes in the perception of 'normal' weather. Reconstructions can be verified by using the atmospheric circulation as a frame of reference for individual reports. Depression tracks are temperature-sensitive. This is illustrated by the more northerly track in summer, initiated by the poleward retreat of polar ice. It follows that depression tracks will be further north in warm periods, promoting drier summers (Fig. 1). Lamb has drawn attention to the care with which medieval water supply channels were constructed in Alpine valleys. Conversely, bridges dating from medieval times over some rivers in the Mediterranean have longer spans than the modern river width would require.

Evidence from dating techniques: assessing the global distribution of the MWP

Dating techniques have provided confirmation of archival evidence. In Europe, radiocarbon dates of wood found in moraines indicate a glacial minimum before 1200 AD, and it is known that tracks made over several Alpine passes were later overrun by glaciers. Ladurie concluded that a glacial advance in central Europe had begun by the early 1200s.

Elsewhere, evidence for warmth has been less conspicuous than increased supply of moisture. This could reflect the fact that agriculture in lower latitudes might be expected to be more sensitive to scarcity of moisture. In the drier western areas of North America, archaeological evidence indicates productive agriculture on land that is now arid. The Indian subculture of New Mexico and Arizona, for example, developed field terracing and irrigation.

A recent advance is the analysis of dendrochronologies that extend over a thousand years. The Laboratory of Tree Ring Research at the University of Arizona has analysed carbon isotope ratios in wood from bristlecone pines (*Pinus longaeva*) growing in the White Mountains region, which indicate a moist anomaly from 1080 to 1129. This is thought to represent the wettest period of the past thousand years. Rings from trees in the Sierra Nevada region indicate that temperatures exceeded late twentieth-century values between 1100 and 1375. A tree-ring chronology in the southern Canadian Rockies shows a relatively early temperature peak from 950 to 1100 AD, followed by a deterioration in which glaciers are known to have overrun mature forest. This early onset of the MWP at high latitudes is consistent with the results of Greenland ice cores that show rapid cooling as early as 1130 to 1190.

Evidence for warmth can be found in many parts of the world, especially at high latitudes. There is a bias to the northern hemisphere, but as dating techniques become more widely applied—and as interest in the event develops—the global distribution of climate anomalies is becoming clearer. A warming in the twelfth century was detected in a tree-ring chronology for Tasmania. Documentary evidence of citrus cultivation in China

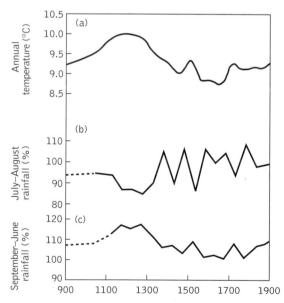

Fig. 1. Reconstructions of (a) annual temperature for central England (°C); (b) July and August rainfall for England and Wales; (c) September–June rainfall for England and Wales; ((b) and (c) expressed as percentages of the 1916–50 averages). (Source: H. H. Lamb (1965).)

suggests that a peak in warmth occurred in the thirteenth century. Hughes and Diaz, at the University of Arizona, do not support the notion that the MWP was a global event. This is contradicted by radiocarbon dating of glacial deposition by Grove and Switsur. There is, however, little doubt that the period represents the last occasion when substantial parts of the world were warmer than in the late twentieth century.

JULIAN MAYES

Further reading

Hughes, M. K. and Diaz, H. F. (1994) Was there a 'Medieval Warm Period', and if so, where and when? *Climatic Change*, **26** (2–3), 109–42. (See also other papers in the same issue.)

Lamb, H. H. (1965) The early Medieval warm epoch and its sequel. *Palaeogeography, Palaeoclimatology, Palaeoecology*, **1**, 13–37.

Le Roy Ladurie, E. (1972) *Times of feast, times of famine: a history of climate since the year 1000*. Allen and Unwin, London.

Grove, J. M. and Switsur, R. (1994) Glacial geological evidence for the Medieval Warm Period. *Climatic Change*, **26** (2–3), 143–69.

melt compositions *see* IGNEOUS MELT COMPOSITIONS

mesosphere The layer of the atmosphere in the altitude range 50 to 80 km is the mesosphere. Because *meso* is Greek for 'middle', the name is not really appropriate. All significant weather occurs in the troposphere at altitudes below about 15 km, and the mesosphere accounts for less than 0.1 per cent of the mass of the atmosphere.

There are no significant sources of heat in the mesosphere as there are in the stratosphere below and the thermosphere above. At the lower boundary of the mesosphere the temperature is about 0 °C; slightly higher in summer and slightly lower in winter. Absorption of ultraviolet radiation in the stratosphere produces heating below the mesosphere, which leads to this relatively high temperature. Some of this heat is transferred into the lower mesosphere and produces weak vertical motion throughout the mesosphere. As the air rises it cools, and a temperature decrease with height is observed throughout the mesosphere. The decrease in temperature is not, however, as great as would be expected from overturning air and it is about half of the decrease observed in the troposphere. This implies that radiative processes also contribute to the temperature structure of the mesosphere.

At the upper boundary of the mesosphere, the *mesopause*, the temperature is between –110 °C, in summer, and –60 °C, in winter. In summer the mesopause is the coldest part of the entire atmosphere. The low temperatures combined with weak updraughts are sufficient to form tenuous clouds in this region. These clouds are so thin that under normal circumstances they cannot be seen. At twilight, however, there is a period when the lower atmosphere is in the shadow of the Earth while the upper parts of the atmosphere are still in sunlight. At these times the clouds are visible and are therefore referred to as *noctilucent clouds*. The best viewing conditions

are at middle latitudes near the summer solstice, when the mesospheric temperatures reach their lowest values and the twilight hours are at their longest.   CHARLES N. DUNCAN

Further reading

Wallace, J. M. and Hobbs, P.V. (1977) *Atmospheric science: an introductory survey*. Academic Press, London.

Mesozoic The term 'Mesozoic' (Greek: middle life) was first used by Charles Lyell in his pioneering book *Principles of geology* (1830) to distinguish the rocks formed between the end of the Palaeozoic era and the beginning of the Tertiary (Cenozoic). It was thus roughly equivalent to the old Secondary category of the Italian classifications, and it included the rocks of the Triassic, Jurassic, and Cretaceous periods. Mesozoic rocks occur on all the continents, the Cretaceous being the most widespread and with enormous areas of outcrop in the northern continents and in Australia.

The fossils of the Mesozoic lie between two major extinction events ('catastrophes', as some of Lyell's predecessors called them). At the end of the Permian period, about 90 per cent of the marine biota seem to have died off, perhaps suddenly as a result of low global sea level and a profound change in atmospheric composition. This took place about 250 million years ago. A further mass extinction occurred at the end of the Cretaceous period, some 185 Ma later when, many geologists believe, a giant meteorite impact may have been the critical cause.

The biostratigraphical division of the Mesozoic rocks is primarily based upon common marine fossils: ammonoid and belemnite molluscs. The living world of the Mesozoic witnessed the reign of the mammal-like and other reptiles, including dinosaurs, the first appearance of the mammals and birds, the rise of many extant marine invertebrate families and of conifers and flowering plants.

At the beginning of the era the continents were gathered into a single supercontinent, Pangaea, but while moving northwards across the Equator and rotating during Mesozoic times this supercontinent broke up into the major present-day continental masses. Climatic conditions over much of Pangaea were markedly arid to begin with, but in Jurassic and Cretaceous times the climate grew to be predominantly warm and continental to equable. Perhaps associated with the continental movements and with processes in the mantle beneath was the enormous extrusion of basalts in several continents, and prodigious volcanic activity in the ocean basins. Activity in the mantle may also have been responsible for the changeable nature of the Earth's magnetic polarity, which swung from reverse early in the Triassic to normal and back to reverse many times throughout the rest of the era.

D. L. DINELEY

Further reading

Hailwood, E. A. and Kidd, R. B. (eds) (1993) *High resolution stratigraphy*. Geological Society Special Publication No. 70.

Moulade, M. and Nairn, A. E. M. (1978) *Phanerozoic geology of the world. Mesozoic A.* Academic Press, London.

Pomerol, C. (1982) *The Cenozoic Era: Tertiary and Quaternary.* (Trans. D. and E. E. Humphries.) Ellis Horwood, Chichester.

Messinian salinity crisis

Messinian salinity crisis The uppermost stage of the Miocene Series (5–25 million years (Ma) old), the Messinian, in its type area in Italy, consists very largely of salt and gypsum. In the rare shale beds there are a few diatoms, ostracods, spores, and pollen. Throughout the rest of the Mediterranean lands this stage appears to be composed of either the thin marginal deposits of a shallow marine basin or shallow-water, brackish-water sediments. However, in 1970 the drill ship *Glomar Challenger*, operating in the western Mediterranean, was able to core down through several hundred metres of pelagic Pliocene and Holocene beds into very thick evaporites, indicative of subtidal to supratidal conditions. It was unexpected to find rocks formed by evaporation of sea water in a deep basin currently occupied by the sea, but they have now been found in all the major basins of the Mediterranean, and they accompany a drastic drop in fossil numbers—a faunal crisis. The origin of this immense body of evaporites continues to intrigue us.

Salt deposition seems to have begun with the isolation of the Mediterranean Basin from both the Atlantic and the Indian (former Tethyan) oceans at the end of the deep-water Tortonian stage, about 6 Ma ago. The next 500 000 years or more saw the rapid shrinking of the water mass—first with depletion of oxygen, as witnessed by thin, black shale, then with the deposition of the great series of evaporites. One of the scientists aboard, the *Glomar Challenger*, the Chinese geologist resident in Switzerland, K. J. Hsu, and his co-workers proposed that the rapid evaporation of the trapped basin waters effectively took the water level down by a few kilometres. The surface area of the enclosed waters possibly diminished by as much as 2000 km², and this Mediterranean basin would have closely resembled the present Dead Sea except that it was much larger and much lower, and hence hotter. There is strong sedimentological evidence that the evaporites were precipitated in sabkhas or desert salt lakes. Also at this time, subaerial canyons were cut by rivers such as the Rhône draining from Europe into the basin; similar deep erosion was effected by the Nile in Egypt. The Alpine lakes of northern Italy, also, are now thought to occupy sites of headward erosion during Messinian time.

The precipitation of the million or more cubic kilometres of salts during the Messinian may have caused a lowering of world ocean-water salinity by 2 parts per million (p.p.m.). This in turn may have raised the freezing point of sea water sufficiently to allow the formation of an Antarctic ice cap, and a lowering of sea level served further to isolate the Mediterranean Basin. At the end of Messinian time a sudden influx of water produced a series of brackish and freshwater lakes, linked perhaps to the Paratethys Sea in the east. Salt precipitation ceased and the evaporite beds were sealed over with thin, shaly coverings. The end of the salt basin came when the Straits of Gibraltar, and perhaps other channels in what is now Spain, were breached by a rise in sea level as the Pliocene period began. The Messinian salts were effectively protected by the overlying argillaceous capping beds from solution by the Pliocene waters. Whether or not the rise of Atlantic sea level produced a gigantic cascade of Atlantic waters into the basin to restore it to world sea level and pelagic deposition is not certain, but the possibility of such a gigantic waterfall operating over thousands of years is by no means to be ruled out. Perhaps man's distant ancestors were able to watch it from an viewpoint near the Straits of Gibraltar?

Once world sea level had been restored in the Pliocene epoch a full succession of faunal communities was soon re-established in the Mediterranean Sea. There was little in the Messinian stage on land to suggest what a fantastic set of events had taken place in the deeper regions of the western Mediterranean Basin.

D. L. DINELEY

Further reading

Adams, C. C. *et al.* (1977) The Messinian salinity crisis and evidence of Late Miocene eustatic changes in the world ocean. *Nature*, **269**, 383–6.

Drooger, C. W. (ed.) (1973) *Messinian events in the Mediterranean.* North-Holland, Amsterdam.

Hallam, A. (1981) *Facies interpretation and the stratigraphic record.* W. H. Freeman, Oxford.

Hsu, K. J. *et al.* (1977) History of the Mediterranean salinity crisis. *Nature*, **267**, 399–403.

metallogenic provinces

metallogenic provinces Deposits containing high concentrations of valuable metals are not distributed randomly over the Earth's surface. In certain areas of the world, a higher density of deposits of a particular metal or group of metals is a conspicuous feature. Such regions delineate metallogenic provinces, which represent restricted areas of the crust where geological processes were suitable for the concentration of metals to form a group of ore deposits of a similar type. Some of these deposits were formed over a relatively short period of geological time: for example, the Iberian pyrite belt of southern Spain and Portugal contains over 80 known sulphide deposits that were formed in the Devonian to early Carboniferous. Formation over a short period of geological time is not, however, a strict requirement for the delineation of a metallogenic province. In the copper provinces of the southwestern United States and South America, numerous porphyry deposits of a comparable type and metal association are found, yet these have formed over a wide interval of geological time.

Metallogenic provinces containing concentrations of ore deposits of similar characteristics can also be found at different locations on the Earth. These may have at one time composed a larger single province that has been broken up by the processes of plate tectonics. They may alternatively have formed in different regions but by similar geological processes. For example, an unusually high density of vein-hosted gold deposits is found in certain areas of Precambrian shield rocks in Canada, Australia, Brazil, and southern Africa. The remarkable similarities between these deposits indicate that they were formed by identical processes. Metallogenic provinces can also overlap in areas where concentrations of more than one type of ore deposit or metal association occur. In the Precambrian shield rocks mentioned above, concentrations of volcanic-hosted *massive sulphide deposits* commonly overlap areas where the vein gold deposits are found, even though these two types of deposit formed at different times and by different processes.

The metal contents in common crustal rocks are too low to be mined economically. The generation of an ore deposit typically requires that metals be derived from a large volume of source rock and then concentrated many times into a relatively small volume of host rock. This is accomplished by a combination of geological conditions and processes that are active during the mineralizing event. These processes are in turn controlled by regional geological characteristics, such as rock type, structural discontinuities, subsurface fluid flow, temperature, and pressure. The regional tectonic setting determines the make-up of these features, and it is readily apparent why such concentrations of particular ore deposit types exist. Metallogenic provinces could thus be said to be a manifestation of the regional control of geological processes by tectonic setting.

Metallogenic provinces are delineated by the occurrence of numerous concentrations of mineralization with similar char-

acteristics and metal association. These concentrations may include economic orebodies, subeconomic deposits, and small showings. Metallogenic provinces have no fixed size and may always be extended or confirmed by the discovery of new mineralization. They provide a valuable first step to the exploration geologist in the search for further mineral resources.

BRUCE W. MOUNTAIN

metamorphic core complex The term 'metamorphic core complex' was first introduced in the early 1970s to describe occurrences of metamorphic and igneous rocks within North American Cordillera that have a very distinct structural setting. Metamorphic core complexes are large domal features that consist of three elements: a gneissic core, overlain by a low-angle extensional detachment (with dips between 0° and 20°), in turn overlain by supracrustal rocks. The detachment (also termed décollement) and the gneissic core are domal or broadly antiformal in shape. The gneissic core is commonly referred to as the *lower plate* or *footwall* and the supracrustal rocks as the *upper plate* or *hangingwall* (Fig. 1). The gneissic core is normally composed of high-grade gneisses, which commonly contain eclogites. The rocks are older than the detachment above them, and Precambrian gneisses can thus be involved in a Tertiary core complex. However, syn-extensional granitoid plutons commonly occur in the lower plate and may be deformed by the extensional detachment (Fig. 1). The upper plate can comprise sediments, basement slivers, and syn-extensional volcanic and sedimentary rocks; the sediments commonly include continental basin infill. High-angle normal faults commonly occur in the upper plate, rooted into the low-angle detachment. The extensional

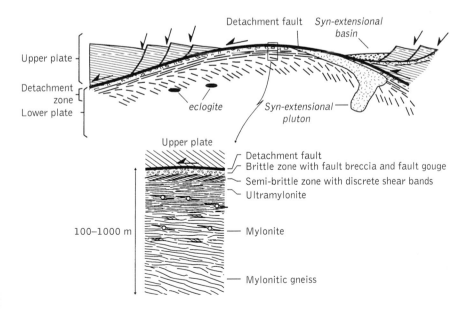

Fig. 1. Cross-section through a typical metamorphic core complex, showing the most important features. Features that are not diagnostic are shown in italics. The enlargement of the detachment zone shows the structures that normally occur in an extensional detachment zone.

669

detachment itself is usually a knife-sharp contact, underlain by a wider zone of brittle deformation containing fault gouge, fault breccia, and chlorite breccia. Below this there is an even wider zone of mylonite and mylonitic gneiss. This zone can be several kilometres thick.

Despite the low angle of the detachment, it differs from a conventional thrust fault because it puts younger, low-grade rocks over older, high-grade ones. The metamorphic core complex is thus formed as mid-crustal rocks are exhumed by tectonic unroofing in an extensional setting. In the classic examples of Utah, Nevada, and Arizona, the detachments are of mid-Tertiary age, and there is a Tertiary overprint on the high-grade metamorphic cores, for example resulting in local mylonite bands. The extent to which metamorphism or magmatic activity or infiltration of meteoric water accompany the development of the detachment and the emplacement of the core complex have been subjects of much controversy.

More than 30 core complexes are now known to occur in the North American Cordillera in a zone stretching from Arizona to the Canadian Rockies, and hundreds more have been recognized in orogenic belts elsewhere; for instance, in the Aegean, Turkey, the European Variscan, Papua New Guinea, and Egypt. They range in size from about 500 km² in the Aegean to the vast Western Gneiss Region (about 35 000 km²) in the Scandinavian Caledonides of west Norway.

Core complexes have been formed throughout the Earth's history: Pliocene–Pleistocene core complexes (possibly still active) occur in the D'Entrecastaux islands off Papua New Guinea, and Archean core complexes have been described in the Slave Province of the Canadian Shield and in the granite–greenstone belts of the Pilbara in Western Australia. It is likely that many more metamorphic core complexes remain to be discovered.

The detachment fault and the underlying shear zone are the result of a progressive evolution. The deep and hot part of the shear zone (at a depth of 20 km or more) initially deforms in a very ductile manner, resulting in a wide zone of mylonitic gneiss. During further extension, the lower plate is pulled towards higher regions and brought against the cooler upper plate. Ductile deformation is localized in narrow shear zones, cross-cutting the mylonitic gneiss. Continuing extension cools the shear zone further and drags the rocks through the brittle–ductile transition; the ductile structures are locally overprinted by cataclasis and brecciation. This overall retrogressive evolution results in a detachment zone in which ductile and brittle structures occur close together, but in which the latter will always have overprinted the former.

As mentioned above, the extensional detachments juxtapose younger, low-grade rocks on top of older, high-grade rocks. Mineral cooling ages (e.g. K–Ar and Ar–Ar ages) in the lower plate tend, however, to be *younger* than in the upper plate, because the lower plate cools later than the upper plate. A significant metamorphic break commonly occurs across the detachment, with low-pressure rocks in the upper plate and high-pressure rocks in the lower plate. Metamorphic core complexes are thus also effective vehicles for the exhumation of high-pressure rocks.

The recognition that metamorphic core complexes and their low-angle detachments were the result of continental extension was hampered for a time because rock mechanics predicts that it is implausible for extensional faults to form at such low angles. In addition, very few earthquakes have been recorded that had their origin in low-angle extensional faults. This apparent contradiction was explained when it was suggested that high-angle normal faults may rotate late during their evolution into low-angle normal faults, either by 'domino-faulting' or by the rolling hinge model, whereby a steep fault is dragged towards the surface and bent into a lower dip. Many observations of core complexes are, however, incompatible with these models, and it appears that many low-angle detachments were initiated *and* developed along low-angle faults: a typical example of theory being incompatible with geological observations. Alternative explanations for initiation and development along low angles have invoked high fluid pressures along the detachments or pre-existing anisotropies, as represented by the mylonitic gneiss, that were produced by ductile deformation deeper in the crust. These two mechanisms (possibly in combination) may create a weakness along which brittle movement can take place, even at low angles and with little seismic activity.

Despite the fact that continental extension often leads to subsidence, metamorphic core complexes are commonly elevated with respect to their surroundings. The high topography of individual core complexes (the D'Entrecastaux islands have an elevation of 2.5 km above sea level, but are only 20–30 km across) is the result of isostatic rebound of the lower plate as the overburden of the upper plate is progressively removed. The high topography of larger regions that contain a series of core complexes, such as the North American Cordillera, may, however, be better explained by removal of dense lithospheric mantle underneath the orogenic belt that resulted in uplift of the region as a whole and may also have been the cause of extension.

Metamorphic core complexes are sites of substantial continental extension and thinning of continental crust. All metamorphic core complexes occur in orogenic settings, that is, in sites of previously over-thickened crust, and the extension always post-dates the thickening. This post-thickening extension can be explained by orogenic (or extensional) collapse or by 'subduction roll-back'. The extension that produces core complexes is not necessarily related to continental break-up, although later continental rifting may reactivate the extensional detachments, as may have been the case in west Norway during the opening of the North Atlantic.

M. KRABBENDAM and BRUCE W. D. YARDLEY

Further reading

Coward, M. P., Dewey, J. F., and Hancock, P. L. (eds) (1987) *Continental extensional tectonics.* Geological Society of London, Special Publication No. 28.

Crittenden, M. D., Coney, P. J., and Davis, G. H. (eds) (1980) *Cordilleran metamorphic core complexes.* Memoir, Geological Society of America No. **153**.

Hill, E. J., Baldwin, S. L., and Lister, G. S. (1992) Unroofing of active metamorphic core complexes in the D'Entrecasteaux Islands, Papua New Guinea. *Geology*, **20**, 907–10.

metamorphism, metamorphic facies, and metamorphic rocks

'Metamorphism' is a term for the alteration that takes place as a rock, formed originally in an igneous or sedimentary environment, recrystallizes to produce a metamorphic rock as a result of changes in temperature, pressure, or the availability of fluids. The grains that make up metamorphic rocks have typically recrystallized during metamorphism, and commonly also include new mineral species formed by metamorphic reactions. Metamorphism normally takes place in a near-solid state: pore fluids play a vital kinetic role in many instances, but are only a very minor component of the rock. As a result, most metamorphic rocks retain some of the characteristics of the parent material, such as bulk chemical composition or bedding. Extreme metamorphism can lead to the onset of partial melting and the formation of migmatites.

Metamorphic changes are of two types: phase changes (crystallization), which reflect the depth and temperature at which metamorphism has taken place, and textural changes (recrystallization), which are particularly sensitive to strain accompanying metamorphism, and affect all rock-types, irrespective of how chemically reactive they are. For example, pure quartzites or pure calcite limestones do not undergo reactions under most metamorphic conditions because quartz and calcite are stable over a very wide range of crustal conditions. They may, however, experience grain growth or other forms of recrystallization, and may develop mineral orientation fabrics that record much information about the deformational history of the rock during metamorphism. Interbedded clay-rich sediments (pelites) are, however, very reactive during heating, because the original clay minerals are of very limited stability, and so develop a sequence of metamorphic minerals that provide a measure of the temperature and pressure at which metamorphism took place.

Types and settings of metamorphism

Most metamorphism takes place as the result of the progressive burial and accompanying heating of sedimentary sequences in deep basins, accretionary prisms, and subduction zones, or as the result of heating around igneous intrusions. These two types of setting are not mutually exclusive, but are the basis for the distinction between *regional metamorphism* and *contact* (or *thermal*) *metamorphism*. Regional metamorphism has affected large tracts of rock in orogenic belts, and heating is typically accompanied by deep burial and ductile deformation. As a result, the rocks produced are often strongly foliated slates, schists, or gneisses. Pre-existing crystalline basement rocks formed in much

earlier orogenic events can also be caught up in younger regional metamorphism of their cover sequence: for example, the Hercynian massifs of the Alps. Magmas may provide a component of the heating of regional metamorphism, but the classic contact metamorphism caused by the heat of intrusions is unaccompanied by any deformation. Other types of metamorphism are generally rather localized. *Hydrothermal metamorphism* is caused by the action of circulating hot water, leading to hydration of any original igneous material and growth of cements and veins of minerals stable in the presence of hot, pressurized water (typically at 70–350 °C). *Cataclastic metamorphism* is the result of intense deformation with little additional heating, and is localized in fault zones. *Impact metamorphism* takes place at sites of major meteorite impacts; its effects range from melting or even volatilization of the original surface rocks, through the production of dense minerals during the passage of intense shock waves, to the formation of shock-induced deformation structures.

Although it is not possible to see active metamorphism taking place in the same way that we can watch some igneous or sedimentary rocks being formed, hydrothermal metamorphism is occurring near the surface today. For example, in high-temperature geothermal fields, such as those that are exploited for power in Iceland, Italy, New Zealand, and elsewhere, volcanic glass and high-temperature basalt minerals are being actively converted to clays, chlorite, zeolites, epidote, and other minerals that are more stable in the cooler, but wet, environment of the geothermal system at depths of only a few hundred metres. In these settings, drilling into the geothermal system provides both samples and detailed information about the conditions of metamorphism.

Depth and temperature of metamorphism

One of the most interesting things that we can learn from the study of metamorphic rocks is the depth and temperature at which their constituent minerals have grown. This tells us not only about the depths to which they have been buried (and values of about 100 km for the burial of some extreme rock-types accord well with the depth to which continental crust has been reported from modern geophysical measurements), but also about the temperature gradient when they formed. Temperature gradients can be diagnostic of particular tectonic settings, but can also have changed significantly during the early history of the earth. Geologists and chemists at the beginning of the twentieth century had recognized not only that certain minerals were indicative of higher or lower temperatures of metamorphism but also that it was associations or *assemblages* of minerals, rather than the occurrence of specific individual minerals, that best constrained the conditions under which a rock had formed. V. M. Goldschmidt, and later P. E. Eskola, pioneered the application of the principles of chemical equilibrium to metamorphic mineral assemblages, and this has remained one of the cornerstones of metamorphic petrology. If it can be demonstrated that mineral assemblages formed in equilibrium for the prevailing conditions of pressure (a simple function of depth of burial) and temperature, then laboratory experiments or thermodynamic calculations can in principle be used to quantify those conditions. In practice it has only been in the latter third of the twentieth century that the experimental and latterly the thermodynamic database for metamorphic minerals has been adequate for the purpose in all but the simplest cases, but *geothermobarometry*—estimating the conditions of formation of a metamorphic assemblage—has now become relatively routine.

Metamorphic grades, zones, and facies

The principal reason for believing that the coexisting metamorphic minerals in a rock (i.e. the metamorphic mineral assemblage) usually form near chemical equilibrium is the observation that, over large areas, rocks of similar chemical composition contain the same assemblage of minerals. In areas such as contact metamorphic aureoles around a pluton, these assemblages may vary systematically towards the heat source in a way that reflects increasing temperature of crystallization. There are a number of ways in which the qualitative variation in the conditions of metamorphism between different rocks can be described. Most generally, *metamorphic grade* is a useful and rather loosely defined term, used to distinguish rocks that have undergone more extreme conditions of temperature or pressure, or both (high-grade metamorphism), from those that have recrystallized under less extreme conditions (low-grade metamorphism).

Mineral assemblages reflect the initial composition of the rock as well as the conditions of metamorphism, but for any one rock-type it is normally possible to recognize distinct metamorphic zones characterized by a different mineral or assemblage of minerals. A *metamorphic zone* is a mappable region of rocks in which a particular mineral or association of minerals is present within specific compositions of rock. Metamorphic zones were originally often defined on the basis of the appearance of a specific mineral, known as an *index mineral* (see *regional metamorphism*). This index mineral will usually survive to higher metamorphic grades, in the presence of new index minerals indicative of higher metamorphic grades. For example, biotite is the index mineral of the biotite zone, in which it appears at relatively low grades of metamorphism. However, biotite continues to be present in a wide range of metamorphic rocks formed at all but the very highest grades. For this reason, metamorphic zones are generally defined, where possible, by the occurrence of an assemblage of coexisting minerals, rather than on the basis of any one phase; for example, the staurolite zone is characterized by the presence of staurolite in rocks that contain muscovite and quartz but do not contain an aluminium silicate polymorph. Metamorphic zones have been most widely recognized in mica-rich rocks that were originally clay-rich sediments, but zones can also be identified in a number of other rock-types, notably impure limestones and metamorphosed volcanic rocks. The boundaries that can be mapped between zones by recording the distribution of mineral assemblages are known as *isograds*; they represent surfaces of constant metamorphic grade. Metamorphic zones are generally specific to one particular rock-type; for example, only aluminium-rich pelites normally develop staurolite in the staurolite zone, but some index minerals are more widespread.

In most successions of metamorphic rocks, those with reactive lithologies, such as mature pelites, make up only a small proportion of the rock mass. As a result, while it may be possible to define metamorphic zones precisely in a few places, the zones and their associated isograds cannot be traced through the intervening tracts of ground. The concept of *metamorphic facies* was introduced by Eskola in order to provide a more general indication of conditions of metamorphism that would be more widely applicable than precisely defined zones. A metamorphic facies is characterized by an assemblage or sub-assemblage of metamorphic minerals, indicative of a specific range of pressure–temperature (*P–T*) conditions of metamorphism. In general, metamorphic facies have been defined on the basis of the assemblages of metamorphosed basic igneous rocks (metabasites), but equivalent assemblages in other rock compositions are well known. It is interesting to note that at very low grades of metamorphism, where the original igneous mineral assemblages are far removed from their original conditions of formation, a number of discrete zones can be recognized in metamorphosed volcanic rocks, but not in any other rock-types. These zones do not therefore have the status of facies. The approximate ranges of pressure (*P*) and temperature (*T*) over which the assemblages of the various facies develop are shown in Fig. 1. The characteristics of the main metamorphic facies can be summarized as follows:

Facies of regional metamorphism (see *regional metamorphism*). The most widespread metamorphic facies are those typical of regional metamorphic terrains. The lowest-grade facies, i.e. the zeolite facies and the prehnite–pumpellyite facies, are characterized by the occurrence of hydrous minerals developed from the breakdown of volcanic glass and fragments of igneous minerals, especially in volcanogenic sediments. Massive volcanic rocks are usually too impermeable to undergo significant hydration at these grades, and so may show only local alteration, while siliciclastic sediments undergo normal diagenetic processes at zeolite facies temperatures (below *c.* 250 °C), and with further heating evolve towards greenschist facies assemblages (white mica and chlorite) without developing any characteristic assemblages at lower grades (e.g. most slates). Increasing regional metamorphic grade is generally accompanied by thorough recrystallization of all but the most massive igneous bodies, with the formation of calcic amphiboles whose chemistry reflects the conditions of formation. Both metabasites and metasediments become progressively dehydrated through the greenschist and amphibolite facies. Mineral abundances generally change progressively, but garnet appears in the upper greenschist facies, while chlorite and muscovite become rarer. Temperatures in the uppermost amphibolite facies are

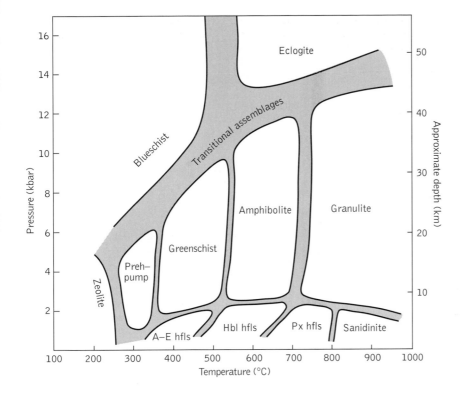

Fig. 1. Pressure–temperature diagram showing the approximate conditions in which the assemblages of the principal metamorphic facies are stable. (Modified from Yardley 1989 and Yardley *et al*. 1990). Hfls, hornfels; A–E, albite–epidote; Hbl, hornblende; Px, pyroxene, Preh–pump, prehnite–pumpellyite.

sufficient for melting to take place in dehydrating metasediment, to produce migmatites. Further metamorphism in the granulite facies results in the formation of minerals that are also typical of igneous rocks (e.g. pyroxenes), although some hydrous minerals, notable biotite and hornblende, can persist.

In some regionally metamorphosed terrains, distinct assemblages are present that are indicative of unusually high pressures. In the blueschist facies, the sodic-amphibole glaucophane is diagnostic, most characteristically occurring with lawsonite (a calcium aluminium silicate). In pelitic rocks, biotite is rare or absent, and garnet appears at relatively low grades. Low-grade blueschists may contain aragonite rather than calcite. Eclogite-facies metamorphism is typical of deeply subducted or overthickened crust, and is notable for the absence of plagioclase (see *eclogite*). In recent years, a number of eclogites have been found to contain evidence for ultra-high-pressure metamorphism, characterized by the presence of the dense silica polymorph coesite, which is preserved as inclusions in garnet; in a few instances, diamonds have also been reported.

Facies of contact metamorphism. The assemblages produced in contact metamorphic aureoles have many similarities to regional assemblages, but dense phases such as garnet are usually absent, while less dense phases such as andalusite and cordierite are widespread in rocks of suitable composition.

Textures of metamorphic rocks

Metamorphic rocks are distinctive for their textures as well as for their mineral assemblages. Regionally metamorphosed rocks commonly recrystallize in response to deformation as well as to changes in pressure and temperature, although the two driving forces can operate together. As a result of the deformation, minerals are commonly aligned in a way that reflects the finite strain over the period during which they recrystallized or grew. Most commonly, platy minerals such as micas are aligned perpendicular to the maximum shortening direction to produce a *schistosity*. In intensely deformed rocks, many minerals can in fact form aligned tabular grains: quartz, feldspar, and calcite for example. Elongate mineral grains, such as those of amphiboles, may also be aligned in the maximum extension direction, giving rise to a *lineation*. Many regionally metamorphosed terranes have experienced complex histories of deformation, and fabrics of aligned metamorphic minerals can locally preserve traces of early deformation episodes where they have not been obliterated by later events.

Although contact metamorphism can be accompanied by deformation, many metamorphic aureoles around plutons develop by static recrystallization (i.e. without strain), and so the minerals grow in random orientations, giving rise to a tough rock that does not split in any preferred manner: a *hornfels*. Hydrothermal metamorphism is likewise commonly without deformation, and often results in the formation of *pseudomorphs*, in which the patterns of pre-existing minerals or textures are preserved in the shape, distribution, or compositional zoning of newly formed metasomatic minerals.

Many of the distinctive metamorphic minerals that provide indicators of metamorphic grade (e.g. garnet) do not readily align into fabrics, but commonly grow as crystals that are relatively large compared to the rock matrix. Such grains are termed *porphyroblasts*. It is common for porphyroblasts to contain inclusions of minerals that were present in the matrix when they grew, and in this way they preserve information about the history of metamorphism. Such inclusions may be aligned, preserving evidence of the fabric in the rock when the porphyroblast grew. An important development in the 1950s and 1960s was the recognition of how such textures could be used to deduce the relative timing of burial, heating, and deformation in the development of an orogenic belt.

Metamorphic processes and rates

In the early days of geology, it was assumed that crystalline metamorphic rocks were inevitably older than sedimentary strata, and that their formation required long periods of geological time. It is, however, now known that although the oldest rocks on earth have experienced metamorphism, there are also much younger metamorphic belts exposed at the surface of the planet, some affecting rocks of Cretaceous age or younger. The youngest belts doubtless remain at depth and are continuing to evolve today. Geochronology indicates that while many metamorphic terranes have experienced long histories of crystallization, spanning hundreds of millions of years, individual metamorphic cycles probably have durations of the order of a few tens of millions of years. Progressive burial and metamorphism of sedimentary sequences consumes large amounts of heat, for the reactions are strongly endothermic. It follows from simple heat-flow considerations that heating rates on a regional scale will be of the order of a few degrees to a few tens of degrees per million years. As geochronologists developed techniques for dating metamorphic index minerals themselves, it became possible to address this question directly, and in a pioneering study, K. W. Burton and R. K. O'Nions showed in 1991 that garnet-bearing rocks from the Norwegian Caledonides had indeed been heated up at close to 10 degrees per million years.

Nomenclature of metamorphic rocks

It is impossible to have a rigidly defined system of nomenclature for metamorphic rocks, partly because of the complex history of events that give rise to them, and partly because of the range of purposes for which the name is required. Metamorphic rock names may reflect any or all of the following: the nature of the parent rock; the grade of metamorphism and the minerals present; the texture of the rock. There are also a few special rock names which encompass several of these factors.

The simplest metamorphic rock names are obtained by adding the prefix *meta-* to the original rock type, as in *meta-gabbro, metabasite, metagranite*. For most metasediments, however, there are specific names: *marble* for metalimestone, or *pelite* for metaclay. These may be qualified by mineral names to indicate the distinctive members of the mineral assemblage, either based on abundance, to indicate rock composition and aid stratigraphical correlation, or based on the distinctive association that best defines the *P–T* conditions of metamorphism. In high-grade gneiss terranes, it can be impossible to deduce much about the nature of the parent rock-type in detail, and the term *paragneiss* is then used to denote a rock with a presumed sedimentary parent; *orthogneiss* implies an igneous origin. High-grade metamorphic rocks can be very difficult to distinguish from igneous rocks that crystallized at depth.

Textural terms are also an important component of metamorphic rock names, because the texture dominates the appearance of the rock in hand specimen. Foliated rocks may be slate, phyllite, or schist according to grain size. *Gneiss* is variously used for a uniformly coarse-grained high-grade metamorphic rock lacking any marked foliation and for a banded high-grade rock in which the foliation is defined by alternating bands of distinct composition, rather than by mineral alignment. The term *granulite* can also be applied to many high-grade rocks; it carries specific connotations of formation under conditions of granulite-facies metamorphism.

Specific rock names that record both the parent rock and its mineralogy are relatively rare, but include *eclogite, blueschist*, and *amphibolite*. These are all varieties of metabasite; eclogite is dominated by sodic pyroxene and garnet, and lacks plagioclase; blueschist is named for a slaty blue-grey colour resulting from the presence of glaucophane; and amphibolite denotes a rock dominated by hornblende and plagioclase. There is a complex nomenclature for rocks of the granulite facies that includes a number of traditional names that lack any rational basis and appear to be applied to rocks of both metamorphic and igneous origin. The most commonly encountered is *charnockite*, a coarse-grained granitic rock containing both orthopyroxene and potassium-feldspar.

BRUCE W. D. YARDLEY

Further reading

Spry, A. (1969) *Metamorphic textures*. Pergammon Press, Oxford.

Yardley, B. W. D. (1989) *An introduction to metamorphic petrology*. Longman, Harlow.

Yardley, B. W. D., MacKenzie, W. S., and Guildford, C. (1990) *Atlas of metamorphic rocks and their textures*. Longmans, Harlow.

metasomatism Metasomatism is a type of metamorphism caused by reaction between the minerals in the original rock and hot fluids that have passed through it. These fluids flush specific elements in or out of the rock, thereby changing its overall chemical composition, which thus distinguishes the process from other kinds of metamorphism.

The usual situation in which metasomatism occurs is when igneous rock, especially a large body such as a granite pluton, is intruded into water-bearing sediments. The heat of the igneous body initiates convection of the water, which is heated as it is drawn into the body and carries heat away after it becomes warm and buoyant, only to be replaced by cooler water drawn in from the sides. Such a hydrothermal circulation system can survive for hundreds of thousands of years. The consequent dissolution and precipitation of chemical elements by the hydrothermal fluids can be responsible for major redistribution of elements, and in some cases their concentration into economic ore deposits. For example, hydrothermal circulation and metasomatic redistribution of elements around igneous intrusions in the Andes is responsible for the world's largest copper deposits, and the 'china clay' of Cornwall is the metasomatized remains of granite that has been thoroughly rotted to leave a residue of the clay mineral kaolinite.

DAVID A. ROTHERY

metazoans Metazoans are complex multicellular organisms and comprise almost all of what are generally considered to be animals. They show an enormous diversity and encompass twenty-nine phyla, all of which, apart from the Chordata, are considered to be invertebrates. The metazoans had their origins in unicellular eukaryotic ancestors (single-celled but with a nucleus), but it is currently not possible to say exactly when that occurred. The earliest eukaryotes might be as old as 1600 million years (Ma), whereas the oldest known metazoans are about 620 Ma, thus leaving about 1000 million years for the development of the metazoans.

The earliest good evidence for the presence of metazoans is provided by the Ediacara fauna, which was described from Ediacara, south east Australia, in the late 1940s. Although it was originally thought to be Cambrian, the persistence of this fauna below shelly fossils of early Cambrian age led to its being correctly assigned to the Precambrian. This entirely soft-bodied fauna appears to have existed worldwide from 620 to 550 Ma. Obvious features of the fauna include concentration in shallow-marine environments, an absence of mobile predators and scavengers, a dominance of coelenterate-like animals (modern corals and jellyfish), and large size; the leaf-like *Dickinsonia* for instance can be up to a metre in length. These organisms have traditionally been compared to younger Palaeozoic and even Recent animals, with the result that they have all been placed in modern taxa. This approach has been criticized by Adolf Seilacher of Tübingen University, who has pointed out that many of the organisms in the Ediacara fauna show a unique organization and no close affinity with Recent invertebrates. Characteristically they are relatively flat with complex surface relief, which provides a low ratio of body volume to surface. This morphology may have allowed gases and nutrients to diffuse through the body surfaces without the need for complex respiratory or digestive systems. This group of metazoans, termed Vendozoa by Seilacher, became extinct before the end of the Precambrian and were replaced by the Tommotian fauna, small shelly fossils which are particularly numerous at the base of the Cambrian.

The reason for the extinction of the Vendozoa is unclear; it was, however, followed by a spectacular diversification of invertebrates during the Cambrian. The development of hard-parts formed of calcium carbonate, silica, and calcium phosphate is clearly important to this diversification, not least because it greatly enhanced preservability and thus provided an extensive fossil record for the first time. The earliest organisms had composite skeletons composed of small elements that disarticulated after death. These somewhat enigmatic organisms were succeeded by complex invertebrate communities after about ten million years. Dominant elements of this Cambrian fauna were trilobites, inarticulate brachiopods, molluscs, echinoderms, and archaeocyathids. This community of deposit- and suspension-feeders was replaced in the Ordovician by the Palaeozoic fauna, in which articulate brachiopods, crinoids, and bryozoans dominated. The Palaeozoic fauna, consisting mainly of upper-level suspension-feeders, suffered badly during the Permian–Triassic extinction event and was replaced by a more complex modern fauna in which bivalves, gastropods, corals, and echinoids dominate, and which is present today.

DAVID K. ELLIOTT

meteorite and comet impacts on Earth The recognition of the results of impacts upon Earth has come about as a relatively recent, and at first esoteric, study. At the end of the first quarter of the twentieth century the number of what are termed 'certain' impact craters could be counted on the fingers of one hand, and even by 1950 two hands would have sufficed for the finger count. However, in the decade 1960–70, the number of 'certain' impacts increased dramatically, and by the early 1990s the number reached 130. These craters range in size from a few tens of metres in diameter to major impact craters up to about 200 km across, and they range in age from Recent back to Precambrian times. The reasons for the sudden increase in the accredited number of 'certain' impact features are twofold. First, experiments demonstrated that impact craters could be significantly more complex than was expected. Secondly, it was realized that the stress generated by major natural impacts was so high that it could produce, among other effects, vaporization, melting, and the generation of stress-induced changes in minerals (for example, quartz to coesite, a high-pressure polymorph).

In the period immediately after the Second World War, several nations carried out explosive tests to establish the effectiveness of megaton nuclear devices; for example, the Bikini Test in the Pacific. Of these tests, it was those carried out at Suffield, in Alberta, which were the most important for the study of natural impact features. The importance of the Suffield results, which were reported by G. H. S. Jones, is

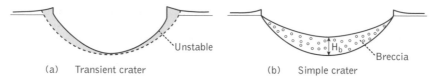

Fig. 1. (a) Section across a transient crater formed by a 20-tonne TNT explosion; (b) section across a crater formed by a 100-tonne TNT explosion. (From Report SSP 177 [unclassified], Defence Research Establishment, Suffield, Canada, by courtesy of G. H. S. Jones.)

related to the physical properties of the sediments on which the tests were conducted. The test site was set on a plain which is the base of a dried-out glacial lake. These Recent sediments consist of uncompacted and uncemented sands, clays, and silts, and they have a strength of the order of one thousandth that of many strong metamorphosed basement, or igneous, rocks. Also, the water-table is about 8 m below the surface.

The explosive tests conducted at Suffield from 1960 until 1970 were mainly in the range from 20 to 500 tons of TNT. A feature of these explosive experiments, especially those of high tonnage, was that the sites were carefully surveyed and numbered, sand-filled, cans were set at specific depths within the sand columns. After an explosion, the site was resurveyed and sectioned by deep trenches so that the geometry of various structures and ground movement that resulted could be accurately established.

The crater created by a 20-ton TNT explosion had a diameter of 23 m and a depth of 4.5 m; 100-ton explosions gave rise to craters with diameters of 36 m and depths of about 8.5 m. It was realized that the production of such experimental craters resulted from two phases of development (see Fig. 1a, b). These consisted of an initial *transient crater*, which then rapidly developed by slumping of the walls and infilling of the lower regions of the transient crater to give rise to a wider and shallower *simple crater*.

As M. K. Hubbert showed in 1937, when model experiments are conducted, meaningful results are attained only when the strength of a model is appropriately scaled (i.e., reduced) with respect to the full-scale event being investi-

gated. Explosions in the 20–100-tonne range produced excellent models of natural craters with peripheral recumbent folds (Fig. 2a, b), which are a feature of natural craters such as the Barringer (or Meteor) Crater, Arizona and other simple craters up to a diameter of about 3.5 km. Until explosive tests were conducted at higher energies, it was generally considered that the simple crater was the only manifestation of natural impacts. A series of 500-ton TNT explosions completely changed this viewpoint.

The first 500-ton experiment, code-named 'Snowball', had a hemispherical configuration for the bomb. The resulting crater was 88 m in diameter. There was a central uplift in the crater as well as a variety of fractures and folds external to the crater (Fig. 2a). Beyond the crater, the rim marked the development of a flat-lying recumbent fold (F1). Outside this fold, and in some instances partly obscured by it, circumferential vertical (normal) faults defined a peripheral graben (*PG*, Fig. 2b). Minor folds were seen in the macadam survey paths leading to the bomb at ground zero (GZ). Vertical radiating fractures (*R*) were also developed well beyond the crater rim. The graben faults and vertical radiating fractures formed pathways for the upward migration of water-bearing sands, muds, and silts that formed mud flats or sedimentary volcanoes. The final structures, revealed by deep trenching, were a series of shear planes (*Thr*) which dipped inward towards ground zero at about 25°. These features originated as thrusts, but were in some instances transformed into low-angle normal faults by slumping of the sediments into the crater.

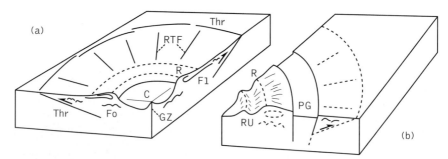

Fig. 2. Block diagrams illustrating structures associated with craters formed by 500-tonne TNT explosions: (a) central uplift crater ('Snowball' test); (b) ringed uplift crater ('Prairie Flat' test). C, central uplift; F1, flat-lying recumbent fold; Fo, folds; GZ, ground zero; PG, peripheral graben; R, rim; RTF, radial (vertical) tensile fractures; RU, ringed uplift; Thr, thrusts.

A later experiment, code-named 'Prairie Flat' used a spherical configuration for the bomb, which merely touched the horizontal target rocks. The 'Prairie Flat' and similar 500-ton experiments produced many of the features external to the crater that were formed by 'Snowball'. Within the crater, however, the morphology was completely different, in that, instead of a central uplift, the crater contained concentric ringed uplifts (*RU*, Fig. 2b).

Many of the features shown in Fig. 2 had been observed on the Moon, but their significance was, until the time of these experiments, not understood. It was, however, soon realized that the Snowball and Prairie Flat features were the first man-made Moon craters on Earth. The pioneers who were studying Earth impacts immediately recognized that many of structures shown in Fig. 2 were diagnostic features that made it possible to identify 'certain' natural impact features on Earth.

An example of an impact feature with a central uplift is that of Mistastin Lake in Labrador, Canada, which is 28 km in diameter. Manicouagan, in Quebec, Canada, is a structure 100 km in diameter with a surrounding peripheral graben (Fig. 3a). The central areas exhibit concentric ridges and include a layer of impact melt 100 m thick. An unnamed ringed uplift structure, 50 km in diameter, in Mauritania, Africa, is shown in Fig. 3b.

The Moon exhibits a much higher concentration of impacts than can be seen on Earth. This can, in part, be attributed to the fact that the Moon is devoid of atmosphere. Because the Moon has no atmosphere it has not experienced the intense weathering, erosion, and sedimentation that characterizes the Earth's history. On the Moon, old small craters have been covered by ejecta, or lava flows, created by later

impact events. Otherwise, even very old craters remain largely unaltered from the date of impact.

On Earth, continental erosion, over a period of perhaps hundreds of millions of years, may completely destroy evidence of even moderately large impact structures: sedimentation can completely obscure impacts of moderate size. At the time of writing, five such buried impact structures have been reported in the literature, of which only one is classified as certain. However, as J. W. Norman has indicated, the features that may remain to indicate the existence of such an impact are traces of arcuate grabens in relatively young flat-lying sediments which cover Precambrian basement.

It usually requires considerable effort to prove that a given terrestrial feature is the result of an impact. When such an impact occurs in a deep ocean, it becomes even more difficult to detect. This, of course, is further compounded by the fact that oceanic lithosphere is subducted. Very little of the present-day ocean floor dates back more than 150 Ma; only the continents can therefore show records of impacts back into the Precambrian.

The Earth's atmosphere has a further protective influence. From the bottom of the exosphere (the outer region of the upper atmosphere) the atmosphere has a thickness of 400 km. Meteoritic or cometary material usually approaches the Earth's surface at an acute angle, so that the apparent thickness of the atmosphere can exceed 1000 km. Small bodies that enter the upper atmosphere are heated by the frictional effect of encountering a progressively more dense atmosphere. The temperature in such bodies rises rapidly, so that small particles vaporize, giving rise to 'shooting stars', which are usually 'consumed' at altitudes of about 45 km above the Earth's surface. In order to impact upon Earth, such bodies have to

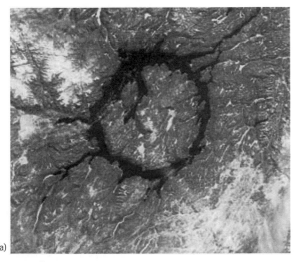

(a)

(b)

Fig. 3. (a) The Manicouagan impact structure, with peripheral graben. (b) A probable ringed-uplift impact structure, 50 km in diameter, in Mauretania, Africa. (From a NASA satellite image).

exceed certain dimensions; and these dimensions depend upon the composition of the body. For spheres composed wholly of ice, stone, or iron to impact upon Earth they must have diameters in excess of 150, 60, and 20 m respectively. If such bodies were to strike the Moon's surface they would, depending upon their velocity, give rise to craters with diameters of 0.25 to 2.5 km. Hence, craters on Earth of such diameter are under-represented relative to those on the Moon.

However, even if such bodies do not quite reach the Earth's surface, they are still capable of causing considerable destruction. For example, in late June 1908 what is thought to be anice body exploded in the remote area of Tanguska in northern Siberia with an estimated energy of a million tonnes of TNT at a height of 7 km above the Earth's surface. The air blast flattened trees over an area of about 2000 km². There were no human casualties, except for two brothers who experienced shock when their tent was blown away. Had such an event taken place above one of our major cities the ensuing devastation would have concentrated the minds of Earth scientists on our greatest natural hazard.

There is one other major problem associated with identifying very large impacts, whether terrestrial or oceanic. The Moon is cold. Nevertheless the energy of a very large impact is sufficient to generate a volume of melt that almost fills the crater. On Earth, hot asthenosphere, near its melting point, rises to within a few kilometres of the surface of the ocean ridges and only rarely has a thickness in excess of 100 km.

The Ontong–Java plateau basalt, which dates from about 120 Ma, has an area equal to about one-third that of the USA. It is formed by an estimated volume of $5–7 \times 10^7$ km³ of basalt emplaced in about a million years. Conventional thinking has great difficulty in explaining how such a vast feature could develop so quickly. The age of the ocean floor surrounding this extrusion is only 20–30 My older than the eruption event. Hence one may infer that when the eruption occurred it was set quite close to the ocean ridge, where the oceanic lithosphere would have been only 30–50 km thick. Let us assume that a major impact occurred in such an area, near the ocean ridge, and that it gave rise to the development of a 'transient crater' (Fig. 1a) with a diameter of 300 km and a depth of 100 km. The crater would excavate a large volume of asthenosphere (at a temperature of 1300–1500 °C), which would experience a great reduction in confining pressure so that pressure-release melting would immediately take place.

Similarly, the asthenosphere forming the limits of the transient crater would also experience a significant decrease in confining pressure, so that much of the asthenosphere would also melt *in situ* and be moved towards the surface as a central, or ringed, uplift crater developed. According to H. J. Melosh, the volume of melt created directly by the impact, coupled with the volume of mantle material already at a suitable temperature that would melt when the static, pre-impact pressure was released, would greatly exceed the

volume of the crater. The result would be that obvious evidence of the impact event would be concealed beneath an extensive sheet of extruded basic volcanic rock.

Such sheets of basic volcanic rock also occur in continental areas. Indeed, it has been argued by J. Negi and his colleagues, from geophysical evidence, that the flood basalts of the Deccan, which have an area about one-sixth that of the Indian peninsula and a volume in excess of a million cubic kilometres, were emplaced in a period of about a million years as the result of a major meteoritic impact.

There are about sixty known major oceanic plateau basalts and eight continental plateau basalts which have developed within the past 250 My. If most, or all, of these features are the result of impacts which gave rise to craters with diameters well in excess of 100 km, such events would profoundly affect the geological record.

The large 'certain' impact structure of Chicxulub, Yucátan, Mexico, which is 180–200 km in diameter, is now widely held to be the impact which was a major contributary cause in the demise of the dinosaurs at the period marked by the Cretaceous–Tertiary (K/T) boundary. It is interesting to note that the initiation of the Deccan Traps occurred within about a million years of the K/T boundary, so that this event would also have contributed to the problems of the dinosaurs.

The conventional view on the rapid development of large-volume eruptions of basalts is based on assumptions about plumes which come from deep in the mantle. Other models which have been proposed to explain the large volume of melt, required to provide the extrusive basic rock require one to assume that thinning of the lithosphere takes place by instantaneous stretching. Such thinning cannot be explained by mechanism associated with plate tectonics. However, near-instantaneous removal of lithosphere, with the attendant generation of pressure-release melting of high-temperature asthenosphere, is inherent in the major impact model.

How large and how frequent can we expect such impacts to be? Even in the 1980s it was generally considered that impact events on planets in the inner Solar System were from meteorites that originated in, but were dislodged from, the asteroid belt and crossed the Earth's orbit. By August 1994, 296 such 'Earth-crossers' had been observed. It is estimated that there are in all about 2000 such bodies. In recognition of the potential dangers of impact events to mankind, an Earth and Space Watch was introduced in the late 1980s. Observers have noted that not only are there 'Earth-crossers', but there are also 'Earth-approachers'. Although only a small number of such Earth-approaching bodies have been seen, they have prompted observers to suggest that the Earth itself is within its own asteroid belt. Some alarm resulted, for example, when asteroid 1997 XF11, with an estimated diameter of about 1.5 km (1 mile), was sighted in December 1997 and calculations indicated that its orbit would bring it close enough to the Earth for there to be a risk of impact in October 2028. More than a hundred potentially hazardous asteroids (PHAs) have been listed. They are all in orbits that intersect the

Earth's orbit and they are all large enough to cause widespread damage in an Earth impact.

Comets originate in the Oort Cloud, which lies beyond the orbit of the outermost planet. Most of the bodies in this cloud, which are numbered in billions, are composed of ice. Some have a core of rocky blocks and fragments cemented together by ice, which also forms a carapace. From time to time these bodies become dislodged from the Oort Cloud. Some are lost to outer space while others enter recurring orbits around the Sun. As the comets pass close to the Sun, the ice vaporizes and is blown away by the solar wind. After a number of passes around the Sun the comet becomes largely or completely de-iced and so resembles an asteroid.

It has been argued that, at any one time, there will be at least 4000 such comets or de-iced comets which cross or closely approach the Earth's orbit. Less than 40 such cometary bodies are recorded. One may therefore infer that, because of their intrinsic high velocity as they approach the Sun, these bodies will probably be more dangerous than the meteoritic Earth-crossers and Earth-approachers. In consequence, most craters on Earth with a diameter up to 25 km are considered to be the result of impacts by comets, and almost all impacts with a diameter of more than 100 km are the result of comets.

How frequently are major impacts likely to strike the Earth? V. Oberbeck and his co-workers have derived a relationship between the number and size of craters from the number of certain impacts which have been recorded in the ancient tectonically stable areas of North America and Europe and a power-law relationship between the number and diameter of craters greater than a specific value (derived from lunar data) (see Table 1).

Table 1 Frequency of impacts of varying sizes

Number of impacts per 2000 Ma		Average time interval (Ma) between impacts of a given size
> 750 km diameter	3	670
> 500	6	400
> 100	110	18
> 20	2003	1
> 10	6974	287 000yr
> 5	24 283	82 000
> 1	1 430 000	1400

From Oberbeck et al. (1993)

The estimates of number and frequency of impact given in Table 1 are likely to be conservative. On the basis of these and other estimates, the inherent dangers to mankind and also to individuals have been assessed. At present, the probability of an individual being killed as the result of a relatively minor impact event is estimated to be the same as the probability of being killed in a commercial flying accident, namely 1 in 20 000.
                                                    NEVILLE J. PRICE

Further reading

Hodge, P. (1994) *Meteorite craters and impact structures of the Earth.* Cambridge University Press.

Melosh, H. J. (1989) *Impact Cratering.* Oxford University Press, New York.

Oberbeck, V., Marshall J., and Aggarwal H. (1993) Impacts, tillites and the breakup of Gondwanaland. *Journal of Geology*, **101**, 1–19.

meteorites *see* ASTEROIDS AND COMETS

methanogenesis Methane (CH_4) is formed by a unique group of anaerobic bacteria belonging to the Archaea, the methanogens. Methanogens are among the most difficult bacteria to grow artificially, for they require strict anaerobic conditions and can be killed even by traces of oxygen. Despite this, they are widespread in the environment, producing about 10^{12} g methane per year: much more than abiological sources (such as volcanoes, vehicle emissions, and natural gas). Methanogens require the activity of other bacteria, both to remove oxygen and to supply their simple substrates, which are of low molecular weight. Hence they occur in interacting bacterial communities within anoxic habitats such as sediments, swamps, paddy fields, sewage digesters, landfill sites, the rumen of cows and sheep, and the guts of insects (e.g. termites) and animals. In these environments methanogenesis can be the major route by which organic matter is mineralized. Methanogenesis can also occur in oxygen containing environments such as forest and grassland soils, but here it is restricted to anoxic microenvironments (e.g. within soil crumbs). A unique environment for methanogenes is provided by geothermal and hydrothermal vents, where they are able to grow at temperatures up to 110 °C, using thermogenic hydrogen for energy.

The most common mechanism of methane formation is from hydrogen and carbon dioxide:

$$4H_2 + CO_2 \rightarrow CH_4 + 2H_2O.$$

Alternatively, acetic acid can be fermented:

$$H^+ + CH_3COO^- \rightarrow CH_4 + CO_2.$$

A few other substances, such as methanol, methylamine, and ethanol (ethyl alcohol) can also be metabolized by methanogens. In co-culture with other bacteria, such as acetogens, a greater range of substrates is used, including hydrocarbons.

In the marine environment, methanogenesis is restricted by competition with anaerobic sulphate-reducing bacteria, which use similar substrates.
                                                    R. JOHN PARKES

Further reading

Oremland, R. S. (1988). Biogeochemistry of methanogenic bacteria. In Zehnder A. J. B. (ed.) *Biology of anaerobic microorganisms*, pp. 641–705. John Wiley, New York.

MHD waves *see* MAGNETOHYDRODYNAMIC WAVES IN THE EARTH

Michell, John (1724–93) The astronomer and geologist John Michell made pioneering contributions to the study of earthquakes and of stratigraphy. After graduating at Queens' College, Cambridge, he was elected to a Fellowship there. He was subsequently elected a Fellow of the Royal Society in 1760 and became Woodwardian Professor of Geology in 1762.

In his *Conjectures concerning the cause and observations upon the phenomena of earthquakes*, published in 1760, Michell put forward the idea that steam generated at depth by the combustion of coal provided the energy source for earthquakes. In this publication he displayed a relatively wide knowledge of British stratigraphy and recognized that beds can show little change when traced laterally for appreciable distances. Michell was one of the first to explore systematically across country, tracing outcrops of strata over wide areas and realizing that specific rock formations may produce characteristic landforms. He later compiled a table of English strata from the Coal Measures to the Chalk, which was published postumously.

Michell left Cambridge in 1767 and spent the rest of his life in Yorkshire. In his last years he devised a method for weighing the Earth by using a torsion balance of his own invention, but it was left to W. H. Wollaston and Henry Cavendish to carry out the experiments after his death.

D. L. DINELEY

micropalaeontology Micropalaeontology is the study of microfossils, a microfossil being any fossil that is best studied by means of a microscope. In practice, however, the term 'micropalaeontology' is usually applied only to studies of

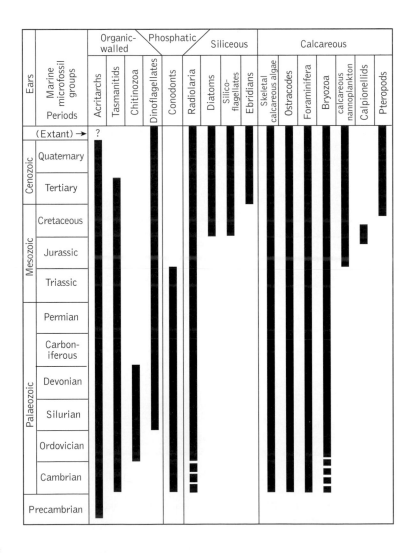

Fig. 1. Time-ranges of various groups of protozoans of partiaular geological interest.

microfossils with mineral walls (e.g. foraminifera, ostracods). The investigation of microfossils with organic walls (e.g. pollen grains, spores, acritarchs) is regarded as the subject of a separate discipline called palynology. The material studied by micropalaeontologists includes dissociated fragments of larger organisms, whole organisms of microscopic size, or embryonic or immature forms of larger fossil organisms.

In order to study any microfossils which might be present within a sample, they need to be extracted from the rock. Various procedures are used for this purpose, the technique used depending on the nature of the rock and the microfossils. Essentially each method involves disaggregation of the rock, followed by the separation of the microfossils from any remaining rock residues. The character of the rock will determine the process by which disaggregation is achieved. Soft muds, silts, and sands can simply be dried thoroughly and then soaked in water or, more effectively, in solutions of washing soda, water softener, or a suitable detergent. Even partly indurated rocks can be broken down by this method. The process is speeded up if the solutions are gently boiled for a short time. Non-calcareous microfossils can be extracted from indurated limestones by dissolving the rock matrix in formic or acetic acid. The residue obtained from disaggregation will normally comprise a mixture of various microfossils and insoluble inorganic debris. The microfossils can be separated from the debris by various methods, such as sieving, decanting, and the use of various heavy liquids. When calcareous microfossils are enclosed in an indurated limestone, disaggregation may be impossible without damaging the fossils. In such cases, it is necessary to use thin sections or stained acetate peels.

The study of microfossils has become a major tool in geological investigation, both in academic research and commercial laboratories. It is valuable because most sedimentary rocks yield microfossils of some kind (Fig. 1). Their small size and generally great abundance mean that only small volumes of rock need to be processed to obtain representative microfossil assemblages. They are therefore particularly valuable in the study of drill cores or drill chips. Microfossil assemblages can be used in a variety of ways, including biostratigraphic dating, determining of the depositional environment of the host rock, and interpreting the burial history or thermal history of the rock.   ALLAN N. INSOLE

Further reading

Bignot, G. (1985) *Elements of micropalaeontology.* Graham and Trotman, London.

Brasier, M. D. (1980) *Microfossils.* Allen and Unwin, London.

middle-latitude tropospheric circulations People

who live in latitudes between about 30° and 65° are accustomed to much greater changes in the weather from day to day than those who live in other regions. This region is characterized by weather systems which develop quickly and

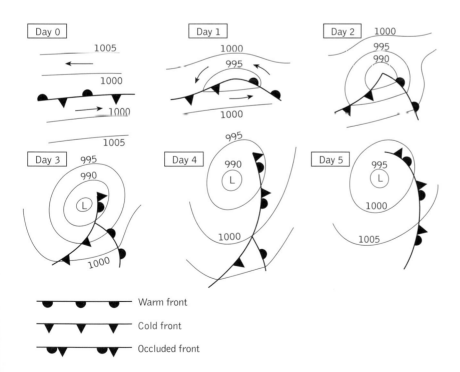

Fig. 1. The life cycle of a polar front depression. L, low-pressure area. Atmospheric pressures are shown in millibars.

681

move quickly. Anticyclones (high-pressure systems) and depressions or cyclones (low-pressure systems) are responsible for most of these changes in weather. Most of the particularly dramatic changes in weather are associated with weather fronts, which are an integral part of depressions.

Horizontal variations in temperature are greater in middle latitudes than anywhere else in the troposphere. Bounded at low latitudes by the warm tropical air, and at high latitudes by cold polar air, there is a strong north–south temperature difference. The region of greatest temperature difference is the polar front. It is at this front that depressions usually develop. The depressions gain energy from the temperature difference. This is an unstable process. When the temperature gradient is strong and a depression begins to evolve, the development is rapid and, as it grows, it circulates air to remove the temperature gradient. When the depression is at its most vigorous the temperature gradient in that region is at its weakest. When it is past its peak the depression loses its energy gradually through friction, and the temperature gradient builds again until another depression develops. Because the temperature gradient is usually not completely removed, depressions can develop in quick succession, forming one after another. The whole process takes about five to six days (Fig. 1), during which the depression has been moving, usually from west to east, as well as developing.

At the beginning of the process the polar front separates a warm air mass from a cold air mass (day 0). A depression usually starts to develop when a region of divergence passes over the front at higher altitudes. This region of divergence removes some air from above the front and causes a drop in atmospheric pressure at the Earth's surface so that a small low-pressure area develops (day 1). Air circulating round the low-pressure area pushes the warm air ahead of the depression while the cold air sweeps in behind the depression (day 2). The front ahead of the depression is a warm front because warm air is replacing cold air. The cold front behind the depression moves faster than the warm front and eventually catches up with it. This happens first near the centre of the depression, where the fronts are already close together (day 3). As the cold front reaches the warm front it forms an occluded front. At this stage the pressure at the centre of the depression has reached its minimum value and the centre of the depression begins to move more slowly while the fronts move further ahead (day 4, Fig. 2). Now that the development processes have ceased, friction starts to dominate and the central pressure of the depression starts to increase while slowing further and turning slightly poleward (day 5). The cold front trailing behind the depression is now the only remnant of the original polar front, and further depressions can develop on this front (day 6).

This view of the life cycle of a depression is easily illustrated on a weather map but is essentially a two-dimensional view of a three-dimensional process. The fronts are not lines but sloping surfaces; the lines are the places where the sloping surfaces meet the Earth's surface. The air not only moves around

Fig. 2. A mature depression as seen from a weather satellite.

the depression but moves upwards at the fronts to produce cloud and precipitation. At upper levels the wind speed increases with height. This is a result of the equator–pole horizontal temperature gradient, which requires that the wind speed increases with height. The maximum wind speed occurs close to the tropopause at about 12 km. This core of high wind speeds is the jet stream, from which commercial aircraft sometimes benefit as a tail wind on west-to-east journeys.

Depressions can sometimes develop much more rapidly than usual (explosive cyclogenesis), producing a drop in central pressure of perhaps 50 millibar (hPa) in 24 hours. A typical pressure change is about 30 to 40 millibars decrease in 48 to 72 hours.

High-pressure systems (anticyclones) form between depressions. These are part of the same dynamical system as the depressions themselves and the increase in pressure is produced by convergence aloft in the same way as the depression is produced by divergence aloft. A more significant type of anticyclone is a 'blocking' anticyclone, which develops in one location and moves very little, while forcing depressions to alter their track around the block.　　CHARLES N. DUNCAN

Further reading

Ahrens, C. D. (1994) *Meteorology today* (Chapter 13). West Publishing Co., St Paul, Minnesota.

Palmen, E. and Newton, C. W. (1969) *Atmospheric circulation systems.* Academic Press, London.

mid-ocean ridges The longest linear, uplifted features of the Earth's surface are to be found in the oceans. They are giant submarine mountain chains with a total length of more than 60 000 km, are between 1000 and 4000 km wide, and have crests that rise 2–3 km above the surrounding ocean basins, which are 5 km deep. The average depth of water over their crests is thus about 2500 m.

These features are the mid-ocean ridges, famous now not only for their spectacular topography, but because it was with them, in the early 1960s, that the theory of ocean-floor spreading, the precursor of plate tectonic theory, began, and we now know that it is at these mid-ocean ridges that new lithosphere is created.

Similar ridges occur at the margins of oceans; the East Pacific Rise is an example. There are other spreading ridges behind the volcanic arcs of subduction zones. These are usually termed *back-arc spreading centres*.

The first ridge to be discovered, the Mid-Atlantic Ridge, was found during attempts to lay a submarine cable across the Atlantic in the mid-nineteenth century.

Although many of the earlier supporters of continental drift (Arthur Holmes, for example) had suggested that the mid-oceanic 'swells' and volcanic islands such as Iceland were the result of ascending and laterally spreading convection currents, it was not until the early 1960s that R. S. Dietz and H. H. Hess proposed the mechanism of 'sea-floor spreading' to explain continental drift. The theory suggested that continents moved in response to the growth of oceanic crust between them. Oceanic crust is thus created from the mantle at the crest of the mid-ocean ridge system, a volcanic sub-marine rise. In the Atlantic, this rise sits in the middle of the ocean between continents, which are symmetrically distributed on either side. The margins of these continents had long been thought by proponents of continental drift to have once been joined together.

Confirmation of the hypothesis of sea-floor spreading came with the discovery by F. J. Vine and D. H. Matthews that magnetic anomalies across the Mid-Atlantic Ridge were symmetrical on either side of the ridge axis (Fig. 1). The only acceptable explanation for these magnetic anomalies was in terms of sea-floor spreading and the creation of new oceanic crust. When lava cools on the sea floor, magnetic grains in the rock acquire the direction of the Earth's magnetic field at the time of cooling. Studies of lava flows on land show that from time to time the Earth's magnetic field has reversed its direction (that is, the north and south magnetic poles have changed places). The upper oceanic crust thus shows alternate normal and reversed magnetization.

The anomalies found across the Mid-Atlantic Ridge could, moreover, be matched with similar anomalies that had been discovered in Iceland and other parts of the world where young volcanic rocks could be dated. Given that the distance of each anomaly from the present rifted ridge crest could be measured, and there was now a time for its creation, the *rate* of ocean floor spreading could be calculated. Over the next decade, as magnetic anomalies were detected in all the oceans and a magnetic stratigraphy was formulated, a picture of oceanic ages throughout the world was gradually built up.

The reason why the ridges are elevated above the ocean floor is that they consist of rock that is hotter and less dense than the older, colder plate. Hot mantle material wells up beneath the ridges to fill the gap created by the separating plates; as this material rises it is decompressed and undergoes partial melting. Geophysical data indicate that melting can occur at depths of more than 100 km and over a broad region perhaps several hundreds of kilometres across. The molten material from the mantle migrates upwards along boundaries between crystals or along channels to accumulate in magma chambers beneath the ridge. Magma will occasionally escape from these chambers to intrude the rocks above as dykes, and may erupt at the ocean floor to form volcanoes and lava flows. The settling of crystals in the magma chambers can result in the formation of gabbros and other cumulate rocks.

At the ridge crest the lithosphere may consist only of oceanic crust, which is basaltic in composition. As it moves away from the spreading centre, it cools and contracts and the peridotitic (olivine-rich) mantle component grows rapidly (Fig. 2). The contraction leads to an increase in water depth. A simple relationship thus exists between depth and age (Fig. 3), and, at least for the first 60 Ma (60×10^6 years), the mean depth of the oceanic crust is proportional to the square root of its age. After that time the ocean floor deepens less quickly, probably because the lithosphere begins to receive a significant amount of heat from below.

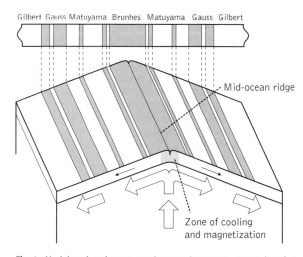

Gilbert Gauss Matuyama Brunhes Matuyama Gauss Gilbert

Mid-ocean ridge

Zone of cooling and magnetization

Fig. 1. Model to show how magnetic anomaly patterns are produced at the crest of a mid-ocean ridge by accretion of new volcanic rock followed by bilateral symmetrical spreading away from the ridge axis. The magnetic polarity timescale is shown at the top. (From Kennett, (1982).)

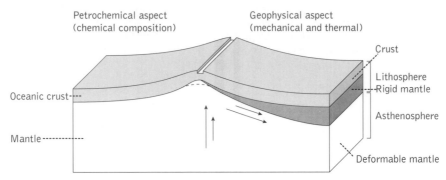

Fig. 2. The nature of crust/mantle and lithosphere/asthenosphere across a mid-ocean ridge. The terms 'crust' and 'mantle' refer to the petrographic aspects (chemical and mineralogical composition). Note that their relative thicknesses do not change away from the ridge crest, i. e. oceanic crust is still only 5–6 km thick under ocean basins. The terms 'lithosphere' and 'asthenosphere' refer to the geophysical aspects (physical structure) based on differences in temperature and mechanical behaviour. Thus the lithosphere consists of the outermost parts of the Earth (up to 100 km thick) that includes the crust and upper mantle. It is distinguished by its ability to support surface loads, like volcanoes, without yielding. (From Nicholas (1995),)

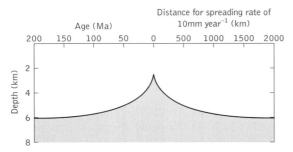

Fig. 3. Theoretical relationship between water depth and age for a spreading rate of 10 mm year⁻¹ (From Kearey and Vine (1996) *Global tectonics.*)

Over the past 30 years a more detailed picture has gradually been built up of the internal nature of mid-ocean ridges and their composition in terms of rock-types (Fig. 4). This was achieved initially by comparison with the supposedly equivalent ophiolite complexes exposed on land, especially in the Troodos Mountains of Cyprus and in Oman (see *ophiolite sequences*). Later, the models were confirmed by a combination of deep-sea drilling on the ridges (down to about 2000 m) and by the physical examination by submersibles of exposed escarpment faces at fracture zones and collection of rock samples at 4500 m depth, where even peridotites are exposed at the surface.

Spreading rates (that is, half the rate of separation from the axis) are not the same throughout the mid-ocean ridge system

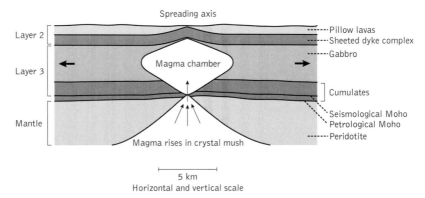

Fig. 4. Diagrammatic cross-section of the structure and composition of the crust and upper mantle beneath the crestal region of a mid-ocean ridge, showing how hot asthenospheric material ascends buoyantly to extrude pillow lavas and intrude sheeted dykes of Layer 2 (Layer 1 is the non-existent sedimentary cover). The gabbroic Layer 3 develops from crystallization of the axial magma chamber. (From Kearey and Vine (1996).)

but vary considerably from a few millimetres per year in the Gulf of Aden to 1 cm year^{-1} in the North Atlantic near Iceland and 6 cm year^{-1} for the East Pacific Rise. This variation in spreading rates appears to influence the ridge topography. Slow-spreading ridges, such as Mid-Atlantic Ridge, have a pronounced rift down the centre, producing an axial valley at the ridge crest and a high relief across the ridge. Fast-spreading ridges, such as the East Pacific Rise, lack the central rift and have a smooth topography. In addition, spreading rates have not remained constant through time. There appear to have been periods in the past when they were either faster (perhaps 18 cm year^{-1}) or slower. If this is so, there may have been at least two consequences. If the Earth did not expand, then, at times of rapid spreading, loss of lithosphere by subduction must also have increased; secondly, the increase or decrease in the total volume of the mid-ocean ridge system would have had a major effect on global sea level. For example, it appears that the increase in volume of ridges in the late Cretaceous resulting from an increase in spreading rates led to worldwide transgression at that time, followed by regression during the early Tertiary when rates of spreading were reduced.

The prime reason for the differences in spreading rates is that the slow-spreading ridges are fed by small and discontinuous magma chambers, thereby allowing for some magma differentiation and eruption of a comparatively wide range of basalt types. Thus the ridge crest consists of numerous very small hills. Fast-spreading ridges have large, continuous magma chambers that generate comparatively homogeneous magmas. Because of the higher rates of magma discharge, sheet lavas are more common.

The differences are also apparent in the nature of the sediment cover. The fractured topography of slow-spreading ridges yields highly localized and lenticular sediment patterns, and these rest on a variety of basic and ultrabasic rocks (Fig. 5). Talus breccias (accumulations of coarse angular material) abound, derived from the exposed fault scarps, a 50–85 m thick breccia having been found within an axial/median rift, itself 2 km deep. Slightly further away, the still-unfilled rift valleys receive a variety of fine-grained sediments, including volcaniclastic sediments derived from the break up of volcanic rocks. Pelagic sediments form on the highs and these may be swept off them to accumulate in the lows as calcareous turbidites, partly filling the valleys with several hundreds of metres of sediment. Eventually, as the ridge sinks to the depths of a typical basin plain, if this is within the reach of land-derived terrigenous sediment, the rugged topography is progressively covered up and smoothed so that it becomes an abyssal plain (see *ocean basins*).

Fast-spreading ridges, with their subdued topography, lack the axial rift, and volcaniclastic sediments are rare (Fig. 6). Instead they are covered by patches of brown sediments rich in iron, manganese, and a host of other metals. Passing away from the ridge crest, carbonate ooze at first accumulates, but as the ridge deepens and passes below the CCD (carbonate compensation depth) the pelagic sediments change to siliceous oozes and pelagic clay.

Although mid-ocean ridges appear at first sight to be continuous features within the oceans, on closer inspection this is clearly not so. They are all broken into segments by transverse fractures that displace the ridges by tens or even hundreds of kilometres. Fractures are narrow, linear features that are marked by near-vertical fault planes (Fig. 7). Along these exposed fault scarps, 500 m or more high, not only can the nature of oceanic crust be examined (see below), but complex patterns of fault breccias, talus, and lavas can be discerned. In addition, because the fault lines are subject to complex stress patterns, suffering both elements of extension and compression, they have many of the features of transtension and transpression that can be seen in strike-slip zones on land. Small, very deep, elongate basins also develop.

These transverse fractures show a spectrum from primary global lineaments that form plate boundaries, along which ocean-floor spreading patterns were initiated, to secondary transform faults that were an inevitable consequence of ocean-floor spreading.

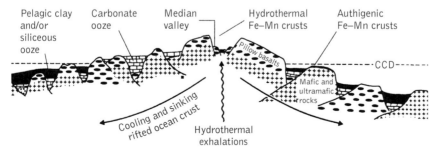

Fig. 5. Sediment distribution on a slow-spreading, rifted mid-ocean ridge typical of the Atlantic Ocean. Notice the great variety of exposed igneous rocks and sediment types, and that the mineralized crusts change from hydrothermal to authigenic (formed in place) as they move away from the vents in the median valley. CCD, Carbonate compensation depth.

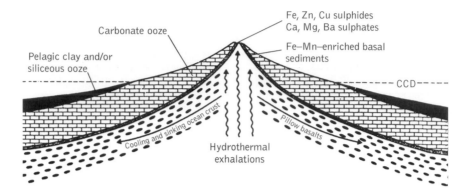

Fig. 6. Sediment distribution on a fast-spreading, smooth mid-ocean ridge typical of the Pacific Ocean. Notice the enhanced development of iron–manganese-enriched basal sediments, in part derived from oxidation of sulphides. CCD, Carbonate compensation depth.

(From Jenkyns, H. (1986) in *Sedimentary environments and facies* (Reading H. G. ed.). Blackwell Science Ltd, Oxford. Based on Garrison (1974) and Davies and Gorsline (1976).)

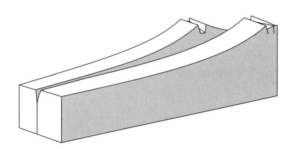

Fig. 7. Sketch showing the differential topography and development of fault scarps along a transform fault perpendicular to the axis of a mid-ocean ridge. Notice that the heights of the ridge crests either side of the fault are very different, suggesting that the normal depth–age relationship does not apply. (From Kearey and Vine (1996).)

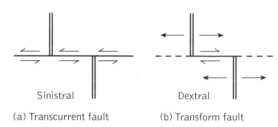

Fig. 8. Comparison of (a) a transcurrent fault which causes a sinistral offset along a vertical fault plane which must stretch to infinity beyond the ridge crests with (b) a transform fault which is active only between the offset ridge crests and where the relative movement of the lithosphere on either side is dextral. (From Kearey and Vine.) (1966).

Although the term 'transform' is frequently used for what should be called fracture zones or strike-slip faults (e. g. continental transforms), in its true, limited sense, as defined by J. Tuzo Wilson in 1965, it refers to those portions of the fracture zone that are active and lie between the offset ridge crests. Away from the ridge crests, on either side, there is no movement between the blocks (Fig. 8).

Undoubtedly the most spectacular discovery in recent years has been the presence of extremely hot springs resulting from extensive hydrothermal activity within the ridges, and the realization that these quite extreme environments can yield unique ecosystems consisting of prolific indigenous faunas of crabs, giant clams, and tube worms. These ecosystems are not only quite unlike any other systems known in the world, but each is unique, all clearly evolving independently.

That the waters and sediments atop ridges are hot and sulphurous has been known for a long time. Temperatures of hydrothermal mineral-rich solutions are as high as 350 °C and pH is as low as 4 (acid) on the East Pacific Rise. The solutions are precipitated as a range of columnar structures, called chimneys, built of various sulphide and sulphate minerals. Two basic types of chimney exist: the high-temperature 'black smokers' growing typically at about 8 cm a day, and belching clouds of finely disseminated pyrrhotite, sphalerite, and pyrite as hydrothermal plumes, which can be detected many kilometres from the vent. The cooler vents (up to 300 °C) are the 'white smokers', which emit particulate amorphous silica, barite, and pyrite.

Black smokers have a characteristic concentric zonation. White smokers are chimneys in which hydrothermal precipitation has sealed the orifices and decreased the permeability of the chimney walls, thus allowing more intra-chimney cooling by sea-water entrainment and heat conduction. These less active chimneys sustain the most prolific growth of tube worms, which may become so densely packed that they form a spherical mass of white tubes called 'snowballs'. Ultimately chimneys become mechanically unstable and collapse to form a mound of chimney talus on which chimney growth begins again. Sea-floor sulphides are chemically unstable in modern sea water, and if hydrothermal activity ceases they oxidize rapidly to form ochreous deposits dominated by hydrated iron oxides, some reacting with silica to form iron-rich smectites.

The hydrothermal fluids that emanate from mid-ocean ridges are sea water that has circulated through the oceanic crust within a convection cell and leached metals from the rocks along its flow path. The metal content of the fluids thus reflects the trace-element composition of the subjacent source rocks which, in mid-ocean ridges, are basaltic. The metals are copper-rich, with zinc as well, and contrast with spreading sites in back-arc basins with their bimodal volcanism of both acid and basic material which are relatively lead-rich (see *island arcs*).   HAROLD G. READING and N. C. MITCHELL

Further reading

Nicholas, A. (1995) *The mid-oceanic ridges: mountains below sea level*. Springer-Verlag, Berlin.

Kearey, P. and Vine, F. J. (1996) *Global tectonics*. Blackwell Scientific Publications, Oxford.

Kennett, J. P. (1982) *Marine geology*. Prentice-Hall, Englewood Cliffs, N. J.

migmatite

'Migmatite' is a textural term that is used to describe a coarse-textured metamorphic rock, usually of high grade, that contains pockets of crystallized melt, either in the form of small lenses or large sheets. The term was introduced by J. J. Sederholm in 1907. Migmatites have long been thought of as having been formed near the boundary between metamorphic and igneous conditions. For example, a migmatite could represent a gneiss in which some pockets of melt had begun to form. Some migmatites have now been shown to be the source regions of granite, from which granite melts have been extracted, leaving an unmelted residue containing a few patches of trapped melt.

Migmatites find uses as ornamental building stones, where their striped or patchy appearance is displayed to good effect, especially on polished surfaces.   DAVID A. ROTHERY

Milankovich, M. (1879–1958)

Milutin M. Milankovich, a Serbian mathematician, developed the idea that the Earth's rotational wobbles and orbital deviations have combined to affect in a cyclic way global climatic changes. He calculated that every 40 000 years circumstances would combine to vary the pattern of solar radiation, lowering temperatures and bringing about increased snowfall in high latitudes. The idea

was primarily developed to account for the Pleistocene record of ice ages and it has led to a proliferation of writings on the topic. Milankovich's basic contribution was concerned with three items for investigation: the problem of the celestial mechanics, the question of how much solar energy arrives at the surface of the Earth, and, finally, what the climatic consequences of this energy supply are.

Milankovich was appointed to a chair in the University of Belgrade in 1909 and taught mechanics, physics, and celestial mechanics. His output of research was large and in several fields. His major contribution on climatic cycles was, however, published in 1941 by the Royal Academy of Serbia, while war was engulfing Europe. It was not until several years later that his ideas were brought to bear upon the growing mass of data on the Pleistocene.

Since then, climate-influenced sedimentary cycles with periodicities compatible with those predicted in Milankovich's calculations have been claimed for parts of the stratigraphic column as far back as the mid-Palaeozoic. Milankovich cycles are likely to have been operative since very early in geological history. Their parameters may have changed and their effects may have been modified by other geological factors, but the value of the recognition of the basic mechanism by Milankovich remains.   D. L. DINELEY

Milankovich cycles and climate change

The Milankovich theory of climate change was first advocated by James Croll in the late nineteenth century and later elaborated by Milutin Milankovich (1879–1958), a Serb mathematician. The mathematical basis of this astronomical theory of the ice ages was substantially refined by André Berger. More recently, Berger has provided highly detailed information on past insolation variations.

The Milankovich theory of climate change is based on the premise that the alternating cold and warm periods of the Quaternary were principally due to changes in the nature of the Earth's orbit around the Sun. One important assumption of the Milankovich theory of orbital changes in that there has been no absolute annual change in the amount of incoming solar radiation. Long-term changes in the Earth's orbit are believed to cause a redistribution of insolation across both hemispheres, and these changes, in turn, lead to changes in climate.

Milankovich orbital processes

At present, the Earth has an elliptical orbit around the Sun. The Sun is not located at the centre of the ellipse but at one of the foci (Fig. 1). During the winter solstice in the northern hemisphere, the Earth is at one end of its elliptical orbit when it is nearer the Sun (in perihelion) and thus receives greater heat. In contrast, the northern hemisphere summer solstice is characterized by a position in the elliptical orbit that is more distant from the Sun (in aphelion). The principal cause of the seasons is the fact that the Earth's axis is tilted at an angle that

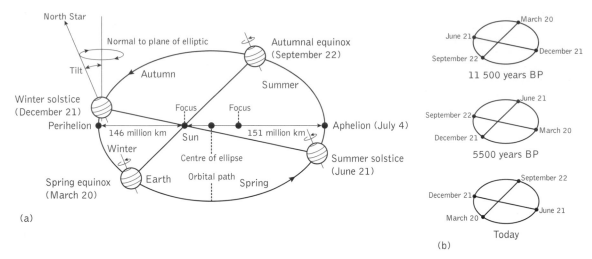

Fig. 1. (a) Geometry of the Sun–Earth system. The Earth's orbit, an ellipse with the Sun at one focus, defines the plane of the ecliptic. The Earth moves around its orbit in the direction of the arrows while spinning about its own axis, which is tilted to the plane of the ecliptic at *c.* 23.5° and points toward the North Star. (b) Precession of the equinoxes causes the position of the equinoxes and the solstices to shift slowly around the Earth's elliptical orbit. During the late glacial, the winter solstice occurred near aphelion, whereas today it occurs at the opposite end of the orbit near perihelion.

is not perpendicular to the plane of its orbit. During the course of its annual orbit, the tilting of the Earth results in the seasonal heating of each hemisphere. The timing of perihelion at present coincides with the period when the southern hemisphere is tilted towards the Sun. Conversely, aphelion coincides for the northern hemisphere in summer, when the Earth as a whole receives approximately 3.5 per cent less solar radiation than the annual mean (as measured at the outer edge of the Earth's atmosphere).

Throughout time, there has been a continual change in the eccentricity of the elliptical orbit. Marked changes in the axial tilt (obliquity) of the Earth have also taken place. In addition, there also have been changes in the timing of perihelion and aphelion with respect to seasonal changes on Earth (the precession of the equinoxes).

Changes in orbital eccentricity

Changes in orbital eccentricity have varied with time from a circular orbit (when perihelion and aphelion are identical) to maximum eccentricity, when the values of incoming solar radiation may have varied by as much as 30 per cent between perihelion and aphelion. The periodicity of this cycle is 95 800 years, during which time the Earth alternates from a circular orbit to a highly eccentric orbit and back again to a circular orbit. It should be stressed that changes in orbital eccentricity do not cause any change in the amount of solar radiation reaching the Earth during summer or winter, nor any change in the total annual heat received by either hemi-

sphere. Instead, the effect is to increase the contrast in seasonality in one hemisphere and reduce it in the other. Croll believed that when this contrast was at its maximum it would cause increased snowfall in the northern hemisphere during winter. The increased global albedo (the fraction of the incident radiation that is reflected from the surface) resulting from a widespread snow cover might then modify the climate of the succeeding seasons and, in this way, initiate ice ages.

Changes in inclination (obliquity)

Changes in the inclination of the Earth's axis have varied between extreme values of 21.39° and 24.36° (the present value is 23.44°) with a periodicity of 41 000 years. Increases in axial tilt result in a lengthening of the period of winter darkness in polar regions. They also result in changes in the seasonal range of latitude in which the Sun occurs overhead. Changes in obliquity therefore cause significant changes in the amount of solar radiation received at high latitudes but do not greatly affect the amount of incoming solar radiation at low latitudes. Because changes in axial tilt are equal in both hemispheres, the resulting changes in incoming solar radiation are the same in both hemispheres.

Precession of the equinoxes

Variations in the timing of perihelion and aphelion are caused by a 'wobbling' in the Earth's axis of rotation as it rotates around the Sun. Over a period of 21 700 years, the axis slowly swings in a conical manner around a line perpendicular to the orbital plane (Fig. 1). During this period of time, the northern hemisphere is tilted towards the Sun at successively different points in the Earth's orbit. At present, this takes place during summer when the Earth is in aphelion. However, approximately 11 000 years ago during the Younger Dryas time period, the Earth was tilted towards the Sun in midsummer during perihelion. In theory, therefore, northern hemisphere winters during the Younger Dryas were much colder and longer than at present, and summers were shorter and warmer.

Combined effects

The solar radiation received in low-latitude areas is principally affected by variations in eccentricity and precession of the equinoxes. By contrast, higher latitudes are mainly affected by changes in axial tilt (obliquity). The combined influence of changes in eccentricity, obliquity, and precession of the equinoxes produces a complex pattern of variations in insolation (Fig. 2). These are indicative of increased seasonal contrasts in one hemisphere and diminished contrasts in the other. It is not, however, known whether, during certain time periods in the Quaternary, insolation variations in the high latitudes of the northern hemisphere may have induced similar environmental changes in the southern hemisphere. Similarly, it is not known whether, on certain occasions, insolation changes in the high latitudes of the southern hemisphere (particularly in Antarctica) may have led to a climatic response in the northern hemisphere. Some scientists have even argued that climate changes in both hemispheres have taken place in an approximately synchronous manner.

A critical factor that affected rates at which ice sheets built up and decayed in the middle latitudes during the Quaternary was the seasonal temperature gradient between low and high latitudes—known as the insolation gradient. During periods when the insolation gradient from the Equator to the pole was high, meridional (north–south) circulation would have been increased, thus increasing the rate at which snow-bearing precipitation was delivered to high latitudes.

It has often been argued that the observed patterns of Quaternary environmental changes reflect well-defined cycles of Milankovich insolation. For example, it is generally agreed that the timing of the principal glacial–interglacial cycles during the Quaternary reflects the influence of the 96 000-year eccentricity cycle and that the last glacial–interglacial cycle, approximately corresponding to the Late Quaternary, was principally due to this process. For shorter timescales, scientists have proposed that climate changes during the last 100 000 years are indicative of the occurrence of the 41 000-year obliquity cycle. Similarly, attention has been drawn to the likelihood that some of these changes may have been related also to a 21 000-year precessional cycle.

Milankovich cycles and the ocean sediment record

Milankovich cyclicity and its relationship to climate change has been given increased credence in recent years by the similarity between the oxygen isotope curve derived from sediments on the ocean floor and the Milankovich curve. For a long time the oxygen isotope curve was considered to represent long-term changes in temperature, and, because the Milankovich curve was similar to the oxygen isotope curve, it was argued that these temperature changes were related to Milankovich cycles. However, in recent decades, persuasive arguments have been made proposing that the oxygen isotope curve derived from the ocean sediment record is a record of changes in the isotopic composition of sea water over time, and is therefore a proxy record of palaeoglaciation. Scientists have accordingly sought to explain past global changes in ice volume as a result of Milankovich cyclicity. Thus, the present view is that Milankovich cycles have been the prime driving force behind the many complex ice ages that have taken place during the Quaternary.

Notwithstanding this recognition, it should be noted that Milankovich cyclicity cannot be used as an explanation for the pattern of glaciation that has taken in any particular region of the world. Milankovich cycles can be related only to aggregate changes in global ice volume. Furthermore, there are many complex climate changes that have taken place in the past that have occurred at timescales shorter than 10 000 years. For example a well-known dramatic period of ice accumulation appears to have taken place during the Younger Dryas,

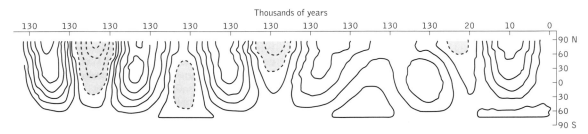

Fig. 2. Mid-month values of daily insolation for July in thousand-year intervals before 1951 AD. The shaded areas indicate areas where the values are negative. Note the relatively rapid mid-month insolation between 130 000 and *c.* 65 000 BP and the relatively small changes between *c.* 65 000 BP and the present (see Dawson 1992).

mineral zoning The term 'mineral zoning' refers to the systematic spatial variation in the composition of a single mineral grain. Concentric zonation reflects compositional variations perpendicular to the crystal surfaces, whereas sector zonation arises from compositional variations between crystallographically different faces. Growth zoning originates during mineral growth and implies temporal variations in the physical or chemical conditions near the surface of the growing mineral grain. Such variations can be triggered by the growth process itself or by unrelated external forcing. Growth-zoned minerals may occasionally show concentric oscillatory zoning, where the composition of the mineral grain fluctuates periodically or non-periodically from its core to its rim. Diffusion controlled changes in the mineral composition imposed after the cessation of mineral growth may give rise to diffusion zoning.

Mineral zoning arises in a variety of geological situations and many common minerals are frequently zoned. Some well-known examples are plagioclases from magmatic rocks, garnets from metamorphic rocks, and carbonates from sedimentary rocks. The study of zoned minerals has given valuable insights into a variety of geological processes, including the mixing of magmas, the thermal evolution of metamorphic terrains, and the dynamics of hydrothermal systems.

BJORN JAMTVEIT

mineralogy, medieval *see* MEDIEVAL MINERALOGY AND FIGURED STONES

minerals and mineralogy A mineral can be defined as a naturally occurring homogeneous solid, inorganically formed, with a definite chemical composition or a definite range of composition, and an ordered atomic arrangement. The study of such substances—their properties, occurrence, associates, and origin—is the science of mineralogy.

Strict application of this definition would exclude opal, which is amorphous though perhaps becoming crystalline with time; native mercury, which is liquid; synthetic gems, such as ruby and emerald, because they are not naturally occurring; perhaps not even the aragonite of shells since it is made by an organism. It would certainly not include copper sulphate grown on a substrate of basalt and sold on the slopes of Etna. It would include malachite on Roman copper slag in dumps on Cyprus and laurionite on fifth-century Greek lead slag at Laurium, since both are natural growths.

Environments

Of the 3700 minerals discovered so far, most are rare, sometimes merely a thin coating on a rock slab at one isolated locality, while others, such as quartz, feldspars, and calcite, are of worldwide distribution. Some exist only as microscopic crystals; other crystals may weigh several tonnes. Minerals are formed in a wide variety of environments: opal from water percolating down near-surface joints; gypsum from warm saline waters; garnet in solid rocks 25 km below the surface at a temperature of 500 °C; olivine from magma at 1100 °C, either on the surface or deep below the crust; galena formed in veins from hydrothermal waters at depth; coesite formed (almost momentarily) at several hundred degrees Celsius and at very high pressure in meteorite impact craters.

The oldest known mineral is a zircon 4300 million years old, found in a sedimentary rock in Australia and derived from an earlier igneous rock.

Formation

The whole Earth, apart from the liquid outer core and isolated patches of magma elsewhere, is made of minerals; those which are known have formed mainly in the crust, many of them from crustal materials. Eight elements—oxygen, silicon, aluminium, iron, calcium, sodium, potassium, and magnesium—make up 98 per cent of the crust by weight, oxygen being 47 per cent by weight and 93 per cent by volume. Some 20 or so mineral families make up 99 per cent of the crust by weight. The chemical constituents of the Earth are continually being reworked; virtually nothing new is being made. Minerals grow, break down, and release elements for recombination. The growth of a particular mineral depends on the appropriate elements being present in the right concentration under the right conditions (primarily of temperature and pressure) in order for nucleation and continued growth to take place.

The same chemical constituents can be arranged in different ways to give different minerals with different properties according to the physical conditions. The aluminium silicate minerals andalusite, kyanite, and sillimanite are good examples of this phenomenon of *polymorphism*. Better known are two of the different forms of carbon: graphite and diamond, low- and high-temperature and low- and high-pressure forms respectively. Carbon is one of the softest and diamond the hardest mineral known. The various forms of silica (SiO_2) are discussed elsewhere.

Bonding

Atoms in crystals are held together by electrostatic bonding forces, which in about 90 per cent of minerals are ionic or predominantly so. An ion is an atom that has become charged, either by the loss of an electron to become positively charged (a *cation*), or by gaining an electron to become negative (an *anion*). Electron gain or loss can involve more than one electron and result in higher charges. Anions have greater radii than cations, and ionic structures can be regarded as lattices of anions with cations in the spaces. The Earth's crust is effectively a packing together of oxygen anions with cations filling the spaces between. Cations are surrounded by as many anions as the geometry allows and the structure has overall electrical neutrality: the positive and negative charges balance. In calcite ($CaCO_3$), for example, Ca is 2+, C is 4+, O is $(2-) \times 3$, and the formula balances. If cations are of similar size and charge they may replace one another in the structure.

Olivine is a magnesium—iron silicate with a continuously variable composition between two end-members, Mg_2SiO_4, forsterite, and Fe_2SiO_4, fayalite, in a series called a *solid-solution* or *isomorphous series*. The atomic structures are similar and because Mg and Fe are both divalent ions and have very similar ionic sizes they can replace each another with equal facility. The physical properties also vary continuously and regularly between the end-members. The general formula for olivine is $(Mg, Fe)_2 SiO_4$. Many silicate minerals are members of complete or partial solid-solution series.

Physical properties

The physical properties of minerals depend on their chemical composition and structure. Graphite and diamond have the same composition but quite different structures and totally different properties. The carbonates aragonite, strontianite, barite, and cerussite have the same structure (are isostructural), but with the cations calcium, strontium, barium, and lead respectively, which have different atomic weights ranging from 40 to 207 and specific gravities ranging from 2.93 to 6.58.

Form, hardness, and cleavage are described in separate entries.

Colour is due mostly to the presence of the transition elements vanadium, chromium, manganese, iron, nickel, cobalt, and copper, but in part to imperfections in the crystal lattice. Some minerals, such as malachite and turquoise, have quite characteristic colours because of elements (copper in this instance) that are part of the structure and are called *idiochromatic*; others are allochromatic because of variable colours due to minute amounts of impurities in the structure. Quartz and fluorite are good examples; so, too, is corundum, which with trace amounts of chromium is ruby but with iron or titanium is sapphire. Other properties, sometimes quite diagnostic, are lustre, opacity, streak, taste, solubility, magnetism, pyroelectricity, piezo-electricity, and radioactivity. These together with optical properties, X-ray diffraction, X-ray fluorescence, and electron probe analysis enable minerals to be identified.

Classification

Minerals do not fit easily into categories if the aim is to put like things together and assemble them into more or less comprehensive groups. In the early days of mineral investigation, magical or supposed properties and resemblances to other natural objects formed the basis for groupings. Later, a more rigorous approach using physical properties was adopted. Eventually, with the development of chemistry in the late eighteenth century, classification based on chemical composition became more refined. X-ray studies beginning in the early twentieth century provided the basis for groupings using atomic structures. The classification most commonly used nowadays has its origins in the work of Berzelius round about 1800. In this scheme, the major divisions or classes are based on the nature of the anionic groups. Details are set out in works by Dana and his associates. A simplified version by Berry and Mason is given here. Classes: 1, native elements; 2, sulphides (including sulphosalts); 3, oxides and hydroxides; 4, halides; 5, carbonates, nitrates, borates, iodates; 6, sulphates, chromates, molybdates, tungstates; 7, phosphates, arsenates, vanadates; 8, silicates. Further subdivisions are based on chemistry and structures. Silicates are dealt with elsewhere in this volume.

Nomenclature

A few mineral names—gold, beryl, amethyst, gypsum—have come down from antiquity, though not all with their modern meaning. By 1800 there were perhaps less than a hundred valid names but thousands of invalid ones. Rational names provide some information about the mineral; irrational ones very little. Werner in 1783 was the first to name a mineral after a person: prehnite after Count von Prehn. At the present time about 45 per cent of minerals are named after people (braggite, goethite), and 23 per cent after discovery localities (andalusite, vesuvianite); 14 per cent of names refer to chemical composition (chromite, molybdenite), 8 per cent refer to distinctive physical properties (orthoclase, axinite) and 10 are combinations and oddities (peristerite, tantalite). An international body, the Commission on New Minerals and Mineral Names, now validates claims for the discovery of distinctive new minerals and their names. There is only one fixed rule—that the name must end in -*ite*—and a suggestion that the name should not be too frivolous.   R. BRADSHAW

Further reading

Battey, M. H. (1981) *Mineralogy for students*. Longman, Harlow.

MacDonald encyclopedia of rocks and minerals (1988) Macdonald Orbis, London.

mining Mining is the process of extracting solid natural resources from the shallow parts of the Earth's crust. Mining has been carried out by man since it was first realized that materials within the ground could be valuable, whether as fuel, for tools, or as decorative materials in jewellery or for religious use. Mining is known to have taken place for thousands of years; there are early indications of the perceived importance of geology at a number of sites. In the remains of the gold–silver workings at Cassandra in Greece there is evidence that miners dug in search of faulted vein segments at some time prior to 300 BC. Their Athenian contemporaries, faced with the depletion of silver and lead ore at Laurium, recognized the significance of marble near a schist contact and sank more than a thousand shafts through barren rock, some to depths of 100 m, in search of hidden orebodies.

The materials that are recovered by mining operations can be divided into six general categories: coal and low-grade coals; metalliferous materials (non-ferrous); iron, including all types of iron ore; rock, including construction materials such as building stone; precious stones; and evaporite deposits.

The two main methods of extraction are opencast mining and underground mining. Opencast mining entails the excavation of open pits from the ground surface so that the raw materials can be transported from the mine to a processing plant or export point. Underground mining entails sinking shafts to reach the target resource and driving tunnels and adits, either inclined or horizontally.

Whether to extract a resource by opencast or underground mining depends on the depth of the ore or rock below the ground surface and the concentration of the material per unit volume. In historical times, underground mining was carried out at extremely shallow depths because of the lack of mechanical equipment for removing the overburden. As a result of advances in mechanization, underground mining is no longer carried out on an industrial scale at shallow depths. There is no comprehensive formula for the depth of a mine: each mine and its ore are unique. For example, as a result of highly efficient production methods, mines in the Witwatersrand gold fields of South Africa operate at depths in excess of 3000 m where the specific geological conditions are suitable.

The 'bigger, faster, and deeper' requirements of modern mining are often met by scaling up the mining equipment and giving that equipment enough room to work. This 'up-scaling' is generally accommodated easily in open-pit mines, but in underground workings the creation of large caverns brings with it problems of rock mechanics and increased risk of instability.

In subsurface ore mining, headings are driven into new ground, level drifts follow the ore, level crosscuts connect drifts, and vertical or inclined raises connect the workings from level to level. Underground mining systems consist of workings that may be naturally supported, may require artificial support, or may be allowed to collapse as part of a caving method, according to the orebody geometry and the engineering properties of the strata. Because very few mineral deposits are uniform, most mines make use of more than one method.

The three main underground mining methods are stoping, room-and-pillar mining, and longwall mining.

In stoping, ore is blasted or excavated in a vertical or inclined stope and extracted from the base of the stope for removal from the mine. Various methods are used, including sub-level open stoping, longhole stoping, and cut-and-fill stoping, according to the geometry of the orebody and the relative strengths of the orebody and wall rock.

Room-and-pillar mining is best suited to gently dipping and relatively uniform bedded deposits, provided that the deposits are not too deep and the rock is strong enough for the pillars to support the overlying strata. At depths beyond 1000 to 2000 m, hard-rock room and pillar mining methods are considered dangerous because of rock bursts and other results of high stress concentration. The limiting depth in coal mines is much less, in the region of 200 to 300 m, since the rocks of the Coal Measures tend to be weaker. Room-and-pillar mining was used extensively in the UK and Europe for the extraction of building stone, such as the Jurassic oolitic limestones of the Bath area.

Longwall mining began in the coalfields as a method of extracting seams at depths in excess of 200 m. It is by far the most common method of working in European coal mines where the shallower seams have been depleted. Longwall mining is best suited to deposits ranging in thickness from 1 to 2.5 m, dipping at less than 12°, with relatively incompetent rock and a fairly competent floor. The method lends itself to mechanization, especially in coal mines where fast-moving shearing machines move across a 100- to 200-m face and load the coal into conveyors without the need for blasting.

Other forms of mining take place, most notably solution mining (leaching) or brining of evaporite deposits, which is a long-established method. The process is extensively used in the mining of halite in Cheshire.   BRIAN J. McCONNELL

Miocene

Miocene The Miocene epoch is a division of Cenozoic time that was first distinguished by Charles Lyell in his famous book *Principles of geology* (1830) on the basis of the fossil mollusc fauna present in the rocks of the Paris Basin. As the earlier of the two divisions of the Neogene sub-period, it follows the Oligocene epoch and precedes the Pliocene. Today, Lyell's criterion of some 17 per cent of the Miocene mollusc fauna being extant is no longer applied; the Miocene is held to have begun at about 23 Ma and lasted for around 17 million years. It contains three series and six stages, based upon sections in the Mediterranean region. The marine bio-stratigraphy is based upon microfaunas and floras, notably planktonic foraminifera and radiolaria.

In North America the continental Miocene deposits of the western interior have been apportioned to three stages on the basis of their mammalian faunas, which locally are prolific.

The principal feature revealed in Miocene stratigraphy is a continuation of climatic cooling. By about 15 Ma a major ice cap had accumulated on the Antarctic continent and a permanent ice shelf was beginning to form around its shores. The circum-Antarctic current intensified and expanded. Glaciers made their appearance in the southern Andes. On the other side of the world the first significant ice fields were developing in Alaska, and polar sea ice became permanent. In consequence, by 10 Ma world sea level began to fall.

At the end of the Oligocene the Ural sea had disappeared and the Himalayan uplift was taking place. Eurasia was more or less consolidated except that the Mediterranean–Tethyan region was still in motion; the seas became reduced in area, and by 18 Ma the Mediterranean was separated from the Indo-Pacific seas. Africa and Arabia were now in contact with Eurasia and free access between the two was available to land animals. The Mediterranean also became closed off from the Atlantic. In the late Miocene, desiccation lowered the level of the Mediterranean waters to well below that of the Atlantic and diminished their area by as much as 2000 km². The last stage of the Miocene was influenced by global environmental changes, probably sharp climatic fluctuation which regulated sea levels up and down. A eustatic rise of Atlantic sea level breached the Gibraltar barrier to flood the western Mediterranean evaporite basin.

Miocene littoral and sublittoral deposits on the western side of the Atlantic show the continuation of the transgression begun in earlier times, while the Gulf of Mexico continued to subside under a growing load of sands and silts.

Around 14 Ma there was a sharp increase in volcanic activity in many regions, especially in the Central American and Pacific regions. The Hawaiian hot-spot was notably active, producing basaltic flows and ashes in the islands.

In the oceans, the foraminifera were flourishing as earlier in the Cenozoic, and underwent a successful radiation to reach a peak of species diversity around 15 Ma. Thereafter they went into decline. The calcareous nannofossils, too, steadily declined in diversity until at the end of the Early Miocene only about 35 species are known. They made a short and slight comeback in the Middle Miocene.

The land flora was susceptible to minor fluctuations in temperature, and the cooling episodes were felt even in the tropics. Coniferous forests spread across the high northern latitudes, and grasslands spread in the lower latitudes. Several modern plant families made their first appearance.

In the history of the mammals the Miocene is important for the diversification of the primates around 21 Ma and the emergence of the early hominid *Ramapithecus* some 7 Ma later. It was around 14 Ma that many land mammals of their kind that had been common in the earlier Cenozoic were lost and new grazing types became abundant. As the climate became cooler, grasses spread and provided habitats for new grazing animals. Towards the end of the Miocene, modern cats and the first elephants arrived on the scene.

D. L. DINELEY

Mississippian *see* CARBONIFEROUS

Mohorovičić discontinuity (Moho)

The Earth's internal structure can be thought of as a series of concentric shells, each of which transmit seismic waves at a different speed. The boundaries between shells are marked by seismic discontinuities at which the speed of transmission of seismic waves makes a sudden jump. Distinct seismic discontinuities occur between the inner and outer core and between the outer core and the mantle, and these reflect major changes in composition. There are two well-marked seismic discontinuities within the mantle, which are believed to represent depths at which the pressure is great enough to cause minerals in the mantle to adopt a denser structure (a phenomenon described as a phase change) although there is no difference in overall chemical composition on either side of the discontinuity. The shallowest seismic discontinuity that is global in its extent is the one known as the Mohorovičić discontinuity (or Moho for short) after the talented Croatian meteorologist and geophysicist Andrija Mohorovičić (1857–1936) who, in 1909, was the first to recognize it.

The Moho represents the base of the Earth's crust, and represents a difference in composition rather than a phase change. Both crust and mantle are made of silicate rock and the compositional difference between them is slight compared to the difference between mantle and core. Even so, it is sufficient to allow compressional seismic waves (P-waves) to travel at about 8 km s^{-1} in the upper mantle, as compared with only 6 or 7 km s^{-1} in the lower crust. The effect of this is to allow seismic waves from a crustal earthquake to reach a sufficiently distant detector faster by passing through the base of the crust and into the mantle before being refracted back up through the crust towards the detector than waves that have travelled directly through the crust to the detector. Mohorovičić proved this by analysing signals collected by seismometers at the Zagreb observatory and elsewhere, that had emanated from a destructive local earthquake.

Mohorovičić's technique relied on refraction of naturally occurring seismic waves at the crust–mantle boundary, but the Moho can be pinpointed more precisely by seismic reflection. The conventional method of mapping the Moho along the line of a seismic traverse is to detonate powerful explosive charges just below the surface and to record the reflections by an array of detectors known as geophones. An alternative seismic reflection technique capable of detecting the Moho and resolving its fine-scale structure is to use a truck-mounted vibrator as a source; this has the advantage of providing a pure signal of known (and variable) frequency.

Continental crust is thicker than oceanic crust, so the Moho is usually at a depth of about 35 km below the continents, though it can be as shallow as 25 km where the crust has been thinned and stretched or as deep as 90 km below major mountain belts. The Moho is only about 7–10 km deep below oceanic crust, which is thinner than the continental variety.

The Moho was formerly assumed to represent a sharp discontinuity, but detailed seismic studies have revealed that in at least some places it is complex and probably layered, with peridotite from the mantle interleaved with gabbro (below the oceans) or diorite (below the continents) over a depth range of perhaps a few hundred metres. In collision zones the Moho may also be duplicated by compressional faulting.

It is a common misconception that the Moho marks the base of the Earth's tectonic plates, but this is not so. The Moho is a firmly welded interface across which crust and upper mantle are held together. The crust and uppermost mantle together constitute the Earth's lithosphere, its outer strong and rigid layer, which is able to move across the deeper mantle (the asthenosphere) because the pressure and temperature are such that the latter is relatively weak. Because of the great depth to the Moho, erosion is nowhere able to expose it at the Earth's surface. However, in rare instances a slice of oceanic crust and upper mantle, known as an ophiolite, has escaped subduction during ocean closure and has instead been thrust over the edge of a continental plate (a process described as obduction). Here the 'fossil' Moho within the ophiolite can be examined (Fig. 1).

Fig. 1. A 'fossil' Moho revealed in the Oman ophiolite: a piece of oceanic lithosphere of Cretaceous age now sitting on top of the eastern edge of Arabia. Dark peridotite representing the mantle (lower right) is overlain by pale gabbro (upper left) from the crust. Note the metre- to centimetre-scale interlayering of gabbro and peridotite near the actual contact.

In such places the mantle is revealed mostly as varieties of peridotite such as harzburgite (the main minerals of which are olivine and orthopyroxene) or lherzolite (main minerals as for harzburgite with the addition of clinopyroxene), which are evidently residua from which the basaltic magma that has gone to make the oceanic crust has been extracted. The contact between the harzburgite or lherzolite mantle and the overlying rocks is described as the 'petrological Moho', because it divides rocks that are petrogenetically part of the mantle from rocks that are petrogenetically part of the crust. Most of the deep oceanic crust is gabbroic in composition, but there are layers, especially near its base, consisting of varieties of peridotite such as wehrlite (main minerals olivine and clinopyroxene) that are not residual but instead have crystallized from magma that has had the remaining melt extracted (perhaps squeezed out) before it was fully solidified. Petrogenetically peridotite of this type belongs to the crust, but seismic waves would travel just as fast through it as through any other peridotite. If there is a well-marked interface between overlying gabbros and underlying magmatic peridotites, this is described as the 'seismic Moho', because it corresponds to the Moho that can be detected *in situ* round the globe by seismology.

The only seismic data we have for any planetary body other than the Earth comes from a network of seismometers installed on the Moon by the *Apollo* landings. This network was able to use seismic refraction calculations to show that there is a clear crust–mantle interface beneath the near side of the Moon, at a depth of about 60 km between the *Apollo* 12 and *Apollo* 14 sites and about 75 km near the *Apollo* 16 site. By analogy with the Earth, this interface is usually referred to as the lunar Moho. Doubtless seismometers on Mars will before long determine the depth to the martian Moho.

In 1957 the main civilian science funding agency in the United States of America, the National Science Foundation, began a flirtation with the idea of attempting to drill a borehole right through the Earth's crust and into the mantle. The scheme acquired the splendid name of Project Mohole, and began promisingly. It was realized that because the oceanic crust is so much thinner than continental crust, the best chance of success would be to drill the 'Mohole' through oceanic crust. Despite the difficulties of drilling from a platform floating on some thousands of metres of water, this would be easier than on-land drilling of the much deeper hole required to reach the Moho below continental crust. In 1961, using a drilling rig mounted on a converted naval barge, a trial hole reached basalt at a depth of 200 m below 3.5 km of water off the coast of Mexico. After this promising start Project Mohole ran into political difficulties and it was terminated in 1966 amid an aura of financial scandal. However, the techniques and expertise that had been developed in Project Mohole sowed the seeds for the international Deep Sea Drilling Project (DSDP) and Ocean Drilling Project (ODP) in the following decades, which although never attempting to reach the Moho itself, have provided a wealth of information about the ocean basins and their history by drilling less ambitious holes at well-chosen sites around the globe.   DAVID A. ROTHERY

Further reading
Brown, G. C. and Mussett, A. E. (1993) *The inaccessible Earth* (2nd edn). Chapman and Hall, London

Mohs' scale of mineral hardness
The hardness of a mineral is mainly controlled by the strength of the bonding between the constituent atoms and partly by the composition and size of the atoms. It is a measure of the resistance of the mineral to abrasion and scratching.

The element carbon exists in two distinct forms—graphite and diamond—which have identical compositions but different structures, according to their conditions of crystallization, yet graphite is one of the softest minerals known whereas diamond is by far the hardest and, indeed, is the hardest substance yet discovered.

In 1822 the Austrian mineralogist Friedrich Mohs proposed a scale of relative hardness based on fairly common minerals (Mohs' scale). The 10-point scale is: 1, talc; 2, gypsum; 3, calcite; 4, fluorite; 5, apatite; 6, orthoclase; 7, quartz; 8, topaz; 9, corundum; 10, diamond.

Each mineral will scratch the ones below it (i.e. with a lower number) on the scale and will be scratched by those above. The intervals are not equal; 9–10 is much greater than 1–9. (On the Rosiwall absolute hardness scale diamond is 140 000; corundum, 1000; topaz, 175; and gypsum, 1.25.) On Mohs' scale a finger nail is 2.5; a copper coin 3.5; a knife blade 5.5; and a metal file 6.5–7.

Hardness is a fairly useful diagnostic property for use in field identification of minerals.   R. BRADSHAW

molasse
Molasse is a dialect word used by French-speaking farmers in western Switzerland to describe soft, friable sandstones. It was subsequently used by Studer in 1825 to describe the entire Oligo–Pliocene sedimentary sequence of the central Swiss Plateau and adjacent areas. The sequence consists of approximately 7000 m of sediments which accumulated in a subsiding foreland basin to the north of the Alps, during the later phases and after the main mountain-building movements. The sediments were deposited under alternately freshwater and marine conditions. The Lower Marine Molasse consist of marine turbidites and shallow-water deposits, and the Upper Marine Molasse of shallow-marine, tidally influenced, sediments. The Lower and Upper Freshwater Molasse consists of alluvial-fan deposits, derived from the Alps, which pass into fluvial deposits that were transported longitudinally along the basin. Whereas in the Lower fluvial interval they were transported to the north-east, in the Upper fluvial interval they were transported to the south-west.

Since Bertrand's employment of the name of 1847, *molasse* has been widely used outside Switzerland to describe an assemblage of sedimentary rocks thought to have accumulated in a tectonic environment comparable to that north of the Swiss Alps.   G. EVANS

molluscs
The molluscs are an extremely diverse and numerous, mostly marine, invertebrate phylum, second only to arthropods in numbers of species. They include a range of organisms, both living and fossil, that seem at first sight to be so different as to be unrelated. These range from slow-moving herbivores, scavengers, and predators, such as chitons and slugs and snails, to suspension-feeders such as tusk-shells and bivalves, and their extinct ancestors the rostroconchs, and the most complex and highly organized group the cephalopods, which includes squids, octopus, and the fossil ammonoids and belemnoids.

Although no one character is unique to every mollusc, there is a suite of characters that serves to unite the disparate members of this phylum. Characteristically, molluscs are free-living organisms with a calcareous shell secreted by a sheet of tissue termed the mantle, which in turn encloses a space termed the mantle cavity. The mantle cavity surrounds the viscera of the animal and opens externally, allowing the discharge of waste products and the free circulation of water across the gills. It is the variable development of these characters in the different classes that is used to separate them.

Molluscs are biostratigraphically important animals and are widely used for zoning marine rocks from the Upper Palaeozoic to the Cenozoic.   DAVID K. ELLIOTT

moment of inertia and precession
Moment of inertia is a fundamental property in rotational mechanics. It is the ratio of torque to angular acceleration and is analogous to the role of mass in linear mechanics (ratio of force to linear acceleration). For solid bodies it is a measure of the distribution of mass. The greater the concentration of mass towards the centre of a sphere the smaller is its moment of inertia. In the 1930s K. E. Bullen recognized that a value for the moment of inertia of the Earth was available from astronomical observations of the motions of the Earth and Moon. Using this information he was able to develop the first realistic models of the Earth's density structure (see *density distribution within the Earth*).

The moment of inertia of a point mass m about an axis at distance r from it is mr^2. For a solid body with a known distribution of density the moment of inertia about any axis can be obtained by integration of this formula. We are normally interested in moment of inertia about an axis through the centre of mass, and we can compare the Earth with two simple standard results. For spheres of mass m and radius r:

Hollow spherical shell $I = (2/3) \, mr^2$
Uniform solid sphere $I = (2/5) \, mr^2$
Earth $I = 0.330695 \, mr^2$

To gain an idea of the mass concentration needed to give the terrestrial value, consider a model Earth with a uniform mantle of density ρ and radius r surrounding a uniform core of radius $0.546r$ and density $f\rho$. We would then require $f = 3.01$, or, in other words, a core with a density about three times greater than that of the mantle. Of course, neither the mantle nor the core of the real Earth is uniform, but this radius ratio is the observed one, and we can thus see that there must be a strong concentration of density towards the centre.

To measure the moment of inertia of a body we need to observe the angular acceleration produced by a known torque. In the case of the Earth the only torques of sufficient magnitude to cause observable effects are those from gravitational interactions with the Moon and Sun. If the Earth had perfect spherical symmetry then no such torques would be possible because all gravitational interactions with a sphere are equivalent to interactions with an equal mass at its centre. Thus the lunar and solar torques on the Earth act on the asymmetry of the mass distribution, specifically the equatorial bulge. They act in a direction that tends to pull the bulge into alignment with themselves.

The geometrical problem is represented in Fig. 1. At the equinoxes, on the y-axis, the Sun is directly above the Equator and no torque arises. The torque is greater when the Sun is at A or B on the x' axis at an angle $\theta = 23.5°$ to the equatorial plane. The Earth then experiences a torque tending to pull the equatorial (x) axis towards x'. The sense of this torque is the same when the Sun is at both of the positions A and B. Thus the solar torque oscillates in strength between the maximum and zero, but never reverses in sign. It is a torque that switches on and off in a semi-annual cycle. There is similarly a torque due to the Moon, which oscillates in strength in a semi-monthly cycle.

When a rotating body experiences a torque that tends to turn its axis of rotation, the response is a shift of the axis in a perpendicular sense. This is precession. The rotational axis does not turn towards the pole of the ecliptic, that is, the normal to the orbital plane ($Ox'y$ in the figure), but precesses about it. It describes a cone of semi-angle 23.5°, about the ecliptic pole, completing the cycle in 25 730 years.

The lunar and solar precessional torques are proportional to the difference between moments of inertia of the Earth about the rotational axis (C) and an axis in the equatorial plane (A). There is thus a torque proportional to ($C-A$) which acts on the Earth's rotational angular momentum, $C\omega$, where ω is the angular speed of rotation. The rate of change in the direction of the rotational axis is the ratio of torque to angular momentum and so gives a value for the precessional constant. The torque on the Earth exerted by any external body is balanced by a torque on that body, causing a precession of its orbit. The effect is well observed for the Moon but is now more precisely seen in the orbits of artificial satellites. The

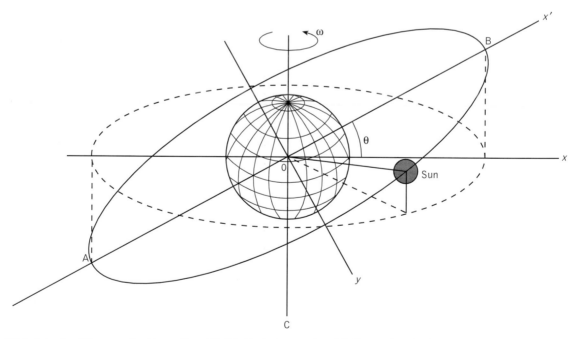

Fig. 1. Geometry of the precessional torque. The orbit of the Sun is shown as the solid ellipse in plane ($Ox'y$). The broken ellipse is the projection of this orbit on to the equatorial plane (Oxy).

nodes of an orbit, that is the points at which it crosses the equatorial plane, slowly regress round the Equator and the rate gives a measure of the asymmetry of the Earth's mass. We can thus obtain a value for the axial moment of inertia, C, of 0.330695 Ma^2, M being the mass of the Earth and a its radius.   FRANK D. STACEY

Further reading

Stacey, F. D. (1992). *Physics of the Earth* (3rd edn). Brookfield Press, Brisbane.

monsoonal currents Monsoon systems constitute the largest and most energetic disturbances of the atmosphere. Not surprisingly, monsoons have assumed historical importance, bringing, for example, the coast of East Africa under the influence of Arabic culture through the reliability of seasonal wind systems.

Monsoons arise from the differential heating and cooling of the land and sea surface. At the most basic level they are akin to land and sea breezes. Three simultaneous criteria define a monsoonal climate:

(1) prevailing wind direction shifts by at least 120° between winter and summer;
(2) the average frequency of prevailing wind directions in winter and summer exceeds 40 per cent; and
(3) the mean resultant wind speed is at least 3 l m s^{-1}.

These conditions reach their largest expanse over Asia. At its height, the Asian monsoon dominates much of the global tropics. In summer, the Tibetan plateau and hence the overlying atmosphere is heated by solar radiation, leading to a reduction in surface pressure. The low surface pressure induces low-level flow of warm, moist air from the surrounding ocean to the land. Convergence and ascent lead to intense precipitation over India and the slopes of the Himalaya during the summer months. Release of latent heat in the rising air further heats the air column, thereby driving the system further. The summer monsoon begins with marked changes in temperature, air pressure, and an increase in wind flow from April to May. This coincides with the movement of warm surface waters from the central Arabian Sea and the Bay of Bengal. Simultaneously, a longitudinal surface pressure gradient extends from the south Indian Ocean anticyclone (high pressure) across the Equator into the south Asian continent. The typical zone of low pressure at the Equator disappears. Inflow into the heated Asian continent occurs as three main airstreams: over the Arabian Sea, the Bay of Bengal, and the South China Sea. The moist yet cool south-west airstreams undercut the dry and hot continental air, leading to vigorous and deep convection. There has been controversy over the importance of cross-equatorial transport of water vapour for the moisture budget of southern Asia. More recently, the role of the low-level East African Jet

(a)

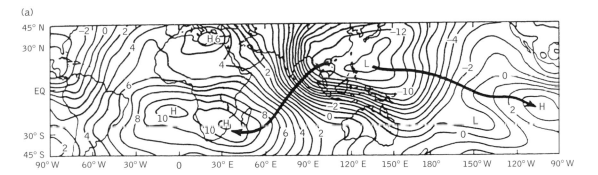

(b)

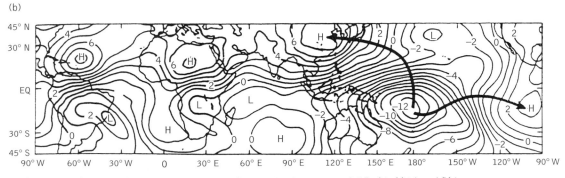

Fig. 1. Divergent circulations as shown by velocity potentials (non-rotational component of airflow) in (a) July and (b) January.

(over the Arabian Sea) in moisture import has been emphasized. The surface pressure trough extends from the desert regions of Iran and Pakistan across northern India. The principal belt of monsoon rainfall tends to migrate along a belt of roughly 500 mm of rainfall which moves from Indonesia to the foothills of the Himalaya.

Unlike the typical tropical Hadley cell, which is characterized by north–south overturning, much of the air flow in the summer monsoon is oriented east–west (Fig. 1a). The extent of this flow is massive, dominating much of the tropical atmosphere. As a result of the change towards a more negative radiation balance (long-wave loss exceeding short-wave solar radiation gain) and cold air advection in the westerlies, the seasonal cooling of the land surfaces of south Asia and the surrounding seas leads to the onset of the boreal winter phase of the monsoon. It is thought that the extensive cooling of the Asian landmass is more important than the more regional heating over Australia. As a result of the cooling, anticyclonic vortices develop over land, the Arabian Sea, and the Bay of Bengal by October, while the major monsoon convective zone migrates from the summertime position over north-eastern India to the maritime continent of Borneo and Indonesia. This occurs despite the much smaller area of land in the southern hemisphere compared with its Asian counterpart. When fully developed, the winter monsoon convection drives a gigantic planetary-scale overturning motion in the east–west and north–south direction (Fig. 1b). Between November and March, north-easterlies blowing out from the surface high pressure over southern Asia sweep much of the northern Indian Ocean, including the Arabian Sea, the Bay of Bengal, and the South China Sea. These airstreams then turn northwards to the north of the Equator and meet the southern hemisphere south-easterlies in a trough zone south of the Equator.

In the upper atmosphere during this season, westerlies dominate in the upper troposphere over most of the monsoon region. The Subtropical Westerly Jet is found to the south of the central Asian mountain massifs, in contrast to the boreal summer when the Tropical Easterly Jet occupies the upper troposphere to the south of the Himalayas. The fundamental difference between the winter and summer monsoon is that convection is based over land in July but over land and sea in January. There is, however, an important difference in the moisture supply between the two systems The daily cycle of winter monsoon convection over Indonesia is influenced by land–sea breeze effects, whereas the diurnal variation of the continental monsoon has continental features, for example, dry convection over Tibet. The divergent component of the equatorial winter monsoon is also much stronger. Overall, the boreal summer monsoon is much more vigorous than its winter counterpart, except over East Asia. This, together with the hemispheric asymmetry in land–sea distribution in the monsoon region, points to the role of hemispheric asymmetry in surface heating as the main driving mechanism of the monsoon.

Much can be learned by considering some of the basic energetics of the monsoon. Descent over Asia in January leads to a gain in geopotential and sensible energy which is huge compared with the small radiative heat and oceanic heat import. Latent heat from precipitation is non-existent, this form of energy being exported in the form of evaporative fluxes largely off the ocean. The gain in radiative heat input at the top of the atmosphere in July between 0° and 30° S is small. Added to this, heat is imported into the southern tropical Indian Ocean by the southward-directed currents. Both sensible and latent heat losses from the southern tropical Indian Ocean are far higher than the radiative heat gain. The latitudinal temperature gradient in July provides an important energy source in the form of available potential energy, providing the monsoon in this season with higher kinetic energy.

Much research effort in the past two decades has been invested in numerical modelling of the Asian monsoon. Since these models allow for changes in one variable at a time, the role of mountains in maintaining the south Asian low-pressure system, or the importance of sea-surface temperature anomalies in the Arabian Sea, have, for example, been investigated.

A remaining problem in monsoon studies concerns intraseasonal oscillations. In the course of the south-west monsoon, there are periods when the monsoon trough shifts northwards to the foot of the Himalayas, and rains decrease over much of India except along the slopes of the Himalayas. This is called the 'break in the monsoon'. These breaks are most frequent in July and August and last for 2 to 3 weeks. Recent studies emphasize the propagation of low-frequency modes and middle-latitude interactions. There has similarly been a great deal of research aimed at predicting monsoon onset as well as understanding the interannual variability of the monsoon. A concerted effort to address such problems saw the initiation of the GARP (Global Atmospheric Research Programme) Monsoon Experiment (MONEX) conducted from 1978 to 1979. Initially staged in Malaysia during the winter of 1978–9 for the study of cold surges, rain-producing waves, and vortices, the entire monsoon trough and the planetary-scale structure of the monsoon were eventually examined. It is thought that the mode of migration of the region of heating from northern Burma in the month of May to the foothills of the Himalayas in June is a critical factor in determining the onset date.

It has long been known that the Asian monsoon is related to El Niño–Southern Oscillation (ENSO) events, which feature the interannual oscillation of air pressure and sea surface temperatures across the global tropics, particularly the Pacific. Indeed, in the 1920s that Sir Gilbert Walker uncovered this see-saw of pressure in an attempt to predict the Asian monsoon. Generally, higher sea-level pressure at Darwin is accompanied by lower monsoon rainfall in India. Research into this relationship continues.

The criteria outlined above are by no means confined to the atmosphere overlying Asia and Indonesia. Africa also has an important monsoon circulation. During the boreal winter, for example, the equatorial surface pressure trough stays closest to the Equator over Africa, while the Harmattan winds, a stream of dry desert air (trade winds) sweep far southwards, confining rains to a belt extending from the Gulf of Guinea eastward into the interior of the continent. During the boreal summer, the surface pressure trough shifts northwards, so that a cross-equatorial moist and cool monsoon airstream penetrates into the continent.

Work by Neil Ward has shown that the monsoon systems of the globe, including Asia and Africa, may be linked in an integral system, such that in some years a tropic-wide mode is triggered, as rainfall throughout the global monsoon system is above. These modes are thought to be oscillations internal to the atmosphere that are excited by sea-surface temperature conditions in the Pacific.   R. WASHINGTON

Further reading
Grotjahn, R. (1993) *Global atmospheric circulations: observations and theories.* Oxford University Press.

Moon The Moon is a familiar object in our sky because it is our nearest neighbour and is the only natural object to orbit the Earth. It is smaller than the Earth, but has all the important attributes of a terrestrial planet except that of being in independent orbit about the Sun. Some basic data are presented in Table 1.

Something that adds to the Moon's familiarity is that it always presents the same face towards us. This is because the Moon rotates exactly once per orbit, in what is described as *synchronous rotation*. This is a phenomenon that is also exhibited by most of the satellites of the giant planets. It is brought

Table 1 Basic data for the Moon

Equatorial radius	1738 km
Mass (relative to Earth)	0.0123 (7.35×10^{22} kg)
Density	3.34 g ml^{-1}
Surface gravity	1.62 m s^{-2}
Rotation period	27.32 days
Axial inclination	6.67°
Distance from Earth	384 400 km
Orbital period	27.32 days
Orbital eccentricity	0.05
Composition of surface	Rocky
Mean surface temperature	1 °C
Composition of atmosphere	argon, helium, atomic oxygen, sodium, potassium
Atmospheric pressure at surface (relative to Earth)	2×10^{-14}

about because tidal forces between the planet and its satellite slow the satellite's rotation until it matches its orbital period.

Although the Moon rotates once per orbit, with patience it is possible to see about 59 per cent of its surface from the Earth. One reason is that the Moon's proximity is such that the Earth's rotation results in slight but significantly different points of view from a single site across a 12-hour period. A similar change in perspective could be gained by moving from the North Pole to the South Pole. The other reason is that although the Moon rotates at an exactly constant rate its progress round its slightly non-circular orbit varies slightly in accordance with Kepler's second law of planetary motion. It thus travels faster when its slightly elliptical orbit brings it closer to the Earth. As a result, if seen from an imaginary point at the Earth's centre the Moon would appear to nod to and fro as it moved round its orbit, so that sometimes the observer could see slightly round its western limb (edge) and sometimes round its eastern limb. However, the terrain seen near the edge of the lunar disc is never well seen, because we only ever see it obliquely, and there remains 41 per cent of the lunar surface that can never be seen at all from Earth.

Missions to the Moon
The Moon was the first extraterrestrial target for space missions. Probes have been directed towards it since almost the very dawn of the space age (Table 2), and it was the main focus of the 1960s–1970s 'space race' between the USA and the then Soviet Union. In the end, only NASA (the National Aeronautics and Space Administration) attempted to put people on the Moon, and the six successful *Apollo* landings brought back at total of 382 kg of lunar rocks. Radiometric dating of these samples, together with 0.3 kg collected from other sites by unmanned Soviet sample return missions, enabled us to calibrate the cratering timescale (based on the number of impact areas per unit area) that is used for dating unsampled areas of the Moon and other surfaces in the inner Solar System. Geochemical analysis of these samples was immensely important in developing our current level of understanding of the Moon's origin and history.

After the budget for the *Apollo* programme was terminated in 1972, there was little further effort in lunar exploration until 1994, when the *Clementine* probe went into lunar orbit and collected a wealth of previously unknown information about the topography, crustal thickness, and variations in crustal composition across the whole lunar globe. This was followed in 1998 by another orbiter, *Lunar Prospector*, which provided even more insights and discoveries, such as the existence of ice dispersed within the regolith at both poles. The European Space Agency and Japan each plan lunar missions for the early years of the twenty-first century.

Origin and history of the Moon
The Moon apparently owes its origin to a giant impact between a Moon-sized or Mars-sized body and the primitive Earth towards the end of planet formation 4.5 billion years (4.5 Ga) ago. This was the last of several giant impacts by

Table 2 Some of the successful and anticipated missions to the Moon

Name	Description	Date
Luna 2	Impact with surface	September 1959
Luna 3	Flyby, images of far side	October 1959
Rangers 7–9	Images at close range prior to impact	July 1964–March 1965
Luna 9	Unmanned landing, pictures from surface	February 1966
Luna 10	First probe to orbit Moon	April 1966
Surveyor 1, 3, 5–7	Unmanned landings, images from surface	June 1966–January 1968
Lunar Orbiter 1–5	Images from orbit	August 1966–August 1967
Luna 11–12	Images from orbit	August–October 1966
Apollo 8	First manned orbits	December 1968
Apollo 11, 12, 14–17	Manned landings, geological and geophysical studies, sample return	July 1969–December 1972
Luna 16, 20, 24	Unmanned sample returns	September 1970–August 1976
Luna 17, 21	*Lunokhod* unmanned surface rovers	November 1970, January 1973
Galileo	Flyby *en route* to Jupiter, compositional mapping	December 1992
Clementine	High-resolution multispectral imaging and laser altimetry from orbit	January–March 1994
Lunar Prospector	Geophysical and geochemical mapping from orbit	January 1998–July 1999
SMART 1	ESA lunar orbiter	Late 2001
Lunar A	Japanese lunar orbiter and seismometers	2003
Selene 1	Japanese lunar orbiter and lander	2003

Luna missions were launched by the Soviet Union; the others are NASA missions unless specified.

which the Earth grew to its current size, but it was unusual in that much of the debris from the fragmented impactor plus some ejecta from the Earth itself ended up in orbit about the Earth, where it collected together to form the Moon. Once the last giant impact in the inner Solar System had occurred, there were just four surviving terrestrial planets (Mercury, Venus, Earth, and Mars). The total is five if you count the Moon, which is a planet in a geological sense (because of its size and character) though not in an astronomical sense (because it orbits the Earth rather than the Sun).

The subsequent evolution and further internal differentiation of these bodies was driven mainly from within, and changed the chemical composition of the outermost part of the mantle sufficiently to merit its distinction by the term 'crust'. However, all these bodies were subject to one very important external influence in the form of bombardment by 'junk' left over from the origin of the Solar System, including the surviving small planetesimals, which scarred their surfaces with impact craters. These are most easily studied on the Moon, where the lack of erosion means that once formed a crater is likely to survive until obliterated by a younger crater. Crater formation has continued up to the present day, but its intensity waned to something like its present level about

3.9 Ga ago, by which time most of the original debris has been mopped up during what is referred to as the 'late heavy bombardment'. Counting the density of impact craters on a planetary surface is virtually the only way we have of estimating the age of the surface (that is, how long the material now forming the surface has been there). The basis of this cratering timescale is that older surfaces have had time to accumulate more craters per unit area than younger surfaces. Rock samples collected from the Moon and then dated in the laboratory (by means of their natural radioactivity) enable us to put numerical values to these ages, which would otherwise be on a purely relative scale.

Polar ice and the atmosphere

The *Clementine* mission returned our first clear views of the lunar poles, showing sites in particular near the south pole that are permanently in shadow, and could therefore be places where ice might accumulate. *Clementine*'s simple radar gave the first indications that frozen water is present there, and this theory was dramatically backed up by measurements by *Lunar Prospector* that show a reduction in the average speed of neutrons (produced by cosmic radiation) over both lunar poles. The only reasonable explanation of this seems to be collisions between neutrons and the hydrogen atoms within ice molecules, and it now seems that there may be as much as three billion tonnes of ice mixed with the regolith at each pole. This is not a lot in terms of the size of the Moon (it is the equivalent of a 1.5 km ice cube at each pole), but it could be ample to supply the immediate needs of human habitation on the Moon.

The Moon's gravity is too slight for it to be able to retain a gaseous envelope, so its atmosphere is tenuous in the extreme. Atoms that have been detected in the Moon's atmosphere are listed in Table 1. These are all light enough to escape to space, and what has been detected must be a steady-state mixture that is continually replenished. Presumably these elements are either supplied continually by micrometeorites or are liberated from the surface under the influence of solar radiation or meteorite impact.

Although water has not been detected in the atmosphere, it is likely that some or most of the Moon's polar ice derives from molecules of atmospheric water (supplied by cometary impact or outgassing) that wandered into the cold polar regions and became incorporated into the surface, thus avoiding the more usual fate of escaping or becoming split into hydrogen and oxygen atoms by radiation.

The interior

The Moon is the only planetary body other than the Earth for which we have any seismic data that can tell us about its interior. Four seismic stations were established at the *Apollo 12*, *14*, *15,* and *16* sites and these continued to send back data until September 1977, when operations were terminated for budgetary reasons. The Moon is seismically very quiet compared to the Earth, and moonquakes are many orders of magnitude less powerful than typical earthquakes. The *Apollo* seismometers also detected the vibrations from meteorite impacts (including one from the far side that gave crucial data on the maximum possible size of the core) and deliberate crashes of various expended units of the *Apollo* spacecraft.

The picture that has emerged from combining *Apollo* seismic data with topographical and gravitational mapping by *Clementine* and *Lunar Prospector* is that the Moon's crust is on average about 70 km thick. However, it exceeds 100 km in some highland regions and falls to about 20 km thick below some impact basins. The crust is thicker on the far side, with the result that the Moon's centre of mass is offset from its geometric centre by about 2 km towards the Earth.

The mantle is rigid to a depth of about 1000 km, which is also the greatest depth for the source of moonquakes. This depth therefore marks the base of the lunar lithosphere, below which it is probably slowly convecting. Like the Earth's mantle, that of the Moon is approximately peridotite in composition, though there are signs that it varies in detail from region to region.

The Moon's core is between about 220 and 450 km in radius (so the core–mantle boundary is at least 1290 km depth), and is probably iron-rich and solid. It makes up less than 4 per cent of the Moon's total mass, a much smaller fraction than for any of the other terrestrial planets. This can be explained by computer models of the giant impact in which the Moon was born, which show the core of the impacting body accreting on to the core of the proto-Earth, whereas the Moon formed from the mixture of fragments of the mantles of the two bodies that were thrown into space. The fact that no magnetic field is generated in the Moon's core today shows that it must be solid. There are, however, local weak remanent magnetic fields associated with some of the Moon's major impact basins that suggest that part of the core may still have been liquid and generating a global field when these formed 3.6 Ga ago.

The surface

Look at the Moon even with the unaided eye, and you will see that it has dark patches on a paler background. This simple observation picks out the two distinct types of crust on the Moon. The paler areas are the lunar highlands, and the darker

Fig. 1. *Left*: An *Apollo* astronaut's footprint in the lunar regolith. *Right*: A large fractured boulder flung out by a relatively recent impact and resting on the surface of the lunar regolith. The astronaut in the lower left provides the scale.

areas are the lunar 'seas' or maria (singular: *mare*). On both the highlands and the maria there is very little bedrock exposed at the surface, because it is buried by several metres of regolith composed of ejecta from crater-forming impacts at all scales. In detail the regolith consists of a mixture of rock fragments, crystal fragments, and droplets of glass (which are congealed globules of melt produced by impacts). This forms a 'soil' capable of bearing the imprint of a booted foot (Fig. 1, left), but there are also scattered boulders of a variety of sizes (Fig. 1, right).

The highlands are the first lunar crust to have formed, probably by crystals rising to the surface of the 'magma ocean' that is thought to have covered the young Moon as a result of the heat generated as the Moon-forming fragments collided. Because of the preponderance of the mineral anorthite (a calcium-rich variety of the mineral plagioclase feldspar) this rock type is called anorthosite. The oldest anorthosite sample collected from the Moon has been dated at 4.5 Ga, but the highland surface suffered such heavy bombardment (until the rate of impact cratering declined towards its current level) that most highland rock samples consist of fragments of rock that were welded together between about 4.0 and 3.8 Ga ago.

The high degree of cratering suffered by the highland crust is notable in the view seen in Fig. 2. This also illustrates the dependence of crater morphology on size. Craters less than about 15 km across are simple bowl-shaped depressions with raised rims, but the largest crater in this view has the terraced inner walls and central peak that are characteristic of craters

larger than this. At yet larger diameters the central peak is replaced by a ring of peaks. The largest impact structures of all are multi-ringed impact basins dating from 4.0 to 3.8 Ga ago. At 2500 km in diameter, the south pole Aitken basin, which lies on the far side, is the largest lunar example, but there are several others, notably the Imbrium basin, occupying the north-west of the Moon's near side.

All the Moon's multi-ringed impact basins are older than the Moon's second kind of crust, consisting of basalts that have flooded low-lying areas to form the lunar maria. Many of the maria occupy multi-ringed impact basins, and it was once thought that the melting to produce the mare-filling basalts was a direct response to the basin-forming impacts. However, the ages of samples of mare basalt that have been returned to Earth range from 3.8 to 3.1 Ga; so clearly the magma was from unrelated sources and ended up in the basins merely because it took advantage of the easiest routes to the surface. Inspection of detailed images of some mare areas that have yet to be sampled suggest that the mare-forming eruptions did not cease until about a billion years ago. The only extensive mare regions are on the lunar near side, and most of the far-side features (even the south pole Aitken basin) have escaped mare flooding, presumably because of the greater thickness of far-side crust.

The mare basalts probably came into place as series of extensive lava flows, amounting to some hundreds of metres in total thickness. In some places the margins of individual lava flows can still be seen, and elsewhere there is spectacular evidence of the lava having flowed in tubes, which are

Fig. 2. A cratered region on the lunar far side, photographed from orbit by *Apollo 11*. The largest crater is about 80 km in diameter.

Fig. 3. Progressively more detailed views of the area where *Apollo 15* landed in July 1971. *Top left*: area 200 km wide (*Apollo 15* metric camera). *Top right*: more detailed view of the central portion of Hadley Rille. The largest crater is 2 km in diameter. Above right of the largest crater on the rim of the rille is a crater 300 m in diameter named Elbow. *Below*: photograph looking north-west along Hadley Rille, taken from the rim of Elbow crater. The astronaut is standing beside the Lunar Roving Vehicle, which had brought him and his companion more than 3 km from the landing site.

revealed by later collapse of their roofs. A good example of this is seen in Fig. 3, which shows Hadley Rille, the main focus of the *Apollo 15* surface explorations. Most of the top left view shows the Apennine mountains, which are highland crust forming the rim of the Imbrium basin (a large near-side multi-ringed impact basin). Hadley Rille emerges from an unseen source at the edge of the mountains, and snakes across the much flatter mare basalt terrain, appearing to have acted as a channelway that fed the later stages of mare flooding. It probably became roofed over by chilled and solidified lava, and this insulating cap enabled a vast quantity of lava to continue to flow through the tube without congealing. The rille is now visible because the roof of the tube collapsed after it became drained of lava.

Living on the Moon?

The projected date for a permanent lunar base seems to have been receding into the future ever since the demise of the *Apollo* programme. However, unless we give up space exploration entirely it is bound to happen eventually. One major objection to lunar habitation, the lack of water, has been removed by the discovery of substantial quantities of ice near the poles. No convincing reasons have yet emerged for basing industries on the Moon, but apart from the advantages for lunar exploration itself the far side would offer an excellent place from which to do radio astronomy, shielded from the increasing interference from cellular telephones and the like.

DAVID A. ROTHERY

Further reading

Spudis, P. D. (1996) *The once and future Moon*. Smithsonian University Press, Washington D.C.

Rothery, D. A. (2000) *Teach yourself planets*. Hodder and Stoughton. London

morphotectonics Morphotectonics combines shape and tectonic activity to explain landforms mainly at continental and sub-continental scales, whereas structural geomorphology considers the passive influence of geologic structure on landforms. Morphotectonics studies the direct effect of Earth movements on landform evolution. It includes the study of mountains, erosion surfaces, drainage patterns, and sediments on land and offshore.

Before the 1960s, morphotectonics concentrated on, for example, the influence of faulting on landforms, and uplift as a mechanism for raising large areas, which then eroded to form erosion surfaces. These have now been joined by studies of the relationships between global tectonics and geomorphic evolution. The morphotectonic development of landforms takes place on the same timescale as plate tectonics. C. D. Ollier has proposed the term 'evolutionary geomorphology' for this part of Earth science.

Morphotectonics requires input from at least three branches of Earth science: geophysics, geology, and geomorphology. Until the 1980s, tectonics was largely dependent on the first two, and the directly observable evidence of landforms was neglected. Landforms are, however, much easier and cheaper to study than the deeper parts, as observations are direct and not made by interpretation of distant geophysical signals.

The contrast between active and passive plate margins is one way to subdivide the study of morphotectonics. Along many active margins, plates converge, leading to subduction, uplift, and formation of mountains such as the Andes and the Himalayas. Passive margins are where continents or cratons have rifted to form rift valleys and then trailing edges. They are also often the location of highlands, such as the Appalachians in the North America, the Drakensberg in Africa, and the Eastern Highlands of Australia.

The orogenic (mountain-building) processes that thrust and fold rocks are very different from the vertical uplift that provides the elevation for erosional development of mountains. When large areas are uplifted high enough, plateaux or mountains are formed; for example, the Andes, often cited as an example of mountains formed by subduction and associ-ated folding. However, the Andes appear to have formed by vertical uplift and faulting followed by intense erosion, on both sides, during the latter half of the Tertiary. Vertical uplift and warping of large areas are also features of the Himalayas, demonstrated by convex river terraces along the major rivers. The fact that the same major rivers rise on the Tibetan Plateau and flow through the highest mountain range on Earth is important morphotectonic evidence that the rivers existed before uplift of the Himalayas.

Figure 1 gives an example of major morphotectonic features of a passive margin and their interpretation. Inland areas are maintained as a palaeoplain that was formed before continental break-up. The palaeoplain, which is little eroded and retains ancient drainage lines, is equated with the basal unconformity. Erosion of the landmass is concentrated in the near-coastal zone, which is backed by a great escarpment (see *escarpments*). The great escarpment is embayed by major valleys, and a belt of mountains lies between it and the coastal plain. The mountains do not rise above the level of the projected palaeoplain. Outlying plateaux and coastal facets are remnants of the palaeoplain. A warp axis in the zone of maximum uplift forms a drainage divide, and may be revealed by drainage modification. In this diagram, drainage before formation of the passive margin was in the direction from sea to land. The great escarpment retreats and may reach or pass the warp axis.

Exotic terranes have important morphotectonic implications, because they bring together continental fragments that have very different geomorphic histories. If this is the case, as has been suggested for New Guinea, it may be useless to try to match erosion surfaces or river patterns, even in neighbouring valley sides, if they have evolved in quite different ways.

Morphotectonics can make an important contribution to theories of continental evolution by placing constraints of

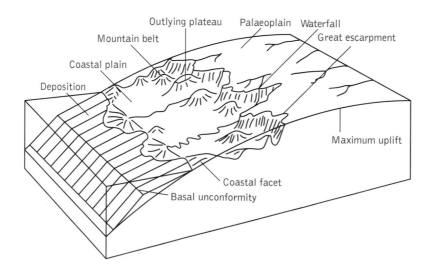

Fig. 1. Block diagram of the major morphotectonic features of a passive margin. See text for explanation. (After Ollier and Pain (1996, *Geomorphology*).)

age and processes on models developed using geology and geophysics.   COLIN F. PAIN

Further reading

Ollier, C. D. (1981) *Tectonics and landforms*. Longman, London.

mountain-building (orogenesis)

Mountains are created by processes that result in either local or regional elevation of the land surface. The processes responsible may be either primarily volcanic or primarily tectonic in origin, although tectonic mountain-building processes typically also involve volcanism or magmatism. The term 'mountain-building' or 'orogenesis' is usually applied by geologists to the creation of substantial belts of mountains ('mountain belts' or 'orogenic belts') arising from plate-tectonic processes. Orogenesis is contrasted with the term 'epeirogenesis', which is applied to the vertical movements involved in the formation of plateaux and basins in stable plate interiors. Although vertical uplift is an essential component in the creation of a mountain belt, it is the origin of the uplift that is important in the study of orogenesis.

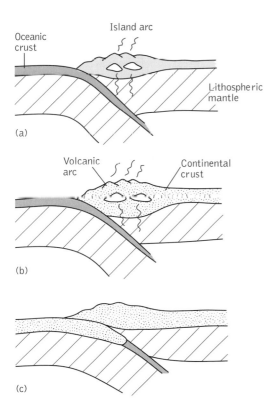

Fig. 1. (a), Island arc formed above an intra-oceanic subduction zone; (b), orogenic belt formed above a subduction zone at a continental margin; (c), orogenic belt formed by continent–continent collision.

In terms of plate-tectonic theory, the major mountain belts (orogenic belts) of the Earth are situated at convergent plate boundaries, and are of two main types. The first type is found on the upper plates of subduction zones, and is formed by the addition of large amounts of igneous material in the form of major intrusions such as granite batholiths within the crust and volcanic extrusions at the surface. This process thickens the crust and elevates the surface. Magmatic belts formed in this way occur either within oceanic crust, where they form arcuate belts of partly submerged mountains known as island arcs (Fig. 1a) or along continental margins (Fig. 1b), such as the Andes along the western margin of South America. There are many examples of island arcs around the northern and western borders of the Pacific Ocean (e.g. the Aleutian and Marianas arcs) and locally in the western Atlantic Ocean (the Caribbean and Scotia arcs).

The second type of mountain belt is produced by continent–continent collision, where a piece of continental crust situated on a subducting (lower) plate meets the continental crust of the opposing (upper) plate (Fig. 1c). As the lower plate subducts beneath the upper, the continental crust of the lower plate underthrusts that of the upper, but, because of its buoyancy, cannot be subducted. Thus, when all the oceanic lithosphere of the lower plate has been subducted, the convergent motion of the two plates gradually decreases and eventually comes to a halt. However, during the underthrusting, a zone of crustal overlap has been produced, which may be about 300 km or more in width, resulting in a belt of continental crust that may be double its original thickness. Because of the buoyancy of the crust, this process also results in elevation of the surface, adding as much as 8 km of vertical relief.

The best current example of a continental collision belt is the Himalayas, which formed as a result of the collision of India with Asia. The two continents first came into contact about 30 Ma ago but subsequent convergence has continued at a slower rate until the present day. The main uplift nevertheless occurred about 20 Ma ago, during the Miocene period.

Because continental collision belts evolve from a subduction to a collision stage, they typically show structural and magmatic features associated with subduction (e.g. magmatic arcs and trench-fill sedimentary sequences) overprinted by those formed during collision (underthrusting and crustal thickening). Because the uplifted belt is rapidly eroded, mid- to lower-crustal sections are soon exposed, and older orogenic belts typically contain zones of uplifted metamorphosed basement belonging to lower crustal levels of the upper plate. Crystalline basement complexes formed in this manner are typical of old orogenic belts such as the Caledonides of the northern British Isles and the Appalachian belt of eastern North America.   R. G. PARK

Further reading

Duff, D. (ed.) (1993) *Holmes' principles of physical geology* (4th edn), Ch. 31. Chapman and Hall, London.

mountain geomorphology Mountains have been defined in various ways. Peattie in his classic book *Mountain geography*, published in 1936, suggested several subjective criteria by which to define a mountain. He believed that mountains should be impressive, they should enter into the imagination of the people that live in their shadow, and that they should have individuality. He cites Mount Etna and Mount Fuji (Fujiyama) as examples. A more objective and common definition is based on altitude. In Britain and America, for example, elevations greater than 300 m above sea level (a.s.l.) are frequently considered as mountains, whereas geographers in continental Europe believe that mountains should be at least 900 m a.s.l. If the former definition is used, we would have to include such regions as the Great Plains of North America, and the plains of central Asia and eastern and southern Africa, which are above 1000 m in elevation. These regions can hardly be described as mountainous. If we use the 900 m definition, however, only the Alps, Pyrenees, Caucasus, Himalayas, and Andes would qualify to be called mountains. In stark contrast, there are regions such as parts of highland Scotland which rise from sea level to only a few hundred metres, but with their steep slopes, glacially carved valleys, snow-capped peaks, severe weather conditions, and sparse vegetation clearly have the appearance of high mountains. Clearly, there are problems with defining what constitutes a mountain. Most geomorphologists would generally agree that a mountain is a topographic high with a large relative relief (e.g. greater than 300 m); that is, the difference between the highest and lowest elevation in a region. A mountain has steep slopes (greater than 10–30°) and has great climatological, vegetational, geological, and hydrological contrast within short distances. Furthermore, most mountains are glaciated or have been glaciated during the recent past, and they contain many landforms that have been formed by ice and snow. Mountains inspire the imagination of the people who live and work in them, and they are usually very beautiful.

Although a mountain can be a single isolated feature (for example, Mount Kilimanjaro in equatorial Africa), most mountains occur as a succession of mountains or narrowly spaced mountain ridges: a mountain range. The Transverse Ranges of southern California are a classic example. A mountain range may form part of a mountain chain which persists for hundred or thousands of miles. The Mediterranean mountain chain of southern Europe is a good example. A group of mountains of irregular shape may also exist; such a group is known as a mountain mass. The Tibetan Plateau and bordering mountains constitute the world's largest mountain mass. When mountain ranges, chains, and masses span continents they are called *mountain systems*: the Alpine–Himalayan system and the Andes–Western Cordillera of North America are the best examples.

Geomorphologically, mountains are very complex and varied. Geomorphologists require a knowledge of geology, climatology, biology, hydrology, and anthropology to understand the dynamics of the Earth-surface processes and the for-

mation of landforms in mountain regions. We shall consider each of these in turn.

The underlying geology and tectonic setting is the ultimate control on the creation of positive relief and the formation of mountains; as mountains form they are subsequently shaped by the erosive power of Earth-surface processes. Most young mountains (less than 200 million years old) are globally distributed along zones that are generally a reflection of the tectonic setting. They occur most commonly along the collision zones between tectonic plates. These collision zones are known as orogenic belts, and the mountains are referred to as orogens. Mountain belts produced by the collision of oceanic and continental plates are known as Cordilleran-type orogenic belts. These include the Western Cordillera and Andean mountain belt, which formed and are still forming as a consequence of the oceanic Pacific plate colliding with, and being subducted under, the continental plates of North and South America, respectively. As oceans close, the oceanic–continental collision may progress to become an interaction between two continental masses. This results in mountain belts of the Alpine type. These include the Alpine–Himalayan belt, where the African and Indian continental plates are colliding with the Eurasian continental plate. Where an oceanic plate is being subducted under another oceanic plate, as for example, in the western Pacific on the margins of east Asia, volcanic island arcs are forming. Mount Fuji is a classic example of such a mountain that has formed within a volcanic island arc. Isolated mountain masses also occur on hot-spots, where anomalously hot magma rising from the Earth's interior towards its surface causes volcanic activity that produces high volcanic mountains. Mount Kilimanjaro is a classic example, rising to 5895 m from a high plain of a little over 1000 m above sea level. Other mountains may be uplifted along zones of compression in major strike-slip fault systems. The Transverse Ranges of southern California along the San Andreas fault system and the Gobi Altai Mountains in Mongolia along the Gobi-Tien Shan fault system are examples of these transpressive-type mountains.

Young mountains are seismically active and may also be volcanically active. These active tectonic processes may also influence the geomorphology of the mountain. Earthquakes, for example, may initiate landslides, and volcanoes extrude lava or may collapse in on themselves, or fail by landsliding, to produce new landforms almost instantaneously.

Other mountain ranges are remnants of former orogenic belts. These are usually more than 200 million years old. They include mountain ranges such as the Appalachians of the United States, the Caledonides of Ireland, Britain, and Scandinavia, and the Urals in Russia.

Figure 1 shows a model of a composite orogenic belt that can be used as a mechanism for organizing observations of real mountain belts and comparing them with others. This model assumes that an orogen is the final result of the collision of two continents. It is most similar to the ancient Appalachian–Caledonide that stretches along eastern North

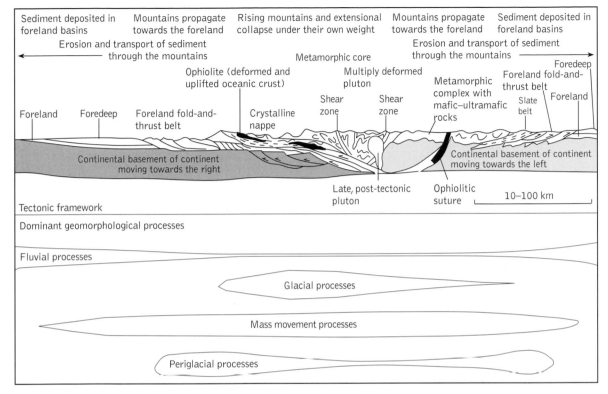

Fig. 1. Schematic cross-section across a model composite orogenic belt, showing the main geological and geomorphological characteristics. The main geomorphological processes are shown below the cross-section to illustrate their geographical importance within an idealized mountain belt. The variation in the thickness of each vertical bar is relative to the magnitude and frequency of the process.

America, Scotland, and Norway, and the younger Alpine–Himalayan belt. The orogen is characterized by:

(1) an outer foredeep or foreland basin;

(2) a foreland fold-and-thrust belt;

(3) a crystalline core zone that includes: sedimentary rocks and their basement; volcanic and igneous rocks and associated sediments; metamorphosed ocean crust (ophiolites); gneissic terranes with abundant ultramafic bodies; and granitic batholiths;

(4) rectilinear (high-angle) fault zones.

The tectonics and geology described above essentially control the distribution of elevations within mountains. This, in turn, controls the development of drainage patterns, and it influences climate, which, in turn, controls vegetation and soil development. The altitudinal distribution of Earth-surface processes is controlled by all these factors. Figure 2 illustrates the linkage between these factors by showing the altitudinal distribution of environments and processes in a typical high mountain.

Mountains control climate on global to microclimatic scales. On a global scale, the highest and largest mountain chains may block or deflect global atmospheric systems. The Alpine–Himalayan chain, for example, deflects the mid-latitude westerly jetstreams, and the Andes block maritime air masses from reaching the easternmost ranges, which creates the hyperarid mountain environments of the Atacama Desert. Mountains may also help to drive monsoonal systems by thermally induced pressure gradients. The South Asian summer monsoon is produced in this way, air masses being forced by the heating effect of the Tibetan Plateau. On smaller scales, climatic contrasts exist because of altitudinal differences. Temperature decreases with increasing altitude, by approximately 6.5 °C km^{-1}. Pressure also decreases with altitude, which further influences meteorological processes. The decrease in temperature with elevation allows permanent snow fields and glaciers to exist in high mountains. The altitude at which these forms is also a function of latitude; for example, the snow line at latitude 60° S is at sea level and rises to a maximum altitude of 6000 m at latitude 23° S. The size of

709

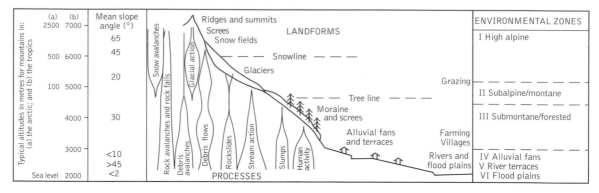

Fig. 2. The altitudinal variation of geomorphological processes and landforms for typical high mountain environments. The variation in the thickness of each vertical bar is relative to the magnitude and frequency of the process.

the mountain system also affects the snow-line altitude. Generally, the larger the mountain mass, the higher the snow line. Precipitation also varies altitudinally, becoming progressively greater with rising altitude until the rising air masses have lost their moisture, which then results in a decrease in precipitation with altitude. On even smaller scales, slope aspect may affect the amount of solar radiation a slope receives; for example, south-facing slopes are warmer than north-facing slopes. On north-facing slopes, this may increase the persistence of snow fields into the spring or summer, and it allows glaciers to become larger than on the equivalent south-facing slopes. Proximity to a lake or glacier may have a cooling effect on, or may increase the windiness of, a particular valley. Furthermore, the world's climate has changed drastically over geological times, altering the distribution of climatic belts both altitudinally and geographically through all the world's mountains. The most dramatic climate changes have occurred during the past 2 million years when the world entered the present ice age. During this time the world's climate has fluctuated rapidly from glacial times, when ice sheets were extensive and sea ice and permafrost were widespread in middle and high latitudes, to interglacial periods, as at the present time, when ice sheets, sea ice, and permafrost are much less extensive.

All these climatic contrasts, on their various scales, help to control the nature and distribution of Earth-surface processes throughout mountainous regions. They also control the distribution of animal and plant life, soils, and even human occupancy. These, in turn, affect the distribution of Earth-surface processes that ultimately help to shape mountain landforms. Given this framework, it is easy to appreciate the complexity of mountain landscapes and the processes that operate in them and shape their landforms. Nevertheless, it is possible to make some generalizations about the nature of geomorphological processes and the landforms that characterize mountain environments. Glacial, periglacial, fluvial, and mass-movement processes dominate in mountain environments. Other processes that are, however, important on local

scales include aeolian (wind), lacustrine (lake), and pedogenic (soil) processes.

In the highest mountains, where average annual temperatures are well below 0 °C, permanent snow fields and glaciers exist. Glaciers may be of many different types, ranging from ice caps (e.g. the Agassiz ice cap on Ellesmere Island in the Canadian Arctic) and long valley glaciers (the Khumbu glacier flowing from the slopes of Everest; Fig. 3), to cirque glaciers that fill small basins in high mountain slopes (e.g. Cirque Mountain in Labrador, Canada). There is much debate about the amount of erosion occurring as a result of glacial activity in mountainous regions. However, given the steep slopes, and the generally high snow falls and rapid ice formation, as well as the high ablation rates, glacial movement in mountainous regions is fast and erosion rates are probably among the highest of all glacial environments. Mountain glaciers also tend to be major transporters of debris and sediments. The sediment falls on the glacier surface by avalanche processes and is transported as the glacier flows, eventually to be deposited when the sediment reaches the edge of the glacier or when the glacier melts. These sediments form impressive landforms called moraines (Fig. 3). In many mountain areas, multiple series of moraines are present, recording past changes in climate, from glacial to interglacial times. Figure 4 shows the fluctuations of mountain glaciers during the past 10 000 years.

Periglacial processes, conditions, climates, and landforms (those associated with the immediate margins of former and existing glaciers and ice sheets) are influenced by the low temperature of the ice. Frost action is an important factor in a periglacial climate beyond the periphery of glaciers. Periglacial processes in high mountain regions are dominated by cold-process weathering (cryogenic weathering), which leads to the disintegration of rock as a result of pressures exerted within cracks in the rock as interstitial water (water inside the rock) freezes with falling atmospheric temperatures. This leads to the production of extensive areas of broken rock and large scree slopes. In many high mountains regions, permafrost

Fig. 3. Part of the Khumbu Himalayas in Nepal, looking north-east up the Lhotse Nup Glacier, with Mount Everest (8848 m above sea level) surrounded by clouds in the far distance. These mountains have been uplifted very rapidly during recent geological time (50 million years to present), forming the world's highest mountain range. The mountains have been deeply denuded by glacial, periglacial, and mass-movement processes. Moraines form high ridges around the margins of the Lhotse Nup Glacier. These testify to glacial retreat induced by climate change. In the foreground, Dr. Dougie T. Benn (a glacial geomorphologist) rests against debris that comprises an old moraine which formed during a former glacial advance.

may be present. This is ground that has a temperature below 0 °C for 2 or more years. The upper layer of permafrost (0.1–1 m thick), the active layer, warms in summer and cools in winter. This allows water to change to freeze and thaw seasonally, creating stresses within the active layer that may lead to sorting of stones of different sizes within the substrate, to the formation of 'patterned ground'. Gelifluction (the slow movement of debris downslope) is associated with frost action and is particularly common in mountains where permafrost exists.

Mass movement and fluvial processes dominate in mountain regions that are not high enough for glacial and periglacial processes to operate, or at lower altitudes of high mountains. Many such areas are forested, and human activity is generally more prevalent. In these regions, anthropogenic processes are most active in shaping the landscapes and influencing natural Earth-surface processes. There is much current concern about the environmental impact that human activity has in many mountain regions, and there is pressure for active conservation in threatened areas.

Mass movement is extremely prevalent in mountain environments because of the very steep and long slopes. It can take many forms, including free fall as rockfalls and avalanches, sliding down discrete surfaces as rockslides or mudslides or flowing as debris flows, and the extremely rapid movements (faster than several hundred km h⁻¹) known as flowslides or sturzstroms. Figure 5 illustrates these various types of slope failures and processes in relation to a mountain valley system. Many geomorphologists believe that low-frequency, mass-movement events of large magnitude are probably among the most important factors in shaping mountain landscapes. Other geomorphologists, however, believe that the higher-frequency, smaller processes, such as weathering and fluvial action, are more important in shaping mountains. It is also likely that mass movement is much more common during periods when glaciers are retreating; large moraines and steep valley sides are then unstable because they are no longer supported by the glacier. The landscape then readjusts itself, by mass movement and erosional processes, to a new, geomorphologically more stable, environment.

Many of the world's most important rivers drain from high mountain regions. For example, the Ganges and Indus rivers from the Himalayas, the Huanghe and Yangste from Tibet, and the Amazon from the Andes. Not only do these rivers support vast populations, but they erode the mountains intensely and carry vast amounts of sediment from the mountains to their forelands. Rivers also help to induce other geomorphological processes. For example, undercutting of river banks may induce slope failure. Major floods are also common along many mountain rivers. These may be induced by heavy rainfalls, particularly in monsoon climates, or when rivers have been blocked by landslides or advancing glaciers,

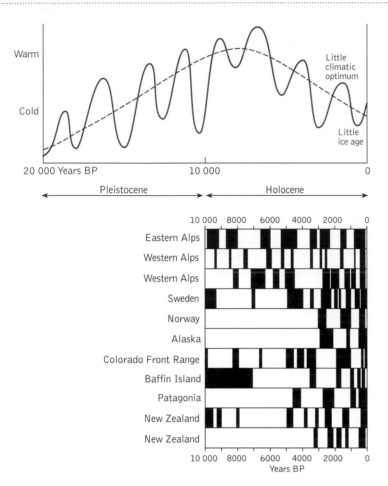

Warm

Cold

Little
climatic
optimum

Little
ice age

20 000 Years BP   10 000   0

Pleistocene   Holocene

10 000 8000   6000   4000   2000   0

Eastern Alps

Western Alps

Western Alps

Sweden

Norway

Alaska

Colorado Front Range

Baffin Island

Patagonia

New Zealand

New Zealand

10 000 8000   6000   4000   2000   0
Years BP

Fig. 4. Schematic variations in relative temperature during the last 20 000 years (solid black curve) and the advance of glaciers (black bars) for selected mountain regions. The dashed curve shows how solar radiation varied because of changes in the Earth's orbit for a latitude of 65° N. (From Pickering, K. T. and Owen, L. A. (1997) *An introduction to global environmental issues*. Routledge, London.)

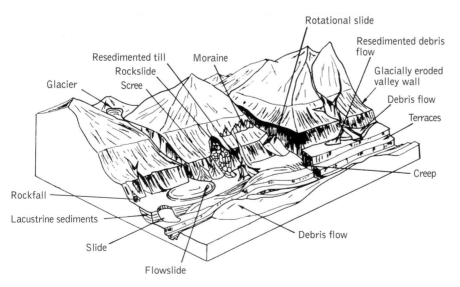

Rotational slide

Resedimented debris flow

Resedimented till

Moraine

Rockslide

Scree

Glacier

Glacially eroded valley wall

Debris flow

Terraces

Creep

Rockfall

Lacustrine sediments

Slide

Flowslide

Debris flow

Fig. 5. Sketch summarizing the variety of mass-movement processes and landforms within a typical mountain environment. (From Owen, L. A. (1991) M. *Zeitschrift für Geomorphologie*, **35**, (4) 401–24).)

producing lakes that have the potential to be breached. When these natural dams burst, the result is often catastrophic. The floods not only kill people and livestock, destroy property and farmland; they also destroy and create new landforms. Many geomorphologists believe that large flood events are are among the most important formative (landform-forming) processes in mountainous environments.

Tectonics, climate, vegetation, hydrology, and geomorphological processes are intricately linked, and understanding the relative importance of each set of factors in the evolution of mountain landscapes and the dynamics of mountain processes is one of the chief goals of most mountain geomorphologists. However, only in relatively recent years have detailed geomorphological studies of individual catchments in mountains begun to elucidate the relative roles of each of these sets of factors. Unfortunately, it is difficult and scientifically unsafe to extrapolate these studies to other areas because of the complexity of mountain environments: no two mountain environments are identical or have the same geomorphological history. Nevertheless, geomorphologists are beginning to understand some of the dynamics of mountain environments.   LEWIS A. OWEN

Further reading

Benn, D. I. and Evans, D. J. A. (1998) *Glaciers and glaciation.* Arnold, London.

French, H. M. (1996) *The periglacial environment.* Longman, Harlow.

Gerrard, A. J. (1990) *Mountain environments: an examination of the physical geography of mountains.* Belhaven Press, London.

Owen, L. A. (1995) Shaping the Himalayas. *Geographical Journal* 117, (2) 23–5.

Price, L. W. (1981) *Mountains and man.* University of California Press, Berkeley.

mud flats Mud flats are gently sloping or flat expanses of fine grained sediment. They form in sheltered environments around the coast and in estuaries, where deposition is favoured by low-energy conditions. Mud flats are made up of fine particles of mineral and organic origin, generally in the silt size range (0.002 to 0.06 mm) or clay size range (less than 0.002 mm). The current velocities required for mud deposition are extremely low, and intertidal mud flats characteristically accumulate sediment when current velocities fall to zero during the slack water period at high tide. Muds are cohesive sediments, which despite their small particle size have a large surface area relative to mass. They are rich in clay minerals, which have an interlayered crystal structure and (normally) a net negative surface charge. The cohesive nature of mud-flat sediments is a result of interparticle attraction between these clay minerals. Particles cohere (or flocculation occurs) when the repulsive surface charges on the clay minerals are suppressed in a weak electrolyte such as sea water. Flocculation is reinforced by organic cohesion, where particles are bound by mucus secretions or biogenic pelletization. The cohesive

nature of mud means that after deposition it is relatively difficult to erode, aiding the build-up of sediment on the mud flats.

Owing to the fine nature of mud-flat sediments, the spaces between the individual sediment particles are extremely small, maximizing water retention as the tide retreats and minimizing the penetration of oxygen into the sediment. The high organic-matter content of mud-flat areas (as much as 5 per cent of the sediment may be organic carbon) provides a rich food source both for burrowing and grazing animals and for bacteria which consume the organic material. A consequence of limited oxygen penetration into the sediment is that the bacteria present quickly use up the available oxygen to break down the organic material. Under these conditions, breakdown of organic matter continues using oxygen from other sources, particularly the sulphate ion (SO_4^{2-}). It is this sulphate reduction which gives mud flats their characteristic 'rotten eggs' smell, produced when SO_4^{2-} is reduced (or used as an electron acceptor) leading to formation of hydrogen sulphide, H_2S, and HS^-. In organic-rich, extremely fine-grained mud-flat areas, sulphate reduction may occur only a few millimetres below the sediment surface. This produces a clear colour change in the sediment from light brown to grey-black (the grey-black coloration being caused by precipitation of iron sulphides).

In comparison with the salt-marsh environments that may form on upper mud-flat areas, the mud flat proper is poorly

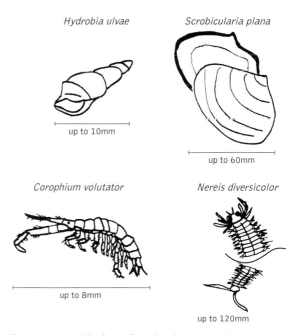

Hydrobia ulvae

up to 10mm

Scrobicularia plana

up to 60mm

Corophium volutator

up to 8mm

Nereis diversicolor

up to 120mm

Fig. 1. Typical mud-flat fauna. (Based on drawings in J. M. Baxter (1990) *The Mitchell Beasley pocket guide to the seashores of Britain and Northern Europe.* Mitchell Beasley, London.)

vegetated. Seaweeds may be found where solitary pebbles or beds of molluscs provide a stable anchoring point, but only a few flowering plants have adapted to truly marine conditions, all of which are in the genus *Zostera* (Eel Grass). This genus was formerly widespread on mud flats along the lower shore and below the low tide mark. In the 1930s, however, the *Zostera* beds were almost completely wiped out by disease, although there has since been a degree of recovery.

The mud-flat fauna is governed to some extent by sediment grain size. Areas exhibiting mixed sand and mud constitute a relatively stable substrate in which typical sand-living animals, such as sand gapers, lugworms, and razor shells, dwell. Mud-flat areas proper are characterized by an unstable substrate in which oxygen-free conditions and the fine nature of the sediment combine to clog the breathing and feeding mechanisms of many species. As a result, relatively few species have adapted to mud-flat conditions, but those animals that are present are often found in vast numbers because of the rich supply of food in the highly organic sediment. Typical mud-flat fauna in northern Europe (Fig. 1) include the deposit-feeding bivalve *Scrobicularia plana*, the snail *Hydrobia (Peringia) ulvae*, the amphipod *Corophium volutator*, and the ragworm *Nereis diversicolor*. Large mussel beds are also common in many areas. Burrowing animals may act to increase or decrease the stability of mud-flat areas; mucus secretions may bind the sediment and make it less susceptible to erosion, whereas sediment stirring (bioturbation) may mix and re-suspend sediment. Mud-flat fauna provide a major food source for waders and wildfowl and, as such, these environments may be host to thousands of overwintering birds. The ecological importance of mud flats has been recognized by the designation of some as international sites, but despite their obvious conservation value, they remain threatened in many areas by pollution, over-exploitation, and land reclamation.   ANDREW B. CUNDY

Further reading

Davis, R. A. Jnr. (ed.) (1985) *Coastal sedimentary evironments* (2nd edn). Springer-Verlag, Berlin.

mud and sand volcanoes

Mud and sand volcanoes are positive topographic features that have many of the aspects of normal phreatomagmatic volcanism resulting from explosive interaction between hot magma and water. They are fed from a chamber of mud or sand through clastic dykes or spines so that they extrude at the surface. Mud volcanoes (also referred to as mudlumps or mud/shale diapirs) range in diameter from less than a metre to several hundred metres. They are associated particularly with places where sedimentation rates are high, where there is a large supply of mud, and where methane generates in the near-surface sediments. In front of the present Mississippi delta, for example, suddenly emergent mudlump islands are a hazard to shipping. Others are well imaged on seismic sections across the Barbados accretionary prism, where tectonic compression is thought to have been an important trigger for the migration of high-pressure pore waters, muds, and gases.

Sand volcanoes range in diameter from a few centimetres to several metres. When seen on the surface of bedding planes, they are more or less circular with a central vent and crater and a surface of radially arranged sand-flow lobes with rounded edges. They are the result of liquefied sand being extruded through a local vent to the surface and reflect the release of pressure from a liquefied layer following a shock, perhaps caused by a seismic event or storm waves at the surface.   HAROLD G. READING

muds and mudstones

Mud and its indurated (i.e. hardened) equivalent mudstone are the most common sedimentary rocks in the Earth's crust. The terms 'mud' and 'mudstone' are loosely used for sediments in which the majority of the particles are less than 63 μm (micrometres) in diameter. The terminology is more precisely defined in Table 1. Silts or siltstones have physical properties more comparable to those of sands than mudstones. Very fine-grained sediments in which more that 66 per cent of the grains are less than 2 μm are called clays or claystones. When subjected to severe compression on burial, some muds become fissile and are then termed shales.

Muds and mudstones are composed of particles produced by the breakdown and alteration of silicate minerals together with silica, calcium carbonate, and organic matter. The particles less than 2 μm in diameter that dominate their physical properties consist mainly of clay minerals. These minerals are hydrated aluminosilicates consisting of sheets

Table 1. Classification of mudstones

Percentage of clay-size constituents ($<$ 63 μm)		0–32	33–65	66–100
Non-indurated	Beds $>$ 10 mm	Bedded silt	Bedded mud	Bedded clay-mud
	Laminae $<$ 10 mm	Laminated silt	Laminated mud	Laminated clay-mud
Indurated	Beds $>$ 10 mm	Bedded siltstone	Mudstone	Claystone
	Laminae $<$ 10 mm	Laminated siltstone	Mud-shale	Clay-shale

(Adapted from Potter *et al.* (1980), Table 1.2)

of alumina and silica with interchangeable cations and considerable amounts of combined or loosely bound water. Clays are produced by weathering in the area that was the source for the muds, and their composition depends upon the character of the source rocks, the type of weathering they undergo, and changes produced during transport, deposition, and burial.

When muddy sediments are composed dominantly of calcium carbonate they are called *carbonate mudstones*. Mixtures of siliciclastic mud and carbonate mud are termed *marl*.

Muds are richer in organic matter than most other sediments. This is partly because fine-grained organic matter has hydrodynamic properties similar to those of clays, but also because it is preferentially preserved in these impermeable sediments. On burial, organic matter is broken down to yield oil and gas; hence some mudstones are major source rocks for hydrocarbons.

Muds are transported in suspension by surface waters and are widely dispersed from their source area. They settle in areas of low turbulence ranging from river flood-plains, lakes, and sheltered coastal areas to the deepest parts of the ocean.

When initially deposited many muds contain up to 80 per cent water, partly within and also partly bound to the clay minerals, as well as infilling primary voids. Muds are thus very reactive to physical disturbance or differential loading, and they slump and flow easily when subjected to stress. They sometimes flow upwards through overlying sediments to form mud-diapirs. Compaction and subsequent loss of water produce a deposit which is only a fraction of its original thickness. When sedimentation is very rapid, as in a delta, the water is sometimes prevented from escaping and it is sealed in the sediment column to produce zones of very high pore-pressure. These zones are a hazard when penetrated by subsequent drilling for, say, oil.

Muds and mudstones are important as raw materials for industry: for example, bricks, cement, ceramics, and drilling muds. They are a major source of the natural mineral cements of sandstones when their contained water is driven into adjacent porous and permeable beds during compaction, and they are also regarded as the major source of mineralizing solutions. Mudstones are the main source of oil and gas, although by the time the oil or gas is extracted it has usually migrated to porous rocks such as sandstones of limestones. Muds and mudstones give rise to many problems in civil engineering because they are weak and shrink or swell on being dried or wetted.

G. EVANS

Further reading

Potter, P. E., Maynard, J. B., and Pryor, W. A. (1980) *Sedimentology of shale*. Springer-Verlag, New York.

Murchison, Sir Roderick (Impey) (1792–1871)

In his day Roderick Impey Murchison was one of the foremost men of science in Europe. He was from a Scottish landowning family, and after serving in the Peninsular War he returned to Britain to the life of a country gentleman. In 1824, however, he settled in London to take up a vigorous interest in science. In 1825 he became a prominent member of the Geological Society and a year later was elected to the Royal Society. His private income enabled him to travel on geological excursions virtually every year, visiting much of Europe in pursuit of stratigraphical successions reaching further and further back in geological time.

He spent the next 40 years studying the Palaeozoic rocks of Britain and Europe and laying the foundation of the present stratigraphical column. From the Welsh Borderland he established the Silurian system (1835) and with Professor Adam Sedgwick of Cambridge he turned to south-west England to examine the thick succession of rocks from which H. T. De la Beche had recently obtained coal measure plant fossils. The result was the Devonian system (1840). Visits to Russia at the invitation of the Czar at this time led to the founding of the Permian system, based on his observations in Perm Province near the Urals. In the 1830s Murchison and Sedgwick found themselves in dispute about the base of the Silurian and over the years this developed into a bitter quarrel.

Relationships with De la Beche, who had become the head of the new Geological Survey were similarly strained, and Murchison now grew increasingly rigid and intolerant of scientific innovation. He was a founder of the Royal Geographical Society and a president of the British Association for the Advancement of Science. Knighted in 1846, he succeeded De la Beche as head of the Geological Survey in 1855. In 1866 he was created Baronet.

Murchison was a splendid example of the gentleman scientist. He did on occasion make serious mistakes, but in military fashion pressed on regardless.

D. L. DINELEY

Murray, Sir John (1841–1914)

Canadian by birth, John Murray attended the University of Edinburgh, where his inclination to oceanography was aroused. His first seagoing work in 1868 took him as far north as 81° N, where he recorded currents, sea and air temperatures, and sea-ice coverage and made a collection of marine organisms. In 1871 and 1872 he assisted in the preparations for the British government's *Challenger* world oceanographic expedition and during its three and a half years' duration he was active in all aspects of its research. On returning to Edinburgh, Murray was given charge of the expedition's collections of biological specimens and, on the death of the expedition director, Sir Wyville Thomson, he was appointed in his place. The '*Challenger* Office' now became an internationally recognized centre for oceanographic research. Murray undertook the compilation and editing of the fifty volumes of the *Challenger Report*, a task which occupied him for some twenty years after the return of the expedition.

Murray's own research interests and publications concerned deep-sea deposits and the formation of atolls. His ideas on the atoll formation differed from those that had recently given out by Charles Darwin. By this time (1880) Murray was also interested in the waters around Scotland, and he set up small laboratories for this work that were to become the basis for the Scottish Marine Biological Association. He subsequently secured funds for bathymetrical study of the freshwater lochs of Scotland. Later he was able to initiate surveys in the tropics and the North Atlantic, the latter being particularly successful. Murray was a tireless and meticulous organizer, highly original and practical in ship and laboratory, and was an international leader in marine biology and oceanography.   D. L. DINELEY

museums and geology The science of geology has been associated with museums and collections since its inception. When geology became a recognizable discipline in the early nineteenth century the principal interest was in stratification. At this time William Smith, a relatively plain and unlettered man, utilizing his collection as a physical model of the strata as he knew them, proved the correlational value of fossils without recourse to complex taxonomies. More than a decade later, Georges Cuvier, who in Paris had established the world's most important museum of natural history, independently reached the same conclusions with the help of Alexandre Brongniart. But here he founded this correlation on a superior understanding of zoology and in the process laid the foundations for palaeontology. Modern mineralogy also developed as a discipline of the cabinet.

These examples of the geological museum in its nascency give hints of aspects that remain important in its modern role. In some respects the museum can be viewed as a three-dimensional publication, but with the difference that the collection is never a fixed entity: it is always being developed and can be manipulated in various ways. Many museum collections, such as those gathered by geological survey organizations, are so vast and data-rich that to explore them is akin to fieldwork but with the advantage of being able to compare specimens that were collected from distant localities and at various times. There are also facilities for the scientific preparation and analysis that may be necessary to provide basic information or identify species.

Collections are also a flexible resource for research and communication. William Smith, for example, believed that the pursuit of geology did not require an understanding of 'hard names'. Collections can talk to the public in ways that are not available to other media: the power of the real object in itself is very persuasive. Yet museums are also places where complex matters such as systematics, biomechanics, and crystallography are investigated. Much of this information is of fundamental importance, underpinning other more complex concepts and interpretation.

The concept of the geological museum is undoubtedly a product of its history: it is defined by the type of geology that was being undertaken in the early nineteenth century and the ways in which governments developed the institutional infrastructure of science. Museums remain wedded to the materials that were then the building-blocks of the science: fossils, minerals, and rocks; materials that remain at the heart of amateur activity and provide a bond between the professional and the amateur. In addition, collections may hold samples from industrial processes or scientific analysis, drill cores, meteorites, fakes and forgeries, archives of notebooks and correspondence, and historical geological equipment. Museums are, however, much more than collections, of equal importance is the expertise and equipment they hold. The museum geologist may be the only publicly accessible scientist or geologist in an area. Many museums have laboratories where fossils are extracted, minerals analysed, rocks sectioned, and conservation undertaken. They are, in addition, centres for expertise in the care of collections and science communication.

It would be wrong, however, to view geological collections as simply the products and materials of science. Inevitably over time they also become the products of history and are capable of saying much about how geological investigations were undertaken in the past. That past shows that these objects had a wide range of cultural roles in, for example, patronage, individual and civic rivalry, reputation-building, and the local economy. Today they reveal much about scientific method and scientific culture. They are as likely to have arisen from an amateur field trip, a house extension, a holiday excursion, or a piece of research. Indeed, much that is held in the world's museums has yet to come under scientific scrutiny, and such collections provide a unique resource for future discovery. The first conodont animal, for example, was discovered in a museum where it had lain unnoticed for decades.

In varying degrees museums have changed with the times. It is, however, difficult to predict the content of any museum. A small multidisciplinary local museum may have a geological collection of international importance and innovative displays. Other larger and more prestigious organizations may have little more than a systematic arrangement. Such things are determined by who that museum sees as its audience, as well as by its governance and funding. Museums are run by universities, local and national governments, and independent trusts, and each views them differently.

Natural history museums have been at the cutting edge of attempts to make museums more accessible: to communicate more effectively. Academic purists no longer hold sway, and galleries now roar to the sound of robotic dinosaurs and sparkle with minerals presented as works of natural art. Museums have recognized that changing their visitors' attitudes to the subject (inspiring them) may be more important than imparting facts. Much of their inspiration comes from viewing real specimens. At the National Museum of Wales in Cardiff and Sharjah Natural History Museum in the United Arab Emirates real objects are placed to the fore with rich multimedia displays in support (Fig. 1). Other museums, such

Fig. 1. A mammoth in the Evolution of Wales gallery at the National Museum of Wales. Staff from this museum created an identical gallery at Sharjah Natural History Museum in the United Arab Emirates. (Photo: T. Sharpe.)

as the Senckenburg Museum in Frankfurt, allow fabulous fossils, such as those of Solnhofen and Messel, to speak for themselves. At the Dinosaur National Monument in Utah, the dinosaur exhibits are *in situ* and constantly under excavation. More and more museums throughout the world are introducing discovery centres where visitors can handle and examine real specimens assisted by trained staff. A display of an unusual type is the Géodrome, 120 km south of Paris. Here, 800 tonnes of rock and mineral specimens from various parts of France are set out in a 'geological garden' in the shape of the country.

There are other museums, such as the Natural History Museum in London and the National Museum of Natural Science in Taichung, Taiwan, where the pursuit of communication objectives has moved the museum away from displaying large quantities of real objects. Here interactivity and carefully constructed didactic exhibits, which owe much to the science centre movement, fill the galleries. If the display of fewer specimens is seen as a loss, there is consolation in the fact that large-scale geological features can now be represented in facsimile. In the Hall of Planet Earth at the American Museum of Natural History, for example, latex casts taken from the Grand Canyon, the San Andreas fault, Vesuvius, the Swiss Alps, and Siccar Point in Scotland are on display. The

scope of the displays is also becoming wider: the atmosphere, the oceans, and environmental issues are now usually included as well as the solid Earth. But, regardless of the general approach, one theme is prominent in museum galleries throughout the world and that is dinosaurs.

In addition to exhibitions, museums provide a wide variety of public programmes. The Royal Tyrrell Museum in Alberta, for example, gives visitors the opportunity of taking part in dinosaur excavations. The Natural History Museum in London arranges laboratory visits, tours, and other activities. Formal and informal education programmes for children and adults exist in many museums and in many instances form a link between the museum and work in the field. Where a geologist is available, a museum will generally provide support for local geological activities, but not all museums with geological collections have geological staff.

Many geological specimens will rapidly deteriorate without proper care, but volunteers should be aware that hasty or ill-informed curation can be even more destructive. There are relatively few national or international groups concerned with such issues; the Geological Curators' Group of the Geological Society in the UK is probably the best known.

SIMON J. KNELL

music and the Earth sciences The connections between music and the Earth sciences are various, although some of them exist only in the mind. No one, for example, has ever heard the music of the spheres, although this concept of heavenly sound is as old as the Greeks and was still extant in the seventeenth century, when Sir Thomas Browne wrote, 'For there is music wherever there is harmony, order, and proportion; and thus we may maintain the music of the spheres.' Pythagoras saw a correlation between the ratios of consonant musical intervals and the ratios governing the heavenly bodies. He thought that as the planets moved they generated sounds that were determined by their velocities and their distances from the Earth. In the *Timaeus*, Plato maintained that the music of the spheres, sung by sirens, was audible but went unnoticed by humans because they had heard it from birth. Certain composers, undeterred (or perhaps even inspired) by its elusive character, have attempted to represent this music in their own compositions, as, for example, in the tableau *L'armonia delle sfere* (1589). Even in the twentieth century the music of the spheres was a significant concept for some composers, Hindemith among them.

Modern planetary science has no place for the music of the spheres. We know that the solid Earth 'rings' like a great bell in response to deep earthquakes, but these low-frequency vibrations hardly qualify as musical sounds. Physics may, however, reintroduce even more fundamental vibrations in the form of string theory, according to which matter is simply the harmonies created by vibrating strings of subatomic length. The theoretical physicist Michio Kaku has, indeed, commented that the 'universe itself, composed of countless vibrating strings, would then be comparable to a symphony'.

Poets have credited the oceans and the atmosphere with their own music. Byron wrote of 'the deep sea and music in its roar'; Masefield of 'the wind's song'. The wind itself is usually kept out of concert halls, but wind machines have been used, for example, by Richard Strauss and by Vaughan Williams in some of their orchestral works, and other composers have used recordings of the wind and the sea as elements in composition. Recordings of waves breaking gently on beaches have also been issued commercially to provide soothing background sound.

The aeolian harp uses the wind to generate musical sounds. It is of great antiquity (there are references to it in Homer), and takes its name from the Greek god Aeolus, the Keeper of the Winds. The instrument typically consists of a set of 12 strings and a soundbox. The strings, of catgut or wire, are all of the same length but of different thicknesses, and are all tuned to the same note. The harp is placed where the wind can blow over it, causing the strings to vibrate. The pitches generated are those of the natural harmonic series (with frequencies one-half, one-third, one-quarter, etc. that of the fundamental note of the string). The harp thus sounds a chord, which contains more notes (at higher frequencies) as the wind speed increases. The harp became particularly fashionable in Europe in the late eighteenth century and the earlier part of the nineteenth century. It is also known in the far East.

'Singing sands', also called 'musical sands' or 'booming dunes', are dune sands that emit sounds variously described as 'booming', 'singing', or 'wailing'. They are known in Arabia, the Atlantic coast of North America, and in the British Isles (Dorset and the Hebrides). The explanation put forward by C. Carus-Wilson was that the sound was produced by 'the rubbing together of millions of clean and incoherent grains of quartz, free from angularities or roughness'.

One curious natural source of musical sound is the 'blowing stone', a naturally perforated block of sarsen that can be blown to produce a wailing sound. Sarsens are boulders of sediment that occur in south-east England. They are typically made up of sand grains or flint (another form of silica) held together by a siliceous cement and are thought to have formed under tropical conditions in Tertiary times and to have been broken down subsequently by frost action and mass movement. The blowing stone at Kingston Lisle in Oxfordshire, about ten miles east of Swindon, was according to legend used by King Alfred to summon the Saxons to fight the Danes. More recently it at least afforded gainful occupation for a village boy, who in 1909 was paid 6*d* (2¹/₂p) a week to demonstrate it to visitors.

Stones have found occasional use in percussion instruments. Some of these *lithophones*, as they are termed, are of great antiquity. 'Ringing stones' have been found in the Far East, in Samoa, and in southern India. In China, stone chimes are known from about 2300 BC. The best-known type is the *pien ch'ing*. This takes the form of 16 L-shaped stone slabs hung in a frame, which are struck with wooden mallets or padded sticks. In some instances the slabs are of jade (a precious stone consisting of jadeite, a green pyroxene, or nephrite, an amphibole). Korean music has also used sets of L-shaped stones, the *p'yŏn'gyong*, tuned chromatically. 'Sonorous stones' are found in many parts of the world: in Ethiopia (where slabs of stone are used in place of church bells), in Venezuela, and in southern Europe. Rock gongs (sets of resonant boulders) are said to be still in use in Nigeria. Two lithophones are preserved in Keswick, in the English Lake District. One of these, the 'rock harmonicon', has five chromatic octaves of stone slabs, which came from Skiddaw, a few kilometres away (Fig. 1). It dates from 1840 and was played on two occasions before Queen Victoria. Another instrument of

Fig. 1. The nineteenth-century 'rock harmonicon', now preserved in a Keswick museum.

the same period is the *lithokymbalom*, built by Franz Weber, which was displayed in Vienna in the nineteenth century. It had slabs of alabaster (a fine-grained variety of gypsum). Western composers have generally found little use for lithophones, but Carl Orff specified a *Steinspiel* in several of his works. The *Steinspiel* consists of a set of stone bars of various sizes, which are struck with beaters. There is an obvious resemblance here to the rock harmonicon.

Geological phenomena have provided the impetus for many works in the musical repertoire. Haydn depicts an earthquake at the end of his *Seven Last Words* (written for performance on Good Friday). The same composer's opera *Il Mondo della Luna* ('The World of the Moon') takes us into the realm of planetary geology, although Goldini, his librettist, had to rely entirely on his imagination for the text. Other eighteenth-century composers also set the same libretto. For many geologists Mendelssohn's overture *The Hebrides (Fingal's Cave)* evokes images of the Isle of Staffa and its columnar basalt. Among twentieth-century works, Milhaud's *La Création du Monde* (music for a ballet), and Richard Strauss's *Alpensinfonie* spring to mind. American geology has been depicted in Ferde Grofé's *Grand Canyon* suite (1931) and by Alan Hovhaness in an orchestral work, *Mysterious Mountain*, and his *Mount St Helens* symphony. In a different vein, Duke Ellington's suite *The River* conjures up rapids, whirlpools, and meanders.

The oceans and the atmosphere feature prominently in western music. Handel depicted hailstones in the plague choruses in his oratorio *Israel in Egypt*. John Mundy's fantasia *Faire Wether* (surprisingly) portrays thunder and lightning. No doubt because of its greater sonic interest, bad weather continued to be portrayed in the music of later centuries: the storms in Beethoven's Pastoral Symphony and in the *Four Sea Interludes* from Britten's opera *Peter Grimes* are familiar examples; Liszt's symphonic poem *Les Préludes* includes a tempest as its third movement. Arnold Bax's tone poem *Tintagel* portrays breakers beating against rocky cliffs. Debussy's Nocturnes and his three symphonic sketches *La Mer* present the sea and the atmosphere in various moods, mostly calmer. Many other examples could be cited, although it has to be added that some titles that sound promising in this respect prove on closer examination to have no connection with the Earth sciences.

Geological materials can be of musical significance in concert halls and other auditoria. There is a noticeable acoustic difference between a church with a wooden roof and a similar building with a stone vault, as many choristers will testify. Rocks are generally very efficient reflectors of sound waves and thus contribute significantly to reverberation; slabs of polished marble in particular have mirror-like properties in this respect.

Geological factors do not merely affect buildings; ethnomusicologists consider that they have been of far-reaching consequence in the musical development of entire continents. The four largest rivers in Africa—the Nile, the Congo, the Niger, and the Zambezi—are all affected by rapids, cataracts, and waterfalls because of the geology of the continent, and navigation is consequently impeded. As a result, the tribal cultures remained independent, which profoundly affected the early development of African music. South-east Asia has similarly been affected by the geology of the region, where the civilizations of the mainland countries have been isolated from each other by mountains and jungle. Musical development consequently took place independently in each region. Palaeoclimate had an important effect in North America, where the ancestors of the modern American Indians migrated, bringing their music with them, from what is now Siberia across the Bering Strait during the last glacial period, about 20 000 to 35 000 years ago.

D. L. DINELEY and B. WILCOCK

Further reading

Arnold, D. (ed.) (1983) *The new Oxford companion to music* (2nd edn), 2 vols. Oxford University Press.

Sadie, S. (ed.) (1980) *The new Grove dictionary of music and musicians*, 20 vols. Macmillan Publications Ltd, London.

mylonite A mylonite is a strongly foliated rock that has undergone intense ductile deformation, with accompanying reduction in grain size. Mylonites occur in zones of high strain, or shear zones, which may vary in thickness from a few millimetres to several kilometres. They were originally interpreted as forming by brittle break-up of original grains (*cataclasis*), but it is now recognized that mylonite textures result mainly from *syntectonic recrystallization*, in which large original grains become strained, leading to the nucleation of small unstrained grains which then grow at the expense of the parent grain. Quartz and carbonate are particularly susceptible to syntectonic recrystallization in this way; less readily recrystallized minerals such as feldspar or garnet often crack and break up as the ductile matrix deforms around them, but remain as relatively coarse *porphyroclasts*, albeit smaller than their original grain size. Many minerals will undergo some syntectonic recrystallization if subjected to sufficient strain at appropriate temperatures. For example, the quartz in granite mylonites may be fully recrystallized into long-drawn-out fine-grained 'ribbons' that wrap round porphyroclasts of feldspar whose outermost part has itself recrystallized to a fine-grained *mortar texture* of polygonal grains.

A number of distinct terms are used for mylonites that have experienced various amounts of strain and recrystallization, as indicated by the proportion of porphyroclasts that remain. *Protomylonites* retain more than 50 per cent of original grain fragments as porphyroclasts, whereas *mylonites* in the strict sense have 10–50 per cent of porphyroclasts only. In cases of extreme deformation, *ultramylonite* may be produced, in which less than 10 per cent of porphyroclasts remain. Ultramylonites are usually restricted to narrow zones within a mylonite. The term *phyllonite* is sometimes used for a mylonite dominated by sheet silicates.

The fabric that is produced as a mylonite is deformed and recrystallizes usually contains characteristic elements that reflect the movement during deformation. Thus there is commonly a stretching lineation visible on fabric planes that indicates the movement direction, at least during the final stages of fabric development. The progressive development of the mylonite fabric also commonly leads to shearing around porphyroclasts with a distinct asymmetry that reflects relative movement.

Mylonites can form in most rock types over a range of metamorphic conditions. Although some mylonites, especially those in igneous bodies, are associated with prograde metamorphism, there are also many examples of mylonites formed during retrograde metamorphism, typically under greenschist facies conditions, of higher-grade rocks. In such instances, the ductile deformation leading to the development of the mylonite is coupled to infiltration of water from an external source that both enables syntectonic crystallization to occur and leads to hydration of the original assemblage.

BRUCE W. D. YARDLEY

Further reading

Passchier, C. W. and Trouw, R. A. J. (1996) *Micro-tectonics*. Springer-Verlag, Berlin.

Sibson, R. H. (1977) Fault rocks and fault mechanisms. *Journal of the Geological Society of London*, 133, 191–213.

myths and the Earth sciences Myth: 'a purely fictitious narrative usually involving supernatural persons … and embodying some popular idea concerning natural or historical phenomena … a traditional story originating in a pre-literate society dealing with supernatural beings, ancestors or heroes … set in the remote past in the other world or an earlier world … their principal characters are gods or animals'.

Vitaliano (see further reading, below) believes that a clear distinction between myths and legends is impossible and that the two are used more or less interchangeably; she uses the term *geomythology* in a broad sense to refer to any geologically inspired folklore, regardless of its origin. Geomyths, then, are the attempts by ancient peoples or more recent unsophisticated people to explain natural phenomena. The earliest are those concerned with origins: of the Earth, the Solar System, the heavens, the gods, and mankind. They are early science, an explanation of the observable near and distant environment.

From the very many instances now known, only a few examples are given below; a wide range of other examples could be quoted from aboriginal peoples of all continents.

Hesiod's *Theogony* (eighth century BC) is one of the earliest accounts. Three Muses appeared to him and urged him to tell how gods and Earth and seas and stars and the heavens came into being out of chaos. The gods performed valiant deeds, but they fought among themselves with much groaning and shaking, thunder, and lightning. The old gods were conquered and banished into the nether regions. Zeus prevailed, even

over the terrible many-headed monster Typhoeus, and reigned over heaven and Earth for ever.

The Book of Genesis gives another view of creation and later of the great flood of Noah—the Deluge—which, because God was angry with the wickedness of men, killed off all living things apart from those that were with Noah in the ark. This story, based at least in part on the Babylonian Gilgamesh Epic, and perhaps on a much earlier Sumerian one (*c.* 3400 BC) was almost certainly based on a folk memory of localized disastrous floods in the Tigris–Euphrates valley. Genesis, however, goes further and has the flood waters cover the highest mountains; and it is here that it becomes a myth that had a profound effect on 'geological' thinking for hundreds of years—and even today.

Early views of the Earth, for example, those of Thales (sixth century BC), had it as a disc floating on water or variously supported by a giant tortoise, a pig, or eight elephants (among many others); movements of these supporters caused the Earth to shake and produce earthquakes. Other earthquake generators were gods who shook the Earth to indicate their displeasure at and to punish men for their sins; demons who lived inside the Earth and in their fury shook it; a giant burrowing mole; vapours and winds rushing through internal caverns.

Greeks from the Aegean with folk memories of the huge Santorini eruption and knowledge of minor eruptions in the area were impressed by the larger and more continuous activity on Etna and Stromboli and the Aeolian Islands. They believed that Typhoeus was buried under the region between Etna and the Phlegraean Fields, that his turning on an uneasy bed caused tremors, and that the belching of the volcano was due to the gasping of the imprisoned monster. The Greek god Hephaestus and the Roman Vulcan were said to have a subterranean workshop below Etna in which they forged thunderbolts for Jove and weapons for other gods and heroes. Their fires resulted in the volcanic eruptions.

Perhaps the most famous of volcanic myths concerns the activities of the Hawaiian goddess Pele, who fled from Tahiti with a magic digging tool and began digging in the furthest north-west of the Hawaiian islands, Kauai, without result. She visited each island in turn with varying degrees of success, finally reaching Hawaii, where she dug the pit of Halemaumau or Pele's Fire Pit, the most spectacular of the region's volcanoes, which was in continuous eruption from 1823 to 1924. Pele is a most temperamental deity who sends out floods of lava when she is angry and causes tremors when she stamps her foot. Even today she is feared by the locals, who offer food to her to try to stop advancing lava flows.

Topographical features figure largely in geomyths. Gibraltar and Ceuta (the gateway to the Mediterranean) are known as the Pillars of Hercules, erected as a record of his labours and the limits of his wanderings. He is said either to have excavated the straits, thus allowing the Atlantic to flow into the Mediterranean, or to have to built up the features in order to prevent Atlantic monsters getting in. Huge blocks of rock (so-called 'erratics') scattered in large numbers over the north German plain were said to be missiles hurled at each other by warring giants. The Giant's Causeway with its splendid basaltic columns was believed to be the landward end of a causeway built by the giant Fionn MacComhal to get at his hated rival Fingal, who had built something similar on the island of Staffa. The hole made by another irate giant who picked up a large chunk of Northern Ireland is now Lough Neagh, and the lump that he hurled into the Irish Sea is now the Isle of Man. A carline (witch) undertook to carry a large hill from Ayrshire to Ireland, but she dropped it on the way to form what is now Ailsa Craig in the Firth of Clyde. The grooves on the sides of the Devil's Tower in Wyoming were thought by the Indians to have been made by the claws of bears trying to scale the sides to get at escaping braves.

On a smaller scale, rounded pieces of amber were ascribed to the tears of the sisters of Phaeton as they mourned his tragic death. R. BRADSHAW

Further reading

Vitaliano, D. B. (1973) *Legends of the Earth*. Indiana University Press. Bloomington.

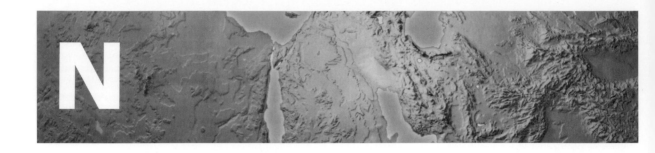

N

Nalivkin, D. V. (1889–1982) One of the great names in the geology of the Soviet Union is that of Dimitri Vasilievich Nalivkin, a man who not only carried out stratigraphic mapping in many parts of Russia and neighbouring Asian republics but was also largely responsible for the Geological Map of the USSR. Born in St Petersburg, the son on a mining engineer, he entered the Mining Academy there in 1907 and was associated with it for the rest of his long life. While still a student he began to teach there and was involved in the Institute's fieldwork in the Caucasus and in Central Asia. One of his earliest studies was of the Devonian brachiopod faunas of the Fergana Valley in Khirgizia. The Devonian System remained a favourite topic of study throughout his life.

Nalivkin graduated from the Mining Institute in 1915 and was immediately appointed a leader of the Pamir Expedition, organized by the Russian Geographical Society. He produced a major work on the geology of the Pamir Range and associated mountain belts in Central Asia in 1925. The many new ideas and concepts it contained won immediate recognition for Nalivkin as a stratigrapher and structural geologist of great acumen. He retained his affection for the central Asian republics, later helping to set up new academic institutes in Turkmenia and Khirgizia.

Early on his career he became a member of the Geological Committee of Russia and was elected an officer of it in 1917. He remained with the Committee for more than sixty years, directing many programmes of research in stratigraphy and palaeontology, regional geology, and sedimentary rocks. Resources such as petroleum, coal, and bauxites, associated with bedded rocks were targeted under his leadership.

Perhaps the greatest of his labours, the creation and editing of the index geological maps of the USSR and adjacent parts of the world, aroused much international acclaim. Then came his Geological Map of the USSR on a scale of 1 : 2 500 000, for which he was awarded a Lenin Prize in 1957. Meanwhile, his research in the Devonian and Lower Carboniferous stratigraphy of the USSR continued. In 1955 he prompted the creation of the Interdepartmental Stratigraphical Committee of the USSR and became the editor-in-chief of a major series on the stratigraphy of that country. No one has had a greater influence upon the progress of regional mapping and stratigraphic practice in the territories that constituted the USSR.

Nalivkin received may national and international awards, and was held in great affection by his students and colleagues. He was still active when he died at the age of 93.

D. L. DINELEY

natural gas The term 'fossil fuels' is commonly used for tar sands, oil, gas condensates, natural gas, coal, and oil shale, and their products. Natural gas is a petroleum and one of the fossil fuels.

Petroleums have a wide range of compositions but can be classified into three major types depending on the relative proportions of gaseous hydrocarbons (C_1–C_5) and components with more than six carbon atoms (C_{6+}). Natural gases are petroleums that are in the gas phase in the subsurface and are composed of hydrocarbon mixtures dominated by methane, with contributions from ethane, propane, butanes, and the pentanes, and traces of inorganic gases, principally carbon dioxide and nitrogen. Gases by definition contain no C_{6+} components. Natural gases can exist in the reservoir as a gas phase or in solution in a liquid petroleum (see *petroleum*). Natural gas does not condense at standard temperature and pressure and is usually dominated by methane.

Natural gas contains other compounds than hydrocarbons, including significant quantities of gases such as carbon dioxide, nitrogen, and hydrogen sulphide and traces of inert gases such as helium. These non-hydrocarbon gases can be the dominant components in some reservoirs. Gases high in carbon dioxide are usually associated with basins, such as the Pattani basin in Thailand, that have high geothermal gradients and deep carbonates.

Natural gas has the same origin as the other forms of petroleum and natural gas, and can be thought of as a petroleum with an extreme composition. Most natural gases are derived from the action of heat on organic matter in sediments, and all types of kerogens (insoluble organic matter in sediments) generate natural gas. Most commercial accumulations of natural gas are associated with source rocks that have been heated to temperatures of over 160 °C. Coals do not generate as much gas per weight of organic matter as organic matter in oil source rocks, but they are prolific source rocks for natural gas since they are rich in organic matter and rarely expel any generated oil, which remains in the source rock and is con-

722

verted to natural gas at higher temperatures. Most natural gases migrate and accumulate in reservoirs in the same types of rocks, and by the same processes, as other petroleums. Coals may act both as a source and reservoir for natural gas. This is because the methane is readily absorbed in the coal matrix and may be desorbed by reducing the pressure in the coal by pumping out the water. The natural gases in the San Juan basin in the USA are principally of this type.

Natural gas is also derived from the thermal breakdown of crude oil at temperatures above 160 °C, either in reservoirs or from oil remaining in a source rock. Natural gas may also be generated by biological activity in shallow sediments when bacteria reduce carbon dioxide to methane to produce so-called 'biogenic methane'. Natural gases derived from biogenic methane can also be produced during the biodegradation of oil. This is termed 'secondary biogenic gas'. Natural gases may also be bacterially altered; this has happened in several of the large North Sea accumulations that have large gas caps, such as the Frigg and Troll fields. Usually when this happens bacterial activity destroys the ethane and propane in the gases, leaving a gas in which methane predominates.

The reserves of natural gas greatly exceed those of crude oil.

STEVEN LARTER

Neo-Darwinism *see* EVOLUTION (DARWINISM AND NEO-DARWINISM)

Neogene
The Neogene is the later of the two sub-periods that comprise the Cenozoic era. It includes the Miocene and Pliocene epochs and is succeeded by the Quaternary. Thus it extends from about 23 Ma to about 1.6 Ma. Some authors have included the Pleistocene and Holocene epochs (divisions within the Quaternary) within the Neogene, but this convention has not been followed by the International Commission on Stratigraphy. Charles Lyell's original subdivision of Tertiary time (1833) was made on the basis of the percentage of modern species in the fossil molluscan faunas. He found a big difference between the figures for the Eocene (3.5 per cent) and the Miocene (17 per cent). (The Oligocene had yet to be identified.) Lyell's original Eocene division comprised the entire sequence that today we would call the Palaeogene, making it a very large unit. However, recent re-examination of the percentages to be found at different levels in these rocks reveals that less than 5 per cent of modern gastropod species were present at the opening of the Miocene.

Lyell's concept of the Miocene was of a moment in geological time when some 17 per cent of the molluscan species were modern. Even in the type-area of the Paris basin it would be difficult to apply this criterion to the definition of a specific body of rocks. The attempt to repeat the kind of measurement that Lyell carried out gives no encouragement to use palaeontological methods of this kind as means of defining a global criterion for the base of the Neogene. If (despite the fact that it is only a sub-system) the Neogene is to be defined in a similar way to the main stratigraphical systems, the first appearance

of a species in an unbroken succession and within an established lineage is needed. With the growing provincialism of marine faunas during the Cenozoic this becomes the more difficult, and the search for a suitable species is not completed yet. Palaeomagnetic, stable isotope, and radiometric criteria may prove to be of much greater immediate assistance in correlation than have biostratigraphical data.

D. L. DINELEY

neotectonics and seismotectonics
Neotectonics, the study of recent plate tectonic movements and their local effects, and *seismotectonics*, the study of recent earthquakes and their relationship with plate tectonics, are very closely allied fields. At a local scale, the relationships between earthquake recurrence time and magnitude, fault movement and growth, sedimentation and erosion patterns, the evolution of drainage systems, and landscape development are used to build up a model of the recent geological history. This history will reflect the larger-scale plate-tectonic movements affecting the region, and also climate and sea-level changes.

Seismotectonics is particularly concerned with the distribution in space and time of recent earthquakes, and with their magnitudes. For earthquakes that have occurred in the decades since the deployment of the World Wide Standardized Seismograph Network (WWSSN) in 1961, this information is readily obtained, and less reliable information is available for the earlier part of the twentieth century. The parameters of older earthquakes can to an extent be deduced from historical or archaeological records (the field of *palaeoseismology*). In some instances, the dates of past movements on faults can be accurately determined by carbon-dating organic-rich sediments in old river channels that cross the fault.

PETER CLARKE

Further reading
Vita-Finzi, C. (1986) *Recent Earth movements: an introduction to neotectonics*. Academic Press, London.

neptunian dykes
A neptunian dyke is a sheet of sediment or sedimentary rock, usually vertical or subvertical, which cuts across any other rock type and, at first glance, resembles an igneous dyke. It is, as Darwin remarked about an occurrence in Patagonia, a 'pseudo-dyke'.

Neptunian dykes are formed by the infilling of open cracks or fissures by sedimentary material—minerals or rock fragments—falling from above the fissure. Examples up to 10 m wide and several hundred metres long have been recorded. The dykes usually run parallel to the tectonic grain of the country. The fissures may be joints which have opened once so that the sides match up and the infill is either horizontally bedded or chaotic, or they may be, either wholly or in part, solution features developed along joints, principally in karstic terrains. Multiple opening of the fracture may have occurred, in which case the infill may be laminated parallel to the sides, the youngest material being in the middle. The sediment may be emplaced either in a terrestrial environment, particularly a karstic one, or on a sea floor after the drowning of a karst landscape.

A significant feature is that a neptunian dyke may preserve within it sediments and fossils not known elsewhere in the region.   R. BRADSHAW

See CLASTIC DYKES

New Red Sandstone

The New Red Sandstone—or more simply the 'New Red'—was recognized by the early geologists in Britain as a prominent post-Carboniferous formation. Thus it was separated from the lithologically similar Old Red Sandstone, which we now recognize as the continental facies of the Devonian. In north-east England, the New Red Sandstone can also be seen to rest on the (Permian) Magnesian Limestone; so it is equivalent to the red beds in Germany (the *Germanic* facies) which were incorporated into the Triassic system by Friedrich August von Alberti in 1834.

The Bunter, Muschelkalk, and Keuper Series of Germany are represented in Britain by non-marine red beds occurring in south-west England and South Wales, the English Midlands, and the Cheshire basin; on the eastern side of the Pennines from Nottingham to Durham; and in coastal parts of Scotland. At their thickest, in the Cheshire basin, they reach about 1500 m. In their upper parts there may be considerable layers of salt.

New Red Sandstone deposition was a continuation of the continental regime that followed the late Carboniferous earth movements. These movements deformed the area of the British Isles with thrusting and folding in south-west England and with uplift and faulting elsewhere. The new uplands were immediately subject to subaerial weathering and denudation, which persisted into the Mesozoic era. Thus, in Permian and Triassic times much of the British Isles was given over to local fault-bounded basins and intervening uplands. The climate was hot and arid and the drainage episodic. Early New Red Sandstone deposits include breccias and locally derived clays, and fluvial roundstone conglomerates. Thick variable sandstone and siltstone formations follow, overlapping progressively on to the margins of the original basins. They are in turn followed by monotonous red siltstones with evaporites, which have been correlated with the Keuper Series in Germany. These siltstones are topped by passage beds in the uppermost stage of the Triassic and marine strata of the basal Jurassic.   D. L. DINELEY

Further reading

Audley-Charles, M. G. (1992) Triassic. In Duff, P. McL. D. and Smith, A. J. (eds) *Geology of England and Wales*. Geological Society, London.

Sellwood, B. W. (1978) Triassic. In McKerrow, W. S. (ed.) *The ecology of fossils*. Duckworth, London.

Smith, D. B. (1992) Permian. In Duff, P. McL. D. and Smith, A. J. (eds) *Geology of England and Wales*. Geological Society, London.

nickel deposits *see* IRON–NICKEL DEPOSITS

Niggli, Paul (1888–1953)

Niggli made fundamental contributions in a number of fields in geology: crystallography, mineralogy, igneous and metamorphic petrology, and, to a lesser degree, ore deposits. He was also a great administrator and a good teacher, interested in providing a balanced programme that also included fieldwork. He wrote a number of influential textbooks and was for twenty years editor of the *Zeitschrift für Kristallographie*. After graduating in Zürich he held posts in Zürich, Leipzig, and Tübingen and in 1920 he succeeded Grubenmann, with whom he collaborated for many years, to the chair of geology at Zürich. A year (1913) at the Geophysical Laboratory in Washington was the beginning of a long association with N. L. Bowen. Although his first work was on schists, much of his early work was on crystallography, on atomic structures, crystal chemistry and physics, and the morphology of crystals.

In igneous geology Niggli has been called an 'extreme orthodox magmatist'. He laid great emphasis on the physical chemistry of magmas and the use of phase diagrams to explain relationships. He devised 'Niggli numbers' and 'norms' as a method of calculating chemical relations and establishing classifications. He was deeply and personally involved in the controversy over granitization and transformism, insisting that granite magmas were the result of differentiation of basic parents, and was against widespread diffusion. In his opinion, differentiation of magmas led to volatile end-products that gave rise to hydrothermal alteration of existing igneous rocks and ultimately to ore deposits.

Niggli followed Grubenmann in his concept of depth zones in metamorphism that reflected increase in temperature and pressure with depth and were characterized by specific mineral associations and rock fabrics. He regarded metamorphism and deformation as continuously related processes and found it impossible to accept migmatites as metamorphic rocks.

His *Gesteine und Minerallagerstätten*, published in 1948 and later translated into English as *Rocks and mineral deposits* (1954), was a digest of much of his earlier work, taking petrological science from the level of crystal structure to mineralogy and on to regional and global scales.

R. BRADSHAW

nuclear energy

In 1938, while working with his colleague Fritz Strassman at the laboratory in Berlin, Otto Hahn noted a special property of the element uranium-235 (^{235}U). When bombarded by slow neutrons the nucleus of this atom split into two parts of similar mass with an emission of prodigious amounts of energy, a process known as 'nuclear fission' (Fig. 1). At the time the discovery was regarded as little more than a scientific curiosity, and Hahn and Strassman reported their findings in the open literature.

The fission process

It did not take long for scientists in various parts of the world to appreciate the significance of the discovery. Apart from the release of large amounts of energy and the production of fission fragments, two or three neutrons are also emitted during the fission of ^{235}U. Normally present only within their parent nuclei, these escaping neutrons, with a diameter of

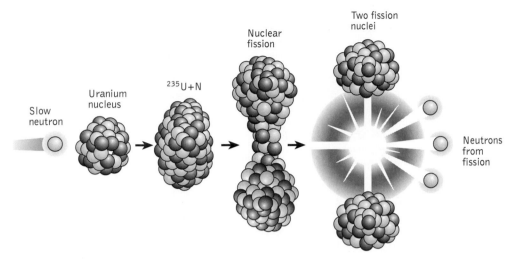

Fig. 1. The process of uranium fission.

only 10^{-14} metres and carrying no electrical charge, can wander freely through materials. On their scale, a solid is mainly open space but should they chance upon an atomic nucleus they may often merely bounce off. Occasionally, though, the nucleus may be entered, making it unstable and provoking radioactive change. To illustrate this process, Otto Frisch recalled that the Danish scientist Niels Bohr had suggested that the atomic nucleus may be regarded as a globular water droplet wobbling to and fro. In the case of the large uranium nucleus it could even wobble into a dumb-bell shape, from which it might occasionally split into two distinct

smaller nuclei of slightly unequal size. After the disintegration of ^{235}U, the neutrons released may strike other ^{235}U nuclei, causing them also to undergo fission, leading to more energy release, and the emission of yet more neutrons, which can then cause further fission, and so on. The possibility therefore existed of establishing a self-sustaining chain reaction, rapidly multiplying the individual atomic-scale energy releases into a huge energy release on a macroscopic scale.

Other fates may, however, befall the neutrons released during fission of ^{235}U. In its natural state uranium consists principally of nuclei of two isotopes: ^{235}U (0.7 per cent) and

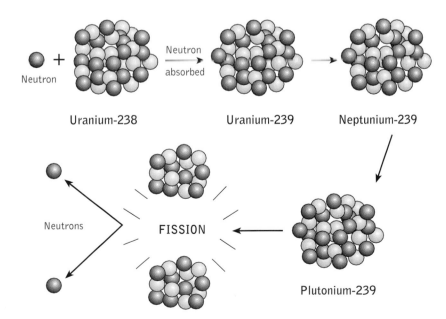

Fig. 2. The uranium-238–plutonium-239 cycle.

^{238}U (99.3 per cent). It is thus highly probable that a neutron may be captured by the more abundant ^{238}U nucleus, which does not disintegrate but undergoes radioactive change. This does not represent a total loss, because the capture process causes it to transmute to a useful new element. Adding a neutron, which has one unit of mass, to ^{238}U gives ^{239}U, which decays rapidly to a new element by emitting a beta particle. As beta decay does not change the atomic mass, the isotope of the new element number 93, neptunium, also has mass 239. Neptunium-239 is radioactive and undergoes beta decay with a half-life of 2 days to yet another new element, number 94: plutonium (Pu). Again, the isotope with the same mass, 239, is produced.

The significance of this chain of events, illustrated in Fig. 2, is that the isotope ^{239}Pu is capable of undergoing fission, and so can be used in its own right as a nuclear fuel. We therefore have a mechanism whereby the non-fissile material ^{238}U can be converted into a fissile material, ^{239}Pu. We thus have the possibility of using the whole of natural uranium for energy production.

The nuclear reactor

Having seen how nuclear energy can, in principle, be produced through the process of uranium fission, it is important to see how this can be achieved in practice in a nuclear reactor. It is possible to establish a self-sustaining chain reaction using natural uranium as a fuel, and most of the early reactors did so. To give the neutrons a better chance of meeting fissile nuclei, rather than getting lost or captured without fission, later designs used 'enriched' uranium in which the concentration of ^{235}U is above the natural level. Fuel elements are therefore constructed as rods of either natural or enriched uranium in either metallic or oxide form, depending on the type of reactor. The fuel rods are contained or 'canned' in some suitable material to prevent the fission products, some of which are gaseous, from seeping into the coolant stream. The choice of canning material is governed by its properties with respect to neutrons and by its strength and corrosion resistance. It is essential that it does not absorb too many neutrons, since this would impede the perfor-

mance of the reactor. In some designs an alloy of magnesium is used; in others, an alloy of zirconium; and in yet others, stainless steel.

Neutrons are emitted from the fission process at high speeds and are better able to induce further fissions at much slower speeds. An important component of a reactor is therefore the 'moderator', within which the neutrons bounce to and fro off many nuclei without being absorbed, gradually losing energy of motion. It is essential that the moderator's atoms are very light and take up as much energy as possible to slow the fission neutrons to speeds approaching those of ordinary gas molecules, about 1.5 km per second. Some reactors use carbon as their moderating material in the form of graphite. Others use normal or 'heavy' water. The core of a reactor thus consists of a block or tank of moderator into which fuel rods are inserted.

To control the reactor (the equivalent of turning up and down a flame on a gas cooker) control rods made of boron or cadmium are inserted near the fuel rod to absorb neutrons. When the fuel rods are fully in they mop up so many neutrons that the chain reaction cannot be maintained and the reactor stops. The energy from the fission process appears as heat in the fuel rods, and this is extracted by passing a coolant over them which absorbs the heat and transfers it to a boiler in which steam is generated. From then on the hardware is identical to that of a conventional coal- or oil-fired power station. The steam is passed through turbines in which the heat energy of the steam is converted into mechanical energy of rotation. The turbines drive electrical generators which feed the public electricity grid.

The coolant can be a gas or a liquid. Some designs of reactor use carbon dioxide gas, others use water. In order to get a sufficiently high rate of heat transfer from the fuel rods, the coolant has to be pushed over them at a pressure which is many times higher than atmospheric pressure. In the pressurized water reactor (PWR), shown schematically in Fig. 3, the water is held under very high pressure in completely liquid form.

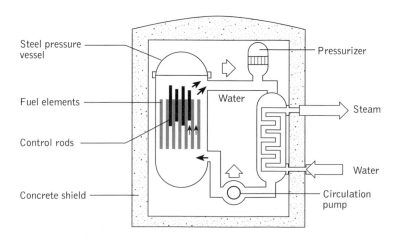

Fig. 3. A schematic representation of the pressurized water reactor.

Table 1. Types of nuclear reactor

Reactor type	Fuel	Fuel cladding	Moderator	Coolant	Pressure vessel
Thermal reactors (slow neutrons)					
Advanced gas-cooled reactor (AGR)	Enriched uranium oxide	Stainless steel	Graphite	Carbon dioxide	Prestressed concrete
Candu reactor (CANDU)	Natural uranium oxide	Zirconium	Heavy water	Heavy water	Steel
Boiling water reactor (BWR)	Enriched uranium oxide	Zirconium	Light water	Boiling water	Steel
Light water reactor (LWR)	"	"	"	Light water	"
Pressurized water reactor (PWR)	"	"	"	Light water	"
Magnox reactor (MAGNOX)	Natural uranium metal	Magnesium	Graphite	Carbon dioxide	Steel or prestressed concrete
Steam-generating heavy water reactor (SGHWR)	Enriched uranium oxide	Zirconium	Heavy water	Light water (boiling)	Zirconium pressure tubes
High-temperature reactor (HTR)	Enriched uranium carbide	Silicon carbide	Graphite	Helium	Prestressed concrete
Fast reactors (fast neutrons)					
Liquid metal fast breeder reactor (LMFBR)	Mixed uranium oxide–plutonium oxide pellet	Stainless steel	–	Sodium (liquid)	–

The main types of reactor used for generating electricity are listed in Table 1. In the thermal type of power reactor, the neutrons are slowed to become thermal neutrons, and the number of fissile nuclei is less than the number consumed in fission; the 'breeding' process is therefore not self-supporting. In the fast breeder reactor, the fuel is enriched up to 25 per cent and the reactor can be run on fast neutrons alone.

The radiation from the core of a working reactor is intense and consists of escaping fission neutrons and gamma rays emitted from products. Fission products also emit alpha and beta particles, of course, but as these have a relatively short range they are reabsorbed in the core of the reactor. To protect operators from this intense radiation, the core of the reactor is surrounded by a reinforced concrete shield, typically about 3.5 m thick.

The natural reactor at Oklo

The ratio of the isotopes ^{235}U to ^{238}U in naturally occurring uranium deposits is virtually constant. It has been determined very accurately as 0.7202 ± 0.0006 per cent. It is also the same for the traces of uranium found in samples of lunar material collected by the *Apollo* missions and in meteorites.

In May 1972 however, the French scientist H. Bouzignes recorded a ^{235}U/^{238}U ratio of 0.7171 ± 0.0007 per cent from a uranium sample extracted from a mine at Oklo in the West African state of Gabon. Subsequent core samples revealed lower values. One was even as low as 0.296 per cent. At first it was difficult to understand why the ^{235}U content was so low in minerals from this source, but a possible clue was to be found in the associated deviation from natural values of the isotopic ratios of other elements, such as neodymium (Nd). Natural neo-

dymium contains 12 per cent neodymium-143, while in Oklo samples containing 24 per cent ^{143}Nd were identified. The only plausible explanation for this is that long ago the missing ^{235}U fissioned, producing ^{143}Nd among other fission products. The volatile fragments such as iodine and the noble gases would have escaped and the easily soluble ones would have been washed away. Of special interest, though, particularly in the context of waste disposal, is the fact that that analyses showed that elements such as plutonium and the rare earths had not moved from their place of origin.

It is amazing that the random forces of nature could have conspired to create the conditions necessary for a nuclear reactor, given the degree of precision engineering needed in modern-day facilities. A detailed analysis of the isotopes of other elements in the Olko deposits, including rubidium-87 and strontium-87, revealed them to be 1740 million years old when, by virtue of their respective half-lives, the ratio of the uranium isotopes would have differed markedly from the present natural ratio. At that time the ^{235}U content of natural uranium would have been 3.0 per cent, a sufficiently high level in the absence of neutron-absorbing materials and the presence of water to support a spontaneous continuing reaction.

Nuclear fusion

The naturally occurring atomic elements may be arranged in a series from the lightest (hydrogen, atomic number 1) to the heaviest (uranium, atomic number 92). Their nuclei are more tightly bound at the centre of this series, where the precious metals and the elements used in everyday manufacture occur. Thus at the heavy end of the series there is a tendency to split into smaller fragments, releasing energy; and by the same

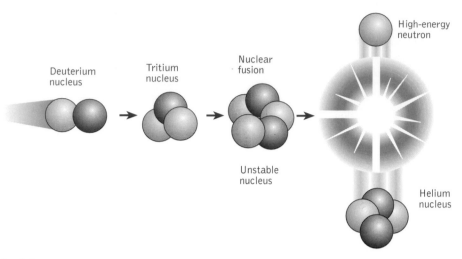

Fig. 4. The nuclear fusion process.

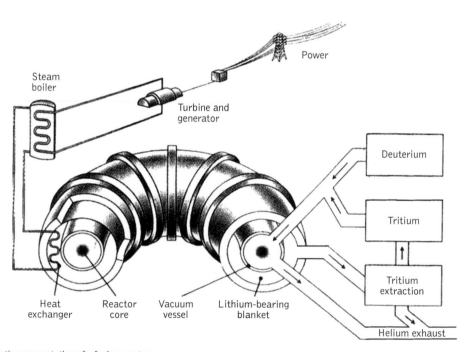

Fig. 5. Schematic representation of a fusion reactor.

token it would seem likely that if nuclei of the lightest elements could be made to join or fuse together to form heavier ones, then large amounts of energy would also be produced. The latter process, illustrated in Fig. 4, is called nuclear fusion, and it is from this process that the Sun and stars derive their energy.

For fusion reactions to occur, the fuels, which are gases, must be heated to very high temperature. The atoms then become ionized, freeing the electrons which orbit the nucleus. This mixture of randomly moving electrons and nuclei is called plasma. In this state, at temperatures around 100 million °C, abundant fusion reactions occur between deuterium and tritium nuclei. These two gases, which are different forms of hydrogen, were used in the first fusion reactors.

In Europe a joint undertaking was set up to construct and operate a fusion reactor known as the Joint European Torus (JET). The machine began operation in June 1983. A schematic representation of a fusion reactor is given in Fig. 5. As plasma is a mixture of positive and negative particles, magnetic fields may be used to contain it and prevent the particles hitting the wall of the containing vacuum vessel.

During a JET experiment, a small quantity of gas is introduced into the doughnut-shaped vacuum vessel—the Torus. The gas is heated to form a plasma by the passage of a large electric current of up to 5 million amperes (5 MA). This plasma current also produces a magnetic field which combines with a second field produced by 32 D-shaped coils equally spaced around the vacuum vessel. This complex system of magnetic fields, called a TOKAMAK (a Russian acronym for toroidal magnetic chamber), provides a cage that prevents the hot plasma from hitting the walls of the vacuum vessel. A set of six hoop coils around the outside of the machine produces the magnetic field that shapes and positions the plasma centrally in the torus.

Fusion reactor

Should the JET experiments ultimately lead to the construction of a commercial reactor, the neutrons produced during the fusion reactions would be captured by a blanket surrounding the plasma region. This blanket would almost certainly contain lithium to produce the tritium needed in the reaction. It is also envisaged that the neutrons so captured would heat up the blanket to temperatures in the region of 500 °C so that steam could be raised to drive turbines and generate electricity in the normal manner.

GEOFFREY C. ALLEN

nuclear winter scenarios In 1983 the journal *Science* published a paper (known from the initials of its authors surnames as TTAPS) which suggested that a nuclear war could lead to a widespread cooling of the climate because the amount of radiation penetrating to the Earth's surface would be reduced as a result of smoke and dust (generated by fires)

blanketing the atmosphere. Darkness could result over large areas with land temperatures being reduced to between −15 and −25 °C, which would have catastrophic biological consequences. In what is called the 'baseline scenario' of a nuclear war (10 000 explosions with a total yield of 5000 megatons) it is supposed that 1000 million tons of dust would be created, 80 per cent of which would find its way into the stratosphere. The atmosphere would be loaded with vastly more fine dust than even the largest volcanic eruptions in historical time have produced, and the dust clouds could therefore be expected to last longer and have a more protracted and drastic effect on the climate.

Some scientists believe that large horizontal and vertical temperature gradients caused by absorption of sunlight in smoke and dust clouds might induce strong monsoonal-type circulations leading to heavy snowfalls in some continental areas. Other authorities suggest that owing to reduced heating over the continental land masses in the northern hemisphere, the monsoon rains would be greatly reduced. It has also been suggested that a consequence of the extensive freeze would be the generation of large quantities of water vapour which would condense and fall as heavy rain. This would tend to wash out the suspended soot and dust from the atmosphere. Wash-out was not allowed for in the TTAPS model. The ozone layer might be damaged by the release of nitrogen oxides from nuclear explosions.

The original predictions came from a simple computer model and, as more realistic three-dimensional models were run, the more apocalyptic claims fell away. While the climate models that have been developed assume that nuclear explosions would result in a uniform smoke distribution covering much of the northern hemisphere, critics have pointed out that patchiness in the smoke cover could be very important. It was suggested by the late 1980s that if a war happened in July (the worst month) temperatures would fall by only a few degrees for up to a month instead of months of freezing darkness.

'Nuclear winter' is a hypothesis, and the detailed reasoning behind it is still under discussion. While the magnitude of the effects is not agreed, it can be argued that the policy implications that have resulted from the wide dissemination of such ideas might have helped to deter the initiation of a nuclear strike. Nuclear winter makes the climate consequences of a nuclear first strike unacceptable, even if there are doubts as to whether the effect would not be as severe as was originally suggested.

ALLEN PERRY

Further reading

TTAPS (1983) Nuclear winter: global consequences of multiple nuclear explosions. *Science*, 222, 1283–92.

Harwell, M. A. (1984) *Nuclear winter*. Springer-Verlag, New York.

nummulites *see* FORAMINIFERA AND OTHER UNICELLULAR MICROFOSSILS

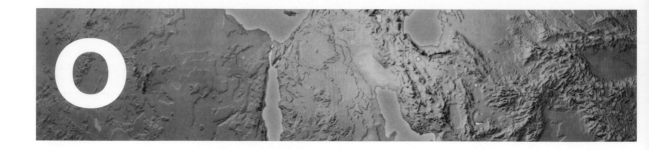

obduction In the late 1960s and early 1970s it began to be accepted that ophiolites (consisting of basaltic lavas, overlying sheeted dykes, gabbro, and peridotite) are slices of oceanic crust and upper mantle that have somehow found themselves exposed on land. Most of the world's oceanic lithosphere is destroyed at subduction zones, below island arcs or Andean continental margins. In those places the oceanic plate descends below an overriding plate and, after providing the stimulus for volcanism on the overriding plate, is eventually absorbed back into the deep mantle. Although it was not clear how or why those particular slivers of ocean floor represented by ophiolites have escaped subduction, a term was required to denote the phenomenon. To meet this need, in 1971 the American geologist Robert G. Coleman introduced the word *obduction*. The processes of obduction are still not understood, and it is apparent that no two ophiolites have been emplaced in exactly the same manner. However, the term 'obduction' has remained a useful counterpart to subduction in the geologist's vocabulary.

Most ophiolites are found in zones of collision or at least convergence between tectonic plates, and they are commonly the only remnants of the floor of a vanished ocean that formerly separated two continents, and which are now firmly sutured together. Small ophiolites along the Alpine–Himalayan mountain chain are of this sort.

It is not surprising that ophiolites are generally small and incomplete, because there is a considerable problem associated with achieving obduction. This arises from the fact that oceanic crust is denser than continental crust. During convergence between two tectonic plates the plate bearing denser oceanic crust is thus liable to be forced below the one bearing less-dense continental crust. It was particularly hard to imagine how obduction could occur when ophiolites were generally thought to be ocean-floor generated by sea-floor spreading at mid-ocean ridges in major ocean basis, because by the time the ocean closed the ocean floor near the edge of the continent would be something like a hundred million years old, and therefore would have become cold and particularly dense. However, it is now recognized that ophiolites are generally younger than most of the main vanished ocean basin with which they are associated, and that they probably began their existence as ocean floor created in marginal basins

above subduction zones (Fig. 1). This is backed up by a variety of geochemical evidence.

The chances of marginal basin lithosphere being obducted over an adjacent continental margin during a collision event are greater than for normal ocean crust for a number of reasons. First, the young marginal basin ocean flow will be hot, and therefore less dense and more buoyant than most ocean floor. Although it will still be denser than continental crust, the density contrast will be less. Secondly, marginal basins form above subduction zones, so when eventually the major ocean basin closes completely continental crust will begin to be subducted below the marginal basin, which could be the first step in the obduction process. Thirdly, water escaping upwards from the downgoing slab can react with the olivine in the mantle below the marginal basin, turning it into a new mineral, a hydrated magnesium silicate called serpentine. The resulting rock, called serpentinite, has a much lower density than the original peridotite, thereby increasing the relative bouyancy of the future ophiolite — and with this its chances of obduction. Another helpful factor is that marginal basin crust is usually only about two-thirds the thickness of typical ocean crust; so if obduction is going to occur it will be easier to slice off a complete section penetrating through this crust and into the upper mantle.

Once continental crust has begun to be forced below a marginal basin, and if continent–continent collision follows, some at least of the marginal basin ocean floor is likely to find itself sandwiched between two masses of continental crust and can be brought to the surface by the processes of uplift and erosion that follow continental collision. This cannot, however, explain the emplacement of the largest ophiolites, which tend to lack the signs of high-pressure metamorphism that would affect rocks that had been taken down to the depth implied by such a process. For example, the world's largest and best-exposed ophiolite, the Oman or Semail ophiolite in Arabia, shows no sign of collision-related metamorphism. It was formed by marginal basin sea-floor spreading in the mid-Cretaceous, and within about 20 million years had been obducted over the edge of the Arabian continent as a result of the closure of the Tethys Ocean that formerly separated Arabia and India from the main Asian continent. However, that part of Arabia was never in direct collision with Asia. The

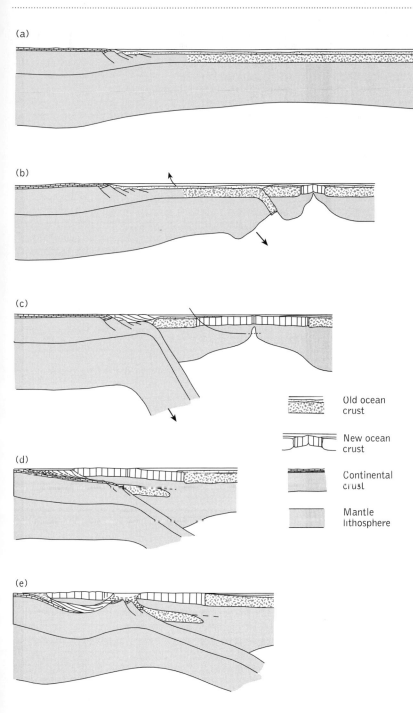

(a)

(b)

(c)

Old ocean
crust

New ocean
crust

Continental
crust

Mantle
lithosphere

(d)

(e)

Fig. 1. Stages leading to obduction of an ophiolite. (a) A passive continental margin with adjacent ocean floor. (b) Subduction begins within the ocean in a direction away from the continent. New ocean floor is generated in a marginal basin above the subduction zone. (c) The continental margin reaches the subduction zone, blocking further motion. A new thrust develops where the new ocean crust joins the old. (d) New ocean crust is displaced across the jammed subduction zone and on to the continent. This is the act of obduction. Some sediment is pushed in front of the advancing ophiolite, other sediment is overridden. (e) Postobduction uplift and erosion isolates the ophiolite from the remaining ocean floor. The time between (a) and (e) is about 30 million years. (This model is adapted from one proposed to explain the obduction of the Oman ophiolite, in which (a) was about 105 million years ago and (e) about 75 million years ago.)

only metamorphism associated with the obduction of this ophiolite is seen in a few slivers of 'metamorphic sole' immediately below the ophiolite and consisting of volcanic and sedimentary rocks metamorphosed to greenschist and amphi-

bolite grade. These appear to represent the first rocks across which the ophiolite was obducted; they were metamorphosed because the base of the ophiolite sheet was still hot. The ophiolite subsequently passed over a full sequence of continental margin sediments. In some places it overrode them; elsewhere it pushed them along in front, in both cases compressing them by folding and faulting. The Oman ophiolite appears to have reached its final position by sliding downhill, riding on a mélange of broken blocks of marine sediments. To allow all this to happen, the leading edge of Arabia must somehow have been depressed, although it has now rebounded back above sea level, despite the extra weight of a ophiolite 10 km thick on top of it.

The Bay of Islands ophiolite of Newfoundland, Canada was obducted over a continental margin in a manner similar to the Oman ophiolite, about 460 million years ago during the Ordovician, which was within about 40 million years of its origin as ocean floor. In this case the associated event was a genuine continent–continent collision when a proto-Atlantic ocean, referred to as the Iapetus Ocean, closed roughly along the line of the Appalachians. By contrast, an ophiolite along the north coast of eastern Papua New Guinea is still attached to the ocean floor of the Solomon Basin from whence it came, and may still be in the process of obduction. Here the Papua New Guinea continental crust appears to be underthrusting the adjacent marginal basin and obduction of the ophiolite is occurring in an uphill direction. Another well-studied ophiolite, the Troodos massif forming the core of the island of Cyprus, appears not to have been obducted over continental crust at all. Instead it has been uplifted as a result of extreme serpentinization of the underlying mantle peridotites. Macquarie Island, 1000 km south-west of New Zealand, is another ophiolite that appears to be ocean floor that has been uplifted without the intervention of obduction.

DAVID A. ROTHERY

Further reading

Coleman, R. G. (1977) *Ophiolites: ancient oceanic lithosphere?* Springer-Verlag, Berlin.

ocean basins

The enormous depths of the ocean floor have been known since the days of the *Challenger* expedition in the nineteenth century. However, the methods used to determine its depth (i.e. sending a weight attached to a line down to the ocean floor 8 km or more below) meant that only a few random measurements could be made. When these spot heights were used to construct contour maps, the ocean floor looked extremely smooth. It was not until echo sounding could be used to give a continuous profile that the ruggedness of the ocean floor was appreciated. These echo sounders showed that not only were some parts of the ocean floor, such as the mid-oceanic ridges, extremely rugged, but, equally surprising, that other parts, such as some basins, were extremely smooth. Another noticeable feature was that much of the ocean floor reaches a depth of about 5–6 km, and, outside the narrow, elongate deep-sea trenches (see *island arcs*), is no deeper. Now that we know that oceans are formed by the creation of new lithosphere at mid-ocean ridges and that this lithosphere progressively cools and sinks, it is clear that the depth of the ocean

floor is limited, unless some other factor is involved. This may be a subduction zone that produces very deep trenches or intraplate volcanism that forms oceanic plateaux and raises the level of the ocean floor significantly above its 'normal' (5–6 km) depth by the creation of thickened crust.

Ocean basins are defined as those basins that are underlain by oceanic crust. They have many different forms, large or small, rugged or smooth, and, as with their complementary ridges, hills, and plateaux, have quite diverse origins.

The most spectacular ocean basins are the abyssal plains, a specific type of basin plain located on true ocean floor where any original abyssal hills have now been submerged by the infilling of the deeper areas by sediment. They range in size from very small ones, such as the Alboran in the Mediterranean (2600 km^2) through the Madeira (54 000 km^2) in the North Atlantic and the Angola in the South Atlantic (1 000 000 km^2) to the Enderby in the Antarctic (3 703 000 km^2), some thousand times larger than the smallest ones. Because the supply of sediment from the land is an essential condition of their formation, abyssal plains are abundant in the Atlantic and generally absent from the Pacific, where subduction-related trenches and marginal (back-arc) basins entrap most terrigenous sediment except in its north-east corner, adjacent to the North American continent. Their importance around the Antarctic is a reminder that the Antarctic continent supplies enormous volumes of sediment to the oceans, because ice is such an efficient agent of erosion.

Since the principal control on their formation is the blanketing of an original rugged topography by sediment, abyssal plains are distinguished on the basis of:

(1) sediment composition (e.g. terrigenous vs. carbonate);
(2) whether they are open or closed (i.e. whether sediment-supplying turbidity currents bank up within them or pass out from an outlet;
(3) depth, i.e. distance above or below the carbonate compensation depth (CCD);
(4) the volume of sediment supplied compared with the area of the basin; and
(5) the size of the surrounding drainage area as compared with the basin area.

The most intensively studied abyssal plain in the world is the Madeira Abyssal Plain (Fig. 1). (It was once considered as a possible site for the disposal of radioactive waste.) It is over-supplied because the area from which sediments are drained (which includes part of the north-west African continent) is 50 times larger than the basin area. This contrasts with the large under-supplied abyssal plains in the western North Atlantic, where the ratio is an order of magnitude less. The basin is fed by turbidity currents from three distinct sources, emplaced about every 20 000 to 30 000 years. Turbidites make up about 90 per cent of the basin sediment (600 km^3 over 300 000 years), the largest having a volume of 190 km^3. Pelagic sediments comprise only 10 per cent of the total sediment volume.

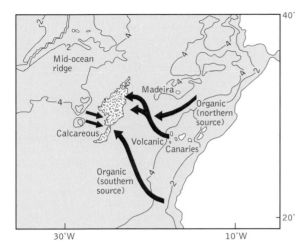

Fig. 1. Map of the Madeira Abyssal Plain at 5400 m depth, showing sources of turbidites: (1) volcanic-rich, derived from the oceanic islands of Madeira and Canaries, from which enormous slides have emanated; (2) organic-rich, derived from the African continental margin; and (3) calcium carbonate-rich, derived from seamounts to the west. The third group is of very minor importance. (From Weaver *et al.* (1992) *Marine Geology.*)

Dominating the Pacific Ocean are the rugged ocean basins still displaying their original topography of lows and abyssal hills. In these ocean basins, pelagic sediments dominate and, instead of just filling the lows, as in abyssal plains, tend to drape the existing topography with a blanket of sediment. The rate of sedimentation and thickness of the blanket depend primarily on the productivity of the overlying waters. Since most of the basins are below the CCD, the sediments comprise red clay, radiolarian ooze, and diatomaceous ooze. Ferromanganese nodules are common, as are whales' earbones and sharks' teeth. The smooth blanket is often disturbed by local resedimentation from topographic highs into lows. Where the ocean floor is not below the CCD the calcareous skeletal matter is dominated by foraminifera and nanno-fossils. One of the interesting features of deep-sea drilling is the discovery of a marked cyclicity in carbonate productivity, terrigenous dilution, or carbonate dissolution. The cycles have been matched to the Milankovich timescale, suggesting that there were cyclic changes in CCD and the glacial supply of terrigenous material.

In addition to the relatively large basins in the principal oceans, small ocean basis also occur where incipient or local-ized ocean-floor spreading takes place. Examples are the Red Sea, the Gulf of California, and some Mediterranean basins; others occur in back-arc settings, particularly around the Pacific, such as the Japan Sea. These small ocean basins are characterized by having a large supply of terrigenous sedi-ment and being above the CCD. Pelagic sediments occur only on the highs and in parts of the basins shielded from the clastic sediments. Hemipelagic sediments (mixed terrige-nous and pelagic sediments) abound.

HAROLD G. READING

Further reading

Weaver, P. P. E. and Thomson, J. (eds) (1987) *Geology and geochemistry of abyssal plains.* Geological Society of London Special Publication No. 15.

Weaver, P. P. E., Rothwell R. G., Ebbing, J., Gunn, D., and Hunter, P. M. (1992) *Marine Geology*, **109**, 1–20.

ocean chemistry The oceans comprise 98 per cent of the free water available at the Earth's surface. They also represent a dynamic system which is continuously undergoing chemical exchange at their boundaries, as well as cycling their con-stituents internally. When the first scientific studies of the oceans began approximately 100 years ago with the *Challenger* expedition, only a few major constituents of the ocean's com-position could be determined. Of the positively charged salts (cations), these were found to be principally sodium, magne-sium, potassium, and calcium, while of the negative (an)ions, the principal constituents were chloride, sulphate, and bicar-bonate. These components dominate the total concentration of dissolved salts in the oceans, referred to as *salinity*, which is always close to 35 g solids per kilogram of sea water. Further, the ratio of these various elements to one another is always constant throughout the oceans—a feature referred to as Dittmar's law. An important consequence of this is that the oceans' salinity can often be obtained simply by measuring the conductivity of a sample of sea water. This can be used routinely to determine automatically the total salinity of any portion of the ocean to within 1 mg kg^{-1}.

The continents are a principal source of chemical input to the oceans. Physical weathering of rocks can lead to the production of fine-grained particles which can be transported to the oceans either by rivers (15×10^{15} g yr^{-1}) or by aeolian (wind-borne) transport (1×10^{15} g yr^{-1}). In the case of riverine transport, it is estimated that some 90 per cent of the total solid-phase load is deposited in estuaries and the submarine canyons that represent the offshore submerged extensions of these landforms. Thus, the net supply of riverine solids to the deep open oceans is not significantly greater than that from aeolian transport. Dissolution of atmospheric carbon dioxide into rainwater can also result in the formation of very weak solutions of carbonic acid, which also plays an important role in chemical weathering (4×10^{15} g yr^{-1}). As well as providing a significant dissolved flux of elements to the oceans, chemical weathering reactions also give rise to an important set of residual products—the clay minerals. Of note is that the primary cation of the ocean is sodium (Na^+) because potassium (K^+) is held within the structure of clay minerals. Although there is a steady influx of material into the world's oceans every year, through the world's rivers, the composition of the oceans, as recorded from marine sediments, has not changed significantly for the past 600 million years. This is because the rate at which material is added to the oceans each year is matched precisely by the rates at which the same material is deposited to the underlying sediments of the deep-sea floor, which, in turn, are recycled to the continents through uplift and metamorphism during plate-tectonic cycles. Thus, the oceans are said to be at steady state.

The only one-way process that has been identified in ocean chemistry is the degassing of volatile elements and compounds from the Earth's mantle at mid-ocean ridges—the approximately 60 000-km-long chain of submarine volcanoes which circumscribes the Earth, demarcating the boundaries between the various tectonic plates. All along these boundaries, volatiles escape from where they have been locked within the Earth's interior, typically as a result of hydrothermal activity. This is the process by which cold sea water seeps down into the fresh, hot volcanic rocks of the mid-ocean ridge crest and undergoes a series of progressive reactions. Eventually, the modified sea water becomes buoyant and is driven back upward through the rock and is expelled from the sea bed (through *black smokers*) as a hot (350 °C), anoxic, hydrogen-sulphide-bearing fluid which is also extremely enriched in a number of minerals, e.g. iron, manganese, copper, zinc, and even silver, platinum, and gold. These submarine hydrothermal vents, which were discovered only in the 1970s, appear to represent a significant input of chemicals to the oceans, comparable in importance to the physical and chemical weathering of the continents.

There are many chemicals present in sea water at concentrations less than 1 mg kg^{-1} which do not significantly affect the total composition of sea water but which remain oceanographically useful. These are described as trace constituents. Marine geochemists can exploit the distributions of these tracers to investigate physical, chemical, and biological processes operating in the world's oceans. For example, measurement of sea-water distributions of chlorofluorocarbons (CFCs), which have been released to the atmosphere since the beginning of the 1950s, can reveal important information about how chemicals exchange between the atmosphere, the surface layers of the ocean, and the deeper waters below the depth of wave mixing. The anthropogenic release of short-lived radionuclides through atmospheric testing of nuclear weapons during the 1950s and 1960s has also been used, in a similar manner.

Another example is the study of the constituents referred to as the micro-nutrients in sea water—dissolved nitrate, phosphate, and silicate. These components, together with calcium and carbon, form the basic building blocks for all life in the ocean, with a fixed composition in plankton (microorganisms) of the form $C_{120}N_{15}Ca_{40}Si_{50}P$. While sunlight, carbon, and calcium are always present in abundance in the surface ocean, silicate, phosphate, or nitrate—and sometimes more than one of these compounds—are typically exhausted during photosynthesis and the growth of marine micro-

organisms. This process is known as primary productivity, but the small plankton responsible for this chemical uptake are short-lived and soon fall into the deeper, colder, and darker portions of the oceans where both the soft parts (organic tissue) and hard parts (shelly material) of their bodies decay and are recycled into solution. Study of the dissolved nutrient distributions of the present-day oceans thus provides geochemists with information about what is limiting marine plankton growth throughout the surface oceans and also yields important information about the recycling processes which are occurring at depth. One of the most important processes deduced from such study is the deep-ocean conveyor belt theory. Cold surface water near the northernmost extremes of the Atlantic is cooled during formation of the Arctic ice cap and sinks to the sea bed, where it flows south along the length of the Atlantic Ocean. In the very southernmost Atlantic it meets Antarctic bottom waters which have formed in a similar way around the Antarctic ice pack and are flowing northwards. The two water masses combine and then both flow eastwards, past the southernmost tip of South Africa and around the Southern Ocean, with tongues of deep water also flowing North along the sea bed into the Indian and Pacific Oceans. As organisms are continuously raining down into the deep ocean from the productive sunlit waters above, this pattern of deep-water flow is reflected in the build-up of dissolved nutrients observed from the Atlantic to the Indian and Pacific Oceans. Measurement of the decay of the radioactive form of carbon (^{14}C), which is produced in the Earth's atmosphere and therefore can enter the deep ocean only through entrapment in the surface ocean, has been used to determine the rate of the mixing process of this deep-ocean conveyor. This has confirmed that, on average, the deep waters in the North Atlantic Ocean have an apparent age which is younger than the deep waters of the northern Pacific. Further, on average, the transit time for a parcel of water to travel from its point of sinking in the northern North Atlantic Ocean until it returns to the surface in the north-eastern Pacific Ocean is approximately 1500 years. Of course, upwelling continues progressively throughout the mixing-time of the deep oceans, just as return flow of surface waters towards the poles must take place across the entire surface of the oceans. Nevertheless, it is useful to remember that any effect introduced to the deep oceans today will not work its way through the system for at least another thousand years.

A major emphasis in the study of ocean chemistry is to understand how it relates to processes in modern-day climate and, thus, to understand how past episodes of global climate change might be preserved in deep-ocean sediments. We have already discussed how distributions of dissolved nutrients can be used to monitor biological processes throughout the world's oceans. With improving sample handling and analytical techniques, the 1970s and 1980s also brought the recognition that various trace metals exhibit nutrient-like behaviour. Unlike the nutrients themselves, however, these various trace metals are also preserved in deep-sea sediments accumulating on the ocean floor. For a number of these tracers, it has recently emerged that the concentrations or ratios of various elements and their isotopes (e.g. cadmium, barium, boron) preserved in modern-day shell fragments are in direct proportion to the composition of the sea water from which those shells have been precipitated. Thus a whole new field of research, paleoceanography, has now sprung up, which is dedicated to examination of the record laid down in deep-sea sediments, in a bid to reconstruct past variations in the distribution of the oceans. Important tracers that can be studied in this way include dissolved phosphate and silicate concentrations and pH—a measure of the relative acidity or alkalinity of sea water. Modern-day sea water is typically close to pH 8.5, indicating a mildly more alkaline solution than neutral waters (pH 7). Knowledge of the past distributions of these tracers immediately yields new insight into past oceanic behaviour, such as oceanic circulation patterns during previous glacial and interglacial periods of Earth's history.

The oceans cover more than 70 per cent of the Earth's surface. It is clear, therefore, that any major perturbations in atmospheric compositions—be they natural or anthropogenically induced—will be buffered significantly by exchange processes between the atmosphere and the surface of the oceans. Studies of air–sea exchange are increasingly being linked with studies of the records preserved in past sediments, the key concept being that understanding the past should provide a key to predicting the future. Only if we can understand how natural processes have affected the Earth in the past, it is argued, can we successfully anticipate how man-made processes might alter our natural environment in the future. A recent example is the so-called iron hypothesis. This working theory predicts that during past ice ages increased atmospheric aridity led to a higher flux of wind-borne aeolian dust to the surface layers of the deep ocean. The iron supplied by this increased flux of particulates, the argument goes on, would have helped to stimulate increased primary photosynthetic productivity across significant portions of the world's oceans and, hence, a significant proportion of the entire Earth's surface. This would simultaneously have led to an important drawdown and removal of gaseous carbon dioxide from the Earth's atmosphere as the carbon was fixed into both the soft organic tissue of the marine phytoplankton and also their calcium carbonate shells. If proved correct, such a mechanism might one day lead to purposeful seeding of some part of the world's oceans to help alleviate the progressively increasing levels of atmospheric carbon dioxide that are experienced through increased burning of fossil fuels dating back to the start of the industrial revolution.

On a somewhat shorter timescale, more information is urgently being sought to understand the fate of the organic wastes and toxic metals which are being delivered daily to our coastal seas through discharge from rivers. With the ever-increasing outputs from both established and developing industrial nations, the race is on to collect as much information as possible from the world's few remaining unaffected coastal environments. Only when the nature of the unaffected environment has been properly established can the true effects of industrial pollution be properly judged. There is therefore a pressing need for an improved understanding of basic chemical oceanographic processes upon which future governmental decisions of both national and international importance can be based. This was perhaps most clearly exemplified by the controversy over—and eventual abandonment of—the plan to dispose of the Brent Spar oil installation in the North Atlantic Ocean in 1995. Perhaps for many of the wrong reasons, research into ocean chemistry looks set to receive continuing widespread attention well into the twenty-first century.

CHRISTOPHER R. GERMAN

ocean currents Ocean currents are analogous to atmospheric winds. The general circulation of the world ocean is driven partially by atmospheric winds and partly by the thermohaline effects across the world ocean (see *sea ice and climate*). The ocean currents are the movement of water resulting from these two driving forces. The world ocean has a complex network of shallow ocean currents (Fig. 1) which are dominated by the prevailing winds; the larger-scale circulation is controlled principally by longer-term thermohaline forcing. The currents generated by prevailing winds derive from a transfer of momentum (through friction) from the atmosphere to the ocean; the currents do not, however, follow the prevailing wind direction. The water, once moving, is deflected by the Coriolis force to the right of the wind in the northern hemisphere, and to the left of the wind in the southern hemisphere. The resulting direction of the current is at approximately 45° to the right of the wind in the northern hemisphere (to the left in the southern hemisphere) and is called the Ekman drift, after the Swedish mathematician who first formulated the problem in 1905.

The general circulation of the two largest ocean basins is very similar; there is a clockwise gyre in the northern hemisphere of both the Pacific and the Atlantic and an anticlockwise gyre in the southern hemisphere. In both hemispheres the gyres are anticyclonic (i.e. opposite to the rotation of the Earth) and are especially strong in the mid-latitude regions. The circulation of the Indian Ocean is slightly more complex because of the seasonal wind changes which accompany the monsoon. In this region from May to September the winds typically blow from the south-west, whereas from November to March they reverse to blow from the north-east. The seasonal change in wind direction results in a seasonal change in the surface currents. The high-latitude currents of the polar seas are discussed in the entry on *high-latitude ocean currents*.

Another important observation is that there are strong narrow currents at the western edge of all the ocean basins. This western intensification drives some of the strongest currents, and these were among the first currents of the oceans to be surveyed accurately. By far the most studied of these western boundary currents is the Gulf Stream in the North

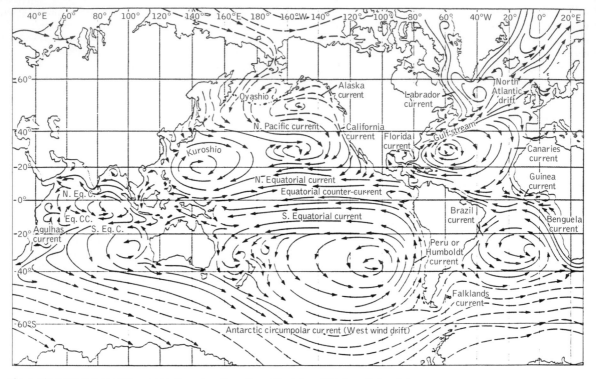

Fig. 1. The ocean currents of the world in December. (From *Ocean Circulation*, 1989. Open University.)

Atlantic. The prevailing west to south-west winds across the North Atlantic are responsible for extending the Gulf Stream and giving the British Isles a temperate maritime climate. The other major western boundary currents are the Agulhas on the east coast of South Africa, which was so important to the tea clippers of the nineteenth century, the Kuroshio on the east coast of Japan, and the East Australia Current. In contrast, eastern boundary currents on the east of the ocean basins tend to be broad and slow.

Changes in the atmospheric windfield can be responsible for rapid and dramatic changes in the ocean currents. An example of this is the periodic extreme climate event known as El Niño.   MARK A. BRANDON

Further reading

Pickard, G. L. and Emery, W. J. (1982) *Descriptive physical oceanography*. Pergamon Press, Oxford.

Ocean Drilling Program
The Ocean Drilling Program (ODP) is an international partnership of scientists and research institutions organized to explore the structure and history of the Earth beneath the ocean basins. Its infrastructure cost of some $45 million per annum is funded by the US

National Science Foundation together with contributions from 20 non-US partner countries. The scientific operations are carried out from Texas A&M University in the USA. The overall planning and programme advice is provided by the Joint Oceanographic Institutions for Deep Earth Sampling (JOIDES), an international group of institutes and universities. The JOIDES structure incorporates a series of managerial, scientific, and technical panels staffed by representatives of the contributing countries. The legacy of ODP includes hundreds of kilometres of drill core, the JANUS database containing many gigabytes of electronically stored shipboard measurements, and thousands of scientific and technical articles in its own volumes (*Proceedings of the Ocean Drilling Program*) and in the outside literature. The unused core is currently stored in purpose-built repositories in the US and Germany.

ODP is the latest of several international scientific ocean drilling projects. Their inspiration was the Mohole Project of the early 1960s. This was an ambitious but unfulfilled plan to drill through the oceanic crust of the Eastern Pacific to reach the crust–mantle boundary (the Moho) at one of its most accessible points. It led to the Deep Sea Drilling Project (DSDP), which began in 1968 as a US initiative aimed at the global reconnaissance of the ocean basins. DSDP ran its own drilling ship, the *Glomar Challenger*, which was purpose-built for the scientific drilling of the deep ocean floor. In 1974, DSDP became an internationally funded programme for what is known as its International Phase of Ocean Drilling (IPOD). DSDP ended in 1983, when the *Glomar Challenger* reached the end of her capabilities. ODP was set up in 1985 with its new ship, the *JOIDES Resolution*.

The *JOIDES Resolution* is a commercial drilling vessel 143 m long, which has been extensively modernized and adapted for scientific drilling (Fig. 1). It can drill in water depths in excess of 8000 m and carries over 9000 m of drill pipe on board. It has twelve computer-operated thrusters which enable the ship to maintain position to within 1 m even in heavy seas. A 400-ton heave compensator keeps the drill string stable relative to the sea floor. The feature that sets it apart from commercial drilling vessels is its seven-storey stack of laboratories where scientists can process and analyse each 9.5 m (maximum) length of core and interpret measurements made *in situ* within the hole. The laboratories contain space for studies of sedimentology, palaeontology, petrology, geochemistry, palaeomagnetism, and physical properties and contain equipment that includes X-ray diffraction, X-ray fluorescence, and a cryogenic magnetometer.

Fig. 1. The ODP drilling ship, *JOIDES Resolution*. The ship is 143 m (470 feet) long and 21 m (70 feet) wide with a derrick that rises over 60 m (200 feet) above the water-line. (Photograph by courtesy of Joint Oceanographic Institutions, Washington, DC.)

Ocean drilling is organized into 'Legs', each of which usually lasts about two-months and comprises a single scientific project. By the time ODP ends in 2003, DSDP and ODP will together have completed more than 200 Legs of drilling. The deepest hole at the time of writing, in Eastern Pacific oceanic crust, exceeds 2 km.

The science carried out by ODP is currently driven by its Long Range Plan entitled *Understanding our Dynamic Earth through Ocean Drilling*. This plan defines two major scientific themes: dynamics of the Earth's environment and dynamics of the Earth's interior. Within each theme are a series of specific sub-themes within which ocean drilling can make a major contribution. For the dynamics of the Earth's environment, these include: understanding the Earth's changing climate; causes and effects of sea-level change; and sediments, fluids, and bacteria as agents of change. For the dynamics of the Earth's interior, the sub-themes include: transfer of heat and materials to and from the Earth's interior; and investigating deformation of the lithosphere and earthquake processes.

Studies of ODP core have shown that bacteria may live in deep-sea sediments at least a kilometre below the sea floor. The Long Range Plan includes a pilot programme to explore this deep biosphere in more detail. It also includes the developments in drilling and logging technologies and in down-hole experimentation that are needed to achieve its scientific objectives. Some of these developments may lead to the longer-term goals of deep penetration of the ocean crust and continental margins and the creation of sea-floor observatories.

The scientific achievements of the Ocean Drilling Program are too many to list here. They have led to the continuing refinement of our understanding of global tectonics, mantle dynamics, global element fluxes, ocean history, glacial history, sea-level change, climate change, global hazards, and the Earth's resources. Examples of more recent 'high-profile' ODP Legs include: Leg 164, which drilled a deposit of gas hydrate, a potential new fossil fuel resource; Legs 165 and 171B, which drilled the K–T (Cretaceous—Tertiary) boundary, with its evidence for the global consequences of a large meteorite impact; Legs 158 and 169, which drilled actively forming massive sulphide deposits and provided new insights into the process of ore genesis; and Leg 169S, which drilled finely laminated sediments containing the best record so far of past global climate changes.

The scope for scientific discovery by ocean drilling has evolved with time. The greatest achievement of DSDP was probably the 'ground-truthing' of the plate-tectonic hypothesis by providing key information on the age of the ocean basins and the processes that take place at plate boundaries. By contrast, the greatest achievement of ODP so far is probably the development of an astronomically tuned geological timescale, which now provides the most reliable absolute method of dating yet available for the Neogene (23.3–1.64 Ma). This has been achieved by matching the cyclic variations in physical and chemical properties down ODP core to the quasi-cyclic variations in Earth—Sun orbital geometry known as Milankovich cycles. This timescale has formed the foundation of many other achievements of ODP, such as the construction of ultra-high-resolution climate records from the Earth's past.   JULIAN A. PEARCE

Further reading

25 Years of Ocean Drilling. *Oceanus* **36**, 1993. Woods Hole Oceanographic Institution, Wood's Hole, Mass.

oceanic density and pressure

The density of sea water is a complex and non-linear function of temperature, salinity, and pressure. (This can be seen in Fig. 1 of the entry on *ocean water*, where the lines of equal density do not follow straight line.) At temperatures above 20 °C, temperature and salinity generally have an approximately equal control on the density of sea water. As temperatures decrease, the control of density is passed to salinity. As temperature falls, the maximum density of fresh water is reached at 3.98 °C; above and below this temperature the density of fresh water is lower. In the ocean, however, the addition of salt to the system alters the behaviour and when the salt content is greater than 24.7 practical salinity units (PSU) (as in virtually all of the world's oceans) the density of the sea water increases with salt content as temperatures approach the freezing point.

Density is expressed in kilograms per cubic metre. For typical ocean depths of up to 4 km the density typical ranges between 1020 and 1030 kg m^{-3} (kilograms per cubic metre). Oceanographers use a unit σ_t (pronounced sigma-t), which is the density minus 1000 kg m^{-3}. Throughout the ocean σ_t generally varies between 20 and 30 kg m^{-3}. For 80 per cent of the worlds ocean water σ_t ranges between 26 and 28.5 kg m^{-3}. The temperature and salinity values used for the calculation of σ_t are called the *in situ* values, and this introduces a problem. The compression of sea water under pressure is significant. Consider a volume of sea water with a salinity of 35 PSU and a temperature of 5 °C at 4000 m, which is raised to the surface adiabatically (i.e. without exchange of heat with its surroundings). As pressure decreases, the volume will increase and the temperature will fall, in this example to 4.56 °C. The temperature of 5 °C is the temperature value *in situ*. To take into account the variation of temperature with pressure we calculate the density using the temperature that the volume of water would have at the surface (4.56 °C); this is called the potential temperature (θ).

When density is calculated using the potential temperature it is termed the potential density σ_θ (pronounced sigma theta). The potential density does not have to be calculated relative to the surface. If we use potential density for comparing the density of water in the deep ocean, it is usual to calculate the potential density at a comparable pressure level. For example, if we were comparing the densities of water at a depth of 3000 m it would be usual to calculate the potential density relative to 3000 m, which would be called σ_3 (sigma 3), or even $\sigma_{2.5}$, which would be relative to 2500 m; and so on.

For a particular depth in the ocean the hydrostatic pressure is equal to the height of the water column in metres above our point, multiplied by the density of the water and the acceleration due to gravity. As a first approximation it is usual to assume the acceleration due to gravity as 10 m s^{-2} and the density of sea water as 1000 kg m^{-3}; pressure in Newtons per metre squared is then 1000 times the depth. In reality the pressure is greater than this, and a uniform depth will not have equal pressure throughout the ocean as it is dependent on the density of water above the depth. A CTD (conductivity, temperature, depth) measurement usually shows the density increasing with depth except in special circumstances.

The surface density of water also varies with latitude (Fig. 1) because of the different temperatures and salinities encountered in different climatic zones. Surface density is greatest in the Southern Ocean south of the Antarctic Circumpolar Current, decreasing to a minimum just north of the Equator before rising again with latitude. The surface density then falls towards and under the permanent ice cover of the Arctic because of the decrease in sea-surface salinity due to melting of sea ice and freshwater input from rivers. This variation has important consequences for the interiors of the oceans. The dense surface water at high latitudes can descend along density surfaces into the interior of the more temperate oceans and can be tracked using temperature, salinity, and potential vorticity. Changes in the surface waters

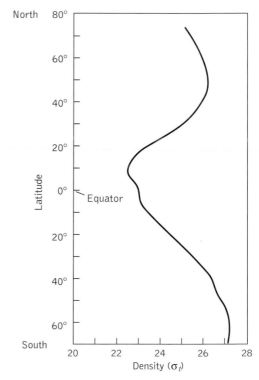

Fig. 1. The variation of surface density with latitude averaged for all oceans. (After Pickard and Emery (1982).)

at high latitudes can therefore be responsible for determining the water properties of the interior of the oceans, making the polar oceans important in climate change.

MARK A. BRANDON

Further reading

Pickard, G. L. and Emergy, W. J. (1982) *Descriptive physical oceanography*. Pergamon Press, Oxford.

oceanic dissolved constituents

In marine chemistry, dissolved species are generally defined as those which will pass through a 0.45 μm (micrometre) filter. This division between dissolved and particulate species is made for convenience and is somewhat arbitrary, since many colloidal (coalesced) forms are included as dissolved species by this definition. Most naturally occurring elements have been measured as dissolved species in sea water, and show a vast range of concentrations. The most prevalent elements are termed 'major constituents', and occur at concentrations greater than 1 part per million (ppm). These elements account for over 99.9 per cent of the total dissolved salts in sea water. The remainder consists of minor and trace constituents. The total concentration of an element in sea water depends on its natural availability (its abundance in the Earth's crust) and its mean oceanic residence time. Dissolved elements in the oceans are subject to long-term recycling processes where their input (mainly from river run-off, atmospheric deposition, and hydrothermal activity) is balanced over geological time-scales by output to sediments (the so-called steady-state ocean). These sediments are then recycled to the continents by sea floor spreading and mountain-building processes, which the cycle begins again. The residence time is the average time an element spends in the ocean before it is removed to the sediment. Residence times range from a few hundred years or less to tens of millions of years, and depend on the element's reactivity in sea water. Highly reactive elements are rapidly scavenged to the sediments by inorganic removal processes (interactions with sinking particles or the sediment itself) and organic removal processes (uptake by organisms and settling organic material). In contrast, unreactive elements tend to stay in solution and so have long residence times.

Dissolved gases

Gases enter or leave the ocean across the air–sea interface, and are transported through the oceans by physical processes. In sea water, some gases (e.g. nitrogen and the inert gases) show conservative behaviour, which means that their concentrations are not significantly changed by the biological or chemical reactions that occur in the oceans. The distribution of these gases is controlled by the mixing of water bodies (physical transport) and the effects of temperature and salinity on their solubility. Other dissolved gases, such as oxygen and carbon dioxide, show non-conservative behaviour because of their involvement in biological and chemical sea water reactions. Dissolved oxygen is generated by photosynthetic processes in the surface layers of the ocean, and consumed during the breakdown of organic matter at greater depths. Carbon dioxide, CO_2, is fixed during biological production, and its removal or release exerts a strong control on the alkalinity of sea water.

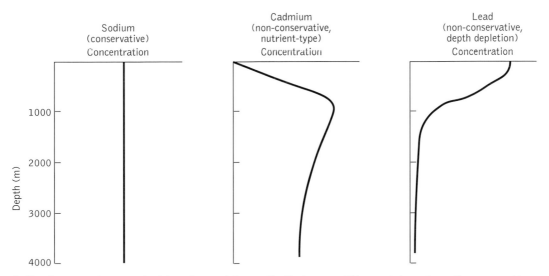

Fig. 1. Profiles of concentration versus depth for sodium, cadmium, and lead in the ocean. Differences between the profiles are caused by biological and chemical processes. (After Chester (1990).)

Major constituents (salinity-contributing elements)

The major constituents of sea water are chlorine, sodium, magnesium, sulphur, calcium, potassium, bromine, strontium, barium, and fluorine. These elements are operationally defined as being 'conservative' in sea water. Although these constituents (and particularly calcium and strontium) may be involved in chemical and biological processes, such processes generally have little effect on the overall concentration of the element. Under most conditions, therefore, the concentrations of major constituents are affected only by physical processes, and they are present in almost constant proportions. This means that while salinity may vary across the oceans, the ratio of the concentrations of these elements to total salinity is almost constant. This is termed constancy of composition. The element silicon, which is present at average concentrations greater than 1 ppm, is classified as a minor constituent because of its involvement in biological processes and strong non-conservative behaviour.

Minor constituents

Most of the minor constituents of sea water are non-conservative, and so their concentrations vary significantly from area to area. This has little effect on overall salinity, however, because of their low concentrations. Involvement in chemical and biological reactions significantly affects the vertical distribution of minor constituents. Typical oceanic concentration v. depth profiles for sodium, cadmium, and lead are shown in Fig. 1. Sodium (conservative) shows no change in concentration with depth. Cadmium s non-conservative, and exhibits labile nutrient-type behaviour. Nutrient elements are those that are essential for the growth of organisms, mainly nitrates, phosphates, and silicates. Cadmium shows similar behaviour to these nutrients, being depleted in surface waters by phytoplankton uptake, and regenerated at depth by decomposition of sinking organic material. Lead shows a profile that is enriched in surface layers by atmospheric input. The dissolved lead is rapidly scavenged by particles, causing a decline in concentration with depth.

Organic material

Dissolved organic matter (DOM) present in sea water is largely inert and has been termed *Gelbstoff* (because of the yellow colour it gives to sea water). DOM is a complex mixture of terrestrially derived oil residues and phytoplankton products.   ANDREW B. CUNDY

Further reading

Chester R. (1990) *Marine geochemistry*. Chapman and Hall, London.

Open University (Oceanography Course Team) (1989) *Seawater: its composition, properties and behaviour*. Oceanography Series. Pergamon Press, Oxford.

oceanic islands
Oceanic islands are best defined by geography rather than geology: as islands in the ocean basins, away from the margins of continents and certainly not within them. This excludes Guernsey and Jersey, the islands of the

Table 1. A genetic classification of oceanic islands

Level One	Level Two
Plate-boundary islands	Islands at divergent plate boundaries
	Islands at convergent plate boundaries
	Islands along transform plate boundaries
Intraplate islands	Linear groups of islands
	Isolated islands
	Clustered groups of islands

English Channel; it excludes Tasmania and Madagascar because of their clear continental connections, both geographical and geological. It includes all islands within the ocean basins and thus it includes both New Caledonia and the Seychelles, despite their continental origins. In the Atlantic, it includes the Canary and Cape Verde groups—favoured sites for an Atlantic Atlantis—together with most islands in the Caribbean. It includes the large islands of Japan, the masses of romanticized islands in the south and west Pacific; the rumbling volcanic islands of Hawaii, the Azores and Galapagos groups, and those in South-East Asia; and the monolithic outposts of land in the ocean basins whose names strike chords in the hearts of everyone curious about world geography: Pitcairn, Easter Island (Rapanui), Heard Island, Kerguelen, St Helena, Ascension, and Juan Fernandez.

A genetic classification of islands is the most useful type of classification for geoscientists. This divides islands according to their association with either plate-boundary or intraplate locations and thence according to the boundary type or pattern, respectively (Table 1).

There are few islands associated with divergent plate boundaries for, although these are sites of active submarine volcanism, they are also sites of normal faulting and rifting associated with divergence. Iceland is an exception to this general rule; it forms an island because it coincides with a hot spot (mantle plume) as well as lying on the Mid-Atlantic ridge.

Islands along convergent plate boundaries are distinguished by their tectonic instability. Along the *frontal arc* (on the overriding plate nearest the trench) islands experience rapid uplift: islands like Nias on the Sunda frontal arc in South-East Asia are rising by as much as 1 mm/year; rates are higher elsewhere. Behind the frontal arc is the *volcanic arc*—a line of active volcanoes, andesitic or basaltic depending on the nature of the magmatic 'mix' beneath. Classic examples are found in the Caribbean and western Pacific; many are inactive, that of 'Ata in the Tonga volcanic arc is shown in Fig. 1. Many islands on downgoing plates at convergent boundaries have been conspicuously elevated as the result of lithospheric flexure: Christmas Island in the eastern Indian Ocean and Niue in the Pacific are examples.

Islands can form along transform plate boundaries where there is significant convergence across them. Examples include Cikobia in Fiji and Clipperton in the eastern Pacific.

Fig. 1. The island of 'Ata, the southernmost in the Tonga volcanic arc, south-western Pacific. (Photograph used with permission from the Government of Tonga.)

The best understood islands in intraplate situations are the linear chains, formed as a plate moves across a fixed *hot-spot*, a place where magma from the asthenosphere leaks to the surface. The Hawaii–Emperor island–seamount chain is the classic example; other chains are known, or have been proposed, for other parts of the oceans.

Isolated islands have a variety of origins; most appear to have been severed from mid-ocean ridges, others may mark nascent hot-spot chains. The origins of island clusters are unresolved; most authorities regard their volcanism as the product of fissuring produced by intraplate stress; the Galápagos and Cape Verde island groups are examples.

An important aspect of this classification is that it is based on island origins. By far the majority of oceanic islands have a volcanic origin but many now appear to us as atolls or high limestone islands whose igneous foundations have been swathed in great thicknesses of limestone. Atolls and high limestone islands are but two examples of how oceanic islands can develop: through subsidence in the case of mid-ocean atolls; through subsidence then uplift in the case of many high limestone islands.

Oceanic islands are of potential interest to all manner of geoscientists. They can inform us about the origin and ongoing development of many parts of the ocean basins, about former and current plate movements, about upper-Earth rheology, and about postglacial sea-level rise, among other processes. Oceanic islands are also of great interest in their own right. Charles Darwin stands out among nineteenth-century visitors to oceanic islands for the insights he reported into island origins and distribution. In the early decades of the twentieth century, Reginald Daly demonstrated his aptitude for independent thinking by studying a great number of oceanic islands across the world's oceans, amassing data on volcanic activity and a recent worldwide fall of sea level. W. M. Davis emulated Daly, largely in order to challenge his ideas, particularly about atoll origins, by visiting a large number of islands, a study which culminated in his 1928 classic, *The coral reef problem*.

Few Earth scientists carried on the tradition established by Daly and Davis until the 1960s, when people like Tuzo Wilson and H. W. Menard demonstrated how marine geologists should integrate ocean-floor and oceanic island data.

Arthur Bloom, in a seminal paper in 1967, showed how oceanic islands were by far the best places to interpret the record of past sea levels, since they acted like 'dipsticks' in the oceans. French workers in particular have followed up Bloom's lead, as have I in both the Atlantic and Pacific.

Pioneering work by John Chappell on the Huon Peninsula in Papua New Guinea led to the realization that the tectonic component of an observed shoreline displacement must be known before its eustatic (sea-level) component can be isolated. This was accompanied by and, indeed, led to an upsurge in studies of tectonically active oceanic islands; for example, in Vanuatu, on Barbados, on the Indonesian island of Atauro, and in various parts of Japan. This in turn stimulated interest in oceanic island tectonics. Many oceanic islands are in the coral seas and thus record slight and/or long-term changes of level with uncommon precision. Much recent work in the south and west Pacific has focused on the nature of uplift along the region's island arcs: either aseismic (vertical creep) or coseismic.

Interpretation of oceanic island landscapes is potentially of great value, particularly for bridging the gap between volcanic extinction of a particular island and the period of recorded human observation. Oceanic island landscapes also have considerable potential for testing models of landscape change, owing to the simplicity of these landscapes (and the controls on their development) compared to continental landscapes.

Throughout the history of the Earth sciences, oceanic islands have been considered of little importance because they are commonly small and remote from the traditional centres of intellectual enquiry. Since the rise of marine geology, it might have been expected that oceanic islands would become a more pressing topic for study, but most marine geologists are interested only in what is below the sea surface and most terrestrial geologists remain secure in their continental strongholds. This will change in the future as oceanic islands become more accessible and better known, but at present, for many Earth scientists, they remain at the frontiers of understanding. PATRICK D. NUNN

Further reading

Darwin, C. R. (1844) *Geological observations on the volcanic islands and parts of South America visited during the voyage of HMS 'Beagle'*. Smith, Elder, London.

Menard, H. W. (1964) *Marine geology of the Pacific*. McGraw-Hill, New York.

Menard, H. W. (1986) *Islands*. Scientific American Books, New York.

Nunn, P. D. (1994) *Oceanic islands*. Blackwell, Oxford.

Nunn, P. D. (1998) *Pacific Island Landscapes*. The University of the South Pacific, Suva.

oceanic salinity Salinity is a fundamental property of sea water, and is a measure of the saltiness of the oceans. It is theoretically defined as 'the total amount of solid material dissolved in a kilogram of sea water when all the carbonate has been converted to oxide, all bromine and iodine have been replaced by chlorine, and all organic matter has been com-

pletely oxidized'. For most practical purposes this is simplified to the weight of total solids dissolved in a quantity of sea water.

The most abundant solutes in sea water are chloride (Cl^-), sodium (Na^+), magnesium (Mg^{2+}), sulphur (SO_4^{2-}), and calcium (Ca^{2+}) (see *sea water*). As a group, the eleven most abundant elements account for over 99.9 per cent of the total mass of solutes in sea water, with Cl^- and Na^+ alone making up 86 per cent. As long ago as 1819, Marcet suggested that the major solutes in sea water occur in essentially constant proportions to each other, regardless of their absolute concentration. In 1884 Dittmar reported results from the chemical analysis of 77 samples of sea water collected during the five-year circumnavigation of the globe by HMS *Challenger* (1873–7), which generally supported Marcet's observation. This relationship between the major ions became known as the Marcet principle. It occurs because the rate at which sea water mixes through the oceans is sufficiently great with respect to the chemical processes acting to remove or add the major ions. Thus, the addition or loss of water alters only the concentration of the ions, and hence salinity, but does not change the total amount of salt in the ocean.

This constancy allows the salinity of sea water to be inferred from the analysis of any of the major ions. Historically this entailed the determination of the chloride concentration of sea water, by titration with silver nitrate, and scaling up to salinity using the ratio of Cl^- ion to the other major ions, expressed as a parts per thousand concentration. This procedure has since been replaced by techniques that measure electrical conductivity. In seawater, dissolved salts ionize forming a solution containing charged particles that are responsible for most of its conductivity. The amount of dissolved salt thus controls the current that passes and so allows determination of salinity. The greater convenience and higher precision of these measurements has led to a new definition of salinity which relates to the measured conductivity ratio of a standard sample of sea water at a given temperature and pressure. Salinities measured by this method are known as practical salinities, and because they are based on the conductivity ratio they have no dimensions. Marine scientists consequently now refer to a sea water having a salinity of, for example, 32 rather than 32 parts per thousand.

The salinity of the world's water masses varies somewhat from the average of 35, although 99 per cent of them are in the range 33 to 37. The chief processes that bring about changes in the salinity of sea water are those that lead to a loss or addition of fresh water. Increases in salinity are caused by evaporation and salt exclusion during ice formation in polar regions. Decreases in salinity result from inputs of fresh water through precipitation, river and groundwater run-off, and melting ice. These processes affect only the surface layer of the oceans; once the water body has left the surface, changes to its salinity can occur only through physically mixing with another water mass of different salinity.

Salinity is most variable in the coastal region because of the influence of river and groundwater run-off. In the open

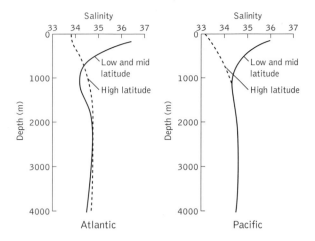

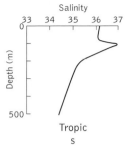

Fig. 1. Latitudinal variation in salinity with depth. (From Riley, J.P. and Chester, R. (1971) *Introduction to marine chemistry*. Academic Press, London.)

ocean the distribution of surface-water salinities tends to be latitudinally controlled by the relative importance of evaporation and precipitation. Salinities in each ocean show the same features, with a minimum in the equatorial regions, where light winds combine with high rainfall to produce a salinity minimum of about 34. In tropical latitudes between about 20° to 30° N and 15° to 20° S, maximum values of about 37 occur where the stronger trade winds produce high evaporation and low precipitation rates. At higher latitudes, salinity again decreases owing to increased precipitation and low evaporative losses, a feature intensified in the Arctic seas by melting ice in the summer months. Greatest variations in salinity are found in enclosed seas; high salinities, exceeding oceanic values, are found in the Mediterranean (37–40) and the Red Sea (40–41), both of which have high rates of evaporation and little fresh-water input. In contrast, low salinities occur in the Baltic, where high riverine inputs dilute the sea water, producing a relatively brackish environment.

In addition to latitudinal variation, sea-water salinity also varies with depth (see Fig. 1) At low and mid latitudes there is a clear salinity minimum between about 600 m and 1000 m, followed by increasing salinities down to about 2000 m. At higher latitudes this minimum is not present, and salinity increases from the surface down to about 2000 m. In the tropics a pronounced maximum occurs at around 100 m, near the top of the thermocline. Deep waters tend to have lower salinities than the surface waters at low and mid latitudes, reflecting their origin as submerged surface polar water. For all ocean water below about 4000 m the salinity remains fairly constant at 34.6 to 34.9.   IAN R. HALL

Further reading

Open University (Oceanography Course Team) (1989) *Seawater: its composition, properties and behaviour*. Oceanography Series. Pergamon Press, Oxford.

Libes, S. M. (1992) *An introduction to marine biogeochemistry*. John Wiley and Sons, New York.

ocean-ridge geochemistry The outer solid skin of the Earth, called the *crust*, consists of the basaltic *oceanic crust*, which is relatively dense and thin (~6 km), in contrast to the granitic *continental crust*, which is relatively less dense and is thick (~35 km). The rocks of the continents preserve a rich record of plate collisions and other events over 4 billion years of Earth history. In contrast, the present ocean basins are no older than about 200 million years. These striking differences are due partly to differences in their properties: for example, being denser, oceanic crust sinks or *subducts* easily into the mantle during plate collisions. Continental crust, being less dense, floats on the underlying mantle rocks and is only rarely subducted. As a result, oceanic crust is produced and recycled at a rapid rate, whereas the continents, though repeatedly battered by collisions and rifting, remain.

The oceanic crust is produced by volcanic processes along the globe-circling system of *mid-ocean ridges* (Fig. 1) and consists of three main layers. The topmost is sediment cover that thickens as ocean crust ages and spreads away from the axis of the oceanic ridge. The middle layer comprises basalt lava flows underlain by their frozen feeders, termed *dykes*. The deepest layer is gabbro, coarse-grained but chemically equivalent to basalt, which forms when magma cools and crystallizes slowly. This layered crust rests on rocks (mainly peridotite) of the Earth's mantle. The moving ocean *lithosphere* consists of the crust and a thickness of mantle that varies from near zero at the hot, elevated ridge axis to 75–100 km for old (>60 million years), cold, and deep ocean floor.

The geochemical characteristics of ocean-floor basaltic lavas provide clues to the formation of oceanic crust. Since ocean crust accounts for about two-thirds of the surface on the Earth, its formation is a major Earth process. Also, the chemistry of ocean-ridge basalts enables us to infer the nature of the Earth's mantle and the processes occurring there. All basalts are thought to be derived by partial melting of the solid mantle. However, on the continents basalt melts must

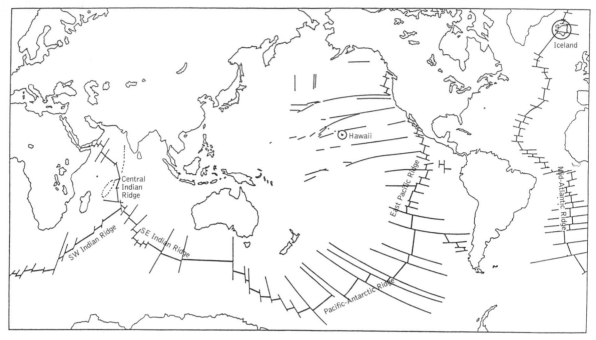

Fig. 1. The global ocean ridge system and fracture zones; the Hawaii and Iceland hot-spots are circled. (Adapted from Floyd (1991).)

rise through thick granitic crust, which can melt and thereby contaminate the magma. Within the ocean crust contamination of this type is minimal, and ocean basalts can thus be used to cast light on the chemical nature of the Earth's mantle and the processes that take place in it. This, in turn, is useful for developing ideas about the Earth's early history of *accretion*, the early development of its layered structure, and the process of slow convection that cools the Earth's interior and drives plate motion. Young ocean-ridge basalts can also be compared with ancient examples to learn about gradual changes within the Earth. During plate collisions, most ocean crust is subducted back into the Earth's mantle; however, small slabs may be preserved in ophiolite complexes on the continents. These complexes, in some cases over 3 billion years old, can be compared with modern ocean crust to tell how the Earth has changed through time.

Samples of ocean-ridge basalt were collected early in this century, for example by J. D. H. Wiseman on the John Murray Expedition of the British Museum. Most of the major oceanic ridges were not, however, sampled until the 1950s and early 1960s. Modern geochemical study of the oceanic ridges dates from 1965, when A. E. J. Engel and others at the Scripps Institution of Oceanography showed that, compared to the basalts of oceanic islands, ridge basalts are chemically unusual in having very low abundances of alkali and other elements that are not easily accommodated in the lattice sites of common silicate minerals. For this reason, ocean-ridge

basalts, or MORBs (mid-ocean ridge basalts), are usually referred to as being 'depleted' in the *incompatible* or excluded elements (for example, potassium, rubidium, strontium, barium, and uranium) and the light rare-earth elements (lanthanum and cerium).

Also in 1965, Engel and two geochemists, M. Tatsumoto and C. Hedge of the US Geological Survey, showed that ocean-ridge basalts have unusually low ratios of the isotopes $^{87}Sr/^{86}Sr$. This work and subsequent studies using other isotope systems (lead, neodymium, hafnium, helium, etc.) has shown that the mantle is not chemically homogeneous. Instead, it consists of chemically and isotopically distinct regions with different evolutionary histories, all of which are capable of melting to produce various types of oceanic basalts. A present and future challenge is to map the chemistry of the mantle and to relate distinct subregions to particular components, such as old ocean crust and sediments, known to be recycled into it by subduction.

Because of the divergence of plates at oceanic ridges, solid mantle material flows passively upward to fill the gap that would otherwise be produced. This hot mantle material begins to melt because of the decompression (lowering of pressure) upon ascent, at a depth determined mainly by the temperature of the mantle and its composition. Melting and volcanism will occur wherever hot mantle material upwells: at ocean ridges or within plate interiors. Hot-spots, a major form of within-plate volcanism, are thought to result from narrow buoyant columns of hot mantle material (*mantle plumes*) rising from great depth. While in some cases (for example, Iceland) hot-spots coincide with ocean ridges, many (such as Hawaii) are far from plate boundaries (Fig. 1). Unlike the moving plates, hot-spots seem to be immobile and have been compared to fixed blow torches held under slowly moving sheets of wood, building long island and sea mount chains on the moving ocean lithosphere.

Chemically, the mantle source of hot-spot volcanic rocks is distinct from that of MORBs, but the two can mix. J.-G. Schilling at the University of Rhode Island and others have shown that at Iceland and other ridge hot-spots the deeper plume mantle source mixes with, and is diluted by, the MORB source. In the North Atlantic, the chemistry of lavas at the Mid-Atlantic Ridge is strongly influenced by the Iceland and Azores plumes. In the Pacific and Indian oceans, plume effects are less dominant. Instead, the mantle seems to contain small, scattered wisps of plume mantle that melt beneath the ridges giving rise to 'enriched' MORB (E-MORB), which occurs sporadically along the East Pacific Rise and the Indian Ocean ridges, far from hot-spots. Interestingly, the chemical and isotopic fingerprints of E-MORB can usually be linked to any of four or five distinct mantle components identified in nearby or distant plume-related lavas.

Ocean-ridge lavas far from plumes and unaffected by wisps of plume mantle are called 'normal' MORB (N-MORB) and are surprisingly uniform in composition. Small chemical variations do occur, however, not only locally at a given ridge but also regionally among various segments of the global system. Several causes for such chemical variation in the major elements (silicon, aluminium, titanium, magnesium, iron, calcium, and sodium), minor elements (manganese, potassium, and phosphorus) and dozens of trace elements have been identified on the basis of numerous lines of geological, geochemical, and geophysical data. One ubiquitous effect is from the cooling and crystallization of melt after it is produced by partial melting of rising upper mantle. Because basalt melt is very buoyant, it can readily escape upwards from the region of melting and, in doing, it can encounter much cooler environments. The ensuing heat loss causes crystals to grow, changing the chemistry of the remaining liquid. Virtually all ocean-ridge basalts have been affected by this process, which can occur in the mantle or in shallow (1–6 km) magma chambers in the crust.

Other chemical changes may result from differences in composition of the mantle source, the depth of melting, and the amount of melting that takes place. These effects have been studied in the laboratory by melting solid rocks under controlled conditions. A complication, however, is that a change of melting depth might have the same chemical effect on resulting melt as, for example, a change of source composition or a difference in the physical style of melting. Thus assigning a unique cause or combination of causes to chemical differences among MORB lavas is a difficult challenge.

A fundamental control on the major element chemistry of MORBs seems to be mantle temperature. As has been shown by E. Klein and C. Langmuir of Columbia University, there is a strong and self-consistent association of MORB chemistry, the depth of ocean ridges, and the thickness of the ocean crust. Hotter mantle begins melting at a greater depth, producing more total melt and thus a thicker crust. This is consistent with earlier theoretical calculations by D. P. McKenzie at Cambridge University.

While differences in mantle temperature and initial melting depth appear to explain many observations, there are other important factors, such as chemical differences among various mantle sources, the effects of which on melt chemistry are not yet well understood. In addition, systematic chemical differences between the lavas of slow- and fast-spreading ridges, found by Y. Niu and R. Batiza at the University of Hawaii, may point to a basic difference in the style of mantle upwelling and melting with plate speed. Much has been learned about the geochemistry of oceanic ridges, but much still remains to be discovered and explained.

RODEY BATIZA

Further reading

Floyd, P. A. (ed.) (1991) *Ocean basalts*. Blackie and Son, Glasgow.

Phipps Morgan, J., Blackman, D. K., and Sinton, J. M. (eds) (1992) *Mantle flow and melt generation at mid-ocean ridges*. Geophysical Monograph 71, American Geophysical Union.

ocean temperatures Sea temperatures are measured in a variety of ways and to varying degrees of accuracy. Temperatures at the surface are normally measured using a thermometer immersed in a canvas bucket, or in the engine inlets of ships (where water is drawn from just below the surface). Since the Second World War a mixture of uninsulated and insulated buckets, engine inlets and, latterly, hull sensors has been in use. Techniques have been devised to correct historical temperature measurements to remove biases and to derive homogeneous global and hemisphere time series that extend back to 1856. Systematic errors may occur in the data set because of developments in instrumentation and procedures, and quality control checks are required to homogenize the data. Today, estimates of sea temperature are increasingly becoming available as a result of remote sensing from satellites.

Annual, seasonal, and monthly mean data, especially of sea-surface temperature (SST), are readily available in atlas form. However, different amounts of data have been used, different numbers of years included in the averages, and different methods of computation employed to prepare the presentations; comparability from one atlas to another is thus often difficult, and evidence points to considerable differences between sources.

The vertical temperature structure in the ocean to a depth of about 450 m is normally investigated using an instrument known as a *bathythermograph*. This is a bomb-shaped device containing a thermistor at its nose and a spool of thin copper wire attached to a deck unit which records variations in tem-

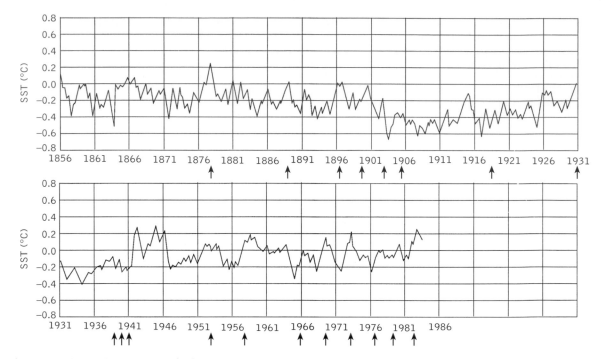

Fig. 1 Seasonal sea-surface temperature (SST) from 1856, with respect to 1951–60 means, averaged over the globe. Arrows mark October–December of El Niño years.

perature as a function of depth. The oceans are heated in their surface layers by contact with the atmosphere, and temperature normally decreases with depth, usually in a series of steps, which are known as *thermoclines*.

Charts of mean SST show that the expected variations of SST with latitude are much modified by ocean currents. Sea temperatures in the southern hemisphere are generally somewhat lower than those in the northern hemisphere because of differences in the character of the prevailing winds and the effects of the large ice-covered Antarctic continent. The highest recorded sea-surface temperature was 35 °C, in the Persian Gulf. Minimum SST may be defined as the freezing point of sea water, which is a function of salinity. The average position of the maximum oceanic surface temperature, the oceanographic thermal equator, is located at about 5–10°N.

The annual range of SST over most of the oceans of the globe is less than 5–10 °C, but annual ranges are higher in land-locked seas and in continental-shelf waters, approaching 20 °C near Korea. By portraying SST as deviations from the average of each geographical latitude, anomalous warmth and coldness is brought into focus. The waters north of the British Isles are more than 9 °C warmer than is normal for the latitude, the largest positive anomaly in any ocean.

Advection by currents, heat exchanges between sea and atmosphere, mixing by wave action, and convective stirring,

can all lead to day-to-day changes of sea temperature, which in many areas of the ocean can amount to as much as 1.5 °C over a 2-day period. If several factors act together to change temperature in the same direction, the magnitude of these short-period changes of temperature can equal the total annual range. In general, water temperature decreases with increasing wind speed, the amount varying seasonally but being greatest in winter and spring, and probably being caused by increased turbulent mixing of the water body.

The SST is a key parameter in modelling ocean–atmosphere interaction and understanding climatic changes. In addition, the planning of oceanographic expeditions designed to study phenomena such as upwelling and oceanic productivity require such data. Historical SST data sets reveal significant fluctuations over the past century and a half. Global minimum temperatures occurred about 1905–10 and maxima about 1960–65, with a worldwide temperature fluctuation of about 0.6 °C. Renewed warming appears to have set in after the early 1970s. The SST maximum appears to lag behind that of the air temperature by about 15 years, but in general it is of a magnitude and consistency to match that of global air temperature.

Changes in ocean climate seem to be linked closely with large-scale atmospheric changes, and this is particularly true in the Pacific where large-scale SST warmings known as El Niño events occur on average every 4–6 years and are thought to have worldwide implications for climate. El Niño may give rise to departures from long-term averages amounting to 6 °C and persisting for several months, and, as Fig. 1 shows, the scale of these anomalies generally leads to peaks in the global temperatures series. ALLEN PERRY

Further reading

Meteorological Office (1990) *Global ocean surface temperature atlas*. UK Meteorological Office and MIT joint project.

Wuethrich, B. (1995) El Niño goes critical. *New Scientist*, 4 February, 32–5.

ocean water Across the Earth's surface the sea water in each ocean basin has different gross physical and chemical properties. The most important properties for defining a particular element of sea water are its potential temperature and salinity. Potential temperature is the temperature that a volume of water would have at a standard reference level (usually the surface), and it is used instead of the temperature *in situ* (i.e. what is actually measured) to calculate density; this takes into account the increase of temperature *in situ* with pressure through adiabatic heating (i.e. without exchanging heat with the surroundings). The two properties are conservative: that is, they are altered only at the boundaries of the oceans (e.g. the surface). Away from the surface, the potential temperature and salinity of a volume of water can be significantly altered only by mixing with a volume of water with different potential temperature (T) and salinity (S) characteristics.

The differences in T and S across the surface of an ocean are due principally to the variation in climatic conditions around the world. The water in the Mediterranean Sea provides a good example. The Mediterranean is an enclosed basin and the ambient climate is relatively dry and hot. The warm atmospheric temperatures result in an excess of evaporation of the surface waters over precipitation, which in turn increases the salinity of the Mediterranean surface waters. Therefore, compared to the North Atlantic waters, the Mediterranean waters are high in both T and S. The excess of T and S increases the density of the Mediterranean water, and at the Straits of Gibraltar the water sinks below Atlantic water as it enters the North Atlantic. Once away from the surface, the water can be tracked throughout the North Atlantic by its T and S values because they are conservative.

There are many areas of the world where uniform climatic conditions create a large volume of water with uniform T and S characteristics, this is called a water mass. The variation in climatic conditions around the world enables us to map the geographical distribution and subsequent movements of water masses. Temperature and salinity are not the only properties we can use to track a water mass. We can also use the oxygen and the varying amounts of nutrients such as phosphates and nitrates dissolved in the water. We must be careful, however, for the nutrient concentration and oxygen are non-conservative properties and are depleted by biological activity. Modern techniques have been developed using man-made tracers such as chlorofluorocarbons (CFCs) which are used in refrigeration

and as propellants in aerosols. The chlorofluorocarbons have received widespread news coverage as they are thought to be responsible for the destruction of the Earth's ozone layer, but they are also a useful oceanographic tool. The concentration of

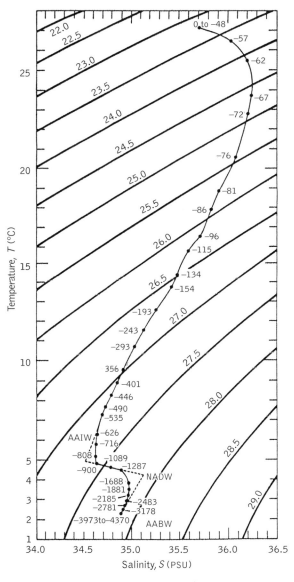

Fig. 1. A potential temperature salinity (T and S) plot of a CTD cast in the tropical North Atlantic at 5 °N, 25 °W. The numbers next to the curve are depths in metres; contours of density are also shown. The water types marked are NADW, AAIW, and AADW. The straight line between AAIW and NADW implies mixing between the two water types. The same has happened between NADW and AABW. (From J. R. Apel (1987) *Principles of ocean physics*. Academic Press, London.)

CFCs in the atmosphere has increased almost exponentially since the 1930s but different chlorofluorocarbons have been released at known different rates. As different CFCs are absorbed at different rates by sea water, it is possible to 'age' the ocean water by calculating the ratio of different chlorofluorocarbons: that is, to calculate roughly when the water was last in contact with the atmosphere during the past fifty years or so. Using CFC ratios, oceanographers have determined that there has been a decrease in deep water formation in the Greenland Sea. Using other tracers, such as carbon-14, some deepest water can be 'aged' at thousand of years.

To make a point measurement of the water characteristics, oceanographers use an instrument called a CTD. The acronym CTD denotes conductivity, temperature, and depth, and the instrument measures these three parameters. From the conductivity of the sea water we can calculate the salinity. The CTD is lowered from a ship, usually down to full ocean depth, which in the case of water 4000 m deep (roughly the average ocean depth) can take as long as 4 hours. The resulting temperature and salinity profiles can be difficult to interpret, but by far the clearest way to look at the vertical temperature and salinity variation is by plotting the potential temperature against salinity.

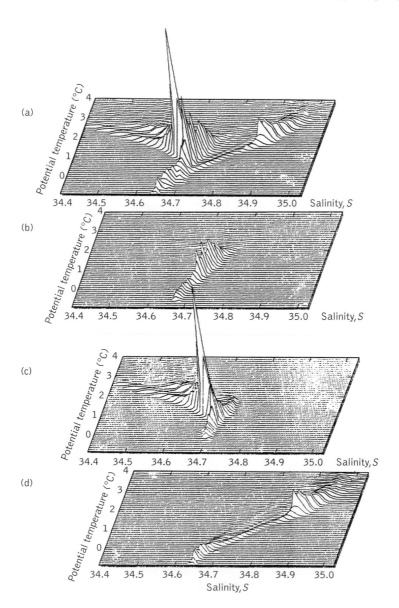

Fig. 2. A potential temperature–salinity–volume plot of the world oceans for temperatures less than 4 °C: (a) the world ocean; (b) the Indian ocean; (c) the Pacific ocean; (d) the Atlantic ocean. (From Worthington, in Warren and Wunsch (1981).)

Temperature–salinity plots are a common way of interpreting CTD profiles. Because density is a function of temperature and salinity it, too, appears on the plot. It was realized in the 1930s that mixing in the ocean will preferentially take place along lines of equal density across the ocean, rather than along lines of equal depth. Once formed under particular climatic conditions in a particular region, sea water can descend into the deep ocean along density surfaces. As sea water from the high latitudes is cold, relatively fresh, and more dense than the warm water at tropical latitudes, the water from high latitudes descends beneath the more temperate waters. This is one way in which the deep ocean can communicate with ('ventilate') the atmosphere.

Water types of different densities from layers throughout the depth of an ocean.

A CTD measurement from the central Atlantic Ocean plotting potential temperature against the salinity (Fig. 1) shows that the water is structured in such a way that there is a layer of water from the central Atlantic, a layer of water influenced by Mediterranean water, a water type called Antarctic intermediate water (AAIW), then North Atlantic Deep Water (NADW), and finally Antarctic Deep Water (AADW). With such a vertical distribution it is easy to see the value of CFC and other tracer measurements for 'ageing' the water types. Because of the mode of formation of a water mass, each type with its own uniform characteristics will by definition appear as a point in what is called T–S space. When two water types of differing T and S are mixed together, the resulting water will lie on a straight line between the two source masses (Fig. 1). The distance along the mixing line from each source indicates the relative proportions of each water type in the water sampled. The same principle applies for a more complex mixture of three or more water masses.

Another conservative tracer for a water mass is potential vorticity. This is an oceanic expression of the conservation of angular momentum; it is equal to the absolute vorticity of a water parcel plus the planetary vorticity, all divided by the thickness of the layer. As the relative vorticity is much smaller than the planetary vorticity in most situations, the potential vorticity becomes the Coriolis parameter divided by the thickness of the layer. Geographical maps of potential vorticity can be used to trace water masses. For example, Labrador Sea Water is a well-known North Atlantic potential vorticity minimum. This is not, however, as common a technique as temperature and salinity mapping.

Throughout the oceans, temperature and salinity can typically range from −1.9 to 30 °C and from 30 to 38 practical salinity units (PSU). There are specific areas where water falls outside this range, for example, the Red Sea has ocean temperatures of up to 45 °C and a salinity that is occasionally greater than 40 PSU, whereas near rivers the salinity can fall to almost 0 PSU. These areas are exceptions; 90 per cent of the world's ocean waters falls into the temperature range −1.9 to 10 °C and salinity range 34 to 35 PSU.

By dividing the temperature and salinity ranges into small increments it is possible to calculate the volume of ocean water of a particular character. An accurate volumetric census of the oceanic water types was calculated in 1981 by Worthington using this method and setting the temperature increment as 0.1 °C and the salinity increment as 0.01 PSU. A temperature–salinity–volume plot of the ocean waters colder than 4 °C can reveal much about the distribution of the world's water masses (Fig. 2). The largest peak in Fig. 2(a) also appears in 2(c); it corresponds to 26×10^6 km^3 and is water from the Pacific Ocean. The largest peak of 2(b), a volume of 6.0×10^6 km^3, is central Indian Ocean water; this water mass is close in character to Pacific water. The largest peak in 2(d) in the Atlantic corresponds to a volume of 4.7×10^6 km^3. The water with a potential temperature of less than zero which can be seen in all the plots corresponds to Antarctic Bottom Water. That the water appears in all of the ocean basins indicates that the ocean basins are connected in the southern hemisphere.

MARK A. BRANDON

Further reading

Pickard, G. L. and Emery, W. J. (1982) *Descriptive physical oceanography*. Pergamon Press, Oxford.

Warren, B. A. and Wunsch, C. (eds) (1981) *Evolution of physical oceanography*. MIT Press, Harvard, Mass.

offlap, onlap, overlap, and overstep

The terms offlap, onlap, overlap, and overstep describe the relationships of beds one to another within sedimentary sequences, especially at the margins of depositional basins, and between these strata and those that had been deformed before initiation of the basin. They are all related to changes in the nature of contacts between adjacent sedimentary bodies (*see* unconformity).

Offlap refers to the situation where successive wedge-shaped beds in a basin do not extend to the margin of the underlying bed but terminate ('feather-edge') within it. Offlap is typical of marine sedimentary sequences in contracting basins with regressing shorelines. Such situations arise where the land is rising from the sea, where sedimentation is building up new land (as in a delta), or where the sea level is declining globally. In a lake environment offlap will occur in the sediments when the area of the lake is reduced by desiccation or build-up of sediment.

Onlap is the opposite of offlap: the basin of deposition or the sea increases in area while advancing its shoreline over the surrounding terrain. Successively higher and younger onlapping beds extend beyond the former limits of lower and older beds.

Overlap is the areal extent to which successive beds increase over those beneath them, as with beds above an unconformity. The successive deposits in a marine transgression are commonly of greater area than those beneath, and they may be expected to display somewhat different characteristics (facies).

Overstep refers to the truncation of beds lying below an angular unconformity. It indicates the relationship between

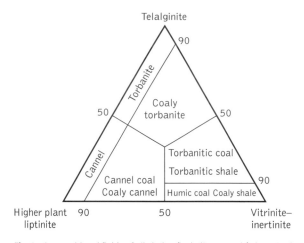

Fig. 3. Compositional fields of oil shales (including cannels) deposited under non-marine conditions where the maceral telalginite is dominant over lamalginite. (From Cook and Sherwood (1991).)

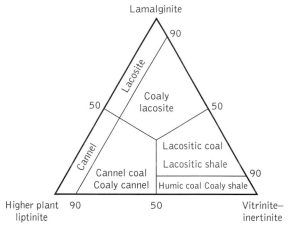

Fig. 4. Compositional fields of oil shales (including cannels) deposited under non-marine conditions where the maceral lamalginite is dominant over telalginite. (From Cook and Sherwood (1991).)

Oil-shale reserves

World-wide resources of shale oil in place are estimated to be in excess of 10×10^{12} barrels, but the level and quality of available information on reserves vary considerably for different countries. Reasonably comprehensive data are available for several countries (Table 1). The resource base of Australia is enormous, with many schemes projected for exploitation; Morocco has large resources, but has made relatively little progress so far in planning for exploitation. Jordan is another country with extensive deposits, these shales being virtually

the only fossil-fuel resource of the country; fortunately they are of quite good quality.

Notable omissions from Table 1 are Brazil, China, Estonia, and the United States of America, all of which have extensive reserves of oil shales but for which less information is available. Of the countries omitted from Table 1, Brazil was reported in 1996 as possessing 0.4 billion tonnes measured and 8.1 billion tonnes estimated of shale oil. Little information is available for China, but output of shale oil is said to be about 60 000 tonnes/year. Total deposits in Estonia, the only indigenous energy source for that country, amount to 5.9 billion tonnes; oil output in 1996 was 343 000 tonnes, of which 184 000 tonnes were exported, the greater part to Denmark. There is again a vast resource in the United States; 'in-place' shale oil in the Green River Formation amounts to 1700 billion tonnes. Recoverable shale oil from the marine black shales of the eastern United States is said to exceed 400 billion barrels.

The future for oil shales

Petroleum, natural gas, and coal are finite fossil-fuel resources. As world population increases and conventional fossil fuels diminish in amount, oil shales may eventually have to provide a substantial part of world energy. At present the costs of mining and processing oil shale are too high to compete with conventional petroleum production, but oil shales may have a brighter future. First, there is an immense resource base; second, the price of crude oil will inevitably rise as supplies decline; third, the geology of many of the major oil-shale deposits is now well understood; and fourth, many oil shales contain potentially valuable by-products (minerals and metals) that can be exploited together with the extractable hydrocarbons.

If the future for oil shales is to improve, an essential requirement is that recovery of the oil should become more

Table 1 Reserves of oil shales in selected countries with substantial deposits (at end 1996)

Country	Proved amount in place (million tonnes) (shale)	Proved recoverable reserves (million tonnes) (oil)	Average yield of oil (kg oil/ tonne)	Estimated additional reserves (million tonnes) (oil)
Morocco	58 800	3083	60	300
Thailand	18 600	1700	43	–
Turkey	1 600	20	50	10
Ukraine	2 674	300	126	6 200
Israel	15 360	600	62	–
Jordan	40 000	4000	100	20 000
Australia	69 200	3651	53	32 374

Data from *World Energy Conference Survey of Energy Resources* (1998).

cost-effective and less environmentally damaging. The simplest method of oil-shale recovery is by opencast excavation or, where the overburden is too thick, by deep-mining methods. The environmental consequences of both methods are serious, as are those from retorting and refining. The problems are widely recognized, but not easily resolved. Attempts have been made to develop potentially less costly methods of recovery *in situ* by circulating hot hydrocarbon gases through oil-shale deposits, but because of technological problems the costs of retorting *in situ* are as great as for conventional surface retorting after excavation. Other technologies, for example catalytic processing and radio-frequency heating, may well develop if interest in oil shales revives because of economic and political circumstances. Three principal factors will determine whether oil shale is to have a significant impact on the energy market in the future: reduction in the costs of producing oil from oil shale; the price of competing fuels, including the costs of methane and synthetic fuels made from coal; and the costs of dealing with environmental issues (oil and water pollution) related to shale-oil production, which currently pose horrific problems for countries such as Estonia and the Russian Federation that are producing shale oil. DUNCAN MURCHISON

Old Red Sandstone

Old Red Sandstone Early in the geological exploration of the British Isles the red-bed non-marine part of the Devonian system, known as the Old Red Sandstone, was recognized as a distinctive unit in the Welsh Borderlands and South Wales, and in parts of Scotland. Some geologists regarded it as essentially part of the sequence that also incudes the Carboniferous, but the classical studies of Sir Roderick Murchison and Professor Adam Sedgwick in England and Wales and the essays of Hugh Miller in Scotland soon established the separate identity of these rocks. Notable within these red beds was the presence of some remarkable fossil fishes.

Sedgwick and Murchison were in 1834 to establish the correlation of the Old Red Sandstone with the marine rocks in south-west England which they named 'Devonian'. The Old Red Sandstone facies has been identified elsewhere in parts of mainland Europe, Spitsbergen, Greenland, and eastern North America as post-orogenic continental basin-fills ranging in age from Late Silurian to early Carboniferous. Local volcanic centres in Scotland contributed lavas and pyroclastic rocks, but for the most part the sediments are clastics. Sedimentary environments range from piedmont fans through alluvial and flood-plain to lacustrine and deltaic-marine.

Fossils are relatively scarce; however, the agnath, placoderm, and osteichtyan fishes are locally common. Less so are fossil invertebrates, including various arthropods, ostracods, and bivalves. Plant fossils are small, primitive, and scarce in the Lower Old Red Sandstone, but rich floras include giant tree ferns in the Middle and Upper Old Red Standstone.

Similar facies with plant and vertebrate fossils are also extensive in parts of Siberia, North and South China, Australia, and Antarctica. Although the global distribution of the vertebrates shows strong provincality in early Devonian times, it was cosmopolitan by the later part of the period. All these regions appear to have been adjacent to, or within, active orogenic areas or uplifts in equatorial to temperate latitudes; and whatever barriers to migration had existed at the beginning of Devonian time, they had been overcome by the end of the period. D. L. DINELEY

Further reading

Dineley, D. L. (1984) *Aspects of a stratigraphic system: the Devonian.* Macmillan, Basingstoke.

McMillan, N. J., Embry, A. F., and Glass, D. J. (eds) (1988) *Devonian of the world.* Proceedings of the Second International Symposium on the Devonian System, Calgary, Canada.

Ziegler, P. A. (1989) *The evolution of Laurussia: a study in Late Palaeozoic plate tectonics.* Kluwer Academic Publishers, Dordrecht.

Oligocene Rocks of the Oligocene series (Greek 'small recent') were deposited over about 13 million years from the end of Eocene time at about 37 Ma to the beginning of the Miocene. It is the latest of the three Palaeogene epochs and contains two stages, the Rupelian/Stampian (below) and the Chattian (above). The series was first named by the German August von Beyrich (1854) when he recognized the deposits of an early Cenozoic marine transgression in northern Germany.

During Oligocene time the fragmentation of the supercontinent Pangaea continued with the doming of the African continent and the initiation of the East African rift valleys with flank volcanoes and central lakes. The Red Sea rift and the separation of Arabia also began. There was some rifting in parts of Europe and volcanism in central France. India was then pressing against the southern flank of Asia and causing the uplift of the Himalayan ranges. At around 30 Ma, Australia separated from Antarctica and began to move northwards. This allowed the circum-Antarctic current to begin, isolating the continent from the warmer waters to the north. The accumulation of Antarctic snow and ice soon led to the beginnings of the continental ice cap. Sea level inevitably soon started to drop with the rapid growth of this ice mass.

A number of major changes occur in both marine and terrestrial faunas at about the Eocene–Oligocene boundary. They include the replacement of entire land mammal faunas; the new arrivals included the cats, dogs, pigs, and bears. The New World monkeys appeared around 35 Ma. The high productivity of the Antarctic Ocean provided for the rise of sea mammals there, especially the whalebone whales. In the oceans, extinctions of large benthic foraminiferal groups, marine molluscs, and the ultramicroscopic nannoplankton took place; major floral changes occurred in Europe where angiosperms (flowering plants) were then dominated by gymnosperms.

The Oligocene was a time of numerous palaeomagnetic reversals, which are of some assistance in correlating deposits of the epoch.

D. L. DINELEY

onlap *see* OFFLAP, ONLAP, OVERLAP, AND OVERSTEP

ophiolite sequence The word 'ophiolite' is derived from the Greek for 'snake-rock'. This name was originally given to outcrops of dark-green shiny rocks composed of a group of minerals called serpentines. Ophiolitic rocks are common in mountain chains, that is, places where two continents have collided. In practice, because they occur in areas of continental collision, ophiolites are usually very deformed and it is rare that a complete sequence of ophiolitic rocks is evident. Study of a number of areas where ophiolites can be seen has, however, led to the construction of an idealized ophiolite sequence, which is shown in Fig. 1. This general ophiolite sequence includes, from bottom to top, sheared garnet lherzolite (thought to be part of the mantle); garnet lherzolite that has been melted; peridotites (formed when olivine and pyroxene minerals crystallize and sink out of a basaltic melt); gabbros (coarsely crystalline material of basaltic composition); sheeted dykes (thin, vertical-layered intrusive basalt sheets); pillow lavas (pillows of basalt formed when lava is erupted into water); and finally, at the top of the sequence, red cherts which are similar to sediments found on the ocean floor. The lherzolites that form the lowest part of the sequence are ultrabasic rocks composed of olivine, clinopyroxene, orthopyroxene, and either spinel or garnet.

In the 1960s, seismic profiles of the oceanic crust were obtained and a comparison was made between the three-layered seismic sequence of the ocean and a generalized ophiolite sequence. The peridotites and gabbros are equated with the thick, lower layer of the oceanic crust (layer 3); the sheeted dykes and pillow lavas with layer 2, and the sediments with the thin upper layer (layer 1): see Fig. 1.

As its name implies, oceanic crust forms the ocean floor. Oceanic crust develops by melting of the Earth's mantle where two plates separate. The ridge where new, hot, buoyant crust is forming is called a mid-ocean ridge. Where two oceanic plates collide, one is forced to sink (subduction) and is recycled into the mantle from which it originally came. How then have ophiolites, if they do represent slices of oceanic crust, been pushed up on to the land (obducted) instead of being subducted?

Measurements of the composition of ophiolites and oceanic crust indicate that ophiolites differ slightly in composition from 'real' oceanic crust. Although some modern oceanic crust is as old as 200 million years, most ophiolites are of about the same age as the mountains in which they are incorporated. These factors have led some people to conclude that ophiolites represent a type of oceanic crust that is formed in association with mountain-building events. As an old oceanic plate sinks, it creates a stretching force in the plate beneath which it is sinking, which sometimes results in exten-

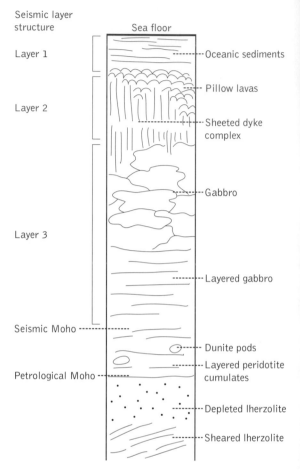

Fig. 1. A characteristic ophiolite sequence representing a vertical sequence through the oceanic crust to the Moho.

sion to form a new ridge with new oceanic crust in a back-arc basin. It is possible that ophiolites represent young back-arc basins that have been caught during the collision of two continents. Other authorities hold that the mid-ocean ridge of the closing ocean, while it is still young, hot, and buoyant, is obducted into the mountains. It seems likely that ophiolites do in fact represent slices of the oceanic crust, but whether it is the back-arc basin or an ordinary mid-ocean ridge that is obducted continues to be a matter for debate.

JUDITH M. BUNBURY

optical mineralogy Optical mineralogy is the study of the optical properties of minerals, their relation to the external crystallography and hence their structure, and their application to the identification of minerals. The principal item of equipment is the polarizing microscope (Fig. 1). This is a compound microscope with the addition of a graduated rotating stage and two polarizing filters: one, the *polarizer,* is below the substage; the other, the *analyser,* is above the objective. An auxiliary lens, the *Bertrand lens,* is fitted below the eyepiece. This account will deal only with transmitted light but reflected light techniques are used in the study of opaque minerals.

The standard technique entails the preparation of a slice of the mineral or rock 0.03 mm thick. At this thickness, most rock-forming minerals are transparent. Mineral grains may also be studied. The slice is usually mounted on a glass slide and embedded in a mount with a refractive index of 1.54.

The optical properties of a mineral are a function of its composition and atomic structure and hence of its symmetry. On the basis of their symmetry crystalline materials can be classified into seven systems. These are, in order of decreasing symmetry, the cubic, hexagonal, tetragonal, trigonal, orthorhombic, monoclinic, and triclinic systems. The cubic system is so highly symmetrical that it is in a class of its own; the other six are divided into two groups: one, the *uniaxial,* consisting of the hexagonal, tetragonal, and trigonal systems, each having a unique symmetry axis of high degree (6-, 4-, 3-fold), the other, *biaxial,* the orthorhombic, monoclinic, and triclinic systems, with lower or no symmetry.

White light can be considered as waves with wavelengths ranging from 390 nanometres (nm) (violet) to 770 nm (red) vibrating perpendicular to the direction of propagation and, in air, in all planes in that direction. Plane polarized light vibrates in only one plane perpendicular to the direction of propagation. In the cubic system the atomic arrangement is so symmetrical that light passing through the substance encounters virtually the same atomic distribution in any direction, and so the substance has the same properties in all directions. Such substances whose optical and other properties are independent of direction are called *isotropic.* Substances in the other six systems are *anisotropic,* that is, their properties are directionally dependent. One important feature differentiating cubic from all other crystals is that light entering cubic (isotropic) crystals is only singly refracted and behaves like light in air, whereas in anisotropic crystals light is, in general, doubly refracted and the two resulting rays are plane polarized in directions at right angles. They travel with different velocities and hence have different refractive indices. The difference between the two indices is the *birefringence.* Except in a very few special cases all sections of anisotropic crystals are birefringent and light passing through them is plane polarized in directions at right angles.

In many microscopes, the vibration direction of the polarizer is (looking down the microscope) 'east—west', while that of the analyser is north—south; with both polars in the system the set-up is called *crossed polars* and no light gets through to the eye so the result is blackness. If the analyser is out of the system we are using *plane polarized light.*

Properties in plane polarized light

A distinctive property is the refractive index (R. I.), which can be measured qualitatively with reference to that of the mount

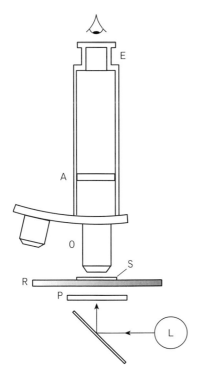

Fig. 1. Optical system of a typical petrological microscope. L, light source, P, polarizer; R, rotating stage; S, slide with thin section; O, objective lens; A, analyser; E, eyepiece lens.

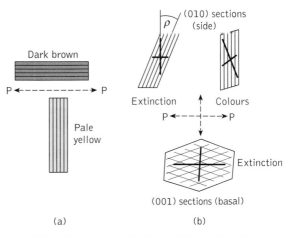

Fig. 2. (a) Pleochroism scheme for biotite. (b) Optics of hornblende: cross-hatching, cleavages; **+,** vibration directions in mineral; ρ, extinction angle; A—A, P—P, analyser and polarizer vibration directions.

755

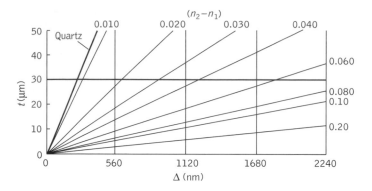

Fig. 3. Michel-Lévy birefringence chart. Thick line is for sections of standard thickness (30 μm). Example: for quartz, $n_2 - n_1 =$ 0.009, $t = 30$ μm $= 3 \times 10^4$ nm, $\Delta = 270$ nm.

(R. I. 1.54) or quantitatively by reference to liquids of known refractive index. If the mineral 'disappears', then the mineral and liquid have matching R. I.s; but if there is a large difference between mineral and mount the mineral will have a strong boundary called the *relief*. On rotation of the microscope stage stage a cubic mineral will show constant relief; if coloured, it will show no change of colour.

In colourless anisotropic minerals the relief varies as the stage is moved because the E—W light from the polarizer passes through different orientations of the mineral. Similarly with most coloured minerals there is a change of colour on rotation of the stage: two colours or tones will be seen for any one section, but if several sections are examined it is found that in biaxial crystals there are three different main colours. This colour variation is *pleochroism* (strictly *trichroism* in biaxial minerals); see Fig. 2. The mineral glaucophane, for example, has a pleochroism scheme of colourless, lavender, and sky blue. The colours can be related to the vibration directions of the mineral.

Properties under crossed polarizers

With crossed polarizers and no mineral on the stage the result is blackness. Similarly, with cubic substances even a complete rotation of the stage leaves the field completely black. In the general case, anisotropic substances will show colours, the *interference colours*; but on rotation through 360° they go black, or *into extinction*, four times. In the extinction position the vibration directions of the mineral are parallel to those of the polarizer and the analyser. If these vibrations are parallel to cleavages or edges in the crystal, then the extinction is called *straight*; if there is an angular difference, then the extinction is *oblique*.

Interference colours are displayed on a chart called the Michel-Lévy chart or Newton's scale of colours; the colours are the same as those produced by an oil slick on a wet surface. They are a function of the retardation between the two rays passing through the crystal, which is determined by the birefringence and the thickness of the slice and given by the formula D $= (n_2 - n_1)\ t$, where D is the retardation, $(n_2 - n_1)$ the birefringence, and t the thickness; D and t are in nanometres (see Fig. 3).

Each mineral has a maximum birefringence and therefore a maximum colour, but in a polygranular aggregate of a single mineral there is a range of colours from black to the maximum because of the variable orientation of the grains with respect to the plane of the slice.

Properties which can be investigated using the polarizing microscope include refractive indices, birefringence, colour, pleochroism, interference colours, interference figures, extinction angles, and the relation of these to the morphology and thus the structure of the mineral. These properties are then used for determinative purposes. R. BRADSHAW

Further reading

Gribble, C. D. and Hall, A. J. (1985) *A practical introduction to optical mineralogy.* George Allen and Unwin, London.

Ordovician The Ordovician System was originated by Charles Lapworth of the University of Birmingham in 1879 to put an end to the uncertainty and confusion regarding the boundary between the Cambrian System of Adam Sedgwick and the overlying Silurian System of R. I. Murchison. Lapworth resolved the long-protracted squabble between the two earlier geologists by grouping the upper formations of Sedgwick's system with the lower of Murchison's into a new convenient unit, based on the succession in the Arenig–Bala area of North Wales. This region was the home of the early British tribe, the Ordovices. Lapworth's creation rapidly won the approval both of British and foreign geologists. They found that fossils distinguished the new system clearly from those above and below.

The Ordovician is thus the second of the three Lower Palaeozoic systems and spans the time interval 510 Ma to 437 Ma. British usage divides the system into five series; North American practice is to recognize only three series. There are 12 named stages with two unnamed intervals. Recent work has advocated drawing the base at the base of the (graptoloid) zone of *Rhabdinopora flabelliformis s. l.* The standard British subdivisions are rather difficult to recognize elsewhere. Intercontinental correlation can to some extent be achieved by graptolites and by conodonts, with biostratigraphic zonal schemes established not only in the British Isles but also in North America, China, and Australia.

The two fossil groups mentioned above are effective for the black shale and carbonate facies, respectively, while in the fine-grained clastic sequences trilobites, brachiopods, and palynomorphs are used in correlation.

During the Ordovician period the South American and African parts of the great Gondwana supercontinent (Gondwanaland) occupied Antarctic latitudes while Eastern Gondwana, made up of Antarctica and Australia, reached northwards almost to the Equator and the Chinese continent. Laurentia, Baltica, and Kazakhstania lay in the equatorial belt with Siberia (Angaraland) close to the north of Baltica (Fig. 1). The Iapetus Ocean between Laurentia and Baltica continued to close towards the north with active continental margins on each side of the now rather narrow band of ocean crust. Vigorous igneous activity made episodic contributions to the continental crust, with island arc volcanicity and intrusions. Other mobile belts occurred in western South America, the Urals area, and parts of north-eastern Russia, eastern Australia, and China.

During the long Cambrian period many of the old land masses had become flat and low-lying from weathering and erosion. Small changes of sea level would have had far-reaching effects on the sizes and shapes of land masses. Ordovician global sea level was initially rather low but a series of transgressive rises spread epeiric seas across much of the interior of Laurentia, and there were similar floodings of other continental regions. Towards the end of the period a marked regression occurred prior to the rise in sea level of the Silurian period. These eustatic movements are thought to be related to the growth and decay of the ice cap present on the Gondwana continent. This glacial mass appears to have been centred upon Saharan Africa; it has left a remarkable array of glacially striated rock surfaces and *glaciogenic deposits*. The South Pole itself was probably occupied by western Morocco.

The shallow tidal epeiric seas of the tropical regions became the sites of carbonate (mainly limestone) depositions, especially over Laurentia. It is possible that virtually all the Canadian shield was covered by Ordovician calcareous sediments. Organic productivity in these waters was immense. Reefs appeared in several flooded continental areas. The volume of carbonate sediment deposited was prodigious and may well reflect an increase in the number of different organisms using calcium carbonate for skeletal structures. New kinds and patterns of animal communities spread across the sea floor. The articulate brachiopods were especially abundant and diverse on Ordovician sea floors. Echinoderms, especially crinoids, also were abundant. The corals underwent a radiation with both rugose and tabulate

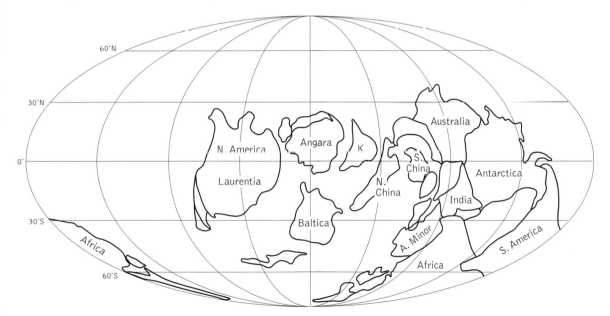

Fig. 1. The world in Middle to Late Ordovician time, about 450 Ma ago. The continents were spread largely throughout the southern hemisphere. North America (Laurentia) was the largest of the several continents to the west of the supercontinental grouping of Gondwanaland. Immediately to the east of Laurentia lay Angarland (Siberia), Baltica (north-west Europe), and the small microcontinent of which southern Britain, Brittany, and Newfoundland were parts.

Kazakhstania (K) and North China were placed off the north-western edge of Gondwanaland where the several blocks of eastern and south central Asia line the long western margins of Australia, India, and Africa. At this time Gondwanaland extended from the northern temperate latitudes to the south polar area. Extensive glaciation took place in what is now the western Sahara. (After Scotese and McKerrow).

forms contributing—with the stromatoporoids—to the building of reefs in many areas within Laurentia, Baltica, Siberia, Kazakhstan, and eastern Australia. The trilobites diversified and multiplied greatly. We know little about the planktonic and pelagic faunas, except for the graptoloids, nautiloids, and conodonts. The graptolites and the conodont animals became abundant, highly diversified, and very widely distributed, making ideal biozone index fossils. The Ordovician nautiloids possessed predominantly straight or slightly curved shells; the so-called Arctic Ordovician Fauna from Canada contains species that may have reached several metres in length.

So plentiful are many of these fossils that we tend to think of the Ordovician seas as teeming with invertebrate life. Vertebrate (fish) remains suggest small primitive armoured forms that might have been benthonic or even burrowing creatures. They have been found in near-shore sandstones in North and South America and in central Australia. Towards the end of the period the active, swimming acanthodian fishes were also present.

The crisis that beset the marine faunas at the end of Ordovician time saw great reductions in the numbers of reef-forming organisms such as the bryozoa, brachiopods, crinoids, the pelagic nautiloids, trilobites, conodonts, and graptolites. Comparatively rapid cooling of the surface waters around the continents is held to be the most probable cause of these extinctions. The ice cap covering the central and western Saharan area was sufficient to bring about the lowering of global ocean temperatures to a point there they caused real harm to the shallow-water communities. Recovery followed the dawn of the Silurian period.

D. L. DINELEY

Further reading

Bassett, M. G. (ed.) (1976) *The Ordovician System, Proceedings of the Palaeontological Association Symposium, Birmingham, September 1974.* University of Wales Press and National Museum of Wales, Cardiff.

Cocks, L. R. M. and McKerrow, W. S. (1978) Ordovician. In McKerrow, W. S. (ed.) *The ecology of fossils*, pp. 2–92. Duckworth, London.

McKerrow, W. S. Dewey, J. F. and Scotese, C. R. (1991) *The Ordovician and Silurian development of the Iapetus Ocean.* Special Papers in Palaeontology 44, 165–76.

McKerrow, W. S. and Scotese, C. R. (eds) (1990) *Palaeozoic palaeogeography and biogeography.* Geological Society of London Memoir No. 12.

ore deposits Ores are naturally occurring rock materials consisting of one or more minerals from which a valuable constituent or constituents can be extracted for profit. The locations in the Earth's crust where these accumulations occur are collectively referred to as *orebodies*, *ore reserves*, or *ore deposits*. The definition of an ore, in the strictest sense, refers only to mineralized rocks that can be profitably mined.

The term is, however, generally applied by geologists to any anomalous accumulation of a valuable constituent or constituents that may have this potential. Whether a deposit can in fact be mined for a profit will depend on such factors as the market price of the constituent, mining costs, taxation, and government environmental policy. The valuable constituents that are extracted from particular ore deposit can include one or more of the metals (e.g., gold, copper, uranium), non-metals (e.g. sulphur, selenium, tellurium), or gemstones (e.g., diamonds and emeralds). The minerals that contain the valuable constituents are called *ore minerals*; barren minerals associated with these ore minerals are *gangue minerals*. The presence of an unusual concentration of ore minerals, as compared with the local host rocks, constitutes mineralization, and rocks are said to be *mineralized* where such a concentration is found.

Many classification schemes for ore deposits have been developed, based variously on morphology, metal association, host rocks, or mode of formation. Many of the older classification schemes relied heavily on the mode of formation. These schemes have consequently become obsolete as ideas about the genesis of certain types of ore deposit have changed. The more recent classifications are likely to be more enduring, since they are based more on host-rock association, with less emphasis on genetic factors. The best classifications are those that provide useful insights into the characteristics of newly discovered mineralization, enabling it to be more efficiently delineated and profitably mined.

Two terms that are still commonly encountered when discussing the classification of ore deposits are *syngenetic* and *epigenetic*. Ores formed at the same time as their host rocks are referred to as *syngenetic*; ores formed after the host rock are called *epigenetic*. There is a general consensus on whether most types of ore deposit are syngenetic or epigenetic, but debate continues about some. Some ore deposit types, such as volcanogenic massive sulphide deposits, can contain both syngenetic and epigenetic mineralization.

The shape of an ore deposit can have a variety of forms that can reveal much about its genesis. Much epigenetic mineralization is found in tabular-shaped zones called *veins*, where, ore and gangue minerals have been deposited from heated solutions in vertical, horizontal, or inclined fractures in pre-existing host rocks. Veins can range in width from microscopic dimensions to many metres. In many instances the density of veining determines the viability of an ore deposit. If many cross-cutting veins are present, they form *stockwork mineralization*, as is common in many porphyry copper deposits. Epigenetic deposits can also occur as tubular bodies or pipes as well as replacement bodies. Here the minerals in the original host rock have been wholly or partially replaced by the ore and gangue minerals, as in skarn deposits. Syngenetic mineralization commonly occurs as beds that lie conformably within the layers of the host rocks and hence are called *stratiform deposits*.

Magmatic deposits

Magmatic deposits are those formed during the crystallization of a magma; they are therefore syngenetic deposits. Two processes can lead to the concentration of valuable commodities: liquid immiscibility and fractional crystallization. Major deposits of nickel, copper, and platinum-group elements are the result of liquid immiscibility, a process in which a sulphur-rich liquid separates from a cooling silicate magma. These metals tend to migrate from the silicate magma into the sulphide. This sulphide liquid is much denser than the surrounding silicate-rich magma, and it settles and becomes concentrated at the base of the magma chamber. Major deposits of this type are found in Sudbury, Canada; Noril'sk-Talnakh, Russia; and Kambalda, Australia, as well as in numerous smaller occurrences throughout the world. Fractional crystallization also occurs during the cooling of a magma. Early minerals, such as the chromium-bearing mineral chromite, crystallize early and are denser than the surrounding magma. They settle and are concentrated in layers within the magma chamber. These de-posits are high-grade sources of iron, chromium, titanium, vanadium, and platinum-group elements.

Another rare type of magmatic ore deposit is that associated with carbonatites, which are unusual igneous rocks composed primarily of carbonate minerals such as calcite, dolomite, and ankerite (a calcium magnesium iron carbonate). These deposits are major sources of tantalum, niobium, rare-earth elements, and vermiculite. Magmas generated at great depths are the ultimate source of diamonds. When these magmas erupt violently near or at the Earth's surface, they generate kimberlites such as are found in South Africa. Kimberlites occur in many parts in the world, but only rarely do they contain enough diamonds for them to be worth mining.

Hydrothermal deposits

Ores generated by the actions of high-temperature fluids are grouped into a large class known as *hydrothermal deposits*. This broad group contains many sub-types relating to how and when they were formed. The term *hydrothermal* refers to the fact that the fluids responsible were composed primarily of hot water containing small amounts of carbon dioxide, sulphur, sodium chloride, and other components. Valuable metals are dissolved in these fluids and are subsequently deposited as a result of changes in the temperature, pressure, or chemical composition of the fluid. Ores that are the direct result of the interplay of high-temperature fluids and intruding magma bodies include pegmatites, skarns, and porphyry-type deposits. Pegmatites are igneous rocks composed of the late differentiates of large magma bodies. As these magmas cool, certain constituents are not fully incorporated in the crystallizing minerals. These become increasingly segregated in the remaining magma, which has a high content of water and silica. This late differentiate is forced into fractures and tabular bodies and cools to form pegmatoidal rocks. Peg-matite orebodies are rich sources of metals such as lithium, beryllium, and the rare-earth elements. Skarn deposits form when magmas are intruded into reactive host rocks such as limestone. The high temperature of the magma and its exsolved fluids causes extensive reactions with the wall rocks and results in the deposition of metals. Many varieties of skarn deposit are found throughout the world and are mined for numerous metals including iron, copper, molybdenum, tin, tungsten, lead, and zinc.

Porphyry-type deposits are the world's primary source of copper and molybdenum. They comprise very large ore-bodies, which normally range from hundreds to thousands of million of tonnes of ore. Gold can also be an important metal in these deposits and can make an orebody economically mineable when it would not otherwise be worth mining. Porphyry copper deposits are formed when a magma body is intruded close to the Earth's surface. The small proportion of water in the magma remains dissolved at depth because of great pressures. As the magma rises, cools, and begins to crystallize, the water can no longer remain in the melt and is expelled to form a high-temperature concentrated brine. It is generally accepted that this brine contains high concentrations of ore metals. When it rises through fractures in the surrounding solid rocks, these metals are precipitated, normally as the minerals chalcopyrite, bornite, and molybdenite. The cooling intrusion also heats the local water contained in fractures in the cooler host rocks, causing this water to expand and become less dense. Hydrothermal convection cells develop, causing further alteration of the rocks and redistribution of the valuable metals.

Although porphyry deposits are intimately associated with an intruding magma body, there are other types of hydrothermal deposits that appear to be formed as a result of magmatic activity, albeit at a greater distance from the magma. Epithermal deposits are a major source of gold, silver, copper, lead, and zinc. The ore minerals of these metals are deposited in veins or by replacement of the host rocks. The term *epithermal* is applied because these deposits form near the Earth's surface from hot fluids at temperatures that are generally lower than those encountered in porphyry-type deposits. These epigenetic deposits can be divided into two sub-types according to the ore metals present and the alteration minerals found in the host rocks. *High-sulphidation epithermal deposits* contain high concentrations of gold, silver, and copper and are formed when gases at high temperature from a nearby cooling intrusion mix with local groundwaters. Rapid chemical changes occur in the groundwater, generating hot fluids that are extremely acidic and cause intense alteration of the host rock. In some instances the alteration is so intense that all that remains of the host is a porous siliceous cap rock. Chemical and isotopic evidence show that a major component of the fluid involved in these deposits is derived directly from the magma, which also appears to be the source of the ore metals. In *low-sulphidation epithermal deposits*, the fluids are less acidic, and alteration is less intense. These deposits also host major concentrations of gold, silver, and copper. They differ from high-sulphidation deposits in that they can also contain mineable amounts of lead and zinc. The source of fluids for low-sulphidation deposits appears to be mainly from local groundwater which becomes heated by a distant magma body. Economic geologists continue to debate whether this magma body supplies ore metals and other constituents to the mineralizing fluid.

Vein deposits that form at greater depths from fluids at higher temperature than those associated with epithermal deposits are termed *mesothermal deposits*. These deposits generally consist of large groupings of veins that extend for great depths in highly altered host rocks. The major commodity of interest obtained from mesothermal deposits is gold, but they are also mined for silver and, in some rare instances, for tungsten. Many of the world's largest gold deposits, such as those in Canada, South Africa, Australia, and India are classified as mesothermal deposits. One common feature of these deposits is that they are emplaced in low-grade metamorphic rocks such as greenschists and were formed by hydrothermal fluids rich in carbon dioxide and sulphur. The exact source of the fluids and metals for these deposits remains a matter for debate. Genetic models calling upon magmatic fluids, metamorphic fluids, and even deep circulating groundwater have been suggested. The source of gold in mesothermal deposits continues to be a matter of intense discussion among economic geologists.

In contrast, volcanogenic massive sulphide deposits provide an example of how scientific investigations have revealed the undisputed origin of a major type of ore deposit. These orebodies are composed of massive accumulations of sulphide minerals that are hosts to high concentrations of copper, lead, zinc, and can also yield mineable concentrations of silver and gold. They are generally tabular in form, but many of them have been disrupted by metamorphism and folding of their host rocks. These are invariably volcanic rocks, including rhyolites, andesites, basalts, and their explosive equivalents, pyroclastic rocks. Scientific debate originally focused on whether these deposits were sea-floor deposits or were formed as a result of replacement of the volcanic hosts by the action of hydrothermal fluids during metamorphism. The discovery of 'black smokers' on mid-ocean ridges in the late 1970s showed the former model to be correct. Black smokers form on the sea floor at spreading plate boundaries, such as the Mid-Atlantic ridge. In these areas of the Earth's crust, magmatic rocks lie only a short distance below the sea floor. The magmas heat sea water contained in the overlying volcanic rocks, causing convection cells to develop. These convection cells continuely circulate the sea water, which becomes heated and reacts with the volcanic rocks, leaching out metals such as copper, lead, and zinc. The high-temperature fluids ultimately reach the sea floor and are expelled. The immediate drop in temperature and mixing with cold sea water causes metal sulphides to be deposited on the sea floor, building up thick mineralized layers in the course of time. These accumulations eventually become covered by further volcanic eruptions and mineralization ceases.

Carbonate lead–zinc deposits are an example of epigenetic deposits for which hydrothermal fluids at rather low temperatures are responsible. The host rocks for these deposits include limestone and dolomite that have undergone dissolution by low-temperature fluids, either before or during the mineralizing event. This karsting provides highly permeable rocks in which the sulphide minerals sphalerite and galena represent the valuable mineralization. The metals in carbonate lead–zinc deposits are thought to have been carried by fluids expelled from deep sedimentary basins adjacent to the carbonate sediments. Debate continues, however, on the source of the sulphide that is necessary to precipitate the metals from the fluids when they reach the host rocks.

Ores formed at the Earth's surface

All the ore deposit types described above are examples of primary mineralization in which the ore metals were carried from a dispersed source and concentrated in a relatively small area of host rocks. At the surface of the Earth, exposure of such types of ore deposit to erosion can lead to secondary mineralization. Many ore minerals are not chemically stable when exposed to air and surface waters. This causes them to decompose, which results in the production of large amounts of acid. Acidic groundwaters with low pH values are produced and react with the primary minerals, releasing more metals. These become concentrated in the water table when the groundwaters move downwards and are neutralized by water–rock reactions forming *supergene mineralization*. Some minerals, such as gold and platinum, are particularly resistant to chemical breakdown; when host rocks rich in these minerals are eroded, they are released unharmed. The action of water carries these into streams and rivers. These minerals have much higher densities than the more common minerals, and this causes them to be locally concentrated in the stream and river deposits. Such placer deposits generated the great gold rushes during the nineteenth century in the Yukon and California.

Residual deposits form when unmineralized common rocks are weathered by acidic surface waters. These waters are generated in areas with warm climates with high rainfall and dense vegetation. Decaying vegetation releases large quantities of organic acid, which descends through thin soils and attacks the bedrock. The acid solutions dissolve out many of the major components of the rock, leaving only the most immobile elements such as silicon, aluminum, and iron. Over thousands of years, thick accumulations of residual deposits can accumulate if erosion is minor. Perhaps the most important example of residual deposit is bauxite, which is, at present, the only economically mineable ore of aluminium. Other residual deposits are mined as sources of iron and nickel.

BRUCE W. MOUNTAIN

ore minerals Ore minerals are economically valuable minerals that through various methods of mining and mineral extraction can be separated from waste and processed to yield valuable commodities. This definition is used, in the strictest sense, by economic geologists and mining engineers to apply only to those minerals that contain the valuable commodity of interest. For example, they do not consider that worthless iron sulphides associated with valuable copper sulphides in a copper deposit are ore minerals. However, the term 'ore minerals' is also encountered in petrography, where it refers to any mineral that is opaque to transmitted light when examined in thin section. Ore minerals are found in several of the main chemical classes into which minerals can be classified, including native elements, sulphides, oxides, silicates, arsenides, phosphates, and carbonates.

The simplest ore minerals are the native metals in which the mineral is composed of a single element. The best-known example in this group is native gold, which is the most common ore mineral of this metal. Other metals for which the native form is an important ore mineral include silver, copper, and bismuth. Two or more metals can also combine to form a group of ore minerals called *alloys*. Electrum, an alloy of gold and silver, is a common mineral in gold–silver deposits, and many of the metals of the platinum group combine with each other, as well as with other metals, to form complex alloys.

The sulphide minerals comprise an important group of ore minerals in which one or more metals combine with sulphur. Simple sulphide minerals are formed when a single metal combines with sulphur. For example, sphalerite and galena (the simple sulphides of zinc and lead), are the world's primary source for these metals. Other rarer simple sulphides are important sources of antimony, arsenic, bismuth, copper, mercury, molybdenum, nickel, and silver. Sulphide minerals in which two metals combine with sulphur are also common ore minerals. Chalcopyrite, a combination of copper, iron, and sulphur, is an important ore mineral for copper and is found in many of its deposits. Other more complex sulphides, formed by the combination of two or more metallic or metalloid elements with sulphur, are called *sulphosalts*. These minerals are less common than the simpler sulphides and are important as ore minerals in only a few deposits.

Many oxide minerals also constitute economically important ore minerals. Simple oxides, in which a single metal combines with oxygen, are the major ore minerals of iron (magnetite and haematite), tin (cassiterite), manganese (pyrolusite), titanium (rutile), and uranium (uraninite). Hydrated oxides, in which water is incorporated into the chemical structure of the mineral, also provide important sources for several metals, including iron (goethite) and manganese (manganite). Bauxite, the major ore of aluminum, is composed of a mixture of hydrated aluminum oxides (gibbsite, diaspore, boehmite).

Although many metals combine with silicon and oxygen to form silicate minerals, these cannot in general be refined economically because of the large amounts of energy required to extract the metals. There are, however, some metals for which silicate minerals are an important source. For example, the major ore minerals for lithium and beryllium (spodumene and beryl) are silicate minerals.   BRUCE W. MOUNTAIN

organic acids in geochemistry Organic acids are integral to many geochemical processes in the atmosphere, soils, sea water, sedimentary basins, and hydrothermal systems. Atmospheric levels of carboxylic acids are generally not more than 3 parts per billion (ppb) in uncontaminated clouds, fog, and rain, but can reach 5 parts per million (ppm) in polluted systems. Natural levels of carboxylic acids in river and lakes are generally not more than 1 ppm, but pollution can raise levels in river water to 1000 ppm. More complex organic acids (humic and fulvic acids) can reach relatively high concentrations in surface water, and in rivers that drain swampy areas the concentration of these acids can be more than twice the concentration of inorganic solutes in the water. These rivers often have acidic pH values because of the abundance of organic acids. Sea-water levels of carboxylic acids are generally more than 1 ppm, but concentrations in microbially active marine sediments can reach 150 ppm. The sources of organic acids in surface and near-surface environments are biological processes of micro-organisms and plants. In addition to natural sources, organic acids enter the environment through industrial processes, or as a result of microbial attack of organic compounds in landfills, petroleum spills, and other locations of organic pollution. It is thought that organic acids are generated at higher temperatures by reactions between water and organic matter, including petroleum. These reactions have been studied experimentally, but much remains to be understood about the production of organic acids by micro-organisms at elevated temperatures in sedimentary basins. Organic acids have been shown to increase the solubility of minerals and form strongly bound complexes and chelates with many metal cations. These complexes can account for a large percentage of some trace metals in rivers, groundwater, and sea water, but are typically of less importance in concentrated sedimentary-basin brines, even though the concentrations of carboxylic acids can approach 10 000 ppm.   E. SHOCK

organic geochemistry Organic geochemistry entails the application of chemical techniques to the study of organic matter and the processes which transform organic matter in the geosphere. Although seemingly a modern discipline with a heavy reliance upon state-of-the-art analytical technology, organic geochemistry was established in the pioneering work of scientists studying the composition of oils and coals in the first half of the twentieth century.

Organic matter in the geosphere is largely biological in origin, and the organic compounds which comprise it are fundamentally based on the element carbon. Most of the carbon on Earth is to be found in the sedimentary rocks of the crust; and of this, about 20 per cent is in the form of organic carbon and the remainder as carbonates. A key concept in organic geochemistry is the global organic carbon cycle. Organic compounds are synthesized by organisms using carbon dioxide or existing organic molecules, and after death their organic matter is a reactive phase in the environment. This organic matter is mostly scavenged or degraded by microbial action, but a relatively small proportion of the organic carbon flux escapes the otherwise efficient carbon cycle, and can be preserved in sediments and sedimentary rocks. Progressive burial and maturation of this preserved organic matter can give rise to petroleum and natural gas, which, if geological conditions are suitable, may accumulate in reservoirs suitable for exploitation. Petroleum geochemistry is an important aspect of organic geochemistry. In the past, petroleum geochemistry has been concerned with the integration of organic geochemistry with other approaches in the exploration for petroleum reserves, but increasing emphasis is now being placed on the composition of petroleum itself, leading to a better understanding of oil generation, migration, and accumulation in reservoirs. Petroleum geochemistry now focuses on the discovery of migration pathways and satellite oilfields and on improved recovery of emplaced petroleum.

The bulk of organic matter in sedimentary rocks is in the form of kerogen: complex polymeric material that is insoluble in inorganic solvents, and difficult to analyse chemically. The precise mechanism of formation of kerogen from the biological precursor material is unclear, but it may entail both the preservation of intact (or only slightly modified) biological polymers and condensation reactions between products of diagenetic decomposition of biological material, perhaps associated with mineral surfaces. Current approaches to determining the detailed chemical composition of kerogen include breaking down the polymeric material by heat or chemical reagents, followed by analysis of the products by gas chromatography–mass spectrometry (GC–MS for short). This analytical technique is of fundamental importance in organic geochemistry, being the standard method for determining the molecular composition of organic matter; it is particularly important in the study of biomarkers. Owing to its relative ease of analysis, the organic solvent soluble fraction of sedimentary organic matter, bitumen, has been studied most widely, although newly developed analytical technology is opening the way for more detailed investigations of total sedimentary organic matter. The molecular composition of bitumen in geological samples has been shown to preserve information of value in the assessment of palaeoceanographic or palaeoclimatic conditions, and for archaeological and palaeontological and taxonomic studies. The molecular composition of the quantitatively more important kerogen will

prove equally, or even more, instructive in the future. The power of molecular organic geochemistry has advanced further with the application of isotopic analysis to individual organic molecules, making them even more specific tools for studying processes influencing organic matter in the geosphere.

While much research in organic geochemistry addresses the fate of biologically derived organic matter in the geosphere, an increasingly important branch of the subject is concerned with understanding the transport and fate of organic pollutants (i.e. man-made organic chemicals) in the natural environment. Another interesting facet of organic geochemistry addresses the origin of life on Earth; and among other approaches the nature and origin of organic molecules in extraterrestrial material (e.g. meteorites) has been of long-standing interest.

It will be apparent that organic geochemistry is a discipline of the Earth sciences which overlaps into the fields of biology and biochemistry (the source and composition of organic matter), microbiology (decomposition of organic matter in sediments), environmental sciences (organic pollutants), planetary science (organic matter in meteorites), oceanography, climatology, and archaeology, among others. Its scope is wide indeed.   PAUL FARRIMOND

Further reading

Engel, M. H. and Macko, S. A. (1993) *Organic geochemistry: principles and applications.* Plenum Press, New York.

Killops, S. D. and Killops, V. J. (1993) *An introduction to organic geochemistry.* Longman, Harlow.

organic matter (kerogen) The bulk (about 95 per cent) of the organic matter held in the Earth's crust is insoluble, amorphous organic material composed of a variety of polymers. This material is often termed 'kerogen'. In young, immature sediments it is rich in hydrogen and contains other elements such as oxygen, sulphur, and nitrogen in lesser amounts. In old, mature sediments it is highly carbonaceous and commonly graphitic, and is very low in hydrogen and elements other than carbon. Whatever its state, kerogen is almost entirely derived ultimately from biological debris and when immature contains component pieces of biomacromolecules and smaller metabolites cross-linked into the kerogen structure by carbon–carbon, carbon–sulphur, and other bonds. Experiments specifically designed to release molecular entities from this heteropolymeric material have shown that the kerogen matrix has protected some components from the diagenetic changes experienced by free unbonded biomarkers in the same sediment. However, the original biosynthetic order incorporated in the molecular components of the kerogen is gradually obscured and destroyed by diagenetic and maturational processes. Deep burial with accompanying rise in geothermal temperature, pressure, and the passage of time eventually erases almost all the original biosynthetic order as the carbonization proceeds. For example, the biosynthetic dominance of odd over even carbon numbers in the C_{30} region of plant wax alkanes is gradually replaced by a smooth distribution at lower carbon numbers as the molecules are broken down through thermally induced carbon–carbon bond breakage. The same maturational processes operating on organic-rich sediments, such as black shales, release major quantities of carbon dioxide, natural gas (methane, etc.), and petroleum.

GEOFFREY EGLINTON

Further reading

Hunt, J. M. (1996) *Petroleum geochemistry and geology* (2nd edn). W. H. Freeman, New York.

origin of life: geological constraints

Irrefutable evidence about how life originated does not exist. Hypotheses regarding its origin, however, are plentiful. Ideas that fuelled research for most of the twentieth century started from the premise that life emerged with a heterotrophic metabolism (that is, one requiring organic carbon) that was able to consume a 'prebiotic' soup of organic compounds that were previously generated by the input of (largely) external energy into the atmosphere. The postulated atmosphere was nothing like that of the present-day Earth, but rather a mixture of reduced gases (methane, ammonia, water, and hydrogen) whose origin was assumed to be consistent with late differentiation of the core after a relatively cool accretion of the planet. The tremendous early success of organic synthesis from reduced gas mixtures using spark discharges and other energy sources strengthened the case for a 'prebiotic' soup into a paradigm for the origin of life.

Many advances in the latter half of the twentieth century challenged this paradigm. The slow differentiation model for the Earth is now largely abandoned in the light of evidence from planetary exploration, dynamic simulations of nebular formation, and isotopic geochemistry of mantle, meteorite, and lunar samples. Prevailing models have the Earth forming through violent accretionary processes that drove simultaneous melting, differentiation, and release of volatiles. Plate tectonics has emerged as a paradigm in the Earth sciences after the conclusive demonstration of the continuing dynamic behaviour of the mantle. Even though there has been extensive recycling of the crust, co-ordinated studies by geologists, palaeontologists, geochronologists, and geochemists have revealed over 4 billion years (4 Ga) of recorded history of Earth processes and life in unprecedented detail. During the 1970s and 1980s, atmospheric chemists showed that ultraviolet photolysis of water vapour yields hydroxyl radicals that rapidly react with methane and ammonia to yield carbon dioxide (CO_2) and nitrogen (N_2). The possibility of a greater ultraviolet flux from the young Sun means that at the relevant time reduced gases were doomed to evanescent lifetimes in the presence of water vapour. As a consequence, methane and ammonia were unlikely starting materials for organic synthesis on Earth in the absence of any form of life. Simultaneously, advances in genetics led to modern molecular phylogenies which reveal that the most recent common ancestor of all life was not likely to have been a heterotroph, requiring its carbon in organic form, or a photosynthetic autotroph, able to utilize inorganic compounds by means of sunlight but rather a chemosynthetic autotroph—the type of organism that synthesizes biomass from inorganic starting materials like carbon dioxide. So, the evidence from life itself demonstrates that the reduction of inorganic carbon oxides to form organic compounds is more fundamental than organic synthesis from methane. These developments require a new set of geologically consistent ideas about how life began.

The early Earth and the geological record of life

The early Earth was a violent place in which any form of life would seem to have been impossible. Accretionary energy provided enough heat, probably enhanced by short-lived radionuclides, to drive differentiation of the core, mantle, and crust. Some theoretical models of these processes indicate the likelihood of a magma ocean extending to considerable depths in the mantle. If the Moon formed through a giant impact, the cataclysmic nature of this process would have vaporized silicate minerals. Using the record of lunar impacts as a guide suggests that the period of bombardment by large objects (100 km or more in diameter) continued for up to 700 million years (Ma) after the Earth had formed, although it is difficult to be sure of the frequency with which such events occurred. Impactors of about 450 km in diameter are estimated to be large enough to vaporize a volume of water equivalent to the present oceans. This would have provided a major perturbation in the evolution of life at the Earth's surface.

If sterilizing events occurred, more than one origin of life was required. At the very least, major impacts generated bottlenecks for evolution through which relatively few organisms could pass. Those that did survive and found themselves at the surface would have been faced with an extremely hostile environment. Not only would the flux of ultraviolet radiation from the young Sun have been unmitigated by an ozone shield, but it may have been many orders of magnitude greater than at present. At the same time, the Sun's luminosity at visible wavelengths was probably feeble by comparison. Nevertheless, the record of microfossils shows that surface or near-surface environments were conducive to life within about 700 million years after the Earth formed. This suggests that life became established very early on as a normal process on the Earth.

Owing to the inadequacies of the geological record, which is often the product of chance preservation of crustal material on an extremely active planet, the timing of events on the early Earth is open to considerable uncertainty. Nevertheless, there are several facts that help in following the transition from the Earth's dramatic origin to conditions that could support life. The oldest known preserved portions of crust are in the Acasta gneiss complex in the Slave province of Canada and are about 4.03 billion years (Ga) old. These rocks are deformed granites (tonalites) that bear indications of having formed through intracrustal reprocessing of mantle-derived magma, juvenile crust, and older, hydrated mafic crust. The inference has been made that this recycling probably occurred in a subduction-related setting. In other words, indications from the oldest known rocks are that normal geodynamic and geochemical processes were at work in the mantle and crust, and had been at work for some time.

Individual zircon grains with ages up to about 4.3 Ga found in younger rocks are the only older terrestrial materials preserved. Trace-element signatures and mineral inclusions in these zircons indicate that they formed from granitic melts, again suggesting that familiar crust-forming geological processes, including subduction and something similar to plate tectonics, were under way less than 300 Ma after the formation of the Earth. It is tempting to conclude that familiar crustal geochemical processes like weathering, sediment accumulation, diagenesis, hydrothermal circulation and alteration, and metamorphism were also active as early as this in Earth history. If so, processes on the early Earth may have been quite similar to processes that are observable at present.

Evidence for life in the rock record does not yet extend back as far as the oldest surviving pieces of continental crust. Nevertheless, there are well-preserved fossils of filamentous microbes from the Apex chert of Western Australia that are about 3.465 Ga old, additional filamentous fossils from the underlying Tower formation (3.556 ± 0.032 Ga) for which the exact provenance is unclear, and filamentous microfossils from the Onverwacht group of South Africa of comparable age (3.540 ± 0.030 Ga). Additional evidence for active biological processes earlier in the rock record comes mainly from carbon isotopic measurements on the metamorphosed sediments of the Isua supracrustal rocks of western Greenland, which are about 3.8 Ga old. The biogenicity of the stable carbon isotopic signatures obtained from these rocks has been argued for many years. Ion microprobe measurements on the scale of individual mineral grains provide supporting evidence for such origins.

Molecular evidence for early life

Developments in molecular biology have made it possible to sequence rapidly biomolecules made of multiple monomers and to extract the evolutionary information in biomolecular sequences. Proteins are made of twenty amino acids, and nucleic acids are made of sequences of four nucleotides. The order of amino acids in proteins is recorded in the genetic code housed in the nucleic acids, DNA and RNA. Over the course of evolution the genetic code changes in response to environmental stresses or through random mutations. The order of amino acids or the order of nucleotides therefore provides the basis for comparisons between biomolecules from diverse organisms that can be interpreted in terms of their evolutionary relatedness. If enough information is available from enough organisms, then a phylogenetic tree can be constructed mathematically that shows the relatedness of the organisms considered.

One example of this type of mathematical model is the universal phylogenetic tree based on the sequences of RNA taken from the ribosomes of various organisms. Ribosomes are the locations in cells where proteins are made. Much of the genetic information is conserved in ribosomal RNA. Errors would otherwise accumulate in proteins to levels that would disable enzymatic function. It follows that ribosomal RNA provides a solid candidate for tracing the long-term changes wrought by evolution. When this is done the tree that emerges has several striking features, including the observation that the shortest and deepest branches in the tree are populated by high-temperature (thermophilic and hyperthermophilic) organisms.

The deepest branches of all are populated by hyperthermophiles that do not conduct photosynthesis, nor are they directly dependent on a source of organic compounds. Instead they thrive by making their own organic compounds from inorganic starting materials. The reaction pathways that these *chemoautotrophs* employ involve the reduction of inorganic carbon compounds to form organic compounds, which include their biomolecules. Analysis of the energetics of these processes shows that biosynthesis of organic compounds in these high-temperature habitats releases energy, making these conditions radically different from the familiar conditions in which we live, but nevertheless much more conducive to life. For example, plants in an oxygenated atmosphere have to perform thermodynamic work against their surrounding chemical systems in order to reduce carbon dioxide (CO_2) with water (H_2O) to form carbohydrates. They do this by harnessing sunlight. Many hyperthermophiles live in habitats where geological processes generate geochemical disequilibria in the C–H–O–N chemical system. As a result, organic synthesis lowers the energy level of the system, avoiding the need for external energy sources. It follows that organisms of the type that populate the shortest and deepest branches of the universal phylogenetic tree are independent of photosynthesis and have no need to be at the surface of the Earth. It is therefore worth considering whether the emergence of life could have been an internal process of the planet. This makes it a geological problem.

Ideas old and new

The concept of a 'prebiotic' Earth popular in many textbooks is substantially based on the success of experiments in chemistry laboratories rather than on observations of natural systems. In consequence, as more is learned about the processes leading to the formation of the Earth, it has become increasingly difficult to connect the features common to popular concepts of a 'prebiotic' Earth with present knowledge of the geochemical, geophysical, and geodynamic constraints on the early Earth. This mismatch is a profound flaw in efforts to apply what has been learned in the laboratory to what may actually happen in nature.

Before research workers had attained their current appreciation of planetary formation through accretionary processes, the view held by many was that life originated in a rather benign environment. Many researches followed Darwin's lead in imagining a 'warm little pond' as the habitat conducive to the origin of life.

The success of abiotic organic synthesis experiments since the early 1950s helped to turn the idea of the 'warm little pond' into a 'prebiotic soup'. In the decades of research that followed the early success of abiotic organic synthesis experiments it was often assumed that there would have been a source for almost any organic compound of interest. This type of thinking continues in popular notions like the RNA-world. The discovery that some RNA molecules have catalytic as well as genetic abilities has lead to the idea that a single type of biomolecule may have predated the DNA–protein dichotomy in modern life. This exciting notion is easy to accommodate in the soup hypothesis, because the thorny issue of the source of the RNA does not have to be addressed.

Although evidence is lacking in the geological record to support the 'soup' hypothesis or the 'prebiotic' Earth, the sketchy nature of that record is often cited in defence of these ideas, usually with the comforting notion that absence of evidence is not evidence of absence. Indeed, lines of evidence from other planetary bodies and objects, such as the wealth of organic compounds in carbonaceous chondrites or the smoggy haze of Titan, are produced to support the soup hypothesis despite the fact that the histories of these objects are quite distinct from that of the Earth. If anything, analysis of carbonaceous meteorites, interplanetary duct particles, pre-solar grains, cometary exploration, and astronomical measurements of the interstellar media and nebulae show that organic compounds are present and even abundant nearly everywhere. Available evidence leads to the conclusion that organic synthesis is a normal process throughout the solar system, but that it appears to be predominantly abiotic and may or may not tell us much about the origin of life.

The search for alternative pathways of organic synthesis consistent with what is known about the formation and early evolution of the Earth is a nascent field of research. Much of the search has focused on hydrothermal systems, which—as the engines that generate many ore deposits—are widely recognized sources of geochemical disequilibria. They are also inevitable consequences of magmatic and impact activity in the presence of liquid water. Early ideas were associated with the discovery of high-temperature submarine hydrothermal systems in the mid-1970s. These were followed by additional arguments and theoretical models that showed the potential for organic synthesis during water–rock reactions hosted in basalt and other sea-floor rocks.

Hydrothermal experiments showed that abiotic organic synthesis was plausible, although many early experiments were conducted under conditions unlike those in hydrothermal systems on the Earth. Extreme examples of mismatch between the laboratory and nature gave reason for some investigators to be highly dubious of the potential for hydrothermal organic synthesis. However, experiments constrained to be more like natural hydrothermal processes show great success in generating and transforming a wide variety of organic compounds. Theoretical models show an enormous potential for organic synthesis from carbon dioxide and hydrogen as hydrothermal fluids mix with sea water, or from carbon dioxide or carbon monoxide as volcanic gases mix with crustal aqueous fluids and the atmosphere. In addition, there is growing recognition of abiotic organic synthesis in various geological materials. Indeed, the success of experiments and models supports the notion that abiotic organic synthesis is a continuing process on the Earth. Perhaps we do not perceive its importance because it is less efficient than biological synthesis or geochemical transformation of biologically generated organic compounds.

The success of hydrothermal organic synthesis experiments opens up the first geologically plausible pathway from abiotic to biotic processes. Facile hydrothermal abiotic organic synthesis could provide a steady source of compounds for condensation and polymerization reactions. Many researchers have shown that construction of biomolecules can be facilitated by adsorption on to mineral surfaces, some of which may be catalytic for reactions involving the reduction of carbon oxides or the polymerization of monomers. Even the RNA-world could be accommodated in hydrothermal systems, given the proclivity of nucleotides for minerals surfaces and their enhanced reactivity at higher temperatures. Favourable geological conditions for the synthesis of organic compounds might also favour biosynthetic pathways, especially if easily formulated substances provided low-tech versions of the catalytic functions now conducted by enzymes. An environment sheltered from impacts and extreme UV radiation, complete with a steady supply of abiotically generated organic compounds, would even be an extremely suitable habitat for an early heterotroph. Although possible, the heterotroph-before-autotroph hypothesis requires a reconfiguration of the deepest branches of the phylogenetic tree or considerable faith that even deeper branches populated by obligate heterotrophs will be identified.

The geochemical drive for organic synthesis in hydrothermal systems permits a distinct departure from the soup hypothesis. Several investigators promote various scenarios in which the earliest metabolic strategy is some form of geochemically derived autotrophy, and aim for a clean break from the historical paradigm by referring to these as 'emergence of life' rather than 'origin of life' models. These alternative approaches require no stockpiling of soup or other inventories of raw materials; they rely instead on normal geological processes like hydrothermal circulation to supply conditions conducive to life and envisage the appearance of simple metabolic processes in response to easily established geochemical potentials. In a sense, these are habitat-based models for the emergence of life—something that has been missing from most origin-of-life discussions.

The emphasis on gradual processes rather than specific events enlivens the realm of experimental and theoretical studies that elucidate the transformation from geochemical to biological processes. In its extreme forms, this approach takes organic synthesis as the means rather than a prerequisite, and views complex biomolecules like DNA and proteins as symptoms of life rather than its cause. Importantly, these models demand that something as ubiquitous on the Earth as life developed in response to perfectly normal geological processes driven by inescapable aspects of the geodynamics of an active planet. The proposal that the emergence of life is a normal process stands in contrast to the largely abandoned origin-of-life paradigm that seems to require special circumstances if applied to the Earth.   E. SHOCK

origins of major fossil groups *see* MAJOR FOSSIL GROUPS: THEIR ORIGINS

orogenesis *see* MOUNTAIN-BUILDING (OROGENESIS)

Osborn, Henry Fairfield (1857–1935) The son of a wealthy Connecticut family, H. F. Osborn is one of the giants of American palaeontology. Even before he had graduated from Princeton University in 1877 he had taken part in a highly productive expedition to Wyoming in search of fossil vertebrates. At Princeton he became a professor of zoology and anatomy but moved in 1891 to New York to be Curator of Vertebrate Paleontology at the American Museum of Natural History, and to an appointment as Professor of Zoology and Anatomy at Columbia University. In these capacities he achieved great things, not only in his scientific research but also by his administrative and organizational prowess. In 1908 he became the Director of the American Museum of Natural History and gave it ultimately the status of the world's leading palaeontological collection.

Osborn's major palaeontological works concerned the early Cenozoic mammalian faunas of the West. In these he was the pupil, friend of, and successor to E. D. Cope. He is particularly associated with work on the giant brontotheres, and he proposed a theory to account for their evolution and extinction. His other great contributions concerned the evolution of the elephants and the horses in America.

As director of the Museum he was responsible for training research staff and, with R. C. Andrews, for palaeontological expeditions to the Gobi desert of Mongolia that were not only spectacularly successful scientifically but also captured popular attention worldwide. In addition, Osborn kept up a steady flow of scientific publications, and in his day was known as 'the Darwin and Huxley of the New World'.

D. L. DINELEY

ostracods Ostracods (technically Ostracoda) are a group of small laterally compressed crustaceans which are enclosed within an exoskeleton or carapace consisting of two valves. The carapace is made of chitin (a hydrocarbon related to cellulose) but is usually reinforced to a greater or lesser extent by calcium carbonate or, less commonly, calcium phosphate. The shape varies but it is usually ovate, kidney-shaped, or bean-shaped. Most ostracods are small, generally 0.3 to 3 mm long, although a few giants are as much as 30 mm in length. The two valves are unequal in size, the larger one partly overlapping the smaller. The valves are hinged along their dorsal edges and are closed by a series of muscles which leave scars on the valve interiors. Ostracods, like other crustaceans, grow by moulting the carapace at regular intervals. During its life each individual thus produces a series of shed carapaces or instars which are graded in both size and morphology. The exoskeletons of some species of ostracods exhibit sexual dimorphism: male and female carapaces differ in size or morphology, or both.

Modern ostracods are classified on the basis of their soft anatomy, particular note being made of the form and number of the appendages and the structure of the reproductive organs, features which are rarely preserved in fossils. Consequently, fossil forms are classified using the characteristics of the carapace, such as overall shape, hinge structure, muscle scar pattern, and ornamentation.

The ostracods first appeared in the early Cambrian but are relatively rare until the Silurian. In later Palaeozoic, Mesozoic, and Cenozoic sequences they can be extremely abundant. They have been widely used biostratigraphically in Jurassic to Pleistocene rock sequences. Most ostracods occupied a restricted, usually benthic, habitat and they therefore tend to have a limited geographical distribution. This makes them of limited use for long-distance correlations. However, where rocks of shallow-marine, brackish, or freshwater origin are interbedded, they are particularly valuable and can be used to correlate marine and non-marine sequences.

Living ostracod species inhabit a wide range of marine, brackish, and freshwater environments. A few species have even adapted to life in humid terrestrial habitats. Their diet includes bacteria, protists of all kinds, living algae and higher plants, and the organic debris that results from the decomposition of plants and animals. Since the majority of Recent genera are present in fossil assemblages as far back as the Miocene and many have close relatives in Mesozoic assemblages, ostracods are widely used in palaeoecological analysis. Even where the biological affinities of a fossil are uncertain, palaeoecology can be inferred from carapace morphology. For example, the carapaces of freshwater species tend in general to be thin and smooth, while those of marine forms are thick and heavily ornamented. There are similar correlations between carapace features and behaviours such as swimming, crawling, and burrowing.   ALLAN N. INSOLE

Further reading

Bignot, G. (1985) *Elements of micropalaeontology.* Graham and Trotman, London.

Brasier, M. D. (1980) *Microfossils.* Allen and Unwin, London.

outcrop, exposure 'Outcrop' and 'exposure' are two terms that are often confused or misused in casual speech. It is, however, useful to retain a distinction between the two. An *exposure* of a particular rock unit is where you can actually see the solid rock unit forming the ground surface. Cliffs, quarries, and areas of bare rock sticking up through the soil are all examples of exposures. On the other hand, the *outcrop* of a rock unit is the complete area where it would be exposed if all the cover by vegetation, soil, windblown sand, glacial deposits, and so on could be removed.

When constructing a geological map, it is the outcrop of each rock unit that a geologist is trying to delineate. This is usually done by making inferences based on the actual exposures, but these inferences can be supplemented by geophysical measurements or by using subtle changes in landform or vegetation that can reflect the nature of the bedrock, even when it is covered to a depth of many metres. The outcrop pattern of a rock unit can be used, in conjunction with measurements of how steeply the strata are dipping, to determine the stratigraphic thickness of beds of rock, and can indicate where folding or faulting have occurred. In planning quarrying or open-cast mining operations, it is important to know the full extent of the outcrop proposed for exploitation. Documenting exposures is useful for indicating where the rock can be conveniently seen and studied, but a map showing just these exposures would rarely give a useful insight into the geological history or an area.   DAVID A. ROTHERY

outlier *see* INLIER, OUTLIER

overland flow The term 'overland flow' refers to the flow of water over a hillslope surface. This can occur in one of two ways. First, it is generated when the rainfall intensity exceeds the surface infiltration rate. The excess rainfall then accumulates on the soil surface in small depressions. Once these depressions are filled, the water spills out and flows downslope as overland flow. Overland flow originating in this way is called *Horton overland flow*. For a given rainfall intensity, the occurrence of Horton overland flow and the rate at which it takes place are functions of the infiltration rate. Many factors influence the infiltration rate, including soil type, vegetation, and land use. In desert landscapes with their thin vegetation covers, shallow soils, and abundant outcrops of bedrock, infiltration rates are low and Horton overland flow is widespread. It is much less common in humid landscapes where thick vegetation and deep soils favour high infiltration rates. Exceptions occur where the natural vegetation cover has been thinned or removed, as in farmland. Destruction of the vegetation permits raindrop impact to seal the soil surface and lower infiltration rates without saturating the soil.

Secondly, overland flow is generated when a rising water table intersects the soil surface. Subsurface water then escapes from the soil and flows downslope over the soil surface, This exfiltrating water is termed *return flow*. That portion of the hillslope over which return flow emerges is saturated, so any rain failing on to it is unable to penetrate the surface and also flows downslope. Direct precipitation on to saturated areas and return flow together constitute *saturation overland flow*. This type of overland flow is rare in desert landscapes but is common in humid ones. It typically occurs on valley floors and concave foot slopes, in valley-head and valley-side hollows, and in areas where the underlying geology directs subsurface flow to the surface. The extent of the saturated area varies both between storms and within storms, and controls the rate of overland flow.

Interrill flow

Overland flow may occur as either interrill flow or rill flow. *Interrill flow*, also known as sheet flow, sheet wash, or slope wash, generally appears as a thin layer of water with threads of deeper, faster flow diverging and converging around surface protuberances, rocks, and vegetation. As a result of these diverging and converging threads, flow depth and velocity may vary markedly over short distances, giving rise to changes in the state of flow. Thus, over a small area the flow may be wholly laminar, wholly turbulent, wholly transitional, or consist of patches of any of these three flow states. Truly laminar interrill flow is rare, since falling raindrops agitate the flow, creating disturbances akin to turbulent eddies. Where the flow is very shallow, these disturbances may significantly alter its hydraulics.

Erosion by interrill flow involves *soil detachment* and *sediment transport*. Soil particles may be detached by a variety of processes, but diffusive ones, such as raindrop impact, frost action, and animal activity, dominate. Raindrop impact is the most widespread of these processes. Raindrop detachment rates are positively correlated with rainfall intensity and soil erodibility and negatively correlated with plant canopy and ground cover. Sediment transport by interrill flow has traditionally been related to measures of flow intensity such as shear stress or flow power. Such relations are valid for most flow conditions. However, where rainfall intensities are high (greater than 50 mm h^{-1}), flow power is low, and the ground surface is gently inclined (less than 5 degrees) and covered with fine sediment (less than 1 mm), sediment transport rates are significantly enhanced by raindrop impact, which agitates the flow and sustains already detached particles in transport.

Rill flow

Rill flow is deeper and faster than interrill flow and is typically turbulent. Rill flow occurs in small hillslope channels, called *rills*. Rills are commonly only a few centimetres wide and deep but grade upward into gullies. The boundary between rills and gullies is necessarily arbitrary, but one that has been widely adopted specifies that gullies are wider than 0.33 m and deeper than 0.67 m. Studies of rill morphology suggest that rill width is positively correlated with soil erodibility, whereas rill spacing is negatively correlated with hillslope gradient. Rill depth usually reflects the degree of rill development or depth to a resistant layer in the soil. Small rills can exist for only short intervals before being obliterated by other slope processes or ploughing. Large rills, on the other hand, may persist for decades, Rills occur naturally on desert hillslopes, particularly where the substrate is easily eroded. They do not form on densely vegetated hillslopes and are found in humid climates only where the natural vegetation cover has been thinned or removed.

Rill flow is typically too deep for raindrop impact to have a significant effect on either soil detachment or sediment transport. Both soil detachment capacity and sediment transport capacity are consequently functions of flow shear stress or flow power. Actual soil detachment in a rill is also a function of the sediment load entering the rill. As the sediment load approaches the transport capacity of the flow, the detachment (i.e. erosion) rate diminishes. Should the sediment load exceed the transport capacity, deposition will occur. Thus, two conditions must be met in order for rills to form: first, the flow must be the dominant soil detachment process and, secondly, the transport capacity of the flow must not be satisfied by the sediment load arriving from further up the slope.

Modern research on overland flow and erosion has focused on the development and calibration of physically based soil erosion models that are used chiefly for soil conservation planning and design. The most sophisticated of these models apply to pasture as well as the farmland, to watersheds as well as hillslopes, and to individual storms as well as many recorded years. These models represent a synthesis of a large body of knowledge and, as such, mark an important milestone in our understanding and application of the hydraulics and erosion mechanics of overland flow.

ATHOL D. ABRAHAMS

overlap and overstep *see* OFFLAP, ONLAP, OVERLAP, AND OVERSTEP

Owen, Richard (1804–92) An early interest in animals and anatomy led Richard Owen to an apprenticeship to a surgeon when he was only 16 in 1820; he was to become one of the most famous anatomists of all time. The Natural History Museum (now part of the British Museum) in London is largely a result of his foresight and energies.

The young Owen matriculated (but did not graduate) at the University of Edinburgh and took up an appointment at St Bartholomew's Hospital in London in 1825. He later set up in practice as a surgeon but soon moved to the Royal College of Surgeons, where he was soon appointed Professor and Conservator of the Hunterian Museum. By this time Owen's

reputation as a brilliant anatomist and zoologist was well known, and he was invited to take charge of the Natural History Department at the British Museum. Far from content with what he found there, Owen began agitating for a new large building to house the many branches of the subject. The outcome was the Natural History Museum much as we see it today, designed by Sir Alfred Waterhouse and built in the 1880s.

Owen's studies on fossil vertebrates were immensely productive. He had described those brought home by Charles Darwin from South America, had carried out a detailed analysis of the remarkable fossil bird *Archaeopteryx* from the Jurassic of Germany, had predicted from the evidence of a small piece of bone the discovery of the giant moa in New Zealand, and had realized that large fossil bones and teeth from Mesozoic rocks indicated an extinct group of reptiles to which he gave the name Dinosauria (terrible lizards). Dinosaur fantasies have given rise to a widespread late twentieth-century industry. Primates ancient and modern fascinated Owen and he became interested in the origin and development of teeth, but it was the dinosaurs and the newly found hominid fossils that brought him so much fame.

Owen was not entirely happy with Darwin's theory of evolution and he was Huxley's antagonist at the famous debate on the origin of species at the meeting of the British Association in 1860. He retired in 1884 and was awarded the KCB.

D. L. DINELEY

oxygen isotope records Oxygen stable isotope records provide a vital method in understanding the Earth's earlier climates. Major long-term variations in $^{16}O/^{18}O$, the ratio of the isotopes oxygen-16 and oxygen-18, in sea water are controlled by variations in ice-sheet volume. They reflect past sea levels, ice volumes, and marine temperatures and can be directly correlated to a range of parameters, including recorded levels of atmospheric greenhouse gases. In the 1940s, Harold Urey at Chicago University argued that although supposedly identical in chemical composition, the stable isotopes of an element do not behave similarly in all chemical processes. When water evaporates more of the lighter ^{16}O isotope is removed by the vapour, enriching the remaining water in ^{18}O. Also, water molecules containing the heavier isotope tend to condense and fall as precipitation more readily than molecules containing the lighter isotope. This results in fresh water having a higher $^{16}O/^{18}O$ ratio than sea water. As ice-sheet volume increases, the amount of ^{16}O removed from the sea increases, thus enriching sea water in ^{18}O.

Urey suggested that any oxygen-bearing substance precipitated from water, for example organic or inorganic carbonate, should have an identical isotopic ratio. Changes in the isotopic composition should be monitored directly by any contemporary precipitates, which may be preserved for example as fossils. Cesare Emiliani in the early 1950s produced the first complete oxygen isotope record from shells of foraminifera.

Oxygen isotope records have been most commonly developed from the carbonate precipitated by planktonic and benthic foraminifera, corals, and mollusca. Ice cores and oxygen trapped in bubbles within ice are also increasingly being analysed for oxygen isotopes.

Analytical results are expressed in relation to the $^{16}O/^{18}O$ ratio of a standard, given as $\delta^{18}O$. A common standard in the past was based on a carbonate from the rostrum of a belemnite from the Cretaceous Pee Dee Formation, South Carolina (PDB). Today, however, Standard Mean Ocean Water (SMOW) is used for it is more widely available and gives more reproducible results. A correction factor can be applied to convert PDB results to SMOW. Samples are processed using a mass spectrometer. The sample is converted into CO_2 and is analysed by comparing its $^{16}O/^{18}O$ ratio to that of a standard gas, which is in equilibration with SMOW.

It is assumed that the organisms which deposited the calcite in their shells did so at equilibrium or constant disequilibrium through time. This may not always be true, for certain species are known to vary their habit during their life cycle and also as a result of evolutionary changes. For example, certain species of benthic foraminifera may be either epifaunal or infaunal, respectively recording ambient benthic-water or pore-water changes. Planktonic foraminifera are also known to migrate through a range of water depths. The ratio of $\delta^{18}O$ between ambient water compared to carbonate precipitates can also be altered by temperature, salinity, and biological influences such as respiration. Measurements from foraminifera cultured in water of a known isotopic value are used to develop our understanding of their degree of fractionation.

Sea level

Global sea level is affected, among other things, by the amount of water contained within continental ice sheets; this in turn is recorded in the $\delta^{18}O$ of marine organisms. It is therefore possible to use the $\delta^{18}O$ record to record former sea-level changes. A $\delta^{18}O$ increase of 1 $^0/_{00}$ (one per thousand) represents a drop in sea level of about 10 m. However, it is impossible to apply a simple linear relation between $\delta^{18}O$ and sea level. Estimates of past sea level based on marine $\delta^{18}O$ values show that discrepancies can arise because of the effect of sea temperature variations and the form in which the ice exists. The formation of sea ice will remove ^{16}O, but will have no effect on the sea level.

During the Miocene epoch a series of sea-level falls related to ice-sheet growth are recorded in the marine $\delta^{18}O$ record. Records for the past 16.5 million years have a series of step-like increases in the $\delta^{18}O$ values, because of the transition from a relatively unglaciated world to one similar to that of today.

Temperature

Temperature is also known to influence the δ^{18} O record; becoming more negative with warmer temperatures. Using the $\delta^{18}O$ of planktonic foraminifera as a record of sea surface temperature (SST) is problematic because (1) these foraminifera respond to environmental changes and vary in their habitat, and (2) temperature changes are often associated with salinity changes, which also have a major effect on the $\delta^{18}O$ record. For benthic records, complications also arise because of variations in the original surface sea water $\delta^{18}O$ during deep water formation, caused by changes in evaporation rates and freshwater input.

Corals

Corals provide the $\delta^{18}O$ records with highest resolution. Their growth rings are deposited monthly, and by counting back from those presently being formed it is possible to produce a monthly record over many decades. Corals in the western Galápagos Islands show a close correlation between $\delta^{18}O$ and SST since 1965, when SST records for this region start. Using this information it is possible to estimate earlier SST values from the coral $\delta^{18}O$ record. El Niño events, characterized by periods of very high SST in the eastern Pacific are clearly visible in the coral $\delta^{18}O$ record. Also an 11-year cycle is seen in the records of $\delta^{18}O$ and the growth rate of corals, supporting the idea that climate variability in the tropics is driven by solar cycles.

Ice cores

Ice cores drilled from the ice sheets of Greenland and Antarctica provide $\delta^{18}O$ records back to 250 000 years ago. Ice core $\delta^{18}O$ records are developed from (1) ice crystals, which produce a record directly comparable to atmospheric dust levels and composition, and (2) O_2 trapped in air bubbles, which gives a contemporaneous record of atmospheric gaseous changes; for example CO_2 and methane. At present the atmospheric $\delta^{18}O$ ($\delta^{18}O_{atm}$) is $+$ 23.5 $^0/_{00}$ relative to SMOW. This difference is known as the Dole effect and is primarily the result of biological isotope fractionation during photosynthesis (the main method of O_2 production), respiration, and hydrological fractionation during evaporation and precipitation. As $\delta^{18}O_{atm}$ and oceanic $\delta^{18}O$ ($\delta^{18}O_{ocean}$), developed from foraminifera, vary at the same time and to the same magnitude during the late Quaternary, it appears that the Dole effect has remained reasonably constant during this period. Therefore, $\delta^{18}O_{atm}$ can be used as a proxy record for ice volume records in a similar manner to that for $\delta^{18}O_{ocean}$. This is very useful, for it provides comparable records of atmospheric CO_2 and ice volume. It appears that atmospheric CO_2 began to increase at least 4000 years before an initial decrease in continental ice volume.

Problems with the ice core $\delta^{18}O$ record include (1) the presence of ice formed elsewhere; (2) changes in temperature between times of ice deposition, commonly due to the elevation of the ice sheet surface as a result of thickening of the sheet; (3) changes in atmospheric circulation, as 'older' clouds lead to precipitation enriched in ^{16}O because of the earlier preferential condensation of ^{18}O; (4) long-term ice sheet deformation.

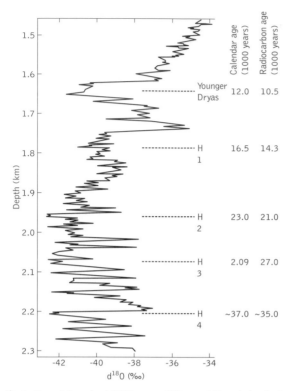

Fig. 1. Summit Greenland GRIP ice-core $\delta^{18}O$ record. H symbols refer to Heinrich events. (After Broecker (1994): *Nature*, **372**, 421–4.)

A section of the Summit Greenland GRIP ice-core $\delta^{18}O$ record is given in Fig. 1. The section is dominated by rectangular climate cycles averaging a few thousand years (ka), termed Dansgaard–Oeschger (D–O) cycles. Packets of D–O cycles characterized by progressively less negative $\delta^{18}O$ measurements are termed Bond cycles and last up 8 ka. They are marked by a Henrich event (H) at their start and end. These cycles reflect the gradual build-up of the ice sheets and their rapid melting (Henrich events). The colder Younger Dryas is clearly marked by more negative $\delta^{18}O$ values and the warmer Holocene by more positive values at the top of the diagram.

Stratigraphy

As the ice volume increases, the $\delta^{18}O_{ocean}$ becomes more positive, and vice versa. This produces a cycle of variations, contemporaneous around the world, recorded down cores within calcite shells. From this a chronostratigraphy was developed; the more positive glacial stages have even numbers and the negative interglacial stages have odd numbers. The stages start in the present interglacial, Stage 1. Similarly the last interglacial period refers to isotope Stage 5. Substages within major stages are defined by letters (or numbers), the lowest at the termination of the stage; for example, Substage 5e is a more negative (warmer) period at the start of interglacial Stage 5.

A composite record based on a large number of $\delta^{18}O$ planktonic foraminifera records has been developed by Prell and others. The ends of stages are characterized by rapid terminations with large changes in the $\delta^{18}O$ values. For example, the termination of the last glacial period, Stage 2, occurred over less than 10 000 years with a change in $\delta^{18}O$ values from + 1.8 $^0/_{00}$ to –2.0 $^0/_{00}$ (for the present interglacial). The past 4.5 million years contains over 152 stages. New records are compared with this standard to deduce the age of the core. Because of gaps in the sedimentary record and varying degrees of intensity between sites, not all stages or complete stages are recorded in all cores.

The $\delta^{18}O$ record peaks every 100 000 years, with superimposed, smaller cycles of roughly 23 000 and 41 000 years. These cycles strengthen the argument for astronomically driven climate cycles. The longer cycles correlate with the Earth's eccentricity cycle; the shorter cycles relate to the Earth's precessional and obliquity cycles. Age models based on counting the obliquity cycles in the $\delta^{18}O$ record of planktonic foraminifera commonly yield ages and average sedimentation rates higher in resolution than those from fossil data.

Isotopic data support earlier climate change theories that the importance of different astronomical cycles varies with latitude. For example, the 41 ka (41 000-year) cycle is more pronounced in cores from high latitudes, while cores from middle to low latitudes have a stronger 23 ka cycle. Cycles also vary over time; for example the 100 ka cycle plays a dominant role only during the past 650 ka years. STEPHEN KING

Further reading

Prell, *et al.* (1986) Graphic correlation of oxygen isotope stratigraphy: application to the late Quaternary. *Paleoceanography*, **1**, 137–62.

Sowers, T. and Bender, M. (1991) The $\delta^{18}O$ of atmospheric O_2 from air inclusions in the Vostok ice core: timing of CO_2 and ice volume changes during the penultimate deglaciation. *Paleoceanography*, **6**, 679–96.

ozone layer Ozone is a blue-green poisonous gas that is composed of three oxygen atoms (O_3). It occurs in trace amounts throughout most of the atmosphere but it is most abundant in the stratosphere. Here, there is a thin layer between 15 and 40 km altitude with a maximum concentration at around 25 km (Fig.1). This ozone layer is believed to be most important to life on Earth because ozone absorbs the most carcinogenic part of the solar spectrum. This fraction of the solar spectrum has the wavelength range 280 to 315 nm and is referred to as ultraviolet-B (UV-B). Indeed, ozone is so efficient that, with the small amounts of oxygen that also occur in the stratosphere, no radiation with wavelengths less than 298 nm manages to reach the troposphere. It is believed that if the ozone layer is depleted, then more UV-B radiation will penetrate into the lower atmosphere and so be detrimental to life on Earth. Some of the many health risks associated with increased UV-B are higher rates of skin cancers and

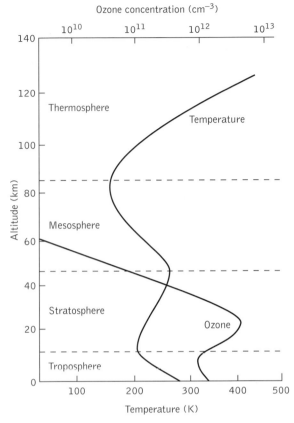

Fig. 1. Vertical distribution of atmospheric ozone. (After Harries, J. E. (1990) *Earthwatch: the climate from space*. Ellis Horwood, Chichester.)

cataracts, and also possible suppression of the human immuno-defence system. In the plant world, growth, yield, and photosynthesis may all be reduced.

Stratospheric ozone distribution

The abundance of ozone is usually measured as the amount of it that exists in an entire atmospheric column rather than the amount at a particular altitude. The unit of measure is a Dobson unit, which is equivalent to 1 milli-centimetre (0.01 mm) atmosphere of column ozone. The amount of stratospheric ozone varies naturally in both space and time. Production of ozone takes place mostly above 25 km altitude by photochemical reactions. From here it moves to the lower stratosphere, where it is stored and redistributed by atmospheric winds. Measurements tell us that ozone is most abundant at the poles and least abundant at the Equator, even though creation of ozone takes place at the Equator and is succeeded by transport to the poles, where it is destroyed. Seasonal change in the distribution of ozone is greatest at the poles; little seasonal change occurs at the tropics. The

maximum amount of ozone at the poles occurs during late winter and early spring. There are, however, distinct hemispheric differences: the maximum occurs nearer the pole and later in the season in the northern hemisphere than in the southern hemisphere. This is probably due to differences in the amount of ozone below 20 km altitude as a consequence of hemispheric differences in the way that lower atmosphere weather systems affect the stratosphere. Short-term changes in stratospheric ozone are linked to weather phenomena that occur in the troposphere. There are also longer-term changes that are of importance. Approximately every 28 months the zonal flow in the tropical lower stratosphere reverses direction (the quasi-biennial oscillation or 'QBO'). There are variations in the amount of ozone linked to the QBO, the solar sunspot cycle, and solar flares. The Junge layer, an aerosol layer of sulphate that exists in the stratosphere, is believed to help in the destruction of ozone. Volcanic eruptions will enhance this layer. This short discussion shows that there are several interlinked processes that affect the amount and distribution of ozone, namely chemistry, dynamics, and radiation.

Photochemical reactions in the upper stratosphere are common because of the influx of short-wave radiation. The lifetime of a chemical is determined by the reaction rate, which itself is temperature-dependent. If the lifetime of the chemical is short compared to the time needed for the atmospheric winds to move the chemical (the transport time), then photochemical equilibrium can occur; this is the case for ozone above 25 km. There is a constant cycle of creation and destruction of ozone, and in the 1930s the first attempt at describing these photochemical reactions was made. It was recognized at the time that a catalyst was needed in the creation and destruction cycles. Since then, further cycles of destruction resulting from catalysts have been identified, most notably those due to hydrogen oxides, nitrogen oxides, and chlorine oxides.

Although the vertical distribution of ozone can be principally explained by photochemical reactions, its geographical distribution relies heavily on dynamical processes. This is because in the lower stratosphere the lifetime of ozone becomes comparable with that of the transport time, and so the atmospheric winds become important. In the stratosphere, because ozone absorbs ultraviolet radiation, the temperature increases with height. This temperature structure means that there are a few vertical motions. Aerosols and gases injected into the stratosphere remain for a long time. The extreme dryness of the stratosphere enhances this feature and there is little opportunity for material to be rained out. As a consequence, large accumulations can occur. The zonal flow is extremely strong, and in winter wavelike structures can occur at high latitudes.

The polar vortex is not only an important feature of the stratospheric winds but also of considerable significance to ozone depletion, which is discussed below. As winter approaches, the long polar night begins and temperatures begin to fall. Centred close to the winter pole, a vortex forms which effectively isolates the stratospheric air over the polar region during the polar night. Temperatures in the vortex can reach below −78 °C in the lower stratosphere. These low temperatures allow the formation of polar stratospheric clouds. Those which form at −88 °C are composed of water-ice, but at −78 °C nitric acid trihydrate clouds form. The vortex that forms over the southern hemisphere pole is much more intense than that over the northern hemisphere. This is because of the position of Antarctica, isolated from surrounding land masses with a circumpolar ocean current that cuts it off from warmer water. In the atmosphere the large-scale planetary waves that occur in the upper troposphere rarely penetrate into the stratosphere over Antarctica, but they frequently disrupt the polar vortex over the Arctic. When the Sun returns in spring the polar vortex breaks down. This happens quite rapidly and is called the final stratospheric warming event. The polar vortex breaks into segments that drift equatorward. The final warming usually occurs in early spring in the northern hemisphere, but it can be very late in the season in the southern hemisphere. The intensity of the polar vortex responds to changes in the QBO. An easterly flow seems to weaken the polar vortex.

Ozone depletion

In the early 1970s concern began to mount that human activity might deplete the ozone layer. At first the chief concern was from the then proposed fleets of supersonic aircraft, such as Concorde. These supersonic planes would fly in the stratosphere, and combustion products such as water vapour and nitrogen oxides would then be deposited there. As mentioned above, the stability of the stratosphere would allow the nitrogen oxide to accumulate and then enhance the catalytic destruction of the ozone layer. The large number of supersonic planes envisaged were never built, but in 1974 came a new worry: that chlorofluorocarbons (or CFCs), a synthetic product, might also be capable of depleting the ozone layer. CFCs had been used commercially as refrigerants since the 1930s, but because they were inert and non-poisonous they were soon used for many other applications, most notably as propellants in aerosol spray cans. By the 1970s they were a measurable quantity in the troposphere worldwide, and because they were so inert in the troposphere they had a long enough lifetime to penetrate into the stratosphere. It was suggested by two scientists, M. J. Molina and F. S. Rowland, that in the stratosphere CFCs could be broken down by ultraviolet light. The chlorine released from the CFC would then be capable of destroying ozone.

Although there were many criticisms of the theory, it was quickly accepted. The level of ozone destruction predicted from models at the time was a 13 per cent decrease in the ozone column in 50 to 100 years. However, even though this initial figure rose to 19 per cent in 1979, successive model predictions down-rated the level of ozone destruction until by 1982 it was less than 5 per cent. CFC use, which had declined after the initial reports, subsequently began to increase again.

Then in 1985 it was reported that surface observations taken by the British Antarctica Survey indicated a substantial thinning of the ozone layer over Antarctica in early spring. Satellite observations from NASA confirmed the report. It is now widely accepted that the ozone hole, as it became popularly known, is caused predominantly by the chlorine from CFCs and other anthropogenic sources. The depletion of the ozone layer first manifested itself over Antarctica because of the wintertime polar vortex. The polar stratospheric clouds that form in the extreme cold provide a surface on which the chlorine from CFCs converts from a stable to a reactive component. Once the sunlight arrives in spring, the chlorine is released and is free to attack the ozone layer. As spring progresses, the vortex breaks up and ozone-rich air can penetrate the region. The extent of the thinning was initially linked to the QBO, being greatest when the vortex was strongest. This, however, appears to be no longer the case, and it has been postulated that the change is due to increasing levels of chlorine. Measurements now indicate significant thinning in the northern hemisphere polar regions and at mid-latitudes.

FRANCES DRAKE

Further reading

Fisher, M. (1992) *The ozone layer*. Chelsea House Publishers, New York.

Roan, S. L. (1989) *Ozone crisis: the 15-year evolution of a sudden global emergency*. John Wiley and Sons, New York.

ozone-layer chemistry
Ozone, a gas composed of three oxygen atoms ($O_{3(g)}$), protects the Earth's surface from harmful ultraviolet radiation from the Sun. More than 90 per cent of the ozone in the atmosphere resides in the upper atmosphere, the stratosphere.

Ozone is formed as a result of the dissociation of molecular oxygen by ultraviolet radiation in the stratosphere to generate oxygen free radicals ($O^•$):

$$O_{2(g)} + \text{energy} \to O^• + O^•.$$

These free radicals may then combine with another molecule of oxygen to form ozone:

$$O_{2(g)} + O^• + M \to O_{3(g)} + M,$$

where M is another molecule which acts simply to remove energy from the system and stabilize the ozone molecule that is generated.

Ozone is destroyed naturally in two main ways:

(1) The ozone formed is rapidly broken up by ultraviolet (UV) light:

$$O_{3(g)} \to O_{2(g)} + O^•$$

generating a molecule of oxygen and an oxygen free radical; or

(2) an ozone molecule combines with an oxygen free radical to form two molecules of oxygen:

$$O_{3(g)} + O^• \to 2O_{2(g)}.$$

Ozone can also be destroyed in other ways by man-made chemicals that have been released into the atmosphere. Most problematic are chlorofluorocarbons (CFCs). Chlorofluorocarbons are volatile compounds, used as refrigerants, which have escaped into the atmosphere. With time these chemicals have reached the upper atmosphere, where they can react with ozone and destroy it. Concern with CFCs, which persist for long periods of time in the atmosphere, grew and led to international regulation in the form of the Montreal Protocol which (as amended in 1990) imposed a complete ban on CFC production by the year 2000. Chlorine gas ($Cl_{2(g)}$ is released into the atmosphere as a component of volcanic gases. If chlorine molecules undergo photodissociation (break-up by light) two chlorine free radicals are generated ($Cl^{\bullet}$) which may then cause the destruction of ozone:

$$O_{3(g)} + Cl^{\bullet} \rightarrow ClO + O_{2(g)}$$

$$ClO + O^{\bullet} \rightarrow Cl^{\bullet} + O_{2(g)}.$$

The net reaction is

$$O^{\bullet} + O_{3(g)} \rightarrow 2O_{2(g)}.$$

Ozone can also be destroyed by reactions with nitrous oxide emissions from aircraft. This, once thought to be a major factor responsible for ozone depletion, is now considered to be less significant than the other mechanisms.

In recent years 'holes' in the ozone layer have been reported to develop each spring over the polar regions. These holes result from the formation of a body of still, cold air over the poles during the winter months. Temperatures within this cold air may drop to below −80 °C. If this occurs, reactions take place which generate forms of chlorine that become active ozone depleters when sunlight returns in the spring.

ELIZABETH H. BAILEY

773

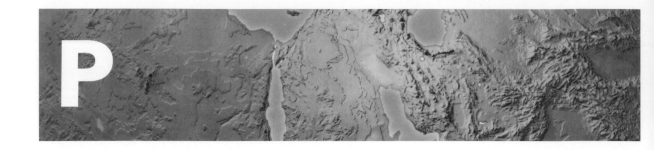

P

palaeobiology *see* PALAEONTOLOGY, PALAEOBIOLOGY

palaeobotany Palaeobotany is the study of fossil plants and includes the reconstruction of both plants and their ecology from fossil evidence. Palaeobotany includes not only macrofossil plants but also microscopic plant material that may be included in *palynology*, the study of fossil spores and pollen and other organic-walled microfossils. In its broadest sense paleobotany includes the study of the fate of plant material from when a plant dies (*biostratinomy*), through transport and burial (*taphonomy*) to fossilization (*diagenesis*). As most fossil plants are fragmented and are found away from their original site of growth, the reconstruction of whole plants is of prime importance. Data on the biology of the plants, together with taphonomic data, can be used in the interpretation of their ecology. Other studies may include the use of plant fossils for palaeoclimatological analysis, for dating and correlation, for palaeoenvironmental analysis, for the study of palaeoatmospheres, paleobiogeography, and evolution. Recent palaeobotanical innovations have involved studies of plant–animal interactions to understand co-evolution; studies on fossil charcoal as evidence of ancient fires; studies on stomata on plant leaves as proxies for ancient CO_2 levels; studies of isotopic and molecular structure of plant fossils, from the study of ancient DNA to the resistant biomacromolecules that allow the preservation of plants as fossils; and the contribution of fossil plants to kerogen formation and to the formation of fossil fuels—coal, oil, and gas.   ANDREW C. SCOTT

palaeoceanography Palaeoceanography is the study of the properties of the oceans over geological time-spans. Approximately 71 per cent of the Earth's surface is now covered by water, and it is clear that the oceans exert a huge influence on all systems on this planet, both marine and terrestrial. Palaeoceanography is critical to our understanding of the evolution of the Earth, its climate, and its biota. A variety of geochemical, palaeontological, and sedimentological proxies recorded in oceanic sediments provide the data from which models are developed for past oceanic conditions: for example ocean circulation patterns, ocean chemistry, and biological activity.

The main contemporary themes of interest and study within palaeoceanography are: (1) oceanic influence on atmospheric carbon dioxide; (2) oceanic heat transfer, for example, surface currents and deep water formation; and (3) the relationship between marine and terrestrial records.

There are two major problems with the palaeoceanographic record.

(1) Because of the subduction of oceanic crust at destructive plate boundaries, little material older than the Tertiary remains. Older deep-sea sediment is preserved when thrust on to continental crust during the closure of oceans; for example, Tethyan material in the Alps.

(2) Bioturbation (disturbance of sediments by organisms) removes short-term fluctuations. Regions with high sedimentation rates provide the most detailed palaeoceanographic records; for example, regions of high productivity such as upwelling zones off Peru and West Africa.

STEPHEN KING

Palaeocene The first epoch of the Cenozoic era, the Palaeocene (also spelt Paleocene) was distinguished by Wilhelm Schimper in 1874. It spans the time from the end of the Cretaceous (65 Ma) to the beginning of the Eocene (54 Ma). Like the other Cenozoic epochs, it was based on a type section in France. There are two stages, a basal Danian and an upper Thanetian/Selandian (two different facies-types).

At the inception of the Cenozoic, Laurasia, like its southern counterpart Gondwanaland, began to split apart. Ocean-floor spreading began to expand the North Atlantic northwards, with a new gulf extending from near Gibraltar to the vicinity of present-day Newfoundland. Canada and Greenland were meanwhile still attached to Europe. Africa and Arabia were joined but Madagascar had separated. India had yet to make contact with the southern margin of Asia. A warm climate prevailed everywhere, with more precipitation over the land than there had been in Cretaceous times. Antarctica had not yet reached its fully polar position and was only just beginning to exert a cooling influence upon the southern hemisphere.

The Palaeocene biota, both marine and terrestrial, contrasted sharply with that of the Cretaceous. In the seas, only the gastropods and bivalvia remained common among the bottom-dwelling mollusca; some nautiloids persisted but the ammonoids were not extinct. On land, new orders of plants, including the grasses and other modern types, flourished, but the deciduous forests were of low diversity. The spread of vegetation over an increasing part of the land surface now took place, assisted by the amelioration of the climate as Pangaea fragmented and new seas came into being. In their turn, the new forests and grasslands constituted habitats, and in the absence of the large reptiles that had been so active in the Mesozoic era, the mammals became the dominant tetrapods, spreading rapidly far and wide and evolving many large species. The contrast between the communities of very small Cretaceous mammals and the new Palaeocene faunas of large—even gigantic—mammals is very striking. South American mammal faunas included prolific numbers of marsupials. Elsewhere the 'archaic' orders of mammals reached prominence, and were not to disappear until Miocene times. D. L. DINELEY

Further reading

Savage, R. J. G. and Long, M. R. (1986) *Mammal evolution, an illustrated guide*. British Museum (Natural History), London.

Taylor, J. (1978) Cenozoic. In McKerrow, W. S. (ed.) *The ecology of fossils*, pp. 328–30. Duckworth, London.

palaeoclimate data from historical sources
Historical palaeoclimate data incorporate a variety of written sources of information on past climates. Many of these records relate to short-term (high-frequency) climatic fluctuations during the most recent past. The geographical distribution of data of this type is highly varied, as is the range of the records preserved. The oldest and longest records include the documented levels of Nile floods, which date back to c. 3000 BC.

Sources of historical data include ancient inscriptions, annals, chronicles, governmental and private estate records, maritime and commercial records, personal papers, and scientific writings. A major difficulty in the utilization of such data is the need to 'filter' and 'calibrate' observer bias in what are typically short-lived, discontinuous records which may overemphasize extremes at the expense of general climatic information. Where possible, the calibration procedure is achieved by the use of instrumentation, providing records that overlap the given proxy historical data set. An additional limitation relates to the importance of discriminating first-hand accounts from those disseminated by rumour or the passage of time; evidence of the latter types is typically exaggerated.

Three main types of historical palaeoclimatic records are recognized:

(1) Specific observations of weather phenomena. These data sources are varied and include specific accounts of climatic phenomena (e.g. days of snow, occurrence of hoar frosts) and climate-related observations, including those that are secondary to our observer's purpose. A notable example of the latter is the collation of weather data included in contemporary paintings covering the period 1400–1967. Researchers collated data on degree of blue sky, visibility, cloudiness, etc. and concluded that paintings of the period 1400–1549 exhibited high levels of blue sky and good visibility, while paintings of the period from 1550 to 1850 were typically darker and exhibited a greater degree of cloudiness.

(2) Records of weather-dependent phenomena. Such phenomena include, for example, droughts, floods, and river-freezing events. Ratios of droughts to floods for a given period have been found to be useful in removing observer bias. As these directly affect local communities on a range of personal and commercial levels (e.g. grain price), such records have been documented and gazeteered extensively.

(3) Phenological records encompass recurrent weather-dependent biological phenomena. Examples of such records include timing of bird migration, tree blossoming, and crop maturation. The historical extents of climate-sensitive plants are also included in this category. One of the best known phenological records comes from the timing of cherry blossom in Kyoto, Japan. The record extends back to the ninth century, early blossom correlating with higher springtime temperature.

Climate records using instrumentation have provided a direct, and for the most part unbiased, record of past climatic parameters over the period of late historical to modern times. Despite the length of some European records, a lack of systematic approaches to measurement has made their detailed interpretation problematic. The longest continuous climatic record (more than 350 years) is that collated at the Radcliffe Meteorological Observatory in Oxford. STEPHEN STOKES

palaeoclimate and magnetism
Rock-magnetic variations within sediments provide continuous proxy records of palaeoclimatic and palaeoenvironmental changes. *Magnetic susceptibility* provides an ideal tool for examining such records, since it reflects the concentration, mineralogy, and crystal size of magnetic minerals. Provided that the relationship between magnetic susceptibility and palaeoclimatic fluctuation can be deduced, it is possible to recover the palaeoclimatic signal preserved by these records.

The compositions of the magnetic minerals deposited within a sedimentary environment depend primarily on the compositions of the rocks in the headwaters of the streams and rivers entering the sedimentary basin. However, their concentration is governed by the hydrological regime of the river systems. During a dry year, water movement will be sluggish, allowing time for oxidation and hydration. In a wet year, swift-flowing rivers will transport immature detritus into the sedimentary environment. Thus, the composition and concentration of magnetic minerals within the sediment will fluctuate in response to climatic variability through time. Successive years will be characterized by strata with higher or lower concentrations of magnetic minerals. Similarly, variations in the magnetic mineral concentrations of deep-sea sediment will also tend to preserve a record of climatic change. However, since deep-sea sediment accumulates extremely slowly, a record with a much lower resolution but on a longer timescale will be preserved. In such environments the composition of the magnetic minerals deposited is unlikely to change significantly. Thus, most observed changes reflect variations in the rate of deposition of non-detrital (and non-magnetic) materials such as carbonates. This dilutes the proportion of detrital grains deposited during warmer periods when carbonate deposition is intensified.

However, a variety of factors influence the magnetic mineral content of a sediment. These include allogenic (external) input (terrigenous, aeolian, volcanic, cosmic, or bacterial); bacterial or inorganic precipitation of magnetic minerals and the effect of diagenetic processes (e. g. cementation) on magnetic mineral preservation. It is thus important to establish the origin of the magnetic assemblages preserved within the sediment. *Magnetite* grains of volcanic origin are typically coarse-grained (more than 1 micrometre: 1 μm) and commonly contain substituting cations, such as titanium, within their crystal lattices. Alternatively, they may occur as elongate inclusions within other mineral grains. Bacterial magnetite,

produced by magnetotactic bacteria, is ultra-fine-grained (less than 0.1 μm), unidimensional, and contains no foreign cations.

An example of the link between magnetism and palaeoclimate is provided by loess deposited during glacial periods. Loess sections exhibit characteristic variations in magnetic susceptibility and show close correlation with oxygen isotope variations, suggesting that they are controlled by climate. In general, primary loess layers have low susceptibility, and interbedded clayey *palaeosol* (fossil soil) layers have high values. The source of the susceptibility signal was originally believed to be subaerial deposition from high-level transport of ultrafine magnetite from distant sources. A constant rate of atmospheric input maintained for thousands of years was assumed, with dilution of the magnetite concentration occurring during cold and dry periods when aeolian silt of low susceptibility was deposited. During warm and humid interglacial periods wind-blown sedimentation rates were low and the dominant magnetic component consisted of the ultra-fine magnetic dust, thus resulting in relatively high magnetite concentration. However, more recent studies, incorporating magnetomineralogical examination and optical or/electron microscope analysis of the sediment, indicate that the higher susceptibility of the soils is a result of authigenesis *in situ* of ultra-fine magnetite during soil formation. The rate of magnetite formation within the soil, and its preservation, depend on the temperature and wetness of the soil. It is therefore pos-

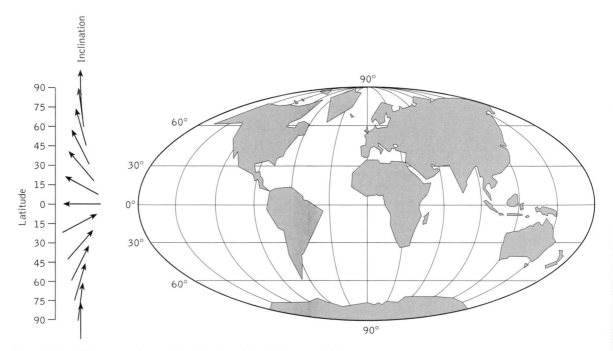

Fig. 1. Relationship between latitude and the inclination of the Earth's magnetic field.

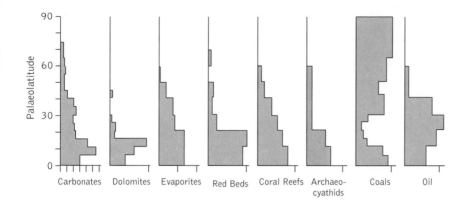

Fig. 2. Relationship between palaeolatitude and palaeoclimatic indicators. (From Tarling (1983).

sible to identify how the magnetic properties of the loess/palaeosol responded to changes in the climate.

Palaeomagnetism also offers an indirect method for determining palaeoclimate based on calculating the palaeolatitude at which the rock was formed. Since the Earth's magnetic field behaves in a similar fashion to a dipolar magnet located at the geocentre, a direct relationship exists between the magnetic inclination and the angular distance from the magnetic pole:

$$\tan (\text{inclination}) = 2 \tan (\text{latitude})$$

Provided that a rock records the inclination of the Earth's magnetic field on formation, it is possible to determine the latitude at which the rock was formed (Fig. 1). The correlation between palaeolatitude, determined from a rock's palaeomagnetic characteristics, and the rock's mode of formation provides a means of gauging the palaeoclimate. A broad range of geological materials exist that can be used to scale the Earth's ancient palaeoclimate (Fig. 2). Though none of these palaeoclimatic indicators is precise enough to define the true axiality of the axial geocentric dipolar mode for the geomagnetic field, the consistency demonstrated by all such indicators with this model implies that palaeomagnetic data can be used to infer palaeoclimatic conditions for specific rock-types.

PAUL MONTGOMERY

Further reading

Thompson, R., and Oldfield, F. (1986) *Environmental magnetism.* Allen and Unwin, London.

Tarling, D. H. (1983) *Palaeomagnetism.* Chapman and Hall, London.

palaeoclimatology

Palaeoclimatology is the study of past climate and its variation in space and time during the period before instrumental measurements began to be made. A key aim of palaeoclimatology over and above the simple documentation of past climates is to elucidate the mechanisms that gave rise to the observed variations. This entails both the collection of proxy records of climatic oscillations and the establishment of a timescale for their occurrence. The sources of proxy data are various: biological (e.g. dendroclimatology, pollen grains, plant and animal macrofossils); terrestrial (e.g. ice cores, glacial and periglacial sediments and landforms, aeolian sediments, lacustrine [lake] deposits, speleothems [chemical deposits found in caves] and pedological evidence); and marine (e.g. floral and faunal abundance and morphology, mineral composition, dust and glacial debris concentrations, and geochemical data), as well as other geological evidence and stable isotopic data. The preserved climatic record is frequently complex, incomplete, and exhibits considerable extraneous 'noise' from non-climatic influences. The palaeoclimatic record does, nevertheless, indicate large shifts from glacial to interglacial conditions, with, in some instances, abrupt shifts between climatic modes.

The most significant global-scale mechanism that is currently invoked to explain glacial–interglacial climatic fluctuations during the past 2.5 million years is embodied in the Croll–Milankovich theory of ice ages. This theory relates the changes in climate to periodic variations in the orbit of the Earth round the Sun which take place at intervals of about 100, 42, and 19–23 thousand years.   STEPHEN STOKES

palaeoclimatology, transfer functions in *see*
TRANSFER FUNCTIONS IN PALAEOCLIMATOLOGY

palaeocurrent analysis

Palaeocurrents are former current directions deduced from sedimentary structures preserved in ancient rocks. The first realization that such data could be obtained and would be useful in paleogeographic reconstructions is usually attributed to Sorby's work of 1839.

Structures produced in sediments by moving fluid, such as ripple-marks and cross-stratification, as well as scour-marks preserved as casts on the bases of beds, indicate the direction of current flow. In fluvial environments this makes it possible to determine the direction of slope down which the sediment was transported (i.e. the palaeoslope). In contrast, cross-stratification of aeolian deposits reflects the distribution of ancient wind systems which were dependent on climatic and topographical controls. In the shallow-marine realm, sedimentary structures are related to waves and currents, and the relationship between these and the trend of ancient shorelines was sometimes complicated and variable. Bottom structures, produced by gravity-driven dense bottom flows, such as groove-casts and flute-casts, when preserved on the base of a turbidite are good indicators of palaeoslope direction.

Careful corrections have to be applied to measurements of sedimentary structures when they are taken from deformed rocks, because of the rotation of the structures during later tectonic deformation.   G. EVANS

palaeoecology
Palaeoecology is the study of the interactions of organisms with each other and with their environment in the geological past. It thus borrows heavily from the modern discipline of ecology although its emphases are different. Modern biologists can sample populations on scales of years and describe environments in detail, while palaeontologists are limited by the data they can collect and generally work on timescales of thousands or millions of years. Despite these restrictions, palaeoecology is important because of its value in helping to reconstruct ancient environments and in understanding fossils as once-living organisms.

The ecosystem is the largest unit of ecological study and consists of a portion of the physical environment, however large or small, together with all the organisms it contains. Within this the habitat is the environment in which the organism lives; its niche is its position in the habitat, including its relationship with other organisms. Study may be at the level of the individual or encompass populations of two or more individuals or communities of two or more species. Modern organisms are grouped according to the role they play in the flow of energy through the ecosystem. Primary producers such as plants convert nutrients and light into organic compounds, which are then consumed by herbivores. In their turn the herbivores are eaten by carnivores or by scavengers and, eventually, all organisms are broken down by bacteria into nutrients that return to the environment.

In order to study the ecology of fossil organisms, palaeoecological inference is used extensively. This uniformitarian approach is based on the assumption that the general laws governing the functioning of organisms have not changed with time. Producers, herbivores, carnivores, and decomposers were present in the past, and although the complexity of ecosystems has increased, the laws governing the flow of energy through them are assumed not to have changed. Some interactions, particularly predator–prey relationships, are independent of inference from modern communities; thus, for example, teeth marks in an ammonite show that it was bitten by a mosasaur. Other relationships require more inference, as in the study of ancient reefs. Modern reefs consist mostly of corals that live in symbiotic relationship (in which both gain an advantage) with algae. The algae need sunlight to photosynthesize and the corals provide the framework to hold them near the surface. Similar reefs can be seen in Cenozoic and Mesozoic rocks and a similar relationship can be inferred.

The functions of various features in fossils can also be inferred by comparison with modern organisms; in many instances, however, these features may be unknown in modern taxa. In order to interpret these so that an understanding of adaptation and mode of life can be developed a coherent methodology such as the paradigm approach is used. In this system optimum structural designs for all possible functions of the feature are developed and the design that appears most similar is considered most likely to be correct.

Other areas of palaeontology are important in the study of palaeoecology. Taphonomy deals with the processes that affect an organism from death to burial, and is necessary for an understanding of the extent to which the preserved organisms accurately reflect the living communities. Ichnology deals with tracks and trace fossils and therefore provides evidence of the soft-bodied members of communities that otherwise leave no record.   DAVID K. ELLIOTT

palaeoenvironment
Reconstructions of ancient geographies are essentially models of the distribution of palaeoenvironments, and the postulated relationships of these environments must comply with the general spatial relationships of modern environments. Such reconstructions are interpretations of the lithological and other properites of the rocks within the area studied. They are limited by the amount and kinds of data available, and they must fit into the geological history of the area.

For the most part, the study of palaeoenvironments is associated with the need to understand the origin of rock formations, especially fossiliferous sedimentary rocks and volcanic rocks, or the conditions under which past floras and faunas existed. Many sedimentary processes are highly discontinuous or episodic, and the rate of deposition of many geological formations is consequently uncertain or hypothetical. The flow of energy within the environments where this occurs is likely to be uneven or discontinuous and may be due to factors originating well beyond the area of study. The (physical) energy regime is important to the biota of any environment and to the transport and deposition of organic material. Thus the type and state of preservation of fossils is significant in interpretation of the palaeoenvironment.

Palaeoenvironmental reconstruction begins with the assembly of all available data. These are then compared with the data of modern well-understood environments and environmental processes. The best available fit is then chosen, but modifications to account for rock properties or features not explicable by this means have to be added and, if possible, tested against the acquisition of new data. Hutton's classic principle that 'the present is the key to the past' has to be applied with due circumspection, not only in regard to the physical and chemical aspects of ancient environments but also where the interpretation of organisms (fossils) in those environments is concerned.   D. L. DINELEY

Further reading

Prothero, D. R. (1990) *Interpreting the stratigraphic record.*
    W. H. Freeman, New York.

Reading, H. G. (ed.) (1996) *Sedimentary environments* (3rd edn).
    Blackwell Scientific Publications, Oxford.

Palaeogene The Palaeogene is the older of the two Ceno-zoic sub-periods that are now recognized. It includes the Palaeocene, Eocene, and Oligocene epochs, spanning the time from the end of the Mesozoic to the beginning of the Neogene, i.e. from 65 Ma to 23 Ma. The term 'Palaeogene' was originally introduced by the Austrian geologist Moritz Hoernes in 1856 and was used as a basic equivalent to the Eocene of Charles Lyell. The term 'Neogene' was employed to embrace the two later stages, the Miocene and the Pliocene. Hoernes, like Lyell, was much impressed with the wealth of the fossil molluscan faunas of Cenozoic rocks and was influenced by their stratigraphical distribution in Eastern Europe. Subsequently the recognition and adoption of the Oligocene Series restricted Lyell's original Eocene, but had no impact upon the biostratigraphical definition of the Palaeo-gene. The Oligocene is today generally recognized as the third and uppermost series within the Palaeogene. The Paleocene, as recognized both in North America and Eurasia, is the oldest series.

There is an appreciable difference between the fossil record of the Palaeogene and its successor, not only among the marine molluscan faunas, but also among the fossil ver-tebrates, principally the mammals. Palaeogene forms include those large herbivores commonly known as the 'archaic mammals', whereas the basic stocks of modern mammalian faunas begin to appear with the dawn of the Neogene; the first hominids are present.

Palaeogene time saw a continuing opening of the North Atlantic and a general disposition of the continents similar to that of today; for example, Australian separation from Antarctica was completed. Climatic changes continued, largely influenced by the growth of the Tethyan–Himalayan ranges.   D. L. DINELEY

palaeogeography Palaeogeography is the study of ancient geographies, using geological evidence to infer the distribution of land and sea, uplands (areas of denudation) and lowlands (sedimentary basins), the vegetational cover of the land, and the past climates that influenced them. On the grandest scale, the study concerns the positions, surface areas, topographies, and geology (palaeogeology) of the continents and the bathymetry of the oceans. Smaller areas of study may concern the palaeogeographical settings of specific sedi-mentary basins or the environments in which certain geo-logical formations originated. Study of the habitats of ancient plant and animal communities falls within the purview of palaeogeography, and the distribution of ancient floras and faunas constitutes palaeobiogeography.

Essential to the description of an ancient geography is the definition of the time-span confining the data. This may be achieved in terms of biostratigraphy (for example, a zone, stage, or epoch) or of the existence of a particular deposi-tional environment (for example, an evaporite basin or a marine delta).

All the geosciences contribute directly or indirectly to palaeogeography because the study of successive palaeo-geographies indicates much about the evolution of the surface of the Earth together with the hydrosphere and the biosphere. The verification of the concept of continental dis-placements from palaeomagnetic studies is an instance of the successful application of geophysical techniques to a basic geological and geographical problem. Palaeobiology con-tributes much to our knowledge of the spread of life throughout different environments and of its interaction with the lithosphere, hydrosphere, and atmosphere during geological time.

Our reconstruction of ancient geographies depends upon an understanding of modern geographical and geological processes. The limitations of 'uniformitarian' reasoning have been much debated of late. It is important to recognize that many ancient geographies may have no exact analogues in today's world. Direct observation of geological processes has been limited to only a few centuries, and the rates at which many geological processes proceed are unknown or uncertain. Nevertheless, it is assumed that certain mineralogical or petrological materials have always been generated in the same fashion, and their presence may be a good indication of similar factors operating in the past. Greensand, coal, rock salt, coral limestone, oolites, and boulder clays are examples of indicators of depositional environments. Climate or water temperature, or both, may be indicated by these materials, especially when considered in connection with geochemical data such as isotope ratios.

Modern deep-ocean studies throw light upon ancient palaeogeographical scenarios. The circulation of ocean currents is influenced by ocean-basin topography and by climate and has an important effect upon world temperature regulation. Similar relationships are presumed to have existed in the past, and the consequences for global or continental palaeogeographies need to be assessed.

Most palaeogeographical data are conveyed on palaeo-geographical maps, which can include a *palaeogeological map* designed to show the outcrop pattern during a particular past time-span. This information may be of economic value in delimiting areas of search for fossil fuels and other natural resources.   D. L. DINELEY

Further reading

Cope, J. C. W., Ingham, J. K., and Rawson, P. F. (1992) *Atlas of palaeogeography and lithofacies.* Geological Society, London.

Prothero, D. R. (1990) *Interpreting the stratigraphic record.*
    W. H. Freeman, New York.

palaeohydrology Palaeohydrology has been defined by Stanley Schumm as 'the study of the quantity, quality and distribution of surface waters in the past'. This broad subject can be divided into three basic approaches: inferential palaeohydrology, classical palaeohydrology, and palaeohydrological modelling.

Inferential palaeohydrology is the hydrological interpretation of sedimentary sequences or landforms, where a change in hydrological conditions is inferred but no attempt is made to estimate its magnitude. This is extremely common in the interpretation of fluvial sequences and lake sediments (inferring depth or volume changes) of all ages. It is also generally a prelude to more formal palaeohydrological work as well as the standard geomorphological method of linking changes in Earth surface processes and rates to changes in the environment such as climate and land use.

Classical palaeohydrology is the estimation of past hydrological parameters from sediments or landforms. One approach is the estimation of palaeoflood discharges, in particular form slack water deposits at tributary junctions and on benches within reaches confined by bedrock (i.e. undeformable channel beds; Fig. 1). The flood slack-water sediments are then dated (e.g. by radiocarbon or luminescence, or by archaeology) and this makes it possible to calculate a return interval for the water level (stage), or flood discharge, if flow can be modelled (e.g. by using slope–area or step–backwater methods). The practical application of this methodology is to calculate to return intervals of floods far beyond the instrumented record.

The second approach in classical palaeohydrology is the estimation of discharge from palaeochannel parameters. This is based upon the well-understood relationships between modern channels and discharge. The most obvious case is the estimation of bankfull discharge from palaeochannel capacity (A) and gradient, using one of the well-known flow equations, such as the Manning's or Darcy Weisbach for velocity (V), and the continuity equation ($Q = VA$) (see *river channel hydraulics*) There are many sources of potential error even when the palaeochannel is well preserved and of a sufficient length. An alternative is to use empirical relationships between channel parameters, such as width or meander wavelength, and a significant discharge (e.g. mean winter flow, or bankfull flow). This method was used by Stanley Schumm in his classic study of two generations of palaeochannels (prior and ancestral streams) of the Murrumbidgee on the Riverine Plain of Australia.

A third method is to estimate flow discharge and velocity from fluvial sediments and structures. Bedforms such as ripples and dunes can for example, indicate flow regime and depth. This approach has been used by John Allen in his classic studies of the Old Red Sandstone, and has been used to estimate the magnitude of the mega-floods associated with the catastrophic drainage of Lake Missoula at the end of the Last Glacial period in North America. In addition, the flow velocity can be estimated from the size of the clasts (particles) moved, or indeed not moved, by the flow, an approach known as 'palaeohydraulics'. Victor Baker was one of the first to use this approach in his classic study of the alluvium at Golden, Colorado. There are, however, many problems with all these methods; defining channel width can be difficult, as can the estimation of channel gradient and the size of the largest entrained particle. Additional problems arise from the state of the channel bed, such as armouring, loose and over-loose bed conditions, and sediment supply. For these and other reasons, a realistic estimate of the level of accuracy attainable is probably only an order of magnitude.

Palaeohydrological modelling is a deductive process in which an estimate is made of the past water balance of an area from hydrometeorological estimates and catchment conditions. It is most obviously applied to modelling the hydrological balance of lakes, in order to estimate precipitation (P) using the standard water-balance equation ($P = R + (E + T) + (DS + DG)$) from runoff ($R$), evaporation ($E$), transpiration ($T$), and changes in surface-water storage (S) and groundwater flux (G). There are, however, problems and there is considerable 'noise' in the records of lake levels (as held by various lake-level databases), principally derived from problems in estimating palaeo-depths and unknown groundwater fluxes. A similar approach has been used for river basins, an example being John Lockwood's estimate that a small catchment in central England could, in the mid-Holocene, have had a 10 per cent lower discharge despite 10 per cent more rainfall as a result of the greater interception and evapotranspiration of a complete forest cover. Palaeohydrological modelling, whether from lakes or river basins provides a bridge between the scale of global climate models and the sedimentary section or borehole record.

There are several other approaches to estimating past hydrometeorological conditions that are closely related to palaeohydrology. One is the analysis and climate interpretation of historical hydrological records, generally of river levels (historical hydrology). A second is the reconstruction of

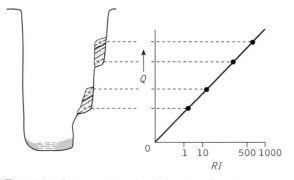

Fig. 1. Palaeoflood estimation using slack-water sediments on benches within bedrock channels to date past levels of flood discharge (Q) and thereby estimate return intervals (RI) for these water levels.

precipitation and stream flow from tree rings (dendro-climatology and dendrohydrology). What all these methods have in common is that they attempt to provide quantitative estimates of past hydrological conditions, which can at least stand as testable hypotheses. The palaeohydrological approach has altered our ideas about hydrological stationarity (the assumption of a constant relationship between flood magnitude and frequency), enabling us to measure, as opposed to estimating by extrapolation, the magnitude of rare events. Palaeohydrological studies have relevance to the flood and drought hazards faced by people in the past, today, and in the future. A. G. BROWN

Further reading

Allen, J. R. L. (1984) *Sedimentary structures: their characteristic and physical basis.* Elsevier Science Publishers, Amsterdam.

Baker, V. R., Kochel, R. C., and Patton, P. C. (1988) *Flood geomorphology.* John Wiley and Sons, New York.

Benito, G., Baker, V. R., and Gregory, K. J. (1998) *Palaeohydrology and environmental change.* John Wiley and Sons, Chichester.

Brown, A. G. (1977) *Alluvial Environments: Geoarchaeology and Environmental Change.* Cambridge University Press.

palaeomagnetism and continental drift

The development of a very sensitive rock magnetometer in the mid-1940s paved the way for much palaeomagnetic research, including measurements on weakly magnetized sedimentary rocks, some as old as Precambrian: more than about 545 million years (Ma) old. In the 1950s, workers from England, including Keith Runcorn, then at Cambridge University, began to acquire palaeomagnetic data from the Phanerozoic rocks of western Europe and North America to test hypotheses of continental drift and true magnetic polar wandering. By 1956, data sets showed systematic discrepancies of about 25° between palaeomagnetic poles inferred for Europe and North America. Thus, the palaeomagnetic database supported the idea of continental drift well before an actual mechanism for drift had been proposed and accepted, a not surprising situation.

The Vine–Mathews, Morley–Larochelle hypothesis of the early 1960s explained marine magnetic anomaly patterns by proposing that oceanic crust was created at mid-ocean ridges. The ocean crust, to a first approximation, faithfully records the direction and polarity of the geomagnetic field. By the late 1960s, a sufficiently detailed geomagnetic polarity timescale for the past few million years of its Earth's history had been compiled and was used to verify that marine magnetic anomaly records described accurately the polarity history of the planet. The paradigm of plate tectonics was finally accepted by the majority of the geoscience community and the use of palaeomagnetic data to interpret continental drift began in earnest.

Palaeomagnetic poles became available for all the continents, and their temporal distribution on the globe defined

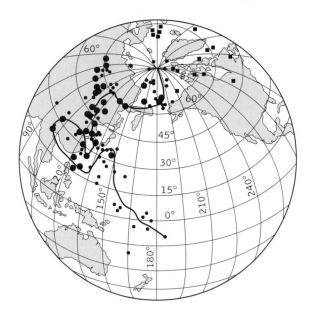

Fig. 1. Phanerozoic apparent polar wander (APW) path (heavy line) constructed for the North American continent on the basis of palaeomagnetic poles (circles) which meet a number of reliability criteria, including (but not limited to) reasonably accurate age of the magnetization, complete laboratory demagnetization, enough samples/sites, adequate averaging of the geomagnetic field, and tectonic coherence of the study area relative to stable North America. Poles meeting a greater number of criteria are indicated with larger circles. For comparison, selected palaeomagnetic poles from rocks of the western borderland of North America are indicated by squares. These data have been interpreted to indicate that, relative to the continent, much of the western borderland has been displaced northward and rotated clockwise before and during amalgamation with stable North America.

paths of apparent polar wander (APW). Each continent exhibited an APW path of different geometry and the conclusion was reached that, to a first approximation, the paths recorded latitudinal shifts and rigid-body vertical-axis rotation of the continents, rather than true polar wander. Increasingly refined comparisons of APW paths from different continents clearly indicated that they recorded the motion in the geological past, often of large magnitude, of what had by then become known as lithospheric plates. Symmetry arguments show that the position of a palaeomagnetic pole relative to a sampled continent does not constrain the palaeolongitude of the continent. However, and very importantly, the distance from a palaeomagnetic pole and the sampled area of the continent, as reflected by palaeomagnetic inclination, provides a direct estimate of the palaeolatitude of the continent.

The APW path for North America in the Phanerozoic (Fig. 1) records, to a first approximation, several global tectonic events. Palaeomagnetic poles for much of the Palaeozoic and early Mesozoic are about 90° from the interior of North America, and are consistent with a near-equatorial position for the continent, as the stratigraphical record indicates, for this period of time. The early Palaeozoic part of the APW path suggests a significant counterclockwise rotation. Poles for the late Palaeozoic and earliest Triassic are nearly identical, implying relatively quiescent motion of the continent; they probably reflect the final closure of Iaepetus and the stabilization of the supercontinent Pangaea. From Middle Triassic to earliest Jurassic times, North America again experienced counterclockwise rotation. At the beginning of the Jurassic, however, a major change in plate motion occurred, because the continents bordering the central Atlantic began to separate. Although there is controversy about the details of the APW path and the ages assigned to particular palaeomagnetic poles, clockwise rotation, ultimately combined with northward translation, became the dominant motion. North America was closest to the pole in the mid-Cretaceous, between about 110 and 80 Ma ago. Since the latest Cretaceous, a slight counterclockwise rotation and southward translation has occurred.

Phanerozoic APW paths have been established for most of the continents, but unfortunately the database for them is not as comprehensive as that for North America. Nonetheless, sufficient palaeomagnetic data exist to reveal many features of the past tectonic history of the continents. For example, the database for India documents its large-magnitude northward drift since the break-up of Gondwanaland, beginning in Early Jurassic time. In contrast to North America, the Mesozoic and Cenozoic parts of the APW path for South America are of exceptionally low amplitude. The limited Palaeozoic data from South America, however, indicate considerable motion.

A fundamental consequence of plate tectonic movements is that continents not only move but also grow with time, by the accretion of fragments of previous continents, oceanic islands, or parts of ocean floors being subducted. While many North American palaeomagnetists were devoted to establishing an APW path by studying rocks in the interior of the continent, the western margin of North America also became the focus of numerous palaeomagnetic studies. Most palaeomagnetic poles from these rocks did not agree with the reference poles of appropriate age from the APW path. Typically, they lay to the right of or to the far side of the APW path. These discrepancies were interpreted by Myrl Beck of Western Washington University and others to indicate, in an absolute sense with respect to North America, clockwise rotation (eastward declinations) and northward translation (gentler inclinations) of much, if not most, of the crust comprising the western margin of the continent. Rocks from Baja California to south and central Alaska yielded data interpreted to suggest the impressive rates at which fragments of crust can be transported before and during final accretion with a larger continent. Just as with the interpretation of APW paths, much debate still centres on the interpretation of the palaeo-

magnetic data; for example, the data for intrusive igneous rocks could be explained by local tilting, rather than latitudinal transport and rotation.   JOHN WM GEISSMAN

Further reading

Butler, R. F. (1992) *Palaeomagnetism: magnetic domains to geologic terranes.* Blackwell Scientific Publications, Oxford.

Van der Voo, R. (1993) *Palaeomagnetism of the Atlantic, Tethys, and Iapetus Oceans.* Cambridge University Press.

palaeomagnetism: local deformation Palaeomagnetic methods can be used to quantify crustal deformation at scales considerably smaller than that of a continent. The palaeomagnetic approach has often proved most powerful in cases where conventional field-based studies provide limited or ambiguous information. Deformation mechanisms of interest can generally be separated into the rotation of crustal blocks about vertical axes (e.g. differential transport of a thrust sheet) and tilting of crustal blocks about horizontal axes. Determination of the magnitude of either rotation about a vertical axis or of tilting about a horizontal axis can be made either in absolute or in relative reference frames. In the former case, representative averaging of the ancient field is required because the data from the deformed region are quantitatively compared with reference, time-average field directions determined from palaeomagnetic poles from the stable part of a continent. Confidence estimates for such determinations are based on the quality of the reference and the observed data. Thus, a large number of independent records of the field must be obtained from the area suspected of deformation. On the other hand, comparison of relative deformation across a region can be made by obtaining data from one single rock unit (e.g. a laterally extensive lava flow) whose palaeomagnetism can be considered to be internally consistent, at least prior to deformation.   JOHN WM GEISSMAN

palaeomagnetism: past intensity of the field A complete description of the geomagnetic field at any site requires information about its direction and intensity. Understanding the ancient geomagnetic field, and its changes with time, requires data on both palaeomagnetic secular variation (directional) and palaeointensity. Data on secular variation from geologically young sediments and data on palaeointensity from igneous rocks and archaeological materials have revealed periodicities in field phenomena. Such palaeointensity determinations are based on the general assumption that the intensity of thermoremanent magnetizations, acquired in cooling from high temperatures, is proportional to the intensity of the ambient field, The laboratory experiments are tedious and complicated, and require that the magnetic mineralogy of the material does not change appreciably during repetitive heating experiments.

The intensity of the field, described in terms of a global dipole moment, has fluctuated considerably over the past several thousands of years. Between about 6000 and 2000 years BP, it increased by almost a factor of two. Both non-dipole and dipole sources appear to contribute, in approximately constant proportions, to variations in field intensity. Despite the fact that sources of geologically short-term variations in the field are not completely known, their effects might be considerable. Changes in the atmospheric ^{14}C (carbon-14) concentration over periods of thousands of years have been attributed to fluctuations in the dipole moment and the resulting effects on the cosmic ray spectrum. Both geomagnetists and workers in core–mantle processes agree that the ultimate demise of the dipolar field and the resulting drastic decrease in field intensity and the increased probability of a reversal of the geomagnetic field are caused principally by fluctuations in motion of the liquid outer core, possibly related to the irregular topography of the core–mantle boundary. The relationship of such phenomena to orbital forcing dynamics is, and will continue to be, of interest.   JOHN WM GEISSMAN

palaeomagnetism and polar wander
Central to understanding the importance of ancient magnetisms in rocks is the concept of the palaeomagnetic pole. In early studies of this subject it was necessary to compare magnetizations of identical age from widespread localities on a continent, and simple inspection of palaeomagnetic directions was inappropriate. The earlier research workers proposed an axial geocentric model for the time-average magnetic field. If the coordinates of the sampling locality were known, directions of magnetization could then be transformed into magnetic poles (strictly termed *palaeomagnetic* poles) that would be expected to give rise to the field at the locality studied. By definition, a palaeomagnetic pole reflects accurately the averaging of the geomagnetic field from a set of independent measurements. The term 'virtual geomagnetic pole' (VGP) was used for a representation of a short part of geomagnetic field time (e.g. the field recorded by a single lava flow or sedimentary bed), regardless of whether the field could be represented as an axial geocentric dipole at that instant in time. Compilations of thousands of VGPs from young igneous rocks worldwide have demonstrated that, to a first approximation, the axial geocentric dipole hypothesis is valid. Symmetry arguments show that the position of a palaeomagnetic pole relative to a sampled continent does not determine the palaeolongitude of that continent. However, and very importantly, the distance from the palaeomagnetic pole to the sampled area of the continent, as reflected by palaeomagnetic inclination, does provide a direct estimate of the palaeolatitude of the area. The palaeomagnetic declination gives an indication of the past orientation of the continent.

One goal of palaeomagnetic research has therefore been to provide a set of well-dated palaeomagnetic poles from rocks that have remained part of the stable continent since their

formation. Understandably, certain time periods have been represented by better and more abundant palaeomagnetic data for some continents than for others. Not all palaeomagnetic poles are dated using isotopic age methods. In the mid-1950s, when the early research workers began compiling palaeomagnetic poles for the various continents, they soon realized that poles older than about mid-Cenozoic time did not coincide with the spin axis, nor was there appreciable agreement among poles of similar age from different continents. In consequence, the tracks of palaeomagnetic poles from each continent were introduced as apparent polar wander (APW) paths because they recorded the motion history of the continents, rather than significant motion of the dipole relative to the spin axis. Accurate APW paths provide a powerful base by means of which ancient motion of the continents, relative to the Earth's spin axis, can be assessed and compared. APW paths provide the only means of assessing such motion prior to the early Mesozoic break-up of Pangea. Inferences based on APW path analysis are limited, however, because the palaeomagnetic pole concept assumes an axial geocentric dipole and the palaeolongitude is undetermined. In addition, APW paths are only as robust as the database, specifically the quality of palaeomagnetic pole determinations and the accuracy and reliability of the age assigned to each pole. True polar wander implies non-axiality of the spin axis and the time-averaged dipole. Although difficult to evaluate, true polar wander has been suggested for parts of the Cretaceous and the mid-Palaeozoic.

                                        JOHN WM GEISSMAN

palaeomagnetism: techniques and remanent magnetization
Palaeomagnetic research has addressed virtually all types of geological materials from the pyroclastic deposits of the 1980 eruption of Mount St Helens to Archaean rocks over 3.4×10^9 years (3.4 Ga) old. The goals of these studies, and the techniques employed, have been very diverse, and consequently the scope of each study has varied. The following is a general discussion of conventional techniques in palaeomagnetism and the types of remanent magnetizations possible in rocks.

Collection of geological materials for study typically follows a hierarchical sequence, in which several independently oriented samples are collected at an individual sampling site, generally a volume of materials considered to be magnetized at essentially the same time (e.g. a single lava flow). Ideally, from 7 to 10 samples are collected from any site, and each sample is divided into one or more specimens for measurement. Numerous sampling sites are established at a locality, and several localities are visited for a thorough sampling of a series of rocks.

Samples are collected in the field using several methods and are oriented using a magnetic compass and, for strongly magnetized rocks, a solar compass, prior to removal from the outcrop. For well-consolidated sedimentary and all crystalline rocks, the preferred method is coring, using a portable field drill. Cylindrical samples, up to 15 cm or so in length, are obtained using a hollow, water-cooled diamond drill bit. In remote and/or environmentally sensitive areas, collection of samples as small oriented blocks (approx. 12 cm or so on a side) is necessary. These blocks are later cored in the laboratory into cylindrical samples. Poorly consolidated rocks require more tedious methods, including carving of samples using non-magnetic tools and encasing them in plastic cubes or quartz glass tubes, impregnating samples with water-glass solution, or encasing samples, *in situ*, with plaster of paris.

In the laboratory, one or more specimens are prepared for measurement from each independently oriented sample, each being marked to preserve its initial orientation. Cored samples are cut using a diamond-tipped, water-cooled blade into cylinders with a length diameter ratio of approximately 0.86, to minimize shape effects. For weakly magnetized materials, it is often important for specimen preparation to be carried out using non-magnetic tools in a low magnetic induction environment.

Determination of the natural remanent magnetization (NRM) requires measurement of both moment and direction, in the absence of an inducing field. Commercially available magnetometers are of either the spinner or superconducting type. Spinner magnetometers utilize the rotating moment of the specimen to alter the current in a nearby fluxgate pick-up coil, of cylindrical or ring design. Superconducting or cryogenic magnetometers consist of one or a set of orthogonally wound superconducting pick-up coils connected to superconducting rings at the temperature of liquid helium. A magnetic moment in the pick-up coil induces a current that is proportional to the magnetic moment normal to the plane of the coil. With three orthogonal coils, the total magnetization can be measured quickly. The main advantage of this magnetometer is the speed with which measurements are made. Magnetometers in most laboratories are linked to a microprocessor for rapid data acquisition, reduction, and analysis.

The NRM in a rock may be very complex, consisting of several discrete magnetizations, each representing a different remanence acquisition during the history of the rock. Because most rocks have been exposed to a normal polarity field for the past 780 000 years at ambient temperatures, they typically contain a viscous remanent magnetization (VRM). Geologically older magnetizations may be of essentially primary origin, acquired at the time the rock was formed, secondary in that they were acquired after the formation of the rock, possibly during a very different field direction polarity, or both. To determine the direction and relative intensity of all components of magnetization present in the rocks, the NRM is demagnetized in the laboratory using, in general, alternating field (AF), thermal, or chemical techniques. In some instances, a successive combination of two of the methods may be employed.

In AF demagnetization, the specimen is uniformly subjected to an alternating magnetic field which decays with time. As the value of the peak alternating field is increased,

the coercivity of an increasing percentage of grains is exceeded, and the magnetization of each grain is effectively randomized as the field alternates during decay. Commercial AF demagnetizers are capable of maximum fields in the range of 1000–2000 oersted (100–200 millitesla inductions), which are usually sufficient to activate assemblages of magnetite grains. The coercivity of haematite exceeds that of magnetite, and hence this technique is of little use on haematite-dominated magnetizations.

Thermal demagnetization entails uniform heating of the specimen and subsequent cooling in a zero field. When the temperature exceeds the unblocking temperature of a magnetized grain, that grain is effectively demagnetized as it

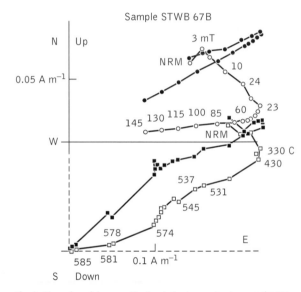

Fig. 1. Examples of demagnetization behaviour, using two specimens, one subjected to alternating field (AF) demagnetization (circles, axes as solid lines), the other thermal (squares, axes as dashed lines), from the same core sample of a gabbro from the Stillwater Complex, Beartooth Mountains, southern Montana (c. 2.7 Ga). For the two orthogonal demagnetization diagrams, the end point of the magnetization vector is simultaneously plotted on to a horizontal (filled symbols) and a vertical (open symbols) plane, and the E–W axis serves as the common axis in both diagrams. Peak alternating field values (in millitesla (mT) or temperatures (in °C) are given beside the projections on to the vertical plane. Magnetization intensity values are given in A m⁻¹ (amperes per metre). In these diagrams, the data are plotted with respect to geographic coordinates; for layered rocks it is preferable in some instances to plot them with respect to stratigraphical coordinates. In both AF and thermal demagnetization, a magnetization of east-north-east declination and moderate negative inclination (trending towards the origin) is removed above about 30 mT in AF demagnetization and about 430 °C in thermal demagnetizations; this magnetization is interpreted as a primary thermoremanent magnetization (TRM) acquired during the initial cooling of the Stillwater Complex c. 2.7 Ga ago.

cannot acquire a magnetization by cooling in a zero magnetic field. This technique reveals the unblocking temperature spectrum of the NRM and can allow for complete demagnetization of all rocks. Maximum unblocking temperatures for magnetite are about 580 °C, for haematite about 680 °C.

Chemical demagnetization has typically been applied to permeable, haematite-cemented detrital sedimentary rocks. Specimens are immersed in concentrated hydrochloric acid for a known time, rinsed in distilled water in a low magnetic field, dried and measured. In principle, the finest grains are haematite pigment removed during early stages of treatment, with coarser material later; magnetite in the form of detrital grains in the rocks is mainly unaffected.

Demagnetization results consist of a set of magnetization directions and intensities obtained after each demagnetization step. Barring any changes in the magnetic mineralogy and contamination by artificial magnetizations, at least 95 per cent of the NRM should be removed in demagnetization to assess all NRM components adequately. A useful approach for inspecting demagnetization data, used in virtually all laboratories, employs orthogonal demagnetization diagrams, first introduced in the early 1960s by J. D. A. Zijderveld and colleagues at the University of Utrecht, where the end point of each magnetization vector is simultaneously projected on to horizontal and vertical planes (Fig. 1), the vertical plane having been folded 90° about a common horizontal axis. The colinearity of data points on both the horizontal and vertical projections usually signifies the removal or isolation of a single vector component magnetization over that range of demagnetization treatment. On the other hand, curvilinear trends of data points on one or both projections indicate that at least two vector components are being removed over similar demagnetization steps, possibly in different proportions between each step. The directions of individual vector components and the orientations of great circles or planes reflecting the simultaneous removal of more than one magnetization, defined by at least three data points and identified by inspection of orthogonal diagrams, are routinely determined using a linear three-dimensional principal components analysis.

The statistical evaluation of magnetization data, as directions or planes, takes several forms. The classic approach,

referred to as Fisherian statistics after the English mathematician who first described their use, assumes circular symmetry about a true mean and that directional data are distributed according to a probability density function. Palaeomagnetic data sets are often bimodal in that both normal and reverse polarities of the field have been sampled and inversion of one part of the population is required. The data can be elliptically, rather than circularly, distributed for several reasons, including inadequate removal of an additional magnetization or incomplete averaging of field variations.

Palaeomagnetic data, with corresponding statistics and acceptance information, are tabulated in any publication. Demagnetization results, particularly in the form of representative orthogonal demagnetization diagrams, are shown. Site mean and formation mean directions, with projected cones of 95 per cent confidence, as well as virtual geomagnetic pole and palaeomagnetic pole projections, illustrate the distribution of data graphically. It is important that enough samples are collected to obtain a statistically well-defined magnetization. The magnetization characteristic of the rocks studied should be found at least at three sites, where seven or more independently oriented samples have been collected from each site. Reverse and normal polarity magnetizations should be shown to be antipodal at a significant level. Finally, the demagnetization characteristics of the magnetization should be consistent over samples from several sites.

For all palaeomagnetic investigations, the geological age of the rocks investigated and the age and origin (or likely origin) of magnetizations characteristic of the rocks should be well determined. Age information may be provided by isotopic age determinations, to as great a precision as possible, and/or biostratigraphical study. The geological stability of characteristic magnetizations can often be assessed from field-based tests (Fig. 2), response to demagnetization behaviour, rock magnetization tests, and directional grouping of magnetization data.

The earliest applied field test entailed sampling deformed sedimentary strata of the same age at several different positions in a fold. Magnetizations acquired before folding are dispersed by deformation, and restoring the beds to the palaeohorizontal, by back-tilt about the present strike of the

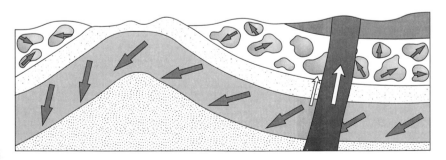

Fig. 2. Fold, conglomerate, and baked contact field tests for the geological stability of a magnetization. A pre-folding magnetization is shown for a rock unit given a light stipple pattern. This magnetization is of random orientation in clasts of unit 2 contained in the conglomerate. Adjacent to the contact with a dyke, all the host rocks contain a magnetization parallel to that of the dyke (open arrow). (Fran Butler (1992).)

rocks (structural correction) should decrease the dispersion of pre-folding magnetizations. Alternatively, magnetizations acquired after folding should increase in dispersion as beds are restored to the horizontal. The test, even under the most ideal circumstances, can only be used to identify a magnetization as pre- or post-folding, not whether is it primary. Recent work on the deformed rocks of mountain belts demonstrated that rocks often acquire a characteristic magnetization during deformation. Interpretation of these magnetizations that are of the same age as the folding is complicated by the possibility of penetrative deformation affecting the magnetization-carrying grains.

Conglomerate tests are applied to mechanically disaggregated clasts of a formation whose palaeomagnetism is under investigation. If clasts contain a magnetization with properties similar to those of the intact information, and if the magnetization is random, then the magnetization predates erosion and transportation. If the magnetization is uniform throughout the clasts, then the clasts, and in all likelihood the formation in question, have been remagnetized since clast deposition.

Contact tests assess the age of magnetization in an intrusive igneous rock relative to that of its host. The host rock is thermally disturbed during intrusion and, ideally, the igneous rock and host at and near the contact cool over a range of magnetization blocking temperatures at the same time, so that the magnetization in the host adjacent to the intrusion is identical to that in the intrusion. If the host rock distant from the intrusion possesses a well-grouped magnetization that is distinct in direction, the overall results provide a positive contact test. A potential drawback to this test is that planar intrusions (i.e. dykes and sills) are excellent conduits for passage of fluids and they, as well as the rocks in immediate contact, may have been chemically remagnetized, possibly long after intrusion.

Many palaeomagnetic investigations are augmented by additional studies. Magnetization carriers should be identified through rock-magnetization experiments. ^{40}Ar/^{39}Ar isotopic age-spectrum determinations may provide high-precision age information on magnetizations acquired in thermal blocking. Scanning electron and scanning transmission electron microscopy can provide fundamental information on the nature and origin of magnetic phases. Low-field magnetic susceptibility, used as a proxy for total magnetic content, is usually monitored between laboratory demagnetization steps, and, as such, is an important parameter in palaeomagnetic work. Anisotropy of magnetic susceptibility (AMS) data are obtained routinely to evaluate whether the rocks studied have a well-developed magnetic fabric. In penetratively deformed rocks, AMS data have in some instances been used to correct magnetizations for the deformation.

Rocks acquire permanent or remanent magnetizations in several ways. In igneous rocks, thermoremanent magnetizations (TRMs) are acquired in subsolidus cooling over a range of blocking temperatures below the Curie temperature of the magnetic phases. For an individual single-domain grain, the time constant associated with the 180° flipping of a magnetization, in cooling over a short temperature interval, increases from about a thousand seconds, or less, to several millions of years, and the magnetization is said to be blocked. With cooling, the probability of a single grain reversing its magnetization decreases even further, because the energy barrier to inversion increases. TRMs residing in fine magnetic particles typically have a great stability with respect to geological time; the oldest of such magnetizations are over 3.0×10^9 years (3.0 Ga) in age. Because TRMs are acquired soon after magmatic magnetic minerals crystallize, they are referred to as primary magnetizations.

Chemical remanent magnetizations (CRMs) are acquired during the growth of magnetic phases, the magnetization is blocked as a result of an increase in grain size. CRM acquisition typically occurs at some point after the formation of a rock; these magnetizations are referred to as secondary, and field-based tests support this assignment. Examples of CRM acquisition include the growth of authigenic magnetic phases in sedimentary rocks, the oxidation of a magnetic phase and replacement by a second magnetic mineral, and the alteration of an iron-bearing silicate mineral into a magnetic iron oxide and other silicate minerals. CRMs do not always faithfully record the direction of the ambient field as the time of magnetization blocking their directions may be a function of both the pre-existing magnetization and the actual process of secondary mineral formation.

Eroded and transported magnetite and haematite grains ultimately comprise part of the sediment in most depositional environments. The magnetic moments of sufficiently fine magnetic particles are readily aligned into parallelism with the ambient field in settling over short vertical distances in an aqueous medium. This results in a depositional remanent magnetization (DRM) which, in a strict sense, may be primary in origin. DRMs in fine-grained sediment deposited rapidly are often exceptionally high-fidelity recorders of the Earth's magnetic field. Depositional effects, such as water current and irregular depositional surface; early post-depositional effects, such as dewatering/compaction-related grain motion and organic activity, and later diagenetic changes can all significantly modify a primary, high-fidelity signal represented by the DRM.   JOHN WM GEISSMAN

Further reading

Butler, R. F. (1992) *Palaeomagnetism: magnetic domains to geologic terranes*. Blackwell Scientific Publications, Oxford.

Van der Voo, R. (1993) *Palaeomagnetism of the Atlantic, Tethys, and Iapetus Oceans*. Cambridge University Press.

palaeontology, palaeobiology

Palaeontology is the study of the record of life through geological time (from the Greek *palaios*, ancient, *ontos*, thing or being, and *logos*, study). The subject has traditionally encompassed the study of fossils (from the Latin *fossilis*, dug up), which are the preserved remains of once-living organisms, in order to reconstruct them and use them to establish evolutionary relationships. This work is also important in the reconstruction of ancient environments, and particularly in biostratigraphy in which the observed distribution of fossils through geological time is used both in dating and correlation. Palaeontology therefore bridges the sciences of geology and biology, derives its principles and methods from both, and supplements both. In recent years, new methodologies such as cladistic methods in phylogeny reconstruction and the paradigm method in functional morphology have been used to answer palaeontological questions, but those questions have remained the traditional ones concerning the appearance of extinct organisms, their way of life, and evolutionary relationships. Palaeobiology is a more theoretical approach in which emphasis is placed on the understanding of processes. It has, therefore, been involved in the study of extinctions, particularly mass extinctions such as those that took place at the Cretaceous–Tertiary and Permian–Triassic boundaries, and also the discussion of evolutionary processes such as punctuated equilibrium.

DAVID K. ELLIOTT

palaeoseismology

Palaeoseismology is the study of prehistoric earthquakes using geological and geomorphological evidence. 'Prehistoric', in this context, is commonly taken to mean the past 500 000 years in contrast to the usual elastic use of this term. Palaeoseismology is of importance in assessing earthquake hazard, especially in regions where the time interval between damaging earthquakes is measured in thousands of years, and thus exceeds the length of our instrumental and historical records.

The two principal methods of palaeoseismological investigation are the analysis of (1) fault scarps and (2) sedimentary strata that have been offset or internally deformed by movements on an earthquake fault. The principle of analysing fault scarps to determine the magnitudes of, and intervals between, past earthquakes rests on the premise that the shape of fault scarp is controlled by denudational processes from the moment that it formed during an earthquake (see *fault scarp*). Where, in a natural outcrop or specially excavated trench, increasingly large offsets can be measured downwards across an earthquake fault it is possible to calculate the time between increments of movement, provided that the strata can be precisely dated. If sedimentary horizons that have been internally deformed by earthquake shaking can be distinguished from those related to other processes, and if they too can be dated, it is again possible to work out when the earthquakes occurred.

PAUL L. HANCOCK

palaeosols, duricrusts, calcrete, silcrete, gypcrete

Palaeosols are soils preserved in the geological succession that were formed in the past on ancient landscapes. They are useful to the geologist in that they indicate that the sediment surface was exposed to the atmosphere at the time of deposition. Furthermore, comparison with modern soils gives valuable information about climatic conditions in the geological past.

Palaeosols are commonly recognized by noticeable colour changes in a sedimentary sequence. They are usually dark in colour because of the presence of organic matter, or reddened because of oxidation of the original sediment. The original stratification of the underlying sediment is disrupted, and in well-developed palaeosols a new horizontal layering is produced. Their macrostructures are similar to those of present-day soils, with pseudo-anticlinal structures and cuspate structures comparable to those produced by the expansion and contraction of swelling clays. They commonly show the grain-size trends that are observed in modern soil profiles, as well as vertical profiles of trace elements which can be related to modern soil profiles formed under different climates. Palaeosols also show micromorphological features with distinctive sequences of mineral redistribution and precipitation. Under ideal conditions, evidence of biological activity, such as rootlets and burrows, can be preserved, but because of their low potential for preservation these features are not always evident. Palaeosols have been found in rocks of all ages and have even been recognized in deposits that have been metamorphosed (the Waterval Onder Formation in South Africa). They are more common in sediments after the later Silurian–Devonian owing to the marked spread of plants since that time.

Terrestrial weathering commonly leads to concentrations of various minerals in the surface materials that have enclosed, cemented, or replaced the original material to form a hard duricrust. They can be formed of aluminium sesquioxide: *alcretes*; iron oxides and hydroxides: *ferricretes*; silica: *silcretes*; calcium carbonate: *calcretes*, or *gypsum: gypcretes*. Of these, calcretes, silcretes, and gypcretes are among the more common.

Because of their hardness palaeosols influence the landscape and commonly cap hills or coat slopes, protecting the underlying sediments from erosion. They are best developed on sub-horizontal or gently sloping surfaces. These features take a considerable time to form, and when well developed indicate a long period of crustal stability at that site; they are consequently very characteristic of stable cratonic areas. They can form a thin skin over rock or reach thicknesses of tens of metres in some areas. There has been considerable discussion about the mode of origin of all these materials. They have generally formed by leaching and precipitation within the soil profile, or by laterally moving subsurface or even surface flow.

Palaeosols are frequently used as palaeoclimatic indicators. Calcrete (caliche) can be found in almost all climatic zones but is best developed in warm areas with low precipitation. Silcrete is found in humid or arid continents; sometimes it is possible to ascertain which from their geochemistry. Gypcrete is characteristic of arid areas.

Over the past few decades there has been an increasing recognition of calcretes and silcretes in the geological column. Calcretes are known in the Precambrian and Palaeozoic, but they become much more common in Devonian and younger rocks. Silcretes are more rare but have also been reported from Precambrian rocks and deposits of other ages, for example, capping aeolian dune sands of Permo-Triassic age in northern England.

G. EVANS

Further reading

Goudie, A. S. (1983) Calcretes. In A. S. Goudie and K. Pye (eds) *Chemical sediments and their geomorphology*, pp. 93–132, Academic Press, London.

Summerfield, M. A. (1983) Silcretes. In: A. S. Goudie and K. Pye (eds) *Chemical sediments and their geomorphology*, pp. 59–92. Academic Press, London.

Waltson, A. (1983) Gypcretes. In A. S. Goudie and K. Pye (eds) *Chemical sediments and their geomorphology*, pp. 133–162. Academic Press, London.

Wright, V. P. (ed.) (1986) *Palaeosols*. Blackwell Scientific Publications, Oxford.

palaeovolcanism and climate change *see* CLIMATE CHANGE AND PALAEOVOLCANISM

Palaeozoic

The Palaeozoic era and rocks are named on account of the fossils of 'ancient life' (from the Greek) within them. The term was first used by Professor Adam Sedgwick in 1838 for his series of somewhat deformed rocks lying beneath the Old Red Sandstone in Wales. It was soon afterwards used by John Phillips to include all the rocks that Sedgwick had described plus the Old Red Sandstone, the Carboniferous, and the Magnesian Limestone of northern England. Today it encompasses the systems from Cambrian to Permian inclusive, and spans geological time from 545 Ma to 250 Ma. Conventionally, the Cambrian, Ordovician, and Silurian systems constitute the Lower Palaeozoic and the Devonian, Carboniferous, and Permian comprise the Upper Palaeozoic.

The base of the Cambrian system is thus also the base of the Palaeozoic erathem. For many years the lowest occurrence of trilobites was taken as the first indication of Palaeozoic life, but the International Commission on Stratigraphy regarded a properly defined and internationally recognized base of the Cambrian system as a priority, and set up a working group to find one. This has now been done (see *Cambrian*).

The lowest Cambrian rocks are not everywhere very fossiliferous, but the record shows that by Mid-Cambrian time there was a very diverse marine ecosystem. Faunal communities, largely of animals lacking hard parts, were superseded by those in which well-skeletonized creatures were prominent. This probably indicates that animals evolved to colonize the more turbulent and shallow reaches of the seas and inland waters during early Cambrian times.

The biostratigraphy of this immense sector of geological time is highly varied, with many fossil groups providing zonal schemes. All modern animal phyla and a large part of the plant kingdom are represented. Habitats ranged from the terrestrial to the bathyal. There were several groups of animals that were highly successful during this era but became extinct before its end. Certain facies in each system can be correlated and dated with great precision on the basis of zonal fossils. The development of terrestrial vegetation, culminating in the great forests of the Carboniferous period, had profound effects upon ecology. Not only were new habitats provided and habitable space on the planet increased, but the spread of vegetation from water to land affected rates of rock weathering, erosion, and sedimentation. Terrestrial tetrapods had appeared well before the end of Devonian time. Marine fossil evidence tells of numerous extinctions, global mass extinction, and geochemical events. At the end of the Palaeozoic era the greatest extinction event of all time took place, providing the classical catastrophe of early nineteenth-century writers.

Palaeozoic geography progressed through an initial dispersal of cratons from a late Proterozoic supercontinent in the southern hemisphere, followed by a progressive regrouping near to the equator by late Carboniferous times. The result was the supercontinent Pangaea with a widespread continental climate. During this long period of time, plate-tectonic margin activity was vigorous and collision tectonics common. At various times parts of the great southern continental area (Gondwanaland) experienced the growth of ice sheets of enormous proportions. Evidence for former glaciation occurs in the Silurian of North Africa and the Devonian of South America, but in the Carboniferous and Permian rocks of all the southern continents there are abundant signs of a very long glacial period, with many phases of growth and decline of an ice mass of unrivalled size. These glacial oscillations constantly affected global sea level and sedimentation at the continental margins.

Palaeozoic volcanicity was conspicuous along active continental margins especially where there was subduction. All traces of Palaeozoic ocean floor are presumed to have been subducted. There were numerous polar reversals, and mantle activity appears to have been an underlying cause of these events.

D. L. DINELEY

Further reading

Hailwood, E. A. and Kidd, R. B. (eds) (1993) *High resolution stratigraphy*. Geological Society of London Special Publication No. 70.

Harland, W. B. *et al.* (eds) (1967) *The fossil record*. Geological Society, London.

McKerrow, W. S. and Scotese, C. R. (1990) *Palaeozoic palaeogeography and biogeography*. Geological Society of London Memoir No. 12.

Moullade, M. and Nairn, A. E. M. (eds) (1991) *The Phanerozoic geology of the world. I. The Palaeozoic*. Elsevier, Amsterdam.

palinspastic maps and sections

A palinspastic reconstruction, which may be in the form either of a map or a cross-section, is a reconstruction of the geometry of an area of folded or faulted rocks as it is thought to have existed at some time in the past, before certain known geometrical changes took place.

The rocks may thus be shown in an unfolded or un-faulted state, with the geometry restored to take account of the displacements produced during the process of deformation.

A special type of palinspastic reconstruction is the 'balanced section', much employed by structural geologists to restore complex folded and faulted belts to their undeformed state. First used in the thrust belts of the Canadian Rocky mountains, the technique is now widely applied to orogenic belts in an attempt to understand their subsurface geometry and to estimate the amount of shortening or extension that a piece of crust has undergone. A section is said to be 'balanced' when several different horizons are restored to the same original length, thus helping to eliminate errors in the restoration process. Understanding the present geometry of complexly deformed areas has a practical value in that the shapes of hidden oil traps or orebodies may, for example, be determined.   R. G. PARK

palynology

The term 'palynology' (from the Greek word *palynos*, 'dust') was first used by Hyde and Williams in 1944. It is now used to include the study of both modern and fossil spores, pollen, dinoflagellates, acritarchs, chitinozoa, fungi, bacteria, and other related marine forms, which are all collectively known as palynomorphs. Although palynology is used to study present-day living micro-organisms, its main application is the study of fossil palynomorphs. Palynology is a subdivision of the broad field of botany.

The underlying assumptions of palynology are no different from those in other areas of palaeontology. They include the familiar trust in the present being the key to the past and a belief that fossils are elements of a continuum of once-lived organisms whose succession was shaped by organic evolution.

Palynology is both a pure and an applied science. Academically it is used to define more precisely the fossil record, making possible the chronostratigraphic (time) subdivision of rock units. It is also used to help in the understanding of the origin of life and the spread of land plants throughout geological time.

The industrial applications of palynology are most evident in the hydrocarbon and coal industries, where it is used to define discrete packages of rock, making possible chronostratigraphic subdivision and correlation of samples from outcrops and boreholes.

Palynological examination of archaeological sites is an important tool in helping to determine the geographical origin of artefacts. This is possible because palynomorphs from one region tend to develop differently from those from another. If two or more types are found at the same site, some transport of material may therefore be inferred.

Microscope slides for palynological analysis were originally prepared from thin sections of rock. Today, rock samples are treated with various acids in order to release the fossil palynomorphs so that they can be mounted on slides for examination, either by transmitted light microscopy or by scanning electron microscopy.

Palynology depends mainly on four characteristics of palynomorphs: (1) they have a much greater resistance to degradation than most other plant parts; this makes them more likely to survive as fossils; (2) their small size, mostly less than 200 microns, means that they are transported and deposited as sedimentary particles by both wind and water; (3) their morphological complexity and rapid evolutionary development make them easy to distinguish and characterize; (4) their production in enormous numbers (up to four thousand in one gramme of coal, for example) facilitates the recovery of statistically significant assemblages.

Palynology is a relatively new member of the palaeontological sciences that crosses traditional disciplines. Palynological studies can be approached from the viewpoint of botany, with the emphasis on the plant relationships, or from the geological perspective, with emphasis on the biostratigraphy. These approaches are not mutually exclusive; indeed, the full comprehension of the science requires knowledge of both fields.

Palynology is a continually expanding science because as new areas of the Earth's surface are explored new forms of palynomorphs are being continually discovered. In addition, new laboratory techniques for releasing the palynomorphs from the host rock are being developed all the time. This expansion will lead to more refined and extensive geological and stratigraphical data, in addition to furnishing a better understanding of the development of the plant kingdom that will provide greater knowledge of plant geography, palaeoclimatology, climatic change, evolution, and palaeoenvironments of deposition.   A. LOY

Pangaea

Early in the twentieth century the German scientist Alfred Wegener postulated that, commencing in the Mesozoic and continuing up to the present, a huge supercontinent, 'Pangaea' (meaning 'all land'), had rifted and the fragmented components had moved apart as a result; this soon came to be generally known as continental drift. His interpretation was that South America and Africa began to separate in the Cretaceous, as did North America and Europe, but North America and Europe had retained contact in the north as late as the Quaternary. The Indian Ocean had begun to open up in the Jurassic but the principal movements took place in the Cretaceous and Tertiary.

Although Wegener's supercontinent was controversial for many years, the general outlines of his interpretation are now widely accepted, although parts of it have had to be modified as more information has become available. Thus no one accepts today that the final separation of North America and Europe took place in the Quaternary. With the use of the magnetic anomaly record of the ocean floor it is possible to trace the opening of the Atlantic and Indian Oceans in considerable detail. The earliest opening apparently took place in the Middle Jurassic, in the central sector of the Atlantic between the present United States and north-west Africa, followed slightly later by opening of the Mozambique Channel

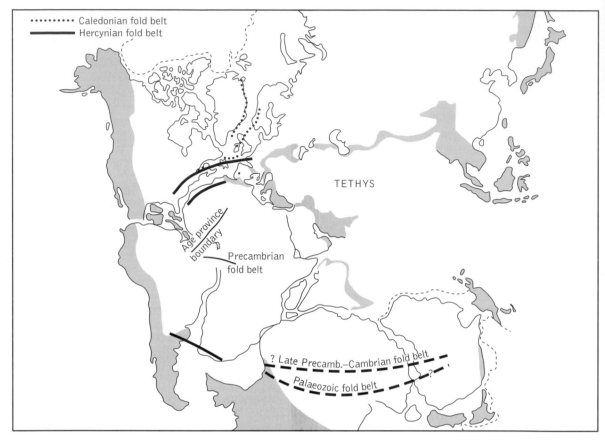

Fig. 1. Reconstruction of Pangaea in Triassic times based on the reconstruction by Alan Gilbert Smith and Jim Briden. The stippling indicates areas affected by Tertiary deformation whose exact position is in consequence uncertain. (After Hallam (1973).)

off East Africa. The principal separation of the continents took place, however, in the Cretaceous, with the South Atlantic commencing its opening, Western Europe separating from North America, and India separating from Australia, Africa, and Antarctica. In the early Tertiary the northernmost sector of the Atlantic, the Norwegian Sea, began to open, and North America became separated from Europe in the Eocene. The final separation of Australia from Antarctica also took place in the Early Tertiary.

Figure 1 shows a modern reconstruction of Pangaea on a Mercator-type projection, showing a number of geological tie-lines between the continents. The reconstruction is much more detailed than Wegener could attempt; it takes advantage of bathymetric and geological information unavailable to him, and uses least-squares computer fits of continental edges. The fit is awkward in two regions. There is some overlap of North and South America, and the Antarctic Peninsula does not align neatly, as the geology would suggest, with the Andean mountain chain of South America. It is believed that the reason for

this lies in a mixture of continental stretching and strike-slip faulting between continental segments. Major modifications also have to be made of such Pangaea reconstructions undertaken in the years soon after the general acceptance of continental displacements and plate tectonics. It has become established within the past two decades that extensive parts of southern and eastern Asia have been accreted to the main Asian continent since the Late Palaeozoic, with a series of small continents progressively moving towards and suturing with the mainland (so-called displaced or exotic terranes). Thus, while western Pangaea was splitting up in the Mesozoic, northeastern Pangaea experienced the reverse process.

Because of a positive hypsometric relationship between continental area and average height, Pangaea must have stood higher above the ocean level than the present dispersed smaller continents. It would also have induced a global monsoonal circulation rather than the zonal atmospheric circulation of today. Although the Mesozoic is generally considered to have experienced a more equable climate than that of the present day, the absence of polar ice caps in the Mesozoic must have resulted in considerable seasonal temperature extremes in the Pangaea climate, and much of the continental interior, far from the ocean, would have been arid in low latitudes.   ANTHONY HALLAM

Further reading

Embry, A. F., Beauchamp, B., and Glass, D. J. (eds) (1994) *Pangea: global environments and resources.* Canadian Society of Petroleum Geologists Memoir No. 17.

Pannotia *see* RODINIA AND PANNOTIA

pans The word 'pan' has at least three meanings in Earth science. In the context of soils it can refer to compacted or cemented subsurface horizons (hardpans, duripans, fragipans, etc.). In the context of geomorphology it can refer either to small weathering pits developed on rock surfaces (equivalent to rock basins or grammas), or to various types of closed depression that occur widely in many of the drier parts of the world. The last of these three meanings is the most prevalent and is the one that is discussed here.

Pans are depressions with no active outlet (i.e. they are closed) that range in size from small depressions, a few tens of metres across and only centimetres deep, to large depressions many kilometres in extent and many tens of metres deep. Many examples display a characteristic kidney-, clam-, chop-, or heart-shaped morphology. On their lee sides some pans have clay dunes or lunettes composed of sandy, silty, clayey, and salty materials blown out from the pan floor. Pans develop in a variety of settings: on susceptible rock surfaces (especially shales and fine sandstones), on desiccated lake surfaces, in inter-dune corridors, and on coastal plains. They are especially widespread in the Pampas, Kalahari, Texan High Plains, and the drier parts of Australia. Solution, tectonics, and annual activity may play a role in their development but deflation by wind is probably the crucial factor in most cases.
                                    ANDREW S. GOUDIE

partial melting and magma production The

igneous rocks observed at or near the Earth's surface are the culmination of many processes within the mantle and crust. Molten rock (magma), produced from an initially solid source, separates and rises to the surface. During this ascent, it may be modified extensively. Although some magmas are generated within the Earth's crust, they are insignificant compared with the enormous volumes generated from within the mantle, and are not discussed here.

Normal conditions in the mantle

The upper mantle of the Earth comprises two parts: a non-convecting lithosphere and an underlying convecting asthenosphere. These distinctions result from temperature differences that control the physical behaviour of mantle rocks; compositionally, the two parts are very similar, consisting of lherzolite (a type of peridotite, an ultramatic rock composed of olivine and pyroxenes). Radioactive heat and the residual heat of core formation are transferred from the Earth's deep interior by convection through the asthenosphere and conduction through the lithosphere.

Under normal conditions, the flux of heat is insufficient to raise mantle temperatures to solidus temperature for lherzolite, that is, the temperature at which it will melt. Increasing pressure with depth, which also increases the solidus temperature, counteracts the effects of increasing temperature. In addition, melting requires significant amounts of extra heat, (in thermodynamic terms, the enthalpy of fusion), to break molecular bonds and cause a phase transition from solid to liquid. In sum, the interior of the Earth is normally not hot enough to melt, and whatever magmas are formed must be produced from previously solid rock.

Melting in the mantle

How then does melting occur? Consensus has been reached that two mechanisms dominate melt production in the mantle: (1) decompression melting caused by the rapid upward movement of volumes of mantle lherzolite; and (2) the lowering of the solidus temperature of mantle lherzolite by the introduction of volatiles. In both cases, these mechanisms produce partial melting of the mantle. Total melting is not possible, that is, raising the temperature of mantle lherzolite to its liquidus (the temperature at which it is completely melted), requires too much heat, even when significant amounts of volatiles are present. It is not possible to retain large fractions of melt within the mantle; when a few per cent of melt are present, the molten rock will separate from the solids and ascend.

Decompression melting

The process of decompression melting was described in a landmark paper by D. McKenzie and M. J. Bickle of Cambridge University in 1988; the following discussion owes much to their work. It is possible to generate magmas of a wide range of compositions from a source material of uniform composition by changing the percentage of material melted and the depth at which melting occurs. For example, the type of melt produced changes with increasing depth from tholeiite (a silica-rich basalt) to alkali basalt to picrite (an ultramafic rock) to komatiite (a lava rich in magnesium). However, to produce melt at great depths (greater than 100 km) requires very high potential temperatures. The next most important control on magma composition is the mineralogy of the source material.

Figure 1 shows the melting interval of anhydrous mantle lherzolite. The movement of heat into and out of an ascending volume of mantle is very slow compared with the rate at which it ascends, and the process may be considered to be adiabatic (that is, no transfer of heat occurs). The temperature of the volume drops as some of its internal heat is used to cause expansion of its constituent minerals as they move to lower pressures, and the batch follows an adiabatic gradient. The amount of heat carried by an ascending volume of mantle can be characterized by its potential temperature (T_p),

791

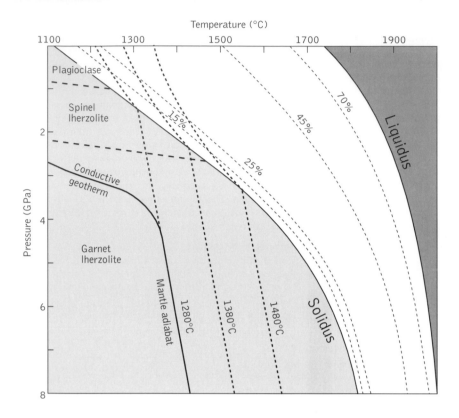

Fig. 1. Melting interval of anhydrous lherzolite, contoured by percentage of melt. Three decompression paths (labelled by potential temperature) are shown by dotted lines. These paths are adiabatic below the solidus, but become strongly inflexed above it as internal heat is consumed during melting. Approximate positions of the different types of mantle lherzolite are indicated by dashed boundaries. For reasonable potential temperatures, most melting occurs within the spinel lherzolite facies. (Modified from D. McKenzie and M. J. Bickle (1988) *Journal of Petrology*, **29**, 625–79.)

which specifies the temperature the volume would reach if permitted to ascend to the surface. Ascent could be due to mantle convection, compositional (density) differences, or extension of the overlying lithosphere.

The batch of mantle ascends adiabatically, whereas the geotherm becomes conductive. The batch moves off the geotherm and continues to ascend until it meets the solidus where melting begins. Along with the loss of heat due to expansion, significant amounts of internal heat are used during melting, accounting for the change in slope of the paths shown. The higher the initial potential temperature of a volume of ascending mantle (the two other curves), the greater the extent to which the batch will melt.

Decompression melting of the asthenosphere dominates environments where the overlying lithosphere is extending: for example the world-girding mid-oceanic ridges, back-arc spreading centres, and continental rifts. The magmas that form oceanic islands and some continental volcanic centres result from pressure release melting of diapirs (buoyant masses of low-density rocks capable of rising through and piercing overlying rocks) rising from deep in the mantle.

Melting by volatile fluxing

Mantle volatiles include water, fluorine, chlorine, and carbon dioxide, and may be present in the mantle as either volatile-

rich minerals (phlogopite, amphibole, magnesite) or a separate fluid. The decompression melting described above assumed a mantle lherzolite lacking volatile-bearing minerals and requires the ascent of mantle lherzolite to produce melting. However, experimental studies of natural lherzolite and various mantle analogues at high pressure have shown that the presence of volatiles or volatile-bearing minerals dramatically lowers the solidus temperature of lherzolite (Fig. 2). Thus, the addition of volatiles alone may be sufficient to cause melting.

While the mantle can retain primordial volatiles from the initial accretion of the Earth, most water and other volatiles in the mantle are recycled, resulting from subduction of near-surface materials. The upper few kilometres of subducting oceanic plates have been altered extensively by the circulation of hydrothermal fluids, which produce volatile-bearing minerals. In addition, there is a thin veneer of sediments covering the subducting plates, which contain large volumes of unbound pore water and water- and carbonate-bearing minerals. As a plate is subducted, much pore water is released early and flushed into the wedge of sediments scraped from the descending plate. However, a significant volume is subducted and released by the breakdown of volatile-bearing minerals in the sediments and altered oceanic crust. Movement of volatiles into the mantle wedge above the subducting plate strongly depresses the lherzolite solidus and allows melting to occur.

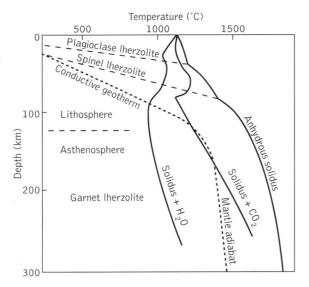

Fig. 2. Effects of the addition of small amounts of volatiles to mantle lherzolite. A mantle adiabat with potential temperature of 1280 °C is shown for reference. The lithosphere – asthenosphere boundary is marked by the inflection in the geotherm. Even small amounts of water ($< 0.5\%$) and carbon dioxide ($< 0.5\%$) strongly depress the temperatures of the solidus, moving it below the geotherm at all depths. However, this effect dominates only in subduction environments. Small addition of volatiles may also be important in permitting melting within the lithosphere. (Modified from B. M. Wilson (1989) *Igneous petrogenesis: a global tectonic approach*. Chapman and Hall, London.)

Segregation and ascent

Melt initially forms at the intersections of unlike minerals, forming isolated blobs of liquid. As the percentage of melt increases, these blobs grow to form an interconnected network, permitting porous flow between grains. The melt has a lower density than that of the surrounding matrix minerals and ascends buoyantly. The density difference between the melt and source rocks decreases with increasing pressure and, at depths greater than 230 km, melts are more dense than the source rocks and cannot rise.

Most melts have very low viscosities at mantle temperature and can easily percolate upwards through the mantle. Basaltic magmas representing as little as 1 per cent melt may easily separate and ascend, as can more alkali-rich melts at melt percentage as low as 0.1 per cent. The ease with which porous flow occurs means that any relatively large volume of melt in the mantle represents the integrated composition of many smaller batches. Source rocks compact during porous flow and it is unlikely that melt can exceed 3 per cent by volume under any conditions. The melt may not be in chemical equilibrium with its matrix because of its easy movement.

The initial stage of melt separation is by porous flow. Other later transport mechanisms include ascent through cracks forced open by melt buoyancy (magma fracturing) and movement through pre-existing conduits. Ascent rates probably range from about 5 to 100 cm year^{-1}. Melts will thus require about 10^5–10^7 years to move from their source to the surface. As a magma ascends through the mantle, the difference in density between it and its surroundings decreases. Ultimately, when it enters the crust, a magma may stall and pool to form a magma chamber.

Mechanisms that modify magma compositions

Once generated, the chemical and isotopic composition of a magma can be modified as it passes through the mantle and overlying crust. As a magma ascends, it cools, losing heat to the conduit walls, and begins to solidify. Minerals crystallize out of the completely molten rock, forming a liquid crystal mixture analogous to that created by high degrees of melting. However, in contrast to the case of melting, temperature–pressure conditions of a crystallizing melt start from the liquidus and move towards the solidus.

Crystal fractionation, which entails removing crystals from a melt as it moves through conduits or resides in a magma chamber, is the most important mechanism for modifying the composition of a magma. Crystals may float or settle because of density differences, be moved by currents within a magma chamber, or stick to the walls of conduits to be left behind during flow. They may continue to react with adjacent melt, constantly re-attaining equilibrium with it, or may become isolated. The latter process, termed fractional crystallization, is generally accepted to be the dominant process. Fractionation shifts the magma composition away from the composition of the crystals; thus, if the crystals are magnesium-rich, the magma will become magnesium-poor.

Contamination by melts from the walls of conduits or the magma chamber can also change the composition of a magma, as can the mechanical incorporation of fragments of the walls. The heat required to melt the wall rock is derived from crystallization of the magma as some of the heat initially absorbed during melting is released during the change from liquid back to solid. The contaminating melt can be of several different compositions but the crustal melts are granitic. A related mechanism is magma mixing, where the magma mixes physically with a melt produced from a distinctly different source.

JAMES H. WITTKE

Further reading

McKenzie, D. and Bickle, M. J. (1988) The volume and composition of melts generated by extension of the lithosphere. *Journal of Petrology*, **29**, 625–79.

particle morphology of sediments
Particle morphology, or form, refers to the sum of the surface characteristics of sedimentary particles, and is an important source of information on the origin of sediments and sedimentary rocks. Processes of weathering, erosion, and transport may all leave distinctive imprints on particles, in the form of

fractures, worn surfaces, and particular surface textures. Careful study of particle morphology can therefore yield valuable data on past processes and environments.

Casual examination of any stone shows that its surface, in all its minute detail, has a very complex form. A full description would clearly be impractical, and Earth scientists have identified a few measurable characteristics which provide the most useful summary. Three types of characteristics may be defined: shape, or the relative dimensions of the particle; roundness, or the overall smoothness of the particle outline; and texture, or small-scale surface features. (In some publications the definitions of 'form' and 'shape' have been used in the opposite senses to those used here.) Roundness is superimposed on form and is a characteristic of particle edges, and texture is superimposed on roundness. Studies of how two or more morphological characteristics vary together can often yield much more information than one characteristic studied in isolation.

The shape of a particle can be visualized as the shape of the smallest rectangular box into which the particle can be fitted.

The three dimensions of the box are referred to as the long (L), intermediate (I), and short (S) axes (or alternatively, the a, b, and c axes, respectively). The shape of the box (which is independent of its size) is then defined in terms of the relative dimensions of the axes. In the 1950s, Edmund Sneed and Robert Folk showed that all possible box shapes can be arranged on a triangular diagram, with three limiting cases:

(1) a cube, with $L = I = S$;
(2) a rod, with $L > 0$, and $I = S = 0$; and
(3) a square plane (or disc), with $L = I > 0$, and $S = 0$ (Fig. 1a).

Several indices, descriptive of particle shape, can be plotted on this diagram without altering its basic geometry. Three of the most useful are: S/L, which separates cubes from rods and discs; $(L–I)/(L–S)$, which differentiates rods and discs; and I/L, which separates rods from discs and cubes (Fig. 1a). Another useful family of indices is based on the behaviour of particles in a fluid. One such index is maximum projection sphericity

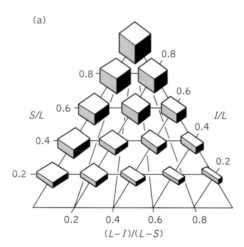

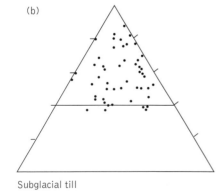

Subglacial till

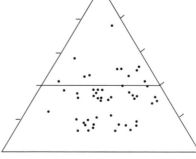

Rockfall debris on ice surface

Fig. 1. (a) Ideal particle shapes shown on a triangular diagram, and (b) shape data from Slettmarkbreen, a small glacier in Norway.

$((S^2/LI)^{0.33})$, which reflects the balance of drag and gravitational forces acting on an immersed particle. Some researchers favour other shape diagrams, such as the two-axis plot devised by the German geologist T. Zingg in the 1930s, in which I/L is plotted against S/I. In fact, the triangular and two-axis plots are equivalent, and one can be transformed into the other by stretching it like a sheet of elastic. The triangular version is preferable because it more accurately reflects the three-way variation of particle shape.

Samples of particles from different environments commonly have distinct distributions when plotted on shape diagrams, reflecting their weathering and transport histories. For example, frost-shattered particles tend to be rod- or disc-shaped, whereas subglacially crushed and abraded particles tend to be more cubic (Fig. 1b). An additional influence is exerted by the joint distribution in the parent material. For example, fissile slates tend to yield much flatter particles than massive granites.

Roundness can be measured in several ways. One method is to measure the radius of curvature of the sharpest edge of the particle, and to compare it with the radius of the smallest circle that can be inscribed within the particle outline. Two disadvantages of this method are that radii of curvature can be very difficult to measure for sharp-edged particles, and the sharpest edge may not be representative of the particle as a whole. Because of this, many geologists favour comparison charts or descriptive tables which define summary roundness categories. Alternatively, computer programs have been produced that give mathematical descriptions of roundness from digitized particle outlines. While effective for sand grains and smaller particles, this method is costly and time-consuming to implement. Roundness is increased by abrasion and chemical weathering processes, which blunt particle edges, and decreased by fracturing, which creates new, unworn edges. Roundness studies can thus assess the relative importance of abrasion, weathering, and fracturing in the history of a particle.

Most studies of texture have focused on particles smaller than sand size. Quartz-grain microtextures have received particular attention, and many distinctive textural features, such as conchoidal fractures and stepped surfaces, have been classified. Early work focused on establishing textures diagnostic of particular environments or processes (e.g. 'glacial' or 'aeolian' textures), but it is now recognized that the origin of the grains (igneous, metamorphic, or diagenetic) is also an important controlling variable. For pebbles and larger particles, surface textures, such as weathering pits and percussion fractures, provide important clues to particle history. In glacial studies, striations and polished facets are often recorded as evidence for subglacial transport.

An additional aspect of particle morphology worthy of attention is the asymmetry of wear patterns. Particles from fluvial or coastal environments are generally worn evenly on all parts of their surfaces, whereas the abraded surfaces of particles that have undergone subglacial transport or deposition are usually concentrated at one end and fracture surfaces at the other. Such asymmetric 'stoss-and-lee' stones reflect the asymmetric distribution of stresses in the subglacial environment, particularly the shearing of debris-rich ice or till over or around the stone. Wear patterns have yielded detailed information on subglacial processes and are also potentially very useful in other environments.

DOUGLAS I. BENN

partitioning In chemistry and geochemistry, the term 'partitioning' is used to describe the extent to which a chemical component will *partition* or distribute itself between two phases that are in contact. In effect, partitioning describes the strength of the preference shown by a chemical component for one phase rather than another. (A phase is a physically distinct form of matter present in a system. For example, a liquid and a solid that has crystallized from it constitute a two-phase system in which the liquid and the solid are separate phases with a *phase boundary* between them.) Thus, in a natural system consisting of an aqueous solution and calcite crystals, ions of strontium or magnesium will not occur in equal concentration in both calcite and solution, but will be unequally partitioned between these two phases according to thermodynamic principles. The direction and strength of partitioning determines how components behave when the system is subjected to change, as, for example, in melting, decompression, or crystallization. Of particular importance in geochemistry is the partitioning of trace elements between minerals and silicate melts in magmatic systems. Because magmatic processes have played a major part in the geochemical differentiation of the Earth, and because much of what is known about these processes derives from chemical analyses of trace elements and their isotopes in magmatic rocks, trace element partitioning is a prerequisite for geochemical modelling. Partitioning is also important in many other geological processes, including the formation of ores, the transfer of contaminants through groundwater aquifers, and mineralogical phase changes in the Earth's mantle.

Partitioning for a given element can be expressed by a partition (or distribution) coefficient, D, which is defined as the ratio of the fractional concentration of the element in one phase to that in another. Partition coefficients vary considerably from component to component and from phase to phase. Partitioning is a thermodynamic process, and D is related to the equilibrium constant, K, for a given chemical reaction. D, like K, is controlled by temperature and pressure and, in the case of multivalent species, by the oxidation state. At a fixed pressure and temperature, partitioning is further controlled by the size of the substituent species (e. g. its ionic radius), its ionic charge, the extent of any complexation (formation of complexes), and the structure of the coexisting phases. For crystal–melt partitioning the structure of the crystal exercises the dominant control, accepting preferentially ions that fit closely into its lattice sites. Plots of mineral–melt partition coefficients against ionic radius show characteristic parabolic patterns, with the maximum located at a radius corresponding to the size of the lattice site at which substitution occurs. Elements that partition preferentially into the melt phase are known as 'incompatible'; those that prefer the solid phases are known as 'compatible'. Solid–solid partitioning requires consideration of the structure of both phases. In some circumstances partitioning is very sensitive to temperature (for example, the partitioning of strontium between coral aragonite and sea water) or pressure (for example, the partitioning of calcium between olivine and clinopyroxene in mantle rocks), making these suitable for geothermometry or geobarometry, respectively.

Partition coefficients can be determined by analyses of coexisting phases in rocks or from laboratory experiments. In both cases partitioning may occur under either equilibrium or disequilibrium conditions according to the temperature and rate of the geochemical process. Disequilibrium effects include zoned crystals, surface adsorption, and rapid crystal growth. Under such conditions partitioning is controlled by dynamic or kinetic factors and may differ considerably from the equilibrium case. At magmatic temperatures most partitioning is assumed to take place under equilibrium conditions.
J. D. BLUNDY

passive margins Continental margins, the linear zones that mark the position of the edge of continental crust, are classified as *passive* or *active* according to whether they are attached to adjacent oceanic crust, as part of the same plate, or detached from it across a plate boundary. Passive margins may be further subdivided according to their maturity, and whether or not they are protected by offshore island arcs. Extensive mature passive margins fringe the Atlantic Ocean, East Africa, India, Antarctica, and the western seaboard of Australia. The best example of an immature passive margin is the Red Sea, only 10 Ma old, an embryonic ocean. As well as providing the setting for much of the world's oil and gas, most

of the developed and developing world is bordered by passive margins. Consequently, oil companies and national governments alike have invested considerably in understanding their development.

Mature passive margins develop during oceanic widening as part of a continuous cycle of ocean opening and widening, subsequent shrinking , continental collision (orogenesis), and continental splitting (rifting). Consequently, they display characteristic syn-rift and post-rift sedimentary successions related, respectively, to the transition from rifting and isostatic subsidence to the cessation of lithospheric extension and the onset of 'passive' thermal subsidence that accompanies the production of new oceanic crust.

Continental rifting commonly continues until the crust has been thinned to practically zero thickness, such that mantle rocks erupt at the surface to form new oceanic crust. Studies of present and former continental rift basins show that once the crust has been thinned to half of its normal 35 km thickness liberation of basaltic partial melts results in extrusive volcanism. Volcanic rocks of Middle Jurassic age mark the point in time in which the North Sea crust was thinned beyond this critical limit. However, further rifting ceased soon after this episode of volcanism, and the basin did not develop into a fully fledged ocean basin.

The onset of ocean opening is signified by two major changes: the cessation of extensional faulting, and the thermal relaxation of thinned, and therefore hot, lithosphere. These changes mark the onset of post-rift deposition. The base of the post-rift sequence is usually defined by a marked, end-rift unconformity, recording the depositional hiatus accompanying the thermal subsidence of a formerly uplifted region of net erosion and non-deposition. Interpretation of boreholes drilled into passive margins reveals a distinct transition from generally land-based syn-rift sedimentation to deepening marine environments of deposition during the post-rift phase. Thermally subsiding basins are always considerably

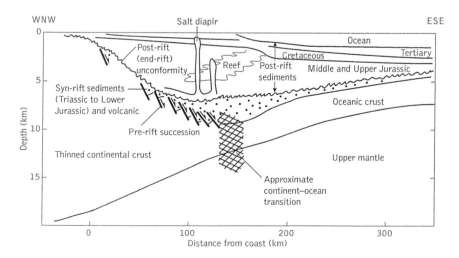

Fig. 1. Crustal cross-section across a passive continental margin, based on observations from the north-east Atlantic margin of the USA. The section shows syn-rift sediments in sub-basins controlled by tilted fault blocks, very thick post-rift sediments, progradation of a Jurassic carbonate reef, and salt diapirs. (Based on Grow and Sheridan (1988).)

larger than the area of their precedent rift basin. Marine flooding of the end rift unconformity, and the deepening of the post-rift marine environment, reflects the areal extent and prolonged nature of thermal subsidence (Fig. 1). Using bore-hole data, particularly from the eastern United States shelf, modelling of the subsidence history of passive margins has revealed that initial thermal subsidence occurs typically at a rate of 30 m per million years, subsequently decaying exponentially until it stabilizes some 150 Ma later.

JONATHAN P. TURNER

peatlands and bogs
Peatlands are biological landforms and can cover large areas, completely changing the pre-existing ecology and hydrology. They are the only plant community to lay down *in situ* a detailed record of their own history in the form of the partially decomposed remains of the plants themselves. Included within these plant remains are stratified records of pollen of their own and surrounding communities, of some invertebrates, of windblown dust, charcoal, and magnetic particles, of volcanic ash, and of past hydrological conditions on the bog surface. These records considerably enhance the scientific and conservation value of peatlands, for it is only by understanding the recent past that we can evaluate fully the present environment.

Of the types of peatlands that form this 'living history book', the *ombrotrophic* or 'rain-fed' *raised bogs* are most distinctive. By the accumulation of peat over the millennia they become raised above the mineral groundwater limit and assume the form of a giant upturned soup plate with a flat top and quite steeply sloping sides. They are distinguished from the other main type of peatland, the *minerotrophic* or 'rock-fed', which occur in valleys and basins or on river flood plains and are nourished by inflowing streams as well as by precipitation.

Raised bogs are dominated, in their natural state, by species of the genus *Sphagnum*, the bog moss. These small plants grow from their tips at a rate of about 2–3 cm in a season. The lower parts of the plants (which, being mosses, have no roots) are shaded because they grow close together, and are soon submerged in the bog water-table, which is usually only a few centimetres below the surface. Here the lower part of what is potentially a long-lived individual dies, but the plant remains are preserved because, in contrast to virtually all other ecosystems, decomposition is extremely slow in the waterlogged, acid, and anoxic conditions. Peat accumulation consequently takes place at a rate of about 100 cm in 1000–2000 years.

By this simple process, involving a small and insignificant plant when viewed in isolation, but a powerful one when growing packed together on a bog surface with sedges and heathers, whole peat landscapes developed during the Holocene, the past 10 000 years. By the infilling of ponds and lakes, and the colonization of ill-drained surfaces such as estuarine flats, raised bogs developed by about 7500 years ago. During succeeding millennia they grew larger through the

process of *paludification*, involving waterlogging of the surrounding area. In low-lying maritime areas, such as the German–Dutch border zone or the central plain of Ireland, vast bog complexes some tens of kilometres across were at their zenith by 2000 BP. Individual raised bogs, such as the three bogs known as Cors Caron, near Tregaron in Wales, which expanded to almost fill the valley of the River Teifi, are impressive landscape features and of international importance to nature conservation.

Blanket bogs also cover huge areas in cold, wet temperate zones such as northern Scotland, coastal areas of Norway, and the tip of South America. They are a mixture of ombrotrophic and minerotrophic bogs, and while they may contain much bog moss they also may be dominated by sedges. Minerotrophic peatlands, such as the valley bogs of the New Forest in southern England, are generally much smaller, being a few tens of metres across but some hundreds of metres long. They commonly have wet-loving trees such as alder and willow growing on the peat surface.

Although peatlands have long been valued for nature conservation on account of their specialized plant and animal life and the birds that are attracted to areas such as the Flow Country of northern Scotland, their value as a scientific archive is immense. In studies of environmental change and palaeoecology the advantages of raised bogs as proxy-data sources are manifold. Since raised bogs derive all their moisture from the atmosphere, there is a relatively direct relationship between the mean water-table of the bog and precipitation. There is similarly a relatively direct relationship between the occurrence and zonation of species on the hummocks and hollows of the bog surface and the mean water-table. By reconstructing the peat-forming vegetation from the plant remains we can therefore reconstruct the changing bog water-table through time, and this is a proxy for past climate. A number of records, particularly from northern Europe, have shown consistent trends with deterioration to wetter, cooler climates during the past 5000 years compared to the warmer climate of the previous 5000 years.

Other results from the proxy-records contained in the peat include the use of pollen analysis, both for the investigation of long-term natural changes, such as the migration of European forest vegetation after the end of the last glacial stage, and for tracing human impact on the wider landscape through forest clearance and the growth of agriculture in prehistory and more recent times. The increase in air pollution consequent upon industrialization has been traced using sensitive techniques of low-level measurement of the concentrations of magnetic minerals which have been trapped during the past few centuries of peat accumulation. The detection and geochemical identification of microscopic shards of volcanic ash (tephra) in peat is a recent development. Tephra from Icelandic eruptions of known historical dates, and from prehistory, have been found in Irish and Scottish bogs, and can be used as synchronous time markers in the peat, linking the other proxy-records precisely in time. Data such as these, and the remarkable preservation of prehistoric human bodies in peat, show what potential there is for unravelling the past in the peat.   KEITH E. BARBER

Further reading

Biodiversity and conservation, Vol. 2, No. 5, 1993; special issue, *Peatlands and people*. Chapman and Hall, London.

Chambers, F. M. (ed.) (1993) *Climate change and human impact on the landscape*. Chapman and Hall, London.

Gore, A. J. P. (ed.) (1983) *Ecosystems of the world 4A. Mires: swamp, bog, fen and moor*. Elsevier, Amsterdam.

pediments Probably more has been written about the nature and origin of pediments than of any other desert land-form, yet to a large extent they remain an enigma. Broadly speaking, pediments are low-angle, more-or-less planar, rock-cut slopes, usually veneered by gravel and separated from fringing hillsides by an angular break of slope. Pediments vary in length from a few hundred metres to several kilometres; their slopes are usually less than 5° and sometimes as little as 0.5°. Individual pediments can coalesce to form extensive surfaces called *pediplains*, commonly several thousand square kilometres in extent.

Just why and how pediment surfaces develop at such low angles is a matter of much debate. Anglo-American geomorphologists tend to the view that pediments are the result of either lateral planation by ephemeral, migrating streams issuing from bordering mountains, or sheet-flood erosion. Both theories have problems. Sheet floods, for example, require a priori a flat surface and can hardly, therefore, initiate one. And streams issuing from a mountainside on to a pediment simply are not capable of reaching much of the upper pediment edge without defying the laws of gravity.

In contrast, German geomorphologists, especially Johannes Büdel, have developed theories based on deep chemical etching followed by the stripping of the weathered mantle. Such theories are problematic in that they imply that desert pediments were once subject to a much wetter climatic regime. In any case, the idea of deep chemical weathering followed by stripping is difficult to reconcile with detailed measurements from Saudi Arabia, which clearly show that long pediments tend to have lower overall slope angles and are also more rectilinear than short pediments, which are usually concave in profile. In other words, pediment slope curvature declines as pediments lengthen. This leads to the possibility that pediments are modifications of vegetated convexo-concave slope systems typical of contemporary temperate climates and were initiated as a result of aridification in the Quaternary. Very short pediments are thus slightly modified convex foot slopes in this model.

Fossil pediments and pediplains are probably quite widespread but are difficult to recognize as they have been modified by recent processes such as glaciation. In Norway, for example, Professor Just Gjessing believes that the broad inter-montane basins forming the Vidda are remnants of preglacial pediplains; and in southern England, Professor G. H. Dury has suggested that the extensive upland surfaces developed on chalk downland are Tertiary pediments modified slightly by periglacial processes.   P. VINCENT

Further reading

Cooke, R., Warren, A., and Goudie, A. (1993) *Desert geomorphology*. UCL Press, London.

Vincent, P. and Sadah, A. (1996) The profile form of rock-cut pediments in western Saudi Arabia. *Journal of Arid Environments*, **32**, 121–39.

pegmatite Pegmatites are extremely coarse-grained igneous rocks. The crystals in a pegmatite are commonly several centimetres across and can be much larger. The processes that lead to the formation of pegmatites also result in the crystallization of rare and valuable minerals. Pegmatites usually form in the final stages of crystallization of an igneous intrusion. The early stages of crystallization involve the nucleation and growth of minerals that contain common elements. Fluids, immiscible droplets of metal sulphides, and elements like boron and fluorine are concentrated in the remaining melt. In the very last stages of crystallization the melt is both watery and rich in rare elements. The high concentration of water in the melt increases the mobility of the elements it contains, and so ions and molecules in the melt can travel farther before they become part of a crystal. Very large crystals can then grow at temperatures of only 100–200 °C over short periods of time. High concentrations of fluorine lead to the formation of the mineral fluorite; a high boron content may similarly lead to the formation of tourmaline. In these late stages of cooling, immiscible droplets of metal-rich substances commonly collect to form economic metal-ore deposits such as gold, silver, galena (lead ore), cassiterite (tin ore), or wolframite (tungsten ore).   JUDITH M. BUNBURY

pelagic carbonates Pelagic carbonates are composed of coccolithophores (plant), foraminifera (zooplankton) and, less commonly, pteropods (marine molluscs). All these organism secrete calcareous tests (skeletons), although pteropod skeletons are composed of a high-magnesium calcite (aragonite). Accumulation of pelagic carbonates occurs throughout the ocean basins above the carbonate compensation depth, or CCD, but predominates within the Atlantic Ocean.

Coccolithophores, commonly called mannofossils, are spherical, photosynthetic plants between 10 and 50 μm (micrometres) in diameter. Coccolithophores secrete minute plates of calcite (coccoliths) between about 2 and 10 μm across, and form coccolith oozes. Intact coccolithophores or coccospheres are occasionally preserved within sediments. Coccolithophores inhabit the euphotic zone (the upper 100 m of the ocean), and commonly exhibit vertical species stratification.

Globally their diversity increases from a minimum in subpolar waters to a maximum in tropical and equatorial waters. Like most phytoplankton, their species distribution is closely linked to water masses. Fossil assemblages can thus help to identify the distribution of past water masses. The shells of colder-water species are generally more robust than their warmer-water counterparts. This often leads to bias in preserved floras, especially at water depths at or near the CCD, for these sediments are primarily affected by thermohaline circulation changes during glacial–interglacial transitions.

Foraminifera (size range from 30 μm to 1 mm) are a diverse group of marine organisms composed of both planktonic and benthonic species. Planktonic species can be subdivided into spinose (Globigerinidae) and nonspinose (Globorotaliidae) forms. The shape of the test (shell) takes a variety of forms as adaptations to maintain buoyancy, and the distribution of species is generally determined by the availability of food and light intensity. Most planktonic forms inhabit the euphotic zone. Thus, areas of increased productivity, typically below latitudes of 60°, are characterized by high numbers of foraminifera. The diversity of planktonic foraminifera decreases with increasing latitude, and the occurrence of specific species is sometimes used to infer the presence or absence of particular water masses.

Benthonic foraminifera are either mobile or sessile, infaunal or epifaunal, and their shape and wall texture differentiate them from planktonic species. Their habitats include both shelf seas and the abyssal plain, and although they occur at all latitudes, their diversity is greatest in the tropics. Individual species show a strong correlation with depth. However, the variety of depth-dependent environmental factors (e.g. light availability, nutrient concentrations, temperature, salinity, oxygen and carbon dioxide contents), often makes it difficult to determine which is the limiting factor for any particular species. Some abyssal plain species are believed to be associated with particular bottom water masses and can be used to chart their palaeoceanographic courses.

Oxygen isotope analysis of planktonic foraminifera tests gives estimates for past sea-surface temperatures and salinities, making it possible to reconstruct palaeo-surface circulation patterns. Isotope data from benthonic foraminifera tests similarly make it possible to reconstruct of bottom water mass history, and therefore the thermohaline structure of the oceans.

Pteropods (planktonic gastropods) form only a minor component of biogenic carbonate sediments. Their thin aragonitic shells come in a variety of shapes, with lengths from 0.3 to 10 mm. Most pteropod species typically inhabit the upper 500 m of tropical and subtropical water masses. In sea water, aragonite is less stable than calcite, and the aragonite compensation depth (ACD) is shallower than the CCD, typically less than about 2–3 km. Pteropods are useful as biostratigraphical markers and water mass indicators, especially in semi-enclosed and marginal seas, such as the Mediterranean Sea and the Red Sea.   R. B. PEARCE

Further reading

Open University (1991). *Ocean chemistry and deep-sea sediments* (2nd edn). Pergamon Press, Oxford.

pelagic environments of the oceans
The pelagic environment is that of the open sea. It is the region inhabited by plankton, which are minute organisms that drift or float at various depths in the water, and by nekton, which are free-swimming organisms. The pelagic organisms that live in the open sea are not only an important constituent of the marine fauna; they also make a significant contribution to the sediments that are deposited on the ocean floor. Benthonic organisms (those that live on the ocean floor) also make their contribution to the ocean-floor sediments.

Pelagic biogenic sediments
Pelagic biogenic sediments consist of the fine-grained skeletal debris of marine planktonic and benthonic organisms. These sediments can be subdivided into those comprised of calcium carbonate and those comprised of opaline silica. The carbonate-rich debris is made up of the remains of *coccolithophores* (plants), *foraminifera*, and *pteropods* (animals); most siliceous components originate from *diatoms* (plants), and *radiolarians* (animals). Pelagic biogenic sediments (containing not less than 30 per cent of biogenic debris), also termed *biogenic oozes*, can be further subdivided according to their principal component; for example, diatom ooze.

Controls on biogenic ooze deposition
The spatial distribution and character of biogenic oozes depend on several factors: (1) the flux of biogenic material from the sea surface to the sea floor, (2) its rate of dissolution before and after deposition, and (3) its dilution by non-biogenic components during deposition.

The initial flux of carbonate and silica to the sea bed is determined by the level of primary productivity within the *euphotic zone* (the upper layer of the ocean penetrated by sunlight), which in turn depends on the availability of nutrients (e.g. phosphorus, nitrogen, silica). Relatively low nutrient levels are characteristic of the euphotic zone, owing to limited vertical mixing with underlying nutrient-rich water. However, the continuous 'rain' of decaying organic matter from the euphotic zone ensures that nutrients build up in the underlying water column. There are high rates of primary production in regions where these nutrient-rich waters upwell, such as coastal upwelling zones (e.g. off the coast of Peru, the Gulf of California) and at sites of oceanic divergence (e.g. eastern equatorial Pacific). These environments favour diatom algal blooms; thus the underlying sediments are typically dominated by diatom oozes. In contrast, pelagic carbonates predominate in areas peripheral to upwelling cells, i.e. where primary productivity is still moderately high. Regions of low productivity (i.e. nutrient-poor waters) are characterized by biogenic sedimentation rates low in carbonate and negligible in silica. Coccolith oozes are typical of such environments; for example, the Black Sea. Dilution of biogenic sediments by non-biogenic material occurs mainly at continental margins (the continental slope and rise). Near-shore terrigenous sedimentation rates usually exceed those of biogenic sediments. With increasing distance from the continental margin, there is a gradation from terrigenous-dominated sediment to pelagic-dominated sediment.

The dissolution characteristics of opaline silica in sea water are very different from those of carbonate. Sea water is everywhere undersaturated with respect to silica. Opaline silica thus starts to dissolve immediately it is exposed to sea water. This dissolution process continues throughout the time it takes for material to settle down through the water column, at the sediment–water interface, and during burial, until such time as the pore waters become saturated with respect to silica, or the material is dissolved.

At ocean water depths of less than 4–5 km, sea water is supersaturated with respect to carbonate, whereas at greater depths it is undersaturated. The depth at which carbonate sediments start to show signs of dissolution is termed the *lysocline*; the depth at which they are no longer preserved is termed *the carbonate compensation depth*, or CCD. Hence, pelagic carbonates are preferentially deposited on topographical 'highs', such as mid-ocean ridges and seamounts. Also, the position of the CCD is not static throughout the ocean basins, but varies both in space and time. The depth of the CCD in modern oceans is controlled by several factors, including temperature, pressure, and the decay of organic matter. As the temperature of sea water falls, the solubility of carbonate increases. Increasing pressure has a similar effect. At depths greater than 4–5 km, the low temperature of sea water and the pressure exerted on it by the overlying water column are sufficient to ensure that it is undersaturated with respect to carbonate. The presence of carbon dioxide further increases the solubility of carbonate. With increasing age, the gradual addition of decaying organic matter to deep water progressively increases its carbon dioxide content. Older deep waters in ocean basins therefore have shallower CCDs than younger deep waters in ocean basins. Glacial and interglacial climatic conditions are characterized by contrasting deep water (*thermohaline*) circulation patterns, and reduced atmospheric carbon dioxide contents. The CCD will thus differ from one climatic state to the next.

The distribution of pelagic sediments

The Atlantic Ocean is characterized by widespread *carbonate oozes*, the Pacific and Southern Ocean by widespread *siliceous oozes*, and the Indian Ocean typically contains a mixture of both sediment types (Table 1). As outlined above, this sediment distribution is determined by the *bottom water circulation*, which controls both the rate of dissolution and the productivity of surface waters through upwelling. By comparing the depth of the CCD in the North Pacific and North Atlantic Oceans, the effect of ageing deep water is clearly illustrated. In the North Atlantic, the deep water circulation is

Table 1. Percentage of deep ocean floor covered by pelagic sediments.

Sediments	Atlantic	Pacific	Indian	World
Calcareous ooze	65.1	36.2	54.3	47.1
Pteropod ooze	2.4	0.1	–	0.6
Diatom ooze	6.7	10.1	19.9	11.6
Radiolarian ooze	–	4.6	0.5	2.6
Pelagic clays	25.8	49.0	25.3	38.1
Relative size of ocean (% of total)	23.0	53.4	23.6	100.0

Note: pelagic clays are not detailed in this entry. (Source: Open University (1991).)

dominated by oxygen-rich *North Atlantic Deep Water* (NADW), which forms from nutrient-poor surface waters in the Greenland and Norwegian Seas. The CCD of the North Atlantic is correspondingly deep (*c.* 5 km). In the Pacific deep water contains a component of NADW, as well as a component of surface Antarctic water. Combined, these two water masses form Antarctic Bottom Water. Thus, the relative age of Pacific deep water is greater than that of NADW, and the CCD is relatively shallow (*c.* 4 km in the South Pacific, 3.5 km in the North Pacific). The shallow depth of the CCD in the Pacific, and the abundance of nutrients in upwelled waters, also contribute to the widespread occurrence of siliceous deposits in this ocean.   R. B. PEARCE

Further reading

Broeker, W. S. and Peng, T. S. (1983) *Tracers in the sea*. Eldigio Publications, Palisades, New York.

Kemp, A. E. S. and Baldauf J. G. (1993) Vast Neogene laminated diatom mat deposits from the eastern equatorial Pacific. *Nature* 362, 141–4.

Kennett, J. P. (1982) *Marine geology*. Prentice Hall, Inc., Englewood Cliffs, New Jersey.

Open University (1991) *Ocean chemistry and deep-sea sediments* (2nd edn). Pergamon Press, Oxford.

pelagic silica

Diatoms and radiolarians predominate in siliceous pelagic sediments. Diatom oozes (those containing 30 per cent or more of diatoms) are preferentially deposited beneath coastal upwelling cells, and in the Southern Ocean beneath the Antarctic divergence. Radiolarian oozes are primarily found in the equatorial Pacific.

Diatoms

Diatoms are photosynthetic single-celled algae, which secrete a test (or frustule) of amorphous hydrated silica, also known as opal or opaline silica ($SiO_2 \cdot nH_2O$). Diatoms vary greatly in size, from about 10 micrometres (μm) to several millimetres in length, and occur either singly or in colonies. Colinial species typically form long chains (e.g. *Skeletonema* spp.) or entangled masses of diatoms held together by 'sticky' threads and spines (e.g. *Thalassiothrix* spp.) known as diatom mats. Colonial forms are commonly preserved in pelagic siliceous sediments; for example, *Chaetoceros* is a common species of coastal upwelling regions, and diatom mats are widely observed in Miocene and Pliocene laminated diatom oozes in the eastern equatorial Pacific.

Diatoms are susceptible to dissolution, both during transit from the sea surface and on the sea bed. Thus, it is the rate of supply of silica that determines the quantity of material preserved in the sediments: high supply rates mean an increase in preserved silica. In areas of low silica supply, diatom assemblages are frequently biased towards dissolution-resistant forms, and so robust species, such as *Coscinodiscus* spp., are often more abundant than fragile species, e.g. *Skeletonema* spp. Like other microplankton, diatoms can be used to infer water mass distributions, although they occur in greatest numbers in the sedimentary record beneath areas of upwelling.

Radiolarians

Radiolarians are a diverse group of open-ocean zooplankton that construct complex skeletons of biogenic silica or strontium sulphate. There are three major groups of radiolaria, the Orders Acantharia, Tripylea, and Polycystina. Only the Polycystina group form skeletons of opaline silica, and are preserved in the sedimentary record. This group can be further subdivided into spherical forms (suborder Spumellaria) and bell- or cap-shaped forms (suborder Nassellaria).

Radiolarians range in size from 50 to 300 μm, and occur singly or occasionally as colonies. Radiolarians show adaptations of the test that have developed to increase their buoyancy; for example, near-surface forms are more delicate, spinose, and ornate than deeper-water forms, which are larger and have shorter and fewer spines.

Radiolarian faunas are most diverse in the near-surface waters (50–200 m) of tropical and temperate regions, although some polar species do exist. The distribution of radiolarians in deep sea sediments is strongly influenced by the undersaturation of the water column with respect to silica. Regions containing well-preserved faunas are believed to occur primarily beneath regions of enhanced productivity. Radiolarians are generally more robust than diatoms and tend to be better preserved, although radiolarian faunas, like diatoms, are also biased towards the more dissolution-resistant species. In palaeoceanography the occurrence of radiolarians is used to delimit the geographical extent of water masses.

R. B. PEARCE

Further reading

Open University (1991) *Ocean chemistry and deep-sea sediments* (2nd edn.). Pergamon Press, Oxford.

Penck, A. (1859–1945), Penck, W. (1888–1927)

In the early part of the twentieth century, when landscape studies in North America, Britain, and France were in the stranglehold of W. M. Davis's cycle of erosion, German geomorphologists, led by Albrecht Penck, and championed by his son, Walther, offered an alternative view of landscape evolution as a response to ongoing crustal movements (tectonism). The basic dichotomy between the Davisian and Penckian schemes is sharply illustrated in their contrasting views on how mountain landscapes evolve. To Davis, the accordance of summits in various mountain ranges represented the remnants of pre-existing erosional surfaces (peneplains) that had been uplifted by short-lived pulses of tectonism and subsequently passively eroded and shaped by progressive downwearing. In contrast, Albrecht and Walther Penck argued that accordant summits, rather than being the inherited 'floors' of previous denudational episodes, were upper limits to which mountains could grow and were limited because maximum uplift rates balanced denudation rates at these elevations. Albrecht Penck's research was undertaken in the European Alps, and it was in providing the first unified framework for understanding the Pleistocene history of this mountain range that he made his notable contribution. His model, developed with E. Bruckner, of four main glacial phases—Gunz, Mindel, Riss, and Würm—was immensely influential, although it has since been superseded by a more complex Pleistocene history determined on the basis of deep-sea core studies. It was Walther Penck, however, working in the European mountain ranges as well as further afield in the Andes, who provided the real (and arguably the only major) contemporary challenge to Davisian theory.

The central tenet of Walther Penck's work was that landforms resulted from the opposing tendencies of exogenetic processes (i.e. weathering and transport) and endogenetic (crustal) processes. In contrast to the faulted basin-and-range topography expounded by Davis, Penck considered folding as the main relief-producing crustal process. Furthermore, at the core of his view of landscape evolution was a vision of slope development that was dominated by mass movement. Penck considered that where rates of denudation are declining, hillslopes evolve through a process of flattening from the base upwards. In effect, each part of the slope profile is replaced by a slope of lower gradient as it retreats, and through this process of 'slope replacement', a broadly concave profile is produced. Penck's inability to illustrate his often complex ideas with convincing diagrams contrasted markedly with the artistic ability and skilful advocacy of Davis, which allowed him to misrepresent Penck's concept of slope replacement as one of 'parallel retreat' of the major part of the slope. With the comparative obscurity of Penck's original German papers, and then his early death in 1927, the general view of his work was long dominated by Davis's misinterpretations. In 1953, Penck's views were somewhat clarified for a wider audience by the English translation of his main work, *Die Morphologische Analyse*, and in the 1970s his deductive reasoning on the nature of mass-movement processes began to be incorporated into emerging process–response models. During the 1980s and 1990s, increasing research into active tectonic processes revealed the remarkable prescience of many Walther Penck's observations on the interplay between crustal movements and surficial processes, and promises a rekindling of interest in his ideas.

IAIN S. STEWART

Further reading

Beckinsdale, R. P. and Chorley, R. J. (1991) *The history of the study of landforms or the development of geomorphology*. Volume 3: *Historical and regional geomorphology 1890–1950*. Routledge, London.

Pennsylvanian *see* CARBONIFEROUS

periglacial landscapes
Periglacial landscapes form under the action of intense frost, often combined with the presence of permafrost. Periglacial landscapes are restricted to areas that experience cold, but essentially non-glacial, climates.

The term 'periglacial' was first proposed by a Polish geologist, Walery von Lozinzki, in 1909 to describe the frost weathering conditions that occurred in the Carpathian Mountains of central Europe. Subsequently, at the Eleventh Geological Congress in Stockholm in 1910 the concept of a 'periglacial zone' was introduced to describe the climatic and geomorphic conditions of areas peripheral to Pleistocene glaciers and ice sheets. Theoretically, this zone was a tundra region which extended as far south as the treeline (Fig. 1a).

For several reasons Lozinzki's original definition is unnecessarily restricting. First, frost action phenomena are known to occur at great distances from ice margins and can be unrelated to ice-marginal conditions. For example, parts of central Siberia and the interior of Alaska, commonly regarded as being 'periglacial' in nature, owe their distinctive nature not to ice sheet proximity but rather to factors such as permafrost or low mean annual temperature. In many areas, the permafrost is relict and unrelated to the present climate, since it is of Pleistocene age. Secondly, although Lozinzki used the term to refer primarily to areas and not to processes, it has increasingly been understood to refer to a complex of geomorphic processes. These include not only the relatively unique frost action and permafrost-related processes but also a broad range of processes which demand neither a peripheral ice-marginal location nor a cold climate for their operation.

Thus, modern usage of the term 'periglacial' refers to a wide range of cold, non-glacial conditions, regardless of proximity

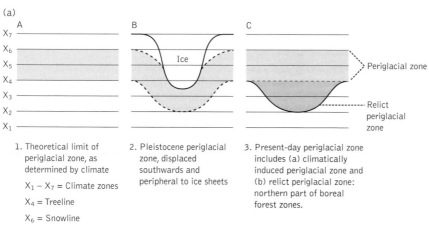

(a)

1. Theoretical limit of periglacial zone, as determined by climate

   $X_1 - X_7$ = Climate zones

   X_4 = Treeline

   X_6 = Snowline

2. Pleistocene periglacial zone, displaced southwards and peripheral to ice sheets

3. Present-day periglacial zone includes (a) climatically induced periglacial zone and (b) relict periglacial zone: northern part of boreal forest zones.

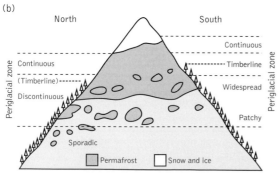

(b)

Fig. 1. Limits of the periglacial domain: (a) high-latitude domain; (b) high-altitude domain.

to a glacier, either in time or space. Periglacial environments occur not only in high latitudes and in tundra regions, as defined by Lozinzki, but also below the treeline. They include therefore, (a) the polar deserts and semi-deserts of the High Arctic and the ice-free areas of Antarctica, (b) the tundra zones of high latitudes, and (c) the extensive boreal forest zones of northern latitudes. To these must be added the high-altitude (Alpine) periglacial zones of middle and low latitudes (Fig. 1b), the largest of which is the Qinghai-Xizang (Tibet) Plateau of China. However, since alpine regions have their own distinctive characteristics, they are not discussed further in this entry.

It is not easy to describe a typical periglacial landscape because there are gradations between environments in which frost processes dominate and where a whole or major part of the landscape is the result of such processes, and those in which frost action processes are subservient to others. There are additional considerations, the most important being that certain rock-types are more prone to frost action than others, and hence, more susceptible to periglacial landscape modification. A second complication is that many areas experiencing periglacial conditions today have only recently emerged from beneath continental ice sheets. These regions cannot be regarded as being in true geomorphic equilibrium. A third problem is that the climate during the Pleistocene fluctuated considerably, ranging from full glacial to full inter-glacial. As a result, periglacial conditions fluctuated in both duration and extent. It follows therefore, that many of the so-called 'periglacial' landscapes of the world should be described more accurately as being 'paraglacial' (i.e. being in the transition from a glacial to non-glacial condition). According to the definition adopted, periglacial landscapes occupy between 5 and 15 per cent of the Earth's land surface.

Processes that are clearly unique to periglacial environments include the formation of permafrost, the development of thermal contraction cracking, the thawing of permafrost (a process termed *thermokarst*), the creep of ice-rich permafrost, and the formation of wedge and injection ice. Other processes, not necessarily restricted to periglacial regions, are important on account of their high magnitude or frequency. These centre around the frost weathering of soils and bedrock, including the disintegration of exposed rock by either frost wedging or a complex of poorly understood physico-chemical (cryogenic) weathering processes. Fundamental to all these processes is the freezing of water and its associated frost heaving and ice segregation.

A number of frost action processes operate in the seasonally frozen layer near the ground surface; they include frost heaving, soil churning, soil creep, stone tilting, upfreezing, mass displacement, and particle size sorting. These processes give rise to various forms of patterned ground, and involuted and disturbed soil horizons. In addition, rapid mass movements may occur in the seasonally thawed layer, or active layer. In regions underlain by permafrost, the relatively slow flow of water-saturated soil is termed *gelifluction*; in areas of

Fig. 2. A tundra landscape of the Low Arctic, Sachs River Lowlands, southern Banks Island, Northwest Territories, Canada (latitude 72° north), showing well-developed ice-wedge polygons and thaw lakes.

seasonal frost it is termed *solifluction*. In contrast to the soil creep of temperate regions, solifluction and gelifluction are relatively rapid processes in periglacial regions and can result in the active development of slopes. However, the most rapid and unpredictable movements are shallow failures restricted to the active layer, sometimes termed '*skin flows*'. These occur upon certain rock-types where high pore-water pressures develop in the active layer consequent upon exceptionally rapid or deep thaw in the spring and late summer, respectively. The melt of an ice-rich zone, which typically occurs near the base of the active layer, contributes to this process.

A common assumption is that periglacial landscapes are distinct. This is probably because certain landforms are uniquely associated with permafrost. Undoubtedly, the most widespread of these are tundra polygons formed by thermal contraction cracking (Fig. 2). Less widespread, but equally distinctive, are ice-cored mounds, or *pingos* (Fig. 3). Other landforms, such as palsas and peat plateaus, are associated with the growth of permafrost, while ground-ice slumps and thaw lakes and depressions result from the melt and erosion of ice-rich permafrost. The creep *in situ* of permafrost, especially as it thaws, may also cause large-scale deformation structures in sedimentary strata. These include uparching beneath valley bottoms (valley bulging) and bending, deformation, and sliding on slopes (cambering and joint widening, or 'gull' formation). These structures are known to occur in middle latitudes where formerly frozen, fine-grained (i.e. ice-rich) strata are overlain by more coherent (i.e. less ice-rich) beds.

A second category of landforms includes those that relate to the intense cryogenic weathering of exposed bedrock. Coarse, angular rock-rubble, termed *blockfields* or '*kurums*', occurs widely over large areas of the polar deserts and semi-deserts, while outcrops of more resistant bedrock form isolated hills or tors (Fig. 4). On a smaller scale, the periglacial

Fig. 3. This pingo is one of over 1350 which are known to exist in the Pleistocene Mackenzie delta. General view, showing its location in a drained lake basin. Note the massive ice-core exposed.

landscape possesses a diversity of patterned ground phenomena, including circles, stripes, and nets. In the tundra and boreal forest zones, by far the most common is the non-sorted circle, or hummock, thought to result from cell-like circulation patterns set up in the active layer. On sloping terrain, stripes are more common.

Gentle undulating relief is also regarded as a characteristic of periglacial landscapes. This is attributed to mass wasting upon slopes and the associated cryoplanation, or flattening, of the landscape. Frost creep, solifluction, and surface wash are the agents of transport.

It must be recognized that the majority of periglacial landscapes possess important glacial or non-periglacial components. Only those landscapes of cold regions which escaped

Fig. 4. A polar demi-desert landscape of the High Arctic, eastern Melville Island, Northwest Territories, Canada (latitude 77° north), developed in sandstone and shale of Mesozoic age.

glaciation throughout the majority of the Pleistocene can be regarded as distinct periglacial landscapes. To these must be added the special case of cold-climate deltas, where sediments are being deposited and permafrost is currently aggrading.

Many landscapes of the cold, non-glacial regions of the world possess some degree of inherited, largely glacial, characteristics. At the meso- and micro-scale, however, the abundance of patterned ground phenomena and frost-weathered features distinguishes the periglacial landscape from others, and the presence of permafrost promotes unique landforms and processes.

HUGH M. FRENCH

Further reading

French, H. M. (1996) *The periglacial environment* (2nd edn). Addison-Wesley Longman, Harlow.

permafrost Permafrost is ground that remains at or below 0 °C for at least 2 years. About one-quarter of the Earth's land surface is underlain by permafrost, including over half of Canada and Alaska, the Tibetan Plateau, and north-eastern Russia. Permafrost that formed during the last glacial period when the sea level was lower is also found on the continental shelf of the Arctic Ocean in the Beaufort, Kara, Laptev, and East Siberian Seas. Continental permafrost regions are usually divided into zones of continuous, widespread discontinuous, and scattered discontinuous permafrost, where respectively over 80 per cent, 30–80 per cent, and less than 30 per cent of the terrain is underlain by perennially frozen ground. In the northern hemisphere, the southerly boundaries of these zones are defined by the mean annual air temperature isotherms of −5, −4, and 0 °C, respectively. The zones not only indicate the spatial extent of permafrost, but also its temporal stability, for short-term climatic changes may lead to its eradication or development in the scattered zone, while, further north, permafrost is continuous in space and time. Alpine, or plateau, permafrost is found at high elevations in mountain regions throughout the world.

Permafrost terrain comprises a seasonally thawed active layer underlain by perennially frozen ground. The thickness of the active layer varies from a few centimetres to several metres, depending on summer conditions and the thermal properties of ground materials. The thickness is greatest in dry sand and bedrock, and least in moist, organic soils. The total thickness of permafrost depends on the annual mean temperature at the base of the active layer, the geothermal flux, and the thermal conductivity of ground materials. The heat flux is equal to the product of conductivity and geothermal gradient, represented by the near-surface temperature divided by the depth to the base of permafrost. Permafrost thicknesses vary from a metre or so at the southerly margins of the scattered permafrost zone to over 1000 m in Arctic Canada and Russia. In the discontinuous zones, snow depth, soil moisture, and soil organic matter content dominate environmental variables responsible for the presence or absence and temperature of permafrost.

The geotechnical significance of permafrost is derived from the occurrence of ground ice, often close to its melting point. The ice may be buried relict glacier ice, or intrasedimental ice formed either by segregation during permafrost development or by intrusion of water under pressure into frozen ground. Ice wedges may form at the base of the active layer after snow melt infiltrates thermal contraction cracks. Characteristically the uppermost layers of permafrost are ice-rich owing to these wedges, and to water incorporated into permafrost since the Early Holocene, when active layers were generally deeper. The presence of such near-surface ground ice renders the terrain liable to subsidence and ponding or to hill-slope failure if vegetation is disturbed and the active layer deepens. Exceptional care is therefore taken to preserve the integrity of permafrost during construction activities.

Since permafrost is a climatic phenomenon, climatic changes lead to an altered permafrost regime. The depth of permafrost development and degradation follows the square root of time since the change occurred. Ice-rich ground over 10 m thick requires centuries to millennia for complete development or thawing, and for other adjustments to permafrost thickness. The response of permafrost temperatures to climate warming is usually rapid until the ground reaches −2 to −1 °C, at which point thawing of pore water has begun. The latent heat required for further increases in ground temperature then slows the warming. The terrain response to climate warming may, however, be rapid, since thickening of the active layer will lead to melting of the abundant near-surface ground ice. This is a significant issue in considering the response of permafrost terrain to potential global warming.

C. R. BURN

permafrost and climate change
Areas of permafrost are usually defined as those regions of the world in which ground temperatures below freezing have existed for a long time (from two years to tens of thousands of years). The thickest accumulations of permafrost in the world presently occur in north-eastern Siberia, where its thickness normally exceeds 800 m (locally reaching up to 1400 m). The growth of permafrost depends upon the radiative heat loss from the ground during winter being greater than the supply of heat to the ground surface during the summer months. Conversely, permafrost degradation may occur when summer heating is greater than winter heat loss. During summer, melting of the permafrost surface, usually to a depth of no more than 1–2 m, results in the development of an active layer, the base of which is known as the *permafrost table*. In the context of climate dynamics, areas of permafrost during winter are characteristically associated with the development of high pressure which occurs as a result of the cooling of the lower atmosphere by the ground surface. In contrast, areas of permafrost during summer are usually associated with the development of low air pressure caused by widespread melting of ice, moisture evaporation, and the convergence of air in the lower troposphere.

At present the most extensive areas of permafrost occur in Arctic Canada and in Russia. During ice ages the growth and thickening of permafrost principally took place in arctic and alpine areas that remained unglaciated. The large ice sheets that formed over Canada and Russia during the last ice age served to insulate the underlying ground surface and may even have promoted decay of permafrost beneath the ice sheet. Not surprisingly, therefore, the areas of the world where permafrost is now at its thickest coincide with those areas that are not only cold at present but also were not covered by ice during the last (and previous) glaciations.

A major feature of atmospheric circulation in Europe and North America during glacial periods was an enhanced zonal wind flow associated with the anchoring of a mid-latitude jet stream south of the major ice sheets. The easterly winds that were generated south of the ice-sheet margins, acting in conjunction with katabatic winds and low air temperatures, led to a negative heat budget at the ground surface and the consequent aggradation of permafrost.

Studies of permafrost history in Russia have provided valuable information on patterns of environmental change between the culmination of the last interglacial (125 000 years ago) and the last glacial maximum (18 000 years ago). Two distinct periods of permafrost growth have been recognized: the first cold period appears to have occurred more than 70 000 years ago; a second phase of permafrost growth culminated between 60 000 and 50 000 years ago. During these periods the southern limit of continuous permafrost may have been displaced 6 to 8° of latitude south of its present position reaching the centre of the Russian Plain, where mean annual ground temperatures may have been as low as −3 °C. The greatest expansion of permafrost, however, took place during the last glacial maximum between 24 000 and 18 000 years ago when permafrost may have been as much as 200 m in thickness in European Russia. The most severe climatic conditions and the thickest permafrost at this time was in north-eastern Siberia, where average annual ground temperatures may have fallen below −15 °C (Fig. 1).

The evolution of permafrost in response to climate warming and cooling is relatively slow. For example, during the Late Quaternary, changes in permafrost thickness took place at a much slower rate than the growth and decay of ice sheets and sea-level changes. Thus, extensive areas of relict permafrost occur beneath sea level on the sea floor of the Arctic continental shelf as a result of the rise in sea level that has taken place during the last 15 000 years. The existence of certain permafrost regions out of phase with the Earth's present climate makes it difficult to assess the response of permafrost to the present global warming. Changes in permafrost thickness arising from global warming reflect changes in the annual heat balance for particular areas. Surprisingly, there have been relatively few studies of the influence of global warming on permafrost. One might expect, for example, that permafrost thinning as a result of increased air temperatures could result in a transfer of melt water to the world's oceans

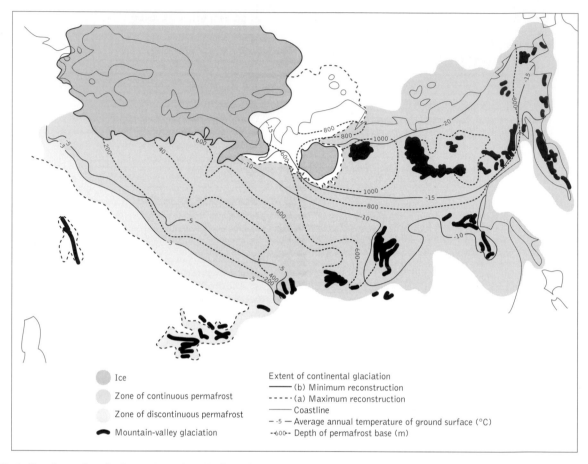

Fig. 1. Map of permafrost distribution during the Valdai (Sartan) glaciation. (Baulin, V. V. and Danilova, N. S. (1984) In A. A. Velichko (ed.) *Late Quaternary Environments of the Soviet Union*, pp. 69–78. Longman, Harlow.)

and a consequent rise in sea level. Yet such analysis has not so far been undertaken and is a notable omission from the Intergovernmental Report on Climate Change. Similarly, permafrost degradation, despite having been described as presently taking place, may also play an important role in the modification of regional climates as a result of changes in rates of air convergence in the lower troposphere during the summer months. It remains to be discovered by the scientific community whether or not the rapid changes in air temperatures experienced over the past two decades have had a profound effect on the stability of permafrost and of the climate over these particular areas. ALASTAIR G. DAWSON

Further reading

French, H. M. (1976) *The periglacial environment*. Longman, London.

Washburn, A. L. (1979) *Geocryology*. Edward Arnold, London.

Péwé, T. L. (1983) The periglacial environment in North America during Late Quaternary time. In S. C. Porter (ed.) *Late Quaternary environments of the United States*, Vol. 1. *The Late Pleistocene*, pp. 157–89. Longman, Harlow.

permeability and porosity Permeability and porosity are the two most important properties that determine the capacity of rocks and sediments to transmit and store groundwater or other fluids such as petroleum. Permeability is a measure of the ease with which fluid will flow through a porous medium under a specified hydraulic gradient. This depends both on the properties of the solid medium and on the density and viscosity of the fluid. Intrinsic permeability depends on the properties of the medium alone. Where the fluid of interest is water, it is common to express permeability in terms of the hydraulic conductivity, which depends on intrinsic permeability and on the physical properties of water. Hydraulic conductivity is defined as the product of the intrinsic permeability and the specific weight of water, divided by the dynamic viscosity. The permeability of a medium can also be determined with respect to other fluids, for example, air permeability.

In the petroleum industry, the conventional unit of intrinsic permeability is the darcy, named after Henry Darcy, who first identified hydraulic conductivity as the constant of proportionality between flow rate and the hydraulic gradient. Sandstones and limestones that are petroleum reservoirs typically have intrinsic permeabilities ranging from 10^{-3} darcies up to about 1 darcy. Groundwater studies generally employ hydraulic conductivity, with units such as metres per second, equal to 100 000 darcies. Hydraulic conductivity can range from about 1 m s^{-1} in coarse gravel aquifers down to 10^{-13} m s^{-1} in unfractured shale confining units.

Porosity is defined as the ratio of the volume of pores, or void space, in a porous material to the bulk volume of the material, which includes both solids and voids. Porosity is frequently expressed as a percentage. It can range from less than 1 per cent in dense crystalline rocks with few joints and fractures to over 70 per cent in clayey sediments. The higher the porosity, the greater storage potential for groundwater. Intu-

itively, one might expect that rocks or sediments with high porosities would also have high permeabilities. Such a relationship can be found where the pore sizes of the rocks or sediments are similar. Permeability is, however, more strongly controlled by pore size than by total porosity. Resistance to flow through a porous medium is generated by friction at the fluid–solid interface. A medium with small pores has a high ratio of surface area to pore volume, and hence presents significant frictional resistance to flow. The result is that while clays may have twice the porosity of sands, the intrinsic permeability of clay is many orders of magnitude smaller than that of sand.

The porosity of sedimentary rocks comprises the intergranular voids that remain from the original sediments, known as primary porosity, as well as later openings such as fractures and joints, known as secondary porosity. Igneous rocks such as basalts may have primary porosity in the form of cooling cracks or vesicles. Most crystalline igneous and metamorphic rocks, however, have only secondary porosity in the form of fractures.

J. BAHR

Permian The Permian System was named by Sir Roderick Murchison in 1841 after the Russian province of Perm. He had been invited by the Tsar to report upon the geology of European Russia and had noted the great outcrop of red rocks

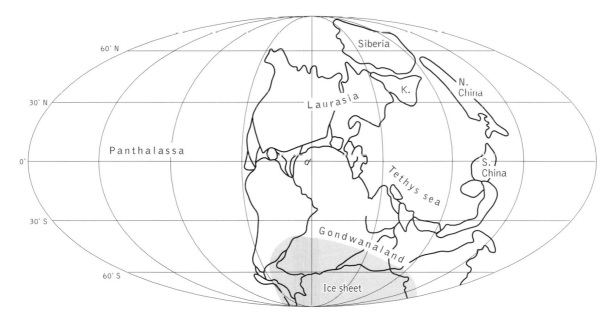

Fig. 1. The world in Late Permian time, about 260 Ma ago. This was a period when almost all the continental parts of the crust had been drawn together into the supercontinent Pangaea. The shallow Tethys sea separated a northern continental mass (Laurasia) from a southern (Gondwanaland). To the east, small elongate continental 'islands', North China, South China, and parts of south-east Asia, lay between Tethys and the great Panthalassa Ocean. K, Kazathstan. Most of

Pangaea was subject to an arid hot climate, but in the south polar region (stippled on map) ice accumulated and spread as one or more great sheets across parts of South America, South Africa, India, Australia, and Antarctica. During this part of the Permian period sea level fell to the point where most of the narrow continental shelf was no longer covered. The consequences for marine life were catastrophic.

lying close to the western Ural Mountains. These rocks follow on the Carboniferous and strata of this age were already known in Germany and Britain as part of the New Red Sandstone. In Perm the formations include fossiliferous red limestones, evaporites, and continental red beds. Murchison's account was based upon beds which are nowadays regarded as only the upper two-thirds of the Permian System. The lower part of the Lower Permian contains marine limestones thought in the nineteenth century to be Carboniferous. Today the System is divided into a Lower Series with four stages, and an Upper Series with two stages. By 1853 the Permian had been recognized over a wide area between the Mississippi and the Colorado rivers in the USA. Other regions in which marine rocks of Permian age were soon to be found include the Himalayas and the high Arctic from Canada to Siberia.

The Permian Period lasted from 290 Ma to 250 Ma. From the Late Carboniferous into the Triassic the Earth was experiencing a long phase of reverse magnetic polarity, which is conspicuous in magnetostratigraphic records.

Permian palaeogeography is that of the single supercontinent *Pangaea*, extending from the equatorial latitudes to the South Pole (Fig. 1). Pangaea resulted from the gathering together by plate collisions of the major continental masses throughout Carboniferous and Permian times. One of the later events in this process was the juxtaposition of Angaraland (Siberia) with eastern Laurussia to produce the Ural Mountains. Sea level sank to an all-time Phanerozoic low, probably in response to the widespread glaciation of much of southern Gondwanaland. The fixation of carbon dioxide in the coals of the Carboniferous and Permian had been the first step in initiating the global change from 'glasshouse' (hot humid) to 'icehouse' (cold arid) conditions. The cold exercised an enormous effect directly in the multiple glaciations of the Southern Hemisphere. This precipitated several waves of extinction, seeing the demise of almost all corals, fusuline foraminifera, brachiopods, bryozoans, and ammonoids. By the end of the period some 90 per cent of all invertebrates had perished.

On land, the mammal-like reptiles first appeared in the early Permian, but by the end of the period most of them, too, were extinct. Permian marine biostratigraphy is based on the fusulinid foraminifera, brachiopods, and goniatitic ammonoids. In South Africa terrestrial deposits contain great numbers of reptiles, which provide a form of biostratigraphy.

D. L. DINELEY

Further reading

Erwin, D. H. (1993) *The Great Palaeozoic Crisis: life and death in the Permian*. Columbia University Press, New York.

Ramsbottom, W. H. C. (1978) Permian. In McKerrow, W. S. (ed.) *The ecology of fossils*, pp. 184–93. Duckworth, London.

Stanley, S. M. (1987) *Extinction*. Scientific American Library, New York.

Sweet, W. C., Yang Zunyi, *et al.* (1992) *Permo-Triassic events in the Eastern Tethys. Stratigraphy, classification and relations with the Western Tethys*. Cambridge University Press.

petrogenesis *see* PETROLOGY, PETROGRAPHY, PETROGENESIS

petrography *see* PETROLOGY, PETROGRAPHY, PETROGENESIS

petroleum The term 'fossil fuels' is commonly used for the materials tar sands, oil, gas condensates, natural gas, coal, and oil shale and their products. Petroleum comprises oil, gas condensate, and natural gas.

'Petroleum' is a term for a series of subsurface fluids, composed significantly, but not entirely, of hydrocarbons, which occur concentrated in reservoirs and can be produced at the surface as oil or gas, or both, and sometimes as liquid condensates. The term 'hydrocarbon' in its strict chemical sense refers to a chemical compound composed solely of hydrogen and carbon. Examples would be methane (CH_4), a gas; decane ($C_{10}H_{22}$), an alkane and a liquid at room temperature and pressure; or phenanthrene ($C_{14}H_{10}$), an aromatic hydrocarbon and a solid at room temperature and pressure. All three occur naturally in solution in oil, and gas condensates; methane is the major component of most natural gases.

Petroleum also contains compounds other than hydrocarbons, including organic carbon compounds that contain oxygen, nitrogen, and especially sulphur. Normally it also contains significant quantities of non-hydrocarbon gases such as carbon dioxide, nitrogen, hydrogen sulphide, and traces of inert gases such as helium. The three commonly described forms of petroleum are natural gas, which does not condense at standard temperature and pressure; gas condensate, which is a gas-phase petroleum in the reservoir but partly condenses to a liquid (the condensate) at the surface; and oil, a petroleum that was liquid in the reservoir but produces a gas (solution gas) and a liquid phase, crude oil, at the surface. Some oils, such as those in tar sands, or very waxy oils from South-east Asia, may be solid or nearly solid at the surface.

Although usually dominated by the hydrocarbons, petroleums may contain up to 80 per cent or more of non-hydrocarbons. Most North Sea oils contain 50–90 per cent of hydrocarbons. Inorganic gases such as carbon dioxide, nitrogen, helium, and hydrogen sulphide are common components of petroleums, ranging in abundance from trace components to major fractions of the fluid.

Composition

Petroleum can be classified by its composition and the relative amounts of various chemical components.

Natural gases contain principally hydrocarbons with from 1–5 carbon atoms. They can exist in the reservoir as a gas phase or in solution in a liquid petroleum. In the reservoir under pressure, crude oil is a mixture of liquid hydrocarbons and non-hydrocarbons. It will contain many thousands of individual components with from 1 to 40 or more carbon atoms per molecule, although compounds with up to 100 or more carbon atoms are found. The largest molecules abundantly present in solution in crude oil will be the asphaltenes, which probably have up to an average of 70 or more carbon atoms, attendant hydrogen atoms, and several oxygen, nitrogen, or sulphur species per molecule. Some asphaltenes may have even larger numbers of carbon and associated atoms in their molecules. Asphaltenes represent small portions of source-rock organic matter dissolved in the crude oil medium. The average size of the molecules in a light crude oil would be about 15–20 carbon atoms per molecule. In most petroleums, even oils, methane is commonly the most abundant individual molecule. When oil is produced, the gas comes out of solution, leaving molecules with from 6 or more carbon atoms (the larger molecules) in the liquid oil.

Petroleums have a wide range of compositions but can be classified into three major types according to the relative proportions of gaseous (C_1–C_5) hydrocarbons and components with more than six carbon atoms (C_{6+}). Natural gases are petroleums which are in the gas phase in the subsurface and which are composed of hydrocarbon mixtures dominated by methane with contributions from ethane, propane, butanes, and the pentanes and traces of inorganic gases, principally carbon dioxide and nitrogen. Gases by definition contain no C_{6+} components.

Gas condensates are petroleums that are the gas phase in the subsurface but contain significant amounts of C_{6+} components (condensate) in gaseous solution. These components

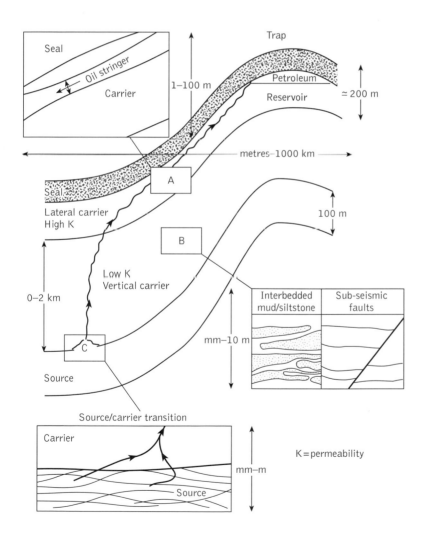

Fig. 1. A petroleum system consists of a source rock containing organic matter (kerogen), vertical and lateral carrier beds, a trap consisting of a reservoir rock, and a seal. Petroleum systems are usually less than 5 km deep; they can be very small laterally, with short migration distances, or range up to very large systems with lateral dimensions up to several hundred kilometres in foreland basins. Source rocks and reservoir rocks are typically about 100 metres thick. Vertical migration distances to a lateral carrier bed may be from 0 to several kilometres, and lateral migration distances can range up to several hundred kilometres. Vertical migration through mudstone carriers may take place through mudstone and interbedded siltstone pores, or through fracture systems.

include hydrocarbons and non-hydrocarbons, with up to more than 40 carbon atoms. The ratio (by mass) of gas (C_1–C_5) to C_{6+} components in gas condensates ranges from 3:1 up to very large values. On coming to the surface, as pressure and temperature are reduced in the petroleum production system, gas condensates undergo retrograde condensation, and a liquid phase (the condensate) condenses from the gas phase.

Oils are petroleums that are in the liquid phase in the subsurface but contain up to 50 weight per cent of gas in solution. According to the relative amount of solution gas, the dissolved gas, the oil is termed 'black oil' or 'volatile oil'. This solution gas exsolves from the liquid during the reduction in pressure and temperature accompanying petroleum production. Oils and natural gases can coexist in contact in the same reservoir. In this case the oil is saturated with gas under reservoir conditions. Where no separate gas phase is present the oil may not be saturated with gas and is termed 'undersaturated'.

The C_{6+} components of oils and condensates from the North Sea petroleum province are typically dominated by saturated hydrocarbons, with aromatic hydrocarbons in abundance; non-hydrocarbons (resins and asphaltenes) usually represent less than 30 per cent of the fluid. Other petroleums contain different proportions of these fractions, but most petroleums are dominated by the hydrocarbon fractions.

Petroleum systems

The elements of a petroleum system include the source rocks, the carrier beds, the trap, the reservoir and the seal (Fig. 1). The various processes acting on a petroleum system to produce an accumulation involve generation of petroleum in the source rocks, migration of petroleum from the source beds through the carrier beds and trapping of petroleum in the trap.

Source rocks and petroleum generation

Source rocks for oil and gas are usually fine-grained sediments, such as mudstones or marls rich in organic matter derived from bacterial and chemical alteration of algae, bacteria, or land-plant debris. Organic sediments, such as coals, can also act as source rocks, principally for natural gas and gas condensate. Where the coals are rich in hydrogen-rich land-plant debris, such as plant exines and resins, coals can also generate liquid petroleums, as is the case in parts of Australia. Source rocks are the site of petroleum generation from insoluble organic matter known as kerogen. Kerogen is the term used for the fractions of organic matter in a sediment that are insoluble in common organic solvents. It is the most abundant form of organic matter in the Earth's crust and is the chief organic component of oil shales, coals, source rocks generating crude oil and natural gas, and other sediments. It represents more, often much more, than 80 per cent by weight of the organic matter in these materials. Biochemical and chemical reactions convert reactive biochemical compounds, such as lipids, proteins, carbohydrates, and lignins, derived

from dead organisms, to kerogen, biochemically unreactive, insoluble material, in shallow unconsolidated sediments. Some soluble organic matter is also present in all sediments, including coals, oil shales, source rocks, and other sediments and contributes in a minor way to any petroleum expelled from the source rock; but most petroleum is generated from alteration of the kerogen. As the source rock subsides in the sedimentary basin, it is heated to temperatures that are typically above 100 °C (Fig. 2). The complex kerogen decomposes thermally to produce oil and gas, which may migrate away from the source rock to produce a petroleum accumulation.

A source rock is thus a fine-grained sediment rich in organic matter that could generate (from kerogen) crude oil or natural gas on thermal alteration in the Earth's crust. The oil or gas could then migrate (move) away from the source rock to more porous and permeable sediments, where ultimately the oil or gas could accumulate to make a commercial accumulation. If a source rock has not been exposed to temperatures of about 100 °C in the Earth's crust, it is termed a *potential source rock*. If generation and expulsion of oil or gas have occurred, it is termed an *actual source rock*. The terms *immature* (not adequately heated) and *mature* (has been heated and generated crude oil and natural gas) are commonly used to describe source rocks.

Many of the major accumulations of oil on Earth were generated from organic-rich marine mudstones or marls with from 1 to 10 weight per cent organic carbon. Examples would include the Jurassic–Cretaceous source rocks of the Middle East fields, the Jurassic Kimmeridge Clay Formation of the North Sea basins, the Cretaceous La Luna Formation of Venezuela and Colombia, the Shublik, Kingak and Torok Formations of the Alaskan North Slope basins, the Miocene Monterey Formation of the California borderland basins, and the Devonian Duvernay, Jurassic Nordegg, and Cretaceous Speckled Shale Formations of the Western Canada sedimentary basin which have variously contributed to the vast Athabasca petroleum accumulation in western Canada. Very organic-rich shales or marls that have not been heated in the crust of the Earth may sometimes be termed oil shales.

Although most of the largest oilfields on Earth have originated from source rocks deposited in marine environments (marine shales), many major accumulations result from petroleum generation in non-marine sediments from, for example freshwater or saline organic-rich shales or coals deposited in lakes. Many of the accumulations of petroleum, both gas and oil, found in South-east Asia were generated in non-marine sediments. Examples of such source rocks generating oil would be the Eocene Shahejie Formation shales of the Liaohe Basin in China. In all cases source rocks were deposited in slowly sedimenting basins with high surface biological productivity and restricted oxygenation of stagnant bottom waters resulting from bacterial consumption of free oxygen in bottom waters and sediments. Under such reduced oxygen conditions, large fractions, up to perhaps 1 to 10 per cent, of the primary biological productivity in the surface

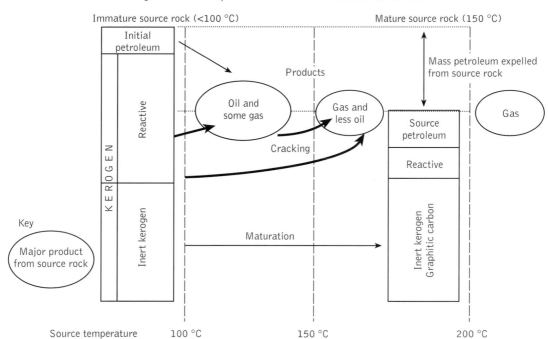

Organic matter in equivalent immature and mature oil source rocks

Fig. 2. Maturation converts immature organic matter in source rocks—principally kerogen—to oil and gas, and residual mature kerogen. The diagram shows the quantities of organic matter remaining in an immature source rock (temperature less than 100 °C) and a mature source rock (temperature: 150 °C). Both boxes are to the same scale, and the difference in heights represents the quantity of petroleum expelled from the source rock. At temperatures below 100 °C the kerogen is unaltered, but at higher temperatures up to 150 °C the reactive parts of the kerogen decompose to oil and associated solution gas and most of the oil and gas are expelled from a rich source rock. At temperatures over 150 °C any oil remaining in the source rock is converted to gas, which is expelled. Perhaps 50 per cent or more of the kerogen may generate petroleum in a good oil-generating source rock, and a large fraction of this would be expelled.

waters may have been preserved as kerogen and related bitumen-soluble components in sediments.

Petroleum generation

Most commercial source rocks contain 1–10 per cent by weight of organic carbon, mostly in the form of kerogen but with 5–15 per cent of the organic matter present being extractable. At temperatures above about 100 °C the kerogen decomposes and is converted to oil and/or gas depending on the type of organic matter in the sediments (typically over the temperature range 100–150 °C). Matter derived from algae and bacteria produces oil and associated gas. Matter derived from the woody parts of plants produces gas, principally at temperatures above, say, 150 °C. In the temperature range 100–150 °C, oil is generated in the crust and is expelled as a result of large pressure (fluid potential) gradients induced in the source rock as a result of the process of conversion of kerogen to oil and resulting compaction of the source rock. Generation and expulsion of oil from a source rock typically takes place at depths of about 3–5 km. Gas generation and expulsion continue at greater depths and at temperatures above about 130–160 °C.

In the thermal alteration of kerogen to oil and natural gas the solid-phase kerogen of high molecular weight becomes broken down to fluid products, such as oil and gas, which may migrate. While the average molecular weight of crude oil may be around 200–400, the mobile phase can contain asphaltenes, which are an operationally defined fraction of crude oil or solvent-soluble organic matter; they consist of material of higher molecular weight similar in bulk composition to source-rock kerogens. Asphaltenes may represent 'kerogen-like' material that is soluble in the fluid phase and may migrate with them.

All kerogens will generate petroleum mixtures on heating in the crust, but A. Pepper of BP suggests that not all kerogens efficiently expel the generated material from the source rock. Thus, most coals generate liquid materials but do not expel them; they retain the material in the coal matrix, where on further burial it becomes 'cracked' to gas, which may be expelled to migrate to a trap. Although all types of kerogen generate some petroleum on exposure to heat in the crust at depths below, say, 3 km (temperatures over 100 °C), not all types of material generate the same amount. Kerogens derived from algae can generate 40–80 per cent, by weight of petroleum components on maturation, whereas kerogens derived from higher plants generate 5–30 per cent, but usually less than 20 per cent by weight of petroleum (Fig. 2, reactive kerogen fraction).

Sediments rich in algal-derived kerogens (i.e. with total organic carbon contents of more than say 3 per cent by weight of the shale) can expel a large fraction of the generated crude oil and natural gas: as much as 80 per cent or more from some shales. As the richness of the sediment declines, the amount of crude oil and natural gas generated decreases and the proportion expelled also decreases. Thus source rocks lean in organic matter tend to be poor expellers of oils. Sediments in which only a small fraction of the kerogen is convertible to oil and natural gas are the poorest expellers of fluid-phase petroleums, and it is generally considered that coals rich in land-plant material are the least effective at expelling oils and typically expel only gas.

'Coal' and 'oil shale' are economic terms for organic-rich sediments. These may also be capable of generation of petroleum in sedimentary basins. In most instances coals principally generate and expel gas on heating in the subsurface. However, some coals derived from algae or from hydrogen-rich parts of plants such as cuticles and spores may generate and expel oil. These types of coals are common in South-east Asia and Australasia.

Although coals can generate the components present in oil and gas, the coal kerogen (principally in the maceral vitrinite) is rich in aromatic structures and is an efficient adsorber of all hydrocarbon and heterocompound components, but especially those components with molecules larger than methane. Most coals thus typically expel only gas. Coal kerogens contain many of the same basic chemical structural elements as algal kerogens, but coals usually have lower hydrogen contents than algal source rocks and generate smaller amounts of petroleum. This combination of a highly adsorbent aromatic base structure, together with a low yield of liquid product on generation typically results in only gas being expelled.

On a worldwide basis, the primary commercial source rocks for crude oil are marine or lacustrine shales rich in algal or bacterial organic matter (Alaska, the Middle East, Venezuela, Colombia, China, the North Sea, the Gulf of Mexico). D. McGregor of BP suggests that coal-bearing sequences are significant oil generators only in very specific and relatively uncommon geological settings. Coals and associated carbonaceous shales are thought to be the primary oil-prone source facies in Australia and an important secondary source in South-east Asia. Australian coals and associated carbonaceous sediments may thus well be oil source rocks in the Gippsland basin area and elsewhere in Australia. In other parts of the world there is no evidence that coals and carbonaceous shales have expelled major quantities of oil.

'Oil shale' is a term for shales that are very rich in organic material (see *oil shale*). They are essentially an organic rich immature source rock which has not been heated to a high enough temperature in the Earth's crust to convert the kerogen into crude oil and natural gas. There are no differences in the nature of the organic matter in crude oil source rocks and in oil shales except that oil shales would typically be very rich in organic matter and would have been prolific crude oil source rocks if they had been heated to temperatures of more than 100 °C in the Earth's crust (matured).

There is a wide variety of source rock types, and their behaviour on maturation varies with the composition of the organic matter, the organic content, and (to a lesser extent) the manner in which the organic matter and associated mineral content of the source rock are arranged. Two extreme end-member types of source rocks can be recognized.

Sediments such as shales rich in organic matter of algal and bacterial origin (1–10 per cent organic carbon by weight) start to generate oil and associated gas at temperatures about 100 °C under normal geological sedimentary basin heating rates of 1–10 °C/My. At about 150–160 °C most of the potential of the kerogen to generate liquid hydrocarbon is exhausted, and above this temperature any oil remaining in the source rock will start to be thermally 'cracked' to shorter-chain compounds and gas. Above 200 °C little further potential for generation exists. With typical geothermal gradients of, say, 33 °C/km, oil generation (and expulsion) starts at about 3 km and peak amounts of oil are generated by 4.5 km. With a typical algal-derived kerogen, upwards of 50 per cent of the kerogen may be converted to oil in this depth/temperature range over a period of from 5 to 50 Ma with a normal heating rate during burial of the source rock. Such source rocks are typified by the upper Jurassic Kimmeridge Clay Formation of the North Sea basins.

Source rocks such as humic coals with a large input of land plant lignin probably generate liquid like petroleum, but because of the high sorbtion capacity of the coal kerogen for bitumen this is rarely expelled from the source rock. At temperatures above about 160 °C humic coal kerogen and any liquid petroleum trapped in the sediment becomes 'cracked' to natural gas or gas condensate, which is expelled. Such source rocks are typified by the Carboniferous coals of western Europe. These have been the source of many major gas fields, including those in the southern North Sea basins, Holland, and Germany.

Migration of petroleum

'Petroleum migration' is the term for the series of processes by which petroleum is transported from its site of generation, the source rock, to the trap. 'Primary migration' is the term for the movement of petroleum from within the source rock to an adjacent carrier bed, at distances typically up to a maximum of a few hundred metres. 'Secondary migration' is the term describing the movement of petroleum from the source rock–carrier bed contact to the trap. A 'carrier bed' is a porous and permeable rock near the source rock through which petroleum flows from source to trap. Carrier-bed rocks are lithologically the same as reservoir rocks, that is, sandstones, limestones, or fractured rocks of all types. In some instances fluid-phase oil and natural gas may migrate through fine-grained rocks such as mudstones. Such is the case in the Gulf of Mexico, where petroleum migrates vertically over several kilometres vertically through mudstones and faults assisted by very high pressures in the deep sediments.

The principal driving force that moves the fluid-phase, crude oil and natural gas, from source to trap in most permeable carrier beds is buoyancy; the fluid-phase crude oil and natural gas in the deep subsurface have densities in the region of 500–800 kg m^{-3}, whereas waters in sediments have densities of over 1000 kg m^{-3}. The oil flows through the pore systems of the carrier beds as rivers of high oil saturation; most of the carrier system does not contain any oil. Even where the reservoir is a fractured source rock, some internal flow (migration) of oil is involved in concentrating the oil to form an accumulation.

The distance from the petroleum source rock to the trap may be as large as many hundreds of kilometres or very short, as where petroleum is trapped in high-permeability fractures within the source rock itself, as in the Bakken Formation oilfields of North Dakota. In the North Sea petroleum basins secondary migration distances are typically of the order of 20–40 km from source to trap, but fields such as the giant Troll field in the Norwegian sector may drain petroleum from source rocks over 100 km from the trap. The rates at which oil and gas move in the subsurface during secondary migration can be very high. Oyvind Sylte at the Continental Shelf Institute in Norway suggests that in the North Sea once generated, petroleum may in some instances flow through the carrier from source to trap in as little as 10 000–100 000 years. The distances can also be impressive: J. Allan and S. Creaney of Esso consider that petroleum from Cretaceous or Devonian source rocks may have migrated over 400–500 km laterally from the source rocks in the Western Canada basin.

The primary migration of petroleum from source rocks is still the subject of much debate, but some understanding of the process has been achieved for the most common types of source rocks: those rich shale source rocks that generate most of the world's oil. Migration of crude oil and natural gas from oil-prone algal source rocks proceeds in a complex manner but is probably related to compaction of the source rock. As the solid kerogen phase converts to a fluid phase (solvent-soluble organic matter), the fluid phase has to bear part of the load of the overlying sediments that was previously borne by the solid kerogen. In consequence, fluid pressures in the source rock rise and the fluid phase is driven out of the source rock through fractures induced during the overpressuring, through the pore system of the source rock, and also partly in solution in the remaining kerogen matrix. Once in the carrier bed (outside the source rock) it is buoyancy that principally causes the oil to flow from source to reservoir; the oil being less dense than the surrounding waters in the rock-pore systems. In most oil source rocks the gas associated with the oil is expelled in solution in the oil.

In secondary migration in the carrier bed, petroleum moves updip in a carrier rock predominantly because of its buoyancy, since it has a lower density than the surrounding formation waters. Petroleums in the subsurface will typically have densities ranging from 800 to 500 kg m^{-3} for oils and as low as 100 kg m^{-3} for gases according to the pressure and temperature. In contrast, formation waters typically have subsurface densities greater than 1000 kg m^{-3}. Regional hydrodynamic gradients driving water flow in the basin may have a very minor role in petroleum migration, but petroleum is considered to move as a separate phase from the waters in basins. In many basins there is in fact little or no water flow associated with petroleum migration.

Theoretical studies by W. England of BP and examination of migration pathways in the field clearly indicate that petroleum does not flow through the entire carrier bed but tends to move in concentrated zones where the petroleum has displaced most of the formation water. Petroleum flow perhaps takes place in as little as 1–10 per cent of the total potential pore space of a carrier bed, the petroleum moving in zones in which it occupies most of the pore space. These petroleum rivers probably have lateral and vertical dimensions a few metres to tens of metres in extent; away from these zones most of the carrier system pore space probably contains only water.

Reservoir rocks and traps

If petroleum is to be produced from the subsurface it must accumulate in rocks that are porous and permeable enough for reasonable volumes of petroleum to be produced at reasonable cost. Most reservoir rocks have permeabilities greater than 1 millidarcy; some have permeabilities of many darcies (1 darcy is 10^{-12} m^2). Most reservoir rocks are sediments such as sandstones, conglomerates, and limestones with porosities typically ranging from 10 to 30 per cent, the petroleum being contained in the pores of the sediments. Petroleum can also accumulate in fractures in limestones, or even within the source rocks themselves. Examples of reservoir rocks include the Middle Jurassic Brent Group sandstones or the Upper Cretaceous fractured Chalk of the North Sea basins, which contain both oil and gas. Reservoired petroleum columns may be as thick as several hundred metres. The petroleum typically fills 50–90 per cent of the pore space, the remainder of the pore space containing formation water trapped when the petroleum entered the field.

In order for petroleum to accumulate in a trap, several criteria must be satisfied. The crude oil and natural gas is trapped in a porous and permeable rock body: a reservoir rock encased above, and sometimes below, by fine-grained rocks that are relatively impermeable to crude oil—the seals. Crude oil (and gas) is kept in the trap by the buoyancy forces acting on it, which tend to move the oil to the surface of the sedimentary basin but are prevented from doing so by the seals. The trap must completely seal an area of reservoir rock; if it does not, no petroleum accumulation will develop. Most seals are fine-grained rocks such as shales or mudstones. The best seals are salt beds, which can even trap gas for long periods of geological time. Lateral sealing is accomplished either by structural deformation of the reservoir rock and seal into a closed anticline or by the reservoir rock changing laterally into a seal rock such as a shale. Faulting of the trap might also place a shale next to a reservoir rock and thus seal the trap.

The oil usually occupies about 50–80 per cent of the pores of the reservoir rock, the remainder of the pore volume being occupied by water. In some accumulations the reservoir can be filled by gas or both gas and oil may occupy the pore volume. Oil reservoirs can occur at any depth within the crust but liquid oil accumulations are rarely found deeper than 5 km, for the temperatures at greater depths may be so high (more than 160 °C) that any oil in the reservoir would be cracked to gas.

Petroleum accumulations are ephemeral features; the petroleum often leaks through the seals with the passage of time. It is thought that most of the petroleum generated during the history of the Earth has leaked away and has been destroyed at the surface by bacterial action and evaporation. Most of the world's existing large petroleum accumulations were generated and migrated within the past few tens of millions of years, and few petroleum accumulations have been found that have contained petroleum for more than 100 million years.

Petroleum alteration

Once in a trap, petroleum can have its composition altered by a variety of processes, which usually reduce the commercial value of the oil and complicate extraction of the petroleum. Five types of processes are involved in petroleum alteration in the reservoir: thermal degradation, biodegradation, thermochemical sulphate reduction, water washing, and asphaltene precipitation.

Thermal degradation or cracking affects liquid petroleum trapped in reservoirs hotter than about 160 °C. Heavier fractions of the petroleum are broken down to lighter fractions, including large amounts of methane and other gases. At temperatures over 200 °C even gases such as propane or butane may be broken down to methane. If degradation is severe a carbon-rich deposit, pyrobitumen, may form in the reservoir, blocking the pores and preventing easy production of the petroleum.

Bacterial action, or biodegradation, affects oils, and, to a lesser extent, natural gases in shallower petroleum accumulations where the reservoir temperature is less than about 70 °C, where there are nutrients such as phosphate ion, and where a chemical oxidant is present in the reservoir waters. Traditionally the oxidant is thought to be molecular oxygen brought

into the reservoir by slowly moving groundwater, but this idea is increasingly being challenged and other oxidants are undoubtedly involved. Oxidation of the hydrocarbons in the petroleum results in the production of a heavier, more viscous oil with lower economic value that is more difficult to refine. Perhaps more than half the oil on Earth has been biodegraded. The largest accumulations of oil are the giant heavy oil accumulations of the Canadian and Venezuelan heavy oil belt. The Alberta heavy oil deposit alone contains over 2500 billion barrels of petroleum which can be produced by mining and extraction of the 'heavy oil'. Although less familiar than the giant accumulations of oil in the middle East, these biodegraded oils will become more economically important in the twenty-first century when the conventional, easily produced, crude oils have been largely consumed.

Thermochemical sulphate reduction is a less common alteration process in which methane, and sometimes other hydrocarbons, in the petroleum react with sulphate ion in the formation waters to produce hydrogen sulphide and remove the hydrocarbon. This lowers the value of the deposit. This alteration process takes place only in reservoirs where the temperature is greater than 140 °C and abundant sulphate ion is present in the water. Gases rich in hydrogen sulphide associated with thermochemical sulphate reduction are common features of petroleums found in deep reservoirs in the Gulf of Mexico and Alberta.

Water washing results from the removal of water-soluble components in the petroleum by slowly moving groundwater in the reservoir. Water washing usually accompanies biodegradation and is commonly associated with shallow petroleum accumulations close to mountain ranges or other elevated terrain where groundwater at a high level can drive water flow in the subsurface.

Asphaltene precipitation results when gas, usually rich in methane, enters an oil reservoir and causes precipitation of the heaviest fractions of the oil. The asphaltenes, with about 70 carbon atoms per molecule, are the largest molecules dissolved in oil. They become insoluble when the oil becomes charged with much gas and they precipitate to form a feature known as a tar mat. Several large oilfields in the world have thick tar mats. The Prudhoe Bay field in Alaska and the Oseberg oilfield in the North Sea both have tar mats more than 10 metres thick at the base of the oil column in the trap. Tar mats complicate the production of the oil. When they are present, asphaltene deposits also form in the production piping, causing production problems and blockages.

The future

Petroleum is a finite natural resource that has been rapidly consumed by man in the twentieth century. As Hubbert showed, when world demand for petroleum rises, petroleum production must pass through a maximum, and eventually the remaining discoveries must be smaller and smaller irrespective of the additional exploration effort. Although gas reserves will last well into the twenty-first century, oil production rates in the USA peaked in the 1970s, and those in the North Sea in the late 1990s. Colin Campbell estimates that world production rates will peak in the decade 2000–10 at above 60–70 million barrels a day. From then on world petroleum production will decline to a level below about 20 million barrels a day. By 2050 other energy sources will be needed to replace the demand. With increased use of heavy oil and tar sands as the lighter lower-viscosity oils are consumed, it is likely that petroleum will form a small part of the energy supply until well into the second half of the twenty-first century. Total recovery of oil will amount to about 1800 billion barrels.   STEVEN LARTER

Further reading

Hunt, J. M. (1979) *Petroleum geochemistry and geology*. W. H. Freeman, New York.

Tissot, B. P. and Welte, D. H. (1984) *Petroleum formation and occurrence*. Springer-Verlag, Berlin.

petroleum industry The petroleum industry barely existed 100 years ago and will probably be largely gone 100 years from now. Nevertheless in this relatively brief period oil and gas have become fundamental to our modern technology-oriented society. The use of oil as fuel can be thought of as a time of growth from about 1860 up to 1975, a time of maturity to the turn of the twentieth century, and a time of decline after that. Although nearly all the world's oil resources were created over a period of about 500 million years, it is a sobering thought that mankind will burn up the vast majority of the recoverable portion of these in less than 200 years.

Proven oil reserves in the world are estimated at around 1000 billion barrels, two-thirds of which are located in the Middle East. With oil production in the region of 75 million barrels per day, if there were neither increases in reserves nor in demand the world's oil would last another 25 years. Alongside this premise stands the fact that only 330 giant oilfields account for around 65 per cent of the world's total oil that has ever been discovered while the other 10 000 or so fields provide the remaining 35 per cent. It is clear that giant fields have been vitally important to the overall health of the petroleum industry by providing its operating base.

Unsurprisingly, giant fields, being large, are relatively easy to find. Most of the 330 cited above were located early in the history of petroleum exploration and over the past 30 years there has been a sharp decline in their discovery rate. Consequently, the remainder of the world's giant fields probably lie in completely unexplored frontier regions such as the Arctic, eastern Russia, central Africa, and underlying the very deep waters of the ocean shelves. Such areas are not only few and far between but are also remote and expensive to exploit.

Meanwhile, the countries of the world with the largest populations, such as China, India, and Indonesia, are relatively undeveloped and have significantly lower energy consumption per capita than that of the developed world. As all see significant economic growth and a surging population (in fact each and every year an additional 80 million or so people are added to the world's population) it would seem that world energy demand, including oil demand, will continue on an upward trend. It is under this scenario of tight supply and escalating demand that the petroleum industry now has to operate.

Fortunately for the industry, available technology is ever improving and this is enabling additional oil reserves to be recovered both from traditional and from frontier areas at reduced cost. Furthermore, huge undeveloped reserves of natural gas have also been located in many remote parts of the world, so it is to both technology and gas that the petroleum industry is turning.

Industry objectives

By far the greater proportion of those working in the Earth sciences earn their living from the petroleum industry. Besides the hundred or so oil companies and twenty or so governments that effectively control the industry today there are several thousand small independent oil companies, service companies, and consultancies that look for opportunities to explore for, develop, and produce oil and gas, to sell new and traditional technologies, and to provide manual and intellectual services. Opportunities exist throughout the world.

There are a number of ways in which oil and gas companies can acquire exploration or production acreage so that they can search for, develop, or produce oil and gas fields:

Directly from governments: interests are obtainable by negotiation or by competitive bidding in licensing rounds. They involve work commitments (acquiring seismic data and drilling wells), cash payments, and taxes on production.

Farm-ins: interests are acquired from other oil companies. They involve the agreement to do work (normally drilling wells) in exchange for an interest in a licence.

Purchase of asset packages: these usually have attendant government work commitments but for overhead or tax reasons are worth more to the buyer than to the seller.

Asset swaps: normally used by companies wishing to improve core asset areas. No money changes hands. Asset swaps are becoming increasingly popular as companies focus their objectives.

Company takeover and merger: where one or both of the companies involved see an economy of scale and so they unite, willingly or unwillingly.

With such activity, acreage is progressively being exhausted so that available exploration areas have started to decline. Together with the escalation of environmental issues and the presence of alternative but expensive energy sources, which will perhaps take an increasing role, so the industry will need to adapt to changing circumstances. Be they the national oil companies such as Pertamina in Indonesia, Petrobras in Brazil, or Aramco in Saudi Arabia; the oil majors, such as

Shell, Exxon/Mobil, BP/Amoco/Arco, Total/Fina, Texaco, and Chevron; or the independents large and small, they will need to develop new ideas and technologies to maintain the world's energy resources and to protect the environment. The petroleum industry has a great responsibility to society.

Many new opportunities have indeed appeared since both the demise of communism and as a result of continued advances in technology brought about by enhanced computing power. However, perhaps unique to the petroleum and mining industries, the risks in pursuing new frontier and technological opportunities are very large, but so are the rewards, and whenever an opportunity comes on the scene there is a rush to acquire the chance to evaluate it. Companies within the petroleum industry need, above all, to understand the risks.

Petroleum industry risk

Oil and gas companies will invest the large sums required for the exploration, development, and production of oil and gas fields only if they can expect a reasonable rate of return from their investment. But every time an area is evaluated, a licence negotiated, a drilling prospect chosen, a well drilled, or a decision on field development made, companies are playing a game of chance. The odds are long, the rewards are high, but there is no assurance of winning.

It is rarely possible, using the data available when decisions on investment are required, to determine the exact value of each of the parameters that affect the ultimate profitability of a field. Thus the determination of risk is fundamental to the industry and the concept of risk has been embedded into the petroleum industry's definition of reserves.

Reserves are *estimated* volumes of crude oil, condensate, natural gas, and natural gas liquids that are expected to be commercially recoverable from known accumulations from a given date, under existing economic conditions, by established operating practices and under current government regulations. Reserves are estimates, and this holds true for so-called proven as well as probable and possible volumes. They can change in magnitude as a result of changes in both technical and economic factors and, because of this, reserves figures are highly subjective. Reserves quoted by a government, a journalist, a geologist, or by the industry itself in a company annual report or press release will all differ. They are influenced by bias.

Risks are nowhere better demonstrated than through an appreciation of the four phases in the exploration and development process of an area.

Phase 1 (original venture)

The first company to enter a newly available exploration area will take a large part of the risk. This new area might be a region (for example, the Arctic), a country (for example, Russia), a sedimentary basin (for example, west of the Shetlands off the UK), a new geological idea (for example, a new reservoir target such as beneath the salt layers in the Gulf of Mexico), or a new technical advance (for example, innovative production systems applied to the excessively deep waters off West Africa). Because of a shortage of knowledge, the risk of failure is high but rewards can be large.

Phase 2 (competitive)

Should Phase 1 companies be successful, those that follow will not only have tougher terms to negotiate but will also have a poorer choice of acreage as the high-reward opportunities, such as giant fields, are rapidly exhausted. On the other hand, they will have learnt from the experiences of the first, and the risks will be lower. However, there have been many instances where a second wave of companies became caught up in a frenzy of activity so that costs escalated and risks were not properly evaluated.

Phase 3 (mature)

Once the discoveries of Phase 1 companies and successful Phase 2 companies are developed, the area then enters Phase 3. It is now mature and the technical challenge has changed. There are still opportunities to be had, but these rely less on luck and more on technical specialities. The competition is intense, and companies with particular skills or experience in the opportunity are better placed for success. Risks are likely to be more carefully gauged and understood. The traditional areas of the North Sea are typical Phase of 3.

Phase 4 (depleting)

After years of production from a proven area, opportunities for new exploration and development become very restricted. In Phase 4 engineering skills come to the fore, for example in redeveloping old fields using enhanced recovery methods and applying innovative production techniques to make small fields economic. Many of the oldest production areas in the world are in this phase of development; for example, the onshore areas of Sumatra and Java in Indonesia, producing basins in western Russia, and many of the producing basins in North America.

Technology

It is clear that, in order to exploit successfully any of the four phases, the petroleum industry must employ a large array of different technologies. These fall broadly into three categories:

those used in exploration, especially the acquisition and processing of onshore and offshore seismic data;

those used in drilling, especially directing the drill bit to the reservoir, acquiring information from a well while drilling, and acquiring information from a well after drilling;

those used in production, especially in enhancing recovery from fields and in the application of production systems to extreme conditions such as very deep waters.

Advances in technology lead, not only to reductions in costs, but also to reduced risk, so that the industry can now find and develop offshore fields, including those deemed to be marginally economic, in as little as 3–4 years.

The acquisition and processing of seismic data dominate exploration. Quality seismic data rely on fast computer processing speeds. Seismic records improved dramatically in the 1990s, and the petroleum industry can now define ever smaller and more complex subsurface hydrocarbon traps. Drilling technology has also improved. Sophisticated directional drilling has revitalized drilling capabilities so that where wells once followed relatively simple paths they can now snake through the subsurface on tortuous routes determined by geological features.

Increasing the recovery rate from existing fields is also a key application for new technological innovation. On a worldwide scale, increasing the recovery rate (the percentage of oil or gas that can be recovered from known in-place reserves) by just 1 per cent will lead to an extra 2–3 years' available consumption. Consequently there is an ever-increasing menu of engineering options for recovery enhancement.

Marginal discoveries are those that would appear to be uncommercial to develop using traditional methods but are perhaps economically attractive should some innovative ideas be applied. Many new technologies serve to make marginal fields commercial, but the most important is in the introduction of new types of production systems. In a similar fashion, exploration and production in deep water have also become feasible as a result of new designs for production system.

As oil reserves become harder to locate, despite technological advances, the industry is also exploiting new markets for natural gas. There are abundant world reserves of gas, it is relatively clean compared to oil, and it is cheaper to produce but, in its natural state, gas requires nearby markets for cost-effective sales by pipeline. The industry is now converting large quantities of this gas to liquefied natural gas (LNG) for export to energy-deficient countries, primarily in the Far East. The LNG trade is growing rapidly together with the design and construction of new regional gas pipeline systems.

Environmental influences

The public perception of non-renewable energy production is a poor one, namely that the act of burning fossils fuels is dirty, it harms the atmosphere, and that operations to produce and transport oil are also environmentally damaging. True or not, for all practical purposes this is what the petroleum industry has to accept. Consequently, the vast majority of oil and gas companies now make preserving the environment a key part of their company objectives.

The real environmental impact of fossil fuels is, of course, very complex, but since the use of fossil fuels may be altering the environment, it is better to prepare for the worst than hope for the best. Regardless of whether or not climate change arising from the burning of fossil fuels is proved, new and stricter environmental policies in the form of CO_2 taxes, energy conservation programmes, and the promotion of renewable alternatives are now factored into petroleum industry planning for the future.

Strict environmental regulations also govern petroleum industry operations. The industry has to be very careful not to upset public opinion, for adverse views are soon communicated to the world's media. For example, there are regulations in most countries as to how redundant installations should be disposed offshore, but the application of environment principles to such structures is now beginning to mean that under no circumstances can they be dumped in the sea.

The other challenge is in the development of alternative fuels. In the short and medium term there would appear to be limited potential for renewable energy sources to affect the petroleum industry's competitiveness as, except in special circumstances, alternative fuels are more expensive. Furthermore natural gas has taken much of their role, particularly in the creation of electricity. Many groups now oppose nuclear power, its environmental impact being deemed to be an even greater threat to populations than burning fossil fuels. Nevertheless in the long term energy supplies will become tight unless other new and cost-effective alternatives are developed in the realms of hydroelectricity, solar, wind and tidal power, and hydrogen and alcohols from crops.

The real challenge for the petroleum industry is to support environmental action plans and proposals as well as alternative energy sources provided they are equitable, cost-effective, and really do prevent environmental harm. The industry will oppose existing or new laws and regulations that do not meet these criteria.

Critical success factors

The petroleum industry thus concentrates on six matters when determining the viability of a project. They differ in importance for different project types—exploration, development, and production—but an assessment of all of them ensures that a factor critical for the success of a project has not been overlooked.

Exploration and development history of an area: in an area where an asset is located or an acquisition proposed this provides a direct gauge of an area's commercial possibilities in a way that is relatively easy to evaluate using data that are relatively easy to come by.

Financial support: enough finance needs to be available to risk, either from internal funds derived from production or from willing investors prepared to support proposed ventures.

Reserves and production rate potential: geologists and engineers should have made a full assessment of a project and have some idea of the risks involved, the traditional, or new technologies that can be applied, and the reserves that might result.

Terms and conditions: there is no point in acquiring potential or actual reserves if the production rates or volumes thought to be present are insufficient to be profitable as a result of heavy taxes, other committed costs, or environmental constraints.

Legislative and political stability: economic assessment of potential exploration and production ventures must not only take into account the impact of local fiscal and contractual arrangements but also the stability of these arrangements.

Infrastructure and market: an export route and market are critical for the success of a project. An obvious point perhaps, but so many exploration projects fail because one or both of these have not been fully considered at the outset.

For most of its history the petroleum industry has driven the world's economic growth and made good profits for itself. In order for it to continue to thrive, companies within it must be aware of impending declines in world reserves, must understand the risks, must develop and apply new exploration and engineering technologies, must improve old technologies, must respond flexibly to new opportunities in traditional and extreme areas, and must have a worldwide reputation for environmental concern.   MICHAEL R. SMITH

Further reading

Hatley, A. (ed.) (1992) *The oil finders: a collection of stories about exploration.* AAPG Bookstore, Tulsa.

Shell Briefing Service(1995) *Energy in profile.* Shell International Petroleum Company Limited, London.

Smith, M. (1996) *Reserves acquisitions: a strategic handbook for oil and gas companies and investors.* Financial Times Energy Publishing, London.

Smith, M. (1997) *The upstream oil and gas industry into the 21st century—opportunities and challenges for the future.* Financial Times Energy Publishing, London.

Yergin, D. (1993) *The prize: the epic quest for oil, money and power.* Simon and Schuster, London.

petrology, petrography, petrogenesis

Petra is the Greek for rock. Petrology is the science of rocks; petrography and petrogenesis are subdisciplines: petrography is concerned with the description of rocks; petrogenesis with the ways in which they are formed.

Three major divisions of rocks are recognized in geology: igneous, metamorphic, and sedimentary. Each of these terms carries petrogenetic connotations. Igneous rocks are those that have formed from molten magma. Metamorphic rocks, as the name implies, are those that have formed by the modification in the solid of an existing rock by the effects of heat and pressure. Most sedimentary rocks are formed at the surface of the Earth by deposition of particles that have been eroded or weathered from pre-existing rocks, which can be igneous, metamorphic, or sedimentary. Some sedimentary rocks are formed of particles deposited by or from a fluid by chemical or biological precipitation.

There are some rocks where more than one set of processes has been active in their formation. For example, igneous cumulates are formed from the accumulation of dense minerals that have crystallized out from cooling igneous melts. Fluid dynamic processes operate upon crystals settling out of a melt in the same way as on sand grains being deposited in water; sedimentary textures can thus be formed in an igneous rock. Another example of an igneous rock with a sedimentary texture is a tuff: a rock formed by the accumulation of volcanic ash. Metamorphic rocks that have formed from igneous or sedimentary rocks can still display relict igneous or sedimentary textures.

Petrography

Petrography, the description of rocks, includes the description of hand specimens and the examination of thin sections under the microscope.

Hand-specimen description includes observation of the colour, texture, grain-size, and mineral content of the rock. Where the rock contains mineral grains it may be possible to determine the hardness, lustre and streak of the minerals. Hardness is measured using Moh's scale of hardness, which is based on a ten-point scale whose points are defined by a series of minerals. Two tests that can be applied when a complete set of test minerals is not available are a fingernail (hardness 2.5) and the blade of a pen-knife (4.5). Streak is the colour of the mark produced by rubbing the mineral across the surface of unglazed porcelain or ground glass. Lustre describes the surface appearance of the mineral.

The use of thin sections greatly aids the description of rocks and the identification of minerals. A thin section is a slice of rock 30 micrometres thick affixed to a glass plate or slide with a resin called Canada balsam. Most rocks and minerals are transparent in slices as thin as this and can be viewed in transmitted light at various magnifications. Many minerals have properties that vary with the direction in which the light passes through the crystal. Rocks are viewed on a petrological microscope in polarized light on a rotating stage so that the various colours and refractive indices become apparent as the stage is rotated. The Canada balsam upon which the rock is mounted has a refractive index of 1.538. Minerals whose refractive index is higher than this appear to stand out from the surface of the slide like hills and are said to have positive relief; minerals of negative relief appear as pits in the slide. Minerals that are not transparent even when sliced so thinly are described as opaque. Opaque minerals are examined in reflected light, i.e. by episcopy. For exact determinations of the mineral composition from thin section, analytical tools such as an ion-probe or an electron microscopy are used. The proportions of matrix and minerals in a rock can be measured by point counting.

Petrogenesis

Igneous petrogenesis is the process or set of processes by which igneous rocks are formed. The basis of the formation of any igneous rock is melting to form a magma. Igneous petrogenesis includes not only the generation of a melt but also the subsequent modification of the magma composition. The petrogenesis of a rock cannot generally be observed directly and must be deduced from petrographic observations of the rock itself or experiments on synthetic materials designed to stimulate the rock-forming process; i.e. experimental petrology.

                                        JUDITH M. BUNBURY

Further reading

Dud'a, Rudolf and Rejl, Lubos (1986) *Minerals of the world.* Spring Books, Twickenham, Middlesex.

Deer, W. A., Howie, R. A., and Zussman, K. (1985) *Rock forming minerals* (2nd edn). Geological Society, London.

Phanerozoic Derived from the Greek and meaning 'apparent or evident life', the term 'Phanerozoic' is given to the long span of geological time (eon) that has passed since the beginning of the Cambrian period, and to the rocks formed during that time (eonothem). Thus it succeeds the span known as the Cryptozoic ('hidden life') Eon, which covers all Precambrian (i.e. Proterozoic and Archaean) time. The terms were invented by the American stratigrapher G. H. Chadwick in 1930, and have been in general use ever since.

It is now known that the later Proterozoic rocks include fossils that are clearly evident and identifiable, but the palaeontology of Precambrian rocks is mainly obscure. Algae and other primitive plants and relatively simple animal body fossils and trace fossils (ichnofossils) of animals indicate that life was widespread and diverse in late Cryptozoic time. Then, however, something of an organic revolution took place at the beginning of the Phanerozoic when hard skeletal structures appeared in many animals for the first time. Fossilization of these forms immediately became more common and ensured the improved record of the biosphere that characterizes the Phanerozoic systems of rocks.

The term 'Precambrian' is still used for Cryptozoic rocks, but tends to be applied loosely to formations lying beneath identified Cambrian strata and of uncertain age.

The beginning of the Cambrian period, and hence of Phanerozoic time, is estimated to be around 570 million years ago. The base of the Cambrian system is defined by international agreement on palaeontological criteria. On similar lines, agreement is being sought for the base of each successive geological system.

Throughout the Phanerozoic systems, fossils provide the best means of correlation and also record the progress of organic evolution over more than 500 million years. The division of the Phanerozoic into chronostratigraphic divisions—eras, periods, epochs and ages—has itself evolved over a period of about 200 years. It is based upon both lithostratigraphy and biostratigraphy and seeks to provide a basis for world correlation and the relative dating of geological events, processes, and materials.   D. L. DINELEY

Further readings

Calder, N. (1984) *Timescale: an atlas of the fourth dimension.* Chatto and Windus, London.

Cowie, J. W. and Brasier, M. D. (eds) (1989) *The Cambrian–Precambrian boundary.* Clarendon Press, Oxford.

phase transformations at depth in the Earth

The disciplines of chemistry, geology, and materials science refer to a 'phase' as a substance of a fixed chemical composition which is mechanically separable from other material of different composition or structure. Accordingly, liquid water, water vapour, and ice are distinct phases with the same composition, H_2O, but with different structures. The particular structure adopted by the material depends on the pressure and temperature conditions it experiences.

Phase changes are fundamental to many biological and non-biological processes on the Earth and in the universe in general. The two major factors contributing to their central role are the discontinuous change in physical properties (deriving from the structural change) and simultaneous emission or absorption of heat. The importance of structure is obvious: a change from a solid to a gas, such as when ice sublimes, dramatically alters density and thus the way H_2O behaves. Equally significant is the heat transfer that generally accompanies a phase transition. The release of heat when atmospheric water condenses to rain, for example, is a critical factor in the evolution of weather systems on the Earth and the gaseous planets of the outer solar system.

Why phase transitions occur inside the solid Earth may be understood by considering the consequences of the increase in pressure and temperature with depth. Pressure increases by 3.5×10^6 bars from the surface to the Earth's centre, and temperature by a factor of 17 (Fig. 1). Different though these ranges may be, they have nearly equivalent effects on the

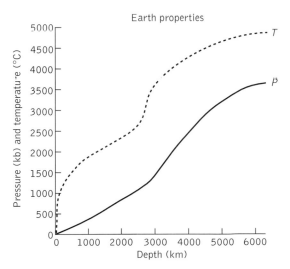

Fig. 1. Pressure (*P*) and temperature (*T*) profiles in the Earth's interior. Pressure increases from 1 bar at the Earth's surface to 3.6 million bars at its centre. In contrast, temperature increases from approximately 25 °C to about 4500 °C, or by a factor of 17 in the absolute temperature scale units (Kelvin, K).

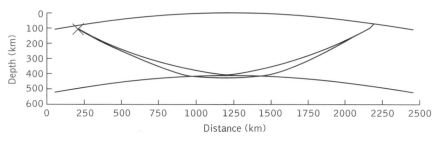

Fig. 2. Sketch of the way in which seismic waves from an earthquake both reflect off and carry through a layer with different elastic properties. Seismic wave speeds generally increase with depth in the mantle of the Earth, which makes seismic waves turn back upwards if they are initially emitted downwards. The reflected path is shorter but in slower material. The path taken by the transmitted energy goes further but deeper. The two waves arrive at a seismic station at slightly different times, permitting the depth of the reflective layer to be inferred. The layer is at 410 km depth in this figure, showing one way in which this discontinuity can be investigated.

thermodynamic factors governing phase transitions. Thus, neither pressure nor temperature is the dominant control upon phase transitions in the Earth's deeper mantle. Because of the corresponding increase in pressure and temperature with depth, a particular phase transition occurs at a characteristic depth. A transition occurring at a lesser or greater depth than expected indicates local temperature or compositional anomalies. Thus, with increasing depth we find a succession of layers of different phases, occasionally disturbed by local influences.

Seismological study of the Earth's interior uses remote sensing tools capable of detecting phase transitions. Elastic properties and densities of rocks determine seismic wave speeds. Each changes upon structural rearrangement. When a seismic wave field meets an elastically different layer, a fraction of the wave reflects off the layer and the remainder passes through it. The seismic wave field of an earthquake, travelling downwards and meeting such a layer, thereby follows two separate paths towards the surface (Fig. 2). A suitably located seismometer would detect two waves, one reflected and one transmitted, arriving at slightly different times. The reflected wave does not travel as far as the more deeply penetrating one, but the deeper one travels faster because rock strength increases with depth, giving rise to higher seismic wave speeds. The interplay of the timing and intensity of these reflected and transmitted arrivals yields information about the depth of the layers and its elastic property contrasts.

Below the crust, seismic wave speeds increase uniformly down to the core, where the change from a dominantly rocky to a liquid metallic composition reduces the wave speeds nearly back to crustal values, and extinguishes shear waves (see *Earth structure*). The notable departures from uniform increases in seismic wave speeds occur in the mantle at 410 and 660 km depth (Fig. 3). The '410-km discontinuity' and '660-km discontinuity' are typically referred to by their nominal depths. At the 410-km discontinuity, there is about a 4 ± 0.5 per cent jump in P-wave speed and a 6 ± 0.5 per cent jump at the 660-km discontinuity. Density also increases by 8 ± 4 and 8 ± 1 per cent respectively. The greater uncertainties

in the density of mantle rocks arise because whole-Earth oscillations of very low frequency principally constrain their values. Densities of rocks are not very sensitive to abrupt changes in rock properties with depth, and the inferred densities depend on other, poorly constrained, properties.

The observed seismic properties of rocks are volume-weighted averages of the properties of its constituent minerals. Thus to understand the seismic properties of the mantle, we can neglect its minor mineral components and focus on the three major ones: olivine, $(Mg,Fe)_2SiO_4$; pyroxene, $(Ca,Mg,Fe)_2Si_2O_6$; and garnet, $(Ca,Mg,Fe)_3Al_2Si_3O_{12}$. With depth, each member of this trio follows a separate path of phase transformation, ultimately resulting in a bimineralic

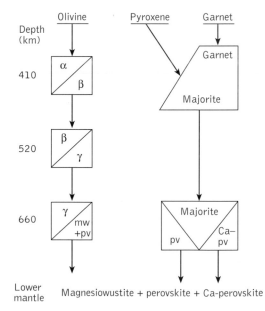

Fig. 3. Successive phase transformations taken on by the major mantle minerals garnet, pyroxene, and olivine.

mixture of magnesiowustite $(Mg,Fe)O$ and perovskite $(Mg,Fe)SiO_3 + CaSiO_3$. The trail of garnet is the most direct and entwines with that of pyroxene. Garnet absorbs pyroxene with increasing pressure (accommodating silicon on the sixfold site normally occupied by aluminium and then breaks down to magnesiowustite plus perovskite at the pressures of the lower mantle. During this latter stage, the Ca-pyroxene (clino-pyroxene) component separates to form a separate Ca-rich perovskite distinct from (Mg,Fe)-perovskite. The more chequered fate of olivine involves three polymorphs with different structures, α (olivine), β (modified spinel or wadsleyite), and γ (spinel or ringwoodite). These are adopted successively with increasing depth in the upper mantle. Ringwoodite eventually becomes unstable and forms magnesiowustite and perovskite. Deeper down, there are no known phase transformations that are seismically detectable. The portion of the mantle just above the core is anomalous, as reflected in a kink in the velocity–depth trend (Fig. 3), but the seismic properties of this region are not globally uniform. The cause of the velocity decrease must be other than a phase transition.

Inside the Earth both temperature and pressure increase with depth, but the surface heat flow varies geographically. Subduction zones, mid-ocean ridges, and hot-spot volcanoes represent the extremes. Beneath these sites, convective downwelling or upwelling influences the temperature profile of the mantle, and thus the phase transformation characteristics. Why this happens is related to the other major effect of a phase transformation: heat. Some phase transformations emit heat while others absorb it. The ones that emit heat will occur at a lower pressure upon a lowering of ambient temperature; conversely, those that absorb heat will occur at a lower pressure in hotter regions. A phase transition is thus a type of temperature sensor, moving deeper or shallower (changing pressure) in response to regional heating or cooling. Seismology provides a probe of the Earth's interior temperature through its ability to detect the changes in depth of the 410- and 660-km discontinuities.

Thorough mapping of mantle temperatures in the vicinity of the 410- and 660-km discontinuities is incomplete. Early results show depth excursions of $c.$ ± 50 km, which translate into lateral temperature variations of around ± 650 °C. Small-scale features in the present discontinuity depth maps are invisible, frustrating much-needed insights into mantle properties and processes at these scales. Further work will reveal features such as mantle plume sizes, the depth of mid-ocean ridge melting, and the detailed temperature structure under continents.

G. HELFFRICH

philosophy and geology Philosophy and the sciences interact in several ways. The metaphysical foundations of whole ways of thinking are affected by and affect the conceptual frameworks of sciences that at different times play a salient role in their time. But there are also mutual influences between philosophy and scientific method to be discerned in the way scientific research in particular fields is conducted. Both these kinds of mutual influence can be seen in the history and practice of the science of geology.

Metaphysics

The thought of any epoch is imbued with certain taken-for-granted principles that are sometimes brought to light in controversies that seem to be on the boundaries between science proper and broad metaphysical abstractions. The deep foundational role of conceptions of time in the world-views of post-Renaissance civilization has been intimately interwoven with geologists' revelations and debates concerning the history of the Earth and the processes that have led to its current state. We can hardly imagine the Darwinian conception of organic evolution in separation from Hutton's expansion of the finite temporality into at least the possibility of immense ages. The two great debates in geology between neptunists or Wernerians (who held that the present state of the Earth is the result of a flood) and the vulcanists, for instance Hutton (who held that the present state of the Earth can be understood only by reference to the action of great heat), and between catastrophists (changes are discontinuous and violent) and uniformitarians, for example Lyell (changes are gradual), were also debates about time. The neptunists and catastrophists were generally finitists, and their opponents were in favour of an open geological time. Cutting across these debates were differences of commitment to an Earth created by natural processes and one brought into being by God. These debates were of far-reaching philosophical import in that they brought novel and disturbing metaphysical issues to the attention of the educated public in a specially dramatic way.

The metaphysics of the early period of scientific geology was part of a larger shift away from a human-centred world-view. The very possibility of volcanists in debate with neptunists, and catastrophists with uniformitarians, depends on the adoption of time-frames very different from those derived from the human life-span. Much the same shift from a human foundation for science occurred in the late eighteenth century in other branches of science; for example, in the creation of temperature scales, such as the Celsius scale, in which the temperature of the human body played no role.

Methodology: the boundaries of the 'scientific'

But what of the methodology of geology as a science? As Rupke has shown, instancing the famous painting of Pegwell Bay by William Dyce as a turning-point, geological studies and landscape painting were closely interwoven. The transition from a representation of a portion of the Earth's surface for aesthetic reasons into an illustration of geological formations occurred around the 1830s. The insight that the state of the visible surface was in need of explanation called into play two different scientific procedures, each of which tried to explain the observed state of the Earth by reference to unobserved states and processes.

There was the question of explanations of the present by reference to past states of the Earth, in the situation that knowledge was needed of processes that had occurred at times when there were no human observers to make records even in principle. Whatever happened in the past was not directly observable in the present. But there were traces. By what inductive assumptions were inferences to the past to be made from the present, to reveal states of affairs which could then, without circularity, be cited in explanations of the very present from which they were themselves derived? How were geologists to justify their 'readings of the rocks'? The status of fossils engendered philosophical debates, and the last-ditch claim of the creationists that they were put there to test our faith is not unheard even today.

From the point of view of scientific method we could look on the debates between vulcanists, neptunists, catastrophists, and uniformitarians as a long-running discussion as to the best inductive principle to adopt in trying to break the circle between the present as an ensemble of traces and the past as their sources. The advent of radioactivity as an inductive principle is technically novel but much in the tradition of geological inductions.

There was, however, another realm of unobserved phenomena about which hypotheses began to have a prime place in geological method. There was the question of structures within the contemporary Earth that are deep below the surface explaining the contours, rock formations, and so on that were directly observable. Philosophers of science generally refer to imagined structures playing this role as 'models'. The creative imagination of geologists is demanded in the construction of models of subterranean processes and structures not currently directly observable. The logic of model-building is more complex than that engaged in the hypothetico-deductive method. Grounds for belief in the representational authenticity of models must be found before their acceptance is widespread. Geology offers one of the most interesting examples of a debate on the viability of a powerful model of a whole-Earth process, namely continental drift and its descendant, plate tectonics. The idea has a long history but its formulation early in the twentieth century by Wegener sparked an acrimonious debate. The arguments are illustrative not only of the logic of model-building but also of the recent insight that belief among scientists is as much a matter of social and psychological forces as it is the result of a cool assessment of evidence. On the one hand Wegener's hypothesis ran into scientific difficulties. As Hallam says, 'the greatest stumbling block to acceptance of Wegener's hypothesis involved the nature of the underlying mechanism.' Wegener's model needed another model, of the engine that drives the drift. On the other hand it provoked the ire of the conservatives that are found in every scientific field. Jeffreys declared it 'unscientific' because it was based on 'qualitative data'. R. T. Chamberlin attacked not so much the theory as the upstart Wegener with 'barely suppressed fury'.

Most sciences have their scandals, moments at which the defenders of an entrenched 'mainstream' turn on an inno-

vator with some ferocity. Geology's moment has come and gone. Who now would reject what we now call 'plate tectonics'? Geology touches philosophy both at the point at which our being as humans in a vast and open cosmos comes into question, and at the more mundane but no less vital point at which the disciplines of scientific method are to be defined.   R. HARRÉ

Further reading

Gillispie, C. C. (1951) *Genesis and geology.* Harvard University Press, Cambridge, Mass.

Hallam, A. (1973) *A revolution in the Earth sciences: from continental drift to plate tectonics.* Clarendon Press, Oxford.

Hutton, J. (1795) *Theory of the Earth with proofs and illustrations.* 2 vols. Edinburgh.

Lyell, Sir Charles (1830–3) *Principles of geology.* London.

Rupke, N. A. (1983) *The great chain of history: William Buckland and the English school of geology.* Clarendon Press, Oxford.

phosphates Phosphate is present in many sedimentary rocks as grains of apatite minerals (calcium phosphate). Rocks that are largely composed of phosphate (phosphorites) are less common but of economic importance as a resource for fertilizers. These phosphorites are formed either directly from a marine source or indirectly through biological uptake.

Marine deposits such as the Upper Cretaceous rocks of North Africa and the Middle East (a leading world resource) form in specific environments. Phosphate in the shallow photic zones is used up by marine organisms, but it will accumulate below this zone if oxygen is sufficiently depleted. This phosphate will form deposits if there is a low influx of sediments, particularly where there are upwelling currents of cold phosphate-rich waters. The phosphate replaces the carbonate in skeletons and lime mud; and phosphate in decaying organic matter can be liberated by bacterial decay. Together this phosphate accumulates as small nodules or crusts and can be seen forming on the ocean shelves today.

Biogenic phosphorites may form where phosphate-rich matter, such as bones, teeth, or excrement, is concentrated and cemented together or dissolved and later precipitated. Guano is made from seabird and bat droppings; it may be up to 30 m thick and is an economic resource in the eastern Pacific. These biogenic deposits may be cemented together into rocks or may be dissolved and re-precipitated within limestones as cements or as nodules.   N. MANN

Further reading

Shelley, R. C. (1988) *Applied sedimentology.* Academic Press, London.

Tucker, M. E. (1981) *Sedimentary petrology: an introduction.* Blackwell Scientific Publications, Oxford

photogeology *see* REMOTE SENSING

phylogeny (morphological and molecular)

'Phylogeny' is the term used to describe the history of evolutionary relationships of organisms. The study of these relationships is termed 'phylogenetics'. Phylogeny is very much interconnected with taxonomy, the science of the classification of organisms, for taxonomy involves the analysis of characters and the ordering of species in taxonomic units that are usually considered to have evolutionary significance. Because of the imperfections of the fossil record, there are large gaps which make it particularly difficult for palaeontologists to reconstruct evolutionary relationships. Although modern biologists can use hard-part and soft-part morphology, biochemical studies, DNA analysis, and even behavioural information in their analysis, palaeontologists are frequently limited to hard-part morphology only. The two main methods for developing phylogenies in palaeontology are evolutionary systematics and phylogenetic systematics or cladism. Both use morphological criteria as their main source of characters, but the approaches in each case are different.

Evolutionary systematics is a more traditional approach in which all available information is used. This consists primarily of an analysis of the distribution of characters among the species concerned. In this approach the number of characters held in common would indicate the closeness or distance of the evolutionary relationship. There are, however, problems with this approach, for it is clear that not all characters are equally valuable in indicating relationships; some similarities may be due to parallelisms, convergences, or reversals. In the past this method may not have been used rigorously enough, with the result that inappropriate characters have been used to indicate the common ancestry of groups. This method also includes stratophenetics, in which the geological age of species is considered to be extremely important. This is an approach that is widespread in palaeontology as a consequence of the perceived distribution of fossils through the geological record and the fact that they have changed through time. However, once again the incomplete nature of the fossil record causes problems and can result in more primitive members of a taxon being preserved at a higher stratigraphical level than more advanced forms. This approach must therefore be combined with rigorous character analysis. Analyses based on evolutionary systematics are represented graphically as dendrograms (family trees), on which a time scale is commonly superimposed (Fig. 1a).

Phylogenetic systematics relies on morphological characters alone, and emphasizes that characters shared by organisms manifest a hierarchical pattern in nature. This method was developed by Willi Hennig, a German entomologist who published his ideas in 1966, and it is now used extensively both in modern biology and palaeontology. Phylogenetic systematics is generally known as cladistics because of the emphasis on recognizing clades, that is, lineage branches resulting from splits in an earlier lineage. Cladistics is based on the view that recency of common origin can be shown by the distribution of evolutionary novelties or shared derived characters. These synapomorphies allow groups to be distinguished from each other, the resulting hierarchy being expressed as branching diagrams termed 'cladograms' (Fig. 1b). It is also accepted that many characters are primitive (symplesiomorphic) and do not contribute to an understanding of close relationships. However, it has to be understood that the system is relative and a character may be derived at one level but primitive at another. For instance, the possession of feathers is a synapomorphy for birds in relation to other reptiles which do not possess this feature. However, it becomes a symplesiomorphy if the relationship of two groups of birds is being considered. In order to decide whether characters are primitive or derived (their polarity), several criteria are used, including comparison with a related group or outgroup, and ontogenetic development of characters. Cladists assume that dichotomous splitting has taken place in each lineage, and they accordingly search for what are called 'sister-groups', which can be denoted by the number of synapomorphies that they share.

Three major cladistic groupings are recognized. (1) Monophyletic groups contain the latest common ancestor plus all, and only all, its descendents (Fig. 1b, A+B). Birds, including *Archaeopteryx*, would be a monophyletic group. (2) A paraphyletic group is one descended from a common ancestor, but not including all the descendents (Fig. 1b, C+D). Thus, although dinosaurs plus birds are monophyletic, dinosaurs alone are paraphyletic. (3) Polyphyletic groups are those containing organisms descended from different ancestors (Fig. 1b, B+D). A modern example of this would be seals, sea lions, and walruses, previously classified together as

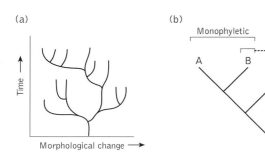

(a)

Time →

Morphological change →

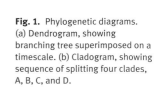

(b)

Monophyletic   Paraphyletic

Polyphyletic

A   B   C   D

Fig. 1. Phylogenetic diagrams. (a) Dendrogram, showing branching tree superimposed on a timescale. (b) Cladogram, showing sequence of splitting four clades, A, B, C, and D.

pinnipeds, but now thought to derive from separate groups of carnivorous land mammals. Cladistics traditionally takes no note of time, and cladograms therefore indicate only the sequence of dichotomous splitting. In some instances cladograms are superimposed on time frames to show the relationship to fossil distributions.

In recent years the results of molecular studies have been used to verify phylogenies developed by the more traditional methods already discussed. This involves comparing the structural details of homologous proteins and other molecules in different species. The methods used to develop a phylogeny are similar to those employed when morphological characters are used. The value of these studies in the study of extinct groups is clearly limited.   DAVID K. ELLIOTT

Further reading

Eldredge, E. and Cracraft, J. (1980) *Phylogenetic patterns and the evolutionary process.* Columbia University Press, New York.

Wiley, E. O. (1981) *Phylogenetics: the theory and practice of phylogenetic systematics.* John Wiley and Sons, New York.

piping

Piping is a form of subsurface erosion, which normally creates open tubes or 'pipes', particularly in soils, unconsolidated sediments, and loosely bonded bedrock. It is predominantly a process of removal of solid particles rather than solution. Piping is favoured by high infiltration rates, pre-existing cracks and voids, an impermeable sub-layer, steep gradients, and erodible materials. It is especially common in soils that have high expansion rates and/or highly dispersible clay minerals, and in locations which experience intense sequences of wet and dry periods.

The wide variety of initiating factors means that piping occurs in almost every climatic region of the world. The largest pipes, often several metres in diameter, have been recorded in semi-arid regions, where heavy convective storms or sudden snowmelt can cause rapid erosional development. Soil piping is, however, probably most widespread in marine, humid, temperate climates, where, although typically less than 25 cm in diameter, the pipes have been found to be hydrologically important, with pipeflow carrying up to half the storm run-off in some upland basins.

Piping erosion has proved to be an economic problem for some farmers, especially in south-western Australia, Arizona, California, and New Zealand. The term 'tunnelling', introduced by soil conservationists in Australia and New Zealand, is essentially synonymous, though some have sought to restrict its use to fluvial enlargement of pre-existing cracks above the water-table, and have sought to limit the process of 'true' piping to the pressurized removal of particles within the saturated zone.   J. A. A. JONES

placer deposits

A placer is a surficial mineral deposit formed by mechanical concentration of mineral particles from weathered rock debris. The concentrating medium can be either flowing air or water. Placer mining, which is the most ancient kind of mining, involves the extraction and concentration of minerals by the use of running water. Placer deposits supply a considerable fraction of the world's production of tin, titanium, gold, platinum, diamond, ruby, sapphire, and other mineral commodities. The economically most important minerals concentrated in placers are listed in Table 1.

For placer concentration to occur, a mineral must be resistant to chemical decomposition, and either tough or malleable so that it will not become fragmented during weathering and transport; the mineral being concentrated must also be more dense than the minerals with which it occurs. The common rock-forming minerals, quartz, feldspar, mica, amphibole, and pyroxene, all have densities equal or below 3.3 cm^{-3}, which is less than all the minerals listed in Table 1.

Concentration of dense mineral particles happens because the less dense particles are removed by flowing air or water, leaving behind an enrichment of the more dense minerals. The environments in which concentration can happen are numerous, but can be classified in four categories:

(1) Colluvial-slope deposits are those in which a blanket of weathered rock slowly creeps down-slope; less dense particles are washed down-slope more rapidly than dense particles, and so the dense particles become concentrated at the base of the weathering profile.
(2) Alluvial deposits are those formed in stream gravels or in delta sands by the action of flowing water.
(3) Marine deposits are formed along seashores by sea water flowing in longshore currents.
(4) Aeolian deposits are those concentrated by flowing air, and are found either in shoreline dunes or in desert areas.

Placer deposits can be of any geological age but most are geologically young. Placers of great antiquity are called *paleoplacers*, and one special group of paleoplacers, the gold deposits the Witwatersrand in South Africa, are about

Table 1 Minerals of economic importance recovered from placer deposits

Mineral	Composition	Density[1]
Gold	Au	15–19 g cm^{-3}
Platinum	Pt	14–19
Cassiterite	SnO_2	6.8–7.1
Monazite	$CePO_4$	5
Ilmenite	$FeTiO_3$	4.7
Zircon	$ZrSiO_4$	4.7
Rutile	TiO_2	4.2
Sapphire and ruby	Al_2O_3	4.0
Diamond	C	3.5

[1] Densities vary somewhat because of atomic substitution by different elements, e.g. silver for gold.

2.7 billion years (2.7×10^9 years) old. The Witwatersrand deposits are an enigma because they are much larger than any other placers of any age. In addition to size, there are many other features of the Witwatersrand that are puzzling: for example, the primary source of the gold is not known; details of how the gold was transported to its final resting-place are conjectural; and the fact that the gold is associated with pyrite (FeS_2), and uraninite (U_3O_8), two minerals which oxidize in today's surface waters and do not become concentrated in placers, suggests an environment at the time of formation that was very different from today's environment. For example, the presence of pyrite and uraninite can be explained if the atmosphere and surface waters 2.7 billion years ago contained little or no oxygen, a possibility that is supported by several lines of evidence.   BRIAN J. SKINNER

planetary geodesy

The geodetic study of the Moon and of the other planets in the Solar System has been restricted by the time and expense required for interplanetary missions, but significant discoveries have been made with techniques similar to those used in spaceborne terrestrial geodesy. The gravity field of a planet can be measured by observing its effect on a satellite in low orbit. Accelerations of the satellite caused by the planet's irregular gravitational field will cause a change in the Doppler shift of the radio signals transmitted back to Earth by the satellite. Only the component of gravitational acceleration in the line of sight from the orbiter to Earth will have an effect on the Doppler shift, so this method is often referred to as *line-of-sight Doppler*. Over many orbits, measurements of the line-of-sight Doppler shift can be combined to reconstruct all components of the acceleration and hence the gravity field. This technique has been applied since the 1960s on the NASA *Lunar Orbiter* and the *Mariner* missions to Mars and Venus, and more recently on the *Magellan* mission to Venus, the *Voyager* and *Galileo* missions to the outer planets, and will be used on the *Mars Global Surveyor* programme and the *Cassini* mission to Saturn.

The other major geodetic observable of a planet is its topography, which when combined with the gravity data will allow the planet's internal structure to be modelled and investigated. For the nearer planets, topography at a very coarse scale has been determined using Earth-based radar measurements since the late 1960s. The *Mariner 9* mission to Mars was able to improve on this information using the method of radio occultation. As the spacecraft passes behind the planet (as seen from Earth), the radio signals will be blocked, and the precise time of radio eclipse can be used to give details of the topography. Unfortunately this method is limited in accuracy because of the refracting effect of the planetary atmosphere on radio wave propagation. The *Magellan* mission to Venus used radar altimetry, a much more accurate topometric technique similar to that used by some Earth-orbiting satellites to measure the geoid. A downward-pointing radar is used to determine the height of the satellite above the ground, and this information is combined with the satellite position data derived from line-of-sight Doppler to give the actual topography with resolution of a few kilometres horizontally and a few tens of metres vertically. Because the atmosphere of Mars is thinner than that of Venus and is also transparent to visible light, the *Mars Global Surveyor* will incorporate an even more precise laser altimeter, which works on similar principles but uses the two-way travel time of a laser pulse to determine the spacecraft altitude. This technology has also been used on the *Clementine* lunar orbiter. Laser altimetry gives much more precise measurements than radar, with a horizontal resolution of around 100 m.   PETER CLARKE

planetary geomorphology

Geomorphology is the science of Earth's surface forms, but its concern with landforms is not limited to the Earth alone. A science is a way of reasoning or enquiry, and enquiry into the landforms of one planet, Earth, is readily enriched by reasoning about the landforms of other planets. It was in this spirit that Galileo Galilei pointed his 3.8-cm diameter telescope at the Moon in 1609 to observe the curious circular depressions on its surface. By 1665, Robert Hooke, a contemporary of Sir Isaac Newton, was performing the first geomorphological experiments, comparing the lunar depressions (1) to the cooled surface crust of boiled gypsum, and (2) the impact of musket balls into a clay–water target material. Reasoning by analogy. Hooke realized that alternative hypotheses for lunar craters might involve (1) internal heat melting (and bubbling) surface crusts, or (2) impacts by particles from space. By the time of Grove Karl Gilbert's classic study of lunar craters in 1893, both these geomorphological hypotheses remained viable, either as (1) volcanism, or (2) impact cratering. Gilbert also propelled balls of clay and metal into various target materials. The similarity in morphology to the lunar craters he observed through the Naval Observatory telescope provided compelling evidence for an origin by impact cratering. Gilbert even studied a possible Earth analogue to this process at Coon's Bluff in northern Arizona. Although Gilbert eventually concluded that Coon's Bluff was volcanic, the example of his reasoning process was taken up by later geologists, notably Eugene Shoemaker, who unequivocally established its impact origin in the 1960s. The site is now known as Meteor Crater.

Scientifically, the most interesting aspect of planetary geomorphology is the discovery of new landscapes, particularly as space missions image planetary surfaces by means of remote sensing systems with increasing spatial and spectral resolution. For example, the surface of Venus has been revealed only since the mid-1980s through Russian and American orbiting radar spacecraft. Despite its geophysical similarity to the Earth, in size, density, and general composition, Venus has a very different surface. It lacks plate-tectonic landforms, and is instead dominated by volcanic plains, locally punctuated by upland plateaux that may form by processes of mantle upwelling. Long channels of probable volcanic origin occur on the plains. The longest of these extends for 6800 km, longer than Earth's longest river.

Fig. 1. High-resolution images of small valleys incised into very ancient heavily cratered terrain on Mars (latitude 10° S, longitude 278° W). The imaged scene is about 40 kilometres wide and represents an oblique view towards the south-east (upper right). The images were produced from the *Viking* orbiter spacecraft.

envisaged to provide new data on Mercury, Venus, Mars, and various satellites of the outer Solar System. As with Galileo's first telescopic observations, these explorations will stimulate discussion on the origins of the newly revealed landscapes. Geomorphological studies of these landscapes will employ digital processing of remote sensing images (Fig. 1), computer simulation of geodynamics, and morphometry, and regional mapping. Fieldwork on terrestrial analogues, like Meteor Crater, remains important. Moreover, plans for human exploration may well lead to extraterrestrial fieldwork. Nor are the prospects for planetary geomorphology exhausted by the limits of the Solar System. The existence of extrasolar planets has been confirmed, and space telescopes are envisaged to image their surfaces. The legacy of Galileo, Hooke, and Gilbert will live on in future reasoning by planetary geomorphologists.   VICTOR R. BAKER

Further reading

Cattermole, P. (1995) *Earth and other planets: geology and space research.* Oxford University Press.

Greeley, R. (1993) *Planetary landscapes.* Chapman and Hall, London.

planets *see* GIANT PLANETS

plate tectonics: the history of a paradigm

Through the efforts of James Hutton and his successors the outlines of the geological evolution of the Earth became apparent, yet for 150 years the means by which large-scale crustal change was brought about remained unknown. When Alfred Wegener proposed his continental drift hypothesis in *Die Entstehung der Kontinente und Ozeane* in 1915 he opened a great controversy on how the continents come to be where they are. Orogeny and volcanism had generally been held to be associated with the contraction of the Earth, but the continents and oceans were regarded as primordial features, fixed in their positions since the crust first formed. Wegener's idea demanded great lateral movement of the continents, and it was definitely not appreciated outside Germany. His book provided no mechanism whereby the drift of continents could come about.

Geophysicists were loud in their objections to any lateral movement of continental blocks. At international conferences on continental drift held in 1922 and 1956 there was only a minority in favour of continental drift in view of the perceived difficulties. At that time advocates of continental drifting were Alex du Toit in South Africa and Arthur Holmes in Britain. Between the two world wars du Toit assembled a great mass of supportive data, while Holmes contributed a mechanism of continental transport in the form of convecting currents in the mantle. After the Second World War advances on two scientific fronts took place; palaeomagnetism and ocean floor surveying and sampling. From P. M. S. Blackett's original discovery of palaeomagnetism came the recognition of past polar reversals and the conclusion that either the poles or the continents had moved.

Venus is similar to Earth in its lack of very densely cratered ancient terrains. During the first few hundred million years of Solar System's history, the number and sizes of impactors were much greater than during the past three to four billion years. Remarkably, Mercury, the Moon, Mars, and most satellites of the gaseous outer planets preserve the ancient cratered landscapes that date back to the earliest history of the Solar System. Thus, impact craters are the most ubiquitous of planetary landforms.

On Mars, the ancient heavily cratered terrains are locally dissected by erosional valleys, as shown in Fig. 1. This shows that Mars, like the Earth, once had an active hydrological cycle that ultimately led to running water on its surface. Mars also shows a great variety of aeolian, glacial, periglacial, volcanic, and hillslope landforms. It clearly has had a complex history involving episodes of active erosion as well as tectonism and volcanism.

Future plans for planetary exploration by the United States, Japan, and the European Union should provide continuing opportunities for planetary geomorphology. Missions are

The second front on which great changes occurred was in our knowledge of what underlies the oceans. Improved (American) instrumentation and the deployment of large resources in submarine bathymetry and geophysics led to the discovery of the world's mid-oceanic ridges. Sampling programmes brought up rock specimens from the oceanic plains as well as the ridges. The volcanic rocks virtually everywhere were basaltic or ultrabasic types, and the sediments were nowhere older than the Cretaceous or possibly the Jurassic. It also became apparent that the ridges are prime sites of seismic activity and basaltic eruption. Geophysics showed the basaltic floor of the oceans to be much thinner than the granitic crust of the continents. Deeply subsided volcanic islands (sea mounts) were found in all oceans.

All this led Harry H. Hess, a distinguished geologist at Princeton University, to suggest that a kind of convective movement beneath the crust brought material up along the axis of the mid-oceanic ridges to spread out across the ocean floors and eventually sink into the trenches at the edges of continents. It had meanwhile been discovered that the ocean floors exhibit a remarkable regular striped pattern of variations in magnetic intensity. In places these are offset along lines that run at right angles to the magnetic lineaments, but the reasons for this remained a mystery. Then F. J. Vine and D. H. Matthews of Cambridge University came up with a possible test of the Hess hypothesis. They had in mind that the Earth's magnetic field had reversed direction several times in recent past ages and that when basalt was ejected at the mid-oceanic ridges it had acquired the imprint of the magnetic direction of the day. If this new strip of basalt were then pushed away from the ridge by subsequently erupted material, a record of the changes would appear from the polarities of the strips outwards from the ridge. Something like a large-scale tape recording of palaeomagnetism during the creation of the ocean floor would thus be made. The idea caught on and many geophysicists took to sea with improved magnetometers to chart the palaeomagnetic patterns of the ocean floors and to improve our detailed view of the ridges themselves. The Vine–Matthews hypothesis was confirmed by the new data.

The mid-ocean ridge, the newly found axis of sea-floor spreading, does not meander and curve at random, but is offset by fracture zones which can extend great distances from the ridge itself. Here originate the many earth tremors associated with the ridge, which occur when one side of the fault zone moves relative to the other. The rate of basalt injection may vary from fault-bound sector to sector, causing stresses and fault-zone movements. Differing rates of spreading from place to place along the ridge cause plate growth to be uneven, though the growth on each side of the ridge seems to be similar throughout. Improvements in isotopic dating of the ocean floor basalts identified the ages of the magnetic strips, and revealed more than 170 reversals of the Earth's magnetic field over about 76 million years. Spreading continues: the Atlantic is at present widening by about 5 cm per year.

Close to the ocean trenches the basaltic plate begins to turn down at an angle of about 45° and to descend into the depths. Soft sediment that has accumulated on the surface of the basalt during its passage from ridge to trench tends not to continue down but lingers as a accretionary wedge on the flank of the trench. Heat and pressure above the descending plate cause island-arc volcanicity.

The basic idea behind plate tectonics was first mooted by a Canadian, J. Tuzo Wilson, in the scientific journal *Nature* in 1965. He proposed that the elongate 'mobile belts' of the crust, where mountain ranges, earthquakes, volcanicity, and thrust faulting occur, commonly end abruptly at a *transform fault*. Transform faults link a continuous network of ridges, mountain belts, and trenches which divides the entire outer mechanical shell of the Earth into several large lithospheric plates. Soon afterwards Jason Morgan at Princeton and Dan McKenzie at Cambridge independently envisaged the Earth's surface as being divided into plates or blocks to which there were three kinds of boundary: ocean ridge, where new crust was generated; ocean trenches, where it was destroyed; and transform faults. The plates were thought to be about 100 km thick and to rest on a weak or plastic layer, the asthenosphere. On some plate surfaces there are the continents, not drifting but being moved passively as passengers on their host plates.

During the late 1960s and 1970s the explanation of mountain-building (orogenic) cycles came to be attributed to activity along active continental plate margins or to continental collisions. Overthrusting, volcanism, and plutonic igneous intrusion were identified as originating above the subduction zone where one plate is forced beneath the edge of its neighbour. Palaeomagnetic and tectonic studies now point to ancient world geographies and to the rotation and migration of continents. The expanding and contracting oceans, continental collisions, and fracturing can be modelled with increasing conviction on the basis of the plate-tectonic model. The process is recognizable throughout the Phanerozoic Eon far back into the Proterozoic, and even into the Archaean in a somewhat modified form. The impact has been called a 'revolution in the earth sciences', the arrival of the long-sought 'theory of the Earth': a true paradigm for all Earth scientists.
D. L. DINELEY

Further reading

Hallam, A. (1983) *Great geological controversies*, ch. 6. Oxford University Press.

Kearey, P. and Vine, F. J. (1990) *Global tectonics*. Blackwell Scientific Publications, Oxford.

Muir Wood, R. (1985) *The dark side of the Earth*. Allen and Unwin, London.

Park, R. G. (1988) *Geological structures and moving plates*. Blackie, Glasgow.

plate tectonics, principles The theory of plate tectonics describes tectonic events in terms of the motions of a small number of rigid plates which move relative to one another on the surface of the Earth, interacting and deforming only along their common boundaries. Because it provides an overarching, self-consistent, and global view of tectonic activity, its development in the 1960s ushered in a true revolution in the Earth sciences. Previous *ad hoc* explanations for local and regional events could be seen to be part of a wider, ultimately global, pattern. Moreover, by providing an understanding of the types of tectonic and volcanic activity that are associated with different types of plate boundary, it brought an important predictive power to geology. It was also important in finally providing a physically realistic mechanism to explain the observations of continental drift.

Approximately 12 major plates are recognized, together with a number of smaller ones (Fig. 1). They are outlined by narrow bands of seismic activity, and have almost aseismic interiors (see *seismology and plate tectonics*). It was this observation, which resulted from the setting up of the World Wide Standard Seismograph Network in the 1960s, that led to the idea that the lithosphere is divided into a number of large, rigid plates that are tectonically active only at their boundaries.

Plate tectonics thus relies on the concept of a rigid lithosphere (see *lithosphere*). This layer, which is rheologically defined as the strong outer layer of the Earth, generally includes both the crust *and* the uppermost part of the upper mantle. It is approximately 100 km thick, whereas the major plates are many thousands of kilometres wide; the plates thus

behave as thin shells sliding on the surface of the Earth. The asthenosphere (see *asthenosphere*) is the ductile layer immediately below the lithosphere, and is believed to decouple the plates from the deeper mantle.

An important difference between plate tectonics and earlier ideas of continental drift is that plates may comprise both continental and oceanic lithosphere. One of the earliest objections to continental drift was the absence of a mechanism that would enable the continents to move through what was perceived as a rigid mantle without producing major disruption at the leading and trailing continental edges and in the ocean basins. In plate tectonics, the oceans themselves are part of the plates, moving rigidly with the continents. The motion is facilitated by the presence of the asthenosphere, whose high ductility results from thermally activated creep, a process unknown to the physicists of the early twentieth century.

In its broadest sense, plate tectonics can be thought of as covering two distinct areas: plate kinematics, which restricts itself to describing the movements of plates and the changing geometry of their boundaries, and plate tectonics *sensu stricto*, which deals with the geological consequences of the creation, destruction, and interactions of plates.

Plate kinematics

As stated above, plate boundaries are primarily defined by the seismicity along their boundaries. However, plate boundaries are also marked by distinctive morphological features, narrow

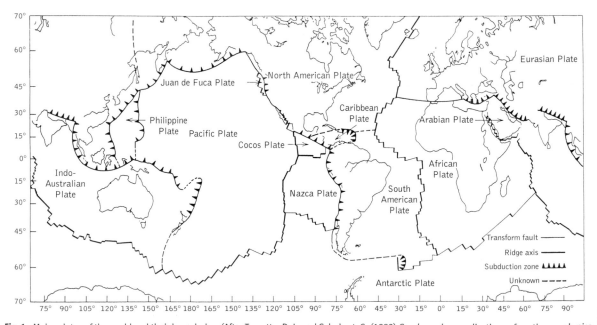

Fig. 1. Major plates of the world and their boundaries. (After Turcotte, D. L. and Schubert, G. (1982) *Geodynamics: applications of continuum physics to geological problems*, Fig. 1.1. John Wiley and Sons, New York.)

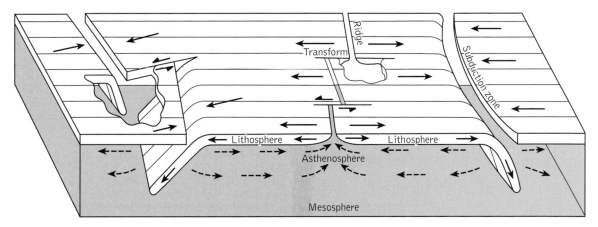

Fig. 2. The three types of plate boundary. (After Kearey and Vine, Fig. 5.1.)

bands of tectonism, and sometimes by individual major faults. The active plate boundary is never more than a few tens of kilometres wide, and may be as narrow as a few kilometres or even less.

Three types of plate boundary are recognized, corresponding to the three types of relative motion (Fig. 2); divergent (also ridge or accretionary) plate boundaries, transform (or conservative) plate boundaries, and convergent (subduction or destructive) plate boundaries. In general, any type of plate boundary can join any other type. Divergent boundaries occur only at mid-ocean ridges, and are the sites of creation of new lithosphere by sea-floor spreading. Transform boundaries are those where plates slide past one another with no change of area; they join one plate boundary to another, and can thus *transform* one type of tectonic boundary (e.g. extensional) to another (e.g. compressional), hence the name. They are found in both continental and oceanic lithosphere. Convergent plate boundaries in the strict plate-tectonic sense occur only at the deep-sea trenches and related subduction zones.

There are also zones of compressional deformation between converging plates in continents (such as the Alpine–Himalayan zone), but these are much broader than normal plate boundaries. This is partly because continental lithosphere is weaker and more easily deformed than oceanic lithosphere, and partly because it is less dense and is therefore not readily removed by subduction. Plate tectonics does not provide a very useful description of such broad continental convergence zones.

Apart from transform faults, plate tectonic theory does not prescribe the precise directions of relative motion at plate boundaries. However, although small degrees of oblique spreading and subduction are relatively common, strongly oblique motions are rare. In the same way, plate tectonics allows asymmetrical spreading (one plate accreting faster than the other). Although temporary asymmetrical spreading is common, the net effect averaged over millions of years is usually approximately symmetrical. The precise reasons for this are not fully understood.

Outlines of the geological consequences of the different types of plate boundaries are given under their individual entries.

Rotation poles

The great strength of plate kinematics is that it can describe plate motions in terms of simple geometry, and hence make precise predictions of relative motions anywhere on the globe. At the heart of this geometry is the concept embodied in Euler's theorem, which states that any displacement of a rigid body on the surface of a sphere can be described in terms of a single rotation about a specified axis. Such axes cut the Earth's surface at pairs of points called 'rotation (or Euler) poles'. Once the poles and the angular rotations are specified, the whole motion is completely determined. Thus the motion of a given plate is specified in terms of its Euler pole and a corresponding angular rotation rate.

In practice, determining the motions of individual plates relative to some common reference frame is not easy; however, determining *relative* motions between pairs of plates is quite straightforward. Such relative motions can also be described in terms of rotation poles (although the pole position must be specified relative to one or other of the plate pair), and such relative rotation poles are the basis of most plate-kinematic descriptions.

Since Euler poles define the directions of the rotation axes, and the angles or rates of rotation define their magnitudes, poles can also be described as vectors. Vector and matrix algebra can then be used to calculate plate motions, greatly simplifying and speeding up such calculations, which can be performed routinely by digital computers.

It is, however, important to distinguish between so-called 'instantaneous poles', which describe motion at an instant only, and 'finite rotation poles', which may describe the net

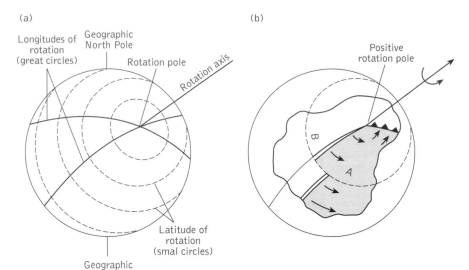

(a)

Longitudes of rotation (great circles)

Geographic North Pole

Rotation pole

Rotation axis

Latitude of rotation (smal circles)

Geographic South Pole

(b)

Positive rotation pole

B

A

Fig. 3. (a) Rotation poles, small circles, and great circles. (b) Two plates, opening on one side of the rotation pole and closing on the other. In practice, Euler poles do not have to lie on plate boundaries. (After Fowler, C. M. R. (1990) *The solid earth: an introduction to global geophysics*, Fig. 2.8. Cambridge University Press, Cambridge.)

result of motion over long periods of time. In descriptions of current plate motions, 'instantaneous' is usually taken to be about the past 1–3 million years. Motions over longer periods can be approximated by successions of so-called 'stage poles', each of which may describe the motion over a period of a few million years.

We can imagine the Euler poles, like the geographic poles, as centres of coordinate grids (Fig. 3). Equivalent to lines of latitude are 'small circles' centred on the poles. Plate relative motions are everywhere parallel to these small circles. Moreover, angular separation rates are constant along a given small circle, and are proportional to the sine of the radius of the circle (measured as a geocentric angle from circle to pole). 'Great circles' (for which the angular radius is 90 °) through the pole are equivalent to lines of longitude, and cut the small circles at right angles.

Measuring plate motions

Plates move at average rates of a few tens of millimetres per year; consequently, relatively indirect methods have generally been used to determine plate motions, although more recently sufficiently precise direct geodetic measurements have been developed. The most common way of determining plate divergent rates is via the linear magnetic anomalies (Vine—Matthews anomalies) produced during sea-floor spreading (see *sea-floor spreading*). Such anomalies mark isochrons (lines of equal age) of crustal creation which have been dated by a variety of methods. Thus, measuring the distance between a recent magnetic isochron on one plate and its conjugate on the other allows the 'instantaneous' divergence rate to be determined. (The commonly quoted 'spreading rate' is the rate at which a single plate accretes—for symmetrical spreading it is half the divergence rate.) Since the spreading

or divergence rate varies with distance from the Euler pole, in principle one can determine the distance to the pole by measuring the spreading rate at several places along the plate boundary. This distance, combined with the linear divergence rate, gives the angular rate of opening about the pole.

The best measure of the direction of relative plate motion is the azimuth (bearing) of transform faults. At these, the relative motion is purely strike-slip (i.e. along the fault direction). Transform faults are readily recognized by their morphology (for example, a narrow linear valley for ridge—ridge transforms along mid-ocean ridges). Transform faults thus follow small circles about Euler poles. A great circle at right angles to a small circle passes through the pole. Thus, if the azimuths of several transforms along a plate boundary are determined, great circles can be constructed normal (at right angles) to them, and should intersect at the Euler pole.

In practice, both spreading rates and transform azimuths are used together to solve for pole positions, sometimes supplemented by other data such as earthquake focal mechanisms (see *seismology and plate tectonics*) to estimate relative motion directions. Euler poles may be calculated for individual plate pairs, for groups of plates, or for the global plate system. The determination of global plate motions performed by C. DeMets and others from Northwestern University, Illinois, provided a remarkably precise and self-consistent description of these motions.

As stated above, an important attribute of plate kinematics is its ability to predict plate motions. An interesting example of this is the possibility of determining convergent rates across subduction zones. Even at ocean—ocean subduction zones, one plate is destroyed, together with the record of magnetic lineations carried on it. Thus there was no direct way of measuring such motion until the development of sufficiently precise geodetic methods. However, the relative motions of the plate pair can be determined by global fits as described above, and then the motion at any point on a plate boundary, including subduction zones, can be calculated from the Euler pole data.

Geodetic methods have now been developed to the level where they can be used to measure plate motions directly. Where plate boundaries exist on land (such as the Mid-Atlantic Ridge in Iceland or the San Andreas Fault in California), standard geodetic methods such as electronic distance measurement can be used at a local scale (over ranges of a few kilometres). On a slightly larger scale of tens to hundreds of kilometres, precise relative determinations of position (to precisions of a few millimetres) can be made by careful use of the Global Positioning System satellite network. Relative positions between widely separated continents can be determined by very long baseline interferometry (VLBI), in which the variation in phases of radio signals from distant quasars is used. Repeat measurements by these methods over times of a few years can now resolve plate motions and give results that, in general, agree well with the more traditional determinations. Palaeomagnetism can on occasion also be used to determine rotation rates for rapidly rotating microplates.

The rotation rates of the major plates about their Euler poles range from about 2.1° per million years for the Cocos—Pacific pair to about 0.1° per million years between Africa and Europe or Africa and Antarctica, and only 0.03° per million years for India—Arabia. Many of the minor (or 'micro') plates rotate much faster than this, at tens of degrees per million years. In terms of linear rates, the fastest plate-divergence rate at present is on the East Pacific Rise between the Pacific and Nazca plates, at about 160 km per million years (or 160 mm per year). At the slow end, the North America—Eurasia plate boundary passes very close to its Euler pole in northern Siberia, where the relative motion becomes essentially zero.

Absolute plate motions

The discussion so far has dealt only with relative motions, which are fairly easy to determine. There is also interest in determining so-called 'absolute' plate motions, in which the motions of all plates are related to some common reference frame. The possibility of doing this arises from the recognition of mantle plumes, which rise as narrow columns of relatively hot rock from deep in the mantle, possibly from the core–mantle boundary. They reach the Earth's surface in so-called 'hot-spots' where they are manifest by clusters of intense volcanic and seismic activity. Well-known examples are Iceland and Hawaii, but there are thought to be many tens of such hot-spots and associated plumes.

Hot-spots leave clear trails on the Earth's surface, which comprise lines of volcanoes or volcanic sea mounts and zones of thickened, volcanically produced crust. The Hawaii–Emperor sea-mount chain, trending north-west from Hawaii, is an excellent example, but there are many other sub-parallel sea-mount trails in the south-western Pacific that have

resulted from plumes. If points along these trails are dated (e.g. by radiometric dating of volcanic products), the relative motion between the plumes and the plate and between given plumes can be determined. We can also calculate the motions of individual plates relative to the average plume motion.

It turns out that the relative motion between plumes is quite small, and significantly less than average relative motions between plates. From this, and the fact that plumes are thought to rise through the mantle, it seems reasonable to assume that the average plume motion relative to the mantle is quite small. If it is assumed that it is zero, plate motions can be given relative to the mantle, and these are referred to as *absolute plate motions*. They can also be described in terms of Euler poles, and are shown in Fig. 4.

Mechanisms and driving forces of plate tectonics

Plate motions are ultimately driven by Earth's heat energy, and they are intimately related to the mantle convection that is driven by this heat. Indeed, one view of plates is that they simply represent the surficial parts of mantle convection cells: as hot, ductile mantle rises to the surface it cools and becomes brittle—a plate—then moves as a rigid block over the surface before being subducted, gaining temperature, and becoming ductile again. Results from seismic tomography suggest that around the Pacific rim, sheets of cold material descend below subduction zones deep into the lower mantle, implying a strong coupling of mantle motion and subducted plates.

However, the coupling is not perfect. For example, there are some parts of mid-ocean ridges (divergent plate boundaries) where it seems that the deeper mantle (below the asthenosphere) may be descending rather than rising. One such is the so-called Australo-Antarctic Discordance south of Australia. Also, some plates, such as Africa, are almost entirely surrounded by ridges and have very few subduction zones on their boundaries. In such instances, a rigid coupling of plates to convection cells would imply the unusual scenario of upwelling along an expanding ring, with a downwelling column inside it. In fact one of the advantages of plate tectonics is that it allows partial decoupling of plate motions from deeper mantle flow via the ductile asthenoshphere.

Another way of looking at the problem of the driving mechanism is to consider the forces acting directly on the plates. There are many possible forces, but among the most important are ridge push, slab pull, trench suction, and mantle drag. Ridge push arises from the tendency of the plates on a mid-ocean ridge flank to slide down the slopes of the wedge of thermally expanded asthenosphere that lies beneath the ridge. Slab pull arises from the negative buoyancy of the subducted plate which tends to drag the rest of the plate down with it. Trench suction is an additional force tending to pull plates together at subduction zones, perhaps as a result of local convection driven by the subduction. Mantle drag is the frictional force between the base of the plate and the underlying asthenosphere.

It is possible to estimate some of these forces, at least approximately, and their relative importance can also be

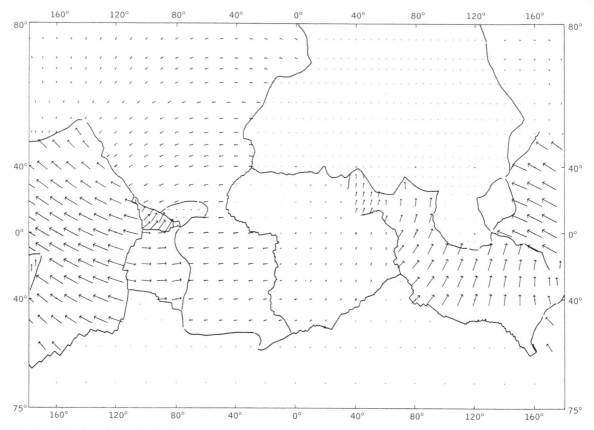

Fig. 4. Map of plate boundaries and absolute motion vectors. Lengths of arrows are proportional to plate speed, the longest arrows (in the western Pacific) corresponding to speeds of about 100 mm per year. (After Fowler, C. M. R. (1990)*The solid earth: an introduction to global geophysics,* Fig. 2.20. Cambridge University Press, Cambridge.)

estimated by considering the observed stresses in plates and inferred absolute plate velocities. The plate velocities are particularly instructive. It is found that absolute velocities are largely independent of the total area of the plate, and so it is unlikely that mantle drag is an overall driving force, as was once thought: in other words, plates do not ride as passive passengers on top of mantle convection cells. However, plate velocities are inversely correlated with the area of continental lithosphere, which suggests that large areas of continent act as a brake (perhaps because such lithosphere is very thick or because sub-continental asthenosphere is rather viscous). The fastest plates are those (mainly in the Pacific) that have large lengths of subduction zone along their boundaries. This implies that slab pull or trench suction, or both, are important driving forces, as has been suggested by calculation. There is a modest correlation with the effective length of ridge on the plate, indicating (again, as suggested by calculations) that ridge push is a driving force, but less strong than trench pull. Observations of intraplate stresses agree with these conclusions.

Tests of plate tectonics

There have been many tests of plate tectonics theory. Its self-consistency, direct measurements of predicted plate motions, earthquake focal mechanisms, and distribution of earthquakes have all played their part in confirming the theory.

ROGER SEARLE

Further reading

DeMets, C., Gordon, R. G., Argus, D. F., *et al.* (1990) Current plate motions. *Geophysical Journal International* **101**, 425–78.

Kearey, P. and Vine, F. J. (1996) *Global tectonics.* Blackwell Science, Oxford.

playas The flattest, smoothest geomorphological surfaces on Earth are playas. Commonly composed of compact silt and clay, sometimes with a surface salt efflorescence, they are often ideal for high-speed car racing as in the case of the Bonneville salt flats in Utah. However, they can also be crossed by giant desiccation cracks that have been known to catch unwary drivers by surprise. But how do playas form, what lies beneath them, and what processes are active on their surfaces?

Playas (*salinas* in South America; *sabkhas* or *sebkhas* in Africa) are found in closed interior basins, or *bolsons*, in arid regions (Fig. 1). In such areas (the western United States is a good example) differential movements of the Earth's crust or other processes have resulted in depressions surrounded by mountains. Under more humid climates in the past, many of these depressions contained perennial lakes. At present, however, water reaching the basin floor evaporates within days or weeks.

Surrounding a typical playa, and sloping towards it, are deposits of gravel called alluvial fans (Fig. 1). The apices of these fans are at the mouths of valleys carved into the mountains. The boundary between the alluvial fans and the playa is normally a few tens of metres wide. Over this distance the slope decreases abruptly.

Role of water under present climatic conditions

Storms in arid regions, although infrequent, are often intense, and the sparse vegetation does little to inhibit run-off or erosion. Consequently, floods coursing out of the mountains typically carry a heavy load of sediment. When these floods reach the toe of a fan, the break in slope instigates deposition of the coarser material—that carried as bed or traction load. Waves of deposition propagate up the fan from this break in slope. (Contrary to some early opinions, there is rarely a break in slope in the stream profile at the fan apex.)

Water continuing out over the playa surface carries with it a quantity of fine sand, silt, and clay in suspension. This sediment is deposited as the water evaporates. Rates of sedimentation may range from a few centimetres to about a metre per thousand years. Because the water fills the lowest parts of the bolson, sedimentation is concentrated there and serves, over the course of millennia, to level the playa surface. This is unlike the situation in perennial lakes where sedimentation also occurs on sloping margins, thus perpetuating variations in depth.

Playa area

Playas typically occupy 2 to 6 per cent of the depositional area in a bolson; alluvial fans make up the remainder. This reflects the size distribution of material supplied by the mountains. Were there no silt and clay in the sediment load, or only enough to fill interstices in the gravel on the fans, there would be no playa. Conversely, geological terrains yielding higher percentages of silt and clay in comparison with coarser material have playas that occupy a larger fraction of the depositional area.

Surface processes

A playa surface may be either 'wet' or 'dry'. Both types of surface may coexist on the same playa. A wet playa is one beneath which the water table is so shallow that water is drawn to the surface by capillary action. As the water evaporates, salts (particularly halite) are precipitated. Initially, this results in a surface coating of delicate white crystals or, occasionally, a puffy dry porous surface. Over time-spans of decades, a layer of silty salt that is centimetres or even decimetres in thickness may develop. Because crystallization of salt in such a layer involves lateral expansion, pressure ridges form. These are normally arranged in a polygonal pattern. The polygons range from one to a few metres across. Flooding of a playa by a rainstorm dissolves the salt and 'resets the clock' on this process.

Dry playas are those beneath which the water table is at some depth. After flooding and desiccation, newly deposited mud on such a playa dries and contracts, forming mud cracks. As the mud contracts, it commonly detaches along thin layers of fine sand that settled out of the water column just before the muds. The area and thickness of the individual mud blocks isolated by such cracks is controlled by the depth to these weak sand partings. Decimetre-scale blocks, for example, may be associated with sand lamellae a few centimetres beneath the surface, while chips a centimetre across may be only a millimetre thick. Because contraction is greatest in clays that settle out last, chips less than a few millimetres thick commonly curl.

As mentioned, giant desiccation cracks, up to a metre wide and metres deep, may also form. These appear to be associated with deep and declining water tables (lowered by pumping) and prolonged periods of drying.

Shallow channels may form on some playas, particularly if the playa surface is sloping. In Death Valley, California, for example, tectonic tilting is taking place faster than deposition can relevel the playa surface. Thus, water reaching the playa continues across it towards the lowest point, and flows in this direction persist long enough to erode and maintain channels that are metres in width and decimetres deep.

Stratigraphy

Under present climatic conditions, the sediment accumulating on playas is normally silty clay with little if any stratification. The lack of stratification is probably a consequence

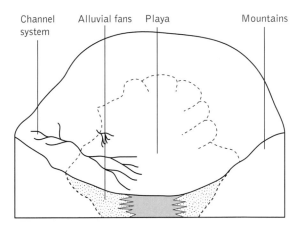

Fig. 1. Sketch of a bolson showing mountains, alluvial fans, and playa. Subsurface stratigraphy is shown schematically.

Channel system   Alluvial fans   Playa   Mountains

of the mixing of old and new sediment whenever the playa is flooded. On playas with channels, overbank flooding may leave thin layers of fine sand, silt, and clay.

Many bolsons, however, contained perennial lakes during the ice ages. Boreholes in these playas reveal a more complicated stratigraphy, consisting of tan, green, and black muds, fine sands, and evaporites (mostly halite, less commonly gypsum, calcite, or other minerals). Near the playa margin, gravels may also be encountered. Study of such stratigraphy can elucidate the history of the lake phases and can yield information on the climate at the time when the sediments were deposited. Black clays, for example, are evidence of anoxic conditions and diagnostic of a perennial lake. Thick evaporites indicate a period of desiccation, and detailed study of them can provide estimates of palaeotemperatures. Interbedded mud and halite layers are characteristic of periodic flooding in a salt-pan environment such as is found in Death Valley today. Where such sedimentary units can be correlated between two or more boreholes, it is sometimes possible to show that the floor of the former lake basin was not as flat as the present playa surface. Correlation of post-Pleistocene beds may also provide information on rates of tectonic deformation.

ROGER LeB. HOOKE

Further Reading

Hooke, R. LeB. and Dorn, R. I. (1992) Segmentation of alluvial fans in Death Valley, California: New insights from surface exposure dating and laboratory modelling. *Earth Surface Processes and Landforms* 17, 557–74.

Jansson, P., Jacobson, D., and Hooke, R. LeB. (1993) Fan and playa areas in southern California and adjacent parts of Nevada. *Earth Surface Processes and Landforms*, 18, 109–19.

Playfair, John (1784–1819)

Although a mathematician of distinction in his own right, John Playfair is more famous as the advocate of the views of his friend James Hutton. They were members of the Edinburgh intelligentsia at the end of the eighteenth century and Hutton had published his epoch-making, but difficult, *Theory of the Earth* (1795). In it he had propounded the principle of constant change and renewal, later known as uniformitarianism. Two years later Hutton died. This revolutionary book was the foundation to which Playfair, who had visited so many geological localities with Hutton, added illustrative examples and a readable gloss, with immediate success. It was a service not only to his old friend but to science itself. Playfair was perfectly capable of adding observations of his own and was a very competent writer and lecturer. *Illustrations of the Huttonian theory of the Earth* influenced, among others, William Smith, W. D. Conybeare, and John Phillips.

Among the important points that Playfair emphasized in *Illustrations* are two worth noting. The importance of unconformity in the geological cycle was singled out as an indication of profound change and of the dimension of geological time. Fossils were recognized as of organic origin and, with

sedimentary features, as clues to depositional environments. Playfair also introduced many words into use in the geological sense for the first time.

A Scot educated at the University of St Andrews, John Playfair was originally to enter the ministry, but showed such ability in mathematics that he took up that discipline and in 1785 became Professor of Mathematics in the University of Edinburgh. He was also a Fellow of the Royal Society of Edinburgh, and friend of many of its most active Fellows. In 1805 he was created Professor of Natural Philosophy.

D. L. DINELEY

Pleistocene

The Pleistocene Series was proposed as a stratigraphic unit by Charles Lyell in 1839, and is generally taken to be that in which the deposits of the Quaternary glaciations occur. These rocks, and their contemporary superficial deposits, thus rest upon the Pliocene and older rock formations. Pleistocene time began about 1.8 Ma ago and lasted until 0.1 Ma ago, after which even younger sediments, or volcanic rocks, of the Holocene epoch were deposited. Lyell had in 1833 described marine sequences now known to belong to the Pleistocene, under the name 'Newer Pliocene', based upon exposures of rocks in south-eastern Sicily. Other similarly young formations were found on the Italian mainland. Today the marine Pleistocene is divided into a lower part with two stages, the Calabrian and the Emilian, followed by the upper part with the stages Sicilian, Milazzian, and Tyrrhenian (uppermost).

The lower Pleistocene is also known as the Villafranchian in Italy, where it occurs as continental formations in the major river and rift valleys of the north. Part of the Villafranchian has been considered as Pliocene, largely on account of its fossil mammal faunas.

Many thousands of metres of drill cores have been obtained from the sediments that have accumulated continuously on the deep-sea floor throughout Pleistocene and Neogene times, and for these a plankton biostratigraphy has been constructed. The fossils used in this biostratigraphy are preeminently the foraminifera. There is also a well-documented oxygen isotope stratigraphy and a similar magnetostratigraphy. By these means a high degree of stratigraphic resolution and correlation is obtained throughout many parts of the world's oceans.

In the previously glaciated parts of northern and western Europe, a Pleistocene chronology of glacial and interglacial deposits is established, and to a degree these can be correlated with those found in North America. Pleistocene history as revealed on the continents of the northern hemisphere is essentially that of climatic fluctuations and is chronicled as a series of glacial and interglacial stages (Table 1). The classical sequence of Middle and Late Pleistocene tills (moraine deposits) and interglacial loess and other sediments was deciphered with reference to the Alps, to the North Sea area, and to North America.

It has to be emphasized that this very simple schedule is far from the expert view of events now held. Climatic fluctuations were indeed far more numerous. In Britain some fourteen to seventeen cold and temperate stages in the last 2.4 Ma are recognized. From the deep ocean sediments we have oxygen isotope records which suggest at least twice that number of major fluctuations during the same time. It seems that many of these fluctuations have so far eluded us on land, and the entire Pleistocene and Holocene timetable of climatic change in this part of Europe is a major research target.

At its greatest extent, the ice covered some 28 per cent of the land surface, in contrast to around 10 per cent today (Fig. 1). A eustatic lowering of sea level followed, perhaps by as much as 100m . During interglacial episodes sea levels were somewhat higher than they are today. Raised beach deposits and submerged forest remains testify to recent changes of sea level but isostatic uplift of the land following the removal of the Pleistocene ice sheets has complicated the story.

Response to the changes in world climate was rapid among the mammals. Species living close to the ice front included woolly mammoth, rhinoceros, caribou, and giant forms of familiar arctic species. In lower latitudes, mastodons and

Table 1. Glacial and interglacial stages of the Pleistocene

Regime	Alps	North Sea	N. America
Glacial	Würm	Weichsel	Wisconsin
Interglacial		Eem	Sangamon
Glacial	Riss	Saale	Illinois
Interglacial			Yarmouth
Glacial	Mindel	Elster	Kansan
Interglacial			Afton
Glacial	Gunz	?	Nebraskan

elephants, giant deer and ox, beavers, dogs and cats, and other familiar species existed in the forests and grasslands. *Homo sapiens* and the extinct neanderthaloid humans were widespread throughout Eurasia. Interglacial intervals were characterized by climates rather warmer than the present, and many animals now present in tropical or subtropical latitudes were to range into the present temperate zones. Britain, for

Fig. 1. At their maximum the Pleistocene ice caps and sea ice shelves in the northern hemisphere covered the area shown. All the major mountain ranges gave rise to glaciers which flowed out from the centres of accumulation into surrounding lowlands.

example, provided habitats for hippos, lions, and hyaenas. Quick as they appear to have been in establishing themselves in the Pleistocene world, very many of the larger mammals soon died out. The Pleistocene extinctions were the last in a series that occurred over about 9 Ma, the earlier affecting mostly the smaller mammals. During the Pleistocene, and especially between about 10 000 and 18 000 BP, the mortality was greatest among the larger animals and in the Americas. It seems to have coincided with phases of deglaciation, and was possibly influenced by human activity.

Dating of Pleistocene organic materials is now very largely achieved by radiocarbon dating, which can be carried out on remains up to 40 000 years old. Radiocarbon dates are usually given in years BP (meaning 'before present'; more precisely before 'AD 1950', a date agreed internationally as a baseline). The degree of accuracy of radiocarbon dating has improved greatly in recent years, but anomalies are frequently found and multiple analyses are carried out to eliminate these wherever possible. Potassium–argon dating is commonly used for the dating of relatively very young igneous rocks. Here the degree of reliability is good enough to attempt dates on material as young as 50 000 years BC. This is in its turn important in corroborating the magnetic polarities in Pleistocene igneous rocks. Magnetostratigraphy is playing an increasingly important role in Pleistocene correlation and chronology. The Middle and Late Pleistocene is marked by the Brunhes Normal Epoch of polarity; the Early Pleistocene is taken up by the Matuyama Reversed Epoch, in which there are three very short reverse polarity events.   D. L. DINELEY

Further reading

Dawson, A. G. (1992) *Ice Age Earth: Late Quaternary geology and climate*. Routledge, London.

West, R. G. (1977) *Pleistocene geology and biology, with especial reference to the British Isles* (2nd edn). Longman, London.

plesiosaurs
Plesiosaurs are one group of tetrapods that returned from the land to a wholly aquatic existence. In the process of doing this the reptilian features that had made them efficient land animals had to be modified in order for them to become efficient aquatic animals. Plesiosaurs first appear in the Triassic, where they apparently developed from a group known as the nothosaurs, which show specializations for propulsion in water, particularly reduction in the limbs and development of the tail vertebrae to support a laterally flattened muscular tail. They probably moved as modern crocodiles do, folding the limbs against the body and progressing by lateral undulations of the tail and body.

Plesiosaurs became numerous and worldwide in their distribution during the Jurassic and Cretaceous periods as they further developed the basic nothosaurian pattern. They were large animals, ranging from 3 to 15 m in length with short stocky bodies, large flippers, and short tails. Although many of them had long necks and small heads, one group, the pliosaurs, had short necks and large heads. Both groups had

long jaws with rows of narrow conical teeth which were adaptations for a piscivorous (fish-eating) way of life. In both forms, the abdominal region was strengthened by the development of additional ventral ribs termed gastralia. The pectoral and pelvic girdles were also greatly expanded ventrally, although the dorsal parts of the girdles were reduced. This adaptation provided large areas for the attachment of the powerful muscles that pulled the flippers backwards and forwards. Analysis of the range of motion of plesiosaur flippers shows that these animals swam in a way that is comparable to that of modern sea lions. These animals create lift by an initial downward and then backward stroke, and can also create lift during the forward feathered part of the stroke. Although sea lions use their front flippers only in propulsion, plesiosaurs had pectoral and pelvic flippers that were almost equal in size and probably used them out of phase to provide continuous motion. It seems probable that the long-necked plesiosaurs hunted by swimming among shoals of fish and using their long necks to dart from side to side and strike at their prey. The pliosaurs, which were faster swimmers, might have pursued and caught larger prey.

One trend seen in the plesiosaurs during the Cretaceous was increase in size. The pliosaur *Kronosaurus* from the Cretaceous of Australia had a skull about 3.5 m long. The plesiosaur *Elasmosaurus* from the Cretaceous of North America had a neck that was twice as long as the body and contained 60 vertebrae. Although the plesiosaurs were a vigorous and widespread group, they became extinct at the end of the Cretaceous together with the dinosaurs and many other groups. However, despite the fact that these animals have no fossil record after the Cretaceous their modern presence has frequently been put forward to explain sightings of the 'Loch Ness Monster' and similar phenomena.   DAVID K. ELLIOTT

Pliocene
In 1833 Charles Lyell proposed to use the rate of extinction of molluscan species throughout what we know as Cenozoic time as a 'palaeontological clock'. Each Cenozoic epoch was characterized by a molluscan fauna which contained both extinct and extant species, the proportion of extant forms increasing with time. The Pliocene was one of his original divisions of time, itself divided into Newer Pliocene with 90 per cent of its mollusc species living to the present, and Older Pliocene with 33–50 per cent of its species still alive. The Pliocene is today defined as the later epoch of the Neogene period, between 1.8 and 5.3 Ma. It contains two stages based on sections in mainland Europe. Pliocene deposits in Britain are limited to thin shelly sands and clays in East Anglia and small patches of sands in southern England.

All evidence points to the Pliocene as a time of continuing cooling of world climate, with major pulses of ice-rafted rock debris being released into the North Atlantic. The warm Gulf Stream and the cold Labrador Current were in operation and climate was highly seasonal.

About halfway through the Pliocene epoch several important tectonic events occurred. The continuing activity of the

African Rift system saw the opening of the Red Sea rift to the point where oceanic crust was generated there for the first time.

Orogenic activity in the Himalayan ranges reached its climax and the Andean cordilleras were given a major uplift, raising the land to about 4000 m. The Panama ithsmus connecting North and South America was established, allowing many of the northern mammals to migrate southwards and displace their counterparts.

Many modern mammals evolved during this epoch: apes, horses, camels, bears, and pigs. Grazing animals spread into the great grasslands of North America, Africa, and Asia. The horse appeared in North America. In Africa evolution took important steps; the Australopithecines were active in the early part of the epoch, and *Homo habilis* and *Homo erectus* appeared at about 2.5 Ma and 1.7 Ma respectively.

D.L. DINELEY

Further reading

Stanley, S. M. (1987) *Extinction.* Scientific American Library. W. H. Freeman, New York.

plumes (mantle) *see* MANTLE CONVECTION, PLUMES, VISCOSITY, AND DYNAMICS

plutonic rocks
The word 'pluton' is derived from the Greek and Latin *Pluto*, the lord of Hades, the infernal depths. Plutonic rocks are igneous rocks that have crystallized at depth. They are always coarse-grained because the magma from which they are derived cooled slowly. Plutonic rocks are formed when magma rising through the crust from its source solidifies before it reaches the surface of the Earth. In its upward journey magma may reach a point of neutral buoyancy at which it ceases to rise; heat loss to the surrounding rocks then results in crystallization. Crystals forming in the melt commonly sink to the bottom of the magma chamber to form a rock called a *cumulate*. The remaining melt is cooler, paler, more viscous, richer in silica, and less dense; it may then continue its rise through the crust. The process of cooling, crystallization, and separation of the melt will continue until the melt either erupts at the surface or experiences heat death, i.e., cools within the crust. Plutons can be seen only where uplift and erosion have unroofed the intrusion. Plutonic rocks can be dense, silica-poor and dark in colour, like most cumulates, or silica-rich and pale in colour like the melt remaining after cumulate formation. Granite and granodiorite are examples of silica-rich plutonic rocks. JUDITH M. BUNBURY

pluvial lakes
A lake is a natural inland body of water in which the water is generally standing. It has various characteristics and can, for example, be classified with regard to its size, shape, source of water, or water chemistry. A *pluvial lake* is one which has experienced large fluctuations in area and volume (i.e. changes in water balance) in response to climatic variations. For a lake to be classed as truly pluvial, changes in lake level should not be affected by any change in base level, nor by being dammed in any way, for example by ice, lava, or moraines. It has been noted that some lakes may have increased in size owing to the melting of ice; however, many pluvial lakes have not had glaciers in their catchments.

Pluvial lakes are so named because they reached their greatest extent during the 'pluvial' (from the Latin *pluvius*, meaning rain) periods of the Quaternary (the Quaternary being the last 1.6 million years). Within this time period there were significant changes in the Earth's climate, one of which was the greater availability of moisture at particular times in certain parts of the world. The term 'pluvial' was first used by Alfred Taylor in 1868 to indicate a period of increased moisture which resulted from greater precipitation or reduced evaporation, or a combination of the two. It was believed that these pluvial periods could be linked simply with the glacial fluctuations of the northern hemisphere, since initial studies in the American south-west indicated such a synchronicity. However, other, more recent studies, in East Africa for example, show that glacial advances in the northern hemisphere were coupled with very dry periods in some regions closer to the Equator. Since the simple temporal linking of the 'glacial–pluvial' theory is no longer acceptable, it is often suggested that the term 'pluvial lake' is not suitable; many therefore prefer the term 'megalake' or 'ancient lake'.

Hydrology and basin form

Pluvial lakes occupied large, hydrologically *closed basins*. The catchments of these basins may have a very large surface area, collecting run-off from a vast region. Groundwater is important to many pluvial lakes, for it is commonly a major source of water for closed basins. As the basins are hydrologically closed, the lake level varies rapidly in response to changes in moisture, whether seasonally or over much longer periods of time. Pluvial periods kept lake levels high for thousands of years at a time and recharged the groundwater of the catchments.

Pluvial lakes were common in areas where geological processes such as tectonism created large basins and depressions. Many of these basins are characterized by an elongated shape, such as those in the south-western USA, reflecting the Basin and Range topography and tectonic regime of the area. Some of the basins are flat-bottomed, and increases in moisture thus result in a vast expansion of surface area with little increase in depth until the whole basin floor is covered. However, some lakes grew to such an extent that they flooded a single catchment and overflowed into adjacent basins. The result was the formation of chains of interconnected lakes, such as the Owens River valley system in California. This is a series of four lakes (Owens, China, Searles, and Paramint) with progressively lower base levels through the system, although each is separated by a ridge so that an increase in water level is required before the lake overflows. At the time of the last glacial maximum (about 18 000 years ago) overflow linked Owens, China, and Searles lakes; although not connected, Paramint contained a lot of water. There was also a lake in Death Valley below this, known as Lake Manly, although this is not believed to have been connected to the Owens valley system.

Evidence of pluvial lakes

Pluvial lakes have been used in the reconstruction of palaeo-climates, using shoreline evidence and analysis of lake sediments to link lake levels to climate. When pluvial lakes have been at the same level for some time, shorelines will form which, as the lake recedes, are left abandoned. Lake stands will be recorded only during periods of lake regression, since a transgressing lake will rework previous shorelines. These relict features can be seen cut into the surrounding uplands of former pluvial lakes and they provide evidence for higher lake levels in the past. For example, shorelines that cut into the Wasatch mountains in Utah, USA are recognized as being the eastern borders of the pluvial lake, Lake Bonneville. In places where substantial parts of shorelines can be seen, they can be mapped to indicate the former maximum extent of the lake. It is also possible to see shoreline features on satellite images, where different vegetation bands can be identified occupying different shorelines.

Where there is no shoreline evidence, the existence of former lakes may be recorded by the lake sediments. The covering of the basin floor during periods of enhanced moisture conditions allows the deposition of lacustrine sediments such as marls and clays. At times of greater evaporation the water in pluvial lakes becomes more concentrated, and more saline deposits, for example, algal limestones, may be laid down. During the more arid periods of the Quaternary the surfaces of some pluvial lake basins were almost desiccated, allowing the deposition of evaporites such as gypsum and halite. If the lake bed became completely dry, material could have been lost through wind erosion or deflation. This could be recorded in the sediments as a section of aeolian deposits.

Location

The south-western United States has one of the greatest concentrations of pluvial lakes in the world. The Basin and Range province, caused by extensional tectonic activity, provided the closed basins and interior drainage required to support pluvial lakes. One of the most extensively studied pluvial lakes is Lake Bonneville. It was the subject of a classic study by G. K. Gilbert (during the 1890s), who mapped abandoned shorelines and reconstructed four past lake stands. The highest shoreline from Lake Bonneville is at 1565 m above sea level (a.s.l.) and is named the Bonneville shoreline. The others include the Provo shoreline (1470 m a.s.l.) and the Stansbury shoreline (1350 m a.s.l.). The fourth is a minor shoreline, 15 m above the present level of 1280 m a.s.l. and is named the Gilbert shoreline. Lake Bonneville was situated around the present Great Salt Lake, Utah Lake, and Sevier Lake (Fig. 1), which today represent about 5 per cent of its

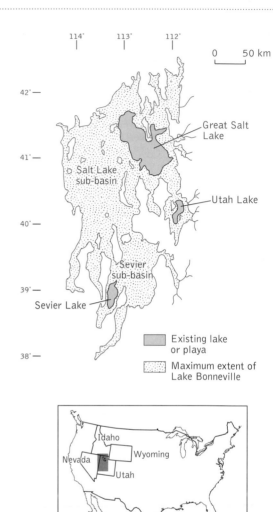

Fig. 1. Lake Bonneville and its remnant lakes, Utah, USA. (After M. A. J. Williams *et al.* (1993) *Quaternary environments*. Edward Arnold, London.)

maximum size. At its greatest extent, Lake Bonneville covered more than 51 000 km² and was more than 330 m deep.

In Africa there is evidence of one of the largest pluvial lakes that is known to have existed: Lake Chad. At present, Lake Chad has a shoreline at 280 m a.s.l., whereas at its maximum extent (known as ancient Mega Chad) the shoreline was 335 m a.s.l. Further east, the valleys of the East African Rift also exhibit evidence of former high lake stands. Lake Abhé, in Afar, reached a depth of more than 150 m and had a surface area of 6000 km². It is recognized that tectonic activity in the rift valleys may have disturbed the lake sediment records and altered the base level. This implies that the recorded lake levels will not reflect only changes in climate. The widespread conformity of the records from numerous lakes indicates, however, that the regional climate was a dominant factor in the fluctuations of these lakes.

The Middle East also has evidence of pluvial lakes. In the Dead Sea Rift Valley there was a large lake called Lisan Lake. The highest shoreline is recorded at a height of 220 m above the present Dead Sea. Again, it is thought that tectonic activity may have altered the base level, but that the climate was a significant factor in lake-level change.

Pluvial lakes today

Pluvial lakes were most commonly found between 30° N and 30° S, in areas that today are arid or semi-arid and where the climate is dominated by high-pressure systems. Owing to the drier nature of the climate in these areas today, former pluvial lakes are usually seen only as shallow saline pools, a fraction of their maximum size, or as predominantly dry playa basins. (*Playa* is a term given to a closed basin or a dry lake bed which has interior drainage, in an arid environment.) The similarity between this definition and that of pluvial lakes is notable, although a playa can form in many different ways, and the basin floor of a pluvial lake is just one example.

The basins left by pluvial lakes have proved useful to humans. Many lacustrine deposits which were laid down during low lake stands, such as chlorides, sulphates, and carbonates, are mined for their commercial value. Deposits from less stable phases of the lake may also be used, such as sands and gravels from the former lake margins. Where the exposed basins are large and flat, they can provide surfaces for highways or airfields. World record-breaking land-speed trials have taken place on the basin floor of Lake Bonneville.

A. COURTICE

Further reading

Goudie, A. (1992) *Environmental change*, (3rd edn). Clarendon Press, Oxford.

Lowe, J. J. and Walker, M. J. C. (1984) *Reconstructing Quaternary environments*. Longman, Harlow.

polar regions, geological exploration of *see* GEO-LOGICAL EXPLORATION OF THE POLAR REGIONS

polar wander and palaeomagnetism *see* PALAEO-MAGNETISM AND POLAR WANDER

pollution control
Pollutants in groundwaters, surface waters (rivers, lakes, and oceans), and the atmosphere are introduced into the environment as a result of many industrial activities. After release these pollutants can be affected by physical, chemical, and biological processes. Mathematical models that are used to attempt to predict the mobility of pollutants by accounting for all three processes are generally referred to as 'fate and transport models'. The majority of processes that affect the fate and transport of elements can be described by rate equations which show that the rate of pollutant reduction is a first-order process. This means that the rate (R) is directly related to the concentration of the pollutant ($[A]$) times a rate constant (k): $R = k[A]$. Few processes are second-order, in which the rate of pollutant reduction also depends on the concentration of a second chemical ($[B]$) that is present in nature: $R = k[A][B]$.

In this context, physical processes are transport processes; they include sorption, volatilization, diffusion, and advection (horizontal flow). Chemical processes are solubility (chemical equilibria), oxidation and reduction (redox), hydrolysis, hydration, and photolysis. Biological processes are primarily biotransformation, biodegradation, and bioaccumulation. It is important to understand the transport processes for toxins because of their effect on living things. We therefore also need to understand the toxicology of each pollutant.

Transport processes

Transport processes can result in the direct dilution of the contaminant as a result of retardation, diffusion, or dispersion.

Sorption refers to the interaction of dissolved constituents (solute) with mineral surfaces in groundwater. The solute is generally a charged ion which can be adsorbed to the surface (see *adsorption*), absorbed into the surface, or undergo ion exchange with ions in the mineral. *Adsorption* results from mineral surfaces having a surface charge in aqueous solutions, the charge being pH-dependent. In general, mineral surfaces are positively charged in acidic waters (low pH) and negatively charged in alkaline waters (high pH). As a result, positively charged metal ions (e.g. Pb^{2+}) adsorb to mineral surfaces at intermediate to high pH. When these ions diffuse away from the mineral surface into the structure of the mineral because of concentration gradients, the process is referred to as *absorption*, not to be confused with adsorption. Dissolved ions can also exchange with ions already present in the structure of a mineral. This process is generally referred to as *ion exchange*, and is particularly important for minerals such as zeolites and clay minerals which have loosely bound cations in channels or interlayers within their structure. These loosely bound ions generate charge balance with the negatively charged structural units composed of silica, aluminium, and oxygen. For minerals that have ion-exchange properties, the maximum amount of inter-layer or channel ion exchange with dissolved solids is generally referred to as the *ion-exchange capacity*.

Sorption of inorganic ions is related to the concentration of the dissolved constituent, the surface area of the minerals, and the pH of the solution. Sorption is generally described by *adsorption isotherms*, which show the amount of a chemical adsorbed as a function of pH or some other chemical variable.

Sorption of organic molecules is primarily on to organic compounds in sediments. Their sorption is hence related to the amount of organic material present and whether the organic compound is hydrophobic (has low water solubility) or hydrophylic (has high water solubility). Hydrophylic organic compounds have low adsorption coefficients and a low bioaccumulation factor (see *biogeochemistry*). It follows that hydrophobic organic compounds have high adsorption coefficients and high bioaccumulation factors.

Volatilization of organic chemicals from water to the atmosphere is particularly important for chemicals that have high vapour pressures and low water solubilities. Organic vapour production is experienced, for example, at filling stations, where obnoxious petrol odours are usually prevalent. In contrast, many of us are familiar with the pleasant smell of a good whisky or brandy, which also results from the volatilization of certain organic molecules.

Volatilization is a first-order rate process and is an important control on the composition of the upper layers of surface waters and groundwaters. Deep down within fluids or aquifers volatilization is not an important transport process.

Diffusion occurs when solutes move under the influence of thermal-kinetic energy down a concentration gradient. Diffusion continues until the concentration gradient is absent.

The diffusion constant varies as a function of temperature, falling by about 50 per cent upon cooling from 25 °C to 5 °C. Diffusion is a relatively slow process and is not of major importance unless water flow is extremely slow. Over geological timescales, however, diffusion can be an important process.

When dissolved solids are carried along with the flowing groundwater, the process is called *advective transport* or *convection*. The amount of solute that is transported is a function of its concentration in the groundwater and the quantity of the groundwater flowing. Groundwater can move at rates that are both greater and less than the average linear velocity. This is caused by differences in pore size, the tortuosity of pore channels, and the differential velocity of fluid across pores caused by drag resulting from the roughness of pore surfaces. Differences in velocity cause mixing to occur along the flow-path. This mixing is called *mechanical dispersion*, and it results in dilution of the solute at the advancing edge of the flow. The magnitude of dispersion is largest in the longitutinal direction of the flow; the dispersion in the transverse direction of the flow is roughly 10 per cent of the longitudinal dispersion. Dispersion is usually described with the advection–dispersion equation. This equation describes the change in concentration with respect to time in one, two, or three dimensions. In general, dispersion calculations do not take account of any retardation processes and are thus worst-case calculations. Reactive transport modelling is a relatively new area of research, which, because of its complexity, is actively pursued by few research groups at present.

Chemical processes

Chemical processes involve a variety of chemical reactions which affect the concentration of dissolved constitutents

along the flow-path. Minerals dissolve in water in a similar way to sugar in a cup of coffee or tea. However, mineral *solubility* is, in general, very low under the near-neutral pH conditions typical of natural aquatic systems. The amount of a mineral that can dissolve (its solubility) depends on the pH of the water, the temperature, the pressure, and, for elements that have more than one oxidation state (e.g. iron, Fe^{2+} and Fe^{3+}), on the amount of oxygen in the water. Mineral solubilities are known from experiments from which thermodynamic data have been derived. In order to use these data it is necessary to assume that the system has reached equilibrium. This takes a very long time (years or decades) at ambient temperatures, but a very short time at high temperatures (a matter of seconds or hours at 1000 °C). Polluted waters at ambient temperatures may not have reached thermodynamic equilibrium. However, by assuming equilibrium conditions it is possible to predict how the system will react in the future; such calculations are therefore of great value for the geochemist when evaluating the effect of chemical reactions on the fate of pollutants in groundwater.

Evaluating chemical equilibria in groundwater systems is done either in terms of solubility or by inferring an alteration from a primary mineral (unstable at ambient conditions) to a secondary mineral (stable at ambient conditions). These mineralogical reactions, whether solubility or alteration reactions, are generally described as proton (H^+) transfer reactions. Oxidation–reduction reactions are generally referred to as electron (e^-) transfer. The solubility of calcite (calcium carbonate, $CaCO_3$) can be simply described by the following reaction:

$$CaCO_3(s) + 2H^+ = Ca^{2+} + H_2CO_3(aq).$$

This reaction shows that in dissolving $CaCO_3$ there is a transfer of $2H^+$ for each Ca^{2+} and that the solubility is dependent on pH (H^+ concentration).

Similarly, the weathering of feldspar ($KAlSi_3O_8$) to form kaolinite ($Al_2Si_2O_5(OH)_4$), one of the most common weathering reactions on the Earth's surface, can be described by

$$2KAlSi_3O_8 + 2H^+ + 9H_2O =$$
$$= Al_2Si_2O_5(OH)_4 + 2K^+ + 4H_4SiO_4(aq).$$

This reaction also requires the transfer of protons, that is, one H^+ for each K^+. Such reactions can be illustrated on activity–activity diagrams on which are plotted the effective concentration (activity) of dissolved ions.

Oxidation and reduction reactions can be described either in terms of oxygen transfer or of electron transfer. For example, the oxidation of the mineral magnetite, ($Fe(II)Fe(III)_2O_4$), to form the mineral haematite, ($Fe(III)_2O_3$), can be represented by the equation

$$2Fe_3O_4 + 0.5\,O_2 = 3Fe_2O_3$$

or by

$$2Fe_3O_4 + H_2O = 3Fe_2O_3 + 2H^+ + 2e^-.$$

These reactions are thus dependent on the amount of oxygen in the system, which can also be represented by the dependence of pH (activity of H^+) and the pe (activity of e^-). In this respect the solubility of minerals can be shown on pe–pH diagrams which describe the solubility of the minerals as a function of pe and pH.

It will be apparent from the preceding discussion that minerals dissolve in water and the mineralogy of the groundwater reservoir will be reflected by the water composition. The composition of the groundwater in turn plays an important role in determining the concentration of pollutants in groundwater.

The effect of chemical equilibrium reactions can be calculated where thermodynamic data are known. These data are generally commonly available for inorganic species. However, the effects of organic ligands (molecules or ions bonded together) on the amounts of dissolved metals (or metal–organic complexes) in water are still poorly characterized and are therefore the subject of active research.

Oxidation of organic chemicals occurs as a result of reaction with oxidants formed during photochemical processes in natural waters. These oxidants include singlet oxygen and alkylperoxyl radicals (monovalent aliphatic hydrocarbon radicals containing O_{2-} groups). The oxidation process in thus a second-order rate process.

Reduction in anaerobic environments occurs by both inorganic and biological processes. An example of such a reduction process is the replacement of a chlorine atom for a hydrogen atom on industrially produced organochlorine chemicals. Rate expressions for this process are still poorly defined for organic chemicals. Oxidation and reduction of inorganic chemicals depend solely on the amount of oxygen in the system.

Hydration is important for some organic chemicals. In this process carbonyl compounds form hydrates which have different properties from those of the parent (unhydrated chemical + water = hydrated chemical). Rate expressions for this process are not well known at present.

Hydrolysis of organic compounds results in the introduction of a hydroxyl group (–OH) into the chemical structure, commonly with the loss of a functional group (–X),

$$RX + H_2O = ROH + H^+ + X^-.$$

This process is a first-order rate reaction. Some organic chemicals show a pH-dependent elimination reaction. These are second-order rate processes in which the hydrolysis reaction is dependent on the concentrations of H^+ or OH^-.

Photolysis is important for organic chemicals in surface waters and in the atmosphere. It occurs when chemicals are bombarded by light of wavelengths of more than 290 nm (ozone filters out shorter wavelengths). Photochemical transformations can occur by direct or indirect photolysis.

Direct photolysis takes place if the chemical absorbs light and then undergoes a transformation reaction from an 'excited state'. These transformations include rearrangement, dissociation, and oxidation. The rate of a photolysis reaction is a first-order process.

Indirect photolysis occurs when substances naturally present in aquatic environments absorb sunlight to form excited chemical species or radicals, which then react with the pollutant chemical. Indirect photolysis is thus a second-order rate process.

Biological processes

Biological processes involve enzyme-catalysed transformation of chemicals and the build-up of chemicals in the food chain.

Organisms require energy, carbon, and other fundamental inputs from the environment for their growth and maintenance. In life they manufacture enzymes which may transform or biodegrade contaminants that have been introduced into the environment. By far the most important biodegradation processes in aquatic and soil environments are carried out by microbes.

The biodegradation rate is a function of microbial biomass and the concentration of the chemical under given environmental conditions. Micro-organisms use the chemical substrate (the substance acted upon by an enzyme) as an energy source. In doing so the biomass is increased. Biodegradation rates, are therefore a function of cell growth rate. If an organic compound is used by a micro-organism as a sole source of carbon, the specific growth rate of the organism is a function of the concentration of the compound.

In the environment, where cell concentration, X, is relatively large, and pollutant concentration is low, microbial populations will not change significantly when the chemical is consumed. The degradation rate under these conditions is thus a 'pseudo'-first-order process. The first-order rate equation can be used under conditions where micro-organisms become acclimatized to the chemical and can actively use it. The acclimatization period is required to induce the organisms to produce necessary enzymes, to develop biodegradation organisms by mutation, to increase the number of microbes to substantial levels, and to use the diauxic (other readily metabolized) substrates. This acclimatization period cannot generally be ignored when a chemical is newly introduced into an uncontaminated environment.

Some pollutants may be biotransformed only when another organic compound is present to serve as a carbon energy source. This phenomenon is known as co-metabolism and is a second-order rate process. Mathematical expressions for such transformations are still poorly defined.

Bioaccumulation of chemicals in living species is especially important for hydrophobic chemicals which can be partitioned into fat and lipid tissues. Bioaccumulation also occurs where inorganic chemicals are partitioned into bones (which are composed of the mineral hydroxyapatite) and bone marrow. Bioaccumulation is usually evaluated in terms of the bioconcentration factor (BCF), which is equal to the concentration of the chemical in tissue (C_t on a dry weight basis) normalized to the concentration of the chemical in water (C_w); $BCF = C_t/C_w$.

Bioconcentration data are complicated by the fact that the concentration of a chemical is usually higher in the fatty tissues of the species than in leaner tissues. The rate of uptake and the time for attainment of equilibration in various organs (and species) also depends on the route of uptake (e.g. whether it is part of the diet or is absorbed through the skin). There is a correlation between the BCF and how hydrophobic the organic chemical is.

Environmental fate and transport

In evaluating the most important fate process for a given contaminant it is important to pinpoint the processes that have the shortest half-lives ($t_{1/2}$), that is, the time required for the removal of one-half of the initial concentration of the chemical. As for radioactive decay, the half-life of a chemical can be calculated if the rate constants (k_1, k_2, k_3, etc.) are known: $t_{1/2} = 0.693/(k_1+k_2+k_3)$.

When considering the fate and transport of contaminants in the environment, each pollutant has to be evaluated individually. The heavy metal lead and the lubrication fluids known as PCBs (polychrorinated biphenyls) provide good examples. Lead is an important pollutant in mining areas where sulphide minerals (such as galena, PbS) are mined, and in many industrial contexts such as metallurgy, paints, and permanent magnets. PCBs were the miracle fluids, used widely in the western world in transformers, turbines, and vacuum pumps for their high thermal stability and dielectric properties. They were also used as plasticizers in paints, plastics, resins, inks, etc. PCBs were used until it became known that they are both highly toxic to most forms of life and are very stable in the environment.

The dominant mechanism controlling the fate of lead is adsorption. Precipitation of lead sulphate ($PbSO_4$), lead carbonate ($PbCO_3$), and lead sulphide (PbS) minerals and bioaccumulation are also important. At low pH values, adsorption and precipitation are not nearly as effective in removing lead from solutions as at higher pH. Lead is therefore much more mobile in acidic waters (such as mine drainage) than at higher pH values. In alkaline and near-neutral waters, removal of lead by adsorption and precipitation occurs relatively quickly. Lead causes genetic changes, foetal malfunctions, and bowel cancer in humans. It is well known that lead causes irreversible retardation in the development of children, especially those exposed to leaded car exhausts. Lead accumulates in bone, because it can partition for calcium in the bone-forming mineral apatite.

Adsorption and volatilization dominate the environmental dynamics of PCBs in natural waters. PCBs do, however, adsorb only to organic compounds within groundwater reservoirs. PCBs can be desorbed from sediments, causing a continuous source of contamination. Dissolved PCBs move with the bulk water flow. PCBs cause growth inihibition and reductions in population in a range of species from algae to birds. In humans PCBs cause cancer of the kidneys and the liver and are known to be mutagens (i.e. capable of changing inherit-

able characteristics), teratogens (i.e. capable of causing malformation), and act as hormone mimickers.

K. VALA RAGNARSDOTTIR

Further reading

Moore, J. W. and Ramamoorthy, S. (1984) *Heavy metals in natural waters*. Springer-Verlag, New York.

Moore, J. W. and Ramamoorthy, S. (1984) *Organic chemicals in natural waters*. Springer-Verlag, New York.

pollution of the natural environment

Pollution alters the natural environment, adversely affecting its use by 'environmental receptors' such as plants, animals, and humans. The natural environment includes three environmental media: air, land, and water. Water includes surface water (wetlands, rivers, lakes and oceans) and groundwater aquifers which flow underground. Pollution contaminates these environmental media, which may then serve as pathways that transfer contaminants to other media or to receptors. For example, industrial waste disposed in landfills may contain contaminants such as solvents, metals, or pesticides that leach through soils to underlying groundwater. The contaminated groundwater might then be used as a human drinking water source or might discharge to a surface water body, adversely influencing aquatic life or humans using the water for recreational purposes.

Some natural events can be pollution sources. For example, volcanic eruptions and forest fires emit harmful gases and particulate matter to the air. The contaminated air is harmful to plants, animals, and humans. In addition, volcanic ash carried through the air and deposited on the land in large quantities adversely affects plant growth and agriculture in the short term. In certain circumstances animals can cause pollution. For example, manure in large concentrations can pollute surface water quality by increasing the organic loading of a water body (see below). This loading reduces dissolved oxygen levels in the water.

The term 'environmental pollution' is, however, more often used to describe the direct or indirect impacts of human activities. The cumulative effects of concentrated human activities can create large-scale or long-term environmental consequences beyond the assimilative capacity of the environment. Land, water, air, and other types of pollution such as radiation, noise, and visual pollution are some direct results of human activities. Indirect pollution like global warming and ozone depletion in the atmosphere are also occurring. Indirect effects are more difficult to prove because they can occur over long periods of time and result from complex chemical and atmospheric reactions. Indirect pollution is the subject of important environmental research.

Water pollution

Pollutants of concern for water include (1) pathogenic organisms and organic matter present in human, animal, medical, and food-processing wastes and (2) chemicals, particulates, and heavy metals present in industrial discharges or associated with soil erosion. Organic wastes are often oxygen-demanding and decrease the dissolved oxygen content of water. Plants and animals in the water rely on this oxygen and are adversely affected by its absence. Bacteria present in organic matter can have adverse effects on human and animal health. Thermal pollution consists of excess heat contained in water discharged from industrial cooling plants or processes. It can increase the temperature of water bodies and also decrease their available oxygen content. Erosion consequent on agriculture, deforestation, construction, or mining operations can deposit particulates like soil into water bodies. Although these particulates may not be inherently toxic, they can impede water flow or completely fill a water body.

Storm water run-off from roads, industrial units, farms, and construction sites also has significant potential for causing water pollution. Contaminated storm-water run-off can contain oil, chemicals, animal waste, and particulates that are transported over land or through drainage systems into surface water bodies. Groundwater pollution generally results from contaminants leaching through soils to underlying aquifers. For example, landfills and underground storage tanks may release contaminants to underground aquifers. These aquifers might then discharge contaminated groundwater to surface water bodies. In rural areas, groundwater pollution poses a significant cause of concern for residents who rely on wells for their drinking-water supplies.

Air pollution

Air pollution includes materials released to the atmosphere that alter the constituents or the ratio of constituents found in normal dry atmospheric air. Types of air pollution include particulate and gaseous emissions and odour pollution. Particulates include dispersed airborne solid and liquid particles larger than single molecules. Particulates are often present in dust, fumes, smoke, mists, and sprays. Gaseous pollutants include wastes or emissions that are gaseous at normal atmospheric temperatures and pressures. These pollutants include ozone, sulphides, nitrogen oxides, carbon monoxide, hydrocarbons, and sulphur oxides. Air emissions can include metal particles, asbestos fibres, or volatile organic compounds.

Land pollution

Land pollution results from a variety of sources, including spills of industrial and household materials and air transport of contaminants deposited on land. For example, lead dust generated from industrial operations can be deposited on land, posing a health hazard to animals or humans through ingestion or inhalation.

A primary cause of land pollution is waste disposal. Open dumps or tips are still used in many parts of the world and pose a risk of disease transmission to humans and animals through pests living on the dump and pathogens or chemicals present in exposed waste. Open dumps are also sources of odour and air-emission pollution. They can contaminate underlying soils and groundwater or contaminate storm water which travels to bodies of surface water.

Because of the problems associated with open dumps, some countries require waste disposal in landfills. Modern landfills incorporate controls to prevent pollution of the environment. For example, they are built in areas where the geology can be expected to reduce potential migration of pollution. They are also equipped with clay or synthetic liners to prevent release of contaminants to soil and groundwater. The waste is compacted and covered regularly to control pest populations and prevent human contact and air releases. Experience has shown, however, that in time contaminants may still leach to underlying soils and groundwater as a result of liner leaks, improper design, or water infiltration. When releases occur, pollution results.

Odour pollution is often associated with air pollution. Examples include strong odours emanating from industrial facilities like paper mills, concentrated vehicle emissions, farms and food-processing facilities, and inadequate sewage facilities.

Air pollution can pose health hazards and damage plant life over time. Humans and animals may breathe contaminated air or ingest plants coated with pollutant deposits. Air pollution can result indirectly from air emission. For example, photochemical smog indirectly results from industrial and vehicle emissions, through chemical reactions occurring in the atmosphere. Similarly, sulphur dioxide emissions from coal- and oil-fired power plants can interact with water molecules present in the air to form acid rain. Acid rain is a threat to plant life and can damage exposed statues and buildings as well.

Other types of pollution

Radiation is generated by the nuclear power industry, medical activities (for example, X-rays), and nuclear weapons production. If it is not properly contained, radiation pollution will occur and affect environmental media and receptors. Large doses of radiation pollution can damage plant and animal life. Radiation might increase cancer risks, and might be associated with genetic mutations. Noise pollution generated by airports, traffic, loud music, construction, and industry can be a nuisance and permanently impair hearing over time. Visual pollution includes billboards, rubbish dumps, and other man-made structures that destroy the beauty of a landscape. Noise, visual, and odour pollution are sometimes evaluated subjectively. One person may consider a piece of music to be art while another considers it noise pollution. One person may enjoy a billboard while another considers it a visual insult.

C. BURIKS

pore-water chemistry Pore waters fill the interstices of sediments and sedimentary rocks. Their chemical composition reflects both the origin of the waters and also the way in which the original composition has been modified by diagenetic (post-depositional) reactions with rock minerals and organic matter.

Scientists aboard the *RRS Challenger* first revealed in 1879 that the composition of pore waters can change very quickly after sediments are deposited. Improved sampling techniques, inspired partly by the Deep Sea Drilling Project (1968–83), had subsequently led to a much greater understanding of the key controls on the chemistry of the pore waters within shallow-buried sediments. Recently deposited sediments are unstable mixtures of terrestrial weathering products and organic matter. Within millimetres of the sediment–water interface, microbes start to exploit the system's inherent chemical energy by sequentially catalysing reactions between organic matter and oxidized phases such as oxygen, nitrate, manganese oxides, iron oxides, and sulphate. The composition of pore waters is changed by the reactions, losing oxygen, nitrate, and sulphate, and gaining iron, manganese, ammonium, bicarbonate, and hydrogen sulphide. Continued metabolism of residual organic matter by methanogenic bacteria results in the production of biogenic methane which, under favourable circumstances, can accumulate in commercial quantities. Pore waters also sensitively record the occurrence of other reactions, such as the dissolution, precipitation, and recrystallization of phosphates, carbonates, and sulphides, during early diagenesis.

The composition of waters deeper within sedimentary basins is known primarily through the drilling activities of petroleum companies. Salinities vary from almost zero to 300 000 milligrams per litre (mg l^{-1}) and are strongly influenced by patterns of fluid flow and proximity to subsurface evaporite deposits. However, the occurrence of highly saline subsurface brines was known substantially prior to the onset of petroleum exploration; historically, brines were a valuable source of salt and are known to have been exploited from neolithic times in Europe and China. Present-day commercial interest focuses on the problems that can occur when pore waters (often known within the industry as formation waters) within petroleum reservoirs mix with waters which are injected into the reservoir to maintain pressure during production. Chemical incompatibility between the fluids leads to the precipitation of mineral scales, which can severely impede oil production.

Chloride, sodium, and calcium are the most common ions in pore waters within deeply buried sediments, with lesser amounts of most elements. These major elements are largely derived from sea water trapped with the sediment at deposition and by the dissolution of buried evaporites. Fluid mixing and reactions with clay and carbonate minerals exert important, additional controls on the composition of pore waters. Ratios of the major cations such as sodium, potassium, magnesium, and calcium are strongly influenced by chemical reactions between waters and rock minerals, as is pH. More detailed analyses—for example, the oxygen and hydrogen isotopic composition of the water and its complement of noble gases—give further insights into the origin and age of the water, and its history of interaction with associated rock minerals. The pore water is thus an integral part of a sediment or sedimentary rock, revealing much about its post-depositional history.　　　　ANDREW C. APLIN

porosity *see* PERMEABILITY AND POROSITY

porphyry copper deposits Porphyry-type orebodies are low-grade, high-tonnage accumulations of mineralized rock associated with intrusive magmatic bodies. They are most commonly mined by open-pit methods, and because of their great size have led to the creation of some of the largest man-made excavations on Earth. Porphyry-type deposits are by far the biggest single orebodies exploited at present; they can range from hundreds to thousands of millions of tonnes of ore. An example is the giant Chuquicamata deposit in Chile, which originally comprised over 9000 million tonnes of mined ore and reserves. The igneous rocks that are intimately associated with these deposits are normally porphyritic (that is, they contain large crystals). They range from intermediate to felsic in composition, and include diorites, monzonites, granodiorites, granites, and tonalites. The most common commodity of interest is copper (hence the name 'porphyry copper deposit') but they are also important sources of molybdenum and gold. Smaller quantities of other metals, including silver, tin, tungsten, and palladium, can also be refined from porphyry copper ores. Mineralization in porphyry deposits in widely disseminated throughout the host rocks and ore grades are very low, but because of their large size, the orebodies can be mined profitably.

Although closely associated with plutonic igneous rocks, porphyry mineralization commonly encompasses large volumes of the surrounding host rocks to the intrusion. These can include igneous, metamorphic, and sedimentary rocks ranging in age from Precambrian to Phanerozoic. The emplacement of porphyry bodies is usually controlled by regional fault structures and zones of fractured rock. The intrusive bodies themselves may be composed of a single intrusion or of multiple intrusions. The intrusions are passively emplaced into the surrounding host by stoping and assimilation. (Stoping is a process in which pieces of country rock are wedged off by magma sink, and are assimilated.) Geological evidence shows that the igneous bodies that generated porphyry ores were emplaced at relatively shallow levels in the crust (less than 4 km) and that they may have provided the magma source for the generation of large volcanoes on the surface which have since been eroded away. Where several intrusions of magma are present, it is common for mineralization to be related to the latest intrusions, which tend to be most differentiated. The presence of phenocrysts in the intrusions indicates that their magmas were partially crystalline when emplaced and that crystallization of the remaining melt occurred rapidly.

The worldwide distribution of porphyry deposits is controlled by orogenic belts, where deformation of the crust is caused by the collision of two of the Earth's tectonic plates. Two types of orogenic belts host porphyry deposits: those created by the subduction of oceanic crust beneath continental crust along a continental margin and those found along island arcs where two oceanic plates are colliding. The numerous deposits located along the west coast of South America are perhaps the best example of a group of porphyry deposits formed by subduction along a continental margin. Those in the Philippines and Papua New Guinea are examples of deposits generated in island arcs formed by colliding oceanic plates. Most deposits are of Mesozoic or Cenozoic age, but some Palaeozoic examples are known, such as those in the Appalachian orogen of Canada and the United States. Older Precambrian deposits are difficult to recognize because of later deformation and erosion, but some rare examples have been found.

The magmas that form porphyry copper deposits are thought to be generated by the melting of subducted oceanic crust. This crust contains a high concentration of water because of the presence of minerals, such as chlorite, that contain water in their structures. When oceanic crust is heated, it dehydrates and then melts to produce magma into which the water can dissolve. The magma is less dense than the surrounding mantle rocks; it rises and penetrates the lower crust, where further melting and assimilation of crustal rocks can occur. These modified magmas can rise up to high levels in the crust. During their ascent, the pressure drops, causing the water dissolved in the magmas to separate—a process referred to as 'first boiling'. The rising magma may also begin to crystallize as it ascends, generating crystals of minerals such as plagioclase. These crystals eventually form the phenocrysts found in porphyritic rocks. The exsolution of water cools the remaining magma and induces more rapid crystallization before it can reach the surface. Further crystal-

lization results in more water being expelled in a process called 'second boiling'. The outer surface of the intrusive body cools more rapidly, forming a carapace of essentially solid rock at temperatures much lower than its centre. This confines the remaining partially molten interior and any water exsolving from it, resulting in a large increase of pressure (Fig. 1a). When the internal pressure builds up to a high enough level, the carapace fractures and the high-temperature fluids are released upwards into the solidified porphyry and its surrounding host rocks (Fig. 1b). The release of these fluids and the concomitant drop in pressure induces crystallization deeper in the intrusive mass, and the cycle begins anew. This cyclic process continues and the carapace and the fractures generated within it migrate downward to greater depths (Fig. 1c). Eventually, confining pressures, caused by the overlying rocks, rise to the point where fracturing cannot occur, and the process ceases.

Although water is an important component released from the magma as it crystallizes, other components can also be expelled from the melt. In particular, chloride salts such as sodium chloride, potassium chloride, and iron chloride also partition between the magma and the aqueous fluid forming brines, which can contain salt concentrations of up to 70 per cent by weight. Trace metals in the magmas may also partition into the brines, causing enrichment of valuable metals in these high-temperature fluids. When the brines are expelled during fracturing, they pass through rock with which they are no longer in chemical equilibrium. This results in an exchange of chemical components between rock and fluid, a process

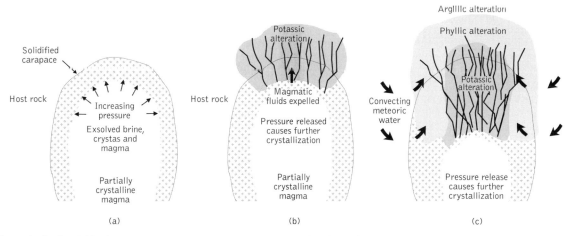

Fig. 1. Idealized model for the generation of porphyry copper deposits. (a) Early-stage crystallization of the outer regions of the intrusion confines partially crystallized magma and its exsolved brine. This causes a build-up of pressure in the intrusion, which is released by fracturing of the carapace and its surrounding host rocks. (b) High-temperature brines are released into the fractured rocks, resulting in potassic alteration and low-grade copper mineralization.

The release of pressure and loss of fluid cause further crystallization deeper into the intrusion. (c) As the influx of magmatic fluids wanes, meteoric waters invade the solidified intrusion, become heated, and react with the rock, causing phyllic alteration. These circulating fluids can also generate argillic alteration of the periphery of the phyllic zone as well as propylitic alteration further from the intrusion.

known as 'hydrothermal alteration'. This exchange causes changes to the fluid as well as to the wall rock. At the same time, cooling of the fluid can occur. Taken together, these two processes induce gradual changes in the physical and chemical properties of the fluid. These changes bring about the precipitation of ore minerals in the fractures, together with other gangue minerals such as quartz and potassium feldspar.

The hydrothermal alteration generated during the formation of porphyry deposits is one of their distinctive features. Its study has led to great advances in the understanding of hydrothermal ore deposits of many types. In 1970, Lowell and Guilbert published a study of the alteration zones of the San Manuel–Kalamazoo porphyry deposit and compared these with the alteration found at 27 other porphyry deposits. This led to the Lowell–Guilbert model for the alteration zonation of porphyry deposits. In this idealized model, alteration zones are centred on the porphyry intrusion and consist of the potassic, phyllic, argillic, and propylitic zones. The central potassic zone is characterized by the presence of potassium-bearing minerals such as orthoclase (a potassium feldspar) and biotite. Other minerals that may be present include anhydrite, chlorite, and sericite. Veins in the potassic zone are filled with minerals similar to those found in the altered wall rock. Fluid inclusion and isotopic evidence shows that potassic alteration is caused by brines which are composed of a high proportion of magmatic fluid and have very high temperatures (400–700 °C). It appears that in the early stages of development these magmatic fluids are expelled upwards and outwards into fractured rock which, at the same time, prevents the incursion of more dilute fluids at lower temperatures from the surrounding wall rock (Fig. 1b).

Phyllic alteration surrounds the potassic core and is distinguished by the mineral assemblage quartz–sericite–pyrite. During its formation, most of the original silicate minerals in the rock are broken down by the hydrothermal fluids and replaced by sericite or clay minerals, or both. These reactions generate large amounts of excess silica, which goes to form quartz. Iron released from the alteration of iron-bearing minerals, as well as iron and sulphur added by the fluid, form pyrite. Fluid inclusion and isotopic evidence show that the fluids responsible for the formation of phyllic alteration were less saline (less than 15 per cent salt), lower in temperature (250–450 °C), and had only a small magmatic component. Phyllic alteration is considered to be the result of the incursion of cooler meteoric water into the porphyry environment. This water becomes heated and rises, reacting at the same time with the host rocks (Fig. 1c). A continuous supply of meteoric water causes convection around the periphery of the intrusion, allowing large volumes of fluid to pass through the rock. As the porphyry cools, less magmatic water is expelled and the convecting fluids can penetrate deeper into the intrusion, overprinting the pre-existing potassic assemblage.

Argillic alteration is normally found on the periphery of the phyllic zone, and is characterized by the presence of clay minerals such as kaolinite and montmorillonite. These minerals typically form when hydrothermal alteration is caused by acidic heated fluids. Most minerals are unstable when in contact with such fluids and break down, releasing most of the metals in their structures. During this process, the rock acts as a neutralizer of the acidic fluid, and metals are exchanged for hydrogen ions. Metals such as silicon and aluminum are less soluble in these fluids and remain as quartz and clay minerals. Isotopic evidence shows that argillic alteration is caused by hydrothermal fluids composed mainly of meteoric water.

The propylitic alteration zone extends outward from the intrusion into less altered host rocks. It is identified by the common occurrence of chlorite and calcite as alteration products of biotite and hornblende in the host. Other minerals that are present in this zone include pyrite and epidote. Propylitic alteration represents the weakest alteration found in porphyry coppers. It is caused by heated convecting meteoric water. The Lowell–Guilbert model represents an idealized porphyry deposit; many variations exist, mainly because of differences in the composition of the host intrusion and its wall rocks.

Ore minerals in porphyry deposits are commonly found in concentric zones around the intrusion, much like the alteration zones. This is not unexpected considering that the ore minerals themselves are alteration minerals and are formed by the same chemical processes that resulted in the hydrothermal alteration. The central potassic core normally contains low-grade mineralization consisting of minor chalcopyrite, molybdenite, and pyrite. These minerals occur in dense microfractures in the altered host intrusion. Along the contact between the potassic and phyllic alteration zones, higher-grade ore is found; this consists of chalcopyrite, molybdenite, and pyrite hosted in microfractures and larger fracture networks. The total sulphide mineral content may be as high as 10–15 per cent; copper concentrations vary from 5 to 10 kg per tonne of ore. Mineralization in the argillic and propylitic alteration zones is typically of low grade and uneconomic.

BRUCE W. MOUNTAIN

post-war specialization in the Earth sciences
see SPECIALIZATION IN THE EARTH SCIENCES AFTER THE SECOND WORLD WAR

Powell, John Wesley (1834–1902)

One of American's great explorers, J. W. Powell planned and led the first boat expedition through the Grand Canyon of the Colorado River. The exploration party consisted of ten men, and their means of transport was four small rowing boats. The boats were launched in the Green River, a tributary of the Colorado, in south-west Wyoming on 24 May 1869, and the expedition emerged from the mouth of the Grand Canyon three months later, on 29 August. On this trip, Powell made the first important geological observations of the geology of the canyon, and demonstrated that it originated by river erosion into rocks that had been slowly elevated.

As a result of his several geological expeditions to the Rocky Mountains, Powell became interested in, and made a special study of, the native peoples of the area and their languages. In order to curate his work with the native peoples, he founded and directed the Bureau of Ethnology within the Smithsonian Institution. Between 1874 and 1879, Powell directed the United States Geological and Geographical Survey of the Territories, jointly carrying out both geological and enthnological field studies in Utah, Nevada, California, New Mexico, and Arizona. During this period, realizing that access to water imposed a limit on development of the western states, he made the first extensive studies of the water supplies available in the arid south-west of the United States. In 1879, the United States Geographical and Geological Survey was incorporated into the United States Geological Survey under the directorship of Clarence King. When King resigned his directorship in 1881, Powell was appointed his successor, carrying out the tasks associated with the directorships of both the Ethnological Bureau and the Geological Survey. He administered both offices until 1894, when he resigned the office of Director of the U.S. Geological Survey in order to devote more time to ethnological studies.

BRIAN J. SKINNER

Precambrian Precambrian rocks are those formed before the beginning of the Cambrian period, about 590 million years ago. The oldest known are dated at rather more than 3800 Ma. They represent about 85 per cent of all geological time. Most are igneous or metamorphic crystalline rocks; many of them have been severely deformed, commonly several times. Their principal outcrops constitute the great Precambrian shields of continental crust upon which later formations were deposited. They provide evidence of the evolution of the Earth and of early life upon it (Fig. 1). Nevertheless, many Precambrian rocks are difficult to interpret, and it is clear that processes no longer operative were once important.

Regional classifications, such as those for the Canadian shield, the Baltic area, or South Africa, identify Precambrian rock units, orogenies, and events including deep burial and metamorphism. Precambrian correlation, classification, and

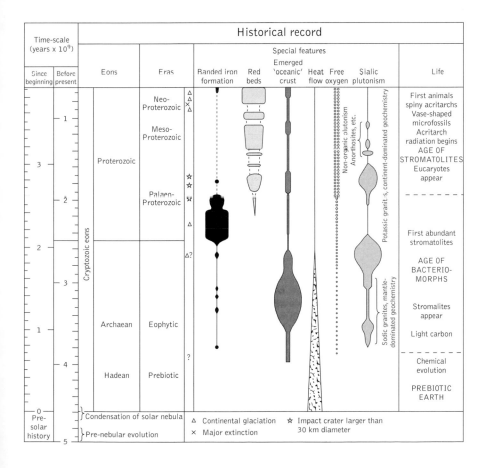

Fig. 1. The immensely long span of Precambrian time saw the development of many of the principal components of the crust, together with a number of remarkable characteristics that were unique to those eons. The most widespread of these are indicated in the figure and they document the chemical and mineralogical evolution of the crust and atmosphere. The evolution of life succeeded a long prebiotic chemical evolution, but only towards the end of Precambrian time does it reveal the advent of animals; prior to then, other forms of life left a relatively long and monotonous, though by no means restricted, record of their existence. (Based on a diagram by P. Cloud.)

847

geochronology are primarily based upon radiometric (isotope) dating. Precambrian time is often referred to as the Cryptozoic (hidden life) Eon, as distinct from the Phanerozoic (apparent life) Eon beginning with the Cambrian Period.

The subdivision of Precambrian time is difficult since the basic principles of stratigraphy (including biostratigraphy) used for the Phanerozoic Eon are difficult to apply. Convention has distinguished two Precambrian eras, the older being the Archaean, extending back to the consolidation of the planet, and the younger being the Proterozoic, which began about 2500 million years ago. However, several specialists recognize also an initial eon, the Hadean or Priscoan, for the 1000 million years or so between the consolidation of the planetary mass and Archaean time around 4000 million years ago, while life on earth began.   D. L. DINELEY

Further reading

Cloud, P. (1988) *Oasis in space: Earth history from the beginning.* Norton, New York.

Goodwin, A. M. (1991) *Precambrian geology: the dynamic evolution of the continental crust.* Academic Press, London.

Nisbet, E. G. (1987) *The young Earth: an introduction to Archaean geology.* Allen and Unwin, Boston.

Plumb, K. A. (1991) New Precambrian time scale. *Episodes*, **14**, 139–40.

Stanley, S. M. (1993) *Exploring Earth and life through time.* W. H. Freeman, Oxford.

precipitation

Precipitation—water falling to the ground—is one of the single most obvious elements of the weather. The presence of precipitation is quickly felt once it occurs, and its total duration in time at any location may consequently seem to be far greater than it actually is. However, even in the wettest parts of the tropics, or in exposed maritime areas in the temperate zones, there is at least as much 'dry' time as there is 'wet', and frequently much more. In truth, very few clouds precipitate.

Precipitation may take a variety of different forms (rain, snow, sleet, etc.); it may show considerable variation in magnitude (intensity), from the heaviest in thunderstorms or tropical cyclones to the lightest drizzle of windward coasts and hills; and it may be fleeting in showers, be continuous, or be intermittent. 'Clouds' at ground level (fog or mist) can also yield precipitation when the wind drives water droplets on to solid obstacles, such as buildings or trees. These general characteristics of precipitation are used in its classification. They also often give vital clues to the processes behind its generation. These processes are still imperfectly understood, and in all of them there is the major physical problem of generating particles, large enough to fall to the ground, from the very tenuous collection of small particles of ice or water which make up clouds. Surviving the fall to the surface is a further problem in the unsaturated air beneath the cloud.

Precipitation is conventionally measured using rain gauges which conform to a standard national pattern. These are read daily, generally at 09.00 hours local time, to the nearest 0.1 mm depth. In many countries such records extend well back into the nineteenth century. Some gauges are also capable of measuring precipitation automatically at much shorter intervals (e.g. every hour). The gauges themselves have to be carefully sited away from obstructions to obtain a 'realistic' measure of amount, but they are subject to error, and each represents only a minute 'sample' of an often huge surrounding area. Precipitation is highly variable at the microscale in both space and time. The reliance on gauge measurement has until recently restricted precipitation measurement to populated land areas, but during the 1980s and 1990s rapid advances were made in the field of radar and satellite 'sensing' of precipitation. Many nations now operate a network of radar installations, enabling good estimates to be made of precipitation occurrence and movement every 15 minutes, and the newest satellite technology provides good estimates of precipitation amount over the two-thirds of the Earth's surface covered by water.

At lower levels, in the tropical latitudes, and during the warmer parts of the year in the temperate zone, most precipitation is in the liquid form—rain or drizzle—once it hits the ground. The distinction between the two is mainly one of size: drizzle drops have diameters of less than 0.5 mm and are close together. Raindrops are equal to, or greater than, 0.5 mm. Intensities greater than 4.0 millimetres per hour (mm h^{-1}) indicate heavy rain, and intensities less than 0.5 mm h^{-1} slight rain; moderate rain fills the gap between these limits. When the rain falls as showers, a different range of intensities is used, because of the nature of 'short, sharp showers': 'violent', greater than 50 mm h^{-1}; heavy, 10–50 mm h^{-1}; moderate, 2–10 mm h^{-1}; slight, less than 2 mm h^{-1}. The definitions for drizzle intensity are more descriptive: 'thick' drizzle 'definitely impairs visibility and accumulates at a rate of up to 1 mm h^{-1}'; moderate drizzle 'causes windows and road surfaces to stream with moisture'; slight drizzle is 'readily detected on face, but produces very little runoff'.

At higher elevations, in high latitudes, and also during the colder part of the year in temperate areas, solid precipitation is more common. Typically this falls as snow, but there is a wide variety of other terms for solid precipitation based on the size and structure of the individual particles. Snowflakes comprise the familiar loose aggregates of ice crystals which often adopt a hexagonal and branched form; there are also snow and ice pellets, snow grains, granular snow (*graupel*), and ice prisms. The distinction between snow and ice is that snow particles are always opaque. The intensity ranges for snow are ten times those for rain: 10 cm of freshly fallen snow is equivalent to 10 mm of rain. At times, formerly solid precipitation will have melted only partially by the time it reaches the ground surface, or snow and rain can fall simultaneously, that is, as sleet.

Another important category of solid precipitation is that of hail. Hailstones are generated by a unique set of processes that operate in cumulonimbus clouds, which contain rapidly rising thermals of warm air and strong down draughts of cold air, through a considerable depth of atmosphere, and are sometimes associated with thunder, lightning, and tornadoes. When looked at in section, hailstones show a series of alternately opaque and clear ice rings. Each opaque layer represents passage through a part of the cloud where freezing was very rapid, trapping air bubbles. Each clear layer represents times when air bubbles were able to escape when freezing was slower. Together, pairs of rings mark the completion of a cycle of vertical movement within the cloud, during which the hailstone is alternately carried rapidly to high levels in the cloud on top of rising thermals, and then downwards within equally strong down draughts. Freezing is very rapid in the upper layers of the cloud and much slower at lower levels.

Precipitation develops within clouds in which there is notable and sustained uplift. Under such conditions the clouds become thick enough for one or more of a variety of processes to operate and cause the precipitation. Mechanisms for uplift are present at the local scale within shower and thunderstorm clouds (cumulus and cumulonimbus), where air is forced upwards over hills and mountain ranges or along locally developed zones of convergence, such as sea-breeze fronts ('forced uplift' or 'forced convection'), and at the larger scale, within weather systems such as temperate-latitude frontal depressions and tropical cyclones. These synoptic weather features can also incorporate local-scale uplift.

Precipitation thus dominantly occurs in parts of the world that are subject to the development and passage of major synoptic weather features within the temperate westerly belts in both hemispheres, in some parts of the tropics where cyclones can form, and in areas prone to sustained daytime surface heating, generating strong thermals, or the windward slopes of higher ground. Such processes may also typically show marked seasonal or diurnal variation, with the more intense thunderstorm or shower-type precipitation generally occurring in mid- to late afternoon, in the summer months in temperate zones, and within the humid tropics. The larger-scale development of temperate depressions favours the winter half of the year, whereas tropical cyclone activity peaks during the hottest months.

These weather systems are efficient generators of precipitation at varying spatial scales. More generally though, the range of conditions that must be satisfied to generate precipitation within a cloud is very exacting. The concentration of water droplets or ice particles within clouds is very low, and their size is extremely small. There is thus far more 'space' between adjacent particles than there is particle volume. This has important consequences for the potential generation of precipitation. Given their spacing, the opportunities for collision between particles are remote. Their size is such that many are more likely to be carried upwards by thermals, rather than

to fall under the influence of gravity; and even if they were to fall from the cloud base they would rapidly evaporate once in the unsaturated air between cloud and ground. Some mechanisms are required that will significantly increase their size (and therefore their mass).

Early theories of precipitation assumed that as a cloud developed and grew in volume, the cloud droplets would simply continue to grow until some of them became large enough to fall as precipitation. For spherical cloud droplets the surface area increases very rapidly for a comparatively modest increase in radius. An increased surface area means that even if copious amounts of water vapour are available in the surrounding air, encouraging initial condensation and droplet growth, the supply is rapidly exhausted. Although the rate of growth for small droplets may be rapid, for large droplets to grow to raindrop size may in theory take days, which is more than the lifetime of the cloud.

There are two basic processes that enable cloud particles to mature into fully fledged precipitation. The first entails collision between cloud particles. This is afforded by the considerable differential horizontal and, particularly, vertical motion within clouds, and the varying rates of fall (or rise) of particles within the cloud according to their mass (i.e. size). Small particles in static air will fall only very slowly towards the ground under gravity. Initially they will accelerate, but they will soon reach a constant terminal velocity when the air resistance around them offsets their downward acceleration. For larger particles this terminal velocity will be reached later, and will be higher. The terminal velocity for a droplet of 0.05 mm radius will be 0.25 metres per second ($m\ s^{-1}$), and that for a 0.001 mm droplet will be 0.0001 $m\ s^{-1}$. For a 2.5 mm raindrop it is 9.1 $m\ s^{-1}$. Smaller particles will also be more easily carried aloft by updraughts; larger ones, less so. Larger particles fall more rapidly towards the surface, the smaller less so. The opportunity for collision between particles is thus considerably increased, with larger particles sweeping up smaller particles in the downward direction and colliding with upward-moving ones. Where collision occurs between liquid droplets, the process is referred to as coalescence. Collision between solid particles, is called aggregation; between liquid and solid it is known as accretion.

The second process that generates precipitation within a cloud is commonly known as the Bergeron–Findeisen process, after the two meteorologists who independently evolved the theory in the 1930s. The process centres on the ability of liquid and solid cloud particles to coexist at temperatures between 0 °C and −40 °C. All cloud particles above 0 °C will be liquid, and all below −40 °C will be solid. The number of liquid (supercooled) droplets decreases below 0 °C. Between −10 and −30 °C there is commonly a mixture of both solid and supercooled liquid particles. Many clouds (particularly those with considerable vertical extent such as conventional clouds) contain a deep layer between these two temperatures. In thunderstorm clouds, the upper layers are below −40 °C, while the lower layers are frequently above freezing point. When significant quantities of supercooled droplets are available, nearby solid ice particles will grow at the expense of the liquid droplets. This occurs because air which is saturated with respect to ice is unsaturated with respect to water. Water will initially evaporate from a droplet and its size will be reduced. Because the air around the ice is now supersaturated, deposition occurs on to the surface of the ice particle, increasing its bulk. The air near the droplet is now once again unsaturated, and the process continues, allowing ice particles to grow to considerable size at the expense of water droplets.

Within typical clouds a combination of collision and Bergeron–Findeisen processes can occur. For 'warm' clouds (exclusively above 0 °C) only coalescence operates. Outside the tropics, warm clouds are shallow, so that only slight rain or drizzle will result. For 'cold' clouds, however, all processes can operate, and the considerable vertical motion within cumulonimbus clouds, with their tops at temperatures below –40 °C, and their lower layers perhaps above 0 °C, affords the opportunity for particles to circulate within them many times over, growing continually, sometimes by collision, sometimes accreting droplets, sometimes by aggregation, and often producing hailstones of a significant size. In addition, the upper parts of the cloud may continually be 'seeding' the higher-temperature parts of the cloud beneath with ice particles, which 'feed' on this raw material. At the same time, new liquid water may be swept up into the same layer, perpetuating the growth of particles by the Bergeron–Findeisen process.

GRAHAM SUMNER

Further reading

Meteorological Office (1972) *Observers' handbook*. HMSO Publications, London.

Sumner, G. (1988) *Precipitation: process and analysis*. John Wiley and Sons, Chichester.

precipitation during Quaternary ice ages During the Quaternary ice ages, changes in the pattern of global atmospheric circulation, as well as changes in the hydrological budget resulted in complex changes in regional patterns of precipitation. In high latitudes the greatest changes arose as a result of the growth of the major ice sheets of the northern hemisphere and also because of the development of sea ice over large areas of ocean. In the southern hemisphere, the Antarctic ice sheet increased in surface area and was also associated with a major expansion of sea ice. In both hemispheres, where ice sheets existed and where there was also a large increase in sea-ice cover there was an associated expansion of the areas affected by permanent high atmospheric pressure. This, in turn, promoted aridity and the displacement of mid-latitude cyclones to lower latitudes. Indeed, certain high-latitude areas were sufficiently arid to be incapable of supporting the growth of glacier ice. In these regions (for example northern Alaska, north-eastern Siberia, and northern Greenland) ice-age aridity was associated with the growth of permafrost.

The expansion of ice sheets in both hemispheres, together with decreased air temperatures, led to complex changes in atmospheric circulation in low latitudes. In many low-latitude areas during ice ages the position of the Intertropical Convergence Zone (ITCZ) was changed, and this led to changes in the distribution of monsoon precipitation. Thus, for example, in Africa during the last ice age, the position of the ITCZ was displaced several hundred kilometres southwards (Fig. 1). This, in turn, led to the failure of the south-west monsoon over the Gulf of Guinea and the displacement of the south-west monsoon over eastern Africa caused by strengthened north-easterly trade winds. By contrast, most areas of Africa during ice-age winters appear to have been dominated by dry northerly and north-easterly winds. These changes in atmospheric circulation led to increased ice-age aridity over most areas of Africa and this is reflected by an ice-age River Nile that scarcely functioned.

Similar climatic changes may have occurred over northern areas of South America during ice ages (Fig. 2). Here, displacement of the ITCZ over the area presently occupied by the equatorial rainforest may have caused a more restricted winter north-east monsoon and pronounced aridity. These changes in tropical South America and in Africa led to a greatly decreased area of equatorial rainforest during the last ice age. This widespread tropical aridity is also illustrated by the occurrence of windblown sediments in areas of present-day equatorial rainforest and by the presence of wind-deposited sediments offshore in sea-bed sediments.

In the southern hemisphere, the northern extension of sea-ice cover to latitude 50 °S resulted in the northward displacement of the Antarctic polar atmospheric front. This change resulted, in turn, in the northward displacement of mid-latitude cyclones over Argentina, southern Africa, southern Australia, and Tasmania. A strengthened cell of Antarctic high pressure also accounted for the northward displacement of mid-latitude cyclones in the southern hemisphere that tracked farther northwards across the southern Pacific, Indian, and Atlantic oceans.

Not all areas were affected by decreased precipitation during ice ages. For example, in western North America the mid-latitude cyclones that were steered along the southern margin of the Laurentide ice sheet resulted, through the combined influence of increased precipitation and decreased evaporation, in the formation of a number of large lakes (for example, Lake Bonneville). Similarly, increased precipitation in areas south of the ice sheet caused the formation of a shallow lake in Death Valley.

ALASTAIR G. DAWSON

Further reading

Dawson, A. G. (1992) *Ice Age Earth: late Quaternary geology and climate*. Routledge, London.

Goudie, A. S. (1983) *Environmental change*. Clarendon Press, Oxford.

Hamilton, A. C. (1982) *Environmental history of East Africa: a study of the Quaternary*. Academic Press, London.

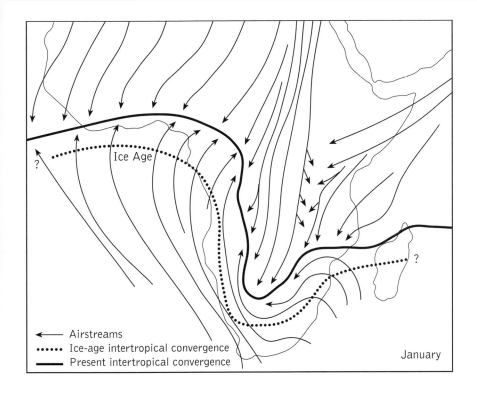

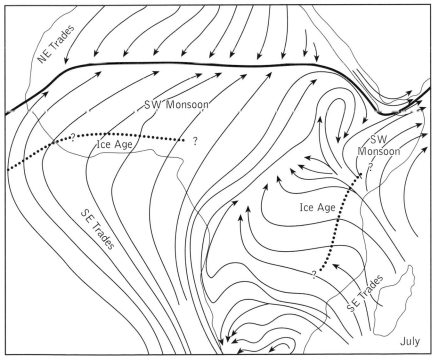

Fig. 1. Mean zones of convergence and airstreams over equatorial and southern Africa in January and July. Hypothetical positions of the Intertropical Convergence Zone at 18 000 BP are also shown (Adapted from Flint (1971) *Glacial and Quaternary geology.* John Wiley, New York and Dawson (1992), *Ice Age Earth; Late Quaternary geology and climate*, Routledge.)

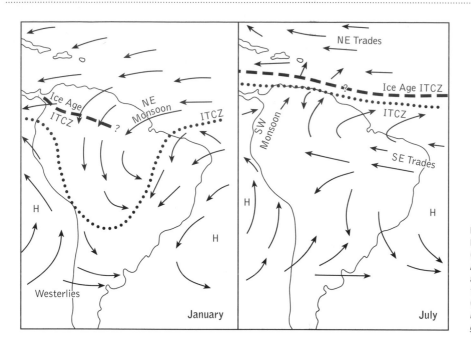

Fig. 2. Mean positions of the Intertropical Convergence Zone (ITCZ) and airstreams over South America during January and July and possible ITCZ positions at 18 000 BP. (Adapted from Flint (1971) and Dawson (1992) *Ice Age Earth: Late Quaternary geology and climate*, Routledge.)

primary deposits *see* GOLD-SILVER DEPOSITS *and* SUPERGENE AND RESIDUAL DEPOSITS

Proterozoic Late Precambrian (Cryptozoic) time is encompassed by the Proterozoic Eon, customarily held to span the interval 2.5 Ga to 590 Ma (2.5×10^9–5.9×10^8 years). Proterozoic rocks rest upon the Archaean and are succeeded by the Cambrian, but the boundaries are problematical. No rock standard has been identified formally as marking the base; at the top of the Proterozoic the Cambrian–Precambrian boundary is identified on the basis of fossils. In the past there have been several different divisions of the later Proterozoic into eras and periods, etc., based upon regional successions. Several of these, for example the Sinian Era, the Vendian Period, and the Ediacarian and the Varangian epochs, have received intensive study on account of the fossils they contain. So far their acceptance by the international community is not unanimous and their usage remains difficult. Generally accepted today are the Palaeoproterozoic (1.6–2.5 Ga), Mesoproterozoic (1.6–1.0 Ga), and Neoproterozoic (1.0 Ga–540 Ma) eras, respectively represented by approximately 20, 25, and 40 per cent of the Precambrian crust. The Subcommission on Precambrian Stratigraphy of the International Commission on Stratigraphy has given names to four periods within the Palaeoproterozoic and three to each of the other eras. The latest of these periods, the 'Neoproterozoic III', awaits a formal name and, at 650 Ma, follows on the Cryogenian, a period characterized by its glacial climate and sediments.

Proterozoic rocks occur in all the great Precambrian cratons of the world. Here they have been deeply eroded and then uplifted so that at outcrop they display features formed originally when deeply buried. Many of the terrains in which they lie exhibit plate-tectonic attributes, especially asymmetric linear orogenic belts, identified by tectonic style and geochronology. They include a wide variety of platformal clastic and carbonate sediments, and basinal clastics as well as numerous volcanics, banded ironstone formations, evaporites, and phosphorites. Their disposition was controlled by the stabilized Archaean continents and long-lasting sedimentary basins, and by cordilleran and collision types of orogeny. Major crustal rifting has also played a part in the evolution of large bodies of Proterozoic sedimentary and igneous rocks. Conspicuous igneous intrusive activity has resulted in gigantic dyke swarms, massive ultrabasic layered sheets, and large granitoid plutons.

Early Proterozoic rocks have been identified in all continents, but the largest outcrops by far are those of the North American–Greenland–northern Europe (Baltic Shield) region. India and Antarctica have the smallest areas of Early Proterozoic rocks. All regions show a great variety of composition and grades of metamorphism, with amphibolite facies orthogneiss–migmatite subordinate to high-grade granulite terrains and low-grade foreland fold belts. The last-mentioned possess distinctive platform–shelf–slope–rise facies that formed on wide stable shelves. Orthoquartzite–carbonate–banded ironstone sequences are very common. Many fold belts include greenstone–turbidite assemblages and they contain the oldest known ophiolites, glaciogenic rocks, and continental red beds. As a result of photosynthesis by plants, the Earth's environment was at that time rich enough in oxygen to support aerobic (animal) organisms.

The biomass was greatly enlarged to include diverse cyanobacteria, stromatolites, and many problematica (structures thought to be of organic origin but of uncertain affinity). The lithospheric plates making up the crust of the Earth were substantial in size and continuously in motion. Their continental parts were involved in a great deal of deformation, both while in passage and during collision with each other. Much underplating of the continents occurred as whole regions of continental crust were driven under their neighbours by subduction. A grand climax occurred at the end of the era, when, as if in response to a sudden surge of terrestrial heat through the crust, there was a global orogeny (1890–1850 Ma).

Mid-Proterozoic rocks occur in greatest mass in North America, with moderately extensive outcrops in South America, Australia, and Antarctica and relatively little elsewhere. Successions of rocks found in the Precambrian fold belts are largely of moderately metamorphosed supracrustal rocks and plutonic granites; those of the stable cratonic areas are of terrigenous sediments, ironstones and volcanics, and thick pelite–carbonate sequences. Continental rifting took place and was accompanied by immense basalt floods, and gabbroic–carbonatite intrusions. The orogenic belts display thick sequences of stacked thrust sheets, together with shear zones. By the Mid-Proterozoic biogenic photosynthetic activity had raised oxygen to about 4 per cent of present atmospheric level (pal) and the earliest eukaryotic organisms appeared. Plate-tectonic activity resulted in the development of a thick stable crust with extensional rifting in response to thermal plumes in the underlying mantle. There were some plate collisions with accretionary arcs adhering to the major continental masses. During the mid-part of the era large masses of rapakivi granites were generated, as were charnockites, mantle-derived anorthosites, and layered gabbros. At the end of the era (around 1.0 Ga) a very widespread major orogeny occurred.

Late Proterozoic rocks dominate the Precambrian geology of the Gondwanaland continents—Africa, South America, Antarctica, Australia, and India. Formations of this age are widespread throughout Eurasia, but are only minor components of the North American shield. They are comprised of extensive spreads of shallow-water marine platform carbonates and arenites, and red beds. Shallow depositional basins extended across enormous areas of the continents, which underwent rifting and the development of huge sediment-filled troughs (aulacogens). Continental red beds are common and there are local evaporites. Although banded ironstones of this age are widespread, they are not as thick as those deposited earlier. Enormous masses of Late Proterozoic glaciogenic deposits are found in all continents, but especially around the North Atlantic. They include many tilloids and associated rocks, indicating an intense and prolonged glacial period with many advances and retreats of the ice. This does not appear to have retarded the development of the biosphere with its ubiquitous stromatolites and widespread cyanobacteria. Towards the end of the

era various groups of soft-bodied metazoa appeared, no doubt aided by the increasing level of atmospheric oxygen (12 per cent pal).

Towards the end of the era plate-tectonic orogenies occurred in which the old stable cratons were broken and rearranged to constitute a single supercontinent. In some of these movements ophiolitic slices were squeezed up from the oceanic crust and large masses of island-arc volcanics were accreted to the continental margins. This great land mass was vigorously eroded and peneplained before the onset of Palaeozoic time with its global rise in sea level.

D. L. DINELEY

Further reading

Goodwin, A. M. (1991) *Precambrian geology: the dynamic evolution of the continental crust.* Academic Press, London.

Nisbet, E. G. (1987) *The young Earth: an introduction to Archaean geology.* Allen and Unwin, Boston.

Plumb, K. A. (1991) New Precambrian time scale. *Episodes,* **14,** 139–40.

Schopf, J. W. and Klein, C. (1992) *The Proterozoic biosphere.* Cambridge University Press.

Protozoa The Protozoa embrace a wide and diverse group of organisms. They are regarded by many as a separate phylum, while others view the group as a 'taxonomic dustbin'. The Protozoa are now classified as a sub-kingdom of the Kingdom Animalia—at least by the protozoologists. This classification, while having many strengths, leaves out many of the more botanical groups that are now separated as the algae. In micropalaeontology many of the protozoan groups, both botanical and animal, are studied as an integral part of the subject and used in both biostratigraphy and palaeoecology. The majority of the organisms included within the sub-kingdom Protozoa have left no fossil record, although many are almost certain to have had a long evolutionary history.

Geologists and micropalaeontologists have a particular interest in those groups that leave behind either organic-walled fossils or calcareous/siliceous skeletons. The important groups are the foraminifera, thecamoebians, radiolaria, silicoflagellates, ebridians, calpionellids, and tintinnids. The other, botanical, microfossil groups that are often included within the Protozoa by geologists include the acritarchs, tasmanitids, chitinozoans, dinoflagellates, haptophytes (calcareous nannofossils), and the diatoms.

Micropalaeontologists often use either the term 'protozoans' or 'protistids' for these groups of unicellular organisms. Because both groups include organisms that have both animal-like and plant-like characteristics, the classification(s) and phylogentic relationships within the groups are beset with complexities. Protozoologists and phycologists tend to use mutually exclusive classifications. Protists arose from the prokaryotes (unicellular forms that lack a nucleus and any other membrane-bound cell organelles such as plastids or mitochondria) in the Late Archaean or Early Proterozoic (between about 3000 and 2000 million years (Ma) ago). Most protist and protozoan groups are parasitic and non-mineralized. We can infer their presence in the geological record only because they must have been an important part of the food chain. The forms that do produce siliceous or calcareous skeletons or tests are very unrepresentative of the group as a whole. Undisputed protozoans are known from the Early Cambrian (550 Ma ago), but strange flask-like objects are known from rocks dating back to 850 Ma.

MALCOLM B. HART

Further reading

Hart, M. B., and Williams, C. L. (1993) Protozoa. In Benton, M. J. (ed.) *The fossil record*, Vol. 2, pp. 43–70. Chapman and Hall, London.

provenance The term 'provenance' is used to describe the source of sediment. The composition of a sediment depends upon the source rock and the changes it undergoes during weathering, transport, and deposition; post-depositional changes also have their effects. Intense weathering of a source rock breaks it down; some minerals are removed and new minerals are produced. Clay minerals commonly reflect the source rock, but their latitudinal distribution in ocean sediments indicates the importance of climatic control during weathering. Sands, except when produced in areas of intense tropical weathering, retain the characteristics of their source even after being transported over long distances. Gravels are modified considerably during transport.

Sorting takes place according to grain-size and density, and in part according to chemistry in the case of clays, and it results in fractionation of the original supply during deposition. Except for the changes undergone by clays, there are only relatively minor changes during burial and diagenesis.

In spite of the various modifications that are undergone by sediments during their history, information about the composition of siliciclastic sediments—mudstones, sandstones, and conglomerates—is of great value in the elucidation of source rocks, even when they have been buried, lost by erosion, or displaced a great distance from their original position.

G. EVANS

pterosaurs Pterosaurs are one of the three groups of terrestrial vertebrates that have evolved flight independently (the other two are the birds and the bats). They were the first to evolve, in the Late Triassic, but also the first to become extinct, dying out at the end of the Cretaceous. Pterosaurs were archosaurs, closely related to dinosaurs, and both groups seem to have evolved from small agile bipedal reptiles in the Middle Triassic.

Pterosaurs are generally divided into two groups. The earliest of these is paraphyletic (it consists of a common ancestor and some but not all of its descendants) and is referred to as the Rhamphorhynchoidea. The later, monophyletic group (having a common ancestor and all its descendants) is called the Pterodactyloidea. In both these groups the forelimb was transformed into a wing support by the enormous elongation of the fourth digit (the 'little' finger). The remaining three digits projected forward from the leading edge of the wing and bore functional claws; medial to them was a small projecting bone, the pteroid, which probably supported a small anterior flight membrane. The wing itself was a tough membrane that was supported internally by elastic threads attached to muscles. The sternum (breastbone) was greatly enlarged to provide an attachment area for the large flight muscles that powered the propulsive stroke, and the shoulder girdle was modified to allow a powerful downstroke. The body was fairly short, but the legs were long and very similar to those of a bird. It therefore seems likely that the pterosaurs were active bipedal animals when on the ground. The bones were hollow and thin-walled, but also pneumatic: that is, they had openings in their walls that allowed air sacs from the respiratory system to enter the bones. This would have had the effect of lightening the skeleton but would also have provided a greater area for gas exchange. The rhamphorynchoids retained a long tail which had a small rudder as its tip. The pterodactyloids lost the tail and also attained great size.

Pterosaurs were originally considered to have been adapted for gliding and were not thought to have been capable of flapping flight; however, more recent analyses of their skeletons indicate that this view is incorrect. The attachment areas for the flight muscles are much larger than would be necessary for a gliding organism, and, in addition, the shoulder girdle is adapted for the forward and downward motion necessary in active flight. The wing appears to have been narrow and gull-like, attached at its base to the side of the body and the upper thigh, but not to the lower leg or ankle. Estimated body weights for pterosaurs suggest that their wing loading was much lower than in modern birds, which would have made take-off much easier and allowed them to have been efficient gliding hunters once airborne.

Teeth and jaw shapes were very diverse; many of the smaller forms had small pointed teeth but the large species were often toothless. The genus *Pterodaustro* from the Lower Cretaceous of Argentina had numerous bristle-like teeth that might have acted as a straining device, which suggests a mode of life similar to that of the modern flamingo. Despite their efficiency in flight and apparent adaptation to a variety of ways of life, the pterosaurs dwindled through the Cretaceous. The last was the largest of all, *Quetzalcoatlus* from the Late Cretaceous of Texas, which had a wingspan of possibly 11 m, making it the largest flying organism that has ever existed.

DAVID K. ELLIOTT

pyroclastic and volcaniclastic rocks Pyroclastic and volcaniclastic rocks are rocks that are formed of rock fragments ('clast', from the Greek *klastos*, broken in pieces), that are flung out of volcanoes. In practice pyroclastic and volcaniclastic rocks include all the products of extrusive volcanic activity except lava flows. A volcaniclastic rock is one containing volcanic material in whatever proportion and without regard to its origin or environment of deposition.

The term 'pyroclastic' is used for classic rock material formed by volcanic explosion or aerial expulsion from a volcanic vent. The clasts in volcaniclastic rocks may range from fine ash to bombs and enormous blocks. The substances known as Pelée's hair and pumice are also grouped with pyroclastic and volcaniclastic rocks. Pelée's hair, a natural product of some volcanoes, is a very fine volcanic glass spun out like candy-floss. Pumice is a foam of volcanic gas bubbles volcanic glass. Some pumice is so filled with gas that it will float on water.

Large ejecta (materials thrown out of a volcano) are usually deposited close to the vent; progressively smaller particles travel farther. The finest-grained clasts are ash. Much ash falls immediately to the ground but often part of it heats the air around it, making it buoyant. When this happens ash clouds can rise up to 10 km or more before they reach neutral buoyancy and spread out to form a thin sheet in the atmosphere. Ashes from such sheets can be blown great distances by wind over long periods of time and over huge areas; indeed, ash occasionally travels round the world. Large quantities of volcanic ash in the atmosphere cause spectacular sunsets. Sometimes hot ash emerging from a volcanic vent rushes in a hot cloud called a *nuée ardente* down the side of the volcano. A hot ash cloud like this erupted from Mt. Pelée on May 1902. The burning cloud was hot enough to melt bottles and to twist metalwork, and it destroyed the town of St Pierre. The heat of a *nuée ardente* welds and streaks out the ash particles, which are then called *fiammé*, to form an ignimbrite. A welded ash without *fiammé* is called a welded tuff. Ash that has been deposited without being welded is simply called a tuff.

Small particles of lava (1 cm across) are called *lapillae* or *scoria*. Scoria is the most common material in most volcanic cones. A typical volcanic cone also contains lava flows and larger pieces of volcanic material called *blocks* and *bombs*. Blocks are lumps of solidified lava that are blown out of the volcano; bombs are portions of lava that are thrown out still molten. Bombs are divided into several types according to their form. Bread-crust bombs are those whose outer surface solidified while the inside was still emitting gas and expanding as a foam. As the inside expanded it cracked the outer shell, making it look like the crust of a loaf of bread. Splatter bombs are those that did not have time to solidify before they hit the ground; they resemble cow-pats.

Volcanic activity is often so violent that it entrains (picks up) pieces of 'country rock', which are not of volcanic origin. These accidental additions to volcanic rocks are called *xenoliths* (from the Greek, 'strange stones'). JUDITH M. BUNBURY

Further reading

Bullard, F. M. (1962) *Volcanoes: in history, in theory, in eruption.* Thomas Nelson, Edinburgh.

Fisher, R. V. and Schminke, H.-U. (1984) *Pyroclastic rocks.* Springer-Verlag, Berlin.

Francis, P. (1993) *Volcanoes: a planetary perspective.* Clarendon Press, Oxford.

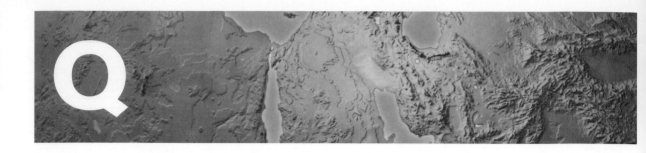

quartz and related minerals

α-Quartz, β-quartz, tridymite, cristobalite, coesite, and stishovite are all polymorphs of silicon dioxide or silica (SiO_2); they have different structures and therefore different symmetries and physical properties and are stable under different conditions of temperature and pressure. α-Quartz is by far the most common and is the form that is stable at the lowest temperature. If heated at atmospheric pressure, α-quartz inverts to β-quartz at 573 °C; β-quartz to tridymite at 867 °C; tridymite to cristobalite at 1470 °C. Cristobalite melts at 1713 °C. Coesite and stishovite form only at very high pressures equivalent to depths in the Earth of 60 km and 250 km, respectively.

α-Quartz is colourless when pure but minute amounts of impurities or lattice imperfections give rise to varieties such as amethyst, cairngorm, rose quartz, and smoky quartz. Silica precipitated from aqueous solution at low temperatures gives cryptocrystalline varieties such as opal, jasper, chalcedony, agate, carnelian, onyx, flint, and chert.

α-Quartz is one of the most common minerals in the Earth's crust and is stable over a wide range of geological conditions. It is an important constituent of granites and related acid igneous rocks, and of a range of metamorphic rocks such as quartzites, schists, and gneisses. α-Quartz is a common constituent of hydrothermal veins where, in cavities, it may grow into large, well-formed crystals. Because of its hardness, lack of cleavage, and resistance to chemical attack it is the most common detrital mineral and is the main constituent of sandstones, greywackes, and related sedimentary rocks.

Quartz is an important industrial mineral component of aggregates, sands, glass, and abrasives. Its piezoelectric properties are of vital importance to the electronic industry and, in order to supply an ever-increasing demand, large quantities are synthesized in the laboratory.

R. BRADSHAW

Quaternary

The Quaternary is the most recent geological period and system of rocks. It is part of the *Cenozoic* era and extends from 1.8 Ma ago to the present time. Its record is extremely varied, being influenced by large-scale climatic oscillations (glacial and interglacial in the higher latitudes, pluvial and interpluvial in the equatorial regions) and by the associated rapid changes of sea level. It is divided into the *Holocene* epoch, dating back to 0.01 Ma and the *Pleisto-cene* epoch from then to about 1.6 Ma ago. Nowadays we have abandoned the notion that the Quaternary represents a separate era following the Tertiary.

The term 'Quaternary' was first used by the French geologist Jules Desnoyers in 1829 to denote the most recent strata. In the following decade the Swiss geologist Louis Agassiz was able to demonstrate from erratic blocks and superficial deposits the previous (but geologically very recent) greater extent of the alpine glaciers. Similar evidence of post-Pliocene glaciation was soon gathered in lowland parts of Europe and North America to confirm the existence of an ice age not long ended. Then early in this century the succession of four Alpine glacial and interglacial deposits was described in the classic study by A. Penck and E. Bruckner. Subsequently, comparable phases in the North Sea region were discovered. In the meantime a fourfold glacial sequence had been recognized in North America. The several glacial phases now known in the southern hemisphere do not correlate exactly with those of the north.

Formal stratigraphic procedures are difficult to apply to the diverse continental and marine deposits of the Quaternary, but climatic criteria appear to be generally the most useful in global correlation. Magnetostratigraphic data also are locally helpful in this connection. From the original concept of four successive glaciations, interrupted by interglacial phases, in the northern hemisphere there has developed a far more complex schedule of climatic changes with attendant complicated stratigraphies.

This realization has come about from studies of the continuous record embodied in marine sediment cores, and of the oxygen isotope ratios of the included foraminifera. The succession of ratios appears to reflect changes in the global ice volume. Cores taken from the world's ice caps also reveal that the carbon dioxide content of the atmosphere has had an important influence upon the timing and intensity of glaciation. The task of reconciling the deep-sea record of climatic changes (perhaps twenty or more cold spells) with the half-dozen or so in the continental record is an urgent one for students of this part of the stratigraphic column. There is nothing to suggest that glacial conditions will not return to the middle latitudes in the near future, continuing the pattern of the past million years or more.

In recent years, there has been renewed interest in the work of the Serbian physicist M. Milankovich, who several decades ago postulated that cyclic changes in the Earth's orbital eccentricity, tilt, and precession are responsible for climatic oscillations. The critical periodicities are estimated as falling at 100 000, 40 000, and 19 000–20 000 years.

The climatic fluctuations have greatly affected biogeography. Mammals in particular evolved to meet the changes, but many suffered extinction. Human exploitation, at least in part, may be responsible for the disappearance of many of the larger animals. Human evolution has been the most distinctive biological feature of the Quaternary. The genus *Homo*, appearing first in Africa, spread rapidly into Eurasia during the early Pleistocene. Migration continued into Holocene time, accompanied by stone-based and more advanced cultures.

D. L. DINELEY

Further reading

Charlesworth, J. K. (1957) *The Quaternary Era.* 2 vols. Arnold, London.

Imbrie, J. and Imbrie, K. P. (1979) *Ice ages: solving the mystery.* Macmillan, London.

Pomerol, C. (1982) *The Cenozoic Era: Tertiary and Quaternary.* (Trans. D. and E. E. Humphries.) Ellis Horwood, Chichester.

Quaternary ice ages, precipitation during *see* PRECIPITATION DURING QUATERNARY ICE AGES

Quaternary ice-sheet disintegration

During the Quaternary the response of ice sheets to climate changes was, for the most part, slow and gradual. Thus, climate warming was normally associated with a rise in the equilibrium line across individual ice sheets which, together with changes in the radiative heat balance, led to changes in the rates of accumulation and ablation. However, at certain times individual ice sheets appear to have undergone catastrophic change. In the relatively few documented cases, those catastrophic changes that affected particular ice sheets took place during ice-sheet melting and were characterised by brief phases of widespread ice-sheet disintegration. Relatively few examples of catastrophic ice-sheet disintegration are known and, of these, limited information is available for three events.

The oldest of the three documented instances is reported to have taken place during the last interglacial period (approximately 125 000 years ago) in Antarctica. It has been proposed by a number of scientists that during this time a large area of the west Antarctic ice sheet disintegrated in a catastrophic manner. It has also been suggested that this catastrophic event led to a global rise in sea level of between 4 and 6 m and that the discharge of icebergs into the southern oceans led to complex oceanographic changes. Some have argued that the West Antarctic ice sheet collapsed partly as a result of increased air temperatures that altered the position of the ice-sheet equilibrium line. Ice-sheet instability may also have been caused by increased sea levels during the last interglacial that undermined the stability of the ice-sheet margin. Much of the West Antarctic ice sheet occurs below sea level and hence the pattern of ice dynamics is greatly influenced by sea level changes that, through changing rates of iceberg calving, alter the mass balance of the ice sheet. The claims that major ice-sheet disintegration took place during the last interglacial, and the associated dramatic sea-level rise, are particularly important since they have relevance to present considerations of global warming, remembering that the latter changes took place in the absence of any human influence.

The second major episode of ice-sheet disintegration is reported to have taken place in Greenland during the last interglacial. Several scientists have proposed that the southern part of the Greenland ice sheet disappeared during the period of maximum interglacial warmth. The evidence on which this argument is based is principally glaciological and geological in nature and, as in the case of the West Antarctic ice sheet, would appear to indicate the capability of ice sheets to disintegrate in the absence of any human influence on climate. Some have argued that the evidence for a global sea-level rise at this time of between 4 and 6 m was in part due to disintegration of the Greenland ice sheet. If this is the case, then the contribution to sea-level rise arising from the collapse of the West Antarctic ice sheet that also took place at this time is correspondingly reduced.

The third, and probably the most unequivocal, episode of ice-sheet disintegration took place over northern Canada approximately 8400 years ago during the decay of the last (Laurentide) ice sheet (see *Quaternary superfloods*). At this time the retreat and thinning of the Laurentide ice sheet was associated with the formation of a series of extremely large ice-dammed lakes along the southern ice-sheet margin (glacial lakes Ojibway and Agassiz). To the north, continued ice decay was associated with the invasion of sea water through the Hudson Strait into the Hudson Basin. Approximately 8400 years ago the ice impounded lake waters breached the ice barrier and drained northwards through the Hudson Strait into the North Atlantic. At this time approximately 70 000–150 000 km³ of water drained from these lakes into the world's oceans. It has been estimated that the catastrophic drainage of water from these lakes would have resulted in a virtually instantaneous rise of world sea level of between approximately 0.2 and 0.4 m within 2 days. Drainage of these lakes was also associated with the disintegration of the Laurentide ice sheet over the Hudson Bay region and the formation of a 'floating ice island'. It has been estimated that by 7000 years ago only small remnants of the Laurentide ice sheets remained.

These three examples, although poorly documented, serve to remind us that during the relatively recent geological past there have been periods of time when ice sheets have disintegrated in a catastrophic manner and have resulted in extremely rapid rates of sea-level rise worldwide. To date, the climatic consequences of these events have not been investigated in detail. At present, particular attention is being directed towards investigations of the stability or otherwise of the West Antarctic ice sheet, there being a fear that this particular ice sheet might undergo catastrophic collapse in the future as a result of global warming. Fortunately, the most recent studies appear to indicate that this possibility of catastrophic ice collapse is not likely, although there are an increasing number of reports of ice-shelf disintegration. Detailed measurements of snow accumulation in western Antarctica over the past two decades appear to indicate that global warming has been associated with increased snow precipitation. This is explained as a result of increased ocean temperatures that have caused increased rates of moisture evaporation from the ocean surface and, in turn, increased amounts of snowfall.　　　　　ALASTAIR G. DAWSON

Further reading

Dawson, A. G. (1992) *Ice Age Earth; Late Quaternary geology and climate.* Routledge, London.

Denton, G. H. and Hughes, T. J. (1981) *The last great ice sheets.* John Wiley and Sons, New York.

Dyke, A. S. and Prest, V. K. (1987) Late Wisconsinan and Holocene history of the Laurentide ice sheet. *Geographie Physique et Quaternaire*, **41**, 237–64.

Quaternary ice sheets and climate

It is now generally agreed that during the Quaternary temperate conditions similar to those of the present existed for only relatively short intervals of time. For the majority of the time, glacial conditions prevailed, and for long time-intervals ice sheets covered large areas of North America, Europe, and

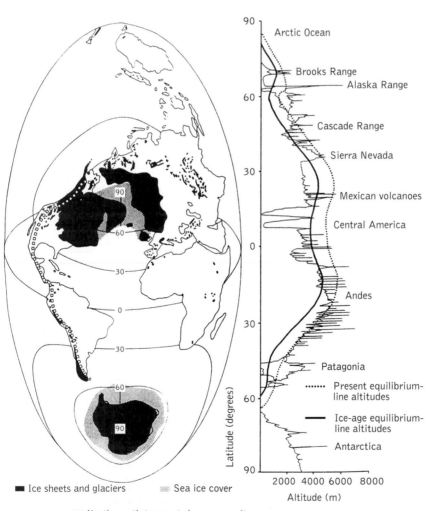

Fig. 1. The distribution of the principal ice sheets and mountain glacier complexes during the last glacial maximum (left). The worldwide lowering of equilibrium-line altitudes during the glacial maximum is depicted along a north–south transect (open boxed line). (After Broecker and Denton (1990).)

■ Ice sheets and glaciers　　　▨ Sea ice cover

▫▫▫ North–south transect shown opposite

Russia. During periods of glaciation the ice sheets in Antarctica and Greenland expanded in size, while in many upland areas a significant lowering of the snow-line (frequently by as much as 700 metres) resulted in the widespread development of alpine glaciers and ice caps (Fig. 1).

During ice ages the various ice sheets played an important part in altering the Earth's climate. The most direct influences were in the northern hemisphere, where large ice sheets in North America and Eurasia resulted in the development of topographic barriers that greatly altered patterns of air flow in the troposphere. Much of what we know of the changing patterns of climate associated with the development of Quaternary ice sheets has been established as a result of the development of numerical models of the Earth's climate during ice ages. This has been supplemented by empirical studies of proxy indicators of climate change (for example biostratigraphical and lithostratigraphical investigations, that have been used to check the accuracy of the mathematical models).

From a climatological point of view, the most important influence of ice sheets on climate arises from the cooling of the overlying air by the surface of the ice sheets. As a result, the air that overlies any ice sheet is stable and results in the formation of permanent high pressure that, in turn, causes

cyclonic depressions to be steered round individual ice sheets. Thus, in the case of the Antarctic ice sheet during the last glaciation, cyclones that formed in the Antarctic Circumpolar Ocean were displaced further northwards and this, in turn, led to increased precipitation in lower latitudes (e.g. across Southern Australia).

In the northern hemisphere, cyclones were similarly steered round the permanent cells of high pressure located above the ice sheets in North America and Eurasia. This process appears to have been associated with a splitting of the mid-latitude jet stream around both the major ice sheets (Fig. 2). Thus, the North American ice sheet was associated with a northern jet stream limb that tracked across northern Alaska and thereafter descended to the south-east to merge with the southern jet stream limb over the North Atlantic Ocean. This limb of the jet stream tracked across the continental United States, where it induced the generation of rain-bearing depressions. Accordingly, during periods of time when an ice sheet existed over North America, much of the continental area of the United States was influenced by increased rainfall and decreased evaporation, the latter a result of lower temperatures.

In Europe there appears to have been a similar split in the mid-latitude jet stream so that a northern jet stream limb

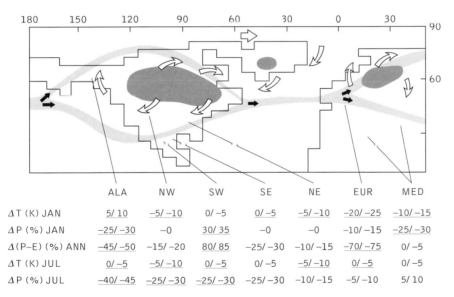

	ALA	NW	SW	SE	NE	EUR	MED
ΔT (K) JAN	5/ 10	−5/ −10	0/ −5	0/ −5	−5/ −10	−20/ −25	−10/ −15
ΔP (%) JAN	−25/ −30	−0	30/ 35	−0	−0	−10/ −15	−25/ −30
Δ(P−E) (%) ANN	−45/ −50	−15/ −20	80/ 85	−25/ −30	−10/ −15	−70/ −75	0/ −5
ΔT (K) JUL	0/ −5	−5/ −10	0/ −5	0/ −5	−5/ −10	0/ −5	0/ −5
ΔP (%) JUL	−40/ −45	−25/ −30	−25/ −30	−25/ −30	−10/ −15	−5/ −10	5/ 10

Fig. 2. Generalized pattern of northern hemisphere circulation for January 18 000 BP. Surface winds are shown by open arrows; jet-stream air flow by continuous stippling and solid arrows. The principal ice-sheet areas are also shown (shaded). Average values of the difference between ice-age and modern temperature (K), precipitation (%), and precipitation minus evaporation (%) are shown for Alaska (ALA), north-western USA (NW), south-western USA (SW), south-eastern USA (SE), north-eastern USA (NE) Europe (EUR), and the Mediterranean (MED). Values are expressed to the nearest interval of 5; a temperature difference of −7° C is thus shown as −5/−10. Underlined values are those that are the most statistically significant. Note the pattern of split jet-stream flow round both the Eurasion and Laurentide ice sheets. (After J. E. Kutzbach and H. E. Wright (1985). Simulation of the climate of 18 000 years BP; results for the North American/North Atlantic/European sector and comparison with the geological record of North America. *Quaternary Science Reviews*, **4**, 147–87.)

tracked across Arctic Russia while a southern section of the jet stream was deflected southwards across the Mediterranean region (Fig. 2). In a similar manner to North America, the southward displacement of the jet stream across Europe resulted in increased precipitation over the Mediterranean.

Further east the behaviour of the jet stream is less well known: it depends largely upon the ice age conditions in the Tibetan plateau. A popular view is that during the last glaciation a large ice sheet formed over the Tibetan plateau and that this ice sheet caused a similar split in jet-stream flow. A contrasting view is that there were relatively small accumulations of ice in Tibet during the last glaciation, since the area was greatly affected by ice-age aridity. This interpretation is consistent with the formation of a single mid-latitude jet stream. This particular controversy is very important from a climatological point of view, since it is of importance for the patterns of ice-age tropospheric air flow over China during ice ages.

The patterns of jet-stream displacement described above were also partly a function of the great thickness that some of the ice sheets attained. In North America, Europe, and Russia the ice sheets were in the region of 3000–4000 m in thickness and, as a result, they presented topographic barriers to air flow within the upper troposphere. The presence of ice sheets (particularly in the northern hemisphere), together with the development of sea ice in the polar oceans, also resulted in an increase in meridional (north–south) temperature gradients. This process in turn led indirectly to increased rates of heat transfer between low and high latitudes, and this resulted in increased rates of air flow. Furthermore, the presence of the ice sheets together with decreased ocean temperatures led to increased aridity at lower latitudes.

An additional factor affecting the patterns of atmospheric circulation round Quaternary ice sheets arises from the fact that in many areas large lakes were created next to the ice-sheet margins. In the case of the Russian ice sheet a series of large lakes were formed along its southern margin. The largest of these, the Pur and Mensi lake complex occupied an area of 1.5 million square kilometres. Numerous large lakes were similarly formed along the southern margin of the Laurentide ice sheet in North America. In addition, numerous large lakes were formed further south of the ice sheet margin, principally as a result of decreased evaporation rates and increased rainfall (for example, Lake Bonneville: see *pluvial lakes*). These lakes are important from a climatological point of view since they provided a supply of moisture for mid-latitude cyclones that tracked across these continental land masses. At present, the significance of these moisture sources for the growth of ice sheets is not fully understood.

The recognition that ice-age conditions were associated with a profound redistribution of global weather systems implies that, over relatively short geological time periods, the dynamics of the troposphere responded to climate change in an unstable manner. The rates at which parts of the atmosphere may be susceptible to such change are at present unknown. However, recognizing that major changes have taken place in the relatively recent past provides a more secure basis from which to predict future changes.

ALASTAIR G. DAWSON

Further reading

Lamb, H. H. (1982) *Climate history and the modern world.* Methuen, London.

Broecker, W. and Denton, G. (1990) What drives glacial cycles? *Scientific American*, January 1990, 43–50.

Dawson, A.G. (1992) *Ice Age Earth: Late Quaternary geology and climate.* Routledge, London.

Quaternary oxygen isotope chronology Unlike

the continental sedimentary record, ocean-floor sediments can often provide an undisturbed record of environmental change for the entire Quaternary. One of the main techniques used for correlation and climatic reconstruction during the past 40 years has been oxygen isotope analysis of calcareous foraminifera within deep-sea cores.

Foraminifera are marine animals which range in size from less than 0.40 mm to 10 cm. Many species secrete a shell of calcium carbonate. When these animals die, their shells preserve a record of the nature of the water, including its oxygen content, during their lifetime. Two isotopes of oxygen are important in this form of analysis: ^{16}O (oxygen-16) and ^{18}O (oxygen-18). The lighter ^{16}O is, on average, about 500 times more common than the heavier ^{18}O. There is a preferential uptake of ^{16}O into the atmosphere during evaporation. During glacial periods the ocean tends to become enriched in the heavier ^{18}O because much of the evaporate containing the lighter ^{16}O is locked within the ice sheets. During interglacial periods the ice caps melt and return the isotopically light water to the oceans, where it mixes rapidly (Fig. 1a).

The difference in the oxygen isotopic composition of the ocean between glacial and interglacial periods is about 1.0‰ (1 part per thousand). This is shown quite clearly by the isotopic composition of the skeletons of the foraminifera in sediments on the sea floor. This isotopic composition depends, in fact, on two main factors: the isotopic composition of the water during life, and the temperature. When the temperature is low there is a greater fractionation of ^{18}O relative to ^{16}O, and the foraminifera shell contains a higher proportion of the heavier ^{18}O than it would if the temperature were higher.

Benthic species of foraminifera are subject to a much smaller range of temperature during their life than planktonic species which float freely in the upper 50 m of ocean, because the temperature of oceanic deep water rarely fluctuates by much. For this reason the changes in the $^{18}O/^{16}O$ ratio of benthic foraminifera are thought mostly to represent ocean isotopic changes which are caused by the changing volumes of continental ice sheets.

The amount of $^{18}O/^{16}O$ is measured during a mass spectrometer on hand-picked foraminifera, usually from the same

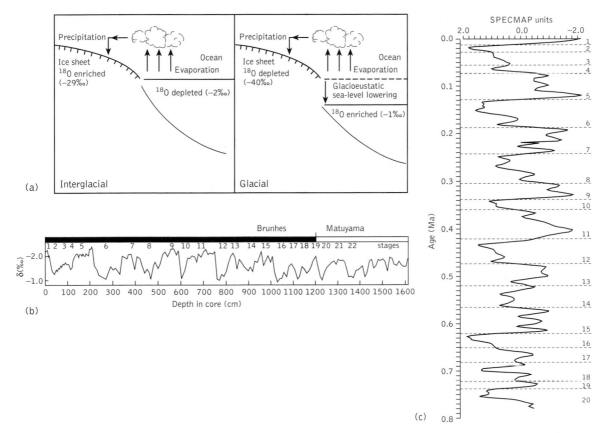

Fig. 1 (a) Schematic diagram showing the variation in isotope composition, ice volume, and sea level between interglacial and glacial periods. (From Dawson, A. G. (1992) *Ice Age Earth: Late Quaternary geology and climate*. Routledge, London., Fig. 2.1.) (b) Oxygen isotope composition of *Globigerinoides sacculifera* in core V28–238. (From Shackleton, N. J. and Opdyke, N. D. (1973) Oxygen isotope and palaeomagnetic stratigraphy of equatorial Pacific core V28–239: oxygen isotope temperatures and ice volumes on a 10^5 and 10^6 year scale. *Quaternary Research*, **3**, 39–55, Fig. 9.) (c) The SPECMAP stack, a calibrated oxygen isotope curve for the past 800 000 years. (From Patience, A. J. and Kroon, D. (1991) Oxygen isotope chronostratigraphy. In *Quaternary dating methods — a user's guide*, (ed. P. L. Smart and P. D. Frances), Quaternary Research Association; Technical Guide, **4**, pp. 199–228, Fig. 10.8.)

species. The result is then related to a standard (for deep-sea cores this is usually a reference carbonate) and the results are expressed as relative deviations in parts per millilitre relative to the standard. For example, a result of –4‰ would mean that the sample was 4 parts per millilitre deficient in ^{18}O relative to the standard.

There are some problems with stable oxygen analysis of foraminifera shells from deep-sea sediments:

(1) Some species of foraminifera migrate through the water column during their life cycle, which means that juveniles and adults of the same species can have different $^{18}O/^{16}O$ ratios.

(2) Physiological differences exist between different types of foraminifera which result in large differences in the $^{18}O/^{16}O$ ratio between species.

(3) Disturbances of the top 20 cm of sediment by burrowing animals can blur the isotopic signal in the record.

(4) Below a depth of between 3 and 5 km, the *carbonate compensation depth* (CCD), the carbonate in foraminifera shells begins to dissolve.

(5) Near to continents and on steep slopes the sediments can be disturbed by terrestrial sedimentation, slumping, or downslope currents.

These factors can be minimized by taking cores with high sedimentation rates from ocean basins above the CCD and by using foraminifera from the same life stage of a single species. Often, however, benthic species may be present in very small numbers relative to planktonic species. It may then be necessary to use mixed assemblages or, alternatively, planktonic species whose ecology and life style are well documented.

861

Oxygen isotope analyses have now been performed on cores from all the main oceans and seas of the world. The most surprising result from these analyses is the degree of similarity in the $^{18}O/^{16}O$ trace between oceans, reflecting the dominance of the signal caused by the changing isotopic composition of the oceans in response to changes in ice volume on the continents. Because of the essential synchronism of the record, once a reference curve has been calibrated for age it can then be used for global correlation.

The usual independent dating control is to use radiocarbon dating in the part of the core that is younger than about 35 000 BP. In addition, the first major magnetic reversal in the core is then identified and assigned an age of about 730 000 BP, assuming that this represents the change in polarity between the Brunhes and Matuyama epochs. Levels in between are then interpolated. Further checks can be employed using other data, such as the Barbados high sea-level stands, which have been dated to about 125 000, 130 000, and 82 000 BP using uranium-series dating of raised coral reefs. These high stands should correspond to peaks in the $^{18}O/^{16}O$ diagram.

One early standard record came from the upper 16 m of core V28–238 which was obtained from a depth of 3120 m on the Solomon Plateau of the western Pacific (Fig. 1b). Nicholas Shackleton of Cambridge University and Neil Opdyke of the

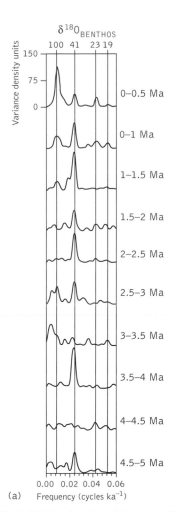

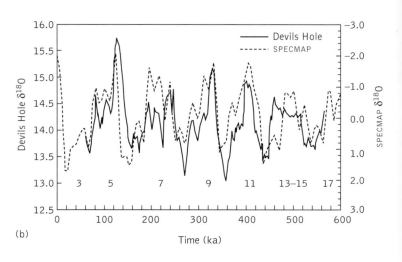

(b)

Fig. 2 (a) Spectral analysis of the past 5 million years in approximately 500 000-year time intervals. Note the change to dominance of the 100 000-year cycle over the past 1 million years. (From Tiedemann, R., Sarnthien, M., and Shackleton, N. J. (1994) Astronomic timescale of the Pliocene Atlantic $\delta^{18}O$ and dust flux records of Ocean Drilling Program site 659. *Paleoceanography* **9**, 619–38, Fig. 5.) (b) Comparison of the SPECMAP and Devils Hole records. (From Crowley, T. J. (1994) Potential reconciliation of Devils Hole and deep-sea Pleistocene chronologies. *Paleoceanography* **9**, 1–5, Fig. 1.)

Lamont-Doherty Geological Observatory published a much-cited paper in 1973 in the journal *Quaternary Research* in which they identified 22 stages covering the Brunhes and Matuyama magnetic epochs. For the upper 230 cm, covering the past 130 000 years, they sampled the core at 5 cm intervals and hand picked samples of the foraminifera for oxygen isotope analysis. They used eight specimens of the planktonic foram *Globigerionoides sacculifera* for each determination. Most horizons were analysed in triplicate, resulting in a precision of ±0.07‰. For the remainder of the core they sampled at 10 cm intervals and analysed in duplicate.

Shackleton and Opdyke interpreted the record in terms of continental ice-volume changes and assigned ages to each stage boundary. The core then became a template for correlation and was subsequently widely used by geologists interpreting the terrestrial record.

In 1976 a further seminal paper appeared in the journal *Science*. The authors were James Hays of Columbia University and the Lamont-Doherty Geological Observatory, John Imbrie of Brown University, and Nicholas Shackleton of Cambridge. They analysed three measures of global climate from two cores, RC11-120 and E49-18, which were located centrally between Africa, Australia, and Antarctica. The indices used were the oxygen isotope record of planktonic foraminifera, an estimate of summer sea-surface temperature derived from analysis of the remains of radiolarians (a group of marine planktonic animals with silica shells), and the percentage of a particular radiolarian, *Cycladophora davisisana*. Hays and his co-authors subjected the data to spectral analysis, which can show if there is any cyclicity in the record, and found that the variation of climate recorded in these cores could be explained by the interaction of three different cycles with periods of 23 000, 42 000, and 100 000 years. These periods are almost identical to the Milankovich cycles of precession of the equinoxes, axial tilt, and eccentricity of the orbit. It was concluded that changes in the Earth's orbit are the fundamental cause of the succession of Quaternary ice ages.

In 1984 a standard reference curve for the past 800 000 years was published (Fig. 1c). Instead of relying on just one curve, the SPECMAP (spectral mapping) curve was constructed by averaging (stacking) oxygen isotope profiles derived from planktonic foraminifera from several cores from low latitudes. John Imbrie and co-workers then used the Milankovich orbital cycles to adjust or 'tune' each curve in order to produce a type section against which other curves could be compared.

With the large number of high-quality hydraulic piston cores retrieved by the Deep Sea Drilling Program, and latterly the Ocean Drilling Program (ODP), spanning the whole Quaternary and beyond, the method of producing a reference curve tuned with respect to the orbital cycles has now been extended back some 5 million years into the Pliocene. Spectral analysis at intervals of approximately 0.5 Ma intervals from ODP core 659 confirms that for much of the Pliocene and Quaternary the axial tilt cycle of 41 000 years seems to have been the dominant factor in forcing climatic change. At about 1 million years ago, however, the 100 000-year eccentricity and 23 000-year precessional cycles become the major climatic forcing mechanisms (Fig. 2).

The SPECMAP model has recently been challenged by a well-dated [18]O record from a terrestrial source, Devils Hole, Nevada. Oxygen isotope measurements have been taken from a core (DH-11) of vein calcite (travertine) 36 cm long and fitted to a timescale based on 21 high-precision replicated uranium-series dates. This record reflects groundwater recharge rather than ice-volume fluctuations and is claimed by Isaac Winograd and his colleagues in the United States Geological Survey to be inconsistent with the Milankovich theory. There is in fact quite good agreement between the two records over the last 500 000 years. There are, however, two intervals of significant difference, around 140 000 and 450 000 BP (Fig. 2b). The discrepancy around 140 000 BP seems to be the most problematic since the uncertainties in the uranium-series dates are only about 2000 years at this point.

One view is that because the main discrepancy is restricted to one time interval it is unlikely that there is a continuous bias affecting one or other of the records. It could arise because the Devils Hole chronology reflects a regional rather than global response to climate change. The arguments about this discrepancy continue, and the outcome will be of paramount importance to our understanding of the mechanisms that force climate change.   B. A. HAGGART

Further reading

Dawson, A. G. (1992) *Ice Age Earth: Late Quaternary geology and climate*. Routledge, London

Quaternary sea-level changes

The Quaternary period (the past 1.8 million years) has been characterized by rapid and large-scale changes in sea level. The extent and nature of these changes has varied around the globe and they have resulted from an interaction between changes in the volume of the world's oceans (*eustatic changes*) and movements of the crust relative to the sea (*tectonic movements*). Since the interaction of these two factors produces a complex response, there is no agreement over the exact nature of sea-level changes during this period.

Causes of sea-level change

Quaternary eustatic changes in sea level result from a number of processes, the most important of which have been changes produced by the growth and decay of Quaternary ice sheets (*glacio-eustatic changes*). As ice sheets grow, water is transferred from the world's oceans on to the land; world sea levels thus fall, perhaps by as much as 150 m. When ice sheets decay, the water is returned to the oceans and the sea level rises. The period has thus been characterized by major fluctuations in sea level.

Nils-Axel Mörner of the University of Stockholm has suggested that major sea-level changes may also have been

(a)

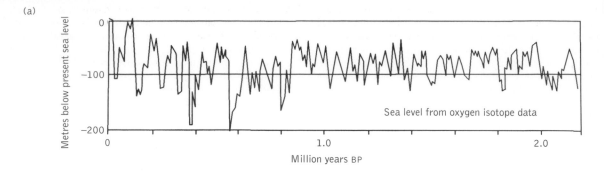

(b)

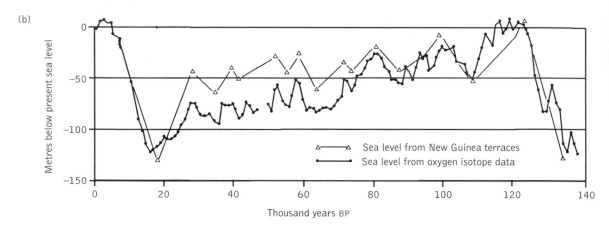

(c)

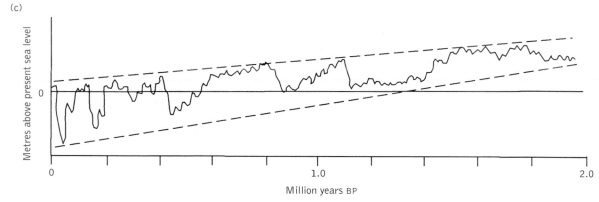

Fig. 1. Sea-level changes during the Quaternary. (a) Oxygen isotope record derived from the Pacific core V28-239. (b) Estimates of sea-level change for the past 140 000 years. (c) The declining sea-level curve derived from Mediterranean data. (After Bowen (1978) *Quaternary geology*, Pergamon; and Dawson (1992) *Ice Age Earth: Late Quaternary geology and climate*, Routledge.)

produced as the ocean's geoid (the surface of the sea) responded to environmental changes. Measurements taken from satellites indicate that the surface of the sea is not uniform, since it shows a variation in altitude from one point to the next. For example, measured from satellites, sea level near Indonesia is some 180 m higher than in the Indian Ocean. These variations in the geoid are the product of variations in the gravitational field and in the shape of the ocean basins. What Mörner has suggested is that as sea level rose and fell in response to the growth and decay of ice sheets, so the ocean basins would have changed shape and gravitational forces would have altered. He concluded that such changes could affect the geoid and thus alter the altitude of the sea surface. He was, however, unable to quantify the magnitude and form such changes may have taken. Other scientists have remained unconvinced about the importance of such processes.

Another eustatic process is the change in the level of the sea induced by changes in the volume of the ocean basins. Increased activity along constructive plate boundaries can increase the size of the mid-oceanic ridges, which in turn displace water and thus raise global sea levels. By contrast, reduced activity along constructive plate margins or mountain-building at destructive plate margins results in reduced global sea levels. Such changes are not as rapid as those associated with glacio- and geoidal eustasy but over a prolonged time period (perhaps longer than the Quaternary) they could have had a major impact on the level of the world's oceans.

More minor eustatic changes are produced by volcanic activity adding new water to the ocean basins or by temperature changes altering the volume of the world's oceans. It has been estimated that if the temperature of the world's oceans were raised by 1 °C, the thermal expansion of the water body would raise sea levels by 20 cm—a relatively minor fluctuation when compared with those produced by glacio-eustatic changes.

Sea-level changes along coastlines are also produced by movements of the land in response to tectonic processes. Some areas of the world (e.g. New Guinea) have experienced extensive uplift during the Quaternary, usually along destructive plate margins. At other sites (e.g. the Mississippi delta) subsidence has dominated, for example, as a consequence of sediment loading or sediment compaction.

More complex tectonic movements have characterized many other coastal areas. In regions that were covered in the past by ice sheets, the crust was depressed during ice-sheet growth and then recovered (rebounded) as the ice sheets decayed. Areas such as Scotland, Scandinavia, and Canada were significantly depressed, and thus the postglacial period has been characterized by rising land and generally falling sea levels. In contrast, many areas at the margins of the ice sheets underwent uplift as material was displaced from beneath the ice. Similarly, many oceanic areas experienced complex movements, since the growth of the ice sheets resulted in reduced crustal loading from ocean water, and thus crustal uplift (*hydro-isostasy*) occurred. As a result, the initial fall in sea level produced by the growth of ice was accentuated by crustal uplift. As the ice sheets melted and world sea levels rose, so the mass of water returned and the crust was lowered, promoting further rises in local water levels.

Nature of Quaternary sea-level changes

A detailed record of Quaternary eustatic sea-level changes has been derived from studying oceanic cores taken at selected sites where there is a continuous record of sedimentation. The tests of foraminifera are extracted from the sediments and oxygen-isotope analysis undertaken. The results (Fig. 1a) are believed to illustrate variations in ice volume through time, and thus they also represent Quaternary variations in the volume of the oceans. As a result such curves have often been used to determine eustatic sea-level changes. This analysis suggests that rapid fluctuations in sea level have occurred during the Quaternary, with 22 glaciations in the past 1.6 million years. It is noteworthy that the curve indicates that prior to 900 000 years ago sea level fluctuated by about 100 m, while since that time the fluctuations have been in the region of 180 m. The current view also suggests that sea level stood near present-day levels during each interglacial.

The oxygen isotope curve has been used to estimate the magnitude of sea-level changes for the past 140 000 years (Fig. 1b). This curve suggests that some 120 000 years ago, representing the last interglacial period, global sea levels were about 5 m higher than at present. Sea level subsequently fell to reach its lowest point at −130 m some 18 000 years ago. This fall in sea level was certainly not a gradual decline; on the contrary, it was marked by major fluctuations. The end of the last glaciation was marked by a very rapid rise in sea level (6000–18 000 years ago), and similar conditions appear to have occurred before the last interglacial (140 000–120 000 years ago).

Some scientists have suggested that Quaternary eustatic sea-level changes were characterized by a series of glacio-eustatic oscillations superimposed on a longer-term fall (Fig. 1c). It has been argued that the oxygen isotope curve reflects only the volume of water in the oceans and it does not illustrate changes in the volume of the ocean basins, which could account for the supposed fall. The proposed curve is noteworthy because it suggests that in the past eustatic sea levels associated with low stands stood at the same level as present-day sea level. Once again, this observation suggests that features of the present coastline may have developed in the past, perhaps under contrasting environmental conditions. This model is, however, based upon raised marine features from the Mediterranean region, and it is considered highly likely that many of these features have been uplifted by tectonic processes.

Sea-level changes in many areas have been dominated by tectonic processes which have been superimposed upon the eustatic fluctuations. One of the best examples is provided from the Huon Peninsula of New Guinea, which is subject to uplift because it lies near a destructive plate boundary. A sequence of coral reefs and deltas rising to altitudes of 400 m illustrate sea-level changes spanning the past 400 000 years. In most instances the coastal features developed when sea levels were high, and once the tectonic component has been extracted from the data, it is possible to estimate local eustatic changes in sea level (Fig. 1b). The resulting eustatic curve is similar to that derived from the oxygen isotope data, although for the past 120 000 years the terraces generally indicate higher sea levels. The data do, however, support the view that eustatic sea level has fluctuated around a constant level and has not been associated with a gradual decline.

Many other areas of the world have been characterized by tectonic uplift. These include Barbados, the east and west coasts of the United States, many sections of the Mediterranean coastline, New Zealand, and the New Hebrides. Other areas have experienced glacio-isostatic movements (Scotland, North America, Scandinavia) and thus have seen rises and falls in the crust, while many oceanic islands have been characterized by subsidence. At one time large sections of the world's coastline were believed to have been stable (e.g. the eastern seaboard of South America, western Australia, the Great Barrier Reef), and although little tectonic movement has occurred such areas are likely to have been influenced by hydro-isostatic movements.

It is now generally accepted that at present it is impossible to construct a truly eustatic curve for the Quaternary period. The oxygen isotope record provides the best approximation for the sea-level changes, but these variations are not entirely controlled by eustatic fluctuations. As a result, one oxygen isotope curve differs from another. Some scientists have suggested that geoidal variations may have been important, but at present the significance of such processes is disputed.

CALLUM R. FIRTH

Further reading

Bowen, D. Q. (1978) *Quaternary geology*. Pergamon Press, Oxford.

Dawson, A. G. (1992) *Ice Age Earth: Late Quaternary geology and climate*. Routledge, London.

Quaternary superfloods During the Quaternary, changes in the distribution of global weather systems and the growth and decay of ice sheets in various parts of the world led to complex changes in river activity. In low latitudes, for example, spatial and temporal variations in monsoonal precipitation led on occasions to periods of time during which severe river flooding took place. In areas close to the former ice sheets, numerous large lakes were impounded against the former ice margins. Overflow from these ice-marginal lakes resulted in greatly increased fluvial discharge through particular drainage basins. The greatest floods, however, took place during periods of ice-sheet melting. On occasions, impounded lake waters breached ice barriers and resulted in some of the greatest river floods ever to have taken place across the surface of the Earth.

Low-latitude superfloods

During the close of the last Ice Age, one of the most significant changes in fluvial activity took place within the Nile drainage basin. It is generally agreed that during the period of the last Quaternary glaciation the Nile drainage basin was extremely arid through much of the time period from 25 000 years BP. Soon after 12 500 years BP, major changes in river activity took place. At this time the waters of Lake Victoria began to overflow into the Nile, while within the Nile Valley in Egypt greatly increased flooding took place, particularly during the time interval 12 000–11 500 years BP. In lower Nubia these floods may have reached as much as +9 m above the flood plain of the time. The floods are thought by several authors to have reflected unusual climatic conditions in sub-Saharan Africa at this time. This period of greatly increased Nile discharge may have been due to the southwesterly monsoon over eastern Africa that had previously failed to traverse the continental landmass during the last glaciation. This major change in low-latitude atmospheric circulation may, in turn, may have been influenced by changes in the position of the Intertropical Convergence Zone. A similar period of monsoonal flooding may also have taken place in west Africa at this time owing to changes in low-latitude atmospheric circulation.

Superfloods related to the melting of the last North American ice sheet

During the beginning of the retreat of the last (Cordilleran) ice sheet in the western part of North America, numerous ice-dammed lakes were produced at various locations along the southern margin of the ice sheet. The greatest catastrophic river floods associated with the drainage of these lakes took place in association with the emptying of glacial Lake Missoula in Washington and Idaho. Several scientists have proposed that glacial Lake Missoula may have repeatedly been breached on as many as forty occasions during the melting of the last ice sheet in western North America, and that on each occasion the floods drained westwards across the Columbia River Basin in Washington and ultimately drained into the Pacific Ocean. Victor Baker has estimated that the peak discharge associated with the floods that drained out of glacial Lake Missoula may have been of the order of 20 million cubic metres per second ($m^3 s^{-1}$). This is an astonishing value for former river discharge: it is about twenty times greater than the average present worldwide run-off of 1.1 million $m^3 s^{-1}$.

Numerous large ice-dammed lakes were also produced along the southern margin of the (Laurentide) ice sheet east of the Rocky Mountains (Fig. 1). The largest of these lakes was glacial Lake Agassiz, which formed during the melting of the last Laurentide ice sheet and between about 12 800 years BP and 8000 years BP fluctuated in size. Throughout much of the time that glacial Lake Agassiz existed, greatly increased fluvial discharge took place through the Minnesota valley into the Mississippi drainage basin and ultimately into the Gulf of Mexico. It has been estimated that the average overflow discharge from Lake Agassiz at this time was between 40 000 and 100 000 million $m^3 s^{-1}$. During periods when Lake Agassiz increased in size owing to ice retreat or when additional flood waters derived from melting ice masses and other lakes emptied into Lake Agassiz, the rate of discharge from Lake Agassiz may have on occasions been as high as a million cubic metres a second.

On several occasions during the melting of the waters of the Laurentide ice-sheet, Lake Agassiz breached ice barriers and drained as a series of catastrophic floods in directions other than through the Mississippi drainage basin. For

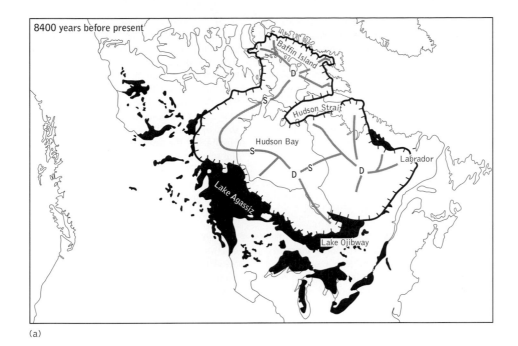

(a)

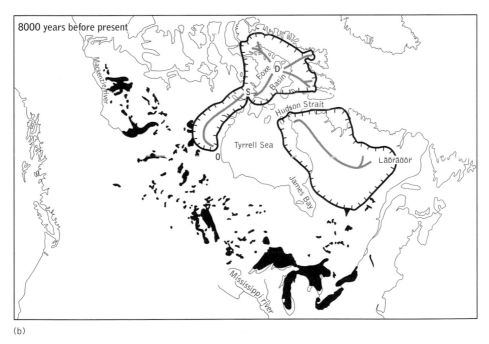

(b)

Fig. 1. Reconstruction of the Laurentide ice sheet for (a) 8400 BP and (b) 8000 BP, showing positions of major ice divides (D) and saddles (S). The distribution of ice for 8000 BP may also include a floating ice island (not shown) over Hudson Bay that was produced during the final disintegration of the Laurentide ice mass and the drainage of glacial Lake Agassiz/Ojibway through the Hudson Strait. (Based on Dyke and Prest (1987) *Géographie Physique et Quaternaire*, **41**, 237–64, and Dawson (1992).)

example, a major flood took place at about 11 000 years BP owing to the eastward escape of lake waters through Lake Superior and the St Lawrence Lowlands into the North Atlantic. Jim Teller has estimated that during each of a series of such flood events, approximately 4 km³ of water drained from the lakes. As the Laurentide ice continued to thin and retreat, a series of newly created lake overflow routes served to transport water out of Lake Agassiz. On each occasion Lake Agassiz alternately filled and drained, and each emptying event caused a catastrophic flood. These floods ended at 8500 years BP when continued ice retreat established a link between glacial Lake Agassiz and neighbouring glacial Lake Ojibway to the east (Fig. 1). Lakes Agassiz and Ojibway became confluent and bordered the southern margin of the ice sheet along approximately 3100 km of its length. Most probably the remainder of the ice sheet was in an advanced state of decay and undergoing widespread stagnation. At this time the waters of glacial Lake Ojibway suddenly breached catastrophically beneath ice in the Hudson Bay area and through the Hudson Strait into the North Atlantic. This resulted in the lowering of the levels of both Lakes Ojibway and Agassiz, in some areas by at least 250 m. It is difficult to estimate accurately the volume of water that must have drained from these lakes and beneath the ice. The figure is most probably between 75 000 and 150 000 km³—representing one of the greatest floods that took place during the Quaternary.

In Eurasia large lakes were dammed along the southern margin of the Russian ice sheet. The development of the very large ice sheet in Eurasia resulted in a complete reorganization of the drainage systems of major rivers such that most of the melt water that drained from this ice sheet was diverted southwards into the Black Sea and ultimately into the eastern Mediterranean. According to M. G. Grosswald, in the west Siberian Lowlands two extremely large lakes, the Pur and Mensi Lakes, occupied a vast area of about 1.5 km². These lakes were comparable in size with those that formed in North America at the close of the last ice age, and the scale of the river discharge associated with water that drained southwards from these lakes was similar to the volumes estimated for melt-water drainage rates in the Mississippi region. The ultimate fate of these large lakes in western Siberia is not known in any detail. It is, however, believed that most of the impounded melt-water in these lakes drained northwards catastrophically as soon as widespread ice-sheet stagnation had taken place during the Early Holocene.

ALASTAIR G. DAWSON

Further reading

Dawson, A. G. (1992) *Ice Age Earth: Late Quaternary geology and climate*. Routledge, London.

quick clays
The term 'quick clay' is used to describe fine-grained, relatively low-density soils that exhibit a dramatic reduction in strength when subjected to either shearing or shaking. The geotechnical term *sensitivity* is used to characterize these materials, and is defined as the ratio of the shear strength in the undisturbed state to the shear strength in the fully remoulded state. Most clays exhibit some drop in strength when sheared, but in the case of quick clays sensitivities well above 50 are quite common, and values above 100 have been reported. Under conditions of rapid shearing the material is transformed in a matter of seconds from a rigid solid to a fluid-like mass. Such unusual behaviour is made possible by the build-up of high pore-water pressure within the moving mass, which itself is caused by a collapse of the low-density material as it is sheared. Since water is virtually incompressible, any reduction in volume caused by structural collapse will cause a large increase in pore pressure. When the pore pressure at a given depth is equal to the total overburden pressure, the frictional forces between the particles are close to zero, and as a consequence the material can behave as a fluid. This is the origin of the 'quick' condition. Although this behaviour can be induced in most soils by adding water and stirring, the crucial difference with quick clays is that a fluid-like state can be induced without increasing the natural water content.

The extreme sensitivity of quick clays renders them very prone to landsliding, especially in Norway and eastern Canada where quick clay deposits are widespread. The term *flow slide* has been coined to describe quick-clay landslides. Flow slides usually begin as a relatively small slump along a steep quick-clay embankment with failure initiated by river undercutting or marine erosion. As the initial slump moves away from the embankment, shear loads are placed on the material directly upslope, triggering a second cycle of failure. This headward progression of failure, known as *retrogression*, may occur at a rate of several metres per minute. Eventually a large horse-shoe-shaped crater of liquefied material is produced, and rapid evacuation of material from the crater may cause either river damming or significant displacement waves in lakes and fjords. The failure process is eventually arrested when the pore pressures within the moving mass start to dissipate by upward drainage of excess water. Decline of pore pressure then allows particles to interlock, thereby restoring frictional strength. Examination of aerial photographs in quick-clay areas of Canada and Norway reveals numerous flow-slide scars of greatly differing ages. As the St Lawrence Valley lowlands are subject to infrequent but large earthquakes, it is possible that many prehistoric flow slides were triggered by seismic shaking.

Given the unusual behaviour of quick clays, a considerable amount of research has been conducted into their origin and geotechnical properties. All major quick-clay deposits accumulated in marine or brackish-water embayments close to the margins of melting ice sheets at the end of the last glaciation, and were then raised above sea level as a result of isostatic uplift in the postglacial period. Given the source of the material, much quick clay does not consist of true clay minerals but rather of glacial rock flour: clay-sized particles of quartz and feldspar produced by glacial abrasion of bedrock. Such particles tend to be negatively charged and mutually repellant in fresh water, but in brackish or salt water the presence of dissolved salts provides swarms of positively charged salt ions which allow aggregation, or flocculation, of fine particles to occur. The small platelets of material tend to align themselves by forming bonds between particle edges and opposing particle faces in a three-dimensional card-house structure. This open, low-density structure favours the retention of large amounts of pore water, and over postglacial time, percolation of fresh groundwater through the deposits has leached out significant amounts of the original salt. As a consequence, the cohesive strength of the card-house structure is progressively reduced over time, although small amounts of cementing material such as iron oxide may be present. When the poten-

tially unstable structure collapses because of shearing or shaking, the pore water is compressed and the 'quick' condition rapidly develops.

MICHAEL J. BOVIS

Further reading

Evans, S .G. and Brooks, G. R. (1994) An earthflow in sensitive Champlain Sea sediments at Lemieux, Ontario, June 20, 1993, and its impact on the South Nation River. *Canadian Geotechnical Journal*, 31, 384–94.

Mitchell, R. J. and Markell, A. R. (1974) Flowsliding in sensitive soils. *Canadian Geotechnical Journal*, 11, 11–31.

Torrance, J. K. (1987) Quick clays. In Anderson, M. G. and Richards, K. S. (eds) *Slope stability*, pp. 447–73. John Wiley & Sons, Chichester.

R

radioactive elements

The phenomenon of radioactivity was not discovered until the late nineteenth century—no doubt because we have no sense-organs for detecting it. Radioactivity cannot be felt, smelled, seen, or heard directly and is detectable only with the aid of mechanical or electronic devices. In about 1880 Henri Becquerel, a Parisian physicist, experimented with the luminescence of crystals of uranyl double sulphate, which is produced by exposure to ultraviolet light. He used his observations and those of the German physicist Wilhelm Conrad Roentgen to demonstrate that these crystals emit an invisible radiation that penetrates black paper and fogs a photographic plate. This work was carried forward by the Polish scientist Marie Sklodowska (later Curie) and Pierre Curie, who discovered that thorium was also an active emitter of penetrating radiation. They then discovered two further elements that emitted radiation, which they named polonium and radium. It was Marie Curie who proposed the term 'radioactivity' on the basis of the emissions of radium. In 1903 the Curies shared the Nobel Prize for physics with Henri Becquerel for the discovery of radioactivity. After Pierre's death, Marie continued her work on the chemistry of radium. In 1911 she received the Nobel Prize in chemistry in recognition of her successful efforts to isolate pure radium. She continued her work on the possible uses of radioactive elements in medicine, but died in 1934 of leukaemia, caused by the radioactive elements that she had studied.

Radioactive decay

The ionizing radiation emitted by radium aroused the curiosity of Ernest Rutherford at the Cavendish Laboratory at Cambridge University. After moving to McGill University in Canada, Rutherford did further work and reported that the radiation emitted by radioactive substances consists of three different components, which he named alpha, beta, and gamma radiation. The alpha component was eventually shown to consist of helium nuclei, and the beta rays were identified as electrons. Only the gamma rays proved to be electromagnetic radiation similar to the X-rays discovered by Roentgen. Rutherford, in collaboration with Frederick Soddy, studied the radioactivity of thorium. This led them to formulate the theory of radioactive decay. They suggested that the atoms of a radioactive element disintegrate spontaneously to form atoms of another element. They proposed that the disintegration is accompanied by the emission of alpha and beta particles and that the intensity of the radiation is proportional to the number of radioactive atoms (N) present. They showed that the change in the number of radioactive atoms present (dN) with change in time (dt) is equal to the decay constant (λ), which represents the probability that an atom will decay in unit time, multiplied by N. Rutherford moved to Manchester University in 1907 and in 1908 received the Nobel Prize for physics in recognition of his work on radioactivity. In Manchester he made further fundamental discoveries on the structure and composition of atoms. The results of his experiments indicated that atoms have very small, positively charged nuclei and that the nucleus is surrounded by orbiting electrons. Rutherford named the positively charged particle of the nucleus the proton in 1919, and a year later he speculated that the nuclei of atoms might contain a neutral particle. The hypothetical particle, the *neutron*, was discovered later on the basis of experiments by Sir James Chadwick and his co-workers. Rutherford's model of the atom was further refined by the Danish physicist Niels Bohr using the principles of quantum mechanics developed by Max Planck and Albert Einstein, and later by Shcrödinger and his co-workers using wave mechanics.

As the radioactive decay series of uranium and thorium was worked out, it emerged that there were several kinds of these elements, which decayed at different rates. It was also found that the atomic weights of the elements are not whole numbers, as had been previously believed, and that lead produced by the decay of uranium had an atomic weight that was different from that of natural lead. These observations led Soddy to suggest that a position occupied by a particular element in the periodic table could accommodate more than one kind of atom. He introduced the term *isotopes* ('same place' in Greek) for the differing forms of the same element. This led to the discovery that neon, for example, is composed of three kinds of atoms which have atomic weights of 20, 21, and 22. This advance relied primarily on the development of the mass spectrograph by F. W. Aston, a young chemist at the Cavendish Laboratory. Aston devoted his life to building increasingly precise mass spectrographs, with which he discovered 212 of the 287 naturally occurring isotopes. He also measured the masses of these isotopes and calculated the atomic weights of elements on the basis of the masses and relative abundances of their naturally occurring isotopes. For this work he was awarded the Nobel Prize in chemistry in 1922.

Geochronology

Alfred Nier became interested in problems of geochronology and the age of the Earth while working with K. T. Bainbridge at Harvard University. In collaboration with G. P. Baxter he began measuring the isotopic composition of lead and evolved dating methods based on the decay of uranium and thorium to lead. The mass spectrometers designed by Nier at the University of Minnesota enabled many other scientists to participate in isotopic research and contributed to the phenomenal growth of isotope geoscience until the end of the twentieth century. As geochronology developed, scientists also became aware that radioactive decay is an exothermic process, in other words, one that gives out heat. The rate of heat production in rocks was described by John Joly in a book entitled *Radioactivity and geology* in 1908. In this book Joly also discussed the measurement of the radioactivity of rocks and the origin of pleochroic haloes.

The ages of minerals and rocks containing uranium were first determined by Rutherford and Boltwood. Rutherford proposed in 1905 that the age of a uranium mineral could be measured by the amount of helium that had accumulated in it. The greatest ages that he obtained were about 500 million years, which provided positive evidence that Lord Kelvin's estimate of the age of the Earth, 40 million years, was in error. In 1907 Boltwood published the first age determinations of three uraninite speciments based on their uranium–lead (U/Pb) ratios. His dates ranged from 410 to 535 million years. The next fifty years saw rapid advancement in our understanding of radioactive decay and how it can be used to determine the ages of rocks. It was Patterson at the California Institute of Technology who in 1956 first established the age of the Earth and meteorites to be 4.54 Ga (billion years) by the use of lead isotope (^{207}Pb, ^{206}Pb, and ^{204}Pb) systematics. Other radiometric dating methods that were exploited in the latter part of the twentieth century included potassium–argon (K–Ar), potassium–calcium (K–*Ca*), rubidium–strontium (Rb–Sr), samarium–neodymium (Sm–Nd), lutetium–hafnium (Lu–Hf), rhenium–osmium (Re–Os), and thorium–lead (Th–Pb), together with cosmogenic radionuclides such as beryllium-10 (^{10}Be) and aluminium-26 (^{26}Al). The study of radioactive isotopes has given Earth scientists insights into the age and evolution of the Solar System, the planets, the Earth, and its moon.

K. VALA RAGNARSDOTTIR

Further reading

Faure, G. (1986) *Principles of isotope geology* (2nd edn). John Wiley and Sons, New York.

radioactive heat production in the Earth

Heat from the Earth's interior is leaking out to the surface at a rate of about 40 000 gigawatts (4×10^{13} W) in total. This 'heat flow' is over 2000 times less than the rate at which heat from the Sun reaches the ground, but the effects of solar heating are limited to the topmost few metres. For example day–night temperature variations amount to less than 1 °C at a depth of 1 m.

The internal, or geothermal, heat from within the Earth is responsible for the fact that temperature increases with depth. In continental crust the rate of increase with depth (known as the geothermal gradient) is about 30 °C km^{-1}. This is why it is so hot in deep mines. Extrapolating the same geothermal gradient to greater depths would mean that the mantle, at 50 km depth, would be hot enough to be molten (about 1500 °C); but we know from seismic studies that the mantle is almost entirely solid, so this cannot be the case. Continuing the near-surface geothermal gradient all the way to the centre of the Earth would imply a temperature of about 20 000 °C (hotter than the surface of the Sun), whereas the actual core temperature is thought to be about 4700 °C.

Evidently the geothermal gradient must be much smaller within the mantle than within the upper crust. There are two reasons for this. One is related to the origin of the heat, and the other is to do with how heat is transported within a planet.

Some of the heat leaking out to the Earth's surface is heat that the Earth was born with, known as primordial heat. The energy released each time planetesimals and planetary embryos collided during the Earth's formation would have made the young planet very hot. One model suggests that the planet was covered with a molten 'magma ocean', but even if this were not the case the Earth must have been cooling down ever since. It is thought that between 20 and 50 per cent of the present-day heat flow is caused by the remains of this primordial heat leaking out to space. This leaves at least half the Earth's heat flow unaccounted for. This heat must be being generated today. A small fraction of this is heat generated by tidal forces, but most of it is radioactive (or radiogenic) heat that comes from the decay of radioactive elements. Many elements have unstable, radioactive isotopes, and when a parent isotope decays to its daughter isotope a small amount of heat is always released.

Only three radioactive elements occur within the Earth in sufficient abundance and release enough heat per decay event to be significant contributors to radioactive heat production. These are uranium, thorium, and potassium. Uranium has two radioactive isotopes (uranium-235 (^{235}U) and uranium-238 (^{238}U)); thorium and potassium have one each (thorium-232 (^{232}Th) and potassium-40 (^{40}K)). Each so-called heat-producing isotope decays at a different rate, so that the total rate of radioactive heat generation has declined and the relative importance of each isotope has varied over time (Fig. 1). Potassium-40 has remained the most important heat-producing isotope throughout the Earth's history, but

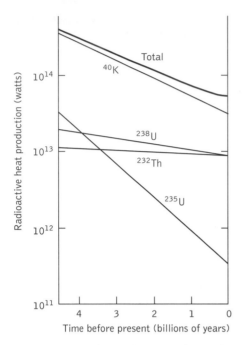

Fig. 1. Graph showing how the rate of radioactive heat production within the Earth has declined over time, as a result of the differing decay rates of the four heat-producing radioactive isotopes—potassium-40, thorium-232, uranium-235, and uranium-238.

uranium-235, which has the fastest decay rate (or shortest half-life), and now makes the smallest contribution of the four, was the second biggest contributor during the Earth's first half-billion years.

Knowing how the bulk of the Earth's heat is generated enables us to explain why the geothermal gradient is much greater in the crust than in the mantle. Uranium, thorium, and potassium are all much more abundant in the crust than in the mantle, because they are all 'incompatible' elements that go preferentially into the magma when rock begins to melt. The mantle-derived melts from which the Earth's crust formed were therefore greatly enriched in these elements. Consequently, about half the Earth's radioactive heat production takes place in the crust, although the volume of the mantle is a hundred times greater than that of the crust.

Heat generation is not uniform within the crust, either. It is greatest in rocks of the granite family, in which the incompatible elements are most highly enriched. This means that the rate of heat generation is greatest in the upper part of the continental crust, where granites are most abundant, and decreases with depth, so that the 30 °C km^{-1} geothermal gradient applies only to the upper crust. Oceanic crust, which is basaltic in composition, is even poorer in uranium, thorium, and potassium; so volume for volume, less heat is produced there, even than in the lower part of the continental crust.

Once into the mantle, the geothermal gradient decreases again, because the mantle is even poorer in heat-producing isotopes.

The other reason why the geothermal gradient is less in the mantle relates to how the heat is transported. Heat gets out through the crust by conduction, and the geothermal gradient is controlled by the rate of heat supply and the thermal conductivity of the rock. The same applies in the upper mantle, but eventually a depth is reached at which the mantle, although still solid, is so hot that it can carry heat outwards by convection. Hot, less-dense mantle rises until it has lost its heat and becomes cooler and more dense, which causes it to sink. Convection is a much more efficient way of transporting heat than conduction, and it acts to keep the geothermal gradient in the convecting part of the mantle much lower than in the rigid, non-convecting uppermost mantle belonging to the Earth's lithosphere.

It is impossible to obtain an accurate estimate of the Earth's present-day rate of radioactive heat production because the total abundances of the heat-producing radioactive elements are poorly known. They can be measured precisely in any single sample taken, but because the crust is so variable in composition it is difficult to estimate a reliable global average. Consequently the rate of heat production even in the relatively well-documented upper crust is uncertain by about 50 per cent either way. A further complication is that we know virtually nothing about radioactive heat production in the Earth's core. Although the outer core seems to be mostly iron, it must have about 10 per cent of a less-dense element in order to explain the speed at which seismic waves pass through it. There are several candidates for this light element, and one of them is potassium. If the light element in the outer core is potassium, then the core could contain more potassium-40 than the entire crust and mantle combined, which would make it a major contributor to the Earth's radioactive heat production.

Although uranium, thorium, and potassium are the only significant heat-producing elements today, when the Earth was very young it probably contained a variety of short-lived radioactive isotopes, most importantly aluminium-26 (^{26}Al) but also chlorine-36 (^{36}Cl), iodine-129 (^{129}I), iron-60 (^{60}Fe), and plutonium-244 (^{244}Pu). These have half-lives so short that they have long since vanished below the limits of detectability. During the Earth's first few million years their decay could, however, have added significantly to radioactive heat generation.
DAVID A. ROTHERY

Further reading

Brown, G. C. and Mussett, A. E. (1993) *The inaccessible Earth* (2nd edn). Chapman and Hall, London.

radioactive waste management Radioactive waste is a hazardous waste substance, the toxicity of which is derived from the radioactive decay of some of its components at a rate exceeding established thresholds. The presence of significant radioactive decay is the only feature distinguishing radioactive wastes from other wastes. They can be in liquid, solid, or gaseous form; flammable or inflammable. They can also have an additional significant chemical toxicity. In this case they are referred to as mixed wastes.

Radioactive wastes are generated at all facilities where radioactive substances (radio-isotopes) are produced or, otherwise, handled directly. These include facilities supporting the industrial, agricultural, and medical application of radio-isotopes, research establishments (including most large universities), and all facilities involving the nuclear fuel cycle. The latter include the mining, milling, and enrichment of uranium, the fabrication of nuclear fuels, their 'burning' in nuclear reactors, and the reprocessing of spent fuel. Radioactive wastes also arise from the decommissioning of nuclear facilities and from the environmental remediation of abandoned sites. Phosphate ore extraction for fertilizer production, as well as drilling for oil and gas, can also result in the abnormal concentration of natural radio-isotopes in waste streams. The latter, however, are not normally classified as radioactive waste, although their radioactivity levels may sometimes reach those of the waste categories mentioned below.

Radiological characteristics

From the point of view of radioactive waste management, three main forms of radioactive decay need to be considered:

(1) alpha (α) decay, during which a heavy, positively charged particle is emitted;
(2) beta (β) decay, during which a light, negatively charged particle is emitted; and
(3) gamma (γ) decay, during which a massless, charge-free particle (photon) is emitted.

These particles have different penetrating powers: the heavier the particle, the less the distance it is able to travel through materials and thus the thinner the shield that is needed to protect oneself from it. Thus anyone could handle without risk a thin-walled metal container of α-emitting waste, but would be in serious danger in front of the same container filled with γ-emitting waste. By the same token, while γ-emitting substances are dangerous upon both internal and external exposure, α- and β-emitting substances tend to be harmful only upon ingestion or inhalation and accumulation in critical organs.

The β- and γ-emitters of interest in radioactive waste management decay typically with a half-life of roughly 30 years or less, while the α-emitters have a much longer half-life, extending to hundreds of thousands of years. *Half-life* is defined as the time needed to reduce the initial amount of the radioactive substance by half. Thus, the presence in the waste of significant concentrations of α-emitters is a relevant criterion for considering their long-term isolation from the biosphere. It must also be noted that, associated with the emission and the slowing down of α-, β-, or γ-particles, there is an energy

transfer which may entail a significant temperature increase in the waste and surrounding medium. Heat generation rate is therefore another important criterion in the management of radioactive waste.

Classification

A certain variability exists in the way national organizations take into consideration the radiological characteristics of waste for classification purposes: this can cause confusion when examining information about waste management programmes and practices or data published in the technical literature. In the context of a packaged radioactive waste ready for shipment to a storage or disposal facility, a traditional but generic and qualitative classification scheme is as follows:

(1) *Low-level waste* (LLW): radioactive waste that does not require shielding during handling and transportation.
(2) *Intermediate- (or medium-) level waste* (ILW): radioactive waste that requires shielding but needs little or no provision for heat dissipation during handling and transportation.
(3) *High-level waste* (HLW): radioactive waste that requires shielding and provision for heat dissipation during handling and transportation. This category includes spent fuel in countries where the latter is not reprocessed.
(4) *Alpha waste*: low- or immediate-level waste with significant contamination from long-lived α-emitting substances. In North America this category of waste is also referred to as TRU (*TransUranic*) waste.

A special category of radioactive wastes, which are not packaged in individual containers for disposal and therefore escape the above classification scheme, are the so-called *uranium* (and/or *thorium*) *mill tailings*. These contain low but long-lived levels of radioactivity and some of the chemicals used in the uranium/thorium separation processes. They are managed by stabilization *in situ* on account of their relatively large volumes.

Finally, some countries have an additional category of *exempt* or *very low-level waste*, that is, wastes with measurable levels of radioactivity, albeit below regulatory concern, which can be disposed of without radiological restriction in approved discharge facilities.

Waste generation and composition

Radioactive wastes arise, for the most part, from the day-to-day operation of facilities where radio-isotopes are handled; that is, operations where parts have to be replaced, samples have to be taken and cleaning has to be performed. The volume, amount, and type of waste vary with each waste-producing facility. However, uranium/thorium mill tailings are generated only during the extraction and refining of the original ore, spent fuel is produced only in nuclear reactors, and HLW only following the reprocessing of spent nuclear fuel. Low-level waste typically consists of paper towels, laboratory glassware, used syringes, rubber gloves, cloth overalls, air cleaning filters. Intermediate-level waste is of a more

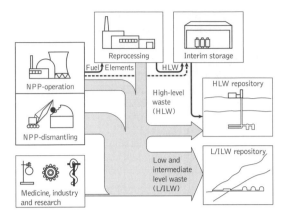

Fig. 1. Example of radioactive waste generation and disposal. The thickness of the arrows reflects the relative volumes of the wastes that are produced.

industrial nature, having an important component of scrap metal, evaporator sludge, spent resins used for the clean-up of contaminated waters, and parts of used radio-isotope sources in medical equipment. High-level waste consists of glass blocks into which the high-level liquid streams from the reprocessing plant are stabilized. Ninety-nine per cent of the total radioactivity produced in the nuclear fuel cycle is concentrated in the spent fuel or, where spent fuel is reprocessed, in the HLW; the remaining 1 per cent is dispersed within the alpha and the low- and medium-level wastes.

In nuclear countries, the facilities of the nuclear fuel cycle produce, collectively, the greatest fraction of the radioactive waste. This general situation is portrayed in qualitative terms in Fig. 1. More quantitatively, and taking the European Community (EC) as an example, for 1991–95 the annual rate of production of radioactive waste (excluding mill tailings) was estimated to be about 81 430 m³/year—75 000 m³ of which (91.5 per cent by volume) was low- and medium-level waste, 6500 m³ (or 8 per cent) alpha waste, and 430 m³ (or 0.5 per cent) high-level waste. Waste generation is expected to increase by twofold after the year 2000 owing to the dismantling of obsolete nuclear facilities, including nuclear power plants. In any event, the volumes to be managed will still be limited with respect to those of other wastes, namely, 1000 million m³/year of general industrial waste in the EC, of which 10 million m³ are classified as being toxic.

Waste treatment

Industrially produced radioactive waste needs to be treated and stabilized for interim storage and final disposal. In particular, liquids are not allowed in disposal containers.

For low- and medium-level waste, as well as for alpha waste, various industrial methods are applied, as needed, to achieve:

(1) volume reduction through incineration, evaporation, centrifugation, chemical separation, and compaction;
(2) waste stabilization through the addition of sorbing additives and filler materials or, for some waste streams, cementation or bitumenization; and
(3) waste packaging in special containers, depending on the waste type.

In the case of high-level waste, the nuclear industry has pioneered the stabilization of the liquid streams through vitrification into borosilicate glass blocks. This technique is now being applied to other hazardous wastes of non-nuclear origin. Alternatively, spent fuel removed from the reactor is already in solid form and needs no additional treatment before disposal, although a packing material is sometimes added in the containers to fill the void space.

Waste storage and disposal

The relatively low volumes of radioactive waste make their management and disposal viable both from economic and environmental points of view. As an example, in France the single 12-hectare disposal facility at La Manche was able to meet the disposal needs of that country's entire production of low- and medium-level waste up to the year 1992. Other useful characteristics of radioactive waste which alleviate the management task are that:

(1) there exists only a limited number of critical radio-isotopes; and
(2) all radio-isotopes decay with time into stable (non-radioactive) substances.

An interim decay period is often useful to lower the γ- and β-activity of most wastes significantly, making them easier to process and to handle. Thus all waste stabilization and disposal activities are preceded by some period of interim storage. Some low-activity wastes are allowed very simply to decay before they are discharged as non-radioactive waste. When it is not feasible to wait for the waste to decay significantly, isolation from the biosphere is practised. This is accomplished by placing the waste within a shallow or deep geological facility, sited and built following an appropriate site characterization programme and environmental impact studies.

Disposal of low- and intermediate-level waste

Until the late 1970s some countries practised both shallow-land and sea disposal of low- and intermediate-level waste, while most other countries have kept the accumulating waste in interim storage. Since the early 1980s sea disposal has no longer been considered an acceptable option, and shallow-land disposal practices have improved with the implementation of engineered barriers. For instance, at the Centre de l'Aube in France, pre-approved, standard waste drums are stacked on each other in concrete cells which are built on a concrete pad a few metres below the ground. The cells are sealed by pouring additional concrete and covering them with impermeable clays to which vegetation-bearing soils have been added. Drainage channels are built within and around the concrete pad, and sampling wells are used in order to monitor potential leaks of radioactive elements. The site will be monitored for roughly 300 years after closure; that is, until the waste decays sufficiently to allow unrestricted access. During that time the information about the source and composition of each waste drum is preserved. Similar disposal concepts have been implemented in other countries, for instance at Rokkasho-Mura, Japan, and El Cabril, Spain.

Deeper geological repositories of low- and intermediate-level waste have also been built and are in operation. They will minimize further the risk of human intrusion, should the continuity of present institutional control be interrupted. Two typical such repositories, at a depth of about 80 m, are the Olkiluoto repository in Finland and the Forsmark repository in Sweden. The repository at Konrad, in Germany, is located in a pre-existing iron mine at the unusual depth of 1000 m.

Disposal of high-level waste, spent fuel, and alpha waste

Geological disposal at depths between 200 and 1000 m is the option normally reserved for waste producing higher levels of heat (HLW and spent fuel) and for long-lived waste (alpha waste). The deeper the disposal, the smaller is the danger of inadvertent human intrusion, the longer is the timescale for geological change, and the less the need for active institutional controls. For maximum passive safety of these deposits, a series of nested barriers is used to prevent the migration of radio-isotopes. These barriers consist of the waste form, the waste containers, earthen materials emplaced around the waste containers, materials sealing the disposal vault and shafts, and the host geological formation itself. Some of the waste containers developed so far have a projected life of several hundreds of thousands of years. Among the identified earthen barriers are bentonite clays, which would swell upon contact with groundwater, ensuring a high resistance to the migration of the waste substances as well as very limited groundwater flow in the proximity of the waste. The host formation would ensure the longest migration (and, therefore, decay) times and the largest volumetric dilution of the waste substances.

No final deep geological repository of HLW, spent fuel, and alpha waste is operating as yet. The most advanced national programmes are at the siting stages of these repositories and only the so-called WIPP facility for the disposal of alpha waste from the defence programmes of the United States is nearing its licensing phase. Following the completion of the conceptual design phase, attention is now being given in many countries to the construction and operation of underground research and development laboratories. These are meant to provide hands-on experience of how a deep repository can be excavated and managed confirmation, *in situ* of the performance of the safety barriers, and demonstration to interested parties of what a repository might look like. The view is taken that the implementation of deep geological disposal is necessarily an incremental process and that the timing of the imple-

mentation will depend not only on technical factors but also on public acceptability. The exact timing required to reach consensus will vary with each country's specific culture, and this will possibly take decades. One factor in favour of long implementation timescales is that, especially for HLW and spent fuel, there is no special urgency for the disposal of the accumulated waste, because the waste volumes are very limited and proven interim storage technologies are available. On the other hand, the disposal task cannot be delayed indefinitely for it behoves the generation that produces the waste to put a safe containment in place, or at least to provide future generations with the means and the technology for its disposal.   C. PESCATORE

radioactivity measurement and surveying The radioactivity of a sample is measured by electrical or photographic methods and is expressed as the number disintegrations observed in unit time. The basic unit for the measurement of radioactivity is the curie (Ci), which is defined as the quantity of any radioactive nuclide in which the number of disintegrations in 3.7×10^{10} per second. For radium, a curie is the amount of radioactive disintegration in 1 g of $^{226}_{88}Ra$. The SI unit of radioactivity is the becquerel (Bq), which is defined as the amount of the radioisotope that gives one disintegration per second. Thus, $1[Bq] = 27 \times 10^{-12}$ [Ci].

Radioactivity needs to be surveyed because the passage of ionizing radiation through biological tissues causes injury, although the symptoms of injury generally become evident much later. The nature and severity of the threat posed by different kinds of particles and radiations depend on the type of particle or radiation, its energy, the duration of the exposure, the parts of the body that are exposed, and whether the exposure is internal or external. The emission energies of alpha, beta, and gamma (α, β, and γ) particles vary as well as the emission range. Beta particles of 0.5 MeV have an emission range of 1 m in air; whereas ? MeV particles have a range of 10 m. When taken up by obstacles, beta particles produce a more penetrative secondary radiation known as *bremsstrahlung*. The best protective barriers are solids of low atomic number such as aluminium, Perspex, and rubber. For example, 2.5 cm of Perspex will protect against beta emission of energy as high as 4 MeV. A beta particle with an energy of 1 MeV will penetrate only 4 mm of human skin. The body-penetrating power of alpha particles is even less. Gamma rays are more difficult to stop because they are not ionic. Their intensity falls away exponentially. Materials of high atomic energy may exhibit photoelectric absorption, the energy of gamma rays is then transferred to an electron, which is then ejected from the atomic shield. Cobalt-60, ^{60}Co, a dangerous product of nuclear fission, emits gamma rays with a mean energy of 1.26 MeV. To reduce their dose by a factor of 10, 4.6 cm of lead is needed; and for each subsequent tenfold reduction a further 4.6 cm of shielding is necessary.

Exposure to radiation is measured in rads. One rad leads to the release of 100 ergs g^{-1} (10^{-5} J g^{-1}) in body tissue. This is also equivalent to exposure to a level of one roentgen. Exposure is also measured in rems. The rem takes account of a 'quality factor', which for the more dangerous alpha particles is 10. Exposure to radiation at the level of 1 rad is rated as 10 rems. The gray (Gy) is the SI unit for absorbed dose, 1 J kg^{-1}. The SI unit for dose equivalent is the sievert (Sv), which is measured in grays times a quality factor for the type of radiation and the weighting factor for the tissue irradiated. The sievert is numerically equivalent to the gray for electrons and for X-rays irradiating the whole body.

Concern about human radiation exposure in the natural and built environment is currently focused on radon, the only radioactive gas. It is formed by the decay of uranium and thorium, and is found in the atmosphere. If inhaled, the gas can cause damage to the lungs. ^{222}Rn has a half-life of 3.82 days. In the open air, the concentration of radon is generally very low. However, in buildings constructed on or out of rocks containing high quantities of uranium, such as granite or bricks, radon levels can rise well above background levels. In the UK the National Radiological Protection Board has defined areas where radiation of over 200 Bq m^{-3} is measured in more than 1 per cent of houses as 'action areas'. In these areas, which are primarily centred around the granites of Devon and Cornwall, special measures must be taken in ventilating houses.   K. VALA RAGNARSDOTTIR

radiolaria *see* FORAMINIFERA AND OTHER UNICELLULAR MICROFOSSILS

radon Radon, atomic number 86, is a natural α-radioactive inert gas produced within the radioactive decay chains of uranium and thorium. Its three natural isotopes, ^{222}Rn, ^{220}Rn, and ^{219}Rn have respective half-lives of 3.8 days, 55 seconds, and 4 seconds. Radon abundance is measured in terms of its radioactivity in becquerels per cubic metre of radon in air, written Bq m^{-3}. ^{222}Rn, commonly referred to simply as radon, is the most discussed isotope.

Since uranium is widely present as a trace element in the ground, radon is ubiquitous in the environment. Radon behaviour has been extensively investigated. Ground-surface radon emissions have been used in uranium exploration and earthquake prediction.

In the late 1980s, high levels of radon were found in domestic houses, and this resulted in substantial surveys of domestic radon levels worldwide. The UK and world average radon exposures are respectively 20 and 42 Bq m^{-3}. Radon is known to cause lung cancer in highly exposed uranium miners. In the general population, radon has been suggested as a partial cause of lung cancer in adults, and of leukaemia in children.   DENIS L. HENSHAW

rainforests and ice ages Rainforests that are rich in plant and animal species occur today in tropical lowlands of

the Earth, up to about 1300 m above sea level. Here temperatures are high with little daily or seasonal variation and precipitation is abundant throughout the year. At higher elevations, and up to about 3000 m in the tropics, montane rainforest is found. This is less diverse taxonomically but contains some species in common with its lowland counterpart.

Until about thirty-five years ago these rainforests were thought to have remained unchanged for a long period of geological time. Their diverse flora and fauna were considered mainly a response to a stable and an equable climatic regime. In particular, the series of Quaternary ice ages, whose harsh climates have lead to marked changes in the composition and distribution of biota (animal and plant life in a region) in temperate regions over the past two million years, were regarded as having had little affect on life in low latitudes, where they caused wetter (pluvial) episodes. However, as John Flenley has stated in his seminal study of the equatorial rainforest, later geological and biological research has shown these ideas to be wrong: it has demonstrated that environments in low latitudes have undergone significant modifications. These were especially marked during Quaternary time, the cold stages of which, while giving rise to extensive glaciations in high and middle latitudes, mainly caused aridity (rather than wetness) in low latitudes. In addition it must not be overlooked that the majority of tropical mountains higher than about 3800 m above sea level were glaciated in the Quaternary. It is in these localities that the environmental perturbations brought about by the last major cold episode, whose ice disappeared at various times between about 15 000 and 9000 years before the present (BP), remain most noticeable.

Climate models and evidence from fossils indicate temperatures 2 to 4 degrees C below those of today in tropical lowlands during Quaternary glacial episodes; mountains at these latitudes experienced a reduction of between 5 and 15 degrees C. It has been predicted that mean annual precipitation amounts fell by 20 per cent during glacial cycles, and that rainfall incidence was more seasonal than at present. As Paul Colinvaux has noted though, a one-fifth reduction in annual precipitation in tropical lowland forest areas (between 2 and 7 m in Amazonia today) would still mean substantial rainfall, and render it possible that the ecological tolerances of a large number of rainforest species were not exceeded. Moreover, as Colinvaux has emphasized, arid conditions were unlikely to have occurred everywhere at low latitudes. Modifications in atmospheric and oceanic circulation patterns would have caused local effects, raising some levels of precipitation and reducing others.

Information concerning the past vegetation of current rainforest areas has come from plant microfossils (pollen and spores) and macrofossils (seeds, leaves, wood) preserved in lake sediments and peat bogs. Most data are from uplands, where climatic cooling lead to the contraction and lowering of vegetation belts, the latter by up to 1500 m. Tropical lowlands have produced fewer fossil records. These have made it possible to infer alternating episodes of forest, within cold stages of the Quaternary, savannah, and shifts in the latitudinal boundaries of these vegetation types by about 5° (see *tropical refugia*).

South America

In South America there is currently a vast expanse of rainforest, mainly in the Amazon Basin and across the Andes on their Pacific side in Columbia and Equador. Vegetation changes in this region over the past few million years have been elucidated largely by T. van der Hammen and his co-workers. A sequence of lake sediments, about 200 m in thickness, from Sabana de Bogotá, 2600 m above sea level in the Colombian Andes and now within montane forest, has produced plant remains which show that Pliocene (Late Tertiary) vegetation in its environs was akin to present-day lowland tropical forest. By the Early Quaternary, Paramo (grassland) vegetation, currently distributed above the montane (mountain) forest, had developed, denoting a cooling of climate. Subsequent changes in the pollen record illustrate shifts between grassland and montane forest as vegetation zones moved up and down in response to climatic fluctuations. Amazonia has yielded little palynological data pertinent to Quaternary vegetation history, but that which exists has been interpreted as depicting the replacement of rainforest by savannah a number of times over this timespan. However, Colinvaux has pointed out that Amazonian ecosystems are currently characterized by a high level of physical disturbance—because of flooding and burning, for example. Such factors would have been operational throughout the Quaternary, and he has suggested that the effects of ice ages have been given undue importance in the explanation of vegetation changes in this region. Pollen evidence gathered by Mark Bush and Paul Colinvaux from lowland Panama probably reflects vegetation changes over the past 150 000 years in this part of Central America. They were unable to demonstrate conclusively the replacement of rainforest by savanna during glacial episodes.

Non-biological evidence of former climatic regimes in tropical Latin America has come from lower Amazonia, north-eastern Brazil, and eastern Colombia, where tropical rainforest now grows in soil developed within unconsolidated rock material—the regolith—that was deposited under arid conditions, purportedly during one or more cold periods of the Quaternary.

Africa

African rainforest is today less diverse than that of other tropical regions. This is perhaps the result of biotic extinctions during arid episodes according to Paul Richards. Compared with other regions, it also lacks major areal extent and is moreover located in two parts, one in the west and one in the east of the continent, both with montane elements.

The Quaternary vegetational history of East Africa is best known. It has been synthesized by a major contributor to its elucidation, A. C. Hamilton. During an ice-age climate about 26 000 to 14 000 years BP, the plant cover of areas now with rainforest resembled that currently growing in areas either with lower precipitation, or with precipitation less evenly distributed seasonally. Between about 14 000 and 8000 years BP, there were botanical changes which culminated in a flora and vegetation analogous to those of the present. This flora and vegetation have remained in existence throughout the last eight millennia.

Non-biological evidence of palaeoclimates in Africa includes sand over a large part of the terrain beneath present rainforest in the Congo. This sand is thought to have been redistributed from the Kalahari Desert during colder and drier conditions in the Middle Quaternary, and to have caused a substantial reduction in forest cover.

Asia and the Pacific

Indo-Malesia is currently well endowed with tropical rainforest, within which montane types are abundant. The floristic composition is similar in Malaya, Indonesia, the Philippines, and Papua New Guinea, but differs in southern India and Sri Lanka. The tropical Pacific islands posses an impoverished flora of Malesian character. The lowering of montane vegetation belts in New Guinea during the last ice age was broadly synchronous with that of South America and Africa. The vegetational data from upland New Guinea suggest that between about 18 000 and 16 000 years BP, temperatures there were up to 11 degrees C below those of today, and that the latter were restored 9000–6000 years ago. Sites at lower elevations in Sumatra confirm these trends and imply that 11 000–8000 years BP temperatures were 2 to 4 degrees C below those of the present.

Australia

The tropical rainforest seen today in northern Australia has floristic affinity with that of South-East Asia. The greatest continuous extent of this rainforest is in north-east Queensland. At Lynch's Crater, within rainforest on the Atherton Tablelands in this region (17 °S), the palynological record from a swamp studied by Peter Kershaw has provided a vegetation history spanning approximately the past 200 000 years (Fig. 1). Over this time-span, rainforest with a complex composition has dominated the area around Lynch's Crater during interglacial stages, when moisture has been plentiful. Between about 165 000 and 126 000 years BP, Araucarian rainforest (of which the gymnosperms *Araucaria* and *Podocarpus* were significant constituents) developed in a drier climate. The palynological evidence suggests that the transition from wetter to drier conditions was abrupt. From about 63 000 to 26 000 years ago (within the last cold stage), Araucarian forest was paramount in the vegetation once again. This forest was gradually replaced by sclerophyll woodland composed of trees with leathery leaves (in which *Casuarina* and *Eucalyptus* were of prime importance) from about 38 000 years BP. Such woodland dominated until about 9000 years BP, when complex rainforest reappeared. A change to sclerophyll

Reduced rainfall in north-eastern Queensland was probably the result of diminished evaporation from the ocean consequent upon lower global temperatures and eustatic changes (in sea level) during ice ages. A fall (in sea level) associated with ice accumulation would have exposed the Sahul Shelf north-west of Australia. The moisture which currently sustains the rainforest is mainly provided by north-west winds, the efficacy of which would have been reduced as they traversed land rather than water. The same effect would have been experienced by the south-east trade winds, which contribute to the climate of the eastern seaboard of Australia.

R. L. JONES

Further reading

Flenley, J. R. (1979) *The equatorial rain forest: a geological history.* Butterworth, London.

Hamilton, A. C. (1982) *Environmental history of East Africa: a study of the Quaternary.* Academic Press, London.

Williams, M. A. J., Dunkerley, D. L., De Dekker, P., Kershaw, A. P., and Stokes, T. (1993) *Quaternary environments.* Edward Arnold, London.

rainsplash Rainsplash is a slope-erosion process induced by the impact of raindrops on the soil surface. As a raindrop hits the soil surface, it becomes compressed under its own weight and spreads laterally. This spreading causes the development of shear on the soil, and detachment of soil particles from the surface. These particles can become entrained in water droplets from the original raindrop as they rebound from the surface. The droplets and particles travel in a parabolic path, usually for a distance of less than 1 m. The travelled distance depends on the size of the soil particle and raindrop and the angle of rebound, which is related to the soil texture and moisture content. The transport paths are usually symmetrical about the original impact, so that the pattern of movement on a hill slope is diffusive, although there is a net downslope movement. Generally, rainsplash affects soil particles less than approximately 20 mm in diameter but rates are much faster for smaller particles. Fine sand-sized particles are most rapidly moved, because silt and clay particles show more cohesion. Vegetation cover is an important control, with plants close to the ground surface protecting the soil from raindrop impact. Taller vegetation canopies can enhance erosion by concentrating water on leaves and producing larger raindrops. Such drops falling from the top of forest canopies can reach terminal velocity and, therefore, impact with an energy equivalent to that of unimpeded rainfall.

Rainsplash is a relatively slow erosion process because of the low energy input from raindrops and the low competence of rebounding drops to transport soil particles. Rates of erosion are an order of magnitude lower than when run-off occurs, and will often show a decline through a storm, either because the available loose soil particles have been exhausted or consolidated, or because water ponding at the surface protects the soil from raindrop impact. However, rainsplash is an

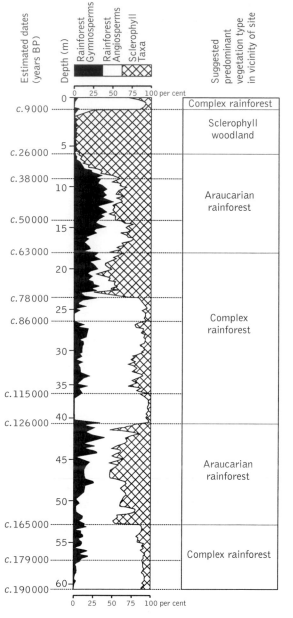

Fig. 1. Summary pollen diagram from Lynch's Crater, north-eastern Queensland, Australia. (From Kershaw (1986) *Nature*, **322**, 47–9, and Williams *et al.* (1993).)

woodland from Araucarian forest did not occur within the first period of drier conditions. This fact, together with the gradual nature of the modification, lead Kershaw to postulate that climate alone did not cause it. Fire was thought to have played an important role in the change.

important process in detaching particles which are then transported by shallow, unconcentrated surface flows, as such flows are generally not capable of detaching particles themselves. Slopes with infiltration rates too high to develop run-off may be dominated by rainsplash and will tend to develop convex profiles. JOHN WAINWRIGHT

raised shorelines Shorelines are surfaces formed at the interface between the land and a water body (lakes or coasts), and relate to the landward margin of a wide variety of geomorphological features (sand or shingle beaches, erosional platforms, estuarine mudflats, mangrove swamps). The term 'raised shoreline' has been used to describe relict shoreline features which are now located above the modern water level. Such features tend to become displaced because of global changes in sea level (eustatic) or because of tectonic movements of the crust.

Classic examples of tectonically displaced raised shorelines are to be found fringing the coasts of formally glaciated areas in Scotland, Scandinavia, and North America. They consist of a wide variety of relict marine features that occur above present sea level as a result of glacio-isostatic uplift. The altitude of the shoreline fragments usually increases as one moves towards the centre of the former ice-sheet, since the thickness of ice in these areas produced greater displacement of the crust. Other examples of tectonically displaced shorelines are found in the coral terraces of the Huon Peninsula, New Guinea.

An example of a eustatic raised shoreline fragment is the Patella raised beach on the Gower coastline of South Wales. Here shingle deposits lie beneath tills and periglacial deposits associated with the last glacial advance. The well-cemented shingle lies on a rock platform that stands between 3 and 5 m above the present beach. It is proposed that when it formed during the last interglacial (125 000 years ago) sea level was 5 m higher than today. CALLUM R. FIRTH

Further reading

Bowen, D. Q. (1973) The Quaternary deposits of the Gower, South Wales. *Proceedings of the Geologists' Association*, **84**, 249–72.

Smith, D. E. and Dawson, A. G. (1983) *Shorelines and isostasy*. Academic Press, London.

rates of geological processes Rates of geological processes (R) are often assessed through the analysis of extent of the process (P) divided by the duration of time over which that process has occurred (t):

$$R = P/t$$

For example, the rate of erosion can be estimated by measuring the depth of erosion of a hill slope, and dividing it by the length of time over which that erosional process may have operated (e.g. the length of time the hill slope has been exposed to weathering). Rates of long-term average erosion estimated using such an equation vary between

10^0 and 10^3 mm/ky. Rates of weathering of silicates estimated by such an analysis (e.g. based upon observations of depletion of feldspars or hornblendes as a function of age of a soil) vary between 10^{-20} and 10^{-17} mol cm^{-2} s^{-1}, where rates are normalized per surface area of mineral. Such a rate estimate yields a *time-integrated* estimate, which may be much slower than the *instantaneous* rate, since many geological processes are intermittent, rather than constant in nature. Thus, rates for a process when calculated for a short time interval are generally higher than rates calculated over a long time interval.

Alternatively, where flux (Q) cannot be measured directly, but gradients in the driving potential $\Delta z/\Delta x$ can be measured, the rate can often be calculated from a simple phenomenological equation, if the constant k can be measured:

$$Q = -k\, \Delta z/\Delta x$$

For example, the rate of conductive heat loss at any location can be calculated by simultaneously measuring the thermal conductivity (k) and the gradient in temperature with respect to depth in a borehole ($\Delta z/\Delta x$). Thus, the areally weighted global mean conductive rate of heat loss is estimated on the order of 90 mW m^{-2}, on the basis of compiled point measurements of thermal conductivity and temperature gradient worldwide.

In geochemical systems operating under ambient conditions where the flux of some chemical component can be directly measured, a formula such as the following can be used to assess kinetics:

$$R = C\, Q$$

where C is the concentration of some element (e.g. Si released during the weathering of feldspar) and Q is the flux of water out of the system (e.g. the flux of water containing C moles of sodium per litre). Such a rate estimate yields a value that is time-integrated (over the period of measurement of C and Q) and area integrated (over the soil or watershed). Often, for rates of weathering, the flux is normalized by the area of the watershed or the surface area of the reacting minerals. For example, rates of weathering of feldspars and hornblende in soils vary between 10^{-19} and 10^{-16} mol cm^{-2} s^{-1} where the rates are normalized by mineral surface area. Laboratory measurements of dissolution of similar minerals under similar conditions characteristically reveal significantly faster rates of dissolution: 10^{-17} and 10^{-15} mol cm^{-2} s^{-1}.

Comparison of the rates of geological processes calculated using different models, and integrated over varying temporal and areal extents for different locations, are often discrepant. However, relative rates are often more well-behaved. For example, the relative rates of dissolution of silicate minerals in the laboratory under ambient conditions are generally similar to the relative rates of dissolution measured in the field for comparable chemical systems. The relative rates of dissolution of silicate minerals roughly follow the Goldich weathering series, in which the rates of dissolution increase with the decreasing connectivity or Si/Al ratio of the silicate: quartz < albite < amphibole < pyroxene < olivine. Laboratory and field studies are also in agreement concerning the fast rates of dissolution of evaporite minerals, including calcite.

The measurement and understanding of absolute and relative rates of geological processes are an important area of active research.

S. L. BRANTLEY

Further reading

Berner, E. K. and Berner, R. A. (1987) *The global water cycle: geochemistry and environment.* Prentice-Hall, Englewood Cliffs, New Jersey.

Gardner, T. W., Jorgensen, D. W., Shuman, C., and Lemieux, C. R. (1987) Geomorphic and tectonic process rates; effects of measured time interval. *Geology*, **15**, 259–61.

Pollack, H. N., Hurter, S. J., and Johnson, J. R. (1993) Heat flow from the Earth's interior: Analysis of the global data set. *Reviews of Geophysics*, **31**, 267–80.

Summerfield, M. (1991) *Global geomorphology: An introduction to the study of landforms.* Longman, Harlow.

White, A. F. and Brantley, S. L. (eds) (1995) Chemical weathering rates of silicate minerals. *Mineralogical Society of America Short Course*, v. 31.

rates and trends in evolution

Evolution is change through time, and hence it is convenient to divide its patterns into rate and direction of change. Rate is defined as the amount of change over a period of time. In graphical terms, this corresponds to the slope of a line tracing the phenomenon through time. If the line is straight, the rate is constant; if the line is curved, then the rate is changing. Direction is the tendency for a phenomenon to increase or decrease persistently over a period of time. A common word for such a tendency is 'trend'.

Evidence from many fossil studies has given rise to a major debate over whether evolutionary rates are generally rapid or gradual. Darwin originally argued for 'gradualistic evolution'. Work by Stephen J. Gould of Harvard and Niles Eldredge of the American Museum of Natural History indicates, however, that a different process, termed 'punctuated equilibrium', is more common. This process is one in which very little change takes place during much of a species' existence, but rapid brief bursts of change punctuate this stasis or equilibrium. It is thought that the change is forced by rapid evolution in small isolated populations during periods in which their range has been reduced by changing environmental conditions. A survey of fossil lineages in which rates of evolution have been studied shows punctuated rather than gradualistic patterns in the majority of cases. There are, however, problems that impede a clear interpretation of the data. In particular, there are gaps in the fossil record and there is disagreement over what is meant by 'rapid'. As the fossil record is very incomplete, we do not have a continuous record of any species. If intermediate forms are missing, a gradual rate of evolution may appear to be punctuated. This is exacerbated by the difficulty of precisely dating sedimentary rocks: we cannot tell the length of the gaps. If they were short, then it could be assumed that evolution really was rapid. Rapid evolution can be shown in some modern examples, as in Death Valley, where small lakes that have been separated for only 20 000 years now harbour different species of fish. It is also clear that evolutionary rates can be very low in some lineages, for it is possible to find modern forms, termed 'living fossils', that differ little from fossil forms. Sequoia trees have changed little in the past 70 million years, and the inarticulate brachiopod *Lingula* has changed little in over 350 million years.

Trends can be seen either as lineage trends in evolving lineages, or biosphere trends affecting the evolution of life as a whole. Lineage trends show directional tendencies in various traits and organs that change through time, and result from consistent directional selection. Thus, the evolution of the horse shows correlated changes in traits resulting in increases in tooth and body size, and reduction in the number of toes from four to one that occurs over a period of 60 million years. This is apparently due to the adaptation of the horse to the spreading grasslands of the Cenozoic. Biosphere trends can be seen in the increasing diversity and size of organisms through time, which indicates a steady increase in their complexity.

DAVID K. ELLIOTT

Recent *see* HOLOCENE/RECENT

recent climate changes

A climatic trend can be distinguished from a fluctuation, since trends are unidirectional and usually have a minimum duration, such as twenty years. The recent past is usually regarded as being within the period of instrumental observations, from which detailed information about climatic trends may be derived.

Long-term instrumental records of 200–300 years' duration are mostly restricted to Europe. Knowledge of the climate of the eighteenth century, when pioneer instrumental observations were being introduced, can be augmented by scrutiny of surviving weather diaries. For Britain, a data-series giving mean monthly temperatures for central England back to 1659 was calculated by Professor Gordon Manley in the 1970s. The series starts with several decades of warming associated with the end of the Little Ice Age, a trend that has resumed since the 1970s.

The construction of an aerial data-set as the central England temperature (CET) series involves reconciling an uneven spatial and temporal distribution of observations with the additional value of a long-period series. At the global scale, this is resolved by calculating temperatures at a regular pattern of grid-points. This requires the analysis of temperature observations made by shipping, from which sea surface temperature estimates can be derived, It is essential to make careful allowance for changes in observational practice over time, to avoid introducing non-climatic 'noise' into the series.

Global temperatures show an increasing trend from the mid-nineteenth century; the warmest year on record from

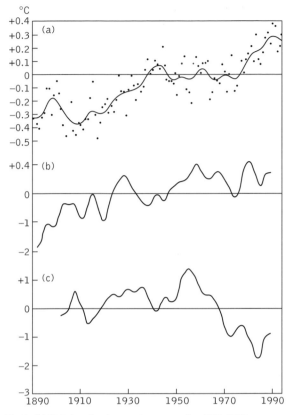

Fig. 1. (a) Global surface temperature anomalies 1890–1994, expressed as deviations from the 1951–80 average with a 21-point binomial filter; (b) regional annual precipitation anomalies for the former USSR; (c) regional annual precipitation anomalies for the African Sahel. Units for (b) and (c): standard deviations from the mean of the whole period. After Folland *et al.* (1990).

1854 to 1999 was 1998 (see Fig. 1 and *global warming*). This was punctuated by cooling in the 1950s and 1960s (mostly in the northern hemisphere), but warming has resumed at an unprecedented rate since the late 1970s in most (but not all) parts of the world. While this period has witnessed an undoubted strengthening of the greenhouse effect, the 0.5 °C warming observed over the past hundred years is regarded as being just within the range of natural variation. A small part of this warming (estimated at 0.05 °C) has been attributed to an increased urban warming effect, but it is thought that the urban effect on the CET since the 1940s is roughly 0.1 °C.

An important diverging trend since the 1960s is the cooling of the North Atlantic and adjacent coastal land masses, such as Greenland, Iceland, and Scandinavia. This has resulted in increased temperature gradients between 50° and 65°N, linked with a recent upturn in the vigour of the westerly airstream over Europe, especially in winter. This has amplified the warming of winters in parts of Europe by encouraging the advection of mild, maritime air.

Variations in solar output have the potential to exert a more widespread influence upon temperature. While the Little Ice Age of the sixteenth to eighteenth centuries coincided with a reduction in solar activity (as revealed by the Maunder Minimum of sunspot observations), it is thought that any future diminution of solar irradiance would be insignificant in comparison with the radiative effects of the enhanced greenhouse effect.

It has been noted that night minimum temperatures are rising faster than day maximum temperatures, especially in the northern hemisphere. This is believed to be caused by the cooling effect of the backscattering of solar radiation in the troposphere by sulphate aerosols emitted from industrial sources. It is possible that this sulphate aerosol cooling has offset up to one-third of the greenhouse warming. Consequently, action to control sulphur emissions (in combating acid rain) might inadvertently hasten further global warming.

Sulphate particles can also affect temperature trends in the stratosphere as a result of volcanic eruptions. Major eruptions—such as El Chichón in Mexico in 1982 and Mount Pinatubo in the Philippines in 1991—can produce a sulphur-rich stratospheric dust cloud that can affect global temperatures for one to two years following an eruption. This can lead to a temporary reversal of warming, as in 1992 and 1993. Conversely, the periodic El Niño events of the equatorial Pacific can provide temporary warming. This arises from the weakening of the easterly trade winds, interrupting the upwelling of relatively cool, nutrient-rich water off the South American coast. Anomalous low pressure over South America (giving floods) and high pressure over the western Pacific can initiate drought in Australia and South-East Asia (the Southern Oscillation). An unusually persistent El Niño event began in 1991, raising speculation of a relationship with global warming.

Trends in rainfall generally have greater spatial and temporal variability than those of temperature. Since the end of the mid-twentieth century, rainfall over the low latitudes of the northern hemisphere has declined, most notably in the Sahel region of Africa. By contrast, high-latitude rainfall has increased (see Fig. 1). In the mid-high latitudes, rainfall is influenced by the vigour of the westerly circulation and depression tracks. Increased rainfall over north-west Britain since the late 1970s, especially in winter, is in marked contrast to a reduction in summer rainfall that has affected much of Europe over the same period. These changes are linked to a northerly shift of depression tracks, believed to be in response to warming, illustrating the links between temperature, rainfall, and the atmospheric circulation.   JULIAN MAYES

Further reading

Bradley, R. S. and Jones, P. D. (eds.)(1995) *Climate since AD 1500.* Routledge, London.

Folland, C. K., Karl, T., and Vinnikov, K. Ya (1990) Observed climate variations and change. In Houghton, J. T., Jenkins, G. J., and Ephraums J.J. (eds), *Climate change: the IPCC scientific assessment*. Cambridge University Press.

Hulme, M. (1994) Historic records and recent climatic change. In Roberts, N. *The changing global environment*. Blackwell, Oxford, pp. 69–98.

Marsh, T. J., Monkhouse, R. A., Arnell, N. W., Lees, M. L., and Reynard, N. S. (1994) *The 1988–92 drought*. Institute of Hydrology/British Geological Survey, Wallingford.

recharge and the hydrological cycle

Recharge is the process by which atmospheric water is added to groundwater in the hydrological cycle. The process entails the addition of water to the land surface by rainfall, melting of snow, or overland flow, followed by infiltration. The rate of recharge can be much smaller than the rate of infiltration because evaporation from shallow pores in the soil or transpiration by plants can return water to the atmosphere before it makes its way to the water-table. Recharge is one of the most difficult components of the hydrological cycle to quantify. It is often estimated indirectly as a water balance residual by subtracting estimated rates of runoff and evapotranspiration from precipitation. In regions of humid climate, annual recharge estimated by this method can equal or exceed 30 per cent of annual precipitation. In arid regions, where potential evapotranspiration can exceed annual precipitation, the estimated recharge is often negligible. Tracer studies have demonstrated that such estimates for arid regions are subject to considerable error and that the actual recharge, while generally low, can be highly variable and locally significant.　　　J. BAHR

recycling

As the world's resources diminished, a new policy of reusing produce made of primary raw materials was adopted in the Western world in the last quarter of the twentieth century. The policy is now to recycle and reuse waste to the maximum possible extent, to reduce the quantity of unrecoverable waste, and to dispose of as safely as possible any unrecoverable waste. The composition of Western domestic waste is 20–45 per cent paper and board, 15–30 per cent kitchen (animal and vegetable) waste, 6–13 per cent glass, 3–8 per cent plastics, 4–11 per cent metals, 3–5 per cent textiles, and 3–5 per cent dust and ashes. In Europe, the recycling policy is most advanced in countries such as the Netherlands and Switzerland, which have little disposable land for landfilling. The European Union Directive on Recycling dictates that by 2001 member countries should recover at least 50 per cent of their waste and recycle at least 25 per cent. According to 1995 figures the European glass recycling rate is highest in Switzerland (85 per cent), the Netherlands (80 per cent), and Germany and Norway (both 75 per cent). The UK is at the bottom of the list of EU countries, recycling only 27 per cent of glass. For steel packaging (1995 figures), Germany has the highest recycling rate in the EU (67 per cent), closely followed by the Netherlands (58 per cent). The UK, on the other hand, recycles only 16 per cent, of steel packaging. The end use of recovered materials is very varied. Paper and board are recycled to produce packaging materials, tissue, toilet paper, printing paper, and writing paper. Glass is recycled to make bottles, flooring materials, and materials for road-filling and road-surfacing. Metals are recycled according to their composition. Iron-based tins are taken to steelworks, and aluminium cans and foil are a primary aluminium substitute. Plastics (PVC: polyvinyl chloride) are recycled to make PVC tubes and pipes as well as fleece fabric. Textile waste is used to make wiping cloths and, more recently, fashion clothes. Note that the recycling of glass and aluminium saves energy. More and more countries now have special arrangements for the collection of wastes; consumers then have to sort their waste into kitchen waste, glass, metals, plastics, paper, and textiles. This practice will increase in the future as the Earth's resources continue to diminish. Some EU countries have stringent regulations for the reuse of packaging materials to minimize waste. For example, Denmark has banned all metal-based drinking cans; drinks, from beer to soft drinks, are sold in bottles, which are returned to the factories and reused. Similarly, Germany has a policy for the reuse of glass used for packaging milk, yoghurt, juice, beer, mineral water, and other common commodities.

K. VALA RAGNARSDOTTIR

recycling, crustal
see CRUSTAL COMPOSITION AND RECYCLING

redox equilibria

Redox is short for *red*uction–*ox*idation. Initially oxidation meant bonding with oxygen and reduction meant parting with it, but later the definitions were broadened to mean, respectively, donating and accepting bonding electrons. *Redox equilibrium* is a state in which such a reaction becomes macroscopically stationary. If chemical energy is at a minimum, the equilibrium is *stable*; if chemical energy is not at a minimum but the reaction is blocked by an energy barrier, the equilibrium is *metastable*.

Redox stratification of the Earth

Redox equilibria played a vital role in the origin, evolution, and characterization of the Earth. Iron and oxygen are the most abundant elements in the Earth, each making up about one-third of the total mass. The Earth's redox history evolved around how the two elements combined or parted. It is a generally accepted view that the hydrogen-dominated Solar Nebula condensed to reduced dust particles that included metallic iron, magnesium-rich silicates, and reduced forms of carbon, and that these particles gravitationally attracted each other and grew (accreted) into rocky proto-planets. Thus the proto-Earth probably had an interior in which metallic iron was in redox equilibrium with other minerals on a local scale. Geophysical evidence indicates that the present Earth is redox-stratified (that is, it has a solid iron–nickel inner core, a molten outer core of iron with some light element(s), a magnesium-rich silicate mantle containing ferrous iron, and a silica-rich crust containing ferrous and ferric iron). Nearly three-quarters of the iron exists as metal in the core, and most of the oxygen exists in the mantle and crust. Both the Earth and the Moon were probably melted in the late stage of (or soon after) accretion. It is now increasingly accepted that a giant impact of another planetary body split the proto-Earth into the present Earth and Moon, tilted the Earth's rotational axis, and melted the planets. A very rapid accretion, which resulted in trapping a substantial fraction of the heat of accretion, could also have melted the planets. The redox stratification probably represented an approach, facilitated by global melting, to the redox equilibrium in a planetary-size continuous system with pressure, adiabatic temperature, and gravitational potential gradients. The gradients were created by the growing mass and, hence increasing gravity of Earth. In the Earth, a lower energy state (i.e. approaching equilibrium) was achieved by releasing gravitational potential energy (by moving iron down and oxygen up) rather than by releasing chemical energy by bonding iron and oxygen.

Redox state of the mantle

A convenient thermodynamic parameter for studying redox relations of rocks is *oxygen fugacity* (abbreviated fO_2). Fugacity, a thermodynamic term introduced by G. N. Lewis, expresses the escaping tendency of a chemical species in any state of aggregation (solid, liquid, or gas). The condition for chemical equilibrium in a *uniform system* (i.e. one with no gradient) is that the fugacity of any component must be identical in all phases (minerals, melt, gas). The local environment in which a rock formed or re-equilibrated determines the intrinsic fO_2 of natural samples. In the 1960s the dominant view held by petrologists was that the redox state of pristine basaltic magma reflected that of the source region (i.e. the upper mantle) because magma behaved as a closed system until it was extruded. The fO_2 of basalt, deduced from mineralogical studies of magnetite (Fe_3O_4)–ilmenite ($FeTiO_3$) pairs, direct measurements in volcanic gases, and the ferric/ferrous ratios of volcanic rocks, happened to be close to that of a mineral mixture called the fayalite (Fe_2SiO_4)–magnetite (Fe_3O_4)–quartz (SiO_2) fO_2 buffer (abbreviated as FMQ). More recent data obtained by methods that are independent of magnetite or volcanic gas, however, point to a more reduced mantle. A view that is gaining acceptance is that the fO_2 of the mantle is governed by the redox equilibria of the carbon (C)–oxidized carbon buffer (CCO). Relevant oxidized carbon species are carbon dioxide (CO_2) and carbon monoxide (CO); the latter is negligible at mantle pressures. Having neither iron nor magnetite that would buffer the fO_2 with ferrous silicates, and having low ferric/ferrous ratios (0.05–0.1), the iron–oxygen system of the upper mantle rocks has little ability to buffer the fO_2 even though the total of iron oxides is nearly 10 weight per cent (Fig. 1). Carbon has a large redox capacity per unit mass (it can reduce 18.6 times its weight of Fe^{3+}). Carbon was trapped in the mantle either as graphite or diamond when the mantle solidified; carbon-

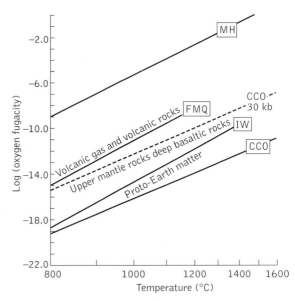

Fig. 1. Plot showing how the logarithm of oxygen fugacities of some standard redox buffers changes as the temperature changes at atmospheric pressure (bold solid lines). When pressure increases, these values all increase at relatively small rates except for the gaseous CCO buffer, which shows a roughly exponential rate of change with pressure (the dotted line shows its log fO_2 at 30 kilobars, equivalent to about 100 km depth). NB. The temperature axis (degrees Celsius) is spaced in linear proportion to the reciprocal absolute temperature so that the plots become linear. The notations show approximate fO_2–temperature ranges of various Earth rocks. The labels in boxes specify the redox buffers as follows: CCO, carbon–carbon oxides; IW, iron–wüstite (FeO); FMQ, fayalite–magnetite–quartz (see text); and MH, magnetite–haematite.

bearing minerals are found in fragments of rocks brought up from the mantle in CO_2^- driven explosive pipes (kimberlite pipes). The fO_2 of the CCO reaction becomes higher with increasing pressure and lower with increasing temperature relative to those of iron–oxygen reactions. In the early Earth the counterbalance of the pressure and temperature effects was such that, when the mantle was molten and CO_2 could ascend in accordance with the barometric principle, diamond could reduce FeO (ferrous oxide) to metallic iron and generate CO_2, in the core region, whereas CO_2 could oxidize metallic iron and precipitate carbon in the upper mantle. The carbon–oxygen system thus probably provided a chemical mechanism for transporting oxygen upward and establishing a redox gradient in a mantle that was increasingly oxidizing outward.

Oxygen fugacity of magma

Basaltic magma originates by the partial melting of the upper mantle. As the magma crystallizes in the crust, the iron concentration in the remaining melt may increase. In 1955

G. C. Kennedy proposed that iron-enrichment occurs in basaltic magmas with relatively low fO_2. Layered rocks crystallized in the deep crust from a water-poor magma (tholeiite basalt) typically show the iron-enrichment trend. They are known to contain graphite of non-biogenic isotopic signature and compressed CO_2 fluid inclusions, they generally have low ferric/ferrous ratios, and they show fO_2 close to that of carbon oxidation at the pressure and temperature of crystallization. Rocks crystallized from water-rich magma (e.g. alkali basalt, calc-alkaline basalt) do not show a clear iron-enrichment trend, do not contain graphite, generally have higher ferric/ferrous ratios, and show fO_2 higher than that of the FMQ buffer. Lunar basalt was water-free, was very reduced, and followed an extreme iron-enrichment trend; it crystallized out metallic iron upon extrusion (probably as a result of reduction by carbon under low pressure).

Water by itself is not oxidizing in a closed system. Except for its lower degree of partial melting (hence, a higher content of volatile matter), alkali basalt is thought to have an origin similar to that of tholeiite. The escape of hydrogen (H_2), the lightest and the most mobile gas, explains the connection between the extent of oxidation of basaltic magma and its initial water content. In magma, H_2O first reacts with carbon, producing H_2, CO_2, and CO, and then with ferrous iron, producing H_2 and ferric iron, if H_2 escapes. The higher the initial water content, the higher are the H_2O and H_2 fugacities, favouring loss of H_2 and hence oxidation of magma. Magnetite crystallizes out of magma when fO_2 becomes a little higher (up to about 1 logarithmic unit) than that of the FMQ buffer. Compared with water-poor magmas, water-rich magmas have greater H_2 loss, and thus are oxidized sooner; magnetite therefore precipitates earlier, depleting the magma of iron. Crystallization of magnetite also buffers the fO_2 increase, because it removes twice as much ferric ion as ferrous ion from the magma. Other magmatic factors also begin to curb the rate at which fO_2 increases. The higher the fO_2, the lower is the fH_2, slowing down the H_2 loss. The change in the logarithm of magmatic fO_2 is similar to that of the logarithm of the Fe^{3+}/Fe^{2+} ratio; so at a steady oxidation rate the fO_2 increase slows down as the ratio approaches unity, which occurs near the fO_2 of magnetite crystallization. The fugacity of sulphur dioxide (SO_2) becomes high enough for the gas to start escaping. The loss of SO_2 counters the loss of H_2 because sulphur exists in magma dominantly in the -2 valent state; the loss of S^{2-} and $2O^{2-}$ ions as SO_2 requires reduction of $6Fe^{3+}$ to $6Fe^{2+}$ to maintain electrical neutrality. If they are not oxidized further by air, many basaltic volcanic rocks end up having fO_2 in this self-regulatory range. The sequence of degassing outlined above indicates that when magma intrudes into the upper crust either H_2 or CO_2 should be the first component to increase in the volcanic gas output. By the time SO_2 increases, the magma has either already started to extrude or is ready to do so. Continuous monitoring of these gases at volcanoes might help to predict imminent volcanic eruptions.

Early atmosphere and iron sediments

The early atmosphere of the Earth is believed to have resembled volcanic gas minus the part condensed as, or dissolved in, the ocean. It is probable that the atmosphere was largely devoid of free oxygen and high in CO_2, and that rainwater was non-oxidizing and moderately acidic. Rock weathering must have been very different in the early Precambrian because iron is soluble in such water and can easily be transported. The oldest (2.2 Ga; 2.2×10^9 years) continental red sediments that contain *haematite* (Fe_2O_3) indicate the time of change to an oxidizing atmosphere and no more transport of iron to the sea. Iron-rich sediments of Early to Middle Precambrian age, called *banded iron formations* (which are major resources of iron today), consist of rhythmically banded iron minerals (oxides, carbonate, and silicates) and chert (fine silica rock), and are believed to have been chemically deposited in shallow sedimentary basins. The largest banded-iron formations are mostly older than 2.2 Ga. Whether haematite and magnetite (or their hydrated equivalents) were of primary sedimentary origin and, if so, how they could be precipitated when free oxygen was presumably scarce, are being debated. Magnetite can be formed inorganically without free oxygen by means of spontaneous decomposition (entailing loss of H_2) of unstable ferrous hydroxide, if the latter is precipitated by neutralization with slightly alkaline sea water. Haematite cannot be made this way. It has been suggested in support of a biological input that oxygen was generated only in specific shallow-water basins by localized, slowly evolving photosynthetic organisms, and abundant ferrous iron in sea water consumed all the oxygen that was generated.

Life and redox catalysis

At the temperatures prevailing in the near-surface environment, some inorganic redox reactions are so slow that stable redox equilibria cannot be achieved, even in geological time. The emergence of life, which was perhaps earlier than 3.5 Ga, depended on the persistence of metastable redox conditions. Except for photosynthetic types, organisms may be regarded as self-replicating catalysts for reactions between metastably coexisting redox pairs. Some microbiologists believe that the most primitive organisms were of the sulphur-metabolic type. Below about 200 °C, the sulphate ion (SO_4^{2-}) is not readily reduced inorganically in aqueous solution and can coexist metastably for eons with hydrogen, hydrogen sulphide, ammonia, and other abiogenic reducing matter. The sulphur-metabolic organisms are thought to have catalysed the redox reaction by providing an alternative reaction path, thus obtaining life-sustaining energy. The earliest photosynthetic organisms might also have used similar sulphur-based mechanisms, but eventually oxygen-based types dominated. Oxic photosynthesis is a process of chemically storing the solar energy in the form of metastable coexisting biological matter and free oxygen. This is possible because O_2 is not a quick oxidizing agent at the Earth's surface temperature; its double bond, O=O, requires a high activation energy to break apart.

Consequently, air cannot quickly oxidize reducing matter that we call food or fuel, even in a moist environment. Aerobes and animals exploit this metastability and extract metabolic energy by catalysing the redox reaction. It is estimated that only 0.1 per cent of biological matter escapes near-surface biotic oxidation, but the total accumulation over geological time is substantial. Fossil fuels (e.g. petroleum, coal, natural gas) form from this buried fraction (mainly oxidation-resistant lipids, lignin, humic matter, and kerogen). *Fossil fuel maturation* is a spontaneous, slow, and tenuous process of removing oxygen as stable H_2O and CO_2 from buried biological matter, ultimately forming methane and graphite.

Near-surface redox reactions

The evolution of the oxygen-rich atmosphere, upon proliferation of photosynthetic organisms, changed the near-surface redox processes completely (Fig. 2). Rocks and minerals formed at depth, and erupted volcanic rocks are far removed from redox equilibrium with the present atmosphere, and redox reactions take place constantly. Ferric oxides (haematite, goethite) form by the oxidation of ferrous silicates, carbonates, magnetite, and iron sulphides in aerated environments and, being nearly insoluble, tend to remain in place. The oxidation of pyrite and marcasite (both FeS_2) creates excess sulphuric acid which facilitates dissolution and transport of various oxidized heavy metals, including toxic cadmium, mercury, and arsenic. These disulphides are very common in metal- and coal-mine waste, and uncontrolled water drainage in such areas can consequently severely harm life. There are, however, two sides of the coin. Many metal sulphide deposits, particularly those in arid to semi-arid regions, are enriched in copper and silver near the ground-water table by the process of *supergene enrichment*, which converts lean primary deposits into mineable resources.

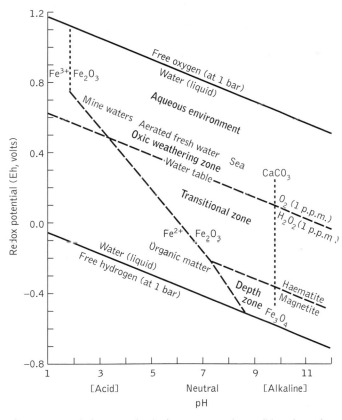

Fig. 2. Redox potential (Eh, or ORP) versus acidity (pH) diagram showing how aqueous redox conditions change from the subaerial zone (O_2-rich) to the depth zone (O_2- absent). Magnetite (Fe_3O_4), ferrous silicates, and metal sulphides, which are common in rocks and mineral deposits of volcanic origin, are not stable in the presence of trace-free oxygen and are readily oxidized to haematite (Fe_2O_3), which imparts a characteristic red colour. The activity of dissolved iron (Fe^{2+} or Fe^{3+}) delineating the haematite and magnetite fields on the low pH side is taken as 10^{-6} molal. The maximum pH is taken as that of calcite ($CaCO_3$) in pure water. Biogenic organic matter and iron sulphides, common in sediments, also consume free oxygen and produce carbon dioxide, water, and sulphate ion (SO_4^{2+}).

Rainwater percolating through oxidized zones of such deposits is acidified by the excess acid, dissolves oxidized copper- and silver-bearing minerals, and transports the metals downwards. When the solution descends below the water table, where free oxygen becomes scarce and acidity decreases, electropositive (i.e. less oxidizable than H_2) copper and silver replace iron and other electronegative metals in primary sulphides. Sometimes metallic silver and copper are precipitated, possibly because of the presence of reducing matter or electrochemical redox cell effects. Toxic metals could also be precipitated similarly under such low acidity and reducing conditions. There is much evidence that the near-surface oxidation and replacement of metal sulphides occurs in conditions more oxidizing than those indicated by stable redox equilibria because of the reaction-path-controlled metastability of metal sulphides. Research on actual reaction mechanisms is thus important for the understanding of near-surface redox processes.

Redox mixing

In the deeper parts of the continental crust, along convergent plate boundaries, and around mid-oceanic ridges and hotspots, reduced magmatic matter, oxidized near-surface matter, and reducing biogenic substances mix and react in complex ways. Redox equilibria of diverse types are involved in processes such as the alteration of sediments by magmatic heat and volatile material, the hydration of rocks of magmatic origin, the deposition and alteration of manganese deposits, the near-surface migration of uranium and vanadium, and the precipitation of metal sulphides in volcanic areas. There is much to learn about the redox equilibria of our planet.

M. SATO

Further reading

Cloud, P. (1988) *Oasis in space: Earth history from the beginning.* W. W. Norton, New York.

Smith, P. J. (ed.) (1986) *The Earth.* Equinox, Oxford and Macmillan, New York.

reefs *see* CORAL REEFS

regional metamorphism
In the most widespread examples of metamorphism, metamorphic rocks occur through an extensive terrane, and the metamorphism is not linked to a specific focus such as an igneous intrusion. Instead it reflects elevated temperature and pressure values resulting from burial within the crust. Regional metamorphism is almost invariably accompanied by penetrative deformation, giving rise to folding and the development of pervasive fabrics: *cleavage* at very low grades where the aligned phyllosilicate grains are too small to be individually visible; *schistosity* at higher grades where the grains are coarser. It is now recognized that virtually all but the lowest-grade regionally metamorphosed terrains have experienced multiple events of folding and fabric development. Petrographic analysis of rock fabrics in thin section makes it possible to determine the relative timing of the growth of metamorphic minerals and the development of structural fabrics. Modern work in metamorphic petrology uses such information in two ways: as a basis for applying single crystal dating techniques to date specific episodes in the metamorphic evolution of a terrain, and as a way of relating particular deformation events to a particular tectonic setting through knowledge of pressure–temperature (P–T) conditions, and hence the depth and the heat-flow regime, when the appropriate minerals equilibrated (see also *metamorphism, metamorphic facies, and metamorphic rocks*).

Some very low-grade metamorphic belts have experienced *burial metamorphism*, a form of regional metamorphism in which new mineral growth is accompanied only by open folding, and pervasive fabrics are not developed. Well-preserved fossils occur in the zeolite facies Triassic rocks of Southland, New Zealand, where Coombs first described the phenomenon in the 1950s. The temperatures and pressures during burial metamorphism are more typical of those in deep sedimentary basins, and the distinction between diagenesis and burial metamorphism is primarily in the rock-types affected. Under the same conditions, sediments composed mostly of volcanic material are much more reactive than normal continental clastic sediments rich in quartz. The changes in siliceous sediments will be confined to the cement, which will recrystallize, but the igneous components in the sediments of volcanic origin will break down and new hydrous phases, such as zeolites, will develop.

Metamorphic facies series. It was inherent in P. E. Eskola's original definition of metamorphic facies (see *metamorphism, metamorphic facies and metamorphic rocks*) that mineral assemblages could be indicative of different pressures and temperatures of metamorphism, but the implications of this for the recognition of thermal gradients in the past were fully appreciated only by Akiho Miyashiro in 1961. He recognized that many metamorphic belts of different ages around the world exhibited one or other of the following sequences of metamorphic grade in going from low to high grade:

prehnite–pumpellyite → blueschist → greenschist or amphibolite

greenschist → amphibolite → granulite

Miyashiro used the term *facies series* for these characteristic associations of facies and pointed out that they distinguished contrasting baric (pressure-) types of metamorphism; in the first instance high pressures were attained at relatively low temperatures, indicative of a low heat-flow setting, whereas in the second instance temperatures rose rapidly at shallow levels, indicating high heat flow. The classical metamorphic zones known from the work of George Barrow are representative of an intermediate pressure association corresponding approximately to a normal crustal thermal gradient. Miyashiro also believed that belts of contrasting facies series and of the same age commonly occurred together. He called this association *paired metamorphic belts*. With the advent of plate-tectonic theory, Miyashiro was quick to recognize the tectonic

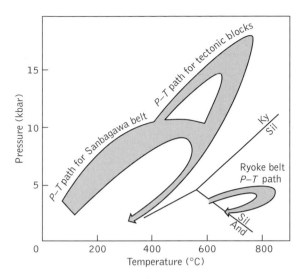

Fig 1. Pressure–temperature diagram showing the pressure–temperature–time paths followed by the rocks of the Sanbagawa (high-*P*) and Ryoke (low-*P*) belts of Japan. *Ky*, kyanite; *Sil*, sillimanite; And, (From Brown, M. in Treloar and O'Brien (1998), pp. 137–69.)

significance of the different facies series. The low heat flow associated with high-pressure metamorphism and the development of blueschists is indicative of a subduction zone setting; the elevated heat flow and associated magmatism associated with low-pressure metamorphism is usually indicative of the roots of a volcanic arc.

Modern work, while building on Miyashiro's recognition of distinct facies series, has emphasized the distinction between the apparent progression of metamorphic conditions that is deduced from studying the conditions of formation of a sequence of zones across a metamorphic belt, and the actual history of changing pressure and temperature experienced by a particular rock. The trend of *P*–*T* conditions indicated by the peak metamorphic conditions of each zone is termed the *metamorphic field gradient*, while the history of an individual rock is known as the *P*–*T*–*t* (pressure–temperature–time) *path*. Perhaps the classic example of paired metamorphic belts, described by Miyashiro, comes from south-western Japan, where the late Mesozoic high-pressure Sanbagawa belt is flanked to the north-west by the low-pressure Ryoke belt, of similar age. Miyashiro's original interpretation of these as representing subduction zone and arc root settings respectively interprets them as a snapshot in time in the development of the Pacific margin, whereas more recent work recognizes the rocks of each as having experienced a prolonged metamorphic evolution. Figure 1 contrasts the *P*–*T*–*t* paths for the Sanbagawa and Ryoke belts, showing their complex individual histories prior to their final juxtaposition, which undoubtedly involved a considerable strike-slip movement component. The

metamorphic field gradients for each belt are a very feeble reflection of this complex history. However, it is still true that ancient metamorphic belts with complex histories involving the juxtaposition of multiple terranes may nevertheless have individual elements with distinctive histories pointing to a particular tectonic setting for their metamorphism.

BRUCE W. D. YARDLEY

Further reading

England, P. C. and Richardson, S. W. (1977) The influence of erosion on the mineral facies of rocks from different environments. *Journal of the Geological Society*, **134**, 201–13.

Miyashiro, A. (1961) Evolution of metamorphic belts. *Journal of Petrology*, **2**, 277–311.

Miyashiro, A. (1972) Pressure and temperature conditions and tectonic significance of regional and ocean floor metamorphism *Tectonophysics*, **13**, 141–59.

Treloar, P. J. and O'Brien, P. J. (eds) (1998) *What drives metamorphism and metamorphic reactions?* Geological Society Special Publication No. **138**.

regression and transgression The most obvious geomorphological boundary on the Earth's surface is the shoreline, a line of demarcation that divides land from water, be it of the sea, of a lake, or a lagoon. A shoreline is thus a thin line within a coast which embraces a broad zone that reaches from the landward limit of marine influence to the seaward limit of alluvial and shoreline processes. Shorelines may be indicated by pebbly or sandy beaches, mudflats, rocky cliffs, or reefs.

To geologists shorelines have a particular importance because they are relatively easy to identify in rock successions: they coincide, not only with changes from strata with dominantly marine biota to strata with freshwater biota, or none, but also with many unconformities, and their migration demonstrates the clearest evidence of sea-level changes.

Seaward migrations of the shoreline are known as *regressions*. Landward migrations of the shoreline are known as *transgressions*. The causes of such shoreline migrations are manifold. Essentially they are the result of the interaction between accommodation space, that is the space available for potential sediment to accumulate, and sediment supply. When sediment supply exceeds the space available for it to accumulate, then regression occurs. When the space available for sediment accumulation exceeds sediment supply, then transgressions occur. Sediment is supplied to the shoreline by rivers, mainly via deltas, by longshore and coastal current drift, and by biogenic, mainly carbonate production. Accommodation space varies as sea level changes because of eustatic (worldwide) changes due, for example, to the waxing and waning of polar glaciations, or to more localized changes due to tectonic movements, glacio-isostasy or changes in intra-plate stress.

Most regressions (Fig. 1) occur when the supply of sediment exceeds the space in the sea or lake available for it to be

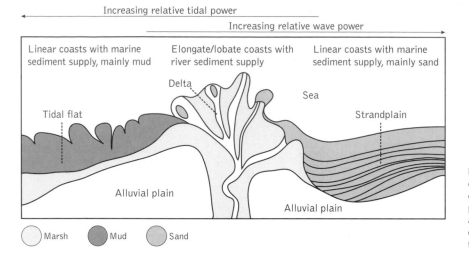

Increasing relative tidal power

Increasing relative wave power

Linear coasts with marine sediment supply, mainly mud

Elongate/lobate coasts with river sediment supply

Linear coasts with marine sediment supply, mainly sand

Delta

Sea

Tidal flat

Strandplain

Alluvial plain

Alluvial plain

Marsh   Mud   Sand

Fig. 1. Various types of regressive coastline, the nature of which depends primarily on the relative power of tides and wave action and whether sediment, either mud or sand, is supplied from rivers or from the sea.

accommodated, so that the land builds out into the sea. If sediment is supplied mainly from a river, then a delta is created. The type and shape of the delta varies according to the degree of reworking of the sediment by tides or waves. Where these processes are small, as in river-dominated deltas such as the Mississippi delta, then an elongate delta is created. Lobate deltas, such as that of the Rhône, reflect substantial reworking by waves. Tidal and wave influenced deltas, such as the Niger delta, have tidal channels that penetrate well on to the delta plain. Tide-dominated deltas, such as the Ganges-Brahmaputra delta, have mouths composed of funnel-shaped estuaries.

Linear, prograding coasts are the consequence of a substantial supply of sediment derived by transport along the coast by longshore drift and coastal currents. If the sediment is primarily sand, then a series of beach ridges develop that build progressively seaward as progradation takes place. Where the sediment is primarily mud, typically downcurrent from a major delta such as that of the Amazon or Orinoco, then the mud flats build out into the sea, forming a rather diffuse shoreline. Sandy coasts tend to enhance the power of waves and storms at the shoreline relative to tidal power, because the shelf tends to be steeper and waves can easily reach the shoreline. Muddy coasts, on the other hand, have much gentler

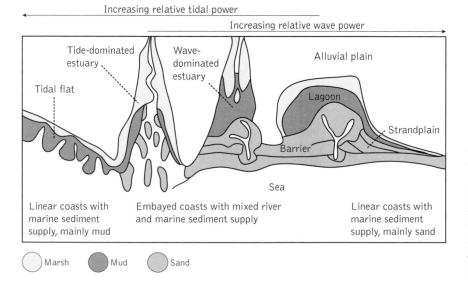

Increasing relative tidal power

Increasing relative wave power

Tide-dominated estuary

Wave-dominated estuary

Alluvial plain

Tidal flat

Lagoon

Strandplain

Barrier

Sea

Linear coasts with marine sediment supply, mainly mud

Embayed coasts with mixed river and marine sediment supply

Linear coasts with marine sediment supply, mainly sand

Marsh   Mud   Sand

Fig. 2. Various types of transgressive coastline, the nature of which depends primarily on the relative power of tides and wave action and whether the sediment, supplied by longshore and coastal current drift, is mainly mud or sand. River supply affects only the inner parts of estuaries. Estuaries are normally located where a deep valley had been incised during the previous fall of sea level.

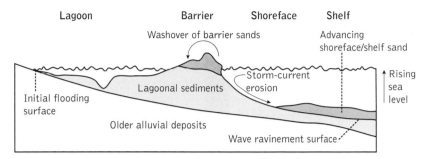

Fig. 3. A profile through a typical barrier-island/lagoonal system that forms during a transgression. As sea level rises, older alluvial sediments are overlain unconformably by lagoonal sediments, which in turn should be overlain by barrier sands, particularly those formed by storm washover events. However, storm-current erosion at the shoreface normally removes most of the barrier sands, leaving another unconformity, the wave ravinement surface.

profiles, and wave and storm action is dampened offshore, seldom reaching the shoreline. In addition, a similar tidal range will have a much greater effect on a gently dipping muddy shelf than on the steeper shelf typical of sandy and gravelly coasts.

Some regressions are caused, not so much by an increase in the sediment supply, but by a rapid fall in sea level, called a *forced regression*. For example, in northern Canada and Scandinavia glacio-isostatic rebound at a rate of up to 250 m in 10 000 years produces a series of parallel beach ridges that form seaward at progressively lower levels as the shoreline retreats.

During a transgression (Fig. 2), as the sea advances upon the land and drowns it, the shoreline retreats. The relative rise of sea level causes a reduction of the gradient of the adjoining coastal plain and sediment supply from the land is reduced.

Estuaries form along indented coasts, especially where rivers have previously entered the sea and have cut deep, incised valleys into the coast. Sediment supply from the land is progressively reduced. The sea floods into the old river valley bringing sediment from the sea. The nature of the estuary differs according to whether tides or storms and waves are the more powerful offshore and shoreline process. Where tidal range is high, as we can see today on either side of the English Channel, or in the Bay of Fundy in eastern Canada, then a funnel-shaped estuary open to the sea forms during transgression, a typical feature being tidal sand bars. Where offshore wave and storm action dominates the shoreline, then a partial barrier, composed of sand develops at the mouth of the estuary, water entering its mouth only through a tidal inlet. Mud predominates in the centre of such estuaries, in contrast to the sand bars that proliferate in tide-dominated estuaries.

Where the coast was previously linear and sand is the dominant sediment, then a beach initially develops. As sea level rises, accommodation space increases and lagoons form above the old alluvial coastal plain and behind the beach, which now becomes a gradually deepening barrier to the lagoon. Typical transgressive coasts are therefore composed of barrier island/lagoonal systems, well known from the south-eastern coast of the USA (Fig. 3).

During transgressions much less of the sediment is preserved than during regressions, because erosion at the shoreface or by tidal channels removes some portion of the

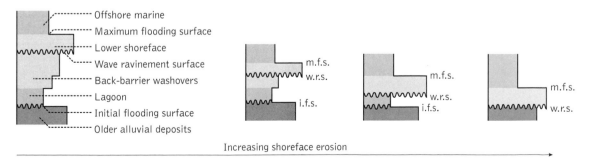

Fig. 4. Sequences preserved as a result of transgression or with differing degrees of shoreface erosion by waves. If tidal inlets, which themselves channel deeply into the barrier, are also present, then the degree of erosion may be even greater and even less of the sequence preserved.

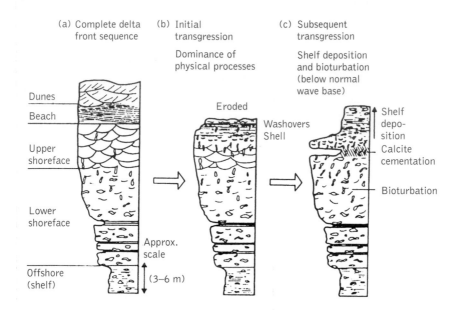

(a) Complete delta front sequence

(b) Initial transgression

Dominance of physical processes

(c) Subsequent transgression

Shelf deposition and bioturbation (below normal wave base)

Dunes

Beach

Upper shoreface

Lower shoreface

Offshore (shelf)

Approx. scale

(3–6 m)

Eroded

Washovers Shell

Shelf deposition

Calcite cementation

Bioturbation

Fig. 5. Delta-front sequences in the wave-dominated Cretaceous Miguel delta system, Texas: (a) complete delta-front progradation sequence; (b) modification of the sequence during the ensuing transgression; (c) final, preserved sequence. (After Wise, B. R. (1980) *Wave-dominated delta systems of the Upper Cretaceous San Miguel Formation, Maverick Basin, South Texas.* Bureau of Economic Geology, Report of Investigations, **107**. Austin, Texas.)

previously deposited sediment deposited on the coastal plain in lagoons or at the coastal barrier. Hence unconformities tend to predominate, and a number of erosive surfaces are present (Fig. 4). The first of these transgressions marks the transgression of the lagoon over the coastal plain (sometimes known as the *brackish transgression* or *initial flooding surface*); the second transgression marks erosion at the shoreface, giving a wave ravinement surface below the shoreface sediments. If transgression continues further, these are overlain by offshore sediments, perhaps deeper-water pelagic sediments that indicate the time of maximum transgression or maximum flooding surface.

In the rock record a regression is generally documented by a progradational sequence (Fig. 5a) that coarsens upwards from offshore mudstone, through siltstone to various types of sandstone, deposited as shorefaces, as beaches, as tidal sand bars, or in deltas, the nature of the deposit depending on the dominant coastal process. Where the cause of regression is a rapid fall of relative sea level, then there may be a sharp break between overlying coastal sediments and the underlying offshore ones.

Transgressive sequences are characterized by erosive surfaces with much of the sequence missing, bioturbation, cementation, and consequent condensation of the sequence (Fig. 5b, c).   HAROLD G. READING

Further reading

Carter, R. W. G. and Woodruff, C. D. (eds) (1994). *Coastal evolution: Late Quaternary shoreline morphodynamics.* Cambridge University Press.

remanent magnetization *see* PALAEOMAGNETISM: TECHNIQUES AND REMANENT MAGNETIZATION

remote sensing
At the roots of geology is the need to visualize how layers and irregular masses of rock relate to one another and extend into three dimensions, particularly where they are hidden from view beneath the surface. Geologists have traditionally pieced together such basic models by marking occurrences of different rock-types on topographic maps and then estimating how the outcrops link together on the land surface. Matching such a geological map with the shape of the land surface gives an idea of three-dimensional geology, which can be extended beneath the surface by using some simple geometry. All this requires much arduous, painstaking, and sometimes confusing work. Looking vertically downwards on a region from an aircraft or an orbiting satellite provides far more information about these geological attributes than one could ever hope to discover on the ground during a reconnaissance mapping programme.

Photogeology: image interpretation

Interpreting views from above speeds up and clarifies geological mapping, and adds otherwise inaccessible information to the generation of ideas about the disposition of rocks and to understanding some geological processes. That, in a nutshell, is what geological remote sensing is all about.

Downward-looking aerial photographs are the stock-in-trade of any kind of modern mapping. Aerial surveys entail taking photographs at intervals along a regular pattern of flight lines, so that every part of the surface appears on at least two frames in a sequence. The detail that appears on the photographs, even from high altitudes, is very fine, usually picking out objects a metre or less across. Equally important, in a similar fashion to the way in which our left and right eyes each record a slightly different view of the scene before us, and thereby allow us to perceive the world in three dimensions, successive aerial photographs incorporate this parallax shift because of the lens geometry and the position of each exposure. Whereas our eyes are about 7.5 centimetres apart, the separation between adjacent aerial photographs is at least several hundred metres. A viewing device called a stereoscope presents the photographs to the left and right eyes, effectively increasing the eye separation to a ridiculous degree. Instead of being completely unable to judge relief by simply looking out of an aircraft at high altitude the result is an amazingly exaggerated stereoscopic impression of topography. Details of rock variation and structure in well-exposed terrain are seen in their full context, and relatively little geological knowledge and experience are needed to make a useful geological map.

Photographs are limited to a narrow range of the electromagnetic spectrum—the visible region and the infrared at slightly longer wavelengths than visible red—that has its origin in the Sun's radiation and is reflected to various degrees by the surface. Apart from a few rare types, even the most pristine rock surfaces appear to us as shades of grey, with tinges of yellows, reds, and browns that result from various amounts of iron oxides and hydroxides. Naturally weathered rock surfaces assume this limited range of colour, with some additional confusion caused by varying cover by lichens. Much of the land surface exposes no rock at all, and is dominated by soil blankets in arid areas and by vegetation where conditions are more humid. The visible range is therefore of little help in enabling different rocks to be discriminated from a distance to the degree that is possible for the field geologist. But a limited differentiation of lithologies is possible.

A photogeologist uses the different ways in which various rock-types respond to weathering and erosion under the prevailing climatic conditions to distinguish them. Some rocks that are compact and hard, such as many crystalline igneous and metamorphic rocks, and carbonates and some sandstones among the sedimentary rocks, resist being worn down, even by the action of glaciers. So they tend to form uplands, and ridges if they are stratified. Less resistant rocks, such as shales, some schists, and all poorly consolidated sediments, form lowlands and vales parallel to the regional strike. It is these ridges and vales formed by variably resistant, stratified rocks that provide the key information from which geological structure can be worked out. Dip slopes and scarps stand out very clearly on stereoscopic photographs and allow accurate estimation of dip angle and direction for strata, as well as the shape of folds. Given a dip direction in a sequence of stratified rocks, it is a simple matter to establish a stratigraphical sequence. Even on a single image the direction and amount of dip can be roughly estimated by using the 'Law of Vs' if the topography includes well-defined valleys and ridges. Faults usually weaken rocks by shattering them to some extent, and some linear valleys are controlled in this way, particularly if the faults assume a steep angle of dip. Careful scrutiny of photographs for truncations of strike and for repetition or omission of strata in a stratigraphical sequence is a means of detecting faults with gentle angles of dip and establishing their sense of movement.

Depending on several factors, of which permeability is the most important, different rocks tend to have variable densities of streams and rilles developed upon them. The shapes in cross-section of individual drainage patterns vary according to lithological differences in resistance to erosion on a variety of scales. These two factors control the topographic textures that develop on different rock-types. The disposition of drainages, interfluves, and ridges in an area contributes to a wide variety of patterns seen on images. Such organization of topography relates to a host of factors. Among these are the spacing of compositional banding and joints, the intensity of foliation, and the presence of landforms, such as volcanoes, associated with the original formation of a rock.

So, geological interpretation of aerial photographs, and those taken from space, depends mainly on assessing erosional resistance, textures, and patterns, as well as the limited differences in colour or grey-tone of rocks in the visible part of the spectrum. This simple 'toolkit' of interpretative methods forms a framework for the geological interpretation of all remotely sensed images, including these that are captured by non-photographic means and use non-visible radiation.

Modern developments

Since the late 1960s scientists have developed a range of solid-state detectors, akin to transistors and capacitors, that are sensitive to electromagnetic radiation extending well beyond the wavelengths of visible light. These make it possible to detect radiation that is emitted by the Earth as well as solar radiation reflected by surface materials. It is in the infrared region where minerals, plants, rocks, and soils exhibit their greatest differences. The spectral variability of natural materials has been known for a long time from laboratory studies. The new detectors, together with instruments that focus radiation from distant surfaces on them and developments in information compression, transmission, and storage, enable entirely new kinds of imagery to be produced and put to use.

Instead of using photosensitive material in a film emulsion to record 'one-off' images through a lens, the new devices employ optical–mechanical methods to build up images during the time it takes a platform—an aircraft or a satellite—to pass over an area. One method uses a back-and-forth sweeping mirror to direct radiation reflected or emitted from a narrow strip of ground beneath on to a detector for each of several ranges of wavelength (line scanners). Another sweeps long arrays of thousands of identical detectors tuned to such wavebands across the ground (pushbrooms). In a line scanner the continuously varying electronic response from the detectors is divided in equal time intervals into segments that correspond to small rectangular patches of the surface. At any instant the detectors in a pushbroom respond proportionally to radiation from patches along a line on the ground. Images are built up of thousands of these lines made up of picture elements (pixels) that represent the patches. To simplify recording and transmission, and also later processing, the signal for each pixel is converted to a digital number in the 0 to 255 range encoded in binary notation. The image data no longer need a medium in order to be preserved; they are simply transmitted and recorded on magnetic tape. In this form countless copies can be made, and, more importantly, images can be manipulated by computer in many appropriate ways.

Pictures made up of mathematically perfect rectangular arrays of pixels can be reassembled on a video monitor—which itself uses pixels—and the many different wavebands collected at the same time are registered exactly with one another. The last point means that data from any three wavebands (and often more, in various mathematical combinations) can be used as the red, green, and blue controls for a video display, thereby producing a colour image. However, unless the bands correspond to the three divisions of visible radiation, the picture is a colour rendition of the completely invisible; a false-colour image. How this is composed depends on the objective of the investigator, for the spectra of materials reveal differences in many different wavelength regions.

The visible region of rock spectra is dominated by the main colouring agent in rocks, iron, in the form of oxides and hydroxides. These minerals, either in the matrix of sediments

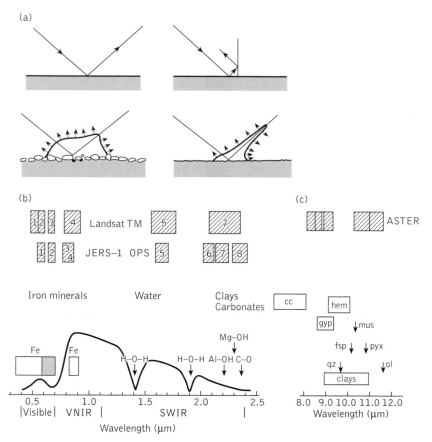

Fig. 1. (a) How radar waves interact with surfaces. Radar imaging requires illumination from the side for technical reasons. The roughness of the surface, relative to the radar wavelength, results in different kinds of reflection or scattering. From top left clockwise: a smooth surface reflects the radar like a mirror, no radiation returns to the detector, and the surface appears dark on the image; a smooth, vertical surface causes all energy to return and gives a bright point on the image; a slightly rough surface results in some energy returning along the radar paths; a rough surface scatters a high proportion of energy back to the detector. The rougher the surface, the brighter it appears on the image. (b) The positions of the main absorptions by minerals in the reflected part of the spectrum relative to a reflectance spectrum for vegetation and the bands recorded by two remote sensing instruments (the US *Landsat* Thematic Mapper and the Japanese Fuyo–1O PS system). (c) Positions of suppressed emission associated with rock-forming minerals in the thermally emitted spectrum. The spectral bands are those planned for the Advanced Spaceborne Thermal Emission and Reflection Radiometer (ASTER) to be carried by the first satellite in the Earth Observation programme. VNIR, very near infrared; SWIR, shotwave infrared; cc, clays, carbonates; gyp, gypsum; hem, haematite; fsp, feldspar; mus, muscovite; ol, olivine; pyx, pyroxine; qz, quartz.

or in a weathering veneer, absorb more blue and green (0.4–0.6 micrometres (μm) than red (0.6–0.7 μm) in sunlight (Fig. 1b); that is the reason for the dominant reds, oranges, yellows, and browns of most bare rock in arid regions. A natural-colour image therefore adds only the parameter of redness to overall reflectivity, texture, and pattern as 'tools' for discrimination of rock-types without the aid of a close-up view, a hammer, and a microscope. The influence of iron minerals extends a little way into the infrared, where processes at the molecular level produce a broad absorption around 0.9 micrometres (μm).

Infrared imagery

In what is known as the short-wave infrared or SWIR, between 2.1 and 2.4 μm, the spectral possibilities become more promising. Molecular interactions with radiation that involves Al–OH, Mg–OH, and C–O bonds result in a number of narrow absorption features, the exact position of which depends on the specific mineral molecule (Fig. 1b). Measurements in this region can therefore be used to detect minerals such as micas, clays, and many magnesium-bearing silicates containing the OH$^-$ radical, such as chlorite, talc, and serpentine, and carbonates, provided that the remotely sensed data are divided into narrow wavelength intervals. Whereas the SWIR region is the focus for much geological remote sensing, the spectral features involved are mainly governed by minerals produced by weathering, not by rock-forming minerals. Distinction between different rocks on this basis usually matches field division very well, but direct identification of rock-types, other than carbonates, is not achievable without field confirmation.

Beyond 2.4 μm, radiation from the Earth becomes dominated by that which is emitted by the surface because of its temperature, rather than by reflected solar energy. As in the SWIR, molecular processes govern the intricacies of thermal spectra, but in this case the dominant processes are those in rock-forming minerals, principally the silicates. Quartz, feldspars, micas, amphiboles, pyroxenes, and olivines have quite different thermal emission spectra; so remote sensing that measures energy emitted in several narrow thermal wavebands potentially provides a means of direct lithological mapping (Fig. 1a). Several factors, such as soil and vegetation cover, and technical problems temper the apparently revolutionary potential of thermal mapping in geology. Several instruments are, however, now being deployed to test the possibilities and are planned to become part of the international Earth Observation programme aboard unmanned satellites (1998 onward).

Although vegetation is one of the curses sent to try geologists, differences in rock and soil on which plants grow subtly affect vegetation, either by encouraging the growth of different floras on different substrates or because underlying chemical factors affect the health of plants. In terms of reflection of solar radiation, plants are unique. As well as being green, because chlorophyll absorbs red and blue, plants have evolved a cellular mechanism that prevents them from overheating by reducing the amount of radiation that they absorb. This dominates their infrared spectra, where wavelengths beyond 0.75 μm are efficiently reflected away. The efficiency varies from species to species, and also depends on the health of the plant. Water molecules in plant cells also absorb radiation at wavelengths around 1.4 and 1.9 μm, and their moisture content can accordingly be assessed. Consequently, by imaging the visible and near-visible infrared it is possible to detect botanical variations and anomalies that are indirectly related to underlying geology. This provides a remote-sensing approach that can be particularly useful in exploration for minerals, hydrocarbon seepages, and water resources. The peculiar reflectance of infrared radiation by plants forms the main source of remotely sensed information about the biosphere on land.

Radar systems

In the microwave part of the electromagnetic (EM) spectrum (wavelength less than 1 mm) natural energy emissions are too low for useful remote sensing. Instead, imaging radar systems are used to illuminate the surface with artificial microwave energy, which can penetrate cloud cover. At such long wavelengths the interactions between radiation and the surface are dominated by the attitude of the surface in relation to the platform and by small-scale surface roughness (Fig. 1a). The information conveyed by radar images is consequently, very different from that obtained at shorter wavelengths. Radar is particularly good at showing up variations in surface texture that depend partly on the small-scale responses of rocks to weathering and erosional processes, and also on the nature of any vegetation canopy. Since radar images can be produced only by illumination from the side, large-scale features in the topography are exaggerated by highlighting and radar shadows. This makes radar images especially effective in outlining geological structures.

Digital image processing

As mentioned above, most modern remotely sensed images are in digital form, with energy levels expressed as arbitrary integers from 0 to 255 (1 byte) in binary code for each of millions of pixels. Digital image processing centres on information in this format. It employs a large and growing range of image transformations aimed at removing defects introduced by imaging instruments, registering images to map projections and map data, modifying or enhancing contrast, combining different kinds of data arithmetically, enhancing or suppressing spatial attributes in images, such as textures, patterns, and shapes, and extracting a wealth of statistical information from image data. The last forms the basis for automatic division of materials at the Earth's surface into different spectral classes. This is most often used in studying vegetation but is potentially useful in geological mapping.

As well as the information gathered in image form, Earth scientists now have access to other information on geographically varying properties of the Earth's surface and its interior, usually at isolated points or along survey lines. These data include topographic elevation, depths to rock boundaries, measures of magnetic and gravitational field strength, and the concentration of chemical elements in soils and stream sediments. By assuming that such isolated measurements are part of a natural continuum that can be represented mathematically by a three-dimensional surface, these data can be transformed into image form by interpolation. Although not remotely sensed in the strict sense, such representations can be enhanced and manipulated in exactly the same way as digital images. This opens up an enormous range of possibilities for comparing every aspect of the Earth's properties rapidly, conveniently, and in a common geographic context. In this way otherwise unachievable correlations and coincidences become visible, so that Earth scientists can use them in modelling natural processes. Such geographical information systems lend themselves to exploration for many physical and biological resources, and environmental management and monitoring, as well as for more academic pursuits.

S. A. DRURY

Renaissance in Europe and geological ideas

The Renaissance in Europe was a remarkable period of artistic, cultural, and intellectual activity. It was also a time of geographical exploration, and it saw the beginnings of modern science and technology. Following the Middle Ages, it lasted from the fourteenth to the sixteenth century. A new curiosity about the Earth and its history, and the revolutionary ideas of Vesalius and Copernicus, Kepler, and Galileo focused the attention of the Catholic Church on these scholars.

Two men who escaped such attention were Georgius Agricola (1494–1555), who lived in Germany, and the Swiss, Conrad Gesner (1516–65). Agricola (christened Georg Bauer) was one of the most important figures in the early history of the geological sciences. He studied in Germany and Italy and spent some years as a physician in the mining town of Joachimstal, Bohemia. He was an enormously industrious and influential writer, and is best known for his six books on geological subjects. Of these, the most famous is *De re metallica* (1556), which deals with mining and smelting. His *De natura fossilium* (1546) has been called the first mineralogy textbook; it gave a new classification of minerals, including many that were previously undescribed, based on their physical properties. Agricola disregarded many of the popular beliefs about minerals and fossils. He encountered some problems with fossils: he acknowledged that shells occur in certain solid rocks, but he thought that many others simply grew within the crust of the Earth. Agricola's great contribution was the descriptions he gave of many metallic minerals and their modes of occurrence. He has been called, with reason, the Father of Mineralogy.

Conrad Gesner is known principally as a zoologist, but his book *De verum fossilium, lapidum, et gemmarum figuris* (*On fossils, stones, and gems*) is a veritable inventory of geological specimens. The term 'fossil' was used here to mean any object dug up, including minerals and inorganic materials as well as fossils in the modern sense. Gesner presented a classification of all these, based on the simplicity or complexity of their forms. Many of his classes refer to fossils and to their similarity to living forms. One of the categories he distinguished was the Glossopterae, or tongue stones, actually the fossil teeth of a Cenozoic shark, which Steno was to discuss in Italy (1669). Reference by Gesner to the 'virtues' or supernatural characteristics of minerals was minimal.

The true nature of fossils was first grasped in Italy and France. Leonardo da Vinci (1452–1519) gave an account in 1508 of marine invertebrate remains found far from the sea. He rejected the ancient theories of a mysterious 'plastic force' or shaping power that constructed these objects in the Earth. To do so was extremely daring at that time, but his rationality prevailed. He acknowledged that the fossils might be the remains of creatures left behind by the sea, but they were not the victims of the catastrophic Deluge of the Scriptures. Leonardo never managed to publish his thoughts on the preservation of fossils, but his notebooks show that his interpretation of them was correct. Independently of Leonardo, Girolamo Frascatoro in Verona arrived at a similar view on the origin of fossils (1517), but did not follow it up. Bernard Palissy, writing in 1580 in Paris, also believed that fossils were organic remains that had somehow become embedded in the crust.

The origins of strata and structures in bedded rocks were discussed perceptively by Nicolaus Steno in Florence in 1669. Like Agricola, he recognized different kinds of mountains—volcanic mountains and erosional mountains—but added block or fault mountains as an extra category. Leonardo had allowed only for mountains to be produced by erosion of the original surface; Agricola was persuaded that most mountains were erosional, but allowed for volcanic and earthquake origins.

Several other, somewhat later, scholars were to consider the nature of the Earth, but all seem to have lacked any concept of terrestrial dynamics and heat, other than what was conveyed in Scripture. This was to come only in the Age of Enlightenment.

D. L. DINELEY

Further reading

Adams, F. D. (1938) *The birth and development of the geological sciences*. Dover Publications, New York.

Wendt, H. (1970) *Before the Deluge*. (Trans. R. and C. Winston). Victor Gollancz, London.

reptiles Reptiles, together with birds and mammals, constitute a single monophyletic group (an ancestor and all its descendants) that is called the Amniota. Amniotes are characterized by the presence of an amniote egg, that is an egg with a hard outer covering that could be laid on land and that contained a membrane, the amnion, that surrounded the embryo in a fluid-filled sac. This feature distinguishes them from the amphibians, which have to lay their eggs in moist conditions and whose young typically hatch out at an immature or larval stage. The organisms that we generally consider to be reptiles,

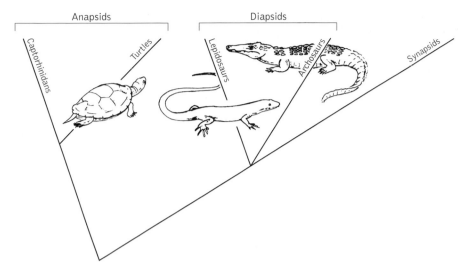

Fig. 1. Cladogram of relationships for the Reptilia.

the lizards and snakes, crocodiles, and turtles, together with fossil groups such as the dinosaurs, form a paraphyletic group (an ancestor and some but not all of its descendents), since the birds and the mammals are not included (Fig. 1).

The development of an amniote egg appears to have occurred in one group of small anthracosaurs (labyrintho-dont amphibians) during Carboniferous time. This was a major step forward, for it freed tetrapods from their dependence on water and allowed them to colonize the land. The earliest reptiles are known from Carboniferous deposits in Nova Scotia, where they occur in the hollow stumps of fossil trees which apparently acted as traps for these small animals. Genera such as *Hylonomus* were small and superficially lizard-like in appearance. Although no fossil eggs have been found with them they are known to be reptiles because of the presence of a suite of advanced features of the skeleton. These features indicate that the animals in question were now fully terrestrial, a step that could have been taken only after the development of the amniote egg. The developments include a change in skull shape from broad and flat to tall and narrow, which brought the eyes more to the side of the head, and changes to the jaw musculature that improved the mechanical efficiency for maximizing force at all stages as the jaws were closed. In the limbs, the main bones elongated and became more slender, while some of the bones of the wrist and ankle fused to provide additional strength. In addition, the attachment between the pelvis and the vertebral column enlarged to provide additional strength in response to an increase in the amount of force generated by the hind limbs. It is from this group of primitive reptiles known as the captorhinidans that the major lineages of reptiles developed in the Late Carboniferous.

The major reptile groups are classified according to the pattern of openings in the skull roof behind the orbits. These openings developed to give increased area for the attachment

and bulging of jaw muscles. In the captorhinidans and their relatives the turtles, no openings are present behind the orbits and they are referred to as being anapsid. Anapsids were important as an ancestral group for the reptiles but their only modern representatives are the turtles, which have a history going back to the Late Triassic. Well-preserved skeletons of this age from Germany and the western USA show that most of the turtle characteristics had already been developed by that stage. In the Triassic *Proganochelys*, teeth were already absent from the jaws and the body was protected by a heavy shell. During the Late Triassic to Late Jurassic the two modern groups of turtles were developed. The pleurodires, in which the neck bends sideways when the head is pulled into the shell, were widely distributed in the Cretaceous but are now restricted to the southern hemisphere. The cryptodires, in which the neck bends vertically, have become important since the Cretaceous, evolving along many lines of adaptive radiation, and they now have a worldwide distribution.

The synapsids are forms in which only one opening is developed behind the orbit. These are represented initially by the mammal-like reptiles, a group that was diverse and important in the Permian and into the Triassic, but was then eclipsed by the dinosaurs. Ultimately this line produced the mammals, which came into their own at the end of the Cretaceous after the demise of the dinosaurs.

The major reptilian group is the diapsids, in which there are two openings behind the orbit. They include the marine plesiosaurs and ichthyosaurs and also the two major groups of archosaurs ('ruling reptiles') and the lepidosaurs, which include snakes and lizards. The major archosaur group was the dinosaurs, which dominated the terrestrial scene during the Mesozoic. In addition, their relatives the pterosaurs and the crocodiles demonstrate the diversity of this group. The crocodiles originated in the Triassic and diversified during the Mesozoic, developing from lightly built terrestrial carnivores to the aquatic predators that we know today. Some of the later Mesozoic forms were indeed formidable; *Deinosuchus* from the Cretaceous of Texas was about 15 m long and probably preyed on dinosaurs.

Lizards developed at the same time as the dinosaurs, but because of the great difference in size there was probably little competition between them. The main modification in the lizard skeleton is the loss of the lower edge of the lower of the two openings behind the orbit; this gave greater flexibility to the jaw. Although lizards are generally small animals the modern komodo dragon is up to 3.5 m long and is an effective predator on mammals. During the Cretaceous the mosasaurs, a group of large marine lizards, developed. These were 9 to 12 m in length, had paddle-like feet, and propelled themselves with a laterally flattened tail. Snakes developed from limbless lizards during the Early Cretaceous and have developed specializations of the skull to allow them to ingest large prey and to inject poison in some cases. Like all the members of the Reptilia, they are a vigorous and diversified group occupying many different habitats.   DAVID K. ELLIOTT

Further reading

Carroll, R. L. (1988) *Vertebrate paleontology and evolution.* W. H. Freeman and Co., New York.

Colbert, E. H. and Morales, M. (1991) *Evolution of the vertebrates.* Wiley-Liss, Inc., New York.

residence times *see* GEOCHEMICAL RESIDENCE TIMES

retardation in groundwater The term 'retardation' applied to groundwater refers to the slow apparent velocity of some dissolved constituents, or solutes, as compared with the velocity of the water itself as it moves through the rocks below the surface. Retardation results from reactions through which solutes become attached to, or incorporated in, the immobile solids of an aquifer. These reactions include surface processes such as ion exchange or sorption (which are particularly important for metals and other cations) and partitioning into solid organic carbon (which is particularly relevant for dissolved organic compounds). If the reactions are irreversible, the mass of solute in solution is permanently reduced, or attenuated. The dissolution and precipitation of mineral phases, while reversible, generally leads to attenuation rather than to retardation of solutes.

The *retardation factor* is defined as the ratio between the advective (horizontal) velocity of water and the average velocity of a reacting solute. Estimates of retardation factors are usually based on laboratory measurements of an equilibrium partition coefficient, which is determined as a ratio of the concentration in the solid phase to the concentration in solution. Such estimates may not compare well with field observations of retardation if the reaction does not reach equilibrium or if the equilibrium partition coefficient varies with concentration.   J. BAHR

Richter, Charles (1900–85) Charles Richter spent his career studying earthquakes, both worldwide and along the faults of southern California, where he was a professor at the Seismological Laboratory of the California Institute of Technology. He is best known for devising the famous 'Richter Scale' of earthquake magnitudes, a universal measure for comparing widely separated earthquakes. Since earthquakes can range from subtle tremors to fault ruptures that extend hundreds of kilometres, Richter chose a logarithmic scale, calibrating the seismic waves from southern California tremors as recorded by standardized seismographic equipment. At the present day, nearly all research seismometers record digital signals that are stored on computer. In Richter's day, seismic waves were recorded by a swinging pen on a sheet of paper that rotated on a cylindrical drum. The 'Richter magnitude' was computed by measuring the pen's swing in millimetres, taking its logarithm, and applying a correction for the distance between station and earthquake. Working in the 1930s and 1940s with Beno Gutenberg, another professor at the Seismological Laboratory, Richter extended his scaling algorithm from near-by to worldwide earthquakes. Later refinements have tied the Richter scale to the actual energy released by an earthquake. Each unit on the Richter scale represents a thirty-fold increase of energy. The largest earthquake since the inception of the scale occurred on 22 May 1960 along the coast of Chile, with estimated magnitude 9.5.

As a public spokesman for seismology and earthquake hazard mitigation, Richter often showed an irascible personality. His renowned 1958 textbook *Elementary seismology* contains many pithy comments and admonitions against sloppy research methods. Ironically, many of his personal records were destroyed in a house fire that followed the Northridge earthquake of 17 January 1994 (magnitude 6.6) north of Los Angeles.   JEFFREY PARK

ring complex A ring complex is a set of circular igneous intrusions around an igneous centre. A typical ring complex (Fig. 1) is generally composed of ring dykes and cone sheets. Ring dykes are thick, nearly vertical igneous bodies that form concentric circles around a central intrusion. Ring dykes are associated with a mechanism of emplacement called cauldron subsidence. If a circular disc of country rock subsides into a magma chamber, molten rock rises around the edges and fills the space above it. When the rock has solidified it forms an igneous body with the shape of an inverted cauldron. The base of the cauldron appears as a circular intrusion; its sides forms a ring dyke. The Sabaloka complex in the Sudan provides a textbook example of a cauldron subsidence with a caldera that is 25 km across at its widest point. Thanks to erosion by the River Nile the caldera and associated ring dykes are well exposed.

Cone sheets are usually thinner than ring dykes, and although they are also concentric to the intrusion they have the general form of a set of inverted cones, the apices of which converge at a point beneath the intrusion. This point is gen-

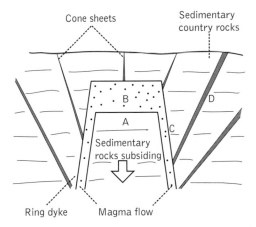

Fig. 1. Cross-section through a ring complex. The sedimentary rocks (A) subside to create space (B) for the magma that intrudes along ring dykes (C); the intrusive event is also accompanied by the formation of cone sheets (D).

erally thought to be in the magma chamber that fed the igneous intrusion. The circular form of these intrusions results from stresses set up in the crust as the magma body forced its way upwards. Ring dykes and cone sheets are often complemented by a set of vertical dykes disposed radially to the igneous body. Classical examples of cone sheets and ring dykes are found in the Tertiary igneous comples of western Scotland.   JUDITH M. BUNBURY

river channelization The practices, in recent centuries, of river engineering for the purposes of flood control, improvement of drainage, maintenance of navigation, and realignment, here collectively referred to as *channelization*, are among the most widespread and obvious direct impacts on river channels in many developed countries. In such regions, the introduction of machinery such as draglines and bulldozers during the past 150 years or so, and the increased demand placed on flood-plain lands for agricultural or urban development, have led to an intensive period of channel modification. In some lowland catchments in Europe, for example, more than 95 per cent of river channels have been directly modified through channelization.

The scientific literature on channelization has tended to focus on the adverse impacts associated with channel

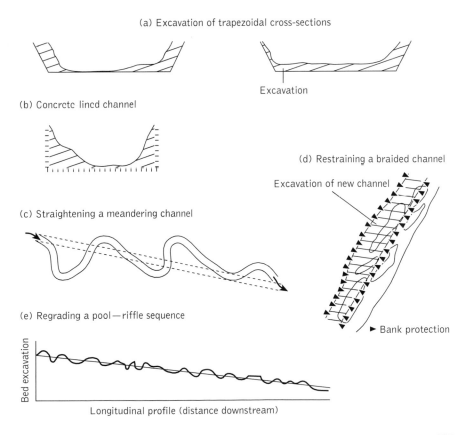

Fig. 1. Aspects of channelization procedures.

widening, deepening, straightening, and bed and bank protection. The planning and design of projects that substantially alter an existing channel (Fig. 1), is not an easy task because rivers are naturally dynamic. Designs have frequently been based on previous experience and intuition. If a design is based on inadequate data, there may be substantial adjustment within the channelized reach, and sometimes upstream and downstream of this, possibly leading to costly and recurrent maintenance. By altering one or more of the interdependent hydraulic variables of slope, width, depth, roughness, or size of the sediment load the existing equilibrium is disrupted, necessitating adjustments of other hydraulic variables as a channel attempts to attain a new state of equilibrium. For example, a straightened river channel (Fig. 1) may immediately react to an increase of slope by increasing sediment discharge through erosion of the bed and banks. Eventually erosion may widen the channel, with a corresponding reduction of velocity, and the adjusted cross-section may then be more efficient in dissipating energy. Case studies documented in the scientific and engineering literature indicate a range of adjustments to river channelization in different environments.

An understanding of the physical impacts of channelization is essential in anticipating interpretation of biological effects of channel change, because altering the original dimensions, shape, and substrate of a channel may cause considerable disruption to a river ecosystem. Pools may be lost, river flow may become more uniform in character, and the channel bed may be of a homogeneous substrate throughout the affected reach. Such changes may reduce habitat diversity and niche potential, and the quality and functions of species occupying the river system may alter. As a consequence, there is now a considerable amount of literature concerned with impacts of channelization on macroinvertebrates, fish and fisheries, plants, birds, and mammals.

With increased environmental awareness of the potential adverse consequences of channel regulation there have been attempts to devise more acceptable practices, based on the need to work with a river channel rather than against it and to minimize the aesthetic degradation of a river and its environs. These practices have been tried, and in a few cases evaluated, for at least twenty years in the USA and Europe. Practices include installing deflectors in channelized reaches to mitigate the effects of widening by narrowing the low-flow channel to a more appropriate width. Natural features such as pools, riffles, substrate, and non-uniform channel geometries have also been created afresh. Such rehabilitation techniques have sometimes been used successfully to accelerate biological recovery of channelized streams.

Many river channels that have been adversely regulated may tend to recover naturally, but usually only in the absence of maintenance. For example, channels that have been straightened, but which are no longer required to be constrained, have occasionally, in higher-energy environments, regained their original meandering course. Such recovery is probably the most pragmatic option for the return of natural attributes

to many modified watercourses, although this needs to managed, particularly with respect to the hydraulic implications of allowing a channel to adjust within an artificially designed section. A more extreme option is restoration, effectively the antithesis of channelization. Although restoration may be taken to encompass all rehabilitation techniques, 'full restoration' involves a structural and functional return to a pre-disturbance condition. There are examples from Denmark, England, Germany, Florida, and California of re-meandering of previously straightened watercourses and re-establishing the hydrological link between a channel and its former flood-plain. Restoring rivers to a pre-disturbance condition is more an ideal than a practical goal; because ecosystems are dynamic, restoration to the present condition of a comparable undisturbed system may be a more appropriate goal. Monitoring and evaluation of the performance of these projects is needed. 'Full restoration' of river channels may have relatively limited application, especially where other demands on rivers have to be met, such as adequate flood protection and continued drainage of agricultural land.

A. BROOKES

Further reading

Brookes, A. (1988). *Channelized rivers: perspectives for environmental management.* John Wiley and Sons, Chichester.

Brookes, A. and Shields, F. D. Jr (eds) (1996) *River channel restoration.* John Wiley and Sons, Chichester.

river management River management refers to human alteration of natural channels, or of the flow in those channels. In practice, alteration of the flow regime in alluvial channels usually results in corresponding changes in channel morphology (see *alluvial channels*). Humans have been attempting to manage rivers since at least the beginning of written history. The oldest constructed dam so far discovered dates to c. 2900 BC in Wadi El-Garawi, south of Cairo, Egypt. Li Ping initiated an extensive system of irrigation canals and flood-control structures more than 2100 years ago in the Szechuan region of China. Levée construction along the Yodo River in Japan began in the fourth century AD.

Present-day river management has two basic goals: the reduction of channel-related hazards such as overbank flooding or bank collapse, and the enhancement of channel characteristics for human uses, such as navigation or water supply.

Management of flow regime is a vital aspect of river management. Flows may be diverted from a channel, or transferred between channels, for purposes of flood control, agricultural irrigation, municipal water supply, navigation, or commerce. Between 1988 and 1992, the number of dams on the world's rivers exceeded 10 000, with major reservoirs storing more than 5000 km^3, or 12.5 per cent of annual global land run-off to the oceans. At present, dams greater than 15 m high are being completed throughout the world at a rate of approximately 500 per year. It is estimated that more than

60 per cent of total streamflow in the world is now regulated by dams and reservoirs used for flood control, hydroelectric power generation, water storage, and recreation. The most common effect of reservoir-regulated flow is a decrease in flood peaks and an increase in the uniformity of base flow throughout the year. In turn, these changes alter channel morphology, and aquatic and bankside (riparian) biota. Flow regulation may also be used to reduce overbank flooding by increasing channel capacity through the use of dredging or levées, or by temporarily storing flood waters in detention basins.

River management may also seek to control downstream movement of sediment. Sediment-control efforts may focus on the introduction of sediment from the hill slopes, valley bottoms, or channel banks. Such control utilizes afforestation, sediment-retention (sabo) dams on tributary channels, hill-slope terracing, and bank protection with loose, broken stone (rip-rap). Sediment control may also be implemented using dams and reservoirs, or stilling ponds, along the main channel.

Management of channel geometry may focus on lateral or vertical channel stability, or on both. Methods used to prevent channels from shifting laterally include *channelization*, which generally involves straightening a meandering channel and thus eliminating meander migration and cutoffs (see *river channelization*), and rip-rap, plating of riparian vegetation, or some means of increasing bank resistance. Methods used to enhance vertical channel stability, or to prevent either incision or aggradation, include grade-control structures that serve as local base levels (see *base level*), energy dissipators that increase channel boundary roughness, and rip-rap along the channel bed. At the extreme, the bed and banks of a channel may be covered continuously with concrete.

River management may also focus on the biotic components of river environments. Large woody debris may be removed to facilitate navigation or fish migration. Beavers may be removed from a channel to reduce overbank flooding. Riparian vegetation may be planted to increase the stability of the banks, enhance overbank sedimentation, or slow the passage of flood waters. Conversely, riparian vegetation may be removed to facilitate flood passage. Flow of water and sediment may be altered to favour a desired fish species, or to enhance a recreational resource.

River management may take the form of an integrated approach encompassing many different methods applied throughout an entire watershed. Such an approach is more common among large, governmental entities, as exemplified by the Tennessee Valley Authority. This project, established in 1933, includes 19 reservoirs for power, navigation, and flood control on the Tennessee River and its tributaries. River management may also be applied by a single individual dumping old tyres or cars into a channel in an attempt to reduce bank erosion along the outside of a meander bend. The majority of river management projects, particularly the larger ones, have produced unanticipated negative side-effects. This results in part from our limited understanding of the complexities of channel processes. In 1994 Schumm summarized three types of misperceptions:

1. a perception of stability, which leads to the conclusion that any change is not natural;
2. a perception of instability, which leads to the conclusion that change will not cease; and
3. a perception of excessive response, which leads to the conclusion that changes will always be major.

In general, limited data collection for short periods can fail to provide knowledge of long-term change and the potential for possibly catastrophic changes of river characteristics. To anticipate and to predict river behaviour, both the past and the present records of river characteristics must be combined in a geomorphological–engineering approach to river management.

Any large, managed river in the world illustrates the consequences of our limited understanding of river dynamics. The 12 major dams along the Colorado River of the western United States have so altered the flow regime that six of the river's 13 endemic fish species are now threatened or endangered, a problem not anticipated when the dams were built. Estimates of channel degradation expected to result from construction of the High Aswan Dam on the River Nile ranged from 2.0 to 8.5 m, but 18 years after the dam became operational, maximum degradation had stabilized at 0.7 m. As both scientists and the public in general become aware of these unfortunate consequences of river management, there is a growing demand for management that aims, as far as possible, to restore natural channel characteristics and functions.

Further reading

Brookes, A. (1994) River channel change. In. Calow, P and Petts, G. E. (eds) *The rivers handbook: hydrological and ecological principles*, Vol. 2, pp. 55–75. Blackwell Scientific, Oxford.

Petts, G. E. (1989) Historical analysis of fluvial hydrosystems. In Petts, G. E. Moller, H. and Roux, A. L. (eds) *Historical change of large alluvial rivers: Western Europe*, pp. 1–18. John Wiley and Sons, Chichester.

Schumm, S. A. (1994) Erroneous perception of fluvial hazards. *Geomorphology*, **10**, 129–38.

Schumm, S. A. and Winkley, B. R. (eds) (1994). *The variability of large alluvial rivers.* American Society of Civil Engineers Press, New York.

Walling, D. E. (1987) Hydrological processes. In Gregory, K. J. and Walling, D. E. (eds) *Human activity and environmental processes*, pp. 53–85. Wiley and Sons, Chichester.

ELLEN E. WOHL

rivers A river is a natural stream of water that flows from a higher to a lower elevation across a land surface. Rivers comprise only 0.0001 per cent of the water at the Earth's surface. This is a small but very important fraction, since rivers transport most of the water as well as dissolved and suspended matter from the continents to the oceans.

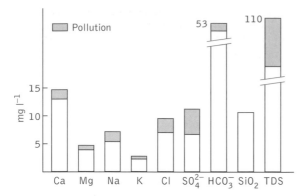

Fig. 1. Chemical composition of world river water, actual and 'natural'. Unshaded, 'natural' values, corrected for pollution (shaded). TDS, total dissolved solids. (Modified from Berner, E. K. and, R. A. (1996).)

Rivers result from the *run-off* of water from the land surface. This run-off is derived from water that reaches the land surface in the form of *precipitation* (rain, hail, sleet, snow, etc.). Some of the water that falls on the land surface is lost by evaporation, some of it forms surface flow, and some of it sinks into the ground to become groundwater. In order to have river run-off, precipitation must exceed evaporation. There are two major belts on the Earth where precipitation exceeds evaporation. One is near the Equator, where there is both high rainfall and high evaporation but where rainfall exceeds evaporation. This results in large rivers such as the Amazon in South America, and the Congo in Africa. The second belt is in the temperate zone, which is characterized by generally adequate rainfall and lower evaporation rates. Two major rivers originating in this area are the Yangtze-kiang in Asia and the Mississippi in North America.

The composition of major dissolved solids in river water is affected by atmospheric sea salts, evaporation, chemical weathering of rocks, temperature, the rate of water circulation, relief, vegetation, and pollution. Small rivers are more diverse in composition than large rivers: the size of a river catchment tends to moderate variations by integrating different environments. The chemical composition of world average river water, both actual and 'natural', is given in Fig. 1. The 'natural' figures are corrected for pollution. Calcium is the most abundant cation and bicarbonate is the most important anion. The total content of dissolved solids of the unpolluted world average river is about 100 mg l^{-1}. Of the major elements, pollution contributes most to sulphur, chlorine, and sodium.

Sea salt in the atmosphere affects the chemistry of surface waters, particularly when they drain from crystalline rocks near the ocean. The concentration of sea salt in precipitation decreases rapidly inland and with increasing altitude. The atmospheric contribution of sea salt as a percentage of actual world average river water is shown in Table 1.

Of the rain that falls on the continents, more than half is returned to the atmosphere, either by direct evaporation or by transpiration through plants. The net effect of evaporation on river water is to remove pure water from solution, so that the concentrations of all dissolved components tend to increase. Although evaporation takes place in all climates, it is only in relatively arid regions that it is of great importance for the composition of river water. The average concentrations in river water are about 20 times those in rain. The increase in concentration due solely to evaporation is only twice that of the concentration in rainwater. The difference is primarily due to contributions from rocks and soil during chemical weathering, and to pollution.

Rocks can be arranged according their tendency to chemical weathering or dissolution in increasing order as follows: granites, gneisses, mica-schists, gabbros, sandstones, volcanic

Table 1 Sources of major elements in world river water (percentages of actual concentrations)

| Element | Atmospheric cyclic salt | Weathering | | | Pollution |
		Carbonates	Silicates	Evaporites	
Ca^{2+}	0.1	65	18	8	9
HCO$_3^-$	≪ 1	61	37	0	2
Na$^+$	8	0	22	42	28
Cl$^-$	13	0	0	57	30
SO$_4^{2-}$	2	0	0	22	54
Mg^{2+}	2	36	54	≪ 1	8
K$^-$	1	0	87	5	7
H$_4$SiO$_4$	≪ 1	0	99+	0	0

Other sources of river SO$_4^{2-}$: natural biogenic emissions to atmosphere delivered to land in rain, 3 per cent; volcanism, 8 per cent; pyrite weathering, 11 per cent. (From Berner, E. K. and Berner, R. A. (1996).)

rocks, shales, serpentines and amphibolites, carbonates, and evaporites. The effect of the chemical weathering of rocks on the composition of river water is to integrate the chemical weathering tendency of each rock-type, its areal extension, and the amount of runoff from each rock-type. The contributions of the three main rock-types, carbonates, silicates, and evaporites, are shown in Table 1. The contribution of evaporite rocks is significant despite the fact that they cover only 1–2 per cent of the surface of the continents. This is due to the rapid dissolution of evaporites. Carbonates dominate world river chemistry. They dissolve rapidly, they are extensive, covering about 16 per cent of the surface of the continents, and they are located where runoff is abundant, primarily in the temperate regions. The bicarbonate (HCO_3^-) derived from weathering comes from two sources. One is the carbon in carbonate minerals, such as calcite and dolomite. The other is the result of the reaction of carbon dioxide dissolved in soil water and groundwater with carbonate and silica minerals. The carbon dioxide is derived almost entirely from the bacterial decomposition of organic matter in soil. Since the organic matter in soil gains its carbon from the atmosphere by photosynthesis, the ultimate source of much of the HCO_3^- in river water is the atmosphere. Of the 61 per cent of HCO_3^- derived from weathering of carbonates, 34 per cent comes from calcite and dolomite, and 27 per cent from soil CO_2. For silicates, however, all 37 per cent is derived from soil CO_2 (Table 1).

The dissolution rates of rock-forming minerals and their solubility are temperature-dependent: the higher the temperature, the faster the dissolution. The solubility of silicates increases with increasing temperature, but the solubility of carbonates decreases with increasing temperature. The weathering of silicate rocks is more temperature-dependent than the weathering of carbonates.

Great physical relief results in intense physical erosion and rapid exposure of fresh reactive rock for chemical weathering, but it provides less contact time between water and rocks. Because the removal of weathered material by transport is rapid with high relief, chemical weathering in the soil is incomplete, and soils are thin. With low relief, chemical weathering is much more complete and thick soils develop over the underlying rocks so that variations in rock-type become less important for the composition of rivers draining the soil.

Vegetation has two opposing effects on the composition of rivers. It increases chemical weathering by supplying CO_2 and organic acids to soil waters. However, vegetation also prevents physical weathering by stabilizing the soil and thus decreasing the exposure of bedrock to chemical weathering. The effects of rock-type, temperature, relief, and vegetation on the composition of river water are often hard to separate in a particular area.

Pollution can be an important source of some ions and major elements, such as sulphate (SO_4^{2-}), chlorine (Cl), and sodium (Na) (Table 1). Population density, type and state of industry (as reflected in energy consumption per capita), and land use affect river chemistry. In general, the major element chemistry of rivers draining highly populated areas is subject to significant variation.

The three major nutrients in river water are carbon, nitrogen, and phosphorus. Carbon in the form of carbon dioxide (CO_2) is generally abundant and readily available. Nitrogen gas (N_2), the most abundant atmospheric gas, must be 'fixed' or combined with hydrogen (as NH_4^+) or oxygen (as NO_3^-), in order to be used by plants and organisms. Nitrogen is fixed by micro-organisms and lichens; 'fixed' nitrogen is dissolved in precipitation; and nitrogen is 'fixed' industrially to produce fertilizers. The concentration of phosphorus in rainfall is low. The chief sources of phosphorus in river water are the weathering of rocks and the leaching out of fertilizers from agricultural land. In general, human activities have raised the levels of dissolved phosphorus and nitrogen to twice their natural values. In Western Europe and North America, however, concentrations of dissolved phosphorus and nitrogen in rivers have increased by factors of 10 to 50. Within individual watersheds, the pollution of rivers is proportional to the population density and the energy consumption.

The trace elements in river waters can be derived from the weathering of rocks, from volcanic eruptions, or, increasingly, from human activities. The processes controlling the concentration of trace elements in river waters are poorly understood. Even the average natural concentrations of dissolved trace elements in river waters are highly debated. Samples can easily be contaminated, detection limits of the analytical methods are sometimes higher than the natural levels, and few major pristine rivers have been studied. The reported average natural concentrations have been declining in recent decades, primarily because of improved sampling and analytical techniques. Considerable amounts of trace elements in river waters are adsorbed on to suspended inorganic and organic material in the water.   SIGURDUR REYNIR GISLASON

Further reading

Berner, E. K. and Berner R. A. (1996) *Global environment: water, air, and geochemical cycles*. Prentice-Hall, New Jersey.

roads and pavements In their simplest form, roads can be made by clearing vegetation and perhaps also shaping the ground surface to aid the run-off of surface water. Such 'dirt roads' are the most common type of road found worldwide. The soil may be treated to stabilize it, but generally undergoes only periodic shaping or 'regrading'. For all-weather traffic, and for the movement of heavy vehicles, a road structure or *pavement* is usually required. Its form depends on the properties of the soil on which it is built, the *subgrade* or *formation* (the ground beneath the road), and on considerations of cost.

The pavement distributes the traffic load to the subgrade, so that the soil undergoes only small, acceptable levels of deformation, thus ensuring that the pavement itself is not excessively deformed or cracked. The pavement also helps to protect the subgrade from the weather; for example, by preventing water from reaching the subgrade and softening it. The main structural component of the pavement is the road-base. Beneath the road-base are the sub-base and capping layer, which provide a firm platform for the construction of the road. Above the road-base are the surface courses on which the traffic runs. If the surfacing and road-base are of granular material bound with bitumen, the pavement is described as 'flexible'; if either layer is of concrete, the pavement is 'rigid'.   T. COUSENS

rock Although it may seem obvious what rock is, formulating a precise definition is not straightforward. Rock may be defined as any hard solid substance naturally occurring in the crust or mantle of a planet. Rock is usually composed of several different minerals, although a few varieties of rock consist of just a single mineral species. Rocks are classified into three main groups according to their mode of origin. Igneous rocks derive from material that has been melted (Fig. 1). Metamorphic rocks are produced by recrystallization of pre-existing rocks under the influence of heat or pressure, or both, but without melting (Fig. 2). Sedimentary rocks are formed from grains or particles that have been deposited, or from minerals that have precipitated from solution in a body of water (Fig. 3).

Most rocks on the Earth are silicates, in that their most abundant minerals contain the silicate group (silicon bonded to oxygen) in their chemical formulae. The most notable exceptions are limestones, which are mainly calcium carbonate, and evaporites, which consist of salts such as gypsum (calcium sulphate) and halite (sodium chloride) that result from the evaporation of saline water.

An unconsolidated deposit such as sand or loose ash from a volcanic eruption is not usually regarded as rock unless and until it becomes hardened (lithified) in some way. The most common ways for this to occur are when groundwater passes

Fig. 2. Gneiss, a metamorphic rock consisting of quartz, feldspar, and biotite (black). Because the assemblage of minerals grew in the solid state under conditions of differential stress and high temperature, gneiss, like most metamorphic rocks has a layered or foliated fabric. Note that the fabric is a product of the way the metamorphic rock formed and is not inherited from the pre-existing rock from which the gneiss was formed.

Fig. 3. Coquina, a bioclastic sedimentary rock consisting of broken fragments of shells and other biological debris cemented together by calcium carbonate.

Fig. 1. Granite, a coarse-grain igneous rock composed of quartz (grey), feldspar (white), and biotite (black). Note that the grains are randomly distributed, giving the rock a typical igneous texture. The coarse grain size indicates that the magma from which the granite formed, cooled, and crystallized very slowly.

through it and a mineral 'cement' is precipitated from solution to bind the grains together, or when the deposit is buried to a great enough depth for pressure to force the grains together into a solid mass.

Moving beyond the Earth to the other terrestrial planets, we find that the rocks there are also dominantly silicates. However, on most of the satellites of the outer planets and on Pluto the rigid material forming the surface, which by analogy should be called rock, is a mixture of ices (though it is unlikely to be pure H_2O) and salts.   DAVID A. ROTHERY

Further reading

Fry, N. (1984) *The field description of metamorphic rocks*. John Wiley and Sons, Chichester.

Thorpe, R. and Brown, G. (1985) *The field description of igneous rocks*. John Wiley and Sons, Chichester.

Tucker, M. E. (1982) *Sedimentary rocks in the field* (2nd edn). John Wiley and Sons, Chichester

Press, F. and Siever, R. (1986) *Earth* (4th edn), especially chapter 3, W. H. Freeman, New York.

rock coasts A large part of the world's coasts is rocky, and even many sand and stony beaches are underlain by shore platforms and backed by marine cliffs. Although our ability to identify and measure the processes that operate on rock coasts has improved, we are still largely ignorant of their precise nature and relative importance. The acquisition of quantitative data has been hindered by the very slow rates of change, the importance of high intensity–low frequency events, the exposed and often dangerous environments for wave measurement and subaqueous exploration, and the lack of access to precipitous or heavily vegetated cliffs. However, even a complete understanding of contemporary erosive processes could not explain the development of slowly changing coasts that frequently retain evidence of environmental conditions that were quite different from those of today.

Mechanical wave erosion usually dominates in storm and vigorous-swell wave environments, and although waves are generally much weaker in polar and tropical regions, they play an important role in removing the products of weathering. Rocks are dislodged by water hammer, high-shock pressures generated against structures by breaking waves, and, probably most importantly, by air compression in joints and other rock crevices. These processes require the alternate presence of air and water, and are therefore most effective in a narrow zone extending from the wave crest to just below the still-water level. Abrasion occurs where rock fragments or sand are swept over rock surfaces or swirled around in potholes. Although it is not as closely associated with the water level, the effectiveness of abrasion rapidly decreases below the surface. Most mathematical models suggest that waves exert the greatest pressures on vertical structures at, or slightly above, the water surface—where the most important mechanical wave erosional processes also operate. The level of greatest wave erosion on a rock coast must therefore be closely associated

with the elevation most frequently occupied by the water surface. This is at, or close to, the neap high and low tidal levels, and wave action is increasingly concentrated between the neap tidal levels as the tidal range decreases.

Cliff erosion also occurs through weathering, and removal of the debris by mass movement and wave action. Weathering is particularly important in high and low latitudes, where there are suitable climates and weak wave environments. It can also play a significant role in sheltered areas within the vigorous-wave environments of the middle latitudes, however, and it prepares rocks for eventual dislodgement and removal by strong waves in more exposed areas. Cool coastal regions provide almost optimum environments for frost action. Rocks in intertidal and spray and splash zones are able to attain high levels of saturation, and because of freezing and thawing caused by the fall and rise of the tide, they experience many more frost cycles than further inland. Alternate wetting and drying in the spray and intertidal zones also creates suitable conditions for many chemical and salt weathering processes. These assume dominant roles in some warm-temperature and tropical regions, but they are probably only important in sheltered areas within cooler, storm-wave environments. In addition to their erosive role on cliffs and shore platforms, chemical and salt weathering contribute to the development of tafoni and honeycombs, and to the suite of processes collectively referred to as water-layer levelling.

There is continuing debate over the processes operating on limestone coasts. The sharp pinnacles, ridges, grooves, and circular basins that are characteristic of coastal limestones in the spray and splash zones are similar to the karren formed by fresh water on land. Although surface sea water is normally saturated or supersaturated with calcium carbonate, solution could occur in rock pools at night, when the carbon dioxide produced by faunal respiration is not extracted by algae during the hours of darkness. *Bio-erosion*, the removal of the substrate by direct organic activity, is probably most important in tropical regions, where waves are generally fairly weak and an enormously varied marine biota live on coral, aeolianite, and other calcareous substrates.

Coastal scenery is the product of a combination of elements, including: the morphology of the hinterlands; present and past climates; wave and tidal environments; changes in relative sea level; and the structure and lithology of the rocks. The occurrence of small bays and headlands often reflects the rather subtle influence of rock structure, including variations in joint density, orientation of discontinuities, and thickness, strike, and dip of the beds. Large bays and prominent headlands are more likely to develop where there are differences in the resistance to erosion of rocks of contrasting type. Because of the effect of wave refraction, the plan shape of crenulated coasts can attain an equilibrium state. This would occur where resistant rocks on exposed headlands are eroded by higher waves at the same rate as weaker rocks are eroded by lower waves in the sheltered bays. The plan shape would then be maintained through time as the coast retreated landwards.

Small bays, narrow inlets, caves, arches, stacks, and related features generally result from accelerated erosion along structural weaknesses, particularly bedding, joint, and fault planes. They form in rocks that have well-defined and well-spaced planes of weakness, yet are strong enough to stand as high, near-vertical slopes, and as the roofs of caves, tunnels, and arches. They are therefore uncommon in weak or thinly bedded rocks with dense joint systems.

Coastal slopes tend to be greater in the vigorous, storm-wave environments of the mid-latitudes than in high latitudes or the tropics, where frost and chemical weathering, respectively, are more important. Geological structure and rock type are at least as significant as climate and wave regime in determining the relative efficacy of marine and subaerial processes and the shape of cliff profiles. Composite cliffs consist of two or more major slope elements. They include multi-storied cliffs, with two or more steep faces separated by more gentle slopes, and bevelled cliffs with a convex or straight seaward-facing slope above a steep, wave-cut face. Some composite profiles reflect geological differences, but high cliffs in resistant rocks have been attributed to Quaternary changes in climate and sea level.

The erosion of sea cliffs is episodic, site-specific, and closely related to prevailing meteorological conditions. Cliff recession rates have been measured in many parts of the world, although much of the data are little more than rough estimates. Reported rates vary from virtually nothing up to 100 m year^{-1}, but it is difficult to identify patterns that can be correlated with variations in rock type or wave and tidal conditions.

Some workers believe that the susceptibility of calcareous rocks to bioerosion and possibly chemical solution distinguishes limestone coasts from those in other types of rock. Nevertheless, exposure to wave action, rock structure, and other factors are responsible for considerable differences in limestone coasts between and within climatic regions. Deep, narrow, horizontal notches usually develop in warm seas where low waves and tidal ranges concentrate the erosive processes within a narrow range of elevations. As tidal range increases, a single notch marking mean sea level may be replaced by one at the high and another at the low tidal levels.

One or more gently sloping erosional terraces occupy the hinterlands of many rock coasts. There are wide planation surfaces along tectonically stable coasts, and flights of emergent marine terraces, ranging from a few metres to several kilometres in width, and up to hundreds of metres above sea level, along tectonically active coasts. Although the most effective erosional processes generally operate at the water level, most workers have assumed that these wide terraces were cut by waves or currents at considerable depths beneath the water surface. Using a mathematical model to simulate terrace development, Trenhaile found that wave erosion within the intertidal zone could produce surfaces between 1 and 3 km in width during a single glacial–interglacial cycle, and much wider terraces at the end of five cycles. This model provided a possible explanation for the way in which very wide marine erosion surfaces are formed, and it supported the contention that they were produced intertidally, particularly when sea level was rising rapidly during periods of deglaciation.

A. S. TRENHAILE

Further reading

Sunamura, T. (1992) *The geomorphology of rocky coasts*. John Wiley and Sons, Chichester.

Trenhaile, A. S. (1987) *The geomorphology of rock coasts*. Cavendish–Oxford University Press, Oxford.

Trenhaile, A. S. (1997) *Coastal dynamics and form*. Cavendish–Oxford University Press, Oxford.

rock coatings
Approximately one-sixth of the Earth's land surface consists of 'bare' rock surfaces. Yet the true mineral faces of the rock are rarely seen. Instead, a variety of paper-thin accretions coat rocks in all terrestrial environments. Since weathering includes the formation of new compounds that are more in equilibrium with the environment at and near the Earth's surface, rock coatings fall under the interdisciplinary field of weathering. Intellectual curiosity about the physical and chemical characteristics, origin, geography, and utility of these encrustations has spawned over 3000 scientific papers. A wide variety of rock coatings is found on landforms at the Earth's surface.

Rock varnish
Rock varnish is the most studied rock coating. Although it is found in all terrestrial settings, it is most easily recognized on the plentiful bare rocks seen in deserts (hence another common name is 'desert varnish'). Also, the physical and chemical stability of rock surfaces in deserts allows sufficient time for this slow-growing accretion to form. It is structured much like a brick wall, but its thickness (ranging from less than 5 micrometres (5 μm) to almost 500 μm) is typically less than 30 μm. The clay materials that comprise the bulk of the structure (about 50 to 70 per cent) are cemented to the rock by hydroxides of manganese and iron (15–50 per cent). The enigma of varnish is its great enrichment in manganese, typically over 50 times compared with the adjacent environment (soils, underlying rock, dust).

The debate over the origin of rock varnish revolves around the source of the manganese and the mechanisms of its enrichment. Until the late 1970s and the more common use of scanning electron microscopy, the majority view was that the manganese and the varnish were derived from the underlying rock. Now, investigators see a clear morphological varnish–rock boundary, together with other evidence that shows varnish to be an accretion. Although most investigators now agree on an external origin, there are still two hypotheses on the mechanism of manganese concentration. The abiotic hypothesis holds the small changes in pH can concentrate manganese by geochemical processes. The biotic hypothesis holds that bacteria, and perhaps other micro-organisms, concentrate manganese—an idea supported by culturing experiments and direct observations of bacterial enhancement of manganese.

Silica glaze

The term 'silica glaze' covers a broad category of rock coatings that are dominated by amorphous silica with variable amounts of aluminium and iron. They are usually less than 200 μm thick, with a clear white to orange shiny lustre, but they can be darker in appearance. Silica glazes have been noted in warm deserts, in cold deserts like Antarctica, on dry tropical islands, along tropical rivers, in mid-latitude humid temperature settings, and in various archaeological contexts. Silica glazes probably precipitate from soluble Al–Si complexes $[Al(OSi(OH)_3)^{2+}]$ that are released from the weathering of phyllosilicate minerals (e.g. micas).

Iron films

Rusty-coloured rocks are readily recognized as 'iron oxides' and are often readily dismissed once such a label has been given. Once iron films are examined in detail, they display a wide variety of characteristics, and different types of iron films occur in very different environmental circumstances. For example, subaerial rock surfaces in hyper-arid deserts host iron films characterized by clay minerals cemented with about 10 per cent iron. Subaerial dolerite rock surfaces in the Dry Valleys of Antarctica are rimmed by iron oxyhydroxides that form both an accretion and a weathering rind over a millimetre thick. Rocks in acid streams in Arctic and alpine settings are often impregnated with iron hydroxides that can physically separate pieces of the rock, much like salt or frost weathering.

Other rock coatings

Other rock coatings include biofilms, which are organic coatings, such as lichens, moss, fungi cyanobacteria, and algae. Carbonate skins are composed primarily of carbonate, usually calcium carbonate, but the carbonate is sometimes combined with magnesium. They can occur as tufas, caliche exposed by rock spalling, or even as coatings on urban buildings. Case-hardening agents represent the addition of cementing agents to rock-matrix material; the agent may be manganese, sulphate, carbonate, silica, iron, oxalate, organisms, or anthropogenic. Dust films are a light powder of clay- and silt-sized particles that adhere to rough surfaces and rock fractures. Heavy metal skins form on rocks downstream from mine tailings; they can have high concentrations of copper, lead, zinc, manganese, iron, cadmium, mercury, and other heavy metals. Nitrate crusts of potassium and calcium nitrate are most often found in caves and rock shelters in limestone areas.

Oxalate crusts are typically composed of calcium oxalate and silica with variable concentrations of magnesium, aluminium, potassium, phosphorus, sulphur, barium, and manganese. They are often less than a millimetre thick and can be found forming near or alongside lichens. Phosphate skins are comprised of various phosphate minerals (e.g. iron phosphates or apatite) that are mixed with clays and sometimes manganese. Polish films form where aeolian and glacial abrasion produce micrometre-scale films that reflect the composition of the underlying rock. Sulphate skins are the superposition of sulphates (e.g. barite, gypsum) on rocks; these are distinguished from gypsum crusts in the soils literature.

Applications

Geomorphologists study rock coatings in part because of the possibility of using them as indicators of time or environmental change. Two developments stand out. The first is the radiocarbon dating of organic matter trapped under rock coatings such as silica glaze; the organic compounds may be directly underneath the coating or encapsulated in the weathering rind of the host rock. Radiocarbon dating of organic compounds trapped by rock coatings has been used experimentally in the dating of such features as alluvial fans, colluvium, and landslides, as well as archaeological rock engravings. Another development is the study of microlaminations; various different layers in rock varnish have been used to correlate rock surfaces in Death Valley (California) and Tunisia (North Africa) with dry and wet episodes of climatic change.

RONALD I. DORN

Further reading

Dorn, R. I. (1998) *Rock coating*. Elsevier, Amsterdam.

Konhauser, K. O., Fyfe, W. S., Schultze-Lam, S., Ferris, F. G., and Beveridge, T. J. (1994) Iron phosphate precipitation by epilithic microbial biofilms in Arctic Canada. *Canadian Journal of Earth Science*, **31**, 1320–4.

Liu, T. and Dorn, R. I. (1996) Understanding spatial variability in environmental changes in drylands with rock varnish microlaminations. *Annals of the Association of American Geographers*, **86**, 187–212.

Robinson, D. A. and Williams, R. B. G. (1992) Sandstone weathering in the High Atlas, Morocco. *Zeitschrift für Geomorphologie*, **36**, 413–29.

rock-forming silicate minerals About one-third of all known mineral species are silicates. Many of them are rare, but a few families—the olivines, pyroxenes, amphiboles, micas, clay minerals, feldspars, quartz, and garnets — make up some 95 per cent of the Earth's crust. According to on one particular model, the mineralogy of the upper mantle is probably 37–51 per cent olivine, 26–34 per cent orthopyroxene, 12–17 per cent clinopyroxene, and 10–14 per cent garnet. Since the mantle extends down to 2900 km it is clear that the most abundant mineral on Earth is olivine or its equivalent.

Silicates crystallize under widely differing physico-chemical conditions ranging from those at the base of the mantle to low-temperature, low-pressure, watery environments in the pores of sandstones. They may have highly variable complex compositions; on the other hand, many minerals are made from the same restricted range of compositions but under different conditions.

Silicate minerals are classified primarily on the basis of their crystalline structure. The essential basic unit is the silicon–oxygen tetrahedron with one silicon atom surrounded

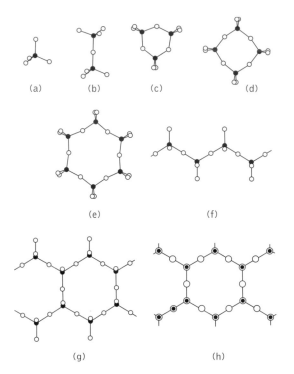

Fig. 1. Structures of the silicates. (a) island or neso (SiO_4); (b) group or soro (Si_2O_7); (c) ring or cyclo 3-member (Si_3O_9); (d) ring or cyclo 4-member (Si_4O_{12}); (e) ring or cyclo 6-member (Si_6O_{18}); (f) single chain or ino (SiO_3); (g) double chain or ino (Si_4O_{11}); (h) sheet or phyllo (Si_4O_{10}). Open circles are oxygen; filled circles are silicon. Other cations omitted.

by and strongly bonded to four oxygen atoms located at the apices of the tetrahedron to give an $(SiO_4)^{4-}$ anionic unit. The SiO_4 units may be present as isolated tetrahedra linked by cations or they may progressively 'polymerize' by sharing one or more oxygen atoms with adjacent tetrahedra so that ultimately all oxygen atoms are shared to give a framework structure (Fig. 1).

Olivines

'Olivine' is the general name given to a group of magnesium–iron silicates with a composition that is mainly $(Mg,Fe)_2SiO_4$ with minor amounts of nickel, manganese, and calcium. It is a continuous solid-solution series from forsterite (Mg_2SiO_4) to fayalite (Fe_2SiO_4). There are names for intermediate members of the series, but it is now more usual to quote compositions in terms of the percentage of forsterite present; e.g. Fo_{20}.

The melting and crystallization relations of dry olivine under atmospheric conditions are shown in Fig. 2a. The upper curve is the *liquidus*, the lower one the *solidus*; above the liquidus olivine is liquid, below the solidus it is solid, and between the two curves olivine and liquid coexist. Consider a liquid of Fo_{50} composition at 1900 °C (point A on the

diagram), and let it cool. The cooling liquid follows a vertical line until it meets the liquidus (at point B) at a temperature of about 1675 °C. It then begins to crystallize, but the composition of the first crystal to form is not Fo_{50} but Fo_{80}. This figure is derived by drawing a horizontal line from B to the solidus and then dropping a perpendicular to the base at C. As the temperature falls, the composition of the crystal is always richer in magnesium than the liquid in which it is growing. At 1410 °C all the liquid has crystallized, and if the cooling has been slow the result is a homogeneous olivine of Fo_{50} composition. If, however, at some stage the olivine crystals are removed, the liquid will have been depleted in magnesium. This process of *fractional crystallization* exemplified here by the olivines is important in the derivation of a variety of magmas from a single homogeneous parent.

Olivine has orthorhombic symmetry, a hardness of 6.5 and no marked cleavage. Its density varies with composition, and its colour becomes a deeper green as the iron content increases; it may then be the gemstone *peridot*. Olivine is prone to alteration: mainly to serpentine, sometimes to chlorite.

Olivine is common as phenocrysts in basalts, as a major component in some dolerites and gabbros, and as the dominant mineral of ultramafic rocks such as peridotites and dunites. Fayalite occurs in acid lavas.

Pyroxenes

There are some 21 species in the pyroxene group, many of them rare. The pyroxenes are anhydrous single-chain silicates with the chains parallel to the z axis of the crystals. The chains are linked by cations, predominantly calcium, iron, and magnesium. The general formula is $X_{1-p} Y_{1+p} Z_2 O_6$, where p can have values between 0 and 1; X can be calcium or sodium, Y can be manganese, iron (Fe^{2+} Or Fe^{3+}), magnesium, aluminium, chromium, or titanium; and Z can be silicon or aluminium. A wide range of substitutions is possible.

Figure 2b shows the range of compositions of the common pyroxenes. Calcium-rich species in the wollastonite area are not pyroxenes but pyroxenoids. There are three principal series: (1) $Mg_2Si_2O_6$–$Fe_2Si_2O_6$, a complete solid-solution series from enstatite to ferrosilite. Members with less than 5 per cent $Ca_2Si_2O_6$ are orthorhombic; all other pyroxenes are monoclinic. Enstatite occurs in plutonic basic and ultrabasic rocks such as peridotite and norite, and also in some high-grade metamorphic rocks. (2) Diopside–hedenbergite, a complete solid-solution series formed mainly by the metamorphism of siliceous dolomites and impure iron-rich carbonates. (3) Augite, with a more complex composition including aluminium, is a very common phenocryst in basalts and is the commonest pyroxene.

Pyroxenes have two good vertical cleavages intersecting at 87°, variable, mainly dark, colours, a hardness of about 6, and are prone to alteration to hornblende, chlorite, and clay minerals.

Some members of the pyroxene family are used as geothermometers and geobarometers.

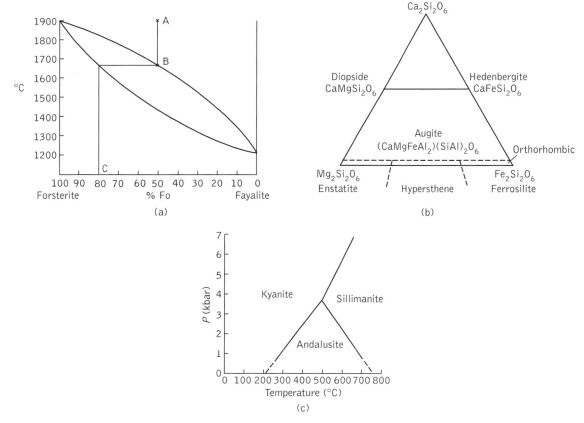

Fig. 2. (a) Melting–crystallization relations of the olivines at atmospheric temperature and pressure. (b) Composition of the pyroxenes. (c) Pressure–temperature relations of the aluminium silicates.

Amphiboles

Some 60 species make up the amphibole family of double-chain silicates. The chains are bonded by chains of cations, all chains being parallel to the vertical axis. In contrast to the pyroxenes they all have hydroxyl (OH) and several have fluorine or chlorine in their structure; there is a wide and complex variation in chemical composition and sometimes extensive solid solution. The amphiboles are stable over a wide range of physical conditions in both igneous and metamorphic rocks.

The general composition of the amphiboles can be written as $W_{0-1}X_2Y_5Z_8O_{22}(OH, F, Cl)_2$, the W cations being Ca, Na, or K; the X cations Ca, Mg, Fe, or Mn; the Y cations Fe, Ti, or Al; and the Z cations being Al or Si. With a few exceptions amphiboles are monoclinic with two good prismatic vertical cleavages intersecting at angles of 56° and 124° (another feature that distinguishes them from pyroxenes). They are usually prismatic or fibrous, elongated parallel to the vertical axis, with a hardness of 5.5–6, a specific gravity of 2.9–3.5, and a wide range of colours varying with composition.

Hornblende covers three series with a range of compositions in which sodium and aluminium are important. Common hornblende is the most widespread of the amphiboles, with a dark green to black colour. It is present in most major groups of plutonic igneous rocks, especially the calc-alkali intermediate ones. It is common in rocks formed by the medium to high-grade metamorphism of impure limestones and basic igneous rocks to give amphibolites.

The tremolite–actinolite series is non-aluminous and varies in composition from Ca_2Mg_5 to $Ca_2(MgFe)_5$. These minerals are white, becoming green with increasing iron. They are usually fibrous; there is a densely felted variety, nephrite, formed by low- to medium-grade metamorphism of siliceous dolomites with additions of iron.

Glaucophane–riebeckite are sodium-bearing, the main substitutions being between magnesium, aluminium, and iron (Fe^{2+} and Fe^{3+}). Their colour varies from pale blue to lavender blue to blue-black. One of the varieties is the dangerous blue asbestos, crocidolite. Glaucophane and riebeckite are found in low-grade schists of basic igneous parentage; crocidolite

907

occurs in medium-grade metamorphosed ironstones. Riebeckite is also found in granites and syenites.

Micas

Of the 30 or so micas, only biotite, muscovite, and phlogopite are common and widespread. The micas are sheet silicates in which oxygens at three corners of the tetrahedron are shared with neighbouring tetrahedra, leaving one free oxygen. Each sheet of tetrahedra has hexagonal symmetry; two such sheets are linked by cations to give three-layer units, which are usually linked by potassium atoms with weak bonding. This results in perfect cleavage parallel to the basal plane of the minerals that gives flexible and elastic sheets.

The general composition of the group can be written as $W_{0-1} Y_{2-3}(Z_4 O_{10}) (OH,F)_2$ the main atoms in the W position are K, Ca, Na, those in Y are Al, Fe, Mg, Li, and those in Z are Al or Si (usually Al Si). All members contain hydroxyl (OH); some have fluorine. They crystallize in the monoclinic system as hexagonal books with perfect basal cleavage and a platy habit; hardness varies from 2 to 4 and specific gravity from 2.70 to 3.30 according to composition, as does colour from colourless to violet, bright green, silvery, and dark brown to black.

Biotite, $K(Mg\ Fe)_3(Al\ Si_3 O_{10}) (OH)_2$, is the most common mica. It occurs in most igneous rock groups but is mainly in intermediate plutonics and in pelitic schists and gneisses; it is an index mineral in low-grade schists of regional metamorphic origin.

Phlogopite, $K\ Mg_3 (Al\ Si_3 O_{10}) (OH)_2$ occurs in kimberlites and metamorphosed siliceous dolomitic limestones.

Muscovite, $K\ Al_2(Al\ Si_3 O_{10}) (OH)_2$, is an important constituent of pelitic schists and gneisses, of some granites and granite pegmatites, and of greisen. *Sericite* is fine-grained muscovite formed by alteration of such minerals as feldspar, cordierite, and sillimanite.

Lepidolite, $K(Li\ Al)_3(Al\ Si_3 O_{10}) (OH, F)_2$ is found in pegmatites and is a source of lithium.

Glauconite is close in composition to muscovite but has some iron, magnesium, sodium, and calcium. It is forming at present in marine environments where sediment supply is limited, and so is authigenic. Because it contains potassium it can be dated by radiometric techniques and can thus give the age of the sedimentary rock directly.

The main use of micas, particularly muscovite and phlogopite, is as sheets for electrical and thermal insulation, and as windows in stoves and furnaces; the variety vermiculite, when treated, is used for insulation blocks and for horticultural purposes.

Feldspars

Feldspars are the most important group of minerals, making up 60 per cent of the rocks of the Earth's crust. They are framework aluminium silicates of potassium, sodium, calcium, and (minor) barium with all the tetrahedral oxygen atoms shared; this would theoretically make the composition SiO_2, but in most feldspars one or two in four silicon atoms are replaced by aluminium. This substitution results in a charge imbalance that is offset by the addition of the cations sodium, potassium, or calcium, which fit into spaces in the structure. The compositions can be represented on a triangular diagram (Fig. 3a), in which the apices represent the three end-members orthoclase, albite, and anorthite. Nearly all feldspars so far analysed plot within the shaded area, and it can be seen that they form two series, both of which are solid-solution series: orthoclase to albite, the alkali feldspars, and albite to anorthite, the calc-alkali or plagioclase feldspars. There are considerable structural variations in the feldspars according to their temperature of formation, which influences the 'ordering' of the aluminium and silicon atoms and the composition.

Alkali feldspars. At temperatures above about 660 °C complete solid solution occurs between the two alkali feldspar end-members sanidine and high albite, but below this there is a break and two feldspars of different composition occur together. The result is an intergrowth of the two, usually with thin lamellae of albite in a host of orthoclase or microcline; this intergrowth is called *perthite* if visible to the naked eye or *microperthite* if visible only under the microscope. This differentiation within one crystal is called *exsolution* and takes place in the solid state.

Potash feldspars. The potash feldspars are polymorphic: sanidine is the monoclinic high-temperature form, microcline the triclinic low-temperature form, and monoclinic orthoclase an intermediate form. Sanidine is found only in quickly cooled lavas such as trachytes; microcline occurs in slowly cooled granites. The potash feldspars are usually pale coloured: glassy clear, white, pink or, rarely, bright green (as the variety amazonite); their hardness is about 6, their specific gravity 2.56–2.71. There are two good cleavages parallel to (001) and (010), at right angles in orthoclase and sanidine and at nearly 90° in microcline. Interpenetrant twinning on the carlsbad law is characteristic of sanidine and orthoclase (Fig. 3d).

Plagioclase. There is complete solid solution among the members of the plagioclase series at high temperatures, but complicated structures and a somewhat incomplete series exist at lower temperatures. This important series is arbitrarily divided into six distinct minerals easily recognized under the microscope but not in hand specimen. The boundaries and names are given in Fig. 3a; compositions are given in terms of the proportion of anorthite and are written, e.g., An_{75}. The melting–crystallization relations of the plagioclases are shown in Fig. 3b. The similarity with those of the olivines is obvious; under atmospheric conditions anorthite melts or crystallizes at 1540 °C, albite at about 1100 °C; a plagioclase of An_{50} composition will begin to melt at 1280 °C but will not be completely molten until 1440 °C. By analogy with the olivines it will be seen that the first crystals to form will be rich in calcium, thus enriching the liquid in sodium. Zoning often results in crystals having centres rich in calcium and rims rich in sodium. Plagioclases are triclinic, white or pale green, with hardness 6–6.5 and specific gravity 2.6–2.75. They twin on a variety of laws, the most common being the albite law in which the twins are multiple, parallel to (010), and visible as

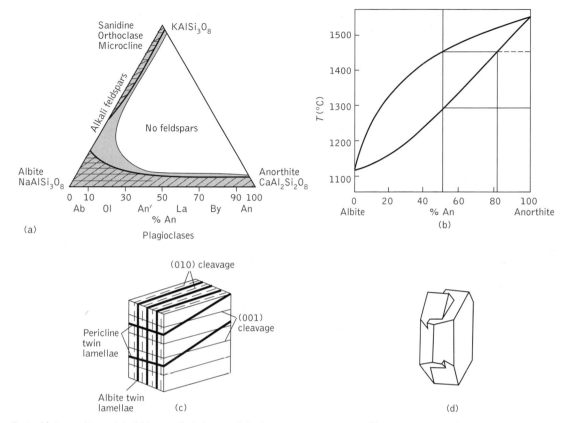

Fig. 3. (a) Compositions of the feldspars. Shaded area, solid solution at high temperature; ruled area, solid solution at low temperature. Ab, albite; Ol, oligoclase; An', andesine; La, labradorite; By, bytownite; An, anorthite. (b) Melting relations of plagioclase under atmospheric conditions. (c) Twinning and cleavage in plagioclase. (d) Interpenetrant carlsbad twin of orthoclase.

bands in hand specimen; pericline twins are also multiple (Fig. 3c).

Plagioclases are common in most kinds of igneous rocks. They range from bytownite–labradorite in basalts and gabbros, andesine–oligoclase in diorites and andesites to albite in granites and rhyolites. They are major constituents of metamorphic rocks such as pelitic schists and gneisses and amphibolites; the composition becomes more calcium-rich with increasing temperature of formation. Pure anorthite is rare in igneous rocks but is found in calc-silicate rocks formed by the metamorphism of impure siliceous limestones and marls.

Feldspathoids

The feldspathoids are a group of comparatively rare anhydrous aluminous framework silicates of sodium and potassium; there are large cavities in the structure, which can accommodate K, Na, CO^{2-}_3, and SO^{2-}_4 ions. The feldspathoids are similar to the feldspars but contain less silica; they are unstable in the presence of quartz and are characteristic of undersaturated (silica-poor) rocks.

The three commonest feldspathoids are leucite, nepheline, and sodalite.

Leucite (K Al Si$_2$O$_6$) is tetragonal (but pseudocubic) at temperatures below 625 °C; cubic, icositetrahedral, above 625 °C. It is unique among silicates in being less dense than the liquid from which it crystallizes. It occurs in lavas and hypabyssal intrusions but not in plutonic rocks. A well-known locality is that of Mount Vesuvius.

Nepheline (Na Al Si O$_4$) is hexagonal. It occurs in lavas and some plutonic rocks, such as syenites.

Sodalite (Na$_8$Al$_6$Si$_6$O$_{24}$) (Cl)$_2$ is cubic, dodecahedral. It occurs in nepheline-syenites. Lazurite, the deep blue variety of sodalite with sulphur replacing some of the chlorine, is dominant in lapis-lazuli.

In general the feldspathoids are pale-coloured, with a hardness of about 6 and specific gravity about 2.5.

Garnets

Garnets are a complex group of island silicates characteristic of metamorphic rocks, rare in igneous rocks such as rhyolites

909

and granites and also in kimberlites, serpentinites, and eclogites, and as detrital grains in sediments. The general formula is $X_3Y_2(SiO_4)_3$, in which X can be magnesium, iron, manganese, or calcium and Y can be aluminium, iron, or chromium. The garnets are assigned to two main groups: pyralspites, consisting of pyrope, $Mg_3Al_2(SiO_4)_3$, almandine, $Fe_3Al_2(SiO_4)_3$ and spessartine, $Mn_3\,Al_2(SiO_4)_3$; and ugrandites, consisting of grossular $Ca_3Al_2(SiO_4)_3$, andradite, $Ca_3Fe_2(SiO_4)_3$ and uvarovite, $Ca_3Cr_2(SiO_4)_3$. There is much compositional variation within the groups but little between them. They are stable over a wide range of temperature and pressure. For example, in the Barrovian metamorphic complex of the Scottish Highlands they occur in rocks that crystallized over the range 350–650 °C and up to 12 kbar pressure. The garnets crystallize in the cubic system, with rhombdodecahedra and icositetrahedra as the common forms; they have poor cleavage, a hardness about 7, specific gravity from 3.4 to 4.6, and highly variable colour: dark red to almost black, colourless, brown, green, greenish yellow, rose red, yellow, orange, red-brown, or emerald green, according to the composition.

Almandine is the most common garnet. It occurs in middle- and high-grade schists and gneisses derived from pelitic rocks; it is the index mineral of one of the Barrovian zones and can also be found in contact-metamorphic aureoles. Grossular results from both regional and thermal metamorphism of impure limestones; spessartite and andradite are mainly found in metasomatic skarn deposits; pyrope occurs in kimberlites and eclogites; uvarovite is rare and is found in serpentinites, usually associated with chromite, and rarely in limestone skarns.

The main uses of garnets are for abrasives and grinding materials; they also provide semi-precious gemstones such as rhodolite, demantoid, and grossular.

Aluminium silicates

The aluminium silicate minerals are an important group of island silicates occurring almost exclusively in metamorphic rocks derived from pelitic protoliths. Three of them, andalusite, kyanite, and sillimanite, form one of the best-known examples of polymorphism with a composition $Al_2Si\,O_5$ or $Al_2O_3.SiO_2$. Figure 2c shows the possible stability relations. Sillimanite is the high-temperature form and cannot exist stably below 500 °C; kyanite is stable at high pressure, and andalusite at low pressure. At temperatures above 1300 °C sillimanite converts to mullite plus silica liquid. (Mullite is not a polymorph because it has a different chemistry, $Al_6O_5(SiO_4)_2$ or $3\,Al_2O_3\,2\,Si\,O_2$.)

Andalusite is orthorhombic, crystallizing as elongate square prisms which at low temperature have a conspicuous cross-shaped arrangement of dark inclusions giving the variety chiastolite. There are two good vertical cleavages almost at right angles. The hardness is 7.5. The colour is usually grey or white, occasionally violet. Andalusite is mainly a contact-metamorphic mineral, but it also occurs in regional schists formed at low pressure.

Kyanite is triclinic with three cleavages and variable hardness (4–7 on a single face). In colour it ranges from pale green to blue to colourless. It forms good porphyroblasts, often bent, in schists and gneisses and is an index mineral in the Barrovian zonal system.

Sillimanite usually grows as elongate orthorhombic prisms but often as felted laths when it is known as *faserkiesel*. It is grey- or greenish-white with a hardness of 6 to 7 and a single good vertical cleavage. It is the index mineral of the highest grade of regional metamorphism.

All three minerals alter to shimmer aggregates of sericite; all three are used in the preparation of refractories.

Staurolite has a similar structure to kyanite but has a high iron content and contains some magnesium. It is pseudo-orthorhombic, short or long columnar, with hardness 7.5. In colour it varies from yellowish to dark reddish brown. Staurolite is well known for its interpenetrant twins, the two parts being related at angles of 60° or 90°; the latter are known as 'fairy crosses'. Staurolite is characteristic of medium- to high-grade regional metamorphism of argillaceous parents rich in iron.

Zeolites

The zeolites are a group of 40 or so hydrated aluminosilicates with a very open framework structure of $(SiAl)\,O_4$ tetrahedra in which sodium and calcium, more rarely potassium and barium, strontium, and water molecules (H_2O) fit into the interconnecting cavities and channels. The water can be driven off as a continuous and partially reversible process without damaging or disrupting the structure; the 'cavity' atoms are deposited in the holes. There are two main groups: (1) fibrous with chains of tetrahedra in the framework and represented by natrolite $Na_2(Al_2\,Si_3\,O_{10}).2H_2O$ with acicular habit; (2) equant or cagelike, such as chabazite $Ca_2(Al_2Si_4O_{12}).6H_2O$.

Because of their open structure zeolites have a low specific gravity ranging from 1.9 to 2.4, a hardness of 3.5–5.5, and are usually colourless or white when pure. They are, in general, late-stage minerals deposited from solutions in vesicles or joints in lavas, but phillipsite is a neoformational mineral that is being formed at present in deep-sea sediments at the water–sediment interface at a temperature of about 4 °C. Zeolites can also be produced by low-grade metamorphism of volcanic glass or tuffaceous sediments.

Zeolites are used as molecular sieves to remove large ions or complexes from solution, i.e. as absorbents, dessicants, and purifiers; in catalytic cracking in the petroleum industry; as fillers in papers; and as cationic exchangers in water softening. Many zeolites are now made synthetically.

Quartz and the clay minerals are dealt with in separate entries.

R. BRADSHAW

Further reading

Deer, W. A., Howie, R. A., and Zussman, J. (1966) *An introduction to the rock-forming minerals.* Longman, London.

Schumann, W. (1992) *Rocks, minerals and gemstones.* Harper Collins, London.

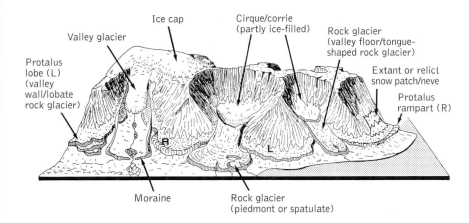

Fig. 1. Rock glaciers and associated features. (After Hamilton and Whalley (1995).

rock glaciers Rock glaciers are alpine features best defined by their morphology and appearance, since there is debate about their origin and significance. They are elongate tongues of rock rubble which flow slowly downhill. The surface velocity is generally less than 1 m year^{-1}, although some with greater velocities (> 5 m year^{-1}) have been described. If no movement can be detected, rock glaciers are generally referred to as 'relict'; if highly vegetated with subdued features, as 'fossil'. Flow-related features are commonly seen as ripples on the surface, although it is not known whether these extend to any depth. The length is typically about l km, but many shorter examples exist and some may be up to 3 km long; their width is generally a few hundred metres. The features typically have their heads in corries (cirques) in which there may be a small glacier, but this is not always seen. The elongate forms are best considered as being rock glaciers proper, although the term has also been used for features that are broader than long and which (typically) have their upper sections against cliffs or scree. It has been suggested that the latter features are best termed 'protalus lobes' (see Fig. 1).

There are three main theories on the origin of rock glaciers. One suggests that they are glacially derived with a veneer of weathered rock debris over glacier ice. The debris protects the ice from melting and sustains a thin body of ice that would otherwise have melted away. Because this glacier core is thin, it flows only slowly. A second 'permafrost model' explains the slow movement as creep of ice dispersed within scree and finds that a glacial derivation is not necessary. It does, however, demand permafrost conditions (usually seen as a mean annual air temperature of less than −1.5 °C) for the formation of the ice and its continued creep. The third model suggests that some rock glaciers (or protalus lobes) are formed by sudden, perhaps catastrophic, failure of scree slopes, but this view is not widely held as a general model. Rock glaciers have been used as indicators of permafrost (or past permafrost if relict features are considered) but this applies only if the permafrost model is valid. Until recently,

the origin and flow mechanisms were generally attributed to the permafrost model. Although traces of glacier ice could be seen in some rock glaciers, permafrost conditions were considered to be the only way in which the features could exist. Observations of glacier ice down the length of some rock glaciers have now been established and thus at least some rock glaciers are known to have glacier ice cores. It is possible that modern dating and isotopic techniques will allow ice from such rock glaciers to provide a climatic record. The full implications for recognizing climate change through rock glaciers still require investigation.

W. BRIAN WHALLEY

Further reading

Giardino, J. R., Shroder, J. F., and Vitek, J. D. (eds) (1987) *Rock glaciers*. Allen and Unwin, London.

Hamilton, S. J. and Whalley, W. B. (1995,) Rock glacier nomenclature: A reassessment. *Geomorphology*, **14**, 73–80.

Martin, H. E. and Whalley, W. B. (1987) Rock glaciers: a review, Part 1. *Progress in Physical Geography*, **11**, 260–82.

Whalley, W. B. and Martin, H. E. (1992) Rock glaciers: II models and mechanisms. *Progress in Physical Geography*, **16**, 127–86.

rock mechanics Rock mechanics is the theoretical and applied science of the mechanical behaviour of rock and rock masses; it is the branch of mechanics that is concerned with the response of the rock and rock mass to the force fields of their physical environment. It is an extremely complex subject that has only since the 1950s become a science, and it remains a blend of advanced analytical techniques and 'rules of thumb'.

Throughout history, underground openings have been used as shelter for both the living and the dead. Man has searched avidly for minerals, has tunnelled for storage, for water, to make roads, and to carry out sieges or escape from them. The importance of underground space is increasing as man explores the options for new uses, such as the disposal of radioactive waste or underground storage. Civil engineering projects of all types can entail rock excavation, whether at the surface or at depth.

Several inherent complexities render the scientific aspects of rock mechanics noteworthy. These include rock fracture, size effects, tensile strength, and the effects of groundwater and weathering. Rock masses are generally heterogeneous, containing intact materials of varying engineering properties as well as joints, faults, folds, and other features. This inherent variability makes engineering in rock different from engineering design within most other media.

Civil engineering projects to which rock mechanics is applicable can be divided into seven categories.

Foundations. Rock is usually an excellent foundation material, but near-surface rock can be heavily weathered and fractured as a result of stress-relief. It is important to determine the competence of the rock mass to bear the required load at the acceptable levels of deformation or settlement.

Rock slopes. Rock slopes can fail in various ways: by plane, wedge, direct, or flexural toppling. The potential for failure in any of these modes can easily be identified by rock-mechanics methods.

Shafts and tunnels. The stability of shafts and tunnels depends on rock structure, *in situ* stresses, stress regime modification, groundwater flow, and construction technique. The fundamental rock-mechanics principles provide the basis for adequate design and mitigation against failure.

Caverns: use of underground space. The risks of underground instability increase as the type of excavation moves from a generally circular tunnel excavation towards a non-circular cross-section and as the size of underground opening increases. Jointing and inhomogeneity within the rock mass become increasingly important as the span of the excavation increases.

Mining. There is a huge range in the geometry of mining operations, but in all cases the mining methods are designed to extract the mineral with the minimum amount of artificial support and in some instances with the minimum amount of natural support.

Geothermal energy. In extracting geothermal energy from hot dry rock, cold water is pumped down into the rock mass to pass through fractures and exit from a borehole or set of boreholes. The optimal configuration for a production system depends on the interactions between rock joints, *in situ* stress, water flow, temperature, and time.

Radioactive waste disposal. The primary aim of radioactive waste disposal is to isolate waste so that unacceptable levels of radiation do not affect the biosphere. Predicting the safety of a repository requires an understanding of all aspects of the containing rock mass and in addition the ability of the rocks to adsorb radiation.

The two main engineering processes to which rock mechanics applies are excavation and support. The design in both cases relies heavily on an adequate quantification of the properties of the rock mass. Quantitative classification of rock masses provides a rapid means of assessing their quality, comparing qualities, and assessing support requirements. Classification applied on a routine basis can be of great value in mines and other underground projects, and has been shown to be of considerable economic benefit. The two most commonly used applications are the so-called Q system and the Geomechanics Classification System. The Q system is based on rock block size, joint shear strength, and confining stress. The Geomechanics Classification, which derives a rock mass rating (RMR), is based on rock material strength, rock quality designation (RQD), joint spacing, joint roughness and separation, and groundwater. Both systems and a number of variations of these basic approaches are now widely used as the basis for the quantification of rock-mass properties in underground excavation. BRIAN J. McCONNELL

rock-slope stability
Rock-slope stability describes the tendency of a rock face, either naturally formed or man-made, to resist failure. Failure encompasses both mass movements in which large volumes of rock slip and smaller-scale displacements such as minor rock falls. All types of rock-slope failure are important for they can have a significant impact on both property and life either on the slope or at its base. The study of rock-slope stability has many applications, including existing natural slopes and man-made slopes, such as those for road construction work, mining, quarrying, and railway cuttings. Rock-slope stability is important both in terms of cost before and during construction and in terms of safety during and after construction.

Instability caused by mass movement of rock along a plane occurs in three ways: plane failure, wedge failure, and circular failure (Fig. 1). Plane failure occurs where a single plane

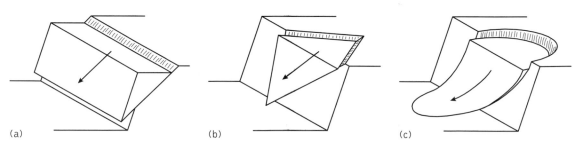

Fig. 1. Primary slope-failure mechanisms: (a) plane failure; (b) wedge failure; (c) circular failure.

(usually a bedding plane or a discontinuity such as a fault line) intersects a rock face. This plane of weakness may cause failure of the rock face because of the lack of frictional resistance. Wedge failure is more common. It is caused by the intersection of two separate planes, for example, two joint sets or a bedding plane and a joint set. The face and the two planes create a wedge that will tend to slide outwards towards the face if the line of intersection of the planes dips towards the face. Circular failures are most common in softer materials such as clays and weak rock. Geometrical methods (stereographic projections) are used to determine whether the face – discontinuity geometry could contribute to slope instability.

Failure can also occur as a toppling or as a ravelling failure, which entails the movement of small pieces of rock to form a scree at the base of a slope. Flexural toppling occurs in slopes where tall blocks of rock separated by steeply dipping discontinuities bend forward and then fracture. Heavily jointed columns of hard rock may suffer block toppling owing to the push of columns behind them. It is not possible to determine a factor of safety for failures of this type.

Several factors need to be taken into account in determining the likelihood of a slope to fail. These include the regional geology, which gives a guide to the general trend of the slopes in the area. This is a good starting-point for determining the likelihood of failure, since it provides a basis upon which to focus a more detailed study. The second stage in assessing a potential slope-stability problem entails the mapping of exposed structures. This calls for extensive field work and depends heavily on the experience of the individual carrying out the mapping exercise, for surface exposures are usually only intermittently visible and are often covered by broken rock material, vegetation, and soil. The field mapping of exposures should provide information on the geology in terms of the type of rock present, information on the dip angle and the direction of dip of the bedding planes or joint surfaces and information on the pattern of jointing in the rock and the frequency of joints with respect to the size of the slope being assessed. All this information can be amalgamated to give a description of the geometry of the rock mass behind the existing or proposed face and indicate any tendency to fail.

The detailed design of a new rock face will entail a ground investigation to determine the geology, the bedding thickness, the discontinuity (fracture) spacing, the ground water conditions, evidence of previous failure or movement, and the bedding dip and fracture orientation. Most of this information is normally obtained from cored boreholes.

The shear strength of a potential failure surface or set of failure surfaces will also need to be determined. This should ideally be determined for the conditions as they exist in the ground, whether the potential shear surface comprises a rock-to-rock contact or a discontinuity with an intermediate filling with properties different from those of the rock faces on either side. The shear strength must be determined as accurately as possible, since relatively small changes in shear strength can have a significant bearing on the safe height or

angle of a slope. The geologist will need to take into account the fact that the shear strength can be affected by the presence of water and also by the surface roughness of the two shear planes that are in contact with each other.

The likelihood of a slope failure is assessed by calculating a factor of safety for the slope in question. The factor of safety is defined as the ratio of the total force available to resist sliding to the total force tending to induce sliding. When the forces resisting sliding and inducing sliding are equal, the factor of safety is 1; when the slope is stable, the factor of safety will be greater than 1.

The stability of rock faces can be improved by a number of methods, including rock anchoring, rock bolting, the construction of retaining structures, meshing, and the installation of slope drainage. The construction of rock-catch fences can also be used to mitigate damage or injury caused by minor rock falls.

BRIAN J. McCONNELL

Further reading

Hoek, E. and Bray, J. W. (1981) *Rock slope engineering*. Institution of Mining and Metallurgy, London.

Rodinia and Pannotia
The names 'Rodinia' and 'Pannotia' denote two supercontinents that are postulated to have existed in Mesoproterozoic and Neoproterozoic times respectively. Evidence of their existence has been found in rock successions and structures in the present Precambrian shields. Their origins, long histories, and ultimate break-up are a record of plate-tectonic events over a span of more than 2000 million years (Ma). Thus they predate Pangea by several hundred million years and each may represent the culmination of a major plate-tectonic cycle in Earth history.

Rodinia was the older supercontinent (Fig. 1). It was a remarkably stable mass that resulted from scattered continental nuclei being drawn together and fused by collision from about 2200 Ma. The principal orogeny within this process was the global Grenvillian. It produced a supercontinent extending from about 60° N to the south polar region. Southward migration rotated the greater part of this landmass into the southern latitudes, whereupon a long and very intense continental glaciation occurred. Thick and very widespread tillites resulted not only in the present Gondwana continents but also in North America and Europe.

In early Neoproterozoic time partial rifting of Rodinia began and there was wide relocation of the numerous continental shields to produce a new assembly in the supercontinent Pannotia. This was accompanied by the initiation of an ocean that was to become truly vast, the Pacific. Pannotia occupied much of the southern hemisphere with a deep narrow gulf separating an 'East Gondwana' half from a 'Laurentia' part. Glacial activity continued in the south.

Pannotia (Fig. 2) began to break up some 250 Ma later, its component shields becoming separated by major fractures and the development of new oceanic spreading centres and subduction zones. The continental drifting initiated then was

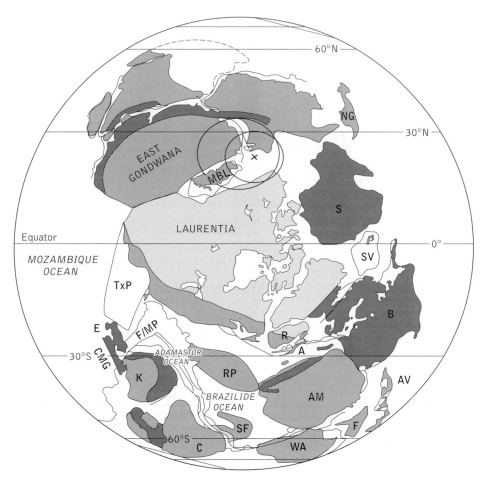

Fig. 1. Reconstruction of the hypothetical supercontinent Rodinia formed by the assembly of older cratons in the global Grenvillian orogeny. Small cross marks palaeomagnetic pole (with 95 per cent confidence); A, Arequipa Massif; AM, Amazonian craton; AV, Avalonia; B, Baltica (Russian craton); C, Congo craton; CMG, Coatsland–Maudheim– Grunchonga province of East Antarctica; E, Ellesworth–Whitmore mountains block (in Pangaea position); F, Florida (in pre-Pangaea position within Gondwanaland and including the Rockall Plateau and adjacent north-west Scotland and north-west Ireland); F/MP, Falkland–Malvinas Plateau; K, Kalahari craton; MB, Marie Bird Land; NG, New Guinea; R, Rockall Plateau with adjacent north-west Scotland and north-west Ireland; RP, Rio de la Plata craton; S, Siberia (Angara craton); SF, São Francisco craton; SV Svalbard block (Barentia); TxP, hypothetical Texas Plateau; WA, West Africa craton. (After Dalziel (1997).)

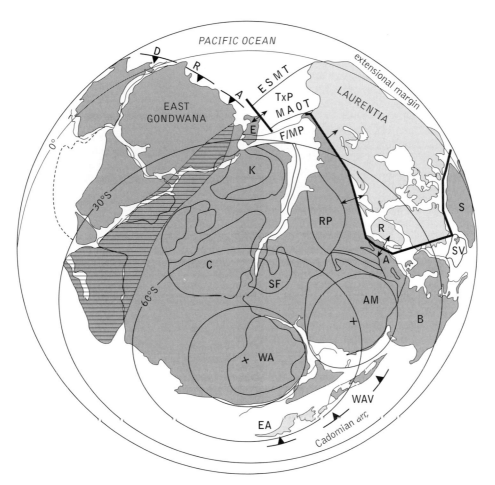

Fig. 2. Reconstruction of the hypothetical Pannotia supercontinent that may briefly have existed at the close of Precambrian time and the beginning of the Cambrian period. l, Laurentia; g, Gondwanaland; barbs, upper plates of subduction zones (Cadomian) and incipient subduction zones (Delamerian–Ross arc, D-R-A); thick lines, incipient mid-Iapetus ridges; diagonal lines, East Alpine collision orogen involving East and West Gondwanaland; EA, east Avalonia (southern British Isles); ESMT, hypothetical Ellesworth–Sonova–Mojave transform; MAOT, hypothetical Malvinas–Alabama–Oklahoma transform; R, Rockall Plateau with adjacent north-west Scotland and north-west Ireland; TxP hypothetical Texas Plateau; WAV, West Avalonia (Nova Scotia, Avalon platform, etc. For other symbols, see Fig. 1. (After Dalziel (1997).)

continued into the Palaeozoic era with accompanying eustatic rise in sea level and global transgressions. A new tectonic–eustatic cycle had begun and with it major changes in the biosphere were to take place.   D. L. DINELEY

Further reading

Dalziel, I. W. D. (1997) Neoproterozoic–Paleozoic geography and tectonics: Review, hypothesis, environmental speculation. *Geological Society of America Bulletin*, **109**, 16–42.

Rossby waves

Carl-Gustav Rossby (1898–1957) was a Swedish meteorologist who worked for much of his life in the USA. In the 1930s he proposed many of the theoretical ideas which led to an understanding of the large-scale dynamics of the atmosphere.

Imagine flying in a balloon at a height of about 10 km in the atmosphere. The balloon would be carried along by the wind at that level. If the site from which the balloon was launched was at 50° N, then the balloon would normally be blown from west to east, since that is the prevailing wind direction at that latitude and height. However, the balloon would not move round the Earth at exactly 50° latitude but would meander north and south of this latitude for long periods of time. It is as if the balloon were caught in waves which at some places would tend to move it southward for a while and would then later move it northward. As the balloon moved round the Earth, eventually returning to approximately where it started, it might have moved south and north about two or three or four times on these waves. The waves are enormous, each one spanning perhaps 5000 to 10 000 km. An even more surprising aspect of these waves is that they are almost stationary. We expect waves, for example those at the sea shore, to be moving, but it is also possible for the wave pattern to be stationary although air is moving through the waves and a balloon carried by the air would follow the wave pattern. If our hypothetical balloon were to begin a second journey round the Earth, and the waves were exactly stationary, then the balloon would follow the same wave pattern at the same locations above the Earth.

Rossby was able to give a simple and elegant explanation for these waves, which had been observed in the atmosphere. Under certain simplifying assumptions it is possible to assume that the absolute vorticity (a combination of the spin of the air and the spin of the Earth) is conserved, that is, it will not change over a period of time. As air moves towards the poles the contribution from the Earth's spin increases. If the sum of the air's vorticity (known as the *relative vorticity*, since it is measured relative to the Earth) and the Earth vorticity is to remain constant, then the vorticity of the air must decrease. This causes it to turn away from the pole and move southward. The reverse process operates as the air moves towards the Equator; so it turns poleward again. Since this motion is superimposed on a general west-to-east motion, the air moves through a wave-like pattern.

More recently the theoretical understanding of these waves has been taken much further by removing some of Rossby's original simplifying assumptions and extending the waves to three dimensions so that their vertical structure and motion can be explained. These more complex waves have, however, retained the name 'Rossby waves'. Our understanding of the structure of Rossby waves has been valuable in explaining some observed climatological patterns which show that some regions of the world tend to be warm when others are cold, or that some parts are dry when others are wet.

CHARLES N. DUNCAN

Further reading

Holton, J. R. (1979) *An introduction to dynamical meteorology.* Academic Press, London.

rotation and orbit of the Earth

To say that the Earth rotates on its axis once per day and completes one orbit of the Sun each year is to encapsulate but also to simplify the situation. In making such a description, the main naïvety is the overlooking of a fixed reference frame. From a vantage point far above its north pole, the Earth would be seen to rotate on its axis in an anticlockwise fashion, and to be moving round the Sun in the same direction (Fig. 1).

By the time the Earth has rotated exactly once, its orbital motion of nearly 30 kilometres per second has carried it

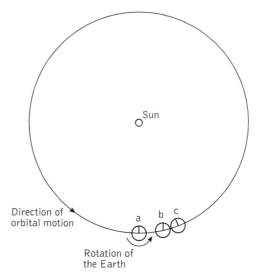

Fig. 1. Schematic view, *not to scale*, to show the relationship between the Earth's orbital motion and the solar and sidereal days. Seen from above the Earth's (and the Sun's) north pole the Earth rotates anticlockwise. The orbital motion of the Earth (and of all the other planets) is also anticlockwise. The Earth at position *a* represents an arbitrary starting time, which is noon at a particular point on the Earth. After exactly one sidereal day the Earth has completed exactly one rotation, but its orbital motion has carried it to position *b* and it is not yet noon at our reference point on the Earth's surface. Noon does not occur until the Earth reaches position *c*, which is one solar day after the start.

approximately 1/365th of the way round its orbit. The result of this is that after exactly one rotation of the Earth, the Sun (seen from the Earth) is no longer in quite the same direction as it was before. Because rotation and orbital motion are in the same sense, it actually takes a fraction more than a complete rotation to complete a day, as defined by the time between successive noons (when the Sun is at its highest point in the sky) at a single point on the Earth's surface. Noon to noon is exactly 24 hours (this is the basis of the definition of the unit of time we call an hour), and this period is referred to as a 'solar day'. The slightly shorter period in which the Earth completes an exact rotation (a 'sidereal day') is approximately 23 hours 56 minutes 4 seconds. An effect of this is that seen from a location on the Earth's surface a given star will rise 3 minutes 56 seconds earlier each day, which is why the stars visible in the evening sky change gradually through the year.

Seasons are caused by the fact that the Earth's rotation axis is not at 90° to its orbital plane, but is inclined at $23\frac{1}{2}°$ and (on a human timescale) points to a fixed direction in space. On about 21 June each year the north pole is tilted as close towards the Sun as it ever gets, and the Sun thus reaches its farthest north point in the sky, which corresponds to the summer in the northern hemisphere and to winter in the southern hemisphere. This is the northern summer solstice, when northern latitudes experience their maximum duration of daylight. Conversely, on about 21 December the north pole is pointed away from the Sun, and the northern hemisphere experiences winter (and the shortest hours of daylight) while the southern hemisphere has summer. This is the northern winter solstice. A quarter of an orbit between these dates, the axial tilt is sideways to the Sun and day and night are of equal length at all places on the Earth; these equinoxes occur on about 21 March and 21 September each year. Calendar dates of events vary slightly because there is not an exact number of solar days in a year. The exact figure is 365.25636 solar days per year, or strictly speaking, per 'sidereal year'. This is close

enough to $365\frac{1}{4}$ for the calendar to be kept in step with the seasons by declaring the 'civil year' to be normally 365 days but 366 days every fourth year (when the year number is divisible by four). These 366-day years are 'leap years'. Our current system of declaring century years to be leap years only if divisible by 400 was introduced by Pope Gregory XIII in 1582, and is known as the Gregorian Calendar. This compensates for most of the 0.00636 of a day difference between the year length and the $364\frac{1}{4}$ day approximation, leaving a discrepancy of only 7 hours 16 minutes per thousand years.

An additional complication is that the Earth's orbit is not quite an exact circle, but an ellipse. At its nearest point to the Sun, *perihelion* (which falls on approximately 3 January), the Earth–Sun distance is 147.1 million km; the distance reaches a maximum of 151.1 million km at *aphelion* on about 4 July. The speed of a planet's orbital motion is greatest at perihelion and least near aphelion, but its rotation is at a constant rate. The Sun therefore usually appears slightly out of position in the sky as compared with the fictitious 'mean Sun' required for regular timekeeping; a sundial will consequently read 14 minutes slow in February and 16 minutes fast in November.

The Earth's orbit and rotation are not constant in the long term, but are affected by factors such as the pull of the Sun and Moon on the Earth's equatorial bulge and, to a much lesser extent, by the gravitational attraction of the other planets. The direction of tilt of the Earth's rotational axis traces out a complete circle in the sky over a period of 22 thousand years; the amount of tilt varies between 21.8° and 24.4° over a period of 40 thousand years; and the shape of the orbit becomes slightly more and slightly less elliptical over a period of 110 thousand years. The combined effect of these cycles is probably an important factor in driving global climate change. It takes much longer for the rate of rotation to change. The main cause is the tidal pull of the Moon, which has slowed the day length by about two hours over the past 400 million years.

DAVID A. ROTHERY

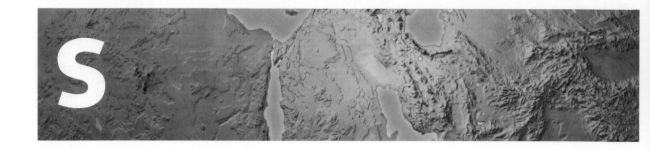

salinity, oceanic *see* OCEANIC SALINITY

salt marshes Salt marshes are areas of land covered mainly by salt-tolerant (halophytic) plant communities which are regularly flooded by the rise and fall of tides. Salt marshes occur in mid and high latitudes along intertidal shores world-wide, being largely replaced by mangrove swamps in tropical and sub-tropical regions. Salt-marsh development requires a net accumulation of mainly fine-grained sediment to provide a stable substrate; so some degree of shelter from wave action is required. Such shelter may be found in estuaries, in embayments, and on sections of open coast which are protected by wide intertidal flats and barrier complexes. Large expanses of salt marsh have often developed in these areas and have an important role in protecting the coastline from erosion by extreme storm events. Salt marshes are also valuable wildlife habitats, providing roosting and feeding sites for birds and acting as a nursery site for many economically important fish and shellfish. The halophyte communities which cover the marshes are biologically interesting in their own right and are extremely productive, exporting large amounts of organic material to adjacent waters.

The dominant organisms found in salt marshes are rooted plants, originally of terrestrial origin, which have adapted to the stresses imposed by the intertidal environment, such as high salinity and periodic tidal inundation. Colonization of sand or mud flats by vegetation can begin only when sediment has accumulated to a sufficiently high level in the tidal cycle for halophyte plants to survive and reproduce. Initial colonization is carried out by pioneer species such as *Salicornia* (glasswort) and *Spartina* (cord-grass). Once vegetation is established, the salt marsh may rapidly grow vertically as it traps and intercepts more incoming sediment, and organic matter (produced by the vegetation itself) is added to the marsh surface. This raising of the salt-marsh surface leads to an increase in drainage and a decreased submergence in salt water. Conditions are thus made suitable for the arrival of less hardy species which may displace the initial pioneers. This results in a zonation of the vegetation, pioneer species giving way to a more terrestrial-type vegetation near the landward margin of the marsh. An individual marsh may therefore develop a series of morphological zones as it matures, open

mud flats with scattered clumps of pioneer vegetation giving way to a zone of continuous vegetation dominated by pioneer species (the low-marsh), which is subject to adjacent estuarine or marine conditions. This grades into a zone which is less frequently flooded and more species-rich (the high-marsh), dominated by plants that are more terrestrial in nature. A generalized transect across a typical North European salt marsh is shown in Fig. 1.

Salt marshes exhibit a number of characteristic geo-morphological features. Cliffs may occur at the seaward edge of the marsh because of wave erosion. Creeks or marsh drainage systems are often well developed, and provide important pathways for energy and material transfer between the marsh and surrounding areas. To a certain extent the initial pattern of marsh drainage is inherited from the original tidal flat on which the marsh formed. A dendritic (or branched) creek pattern may develop which is the most efficient in terms of dissipating incoming wave and tidal energy by friction. *Pans* are natural depressions in the marsh which retain water even at low tide. They may form when the outlet of a tidal creek or depression is blocked by vegetation or accumulating sediment.

The behaviour of relative sea level is a major driving force behind the evolution of salt marshes in the medium to long term. A falling sea level may allow salt marsh progradation, whereas a rising sea level may cause landward migration. Studies of salt marshes on the western Atlantic coast have shown that, where sediment supply or organic production is sufficiently high, salt marshes are able to keep pace with rising sea levels. In such areas the vertical growth of salt marshes

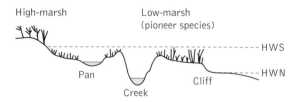

Fig. 1. Transect across a typical north European salt marsh. (HWS, high water, spring tides; HWN, high water, neap tides. (After Boaden and Seed (1985).)

provides a measure of the rate of sea-level rise. In other areas, such as the Mississippi delta and parts of southern England, marshes are not successfully maintaining their elevation and are eroding and dying off as sea level rises. Other salt marsh areas are threatened by land reclamation and pollution. In recent decades, however, the ecological importance of salt marsh areas has been recognized and many are now designated conservation areas.   ANDREW B. CUNDY

Further reading

Allen, J. R. L. and Pye, K. (1992) *Saltmarshes. Morphodynamics, conservation and engineering significance.* Cambridge University Press.

Boaden, P. J. S. and Seed, R. (1985) *An introduction to coastal ecology.* Blackie, Glasgow.

Frey, R. W. and Basan, P. B. (1985) Coastal salt marshes. In Davis, R. A., Jnr (ed.) *Coastal sedimentary environments* (2nd edn), pp. 225–301. Springer-Verlag, New York.

saltation
Saltation (from the Latin '*saltare*') is the process of particle transport by small bounces or leaps along a surface. Saltation occurs in both water and air, although there are some significant differences in the processes in the two media. In water, direct fluid lift and drag forces are responsible for entraining a grain into the fluid, whereas in air entrainment can also result from grain collision. The initial launch trajectory will be steep, though the angle of descent will be far more gentle. The impact of the saltating grain with the surface will be violent in air, normally resulting in multiple grains, up to six times the diameter of the impacting grain, being launched into the airstream. The force at which a grain is set in motion (the *entrainment threshold*), as well as the length and height of its subsequent leap, will be proportional to the shear velocity of the fluid and the size of the grain. In air the entrainment threshold will also be a function of grain impaction and configuration of the sediment surface. In water the impact of a saltating grain on a surface is far less violent, and does not normally result in further grains being launched into the fluid. This is because of the much lower density ratio between the fluid and the grain, and the greater viscosity of the water.
                                                    T. LINSEY

sand and gravel resources
Sand and gravel are geological materials of great commercial value that tend to be taken for granted. They are familiar as piles of material in the builders' yard, and their significance is easily overlooked. If they ever think about such materials, most people probably regard them as boring, and certainly unexciting, and overlook their significance for modern life. Gravels are, however, important to the aggregates and construction industries for making concrete and for road-building. They are also used, to a more limited extent, for railway ballast. Sand is in demand as an abrasive, for mortar, for making glass, as a filter medium in the treatment of water supplies and sewage, and in foundries to make moulds for casting.

Materials such as sand and gravel should ideally be close to the places where they are to be used. Needless to say, this is not always the case; and where it is, there is inevitably competition between extraction and the use of land for other purposes. There is an imbalance between demand and supply of sand and gravel in many countries. The main sources of gravel are unconsolidated superficial or drift deposits of Quaternary age. These may be of fluviatile, glacial, periglacial, lacustrine, aeolian, or marine origin. The estuaries of the River Thames and the River Trent provide large quantities of sand and gravel in the United Kingdom. Elsewhere, beaches and marine sands are used. Marine aggregates are an important source in northern Europe, the United States, and Japan, although some would question whether a relatively scarce resource such as shingle should be used for aggregate when crushed rock would be equally suitable.

Unconsolidated materials such as sands and gravels are usually extracted using mechanical excavators, shovels, loaders, floating cranes, or dredgers. Marine sands and gravels are excavated from the sea bed by dredgers.

After excavation, the material is sorted into various grades according to size, using vibrating screens or by hydraulic methods, scrubbed to remove adhering particles of silt or clay, and then treated to remove water.

In concrete, the role of gravel, as for any other aggregate material, is to provide bulk and strength. The quality of the material is therefore important. Sampling and testing are essential to establish its mechanical and other properties before use. Low porosity is, for example, required for road paving and for concrete. Chemical properties can also be important; surface spalling and serious staining can result if only one particle in about three thousand contains a significant amount of iron sulphide. Opaline silica, if present, can cause expansion and cracking. Sands are subjected to comparable tests.

                        K. VALA RAGNARSDOTTIR AND B. WILCOCK

Further reading

Manning, D. A. C. (1995) *Introduction to industrial minerals.* Chapman and Hall, London.

sand and sandstone
From a geologist's point of view, the word 'sand' refers to a sediment containing 50 per cent or more of grains with a diameter between 0.063 and 2.00 mm, regardless of their composition; for example, quartz, skeletal debris, volcanic rock fragments, etc. However, it is also commonly understood, unless there is a qualifying adjective, that 'sand' refers to deposits formed of silicate minerals, of which quartz is the most abundant. Sand is part of a large group of sediments derived from the breakdown of silicate rocks which are termed detrital, clastic, terrigenous, or, more usually today, siliciclastic deposits. Sandstone is indurated (hardened) sand, composed of siliclastic grains, 50 per cent or more of which are of sand size, and are bound together by chemically precipitated cement or a recrystallized matrix of finer sediment.

Sand and sandstone are second to muds and mudstones in abundance, and form approximately 10–15 per cent of the total sediments of the Earth's crust. The size and degree of homogeneity of the population of grains is usually used to qualify the terms sand and sandstone; for example, very coarse, coarse, medium, fine, and very fine. A measure of the central tendency of the grain-size distribution is used in general description and statistical analysis; for example mean, median, mode, or modal size. The spread of the population of grain sizes is described by the term *sorting*. A well-sorted sediment has a small range of grain sizes, and a poorly sorted sediment the reverse. Sorting is the opposite to the term *graded*, with which it is sometimes confused. A well-graded sediment is the reverse of a well-sorted sediment and has a wide variety of grain sizes. Statistical terms such as skewness and kurtosis are also employed to describe the asymmetry of the grain population and the relationship of the central and peripheral parts of the grain-size distribution. Although these grain-size properties are initially dependent upon the source, they are mainly a reflection of the hydrodynamics of the environment in which the sand is deposited, and hence they can be used to determine conditions in the environment of deposition. Coarse-grained sands are characteristic of deposition in high-velocity regimes, whereas fine-grained sands are deposited in low-velocity regimes; the mean grain size of the sediment indicates the flow strength. Sands deposited in high-energy environments, such as beaches or dunes, or environments of stirring by persistent current or wave action, are usually better sorted than sands deposited by a flash flood on, say, an alluvial fan. Winds blowing over a beach remove only the finer-grained sands and leave a deposit with an increased coarse fraction; that is, a negatively skewed fraction, whereas aeolian sand lacks coarser material but may have a tail of fines; that is, it is positively skewed.

The geometry or shape of individual grains appears to be a property inherited from the original crystals of the silicate minerals in the source rock. In contrast, the sharpness of the edges of the individual grains or their angularity (i.e. roundness) records the history of the grain during transport and, to a lesser extent, during diagenesis. Constant collision of grains during transport smooths the sharp edges of the crystals and grains that were released during weathering. Wind is particularly effective in rounding grains, because of the greater force of impact in lower-density air compared to higher-density water. Nevertheless, the rounding process is very slow, and the famous Dutch geologist Ph. H. Kuenen claimed, on the basis of experimental work, that a cube of quartz would have to be transported over a distance equivalent to travelling several times around the Equator to produce a well-rounded grain, and, furthermore, that wind transport would need to be involved at some time during this process. The surface texture of an individual grain records the history of transport and deposition, although this can be modified or altered during diagenesis. Weathering, transportation by wind, and ice impart characteristic patterns of grooves, pits, and depressions which are revealed when viewed using an electron microscope.

The main components of siliciclastic sands and sandstones are quartz, feldspar, fragments of fine-grained rocks (lithic grains), mica, and accessory minerals. The last-named are often called 'heavy minerals', since they are usually denser than the more important constituents. They are usually present only in small amounts but may be concentrated by hydrodynamic processes to form accumulations which are sometimes of great economic importance—placer deposits. In addition, they can be very useful as tracers of the source of a sediment.

Sedimentologists have for many years been attracted by the possibility of using the composition of sand grains to determine the source rock and hence, in some situations, the tectonic evolution of the source area, a subject we can call *palaeogeology*. Initially, the more varied heavy minerals were used and a vast literature on this subject accumulated in the early part of the twentieth century. Little work was done on the apparently less promising major components, except by a few isolated workers. However, pioneer studies were initiated by Krynine in the United States to relate the relative abundance of the more common minerals of sands and sandstones to the geological character and evolution of some areas. This interest has developed enormously during the past few decades, largely because of the advent of plate-tectonic theory. It has become apparent that not only have sources of sediments, at various times in geological history, become buried or removed by erosion, but also that parts of the Earth's crust have been displaced, during plate-tectonic movements, vast distances from their previous positions. The record of their former presence is reflected in the composition of the sands they shed during erosion into adjacent basins; and the record left in sandstones is one of the few permanent records of the proximity of a particular source-land.

Studies of sands from the Mississippi River by Rittenhouse in the 1930s showed that sediments retained their source characteristics with only minor changes in mineralogy, even when transported thousands of miles. Later work by Potter and his students in the United States and South America has confirmed that this appears to be generally true for most large rivers, which are, in temperate regions, the most important suppliers of sand to sedimentary basins. These geologists also demonstrated that in tropical regions sands show considerable changes in composition during transport. Furthermore, sorting during transport and deposition according to size and density also fractionates the initial population into several populations of varying composition, and this has to be taken into account if reconstructions of source-lands are to be attempted. Diagenetic changes can alter the composition of the sands after deposition; for example, unstable components, such as feldspars and volcanic fragments, may be altered. Petrographic work will sometimes reveal the composition of the original grain, although this is not always possible.

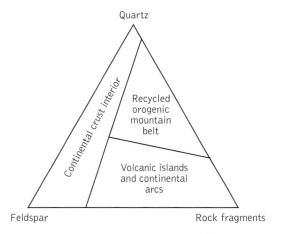

Fig. 1. Composition of sands and sandstones from various sources. (After Siever (1988) *Sand*. Scientific American Publications.)

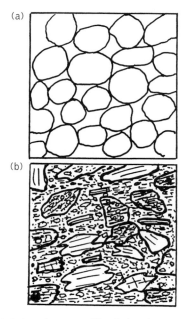

Fig. 2. (a) Typical arenite texture. (b) typical wacke texture.

The detailed understanding of the relationship between the composition of a sand or sandstone and the nature of the source rock, the weathering it undergoes, and the effects of transport, deposition, and diagenetic changes is still in its infancy. It has, however, been claimed by Dickinson and his co-workers in the United States that the influence of the source is, in most sands and sandstones, strong enough to persist through the vicissitudes of a grain's history. The relative percentages of sand grains composed of quartz, feldspar, volcanic fragments, etc. can be used to distinguish source-lands such as the stable cratons of continental interiors; continental and island arcs; and collisional mountain belts (Fig. 1). The study of the accessory minerals, which had gone out of fashion except in a few countries, notably the Netherlands, has received renewed attention. The introduction of the electron probe and the ability to date individual grains has increased the chances of identifying source rocks of sediments.

Sands undergo relatively small physical changes when buried except for an increase in bulk density that results from tighter packing of grains under overburden pressure. Sands buried to over 3000 m in the Gulf coast of the United States still retain porosities close to those of recently deposited sand. However, those containing large amounts of unstable lithic fragments or matrix show greater porosity reduction on loading owing to deformation of the grains and water loss from grains or matrix.

The most important diagenetic change in the conversion of sand to sandstone is the precipitation of mineral cement between the grains. Silica, calcium carbonate, and iron oxides are the most important, but in many deposits precipitated clay cements are common. Several generations of cement are present in some sandstones, sometimes replacing one another. During deep burial, overburden pressure causes solution at points of contact between grains, a process known as *pressure solution*, to produce sutured, interdigitating contacts between

grains or their overgrowths. Although a considerable amount of the cement may be produced locally by solution of carbonate skeletal debris, siliceous organisms, volcanic material, and pressure solution, most must come from solutions driven into the sand from adjacent fine-grained mudstones during compaction and from metoric waters draining down into a sedimentary basin. The fine-grained matrix usually recrystallizes during diagenesis. Occasionally, however, aggressive waters from adjacent organic-rich sediments dissolve earlier cements and produce horizons of enhanced porosity and permeability in the subsurface. Although there are some striking examples of solution of some minerals by incoming fluids (*intrastral solution*), this is not a process of major significance.

Sandstones are classified into two main groups: *arenites* and *wackes* (Fig. 2). Arenites are siliciclastic sediments of sand size with less than 15 per cent of fine-grained material between the framework of sand grains (Fig. 2a). They are usually infilled to a varying degree by a precipitated mineral cement. Wackes are sand-sized sediments which contain more than 15 per cent matrix. The matrix has generally recrystallized to bind the sand grain together, to form a indurated rock (Fig. 3). Interpreting the origin of the matrix of such rocks has been a subject of great controversy because they may be partly primary, but in some sandstones may have been introduced later by infiltration of fine-grained materials or by the breakdown, compaction, and recrystallization of unstable sand grains. *Greywacke* (or *graywacke*) is an old term still sometimes used for a particular type of wacke especially in grey-coloured turbidite sequences.

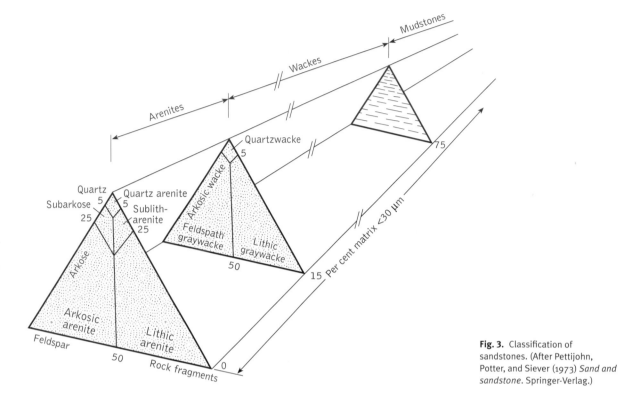

Fig. 3. Classification of sandstones. (After Pettijohn, Potter, and Siever (1973) *Sand and sandstone*. Springer-Verlag.)

The term *arkose* or the prefix 'arkosic' are used for feldspar-rich sands or sandstones. The concept of *maturity* is often applied to sandstone description: those sandstones that are composed of well-sorted and well-rounded grains are said to be *texturally mature*, implying considerable transport; a *mineralogically mature* sediment is composed only of stable quartz with no unstable components, implying that intensive weathering has occurred in the source area or that the deposit has undergone many cycles of sedimentation. Some deposits may be both texturally and mineralogically mature.

Sand is transported by traction currents; it slides, rolls, and saltates along the surface. Fine and very fine sand are also transported in suspension. Sand can be fashioned into a wide variety of bedforms by the transporting fluid, to produce cross-stratified sediments which are invaluable in attempts to ascertain the location of a source area via this palaeocurrent indicator. Such phenomena can also be used to reconstruct past environments of deposition and palaeogeographies.

Sand is deposited in a wide variety of environments, extending from terrestrial settings such as fluvial and desert environments, through coastal beaches and dunes to shelf seas and the turbiditic deposits of the deepest parts of the oceans.

Sands and sandstones are important as raw materials for the construction industry (building stones, building sand, filter-bed sand), as well as for moulding sands and glass sands. In addition, places deposits in sands are often worked for industrial minerals (cassiterite, titanium ores, iron ores), as well as for semi-precious gemstones and precious minerals such as gold, diamonds, and emeralds. Because of their porosity and permeability, sandstones act as hosts for water, and they are important aquifers and reservoir rocks for oil and gas (e.g. the Permian sands of the North Sea). Valuable but small deposits of uranium and copper were precipitated in some geologically ancient sandstones.

G. EVANS

Further reading

Dickinson, W. R. and Suczek, C. A. (1978) Plate tectonics and sandstone composition. *American Association of Petroleum Geologists Bulletin*, pp. 2164–82.

Pettijohn, F. J., Potter, D. E., and Siever, R. (1973) *Sand and sandstone*. Springer-Verlag, Berlin.

Siever, R. (1988) *Sand*. Scientific American Library, New York.

SAR interferometry *see* SYNTHETIC APERTURE RADAR

scale in geomorphology Scale is a concern in any geomorphological study because landforms of different spatial scales are produced by different processes, are described by different variables, and evolve over different time periods. Even though a landscape can be viewed as a nested hierarchy in which the larger landforms are made up of smaller ones, investigating landforms at different scales typically requires changing the research focus, since the functioning and evolution of large-scale landforms often cannot be understood by simply integrating the behaviour of the smaller ones. For example, at the smallest scale of investigation, the microscale, the entrainment of sediment is controlled by turbulence which lifts particles from the stream bed into the main body of the flowing water where the particles are swept downstream. Such processes are investigated using detailed analyses of the three-dimensional, turbulent flow structures which evolve over time-spans measured in fractions of seconds. Even though at a larger scale the formation of a meandering channel pattern is the direct result of sediment entrainment and transport, to study a river successfully at this larger scale requires information about discharge and sediment transport over periods of years. Thus, the increase in the spatial scale of the problem requires a shift in research focus, from the short-term, microscale, flow structures controlled by turbulence, to the longer-term sediment yield and discharge regime controlled by conditions in the contributing drainage basin. Such changes in focus occur over the full scale range of geomorphological interest. Generally, at the megascale, the largest scale of investigation, landforms result primarily from endogenic processes associated with plate tectonics. With decreasing spatial scale, the balance between endogenic and exogenic processes shifts towards a predominance of exogenic processes at the smallest or microscale. Because the methodology for a study at one scale is not necessarily suitable for studies at other scales, and because conclusions drawn from a study at one scale do not necessarily apply to other scales, any geomorphological investigation requires careful identification of the spatial and temporal scales of the problem to allow selection of the relevant processes and variables.

Spatial and temporal scale are directly related. Because the characteristic features of large-scale landforms evolve over long periods of time, studies of these landforms generally need to consider long periods of time, frequently resulting in an emphasis on the historical aspects of landform evolution. Conversely, small-scale landforms evolve rapidly so that their evolution and functioning can be understood by considering only the most recent time interval. In consequence, studies of small-scale landforms tend to de-emphasize the landforms' history, and generally focus on the nature of the processes shaping these landforms.

An additional scale-related issue is that at small spatial and temporal scales data may be collected from many individual landforms and during many events, so that meaningful generalizations and statistical analysis of the data are possible. At larger scales, the smaller number of individual landforms and the longer time-span to be considered limit the number of individual landforms and events that can be studied. Thus,

generalization is virtually impossible at the megascale, and investigations tend to be case studies.   DIRK H. DE BOER

Further reading

Thorn, C. E. (1988) *Introduction to theoretical geomorphology*. Unwin, Boston.

sea-floor spreading
Sea-floor spreading is the process whereby new oceanic lithosphere is created as a pair of plates diverges at a mid-ocean ridge spreading centre. It was originally proposed by R. S. Dietz in 1961, and was an important component of the theory of plate tectonics that developed shortly afterwards.

At a mid-ocean ridge, two plates diverge, probably driven mainly by the pull of subducting slabs and the gravitational sliding of the ridge flanks (see *plate tectonics*). As the plates separate, ductile upper mantle rises from the underlying asthenosphere, and undergoes adiabatic decompression (decompression without loss or gain of heat) and consequent partial melting. The melt rises toward the surface, where it solidifies to form new oceanic crust, while the residual ductile mantle is drawn along with the diverging plates. As the residual mantle cools, it becomes more brittle and is effectively accreted to the lithosphere, gradually thickening it.

The newly formed oceanic crust (especially the extrusive basalts at its top) acquires a thermal remanent magnetization as it cools; and as the Earth's magnetic field episodically reverses direction, strips of crust are formed with alternating polarities of magnetization. These give rise to alternating positive and negative magnetic anomalies that are symmetrical about the spreading centre and can be used to date the oceanic crust, as was first suggested by Vine and Matthews in 1966.   ROGER SEARLE

sea ice and climate
Ice which forms from the freezing of sea water is called *sea ice*. The presence of salt in water has the effect of reducing the freezing point below that of fresh water, and there is a standard empirical formula which tells us that the freezing point will be lowered by approximately 0.05 °C for every 1 PSU (practical salinity unit) of salt in the water. The freezing point also decreases with increasing pressure. For the typical surface salinities in the world ocean the freezing point is approximately −1.9 °C. The presence of salt also increases the density of the water, and for salinities greater than 24.7 PSU (which is much lower than typical ocean surface salinities) the density of sea water increases until it reaches the freezing point. This means that as the water is cooled it will undergo thermal convection and descend, forcing deeper warmer water to the surface. Therefore, unless the water column is stratified, to form ice, the entire water column has to be cooled to the freezing point. With a salinity stratification only the surface waters have to be cooled. This is the situation found in the Arctic Ocean where there is a surface layer of lower salinity above what is termed the Arctic halocline. This means that just the upper 30–50 m of the water column has to be cooled for ice to form.

The initial form of sea ice is a slurry of ice crystals called *frazil ice*, which under various conditions grow together to form many different and complex ice forms. As the frazil ice crystals coalesce, some salt from the sea water is trapped within the crystal matrix. This makes the ice slightly saline (up to 8 PSU), but most of the salt is rejected into the surface layers of the ocean. This salt rejection causes a local density increase in the surface water, which can lead to local convection and under certain conditions to deep convection. There is a standard but complex set of descriptive terms for types of sea ice which has been developed by the World Meteorological Organization, but by far the most common distinction is between first-year and multi-year ice. First-year ice is generally considered to be new ice up to approximately 750 mm thick, which can be reached in one year's growth. Multi-year ice is first-year ice which has survived one or more summers, increasing in thickness each winter up to 5 m thick. The quantity of salt trapped in the crystal matrix of multi-year ice is less than that of first-year ice (3–5 PSU), since the surface of the ice can melt in summer, flushing salt from the crystal matrix.

The different geography of the Arctic and Antarctic has a considerable effect on the type of ice within each region. The central Arctic Ocean is an enclosed basin and retains an extensive covering of permanent multi-year ice throughout the year. At its minimum the ice covers an area in September of approximately 9×10^6 km², rising to a maximum of approximately 16×10^6 km² in March, as the ice expands from the Arctic Basin to cover the Hudson Bay, Baffin Bay, the Canadian Archipelago, and the Kara Sea. Large portions of other peripheral seas are also covered, such as the Greenland Sea, Barents Sea, and smaller areas of the Labrador Sea and the Sea of Japan. In contrast, Antarctica is surrounded by open ocean and the permanent sea ice cover is restricted to small regions of the Weddell Sea, the Bellingshausen Sea, and the Amundsen Sea. Because of the smaller regions of permanent ice cover, Antarctic sea ice has a much smaller percentage of multi-year ice. At its minimum in February the ice covers approximately 3.8×10^6 km² but the comparative increase in area in winter is much greater with the ice expanding from the continent to cover an area of almost 19×10^6 km². The global sea ice coverage can best be observed by satellites with passive microwave sensors such as NASA's *Nimbus 7* and *Nimbus 5*. These sensors can show the interannual variation of the area covered by Arctic and Antarctic sea ice (Fig. 1).

To calculate the volume of water stored in sea ice we must know its thickness. This is much more difficult to measure than the extent of the sea ice. Most thickness measurements from the Arctic ocean have been obtained from nuclear submarines using upward-looking sonar. The current measurements of thickness of the Antarctic sea ice have all been made by drilling holes. In the early 1990s submarine-derived ice thicknesses showed that sea ice north of Greenland was thinning drastically, but this was shown to be the product of

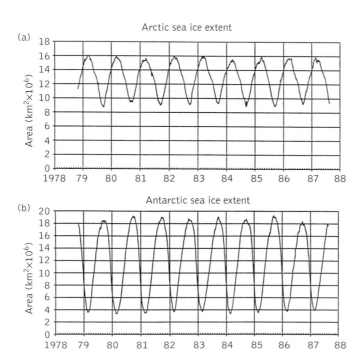

Fig. 1. Sea ice extent from passive microwave satellite measurements of (a) the Arctic, (b) the Antarctic. (From Gloersen, P., *et. al.* (1992) *Arctic and Antarctic sea ice, 1978–1987: Satellite passive-microwave observations and analysis.* National Aeronautics and Space Administration Report NASA SP-511.)

meteorological effects rather than a genuine reduction of ice. The water stored in sea ice is negligible in comparison with the volume of water stored in the world's glaciers and the ice sheets of Greenland and Antarctica. However, sea ice covers a vast area of the world's surface. In the Arctic winter, sea ice can occupy almost 5 per cent of the entire northern hemisphere and in the austral winter the Antarctic sea ice can occupy as much as 8 per cent of the southern hemisphere. On a global scale approximately two-thirds of the ice-covered areas of the Earth are covered by sea ice, which exerts a strong effect on climate.

At the most simple level the ice shuts off the exchanges of momentum, heat, and mass between the atmosphere and the ocean. It thus tends to stabilize the climate system, reducing the heat loss from the ocean in winter. The ice cover also reflects a significant amount of the incident solar radiation (up to 55 per cent for bare sea ice and as much as 90 per cent for snow-covered sea ice) and so provides positive feedback in the climate system. If the ice extent increases, more solar radiation is reflected and ice extent can increase further. If the ice extent decreases, the opposite applies. The ice cover in the polar oceans is not continuous, nor is it complete. Sea ice is a complex and flexible medium and it is moved and broken and piled up by a combination of winds, ocean currents, and (at the edge of the ice pack) wave energy. These forces move the ice to create open water, even in winter, in the form of linear cracks called *leads* and large areas of persistently open water called *polynyas*. The winter open water, although small in area (up to 2×10^6 km^2 in the Arctic and 4×10^6 km^2 in the Antarctic), is of great importance to the heat balance of the polar oceans: about half the total heat exchange takes place through them. Polynyas have also been called ice factories: the open water creates considerable amounts of ice, and salt rejection results. In the 1970s there was a persistent area in the Antarctic where the Weddell Sea Polynya formed which was thought to be responsible for forming deep water. This polynya has, however, been absent since the late 1970s, although the area still shows a reduced ice concentration.

Oceanographers now believe that the world ocean circulation is linked through a series of strong currents driven by deep water formation in the polar seas and heating of water in the tropical seas of the central Pacific Ocean. This circulation is called the 'ocean conveyor belt' or 'thermohaline circulation'. ('Thermohaline' because the circulation is forced by heat and salt.) The formation of sea ice and salt rejection are probably the principal force driving the deep water formation and the cold end of the conveyor belt, and are also thought to be important in the exchange and absorption of carbon dioxide from the atmosphere to the ocean. In the Greenland Sea the ocean loses heat to the atmosphere, rapidly forming a distinctive sea-ice feature in the central Greenland Sea called the 'Odden' (Norwegian for headland). The associated salt rejection from the Odden can drive deep convection. Over thousands of years this continued forcing circulates the cold dense waters of the Arctic Ocean to the Pacific Ocean via deep currents through the Atlantic and Indian Ocean, returning warm waters to the Arctic. There is much evidence from ice cores and pollen records to suggest that there have been rapid variations in the Earth's climate of several degrees Celsius in fifty or so years in the recent geological past, which have been linked to changes in the pathways of the conveyor belt or even in the modes of the conveyor-belt circulation. These changes could have been initiated by modifications in the sea-ice cover.

The ice extent in the global ocean (Fig. 2) would appear to be fairly consistent from year to year. There are, however, regional differences. The Arctic sea ice extent has decreased by roughly 2 per cent in the 8.8 year period of Fig. 2, whereas the Antarctic sea ice extent shows no statistically significant change. Because of the complex link between ocean currents, ice formation, and deep convection, the key issue in any reduction in ice extent is the location.

The Odden ice tongue has been known to exist in the Greenland Sea for at least the past two hundred years from whaling records and has been the subject of much recent research. Peter Wadhams of the Scott Polar Research Institute has alerted the scientific community to the reduction in size of the Odden ice tongue over the last 10 years and its total absence in 1994 and 1995. This reduction has in turn been linked to the depth of convection and resulting deep-water

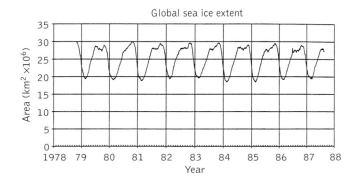

Fig. 2. Global sea ice extent from passive microwave observations. (From Gloersen, P., *et. al.* (1992) *Arctic and Antarctic sea ice, 1978–1987: Satellite passive-microwave observations and analysis.* National Aeronautics and Space Administration Report NASA SP-511.)

formation in the Greenland Sea; whereas convection once reached down to almost 4 km it now extends down only to 1 km, and there has been no formation of Greenland Sea Deep Water since at least 1980. The reduction of deep water formation may not necessarily mean that the conveyor belt has stopped at the polar end. What it is important to remember is that although we can measure sea ice extent and the changes relatively easily we cannot measure the volume; changes in sea ice extent could be linked to as yet undetermined, but more important, thickness changes in the global sea ice cover.   MARK A. BRANDON

sea-level changes and palaeoclimate Sea level is
intimately linked to climate. At the maximum of the last ice age, some 20 000 years ago, large ice caps were present on the great northern hemisphere continents. At this time the Laurentide and Cordilleran ice sheets merged, covering over half the land area of northern North America with over 30 million km³ (cubic kilometres) of ice. The Antarctic ice sheet, which today has a volume of about 24 million km³, expanded to 38 million km³, and other continental ice sheets on Scandinavia, Greenland, and Iceland also grew. This massive build-up of ice resulted in an estimated eustatic sea-level fall of about 120 m below present levels. The immense Laurentide ice sheet contributed about half to this total, and Antarctica about 20 m. The British ice sheet, which reached as far south as north Norfolk, the Midlands, and South Wales, was small by comparison. It had a volume of only 0.8 million km³ and contributed about 1.0 m to global lowering of sea level.

Ice and sea level are also of utmost concern today. If all the remaining continental ice were to melt, the resulting rise in sea level would be about 90 m, with the Antarctic ice sheet contributing about 65 m. It is clear, therefore, that the major determinant of global (eustatic) sea-level change over the past 2 million years is the amount of ice present on the continents. This suggests a direct link with climatic change. There are, however, many other factors, operating on a variety of scales in time and space, that influence the exact level of the sea, and not all of them are linked to climatic change. These other factors can conveniently be divided into those that affect ocean basin volume (such as sea-floor spreading and sediment infill), those that affect ocean water volume (such as the amount of water stored on land), those that affect the geoid (which is the surface of the undisturbed ocean), and finally dynamic sea-level changes which can have a meteorological, hydrological, or oceanographic origin.

The direct link between climate and sea level during the Quaternary can be illustrated by comparing sea-level records with other information on climatic change. It is generally accepted that the oxygen isotope record, which is recovered by analysing benthonic or planktonic foraminifera from deep-sea cores, chronicles the history of worldwide continental ice volume. It should also therefore act as a first approximation to glacio-eustatic sea-level change. Figure 1 shows an attempt from 1987 by N. J. Shackleton of Cambridge to compare the oxygen isotope and sea-level record. Sea level was estimated using data from the remarkable 'staircase' of raised coral reefs on the Huon Peninsula of New Guinea. In this area more than 20 coral reef complexes have been uplifted during the past quarter of a million years. The 'eustatic' sea-level curve from New Guinea shown in Fig. 1 is derived from uranium-series dating of coral and has been corrected for the effects of land uplift. The oxygen isotope record used for comparison was

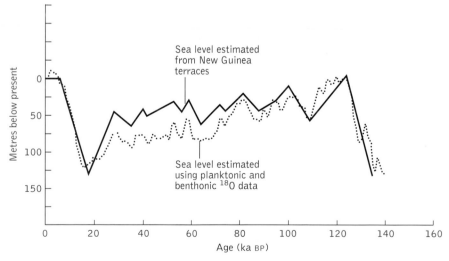

Fig. 1. Sea-level history of the Huon Peninsula, New Guinea compared to sea level estimated using planktonic and benthonic oxygen isotope data. (Source: Shackleton (1987) *Quaternary Science Reviews*.)

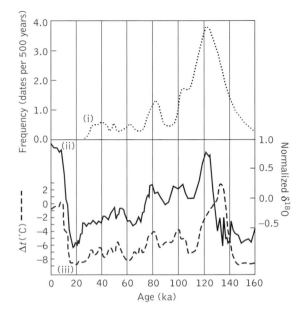

Fig. 2. Weighted frequency distribution curve for 320 uranium-series dates on high-stand coral reef terraces from over 30 areas worldwide (i), compared to the normalized, stacked, deep-sea oxygen isotope record with a chronology tuned to orbital variations (ii) and a reconstructed temperature record based on oxygen isotope data from the Vostok ice core (iii). (Source: Smart and Richards (1992) *Quaternary Science Reviews*.)

combined from two cores, V19-30 from the eastern Pacific, where benthonic foraminifera were analysed, and RC17-177 from the western Pacific, which was based on planktonic foraminifera.

Both sets of data show good agreement. In particular the high sea-level stands at 123 000, 105 000, and 80 000 years BP can be equated with marine isotope stages 5e, 5c, and 5a. In both records, stage 5e shows eustatic sea level approximately 20 m above that attained in stages 5c and 5a. Both curves show a gradual decline to the last glacial maximum at 10 000–20 000 BP and a large rise thereafter. The agreement between 80 000 and 30 000 BP is not so good, which is interesting in the light of the rapid alternations of climate from glacial to interstadial recorded in the Greenland ice cores from this period.

Using the extremes of the oxygen isotope record for the past 750 000 years, Shackleton suggested that marine isotope stages 12 and 16 record significantly more global ice volume and hence sea-level lowering than the last glacial maximum (stage 2). In addition, interglacial stages 7, 13, 15, 17, and 19 did not attain Holocene (stage 1) oxygen isotope values, suggesting that during these periods sea level did not reach its present level. The extreme values of stages 1, 5e, 9, and 11 were considered so similar that it was not possible to suggest that any one was significantly higher than any of the others.

A more recent study from 1992 by Peter Smart and David Richards (Bristol University) used a compilation of the frequency distribution of 320 published uranium-series dates on corals from Quaternary marine terraces relating to high sea levels in more than 30 areas worldwide (Fig. 2). They found that since the peak of marine isotope stage 5e at 123 000 BP there were six other maxima recorded, at 102 500, 81 500, 61 500, 50 000, 40 500, and 33 000 BP. These dates should not be seen as exact—there is a degree of uncertainty about each one—but when they are compared with independent records of climate change the degree of similarity is evident. Figure 2 shows the frequency curve for the coral terraces compared with the normalized deep-sea oxygen isotope record and also a temperature reconstruction based on oxygen isotope records from the Vostok ice core in Antarctica.

There is therefore convincing evidence that, over larger time and space scales of the Quaternary, the main determinant of sea level has been continental ice volume, which is influenced by climatic change. When smaller time and space scales are investigated, however, the link is obscured by the many other factors that can influence the position of sea level.

During the 1970s, for instance, there was much debate over the correct way to draw and interpret Holocene sea-level curves from areas that had undergone little uplift since deglaciation. Figure 3 shows two sea-level curves of mean tide level from north-west England and the Bristol Channel. There is no doubt that the major rise in sea level is due to melting of the large continental ice caps since the last glacial maximum. However, looking at the fine detail, one curve is smooth, the other shows marked oscillations of up to 1 m amplitude and a few hundred years' duration. Both are based on essentially the same sort of information; radiocarbon dated horizons from peat horizons

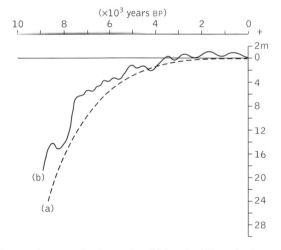

Fig. 3. Holocene sea-level curves from (a) the Bristol Channel and (b) north-west England. (Source: Kidson (1982) *Quaternary Science Reviews*).

927

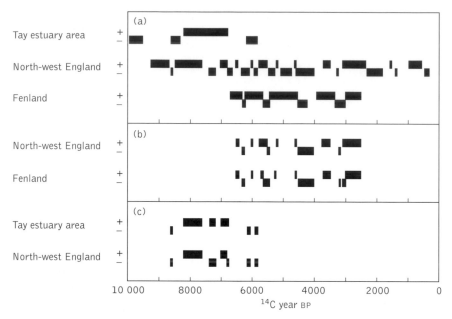

Fig. 4. Regional tendencies of sea-level movement indicated by correlation of data from north-west England, the Fenland, and the Tay estuary: (a) regional chronologies of positive (**+**) and negative (**–**) tendencies; (b) regionally significant tendencies between north-west England and the Fenland; (c) regionally significant tendencies between the Tay estuary and north-west England. (Source: Shennan, Tooley, Davis, and Haggart (1983) *Nature*, **302**, 404–6.)

interleaved with marine clay layers. The main area of disagreement concerned how to interpret peat (terrestrial) overlying clay (marine). One school suggested that given the complexity of the coastal environment it was possible that this could be achieved with no absolute fall in sea level, and it was therefore correct to draw a smooth curve through the data points. The other school suggested that peat overlying clay probably represented absolute falls in sea level, by inference caused by small changes in eustatic sea level. The essential question is whether it is possible to detect the influence of small-scale climatic changes given the resolution of the data, when errors in altitude on each point can be as much as ±1 m and the age can be determined only to ±100 years.

Some of the methodology and operational definitions of Holocene sea-level research were improved by Ian Shennan of Durham University. Defining dated points in a consistent manner, he looked at the chronology of sea-level movements between the 'uplifted', 'stable', and 'subsiding' regions of the Tay estuary in Scotland, north-west Lancashire, and the Fenland. He demonstrated, using simple sequence matching that regionally significant events could be detected and matched in these areas (Fig. 4).

With improvements in the climatic proxy record and with the use of comparable methodologies, it may soon be possible on smaller time and space scales to separate the climate signal from the noise caused by palaeotidal changes, compaction of sediments, and variations in coastal processes, sediment supply, and sedimentation rates.   B. A. HAGGART

Further reading

Devoy, R. J. N. (ed.) (1987) *Sea surface studies: a global view.* Croom Helm, London.

Smith, D. E. and Dawson, A. G. (eds)(1983) *Shorelines and isostasy.* Academic Press, London.

sea-level changes, Quaternary *see* QUATERNARY SEA-LEVEL CHANGES

seamounts Very large areas of the ocean floor are covered by seamounts, submarine mountains that may rise more than 1000 m above the ocean floor. However, the term 'seamount' means no more than a submarine mountain; there are several types of seamount and they have many origins.

Some are relatively small. If they are flat-topped they are known as guyots (see *guyots*). Small seamounts occur all over the Pacific Ocean. Most of them are essentially intraplate volcanoes not apparently associated with hotspots or obvious tectonic features. Others are larger and make up large portions of aseismic submarine ridges and oceanic plateaux, frequently lying above hotspots. For example, some 100 seamounts make up the submarine part of the Hawaiian–Emperor Chain of volcanoes and sunken extinct islands that extend nearly 6000 km across the North Pacific Ocean (see *aseismic submarine ridges and oceanic plateaux*).

In this case the term 'seamount' may, in reality, simply represent a particular stage in the evolution of a major volcano: in the Hawaiian Chain individual volcanoes pass through a series of stages from an early shield-building stage that progressively emerges from below sea level, to one, as magmatic activity diminishes, of subsidence, erosion and perhaps reef-building, before eventually subsiding to become a seamount. A significant feature of these large seamounts is that, with marginal slopes of 5° to 25°, they are unstable and may therefore break off at intervals to yield enormous submarine slides that have catastrophic effects in the oceans and on their margins (see *tsunamis*).

There are also seamounts associated with convergent margins and island arcs. These are characteristically formed of polygenetic intermediate volcanoes and may include both basaltic and more silicic volcanoes that have emerged during various times in their life history.

Not all seamounts are the result of volcanic activity. Relatively shallow ones occur on continental margins where there is a submarine topography of shallower seamounts, banks, and plateaux interspersed with deeper-water basins. These are the consequence of block-faulting caused by rifting in extensional regimes and the build-up of carbonate platforms that subside to be covered by pelagic sediments in the same way as with volcanic edifices.

The sedimentary record of seamounts is very varied, both in type and rates of sedimentation. The sediments may indicate shallow-water currents, and more than 100 m of shallow-water sediments have been recorded. Reefs were particularly abundant. At times seamounts were emergent and were eroded and subaerial sediments formed. In contrast there were periods when only pelagic sediments could form; these indicate deeper water, and there is commonly a fluctuating succession of silica-rich, carbonate-rich, and phosphatic sediments indicating a complex history of changes in oceanic waters and depth relative to the CCD (calcite compensation depth).   HAROLD G. READING

sea-surface temperatures (SSTs) and climate

The atmosphere interacts with its boundary layer, which on Earth is 30 per cent land and 70 per cent ocean. Not surprisingly, interaction between the atmosphere and the ocean is important both for the climatology and the variability of weather and climate. This interaction has only recently acquired the status it deserves in climatology, as is evidenced by the now completed long-term international collaborative study entitled 'Tropical Ocean Global Atmosphere' (TOGA).

Interaction between the oceans and the atmosphere occurs through the transfer of heat, mass, and momentum. Transfer can occur in either direction. In the case of heat, both latent heat resulting from evaporation and sensible heat are important, and both depend on factors such as wind speed, the sea-

surface temperature (SST), and the temperature of the overlying air. SSTs may result from ocean processes such as advection of cold or warm water in ocean currents, from the influence of the radiation budget (which largely depends on latitude, season, and cloudiness), or from the action of the atmosphere itself through wind-driven evaporation.

Despite this complexity, regions of climatological interaction between the ocean and atmosphere can be readily identified. The cold ocean currents on the western side of continents, for example, are associated with desert conditions, examples being the Benguela Current and the Namib Desert, the Humboldt Current and the Atacama Desert as well as the western margins of North Africa, North America, and Australia. If seen as an ocean-to-atmosphere process, these cool waters stabilize the overlying atmosphere by cooling the near-surface air. Convective rainfall is thus unlikely. The atmosphere also helps to maintain the cool water since alongshore or offshore winds driven by the resident subtropical anticyclones in these locations drive upwelling regimes in the ocean which ensure a supply of cold, deeper water.

The exchange of heat from the ocean to the atmosphere is important in driving the large convective regions of the tropics, particularly over Indonesia where the SST is some 30 °C. Here the flux of latent and sensible heat from the ocean to the atmosphere is sufficient to generate convection throughout the depth of the atmosphere. This response is possible since the relationship between air temperature and the amount of water vapour the air can hold (and hence the subsequent release of latent heat when clouds are formed) is highly non-linear. In the eastern tropical Pacific, on the other hand, where SSTs may be half those near Indonesia, skies are clear and subsidence in the atmospheric column is dominant.

Differences between land and sea temperatures are vital in driving the monsoon circulations of the globe. Water responds more slowly than land to heating since it allows heat to be mixed more easily, it delivers radiation through a depth of several metres, it is more reflective, and evaporation cools the surface. Seasonal heating and cooling of oceans lags several months behind that of land and is of a smaller amplitude. These differences are revealed in the monsoon circulations. Similarly, during winter the high-latitude oceans are characterized by low-pressure centres, with the Aleutian and Icelandic lows centred in the northern margins of the Pacific and Atlantic Oceans, with a high-pressure cell over Asia. The picture reverses in summer, with low pressures over land.

Much of the interannual variability in the tropical atmospheric circulation and rainfall is related to variations in SST. For some time this connection was overlooked, largely because variations in sea-surface temperature were regarded as being too small to influence the atmosphere, but also because global data sets of historical SSTs had yet to be developed.

The best-known example of air–sea interaction is the *El Niño–Southern Oscillation* (ENSO) phenomenon. The atmospheric component was identified in the 1920s, but it was not until the late 1960s that Bjerknes argued for ENSO as a fully coupled ocean–atmosphere system. It is now regarded as a fairly simple, yet robust system. Nevertheless, the issue of whether the SST anomalies cause the rainfall anomalies or simply reflect changes in the atmospheric circulation has often been raised. In part, this question has been examined in general circulation model (GCM) experiments forced by SST changes. The most comprehensive to date have been undertaken at the UK Meteorological Office, where an atmosphere-only GCM has been forced with observed SSTs from 1903 to 1993. These experiments show that much of the variability of the tropical atmosphere may be explained by variations in tropical SST, particularly those in the tropical Pacific. The mid-latitude atmosphere, on the other hand, appears to be highly chaotic, varying independently of SSTs.

An important application has resulted from the understanding of ocean–atmosphere interaction. Since changes in SST take place slowly, and since the atmosphere in much of the tropics is linked with SSTs, it is possible to predict seasonal rainfall categories for several parts of the tropics. The Sahel, eastern and southern Africa, and north-east Brazil are among the regions in the tropics where seasonal rainfall forecasts with leads of several months are prepared. The Sahel, for example, is sensitive to SSTs in the Pacific and the Atlantic. The recent decade-long drought in the Sahel has been explained in terms of the changing SST structure in the Atlantic, whereby positive temperature anomalies in the southern part of the Atlantic led to dry summers in the Sahel. This change in the Atlantic may reflect changes in the thermohaline circulation, which is the largest mode of oceanic circulation. Tropical cyclone frequency in the tropical Atlantic is linked to ENSO events and Sahel rainfall amounts. Southern Africa rainfall, on the other hand, is sensitive to variations in sea-surface temperatures in the Pacific and Indian Oceans. Grain yield in southern Africa is related to the pattern of temperatures in the eastern tropical Pacific, half way round the globe! On even longer timescales, the cooling of the global oceans during ice ages is thought to have had a profound effect on the global atmospheric circulation. Included in these timescales is the role of the thermohaline circulation mentioned above.

R. WASHINGTON

Further reading

Open University (1989) *Ocean circulation*. Pergamon Press, Oxford.

sea water At first glance, water seems an ordinary substance; this is not, however, the case. In fact it is only because of the very unusual physical and chemical characteristics of water that such diverse life forms have been able to evolve on land and in the sea. As we all know, water is composed of two atoms of hydrogen bonded to a single oxygen atom. The geometry of these bonds results in a water molecule in which both hydrogen atoms are aligned on one side of the oxygen atom. Because of the way in which the electrons are shared by the atoms, oxygen carries a small net negative charge and hydrogen a net positive charge. Consequently, when water

molecules meet one another they tend to join together, held by the attraction of the positively charged hydrogen atoms to the negatively charged oxygen atom of the neighbouring water molecule (in much the same way that the opposite poles of bar magnets are attracted to each other). This type of bond between molecules is known as a hydrogen bond. It results in water having a highly stable molecular arrangement, which has important implications. Under ordinary conditions, water is the only substance on the planet that naturally occurs simultaneously in all three phases: solid, liquid, and gas. Were it not for the great stability of the hydrogen bonds, the boiling point and freezing point of water would be considerably lower and water would occur on Earth exclusively in the gaseous phase.

Approximately 71 per cent or 361×10^6 square kilometres (km^2) of the Earth's total surface is covered with water. Each year about 3.6×10^{14} cubic metres of water evaporate from the surface of the oceans. About 90 per cent is returned to the oceans as rainfall. The remaining 10 per cent is transported over the continents, where it falls mainly as rain before eventually returning to the sea as river and groundwater run-off. This transport of water forms the basis of the hydrological cycle. Because the rates at which water is removed from and supplied to the ocean is equal, the volume of water in the ocean remains constant over time. Water is known as the universal solvent because of its unique ability to dissolve, to varying extents, the majority of the known elements. During its passage over land, water erodes vast amounts of continental rock. Each year approximately 50×10^6 tons of material are removed from the continents and transported to the oceans as dissolved salts and suspended particles. The bulk of this material falls to the sea floor and accumulates as thick layers of sediment close to the continental margins.

Table 1 Concentrations of the eleven most abundant constituents in sea water

Constituent	Ion symbol	Parts per thousand by weight (g/kg)	Percentage of dissolved material
Chloride	Cl^-	18.980	55.05
Sodium	Na^+	10.556	30.61
Sulphate	SO_4^{2-}	2.649	7.68
Magnesium	Mg^{2+}	1.272	3.69
Calcium	Ca^{2+}	0.400	1.16
Potassium	K^+	0.380	1.10
Bicarbonate	HCO_3^-	0.140	0.41
Bromide	Br^-	0.065	0.19
Borate	$H_3BO_3^-$	0.026	0.07
Strontium	Sr^{2+}	0.008	0.03
Fluoride	F^-	0.001	0.00
Total		34.447	99.99

(From: A. Allaby and M. Allaby (1990) *The concise Oxford dictionary of Earth sciences*.)

The average concentrations of the eleven most common constituents found in sea water are shown in Table 1; together they account for about 99.9 per cent of the total dissolved material in sea water. A significant feature of sea water is that while the total concentration of the more abundant components varies with location, the proportions of these dissolved salts remain remarkably constant throughout the oceans and appear to have done so over the past 600 million years. The total concentration of these dissolved salts determines the saltiness or salinity of sea water (see *oceanic salinity*).

Together with the major elements found in sea water, nearly all the remaining naturally occurring elements are also present in smaller quantities. In fact, several of these minor constituents, in particular nitrogen, phosphorus, silicon, zinc, copper, and iron, are important for the growth of many marine organisms. The variety of different element reactivities is highlighted by the span of the estimated average time each element spends in the ocean before being permanently incorporated in the sediment of the sea floor. These estimates, known as residence times, range over six orders of magnitude from sodium at about 2.8×10^8 years down to aluminium at about 100–200 years. The longest residence times are found for the major ions in sea water. These times are far longer than the average time it takes to complete a single mix of the oceans (about 1600 years); this emphasizes the lack of reactivity in sea water shown by these elements. The shortest residence times are found for elements such as aluminium, chromium, titanium, and iron, which are highly particle reactive and so are quickly removed by sinking particles. A large number of the minor constituents of sea water that are active in biological processes have intermediate residence times (about $5–50 \times 10^3$ years); these times are still long with respect to the oceanic mixing time. As a result, these elements undergo numerous internal cycles before being permanently removed to sediments. Such elements are removed in the surface waters and transported down through the water column by biological carriers. In deeper waters these elements are regenerated into solution from decaying organisms and are subsequently transported back to the surface in upwelling zones.

Besides the solids dissolved in sea water there are also gases, which in order of their relative abundances are nitrogen, oxygen, and carbon dioxide. By volume dissolved nitrogen accounts for about 64 per cent of the total gases dissolved in sea water. This nitrogen is biologically unimportant, as most organisms cannot directly make use of it. Sea water contains a substantially higher concentration of dissolved oxygen than the atmosphere (34 per cent as opposed to 21 per cent). The supply of oxygen to the surface ocean reflects a balance between the input of oxygen across the air–sea boundary and involvement in biological processes. During photosynthesis, marine plants consume carbon dioxide and produce oxygen gas. Since photosynthesis occurs only in the illuminated surface waters, the only source of oxygen in the deeper ocean for use by deep-living aerobic organisms is the sinking of relatively oxygen-enriched surface sea water. The concentration of carbon dioxide (about 1.6 per cent), which is some fifty times greater than in the atmosphere, is also exchanged between the atmosphere and ocean at the air–sea boundary. Once dissolved in sea water, carbon dioxide enters a series of complicated reactions which prevent sudden change in the acidity or alkalinity of sea water. This regulation is known as buffering and it maintains the pH range of ocean water between 7.5 and 8.5. Carbon dioxide is also transferred to the biosphere by the photosynthesis of marine plants and production of carbonate shells by marine plants and animals. Carbon dioxide therefore has importance in both the biological and chemical cycles of the oceans. The role of the oceans as a consumer of excess atmospheric carbon dioxide has generated considerable interest in recent years as our appreciation of the so-called 'greenhouse effect' has increased. The other remaining gases in sea water make up only about 0.5 per cent of the total amount of gases in water.

Temperature and salinity are of particular interest to oceanographers, because combined with pressure they determine the density of sea water. This property is important, for it can be used to identify and trace characteristic water masses. The difference between adjacent water bodies allows physical oceanographers to calculate rates of water movement in the deep sea. The temperature of surface oceanic sea water is determined primarily by the latitudinal variation in solar insolation, which is related to the varying angles at which the Sun's radiation strikes the Earth's surface. The temperature tends to decrease away from the Equator towards the poles from about 30 to –2 °C. Because of its salt content, sea water does not freeze until about –2 °C. There is also vertical temperature variation in the water column, with warm surface water separated from the main body of the colder deep ocean by the thermocline, a zone of rapidly decreasing temperatures that extends from the base of the surface mixed layer down to 1000 m at some locations. Temperatures in water deeper than 1000 m are much lower that at the surface and range between –1.8 and 5 °C. The heat energy balance of the atmosphere and ocean remains in equilibrium despite short-term and local variations. This is achieved by interaction of the circulating sea water with the overlying atmosphere. Warm surface water is carried from the low latitudes to the higher polar latitudes as a surface current. In polar regions net loss of heat occurs through radiation from the ocean to the atmosphere. The decrease in temperature increases the density of the sea water. The resulting water mass then sinks to the bottom and causes the deep water to be pushed along horizontally at a depth of equal density, creating a bottom current which moves at velocities of about 1 to 0.1 mm s^{-1} towards the Equator. This cyclic flow of water initiated by the cooling of surface sea water in the polar regions is known as thermohaline circulation, and is the principal example of a process called advection. Sea water motion is the combination of two processes: advection and turbulence/diffusion. Advection is a large-scale transport mechanism that involves the net movement of water from one location to another, either horizontally or vertically, whereas turbulence is essentially random mixing of water molecules caused by diffusion, in which there is no overall net transport of water, only the exchange of properties. Advection occurs at a much faster rate than turbulent mixing, and the sinking water masses therefore maintain the salinity and temperature signatures acquired from their last period at the sea surface. These characteristic signatures have been used to trace the transport paths as water masses follow the thermohaline-driven circulation. Turbulent mixing with adjacent water masses will, however, eventually homogenize the waters and destroy these unique salinity and temperature signatures.

The speed at which sound waves travel in sea water ranges between about 1400 and 1500 metres per second (m s^{-1}); this is some five times faster than the speed of sound in air. The speed of sound waves in sea water is related to its density and is therefore affected by any slight changes in salinity, temperature, and pressure. For every unit increase in salinity, the speed of sound increases by 1.5 m s^{-1}. While a 1 °C increase in the temperature causes a rise of about 4 m s^{-1} in the speed of sound, increasing pressure causes speed to rise at a rate of 18 m s^{-1} for every 1000 m of depth. The propagation of sound waves in sea water has many applications in oceanography. One of the most important has been the use of precision echo-sounders on ships. These employ a sound pulse sent from a transmitter which is reflected from the bottom to an underwater receiver, or hydrophone, carried by the ship. If the speed of the sound wave through the water is known, the time interval between the pulse being sent and the reflection received can be used to calculate the depth. Accurate depth measurement has enabled oceanographers to construct detailed bathymetric charts of the ocean floor.

The passage of light through sea water is unaffected by salinity, temperature, or pressure. About 60 per cent of the total light energy entering open-ocean sea water is absorbed in the first metre; only 1 per cent is left at a depth of about 150 m, and no sunlight penetrates below 1000 m. The typical blue colour of the open ocean is due to the preferential scattering of the shorter-wave blue component of sunlight by water molecules and suspended particles. The short-wave blue light is the most penetrating, and therefore the colour seen is dominantly blue. Greater colour variation is seen in coastal sea water because of the pigments of photosynthetic plankton that are present in great abundance and the dissolved organic substances in freshwater run-off, which tend to emphasize the yellowish-green colours.

IAN R. HALL

Further reading

Open University (1995) *Seawater: its composition, properties and behaviour.* (Oceanography Series). Pergamon Press, Oxford.

Libes, S. M. (1992) *An introduction to marine biogeochemistry.* John Wiley and Sons, New York.

Sederholm, Johannes Jakob (1863–1934)

Sederholm fought against illness all his life; at university in Helsinki he chose to study geology as a subject that would enable him to work out of doors. After further studies at Stockholm and Heidelberg he was appointed to the Geological Survey of Finland, where in 1893 he became Director, a post he held for forty years. It was Sederholm's task to try to elucidate the complex rocks of the Finnish Precambrian basement. He initiated a programme of mapping on a variety of scales from 1 : 10 to1 : 10 000 and showed that a number of belts were characterized by a distinct geological history. During the years from 1899 to 1925 many maps and descriptions of these areas were published, thereby establishing a basis for much later research. Many of the gneisses in such areas were largely of mixed aspect, with bands of metamorphic rock interleaved with others of more granitic nature. These Sederholm termed *migmatites*. He considered them to be the result of intrusion of granite magma into high-grade metamorphic rocks at depth, the granite formed partly by the action of emanations on the older rocks and partly by anatexis. He thus became involved in, and made most important contributions to, the granitization controversy. He was the first to establish the relation between depth of burial and intensity of metamorphism and the growth of new minerals. His fieldwork was backed up by detailed observations of textures and fabrics of metamorphic rocks and he gave to geology a great many textural terms of Greek derivation. He published papers in five languages.

On the basis of detailed mapping and many analyses Sederholm compiled an average composition of a deeply eroded piece of the Finnish crust, giving weighted proportions according to the surface area visible.

Sederholm was not just a geologist. He was a member of the Finnish Diet, for whom he made several missions to the League of Nations. He was also twice chairman of the Economic Society of Finland, and wrote in 1915 *The science of work.* He was deeply interested in many aspects—social, political, and economic—of the Americas and made several visits to Canada and the USA.

Seven of Sederholm's most influential works were published in book form in 1967 under the title *Selected works: granites and migmatites.*

R. BRADSHAW

Sedgwick, Adam (1785–1873)
Adam Sedgwick left his native Yorkshire in 1808 at the age of 23 to enter Trinity College, Cambridge, as a scholar in mathematics. He was elected to a Fellowship in 1810, ordained in 1817, never married, and remained at Trinity all his life. Sedgwick was clearly an exceptional scholar and in 1818 was appointed Woodwardian Professor of Geology, never having given a geology lecture in his life. He soon began to do so and very popular lectures they were, influencing a long succession of geology students at Cambridge. Sedgwick was a co-founder of the Cambridge Philosophical Society and by 1830 was a Fellow of the Royal Society.

Sedgwick at once began his geological travels and made contact with the growing number of geologists at home and abroad. He began with a study of the Permian and Triassic rocks of north-eastern England, but soon became fascinated by the thick sequences of older Palaeozoic rocks in the Lake District and North Wales. A trip to the Alps with R. I. Murchison aroused his interest in tectonics and he was inclined to a somewhat catastrophist view of structural geology. Although never happy with rigid uniformitarianism, he did grasp the magnitude of geological time. His work with Murchison in Devon and Cornwall was happy (1831), but their friendship later suffered in the dispute about the base of the Silurian system. In 1851 he was presented with the Geological Society's Wollaston Medal, its highest award, for his research into Palaeozoic rocks, and the Royal Society bestowed a Copley Medal upon him in 1863.

In later life Sedgwick became rather remote and he rejected Darwin's work on evolution. He was keen to establish a geological museum at Cambridge. It now commemorates his name as the Sedgwick Museum. He died in Cambridge in 1873.   D. L. DINELEY

sedimentary basins
Sedimentary basins are areas of the Earth's crust with a long history of subsidence and within which a thick sequence of sediments has accumulated. They are the dustbins of continental weathering, erosion, and transport, and their sedimentary successions record the evolution of the Earth's crust in a particular part of the world. Each basin has a stratigraphy controlled partly by local conditions and also by some large-scale worldwide events, such as tectonic activity, climate changes, and eustatic sea-level change. The separation of local events from those of worldwide extent makes possible the correlation of individual basins and the construction of a general synthesis. Undoubtedly the main control is tectonic, because tectonic events produce areas of subsidence that become filled with sediments. Continued tectonic activity, taking place while the basin is being filled, will affect the distribution and type of sediment. In some instances, for example in the North Sea, parts of one basin later become elevated during tectonic inversion (uplift rather than subsidence) and supply sediment to adjacent areas. Likewise, a basin that was formerly on the margin of a mountain belt (for example a foreland basin) may become involved in the mountain belt as the area affected by shortening deformation increases with time and the growth of the belt.

An excellent example of the tectonic control of basin development is shown by modern large fluviomarine deltas which are sited in tectonic depressions. Deltas need large supplies of siliciclastic sediment for their maintenance. Such huge volumes of sediment need large drainage networks, phenomena which are not haphazardly distributed over the Earth's surface but are developed in large tectonic depressions. For example the Tigris and Euphrates flow into the Mesopotamian Plain between the stable shield of Arabia and the orogenic arc of the Zagros Mountains to form a large delta at the head of the Arabian (Persian) Gulf.

The study of sedimentation in the 1950s greatly increased our understanding of how sedimentary basins are filled, and

during the past few decades knowledge of diagenesis and the evolution of interstitial waters has enlarged this understanding. The advent of sequence stratigraphy since the 1970s, building on the early ideas of Sloss in the USA, and those developed by Vail and his colleagues in Exxon, who used new data gained by seismic reflection surveys, led to a re-examination of stratigraphical successions and the introduction of a plethora of new terms and ideas, many of which cast a mystique over phenomena already well understood.

An important attribute of the 'new' sequence stratigraphy is that of eustatic (that is, worldwide) sea-level change and its influence on stratigraphical successions. The new ideas were successful when applied to passive continental margins but they were less successful when applied to destructive or convergent margins, where the effects of local contractional tectonics often overwhelmed other more subtle, but worldwide, controls. Appreciation of this limitation has tempered some earlier views.

A considerable advance was made in the understanding of sedimentary basins by geophysical studies on the eastern margin of the USA by Drake, Ewing, and Sutton in the 1950s. Basins had been envisaged as always being concave-downwards depressions which produced sedimentary fills of similar geometry. However, the studies by Drake and his colleagues revealed two thick pods of sediment underlying the shelf slope and rise, where the contemporary surface of deposition is subhorizontal or convex (Fig. 1c, d). It was the synsedimentary subsidence of the continental margin which produced a downward concave structure. The shape of a sedimentary basin fill on a geological cross-section depends upon the attitude given to the bounding surfaces. In a deltaic succession (Fig. 1a) if the confining upper surface (T_2, Fig. 1a) is drawn horizontally, the deltas appear as a series of convex-downwards basins containing thick sedimentary sequences separated by 'highs' preserving thin sedimentary sequences (Fig. 1b). In reality, however, the deltas were mounds, with a certain amount of downwarping beneath them, separated by thin continental shelf sediments.

For an understanding of the evolution of a basin, the history of subsidence of the basin floor is necessary. This necessitates correction of water depth with time and absolute sea-level changes, compared to those of today. The sedimentary thicknesses have to be corrected for compaction, so that the initial depositional thicknesses can be calculated.

The water depth, or *palaeobathymetry*, of the sediments can be approximately inferred from facies and geochemical indicators, but mainly from the microfauna. Determination of the relative importance of tectonic subsidence of the basin floor in relation to that caused by sedimentary loading is achieved by a calculation procedure known as *backstripping*. The economic potential of a basin not only requires understanding of the burial history of the sediment fill but also its thermal history. Temperature changes accompany subsidence; *thermal maturation* is controlled by temperature changes and the time over which they act, pressure changes being relatively unimportant. The properties used to unravel temperature

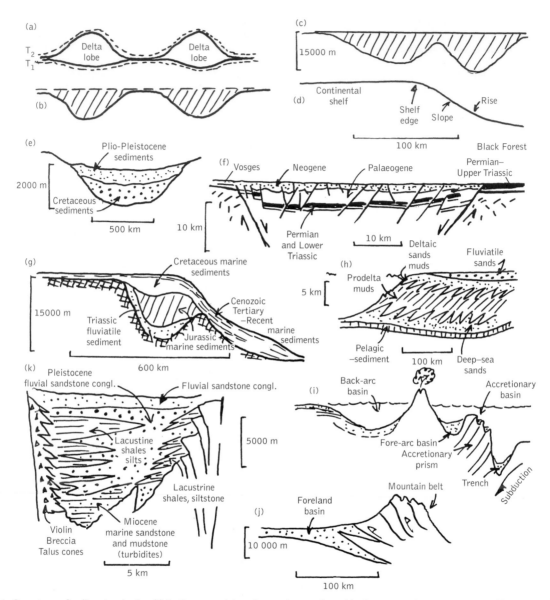

Fig. 1. Some types of sedimentary basins. (a) Section across lobes of a prograding delta; (b) sedimentary pattern (a) if T_2 is plotted horizontally; (c) thickness of sediments under continental margins of the eastern USA; (d) actual topography of continental margin of the eastern USA; (e) section across Lake Chad basin; (f) cross-section across the Rhine rift basin; (g) cross-section across a typical passive margin: the east coast of the USA; (h) cross-section across the Benue Trough; (i) cross-section of a typical active margin basin; (j) cross-section of a typical foreland basin; (k) cross-section across Ridge Basin, California. ((d) after C. L. Drake *et al.* (1959) Continental margins and geosynclines: the east coast of North America north of Cape Hatteras. *Physics and Chemistry of the Earth*, **5**, 110–98; (e) after K. Burke (1976) The Chad Basin; an active intracontinental basin. *Tectonophysics*, **36**, 197–206; (f) after N. H. Woodcock (1981) The deformation of rocks. In *The Cambridge encyclopedia of Earth sciences* (ed. D. G. smith), pp. 189–200. Cambridge University Press; (g) after R. E. Sheridan (1974) Atlantic continental margin of North America. In *The geology of continental margins* (ed. C. A. Burke and C. L. Drake), pp. 391–407. Springer-Verlag, New York; (h) After K. J. Weber (1971) Sedimentological aspects of oilfields of the Niger delta. *Geologie en Mijnbouw*, **50**, 559–76; (k) after J. C. Crowell and M. H. Link (1982) Ridge Basin, Southern California. In *Geologic history of Ridge Basin, Southern California* (ed. J. C. Crowell and M. H. Link), pp. 1–4. Society of Economic Paleontologists and Mineralogists, Pacific Section.)

changes are the state of the organic matter and the changes undergone by the clay minerals. The transformation of various clay minerals, such as the crystallinity state of the contained illite, its percentage in mixed clays, and the type and amount of smectite, are all useful in this respect.

Although sedimentary basins have been studied for many decades, and have become increasingly well understood because of the three-dimensional information provided by deep drilling by oil companies since the 1930s, it was not until the development of plate tectonics in the 1960s that they were placed in a coherent tectonic framework.

The mechanisms that produce sedimentary basins are lithospheric stretching, flexuring of the oceanic–continental lithosphere, and strike-slip faulting. Sedimentary basins develop within plates (*intraplate basins*) or at their edges (*interplate basins*). The former can develop on both continental and oceanic crust; the latter can form at divergent margins (passive margins) where new oceanic crust is being created, within stretched parts of continents; on convergent margins (destructive margins) where oceanic crust is being subducted, at margins where two continents collide, or at conservative margins where plates are moving past one another.

Intracontinental sag basins are important sites of sedimentation. The slowly subsiding sag basin on the continental crust of the Chad Basin of north-east Africa (Fig. 1e) has accumulated several thousand metres of siliciclastic sediment over an area of 600×1000 km^2 since the Cretaceous, and has been estimated to be sinking at 2 cm per 1000 years. In contrast, the Michigan Basin in the USA has been the site of deposition of 3.5 km of carbonate and evaporitic sediments since the Palaeozoic.

Rifting, via normal faulting, of the crust has produced large sedimentary basins in various settings. The East African rift basins developed on continental crust and are filled with up to 2000 m of freshwater lake sediments, evaporites, and associated volcanic rocks. Lake Baikal is underlain by continental crust and up to 5 km of mainly fluvial, swamp, and shallow lake sediments overlain by continental sediments deposited since the Oligocene. The Rhine rift is another well-known example of an intracontinental rift, (Fig. 1f). The Aegean region and Basin and Range in western USA are characterized by abundant grabens (rifts) within areas experiencing horizontal stretching. Each, generally isolated, basin in these areas is a sub-basin within a larger area of subsidence.

Rifts in oceanic crust also produce sedimentary basins covering hundreds of square kilometres which contain hundreds of metres of turbidites derived from the bordering ridges, but, unlike continental margin turbidites, they are composed of re-deposited pelagic sediments.

Ocean-floor spreading is one of the most important modes of basin development. The early history of spreading is characterized by, as in the present Red Sea, alluvial fans and evaporites which are ultimately succeeded in the geological record by deep-water sediments. The separated flanks of such newly opened oceans gradually develop into passive margins, asymmetric downwards, which become the sites of thick wedges of coastal and open-marine sediment (up to 17 km thick) passing seawards into slope, rise, and deep oceanic deposits (Fig. 1g). Where the sediment supply was abundant the basins are dominated by siliciclastic sediments, but elsewhere carbonates accumulated on rimmed shelves and carbonate ramps.

Where the opening of an ocean was aborted, deep embayments cut into continental margins, which could have become new oceans, and are filled with sediment (aulacogens or failed rift systems). The Benue Trough on the west African coast is a good example of an aulacogen (Fig. 1h). The trough, into which the rivers Benue and Niger, are still pouring sediment, is filled with approximately 10–15 km of deep-water pelagic sediments of Cretaceous and Early Palaeogene age, overlain by deep-sea sands, mudstones, and shallow-water sands of the Neogene–Recent Niger delta, covering an area of hundreds of thousands of square kilometres. The North Sea Basin is also a failed rift system and is filled with materials grading from continental desert and lake sediments fully marine sediments deposited in a variety of environments from deltaic to deep basin.

A variety of basins develop on active plate margins where oceanic crust (Fig. 1i) is being subducted beneath other oceanic crust or beneath continental crust (Fig. 1i). Most of the adjacent oceanic floor of the subducting plate is the site of limited pelagic sedimentation because siliciclastic material from the continent is trapped in basins located close to the subducting zone. When these become filled, however, sediment is spread by turbidity currents over the deeper ocean floor as, for example, off the north-western coast of the United States. Deep subduction-zone-related trenches are filled with only a modest thickness of sediment produced by slumping from the walls or turbidity currents flowing along the axis of the trench. The sediment is scraped off the surface of the subducting oceanic plate to produce a thick stack of thrusted deep-water sediments, called *accretionary prisms*. Small basins develop on this accreting wedge of sediments (*accretionary basins*), which are filled with turbidites, debris-flow deposits, and volcanic rocks. Large basins, thousands of square kilometres in area, develop between the volcanic arc generated by the subducting oceanic plate and the seawards trench (*fore-arc basins*). Off Japan typical fore-arc basins are 200×50 km^2 in area and contain a depth of 5 km of sediment, much of it vast quantity of volcanic detritus derived from the adjacent arc. In such basins volcanic-rich turbidites and muds grade into a variety of shallow-water sediments. The volcanic arcs enclose large, deep-water back-arc basins developed by back-arc spreading, sometimes accompanied by strike-slip faulting. Small ocean basins are well developed off eastern Asia where the Japan Sea and other basins cover thousands of square kilometres. They contain a similar range of sediments to those in passive margins on the continental side—deltaic, shallow-water shelf, slope, and pelagic sediments—but lack evidence of geostrophic currents. Where there are large deltas on the continental flank (e.g. off Burma), they may contain up to 17 km of sediment, passing into turbidites form the enclosing arcs.

Where there has been ocean–continent or continent–continent collision, basins develop which are filled quickly with vast thicknesses of sediment derived from the nearby mountain chains that were formed by the collisions (Fig. 1j). *Foreland basins*, a term introduced by Dickinson in 1974, lie between a mountain chain and its adjacent craton (Fig. 1j). Where they lie adjacent to the outer part of a mountain belt (*orogenic arc*) produced by continent–continent collision (e.g. the Swiss plain or Molasse Basin, or the Indo-Gangetic plain between the Himalayas and the cratonic mass of peninsular India) they are termed *peripheral foreland basins*. The other type of foreland basin, *retro-arc basin*, is situated on the continental side of a magmatic arc where there has been ocean–continent collision, subduction, and formation of a mountain chain (e.g. the thick sequence of Mesozoic–Cenozoic sediments adjacent to the Rocky Mountains). Foreland basins initially develop ahead of the outermost thrust sheets of mountain belts, but eventually their sediments become involved in younger thrust sheets that ride over the proximal edge of the foreland basin. Sometimes such basins are carried on the backs of thrust sheets. Orti and Friend have called such basins *sheet-top basins* or *piggy-back basins*.

Strike-slip faulting can produce a series of *strike-slip* or *pull-apart basins*. They vary enormously in size from small pond-sized features, through those the size of small lakes (e.g. the Bovey Tracey Tertiary basin in south-west England, developed along the Sticklepath fault), to those associated with major transform faults (e.g. the San Andreas Fault system of the western USA), which may cover an area of thousands of square kilometres. The evolution of such basins is extremely complicated. Strike-slip faults are rarely straight, and areas of localized compression (i.e. *transpression*) alternate with zones of extension (i.e. *transtension*). Fault movements on the basin margin can be both vertical and horizontal and can result in rapidly changing tectonic source areas for sedimentation. Pull-apart basins are dominated by either continental or marine sediments, according to the setting and the local climate, and are usually the sites of thick piles of conglomeratic sediments. Famous examples of pull-apart basins include the Dead Sea Basin, which is filled with non-marine carbonate rocks and evaporites, with the deltaic sediments of the Jordan River at its north-western extremity. A thick sequence of marine turbidites passing into alluvial fan and lake deposits fills the Pliocene Ridge Basin of California (Fig. 1k) and records the changing conditions of a pull-apart basin associated with the San Andreas fault system (Fig. 1k). Adjacent basins off the California coast are developed in the same area of extensive strike-slip movement and are filled with thick piles of turbiditic sands and muds, hemipelagic muds and large quantities of slumped sediments. Basins of this type, which are either remote from land or are enclosed by raised blocks that cut off the supply of sediment, are areas within which only fine-grained, land-derived sediment or pelagic diatomaceous oozes can accumulate (e.g. the Santa Barbara Basin). Because they are isolated from deep oceanic current circulation, they become anoxic and organic-rich sediments can accumulate.

The concept of a *geosyncline*, introduced by James Hall and named and modified by J. D. Dana, dominated pre-plate-tectonic ideas on the origin of thick piles of sediments. A geosyncline was conceived of as a long, linear trough in which a great thickness of sediment accumulated. The sediments of many of these troughs were thought to be folded during orogenesis to produce mountain chains. Many different types of geosynclines were visualized and elaborate classification schemes were proposed by Kay and others in the USA. Robert Dietz in the USA, and later Tuzo Wilson in Canada, attempted in the 1960s and later to explain the evolution of such large sedimentary basins in terms of modern continental margins and seafloor spreading. Today, the term 'geosyncline' is archaic and has been replaced by terms related to plate tectonics.

The understanding of sedimentary basins is achieved by the application of many techniques. Detailed study of the sediments and their included fauna and flora, *facies analysis*, makes it possible to reconstruct the various sedimentary environments that existed during the evolution of a basin and to reconstruct past geographies—*palaeogeographies*. The study of directional sedimentary structures, *palaeocurrent analysis*, can indicate the position of a source area, and petrographic studies can determine of the type and geographical origin of a source rock, that is, the *provenance* of a sediment, and especially pebbles in conglomerates. The micro- and macrofauna included in a sediment ca be used to determine water depth as well as the salinity and sometimes the temperature in which the sediments accumulated. The burial history of the sediments can be ascertained by the study of their varying thickness; and the petrography of the sediments reveals their diagenetic history and the movement of meteoric and pore waters through the basin. The stratigraphy of a basin fill, combined with a detailed analysis of its sedimentary successions, permits the tectonic evolution of the basin, as well as the effects of worldwide changes in sea level and climate, to be assessed, and allows correlation of successions between basins. Understanding basin evolution is essential for the economic exploitation of sediments for water, oil, gas, and minerals.

G. EVANS

Further reading

Allen, P. A. and Allen, J. R. (1990) *Basin analysis, principles and applications.* Blackwell Scientific Publications, Oxford.

Mitchell, A. H. G. and Reading, H. G. (1986) Sedimentation and tectonics. In Reading, H. G. (ed.) *Sedimentary environments and facies*, (2nd edn), pp. 471–519. Blackwell Scientific Publications, Oxford.

Reading, H. G. (1982) Sedimentary basins and global tectonics. *Proceedings of the Geological Association*, 93, (4), 321–50.

Drake, C. L. *et al.* (1959) Continental margins and geosynclines: the east coast of North America north of Cape Hatteras. In *Physics and Chemistry of the Earth*, V. 5, D110–198. Pergamon Press, Elmsford, N.Y.

Burke, K. (1976) The Chad Basin: an active intracontinental basin. *Tectonophysics*, **36**, 197–206.

Woodcock, N. H. in Smith, D. G. (ed.) (1981) *The Cambridge Encyclopaedia of Earth Sciences*. Cambridge University Press.

Weber, K. J. (1971) Sedimentological aspects of oilfields of the Niger delta. *Geologie en Mijnbouw*, **50**, 559–76.

Sheridan R. E. (1974) Atlantic Continental Margin of North America. In Burke, C. A. and Drake, C. L. (eds) *The geology of continental margins*. Springer-Verlag, New York.

Crowell, J. C. and Link M. H. (1982) Ridge Basin, Southern California. In Crowell, J. C and Link, M. H. (eds) *Geologic history of Ridge Basin, Southern California*, 1–4 Society of Economic Paleontologists and Mineralogists Pacific Section.

sedimentary geochemistry Sedimentary geochemistry applies chemical principles and techniques to decipher the history of sedimentary rocks from their formation as sediments through lithification to form rocks, to eventual erosion to form the components for new sediments. Sedimentary geochemists also model this sedimentary cycle to better understand elemental cycles, such as the carbon cycle, and the fate and transport of chemical compounds, including pollutants, in soils, surface waters and groundwater.

During formation, sediments develop primary chemical signatures reflective of their origins. These signatures include the mineralogy, and the isotopic and trace-element compositions of minerals. After deposition, sediments are subjected to a number of changes which can add new minerals, dissolve old minerals, and alter the organization of components. These processes also cause the loose sediment grains to be bound together to form sedimentary rocks. These processes, collectively termed *sedimentary diagenesis,* occur through interactions between sediment grains and the waters in contact with them. Consequently, a large part of the discipline of sedimentary geochemistry is concerned with study of the chemistry of diagenetic waters.

A number of clues are present in sedimentary rocks that a sedimentary geochemist can use in reconstructing rock history and diagenetic fluid chemistry. Rocks may be analysed to determine their bulk chemistry or mineralogical compositions, or the chemical compositions of individual constituents may be determined separately. In addition, during growth, some minerals incorporate small pockets of liquid, termed *fluid inclusions,* and these too can be analysed.

Materials making up sedimentary rocks can be divided into three general categories: clastic minerals, produced by weathering of precursor rocks; minerals that are chemically or biochemically precipitated, including carbonates, evaporites, sulphides, oxides, and phosphates; and organic matter, including kerogen, coal, and petroleum.

Clastic sediments

Clastic rocks are composed predominantly of those minerals most resistant to weathering; these are also the igneous minerals which form at the lowest temperatures in Bowen's reaction series: quartz, feldspar, and mica. Classic rock groups include sandstones, siltstones and shales. The more soluble minerals in source rocks dissolve, contributing chemicals such as potassium, calcium, sodium, iron, magnesium, and silicon to surface waters and sediment pore waters. These chemicals often subsequently reprecipitate during diagenesis to form new minerals as cements in sediment pore spaces. Common cements include clay minerals such as kaolinite, montmorillonite, or illite; quartz or chalcedony; iron oxides such as haematite; or calcite. Diagenesis may also lead to dissolution of detrital minerals after the sediment has lithified, creating pockets of porosity which may subsequently be filled by petroleum or water.

The geochemistry of clastic rocks focuses on determining the source rocks of weather-resistant grains and on the geochemical conditions that lead to cement formation and diagenetic dissolution. Bulk-rock geochemistry is often determined and usually correlates well with mineralogy. For sandstones, major-element geochemistry also correlates well with tectonic setting because of differences in igneous and metamorphic source rocks that are present in ocean island arcs and passive or active continental margins.

Precipitated sediments

Carbonates

Carbonate sediments are composed predominantly of fragments of the skeletal hard parts of marine organisms, including corals, clams, and other molluscs, and calcareous algae. The hard parts are composed of calcite and aragonite, two mineral polymorphs of calcium carbonate. The calcite may contain some magnesium substituted for calcium to produce low-magnesium (less than 5 per cent substitution) or high-magnesium (5–20 per cent substitution) calcite. High-magnesium calcite is characteristic of organisms in tropical to subtropical marine environments and hypersaline settings; low-magnesium calcite of organisms in freshwater or temperate marine environments. Aragonite is mostly restricted to tropical and subtropical marine and hypersaline environments. After these grains are deposited as sediments, these same minerals are also commonly precipitated in marine environments as cements between sediment particles. In terrestrial, freshwater settings, inorganically precipitated low-magnesium calcite forms in subsurface aquifers, coastal zones where sea water mixes with groundwater, hydrothermal springs, and cave deposits. Dolomite also forms in very small quantities in some modern sediments, apparently as a product of early diagenetic alteration, but it is abundant in ancient carbonate rocks. Siderite (iron carbonate), magnesite (magnesium carbonate), strontianite (strontium carbonate), rhodochrosite (manganese carbonate), and smithsonite (zinc carbonate) may form as minor secondary phases in some sedimentary rocks.

By comparison with clastic rocks, carbonates are very reactive and commonly undergo substantial mineralogical and textural changes during diagenesis. When shallow marine sediments are exposed to fresh water, high-magnesium calcite and aragonite are readily replaced by low-magnesium calcite, often with the structure of the organism's hard parts still preserved. These minerals may also dissolve without being replaced, resulting in considerable increases in rock porosity.

The nature of the organisms whose hard parts are preserved in carbonate rocks helps geologists to identify the type of environment that was the source of the sediment. After deposition and diagenetic alteration, mineralogy and trace-element and isotopic chemistry are keys to deciphering rock history and the chemistry of diagenetic fluids. Ancient carbonate rocks have mainly been converted to low-magnesium calcite or dolomite, and so one of the first tasks of a carbonate geochemist is to reconstruct the original mineralogy of the constituent cements and grains and the chemistry of the waters causing the mineral alteration. For the skeletal fragments, the mineralogy of similar or equivalent modern organisms is assumed for their ancient counterparts. However, some organisms do not have surviving examples, and their original mineralogy is inferred from studying the geochemistry of their fossil hard parts. One clue for determining the mineralogy of an original cement is the shapes of crystals, which are commonly preserved after mineral alteration: aragonite tends to form long, needle-like crystals, high-magnesium calcites tend to form spiky dog-tooth shapes, and low-magnesium calcites tend to be equant, blocky crystals. In addition, the replacement is usually not complete; small domains of aragonite or high-magnesium calcite can be found by electron microscopy in low-magnesium calcite or dolomite crystals. Also, aragonite tends to have much higher amounts of strontium as a trace element in the calcium carbonate. Low-magnesium calcite will thus have higher strontium values when it has replaced aragonite than when it replaces high-magnesium calcite.

Other precipitated rocks

Evaporites are formed by precipitation of minerals as a body of water evaporates. Evaporite deposits formed from sea water have a characteristic mineral sequence, beginning with calcite, gypsum, and halite, followed by potash salts. In lakes on continental crust, such as the Great Salt Lake, or the hypersaline lakes of the East African Rift Valley, the mineral salts are different, and may include trona, mirabilite, glauberite, epsomite, and natron, as well as gypsum, halite, and carbonate minerals. Evaporites also form from early or late-stage diagenetic processes in the subsurface. Geochemists often have the task of determining whether evaporite minerals are primary or diagenetic. Isotopes, trace elements, and analysis of fluid inclusions are valuable approaches for differentiating between the two.

Chert is another precipitated sedimentary rock. It is composed of various polymorphs of silicon dioxide: chalcedony, opal, and quartz. Chert may derive from shells of microscopic silica-precipitating marine organisms: radiolaria and diatoms, and some sponges. Chert (and flint, which is a distinctive form of chert found in the chalk of north-western Europe) also forms diagenetically as stringers and nodules in limestones, or as a replacement for carbonate mineral fossil fragments. Trace elements are common in chert-forming minerals and these give chert some spectacularly colourful banding, creating the complex patterns found in chert nodule geodes and jewellery stones. Jasper, for example, is red chert, coloured by trace amounts of iron. Trace elements and oxygen isotopes are useful geochemical indicators of the chemistry of the waters from which the chert precipitated.

Iron-rich sedimentary rocks are particularly abundant in Precambrian sequences, but were also formed during the Early Palaeozoic, the Jurassic, and Cretaceous. They can be divided into iron formations, which are commonly banded, consisting of magnetite and haematite interbedded with chert, and are primarily of Precambrian age; and ironstones, which are Palaeozoic and Mesozoic in age, commonly consisting of goethite and haematite associated with carbonate minerals, and usually are massive or oolitic. There are no modern analogues for iron-rich sedimentary rocks, and their origins remain enigmatic. Today, iron is deposited as disseminated sulphides and oxides in carbonate and clastic rocks, and as ferromaganese nodules on the sea floor. The chemistry of iron depends on the amount of dissolved oxygen present and the acidity of solutions; precipitation of iron as oxides or hydroxides occurs rapidly unless acid, anoxic (or reducing) conditions are present. Thus, identifying the geochemical mechanism that kept iron in solution long enough to accumulate it in these enriched ancient deposits remains one of the classic problems in sediment geochemistry; also puzzling is why this mechanism is not operating today.

Phosphorites are sedimentary rocks containing more than 15–20 per cent P_2O_5, present as the mineral apatite, often concentrated in carbonate or chert host rocks. The remains of fish skeletons, shark teeth, nodules, pellets, phosphatized carbonate fossil fragments, and ooids are common constituents. Phosphorites are presumed to originate as marine deposits in zones of upwelling currents, but there is, again, a need to find a mechanism to explain how these high amounts of phosphate (up to 40 per cent in some ancient deposits) become concentrated from the low levels present in sea water. Most geochemists believe that this concentration process occurs diagenetically in the pore waters of ocean-floor sediments. Measurement of oxygen and carbon isotopes, as well as trace-element geochemistry, can be used to constrain the compositions of diagenetic fluids.

Carbonaceous sedimentary rocks contain at least 10 per cent organic matter. Two types of organic matter accumulate in sediments: humus and sapropel. Humus is produced from the decay of plant organic matter in soils. In soils, humus is usually only a few per cent of the total material, but in environments where there is little oxygen, such as bogs or swamps, enough humus is preserved to create peat, which is over

40 per cent organic. Sapropel is organic matter produced predominantly from the decay of phytoplankton and zooplankton in marine, estuarine, or lake environments. With increasing burial and temperature, diagenetic processes change organic matter in sedimentary rocks into kerogen, the precursor of petroleum. Organic matter consists of carbon, with much smaller amounts of hydrogen, oxygen, nitrogen, and sulphur. The relative amounts of these elements in kerogen are related to the nature of the source material. Oxygen, carbon, and hydrogen isotopes can also be useful indicators of the diagenetic history.

Thermodynamics

Thermodynamics enables geochemists to determine whether a mineral, in contact with water of a specified composition, will precipitate or dissolve. This is dependent on the quantity of the mineral's constituent ions that are present in solution. A simple experiment with table salt (sodium chloride, chemically the same as the mineral halite) and a glass of water illustrates the concept. If salt is added to the water, it dissolves initially, contributing sodium and chloride ions to solution. If more and more salt is added, a point is eventually reached when no further dissolution occurs, and salt crystals remain on the bottom of the glass. At this point, the water is said to be exactly saturated or at equilibrium with respect to sodium chloride. Before equilibrium was reached, fewer sodium and chloride ions were present in the water than the amounts needed at equilibrium, and the solution was undersaturated with respect to sodium chloride. Similarly, if more sodium and chloride ions were present, the solution would be supersaturated with respect to sodium chloride. The amounts of sodium and chloride ions in solution at equilibrium can be measured to express a quantity called the solubility constant. The value of the solubility constant is in fact constant only for a specific set of geochemical conditions. If the glass of water at equilibrium with the salt at room temperature were heated, the salt at the bottom of the glass would dissolve further, coming to a new equilibrium concentration at the higher temperature; if the water were cooled, some of the sodium and chloride ions would start to precipitate, forming new salt crystals until a new equilibrium was established. If further experiments were done by putting the salt solution in a pressure cooker, or by adding other compounds to the water, the dependence of the solubility constant on pressure, and on the amounts of other ions present in solution could also be demonstrated.

The situation for the geochemistry of natural waters is akin to having a glass of water with many dissolved ions in it and many minerals in contact with it. To determine whether the water is supersaturated, saturated, or undersaturated with respect to the minerals present, a complete water analysis must be done, and the temperature and pressure of the environment must be known. From extensive databases of solubility constants and their dependence on bulk solution chemistry, geochemists can determine which minerals are dissolving or precipitating. These reactions not only determine the types of alteration occurring in the sediments, but also indicate which minerals are influencing the compositions of the diagenetic fluids. For the sedimentary geochemist, however, the situation is similar to having a suite of minerals in the bottom of the glass, but no water present. The goal is to use the minerals to work out what the water composition was at an earlier stage.

Thermodynamics can also be used to assess the stability of minerals relative to each other, under a given set of geochemical conditions. Whether one mineral is more stable than another is determined by the energy associated with the reaction in which one mineral is converted into another, and by the amount of disorder in the system. Reactions tend to proceed to a state of lowest energy and maximum disorder.

From thermodynamic calculations of mineral solubility and mineral conversion reactions, diagrams can be constructed which show stability fields for various minerals. The axes for these diagrams are various geochemical parameters, such as temperature, ion concentrations in solution, or concentration ratios of solution. A particularly important type of stability diagram plots the acidity of solutions, measured as pH, against the amount of available oxygen in solution, measured as Eh, oxidation–reduction, or redox, potential. Eh–pH diagrams are used extensively in sedimentary geochemistry to predict the behaviour of metals in water, and the formation of metal-rich sedimentary rocks, including iron formations.

Kinetics

Thermodynamics predicts the direction and magnitude of a reaction, but the rate of reaction and the pathway by which it occurs is the subject of kinetics. A geochemist once compared thermodynamics and kinetics by use of a feather allowed to drop to the floor. Thermodynamics described its initial state, resting in his hand, the fact that it would fall downward, not upward, and its final state, resting on the floor. But the complex pathway of fall, with all of the feather's various floating motions, is described by kinetics.

Sedimentary geochemists study the kinetics of mineral reactions in order to understand why thermodynamically predicted states do not occur. A classic example is the case of dolomite. Dolomite, from a thermodynamic standpoint, is more stable than calcite or aragonite, and modern sea water is hundreds of times oversaturated with respect to dolomite. However, dolomite is rarely, if ever, precipitated directly from sea water, nor does it replace calcite or aragonite except in special environments (see *dolomite*). Kinetic studies provide some answers: dolomite precipitation rates are extremely slow compared with calcite and aragonite growth rates. Thus, calcite and aragonite, competing for the same constituent ions in solutions, end up predominating over dolomite, in spite of its thermodynamically favoured status.

Isotope geochemistry

Ratios of stable isotopes in minerals are important clues for deducing the chemistry of the water and environment in which the minerals have been precipitated, but the interpretation of these isotopic signatures can be extremely complex. Stable isotopes are atoms of an element which have the same number of protons but different numbers of neutrons in their nuclei. These extra neutrons do not, however, make the nuclei unstable, and so they do not decay radioactively. Sedimentary geochemists commonly measure stable isotopic ratios for carbon, oxygen, strontium, hydrogen, and sulphur in a variety of minerals.

Because of differences in their chemical behaviour, isotopes of an element are present in minerals in ratios that are different from those that are found in the waters from which they were precipitated. The degree of difference, termed the *isotopic fractionation factor* for a mineral, depends on the temperature of precipitation, but is independent of most other environmental conditions. This means that isotopic ratios in minerals can be used to determine the isotopic ratio of the water if the temperature of precipitation is known; conversely, the temperature can be determined if the isotopic ratio of the water is known.

Carbon

About 99 per cent of the carbon atoms on Earth have an atomic weight of 12 (^{12}C), their nuclei consisting of six protons and six neutrons; there is also, however, about 1 per cent of carbon atoms that have extra neutrons, giving them atomic weights of 13 (^{13}C). Most of the carbon in the world is found in two places (i.e. reservoirs): carbonate minerals and organic matter. Carbonate minerals preferentially incorporate more ^{13}C, giving them higher ^{13}C/^{12}C ratios than the waters from which they are precipitated. Plants selectively take up ^{12}C, and so any organic matter in rocks or sediments from plants or phytoplankton (or from any animals that eat plants or phytoplankton) has less ^{13}C than would be expected if plants took in both isotopes indiscriminately from their carbon sources. Thus, carbon is present in one reservoir of mineral or inorganic carbon that is relatively enriched in ^{13}C, and in one of organic carbon that is relatively depleted. The ^{13}C/^{12}C ratios of carbon-containing ions in water reflect the amount of interaction of the water with these organic and inorganic reservoirs: if decay of organic matter is occurring (as in swamps and wetlands), waters will have relatively low ^{13}C/^{12}C ratios; if waters are dissolving carbonate minerals (as in aquifers near recharge areas), they will have relatively high ^{13}C/^{12}C ratios. Any carbonate minerals that are subsequently precipitated from these waters will also have relatively low and high ^{13}C/^{12}C ratios, respectively, offset from the water values by their fractionation factors at the temperature of precipitation.

Oxygen

Oxygen isotopes give different information about the source the water from which minerals have formed. Oxygen has two important isotopes: ^{16}O (99.76 per cent) and ^{18}O (0.20 per cent). When water, H_2O, evaporates, ^{16}O is preferentially removed, leaving residual water enriched in ^{18}O; thus, rain-water has lower ratios than the sea water from which it evaporates; and rain at progressively higher latitudes, formed from multiple evaporation–precipitation cycles, has progressively lower ratios of ^{18}O/^{16}O. Thus freshwater carbonates show a relative depletion in ^{18}O relative to marine carbonates, and there is a progression with latitude in the signatures of freshwater carbonates. However, changes in temperature also produce differences in the oxygen isotope fractionation factors for minerals, making if difficult to separate these effects.

Oxygen isotope ratios can aid in determining sources for detrital particles in clastic sedimentary rocks and can be used to identify diagenetic waters from which clays, silica, and carbonate cements precipitated.

Hydrogen

Hydrogen has two important stable isotopes: ^{1}H, making up 99.984 per cent and ^{2}H, also known as deuterium or D, making up 0.016 per cent. As with oxygen, when water evaporates, the lighter isotope, ^{1}H, is preferentially removed, leaving the residual water relatively enriched in the heavier isotope. Hydrogen isotope ratios are measured to aid in determining the origin of diagenetic waters. They can also be measured in clays or other mineral that contain hydroxyl (OH) groups.

Strontium

Strontium ratios (^{87}Sr/^{86}Sr) can also be used to trace water sources. As noted above, carbonate minerals incorporate strontium in trace amounts. Higher ratios indicate that formation waters have been in contact with clays or other minerals containing radioactive rubidium, which decays to produce stable ^{87}Sr. Carbonates forming from deep groundwaters in sedimentary basins may have high ratios. Sea water appears to have a constant ratio today, but differences in the rate of interaction of sea water with basalt at mid-ocean ridges have produced changes in the sea water ratio over geological time. In turn, these changes are reflected in the ^{87}Sr/^{86}Sr ratios of marine carbonates.

Sulphur

Sulphur has two isotopes of importance to sedimentary geochemistry, ^{34}S and ^{32}S, representing 4.2 per cent and 95 per cent of total sulphur, respectively. In small percentages, metal sulphide minerals are common in both carbonate and clastic rocks. They form when sediment pore waters become depleted in oxygen, which is used up as bacteria break down organic matter. Some bacteria can use sulphate as a substitute for oxygen, and this process causes the sulphate to be transformed to sulphide. If metal ions are presented in solution, the sulphide ions readily precipitate out, forming minerals such as pyrite within the pore spaces of sedimentary rocks.

Trace-element geochemistry

The presence of trace amounts of some elements can be used to determine the conditions of formation of precipitated minerals or the source rocks of detrital minerals. For carbonate minerals, the amounts of strontium and magnesium in calcite are important indicators for the alteration of precursor aragonite or high-magnesium calcite. The amount of magnesium incorporated in calcite also apparently increases with increasing temperature. Iron and manganese concentrations in carbonate cements can be determined qualitatively by the use of a cathodoluminescence microscope, or quantitatively by a microprobe. Incorporation of manganese and iron indicates that reducing conditions prevailed when the cements were precipitated. In clastic rocks, various trace elements are used to match source rocks with detrital quartz and feldspar grains. Cathdoluminescence of quartz, which measures trace-element concentrations and lattice defects, may also be diagnostic. Trace elements in authigenic clays can also be used to constrain diagenetic water compositions.

Trace elements and rare-earth elements (REE) (elements 57 to 71 in the periodic table) can also be used to better understand Earth surface processes and the operation of the sedimentary cycle. These elements occur in specific concentration ratios in different types of rocks: sediments and sedimentary rocks are depleted in some elements but enriched in others relative to igneous and metamorphic rocks. Also, specific elements are concentrated in different types of sedimentary rocks. This geochemical 'sorting' is the result of differences in the chemical properties of elements that govern their behaviour during the processes of weathering, transport, deposition, and diagenesis that make up the sedimentary cycle.
E. BURTON

Further reading

Drever, J. I. (1997) *The geochemisty of natural waters*. Prentice-Hall, Englewood Cliffs, New Jersey.

sedimentary structures

sedimentary structures Fluids transporting sediment particles, their subsequent movement within sediments, the action of gravity, and the effects of biological and chemical changes in the deposited sediment: all these factors control the architecture of the resulting deposit. The operation of any of these processes leads to the formation of sedimentary structures, which are the visible signs of the various processes as preserved in the rock record.

Sedimentary structures range from centimetres to kilometres in scale. They provide invaluable information about the mode and direction of transport and deposition, as well as subsequent changes. Because particular processes are characteristic of different environments of deposition, the study of sedimentary structures makes it possible to reconstruct environments of deposition and hence *palaeogeographies*.

Episodes of sedimentation produce layers or strata which vary in thickness from laminae (less than 1 cm thick) to beds (more than 1 cm thick). Varying current conditions produce mixed deposits: inter-laminated sediments of unlike type, for example mud and sand (i.e. *heterolithic stratification*) (Fig. 1, a, b, c). When sand is being transported by ripples and current velocity falls, the sands often become cloaked by a thin film of mud. Remobilization of the ripples by a new current leads to erosion of the mud except in the ripple-troughs, where it remains. The resulting structure is termed *flaser bedding*. More persistent mud layers are deposited under lower-energy conditions during longer periods of low-velocity currents. *Wavy bedding* then results. Where mud is the dominant sediment and sand is supplied spasmodically, the muddy sediment encloses linked or isolated sand ripples and the resulting structure is called *lenticular bedding*.

Sediments that slump under water can develop into a mixture of sediment and water that flows downslope as a density current along lake or sea floors. Various types of flow develop: *turbidity currents* where the sediment is supported mainly by the upward component of flow; *fluidized-sediment flows* where the sediments are supported by water escaping from between grains as they settle; *grain flows* where sediment is supported by grain-to-grain collision; and *debris flows* where grains are supported by a mixture fine sediment and water. Turbidity currents produce upward-fining beds, expressed by *graded bedding*. The beds show a sequence of structures produced by the waning flows. Such graded beds commonly display a *Bouma sequence*, named after the Dutch geologist who first recognized the phenomenon (Fig. 1, d(i)). Other mechanisms produce distinctive sedimentary sequences containing characteristic structures (Fig. 1, d (ii), (iii), (iv).

A unidirectional current of water moving over sand moulds the sediment into a series of bedforms which migrate downcurrent. The size pattern and sequence depend upon the magnitude of stress exerted by the fluid and the grain size of the sediment. As the velocity increases over a bed of initially smooth, fine-medium sand, the sediment is fashioned into small ripples (from one to a few centimetres high with wavelengths of decimetres). At first these are regular crested ripples which become linguoid as the current velocity increases. Larger ripples (sometimes called *dunes* or *megaripples*), which again are initially straight-crested and then develop irregular forms, subsequently give rise to a flat bed (*plane-bed*) before finally becoming an undulating bed of *antidunes*. (Antidunes, unlike ripples, migrate upstream as sediment is eroded on their downstream (lee) slopes and is deposited on the upstream (stoss) slope of the next antidune downstream. The latter sometimes develop into beds or irregular *chutes* and *pools*. In very fine sands, the sequence of bedforms is small ripples, plane beds, antidunes, whereas in coarse sands and gravels the sequence is plane beds with irregular scours parallel to the current and then large ripples and plane beds.

As ripples migrate, sediment moving up the gentle stoss slopes is deposited on the steeper lee slopes to produce inclined cross-stratification dipping in the direction of flow. In long crested ripple marks of all sizes this produces a series of cross-stratified sediments with individual sets representing the passage of a single bedform. The sets may have non-erosional or slightly erosional contacts with underlying sets (*tabular cross-stratification*) (Fig. 1, e (iv)). Short crested ripples usually have scour-pits developed on their downcurrent side into

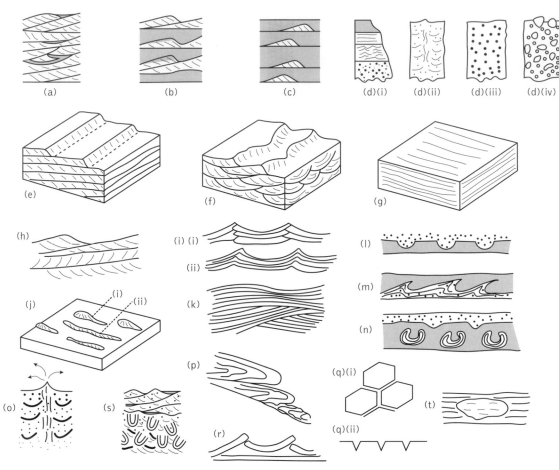

Fig. 1. Sedimentary structures. a, flaser-bedding (in section); b, wavy-bedding (in section); c, lenticular bedding (in section); d, (i) Turbidity current deposit-graded bedding; (ii) fluidized flow deposit; (iii) grain-flow deposit; (iv) debris-flow deposit (each shown as if encountered in a borehole); e, tabular cross-stratification; f, festoon cross-stratification; g, plane bedding with current lineation exposed on the top of the uppermost bed; h, herringbone cross-stratification (in section); i, (i) symmetrical ripple cross-stratification; (ii) slightly asymmetrical ripple cross-stratification (both in section); j, bottom structures as seen on the underside of a bed: (i) flute casts; (ii) groove casts; k, hummocky cross-stratification (in section); l, load casts (in section); m, flame casts (in section); n, pseudo-nodules (in section); o, dish-structures, pipe structures and sand volcanoes (in section); p, slumps (in section); q, desiccation cracks, (i) in plan; (ii) in section; r, tepee structure (in section); s, Bioturbated horizon in flaser bedded deposit (in section); t, chemical nodule (in section). (Sketched from various sources, including: Harms, J. C. (1982) Society of Economic Paleontologists and Mineralogists, Tulsa, Oklahoma; Reineck, H. E. and Singh, I. B. (1973) Springer-Verlag, Heidelberg; Collinson, J. D. and Thompson, D. B. Allen and Unwin, London (1982); Middleton, G. V. and Hampton, M. A. (1973) Society of Economic Paleontologists and Mineralogists (Pacific Coast Section).)

which the cross-stratified sediment advances in a series of spoon-shaped depressions infilled with cross-stratified sands which conform to the scour surfaces. Individual sets of cross-strata fill the scour surface (*trough cross- stratification* or *festoon cross-bedding*: Fig. 1, f). The axis of an individual set forms parallel to the current. Small-scale ripples, under decelerating flow, migrate to form *ripple-drift cross-stratification*, the angle of climb depending upon the rate of sediment deposition. Some large-scale ripples have counter flows in the shelter of the ripple which produce small-scale ripples that move upstream at the base of the lee slope.

Flat beds produce flat laminated deposits (*plane bed lamination*: (Fig. 1, g), with linear streaks or ridges only a few grains thick aligned in the direction of the current flow, giving rise to a structure called *primary current lineation*. Problems arise in the interpretation of plane beds in very find sands, since a very similar structure can be produced when sediment is falling out of suspension against a background of slowly moving currents. However, the presence of primary current lineation is usually sufficient to distinguish one from the other. Changes in velocity, sediment supply, or even exposure lead to the modification of the ripple surfaces and truncation of cross-stratified units to produce so-called *reactivation surfaces*. Reversing tidal currents are usually invoked to explain reversals of dip in succeeding cross-stratified units (*herringbone cross-stratification*: Fig. 1, h).

Oscillatory flow under gravity-driven waves produces ripples which show laminae dipping away from each side of the crest to give a chevron pattern (*oscillatory ripples*) (Fig. 1, i (i)). As waves commonly show variable degrees of asymmetry in shallow water the resulting ripples are asymmetric and show an asymmetric internal structure (Fig. 1, i (ii)), similar to those formed during unidirectional flow. Air flowing over sands produces ripples which are superficially like aqueous ripples but they have lower height/wavelength ratios and long crests, and are slightly asymmetric. In strong winds they are destroyed and a plane bed develops. *Aeolian ripples* are formed by saltating grains and the wavelength is approximately that of the *saltation path* of the grains. These ripples commonly do not show an internal structure of cross-laminae. Ripples of coarser sand and gravel produced by surface creep of grains show a coarse cross-stratification. Aeolian sand transport fashions sand into large-scale bedforms which often have complicated internal structures (dunes and draas).

A fluid together with its entrained load moving over a cohesive bed erodes longitudinal *furrows or grooves* when the stress exceeds the critical erosion velocity. At higher velocities spoon-shaped scours develop, with their blunt end on the upstream side (*flutes*). Transported fragments scour the bottom to form similar grooves and pits. These, when infilled with sediment, can be seen on the base of the succeeding bed, usually sand, as casts (*groove casts* and *flute casts*) (Fig. 1 j). The former record the sense of direction of current flow, and flutes also demonstrate its polarity from their asym-metry.

A curious structure found in coarse silt and fine-grained marine sands that has attracted a great deal of interest in the last few decades is *hummocky cross-stratification*. It consists of very gently inclined medium to large-scale cross-stratification with the laminae in each set parallel or nearly parallel to a lower, gently dipping erosion surface (Fig. 1, k). Hummocky cross-strata are fan-shaped and dip in all directions. When exhumed during later erosion they form an irregular surface of unorientated hummocks and depression 10–50 cm high, spaced a metre to a few metres apart. Units with such structures commonly show an erosional base cut in fine-grained deposits and are succeeded by oscillation-rippled sand and bioturbated sediments. They are thought to be produced by short-lived events such as storms.

Sand deposited on soft muds with a high water content sinks into them because of their greater density; structures of all sizes at the sand–mud interface show downward protuberances (*load casts*) (Fig. 1, l). These appear to develop in some instances as the sand is being deposited, and protrusions of mud occasionally project into the sand and bend in the direction of transport. These are *flame structures* (Fig. 1, m). Sometimes, the sand becomes detached from the main sand unit and sinks into the mud to form *sand balls* or *pseudonodules* (Fig. 1, n), which resemble true chemical nodules. They probably form by normal loading but they can be triggered by shock waves (e.g. earthquakes) which liquify the mud. In large deltas during times of rapid deposition of nearshore sediments over pro-delta sediment, muds flow upwards from hundreds of metres depth to regions of lesser stress, to form mud-diapers or mud lumps on the surface around delta river mouths. Slow creep of pro-deltaic muds under the load of the delta-top sands can produce a radial pattern of grabens and horsts and an accompanying tangential pattern of growth faults.

Water-saturated sands are sometimes disturbed by fluid flow or an object being dragged over the surface. In this way *overturned cross-stratification* can be formed. Convolutions can form the direction of flow in interlaminated sand or mud beds (*convolute bedding*). When sands have been deposited very quickly, the particles do not adopt their most stable packing, and the load is carried by interstitial water as well as by the sediment framework; shock waves or pressure differences due to passing waves can dewater the sand; water escaping upwards produces vertical pipes which are vertically stratified (*pipe structures*), and where they emerge on the sediment surface *sand volcanoes* (Fig. 1, o) can be formed. Between the pipes, the sediment is deformed and forms downward protrusions of sand with the original stratification retained but distorted (*ball and pillow structure*) (Fig. 1, n). Fine-grained sediment is sometimes concentrated to produce upwards concave structures (*dish structures*) (Fig. 1, o).

Large-scale disturbances resembling those produced by tectonic deformation can occur in sediments. Many of these phenomena are slumps. They differ from tectonic structures in that they are usually confined between undisturbed sediments. The distinction is not, however, always easy, since slumps and tectonic folds can have similar orientations. Sediments deposited on slopes of even 1° slump when the shear strength exerted by gravity exceeds the shear strength of the sediment. The shear stress can be increased by the dumping of new sediment on top of a mud, and the shear strength decreased by increased pore pressure or fluidization caused by sudden sediment loading, storm waves, tsunamis, or earthquakes. Such disturbances vary from gentle folds with small thrusts, when the beds retain coherence, to those in which the sediment completely loses its coherence and forms a *sediment gravity flow*. A sheet of mud and sand behaving chaotically can develop from such movements. Sheets of this kind are known as *olistrostromes*. They contain isolated clasts, called *olistoliths*. Olistrostomes can be very extensive, covering hundreds of square kilometres. Almost all the thick and extensive examples have been triggered by earthquakes. Slumps (Fig. 1, p), many of them covering hundreds of square kilometres, are common on basin margins, on delta slopes, and along modern continental slopes, particularly those of active margins.

Loss of water by desiccation, and shrinkage on exposure of sediments to the atmosphere, produce one of the most widely used and least ambiguous pieces of evidence for former exposure in ancient deposits: *mudcracks* and *desiccation cracks* (Fig. 1, q). The polygonal pattern of cracks often becomes filled with sediment and they are thus preserved as downward V-shaped projections filled with sediments from the succeeding bed. Muds sometimes lose water even when permanently submerged, to produce a rather similar pattern of cracks called *syneresis cracks*, which are less regular but can be difficult to distinguish from subaerial mudcracks. Crystallization of cements by salt and evaporites in subaerial or very shallow water, and by calcium carbonate in waters of variable depth, produces horizontal expansion in surface sediments to produce a polygonal pattern of slabs (several metres or tens of metres across) with upturned edges (*tepee structures, pseudo-anticlines, gilgai*: Fig. 1, r).

Bottom-living organisms leave traces—trace fossils—and usually have a profound effect on the sedimentary structures of a deposit. The intensity of this type of disturbance, or *bioturbation*, (Fig. 1, s) and the relative proportions of physically and biologically produced structures provides useful information of the environment and rate of deposition. Whereas the main effect of burrowing organisms is to destroy stratification; other organisms produce delicate stratification by trapping fine-grained sediment to produce *stromatolitic bedding*.

Many sediments contain cemented patches—*nodules*—formed by chemical precipitation of a mineral cement developed around local centres of chemical activity, such as an organic fragment, or a zone of increased permeability such as a burrow (Fig. 1, t).   G. EVANS

Further reading

Allen, J. R. L. (1985) *Principles of physical sedimentology*. Allen and Unwin, London.

Collinson, J. D. and Thompson, D. B. (1982) *Sedimentary structures*. Allen and Unwin, London.

Leeder, M. R. (1982) *Sedimentology process and product*. Allen and Unwin, London.

sedimentology, sediments, and sedimentary rocks

Sedimentology is the study of the nature and origin of sediments. It includes the production, dispersal, and deposition of sediments as well as their conversion into rocks—sedimentary rocks.

Sediments are the products of reactions between the surface rocks of the Earth's crust and the surrounding envelope of water, air, organisms, and plants. They are accumulations of crystals or mineral grains that were deposited in layers from a fluid (water or air) under normal pressures and temperatures at the Earth's surface. Some sediments consist of particles derived from the breakdown and alteration of earlier rocks; these, technically termed terrigenous detrital, clastic, or siliciclastic deposits, include sandstones, silts, and mudstones. Other sediments form by precipitation or biological extraction of calcium carbonate; they are termed limestones. Chemical precipitation produces other deposits, such as evaporites, ironstones, and phosphates; and the accumulation of organic matter results in the formation of coal, oil shales, and hydrocarbons.

Sediments form in many environments, ranging from terrestrial situations to the deepest parts of the ocean. The analysis of sedimentary sequences and their associated floral and faunal remains can make it possible to reconstruct the changing conditions of the Earth's surface in the geological past.

Sediments are of great economic importance as raw materials (coal, ironstone, building materials) as well as containing, in the voids between grains, valuable accumulations of water, oil, gas, and economic minerals.   G. EVANS

seismic anisotropy

Seismologists usually make the initial assumption that the material through which a seismic wave travels is mechanically isotropic (see *seismic waves, principles*). This means the speed of travel is independent of the direction of travel. This assumption greatly simplifies mathematical treatments and is acceptable for most applications. Almost all minerals are anisotropic, but since the individual mineral grains which combine to form a rock are usually randomly oriented, the end result is isotropy on the scale of the wavelengths of typical seismic waves (metres to hundreds of metres). There are, however, many situations in which significant anisotropy occurs, and attempts to measure and understand anisotropy and treat it mathematically are increasingly common. Anisotropy is usually expressed as a percentage representing the variation between the fastest and slowest velocity through a material. Although individual minerals can be highly anisotropic, an anisotropy of 10 per cent would be high for a large body of rock.

It has long been known that foliated rocks, such as gneisses, in which mineral alignments have been produced by the metamorphic effects of pressure and temperature, are anisotropic. Rocks such as shale, which are composed of elongate grains, are also anisotropic because these grains align during the process of sedimentation that forms the rock.

An early significant observation of large-scale anisotropy occurred when seismic refraction measurements (see *controlled source seismology*) in oceans showed that the P-wave velocity of the upper mantle (Pn) was consistently higher for profiles recorded perpendicular to an oceanic spreading centre (i.e. parallel to the direction of spreading or plate movement) than for profiles recorded parallel to the spreading centre.

These measurements have been confirmed by numerous subsequent studies and are attributed to the alignment of olivine crystals in the mantle lithosphere because of flow during the formation of the oceanic plate at the ridge. Similar observations are uncommon for continental areas because there are fewer good determinations of the P-wave velocity in the upper mantle. Anisotropy has, however, been reported in the Basin and Range province of western North America and in the eastern arm of the East African rift.

On a smaller scale, stresses in the Earth can cause bodies of rock to fracture in a consistent manner. If these fractures are aligned, the rock will be anisotropic with the fast direction perpendicular to the fractures and the slow direction parallel to them. In ideal cases, these observations can provide information on the state of stress in the Earth.   G. R. KELLER

seismic body waves Seismic waves of several types are generated when a seismic source such as an earthquake or explosion is energized. The waves that penetrate deeply and represent short pulses (wavelets) of propagating energy are referred to as body waves. In a homogeneous and mechanically isotropic medium, two types of body waves are generated (see *seismic waves: principles*). The fastest ones are called P-waves (primary waves) and are simply sound waves. The P-wave is thus a series of compressions and rarefractions. As a P-waves passes a point in the Earth, the material at this point vibrates back and forth in the direction in which the wave is travelling (Fig. 1a). The S-waves (secondary or shear waves) move more slowly and occur only in solids where there

is an elastic response that resists shearing stresses. For S-waves, the motion at a point in the Earth is perpendicular to the path along which the wave is travelling (Fig. 1b). In mathematical treatments, it is convenient to 'decompose' the S-wave into a vertically polarized component, S_V, and a horizontally polarized component, S_H. If the material through which the wave is travelling is mechanically isotropic, these components travel together and are not really separate waves; if, however, the material is anisotropic, these components separate and travel at different speeds. If separate S-wave arrivals are observed, we call this *shear wave splitting*. This is an important way in which anisotropy in the Earth is detected (see *seismic anisotropy*). The P-wave and S-wave velocities of a material (V_P and V_S respectively) are important physical constants which describe its elastic properties (see *seismic properties of rocks*). In most solids, the ratio of V_P to V_S is about $\sqrt{3}$. An important property of fluids is that $V_S = 0$: in other words, S-waves are not transmitted by fluids. The travel of seismic body waves through the Earth is most easily visualized in terms of ray paths (see *seismic ray theory*). A ray path is the trajectory that a small packet of seismic energy follows as it travels through the Earth. At any point in time, the ray path is perpendicular to the wavefront, which is spreading out in three dimensions. As the seismic velocity changes, the ray path is bent according to the law of reflection and refraction (Snell's law)—very much as light is reflected and refracted. Strong discontinuities in seismic velocity (which are usually related to changes in rock type) produce interfaces which reflect seismic body waves like a mirror and refract them like a

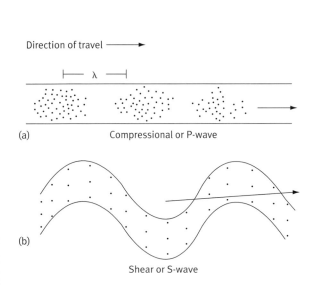

Fig. 1. Cartoons illustrating P- and S- body waves. λ is the wavelength.

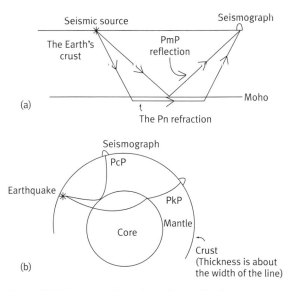

Fig. 2. (a) Reflected and refracted seismic rays; (b) reflection and refraction of body waves passing through the Earth.

lens (Fig. 2a). The body waves detected by seismographs relatively near a source can be processed to form an image of the earth's crust that is similar to an X-ray image. When recorded at relatively large distances from the source, body waves reflected and refracted from the base of the Earth's crust (i.e. the Moho) and other deep interfaces can be analysed to determine structure on a lithospheric scale (hundreds of kilometres) (Fig. 1b). Body waves generated by earthquakes of moderate size travel through the mantle and may be reflected from the core–mantle interface or may travel through the core to emerge on the other side of the Earth (Fig. 2b). Body waves commonly observed to travel along particular paths are referred to as *phases* and are given labels, such as PcP, PkP, Pn, and PmP in the case of P-waves; or ScS, etc. in the case of S-waves (Fig. 2b). An early contribution in the field of earthquake seismology was the observation that S-waves do not travel through the outer core, thus establishing that it is liquid. The phase SkS therefore involves the conversion of the S-wave energy at the core–mantle boundary to form a P-wave which travels through the core before being converted back to an S-wave on the opposite side of the core.

Another important application of the study of body waves is in the study of earthquake sources. Earthquakes radiate seismic energy in a complex pattern which reveals much about the nature of the source. Observations of the amplitude and shape (wave-forms) of body waves at various distances and azimuths are used to study the radiation patterns and the earthquake source.

Most applications of seismology to practical problems in, for example, petroleum exploration and engineering geology employ P-waves (although S-waves can also be employed). The reflection technique requires complex computer processing to form an image of subsurface features; in the refraction technique, patterns of arrivals are analysed to derive Earth structure. Each technique has it strengths and weaknesses (see *seismic exploration methods*).   G. R. KELLER

Further reading

Bullen, K. E. and Bolt, Bruce A. (1985) *An introduction to the theory of seismology.* Cambridge University Press.

Lay, T. and Wallace, T. C. (1995) *Modern global seismology.* Academic Press, San Diego.

seismic exploration methods
Ever since the early textbooks were written on the applications of geophysics, seismic exploration methods have been divided into the reflection and refraction techniques. In each technique, a number, from at least 12 to as many as hundreds, of readily portable seismometers (geophones) are laid out along a line (or lines in the case of 3-dimensional seismic reflection studies), and a number of seismic sources are energized in sequence (Fig. 1a). The methods used for recording the signals from these geophones can vary greatly, as can the character of the seismic source. Because the early practice was to

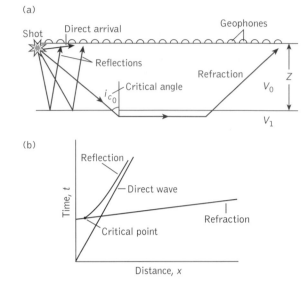

Fig. 1. Principles of seismic refraction and reflection surveying: (a) cross-section showing shot, geophones, direct arrival, reflections, and refraction. V_0 and V_1 are the respective velocities in the layers above and below the interface at depth z; i_c is the critical angle. (b) Graph showing time-distance relationships for direct, reflected, and refracted waves; for the direct wave, $T = x/V_0$; for the reflected wave, $T = 2z/V_1$ at $x = 0$; for the refracted wave,

$$T = \frac{x}{V_1} + \frac{2z\sqrt{V_1^2 + V^2}}{V_1\,V_0}.$$

use small explosive charges, the source is consequently thought of as a shot that generates a distinct package of seismic body wave energy (see *seismic body waves*): a wavelet which travels by various ray paths through the Earth and eventually returns to the surface to be recorded by the geophones. The travel times and amplitudes of these wavelets can be interpreted to reveal the subsurface structure.

The reflection technique is dominant because it can produce a very detailed image of the subsurface structure by recording nearly vertical travelling waves as they are reflected from discontinuities in the Earth (Fig. 1a). The refraction technique, on the other hand, targets waves which have a large horizontal component to their travel path. Earth structure is then deduced from computer modelling of the observed data by employing techniques such as ray tracing (see *seismic ray theory*).

This division into two techniques has historically made sense because of the different approaches that are used to interpret the data and because of the limitations imposed by instrumentation. However, in either case, body waves are employed and both reflected and refracted waves are recorded and used. Either technique is very flexible and can be employed in near-surface studies for engineering and envi-

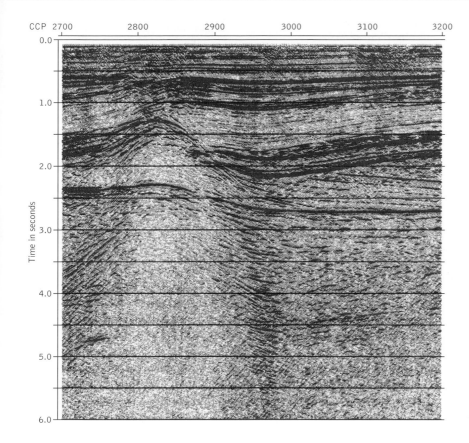

Fig. 2. Example of a seismic reflection image of a structure in the Earth.

ronmental applications or in studies of deep Earth structure (see *controlled-source seismology* and *deep seismic reflection profiling*). If the number of geophones that can be deployed is limited to, say, 40, the choice is either to place the instruments close together, as is done in the reflection technique, or spread them apart, as is done in the refraction technique. Thus, the perception is that the reflection technique produces a useful image of the Earth but has limited resolution for determining velocity, while the refraction technique measures velocity well but can produce only simplistic models of the Earth. However, with advances in instrumentation that allow hundreds or even thousands of geophones to be deployed, and with the development of new theoretical approaches, this division is becoming blurred. Because waves travelling by a wide variety of travel paths can all be densely recorded, the advantages of both ways of treating the problem can begin to be exploited simultaneously.

The basic geophysics of the two approaches, as well as the interrelations between them, can be illustrated by deriving the travel-time expressions for the case of a layer with velocity V_0 over another layer with a higher velocity V_1 (Fig. 1a). The wave that travels essentially horizontally and at velocity V_0 is called the *direct arrival*. The only relationship that is involved in deriving the travel-time expression for this arrival is that the distance travelled equals the velocity multiplied by the time ($x = vt$).

Thus, when plotted on a time (t) versus distance (x) diagram (Fig. 1b), the direct arrival is a line whose slope is $1/V_0$. The refraction technique focuses on what are called *critically refracted* waves. The ray path of such a refraction hits the interface at an angle (the critical angle, $i_c = V_0/V_1$) such that the wave is bent to be horizontal at the interface according to Snell's law (see *seismic waves, principles*). The wave then travels just below the interface at velocity V_1. Thus, at a certain point, called the *critical distance*, this arrival will be the first because it is travelling at a higher velocity than the direct arrival. A simple geometrical derivation shows that the equation for this arrival's travel time is that of a straight line with a slope of $1/V_1$ and with an intercept which is related to the depth of the interface. The slopes of the arrivals and the intercept are measured from data plots. Using these values, the depth to the interface and the velocities can readily be calculated. Additional layers, or layers which are not horizontal, lead to equations that are algebraically more

947

complex but are still conceptually simple. The presence of dip introduces an additional unknown into the problem. This is handled by locating a seismic source at each end of the array of instruments and recording what is called a reversed profile.

The traditional applications of the refraction technique include determining the depth to bedrock so that foundations can be properly designed for construction projects, searching for the water table, and determining the thickness and velocity of near-surface material (the weathered layer).

The reflection technique is expensive because many more sources and geophones are typically employed than in the refraction technique. Because of its cost, this technique is mainly employed in exploration for hydrocarbons. In addition, the data gathered during a seismic reflection survey require a great deal of computer processing in order to form an image of the subsurface structure (*see seismic signal processing*); an example is shown in Fig. 2. In a reflection study, the geophones are usually deployed so that the maximum distance from the source is comparable with the depth of the target. The ray paths the first hit the interface at angles less than the critical angle. The reflected arrival time (Fig. 1) becomes closer and closer to that of the refracted arrival when *x* has large values, because their travel paths become very similar. The intercept of this arrival is simply twice the depth divided by V_0, and is called the *vertical incidence travel time*. The equation of this curve can be shown to be that of a hyperbola whose curvature is inversely related to the value of V_0. The measurement of velocity employing reflection data depends on analysing this curvature. The velocity derived from reflection data is an average for all the materials between the reflector and the surface, but the exact value of this average velocity depends on the geometry of the receivers and seismic sources and on the complexity of the structure. This velocity value may be called the average, rms (root mean square), or stacking velocity. A major aim is to measure the velocity of the material located between two reflectors. This value is called the *interval velocity*. Its measurement becomes more complex as the complexity of the structure being investigated increases.

A smaller-scale application of seismology to exploration problems is well logging. The seismic instrument used in this technique has been miniaturized so that it can be lowered down a borehole. The measurement of seismic velocity in boreholes is usually called sonic logging; it typically consists of a refraction survey in which the altered rock along the borehole wall is the first layer and the unaltered rock is the second (Fig. 3). However, because of its utility in petroleum exploration, this technique is becoming more and more sophisticated. Another borehole technique is cross-hole tomography (see *seismic tomography*). A seismic source is lowered into a borehole and seismometers are lowered into another borehole or other boreholes. The travel times between the sources and receivers can be used to detect anomalies in velocity structure between the boreholes.

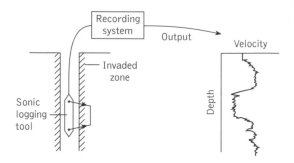

Fig. 3. Principle and practice of downhole sonic logging.

In using either the reflection or the refraction techniques, the geophysicist tries to design the seismic survey so that it can be tied to independent points of control such as other geophysical surveys, boreholes, and places where the strata of interest are exposed (outcrops). This adds significantly to the ability to interpret the seismic data with confidence. In seismic reflection surveys especially, synthetic seismograms are used to predict the Earth's response based on sonic logs from boreholes. The patterns of reflections caused by particular strata as seen in the 'synthetics' are used to identify these same strata in the seismic reflection data. This pattern can then be followed across the seismic reflection record section; changes in it might be related to various physical changes in the strata. In the refraction technique, an Earth model is derived by modelling the travel times of phases which are interpreted as having followed certain ray paths through the Earth. This model is then often refined by using synthetic seismogram modelling to match the wave forms of these phases.

Inversion is a term used to refer to a number of techniques for deriving a velocity model of the Earth directly from the data with a minimum of human intervention. This approach minimizes the effect of bias on the part of a human interpreter and can give quantitative estimates of the uncertainties in the model that is derived. In reflection studies, the goal is to recover the reflectivity of the Earth and use it to arrive at a solution for the velocity structure.   G. R. KELLER

Further reading

Telford, W. M., Geldart, L. P., and Sheriff, R. E. (1990) *Applied geophysics.* Cambridge University Press.

seismic properties of rocks
The seismic properties of rocks are usually thought of in terms of P-wave velocity, V_p, and S-wave velocity, V_s, which are both functions of density (see *seismic waves, principles*). A knowledge of V_p, V_s, and density can tell us a great deal about the composition and physical state of materials in the earth. The Earth is, however, much more complex than this simple approach would indicate. Laboratory studies on individual mineral crystals are

capable of great detail and precision. The situation is very different for measurements based on the analysis of body and surface waves travelling through the Earth. It has, for example, commonly been the practice to assume that materials are mechanically isotropic so that only two elastic constants are required. The constants measured are usually V_p and V_s. Modern studies are often able to go beyond this simple assumption but usually stop at determining V_p, V_s, density, Q, the quality factor (assumed to be a constant), and anisotropy expressed as a percentage. As a practical matter, this amounts to measuring from three to five elastic constants out of a possible 21, and recognizing that observable differences in velocity exist which are a function of the direction of travel.

Seismologists desire to measure physical properties to the greatest level of detail possible. Their aim is to determine not only the composition of the Earth but also its physical state. The question of composition is complex because rocks are combinations of minerals which each have their own seismic properties. In general, the velocities of the crystalline rocks which constitute most of the Earth's lithosphere are inversely proportional to the silicon dioxide content and directly proportional to the content of iron and magnesium oxides.

Many factors, such as porosity, pressure, and temperature, affect the seismic velocity of a particular rock. Figure 1, which is a compilation of typical values, illustrates some relationships which are generally true. Unconsolidated sediments can have very low velocities. Clastic sedimentary rocks (sandstone and shale) are 'slower' than carbonate sedimentary rocks (limestone and dolomite). For typical crystalline rocks,

granite (felsic) is slower than gabbro (mafic), which is slower than peridotite (ultramafic). These relationships are generally reflected in the velocity structure of the Earth's lithosphere, in which the granitic upper crust generally has higher velocities than the sedimentary rocks which cover it but has lower velocities than the more mafic lower crust. The ultramafic mantle, which is very rich in iron, has even higher velocities.

Such factors as the degree of alignment of particular minerals govern properties such as the level of anisotropy. For example, an early large-scale observation of anisotropy was provided by P-wave velocities measured in the mantle near ocean spreading centres. Here, alignment of the highly anisotropic mineral olivine is sufficient to cause velocities measured from refraction profiles that are recorded in the direction of spreading to be higher than those for profiles recorded parallel to the spreading centre. This alignment is particularly interesting because it is related to the flow which reflects plate movements. As a result there is much interest in measuring anisotropy elsewhere in the mantle. Some metamorphic rocks and shales have mineral alignments which are clear in hand specimens and which are reflected in significant anisotropy, but when viewed on a larger scale most crustal rocks do not appear to be very anisotropic. However, detailed velocity measurements, such as those made by logging devices which are lowered down boreholes, often reveal significant anisotropy in individual rock units, which is sometimes a result of the alignment of fractures.

Other properties of rocks can dominate composition, especially in the sedimentary rocks found near the Earth's surface. Porosity (percentage of pore space) is particularly important. Burial at depth, giving rise to high overburden pressures, causes sedimentary rocks in particular to compact. In the process, their porosity decreases and their seismic velocity increases correspondingly. For example, pressure due to burial, accompanied by cementation, causes sand and mud to evolve into sedimentary rocks (sandstone and shale respectively), and their velocity usually increases as this evolution proceeds. Increasing age also causes the velocity of sedimentary rocks to increase because many processes that take place in the Earth tend to decrease the porosity. Crystalline rocks also exhibit an increase in velocity with increasing pressure, as micro-cracks in the rock close, reducing their porosity. In the mantle, phase transformations of olivine result in more compact crystal structures which are believed to cause the seismic discontinuities that are consistently observed at several depths on a global basis. Temperature is another factor that affects the seismic properties of rocks. If other factors are held constant, an increase in temperature generally causes a decrease in seismic velocity. Since temperature increases with depth in the Earth, we might expect seismic velocity to decrease with depth. Except in small regions, this is not, however, the case because in general the temperature effect is offset by the effects of pressure and composition.

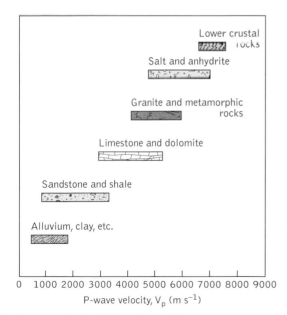

Fig. 1. P-wave velocities (V_p) for various rock types.

In many cases instances it is important to know V_p, V_s, and density (ρ) in order to study some phenomenon; for example, the variation in the amplitude of a wave reflected from a discontinuity in the Earth with the angle at which the wave is approaching. We may, however, be able to measure only V_p. Since we have good values for Poisson's ratio (σ) for typical rock types, we can calculate V_s from a standard equation.

This relationship becomes very useful when we have independent information on density from gravity measurements or logging devices which measure the density of rocks penetrated by a borehole. In addition to Poisson's ratio, there is a good correlation between V_p and density. This relationship is based mostly on measuring the density and V_p for many different rock types. It has been presented in several forms.

These relationships are useful, but only approximate. One should not lose sight of the fact that many types of rocks share the same P-wave velocity, V_p (Fig. 1). More independent measurements of other seismic properties are thus highly desirable so that the type of rock and its physical state can be better determined.

The attenuation of seismic waves as they travel through the Earth is a complex subject which is the target of much research. In general, we find that the amplitude of a seismic wave dies off in proportion to the factor $e^{-\alpha x}$, where x is the distance travelled and α is the attenuation coefficient. The attenuation coefficient is a function of frequency, and it becomes larger as the frequency increases. This preferential attenuation of higher frequencies is a fundamental limitation in seismology that restricts our ability to resolve (i.e. see) deep structures because we need the short wavelengths associated with high frequencies (*see* seismic waves, principles). It would be a major task to tabulate α for many materials as a function of frequency, so it is fortunate that α is an approximately linear function of frequency for typical seismic waves. This relationship leads to the quality factor, Q, being a constant. The quality factor is thus a convenient measure of the degree to which a material absorbs seismic energy; high values of Q mean that attenuation is low. An early large-scale observation was that the cold down-going slabs in subduction zones had high Q values while the hot volcanic arcs above the subduction zones had low Q values.   G. R. KELLER

Further reading

Telford, W. M., Geldart, L. P., and Sheriff, R. E. (1990) *Applied geophysics*. Cambridge University Press.

seismic ray theory The easiest way to visualize the propagation of body waves through the Earth is by considering ray paths and wave fronts (Fig. 1). A wave front is an imaginary line that connects all the points representing a particular peak or trough in a wave, somewhat like the ripples which form when a stone is thrown into a pond. As a wave travels, the wave fronts, which connect points sharing the same type of motion at a particular time, spread out and change shape

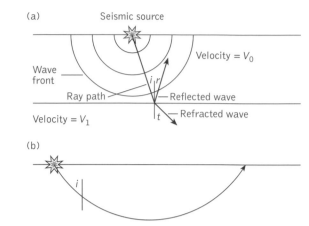

Fig. 1. (a) Seismic ray paths and fronts in two media of contrasting seismic velocities; i, angle of incidence; r, angle of refraction; t, angle at which seismic ray is refracted at the interface between the two media; V_0, V_1, seismic velocities in the two media. The angles of incidence and refraction are related to the velocities in the two media by the equation $\sin i/V_0 = \sin t/V_1$. (b) An example of an arcuate ray path resulting from a continuous variation in seismic velocity. The ray parameter p is equal to $\sin i/V$, where i is the angle of incidence and V is the seismic velocity. V is equal to $V_0 + kz$, where V_0 is the seismic velocity at the surface, k is a constant, and z is the depth.

as the velocity in the material the wave is traversing changes vertically or laterally. A ray starts at the source and is perpendicular to the wave front as it travels. If the material through which the wave travels has a constant velocity (Fig. 1a), the wave fronts in this material are circles and the rays are straight lines. (Remember the stone in the pond.) If the velocity changes in a gradual fashion, the wave front changes shape and the rays become smooth curves. There may, for example, be a linear increase of seismic velocity with depth. This is approximately the case in many sedimentary basins because of compaction, and in parts of the mantle because of the interplay of pressure and temperature. The rays are then arcs of circles (Fig. 1b). The trajectory of the ray is called the *ray path* and can be thought of as the path that a small packet of the energy released by the source has travelled. The ray approach in seismology is similar to that which is used in physical optics, except that it is a seismic wave, not light, that is travelling. For example, when a ray representing a P-wave intersects an interface between materials with different seismic velocities (Fig. 1a), some energy is reflected and some is refracted (i.e. the ray is bent according to Snell's law), and some is converted to reflected and refracted S-waves (see *seismic waves, principles* and *seismic body waves*). A particularly useful relationship, from Snell's law, is that along the ray path $\sin i/v$ (where v is the velocity and i is the inclination of the ray, normally measured from the vertical) is a constant called the ray parameter, p (Fig. 1b). When geophysicists are interpreting seismic data, waves travelling along many different ray paths are employed, and together they reveal the structure through which they have travelled.

If the calculation of travel time along a ray path between a particular source and a receiver (seismograph) is all that is desired and the structure being studied is fairly simple, such simple geometrical approaches are adequate. However, complex structures and the desire for amplitude information require more complex treatments. Theoretical treatments of seismic wave propagation address this problem by using two basic approaches. One is called wave theory (see *seismic wave theory*). In this approach, the wave equation of classical physics is derived in some simplified form, and exact or numerical solutions are sought. The advantage of this approach is that the entire wave field emanating from a seismic source can be considered. Wave theory can provide solutions which include both body waves and surface waves and can provide a good picture of how the amplitude and shape of a seismic wave change as it travels. The computational requirements are, however, extreme, and for most applications it is not practicable to employ wave theory on a routine basis. The second theoretical approach is called geometric ray theory. It employs a useful approximation in which the seismic body waves recorded at a particular location are treated as individual packets of seismic energy (wavelets). The travel paths of these wavelets can be traced through the Earth, and their interaction with the materials through which they travel can be calculated provided that a number of restrictions on the complexity of the Earth are taken into account. The approximation yields what is known as the Eikonal equation, which can be thought of as a simplification of the wave equation. Solutions of the Eikonal equation are not therefore exact solutions to the wave equation but are useful if the rate of change of velocity in the Earth is small relative to a certain wavelength. For a given mathematical model of the Earth, ray theory will yield very accurate values of the travel time for a particular body wave; however, the accuracy of the amplitude and shape of the seismic wave calculated are dependent on the degree to which the actual Earth structure is as simple as must be assumed.

G. R. KELLER

Further reading

Lay, T. and Wallace, T. C. (1995) *Modern global seismology.* Academic Press, San Diego.

Bullen, K. E. and Bolt, B. A. (1985) *An introduction to the theory of seismology.* Cambridge University Press.

seismic reflection surveying

In seismic reflection surveying, seismic waves are propagated through the subsurface and the travel times are measured of those waves that return to the surface after reflection at geological surfaces. Knowledge of the velocities with which the subsurface transmits the seismic waves makes it possible to convert the travel times into the depths to the surfaces. The geological structure can then be determined.

Seismic reflection is the most widely used of the geophysical techniques because it is the only method that provides a representation of the subsurface structure that is sufficiently detailed to identify potential hydrocarbon traps.

The method is analogous to that used by echo-sounders to determine water depth from the arrival times of sonic pulses reflected from the sea bed, but it uses waves of much lower frequency. The technique relies on the fact that the behaviour of seismic waves is similar to that of light in that they are reflected (and refracted) when they meet a boundary between

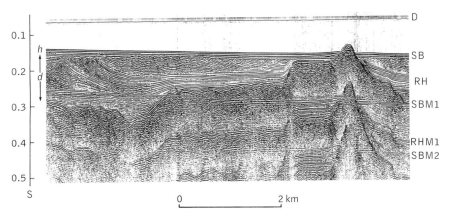

Fig. 1. Seismic reflection section from the Gulf of Patras, Greece, showing Holocene hemipelagic sediments (*h*) and deltaic sediments (*d*) overlying an irregular erosion surface (rockhead, *RH*) cut into older rocks. *SB*, sea-bed reflection; SBM1, SBM2, first and second multiples of the sea-bed reflection generated by reflections within the water layer; RHM1, multiple of *RH* reflection. (From Kearey, P. and Brooks, M. (1991) *An introduction to geophysical exploration* (2nd edn). Blackwell Scientific Publications, Oxford.)

media with different seismic velocities. In general, the seismic velocities of rocks increase as the porosity decreases and the density increases.

Seismic waves are detected by instruments known as *geophones* when used on land or *hydrophones* in water. They essentially convert the seismic wave motion into an electrical signal.

Seismic reflection may be used at many different scales. Thus, the top few tens of metres of the subsurface might be studied using seismic waves generated by blows from a sledgehammer, while penetration through the continental crust can be achieved using more potent sources. Most surveys are undertaken using non-explosive sources such as the air gun, which rapidly discharges high-pressure air into the water column and produces a bubble that oscillates in the same way as an explosion. Sources of this type are more convenient, can be fired again rapidly, and provide an exactly repeatable source 'signature' that can be utilized by certain modern data-processing techniques. Indeed, computer-based data processing is commonly used to enhance the reflected waves at the expense of various types of noise.

The interpretation of seismic reflections is often confused by the presence of multiple reflections: events that have been reflected more than once before reaching the detector. These 'multiples' can be removed by some processing techniques.

Two-dimensional seismic profiles are built up by moving the source and an array of detectors along a line. More complex arrangements enable three-dimensional coverage of an area to be obtained. Traces for each shot-detector set-up are reproduced closely side by side so that the seismic reflections appear to blend into a continuous band and the section so produced mimics a geological section. An example of such a section is shown in Fig. 1. However, further processing is needed to convert the two-way reflection time of the y-axis into depth, using the velocities of the strata, and it may also be necessary to correct for the non-vertical reflection path that would obtain if the reflecting surface had significant dip.

P. KEAREY

seismic signal processing

seismic signal processing For many years, seismograms consisted only of a paper record of the trace drawn by a pen or light-beam. This trace was an analogue record of the motion of the Earth at that point. That total motion consisted of more than the seismic signals which were of interest, but little could be done to separate interfering motions (i. e. noise) from the signal. As the computer revolution unfolded, seismology experienced a digital revolution. As a result, voltage variations from the seismometer that represent the ground motion could be fed into an electronic device called an analogue to digital converter, and the seismogram became a sequence of numbers which represented the amplitude of the ground motion sampled at some constant time interval, often a few milliseconds. This digital record, the signal, is a time series and is ideally suited for mathematical procedures which improve its utility. These procedures are generally known as signal processing, or conditioning, and are quite varied and sometimes complex. The digital revolution occurred first in the seismic reflection industry, which focuses on the exploration for hydrocarbons. Later, earthquake seismographs were developed that recorded digitally, and today virtually all modern seismic recordings are digital and thus involve some sort of signal processing. A simple example of *seismic signal processing* occurs in engineering and environmental applications when a hammer is used as a seismic source. An individual impact of the hammer is too weak to provide a signal strong enough to be useful. However, a usable record can be obtained by a simple signal processing scheme, which begins by storing the digital signal from the first impact in the memory of the computer which is part of the recording system. The digital record that represents each subsequent impact is first summed with the record in the memory, and the sum is then stored as the new record. The instrument operator examines each sum, and the process continues until a suitable record is obtained. At the other end of the spectrum of complexity, there are three-dimensional seismic reflection surveys where each seismic trace in the resulting ultimate image represents tens of thousands of individual seismic signals which have been summed in a variety of complex processes. The signal processing effort required to produce a good three-dimensional image in fact rivals the intensive field effort required to collect the actual data.

The most basic goal of *seismic signal processing* can be expressed simply as increasing the signal to noise (S/N) ratio. Noise is simply any ground motion which is not of interest. In digital terms, S/N is just the ratio of the number representing the amplitude of the signal to the number representing the amplitude of the noise. Another basic goal of signal processing is to increase resolution, which is the ability to recognize individual reflected or refracted seismic waves (see *seismic exploration methods*) as distinct arrivals or wavelets. Processing techniques that address this issue try to make the individual wavelets as compact as possible and to remove interfering waves which have travelled from the source to the receiver (seismograph) along any indirect path. In seismic reflection studies, a process called *migration* is usually the most complex processing step. Here the goal is to unravel the effects of complex subsurface structures and, in essence, sharpen the focus of the image which is the ultimate product of the processing effort.

Seismic signal processing is a complex subject, and processing of modern seismic reflection surveys is one of the most sophisticated and intensive computer applications that exists. However, a relatively small number of basic principles form the basis of most processing operations. Perhaps the most basic consideration is the application of Fourier analysis (named after J. B. J. Fourier, a prominent French mathematician). This common mathematical tool is very useful in seismic signal processing and has been the driving force behind many advances in this field. When viewed as a relationship between amplitude and time, a seismogram is a well-

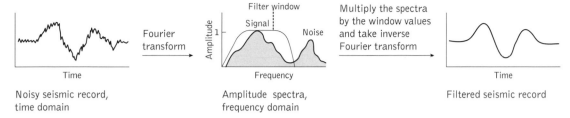

Fig. 1. Idealized representation of time domain, frequency domain, and filtering during seismic signal processing.

behaved mathematical function. Thus, a seismogram can be represented as the sum of a series of time-shifted sine waves with varying frequencies and amplitudes. If one chooses enough different sine waves (frequencies), their sum approximates the seismogram to an arbitrary degree of precision. When one views the seismogram in terms of the relative amplitudes (amplitude spectra) and time shifts (phase spectra) as a function of frequency (Fig. 1), one is analysing the frequency domain representation of the signal. This representation is obtained through a mathematical process called the Fourier transform, and it is a very intuitive way of representing the signal for many applications. For example, frequency filtering becomes simply a process wherein the desired frequencies are retained (i. e. multiplied by 1 in the frequency domain) and the undesired frequencies are multiplied by zero. After doing these multiplications, an inverse Fourier transform provides the seismogram with the undesired frequencies removed (Fig. 1).

When high frequencies are undesired, one employs a low-pass filter. A high-pass filter attenuates low frequencies, and a range of frequencies is passed by a band-pass filter. Modern seismic studies usually place a premium on obtaining broad-band data, which refers to the frequency domain where the desired signal would contain a broad range of frequencies. A broad-band signal has numerous advantages, including superior resolution. This fact can be illustrated by looking at the opposite extreme, which is a single sine wave containing, by definition, only one frequency and which is thus narrow band. A sine wave travelling through the Earth would be reflected and refracted, producing a sinusoidal seismogram on which no distinct arrivals could be recognized. This lack of resolution is the opposite of what is desired.

Another basic principal of signal processing is that of the linear system. Once the seismic wave has been generated at the source, a linear system allows one to model mathematically the processes that represent modifications of the wave as simple operations called convolutions. The main processes are the effects of travel through the Earth and the effects of the response of the seismograph system. These operations are independent of the order in which they are applied in both the frequency domain or the time domain. Often the goal of the seismologist is to model the response of the Earth by cal-

culating seismograms that match those observed (see *synthetic seismograms*). By using the linear system concept, the seismic source and the response of the recording system can be separated from the response of the Earth so that it can be modelled.

A final basic concept is the stacking (summing) of seismograms to produce a result in which the S/N ratio is improved. The idea is that a signal of a given amplitude can be obtained by recording one big seismic source or many smaller ones and summing the results. The simple summing of a series of weak signals from a single source location at one receiver location is one example. Another is common mid-point (CMP) stacking, which was invented by the seismic reflection industry in the 1960s and is illustrated in Fig. 2. Here, seismic reflections travelling along different ray paths but sharing a common mid-point between a series of sources and receivers are analysed. These arrivals are not aligned in time initially. They can be aligned after making time shifts based on determining an average velocity of the material above the reflector. When then summed, a great improvement in data quality (better S/N) is attained. In modern reflection surveys, it is not uncommon to obtain over 200 seismograms that share the same CMP. The S/N ratio improves with approximately the square root of the number of traces. For example, 16 traces must be summed to obtain a fourfold increase in S/N.

The seismic signal processing procedures employed by the seismic reflection industry are by far the most complex of any seismological technique. However, two of the basic steps are common in most applications. The first is simply bookkeeping, in which computer files representing seismograms must be organized so that it is clear when and where they

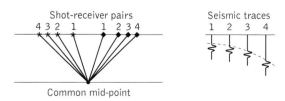

Fig. 2. Principles of common mid-point recording during seismic signal processing.

were recorded. Getting the data into a standard format that is easily recognized is also part of this procedure. This step is not as trivial as it seems when one realizes that hundreds of seismographs and seismic sources spread over large distances may be involved. The second step is harder to generalize, but would include some basic filtering, a first look at the data in some organized form, and perhaps some stacking of the seismograms. The last step entails sophisticated operations that apply to seismic reflection techniques and their goal is producing an image of the subsurface.　　G. R. KELLER

Further reading

Telford, W. M., Geldart, L. P., and Sheriff, R. E. (1990) *Applied geophysics.* Cambridge University Press.

Yilmaz, O. (1987) *Seismic data processing.* Society of Exploration Geophysicists, Tulsa.

seismic stratigraphy

Seismic stratigraphy (seismostratigraphy) involves the construction and study of stratigraphic sections from seismic records, and it has been largely developed by the exploration staff of petroleum companies in the search for suitable oil and gas reservoir rocks and structures. In principle the technique requires the generation of a seismic shock and the timing of its reflections from surfaces such as bedding planes or reflective horizons at depth within the Earth back to receivers at the surface (Fig. 1). The shock may be provided by a charge of explosive, or by a mechanical 'thumper'. The reflected tremors are detected by an array of geophones set along the required profile. This may be many kilometres long and the depth from which the seismic reflections are collected may amount to hundreds—or even thousands—of metres. The techniques have been adapted for use at sea for study of the continental shelves.

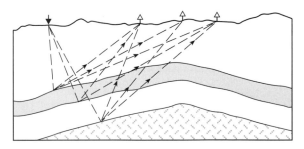

Fig. 1. Seismic exploration of deep structures and stratigraphy involves the generation of a shock, such as an explosion or a heavy blow, at one point at the surface and the timed reception of its reflection from below by geophones at other points. The reflecting surfaces are the contacts between different rock units, and the time taken for the reflected shock waves to return to the surface is an indication of the depth of each reflecting surface.

Seismic reflection profiles can be shot over all kinds of topography or terrain. Generally speaking, the shock waves are transmitted faster through denser material, but their strength diminishes the farther they travel. Essentially, they provide data from the presence of reflecting surfaces in the rock column. These occur where a sharp change in the density of the column occurs. Stratification and the attitude of surfaces in the section are thus revealed. For example, much of the geology of the North Sea and the offshore continental shelf around the British Isles has been deciphered by use of this technique. The strongly developed stratification of many of the Mesozoic and Cenozoic rocks is clearly shown, and their displacement at fault zones or truncation by unconformities gives good two-dimensional control in the modelling of oil-field structure.

The depth of penetration depends upon the strength of the explosive charge and the spacing of sufficiently effective geophones. Depths of six or more kilometres have been commonly attained. Salt domes and other disruptive bodies of rock can be discerned at a glance, and it was this that spurred the early exploration of the North Sea when the family of deep salt domes beneath Holland and Germany was found to extend beyond the shoreline. The seaward extension of deltaic sands has been discovered by this method, as in the Gulf of Mexico, and off the Niger delta.

Around Britain a series of deep seismic traverses has revealed many of the structures produced during the Caledonian orogeny and other earth movements that are continued out over the continental shelf.

One of the important consequences of seismic stratigraphic exploration in the USA has been the recognition by Exxon geologists of widespread unconformity-bounded sequences on the seismic profiles available to them. The profiles were correlated to known borehole records, and this made it possible to allocate the succession to chronostratigraphic units and thereafter be correlated with other sections and sequences on a continent-wide scale.

In the relatively stable continental interiors the technique has been used, not only to explore for oil or other resources, but also to furnish data about the deep structure. One such venture based at Cornell University has revealed that over thousands of square kilometres the Appalachian Mountains are constructed of shallow, thin thrust-slices, rather than deeply rooted upthrust blocks, as was previously believed.　　D. L. DINELEY

Further reading

Sherriff, R. E. (1980) *Seismic stratigraphy.* IHRDC Publications, Boston.

seismic surface waves

In addition to P-waves and S-waves which penetrate deeply into the Earth (see *seismic body waves*), a seismic source such as an explosion or earthquake also generates waves which travel along the surface of the Earth. The most commonly observed waves are called

Rayleigh waves, after Lord Rayleigh who predicted their existence in 1887. These waves are analogous to waves travelling across the ocean. A swimmer is not only pitched up and down, but also to and fro as a wave passes. The actual movement of the swimmer describes an ellipse. In a Rayleigh wave, the motion of a point in the Earth as the wave passes is also elliptical (Fig. 1a). The motion of waves in the ocean dies out quickly with depth, and this is also the case with Rayleigh waves. On the land, these waves spread out in only two dimensions, compared to three for body waves. Thus, their amplitudes diminish with distance more slowly than do those of body waves. Surface waves are usually a prominent feature on seismograms; when particularly large earthquakes occur, the surface waves can propagate so as to encircle the Earth several times. The mathematical treatment of surface wave propagation is complex, but Rayleigh waves can be thought of as arising from the constructive interference of multiply

reflected P-waves and S-waves (S_v) travelling in a vertical plane. Seismic surface waves of the second type are called Love waves, after A. E. H. Love, who developed a theory for their existence early in the twentieth century. These waves can be thought of as the constructive interference of multiply reflected S-waves whose particle motion is horizontal (S_H). In each case, the depth to which the waves penetrate is a function of their frequency. Lower frequencies have longer wavelengths, because of the relationship $v = f\lambda$ (where v = velocity, f = frequency, and λ = wavelength), and penetrate deeper into the Earth. Since velocity generally increases with depth in the Earth, the lower frequencies pass through material with higher average velocity. This gives rise to a major attribute of surface waves, which is that they are dispersed (i.e. velocity is a function of frequency). Thus, surface waves as seen on seismograms are not compact wavelets like body waves, but appear as long wave trains in which the frequency slowly increases (Fig. 1b). Surface waves are used to determine a general picture of Earth structure through the analysis of their dispersion. Since different regions of the Earth have different distributions of velocity with depth, each region is characterized by different dispersion curves (plots of the variation of velocity with frequency, f, or period, T, where $T = 1/f$). A typical dispersion curve for a continental region is shown in Fig. 1c. As shown in the plot, two velocities are associated with the propagation of surface waves. One is phase velocity, which is the velocity at which a particular feature on the wave train (i.e. a peak) or frequency component travels. The other is group velocity, which is the velocity at which a packet of energy or band of frequency travels. There is a mathematical relationship between these two velocities, but they are often treated independently in efforts to determine Earth structure from surface waves.

The analysis of the dispersion of surface waves has led to many important discoveries about the Earth. For example, the profound difference between the deep structure of the oceans and the continents, the existence of a widespread low-velocity layer in the upper mantle, and regional differences between the deep structure of continental features such as mountain belts and ancient cratons were all established on the basis of dispersion of surface waves. Although surface-wave dispersion yields average models for the region between two seismograph stations, it is capable of detecting low-velocity zones, and it is most sensitive to shear-wave velocity variations. These attributes are both advantageous to the seismologist in that they complement body-wave studies. Surface-wave studies are not limited to the analysis of deep structures. High-frequency surface waves can be used on a small scale in engineering studies. Because they are so near the surface, explosions generate surface waves strongly. This fact is a nuisance in seismic reflection studies (see *seismic exploration methods*) and requires special measures to attenuate the large surface waves. On the other hand, part of the method for detecting nuclear explosions is based on the large surface waves they generate. In contrast, deep earthquakes can be

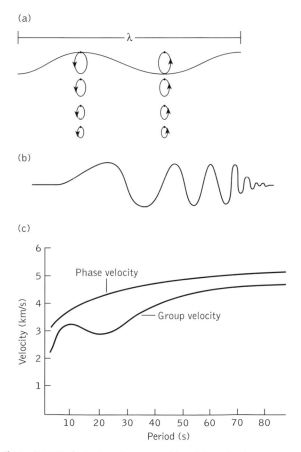

Fig. 1. Aspects of seismic surface waves: (a) particle motion in a Rayleigh wave; (b) a dispersed surface wave; (c) typical dispersion curves for a Rayleigh wave.

recognized by the weak surface waves they generate. In general, the analysis of surface waves is a useful and cost-effective tool for determining Earth structure and seismic source characteristics. The Earth models derived from the analysis of surface waves are, however, more generalized than those obtained using most other seismic techniques.

G. R. KELLER

Further reading

Bullen, K. E. and Bolt, B. A. (1985) *An introduction to the theory of seismology*. Cambridge University Press.

Lay, T. and Wallace, T. C. (1995) *Modern global seismology*. Academic Press, San Diego.

seismic tomography Seismic tomography is a relatively new method of determining the velocity structure of the Earth. In this method, an image of an object (a heterogeneity), is formed by analysing the travel times of many waves that have passed through the target area for which the image is sought. Some reference Earth model (a velocity structure) is assumed; variations from this model are determined and are typically expressed as percentage of changes in velocity. This approach is similar to that used in medicine to form images from X-rays or ultrasound such as the CAT scan (computerized axial tomography) and is fundamentally different from the traditional approaches used in seismic reflection and refraction studies. Unfortunately, tomographic studies of the Earth employ seismic waves which bend as they travel, and the distribution of seismographs and seismic sources is usually less than ideal. For example, earthquakes are concentrated in

distinct regions in the Earth and few seismographs operate in oceanic areas. However, recent advances have lead to the installation of many new, high-quality seismography stations and the ability to process large quantities of data. In addition, a particular region can be targeted by installing portable seismographs for a limited period of time. Also, as in other geophysical techniques, most developments can be applied on either a local or global scale. Thus, seismic tomography is an active area of research that is producing interesting new images of the velocity variations in the Earth.

The basic concept behind seismic tomography is that the travel time of a body wave travelling through the Earth can be expressed by an integral equation which in effect states that the time of travel is equal to the distance divided by the velocity. If the travel time for an individual seismic wave is different from what we expect from the reference Earth model, we say that there is a *travel-time residual*. At this stage we would know only that there was a velocity anomaly somewhere along the ray path that the wave had travelled. The essence of seismic tomography is to employ a large number of such observations for waves that have travelled by a variety of ray paths and to locate the position of the velocity anomaly. There are many mathematical approaches to this complex problem, but the most commonly employed one that is to approximate the Earth by a 3-dimensional mesh of blocks. The travel-time integral in the equation then becomes the summation of the travel times through each block intersected by the ray path. Many observations lead to a large system of equations in which the unknowns are the velocities of the blocks. The quality of the solution depends on the number of observa-

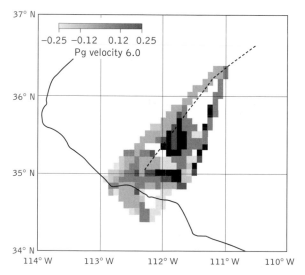
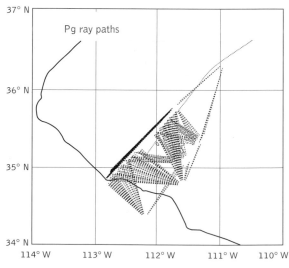

Fig. 1. (a) Seismic tomographic map of a volcanic field south of the Grand Canyon in Arizona. Variations in seismic velocity from 6.0 km s⁻¹ are shown. The upper crustal refraction, Pg, was employed. (b) Ray-path diagram for the same area.

tions and their accuracy, the distribution of ray paths, and the degree to which the block model actually approximates to the true velocity structure.

A typical example is shown in Fig. 1a. This is from a study conducted in a volcanic field south of the Grand Canyon in Arizona. The velocity variations are shown by changes of tone; dark tones represent high velocities. A zone of high velocity was detected under the volcanic centre in the field. The ray-path coverage (Fig. 1b) must be shown because blocks that are not sampled obviously do not have their velocity determined.

Other approaches to this problem are being investigated. In particular, spherical harmonic expansions have been used to look at global-scale velocity variations. A great deal of research is also being undertaken to study more than just travel-time pertubations and isotropic velocity structures (see *seismic anisotropy*).   G. R. KELLER

Further reading

Iyer, H. M. (1989) Seismic tomography. In James, D. (ed.) *The encyclopedia of solid Earth geophysics*, pp. 1133–51. Van Nostrand-Reinhold, New York.

Lay, T. and Wallace, T. C. (1995) *Modern global seismology*. Academic Press, San Diego.

Nolet, G. (ed.) (1987) *Seismic tomography*. D. Reidel Publishing Co., Dordrecht.

seismic wave theory Theoretical treatments of seismic wave propagation pose complex problems whose mathematical solutions can be approached in two basic ways. One is by employing ray theory, which is based on a useful approximation in which the seismic body waves recorded at a particular location are treated as individual packets of seismic energy or wavelets. The travel paths of these wavelets can be traced through the Earth, and their interaction with the materials through which they travel can be calculated provided that a number of restrictions on the complexity of the Earth are taken into account. For a given mathematical model of the Earth, ray theory will yield very accurate values of the travel time for a particular body wave. The amplitude and shape calculated for the seismic waves are, however, only approximate because this approach is based on what is known as the Eikonal equation, not on the true wave equation.

The other (and ultimate) theoretical approach to analysing seismic wave propagation uses wave theory. Here the wave equation of classical physics is derived in a somewhat simplified form, and exact or numerical solutions are sought. The advantage of this approach is that the entire wave field emanating from a seismic source can be considered. For example, the widely used reflectivity technique is based on an exact solution of the wave equation for the case of an Earth which consists of horizontal layers. This is valid because the Earth is approximately layered in many places and because the method takes into account the full complexity of wave phenomena, including surface waves. This approach cannot,

however, treat truly realistic Earth models. To treat complex Earth models, we must rely on numerical solutions to the wave equation, which are usually based on the finite difference technique. Such solutions still entail some simplifying assumptions. They are extremely computer-intensive, and thus cannot be employed on a routine basis.

The physical difference between ray theory and wave theory boils down to the fact that ray theory considers only the effect of the material that the infinitely thin ray encounters. Thus, a major discontinuity could be missed by millimetres and have no effect on the ray path. Wave theory recognizes that there is a finite portion of the wave front that interacts with the Earth as the wave propagates. The width of the Fresnel zone in which discontinuities (i.e. interfaces between beds, faults, etc.) and variations in velocity significantly affect a particular point on the wave front is related to the travel time of the wavefront, the average wave velocity, and the dominant frequency in the wave.

It is convenient to think of the Fresnel zone from the standpoint of a reflecting interface. In this case, we must realize that it is not a point on the reflector that affects a particular ray but that, instead, a zone on the reflector is involved. The main area on the reflector influencing its reflective properties is near the point at which the ray path intersects it, but adjacent portions of the reflector affect the nature of the reflected wave (Fig 1a).

Many aspects of wave propagation can be mathematically treated only by using wave theory. An example would be the diffractions that result when a wave front encounters a sharp

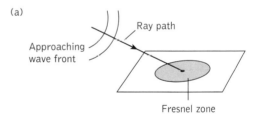

(a)

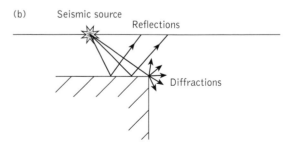

(b)

Fig. 1. (a) Cartoon showing a wave front and its associated Fresnel zone; features in the Fresnel zone, not just the point at which the wave front intersects it, affect the reflected wave front. (b) Multiple diffractions from a 'sharp-corner' discontinuity within the Earth; these are an example of Huygen's principle, which states that each point on a wave front acts as a secondary seismic source.

corner such as a fault (Fig. 1b). Diffractions arise because the corner acts as a seismic source in accordance with Huygen's principle, which states that each point on a wave front acts as a secondary seismic source; the wave propagates by the constructive interference of these secondary sources. Another example is the fact that a discontinuity in a reflecting interface will not appear clearly in the reflected wave front unless the discontinuity is wider than the width of the Fresnel zone. The point is that seismic waves are not only affected by the features directly on the ray path but are also affected by features with a circular zone whose diameter is a Fresnel zone.

G. R. KELLER

Further reading

Lay, T. and Wallace, T. C. (1995) *Modern global seismology.* Academic Press, San Diego.

Bullen, K. E. and Bolt, B. A. (1985) *An introduction to the theory of seismology.* Cambridge University Press.

Aki, K. and Richards, P. G. (1980) *Quantitative seismology.* W. H. Freeman, San Francisco.

seismic waves: principles

When stress builds up in the Earth to such a level that rapid slip occurs on a fracture (i.e. an earthquake takes place) or when an explosion or mechanical device is used to initiate a seismic disturbance artificially, a complex field of seismic waves is generated. This wave field propagates much like the waves that travel away from the point at which a stone is thrown into a pond (see *elastic wave propagation*). In the Earth, however, even ignoring the non-elastic effects near the source, the situation is much more complex because several types of waves are generated. Also, the Earth is composed of many differing bodies and layers of rock which are heterogeneous at many different scales, and the propagation is in three dimensions. Furthermore, discontinuities in the Earth reflect and refract some of the propagating energy, thus changing its direction of travel, and the Earth simply absorbs some of the seismic energy, effectively converting it to heat. Treating the full range of these complexities mathematically is a major challenge which is the topic of much continuing research.

It is usually sufficient to treat the seismic energy of the source as being divided into three types of waves. These are: (1) P-waves, which are the fastest and are simply sound waves resulting in vibrations in the direction in which the wave is travelling (see *seismic body waves*); (2) S-waves, which typically travel at about 60 per cent of the speed of P-waves and result in vibrations perpendicular to the direction in which the wave is travelling (see *seismic body waves*); and (3) surface waves, which typically travel at about 90 per cent of the speed of S-waves and are analogous to waves travelling on the surface of the ocean (see *seismic surface waves*). These waves travel by different paths, as shown in Fig. 1a. As observed at a point on the Earth's surface they produce a seismogram (a record of ground motion produced by a seismograph) as shown in Fig. 1b.

A number of simplifying assumptions are made in order to make seismic theory mathematically tractable. The first is usually that the observer is far enough from the seismic source for the actual ground displacements to be small enough to be within the elastic limit of rocks: in other words that the seismic waves travel without causing rock breakage and that the Earth returns to its undisturbed state after they have passed. This assumption arises from Hooke's law, which states that there is a linear relationship between stress and strain until the elastic limit is exceeded (i.e. until fracture, when permanent deformation occurs). Hooke's law is discussed in more detail below. A second assumption is that the Earth is composed of materials which are either homogeneous at some scale or whose properties change slowly with respect to the wavelength of a particular seismic wave. If one thinks of a snapshot of a seismic wave at a particular moment in time, as shown in Fig. 1c, the wavelength λ is the distance between two adjacent peaks in the wavelet (seismic pulse). This quantity is a very useful measure of scale because inhomogeneities in the Earth which are small relative to a wavelength (one-quarter of a wavelength is commonly used as a relative standard of what is 'small') can, for most purposes, be thought of as being averaged out by the seismic wave. Since the Earth naturally absorbs higher frequencies (i.e. the more rapid oscillations generated by the seismic

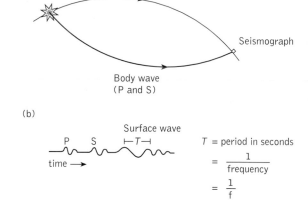

Fig. 1. Propagation of seismic waves. (a) Travel paths of surface waves and body waves; λ is the wavelength. (b) Diagrammatic representation of a typical seismogram; T is the period. If T is expressed in seconds, then $T = 1/f$, where f is the frequency in Hertz. (c) Snapshot of a typical seismic wave in time; λ is the wavelength.

source), the wavelength can be thought of as slowly increasing as the wave travels through the Earth. Thus, the minimum size of a feature that is a directly detectable discontinuity also increases with time. One of the difficulties in applying seismology to study Earth structure (see *controlled-source seismology* and *deep seismic reflection profiling*) is appreciating that seismic body waves travelling through the Earth usually have longer wavelengths than we would prefer, thus limiting our ability to 'see' features at depth clearly. For example, a P-wave travelling at a velocity of 6 km s^{-1} at a depth of about 10 km in the Earth's crust might have a maximum frequency of about 10 Hz, and its wavelength would be about 600 m. Clearly, the inhomogeneities represented by the individual mineral grains, which constitute a rock such as granite and have dimensions of up to a centimetre or so, are not an important consideration. Instead, we can think of this granite as a rock having seismic properties which are the average of its constituents. Discontinuities which are large with respect to a wavelength reflect and refract seismic waves according to the classic laws of optics which are discussed below. The third assumption usually made is that the materials through which a seismic wave travels are mechanically isotropic, which means the seismic velocity is independent of the direction of travel through the material. Mechanical anisotropy adds considerable complexity to the treatment of seismic waves but has been the object of much research interest (see *seismic anisotropy*). Mapping regions of significant anisotropy can provide valuable information on such factors as flow or stress.

Materials which obey Hooke's law are, by definition, elastic. Since seismic waves represent small strains (deformations), we can for most purposes assume that Earth materials are perfectly elastic. This means that once the stress (force per unit area) causing the strain is removed, the material returns to its original state. In its simplest form, the linear relationship which expresses Hooke's law is $s = kx$, where s is the stress, x is the strain, and k is an elastic constant. Several elastic constants are commonly employed and can be measured in a laboratory. One is Young's modulus, which describes the elongation (or shortening) a sample of material experiences when it is pulled (or pushed) by a force applied in a single direction. Another is the bulk modulus, which describes the volume change a sample undergoes when it experiences hydrostatic compression. Yet another is the shear modulus, which is a measure of the material's resistance to deformation by a shearing stress (i.e., one applied across a face of the sample in a laboratory apparatus). Fluids have no ability to resist shearing, so they have a shear modulus of zero. Poisson's ratio is the ratio of the unit lateral contraction to the unit longitudinal extension when the material is stretched. It can be determined by measuring the ratio of the elongation to 'necking' when the material is stretched in a single direction. This ratio relates V_p and V_s; when it has its typical value of 0.25, $V_p = \sqrt{3} V_s$.

If we can assume that a material is homogeneous and isotropic, only two elastic constants are needed to describe its

elastic behaviour because there are algebraic expressions from which one can calculate other constants given any two. In this case, it is also relatively easy to show that two, and only two, body waves (P and S) travel through the material and that their velocities are given by two simple equations.

Anisotropy introduces considerable complexity to the problem. Three body waves exist in this case, and the fastest one (the quasi P-wave) is, for example, no longer analogous to a simple sound wave. In the least isotropic natural material, 21 elastic constants are required to describe its behaviour. In the most general case, the expression for Hooke's law becomes a tensor equation involving stress and strain tensors of rank 2 and a tensor of rank 4 involving these 21 elastic constants.

When seismic waves encounter an interface between two materials with distinctly different seismic properties, some of the energy is reflected from the interface and some is transmitted through it (Fig. 2). Snell's law is a simple relationship which predicts the angles involved. An obvious but important relationship in seismology (and in billiards) is that the angle of reflection equals the angle of incidence.

Some P-wave energy incident on an interface can be converted to S-waves and *vice versa*. The partition of energy can be calculated by setting up a boundary-value problem. The result is an algebraically complex set of equations derived by Knott and Zoeppritz at the turn of the century. These equations not only reveal the relative strengths of the waves reflected from and transmitted through the interface but also reveal the extent to which energy conversion has occurred. These converted waves are particularly interesting, as are the amplitude variations that occur as the angle of incidence changes. From an applied perspective, the amount of energy reflected from an interface is of particular interest (see *seismic exploration methods*). This is normally only a few per cent, and the larger the contrast in seismic properties the larger the amount of reflected energy.

The amplitude of a seismic wave also naturally diminishes as the wave travels through the Earth. As the wave front travels, it expands, and thus the energy is spread over a larger and larger surface area. This effect is called *geometrical spreading* and it

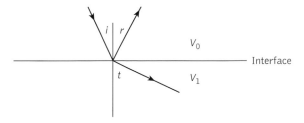

Fig. 2. Reflection and refraction of a seismic wave in accordance with Snell's Law; *i*, angle of incidence; *r*, angle of reflection; *t*, angle at which seismic ray is refracted at the interface between the two media; V_0, V_1, seismic velocities in the two media.

results in amplitude being inversely related to the distance travelled. Another effect on amplitude is *anelastic attenuation*, which refers to a number of phenomena which cause a material to absorb energy. This effect is strongly dependent on frequency: high frequencies are most strongly attenuated. This fact is a fundamental limitation in seismology which limits resolution because high frequencies are needed to 'see' features that are small relative to a wavelength. For typical seismic frequencies, anelastic attenuation is an approximately linear function of frequency which results in Q (the Quality Factor) being a constant. This factor is the quantity most often used to describe the anelastic attenuation in rocks. High-Q materials transmit seismic waves efficiently (i.e. they ring like a bell) while low Q materials absorb considerable energy. G. R. KELLER

Further reading

Aki, K. and Richards, P. G. (1980) *Quantitative seismology*. W. H. Freeman, San Francisco.

Bullen, K. E. and Bolt, Bruce A. (1985) *An introduction to the theory of seismology*. Cambridge University Press.

Lay, T. and Wallace, T. C. (1995) *Modern global seismology*. Academic Press, San Diego.

seismograms, synthetic *see* SYNTHETIC SEISMOGRAMS

seismology, controlled-source *see* CONTROLLED-SOURCE SEISMOLOGY

seismology and plate tectonics

Earthquake seismology may be divided into seismicity—the study of earthquake distribution and mechanisms—and seismology *sensu stricto*—the use of earthquakes to probe the interior structure of the Earth. Both branches have had a profound influence on the development of plate-tectonic theory.

Seismicity and plate boundaries

One of the fundamental tenets of plate tectonics is that plates are rigid and undeformable. This idea arose largely from the observed distribution of earthquakes around the world, which became readily apparent in the 1960s after the setting up of the World-Wide Standard Seismograph Network (WWSSN). Global maps of earthquakes (Fig. 1) show that most events, especially in the oceans, are concentrated in relatively narrow bands, with broad aseismic areas between them. The narrow banks of seismicity are taken to indicate active plate boundaries, while the aseismic areas are the rigid interiors of the plates. Details of the seismicity of different types of plate boundary are considered below.

Earthquake mechanisms

As well as marking the positions of plate boundaries, earthquakes can be used to determine the nature of faulting, the major stress directions, and sometimes the directions of relative plate movement. This is done by determining the 'fault-plane solution' or 'focal mechanism'.

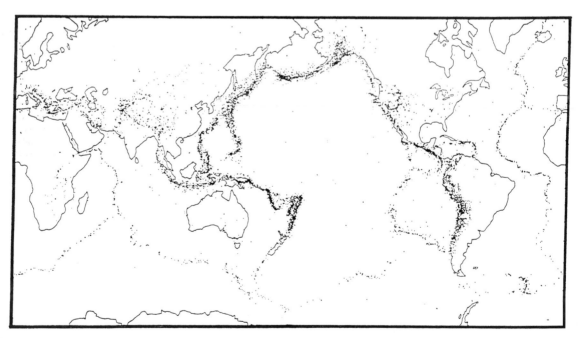

Fig. 1. Seismicity of the world. Each dot represents one epicentre. (After Barazangi, M. and Dorman, J. (1969). World seismicity maps compiled from ESSA, Coast and Geodetic Survey, epicenter data, 1961–1967. *Bulletin of the Seismological Society of America*, **59**, 369–80.)

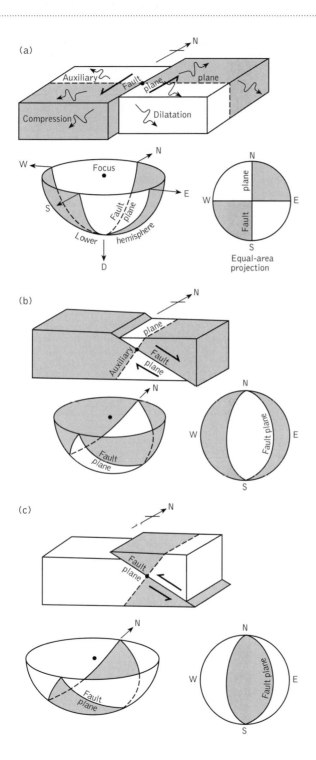

Fig. 2. Presentation of earthquake focal mechanisms in 'beachball' diagrams for (a) a strike-slip fault, (b) a normal fault, and (c) a thrust fault. In each part, a block diagram shows the fault motion. Quadrants in which the first motion is compressive (towards the receiver or away from the epicentre) are shaded. In (a) the arrow-headed wiggles indicate how seismograms of the first motion would appear in each quadrant, assuming that compression is indicated by an upward wiggle. The lower focal hemisphere is shown as an open bowl below, with its quadrants shaded according to the motion of earthquake rays passing through them. The circle shows the equal-area projection of this hemisphere on to a horizontal plane, which is the conventional representation for focal mechanisms.

A relatively simple way of doing this is to measure the direction of first motion recorded by P-waves on a seismograph. In P-waves, the ground motion is back-and-forth along the ray path. If one considers an earthquake at a strike-slip fault, for example (Fig. 2a), the ground around the fault can be divided into four quadrants, according to whether its initial direction of motion is toward or away from the epicentre. At the lines separating the quadrants the initial motion is zero. In fact zero motion is also experienced on the vertical extensions of these lines, which are called 'nodal planes'. Similar divisions into quadrants can be made for faults in any orientation. For simple fault models the nodal planes are always perpendicular to each other.

A seismograph in a quadrant with first motion away from the epicentre will itself experience a first motion away from the epicentre (i.e. towards the seismograph station). Such motion tends to compress the ground it moves into, and this quadrant is thus referred to as a compressional quadrant. In contrast, quadrants in which the first motion is toward the epicentre are dilatational, and first motions on seismographs in these quadrants will be away from the station.

First motions are conventionally plotted on a hemisphere just below the hypocentre (or focus), at the position where the seismic ray crosses the hemisphere. The two best-fitting orthogonal nodal planes are then fitted to the plotted data and the compressive quadrants are shaded, producing a 'beachball' diagram. The axes of maximum compressive and tensional stress run through the centres of the dilatational and compressive quadrants, respectively. One cannot tell from first motions alone which is the actual fault plane. Other information, such as the strike of a mapped fault, observed breaks in the ground, or alignments of aftershocks has to be used.

More recently, new methods have been developed for determining focal mechanisms, including modelling the complete radiation patterns of both P- and S-waves.

As shown in Fig. 2, different patterns of compressions and dilatations are characteristics of strike-slip (horizontal), normal (extensional), and thrust (compressional) faulting. These vary at different plate boundaries.

Divergent boundaries

The narrowest seismic zones are associated with the crests of the mid-ocean ridges, which contain divergent plate boundaries. This association was one of the prime pieces of evidence showing that ridges are tectonically active, which led to the theory of sea-floor spreading, itself an important component of plate tectonics. Global seismic networks such as WWSSN resolve the epicentres of these mid-ocean ridge earthquakes to about 20 km, and show that the bands are normally little wider than this. More detailed local studies, for example using local networks on ocean islands or ocean-bottom seismographs, show that mid-ocean ridge seismicity commonly occurs in clusters only a few kilometres across. These can in some instances be seen to be associated with individual faults or inferred dyke-intrusion zones and fissure eruptions. Such activity rarely occurs over a zone more than a few kilometres wide, although occasional earthquakes may occur tens of kilometres from ridge axes.

Earthquakes at divergent boundaries are virtually entirely confined to depths of 8 km or less which may be taken to define the depth of the brittle–ductile transition or, effectively, the base of the zero-age lithosphere.

Fault-plane solutions (Fig. 3) show that divergent boundaries are mainly characterized, as expected, by extensional,

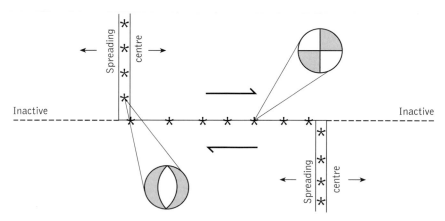

Fig. 3. A ridge–ridge transform fault offsetting two mid-ocean ridge spreading centres. Motion on the fault (half arrows), and therefore seismic activity—(stars), should be confined to the region between the spreading centres. The sense of motion is shown by the full arrows.

'Beachball' symbols (compressive quadrants shaded) show typical observed earthquake focal mechanisms, as predicted by plate tectonics: dextral strike-slip motion on the transform, and extensional normal faulting on the adjacent spreading centres.

normal fault mechanisms. Almost all earthquakes here are of moderate magnitude, usually less than magnitude 6, because the young, weak lithosphere and predominantly extensional faults do not support very large accumulations of stress before breaking or slipping.

Transform boundaries

Along many mid-ocean ridges, especially those opening at higher spreading rates, such as the East Pacific Rise, most seismicity is concentrated not on the divergent boundary itself but on the transform faults (conservative, strike-slip boundaries) which offset the divergent boundaries at intervals of tens to hundreds of kilometres. Here, detailed studies have shown that most seismicity is centred on a single strike-slip fault representing the very narrow plate boundary.

The concept of transform faults is another critical aspect of plate-tectonic theory. The idea was introduced by the Canadian geophysicist J. Tuzo Wilson in 1965. He defined transform faults as a class of strike-slip fault that connects two belts of tectonic activity; in plate tectonics, transform faults are thus themselves one type of plate boundary. Their importance is that they show how major transcurrent faults can terminate by *transforming* into different types of plate boundary.

The transform faults that offset mid-ocean ridges connect two similar types of tectonic activity (two spreading centres or divergent boundaries). If they are true Wilsonian transform faults, they should be active only between the offset spreading centres, and not on the prolongation of their trend beyond these centres (Fig. 3). The distribution of earthquakes on ridge–ridge transform faults shows that this is the case. Note also that the sense of motion on a transform fault is the opposite of that required on a simple transcurrent fault offsetting a ridge in the same direction. This is also confirmed by observed focal mechanisms.

Transform faults are not confined to the oceans, but occur on land in areas such as California (the San Andreas fault complex), Turkey (North and South Anatolian faults) and Asia (Altyn Tagh fault), and elsewhere. Many of these continental fault systems are quite complex, with numerous secondary anastomosing and splaying faults, many of which may be active simultaneously. Many of these complexities are reflected in the distribution of seismic activity, as shown for example by the very detailed seismic maps produced in California.

Continental transform faults can generate larger earthquakes than those on mid-ocean ridges, magnitudes of about 7 or more being not uncommon (e.g. the North Ridge earthquake, California 1994). These larger-magnitude events normally occur on faults with a component of thrusting, where larger stresses can build up.

Seismicity on continental transform faults extends to depths of about 15 km, where much of the deformation appears to change from brittle to ductile.

Subduction zones

A third vital concept in plate tectonics is that of the subduction zones. If, as palaeomagnetic and other studies indicate, the Earth's radius is not increasing rapidly, plate created at divergent boundaries must be destroyed elsewhere. Subduction zones are where this happens. There, plates descend at an angle into the mantle. The traces of these dipping, subducted plates are marked by zones of earthquakes (Fig. 4) called *Benioff* or *Benioff–Wadati zones* after the Californian and Japanese seismologists who first recognized them.

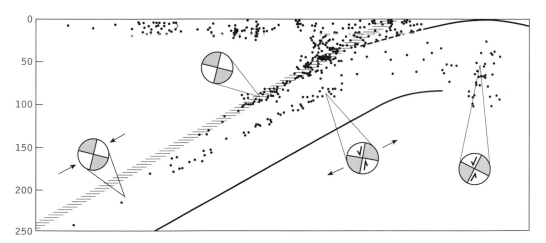

Fig. 4. Vertical cross-section through the shallow part of a subduction zone, showing the distribution of earthquakes and schematic focal mechanisms projected on to a vertical plane through the Benioff zone. (Hypocentres after Hasegawa *et al.* (1978) *Geophysical Journal of the Royal Astronomical Society*, **54**, 281–96.)

Benioff zones normally dip at about 45°, although the angle ranges from about 20° to almost vertical. The earthquakes can extend down to the base of the upper mantle at 670 km, and the epicentres of the deepest can thus be several hundred kilometres horizontally away from the plate boundary (which is usually marked by an oceanic trench) at the surface. This is what accounts for the broadening of epicentral zones around the Pacific rim (Fig. 1).

Where the subducting plate bends down into the subduction zone, most earthquakes are tensional (with normal fault mechanisms), and are thought to result from the tensional stresses developed in the upper part of the plate as it bends. These earthquakes extend to about 25 km depth. At the plate boundary, and for a few tens of kilometres beyond it on the side of the overriding plate, earthquakes are mainly compressional. They have thrusting mechanisms reflecting the contact between the two plates and the compressional deformation of the overriding plate. Deeper earthquakes make up the Benioff zone itself.

Detailed studies using local seismograph networks have resolved two parallel bands of dipping earthquakes in the Benioff zone (Fig. 4). The uppermost corresponds to the upper surface of the subducting slab, while the lower lies within the slab. Those earthquakes in the upper band have compressional mechanisms and are thought to be related to the *un*bending of the slab, combined with the effects of the frictional resistance of the surrounding asthenosphere. Those in the lower band reflect internal deformation of the descending plate. Events in the shallower part of the lower band are tensional, but at greater depths these intraplate events become compressional. Their precise cause is still unknown, but suggested mechanisms include extension of the slab under its own weight (in the upper part), compression due to increasing subduction resistance (in the lower part), and the volumetric effects of mineral phase changes.

The largest-magnitude earthquakes found on Earth occur within subduction zones, mainly in the shallow region of thrusting, where very large stresses can build up. The largest recorded earthquake (Good Friday 1964 in Alaska, magnitude 9.2) occurred on the Aleutian subduction zone.

Continental collision zones

The broadest zones of seismicity occur within continental collision zones such as the Himalayas, Tibet, and the surrounding region (Fig. 1). This reflects the complex nature of tectonic boundaries in such areas, where plate tectonics is least applicable. Here, the lithosphere, which is thicker but weaker than oceanic lithosphere, is broken into large numbers of tectonic blocks delimited by active faults in zones many hundreds or even thousands of kilometres wide. Although individual earthquakes still lie on discrete faults, the plethora of such features produces a complex pattern. Since these small tectonic blocks may all be in motion relative to each other, the full range of thrust, strike-slip, and tensional mechanisms can occur in different parts of the zone.

Most earthquakes in continental collision zones are relatively shallow, being confined to the thickness of the rigid lithosphere: 100 km or so. However, sinking lithospheric slabs may exist this depth, either as the remnants of the subduction zones that led to the collision, or as recently delaminated blocks of lower lithosphere. These may be marked by deeper earthquakes extending hundreds of kilometres below the surface.

As with subduction zones, continental collision zones can support large stresses, and so large-magnitude earthquakes are also found here.

Seismology and the lithosphere

As well as defining the positions of plate boundaries and the nature of the motion and deformation across them, seismology has provided important information on the nature of lithospheric plates and their surrounds.

The lithosphere is defined as a high-strength, rigid layer, and seismology has been able to estimate its thickness. Study of P-wave and S-wave travel times worldwide has shown that the velocities of these waves exhibit distinct minima at depths of around 100–200 km. This feature is almost ubiquitous, and is referred to as the low-velocity zone (LVZ). All else being equal, a lower seismic velocity indicates that a material is nearer to its melting point. The LVZ is thought to correspond to mantle material that is close to its melting point and is thus significantly weakened. In consequence, the LVZ is often approximately equated with the asthenosphere (the weak layer below the lithospheric plates), the strong plates corresponding to the slightly higher velocities above.

The thickness of the plates themselves can be estimated by using seismic surface waves. Like water waves on the ocean, their particle motion penetrates a fraction of a wavelength into the Earth; deeper for longer wavelengths. Surface waves have a range of wavelengths from tens to hundreds of kilometres. Their velocities reflect seismic velocity in the Earth at the depths to which they penetrate. The longer surface waves are thus sensitive to the Earth's properties near the base of the lithosphere. To fit the observed surface-wave velocities requires models with a high-velocity 'lid' (the plate) overlying a lower-velocity channel (the asthenosphere). It is found that the lid thickens with increasing age of the plate in just the way that is predicted by thermal models of plates.

ROGER SEARLE

Further reading

Isacks, B. L., Oliver, J., and Sykes, L. R. (1968) Seismology and the new global tectonics. *Journal of Geophysical Research*, 73, 5855–900.

Kearey, P. and Vine, F. J. (1996) *Global tectonics*. Blackwell Science, Oxford.

seismology, strong-motion *see* STRONG-MOTION SEISMOLOGY

seismostratigraphy *see* SEISMIC STRATIGRAPHY

seismotectonics *see* NEOTECTONICS AND SEISMO-TECTONICS

self-compression The Earth is held together by its own gravity, which pulls it into its nearly spherical shape and compresses the interior. There is a general increase in density with depth in the Earth (see *density distribution within the Earth*) and this self-compression is one of the causes.

Large regions of the Earth's interior, notably the lower mantle and outer core, appear to be uniform in composition and crystal structure, or nearly so. Within these regions self-compression is the only cause of density variation. The magnitude of the effect can be seen in Table 1, which gives

Table 1. Density variation in the mantle and core

Region	Depth (km)	Pressure (GPa)	Density (k gm^{-3})
Lower mantle	Extrapolated	0	3 990 ($T \approx 1800$ K)
	670	24	4 381
	2890	136	5 566
Outer core	Extrapolated	0	6 330 ($T \approx 2000$ K)
	2890	136	9 903
	5150	329	12 166

densities at the top and bottom of each of these regions. Also listed are the calculated densities which the materials would have at zero pressure if their crystal structures survived complete decompression. These numbers show that at the Earth's centre compression causes nearly a twofold increase in the density of core material (which is believed to be mainly iron).

FRANK D. STACEY

sequence stratigraphy Sequence stratigraphy is the subject that is concerned with the large-scale, three-dimensional arrangement of sedimentary strata and with the major factors that influence their geometrical relationship. These factors are sea-level change, contemporaneous fault movements, and supply of sediment. The basic conceptual unit, a 'depositional sequence', is a package of sedimentary strata bounded above and below by major surfaces. These surfaces are erosion surfaces ('unconformities') along some of their length, and are technically defined as the 'sequence boundaries'. The geometrical relationships within and between depositional sequences are observable from geophysical data, particularly seismic reflection profiles, which represent an acoustic image through a rock sequence. Because seismic reflection profiles are an important tool in hydrocarbon exploration, sequence stratigraphy as a discipline was first developed extensively within the oil industry. Depositional sequence geometries may also be observed in mountainside exposures of relatively undeformed strata, or by piecing together evidence from exposures on a much smaller scale.

A simple depositional sequence for particulate (detrital) sediment is shown in Fig. 1. Sediments laid down in different depositional environments are distinguished by different tints, and the sequence boundaries are also indicated. The geometries have been related to changing sea level because sea level exerts a strong influence on the heights to which sediment may accumulate. At the sequence boundary, sediment accumulated only in the most distal locations, and hence it is inferred that sea level was falling or at a low stand. At the opposite extreme, in the middle portion of the depositional sequence, deposition occurred near the source, and it is inferred that sea level was rising or at high stand; at the same time more distal locations may have become completely starved of sediment, and the surface generated as a result has been called a 'maximum flooding surface'.

The proximal region of the sequence boundary will be characterized by erosion of the underlying strata and may include features such as river valleys cut into previously deposited marine strata below. Conversely, the distal region of the sequence boundary may be represented by an increased abundance of sedimentary debris eroded from more landward sites. The maximum flooding surface has a different character. The more distal portion of the maximum flooding surface represents, like the sequence boundary, a depositional break, but unlike the sequence boundary it develops at the far end of the sediment transport path as a result of sediment starvation, and may be characterized by slowly accumulating marine deposits containing an abundant pelagic fauna and well-developed mineralization of the sea floor.

One school of thought prevalent in the early development of sequence stratigraphy was that the sea-level changes responsible for generating depositional sequences were global in extent. Thus, by recognizing the key surfaces of depositional sequences in well-dated sedimentary successions, composite global sea-level curves could be constructed for Phanerozoic time. There are, however, many soundly based objections to the notion of global sea level change as the dominant control on depositional sequence geometries.

Sea-level change taking place in a specific place on the Earth is also the result of localized subsidence and uplift. The rates of subsidence or uplift in many geological settings are likely to have been comparable to, or in excess of, global sea-level change. This is particularly the case during times in Earth history when there were no large, land-based polar ice caps, as during the Mesozoic. Additionally, changes in the rate at which sediment was supplied to a basin would alter significantly the time of formation of a sequence boundary or maximum flooding surface. For example, where there was an abundant supply of sediment, the shoreline would have begun to move seawards much earlier than it would when sediment was scarce. Furthermore, movements along major faults would have affected not just subsidence and uplift but also the local supply of sediment. Thus, the extent to which global sea-level changes may have influenced sequence stratigraphy will have depended both on time and place.

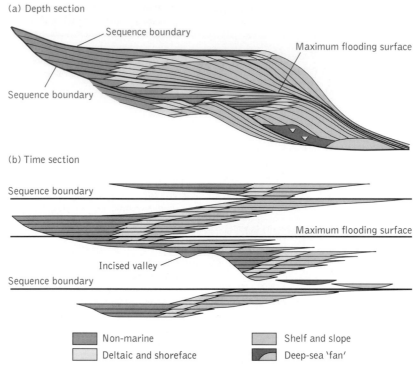

(a) Depth section

Sequence boundary

Maximum flooding surface

Sequence boundary

(b) Time section

Sequence boundary

Maximum flooding surface

Incised valley

Sequence boundary

Non-marine

Deltaic and shoreface

Shelf and slope

Deep-sea 'fan'

Fig. 1. An idealized depositional sequence. Sediment is supplied from the left of the figure. The vertical axis for (a) is sediment thickness, and is greatly exaggerated. Near-horizontal lines represent former positions of the land surface and sea floor. The deposits of different sedimentary environments migrate landward and seaward primarily in response to changes in sea level. The sea surface would have intersected the sediment surface at about the interface between the 'non-marine' and the 'deltaic and shoreface' depositional environments. The vertical axis for (b) is geological time. (Modified from Haq *et al.* (1987) *Science*, **235**, 1156–66.)

Carbonate systems (limestones and dolomites) may be expected to have responded differently to sea-level changes because the sedimentary grains were produced mainly *in situ* rather than having been transported from a hinterland. Many carbonate-producing processes require warmth and light in a shallow-water setting, and are particularly sensitive to changes in nutrient supply. As a result, the large-scale geometrical relationships of carbonate systems are very different from those of detrital systems. During relative sea-level rise carbonate production would generally have been extremely effective; large volumes of shallow-water carbonate grains would accumulate, and enough would be produced for export of carbonate grains to deep-water settings. During falls in sea level, large regions of potentially productive shelf would be exposed and sediment production effectively shut down, so that little sediment would have been able to accumulate in either shallow-water or deep-water settings. It is clear also that in the past some sea-level rises have been associated with increases in nutrient supply that have led to the demise of the carbonate sedimentary system, to form a 'drowning unconformity'. The lower depositional angles of overlying detrital sediments may create stratal geometries that mimic those expected with extreme sea-level fall in purely detrital systems.

STEPHEN HESSELBO

shallow-marine sedimentary deposits It is on the continental shelves where *shelf, epeiric, neritic,* or *epicontinental* sediments are deposited. The water depth is here usually less than 200 m, and the width of the shelves varies from tens to hundreds of kilometres. During times of higher sea level the shelves were much more extensive than they are today, and they covered vast areas of the continental crust. Great thicknesses of marine sediments in the geological record were deposited in this environment.

Sediments derived from the continents are dispersed by a wide variety of processes. Waves stir the sediment over the whole shelf during storms, but during quieter weather only the shallower inner parts are affected. Currents, either locally wind generated, or those forming part of the world-wide pattern of oceanic currents, are important in dispersing silt and clay and occasionally are sufficiently strong to move sand. Some shelves are vigorously stirred by tidal currents capable of moving sediment of almost all sizes. Storm-driven currents, in spite of their relative infrequency, are important in transporting sediments of a wide variety of grain sizes. Gravity-driven density currents, produced by storms that cause vast quantities of sediment to be suspended, are important in some parts of the inner shelf. It has also become apparent that internal waves within the water column can disturb sediments of the outer shelves.

The siliciclastic sediment of the shelves is supplied from adjacent land masses or by reworking of existing sea-floor sediment. Siliciclastic sediment is supplemented by skeletal debris of biological origin or by biochemically extracted calcium carbonate. In areas where the supply of siliciclastic sediment is meagre, carbonate (limestone) production comes dominant. Furthermore, upwelling of water along some shelves produces high organic productivity. Reaction between shelf sediments and shelf waters produces new minerals, such as glauconite, and elsewhere phosphate is precipitated from shelf waters to coat or replace earlier sediments.

On many modern shelves, mud escaping from adjacent continents forms only narrow belts parallel to the coastline where the supply is low or where it has been trapped in local estuaries or lagoons (Fig. 1, c). However, around large deltas where large amounts of mud reach the coast this is dispersed by marine currents to cover the whole shelf to produce extensive mud-sheets or mud-blankets which cover thousands of square kilometres (Fig. 1, d). The extensive shelf mudstones and shales with open marine fossils, which are so common in the geological record, were produced in this way. At present, shelves have only recently been covered to their present depth by water during the last rise of sea level (the *Flandrian trans-*

gression in north-west Europe). The present processes have been operational for only a few thousand years and the supply of siliciclastic sediment has had insufficient time to cover the shelves entirely except where they are narrow.

Today, many shelves are covered with a variety of shallow-water sediments that were deposited when sea levels were lower and when large parts of the shelves were exposed. These sediments are texturally and compositionally anomalous and have been inherited from a previous cycle of deposition: they are *relict sediments*. In most places they have, however, been modified by modern processes by the addition of some modern mud and skeletal debris from modern organisms living either in the sediment or in the water column above. They have appropriately been named *palimpsest sediments*: like an old manuscript which has been reused, they bear the impression of both older and contemporary events. In such areas of slow sedimentation, or sometimes lack of sedimentation, the mineral glauconite and phosphates develop surfaces which are equivalent to diastems in the stratigraphic record. Where shelves are remote from sources of siliciclastic mud, these earlier deposits are gradually enriched in skeletal debris and ultimately form extensive sheets of skeletal limestones, of which many occur in the geological record. Thus although modern shelves are different because of the small amount of time for which much of them have been covered by sea water or covered by water of the present depth. The sequence of deposits forming today from nearshore to the shelf edge is similar to that seen in ancient sediments: sand, muds, and skeletal sands. The ancient sediments were recognized by early geologists as indicating a coast–shelf edge succession.

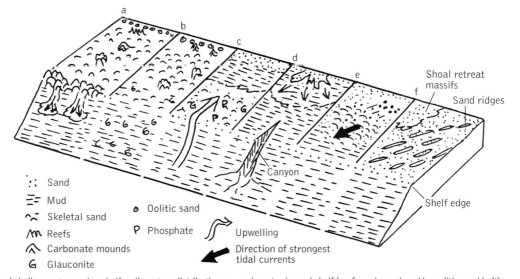

Fig. 1. Typical shallow-water marine shelf sedimentary distributions: *a*, carbonate-rimmed shelf (reef may be replaced by oolitic sand belt); *b*, ramp carbonate shelf; *c*, siliciclastic shelf with limited sediment supply; *d*, siliciclastic shelf with abundant sediment supply; *e*, tidal-dominated siliciclastic shelf; *f*, storm-dominated siliciclastic shelf.

Extensive sheets of sandstones containing marine fossils, which obviously were not of shallow-water origin, puzzled geologists for many years. However, work pioneered by Stride in Great Britain on tidally dominated seas and later by Swift and his co-workers on storm-dominated shelves in the USA eventually produced an explanation. In seas dominated by strong tidal currents, a contemporary mobile sediment cover is produced by the reworking and remobilization of sediments that were deposited when the sea level was lower. This sediment cover is available even where is no contemporary supply of siliciclastic sediment from the adjacent land (Fig. 1, e). Where tidal currents are strong, the sediment has been completely removed or has merely left a sheet of gravel (*lag gravel*), which is usually thin and rich in coarse worn skeletal debris or has been fashioned into large-scale ripples expressed by cross-stratified gravels. In areas of lower current speed, extensive *sand sheets* develop which often cover thousands of square kilometres of the sea floor. The sand is moulded into large-scale ripples (so-called sand waves) with smaller ripples of various sizes to produce extensive sheets of cross-stratified (metre-scale) marine sands. Although the tidal currents are bimodal, the asymmetry of the bedforms and the direction of dip of cross-stratification record the direction of the strongest flow, and hence of sand flux. Sets of cross-stratified sands are sometimes covered with thin mud lenses deposited at times of lower velocity in the tidal cycle. They occasionally show a set of cross-strata recording the contrary flow direction. Bioturbation is rare in these deposits. In areas where current speeds are lower, the sand waves decrease in size and pass into smaller-scale ripple cross-stratified sands with *mud-drapes* and some bioturbation. These in turn pass into areas where tidal current are sufficiently low for deposition of mud over thousands of square kilometres. Sand rarely reaches these areas except during storms, when thin sand layers are formed either by density current material in suspension or by locally reworked and concentrated coarse elements. The sediments are often extensively bioturbated and contain a good marine fauna.

Elsewhere in tidal seas the sand has been piled up into elongate ridges (*sand ridges*) generally parallel to the marine tidal flow. They are asymmetric in cross-section and migrate in the direction of the steeper face. They consist of complex cross-bedded sands commonly enclosing thin wisps of muddy sediment.

In seas which experience intensive storms, currents are generated which produce sand sheets similar to those found an tidally dominated shelves (Fig. 1, f). *Linear sand ridges* cover vast areas of the shelf in some areas, such as the east coast of the USA. Initially they were thought to be drowned coastal barriers; but it is now known that they developed when storm-generated wind-driven currents initiated erosional scours oblique to the coastline and built ridges to seaward, which were then modified by shoreward approaching waves. The connection between the sand ridges and the shoreline is gradually severed and they become isolated on the shelf, where they continue to be modified by waves and storm currents. The latter flow parallel to the ridges and develop secondary currents which sweep sand on to the ridge crests. Waves rework the crests to produce a coarse-grained cap to these features. They have a complex internal structure with cross-stratified sands with mud drapes rather like those of tidal sand ridges, and some horizons of hummocky cross-stratification. Other large sand bodies (*shoal retreat massifs*) have developed in storm-dominated seas and they form major sand accumulations in the inner shelf areas.

The gradual evolution of such storm-dominated shelves produces extensive fields of storm-generated ridges maintained by the violent storm currents even when they are under hundreds of metres of water. As with tidally produced sand fields, they also result in sheet-sands of cross-stratified sands resting unconformably on the scoured surface of the underlying sediments.

Although most normal oceanic currents are capable of transporting only fine-grained sediment in suspension, Fleming showed that in some areas, such as off the East African coast, they can transport sand. Here the Agulhas current has moulded the shelf sediment into large bedforms (17 m high) with steep lee faces, and these bedforms are likely to produce cross-stratified sands dipping in the direction of flow of the oceanic current if preserved in the geological record.

Where there is a lack of supply of siliciclastic sediment the shelves are covered with the remains of skeletal organisms (Fig. 1). On inner shelves, benthonic remains are dominant, those of planktonic and nektonic organisms (mainly foraminifers) increasing as the water deepens. The latter may become dominant in some areas to result in fine-grained pelagic limestones having many similarities to chalk. In tropical areas, blooms of diatoms lead to the direct precipitation of carbonate mud (*whitings*). The resulting sediment cover produces widespread skeletal limestones with increasing amounts of finer-grained carbonate in less turbulent, usually deeper water (*carbonate ramps*) (Fig. 1, b). Extensive sheets of limestones preserved in the geological record were formed in this way.

Reefs and *carbonate-mounds* are characteristic of some tropical shelves, where they form isolated patch reefs or, in some areas, well-defined barrier reefs on the shelf edge (*rimmed platforms*) (Fig. 1, a); e.g. the Great Barrier Reef of Australia and reefs off Belize.

Upwelling of cold oceanic nutrient-rich waters in many shelves (e.g. off south-west Africa and Peru) leads to extensive plankton blooms. The bottom waters, which are normally well oxygenated, are depleted in oxygen and organic-rich muds accumulate, as well as phosphates.

G. EVANS

Further reading

Dalrymple, R. W. (1992) Tidal depositional systems. In Walker, R. G. and James, N. P. (eds) *Facies models*, pp. 195–219. Geological Association of Canada.

Johnson, H. D. and Baldwin, C. T. (1986) Shallow siliciclastic seas. In Reading, H. G. (ed.) *Sedimentary environments and facies* (2nd edn), pp. 229–82. Blackwell Scientific Publications, Oxford.

Sellwood, B. W. (1986) Shallow-marine carbonate environments. In Reading, H. G. (ed.) *Sedimentary environments and facies* (2nd edn), pp. 283–342. Blackwell Scientific Publications, Oxford.

Walker, R. G. and Pint, G. (1992) Wave and storm dominated shallow marine systems. In Walker, R. G. and James, N. P. (eds) *Facies models*, pp. 219–38. Geological Association of Canada.

shear zones, deformation bands, and kink bands

Under certain circumstances, deformation in rocks produces a localized zone within which the rock material undergoes a rotation with respect to the walls of the zone. This process is known as simple shear. Such structures are analogous to faults in the sense that they accommodate a displacement of one wall with respect to the other, but differ in that the displacement is distributed through the zone rather than being confined to a planar fracture. Structures formed in this way can occur on all scales from fractions of a millimetre to many tens of kilometres across.

Deformation bands

Zones of simple shear strain within single crystals on a microscopic scale are termed 'deformation bands'. These are tabular zones whose crystallographic orientation differs from that in the adjoining parts of the crystal. They are commonly observed in quartz crystals, where they show up in polarized light as alternating dark and light bands, forming a feature known as 'undulatory extinction'.

Shear zones

Shear zones are localized zones of mainly ductile deformation within rocks at scales of between millimetres and kilometres across. The smallest shear zones, although only a few millimetres across, are usually many centimetres or even metres in length, and traverse many crystals, so that they represent structures at least an order of magnitude larger than deformation bands. The largest shear zones, found typically in regions of Precambrian basement, are tens of kilometres across and hundreds of kilometres long.

The simplest type of shear zone is illustrated in Fig. 1a. It is a tabular zone of ductile deformation enclosed between two opposing blocks of undeformed rock, and is produced by a lateral displacement of one wall of the shear zone relative to the other (that is, simple shear). In this case the displacement is distributed uniformly across the zone. Such zones are ductile analogues of faults, and, like faults, may have varying orientations and sense of movement. For example, vertical zones exhibit strike-slip (transcurrent) displacement, or shear, and inclined zones can exhibit either a normal or a reverse sense of movement, produced by extensional or compressional stresses, respectively.

In Fig. 1, the deformation (strain) within the zone is distributed evenly across the zone but exhibits an abrupt change at its margins. In most shear zones, however, especially those affecting previously undeformed rocks, the strain increases gradually through the zone, reaching a maximum in the central part of the zone (Fig. 1b). With large displacements, and consequently large strains, the deformed rock in the central part of the shear zone possesses a strongly developed planar and linear fabric that has been rotated into near parallelism with the walls of the shear zone, and can be used to indicate the shear direction, that is, the direction of displacement along the zone.

Shear zones are often more complex than the simple models of Fig. 1. They can, for example, incorporate an element of compression or extension across the zone, or the wall rock may also be deformed, either before the development of the shear zone or contemporaneously with it. In each of these cases, the geometry of the strained rock within the zone will be different from that of Fig. 1, and more difficult to interpret.

Movements on shear zones represent the characteristic method of accommodating displacements at intermediate to deep levels of the crust and uppermost mantle. Major fault displacements at upper crustal levels are thought to transfer downwards into shear zones (Fig. 2). For example, the San Andreas fault zone of the western USA, which comprises a network of branching sub-parallel faults, between 100 and 400 km wide, with a total displacement of perhaps 1000 km, is probably replaced at depth by a broad shear zone in which the displacement is taken up by ductile deformation. Evidence for the nature of deep-level displacements comes from the study of Precambrian gneiss terrains, such as those of southern Greenland and north-west Scotland, where shear zones on

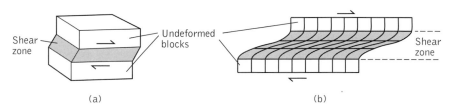

Fig. 1. (a) Idealized shear zone produced by the relative displacement of two undeformed blocks. (b) Varying shear strain within a shear zone, shown by the deformation of originally perpendicular lines. Note that the maximum deformation is in the centre of the shear zone.

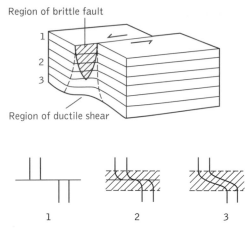

Region of brittle fault

Region of ductile shear

Fig. 2. Transition from brittle fault to ductile shear zone with increasing depth.

every scale are exposed. In south-west Greenland, a thrust-sense shear zone 50 km wide forms the boundary between the Nagssugtoqidian orogenic belt to the north and the Archaean craton to the south. In the Lewisian complex of north-west Scotland, several large transcurrent shear zones occur. Among these is the Canisp shear zone in the Assynt area, which is about 1 km wide and has accommodated right-lateral displacement.

Kink bands

Kink bands are a special type of shear zone, occurring in rocks that already possessed a strongly layered or laminated structure before the development of the kink band (Fig. 3). They are small-scale structures, typically of the order of millimetres to centimetres in width, and are defined by sharp boundaries where the orientation of the pre-existing laminar structure changes abruptly. By analogy with faults, kink bands may display either a normal or a reverse sense of movement on inclined bands and transcurrent movement on steeply dipping bands.

Kink bands are commonly observed in conjugate pairs (as illustrated in Fig. 3), and can lead with increasing strain to the

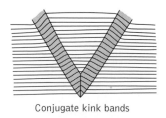

Conjugate kink bands

Fig. 3. Conjugate kink bands in laminated rock.

development of special types of folds known as box folds and chevron folds. Such folds form when two or more kink bands merge to form continuous zones of deformation. However, although they are in some ways analogous to folds, kink bands are formed by a simple shear mechanism and represent an alternative shortening mechanism to continuous folding. They are formed under special conditions where the deforming material consists of a set of strong thin layers separated by weak zones of easy slip. These slip horizons are essential for rotation to occur within the kink bands. Rocks with slaty cleavage or well-developed schistosity are the most likely to be affected by kink bands.

R. G. PARK

Further reading

Park, R. G. (1997) *Foundations of structural geology* (3rd edn), chapters 3, 10. Chapman and Hall, London.

Shepard, F. P. (1897–1985)

Francis Parker Shepard was one of the most productive of the pioneer marine geologists and sedimentologists of this century. He was born in Massachusetts and, after graduating, moved to Illinois. From there he went to the Scripps Institute of Oceanography at La Jolla, southern California, to which he was attached for the rest of his life. Shepard was, to all intents and purposes, the only seagoing American marine geologist from the 1920s until the 1940s. He was author of several books and many scientific papers, mainly concerned with marine geology and research; he was also an editor of the journal *Sedimentology*.

His experience extended to many parts of the continental shelves, to the slope and the deep ocean plains. Among his major achievements was the surveying of the great submarine canyons cut into the continental slope off the American coast, and the monitoring of the bottom currents on the continental shelf and slope. He postulated that the canyons had initially been cut by rivers during a lower stand of the sea in the Pleistocene epoch. During a cruise of the research vessel *E. W. Scripps* in 1951, Shepard was a member of the team which attempted to generate a turbidity current at the head of the Scripps Submarine Canyon. Charges were exploded in the sediment at the top of the canyon, but no turbidity current was detected moving down the canyon.

Because of the need for information about the sea floors during the Second World War Shepard became a principal investigator in marine geology for the US Navy. He continued to participate in work of national importance for the remainder of his career.

In 1951 Shepard was appointed Director of the American Petroleum Institute's major project on sedimentation in the northern part of the Gulf of Mexico. The work continued until 1958.

The Geological Society of London awarded him its highest honour, the Wollaston Medal, for his pioneering work in marine geology and sedimentology.

D. L. DINELEY

Shoemaker, E. M. (Gene) (1928–97) Eugene Merle
Shoemaker was born in Los Angeles, California, in 1928. He discovered that meteorite and cometary impact is a major process in the Solar System. This discovery revived the largely discarded geological concept of catastrophism, which holds that short-term catastrophic events help to shape the surface of the planets and on Earth have contributed to species extinction, and thus to evolution. For this concept, Gene Shoemaker will be remembered as one of the more important geologists of the twentieth century, and as the father of modern planetary geology.

The recognition of impact as a significant geological process allowed the establishment of the widely held hypothesis that a major meteorite impact caused the extinction of the dinosaurs and many other life forms, and that major impacts threaten the Earth in the future: a matter of concern today to civil defence authorities.

Gene Shoemaker was heeded by both sides during the Cold War, when he warned that a large impact could be mistaken for a nuclear explosion. His message that impact is a continuing process was graphically demonstrated to the world in 1994, when the comet Shoemaker–Levy, discovered by Shoemaker, his wife and co-worker Carolyn, and David Levy broke into pieces that crashed spectacularly into Jupiter.

Shoemaker was concerned with both the process of impact and the population and flux of objects in the Solar System (asteroids, comets, and meteoroids) that can impact planets, especially the Earth. To this end, he studied the geology of, and counted the density of, impacts on the Earth and the Moon, and looked for comets and asteroids through the telescope and calculated their orbits. As a team, he and Carolyn discovered 32 comets and 1125 asteroids, which is a record.

He realized that the geology of the lunar surface could be interpreted in terms of stratigraphy, that is in terms of sequence and correlation, one layer upon the other, and very early recognized the importance of impact and volcanism in surface processes. He began the systematic mapping of the Moon by telescope. By measuring the density of impact craters and making assumptions about impacting body fluxes, he established a method of determining relative age and approximate absolute age for the lunar surface. His methods of planetary geology are used today in studying other bodies in the Solar System.

Many honorary doctorates, awards, and medals deservedly came to Shoemaker, the most important being the US National Medal of Science (the highest scientific award in the USA). He died in 1997 at The Granites, Northern Territory, Australia. He leaves as a legacy a world-wide public whom he interested in planetary geology, and many scientists whom he either taught or otherwise influenced through his work.

ROBIN BRETT

shore platforms Shore platforms are prominent features
of rock coasts. Two main types have been distinguished. Gently sloping platforms (1–5°), which are sometimes called ramps, usually extend from the cliff base to below the low tidal level, without any major break of slope (Fig. 1a). Subhorizontal platforms, which can occur above, within, or below the intertidal zone, generally terminate abruptly in a low-tide cliff or ramp (Fig. 1b). Much of the literature has emphasized the occurrence of subhorizontal platforms in Australasia and Hawaii, and of sloping platforms in Britain, the north-eastern USA, and elsewhere in the northern Atlantic.

Most workers accept that mechanical wave action is the main erosive mechanism operating on the sloping platforms in the North Atlantic and other vigorous-wave environments. It has been argued, however, that, because of variations in wave intensity, and in the level at which they operate, subaerial weathering is essential for the formation of horizontal surfaces. In Australasia and other warm temperate and tropical-swell wave environments, horizontal platforms have therefore been attributed to cliff weathering or the modification of wave-cut ramps by weathering processes. There is no consensus about the way in which weathering produces horizontal platforms. Critical examination of the various theories that have been put forward over the past 100 years shows that most attribute a surprisingly major role to mechanical wave erosion, although this is rarely explicitly stated.

The geographical distribution of platform types has been generally thought to reflect differences in climate and wave regime. Most of the century-old debate on their origin, however, has been conducted in almost total ignorance of the fundamental role of tidal range. Although geological and other local factors may produce sloping and horizontal platforms along a coast, the former are most common in areas with high tidal range and the latter in areas with low tidal range. Along the storm-wave coast of eastern Canada, for example, horizontal platforms occur where the spring tidal range is between 2 and 2.5 m, and sloping platforms, with gradients up to about 5°, where the tidal range is almost 15 m. This suggests that horizontal platforms can be cut by waves where there is a small tidal range, but it does not rule out the possibility that they can also be produced by chemical, salt, or frost weathering, or by coastal ice. Nevertheless, any attempt to attribute a primary role to these mechanisms must explain how they are controlled by the tidal range.

Geological factors are responsible for marked variations in the shape of individual platform profiles, although over large areas their effect is often limited to providing local variations in platform morphology about regional means which are determined by tidal range and other aspects of the morphogenic environment. Geology influences the development and morphology of shore platforms in many ways:

1. The nature, intensity, and efficacy of the erosional mechanisms are governed by the structure, lithology, and mineralogy of the rocks.

(a)

(b)

Fig. 1. (a) Gently sloping (2.5–3.5°) platform in conglomerate sandstone, where the tidal range is very large. Bay of Fundy, Nova Scotia, Canada. (b) Horizontal platform where the tidal range is modest, in sedimentary rocks. Western Newfoundland, Canada.

2. Geological factors help to determine whether the foot of the cliff and the back of the platform are abraded or protected by accumulating debris and beach sediment.

3. The surface roughness of shore platforms in sedimentary rocks is strongly influenced by the dip and strike, bed thickness, joint density, and variations in the strength of the beds. Platform surfaces are usually coincident with the more resistant members of horizontal or gently dipping strata, but corrugated (washboard) relief often develops in steeply dipping rocks.

4. The age of coastal features, and the possibility that platforms are at least partly inherited from periods when sea level was similar to that of today, increase with the strength of the rocks.

Early models of platform development were descriptive and were usually structured within a cycle of erosion. More recently, mathematical modelling has suggested that platform width and gradient are maintained through a balance in the rates of erosion at the high and low tidal levels. This is consistent with the changes that occur in wave strength and erosion rates at the high tidal level in response to changes in platform width and gradient. Modelling has also supported the argument that the shape of platform profiles, and the close relationship that exists between platform gradient and tidal range, reflect the way in which wave energy is distributed between the high and low tidal levels. Simulated platforms attained equilibrium when erosion occurred at the same rate at all points along their profiles. This was accomplished by the platform gradient varying along a profile in such a way as to compensate for differences in the frequency of tidally controlled wave attack.

In many areas, erosional processes may be modifying platforms and other elements of rock coasts that were formed, at least in part, in the past. This hypothesis is consistent with the palaeo-sea-level record, which shows that interglacial sea levels were similar to those of today on a number of occasions during the Middle and Late Pleistocene. Inheritance is most likely to have occurred on tectonically stable coasts, and in resistant rocks where present rates of erosion are too low to account for the formation of wide shore platforms since the sea reached its present level. For example, on the southern coast of New South Wales, uranium–thorium (U/Th) dating of ferruginous and calcareous crusts and thermoluminescent dating of associated sediments show that a subhorizontal platform, which is awash at high tide, was formed during a period of higher sea level in the last interglacial stage. It was buried under soil during the last glacial stage, and then exhumed and partly modified by wave erosion in the Holocene.

There is less justification for assuming that inheritance has necessarily occurred in areas of weaker rocks. Even if the platforms were inherited from ancient surfaces in, or slightly above, the modern intertidal zone, till covers, raised beaches, structural remnants, and other evidence would probably have been removed by fairly rapid wave erosion at the present level of the sea. Relationships between platform morphology and aspects of the morphogenic environment are often lacking in resistant rocks. Their occurrence in weaker rocks, however, suggests that even if these platforms were partly inherited, they have been substantially modified by contemporary processes.
A. S. TRENHAILE

Further reading

Sunamura, T. (1992) *The geomorphology of rocky coasts.* John Wiley and Sons, Chichester.

Trenhaile, A. S. (1987) *The geomorphology of rock coasts.* Cavendish–Oxford University Press.

Trenhaile, A. S. (1997) *Coastal dynamics and form.* Cavendish–Oxford University Press.

shoreline (littoral) sedimentary deposits

The character of the shoreline and its associated shallow-marine deposits is controlled by the degree of exposure to waves and associated currents and the range and strength of the tides. A spectrum ranging from high-energy wave-dominated to low-energy tidally dominated shorelines, with many intermediate types, is found around the world's coastlines.

Shoreline sediments can be composed of siliciclastic gravels, sands, and muds, but where the supply of such sediment is limited, they may consist of skeletal debris or limestone (carbonate) muds, ooids, or peloids (grains resembling pellets), especially in warm tropical seas.

Exposed, wave dominated coastlines (Fig. 1a) are fringed by sand or gravel beaches and sand dunes. These features commonly form coastal barriers enclosing lagoons and marshes on the adjacent coastal plain, or the sediments abut directly against cliffs. Coarser sediment derived from river mouths, cliff erosion, or the adjacent sea floor is trapped close to the shoreline by waves and associated currents, and escapes seawards only during storm periods. Finer-grained sediments cannot remain in this high-energy environment and are winnowed out and dispersed seawards to accumulate in the near-shore zone, or are spread more generally over the shelf by currents.

These processes produce a seawards-fining belt of sediments parallel to the coastline. The near-shore sediment affected by normal waves, that is, sediment which is above fair-weather wave base, is usually composed of ripple cross-stratified fine and very fine sands with some layers of plane-bedded sands or hummocky cross-stratification. These sediments are often bioturbated by an infauna that lived in them. They pass landwards, with increasing grain size, into sands with larger-scale ripple cross-stratification and ultimately into cross-bedded sands and plane-bedded sands formed by breaking waves. Above low water mark a low tide terrace can develop in tidal seas where cross-stratified sands accumulate. This is succeeded by plane-bedded sands dipping gently seaward, which are produced by the swash and backwash of the waves on the beach face. In tidal seas additional wave-formed bars of cross-bedded sands form on the beach face. Above the normal limits of wave action, a flat terrace

(a)

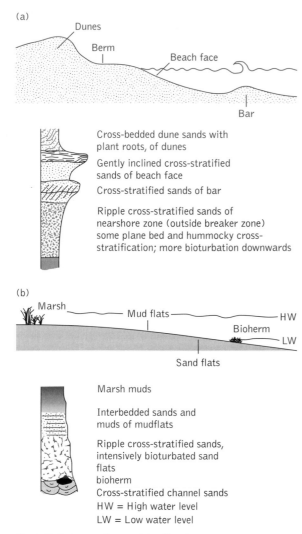

Cross-bedded dune sands with plant roots, of dunes

Gently inclined cross-stratified sands of beach face

Cross-stratified sands of bar

Ripple cross-stratified sands of nearshore zone (outside breaker zone) some plane bed and hummocky cross-stratification; more bioturbation downwards

(b)

Marsh muds

Interbedded sands and muds of mudflats

Ripple cross-stratified sands, intensively bioturbated sand flats
bioherm
Cross-stratified channel sands

HW = High water level
LW = Low water level

Fig. 1. Typical prograding coastal sedimentary sequences: (a) wave-dominated; (b) intertidal flats. Also shown are characteristic shoreline profiles for the two environments.

(*berm*) is underlain by flat-bedded or gently landwards-inclined sands and coarse concentrates of skeletal debris from off shore. Although some infaunal worms, molluscs, and crustaceans inhabit the sediments on such coastlines, physically produced sedimentary structures dominate. Wind transports sand from the berm to produce dune accumulations of steeply cross-stratified sand with some displaced marine fauna in addition to the roots of plants and the burrows and trails of terrestrial organisms.

On coastlines sheltered from strong wave action, particularly in areas of high tidal range, broad sandflats, mudflats, and marshes occur between the tidal limits (Fig. 1b). Tidal currents decrease in velocity, competency, and capacity as they advance over the intertidal zone and deposit a sequence of sediments which becomes finer from low to high water mark.

Ripple-stratified sands, mobilized by tidal currents and small waves, with thin lenses of mud (flaser bedding) deposited during quiescent periods, accumulate on the sandflats. These sands are colonized by a burrowing infauna of molluscs, worms, and crustaceans that disturb (bioturbate) the sediment. They are succeeded landwards by interstratified muds and sands of the mudflats (with wavy and lenticular bedding), which are less bioturbated and commonly have coatings of cyanobacteria that develop stromatolitic structures with abundant desiccation cracks produced during periods of prolonged exposure. The upper parts of the intertidal zone are colonized by salt-tolerant (halophytic) plants which trap sediment to produce muddy deposits with thin sand laminae deposited during storms (lenticular bedding). The sediments are intensely disturbed by plant roots and burrowing organisms. In arid areas, plants are replaced by evaporitic crusts. At the extreme tidal limits in wet areas, organic production may exceed sediment supply and peaty organic sediments may then accumulate.

Coastal progradation produces strand plains of upwards-coarsening sand-sheets on exposed coastlines, and upwards-fining sequences on sheltered coastlines (Fig. 1); both are commonly preserved in the geological record.　　G. EVANS

Further reading

Ginsburg, R. N. (ed.) (1975) *Tidal deposits: a case book of recent examples and fossil counterparts.* Springer-Verlag, Berlin.

Reinson, G. E. (1984) Barrier islands and associated strandplain systems. In Walker, R. G. (ed.). *Facies models* (2nd edn), pp. 119–40. Geoscience Canada.

silcrete *see* PALAEOSOLS, DURICRUSTS, CALCRETE, SILCRETE, GYPCRETE

Silurian The Silurian System is the third and youngest of the three Early (Lower) Palaeozoic systems; its formations came into existence over a time-span of about 25 million years, beginning at about 443 Ma. It was first mentioned by Sir Roderick Murchison in 1835, after several summers of fieldwork in Wales and the Welsh Borderland. He did not publish a detailed description until 1839, when it appeared in his monumental book *The Silurian System*, based on his discoveries there and elsewhere in Britain. Murchison was widely travelled and quick to prepare and publish his findings. *The Silurian System* is a remarkable account of the strata he had encountered and the system he now espoused. It is named after the ancient celtic tribe, the Silures, inhabitants of the Welsh Borderlands in Roman times. Originally it included rocks lower than those now in the basal Silurian series, which were held in dispute by Adam Sedgwick, Professor of Geology at the University of Cambridge, to be part of his Cambrian System. These two able geologists and formidable gentlemen had been good friends, but the dispute very effectively spoiled their relationship.

The controversy was resolved to general satisfaction by Charles Lapworth, who in 1879 allocated the lowest part of Murchison's Silurian and the uppermost part of Sedgwick's Cambrian to the (new) Ordovician System. The name was not officially recognized by the International Geological Congress until 1960, many European geologists having until then preferred Gotlandian. The Swedish island of Gotland was well known for the fossiliferous limestones it displays.

Modern usage divides the system into lower and upper halves, each with two series. The lower three series are divided into stages based upon basal boundary stratotypes that are well-exposed beds in Scotland, South Wales, and the Welsh Borderland. The youngest series (Přídolí), based upon a section in the Bohemian region of the Czech Republic, was added to the Murchison's Llandovery, Wenlock, and Ludlow Series in 1984. In Britain the Downton Series was by many authors regarded as the uppermost representative of the Silurian System, and has subsequently proved to be in part equivalent to the Přídolí. The base of the system is now drawn in the stratotype at Dob's Linn in Scotland at the base of the biozone of the graptolite *Parakidograptus acuminatus*. This

section is in the graptolitic facies, and additional reference sections in other characteristic facies are needed. They exist in North America, Western Europe, China, and Australia. Some 32 graptolite biozones have been recognized, while in the other facies conodonts, chitinozoa, and, to a lesser extent, brachiopods also are of value.

Rocks of the Silurian system occur on all continents, except (so far as is known) Antarctica. Major outcrops occur in North America, north-west Europe, North Africa, and central and eastern Asia. They include broad outcrops of carbonate rocks (mainly limestones) and some evaporites in the mid-west part of North America and the Russian platform, and also in China and Australia. Extensive belts of clastic sedimentary rocks and volcanics, some of the very great thickness, occur in the eastern and western parts of North America, western Europe, central Asia, many parts of China, and eastern Australia. The Silurian continents were distributed as a major agglomeration of ancient continental shields in the high southern latitudes—Gondwanaland—and a chain of separate shields (Laurentia, Baltica, Kazakhstania, and China) in the equatorial region, together with the large Siberian land mass (Angaraland) lying in the middle latitudes north of Baltica (Fig. 1). Plate motions were unceasing and significant, with active island arcs and subduction zones along several continental margins. This must have been particularly spectacular with the closure of the

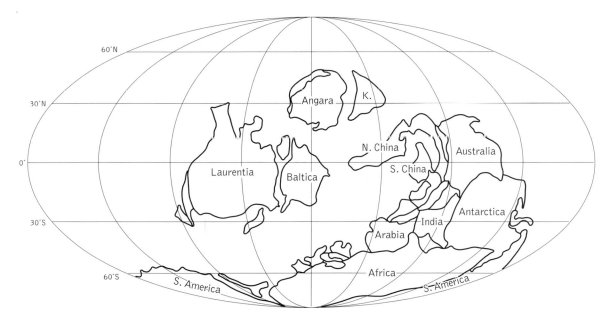

Fig. 1. The world in Middle Silurian time, about 420 Ma years ago. At this time most of the land was in the southern hemisphere. The converging continents of Laurentia (North America), Greenland and Baltica (north-west Europe), and Avalonia (Brittany, Newfoundland) gave rise to the new Laurasia. To the north lay Angaraland (Siberia) and Kazakhstania (K.), and to the east North China, South China, and northern parts of Gondwanaland. In the vicinity of the South Pole glacial ice spread over wide areas, now in South America and West Africa. (After McKerrow and Scotese (1990).)

Iapetus Ocean that took place at the collision and suturing of Laurentia with Baltica. It was the culmination of much volcanism and the growth of great wedges of deep-water sands, like submarine deltas on the margins of the two facing land masses. Similar active belts with much volcanic and clastic sediment accumulation also existed on the eastern margin of Australia, in China– Mongolia, and in Kazakhstan.

There is good evidence of former glacial activity in Gondwanaland, though it appears to have been less widespread than it was in the Ordovician period. World sea level rose significantly during early Silurian time, its later behaviour being locally obscured by tectonic activity. Nevertheless, a eustatic rise is thought to have persisted throughout Silurian time, and the level was generally high all the time. In the tropical waters that flooded the margins of the continents, carbonate deposition and the vigorous growth of coral– stromatoporoid–algal reefs in mid- and late Silurian ages were widespread. In some associated shallow coastal basins evaporite deposits include important masses of halite (rock salt).

Life in the sea witnessed a growing diversity of bottom-dwelling creatures, mostly brachiopods, bryozoans, trilobites, corals and echinoderms, with minor numbers of gastropods, bivalves and other animals. Depth-related 'communities' of Silurian brachiopods and other forms can be found in Britain. Generally these faunas were cosmopolitan throughout the tropical and temperate latitudes. There were, however, two geographically distinct realms of benthos. The southern, cool, Malvinokafric realm includes the distinct shelf communities with the brachiopod *Clarkeia*, from South America and west and southern Africa. In the Siberian–Mongolian region, a shelf fauna with the distinctive *Tuvaella* is also thought to represent a cool setting.

The calcareous algae, stromatoporoids, tabulate corals, and rugose corals became conspicuous reef-builders through-out the tropical seas. The plankton and nekton included graptoloids and many kinds of microfossil organisms and vertebrates. The latter included the primitive ostracoderms and the acanthodians, which were marine creatures early in the period, but which spread to fresh waters before it ended.

Land plants had made their appearance in the early Silurian, as the presence of fossil spores indicates, and vascular plant fossils are widespread from the Mid-Silurian onwards. Fossil soil deposits confirm that algae and, no doubt, other primitive plant life, were spread throughout the warmer latitudes. Scorpion-like eurypterids and other arthropods established themselves in fresh waters during late Silurian time, and might even have been capable of passing considerable time out of the water. All in all, the evidence shows that terrestrial life was making progress, with a variety of plants living close to the water or in areas of suitable climate.

D. L. DINELEY

Further reading

Bassett, M. G., Lane, P. D., and Edwards, D. (eds) (1991) *The Murchison Symposium: proceedings of an international conference on the Silurian System*. Palaeontological Association Special Paper No. 44.

Cocks, L. R. M. and McKerrow, W. S. (1978) Silurian. In McKerrow, W. S. (ed.) *The ecology of fossils*, pp. 93–124. Duckworth, London.

Holland, C. H. and Bassett, M. G. (eds) (1989) *A global standard for the Silurian System*. National Museum of Wales Geological Series 9, Cardiff.

McKerrow, W. S. and Scotese, C. R. (eds) (1990) *Palaeozoic palaeogeography and biogeography*. Geological Society of London Memoir No. 12.

silver deposits *see* GOLD, SILVER, AND COPPER DEPOSITS

sinkholes Sinkholes (French, *dolines*) are topographically closed depressions created by removal of materials downwards through their floors. Loss downwards differentiates them from other types of closed depressions, such as impact craters or deflation hollows that form when material is carried out over the enclosing rims. Four main processes create sinkholes: dissolution of soluble rocks such as salt, gypsum, limestone, and dolostone directly downwards from the surface; collapse of consolidated or partly consolidated rocks into underlying cavities formed by dissolution, compaction, piping, mining, etc.; piping of silts, sands, and gravels grain by grain into smaller cavities (analogous to emptying a sandhopper); and subsidence, the slower settling or sagging of strata into cavities without the large ruptures or total fragmentation that occur in collapses. The term 'suffosion' is also used for erosion by the slumping of unconsolidated sediments at the surface into solution cavity below.

The basic shapes of sinkholes are bowls, funnels, or cylindrical shafts, but all of these may be distorted by geological effects, such as elongation along a fracture or orientation down tilted strata. Dimensions range from less than a metre in diameter and depth for the smallest, fresh collapse or suffosion features, up to kilometres in length and hundreds of metres in depth for ancient dissolutional landforms. The largest dimensions and greatest variety of features are found on the soluble (karst) rocks, particularly limestones. Most large karst sinkholes are of composite origin, initiated by surficial dissolution but enhanced by periodic collapses of bedrock and the suffosion of any insoluble residuum. Blockage of their basal drains may create *pond dolines*. Lateral expansion of the rims often leads to amalgamation into *compound sinkholes* (*uvalas*, Slovene). Around the boundaries of karst outcrops, elongated *stream sinks* may form where streams flow in from other rocks. The outcrops themselves may come to be completely occupied by sinkholes, creating *polygonal karst*. Densities greater than 1000 sinkholes per square kilometre may be counted on very soluble rocks such as salt or gypsum. On thick limestones where the water-table is deep, large solutional sinkholes may be punched downwards to create 'eggbox topography' with sharp residual peaks scattered along the doline rims and a local relief of hundreds of metres. This is known as *fengcong* ('peak forest cluster') in southern China, where the greatest karstlands are found. They are the inspiration of much classical Chinese painting.

Sinkholes are a significant hazard. Thousands of collapse and suffosion features develop rapidly and quite naturally each year, destroying property. However, tens of thousands more are induced by human activity, chiefly mining, improper design of urban drainage, overpumping of aquifers that removes the buoyant support of water, or inadequate land filling that allows compaction cavities to form. Heavy trucks and buses, pitheads, high-rise buildings, and swimming pools have all foundered abruptly in recent decades, often with loss of life. Everywhere in the world sinkholes are looked upon as God-given sites for the convenient disposal of raw sewage and other refuse; as they are commonly the principal sources of recharge for aquifers as well, this practice ensures widespread water pollution, especially in the karst regions.

D. C. FORD

Further reading

Ford, D. C. and Williams, P. W. (1989) *Karst geomorphology and hydrology.* Chapman and Hall, London.

Waltham, A. C. (1989) *Ground subsidence.* Blackie, Glasgow.

slate
In popular usage the word 'slate' refers to any sort of natural roofing material, including those composed of thin-bedded, flaggy, sandstones and limestones such as the Cotswold Slates of Jurassic age that cover so many buildings in Oxford. However, to a geologist these 'slates' are simply *flagstones*, unmetamorphosed sedimentary rocks that can be split into sheets thin enough to be used for roofing buildings.

True slates do not necessarily split along bedding planes. Instead, most of them split along planes that are inclined at an angle, sometimes a very high angle, to the bedding planes. These planes, referred to as *slaty cleavage*, are caused by the rearrangement of minerals such as mica, chlorite, and clay within the rock. It is along these planes that the rock splits into the characteristic roofing slates that we know from so many parts of Europe. These rocks were affected by the Caledonian orogeny, in which fine-grained shales composed of mud and volcanic ash were transformed by low-grade metamorphism and pressure into the variously coloured grey, purple, and green patterned slates. The colour banding indicates the original bedding of the rocks. As the intensity of metamorphism increases, so slates are transformed first into phyllites and then into schists.

HAROLD G. READING

slope stability and landslides
All slopes are potentially unstable. Landslides occur in any rock type and represent a single dramatic event in the slow degradation of a slope by weathering and gravity forces. The tendency of a slope to move is described as the *instability* of the slope, whereas failure describes the actual mass movement, which may occur along well-defined planes with large catastrophic displacements or by slow, infinitesimal, grain-by-grain movement, as in creep. Failure planes may be shallow and restricted to near-surface layers or may be deep and displace large masses of soil and rock; the movement can be controlled by pre-existing geological zones of weakness or bands of different lithology.

Slopes can be either natural or man-made; they include hill and valley slopes, scree slopes, embankment and earth dam slopes, cuttings, the sides of deep excavations, and spoil heaps. All these slopes are likely to degrade very slowly over time but may be triggered into more dramatic failure by a variety of natural and man-made processes.

A landslide occurs when the shear stresses within the slope exceed the shear strength of the soil, either of which may have changed to cause failure. The stresses within a slope are generated by the weight of the material of which it is made, by stress release as overburden and side stresses are removed during slope formation, by the presence or movement of groundwater, and occasionally by seismic disturbance.

The stability of a slope is controlled by the characteristics of the material of which the slope is made, the slope gradient, and by the geological structure. Vertical cliffs may be quite stable if made up of massive rock; they may also be stable if the bedding of the rock and any fractures in it are vertical, horizontal, or dipping into the face. Fractures dipping out from the face will, however, cause instability and frequent rockfalls. The orientation of joints and bedding can frequently be seen to result in asymmetric road cuttings, in which the angle of slope depends on whether the inclined joints or bedding dip out into the cutting or back into the face. In contrast, some clay slopes are unstable at gentle angles of less than 10°, especially if past slippage has imparted discrete shear zones in the clay fabric, where the alignment of clay mineral platelets results in very low shear strengths.

Large rock failures tend to occur along bedding planes, joints, or a combination of the two, as in wedge failures. According to the orientation of the failure plane and geometry of the blocks, the blocks may slide or topple. Rotational slips, where the failure surface forms a circular arc or a non-circular curve in section, tend to occur in homogeneous soils; translational slips have a planar failure surface subparallel to the ground surface; and compound slips have both curved and planar portions of the failure surface. Both types tend to be controlled by the presence of layers of contrasting strength (Fig. 1).

The influence of water has proved to be the single most important factor in the assessment of slope instability and landslides. Water increases the weight of soil and rock; it will soften or weaken clays, and can widen fractures and joints upon freezing because of the increase in volume associated with the formation of ice. Heavy rains or rapid snow-melt cause a sharp rise in pore water pressures, leading to a decrease in effective stress and therefore a sudden decrease in

977

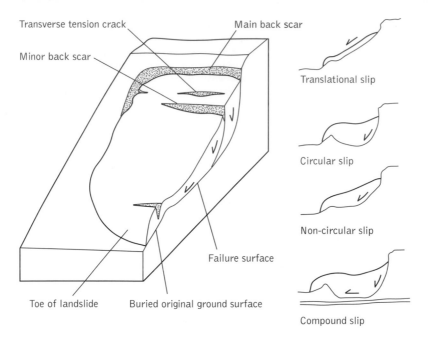

Fig. 1. Clay landslide morphology.

shear strength; consequently, many slope movements show a seasonal variation. The infiltration of water results in the leaching or alteration of soluble minerals; for example, the leaching of calcareous particles from clays can lead to a dramatic reduction in the shear strength across pre-existing shear surfaces as the interlocking, granular calcareous material is removed to leave a smooth microfabric of aligned clay mineral platelets. Infiltrating water may also lead to cation exchange in the clay minerals, with a consequent change in shear strength.

Unstable material moves downslope by a variety of mechanisms, ranging from very slow creep to the toppling of blocks or the rotation of large soil masses. *Solifluction* is the slow downslope movement of rock and soil debris under saturated conditions; the material may even move on very shallow slopes, resulting in a large accumulation of soliflucted debris, or head, at the valley bottom. The zone of movement is generally only to shallow depth, perhaps extending to 4 m, and most commonly represents the summer thawed surface layers in cold climates, where downward infiltration of water is prevented by permafrost. Solifluction deposits consequently mantle many slopes in active glacial areas or where periglacial conditions existed in the geological past.

A flow slide is the rapid movement of soil or rock particles where the material behaves as a liquid and can travel large distances over relatively flat ground. A flow slide may be caused by liquefaction. In liquefaction there is a total loss of strength in a material with an open structure after it has been disturbed and the particle fabric destroyed. The effective stress is then reduced to zero and all the load is transferred to the pore water. The flow ceases when sufficient drainage has occurred to recover the effective stress. This is especially likely to occur in 'quick' or sensitive clays, where collapse of an open clay mineral fabric allows liquefaction to take place which is maintained by the fine grain size and low permeability. Alternatively, a flow slide may be caused by the fluidization of particles. This occurs where a mass of rapidly moving soil and rock particles continually bounce off each other with no permanent grain contact; the whole mass then behaves as a fluid, even though there may be little or no water present.

Other causes of slope instability include the removal of the toe of a slope or loading the head of a slope, both of which may be the result of man-made and geological factors. Material at the toe of a slope increases the slope's natural resistance to movement, but this material may be removed by rivers, glaciers, or coastal erosion, or by the excavation of soil for road cuttings or building developments. Alternatively, the head of a slope may be surcharged by the addition of material, which may include lava and volcanic ash or the dumping of waste material or the construction of buildings or roads. Vibrations caused by earthquakes, explosions, or rail or road traffic can also temporarily increase stresses within slopes, resulting in landslides.

Clay is generally the most unstable slope-forming material, and its stability may be time-dependent. Vertical slopes can be cut in a clay which will remain stable in the short term by virtue of the internal friction between the clay particles, their cohesion, and water suction. With time, however, drainage occurs, resulting in a reduction of water suction and cohesion, degradation, and eventual collapse. Exposure at the surface results in physical and chemical weathering and in stress release, with the formation of stress-relaxation joints and fissures which allow the further infiltration of water.

Mass failure of the slope in a rotational or translational slide will result in the formation of a discrete shear zone where the moving mass has passed over undisturbed strata beneath. In clay slopes, the clay mineral platelets realign themselves parallel to the direction of movement, giving a very low angle of internal friction between the grains, and the slope is said to be at residual strength. Such slopes are particularly susceptible to reactivation, especially by removal of the toe, head loading, or elevated pore-water pressures within the underlying material. Shallow shear surfaces are formed, especially under periglacial conditions, owing to the mobilization of the saturated upper layers.

The tendency of a slope to fail is quantified by calculating a factor of safety, which is defined as the ratio of the forces resisting movement to those promoting movement. Thus, if the factor of safety is less than one, the slope is unstable and likely to move. There are several methods of finding the factor of safety by considering the slope along a two-dimensional section and dividing the slope into sections or slices to assist in calculation. The choice of method depends on the morphology of the slope, the time available, and the effort and expense that can be afforded in calculation. The accuracy obtained will depend on the certainty of material parameters such as the shear strength or weight of material and knowledge of the slope morphology. New slopes may thus be designed with a factor of safety much greater than one to allow for uncertainties and assumptions in the analysis.

The simplest forms of analyses consider the slope to be of infinite length; they are therefore applicable where the length of the slipped mass is considerably more than its depth, as in long translational slides. Non-circular slips are mainly analysed by examining and summing the forces acting on the shear plane for each slice in a cross-section of the slide. When the slip surface can be approximated to a circle in section, this method can be modified by establishing moments of equilibria about a centre point of rotation. The ratio of the forces or moments resisting failure to those promoting failure, summed for all the slices, gives the factor of safety. The effectiveness of any one method depends mainly on the assumptions made about forces between slices; the more rigorous methods use both forces and moments in their solution. Some of the methods are iterative and entail repetitive calculations to find the lowest factor of safety; such methods are therefore particularly amenable to solution by computer.

The selection of appropriate parameters from a comprehensive site investigation detailing the geology, topography, and groundwater distribution is critical to a successful stability analysis. Trial pits may be dug to search for existing failure surfaces, and boreholes may be used to establish the geological succession and the groundwater regime. Samples may be removed for the measurement of properties such as the density and shear strength of materials, and the results may be included in a slope-stability analysis. Slopes may be monitored over a period of time to assess movement rates, groundwater influence, or seasonal variation. Such monitoring may call for detailed surveying against fixed reference points, the installation of inclinometers in boreholes within the slipped mass, or crack-dilation measurements across rock fractures.

If a natural hill or valley slope is to be excavated or built upon, it is imperative to discover whether or not the slope has already failed in the past. It can then be decided whether peak strength or residual strength parameters are appropriate. Evidence of recent slippage may be provided by fresh topographic expression, highly disturbed topsoil, tension cracks, fresh back scars and toe bulges (Fig. 1). Many older slips initiated during the Quaternary may not be so evident, and their current topographical expression may be masked by erosion or agricultural activities. Underlying failure surfaces will, however, remain; they will be clearly evident, particularly in cohesive soils, and will be identifiable in a trial pit as the soil is excavated slowly either by hand or with an mechanical excavator.

Localized, unfavourable slope conditions may nearly always be accommodated by engineering design, limited only by cost, but larger slopes cannot usually be held successfully. Soil slopes may generally be stabilized, either by modifying the slope profile or by draining the slope material. The slope profile may be modified by reducing the gradient or by cutting berm ledges to redistribute the stresses; these ledges will also serve to catch small failures. Alternatively, the addition of load to the toe of a landslide can be effective, particularly in circular slips; it can take the form of an earth bank, rock fill, or a concrete block.

Drainage of the slope is very effective and economical because of the overriding influence of pore-water pressures in promoting instability. Water may be prevented from infiltrating the ground by the installation of surface water drains to divert run-off away from the back scar area and to drain surface layers within the slipped mass. Deep drains are very effective; they take the form of mined adits or sub-horizontal boreholes gently inclined to drain groundwater out from the slide. In addition, drainage measures installed behind the slope can reduce the groundwater flow from the ground behind into the slide at depth.

Rock slopes can be hazardous because of the periodic fall of loose blocks and debris, but these can be caught in rock-trap ditches at the base of the slope or fences on the slope. Alternatively, geotextile or wire mesh, bolted to the surface, can be deployed to cover the loose debris, catch small stone falls, and provide a stable surface for plants. Slopes in weak rock can also be stabilized by concrete sprayed over a reinforcing mesh that has been bolted to the face, or by the construction of concrete walls with drainage holes to prevent the build-up of pore-water pressure behind (Fig. 2).

Rock slopes can generally be stabilized by modifying the slope profile or by anchoring and supporting the existing

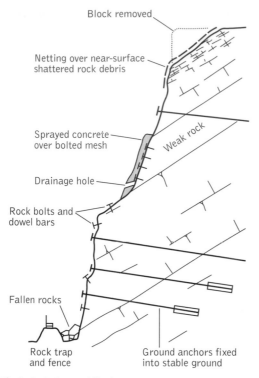

Fig. 2. Rock slope stabilization.

Labels on figure:
- Block removed
- Netting over near-surface shattered rock debris
- Weak rock
- Sprayed concrete over bolted mesh
- Drainage hole
- Rock bolts and dowel bars
- Fallen rocks
- Rock trap and fence
- Ground anchors fixed into stable ground

profile. The slope can be modified by the construction of berm ledges to redistribute stresses and catch minor rockfalls, or by removing hanging blocks and wedges. Individual unstable blocks may be secured by rock bolts, which are short steel bars passed through the block and fixed into the rock mass behind. They are put under tension to increase the stress across individual joints. Dowel bars are similar steel bars installed across joints to provide resistance to shear, but are not pre-tensioned.

Rock and soil slopes may be stabilized by more substantial ground anchors constructed to oppose the driving force of the landslide. The anchors consist of multiple steel cables up to 40 m long, installed in inclined boreholes through the landslide material into stable ground beneath. They are bonded in resin at depth and the applied tension opposes the driving force of the landslide and closes rock fractures.

CHRISTOPHER McDONALD

slopes *see* CLIFFS AND ROCK SLOPES

Smith, William (1769–1839) The 'Father of English Geology', William Smith, was born in 1769 at Churchill, Oxfordshire, and received a village school education. Apprenticed to a surveyor, he soon mastered the business. He was

interested in the local rocks, and in 1791 set up for himself in the Somerset coalfield, where he was engaged in underground mine surveys and calculations of coal resources. He was soon surveying for a proposed canal and this gave him a further chance to note the disposition of the local strata. He collected fossils and marked the facies changes of formations as they were traced across country. By 1796 he had conceived the idea of identifying the rock units by the fossils they contain, irrespective of lithology. In nearby Bath he discussed his ideas with friends and studied the soil maps that accompanied some of the County Agricultural Reports. By about 1793, he was probably making similar maps to indicate rock outcrops. He coloured each formation distinctively and later gave a list of the characteristic fossils of each. The succession around Bath was published in 1801. Meanwhile he was travelling incessantly around the country as a surveyor, and took every opportunity to record the rocks he saw.

By 1801 he had completed a small geological map of England and Wales, which shows seven major rock units. During the following decade he made several other maps and increased the number of mapped formations. A map-publisher friend undertook to produce Smith's *magnum opus*, the epoch-making, *A delineation of the strata of England and Wales with part of Scotland …* (1815). The scale was 5 miles to the inch, the topography newly improved, and some 20 or more formations were listed. A short memoir came with the map, and also a table of formations found in the various counties. Then came his *Strata identified by organic remains* (1810–19), with engravings of the fossils. He also produced a series of sections, showing his interpretation of the geological structure.

In 1819 he was at work again in northern England, eventually settling in Scarborough among congenial clients and friends. In 1831 the Geological Society of London awarded its first Wollaston Medal (still its most prestigious gift) to William Smith. He continued in semi-retirement in Yorkshire. An essentially practical geologist, he had virtually invented the geological map and set stratigraphy upon a course that still runs. He died in 1839.

D. L. DINELEY

soft-sediment deformation Soft-sediment deformation includes an enormous range of structures formed by the movement of masses of unlithified sediment relatively close to the surface of sedimentation. The resulting structures may resemble not only those formed by tectonic folding and faulting, but also those formed by volcanic activity, e.g. dykes, sills, and volcanoes (see *clastic dykes* and *mud and sand volcanoes*).

Both lateral and vertical movement of sediment may be involved, at a great variety of scales. Some structures clearly formed by lateral downslope movement at the surface. For example, slump-folds, bounded by undisturbed sediments, and ranging from less than a metre to hundreds of metres in thickness, can occur in any environment from river bank margins to continental margin slopes. In ancient sediments slump-fold axes can be measured to determine the palaeoslope. Syn-sedimentary faults, ranging from centimetre-sized fractures to major slump scars, may also be used as palaeo-slope indicators.

Other soft-sediment deformation structures are associated with the escape of water (dewatering) and sediment loading (i.e. a relatively dense bed sinking into a less dense bed) or escape of pore water from below due primarily to lique-faction. These structures include dish and pillar structures and convolute bedding.   HAROLD G. READING

soil classification and description in engineering geology

In civil engineering the term 'soil' is used for any cemented or weakly cemented accumulation of detrital mineral particles and rock fragments formed by the erosion and weathering of rock. It can generally be excavated easily and has very different properties from the underlying rock head (the surface between the bedrock and unconsolidated material). A soil may typically contain some organic matter, especially in near-surface layers, and may also have void spaces between the particles which will contain water, or air, or both.

Soils are formed by many processes. Physical weathering processes include the action of wind, water, and glaciers, or disintegration by alternate freezing and thawing; chemical weathering processes include the leaching of soluble materials and the accumulation of insoluble residues; other processes include the incorporation of organic matter; the downward movement of fine particles; and disturbance by root penetration, animal burrowing, and desiccation.

If the products of weathering remain at their original location, they constitute a *residual soil*; alternatively, if the products are transported by gravity, wind, water, or glaciers and are deposited in a different location, they constitute a *transported soil*. Soil may be termed after their mode of transport and deposition; for example, *drift* is a glacial or fluvioglacial deposit left after the retreat of glaciers and can vary considerably in composition and thickness, *colluvium* is slope debris that has moved downslope, largely by gravity alone, and includes head and scree.

The soil profile, shape, size and arrangement of the particles, soil strength, water content, the degree of cementation, the chemical composition, and potential changes in composition are frequently recorded by the engineering geologist as they present a range of problems when considering the design of foundations, the settlement of buildings, the stability of slopes, or the quarrying of material.

Soil description must be distinguished from soil classification. The description indicates the characteristics of both the soil material and the soil mass *in situ*. Such a description should include reference to the particle size distribution of the soil, plasticity, colour, texture, and mineral composition. Description of the soil mass is best done in the field or on undisturbed samples and would refer to the soil strength *in situ* with any details of bedding, weathering, and discontinuities. Soil description is particularly important when the soil is to be used in an undisturbed condition, perhaps for supporting foundations. Soil classification is the allocation of a soil to one of a limited number of soil groups using parameters that are independent of its condition *in situ* and provides a common language for the exchange of information. Soil classification is therefore more useful when the soil is to be disturbed prior to use, perhaps in a dam or embankment, resulting in changes in its water content, fabric, and texture.

Soils having properties that are influenced mainly by clay- and silt-sized particles are referred to as fine-grained soils. Those with properties that are influenced mainly by sand- and gravel-sized particles are referred to as coarse-grained soils.

Engineering soils are characteristically divided into cohesive and non-cohesive types. Cohesion between particles is caused by the presence of clay minerals which, even when they are present in relatively small quantities, usually exert a considerable influence on the properties of a soil, an influence which is sometimes disproportionate to their percentage by weight. In general terms, a soil is considered to be cohesive if the particles adhere after the soil is wetted and subsequently dried, and if significant force is then required to crumble the soil: this definition excludes those soils in which particles adhere when wet due to surface tension.

A convenient way of separating the cohesive and non-cohesive soils is the plastic limit test introduced by Atterberg in 1911. Very simple index tests have been shown to be invaluable in indicating both the proportions of cohesive/non-cohesive materials and the nature of the clay minerals. According to its water content, a soil may exist in the liquid, plastic, semi-solid, or solid state. For a soil to exist in the plastic state, the water content must be sufficient for the particles to slide freely past each other but the magnitude of the net interparticle forces must be enough for cohesion or an overall attraction to be maintained between them. Thus, the plastic limit is defined as the minimum moisture content at which a soil may be rolled by hand on a glass plate into a cylinder 3 mm in diameter; non-cohesive soils will crumble before reaching this stage.

Most fine-grained soils exist naturally in the plastic state. The upper and lower water contents at which plastic behaviour is observed are the liquid limit (WL) and plastic limit (WP), respectively, and these limits are widely used in soil classification. The liquid limit is also determined by arbitrary but standardized test procedures. The liquid limit is defined as the minimum moisture content of the soil at which it will flow under its own weight. It is assessed by dropping a standardized steel cone into a creamy paste of the clay sample; the moisture content at which the cone penetrates the sample by 20 mm is defined as the liquid limit. Similarly, the shrinkage limit is the water content at which further loss of water from the soil will not cause a further reduction in volume.

The plasticity index is defined as $WL - WP$ and represents the range of moisture contents over which the soil behaves in a plastic manner. Soils with a high plasticity index tend to have a high clay fraction and be prone to swelling on the uptake of water. Conversely, soils with a low plasticity index may be susceptible to heave owing to the formation of ice beneath the surface.

Non-cohesive soils include silty, sandy, and gravelly soils in which there is no cohesion except that derived from any minor amount of clay matrix or from water tension between particles. The strength of these soils is derived from any partial cementation and the internal friction provided by the interlocking nature of the grains. Locked sands are a particular type of non-cohesive soil in which the particles have undergone pressure solution at their point contacts, giving a highly interlocking grain structure and a high strength until they are disturbed and the structure is lost.

The properties of non-cohesive soils can be assessed *in situ* by the standard penetration test (SPT). The test is performed at the required depth in a site investigation borehole by driving a 50 mm split tube sampler 150 mm into the soil. A 65-kg hammer is then used to drive the sampler a further 300 mm and the number of blows required is referred to as the standard penetration resistance (N). The test is very simple and is widely carried out in coarse soils as the value of

N relates to the properties of the soil, particularly the density and bearing capacity. It must, however, be used with caution in clayey soils. The split barrel enables samples of the soil under test to be recovered, but in very coarse soils a conical shoe may be fitted to the sampler without affecting the results obtained.

Another test frequently undertaken *in situ* on non-cohesive soils is the cone penetration test (CPT). This entails driving a conical probe into the soil at 20 mm s^{-1} followed by a concentric outer sleeve. The ratio of the resistance against the sleeve to the resistance against the cone provides information on soil type and packing arrangement and is used to estimate bearing capacity and settlement.

Soils are classified according to grain size and index limits, and standard terminology is used for effective communication between engineering geologists and other engineers (Fig. 1). If the plasticity index and liquid limit have been determined in the laboratory, then the soil is represented by a point on the chart. Soils plotting below the A-line are silts (M) and those above the A-line are clays (C). In turn, the plasticity of the soil may be described as low, intermediate, high, very high, or extremely high by reference to the liquid limit.

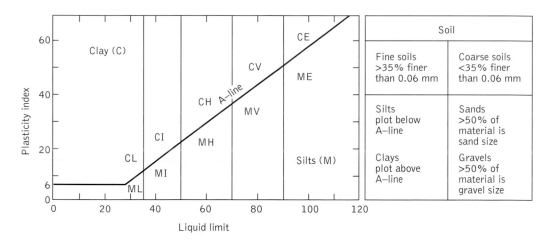

Fig. 1 Classification of engineering soils. Clays (C) plot above the A-line; silts (M) below it. The plasticity of the soils is denoted by the second letter: L, low; I, intermediate; H, high; V, very high; E, extremely high.

The boundary between fine soils and coarse soils is placed at the point where 35 per cent of material is finer than 0.06 mm. The choice of this dividing point reflects the overriding engineering importance of the finer grains and the particular properties of the clay minerals.

Soils are described according to the dominant particle-size fraction, which may be qualified by the presence of other size fractions or may give an indication of the particle-size distribution, e.g. a gravelly sand or a sandy silty clay. Additional information may also be given about the presence of cobbles and boulders or organic matter. Well-graded soils have no excess of particles in any one size range. Conversely, poorly graded soils either have a high proportion of particles in one size range in a uniform soil, or have large and small particles present with a low proportion of intermediate sizes in a gap-graded soil.   CHRISTOPHER McDONALD

soil development To produce the three-dimensional organization of a well-developed soil requires the relatively consistent application of a complex suite of physical, chemical, and biological processes (see *soils*). The fact that basic sets of soils all over the world contain recognizable horizons implies that some processes are common to the development of all soils. As soils are open systems, these processes can be grouped into those that result in inputs, outputs, transfers, and transformations. The main inputs or additions to the upper parts of soils are organic matter and the contained elements from surface vegetation and water from precipitation; this water can also contain dissolved elements. There may be an additional input of particulate matter to the soil by wind and water action and by mass movement processes, such as soil creep. Energy input is provided by incoming solar radiation, and the combined temperature and moisture status determines the nature and

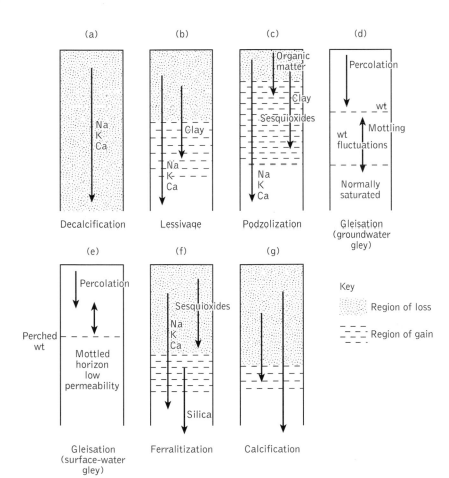

Fig. 1. Some of the more important soil-forming processes (see text for details). Ca, calcium; K, potassium; N, nitrogen; wt, water-table.

983

rate of operation of most soil processes. Weathering of bedrock can also be considered as an addition or input, although some workers regard it mostly as a transformation. Outputs from the soil system include the downslope movement of soil particles and water, throughflow, deep percolation, leaching, and the uptake of water and nutrients by roots. Leaching is the downward movement of water through the soil which results in the removal of water-soluble minerals.

Soil development is mainly the result of transfers and transformations. Transformations include the great range of organic compounds that form during the decomposition of organic matter and the weathering of primary and secondary minerals. Transformation of energy can also occur. Transfers are probably the most important group of processes; they include the redistribution of energy and the internal reorganization of matter. There is the translocation of iron, clay, humus, and hydrated ions; ion exchange; and the diffusion of gases. *Eluviation* is the movement of soil material through the soil zone, resulting in depletion in certain horizons and accumulation in others. *Illuviation* is the precipitation or accumulation of material leached from upper horizons within what is known as the B horizon (which is usually in the middle of the soil sequence). Material can also move up the soil profile, for example by capillary rise of dissolved ions and mixing by soil fauna. In areas subject to freeze–thaw activity, frost heave (cryoturbation) will create a similar effect.

As a soil develops, distinctive horizons become apparent. These are essentially created by processes of transfer and transformation, as can be illustrated by examining a number of specific examples. *Decalcification* (Fig. 1a) is the result of the leaching of calcium, but the process may be balanced by the release of calcium by the weathering of bedrock. If the production of nutrients by the weathering of bedrock is small, and if the rate at which cations are replaced from vegetable litter is poor, the effect of leaching may not be balanced. Under such conditions the stability of soil aggregates will decrease and clays will break down and deflocculate. *Lessivage* (Fig. 1b) then occurs; this is the movement of silicate clays in colloidal suspension within the soil profile without any change in chemical composition. The soils that are then produced are known as acid brown earths, *sols lessivés*, or luvisols in the FAO–UNESCO classification (see *soils*).

Podzolization (Fig. 1c) occurs in freely drained soils on siliceous parent materials, usually under a cool, temperate climate where the vegetation provides a litter deficient in bases that promotes the formation of raw acid organic horizons. Under such conditions base cations such as Na^+ and Ca^+ are leached rapidly and are not replaced by other processes; in consequence, clay minerals break down and hydrous oxides of iron and aluminium are mobilized and removed from the upper horizons. This results in the bleached eluvial horizon characteristic of podzols. Some of this mobilized material accumulates lower down in the profile to produce a diagnostic horizon enriched in aluminium and iron. The iron can become indurated to form an iron pan.

Waterlogging in the soil, which can be the result of a perched water table or high groundwater levels, will alter the balance between oxidative and reducing conditions (Fig. 1d, e), creating a mottled effect in the zone where the water table fluctuates. The mottling is created by the alternation of ferrous and ferric (oxidized) iron compounds. In the humid tropics, where intense weathering is occasioned by high temperatures and high water availability, the process of *ferralitization* produces a red soil as a result of the dehydration of iron compounds (Fig. 1f); this reddenning is known as *rubefaction*. During this process, iron and aluminium compounds are removed from upper horizons and redeposited lower down in the soil profile.

In semi-arid environments the amount of precipitation may be insufficient to produce complete leaching of solutes. Sodium and potassium may be removed but calcium is often precipitated as calcium carbonate, a process known as *calcification* (Fig. 1g). Under more arid conditions only the more mobile ions of sodium and potassium are dissolved. These, however, are soon precipitated and may also be drawn upwards by intense surface evaporation to be precipitated near the surface. This is the process of *salinization*, which produces a group of soils known as *solonchaks*.

Soil development is not necessarily a gradual sequential process. A change of climate or of vegetation cover will affect the processes and may lead to a complex soil with one type superimposed on another. Excessive erosion of soil and excessive burying (known as retardant upbuilding) may lead to a regression of soil development. Soil is a complex entity and its development is consequently complicated.

JOHN GERRARD

Further reading

FitzPatrick, E. A. (1980) *Soils: their formation, classification and distribution.* Longman, Harlow.

Ross, S. (1989) *Soil processes: a systematic approach.* Routledge, London.

soil erosion Soil erosion, the loss of soil from the surface of the Earth, is principally due to the action of water, wind, and mass movement, alone or sometimes in combination. Other agents that contribute to erosion are animals and humans, for example, in footpath erosion.

Erosion is a natural process generally occurring at low rates. Under woodland or grassland, rates are less than one tonne per hectare per year. Human-induced or accelerated erosion may be a serious environmental problem which threatens the ability of communities to grow crops or graze animals. It may also damage property or pollute water some distance from the site of the erosion (Fig. 1). The most famous example is the Dust Bowl of the American Great Plains in the 1930s, particularly between 1933 and 1938. These were drought years; but ploughing of grazing land for wheat farming led to severe wind erosion across a huge area and the migration westwards of hungry and dispossessed farmers (as depicted in Steinbeck's *The grapes of wrath*). In

Fig. 1. Severe erosion on a field drilled with winter wheat, Rottingdean, South Downs, England, October 1987. Many gullies cut down to the chalk beneath a silty stony rendzina soil about 200 mm thick. The valleys in this region are normally dry. Fans of chalk and flint stones can be seen in the valley bottom. Erosion occurred as a result of a storm of about 60 mm in 12 hours. Rates of erosion on the field were about 200 tonnes per hectare over a field of 12 hectares. Downvalley a housing estate was flooded with soil-laden water and about £500 000 worth of damage occurred. (Photograph by John Boardman.)

Amarillo, Texas, during one month at the height of the period there were 23 days within each of which there were at least 10 hours when windblown soil was in the air, and one in five storms gave rise to zero visibility.

Blaikie has argued that the explanation for soil erosion is usually to be found in social and political relations in society, especially factors concerning land ownership. Such factors explain why unsuitable land is farmed and management decisions are taken which increase the risk of erosion occurring. Physical factors such as slope, soil type, vegetation cover, and the amount, timing, and intensity of rainfall and wind influence how much erosion occurs.

Soil erosion is a potentially serious problem because rates of soil formation are low, depending as they do on the weathering of rocks, the deposition of sediment, and the accumulation organic material from animal and vegetable life. Rates of formation are estimated to be less than one tonne per hectare per year (t ha^{-1} year^{-1}). Rates of soil erosion can locally exceed 100 t ha^{-1} year^{-1}; average rates are much lower, yet still frequently exceed rates of soil formation. Although rates of erosion are commonly quoted for various countries, there are problems with such figures. Many are obtained from small-plot experiments which poorly replicate field conditions. Others are based on the sampling of sediment carried away by rivers; this tells us little about amounts of erosion on fields within the river basin.

Soil erosion by water appears to be more extensive than that by wind. The characteristic forms of water erosion are rills and gullies. Rills are small, temporary channels that can be removed by cultivation, for example, by a plough. Gullies are more substantial and, unless filled in, will become permanent features of the landscape. In semi-arid landscapes gullies may be large: in the American south-west they are known as arroyos. Their development has been attributed to landscape destabilization resulting from the introduction of cattle in large numbers in the late nineteenth century. Climatic change or wetter periods have also been implicated. Landscapes which have suffered extreme dissection by gullies are known as badlands. Examples occur in Cappadocia (Turkey), Alberta (Canada), Swaziland, and southern Spain. These forms have probably developed over thousands of years.

The impact of raindrops on bare soil causes splash erosion. This is important since it disaggregates soil particles and allows them to be carried away by sheet flow (interrill erosion). Splash itself moves soil particles only short distances.

Wind and water erosion remove the most valuable part of the soil, the organic-rich upper horizon. This leads to reduction in its water-holding capacity and in the stability of soil aggregates. Fine-grained soil particles (clay and silt) and organic matter are preferentially removed by run-off. Soil nutrients are commonly attached to the finest particles, and their removal thus reduces soil fertility. Over long periods of time soils then become coarser and in some cases more stony. Erosion consequently results in thinning of the soil, and many experiments have shown that loss of soil leads to lower crop yields. This can be compensated by addition of fertilizers; these, however, may be costly, and excessive use of them can damage freshwater habitats and aquifers. Crops may be damaged by the abrasion of moving particles carried by the wind or buried by deposits of soil.

Overgrazing by animals is a common cause of erosion. Semi-arid areas are especially at risk because of seasonal or periodic drought. The loss of vegetation cover exposes the land to erosional processes. Overgrazing by goats and sheep has led to erosion in the Mediterranean basin both in the past and at the present day. Erosion is therefore an important component of desertification, which is defined as 'land degradation in arid, semi-arid, and dry sub-humid areas resulting mainly from adverse human impact'. Adverse human impact on vegetation is also included in this concept.

The impact of mass movements (landslides, debris flows, etc.) is difficult to estimate because they tend to occur irregularly. They are of particular importance in upland or montane areas, such as parts of China, the Himalayas, and Tanzania. The extent to which such occurrences are due to human action as opposed to natural forces is in dispute. In the Himalayas there is strong evidence that rates of erosion are high because of steep slopes and high rainfall rather than farming and deforestation.

Soil erosion is not exclusively a modern phenomenon. In many societies population pressure has led to the cultivation of unsuitable land and erosion. In Greece, forest clearance and population increases led to severe soil erosion in the period 4000 to 3000 BC. On the southern chalklands of Britain, Neolithic and Bronze Age clearance of wooded landscapes and the introduction of arable farming gave rise to substantial erosion.

Soil erosion affects the quality of the soil and therefore the success of farming: these are known as on-farm impacts. From the point of view of short-term economics, the off-farm impacts appear to be much more serious. These include the pollution of freshwater habitats by soil and agricultural chemicals; flooding and damage to property by soil-laden water; and the silting up of dams, reservoirs, and river courses.

The distribution and the rate of soil erosion are predicted by compiling erosion hazard maps and by mathematical approaches, most commonly the Universal Soil Loss Equation (USLE):

$$\text{Soil loss} = R \times K \times LS \times C \times P,$$

where R is an index of rainfall erosivity, K is an index of soil erodibility, LS is an index of slope length and steepness, C is an index of crop cover, and P is an index of conservation measures.

The terms in the USLE may be quantified and an estimate of soil loss obtained for a specific situation. Although the USLE is useful for identifying the factors that control erosion, it is poorly tested and there is little evidence that it gives reliable estimates of erosion on field-size areas. New models for estimating soil loss are being developed: these are based on detailed simulation of the processes of erosion.

Although understanding and predicting erosion is often imperfect, many societies have developed methods of preventing soil loss, such as the terracing of slopes. If well maintained, terraces are effective but they are costly to construct and constrain mechanized agricultural operations. Soil conservation techniques are difficult to introduce to societies where the main issue is producing the next harvest.

JOHN BOARDMAN

Further reading

Bell, M. and Boardman, J. (1992) *Past and present soil erosion.* Oxbow Books, Oxford.

Blaikie, P. (1985) *The political economy of soil erosion in developing countries.* Longman, Harlow.

Morgan, R. P. C. (1986) *Soil erosion and conservation.* Longman, Harlow.

Pimental, D. (ed.) (1993) *World soil erosion and conservation.* Cambridge University Press.

soil fabric and structure

The term 'soil fabric' describes the arrangement, shape, size, and size distribution of the mineral constituents in a soil. On a small scale, the soil microfabric can be studied by looking at undisturbed samples under a scanning electron microscope or in thin section. On a large scale, the soil macrofabric can influence the engineering behaviour of the soil and should be noted in a soil description; for example, thin lenses of impermeable clay in a permeable sand would be significant, as would the inclusion of organic material, the presence of fissures or root holes within a clay, or the preferential growth of secondary minerals on soil laminations. A macrofabric may not be evident in a small sample taken for laboratory testing and caution must therefore be exercised when interpreting test results (in permeability testing, for example).

In non-cohesive soils, i.e. soils without a significant content of clay minerals, the principal characteristics of the fabric are the particle size distribution, particle shape, particle orientation, and the packing arrangement. For example, soils with well-rounded, equal grains have a more open structure than similar soils with angular grains, and soils with a range of particle sizes may be more dense because smaller grains fit between bigger ones. A *grain-supported fabric* describes a soil where the larger particles touch, with finer sediment filling the spaces between; conversely where there is enough finer sediment for the larger particles to be separated, the fabric is described as *matrix supported*. *Locked sand* is a term used to describe sands that have undergone pressure solution and have interpenetrative point contacts, giving a much stronger fabric than might otherwise be expected, although lost irreversibly upon disturbance.

In cohesive soils, the particular properties of clay minerals and the forces of repulsion and attraction acting between adjacent clay particles have important influences on the soil fabric. Clay mineral platelets carry a residual negative charge and attract cations present in the pore water to form a cloud around the particle, in what is termed the double layer. The cations are not held strongly and the equilibrium may be readily upset if the nature of the water changes and other cations are introduced, since some cations have a greater affinity for the layer than others. Those with a greater affinity tend to replace those with less, by a process called *cation exchange*, resulting in a change in the thickness of the double layer.

Dispersed clay mineral structure

Flocculated clay mineral structure

Natural clay deposit structure

Fig. 1. Clay structures.

Attraction between adjacent clay particles is due to short-range van der Waals forces, which decrease rapidly with increasing distance between particles. On the other hand, repulsion occurs between the like charges of the double layers. If the layer of cations is thin and the attractive van der Waals forces dominate over the repulsive forces between cations, then the particles will orientate themselves with edge-to-face orientation (positive to negative). This is termed *flocculation* and the soil is said to exhibit a flocculant structure. In a marine depositional environment where there is a high concentration of cations, adsorbed layers are thin and the clay minerals tend to settle out of suspension with this structure. This is in contrast to lacustrine clays deposited in a freshwater environment which settle in a face-to-face orientation because of a net repulsion. This is termed a *dispersed structure*.

Natural clays invariably contain a mixture of various types of clay minerals and larger particles of more inert minerals such as quartz, leading to very complex structural arrangements. Individual platelets generally form aggregations of particles with a face-to-face orientation which combine with other aggregations in a structure determined by the depositional environment (Fig. 1). The property of cation exchange in clays is used in a common agricultural process whereby the dispersed mineral fabric of a heavy clay soil is flocculated by the additional of calcium (lime) to improve the workability. The clay minerals take up the calcium in preference to the existing sodium, and the rearrangement of the particles creates a more pervious, open structure.

The engineering significance of a soil fabric can be assessed by measuring the *sensitivity* of a soil. This is the ratio of the strength of a soil *in situ* to the strength of a remoulded sample at the same moisture content. Many clays are relatively insensitive to remoulding, with a sensitivity that is normally less than 4, but some clays have a sensitivity exceeding 8 and are termed 'quick clays'. Some marine clays that have had all the salt removed from the pore spaces by percolating non-saline groundwater may have sensitivities up to 500 and require extreme care to avoid disturbance.

CHRISTOPHER McDONALD

soil geomorphology 'Soil geomorphology' is the general term for studies that attempt to describe and explain relationships between soils and landforms. It includes analysis of the way in which landform (*geomorphological*) processes interact with soil-forming (*pedological*) processes. Hence it is sometimes called *pedogeomorphology*. Geomorphological processes and soil-forming processes interact in the landscape, especially where the movement of soil and water is involved. Many geomorphological processes are also pedological processes; in consequence soil patterns and landscape elements often coincide, and a knowledge of the one allows predictions to be made of the other.

Geomorphological processes, both erosional and depositional, create distinctive landforms which have a great influence on soil types and soil distribution. For example, a valley which has been glaciated will exhibit erosional landforms where the soil has been stripped down to bedrock, and depositional landforms where material has been dropped and piled up by the ice. The development of soils will vary according to the nature of these landforms. Landscapes shaped by rivers are relatively similar worldwide in terms of both landforms and processes, and soil materials and drainage conditions share sharp boundaries in relation to distinctive fluvial landforms such as terraces, flood plains, and levees. Soils can also be used as indicators of long- or short-term landform change and may also, under certain circumstances, indicate the extent of past climate or environmental change.   JOHN GERRARD

Further reading

Gerrard, J. (1992) *Soil geomorphology*. Chapman and Hall, London.

soils It is possible to define soil in many ways. A useful, well-established, and all-embracing definition proposed by J. S. Joffe in 1945 is that soil is a natural body consisting of layers or horizons of mineral and/or organic constituents of variable thickness which differ from the parent material in their morphology, physical, chemical, and mineralogical properties and their biological characteristics. A soil can be described in terms of its *profile*, which is the vertical arrange-

ment of the soil horizons down to parent material. The fact that soils can be grouped into broad classes with similar properties suggests that all soils are influenced by a universal set of factors. These are known as *state factors*.

Factors of soil formation

Soils develop as a result of the interplay of five factors of soil formation: parent material, climate, organisms, topography, and time. These were incorporated by Hans Jenny in 1941, in a state factor equation:

$$S \text{ or } s = f(c, o, r, p, t)$$

where S denotes the soil, s is any soil property, c is the climate factor, o is the organism or biotic factor, r is the relief or topography factor, p is the parent material, and t is the time factor. Jenny defined these factors in the following terms. The climate factor represents the regional climate, with precipitation and temperature being considered as separate functions. The organism or biotic factor is essentially vegetation and is the summation of the plant matter reaching the soil. The topographic factor includes the shape and slope of the landscape, the aspect of the slope and the height of the water-table, the latter usually being related to topography. Parent material includes both weathered and unweathered material from which the soil formed. This includes material both *in situ* and transported and may also include a pre-existing

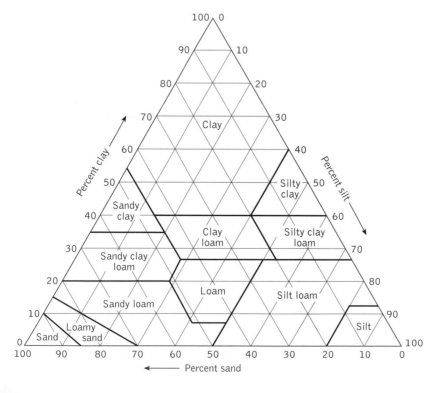

Fig. 1. Triangular diagram of soil texture classes.

soil. Time is the time elapsed since the deposition of material, the exposure of material at the surface, or the formation of the slope.

A number of attempts have been made to show that some factors are more important than others. In the early days of soil science, rock type was thought to be the most important factor. Early Russian workers stressed the importance of climate, and many soil classification schemes and world soil maps were based on climate characteristics. Each factor is, however, essential; none can be considered generally to be more important than any other, although locally one factor may exert a strong influence (*see* soils and topography). These factors define the 'state' of the soil system and provide the framework within which soil processes operate.

Soil characteristics

Undisturbed soil is a mixture of organic and inorganic solid particles and interconnected voids containing varying amounts of soil, water and gases. The appearance of a soil is influenced by a large number of characteristics, of which only the more significant ones can be examined here. Colour is often the most obvious characteristic and may indicate some of the processes that are, or were, operating within the soil. Dark colours near the surface usually indicate an accumulation of organic matter. Yellow-brown to red colours generally indicate the presence of iron (ferric) oxides, and greyish colours are usually due to ferrous iron compounds formed under reducing (gleying) conditions. White or light grey colours characterize a horizon from which leaching has removed oxides or hydroxides of aluminium or iron.

The textures of soils reflect the proportion of sand, silt, and clay sizes within that portion of an inorganic soil fraction that is less than 2 mm. The relative combinations of sand, silt, and clay are the formal basis of the various soil textural classes (Fig. 1). Texture will affect processes operating within the soil, and will affect chemical exchange because surface area per unit volume increases greatly as particle size decreases. Organic matter, which is usually concentrated near the surface, ranges from undecomposed plant and animal tissue to *humus*—a relatively resistant mixture of brown and dark-brown amorphous and colloidal substances modified from the original organic matter by various soil organisms. Carbon usually makes up over one-half of organic matter and the carbon–nitrogen ratio is a good indication of the amount of decomposition of the original organic material. The ratio is high (greater than 20) in plant tissue and low (less than 10) in humus. Soil organic matter increases the water-holding capacity of soils and the cation-exchange capacity. Carbon dioxide is released during the formation of humus, which aids the formation of carbonic acid, increases soil acidity, and enhances weathering.

Soil structure reflects the way in which individual particles are aggregated together. The individual aggregates, called *peds*, are classified into a number of basic types (Fig. 2). Texture can be an important control here; clay content is important in the creation of blocky, prismatic, and columnar structures. The

Structure

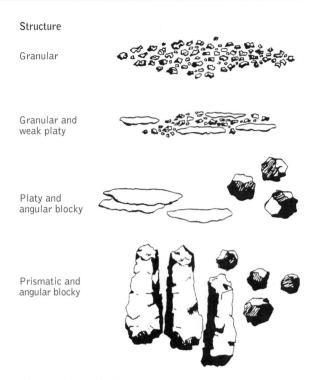

Granular

Granular and weak platy

Platy and angular blocky

Prismatic and angular blocky

Fig. 2. Main types of soil structure.

movement of water through the soil and surface erosion are very much affected by structure. Some aggregates are unstable in water, breaking up and allowing clay to be removed by percolating water; some of this clay may be redeposited as clay skins on ped surfaces lower down the profile.

Water is an important constituent of soils. It is held in the soil by adhesive forces between water molecules and organic and inorganic particles, and by cohesive forces between adjacent water molecules. Moisture content is usually expressed as a percentage of the oven-dry soil weight, but it can also be related to its availability to plants. When the forces holding water films on soil-particle surfaces equal the forces of downward gravitational pull, the soil is said to be at *field capacity*. Water is removed easily from a soil by evaporation and transpiration through the vegetation. As more water is removed from the soil, water films become thinner and are held by ever-increasing forces of attraction. Eventually, the water is held so strongly that roots cannot extract it. The water content at which this occurs is called the *permanent wilting point*. Water retention is largely related to the organic matter and clay contents, whereas water movement is influenced by bulk density, porosity, and permeability; *bulk density* is the weight of soil per unit volume; *porosity* is the percentage of voids to the total volume of soil; and *permeability* is a measure of the ease with which water can pass through the soil.

Soil acidity, or pH, is an important characteristic and normally varies from 5 to 9. A value of 7 indicates neutral conditions, below 7 indicates acid conditions, and above 7 indicates alkaline conditions. Soil acidity is largely a function of organic matter and the type and amount of cations. Large amounts of organic matter tend to produce acid conditions unless counterbalanced by basic cations. Hydrogen and aluminium ions are largely responsible for soil acidity; aluminium is released by the weathering processes of hydrolysis and solution, and the production of free hydrogen ions increases the acidity. pH tends to decrease as rainfall increases because leaching of basic cations is then increased.

Soil horizons

The soil characteristics discussed above are not distributed randomly in the soil profile but are generally organized to produce a definite vertical and lateral structure to the system. *Horizons* are layers differentiated vertically in the soil body that differ in their physical, chemical, and biological attributes. The distinctiveness of soil type is usually related to the properties of its horizons, and horizon designations are an element in the definition of soil units and in the description of representative profiles.

A consistent set of master horizons, designated by capital letters, H, O, A, E, B, C, and R, with distinctive characteristics can be described (Fig. 3). The upper part of a soil is usually an organic horizon (O), formed or forming from accumulations of organic matter deposited at the surface. It may be subdivided into fresh litter (L) deposited during the most recent phases of litter fall and the partly decomposing litter (F) remaining from earlier periods of litter fall; this is also known as the *fermentation layer*. In the next layer (H) the organic matter is well decomposed and original plant structures cannot be discerned.

The A horizon is the first dominantly mineral horizon, in which humified organic matter is mixed with the mineral fraction. The organic matter is either distributed as fine particles or occurs as coatings on the mineral particles. A horizons are usually quite dark in colour because of the organic content. An E horizon, which usually underlies an O, H, or A horizon, is an 'eluvial' horizon (see *soil development*), having lost compounds of iron and aluminium (sesquioxides) and silicate clay by leaching and translocation. It contains a lower organic content than the A horizon and is usually paler in colour. The B horizon is commonly a mineral horizon characterized by weathering of the original parent material *in situ*. It may possess an 'illuvial' (see *soil development*) concentration of silicate clay, iron, aluminium, or humus, alone or in combinations. It is also characterized by the release of sesquioxides by weathering and is the zone where large-scale granular, blocky, or prismatic structures are formed. B horizons are the most variable of soil horizons. The C horizon is the parent material from which the soil has been derived. It lacks the properties of A and B horizons but includes weathering, as indicated by mineral oxidation, accumulation of silica, carbonates, or more soluble salts, and is often gleyed. The R horizon is usually hard, unweathered bedrock.

It is sometimes necessary to qualify the master horizon description with a suffix, or suffixes, to indicate additional properties of a horizon. Thus Ap indicates that the A horizon has been mixed by ploughing, Ag signifies gleying, and Ah denotes an accumulation of organic matter. Suffixes are used extensively in the B horizon to represent alteration by weathering *in situ* (Bw), illuvial concentration of clay (Bt) and sesquioxides (Bs), or a cemented layer (Bm). In this way a detailed description of a soil profile is also an interpretation of the processes that have shaped the profile. The nature and arrangement of horizons also form the basis for most soil classifications.

Soil classification

It is necessary to stress that there is no international agreement about the classification of soil types and there are at least ten different systems in use in various countries. Early Russian workers established that there was a close relationship between soils and vegetation and between soils and climate. This led the famous Russian soil scientist Dokuchaev to propose a classification based on broad vegetation zones, such as boreal, taiga, forest–steppe, steppe, etc. Such soils were called *zonal soils*, but it was soon apparent that soil types were not completely uniform over these broad areas. Because of this, soils that were formed as a result of the influence of a specific local factor, such as parent material or topography, were called *intrazonal soils* and young or poorly developed soils, such as those developed on river alluvium, were termed *azonal* soils. The division into zonal, intrazonal, and azonal soils was the basis for the soil classification used in the USA until 1960. It contains soil types, such as podzols, chernozems, brown earths, chestnut soils, etc., familiar to many people.

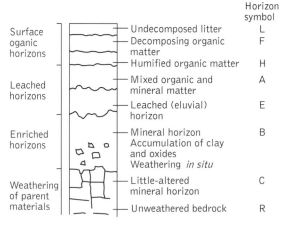

Fig. 3. Idealized soil profile showing main soil horizons. Note that all the horizons could never occur together.

Because it was based on zonal concepts this classification was very attractive to soil mappers, geographers, and ecologists. But such a scheme has severe limitations. Soils, such as podzols, which are zonal in some parts of the world might be intrazonal in other areas. Also, the system was based on environmental factors rather than on the essential characteristics of the soils. This type of classification has now been superseded by the United States Soil Conservation Service's Soil Taxonomy scheme and by an FAO–UNESCO classification.

The Soil Taxonomy scheme has six levels of categorization: order, suborder, great group, subgroup, family, and series. There are ten orders, differentiated by the presence or absence of certain diagnostic horizons or features that demonstrate which soil-forming processes have been dominant. It is a complicated system which often requires detailed measurement of the soil profile and quite sophisticated laboratory determinations, but it is a flexible system which is being adapted continuously. The FAO–UNESCO scheme was developed for use as the basis for the production of a world soil map at a scale of 1 : 50 000 00, which was completed in 1981. In this scheme, soils were divided into 26 major groups at the first level of the classification and into 106 soil units at the second level. This classification is also based on observable or measurable attributes of soils. The scheme was revised in 1988, and the number of major groups was then increased to 28 and the number of soil units to 153. These changes demonstrate the fluid nature of soil classifications and the great variability of soil types, which reflects the interaction of the soil-forming factors discussed above.   JOHN GERRARD

Further reading

FAO–UNESCO (1974) *Soil map of the world*, Vol. 1. FAO–UNESCO, Paris.

FitzPatrick, E. A. (1980) *Soils: their formation, classification and distribution*. Longman, Harlow.

USDA (1975) *Soil taxonomy*. Agricultural Handbook No. 18. USDA, Washington, DC.

soils and topography

Relief or topography is one of the five main factors that influence the way soils develop. The influence of topography on soil formation operates through three main elements: slope angle, slope position, and altitude. Slope angle influences processes such as the movement of the soil and water. Slope position governs whether soils are net receivers or net losers of soil and water. Altitude affects climate, which is also an important soil-forming factor. Areas of great relative relief, such as mountain areas, often experience different climate characteristics from lower to upper slopes, which are usually expressed in terms of different rainfall and temperature regimes. This generally results in a vertical zonation of soil types, which can correspond to a zonation of vegetation types. Thus, a typical sequence on tropical mountains is from lowland tropical forest, through submontane, montane, and subalpine vegetation zones to the alpine zones of grasses and shrubs.

It has been known for some time that soils vary systematically along a transect from the top to the bottom of a hill slope. The effect of topography is seen best on slopes developed on a single rock-type; it may otherwise be difficult to differentiate the effect of topography from that of rock-type. Soil differences are usually related to changes in soil moisture and availability of water. In any explanation of the way in which soils vary with topography, emphasis is placed on the difference between freely drained upper parts of hill slopes and imperfectly to poorly drained lower portions.

There is a continuum between those parts of a slope where the influence of soil moisture is at a minimum and those parts where maximum influence of soil moisture is felt. Slope steepness is one of the most important factors affecting soil moisture, because it influences the balance between the amount of water that infiltrates into the soil and the amount of water that runs off as surface flow. On steep slopes relatively less water infiltrates, thus reducing percolation and the intensity of leaching processes within the soil. On steep slopes more water within the soil will also flow downslope as 'through flow'. If less water infiltrates, more will flow across the surface, perhaps resulting in soil erosion. Soluble minerals are leached from soils on upper slopes, move down the slope, and are often deposited at the foot of the slope (Fig. 1). This leads to different trends in soil properties on upper and lower slopes. Upper slopes are associated with processes of removal of soil and water, whereas lower slopes are associated essentially with deposition and accumulation. Some workers have made the distinction between 'non-cumulative' soils on upper slopes and 'cumulative' soils on lower slopes. The operation of such effects across the entire hill slope produces a series of systematic changes in soil properties and soil profiles. The greatest concentrations of certain soil properties, such as the amount of organic matter present, would be expected on the gentler areas at the top and bottom of hill slopes.

Variation in soil colour often provides a clue to the processes that are operating. Colour changes are especially prominent on many tropical hill slopes. Upland, well-drained soils are reddish-brown, the colour indicating the presence of non-hydrated iron oxide. Drainage is slower on middle and lower parts of hill slopes, partly because of moisture seeping downslope from upper soils. Middle- to lower-slope soils remain moist longer and dry out less frequently and less completely. This causes the iron to become increasingly hydrated, and the red colour changes to brown or yellow. Drainage is poor on the lowest slopes and part, or all, of the soil profile may be waterlogged, leading to chemical reduction of iron. Under waterlogged conditions bacteria obtain their oxygen from oxygen-containing compounds, which are then reduced to other compounds (see *soil development*). Such waterlogged soils are usually bluish-grey, greenish-grey, or neutral grey in colour, although if the water-table fluctuates, alternate oxidizing and reducing conditions will lead to the formation of red and grey mottles.

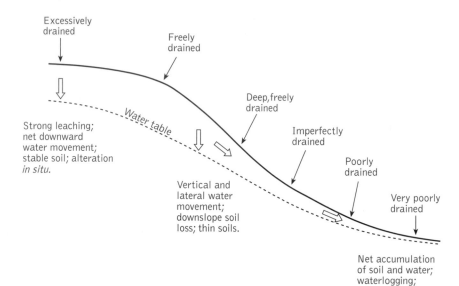

Excessively drained

Freely drained

Deep,freely drained

Imperfectly drained

Poorly drained

Very poorly drained

Water table

Strong leaching; net downward water movement; stable soil; alteration *in situ*.

Vertical and lateral water movement; downslope soil loss; thin soils.

Net accumulation of soil and water; waterlogging; chemical reduction.

Fig. 1. Processes that lead to the differentiation of soil profiles on hill slopes.

The recognition that there are clear patterns of soils on slopes and that these patterns repeat themselves on similar slopes in similar environments led to the formulation of the *catena concept*. A catena is a grouping of soils which are linked in their occurrence by conditions of topography and are found in the same relationships to each other wherever the same conditions are met. The concept was first developed in the 1930s by G. Milne, working in East Africa, and was used as a way of mapping soils over wide areas. The assumption made was that where slope patterns were similar, soil patterns would also be similar. Since then catenas have been recognized in a variety of areas and under a variety of climatic conditions, an indication of the strength of the relationship between soil and topography. JOHN GERRARD

Further reading

Birkeland, P. W. (1984) *Soils and geomorphology*. Oxford University Press, New York.

Gerrard, J. (1992) *Soil geomorphology*. Chapman and Hall, London.

soil stabilization
Soil stabilization is any process which improves the physical properties of a soil, such as increasing the shear strength, bearing capacity and the resistance to erosion, dust formation, or frost heaving. For some operations it is of a temporary nature only (e.g. in reducing air loss in compressed air tunnelling), but for most it is a permanent part of the construction. The stabilization methods used are divided into mechanical, chemical, and electrochemical methods.

Mechanical methods are those in which the compaction or bulk density is increased by dynamic or vibro-compaction. Dynamic compaction is usually affected by mechanical means such as rolling, tamping, or vibrating the soil. In the construction of road bases, runways, earth dams, and embankments, the soil is placed in layers of specified thickness. Each layer is then subjected to a specified amount of compactive effort. Vibroicompaction is a process in which a vibratory poker is placed into a granular material causing compaction. Under good conditions in granular materials a cylinder zone as much as 3 m in diameter can be compacted in a single penetration. Vibratory techniques are not used in biodegradable, heavily obstructed, chemically unstable, or potentially combustible materials, since penetration may create paths through which leachates and landfill gases may leak.

Chemical stabilization in engineering is not new, but its use has become much more widespread since the 1960s. Initially only lime and cement suspensions were used. These are still popular, since they remain the cheaper methods. Lime stabilization involves the formation of strong bonds between the clay minerals and other soil particles and is therefore ineffective in granular soils. Three processes appear to be operative: (1) rapid replacement of exchangeable cations by calcium with increased interparticle bonding and a more flocculated fabric; (2) slower attack on the edges of the clay minerals by alkaline solutions in which the solubility of silicon is increased, with the possible formation of new silicates similar to those in cement; and (3) crystallization of calcite to form additional interparticle bridges. Chemical stabilization can also be achieved by relatively expensive organic agents which are either water-repellent, oily, or bituminous compounds or cementing resins. The application of the chemicals to the ground is usually performed by injection into the ground. Up to 150 chemicals are now used commercially in soil stabilization.

Freezing of the ground provides a method of temporary soil stabilization. This is achieved by circulating liquid nitrogen through the ground in a set of access tubes installed either from the ground surface or within a tunnel. The liquid boils and the gas may be used further or exhausted to the atmosphere.

Electrochemical methods of stabilization involve the reduction of water content by electro-osmosis. An electrical potential is applied to the soil across two electrodes, to which ions migrate carrying water with them away from the area of treatment.   M. HIGGINS

solar activity and sunspots
The Sun provides most of the extraterrestrial energy which reaches the Earth's surface. The average annual energy reaching the Earth from the Sun is approximately 1400 W m^{-2}. Average surface temperatures on the Sun are estimated to be of the order of 6000 °C. Earth-based observations of the lower latitudes (30° N–30° S) of the Sun's photosphere reveal dark patches or spots; such observations have been made since the seventeenth century. The so-called 'sunspots' vary in size, abundance, and duration. At any one time there may be as many as 20 or 30 such sunspots which can range in size from 10^3 to 2 × 10^5 km in diameter. Each sunspot consists of a dark central umbra, which is estimated to have a temperature in the region of 4000 °C, and a surrounding penumbra, at a temperature in the region 5000 °C.

Sunspots show pronounced cyclic behaviour, the major fluctuations having an approximately 11-year period. Maxima of successive 11-year periods show a periodic variation in intensity which reflects a periodicity of around 90 years (termed Gleissberg cycles). The cycles do not exhibit an exactly 11- or 90-year periodicity; power spectra calculated for sunspot time-series consequently exhibit broad rather than distinct peaks. An observed periodicity of 22 years (termed the Hale cycle) may be due to the coupling of adjacent 11-year cycles, which are magnetically oppositely polarized. Evidence has also been presented for the existence of a 200-year cycle. During the 11-year cycle, the abundance of sunspots and their size fluctuate in a more or less regular manner. At the start of the cycle, spots appear in both hemisphere at about 35° from the solar equator. As the cycle proceeds, the initial spots fade and new, more numerous spots occur at lower latitudes. By the end of any given cycle the sunspots are concentrated at latitudes around 5° from the solar equator.

Sunspots are typically measured using the Wolf Relative Sunspot Number. This is a measure of the number of groups of spots combined with the number of individual spots, calculated on a monthly basis. Sun-spot activity varies from zero, as was observed during the Maunder Sunspot Minimum which occurred synchronously with a period of global cooling from AD 1525 to AD 1850, up to about 200 in the highly active period which peaked in 1957. An instrumental record of sunspot cycles now extends from 1750. It includes 22 sunspots cycles, and indicates that the overall variation of energy during a cycle is of the order of a few parts per thousand (i.e., less than 1 per cent).

Associated with sunspots are brighter areas called *faculae*. The total solar radiance, or solar energy flux, is related to the sunspot cycle in the following manner:

$$E = 1366.82 + 7.71 \times 10^{-3} \, S_n \, \text{Wm}^{-2},$$

where E is the solar energy flux reaching the Earth and S_n is the number of sunspots. The fact that solar output rises with increasing sunspot numbers suggests that variations in faculae brightness are a major factor in determining total solar energy flux. Instead of exhibiting a uniform change in brightness across the entire solar spectrum, variations are observed to be most marked in the UV range.   STEPHEN STOKES

Further reading
Bradley, R. S. and Jones, P. D. (1992) *Climate since A.D. 1500.* Routledge, London.

solar system abundances of the chemical elements
Relative abundances of the chemical elements in the solar system span some 12 orders of magnitude (Fig. 1). The two lightest elements, hydrogen (H) and helium (He), together constitute more than 99 per cent of the atoms and the mass of the solar system. At the opposite end of the abundance scale are the rarest heavy elements, such as tantalum (Ta), rhenium (Re), thorium (Th), and uranium (U). Even elements as familiar as oxygen (O), iron (Fe), copper (Cu), and silver (Ag) are 10^3–10^{11} times less abundant than hydrogen.

The most accessible solar system materials are the rocks of the Earth's crust and upper mantle. Elemental abundance patterns in some terrestrial rocks do partially mimic those of the solar system as a whole. In general, however, igneous, metamorphic, and sedimentary processes operating over the 4.6 billion years of Earth history have strongly fractionated the chemical elements within the Earth. Determination of the overall chemical composition of the solar system requires information from other, less evolved members of the solar system. The principal data come from chemical analyses of meteorites, particularly the chemically primitive meteorites known as carbonaceous chondrites. Supplementary data are provided by atomic absorption spectra of the outer layer of the Sun.

The abundance patterns of the chemical elements record nucleosynthesis (production of the elements) in the Big Bang, in pre-solar system stars, and in the stellar supernova that triggered the formation of the solar system. The solar system abundance curve is dominated by two features: exponentially decreasing abundance with increasing atomic number, and sawtooth-like alternation of elements of even and odd atomic

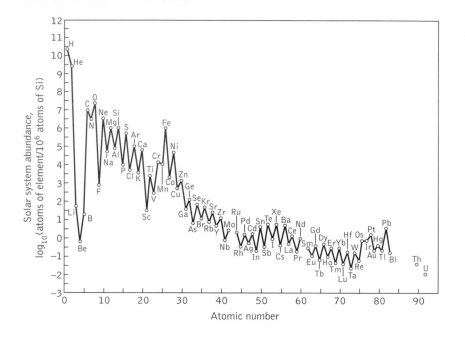

Fig. 1. Relative abundance of the chemical elements in the solar system, expressed as atoms of each element per million atoms of silicon (Si). Elements with atomic numbers 43, 61, 84–89, and 91 have no stable or long-lived isotopes, and therefore have vanishingly small abundances.

number. Heavier nuclei were produced by stepwise fusion of lighter nuclei and the addition of neutrons to lighter nuclei. Progressively heavier elements required more steps and therefore generally were produced less abundantly. Elements of even atomic number have more numerous stable isotopes than do odd-numbered nuclei, and are therefore more common. The iron-group elements (Mn, Fe, Co, Ni) are 10 to 1000 times more abundant than adjacent elements because they have the most stable nuclei, with the greatest binding energy per nuclear particle. The light metals lithium (Li), beryllium (Be), and boron (B) are grossly underabundant because they are consumed, rather than produced, in stars. Slightly elevated relative abundances of the heavy metals osmium (Os) through bismuth (Bi) are related to their nuclear stability and nucleosynthetic processes, as are other second-order features of the abundance curve.

Further reading

Cox, P. A. (1989) *The elements; their origin, abundance, and distribution.* Oxford University Press.

G. B. HAXEL

solar system *see* AGE AND EARLY EVOLUTION OF THE EARTH AND SOLAR SYSTEM

Solnhofen lithographic limestone fauna The
Solnhofen lithographic limestones are Late Jurassic in age and occur beneath a large part of Bavaria, southern Germany. They consist of extremely fine-grained evenly bedded limestone, formerly quarried for use in printing. The environment in which the limestones were deposited was a series of marine lagoons restricted by barrier islands and a coral–hydrozoan barrier reef. High rates of evaporation coupled with restricted exchange of water with the open sea resulted in hypersalinity and the development of density stratification. The bottom waters of these lagoons were probably stagnant and provided an environment hostile to life. More than 600 species of organisms are preserved in the limestone. They represent a variety of different environments, including open marine, reef, lagoon, and terrestrial habitats. Most of the specimens were probably washed into the hypersaline lagoons during storms and then preserved by rapid burial as a result of suspension fallout of carbonate particles. Soft parts are often preserved, including the wing membranes of pterosaurs (flying reptiles) and the muscle tissues and intestines of fish. Many of the specimens show contraction and coiling that is probably a post-mortem feature caused by dehydration in the brine and subsequent contraction of ligaments.

The invertebrate fauna includes many planktonic (floating) forms, particularly jellyfish and the stemless crinoid *Saccoma*, and also nektonic (swimming) organisms such as cephalopods. Many arthropods are also present including the horseshoe crab *Mesolimulus* and the crayfish *Mecochirus*. In some instances these arthropods were still alive when they were washed into the lagoons and were sufficiently tolerant of the salinity change to survive for a short while before being preserved at the end of their trails. Many species of flying insects such as dragonflies are also present, apparently washed into the lagoons from the land by seasonal rainwater floods.

The vertebrate fauna consists mostly of fish, including sharks and rays, but also terrestrial organisms such as dinosaurs, pterosaurs, and the earliest bird, *Archaeopteryx*. The small theropod (carnivorous dinosaur) *Compsognathus* is known only from a small amount of skeletal material. Some pterosaurs, such as the small form *Pterodactylus*, are preserved with the outlines of the wing membranes and tail rudders, which have been of great value in their reconstruction. The most important fossils are those of *Archaeopteryx*, an animal that shows a mixture of bird-like and reptilian features. Although the skeleton is essentially reptilian and features a long bony tail, a theropod pelvis, hands with functional fingers, and a reptilian skull with teeth, its feathers were preserved in the fine-grained lithographic limestone, leaving no doubt as to its avian affinities; study has shown that these were flight feathers. Only six specimens have been collected since the first was found in 1861, making this the rarest and certainly most valuable fossil in the Solnhofen Lagerstätte (fossil accumulation). As the *Archaeopteryx* and pterosaur skeletons are generally complete, it seems unlikely that they were washed into the lagoon from the land. A more probable explanation is that they were caught in flight by storms, blown into the ocean and drowned. Sinking to the bottom they would have been rapidly covered by the fine-grained carbonates that preserved the details of their soft parts.

DAVID K. ELLIOTT

Sorby, Henry Clifton (1826–1908)

Henry Clifton Sorby was one of the most original, productive, and innovative of nineteenth-century researchers in the Earth sciences in Britain. He came from a wealthy family in Sheffield, and devoted his private fortune to a life of scientific investigation.

Sorby's most influential work was carried out between 1849 and 1864, when he applied the techniques of microscopy to geology and metallurgy. He rapidly developed the techniques of thin-section petrographic analysis and the use of the polarizing microscope.

There followed papers on slaty cleavage, the micropalaeontology of the Chalk, and other sedimentary phenomena. Sorby initiated the study of fluid inclusions in crystals, and from them determined the temperatures at which granites had crystallized. He investigated the role of water in magmas before N. L. Bowen in the USA. Etching minerals, meteorites, and metals by acids, he revealed their textures and histories.

In 1864 he experimented in spectrum analysis, combining the use of microscope and spectrometer, thus initiating a valuable analytical method. Then in 1878 he set off in another direction. With a 75-ton yacht as a floating laboratory, he spent summers cruising off the eastern coast of Britain, studying water and sediment movements, marine biology, and ecology.

Sorby published only a small part of all his findings, but his papers proved to be a great stimulus to geologists. Sedimentology, in particular, owes much to his curiosity and his methodical investigation of sediment particle dynamics and sedimentary rocks.

D. L. DINELEY

specialization in the Earth sciences after the Second World War

Improvements in transport, in communications, and in instrumentation were almost immediate effects of the cessation of the Second World War. Technological advances made during the hostilities were put to peaceful uses, especially in the USA. New requirements in natural resources, especially fossil fuels and building materials, were felt almost worldwide at this time, and economic needs were a spur to the development of both laboratory and field studies in the Earth sciences. The petroleum companies led the way in exploration geology and exploitation engineering, closely followed by the mining companies, mostly operating in other kinds of terrain. New service companies offering aerial photography, geophysical prospecting and analytical facilities were soon to appear in the developed countries for engagement in petroleum and mineral exploration. Somewhat later, computing services and information technology were to enhance the pace of research and development, and by the 1980s satellite photography and remote sensing were available on a commercial basis. The space programmes of NASA, the Soviet Union, and Europe prompted interest in lunar and planetary geology.

Meanwhile, both academic and commercial interests had given an enormous impetus to marine studies, as had strategic needs and, for the West, the expansion of Soviet sea power. Canada and the USA were very active in surveying and investing in the defence development of Alaska, the Canadian Arctic archipelago, and Greenland. Newly independent and developing countries were keen to attract investment in the exploitation and utilization of their natural resources. The market for qualified Earth scientists was large and international, and so there was a significant impetus to academic institutions to expand their teaching and research capabilities. All of this was favourable for the development of new fields within the Earth sciences. Expansion of universities and other research and teaching institutions began immediately after the war and was well under way in the early 1950s. By the mid- to late 1960s a further important factor came into play. This was a veritable revolution in the Earth sciences occasioned by the theory of plate tectonics. It provided a tremendous stimulus in virtually every branch of geological science, and it prompted the geological reconnaissance or re-examination of many parts of the world.

As a result of the general post-war stimulus and acceptance of plate-tectonic theory, there was a growing need for interdisciplinary studies and the breaking down of perceived distinctions between different members of the Earth science community. In the late 1970s and 1980s this was accentuated by growing environmental consciousness, emphasis upon the limits of non-renewable resources, and the effects of pollution. By this time technological innovations had greatly improved the speed, number, and effectiveness of both investigations and communications. Scientific capabilities had outstripped social and international competence in the use of science for the well-being of the planet. Basic science has extended its attractions while the applied Earth sciences have found increasing demands made upon them.

In the laboratory one of the most fruitful post-war areas of growth has been that of sedimentology. Both carbonate and clastic rocks have been subjected to new techniques of study, including ultramicroscopy, ion-probe analysis and mass spectrometry, and other forms of rapid analysis. The dynamics of clastic sediment transport and deposition have yielded to laboratory study. Geochemical and isotope studies have been of use in understanding the genesis, diagenesis, and geochronological aspects of the evolution of sedimentary rocks. Organic geochemistry and microbiological studies have been undertaken in modern sedimentary environments and on ancient sediments. The role of organisms in generating sediment, and in leaving signs of their behaviour (as trace fossils) within it, have been notable areas of special study. The modelling of sedimentary facies and sedimentary environments has prospered, and sedimentary tectonics has developed as a discipline linking tectonic behaviour to the provenance and rate of supply of sediment types.

Experiment has become a widespread methodology in the study of mobile materials, under both atmospheric and aquatic conditions, and higher temperatures and pressures. Physical and chemical processes in crystalline rock genesis, volcanism, and mineral emplacement have been elucidated, and research in these areas continues to expand.

Palaeobiology has seen the spectacular expansion of knowledge of microfossils, with emphasis upon their use in biostratigraphy. Conodonts, palynomorphs, and nannofossils, in particular, have contributed to global biostratigraphy and correlation. Cladistic analysis has influenced evolutionary thinking and has been employed in studies of taxonomy and biogeography, particularly in response to the impact of plate tectonics. Palaeoecology has emerged as a significant part of palaeobiology, relating organisms to habitats of the past. More recently, taphonomy has become a topic of wide interest, especially since the discovery of new, exceptionally preserved fossil faunas.

In structural geology the interpretation of mesoscale structures and microstructures (petrofabrics) was a rapidly developing theme and has been joined by those of large-scale tectonics and, most recently, by neotectonics—that is, the study of structures that are young or still forming.

A topic of importance in correlation and palaeogeography has been palaeomagnetism, perhaps the most far-reaching branch of geophysics since the Second World War. It has been matched in significance by the development of isotope geochemistry as an invaluable aid to geochronology and palaeoenvironmental determinations.

Geophysical prospecting and exploration, for which the petroleum companies provided a lead, is now valuable to academic and applied geology alike. An important development, seismic stratigraphy, now provides data over both land and sea that bears upon the Mesozoic and Cenozoic evolution of the continental shelves.

Expanding specialist areas of applied geology have included hydrogeology, soil mechanics, rock mechanics, and engineering geology. More recently, studies of both the supply of extractable resources and waste disposal have greatly increased in number, detail, and size. In all these fields modelling is now an essential part of assessment programmes. The roles of statistics, data handling, and computing have grown steadily over the past 30 years or more, and all branches of the Earth science tree are becoming more quantitative, pure and applied alike.   D. L. DINELEY

species concept The definition of a species is a problem that has exercised the minds of many evolutionary biologists. Although we can use the term rather loosely for groups that are visibly distinct because directional selection causes accumulating changes in traits through time, a more objective definition is needed. It is particularly important to know whether species really exist as natural units or whether they are merely divisions that are useful for taxonomists. In biology two meanings of species are recognized, one of which is reproductive and the other morphological or shape-related.

The reproductive species concept states that species are communities of interbreeding organisms. One definition put forward by Ernst Mayr stated that species are 'groups of actually or potentially interbreeding natural populations that are reproductively isolated from other such groups.' This definition of reproductive isolation is important because many closely related species may look similar yet be unable to interbreed. Isolating mechanisms that develop to cause reproductive isolation might be behavioral changes that prevent mating occurring, or postmating changes that prevent production of fertile offspring. Isolation could also occur by geographic separation, as when individuals are blown or swept out to islands, or left behind as isolated populations, when ranges are reduced by environmental changes. It is processes of this kind that lead to speciation events in which new species are developed. This can occur in a non-branching fashion, in which one species is gradually transformed into a new species, or in a branching fashion, in which separate populations undergo different directional selection.

Morphological species are defined by their appearance, so that individuals that look similar are assumed to be members of the same species. This concept has been tested by determining that natives in New Guinea recognize the same species of birds that are recognized by Western taxonomists. These groupings hence have some reality in nature and they can be recognized without using the reproductive concept. Morphological species are important in palaeontology, for interbreeding ability does not fossilize. Palaeontologists therefore have to define species solely according to their anatomy. This, however, raises additional problems that are generally known as the species problem in palaeontology. Modern biologists deal with groups of organisms in the present, but palaeontologists have to deal with the added dimension of time. They thus not only compare populations with others that existed at the same time; they also compare them with both ancestral and descendent populations. There is no way of eliminating this problem, although the widespread gaps that are present in the fossil record alleviate it to some extent. Only short segments of a particular lineage will usually be preserved, and the large gaps between these segments provide convenient locations for species boundaries. Examples of continuous preservation of lineages are, however, known. One has been studied by Peter Sheldon of Cambridge University, who found evidence for gradualistic evolution in Ordovician trilobites over a period of three million years. The presence of intermediate morphologies made taxonomic subdivisions of the lineages impossible.   DAVID K. ELLIOTT

spits A spit is a landform created by the extension of sediments from the shoreline into a water body (lake or sea). Spits commonly occur near the mouths of estuaries or embayments, where the line of the coast turns inland, the spit continuing the former coastal trend. As a result, many spits form a natural defence against coastal inundation.

Although spits are found in lacustrine settings, the majority are associated with coastal areas. They usually consist of sand or shingle that has been transported by wave action. The formation of a spit thus depends on a source of sediment and sufficient wave energy to transport the material towards and then across the mouth of the embayment. The direction of dominant wave attack in comparison to the available sediment load is crucial. Where sediment supply is lacking or wave energy greater, the resulting spit will tend to turn into the embayment rather than continue the line of the original coast.

The distal end of a spit commonly displays alterations in alignment, known as *recurves*, which usually curve into the embayment. Recurves usually develop as a result of wave refraction in deeper water. The transport of sediment, and hence the development of a spit, reflect the changing direction of wave movement (e.g. Blakeney Point, Norfolk). In some circumstances, however, recurves reflect the complex wave environment at the site (e.g. Hurst Castle Spit, Hampshire) with the dominant direction of wave action altering along the spit.

Many instances spits are both prograding into the embayment and are being rolled onshore at the same time. As a consequence the modern shingle ridge truncates former recurves and such patterns can be used to reconstruct the evolution of the spit. Many spits also display cyclic phases of evolution, perhaps reflecting cyclic patterns of sediment supply. When coastal sediments abound, the spits tend to prograde, both in length and width. If the sediment supply diminishes, the wave processes continue to move material along the feature; as a result, progradation at the distal end results from cannibalization of the feature at its proximal end. This can result in breaching of the feature at its proximal end and ultimately detachment from the mainland. The subsequent input of a fresh supply of sediment results in the reattachment or regrowth of the spit. Many of the world's spits are currently showing signs of cannibalization. This may be due to the fact that they developed as a result of abundant sediment supplies

derived from Quaternary sea-level changes. Now that this sediment has been fully utilized, the spits are decaying to a more stable form.

At certain sites two spits appear to be growing towards each other across the embayment. At some of these sites (e.g. Poole Harbour, Dorset) this is a product of the coastal configuration, the dominant wave action being in opposite directions along each side of the embayment. In other cases, a major spit has developed along the coast as a result of dominant wave activity (e.g. Orford Ness, Suffolk); the shelter the feature provides allows a second smaller spit to develop on the other side of the embayment. This second spit grows in the opposite direction.   CALLUM R. FIRTH

Further reading

Bird, E. F. C. (1984) *Coasts*. MIT Press, Cambridge, Mass.

Carter, R. W. G. (1992) *Coastal environments*. Edward Arnold, London.

sponges and related fossils Sponges are simple multicellular organisms intermediate in grade between protozoans and metazoans and hence are often called parazoans. They are sedentary aquatic filter-feeders, using flagellae on the cells to create currents that pass into the organism through multiple external openings; these currents then circulate before exiting through a common opening. About 1500 genera of modern sponges are known, of which about 80 per cent are marine. Sponges have an internal skeleton composed of needle-like elements called spicules, which can be composed of calcium carbonate, opaline silica, or fibres of the scleroprotein spongin. Although relatively rare in the fossil record, the presence of isolated spicules shows that sponges existed at least as far back as the Cambrian. The sponges form the major part of the phylum Porifera, but there are several groups of sponge-like organisms that are found as fossils and are generally included within the phylum. These include stromatoporoids, which formed massive calcareous skeletons, and archaeocyathids, which were conical and also calcareous.

Sponges are classified on the shape and composition of the spicules and also on their external morphology. These characters are used to divide them into five major classes. Demospongea are the dominant modern group with 600 genera; they have a skeleton formed of spongin or silica, or a mixture of the two. Hexactinellids are deep marine forms with siliceous skeletons that can be solidly fused. Sponges with calcareous skeletons are included within the Calcarea, an exclusively marine shallow-water tropical group. Sclerosponges are a minor group with a skeleton of massive calcite and spicules, described from modern forms only in the 1960s but with a record that extends from the Ordovician. The stromatoporoids have a laminated calcareous skeleton with no spicules; they became extinct in the Cretaceous. In addition to these groups the archaeocyathids (which are sometimes treated as a phylum in their own right) were exclusively Cambrian and had a calcareous skeleton in the form of a porous inverted cone.

Although sponges in general show great ecological variation and adaptability, individually they are restricted environmentally by temperature, salinity, currents, turbidity, and the substrate. Because of this they show marked provinciality when viewed worldwide. Together with their incomplete fossil record this makes them generally unsuitable for use in biostratigraphic zoning, although they sometimes have value on a local scale. One exception is the archaeocyathids, which occur worldwide in the Cambrian, reaching their acme in the Lower Cambrian, where they are stratigraphically useful.

The main geological significance of sponges is as a source of biogenic silica that can contribute to the formation of chert; they have also been important as reef-formers through time, although generally in a subsidiary role to other organisms. During the Cambrian, archaeocyathids formed shallow-water reefs in association with stromatolites; during the Silurian and Devonian stromatoporoids formed reefs in association with tabulate corals. Although sponges continued to be an important element of reef structures into the early Mesozoic, their importance subsequently declined. This was probably due to the rise of the reef-building modern corals, which grow more quickly and can thus colonize space more rapidly.

DAVID K. ELLIOTT

springs

A spring is a location where groundwater flows naturally, that is, discharges to the land surface or into a surface of water body such as a stream, a lake, or even the sea. The source of this discharge may be a water-table aquifer in which the water-table intersects the land surface. Springs associated with water-table aquifers are usually found in valleys or lowlands, where they provide water to wetlands and base flow to streams. Springs can occur in uplands if zones of perched water extend laterally to cliff faces or other steep slopes. Springs are also found in places where a confined aquifer is connected to the land surface by a permeable fracture or fault zone. If the confined aquifer is at great depth, where temperatures are naturally elevated, or is connected to heat sources associated with recent volcanic activity, the springs can discharge hot water, often of high salinity. Spring water is often assumed to be unusually pure and healthful. Although spring water from confined aquifers may be less susceptible to shallow sources of contamination than surface water, the fact that water comes from a spring is no guarantee of high quality.

J. BAHR

stalactites and stalagmites

Stalactites and stalagmites are secondary carbonate deposits found in caves in limestone areas. They can be classified together as speleothems (from the Greek words for 'cave' and 'deposit'). Stalactites hang down from the ceilings of caves while stalagmites grow up from the floor. The majority of stalactites and stalagmites are composed of calcite (calcium carbonate), although in certain circumstances they may consist of other minerals, such as aragonite or gypsum. They are deposited from supersaturated waters and their formation is favoured during periods of relative warmth and wetness. Therefore, their rate of growth is controlled climatically. In addition to climatic controls, local chemical and hydrological factors may determine whether one individual speleothem grows faster or slower than another. Estimated rates of stalagmite growth range between 0.2 and 36 cm per 1000 years, while stalactites may grow significantly faster. Some stalactites and stalagmites contain annual bands, analogous to tree rings, which are in some cases visible to the naked eye and in others visible only under ultraviolet light. In the future, it may be possible to reconstruct interannual climate variations from stalactites and stalagmites in the same way that tree rings have been used. The age of stalactites and stalagmites can be determined precisely by measuring certain radioactive isotopes of uranium and thorium. When a stalactite or stalagmite is deposited it contains a large amount of uranium and no thorium because of differences in the chemical behaviour of these elements. Uranium decays radioactively to produce thorium-230 at a known rate. By measuring the amount of thorium-230 that has been produced relative to the amount of uranium, an age can be calculated. These ages have provided geologists with insights into climate change over the past 350 000 years (the limit of this dating technique), because periods when there was minimal formation of stalactites and stalagmites can be interpreted as being relatively cold and dry. Stalactites and stalagmites from caves which are now submerged beneath the sea demonstrate that sea levels have changed in the past. By dating such stalactites and stalagmites (which can form only when they are exposed by a fall in sea level), constraints can be placed on the extent and timing of these sea-level changes. Stalactites and stalagmites can also be used to provide constraints on rates of landscape evolution by dating valley incision or tectonic uplift. In addition, ages from stalactites and stalagmites associated with archaeological and palaeontological material from cave sites may provide valuable chronological control on that material. It may be possible to obtain further information about past climates by determining the ratios of different oxygen isotopes in the calcite. Although stalactites and stalagmites are scientifically valuable, it is also important to conserve them as they are an integral and beautiful component of the cave environment.

M. S. ROBERTS

Steno, Nicolaus (1638–86)

A Dane of great intellect, Neils Stensen (1638–1686) had studied in Copenhagen and Leiden before entering the service of the Medici in Florence and had acquired a formidable reputation as an anatomist. In 1666 he was presented with the head of a giant shark for dissection and noted the similarity of the teeth to fossil sharks' teeth from rocks near by. From this Steno, as he was now known, hit upon the concept of fossils as preserved organic remains. He also began to travel about Tuscany looking at strata and collecting fossils. His commitment to geology overtook his many other interests and he produced a summary volume describing the origins of fossils, the implications of fossil marine shells in strata in the Tuscan hills, and a few other geological phenomena. This, his *Prodromus* (1669), was to herald a major work on geology, and it led Steno to sense the conflict between Genesis and his own careful observations. He was able to formulate what amounted to the law of stratal superposition. The formation of mineral deposits and metalliferous vein ores in recognizable crystalline forms took his attention and were carefully described in scientific fashion.

Nevertheless, he was converted to Catholicism, and incidentally calculated that Noah's flood took place about 4000 years ago. The *Dissertation* forecast in the *Prodromus* was never finished. Steno recognized that mountains may be formed by volcanic activity and by erosion, and that collapse into subterranean cavities may cause faulting. He clearly had some idea of the way in which geological processes gave rise to different landscapes.

Steno eventually took holy orders, was appointed a bishop, and spent the remainder of his life in peripatetic poverty.

D. L. DINELEY

Stille, Wilhelm Hans (1876–1966)

By the end of the nineteenth century much of the outline geology of the continents was known, but explanations for their configuration were few and vague. Wilhelm Hans Stille became one of the outstanding investigators of tectonics and the principal collator of the history of global tectonic events. In 1913 he was appointed to the chair of geology at the University of Göttingen, but moved to the University of Berlin in 1932. During his long career he was a much-admired teacher, who published over 200 papers and directed more than 100 graduate students.

Stille's principal creation was a table of some 50 orogenic phases, each thought to be essentially synchronous globally, within Phanerozoic time. These phases were small pulses within tectonic eras—Assyntic, Caledonic, Variscic, and Neoidic/Alpidic. Each one was thought to have achieved the stabilization of a portion of crust and enlarged the continents. The eras were named after stages in the consolidation of the European continent, giving rise to Ur-, Palaeo-, Meso-, and Neo-europa. In the light of plate-tectonic theory, however, we no longer hold to the notion of such globally correlatable mountain-building events.

Within the crust, Stille identified orthogeosynclinal belts with marginal oceanic low cratons (stable areas) and high continental cratons. He was to develop the geosynclinal concept, a geological synthesis that was useful until the 1960s, when plate tectonic theory provided a better and more rational explanation for the origin of global tectonics.

Stille established the relationships between large-scale crustal features of the Earth, but he was unable to provide a mechanism to explain their disposition. Ironically, perhaps, it was not until the year of Stille's death that ocean-floor spreading was established as the mechanism driving what he had called the geosynclinal cycle.

D. L. DINELEY

stone decay and conservation *see* URBAN STONE DECAY AND CONSERVATION

storm surges

'The deadliest tropical cyclone in history' on 12 November 1970 in the Bay of Bengal resulted in a storm which elevated water levels by 9 m and caused extensive flooding in the Ganges delta. An estimated 300 000 people lost their lives and 4.7 million people were affected. The effects were disastrous, but such storm surges are not uncommon, and their frequencies and intensities have changed through time.

Storm surges are set up by changes in atmospheric pressure and associated wind stress. They are particularly severe in shallow seas, and occur both in the tropics and in middle

Fig. 1. Hurricane 'Gloria', Long Island, 1985 (from *Time* (1987), **180** (6), 53.)

latitudes in both hemispheres. Tropical storm surges are generated over the oceans between 7° and 25–30° latitude where the sea-surface temperature is at least 26 °C. They are known as typhoons in the western North Pacific, as hurricanes in the eastern North Pacific, Caribbean, and Atlantic (for example, Hurricane Gloria, 1985, Fig. 1), and as tropical revolving storms in the Indian Ocean, including the Bay of Bengal and Arabian Sea, and around Australia.

Strong onshore winds associated with the passage of a deep depression cause sea water to pile up along the coast, generating a positive surge, the height of which is increased by low pressure; whereas offshore winds and rising pressure produce a fall in water levels and a negative surge. A fall of atmospheric pressure of 1.005 millibars (mb) will result in a rise of sea level of 1 cm. The disastrous Moroto Typhoon of 9 September 1934 that affected the Osaka area of Japan was set up by an intense low pressure of 954.3 mb and maximum mean wind speeds onshore of 42 m s^{-1} which produced a surge of almost 3 m.

A tropical revolving storm will have an 'eye' of 20–50 km diameter where pressure is at its lowest, and here sea level may rise as much as 4 m. Most (70 per cent) of all tropical revolving storms occur in the northern hemisphere, and of these 97 per cent are recorded in July to November. An increase in the area where the sea-surface temperature is 26 °C or more will result in an increase in the incidence of tropical revolving storms, which will penetrate middle latitudes more frequently.

In middle latitudes, an increase in temperature gradient between high and middle latitude surface water masses will also lead to an increase in storm surge frequency. Such increases were characteristic of the Little Ice Age from the sixteenth to the eighteenth centuries, for which there are many historical records. The ships' logs of the Spanish Armada graphically illustrate the great storms of 14–18 August 1588 that scattered the fleet; surface wind speeds of 20–35 m s^{-1} can be estimated from these records. Storm surges were particularly frequent in the seas around the British Isles between AD 1570 and 1720, beginning with the 'All Saints Flood' of 11–12 November 1570, which was one of the greatest North Sea storms recorded, and during which up to 400 000 people are estimated to have drowned, particularly in The Netherlands.

Improvements in sea defences in north-western Europe have meant that loss of life has been minimized, but during the infamous 1953 storm surge in the North Sea on the night of 31 January to 1 February, 1800 people were drowned in the The Netherlands and 307 in England, particularly on Canvey Island in the Thames estuary (Fig. 2). Gusts of wind exceeded 100 knots (31 m s^{-1}) and atmospheric pressures were as low as 968 mb. The surge added 2.2–3.0 m to the predicted astronomical tide levels. At the time, the recurrence interval of this storm surge was calculated as 1 in every 200 years, but the extreme water level recorded then has been exceeded in eastern England no less than 12 times in the past 40 years. Hubert Lamb has recorded the great storms and associated storm surges in the North Sea and adjacent areas for the period AD 1509–1990, and summarized the meteorological conditions inferred from the historic records. A similar compilation and analysis has been undertaken by Yoshito Tsuchiya in Japan; for the Osaka area he has established the frequency of storm surge disasters, and has noted that since AD 700 the average interval between major storm surges is about 150 years, which is close to the recurrence interval for the disastrous Moroto Typhoon of 1934.

Neither the instrumental record nor the documentary record of storm surges is long enough to make it possible to calculate, with any accuracy, the recurrence interval of extreme water levels and the extent of flooding associated with storm surges. The record can be extended by identifying and dating storm surge signatures in the sediments of coastal

Fig. 2. Flooding associated with the storm surge on Canvey Island, England, 1953 (from *Geographical Journal*, **119** (3), 289.)

lowlands or staircases of gravel beaches in formerly glaciated areas. By dating shells deposited during storm surges in the Dutch sand dunes, Saskia Jelgersma has demonstrated a progressive increase in the height of extreme water levels associated with storm surges since 2450 BC. In coastal freshwater lagoons, isolated peaks of marine planktonic diatoms bear witness to storm surges that have taken place during the past 10 000 years, since the beginning of the present interglacial (the Holocene).   M. J. TOOLEY

Further reading

Lamb, H. H. (1991) *Historic storms of the North Sea, British Isles and northwest Europe.* Cambridge University Press.

Tooley, M. J. and Jelgersma, S. (ed.) (1992) *Impacts of sea-level rise on European coastal lowlands.* Blackwell, Oxford.

Tsuchiya, Y. and Kanata, Y. (1986) Historical study of changes in storm surge disasters in the Osaka area. *Natural Disaster Science,* **8** (2), 1–18.

strain and stress analysis

Strain analysis is a subdiscipline of structural geology concerned with the measurement of the shape changes (strains) associated with geological deformation. Stress analysis (or palaeostress analysis) is a closely allied field devoted to the estimation of past stresses (palaeostresses) from observed rock strains.

The calculation of geological strain involves the comparison of the original and deformed shapes of objects (*strain markers*) enclosed within the rock. Examples of strain markers are fossils and the outlines of sedimentary particles. Most useful are objects with known original dimensions. For example, the original length of certain graptolites can be estimated by counting the thecae present. A measure of length change is the *stretch*; the ratio of new to old lengths. Such measurements commonly reveal that the value of stretch varies according to the direction of the marker being measured. This accords with the concept of the *strain ellipse*; a representation of the state of strain, which considers the final form adopted by an original sphere with unit radius. The results of strain analysis are usually expressed in terms of this ellipsoid, which can be completely described by six quantities: the stretches in the directions of the principal axes ($S_1 \geq S_2 \geq S_3$ (and three angles defining the orientation of those axes. Ooids, that is grains in limestones with near-spherical original shapes, are examples of strain markers which allow direct estimation of strain ellipsoid parameters, though even here the usual lack of information on sizes means that only ratios of

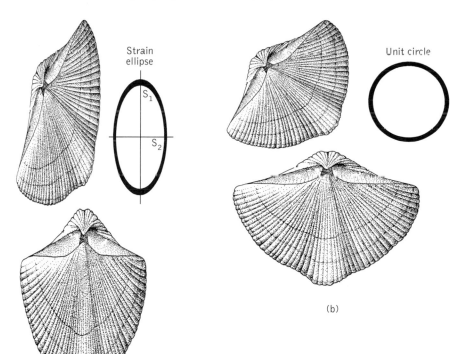

(a)

(b)

Fig. 1. (a) The deformed shapes of strain markers (e.g. deformed fossil brachiopods) allows the calculation of the strain ellipse, with principal axes S_1 and S_2. (b) From a knowledge of the strain ellipse, the original pre-deformation form of objects can be reconstructed.

the principal stretches (R_s) can be determined. Only where *principal stretches* are known can volumetric strain be assessed. A large number of different techniques, many of ingenious design, have been developed for the treatment of natural strain gauges. An example is provided by the deformed fossil brachiopods shown in Fig. 1, which record changes in the length of lines or the modification of angles.

The application of strain analysis is limited by the availability of suitable markers. Furthermore, to give meaningful results the determination of strain should be carried out on samples where the strain is constant in both magnitude and direction (*homogeneous strain*). In practice, however, *heterogeneous strain* is the rule, brought about by differences in mechanical properties between components of the rock, e.g. larger grains versus matrix, lithological changes between beds. In many instances, there may be a competence contrast between strain markers and the rock enclosing them, leading to biased estimates of the bulk strain of the rock. In addition, angular strain gauges lose their sensitivity at some of the high levels of strain commonly found in rocks (R_s greater than 20 : 1). Some structural features used for the computation of strain (e.g. mineral growths in the pressure shadows around porphyroclasts) are of secondary origin and therefore do not record the total or *finite strain* experienced by the rock.

In spite of these problems, strain has been estimated in a wide range of structural environments. Such analyses have provided invaluable information on displacements in deformed rocks (e.g. the amount, direction, and sense of motion within shear zones) and greater understanding of the significance of certain structures (e.g. slaty cleavage, arcuate fold belts). Results from strain analyses can also be used to reconstruct the pre-tectonic geometry (e.g. to compute original thicknesses of rock units in metamorphic regions, and thus to construct balanced cross-sections).

The derivation of palaeostresses poses an altogether more daunting task, because a concept equivalent to the finite strain ellipsoid does not exist for stress. Stress conditions during deformation are likely to have been transitory, changing in magnitude and direction with time. Attempts to estimate stresses in deformed rocks almost invariably entail the measurement of strains coupled with making assumptions about the stress–strain relationship for the rock materials involved. *Fault-slip analysis*, sometimes called *striation analysis*, interprets field data from fault orientations and slip directions in terms of a 'bulk' stress state for the rock volume containing the faults. These methods, which yield estimates for the orientation of the three principal stress axes (σ_1, σ_2, and σ_3) and a parameter expressing the ratio of the differences of the principal stress values, assume that fault slip directions measured from lineations on each fault plane are parallel to the direction of resolved shear stress on the plane of the fault. Similar methods of stress analysis have been applied to microscopic deformation features in crystals, e.g. deformation twins. The orientation of brittle structures such as tensile and shear fracture are used to infer stress orientations at the time of fracture

formation. *Tensile fractures* are oriented normal to the least compressive stress axis, σ_3. *Stylolites*, irregular pillar-and-socket surfaces produced by a pressure solution deformation mechanism, are frequently used as stress direction indicators. The greatest compressive stress axis (σ_1) is assumed to be parallel to the columns. The magnitudes of principal stresses can be inferred from experimentally derived data on brittle strength of rocks. Unfortunately, this is complicated by the fact that rock strength (the principal stresses at failure) is a function of the fluid pressure in the rock pores. Stress magnitudes may be more reliably estimated from intracrystalline deformation structures, e.g. mechanical twinning.

RICHARD J. LISLE

Further reading

Angelier, J. (1994) Fault slip analysis and palaeostress reconstruction. In Hancock, P. L. (ed) *Continental deformation*, pp. 53–100. Pergamon Press, Oxford.

Lisle, R. J. (1994) Palaeostrain analysis. In Hancock, P. L. (ed.) *Continental deformation*, pp. 28–42. Pergamon Press, Oxford.

Ramsay, J. G. and Huber, M. I. (1983) *The techniques of modern structural geology. Vol. 1: Strain analysis.* Academic Press, London.

strategic minerals A strategic mineral is usually defined as one that is vital to the security of a country but has to be obtained largely or entirely from foreign sources because the supplies available within the country concerned would not be adequate in a time of national emergency. A variety of minerals that are vital for defence, for the aerospace industry, for energy supplies, and for transport have at particular times been listed as of strategic importance. During the Second World War, chromium- and tin-bearing minerals, quartz crystals, and sheet mica were among the minerals placed in this category. During the war period, the United States, for example, had to import 65 critical minerals from 57 countries; 27 of these minerals came from foreign countries.

Commercial deposits of valuable minerals are not evenly distributed round the world: they tend to be concentrated in a small number of localities, and some are obtained from only one or two sources. The balance between supply and demand can be disturbed in various ways. The exhaustion of a deposit, the discovery of new reserves, civil war, a change in government, or a new trade agreement can affect supplies. Substitution, recycling, and technological developments can affect demand. Thus, the so-called 'Eighty-day War' in Zaire in 1987 disrupted production of copper and cobalt there; and the opening-up of trade with China made supplies of graphite and antimony available to the United States in the 1980s. On the other side of the equation, part of the demand for tin has been met by recycling since the Second World War; and sources of uranium, thorium, and zirconium became important with the development of atomic energy.

An interesting case history is that of niobium, which was of little commercial interest until new uses made it important in the second half of the twentieth century. Sources of niobium are few, and more than 80 per cent of the world's supplies from mining come from Brazil. More than 90 per cent of niobium comes from pyrochlore, an oxide mineral with a complex formula containing thorium, uranium, titanium, tantalum, and zirconium as well as niobium and various other elements. Pyrochlore occurs in carbonatites, but is mined directly from these rocks only in Quebec in Canada. It is also found in some igneous rocks, but not in commercial quantities. The Brazilian pyrochlore comes from lateritic soils overlying a carbonatite. In the former USSR, niobium has been obtained from loparite, a mineral of the perovskite group, but this is a more expensive source.

Niobium has many uses in industry: as an alloying agent in high-strength and corrosion-resistant steels; in 'superalloys' used in jet engines and other equipment operating at high temperatures; in cutting tools, some superconductors, and optical components. The United States and the European Union are major importers of niobium, and the EU has sponsored research into the development of mining a pyrochlore resource at Lueshe in north-eastern Zaïre, which has been worked since 1987. This deposit was seen as a potential source that would provide for the needs of Europe, but a number of geological problems had to be addressed for production to be fully developed. G. MORTEANI and B. WILCOCK

Further reading
Dennen, W. H. (1989) *Mineral resources: geology, exploration, and development*. Taylor and Francis, New York.

stratification A primary depositional feature of many layered rocks is *stratification*, or bedding. The bedding planes that define the individual layers are commonly more or less smooth, parallel to each other, and roughly horizontal when formed. Bedding planes that were originally smooth can subsequently be disturbed physically by uneven loading or by movement under gravity, or both. They can also be deformed by the uneven action of cementation or other chemical reactions, and modified by the activity of organisms moving over or on them, producing trace fossils. Each bedding plane represents a hiatus or change in the sedimentation process. Today, most sedimentation is viewed as a discontinuous build-up of material with much reworking of the particles before they finally come to rest.

Thus, stratification defines bed thickness, which may give important information about the hydrodynamic or other kind of event responsible for the deposition of the original sediment. Between major bedding planes, small-scale surfaces that are parallel or nearly parallel to them, such as laminae, may occur. A *stratum* is generally regarded as a layer of rock 1 cm or more in thickness; a *lamina* is similar but less than 1 cm in thickness. There are various scales of bed thickness, ranging from thinly laminated to very thickly bedding (over 100 cm) and *megabeds* at more than 100 m. The sediment lying between two bedding planes can generally be regarded as the product of a single depositional event, with no sig-

nificant hiatus. Thick but otherwise structureless beds can be the products of long, unbroken phases of slow deposition, or they can reflect the rapid accumulation of heavy loads of sediment. They are commonly produced where animals have burrowed and thus reworked a bed that was formerly internally laminated or bedded.

In addition to normal or planar bedding, other forms may be distinguished: tabular, wedge, lens, or irregular, in addition to cross-bedding, graded bedding, and rhythmic bedding. *Cross-bedding* is where laminae or closely spaced bedding planes are transverse to the main planes of stratification; *graded bedding* is where particle size decreases with height above the base of the bed; and *rhythmic bedding* is produced where different lithologies are repeated in a regular fashion throughout a sequence of strata.

Stratification may be deformed penecontemporaneously by various agencies (i.e. almost as soon as the sediment is deposited), before the rock is lithified. It is often the most widespread and constant primary sedimentary structure that any geological formation displays, and it provides the framework within which all the others are created.

D. L. DINELEY

Further reading
Collinson, J. D. and Thompson, D. B. (1982) *Sedimentary structures*. Allen and Unwin, London.

Tucker, M. E. (1982) *The field description of sedimentary rocks*. Geological Society, London.

stratigraphy The definition of stratigraphy is simple: it is the study and description of rock successions. Its more philosophical aspects, especially those concerned with the extent to which the physical nature of the record can be used to mark the passage of geological time, are less easy to grasp and have been much debated. Stratigraphy seeks to interpret rock successions as sequences of events in the history of the Earth. Many different field methods and laboratory techniques contribute to stratigraphical investigations, while, in its turn, stratigraphy provides the basis for basin analysis, palaeogeography, and the geology of petroleum and other resources. Stratigraphy is the heart of much historical geology, because it reveals the order in which past events and processes occurred. The Cryptozoic part of the geological record, with its largely unfossiliferous and commonly highly deformed rocks is much less susceptible to the techniques of stratigraphy than is the Phanerozoic (fossiliferous) part.

Stratigraphy figures large in the early history of the Earth sciences. Between the Renaissance and the Age of Reason the foundations of geology were laid in western and central Europe, where stratified rocks were ubiquitous. Nicolaus Steno's mid-seventeenth century recognition of the law of superposition and other fundamentals of what we might call field relationships was soon followed by classifications of rock types grouped into successions. These ultimately became the 'stratigraphical column' in use today. The English civil engineer William Smith (1769–1839), sometimes referred to as the 'Father of Stratigraphy' laid the foundations of biostratigraphy and produced a geological map of England and Wales in which rock units correlated by their fossil contents were depicted. This, the first modern geological map, first appeared in 1815. Throughout the nineteenth century many geologists were to investigate their local geology, and the geological map of Europe began to take shape. A primary objective almost everywhere was to establish local stratigraphical successions. This brought many new formation names into the literature. Correlation with the aid of fossils was highly successful and made it possible to begin building the stratigraphical column to the point where it seemed to represent every moment of Phanerozoic time, if not some Proterozoic time too. Charles Lyell in his *Principles of geology* in 1833 included a columnar diagram in which the main divisions of rocks and the corresponding periods of Phanerozoic time were represented. He was among the first to appreciate—indeed to demonstrate—the magnitude of the time involved. He argued that a vast

period of time was necessary on two counts: the uniformity of the laws and processes that affect the accumulation of stratified rocks, and the rates of these processes.

Disputes and disagreements about the relative positions of newly described sections and formations, and the terminology to be employed, abound in the literature of the time. One of the most active and important scientific bodies that sprang up early in the nineteenth century was the Geological Society of London, which provided a forum for the presentation and discussion of new facts and ideas.

Today's stratigraphical column has evolved as knowledge of the rocks has improved, yet it still contains inconsistencies that derive from the divisions employed more than a century ago. The stratigraphy of the more recent geological periods is more accessible than that of very ancient times, and the problems encountered in the higher (i.e. younger) parts of the stratigraphical column tend to be different from those found below. As research continues and our methods of geochronometry improve, some of the problems may be resolved,

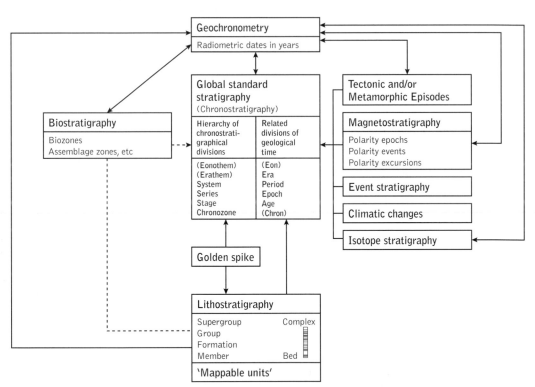

Fig. 1. Various categories of stratigraphy and the links between them. The basic data are acquired from mapping or drilling practice; correlation and, possibly, biostratigraphy follows. The other procedures and techniques are means by which chronostratigraphy is supported. They are becoming increasingly detailed and quantitative. Biochronology is the integration of radiometric data with biostratigraphy. The 'golden spike' signifies the selection of a section and a point within it at which a geological datum is designated. Event stratigraphy is the ordering of events, rather than the rocks and their fossils. Isotope stratigraphy is based upon the proportions of isotopes of elements such as strontium, and is particularly useful in dealing with marine sedimentary rocks. (After C.H. Holland (1978) *Lethaia* **11**, 85–90; (1986) *Journal of the Geological Society* **143**, 3–21.)

but it remains a truism that the older rock formations, and especially those that have been severely deformed, are the more difficult to date and correlate.

Stratigraphy, then, was originally concerned with rocks, that is, lithostratigraphy; it soon afterwards included the ranges of fossils throughout the successions, biostratigraphy; and the more abstract chronostratigraphy. The techniques of magnetostratigraphy make use of the stratigraphical variations in the magnetic properties of rocks as a basis for correlation. The dating of rocks in years ('absolute dating') is the science of geochronometry, which today usually entails isotope analysis. Figure 1 illustrates links between different stratigraphical approaches, some of which are discussed below in more detail.

Lithostratigraphy, concerned with the definition and naming of rock units, relies upon the lithological characteristics and geometry of the rock bodies. The basic unit is the *formation*; it is commonly regarded as the smallest mappable unit, though there are many exceptions. Formation boundaries may be sharp or gradational, and thicknesses vary from a few metres to several hundreds. Within the formation local variations may distinguish one or more *members*, each of which may consist of several *beds*, the bed being the smallest recognizable formal unit of all. Several contiguous formations may be distinguished as a *group*; for reasons of local convenience or genetic similarity several groups together may be considered as *supergroup*.

Biostratigraphy is, as we have seen, the use of fossils for correlation and the organization of strata into units on the basis of their fossil content. Each basic unit is a *biozone*, identified by a characteristic *taxon* (or *taxa*) of fossil (or fossils). *Sub-biozones* may be distinguished on the basis of small faunal or floral variations within a biozone. Many biozones may also be called *range biozones* because they exhibit an aspect of the total range of the index species. Different are *assemblage biozones*, which are characterized by unique assemblages of taxa, several of which have different stratigraphical ranges from the others. These biozones are commonly of restricted or local occurrence.

Chronostratigraphy is the unifying construct that defines (ideally by international agreement) boundaries for systems, series, and stages (*see* Appendix 1a). Its ultimate product is the global stratigraphical column in which every moment of Phanerozoic time is represented by strata which by their contained fossils are distinct from those above or below. It is a compilation that was begun two centuries or more ago and is still imperfect, if not incomplete. So much confusion has been generated in the past by incorrect correlations, imprecise definitions of stratigraphical units, and other inconsistencies that a regular procedure has been devised for defining stratigraphical units. This procedure involves the selection and definition of a *Global Stratotype Section and Point (GSGP)* to fix such a boundary. Because most fossiliferous formations are of marine origin, attention has been focused upon these strata in the belief that the most continuous and fullest record of the evolution of life (and hence of the passage of Phanerozoic time) has been recorded in the seas. Other habitats and

Table 1. Chronostratigraphic divisions and time divisions

Chronostratigraphic divisions	Corresponding divisions of geological time
Eonothem	Eon
Erathem	Era
System	Period
Series	Epoch
Stage	Age

environments are prone to many disruptions and variations, and they consequently present difficulties in establishing comparable standards.

The procedure used for establishing a GSGP entails collection of all possible biological and other information about the rocks in an unbroken stratigraphical section and the selection of a point in the rock sequence that can be defined by the fossils around it. Correlation elsewhere with this point has to be achieved as accurately as possible, using all the available data. GSGPs are used to define the base of a standard *chronostratigraphic unit*. In effect this also defines the top of the underlying unit. The section containing the point is known as the stratotype section, and the point is said to be designated by an imaginary 'golden spike'. All this is achieved by investigation and agreement under the aegis of the Commission on Stratigraphy (and its Subcommissions), an offspring of the International Union of Geological Sciences.

The chronostratigraphic hierarchy of divisions corresponds to divisions of time, as listed in Table 1.

Most of the names in use for the series and stages refer to the place where the type section is located. Ultimately, it is hoped, a global stratotype section and point will be designated for every stage.

When the detailed biostratigraphic division of a set of rocks is undertaken, zones may be established at the outset which eventually prove to be of great value in very wide geographical, if not global, correlation. These zones have been called *chronozones*, the smallest chronostratigraphic unit. They are exemplified by the ammonite biozone stratigraphy of the Mesozoic, but so far few have been established for the Palaeozoic erathem.

Geochronometry is the application of techniques which determine the ages of rocks in years. Many of these techniques have been concerned with assessing the rates of geological processes and using these values to obtain the age of their products at specific sites. The study and analysis of sections of annual clay and silt laminae (varves) from Quaternary lakes has had some success, and the measurement of the growth rings of long-lived trees has led to the establishment of dendrochronology as useful in archaeological and Holocene studies. Radiometric or isotopic dates, however, are the most favoured dating techniques; radiocarbon dating refers to isotope analysis of carbon from biological materials not older than about 140 000 years.

High-resolution stratigraphy is the term applied to the kinds of precise stratigraphical resolution that may be possible on the basis of microfossil biostratigraphy, magneto-stratigraphy, carbonate analysis, and oxygen or other stable isotope stratigraphy. It has great value in deciphering the sedimentary record where it reflects specific geological events (event stratigraphy), and it has been much influenced by fine-scale stratigraphical studies of ocean sediment sequences. These approaches have been singularly successful where Recent and Cenozoic sediments are concerned, and they are also increasingly applied to Mesozoic and Palaeozoic successions. Some of the studies have included attempts to establish rates of geological processes such as erosion, sediment transport, and organic evolution.

Ecostratigraphy, according to the American pioneer in this field, A. J. Boucot, covers 'the evolutionary, ecologic, biogeographic, stratigraphic correlation and basin analysis consequences of the basic fact of biofacies'.

The study of regional stratigraphy can make it possible to recognize the uneven and geographically varying rates at which sedimentary basins subside and are filled and to reveal the erosional history of upland areas. The constant shift and evolution of drainage, glacial, and other physiographical systems may be indicated, and, most significant of all, the advance and retreat of the shoreline can be measured by stratigraphy. By these means, too, relative rates of geological processes can be demonstrated. During sedimentation certain distinctive beds may be formed time and time again in a regular pattern, and we refer to deposition of this type as *cyclic sedimentation* (see also *cyclicity*).

Because the distribution of many important economic raw materials, especially the hydrocarbon fossil fuels, is controlled by stratigraphy, this branch of the Earth sciences has been vigorously pursued by applied geologists. The first petroleum well in the United States was sunk in the mid-nineteenth century, and within a few decades companies were employing geologists in the search for petroleum deposits. Since then the industry has invested immense capital sums in exploration work in many parts of the world. The job of stratigraphers in the petroleum industry is to locate the strata that generate and retain commercial quantities of oil and gas. Coal, evaporites, and many other important materials also occur as stratified deposits in association with specific lithologies and under special stratigraphical conditions. To assist in the exploration for these materials at depth, techniques of rock sampling in boreholes, and the measurement of the geophysical properties of stratigraphical formations have reached a very high degree of sophistication. Some of these techniques have been employed also in purely academic research in stratigraphy, with far-reaching effects of the interpretation of the stratigraphical record. D. L. DINELEY

Further reading

Blatt, H., Berry, W. B. N., and Brande, S. (1990) *Principles of stratigraphic analysis*. Blackwell Scientific Publications, Oxford.

Briggs, D. E. G. and Crowther, P .C. (eds) (1990) *Palaeobiology*. Blackwell Scientific Publications, Oxford.

Dineley, D. L. (1992) Stratigraphy and dating. In *Encyclopedia of Earth Systems Science*. Academic Press, New York.

Prothero, D. M. (1989) *Interpreting the stratigraphic record*. W. H. Freeman, New York.

Schoch, R. M. (1989) *Stratigraphy: principles and methods*. Van Nostrand Reinhold, New York.

stratosphere The stratosphere is the atmospheric layer between the troposphere and the mesosphere and extends from about 12 km altitude to about 50 km. Within this layer the temperature rises with altitude to a maximum of about 0 °C at 50 km. A region in the atmosphere in which temperature increases with height is very stable in relation to vertical movement. Even large thunderstorms, which can produce updraughts of several metres per second, cannot penetrate far into the stratosphere. *Stratosphere* literally means 'layered sphere', and air in one layer remains in that layer without significant upward or downward motion. This contrasts strongly with the troposphere, where vertical overturning of air is the norm. In the troposphere, any small dust particles in the air are swept around rapidly and can fall back to the surface in a matter of days or weeks. In contrast, dust particles injected by very large volcanic eruptions can reside in the stratosphere for a year or more before slowly falling back to the troposphere. Gases that reach the stratosphere can remain there for many years, particularly if they are chemically inert.

Although 90 per cent of the mass of the atmosphere lies below the stratosphere, and the chemical composition of the stratosphere is almost exactly the same as the troposphere, the stratosphere has enormous significance because it contains minute amounts of ozone. Even at 25 km altitude, where the maximum concentrations of ozone are to be found, only about ten molecules in every million are ozone. However, the temperature structure of this part of the atmosphere, the radiation received at the Earth's surface, and, ultimately, the fate of many living things depend crucially on these ozone molecules.

Ozone absorbs ultraviolet (UV) light from the Sun. Ultraviolet light has a wavelength of less than 0.39 μm (micrometres). What happens to UV light that falls on the top of the atmosphere depends on its wavelength. If the wavelength of the UV light is less than 0.246 μm, it is capable of splitting oxygen molecules (O_2) into their two component oxygen atoms (O + O). Nearly all of this very short UV light is absorbed in the thermosphere at altitudes above 80 km, but a very small amount reaches the stratosphere and separates oxygen molecules there (Fig. 1). The odd oxygen atoms (O) can then combine with some of the many oxygen molecules (O_2) to form a molecule with three oxygen atoms: ozone (O_3). Some of the oxygen atoms can combine with other single oxygen atoms to form O_2 molecules, but, because oxygen molecules are much more common, ozone is more likely to be formed. These reactions in which oxygen com-

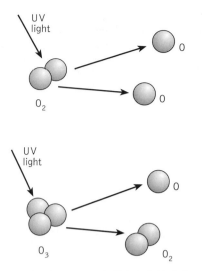

Fig. 1. Oxygen molecules are separated into individual atoms by ultraviolet light.

bines to form either O_2 or O_3 cannot occur without another molecule also being involved. The role of this extra molecule, which can be of any kind but is most likely to be a nitrogen molecule, is to carry away excess energy produced during the combination. UV light at wavelengths less than 0.31 μm can break up ozone molecules into a single oxygen atom and an oxygen molecule ($O + O_2$). The net effect of all these processes is that ozone is being formed and destroyed continuously. A fine balance has been achieved during the evolution of the atmosphere, resulting in a layer in which a small amount of ozone is concentrated (if such a word can be used for concentrations of ten parts per million) in a layer between 20 and 30 km above the Earth's surface. UV light is absorbed in the process that leads both to the formation and destruction of ozone. The energy carried by that light is left in the stratosphere where the light is absorbed, and it is this energy that is responsible for the increase in temperature with height above the troposphere that is observed.

Why should the ozone layer exist at exactly the height it does? The same delicate balance that maintains the reactions which produce and destroy ozone is the reason why the ozone layer peaks at 25 km altitude. At greater altitudes there is more UV light, and more single oxygen atoms are produced. However, the air is less dense and the probability of three molecules colliding to form ozone is less than the probability that an ozone molecule will combine with a single oxygen atom to produce two oxygen molecules ($O + O_3 \rightarrow 2O_2$). This both destroys ozone and removes some of the single oxygen atoms needed to form ozone. At altitudes below the ozone peak very little UV radiation is found because it is absorbed by the ozone above. As a result, both creation and destruction of ozone below the peak in the ozone layer occur more slowly

than above, and the observed distribution of ozone at these levels is determined largely by atmospheric motions.

The reactions which maintain the equilibrium of the ozone layer are more complicated than this relatively simple explanation because more minor constituents in the air are involved in reactions. However, before the late 1970s, all observations of the ozone layer suggested that it was in a stable equilibrium state in which the ozone production processes were balanced exactly by ozone destruction.

The ozone hole

Human activities have disturbed the delicate balance which maintains the ozone layer. CFCs (chlorofluorocarbons) are still used in refrigerators and as propellants for blowing plastic foam insulation and fast-food containers. They were in the past widely used as propellants for spray cans. Although their use in spray cans has largely ceased, the CFCs in the stratosphere have a very long residence time, and much of the CFC currently in the stratosphere came from spray cans.

In the early 1970s it was suggested that CFCs could destroy ozone, but it was not until the late 1970s that ozone depletion was actually observed by scientists at the British Antarctic Survey. For many years they had routinely made measurements of the amount of ozone at different heights over Antarctica. A detailed study of their records showed that in September and October (spring in Antarctica) the total amount of ozone is depleted by as much as 40 per cent. At

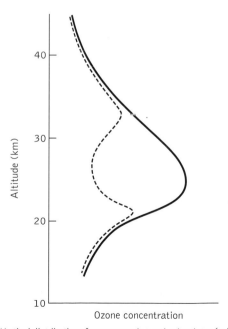

Fig. 2. Vertical distribution of ozone over Antarctica in winter (solid line) and spring (dashed line), before and after development of the ozone hole.

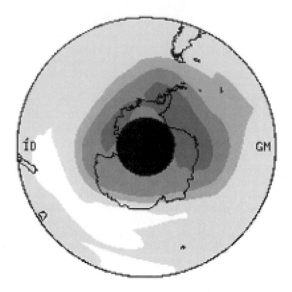

Fig. 3. Ozone distribution in the southern hemisphere in spring observed by the Upper Atmosphere Research Satellite. The black central ring is a region which the satellite cannot observe. The darkest grey area has less than half the ozone of the lightest grey area.

some levels it is almost completely destroyed (Fig. 2). Satellite observations have now established that the ozone depletion is mainly limited to an area slightly larger than the Antarctic land mass (Fig. 3), but depletion on such a large scale is not observed elsewhere. Why should this happen over Antarctica and why in spring? The atmospheric circulation in the Antarctic stratosphere is marked by a belt of high winds at about 50 °S which defines a region within which air is relatively isolated from the rest of the atmosphere. During the Antarctic winter this region is in darkness and cools to temperatures as low as −85 °C. Under these conditions diffuse ice clouds, known as polar stratospheric clouds (PSCs), are formed. Chemical reactions occurring on the surfaces of the ice particles in these clouds allow chlorine molecules (from CFCs) to become fixed and stored in the clouds. This large reservoir of chlorine is released in the spring when sunlight returns, temperatures rise, and the clouds evaporate. The sudden release of the stored chlorine allows rapid destruction of ozone to occur and the ozone hole is formed. Over a period of a few months the usual equilibrium is regained and the hole is removed.

Although the ozone hole exists for only part of the year and over an inaccessible part of the world, it is natural to ask if the same process could occur elsewhere. Less dramatic, but nonetheless significant, ozone depletion has been measured in the southern hemisphere latitudes outside the belts of high winds. This could lead to increased ultraviolet radiation reaching the surface in places such as Australasia and South America. In the northern hemisphere most research has concentrated on the Arctic. A belt of high winds similar to that in the southern hemisphere also exists in the northern hemisphere, but it is not a complete ring; the effect of oceans and continents is greater in the northern hemisphere. Breaks in the belt of winds allow air from outside the polar regions to mix with polar air. This results in higher temperatures than are observed over Antarctica and only a few, relatively isolated, polar stratospheric clouds form. Some depletion of ozone is observed in the northern hemisphere, though not on the same scale as the Antarctic ozone hole.

CHARLES N. DUNCAN

Further reading

Ahrens, C. D. (1994) *Meteorology today.* West Publishing Co.

stress analysis Stress analysis is the process used to estimate the distribution of forces throughout a mass of material. In the geosciences, stress analysis can be used to ascertain whether a material is approaching, or has reached, a state of limiting equilibrium (or failure); if the appropriate stress–strain relationships are known, it can also be used to calculate the resulting deformations. Calculations of this kind are vital in many areas of engineering geology.

A stress is a force divided by the area over which it acts. If, for example, a vertical cylinder of rock, with a horizontal cross-sectional area a, is subjected to a vertical compressive force N, which will tend to reduce its height, the cylinder is then under a compressive normal stress (that is, a compressive stress acting parallel to its axis) of N/a (Fig. 1). The converse is the application of a tensile normal stress—one that tends to pull the ends of the cylinder apart—which will obviously tend to lengthen the cylinder. The customary units for stresses are newtons per square millimetre (N mm^{-2}) and kilonewtons per square metre (kN m^{-2}). Two stresses will usually act on any

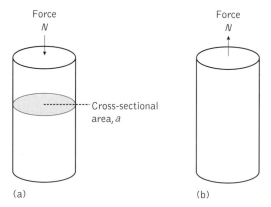

Fig. 1. (a) A compressive force N applied to a vertical cylinder of rock with horizontal cross-sectional area a; the resulting stress is N/a. (b) The same cylinder subjected to a vertical tensile force N; the resulting tensile force is $-N/a$.

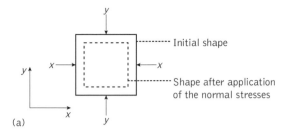

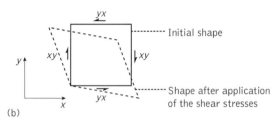

Fig. 2. Normal and shear stresses: (a) positive normal stresses and the associated deformations; (b) positive shear stresses and the associated deformations. In each case solid outlines show the initial shape; dashed lines show the shape after the application of the stresses. Full arrows denote normal stresses; half arrows denote shear stresses.

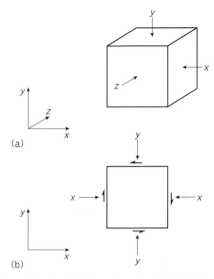

Fig. 3. Three-dimensional and two-dimensional states of stress. (a) General three-dimensional state of stress. Similar stresses act on the hidden sides of the cube. (b) Two-dimensional state of stress. Positive stresses are shown.

chosen plane in a mass of material. One of these is the normal stress, as described above, which acts at right-angles to the plane; the other is the shear stress, which acts along the plane, tending to distort the material as shown in Fig. 2. (It should be mentioned that in many textbooks on stress analysis tensile normal stresses are treated as positive, and compressive normal stresses are treated as negative. The opposite convention is, however, usually adopted for geomaterials (rocks and soils), because the tensile strengths of these materials tend to be low in comparison to their compressive strengths.)

Many problems in stress analysis can be analysed in two dimensions. Two normal stresses and one shear stress then need to be specified to define the state of stress at any specified point in the material (Fig. 3). More generally, stress states exist in three dimensions. Three normal stresses and three shear stresses will then need to be considered, as shown in Fig. 3a.

At any point in a continuous mass of material the stresses will change as the direction of the plane that is being considered is varied. These changes can conveniently be shown in graphical form by a *Mohr's circle of stress*. In practice, numerical methods, such as finite elements, are commonly used to carry out stress analysis.

In soils, which consist of uncemented, or weakly cemented, particles, there are voids between the particles. These are generally interconnected and are filled with liquid, usually water, or gas, such as air. Any stress applied to the soil may be taken, wholly or in part, by the pore fluid phase. Assuming that the fluid has no shear strength, the contribution of the fluid phase will then be to help support the normal stress. Any total normal stress will therefore be taken by a normal stress carried by the assemblage of soil particles, the effective normal stress, and the normal stress in the pore fluid. The total normal stress will therefore be the sum of the effective normal stress and the normal stress in the pore fluid. This is known as the *principle of effective stress*. If the soil is fully saturated, that is, if the pore space is completely filled with water, the normal stress in the pore fluid is simply the pore water pressure. If the voids are partially filled with water, the expression can be amended. If the soil has an appreciable air content, then the total normal stress is often taken as zero. Soils can thus be analysed in terms of total or effective stress. The total stress is often used for convenience, although as the shear resistance of soil derives from the interaction of the particles, the effective stress should be used.   T. COUSENS

stress field of the Earth The solid part of the Earth (the crust and mantle) has a short-term strength (over periods of seconds or minutes) that allows the shear waves produced by earthquakes to travel through the Earth. However, over longer periods of time (years) much of the Earth's mantle behaves as a weak medium, capable of flow, thus enabling convection currents and plate movements to take place.

The outer, strong, layer of the Earth, known as the lithosphere, has a mean thickness of around 100 km, and consists of the crust and uppermost part of the mantle. It is defined on

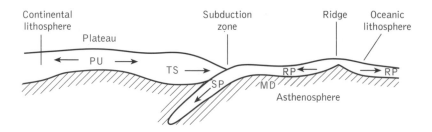

Fig. 1. Main sources of stress in the lithosphere. RP, ridge push; SP, slab pull; TS, trench suction; MD, mantle drag; PU, plateau uplift.

the basis of its strength in relation to the underlying weaker asthenosphere. Because of its relative strength, the lithosphere can support substantial stresses, and forces applied to one part of the lithosphere can be transmitted through considerable distances.

Stress systems affecting the lithosphere are divided into renewable and non-renewable types. Renewable stresses are those that persist because of continued re-application of the causative forces despite continual dissipation of the resulting strain energy. The main sources of renewable stress are plate boundary forces and surface loading of the crust (e.g. by mountain ranges and high plateaux). Non-renewable stresses, resulting, for example, from thermal expansion or contraction or from bending of the lithosphere, do not make a significant contribution to the overall state of stress.

The main types of force acting at plate boundaries are slab pull, trench suction, ridge push, and mantle drag (Fig. 1). The slab-pull force acts on a subducting plate and results from the negative buoyancy of the cool, dense sinking slab. The trench-suction, or subduction-suction, force acts on the upper plate of a subduction zone because the upper plate is fixed to the downgoing slab. The ridge-push force is caused by the lateral spreading effect of the mass of warm, less dense, mantle ma-

terial underlying the ocean ridges. Mantle drag acts on the base of a plate and, according to the direction of flow in the asthenosphere, may either assist or counteract the plate movement. Significant stresses are also produced where a large topographic load, such as a mountain range or high plateau, is supported by material of lower density at depth. These features act in the same way as ocean ridges, exerting horizontal extensional stresses across the uplift and compressional stresses on the adjoining lithosphere.

Estimates of the magnitude of these forces are in the range 0–50 megapascals, acting over the whole thickness of the lithosphere. Over time, however, stress decay in the lower lithosphere causes the stress in the upper lithosphere to be amplified to much higher levels, up to several hundred megapascals, which is enough to cause fracturing in the strongest part of the crust.

Estimates of the current stress field in the upper crust are derived from studies of earthquake source mechanisms and by direct measurement using *in situ* techniques, such as induced borehole fracturing and the use of strain gauges. Data are available for many parts of the world, but are very unevenly distributed; most measurements have been made in the USA, in western Europe, and in Central Asia. It is clear,

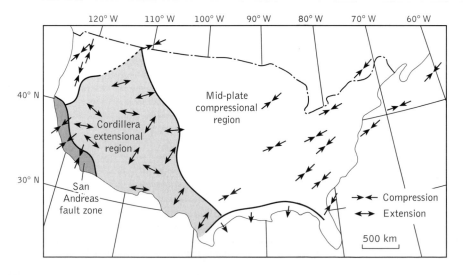

Fig. 2. Stress orientations in the coterminous USA. (After Zoback, M. L., and Zoback, M. (1980) *Journal of Geophysical Research* **85**, (B11) 6113–56.

however, that the horizontal stress orientation and magnitude are uniform over large areas, and that the stable continental and oceanic interiors are in a state of general compression. Localized extension affects continental and oceanic rifts and uplifted plateaux.

Analysis of the upper-crustal stress pattern in the USA (Fig. 2) shows that the stable central and eastern part of the country is characterized by uniform NE–SW to ENE–WSW compressive stress, but that west of this region a wide zone consisting of the Rocky Mountains and Basin-and-Range Province is characterized by horizontal extensional stress varying from NE-SW to NW-SE in orientation. Near the San Andreas fault, along the western continental margin, the stress state changes again to NE–SW compression. The stress state in the extensional regions is consistent with that expected from uplifted plateaux and can be transmitted into the stable continental interior as a compression in an ENE–WSW direction, approximately perpendicular to the boundary of the uplift. However, the NE–SW directions in the eastern USA are not obviously related to Atlantic ridge push.

The stress field in western Europe is also fairly uniform and is characterized by NW–SE maximum horizontal stress. This direction is consistent with a compromise between the ESE-directed ridge push from the North Atlantic in the west and a northwards push from the Alpine front.

It has been pointed out that there is a worldwide correlation between the directions of the stress axes and those of the absolute plate velocities. This suggests that the net plate boundary forces responsible for moving the plates also dominate the stress distribution in the plate interiors.

R. G. PARK

Further reading

Park, R. G. (1988) *Geological structures and moving plates*, Chapter 2. Blackie, Glasgow.

strike-slip tectonics of the continents

Ever since the San Francisco earthquake of 1906, in which up to 7 m of lateral displacement was measured, the social and economic consequences of such movements have been known to the world. Earthquakes associated with lateral movement on land are devastating to human populations. Most of the major ones, and many smaller ones, are the result of strike-slip faults on the continents. Some, such as the San Andreas fault system of California and the Alpine Fault of New Zealand, although they involve continental rocks, are essentially plate margins. Others are associated with continental collision and thus today occur over a very wide belt extending from the lands surrounding the Mediterranean (the Alps and North Africa) through Turkey, the Middle East, and into the Himalayas, China, and the Indian subcontinent. The belt is last seen in South-East Asia. The abundance of strike-slip faults in the Middle East explains why there are so many vivid descriptions of the violent destruction of cities in the Old Testament.

Although the effects of strike-slip movements are now well known, it took many decades before the geological establishment came to appreciate the length of such faults. For example, it was not until 1942 that it was realized that the Alpine Fault of New Zealand was a continuous feature 600 km long, and the idea that displacement along it could be hundreds of kilometres (490 km being postulated) had to await the 1950s. Similarly with the San Andreas Fault, where, although it had been argued in 1926 that there was 40 km of displacement, it was 1953 before a 640 km post-Jurassic displacement was postulated. In 1958 evidence was accumulated for about 100 km of sinistral movement for the Dead Sea rift.

The 1950s were thus the decade when most hypotheses of global tectonics were based on concepts of lateral movement along strike-slip faults, although terms such as wrench faults or megashears were more commonly used. Moody and Hill, for example, saw the Earth as a worldwide system of wrench faults, thought to be the result of north–south global compression. Carey explained his expanding Earth by movements along megashears.

Although small-scale pre-Mesozoic strike-slip faults had been recognized since the start of the twentieth century, it took the work of W. Q. Kennedy on the Great Glen fault of Scotland to demonstrate to geologists that a major lateral displacement had occurred. He matched two apparently similar granite intrusions on either side of the fault and argued that they had been displaced sinistrally about 100 km. Interestingly, as so often happens when a new quite convincing hypothesis is postulated which is later shown to be true, it now appears that the two granites are not the same. It is now thought that the Great Glen fault, and its continuation into the North Sea has had a very complex geological history of movements, both sinistral and dextral, acting therefore as a fundamental fault or lineament (see *lineaments*). Although Kennedy's evidence is now discredited, the hypothesis, that large-scale lateral movement along strike-slip faults can occur in older rocks, still stands.

During the period of plate-tectonic euphoria when most Earth scientists, especially geophysicists, thought that global tectonics could be explained by a simple combination of crustal creation by ocean-floor spreading and loss by subduction, that is, by a simple two-dimensional model, strike-slip movements were relegated to a minor role. Many global tectonic maps of the early 1970s failed to show any strike-slip faults, even the San Andreas fault. There were, however, always persistent believers in the importance of strike-slip faults, many of them working in the research laboratories of the Esso Petroleum Company, and from the 1980s onwards a more balanced view has prevailed. Strike-slip movements are now recognized as a major feature of global tectonics, especially on continents and in the later stages of continental collision in ancient orogenic belts.

When seen on a map, strike-slip faults are characterized by a linear or curvilinear principal displacement zone, along which there is a great variety of features (Fig. 1) resulting

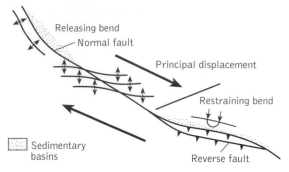

Fig. 1. Map view of an idealized dextral (right-handed) curvilinear strike-slip fault showing alternate zones of extension at the releasing bend and compression with folding and thrusting at the restraining bends. Note that basins can also develop adjacent to the restraining bend. (After Christie-Blick and Biddle, in Biddle and Christie-Blick (1985), pp. 1–35.)

from the orientation of the fault relative to the regional shear direction. Areas of extension, subsidence, and basin formation alternate with areas of compression and folding, thrusting, and uplift, sometimes involving basement rocks at the surface.

The orientation of structural features such as folds and faults is not random in strike-slip belts (Fig. 2). Experiments using clay-cake models in the 1970s, borne out by studies of geological examples, show that if the movement along a

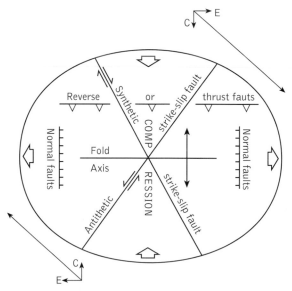

Fig. 2. Idealized structural pattern resulting from a NW–SE dextral (right-lateral) simple shear couple, under ideal conditions. (From Reading (1980), based on Harding (1974) *Bull. AAPG.*)

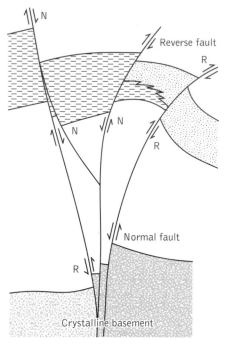

Fig. 3. A transverse profile through an idealized strike-slip fault or 'flower structure'. Note: (1) basement is involved, showing both normal and reverse separation; (2) there are abrupt changes of thickness and facies across the faults; (3) a single fault may show both normal and reverse separation in the same profile and there may be variable magnitude and sense of separation for different horizons offset by the same fault. It should also be remembered that if one were to see other profiles along the length of the strike-slip fault, each profile would show different magnitudes, dip directions, sense of separation, and sedimentary histories. (From Christie-Blick and Biddle, in Biddle and Christie-Blick (1985), pp. 1–35.)

strike-slip fault is known, one can then predict the orientation of folds and faults. Alternatively, by measuring their orientation, one can predict the sense of movement along the fault. In profile (Fig. 3), there is commonly a subvertical fault zone that ranges from braided to upward-diverging. These fault zones typically consist of a relatively narrow, subvertical displacement zone at depth that passes upwards into a zone of braided splays that diverge and rejoin both upwards and laterally. These upward-diverging splays, known as 'flower structures', are recognized particularly by petroleum geologists, for they are well seen on seismic sections. Although many strike-slip faults penetrate igneous and metamorphic basement rocks, others terminate at depth against low-angle detachment faults that may be located entirely within the sedimentary section. For example, the classic Vienna strike-slip basin is situated on a detached nappe of the Carpathian flysch belt.

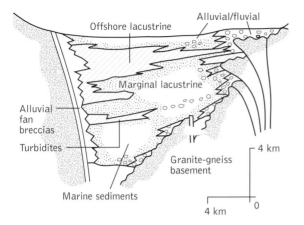

Fig. 4. Cross-section through the asymmetric Pliocene Ridge Basin of California, one of many such basins associated with the San Andreas fault system. Note the range of sedimentary facies, the lateral restriction of facies (very coarse alluvial conglomerates passing rapidly into fine-grained lacustrine muds), and enormous thickness of sediments; the vertical and lateral scales are the same. (After Reading (1980).)

The basins that develop along strike-slip faults are all relatively small, especially when compared with rift basins (see *extensional tectonics in the continents*). They tend to be about 20 by 50 km in area. They are also relatively short-lived, because, before long, extension changes to compression. They are then filled as rapidly as they were formed, since there is an abundant source of sediment close at hand. Another important feature is that, since they open progressively from one end, their fill is asymmetric along their length, as well as across.

Some strike-slip basins at continental margins are filled by marine sediments, as in the Californian Borderland basins such as the Pliocene Ventura Basin, where the first North American turbidites were recognized in 1950 by Kuenen and Migliorini. Most continental strike-slip basins are filled by lacustrine sediments. In the Californian Ridge Basin (15 by 40 km in area) relatively humid conditions gave rise to over 9000 m of coarse-grained marginal breccias and conglomerates deposited as alluvial fans on the western margin, where the major fault existed at the time of sedimentation. This contrasts with better-sorted, laterally persistent, finer-grained marginal sediments on the opposite margin, where large alluvial fans and deltas extend further into the basin (Fig. 4).

Very different sediments are found in the Dead Sea, which is the classic on-land strike-slip (pull-apart) basin (Fig. 5). It was shown by Quennell in 1958 to have been formed by side-stepping of the Dead Sea fault as the Palestine plate to the west moved sinistrally with respect to the Arabia plate to the east. Curvature of the fault to the north, in the Lebanon, results in uplift and run-off of detrital sediment, but this is limited in this arid area; the River Jordan carries very little sediment, and sedimentation in the Dead Sea is dominated by evaporites with only marginal alluvial fans that barely reach the lake itself.

The basins described above form along the line of major strike-slip faults that, although they occur on continental crust, do, at least in part, form plate boundaries. In north-eastern China, however, there appears to have been an area of continental crust about 500 km wide that was subject to a north-easterly trending, sinistral tectonic regime, continuing perhaps to the present day. A transtensile regime was followed by transpression, leading to the creation of some 60 elongate

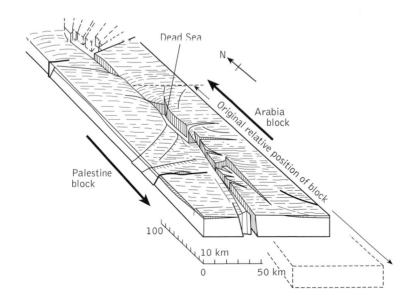

Fig. 5. The Dead Sea pull-apart rift, formed by sinistral strike-slip motion of about 100 km at a rate of 6–10 km per million years, with opening and sedimentation progressing from Miocene times in the south to the present-day Dead Sea in the north. (From Quennell (1958).)

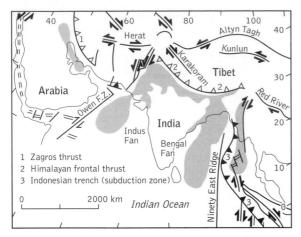

Fig. 6. Map showing continental collision of the Indian and Arabian plates with Asia and the associated pattern of strike-slip faults. (From Reading (1982).

north-east-trending Late Mesozoic basins (30 by 150 km) filled by lacustrine sediments with abundant coals.

Continental crust responds to continental collision in three principal ways: by thrusting, by uplift, and by strike-slip movements. Strike-slip movements increase in importance as collision progresses. Theoretical models of continental collision have suggested that the shapes of the colliding blocks, especially the overriding one, govern the direction and sense of movement along faults as blocks adjust to the pressures. To some extent this is true. For example, the impact of the Indian plate on Asia as it forms the Himalayas is in part resolved by lateral translation at the margins of the Indian plate with sinistral movement along the borders of Pakistan, where the Quetta earthquake struck in the 1930s, and dextral movement along the faults of South-East Asia (Fig. 6). Although faults in Tibet respond to some degree in similar ways, the position of the faults, and in some instances the sense of movement, are governed more by the position of an existing fundamental fault or lineament and the existence of innumerable mosaics of existing blocks (microplates), both large and small.

HAROLD G. READING

Further reading

Ballance, P. F. and Reading, H. G. (ed.) (1980) *Sedimentation in oblique-slip mobile zones*. International Association of Sedimentologists Special Publication, 4. Blackwell Science Publications, Oxford.

Biddle, K. V. and Christie-Blick, N. (eds) (1985) *Strike-slip deformation, basin formation, and sedimentation*. Society of Economic Paleontologists and Mineralogists Special Publication 37.

Nilsen, T. H. and Sylvester, A. G. (1995) Strike-slip basins. In Busby, C. J. and Ingersoll, R. V. (eds), *Tectonics of sedimentary basins*, pp. 425–57. Blackwell Science, Cambridge, Massachsetts.

Quennell, A. M. (1958) The structural and geomorphic evolution of the Dead Sea Rift. *Quarterly Journal of the Geological Society, London*, **114**, 1–24.

Reading, H. G. (1980) Characteristics and recognition of strike-slip systems. In Ballance, P. F. and Reading, H. G. (eds), *Sedimentation in oblique-slip mobile zones* pp. 7–26. International Association of Sedimentologists, *Special Publication* 4. Blackwell Science Publications, Oxford.

Reading, H. G. (1982) Sedimentary basins and global tectonics. *Proceedings of the Geologists' Association*, **93**, 321–50.

stromatolites Stromatolites are columnar or mound-shaped, finely layered, rock structures that are found in rocks up to 3500 million years old. Their origin was, and to some extent still is, controversial, but studies of modern stromatolites, such as those found in Shark Bay, Western Australia, indicate that they are formed by the activities of cyanobacteria (blue-green algae) in shallow-water marine environments. These bacteria form a living mat of intertwined filaments at the surface of the stromatolite, which also harbours numerous other types of micro-organisms. Light energy is required to power cyanobacterial photosynthesis, however, sediment particles tend to wash onto the surface of the stromatolite, gradually blocking the light and decreasing the light energy available. The microbes are phototactic and will move up through the accumulated sediments to avoid being totally deprived of light. A new growth surface is thus developed periodically and the sediments accumulate as a series of fine laminations. There is also the possibility that some stromatolites may not be of biological origin.

Stromatolites form the oldest known reefs, over 2000 million years old, some of which were up to 450 m thick. Although abundant and widespread in the Proterozoic, where they are biostratigraphically useful, they declined sharply in the Late Precambrian. Since then predation by Metazoa, together with competition from red and green algae and other marine plants, have confined stromatolite-builders mostly to hypersaline and freshwater environments. DAVID K. ELLIOTT

strong-motion seismology Since the 1970s, a branch of earthquake seismology (see *earthquake seismology*) has evolved, focusing on the assessment of ground shaking in strong earthquakes. Ground shaking is the most significant hazard generated by earthquakes because of the resulting potentially severe and widespread damage (see *earthquake hazards and prediction*). As a result of attempts to understand, predict, and mitigate this hazard, strong-motion seismology emerged as a significant application of the science, particularly in the 1990s.

It has long been known that (1) the severity of ground shaking increases with increasing earthquake magnitude; (2) ground shaking decreases in strength or attenuates as the distance from the epicentre of the earthquake increases; (3) there are regional differences in the rate at which ground motions attenuate with distance; and (4) geological site conditions (e.g. soil versus rock) can have a significant impact on ground shaking.

Site conditions may be the most significant factor affecting ground motions, at least in terms of earthquake damage. In particular, the amplification of ground motions by soil and unconsolidated sediments has resulted in major earthquake losses. One of the most notable recent examples was the 1985, magnitude (M) 8, Michoacan earthquake, which caused 10 000 deaths and resulted in the collapse of numerous highrise buildings in Mexico City more than 350 km away from its epicentre in the Pacific Ocean to the west.

In a sense, strong-motion seismology has its roots in earthquake engineering and the strong-motion instrument, or accelerograph. In an attempt to predict future ground shaking, earthquake engineers such as Harry Seed of the University of California at Berkeley began analysing accelerograms (strong-motion recordings) of actual earthquakes in an effort to develop empirical relationships between earthquake magnitude, distance, site geology, and some specified ground-motion parameter.

In the United States, strong-motion instruments are generally abundant in California. A few countries which also have notable strong-motion instrumentation programmes include Japan, Italy, Taiwan, and Mexico. Elsewhere, strong motion instruments are rare. As a result of the recent occurrence of several large California earthquakes, such as the damaging 1994, M 6.7, Northridge event and the 1999 Chi Chi, Taiwan earthquake the strong-motion database has increased dramatically in size and in quality. Strong-motion instruments have not only been useful in providing observations of ground motion in the free-field (away from any structures) but also in showing the structural response of buildings, dams, and bridges. Analyses of such records have thus improved understanding of structural response and, hence, enabled engineers to improve seismic design.

Based on the analyses of the strong-motion database, numerous attenuation relationships have been developed, principally for peak ground acceleration of different geographical regions and, more recently, different styles of earthquake faulting (see *earthquake mechanisms and plate tectonics*). The attenuation of ground motions is due to the geometrical spreading of seismic waves and the loss of their energy that results from material damping or absorption of energy by the Earth's crust and mantle.

In the 1990s, theoretical approaches were also developed to predict earthquake ground shaking. These approaches, consisting principally of numerical modelling techniques, which have been validated with empirical data, provide an important tool for strong-motion seismologists because of the general lack of strong-motion data, particularly for large earthquakes at short distances. In this regard, ground motion amplification due to soils has been an important issue in strong ground motion prediction and the focus of considerable research. The near-source aspects of ground shaking

and two- and three-dimensional effects, such as might be due to topographic features and basin geometry, have also been modelled numerically to assess their contributions to the damage that resulted from earthquakes such as the 1971, M 6.7, San Fernando, California and 1995, M 7.0, Kobe, Japan earthquakes.   IVAN G. WONG

structural geology Structural geology is the branch of geology that deals with the description and interpretation of the structure of rocks. It is concerned with the whole range of scales from the regional to the microscopic.

At the simplest level, structural geology describes the geometric relationships between and within rock units. Thus, the geometry of individual units can be planar, folded, or faulted (laterally offset along a break in the unit). Adjacent units may be in conformable contact (that is, they are essentially parallel and continuous) or unconformable (that is, with an angular or temporal discontinuity between them).

In the field, the geometry of a rock unit is characterized by determining its strike (the azimuth of a horizontal line) and dip (the steepest gradient, which is at right angles to the strike line) at several points. When a geological map is available, details of the structure can be obtained by measurements from it.

A large part of structural geology is concerned with inferring the tectonic (deformational) history that has given rise to a particular structure. This may have involved one or more episodes of uplift or subsidence, faulting, folding, intrusion, erosion, or non-deposition.

Small-scale and microscopic observations can be as important as macroscopic ones. For example, slickensides (small grooves and ridges) on a fault surface can reveal the direction of motion on the fault; the way in which mineral grains are deformed can indicate not only the direction of deformation, but often the rheological regime (brittle or ductile) and the type of stress field in which it occurs.

At a regional scale, structural observations are often complemented by palaeomagnetic studies, which can reveal the ancient orientations of the rock units.   ROGER SEARLE

stylolites It is a fundamental axiom of geology that bedding planes indicate changes in conditions of deposition at the sedimentary surface. However, not all the obvious 'bedding' surfaces are true depositional surfaces. In tectonically deformed regions the most obvious stratification may be cleavage surfaces (see *slate*): jointing, too, can often obscure true bedding. In undeformed regions, especially within rocks that are relatively homogeneous such as limestones, dolomites and quartzitic sandstones, a type of 'pseudobedding' can develop.

The process by which this forms is pressure solution, nowadays commonly referred to as pressure dissolution. Because crystalline solids under stress are more soluble than less stressed or unstressed portions, they may be dissolved at points of contact between mineral grains where there is an increased local pressure. This dissolution leads to mass transfer

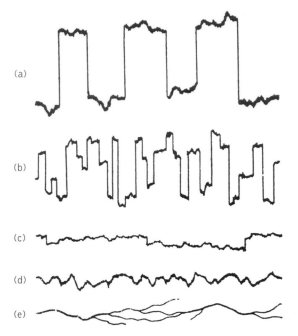

(a)

(b)

(c)

(d)

(e)

Fig. 1. Sketches of pressure–dissolution surfaces. (a) and (b) show columnar types of stylolites; (c) and (d) show serrated surfaces of stylolites; (e) shows a serrated undulatory surface that is a dissolution seam. (From Railsbeck, L. B. (1993) *Journal of Sedimentary Petrology*, **63**, 513–22.)

through an aqueous phase, either by diffusion or bulk flow, and can result in precipitation in less stressed domains.

Both tectonic stress and gravitational loading by overburden can produce pressure dissolution. Since carbonates are more soluble than quartz, pressure solution induced by gravitational loading tends to start at very shallow depths in limestone and dolomites; in quartzites it occurs at depths of 1000 m or more.

The surfaces produced by pressure dissolution may be either smooth and undulatory (*dissolution seams*) or serrated (*stylolites*), with gradations between the two (Fig. 1). The former tends to occur in clay-rich rocks, the latter in clay-poor rocks. Although there has been a long evolution in the definition of the term stylolite, it is generally taken today to refer to a serrated interface between two rock masses that have a sutured appearance in section normal to the plane of the stylolite. These suture-like seams are marked by the presence of insoluble minerals left behind after solution of the more soluble constituents. These insoluble minerals may be clays or other silicate minerals, organic or carbonaceous material, or even iron minerals. The amplitude of the suture is much greater than the diameter of the transected grains. It cuts indiscriminately through the rock fabric, across grains, cement, and matrix; it may truncate fossils, ooliths, veins, and other stylolites.

Stylolites that are the result of overburden pressure run approximately parallel to the bedding. Those that are caused by tectonic forces may be oblique or even perpendicular to bedding. Where one cuts another the relationship can be used to unravel the diagenetic and tectonic history of the rock. Stylolites serve not only as permeability barriers but also as conduits to fluids. They thus have a considerable effect on later diagenetic processes such as dolomitization and the migration of hydrocarbons. HAROLD G. READING

subduction zones Subduction zones form where an oceanic lithospheric plate is in collision with another plate, either oceanic or continental. The density of oceanic lithosphere is similar to that of the asthenosphere, and it can thus fairly easily be pushed down into the uppermost mantle. Subducted lithosphere remains cooler, and therefore denser, than the surrounding mantle for many millions of years; so once started, subduction tends to continue, driven in part by the weight of the subducting slab.

Palaeomagnetic studies have shown that the Earth's radius has not grown substantially since the Mesozoic, and certainly not enough to accommodate the amount of new oceanic lithosphere that has been created at mid-ocean ridges in the past 200 Ma. In plate-tectonic terms, subduction zones allow lithosphere to be 'consumed', thus balancing its production at ridges.

The most persuasive evidence for the existence of subduction zones is the narrow Benioff zones of earthquake epicentres dipping away from deep-sea trenches (see *seismology and plate tectonics*). The subducted slab can also be imaged by seismic tomography, which shows it as a zone characterized by high seismic velocity and low seismic attenuation, both of which can be attributed to its low temperature compared to that of the surrounding mantle.

Most subduction zones are marked at the surface by arcuate deep-sea trenches (see *deep-sea trenches*) and include an arc of active volcanoes that lies parallel to and several hundred kilometres behind the trench, above the dipping slab. Trenches and volcanic arcs tend to be convex toward the underriding plate. Geochemical analysis of arc lavas shows that their magmas originated at about the depth of the subducted slab immediately beneath them.

Ocean–ocean subduction zones

There are two kinds of subduction zones, depending on whether one or both of the colliding plates is oceanic. Ocean–ocean subduction zones are slightly simpler, and will be considered first. Figure 1a shows a schematic cross-section across a typical ocean–ocean subduction zone, although not all the features shown are always present. The various features and their origins are as follows, described in order of a traverse from the subducting plate through the subduction zone to the overriding plate.

A little over 100 km 'ahead' of the trench, on the subducting plate, there is a low bulge in the sea floor called the trench

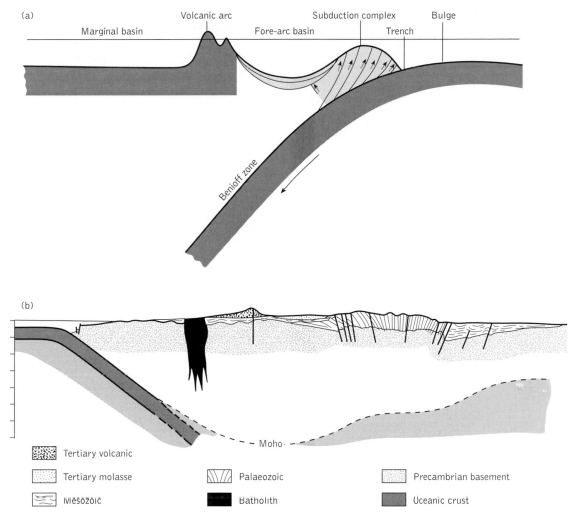

Fig. 1. Cross-sections through (a) a typical ocean–ocean subduction zone and (b) a typical ocean–continent subduction zone. (Modified from Kearey and Vine, Figs 8.2 and 9.3.)

outer rise. This is formed as a result of elastic bending of the plate as it is subducted.

One of the major surface features of the subduction zone is the trench, the bottom of which marks the boundary between the two plates. The outer slope of the trench is fairly gentle, but is usually marked by active normal faulting which breaks the brittle upper part of the plate as it bends down. The trench floor is commonly flat and covered by sediments (turbidites) that have slumped off the walls.

The trench inner slope is usually steeper than the outer slope. It usually consists of sediments scraped off the subducting slab to form an accretionary wedge or prism, sometimes called the subduction complex. The accretionary prism can contain repeated slices of sediment transferred from one plate to another in this way. The degree of development of the accretionary prism is very variable, but it can build up to produce a significant ridge, which may even appear above sea level. In the Lesser Antilles, for example, the islands of Trinidad, Tobago, and Barbados are actually the top of the accretionary prism.

Most subduction zones contain a volcanic arc, parallel to the trench and situated about 150–200 km away from it on the overriding plate. Many of the volcanoes stand above sea level to form a volcanic 'island arc'. The Lesser Antilles islands from Grenada to St Kitts represent the volcanic arc there. Arc volcanism results mainly from the melting of the asthenosphere immediately above the descending slab, together with the subducted sediments. Melting is facilitated by the large amount of water carried down as pore water and in hydrated minerals; this water lowers the mantle melting temperature.

Between the volcanic arc and the accretionary prism (where developed) is a 'fore-arc basin'. This may contain sediments derived from the volcanic arc and possibly also from the accretionary prism.

Beyond the volcanic arc there is often a 'back-arc basin' or 'marginal basin'. Most mature ocean–ocean subduction zones contain a back-arc spreading centre, which results from tensional forces on the overriding plate generated by the subduction process itself. Back-arc spreading centres are similar to mid-ocean ridges, and produce new oceanic lithosphere in back-arc basins by sea-floor spreading. However, the presence of water from the subducted lithosphere results in magmas that are chemically distinct from those produced at mid-ocean ridges. There are many back-arc basins in the western Pacific, including the Japan Sea and a string of basins from the South China Sea to the South Fiji and Lau basins.

Ocean–continent subduction zones

In many respects ocean–continent subduction zones (Fig. 1b) are similar to ocean–ocean ones, and can include an outer rise, trench, subducted slab, accretionary prism, fore-arc basin and volcanic arc. However, the arc magmas are here intruded through a thick continental crust, which may modify their chemistry. Perhaps more importantly, the weak continental lithosphere is readily deformed by compressional stresses so that extensive folding and thrusting may occur both in the fore-arc and in the back-arc regions. This can lead to the formation of complex mountain belts such as the Andes (which flank the Peru–Chile subduction zone), composed of both folded and faulted sediments and arc volcanoes. While there is evidence of back-arc extension in such belts, full sea-floor spreading does not seem to occur.

The accumulation of accretionary prisms and arc volcanic rocks at continent–ocean subduction zones represents one of the main mechanisms for the accretion of continents. For example, western North America and eastern Asia display well-developed concentric bands of crust, ageing towards the cratonic interior, that are thought to have been accreted in this way in successive episodes of subduction. R. C. SEARLE

Further reading

Kearey, P. and Vine, F. J. (1996) *Global tectonics*. Blackwell Science, Oxford.

submarine ridges, aseismic *see* ASEISMIC SUBMARINE RIDGES

submarine sediment slides With the detailed seismic surveys of the ocean floor that have taken place during the past three decades, there has developed an awareness of the complex geomorphological processes that take place on the continental shelf, slope, and abyssal plains the world's oceans. The surveys that have taken place so far have shown that most of the largest 'landslides' (really submarine slides) in the world are located on the sea floor. Most of these sediment slides are located in areas of crustal subduction, but many are on the flanks of submarine volcanoes. Although scientific studies of submarine slides have been relatively few, the information available appears to indicate that those slides that occur on the sea floor have a wide age range. Many of the largest features are composed of unconsolidated Quaternary sediments that have been subject to complex mass-movement processes. The processes by which slides are generated are poorly understood. In many instances it has been argued that the sediment slides are generated by offshore earthquakes and submarine faulting, which trigger the initial sliding. In other cases slides have been related to the release of methane gas (clathrates) from sea-bed sediments. In some instances it has been proposed that the slides themselves have been capable of generating earthquakes. A well-known submarine slide is the Grand Banks slide and turbidity current, which took place in November 1929 after a large earthquake. In this instance the slide sediments broke a series of submarine cables, and analysis of the precise times of the cable breakages made it possible to calculate the velocity of the slide. Heezen and Ewing demonstrated that the Grand Banks slide was associated with maximum velocities in excess of 50 knots (93 km per hour).

One of the world's biggest submarine sediment slide complexes is located in the Storegga area in the Norwegian Sea west of Norway (Fig. 1). Three slides are present. The oldest, known as the First Storegga Slide, involved the movement of approximately 3600 km^3 of unconsolidated Pleistocene sediments. This slide has been dated by oxygen isotope analysis to approximately 28 000–30 000 years before the present. The Second Storegga Slide involved the movement of approximately 1700 km^3 of Quaternary sediments and appears to have taken place approximately 7000 years ago. The Third Storegga Slide appears to have taken place shortly after the second slide; it involved the transport and redeposition of some sediment already subjected to transport during the second slide. The Second Storegga Slide is of particular interest because it appears to have generated large tsunami waves that were propagated across the North Atlantic Ocean and caused severe coastal flooding along most coastlines of north-western Europe.

Despite the detailed sea-floor surveys that have been undertaken, many areas of the ocean floor are still unmapped. In an era characterized by intensive offshore oil exploration it is extremely important to identify those areas of the continental shelf and slope that have been subject to slide activity, since these areas are potentially unstable and susceptible to slope failure. ALASTAIR G. DAWSON

Further reading

Bugge, T. *et al.* (1987) A giant three-stage submarine slide off Norway. *Geo-Marine Letters*, 7, 191–8.

Prior, D. G. (1978) Submarine slides. In Voigt, R. (ed.) *Rockslides and avalanches*. Amsterdam.

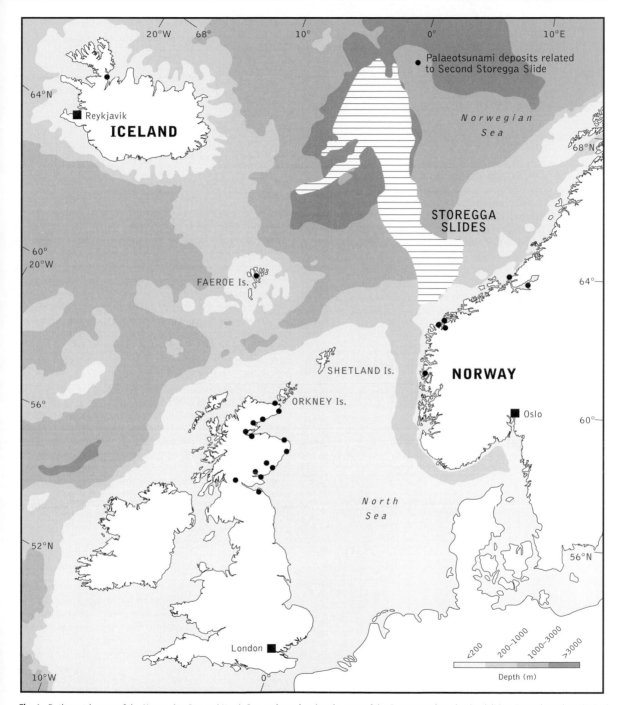

Fig. 1. Bathymetric map of the Norwegian Sea and North Sea regions showing the area of the Storegga submarine landslides. Dots show the principal locations where coastal deposits exist that have been attributed to deposition by the tsunami generated by the Second Storegga Slide. (After Bugge (1983) and Bugge *et al.* (1987).)

subsidence and uplift

The terms 'subsidence' and 'uplift' refer, respectively, to descent and elevation of the Earth's surface relative to the geoid. Rates of vertical movement across the Earth's surface vary from 100 mm per year to 10 m per Ma, the greatest rates being recorded from the plate margins, where the driving forces are at their strongest. Three main processes are essential to the understanding of mechanisms of large-scale subsidence and uplift: thermal perturbations of the lithosphere, isostasy (mass balancing), and flexure.

Perturbations of the normal temperature field of the crust (about 30 °C temperature increase per kilometre depth) cause changes in its buoyancy, and uplift or subsidence at the surface is the direct result. Imagine that a volume of crust is stretched, and therefore thinned, resulting in the bunching together of the isotherms and a consequent increase in geothermal gradient. The immediate effect of an extensional phase will be for the thinner, and therefore hotter and more buoyant, crust to undergo doming, the centre of the dome being sited over the most attenuated crust. Thermal uplift occurs today in the Afar region of Ethiopia, where extension at the junction of the Red Sea and the East Africa rift system has produced a domed region approximately 1100 km in diameter.

Conduction of heat between the heated lithosphere and its adjacent thermally normal lithosphere will gradually return the lithosphere to thermal equilibrium. This will cause the uplifted region to subside thermally, resulting in a sedimentary basin located above the locus of thermal uplift. The phase of heating during initial thermal uplift, and subsequent subsidence and burial, is responsible for creating the necessary conditions of heat and temperature that led to the expulsion, and migration into suitable reservoir rocks, of most North Sea oil.

Lithospheric extension generates thinned crust which is lighter than the mantle it replaces. Consequently, the mass of a column through part of the lithosphere which underwent stretching will be greater than the mass of a similar column through an adjacent region whose thickness remains unchanged. The thinned crust will subside isostatically until the mass of the two columns overlying a notional equilibrium depth is equal. Conversely, the central parts of most mountain ranges comprise rocks which, although now exposed at the surface, formed at great depths, often as deep as the base of the crust itself. An appreciation of the concept of isostasy allows us to see how an initial episode of uplift and erosion will lead to more isostatic uplift, and hence more erosion. Continuation of this process over several millions of years will bring deep-seated rocks to the surface in a process known as *exhumation*. For example, at the foot of Nanga Parbat, in the Pakistan Karakoram, rocks which were formed about 2 Ma ago at a depth of about 35 km are today exposed at the surface, thereby giving a time-averaged exhumation rate of nearly 20 mm per year.

The stiffness of the lithosphere means that it can partially support the extra mass represented by a load such as a mountain belt or volcano. Volcanos are a special case because they constitute a point load, rather than the more common line loads represented by linear features such as mountain belts. Studies of ocean-floor bathymetry around Hawaii, a major active volcanic complex, reveal an oceanic 'moat' recording flexural subsidence in response to the extra mass represented by the weight of the volcano.   JONATHAN P. TURNER

subsurface flow and erosion

Subsurface erosion of the land can produce some spectacular features, but it has often been regarded as unusual. Until quite recently, theorists and modellers of both landscape evolution and storm-water run-off have concentrated on surface processes, following in particular the lead given by the American hydrologist Robert E. Horton in the 1940s. Hortonian theory maintained that erosion and storm run-off in humid regions are caused solely by overland flow, which occurs after rainfall rates exceed the infiltration capacity of the soil surface. According to Horton's premise, stream channels begin where the erosional power of this overland flow exceeds the erosional resistance of the soil surface.

Evidence has grown over the intervening decades to support an alternative view: that subsurface flow plays an important and sometimes even dominant role in perhaps the majority of environments. One of the most widespread and important roles that subsurface flow can play is in the formation of saturation overland flow. It is now realized that Horton's idea that overland flow is generated all over the river basin is rarely supported by observations, except perhaps after heavy storms in the desert. Surface flow normally begins in just a few areas, especially dips or hollows near the stream. If the storm lasts long enough, these areas may extend upslope away from the stream, but they rarely cover the whole basin.

Subsurface flow tends to create this localized saturation (or saturated) overland flow in two ways:

(1) soils in the hollows and bank-side areas are wetter at the beginning of the storm because water has seeped downslope into these places beforehand; and

(2) subsurface stormflow carries water that has infiltrated the soil upslope rapidly downhill and into these places.

The longer the storm lasts, the more time there is for water to seep downslope from further and further upslope to the edge of the saturated areas, and so the saturated area expands.

This saturation overland flow therefore commonly contains water from three sources:

(1) 'old' water left over from previous storms that may be displaced by the 'new' water that has entered the soil during the current storm;

(2) 'new' water that has infiltrated during the current rainstorm; and

(3) rain that has fallen on the saturated areas and has not been able to infiltrate the soil.

The first two are known collectively as *return flow*. The proportions of each vary from place to place and storm to storm.

Subsurface flow can also deliver water directly into the stream channel either as diffuse seepage or as pipeflow emerging from the stream bank. Diffuse seepage can occur on a wide front, especially near to the channel where the water-table joins the stream. This flow may be either beneath the water-table, as 'saturated flow', or above it, as 'unsaturated flow'. Flow is much faster in saturated soil, but it still tends to be rather slow, typically moving at rates of barely 1 mm per second. Flow may be more efficient in *percolines*, which are localized lines of extra deep soil which concentrate drainage in some soils. Saturated flow persists for longer there, and the concentrated flow may winnow away the finer soil particles, increasing the permeability. Persistent saturation increases weathering and increased erosion may also lower the surface, capturing more surface water. Percolines can grade downslope into *seepage lines*, where return flow commonly emerges or *exfiltrates*. Theories of hillslope evolution still generally need to incorporate the spatial diversity in slope development that this linearly localized weathering and erosion create.

Flows are very much faster in soils that have connecting voids called *macropores*. Macropores may be cracks, caused by desiccation or mass slumping, cylindrical holes (mainly biotic), vughs (between these two types), or water-sculpted soil pipes (see *piping*). The larger macropores may extend for long distances, as water erodes and enlarges the segments that run downslopes. Networks of soil pipes up to 750 m long have been reported. These act as 'soil springs' and have been found to carry half of all the storm flow reaching the stream in a British upland basin. Rainwater can reach these pipes within an hour or so, even though they may be 50 centimetres below the surface. It seems that smaller macropores and 'blowholes' at the surface allow rainwater to infiltrate the soil very rapidly, causing the water-table to rise and then drain through the pipes.

While flow through the fine micropores of the soil matrix is essentially non-turbulent or laminar flow, flow through macropores can be turbulent and erosive. Typical flow rates in macropores range from 5 mm per second up to 0.5 m per second in pipe flow. The exact threshold at which erosion begins is highly variable and difficult to determine, because it depends on the nature of both the flow and the soil; individual, loose (or 'dispersed') clay particles may be transported through pores a mere 10 micrometres (μm) in diameter, whereas aggregated soil particles may be eroded only in the occasional, more extreme flows, even in pipes of 100 mm diameter.

All forms of subsurface flow within the soil or regolith are now commonly referred to as *through flow*. The older term, *interflow*, is still occasionally used synonymously, although it originated among engineering hydrologists to refer to water that arrives towards the end of stormflow in the stream after following a deep subsurface route, typically thought of as through bedrock. In some cases, subsurface flow through the soil is distinct from subsurface flow through bedrock, especially where the surface bedrock is impermeable. At other times, the two may be connected, especially near stream channels. The term *groundwater flow* is generally restricted to deep seepage in the saturated zone of the bedrock and its subsequent contribution to stream flow, primarily because this is of greater economic importance as a source of public water supply. Nevertheless, in reality soil water and this groundwater *sensu stricto* often interact and mix. In broad terms, all subsurface water is groundwater.

Deep groundwater flow exploits joints and other planes of weakness in rocks, just like through flow, and can develop spring networks. This reaches its most spectacular expression in the underground rivers of a karst landscape. Whereas karst is predominantly formed by solutional erosion, piping processes are essentially mechanical. Karst-like caves, sinks, and collapse features formed in non-calcareous bedrock by piping processes have been called *pseudokarst*.

Spring sapping occurs where piping processes extend the head of a channel or gully or develop tributary channels. This may occur as a general washing out of the subsoil or by enlargement of a pipe until the roof collapses. Gully networks with stubby tributaries or blind valleys meeting at high angles are typical of surface channels that have been formed by subsurface erosion. Many channels are, however, indistinguishable from those formed by overland flow, whether or not surface flow has played any part in their development. The extent to which subsurface erosion is responsible for developing surface channels is still unknown, but it seems reasonable to suggest that subsurface erosion is far more important than has been supposed in humid areas with a good vegetation cover that restricts surface erosion.

As with channel initiation, subsurface erosion is often found to be inextricably linked with surface processes in many instances of severe soil erosion. Rills and gullies may stimulate subsurface erosion by creating a hydraulic gradient within the soil towards the new channel. At the same time, piping may be creating the rills.

J. A. A. JONES

Suess, Edouard (1831–1914)

One of the great names in late nineteenth-century geology, Edouard Suess lived for most of his life in Vienna. He was a natural academic of great charm and energy. After graduation he worked at the Museum of Natural History in Vienna. During an ascent of the Dachstein in the eastern Alps he was so impressed by the view that he soon began a scientific study of landforms and the geological structures that underlie them. This was to be the main theme of his research for many years. Suess was interested in fossils, having been involved with classifying brachiopods while at the museum. To these studies he added others concerning graptolites, ammonoids, and mammals.

He gained a professorship in the University of Vienna in 1861, and on becoming President of the Imperial Academy in 1898, he transformed it, too, into a highly efficient scientific body.

Suess was the first to recognize the apparently symmetrical arrangement of structures within the various Alpine mountain ranges of Europe. He distinguished stable forelands and hinterlands, postulated sequences of tectonic events in orogenic belts, and the relative youth or antiquity of mountain landscapes. His influence upon Alpine geologists was huge. He was the first to develop the concept of the geosyncline, though he did not use the word. He also paid attention to the importance of global sea level throughout geological history, and suggested that change in its level resulted from deformation of the ocean floor and from build-up of sediment. He distinguished continental crust (which he called sial) from oceanic crust (sima) on the basis of their overall composition. Many such concepts were discussed in his three-volume masterpiece *Das Antlitz der Erde* (*The face of the Earth*, 1883–1909).   D. L. DINELEY

sulphate minerals

Sulphate minerals are formed in a variety of environments, mainly at low temperatures. Of the more than one hundred species, many are rare and only one, gypsum, is of worldwide abundant occurrence; anhydrite is abundant at depth and barite (baryte) is fairly ubiquitous but generally occurs only in small amounts.

The basic structural unit is the sulphate group $(SO_4)^{2-}$ linked by metal ions or by water molecules. In one anhydrous orthorhombic group, barite $(BaSO_4)$, celestite $(SrSO_4)$, and anglesite $(PbSO_4)$ have the same structure, but anhydrite $(CaSO_4)$, because of the smaller calcium ion, has a different structure. Some sulphate minerals, for example jarosite, contain the hydroxyl (OH) ion; others, such as the rare epsomite and gypsum $(CaSO_4 \cdot 2H_2O)$, contain the water molecule.

Gypsum and anhydrite are closely related; both are calcium minerals and both are formed early (after carbonate) in the sequence of minerals when sea water is evaporated or in sabkha environments, as in the present-day Persian Gulf. Anhydrite is not precipitated below 42 °C unless the salinity of the water is high. Gypsum is generally formed at the surface and is stable down to about a kilometre or so, where it inverts to anhydrite. On uplift, or with the ingress of water, anhydrite turns into gypsum. In this process there is a 61 per cent increase in volume, which often results in intense deformation of the evaporite bed. Under certain conditions the growth of gypsum, resulting from the oxidation of pyrite, the formation of sulphuric acid, and its reaction with the carbonate of shelly material in clays and shales, can cause the heaving of floors in buildings.

Gypsum of the variety known as selenite is monoclinic; its hardness is 2 on Mohs scale and it has one perfect cleavage yielding flexible but not elastic laminae; other, rarer cleavages may result in rhombohedral masses. Twin crystals are common; they show swallow-tail or arrow-head forms and more rarely cruciform growths, and may grow to 3–4 m long. Gypsum grows in four different habits: selenite as just described; alabaster, variously coloured in nodular granular masses; satin spar, which is usually white and finely fibrous;

and desert rose, like selenite in form but enclosing many quartz grains. Gypsum has many commercial uses: as a retarder in Portland cement, as a filler in paper, textiles, and paints, in the manufacture of sulphuric acid and fertilizers, and in the production of plaster of Paris.

Barite is primarily a gangue mineral in hydrothermal veins but is also widespread in nodules, veins, and lenses in limestones, sometimes growing with a cockscomb habit. Most commonly white or colourless, its density makes it an important constituent of drilling muds; it is also used as a filler in paints, in the sugar industry, and as the primary source of barium.

Celestite (celestine) has similar properties to barite, but usually has a characteristic pale blue colour. It occurs as an evaporite mineral associated with gypsum and halite, in veins of hydrothermal origin, or in veins and cavities in limestones; the strontium derives from the alteration of aragonite to calcite. It is used for its red colour in pyrotechnics, in paints, and in the refining of sugar beet.

Anglesite is fairly rare; it commonly forms white granular skins around silvery-grey galena as a result of the alteration of galena in the zone of oxidation.   R. BRADSHAW

sulphide deposits *see* MASSIVE SULPHIDE DEPOSITS

sulphide minerals

Minerals in which sulphur is combined with one or more metallic elements are classified as the sulphide minerals. Sulphides are typically metallic grey to brass coloured, and are relatively soft when compared with silicate or oxide minerals. Pyrite, FeS_2, a compound of iron and sulphur, is one of the most common sulphides and is found as a minor component in many rocks. It is also an important accessory mineral in most mineral deposits formed from high-temperature fluids. Its golden metallic colour and common occurrence in gold deposits has lead it to be mistaken for gold; hence its colloquial name of 'fool's gold'. Other sulphides constitute the most important hosts of several elements that are extracted from ore deposits. For example, sphalerite (zinc sulphide, ZnS) and galena, (lead sulphide, PbS) are the chief ore minerals of zinc and lead, respectively. Other rarer simple sulphides are important sources of cobalt, nickel, copper, molybdenum, cadmium, mercury, antimony, and arsenic. Sulphides in which two metals combine with sulphur are also common and two of these minerals, chalcopyrite $(CuFeS_2)$ and bornite (Cu_5FeS_4), in which copper combines with iron and sulphur, are the most commonly encountered minerals in copper deposits. Minerals in which a metal and sulphur bond with a metalloid element, usually arsenic or antimony, form a sub-group of the sulphide minerals known as sulphosalts. These are accessory phases, commonly subordinate to the simple sulphides, in many types of mineral deposits. Sulphides are relatively unstable minerals in the surface environment and when exposed to oxidizing groundwaters resulting from mining activities, they degrade, generating high concentrations of acid: hence the term 'acid mine drainage'.   BRUCE W. MOUNTAIN

superficial deposits Superficial deposits consist of the unconsolidated material that obscures much of the outcorp of solid bedrock in many parts of the world. They include such features as sand dunes, sandy beaches, river gravels, glacial debris, and soil. Most geological maps omit superficial deposits in order to show the underlying geology more clearly. In the British Isles some geological maps are published in two editions, a 'Solid' edition showing bedrock only, and a 'Drift' edition that includes superficial deposits deposited by rivers and glaciers, and shows the bedrock only where such cover is thin or absent (although soil is ignored).

By their very nature, superficial deposits are young, unconsolidated, and prone to erosion. Only a small fraction of superficial deposits is therefore likely to survive long enough to be buried and turned into true rock by the processes of lithification.   DAVID A. ROTHERY

supergene and residual deposits Mineralization formed deep in the Earth's crust can be exposed to the surface environment by uplift and erosion of its host rocks. The high oxygen content of the Earth's atmosphere and the presence of water result in chemical reactions that break down ore minerals that are not chemically stable under surface conditions. Sulphide minerals are particularly susceptible to chemical weathering, which results in the release of their contained metals and sulphur. The oxidation of the sulphur produces large quantities of acid which then promotes further breakdown of other more stable minerals. In the air-saturated zone near the surface, valuable metals can be dissolved into groundwater because of oxidizing conditions in this region of the soil. When the groundwater percolates down to the water-table, conditions become more reducing and the metals can no longer remain dissolved and are precipitated as supergene minerals. Over time, large accumulations of supergene minerals can form and constitute a supergene deposit. In some instances sparse mineralization that is uneconomical to mine can be concentrated by supergene processes into mineable ore; supergene deposits are commonly underlain by such primary mineralization. Common supergene ore minerals include those of copper (chalcocite, covellite, native copper), lead (anglesite), and zinc (smithsonite). Carbon dioxide can also combine with metals to form spectacular encrustations of colourful minerals such as azurite and malachite, both carbonates of copper. Other metals that are concentrated in supergene deposits include gold, silver, nickel, manganese, and uranium. Surface exposures of supergene deposits are unusually rich in iron hydroxide minerals, imparting a red coloration to the surface that contrasts strongly with surrounding unmineralized rocks. These exposures, commonly referred to as *gossans*, have been useful indicators of subsurface mineralization to prospectors and exploration geologists.

Supergene deposits are formed by the weathering of earlier mineralization, but there are other types of ore deposit that are formed by the weathering of ordinary unmineralized rocks. Residual deposits form when intense weathering removes much of a rock's components and concentrates a few of them to very high levels. Bauxites are ores of aluminium generated by this process. Bauxites form in areas of well-drained terrain that has low to moderate relief and has been geologically stable for long periods of time. A tropical climate and intense vegetation favour their formation over suitable host rocks. In Jamaica, extensive bauxite deposits are found overlying limestone and dolomite. These resulted from the dissolution of the carbonate component of the host rock by organic acids that were generated by decaying vegetation. The remaining clay component from the limestone can be further altered by these acids and, over time, it is converted to aluminium oxyhydroxide minerals such as gibbsite, diaspore, and boehmite. In some other locations, thick accumulations of bauxite were formed by the weathering of volcanic rocks. A major source of iron, called *iron laterite*, is a type of residual deposit generated by the intense weathering of iron-rich rocks such as mafic and ultramafic volcanic rocks. During weathering, acid solutions remove much of the rock-forming elements and, because mafic and ultramafic rocks are poor in aluminum, the remaining laterite is highly enriched in iron oxyhydroxide minerals such as haematite and goethite. If an ultramafic rock rich in nickel undergoes such a process, a nickel laterite is formed. These nickel laterites are important resources of this metal in the South Pacific and South America. Other residual deposits are sources of chromium, titanium, rare-earth elements, and even gold.

BRUCE W. MOUNTAIN

superposition principle It is a common observation that rocks occur in layers; that is, *strata* or beds. Most of these layered or stratiform rocks are the consolidated products of sediments or volcanic particles deposited in bodies of water or from the air. As each layer was deposited it must have had a solid base upon which to form, and is necessarily younger than that base. So in any undisturbed succession of bedded or layered rocks the lower rocks are older than those above. Even where the rocks have been greatly disturbed, contorted, or metamorphosed, many clues as to the correct 'way up' or direction of younging within a sequence may survive.

All geologists accept this without question, but the principle was not recognized until the mid-seventeenth century. This milestone in the history of the Earth sciences is credited to a young Danish physician, Niels Stensen (1638–86), who took service in the court of the Grand Duke of Tuscany. He is usually known today by the name of Nicolaus Steno. In Italy he was able to travel and observe closely the tilted fossiliferous strata of the Tuscan countryside. Taken with the similarity of the fossils to modern shells and sharks' teeth, he deduced that the rocks had once been soft unconsolidated sediments on the sea floor, and had subsequently been hardened and uplifted from their original sites to their present positions. He stated this revolutionary new idea in his famous book *De solido intra solidum naturaliter contento dissertationis prodromus* (*Preamble to a dissertation on a solid naturally enclosed within a solid*). It was an explanation of fossils, in particular, the tooth of a fossil shark, but Steno went on to deal with sedimentary strata. He came up with three 'laws':

(1) *Original horizontality*. Sediments deposited on a solid base must originally have been horizontal or they would have slithered to a lower point. Rocks that lie at an angle must have been tilted after the sediments were consolidated.

(2) *Original continuity*. Although we now see discontinuous outcrops of the same layers across the topography, these deposits were originally continuous sheets and their once laterally continuous extent is marred by erosion or by overlying material.

(3) *Superposition*. Each layer must have been deposited on top of an existing base or solid substrate; so it must be younger than the layer beneath.

Steno must have wondered how long this process of change from soft sea-floor sediment to tilted strata up in the hills had taken, but he had no means of estimating this. In the Italy of his day, any disagreement with the time allowed by Scripture would have been dangerous.

Superposition enables us to determine the relative ages of strata in a succession. There are, however, circumstances in which the beds have been tilted so much that they are inverted, and this has given rise to problems in determining which beds are the younger. Today a considerable number of clues may be recognized to give the 'way up' of rocks in a tilted or overturned section. The methods used for sedimentary rocks are rather different from those in once-molten igneous material, but many of them rely upon the effects of gravity upon material that was once fluid or deposited from water.

Fossils may also give valuable information about which beds are the younger, since biostratigraphy is based upon the sequence of fossils in undoubtedly 'normal' rather than overturned or inverted successions. Just how valuable is the contribution of fossils to establishing the relative ages of rocks was demonstrated at the end of the eighteenth century when William Smith, a civil engineer, drew up his table of strata around the city of Bath with 'their respective organic remains', and again in 1815 with the production of his geological map of England and Wales. He showed that specific fossils characterize rocks of specific ages and that there is a natural and invariable succession of fossils throughout the geological column. Superposition (that is, relative age) can be confirmed by palaeontology.

D. L. DINELEY

Further reading

Prothero, D. R. (1989) *Interpreting the stratigraphic record.* W. H. Freeman, New York.

surging glaciers

Surging, or surge-type, glaciers are those that undergo periodic phases of rapid flow, punctuating longer intervals of stagnation. Fast flow may last from a few months to a few years, and the inactive phase from between about 20 and 200 years. Velocities at the ice surface can reach up to 50 m day^{-1} during a surge, and the ice becomes heavily crevassed (Fig. 1). These high velocities are often associated with rapid advance of the glacier terminus. One glacier in Svalbard, the island archipelago north of Norway, advanced by 20 km along a 30 km-wide front in less than 2 years, thus earning the name Bråsvellbreen, that is, rapid-growth glacier. Large amounts of ice are transferred from an upper or 'reservoir' area to a lower or 'receiving' area during surges (see *glaciers and glaciology*).

Fig. 1. The heavily crevassed surface of a surging glacier in Spitsbergen.

The high velocities recorded during the active phase of the surge cycle are linked to the presence of lubricating water at the glacier bed. It is clear from investigations of the subglacial hydrology of surging glaciers that reorganization of this drainage system controls the onset and termination of rapid flow. At Variegated Glacier in Alaska, for example, little melt-water escapes from the glacier during a surge, whereas the rapid shift to low velocities at the termination of a surge is linked closely with the draining of large quantities of sediment-rich water from the glacier.

Only a few per cent of all glaciers are known to surge, and the geographical distribution of surge-type glaciers is remarkably non-random. For example, Alaska, the Yukon, Iceland, the Russian Pamirs, and Svalbard all contain relatively large numbers of surging glaciers, whereas there are very few in the Russian and Canadian Arctic archipelagos. No surges have been observed in Antarctica. Both small and large glaciers surge, including ice-cap outlet glaciers of over 1000 km^2 in area. Glaciers which have undergone surges in historical times, and at which no direct observations of the active phase have been made, are identified by characteristic series of looped medial moraines on the ice surface, which can easily be recognized from aerial photographs and satellite images.

Surges of glaciers ending in fjords can pose a significant natural hazard. Not only do they calve large numbers of icebergs into the fjord waters during the active phase, making navigation hazardous, but they may also advance across fjords to form a glacier dam. Water levels rise and flooding may follow, together with a shift towards less saline water with potentially disastrous consequences for local marine life.

JULIAN A. DOWDESWELL

Further reading

Dowdeswell, J. A., Hamilton, G. S., and Hagen, J. O. (1991) The duration of the active phase on surge-type glaciers: contrasts between Svalbard and other regions. *Journal of Glaciology*, 37, 388–400.

synoptic climatology

Synoptic climatology is the branch of climatology that concentrates on the links between atmospheric circulation and the weather and climate conditions experienced in a given area or at a given location. The subject is a broad one and may concentrate either on a single climate variable, such as precipitation, or on groups of measures, involving temperature, sunshine, etc., but the link is always made between these manifestations of atmospheric circulation and the circulation itself. Synoptic climatological studies are, by their nature, empirical in approach, utilizing present and past observational data from climatological stations, satellites, etc. Such empirical approaches, using 'real' data, demand the careful use of statistical techniques to draw out significant findings. As computers have become more powerful, the opportunity has arisen for analyses to be carried out using much larger volumes of data and more sophisticated statistical techniques. It is not, of course, sufficient merely to prove that a correlation exists between circulation and weather and climate: the cause must also be determined. In this context, synoptic climatology has made a contribution to the understanding of atmospheric processes that give rise to particular weather conditions which is complementary to the more theoretical process studies conducted by meteorologists and other climatologists.

Synoptic climatological studies vary considerably in their scale of approach. At the broadest, global, scale, for example, links have been made between circulation changes associated with the El Niño phenomenon and Australian droughts and Asian monsoon rainfall. Such links are referred to as 'teleconnections'; they provide valuable information that aids the prediction of seasonal weather at the regional scale (e.g. for Australia or India). Regional and meso-, or local-scale studies, however, form the basis for much research in synoptic climatology. These contribute to the science of weather prediction by helping to provide a general picture of, for example, precipitation distributions over an area, when the surface atmospheric circulation is of a particular type: what for example, is the expected distribution of rainfall over northern England when a showery north-westerly air mass dominates? They can also help in the longer term in assessing water resource potential. Similarly, global warming may produce notable regional changes in the surface circulation. Knowing the precipitation patterns associated with different types of circulation, and the predicted changes in the frequencies of each circulation type, will help in providing estimates of future changes in the distribution of precipitation.

A basic theme of synoptic climatology is the ability to classify atmospheric circulation. One of the first such classifications was evolved by Professor Hubert Lamb in the 1950s and 1960s. This classification is still in use today, and forms the basis of many current synoptic climate studies. Similar classifications have been derived for many other parts of the world. They focus on the identification of the most frequent atmospheric circulation types. An assumption is made that for many of these types distinct surface weather patterns will develop, and statistical studies are carried out to test whether this is so. This approach to classification is, however, subjective, deriving as it does from the precomputer era. Computer technology now makes possible the more objective classification (*typing*) of circulation types, based purely on a grid of numerical values, rather than visual inspection of surface synoptic charts (weather maps).

GRAHAM SUMNER

Further reading

El Kadi, A. K. A. and Smithson, P. A. (1992) Atmospheric classifications and synoptic climatology. *Progress in Physical Geography*, 16, 432–55.

Lamb, H. H. (1972) British Isles weather types and a register of the sequence of circulation patterns, 1861–1971. *Geophysical Memoir*, 116, Meteorological Office.

Perry, A. (1983) Growth points in synoptic climatology. *Progress in Physical Geography*, 7, 90–6.

Smithson, P. A. (1986) Synoptic and dynamic climatology. *Progress in Physical Geography*, **10**, 100–10.

synthetic aperture radar Ordinary radar works by measuring the precise time between the transmission of a radio pulse and the return of its echo from a distant object. In this way, a very precise distance (range) measurement can be achieved, but the angular (azimuthal) resolution is limited by the physical size (aperture) of the antenna. Synthetic aperture radar (SAR) overcomes this limitation by combining the radar echoes received at several small fixed antennae or one small moving antenna to simulate a single antenna with a large effective aperture. The latter method allows SAR to be used as a detailed imaging tool from a small platform such as an aeroplane or satellite. The image must be reconstituted by combining the radar returns some time after measurement, since each pixel in the image depends on several sets of returns as the antenna progresses along track. The reconstitution method relies on the Doppler effect: echoes from points ahead of the platform are shifted to higher frequencies, whereas returns from points behind the platform are shifted to lower frequencies. Because microwave SAR is mostly insensitive to cloud cover and does not require solar illumination, SAR instruments have been placed on several space-borne systems, including ERS-1, ERS-2, RADARSAT, the Space Shuttle, and the now-defunct SEASAT and JERS-1. SAR has been used by the Magellan space mission to map the surface of Venus, which is hidden at optical frequencies by the dense atmosphere.

A further development of SAR is *interferometric SAR* (InSAR). SAR images contain information about the amplitude and phase of the radar return from each pixel. If two images of an area are taken with the same SAR system, the difference in the phase of a pixel from one image to the next will depend on the ground deformation in that pixel. There may also be a phase change dependent on the ground elevation, because of the stereoscopic effect that arises if the images are not taken from exactly the same track. In the known absence of ground movement, or if the images are taken simultaneously, this latter effect can be used to construct *digital elevation models* (DEMs) of wide areas. If the ground elevation is known sufficiently accurately for its effect to be removed, surface movements can be detected which are of magnitudes somewhat smaller than the wavelength used to form the radar pulse. In the case of ERS-1 and ERS-2, which have a radar wavelength of 56 mm, movements of a few millimetres can be resolved. This level of sensitivity makes InSAR a readily feasible method of observing landslides, glacier movements, earthquakes, and deformation associated with volcanic eruptions. Sensitivity is limited by seasonal and climatic effects: the radar return is affected by groundwater movement and the growth of vegetation, and to a lesser extent by atmospheric variability. Under optimum (desert) conditions, it may be possible to use InSAR to monitor the still smaller deformational signals associated with inter-seismic and post-seismic fault movements.

PETER CLARKE

Further reading

Curlander, J. C. and McDonough, R. N. (1991) *Synthetic aperture radar: systems and signal processing*. Wiley Interscience, New York.

synthetic seismograms In many applications, a seismologist is content to derive information about the subsurface region from calculations utilizing only the travel times of the observed seismic waves. This approach is adequate for many situations and is relatively simple computationally. However, it does not consider fully the interaction of the seismic waves with the materials and structures they encounter as they travel through the Earth. These interactions change the amplitude and shape of the seismic waves, and understanding them can reveal much more about the Earth than simple travel-time calculations. In order to interpret these interactions, one must employ complex theoretical treatments to model the effects of the Earth on seismic waves travelling through it. One of the chief aims of these treatments is to calculate a theoretical response for a particular seismic source at a particular place on, or even in, the Earth. This theoretical response can be displayed as a synthetic seismogram, which is compared with observed data. The model of the Earth employed in the calculation is then changed until the synthetic and observed seismograms match in appearance.

The theoretical treatments of seismic wave propagation used to derive the synthetic seismograms treat this complex problem in two basic ways. One is by employing ray theory (see *seismic ray theory*), which is based on a useful approximation in which the seismic body waves (see *seismic body waves*) recorded at a particular location are treated as individual packets of seismic energy (wavelets). The travel paths of these wavelets can be traced through the Earth and their interaction with the materials through which they travel can be calculated, provided that a number of restrictions on the complexity of the Earth are taken into account. For a given mathematical model of the Earth, ray theory will yield very accurate values of the travel time for a particular body wave. The amplitude and shape calculated for the wavelet are, however, only approximate. Because the calculations for even complex Earth models are fairly straightforward on modern computers, this is the method most commonly employed to calculate synthetic seismograms.

The other theoretical approach is best suited to the calculation of synthetic seismograms and is called wave theory (see *seismic wave theory*). Here, the wave equation of classical physics is derived, at least in a somewhat simplified form, and exact or numerical solutions are sought. The advantage of this approach is that the entire wave field emanating from a seismic source can be considered; the calculations are, however, very complex, and exact mathematical solutions can be derived only for simple classes of Earth models. For example, the reflectivity technique is a commonly employed and mathematically elegant method which provides complete wave-theoretical seismograms for the restricted case of a horizontally layered Earth. For more complex Earth structures, a number of theoretical short cuts have been developed for specific applications, but even with these simplifications the calculations are very time-consuming on even the fastest computers. For truly complex Earth structures, numerical techniques such as finite difference methods must be employed, and these calculations are extremely computer-intensive. Wave-theoretical synthetic seismograms are thus used primarily by academic researchers and in applications where the extra expense can be justified. G. R. KELLER

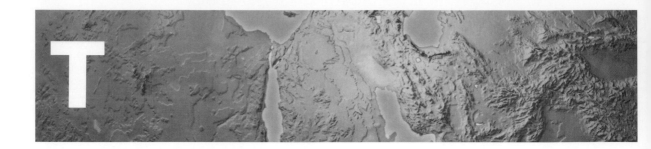

tafoni Tafoni (singular *tafone*) are typically cavernous weathering features which can be up to several metres in height, width, and depth. They have arch-shaped entrances, concave inner walls, overhanging upper margins (visors), and smooth, gently sloping debris-covered floors. These landforms are found in many parts of the world, particularly in semi-arid and coastal environments, and in a variety of topographic situations. They also develop in a variety of rock types but more especially in medium- to coarse-grained lithologies including granites, sandstones, and limestones.

The origin of tafoni is complex and appears to involve a combination of different processes operating on different parts of the landform, although the marked similarity of form suggests a similarity of origin and development between regions. The cavernous hollows appear to result from hydration, dissolution, and re-precipitation of calcareous and saline cements, wetting and drying, rain-wash, and wind deflation, while the smooth and often rounded outer surfaces appear to develop through near-surface cementation, thermal expansion and contraction, and formation of surface varnish. Although many tafoni are undergoing active development today, this is not always the case and some appear to be relics inherited from a different weathering environment in the past.

A. MELLOR

taphonomy Taphonomy is the study of the processes that affect organisms between death and incorporation in the sediment. It is important to an understanding of the extent to which fossils actually represent, in abundance and distribution, original living communities and, therefore, of fundamental importance to palaeoecology, which deals with what happens to organisms during their life. Taphonomy helps in understanding how fossil assemblages differ from the original community from which they were derived, and also provides insight into the environments in which the organisms lived and were preserved. In addition, it can be important in helping to establish the extent to which fossil assemblages were formed rapidly or as a result of the accumulation and mixing of distinct communities over a period of time. This has considerable significance in biostratigraphic studies.

Taphonomy is generally divided into three main areas of study; necrolysis, biostratinomy, and fossil diagenesis. Necrolysis deals with the processes that act on an organism upon its death and shortly afterwards, and is thus concerned mostly with decomposition and disaggregation. In this area much experimental work has taken the form of aktuopalaeontology, in which modern organisms are observed in natural conditions as they decay. An example of this is the work by Roy Plotnick of the University of Chicago, who buried shrimp in marine sediments and was then able to monitor their destruction by scavengers and the disturbance of their remains by infaunal burrowers. Biostratinomy encompasses the sedimentological processes that form the fossil assemblage, and therefore includes the processes of transportation, deposition, and compaction. In this area, studies of the extent to which marine organisms show characteristic disarticulation, fragmentation, abrasion, sorting, and orientation have been used to develop the concept of taphofacies in which particular environments characteristically produce, and can therefore be recognized by, particular levels of damage to the fossils. High levels of disarticulation and breakage are generally correlated with high levels of transportation and water energy, while rapid burial in quiet conditions can lead to accumulations of well-preserved organisms, termed fossil Lagerstätten. Fossil diagenesis can lead to the preservation of fossils, as in the permineralization of plant material by resistant minerals, but may also result in their loss if the original material is dissolved. This can be a source of diagenetic bias if material of one composition is removed while other compositions are preserved.

Much research on taphonomy emphasizes taphonomic loss, in which the differences between the original living community and the preserved remains are viewed as being detrimental. From this perspective, taphonomic processes can be viewed as a series of screens or filters that selectively reduce information; the goal for the palaeontologist is then to recognize and compensate for the lost information in order to reconstruct ancient communities. A different approach, termed taphonomic gain, seeks to recognize the agents of taphonomic loss in order to gain additional knowledge of the ancient environment. This has been particularly fruitful in studies of predation, making it possible to identify predators by the characteristics of the preyed-on remains, even though the predators themselves were not preserved.

DAVID K. ELLIOTT

tar sands Tar sands are natural accumulations of extra-heavy oils, occurring in reservoirs in porous rocks such as sandstones, which are not possible to extract using conventional crude oil production techniques. Extra-heavy oils have been defined as having a viscosity greater than 10 000 centipoises (mN s m^{-2}) under reservoir conditions and a specific gravity of greater than 1.00 g cm^{-3}. Although some heavy oils may be unaltered immature oils, most are thought to be derived from normal crude oils that have been altered by biodegradation, water washing, oxidation, or loss of volatiles. These processes are more likely to occur in reservoirs near the surface.

Compared with conventional oils, heavy and extra-heavy crude oils have increased relative amounts of resins and asphaltenes and decreased amounts of saturated and aromatic hydrocarbons. These undesirable compositional features result in the oils being less valuable than conventional crude oils, which, together with their high extraction costs, make many accumulations uneconomic to exploit at present. However, the amount of in-place reserves of heavy oil in tar sands is enormous (450–1000 billion m^3). The principal accumulations are Venezuela, Canada, the United States, and Russia. Furthermore, accumulations of tar sands are often very large when compared to accumulations of conventional oils. For example, the in-place oil in the Athabasca tar sand field in Alberta, Canada is 140 billion m^3, which is three times the volume of the world's largest conventional field in Saudi Arabia.

D. M. JONES

Further reading

Tissot, B. P. and Welte, D. H. (1984) *Petroleum formation and occurrence* (2nd edn). Springer-Verlag, Berlin.

taxonomy and taxonomic hierarchies

taxonomy and taxonomic hierarchies Taxonomy, the science of classification of organisms, entails placing them in taxonomic units and developing an understanding of the characters that are used to distinguish them. The taxonomic system used is a binomial system, first developed by Linnaeus in 1758, in which each different organism is given a two-word name, the first of which is the generic name, the second the species name. Thus humans are called *Homo sapiens*. Above this level, organisms are grouped into larger and larger taxonomic units as follows: family, order, class, phylum, kingdom. Although these levels cannot be defined further than as groupings of the units below or subgroups of the units above, the phylum can be seen as a large grouping of organisms with a few basic characters in common, as in the Echinodermata (starfish, sea-urchins, crinoids, etc.), all of which have a calcareous exoskeleton, pentameral symmetry, and an internal water vascular system. When Linnaeus produced his original scheme he divided all organisms into two kingdoms, Plantae (plants) and Animalia (animals). Since then it has been necessary to add Fungi and, for microscopic unicellular organisms, Monera for those that have no nucleus and Protista for those that do.

The Linnaean system was developed for use with modern organisms but it is also applied to fossils, which do not always fit comfortably into this scheme. In principle, modern organisms provide inexhaustible taxonomic information, morphological, biochemical, and behavioural, whereas fossils provide only limited information. The limitations are due to the fragmentary nature of fossils and the lack of preserved soft parts; molecular taxonomy, increasingly used on modern organisms, cannot therefore be applied. In addition, the relationship of separate parts of disarticulated animals might not be recognized, and they may be given form-generic status.

An example of such a situation would be the conodonts, in which the elements comprising the feeding apparatus of the otherwise soft-bodied animal are rarely found in association, and thus may receive separate generic and specific names. Whole plants are also rarely found in the fossil record, and separate names can be given to separate parts of the plant (leaves, spores or pollen, bark, etc.). An additional problem is posed by trace fossils, which are sedimentary structures produced by any organism and thus reflect behaviour rather than the morphology of the trace-maker. Although trace fossils can rarely be attributed to a particular organism, a modification of the Linnaean system is used to classify them. Traces are given ichnogenera and ichnospecies names based on morphological characters, but above that level they are grouped into a series of behavioural categories.

As taxonomic study proceeds on the assumption that evolution has taken place, it is much involved with phylogenetics, which deals with the description of evolutionary relationships. This is perhaps more important in palaeontology than in the study of modern organisms because of the added dimension of time that fossils bring to taxonomic studies. The existence of the fossils provides the direct evidence for the timing of the branching events that gave rise to the major taxonomic groups.

DAVID K. ELLIOTT

tectonic geomorphology

tectonic geomorphology Tectonic geomorphology is an exciting new subdivision of geomorphology that is concerned with active tectonic processes, including earthquakes resulting from faulting, that have produced some of the most spectacular mountain scenery in the world. Let us first define some basic terms. The term *tectonic*, from a geological viewpoint, refers to those structures and processes that are associated with the deformation of the Earth's crust as a result of deep-seated processes. On the other hand, *geomorphology* is concerned with the evolution of the landscape, special attention being paid to those processes that produce, modify, or destroy landforms. Putting these together, *tectonic geomorphology* is defined as the study of landforms produced by tectonic processes, such as faulting, tilting, folding, uplift, or subsidence. Another view of tectonic geomorphology is that it is the application of geomorphology to the study of tectonic problems. One of the most significant tectonic problems of relevance to society today is to obtain a better understanding of earthquakes and their effects. For example, we would like to know how tectonic processes produce hills and mountains and at what rates so that we can estimate how often large earthquakes may occur and what their consequences at the surface of the Earth are likely to be. When we bring in the human dimension, we are talking about what Robert Wallace, a scientist with the United States Geological Survey, defines as *active tectonics*, which refers to those tectonic processes that operate on a timescale of significance to human society. This period of time ranges from several decades to several hundred years, which is the time period over which we plan the lifetimes of structures such as buildings, bridges, and critical facilities, including dams and power plants. It is, however, important to understand that if we are to study and make estimates concerning tectonic processes over decades to hundreds of years, we will have to study a longer period of time. This is because geological structures such as faults may move at intervals of several hundred to several thousand years, or longer. For example, if you are interested in the earthquake hazard of the Los Angeles Basin, which experiences a moderate to large earthquake almost every decade, then you would need to consider the pattern and processes related to tectonic activity that has produced the basin, as well as the surrounding mountains. The mountains are the result of tectonic processes and geomorphological processes that have been going on for several million years, and there are numerous faults that have, over a period of hundreds of thousands of years or longer, caused uplift, folding, and other deformation at the surface of the Earth. If we are to evaluate and eventually predict the activity of a given fault, then we must be able to determine the history of faulting and put this in the context of the evolution of the entire region.

On a regional scale, observations and measurements from maps and aerial photographs, together with fieldwork, are used to evaluate geomorphic indices of active tectonics that are useful for establishing relative tectonic activity. Regional or reconnaissance work can also help to identify locations where more detailed field evaluation may yield important information at the local level. Finally, at the regional level, mathematical models may be developed to establish relationships between topography, geology, and active tectonics.

Geomorphic indices of active tectonics have been developed with the objective of establishing relative tectonic activity, and identifying those locations where more detailed work may yield valuable information concerning rates of tectonic processes and recurrence intervals for seismic events. This endeavour was pioneered in the 1970s by William Bull at the University of Arizona, who developed several important indices including 'mountain-front sinuosity' and 'ratio of valley-floor width to valley height'. The former is defined as the ratio of the length of a mountain front, as measured from an aerial photograph or topographic map or other method, to the straight-line length along the mountain front. Mountain-front sinuosity therefore reflects a balance between the tendency of erosional processes to produce an irregular or sinuous mountain front and the effect of vertical active tec-

tonic movement on steeply dipping, range-bounding faults, which tends to produce a relatively straight front. Low values of the indices thus correlate with relatively high rates of uplift along faults bounding mountain ranges, compared to others with greater sinuosity and lower rates of tectonic activity. The ratio of valley-floor width to valley height is used to differentiate between broad-floored valleys with relatively high values of the index and V-shaped valleys with relatively low values. Those valleys that are V-shaped tend to be those associated with active downcutting in response to uplift, and are therefore more likely to be associated with active tectonics at the mountain front than are broad valleys with relatively high values of the index.

Another geomorphological index that has proved valuable in reconnaissance studies of tectonic geomorphology to establish relative rates of tectonic processes is the 'stream-gradient index' developed by John Hack of the United States Geological Survey in the 1970s. The index crudely correlates with available stream power, which is defined as being proportional to the product of the discharge of water and the energy slope of the river. The index is related to available stream power because the energy slope may be approximated by the slope of the channel bed, and total upstream channel length is proportional to discharge. In practice, the index is evaluated at a number of locations from topographic maps, and the values are then contoured. The index is sensitive to rock resistance and other processes, including tectonic activity, which may influence the gradient of a stream. High values of the stream-gradient index may thus be related to locations where a stream flows across resistant rocks with a consequent steep gradient, or to areas that have experienced recent vertical tectonic movements that have increased the slope of the stream bed.

A valuable tool in evaluating the tectonic geomorphology of a region is to characterize the assemblage of landforms present. For example, there is a characteristic suite of landforms associated with strike-slip faulting, including offset or deflected streams, linear valleys, sags, pressure ridges, and topographic benches and fault scarps. As a second example, consider alluvial fans as indicators of active tectonics. Early work in the 1960s by William Bull demonstrated that many alluvial fans have radial profiles that are segmented into a series of relatively straight segments that make up the overall concave profile of the fan. Bull was able to demonstrate that fan segments are of differing ages, and their arrangement is often related to tectonic activity at the mountain front. For example, if the rate of uplift of the mountain front is relatively high compared to the rate of stream-channel downcutting in the mountain, then deposition on the fan tends to be near the mountain front, and that is where the youngest fan segment is most likely to occur. On the other hand, if the mountain block and alluvial fan have been tilted in the downstream direction, or if tectonic processes are such that the rate of uplift of the mountain front is less than or equal to the rate of downcutting of the stream in the mountain, deposition on the fan is shifted downstream and young fan segments are found well away from the mountain front.

The completion of reconnaissance or regional geomorphic evaluation that leads to the identification of promising sites where more detailed work is likely to yield results to determine the rates of tectonic processes is followed by evaluation at a more local scale. The tectonic framework is established and measurements are made to determine the deformation of landforms. Establishing the tectonic framework is based on recognizing the various types of tectonic features and structures that are present. For example, the tectonic framework comprising a fold-and-thrust belt commonly consists of a series of anticlines and synclines that respectively form linear hills and valleys arranged *en échelon*, that is parallel to one another but oblique to the zone containing item. Similarly, the tectonic framework associated with large-scale crustal extension is the presence of tilted fault blocks and of horst and graben structures. Measurement of deformation is relatively easy. Stream channels may be offset, marine terraces may be faulted, uplifted, tilted, or folded, and alluvial fans may likewise be faulted, tilted, or folded; and deformation can be measured in the field or from topographic maps. What is most difficult in tectonic geomorphological studies is to establish the late Pleistocene–Holocene chronology. In order to complete a successful study of this type, the rates of the processes should be calculated (or estimated). If no chronology has been established, then it is impossible to estimate rates. Relative chronology is not sufficient, and so we must look to a variety of other methods, including such techniques as dendrochronology and numerical dating, utilizing radioactive isotopes.

An important aspect of tectonic geomorphological evaluation is the construction of process–response models in conjunction with rates of active tectonics. These models are broadly defined so as to include the integrative investigation of deformation of Earth materials and landforms with the late Pleistocene–Holocene chronology. When this is accomplished, it is possible to predict future changes of the landscape and derive rates of tectonic processes, such as rates of tilting, uplift, or displacement caused by faulting. One of the most successful process–response models has been the evaluation of fault-scarp morphology as an indicator of active tectonics. When faulting ruptures the ground and produces a scarp in alluvial material, the scarp (i.e. steep slope) changes with time as the crest becomes more rounded by erosion and the toe is buried by deposits. The changes following the faulting event have been quantified mathematically, and so it is possible to estimate the time since faulting took place by examining the geomorphology of the fault scarp and knowing something about the climate of the region. Other process–response models have been developed for evaluating offset stream channels along strike-slip faulting. This work has been pioneered by Kerry Sieh at the California Institute of Technology. Careful evaluation of the stream channel deposits that have been displaced by strike-slip faulting, coupled to numerical dating (by the carbon-14 method) have yielded rates of slip along the San Andreas fault in California. As a final example of process–response models, consider the study of uplifted marine terraces. Palaeo-shoreline angles delineate past sea levels. Coupling of the geomorphology of dated marine terraces and elevation of shoreline angles to palaeo-sea-level curves allows rates of surface uplift in coastal areas to be estimated.

Some of the interesting and significant research questions being addressed by process–response models and quantification of tectonic processes are:

(1) What is the recurrence interval between seismic events for a specific fault?
(2) Which faults are responsible for the greatest earthquake hazard?
(3) Are surface rates of uplift, tilt, and faulting constant through time?
(4) How does the landscape respond to varying rates and types of tectonic processes?

Answers to some of these fundamental questions are critical in our efforts to reduce the hazards resulting from tectonic processes, particularly earthquakes. To this end, tectonic geomorphology is becoming an interdisciplinary field in the Earth sciences that incorporates aspects of geomorphology, hydrology, geophysics, and structural geology. The integration of these fields is producing developments in tectonic geomorphology and the construction of process–response models to help in our evaluation of how tectonic processes modify the surface of the Earth.

E. A. KELLER

Further reading

Bull, W. B. (1991) *Geomorphic response to climate change*. Oxford University Press, New York.

Keller, E. A. and Pinter, N. (1996) *Active tectonics*. Prentice Hall, Upper Saddle River, New Jersey.

Wallace, R. E. (ed.) (1989) *Active tectonics*. National Academy Press, Washington, DC.

Yeats, R. S., Sieh, K., and Allen, C. R. (1997) *The geology of earthquakes*. Oxford University Press, New York.

tectonics Most processes causing or associated with deformation of the Earth's crust or lithosphere can be described as tectonics. A *tectonic province* is distinguished by having a coherent history of deformation different from that of neighbouring regions.

Tectonic structures such as faults, folds, and cleavage are the result of deformation of pre-existing rock, as opposed to features generated during the original deposition or formation of the rock (such as cross-bedding in a sedimentary rock or joints caused by cooling and contraction in an igneous rock). Superficial events like landslides and slumping are excluded from the definition, even though they may generate faults or folds.

Deformation on a regional scale in the form of compression, extension, or uplift has long been attributed to 'tectonic forces', but it was not until the development of the theory of plate tectonics in the late 1960s that the underlying causes of most tectonic deformation began to be understood. Plate tectonics is now widely accepted, and recognizes that the Earth's rigid outer shell, the lithosphere, is split into a number of plates. These glide over a weak zone in the mantle known as the asthenosphere, and the relative motion between plates causes most large-scale tectonic structures. Plate-tectonic models based on the motion of rigid plates over the surface of a sphere explain tectonic features in the oceanic crust very well. However, where plate boundaries pass through continental crust they are complicated, and although simple plate-tectonic theory can explain large-scale structures in the continents in a general sense it often fails to predict the detailed structure.   DAVID A. ROTHERY

temperature *see* ATMOSPHERIC TEMPERATURE

terraces, coral *see* CORAL TERRACES

terrestrial planets and other Earth-like bodies

The Earth is just one of many planets and planet-sized bodies orbiting the Sun. Information from space missions, most spectacularly in the form of close-up pictures, has removed these planetary bodies from the realm of astronomy into the purview of geologists and meteorologists. Other Earth-like bodies display all the processes with which we are familiar on the Earth, but blended in different proportions and therefore having different net results, and their study helps considerably in understanding our own planet. By way of analogy, suppose you were trying to study the life cycle of a tree. Imagine how poor and incomplete an appreciation you would have of this tree in particular, and of trees in general, if you only ever looked at a single specimen—say an oak growing in a sheltered meadow—and did not realize that its way of growth would be entirely different were it on a windy hillside or in a dense forest. Consider also how wrong you would be to assume that essential attributes of all trees were (like the oak)

to shed their leaves in autumn and to have no defence against browsing animals, because you had never looked at an evergreen pine tree or a thorn bush.

The bodies most like the Earth are referred to as the *terrestrial planets*. Traditionally there are four of these. In sequence outwards from the Sun they are Mercury, Venus, Earth, and Mars. These planets share the fundamental characteristic of being large, rocky bodies with a dense, iron-rich, core. The Moon, although a satellite of the Earth, and therefore not a planet in the astronomical sense, has so much in common with them that geologically it can be regarded as a fifth terrestrial planet. Io, the innermost large satellite of Jupiter, belongs to the same class as the Moon. The tally of Earth-like rocky bodies thus stands at six (Table 1).

The giant planets (Jupiter, Saturn, Uranus, and Neptune) are sufficiently unlike Earth to be discussed in a separate entry (see *giant planets*), but there are 19 other bodies in the solar system large enough to show (to varying degrees) Earth-like attributes in terms of internal differentiation into core and mantle, and surface processes of volcanism, fault movements, and cratering. These bodies tend to be found further from the Sun than the terrestrial planets and have icy mantles surrounding rocky cores. Although compositionally different from the Earth, the geological processes in their icy outer layers are close analogues to those in the terrestrial planets. Most of these icy bodies are satellites of the giant planets, but Pluto and its moon Charon also belong to this class (Table 2).

The other solid matter in the Solar System consists of objects with insufficient gravity to pull themselves into spherical shapes and generally too small to show internal differentiation. These are rocky asteroids, ranging from 900 km across downwards, and icy bodies less than about 400 km across. They, too, are sufficiently unlike the Earth to be described in a separate entry (see *asteroids and comets*).

The Moon

The most appropriate Earth-like body with which to begin a review is the Moon, which is the closest body to the Earth and the only one from which samples have been collected. Like the Earth, the Moon's main source of internal heat is the decay of

Table 1. Earth-like rocky bodies

Name	Distance from Sun (millions of kilometres)	Diameter (km)	Mass relative to Earth	Density (tonnes m⁻³)
Mercury	57.9	4878	0.0553	5.43
Venus	108.2	12104	0.815	5.25
Earth	149.6	12756	1.000	5.52
Moon	149.6	3476	0.0123	3.34
Mars	227.9	6786	0.11	3.95
Io	778.3	3630	0.0149	3.57

The Earth's mass is 5.98×10^{24} kg. The Moon is the Earth's only natural satellite, and Io is a satellite of Jupiter.

Table 2. Earth-like icy bodies

Name	Satellite of	Diameter (km)	Mass relative to Moon	Density (tonnes m⁻³)
Europa	Jupiter	3138	0.653	2.97
Ganymede	Jupiter	5264	2.02	1.94
Callisto	Jupiter	4800	1.47	1.86
Mimas	Saturn	396	0.00052	1.17
Enceladus	Saturn	502	0.00109	1.24
Tethys	Saturn	1048	0.0104	1.26
Dione	Saturn	1108	0.0143	1.44
Rhea	Saturn	1524	0.0339	1.33
Titan	Saturn	5150	1.83	1.88
Iapetus	Saturn	1436	0.026	1.21
Miranda	Uranus	472	0.00102	1.35
Ariel	Uranus	1158	0.0184	1.66
Umbriel	Uranus	1192	0.0173	1.51
Titania	Uranus	1580	0.0474	1.68
Oberon	Uranus	762	0.0397	1.58
Proteus	Neptune	418	0.00054	1.1
Triton	Neptune	2700	0.291	2.08
Pluto	(Sun)	2320	0.191	2.1
Charon	Pluto	1270	0.015	1.3

With the exception of Pluto these are all planetary satellites, and it is convenient to compare their masses with that of the Moon (which is 7.35 × 10²²kg). Europa has only a thin icy shell and is intermediate in character between the typical icy bodies and the rocky bodies listed in Table 1.

radioactive elements. The total rate of heat production within a rocky planetary body depends on its mass, but the rate of heat loss depends on its surface area. The smaller the body, the greater the ratio of its surface area to its mass, and so the faster it loses its heat. Having less than an eightieth of the Earth's mass, the Moon has lost so much of its internal heat that its interior is rigid down to a depth of about 1000 km, and there has been little volcanism or other geological activity on its surface for the past 3 billion years. Consequently, the Moon retains a record of ancient events whose traces have been almost entirely erased from the Earth's surface by erosion, volcanism, deformation, and burial.

The oldest regions of the Moon's surface are the lunar highlands (Fig. 1). These occupy just over half of the Earth-facing hemisphere and most of the far side. The most obvious feature of the highlands is that they are completely covered by impact craters, ranging downwards in size from hundreds of kilometres across. Once thought to be volcanic in origin, it has now been proved that virtually all lunar craters are the result of impacts by meteoritic debris hitting the surface at speeds of a few tens of kilometres per second. In such an event, a shock wave is generated at the point of impact, which propagates radially to excavate a crater with a diameter about thirty times that of the impacting body, and distributes fragmentary ejecta over the surrounding area.

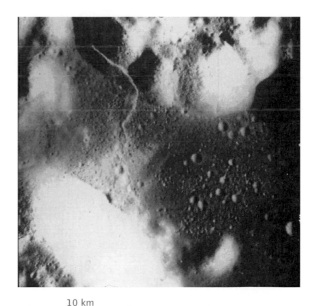

10 km

Fig. 1. A lunar highland area including the landing site for *Apollo* 17. Only relatively small impact craters appear in this view, the largest being about 2 km across. (Image by courtesy of NASA.)

The Moon has several multi-ringed impact basins up to 1000 km or so in diameter that were caused by particularly large impacts. All those on the Earth-facing hemisphere have been flooded by basaltic lava flows. However, the largest one, the South Pole–Aitken basin (which is on the far side), escaped this fate. The lava-covered regions, which in places have spilled beyond the limits of the basins, make the familiar dark patches on the face of the Moon and are referred to (inappropriately) as the lunar seas or *maria* (Latin, *sea*; singular *mare*).

That the maria are younger than the surrounding highlands can be demonstrated by two simple observations. The first relies on the basic geological *principle of superposition*: the lavas of the maria can be seen to overlap on to the highlands. The second is that any given area of maria has considerably fewer craters than an equivalent area of highlands, and (because impact cratering is essentially a random process) a region with a consistently lower crater density must be younger than a region with a higher density of craters.

Radiometric dating of samples from the Moon has made it possible to put the above deductions on a more refined basis. The lunar maria were created by series of lava eruptions lasting about half a billion years and terminating about 3.1 billion years ago. The lunar highland surfaces, despite having much greater densities of impact craters, are not much older than the oldest maria; they date back to about 3.9 billion years ago. It is believed that the cratering we can see on the lunar highlands records the tail end of the cataclysmic bombardment inherited from the birth of the Solar System (4.6 billion years ago), which obliterated the traces of the highland's previous history. The craters that appear on the lunar maria (and the youngest craters in the highlands) result from occasional impacts by small asteroids and comets that hit the Moon (and, inevitably, the Earth) during subsequent time. On bodies other than the Moon, where we have no radiometric dates, counting the number of craters in a given area and relating these to the calibrated lunar catering timescale is the best way we have of estimating the age of the surface.

The Moon has no atmosphere to shield it from smaller cosmic debris, and as a result it is cratered on all scales down to the microscopic. Because there is no erosion to remove the ejecta thrown out during crater formation it remains where it falls until redistributed by the arrival of a fresh impactor. Consequently, in most places the lunar bedrock is buried by at least several metres of fragmental and dusty debris constituting the lunar 'soil' or '*regolith*' in which the astronauts left their footprints during the *Apollo* Moon landings of 1969–72.

As well as bringing samples back, the astronauts were able to deploy various geophysical instruments on the Moon's surface; in particular, seismometers to record the vibrations from 'moonquakes' that have given us a clearer picture of the Moon's interior than we have of any planet other than the Earth. From these data it has been deduced that the Moon's core occupies no more than about 2 per cent of its volume.

500 km

Fig. 2. Part of the planet Mercury, including the sunlit half of the 1300 km diameter Caloris multi-ringed impact basin on the left. Many impact craters of 100 km diameter and less are visible. (Image by courtesy of NASA.)

Mercury

Mercury is superficially the most Moon-like of the other terrestrial planets, but is particularly dense for its size, implying that it has a relatively large iron core occupying about 40 per cent of its volume. Only one space probe has visited Mercury, *Mariner* 10, which made three fly-bys of the planet in 1974 and 1975. The pictures sent back by this mission show a heavily cratered surface (Fig. 2), which, as we know from lunar experience, shows that there has been little geological activity to erase the scars of the intense cratering that occurred during the first half-billion years of the history of the Solar System.

Venus

In contrast, Venus is the most Earth-like of the other planets, in that its size, mass, and density are each only slightly lower than those of the Earth. Its history has, however, been very different. The most obvious difference is that Venus's atmosphere has evolved in an entirely different manner from that of the Earth. It is mostly carbon dioxide, with a surface pressure more than 90 times that at sea level on the Earth. Sulphur dioxide droplets high in Venus's atmosphere hide the planet's surface behind a perpetual cloud layer that reflects sunlight so strongly that Venus outshines the brightest star in Earth's sky, when it is visible as the evening or morning 'star'. Although Venus's atmosphere reflects most of the sunlight, it traps enough solar heat by the greenhouse effect to give the planet a surface temperature of a searing 460 °C.

The surface of Venus has been seen by a series of seven Soviet probes that soft-landed on Venus between 1985 and 1992. Pictures from these landers show a rather slabby surface. Crude chemical analyses identify the rock type as resembling the Earth's ocean floor rather than the Earth's continents. More important than these isolated snapshots are the detailed images of virtually the entire globe obtained by radar from spacecraft in orbit about the planet, notably the American probe *Magellan* (1990–93). These show that Venus has a bewildering variety of landforms, such as Hawaii-like shield volcanoes; vast fields of lava flows; long, narrow channels apparently carved by flowing lava; crumpled belts of fold mountains resembling the Appalachians on the Earth; fields of sand dunes blown by dense but sluggish winds; and regions of intensely fractured terrain (Fig. 3).

No present-day internally driven activity has been observed on Venus. Moreover, impact craters are not uncommon on Venus, and are distributed randomly across the globe in numbers suggesting an age of about 500–800 million years.

50 km

Fig. 3. A radar image of a heavily deformed region on Venus (Ovda Regio). (Image by courtesy of NASA.)

There are no identifiable tracts of younger terrain with significantly lower crater density. The entire surface is therefore relatively old compared to that of the Earth, where ocean floor is continually being created by sea-floor spreading and destroyed at subduction zones, and where areas of continental crust are resurfaced by volcanic eruptions, erosion of actively uplifting mountains, and burial by deposition in low-lying areas. This contrast between the two planets is surprising, because they should have virtually the same radiogenic heat supply and be losing heat at roughly the same average rate. Outward heat transfer in the Earth is ultimately responsible for all the processes listed above, and yet they seem not to be occurring today on Venus. One explanation is that Venus behaves in an episodic fashion. Most of the time (as at present), heat is trapped below the planet's rigid outer shell (or *lithosphere*). As the interior part of the mantle warms up, it becomes progressively more buoyant in relation to the colder and denser exterior. Eventually the entire surface founders in a cataclysmic event, leading to volcanic resurfacing on a global scale, followed by tectonic deformation of some areas as the new surface settles down. If this last happened 500–800 million years ago, it would explain what we see today on Venus, with the youngest volcanic features representing the waning phase of activity as the new lithosphere thickened and strengthened. Perhaps we shall have to wait only about another 100 million years before it all happens again.

Mars

Although small, Mars displays clear signs of a wide range of geological processes. Its southern hemisphere is mostly ancient, heavily cratered terrain, whereas its northern hemisphere consists of lower-lying plains that are probably of volcanic origin but are much modified by the deposition of sediment derived from the southern highlands. Although the theory that the northern hemisphere was occupied by a shallow sea some billions of years ago remains controversial, it is undeniable that torrents of water once debouched into it from the southern highlands, fed by channels that are clearly visible today.

At a mere 6 millibars, the present atmospheric pressure on Mars is now too low to permit liquid water to exist, and the martian channel systems are the clearest evidence in the Solar System for a major global climatic change. The only known water on Mars today resides in the residual ice caps at either pole, which are surrounded by seasonal carbon dioxide ice that condenses out of the atmosphere in autumn and sublimes again in spring. There may, however, be an even greater volume of invisible water ice trapped in a permafrost layer below the surface in many parts of the globe. Indeed, although some of the martian channels form dendritic networks characteristic of having been fed by surface run-off resulting from rainfall, others have morphologies consistent with headwards sapping by melting of subsurface ice.

10 km

Fig. 4. The wall of part of the Valles Marineris system on Mars, showing many signs of collapse, with some impact craters on the plain beyond. (Image by courtesy of NASA.)

Water also occurs in hydrous minerals, such as the alteration products found within rare groups of meteorites whose martian origin is now well established. The extreme fractionation of stable isotopes of carbon that is consistent with biological activity in parts of these meteorites, the array of organic molecules associated with these, and ambiguous bacteria-like shapes seen in scanning electron microscope images provide tentative evidence that primitive life once existed on Mars. As Mars clearly once had much more water than today and (inevitably) hot, mineral-rich springs associated with volcanic activity, and bearing in mind that life on Earth seems to have begun in comparable conditions, the proposition that life could have originated on Mars is entirely reasonable. This being so, bacteria-like organisms could have survived there to this day within the polar caps or in the subsoil.

Turning to less controversial aspects of martian geology, sediment transport today is largely by wind. Seasonal dust storms frequently obscure our view of the surface, and there are extensive fields of dunes. These are particularly widespread around the northern polar cap, where longitudinal dunes predominate, although crescent-shaped barchan dunes are also found.

Major volcanoes occur in several regions of Mars. The oldest (dated at about 3.1–3.7 billion years ago on the basis of impact crater counting) are very gently sloping structures called *paterae*, hundreds of kilometres in diameter but only a few kilometres high, and usually with a caldera at the summit a few tens of kilometres in diameter. They are most likely to be products of explosive volcanism, and their flanks are interpreted as deposits of volcanic ash. Younger volcanoes have forms similar to shield volcanoes on the Earth and have clearly been constructed mainly by eruption of basaltic lavas. Most of them are concentrated in the Tharsis volcanic province, an upwards bulge in the martian crust some 3000 km across that locally obscures the distinction between the high southern and low northern hemispheres. The youngest of the Tharsis shield volcanoes is Olympus Mons, which with a height of 24 km and a diameter of over 600 km is the largest volcano in the Solar System (Earth's largest is Hawaii, whose summit is 9 km above the floor of the Pacific Ocean). Olympus Mons is hard to date, but may have most recently been active only about 200 million years ago.

Another remarkable feature on Mars is Valles Marineris, a canyon system 4000 km long that runs eastwards from the Tharsis bulge before draining northwards into the northern hemisphere lowlands. The Valles Marineris system probably owes its origin to tectonic fracturing, but has subsequently been modified by water flow and landslips (Fig. 4). It reaches 7 km in depth and over 200 km in width and would dwarf the Earth's Grand Canyon as thoroughly as Olympus Mons would dwarf Hawaii.

Io

Io's size, mass, and density (Table 1) would lead one to expect it to be a Moon-like body. However, whereas volcanic activity ceased on the Moon several billion years ago, Io still has about a dozen volcanoes erupting at any one time. This is because of tidal heating, caused by Jupiter's powerful gravitational field, which raises tidal bulges several kilometres high on Io's Jupiter-facing and anti-Jupiter hemispheres. The changing stresses on these tidal bulges as Io orbits Jupiter heats Io's interior about a hundred times faster than does radioactive decay. Io has a strong magnetic field, suggesting that heating of its interior has not merely made it possible for Io to develop a fully differentiated structure but also that its iron-rich core is fluid and therefore molten.

No impact craters are known on Io, which is to be expected on a body where the global average rate of resurfacing by volcanic deposits is apparently well in excess of 1 mm a year. Instead, Io's surface is dominated by volcanic features, notably gentle shield volcanoes with summit calderas several tens of kilometres across and surrounded by radial lava flows (Fig. 5). The strong yellowish colour of Io was formerly thought to indicate that many of these flows had been formed by molten sulphur, but it is now accepted that these flows are conventional silicate flows (perhaps basaltic) and that the colours are a result of surface coatings by sulphur and sulphur oxides. A major contrast with volcanism on the Earth is that whereas on Earth the most abundant volatile escaping from erupting volcanoes is water vapour, on Ió this role is taken by sulphur dioxide and sulphur. Violent escapes of volatiles at eruption sites on Ió drive pyroclastic eruptions in the form of eruption plumes that reach up to 300 km in height and spread, umbrella-like, over regions in excess of 1000 km across.

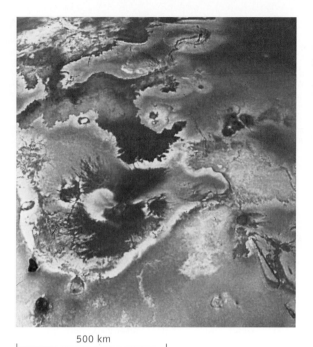

Fig. 6. 'Rafts' of thick ice separated by newer tracts of thinner ice on Europa. The ridge and groove patterns on the rafts can be fitted together to see how the rafts must have moved since they broke apart. (Image by courtesy of NASA.)

Fig. 5. Part of Io's surface. Young lava flows are relatively dark, except where bleached by condensation of gaseous exhalations. (Image by courtesy of NASA.)

Jupiter's other satellites

Jupiter has three other large satellites, orbiting beyond Io. The closest is Europa, which is transitional in nature between the terrestrial and icy classes of bodies. It has a cold, icy surface, but its high density shows that the ice can be no more than about 100 km thick. Below this is a thick icy mantle, probably with a small iron-rich core in the centre.

Europa's exceptionally complex surface is marked by multiple generations of fractures and curved ridges that appear to indicate episodes when the icy shell was broken into numerous jostling platelets (Fig. 6). Unlike Jupiter's other icy-surfaced satellites, Europa has few impact craters, which means that its surface must be very young. This is consistent with the amount of tidal heating to be expected in the upper part of Europa's rocky mantle. It is uncertain whether the ice today is solid all the way to its base, or whether it floats on a layer of water. Local, if not global, melting at depths of only a few tens of kilometres in recent times seems highly likely in view of the young surface and the identification of over-lapping flow-like features probably formed by the extrusion of partly crystallized icy brines.

Speculation abounds about the interactions that could occur within Europa where the ice or water is in contact with the underlying rock. In view of the tidal heating, some kind of hydrothermal circulation seems inevitable, with water being drawn through the rock and expelled at nearby warm vents as a brine charged with chemicals dissolved from the rock. On Earth's deep ocean floors, comparable hydrothermal vents such as black smokers are known to host self-contained ecosystems, and it is probable that heat-tolerant bacteria that derive their energy through chemical reactions could survive if transplanted from a black smoker to a vent site below Europa's ice. Whether life has actually begun there remains unanswered as yet, but the chances are good given that the origin of life on Earth is now attributed to hot vents.

Ganymede and Callisto are true icy bodies, with ice extending to depths of more than half their radius. Magnetic and gravity measurements suggest that Ganymede's rocky interior has a fluid iron core within it. Callisto, the outermost of Jupiter's large satellites, has a heavily cratered surface (Fig. 7) bearing no clear traces of internally driven geological processes. On the other hand, Ganymede has tracts of dark, heavily cratered ancient terrain broken by belts of paler, younger terrain. The pale terrain displays a surprising pattern of ridges and grooves, occurring on a variety of scales, and apparently recording some blend of tectonism and icy volcanism. Cross-cutting relationships demonstrate that the pale, grooved terrain was developed in multiple stages, but the large number of impact craters on even the youngest parts show that these events concluded long ago (probably 3–4 billion years ago). With surface temperatures considerably below −100 °C, the ice is far too cold to flow like a glacier, it behaves to great depth mechanically like the Earth's lithosphere, and large impact craters retain their shapes without slumping.

Other icy satellites

The large satellites of Saturn, Uranus, and Neptune are essentially icy bodies with small, rocky cores. Many are heavily

5 km

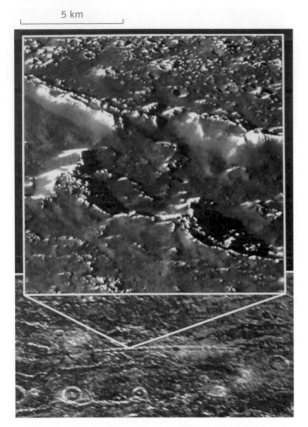

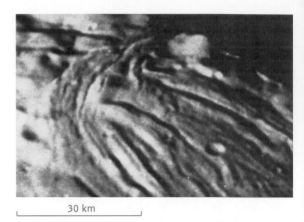

Fig. 8. A part of Miranda, including (top centre) what appears to be an icy 'lava' flow, which was probably formed by eruption of a viscous ammonia–water melt. (Image by courtesy of NASA.)

Fig. 7. Close-up of part of Callisto, revealing details as small as 100 m across. The main image is set in context within the wider area shown below, revealing that it shows part of a chain of overlapping craters. Such crater chains are thought to be produced by the serial impact of fragments of tidally disrupted comets. (Image by courtesy of NASA.)

cratered, but those that have experienced significant heating (presumably tidal in origin) at times during the past 4 billion years have younger surfaces and are therefore less heavily cratered and preserve evidence of internally driven geological processes. An important factor governing the volcanology of these icy bodies is that their ice is believed to be contaminated with ammonia, methane, carbon monoxide, and other species that would have condensed out of the solar nebula at the low temperatures prevailing in the outer Solar System. These mixed ices melt in exactly the same way as mixtures of silicate minerals found in ordinary rock. Mineralogically, an ice formed from water and ammonia will consist of crystals of two distinct compositions: pure water-ice (H_2O) and ammonia hydrate ice ($H_2O.NH_3$). On heating, this ice will begin to melt at −97 °C, well below the temperature at which melting would occur in either pure water-ice or pure

ammonia hydrate ice, yielding a melt consisting of about two parts water and one part ammonia. Unless the starting material was particularly rich in ammonia, the solid residue would be crystals of pure water-ice. Melting behaviour of this type is described as *partial melting* and is important because it allows melts to be generated at low temperatures and, by separating the first-formed melt from the solid residuum, it is a way of generating an evolved crust that is chemically distinct from the underlying mantle. Extrusions of viscous ammonia–water melts are most clearly seen in the form of icy lava flows (long-since solidified) on Miranda (Fig. 8).

There are many features of note other than icy volcanism. Several icy satellites are marked by fracture systems demonstrating phases of tectonic activity. Two have atmospheres. Titan, a particularly large satellite of Saturn, has a dense atmosphere consisting mostly of nitrogen but made opaque by clouds of methane and heavier hydrocarbons, not greatly dissimilar, perhaps, to the atmosphere of the primitive Earth. Triton, Neptune's largest satellite, has a much more rarified atmosphere composed essentially of nitrogen that sublimes from the seasonally sunlit polar cap of nitrogen ice (which overlies the methane-rich permanent surface) and condenses on the opposite polar cap in darkness.

Pluto and Charon

Most icy satellites probably formed in orbit around their planets, but Triton's retrograde orbit suggests that it formed separately and was captured by Neptune later. Pluto appears to be a Triton-like body that has evaded capture, although it has its own satellite, Charon. As a pair, Pluto and Charon are much more evenly matched than the Earth and Moon, and are the closest approximation in the Solar System to a double planet.

DAVID A. ROTHERY

Further reading

Beatty, J. K., Petersen, C.C., and Chaikin, A. (eds) (1999) *The new Solar System* (4th edn). Sky Publishing Corporation and Cambridge University Press.

Rothery, D. A. (1999) *Satellites of the outer planets* (2nd edn). Oxford University Press, New York and Oxford.

Rothery, D. A. (2000) *Teach yourself planets.* (2nd edn). Modder and Stoughton, London.

Tertiary The stratigraphical term 'Tertiary' is one of the oldest still in use, though it has in fact been superseded largely by 'Palaeogene' and 'Neogene' combined or by 'Cenozoic' which embraces a slightly longer time-span. It was originally coined by Giovanni Arduino (1714–95), a professor at the University of Padua. In 1758 he recognized three major divisions in the rock succession of the Apennine mountains of central Italy, naming them Primitive, Secondary and Tertiary. The last of these was a group of very fossiliferous sedimentary rocks, not very strongly indurated and hence sharply different from the hard strata of the Secondary and the crystalline rocks of the Primitive categories. It is the only one of Arduino's terms to survive, but its meaning is now different from the original.

Charles Lyell (1833), writing in London, adapted the word for the post-Mesozoic strata of western Europe; his 'Tertiary Period' was subdivided into Eocene, Miocene, and Pliocene (Older and Newer), and was followed by his 'Recent Period'. Lyell's division of this new unit was based on a new concept: that of dividing the Tertiary on the basis of the percentage of living species to be found among the molluscan fossils at each level. The fossils had been carefully collected by the French conchologist Paul Gerard Deshayes (1797–1875), and Lyell was able to reach a conclusion similar to that of Deshayes, on the basis of about 8000 species allotted to over 40 000 specimens from the Paris basin. Lyell was really intent upon establishing the use of molluscan fossils as a sort of palaeontological 'clock'. By finding the percentage of modern species in a sample from anywhere within the Tertiary rocks, a geologist would, he thought, be able to establish the relative age of the sample. Subsequent events were to prove that local geologists found it anything but easy to put the concept to work. Disputes and mistaken correlations were soon all too common. It was a nice idea, but unfortunately the rates of extinction and replacement are neither uniform nor constant. The process is decidedly episodic rather than smoothly gradual. Recent work on these rocks and fossils makes out a rather different picture of the taxonomy at each stage, and much of the fossil mollusc fauna of the Paris basin is restricted to that region. Lyell's terms have by now all been redefined, but remain in use as part of the Cenozoic.

Invertebrate fossils apart, the Tertiary strata to which so much attention was given by Lyell and his contemporaries are also remarkable for their mammalian remains and also,

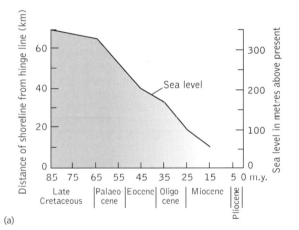

(a)

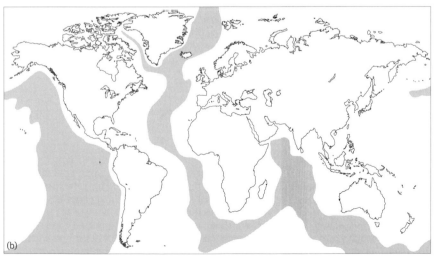

(b)

Fig. 1 Throughout post-Cretaceous time world sea level has been falling. One calculation of the rate and timing of this fall, based upon both stratigraphy and the record of ocean-ridge activity, is shown in (a). The process began towards the end of Cretaceous time. The graph seems to support the idea that marine regressions take place more quickly than transgressions. Whether or not ocean-ridge growth and subsidence are entirely responsible for these rates is debatable. (b) Areas of ocean floor (grey tint) created in the last 75 Ma, i.e., during a time period only slightly longer than the Tertiary period. Older ocean is shown uncoloured.

locally for their teleost fishes. The mammals include many marsupials and large placental forms which arrived on the scene abruptly after the extinction of the dinosaurs. The remains of mammals are, in fact, so numerous and stratigraphically sensitive in parts of the equivalent continental rocks in North America that biostratigraphic mammal zone successions have been drawn up there.

The *Global Stratigraphic Chart* published by the International Union of Geological Sciences (1989) uses the term 'Cenozoic' for all the post-Cretaceous part of the column. Arduino's Primitive (Primary), Secondary, and Tertiary categories have all dropped from modern use. The term 'Quaternary' remains as a strange relic of past geological debate, but it is well defined and understood, despite the fact that some geologists wish to make it redundant. Ironically, it was never used by Lyell. Today 'Tertiary' is being replaced by the terms 'Palaeogene' and 'Neogene', which denote two sub-periods that are justified on the basis of both physical and biostratigraphy in a way that would have won Lyell's approval.

During the Tertiary times the ocean ridges diminished in volume to a marked degree, and this has probably been responsible for most of the withdrawal of ocean waters from continental areas. Tertiary stratigraphy is thus dominated by regression and fall of sea level. This is shown in Fig. 1a; areas of sea floor created during this time are shown in Fig. 1b.

D. L. DINELEY

Further reading

Pomerol, C. (1986) *The Cainozoic Era; Tertiary and Quaternary.* (Trans. D. and E. E. Humphries.) Ellis Horwood, Chichester.

Savage, R. J. G. and Long, M. C. (1986) *Mammal evolution; an illustrated guide.* British Museum (Natural History), London.

Tethys

The great Austrian geologist Eduard Suess pointed out in 1893 that an inferred Jurassic marine realm extending from the Alps to the Himalayas had originally been an ocean, and that its compressional obliteration between the northern Angaraland (a shield area east of the Urals) and southern Gondwanaland had generated the Alpine–Himalayan mountain chains. To underline its oceanic character Suess called this sea 'Tethys', after the sister and wife of Okeanos, the god of the ocean in ancient Greek mythology. Shortly afterwards Suess extended the age of Tethys down to the Triassic.

The notion of Tethys turned out to be one of the most important and durable concepts that modern geology has inherited, but it has been interpreted in a variety of ways. Palaeobiogeographers have considered it to be an equatorial seaway extending from Central America to South-East Asia, characterized by a distinctive fauna. For stratigraphers, Tethys implied a certain collection of mainly Mesozoic marine facies realms. Tectonicians were divided into two camps. Some considered it a 'fixed' geosyncline that had existed since the later Proterozoic, which had diminished in size progressively in the Phanerozoic and was eliminated finally during the 'Alpine orogeny'. Others saw it as a rather narrow 'mobile' marine realm caught up between the drifting continental rafts of Laurasia and Gondwanaland. In plate-tectonic reconstructions of Pangaea Tethys is seen as an ocean wedging out westwards as the Americas are approached.

In the most modern tectonic interpretation, the Turkish geologist Celal Sengör has recognized the former existence of another, older Tethys (Palaeo-Tethys), a contemporary of Pangaea. Its closure formed the Cimmeride orogenic system, which is distinct from, but largely overprinted by, the Alpide orogenic system, a product of the demise of the 'classical' Tethys.

ANTHONY HALLAM

Further reading

Sengör, A. M. C. (1985) The story of Tethys: How many wives did Okeanos have? *Episodes*, **8**, 3–12.

tetrapods

'Tetrapods' (four-footed animals) is the term used to designate land vertebrates in which the fins of fish have been transformed into front and hind limbs. This involved a major advance for vertebrates that invaded a new environment very different from the one in which they had been living. This invasion gave rise initially to the amphibians, and subsequently to the amniotes, which include the reptiles, birds, and mammals. Amphibians retained an aquatic larval stage, but amniotes became totally terrestrial by developing a hard-shelled egg within which a relatively waterproof membrane (the amnion) enclosed the embryo in fluid.

Tetrapods are thought to have been derived from one of the groups of bony fish (Osteichthyes) at some time during the Devonian. The generally held view is that the osteolepiforms, an entirely fossil group of bony fish form the sister group (closest relatives) of the tetrapods. An opposing view that the dipnoans (lung-fish) are the closest relatives is not strongly supported at present. Osteolepiforms are known from the Lower Devonian, which suggests that tetrapods must have a similar antiquity; the earliest fossil tetrapods occur, however, in the Upper Devonian of East Greenland. Material collected from this area in the 1930s represents three separate genera, of which *Ichthyostega* is the best known. This is undoubtedly the most primitive tetrapod so far discovered. It retains many fish-like characters, including a fish-like tails, gills and sense-organs. Other tetrapods of similar age from Russia show that these organisms were already diverse and widely distributed by this time.

Movement on to the land required the modification of several important body systems to solve problems of support, locomotion, respiration, dessication, reproduction, feeding, and the functioning of sensory systems. Support and locomotion were developed by the modification of the paired fins, pelvic and pectoral girdles, and the vertebral column to provide limbs strongly connected to an interlocking vertebral system. The major bones of the limbs are already present in the fin supports of the osteolepiforms. Lungs were apparently already present in the osteolepiforms and were merely enlarged at the expense of the gills as terrestrialization took place. Dessication, or water loss, was prevented by the development of the dermal scales of fish, and the eyes were protected by lids and the development of tear glands. The main sensory system of fish is the lateral-line system, which detects vibrations in the surrounding water. This system does not operate effectively in air and was replaced by one in which a small bone, the stapes, moved into a position between a tympanic membrane (ear-drum), at the back of the skull, and the inner ear.

Tetrapods are traditionally considered to have developed in freshwater conditions and to have moved on to the land during arid periods when they needed to move from one body of water to another. A more recent view is that tetrapods originated in marine conditions. This is suggested by the fact that osteolepiforms lived mostly in marine conditions. In either case, the move on to land might have resulted from the need for juvenile osteolepiforms to escape from predators in the water. An additional influence may have been the presence of a great diversity of animal food, particularly arthropods, on the land at that time.   DAVID K. ELLIOTT

textures of rocks The texture of a rock shows the relationships between its constituent grains and crystals. A description of its texture includes both grain sizes and the patterns made by the grains. The texture of a rock often reveals information about the processes that formed it—its petrogenesis.

Igneous textures

Igneous textures provide a record of the cooling history of the rock. Magma (molten rock) is a liquid composed of silicates with small amounts of dissolved gases. As a magma cools, the silicates tend to form crystals. If cooling is very rapid, there is no time for crystals to form and the magma is quenched as a glass. Slower cooling leaves time for all the magma to form crystals and the result is a crystalline solid: fine-grained if cooling was relatively rapid, coarse-grained if cooling was slow. Some basalts cool slowly underground and then erupt and are quenched. Crystals that are larger than average (called phenocrysts) form during slow cooling and become set in a glassy or fine-grained matrix if quenching occurs. This type of porphyritic rock has a texture that is diagnostic of two stages of cooling. Just as bubbles on the surface of a cup of coffee tend to clump together, so do the crystals in a rock.

Some igneous rocks have different names depending upon the grain size (which is, of course, dependent on the cooling history). Although basalt, dolerite, and gabbro all have the same composition, a basalt is fine grained whereas a dolerite, which has cooled more slowly, is coarser grained. Gabbros have cooled even more slowly and have large grains (more than 1 mm across); they are plutonic rocks. In coarse-grained rocks a mixture of different crystals can be visible. Crystals that are entirely enclosed by others are interpreted as having formed earlier than the crystals that surround them. The order of crystallization of minerals reflects part of the cooling history. Some early formed crystals can become unstable and start to dissolve. These dissolved crystals can be recognized because the normally straight sides and sharp corners of the crystal are partly eaten away, forming a resorption texture.

Metamorphic textures

Metamorphic rocks are those that have been altered in the solid state by heat, pressure, or movement of fluids. Heating and compressing a rock can mean that a different combination of minerals becomes stable. The results of these changes are often spectacular and result in the formation of minerals that are rare in igneous rocks. Garnets, for example, are a common result of metamorphism. When one mineral changes into another new mineral it may pseudomorph the old mineral. A pseudomorph is a mineral that has a crystal shape that is not its own. For example, the metamorphic rocks of the Pyrenees contain graphite (a form of carbon like pencil lead) in the shape of octahedral diamond crystals. The diamonds formed by the heat and pressure of metamorphism have changed back to the low-temperature form (isomer) of carbon, which is graphite. These are graphite pseudomorphs after diamond formed by retrogressive metamorphism (that is, metamorphism that takes place when the temperature and pressure are falling). When a metamorphic mineral grows in its own shape it either pushes aside the surrounding minerals to form a porphyroblast, or it surrounds them to give rise to a poikilitic texture. In places where rocks are sheared, minerals that deform by brittle deformation are broken into long strings; minerals that deform under these conditions by plastic deformation are streaked into long rods. The rock then possesses a shear fabric (or texture).When spherical objects such as pebbles have been plastically deformed the arrangement of their long axes can be used as strain indicators. Where much shearing and brittle deformation have occurred, the grain size is reduced and the grains are angular and jumbled; the resulting rock is called a cataclastite.

Sedimentary textures

The depositional processes of sedimentary rocks reflect the fluid dynamics of the particles in the fluid (wind or water) that deposited them. Particles that are deposited close to their source are usually angular and poorly sorted (the grains are of mixed size). Particles that have been transported a long way become rounded and well sorted (the grains are of identical size). Fast-moving currents are able to transport large particles to form coarse-grained sediments, whereas slow currents can carry only smaller particles. As a current slows, progressively smaller particles settle out and form fine-grained sediment. When sediments are buried, the increasing pressure and temperature can result in the dissolution of minerals and new growth elsewhere in the rock. Minerals deposited in the pore spaces of a rock commonly cement together the clastic grains. These processes in which sediments are altered and compacted are known as *diagenesis*. Metamorphism could be thought of as an extreme form of diagenesis, but diagenetic processes are generally distinguished from metamorphism and are considered in terms of the sedimentary rocks on which they act.   JUDITH M. BUNBURY

thermal metamorphism *see* CONTACT METAMORPHISM
(THERMAL METAMORPHISM)

thermobarometry Igneous and metamorphic rocks bear
witness to processes that occurred under conditions of pressure and temperature much higher than those at the Earth's
surface. Indirect techniques that use the compositions and
assemblages of minerals in rocks to determine the temperatures and pressures at which they formed are called
geothermometry and *geobarometry* respectively.

Although there are many different geothermometers and
geobarometers, most rely on equilibrium chemical reaction(s)
between two or more phases (e.g. minerals) in a rock. Because
the extent of chemical equilibrium is determined by kinetic
factors, such as solid-state diffusion, reactions become inhibited below a threshold (or 'closure') temperature. Rocks thus
commonly preserve chemical or mineralogical evidence for
the highest temperature that they have experienced. The most
widely used geothermometers depend on exchange reactions
between two components (e.g. Fe^{2+} and Mg^{2+}, or ^{16}O and ^{18}O)
whose kinetics are sufficiently sluggish to preclude re-equilibration during cooling. Geobarometers typically employ net
transfer reactions in which one or more phases is consumed
or produced. The ability of thermobarometry to determine
the magmatic or peak metamorphic temperature and pressure
of a rock depends on the cooling history of the rock and the
extent of subsequent re-equilibration. Ideally, a number of
geothermometers and geobarometers are applied to a given
rock in an attempt to counter these effects. Careful consideration must also be given to textural evidence for equilibrium
between the phases of interest.

Thermobarometers require thermodynamic calibration of
the equilibrium constant for a particular reaction or set of
reactions. In general, geothermometers depend on reactions
with large enthalpy (heat) changes and barometers on reactions with large volume changes, so as to maximize the sensitivity of the equilibrium constant to changes in temperature
or pressure. In nature there are far fewer reactions suitable for
geobarometry than geothermometry. Thermodynamic data
used to calculate the equilibrium constant may be derived
calorimetrically or empirically. In either case a reliable
description of activity–composition relationships in the
relevant minerals or phases is essential in determining the
equilibrium constant, and hence pressure or temperature,
from natural mineral compositions.

In practice, thermobarometry entails the precise determination of mineral compositions in a rock, typically using
the electron microprobe. Common exchange geothermometers include: Fe^{2+}–Mg between garnet and biotite (useful
for metapelitic schists); Ca–Mg exchange between clinopyroxene and orthopyroxene (basic and ultrabasic igneous rocks);
Fe–Ti exchange between ilmenite and magnetite solid solutions (acid and intermediate volcanic rocks); ^{16}O–^{18}O isotopic
exchange between magnetite and quartz (many igneous and
metamorphic rocks); NaSi–CaAl exchange between plagio-

clase and amphibole (amphibolites and some igneous rocks);
and Ca–Mg exchange between calcite and dolomite (metamorphosed limestones). Geobarometers for use with
metapelitic rocks use net transfer reactions in the assemblages
plagioclase–garnet–Al_2SiO_5–quartz; or garnet–rutile–Al_2SiO_5–
ilmenite–quartz. MgSi–Al_2 exchange between garnet and
orthopyrozene can be used as a geobarometer in basic and
ultrabasic rocks. Many continuous reactions offer potential
for relative thermobarometry: for example, during growth of
zoned crystals. Some discontinuous reactions can be used to
estimate minimum or maximum pressure or temperature; for
example, the stability of graphite and diamond, or the Al_2SiO_5
polymorphs. Related techniques use exchange reactions to
constrain the partial pressures of oxygen and water during
rock formation.   J. D. BLUNDY

thermosphere When the various layers of the atmosphere
are described in terms of their temperature structure, the
thermosphere is the uppermost layer. It extends from about
80 km to the top of the atmosphere. *Thermos* is Greek for
'hot', and this is the hottest part of the atmosphere. At 80 km
the mesopause is the coldest part of the atmosphere, but as
altitude increases temperature also increases. In the absence
of a heat source temperature will decrease with height. The
heat source in the thermosphere is radiation and energetic
particles in the solar wind. The Sun emits not only sunlight
but also ultraviolet light, X-rays, and a stream of charged
particles, electrons, and ions. These bombard the upper part
of the atmosphere and if they are sufficiently energetic they
can disrupt the atomic and molecular structure of the
atmospheric gases.

An individual molecule may respond in several different
ways when it absorbs radiation or collides with an incoming
particle. It may absorb some of the energy in a collision so
that the incoming particle moves off with less energy. It may
absorb radiation and change its internal energy states. It may
absorb so much energy that the forces binding the molecule
together can be overcome and the molecule can split into its
component atoms. All these processes can and do happen in
the thermosphere and all of them contribute to an increase
in the energy of the atmospheric gases which is equivalent to
an increase in temperature. Because these processes can result
in ionization, this region of the atmosphere is also known as
the ionosphere.

At lower levels in the atmosphere the predominant gases
are nitrogen (78 per cent by volume) and oxygen (21 per cent
by volume). (These figures neglect water vapour, which varies
between 4 per cent and negligible amounts.) Both oxygen and
nitrogen exist in molecular form, that is, each molecule is
formed from two atoms, N_2 and O_2. In the thermosphere both
gases suffer changes through collisions and absorption of
radiation. Oxygen is readily split into its atomic state so that
each molecule becomes two oxygen atoms which have more
energy than the complete molecule. The molecular bonds of
nitrogen are stronger than those of oxygen, and it is therefore

not separated, but it can have electrons removed to become ionized. In the lower parts of the thermosphere, molecular nitrogen is still the most common constituent, but atomic oxygen (O) is more common than molecular oxygen (O_2). At altitudes greater than 180 km, atomic oxygen becomes more common even than molecular nitrogen. This is partly because all the molecular oxygen is now in atomic form. It is also due to the fact that molecular nitrogen is much heavier than atomic oxygen and gravity is stratifying the heavier species at lower levels, leaving only lighter species at higher levels. Gravity does not have this effect below the thermosphere, because turbulence is sufficient to ensure homogeneous mixing of all the gases. However, at 100 km altitude the distance that an atom or molecule will move before colliding with another is, on average, 0.1 m. This is a hundred million times farther than at the Earth's surface. At 180 km this distance has increased to 100 m between collisions.

CHARLES N. DUNCAN

Further reading

Ahrens, C. D. (1994) *Meteorology today*. West Publishing Co.

McIlveen, J. F. R. (1986) *Basic meteorology*. Van Nostrand Reinhold, New York.

Thomson, William (Lord Kelvin) (1824–1907)

William Thomson became the foremost British physicist of his day, and made important contributions to Earth science. He took up the chair in natural philosophy at Glasgow in 1846 and occupied it for 53 years.

His interest in thermodynamics prompted his idea that the dissipation of heat from the Solar System could provide a clue to the age of the system, including that of the Earth. He was aware of the temperature gradient within deep mines and calculated a terrestrial heat loss. It was possible to calculate the time needed for this from the point when the crust of the Earth solidified. His figure of about 100 million years (Ma) was quite beyond anything previously suggested and caused prolonged controversy. It outraged the adherents of uniformitarianism, but it gave many of them pause for thought. At that time most geologists were content with less than 100 Ma for the age of the planet, and indeed for the Sun also.

Kelvin summarized his ideas in this field in 1897, concluding that the Earth had been habitable for as little as 20 Ma, certainly not longer than 40 Ma. Clarence King in the United States estimated 24 Ma that same year. All this was to dismay Darwin, who needed a longer span of time for natural selection to bring about the living world of today. At the time of this debate scientists were ignorant of the generation of heat within the Earth by radioactive elements (radiogenic heat). When the contribution made by this hitherto unsuspected source of heat was later recognized, it became apparent that the Earth must be much older than 100 Ma.

Thomson was knighted in 1866 and raised to the peerage as Lord Kelvin of Largs in 1892. He retired in 1899 and died in 1907. He is buried in Westminster Abbey.   D. L. DINELEY

tornadoes

Tornadoes are small and short-lived but highly destructive phenomena. Because of their severe nature and small size, comparatively little is known about their detailed structure: measurement and observation within them is difficult. They comprise elongated funnels of cloud which descend in a snake-like fashion from the base of a well-developed cumulonimbus cloud, eventually making contact with the ground beneath. They represent a zone of marked convergence into the cloud base generated by the intense updraughts that feed the cloud system, and are manifested as a narrow columnar zone of upwardly rotating violent winds, perhaps exceeding 100 metres per second. The vortex core is at very low atmospheric pressure, so that cloud develops within the funnel, and pressure gradients can approach an estimated 25 mbar (hPa) per 100 m. This compares to the most extreme pressure gradients of about 20 mbar per 100 km around a larger-scale cyclone.

A tornado's impact as a hazard is extreme. There are three damaging factors at work. First, the winds are often so strong that objects in the tornado's path are simply removed or very severely damaged. The pronounced rotational element also tends to twist objects from their fixing, and strong uplift can carry some debris upwards into the cloud. More typically, the heavier material is thrown tangentially outwards by the system's rotational winds: the second source of hazard. The third source of damage is the very low atmospheric pressure near the vortex centre. When a tornado approaches a building, external pressure is rapidly reduced, and unless there is a nearly simultaneous and equivalent decrease in internal pressure, by opening windows or doors, the walls and roof may explode outwards in the process of equalizing the pressure differences.

Many tornadoes have a short life. The convectional activity which creates the source cloud is itself highly variable, and a single cloud can spawn a number of different tornado vortices, either simultaneously or in sequence, beneath different areas of the cloud, as parts of it develop and decay. Movement is generally with the parent cloud, perhaps with the funnel twisting sinuously across the ground beneath. Once contact with the ground is made, the track of a tornado at ground level may frequently extend for only a few kilometres, though there are examples of sustained tracks extending over hundreds of kilometres. The diameter of the funnel is rarely more than 200 m; track length and width are consequently modest, and the probability of being affected is comparatively remote—though this is no consolation to those afflicted.

Tornadoes, being associated with extreme atmospheric instability, show both seasonal and locational preference in their incidence. 'Favoured' areas are temperate continental interiors in spring and early summer, when insolation is strong and the air may be unstable, although many parts of the world can be affected by tornado outbreaks at some time or another. The Great Plains of the USA, including Oklahoma, Texas and, of course, Kansas, have by far the highest global frequency, and are particularly prone at times when cool, dry air from the Rockies overlies warm, moist 'Gulf' air.   GRAHAM SUMNER

Further reading

Eagleman, J. R., Muirhead, V. U., and Willems, N. (1975) *Thunderstorms, tornadoes and building damage*. Lexington Books, London.

tors Tors are steep-sided residual hills. Their morphology varies but they have been defined as 'irregularly prominent masses of rudely shaped rock fragments' and as bare rock outcrops, usually of monumental form and 'about the size of a house', commonly bounded by near-vertical fractures, and boldly fissured by widely spaced joints. They are characteristic of the granitic uplands or moors of south-western England and of other Hercynian (Variscan) massifs in western and central Europe. They are developed in sandstones in the English Pennines.

In most tors the granite is subdivided by orthogonal fracture sets, and the residual landforms are blocky, but at a few sites convex-upward slopes are associated with arcuate sheeting fractures. Near the surface, the granite is also commonly subdivided by closely spaced subhorizontal partings, known as 'pseudobedding', which delineate discontinuous slabs or attenuated lens-shaped masses. Rock basins are well developed on flattish upper surfaces.

Many, if not most, of the tors of south-western England have evolved in two stages: the first involves differential fracture-controlled subsurface weathering, which results in the isolation of a compartment of massive fresh rock surrounded by weathered granite or 'grus'; and the second involves the erosion of the grus, leaving the massive compartment in positive relief. Whether the differential subsurface weathering took place in humid tropical conditions or involved frost action is debated, as is the nature of the agencies responsible for eroding the weathered bedrock, though solifluction must have played a major role. Tors probably represent the cores of larger massive compartments, and are the last remnants surviving after prolonged weathering and erosion: hence their comparatively small size, and the French name, *inselbergs de poches*.

Tors have their congeners in other parts of the world, for example in Zimbabwe, where they are known as 'castle koppies', a type of inselberg or isolated hill. The general terminology is further confused because in the Antipodes especially, but also in Africa, some workers have used the word *tors* for what are properly termed boulders. The term *tor* is of ancient origin. It is Cornish and is comparable to the Anglo-Saxon *torr*, Welsh *twr*, and the Latin *turris*, in each instance denoting a tower. The word has long been used in south-western England in the sense of a small steep-sided residual hill on the moors. A boulder, on the other hand, is defined as a detached rock mass, somewhat rounded or otherwise modified by weathering or transport, and larger than a cobble. The minimum diameter is 256 mm (10 inches). Boulders with a diameter of more than 30 m are recorded, but most are 1–2 m in diameter. Some stand in isolation, others in disordered groups (the *compayrés* of French authors).

Tors and boulders are thus distinct and different forms. The tors of south-western England and the Pennines are similar to the castellated inselbergs or castle koppies of other parts of the world. They ought to be recognized as such and the word *tor* retained as a regional term for the same form.

C. R. TWIDALE

Further reading

Linton, D. L. (1955) The problem of tors. *Geographical Journal*, 121, 470–487.

Twidale, C. R. (1982) *Granite landforms*. Elsevier, Amsterdam.

Worth, R. H. (1953) In Spooner, G. M. and Russell, R. S. (eds) *Dartmoor*. David and Charles, Newton Abbott.

trace elements A trace element is defined as a component of a rock occurring in concentrations of less than 0.1 wt% (weight per cent). The term is also used in mineralogy for elements that are not essential components of a mineral but are present in small concentrations (usually significantly less than 1.0 wt%), either accommodated in the crystal structure or adsorbed on the surface of a mineral. Trace elements are incorporated into the crystal structure of many rock-forming minerals by substituting for major elements. Adsorption of trace elements is most commonly observed on the surface of oxides, hydroxides, and clay minerals in sedimentary, diagenetic, and hydrothermal environments.

Trace elements have gained increasing importance in the Earth sciences because their geochemical behaviour is very distinct and characteristic of both geochemical processes and specific geological environments. They are therefore classified according to their geochemical behaviour. Two different classifications are widely used to distinguish between several groups of trace elements. The first of these is based on the position of an element in the periodic table. Because the chemical properties of an element are largely determined by its electronic configuration, elements with similar configurations are expected to display a similar geochemical behaviour. Four main groups are usually distinguished in this way: (1) rare earth elements (REE), comprising the lanthanides and yttrium; (2) actinides, with thorium and uranium as the most important naturally occurring elements in the group; (3) platinum group elements (PGE), comprising the elements ruthenium, rhodium, palladium, osmium, iridium, and platinum; and (4) transition metals of the first transition series (scandium, vanadium, chromium, manganese, cobalt, nickel, copper, and zinc). The second, more empirical classification of trace elements is based on observations of their geochemical behaviour in magmatic systems. During the crystallization of minerals from a cooling melt, some trace elements, the so-called *compatible elements*, exhibit a preference for the solid phase, whereas others tend to remain in the liquid phase. The latter are known as *incompatible elements*, because they are not easily incorporated into the structures of most crystallizing minerals. As crystallization proceeds, they will thus become more and more concentrated in the melt, from which they will be removed eventually either by crystallization of minerals that are able to accommodate them into their structure or by fractionation into a fluid phase that escapes from the magmatic system. During such late-stage magmatic processes trace elements may form mineral species on their own (e.g. REE minerals), which sometimes are concentrated locally in economic ore deposits. The incompatible elements are often further subdivided because of differences in the ionic potentials (ratio of charge to ionic radius) of their cations. Cations with high ionic potentials show geochemical behaviour that is significantly different from that exhibited by cations with low ionic potentials. The former belong to the group of high-field-strength elements (e.g. zirconium, hafnium, uranium, thorium, niobium, tantalum, and the rare-earth elements); the latter to the so-called large-ion lithophile elements (e.g. rubidium, caesium, strontium, and barium). Some trace elements, such as uranium, thorium, lead, rubidium, strontium, samarium, and neodymium, are not only important for characterizing geochemical processes and environments, but are also essential in the determination of the absolute age of a rock by means of the abundances of their isotopes.

The determination of trace-element concentrations (which are usually of the order of parts per million or parts per billion) in rocks and minerals generally requires elaborate sample preparation and advanced analytical techniques (e.g. neutron activation, isotope dilution, inductively coupled plasma spectroscopy). Many of these techniques are now widely used in environmental geochemistry, particularly in order to recognize contamination of groundwater and soil by trace elements (e.g. lead, cadmium, nickel, and zinc), but also in order to monitor environmental conditions in the vicinity of incinerators and disposal sites for toxic waste. RETO GIERÉ

trace fossils and ichnology

Trace fossils are biogenic sedimentary structures produced by the activity of organisms and, therefore, reflect their behaviour. They include tracks, trails, burrows, and borings but not body fossils (the actual remains of organisms). They can be found in sedimentary rocks that do not contain body fossils and may thus provide the only record of life, and they may also provide the only record of soft-bodied organisms that are not normally preserved. One important aspect of trace fossils is that they are almost always preserved in place, since reworking would result in their destruction. They can, therefore, always be related directly to the sediments in which they are found. The study of trace fossils is called ichnology (from the Greek *ichnos*, a footprint or track) and is principally concerned with gleaning information that is of use in understanding and reconstructing palaeoenvironments.

Classification of trace fossils is difficult because they represent behaviour rather than body remains. It is also usually difficult to relate traces to the organisms that formed them, for organisms are rarely found with the trace, and similar traces may be left by different organisms. Despite the fact that traces are not organisms the Linnaean binomial scheme is used to name them, but only to the level of ichnogenus and ichnospecies. Above that a broad behavioural grouping is used in which six basic categories are recognized. Domichnia (dwelling structures) consist of living burrows made by organisms that fed outside the dwelling rather than from the surrounding sediments. Fodinichnia (mining structures) are burrows formed by animals that feed on the sediment. If these structures are actively back-filled as the organisms move through the sediment distinctive laminations termed spreite can be formed. Pascichnia (grazing traces) are formed by animals feeding on a surface; they normally consist of highly organized patterns resulting from efficient search procedures. Repichnia (locomotion traces) are formed by animals during travel and can represent a variety of activities, including migration or searching for food. Cubichnia (resting traces) are formed when an animal comes to a temporary stop and

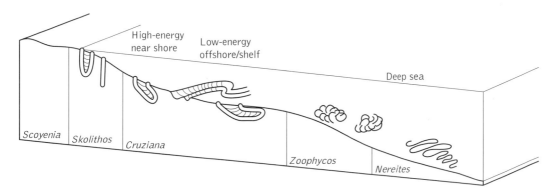

Fig. 1. Terrestrial to offshore profile showing distribution of trace fossils and ichnofacies.

buries itself in the sediment. This will produce an impression of the ventral surface of the animal and can represent hiding behaviour either on the part of prey or predator. The final category is fugichnia (escape burrows), which are formed by animals as they move upward through sediment that has been rapidly deposited over them.

One of the most important uses of trace fossils has been their grouping into recurring ichnofacies (traces reflecting a particular sedimentary environment). This is linked to a particular set of environmental parameters which are generally correlated with depth. These parameters include temperature, light level, salinity, and oxygen level as well as bottom conditions. The ichnofacies concept has been particularly valuable in recognizing shallow-marine facies and has the advantage of being independent of time, because particular environments, and therefore behaviour, persist through time even though the trace-formers may change. Each ichnofacies is named after a typical trace fossil and is characterized by the common occurrence of a number of others (Fig. 1). The *Scoyenia* ichnofacies consists of terrestrial traces and is not well known; however, the *Skolithos* ichnofacies is associated with high-energy, shallow marine environments and consists of vertical burrows, commonly U-shaped, made by suspension feeders. The *Cruziana* ichnofacies indicates deeper conditions off shore and characteristically consists of traces made by mobile sediment-feeders (*Cruziana* is formed by trilobites). The *Zoophycos* ichnofacies occurs below storm-wave base and into fairly deep water and consists of grazing and feeding structures produced by deposit-feeders. The deepest-water facies is termed the *Nereites* ichnofacies and is characterized by complex horizontal grazing traces. Additional ichnofacies are recognized for specialized environments and include the *Trypanites* ichnofacies for lithified marine substrates, which are often useful in the recognition of unconformities, hard grounds, and other omission surfaces.

Trace fossils can be useful in the recognition of sedimentary characteristics and can also be involved in depositional processes. Sediment cohesiveness may be indicated by the type of burrows formed in it. Lithified sediments will show borings, while at the other extreme poorly cohesive sediments may contain burrows lined with pellets or grains to prevent collapse. Rate of sedimentation can be indicated by trace abundance; low abundance indicates rapid sedimentation, whereas abundant traces may indicate a decrease in depositional rate. Acting as agents of deposition, burrowing organisms can affect stratification, causing shell-lag layers for instance, and may also sort or disorganize sediments and cause chaotic textures generally recognized as bioturbation.

DAVID K. ELLIOTT

Further reading

Boardman, R. S., Cheetham, A. H., and Rowell, A. J. (eds) (1987) *Fossil invertebrates*. Blackwell Scientific Publications, Oxford.

Frey, R. W. (ed.) (1976) *The study of trace fossils*. Springer-Verlag, Berlin.

tracers and groundwater Tracers of groundwater flow include both naturally occurring and artificially introduced dissolved materials, or solutes. Monitoring of tracers in wells or springs can provide information on directions and rates of flow as well as on the network of permeable features, such as fractures, that form the pathways of groundwater movement. Many tracer studies in limestone aquifers with significant solution-enhanced fractures or caves have employed fluorescent dyes that can be detected at low concentrations. Such studies have demonstrated connections between wells and springs located hundreds of metres apart, with estimated flow velocities exceeding many kilometres per day. At the other extreme of flow velocities, radioactive isotopes such as tritium (^{3}H) and chlorine-36 (^{36}Cl), which were introduced to the hydrological cycle through nuclear weapons fallout, have been used to trace infiltration of water through the unsaturated zone at rates of millimetres per year. Experiments involving the injection of water tagged with dissolved chloride and bromide salts have been used to develop an improved understanding of dispersion in aquifers and to characterize sites proposed for radioactive waste disposal. J. BAHR

Further reading

Bonacci, O. (1987) *Karst hydrology*. Springer-Verlag, Berlin.

transfer functions in palaeoclimatology

Transfer functions are frequently employed in Quaternary palaeoclimatology because they provide an important method of reconstructing past climatic conditions. Transfer function equations are derived by using multivariate statistical techniques to define the relationship between biological or chemical entities and a physical parameter (which is usually an environmentally sensitive variable such as temperature) from within a large and diverse modern-day calibration data set. The application of these equations to geological palaeodata enables a quantitative estimate of the physical parameter to be calculated for the past. Transfer functions have been constructed for numerous faunal and floral biological groups, including beetles, coccoliths, foraminifera, molluscs, and pollen. Transfer functions therefore facilitate the investigation of biosphere–climate interactions in both the terrestrial and marine realms and support palaeoclimatic reconstruction on a global scale.

The underlying rationale behind the transfer function approach is that biological organisms are preferentially adapted to a particular set of ecological and environmental conditions and that the distribution and composition of species assemblages can therefore be related to certain variables within the physical environment. The successful application of a transfer function is, however, also dependent on certain assumptions being satisfied: (i) the distribution of the biota has to be closely linked to the predicted environmental variable; (ii) the modern calibration data set must provide an accurate representation of the species–climate relationship; and (iii) the ecological requirements of the various taxa being studied cannot vary appreciably in the time interval under consideration.

Transfer functions can be calculated from qualitative, semi-quantitative, or quantitative data, although the greater objectivity and precision associated with quantitative data make them the most suitable for multivariate statistical analysis. Similarly, variations in the composition of species assemblages, rather than the distribution record of a single taxon, are preferable since they can better account for the complexity of factors that may affect the ecological response of an individual species and hence improve the precision of the transfer function.

The Imbrie–Kipp transfer function approach is a commonly applied method; this technique was used by members of the CLIMAP project to calculate past variations in ocean surface temperature using data for planktonic microfossil assemblages (i.e. data for, foraminifera, coccolithophorida, and Radiolaria) from deep-sea cores. The transfer function is calculated in two steps: (i) a factor analysis of individual species abundance data from the core top calibration data set to generate ecological groupings or assemblages; and (ii) a regression analysis of the assemblage factor loadings with modern sea surface temperature to produce a set of temperature equations which can subsequently be applied to fossil assemblages.

A number of alternative methods for calculating transfer functions also exist. For example, the modern analogue technique employs similarity coefficients (e.g. squared Euclidean distance) to evaluate the differences between the palaeoproxy data and the modern calibration dataset. The palaeo-environmental estimate is calculated as a simple or a weighted average of the climatic values from the most similar modern subset. Transfer functions derived using the mutual climatic range method, which are also determined using similarity coefficients, differ in that they are based on qualitative presence or absence of data.　　MARK R. CHAPMAN

transform plate margins In plate tectonics, a transform fault is a strike-slip fault extending throughout the lithosphere and joining any two other plate margins. Transform faults are common in the oceans, where they are mostly ridge–ridge transforms offsetting mid-ocean ridge spreading centres. However, trench–trench transforms (such as that joining the southern Chile Trench to the northern Scotia Trench) are quite common. The famous San Andreas fault system is a transform fault joining a ridge at its southern end to a trench and a transform fault (at the Mendocino trench–transform–transform triple junction) at its northern end. As with most plate boundaries, transform margins tend to be more complex where they run through weak continental lithosphere, compared with their counterparts in strong oceanic lithosphere.

Ocean transform plate margins

Ridge–ridge transform faults typically occupy a zone a few tens of kilometres wide, within which there is usually a single active strike-slip fault which is the actual plate margin. The transform fault normally lies within a narrow, linear valley which runs along the plate margin and may be up to several kilometres deep (Fig. 1). The trend of such transform valleys is one indicator of the relative direction of plate movement.

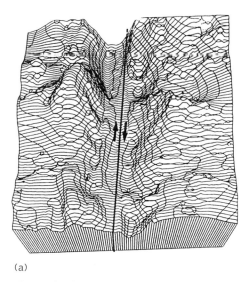
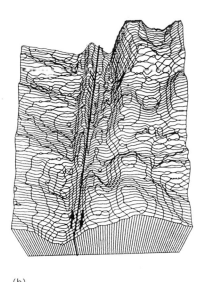

(a)　　　　　　　　　　　　　　　　　　　(b)

Fig. 1. Topography of the Kane transform fault, Mid-Atlantic Ridge, 24 °N, showing views (a) westward and (b) eastward from the mid-point of the 150-km-long transform. Vertical exaggeration 10 : 1. (Modified after R. A. Pockalny *et al.* (1988) *Journal of Geophysical Research*, **93**, 3179–93.)

The active strike-slip fault is usually indicated on the sea floor by a narrow, linear furrow a few tens of metres wide and a few metres deep. The furrow is commonly associated with fault breccia containing fragments from all levels in the crust and uppermost mantle.

Transform valleys probably have their origin in several different processes. The oceanic crust is known to be quite thin, possibly as little as 2 km thick or less, at transform faults, and the isostatic effect of this thin crust would create a valley. There may be minor extension across the transform, leading to normal faulting and graben formation. Dynamic forces associated with the sea-floor spreading process may also contribute to some of the topography. Parts of some transform valleys can reach over 6 km below sea level, and are thus some

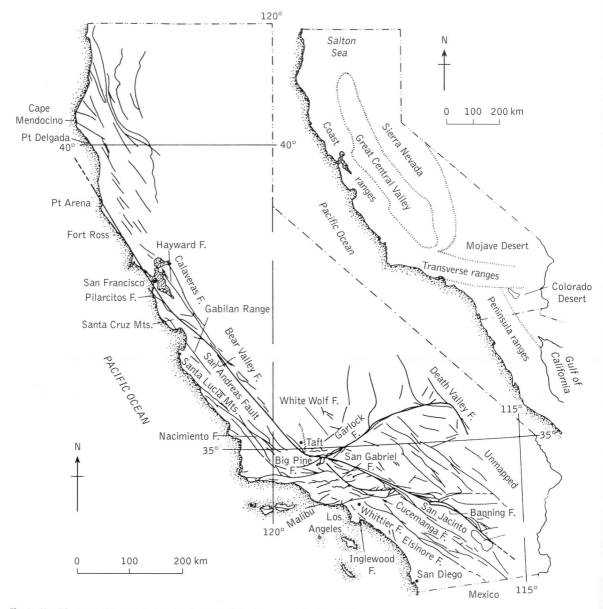

Fig. 2. Simplified map of the San Andreas Fault System (After Kearey and Vine, Fig. 7.8.)

of the deepest parts of the ocean basins outside deep-sea trenches.

The flanks of transform valleys are commonly faulted, and these faults expose a variety of rock types from the crust and upper mantle. Some authors have tried to infer a crustal stratigraphy from them, but in view of the very limited amount of sampling and observation available, such attempts should be treated with caution. It is clear nevertheless that transforms are sites where deep-seated rocks, such as serpentinized peridotite (derived from the upper mantle), are relatively common. Such normally deep-seated rocks may be exposed in transforms either by faulting or by diapirism, or simply because they constitute a significant component of the oceanic crust formed at ridge–transform intersections.

Transform valleys, especially those of larger-offset transforms, are usually closely flanked by transverse ridges that run parallel to them. Various models for the origin of such ridges have been proposed, including dynamic forces, faulting, excess volcanic construction, and buoyant uplift as a result of the hydration of upper-mantle peridotite to form low-density serpentinite.

Much of the relief generated in the transform fault zone is preserved in the inactive traces of ridge–ridge transforms, referred to as fracture zones. In addition to this relief, each flank of the fracture zone will subside to a different depth, according to its age (see *mid-ocean ridges*), resulting in differential depth across the fracture zone. This can lead to spectacular escarpments, hundreds or even thousands of metres high, that extend right across ocean basins. The great fracture zones of the northern Pacific—Mendocino, Murray, Molokai, Clarion, and Clipperton—are excellent examples, standing out on global maps of sea-floor topography and satellite gravity.

Continental transforms

Examples of major continental transform margins are the San Andreas Fault Zone, which separates the Pacific and North American plates in California, the Dead Sea Fault between the African and Arabian plates north of the Red Sea, the Alpine Fault which bounds the Indo-Australian and Pacific plates in New Zealand, and the North and South Anatolian Faults in Turkey.

Because continental lithosphere is weaker, more readily deformed, and less homogeneous than oceanic lithosphere, continental transform plate margins tend to be somewhat more complex. The San Andreas Fault Zone is actually a complex fault system that is 200–300 km wide in places (Fig. 2). Resistant geological structures cause the main fault zone to bend in places so that compressive or mixed strike-slip and compressive (so-called 'transpressive') zones develop; secondary strike-slip and thrust faults branch off the main strike-slip zone; and palaeomagnetic results indicate that there are many small-scale fault blocks that have undergone significant rotations during movement along this plate margin. Although the San Andreas is the main through-going

fault, it has accounted for only about 300 km of an estimated total 1500 km motion on the system, the rest presumably being taken up by the secondary faults and by folding.

Seismicity in the San Andreas system is confined to about the uppermost 15 km, below which deformation is considered to be ductile. Measured heat flow on the San Andreas fault is much lower than would be expected for a frictional fault, and local compressive stresses are oriented at right angles to the fault. This suggests that the fault is very weak, a situation that might be due to the presence of pervasive high-pressure fluids in the fault zone.

The topography of continental transform faults is generally more complex than that of their oceanic counterparts. Even where there is a single primary strike-slip fault, it may be associated with quite complex topography. Minor vertical movement on the fault can give rise to small scarps aligned along it; juxtaposition of different lithologies can lead, through differential erosion, to a height differential across the fault; erosion of brecciated rock in the fault zone can produce a linear valley along the fault line; streams, depositional fans, and other topographic features may be displaced by the fault. If the strike-slip fault contains a bend (resulting perhaps from inhomogeneities in the continental lithosphere), leading to a component of compressive stress, this is referred to as a *restraining bend*, and may lead to folding, thrusting, and elevated topography. A bend in the opposite sense is a *releasing bend*, leading to extension, subsidence, and the formation of graben and basins. Such topographic features are important in controlling erosion and deposition along transform plate margins, and are consequently of importance in hydrocarbon exploration.

The deep structure of continental transforms is of interest. Seismic observations of the San Andreas system suggest that in some places the strike-slip fault observed at the surface may penetrate only to about 8 km, whereas there is evidence of structural boundaries in the deeper lithosphere offset laterally from this. The transform margin may therefore occur in slightly different locations at different depths, these different active zones perhaps being connected by horizontal detachment surfaces. Whether this is a common feature of transform plate margins is not yet known.   ROGER SEARLE

Further reading
Kearey, P. and Vine, F. J. (1996) *Global tectonics*. Blackwell Science, Oxford.

transgression *see* REGRESSION AND TRANSGRESSION

transpression *see* TRANSTENSION AND TRANSPRESSION

transtension and transpression In 1971, W. B. Harland, who had for many years worked in Spitsbergen where a major transcurrent lineament crosses the country, pointed out that it is seldom possible to describe the horizontal movement of tectonic blocks and plates in terms of three simple tectonic regimes of *extension, transcurrence* (= strike-slip or wrench), and *compression*. Movement between blocks is normally intermediate between these extremes, and strike-slip motion is therefore more commonly either divergent or convergent.

Harland termed a combination of transcurrence and extension *transtension* and a combination of transcurrence and compression *transpression*, terms that are synonymous with *divergent strike-slip* and *convergent strike-slip*. Studies of both plate margins and continental fault systems at the present day suggest that most systems do have combinations of more than one vector; the concepts established by Harland are therefore important.

Transtensile regimes are marked by normal faulting, the formation of small sedimentary basins, and volcanicity. Transpressive regimes are marked by thrust faulting, reverse faulting, folding and uplift, and the formation of sedimentary basins adjacent to the uplifted fault.

Combining the two regimes into a cycle of sedimentation and orogenesis, H. G. Reading suggested in 1980 that one could detect a *strike-slip cycle* within continental crust comparable to, but on a smaller scale than, the Wilson Cycle of sea-floor spreading → subduction → continental collision. The Reading Cycle comprises three phases.

(1) A phase of *transtension* leading to the formation of small basins, as in present-day California both offshore and on land. If extension is substantial, oceanic crust may form, giving rise to ophiolites and hydrothermal deposits as at the spreading ridges of the present Gulf of California.

(2) A phase of basin filling when the transtensional phase overlaps with transpression. Sedimentation is both substantial and rapid, and is dominated by slides, debris flows, and turbidity currents. As in all strike-slip basins, the margins may have a complex pattern of extensional faults, thrusts, and folds. Filling of the basin is very rapid, since both basin filling and uplift of the adjacent margin are equally rapid.

(3) A phase of *transpression* succeeds the basin fill phase. The basin fills up and is perhaps uplifted above sea level. Earlier, more marine, sediments pass up into lacustrine fluvial sediments. Compressive structures, especially thrusts, now dominate and the continued uplift leads to erosion of the deformed sediments, and their collection in adjoining basins.

HAROLD G. READING

Further reading

Ballance, P. F. and Reading, H. G. (eds)(1980) Sedimentation in oblique-slip mobile zones. *International Association of Sedimentologists Special Publication* **4**.

Biddle, K. V. and Christie-Blick, N. (eds)(1985) Strike-slip deformation, basin formation, and sedimentation. *Society of Economic Paleontologists and Mineralogists Special Publication* **37**.

Nilsen, T. H., and Sylvester, A. G. (1995) Strike-slip basins. In Busby, C. J. and. Ingersoll, R. V (eds) *Tectonics of sedimentary basins*, pp. 425–57. Blackwell Science, Cambridge, Mass.

travertine, tufa When water becomes supersaturated in dissolved calcium carbonate, solid calcium carbonate can grow by precipitation from the water. This solid build-up is a form of 'chemical limestone' (as opposed to biogenic or clastic limestone), usually consisting of the calcite variety (or polymorph) of calcium carbonate, but sometimes as the alternate form aragonite.

Rock formed in this way is usually referred to as *travertine* or *tufa*. The terms are sometimes used interchangeably, but it is more useful to distinguish dense varieties as travertine and porous or spongy varieties as tufa. Calcium carbonate precipitated near the water-table within soils or gravels that eventually acts to cement the formerly loose deposit together is of comparable origin but is not usually referred to as either travertine or tufa, the term *calcrete* being preferred to describe the cemented deposit as a whole.

The most common ways for travertine to form are by evaporation of fresh water, and around hot springs (as, for example, at Mammoth Hot Springs in Yellowstone Park, Wyoming). Perhaps the best-known locations of travertine formation are caves in limestone regions. Where water rich in dissolved calcium carbonate drips slowly from the roof a downward-hanging travertine stalactite grows, and where drops persistently hit the floor an upward-pointing stalagmite grows. Cave walls down which water habitually runs may become covered with sheets of travertine flowstone.

Travertine is often attractively banded and takes a good polish. It has therefore long been a favoured ornamental stone. For example, travertine quarried near Tivoli was used as the facing stone for the Colosseum in Rome.

Tufa usually gets its porous structure because of the decay of algae or other plant material around which the material was initially precipitated. It is sometimes called calcareous sinter, by analogy with siliceous sinter, which is a comparable deposit formed in a similar way by precipitation of opaline silica. Most of the world's so-called petrifying springs deposit a sintery encrustation on objects immersed in them. In other situations, algae play a direct role in encouraging precipitation to form algal mats or mounds (stromatolites), many of which are made of tufa bound by algal filaments.

Tufa's porosity makes it a soft and easily cut variety of stone and it has found many uses. Notable among these is its use to line the Aqua Appia underground aqueduct, 16 km long, built in 312 BC to supply water to the city of Rome.

The chemical reaction by which travertine and tufa form may be summarized as:

$$Ca^{2+} \text{ (in solution)} + 2(HCO_3^{2-}) \text{ (in solution)} \rightarrow$$
$$CaCO_3 \text{ (solid)} + H_2O + CO_2 \text{ (in solution)}.$$

The bicarbonate ion, (HCO_3^{2-}), is soluble in water, but if either water (H_2O) or carbon dioxide (CO_2) is removed the effect will be to drag the reaction to the right and cause calcium carbonate to be precipitated. Carbon dioxide can be removed by plant metabolism or driven out of solution by heat (it is less soluble in warm water than in cold water), and these conditions give rise to the formation of tufa at hot springs and in stromatolites. Water can be lost by evaporation. This produces stalactites and related deposits in underground caves. It can even cause a canopy-like feature to grow over a waterfall, which may completely bridge the river to create a natural tunnel referred to as a tufa cave.   DAVID A. ROTHERY

trenches, deep-sea *see* DEEP-SEA TRENCHES

Triassic

The Triassic is the oldest of the three Mesozoic systems, with rocks deposited over a span of about 45 million years, beginning at around 250 Ma. Its name reflects the three distinctive formations that were grouped together by August von Alberti, writing in 1834. He was employed as an official in the German salt-mining industry and spent many years from 1815 studying the geology of south Germany and Austria. The three formations that held his attention are not very fossiliferous but are highly distinctive and widespread. Their names, Bunter, Muschelkalk, and Keuper, have been widely used as series names, but they are not strictly equivalent to three equal series. Marine Triassic rocks are found in many parts of the present Arctic, for example, in Canada, East Greenland, Spitsbergen, and parts of north-eastern Russia. They also occur in a belt running from Spain eastwards throughout the Middle East to Iran, the Himalayas, India and Indonesia, China, and Japan, while patches crop out also in Australasia. Major outcrops extend along the western coastal regions of North and South America.

The base of the system and the subdivision into six stages was originally recognized in the marine facies of the Southern Alps. The base of the Triassic system is also the base of the Mesozoic and reflects a profound change in fossil biota. Biozones are based upon the ammonoid succession and the ceratite ammonoids are typical of the Triassic. Since the original descriptions of the sections in the Southern Alps were made, several more complete and fossiliferous ammonoid successions have been discovered in the USA, Canada, and Asia. Today, the system is divided into 23 ammonoid biozones in the Alpine–Himalayan region and 39 in North America. The base of the system is now formally defined as the base of the zone of the ammonoid *Otoceras woodwardi* in the Tethyan (or Alpine–Himalayan) realm. There is a proposal to divide the original basal stage, the Scythian, into four stages, and perhaps to drop the use of the name Rhaetian for the top stage.

Triassic palaeogeography is dominated by the supercontinental mass Pangaea (Fig. 1). It stretched from arctic to antarctic latitudes as a great but varied continent upon which many sedimentary basins and fault-controlled rifts were to develop. The oceans were reduced two in number; the larger by far was the enormous Panthalassa Ocean, roughly equivalent to the Pacific of today, while the smaller Tethys Ocean lay as a gigantic bight on the eastern side of Pangaea. Numerous small archipelagos of volcanic islands lay off the western seaboards of Pangaea. These were a consequence of the active subduction of the oceanic crust under the continent throughout this period. Thick volcanic Triassic rocks are known from the coastal region from Alaska to California and beyond. On the western margin of South America similar features occur.

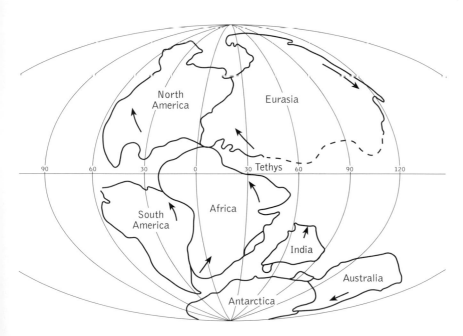

Fig. 1. The Triassic world, about 220 Ma ago. Pangaea was then beginning to split into the continental units of the present globe. The two supercontinental masses, Laurasia in the north, and Gondwanaland in the south, are now distinct, with the Tethys bight or gulf separating them. New sea floor was being generated between Africa and the two Americas, and between Africa, India, and Antarctica. The small arrows indicate the directions in which the continents were rotating as they became separated. (Based on an early map by Dietz and Holden.)

Reefal islands occurred in the tropical Tethys. Palaeo-magnetic evidence suggests no significant motion of Pangaea across the latitudes, but small land masses such as South-east Asia were accreted to the eastern margin of the Laurasian segment. It was essentially a quiet time in terms of orogenic activity but flood basalts on the grand scale were erupted at a number of times and places. In Siberia, basalts were erupted in a series of flows that span the Permo-Triassic boundary, while in southern Africa gigantic quantities of basalts form part of the very extensive Karroo System. Smaller volumes were erupted in eastern North America, and at the end of the period a phase of basalt flow emission in North Africa persisted into the Jurassic. Many parallel sets of Triassic dolerite dykes occur in the coastal regions of eastern North America, West Africa and north-eastern South America. All these occurrences are associated with phases of regional tensional stress affecting much of Pangaea. This stress was to lead to the break-up of the land mass, first appearing in the vicinity of the South Atlantic and the Indian Ocean.

In early Triassic time sea level was very low but soon began to rise, giving the Mid-Triassic or Muschelkalk transgression. Evidence of transgression is very strong in eastern Greenland, China, and Western Australia. The Tethyan Ocean began to flood the (northern) European part of the supercontinent, reaching a maximum in the late Triassic Norian stage. Something of a standstill or regression then held until the end of the Triassic period.

The climate seems to have been largely equable and warm to hot. Nowhere has evidence of ice caps been found. Vast areas of the land surface were arid deserts; in consequence Triassic evaporite deposits occur in many places. These salt beds include some of the world's largest. Elsewhere strong monsoonal regimes were established with abundant precipitation, which enabled coal-forming forests to grow in eastern Laurasia and in Gondwanaland (China, Australia, eastern Russia). Round about these regions there were widespread continental (red-bed) sedimentary basins. Around the margins of the Tethyan Gulf the Tethyan marine facies, now seen in the regions of southern Europe and the Mediterranean lands eastwards to the Himalaya, consists of carbonates and subsidiary clastics with local reefs and shell banks.

The first half of the Triassic period was of reverse palaeomagnetic polarity; the second mainly of normal polarity. Some writers have suggested that the change was linked to processes in the mantle which also prompted the eruption of the flood basalts and the expansion of the ocean basins.

Great changes in the living world were set in motion at the beginning of Triassic time. The end-Permian extinction caused the loss of some 83 per cent or more of all marine genera. Only about 13 per cent of the continental shelves were covered by the sea, probably as a result of the expansion of the ocean basins, and the marine faunas there seem to have been strongly cosmopolitan. Bivalves appear to be the most important and numerous macrofossils, but new groups of corals, bryozoans, and echinoderms arose. Brachiopods were now only a minor element in the benthos, but the decapod crustaceans made their first appearance. The ammonoid molluscs and the conodonts are significant in Triassic biostratigraphy but became extinct by the end of the period. Land floras were remarkably uniform everywhere and were given over very much to gymnosperms, horsetails, and ferns.

Among the vertebrates, the elasmobranchs and the chondrostean and holostean fishes were conspicuous in the seas, but the crossopterygians and dipnoi were comparatively rare. Among the aquatic reptiles, ichthyosaurs and plesiosaurs arrived to make a strong showing and probably swam in small schools in many parts of the oceans, feeding on the abundant molluscs.

Terrestrial vertebrate faunas were involved in spectacular changes. The labyrinthodont amphibians, the most widely distributed and common of Triassic vertebrates, had almost completely died out by the end of the period. Increasingly arid environments limited their breeding grounds, while the reptiles were not so handicapped. The newly evolved mammal-like reptiles were important throughout the first two-thirds of Triassic time but were then largely replaced by a dynamic group, the archosaurs, from which sprang the crocodiles, dinosaurs, and pterosaurs. By the end of Triassic time the mammals, tiny but vigorous, had arrived.

It has long been recognized that the Triassic was a period of crises and extinctions, and in recent years there has been some debate as to their number and timing. In the seas, great changes took place; 58 families of cephalopods, 15 families of gastropods, 12 of brachiopods, and 8 each of bivalves and sponges died out. Thirteen families of marine reptiles also went. On land, extinctions accounted for 35 families of insects, 8 of freshwater bony fishes and the 8 thecodont reptile families. Depletion of previously well-established tetrapod stocks in the late Triassic made way for the reptile newcomers mentioned in the last paragraph. What effects these had upon the contemporary mammals is not known, but fur, warm blood, omnivorous feeding habits, and great fecundity may have given them some advantage over the reptiles. It was not until the end of Cretaceous period, however, that the day of the mammals was nigh.

Overall there now seems to be convincing evidence of three extinction events in the Triassic: at the end of the Early Triassic (Scythian), in the Middle to Late Carnian (early Middle Triassic), and at the Triassic–Jurassic boundary (205 Ma ago). Climatic changes are generally held to be the cause of these extinctions, with closely linked floral changes, and with marine regression. The proponents of meteorite impact as a cause of extinction have not been able to advance so convincing an argument for the end-Triassic extinctions as for the end-Cretaceous event. However, at least one large meteorite impact crater—that at *Manicuagan* in Canada with a diameter of 70 km—is known, and the impact effects from this would have had consequences for terrestrial animals in the region, if not throughout all Pangaea.

D. L. DINELEY

Further reading

Sweet, W. C., Yang Zunyi, *et al.* (1992) *Permo-Triassic events in the Eastern Tethys. Stratigraphy, classification and relations with the Western Tethys.* Cambridge University Press.

Tozer, E. T. (1984) *The Trias and its ammonoids: the evolution of a time scale.* Geological Survey of Canada Miscellaneous Report No. **35**, Ottawa.

Zapfe, H. (ed.) (1974) *The stratigraphy of the Alpine–Mediterranean Triassic.* Springer-Verlag, Berlin.

trilobites and related fossils
The trilobites are among the best-known fossil examples of the phylum Arthropoda, an extensive group of organisms in which the body is segmented and has a chitinous exoskeleton and numerous jointed legs. The arthropods are the most successful living organisms, comprising almost 75 per cent of all described animals, and trilobites are the earliest known examples, occurring in rocks of Early Cambrian age just above the earliest Cambrian faunas. This entirely marine group was widespread and diverse throughout the Palaeozoic and became extinct in the Permian. They are important geologically, for they are widespread and numerous in Palaeozoic rocks and their rapid evolution is easy to detect in their distinctive morphological features, making them valuable for biostratigraphic correlation. A number of other arthropod groups are present in the Palaeozoic, although their relationship to the trilobites is unclear. Some of these, insects and crustaceans, are dealt with in separate entries. One major group, the eurypterids, will, however, be considered here.

Morphology

Trilobites take their name from the division of the exoskeleton into three longitudinal portions which is clearly seen in most species (Fig. 1). The axis, or axial lobe, extends from near the anterior margin to the posterior extremity of the body and is flanked on either side by the pleural regions. The most anterior section of the dorsal exoskeleton is termed the head or cephalon, and this bears a central hump or glabella that is flanked laterally by the paired compound eyes. A thin lineation, the facial suture, runs from the postero-lateral margin of the cephalon forward and around the eyes before terminating at the anterior margin. This follows different courses in different trilobite species and is a weak area of the skeleton that apparently split during ecdysis to facilitate moulting of the exoskeleton. Ventrally, the mouth of the trilobite was protected by a flat plate, the hypostome. Behind the head are a series of thoracic segments, which vary in number in different taxa. Each of these is articulated with its anterior and posterior neighbour, allowing the animal a considerable degree of flexibility, in some instances even allowing it to roll up completely. The most posterior section of the animal is termed the pygidium, a flat plate of fused segments similar to the thorax. Ventrally, the appendages are very rarely preserved since they were apparently not well mineralized. However, unusually preserved specimens show the presence of paired antennae anteriorly, followed by paired two-branched appendages under the cephalon and one pair for each segment through the thorax and the pygidium. The appendages are divided into a gill branch and a walking leg, thus combining the functions of locomotion and respiration.

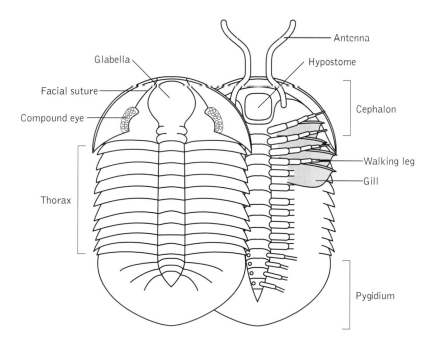

Fig. 1. Typical trilobite showing major morphological features.

Taxonomy

Arthropods are extremely numerous and show an amazing diversity of morphology. Because of this a number of different taxonomic schemes have been raised for them and there is still much controversy over whether they were derived from single or multiple ancestors. In a series of papers in the 1970s S. M. Manton championed the view that 'arthropodization' occurred several times. She therefore divided arthropods into four phyla: Uniramia, which includes the millipedes, centipedes, and insects; Crustacea, which includes lobsters and crabs, and also ostracods and barnacles; the Chelicerata, composed of spiders and scorpions as well as the extinct eurypterids; and finally the Trilobita. In other schemes the phylum Arthropoda is retained and the major taxa are given subphylum or class status.

The trilobites are divided into nine orders according to the morphology of the dorsal exoskeletons, and in particular details of the cephalon. The important forms are the Agnostida, very small blind trilobites with only two thoracic segments, and the Redlichiida, large forms with many lateral spines that are characteristic of Lower Cambrian rocks. The Corynexochida were an extensive and rather heterogeneous group that lasted until the Middle Devonian. The Phacopida were Ordovician to Devonian trilobites characterized by large well-developed eyes. The Ptychopariida are a diverse group characterized generally by a forward-tapering glabella. The Proetida were the last of the trilobites, holding on until final extinction in the Permian.

Ecology

Evidence for the life habits of trilobites comes both from their morphology and from the trace fossils that they formed in the sediment. The traces are of several distinct types and appear to record a variety of activities. Some show a series of V-shaped markings indicating movement on the surface; others consist of elongated bilobed trails, convex down and with oblique striations formed by the legs, which are probably related to burrowing activity. Short ovoid markings of a similar structure seem to indicate 'resting' traces excavated by the animal near the surface, although some reported from the Lower Cambrian of Sweden are related to the burrows of priapulid worms and might record hunting activity by the trilobites.

In general, trilobites appear to have been benthonic organisms feeding on organic detritus on the sea floor, but some may have captured small soft-bodied organisms which would have been passed forward to the mouth by the limbs. Forms that carried out this type of benthonic feeding were oriented parallel to the surface and may have been able to form a seal with it, below which the legs would churn up the sediment and remove organic particles. Some trilobite lineages show adaptations that are consistent with a pelagic or swimming lifestyle. These include large cephalons with extremely large eyes, and an overall morphology that seems maladapted to movement on the sea floor. By analogy with modern amphi-pods it is thought that these animals may have swum on their backs and fed on suspended particles. It has also been suggested that agnostid trilobites might have developed a pelagic mode of life, for some of them are very widely distributed, the same species occurring, for instance, in the Middle Cambrian of Scandinavia and western North America.

Trilobite eyes

Trilobites characteristically possess well-developed compound eyes. These represent the first visual systems known and are thus of particular interest. Most trilobite eyes are holochroal, having many round or polygonal lenses in close contact covered by a single cornea. These are the oldest eye-types known and persist until the extinction of the trilobites in the Permian. The lenses are commonly long prisms formed of calcite with the crystallographic axis parallel to the light path to reduce birefringence (splitting of light). A second type of eye, termed schizochroal, is found only in phacopids; it consists of an array of large lenses separated from each other. Work by Euan Clarkson of the University of Edinburgh has shown that the strongly biconvex lenses were constructed in two parts separated by a complex surface, a structure that would tend to improve focusing ability. Such an array may have given the animals sharp stereoscopic vision through 360° and could also have provided them with improved vision in the dark.

Geological history

Trilobites appeared in the Early Cambrian with no obvious ancestral forms, although it has been suggested that arthropod-like forms in the Proterozoic Ediacara Fauna may be on the ancestral line. Their mobility allowed them to take advantage of available habitats and they developed rapidly during the Cambrian; all the orders except the proetids appeared during this period. This trend was reversed, however, during the Ordovician, probably through increased competition from other benthonic forms and the rise of predatory cephalopods. The redlichiids became extinct before the end of the Cambrian; several other groups, including the agnostids, did not survive the Ordovician. A gradual decline then set in until the extinctions of the Middle and Late Devonian disposed of the majority of the trilobite groups. Only the proetids continued until the end of the Permian, when they died out during the major extinction event that affected all the shallow-water marine invertebrate groups. Because of this history, trilobites are most useful in the biostratigraphy of the Cambrian on a worldwide scale, and in the Ordovician and Devonian locally.

Eurypterids

Eurypterids are arthropods of the phylum Chelicerata that have a fossil record ranging from the Ordovician to the Permian, reaching their acme during the Silurian and Devonian. They occur in marine, brackish, and freshwater rocks and appear to have been active benthonic predators that were also capable of swimming. The animal had a large head region or prosoma (Fig. 2) with prominent compound eyes dorsally, and six pairs of appendages ventrally. The first pair were

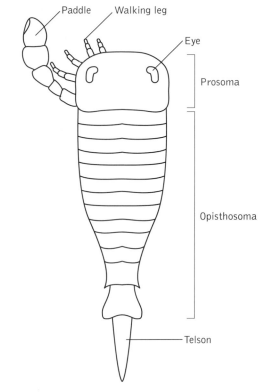

Paddle   Walking leg

Eye

Prosoma

Opisthosoma

Telson

Fig. 2. Typical eurypterid showing major morphological features.

adapted for feeding, the next four were walking legs, and the most posterior pair formed large swimming paddles. Behind the head was a long body or opisthosoma of 12 segments terminating in a stout telson. Analysis of the sixth appendage has shown that the blade moved from a vertical orientation during the propulsive stroke to a horizontal orientation during the recovery stroke, an arrangement necessary for efficient swimming. There is, however, evidence from fossil trackways (a type of trace fossil) that suggests that some euryptids may have been able to walk on land. Most eurypterids were probably predators on other invertebrates and possibly fish. Although most were quite small, the largest, at almost 2 metres long, were the largest arthropods that have ever existed.

DAVID K. ELLIOTT

Further reading

Boardman, R. S., Cheetham, A. H., and Rowell, A. J. (eds). (1987) *Fossil invertebrates*. Blackwell Scientific Publications, Oxford.

Clarkson, E. N. K. (1979). The visual system of trilobites. *Palaeontology*, 22, 1–22.

Manton, S. M. (1973). Arthropod phylogeny—a modern synthesis. *Journal of the Zoological Society of London* 171, 111–30.

tropical cyclones Tropical cyclones are the single most damaging weather events anywhere in the world, with, on average, between 4000 and 5000 deaths attributable per tropical cyclone event. They devastate wide areas, sometimes with extreme loss of life and considerable damage to property. Winds can exceed 75 metres per second (m s⁻¹) (250 kilometres per hour); rain is torrential, perhaps yielding more than 250 mm in less than a day; and sea level is locally elevated by low atmospheric pressure and driving wind. These effects are commonly sustained for many hours, and the typical lifetime and track of such systems means that many communities may be affected within a few days by the same system.

The nature of the impact, split between loss of life and injury on the one hand and damage to property on the other, depends broadly on the community afflicted. The technological expertise, degree of preparedness, and ability to recover in countries such as Australia or the USA ensure that loss of life and injury are limited by effective warning, evacuation, and protection measures. On average fewer than a hundred deaths occur per tropical cyclone event in either of these countries. On the other hand, the cost of damage to property is frequently extreme. Hurricane 'Hugo' in September 1989 caused nearly US$10 billion damage in the USA and Caribbean, a large proportion of which was in the USA. In poorer countries, such as Bangladesh, it is loss of life which is more significant: during the Bangladesh cyclone of November 1970 more than 200 000 people lost their lives; in May 1990 a cyclone hit Andra Pradesh in India, killing 514 people, but directly affecting a further 6.5 million people in one way or another; hurricane 'Hugo' left 90 per cent of the population of Montserrat homeless. Death rates are commonly measured in thousands or tens of thousands per event.

Tropical cyclones are known by different names in different parts of the world (Fig. 1): hurricanes in the tropical North Atlantic and Caribbean; typhoons in the north-eastern Pacific; and simply as tropical cyclones in the south-western Pacific and the Indian Ocean. In some areas local names apply: for example, *papagallos* in the eastern North Pacific, *baguios* in the Philippines, and *travados* near Madagascar. The nature of tropical cyclones, however, is the same in all areas, and individuals are identified by name, alternating between female and male: Annie, Bob, etc. In a typical year the north-western Pacific area experiences the largest number, typically about 25 to 30 (Fig. 1), although the year-to-year variation is high, as in all areas of occurrence. Such variation has been linked to the occurrence of the ENSO (El Niño, Southern Oscillation) phenomenon. The north-eastern and south-western Pacific (Australian) areas each experience between ten and fifteen a year; the Caribbean commonly suffers less than ten. They also occur in the Indian Ocean: in the south-west, near Madagascar and southern Africa; and in the north, west of the Indian subcontinent and in the Bay of Bengal. In all Indian Ocean areas the annual incidence is generally much less than ten. Tropical cyclones also exhibit a strong seasonal-

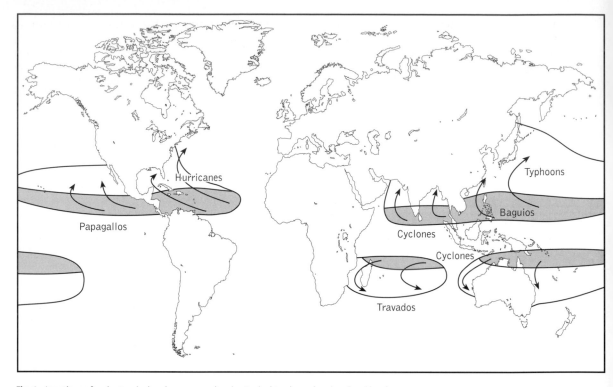

Fig. 1. Locations of major tropical cyclone areas, showing typical tracks and regional and local names.

ity; most of them occur during the late summer months of the respective hemisphere: July to October in the northern hemisphere; January to March in the southern.

There has until recently been comparatively little understanding of they origins, occurrence, and behaviour of tropical cyclones. Significant advances in our knowledge of them, and improvements in forecasting, have been made only since the deployment of weather satellites in the 1960s. Tropical cyclone systems develop over parts of the tropical oceans, drawing on heat and moisture from the ocean surface. Observation is difficult in these areas without satellite surveillance. Their effects are most noticed, however, once they reach land, by which time they may usually have reached their maximum severity. This fact emphasizes the need for constant satellite monitoring of their birth, development, and track.

Tropical cyclones form from pre-existing disturbances. Their life history is particularly well known for the tropical north Atlantic, where much research effort has been concentrated. In this region, many develop from so-called 'easterly waves', some of which also derive from pre-existing squall-lines over West Africa. In any part of the tropics subject to their occurrence, any pre-existing organized mass of cloud yielding precipitation is carefully monitored at the appropriate time of year. A tropical cyclone is born when the sus-

tained wind speed exceeds hurricane force (33 m s^{-1}) and there is a definite cyclonic circulation about a central core, with a central pressure of less than 990 millibars or hectopascals (hPa). At lesser sustained speeds such organized systems are termed 'tropical storms' or 'tropical depressions'. Central pressures can reach well below 950 mbar, similar to some of worst temperate depressions, though the pressure gradient is confined within a much smaller radius: rarely greater than about 300 km. Over the open ocean such low atmospheric pressures elevate the ocean surface to perhaps four metres above normal, and the wind will easily whip up waves well over 10 m high. This means that a tropical cyclone reaching land will, as well as bringing with it destructive winds and flood-inducing rains, commonly produce serious marine flooding in low-lying coastal areas.

The seasonal peaks in their occurrence, together with the location of the major areas of development, provide clues to the origin and development of cyclones as weather phenomena. The incidence of tropical cyclones is high when ocean surface temperatures exceed 27 °C: they draw their energy from the heat of the ocean surface. This same ocean, of course, ensures a continual supply of moisture. An active cyclonic (rotating) circulation can, however, develop only where there is sufficient rotational energy derived from the Earth's surface.

Tropical cyclones are unable to form within about five degrees latitude of the Equator, where the Coriolis effect is very small: their main source areas lie between about 5° and 15° latitude. In either hemisphere these features possess an inward spiral of strong winds at the surface (counter-clockwise in the northern hemisphere; clockwise in the southern hemisphere), introducing heat and moisture to the system, rising within the cyclone to feed an outward spiral, linked to an upper anticyclone aloft. Both horizontal and vertical aspects of the circulation are important to its maintenance as an active weather feature. Should there be a pronounced pre-existing flow in the region, its ability to form will, however, be severely restricted. Such circumstances commonly prevail near the Indian subcontinent at the height of the monsoon, so that, for example, very few cyclones occur in the Bay of Bengal during July and August.

Both heat and moisture are important to the sustenance of a tropical cyclone. Once taken into the system, moisture plays an important part in maintaining the cyclone's continued development. The hot and very moist air fed in near the surface, once uplifted, is cooled, forcing condensation of excess moisture and adding further to cloud volume. In the process, massive quantities of latent heat are released, adding still further to the heat input to the system as a whole, reinforcing the upper outflow by expanding the air, and drawing in more air at the surface by enhancing uplift, so that the system is well on the way to becoming self-perpetuating. This also ensures that tropical cyclones, in contrast to other organized tropical disturbances, possess a 'warm core', with oppressively high temperatures and humidities, near the centre.

The strongest winds and the heaviest rain occur close to this central zone. Strong winds and heavy rainfall are also associated with organized spirals of cloud, frequently containing clusters of severe thunderstorms, often with localized tornado development. The spiral bands appear to migrate outwards through the system, decreasing in intensity and spreading horizontally as they move, commonly ending up well away from the main system as pre-cyclone squall-lines. The source for the spirals appears to be near the centre of the system, near to the 'eye'. The eye is one of the main tell-tale signs that a system has achieved cyclone intensity. It is an almost totally cloud-free area at the centre of the vortex, with near-calm winds and brilliant sunshine. The eye may be from one to forty kilometres in diameter, and is surrounded by a wall of towering cumulonimbus clouds. Although tropical cyclones are dangerous simply because of the associated severe weather, the occurrence of the eye itself, ironically, has often compounded cyclone impact. The appearance of bright sunshine and the arrival of calm conditions may sometimes be seen as the end of the storm rather than a mere lull. Anyone venturing outside in the eye would, however, do well to consider that once it has passed, torrential rain and violent winds will return—from the *opposite* direction. Any debris loosened during the arrival of the storm will then be carried back in the opposite direction by winds on the other side of the system.

Tropical cyclones rarely outlast a week, although they may have travelled thousands of kilometres during that time (Fig. 1), initially moving broadly with the prevailing westward flow that occurs in the tropics, but subsequently gradually 'recurving' polewards around the subtropical anticyclones, perhaps into temperate latitudes. Subsequently their track might take them back eastwards across the temperate ocean, now as severe examples of temperate depressions, towards, for example, Europe. A former Caribbean hurricane ('Charlie') wrought havoc in western Britain in August 1986, though it is important to stress that it was not a hurricane by this stage: it had lost all the characteristics by which we define such a feature. The tracks of many individual tropical cyclones are, however, notoriously haphazard and unpredictable. They may remain nearly stationary for some days, move round in a circle, and may often reverse their former track. Once they make landfall, however, they generally begin to lose their identity, since the source of moisture over the ocean surface is removed. They can nevertheless still take torrential rainfall many hundreds of kilometres inland. Tropical cyclone 'Bobby' carried copious amounts of rainfall south-eastwards across the desert areas of Western Australia in February 1995. A cyclone can also track along a coastline, drawing on the adjacent ocean. The eastern seaboards of North America, Australia, and Asia can be particularly vulnerable if this occurs. In June 1972, hurricane 'Agnes' tracked up the eastern seaboard of the USA and retained its tropical identity as far north as the New England states, carrying torrential rain and strong winds over many populated parts of the country as it did so.

GRAHAM SUMNER

Further reading

Eden, P. (1988) Hurricane 'Gilbert'. *Weather*, 43 (12), 446–8.

Rappaport, E. N. (1993) Hurricane 'Andrew'. *Weather*, 49 (2), 51–61.

Simpson, R. H. and Riehl, H. (1981) *The hurricane and its impact*. Basil Blackwell, Oxford.

tropical karst Karst landscapes in the tropics are among the most spectacular anywhere in the world. Although the formative processes, primarily dissolution of carbonate rocks, are essentially the same as in other climatic zones, the higher temperatures, humidities, and rainfall, particularly in the humid tropics, accelerate and accentuate karst development (see *karst*).

High temperatures favour chemical erosion, and high humidities and rainfall ensure that abundant water is available for the dissolution process. Moreover, the warm, humid conditions encourage high levels of biological activity, particularly in the soil, generating the production of carbon dioxide, which is critical to the dissolution process in the near-surface, or epikarstic zone. Intense tropical rainfall may also result in rapid surface erosion and gullying by short-lived streamflow. In some tropical areas, surface weathering crusts develop through dissolution and recrystallization, a process termed *case-hardening*. This renders the rock surface less porous and increases its resistance to erosive stresses. Elsewhere, springs and waterfalls are associated with the precipitation of sheets and dams of tufa or travertine.

Extensive areas of tropical karst occur in southern Mexico, Central America, the Caribbean, in South-East Asia, and in southern China, where (in the provinces of Yunnan, Guizhou, and Guangxi) there is over 500 000 km^2 of the most impressive karst in the world. There are also significant tropical karst areas in South America, Madagascar, the Middle East, New Guinea, and northern Australia.

The most diagnostic elements of tropical karst terrain are large, more-or-less enclosed and polygonal depressions, sometimes termed *cockpits* because of their resemblance to cockfighting arenas, and systems of dry valleys, commonly dismembered by subsequent depression development. Bordering these negative relief features are sinuous ridges and interconnected or isolated residual hills, generally known as *cones* or *towers*. Although these hills may be striking, sometimes exceeding 100 m in height and with sheer cliffs, it is the valleys and depressions that are the active foci of the landscape, often channelling surface water into underground drainage systems. The enclosed depressions are complex forms of dolines or sinkholes, which act as centripetal drainage basins (see *sinkholes*). Although dissolution is the dominant process in their development, collapse of depression bases over subterranean voids may occur. For example, *cenotes* of the Yucatán Peninsula of Mexico are collapsed depressions, in the base of which groundwater flow is exposed.

Valleys in tropical karst are also highly variable. Steep-sided gorges develop where large rivers, rising in adjacent terrain, traverse karst belts, but many smaller rivers disappear underground at the end of blind valleys. Springs near the periphery of a karst area may supply rivers draining out of the karst via pocket valleys and some rivers may enlarge valleys and deposit alluvium within the karst to produce the large, flat-floored features known as *poljes*.

The classification of tropical karst is highly complex, with a tortuous terminology derived from several languages. The Chinese recognize a distinction between *fenglin*, or peak forest karst, in which residual hills are solitary and isolated; and *fengcong*, or peak cluster karst, in which numerous residual hills share a common bedrock base. Quantitative indices of tropical karst terrain have been developed, but are not employed widely.

In combination with the large landforms, a wide variety of smaller-scale features produced by dissolution are found. These features are generally called *karren* (see *karren*). Many of them are small and of eccentric morphology, but others are larger and mushroom- or spire-shaped. Among the most impressive of the latter are the pinnacle karsts, or stone forests of southern China, Borneo, and Madagascar, where the steep, fluted limestone spires may exceed 50 m in height.

Integral to tropical karst are some of the world's largest and most spectacular cave systems, the formation and evolution of which is favoured by the large volumes of water available underground (see *caves*). The caves take on a wide variety of forms, but they include some of the largest active river passages and several of the world's largest cave chambers. The largest of the latter, the Sarawak Chamber in Good Luck Cave, Sarawak, Malaysia, has horizontal dimensions of over 600×400 m, with a roof span of 300 m, a floor area of nearly 163 000 m^2, and a volume of some 12 million m^3. Other massive cave chambers occur in Belize, Central America, and in Oman. Many tropical karst caves are extensively decorated with recrystallized deposits of calcite and other minerals.

Although karst in the humid tropics attracts the greatest attention, karst in arid and semi-arid tropical regions is also significant, in part because it illustrates the importance of climatic and environmental change in karst development. Tropical karst is developed on rocks ranging in age from the Palaeozoic (more than 250 million years old) to the Cenozoic (less than 65 million years old), but it is clear that it has often developed under changing environmental conditions, both at the surface and underground, with considerable inheritance of characteristics from karst terrain developed under earlier conditions.

Studies of tropical karst have influenced the understanding of karst in general. For example, in the early twentieth century, drawing on tropical experience, Alfred Grund developed models of groundwater circulation and landscape evolution in karst regions. The Chinese traveller Xu Xiake (1586–1641), who journeyed extensively throughout the tropical karst of southern China, was one of the first to appreciate the singular character of karst landscape.

Karst landscapes in the tropics are significant in terms of regional water supplies, as archaeological and cultural reserves, and as reservoirs for plants and animals. They are also increasingly important in the context of tourism. At the same time, they are inherently fragile, subject to severe water contamination, soil erosion, and environmental degradation. Increasing numbers of tropical karst landscapes are being designated as protected areas, but pressures upon them from agriculture, industry, and other sources are also increasing, and their conservation is of paramount importance. Notable protected areas of tropical karst include the Gunong Mulu National Park in Sarawak and the Lunan Stone Forest in Yunnan Province, China.   MICHAEL DAY

tropical landforms The tropics differ from other parts of the world in that their seasonality is determined by water availability, not by temperature. There is sufficient heat for biological activity throughout the year, but whether or not that activity occurs depends on the distribution of rainfall. In the wettest areas, water is sufficient throughout the year to support the evergreen tropical rainforest in which continuous biotic activity gives rise to rapid turnover of decaying plant material. Away from these humid areas seasonal water shortages gives rise to savannah, sahel, or desert biomes in which biotic activity is reduced or virtually nil in the dry season.

The work of geomorphological processes on the landscapes of the tropics is thus moderated by the vegetation cover: completely in the areas of undisturbed tropical rain forests and less and less as the arid areas are approached. However, the magnitudes and frequency of rain events in the humid tropics can be such that geomorphological change is rapid, particularly in the forested areas on steep slopes affected by tropical cyclones. In all these areas, there is great variability from month to month and year to year, especially when El Niño and La Niña events occur.

High rainfall, high temperature, and high biological productivity lead to high rates of biogeochemical processes, particularly decomposition of organic matter and rock weathering. In the humid tropics, the dissolution and removal of chemical elements is thus more complete than elsewhere. The clays in weathering profile are the simplest, with few soluble elements. In places, even the true clay minerals disappear and the residual soils are instead dominated by iron and aluminium oxides. In stable areas, over long periods of time, these residual soils may develop ferricrete or bauxite duricrusts on deeply weathered rotted profiles. The existence of both the deep weathering profiles (deep *regolith*s) and the duricrusts depends on landscape stability and parent material.

The tropics cover a wide range of rock types and tectonic conditions, from highly active areas such as New Guinea, the Philippines, and Central America, to stable shield areas such as eastern Brazil, central and southern Africa, and the Deccan. In the earthquake-prone unstable areas, landsliding is so frequent that only skeletal soils remain on some sandstones and mudrocks. Weathering profiles here are rarely more than 2 m deep, whereas more than 120 m of regolith occur at places in Africa and Australia. These elements provide a fundamental contrast of tropical landscapes, from the landslide-prone steep and dense valley networks of the feral relief of New Guinea and parts of Central America to the old, deeply weathered plains of stable shield areas. This contrast was expressed at the beginning of the twentieth century in the writings of the pioneer German geomorphologists, Behrmann in New Guinea and Thorbecke in East Africa. Contrasts of this type can be found within relatively short distances. Much of the island of Borneo is made up of cuestas of Tertiary sandstones with intervening areas of weak mudstones within shallow weathering profiles, but in the west of the island the igneous rocks of the Thai–Malay Peninsula and Bangka and Billiton Islands Mesozoic intrusions occur, with deep weathering profiles on the granites. Landform evolution in the tropics is dominated by these spatial variations in conditions and by the way that conditions have changed during the Quaternary.

Landforms in the tropics are an expression of the balance between rate of rock rotting and the formation of regolith on the one hand, and the rate of removal and subsequent deposition of these products of weathering on the other. Overall, the landforms are governed by the interrelationships between climate, lithology, and past landscape history. Few landforms can be said to be uniquely tropical, because the weathering and transport processes associated with running water are found in other climates, but some are much more likely to be found in the tropics and others evolve more completely in the tropics. The more complete development may be due to the continuous landform evolution and fluvial conditions that have taken place, not only through the Quaternary, but probably through much of the Tertiary in some of the stable shield areas of tropical regions. The Quaternary may have produced fluctuations in moisture regimes, with periods of longer or shorter dry seasons, but basically the dominant landform processes were related throughout the time to chemical weathering and the work of running water. Such continuity is a dominant force over about 70 per cent of the tropics, which have escaped orogenic activity since the early Tertiary. This has meant that large areas of the tropics are covered by extensive plains, of relatively low relief with relatively deep regolith cover. Such plains exist outside the tropics, particularly in the continents of the southern hemisphere and in parts of the northern hemisphere, but some scholars argue that they were the product of evolution under early warm, palaeotropical climates in the early Tertiary. They are not, however, uniquely tropical.

Only two types of landform—the coral reefs and the mangrove swamps—are uniquely tropical and even the latter can be found outside tropical latitudes along the coasts of New South Wales and the North Island of New Zealand. They are both biogenic landforms resulting from biotic activity regulated by temperature and salinity.

The shores of continental mainlands and many islands in the tropics are fringed with reefs or other formations built of the bodies of coral polyps, tiny marine animals that secrete external skeletons of calcium carbonate. The polyps live in shallow tropical waters and can build up reef formations on any stable coastal platform or substrate. The coral reefs of the Bahamas and southern Florida are built on such a platform. The Australian Great Barrier Reef is a huge complex of many individual reefs, irregular coral masses, and some islands on a shallow bedrock platform.

In the oceans, characteristic fringing reefs develop around the volcanic islands surmounting hot spots, as in Hawaii, or the mid-oceanic ridges. When a volcano first grows up to the ocean surface, a reef may become established. If later the volcano sinks, the reef may continue to build on itself, growing upwards just below the sea surface until it forms a barrier reef some distance out from the volcanic island. Further subsidence of the volcano may leave only the coral reef above the sea surface, forming an *atoll*. The term 'atoll' implies a ring-shaped structure. However, a perfect ring rarely occurs. An atoll usually consists of a string of closely spaced coral islets separated by narrow channels of water. Each individual islet is called a *motu*.

Mangrove swamps are the tropical equivalent of coastal salt marshes. Mangroves tend to occur equatorwards of 30° in the northern hemisphere, and from slightly higher latitudes in the southern hemisphere, with salt marshes towards the poles. Mangroves develop at river delta fronts, as in the case of the Sundarbarns of Bengal and Bangladesh, and along major river estuaries, as in Auckland and Sydney harbours, or behind coastal barriers, as on the north of Borneo. Most importantly, they form extensive prograding areas along sheltered coasts, as on the sides of the Malacca Straits between Sumatra and the Malay Peninsula.

The supply of sediment and the activity of mangrove stilt roots lead to the accumulation of muds that form mangrove-covered islands, inundated at high tide, but with the trees standing clear of the water. If the balance of sediment supply and tidal action is appropriate, the succession of the mangrove plants moves seaward. Landward of the mangrove there commonly develops a pandanus swamp forest. These formations are commonly the seaward part of coastal swamplands that eventually develop into deep peat formations. Further inland they can give way to the vast freshwater swamp forest formations characteristic of many tropical lowlands, such as the flooded forest of the Amazon and the wetlands of the Congo Basin. Such forested wetlands deserve more attention as significant alluvial landforms and as present-day analogues of the formation of coal deposits.

Landforms that are typically, but not uniquely, tropical are those associated with the deep weathering and efficient debris-removal systems of the tropics. The most characteristic such assemblage is the extensive plain of a deeply weathered regolith surmounted by isolated bare rock hills, or *inselbergs*. An inselberg has been variously defined as a 'small, island-like hill or mountain rising sharply above a surrounding pediment or alluvial fan' (Strahler and Strahler), an 'isolated summit rising abruptly from a low relief surface' (McKnight), a 'stark erosional remnant standing above surrounding terrain as exemplified by Uluru (Ayers Rock), Australia' (Christopherson), or 'a steep-sided residual hill that has arisen from the operation of scarp retreat pediplanation' (L. C. King).

One investigation of research on inselbergs found that 40 per cent of studies came from savannah environments, 32 per cent from arid or semi-arid areas, 12 per cent from humid continental and subarctic climates, and 6 per cent each from humid tropical, subtropical, and Mediterranean zones. Even though such a count reveals something about where scientists are based and work, it does show that inselbergs are by no means a uniquely tropical landform.

The level, and often extensive, plains above which the inselbergs rise evolve by ground surface lowering and the retreat of steep slopes. The combination of processes associated with heavy seasonal rainfall and deep chemical weathering has been termed *etchplanation*. Etchplains result from the vertical movement of infiltrating water down through previously weathered rock to the basal surface weathering to rot further fresh rock material and so enhance the depth of weathering. If, at the same time, processes on the land surface are removing material, the whole land surface may be lowered, while the regolith depth is maintained. This process of parallel stripping of the land surface and deepening of the regolith was identified in East Africa, but was found to be equally operative in Brazil and southern India. Working in the latter area, Julius Büdel found four circumstances to be indicative of etchplanation:

(1) weathered rock thicknesses of 3–30 m;
(2) homogeneity of the regolith, indicating a steady, dynamic evolution;
(3) a sharp transition from weathered to fresh rock over wide-jointed, quartziferous rocks, indicating intense weathering; and
(4) the widespread existence of a deeper zone of decomposition up to as much as 200 m thick.

This combination is seldom active at the present day outside the tropics, and thus these inselberg-dominated plains may be considered a characteristic tropical landform assemblage.

The deep weathering can also result in the duricrust formations mentioned earlier. Over periods of time, one level of duricrusted surface may be dissected and a landscape dominated by flat, duricrust-capped residuals may be created. Where the duricrusts are bauxite or ferricrete the residuals almost certainly developed under tropical or subtropical conditions, whereas silcrete- and calcrete-capped residuals are characteristic of drier and Mediterranean climates.

The feral relief of tropical steeplands is not quite so characteristic, for it is found in other areas free of extensive Quaternary glaciation, such as the North Island of New Zealand. Nevertheless, steeplands of landslide-controlled slopes found in the wettest areas of the tropics, particularly in areas of both earthquake and tropical cyclone activity, are some of the most geomorphologically active, most rapidly changing natural landscapes on Earth.

Equally typically tropical are the karst landforms of the 'botanical hothouse extreme' (Jennings) characterized by tower karst hills, cone (or cockpit) karst, and by smaller-scale 'knife-edge' karren formations (see *tropical karst*). Tower karst hills are the inselbergs of limestone landscapes, standing up with white, exposed, vertical or overhanging limestone walls from surrounding jungle-covered plains. Huge caverns may exist inside the hills. Cone karst is a highly dissected doline landscape, with closely spaced sinkholes resulting from extremely high rainfall. Sometimes a vertical-walled tower karst block has a cone karst summit surface.

In detail, some of these tropical karst massifs are surmounted, not by the karst pavements typical of temperate regions, but by sharp ridges or pinnacles of limestone formed by the solvent action of huge quantities of rain running down joints in the limestone, and so rapidly widening the joints that only the pinnacles are left. In seasonally wet tropical areas, tower karst hills are often residuals from past wetter climates and tend to have their caves partially filled by secondary calcite deposits. Their exteriors have been sculpted into fine solution features (see *karren*).

Tropical landforms thus show the effects of the interaction between climate and lithology, as mediated by biotic systems. The most characteristic features, apart from the temperature-controlled biogenic landforms, are those where continuity of landform evolution and tectonic stability are greatest; but the most problematic for people are often those where both tectonic and climatic forces are the most active and violent.   IAN DOUGLAS

tropical refugia
Tropical refugia are patches of vegetation and their faunal inhabitants which remained unchanged while others were modified during Quaternary ice ages. The refuges were mainly areas of lowland rainforest that survived when savanna became more widely distributed in colder and drier episodes (see *rainforests and ice ages*). They were not places to which plants and animals retreated to shelter from hostile conditions.

Tropical refugia have been identified by means of a combination of biogeographical and palaeoecological data. Biogeographical studies have demonstrated the current existence of rainforest areas with high levels of *diversity* (richness) and *endemism* (localized occurrence) among their component species. Populations of these species are normally characterized by a marked degree of morphological similarity. Between these areas are contact zones containing fewer species, whose members exhibit more differences of form. Information about past ecology has come mainly from examination of the composition of superficial deposits beneath rainforest and the pollen content of Quaternary sediments.

At present, species diversity in areas considered to have been tropical refugia is perpetuated by a genial climate, varied environment, abundant biological productivity, and a substantial amount of competition and predation within the biota. These factors produce smaller niche breadths and higher niche overlaps in the species.

However, even if such mechanisms have been operational in the past, it is possible that current species richness had another contributory cause. This has been hypothesized as evolutionary activity associated with the persistence of refugia in the tropics during cold stages of the Quaternary. While the notion of such localities has been in existence for a considerable time, explanation of the Theory of Tropical Refugia is generally credited to Jürgen Haffer in 1969. This theory neither suggests that speciation has been confined to refuges, nor that the entire complement of present-day species has evolved during the past 2 million years. However, it does propose that the Quaternary has witnessed the most recent, and possibly the most fruitful, of a series of evolutionary episodes which have occurred over geological time. The theory is underpinned by evidence that the climatic cycles of the Quaternary have influenced biogeography in the tropical as well as the temperate part of the Earth. As Ghillean Prance observed in 1982, the major point is to what

extent did oscillations in the extent of tropical forest during the Quaternary affect speciation? He also pointed out that analogous families and genera of tropical rainforest plants to those of today had developed by the Early Tertiary (about 60 million years ago) and that over this time-span, episodes of continental drift and suturing have influenced both the distributional and evolutionary strategy of biota.

Haffer stated in 1982 that there are three potential results of a contraction of forest cover into refugia:

(1) species will become extinct because they are unable to resist heightened competitive and predatory pressures in a reduced area;
(2) they will continue to exist without change; or
(3) they will survive and evolve into new subspecies and species by means of *genotypic* (genetic constitutional) and *phenotypic* (physical) modifications.

A combination of outcomes is likely to occur in many cases.

The most prevalent method of speciation is that brought about by the presence of barriers to reproductive contact between members of the same plant and animal species. Such barriers are some kind of geographic feature which isolates parts of populations, and leads to *allopatric* (geographical) speciation. Refuge theory suggests the development of ephemeral habitat barriers which have caused breaks in the distribution (range) of species. Thereafter, separated parts of the populations evolved in isolation. Haffer pointed out that certain animals (small mammals, for example) can speciate in under 10 millennia. Such a period is considerably shorter than those that have been assigned to cold stages of the Quaternary, and would have enabled new species to emerge. Additionally, a certain amount of *sympatric speciation* (involving reproductive isolation between segments of one population) and *parapatric speciation* (whereby a population enters a new habitat but the barriers between it and its parent population are related to gene flow rather than physical conditions) probably took place in refugia, as did some evolution of different kinds of plants and animals arriving by dispersal from long distances away.

The theory also states that re-expansion of the tropical rainforest flora and fauna, mainly at the expenses of that of savanna, during intervening temperate stages of the Quaternary will have led to contact with populations of other refugia. The amount of species interaction in these contact zones will have varied, depending upon the degree of differentiation achieved by particular plants and animals. In this context, biota evolve at differing rates and the duration of episodes of isolation has been unequal. A sexually and ecologically discrete species which evolved in a refugium might extend its range over a considerable part of the newly enlarged habitat. During the next episode of inhospitable climate, both this distribution and that of its parent species would undergo additional fragmentation. As John Flenley has stated, the amalgamation of formerly separate patches of tropical rainforest in which genetic isolation had taken place would produce groups of closely related species. Such occurrences are frequent in this biome today.

It is worth noting that the effects of a colder and drier climate would have had different biogeographical and evolutionary consequences at altitude than over latitude. On mountains, the downward movement of life zones during the ice ages lessened the efficacy of the barriers between them, facilitated the dispersal of organisms, and militated against evolutionary change. More effective isolation will have been achieved during interglacials, when upward migration of zones took place and species differentiation was fostered.

In South America, rainforest seems to have survived cold episodes beside major rivers and in several limited areas of tropical lowland (Fig. 1). Such localities have the most abundant precipitation today, and have relief that has endured with little change since the Late Tertiary. They are likely to have had the highest rainfall and most fertile soils during colder and drier phases. Thus they probably were recurrent refugia, a phenomenon that would have enhanced the differentiation of certain of the species populations within them. The floral diversity of Amazonia has been attributed to refugia by Prance, while Haffer has interpreted its avifaunal richness in the same light. Other animals (butterflies and lizards, for example) possess similar patterns of species richness and endemism (Fig. 1), which can be explained in a similar way.

In Africa, sectors of coast in Upper Guinea, Cameroon and East Africa, together with parts of Gabon and Zaïre, are thought to have had rainforest refugia. Distributional data on amphibians, reptiles, mammals and birds demonstrate centres of species diversity and endemism. There is currently a distinct western and eastern distribution of rainforest in Africa. According to R. E. Moreau, links between the two areas during former periods with higher precipitation may afford an explanation for the pattern of bird speciation in Africa. It is interesting to note that Paul Richards has suggested that the diversity of African rainforest was reduced as a result of arid episodes.

Donald Walker has speculated that the diversity of Sunda (mainly Malaya, Sumatra and Borneo) and New Guinea rainforests in South-East Asia was not significantly increased during arid episodes of the Quaternary. In Sunda, niche subdivision, and in New Guinea environmental differences promoting rapid allopatric speciation, may be responsible for current diversity levels.

Doubts have been expressed concerning the validity of the refuge theory. Some of its opponents contend that loci of current species diversity in tropical rainforest could be explicable in terms of existing ecological (demographic and physical, for example) factors. Contact zones would then relate to gradients of these factors. Refugia should be assessed against the backdrop of this and other hypotheses regarding evolution in tropical lowlands. Alternatives include the island and river theories. The former states that modifications to the geography of the Earth during the Tertiary caused the separation of substantial populations which evolved into species that are extant. The latter propounds the view that the development of large rivers (such as the Amazon) early in the Quaternary led to the fragmentation of hitherto widely distributed populations, thereby bringing about speciation on opposite banks.

R. L. JONES

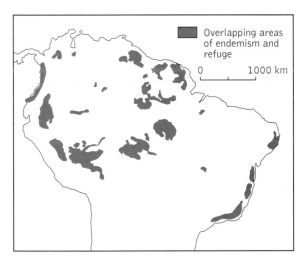

Fig. 1. Overlap of areas of endemism in tropical South American plants, butterflies, and birds with palaeoecological evidence for forest refuges. (From Brown, K. S. (1987) Conclusions, synthesis, and alternative hypotheses. In *Biogeography and Quaternary history in tropical America* (ed. T. C. Whitmore and G. T. Prance), pp. 175–96. Clarendon Press, Oxford.)

Further reading

Bradley, R. S. (1985) *Quaternary palaeoclimatology: methods of palaeoclimatic reconstruction.* Allen and Unwin, Boston.

Brown, K. S. (1987) Conclusions, synthesis and alternative hypothesis. In Whitmore, T. C. and Prance, G. T. (eds) *Biogeography and Quaternary history in tropical America,* pp. 175–96. Clarendon Press, Oxford.

Flenley, J. R. (1979) *The equatorial rain forest: a geological history.* Butterworth, London.

Haffer, J. (1969) Speciation in Amazonian forest birds. *Science,* **165**, 131–7.

Haffer, J. (1982) General aspects of the refuge theory. In Prance, G. T. (ed.) *Biological diversification in the tropics,,* pp. 6–24. Columbia University Press, New York.

Moreau, R. E. (1966) *The bird faunas of Africa and its islands.* Academic Press, New York.

Prance, G. T. (ed.) (1982) *Biological diversification in the tropics.* Columbia University Press, New York.

Richards, P. W. (1973) Africa, the 'odd man out'. In Meggers, B. J., Ayensu, E. S. and Duckworth, W. D. (eds) *Tropical forest ecosystems in Africa and South America,* pp. 21–6. Smithsonian Institution Press, Washington.

troposphere The troposphere is the part of the atmosphere in which all 'weather' occurs: clouds, rain, snow, hurricanes, tornadoes, or fronts. The name comes from the Greek *tropos*, 'turning'. In this part of the atmosphere turbulent overturning of air is dominant and causes mixing vertically as well as the horizontally.

The atmosphere is divided into layers or 'spheres' according to the different properties found in each layer. If the layers are characterized by the rate at which temperature changes with height, then the atmosphere is split into four layers: the troposphere, stratosphere, mesosphere, and thermosphere. If the layers indicate the chemical composition, they are referred to as the homosphere and the heterosphere, according to whether the composition is homogeneous or heterogeneous. Below about 80 km altitude the atmosphere is homogeneous: it has the same chemical composition at all locations. Other layers given special names are the ionosphere, where some of the atoms and molecules are electrically charged, and the exosphere, where some molecules escape from the gravitational attraction of the Earth and move off into space.

The troposphere is a layer in which, on average, the temperature decreases with height. It extends from the Earth's surface to the tropopause, the name given to the boundary between the troposphere and the stratosphere (Fig. 1). In the stratosphere the temperature is constant at lower levels and begins to rise at greater altitudes. The tropopause can be defined as the place where the temperature stops decreasing with height and becomes constant. The altitude of the tropopause is not constant. It varies from place to place and from time to time. It is usually higher in the tropics, where it is about 16 km, than near the poles, where it is about 6 to 8 km. At middle latitudes, between about 40 and 60° N and S, the tropopause lies between about 10 to 12 km. Its height varies from day to day, depending on the large-scale movement of air within the troposphere.

The average rate at which the temperature decreases with height in the troposphere is 6.5 degrees Celsius in every kilometre. Since this is a decrease in temperature it is usually referred to as the *lapse rate*. If we were to measure the lapse rate on any particular day, at an arbitrary location, it would probably not be 6.5 degrees Celsius per kilometre. We would find that the observed lapse rate would almost never exceed 10 degrees Celsius per kilometre but would sometimes be very small. In fact, the lapse rate can often be zero, in what is known as an isothermal layer (since the temperature in the layer is constant). The lapse rate can also be negative, indicating that the temperature is increasing with height. This is called an inversion layer, since it is an inversion of the usual decrease in temperature with height. Inversion layers and isothermal layers are usually not very deep: rarely more than a few hundred metres.

Because air is constantly overturned in the troposphere, it is worth considering the changes that are forced on the air as it moves. In order to imagine these changes consider a 'parcel' of air: a mass of air which will move about without mixing with its surroundings as if it were enclosed in an invisible, totally flexible balloon. If this air parcel lies near the Earth's surface it will gain or lose heat according to whether it is colder or warmer than the surface. The surface itself gains heat during the day by absorbing sunlight and loses heat at night when there is a net heat loss through infrared radiation to space.

The air parcel is most likely to be warmed during the day. (Some air parcels will be warmed more than others, because some parts of the surface will be warmed more than others.) As the air parcel is warmed, it expands: the greater the heating, the greater the expansion. Because the air in the parcel now occupies more space, it has a lower density than the relatively cooler air parcels. The less dense air parcels will rise. This rising air is part of a *thermal*, a rising current of air which develops over parts of the surface that are warmer than their surroundings.

Because the atmospheric pressure decreases with height, the rising air parcel encounters air around it which has a lower pressure. As the parcel rises, the pressure inside it will tend to become the same as the pressure outside. This causes the air parcel to expand further. Expansion requires work to be done against the surrounding air. At the surface the energy for the expansion came from the heat provided by the warm surface. In the atmosphere there is no such source of energy. The energy

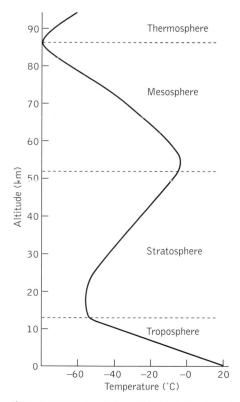

Fig. 1. Temperature variation with height in the atmosphere, averaged over all seasons and all locations.

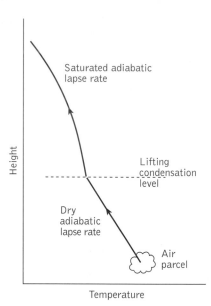

Fig. 2. Variation of temperature with height of a parcel of air as it rises through the point where it becomes saturated.

needed for expansion has to come from the heat within the air parcel. As the air parcel rises it thus cools as it expands. This can be explained in physical terms by the first law of thermodynamics, through which we can calculate the rate at which the air parcel cools. The cooling rate is 10 degrees Celsius for every 1 km of ascent. This is referred to as the *dry adiabatic lapse rate*: 'dry' because no water has condensed in the air parcel; 'adiabatic' because all the energy for expansion comes from the air parcel itself; 'lapse rate' because the air parcel cools as it rises.

Because the rising air is cooling at a rate greater than the commonly observed lapse rate, the air parcel will eventually cool until it has the same temperature as the surrounding air. It will also have the same pressure and density. It will then no longer have any buoyancy, and it will stop rising; see Fig. 2.

The rising air will contain some water vapour and will often be unsaturated as it begins to rise. As it cools it will eventually reach a temperature at which it becomes saturated. Further rising will cause condensation of water into drops. This is the *lifting condensation level*, which marks the base of clouds formed in the rising air. As the air rises further, more water condenses. As the water condenses it releases latent heat. The rate at which the air cools as it rises is less after condensation starts, since the latent heat that is released warms the air slightly. The amount of warming depends on how much water vapour condenses. The lapse rate under these conditions, the *saturated adiabatic lapse rate*, is less than the dry adiabatic lapse rate. It varies from place to place and from time to time, but is about 6.5 degrees Celsius per kilometre on average—the same as the average lapse rate observed in the troposphere.

CHARLES N. DUNCAN

Further reading

Ahrens, C. D. (1994). *Meteorology today*. West Publishing Co.

McIlveen, J. F. R. (1986) *Basic meteorology*. Van Nostrand Reinhold, New York.

tsunamis The word 'tsunami' is derived from the Japanese words meaning 'harbour wave'. Often described as tidal waves, although they have nothing to do with tides, tsunamis are in fact generated by offshore earthquakes, submarine landslides, and occasionally by undersea volcanic activity. In each case, water disturbance is created by large-scale underwater displacement of sediment or rock on the sea bed, usually as a result of a fault or a landslide. The initial water movement is often characterized by a rapid drawdown or lowering of the sea surface at the coast as the water moves into the area of sea-bed displacement. Thereafter, large kinematic waves are propagated outwards from the zone of sea-bed disturbance, travelling across the ocean at very high velocities, often in excess of 450 km per hour, and possessing very long wavelengths and periods. At the coast, the flood level or run up associated with a tsunami is partly a function of the dimensions of the propagated waves, but it is also greatly influenced by the topography and bathymetry of the coastal zone. The waves can reach considerable elevations and can cause widespread destruction and loss of life.

The geological processes associated with tsunamis are poorly understood. Remarkably, most Earth science textbooks do not consider the geology and geomorphology of tsunamis. Tsunamis, however, produce unique coastal landforms that are characterized by the effects of erosion and deposition of large magnitude. They frequently deposit boulder accumulations, and in many coastal areas, run-up processes commonly result in the deposition of continuous and discontinuous sheets of sediment. Many earthquake-generated tsunamis are also associated with the flooding of coastlines caused by subsidence associated with the earthquakes (coseismic subsidence). On the sea bed, tsunamis can also result in the remobilization of sediments and the formation of distinctive strata (so-called tsunamites).

Until recently, knowledge of the long-term frequency of tsunami flooding along individual coastlines was based simply on whether or not the occurrence of former tsunamis was described in historical records. Thus, for example, the record of documented tsunamis for Hawaii extends only to 1850 AD. Before this date, the return frequency of tsunamis is not known, since no historical record exists, yet many tsunamis are likely to have occurred. In recent years, it has proved possible to investigate the occurrence of prehistoric tsunamis for individual areas because many 'palaeotsunamis' have been associated with the deposition of sediment in coastal areas subject to flooding. In this way it has been possible to calculate the number of tsunamis that have taken place in particular areas over timescales of 10^4–10^5 years. For example, geological research in Japan has provided a chronology of past tsunami flooding for the last 4000 years.

During the early 1990s a number of large tsunamis occurred in various parts of the world. Studies of the geomorphological processes associated with these floods have provided a unique opportunity to understand processes of tsunami erosion and deposition. The most important tsunamis that have taken place during this period include a major tsunami that struck the Pacific coastline of Nicaragua and Costa Rica during 1991 and a very destructive tsunami on the island of Flores, Indonesia on 12 December 1992. More recently, destructive tsunamis have struck on two occasions in Hokkaido in northern Japan and in the neighbouring Kurile Islands, Russia, as well as on three occasions in Java, Indonesia, and in the Philippines. Data on coastal flooding resulting from these tsunamis has bee used to develop models of individual tsunamis. In most instances, this type of numerical modelling has been used to reconstruct the properties of the offshore earthquakes and sea-bed faulting that generated the tsunami waves. In general, these studies have shown that the observed flood run-up at the coast is much greater than the run-up values predicated by the mathematical models.

Not all tsunamis are generated by offshore earthquakes. As mentioned above, many are triggered by submarine sediment slides. One of the world's largest areas of submarine slides is located west of the coast of Norway in the Norwegian Sea. This area, known as the Storegga area, has been the site of three exceptionally large underwater slides during the past 30 000 years. Each of these submarine slides is believed to have generated an exceptionally large tsunami. The best known of these is a slide that took place approximately 7000 years ago and involved the movement of approximately 1700 km³ (cubic kilometres) of debris from the continental slope to the abyssal plain east of Iceland. The tsunami generated by this slide produced flooding up to 10 m above former sea level along parts of the west coast of Norway; it also caused severe flooding along the northern and eastern coastlines of Scotland and as far south as the present location of Amsterdam.

What was probably the world's largest tsunami took place in the Pacific basin approximately 105 000 years ago. This tsunami was generated by a large underwater slide south of the island of Lanai in Hawaii. The tsunami is presently believed to have caused flooding up to approximately 360 m above former sea level on Lanai. It has also been proposed that the tsunami waves reached up to 20 m, above sea level along the New South Wales coastline of eastern Australia.

Recognition that underwater slides are capable of generating large tsunamis has, not surprisingly, rendered the task of mathematical modelling of tsunamis much more difficult. More importantly, it has made it very clear that it is a mistake to assume that all major tsunamis are solely the products of offshore earthquakes. The clearest example of this is the well-known meteorite impact that took place in central America approximately 65 million years ago (the so-called K/T impact) that is commonly associated with the extinction of the dinosaurs. A number of Earth scientists believe that this meteorite impact also generated an extremely large tsunami that caused severe flooding throughout many coastal areas. Some scientists have gone so far as to suggest that this tsunami might have caused flooding throughout the entire world at this time.

As a result of the loss of life and damage to property caused by tsunamis in the Pacific, attempts have been made to develop a tsunami warning system, the purpose of which is to alert the public in advance of the arrival of individual tsunamis. Tsunami warning systems did not exist for the Pacific prior to the Aleutian trench earthquake of 1 April 1946. A tsunami warning system network was eventually established by the United States Coast and Geodetic Survey with its headquarters on the Island of Oahu, Hawaii. The centre, known as the Pacific Tsunami Warning Centre (PTWC), became operational in 1948 and is now linked to over 30 seismological stations throughout the Pacific Basin. These provide data on Pacific earthquakes whose magnitude and epicentres make them tsunamigenic (capable of producing tsunamis). Once an earthquake has taken place, the PTWC issues a 'tsunami watch' to all receiving stations. In addition, the initial registration of a tsunami on tide gauges is relayed to the PTWC and, once the estimated times of tsunami arrival are computed for different coastal areas, a tsunami warning is issued.

The accuracy of PTWC tsunami warnings is exemplified by the Chilean earthquake and subsequent tsunami of 21 May 1960. Once the earthquake epicentre had been determined and tide-gauge data analysed, it was predicted that the tsunami would reach the Hawaiian islands 14 hours 56 minutes after its generation off the Chilean coast. The prediction was that the first wave would strike Hilo, Hawaii at 9.57 p.m. It arrived one minute later.

Because tsunamis are less frequent outside the Pacific region, it might be argued that the cost-effectiveness of separate tsunami warning systems for other areas is very low. Until recently, this was the case in Portugal. On 1 November 1755 AD, one of the highest-magnitude earthquakes documented for Europe led to the complete destruction of the city of Lisbon, as well as causing widespread damage and loss of life across a coastal zone stretching from northern Portugal to southern Morocco. In Lisbon alone, some 50 000 people were killed. Most of them were drowned by the resulting tsunami, which reached a height of 17 to 20 m in many areas. In recent decades, two minor tsunamis have struck the coast of Portugal, one on 25 November 1941 and the other on 28 February 1969. The Portuguese authorities, aware of the risk posed by tsunamis, subsequently decided to instal seismometers and wave recorders west of Portugal with the aim of providing advance warning of any future tsunami comparable to the one that struck Lisbon in 1755. In most other areas of the world (apart from the Pacific), however, coastal populations have no protection from tsunamis. They rely instead on the expectation that no tsunamis will ever strike the particular coastlines in which they live.

ALASTAIR G. DAWSON

tufa *see* TRAVERTINE, TUFA

tunnels The earliest recorded tunnels are the *quanats* found in the Middle East and North Africa. These tunnels, some dating from 4000 BC, conducted water from springs in the foothills to communities in the desert. The use of tunnels, some of 60 km length, made it possible to even out gradients and prevented the evaporation that would have taken place in an open watercourse. Many *quanats* operate to this day.

The Romans developed tunnelling further, and many Roman water-supply systems used a combination of aqueducts and tunnels to negotiate uneven terrain at a constant gradient. The Romans also developed tunnelling for military purposes, either by breaking through behind enemy defences or by undermining fortifications to cause their collapse.

Much early tunnelling was carried out in association with mineral mining, either to provide access to the body of ore or to provide drainage as workings became deeper. Excavation was carried out using pick and shovel. Harder rocks were broken down by 'fire-setting', a process in which the rock face was heated by burning faggots against it, and then quenched with water, causing the rock to fracture.

Modern tunnelling developed with the canal age, which required tunnels of relatively large cross-section to pass beneath the watersheds between valley systems. Arched brickwork linings were developed and the use of gunpowder eased excavation in hard ground. Railways created a further demand for tunnelling, and by the mid-nineteenth century the construction of urban sewers, which now form the greatest length of tunnels constructed, began in earnest.

The nineteenth century saw the development of compressed air tunnelling, in which the entire tunnel was pressurized in order to hold back groundwater and support the face. The elder Brunel (Marc Isambard) also developed the tunnelling shield, a steel cylinder from within which excavation could safely take place. These techniques enabled softer ground to be tackled.

Machines to excavate tunnels began to appear in the latter half of the nineteenth century. A machine drove over a mile of a Channel tunnel in the late 1800s before the work was stopped by political problems. Preformed linings for tunnels were first developed using cast iron. Later, concrete segments which could be bolted together to form a circular lining were used to provide support to ground and watertightness.

Modern tunnelling can take place in virtually any ground conditions. Machines 15 m in diameter bore through the hardest rock for two-lane highways. Sealed-face machines tackle soft silts, sands and water pressure beneath river crossings. Remotely controlled machines enable tunnels below man-entry size to be constructed, and at the smaller sizes the boundary between tunnelling and drilling techniques is becoming increasingly blurred. The growth of modern cities has greatly increased the demand for tunnelling for road transport and metro systems and for water, gas, electricity, and telecommunications routes. The high land costs of alternative routes more than offset the additional costs of tunnelling.

Nowadays tunnels are mainly lined with concrete segments or with concrete pumped in between the excavated ground and internal shuttering. Any voids between the lining and the ground are filled by pressure grouting to improve watertightness and ensure that the loading on the lining is even. Excavated material is removed by rail or, in some systems, by mixing the spoil with a bentonite clay slurry, which is then pumped to the surface. After settling, the slurry is reused. Lining segments are brought in by rail and erected using mechanical handling systems. Excavation is carried out either by a cutting head the size of the finished tunnel or by boom cutters which can be operated over the complete tunnel face by the driver. In more variable rock conditions, drill and blast techniques are used for excavations. Steel arches or sprayed concrete keep the tunnel stable until a final lining is put in place.

The Channel Tunnel, with 102 km of main tunnels and 51 km of service tunnels, achieved progress rates of 400 m a week on each tunnel face. Rates will continue to improve as machines and materials are developed further.

HAMISH J. ORR-EWING

type locality, type section A *type locality* is the place where a *type section* or *stratotype* is to be found—the stratotype being the best available exposed section of accessible undisturbed strata of a described rock body. In particular, it is a section that reveals the contacts with formations above and below, and possesses characteristics, including fossils, that are common throughout the geographical extent of the rock unit. In setting up the formal naming, description, and correlation of a rock formation, the geologist will refer to the type section as the principal standard to which he or she can compare and correlate other exposures. Thus the description of the type section has to be as complete as feasible; its location and extent must be accurately fixed; and, if possible, access to it by other geologists must be ensured. Lists of fossils and the beds from which they come are critically important for purposes of biostratigraphic identity and dating.

Where a type section is established in the subsurface region (as, for example, in a borehole) the log of the appropriate part of the hole, together with cored material or drill chips, must be described. Geophysical survey data may also be critical in distinguishing the formation from other rock bodies.

The term *type locality* is also applied to the location from which a particular fossil, rock, mineral, or phenomenon was originally described. It is especially important in the accurate naming of fossils. At type localities for fossils the nature of the enclosing rock is important, not only for biostratigraphic reasons but also for a better understanding of possible palaeoecological factors.

D. L. DINELEY

Further reading

Whittaker, A. *et al.* (1991) *A guide to stratigraphical procedure.* Geological Society Special Report No. 20. Geological Society, London.

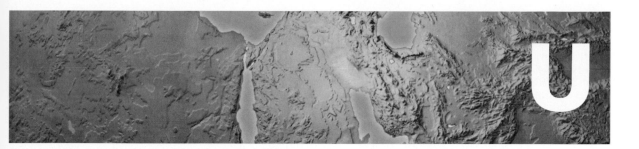

unconformity An unconformity is a large gap in the geological record that is formed when deposition of sediment ceases for a considerable time. The most spectacular and best-known examples are called *angular unconformities* (Fig. 1a): they illustrate the folding and faulting of an older sedimentary record, its planing down by erosion, and the deposition of a younger sequence that truncates the old one. Wherever sedimentary sequences have been studied they are interrupted by such gaps. The lost intervals range in magnitude from extremely long time periods to geologically very short ones. Thus the time gap represented by a given unconformity varies laterally. It commonly decreases away from an angular unconformity towards the main or central part of the sedimentary basin which received the material eroded during the development of the unconformity. In a situation of this kind the angular unconformity passes down dip, that is, basinwards, into a correlative conformity somewhere in the middle part of an uninterrupted sedimentary sequence.

Before discussing the concept of unconformity in more detail it is important to realize that the prime concern of geologists on first entering a new area, in addition to examining and describing the types of rock present, is to establish the geological succession or sequence. The geologist begins by mapping the rock exposures and recording the structural disposition of the rock layers (strata), and their angle of inclination (dip). From observations like this it is usually possible to elucidate the relationships between adjacent layers of rock and their arrangement in vertical sequence, thus giving a first approximation to the three-dimensional arrangement of the succession of rocks.

In addition, the geologist studies the nature of the stratification. For example, some bedding planes (surfaces that separate strata) are caused by vertical changes in sediment type. Many bedding planes, however, represent pauses in sedimentation, possibly accompanied by some local erosion; such breaks are commonly described as diastems or non-sequences. These relatively short interruptions in sedimentation, representing only a brief interval of time with little or no erosion before the resumption of deposition, are, of course, also gaps in the sequence; they may include the down-dip, basinward, conformable equivalents of unconformities. These very short breaks are those that would occur *within* a particular sedimentary environment, unlike those related to

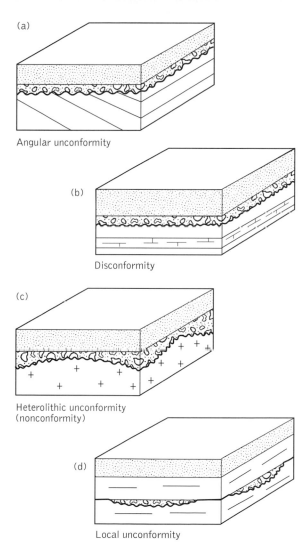

Fig. 1. Types of unconformity: (a) angular unconformity, (b) disconformity, (c) heterolithic unconformity (nonconformity), (d) local unconformity. (Modified from Tomkeieff (1962) and from Bally (1987).)

an unconformity, which may be associated with a major change in environment.

Thus, an early task of the geologist is to distinguish the individual types of rock exposed in an area and to work out their chronological succession: that is, their sequence in time. This elucidation of the stratigraphy begins by separating the succession into its different rock-types or formations; for example, limestone, overlain by shale, overlain in turn by sandstone. This type of stratigraphy is known as rock-stratigraphy or lithostratigraphy. Additional stratigraphical study may work out the distribution of fossils (biostratigraphy) present in the succession so as to facilitate or further refine correlations with strata in adjacent areas.

The present-day concept of stratigraphical sequence was developed from principles first applied in the continental interior of the United States of America using information from surface outcrops. There, rock-stratigraphical units of higher rank than Group (i. e. approximately at System level) are traceable over large areas of the continent and are bounded by unconformities of interregional scope. It is now apparent that the description and mapping of unconformities are the first steps in understanding the regional and super-regional geological history and development of geological provinces and basins.

In addition to the observation of unconformities at outcrop, present-day high-resolution subsurface data now make it possible to recognize unconformities in seismic reflection profiles and also in geophysical well-log data. Most importantly of all, these unconformities (and indeed their correlative conformities) are used as the boundaries of stratigraphical units in the establishment of what is now termed sequence stratigraphy. The methodology of sequence stratigraphy has greatly augmented stratigraphical practice. It is found that such stratigraphical units and their bounding, regional unconformities provide an excellent means of mapping the distribution in time and space of local and regional stratigraphical and tectonic events, and of elucidating and describing the development of geological provinces and basins through time.

The unconformity-bounded units known as sequences are defined formally as relatively conformable successions of genetically related strata bounded at top and bottom by unconformities or their correlative conformities. The sequences are interpreted as depositional sequences which are described from seismic, borehole, and outcrop data. As a depositional sequence is determined by the single objective criterion of the physical relations of the strata themselves, it is useful in establishing a comprehensive stratigraphical framework rather than being primarily dependent on the more subjective criteria used in determination of rock-type, fossils, depositional processes, etc. that vary within a given sequence.

Once the local stratigraphy has been determined and mapped so that geological events have been organized into a reasonable geological history, the next step is to compare and correlate with adjacent areas so that geological understanding of the wider region is facilitated. To achieve this, the rocks of the local area must be equated with those of other areas by correlating laterally equivalent rock-types and establishing their ages in terms of the geological timescale. The time framework is based upon the geological systems (e.g. Carboniferous System, Jurassic System), which, in their type areas at least, are well developed in terms of thickness, etc. Each system lies between older strata below and younger strata above, and the systems are contiguous; that is, they do not overlap in time. Thus any geological system was deposited during a discrete portion of geological time. The boundaries of many of the systems were originally chosen at important unconformities, although as research proceeded and more geological data accumulated it became clear that these gaps could be represented by complete stratigraphical successions in other places. The time equivalents of the geological systems are the geological periods, which constitute the standard of time measurement in the geological sciences.

Returning to the idea of gaps in the sequence, it is possible to define an unconformity as a surface of erosion or non-deposition separating younger strata from older rocks and representing a significant hiatus or gap in the geological succession. The dip-discrepancy type of unconformity now known as an angular unconformity (Fig. 1) was the first type discovered in the late eighteenth century. By the early twentieth century, however, several types of unconformity other than the angular discordance type were being recognized as a result of further accumulation of data. One example is what is now known as a disconformity (Fig. 1b), which is characterized by a break between two sets of parallel strata recognized on a regional scale (that is, there is no local discrepancy in dip). Also recognized is what was originally known as a heterolithic unconformity, which is a contact between two rock-types of wholly unlike origin, such as sedimentary rocks overlying granite. This type of unconformity is now sometimes known as a nonconformity. A fourth major category is the local unconformity, that is, a depositional break on either side of which the beds might be parallel to each other, and where the unconformity appears to be only of local extent or significance. From the above it will be seen that unconformities are classified on the basis of the structural relationship between the underlying and overlying rocks and thus are of tectonic or structural significance as well as of time significance.

It will also be apparent that in an area of low dip or horizontal strata it is possible to project the known surface geology into the deep subsurface only to a limited extent: at most to the top of the next lower (usually concealed), underlying unconformity. In some areas boreholes may provide additional constraining information on the thickness of the topmost (i.e. surface-mapped) conformable sequence, or perhaps even by penetration of an underlying and concealed unconformity. However, the fact that borehole information provides data only for the close proximity of the borehole section itself usually precludes too wide or regional an application of the results unless many boreholes are available in a small area.

Primarily as a result of extensive oil exploration during the past two or three decades, geological science has seen the advent of high-quality seismic reflection profiles, which can be accurately calibrated geologically by downhole geophysical logs. These techniques provide a much wider and more accurate image of the regional concealed deep subsurface geology than hitherto. In addition, the new methods have facilitated a much fuller understanding of unconformities, and especially of the associated genetically related sequences that are present between them.

The expense of modern drilling is such that continuous coring is impractical. The geologist thus turns to downhole geophysical logging to provide the data in addition to cuttings samples that are required for sequence-stratigraphical studies. These so-called 'electric' logs, which are acquired via an instrument called a sonde, measure various physical properties of the subsurface rocks and formations. Many of the log suites acquired are of great use and relevance to the interpretational methods discussed here. For example, the sonic or acoustic log measures the speed of sound in the concealed rock formations and enables detailed geological calibration of seismic reflection data in terms of depth below the surface and, under the right conditions, rock-type. Microresistivity measurements are used to prepare what are known as dipmeter data and allow the determination of structural dip (and thus the location of unconformities in the concealed subsurface) as well as the location of geological faults and the presence of some sedimentary features such as cross-bedding.

Other types of electric log data allow the detailed determination of a 'high-fidelity' record of the rock-stratigraphical section penetrated by the borehole. Furthermore, the log data allow accurate recognition of patterns (sometimes known as electrofacies) which are related to fining-upwards or coarsening-upwards sequences and thus present information of direct use and relevance to sequence interpretation. In fact the geophysical log data make possible the direct recognition of individual unconformities in the uncored borehole section itself (via dipmeter data), or provide evidence of local or regional unconformity by means of correlation and comparison of log-derived stratigraphical and sequence information between individual, widely spaced boreholes.

Perhaps more than anything else it has been the introduction of sophisticated seismic reflection interpretation techniques which has allowed further extension and application of the concept of depositional discontinuities or unconformities between stratigraphical sequences. The seismic reflection technique is rather akin to echo-sounding. Energy, usually from an explosive or vibrating source at the surface, is put into the Earth's crust so that where there is a change in rocktype (known as an acoustic impedance contrast) some of the downgoing waves are reflected back upwards towards recording devices (geophones or hydrophones) at the surface. After processing, the data are presented as seismic reflection profiles

which represent the stratal layers in the Earth's crust. The representation looks rather like a geological cross-section except that the vertical dimension is given in two-way travel time rather then depth. However, by incorporating relevant velocity information into the section it is possible to transform the vertical axis of the seismic section from time to depth. It is important to remember that the vertical resolution of the method is coarse (say, tens of metres), as compared with exposed rock sections (observable by the naked eye down to millimetre thickness) and geophysical log data (resolvable down to decimetre thickness or less).

The objective of seismic sequence analysis is to interpret depositional sequences on seismic sections by identifying discontinuities (or unconformities) using reflection termination patterns. Two types of patterns (onlap and downlap) occur above the discontinuity; three types of patterns (truncation, toplap, and apparent truncation) occur below the discontinuity. Sequence boundaries are characterized by what geologists call regional onlap and truncation. Such depositional sequences (see Fig. 2a) are defined as stratigraphical units composed of relatively conformable successions of genetically related strata bounded at top and bottom by unconformities and their correlative conformities. Seismic reflection data give a clear and unequivocal demonstration that unconformities at basin margins or in mid-continent situations pass basinwards into correlative conformities. The results provide a useful reminder that the onshore regions (geological 'highs') are in fact atypical stratigraphically and are thus not the best places to define type sections.

A depositional sequence is of great interest to geologists because it was deposited during a given interval of geological time limited by the ages of the sequence boundaries where they are conformable, although the age range of the strata within the sequence may differ from place to place where the boundaries are unconformable (see Fig. 2). The hiatus represented by the unconformable part of a sequence boundary generally is variable and may range from about a million years to hundreds of millions of years. In addition, unconformities are of considerable importance in time stratigraphy (chronostratigraphy) because strata above an unconformity are everywhere younger than rocks below it. The conformable part of a sequence boundary is usually thought of as being effectively synchronous because the hiatus is so small as to be virtually immeasurable. It is clear that a depositional sequence may have more significance in geological history and associated stratigraphical studies than a unit bounded only by synchronous surfaces that are chosen arbitrarily. This is because a sequence represents a genetic unit that was deposited during a single episodic event, whereas the arbitrarily chosen unit may span two or more incomplete portions of genetically related units, and may therefore not accurately display the depositional history.

The definition of an unconformity surface as representing a *significant* gap in the geological succession begs the question of the amount of time involved. Angular unconformities

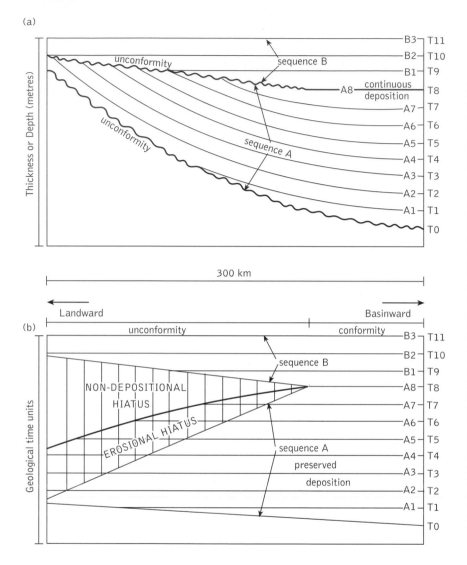

(a)

Thickness or Depth (metres)

unconformity

sequence B

B3 — T11
B2 — T10
B1 — T9

A8 — continuous deposition — T8

A7 — T7
A6 — T6
A5 — T5
A4 — T4
A3 — T3
A2 — T2
A1 — T1
T0

unconformity

sequence A

300 km

Landward   Basinward

(b)

Geological time units

unconformity   conformity

sequence B

NON-DEPOSITIONAL HIATUS

EROSIONAL HIATUS

sequence A

preserved deposition

B3 — T11
B2 — T10
B1 — T9
A8 — T8
A7 — T7
A6 — T6
A5 — T5
A4 — T4
A3 — T3
A2 — T2
A1 — T1
T0

Fig. 2. Basic concepts of a depositional sequence and bounding unconformity. (a) Stratigraphical sequence A rests unconformably (nonconformably) on igneous rocks and is overlain by stratigraphical sequence B. The bounding surface between sequences A and B at the landward (left-hand) side of the diagram is an unconformity that passes basinwards (right-hand side of the diagram) to a correlative conformity at the level marked A8, where sediment deposition has been continuous. Sequence A is truncated by sequence B, which in turn shows an onlapping relationship against sequence A. Where units of strata are missing, hiatuses are evident. (b) Generalized time-stratigraphical (chrono-stratigraphical) section representing the sequence-stratigraphical relations shown in Fig. 2 (a). Missing strata at the unconformity surface are shown by vertical lines and illustrate types of hiatus. Some of sequence A was deposited but later removed by erosion (an erosional hiatus) as the base level changed; the earlier parts of sequence B were not deposited at all in the landward areas (non-depositional hiatus) owing to non-deposition. T_0–T_{11} are time divisions. (Modified from Eicher (1976).)

(Fig. 3) can represent many millions, even hundreds of millions, of years; some estimate is usually possible from outcrop studies using fossil evidence, or from regional geological studies. With regard to smaller time-spans, of great interest and relevance in this connection is the magnitude of a hiatus along a sequence boundary, which is analogous to the length of time taken to deposit a sequence. Estimates of the order of magnitude depend essentially on the vertical resolving capability of the method used. A sequence determined primarily from seismic reflection data may encompass a geological time-span of a few million years. A well-log sequence may be recognized that is much smaller than that determined from seismic data and may encompass a million years to hundreds

of thousands of years. Very small-scale stratification, bedding, or laminae may be used to determine a hiatus of very short duration, but such examples are likely to fall below what is normally considered to be a significant hiatus and therefore not to fall within the definition of an unconformity.

ALFRED WHITTAKER

Further reading

Bally, A. W. (1987) *Atlas of seismic stratigraphy*. American Association of Petroleum Geologists Studies in Geology, No. 27.

Eicher, D. L. (1976) *Geologic time*. Prentice Hall International Inc., London.

Fig. 3. The horizontal Dolomitic Conglomerate (Triassic) rests unconformably on dipping Old Red Sandstone (Devonian) at Portishead, Somerset, England.

Tomkeieff, S. I. (1962) Unconformity—An historical study. *Proceedings of the Geologists Association*, 73, (4), pp. 383–416.

Whittaker, A., Holliday, D. W., and Penn, I. E. (1985) *Geophysical logs in British stratigraphy*. Geological Society of London, Special Report No. 18.

Whittaker, A., Cope, J. C. W., Cowie, J. W., *et al.* (1991) *A guide to stratigraphical procedure*. Geological Society of London, Special Report No. 20.

Whittaker, A. (1998) Borehole Data and Geophysical Log Stratigraphy. In Doyle, P. and Bennett, M. W. (eds) *Unlocking the Stratigraphical Record – Advances in Modern Stratigraphy*, pp. 243–73. John Wiley and Sons, Chichester.

Whittaker, A. (1998) Principles of Seismic Stratigraphy. In Doyle, P. and Bennet, M. W. (eds) *Unlocking the Stratigraphical Record – Advances in Modern Stratigraphy*, pp. 275–98. John Wiley and Sons, Chichester.

This account is published with the approval of the Director, British Geological Survey (NERC).

uniformitarianism *see* CATASTROPHISM AND UNIFORMI-TARIANISM

uplift *see* SUBSIDENCE AND UPLIFT

urban geology Urban geology can be defined as all aspects of the geology of the built environment; it is in essence a branch of environmental geology. Urban growth is a feature of all the industrialized countries, where it is common for up to 80 per cent of the total population to be concentrated in urban centres; and growth rates for Africa and southern Asia are increasing. Given that the Earth's population will in future be concentrated into 'mega-cities', with the attendant problems associated with the concentration of the human race into a small part of the Earth's land surface, an increasing focus for geologists will be the urban environment. Urban geologists have three broad tasks: to manage the provision of mineral, construction, water, and conservation resources; to provide appropriate geological information for construction, engineering, and waste-management projects; and to manage and mitigate the natural and human-induced hazards that threaten an increasingly concentrated and therefore vulnerable urban population.

Geology and planning in the urban environment

The local geology has always affected the way in which urban centres develop. For example, geology has historically had a major role in the siting and development of urban centres in the majority of ancient cities in Europe, Asia, and the Americas, which were originally located to take advantage of some geological or geomorphological feature for defence or commerce. Early cities were also located to make use of abundant natural resources such as water, construction materials, or economic minerals (Fig. 1a). The same became true of the more recently founded North American cities, as the European settlers of the 'New World' found suitable locations for the provision of transport or resources. Urban growth in the late nineteenth and twentieth centuries followed at least three distinctive patterns, mostly associated with the constraints of the geological environment: lateral spreading; building upwards; and utilizing or creating underground space.

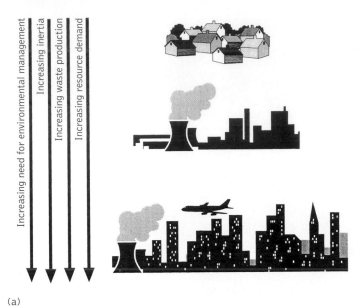

Strong geological influence: location controlled by geology—water, building materials, soils, defensive landforms—building materials and urban fabric reflect local geology.

Strong regional influence: growth of industrial urban areas strongly influenced by geological resources; geology visible in building materials; engineering geology plays an important role in development; it is important to note that urban growth may also be stimulated by non-geological factors such as economic, commercial, and political constraints.

Local geological influence weak; urban inertia strong; regional geology has important role in determining resource supply (drinking water and aggregates) and waste disposal; engineering geology may play an important role in urban redevelopment/reclamation of industrial areas.

(a)

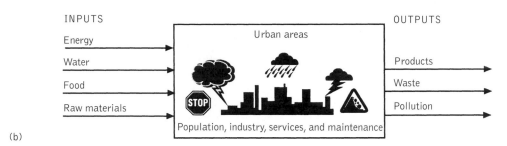

(b)

Fig. 1. (a) A model of the role of geology in urban development. Initial urban settlements are strongly influenced by local geological conditions, an influence that declines through time. Geological control continues, however, through the need for the provision of resources, engineering geology, and waste management. (b) The 'urban machine', which draws inputs derived from the geological environment (energy, water, food from agriculture, and building materials) and produces wastes to be disposed of in the geological environment. (From Bennett, M. R. and Doyle, P. (1996) *Environmental geology*. John Wiley, Chichester.)

Lateral spreading is a common feature of most cities. It can be achieved in many different ways, usually coupled with the development of transport links. Geological constraints on lateral growth are commonly associated with the geomorphology and engineering properties of the region; the initial foundation of the city will usually be in the optimum area for development, and further development will be on flood plains, slopes, coastal regions, and so on, which can often increase the impact of natural hazards (Fig. 1a). Lateral growth may also be constrained by proximity to resources, because quarry or mine activity may become included within city boundaries, thereby sterilizing future exploitation of a resource. The pressure for this type of growth often threatens belts of agricultural, recreational, and conservation import-ance that are commonly maintained around cities. The use of old industrial sites and other contaminated 'brown-land' areas can relieve pressure on such resources ('green land'), although problems can arise from such activities as mining. Transport corridors are also an important factor in the lateral growth of large towns and cities. The building of extensive road and rail systems, often with the need for adequate foundation works and for adequate hazard protection, is a major undertaking.

Building upwards, thereby creating 'cities in the sky', which was a utopian vision for many architects and town planners of the middle part of the twentieth century, is largely dependent on geological factors. The construction of tall buildings and their safety depend of the nature of the foundation works, which are themselves ultimately controlled by the bedrock and soil conditions.

The utilization of underground space has always been a consideration in urban areas, particularly in the development of underground infrastructure and transport, and this is obviously subject to geological control. Many ancient cities have extensive underground tunnels that have at times been used for habitation, and most modern cities have an extensive network of underground infrastructure, railways, etc., which require adequate geological input. There has been much interest in a number of North American cities in the use of underground space, especially in old mining regions, for industrial, commercial, and even educational uses.

The urban machine

The concentration of people into urban centres produces a kind of machine that can strain the geological resources of the environment (Fig. 1b). This 'urban machine' consumes inputs, which are mostly derived from the natural and geological environment. These inputs are water, locally derived from springs and aquifers, or as part of a reservoir and transfer system; raw materials, in the form of mineral resources for industry and construction, traditionally locally derived, but now, with the development of transport links, often obtained globally; food resources, derived from farming of agricultural lands, but also from effective transport links with other countries; and energy, derived from the use of the Earth's physical resources. The machine produces outputs, in the form of products of industry and commerce created from the raw materials imported into the city; waste, in the form of worn-out materials, by-products from industry, and day-to-day wastes from domestic, commercial, and industrial sources; and pollution, caused when poor waste management overloads the ability of natural atmospheric, land, and water systems to recycle and redistribute the waste gases, solids, and liquids produced by industry, commerce, and domestic activities. Like any machine, the urban system needs regular maintenance, particularly in the upgrading of its buildings and infrastructure of utilities and transport. All aspects of the efficient running of the urban machine rely in some way on the nature of the urban geological environment (Fig. 1b).

Urban geology and the urban machine

The efficient running of the urban machine depends on strategic planning, both in terms of recognizing opportunities for development and in identifying constraints. Geologically speaking, typical opportunities reflect the supply of inputs to the system and low maintenance costs; they would include favourable locations and ground conditions; the presence of natural resources, such as water and minerals; and the absence of significant hazards, whether natural or human-induced (Fig. 1b). Most of the constraints that limit the success of the city and the level and extent of its output are a function of the same set of factors: unfavourable locations and ground conditions; the absence of natural resources; and the presence or growth of

significant hazards. Urban geological planning therefore needs to recognize the balance between these opportunities and constraints in the development of a town or city. Most of them are a function of the human interaction with the geological environment. There are four aspects to this interaction: the provision of mineral resources, including water for industry and construction; the assessment of ground conditions for transport links, urban expansion, urban redevelopment, or the creation of new urban areas; the disposal and management of the waste products generated by urban and industrial areas and their utilities, such as power stations; and the identification of threats from natural hazards and their mitigation.

Essential to effective planning is the availability of accurate, understandable, and pertinent data on urban geology that decision-makers can use in making informed choices. The provision of geological information as a means of identifying both the opportunities and the constraints for development is an essential part of urban planning. Unfortunately, geological information is an underused resource in planning, despite the fact that the majority of cities owe their historical development to geological factors.

The planning process is complex, but an effective urban strategy will involve land-use planning to guide development and appropriate use of land areas; the identification of specific uses for designated sites; and the regulation and granting of permissions for development. In all these stages it is essential to be able to make informed judgements; the provision of appropriate geological information about the urban environment is therefore essential. Such information can help in avoiding costly remedial work to foundations and other calamities. Information is typically needed on opportunities for development in the form of the nature and extent of mineral, soil, water, and undeveloped land resources; the nature of constraints on development, such as the assessment of contaminated land, geotechnical information about ground conditions, and the nature of the hydrology; and sources of expert guidance where necessary. This will be achieved by the collection of data from desk surveys, field surveys and monitoring, terrain analysis, and the presentation of these data in a form that is accessible to all, usually in a set of thematic maps displaying the geology, geomorphology, resources, hazards, and so on.

In considering new urban areas, planners need to realize that a given land area will not simply be a 'blank piece of paper' on which they can 'draw' their town or city. Instead, the landscape is a function of numerous geological factors, including the nature and form of the bedrock and the processes that form its land surface and hydrogeology. The role of the urban geologist is to give advice on the nature of these conditions and to consider the impacts of prevailing conditions in the future, particularly with regard to hazard mitigation and the effects of urban growth. In sum, it is important to identify those elements in the geological environment that provide opportunities for, and constraints to, urban development.

PETER DOYLE

1073

Further reading

Bennett, M. R. and Doyle, P. (1996) *Environmental geology. Geology and the human environment*. John Wiley and Sons, Chichester.

McCall, G. J. H., De Mulder, E. F. J., and Marker, B. R. (eds) (1996) *Urban geoscience* A. A. Balkema, Rotterdam.

urban stone decay and conservation

Urban stone decay concerns many professions, each with different views on what causes damage and what should be done about it. What is undisputed, however, is that stone decay exacts enormous costs through structural damage, loss of cultural heritage, cleaning, replacement, and conservation. It is also no longer restricted to industrialized Europe and North America, but is increasing in global significance as urbanization spreads.

To succeed in preventing or treating stone decay the processes responsible must be understood. Failure to do this can lead to inappropriate strategies which exacerbate original problems or trigger new decay. In this context, the first point to note is that building stone is subject to the same weathering processes as natural outcrops (see *weathering*). However, placing stone in a building exposes it to additional influences. Some, such as surface polishing, may retard weathering, but overwhelmingly, urban exposure accelerates decay. Sometimes this decay takes place through new processes associated with, for example, specific pollutants, but anthropogenic processes can also combine with and reinforce natural (climatic) weathering. This additional complexity is traditionally ascribed to the effects of atmospheric pollution, particularly 'acid rain'. Other factors include the physical loading of stone, artificial microclimates created on complex building façades, and treatments carried out in the name of cleaning, weatherproofing, and conservation.

The most obvious decay process is solution—although this is primarily restricted to limestones or stones with a calcium carbonate cement. Accelerated solution is principally associated with the conversion of atmospheric sulphur dioxide, from the burning of fossil fuels, to sulphuric acid in rainfall. This dissolves calcium carbonate by converting it to more soluble calcium sulphate (gypsum). However, this process is additional to natural or 'karstic' solution by rainfall. Sulphation also occurs on surfaces protected from rainfall through reactions between moist stonework and atmospheric gases and through concealed deposition from dew, mist, fog, and frost. In sheltered areas, gypsum can crystallize on stone surfaces and, by incorporating flyash and other windborne debris, forms 'black crusts' and growths. On rainwashed surfaces gypsum is either completely removed in surface run-off, washed into the stone, or washed over and into adjoining stone and brickwork.

It is this incorporation of a salt (calcium sulphate) into stonework that is largely responsible for urban stone decay. Salt within pores and fractures exerts expansive forces through crystallization, differential thermal expansion, and hydration. Hydration and dehydration, in particular, are controlled by subtle changes in temperature and humidity, and salt weathering exploits a wide range of climatic conditions to be almost continuously at work in urban environments. Salt weathering also affects non-calcareous stone through salts formed from lime-based mortars, material washed or blown in from adjacent stone, and flyash containing gypsum. In addition, salt comes from the upward movement of salt-rich groundwater, road de-icing, and marine aerosols. Thus, although gypsum is the dominant urban salt, others, especially halite (sodium chloride), are present and salt weathering does not have to be associated with human intervention.

Salt accumulation near a stone surface can produce granular disintegration, but accumulation at depth causes blistering and scaling or multiple flaking. Salt also tends to concentrate in and enlarge any depression protected from rainwash. In this way the stone becomes honeycombed and eventually larger, cavernous hollows can form. Increased salinity also enhances the solubility of quartz and can further encourage the disaggregation of many sandstones. Other chemical changes in sandstones include the mobilization of iron to form hard outer skins. If these skins are breached, the weakened stone behind can rapidly be hollowed out.

Solution, disaggregation, and the visual impairment of buildings are also caused by lichen, fungi, algae, and bacteria which colonize surface and subsurface layers of stonework. Other damage results directly from human actions, from metal fixings that corrode or react with stone, and from inappropriate combinations of different types of stone. Ironically, damage can also ensue from procedures, such as cleaning, designed to mitigate decay.

Stone cleaning can be carried out by abrasives, water jets or mist sprays, solvents (especially acids) applied either as washes or as poultices, physical removal of crusts, herbicides, and the vaporization of grime by lasers. Damage from these treatments includes excessive surface loss where abrasives are harder than the stone, washing of salts into stonework, relocation and activation of salts already present, mobilization of iron leading to discoloration, weakening of cements, exposure of weakened stone behind outer crusts, destruction of original patinas, and biological colonization of cleaned stone.

Once cleaned, stone can be protected and stabilized by a wide range of consolidants, but if wrongly applied they may trap moisture and salts by preventing the stone from breathing and can lead to discoloration. Most conservation is carried out following the enactment of clean-air legislation. However, even after pollution has been reduced, stone can continue to decay because of inherited contaminants and stresses—the so-called 'memory effect'. Some inherited effects are created before stone is put in place, by processes such as 'curing' at quarries and by different cutting and finishing techniques. After construction, stone can also store up legacies of salt and other pollutants, weaknesses caused by previous weathering, and the effects of earlier cleaning and conservation.

Cleaning and conservation must clearly be matched to individual buildings, stone types, environments, and pollution histories. Whether or not to clean is not, however, a purely scientific decision. Much cleaning is for aesthetic reasons, but there are equally cogent arguments for allowing buildings to reflect their history. Cleaning can never recreate the original surface conditions of a stone; 'memory effects' mean that replacements will never perfectly match adjacent stonework, and many buildings were designed specifically to mellow with age. It is because of these problems that cleaning is so controversial.

BERNARD J. SMITH

Further reading

Amoroso, G. G. and Fassina, V. (1983) *Stone decay and conservation*. Elsevier, Amsterdam.

Ashurst, N. (1994) *Cleaning historic buildings*, Vol 1 and 2. Donhead Publishing, London.

Schaffer, R. J. (1991). *The weathering of natural building stones*. HMSO London. (Reprinted from 1932 original and available through the Buildings Research Establishment, Garston, Watford, WD2 7JR.)

Smith, B. J. and Warke, P. A. (eds) (1996) *Processes of urban stone decay*. Donhead Publishing, London.

Urey, Harold Clayton (1893–1981) H. C. Urey is
best known for the discovery of heavy hydrogen (deuterium), for his part in the development of the atomic bomb, and for his fundamental contribution to a widely accepted theory of the origin of the Earth and planets. He was awarded the Nobel Prize for Chemistry in 1934 for his work on deuterium.

He entered the University of Montana to study Latin but soon switched to chemistry and later held appointments in several American universities, during which time (1923–4) he did research on the theory of atomic structure with Niels Bohr in Copenhagen. He was professor at Columbia from 1929 until 1945. His work there on heavy isotopes led to his appointment in 1940 as Director of the Columbia Research Project , which was part of the Manhattan Project. The particular contribution of his group was in the separation of fissionable uranium-235 from uranium-238. From 1945 until 1958 he was Professor of Nuclear Studies and, later, Professor of Chemistry in the University of Chicago. Following up a suggestion that the atmosphere of the early Earth might have been rich in ammonia, methane, and nitrogen, one of his team at Chicago irradiated such a mixture, passed an electric discharge through it, and produced amino acids. This experiment opened up the possibility that life on Earth might have had an inorganic origin. It was also at Chicago that studies of the uptake of stable isotopes of oxygen by shells showed that temperature is an important factor and led to the use of the $^{18}O/^{16}O$ ratios of fossil shells as indicators of palaeo-temperatures.

Studies of the relative abundances of elements in the Earth's crust led Urey to propose that the Earth and other planets were not torn off the Sun as molten balls but originated as cold accretions of gas and dust from a disc rotating round an already formed Sun. This theory was encapsulated in his book *The planets: their origin and development*, published in 1952.

From 1958 until his death Urey was Professor Emeritus at the University of California at San Diego.

R. BRADSHAW

USA, western *see* GEOLOGICAL EXPLORATION OF THE NORTH AMERICAN WEST

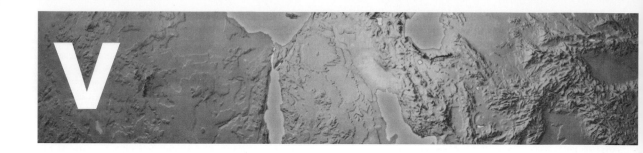

valleys A true valley is a linear depression on the surface of a planet. It is generally longer than it is wide and its floor slopes downwards towards a junction either with another valley or a plain. On Earth, valleys occur both on dry land and beneath the sea; there are also valleys on the Moon, Mars, and Venus. One of the most striking features of valleys is that they occur in sets, or networks. To a certain extent, the nature of the network gives an indication of its origin. Those networks which are basically created by the forces within the crust of a planet that are linked with earthquakes and volcanoes (tectonics) tend to have very straight sections, joining at right angles. The main valley is also commonly at a higher elevation than its tributaries. An excellent example of this is the Mid-Atlantic Ridge on Earth (part of a 60 000 km valley system which circles the globe at the 'constructive' edges of the principal lithospheric plates, where new sea floor is being created): the central valley and its tributaries look rather like an irregular zip fastener on top of a bulge. In contrast, the valley networks which are due to some substance (usually, but not always, water) flowing over the surface have a generally tree-like structure: a large number of small, shallow depressions join together, at angles of 90° or less, to create fewer, larger and deeper valleys and, ultimately, a major—or 'trunk'—valley, which is at a lower elevation than its tributaries.

In most instances valleys are probably formed by a mixture of tectonic and surface processes. That is particularly likely to be the case where the depressions are old. There is evidence from Australia that some valley networks are about 500 million years old (or one-ninth as old as the Earth itself), and the great martian valleys are almost certainly older. On the other hand, new valleys may form on Earth literally overnight as a result of earthquakes, the collapse of ground over caves and tunnels, the release of flood waters from a lake, or just very heavy rain falling on bare ground. Whatever the original cause, any linear depression is likely to deepen over time and act as a route that channels the flow of liquids and solid debris. This tendency for valleys to persist on planetary surfaces and to be used as lines of movement by wind, water, ice, and sediment has made it impossible to develop simple theories about how these networks form. However, since, on Earth, most valleys contain streams, the question of the relationship between valleys and streams has been a matter of tremendous interest for the past 250 years.

In the late eighteenth century, when the modern study of the Earth's surface began in earnest in Europe and North America, the reports of scientific travellers showed that the great mountain belts of the world enclosed large, long valleys. Many of these mountain-and-valley system—particularly the Andes—ran approximately parallel to sea coasts and contained active volcanoes. As Charles Darwin experienced when he was in Chile in 1835, violent earthquakes caused the whole mountain range to move upwards (or downwards). In cases like this, the valley networks seemed to be principally related to the mountain-building forces, and their flat floors were often thought to be caused by sea water draining away as the land was shunted upwards by successive earthquakes. The fact that a large number of valleys contained no active rivers seemed to support the idea that running fresh water simply 'colonized' the network from time to time.

While all the processes described above are undoubtedly important, the modern view is that most tree-like valley networks, including those that extend under the sea and are known as submarine canyons, are primarily the work of the flow of material over thousands or millions of years. Flow can vary from time to time and from place to place. In high mountains of Earth, glacier ice has been a key feature of valley development over the past two or three million years. Deep valleys have been detected under the great Antarctic ice cap: the ice flows much faster along them than over the intervening higher ground. Below the ocean, great slurries of sediment flow down the canyon systems to the abyssal plains of the ocean floors. In deserts, and on Mars, fine sand and silt are funnelled down valleys by wind and may even carve new systems of ridges and depressions in solid rock. But certainly on Earth, and possibly on Mars, it is liquid fresh water which has had the chief role in turning the ground surface into an orderly hierarchy of valleys (or drainage basins). Some of those valleys may be oriented in directions that betray the existence of a major fault trace (or merely the very minor fracture we call a joint) in the underlying rocks. Parts of some networks—dry valleys—do not contain flowing streams: this may be because the climate has changed since the valley was cut (much of the Earth was cooler and wetter during the glacial advances of the past two million years); because the water has drained into underground caverns as in limestone terrains; or because human beings have cut down trees or piped water to their settlements. Alternatively, the size of the valleys may seem far too large for the tiny streams that flow in them—these 'misfits' may also have been produced by the changes that, carried to extremes, will produce dry valleys.

All in all, the planetary surfaces we know have an enormous number of valley systems. Many of these systems are very old and, we think, have experienced a great range of processes. The valley networks on Earth may show traces of all those formative agencies, but the key processes at work to modify today's features are tectonics and running fresh water.

BARBARA A. KENNEDY

Further reading

Darwin, C. (1845) *The voyage of the* Beagle (Everyman Edition, 1959). Dent, London.

Greeley, R. (1987) *Planetary landscapes.* Allen and Unwin, Hemel Hempstead.

Schumm, S. A. (1977) *The fluvial system.* John Wiley and Sons, Chichester.

Variscan orogeny The Variscan is a late Devonian (Frasnian) to late Carboniferous (Stephanian) orogeny caused by the collision of Gondwana and associated microcontinents with Laurussia during the gradual assembly of the supercontinent Pangaea. Variscan metamorphic and intrusive complexes comprise the basement of much of central and western Europe (Fig. 1). These rocks do not, however, crop out as a continuous mountain range but in isolated inliers, many of which were uplifted during the Tertiary in response to Alpine movements. The term 'Hercynian' is used within the romance languages as a synonym for 'Variscan', but may cause confusion because German literature describes the post-Variscan fault trend as 'herzynische.'

The eastern Variscides are classically divided into three zones (Fig. 1). A northern Rheno-Hercynian zone, of low metamorphic grade, with mainly Devonian sediments and few Permian granites, is bounded to the north by north-verging thrusts that override Stephanian coal-bearing foreland basins. The central Saxothuringian zone was a source of the Late Devonian to Early Carboniferous flysch of the Rheno-Hercynian zone. A high-grade gneissic core is thrust northwards over lower-grade lower Palaeozoic sediments. Granites are common. The southern Moldanubian zone represents the internal zone of the orogen comprising high-grade metamorphic rocks, commonly Precambrian, and abundant (80 per cent) late Devonian to late Carboniferous granites and migmatites. However, low-grade Lower Palaeozoic sediments occur locally. In the south of this zone structures *verge* southward; that is, the folds are inclined or overturned in a southerly direction. Further south, basement inliers within the Alpine nappes and the Pyrenees contain Cambrian to Devonian sediments affected by pre-Stephanian metamorphism with syn- and post-tectonic granites. The south Por-

tuguese, Ossa–Morena, and Central Iberian zones are broadly correlated with the Rheno-Hercynian, Saxothuringian, and Moldanubian zones respectively. An important feature of all the Variscan zones is that they have a basement deformed during the Pan-African Cadomian orogeny (600–550 Ma) which is typical of Gondwana. The Saxothuringian and Moldanubian zones also contain Ordovician faunas and glacial deposits typical of Gondwana.

Recent studies have refined this zonal division. Major shear zones containing basic and ultrabasic igneous rocks are interpreted as suture zones (Fig. 1) in which subduction of intervening oceanic crust has produced high-pressure eclogite-facies metamorphism. The Devonian Lizard–Rhenish–Ossa–Morena suture separates the Rheno-Hercynian zone, attributed to the Avalonian microcontinent, and the Saxothuringian zone. The Münchberg–Tepla suture, containing Cambro–Ordovician oceanic basalts that record Siluro-Devonian subduction, marks the southern limit of the Saxothuringian zone. Structures in both of these sutures *verge* to the north, and so are related to a northern or Rheic ocean. The Moldanubian zone is divided by the Massif Central–Coimbra–Cordoba suture. This zone is typified by Cambro-Ordovician rift-related volcanism and Ordovician to Devonian subduction attributed to a southern proto-Tethys ocean separating the Armorican– Barrandian microcontinents to the north from Gondwana. The geological similarities of terranes on either side of this suture indicate that this ocean was not very wide.

Deep seismic sections of the Variscan confirm that the sense of vergence changes across the orogen (Fig. 1). This bilateral thrust symmetry is further complicated by the arcuate trend of all the major structures attributed to the collision of 'indenters', rigid blocks whose margins are marked by zones of strike-slip. Much of western Europe is typified by dextral strike-slip marking the collision of the Iberian and north Brittany blocks, whereas sinistral strike-slip occurs on the western margin of the Iberian block (Fig. 1).

The tectonic evolution of the Variscan orogeny can be summarized as follows. Cambro-Ordovician rifting of Gondwana led to the formation of microcontinents. Avalonia drifted rapidly northwards, opening the Rheic ocean until it collided with Laurentia in the Silurian during the Caledonian orogeny. Armorica–Barrandia failed to migrate far and was separated from Gondwana by a narrow proto-Tethys ocean. Both oceans were closed by Silurian and Devonian subduction producing high-pressure metamorphic rocks. Continental collisions with associated flysch sedimentation began by the late Devonian. Final closure took place in early Carboniferous times with the Gondwana margin indenting the orogen. The opposed structural vergences (Fig. 1) suggest that subduction was directed beneath both margins of the central Armorican-Barrandian microcontinents. When subduction ceased, the down-going oceanic slabs became detached, leading to the emplacement of basic magma in the orogenic core. This was associated with high-temperature metamorphism, which produced migmatites

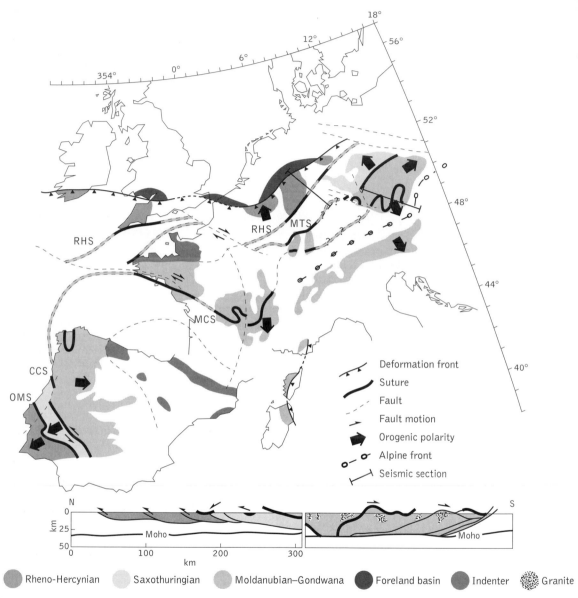

Fig. 1. The Variscan orogeny: map and seismic section. CCS, Coimbra–Cordoba suture; MCS, Massif Central suture; MTS, Münchberg–Tepla suture; OMS, Ossa–Morena suture; RHS, Rheno–Hercynian suture.

and S-type granites of sedimentary origin. The resultant topographic uplift of the orogenic core caused the deformation to spread north and south to produce thrust fronts and associated foreland basins by the late Carboniferous. Collapse of the rapidly spreading hot, weak orogen was followed by the formation of early Permian basins.   PAUL D. RYAN

Further reading

Matte, P. (1991) Accretionary history and crustal evolution of the Variscan Belt in Europe. *Tectonophysics*, **196**, 309–37.

Franke, W. (1989) Tectonostratigraphic units in the Variscan belt of central Europe. *Geological Society of America Special Paper* **230**, 67–90.

Windley, B. F. (1995) *The evolving continents*, pp. 168–86. John Wiley and Sons, New York.

varves Varves represent rhythmic accumulations of fine sands, silts, or clays that are deposited annually in couplets. The relatively coarser-grained layers alternate with layers of finer-grained sediments. Deposition of varved sediment is always associated with cycles of annual deposition, usually (although not always) within ice-marginal lakes. Because varves are deposited annually, they have been used in Quaternary studies as a means of relative dating.

The annual deposition of varved sediment is believed to take place as a result of two distinct seasonal depositional processes. During winter, when the lake is covered in ice, the finest-grained sediment is deposited out of suspension on to the floor of the lake. During spring and summer, when there is no surface cover of ice and when glacial meltwaters enter the lake, coarser material is deposited on the lake floor, and this forms the annual sediment couplet. The use of varves as a dating technique was pioneered by the Swedish scientist de Geer, who first investigated late- and post-glacial Swedish varve deposits in the late nineteenth century. De Geer was the first scientist to present detailed measurements of annual varve sequences, and he was also the first to demonstrate spatial correlation between different varve sequences. He established a varve chronology for Scandinavia that extends back to approximately 13 000 years before the present.

The deposition of varves need not always take place in ice-marginal lakes. In some areas, chemical precipitation of sediments can take place on the floors of lakes on an annual basis. In other lakes, seasonal changes in the deposition of organic detritus can result in the production of organic varves. Similarly, seasonal variations in the deposition of certain microfossils (e.g. diatoms) may result in the production of annually laminated sediments. Annual varves are also deposited in marine environment, although here the processes of sedimentation are complicated by clay flocculation.

ALASTAIR G. DAWSON

Further reading

De Geer, G. (1912) *Geochronology of the last 12,000 years.* XI International Geological Congress, Stockholm, **1**, 241–53.

Lowe, J. J. and Walker, M. J. C. (1984) *Reconstructing Quaternary environments*. Longman, Harlow.

vegetation and climatic change Climate is one of the dominant controls on vegetation; this domination is clearly illustrated by comparing maps of global climate with the world's major vegetation communities. Ten major vegetation communities, or biomes, are recognized on the terrestrial

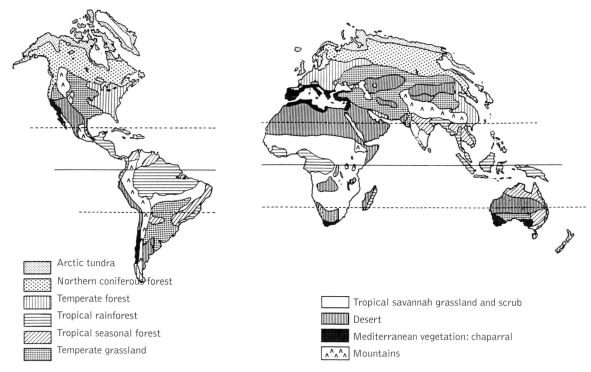

Arctic tundra
Northern coniferous forest
Temperate forest
Tropical rainforest
Tropical seasonal forest
Temperate grassland

Tropical savannah grassland and scrub
Desert
Mediterranean vegetation: chaparral
Mountains

Fig. 1. Distribution of the world's major terrestrial biomes. (Source: Cox, C. B. and Moore, P. D. (1985), pp. 54–5.)

surface of the Earth: Arctic tundra; northern coniferous forest (taiga); temperate forest; tropical rainforest; tropical seasonal forest; temperate grassland; tropical savannah grassland and scrub; and chapparral, which corresponds to those regions of the world that possess a Mediterranean climate; and mountains (Fig. 1). The distribution of these major biomes correlates extremely well with global patterns of climate based on a combination of factors such as temperature, precipitation, and the intensity of isolation. Hence it is concluded that the two are interrelated in some way.

This interrelationship can be shown by looking at the present geographical range of plant communities and their physiological variables, such as the length of the growing season, temperature range, and precipitation. In the late 1960s and early 1970s, Reid Bryson and co-workers demonstrated the close association between biomes and weather systems on a subcontinental scale. Their investigations revealed that there was a good statistical correlation between the mean frontal positions of meso-scale airmass boundaries and the modern distribution of biotic communities. In northern Canada, the position of the boreal forest coincides with the area over which continental and Arctic air masses dominate in winter and over which the Pacific and tropical airmasses are present during the summer months. Bryson found that similar patterns exist for the areas of the North American continent occupied by tundra, the prairies, the eastern deciduous woodlands, and the south-eastern evergreen forests, all of which appear to be strongly associated with different and areally well-defined air masses. Thus, the correlation between climatic variables such as air masses and vegetation communities occurs on both a subcontinental and a regional scale.

The clear correlation between climate and vegetation becomes more obscure at a local scale. At this scale, the distribution of plant communities and individual plant species can be controlled by a variety of environmental factors, which include non-climate variables such as soil type, topography, and factors related to the soil, as well at some climatic variables like precipitation and temperature. Many ecologists, such as Margaret Davis and John Birks, argue that individual plant species respond differently at a finer scale of resolution to non-climatic and climatic variables, and the effects of both combine to structure vegetational communities. However, temperature, rainfall, evaporation, water, and sunlight can be the dominant controls in explaining the geographical distribution of a plant, and this can also be demonstrated at a species level. For example, the grey hair grass (*Corynephorus canescens*) occurs throughout central and southern Europe, reaching its northern limit in the British Isles and southern Scandinavia. This northern limit corresponds to the 15 °C July mean isotherm. Marshall suggests that climatic factors restrict the distribution of the grey hair grass to the south of this isotherm because the grass cannot germinate or flower in areas affected by low temperatures. The study of modern ecology and environmental tolerances of plant communities

and plant species enables ecologists and biogeographers to determine how far climate can influence geographical distribution.

Studying the development and distribution of past vegetation has proved valuable for understanding the relationship between climatic change and vegetation. Pollen analysis has been a productive technique for reconstructing past vegetational communities because every species of plant produces a morphologically distinctive pollen grain or spore. Once released by their host plant, pollen grains and spores can be transported over wide areas and are ultimately deposited and preserved in a variety of sediments. Sediments such as peat and lake deposits that are acidic and anaerobic preserve pollen and spores particularly well. By employing pollen analysis, scientists have been able to reconstruct past vegetational communities and analyse how those vegetation patterns have changed over time. Fossil pollen records have, however, been used extensively as a source of information for inferring past changes in climate. If a circular argument is to be avoided in assessing the role of climate in determining the structure of plant communities, it is therefore necessary to have a method of establishing the nature of past climates that is independent of the pollen record. Fortunately, we can also gain an idea of temperature change during the recent geological past by using a range of proxy sources (e.g. oxygen and deuterium isotope records from ice cores) to reconstruct past temperature curves. Another source of proxy data is that of fossil Coleopteran (beetle) remains, which, like pollen grains, are also preserved in sediments. Many species of Coleoptera have a very specific range of tolerance to temperature. Their occurrence and preservation in sediments deposited during the recent geological past thus provides an indication of the temperature conditions when the sediment was deposited. By looking at both the pollen and Coleoptera records, the relationship between past climate change and vegetational history can thus be established. A chronology of vegetation history and climate change has been established by radiocarbon dating of the sediments in which the pollen grains and Coleoptera are preserved.

One of the most valuable ways of studying the relationship between vegetation and climate change is to analyse how vegetation patterns develop as the Earth moves from a period of glaciation (a glacial period) into warmer deglaciated conditions (an interglacial period). The last glacial–interglacial transition occurred between 14 000 and 9000 years before the present (BP), when the last glacial period ended and the present interglacial, known as the Holocene, commenced. As part of an international project, known as IGCP–253, a sub-project called the North Atlantic Seaboard Programme (NASP) was initiated to investigate the environmental changes that accompanied the termination of the last glaciation in north-west Europe. The preservation of deposits that contain microfossils such as pollen during this transition has provided a unique opportunity of studying how vegetation responded to what can be described as a relatively abrupt change in climate. Research in the UK, primarily by Professor Russell Coope of Birmingham University, has provided temperature curves, based on Coleoptera records, for the last glacial–interglacial transition. This makes it possible to compare the vegetation record, as depicted by pollen analysis, with a proxy climate record. A brief summary of the climate and vegetation changes during this time-period given below illustrates the strong interrelationships between them (see Fig. 2).

Between 14 000 and 13 000 years BP, glacial conditions prevailed in the British Isles. Vegetation was dominated by open ground communities containing Gramineae (grasses), *Rumex* (docks) and *Artemisia* (mugworts). The climate appears to have improved rapidly after this in the British Isles, and a period known as the Late Glacial Interstadial occurred between approximately 13 000 and 11 000 years BP. Initially, between 13 000 and 12 000 years BP, July mean sea-level temperature averaged 18 °C, before gradually falling to below 12 °C by 11 000 years BP. The amelioriation in climate allowed a scrub vegetation to develop, and this was followed by woodland in many areas of England, Wales, and Scotland. The scrub vegetation appears to have been characterized by pioneer trees such as *Betula* (birch) and *Juniperus* (juniper). In southern areas of the British Isles, conditions were suitable for the development of woodlands containing *Pinus* (pine) by about 11 500 years BP. The transition from glacial to interglacial conditions does not, however, appear to have been smooth. Every so often evidence for a sudden deterioration in climate can be identified in the fossil record. These sudden reversals in climatic conditions are known as 'revertance' episodes. One such revertance episode seems to have taken place throughout large areas of north-west Europe, including the British Isles, between approximately 11 000 and 10 000 years BP (the Younger Dryas). These revertance episodes were accompanied by a change in vegetation. During the Younger Dryas in the British Isles, vegetation was characterized by a cold-loving tundra and low alpine scrub, typically comprising species such as Gramineae, Cyperaceae (sedges), Caryophyllaceae, Cruciferae, *Rumex*, and, in upland areas, *Selaginella* and *Lyocopodium* (both clubmosses). These correspond with a phase of much lower temperatures; the mean July temperature between 11 000 and 10 000 years BP is estimated to have been below 10 °C. The Holocene period began at approximately 10 000 years BP, and it is marked a vegetational succession from tundra and low alpine scrub to closed mixed deciduous woodland. This vegetational succession correlates closely with the development of warmer climatic conditions. First, the open ground communities are gradually replaced by *Betula* and *Juniper*, which are thought to have formed an open woodland. This is then followed by the arrival, in sequence, of the tree species that characterize a mixed deciduous woodland. By about 9000 years BP, *Pinus* and *Corylus* (hazel) became established in the British Isles, closely followed by *Quercus* (oak) and *Ulmus* (elm). A more diverse woodland community developed as climate warmed, allowing more thermophilious taxa such as *Alnus* (alder), *Tilia* (lime), and *Fraxinus* (ash) to establish themselves between 8500 and 6000 years BP. By then, climatic conditions in the British Isles are thought to have reached their optimum, highlighted by the mean July temperature which had risen to over 16 °C by 9000 years BP. Thus, by using pollen and Coleoptera data a proxy record of vegetation and climate change can be inferred, and these data suggest that there was a strong degree of correlation between climatic change and vegetational development in the past.

One of the main controversies that exists in the study of the relationship between vegetation and climate concerns the degree to which plant communities are in equilibrium with the prevailing climatic conditions, and how quickly vegetation responds once a change in climate takes place. In theory, if the vegetation response rate occurs relatively slowly once climate has changed, it is plausible that for a certain length of time the relationship between climate and vegetation is out of equilibrium (or is in disequilibrium). The time it takes for the vegetation to readjust to new climatic conditions is known as the lag time. Margaret Davis suggests that the differential response time of plants to a climatic change that results in a lag time results from inherent differences in life-history characteristics, such as the lifespan of individual plants, the dispersal capabilities of a plant, and its intrinsic rate of population increase. Thompson Webb III and others have studied in detail the response time of vegetation to climate changes. Webb suggests that the rate at which vegetation responds to a change in climate varies over different timescales. The time required for vegetation to adjust and then reach a new equilibrium in response to a climatic change on a timescale of 10^4 years is so far unknown. On much shorter timescales of 500 to 1000 years duration, however, vegetation responses or lag times appear to be relatively short.

From our understanding of vegetation and climate interactions, it would appear that climate is the major driving force behind colonization and establishment, or removal, of vegetation within any area. However, on a local scale, site conditions must also be suitable if a plant is to establish itself. Physical factors such as soil development are important; the soil must provide nutrients in order for a plant to survive. Other factors, including parent material, topography, drainage, and the amount of disturbance, can contribute to the fertility of a soil and its suitability for plant growth. Furthermore, geomorphological changes can also influence the structure of plant communities. Suitable climatic conditions do not automatically mean that a plant will colonize and establish itself within an area. Natural barriers, such as a mountain range or an ocean, can prevent the dispersal and spread of a plant. Sometimes the distribution of a plant is dependent on a chance event or it is advertently or inadvertently modified by human activity. Biological interactions, such as competition or disease, can play a role in structuring plant communities. Thus, it is important to remember that vegetation, especially at a species level, can respond differently and vary over different

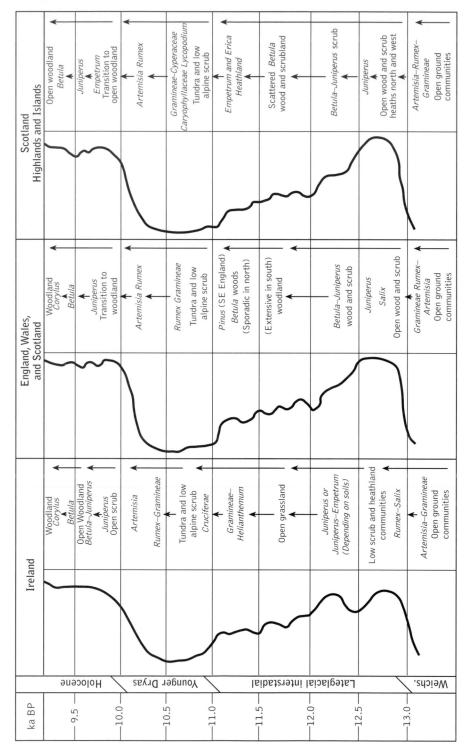

Fig. 2. A summary of the climatic reconstructions (based on Coleoptera) and vegetational history for the Late glacial period in the British Isles and Ireland. (Adapted from Walker *et al.* (1994).)

timescales to climatic and non-climatic variables. Many of the non-climatic factors that can alter the vegetational structure of the landscape could also, in theory, contribute to the lag time in the response to a rapid climatic change. Non-climatic factors might also explain why the distribution of certain plant species does not always correspond to their potential geographical range as defined by climatic variables. Webb has argued that the influence of non-climatic factors to create a lag in vegetation is in fact dependent upon the time-scale over which a climatic change takes place. Many non-climatic processes can limit the total response of vegetation to rapid climate changes of very short duration, say between ten and a hundred years. Webb suggests that when the timescale is extended to 1000 years or more, only very slow rates of plant dispersal or soils with development rates of greater than 1000 years will hinder the readjustment of vegetation communities to new climatic conditions.

Climate, however, might have some degree of control over some of the biological and physical factors that influence the structure of vegetation communities. For example, J. S. Clark has shown that the frequency of fire disturbance of the western margin of the Hemlock–White pine–northern hardwood forest region of north-western Minnesota can be associated with changes in climate. Clark's research shows that during periods of warmer, drier climate, the periodicity of the fire regime of the forest occurred over a much shorter timecycle when compared with periods of cooler and moister climatic conditions. During the dry and warm fifteenth and sixteenth centuries AD, the recurrence interval of major forest fires was approximately 33 to 44 years. A change in climate to cool, moist conditions took place between 1600 and 1864 AD (a period known as the Little Ice Age) and correlates with a doubling of the fire recurrence interval to approximately 88 years.

The relatively fast rate of climate change compared to vegetational response can mean that in certain areas some plants become disjunct from their main areas of distribution. These isolated pockets of vegetation are known as climatic relicts (also refugia); their presence is simply explained by the former presence of more suitable climatic conditions. An example of a climatic relict is the strawberry tree (*Arbutus unedo*). The strawberry tree has a disjunct distribution in western Europe, having its main centre of distribution in the Mediterranean region and scattered pockets in western France and western Ireland. Cox and Moore have suggested that the occurrence of the strawberry tree in western Ireland resulted from the relatively warm climatic conditions which occurred after the retreat of glaciers at the end of the last ice age. This allowed plants that are adapted to a climate of more Mediterranean type to expand their range northwards along the coastlines of western Europe. Climatic conditions were suitable for the northward spread of the strawberry tree because close proximity to the sea and the influence of the warm Gulf Stream provides western Ireland with a wet, mild and frost-free climate. Since then, the population of strawberry trees in

western Ireland has become isolated as a result of a rise in sea-level and a deterioration in climatic conditions but the tree population has so far continued to survive in an area that is outside its normal Mediterranean biome.

Predicting future climate change and how it will affect the pattern of vegetation is an extremely difficult task. However, the need to understand the relationship between climate change and vegetation is growing in importance as human intervention in the environment appears to be upsetting the natural balance of gases in the atmosphere. In particular, there is a strong body of data to suggest that the atmospheric concentrations of the greenhouse gases—those gases that are responsible for absorbing longwave radiation in the troposphere—are increasing and therefore raising the temperature of the troposphere and the Earth's surface. The International Panel for Climate Change (IPCC) published data in 1990 and 1992 which suggests that human activity is primarily responsible either for increasing the processes that release otherwise inert forms of greenhouse gases (e.g. the burning of fossil fuels) or for destroying the sinks (e.g. the destruction of tropical rainforests). They confidently predicted that over the next century the result of increasing greenhouse gas concentrations in the troposphere will result in a change in climate, and especially an increase in global temperature. By using general circulation models, climatologists predict that surface air temperature will increase globally between 1.5 and 4.5 °C if atmospheric levels of carbon dioxide double. A rise in temperature of this kind would have a significant effect on the distribution of vegetation.

There is still, however, a great deal of uncertainty about the extent to which climate, and hence vegetation, will change on a regional and local scale. Most climatologists believe that climatic variables such as the distribution and levels of precipitation and soil moisture will change as a result of increased climatic warming. One example where ecosystem development and vegetation patterns will be altered if global temperatures increase is the Arctic tundra. W. Dwight Billings and Kim Moreau Peterson describe the Arctic tundra ecosystem as one which is dominated by plants that are unique in being able to grow and reproduce at temperatures just above freezing. This means that plants that occupy the tundra can complete successfully with more thermophilious plant species. If, however, the IPCC's predictions prove to be correct, and over the next century the Earth experiences global warming, this could have a disastrous effect on the tundra ecosystem. First, a rise in temperature in the Arctic would cause some permafrost to melt and allow more woody and thermophilious species to invade and displace the present plant communities. Kullman has demonstrated that birch (*Betula pubescens*) expanded into areas of the Swedish tundra in response to warming during the first half of the twentieth century, illustrating how the diversity, range, and composition of plant communities may change if our present climate continues to become warmer. Secondly, a warmer climate could also alter the global carbon balance by melting frozen peat.

Carbon compounds (either carbon dioxide or methane) could then be released and further enhance the concentration of these greenhouse gases in the atmosphere. Research into the effects of global climatic change on forests by Jerry F. Franklin and his colleagues also suggests that the pattern of forest composition and distribution will change as temperature increases. At present, the dense coniferous forests of the Pacific coast of north-western North America are dominated by species like the Douglas fir (*Pseudotsuga menziesii*), western hemlock (*Tsuga heterophylla*), western red cedar (*Thuja plicata*), and the Pacific silver fir (*Abies amabilis*). The structure and distribution of these forests is controlled by different moisture regimes and natural disturbances, especially fire and windstorms, which are also thought to be climatically driven. Results from models of climate change for this region suggest that the climate will become significantly warmer and drier. It is thought that the main impact on climatic conditions will be a change to the present moisture regime: an increase in mean temperature by up to 4 °C, while precipitation levels remain unaltered. This would result in increased potential evapo-transpiration and could lead to shift in the major forest vegetation zones in this area. Sites currently supporting communities characterized by the mountain hemlock are expected to be replaced by a forest zone dominated by the western hemlock. Elevational shifts of certain forest zones are also predicted. For example, it is predicted that on the western slopes of the central Oregon Cascade mountain range that those forests tolerant of drier conditions, such as the area of forest dominated by Douglas Fir, would increase dramatically with a temperature rise of +2.5 °C, while the western hemlock forest zone would contract. However, Franklin and his co-workers suggest that it could be the indirect effects of climate change, altering natural disturbance regimes like the periodicity of fire, storms, and disease, that could force many of the changes to the forest communities in this region of the United States. The results of climate models do, however, illustrate the potential effect of changing climate on future vegetation patterns. Notwithstanding the uncertainties and possible errors contained within these climate models (especially the gaps in our knowledge of the mechanics of the climate system and how it interacts with the oceans and land), we can expect major changes in the pattern of global vegetation to take place in response to future changes in climate.　　　T. MIGHALL

Further reading

Peters, R. L. and Lovejoy, T. E. (eds) (1992) *Global warming and biological diversity.* Yale University Press, New Haven.

Cox, C. B. and Moore, P. D. (1985) *Biogeography: an ecological and evolutionary approach.* Blackwell Scientific Publications, Oxford.

Delcourt, H. R. and Delcourt, P. A. (1991) *Quaternary ecology: a paleoecological perspective.* Chapman and Hall, London.

Walker, M. J. C., Bohncke, S. J. P., Coope, G. R., O'Connell, M., Usinger, H., and Verbruggen, C. (1994) The Devensian/Weichselian Late-glacial in northwest Europe (Ireland, Britain, north Belgium, The Netherlands, northwest Germany). *Journal of Quaternary Science,* 9 (2), 109–18.

veins Veins are sheet-like, usually discordant, bodies formed by the infilling by minerals of fractures, such as joints or fault planes, bedding planes or solution cavities. They are epigenetic, that is, they were formed after the rocks in which they occur, and can be found in almost all types of rock. They are highly variable in width, ranging from a few millimetres to several metres, and may extend for considerable distances both vertically and horizontally. The walls, especially of secretion veins, may be parallel but are more commonly branching and display pinch-and-swell structure which depends on the lithology of the host rock and the nature of the fracture-forming process. Veins commonly display a regularity of pattern over wide areas (the vein system) which is related to that of the fractures or joints produced by extensional, shear, or hybrid stress systems.

The simplest veins are those formed by lateral secretion of material that has migrated out of the enclosing rock from the immediate vicinity of the vein. They thus reflect the nature of the host rock and are usually monomineralic. Quartz is the dominant mineral in veins in siliceous rocks, calcite in limestones, and gypsum in gypsiferous sediments. The minerals in these simple veins commonly show growth fibres whose orientation is related to the origin of the fractures that they occupy. For example, in extensional situations the fibres are usually perpendicular to the vein walls, but along shear fractures the fibres are oblique to the walls, indicating the orientation of the opening of the fracture; they may be curved, reflecting a change in orientation of the openings as crystallization proceeded.

Most veins, however, contain minerals deposited from solution migrating into the system from outside the area, either from above (*supergene*) or below (*hypogene*), or sometimes from both directions. The hypogene veins are deposited from water at temperatures varying from about 100 °C to 500 °C and are called *hydrothermal veins.* Here, too, the veins may be monomineralic, commonly quartz or calcite, but more usually are polymineralic with a mixture of metalliferous sulphide or oxide (ore) minerals and gangue minerals. The gangue minerals are usually of limited or no commercial value, but some, such as fluorite or barite, may be valuable. The minerals seen at or near the surface may not be those originally deposited but may instead be the result of alteration by near-surface oxidation processes.

These hydrothermal veins are often complex and banded, revealing a long history of repeated opening and reflecting the changing composition of the fluids with time. The walls are the nucleating surfaces with the first minerals growing on them; ore and gangue minerals may grow sequentially and symmetrically from the walls inwards producing *crustification veins.* Cavities or vugs in the centre of the veins make it possible for minerals to grow into free space forming large and well-formed crystals.

Veins may be zoned not only internally but also on a regional scale, their mineralogy changing both laterally and vertically away from the source body (usually a granite) as a result of declining temperature of the mineralizing fluids. A good example of this zoning and of the pattern of the vein systems is found in south-west England.

The temperature of formation of the minerals and the composition of the fluids can be determined by studying fluid inclusions, which are small bubbles of liquid trapped on the surface of the growing minerals at the time of formation. The origin of the vein-forming fluids—whether magmatic, meteoric, or connate—may also be determined from a study of the oxygen isotopes of the inclusions.   R. BRADSHAW

Further reading

Park, C. F. and MacDiarmid, R. A. (1975) *Ore deposits.* W. H. Freeman, San Francisco.

Vening-Meinesz, F. A. (1887–1966) Felix Andries

Vening-Meinesz was the foremost exponent of gravimetric surveys and student of the implications of gravity values of the twentieth century. In 1910 he joined a gravity survey of The Netherlands: one of the earliest such surveys. To solve the problem of unstable support for instruments at gravity stations where the ground was soft and insecure, he produced a new mounting for gravimeters which eliminated unwanted vibrations, and by 1921 gravity had been measured at 51 stations. The next step was to attempt to take gravity readings at sea.

Surface vessels being unsuitable, even in the calmest sea, he used submarines around the island arcs of the Dutch East Indies, measuring the great negative anomalies associated with the ocean trenches. By 1923 his early results were published and he offered an explanation: thick masses of light sediment in the depressions in the crust. He postulated that eventually the crust would buckle under the weight of these sediments, causing the sides of the depression to draw closer and squeeze the infilling into folds and nappes as elongate orogenic belts. His later work confirmed these ideas.

Vening-Meinesz also considered areas that are rising but are without signs of crustal buckling, e.g. those from which ice caps had recently melted. He was convinced that there are limits to lateral flexure, and that the continents are firmly fixed in position. He therefore opposed continental drift, but believed that where convection currents in the mantle turned towards the core, orogenic buckling took place at the surface (Vening-Meinesz belts). These were noteworthy contributions to the debate about continental drift, orogeny, and crustal structure that was gathering pace in the mid-twentieth century.   D. L. DINELEY

Vernadsky, Vladimir Ivanovich (1863–1945) Ver-

nadsky, a Russian mineralogist and the country's most famous geochemist, was born and educated in St Petersburg. During his years as a professor at Moscow University (1891–1911) Ver-

nadsky built a chemical and mineralogical laboratory that, for its time, was considered to be among the best equipped in the world. He transformed mineralogical studies in Russia and became one of its first scientists to specialize in radioactive minerals, a specialty that became strategically valuable late in his life when the Russians instituted a crash programme to build an atomic bomb and called on Vernadsky for help in identifying ores from which uranium might be recovered.

Vernadsky was one of the first to investigate the involvement of life in geological processes. This work led to the publication of his influential volume *La biosphère* in 1929. In subsequent publications Vernadsky developed his ideas on geochemistry and biogeochemistry, and laid the foundation for the explosion of geochemical studies in the latter half of the twentieth century.   BRIAN J. SKINNER

vertebrates The vertebrates are of interest to us because the human taxonomic group is included within them. Vertebrates are a subphylum of the phylum Chordata and retain the chordate characters of a notochord (a longitudinal stiffening rod), a tubular dorsal nerve cord, and gill slits (lost in the adult stage of land-dwelling vertebrates). However, in addition, vertebrates have developed a brain and an internal bony skeleton that have enabled them to become far more successful and widespread than their chordate relatives the urochordates and cephalochordates. These lower chordates are marine filter-feeders, and it is generally accepted that the cephalochordates, which are rather fish-like in appearance, are the closest relatives of the vertebrates although not their direct ancestors. The presence of the cephalochordate *Pikaia* in the Middle Cambrian Burgess Shale suggests that vertebrates were already present then, and this has recently been substantiated by the discovery of vertebrates in the Middle Cambrian of Chengjiang, southern China. This seems to indicate that the earliest vertebrates did not possess bone. However, this is present in fossil forms from the Ordovician of North America, Australia, and Bolivia. Sufficient articulated material exists to show that these animals were agnathans (jawless vertebrates), with a bony exoskeleton, and they appear to have inhabited shallow-marine environments.   DAVID K. ELLIOTT

vertical air motion In contrast to horizontal air motion (i.e. wind), vertical motion is both difficult to measure and is typically of very small magnitude. As a weather-producing agent, however, it is the key to differences between sunny conditions or cloud and rain. Air may ascend or descend for a variety of reasons. It may be forced to rise and sink over mountains or through convergence or divergence of horizontal airflow. Changes in the horizontal velocity of wind with height (wind shear) may also induce vertical motion. The most significant vertical motion develops under conditions of thermodynamic instability, where semi-isolated parcels of air rise under buoyancy with respect to the surrounding atmosphere. This form of ascent may readily be seen in cumulus clouds, and is termed convection. The degree of uplift from forced ascent is determined by the thermodynamic state of the atmosphere.

Convection in the atmosphere, in its simplest form, results from the atmosphere being heated from the surface upwards. Solar radiation passes through the Earth's atmosphere without raising its temperature. The Earth's surface is thus the first layer to be heated. On average, the temperature in the troposphere decreases with height. The rate of this decrease changes from place to place, and from one moment to the next, as a function of the degree of surface heating, the nature of the surface, the presence of temperature advection aloft, and so on. This rate of temperature decrease, the environmental lapse rate, must be measured to be known; this is done routinely by balloons carrying instruments called radiosondes.

Parcels of air, or *thermals*, which are isolated from the surrounding atmosphere, cool at a predetermined rate known as the *adiabatic lapse rate*. Since pressure decreases vertically in the atmosphere, thermals move from a region of higher pressure near the surface to lower pressure, a change which allows them to expand. Expansion means that work is being done on the surrounding atmosphere. That work is paid for by a loss of internal energy, or temperature. As the temperature is lowered, so the ability of the air to hold water in gaseous form decreases. If a critical temperature, known as the *dew point*, is reached, condensation occurs and clouds form. Similarly, air which is sinking contracts. The atmosphere is doing work on the parcel of air, which gains internal energy, expressed by a temperature increase. Water evaporates back into vapour form and the cloud disappears. This rate of change of temperature with height is fixed at 10 °C km^{-1} in dry air and is known as the *dry adiabatic lapse rate* Following condensation (i.e. above cloud base), latent heat of condensation is released into the parcel of air. Latent heat is released because water changes state from a gaseous form (water vapour) to a liquid form (cloud water droplets), which corresponds to a change to a lower energy level. This release of latent heat acts to slow the rate of cooling from 10 °C km^{-1} to an average of 6 °C km^{-1}, a rate of change of temperature known as the *saturated adiabatic lapse rate*. It is important to point out that in the case of saturated air (above cloud base) this rate of change of temperature with height is not fixed, since it depends on the amount of water vapour initially in the air and on the temperature of the air itself. A warm, moist air mass will cool very slowly after condensation, since the release of latent heat following condensation will be large. A cool, dry air mass, if lifted to condensation level, will cool at a rate not much slower than that of dry air. In other words, in cool, dry air the adiabatic lapse rate and the saturated adiabatic lapse rate are closer together.

The stability or instability of an air mass depends on whether the isolated rising parcel (thermal), which cools at a fixed rate determined by the dry adiabatic lapse rate to condensation level, and by the saturated lapse rate thereafter,

finds itself at any moment to be warmer or cooler than the surrounding air. If warmer, the parcel is less dense and buoyancy forces will force it to rise further. The air is said to be *unstable*. If cooler, it will be more dense and will sink. The air is said to be *stable*. When air is stable, there is no convection: air which is displaced up or down by mountains, for example, will return to its previous altitude. Unstable air is associated with convection.

These relationships are shown graphically in Fig. 1a, the environmental lapse rate, or measured lapse rate, being represented by the line AB. Temperature decreases rapidly from the surface. This may result for a variety of reasons. Cold air may have been advected aloft, or the surface may have been intensely heated. The profile of line AB will change with time. Parcels of air rising from the surface will cool at a predetermined rates, given by the line AC. In this case we assume that no condensation has occurred and that cooling is 10 °C km^{-1} (dry adiabatic lapse rate). We shall also not concern ourselves with what causes the parcel to rise just yet. We can determine whether the atmosphere is unstable or stable by comparing the temperature of the ascending parcel (line AC) with that

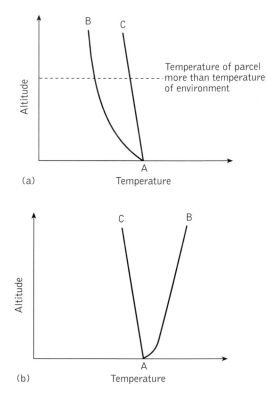

Fig. 1. Temperature profiles: (a) of an ascending air parcel (AC) and that of the surrounding air, showing the environmental lapse rate (AB); (b) showing forced parcel ascent (AC) and environmental lapse rate (AB).

of the surrounding atmosphere (line AB) at any common altitude. At all altitudes, the temperature of the parcel exceeds that of the surrounding atmosphere. The parcel rises by its own buoyancy.

Figure 1b illustrates a stable case. Here, the environmental lapse rate, AB, increases with height, a condition known as a temperature inversion. Forced parcel ascent will occur along line AC, the dry adiabatic lapse rate. Any ascent by the parcel

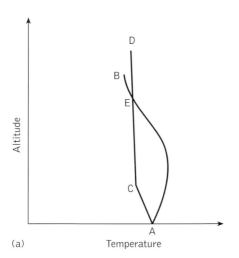

(a)

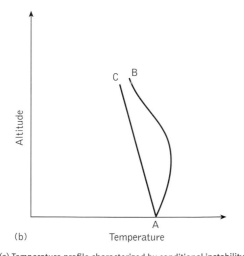

(b)

Fig. 2. (a) Temperature profile characterized by conditional instability. The environmental lapse rate is shown by line AB. The air parcel cools initially at the dry adiabatic lapse rate (AC) and then at the saturated adiabatic lapse rate (CD). Above point E the parcel of air becomes warmer than the surrounding air. (b) Temperature profile showing an identical environmental lapse rate to that of Fig. 2a, but here the air is extremely dry. The air parcel is always cooler than the surrounding air.

will lead to the parcel becoming cooler and more dense than its surroundings. It will therefore sink back to its starting point.

A further type of instability needs to be considered, that of *conditional instability*. Here, rising air may become unstable if there is enough moisture for condensation to occur, or rather if the air is sufficiently close to saturation for condensation to occur. Condensation leads to latent heat release, which further warms the parcel of air and may render it less dense than its surroundings. This is illustrated in Fig. 2a. Line AB represents the environmental lapse rate. Ascent by the parcel occurs at two rates. Initially the parcel cools at the rate shown by AC, the dry adiabatic lapse rate. Condensation of the moist air occurs at point C and further ascent (CD) takes place at a slower rate of cooling, the saturated adiabatic lapse rate. Above point E the parcel becomes warmer than the environment. The atmosphere is classed as conditionally unstable. Figure 2b shows an identical environmental lapse rate. The difference here is that the air is extremely dry. Ascent by the parcel at the dry adiabatic lapse rate is given by line AC. The dry air is unable to cool to condensation level (*dew-point temperature*) throughout the ascent. Since the dry adiabatic lapse rate is a faster rate of cooling, the parcel remains cooler than the surrounding atmosphere and the atmosphere is stable. Conditional instability therefore depends on the amount of moisture in the parcel.

Ascent forced by mountains, convergence in a low-pressure system, or even wind shear may thus be enhanced if the atmosphere is in an unstable state. Typically, forced ascent is needed to trigger conditional instability. Evidence of this is the tendency for thunderstorms to develop near mountains or in low-pressure systems. The forced ascent provides the energy to raise the air to the altitude where instability is released. Subsequent energy release occurs as a result of the differences between the parcel and its surroundings.

An assumption thus far is that rising air is isolated from the environment. In reality, mixing, known as entrainment, between the parcel and the environment does occur, but is less prevalent than would be expected. Clouds, for example, tend to act as conduits for moist, warm surface air to reach higher levels with minimal mixing.

Differences between rising and sinking air are worth noting. Ascent is often concentrated inside cloud turrets, with values reaching 30 m s^{-1} in vigorous deep thunderstorms. Descent is often very slow, of the order of a few centimetres per hour, and tends to occur over much larger areas. Evidence of slow, large-scale descent is provided in the desert areas of the subtropics, such as the Sahara, which cover vast expanses of the globe.   R. WASHINGTON

Further reading

Meteorological Office (1991) *Meteorological glossary*. HMSO Publications, London.

very-long-baseline interferometry (VLBI) The

basic principle of very-long-baseline interferometry (VLBI) is that the difference in time between the arrival of a radio signal at a pair of radio telescopes is a function of the length and orientation of the line joining the two. Because the radio sources used as targets for VLBI are very far away (and thus independent of the Earth) and move very little relative to each other, VLBI is a good means of acquiring information about variations in the Earth's rotation rate and orientation in space as well as the positions of the radio telescopes. The S- and X-band (2–8 GHz) microwave signals that are received are very weak, and large antennae (10–100 m) are normally required, but data from smaller (4–5 m) transportable antennae can be combined with data from the larger ones. Although radio interferometry has been possible since the 1930s, the main advance in the field came in the 1960s when the development of precise atomic clocks removed the need for physical communication between the telescopes. The latter introduced additional delays into the signal path and limited the length of the baseline. Nowadays, the incoming signal is mixed (*heterodyned*) with a pure harmonic signal from a stable local oscillator (usually a hydrogen maser which can provide frequency stability of one part in 10^{14}) and the resulting beat frequency is time-tagged and recorded for later analysis. This process is carried out simultaneously for many frequency bands spanning the range of interest.

The analysis of VLBI observations is complicated by several factors. First, the incoming signal is delayed (and to some extent refracted) by the Earth's ionosphere and troposphere (the ionized outer layer and the lower layer of the Earth's atmosphere respectively). Ionospheric delay is *dispersive*; that is, it affects different frequencies of a signal in a systematically different way. The effect can be corrected by using measurements at more than one frequency. Tropospheric delay is harder to correct, because it is non-dispersive. The 'dry' component (neglecting the presence of atmospheric water vapour) varies mostly with local surface pressure and temperature and can be modelled. However, the 'wet' component, which depends on water vapour content along the signal path, cannot easily be estimated, although it is smallest for observations at high angles of elevation. Secondly, the raw measurements are subject to a *phase ambiguity*. Because the incoming oscillation repeats each cycle, there will be an unknown integral number of cycles which must be added to the phase-delay measurement. Unless this integer can be estimated (which would require a very good pre-existing knowledge of the length of the baseline), this difficulty must be overcome by using the *group delay*, the change of phase delay with frequency, which has no such ambiguity but is less precisely determined. Thirdly, errors in the local oscillators and clocks must be considered. Finally, because radio telescopes are large structures, relating the actual point of measurement to a ground marker may be difficult, and the point of measurement may vary with frequency and will shift as the telescope deforms in response to orientation, wind, and temperature changes. Despite these difficulties, VLBI measurements of position can achieve accuracies of a few centimetres, which is sufficient for measuring very small changes in tides, Earth rotation rates and wobble, as well as plate-tectonic movements over several years.　　　　PETER CLARKE

VLBI *see* VERY-LONG-BASELINE INTERFEROMETRY

volatilization 'Volatilization' is a term used to describe mass transfer from the liquid (condensed) phase to the vapour (gaseous) phase. This includes evaporation of river, lake, and soil water into the atmosphere, making the air humid and forming clouds. This process drives heat transfer at the Earth's surface. Sea water is primarily evaporated near the Equator (requiring heat). It is then transported by atmospheric currents to northern and southern latitudes, forming clouds and eventually rain (generating heat).

Volatilization is also an important phenomenon controlling the fate of many low-boiling chemicals, such as the hydrocarbons in petroleum (gasoline) and organic solvents (e.g. trichloroethylene). These chemicals are also described as having high vapour pressures. The hydrocarbons can be smelled at petrol stations, and trichlorethylene at the dry cleaners. Other less volatile chemicals such as polychlorinated biphenyls (PCBs) are reported to be found in the body fat of polar bears, indicating that the PCBs were transported to northern latitudes by atmospheric transfer. Water–air exchange is important for various gases such as oxygen (O_2) and hydrogen sulphide (H_2S), which potentially interact chemically or through microbial mediation with organic compounds. Organic compounds and gases dissolved in water must diffuse to the air–water interface in order to escape from the water into the atmosphere.　　　　K. V. RAGNARSDOTTIR

volcanoes and volcanic rocks Volcanoes have been of concern to mankind since prehistory. For example, the fertile soils that form from volcanic material have encouraged men to farm on their slopes since agriculture began. Some volcanic products such as obsidian were central to the Middle and Near-Eastern neolithic trading industry from 12 000 BC. Vesuvius, a volcano in Italy, was planted with vines right up to the crater rim when the volcano erupted in AD 79, to drown the nearby city of Pompeii in ash.

Volcanoes are named after the Roman god of fire and metalworking, Vulcan. Vulcano itself is a small volcanic island in the Aeolian Islands in the Tyrhennian Sea. Volcanoes are areas where magma (molten rock) erupts (breaks through the surface of the Earth). Strabo (b. c. 60 BC), a geographer who travelled extensively in Asia Minor, interpreted volcanoes as burns where the ground had been scorched and blistered by escaping tongues of subterranean fire. In the nineteenth century, observations of lava flows in the Massif Central of France led an early school of geologists, the plutonists, to propose that all rocks were formed by volcanic activity. However, a German group, the neptunists, whose area of study was the sediments of Germany, were strongly in support of the hypothesis that all rocks were formed under the sea by sedimentation. It was not until a few mutual field trips had been undertaken that the two schools were able to come to an agreement whereby both methods of rock formation were admissible. The study of volcanoes and volcanic rocks is called volcanology or vulcanology.

The behaviour of magmas, and hence the type of volcanic edifice that they form, is dependent upon their composition and the volume of lava erupted. Gas-rich lavas, particularly those that are of high viscosity, form large quantities of cinders and ash which collect in a cinder cone. Where gas-free, or low-viscosity, lava erupts, the absence or ease of escape of gas means that the eruption is quiet and the volcanic landscape is dominated by lava flows. This type of eruption forms a shield volcano. The volcanic terrain of Iceland is dominated by shield volcanoes. The composition and volume of the volcanic rocks is dependent upon the magma-generating processes, which are the result of plate-tectonic activity. Specific tectonic environments are thus associated with particular types of volcanic edifice. The most abundant volcanoes on land are found round the margins of the Pacific Ocean where plates are being destroyed by subduction, a process commonly associated with volcanic activity. These chains of volcanoes are sometimes called the circum-Pacific Ring of Fire.

Volcanoes

Most magmas erupt at about 1200 °C, at which temperature they glow red; hence the popular idea that volcanic material is aflame. Gases that are dissolved in the magma until the time of eruption bubble out (*exsolve*) as the magma approaches the surface of the Earth. The material that erupts out of the ground is a mixture of molten rock, solid rock, and gas. Figure 1 is a diagram of a volcano, showing the main features of an eruption. The volcanic liquid is called a magma while it is still inside the ground and a lava after it has been erupted. Magma is fed to the eruption from a magma chamber. The pressure of the rock overlying the magma chamber forces the liquid out of the ground. The hole from which lava erupts is known as a vent. While the melt is resident in the magma chamber it cools, crystals form in it and bubbles rise out of it. Many of the crystals remain suspended in the melt while the remainder sink to the bottom of the magma chamber to form a cumulate rock. When irregularities on the roof of the chamber overflow with bubbles there is a rush of gas through the vent. Experimental work on bubbles collecting in magma-chamber-shaped tanks of water or syrup show that bubbles forming, collecting, and escaping in this way can lead to periodicity in volcanic eruptions.

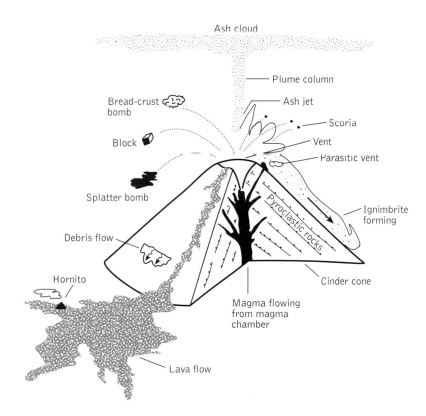

Fig. 1. Principal attributes of a volcano during an eruption. If a strong wind is blowing, the ash cloud will be elongated in the direction of the wind.

If the hot magma reaches the water table before it reaches the surface of the Earth through a vent, the water boils and expands violently, causing an explosion. This type of eruption, called a *phreatic eruption*, is characterized by a volcanic crater that goes beneath the level of the surrounding landscape and is encircled by a wide, low ring of volcanic debris. During a phreatic eruption the early magma is chilled to form a glass, which is broken by the explosion to form angular clasts of *pelagonite*. Where there is no groundwater and the magma is gassy, it travels to the surface and sprays out of the vent in a fire-fountain, the bubbles in it expand, and small vesicular stones called *scoria* (or *lapillae*) are formed. The bubbles in lava are called *vesicles*. If, after the lava has cooled and set, the vesicles are filled with minerals, they are known as *amygdales*.

Small vesicular pieces of volcanic material, *scoria*, collect around the vent to form a cinder cone. As cinders build up near the vent the pile becomes unstable and eventually cinders roll down the slope so that the cone, although it grows larger, maintains its profile. Smaller vents on the flanks of a volcanic cone also form small cones of their own, called *parasitic cones*. Large masses of volcanic material falling down the sides of the volcano are called *debris flows*. The debris flows in the 1980 eruption of Mount St Helens (Washington State) contained enormous volumes of material. Detritus incorporated into these flows included trees and large logging machines that were being used in the vicinity of the volcano at the time of eruption.

If the lava is very viscous (sticky), and there is a great deal of gas present, pumice is formed. *Pumice* is a very pale rock of low density, composed of a foam of glass enclosing gas bubbles. Pumice is of such low density that it can float on water. Submarine volcanoes have been known to erupt so much floating pumice that ships have been stranded in mid-ocean. Sometimes there is so much gas in the magma and the siliceous magma is so viscous that the gas cannot form bubbles quickly enough and the pumice is exploded into Y-shaped sherds. These sherds are blown out of the volcano as ash. Occasionally, volcanic glass is spun out into fine threads like candy floss, called *Pelée's hair*.

Lava flows

When gas from the magma chamber has been exhausted, the lava erupts more quietly to form a broader and lower edifice, a shield volcano. If a suitable crater has been formed, a lava lake develops. The surface of the lava lake freezes while the liquid underneath continues to convect, disturbing it and creating patterns reminiscent of moving tectonic plates. Should the wall of the lava lake be breached, lava escapes and moves down the side of the volcano as a *lava flow*. When the lava flow reaches flat land it spreads out to form a lava field. *A lava field* is composed of large numbers of flows winding through the field. Wherever the side of a lava flow breaks, lava gushes out and starts to form a new flow. Figure 2 shows how a lava flow advances. The molten lava in contact with the air cools quickly to form a skin over the flow. As the lava flows onwards, this skin is left behind as a lava tunnel. Lava tunnels may be several kilometres long, but in many instances the roof of the tunnel collapses to form a trough in the lava field, along which a later lava flow may pass. At the downhill termination of a flow, lava flowing down the lava tunnel from the volcano accumulates behind a flow front which is made of solidified lava. The pressure of the lava behind builds up on the flow front until it breaks and molten lava pours out. When this lava solidifies, it forms a new flow front. Thus, lava flows advance in small steps. The new lava flows over the debris that has fallen down the old flow front to form a type of basal breccia called an agglomerate.

When a lava flow meets trees or other obstructions, it flows round them. The intense heat of the lava causes trees to catch fire. After the lava has solidified the trees are burnt away by the heat of the lava flow, leaving behind tree-shaped holes. The carbonized remains of trees and plants can be used to date recent lava flows by the carbon-14 radiometric dating method.

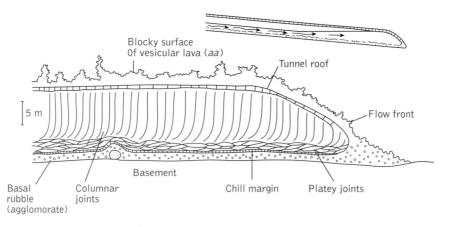

Fig. 2. Cross-section showing phenomena characteristic of the toe of a lava flow.

Blocky surface
Of vesicular lava (*aa*)

Tunnel roof

Flow front

5 m

Basement

Basal rubble (agglomorate)

Columnar joints

Chill margin

Platey joints

As lava flows along a tunnel, gas exsolving from the lava collects immediately beneath the roof of the tunnel. Where there is a hole in the roof, gas bursts out and forms a cinder cone in the middle of the lava flow, called a *hornito*. Eventually the supply of lava to a tunnel ceases and the flow stops advancing. The higher parts of the tunnel are evacuated and lava ponds at the downhill termination of the flow. As the ponded lava cools, it shrinks and cracks. The cracks form perpendicularly to the cooling surface (the top of the flow) unless there are elongated vesicles, which act like the perforations in postage stamps, bending the cracks round them. Many joints in lava flows are vertical cracks which divide the solid lava into polygonal columns. A well-known example of this phenomenon of *columnar jointing* is the 'Giant's Causeway' in Antrim, Northern Ireland.

The surface of a lava flow weathers, particularly in wet climates, to form a rich, reddish volcanic soil, called a *bole*. Although soil may form on the surface, lava flows are usually quite resistant to erosion. Where volcanic material has been extruded on to a soft substrate, the rate of erosion of the substrate can exceed that of the lava flow. Where this is the case, the lava, although having originally flowed down a valley, becomes the site of a ridge. Should more lava flow through the area, it, like its predecessor, will flow down the valleys, with the result that the older lava is topographically higher than the new lava. This is contrary to normal stratigraphical principles, which dictate that older rocks are usually preserved underneath younger rocks.

The propensity of lava to seek out topographic lows means that it often flows into river beds or the sea. Contact with water causes chilling and fracturing of the lava to form a *hyaloclastite*. Where there is an eruption underneath a glacier, hyaloclastites are particularly common. Lava erupting into deep water forms 'pillows'. Initially it is squeezed out like toothpaste from a tube into the water. Then the outside of the lava freezes to form a glassy, vesicle-free skin which inflates like a balloon until the surface is ruptured and a new pillow begins to form. Internally the pillows are characterized by concentric layers of vesicles. While the pillows are still partially molten they are compacted together so that they display convex upper surfaces and downwards-facing cusps at their contacts. Sediments that have settled on the lava are cooked by the heat of the new pillows. Where these sediments contain clay (as would be expected in an oceanic environment) *jasper*, a semi-precious red form of silica, may be formed.

Plinian eruptions

Plinian eruptions are those that are dominated by a plume of hot gas and ash. They are commonly associated with Pliny the Younger, who described them after the eruption of Vesuvius in AD 79. The eruption buried the town of Pompeii in ash, drowned nearby Heraculaneum in lava, and caused the demise of his uncle, the elder Pliny, a natural historian.

Hot gas and ash that are pushed out of a volcanic vent form a jet. The force with which the gas is pushed out diminishes quickly away from the vent, but where the ash and gas are hot they rise through the atmosphere as a plume. As the plume rises it mixes with, and heats, cold air to become more buoyant and rise still further. (If a strong wind is blowing, the plume will be asymmetric.) Under certain conditions (depending on the velocity of the volcanic jet and the size of the vent) the ash cloud may become unstable and collapse down the side of the volcano. The hot cloud of gas and ash ingests air as it rolls down the side of the volcano as a density current. The heat and speed of the cloud are such that the shearing forces within the cloud are large and fragments of pumice and ash are crushed and welded together. Streaks of crushed pumice are called *fiammé* and are diagnostic of an ash deposit called an *ignimbrite*. If the ash deposit has been welded but does not contain fiammé it is called a *welded tuff*. Ash deposits that have no fiammé and have not been welded are called simply *tuffs*. These tuffs may be compacted and turned to stone (lithified) during later geological events. The large quantities of gas (including water vapour) and fine ash particles that are emitted by volcanoes often result in violent rainstorms: water vapour from the volcano and from the atmosphere is nucleated by the fine ash particles to form rain clouds. The deluges of rain upon the volcano slopes, which may be augmented by melting ice, help to mobilize ash and debris flows (*lahars*).

In addition to scoria and ash, other volcanic ejecta include blocks, bombs, and xenoliths. Rocky material formed by the accumulation of large ejecta is classified as agglomerate. Blocks and xenoliths are pieces of material that are broken off the sides of the volcanic feeder as the magma rises from it source. *Blocks* are pieces of volcanic material that have been picked up from the volcanic edifice; *xenoliths* are pieces of non-volcanic country rock. If the magma has risen quickly from the source region of the volcano, the xenoliths may represent country rock from all levels of the crust through which it has travelled. *Bombs* are pieces of molten lava that are thrown out of the volcano. They are named according to their form. For example, *bread-crust bombs* are those whose outer surface has solidified while the inside is still evolving gas and expanding as a foam. As the inside expands it cracks the outer shell, giving it the appearance of the crust of a loaf of bread. *Splatter bombs* are those that do not have time to solidify before they hit the ground. They have a morphology reminiscent of a cow-pat.

Some volcanic vents emit only gases. Water, carbon dioxide, and sulphurous gases are common volcanic gases. Where the vent is in a hollow, the gas collects under certain weather conditions to form a pool of noxious gas. Pools of carbon dioxide have been reported as forming on the Icelandic volcanoes overnight; the casualties are usually livestock that have taken shelter in the hollow. If a gas vent opens into a pool of water, the gas, particularly carbon dioxide (which has a higher solubility at high pressure), dissolves in the water until it reaches a critical point. At this point the gas bubbles out of the water, causing the waters of the pool to overturn, which causes more evolution of gas. The most devastating example of a volcanic gas eruption in recent years was the outgassing of Lake Nyos in Cameroon on 21 August 1986. The waters of Lake Nyos overturned one cold night, releasing an enormous surge of cold carbon dioxide gas. The gas rushed down the valley in a deadly, invisible cloud, suffocating 1700 people in their beds.

After a magma chamber has been exhausted and eruption has ceased, the centre of a volcanic edifice may collapse into the void that represents the former site of the magma chamber. This process of volcanic collapse is called *caldera collapse*. A lake may collect in the caldera until such time as more magma arrives beneath the volcano to end the period of quiescence. Many caldera lakes have islands in the middle, formed where more magma has arrived and erupted in the centre of the lake. A lava dome of this type may be a prelude to further activity. Volcanoes that are prone to erupt but are not doing so at the time of inspection are designated *dormant*, while volcanoes that are not expected ever to erupt again are *extinct*.　　　　　　　　　　　　JUDITH M. BUNBURY

Further reading

Fischer, R. V. and Schminke, H.-U. (1984) *Pyroclastic rocks.* Springer-Verlag, Berlin.

Francis, P. (1993) *Volcanoes: a planetary perspective.* Oxford University Press.

volcanogenic earthquakes
On a global scale the association of earthquakes, particularly those with deep foci, and volcanicity is very close, as is indicated by the fact that both occur at destructive and constructive plate boundaries. In contrast, the earth tremors that are the direct consequence of volcanic activity are small, minor shocks that are spread over a relatively large area.

Before the commencement of a volcanic eruption quite local shocks are commonly registered which are caused by the opening of fissures in the underlying rocks or in the volcanic structure itself. These are due to the sluggish movement of very hot viscous magma several kilometres below the volcanic vent from one storage chamber to another. This motion takes place under great steam pressure through a network of tubes and pipes. In the process, various parts of the surrounding rock become hotter and more strained as the magma pushes through them. This results in fractures in the neighbouring rocks, the strain being relieved by elastic rebound, which is felt as an earthquake.

After a major eruption has ceased, more-or-less circular volcano-tectonic basins appear, termed *calderas*. These form by collapse of the surface rocks owing to partial evacuation of the underlying magma chamber. This caldera collapse induces very complex ground movements that are accompanied by earthquakes.　　　　　　　　　　HAROLD G. READING

volcanoseismicity
Earthquakes are common in volcanically active regions. Many of these events, including virtually all the really large ones, are tectonic earthquakes caused by fault movements. However, smaller seismic tremors are commonly a direct result of volcanic processes, and these are referred to as *volcanogenic earthquakes*. Low-frequency volcanoseismicity, with a maximum frequency of about 5 Hz (5 vibrations per second), is caused by eruptions or violent outgassing at volcanic vents. High-frequency volcanogenic earthquakes with their sources less than about 10 km beneath a volcano resemble tectonic earthquakes, except that they appear to be caused by fracturing associated with the movement of magma through the crust. There is also a type of seismicity peculiar to volcanoes in which vibrations occur at a single frequency (usually about 1 to 5 Hz) and can continue for hours or even days. This is called *harmonic tremor*; it is probably set off by the flow of magma or gas in conduits.

Seismic monitoring is a very effective way of keeping watch on a volcano. If the number of events per day increases, then the volcano may be building up towards a crescendo of activity. Deploying an array of seismometers around a volcano enables the depth of events to be determined. If the depth of high-frequency earthquakes or harmonic tremor is observed to become progressively shallower, this is a strong indication that magma is approaching the surface and can be used to predict the onset of a major eruption.　　　DAVID A. ROTHERY

vorticity
Vorticity measures the rate at which fluids rotate, whether the fluid is bath water flowing out of the plughole or whirling air in a tornado. Even very large weather systems are rotating. In the atmosphere and oceans the spin around the vertical is usually the most significant. In the northern hemisphere anticlockwise spin is defined as positive vorticity, but in the southern hemisphere positive vorticity is clockwise spin. In both hemispheres cyclones have positive vorticity while anticyclones have negative vorticity.

Relative vorticity is vorticity measured relative to the Earth. Since the Earth's rotation implies a spin around the vertical (except at the Equator) the combined relative vorticity and Earth's vorticity is referred to as *absolute vorticity*. The vorticity of cyclones and anticyclones is similar in magnitude to that of the Earth. The vorticity of hurricanes is five to ten times larger, and tornados spin 500 to 1000 times faster.

An important influence on vorticity is convergence. If air with low vorticity converges to occupy a smaller area the vorticity increase. This is analogous to a spinning ice skater increasing her rate of spin as she draws in her arms.

　　　　　　　　　　　　　　　　CHARLES N. DUNCAN

Further reading

Holton, J. R. (1979) *An introduction to dynamical meteorology.* Academic Press, London.

Atkinson, B. W. (ed.) 1981. *Dynamical meteorology: an introductory selection.* Methuen, London.

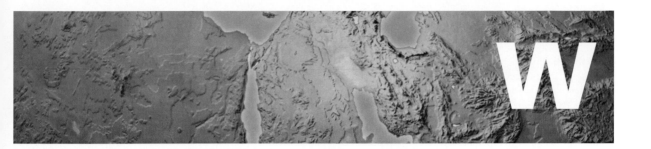

wadis *see* ARROYOS

Walcott, C. D. (1850–1927) Charles Doolittle Walcott
was one of the most powerful and extraordinary of American
scientists. Although not an intellectual innovator, he was a
hard-working investigator in the field and in the laboratory,
producing a steady stream of palaeontological research and he
was also a gifted administrator. His fossil collections, now in
the US National Museum, are an unique international asset.

Walcott's education in upstate New York was modest, but he
was an active collector of natural history specimens. His
special enthusiasm was for trilobites. In 1876 he was appointed
assistant to James Hall, the State Geologist for New York, and
three years later he joined the new US Geological Survey at the
lowest rank of field geologist. By 1883 his abilities as scientist
and administrator led to his appointment as head of palaeontology, and in 1894 he was elevated to be Director of the
Geological Survey. In the interim he had become a world
authority on the Cambrian system and its fossils. His work at
the Survey continued till 1907. Thereafter he was much occupied by the post of Secretary of the Smithsonian Institution.
He was also involved with the birth and growth of the
Carnegie Institute in Washington, later becoming the
Chairman of its Executive Committee. Other duties falling to
him were as President of the National Academy of Sciences
and Vice-Chairman of the National Research Council.

Walcott was in 1909 the discoverer of the famous Burgess
Shale (Cambrian) exquisitely preserved fauna, and began the
descriptions of the hitherto unknown groups of fossils in it.

Despite his life in the corridors of power, Walcott was an
intensely private man. The Earth sciences in the USA
benefited immeasurably from his remarkable abilities.

D. L. DINELEY

Walker circulation In the 1920s and early 1930s Sir
Gilbert Walker, an English meteorologist, published a series of
papers which suggested that the weather in Djakarta
(Indonesia) and Santiago (Chile) was related in such a way
that when the pressure was higher than normal at one place it
was lower than normal at the other. Since these cities are about
15 000 km apart it was difficult at that time to visualize a
mechanism for the connection between these distant locations.

It is now understood that what Walker had discovered was part
of a teleconnection now known as the Southern Oscillation.
Teleconnections are links between one part of the world and
another so that variations at one location are linked to variations at the other. The size of the variations is, however, usually
very small and may be indistinguishable in the normal day-to-
day variations in weather patterns. Modern research shows
that the anti-correlation found by Walker extends over the
whole of the southern tropical Pacific Ocean and acts like a
see-saw pivoting about a line at approximately 170° E. When
the region to one side of this line has higher than normal pressure the other side has lower than normal pressure. The see-
saw moves extremely slowly and somewhat erratically. The
period of oscillation varies between three and seven years.

The large-scale circulation of the atmosphere and oceans
must be responsible for these long-term variations. Over the
tropical oceans the dominant circulation is the Hadley circulation. Air over the warm oceans near the Equator rises
through the troposphere to altitudes up to 16 km and then
moves poleward towards both the north and south, gradually
descending in the subtropical anticyclone belts around 30°
north and south. If water in the west of the tropical Pacific is
warmer than that in the eastern Pacific, then there will be a
relative strengthening of the rising air in the west and weakening of the rising air in the east. Since this implies that more air
is removed upwards in the west and less is removed in the
east, a horizontal flow at low levels from east to west is
required for the circulation to be sustainable. Also an opposite
flow is required at high levels to complete the circulation. The
complete three-dimensional circulation is complex but it can
be visualized as having three important components: the
Hadley circulation in the north–south direction; the effect of
the rotation of the Earth, which means that the winds are
diverted to the right of the direction implied by the Hadley
circulation; and the 'Walker' circulation with air rising in the
eastern Pacific moving eastward at high levels and sinking
over the eastern Pacific to return to the western Pacific near
the ocean surface.

The Southern Oscillation is caused by small periodic variations in the Walker circulation. These are not completely
understood but are related to the ocean circulation, which
periodically produces strong upwelling of cold water in the

eastern Pacific of cold water which then travels westward. This linked atmosphere ocean circulation is also referred to as the El Niño circulation.   CHARLES N. DUNCAN

Further reading

Kumar, A., Leetman, A., and Ji, M. (1994) Simulations of the atmospheric variability induced by sea surface temperatures and the implications for global warming. *Science*, **266**, (5185), 632–4.

Meehl, G. A. (1994) Coupled land–ocean–atmosphere processes and the South Asian Monsoon Variability. *Science*, **266**, (5183), 263–7.

Tziperman, E., Stone, L., Cane, M. A., and Jarosh, H. (1994) El-Niño chaos—overlapping of resonances between the seasonal cycle and the Pacific Ocean–atmosphere oscillator. *Science*, **264**, (5155), 72–4.

waste disposal methods
Methods used for the disposal of wastes are designed to minimize their perceived hazard to the environment. Wastes can be divided into four broad categories: municipal, medical, industrial, and hazardous. Municipal wastes include basic household wastes such as papers, cans, bottles, food scraps, and other debris typically generated by households. Medical wastes result from treatment of patients in hospitals, clinics, and surgeries. Industrial wastes result from the demolition of buildings, roads, and factories, and include other construction debris that is non-hazardous. Hazardous wastes include solids and liquids that are corrosive, flammable, reactive, or toxic. Hazardous wastes also include radioactive materials, which are not discussed here. Waste disposal methods include landfarming, landfilling, incineration, and injection wells.

Landfarming

Oilfield waste, such as petroleum-contaminated soils, can be disposed of in industrial landfills or landfarmed if contaminant levels are within certain limits. Landfarming entails spreading waste over the ground surface in shallow layers. The ground is periodically ploughed to aerate the waste and encourage biodegradation by soil microbes.

Landfilling

Municipal wastes are disposed of in landfills. In the past, landfills have consisted of little more than open dumps located in places of little economic value, such as wetlands, ravines, and gravel pits. In 1993, about 20 per cent of the 1270 hazardous waste sites on the United States Federal Superfund Priority Cleanup list were closed municipal waste sites. No comparable list is available in the European Union. Minimum standards were adopted in 1991 in the USA. regulating the location, design, construction, and operation of municipal landfills to prevent contamination of the groundwater below the landfills. The primary goal of current government regulations in the Western world is to protect groundwater. To achieve this requires extensive monitoring and corrective action in the event of groundwater contamination.

Municipal wastes are covered daily in the landfill, either with soil, to aid in biodegradation, or with synthetic liners, to control odours. A collection system gathers and removes gases (usually a mixture of methane and carbon dioxide produced by biodegration) from the landfill. Gas monitoring is performed at the landfill perimeter. Collection and monitoring of gas are critical to prevent its accumulation in explosive concentrations during the processes of decomposition. Some municipal landfills harness this gas for conversion to electricity.

Solid wastes containing metals or small amounts of organic compounds are also disposed of in landfills. These wastes must be treated according to types of contaminants, before landfilling. Landfills in the USA can no longer accept free liquids. Landfills that are built to accept hazardous waste must undergo a lengthy and strict permitting process that involves public, federal, and state agency interaction. Because of this lengthy process, only a few landfills gain permits each year in the USA.

The bottom and sides of hazardous-waste landfills are constructed of a double-liner system of highly impermeable clay and thick plastic liners. Construction and placement of these liners have to conform to strict engineering specifications to prevent the movement of liquid waste into the groundwater below. Trenches containing gravel and sand surrounding a non-corrosive piping system must be placed in the bottom of the landfill to collect and transport any leachate to a storage area for treatment and discharge. (Leachate is water that percolates through the solid waste and accumulates on top of the liner at the base of the landfill.)

Waste is placed into the landfill in layers or lifts to prevent puncturing of the plastic liner. The location of a particular waste stream within the landfill is recorded to ensure compatibility with other wastes and to identify wastes if future groundwater problems occur. Permits in the USA require groundwater monitoring systems around the landfill to detect any leakage.

After the landfill reaches capacity, the surface is graded and then covered with the same double-liner system. The final clay liner is covered with graded topsoil. The leachate collection system continues to function as before in monitoring leakage through the surface cap.

When a landfill reaches capacity, US federal regulations require that owners and operators have financial assurance to care for the landfill after its closure. Closure requirements mandate continued environmental management of landfills for 30 years after closure.

Industrial wastes are usually landfilled in minimally regulated facilities. Asbestos is an industrial waste exception that must be disposed of according to special regulations and in specially permitted landfills. Because of the airborne fibre problem, asbestos is bagged in specially marked plastic containers, landfilled, and then covered with soil.

Incineration

Solid or liquid industrial wastes with a high organic content or containing chemicals such as polychlorinated biphenyls or other chlorinated hydrocarbons must be incinerated. Medical wastes are also incinerated in specially permitted facilities or commercial hazardous waste incinerators.

Incinerators are designed to destroy gases, liquids, solids, sludges, and slurries at high temperatures. One incineration process, the rotary kiln process, uses a cylindrical shell mounted on its side at a slight angle to the horizontal. As the kiln rotates, the waste moves down the slope, and the organic compounds burn. The kiln or burner is designed to withstand and maintain extremely high temperatures. Wastes are fed into the incinerator in batches or in a continuous stream. Liquid wastes are generally pumped into incinerators through a nozzle that breaks the liquid into fine droplets that burn more easily. Solid wastes may be fed into the incinerator in bulk or in containers using a conveyor or gravity system.

As the wastes are heated, they are converted from solids or liquids into gases mixed with air, and they are then passed through a very hot flame. As the temperature of the gases rises, the organic contaminants begin to break down and recombine with oxygen from the air, forming carbon dioxide and water. Other inorganic compounds may also be formed, according to the waste feed composition.

Temperatures in the combustion chamber generally range from 1000 to 1400 °C. The wastes are usually maintained at these elevated temperatures for less than a second to several seconds. Some incinerators may have a second combustion chamber that maintains higher temperatures than the first chamber. Secondary combustion is used to process chlorinated hydrocarbons such as polychlorinated biphenyls because they are difficult to burn.

Combustion yields two residual products: solid ash and gases. During combustion, most ash collects at the bottom of the combustion chamber but some is carried along with the gases as particulate matter. The composition of the ash depends on the composition of the waste feed. Incineration of highly liquid wastes produces little ash, while the amount of ash produced by combustion of solid wastes can be 10–30 per cent of the original amount of waste. Ash must usually be disposed of in a hazardous waste landfill, regardless of whether the original waste was hazardous or not.

After combustion, the gases produced move through various devices that cool and cleanse them before release to the atmosphere through the incinerator stack. Gases flow from the secondary combustion chamber through the quench chamber, and then through air pollution control devices to remove acid gases and particulates. Fans maintain a predetermined gas pressure and draw rate. The exact number and types of devices used depend on the incinerator and the types of wastes burned.

Complete combustion is not possible. Three factors determine the completeness of combustion in an incinerator: the temperature in the combustion chamber, the length of time the wastes are maintained at the high temperatures, and the degree of mixing of the waste with oxygen to ensure maximum burning. US permit requirements allow for ranges in incinerator operating conditions and mandate continuous monitoring of these conditions.

As part of the energy recovery process to conserve diminishing fossil fuels, hazardous wastes are being disposed of in cement kilns, another type of incineration process. Solvents containing waste are processed to produce a hazardous-waste-derived fuel (HWDF) that is then burned in the cement kiln. The blended HWDF must have a high energy value to qualify for valid energy recovery. Cement kilns are usually of the rotary type and burn at the same temperatures as commercial incinerators. Powdered limestone (90 per cent of the raw material used in cement kilns) chemically scrubs the hot gases as they pass through the preheater of the cement kiln, neutralizing acid gases. The permit process for cement kilns is less strict than for commercial incinerators.

Mobile incinerators were designed for use on sites when it is not economical to transport wastes to an off-site incinerator. Mobile incinerators can be mounted on flat-bed trucks and on ships. Another type of incinerator is the fluidized-bed incinerator, which burns finely divided solids, sludges, slurries, and liquids. The bed consists of an inert granular material, usually sand, that is suspended by pressurized air in a highly turbulent or fluidized state above the combustion chamber floor. Waste in the fluidized bed has direct contact with the bed material, thereby improving the transfer of heat. Combustion gases move out of the combustion chamber for cooling and further treatment. Ash caught in the bed material is eventually removed when the bed material is replaced.

Injection wells

In the USA, industrial wastes that are primarily liquid are usually disposed of in injection wells. Injection wells receiving aqueous wastes can be placed in highly permeable, underground geological formations. These formations are well below 1000 m underground, which is lower than the depth of most aquifers used as sources of drinking water. Before injection, liquid wastes are filtered to remove suspended solids and skimmed for phased organic compounds. Filtration prevents the plugging of the injection formation. If the waste is reactive, it is converted to less reactive compounds before injection. The pH value of the waste may also be adjusted to conform with that of the injection formation.   J. H. BAKER

water and water types Water, the essence of life, is also a fundamental component of the dynamic Earth system; water participates in nearly every geological process. It alters, sculpts, and creates landforms while wearing away mountains. Water is abundant in the atmospheric surrounding the Earth, on the Earth's surface, and within the Earth's crust. More than two-thirds of the Earth's surface is covered with water, as oceans, glaciers, rivers, and lakes. Within the Earth, water occurs as a discrete phase in the void spaces of rocks, is absorbed on to mineral surfaces, is incorporated into mineral structures, and becomes trapped in growing minerals as minute fluid inclusions. Although the depth to which water occurs in the Earth is still debated, deep drill holes encounter zones of aqueous fluids coming from even greater depths. Water may even be carried down into the mantle in subduction zones.

The water molecule

Water, H_2O, is involved in a plethora of Earth processes because of its unique nature and unusual properties. The remarkable qualities of H_2O are exhibited, in part, by its existence in three physical states at the Earth's surface: as solid (ice and snow): liquid (often referred to simply as 'water'); and gas (water vapour, steam). Each state is stable over a range of pressure (P) and temperature (T) conditions, as shown by the phase diagram for the H_2O-system (Fig. 1). For a given pressure–temperature (P–T) regime within the region labelled 'liquid', water exists only in the liquid state: similar considerations apply to regions of solid and gas. The solid region is subdivided because solid water is a mineral, ice, and crystallizes into different structures according to the P–T conditions. Separating these major regions are *phase boundaries* that mark the abrupt transition from one state to another and delineate the marked change in physical properties of H_2O. At P–T conditions on a boundary, two phases coexist in equilibrium. At the surface of the Earth, H_2O is commonly observed along a two-phase boundary: water vapour over ice on glacial fields traces the solid–gas boundary; water ponding on a frozen lake lies along the liquid–solid boundary; and water vapour hanging over a lake represents the liquid–gas boundary (Fig. 1). All three states coexist in equilibrium at the *triple*

point for H_2O near 0.01 °C (273K), 0.006 bar (0.0006 MPa) (Tr. pt, Fig. 1). Cirrus clouds form at triple-point conditions. At °C and surface pressure, liquid water freezes to become solid. Along the solid (ice I–liquid) boundary, increasing pressure causes liquid water to freeze at a lower temperature. Conversely, ice skaters glide effortlessly over the ice as their weight (P) induces melting; beneath their blades a two-phase region develops: a thin film of water over ice. Along the vapour–liquid boundary, water vaporizes at higher temperatures as pressure increases. Increasing T to 100 °C at surface pressure causes water to boil as is crosses the phase boundary from liquid to gas. An example of this transition is commonly seen when steam rises from kettles. Liquid water has a narrow range of thermal stability at the Earth's surface. The two-phase liquid vapour boundary continues until the critical endpoint is attained. For pure H_2O, the critical point is at about 374 °C (647 K), 220 bar (22 MPa) (Fig. 1). Extreme values in many thermodynamic and transport properties of water occur at this point. In natural systems, thermal springs in the mid-ocean ridges emanate fluids near critical-point conditions. At higher temperatures, a discrete vapour and liquid phase no longer exist; instead phase properties vary continuously. Water in this P–T region is referred to as a *supercritical fluid;* aqueous fluids typical of metamorphic and igneous environments.

The *P*–*T* stability of water is also an indication that it has unique bonding characteristics. This uniqueness is derived from the nature of the water (H_2O) molecule, which consists of two hydrogen atoms (H) bonded to one oxygen atom (O) in a V-shaped configuration, with the oxygen atom at the apex

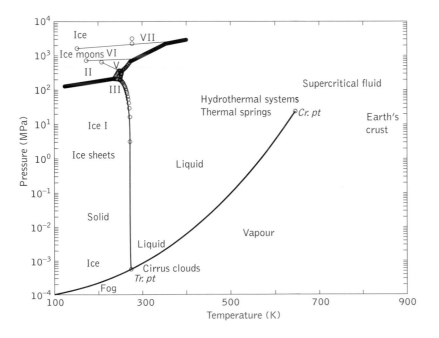

Fig. 1. Phase diagram for the H_2O system showing the pressure and temperature (*P*–*T*) conditions for which each physical state of H_2O is stable: solid (ice), liquid, vapour (gas). Roman numerals denote various structures of ice according to W. Durham. (Ice IV is a metastable phase that exists in the ice V region.) Several additional ice phases have also been found. Lines represent phase boundaries. *Tr. Pt*, triple point, at which all three phases exist in equilibrium; *Cr. Pt*, critical point above which supercritical fluid exists. The *P, T* locations of natural systems are shown in boxes. (Liquid and vapour data from Johnson and Norton (1991), *American Journal of Science*, **291**, 541–648.)

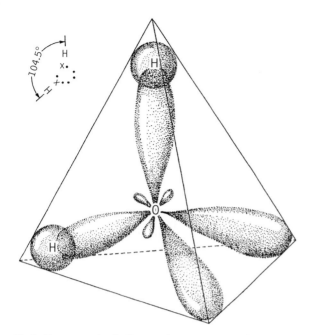

Fig. 2. The water molecule. The central oxygen atom, O is bonded to two hydrogen atoms H, shown as circles. Electron orbitals of the oxygen atom are shown as lobes. The atomic configuration is shown in the upper left.

(Fig. 2). The two H are separated by an angle of about 104.5°. Although the molecule is symmetrical, it has an asymmetrical charge distribution. The oxygen side of the molecule has a local negative charge whereas the hydrogen side has a local positive charge. This non-coincidence of charge centres imparts the property of *polarity* to H_2O: water is a polar molecule. Liquid water is associated because the positive hydrogen atom of one water molecule attracts the negative oxygen atom of another. This polar association is responsible for many of the properties of water, including: low vapour pressure, high boiling-point, and high heats of fusion and vaporization. High energy is required to break the hydrogen bonds holding the polar molecules together. Its capacity to bind both positively and negatively charged ions is also a consequence of it polarity. Liquid water is a universal solvent, behaving as an acid and a base, although it is a poor solvent for non-polar molecules. Water has an abnormally high dielectric constant, which is responsible for its ability to form complexes with other substances.

As mentioned above, solid H_2O, ice, is also a mineral. The most common form of ice (ice I, Fig. 1) crystallizes in the hexagonal crystal system (Fig. 3). This is readily apparent on examination of individual snowflakes. No matter how small and intricate, these ice crystals have symmetrical sixfold shapes. In the crystal structure of ice, each oxygen atom is surrounded by four hydrogen atoms attached by hydrogen bonds

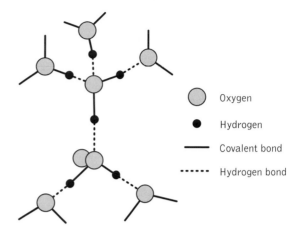

Fig. 3. The crystal structure of ice I. Each oxygen atom is surrounded by four hydrogen atoms. Two are attached by hydrogen bonding, and two by covalent bonds.

and two by electron-pair bonds (Fig. 3). This arrangement results in an open structure and explains the low density of ice. H_2O is also unusual in that this solid form is less dense than its liquid form. On melting, the hydrogen bonds are broken, the open structure collapses, and the water molecules move more closely together because of the intermolecular forces of polarity. This closer packing results in liquid water being nearly incompressible and having a higher density than ice. Thus, ice floats in water (there is a two-phase boundary). The maximum density for liquid water is at about 4 °C, the temperature of ocean bottom waters. Water vapour is much less dense than liquid, and occupies a volume about a thousand times greater than liquid.

Water in geological processes

Erosion, weathering, and alteration processes caused by water at the Earth's surface are observed in every climatic regime. The erosional capacity of water is expressed in features ranging from smooth water worn stones and sculpted rock faces, to enormous sedimentary deposits at the base and sides of mountains. V-shaped valleys carved by moving water are transformed into broad U-shaped valleys by the movement of glacial ice. Silt deposits from floods and boulders far from their source attest to the transport capacities of water. Water is a primary modifier of landscape.

Within the Earth, water is found in all rock-types, as a constituent of minerals or as *interstitial* or *pore water* contained within pore space between mineral grains until the void space becomes vanishingly small at depth. *Groundwater* is a general term used for water contained in the pore spaces of rocks, typically in the upper portion of the Earth's crust. Although this water is a minor component within the crust, it contributes significantly to the chemical, mechanical, and thermal development of the crust because of its superb ability to transfer heat and solutes and to do mechanical work. Water dramatically enhances rates of chemical reactions and mineral transformations. Water is an excellent inorganic solvent. It reacts with both metals and with nonmetals to dissolve rocks. All minerals exhibit some degree of solubility in water. When water flows, it carries the dissolved constituents in solution and redeposits them elsewhere. Ore deposits are the remnants of former metal-bearing water-rich fluids. The migration, emplacement, and destruction of hydrocarbons depend on the movement and presence of aqueous pore fluids. Water alters the mechanical properties of rocks; it reduces rock strength and the effective stress but enhances the ductility of the rock.

Water can also be dissolved within magmas. Upon crystallization, the magmatic water is expelled as bubbles which later form vesicles or microlitic cavities in the rock, or it infiltrates into the surrounding country rocks. These hot igneous intrusions also produce *hydrothermal water* by heating pore fluids contained in adjacent rocks. Because water expands when it is heated, water pressure may increase, if the water is in a confined area, to overcome the effective strength and fracture the rock. *Hydrofracturing* is a method by which water does mechanical work (fractures rock) in response to a driving force and creates passageways through which it can migrate. Hydrothermal waters are excellent transporters of heat and mass, and are responsible for much of the chemical alteration of the crust. If large convection cells of the thermally heated water find pathways to the surface, *geysers* erupt or the fluids emanate at the surface as *hot springs*.

Structural water (H_2O) or hydroxyl ions (OH) may form an integral portion of the structure of a mineral and be necessary for its formation. In some minerals, water occurs in structural cavities and is only loosely bound to the crystal structure. It readily enters and leaves the mineral, as in some swelling clays and zeolites. The 'fire' in opals is due to the reflection of light from water residing in the interstices of the silica spheres. If these minerals are heated, they will dehydrate and lose their *water of hydration*. During metamorphism, as the rocks are subjected to higher temperatures and pressures, hydrous minerals (e.g. muscovite, amphiboles) are exposed to conditions outside their stability range and are converted to more stable anhydrous phases. Water and hydroxyl ions bound within the crystal are released, forming metamorphic aqueous fluid as the minerals devolatilize and dehydrate.

Water types

In addition to the physical state in which it exists, water is described and classified by several features. One of the prevalent classification schemes is based on chemical attributes of water, as suggested by J. I. Drever.

All natural waters contain some quantity of dissolved substances. The total amount of dissolved solids, *TDS* measured in milligrams per litre (mg l^{-1}), provides one basis for the chemical description of water types. (TDS is determined by evaporating a specific quantity of water and measuring the material remaining). The ultimate source of fresh water is unpolluted rainwater. *Meteoric water*, derived from the atmosphere, originates and falls to the Earth as precipitation. Some of this precipitation infiltrates into the ground through porous sediments and becomes groundwater, the most desirable source of drinking water. Pure water is a tasteless, odourless, colourless substance and is not for human consumption. Fortunately, the water molecule has an atomic configuration that solvates rock-forming elements and contaminates itself on the continents into a compositional range on which we thrive. *Fresh water* contains small quantities of solids, less than 1000 mg l^{-1} TDS, and most is potable. Minor amounts of dissolved gases and salts in the water provide the taste. Small quantities of organic compounds can produce an unpleasant taste, and a few hundred parts per million (ppm) of salt will cause water to taste salty. As the TDS increases to *c.* 1000–20 000 mg l^{-1} TDS, *brackish water* becomes too saline for consumption by humans. These waters have lower salinities than sea water, which has an average salinity of 35 000 mg l^{-1} TDS or more. *Brines* are chemically variable but contain extremely high concentrations of TDS, at least ten times the salinity of sea water. Sea water and sedimentary brines are volumetrically more important that fresh waters, but are unfit for human consumption. In areas of active volcanism, some lakes have water which is extremely acidic because of dissolved gases. The range of water composition on which humans can survive is a narrow one, and that range approximates that of average river water.

As water moves and chemical reactions occur, the chemistry of the water changes and its properties are altered. Incorporation of dissolved ions makes water a better conductor of electricity. Continued contamination by metals makes the ores on which we depend.

Specific chemicals dissolved in water or constituting the water also give rise to descriptive terms. *Heavy water* results when the heavy isotope of hydrogen, ^2_1H, deuterium, D, combines with oxygen to form D_2O. These waters occur naturally and are also produced in atomic energy plants, where they are used to slow nuclear chain reactions. *Hard water*, which leaves an insoluble residue and cleans poorly, results from an abundance of dissolved calcium (Ca) with lesser magnesium (Mg). Conversely, *soft water* has lower concentrations of calcium ions and gives the sensation that it cannot be washed off the skin.

The elements of which water is composed, hydrogen and oxygen, both have stable isotopes. The origin of a water sample can thus be determined by use of oxygen and hydrogen isotopes. Meteoric, magmatic, metamorphic, formation, and exotic waters all have definite isotopic signatures that vary in a systematic fashion, as summarized by S. M. F. Sheppard. *Formation water* resides in the pore spaces within deeply buried sediments. *Exotic water* is a catch-all term which indicates that water has been introduced into a system and was externally derived. *Organic water* is water that results from the transformation of organic materials. The use of isotopes as tracers has become important in understanding fluid flow, contaminant transport, and alteration of the crust.

Water holds a special place in several respects. It is used as a standard for the determination of numerous physical constants. It has profoundly influenced the evolution of the Earth. It is the substance that is required to sustain life. It also provides a source of power and for recreation. As our reserves of fresh water diminish and hazardous wastes become increasingly abundant, water quality is now of the utmost concern. The groundwater on which so much of the world's population depends is essentially a non-renewable resource. In most areas infiltration rates are so slow that consumption far outstrips replacement. As demands increase, more and more stress is placed on our planet's ever-dwindling supplies of potable water.

BARBARA L. DUTROW

Further reading

Bentley, W. A. and Humphreys, W. J. (1962) *Snow crystals*. Dover Publications, New York.

Drever, J. I. (1988) *The geochemistry of natural waters*. Prentice Hall, New York.

Fyfe, P. and Thompson, A. B. (1978) *Fluids in the crust*. Elsevier, New York.

Hunt, C. A. and Garrells, R. M. (1972) *Water: the web of life*. W. W. Norton, New York.

Sheppard, S. M. F. (1986) Characterization and isotopic variations in natural waters. *Reviews in Mineralogy*, 16, pp. 165–84.

water-table

The water-table is the surface that divides the zone that is permanently saturated with water (the *phreatic zone*) from the zone above in which the pore spaces are not completely filled with water (the *vadose zone*). The water-table is, in other words, the upper surface of the groundwater or the level below which an unconfined aquifer (a body of permeable rock) is permanently saturated with water. In practical terms, the water-table represents the shallowest depth at which water will flow into an open well.

The word 'table' can be misleading here: the water-table is rarely flat, and it can change its shape and position. The water-table is usually below ground, but where there are streams, lakes, or wetlands, parts of the water-table will coincide with the land surface. In general, the water-table tends to follow the contours of the ground above, but its relief is less marked than that of the ground. The level of the water-table is usually at its highest in the spring and at its lowest in the autumn.

The water-table can also be described as the surface at which the pressure of groundwater is equal to atmospheric pressure. Below the water-table, the pore spaces in the rock are completely filled with water at a pressure that is greater than atmospheric pressure. Between the water-table and the land surface, some water at less than atmospheric pressure is held in pore spaces by capillary forces. Immediately above the water-table is the *capillary fringe* or *capillary zone*, a shallow region in which water can be drawn upon by capillary action.

The capillary fringe is usually not more than a few metres deep, and may be only a few centimetres deep.

A *perched water-table* can exist where there is an impervious layer at a relatively high level and water is reaching its upper surface faster that it is being lost by seeping through the layer and by lateral movement. It is thus possible to have two water-tables at the same point, one above the other. Springs usually mark the contact between the upper surface of the impervious layer and the surface of the ground.

In coastal regions, sea water can infiltrate the rocks, and a salt-water wedge will then form below sea level under the fresh groundwater. (The fresh water remains above the salt water because of the difference between their densities.) If too much water is pumped from the ground in these circumstances, the fresh water will become contaminated with salt and will be unfit for drinking. The rate of extraction will then need to be reduced to allow the supply of fresh water from the landward side to restore the balance.

J. BAHR

weather forecasting

The economic and safety implications of accurate weather forecasting are enormous. Weather-sensitive activities such as flying, road transport, and construction depend on short-term forecasts. Shipping can benefit from longer-term forecasts, while agriculture could use forecasts from a few hours to a month or more.

Since the introduction of computers, weather forecasting has been based on numerical solutions of the equations that govern atmospheric motions. Because the equations are well established and accurate it seems reasonable to assume that reliable predictions can be made.

The process of making a forecast is relatively simple. Observations of the state of the atmosphere are made at thousands of locations around the world; most of these observations are made using radiosondes, instrument packages carried by balloons to heights of about 25 km. These give measurements of temperature, humidity, pressure, wind speed and direction as the balloon rises. Unfortunately, most of the observations are over the Earth's land masses with very few over the oceans. Although ocean areas are well covered by satellite observations, these lack the high accuracy and fine detail of radiosonde measurements.

The observations are interpolated on to a regular grid which covers the entire globe and extends from the surface to about 20 km altitude. Applying the equations of motion it is then possible to calculate the rate at which each weather element will change. From this rate of change it is possible to calculate the value at some small time into the future, say 10 minutes. The process of calculating the rate of change can then be repeated and the forecast can progress another 10 minutes. This can lead eventually to forecasts for periods of between a day and a week. The forecast must be made in small steps because changes in one weather element affect the way other weather elements will change. It cannot be assumed that the conditions on which the calculations are based will remain unchanged for long.

Numerical forecasts are limited to short-term forecasts. Small errors in the initial observations grow during the numerical calculations so that the accuracy of the forecast decreases as the forecast period increases. Most forecasts are of little practical value when the forecast period is more than about one week. This limit to the range of forecasts may be extended by improved observation systems. However, it seems unlikely that it will ever be possible to provide accurate numerical forecasts for periods in excess of about 14 days.

The equations used for weather forecasting are nonlinear. Equations of this sort can exhibit chaotic behaviour. In fact, atmospheric simulation experiments were the first natural systems to reveal their chaotic nature. One consequence of the chaotic equations is that any initial error in the observations of the state of the atmosphere will grow rapidly as the forecast proceeds. The error need not be significant; it may be minute. This has lead to the assertion that a butterfly flapping its wings in Japan may affect the weather in Europe within a week or two.   CHARLES N. DUNCAN

Further reading

Atkinson, B. W. (ed.) (1981) *Dynamical meteorology: an introductory selection*. Methuen, London.

weather fronts Weather fronts are the boundaries between two different air masses. An air mass is a very large body of air which has approximately uniform properties, with little variation in temperature, humidity, visibility, or cloud type. Air masses form over regions of the Earth where air is slow-moving and takes on some of the characteristics of the surface. The main characteristics are the temperature and humidity. These are used to categorize air masses, using the letters c (continental) and m (maritime) to indicate dry and moist air, and T (tropical) and P (polar) to indicate warm and cold air, so that cT (continental tropical) air masses would be warm and dry, forming over regions such as the southern USA or north Africa, while mP (maritime polar) air masses are cold and moist and form over regions such as the northern Atlantic or Pacific oceans. To indicate extremely cold air, usually originating over snow-covered regions, the designation A (arctic) is sometimes used.

Our basic model of the structure of fronts was developed by the 'Norwegian school' of meteorologists based in Bergen, who were very productive in many areas of meteorology after the First World War. It is from that war that the term 'front' comes to indicate a boundary between two very different entities. Although this model is very useful in describing the formation and typical structure of fronts, it is very rare that an individual front looks exactly like this idealized model. Factors such as mountain ranges, coastal regions, the season of the year, or even the time of day can significantly alter the appearance or behaviour of individual fronts.

Fronts are drawn on weather maps as lines with symbols indicating whether they are warm fronts (semicircles) or cold fronts (triangles) (Fig. 1). When a warm front passes a partic-

ular location the air mass will change from cold to warm, while the opposite is true of a cold front. The designation of a front therefore depends on the motion of the front. If, as occasionally happens, a warm front slows down and stops and then moves in the opposite direction, it becomes a cold front. Fronts can also slow down and become stationary fronts. Since the front is not moving it is then neither a cold front nor a warm front. A stationary front is denoted by semicircles and triangles on alternate sides of the front. On a weather map, symbols are shown on the leading edge of a moving front. For stationary fronts the warm front symbols are, however, drawn on the side that would be the leading edge, if the front was moving as a warm front. The cold front symbols would be placed on the opposite side to indicate that if that were the leading edge the front would be a cold front. Occluded fronts are formed from a combination of both warm and cold fronts, and both symbols are accordingly used together.

The lines that represent fronts on weather maps indicate only part of the front. Fronts have a vertical as well as a horizontal structure. They are sloping surfaces, and the line on the weather map indicates where the frontal surface meets the Earth's surface. Fronts always slope so that the colder air, being more dense, is below and the warmer air mass is above. The frontal surface has the remarkable property that it acts almost like a solid, impermeable surface. Air from one air mass does not pass through the frontal surface to the other air mass but rises over the frontal surface or moves along it while remaining within the same air mass.

The polar front is a huge semi-permanent quasi-stationary front which separates warm air in the tropics from cold air in

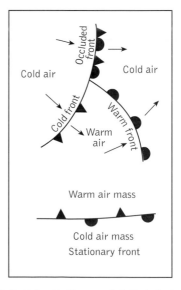

Fig. 1. Symbols used on weather maps to indicate fronts.

polar regions. Mid-latitude cyclones or depressions form on this polar front. As the depressions develop, their circulation bends the polar front and pushes it in such a way that a warm front and a cold front are formed. As the depression develops the cold front moves faster than the warm front and eventually catches up with it, producing an occluded front. The motion of the air around the depression is complicated by the fronts across which the air does not move. The resulting upward motion at fronts causes cooling of the air, condensation of water vapour to produce clouds, and eventually precipitation. The detailed characteristics of each type of front are described below.

Warm front

The warm front (Fig. 2a, right-hand side) has a slope of about 1 in 100. The uppermost part of the front is typically about 10 km above and 1000 km ahead of the region where the frontal surface meets the Earth's surface. Over all of this surface the air in the warm air mass is forced to rise over the frontal surface, and as the air cools water vapour condenses to form clouds. At the highest levels the clouds are composed almost exclusively of ice crystals; at middle levels they contain a mixture of ice crystals and water droplets. The clouds are deepest in the area just ahead of the surface front, and this is where precipitation falls as small raindrops. Embedded within this deep cloud there are commonly regions of more vigorous updraughts that produce heavier bands of rain. The rainfall area associated with warm fronts extends about 300 km ahead of the surface front. As the rain is about 700 km behind the leading cloud at the uppermost edge of the front, and the front moves with speeds in the range 20 to 100 km/hr, it is possible to have several hours' warning of an approaching warm front simply by recognizing the sequence of clouds. This will start with high uniform ice crystal cloud

(cirrus, Ci, and cirrostratus, Cs) followed by a lowering cloud base and darker, thicker cloud (altostratus, As), eventually becoming rain clouds (nimbostratus, Ns). At the same time as the upper level clouds are becoming darker, low-level clouds will become smaller or disappear altogether. The rain precedes the surface front, and so the increase in temperature associated with passage into the warm air mass is coincident with the cessation of rain and usually with the break-up of the dense cloud mass. The wind direction will also veer (i.e. come from a point further clockwise round the compass) as the surface front passes.

Cold front

At a cold front (Fig. 2a, left-hand side) the cold air undercuts the warm air and produces a sloping frontal surface with a steeper gradient than that of a warm front. The gradient at a cold front is typically about 1 in 75 but it is even steeper close to the surface front. The cold air moving behind the front is moving faster than the warm air ahead of the front, and this forces the warm air to rise rapidly at the frontal surface. Deep convective clouds (cumulonimbus) are produced at the cold front which produce heavy precipitation and sometimes hail. Since the clouds form behind the surface front, there is no early warning, as in the case of the warm front, although the towering heads of the cumulonimbus clouds can sometimes be seen from a considerable distance. The precipitation arrives at the same time as the temperature drops and the wind veers. There is often a significant increase in wind speed as well as a change in direction. When the rain stops the clouds usually clear rapidly, leaving the sky almost cloud-free and with excellent visibility. The cold air behind the front is, however, often unstable and showers frequently form in this cold airstream a few hours after the passage of the cold front.

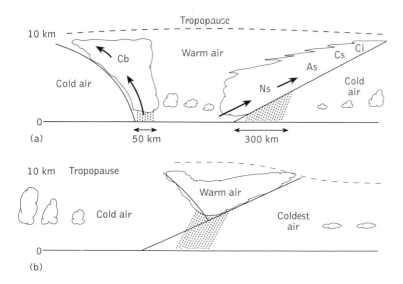

Fig. 2. (a) Vertical cross-section of a cold and warm front showing cloud and precipitation; cloud types; As, altostratus; Cb, cumulonimbus; Ci, cirrus, Cs, cirrostratus; Ns, nimbostratus. (b) Vertical cross-section of an occluded front (warm type).

Occluded front

Occluded fronts are formed when cold fronts meet warm fronts. Two different types of occlusion can form, according to whether the air ahead of the warm front or behind the cold front is the colder. The colder air undercuts the less cold air, and the warm air, which has been in contact with the Earth's surface, is now lifted above the surface. The occlusions are described as warm occlusions (Fig. 2b) or cold occlusions to indicate their similarity to warm or cold fronts. In a warm occlusion the air behind the cold front is relatively warmer than the air ahead of the warm front. This means that the air in the warm sector and the air behind the cold front rise above the coldest air; the precipitation then falls ahead of the surface occlusion, as would be the case at a warm front. In a cold occlusion, the reverse occurs and the occlusion resembles a cold front. In general, occluded fronts are not as vigorous as warm or cold fronts, and the changes in wind speed and direction are not so marked.   CHARLES N. DUNCAN

Further reading

Ahrens, C. D. (1994) *Meteorology today*. West Publishing Co.

McIlveen, J. F. R. (1986) *Basic meteorology*. Van Nostrand Reinhold.

weathering The skin on your face and the paint on your car can suffer from the weather, a process so obvious that 'weathering' is a term used in everyday language. It may take a little longer, but rocks also suffer from weathering, with many interesting geological and practical effects. From the chemical and physical viewpoint, rocks react to a change of conditions. Many rocks are formed deep in the Earth under conditions of high temperature and pressure and in the absence of large amounts of water and air. When such rocks are exposed by erosion at the ground surface they react to the lower pressure by expansion, and to the presence of air and water by chemical alteration of their minerals.

The main effect of pressure release is the opening of cracks in rocks called (*joints*) which let in air and water. The main effect of air is oxidation. Iron oxides and hydroxides are generally red, yellow, or brown and these give the rusty colours of many weathered rocks—a process analogous to the rusting of iron. Water affects minerals by several processes, of which *hydrolysis* (chemical combination with water) is the most important. The most common mineral in the Earth's crust is quartz, silicon dioxide, SiO_2 which is virtually unweatherable in most circumstances. The next most common minerals are silicates, such as mica, feldspar, pyroxene, and the olivines, which break down to clay, the soluble elements they contain being mostly carried away in solution. The weathering products, such as clay and iron oxides, may remain in place or be carried away by erosion. The main work of weathering is to prepare rocks for erosion by breaking them into smaller, softer, and less dense particles; erosion makes the landscape we see around us, although small landforms can result from weathering.

The products of weathering depend on the composition of the original rock. Granite, a plutonic rock formed deep in the Earth, consists mainly of quartz, feldspar, and mica. The quartz remains unweathered, while both the feldspar and mica turn into clay, and the weathering product (sometimes known as *grus*) is a mixture of clay and quartz. Basalt, a characteristic rock of lava flows, consists mainly of feldspar, pyroxene, and a little olivine. All these minerals form clay, and there will be no quartz in the weathering residue. Basalt thus produces heavy clay soils, many of which are abundant in nutrients and provide rich agricultural land. Limestone consists of calcium carbonate in the form of calcite, which is entirely soluble in rainwater (dilute carbonic acid), which dissolves it, producing hollows in the limestone, but nothing else unless the original limestone was impure. Since most limestones contain impurities, a weathered limestone is likely to be covered by a layer of red clay.

Weathered rock can accumulate to a thickness of many metres, and consist of various layers. The succession of layers from ground surface down to fresh rocks is called the *weath-*

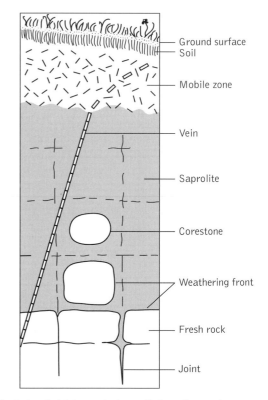

Fig. 1. A typical deep weathering profile formed on granite near Mt. Kosciusko, Australia. Note the large corestone, the vein indicating the extent of saprolite, the mobile zone 2 m thick. Such deep weathering would not occur under the present climate, and so is inherited (in this case from the Mesozoic).

ering profile. A typical deep weathering profile on granite is shown in Fig. 1 and reveals several interesting features. The first surprise is that the weathering does not usually grade slowly into unweathered rock, but instead there is a knife-sharp contact between fresh rock and weathered rock, called the *weathering front*. Weathering penetrates down the joints (cracks) in the granite, isolating the blocks between the joints, which themselves weather into smaller and smaller joint blocks called *corestones*. Each corestone possesses its own sharp weathering front. A second surprise is that the joints, and filled cracks such as quartz veins, retain their original position in the weathered rock, showing that there has been no change in volume of the rock during weathering. Because the products of weathering (clays) are less dense than the original minerals one might have supposed that there would be a volume increase. The fact that volume stays the same indicates that some weathering products have been lost in solution. Weathered rock with no volume change is called *saprolite*. At the top of the weathering profile, rock structure is destroyed by burrowing animals, root growth, hillside creep, and other processes, creating a *mobile zone*. The saprolite, mobile zone, and any surface sediments are collectively termed *regolith*.

Groundwater fills voids in rocks and weathered rocks, and under gravity it naturally accumulates at the bottom of the weathering profile. If there is permanent groundwater, the lower saprolite is satured, and in this condition anaerobic (reducing) conditions prevail. The weathered rock is grey or green, in contrast to the red and brown colours of the oxidized zone above. In such a profile the base of the groundwater corresponds to the weathering front, and the water-table corresponds to the change from green to red, that is, from reducing to oxidizing conditions. On weathered ore bodies the weathering often improves the grades of economic deposits by concentrating desirable elements such as copper around the water-table. Oxidized ore is found above the water-table, and reduced ores (usually sulphides) below the water-table.

Another economic aspect of weathering is the alteration of prepared stones such as building stones and tombstones. If you want immortality, do not choose marble for your tombstone; it is very weatherable. Sandstones can be tough, but many of them flake. Granite starts to crumble in a few hundred years, and often sooner. The most resistant rocks are quartzite and quartz-rich sandstones, and tough fine-grained rocks such as slate.

City buildings suffer excessive weathering because of atmospheric pollution. Enhanced carbon dioxide levels have some effect, but the worst offender is sulphur dioxide, which dissolves in water and acts as a dilute acid. The Houses of Parliament in London are a fine example; as Beresford Pite wrote, 'Ever since the magnesian limestone, of which the new Palace at Westminister is built, was first exposed to the chemical constituents of the London atmosphere, is has been slowly and surely transformed into some sort of heap of Epsom salts.' Air pollution is a modern effect, and many medieval cathedrals and statues that have stoood for centuries have suffered more corrosion in the past 80 years than in all their previous history. The more recently built Taj Mahal remained in fine condition until well into the twentieth century, but lost its shine after three decades of industrial pollution. Two coal-burning power houses, an oil refinery, a large railway shunting yard, and 250 iron foundries are just some of the developments in its vicinity.

Of the numerous clay minerals, two main types are kaolinite and montmorillonite. Kaolinite does not absorb much water, has little capacity to swell, and is not chemically very active. It thus does not hold on to cations, such as calcium or magnesium, that might be useful for plant growth. Kaolinite is, therefore, good for making pottery, bricks, and tiles, and it makes a reasonable foundation, but it is not very fertile. In contrast, montmorillonite has a capacity to absorb water (and release it on drying), with consequent volume changes. This means that it is useless for making pottery or bricks, and makes a bad foundation because as it swells and shrinks with wetting and drying. It produces fertile soils that can retain fertilizers. Granite is a source of commercial kaolinite; in Devon and Cornwall, for example, granite saprolite and other forms of kaolin are extracted commercially and the quartz separated to produce kaolinite of high quality. Weathered shales can provide good brick clays, whereas basalts are more likely to provide good montmorillonite and fertile soils.

Weathering can give rise to a wide variety of landforms. At one end of the scale there are small weathering pits (measured in centimetres), sometimes in groups like honeycomb, mostly formed by disintegration of rock grain by grain. These weathering hollows can grow to larger size, even to caves tens of metres high and deep. Another common feature of weathering is exfoliation or spheroidal (onion-skin) weathering, in which the rock surface seems to weather for a few millimetres and then, when the process is repeated several times, a core develops surrounded by a series of 'shells'.

Many landforms are associated with limestone, producing *karst topography* (see *karst*). Because limestone is soluble, the most important landform of karstic landscape is the cave. Many other landforms, such as solution pipes and widened joints, are formed at the ground surface. In some areas, spectacular tower karst is formed, with many steep-sided hills, like the beautiful landscape of Guilin in China whcih resembles a classical Chinese landscape painting. Everybody knows about stalactites and stalagmites (even if they cannot remember which is which), but few know how they are formed. Soil atmospheres are about four times richer in carbon dioxide than the normal atmosphere. Water trickling through soil thus becomes enriched in carbon dioxide. When it meets limestone, the carbonated water is able to dissolve some calcium carbonate and carry it down. When it reaches a cave, where the carbon dioxide content of the air is roughly the same as in the atmosphere, drips of water diffuse carbon dioxide, and precipitate skin-like veneers of calcium carbonate in the form of the mineral calcite. The calcite skins gradually accumulate to form stalactites (on the roof) or stalagmites (on the floor) of the cave.

Weathering profiles can be very deep; some are hundreds of metres deep. These depths are beyond the influence of weathering processes at the ground surface, but the alteration is still called weathering. To create such profiles takes a very long time, and during that time there must be no more than a little erosion if the weathering products are not to be removed. If erosion exceeds weathering, the weathering profile will be reduced; if weathering exceeds erosion, the regolith profile gets thicker. But imagine what happens when a deep profile on, say, granite is eroded. Erosion can wash away the clay and quartz sand, but the large corestones are too big to be moved, and so accumulate on the surface. This can create a boulder field even though there is not much fresh rock immediately under the boulders. In time the bedrock will be exposed, but because the weathering front is very irregular, solid rock will stick up above the smoother slopes of the regolith as tors, hills, or even inselbergs, such as the Sugarloaf at Rio de Janeiro. This process of landscape formation by deep weathering followed by stripping of the saprolite is called 'etching' or *etchplanation*. It is thought to be very important in parts of Africa, Australia, and India. Localized examples of deep weathering and stripping are also found in cold countries such as Scotland, Sweden, and Finland.

Three climate types that have different styles of weathering are (1) glacial, where water is frozen at least part of the time; (2) humid, where plentiful water is available for weathering; and (3) arid, where water is scarce. In glacial regions frost weathering is dominant. Water that gets into cracks expands on freezing, prising away chunks of rock that are still fresh, having suffered no chemical alteration. Frost weathering produces rugged, sharp landscapes, and the debris released by frost weathering accumulates as scree slopes at the foot of cliffs. In arid regions there is not only a shortage of water, but the most effective water for weathering is usually in the form of dew. There is then the strange phenomenon of weathering being greatest in shady areas. Also in arid regions, salt accumulates and the growth of crystals is effective at breaking up rocks and buildings. Humid areas have the most chemical alteration. The greatest weathering probably occurs in the humid tropics where temperatures are high making chemical reactions more rapid and there is abundant water to take part in chemical alteration of minerals and to carry away the weathering products. Deep, red regoliths and soils are common in the tropics.

Two of the common features of weather, wind and heat, have little effect on weathering. Wind is not a weathering agent; it merely transports weathered debris. (Wind erosion is another matter.) Temperature changes also have little effect, despite the common idea that heating and cooling of rocks is a dominant weathering process, and the common tales of rocks splitting with a noise like a rifle shot. This may happen, but it is very rare.

Some hard crusts (*duricrusts*; from the Latin, *durus* = hard) can develop in association with weathering. The most common is an iron crust, commonly called 'laterite', which consists of concretions of iron oxides and hydroxides. This is generally associated with tropical areas, especially savannahs. *Bauxite*, the main ore of aluminium, also consists of small nodules, called *pisoliths*, and is generally thought to require an extremely wet climate to form. It is forming at the present time in the great bauxite deposit at Weipa in monsoonal northern Australia. A rarer kind of duricrust is silcrete, which is rich in silica. In more arid areas calcrete (consisting of calcium carbonate) is the duricrust. This is a much more mobile and faster-forming material than that of the other duricrusts.

In the past, studies concentrated on the inorganic chemistry of weathering processes. Today, there is much more emphasis on the role of organisms, especially micro-organisms. They are so effective in altering rocks and minerals that they are being harnessed for mineral processing and other commercial roles. Even resistant minerals such as zircon have been found to be etched by bacteria. Iron bacteria are plentiful, and it now seems that aluminium is largely mobilized by micro-organisms. Water from wells dug in deep regolith in Africa proved to be high in aluminium, which may present health problems for the users; and scientists at the Weipa bauxite deposit are convinced that microbiology plays a major role in bauxite formation. Explaining the shape of gold nuggets has long been a problem, because they have neither the size nor the shape of gold fragments released from bedrock by weathering. They seem to grow, indicating that gold is somehow mobile in the weathering environment. Micro-organisms seem to provide the answer. High-resolution electron microscope studies have even revealed the forms of organisms caught in the act of budding, preserved in gold. The Russian scientist Polynov regarded weathering as the pathology of rocks, and ever-increasing understanding of the importance of micro-organisms suggests that he may be right.

Some weathering profiles in Australia and India have been dated by palaeomagnetic techniques and are up to 100 million years old. A kaolin deposit in Sweden has Cretaceous oysters in their growth position on a corestone; it was therefore already weathered, eroded, and exposed on a beach by about 80 million years ago. Other ancient weathering profiles have been dated by overlying lava flows, and prove to be tens of millions of years old.

These great ages mean that the weathering products and landforms that we see today formed long ago when the climate might have been very different. For example, in Great Britain, North America, and Scandinavia many features of the present landscape relate to an ice age that ended only 10 000 years ago. The arid weathering that characterizes much of Australia is a relatively recent process, and features such as sand dunes and salt lakes are almost all less than half a million years old. But in Africa and Australia many weathering profiles date back tens or hundreds of millions of years, when much of the world was warm and wet. This means that we can use weathering profiles and landforms to work out aspects of Earth history in the same way that stratigraphers use strata and fossils. In much of Australia, for example, there was a period of great weathering in the Mesozoic, when very deep profiles were formed, and the Tertiary was a period of stripping of weathered products (etching). Early Tertiary alluvium usually contains no coarse material other than quartz (and gold), because no fresh rock was exposed to erosion. As we near the present, more and more fresh rock is included in the alluvium, and today rock outcrops are common in the landscape. Weathering can thus be studied on the same timescale as plate tectonics, mountain building, biological evolution, and global climatic change.

To come full circle, it is even possible that weathering plays a part in controlling climate. Some think that weathering consumes carbon dioxide, leading to an anti-greenhouse effect and general cooling. Others see a battle between weathering, which locks up carbon dioxide, and plant growth, which on balance releases it. Weathering is of global concern.

C. D. OLLIER

Further reading

Ollier, C. D. (1969) *Weathering*. Oliver and Boyd, Edinburgh.

Selby, M. J. (1985) *The Earth's changing surface*. Clarendon Press, Oxford.

Wegener, Alfred (1880–1930)

Born in Berlin in 1880, Alfred Wegener was to become one of the most influential geological thinkers of the twentieth century with his hypothesis of continental drift. After graduating, he joined a Danish expedition to Greenland for 2 years in 1906. He returned there in 1912 on a second expedition, having been occupied with meteorology and astronomy. In 1924 he moved to a chair in meteorology and geophysics in the University of Graz. Then, in 1930, he was once again on the Greenland icecap where he was to lose his life.

It is said that Wegener got the idea of continental drift from seeing slabs of sea ice splitting apart. As early as 1910 he was persuaded by the correspondence between opposing sides of the South Atlantic that lateral movement of the continents was possible. In 1912 he put forward the idea in two articles in journals, and followed these up in 1915 with his book *Die Entstehung der Kontinente und Ozeane*. It ran to six editions. Two were published posthumously, with translations into English, French, Russian, and Spanish.

Wegener envisaged a Mesozoic supercontinent, which he called Pangaea, and postulated that since its fragmentation the present continents had moved apart. He gave little evidence to substantiate his hypothesis, but went straight into the argument as to how drift had come about. The existing orthodoxy provided no adequate mechanism. He was unconvinced that former land bridges between continents had sunk into the ocean floor, but he was not certain what drove the continents

apart. His book had a mixed reception. Many authorities could see the relevance of evidence to support the idea, but few could suggest how such movement took place. Distinguished geophysicists, British among them, would have none of it.

After the Second World War submarine geophysics revealed the geology of the ocean floor. By the late 1960s the concept of sea-floor spreading was confirmed and the mechanism vital to continental drift was revealed. Wegener was at last vindicated.

D. L. DINELEY

Further reading

Hallam, A. (1989) *Great geological controversies* (2nd edn). Oxford University Press.

Schwarzbach, M. (1986) *Alfred Wegener, the father of continental drift*. Springer-Verlag, Berlin.

Werner, Abraham Gottlob (1749–1817)

Abraham Gottlob Werner is best remembered today as the leader of what became known as the neptunist school of geological thought. In his time he was famous as a teacher of geology at the Freiberg Mining Academy in Germany, where he had previously been a student. His wide-ranging lectures on minerals and rocks generated great enthusiasm and drew students, some of them quite senior people, from afar. His contributions to the classification of minerals and rocks were important and have stood the test of time. In his more general teaching he used the term 'geognosy' to cover topics relating to the solid Earth and the minerals and rocks of which it is composed. He put forward a stratigraphical classification (partly derived from the work of other German geologists) in which four (later five) major units (*Gebirge*) were proposed for the Earth as a whole. According to Werner, the older rocks, including granite and basalt, had been precipitated or deposited from a worldwide primeval ocean—hence the term 'neptunist'. Volcanic activity was, in his scheme, a geologically recent phenomenon. Although it was attractive to many at the time, the Wernerian theory had inherent defects and did not survive.

Even though his neptunist theory was rejected, Werner can be given credit for having discerned that the time at which a rock had been formed was more important than its composition. He also saw the potential value of fossils for correlation, and thus has a claim to have developed fundamental concepts of stratigraphy.

D. L. DINELEY

wetlands

Wetlands are areas that are permanently or periodically inundated by water that support ecological habitats adapted to wet conditions. As defined by an international convention on wetland conservation in 1971, wetlands are 'areas of marsh, fen, peatland or water, whether natural or artificial, permanent or temporary, with water that is static or flowing, fresh, brackish or salt, including areas of marine water the depth of which at low tide does not exceed six metres'. This broad definition covers an extensive range of habitats from coral reefs to upland bogs. Differences in the definitions of wetland environments have often reflected the subjectivity of interested parties and their reason for classification, which has (to the detriment of many wetlands) led to many problems in legislative protection and conservation.

Wetland areas are particularly responsive to processes governing environmental change in adjacent terrestrial and aquatic environments. They are, because of their setting, particularly sensitive to the interaction between terrestrial and aquatic processes (e.g. between land and sea, river channel and floodplain) and they consequently contain many unique habitats.

Interrelated physical, chemical, and biological processes that occur in wetland areas can be viewed as performing a variety of functions. These functions (e.g. the stabilization of sediment by vegetation, the primary production of organic matter) control both internal wetland processes (e.g. nutrient pathways, patterns of vegetation growth) and the relationships between wetland areas and other environmental systems. Wetland areas have had a chequered past so far as human activity is concerned; from the recognition of wetland functions by early human populations as a valuable resource (e.g. for the production of foodstuffs and for defence) in their natural state, to the extensive destruction and replacement of wetland areas for agriculture and industry unrelated to wetland processes.

The physical development of wetland areas is controlled principally by hydrological factors. Wetlands can therefore be broadly divided into two groups: coastal wetlands, including estuaries, lagoons, deltas, and sheltered settings on open coasts (see *mud flats*, *salt marshes*, and *mangroves*), and interior wetlands at the margins of rivers and lakes and areas of restricted drainage or elevated groundwater levels (see *peatlands and bogs*). Although a seemingly obvious distinction, the broad division between wetland areas being influenced by either salt water or fresh water is often complicated by regional and local differences in climate, hydrology, and geomorphology. For example, wetlands situated inland under a semi-arid environment or fed by a drainage basin containing evaporite sequences can display soil salinity and vegetation characteristics typical of tidally influenced coastal settings.

Viewed on a broad scale, the development of wetlands and their ability to respond to changing environmental conditions depend on the inertia of adjacent aquatic and terrestrial systems. However, the ecology of wetland areas, particularly the growth of wetland vegetation adapted to variable hydrological conditions, means that wetlands are particularly affected by shorter-term variations in environmental change and present-day conditions. In coastal and interior wetlands, successive colonization of available areas by adapted wetland species leads to a distinct zonation of habitats and plant types as, for example, from mud flat to salt marsh/mangrove or open inland water to mire and bog environments. Once colonization has sufficiently altered the hydraulic setting, colonizers are usually replaced by successive communities of plants that are less tolerant to previous growing conditions. This process is an important control in determining patterns of wetland establishment and subsequent internal physiographical evolution and biogeochemical processes.

Wetland areas are prime sites for the temporary storage of water in drainage catchments and coastal areas. Surface and subsurface drainage patterns are significantly affected by the growth of wetland vegetation and associated geomorphological changes, such as the development of salt marsh and mangrove creek systems and of valley-bogs. In both interior and coastal wetlands, the absorption of kinetic energy generated by high rates of channel flow or by wave activity is the main physical reason for the use of wetlands in flood mitigation and coastal protection.

The dissipation of energy at the terrestrial–aquatic interface results in wetlands also being preferential settings for the accumulation of organic material and deposition of sediments. The accumulation *in situ* of organic matter in wetland soils often leads to peat formation, driven by a high rate of primary productivity, a lack of physical disturbance, and a reduced rate of anaerobic decomposition. Peats developed in marshes, bogs, and mires reflect the productivity (nutrient status) of the wetland catchment and the dynamic nature of the water-table. Similarly, the deposition of sediment derived from outside the wetland reflects the broader geomorphological and hydrological processes that control the provenance of sediment and its transport to the wetland area. Because deposited wetland sediments are capable of capturing materials from a broad range of sources and timescales, and because they reflect both internal wetland dynamics and external influences, they provide an extremely valuable resource for palaeoenvironmental research (see *peatlands and bogs*).

Wetland areas play a critical biogeochemical role in linking terrestrial and aquatic systems. They act both as temporary sinks between coastal, aquatic, and terrestrial environments and as valuable sources of nutrients and organic material for them. Nutrient cycling within wetland areas is related to biogeochemical interchanges between the overlying water column, wetland vegetation, and accumulated sediment. The uptake, temporary storage, and transformation of nutrients (particularly nitrogen and phosphorous) in wetland soils is dominated by the interaction of biotic processes (e.g. plant growth, organic decomposition, and the associated activity of micro-organisms) and abiotic processes (e.g. pH, oxidation–reduction potential).

Biogeochemical conditions encountered in wetlands are also capable of removing contaminants, such as heavy metals, radionuclides, and organic pollutants, from aquatic systems. Pollutants can be temporarily taken out of the system by plant uptake, but their longer-term removal occurs by adsorption on to organic or minerogenic material and subsequent burial in wetland sediments. The natural effectiveness of wetlands in this process has led to the use of artificial and managed wetlands in wastewater treatment works. Within catchment areas they are used to intercept nutrient–pollutant loads from agricultural–industrial and domestic sources.

Archaeological evidence of early occupation sites in wetland areas has shown that communities were able to take full advantage of wetland resources while adapting to hydrological changes and wetland evolution; the Meso- and Neolithic occupation of coastal lowlands in north-west Europe provides an example. Subsequent human modification of wetland landscapes has been caused both by direct human influences (e.g. peat extraction, reclamation) and by indirect influences (e.g. alteration to drainage catchments). It is only in the past few decades that the perception of wetlands has changed. Formerly regarded as areas of low economic value useful only for reclamation or set-aside conservation, they are now seen as providing a wide range of sustainable socio-economic benefits (e.g. for flood control and coastal protection) as well as being of great ecological and cultural importance. SIMON D. TURNER

Further reading

Williams, M. (ed.) (1990) *Wetlands: a threatened landscape.* Blackwell, Oxford.

Giblet, R. (1996) *Postmodern wetlands; culture, history, ecology.* Edinburgh University Press.

white smokers *see* BLACK SMOKERS AND WHITE SMOKERS

Willis, Bailey (1857–1949) Structural geology in the first half of the twentieth century was much advanced by the work of Bailey Willis, who in his later years made laboratory models of the major structural features found in the Appalachian mountains and the Rockies. After taking degrees in mining engineering and civil engineering and a spell as geologist for the North Pacific Railway, prospecting for coal in Washington State, Willis joined the US Geological Survey. His fieldwork was in the mountainous western States, where he studied the magnificently exposed tectonic and structural features. He produced some sixty publications for the Survey during this work on structural geology.

Willis attracted much attention in 1893 when he produced a masterly account of the mechanics of formation of the Appalachian fold belt. His subsequent work maintained this high standard and his publications show great insight into the three-dimensional characteristics of differently deformed terrains.

In 1915 he was appointed professor of geology at Stanford University in California, where he continued to publish on structural geology and seismology, continental genesis, and rift valleys. Willis's use of a 'pressure box' in modelling the creation of geological structures in material of scaled-down strength to equate to rocks in the crust drew much attention. Opposed to continental drift, he said in 1943 'the theory of continental drift is a fairy tale.'

California is an active earthquake region, and Willis's seismological work was important to civil engineering there. He eventually persuaded the state to improve its building laws, so that earthquake damage and loss of life might be reduced.

D. L. DINELEY

Wilson, J. Tuzo (1908–93) James Tuzo Wilson, a Canadian geophysicist who has been called a 'benign cyclone of science', played a major role in the evolution of the theory of sea-floor spreading and plate tectonics. His interests also included science education, and in his later years he was Principal of Erindale College, Toronto and Director-General of the innovative Ontario Science Centre (1974–85). Always busy and with a broad interest in Earth sciences, as a Princeton student in the 1930s he earned extra money by working up seismic computations for Maurice Ewing at the nearby Lehigh University. His doctorate at Princeton came in 1936.

Returning after the 1939–45 war to the University of Toronto, he became interested in crustal evolution, and in 1946 recognized (simultaneously but independently of J. E. Gill) the major age provinces of the Canadian shield and their boundaries. He was active in Canada's part in the International Geophysical Year of 1957–8 and subsequently was author of *IGY: the year of the new moons* (1961).

In 1963 he believed it likely that the Hawaiian Islands had formed one or two at a time over a fixed hot spot in the mantle (a mantle plume) and then moved away to the north-west to become older and lower with time. Confirmation of this had to wait for palaeomagnetic work by D. H. Matthews and F. Vine of Cambridge. In 1965 Tuzo Wilson became a member of a special study group at Cambridge: Harry Hess, F. Vine, D. H. Matthews, and Sir Edward Bullard, intent upon deciphering palaeomagnetic data from the ocean floors. Tuzo Wilson recognized 'plates' and three kinds of margins, collisions, island arcs, and trenches. His ideas on the relationship between ocean ridges and transform faults led W. Jason Morgan of Princeton to use the term 'plates' for the first time.

By 1968 Tuzo Wilson was convinced that cycles of continental drift had occurred more than once, opening and closing ocean basins; he proposed that suture lines occur where pieces of one continent have become fused to another (e.g. Florida derived from Africa). The outlines of the 'plate tectonics' theory had been drawn.

D. L. DINELEY

wind erosion Removal of soil or sediment by wind can occur either through the action of direct fluid lift and drag forces, or by the impaction of wind-entrained sediment (see *saltation*) on the sediment surface. Wind is also capable of eroding consolidated sediment through the impact of entrained sediment on a surface, to produce ventifacts and yardangs. Eroded sediment can be transported by creep, saltation, or suspension, and where much fine soil or sediment is present, dust clouds can result (see *dust*). Sandy soils are particularly prone to wind erosion, but an increasing percentage of clay in a soil, a greater moisture content, or an enhanced vegetation cover will inhibit the erosivity. Hazards resulting

from wind erosion can induce damage to crops through abrasion and removal of seedlings and damage to soils through the removal of valuable soil constituents. It can also result in a reduction of visibility and other hazards as a result of dust storms. Most wind erosion sites are situated in arid and semiarid regions of the world. However, wind erosion and dust emission have also been reported in more temperate countries, such as New Zealand, Sweden, and the United Kingdom. An instance of particular note was the 'dust bowl' period in the United States during the 1930s.   T. LINSEY

wine and geology Is the connection between wine and geology more than a romantic notion inspired by geologists over long dreamy lunches in a sunny, vine-shrouded landscape? A proposal that there is a real connection was in fact put forward by Peigi Wallace at the International Geological Congress in Montreal in 1972. However, the extent to which this perceived association is a reality remains an issue for serious scientific investigation and debate.

Vines are preserved in the geological record and have certainly existed from early Tertiary times. The origin of wine is, however, unknown. It probably arose from the accidental but fortuitous fermentation of grapes from wild vines. The cultivation of the vine, *Vitis vinifera*, had started by the fifth century BC, and possibly before. The earliest vineyards were selected to provide a sunny aspect, commonly on a hillside slope, a sufficient depth of soil, and access to water for the roots. These basic elements remain central to the siting of vineyards to the present day and are now recognized through the use of the French term *terroir*, a word which has carried some mystique. There is no single meaning for *terroir*, but it is best defined as those physical environmental factors present in a vineyard which regulate the growth of the vine and the production of grapes. *Terroir* includes, therefore, climate (embracing air temperature, soil temperature, and rainfall), topography (embracing aspect, slope angle, and elevation, and influencing microclimate), exposure to sunlight (where topography can have a significant effect), bedrock and superficial geology, soils, and soil moisture and groundwater. Geology is therefore inextricably a part of the concept of *terroir*, whereas, as was put clearly by J. M. Hancock in his Master of Wine notes, the geology of a vineyard is 'only a matter of theoretical importance to the drinker: it is of vital importance to the grower'.

Controlling factors

Three factors appear to demonstrate the closest links between geology and wine: those of landscape; the underlying rock (*sensu lato*) and soils; and the availability of water provided by soil moisture and groundwater.

The contribution made by landscape comprises a number of interrelated components, each of which has an influence on the local climate. The best wines are derived from vineyards that are below an elevation of 500 m, although wine is successfully produced from very much higher altitudes.

Increase in elevation alone can result in a reduction in temperature (0.6 °C per 100 m), an increase in cloudiness, more rainfall, greater risk of frost, and decreased atmospheric carbon dioxide. Slopes facing the sun provide warmer conditions that are beneficial to vines. The soil conditions must, however, be such that heat is stored during the day, generally in pebbles or stone fragments, so that it can be re-radiated during cooler periods at night. By such means the influence of diurnal temperature extremes is ameliorated. Further, the cooler, night-time air which flows down a hillside is replaced by warmer air from above forming a warmer 'thermal zone' on the hillside in contrast to the cooler valley floor below. Such thermal zones are particularly well developed on the sides of isolated hills such as the hill of Corton in the Côte de Beaune and the Montagne de Reims in Champagne. A slope with a easterly aspect also carries the dual advantage of protection from the globally persistent westerly winds and exposure to the early morning rise in temperature that is experienced in the Côte d'Or and the Rhine valley.

During poor vintage years in France, notably in the Bordeaux region and in Burgundy, some vineyards have been able to produce wines of consistently higher quality. Because the other environmental controls on the vines are otherwise constant, and the wines are not blended, this phenomenon is believed to demonstrate the influence of soil type on vines and grapes. The most important properties of the soil and the underlying rock are those that influence the flow and retention of water. The origin of the soil—whether it has been transported or is a weathered rock—and the nature of the original rock-type appear in general to have less direct effect. Darker or stony, pebbly soils are more retentive of heat and can reduce surface evaporation (Fig. 1). Conversely the blinding white calcareous albariza soils of Jerez produce high-quality sherries, possibly because of reflectance qualities in high temperatures. Rocky or stony soils can transmit warmth effectively with depth (and incidentally underlie the vineyards that produce many of the world's best wines). Soil depth is important; vine growth is inhibited over shallow soils, although moderately deep soils are, in general, to be preferred. Steep slopes underlain by stony soils form appealing vineyard sites, but they can be susceptible to rapid soil erosion during storms.

Attractive though the idea may seem, no relationship has been demonstrated between natural soil chemistry, as derived from different rocks, and wine type or quality. Elements that would be released during rock weathering, such as potassium, calcium, magnesium, phosphorus, and sulphur, are significant components (200 to 2000 mg l^{-1}) of grape juice. Of these elements, potassium is particularly widespread as a component of K-feldspar and some clay minerals. Smaller quantities of a range of trace elements such as copper, manganese, iron, and zinc are also present. Organic matter contained in the soil provides a source of nutrients and also assists in creating a water-retaining soil structure.

The ideal soil-water condition would be one that can retain moisture throughout the growing season while being sufficiently free-draining to inhibit waterlogging. The water-table needs to be below the base of the main root system. The soil or weathered rock material should contain enough silt, clay, or organic material to retain water within the partially saturated zone (accessible to soil gases) above the water-table, or to be able to transmit water upwards from the water-table through the capillary fringe. A similar effect can be created by perched water tables formed by silty or clayey layers of low permeability within alluvial soils, or cemented layers in weathered, residual soils. Where water availability is reduced during the season (for example, because of a fall in the water-table), mild water stress can be helpful in inhibiting vine growth during the ripening of the grapes. The mass soil permeability, resulting from pores and fractures, and the slope of the ground-water gradient, must be sufficient to ensure free drainage. Such conditions can be provided in a range of geological situations, notably in low slopes where interlayering within a relatively thin zone of unsorted soils and weathered rock debris with different characteristics (colluvium), and access to water can provide an ideal environment.

Examples of vineyard geology

This consideration of the role of landscape, soil, and water demonstrates that there is no single set of geological parameters that determines the suitability of a vineyard or the response of a particular vinifera to being grown on a specific site. Rather it is an association of geological influences, different in each case, which appear to illustrate a causal link.

Many vineyards are located on valley bottoms underlain by alluvium which can provide deep, free-draining soils of variable grain size. Layering within the alluvium will impede water flow, thereby reducing any decline in storage during the season. The importance of alluvial gravels in providing the correct soil moisture and heat-retention properties is well illustrated by the wines of Médoc in Bordeaux and Marlborough in New Zealand, although irrigation is applied in the latter case.

Chalk and similar soft, bedded limestones provide crumbly soils that are both well drained and have the capacity to retain water in the microporous rock structures. Chalks are commonly aquifers, and water may be drawn up through a capillary fringe that can be several metres in height. However, chalk soils have low fertility unless mixed with clayey or organic material form adjacent formations, as may readily occur in colluvium. Thus Seyval Blanc has been particularly successful in English vineyards located on colluvium derived from the Chalk and the overlying clay-with-flints. These are also the special circumstances that create the natural conditions for the production of the grapes used in champagne.

The Falaises de Champagne form a series of hills east of Paris, the upper part of which is underlain by Chalk of the *Belemnitella quadrata* zone and is capped by a thin layer of Tertiary silty and sandy strata. East-facing vineyards were originally developed on colluvium-covered slopes derived from the Tertiary above and resting on the underlying Chalk. The colluvium was presumably developed under periglacial conditions when the east-facing slopes would have been subject to the most active freeze-and-thaw. The Tertiary contains an impure lignite, the '*cendres noires*', described as the 'black gold of champagne' and containing iron and sulphur derived from the decomposition of pyrite (iron pyrites). This

Fig. 1. Young vines growing in the stones and rocks of Domaine Rimauresq in Côtes de Provence. (Photograph: Cephas/Mick Rock.)

combination of clay, sand, lignite, and chalk provided a thin fertile, free-draining soil with access to groundwater in the Chalk below. It is said that Champagne grapes should come from a vine with its 'head' in the lignite-bearing Sparnacian and its 'feet' in the water-bearing Chalk. There is now rigorous delimitation, on essentially geological grounds, of those areas from which champagne grapes can be produced. Because the soil is thin it has been artificially replenished by the excavation of the Tertiary sediments and lignite on the hilltops, and its redeposition on the Chalk slopes. The soil is further supplemented and fertilized, more prosaically, by ground-up domestic refuse referred to as 'boues de ville'. As Champagne is a blend of Pinot Noir, Pinot Meunier, and Chardonnay grapes, the particular characteristics of an individual vineyard are submerged within the quality of the wine, which is stored in underground tunnels excavated in the Chalk.

The 200-km stretch of Burgundy vineyards strung along the western flank of the north–south Saône valley provides many examples of complex local relationships between soil type and wine quality. These examples can be used to demonstrate the influence of geology on *terroir*, as has been well presented by James Wilson. The east-facing scarp slope of the Côte d'Or has been formed by faulting along the Saône, and further partitioned by extensive small-scale faulting. The slopes are underlain by interbedded, gently dipping Jurassic limestones and marls, which have contributed to colluvial soils spread down the slope. *Grands* and *Premier Crus* red wines are produced from vineyards that form a well-defined zone along the hillside, which is commonly underlain by the *Ostrea acuminata* marl band or its soliflucted products. The limestones, which form the upper part of the scarp, are typically associated with vineyards producing white wines.

Clays are not free-draining and retain water that cannot be efficiently extracted by root systems. In addition, shrinkage during summer drought will cause extensive surface cracking. As a result, clay soils do not make good vineyards unless mixed with other rock materials. The vineyards surrounding Montalcino in Tuscany, from which the grapes used in the production of Brunello (arguably one of Italy's greatest red wines) are grown, are underlain by Tertiary clays interbedded with thin limestones and sandstones that, even when weathered and broken down, provide friable, and sometimes stony, soils. A calcareous component within a clay soil encourages flocculation, and this results in the formation of a crumbed structure that aids drainage. Vineyards on the three different soil types surrounding Montalcino are associated with wines having distinct characteristics.

Weathered igneous and metamorphic rocks generally result in stony, well-drained, and relatively unfertile soils. Granitic soils nevertheless occur in many wine-producing areas, such as Beaujolais, the Rhône, Sardinia, northern Portugal, and the Cape. Soils formed on volcanic debris lining the sides of the Napa Valley in California result in poor yields but intensely flavoured fruit.

The Coonawarra vineyards, which produce some of Australia's finest Cabernet Sauvignon wine, exploit a narrow belt of highly productive terra rossa soil overlying water-bearing limestone, from which it was derived by weathering. Similar red-brown soils with ferricretes (iron-rich crusts) overlying limestone occur in the Southern Vales south of Adelaide, which have a reputation for 'ferruginous' dry reds produced from both Shiraz and Cabernet Sauvignon.

JOHN KNILL

Further reading

Pomerol, C. (1989) *The wines and winelands of France: geological journeys.* Robertson McCarta, London.

Robinson, J. (1994) *The Oxford companion to wine.* Oxford University Press.

Wallace, P. (1972) Geology of wine. *Proceedings of the 24th International Geological Congress,* **6**, 359–65.

Wilson, J. E. (1989) *Terroir: the role of geology, climate and culture in the making of French wines.* University of California Press, Berkeley.

Woodward, John (1665–1728)

Although he was Professor of Physic at Gresham College, John Woodward spent much of his time and resources building up the first great systematic collection of minerals, rocks, and fossils. He also was concerned with the more philosophical and historical aspects of Earth science. He was known as a 'super-diluvialist' because he held that sedimentary strata were the product of a universal deluge during which pre-existing rocks were comminuted, the detritus then settling out according to gravity and entombing the remains of organisms of the time as fossils. A critical part of the process involved the temporary suspension of gravity while the sediments were broadcast before settling. This was described in his *Essay toward a natural history of the Earth* (1695). It prompted widespread interest and critical comment.

Then came *The brief instructions for making observations in all parts of the world* (1696), which laid out the rationale of systematic observation and recording of geological phenomena. A pioneer of its kind where specimens were concerned, it became a model of museum practice for many decades.

While adding to his collection, which ran to thousands of items arranged according to their physical characteristics and provenance, Woodward wrote his *Classification of English minerals and fossils* (1729). This was the best classification of the time and was used for many years.

Woodward's collection was left to the University of Cambridge. His property was auctioned to provide for a lectureship in 'natural philosophy'. The topics of the lectures given were to be based upon Woodward's own views of the Earth, its components and history. From this eventually was established the Woodwardian Chair at Cambridge, while Woodward's specimens became the nucleus of the University's Sedgwick Museum.

D. L. DINELEY

Worldwide Standardized Seismographic Network

One of the most significant advances in earthquake seismology occurred in the early 1960s when a global network of seismograph stations was established. This Worldwide Network of Standard Seismograph Stations, as it was originally called (now referred to as WWSSN), and the associated development of a global interchange of seismographic data between co-operating nations, expanded and improved views of seismicity on a worldwide scale. Many of these observations of global seismicity helped to lay the foundation for the theory of plate tectonics. In large part because of the WWSSN, the quality, quantity, and process of exchanging seismographic data improved dramatically after the 1960s. As a result, the number of earthquake epicentres located increased significantly, and there was a significant improvement in accuracy.

The idea of the WWSSN had its origin in 1959 as part of the United States' project VELA Uniform. The aim of this project was to improve the capability for detecting and identifying underground nuclear explosions on a global scale. The US Coast and Geodetic Survey, under advice from a special committee of the National Academy of Sciences, subsequently designed and installed, generally at existing seismograph stations, a standardized set of instruments and accurate clocks. The instruments included a set of three orthogonally oriented (two horizontal and one vertical), short-period (1 second) seismometers and a similarly oriented set of long-period (15 or 30 seconds) seismometers. The crystal clock, which is crucial in providing uniform and reliable timing between individual stations, was calibrated by a standard radio broadcast. Recording was on paper. At its peak in the early 1970s, there were about 115 continuously recording WWSSN stations distributed throughout most of the world, with the exception of some communist and former communist countries.

More recently, the WWSSN has been gradually replaced by global initiatives that have utilized broad-band digital instruments. In particular, with encouragement from the National Academy of Sciences, efforts began in 1983 to upgrade the WWSSN. As a result, a consortium of universities in the United States banded together to form the Incorporated Research Institutions for Seismology (IRIS). One of the goals of IRIS is to develop a 150-station Global Seismographic Network (GSN). In 1999, the GSN was more than two-thirds complete.

IVAN G. WONG

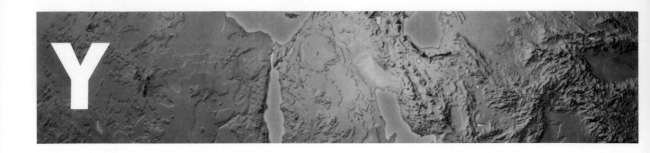

Y

yardangs yardangs are aerodynamically shaped desert land-forms produced by the wind erosion of bedrock or sediments. They tend to be composed of parallel ridges and troughs with the shape of an upturned ship's keel. They range in size from small masses only a metre or two high to great ridges that are several hundred metres high, run for tens of kilometres, and are spaced at intervals of one to two kilometres. They occur in both unconsolidated Pleistocene materials and in resistant dolomites, sandstones, and even gneisses. Their importance in arid landscapes was fully appreciated only when satellite images revealed their great areal extent and regularity of form in areas like the central Sahara, near Luxor in Egypt, and in the Lut Desert of Iran. The ridges run parallel to the local winds and the intervening depressions result from a combination of deflation and sand abrasion achieved by high-velocity winds in areas with a minimal vegetation cover. It is possible that in some instances these processes may be aided by the presence of joints or pre-existing trellis drainage systems.

The availability of remotely sensed images of some of the planets gave a considerable impetus to yardang studies when broadly analogous features to those on Earth were found on Mars. ANDREW S. GOUDIE

Younger Dryas stade The *Younger Dryas* was a cold interval (or stade) that occurred near the end of the last glaciation, immediately prior to the beginning of the present interglacial period (i. e. the Holocene). In geological terms, it was a relatively brief and minor climatic event, although it has attracted a great deal of attention from scientists because of the light it can shed on the causes and mechanisms of climatic change, and the ways in which climate signals in one region can be transmitted to other parts of the Earth.

The Younger Dryas was first recognized in studies of organic sediments preserved in bogs and lake floors in Scandinavia. Pollen grains, buried in the sediments as they accumulated, showed that the succession of vegetation that colonized the land at the end of the last glaciation was interrupted on one or more occasions, and warmth-loving plants were replaced by tundra-type vegetation, such as grasses, sedges, dwarf shrubs, and mosses. The name 'Younger Dryas' was applied to the most recent and pronounced of these late-

glacial cold intervals, and has received the most attention. The name comes from *Dryas octopetala* or Mountain Avens, an attractive white and yellow flower related to the strawberry, which is prominent in the pollen records of those times for lowland Scandinavia. The plant is still common on calcareous soils in some Scandinavian mountain areas, where it can form a distinctive community known as Dryas heath.

The timing of the Younger Dryas was first established using radiocarbon dating and assigned to the approximate interval 11 000 to 10 000 years before the present (BP). The radiocarbon timescale is now known to be in error for the late-glacial period, owing to rapid changes in global carbon

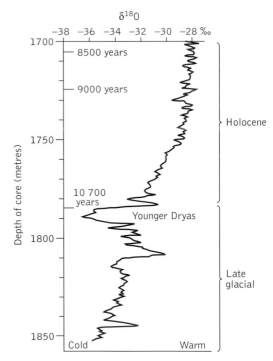

Fig. 1. Part of the oxygen isotope record from the Dye-3 ice core from south Greenland, spanning the Younger Dryas and early Holocene. Low $\delta^{18}O$ values correspond to low temperatures, and vice versa.

reservoirs at that time. The true date of the Younger Dryas has now been established using cores of ice drilled from the Greenland ice sheet. Careful study of annual ice layers that built up near the centre of the ice sheet has established that the onset of the Younger Dryas was about 12 800 calendar years ago, and its termination about 11 600 years ago. The Greenland ice cores provide such a clear and strong record of the Younger Dryas that they can be used as a yardstick against which other types of evidence can be compared (Fig. 1). Reconstructed temperatures, based on the isotopic composition of oxygen locked up in the ice, are about 7 degrees C colder in central Greenland during the Younger Dryas than in the preceding and succeeding warm periods. The ice-core record shows that the cooling at the start of the Younger Dryas was gradual, but the warming at the end was exceedingly rapid, occurring within a few years or decades at most. The rapidity of this climatic change, which is also recorded in marine and terrestrial records from around the North Atlantic, is one of the most intriguing aspects of the Younger Dryas; it is also characteristic of other relatively short-lived episodes during the last glacial period. The Greenland ice cores also record changes in atmospheric composition. Air bubbles trapped within the ice show that during the Younger Dryas concentrations of the 'greenhouse' gases, carbon dioxide and methane, were lower than modern natural levels, and were similar to those of full glacial times. Methane in particular shows a large oscillation, closely following the pattern of temperature change. Finally, precipitation changed during the Younger Dryas, with snow accumulation about half of that for the succeeding early Holocene. This evidence for low precipitation is compatible with the temperature record, since cold air cannot carry as much moisture as warm air.

A less pronounced Younger Dryas cooling might also have occurred in Antarctica, where oxygen isotope records from some cores indicate an interruption or reversal of the warming trend at the end of the last glacial period. At present, it is not certain whether this event occurred in exactly the same time-frame as the Younger Dryas, although synchronous drops in carbon dioxide and methane concentrations, similar to those in the Greenland cores, make it appear very likely.

Studies of microfossils contained in sea-floor sediments show that, in the North Atlantic during the Younger Dryas, cold surface water extended as far south as the northern coast of Spain (Fig. 2). At present, poleward transfer of warm, saline surface water—the North Atlantic Drift—has a major warming effect on western Europe, resulting in January sea-surface temperatures as high as 4 °C off western Britain, at the same latitude as Labrador. The North Atlantic Drift is balanced by the southward transfer of cold water at depth, known as North Atlantic Deep Water (NADW), which forms as the saline surface water cools and sinks near Greenland. This circulation—the 'Atlantic conveyor'—is very important in transferring heat through the world's oceans, and is crucial in maintaining the temperate climate of the North Atlantic region. Geochemical studies indicate that during the Younger Dryas NADW formation either halted or took place in much weaker form further south. The Younger Dryas is much less apparent in other oceans, although there is some evidence for changes in the circulation of North Pacific waters at that time. Ventilation of waters off the coast of California was more efficient than at present, whereas the deep waters of the Japan Sea, which are well ventilated today, became poor in oxygen (anoxic). Additionally, marine microfossil evidence points to lower sea-surface temperatures off parts of South-East Asia.

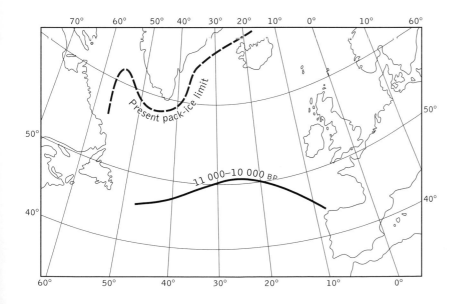

Fig. 2. The position of the oceanic polar front and southern limit of pack ice in the North Atlantic for the Younger Dryas and the present day.

On land, the Younger Dryas is clearly represented in bio-stratigraphic records from the Atlantic fringes of Europe and North America. The severe cooling recorded by changes in vegetation is also well represented by insect remains, particularly beetles (Coleoptera). As shown by Russell Coope at the University of Birmingham, beetles are excellent indicators of past temperatures, since they have well-defined ranges and can migrate rapidly in response to environmental change. Beetle remains in bog and lake sediments indicate that in central England the mean temperatures of the warmest and coldest months were around 10 °C and −17 °C respectively, some 7 and 20 degrees below those of today. The beetle record also shows that temperature increases in Britain at the end of the stade were very rapid: about 1.7–2.8 degrees per century.

In Europe, the Younger Dryas cooling is most pronounced in Britain, southern Scandinavia, northern Germany, the Low Countries, and northern France, and declines in importance to the north, east, and south. On the other side of the Atlantic, the event is clearly recognizable in the maritime provinces of Canada and the north-eastern United States, but is less apparent further west. A short-lived cooling is also indicated by some pollen records from the Pacific coast of North America, as far north as Alaska. Some scientists believe that the Younger Dryas can be recognized in pollen profiles from the northern Andes, Chile, Japan, and Australia, but others disagree.

There is intriguing evidence that the Younger Dryas brought changes in wind circulation and moisture balance to some low-latitude continental areas. Lake levels in East Africa fell, indicating dry conditions and a reduction in the on-land transfer of moist air from over the Indian Ocean. The Tibetan plateau is also thought to have been arid at that time. In contrast, the plateaux of Central China appear to have had a wetter climate than in the preceding and succeeding periods. This wet climate has been attributed to a heightened temperature gradient between a warm land surface and a relatively cool western Pacific, and a consequent strengthening of the summer monsoon. Additionally, the Altiplano in the high Andes also appears to have been wetter at that time.

The climate of the Younger Dryas created the conditions for glacier expansion, especially in north-west Europe. The Scandinavian ice sheet, which had been in retreat until about 12 000 (radiocarbon) years ago, underwent a significant read-

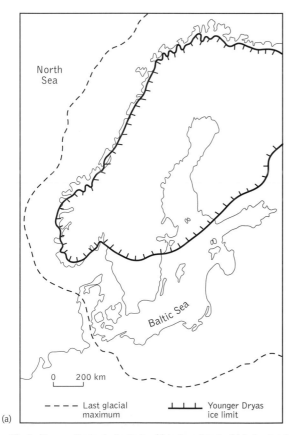

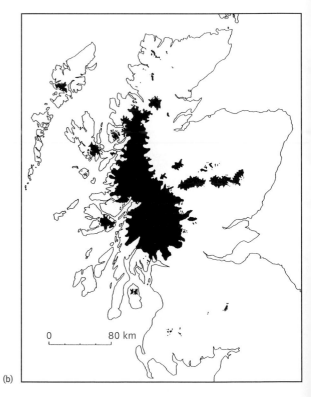

Fig. 3. Younger Dryas glacier limits: (a) in Scandinavia, (b) in Scotland.

vance and formed prominent moraines at the limit of ice advance in Norway, Sweden, and Finland (Fig. 3). Ice cover in Britain had probably been reduced to isolated remnants in Scotland between 13 000 and 12 000 radiocarbon years ago, but during the Younger Dryas a substantial ice field formed over the western Scottish Highlands (locally known as the Loch Lomond readvance) and smaller mountain glaciers became established elsewhere in Britain and Ireland (Fig. 3). Younger Dryas glacier advances of smaller magnitude have been identified in the European Alps, the North American Cordillera, the Andes, and New Zealand, although dating evidence is fragmentary and far from certain in many areas. In some areas, the glacial record hints at climatic changes within the Younger Dryas. In western Scotland, retreat of some glaciers was well under way before the temperature increase that marked the end of the stade, suggesting that deglaciation may have begun in response to a decrease in snowfall. Changes in the abundance of certain pollen types also indicate that the Younger Dryas in north-west Europe consisted of a moist, cold period followed by a more arid (but still cold) period.

Any explanation of the Younger Dryas must account for its varying regional effects, particularly the severe cooling centred on the North Atlantic, and a less pronounced reorganization of oceanic and atmospheric circulation elsewhere. The Younger Dryas is unrelated to the periodic changes in the Earth's orbit around the Sun which modulate the amount of solar radiation reaching high northern latitudes and govern the timing of major glacial episodes (Milankovitch cycles). Instead, the Younger Dryas event appears to have been caused by variations in solar radiation or by internal changes to oceanic circulation, ice-sheet dynamics, and the atmosphere in the North Atlantic region, from where it was transferred to other regions by atmospheric processes.

One interesting model, developed by Wallace Broeker and George Denton, is that the North Atlantic Younger Dryas cooling was due to re-routing of melt water from the retreating North American ice sheet. During the early stages of ice sheet retreat, a huge lake—known as Lake Agassiz after the nineteenth-century Swiss glaciologist—was ponded against the ice margin in what is now southern Manitoba (Fig. 4). At first, the overspill from the lake drained southwards via the Mississippi into the Gulf of Mexico, but about 11 000 (radiocarbon) years ago the retreating ice allowed a new escape route to open up to the east, through the Great Lakes region and along the St Lawrence River. The rate of water flow out of the lake is estimated to have been as high as 30 000 cubic metres per second; this is about one-sixth of the present-day discharge of the Amazon, and represents a huge influx of cold,

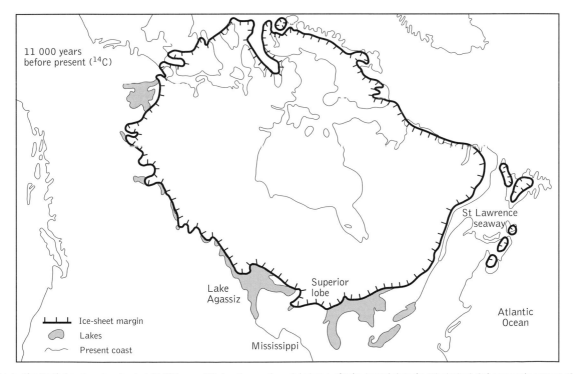

Fig. 4. The North American ice sheet at 11 000 years BP, showing southward drainage of Lake Agassiz into the Mississippi. Subsequently, retreat of the Superior Lobe allowed drainage eastward into the St Lawrence, which persisted throughout the Younger Dryas.

fresh water into the North Atlantic. The resulting reduction in the salt content of Atlantic surface waters might have been sufficient to prevent the formation of NADW, disrupting the Atlantic conveyor and reducing the poleward transport of warm surface water. This effect might have been enhanced by additional fresh water derived from the melting of icebergs, broken from the margins of expanded glaciers in the region. The end of the Younger Dryas coincides with a readvance of an ice lobe in the Lake Superior area, which cut off the eastern outflow from Lake Agassiz, re-routing melt water back down the Mississippi.

Although several lines of evidence appear to support this model, some problems suggest that it may not be the whole answer. First, a second period of drainage from the Lake Agassiz basin to the North Atlantic occurred after 10 000 years BP, yet failed to produced any regional cooling. Second, the Younger Dryas is only one of the several short-lived cold episodes that punctuated the last glaciation in the North Atlantic region; this suggests that they may share a common origin. The Greenland ice-core record for the last glaciation shows that colder than average conditions, known as Dansgaard–Oeschger events, occurred every 2000 to 3000 years during the last glaciation. Each cold event consisted of a gradual cooling followed by a rapid warming; and several culminated in massive discharges of icebergs into the North Atlantic, which are thought to reflect the break-up of advancing ice-sheet margins. The Dansgaard–Oeschger events have been attributed to flow instabilities, or surges, of the North American ice sheet, but recent research suggests that ice margins throughout the region advanced simultaneously during many of the events, and that an oceanic or climatic trigger is more likely. Possibilities include oscillations of the Atlantic conveyor due to ocean–ice-sheet interactions, or some hitherto unknown climatic cycle, possibly driven by changes in solar radiation. It is therefore possible that the Younger Dryas is part of a larger, little-understood pattern, rather than a unique event requiring a special explanation.

Current research is focused on how changes in oceanic circulation and cooling in the North Atlantic region can be transferred to other parts of the world, particularly the south- ern hemisphere. Changes in evaporation and latent heat transport in the North Atlantic might have had 'knock-on' effects on atmospheric circulation, but these are unlikely to have been large or widespread enough to account for the apparently global nature of the cooling. One possibility is that the North Atlantic cooling was amplified and transmitted worldwide by reduced concentrations of carbon dioxide and methane. Short-term fluctuations in such trace gases reflect changes in the amount stored in vegetation, sediments, and the oceans; so the Younger Dryas drop in concentration may be due to changes in biological productivity or oceanic ventilation. Once initiated, the decline in carbon dioxide and methane might have caused additional global cooling as smaller amounts of terrestrial long-wave radiation were retained in the atmosphere. Other proposed mechanisms for amplifying the Younger Dryas cooling include the formation of sea ice in the northern oceans, lowering the planetary albedo and reducing net receipt of short-wave radiation, and the effect of high concentrations of atmospheric aerosols, such as sea-spray and wind-blown silt, held aloft by vigorous air circulation. Resolution of the problems surrounding the origin, extent, and termination of the Younger Dryas is of more than academic interest, as they impinge on a number of mechanisms central to the functioning of today's atmosphere and oceans.

DOUGLAS I. BENN

Further reading

Atkinson, T. C., Briffa, K. R., and Coope, G. R. (1987) Seasonal temperatures in Britain during the past 22 000 years, reconstructed using beetle remains. *Nature*, **325**, 587–92.

Bond, G., Broeker, W., Johnsen, S., McManus, J., Labeyrie, L., Jouzel, J., and Bonani, G. (1993) Correlations between climate records from North Atlantic sediments and Greenland ice. *Nature*, **365**, 143–7.

Broeker, W. S. and Denton, G. H. (1990) What drives glacial cycles? *Scientific American*, January 1990, 43–50.

Lehman, S. J, and Keigwin, L. D. (1992) Sudden changes in North Atlantic circulation during the last deglaciation. *Nature*, **356**, 757–62.

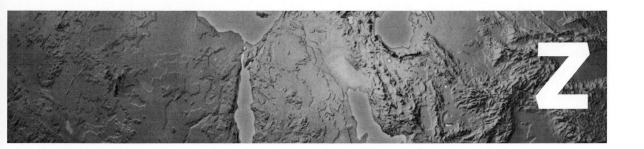

zinc deposits *see* LEAD–ZINC DEPOSITS

zone fossils Zone fossils or index fossils are fossils that characterize a particular time period or biozone. This is the basic unit of biostratigraphy in which fossils are used in the correlation of rock units. Local units can thus be correlated on a worldwide basis. The concept was originated by William Smith, a surveyor, in the latter part of the eighteenth century. Smith realized that although lithology could not be relied upon to carry correlation over long distances the fossils could be used as indices. The potential of different organisms as zone fossils varies, however, and to be most useful they should embody certain characteristics. They must have a high preservation potential, be relatively common, and also be distinctive so that accurate identification is possible. They must have a wide lateral distribution but a short vertical range so that they characterize a short time period over a wide area. It is also very helpful if they are independent of environment, as for example is the case for wind-borne spores and pollen. Although all these characteristics are rarely found in one organism there are a number of groups that are particularly useful in specific time periods. In the Palaeozoic trilobites and graptolites have been used extensively; ammonites are particularly useful in the Mesozoic, and microfossils such as foraminifera are widely used in the Tertiary.   DAVID K. ELLIOTT

zoogeomorphology Zoogeomorphology is the study of the geomorphological effects of animals, including both vertebrates and invertebrates. The geomorphological effects encompass the processes of erosion, transport, and deposition of rock, soil, and unconsolidated sediments. Zoogeomorphological effects specifically involve the movement of rock, soil, or unconsolidated sediment from one location to another, whereas faunal bioturbation involves mixing of rocks and sediment without net transport from a site. The geomorphological effects of animals can be categorized into direct effects involving actual excavational or contructional activities accomplished by the processes of digging for and caching of food, burrowing and nest-building, mound-building (usually in association with burrowing), wallowing, geophagy and lithophagy (animal consumption of soil or rocks), and dam-building by beavers (genus *Castor*); and indirect effects arising through animal influences (trampling,

tunnelling, burrowing, vegetation removal) on surface and subsurface infiltration, soil creep, surface wash, and rainsplash detachment.

Burrowing encompasses features ranging from shallow daybeds and spawning redds to extensive underground tunnel complexes. Burrowing bioturbates surface and subsurface sediment, leads to deposition of sediment in the form of surface spoil heaps and related distinctive landforms such as termite and ant mounds, and can cause surface collapse through complete undermining. Some spoil heaps and burrow complexes are at scales sufficiently large to be discernible on aerial photographs and satellite imagery (e.g. wombat warrens in Australia). Burrowing and excavation of nest cavities by Atlantic puffins led to the undermining and destruction of the island of Grassholm, an 8.9 ha island, off the south-western coast of Wales. Indirectly, burrowing produces profound geomorphological results through its influence on soil texture and structure, fertility, and infiltration capacity; and the resulting changes, both in production of vegetation cover and in surface run-off and erosion.

Mima mounds are earth mounds, approximately 20–30 m in diameter and 2 m high, found throughout the western two-thirds of North America. Seen from the air, they resemble hundreds to thousands of spheres partly buried in the ground. Mima-like mounds have also been reported from the highlands of Kenya in eastern Africa, and Argentina. The mounds are typically composed of unconsolidated fine soil, varying in texture from loamy sands to clay loams. Small stones are more concentrated in mound soils than in intermound soils, but larger stones ranging in size up to cobbles and boulders are commonly exposed in the intermound zones. Cross-sections through mounds reveal a lenticular (biconvex) outline. Five scientific hypotheses for the origin of Mima-type mounds have emerged. These include wind or water erosion, wind deposition, periglacial freeze–thaw dynamics, seismic activity, and soil translocation by fossorial rodents or other animals. The fossorial rodent, or Dalquest–Scheffer, hypothesis states that Mima mounds are the result of soil translocation by burrowing rodents. Objections to the fossorial rodent hypothesis include observations that gophers are widely distributed but mounds are not, and gophers are missing from many mound sites, including those of the type area at Mima prairie, Washington State.

By far the most controversial, and unlikely, geomorphological features attributed to burrowing are the Carolina Bays of the south-eastern Coastal Plain of the United States. These 'bays' are shallow, elliptical, wetland depressions, typically characterized by a long axis orientation from north-west to south-east and a prominent sand rim. Many bays are at least 1 km long, and over 500 000 bays exist from southern Georgia to Maryland. Although current thinking suggests that they were created by south-west–north-east prevailing winds during the Pleistocene, one school of thought attributed their creation to vast quantities of schooling, spawning fish, creating tidal-flat excavations during the Pleistocene.

Trampling by animals can, directly or indirectly, lead to erosion. It is a direct agent of erosion when trampling along the edge of a stream, pond, turf terrace, or erosion pan causes hoof or paw chiselling and sloughing and erosion of banks. Natural animal trails produced by trampling are widespread; examples include trails produced by elephants and hippopotamuses in tropical grasslands, bison in the semi-arid grasslands of the Great Plains, mountain goats in the Rocky Mountains, and caribou in the Arctic tundra of the Barren Grounds of Canada's Northwest Territories. The indirect effects of trampling are difficult to quantify, but can be categorized as the reduction or complete removal of vegetation, and an attendant increase in the bulk density of the soil, leading to lowered infiltration capacity and increased surface run-off.

Animal excavations in search of food sources can result in pit-and-mound topography and the removal of large quantities of sediment. Grey whales feeding on the shelf of the Bering Sea 'slurp' large quantities of ocean-floor sediment from which they extract benthic invertebrates, leaving behind a pitted ocean floor. Sediment is expelled near the surface as large mud plumes. Approximately 16 000 whales feed in the Bering Sea annually, and resuspend a minimum of 120 million m^3 of sediment, which is nearly three times the annual load of suspended sediment discharged into the Bering Sea by the Yukon River.

Wallowing in mud or dust is a common activity among many animals, ranging from small rodents to elephants, African buffalo, seals, and beluga whales. Quantitative data on the effects of wallowing are extremely limited. However, it has been pointed out that nearly every natural depression in the American and Canadian Great Plains has at one time or another been referred to, sometimes in jest and sometimes in all seriousness, as a 'buffalo wallow' resulting from wallowing by American bison. The abrupt cessation of this geomorphological impact, owing to the near-extinction of the bison through overkilling following European contact, must stand as one of the most striking examples of the creation of a disequilibrium landscape anywhere on Earth.

Probably more than any other animal except humans, beavers geomorphologically alter the landscape through their dam-building and related activities. Damming of streams creates appropriate habitat for the animals and may completely alter the local drainage pattern. Hydrological effects of beaver dams include the creation of ponds, diversion channels, and multiple surface-flow paths; the reduction of downstream discharge during dry periods below dams; alteration of discharge during high flow; and alteration of the overall water balance. Beavers may act as agents of karstification; in Ontario, Canada, beaver ponds acted to concentrate water that subsequently disappeared through karstic sink-holes in the base of the ponds. Dams and canals were built by beavers to combat lowering water levels, but only served to further concentrate water and exacerbate the karstification.

Indirect effects associated with animals, particularly with burrowing and tunnelling, include lowering of soil bulk density and increasing soil porosity (with accompanying increases in the moisture-retentive capability of a soil, increasing levels of organic matter, alteration of soil pH, and increasing nutrient levels). The soils of South African ant mounds, for example, have higher infiltrability, and significantly more organic matter, phosphorus, potassium, and nitrogen than soils of intermound areas. Similar results have been described for mound soils of ants in the western USA. The presence of Quaternary stone lines (a differentiated zone in the upper part of soils) has also been attributed directly to pedoturbation by ants.

D. R. BUTLER

Further reading

Butler, D. R. (1995) *Zoogeomorphology – animals as geomorphic agents.* Cambridge University Press.

Appendix 1
Geological timescales

Appendix 1a **Global stratigraphic chart** [Cowie and Bassett 1989]

Left chart:

Eonothem	Erathem	System	Series	Geochronometry Ma BP	Magnetostrat
Phanerozoic	Paleozoic / Upper Paleozoic	Permian	WG		
			Upper	255 / 260 / 270	
			Lower	280 / 290	
			WG		
		Carboniferous / Upper Subsystem	Steph-anian — Gzhelian		Mainly reverse
			West-phalian — Moscovian	300	
			Namurian — Bashkirian	310	
			WG / Serpukov.		
		Carboniferous / Lower Subsystem	Viséan	325	Mixed
			Tournaisian	(342)	
			WG	355	
		Devonian	Upper	375	
			G33P / Middle		Mainly reverse
			GSSP / Lower	390	
			GSSP	410	Mixed
		Silurian	Upper		
			GSSP / Ludlow	424	Reverse
			GSSP / Wenlock	428	
			GSSP / Llandovery		
		GSSP		435	
	Lower Paleozoic	Ordovician / Upper / Cincinnatian	Ashgill		Mainly normal
			Caradoc	446 / (454)	
		Ordovician / Lower / Cham-plainian / Can.	Llandeilo-llanvirn	455 (473) 470	
			Arenig	490	Mixed
			Tremadoc	510	
			WG		
		Cambrian	Upper	(523)	Mainly reverse
			Middle		Normal
			Lower	530 / 550	Mainly reverse

Right chart:

Eonothem	Erathem	System	Series	Geochronometry Ma BP	Magnetostrat
Phanerozoic	Cenozoic	Quaternary	Holocene	(0.1)	Brunches (Magnetochron. 1)
			Pleistocene — Upper		
			WG / Middle		Matu Yama (Magneto-chron. 2)
			WG / Lower		
		GSSP		1.6	
		Neogene	Pliocene — Upper	3.3(3.2)	Gauss (3)
			WG / Lower	5.3 (4.8)	Gilbert (4)
			Miocene — Upper	6.5 / 11(11.3)	
			Middle	(15) (16.5)	Mainly mixed (Magneto-chrons. 5–34)
			Lower	19(21) / 23(23.7)	
			WG		
		Paleogene	Oligocene	37(30) (36.5)	
			WG	34(40)	
			Eocene	39(43.6) / 45(52)	
			WG	53(57.8)	
			Paleocene	59(62.3) / 65(64.4)	
		GSSP		65(64.6)	
	Mesozoic	Cretaceous	Upper	83 / 86 / 88 / 91 / 95	Predominantly Normal
			Lower	107(112) / 114 / 116 / 120 / 128(135)	
			WG	135(140)	Mixed (Anomalies M1–M29)
		Jurassic	Upper	139 / 144 / 152 / 159	
			Middle	170 / 176	Mainly normal
			Lower	180 / 188 / 195 / 201	Mainly Mixed
			WG	205	
		Triassic	Upper	210 / 220 / 230	Mainly normal
			Middle	235 / 240	Reverse
			WG / Lower	250	

Note: Appendix 1a, the 1989 global stratigraphic chart compiled by J. W. Cowie and M. G. Bassett, uses rock-stratigraphic divisions (eonothem, erathem, system, series, stage), whereas Appendix 1b, the Phanerozoic timescale of F. Gradstein and J. Ogg (1996), uses equivalent time-stratigraphic divisions (eon, era, period, epoch, age).

These tables are reproduced as they appear in the original sources, without change. Note that both use American spelling (Paleocene, Paleogene, Paleozoic), not British spelling (Palaeocene, Palaeogene, Palaeozoic).

Note added in proof. A new edition of the International Stratigraphic Chart compiled by Jürgen Remane, A. Fuare-Muret, and G. S. Odin was published by the Division of Earth Sciences, UNESCO, in 2000 for the International Commission on Stratigraphy with the collaboration of all subdivisions of the ICS. The age values given in this chart differ by a few million years from those of Gradstein and Ogg (1996), but these small refinements do not significantly change the magnitudes given here.

GEOLOGICAL TIMESCALE

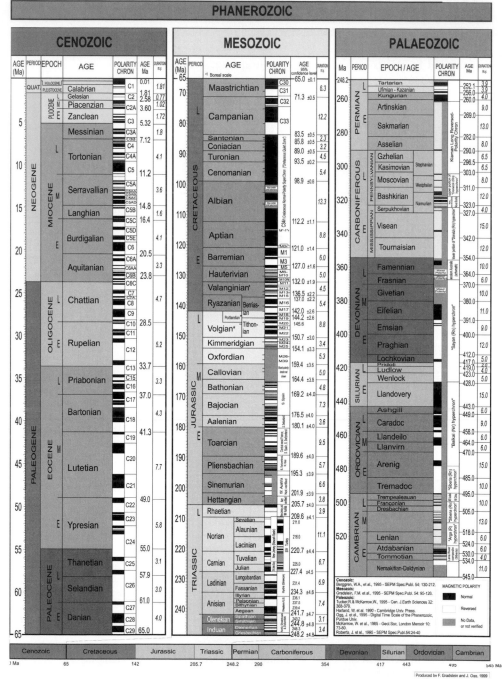

Appendix 2
The Earth and the Solar System

	Mass (kg)	Mean density (g cm^{-2})	Equatorial diameter (km)	Mean distance from Sun (km)
Sun	2.0×10^{30}	1.41	1.393×10^6	—
Mercury	3.3×10^{23}	5.43	4.878×10^3	57.9×10^6
Venus	4.87×10^{24}	5.25	12.10×10^3	108.2×10^6
Earth	5.98×10^{24}	5.52	12.76×10^3	149.6×10^6
Mars	6.42×10^{23}	3.95	6.794×10^3	227.9×10^6
Jupiter	1.90×10^{27}	1.33	142.8×10^3	778.3×10^6
Saturn	5.68×10^{26}	0.71	120×10^3	1427×10^6
Uranus	8.70×10^{25}	1.29	51.2×10^3	2870×10^6
Neptune	1.03×10^{26}	1.64	48.6×10^3	4497×10^6
Pluto	0.01×10^{22}	2.03	2.302×10^3	5954×10^6

Additional data for the Earth

Rotational period: 23 hours 56 minutes 4 seconds
Orbital period: 365.26 days

The Moon

Mass: 7.35×10^{22} kg; mean density: 3.34 g cm^{-3}
Equatorial diameter: 3476 km
Mean distance from Earth: 384.4×10^3 km

Appendix 3
The periodic table

	1 I	2 II		3 III	4 IV	5 V	6 VI	7 VII	8	9 VIII

s

2	3 **Li** RAM: 6.941 P: 0.98 Lithium	4 **Be** RAM: 9.012182 P: 1.57 Beryllium
3	11 **Na** RAM: 22.98977 P: 0.93 Sodium	12 **Mg** RAM: 24.305 P: 1.31 Magnesium
4	19 **K** RAM: 39.0983 P: 0.82 Potassium	20 **Ca** RAM: 40.078 P: 1 Calcium
5	37 **Rb** RAM: 85.4678 P: 0.82 Rubidium	38 **Sr** RAM: 87.62 P: 0.95 Strontium
6	55 **Cs** RAM: 132.9054 P: 0.79 Cesium	56 **Ba** RAM: 137.327 P: 0.89 Barium
7	87 **Fr** RAM: 223 P: 0.7 Francium	88 **Ra** RAM: 226.0254 P: 0.9 Radium

d

21 **Sc** RAM: 44.95591 P: 1.36 Scandium	22 **Ti** RAM: 47.88 P: 1.54 Titanium	23 **V** RAM: 50.9415 P: 1.63 Vanadium	24 **Cr** RAM: 51.9961 P: 1.66 Chromium	25 **Mn** RAM: 54.93805 P: 1.55 Manganese	26 **Fe** RAM: 55.847 P: 1.83 Iron	27 **Co** RAM: 58.9332 P: 1.88 Cobalt
39 **Y** RAM: 88.90585 P: 1.22 Yttrium	40 **Zr** RAM: 91.224 P: 1.33 Zirconium	41 **Nb** RAM: 92.90638 P: 1.6 Niobium	42 **Mo** RAM: 95.94 P: 2.16 Molybdenum	43 **Tc** RAM: 98 P: 1.9 Technetium	44 **Ru** RAM: 101.07 P: 2.2 Ruthenium	45 **Rh** RAM: 102.9055 P: 2.28 Rhodium
71 **Lu** RAM: 174.967 P: 1.27 Lutetium	72 **Hf** RAM: 178.49 P: 1.3 Hafnium	73 **Ta** RAM: 180.9479 P: 1.5 Tantalum	74 **W** RAM: 183.85 P: 2.36 Tungsten	75 **Re** RAM: 186.207 P: 1.9 Rhenium	76 **Os** RAM: 190.2 P: 2.2 Osmium	77 **Ir** RAM: 192.22 P: 2.2 Iridium
103 **Lr** RAM: 260 P: Lawrencium	104 **Rf** RAM: 261 P: Rutherfordium	105 **Db** RAM: 262 P: Dubnium	106 **Sg** RAM: 263 P: Seaborgium	107 **Bh** RAM: 262 P: Bohrium	108 **Hs** RAM: 265 P: Hassium	109 **Mt** RAM: 266 P: Meitnerium

f

57 **La** RAM: 138.9055 P: 1.1 Lanthanum	58 **Ce** RAM: 140.115 P: 1.12 Cerium	59 **Pr** RAM: 140.9077 P: 1.13 Praseodymium	60 **Nd** RAM: 144.24 P: 1.14 Neodymium	61 **Pm** RAM: 145 P: 1.13 Promethium	62 **Sm** RAM: 150.36 P: 1.17 Samarium	63 **Eu** RAM: 151.965 P: 1.2 Europium
89 **Ac** RAM: 227 P: 1.1 Actinium	90 **Th** RAM: 232.0381 P: 1.3 Thorium	91 **Pa** RAM: 213.0359 P: 1.5 Protactinium	92 **U** RAM: 238.0289 P: 1.38 Uranium	93 **Np** RAM: 237.0482 P: 1.36 Neptunium	94 **Pu** RAM: 244 P: 1.28 Plutonium	95 **Am** RAM: 243 P: 1.3 Americium

III

Key

Atomic number · · · · · ·
Relative Atomic Mass · · · · · ·
Electronegativity (Pauling) · · · · · ·
Element · · · · · ·

SYMBOL
00 **Xx**
RAM: 0.0000
P: 0.00
Name

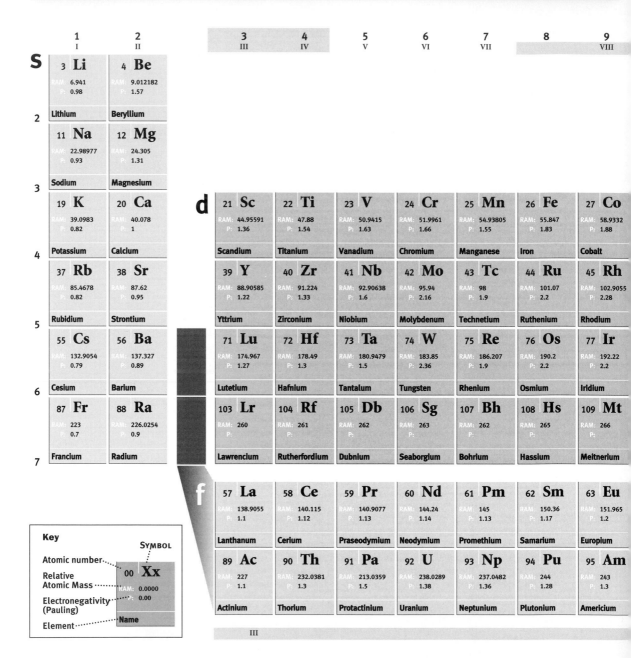

1S

1 **H**	2 **He**
RAM: 1.00794	RAM: 4.002602
P: 2.2	P: 0
Hydrogen	Helium

10	11 I	12 II	13 III	14 IV	15 V	16 VI	17 VII	18 VIII	
			p						

5 **B**	6 **C**	7 **N**	8 **O**	9 **F**	10 **Ne**	
RAM: 10.811	RAM: 12.011	RAM: 14.00674	RAM: 15.9994	RAM: 18.9984	RAM: 20.1797	
P: 2.04	P: 2.55	P: 3.04	P: 3.44	P: 3.98	P: 0	
Boron	Carbon	Nitrogen	Oxygen	Fluorine	Neon	2

13 **Al**	14 **Si**	15 **P**	16 **S**	17 **Cl**	18 **Ar**	
RAM: 26.98154	RAM: 28.0855	RAM: 30.97376	RAM: 32.066	RAM: 35.4527	RAM: 39.948	
P: 1.61	P: 1.9	P: 2.19	P: 2.58	P: 3.16	P: 0	
Aluminium	Silicon	Phosphorus	Sulfur	Chlorine	Argon	3

28 **Ni**	29 **Cu**	30 **Zn**	31 **Ga**	32 **Ge**	33 **As**	34 **Se**	35 **Br**	36 **Kr**	
AM: 58.6934	RAM: 63.546	RAM: 65.39	RAM: 69.723	RAM: 72.61	RAM: 74.92159	RAM: 78.96	RAM: 79.904	RAM: 83.8	
P: 1.91	P: 1.9	P: 1.65	P: 1.81	P: 2.01	P: 2.18	P: 2.55	P: 2.96	P: 0	
ickel	Copper	Zinc	Gallium	Germanium	Arsenic	Selenium	Bromine	Krypton	4

46 **Pd**	47 **Ag**	48 **Cd**	49 **In**	50 **Sn**	51 **Sb**	52 **Te**	53 **I**	54 **Xe**	
AM: 106.42	RAM: 107.8682	RAM: 112.411	RAM: 114.82	RAM: 118.71	RAM: 121.757	RAM: 127.6	RAM: 126.9045	RAM: 131.29	
P: 2.2	P: 1.93	P: 1.69	P: 1.78	P: 1.96	P: 2.05	P: 2.1	P: 2.66	P: 0	
alladium	Silver	Cadmium	Indium	Tin	Antimony	Tellurium	Iodine	Xenon	5

78 **Pt**	79 **Au**	80 **Hg**	81 **Tl**	82 **Pb**	83 **Bi**	84 **Po**	85 **At**	86 **Rn**	
AM: 195.08	RAM: 196.9665	RAM: 200.59	RAM: 204.3833	RAM: 207.2	RAM: 208.9804	RAM: 209	RAM: 210	RAM: 222	
P: 2.28	P: 2.54	P: 2	P: 2.04	P: 2.33	P: 2.02	P: 2	P: 2.2	P: 0	
latinum	Gold	Mercury	Thallium	Lead	Bismuth	Polonium	Astatine	Radon	6

10 **Uun**	111 **Uuu**	112 **Uub**	
AM:	RAM:	RAM:	
P:	P:	P:	
	?	?	7

64 **Gd**	65 **Tb**	66 **Dy**	67 **Ho**	68 **Er**	69 **Tm**	70 **Yb**
AM: 157.25	RAM: 158.9253	RAM: 162.5	RAM: 164.9303	RAM: 167.26	RAM: 168.9342	RAM: 173.04
1.2	1.2	P: 1.22	P: 1.23	P: 1.24	1.25	1.1
adolinium	Terbium	Dysprosium	Holmium	Erbium	Thulium	Ytterbium

Lanthanides

96 **Cm**	97 **Bk**	98 **Cf**	99 **Es**	100 **Fm**	101 **Md**	102 **No**
AM: 247	RAM: 247	251	RAM: 252	257	RAM: 258	RAM: 259
1.3	1.3	P: 1.3	1.3	1.3	1.3	1.3
urium	Berkelium	Californium	Einsteinium	Fermium	Mendelevium	Nobelium

Actinides

Appendix 4 Scientific units and notation, conversion tables, abbreviations

SI units

Base and supplementary SI units

Physical quantity	SI unit	Symbol
length	metre	m
mass	kilogram	kg
time	second	s
electric current	ampere	A
thermodynamic temperature	kelvin	K
luminous intensity	candela	cd
amount of substance	mole	mol
plane angle (supplementary unit)	radian	rad
solid angle (supplementary unit)	steradian	sr

Derived SI units with special names

Physical quantity	SI unit	Symbol
frequency	hertz	Hz
energy	joule	J
force	newton	N
power	watt	W
pressure	pascal	Pa
electric charge	coulomb	C
electric potential difference	volt	V
electric resistance	ohm	Ω
electric conductance	siemens	S
electric capacitance	farad	F
magnetic flux	weber	Wb
inductance	henry	H
magnetic flux density (magnetic induction)	tesla	T
luminous flux	lumen	lm
illuminance	lux	lx
absorbed dose	gray	Gy
activity	becquerel	Bq
dose equivalent	sievert	Sv

Decimal multiples and submultiples used with SI units

Submultiple	Prefix	Symbol	Multiple	Prefix	Symbol
10^{-1}	deci-	d	10	deca-	da
10^{-2}	centi-	c	10^2	hecto-	h
10^{-3}	milli-	m	10^3	kilo-	k
10^{-6}	micro-	μ	10^6	mega-	M
10^{-9}	nano-	n	10^9	giga-	G
10^{-12}	pico-	p	10^{12}	tera-	T
10^{-15}	femto-	f	10^{15}	peta-	P
10^{-18}	atto-	a	10^{18}	exa-	E
10^{-21}	zepto-	z	10^{21}	zetta-	Z
10^{-24}	yocto-	y	10^{24}	yotta-	Y

Ma, Ga. Two non-SI units of time commonly used in the geological sciences are the megayear (one million years), abbreviated to Ma, and the gigayear (one billion or 10^9 years), abbreviated to Ga.

The index or power notation

The index, or power, notation is generally used in science to express large or small numbers. It uses powers of ten. Thus, 100 is 10^2; 1000 is 10^3; 1 000 000 (one million) is 10^6; one billion (a thousand million) is 10^9; and so on. It will be apparent that the index (i.e., the power of ten) is equal to the number of noughts in the number that is represented.

This notation is similarly used to denote decimals by inserting a minus sign. Thus, 0.1 (one-tenth) is 10^{-1}; 0.01 (one-hundredth) is 10^{-2}; 0.001 (one thousandth) is 10^{-3}; and so on.

The index notation is also used in combination with abbreviations for units. For example, 'square metre' is concisely rendered as m^2; 'cubic metre' as m^3. Negative indices are used in place of 'per' or the oblique stroke, as in m s^{-1} (metres per second or m/s); m s^{-2} (metres per second per second). Reference to the table of units in this appendix in conjunction with this note should elucidate any unfamiliar SI unit.

Conversion tables

Length

	metre	centimetre	inch	foot	yard
1 metre	1	100	39.3701	3.28084	1.09361
1 centimetre	0.01	1	0.393701	0.328084	0.0109361
1 inch	0.0254	2.54	1	0.0833333	0.0277778
1 foot	0.3048	30.48	12	1	0.333333
1 yard	0.9144	91.44	36	3	1

	kilometre	mile	nautical mile
1 kilometre	1	0.621371	0.539957
1 mile	1.60934	1	0.868976
1 nautical mile	1.85200	1.15078	1

Area

	m^2	cm^2	in^2	ft^2
1 square metre	1	10^4	1550	10.7639
1 square centimetre	10^{-4}	1	0.155	1.07639×10^{-3}
1 square inch	6.4516×10^{-4}	6.4516	1	6.94444×10^{-3}
1 square foot	9.2903×10^{-2}	929.03	144	1

	m^2	km^2	yd^2	$mile^2$	acre
1 square metre	1	10^{-6}	1.19599	3.86019×10^{-7}	2.47105×10^{-4}
1 square kilometre	10^6	1	1.19599×10^6	0.386019	247.105
1 square yard	0.836127	8.36127×10^{-7}	1	3.22831×10^{-7}	2.06612×10^{-4}
1 square mile	2.58999×10^6	2.58999	3.0976×10^6	1	640
1 acre	4.04686×10^3	4.04686×10^{-3}	4840	1.5625×10^{-3}	1

Volume

	m^3	cm^3	in^3	ft^3	gallons
1 cubic metre	1	10^6	6.10236×10^4	35.3146	219.969
1 cubic centimetre	10^{-6}	1	0.0610236	3.53146×10^{-5}	2.19969×10^{-4}
1 cubic inch	1.63871×10^{-5}	16.3871	1	5.78704×10^{-4}	3.60464×10^{-3}
1 cubic foot	0.0283168	28316.8	1728	1	6.22882
1 gallon (UK)	4.54609×10^{-3}	4546.09	277.42	0.160544	1

Speed

	m/sec	km/hr	mile/hr	ft/s
1 metre per second	1	3.6	2.23694	3.28084
1 kilometre per hour	0.277778	1	0.621371	0.911346
1 mile per hour	0.44704	1.609344	1	1.46667
1 foot per second	0.3048	1.09728	0.681817	1

Mass

	kg	g	lb
1 kilogram	1	1000	2.20462
1 gram	10^{-3}	1	2.20462×10^{-3}
1 pound	0.453592	453.592	1

Density

	kg/m^3	g/cm^3	lb/ft^3
1 kilogram per cubic metre	1	10^{-3}	0.062428
1 gram per cubic centimetre	1000	1	62.428
1 pound per cubic foot	16.0185	0.0160185	1

Force

	N	kgf	dyne	poundal	lbf
1 newton	1	0.101972	10^5	7.23300	0.224809
1 kilogram force	9.80665	1	9.80665×10^5	70.9316	2.20462
1 dyne	10^{-5}	1.01972×10^6	1	7.23300×10^{-5}	2.4809×10^{-6}
1 poundal	0.138255	1.40981×10^{-2}	1.38255×10^4	1	0.031081
1 pound force	4.44822	0.453592	4.44823×10^5	32.174	1

Pressure

	N/m^2(Pa)	kg/cm^2	lb/in^2	atmos
1 newton per square metre (pascal)	1	1.01972×10^{-5}	1.45038×10^{-4}	9.86923×10^{-6}
1 kilogram per square centimetre	980.665×10^2	1	14.2234	0.967841
1 pound per square inch	6.89476×10^3	0.0703068	1	0.068046
1 atmosphere	1.01325×10^5	1.03323	14.6959	1

Work and energy

	J	cal	kWh	btu
1 joule	1	0.238846	2.7778×10^{-7}	9.47813×10^{-4}
1 calorie	4.1868	1	1.16300×10^{-6}	3.96831×10^{-3}
1 kilowatt hour	3.6×10^6	8.59845×10^5	1	3412.14
1 British Thermal Unit	1055.06	251.997	2.93071×10^{-4}	1

Some abbreviations used in the Earth sciences

[This list does not include abbreviations for institutions or organizations or for research programmes.]

AABW	Antarctic Bottom Water
AADW	Antarctic Deep Water
AAIW	Antarctic Intermediate Water
AAS	atomic absorption spectroscopy
Ab	albite (sodium plagioclase)
ACC	Antarctic Circumpolar Current
ACD	aragonite compensation depth
AGCM	atmospheric general circulation model
An	anorthite (calcium plagioclase)
APW	apparent polar wander
aq	(in chemical equations), dissolved in a large volume of water
a.s.l.	above sea level
BIF	banded iron formation
BOD	biological oxygen demand
BP	(years) before present (in practice, before 1950)
c	(in chemical equations), crystalline
c.	circa (about)
CCD	carbonate compensation depth
CCN	cloud condensation nuclei
CFB	continental flood basalt
CFCs	chlorofluorocarbons
CMB	core–mantle boundary
COD	chemical oxygen demand
CPDW	Circumpolar Deep Water
cpx	clinopyroxene
CRM	chemical remanent magnetization
CZM	coastal zone management
DEM	digital elevation model
DOM	dissolved organic matter
DRM	depositional remanent magnetization
DTA	differential thermal analysis
DVI	dust veil index
EAIS	East Antarctic Ice Sheet
EDM	electronic distance measurement
Eh	oxidation potential
EIA	environment impact assessment
ELR	environmental lapse rate
EM	electromagnetic radiation
EMP	estuary management plan
En	enstatite
ENSO	El Niño–Southern Oscillation
EPMA	electron probe microanalysis
Fa	fayalite (Fe_2SiO_4)
Fo	forsterite (Mg_2SiO_4)

g	(in chemical equations), gaseous
Ga	gigayear (10^9 years)
GCM	general climate model
GIS	geographical information systems
GOR	gas–oil ratio
GPR	ground-penetrating radar
GPS	Global Positioning System
GPTS	geomagnetic polarity timescale
GSN	Global Seismographic Network
GSSP	global stratotype section and point
HDR	hot dry rock
HREE	heavy rare-earth element
ICP	inductively coupled plasma spectroscopy
ICZM	integrated coastal zone management
IGRF	International Geomagnetic Reference Field
ITC	International Tropical Confluence
ITCZ	intertropical convergence zone
K–T boundary	Cretaceous–Tertiary boundary
l	(in chemical equations), liquid
LFG	landfill gas
LGM	Last Glacial Maximum
LIL	large-ion lithophile
LNG	liquid natural gas
LPG	liquid petroleum gas
LREE	light rare-earth element
LVZ	low-velocity zone
Ma	megayear, million years
MHD	magnetohydrodynamic (waves)
MIZ	Marginal Ice Zone
MOR	mid-ocean ridge
MORB	mid-ocean-ridge basalt
MS	magnetic susceptibility
MSS	multispectral scanners
MT	magnetotelluric (prospecting method)
MWP	Medieval Warm Period
NADW	North Atlantic Deep Water
NRM	natural remanent magnetization
NRS	New Red Sandstone
OD	(in UK) Ordnance datum (sea-level datum)
OGCM	oceanic general climate model
OIB	ocean island basalt
opx	orthopyroxene
ol	olivine
Or	orthoclase feldspar

ORS	Old Red Sandstone
PCBs	polychlorinated biphenyls
PGE	platinum group elements
pH	measure of hydrogen ion concentration in a solution (and thus of acidity and alkalinity)
Ph	Phanerozoic
ppb	parts per billion (10^9)
ppm	parts per million
PREM	Preliminary Reference Earth Model
QBO	quasi-biennial oscillation
qtz	quartz
REE	rare-earth elements
RI	refractive index
RMR	rock mass rating
RQD	rock quality designation
s	(in chemical equations) solid
SAR	sodium absorption ratio; synthetic aperture radar
SEM	scanning electron microscope/micrograph
SI	Système International d'Unités, SI units
SLAR	side-looking airborne radar
SLIME	Subsurface ChemoLithoautotrophic Microbial Ecosystem
SLR	satellite laser ranging
SMOW	Standard Mean Ocean Water (international standard for isotopic ratios for sea water)
SMP	shoreline management plan
SO	Southern Oscillation

SPCZ	South Pacific Convergence Zone
SST	sea-surface temperature
STEM	scanning transmission electron microscopy
TEM	transmission electron microscopy
TOGA	Tropical Ocean Global Atmosphere
TRM	thermoremanent magnetization
UV	ultraviolet
VLBI	very-long-baseline interferometry
VLF	very low frequency
VRM	viscous remanent magnetization
WPB	within-plate basalt
WWSSN	World-Wide Standardized Seismographic Network
XRD	X-ray diffraction analysis
XRF	X-ray fluorescence analysis

Symbols

$\neq$	not equal to
$\approx$	approximately equal to
$\equiv$	identical to
$\propto$	proportional to
∞	infinity
$>$	greater than
$<$	less than
$\geq$	greater than or equal to
$\leq$	less than or equal to
$\pm$	plus or minus
$\sim$	of the order of
‰	per thousand

Thematic lists

Atmosphere, meteorology
Climate and climate change
Earth sciences
  General
  Historical
Economic geology, applied
  geology, environmental
  sciences
  General
  Engineering geology,
    construction
Geochemistry
  General
  Geochemical processes
  Biogeochemistry
Geological time and
  stratigraphy
  Earth history, stratigraphy
  Divisions of geological time
  Palaeogeography
Geomorphology, glaciology, soil
  science
  General
  Weathering, erosion, and
    deposition
  Coasts
  Glaciers and glaciation
  Rivers and river systems
  Soil science
Geophysics
  General
  Geodesy and surveying;
    gravity surveys
  Earthquakes, seismology,
    and seismic surveys
  Geomagnetism; magnetic
    and electrical methods
    in geophysics
  The interior of the Earth
Hydrology and hydrogeology
Marine geology
Mineralogy
  General
  Mineral species and
    mineral groups
Oceanography
Palaeontology, palaeobiology
  General

Invertebrates
Vertebrates
Petrology: igneous
  General
  Rock-types
Petrology: metamorphic
  General
  Rock-types
Planetary science, the Solar
  System
Plate tectonics, continental drift
Sediments and sedimentology
  General
  Types of sediments
Structural geology
  General
  Faults and faulting
  Rock deformation

Atmosphere, meteorology

acid rain
air pressure
anticyclone
atmospheric convergence and
  divergence
atmospheric electricity
atmospheric high pressure
atmospheric temperature
clouds
cyclone
droughts
dust
evaporation
flash floods
floods
general circulation of the
  atmosphere
gradient wind
high-latitude tropospheric
  circulation
ionosphere
jet streams
low-latitude atmospheric
  circulation
mesosphere
mid-latitude tropospheric
  circulation

ozone layer
ozone-layer chemistry
precipitation
Quaternary superfloods
Rossby waves
storm surges
stratosphere
synoptic climatology
thermosphere
tropical cyclones
vertical air motion
vorticity
Walker circulation
weather forecasting
weather fronts

Climate and climate change

aerosols and climate
Antarctic climate
Antarctic ice cores
Buckland, W.
Chamberlin, T. C.
CLIMAP Project
climate change and deep water
  formation
climate change and lake levels
climate change and laminated
  sediments
climate models
COHMAP project
Croll, James
dendroclimatology
droughts
El Niño
global warming
Greenland ice cores
Gulf Stream
Himalayan–Tibetan uplift and
  climate change
ice ages
ice-age aridity
ice-age theories
ice sheets and climate
Jamieson, T.
Lamb, H. H.
landscape and climate

Little Ice Age
loess deposition and
  palaeoclimate
Medieval Warm Period
Milankovich, M.
Milankovich cycles and climate
  change
nuclear winter scenarios
oxygen isotope records
precipitation during Quaternary
  ice ages
Quaternary superfloods
Urey, H. C.
Younger Dryas stade

Earth sciences: general and historical

General

amateurs in geology
art and geology
Atlantis: fact or fiction?
classical times and the Earth
  sciences
Creationism
Earth sciences
Earth sciences and myth
Earth heritage conservation
fractals in Earth science
fraud in geology
Gaia
geoarchaeology
geoecology
geological controversies
geological humour
geological maps and map-
  making
geological and other Earth
  sciences
geological societies
geological surveys
geoscience and the media
geosphere
information technology and the
  Earth sciences
literature and geology
International Union of Earth
  Sciences

medical geology
medieval mineralogy and
    figured stones
military geology
museums and geology
music and the Earth sciences
philosophy and geology
wine and geology

Historical

Agricola, G.
Agassiz, L.
Airy, G. B.
Alpine pioneers of geology
Barrell, J.
Buckland, W.
Barrell, J.
beginnings of geological
    thought
Buckland, W.
Cuvier, G.
Darwin, C.
De la Beche, T.
Du Toit, A.
geological controversies
geological exploration of the
    polar regions
geological exploration of the
    North American West
geological maps and map-
    making
geological societies
geological surveys
Hall, J. [1761–1832]
Hall, J. [1811–98]
Haüy, R.-J.
Hooke, R.
Hutton, J.
Jamieson, T.
Lamarck, J. B. P. A. de M.
Lapworth, C.
Lyell, C.
medieval mineralogy and
    figured stones
military geology
Murchison, R. I.
Owen, R.
philosophy and geology
Playfair, J.
Renaissance in Europe and
    geological ideas
Sedgwick, A.
Smith, W.
Sorby, H. C.
specialization in the Earth
    sciences after the Second
    World War
Steno, N.
Woodward, J.

Economic geology, applied geology, environmental sciences

General

Agricola, G.
Barrell, J.
building stones
china clay
chromites
coal
containment walls in
    environmental management
Earth resources
ecclesiastical geology
economic geology
elemental associations and ore
    minerals and allied deposits
environmental geology
environmental toxicology
history of the search for
    economic deposits
dust
geochemical mineral
    exploration
geothermal energy
gold–silver deposits
history of the search for
    economic deposits
industrial minerals
iron–nickel deposits
Lamb, H. H.
laterites
lead–zinc deposits
Lindgren, W.
massive sulphide deposits
metallogenic provinces
Minankovich, M. M.
mineral water
mining
Nalivkin, D. V.
natural gas
nuclear energy
oil shale
ore deposits
ore minerals
petroleum
petroleum indistry
placer deposits
pollution control
pollution of the natural
    environment
porphyry copper deposits
radioactive waste management
radioactivity measurement and
    surveying
radon
recycling

river management
sand and gravel resources
strategic minerals
supergene deposits
tar sands
urban stone decay and
    conservation
veins
waste disposal
Werner, A. G.

Engineering geology, construction

aggregates
building stones
dams
deformation
earth fills
engineering geology
ground subsidence
landslides
mine drainage
mining
roads and pavements
rock mechanics
rock slopes and stability
slope stability and landslides
soil classification and
    description in engineering
    geology
soil stabilization
strain and stress analysis
stress analysis
surface water
tunnels

Geochemistry

General

banded iron formations (BIFs)
Berzelius, J. J.
biogenic weathering
black smokers and white
    smokers
brines
carbon cycles
chemical fossils
chemical properties of soils
chemistry of geothermal waters
chemostratigraphy
Clarke, F. W.
cosmochemistry
fluids in the Earth
fluid inclusions
gas hydrates
geochemical analysis
geochemical anomalies
geochemical distribution
geochemical indicator species

geochemical mineral
    exploration
geochemical mobility
geochemical residence times
geochemistry of lakes
Goldschmidt, V. M.
groundwater chemistry
history of geochemistry
hot springs
hydrothermal solutions
iron–nickel deposits
isotope geochemistry
liquefaction
major elements
manganese nodules
mantle and core composition
natural gas
Niggli, P.
ocean chemistry
oceanic dissolved constituents
ocean-ridge geochemistry
oceanic salinity
ozone-layer chemistry
pelagic carbonates
pelagic silica
phosphates
pore-water chemistry
radioactive elements
radon
sedimentary geochemistry
Solar System abundances of
    chemical elements
stylolites
sulphate minerals
sulphide minerals
trace elements
travertine and tufa
Urey, H. C.

Geochemical processes

adsorption
assimilation
biodegradation
biogenic weathering
chemical alteration of rocks
crustal composition and
    recycling
geochemical cycles
geochemical differentiation
geochemistry of geothermal
    cycles
metasomatism
partitioning
redox equilibria
volatilization

Biogeochemistry

bacteria
bacterial isotope fractionation

biodegradation
biogenic weathering
biogeochemistry
biosphere
DNA: the ultimate biomarker
geomicrobiology
life, origin of
methanogenesis
organic acids in geochemistry
organic geochemistry
organic matter (kerogen)

Geological time and stratigraphy

Earth history, stratigraphy

age and early history of the
    Earth and Solar System
Barrell, J.
Buckland, W.
catastrophism and
    uniformitarianism
chemostratigraphy
Cuvier, G.
dating rocks absolutely
De la Beche, H. T.
event stratigraphy
facies
Geikie, A.
geological time
Hall, J. (1811–98)
historical geology
ice ages
isotopic dating
Kuenen, Ph. H.
Lapworth, C.
Lehmann, J. G.
Libby, W. F.
Lyell, C.
magnetostratigraphy
Murchison, R. I.
Nalivkin, D. V.
offlap, onlap, overlap, overstep
outcrop, exposure
palaeocurrent analysis
palaeoecology
palaeoenvironment
Quaternary oxygen isotope
    chronology
rates of geological processes
regression and transgression
Sedgwick, A.
sequence stratigraphy
stratification
stratigraphy
Smith, W.
Steno, N.
superficial deposits

superposition principle
Thompson, W.
type locality, type section
unconformity
varves

Divisions of geological time

Archaean
Carboniferous
Cenozoic
Cretaceous
Devonian
diachronism
Eocene
Hadean eon
Holocene/Recent
Jurassic
Mesozoic
Miocene
Neogene
New Red Sandstone
Old Red Sandstone
Oligocene
Ordovician
Palaeocene
Palaeogene
Palaeozoic
Permian
Phanerozoic
Pleistocene
Pliocene
Precambrian
Proterozoic
Quaternary
Silurian
Tertiary
Triassic

Palaeogeography

Iapetus
Caledonides
Gondwanaland
Laurasia
Messinian salinity crisis
palaeogeography
Pangaea
Quaternary ice-sheet
    disintegration
Rodinia and Pannotia
Tethys
Variscan orogeny

Geomorphology, glaciology, soil science

General

aeolian processes
anthropogeomorphology
applied geomorphology

biogemorphology
catastrophic geomorphology
coral reefs
coral terraces
Davis, W. M., and landscape
    evolution
deep-sea geomorphology
desertification
deserts
drylands
fire as a geomorphological
    agent
Geikie, A.
geomorphological equilibrium
geomorphological systems
geomorphology
Gilbert, G. K.
global geomorphology
King, L. C.
lakes
landscape and climate
landscape evolution
landscape modelling
landscape sensitivity
morphotectonics
mountain geomorphology
peatlands and bogs
Penck, A. and W.
planetary geomorphology
pluvial lakes
Powell, J. W.
scale in geomorphology
tectonic geomorphology
tropical landforms
wetlands
zoogeomorphology

Weathering, erosion, and deposition

alluvial fans and alluvial plains
alluvial channels
avalanches
badlands
debris flows
deforestation and landscape
desert dunes
desert pavement
erosion
erosion surfaces
escarpment
fault scarp
flash floods
Geikie, A.
ground subsidence
hill slopes
hill-slope creep
karren
karst
landslides

laterites
limestone pavements
liquefaction
mud and sand volcanoes
neptunean dykes
overland flow
pediments
piping
playas
quick clays
rainsplash
rock coatings
rock glaciers
saltation
sinkholes
soil erosion
stalactites and stalagmites
subsurface flow and erosion
tafoni
tors
tropical karst
tsunamis
urban stone decay
weathering
wind erosion
yardangs

Coasts

beaches
beachrock
caves
cliffs and cliff slopes
coastal dunes
coastal management
coasts and coastal processes
coral reefs
coral terraces
deltas
estuaries
fiords
lagoons
mangroves
mud flats
oceanic islands
raised shorelines (beaches)
rock coasts
salt marshes
shore platforms
spits

Glaciers and glaciation

blockfields and blockstreams
glacial landforms
glacial rivers
glacial sediments
glacier mass balance and
    glaciology
glaciers and glaciology
glaciotectonics

ice sheets and landscape
   evolution
jökelhlaups
limestone pavements
rock glaciers
surging glaciers

Rivers and river systems

alluvial channels
arid-zone hydrology
arroyos
base level
braided rivers
drainage basins
flash floods
glacial rivers
gullies and gullying
hydrology
meanders and meandering
   rivers
overland flow
palaeohydrology
river channelization
river management
rivers
springs
surface water
valleys

Soil science

bauxite
compaction and consolidation
   of soil
laterites
palaeosols, duricrusts, calcrete,
   silcrete, and gypcrete
pans
peatlands and bogs
permafrost
soil classification and
   description in engineering
   geology
soil development
soil erosion
soil fabric and structure
soil geomorphology
soil stabilization
soils
soils and topography
laterites

Geophysics

General

Coriolis effect or Coriolis force
cratons
Dutton, C. E.
energy budget of the Earth
exploration geophysics
geodynamics
history of geophysics

isostasy
moment of inertia and
   precession
radioactivity measurement and
   surveying

**Geodesy and surveying;
gravity surveys**

aeromagnetic surveying
Bullard, E. C.
figure of the Earth
geodesy and geodetic
   measurement
geomagnetics
Global Positioning System
   (GPS)
gravity measurement and
   interpretation of anomalies
Hess, H. H.
Hooke, R.
inverse methods
planetary geodesy
radioactive measurement and
   surveying
SAR interferometry
synthetic aperture radar (SAR)
Vening-Meinesz, F. A.
very-long-baseline
   interferometry (VLBI)

**Earthquakes, seismology, and
seismic surveys**

Benioff, H.
Bullard, E. C.
controlled-source seismology
deep seismic refraction profiling
density distribution within the
   Earth
Dutton, C. E.
earthquake hazards and
   prediction
earthquake mechanisms and
   plate tectonics
earthquake seismology
elastic wave propagation
Ewing, W. M.
free oscillations
Gutenberg, B.
intraplate seismicity
inverse methods
Jeffreys, H.
Lehmann, I.
Michell, J.
Mohorovičić discontinuity
   (Moho)
palaeoseismology
Richter, C. F.
seismic anisotropy
seismic body waves
seismic exploration methods

seismic properties of rocks
seismic ray theory
seismic reflection surveying
seismic reflection profiling
seismic signal processing
seismic stratigraphy
seismic properties of rocks
seismic ray theory
seismic surface waves
seismic tomography
seismic wave theory
seismic waves; principles
seismology and plate tectonics
strong-motion seismology
subduction zone
synthetic seismograms
Vening-Meinesz, F. A.
volcanogenic earthquakes
volcanoseismicity
Worldwide Standard
   Seismograph Network

**Geomagnetism; magnetic
and electrical methods in
geophysics**

aeromagnetic surveying
Blackett, P. M. S.
controlled-source
   electromagnetic mapping
Curie temperature (Curie
   isotherm)
Dietz, R. S.
electrical techniques in applied
   geophysics
geomagnetic measurement:
   techniques and surveys
geomagnetism: external fields
geomagnetism: main field,
   secular variation, and
   westward drift
geomagnetism: polarity
   reversals
magnetic field (origin of the
   Earth's internal field)
magnetic pole
magnetohydrodynamic waves in
   the Earth (MHD)
magnetostratigraphy
magnetotelluric prospecting
Matthews, D. H.
palaeoclimate and magnetism
palaeomagnetism and
   continental drift
palaeomagnetism: local
   deformation
palaeomagnetism: past intensity
   of the field
palaeomagnetism and polar
   wander
palaeomagnetism: techniques of

remanent magnetization
Wilson, J. Tuzo

The interior of the Earth

Airy, G. B.
asthenosphere
Beloussov, V. V.
density distribution within the
   Earth
Earth structure
Earth tides
Gilbert, G. K.
heat flow in the Earth
Lehmann, I.
lithosphere
mantle convection, plumes,
   viscosity, dynamics
Mohorovičić discontinuity
   (Moho) ˇ
phase transformations at depth
   in the Earth
radioactive heat production in
   the Earth
self-compression
stress fields of the Earth
Thomson, W. (Lord Kelvin)

Hydrology and
hydrogeology

aquifer
arid-zone hydrology
artesian well
contaminant transport
differentiation and dispersion in
   groundwater flow
drainage basins
fluids in the Earth
groundwater
groundwater chemistry
groundwater flow rate
hot springs
hydrological cycle
hydrology
hydrothermal solutions
palaeohydrology
permeability and porosity
recharge and the hydrological
   cycle
retardation in groundwater
springs
storm surges
tracers and groundwater
water-table
water and water types

Marine geology

aseismic submarine ridges
black smokers, white smokers

continental shelf geology
coral reefs
coral terraces
crustal composition and
   recycling
Daly, R. A.
deep-sea geomorphology
deep-sea trenches
deep-water sediments
Dietz, R. S.
Ewing, W. M.
Heezen, Bruce
diagenesis in deep-sea
   sediments
geomagnetism: polarity
   reversals
guyots
Hess, H. H.
island arcs
Kuenen, Ph. H.
oceanic islands
manganese nodules
marine geology
Matthews, D. H.
mid-ocean ridges
Murray, J.
ocean basins
Ocean Drilling Programme
   (ODP)
oceanic islands
ocean-ridge geochemistry
ophiolite sequence
sea-floor spreading
seamounts
Shepard, F. P.
submarine sediment slides
transform plate margins
Wilson, J. Tuzo

Mineralogy

General

Agricola, G.
Clarke, F. W.
cleavage in minerals
crystallographic systematics
crystals and crystal growth
Dana, J. D.
Eskola, P.
gemstones
Haüy, J.
medieval mineralogy and
   figured stones
Miller, W. H.
minerals and mineralogy
mineral zoning
Moh's scale of hardness
Niggli, P.
optical mineralogy
Werner, A. G.

Mineral species and mineral groups

amber
asbestos
bauxites
calcite, aragonite, dolomite
chromites
clay minerals
diamonds
quartz and related minerals
rock-forming silicate minerals
sulphate minerals
sulphide minerals
travertine and tufa

Oceanography

CLIMAP Project
climate change and deep water
   formation
Gulf Stream
Heezen, Bruce
high-latitude ocean currents
Messinian salinity crisis
monsoonal currents
Murray, J.
ocean basins
ocean chemistry
ocean currents
ocean water
oceanic density and pressure
oceanic dissolved constituents
oceanic salinity
oceanic temperatures
palaeoceanography
pelagic carbonates
pelagic environments of the
   oceans
pelagic silica
Quaternary sea level changes
palaeocurrent analysis
sea ice and climate
sea-level changes and
   palaeoclimate
sea-surface temperatures (SSTs)
   and climate
sea water
sea-water chemistry
tsunamis

Palaeontology, palaeobiology

General

Buckland, W.
Burgess Shale fauna
Cope, E. D.
Cuvier, G.
Darwin, Charles
evolution (Darwinian and neo-
   Darwinian)
extinctions and mass
   extinctions
faunal provinces
fossils and folklore
fossils and fossilization
Hall, J. (1811–98)
heterochrony in palaeontology
Hooke, R.
Lamarck, J. B. P. A. de M.
Lapworth, C.
major fossil groups
origin of life
ostracods
palaeoecology
palaeontology, palaeobiology
phylogeny (morphological and
   molecular)
rates and trends in evolution
Solenhofen lithographic
   limestone fauna
species concept
Steno, N.
taphonomy
taxonomy and taxonomic
   heirarchies
trace fossils and ichnology
tropical refugia
Urey, H. C.
Walcott, C. D.
Woodward, J.
zone fossil

Invertebrates

arthropods
bivalves
brachiopods
bryozoans
corals and related fossils
crustaceans
echinoderms
foraminifera and other
   unicellular microfossils
gastropods
graptolites
insects
invertebrates
metazoans
micropalaeontology
molluscs
ostracods
sponges and related forms
trilobites and related fossils
Walcott, C.D.

Vertebrates

Agassiz, L.
Anning, M.
amphibians
birds
chordates
conodonts
Cope, E.D.
Cuvier, G.
dinosaur hunters
dinosaurs
early vertebrates and fish
ichthyosaurs
Leakey, L. S. B.
mammal-like reptiles
mammals
Man (hominids)
Marsh, O. C.
Osborne, H. F.
Owen, R.
plesiosaurs
pterosaurs
reptiles
tetrapods
vertebrates

Fossil plants

fossil plants
palaeobotany
palynology
stromatolites

Petrology: igneous

General

assimilation
Bowen, N. L.
diapirs and diapirism
experimental petrology
geothermometers (geothermal
   systems)
granite–gneiss terrains
Hall, J. (1761–1832)
Harker, A.
igneous melt compositions
   (silicate and carbonate)
igneous rock classification
intrusion types and intrusive
   igneous rocks
lava
magma and magmatism
Niggli, P.
palaeovolcanism and climate
   change
partial melting and magma
   production
partitioning
petrography, petrology,
   petrogenesis
ring complex
textures of rocks
thermobarometry
veins
volcanoes and volcanic rocks

Rock-types

basalt

carbonatites
continental flood basalts (CFBs)
eclogite
glass
granite
hypabyssal igneous rocks
kimberlites
pegmatite
plutonic rocks
pyroclastic and volcaniclastic
    rocks

Petrology: metamorphic

General

contact metamorphism
Eskola, P.
geothermometers (geothermal
    systems)
granite–greenstone terrains
high-grade gneiss terrain
metamorphic core complexes
metamorphism, metamorphic
    facies, and metamorphic
    rocks
metasomatism
Niggli, P.
regional metamorphism
Sederholm, . J.
thermobarometry
Wilson, J. Tuzo

Rock-types

granulite and granulite facies
marble
migmatite
mylonite
slate

Planetary science,
the Solar System

Airy, Sir George
age and early history of the
    Solar System
asteroids and comets
Chamberlin, T. C.
cosmochemistry
Dietz, R. S.
Earth orientation
giant planets
Gilbert, G. K.
Hadean eon
Kuiper, G. P.
meteoric and cometary impacts
moment of inertia and
    precession
planetary geodesy
planetary geomorphology
rotation and orbit of the Earth

Shoemaker, G.
solar activity and sunspots
terrestrial planets and other
    Earth-like bodies

Plate tectonics,
continental drift

aseismic submarine ridges
asthenosphere
Benioff, H.
Blackett, P. M. S.
Bullard, E. C.
continental drift
continental rifts
convergent plate margins
Daly, R. A.
Dietz, R. S.
divergent plate margins
du Toit, A.
earthquake mechanisms and
    plate tectonics
exotic terrane
foreland, foreland basin, and
    hinterland
geodynamics
Hess, H. H.
Holmes, A.
island arcs
Jeffreys, H.
Matthews, D. H.
obduction
ophiolite sequence
palaeomagnetism and
    continental drift
passive margins
plate tectonics: principles
plate tectonics: the history of a
    paradigm
sea-floor spreading
seismology and plate tectonics
subduction zone
transform plate margins
Vening-Meinesz, F. A.
Wegener, A.
Willis, B.
Wilson, J. Tuzo

Sediments and
sedimentology

General

Bagnold, R. A.
cyclicity
diagenesis
Daly, R. A.
diagenesis in deep-sea
    sediments
facies
Kuenan, Ph. H.

neptunean dykes
Niggli, P.
particle morphology of
    sediments
provenance
saltation
sedimentary basins
sedimentary geochemistry
sedimentary structures
sedimentology, sediments, and
    sedimentary rocks
soft-sediment deformation
submarine sediment slides

Types of sediment

banded iron formations
    (BIFs)
black shales
carbonate platforms, reefs, and
    carbonate mounds
chalk
clastic dykes
clays
conglomerate
coral reefs
deep-water sediments
deltaic sandy deposits
desert sedimentary deposits
dolomite
dust
evaporites
fluvial sediments
flysch
gravel and conglomerate
lake sediments
limestones
loess
loess deposition and
    palaeoclimate
molasse
mud flats
mud and sand volcanoes
sand and sandstone
shallow-marine sedimentary
    deposits
shoreline (littoral) sedimentary
    deposits

Structural geology

General

Alpine orogeny
Alpine pioneers of geology
Andes of South America
Chamberlin, T. C.
Cloos, H.
collisional orogeny
cross-sections
Dana, J. D.

deformation of rocks
epeirogeny
extensional tectonics on the
    continents
Earth movements
    (diastrophism)
geodynamics
Gilbert, G. K.
Grabau, A. W.
gravitational sliding
Himalayan tectonics
Hutton, J.
intraplate tectonics
joints and jointing
Lapworth, C.
lineaments
lineation
morphotectonics
mountain-building (orogeny)
Nalivkin, D. V.
neotectonics and
    seismotectonics
obduction
palinspastic maps and
    sections
Sedgwick, A.
Stille, W. H.
structural geology
subduction
subsidence and uplift
Suess, E.
tectonic geomorphology
tectonics
transtension and transpression
Willis, B.

Faults and faulting

active fault
escape tectonics
faults and faulting
fault rock
grabens and rift valleys
growth fault
shear zones, deformation bands,
    and kink bands
soft-sediment deformation
strike–slip tectonics of the
    continents
transtension and transpression

Rock deformation

cleavage and other tectonic
    foliations in rocks
diapirs and diapirism
folds and folding
inversion tectonics
shear zones, deformation bands,
    and kink bands
strain and stress analysis

Index